THE UNIVERSAL ACCESS HANDBOOK

Human Factors and Ergonomics

Series Editor

Gavriel Salvendy

THE UNIVERSAL ACCESS HANDBOOK

CONSTANTINE STEPHANIDIS

CRC Press
Taylor & Francis Group
Boca Raton   London   New York

CRC Press is an imprint of the
Taylor & Francis Group, an **informa** business

CRC Press
Taylor & Francis Group
6000 Broken Sound Parkway NW, Suite 300
Boca Raton, FL 33487-2742

© 2009 by Taylor & Francis Group, LLC
CRC Press is an imprint of Taylor & Francis Group, an Informa business

No claim to original U.S. Government works

10 9 8 7 6 5 4 3 2 1

International Standard Book Number-13: 978-0-8058-6280-5 (Hardcover)

Library of Congress Cataloging-in-Publication Data

The universal access handbook / edited by Constantine Stephanidis.
    p. cm.
  Includes bibliographical references and index.
  ISBN-13: 978-0-8058-6280-5
  ISBN-10: 0-8058-6280-3
    1. Human-computer interaction--Handbooks, manuals, etc. 2. User interfaces (Computer systems)--Handbooks, manuals, etc. 3. User-centered system design--Handbooks, manuals, etc. 4. Computerized self-help devices for people with disabilities--Handbooks, manuals, etc. I. Stephanidis, Constantine. II. Title.

QA76.9.H85U65 2009
004.01'9--dc22
                                        2008048075

Visit the Taylor & Francis Web site at
http://www.taylorandfrancis.com

and the CRC Press Web site at
http://www.crcpress.com

To those who care

Table of Contents

PART I   Introduction to Universal Access

PART II   Diversity in the User Population

PART VI Interaction Techniques and Devices

PART VII Application Domains

PART VIII Nontechnological Issues

PART IX Looking to the Future

Foreword

This new volume in the Human Factors and Ergonomics series represents the current state-of-the-art in a young but rapidly developing and maturing scientific domain—universal access—which addresses principles, methods, and tools to develop interactive technologies that are accessible and usable by all citizens in the information society.

The Universal Access Handbook follows the 2001 publication of the first book dedicated to design for all in human-computer interaction (HCI), *User Interfaces for All: Concepts, Methods, and Tools.*

Since then, the scope of investigation in universal access has broadened, more systematic investigations of users, contexts, and technology diversity in the information society have been carried out, new methodological approaches have been elaborated upon, existing approaches have been embedded in the development of support tools, a wide variety of novel interaction techniques have emerged for supporting user diversity, and a plethora of applications and case studies putting to practice all of these issues have become available. Additionally, awareness and policy have also progressed to the point that now accessibility to the basic technological infrastructure has been recognized as a fundamental human right by many countries in the world and by the United Nations.

This handbook reflects all these recent developments in an effort to consolidate present knowledge in the field of universal access and to open new perspectives for the future. It provides a structured guide to professionals and practitioners working in the field, a comprehensive and interrelated collection of reference articles for academics and researchers, an indispensable source of information for interdisciplinary and cross-thematic study, an important educational tool in an increasingly globalized research and development environment, and a base line for future in-depth studies in the subject matter.

It contains 61 chapters covering the breadth and depth of the subject area, written by 96 authors from 14 countries. Of these individuals, 60 come from academia, 22 from research institutions, and 14 from industry. The book includes 381 figures, 87 tables, and 3575 references.

In summary, this handbook provides a great contribution toward further advancing the concepts and principles of universal access for the benefit of all citizens in the emerging information society. For this, I express my appreciation and extend my congratulations to the editor and the authors of *The Universal Access Handbook.*

Gavriel Salvendy
Series Editor

Preface

Since the 2001 publication of the volume *User Interfaces for All: Concepts, Methods, and Tools,* which was the first and so far unique attempt to edit a book dedicated to a comprehensive and multidisciplinary view of design for all in human-computer interaction (HCI), the field of universal access has made significant progress toward consolidating theoretical approaches, methods, and technologies, as well as exploring new application domains. Universal access refers to the conscious and systematic effort to proactively apply principles, methods, and tools of universal design, in order to develop information society technologies that are accessible and usable by all citizens, including the very young and the elderly, as well as people with different types of disabilities, thus avoiding the need for *a posteriori* adaptations or specialized design. The requirement for universal access stems from the growing impact of the fusion of the emerging technologies, and from the different dimensions of diversity, which are intrinsic to the emergence of the information society. These dimensions become evident when, for example, considering the broad range of user characteristics, the changing nature of human activities, the variety of contexts of use, the increasing availability and diversification of information and knowledge sources and services, and the proliferation of technological platforms.

The Universal Access Handbook is intended to provide a comprehensive and multidisciplinary overview of the field of universal access. It is a collection of 61 chapters, structured into nine parts, written by leading international authorities, affiliated with academic, research, and industrial organizations. The nine parts of this handbook holistically address all major dimensions of universal access, unfolding:

- The historical roots of universal access through the progressive elaboration of diverse and complementary approaches to the accessibility of interactive applications and services
- Current perspectives and trends in the field
- The various dimensions of diversity in the user population, including, but not limited to, various forms of disability
- The various dimensions of diversity in the technological platforms and contexts of use, including trends toward mobile interaction
- The implications of universal access on the development life cycle of interactive applications and services
- The implications of universal access on user-interface architectures and related components
- Required and available support tools for the development of universally accessible applications and services
- Alternative new and emerging interaction techniques and devices to support diversity in user interaction
- Examples, case studies, and best practices of universal access in new and emerging application domains
- Nontechnological issues related to universal access practice, demand, offer, management, and acceptance
- Future perspectives, with emphasis on the role and impact of universal access in the context of ambient intelligence environments

The handbook is targeted to a broad readership, including HCI researchers, user-interface designers, computer scientists, software engineers, ergonomists and usability engineers, human factors researchers and practitioners, organizational psychologists, system/product designers, sociologists, policy and decision makers, scientists in government, industry, and education, and assistive technology and rehabilitation experts, as well as undergraduate and postgraduate students reading in the various relevant scientific fields.

Constantine Stephanidis
Editor

Acknowledgments

This handbook would not have been possible without the dedicated commitment of many people whom I would like to thank. The authors of all chapters are congratulated for their dedicated efforts and collaborative attitude throughout the long process of editing and publishing this volume. My appreciation goes to the book series editor, Prof. Gavriel Salvendy, and the advisory board of this handbook, who have greatly contributed to the highest quality standards of this volume. My sincere appreciation for all their efforts and dedication also goes to the reviewers of the chapters, who are listed below.

At base camp, my gratitude goes to Dr. Margherita Antona for her unwavering dedication to the cause of this project, and for her prominent and unwearied support role throughout the long and demanding editing process. Maria Pitsoulaki deserves a singular mention of appreciation for her commitment and assistance, down to the last detail, during the editing process.

Finally, I express my appreciation to Cindy Carelli and Marsha Pronin of CRC Press/Taylor & Francis, who have supported and facilitated the editorial work during the preparation of this volume.

Constantine Stephanidis
Editor

List of Reviewers

Ray Adams, Middlesex University, United Kingdom
Margherita Antona, Foundation for Research and Technology–Hellas, Greece
Antonis Argyros, Foundation for Research and Technology–Hellas, Greece
Evangelos Bekiaris, Center for Research and Technology–Hellas, Greece
Christian Bühler, Research Institute Technology and Disability (FTB), Germany
Noëlle Carbonell, Henri Poincaré University, LORIA, France
P. John Clarkson, University of Cambridge, United Kingdom
Pier Luigi Emiliani, National Research Council, Italy
Jan Engelen, Catholic University of Leuven, Belgium
Grigori Evreinov, University of Tampere, Finland
Jinjuan Feng, Towson University, United States
Dimitris Grammenos, Foundation for Research and Technology–Hellas, Greece
Vicki L. Hanson, IBM T. J. Watson Research Center, United States
Simon Harper, University of Manchester, United Kingdom
Julie A. Jacko, University of Minnesota, United States
Eija Kaasinen, VTT Technical Research Center, Finland
Arthur I. Karshmer, University of San Francisco, United States
Simeon Keates, ITA Software, Inc., United States
Erkki Kemppainen, National Research and Development Centre for Welfare and Health (STAKES), Finland
Stephen Kimani, Jomo Kenyatta University of Agriculture and Technology, Kenya
Iosif Klironomos, Foundation for Research and Technology–Hellas, Greece
Georgios Kouroupetroglou, University of Athens, Greece
Antonio Krüger, University of Muenster, Germany
Sri Kurniawan, University of California, Santa Cruz, United States

Barbara Leporini, National Research Council, Italy
Clayton Lewis, University of Colorado, United States
Aaron Marcus, Aaron Marcus and Associates, Inc., United States
Mark Maybury, The MITRE Corporation, United States
Klaus Miesenberger, University of Linz, Austria
Alexandros Mourouzis, Center for Research and Technology–Hellas, Greece
Klaus-Robert Müller, Technical University of Berlin, Germany
Stavroula Ntoa, Foundation for Research and Technology–Hellas, Greece
Fabio Paternò, National Research Council, Italy
Lyn Pemberton, University of Brighton, United Kingdom
Enrico Pontelli, New Mexico State University, United States
Roope Raisamo, University of Tampere, Italy
Pei-Luen Patrick Rau, Tsinghua University, China
Patrice Renaud, University of Quebec in Outaouais, Canada
Dominique Scapin, National Institute for Research in Computer Science and Control (INRIA), France
Andrew Sears, University of Maryland, Baltimore County, United States
Randall Shumaker, University of Central Florida, United States
Christian Stary, University of Linz, Austria
Shari Trewin, IBM T. J. Watson Research Center, United States
Jean Vanderdonckt, Catholic University of Leuven, Belgium
Markus Vincze, Technical University of Vienna, Austria
Thorsten Völkel, Technical University of Dresden, Germany
Gerhard Weber, Technical University of Dresden, Germany
Panayiotis Zaphiris, City University London, United Kingdom

Editor

Constantine Stephanidis is a professor in the Department of Computer Science[1] at the University of Crete,[2] Director of the Institute of Computer Science[3]–FORTH,[4] and head of the Human-Computer Interaction Laboratory,[5] the Centre for Universal Access and Assistive Technologies,[6] and the Ambient Intelligence Programme.[7] He is currently a member of the National Advisory Council for Research and Technology of the Hellenic Ministry of Development and president of the Board of Directors of the Science and Technology Park of Crete.[8]

Over the past 25 years, Prof. Stephanidis has been engaged as the Scientific Responsible in more than 40 European Commission and nationally funded projects in the fields of human-computer interaction, universal access, and assistive technologies. In the beginning of the 1990s, he introduced the concept and principles of design for all in human-computer interaction and for universal access to the evolving information society technologies. He has published more than 350 technical papers[9] in scientific archival journals, proceedings of international conferences, and workshops related to his fields of expertise. He is editor-in-chief of the Springer international journal *Universal Access in the Information Society*.[10] He is editor and (co-)author of many chapters of the book *User Interfaces for All: Concepts, Methods, and Tools*,[11] published by Lawrence Erlbaum Associates (2001). From 1995–2006 he was the founding chair of the ERCIM working group "User Interfaces for All"[12] and general chair of its nine international workshops. From 1997–2000 he was the founding chair of the International Scientific Forum "Towards an Information Society for All,"[13] in the context of which he edited two white papers[14] concerning the roadmap and R&D agenda. Since 2001 he has been the founding chair of the International Conference Universal Access in Human-Computer Interaction,[15] and since 2007 he has been the general chair of the HCI International Conference[16] series.

Since the late 1980s, Prof. Stephanidis has been actively involved in activities contributing to the European Commission R&D Policy on Information Society. More recently, he was a member of the FP6 Information Society Technologies Management Committee (2002–2006) and the eAccessibility Expert Group (2002–2005). Since 2006 he has been a member of the eInclusion Group of the i2010 Initiative and the Management Board of the European Network and Information Security Agency (ENISA),[17] having worked tirelessly for the hosting of the European Agency in Heraklion since 2002. He is also a cofounder, in 2004, of the European Design for All

---

1 http://www.csd.uoc.gr/index.jsp?tID=main&sub=1.
2 http://www.uoc.gr/About/index.html.
3 http://www.ics.forth.gr.
4 http://www.forth.gr.
5 http://www.ics.forth.gr/hci/index.html.
6 http://www.ics.forth.gr/hci/cuaat.html.
7 http://www.ics.forth.gr/ami/index.html.
8 http://www.stepc.gr.
9 http://www.ics.forth.gr/hci/publications.jsp.
10 http://www.springeronline.com/journal/10209/about.
11 http://www.ics.forth.gr/hci/publications/book.html.
12 http://www.ui4all.gr.
13 http://ui4all.ics.forth.gr/isf_is4all.
14 http://www.ui4all.gr/isf_is4all/publications.html.
15 http://www.hcii2009.org/thematic_view.php?thematic_id=5.
16 http://www.hci-international.org.
17 http://www.enisa.europa.eu.

e-Accessibility Network (EDeAN)[18] and founder of the GR-DeAN,[19] the corresponding national network. Prof. Stephanidis is senior editor of the EDeAN White Paper,[20] outlining a roadmap for future European initiatives in Design for All, e-Accessibility and e-Inclusion (published in 2006).

Prof. Stephanidis is the scientific coordinator of the European Commission Coordination Action InterLink[21] (2006–2009), which aims to identify and address basic, worldwide research problems in software-intensive systems and new computing paradigms, ambient computing and communication environments, and intelligent and cognitive systems under a human-centered perspective and to define joint basic research agendas for worldwide cooperation in these domains.

---

[18] http://www.e-accessibility.org.
[19] http://www.e-accessibility.gr.
[20] http://www.springerlink.com/link.asp?id=e31p27x51v0mt60w.
[21] http://interlink.ics.forth.gr.

Advisory Board

Ray Adams, Middlesex University, United Kingdom
Christian Bühler, Research Institute Technology and Disability (FTB), Germany
Noëlle Carbonell, Henri Poincaré University, LORIA, France
P. John Clarkson, University of Cambridge, United Kingdom
Pier Luigi Emiliani, National Research Council, Italy
Michael C. Fairhurst, University of Kent, United Kingdom
Vicki L. Hanson, IBM T. J. Watson Research Center, United States
Julie A. Jacko, University of Minnesota, United States
Simeon Keates, ITA Software, Inc., United States
Aaron Marcus, Aaron Marcus and Associates, Inc., United States
Klaus Miesenberger, University of Linz, Austria
Michael Pieper, Fraunhofer, Germany
Jenny Preece, University of Maryland, United States
Pei-Luen Patrick Rau, Tsinghua University, China
Dominique Scapin, National Institute for Research in Computer Science and Control (INRIA), France
Andrew Sears, University of Maryland, Baltimore County, United States
Ben Shneiderman, University of Maryland, United States
Christian Stary, University of Linz, Austria
Gregg C. Vanderheiden, University of Wisconsin–Madison, United States
Gerhard Weber, Technical University of Dresden, Germany
John Wilson, University of Nottingham, United Kingdom

Contributors

Ilia Adami
Institute of Computer Science
Foundation for Research and
  Technology–Hellas
Heraklion, Crete, Greece

Ray Adams
Collaborative International Research
  Centre for Universal Access
School of Engineering & Information
  Sciences
Middlesex University
London, United Kingdom

Elisabeth André
Department of Computer
  Science
University of Augsburg
Augsburg, Germany

Margherita Antona
Institute of Computer Science
Foundation for Research and
  Technology–Hellas
Heraklion, Crete, Greece

Antonis Argyros
Institute of Computer Science
Foundation for Research and
  Technology–Hellas
Heraklion, Crete, Greece

Chieko Asakawa
Accessibility Research
Tokyo Research Laboratory
Kanagawa, Japan

Mahima Ashok
Wallace H. Coulter Department of
  Biomedical Engineering
Georgia Institute of Technology and
  Emory University School of Medicine
Atlanta, Georgia

Haris Baltzakis
Institute of Computer Science
Foundation for Research and
  Technology–Hellas
Heraklion, Crete, Greece

Richard Bates
School of Computing
De Montfort University
Leicester, United Kingdom

Evangelos Bekiaris
Hellenic Institute of Transport
Center for Research and
  Technology–Hellas
Thermi, Thessaloniki, Greece

Nigel Bevan
Professional Usability Services
London, United Kingdom

Marie-Pierre Bonin
Cyberpsychology Laboratory
University of Quebec in Outaouais
Gatineau, Québec, Canada

Stéphane Bouchard
Cyberpsychology Laboratory
University of Quebec in Outaouais
Gatineau, Québec, Canada

Themistoklis Bourdenas
Institute of Computer Science
Foundation for Research and
  Technology–Hellas
Heraklion, Crete, Greece

Christian Bühler
Research Institute Technology
  and Disability
Technical University of Dortmund
Dortmund, Germany

Noëlle Carbonell
Lorraine Laboratory of Computer
Science Research and
  Applications (LORIA)
Henri Poincare University
Nancy, France

Tiziana Catarci
Department of Computer and
  System Sciences
Sapienza University of Rome
Rome, Italy

Sylvain Chartier
School of Psychology
University of Ottawa
Ottawa, Ontario, Canada

P. John Clarkson
Department of Engineering
University of Cambridge
Cambridge, United Kingdom

Michael Donegan
The ACE Centre
Oxford, United Kingdom

Constantina Doulgeraki
Institute of Computer Science
Foundation for Research and
 Technology–Hellas
Heraklion, Crete, Greece

Yael Dubinsky
Department of Computer and
 System Sciences
Sapienza University of Rome
Rome, Italy

Pier Luigi Emiliani
National Research Council
"Nello Carrara" Institute of Applied Physics
Sesto Fiorentino, Italy

Jan Engelen
Department of Electrical Engineering
Catholic University of Leuven
Leuven, Belgium

Grigori Evreinov
Department of Computer Sciences
University of Tampere
Tampere, Finland

Michael C. Fairhurst
Department of Electronics
University of Kent
Canterbury, United Kingdom

Jinjuan Feng
Computer and Information Sciences
 Department
Towson University
Towson, Maryland

Alois Ferscha
Institute of Pervasive Computing
University of Linz
Linz, Austria

Mirko Fetter
Faculty of Media
Bauhaus-University Weimar
Weimar, Germany

Silvia Gabrielli
Fondazione Bruno Kessler
Trento, Italy

Evangelia Gaitanidou
Hellenic Institute of Transport
Center for Research and
 Technology–Hellas
Thermi, Greece

Yannis Georgalis
Institute of Computer Science
Foundation for Research and
 Technology–Hellas
Heraklion, Crete, Greece

Jennifer George
SAE Institute
London, United Kingdom

Paul Gnanayutham
Department of Computing
University of Portsmouth
Portsmouth, United Kingdom

Dimitris Grammenos
Institute of Computer Science
Foundation for Research and
 Technology–Hellas
Heraklion, Crete, Greece

Tom Gross
Faculty of Media
Bauhaus-University Weimar
Weimar, Germany

Gopal Gupta
Department of Computer Science
University of Texas at Dallas
Richardson, Texas

Vicki L. Hanson
Accessibility Research
IBM T. J. Watson Research Center
Hawthorne, New York

Simon Harper
School of Computer Science
University of Manchester
Manchester, United Kingdom

Matt Huenerfauth
Department of Computer Science
Queens College–City University
 of New York
Flushing, New York

Charles Hughes
Institute for Simulation and Training
University of Central Florida
Orlando, Florida

Darin Hughes
Institute for Simulation and Training
University of Central Florida
Orlando, Florida

Julie A. Jacko
The Institute for Health Informatics
Schools of Nursing and Public
 Health
University of Minnesota
Minneapolis, Minnesota

Gunnar Jansson
Department of Psychology
Uppsala University
Uppsala, Sweden

Kristiina Jokinen
Department of General
 Linguistics
University of Helsinki
Helsinki, Finland

Eija Kaasinen
Human Technology Interaction
VTT Technology Research Center
Tampere, Finland

Anne Kaikkonen
Nokia Corporation
Espoo, Finland

Arthur I. Karshmer
College of Professional Studies
University of San Francisco
San Francisco, California

Simeon Keates
ITA Software, Inc.
Cambridge, Massachusetts

John D. Kemp
Powers Pyles Sutter & Verville, P.C.
Washington, D.C.

Erkki Kemppainen
National Research and Development
Centre for Welfare and Health
Helsinki, Finland

Pekka Ketola
Nokia Corporation
Tampere, Finland

Stephen Kimani
Commonwealth Scientific and
 Industrial Research Organisation
Hobart, Tasmania, Australia

Erin Kinzel
Wallace H. Coulter Department of
Biomedical Engineering
Georgia Institute of Technology
and Emory University
School of Medicine
Atlanta, Georgia

Georgios Kouroupetroglou
Department of Informatics and
 Telecommunications
University of Athens
Athens, Greece

Sri Kurniawan
Department of Computer Engineering
University of California, Santa Cruz
Santa Cruz, California

Barbara Leporini
Institute of Information Science and
 Technologies
National Research Council
Pisa, Italy

Clayton Lewis
Department of Computer Science
University of Colorado
Boulder, Colorado

Päivi Majaranta
Department of Computer Sciences
University of Tampere
Tampere, Finland

Aaron Marcus
Aaron Marcus and Associates, Inc.
Berkeley, California

Mark T. Maybury
The MITRE Corporation
Bedford, Massachusetts

Klaus Miesenberger
Institute of Integrated Study
University of Linz
Linz, Austria

Alexandros Mourouzis
Hellenic Institute of Transport
Center for Research and
 Technology–Hellas
Thermi, Thessaloniki, Greece

Michael A. Nees
School of Psychology
Georgia Institute of Technology
Atlanta, Georgia

Stavroula Ntoa
Institute of Computer Science
Foundation for Research and
 Technology–Hellas
Heraklion, Crete, Greece

Maria Panou
Hellenic Institute of Transport
Center for Research and
 Technology–Hellas
Thermi, Greece

Nikolaos Partarakis
Institute of Computer Science
Foundation for Research and
 Technology–Hellas
Heraklion, Crete, Greece

Fabio Paternò
Institute of Information Science
 and Technologies
National Research Council
Pisa, Italy

Helen Petrie
Department of Computer Science
University of York
Heslington, United Kingdom

Ulrike Pfeil
Centre for Human Computer
 Interaction Design
City University London
London, United Kingdom

Enrico Pontelli
Department of Computer
Science
New Mexico State University
Las Cruces, New Mexico

Gilles Privat
Orange Labs
Meylan, France

Roope Raisamo
Department of Computer Sciences
University of Tampere
Tampere, Finland

Pei-Luen Patrick Rau
Department of Industrial Engineering
Institute of Human Factors and
 Ergonomics
Tsinghua University
Beijing, People's Republic of China

Matthias Rehm
Department of Computer Science
University of Augsburg
Augsburg, Germany

Patrice Renaud
Cyberpsychology Laboratory
University of Quebec in Outaouais
Gatineau, Québec, Canada

John T. Richards
IBM T. J. Watson Research Center
Hawthorne, New York

Brigitte Ringbauer
Fraunhofer Institute for Industrial
 Engineering
Stuttgart, Germany

Carmen Santoro
Institute of Information Science
 and Technologies
National Research Council
Pisa, Italy

Anthony Savidis
Institute of Computer Science
Foundation for Research and
 Technology–Hellas
Heraklion, Crete, Greece

Andrew Sears
Interactive Systems Research Center
University of Maryland, Baltimore County
Baltimore, Maryland

Randall Shumaker
Institute for Simulation and Training
University of Central Florida
Orlando, Florida

Eileen Smith
Institute for Simulation and Training
University of Central Florida
Orlando, Florida

Christian Stary
Department of Business
 Information Systems
University of Linz
Linz, Austria

Constantine Stephanidis
Institute of Computer Science
Foundation for Research and
 Technology–Hellas
Heraklion, Crete, Greece

Norbert A. Streitz
AMBIENTE–Smart Environments
 of the Future
Fraunhofer Integrated Publication and
 Information Systems Institute
Darmstadt, Germany

Shari Trewin
IBM T. J. Watson Research Center
Hawthorne, New York

Gregg C. Vanderheiden
Trace Research & Development
 Center
University of Wisconsin–Madison
Madison, Wisconsin

Thorsten Völkel
Department of Computer
 Science
Technical University of Dresden
Dresden, Germany

Bruce N. Walker
School of Psychology
Georgia Institute of Technology
Atlanta, Georgia

Annalu Waller
School of Computing
University of Dundee
Dundee, United Kingdom

Sam Waller
Department of Engineering
University of Cambridge
Cambridge, United Kingdom

Gerhard Weber
Department of Computer
 Science
Technical University of Dresden
Dresden, Germany

Hajime Yamada
Faculty of Economics
Toyo University
Tokyo, Japan

Xenophon Zabulis
Institute of Computer Science
Foundation for Research and
 Technology–Hellas
Heraklion, Crete, Greece

Panayiotis Zaphiris
Centre for Human Computer Interaction
 Design
City University London
London, United Kingdom

Introduction to Universal Access

1

Universal Access and Design for All in the Evolving Information Society

Constantine Stephanidis

1.1 Introduction

The Universal Access Handbook aims at advancing the state of the art in universal access by providing a comprehensive and multidisciplinary overview of the field and unfolding its various dimensions. After some years from its first steps (Stephanidis et al., 1998, 1999), universal access is a continuously growing and dynamically evolving field that accompanies and fosters the evolution of the information society in its current and future advancement and develops and extends its methods, techniques, and tools accordingly. Within a decade, universal access has already scaled up to obtain international recognition, specific research agendas,[1] technical scientific and policy forums and networks,[2] international conferences,[3] and archival journals,[4] while the need is also rapidly emerging to further consolidate the field by establishing technology transfer as well as education and training practices. In such a context, this chapter offers some reflection on universal access as a field of inquiry by providing an overview of progress and achievements so far and discussing

current status and perspectives. This chapter also offers an interpretative key to this handbook by guiding the reader through its 9 parts and 61 chapters in the light of the main dimensions and implications of universal access.

1.2 The Field of Universal Access

The origins of universal access are to be identified in approaches to accessibility mainly targeted toward providing access to computer-based applications by users with disabilities, as well as in human-centered approaches to human-computer interaction (Stephanidis and Emiliani, 1999). The main aim is to prevent the exclusion of users from the information society while at the same time increasing the quality and usability of products and services.

Transcending the traditional view of accessibility and usability, universal access embraces theoretical, methodological, and empirical research of both a technological and nontechnological nature that addresses accessibility, usability, and, ultimately, acceptability of information society technologies (IST) by anyone, anywhere, at any time, and through any media and device (Stephanidis, 2001a).

Universal access puts forward a novel conception of the information society (Stephanidis et al., 1998, 1999) founded on a novel way of addressing the relationship between *techne* and *politeia*, where, on the one hand, technological development highly affects the life of all citizens, and, on the other hand, there is an ethical but also business-driven methodological requirement of informing technological evolution to deeply human-centered

---

[1] See Stephanidis et al. (1998, 1999).

[2] International Scientific Forum "Towards an Information Society for All"—ISF-IS4ALL, 1997–2000, http://www.ui4all.gr/isf_is4all; ERCIM WG UI4ALL, 1995–2006, http://www.ui4all.gr; Thematic Network "Information Society for ALL"—IS4ALL (IST-IST-1999-14101), 2000–2003; European Design for All eAccessibility Network—EDeAN, 2002–present, http://www.edean.org.

[3] Universal Access in Human-Computer Interaction (UAHCI), http://www.hcii2009.org/thematic_view.php?thematic_id=5.

[4] International journal *Universal Access in the Information Society* (UAIS), http://www.springeronline.com/journal/10209/about.

principles. In these respects, the connotation, theoretical underpinnings, and practical results of universal access are informed by, but are also intended to inform, the evolution of the information society as a human, social, and technological construct. Universal access can therefore be viewed as a social-shaping approach in terms of philosophy of technology (MacKenzie and Wajcman, 1985), equally distant from the determinism of both technological enthusiasm and technological nightmare views. In this perspective, it is fundamental that different routes are available in the design of individual artifacts and systems and in the direction or trajectory of innovation programs, potentially leading to different technological outcomes. Significantly, these choices could have differing implications for society and for particular social groups (Williams and Edge, 1996).

Thus, while universal access is targeted to provide a technological substratum for eInclusion in the information society, it also recognizes that "the development of the information society is not likely to be characterized by a linear technological progression, but rather through the often competing forces of innovation, competitive advantage, human agency and social resistance" (Loader, 1998), and that " 'inclusion' must be a process which is the result of the 'human agency' of the many diverse individuals and cultural or national groups who should help shape and determine, and not merely 'access,' technological outcomes" (Henwood et al., 2000).

A direct consequence is the multidisciplinary nature of universal access, evident from the beginning of the field in its conscious effort to ensure a broad scope of research and development activities empowered with new concepts, tools, and techniques from diverse and in some cases dispersed scientific disciplines, technological strands, and socioeconomic and policy perspectives. At the scientific level, this amounts to a need for establishing cross-discipline collaborative views based on synergies among relevant disciplines to bring about a new conceptualization of computer-mediated human activities within the information society (Stephanidis, 2001a).

Another important aspect of universal access is its human-centeredness. In the context of universal access, the study of human characteristics and requirements in relation to the use of IST is of fundamental importance, thus necessitating the contribution of all related scientific disciplines. Universal access goes well beyond current approaches stating the centrality of the human element in the design and development process (Norman and Draper, 1986), as it introduces a new and challenging dimension—the consideration and valorization of human diversity. In the information society and for the vast majority of its applications (e.g., World Wide Web services), the set of users is unknowable (Olsen, 1999). Therefore, in universal access, the consideration of human diversity becomes a *conditio sine qua non*, and the traditional precept of "knowing the users" becomes "knowing diverse user groups and individual users."

The role of technology is equally critical in universal access, as technology is the fundamental provider of the required tools through which humans interact with information artifacts. All technological advances in computing platforms and environments, as well as advances in computing that give rise to new interaction possibilities (in terms of both interaction techniques and domains of application), are potentially relevant to universal access. However, universal access seeks to transcend specific technological manifestations, as the nature of interaction changes dramatically as time goes by and new technologies and trends emerge continuously in the information society. This is evident from the history of computing, which has started with command-line interfaces and has gone through many evolutions, including graphical user interfaces, mobile computing, virtual reality, ubiquitous computing, and ambient intelligence. Therefore, universal access needs to be prepared for new evolutions by elaborating innovative, more fundamental approaches to interaction.

Universal access has a focus on design, as it entails a forward-looking proactive attitude toward shaping new generations of technology rather than short- or medium-term interventions on the present technological and market situation (Stephanidis et al., 1998, 1999). Therefore, innovation in design is invested with a central role in terms of methodological frameworks, processes, techniques, tools, and outcomes. In the context of universal access, *design for all* in the information society has been defined as a general framework catering for conscious and systematic efforts to proactively apply principles, methods, and tools to develop IST products and services that are accessible and usable by all citizens, thus avoiding the need for *a posteriori* adaptations, or specialized design (Stephanidis et al., 1998). Design for all, or universal design, is well known in several engineering disciplines, such as, for example, civil engineering and architecture, with many applications in interior design, building, and road construction. In the context of universal access, design for all either subsumes, or is a synonym of, terms such as *accessible design, inclusive design, barrier-free design, universal design,* and so on, each highlighting different aspects of the concept. Through the years, the concept of design for all has assumed various main connotations:

- Design of interactive products, services, and applications that are suitable for most of the potential users without any modifications. Related efforts mainly aim to formulate accessibility guidelines and standards in the context of international collaborative initiatives.
- Design of products that have standardized interfaces capable of being accessed by specialized user-interaction devices (e.g., Zimmermann et al., 2002).
- Design of products that are easily adaptable to different users by incorporating adaptable or customizable user interfaces (Stephanidis, 2001b).

This last approach fosters a conscious and systematic effort to proactively apply principles and methods and employ appropriate tools to develop interactive products and services that are accessible and usable by all citizens in the information society, thus avoiding the need for *a posteriori* adaptations or specialized design. This entails an effort to build access features into a product starting from its conception and continuing throughout the entire development life cycle.

Finally, while efforts toward technological developments are clearly necessary, they do not constitute a sufficient condition for leading to an information society for all citizens. There are additional requirements for nontechnological measures to assess efficacy and ensure adoption, acceptance, and diffusion of technologies. Socioeconomic and policy issues are relevant to the extent to which they address research and development planning, industrial policy and innovation, assessment of the envisioned products and services, security and privacy, cost factors, diffusion and adoption patterns, standards, legislation, and technology transfer.

The aim of this handbook is to reflect to the largest possible extent the multidisciplinary nature of universal access and at the same time provide a structured path toward unfolding the role and contribution of different research, development, and policy areas in the context of universal access.

1.3 Universal Access Today: An Overview

This section offers an overview of the progress made in the field of universal access as it consolidates itself after some years of intensive efforts and increasing expansion and recognition. The underlying intent is to systematically address, as comprehensively as possible, the various dimensions of universal access, as they hold at present and are reflected in this handbook, and to point out main achievements and prospects.

1.3.1 Theoretical Background

As previously mentioned, universal access emerged through the progressive elaboration of diverse and complementary approaches to the accessibility of interactive applications and services (Stephanidis and Emiliani, 1999). Traditional efforts to provide accessibility for users with disabilities were based on the product-level- and environment-level adaptation of applications and services, originally developed for able-bodied users. These approaches have given rise to several methods, including techniques for the configuration of input/output at the level of the user interface and the provision of assistive technologies. Popular assistive technologies include screen readers and Braille displays for blind users, screen magnifiers for users with low vision, alternative input and output devices for motor-impaired users (e.g., adapted keyboards, mouse emulators, joysticks, and binary switches), specialized browsers, and text prediction systems. Despite progress, assistive technologies and dedicated design approaches have been criticized for their essentially reactive nature (Stephanidis and Emiliani, 1999). Therefore, the need for more systematic and proactive approaches to the provision of accessibility has emerged, leading to the concepts of universal access and design for all. In this context, accessibility refers to the extent to which the use of an application or service is affected by the user's particular functional limitations

or abilities (permanent or temporary), as well as by other contextual factors (e.g., characteristics of the environment). This implies that for each user task of an application or service (and taking into account specific functional limitations and abilities and other relevant contextual factors), there is a sequence of input actions and associated feedback via accessible input/output devices that lead to successful task accomplishment (Savidis and Stephanidis, 2004). In this light, universal access also provides a clear answer to the debate on the relationships between accessibility and usability (see, e.g., Petrie and Kheir, 2007): accessibility becomes a fundamental prerequisite of usability, intended as the capability of all supported paths toward task accomplishment to "maximally fit" individual users' needs and requirements in the particular context and situation of use (Savidis and Stephanidis, 2004), since there may not be optimal interaction if there is no possibility of interaction in the first place.

Along these lines, the historical roots of universal access are addressed in the remaining two chapters of Part I of this handbook from both a European and American perspective. The transition from accessibility to universal access and design for all is illustrated in Chapter 2, "Perspectives on Accessibility: From Assistive Technologies to Universal Access and Design for All," by P. L. Emiliani, through a series of landmark research projects, funded by the European Commission, that have demonstrated the feasibility of the design for all approach in the information and communication technologies (ICT) field. Over the years, suitable technical approaches have been elaborated to design and implement universally accessible interfaces, services, and applications. Looking toward ongoing developments and the emergence of ambient intelligence environments while asking for a more general application of the design for all approach also favors its implementation by making available in the environment the necessary interaction means and intelligence.

New technological achievements will contribute to the blurring of the lines between assistive technologies and design for all in next-generation interfaces. Current dramatic changes in information technologies and human interfaces are creating both new challenges and new opportunities for developing mainstream products that are accessible to and usable by people with disabilities or functional impairments, but in principle also all users. These issues are addressed in Chapter 3, "Accessible and Usable Design of Information and Communication Technologies," by G. C. Vanderheiden.

1.3.2 Diversity in the User Population

In the context of the emerging distributed and communication-intensive information society, users are no longer the computer-literate, skilled, and able-bodied workers driven by performance-oriented motives. Additionally, users no longer constitute a homogeneous mass of actors with standard abilities, similar interests, and common preferences regarding information access and use. Instead, it becomes compelling that designers' conception of users accommodate all potential citizens, including residential users, as well as those with situational or permanent disabilities, but also people of different ages and with different experiences and cultural and educational backgrounds. Therefore, at the heart of universal access lies a deeply human-centered focus on human diversity in all its aspects related with access to and use of ICT. Main efforts in this direction are concerned with the identification and study of various nontraditional target user groups (e.g., disabled, elderly, novice users, etc.), as well as of their requirements for interaction, and of appropriate methods, tools, and interactive devices and techniques to address their needs. Much experimental work has been conducted in recent years to elaborate design guidelines for diverse user groups. Work in understanding human characteristics and needs in relation to interaction has also been facilitated by the functional approach of the International Classification of Functioning, Disability and Health (ICF) (World Health Organization, 2001), where the term *disability* is used to denote a multidimensional phenomenon resulting from the interaction between people and their physical and social environment. This allows grouping and analysis of limitations that are not only due to impairments but also, for example, to environmental reasons.

Part II of this handbook provides an overview of the main issues related to users' diversity in the context of universal access. A general introduction to this topic is provided in Chapter 4, "Dimensions of User Diversity," by M. Ashok and J. A. Jacko. Several dimensions of diversity are discussed, including disabilities and impairments, skill levels, cognitive factors, social issues, cultural and linguistic issues, age, and gender. Each diversity dimension is analyzed with a focus on how differences translate into variations in the use of technology, along with suggestions for how designers can be inclusive in their design by accounting for such differences.

In the subsequent chapters of Part II, various dimensions of user diversity are analyzed in more depth. An overview of the different types of motor impairments that may affect access to interactive technologies, their prevalence across the general population, and their interrelations with the aging process are provided in Chapter 5, "Motor Impairments and Universal Access," by S. Keates. This chapter also overviews common software and hardware solutions for improving access for motor-impaired users and analyzes in detail the effects of motor impairments on key pressing and cursor control.

Various sensory impairments also bring about diverse interaction requirements. Chapter 6, "Sensory Impairments," by E. Kinzel and J. A. Jacko, addresses the general structure and function of the primary human sensory systems—vision, hearing, and touch—that are vital to interaction with technology, as well as some examples of common sensory-specific impairments that may affect interaction. Some of the recent technological developments targeted to enhance the sensory experience of users with impairments are also discussed.

The diversity and complexity of cognitive differences, addressed in Chapter 7, "Cognitive Disabilities," by C. Lewis, also highly affect interaction, and many barriers for access arise for people with cognitive disabilities. Designing technology to

reduce these barriers involves the combination of appropriate interface features, attention to configurability, and user testing.

The current demographic phenomena that lead to an aging society and the degenerative ability changes caused by age determine fundamental differences in the way older and younger persons use ICT. Chapter 8, "Age-Related Differences in the Interface Design Process," by S. Kurniawan, focuses on understanding these changes and accommodating them through aging-sensitive design to mediate differences and considerably improve the use of computers, the Internet, and mobile devices by older persons.

Finally, the linguistic and cultural dimensions of diversity acquire progressive importance as the information society becomes increasingly global. Chapter 9, "International and Intercultural User Interfaces," by A. Marcus and P-L. P. Rau, addresses these issues by proposing an approach to global user-interface design consisting of partially universal and partially local solutions to the design of metaphors, mental models, navigation, interaction, and appearance. By managing the user's experience of familiar structures and processes, user interface design can obtain more usable, useful, and appealing results for an international audience.

1.3.3 Technologies for Universal Access

In the information society, diversity concerns not only users, but also interaction environments and technologies, which are continuously developing and diversifying. The diffusion of the Internet and the proliferation of advanced interaction technologies (e.g., mobile devices, network-attachable equipment, virtual reality, agents, etc.) signify that many applications and services are no longer limited to the visual desktop but span over new realities and interaction environments. Overall, a wide variety of technological paradigms play a significant role in universal access either by providing new interaction platforms or by contributing at various levels to ensure and widen access. Part III of this handbook seeks to offer an overview of these issues.

The World Wide Web and its technologies are certainly a fundamental component of the information society. Chapter 10, "Accessing the Web," by V. L. Hanson, J. T. Richards, S. Harper, and S. Trewin, discusses various challenges and solutions to make the web accessible to all. In the context of the information society, the World Wide Web offers much for those who are able to access its content, but at the same time access is limited by serious barriers due to limitations of visual, motor, language, or cognitive abilities. Current approaches to web accessibility, and in particular guidelines for web and browsers' development, as well as current opportunities and obstacles toward further progress in this domain, are also reviewed.

Another very important and rapidly progressing technological advance is that of mobile computing. Mobile devices acquired an increasingly important role in everyday life, both as dedicated tools, such as media players, and multipurpose devices, such as personal digital assistants and mobile phones. Chapter 11, "Handheld Devices and Mobile Phones," by A. Kaikkonen,

E. Kaasinen, and P. Ketola, describes the specific characteristics of mobile contexts of use, mobile devices, mobile services, and mobile user interfaces and how those characteristics are affected by the demand for universal access. Guidelines for designing and evaluating mobile devices and services for universal access are also offered.

As the information society extends from the real to the virtual world, Chapter 12, "Virtual Reality," by D. Hughes, E. Smith, R. Shumaker, and C. Hughes, describes the basic concepts of virtual reality in terms of specific technologies, physical infrastructures, and applications for accessibility, assessment, and therapy. Included within the discussion are a range of virtual reality concepts, such as augmented reality and mixed reality. Additionally, auditory displays, haptics, and tracking technologies for virtual reality applications are discussed.

Security and integrity of data are of paramount importance in the context of universal access. Chapter 13, "Biometrics and Universal Access," by M. C. Fairhurst, focuses on biometric technologies as a means to verify user access rights. Some of the basic principles underlying the adoption of biometrics as a means of establishing or verifying personal identity are outlined, and approaches are discussed to enhance the reliability and flexibility of biometrics and to ensure their effective implementation.

Agents constitute another enabling technology of universal access. Chapter 14, "Interface Agents: Potential Benefits and Challenges for Universal Access," by E. André and M. Rehm, discusses the use of interface agents to enable a large variety of users to gain access to information technology. As mediators between users and a computer system, agents seem to be ideally suited to adapt to the different backgrounds of heterogeneous user groups. The technological requirements to be met for interface agents to satisfy the requirements for universal access are investigated.

1.3.4 Development Life Cycle of User Interfaces

The notion of universal access reflects the concept of an information society in which potentially anyone (i.e., any user) interacts with computing machines, at any time and any place (i.e., in any context of use) and for virtually anything (i.e., for any task). To reach a successful and cost-effective realization of this vision, it is critical to ensure that appropriate interface development methods and techniques are available. Traditional development processes, targeted toward the elusive "average case," are clearly inappropriate for the purposes of addressing the new demands for user and usage context diversity and for ensuring accessible and high-quality interactions (Stephanidis, 2001b). Under this perspective, universal access affects the entire development life cycle of interactive applications and services. Work in this area has therefore concentrated on design and development methodologies and frameworks that integrate and support design for all approaches, support user interface adaptation, and integrate the consideration of diversity throughout all development phases.

Part IV of this handbook unfolds several aspects of user interface development in a universal access perspective, including user requirements analysis, user interface design, software development requirements, and accessibility and usability evaluation.

Various user requirement analysis methods and techniques present both advantages and potential difficulties in the optimal involvement and usage for diverse user groups, including users with various types of disabilities or in different age ranges. Chapter 15, "User Requirements Elicitation for Universal Access," by M. Antona, S. Ntoa, I. Adami, and C. Stephanidis, provides an overview.

The requirements for designing diversity for end-users and contexts of use, which implies making alternative design decisions at various levels of the interaction design, inherently leading to diversity in the final design outcomes, are discussed in Chapter 16, "Unified Design for User Interface Adaptation," by A. Savidis and C. Stephanidis. To this end, traditional design methods are suboptimal, since they cannot accommodate for diversity. Therefore, there is a need for a systematic process in which alternative design decisions for different design parameters may be supported. Unified user interfaces constitute a theoretical platform for universally accessible interactions, characterized by the capability to self-adapt at run-time, according to the requirements of the individual user and the particular context of use. The unified interface design method is a process-oriented design method that enables the organization of diversity-based designs and encompasses a variety of techniques such as task analysis, abstract design, design polymorphism, and design rationale.

An instantiation of unified user interface design in the area of computer games is presented in Chapter 17, "Designing Universally Accessible Games," by D. Grammenos, A. Savidis, and C. Stephanidis, that discusses its adaptation to the specific domain and its practical application in two design cases.

The last decade has also witnessed the elaboration of a corpus of key development requirements for building universally accessible interactions. Such requirements have been consolidated from real practice in the course of six medium-to-large-scale research projects within a 10-year timeframe and are discussed in Chapter 18, "Software Requirements for Inclusive User Interfaces," by A. Savidis and C. Stephanidis.

In parallel, models for inclusive design processes, methods, and tools that can stimulate and manage the implementation of inclusive design and evaluation methods to support informed decision-making for inclusive design have also been elaborated, and are reported in Chapter 19, "Tools for Inclusive Design," by S. Waller and P. J. Clarkson.

Finally, the last chapter in this part, Chapter 20, "The Evaluation of Accessibility, Usability, and User Experience," by H. Petrie and N. Bevan, introduces a range of accessibility, usability, and user experience evaluation methods that assist developers in the creation of interactive electronic products, services, and environments (e-systems) that are both easy and pleasant to use for a broad target audience, including people with disabilities and older people. For each method, appropriate use, strengths, and weaknesses are outlined.

1.3.5 User Interface Development: Architectures, Components, and Support Tools

Another challenge of universal access concerns the development of methods and tools capable of making it not only technically feasible but also economically viable in the long-term (Stephanidis et al., 1998). In the past, the availability of techniques and tools was an indication of maturity of a sector and a critical factor for technological diffusion. As an example, graphical user interfaces became popular once tools for constructing them became available, either as libraries of reusable elements (e.g., toolkits), or as higher-level systems (e.g., user interface builders and user interface management systems). As development methods and techniques for addressing diversity are anticipated to involve complex processes and have a higher entrance barrier with respect to more traditional means, the provision of appropriate tools can help overcome some of the difficulties that hinder the wider adoption of design methods and techniques appropriate for universal access, both in terms of quality and cost, by making the complex development process less resource-demanding and better at supporting reuse. To support a universal access development life cycle as sketched in Part IV of this handbook, purposeful software architectures, user interface toolkits, representation languages, and support tools have been elaborated, tested, and applied, with the underlying objective to facilitate the adoption and application of universal access approaches and improve ease of development and cost justification. Main achievements in this area are addressed in Part V of this handbook.

A software architecture that supports run-time self-adaptation behavior in the framework of unified user interfaces is presented in Chapter 21, "A Unified Software Architecture for User Interface Adaptation," by A. Savidis and C. Stephanidis. The software engineering of automatic interface adaptability entails the storage and processing of user and usage context profiles, the design and implementation of alternative interface components, and run-time decision making to choose on the fly the most appropriate alternative interface component given a particular user and context profile.

Chapter 22, "A Decision-Making Specification Language for User Interface Adaptation," by A. Savidis and C. Stephanidis, focuses on the decision-making process according to diverse profiles of individual end-users and usage contexts in automatic user interface adaptation. A verifiable language is proposed that is particularly suited for the specification of adaptation-oriented decision-making logic, while also being easily deployable and usable by interface designers.

A novel approach to the development of inclusive web-based interfaces (web content) capable of adapting to multiple and significantly different profiles of users and contexts of use is introduced in Chapter 23, "Methods and Tools for the Development

of Unified Web-Based User Interfaces," by C. Doulgeraki, N. Partarakis, A. Mourouzis, and C. Stephanidis. The unified web interfaces method, building on the unified user interfaces development method, is proposed as an alternative approach to the design and development of web-based applications. An advanced toolkit has also been developed as a means to facilitate web developers in producing adaptable web interfaces following the proposed method.

User modeling is another important area that provides the foundations for design for all, personalization, and adaptation of user interfaces. Chapter 24, "User Modeling: A Universal Access Perspective," by R. Adams, outlines various approaches, implications, and practical applications of user modeling along with the basis for a toolkit of concepts, methods, and technologies that support them. These are discussed in relation to unified user interfaces as an emerging design methodology for universal access.

Model-based approaches are considered very promising in the context of universal access, as they can potentially reduce the complexity of design for all tasks while also facilitating development. As model-based tools capture design inputs and specifications and support their refinement to the implementation level, they can be used along the various phases of development and allow the specification of highly adaptable user interfaces, which are considered key for universal access. Chapter 25, "Model-Based Tools: A User-Centered Design for All Approach," by C. Stary, focuses on model-based tools that support design for all and take into account both user tasks and needs, as well as software engineering principles.

Design for all knowledge also needs to be represented and codified. Chapter 26, "Markup Languages in Human-Computer Interaction," by F. Paternó and C. Santoro, discusses the importance of formalizing interaction-related knowledge in such a way that it can be easily specified and processed and proposes markup languages as an appropriate instrument in this respect. Chapter 26 analyzes how various markup languages are used to represent the relevant knowledge and how such information can be exploited in the design, user interface generation, evaluation, and run-time support phases.

Finally, Chapter 27, "Abstract Interaction Objects in User Interface Programming Languages," by A. Savidis, presents a subset of a user interface programming language that provides programming facilities for the definition of virtual interaction object classes and the specification of the mapping logic to physically instantiate virtual object classes across different target platforms. These constitute a significant step toward universal access, as virtual interaction objects play the role of abstractions over any particular physical realization or dialogue metaphor, thus facilitating the compact development of user interface adaptations.

1.3.6 Interaction Techniques and Devices

A very wide variety of alternative interaction techniques and devices have significant potential to serve and support diversity in user interaction, and recent advances in several domains are crucial to universal access. On the other hand, it is of paramount importance that diversity and human characteristics are taken into appropriate account while developing new interaction devices and techniques, so that diverse user groups and individual users can be provided with the most appropriate interaction means for each application and task. Part VI of this handbook is dedicated to an overview of recent major progress in interaction techniques and devices relevant for universal access, ranging from traditional assistive technologies to advanced perceptual interfaces.

Chapter 28, "Screen Readers," by C. Asakawa and B. Leporini, introduces screen readers as the main assistive technology used by people with little or no functional vision when interacting with a computer or mobile devices. A classification of screen readers is provided, and basic concepts of screen reading technology are described. Interaction using screen readers is described through practical examples. In particular, an overview of two main features is provided for visually impaired people that should be considered when designing user interfaces: user perception and user interaction.

The importance of text entry techniques and tools in the context of universal access is emphasized in Chapter 29, "Virtual Mouse and Keyboards for Text Entry," by G. Evreinov. In this area, the need emerges for elaborating interaction techniques for text entry that are adaptive to personal cognitive and sensory-motor abilities of the users. Progress in this area implies the design of a wide spectrum of text entry systems and the elaboration of novel algorithms to increase their efficiency. This chapter reviews different solutions and principles that have been proposed and implemented to improve the usability of text entry in human-computer interaction. Examples are the virtual mouse and onscreen keyboards.

Speech-based interaction is also a fundamental component of universal access, as it is one of the most natural forms of communication. Effective speech-based hands-free interaction has significant implications for users with physical disabilities, as well as users interacting in mobile environments. Chapter 30, "Speech Input to Support Universal Access," by J. Feng and A. Sears, focuses on the use of speech as input. Its advantages and limitations are discussed in relation to diverse user groups, including individuals with physical, cognitive, hearing, and language disabilities, as well as children. Guidelines for the design, evaluation, and dissemination of speech-based applications are reported.

Natural interaction is an approach to interface design that attempts to empower different users in various everyday situations by exploiting the strategies they have learned in human-human communication, with the ultimate aim of constructing intelligent and intuitive interfaces that are aware of the context and the user's individual needs. Chapter 31, "Natural Language and Dialogue Interfaces," by K. Jokinen, discusses natural language dialogue interfaces. In this context, the notion of natural interaction refers to the spoken dialogue system's ability to support functionality that the user finds intuitive and easy (i.e., the interface should afford natural interaction).

Nonspeech audio can offer an important design alternative or accompaniment to traditional visual displays and can contribute meeting universal access challenges by providing a means for creating more accessible and usable interfaces and offering an enhanced user experience. Chapter 32, "Auditory Interfaces and Sonification," by M. A. Nees and B. N. Walker, discusses the advantages and appropriate uses of sound in systems and presents a taxonomy of nonspeech auditory displays along with a number of important considerations for auditory interface design.

Haptic is another important nonvisual dimension of interaction. Chapter 33, "Haptic Interaction," by G. Jansson and R. Raisamo, provides an overview of basic issues within the area of haptic interaction. An extensive collection of low-tech, high-tech, and haptic displays is presented, with a special focus on haptic interaction for the visually impaired.

Computer vision has recently been employed as a sensing modality for developing perceptive user interfaces. Chapter 34, "Vision-Based Hand Gesture Recognition for Human-Computer Interaction," by X. Zabulis, H. Baltzakis, and A. Argyros, focuses on the vision-based recognition of hand gestures that are observed and recorded by typical video cameras. It provides an overview of the state of the art in this domain and covers a broad range of related issues ranging from low-level image analysis and feature extraction to higher-level interpretation techniques. Additionally, it presents a specific approach to gesture recognition intended to support natural interaction with autonomous robots that guide visitors in museums and exhibition centers.

Hierarchical scanning is an interaction technique specifically suited to people with motor disabilities, ensuring more rapid interaction and avoiding the time-consuming sequential access to all the interactive interface elements. Hierarchical scanning provides access to all the interactive interface elements of a window based on their place in the hierarchy by dynamically retrieving the window hierarchical structure through the use of binary switches as an alternative to traditional input devices (i.e., a keyboard or mouse). Chapter 35, "Automatic Hierarchical Scanning for Windows Applications," by S. Ntoa, A. Savidis, and C. Stephanidis, presents an advanced scanning method that enables motor-impaired users to work with any application running in Microsoft Windows without the need for further modifications.

Eye tracking is a technique-enabling control of a computer by eye movements alone or combined with other supporting modalities. Traditionally, eye control has been in use by a small group of people with severe motor disabilities, for whom eye control may be the only means of communication. However, recent advances in technology have considerably improved the quality of systems, such that a far broader group of people may now benefit from eye control technology. An overview of this technology is provided in Chapter 36, "Eye Tracking," by P. Majaranta, R. Bates, and M. Donegan, where the basic issues involved in eye control are introduced that consider the benefits and problems of using the eye as a means of control. A summary of key results from user trials is reported to show the potential benefits of eye control technology, with recommendations for its successful application.

Brain-body interfaces can be used for communicating, recreating, and controlling the environment by disabled individuals. Chapter 37, "Brain-Body Interfaces," by P. Gnanayutham and J. George, discusses how brain-body interfaces open up a spectrum of potential technologies particularly appropriate for people with traumatic brain injury and other motor impairments.

Significant advances have also been achieved recently in technologies capable of generating sign language animations and understanding sign language input in the context of information, communication, and software applications accessible to deaf signers. In Chapter 38, "Sign Language in the Interface: Access for Deaf Signers," by M. Huenerfauth and V. L. Hanson, challenges in the processing of sign languages are discussed, which arise from their specific linguistic and spatial characteristics that have no direct counterparts in spoken languages. Important design issues that arise when embedding sign language technologies in accessibility applications are also elaborated upon.

Chapter 39, "Visible Language for Global Mobile Communication: A Case Study of a Design Project in Progress," by A. Marcus, reports on long-term attempts to develop a universal visible language and discusses mobile computing and communication technology as a new platform for the use of such a language. The example is a report on the LoCoS language and its application as a usable, useful, and appealing basis for a mobile phone application that provides communication capabilities for people who do not share a spoken language.

Finally, multimodal interaction intrinsic in ambient intelligence is claimed to offer multiple benefits for promoting universal access. Chapter 40, "Contributions of 'Ambient' Multimodality to Universal Access," by N. Carbonell, presents recent advances in the processing of novel input modalities, such as speech, gestures, gaze, or haptics and synergistic combinations of modalities, all of which are currently viewed as appropriate substitutes for direct manipulation in situations where the use of a keyboard, mouse, and standard screen is awkward or impossible. A software architecture for multimodal user interfaces is proposed that takes into account the recent diversification of modalities and the emergence of context-aware systems distributed in the user's life environment.

1.3.7 Applications

Recent years have witnessed a wide variety of developments that exemplify the adoption of universal access and design for all principles and approaches in diverse new and emerging application domains. These developments demonstrate the centrality of the universal access concept toward technologically supported and enhanced everyday human activities in an inclusive information society. The experience accumulated through such concrete applications demonstrates that universal access is more of a challenge than it is a utopia. Part VII of this handbook represents a collection of some relevant case studies.

Digital libraries are crucial for access to information and knowledge as well as education and work in the information society. Chapter 41, "Vocal Interfaces in Supporting and Enhancing Accessibility in Digital Libraries," by T. Catarci, S. Kimani, Y. Dubinsky, and S. Gabrielli, focuses on the involvement of users, and in particular users with disabilities, in the systems development life cycle of digital libraries to support accessibility. A case study is presented that illustrates the use of vocal interfaces to enhance accessibility in digital libraries due to their usefulness in hands-busy, eyes-busy, mobility-required, and hostile/difficult settings, as well as their appropriateness for people who are blind or visually or physically impaired. The involvement of users in the development process has taken place through an integration of user-centered and agile methods.

Online communities have acquired increasing importance in recent years for various target user groups, and in particular for people with disabilities and the elderly. Chapter 42, "Theories and Methods for Studying Online Communities for People with Disabilities and Older People," by U. Pfeil and P. Zaphiris, highlights some of the key benefits of such applications for the disabled and the elderly and points to a number of weaknesses. The key theoretical foundations of computer-mediated communication are addressed, explaining how those could help in studying online social interaction.

Humans have individual cognitive, perceptual, and physical strengths and weaknesses, and universal access aims to build upon the respective user's strengths and to compensate for the weaknesses. Computer-supported cooperative work can contribute to bridging these domains in universal access for groups and communities, particularly toward bringing together users with heterogeneous cognitive, perceptual, and physical abilities. Chapter 43, "Computer-Supported Cooperative Work," by T. Gross and M. Fetter, provides an overview of the field of computer-supported cooperative work from a universal access perspective.

The emerging need for universal access design and implementation methods in the context of e-learning systems is put forward in Chapter 44, "Developing Inclusive e-Training," by A. Savidis and C. Stephanidis. This chapter reports on a consolidated development experience from the construction of training applications for hand-motor-impaired users and for people with cognitive disabilities. In this context, the primary emphasis is put on design and implementation aspects toward accessibility and usability for the addressed user groups.

Along the same lines, Chapter 45, "Training through Entertainment for Learning Difficulties," by A. Savidis, D. Grammenos, and C. Stephanidis, focuses on real-life training of people with learning difficulties. This is a highly challenging and demanding process that can be effectively improved with the deployment of special-purpose software instruments. The development and evaluation of two games are reported, with the main objective of investigating how playing games, and more generally, providing a pleasant and engaging interactive experience, can have a significant role on improving the training of people with learning difficulties.

Enriched documents (e.g., eBooks) that contain redundant alternative representations of the same information aim at meeting the needs of a heterogeneous range of readers. Chapter 46, "Universal Access to Multimedia Documents," by H. Petrie, G. Weber, and T. Völkel, presents a multimedia reading system that meets the needs of a number of print-disabled reader groups, including blind, partially sighted, deaf, hard of hearing, and dyslexic readers. The development of the system is described from an iterative user-centered design perspective, and a set of attributes is established for user personalization profiles that are needed for adapting content and presentation, as well as for adapting and customizing content, interaction, and navigation.

Augmentative and alternative communication supports interpersonal communication for individuals who are unable to speak. Chapter 47, "Interpersonal Communication," by A. Waller, provides an overview of this domain and introduces recent technological advances as examples of how natural language processing can be used to improve the quality of aided communication.

Public access terminals are also omnipresent in the information society. Chapter 48, "Universal Access in Public Terminals: Information Kiosks and ATMs," by G. Kouroupetroglou, examines current barriers in the use of public access terminals, along with user requirements for the elderly, the disabled (visually, aurally, intellectually, and physically), the temporarily or occasionally disabled, and foreigners. Available technologies (both software and hardware) to alleviate the barriers are also discussed, a range of accessibility strategies and prototype-accessible public terminals is presented, and an overview of available specific accessibility guidelines and standards is provided.

Information and communication technologies play an important role in enhancing and encouraging individual mobility in the physical environment. Chapter 49, "Intelligent Mobility and Transportation for All," by E. Bekiaris, M. Panou, E. Gaitanidou, A, Mourouzis, and B. Ringbauer, addresses major design issues emerging from the need to consider the individual and contextual requirements of a far more heterogeneous target group than in ordinary computing. A brief benchmarking of the issues that various population groups face in getting pre- and on-trip information, as well as in actually traveling via various transportation means, is presented. The involved issues are further highlighted by a series of best practice examples.

Chapter 50, "Electronic Educational Books for Blind Students," by D. Grammenos, A. Savidis, Y. Georgalis, T. Bourdenas, and C. Stephanidis, introduces a novel software platform for developing and interacting with multimodal interactive electronic textbooks that provide user interfaces concurrently accessible by both visually impaired and sighted persons. The platform comprises facilities for the authoring of electronic textbooks and for multimodal interaction with the created electronic textbooks. Key findings are consolidated, elaborating on prominent design issues, design rationale, and respective solutions and highlighting strengths and weaknesses.

Making mathematics accessible is also a significant challenge, due to its two-dimensional, spatial nature and the inherently linear nature of speech and Braille displays. Chapter 51, "Mathematics and Accessibility: A Survey," by E. Pontelli, A. I. Karshmer, and G. Gupta, reviews the state of the art in nonvisual accessibility of mathematics. Various approaches based on Braille codes and on aural rendering of mathematical expressions are discussed.

Cybertherapy and cyberpsychology emerge from the application of virtual reality techniques in psychological and psychiatric therapy. Chapter 52, "Cybertherapy, Cyberpsychology, and the Use of Virtual Reality in Mental Health," by P. Renaud, S. Bouchard, S. Chartier, and M-P. Bonin, presents cybertherapy and cyberpsychology through a universal access perspective. It describes the technologies involved and explains how clinical psychology is gaining from these technological progresses. Empirical data from clinical studies conducted with arachnophobic patients are presented.

1.3.8 Nontechnological Issues

As discussed in Section 1.2, universal access is not only a matter of technology. A large variety of nontechnological issues also affect the wider adoption and diffusion of universal access principles and methods in the mainstream, as the information society does not develop independently from the human society but needs to build upon it, while at the same time continuing to shape it. Nontechnological issues related to universal access practice, legislation, standardization, economics and management, ethical principles, and acceptance are addressed in Part VIII of this handbook.

Several frameworks for analyzing and assessing policies and legislation related to accessibility in Europe, the United States, and Japan, as well as the United Nations, are introduced in Chapter 53, "Policy and Legislation as a Framework of Accessibility," by E. Kemppainen, J. D. Kemp, and H. Yamada. Legislative areas that can promote equal opportunities and eAccessibility include nondiscrimination, ICT, privacy and transparency, product safety, public procurement, and assistive technology.

Standards and guidelines are important aspects of accessible design and take a wide variety of forms that serve diverse functions. Chapter 54, "Standards and Guidelines," by G. C. Vanderheiden, outlines the process for the creation of related standards, as well as their impact on developing products that will be usable by people with a wide range of abilities.

Chapter 55, "eAccessibility Standardization," by J. Engelen, continues the discussion of the standardization topic by sketching formal, ad hoc, company-driven, and informal standardization activities in universal design and assistive technology. Standards are intended in this context as a reference instrument for legislation in these domains.

Design for all management is intended as comprising all activities needed for planning, organizing, leading, motivating, coordinating, and controlling the processes associated with design for all. Chapter 56, "Management of Design for All," by

C. Bühler, reflects on management approaches from a business perspective. Particular focus centers on the product level, but the brand and company level need close attention as well. Key business motivations are competitiveness in the global market, shareholder value, and sometimes corporate social responsibility. From a more general perspective, political economics and social considerations are key elements.

Chapter 57, "Security and Privacy for Universal Access," by M. T. Maybury, provides an overview of the security and privacy requirements for universal access. These include confidentiality, integrity, and availability for all users, including those with disabilities and those in protected categories such as children or the elderly. The need for privacy is also addressed in its potential conflict with the need to represent and analyze detailed user properties and preferences to tailor information and interaction to enable universal access. Some important functional requirements to support universal access are reviewed, along with a discussion of the current international legal environment related to privacy.

Finally, Chapter 58, "Best Practice in Design for All," by K. Miesenberger, discusses best practice in design for all, focusing on the context-sensitive and process-oriented nature of design for all, which is often invisible in the final products. This makes the selection of single examples as demonstrators for best practice a difficult task, and the transfer into other contexts not straightforward. Therefore, best practice is outlined as dependent on specific contexts.

1.3.9 Future Perspectives

As the information society evolves continuously, universal access also evolves and expands to address the needs of new technological environments and contexts of use. As a result of the increasing demand for ubiquitous and continuous access to information and services, IST are anticipated to evolve in the years ahead toward the new computing paradigm of ambient intelligence. Ambient intelligence will have profound consequences on the type, content, and functionality of the emerging products and services, as well as on the way people will interact with them, bringing about multiple new requirements for the development of IST; universal access is critically important in addressing the related challenges. Part IX of this handbook looks into the future of universal access from an ambient intelligence perspective.

Implicit interaction, based on the fundamental concepts of perception and interpretation, is a novel paradigm in an "environment as interface" situation. Chapter 59, "Implicit Interaction," by A. Ferscha, overviews the technologies and approaches that make implicit interaction feasible in the near future and discusses application scenarios and their implications.

An overview of the basic concepts, trends, and perspectives of ambient intelligence is presented in Chapter 60, "Ambient Intelligence," by N. A. Streitz and G. Privat. In light of an overall technological frame of reference, a number of constituent ambient intelligence approaches are presented. Furthermore, alternatives are proposed and discussed that characterize the theoretical

and practical challenges to be addressed in this field, and current trends and perspectives that may help overcome these alternatives are discussed.

Finally, Chapter 61, "Emerging Challenges," by C. Stephanidis, concludes the handbook by summarizing current and future challenges in ambient intelligence that emerge from the various universal access dimensions addressed in this handbook.

1.4 Conclusions

Universal access is a young and dynamic field of inquiry that involves, exploits, and affects the development of a very large part of IST and has so far witnessed significant advances in all its main dimensions, as they are presented in this handbook. Currently, universal access is rapidly progressing toward a higher level of maturity, where the consolidation of achieved results and the systematization of accumulated knowledge are of paramount importance for a variety of purposes.

First, as discussed in Section 1.3.9 and thoroughly elaborated upon in Part IX of this handbook, universal access needs to build upon its strength and further evolve to be able to address new challenges that arise as the information society develops and intelligent interactive environments are gradually being created. Second, stronger links between research and industry in the universal access domain need to be established. On the one hand, it is important to encourage industry to become more receptive to universal access and more involved in its adoption and diffusion; on the other hand, research must provide industry with suitable techniques and tools for addressing real practical problems emerging from the consideration of diversity in mainstream product development.

Another critical impediment to the adoption of universal access and design for all principles and methods in practice is the lack of qualified practitioners who understand diversity in the user target population, in the technology, and in the context of use and are able to integrate the related requirements in the development process. Therefore, the need for better preparing present and future generations of scientists, designers, developers, and stakeholders toward developing a more inclusive information society through both academic education and professional training in accessibility, design for all, and universal access is widely felt. Various initiatives are targeted to meet this objective in Europe (e.g., Weber and Abascal, 2006; Keith, 2008).

This handbook aims to establish an important landmark in universal access and provide a useful tool for addressing these challenges.

References

Henwood, F., Wyatt, S., Miller, N., and Senker, P. (2000). Critical perspectives on technologies, in/equalities and the information society, in *Technology and In/equality: Questioning the Information Society* (S. Wyatt, F. Henwood, N. Miller, and P. Senker, eds.), pp. 1–18. London: Routledge.

Keith, S. (2008). Curriculum development in design for all and the use of learning outcomes. *Electronic Acts of the EDeAN 2008 Conference*, 12–13 June 2008, León, Spain. http://www.ceapat.org/docs/repositorio//es_ES//PonenciasEDeAN/keith_dfaspain2008ed1.pdf.

Loader, B. (ed.) (1998). *Cyberspace Divide: Equality, Agency, and Policy in the Information Society*. New York: Routledge.

MacKenzie, D. and Wajcman, J. (eds.) (1985). *The Social Shaping of Technology: How the Refrigerator Got Its Hum Milton Keynes*. Maidenhead, U.K.: Open University Press.

Norman, D. A. and Draper, S. W. (1986). *User Centered System Design: New Perspectives on Human-Computer Interaction*. Mahwah, NJ: Lawrence Erlbaum Associates.

Olsen, D. (1999). Interacting in chaos. *Interactions* 6: 43–54.

Petrie, H. and Kheir, O. (2007). The relationship between accessibility and usability of websites, in the *Proceedings of CHI '07: ACM Annual Conference on Human Factors in Computing Systems*, pp. 397–406. New York: ACM Press.

Savidis, A. and Stephanidis, C. (2004). Unified user interface design: Designing universally accessible interactions. *Interacting with Computers* 16: 243–270.

Stephanidis, C. (2001a). Editorial. *Universal Access in the Information Society* 1: 1–3.

Stephanidis, C. (ed.). (2001b). *User Interfaces for All: Concepts, Methods, and Tools*. Mahwah, NJ: Lawrence Erlbaum Associates.

Stephanidis, C. and Emiliani, P. L. (1999). Connecting to the information society: A European perspective. *Technology and Disability Journal* 10: 21–44.

Stephanidis, C., Salvendy, G., Akoumianakis, D., Arnold, A., Bevan, N., Dardailler, D., et al. (1999). Toward an information society for all: HCI challenges and R&D recommendations. *International Journal of Human-Computer Interaction* 11: 1–28. http://www.ics.forth.gr/hci/files/white_paper_1999.pdf.

Stephanidis, C., Salvendy, G., Akoumianakis, D., Bevan, N., Brewer, J., Emiliani, P. L., et al. (1998). Toward an information society for all: An international R&D agenda. *International Journal of Human-Computer Interaction* 10: 107–134. http://www.ics.forth.gr/hci/files/white_paper_1998.pdf.

Weber, G. and Abascal, J. (2006). People with disabilities: Materials for teaching accessibility and design for all, in *Computers Helping People with Special Needs: Proceedings of the 10th International Conference (ICCHP 2006)* (K. Miesenberger, J. Klaus, W. Zagler, and A. Karshmer, eds.), 11–13 July 2006, Linz, Austria, pp. 337–340. Berlin/Heidelberg: Springer-Verlag.

Williams, R. and Edge, D. (1996). The social shaping of technology. *Research Policy* 25: 856–899.

World Health Organization (2001). *International Classification of Functioning, Disability and Health (ICF)*. http://www.who.int/classifications/icf/site/icftemplate.cfm.

Zimmermann, G., Vanderheiden, G., and Gilman, A. (2002). Universal remote console: Prototyping for the alternate interface access standard, in *Universal Access: Theoretical Perspectives, Practice and Experience: Proceedings of the 7th ERCIM UI4ALL Workshop* (N. Carbonell and C. Stephanidis, eds.), 23–25 October 2002, Paris, pp. 524–531 (LNCS: 2615). Berlin/Heidelberg: Springer-Verlag.

2

Perspectives on Accessibility: From Assistive Technologies to Universal Access and Design for All

Pier Luigi Emiliani

2.1 Introduction

From the perspective of people with disabilities, technological developments have always been concurrently perceived as a potential support to inclusion in society, but, at the same time, also as a challenge to their present situation. The personal computer, for example, was immediately considered as an invaluable new possibility for accessing information, but it needed adaptations for blind and motor-disabled people. Following the development of textual screen readers, the introduction of graphical user interface was considered by blind people as a threat to their recently acquired autonomy in reading and writing. In fact, they had to wait for the development of screen readers for the graphic interfaces to be able to access computers again.

Traditionally, this has been the situation of people with disabilities. They must wait for the technology, even if potentially very promising, to be adapted for their use. The living environment in general, including technology, was also normally designed for the average user and then adapted to the needs of people who are more or less far from "average." Architects then started to think that it might be possible to design public spaces and buildings that are accessible to everyone, even, for example, those who move about in a wheelchair. This approach (design

for all, or universal design) resulted in successful designs for landscapes, which were subsequently documented as guidelines for accessible built environments. It took several years before the approach was able to gather the political support needed for practical application, but the main principles had been developed. Moreover, it turned out that the approach was invaluable not only for disabled people, but also for the population at large. It is only a pity that too many buildings and public spaces are constructed at present whose designers do not take these basic principles into consideration.

This chapter deals with how the concept of designing for all potential users can be generalized outside the original field of architecture, to become applicable and relevant to the information society (i.e., from physical spaces to conceptual spaces). It will be shown that, to design an information society accessible to all, the basic assumptions of design for all as developed in architecture must be reverted. While a single physical space can be designed to be available to all, information environments must be implemented in such a way to be automatically adaptable to each individual user. Therefore, the individual needs of all potential users must be taken into account in constructing the emerging telecommunications and information environment with an embedded intelligence sufficient for automatic self-adaptation to the individual users.

It will be also argued that design for all concepts must be revisited with the evolution of the information society. There is a difference between designing for all an interface with a computer or an application running on it and designing for all the information society as such. It is probably a problem of intelligence that is possible to embed in the system, but certainly it is not limited to only intelligence needed for self-adaptation to the individual users.

However, the design for all approach is not "against" assistive technology, the conceptual and technological environment in which adaptations and add-ons for people with disabilities have been traditionally developed. The variety and complexity of individual situations are such that, at least in the short-to-medium term, it will not be possible and/or economically viable to accommodate all necessary features within the adaptability space of a single product. What will probably be necessary is an expansion of the assistive technology sector toward the use of advanced technology, as was the case 20 years ago when new technology (e.g., voice synthesis and recognition) was primarily applied in the environment of rehabilitation, and a shift from the adaptation of products designed for an average user to an adaptability built in at design time. Probably, the transition from assistive technology to design for all will have to be established upon a careful trade-off between built-in adaptability and *a posteriori* adaptations on the basis of economic and functional criteria.

Even if, conceptually, these developments appear promising, there is limited interest at the level of end-users and also professionals working in sectors related to eInclusion. Probably, people with disabilities think that being embedded in the "for all" concept will reduce interest in their specific problems, while other users, who so far did not have accessibility problems with information and communication technologies (ICT), are not aware of the important changes that the emergence of the information society is bringing about. Professionals, who must cope with problems of users in their present situation, are traditionally suspicious of approaches that, even if interesting, are foreseen to give results in the medium-to-long-term. The main arguments of this chapter are that to guarantee an accessible information society to all users, the design for all approach is the only viable one, and that if needs, requirements, and preferences of all users are taken into account as far as possible in the specification of new technology and corresponding services and applications, this will bring about advantages for all citizens.

2.2 Accessibility versus Universal Access

Presently, there is a shift from accessibility, as traditionally defined in the assistive technology sector, to universal access, due to developments in technology and an increased social interest for people at risk of exclusion, including not only people with disabilities, but any person who may differ with respect to language, culture, computer literacy, and the like. This section deals with the foreseen changes in technology and in the organization of society, as well as with the rationale underlying the concept of universal access.

2.2.1 From Terminals and Computers to the Information Society

In ICT, the issue of accessibility was originally related to people with disabilities. When the interest in the use of information technology and telecommunications for people with disabilities started, the situation was relatively simple: the main service for interpersonal communication was the telephone, and information was distributed by means of radio and television. Computers were mainly stand-alone units used in closed and specialized communities (e.g., those of scientists and businessmen).

In principle, the telephone was a fundamental problem only for profoundly deaf people. For all other groups of people with disabilities, solutions were within the reach of relatively simple technological adaptations. The technology used for implementing the telephone lent itself to the possibility of capturing the signal (electromagnetic induction) and making it available for amplification for deaf people. Even the problems of profoundly deaf people were facilitated by the telephone system itself, when it was discovered that the telephone line could be used to transmit digital data (characters) with suitable interfaces (modems). Radio was an important medium for the diffusion of information. In principle, radio can represent a problem for deaf people. But since amplification is inherent in a radio system, problems occur again therefore only for profoundly deaf people. Television was the first example of a service that used a combination of the visual and acoustic modalities, not redundantly, but for conveying different types of information. Being more complex, television could create more difficulties for people with disabilities, but it had inherent capabilities for overcoming some of the problems. It is evident that television can create problems for blind, visually disabled, and deaf people. On the other hand, the fact that additional information can be transmitted by exploiting the available bandwidth enables support for people with disabilities to be added to the standard service. Therefore, programs can be subtitled for deaf people, and scenes without dialogue can be described verbally for blind people. In addition, text services can be set up (e.g., televideo, teletext), thus solving some of the problems related to the accessing of information by profoundly deaf people.

Television is a simple example of a general situation. An increase in the complexity of a system or service increases the number and extent of problems that such a system or service can create for people who have reduced abilities compared to the majority of the population. At the same time, technical complexity often implies additional features to recover from this unfortunate situation, as well as the possibility of using the same technology in an innovative way to solve problems that have not yet been addressed.

The situation started to change, thanks to the development of computers and technology able to increase the bandwidth of communications channels, which ultimately contributed to creating a completely new environment for communication and access to information, as will be briefly described in the following. From the perspective of the user, the first important innovation was brought about by the introduction of personal computers. Personal computers were immediately seen as a new and very important possibility for supporting people with disabilities in communication and providing access to information. Unfortunately, they were not directly accessible to some user groups, such as blind people and people with motor impairments of the upper limbs. However, the possibility of encoding information, instead of printing it on paper, was immediately perceived as being of paramount importance for blind people. Therefore, personal computers had to be made available to them. Adaptations were investigated, and through the synergy of new transduction technologies (mainly synthetic speech) and specialized software (screen readers), capable of "stealing" information from the screen and making it available to appropriate peripheral equipment, coded information was made available to blind people (Mynatt and Weber, 1994). Blind people could also read information retrieved from remote databases, and write and communicate using electronic mail systems. Adaptations for motor-disabled people (special keyboards, mouse emulators) and for other categories of disabled people were also made available.

It can therefore be concluded that, when the interest in accessibility by people with disabilities and elderly people became more widespread, the worldwide technological scene was dominated by a set of established systems and services. The situation required adaptations of existing systems, which slowly became available with long delays.

Today, after a period of relative stability, the developments in solid-state technology and optoelectronics, which made possible the increase of the available computational power, the integration of "intelligence" in all objects, and the availability of broadband links, and, particularly, the fusion between information technology, telecommunications, and media technologies and industry, are causing a revolution in the organization of society, leading from an industrial to an information society.

The emergence of the information society is associated with radical changes in both the demand and the supply of new products and services. The changing pattern in demand is due to a number of characteristics of the customer base, including (1) increasing number of users characterized by diverse abilities, requirements, and preferences; (2) product specialization to cope with the increasing variety of tasks to be performed, ranging from complex information-processing tasks to the control of appliances in the home environment; and (3) increasingly diverse contexts of use (e.g., business, residential, and nomadic).

On the other hand, one can clearly identify several trends in the supply of new products and services. These can be briefly summarized as follows: (1) increased scope of information content and supporting services; (2) emergence of novel interaction paradigms (e.g., virtual and augmented realities, ubiquitous computing); and (3) shift toward group-centered communication-, collaboration-, and cooperation-intensive computing.

This general trend is exemplified by the shift in paradigm in the use of computers, leading to the present situation made possible by the fusion between information technology and telecommunications (Figure 2.1). As suggested by Figure 2.1, from the early calculation-intensive nature of work that was prevalent in the early 1960s, computer-based systems are progressively becoming a tool for communication, collaboration, and social interaction, which are the main characteristics of the emerging intelligent information environment. From a specialist's device, the computer is being transformed into an information appliance for the citizen in the information society.

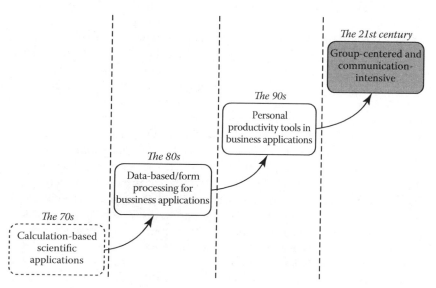

FIGURE 2.1 Paradigm shifts in the use of computers (From Stephanidis, C. and Emiliani, P. L., *Technology and Disability Journal*, 10, 21–44, 1999. With permission.)

To have a basis for discussion, let us consider a possible scenario for the further development of the information society. Almost all experts agree that the information society will not imply the use of an increased number of computers and terminals, such as the ones we are accustomed to use today, or at least this will occur only in small professional activity niches or only for few general activities. According to observatories around the world, the information society is foreseen to emerge and evolve as some form of "intelligent environment." This vision is present not only in Europe, but also worldwide (for example Australia, Japan, and the United States), both at an industrial (as, for example, at Rand, Xerox, Microsoft, IBM, Philips, Siemens, NEC, Fujitsu) and academic level (for example, MIT has an ambient intelligence laboratory and many research projects around it). Apparently, this idea is popular also in the research environment.

This will entail the emergence of a new environment where intelligence is a distributed function, that is, computers will behave as intelligent agents able to cooperate with other intelligent agents on distributed data. Wearable computers, disappearing computers, mobile systems, ambient intelligence, and a variety of technical platforms are some of the expressions emerging from technical discussions. The information society is expected to evolve in the direction of the proliferation of computational systems that integrate a range of networked interactive devices embedded into a physical context (in either indoor or outdoor spaces). These systems will provide hosting for a broad range of computer-mediated human activities and access to a multitude of services and applications. Such systems are based on the distribution of computers and networks in physical environments, and are expected to exhibit increasingly intelligent and context-sensitive behavior. On the user side, it is starting to become clear that the variety of users will increase to the point of including practically all people, since the entire society is supposed to become an "intelligent environment," with a variety of contexts of use, ranging from public spaces to professional environments, from entertainment activities to living environments.

The ambient intelligence (AmI) environment will be populated by a multitude of handheld and wearable "micro-devices," and computational power and interaction peripherals (e.g., embedded screens and speakers, ambient displays) will be distributed in the environment. Devices will range from personal (e.g., wristwatches, bracelets, personal mobile displays and notification systems, health monitors embedded in clothing), carrying individual and possibly private information, to public devices in the surrounding environment (e.g., wall-mounted displays). As technology disappears to humans both physically and mentally, devices will be no longer perceived as computers, but rather as augmented elements of the physical environment. Personal devices are likely to be equipped with facilities for multimodal interaction and alternative input/output (e.g., voice recognition and synthesis, pen-based pointing devices, vibration alerting, touch screens, input prediction, etc.), or with accessories that facilitate alternative ways of use (e.g., hands-free kits), thus addressing a wider range of user and context requirements

than the traditional desktop computer. A variety of new products and services will be made possible by the emerging technological environment, including home networking and automation, mobile health management, interpersonal communication, and personalized information services. These applications will be characterized by increasing ubiquity, nomadism, and personalization, and are likely to pervade all daily human activities. They will have the potential to enhance security in the physical environment, save human time, augment human memory, and support people in daily routines and simple activities, as well as in complex tasks.

A general description of the direction of anticipated technological development can be found in the ISTAG[1] Scenarios (IST Advisory Group, 2003), where a vision of ambient intelligence is offered:

> The concept of Ambient Intelligence (AmI) provides a vision of the information society, where the emphasis is on greater user friendliness, more efficient services support, user empowerment, and support for human interactions. People are surrounded by intelligent intuitive interfaces that are embedded in all kinds of objects and an environment that is capable of recognizing and responding to the presence of different individuals in a seamless, unobtrusive, and often invisible way.

To have an idea of the type of technology, interactions, and services available in the information society, let us now summarize their main characteristics (IST Advisory Group, 2003). First of all the hardware is supposed to be *very unobtrusive*. Miniaturization is assumed to produce the necessary developments in micro and optical electronics, smart materials, and nanotechnologies, leading to self-generating power and micro-power usage; breakthroughs in input/output systems, including new displays, smart surfaces, paints, and films that have smart properties; and sensors and actuators integrated with interface systems to respond to user senses, posture, and environment. Many technologies are conceived as handheld or wearable, taking advantage of the fact that intelligence can be embedded in the environment to support the individual personal system. This means being lightweight, but also readily available. It is taken for granted that people can have with them everything necessary for performing even complex tasks. For example the only communication item (sufficient, e.g., for carrying out navigation, environmental control, and communicating with other people) foreseen in the ISTAG scenarios is a personal communicator (P-Com). Its characteristics are not precisely defined. It does not have a specifically defined interface, but is a disembodied functionality supported by the AmI with different interfaces. It is adaptive and learns from user's interactions with the environment. It offers communication, processing, and decision-making functions. Finally, it must not necessarily be a highly sophisticated piece of equipment, whose performances are limited by size, weight, and

---

[1] IST Advisory Group.

power. The intelligence necessary to support the transduction of information necessary to address the different modalities and to support the user can be in the environment and in the network. In principle, the only limiting factor can be bandwidth.

Then a *seamless mobile/fixed web-based communications infrastructure* is supposed to be available. Complex heterogeneous networks need to function and to communicate in a seamless and interoperable way. This implies a complete integration of mobile and fixed networks, including ultrafast optical processing. These networks will have to be seamless and dynamically reconfigurable.

Dynamic and massively distributed device networks will be in place. The AmI landscape is a world in which there are almost uncountable interoperating devices. Some will be wired, some wireless, many will be mobile, many more will be fixed. The requirement will be that the networks should be configurable on an *ad hoc* basis according to a specific, perhaps short-lived, task, with variable actors and components.

Human interfaces will have to become *natural*. A central challenge of AmI is to create systems that are intuitive in use. This will need *artificial intelligence* techniques, especially dialogue-based and goal-orientated negotiation systems, as the basis for intelligent agents and intuitive human to machine interaction, which are supposed to be multimodal (multiuser, multilingual, multichannel, and multipurpose). It should also be adaptive to user requirements providing context-sensitive interfaces and information filtering and presentation.

Finally, the AmI world must be *safe, dependable, and secure*, considering all physical and psychological threats that the technologies might imply and giving important emphasis on the requirement for robust and dependable software systems components. Various aspects of AmI environments are discussed in Chapters 59 and 60 of this handbook.

2.2.2 From Accessibility to Universal Access

The previously discussed developments are expected to alter human interaction, individual behavior, and collective consciousness, as well as to have major economic and social effects. As with all major technological changes, this can have disadvantages and advantages. New opportunities are offered by the reduced need of mobility, due to the emergence of networked collaborative activities, and by the increased possibility of network-mediated interpersonal communications. However, difficulties may arise in accessing multimedia services and applications when users do not have sufficient motor or sensory abilities. The complexity of control of equipment, services, and applications, and the risk of information overload, may create additional problems.

The abovementioned problems are particularly relevant for people with disabilities, who have been traditionally underserved by technological evolution. Disabled and elderly people currently make up about 20% of the market in the European Union, and this proportion will grow with the aging of the population to an estimated 25% by the year 2030 (Vanderheiden, 1990; Gill, 1996).

Not only is there a moral and legal obligation to include this part of the population in the emerging information society, but there is also a growing awareness in the industry that disabled and elderly people can no longer be considered insignificant in market terms. Instead, they represent a growing market to which new services can be provided. However, due to the foreseen increase of citizens who will need to interact with the emerging technological environment, accessibility can no longer be considered a specific problem of people with disabilities, but of the society at large, if suitable actions are not undertaken.

This final observation is very important—accessibility is not enough. The concept of universal access must be introduced and adaptations fully addressed as a real option for satisfying the eInclusion requirements. Design for all has been mainly introduced in human-computer interaction on the basis of serving a variety of users, that is, addressing users' diversity. The related line of reasoning is that since users are different, and they have different accessibility and usability requirements, it is necessary to take all of them into account in a user-centered design process. However, the emerging environment is much more complex and diversity must be considered from other perspectives. First of all, interaction no longer involved only computers and terminals, but the environment and the physical objects in it. Therefore, it will be necessary to consider a variety of interaction paradigms, metaphors, media, and modalities. Then customers will not have to cope with tasks determined by the used application, but with goals to reach in everyday life, which will be different in different environments and for different users. Additionally goals may be complex not only due to the foreseen merging of functions connected to access to information, interpersonal communication, and environmental control, but also because they may involve communities of users. Finally, the same goal must be reached in many different contexts of use. This gives an idea of the complexity of the involved problems, the limitation of the classical accessibility concepts, and the need for innovative approaches.

In this dynamically evolving technological environment, accessibility and usability of such complex systems by users with different characteristics and requirements cannot be addressed through *ad hoc* assistive technology solutions introduced after the main building components of the new environment are in place. Instead, there is a need for more proactive approaches, based on a design for all philosophy (Stephanidis et al., 1998, 1999), along with the requirement of redefining the role and scope of assistive technologies in the new environment. In such a context, the concepts of universal access and design for all acquire critical importance in facilitating the incorporation of accessibility in the new technological environment through generic solutions.

Universal access implies the accessibility and usability of information technologies by anyone at any place and at any time. Universal access aims to enable equitable access and active participation of potentially all people in existing and emerging computer-mediated human activities, by developing universally accessible and usable products and services. These products and services must be capable of accommodating individual user requirements in different contexts of use, independent of location, target machine, or run-time environment.

Therefore, the concept of accessibility as an approach aiming to grant the use of equipment or services is generalized, seeking to give access to the information society as such. Citizens are supposed to live in environments populated with intelligent objects, where the tasks to be performed and the way of performing them will be completely redefined, involving a combination of activities of access to information, interpersonal communication, and environmental control. Everybody must be given the possibility of carrying them out easily and pleasantly.

This also has an impact on the technological approach to the problem of accessibility. Universal access needs a conscious and systematic effort to proactively apply principles, methods, and tools of design for all to develop information society technologies that are accessible and usable by all citizens, including people who are very young, elderly people, and people with different types of disabilities, thus avoiding the need for *a posteriori* adaptations or specialized design.

2.3 From Reactive to Proactive Approaches

2.3.1 Reactive Approaches

The traditional approach to rendering applications and services accessible to people with disabilities is to adapt such products to the abilities and requirements of individual users. Typically, the results of adaptations involve the reconfiguration of the physical layer of interaction and, when necessary, the transduction of the visual interface manifestation to an alternative (e.g., auditory or tactile) modality.

Although it may be the only viable solution in certain cases (Vanderheiden, 1998), the reactive approach to accessibility suffers from some serious shortcomings. One of the most important is that by the time a particular access problem has been addressed, technology has advanced to a point where the same or a similar problem reoccurs. The typical example that illustrates this state of affairs is the case of blind people's access to computers. Each generation of technology (e.g., DOS environment, Windowing systems, and multimedia) caused a new wave of accessibility problems to blind users, addressed through dedicated techniques such as text translation to speech for the DOS environment, off-screen models, and filtering for the Windowing systems.

In some cases, adaptations may not be possible at all without loss of functionality. For example, in the early versions of Windowing systems, it was impossible for the programmer to obtain access to certain Window functions, such as Window management. In subsequent versions, this shortcoming was addressed by the vendors of such products, allowing certain adaptations (e.g., scanning) on interaction objects on the screen. Adaptations are programming intensive, which raises several

considerations for the resulting products. Many of them bear a cost implication, which amounts to the fact that adaptations are difficult to implement and maintain. Minor changes in product configuration, or the user interface, may result in substantial resources being invested to rebuild the accessibility features. The situation is further complicated by the lack of tools to facilitate ease of "edit-evaluate-modify" development cycles (Stephanidis, et al., 1995). Moreover, reactive solutions typically provide limited and low-quality access. This is evident in the context of nonvisual interaction, where the need has been identified to provide nonvisual user interfaces that go beyond automatically generated adaptations of visual dialogues. Additionally, in some cases, adaptations may not be possible without loss of functionality.

Traditionally, two main technical approaches to adaptation have been followed: product-level adaptation and environment-level adaptation. The former involves treating each application separately and taking all the necessary implementation steps to arrive at an alternative accessible version. In practical terms, product-level adaptation practically often implies redevelopment from scratch. Due to the high costs associated with this strategy, it is considered the least favorable option for providing alternative access. The alternative involves intervening at the level of the particular interactive application environment (e.g., Microsoft Windows) to provide appropriate software and hardware technology to make that environment alternatively accessible. Environment-level adaptation extends the scope of accessibility to cover potentially all applications running under the same interactive environment, rather than a single application, and is therefore considered a superior strategy. In the past, the vast majority of approaches to environment-level adaptation have focused on access to graphical environments by blind users. Through such efforts, it became apparent that any approach to environment-level adaptation should be based on well-documented and operationally reliable software infrastructures, supporting effective and efficient extraction of dialogue primitives during user-computer interaction. Such dynamically extracted dialogue primitives are to be reproduced, at run-time, in alternative input/output (I/O) forms, directly supporting user access. Examples of software infrastructures that satisfy these requirements are the active accessibility technology from Microsoft Corporation, and the Java accessibility technology from Sun Microsystems.

2.3.2 Proactive Approaches

Due to the previously described shortcomings of the reactive approach to accessibility, there have been proposals and claims for proactive strategies, resulting in generic solutions to the problem of accessibility. Proactive strategies entail a purposeful effort to build access features into a product as early as possible (e.g., from its conception to design and release). Such an approach should aim to minimize the need for *a posteriori* adaptations and deliver products that can be tailored for use by the widest possible end-user population. In the context of human-computer interaction, such a proactive paradigm should address the fundamental issue of universal access to the user interface and services and applications, namely how it is possible to design systems that permit systematic and cost-effective approaches that accommodate all users.

Proactive approaches to accessibility are typically grounded on the notions of universal access and design for all. The term *design for all* (or *universal design*—the terms are used interchangeably) is not new. It is well known in several engineering disciplines, such as civil engineering and architecture, with many applications in interior design, building and road construction, and so on. However, while existing knowledge may be considered sufficient to address the accessibility of physical spaces, this is not the case with information society technologies, where universal design is still posing a major challenge. Universal access to computer-based applications and services implies more than direct access or access through add-on (assistive) technologies, since it emphasizes the principle that accessibility should be a design concern, as opposed to an afterthought. To this end, it is important that the needs of the broadest possible end-user population are taken into account in the early design phases of new products and services.

Unfortunately, in ICT there is not yet general consensus about what design for all is, and there is not yet enough knowledge and interest about developments in the information society (apart from what is directly testable in everyday present life). This is particularly strange because design for all by definition can be applicable only to products to be developed, that is, to the future. However, if the development is toward an agreed-upon model (ambient intelligence), since the new society will not materialize in a short time but there will be a (probably long) transition, it makes sense to try and find out the main characteristics of the future generations of technology and services and applications to influence them. This will bring advantages in the short and medium terms as a contribution to innovation in assistive technology, and will hopefully lead to the emergence of a more accessible information society in the long-term.

Some general investigations were carried out during a set of meetings of the International Scientific Forum "Towards an Information Society for All." The result of the activity of this international working group has been published in two white papers (Stephanidis et al., 1998, 1999), as a set of general recommendations and specific suggestions for research activities. An accurate report of the findings of the Scientific Forum, already reported in the white papers, is outside the scope of this chapter. However, a short summary is included that points out the main recommendations useful for the current discussion.

The first set of recommendations is related to the need to promote the development of environments of use, that is, integrated systems sharable by communities of users that allow for richer communications, and the progressive integration of the computing and telecommunications environments with the physical environment. This includes the study of the properties of environments of use, such as interoperability, adaptation,

cooperation, intelligence, and so on; the identification of novel architectures for interactive systems for managing collective experiences, which can facilitate a wide range of computer-mediated human activities; the development of architectures for multiple metaphor environments, adapted to different user requirements and contexts of use; the introduction of multiagent systems and components for supporting cooperation and collaboration, and allowing more delegation-oriented activities; and the individualization and adaptation of user interfaces.

A second group of recommendations is related to the need for supporting communities of users, with emphasis on social interaction in virtual spaces, to enhance the currently prevailing interaction paradigms (e.g., graphical user interfaces [GUIs] and the World Wide Web) and to support the wide range of group-centric and communication-intensive computer-mediated human activities. This includes the study of individual/collective intelligence and community knowledge management; methodologies for collecting/analyzing requirements and understanding virtual communities; access to community-wide information resources; and social interaction among members of online communities.

A third set of general recommendations is connected with the integration of users in the design process and the evaluation of results. It is based on the concept of extending user-centered design to support new virtualities. Detailed recommendations include the identification of foundations for designing computer-mediated human activities, to apply, refine, and extend existing techniques and tools of user-centered design with concepts from the social sciences; metrics for important interaction quality attributes, for measuring different aspects of an interactive system, including additional quality attributes such as accessibility, adaptation, intelligence, and so on; computational tools for usability engineering to automate certain tasks, guide designers toward usability targets, or provide extensible environments for capturing, consolidating, and reusing previous experience; requirements for engineering methods to facilitate the elicitation of requirements in novel contexts of use and different user groups; and protocols for effective user participation in design activities.

A fourth set of recommendations deals with support actions as articulating demand for design for all, supporting the industry, as well as promoting awareness, knowledge dissemination, and technology transfer. These activities are of paramount importance to the real take-up of the technological developments that are needed for the creation of a truly accessible information society. However, their discussion is outside the scope of this chapter.

Technological developments that are considered necessary in contributing to the accessibility and usability of the emerging information society and a list of specific research topics are also reported in Stephanidis et al. (1999). These include suggestions on research activities on design process, methods and tools, user-oriented challenges, input/output technology, and user interface architectures. Even if the list included in the paper was not intended to be exhaustive, it gives a clear indication of the challenges and the complexity of the issues that need to be addressed by the research community to facilitate the development of an information society acceptable for all citizens.

2.4 The Design for All Approach

There are many definitions of design for all (or universal design). As a first definition, let us consider the one that is available in the web site of the Trace Center, a research organization devoted to making technologies accessible and usable: "The design of products and environments to be usable by all people, to the greatest extent possible, without the need for adaptation or specialized design" (Trace Center, 2008).

At an industrial level, Fujitsu has recently published an entire issue of their journal completely devoted to universal design, defined as: designing products, services, and environments so that as many people as possible can use them regardless of their age and physical characteristics (e.g., height, visual and hearing abilities, and arm mobility).

Finally, in the context of a series of European research efforts spanning over two decades, design for all in the information society has been defined as the conscious and systematic effort to proactively apply principles, methods, and tools to develop information technology and telecommunications (IT&T) products and services that are accessible and usable by all citizens, thus avoiding the need for *a posteriori* adaptations, or specialized design (Stephanidis et al., 1998).

These definitions are conceptually based on the same principle, that is, the recognition of the social role of the access to the information and telecommunication technologies, which leads to the need of design approaches based on the real needs, requirements, and preferences of all the citizens in the information society, the respect of the individuals willing to participate in social life, and their right of using systems/services/applications. Furthermore, computer accessibility is gradually being introduced in the legislation of several countries (see also Chapter 53, "Policy and Legislation as a Framework of Accessibility"). For example, in the United States, since 1998, Section 508 of the Rehabilitation Act (U.S. Code, 1998) requires that "any electronic information developed, procured, maintained, or used by the federal government be accessible to people with disabilities." In Europe, the eEurope 2005 (European Commission, 2002) and the i2010 action plans (Commission of the European Communities, 2005) commit the member states and European institutions to design public sector web sites and their content to be accessible, so that citizens with disabilities can access information and take full advantage of the information society. The legal obligation to provide accessible interactive products and services may contribute to the adoption of systematic design approaches under a design for all perspective.

The approach is also in line with the one at the basis of the preparation of the new World Health Organization (WHO) International Classification of Functioning, Disability and Health (ICF) (World Health Organization, 2001), where a balance is sought between a purely medical and a purely social approach to the identifications of problems and opportunities for people in their social integration. When dealing with the problems of people who experience some degree of activity limitation or participation restrictions, "ICF uses the term *disability* to denote a multidimensional phenomenon resulting from the interaction between people and their physical and social environment." This is very important, because it allows grouping and analysis of limitations that are not only due to impairments. For example, people are not able to see because they are blind, or have fixation problems due to spastic cerebral palsy, or are in a place with insufficient illumination, or are driving and therefore cannot use their eyes for interacting with an information system. People may have impairments, activity limitations, or participation restrictions that characterize their ability (capacity) to execute a task or an action (activity), but their performance is influenced by the current environment. The latter can increase the performance level over the capacity level (and therefore is considered a facilitator) or can reduce the performance below the capacity level (thus being considered a barrier). Here the emphasis is on the fact that all people, irrespective of their capacity of executing activities, may perform differently according to different contexts, and the environment must be designed to facilitate their performances.

Even if there is, apparently, a convergence on the conceptual definition of design for all, there is limited interest and sometimes skepticism about it among people working in the social integration of people with disabilities, where the related concepts were firstly explored in ICT. In particular, there is an argument that raises the concern that "many ideas that are supposed to be good for everybody aren't good for anybody" (Lewis and Rieman, 1994). However, design for all in the context of information technologies should not be conceived of as an effort to advance a single solution for everybody, but as a user-centered approach to providing products that can automatically address the possible range of human abilities, skills, requirements, and preferences. Consequently, the outcome of the design process is not intended to be a singular design, but a design space populated with appropriate alternatives, together with the rationale underlying each alternative, that is, the specific user and usage context characteristics for which each alternative has been designed.

If this is the case, then it is argued that this is clearly impossible or too difficult to be of practical interest. However, even if it is true that existing knowledge may be considered sufficient to address the accessibility of physical spaces, while this is not the case with information technologies where universal design is still posing a major challenge, important advances are being made in the development of concepts and technologies that are considered necessary for producing viable design for all approaches, as discussed in the previous section.

Apparently there is a conceptual confusion between the concepts of universal access and design for all (universal design). What is considered important, particularly in the field of disability, is granting people universal access. This is clearly right, but the claim that, therefore, everything that aims to give accessibility to all is design for all is conceptually misleading. Design for all is a well-defined approach, particularly promising due to the developments of the information society, which must coexist at least in the short and medium terms with assistive technology to serve all potential users of ICT systems, services, and applications.

Another common argument is that design for all is too costly (in the short term) for the benefits it offers. Though the field lacks substantial data and comparative assessments as to the costs of designing for the broadest possible population, it has been argued that (in the medium-to-long-term) the cost of inaccessible systems is comparatively much higher and is likely to increase even more, given the current statistics classifying the demand for accessible products (Vanderheiden, 1998).

The origins of the concept of universal access are to be identified in approaches to accessibility mainly targeted toward providing access to computer-based applications by users with disabilities. Today, universal access encompasses a number of complementary approaches, which address different levels of activities leading to the implementation of designed for all artifacts.

At the level of design specifications, for example, there are lines of work that aim to consolidate existing wisdom on accessibility, in the form of general guidelines or platform- or user-specific recommendations (e.g., for graphical user interfaces or the web). This approach consolidates the large body of knowledge regarding people with disabilities and alternative assistive technology access in an attempt to formulate ergonomic design guidelines that cover a wide range of disabilities. In recent years, there has also been a trend for major software vendors to provide accessibility guidance as part of their mainstream products and services. Moreover, with the advent of the World Wide Web, the issue of its accessibility recurred and was followed up by an effort undertaken in the context of the World Wide Web Consortium to provide a collection of accessibility guidelines for web-based products and services (W3C-WAI, 1999). The systematic collection, consolidation, and interpretation of guidelines are also pursued in the context of international collaborative and standardization initiatives. Another line of work relevant to universal access is user-centered design, which is often claimed to have an important contribution to make, as its human-centered protocols and tight design evaluation feedback loop replace technocentric practices with a focus on the human aspects of technology use.

At the level of implementation approaches, the proposed approach, which was first applied in the design of human-computer interfaces and then generalized to the implementation of complete applications, is based on the concepts of adaptability and adaptivity (Stephanidis, 2001a). The central idea is that the variety of possible users and contexts of use can be served only if the systems and services are able to adapt themselves automatically to the needs, requirements, and preferences of every single user. Adaptation must be guaranteed at run time (adaptability) and, dynamically, during interaction (adaptivity). Adaptation to users is now considered an important feature of all systems and services in ICT, even if in most cases this general claim is considered to be satisfied by introducing some form of personalization under the control of the user.

2.5 From Assistive Technology to Design for All: A Historical Perspective

After this general analysis let us now concentrate on an example of migration from assistive technology to design for all, following the evolution and achievements of a series of research projects, the majority of which were funded by European Commission Programs. This research line has spanned across almost two decades and has pursued an evolutionary path, initially adopting reactive, and subsequently advocating proactive, strategies to accessibility.

What is important to notice in this context is the progressive shift toward more generic solutions to accessibility. In fact, with the exception of early exploratory studies, which did not have an RTD development dimension, all remaining research efforts embodied both a reactive RTD component as well as a focus on proactive strategies and methods. The latter were initially oriented toward the formulation of principles, while later an emphasis was placed on the demonstration of technical feasibility.

2.5.1 Exploratory Activities

Early exploratory activities[2] have investigated the possibilities offered by the multimedia communication network environment, and in particular B-ISDN (broadband integrated services digital network), for the benefit of people with disabilities (Emiliani, 2001). To enable the accessibility of disabled people to the emerging telecommunications technology, it was considered essential that the designers and providers of the services and terminal equipment take explicitly into account, at a very early stage of design, their interaction requirements. Several barriers have been identified that prevent people with disabilities from having access to information available through the network. The identified barriers are related to accessibility of the terminal, accessibility of the anticipated services, and the perception of the service information.

To cope with these difficulties, different types of solutions have been proposed that address the specific user abilities and requirements at three different levels:

1. Adaptations within the user-to-terminal and the user-to-service interface, through the integration of additional input/output devices and the provision of appropriate interaction techniques, taking into account the abilities and requirements of the specific user group;
2. Service adaptations through the augmentation of the services with additional components capable of providing redundant or transduced information; and

---

[2] The IPSNI R1066 (Integration of People with Special Needs in IBC) project was partially funded by the RACE Program of the European Commission and lasted 36 months (January 1, 1989 to December 31, 1991).

3. Introduction of special services only in those cases where the application of the two previously mentioned types of adaptation are not possible or effective.

2.5.2 Adaptation of Telecommunication Terminals

Building on these results, the technical feasibility of providing access to multimedia services running over a broadband network to people with disabilities was subsequently demonstrated.[3] Adaptations of terminals and services were implemented and evaluated. In particular, two pairs of multimedia terminals (one UNIX/X-Windows based and one PC/MS-Windows based) were adapted according to the needs of the selected user groups. Special emphasis was placed on the adaptation of the user interfaces, and for this purpose, a user interface and construction tool was designed (Stephanidis and Mitsopoulos, 1995), which takes into account the interaction requirements of disabled users. The tool was built on the notion of separating an interactive system in two functional components, namely the application functional core and the user interface component, thus allowing the provision of multiple user interfaces to the same application functionality. However, for blind users who are not familiar with graphical environments, it was difficult to grasp the inherently visual concepts (e.g., the pop-up menu). Such an observation led to the realization that adaptations cannot provide an effective approach for a generic solution to the accessibility problems of blind users.

These efforts allowed an in-depth analysis of services and applications for the broadband telecommunications environment from the point of view of usability by disabled people, leading to the identification of testing of necessary adaptations and/or special solutions (Emiliani, 2001). This led to the conclusion that if emerging services, applications, and terminals were designed considering usability requirements of disabled users, many of their access problems would be automatically reduced with a negligible expense. One of the conclusions was that, as a minimum, sufficient modularity and flexibility should be the basis of product implementation to allow easy adaptability to the needs, capabilities, and requirements of an increasing number of users.

2.5.3 Adaptations of Graphical Interactive Environments

The subsequent research phase aimed to identify and provide the technological means to ensure continued access by blind users to the same computer-based interactive applications used by sighted users.[4] The short-term initial goal was to improve adaptation methodologies of existing GUIs. Specific developments were carried out through the implementation of appropriate demonstrators enabling access to MS-WINDOWS (PCs) and to interactive applications built on top of the X WINDOW SYSTEM (UNIX-based workstations). The adopted approach to interface adaptation for blind users was based on a transformation of the desktop metaphor to a nonvisual version combining Braille, speech, and non-speech audio. Access to basic graphical interaction objects (e.g., Windows, menus, buttons), utilization of the most important interaction methods, and extraction of internal information from the graphical environment were investigated.

Input operations (e.g., exploration/selection of menu options, etc.) can be performed either by means of standard devices (keyboard or mouse) or through special devices (i.e., mouse substitutes, touch pad and routing keys of Braille device). An important feature of the method is that the entire graphical screen is reproduced in a text-based form and simultaneously presented on a monochrome screen that can be explored by blind users by means of Braille and speech output. Additionally, sounds help navigation and provide spatial relationships between graphical objects. It is important to note that the text-based reproduction facilitates cooperation with sighted colleagues.

A variety of issues related to user interaction in a graphical environment were also investigated, particularly for blind users. For example, different input methods that can be used instead of the mouse were investigated. The problem of how blind users can efficiently locate the cursor on the screen, and issues related to combining spatially localized sounds (both speech and nonspeech) and tactile information to present available information, were examined. Finally, the project addressed the design and implementation of real-world metaphors in a nonvisual form and the development of an optimal method to present graphical information from within applications.

In this context, a first step toward the development of tools aimed at the implementation of user interfaces for all was carried out. The goal of these efforts was the development of innovative user interface software technology to guarantee access to future computer-based interactive applications by blind users. In particular, these projects conceived, designed, and implemented a user interface management system as a tool for the efficient and modular development of user interfaces that are concurrently accessible by both blind and sighted users.

2.5.4 Dual User Interfaces

The concept of dual user interfaces (Savidis and Stephanidis, 1998) was proposed and defined as an appropriate basis for "integrating" blind and sighted users in the same working environment. Figure 2.2 shows the concept of dual user interfaces. A dual

---

[3] The IPSNI-II R2009 (Integration of People with Special Needs in IBC) project was partially funded by the RACE-II Program of the European Commission and lasted 48 months (January 1, 1992 to December 31, 1995).

[4] The GUIB TP103 (Textual and Graphical User Interfaces for Blind People) project was partially funded by the TIDE Program of the European Commission and lasted 18 months (December 1, 1991 to May 31, 1993). The GUIB-II TP215 (Textual and Graphical User Interfaces for Blind People) project was partially funded by the TIDE Program of the European Commission and lasted 18 months (June 1, 1993 to November 30, 1994).

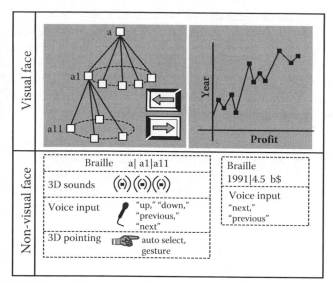

FIGURE 2.2 The concept of dual user interfaces (From Stephanidis, C. and Emiliani, P. L., *Technology and Disability Journal*, 10, 21–44, 1999. With permission.)

user interface is characterized by the following properties: (1) it is concurrently accessible by blind and sighted users; (2) the visual and nonvisual metaphors of interaction meet the specific needs of sighted and blind users respectively (they may differ, if required); (3) the visual and nonvisual syntactic and lexical structure meet the specific needs of sighted and blind users, respectively (they may differ, if required); (4) at any point in time, the same internal (semantic) functionality is made accessible to both user groups through the corresponding visual and nonvisual "faces" of the dual user interface; (5) at any point in time, the same semantic information is made accessible through the visual and nonvisual "faces" of the dual user interface, respectively.

The HOMER user interface management system (UIMS) (Savidis and Stephanidis, 1998) was developed to facilitate the design and implementation of dual interfaces. HOMER is based on a fourth-generation user interface specification language, which supports: (1) abstraction of interaction objects, for instance, representation of objects based on their abstract interaction roles and syntactic/constructional features, decoupled from physical presentation aspects; (2) concurrent management of at least two toolkits, so that any modifications effected on the interface by the user through the objects of one toolkit are concurrently depicted in the objects of the second toolkit; (3) meta-polymorphic capability for abstract objects, for instance, abstract objects can be mapped to more than one toolkit, or to more than one object class within a specific toolkit; (4) unified object hierarchies supporting different physical hierarchies, so that alternative mappings of (portions of) the unified hierarchy to (portions of) physical hierarchies are possible; (5) ability to integrate different toolkits; (6) object-based and event-based model support for dialogue implementation, that is, the dialogue model can be defined either on the basis of the individual objects that participate in it, or on the basis of interaction events that originate

from those objects; and (7) declarative asynchronous control models (e.g., preconditions, monitors, constraints), as opposed to syntax-oriented control models (e.g., task notations, action grammars), or alternative control techniques (e.g., event-based models and state-based methods); the rationale behind the adoption of declarative control models concerns the desired independence from specific syntactic models, which allows for differing models, supported by different toolkits, to be supported.

A nonvisual toolkit to support nonvisual interface development (Savidis and Stephanidis, 1998) was also developed and integrated within the HOMER UIMS. The toolkit was developed on the basis of a purposefully designed version of the rooms metaphor, an interaction metaphor based on the physical environment of a room, and whose interaction objects are floor, ceiling, front wall, back wall, and so on. The library provides efficient navigation facilities, through speech/Braille output and keyboard input. Two different nonvisual realizations of the rooms metaphor have been assembled: (1) a nonspatial realization, supporting Braille, speech, and nonspeech audio output with keyboard input; and (2) a direct-manipulation spatial realization, combining 3D audio (speech and nonspeech), 3D pointing via a glove and hand gestures, keyword speech recognition, and keyboard input. In both realizations, special sound effects accompany particular user actions such as selecting doors (e.g., "opening door" sound), selecting the lift (e.g., "lift" sound), pressing a button or a switch object, and so on.

The HOMER UIMS has been utilized for building various dual interactive applications such as a payroll management system, a personal organizer, and an electronic book with extensive graphical illustrations and descriptions.

2.5.5 User Interfaces for All and Unified User Interfaces

The concept of user interfaces for all (Stephanidis, 2001b) has been proposed, following the concept of design for all, as the vehicle to efficiently and effectively address the numerous and diverse accessibility problems. The underlying principle is to ensure accessibility at design time and to meet the individual needs, abilities, and preferences of the user population at large, including disabled and elderly people.

Collaborative research was conducted[5] to develop new technological solutions for supporting the concept of user interfaces for all (i.e., universal accessibility of computer-based applications), by facilitating the development of user interfaces automatically adaptable to individual user abilities, skills, requirements, and preferences. The problem was approached at two levels: (1) the development of appropriate methodologies and tools for the design and implementation of accessible and usable user interfaces; and (2) the validation of the approach through the design

[5] The ACCESS TP1001 (Development Platform for Unified ACCESS to Enabling Environments) project was partially funded by the TIDE Program of the European Commission, and lasted 36 months (January 1, 1994 to December 31, 1996).

and implementation of demonstrator applications in two application domains, namely interpersonal communication aids for speech-motor- and language-cognitive-impaired users, and hypermedia systems for blind users.

The concept of unified user interface development (Stephanidis, 2001a) was proposed with the objective of supporting platform independence and target user profile independence (i.e., the possibility of implementation in different platforms and adaptability to the requirements of individual users). Unified user interface development provides a vehicle for designing and implementing interfaces complying with the requirements of accessibility and high-quality interaction.

A unified user interface comprises a single (unified) interface specification, targeted to potentially all user categories. In practice, a unified user interface is defined as a hierarchical construction in which intermediate nodes represent abstract design patterns decoupled from the specific characteristics of the target user group and the underlying interface development toolkit, while the leaves depict concrete physical instantiations of the abstract design pattern. The unified user interface development method comprises design- and implementation-oriented techniques for accomplishing specific objectives. The design-oriented techniques (unified user interface design) aim toward the development of rationalized design spaces, while the implementation-oriented techniques (unified user interface implementation) provide a specifications-based framework for constructing interactive components and generating the run-time environment for a unified interface.

To achieve this, unified user interface design attempts to: (1) initially identify and enumerate possible design alternatives, suitable for different users and contexts of use, using techniques for analytical design (such as design scenarios, envisioning and ethnographic methods); (2) identify abstractions and fuse alternatives into abstract design patterns (i.e., abstract interface components that are decoupled from platform-, modality-, or metaphor-specific attributes); and (3) rationalize the design space by means of assigning criteria to alternatives and developing the relevant argumentation, so as to enable a context-sensitive mapping of an abstract design pattern onto a specific concrete instance.

The result of the design process is a unified user interface specification. Such a specification can be built using a dedicated, high-level programming language and results in a single implemented artifact that can instantiate alternative patterns of behavior, at either the physical, syntactic, or even semantic level of interaction. The unified implementation, which is produced by processing the interface specification, undertakes the mapping of abstract interaction patterns and elements to their concrete/physical counterparts.

Unified user interface development makes two claims that radically change the way in which interfaces are designed and developed, while having implications on both the cost and maintenance factors. The first claim is that interfaces may be generated from specifications, at the expense of an initial design effort required to generate them. The second claim relates to the capability of the unified user interface to be transformed, or adapted, so as to suit different contexts of use. For example, in the cases of blind and motor-impaired users, the problem of accessibility of the menu can be addressed through a sequence of steps, involving (1) the unification of alternative concrete design artifacts (such as the desktop menu, the 3D acoustic sphere likely for the nonvisual dialogue, etc.) into abstract design patterns or unified design artifacts (such as a generalized container); (2) a method to allow the instantiation of an abstract design pattern into the appropriate concrete physical artifact, based on knowledge about the user; and (3) the capacity to dynamically enhance interaction by interchanging or complementing multiple physical artifacts at run-time (see adaptivity examples in the AVANTI system, Section 2.6.1).

It follows, therefore, that unified user interface development results in a revised cost model for user interfaces, where there is an initial effort to design, while development cost of alternative versions of an interface and maintenance costs are minimized.

Unified user interfaces and the related design approach are discussed in depth in Chapters 16, 18, and 21 of this handbook.

2.6 Working Examples of Systems and Services Designed for All

The unified user interface design method has been applied and validated in large-scale applications, which have provided both interesting and challenging test beds for the method's application, as well as the opportunity to refine details of its representation, conduct, and outcomes. Two of these applications are briefly discussed in the following. Their main achievement is that they have demonstrated the technical feasibility of the design for all approach. In these projects, the integration of all users has been obtained by implementing systems and services that are adaptable (that is, automatically reconfigurable at run-time, according to knowledge about the user or the user group) and adaptive (that is, able to change their features as a consequence of the patterns of use).

2.6.1 The AVANTI System

The AVANTI system[6] put forward a conceptual framework for the construction of web-based information systems that support adaptability and adaptivity at both the content and the user interface levels (Stephanidis et al., 2001). The AVANTI framework comprises five main components (Figure 2.3):

1. A collection of multimedia databases, which contain the actual information and are accessed through a common communication interface (multimedia database interface, MDI);

---

6 The AVANTI AC042 (Adaptable and Adaptive Interaction in Multimedia Telecommunications Applications) project was partially funded by the ACTS Program of the European Commission and lasted 36 months (September 1, 1995 to August 31, 1998).

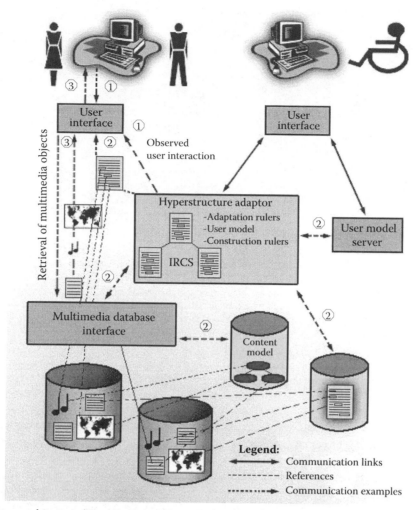

FIGURE 2.3 The AVANTI system architecture (From Emiliani, P. L., *Technology and Disability Journal*, 18, 19–29, 2001. With permission.)

2. A user modeling server (UMS), which maintains and updates individual user profiles, as well as user stereotypes;

3. The content model (CM), which retains a metadescription of the information available in the system;

4. The hyperstructure adaptor (HSA), which adapts the information content, according to user characteristics, preferences, and interests; and

5. The user interface (UI) component, which is also capable of adapting itself to the users' abilities, skills, and preferences, as well as to the current context of use.

Adaptations at the information-content level are supported in AVANTI through the HSA, which dynamically constructs adapted hypermedia documents for each particular user, based on assumptions about the user characteristics and the interaction situation provided by the user model server. The user characteristics that trigger appropriate adaptation types at the content level mainly concern the type of disability, the expertise, and the interests of the user. The resulting adaptations mostly

concern: (1) alternative presentations using different media (e.g., text vs. graphics, alternative color schemes); (2) additional functionality (e.g., adaptive "shortcut" links to frequently visited portions of the system, and conditional presentation of technical details); and (3) different structures and different levels of detail in the information provided. The knowledge about the user and the interaction session is mostly based on information acquired dynamically during run-time (e.g., navigation monitoring, user selection, explicit user invocation), with the exception of the initial profile of the user, retrievable from the UMS, which is acquired through a short questionnaire session during the initiation of the interaction, or retrieved from a smart card if one is available.

The design and development of the AVANTI browser's user interface (which acted as the front-end to the AVANTI information systems) have followed the unified user interface design methodology. The resulting unified interface is a single artifact in which adaptability and adaptivity techniques are employed to suit the requirements of three user categories: able-bodied, blind, and motor-impaired people. Adaptations at the user

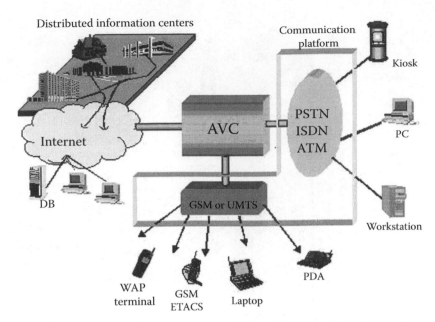

FIGURE 2.4 The PALIO information system (From Emiliani, P. L., *Technology and Disability Journal*, 18, 19–29, 2001. With permission.)

interface are supported through the cooperation of the browser and the user model server.

The categories of interface adaptation supported by the AVANTI UI include (Stephanidis et al., 2001) (1) support for different interaction modalities and input/output devices; (2) automatic adaptation of the presentation of interaction elements; (3) task-based adaptive assistance; (4) awareness prompting; (5) limited support for error prevention; and (6) limited support for metaphor-level adaptation. Additional features that have been included in the AVANTI browser to meet the requirements of the target user categories include adaptive support of multiple interaction metaphors (e.g., desktop application and an information kiosk metaphor), special I/O devices, and extended navigation functionality. Alternative metaphors have been designed for the different usage contexts of the AVANTI system. Furthermore, special-purpose input/output devices have been integrated into the system to support blind and motor-impaired individuals: binary switches, joysticks, touch screens and touch tablets, speech input and output, and Braille output.

2.6.2 The PALIO System

PALIO[7] (Stephanidis et al., 2004) is an open system for accessing and retrieving information without constraints and limitations imposed by space, time, access technology, and so on. Therefore, the system is modular and capable of interoperating with other existing information systems. Mobile communication systems play an essential role in this scenario, because they enable access to services from anywhere and at any time. One important aspect of the PALIO system is the support of a wide range of communication technologies (mobile or wired) for accessing services. In particular, it is possible for users equipped with either a common cellular phone or an advanced WAP phone to access services, wherever they are.

The PALIO system envisages the adaptation of both the information content and the way in which it is presented to the user, as a function of user characteristics (e.g., abilities, needs, requirements, interests); user location with the use of different modalities and granularities of the information contents; context of use; the current status of interaction (and previous history); and, lastly, the technology (e.g., communications technology, terminal characteristics, special peripherals) used.

The PALIO information system consists of the following three main elements (see Figure 2.4):

1. A communications platform that includes all network interfaces to interoperate with both wired and wireless networks;
2. The AVC center, which is composed of the main adaptation components, a service control center, and the communication layers to and from the user terminals and the information services; and
3. Distributed information centers in the territory, which provide a set of primary information services.

The AVC center is the architectural unit that manages diversity and implements the mechanisms for universal access. The AVC will be perceived by users as a system that groups together all information and services that are available in the city. It will serve as an augmented, virtual facilitation point from which

[7] The PALIO IST-1999-20656 Project (Personalized Access to Local Information and Services for Tourists) was partially funded by the EC 5th Framework Program and lasted 30 months (November 1, 2000 to April 30, 2003).

different types of information and services can be accessed. The context and location awareness, as well as the adaptation capabilities of the AVC, will enable users to experience their interaction with services as a form of contextually grounded dialogue: for example, the system always knows the user's location and can correctly infer what is near the user, without the user having to explicitly provide information to the system.

2.6.3 Recent Developments

In an AmI environment, applications are required to continuously follow end-users and provide high-quality interaction while migrating among different computing devices, and dynamically utilizing the available I/O resources of each device. An application experiment addressing these issues is reported in Stephanidis (2003). The developed experimental application is a nomadic music box for MP3 files, providing downloading from a server through the network, local storage, and decoding and playback functionality through a software/hardware player with typical audio control functionality. It exhibits elements of mobile, wearable, and wireless I/O resources, multiple I/O resources employed concurrently, and interface migration with state-persistent reactivation, ability to dynamically engage or disengage I/O resources as those become available or unavailable in mobile situations, and, finally, high-quality interface design to ensure interaction continuity in dynamic I/O resource control. The nomadic music box supports context awareness and adaptation to context, as well as adaptation to the (dynamically changing) available devices, addressing therefore diversity of contexts of use and technological platforms in the DC environment. Additionally, the adopted interface architecture and the dialogue design easily allow catering to additional target user groups (e.g., disabled users such as blind or motor-impaired users). Therefore, the nomadic music box constitutes a first example of prototype application addressing some of the issues raised by universal access in the context of AmI.

Furthermore, more recent developments of universally accessible systems based on design for all approaches and on the unified user interface development methodology are discussed in Chapter 23, "Methods and Tools for the Development of Unified Web-Based User Interfaces," which reports on a framework and the related toolkit for the development of web portals that integrate server-side adaptation capabilities, and Chapter 17, "Designing for Universally Accessible Games," which discusses the application of the unified user interface design method in the domain of electronic games, and presents examples of universally accessible games developed following such an approach.

2.7 Toward Intelligent Interactive Environments

The emergence of AmI environments (see Section 2.2 of this chapter and Chapter 60, "Ambient Intelligence") is likely to require a rethinking of design for all. In fact, in the context of

AmI, the information society is no longer seen as a support to the execution of activities such as accessing and manipulating information or communicating with other people, but is supposed to have an impact on all aspects of social activities. This is clear if the sociopolitical factors at the basis of the deployment of the new technology and application environments as discussed in the ISTAG documents are considered.[8] According to ISTAG, AmI should (1) facilitate human contacts; (2) be oriented toward community and cultural enhancement; (3) help to build knowledge and skills for work, better quality of work, citizenship, and consumer choice; (4) inspire trust and confidence; (5) be consistent with long-term sustainability—personal, societal, and environmental—and with lifelong learning; and (6) be controllable by ordinary people. In essence, the challenge is to create an AmI landscape made up of "convivial technologies" that are easy to live with.

Accordingly, the main high-level design requirements of an AmI environment are that it must be unobtrusive (i.e. , many distributed devices are embedded in the environment, and do not intrude into our consciousness unless we need them), personalized (i.e., it can recognize the user, and its behavior can be tailored to the user's needs), adaptive (i.e., its behavior can change in response to a person's actions and environment), and anticipatory (i.e., it anticipates a person's desires and environment as much as possible without the need for mediation).

It is clear from the previous design requirements that design for all in the information society must provide much more than personalization (adaptability) and adaptivity, which nevertheless are among the required features of AmI. Smart devices and (complex) services are supposed to be embedded in the environment, and must be able to provide support to users only when they need them, anticipating their desires. Moreover, this must occur in any place and context of use and in any moment. This requires the deployment of an infrastructure for supporting ubiquitous connection and computational power, but also of intelligence for identifying the goals of the users and helping users in fulfilling them. Therefore, the environment must not only be filled with intelligent objects (that is, computer-based systems), but must also be able to reason with regard to the goals of the users and to optimize the support in accordance with the resources available.

While many problems related to interaction with the present information society are actually linked to a suitable structuring of information and an accessible human system interface, integration within the AmI environment is much more complex, due to the interplay of different levels (e.g., the physical level with a multiplicity and heterogeneity of intelligent objects in the environment and their need for a continuous and high-speed connection, the level of identification and consideration of the variety of contexts of use, and the level of elicitation of the diversity of user goals and help in their fulfilment).

The AmI environment must be able to seamlessly integrate these three levels. At the lower level, all objects in the

---

[8] See ISTAG online at http://cordis.europa.eu/ist/istag.htm.

environment must incorporate intelligence and must be interconnected and able to cooperate to support the goals of the user. Moreover, the environment must be reconfigurable in real time, to cater for the introduction or removal of components (e.g., objects that users entering the environment may have with them), by remodeling its support as a function of the available resources. This aspect of the intelligent environment is a prerequisite for its development, and the reconfiguration must take into account the variability of the contexts of use and the goals of the users.

At this level, artificial intelligence is crucial in supporting the development of basic technologies considered important in the implementation of AmI. For example, the ISTAG experts write that pattern recognition (including speech and gesture) is a key area of artificial intelligence that is already evolving rapidly. Speech recognition will have a big impact on the miniaturization of devices and the augmentation of objects allowing hands-free operation of personal ambient devices. In the scenarios, the use of voice, gesture, and automatic identification and localization are implicitly used to synchronize systems, so that services are available when people want them. The synchronization of systems is indeed a very important aspect. In the intelligent environment, there may be different users with goals that have different importance and criticality. AmI must be able to decide how to take care of potentially conflicting needs (from the perspective of resources).

At a higher level, the AmI environment must take care of the contexts of use considered as processes, which are defined by specific sets of situations, roles, relations, and entities (Coutaz et al., 2008). Recently, interest on the right definition and role of the contexts of use has grown, when dealing with both the development of user machine interfaces and with the organization and representation of information. For example, in connection with multimodality, it has been argued that the availability of different representations of the same information could be very interesting to avoid problems of context-related accessibility. Typical examples are car drivers who are functionally blind and motor disabled, meaning that they cannot interact with information and communication by using a screen and a pointer. But in AmI, the situation is more complex, because in the ubiquitous interaction with information and telecommunication systems the context of use may change continuously or abruptly, and the same systems or services may need to behave differently in different contexts. A clear example to be considered is the complexity of the situation where a user is carrying out some activity in a room (e.g., in the kitchen) and a second person enters the room. The system must reconfigure itself, not only due to the possible introduction of some additional intelligent component, but also because of the change in the context of use. The system must accommodate the original goals of two persons, and also take into account their interaction, which, for example, may redefine in real time some of their goals or change the time scale of different activities. This requires not only a complete configurability of the system and service at the level of interfaces and functionalities, but also the capacity of

realizing changes in the environment and reasoning as to their impact on the context in which they are used.

When discussing the technology necessary for transforming the scenarios into reality, ISTAG experts (IST Advisory Group, 2003) make a list of its intelligent components. They are:

- Media management and handling, including presentation languages that support "produce once" and "present anywhere" methods and tools to analyze content and enrich it with metadata, and tools for exploiting the semantic web;
- Natural interaction that combines speech, vision, gesture, and facial expression into a truly integrated multimodal interaction concept that allows human beings to interact with virtual worlds through physical objects, and that enables people to navigate the seemingly infinite information which they will be able to access;
- Computational intelligence, including conversational search and dialogue systems, behavioral systems that can adapt to human behavior, and computational methods to support complex search and planning tasks;
- Contextual awareness, for instance, systems that support navigation in public environments, such as in traffic, in airports and in hospitals, service discovery systems that enhance the shopping experience, and context-aware control and surveillance systems; and
- Emotional computing that models or embodies emotions in the computer, and systems that can respond to or recognize the moods of their users, and systems that can express emotions.

These technologies are the necessary building blocks for implementing AmI environments populated by smart artifacts that can adapt to human behavior and to different contexts. The emerging question is whether the list is exhaustive and what the impact of (artificial) intelligence is in really meeting the design requirements. That is, creating an environment where people can reach their goals and in which they do not feel artificial does not create problems of information overload and confusion and is perceived as being worthy of trust and confidence.

Indeed, in addition to objects and contexts of use, there is a higher abstraction level to be considered. Most of the interaction with currently available systems is based on the performance of tasks in a set determined by the application used. In the intelligent environment, the goals of the user are the starting point. They must be inferred by the system and decomposed into tasks that are adapted to the preferences of the individual. This is really the level where intelligence plays a crucial role. The acceptability and uptake of the new paradigm will be essentially dependent on how smart the system is in inferring the goals of the users in the continuously varying contexts of use and in organizing the available resources (intelligent objects in the environment) to help users fulfil them. Moreover, this aspect is also very sensitive from a psychological perspective. For example, the system must be able to deal with the task of inferring the goals of the users without giving them the impression that the system is controlling them (Big Brother), and must be able to support the users without giving them the impression of force. It must "offer" possible solutions, not "impose" them. This requires a lot of ingenuity also on the part of human beings, and appears particularly difficult for a machine. However, it can also be argued that the intelligence in the system does not necessarily have to be artificial, but could and probably will also be in the network of interconnected and cooperating persons.

2.8 Conclusions

Due to the emergence of the information society, which is not conceived as an increased diffusion of computers and terminals as presently available, but as a space populated with interconnected intelligent objects offering people functionalities for communicating, controlling the environment, and accessing information, emphasis is being placed on the problem of granting universal access to the emerging information space, instead of providing accessibility to individual terminals and computers.

This is causing a revision of the traditional ways of using technology for the social inclusion of people with disabilities. In particular, due to the ongoing transition and the possible complexity of the resulting environment, it is commonly accepted that from reactive approaches to inclusion, based on the adaptation of available mainstream technologies with assistive technology, it is necessary to switch to proactive approaches, whereby the needs, requirements, and preferences of all potential users are integrated in the specifications for the design of new technology and its applications. This implies that assistive technology is no more "the technological solution" for inclusion, but one of its components.

This conceptual approach, known as design for all in Europe and universal design in the United States, is shifting the interest of designers from an artificial "average user" to real users in real contexts of use, aiming for an implementation of systems, services, and applications that are usable by all potential users without modifications. This concept, developed in architecture and industrial design, remains valid in the ICT environment, but the implementation strategy and the technical approach must be changed. As a matter of fact, design for all in the context of information technologies should not be conceived (as in architecture or industrial design) as an effort to advance a single solution for everybody, but as a user-centered approach to providing products that can automatically address the possible range of human abilities, skills, requirements, and preferences. Consequently, the outcome of the design process is not intended to be a singular design, but a design space populated with appropriate alternatives, together with the rationale underlying each alternative, that is, the specific user and usage context characteristics for which each alternative has been designed.

However, it is also necessary to define a technical approach for the practical implementation of this general strategy. A possible technical approach for the implementation of designed for all interfaces has been described in the chapter. The description starts from the initial difficulties found in exploratory activities concerned with the adaptation of telecom terminals and graphical interaction environments (GUIs), leading to the definition of a design for all technical approach based on the automatic initial adaptation of the interface when starting interaction (adaptability) and the continuous automatic adaptation as a function of the use (adaptivity), leading to the development, first, of the dual interface concept, and then of the unified user interface concept.

Then the feasibility of the approach is demonstrated through the application of the approach to the development of the interfaces of real web-based applications, both in the classical Internet environment (the AVANTI system) and in the emerging mobile environment (the PALIO system). Finally, its generalization outside the interface implementation to the automatic adaptation of the information contents of the web pages on which the services are based is also shown.

This approach appears particularly suitable in connection with foreseen technological developments. The emergence of ambient intelligence and the deployment of intelligent interactive environments will obviously be instrumental in making available the intelligence that is necessary to grant adaptability and adaptability in a way that is unobtrusive and anticipates the needs of the single user.

References

Commission of the European Communities (2005). Communication from the Commission to the Council, the European Parliament, the European Economic and Social Committee and the Committee of the Regions. *i2010: A European Information Society for Growth and Employment.* Brussels.

Coutaz, J., Crowley, J. L., Dobson, S., and Garlan, D. (2005). Context is key. *Communications ACM* 48: 49–53.

Emiliani, P. L. (2001). Special needs and enabling technologies, in *User Interfaces for All: Concepts, Methods, and Tools* (C. Stephanidis, ed.), pp. 97–114. Mahwah, NJ: Lawrence Erlbaum Associates.

Emiliani, P. L. (2006). Assistive technology (AT) versus mainstream technology (MST): The research perspective. *Technology and Disability Journal* 18: 19–29.

European Commission (2002). *eEurope 2005 Action Plan.* http://europa.eu.int/information_society/eeurope/2002/news_library/documents/eeurope2005/eeurope2005_en.pdf.

Gill, J. (1996). *Telecommunications: The Missing Links for People with Disabilities.* COST 219, European Commission, Directorate General XIII, Telecommunications, Information Market and Exploration of Research.

IST Advisory Group (2003). *Ambient Intelligence: From Vision to Reality.* ftp://ftp.cordis.lu/pub/ist/docs/istag-ist2003_consolidated_report.pdf.

Lewis, C. and Rieman, J. (1994). *Task-Centred User Interface Design: A Practical Introduction.* http://www.syd.dit.csiro.au/hci/clewis/contents.html.

Mynatt, E. D. and Weber, G. (1994). Nonvisual presentation of graphical user interfaces: Contrasting two approaches, in *Proceedings of the SIGCHI Conference on Human Factors in*

Computing Systems: Celebrating Independence, 24–28 April, Boston, pp. 166–172. New York: ACM Press.

Savidis, A. and Stephanidis, C. (1998). The HOMER UIMS for dual user interface development: Fusing visual and non-visual interactions. *International Journal of Interacting with Computers* 11: 173–209.

Stephanidis, C. (2001a). The concept of unified user interfaces, in *User Interfaces for All: Concepts, Methods, and Tools* (C. Stephanidis, ed.), pp. 371–388. Mahwah, NJ: Lawrence Erlbaum Associates.

Stephanidis, C. (2001b). User interfaces for all: New perspectives into human-computer interaction, in *User Interfaces for All: Concepts, Methods, and Tools* (C. Stephanidis, ed.), pp. 3–17. Mahwah, NJ: Lawrence Erlbaum Associates.

Stephanidis, C. (2003). Towards universal access in the disappearing computer environment. *UPGRADE: The European Online Magazine for the IT Professional, Special Number on HCI* (M. P. Díaz Pérez and G. Rossi, eds.), IV.

Stephanidis, C. and Emiliani, P. L. (1999). Connecting to the information society: A European perspective. *Technology and Disability Journal* 10: 21–44.

Stephanidis, C. and Mitsopoulos, Y. (1995). INTERACT: An interface builder facilitating access to users with disabilities. *Proceedings of HCI International* 2: 923–928.

Stephanidis, C., Paramythis, A., Sfyrakis, M., and Savidis, A. (2001). A case study in unified user interface development: The AVANTI web browser, in *User Interfaces for All: Concepts, Methods, and Tools* (C. Stephanidis, ed.), pp. 525–568. Mahwah, NJ: Lawrence Erlbaum Associates.

Stephanidis, C., Paramythis, A., Zarikas, V., and Savidis, A. (2004). The PALIO framework for adaptive information services, in *Multiple User Interfaces: Cross-Platform Applications and Context-Aware Interfaces* (A. Seffah and H. Javahery, eds.), pp. 69–92. Chichester, U.K.: John Wiley & Sons.

Stephanidis, C., Salvendy, G., Akoumianakis, D., Arnold, A., Bevan, N., Dardailler, D., et al. (1999). Toward an information society for all: HCI challenges and R&D recommendations. *International Journal of Human-Computer Interaction* 11: 1–28.

Stephanidis, C., Salvendy, G., Akoumianakis, D., Bevan, N., Brewer, J., Emiliani, P.-L. et al. (1998). Toward an information society for all: An international R&D agenda. *International Journal of Human-Computer Interaction* 10: 107–134.

Stephanidis, C., Savidis, A., and Akoumianakis, D. (1995). Tools for user interfaces for all, in *Proceedings of the 2nd TIDE Congress* (I. Placencia-Porreiro and R. P. de la Bellacasa, eds.), 26–28 April, Paris, pp. 167–170. Amsterdam: IOS Press.

Trace Center (2008). *General Concepts, Universal Design Principles and Guidelines*. http://trace.wisc.edu/world/gen_ud.html.

U.S. Code (1998). *The Rehabilitation Act Amendments (Section 508)*. http://www.access-board.gov/sec508/guide/act.htm.

Vanderheiden, G. C. (1990). Thirty-something million: Should they be exceptions? *Human Factors* 32: 383–396.

Vanderheiden, G. C. (1998). Universal design and assistive technology in communication and information technologies: Alternatives or compliments? *Assistive Technology* 10: 29–36.

W3C-WAI. (1999). *Web Content Accessibility Guidelines 1.0*. http://www.w3c.org/TR/WCAG10.

World Health Organization (2001). *International Classification of Functioning, Disability and Health (ICF)*. http://www.who.int/classifications/icf/site/icftemplate.cfm.

3

Accessible and Usable Design of Information and Communication Technologies

Gregg C. Vanderheiden

3.1 Introduction

Designing accessible information and communication technologies (ICT) has always been challenging. However, the dramatic changes in human interface that are now occurring are creating new challenges, some of which cannot be addressed with old approaches. New types of speech, gesture, and biosensor inputs are being developed. There are also new levels of intelligence, adaptation, and variation in the behavior of interfaces over time. Software is becoming virtual, as is computing. And the introduction of "pluggable user interfaces" changes the definition of "device user interface" from a physical-sensory form to a command and variable form (Vanderheiden and Zimmerman, 2005). About the only thing that is not changing is the human being and the range of abilities and limitations that humans present. However, with the possibility of direct brain interfaces and other direct neural interfaces, abilities and opportunities for human interfaces may be changing as well.

This chapter will negotiate the different facets of accessible interfaces to ICT in a layered fashion starting with user needs, then current techniques and strategies for addressing them. Approaches for addressing both single and multiple disabilities are covered. The chapter will cover access via assistive

technologies, universal design, and pluggable user interfaces. It will also examine how these terms are blurring in ways that are changing them from categories to characteristics. That is, where it used to be possible to sort assistive technologies or techniques into these categories, most devices and technologies in the future will exhibit characteristics of all categories. This will provide advantages but may further complicate things as well, particularly around public policy. This chapter closes with a look at the future of interface as it relates to information and communication technologies, highlighting both the challenges and opportunities.

3.2 Needs of Individuals Experiencing Constraints

This chapter is primarily about individuals experiencing functional limitations due to disability, including those experienced during aging. However, most of the principles for making devices more accessible also solve problems of individuals who do not have disabilities, but who may be experiencing limitations due to some other factor. For instance, in a very noisy environment an individual who ordinarily has no trouble hearing may have great difficulty or find it impossible to hear the auditory output from a device such as a cell phone or ticket vending machine. While driving a car, one needs to operate devices without using vision. Others may find themselves in an environment where it is dark or may be without their glasses and unable to see controls or labels. Table 3.1 provides some parallels between individuals with disabilities and individuals without disabilities who may find themselves experiencing environmental or task-induced constraints. When all of these people are considered, as well as individuals experiencing a wide range of temporary disabilities, it is useful to note that "those with disabilities" are not such a small portion of the population. And when people are considered across their lifetime (rather than looking at a snapshot of the population at any point in time), the result is that most people will acquire disabilities if they live long enough.

Everybody hopes to live well into their sixties, and beyond. Unfortunately, as we age, an ever-increasing percentage of people will acquire functional limitations. In fact, all of us will acquire disabilities—unless we die first. Figure 3.1 provides a glimpse of this effect by plotting out the percentage of individuals with functional limitations as a function of age. If this series is continued, it will be moving toward unity as one increases in age. Figure 3.2 shows that these disabilities include physical, visual, and hearing. In addition, observing the percentages, it becomes clear that people acquire multiple disabilities as they age. Unfortunately, those who are designing the world in which people must live are usually the youngest, most able, and most technically oriented. Perhaps, as an ever-increasing percentage of the population falls in the upper age groups, market pressures may cause designs to take those with disabilities more into account to enable elders to remain productive and to live more independently for a greater portion of their lives.

3.2.1 Profile of User Interface Needs

There are many different ways of looking at user needs. One way is to explore the needs by disability. This is the approach originally taken in studying consumer product accessibility guidelines (Vanderheiden and Vanderheiden, 1992), Guide 71,[1] and many others that are organized by disability or limitations. However, a more useful approach to designers might be to examine user needs by interface component or interface dimension across disabilities. That is, examine the different parts or functions of the human interface individually and look at the impact or barriers experienced by individuals with different disabilities. Cross-disability interface strategies can then be described and understood more easily. Designers can both see aspects of design that would work for multiple disabilities, and identify those strategies that would not create barriers for one disability while solving another. This is particularly important for designing access into mainstream and public devices, where the interface must be usable by all. The Trace Center at the University of Wisconsin first began exploring this approach about 7 years ago and developed a user needs profile based on basic access/use essentials. It should be noted that these are not essentials for individuals who have disabilities, but essential components that must be there for anyone to be able to effectively use an interface. Everyone must be able to *perceive, operate,* and *understand* a product's interface to use the product. It must also be *compatible* with anything that is part of their person (glasses, clothes, or, for people with disabilities, any assistive technologies they must use while using the product). The basic essentials are:

1. Perceive
   To use a product, users:
   1a. Must be able to *perceive any information that is displayed.*
       - This includes information that is displayed passively (labels, instructions) or actively (on displays).
       - It includes both visually displayed information and information delivered in auditory form (usually speech).
       - Includes labels, signs, manuals, text on the product, and information conveyed by symbols on displays, alerts, alarms, and other output.

---

[1] ISO/IEC Guide 71:2001 guidelines for standards developers to address the needs of elderly persons and persons with disabilities.

TABLE 3.1 Parallel Chart: Disability vs. Situation

Requirement	Disability-Related Need	Situation-Related Need
Operable without vision	People who are **blind**	People whose **eyes are busy** (e.g., driving a car or phone browsing) or who are **in darkness**
Operable with low vision	People with **visual impairment**	People using a **small display** or in a high-glare, **dimly lit environment**
Operable with no hearing	People who are **deaf**	People in **very loud environments** or whose **ears are busy** or are in **forced silence** (library or meeting)
Operable with limited hearing	People who are **hard of hearing**	People in **noisy environments**
Operable with limited manual dexterity	People with a **physical disability**	People in a **space suit** or **chemical suit** or who are in a **bouncing vehicle**
Operable with limited cognition	People with a **cognitive disability**	People who are **distracted, panicked,** or under the **influence of alcohol**
Operable without reading	People with a **cognitive disability**	People who just **have not learned to read a specific language**, people who are visitors, people who left reading glasses behind

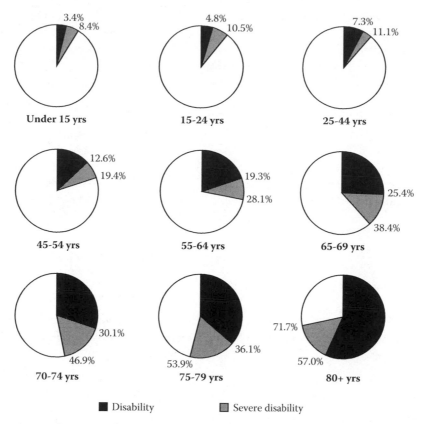

FIGURE 3.1 Prevalence of impairments by age. Pie charts show the percentage of people who have a disability as a function of age. From U.S. Department of Health and Human Services: 2006 National Health Interview Survey (http://www.cdc.gov/nchs/data/series/sr_10/sr10_235.pdf).

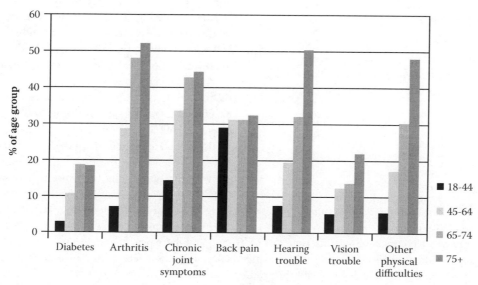

FIGURE 3.2 Disability as a function of age. Physical, sensory, and cognitive disabilities at different ages show that as we age physical, hearing, and visual disabilities rise sharply from small percentages to the 50% range. From U.S. Census Bureau, Survey of Income and Program Participation (June–September 2002).

1b. Must be able to *perceive the existence and location of actionable components.*

- Buttons, controls, latches, etc.
  - Must be able to find them and refind them easily.
  - Must be able to perceive the status of controls and indicators.
  - Includes progress indicators and the status of any switches, dials, or other controls, real or virtual.

1c. Must be able to *perceive any feedback* from operation.

- Includes not only programmed feedback, but natural feedback such as machine sounds that are important for safe and effective use of the device.

2. Operate

To use a product, users:

2a. Must be able to *invoke and carry out all functions* via at least one mode of operation.

- Including daily maintenance and set up expected of users.
- Preferably all of maintenance and set up.

2b. Must be able to *complete all actions and tasks within the time allowed.*

- To effectively compete in the workplace, meeting productivity requirements, etc.

2c. Must be able to *operate without accidentally activating actions.*

2d. Must be able to recover from errors.

- Physical or cognitive errors.

2e. Must have equivalent *security and privacy.*

- If alternate modes are needed, they need to provide equivalent security and privacy.

2f. Must be able to use *without causing* personal risk.

- e.g., seizures, physical injury, etc.

3. Understand

To use a product, users:

3a. Must be able to *understand how to use the product.*

- Including discovery and activation of any access features needed.

3b. Must be able to *understand the output* or displayed material.

- Even after they have perceived it accurately.

4. Compatible with personal technologies

To use a product, users:

4a. Must be able to *use the product in conjunction with any personal technologies.*

- e.g., glasses, wheelchairs, hearing aids, etc.
- For some it would be more efficient if they could use their own personal interface devices with the products they encountered.
- For others, the only way they would be able to use products would be to use specialized input devices

that they would bring with them since it would be impractical to have them built into the products they encounter.

These basic principles were expanded into a profile, including the problems faced by individuals with different disabilities for each of these categories and specific user needs, as part of an online design tool under development at the Trace R&D Center (see Table 3.2 for current version).

In April 2005, these were submitted to the Joint Technical Committee, ISO-IEC Special Working Group on Accessibility (ISO/IEC JTC1 SWGA) where they underwent review, comment, and revision on their way to becoming a JTC1 technical report. The final version from JTC1 is scheduled for release in early 2009.

3.3 Strategies for Addressing User Needs

3.3.1 General Approaches

If someone is not able to use the environment and devices they encounter in daily life effectively, there are three approaches to intervention:

1. Change the individual, so that the person can use the world better as it is found.
2. Adapt the individual products encountered by the person to make them usable by the person.
3. Change the world, so that it is easier for people to use with the abilities they have.

The first approach, changing the individual, is based on a medical model and is a very important strategy. It seeks to increase the basic abilities of the individual through both medical and other rehabilitation strategies. It may include surgery and rehabilitation therapy, but also includes training, the learning of techniques from peers, and in many cases, equipping the individual with personal assistive technologies such as glasses, hearing aids, prostheses, splints, and wheelchairs. In the future, individuals (both with and without disabilities) may also carry around with them specialized interface technologies or devices that are tuned to the individual's need and could serve as personal interfaces to the devices around them (see the pluggable user interfaces discussion later in this chapter). These personal assistive technologies are thought of as extensions of the individual.

The second approach, adapt the individual products encountered, has been around as long as there have been inventive people with disabilities and their inventive friends. This approach basically focuses on adapting the devices around the individual so that they are operable by the individual. This includes, for example, adding tactile markings to a stove or microwave or putting grab bars near the toilet. Adaptations for information and communication technology include special keyboards, screen readers, and enlargers. This category includes products that are developed on a custom basis for individuals, as well as commercially available adaptive or assistive technologies (AT) used with mainstream products to make them more accessible and

TABLE 3.2   User Needs Summary: Trace Center, University of Wisconsin–Madison

Basic	Problems Using Products	User Needs

Users need to be able to PERCEIVE all information presented by the product, including:

Basic	Problems Using Products	User Needs

Perceive static displayed info
-Labels
-Signs
-Manuals
-Text
-Etc.
	People who are blind
	
	• Cannot see (to read)
	  ○ Printed labels on keys, controls, slots, etc.
	  ○ Printed signs near device or instructions printed on device
	  ○ Manuals or other printed material provided with product
	• Cannot access information presented (only) via graphics
	• Cannot find public devices (cannot see where device is or see signs giving location)
	
	People with low vision
	
	• Cannot see (to read) signs and labels:
	  ○ If text is too small for them
	  ○ If contrast with background is too low
	  ○ If text is presented as small raised letters (same color as background)
	  ○ If information is coded with color only (color deficiency)
	  ○ If there is glare (intensity); if they have light sensitivity
	  ○ If there is surface (reflective) glare
	  ○ (many problems same as blindness)
	
	People with physical disabilities
	
	• Often cannot reposition themselves to see information if not in easy sightline
	• May not be able to see due to glare/reflections (and cannot reposition enough)
		Some users with disabilities
		
		• Need to have all static text information required for use provided via speech output or large raised text
		  ○ NOTE 1: Braille is also very useful to people who know it where it is practical to put it on the product, but it would be in addition to speech, not instead of, since most people who are blind do not know Braille, including those who acquire blindness late in life and those who have diabetes, which takes away sensation in the fingertips
		  ○ NOTE 2: Speech output also important for those with cognitive disabilities (see "UNDERSTAND")
		  ○ NOTE 3: Raised text would need to be approx. 3/4 inch high
		• Need to have visual cues provided in auditory form
		• Need sufficient contrast between all printed information and its background
		• Need to have text presented in large easy-to-read fonts
		• Need to avoid surface (reflective) glare
		• Need to have information within viewable range of people in wheelchairs and those of short stature
		• Need to have any information presented in color be also presented in a way that does not depend on color perception

Perceive info presented via dynamic displays
-Screens
-Alerts
-Alarms
-Other output
	People who are blind
	
	• Cannot see what is displayed on visual display units (all types)
	• Cannot determine current function of soft keys (where key function is dynamic with label shown on dynamic display such as LCD)
	
	People with low vision
	
	• Same problems as static text (size, contrast, color) (see above)
	• Glare: from environment or too bright a screen
	• Miss information presented temporarily where they are not looking
	• Sometimes cannot track moving/scrolling text
	
	People who are deaf
	
	• Cannot hear information presented through:
	  ○ Speech
	  ○ Tones
	  ○ Natural machine sounds
		Some users with disabilities
		
		• Need to have all DYNAMIC visual information required for use also provided via speech output
		  ○ NOTE 1: Dynamic Braille displays are very expensive and impractical for inclusion in devices
		  ○ NOTE 2: Speech output is also important for those with cognitive disabilities (see "UNDERSTAND" below)
		  ○ NOTE 3: Raised text won't work for dynamic information
		• Need a means for identifying all keys and controls via speech
		• Need sufficient contrast between all display information (audio or visual) and its background
		• Need to have text presented in large easy-to-read fonts
		• Need to avoid surface (reflective) glare
		• Need to avoid brightness glare
		• Need to have information within viewable range of people in wheelchairs and those of short stature
		• Need to have all auditory information required for use also available in visual and/or tactile form
		  ○ NOTE 1: Tactile presentation only useful for products that will always be in contact with user's body

(Continued)

TABLE 3.2 (Continued)

Basic	Problems Using Products	User Needs

	People who are hard of hearing
	
	• May miss any information presented auditorily because:
	  ○ At a frequency they can't hear
	  ○ Background noise blocks it or interferes with it (including echoes)
	  ○ Too soft
	  ○ Poor-quality speech
	  ○ Speech too fast and user can't slow it down
	  ○ Presented as two different tones that can't be distinguished
	  ○ Information is presented that requires stereo hearing
	
	People with physical disabilities
	
	• Cannot maneuver to see display or avoid glare
	
	People with cognitive disabilities
	
	• Distracted by dynamic movements on screen
		• Need to have auditory events, alerts, etc., be multifrequency so that they can hear it
		• Need sufficient volume (preferably adjustable) for audio output
		• Need to have any information presented in color also presented in a way that does not depend on color perception
		• Need to be able to control the colors in which information is presented
		• Need to be able to control the pitch of information presented auditorily
		• Need to have audio information conveyed by sound pattern, not frequency
		• Need to have audio information conveyed by vibration to use patterns, not frequency or strength
		• Stereo information available in monaural form

Perceive existence and location of actionable components
-Buttons
-Controls
-Latches
-Etc.
(find them and refind them)
	People who are blind
	
	• Cannot determine number, size, location, or function of controls on
	
	  ○ Touchscreens
	  ○ Flat membrane keypads
	
	• Cannot find controls in a large featureless group; cannot be relocated easily even if known to be there
	• Switch or control in an obscure location may not be discoverable even if visible
	• Might touch tactilely sensitive controls while exploring with hands
	• Can be fooled by phantom buttons (tactile) (things that feel like buttons but are not, e.g., a logo, a round flat raised bolt head, a styling feature)
	• Cannot type on a non-touchtypeable keyboard
	
	People with low vision
	
	• Cannot find buttons that don't contrast with background (won't feel where nothing is visible or expected)
	• Phantom buttons (visual) (logos, styling that looks like button when blurred)
	• Cannot locate where the cursor is on the screen
	
	People with physical disabilities
	
	• Often cannot reposition themselves to see controls if not in easy sightline
	
	People with cognitive disabilities
	
	• Do not recognize stylized control as a control
		Some users with disabilities
		
		• Need a means to access all product functionality via tactilely discernable controls
		• Need controls to be locatable without activating control or nearby controls
		• Need sufficient landmarks (nibs, groupings, spacing) to be able to easily locate controls tactilely once they have identified them (per above)
		• Need to have controls visually contrast with their surroundings so they can be located with low vision
		• Need to have any keyboard be operable without sight
		• Need to have controls be in places where they can be easily found with poor or no sight
		• Need to have pointing cursors (on screen) be large enough to be visible with low vision
		• Need to have logos and other details not look like or feel like buttons or controls
		• Need to have controls where they can be seen by people of short stature and those using wheelchairs

Perceive status of controls and indicators (includes PROGRESS indicators)	**All disabilities** • Cannot tell state if the same alternative is provided for different signals **People who are blind** • Cannot tell status of visual indicators (LEDs, onscreen indicators, etc.) • Cannot tell the status of switches or controls that are not tactilely different in different states (or where tactile difference is too small)	**Some users with disabilities** • Need an auditory or tactile equivalent to any visual indicators or operational cues, manmade or natural • Need a visual or tactile indicator for any auditory indicators or operational cues, manmade or natural • Need visual or auditory alternative to any subtle tactile feedback • Need visual indicators to be visible with low vision

TABLE 3.2 *(Continued)*

Basic	Problems Using Products	User Needs
	People with low vision • Cannot read visual indicators with low vision if indicator is not bold • Cannot distinguish between some colors used to indicate status • Cannot see or read small icons for status • Cannot see cursors unless large, high contrast; static harder than dynamic to spot **People who are deaf** • Cannot hear audio indicators of status • Cannot hear natural sounds (e.g., machine running, stalled, busy, etc.) **People who are hard of hearing** • May not hear status sounds due to volume, frequency used, background noise, etc. **People with physical disabilities** • May not have good line of sight to indicators • May not have tactile sensitivity to detect tactile status indications **People with cognitive disabilities** • May not recognize or understand different indicators	• Need all indications that are encoded (or presented) with color to be encoded (marked) in some noncolor way as well • Need large high-contrast pointer cursors • Need sufficient volume and clarity for audio cues • Need alternatives that are different, when different signals are used (e.g., different ringtones, or tactile or visual indicators) • Need indicators and cues to be obvious or explained • Need to have controls and indicators located where they can be seen by people of short stature and those using wheelchairs
Perceive feedback from operation	**All disabilities** • Cannot tell state if the same alternative is provided for different signals **People who are blind** • Cannot see visual feedback of operation **People with low vision** • Cannot see visual feedback of operation unless large, bold ○ Often have hearing impairments as well so cannot always count on audio **People who are deaf** • Cannot hear auditory feedback of operation **People who are hard of hearing** • Often cannot hear auditory feedback of operation due to: ○ Volume ○ Frequency used ○ Background noise ○ Speech feedback not clear or repeatable **People with physical disabilities** • May not be able to feel tactile feedback due to insensitivity or impact of hand or use of artificial hand, stick, splint, etc. to operate the control **People with cognitive disabilities** • Feedback too subtle or not directly tied to action	**Some users with disabilities** • Need visual feedback that is dramatic (visual from 10 ft) • While others need it to be audio or tactile feedback

TABLE 3.2 *(Continued)*

Basic	Problems Using Products	User Needs
Be able to OPERATE the product		

Be able to invoke and carry out all functions (via at least one method)	**People who are blind**
	
	• Cannot use controls that require eye-hand coordination
	   ○ Pointing devices, including mice, trackballs, etc.
	   ○ Touchscreens of any type
	• Cannot use devices with touch-activated controls (can't explore tactilely)
	• Cannot use products that require presence of iris or eyes (e.g., for identification)
	
	People with low vision
	
	• Difficult to use device with eye-hand coordination
	
	People who are deaf
	
	• Many cannot use if speech input is only way to do some functions
	• Cannot operate devices where actions are in response to speech (only)
	
	People with physical disabilities
	
	• Cannot operate devices if operation <u>requires</u> (i.e., no other way to do function):
	   ○ Too much force
	   ○ Too much reach
	   ○ Too much stamina (including long operation of controls with arm extended or holding handset to head for long period unless able to prop or rest arm)
	   ○ Contact with body (so that artificial hands, mouthsticks, etc., cannot be used)
	   ○ Simultaneous operation of two parts (modifier keys, two latches, etc.)
	   ○ Tight grasping
	   ○ Pinching
	   ○ Twisting of the wrist
	   ○ Fine motor control or manipulations (i.e., can't operate with closed fist)
	   ○ Cannot use products that require presence of fingerprints or other specific body parts or organs (e.g., for identification)
		Some users with disabilities
		
		• Need to be able to operate all functionality using only tactilely discernable controls coupled with audio or tactile feedback/display (no vision required)
		• Not requiring a pointing device
		• Need to not have touch-sensitive or very light touch controls where they would be touched while tactilely finding keys they must use to operate device
		• Need alternate identification means if biometrics are used for identification
		• Need alternate method to operate any speech-controlled functions
		• Need to be able to access all computer software functionality from the keyboard (or keyboard emulator)
		• Need method to operate product that does not require:
		   ○ Simultaneous actions
		   ○ Much force
		   ○ Much reach
		   ○ Much stamina
		   ○ Tight grasping
		   ○ Pinching
		   ○ Twisting of the wrist or
		   ○ Direct body contact

Be able to complete actions and tasks within the time allowed (by life, competition, productivity requirements, etc.)	**People who are blind**
	
	• Must use nonvisual techniques that are often slower, requiring more time than usual to read/listen to output, explore, and locate controls etc.
	
	People with low vision
	
	• Often take longer to read text and locate controls
	
	People who are deaf
	
	• May be reading information in a second language (sign language being first)
	• May be communicating (or operating phone system) through a relay/interpreter that introduces delays
	
	People who are hard of hearing
	
	• May have to listen more than once to get audio information
	
	People with physical disabilities
	
	• May take longer to read (due to head movement), to position themselves, to reach, or to operate controls
		Some users with disabilities
		
		• Need to have all messages either stay until dismissed or have a mechanism to keep message on screen or easily recall it
		• Need to have ability to either:
		   ○ Have no timeouts, or
		   ○ Have ability to turn off timeouts, or
		   ○ Be able to set timeouts to 10 times default value, or
		   ○ Be warned when timeout is coming and be provided with ability to extend timeouts except where it is impossible to do so
		• Need to have a way to turn off or freeze any moving text

TABLE 3.2 (*Continued*)

Basic	Problems Using Products	User Needs
	People with cognitive disabilities • May take longer to remember, to look things up, to figure out information, and to operate the controls, all of which can cause problems if: ◦ Information or messages are displayed for a fixed period and then disappear ◦ Users are only given a limited amount of time to operate device before it resets or moves on ◦ Text moves on them while they are trying to read it	
Won't accidentally activate functions	**People who are blind** • Might touch "touch sensitive" controls or screen buttons while tactilely exploring • Might miss warning signs or icons that are presented visually • Might bump low-activation force switch(es) while tactilely exploring **People with low vision** • Might bump low-contrast switches/controls that they do not see **People who are deaf or hard of hearing** • May not detect alert tone and thus inadvertently operate device when unsafe **People with physical disabilities** • Might activate functions due to extra body movements (tremor, chorea) • Might activate functions when resting arm while reaching **People with cognitive disabilities** • Might not understand purpose of control (or control changes due to softkey)	**Some users with disabilities** • Need to have products designed so they can be tactilely explored without activation • Need products that can't cause injury with spasmodic movements • Need to have products that don't rely on users seeing hazards or warnings to use products safely • Need to have products that don't rely on users hearing hazards or warnings to use products safely • Need to have products where hazards are obvious and easy to avoid, hard to trigger
Be able to recover from errors (physical or cognitive errors)	**People who are blind or have low vision** • May not detect error if indication is visual • May not be able to perceive contextual cues (if visual only) to know they did something wrong or unintended (when not an "error" to the device) **People who are deaf** • Will not hear auditory "error" sounds **People who are hard of hearing** • May not hear auditory "error" sounds or be able to distinguish between them **ALL disabilities** • User may not be able to figure out how to go back and undo the error	**Some users with disabilities** • Need a mechanism to go back and undo the last thing(s) they did, unless impossible • Need good auditory and visual indications when things happen so that they can detect errors • Need to be notified if the product detects errors made by the user • Need clear unambiguous feedback when error is made and what to do to correct
Have equivalent, security, and privacy	**People with all disabilities** • Do not have privacy when human assistance is required **People who are blind** • Have more difficulty detecting people looking over shoulder • If no headphone or handset, information is broadcast to others via speaker **People with low vision** • Larger print makes it easier for others to look over shoulder	**Some users with disabilities** • Need ability to listen privately • Need to have product designed to help protect privacy and security of their information even if they are not able to do the "expected" things to protect it themselves

(*Continued*)

TABLE 3.2 (*Continued*)

Basic	Problems Using Products	User Needs
	People who are deaf • May not detect sensitive information being said aloud **People who are hard of hearing** • Louder volume may allow eavesdropping, even with headphones ○ User may not realize volume of audio **People with physical disabilities** • In wheelchair, body doesn't block view of sensitive information like someone standing **People with cognitive disabilities** • Less able to determine when information should be kept private	
Not cause health risk (e.g., seizure, etc.)	**People who are blind** • Cannot see to avoid hazards that are visual • Cannot see warning signs, colors, markers, etc. • If using headphones, they are less aware of surroundings (and not used to it) **People who are deaf or hard of hearing** • May miss auditory warnings or sounds that indicate device failure **People with physical disabilities** • May hit objects harder than usual and cause injury • May not sense when they are injuring themselves **People with photosensitive epilepsy** • May have seizure triggered by provocative visual stimuli **People with allergies and other sensitivities** • May have adverse reactions to materials, electromagnetic emissions, fumes, and other adverse aspects of products they touch or are near	**Some users with disabilities** • Need products that don't assume body parts will never stray into openings or that only gentle body movements will occur around the products (unless required by task) • Need to have products that take into account their special visual, physical, chemical, etc., sensitivities so that they are not prevented from using products except when the nature of the product or task would prevent them (e.g., not by product design)
Be able to efficiently navigate product	**People who are blind** • Often have to wait for unnecessary audio before getting to desired information **People with low vision** • Have trouble tracking cursors on screen **People with physical disabilities** • Have trouble with navigation requiring many repeated actions to navigate **People with cognitive disabilities** • Have trouble with hierarchical structures	**Some users with disabilities** • Need to have alternate modes of operation that are efficient enough to allow them to be able to compete in education and employment settings • Need to control speech output rate • Need ability to preserve their access settings
Be able to UNDERSTAND		
Understand how to use product (including discovery and activation of any access features needed)	**ALL disabilities** • May have trouble understanding how to turn on special access features they need • May have trouble understanding how to operate it if different than standard users	**Some users with disabilities** • Need way to get overview and orient themselves to product and functions/parts without relying on visual presentation or markings on product • Need products to operate in predictable (standard or familiar) ways • Need way to understand product if they don't think hierarchically very well

TABLE 3.2   (Continued)

Basic	Problems Using Products	User Needs
	People who are blind (or have low vision) • Have a more difficult time getting general context for the operation of the product, since they cannot see the overall visual layout or organization • Complex layouts can behave like a maze for someone navigating with arrowkeys **People who are deaf** • English (or the spoken/written language used on the product) may be different from their natural (first) language (e.g., if it is sign language) **People with cognitive disabilities** • Have trouble remembering the organization of a product, its menus, etc. • Have a harder time with any hierarchical structures • Cannot read labels, signs, manuals, etc., due to reading limitations • May have trouble understanding directions, especially if printed • May have trouble remembering steps for use • May have trouble getting it turned on, and therefore active • May be confused by options, buttons, controls, that they do not need or use • Icons and symbols may not make sense to them, and they don't remember • Product may differ from real life experience enough to leave them at a loss • Might have trouble with products that operate in nonstandard ways	• Need to have clear and easy activation mechanisms for any access features • Need interfaces that minimize the need to remember • Need to have language used on products to be as easy to understand as possible given the device and task • Need to have printed text read aloud to them • Need to have steps for operation minimized and clearly described. • Need information and feedback to be "salient" and "specific" rather than subtle or abstract to understand it • Need keys that don't change function • Need cues to assist them in multistep operations • Need to have simple interfaces that only require them to deal with the controls they need (advanced or optional controls removed in some fashion)
Understand the output or displayed material (even after they perceive it accurately; see also **PERCEIVE** above)	**People who are blind** • Output often only makes sense visually. Reading it is confusing (e.g., "select item from list at the right" when they get to it by pressing down arrow) • Have difficulty with any simultaneous presentation of audio output and audio description of visual information (e.g., reading of screen information while playing audio) **People who are deaf** • Reading skills; English may not be primary language (ASL) • Can have difficulty with simultaneous presentation of visual information and (visual) captions of auditory information **People with cognitive disabilities** • May not be able to read information presented in text • Language may be too complex for them • Long or complex messages may tax their memory abilities • Use of idiom or jargon may make it hard to understand • Structures, tabular or hierarchical information may be difficult	**Some users with disabilities** • Need descriptions, instructions, and cues to match audio operation, not just visual operation • Need to have any printed material be worded as clearly and simply as possible • Need to have any printed material read to them • Need to have audio generated by access features not interfere with any other audio generated by device • Need to have visual information generated by access features (such as captions) not occur simultaneously with other visual information they must view (and then disappear before they can read the captions)

(Continued)

TABLE 3.2   (*Continued*)

Basic	Problems Using Products	User Needs
Be able to USE THEIR ASSISTIVE TECHNOLOGIES (in addition)		

Basic	Problems Using Products	User Needs

Ability to use their AT to control the product (not always possible with public devices but common with personal or office workstation technologies)
 	
NOTE: To replace built-in access, AT must allow all of the above basics to be met.
	All disabilities
	• Cannot use their AT to access products if:
	  ◦ Product is in public and they will not have their technology with them
	  ◦ They do not have permission to use their AT with the product
	    • e.g., cannot install AT software on library systems
	
	They are not able to connect their AT to it
	• Cannot use their AT if the device interferes with it
	• Cannot use their AT if they are not easily able to find the connection mechanism given their disability
	• Need to have full functionality of the product available to them via their AT
	• AT is not available for new technologies when they come out
	
	People who are blind
	• Would need all visual information to be available to their AT in machine-readable form via a standard connection mechanism
	• Would need to be able to activate all functionality from their AT (or from tactile controls on the product)
	
	People with low vision
	• Would need all visual information to be available in machine-readable form to their AT via a standard connection mechanism so that the AT could enlarge it or read it
	
	People who are deaf
	• Would need all auditory information to be available to their AT in machine-readable form via a standard connection mechanism
	
	People who are hard of hearing
	• Would need all audio information to be available via a standard connection mechanism that is compatible with their assistive listening devices (ALDs)
	  ◦ Need a standard audio connector to plug their ALD
	  ◦ For something held up to the ear, it should be T-Coil compatible
	
	People with physical disabilities
	• Cannot use products that aren't fully operable with artificial hand, stick, stylus, etc.
	• Need connection point that allow operation of all controls
	
	People with cognitive disabilities
	• Would need all information to be available in machine-readable form to their AT via a standard connection mechanism
		Some users with disabilities
		• Need to not have product interfere with their AT
		• Need to be able to connect their AT
		• Need to have full functionality of product available through their AT if they have to use their AT to access the product
		  ◦ Need to have software use standard system-provided input and output methods
		  ◦ Need to have all displayed text made available to their AT
		  ◦ Need information about user interface elements including the identity, operation and state of the element to be available to assistive technology
		  ◦ All controls need to be operable from AT
		• Need to be able to access all computer software functionality from the keyboard (or keyboard emulator)
		• Need to have all controls work with their manipulators, artificial hands, pointers, etc.
		• Need to have new technologies be compatible with their AT when the new technologies are released

Cross-cutting issues	**All disabilities** • Accessibility is not available in new technologies when they come out • Support services are not accessible (no training or proper communication equipment)	**Some users with disabilities** • Need to have new technologies be accessible when they are released • Need to have support and training services that are accessible

usable by individuals with particular disabilities. These bridging or adaptive technologies are especially important for individuals with severe or multiple disabilities, where building sufficient accessibility into mainstream products is not practical. It is also important in employment settings where the employee with a disability must be able to access a product and use it efficiently enough to be competitive and productive.

The third approach, changing the world, is commonly called universal design (UD) or design for all (DFA). It has also been called accessible design, barrier-free design, and inclusive design. The term universal design was originally coined by Ron Mace, an architect and director of the Center for Universal Design at North Carolina State University. He defined it as follows: "Universal design means simply designing all products, buildings and exterior spaces to be usable by all people to the greatest extent possible" (Mace, 1991).

This definition has served well as a reference point, but has raised concerns among some designers because it sets no practical limits. What is possible is not necessarily commercially viable. As universal design/design for all (UD/DFA) moved from a goal to appearing in social legislation, designers began to fear the implications of such an ideal goal (designing things that everyone can use) if the term was used in a requirements context. For example, building a $2,000 Braille display into every electronic device with a visual display is not generally practical. As a result, some designers began to fight the movement rather than embrace or explore the basic concept.

A debate also surfaced as to whether UD/DFA included compatibility with assistive technology, which many view as key to accessibility of mainstream technologies. This is particularly true with regard to personal assistive technologies, as discussed previously.

To address these issues and create a practitioner's definition of universal design (or design for all), a companion definition was proposed:

> The process of designing products so that they are usable by the widest range of people operating in the widest range of situations as is commercially practical. It includes making products directly accessible and usable (without the need for any assistive technology) and making products compatible with assistive technologies for those who require them for effective access. (Vanderheiden, 2000)

3.3.1.1 No Universal Designs, Only Universal Design

It is important to note the word *process* in the preceding definition. UD/DFA is a process, not an outcome. There are no universal designs. That is, there are no designs that can be used by everyone, no matter how many or how severe their disabilities. Universal design is not the process of creating products that *everyone* can use. It is a process of ensuring that designs can be used by as many people as is practical, and then constantly moving that line as new approaches are discovered and new technologies become available. For some products, this might result in a very narrow

range of users, if it has extremely high user demands (e.g., jet fighters). For other products, it can have an extremely wide range of users. Fare machines, information kiosks, and even voting systems have been designed that can be used by individuals who have low vision, who are blind, who are hard of hearing or deaf, who have almost no reach, who cannot read, or who have various cognitive, language, and learning disabilities (Vanderheiden and Law, 2000; Vanderheiden, 2002). Moreover, they do not require multiple modes of operation, but rather options for operation in the same way that both the keyboard and mouse can be used to navigate windowing environments. Yet no matter how good the universal design, there are individuals who will not be able to operate products directly without the need for some type of assistive technology or alternate interface. Hence, the importance of compatibility with AT to complement direct access.

3.3.1.2 Pluggable User Interfaces

This whole area of personal assistive technologies and universal design has been made more interesting by the recent creation of international standards for pluggable user interfaces. A five-part ISO standard (ISO 24752, adopted in January 2008) describes a standard method for mainstream products to expose their functionality so that they can be directly controlled from personal alternate interface devices. This represents a breakthrough in the ability to design products that can be used by a much broader range of users as they encounter them.

Figure 3.3 shows the typical way an individual interacts with a product. The product has certain needs for information and/or commands from the user. A television, for example, needs to know the channel that the user wants to be watching, the volume that the user wants to set, the various color tints and settings to be selected, the source of the signal (cable, DVD, etc.), and so on. It does not care whether it gets the information by having the user push a button, turn a dial, or pick an item from a menu. However, the television comes with a built-in interface that takes its general device requirements (volume, channel, etc.) and changes them into specific actions that a user must perform. These actions may or may not be easy to perform for an individual with a disability.

Figure 3.4 shows the same device, except that an interface socket has been added that allows the individual to plug a different interface into the device to provide the television with the various types of information it needs, but this time using an interface of the user's choosing. An individual who is blind may choose an all-auditory interface; an individual with a physical disability may choose an interface that only requires sipping and puffing, etc.

The interface socket would provide all of the information necessary for the user's personal interface to be able to construct and present an interface to the user for each device encountered. Whenever the user encounters a device (that supports this standard) in the environment, the personal interface can discover it and download from it all of the information about the product's functions, and any commands or settings needed to operate the device. Optionally the device can also provide hints as to how an

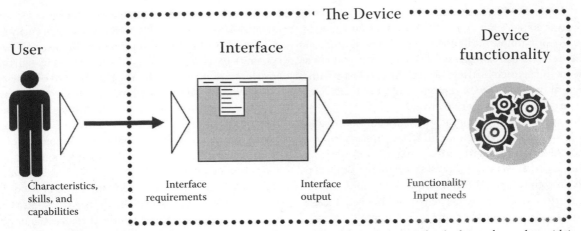

FIGURE 3.3 Typical interface where the only access to the device's functionality is through the interface built into the product with its particular assumptions and requirements regarding the user's abilities.

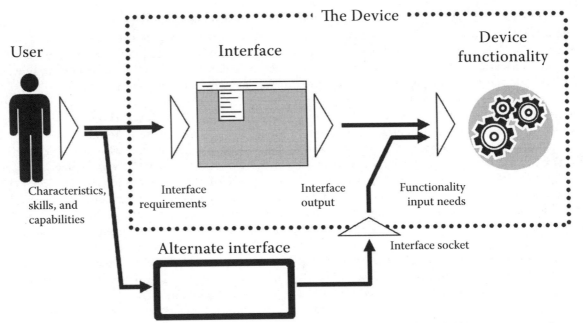

FIGURE 3.4 By adding an alternate interface connection point to the device in Figure 3.3, it is possible for users to connect an alternate interface that better matches their abilities.

interface should be structured. The user's personal interface can then construct an interface for the product that meets the user's needs. If he is blind, it might be auditory. If she has severe paralysis, it might be voice controlled. For a user who is deaf-blind, it might be a Braille-based interface. It should be noted that the pluggable user interface may connect to products without the need for any physical plug. Merely coming near a product may be all that is necessary for an alternate interface to link up to an

interface socket on a product, allowing the individual to control the product using the alternate interface.

Key to this approach are the universal remote console (URC) standards (ANSI-INCIT 2005-389 to 393 and ISO-24752). Unlike products controlled by universal remote controls that need to be programmed, products implementing the URC standards would automatically provide all of the information needed by a URC to construct an interface for a user. In this way, an individual could approach an unknown device and the URC would be able to automatically construct an interface for this device that meets the user's needs and preferences.

Personal pluggable user interfaces like this that the user would carry about are examples of both type 1 (change the user) and type 3 (change the world) strategies working together. They could also be a type 2 strategy (adapt the individual products encountered by the person to make them usable to the person). That would be less common, since it is rarely possible for anyone besides the manufacturer to create an interface socket for a product. It can be seen, however, how the approaches are blending together as technologies advance.

3.3.1.3 Working Together, Blending Together

All of these techniques (change the person, adapt the environment, and change the environment) need to work together to accommodate individuals with the full range of abilities and limitations. In general, individuals with less severe disabilities can be accommodated through universal design/design for all. This includes a large percentage of individuals who are aging. For individuals with more severe or multiple disabilities, it may be more difficult to build interfaces into mainstream products that will work effectively for them. Either adaptive assistive technology, or perhaps personal assistive technology that connects through pluggable user interface sockets, may provide effective access.

More and more mainstream products are being developed that have interfaces that are more naturally flexible. Interface sockets are of interest to mainstream users as much as to people with disabilities. For example, many would like to be able to control their televisions and other devices in their house by simply pulling out their PDA or cell phone. With task-based control and continuing advances in voice recognition in constrained applications, it may soon be possible for mainstream users to control their television by simply telling it what they want to do, rather than having to navigate all of its menus, buttons, and features. ("Turn on *Masterpiece Theatre*"; "Record the 49ers game tomorrow"; "Show me the movies I have recorded"; "Play *Casablanca*"; "Play the DVD I just put in.") Moreover, since the intelligence is in the URC, the same commands would work on devices wherever they are encountered. Thus, changing TVs or changing environments, which can currently confuse many users, including many elders, may no longer be as much of a problem since these people could continue to use their familiar interface (on the URC) to control the new devices in the same manner as the old.

Thus, universal design can take many forms, including design of the main or default interface so that it is more usable, providing the ability to invoke interface modules that would increase the accessibility of the product, and options such as an interface socket to allow users to use alternate interfaces they carry with them. There is even research today looking at direct brain interfaces for both people with disabilities and those without (see Chapter 37, "Brain-Body Interfaces"). Combined with interface sockets, someday people may be interfacing with the devices around them without lifting a finger or having to carry physical interface devices (displays, keyboards, etc.) around with them.

Designing accessible products in the future therefore must look at all three approaches, since different users, environments, and tasks may require different approaches (built-in, adaptive, and alternate, pluggable accessible interfaces). Together with advances in technologies, the combination can provide new options for addressing the full range of functional limitations, including those not well addressed today.

3.3.2 Specific Strategies to Address Needs

There is an extremely wide range of specific techniques that can be used for addressing the needs outlined in Section 3.2. Table 3.3 provides a summary of the basic strategies organized in parallel with the essentials: perceivable, operable, understandable, and compatible. A more comprehensive list of strategies as well as specific techniques that can be useful when implementing them on different technologies can be found at http://www.trace.wisc.edu/resources.

As always, these strategies may be employed in the design of the mainstream product itself, or may be made possible by allowing the connection of adaptive assistive technologies or personal assistive technologies.

3.4 Priorities in Implementation

In looking at accessibility-usability features, it is important to prioritize because of the multidimensional nature of disability (vision, hearing, physical, cognitive) and the large number of individual design techniques or strategies that might be implemented for each dimension. Between 200 to 300 different strategies were identified in a number of design strategy collections for making products more accessible to people with disabilities (and this did not include the large number of different strategies documented in general usability literature). Without a means to prioritize, two behaviors have been observed in interactions with industry.

First, product designers become overwhelmed with the sheer number of different techniques and strategies. Just contemplating building over 100 different strategies into a product causes many to walk away or to approach feature selection (focusing of their efforts) in a somewhat random fashion. Usability tests by themselves are not a solution to the problem of being overwhelmed, since they quickly generate a long list of problems that, in turn, point back to the even longer list of potential solution strategies.

The second behavior observed is a poor prioritization in efforts where features that were first thought of or easiest to implement are chosen rather than strategies or features that are more important. The result is a product that has multiple low-priority features (which are helpful but not essential for access) while lacking key high-priority features that are needed to make the product (or key functions of the product) accessible for the same disability group. This is equivalent to changing the plush carpet in the entire building to a tighter nap to make it easier for wheelchairs to get around, but leaving the steps at the front door and having no elevators to get off of the ground floor.

TABLE 3.3 Basic Guidelines and Strategies for Access to Electronic Products and Documents (July 2001)

Basic Access Guideline	Why	How—General
Make **all information** (including status and labels for all keys and controls) **perceivable** • Without vision • With low vision and no hearing • With little or no tactile sensitivity • Without hearing • With impaired hearing • Without reading (due to low vision, learning disability, illiteracy, cognition, or other) • Without color perception • Without causing seizure • From different heights NOTE: Other aspects of cognition covered below	Information presented in a form that is only perceivable with a single sense (e.g., only vision or only hearing) is not accessible to people without that sense **NOTE:** This includes situations where some of the information is only presented in one form (e.g., visual) and other information is only presented in another (e.g., auditory) **In addition:** Information that cannot be presented in different modalities would not be accessible to those using mobile technologies, for example: • Visual-only information would not be usable by people using an auditory interface while driving a car • Auditory-only information would not be usable by people in noisy environment	**FOR INFORMATION:** Make all information available either in: a) **Presentation-independent form** (e.g., electronic text) that can be presented (rendered) in any sensory form (e.g., visual-print, auditory-speech, tactile-Braille) **OR** b) **Sensory parallel form** where redundant and complete forms of the information are provided for different sensory modalities (synchronized) (e.g., a captioned and described movie, including e-text of both) **FOR PRODUCTS:** Provide a mechanism for presenting all information (including labels) in visual, enlarged visual, auditory, enhanced auditory (louder and if possible better signal to noise ratio), and (where possible) tactile form **NOTE:** This includes any information (semantics or structure) that is presented via **text formatting or layout**
Provide **at least one mode (or set of different modes)** for **all product features** that **is operable**: • Without pointing • Without vision • Without requirement to respond quickly • Without fine motor movement • Without simultaneous action • Without speech • Without requiring presence or use of particular biological parts (touch, fingerprint, iris, etc.)	Interfaces that are input device- or technique-specific cannot be operated by individuals who cannot use that technique (e.g., a person who is blind cannot point to a target in an image map; some people cannot use pointing devices accurately) **In addition:** Technique-specific interfaces may not be accessible to users of mobile devices; for example, people using voice to navigate may not be able to "point" Many individuals will not be able to operate products, such as workstations, with sufficient efficiency to hold a competitive job if navigation is not efficient	Provide at least one mode (set of modes) where… a) All functions of the product are controllable via tactilely discernable controls and both visual and voice output is provided for any displayed information required for operation including labels **AND** b) There are no timeouts for input or displayed information, OR allow user to freeze timer or set it to long time (5 times default or range), OR offer extended time to user and allow 10 seconds to respond to offer **AND** c) All functions of the product operable with: • No simultaneous activations • No twisting motions • No fine motor control required • No biological contact required • No user speech required • No pointing motions required **AND** d) If biological techniques are used for security, have at least two alternatives with one preferably a nonbiological alternative unless biological-based security is required **AND** e) Allow users to jump over blocks of undesired information (e.g., repetitive info; or jump by sections if large document), especially if reading via sound or other serial presentation means. f) Make actions reversible or request confirmation
Facilitate understanding of operation and content • Without skill in the language used on the product (due to poor language skills or because it is a second language) • Without good concentration, processing • Without prior understanding of the content • Without good memory • Without background or experience with the topic	Many individuals will have trouble using a product (even with alternate access techniques) if the **layout/organization** of the information or product **is too difficult to understand** People with cognitive or language difficulties (or inexperienced users) may not be able to use devices or products with complex language	a) Make overall organization understandable (e.g., provide overview, table of contents, site maps, description of layout of device, etc.) b) Don't mislead/confuse (be consistent in use of icons or metaphors—don't ignore or misuse conventions) c) Consider having different navigation models for novice vs. expert users d) Use the simplest, easiest-to-understand language and structure/format as is appropriate for the material/site/situation e) Use graphics to supplement or provide alternate presentations of information f) If phrases from a different language (than the rest of the page) are used in a document, either identify the language used (to allow translation) or provide a translation to the document language

TABLE 3.3 (*Continued*)

Basic Access Guideline	Why	How—General
Provide compatibility with assistive technologies commonly used by people • With low vision • Without vision • Who are hard of hearing • Who are deaf • Without physical reach and manipulation • Who have cognitive or language disabilities	In many cases, a person coming up to a product will have assistive technologies with them; if the person cannot use the product directly, it is important that the product be designed to allow them to use their assistive technology to access the product **NOTE:** This also applies to users of mobile devices and people with glasses, gloves, or other extensions	a) Do not interfere with use of assistive technologies • Personal aids (e.g., hearing aids) • System-based technologies (e.g., OS features) b) Support standard connection points for: • Audio amplification devices • Alternate input and output devices (or software) c) Provide at least one mode where all functions of the product are controllable via human understandable input via an external port or via network connection

The purpose is to map out the dimensions of complexity involved and then to develop simplified and straightforward (as possible) techniques and procedures for addressing or accommodating them. The advice of Albert Einstein is appropriate to remember here: "Everything should be made as simple as possible. But no simpler." Hence the goal is to make this as simple as possible, but not to simplify where that leads to inaccuracy or poor decisions.

3.4.1 First Dimension for Prioritization: Accessibility/Usability

In looking at the usability of a product to different people, there is a continuous range that runs all the way from:

- People who have no problems at all in using all of the functions of a product (usually a small number of people)
- People who have little difficulty with all
- People who have difficulty with some features
- People who have trouble with most features
- People who are unable to use the product at all

Individual product features vary in importance to the overall use of the product. Some features are essential, while others are merely convenient.

The importance of product features, may be evaluated based on the following criteria:

- *Essential*—features that, if they have not been implemented, will cause a product to be unusable for certain groups or situations.
- *Important*—features that, if not implemented, will make the product very difficult to use for some groups or situations.
- *Improving usability*—features that, if they are implemented, will make the product easier to use but do not make a product usable or unusable (except for individuals who are on a margin due to other factors and this small amount of usability pushes them over the threshold).

In looking at this dimension, however, it is important to note that features that may merely improve the usability for some people may be essential to allow use by others. This is especially true for people with cognitive, language, and learning disabilities.

3.4.2 The Second Dimension Affecting Prioritization: Independence versus Reliance on Others

In addition to the accessibility/usability dimension, there is a second dimension that deals with independence versus reliance on others. Everybody depends upon others for some aspects of life. Few people know how to repair a car and television set. Some do not know how to change printer cartridges or clear paper jams or reformat hard drives. In daily life, there are some things people need to be able to do independently and some things that they can depend on others for. In setting usability priorities, this can be taken into account to facilitate decisions regarding expenditure of effort.

For example, it is more important that an individual be able to load their work into the input hopper on a copier and operate the controls to get the required number and type of copies than it is for them to be able to change the toner or clear a paper jam. In fact, in many offices only people trained in clearing paper jams are allowed to do so. Loading new reams of blank paper into the copier generally falls somewhere in between. Similarly, it is more important for an individual to be able to launch and operate programs than it is for him to be able to configure the modem settings. This importance stems not from the technical difficulty of the two tasks but from the fact that one is an activity that is required continuously as a part of daily operations, whereas the other is something that needs to be done only once or that can be planned for and scheduled when there is someone to assist.

Figure 3.5 shows a rough hierarchy based on the need for independence versus reliance on others. The exact order of the items will vary for different types of products and different environments (e.g., the availability of support personnel), but the general order can be seen. This can then be used to set priorities in a resource-constrained or time-constrained product design program.

Functions/features needed for basic use of the product

> 1. Unpredictable, but typically user-serviceable (by "average" user) maintenance or recovery operations
>
> 2. Unpredictable service, maintenance or recovery, typically corrected by support personnel
>
> 3. Predictable or schedulable maintenance that can be delegated to others
>
> 4. Unpacking and initial setup
>
> 5. Repair
>
> Note: The location and availability of support personnel (e.g., in a home office) affect this dimension.

FIGURE 3.5   Priority based on the need for a person to be able to independently accomplish tasks.

3.4.3 A Third Dimension Affecting Prioritization: Efficiency and Urgency

A third dimension to prioritization deals with the need for *efficiency*. If a task is performed only once a day and there is no particular time constraint on its accomplishment (e.g., the person is not trying to disarm an alarm before it goes off), then the relative efficiency of operation is not as critical as in the case of a function that must be used continuously throughout the day. For example, if it takes an individual five times longer to operate the "on" switch on her computer than the average worker, it will not have a major impact on her productivity or effectiveness. In fact, activating the on switch is such a small part of booting a computer that the total time it takes for her to turn the computer on is likely to be only negligibly longer than the time for anyone else to boot her computer. If it takes an individual five times as long to type characters on his computer, however, and he spends the bulk of his day entering information into his computer, the difference in efficiency could be catastrophic. If it takes him five days to get an average day's worth of work done, it would be hard for him to compete in either an educational or work environment. Thus, for the on/off switch, level 1 accessibility may be all that is required. However, for data entry levels 1, 2, and 3 may all be critical on an individual's workstation.

A close parallel to efficiency is *urgency*. If there are situations where a user must do something within a particular time constraint to avoid an adverse situation, then, even if it is rarely done, it may be important to strive for level 2 or level 3 usability to allow the individual to be able to carry out the activity within the time allowed.

The importance that is attached to this dimension is the function of at least three factors.

1. The reversibility of the action
2. The severity of the consequence for failure
3. The ability of the person to adjust the time span to meet increased reaction times

Situations where the result is not reversible or is dire in nature and is also of a type that does not allow for user adjustment or extension (as in some security-related situations) would create the highest priority for providing not only an accessible, but a highly usable interface for the group or situation.

3.4.4 A Pseudo-Priority Dimension: Ease of Implementation

In setting priorities for implementation of usability features in products, a factor that is often used to select features is the ease with which they can be implemented in the product. In this context, *ease* may have many different characteristics, including low or no increase in product cost, low or no increase in development time line, ease in getting clearance from supervisors, minimized impact on other features, minimized impact on testing, minimal impact on documentation, and so on. Often referred to as "low hanging fruit," such features are often very tempting when compared to features that are much more difficult to implement. Although it is always good to look at this dimension, it can result in a belief that five low hanging fruit features are better than one that is more difficult to achieve. This can lead to the implementation of multiple usability features instead of essential accessibility features. Often this occurs with features intended to benefit the same disability group, so that a product may have usability features for a disability group that cannot, in fact, use the product.

Within the essential accessibility features, however, one will also often find either a low hanging fruit or features that would have such mass-market appeal that their "costs" are offset by their market benefit.

3.4.5 Cognitive Constraints: A Unique Dimension

In looking at the dimensions, it is important to note that the cognitive dimension is unique with respect to the other dimensions (see also Chapter 7, "Cognitive Disabilities"). It is possible to make most products usable to individuals with no vision or no hearing and even with severely limited physical ability. However, there are very few products, if any, that are usable by individuals with low cognitive abilities. This is due to the fact that it is possible to translate most types of information between sensory modalities and most types of activities between physical interface techniques, but there is no mechanism for transferring cognitive processing into another domain. While it is true that there are some activities and some types of information for which good strategies do not exist for providing access to by individuals with severe or total visual limitation, severe or total hearing limitations, or severe or total physical limitations, the number of devices and activities that are excluded are much smaller than for severe cognitive limitations. For this reason, strategies for enabling access for people with cognitive disabilities basically look like techniques to facilitate, with each technique that facilitates pushing a few more people over the threshold into the category of individuals who can use a product.

It is also important to note that there are a number of dimensions that are often lumped in with cognitive disabilities, where products can be made accessible. For example, there are strategies that can allow individuals who think clearly but are completely unable to read for some reason to be able to effectively use a very wide variety of products. In this case, the difficulty is not in general cognitive processing or memory, but rather in a specific skill, which is decoding printed information on a page.

The point is just now being reached where there is the computing power and language processing knowledge needed to begin to effectively tackle some of these areas. While some of this specialized cognitive processing may not be appropriate for mainstream devices, approaches like pluggable user interfaces may allow users to carry cognitive orthoses with them.

In the meantime, there is much that can be done to mainstream products to make them easier for people with cognitive, language, and learning disabilities to use, and to make them easier to use by everyone else as well.

3.4.6 Setting Priorities

The suggestion overall is to focus on what is important to people with the full range of disabilities rather than cost or difficulty. Do what can be done and then look for opportunities where difficult or expensive solutions become possible due to other events or discoveries.

Ordering options by difficulty often results in unimportant or even useless (by themselves) features being added (e.g., only half of the provisions needed for access for each disability are included, resulting in a product no one can use). It is also important to remember that what is a usability enhancement for one, may be required for another to be able to use a product, process, or service.

3.5 Impact of Technology Trends on Accessibility in the Future

Going forward it is clear that electronics are being incorporated into practically everything, making it more and more important to be able to access electronic interfaces if one is to be able to live, learn, work, or even move about in one's community.

Recently, a report looking at emerging trends was prepared for the U.S. National Council on Disability. A version was also prepared and submitted to the European eInclusion initiative. Below is an overview and summary of the issues highlighted in the report. The complete U.S. report can be found at http://www.ncd.gov/publications.

3.5.1 Rapid and Accelerating Pace of Technology Advancement

Information and communication technologies are changing at an ever-increasing rate. What used to be multiyear product life cycles have now decreased in many instances to life cycles of less than 1 year. Previous accessibility strategies involving the development of adaptive technologies, or accessible versions of new technologies, are failing due to this rapid turnover (National Task Force on Technology and Disability, 2004). This is exacerbated by the fact that it is not just products that turn over, but the underlying technologies as well. For example, analog cell phones were made accessible just as they were being replaced with digital cell phones. Now some digital phone formats are being phased out in favor of newer technologies (Subcommittee on Telecommunications and the Internet, 2003). This same technology churn, however, is also opening up new opportunities for better assistive technologies and more accessible mainstream technologies.

Many of the changes in technology are evolutionary, but some revolutionary changes are also ahead. Several of these changes may even cause a rethinking of concepts and the definitions of such terms as *disability, assistive technology,* and *universal design,* or how these terms are used.

3.5.2 Technological Advances That Are Changing the Rules

To understand how technological advances can lead to the need to rethink technology and disability funding and policy, it is important to understand just how fundamentally things are changing. Four key technology trends are highlighted here. Opportunities and barriers created by these advances follow.

Some technologies mentioned in the following discussion challenge imagination. Yet, except where indicated otherwise, everything discussed is already commercially available or has been demonstrated by researchers.

3.5.2.1 Trend 1: Ever-Increasing Computational Power plus Decreasing Size and Cost

Computational power is growing at an exponential rate. At the same time, the size of electronic components is shrinking, decreasing product size, power consumption, and cost. Raymond Kurzweil helped to make this growth real to those not used to dealing in exponentials with the following: in 2000, $1,000 could buy a computer that had the computational power of an insect. By 2010, $1,000 will purchase the computational power of a mouse. By 2020, $1,000 will purchase the computational power of the human brain. By 2040, $1,000 will purchase the computational power of all the brains in the human race (Kurzweil, 2001). Kurzweil has also "projected 2029 as the year for having both the hardware and software to have computers that operate at human levels" (Kurzweil, 2006).

Personal digital assistants have shrunk from the size of paperback books to credit card size, and now to a function that runs in the back of a cell phone.[2] Cell phones have shrunk from something just under the size and weight of a brick to cigarette-lighter size, most of which is occupied by the battery. Multiple web servers can fit on a fingernail (sans power supply), and RJ45 (Internet) cable jacks are available that have web servers built directly inside the jack.[3]

Researchers have created gears the diameter of a human hair (Sandia National Laboratories, 1997), motors that are a hundred times smaller than a human hair (Carey and Britt, 2005), and are now exploring tiny cellular-scale mechanisms that would use flagella to move about in the bloodstream (Avron et al., 2004; Svidinenko, 2004). The entire field of nanotechnology is taking off, supported by major federal funding.

Although very expensive technologies are needed to create these devices, the cost per device is dropping precipitously. Sensors that were once hand-assembled are now created en masse, and sometimes even created in a "printing-like" process (Kahn, 2005). The cost of computing drops by a factor of 10 approximately every 4 to 5 years. It is not uncommon to find children's video games that have more computing power than supercomputers of just 10 to 15 years prior. Scientists are now turning to light instead of wires in microchips to keep up with the speed (Paniccia et al., 2004).

This trend toward more computational power, coupled with decreased size and cost, can make possible improved and entirely new types of assistive technology. This trend is also providing capabilities in mainstream technologies that can enable them to more easily and effectively meet the needs of people with disabilities.

3.5.2.2 Trend 2: Technology Advances Enabling New Types of Interfaces

The human interface is one of the most important determinants of whether a technology product can be used by people with disabilities. Advances in interface technology are creating new opportunities for better assistive technologies, more accessible mainstream technologies, and entirely new concepts for controlling both.

3.5.2.2.1 *Projected Interfaces*

Using a projector and camera, companies have created products that can project anything from a keyboard to a full display and control panel onto a tabletop, a wall, or any other flat surface. People can then touch the "buttons" in this image. The camera tracks movements, and the buttons or keys operate as if they really existed (Borkowski et al., 2004). One device is pocket-sized, projects a keyboard onto the tabletop, and allows users to enter data into their PDA by typing on the image of the keyboard on the tabletop (Alpern, 2003).[4] Other projected interfaces use sound waves (Good, 2004).

3.5.2.2.2 *Virtual Interfaces*

Going one step further, researchers have demonstrated the ability to project an image that floats in space in front of a person. With this glasses- or goggle-based system, only the user can see the image floating there (Billinghurst and Kato, 1999, Figure 6). Some systems project the image directly onto the retina (Kollin, 1993). A pocket controller or gesture recognition can be used to operate the controls that float along the display. Motion sensors can cause the displays to move with the user's head, or stay stationary.

3.5.2.2.3 *Augmented Reality*

Researchers are also using this ability to project images to overlay them with what a person is seeing in reality, to create an "augmented reality." One project envisions travelers who can move about in a city in a foreign country by wearing a pair of glasses that automatically recognizes all of the signs and translates them. Whenever foreign travelers look at a sign, they would see a translation of that sign (in their native language) projected over the top of the sign (Spohrer, 1999; Vallino, 2006).

3.5.2.2.4 *Virtual Reality*

Research on ultra-high-resolution displays has a target of being able to display images that appear with the same fidelity as reality. Researchers look forward to the day when the resolution and costs drop to the point that entire walls can be "painted" with display technology, to allow them to serve as "windows," work spaces, artwork, or entertainment, as the user desires. Introducing three-dimensional viewing and displays that work in 360 degrees, researchers have a goal of eventually creating walls or environments that are indistinguishable from reality.

Realistic imaging technologies are already being used in classrooms, primarily (but not exclusively) to teach science. The ability to virtually "shrink oneself" can be used to explore things that would otherwise not be visible or manipulable by humans. The ability to zoom out can provide more global perspectives. The ability to carry out virtual chemistry experiments can allow

---

[2] Xun-chi-138-worlds-smallest-cellphone(2006):http://www.mobilewhack. com/reviews/xun-chi-138-worlds-smallest-cellphone.html.

[3] XPort—embedded ethernet device server (2006): http://www.lantronix. com/device-networking/embedded-device-servers/xport.html.

[4] The I-tech virtual laser keyboard: http://www.virtual-laser-keyboard. com.

students to conduct the experiments that are most interesting or educational, rather than those that are the safest (from poisoning or explosion) or cheapest (not involving expensive chemicals or elements). Time can also be expanded or compressed as needed to facilitate perception, manipulations, or learning (Taubes, 1994). Virtual and augmented reality are addressed in detail in Chapter 12 of this handbook.

3.5.2.2.5 Hands-Free Operation and Voice Control

There are already hands-free telephones. New phase-array microphones have been developed that can pick up a single person's voice and cancel out surrounding sounds, allowing communication and voice control in noisy environments.[5] There are cameras that can self-adjust to track a user's face, allowing face-to-face communication for those who cannot reach out to adjust cameras.[6] Rudimentary speech recognition is available on a $3 chip,[7] and speech recognition within a limited topic domain is commonly used. IBM has a "superhuman speech recognition project," the goal of which is to create technology that can recognize speech better than humans can (Howard-Spink, 2002).

3.5.2.2.6 Speech Output

The cost to build speech output into products has plummeted to the point where speech can be provided on almost anything. All of the common operating systems today have free speech synthesizers built into them or available for them. Hallmark has a series of greeting cards with speech output that, at $3.99, are just 50 cents more expensive than paper, nonelectronic cards. Recently, a standard cell phone that had been on the market for a year received a software-only upgrade and became a talking cell phone, with not only digitized speech talking menus, but also text-to-speech capability for short message service (SMS) messages. The phone, with all speech functionality, is sold for $29, with a service contract.[8]

3.5.2.2.7 Natural Language Processing

The ability of technology to understand people as they normally talk continues to evolve. Although full, open topic natural language processing is a way off, natural language processing for constrained topics is being used on the telephone and soon may allow people to talk successfully to products (see also Chapter 31 of this handbook).

3.5.2.2.8 Artificial Intelligence Agents

Web sites are available that allow users to text chat with a virtual person, who will help them find information on the site.[9] Research on task modeling, artificial intelligence, and natural language are targeted toward creating agents users can interact with, helping them find information, operate controls, etc. (see also Chapter 14 of this handbook). Often the subject of science fiction, simple forms of intelligent agents are reaching the point in technology development where they can become a reality in the home.

3.5.2.2.9 Microprocessor-Controlled User Interfaces

When products are controlled by microprocessor running programs as they are today, they can be programmed to operate in different ways at different times. The use of more powerful processors, with more memory, is resulting in the emergence of new devices that can be controlled in many different ways and can be changed to meet user preferences or needs.

3.5.2.2.10 Multimodal Communication

There is a rapid diversification taking place in the ways people can communicate. Video conferencing allows simultaneous text, visual, and voice communications. Chat and other text technologies are adding voice and video capabilities. In addition, the technology to cross-translate between modalities is maturing (see also Chapter 40 of this handbook). The ability to have individuals talking on one end and reading on the other is already available using human agents in the network.[10] In the future, this ability to translate between sensory modalities may become common for all users.

3.5.2.2.11 Direct Control from the Brain

External electrodes in the form of a band or cap are available today as commercial products for elementary control directly from the brain (Wickelgren, 2003). Research involving electrode arrays that are both external and embedded in the brain have demonstrated the ability to interface directly with the brain to allow rudimentary control of computers, communicators, manipulators, and environmental controls (see also Chapter 37, of the handbook).

3.5.2.3 Trend 3: Ability to Be Connected Anywhere, Anytime—with Services on Demand

New advances will soon enable people to be connected to communication and information networks no matter where they are. People can leave caretakers and still be a button-press away. Everything in the environment will be connected, most often wirelessly, allowing people to think about communication, control, and "presence" in entirely new ways. Individuals who have trouble with wires and connectors will not need them. Network-based services can provide assistance, on demand, to people

---

[5] Andrea electronics headsets (2005): http://www.andreaelectronics.com.

[6] Logitech—leading web camera, wireless keyboard and mouse maker (2006): http://www.logitech.com.

[7] Sensory, Inc. embedded speech technologies, including recognition, synthesis, verification, and music (unspecified date): http://www.sensoryinc.com.

[8] LG VX4500 from Verizon Wireless offers latest in voice command and text-to-speech features (2004): http://news.vzw.com/news/2004/11/pr2004-11-29.html.

[9] KurzweilAI.net (click on Ramona!) (2006): http://www.kurzweilai.net/index.html?flash=1.

[10] Ultratec—CapTel (2006): http://www.ultratec.com/captel.

wherever they are. These advances will create opportunities for whole new categories of assistive technology.

3.5.2.3.1 Wireless Electronics—Connected World

There are already wireless headsets, computer networks, music players, and sensors. New technologies, such as ZigBee, will allow devices that are very small, wirelessly connected, and draw very little power.[11] Light switches, for example, could run off a small 10-year battery and have no wires coming to or from them. People would simply place a light switch on the wall where it was convenient, at a convenient height. Flipping the switch would control the lights as it does now. If someone else needed the light switch in a different place, they would simply move it by pulling it off the wall and replacing it where desired, or placing an additional switch wherever they liked, including on their wheelchair or lap tray.

High-speed wireless networks are also evolving, and costs are dropping. No wires will be needed between televisions, video recorders, or anything else (except sometimes the wall, for power). A person in a power wheelchair could have an on-chair controller connected to everything in the house, and yet still be completely mobile.

3.5.2.3.2 Virtual Computers

Computers may disappear, and computing power will be available in the network. Wherever a person is, he or she will be able to use whatever display is convenient (e.g., on the wall or in a pocket) to access any information, carry out computing activities, view movies, listen to music, and so on. Instead of making each product accessible, things would exist as services and capabilities, which could be accessed through a person's preferred interface (see also Chapter 60 of this handbook).

3.5.2.3.3 Control of Everything from Controller of Choice

New URC standards have been developed that would allow products to be controlled from other devices.[12] Products implementing these standards could be controlled from interfaces other than the ones on the product. A thermostat with a touch screen interface, or a stove with flat buttons, for example, could be controlled from a cell phone via speech, or from a small portable Braille device.

3.5.2.3.4 Location Awareness

Global positioning system (GPS) devices enable people to determine their position when outside and are already small enough to fit into cell phones and large wristwatches. Other technologies, such as radio frequency identification (RFID) and devices that send signals embedded in the light emitted from overhead light fixtures, are being explored to provide precise location information where GPS does not work (see also Chapter 59 of this handbook).

3.5.2.3.5 Object Identification

Tiny chips can be embedded into almost anything to give it a digital signature. RFID chips are now small enough that they are being embedded inside money in Japan.

3.5.2.3.6 Assistance on Demand—Anywhere, Anytime

With the ability to be connected everywhere comes the ability to seek assistance at any time. A person who does not understand how to operate something can instantly involve a friend, colleague, or professional assistant who can see what she is looking at and help work through the problem. Someone who needs assistance if he gets into trouble (and who would currently not be allowed out on his own) could travel independently, yet have someone available at the touch of a button. These assistants could help think something through, see how to get past an obstacle, listen for something, translate something, or provide any other type of assistance and then disappear immediately.

3.5.2.3.7 Wearable Technology

Today there are jackets with built-in music players, with speakers and microphones in the collar (Benfield, 2005).[13] There are keyboards that fold up, and circuitry that is woven into shirts and other clothing. There are now glasses and shoes with a built-in computer that can detect objects within close proximity through echo location and then send a vibrating warning signal to the wearer. The shoes also will use a GPS system to tell the wearer where they are and in which direction she is going.

3.5.2.3.8 Implantable Technology

There are cochlear implants to provide hearing. Heart and brain pacemakers are common. Increasing miniaturization will allow all types of circuits to be embedded in humans. In addition, research is continuing not only on biocompatible materials, but also on biological "electronics."

3.5.2.4 Trend 4: Creation of Virtual Places, Service Providers, and Products

Possibly one of the most revolutionary advances in information and communication technologies has been the development of the World Wide Web. Although the Internet had been around for a relatively long time by the 1990s, web technologies allowed it to be approachable and usable by people in a way not previously possible. It has not only given people new ways of doing things, but has fostered the development of entirely new social, commercial, and educational concepts. It also has allowed for virtual "places" that exist only in cyberspace. This includes virtual environments, virtual stores, virtual community centers, and complete virtual communities. E-travel is allowing people to go places and see things that once were

---

[11] ZigBee Alliance (2006): http://www.zigbee.org.

[12] Myurc.org (unspecified date): http://www.myurc.org.

[13] The raw feed: New jacket sports built-in GPS, MP3, phone (2006): http://72.14.203.104/search?q=cache:TB1l942nXQEJ:www.therawfeed.com/2006/03/new-jacket-sports-built-in-gps-mp3.html.

possible only through books or documentaries. Electronic re-creation can allow people to explore real places, as if they were there, and at their own speed. They could wander in a famous museum, for example. The web also provides an array of products and services that is unmatched in physical stores in most localities.

3.5.3 New Opportunities

Advances in information and communication technology will provide a number of new opportunities for improvement in the daily lives of individuals with disabilities, including work, education, travel, entertainment, health care, and independent living. There is great potential for more accessible mainstream technology with less effort from industry. There is also great potential for better, cheaper, and more effective versions of existing AT, and entirely new types or classes of AT.

3.5.3.1 Opportunity 1: More Accessible Mainstream Products

Some of the changes that will result from mainstream product design are evolutionary continuations of current trends. Other changes will be revolutionary, changing the nature of mainstream technologies and their usability by people with different types of disabilities. Some examples:

3.5.3.1.1 *Potential for More Built-In Accessibility*

Almost everything today, including cell phones, alarm clocks, microwaves, ovens, washers, and thermostats, is being controlled by one or more microcomputers. The increasing flexibility and adaptability that technology advances bring to mainstream products will make it more practical and cost effective to build accessibility directly into these products, often in ways that increase their mass market appeal.

3.5.3.1.2 *Products That Are Simpler to Use*

Although products have been getting progressively more complex for some time now, advances in key technologies such as task modeling, language processing, and constrained voice recognition will soon make it possible to reverse that trend and make products simpler.

3.5.3.1.3 *Interoperability: To Reduce the Need for Built-In Direct Access*

Improvements in connectivity and interoperability will enable individuals with severe or multiple disabilities, who could not operate the standard interface even on universally designed products, to use products via a personal interface device that matches their abilities.

3.5.3.1.4 *Flexible "Any-Modality" Communication*

The trend toward ubiquitous multimodal communication (voice, video, chat) all using the same device, can be a boon for individuals with sensory disabilities, especially individuals who are deaf, hard-of-hearing, deaf-blind, or have speech impairments.

3.5.3.2 Opportunity 2: Better (Cheaper, More Effective) AT and New Types of AT

Technology advances will result in the improvement of current assistive technologies and the introduction of entirely new types of AT. Some of these technologies are realizable today. Some will emerge in the future.

3.5.3.2.1 *Advances in Cost, Size, and Power*

Advances allow for less costly and more effective assistive technologies (AT). More importantly, however, emerging technologies will enable the development of new types of AT, including technologies that can better address the needs of individuals with language, learning, and some types of cognitive disabilities.

3.5.3.2.2 *A Potential for New Intelligent AT*

Previously not possible, opening the door to self-adaptive and environmentally and user-responsive technologies.

3.5.3.2.3 *Translating and Transforming AT*

Takes information that is not perceivable or understandable to many with sensory or cognitive impairments, and render it into a form that they can use.

3.5.3.2.4 *Human Augmentation*

Technologies will enhance some individuals' basic abilities, enabling them to better deal with the world as they encounter it.

3.5.3.2.5 *Losable and Wearable Technologies*

Advances in technology will also reduce the size and cost of products, making them easier to carry, wear, and, in some instances, replace. This can allow the provision of assistive devices (including alternate interface devices) to those who would not have been able to get them in the past out of a concern that they might lose them.

3.5.4 Barriers, Concerns, and Issues

Many of the same technological advances that show great promise of improved accessibility also have the potential to create new barriers for people with disabilities. The following are some emerging technology trends that are causing accessibility problems.

3.5.4.1 Increasing Complexity of Devices and User Interfaces

Devices will continue to become more complex to operate before they get simpler. This is already a problem for mainstream users, but even more of a problem for individuals with cognitive disabilities and people who have cognitive decline due to aging.

3.5.4.2 The Trend toward Digital Controls

Increased use of digital controls (e.g., push buttons used in combination with displays, touch screens, etc.) is creating problems for individuals with blindness, cognitive, and other disabilities.

3.5.4.3 Devices Too Small and Closed to Physically Adapt

The shrinking size of products is creating problems for people with physical and visual disabilities.

3.5.4.4 Closed/Locked Systems

The trend toward closed systems, for digital rights management or security reasons, is preventing individuals from adapting devices to make them accessible, or from attaching assistive technology so they can access the devices.

3.5.4.5 The Trend toward Automated and Self-Service Devices in Public Places

Increasing use of automated self-service devices, especially in unattended locations, is posing problems for some and absolute barriers for others.

3.5.4.6 The Trend away from Face-to-Face Interaction

The decrease in face-to-face interaction and increase in e-business, e-government, e-learning, e-shopping, and so on is resulting in a growing portion of our everyday world and its products and services becoming inaccessible to those who are unable to access these Internet-based places and services.

3.5.4.7 Technology Advancing into Forms Not Compatible with Assistive Technology

In addition, the incorporation of new technologies into products is causing products to advance beyond current accessibility techniques and strategies. The rapid churn of mainstream technologies, that is, the rapid replacement of one product by another, is so fast that assistive technology developers cannot keep pace. Even versions of mainstream technologies that happen to be accessible to a particular group can quickly churn back out of the marketplace.

3.5.4.8 Decreasing Ability of Adaptive AT to Keep Up

To complicate the situation further, the convergence of functions is accompanied by a divergence of implementation. That is, products increasingly perform multiple functions that were previously performed by separate devices, but these "converged" products are using different (and often incompatible) standards or methods to perform the functions. This can have a negative effect on interoperability between AT and mainstream technology where standards and requirements are often weak or nonexistent. Thus, without action, the gap will increase between the mainstream technology products being introduced and the assistive technologies necessary to make them accessible, as will the number of technologies for which no accessibility adaptations are available.

3.5.4.9 Accessibility Rules Being Too Specific to Cover New Technologies as They Emerge

Another concern is that technology advances are causing functions and product types to develop in ways that move them out of the scope of existing policy. For example, in the United States when telephony moved from the public switched telephone network (PSTN) to the Internet (VoIP), the accessibility regulations did not keep pace. The U.S. Federal Communications Commission (FCC) determined that the Internet was information technology, and for some years the telephony access regulations did not apply to VoIP, even though people were using the same phones and the same household wiring to make phone calls to the same people, many of whom were on the PSTN. Although the FCC has recently applied some telecommunications policies to VoIP, VoIP is still not classified as telecommunication.

Internet Protocol Television (IPTV) manufacturers are now talking about including conversation capabilities in their base technologies, again raising the question as to whether telecommunication accessibility will apply to these "phone calls." When accessibility is tied to technologies that become obsolete, often to be replaced by multiple new technologies, the accessibility requirements are often late or deemed not applicable. The shift of education, retail sales, and so on, to the Internet after the Americans with Disabilities Act (ADA) was drafted resulted in the Internet versions of these activities not being specifically mentioned in the law. This is leading some judges to determine that web sites are not places of business as mentioned in the ADA and therefore not covered. This is another example of policy not keeping pace with technology (see also Chapter 53, of this handbook).

3.5.4.10 Open versus Content-Constrained Internet Connections

There is currently debate about whether those who provide Internet connections to a house, or other location, should be able to control the types of information sent to the house, by whom, and at what level of quality connection. If those who provide the connection are allowed to decide what equipment will connect to their systems, or to degrade performance if equipment or software is not from preferred vendors, people who can more easily use other vendors' products, or who must use special equipment, may find their equipment does not work or find its performance is degraded, causing accessibility problems. This problem is exacerbated by the fact that individuals may have to use their technologies from multiple locations and not just from their homes. Unless the Internet operates more like the public road system, where individuals are allowed to take any vehicle that meets safety standards onto the road, rather than having to drive only certain companies' vehicles on certain roads or to certain locations, individuals who must rely on accessible versions of technologies will run into problems.

3.5.4.11 Digital Rights Management

A very interesting subarea in this discussion is digital rights management (DRM). While the need to protect the rights of those who publish authors is critical, the ability to allow access for people with disabilities must be addressed as well. If content is to be locked so that it cannot be copied electronically, then some mechanism for rendering it in different forms should be built into the secure digital media players. For example, if a digital book can be presented visually but the text cannot be read by the operating system (so that assistive technology such as screen readers could read it aloud), then a mechanism within the book player for enlarging it and reading it aloud should be provided. Technologically, this is not a problem, and voice synthesizers with speed control can be, and have been, built into the eBook products directly. A marketing policy, however, whereby publishing companies sell the print (visual access) rights for a book to one distributor but the audio (spoken) rights for the book to another, has created an obstacle. Book player companies have been required to support a bit in their players that, when set by a book publisher, will prevent the voice output option in the book player from functioning. Thus, even though the book reader is capable of reading the book to the blind person, it will not perform that function if the book publisher sets the bit that tells the book reader to not read this book aloud. The same book is also protected, so that it cannot be read by any other technology.

3.6 Conclusion

The needs of people with disabilities are knowable, as are the strategies for providing access to yesterday's technologies. Yet implementation of this knowledge is fairly minimal. This is due in large part to the fact that making products more accessible is not profitable or is not perceived as profitable. Since profit is the underlying motivating force in all product development and deployment, better ways to address this problem need to be found. First, those areas where it is profitable need to be documented. The increasing age of the population coupled with the increasing complexity of products may provide some added impetus to this area. For example, a recent industry survey showed that the rate at which consumers are returning new products has been increasing, with the "no defect found" return rate running 50% to as high as 90%+ (depending on product category) (Sullivan and Sorenson, 2004). These data are for mainstream customers, but the impact of increasing complexity of products on individuals with cognitive disabilities is even greater. And the percentage of elders is increasing, providing a larger market with increasing problems. For those companies for whom a natural market pressure is not sufficient, legislation or regulation may be needed to inject the social concerns into the profit equation. But regulation should not be a replacement for careful research and documentation of the benefits of accessible design for mainstream users.

Even with motivation and action by who design new products to address needs that can be met through universal design/design for all, however, the pace of technology advance will still be a major concern for those who need special access systems. Better methods must be found for creating generic access rather than catch-up patches and adaptations, and these same technological advances are providing some of the keys to doing so. The next decade promises to bring about interesting and revolutionary steps forward in accessibility and information and communication technologies.

Acknowledgments

This work was supported with funding from the National Institute on Disability and Rehabilitation Research, U.S. Department of Education, under grants H133E030012 and H133E040013. The opinions herein are those of the author and not necessarily those of the funding agency.

References

Alpern, M (2003). *Projection Keyboards.* http://www.alpern.org/weblog/stories/2003/01/09/projectionKeyboards.html.

Avron, J. E., Gat, O., and Kenneth, O. (2004). *Swimming Microbots: Dissipation, Optimal Stroke and Scaling.* http://physics.technion.ac.il/~avron/files/pdf/optimal-swim-12.pdf.

Benfield, B. (2005). *Smart Clothing, Convergence, and a New iPAQ.* http://www.pocketpcmag.com/_archives/jan05/European Connection.aspx.

Billinghurst, M. and Kato, H. (1999). Collaborative mixed reality, in the *Proceedings of the First International Symposium on Mixed Reality*, pp. 261–284. Berlin/Heidelberg: Springer-Verlag.

Borkowski, S., Sabry, S., and Crowley, J. L. (2004). *Projector-Camera Pair: A Universal IO Device for Human Machine Interaction.* Paper presented at the Polish National Robotics Conference KKR VIII. http://www-prima.imag.fr/prima/pub/Publications/2004/BSC04.

Carey, B., and Britt, R. R. (2005). *The World's Smallest Motor.* http://www.livescience.com/technology/050412_smallest_motor.html.

Good, R. (2004). *Use Any Surface as Interface: Sensitive Object.* http://www.masternewmedia.org/news/2004/11/25/use_any_surface_as_interface.htm.

Howard-Spink, S. (2002). *You Just Don't Understand!* http://domino.watson.ibm.com/comm/wwwr_thinkresearch.nsf/pages/20020918_speech.html.

Kahn, B. (2005). *Printed Sensors.* http://www.idtechex.com/products/en/presentation.asp?presentationid=215.

Kollin, J. (1993). A retinal display for virtual-environment applications, in *Proceedings of Society for Information Display, 1993 International Symposium, Digest of Technical Papers*, Vol. XXIV, p. 827. Playa del Rey, CA: Society for Information Display.

Kurzweil, R. (2001). *The Law of Accelerating Returns.* http://www.kurzweilai.net/meme/frame.html?main=/articles/art0134.html.

Kurzweil, R. (2006). *Why We Can Be Confident of Turing Test Capability within a Quarter Century.* http://www.kurzweilai.net/meme/frame.html?main=/articles/art0683.html.

Mace, R. (1991). Accessible for all: Universal design. *Interiors & Sources* 8: 28–31.

National Task Force on Technology and Disability (2004). *Within Our Reach: Findings and Recommendations of the National Task Force on Technology and Disability.* http://www.ntftd.org/report.htm.

Paniccia, M., Krutul, V., and Koehl, S. (2004). Intel unveils silicon photonics breakthrough: High-speed silicon modulation [Electronic version]. *Technology@Intel Magazine* 1–6.

Ronald, L., Mace, R. L., Graeme, J., Hardie, G. J., and Place, J. P. (1991). Accessible environments: Toward universal design, in *Design Intervention: Toward a More Humane Architecture* (W. E. Preiser, J. C. Vischer, and E. T. White (eds.), p. 156. New York: Van Nostrand Reinhold.

Sandia National Laboratories. (1997). *New Sandia Microtransmission Vastly Increases Power of Microengine.* http://www.sandia.gov/media/microtrans.htm.

Spohrer, J. C. (1999). Information in places [Electronic version]. *IBM Systems Journal: Pervasive Computing* 38.

Story, M. F., Mueller, J. L., and Mace, R. (1998). *Universal Design File.* Center for Universal Design. http://www.design.ncsu.edu/cud.

Subcommittee on Telecommunications and the Internet. (2003). *Wireless E-911 Implementation: Progress and Remaining Hurdles.* http://energycommerce.house.gov/108/Hearings/06042003hearing947/print.htm.

Sullivan, K. and Sorenson, P. (2004). *Ease of Use/PC Quality Roundtable: Industry Challenge to Address Costly Problems* (PowerPoint slideshow). http://download.microsoft.com/download/1/8/f/18f8cee2-0b64-41f2-893d-a6f2295b40c8/SW04045_WINHEC2004.ppt.

Svidinenko. (2004). *New Nanorobotic Ideas from Adriano Cavalcanti.* http://www.nanonewsnet.com/index.php?module=pagesetter&func=viewpub&tid=4&pid=9.

Taubes, G. (1994). Taking the data in hand--literally--with virtual reality. *Science* 265: 884–886.

Vallino, J. (2006). *Augmented Reality Page.* http://www.se.rit.edu/~jrv/research/ar.

Vanderheiden, G. C. and Vanderheiden, K. (1992). *Accessible Design of Consumer Products: Guidelines for the Design of Consumer Products to Increase Their Accessibility to People with Disabilities or Who Are Aging.* Madison, WI: Trace Research and Development Center. http://trace.wisc.edu/docs/consumer_product_guidelines/consumer.htm.

Vanderheiden, G. C. (2000). Fundamental principles and priority setting for universal usability, in *Proceedings of the ACM Conference on Universal Usability*, pp. 32–38. New York: ACM.

Vanderheiden, G. C. (2002). Building natural cross-disability access into voting systems, in *Proceedings of RESNA 2002 Annual Conference.*

Vanderheiden, G. C. and Law, C. (2000). Cross-disability access to widely varying electronic product types using a simple interface set, in the *Proceedings of the XIVth Triennial Congress of the International Ergonomics Association and 44th Annual Meeting of the Human Factors and Ergonomics Society*, p. 156. San Diego: Human Factors and Ergonomics Society.

Vanderheiden, G. and Zimmerman, G. (2005). Use of user interface sockets to create naturally evolving intelligent environments, in *Proceedings of the 11th International Conference on Human-Computer Interaction (HCI International 2005)* [CD-ROM]. Mahwah, NJ: Lawrence Erlbaum Associates.

Wickelgren, I. (2003). Tapping the mind. *Science* 299: 496–499.

Diversity in the User Population

4

Dimensions of User Diversity

Mahima Ashok and
Julie A. Jacko

4.1 Introduction

User diversity is vital for effective conceptualization and practice of universal access. Failure to consider user diversity during design, development, and testing of applications leads to "technological exclusion" (Gabriel and Benoit, 2003) of some sections in society, preventing them from participating in the dynamic of scientific progress. Technological exclusion refers to a phenomenon whereby a technological environment ignores the presence and needs of the demographic heterogeneity of its users and excludes certain segments of society from benefiting from technological applications. In our globalized world interconnected by technology, failure to consider user diversity in the design and application of technology can lead to a lack of cohesion in society itself. The phenomenon of technological privilege arises when certain users are provided with opportunities to benefit from applications while others are denied the same because of a failure to include their needs during the design process. Hence, technological privilege can add another divisive layer to existing barriers such as class, race, and gender. With an understanding of user differences, designers will be able to incorporate techniques to enable all users to obtain equal advantage when using the system. With this objective in mind, the Web Accessibility Initiative (http://www.w3.org/WAI) was created by the World Wide Web Consortium (W3C). This initiative provides guidelines

to ensure and improve accessibility by taking into consideration various diversity factors. Through the development of guidelines for design, educational and outreach programs, and research operations, the Web Accessibility Initiative aims to make the World Wide Web more usable for all users. Initiatives such as this are required in all areas of technological development, and the widely used World Wide Web provides a good example of such an inclusive application cognizant of diversity.

When studying universal access, it is important to consider not only the nuances of diversity, but also the nuances of every component of the application. For instance, considering again the example of a web site, it is composed of various segments such as text, images, video, audio, scripts, and so on (Freitas and Ferreira, 2003). The designer must take into account these different components, as well as the needs of the users, while composing every facet of a web site design that is universally accessible.

The objective behind the study and understanding of user diversity is to create applications and systems that allow users to access these systems and to prevent their differences from becoming impairments during the use of technology, thereby affecting the quality of interaction. This concept of equal opportunity is vital to the design of usable computing tools, which will benefit all users. This chapter will consider different diversity factors, and how these can affect technological interactions, design, and development. The user diversity schematic is meant

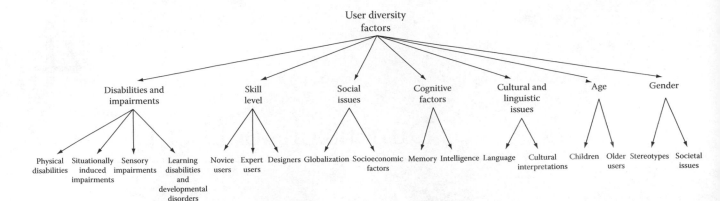

to serve as a guiding tool for this chapter. This schematic represents the various diversity issues and serves as a framework for the content. It shows that diversity is not monolithic, and that there are many different kinds of diversity. Designers, developers, and all members of the human-computer interaction (HCI) community need to understand the heterogeneous nature of diversity, because such an understanding leads to the design of more efficient and usable systems. Factors in user diversity include disabilities and impairments, skill level, cognitive factors, social issues, cultural and linguistic issues, age, and gender. Diversity, a defining feature of the contemporary world, must not become a barrier to access and effective use of technological applications.

4.2 Disabilities and Impairments

Technological tools have become all-pervasive and have found their way into almost every facet of life. For most individuals, these applications increase efficiency, facilitate easier communication, and reduce the time taken for otherwise time-consuming tasks. However, for those users who have certain disabilities, technology may actually reduce efficiency and alter the quality of what can be a positive experience. In this case, technology becomes a barrier by exclusionary design. If the impairments of the users had been considered during the design process, the application could have facilitated or improved communication, and even led to a greater sense of independence among users (Young et al., 2000). Furthermore, designs that lack disabilities-awareness cause, as noted already, a societal divide between the technologically enabled and the technologically excluded. Many established methods of development and evaluation in HCI need to be modified to derive optimal results for technology designs when diverse user populations are considered. For instance, evaluation methods such as iterative user feedback must be suitably modified to ensure valid representation of user groups when dealing with users with learning disabilities (Neale et al., 2003). Designers who consider disabilities during the development process will contribute to a well-integrated society. To be able to design with these concepts in mind, designers must be familiar with the kinds of disabilities and impairments

that may affect the use of technology. It is also important to note the ways in which these disabilities and impairments affect interactions with technology.

Disabilities and impairments take on many forms, such as physical disabilities, situationally induced impairments, and sensory disabilities. Physical disabilities are those that directly affect a user's physical interaction with technology, such as arthritis, multiple sclerosis, and cerebral palsy. Situationally induced impairments are those that arise due to the environment where the user is located, or specific activities that the user is involved in, or is surrounded by, which have special characteristics that affect human-computer interaction. Sensory impairments refer to impairments of the senses such as visual and auditory impairments.

There is an intricate nexus of connections between these terms often used as synonyms: disabilities, handicaps, and impairments (Sears and Young, 2002). The World Health Organization made these distinctions clear in their study published in 1980 (World Health Organization, 2000). These definitions were further clarified in the publication of the World Health Organization in 2000 and are briefly summarized in the following (World Health Organization, 2000; Sears and Young, 2002):

- Health conditions are those that arise because of a disease or injury, such as with multiple sclerosis, joint-related pain due to arthritis, and so on.
- Impairments are those conditions caused when a particular part of the human body begins to function in an abnormal manner or loses its ability to function altogether. For example, people without an arm or with a paralyzed arm suffer from an impairment.
- Disabilities are conditions that prevent a person from completing a task as would have normally been expected. For instance, learning disabilities can hamper communication and can cause difficulties for an individual when interacting with a computing tool.
- Handicaps are defined as those conditions that cause "participation restrictions" in various situations, which could cause problems in transferring information and in communication.

It is clear from these definitions that health conditions can cause impairments, disabilities can result in a handicap, and so on. Therefore, instead of considering the impairments, handicaps, or disabilities in isolation, designers need to understand the user, the environment, and the task holistically in the design process.

4.2.1 Physical Disabilities

Physical impairments interfere with the bodily functions that are necessary for interacting with technology. They may be a genetic condition, or may be caused by trauma, injury, accidents, or illness. This section will consider some specific instances of physical disabilities to illustrate how design transformation to make technology user-centric can assist in improving the quality of experience of diverse users.

Quadriplegia is a disease in which affected individuals are paralyzed in their limbs as a result of spinal damage. Since typical technological interaction requires the use of hands and eyes, quadriplegic users are likely to encounter various problems while engaged in activities such as web browsing (Larson and Gips, 2003). Using a mouse can be extremely challenging and material on the web is often too small to decipher. Due to various vision-related problems, complex visual designs can often make simple tasks very difficult. To overcome many of these difficulties that quadriplegics face, a new browser called WebForward was developed (Larson and Gips, 2003). This browser is equipped with larger buttons that are easier to see and understand, a simple uncomplicated design that reduces the cognitive burden placed on users, and a text reader that reads the text of the page to the user. While designing tools such as these, it should be noted that while these tools help users with impairments and disabilities gain more from the technology, they need not detract from the experience of a user without these disabilities.

Another example of a physical disability that commonly affects many users is arthritis. The Centers for Disease Control has stated that arthritis is the leading cause of disability in the United States. More than 20 million people in the United States have arthritis and by the year 2030, 70 million citizens will be at risk for developing it. Arthritis results in pain, stiffness, and difficulty in moving and performing regular tasks using joints. There are two major kinds of arthritis: osteoarthritis and rheumatoid arthritis. The former, caused by deterioration of cartilage, is the most common form of arthritis in the United States. Rheumatoid arthritis is an inflammatory disease in which severe pains in the joint areas are caused by the weakening of the immune system. All variations of arthritis cause difficulty in movement and debilitating pain in the joints. The inability to freely and painlessly move joints in hands and fingers causes immense difficulty in using a keyboard, mouse, speaker controls, and other commonly used devices. Voice input of spoken commands, in the place of motor-dependent tools such as keyboards, can greatly ease the difficulty faced by users with joint problems. Another possibility for users who have limited or no motor ability (such as those with paralyzed limbs or amputees) is to simulate the functionality of

a keyboard and mouse with head movement applications. These applications recognize head movements as commands and are able to translate them appropriately. Head movement applications can be combined with speech recognition software. Results of a study combining head movements with speech recognition to simulate keyboard and mouse controls were positive (Malkewitz, 1998). A study conducted substituting a head-operated joystick for a regular mouse proved appealing to users and also enhanced usability (Evans et al., 2000).

Cerebral palsy is a neurological affliction that affects individuals from birth. It can result in impairment in muscle coordination, speech, and learning. Up to 4 individuals out of 1000 births are affected by this disease. Individuals with cerebral palsy often find coordinated use of keyboard and other similar input devices extremely challenging, due to lack of muscle coordination. Applications to recognize gestures made by individuals, including facial expressions and movements, can ease the difficulty faced by technology users with cerebral palsy and other diseases that affect muscle coordination, such as Parkinson's disease, multiple sclerosis and muscular dystrophy. Researchers who studied a gesture recognition interface found it to be worth exploring as a method to improve the technology experience of users with cerebral palsy (Roy et al., 1994).

Some of the applications discussed previously, such as speech and gesture recognition software, can provide increased usability and access to users without long-term disabilities or impairments. For example, many users undergo periods in which certain muscles are strained or sprained, or when an arm is broken during an accident or injury. While these users may not otherwise require speech or gesture applications, these tools would be of great benefit during particular times of use.

A more detailed discussion of motor impairments and their impact on access to interactive technologies is provided in Chapter 5 of this handbook.

4.2.2 Situationally Induced Impairments

Temporary states of impairment may be created by the particular contexts in which users interact with technology. For instance, a working environment in which noise level and visual distractions of the environment are extremely high can interfere with the efficient use and navigation through computer-based applications. These impairments, caused by contextual factors influencing the quality of interaction in a negative way, are known as situationally induced impairments (Sears et al., 2003). It is important to understand that context refers to a larger group of factors and not just the physical environment surrounding the user (Sears et al., 2003). Environmental factors can increase feelings of stress, which could impair technology use. Consider the example of an individual working at home when the heating unit is malfunctioning in winter. Due to uncomfortable cold temperatures in the home, the user may suffer from temporary disabling conditions such as mild numbness in the fingers. Cold temperatures also cause a general feeling of physical discomfort leading to a less than optimal computing experience. The same would be true of extreme heat, which could occur in summer when the user works outside. It is also possible for the nature or the number of tasks that the user is working on to become a source of impairment. When a user is performing several tasks at the same time, not necessarily all on the computer, it is possible for attention to be diverted and quality to be affected. A commonly seen example is when a student trying to complete a computerized homework assignment is also on the phone with music playing at the same time. It is likely that, under such circumstances, the capacity to work efficiently on the computer may be so diminished as to justify the term *impairment*.

In a world where computing has become a pervasive phenomenon, there is a steep increase in the variability of environments that users find themselves in. This has also led to increased attention on situationally induced impairments (Sears et al., 2003). Using a personal computer to complete an assignment in a quiet office might produce startlingly different results if the same individual was to use the same computer in an active, noisy kitchen. Another cause of situationally induced impairments is the technology itself. When screens are too small, the user may become vision-impaired in this particular situation (Grammenos et al., 2005). If the volume controls on a music-playing technology do not allow the volume to be loud enough, the user would be unable to use the application as intended. While it would be impossible to come up with all the various situationally induced impairments that a user might be prone to, it is important for designers to be aware of common contextual factors that might affect technology use. An inventory tracking system for use in a noisy warehouse environment should be designed with minimal auditory function, or with attachable headphones that would permit users to block out external sounds. A handheld computing tool should permit for a large-enough screen to allow people to view information comfortably, but not so large that it leads to increased weight, thus making it less portable. Designing with possible contexts in mind helps in the creation of technology that may be widely used and, more importantly, widely well used. Hence, designers must imagine typical and not so typical scenarios of usage for their interactive systems.

4.2.3 Sensory Impairments

Sensory impairments that significantly affect human-computer interaction are those that affect visual and auditory capabilities. Difficulty in seeing and hearing can cause cognitive confusion in the users, as well as prevent them from completing tasks effectively. Statistical estimates show that 1 out of 10 people have a "significant hearing impairment" (Newell and Gregor, 1997). One in 100 people have visual impairments and 1 in 475 people is legally blind (Newell and Gregor, 1997). The 1990 United States Census found that out of the 95.2 million people who are older than 40 years of age, more than 900,000 people are legally blind and 2.3 million suffer from visual impairments. Thus, the extent of sensory impairments and their effect on technology use might be greater than one would expect. It is therefore very important to understand the extent and nature of visual and auditory impairments.

Many forms of visual impairment affect human beings. Age-related macular degeneration (AMD) is one of the leading causes of vision loss in the United States, and understanding the disease and its effects on lifestyle (which includes the use of technology) can help raise the quality of life for millions of individuals. AMD is a disease in which central vision is distorted and lost. The central portion of the retina, known as the macula, undergoes degeneration and destruction, resulting in this severe condition. The "dry" form of the disease is the most common, but the "wet" form is the cause of the most severe kind of macular degeneration. As is clear from the name of the disease, aging is an important risk factor for this disease; most people affected by macular degeneration are over the age of 60 (Prevent Blindness America, 2002). Extensive testing using advanced eye movement tracking applications has shown that those with AMD and those with full sight show different performance levels when using visual search strategies on a computerized tool (Jacko et al., 2002a). The nature of particular features on the display, such as color and icon size, played a strong role in determining performance. Thus, when the nature of impairments and their effect on technology use is studied, it is possible to design systems that enable more and more people to effectively use technology. For example, the use of multimodal feedback to provide cues to users can improve the performance of users with visual impairments by providing a "different sensory feedback" (Jacko et al., 2002b). Multimodal feedback technology would include auditory, visual, and haptic feedback to reinforce the message to be conveyed. For users without visual impairments, a message on the screen would likely be sufficient. However, for users with AMD, when central vision is lost or seriously distorted, audio or haptic feedback can provide useful information. Another example of improving access to technology for those with severe visual impairments is to implement interventions and design strategies into handheld computing that are shown to significantly improve user performance, particularly concerning users with AMD (Leonard et al., 2006). Many of these strategies to make the concept of universal access a reality are low-cost and easy to implement.

Besides AMD, several other visual impairments affect the interaction between a human and technology. Cataract is a disease in which the lens of the eye becomes cloudy and opaque and consequently, vision deteriorates. More than 20.5 million people in the United States suffer from cataract and this disease affects even young people, sometimes from birth. In many situations, loss of vision caused by cataract can be countered by lens replacement surgery (Prevent Blindness America, 2002). For those awaiting surgery, or for those who cannot, for some reason, undergo surgery, technology augmented with systems such as multimodal feedback can be of immense assistance. Glaucoma is another serious eye-related disorder that progresses in stages and finally causes blindness. It is actually a set of diseases that results from increased pressure within the eye, leading to damage in the optic nerve. It is estimated that between 90,000 and 120,000 people have lost their vision due to glaucoma. In diabetic retinopathy, the blood vessels in the retina are affected, which leads to blindness in some diabetic patients. About 5.3 million people in the United States, over the age of 18, have diabetic retinopathy (Prevent Blindness America, 2002). Because the length of time for which a person has diabetes is strongly related to the risk for diabetic retinopathy, most people who develop juvenile diabetes will suffer from some form of diabetic retinopathy during their lives.

Technologists interested in universal access have studied the use of special tools to assist and improve the experience that people with impaired vision have with technology. The use of touch-based interfaces that rely on haptic tools rather than on vision-based feedback has been studied (Ramstein, 1996; Sjöström, 2001). By using Braille displays that can be felt by the hand, blind or visually impaired users are able to read and understand digital information. Adhering to certain guidelines of good design, such as providing reference points on the screen that are easy to identify and not changing reference points very often, can enhance the quality of haptic-based design for the visually impaired (Sjöström, 2001). Some users do not suffer from vision loss but rather from a loss in the ability to distinguish colors. Incorporating additional color palettes in technologies can improve universal access by making the system easier to use by those with color vision deficiencies (Knepshield, 2001).

Loss of hearing is another serious type of sensory impairment that can significantly affect interaction with technology. Hearing loss may be either temporary or permanent, depending on the nature of damage to the ear. In conductive deafness, problems with the ear drum or the bones in the middle ear can cause failure of hearing. It is very possible that hearing will return once proper care has been given to the ear. In nerve deafness, the cochlear nerve is damaged by serious trauma or by other forces. The damage could even be in the brain, resulting in a condition of irreparable hearing loss. There are many reasons for loss of hearing, including genetic predisposition, exposure to noisy environments, accidents, illnesses, and aging. The inclusion of caption-text, text-based descriptions of audio, and the use of simultaneous sign language video transmission can immensely benefit hearing-impaired users of technology (Drigas et al., 2004). Caption-text is now available for most television programs. The topic of situational-induced impairments comes up here, as television viewers (without any permanent hearing impairments) may find it useful to turn on caption-text when children play noisily in the house, or when the telephone rings often, because their hearing is temporarily impaired due to the situation they are in. Therefore, considering sensory impairments, both temporary and permanent, prior to design can provide beneficial results to a large population.

Perceptual design is a design paradigm that humanizes the interaction between the perceptually impaired human operator and technology. Perceptual design defines the human-computer communication in terms of the perceptual capabilities of the individual using the application. Combining knowledge about the abilities of human operators with the performance needs

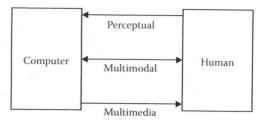

FIGURE 4.1 Flow of information in perceptual, multimodal, and multimedia interfaces. (From Jacko, J.A., Vitense, H.S., and Scott, I.U., Perceptual impairments and computing technologies, in *The Universal Access Handbook*, Lawrence Erlbaum Associates, Mahwah, NJ, 2003.)

of the system enables designers to create applications that consider human needs. Perceptual interfaces prescribe humanlike perceptual capabilities to the computer. Both multimedia and multimodal interfaces offer increased accessibility to technologies for individuals with perceptual impairments (Jackoet et al., 2002c). The flow of information when dealing with perceptual, multimodal, and multimedia interfaces is clearly shown in Figure 4.1.

Sensory impairments are discussed in more detail in Chapter 6, "Sensory Impairments," of this handbook.

4.2.4 Learning Disabilities and Developmental Disorders

In addition to the disabilities and impairments discussed thus far, there are certain conditions that inhibit individuals from processing complex information that is a part of technology use. These conditions include learning disabilities, such as dyslexia, and other disorders, such as autism. Autism is a disorder of the central nervous system, which results in problems with communication, imagination, and social activities. Special care is taken in educating autistic children, and great care is needed in the development of technology for those with autism. Already, technology is being used as an integral part of the educational process for autistic children (Mirenda et al., 2000). Understanding the unique needs of children with autism helps in developing technology that enhances the quality of education (Barry and Pitt, 2006). In fact, it has been observed that using software enhanced with unique sounds and movements increased attention span when used as an educational tool with autistic children (Moore and Calvert, 2000). Incorporating special features into widely used tools, such as the personal computer, can help those with autism. Similarly, taking special care to design tools for those with learning disabilities, such as dyslexia, can enhance education and the quality of daily life. The fundamental concept here from a design perspective is that a clear knowledge of the mental models used by users with these disabilities is required. Such knowledge allows incorporating features into technology that are suitable for these models. Involving end-users throughout the development process is essential to receive useful feedback during design. When involving users with autism, for example, there may be significant challenges in communication during the process, which need to be planned for. Evaluation methods must be specially modified depending on the nature of the group of evaluators and users. For instance, using numerous pictures and designs in questionnaires for those with learning disabilities could improve feedback when compared to using regular heavy-text questionnaires (Neale et al., 2001). The interdisciplinary nature of design is obvious here: universal access technology is not a venture of just software engineers, but a joint effort of members of the educational community, medical community, and technologists. See Chapter 7 of this handbook for more details on this issue.

4.3 Skill Level

4.3.1 Novice Users, Expert Users, and Designers

With the pervasiveness of technology today, many routine activities are now performed electronically. Shopping, banking, information searching, reading, writing, entertainment, are all activities that are now performed easily with personal computers, handheld devices, and even mobile phones. The wide use of technology by a large group of the population has resulted in increased comfort with basic technological tools. However, the level of comfort and the ease of use of technology vary significantly depending on the skill levels of users, and is a valid source of diversity among users. Some groups of users are unfamiliar with technology, particularly older users and those with minimal or no education, and are now required to use computing tools to keep up with the information society we live in. The result is a mix of users with great diversity in technology skill level.

The challenge of designing systems for users who fall within a wide and uneven spectrum of skills can be daunting. This is especially so because designers are typically experts in their respective domains and find it difficult to understand and incorporate the needs of novices. The tendency to homogenize can detract designers from designing with the objective of universal access. As a corrective, it is essential to include the user participation in the design process. A design cycle that does not include feedback from a diverse group of users will result in a design that is not equally accessible by all. Judging the skill levels of users can be more difficult than assessing impairments or difficulties because users who are experts on a particular tool may find a new replacement tool hard to use and understand—this results in a situation where a person who you may think is an expert actually behaves like a novice. Feedback from users with differential skill levels can provide fresh perspectives and new insights.

Including useful help options and explanations that can be expanded and viewed in more detail, consistent naming conventions, and uncluttered user interfaces are just a few ways in which technology can be made accessible by users with less knowledge of the domain and system while at the same time not reducing efficiency for expert users. In fact, these suggestions are guidelines of good design, which will benefit all users, irrespective of skill level.

Users with different levels of skill respond differently to various features in the system, such as feedback methods. A study of older users with varying levels of skill shows that experienced users responded well to all combinations of multimodal feedback while those with less experience showed preference for certain specific combinations (Jacko et al., 2004). In addition, skill level also plays a role in the attitude users have toward using and accepting new technologies. Those who are unfamiliar with using the Internet might resist a web-based service to replace their daily banking chores. To assist in the transition, the service must be built in a way that is easy to understand and maps well to typical mental models of users who perform simple banking tasks.

Skill level differences are often related to differences in age, education, and other factors to make for a complex variable. The example of multimodal feedback described previously is one instance of how age and computer experience play a role in the user's interaction with technology. Another example of the digital divide is the case of people with varying income levels. Inequality in economic status causes inequality in computer access, which in turn causes inequality in computer skill level. This chain of events could lead to a situation where two children of the same age have different levels of comfort using the same word processing tool, because of differences in their home environments. As with all diversity markers, understanding the needs and nature of users is the single most important factor in developing technologies that are easy to use for those with varying levels of skill. Questionnaires handed out during user evaluation must consider technology experience and background as an important parameter. Simple questions regarding how often the user accesses a tool, and so on, can help designers gain valuable insight into the level of experience a user has. Without this understanding, the resulting technology will increase the digital divide and will play a role in furthering the technological exclusion prevalent in today's society.

4.4 Cognitive Factors

Cognition is an intangible quality, which manifests itself tangibly in interactions with other people and with technology. Cognition is the ability of the human mind to process information, think, remember, reason, and make decisions. While all human beings possess some level of cognitive ability, the extent of this ability varies from person to person. This spectrum of variability makes it difficult to define the exact point of cognitive impairment, although it is possible to generally state that there is an accepted level of "normal" cognitive ability (Newell et al., 2002). Levels of cognition that fall below this "normal" level are considered impaired states. Some of the conditions discussed previously, such as dyslexia and autism, create a situation where the cognitive ability of the individual may be different from one who does not have the condition. Some research has shown the potential of technology use with autistic individuals, based on preliminary evidence, that these individuals have more productive interactions with computers than with people (Newell et al., 2002). This section will focus on understanding cognition, how it is measured, how cognitive factors play a role as markers of diversity, and how technology use and design is affected by this.

4.4.1 Memory

The cognitive ability of a human being includes many facets, and one of these is the ability to recall and remember past actions and experiences. This ability is a very important part of cognition. Being able to recall actions from the past enables us to perform them at quicker speeds in the present. Memory is used on a daily basis when performing routine computing tasks. For example, when typing on a keyboard, the location of the various keys in QWERTY format is recalled to type faster. Very often, memory recall in everyday actions may happen unconsciously. Although this happens at a rapid speed, recall helps in finding the keys faster without having to look for their location every time.

It is postulated that there are three distinct kinds of memory: sensory memory, short-term memory (or working memory), and long-term memory (Dix et al., 1998). The first of these, sensory memory, can be compared to a buffer that holds the various bits of information received by the senses. Information is filtered and passed into the short-term memory. As the name indicates, short-term memory serves the purpose of storing information for short intervals of time while the information is processed. Information that is stored for longer periods is moved into long-term memory, which is essentially a collection of memories. Aging is a natural process in which memory is eroded. It also is possible that, as people grow older, the capacity to move information from short-term to long-term memory is also decreased. This could explain why some (not all) older people may require repeated relearning of concepts in technology use before they can independently use a system. Serious conditions such as Alzheimer's disease, as well as various forms of dementia and amnesia, can result in varying degrees of memory loss.

4.4.2 Intelligence

Once information is stored in memory, human beings use their reasoning skills to process and understand information. Great differences are seen in this particular segment of cognitive ability. A 20-year-old man will have better reasoning skills than a 10-year-old boy by virtue of his age, experience, and education. The capacity to process information is vital in technology use where users are constantly bombarded with all forms of data: textual, audio, and video. Sometimes these three forms of data are provided simultaneously, placing a large cognitive workload on the user. The way in which an individual is trained to process information affects her experience with information. For instance, a person whose education has been entirely computer-based will find it easier to use computer tools in the workplace than someone who has used a computer only occasionally. Understanding the educational background of users is

vital during the design of systems, as it can provide insight on cognitive capabilities.

Though "cognition" and "intelligence" are sometimes used interchangeably or in association with each other, the ambiguous nature of the concept of intelligence has made this term a controversial measurement of cognitive ability (Newell et al., 2002). Howard Gardner, a renowned theorist of human intelligence, points out that intelligence is not a single identity but is multipronged. Mathematical and linguistic intelligence are valued and prioritized highly in our educational system, but there are other kinds of intelligence that need to be tapped (Gardner, 1993b). Also, the evolution and progress of human societies will depend on our ability to deploy the multiple levels of intelligence in individuals (Gardner, 1993a). This theory of cognition can be usefully applied to technology design. Some people are especially skilled at processing visual information, while others require detailed text-based explanations. It is not possible to say whether one shows more intelligence than the other. Intelligence, hence, is not a univocal term, but a multifaceted concept, the ramifications of which have immense practical relevance for the technological world. Theories of intelligence need to be translated into the theory and practice of technology design.

The need to standardize this concept of intelligence leads to the idea of the intelligence quotient. The intelligence quotient, popularly known as IQ, is a commonly used scheme for measurement purposes. IQ is a score given to an individual based on performance on a test. Although there has been much debate on the value of IQ scores as a determinant of intellectual prowess, it is one of the most widely recognized and used assessment tools.

Technology plays a major role in improving cognitive abilities and honing various skills. The use of technology with autistic children was discussed earlier. Using computer-based learning tools for children with dyslexia and other learning disabilities can be of immense assistance to both the student and the teacher. These creatively designed tools are structured with complex teaching algorithms and other aids, which help instructors teach dyslexic students. A study of dyslexic undergraduate and graduate students using simple writing software showed that these students uniquely use technology to maximize their cognitive abilities (Price, 2006). The fact that the "simple" application was the one that provided the most benefit shows that the design of technology plays a major role in how useful it will be to end-users.

Ideally, technology is a functional aid to help enhance quality of work and life. However, cognitive impairments can sometimes make technology an impediment rather than a tool of efficiency and quality. For instance, cognitive impairment can decelerate a person's response time and in this situation, technology needs to be adapted to the user's cognitive capacity to help rather than hinder. W. A. Gordon rightly notes that for a cognitively impaired user, the processing time of all "information-laden stimuli" will be considerably increased (Gordon, 2001). Added to this will be the incremental layering of unprocessed information forming a backlog of uncovered territory. Cognitive impairment can result in decreased attention span, and hence any stimulus that is stretched out can be lost on the impaired user.

4.5. Social Issues

4.5.1 Globalization and Socioeconomic Factors

Advances in technology have brought distant and diverse societies closer together. People living thousands of miles apart communicate with each other instantaneously, news is transmitted simultaneously with the occurrence of actual events, and state-of-the-art techniques have helped diasporic lives to connect with each other as an electronic family or community. Globalization has created an environment of rich information and easy communication. However, social issues such as economic and social status pose a serious challenge to universal access. In many parts of the world, only the wealthier segments of society have the opportunity to use technology and benefit from it. Poverty, social status, and meager or nonexistent educational opportunities create barriers to technology access. An overview of globalization and its impact on access to technology enlightens us with an understanding of social challenges for universal access.

The economic structure of the world today is an open channel where barriers in trade are diminishing and organizations from various parts of the planet come together for economic and financial benefit. The "global village" we live in is dependent on information technology not only for simple communications, but also for the dissemination of complex information. Critics of globalization believe that small-scale rural businesses are left behind in the race to establish high-tech corporations around the world. Countries such as the United States tend to develop technologies that are well suited for its own consumers without realizing the varying needs of users in different parts of the world (Marcus, 2002). While proponents of globalization present a strong argument citing the benefits of free flow of trade, finance, and people, others perceive globalization as widening the existing socioeconomic gap in many societies. Designing applications that are equally accessible and equally easy to use for every single socioeconomic group in the world is virtually impossible, but there are lessons to be learned from considering the needs of various social groups. For instance, financial software applications to be deployed in the United States and in Japan need to include translational software to translate commands into Japanese and back to English. Therefore, identifying user demographics within the target populations is important, because it allows designers to refine access parameters.

The use of the Internet has changed fundamental concepts of communication and information access. While the benefits of technology are undoubtedly tremendous, technology in the globalized world has, in some situations, contributed to the rift between the rich and the poor, and the educated and the uneducated. Econometric studies have revealed that a certain level of education, technical education to be precise, is required to receive optimal productivity from the use of technology (Castells, 1999). The two-pronged nature of technology is seen here. On one hand, technology helps greatly in improving the economic status of societies, allowing nations to work with one another, learn from each other, and participate in economic, educational, and political transactions with each other. On the other hand, the fast-paced advances made by a technology-powered globalized world leave behind the poorest of the poor and those who are unable to keep up with these changes. One is comforted by the fact that as technologies gain popularity, they are easier to implement due to sharp decreases in costs. This implies that technological advances will one day become universally affordable.

The realization that technological benefits are available more readily to the educated conveys a simple message regarding the responsibility of designers, developers, engineers, and all those involved in the creation of technology. This team of people creates and distributes technology, and it is critically important for them to be educated in matters of universal access and issues in the diversity of users, including the need to consider designing for the undereducated. Designing for technological literacy must become a top priority.

While globalization deals with the delivery of goods, services, and financial transactions on a global scale, the term *localization* refers to customizing products for specific markets to enable effective use (Marcus, 2002). The process of localizing technology is an important balance to globalization. Included in localization is language translation, changes to graphics, icons, content, and so on (Marcus, 2002). To achieve effective localization of a product, it is necessary to identify groups with similar needs within larger groups of the population. Even though localization is discussed here with respect to social diversity, this concept is applicable to other types of user differences. For example, providing certain audio cues and feedback for users with visual impairments can be compared to a localization process for a particular subgroup of users. An example of localization of technology for a specific social group would be customizing a product for a group of villagers in southern India. India is unique because there are numerous languages and dialects that vary from state to state. Thus, to properly customize a tool for a group in southern India, the developers must be aware of exactly which state the users reside in and which language they speak. In addition, literacy issues would be a major concern when dealing with members of certain rural communities. User interviews and surveys to determine the extent of literacy must be completed, so that the user interface can use the appropriate amount of icons and text-based cues. Furthermore, because of the fact that the users reside in a village, it is likely that their exposure to technology is extremely low. Complex features would be lost on a group that would struggle with basic keyboard functionality. Perhaps in such a situation, speech-recognition software would benefit the users more than regular keyboard entry. The mental models used by people in different parts of the world vary significantly—what appears obvious and simple to one user in California may seem complex and even impossible to decipher to someone residing in a small village in southern India. Understanding users is, as always, vital to the design and development of technological applications for a global audience. This example presents a complex challenge for designing technology for users who have minimal exposure to technology, low literacy levels, and no knowledge of the English language. Nevertheless, even when dealing with English-language users, there are various items that may require localization to be optimally effective and understood. These include changes to address formats, nomenclature, environmental standards, keyboard formats, punctuation symbols, telephone number formats, name formats, icons, symbols, colors, calendar formats, licensing standards, and so on (Marcus, 2002). Fundamental rules such as which side of the road to drive on are different in different countries. A car-manufacturing web site offering online test drives would benefit from localization. An online test drive where the cars ride on the right side of the road would seem quite unusual to a user in the United Kingdom. Certain words, phrases, and even colors have different meanings in different societies. Spellings of words are different as well: examples include behaviour vs. behavior, color vs. colour, etc. Being sensitive to cultural differences is crucial during the design process. The following section will look into cultural and linguistic issues in greater detail.

Class differences in societies, common in earlier centuries, still exist in different parts of the world. Differences in socioeconomic status result in classes of unspoken privileges and denial. Regions in which the more advanced segments of the population are the focus of technological implementations create a situation in which people from all parts migrate to these regions to share in the benefit, while many groups of society from the region itself are neglected (Castells, 1999). This can lead to many consequences, both socioeconomic and political. The same principle applies on the individual level, where differences in education and socioeconomic status can create or contribute to rifts. The cost of technological systems is sometimes prohibitive. For example, the purchase of a basic personal computer may be financially challenging for one family, while posing no financial stress for another. Dealing with, and developing solutions for, socioeconomic conditions is beyond the scope of this chapter. The intention here is to shed light on the fact that, while technologies may be created equal, the ability to purchase, access, and use them is not always equal between all persons and populations. A Harvard study involving telecommunications in Algeria and educational television in El Salvador concluded that new technologies create situations of power concentration as well as a group of "technocratic elite" (Garson, 1995). The concept of an elite group of society who has access to the latest in technological inventions takes us back to the concept of technological exclusion. It is clear that for technology to be accessible to larger groups of the population it is critical not only to be aware of potential individual differences, such as impairments, disabilities, cognitive abilities, and the like, but also of societal differences between segments of people.

4.6 Cultural and Linguistic Issues

Closely related to social issues is the reality of cultural differences. Culture, defined in general, refers to specific habits of everyday living that make us who we are. It is a central factor of human self-definitions, and is crucial to our identities, with which we negotiate our everyday life in the societies we live in. During ancient times, when modes of communication and travel were not technologically advanced, culture was only a matter of geography. However, with the globalized technology shrinking the world and redefining our understanding of the near and the far, home and the world, cultures are not as remote from one another as they used to be. Still, when people visit countries far from their own, their initial surprise at the change in the cultural environment can evoke what is referred to in common parlance as "culture shock." Many residents of the Western world are surprised when they visit countries in Asia, such as India, where many attributes of society, from eating habits to transportation modes, are radically different. Seeing cow-drawn carriages moving routinely through busy streets is an unfamiliar sight for many. In the same way, visitors to the Western world are sometimes amazed by differences in clothing habits—for example, the wearing of shorts by women, even in the most scorching of summers, is taboo in some countries in Asia. Differences in religious practices and beliefs constitute another aspect of culture. Recognizing the importance of cultural diversity and of the need to acknowledge and appreciate it, UNESCO held a convention on the protection and promotion of diverse cultural expressions in 2005. In today's world, we have entered a significant shifting point in the perception, understanding, and experience of cultural epistemology. The inclusion of this knowledge in technology will lead to more inclusiveness and tolerance. Cultural issues relevant to universal access are addressed in Chapter 9, "International and Intercultural User Interfaces," of this handbook.

4.6.1 Language

Language is an integral part of culture and much, as we know, can be lost in translation due to language barriers. For example, many technological applications use English, and this in itself could be a restricting factor for people who do not speak or write the language. Even within the same linguistic group, the usage of language can vary, and certain words or phrases can have many different connotations. In the United States, the word *subway* refers to a popular sandwich brand as well as the underground train system. In England, *tube* is used to refer to these trains. *Elevator* and *lift* both mean the same, but used in the wrong environment can lead to misunderstanding. Abbreviations, spelling, punctuations are all linguistic variables. The connection between language and the layout of text on technical applications is a factor to be considered, since certain languages like English and French lend themselves to shorter representations, while other languages may require longer formats (Marcus, 2001).

4.6.2 Cultural Interpretations

In technology, the design of an application that is not aware of cultural nuances can inhibit access. Differences in culture include interpretations of symbols, colors, movements, phrases, and so on. Aaron Marcus points out the following interesting differences: green is a sacred color in Islam, and saffron yellow is sacred in Buddhism. Reading direction is left to right in North America and Europe, while it is right to left in the Middle Eastern region (Marcus, 2001). In certain Hindu wedding ceremonies, red is considered the festive and appropriate choice for bridal attire, while in Christian ceremonies, the bride wears white. Even within the Hindu culture, red is not uniformly a sacred color. The design of a bridal web site for a primarily Hindu audience with extensive usage of the color white would be inappropriate for this cultural context. Culture-specific notions of the sacred need to be respected during design of technology. Symbols and icons can also mean different things to people from different cultures. In fact, sometimes symbols can be interpreted to have diametrically opposite meanings. Ideas on clothing, food, and aesthetic appeal also vary from culture to culture. These numerous differences make it imperative that designers avoid treating all cultures as the same, but to be sensitive to these differences during the creation of technology. Rather than neutralize cultural and linguistic differences, universal access acknowledges, recognizes, appreciates, and integrates these differences. In theory, this may appear to be a formidable challenge, but investing energy into assessing diversity of users is a valuable effort.

To gain an understanding of the heterogeneity of the human community that is expected to comprise the user group for the technology under consideration, the following guidelines for designers are emphasized:

- Understand the target user population in terms of geographical location, cultural identities, and language usage. Recognize that not all user groups from the same general geographic area speak the same language or hold the same cultural beliefs.
- Ensure that evaluation sessions with users include user groups that are truly representative of the target user population.
- Invest time, effort, and monetary allowances on sessions with users supported by translators and any other assistance that may be required for evaluators to obtain as much useful information as possible from these sessions. This will save considerable time, effort, and money during iterative design, testing, quality assurance, and implementation.

4.7 Age

Age plays a significant role in how a person perceives and processes information. Knowing the age of the target population of a technology product can provide vital clues about how to present information, feedback, video, audio, and so on. The design process becomes more challenging when a wide range of ages is included in the list of potential user groups. While there are some variations in adult users between the ages of 18 and 65, the focus of this chapter is on the two user groups whose age is one of the significant defining factors about them: children (defined as users below the age of 18, but particular focus on younger children less than the age of 12) and the elderly (defined as users over the age of 65). Design for children is a unique realm of study as is design for older users. Older users present a set of challenges that include the fact that they are typically accustomed to performing tasks in a certain way that usually does not include technology. Bringing technology into the picture and requiring that older users adapt to these new systems can be a challenging endeavor.

4.7.1 Children

Children's physical and cognitive abilities develop over a period of years from infancy to adulthood. Children, particularly those who are very young, do not have a wide repertoire of experiences that guide their responses to cues. In addition to this lack of experience, children perceive the world differently from adults (Piaget, 1970). Today, many software applications are developed with the sole purpose of providing entertainment and knowledge to children. Various applications are used to teach children everything from the alphabet to algebra, from shape recognition to grammar. Unlike many applications, which are designed by adults for adults, the design of tools for children poses a special challenge, in that designers must learn how to perceive systems through the eyes of a child. Testing applications with children requires special planning and care. Younger children may experience feelings of anxiety when being asked to perform tasks on an application, or may fear the instructor. If the child is separated from the parent, this may increase feelings of anxiety. Guidelines have been developed to conduct usability testing with children. These guidelines provide a useful framework to obtain maximum feedback from children, while at the same time ensuring their comfort, safety, and sense of well-being. Some of these guidelines are as follows (Hanna et al., 1997):

- The area where the testing takes place should appear colorful and friendly without being overly distracting.
- Preschool-age children have difficulty when asked to use an input device they are unfamiliar with. This situation can be avoided by finding out what device they are comfortable with and having that installed on the system.
- For preschool-age children, it is advisable to set cursor speeds at the slowest possible level.
- Keep fatigue levels in mind and do not schedule long sessions, as even older children will tend to become tired as time goes by.
- Include a representative group of children for the study instead of using one's own children or the children of associates.

- Establish a friendly relationship with children when they arrive.
- To reduce feelings of anxiety, allow parents to be present with young children when needed.

This subset of guidelines makes it clear that usability testing for children is a special process. While the effort to set up these testing procedures may be exhausting at times, the value derived from effective usability testing is immense and is crucial to designing technology for children. The need to involve children in every stage of design, using methods such as cooperative inquiry, is particularly important in the case of children's technology, because for adult designers it is difficult (and often incorrect) to make assumptions about how a child may view or interpret data. Children also tend to view certain specific things differently from adults. Audio feedback may alarm very young children and extremely bright colors and video could easily distract them from the task. When developing systems for very young children, it is important to remember that they may not understand words used regularly to convey information. For example, a picture-based application for toddlers with a Help button on the navigation bar would be futile, since the child may not be able to read the word "help." Complex functionality embedded in applications for children would increase their cognitive workload and result in a frustrating computing experience for the child. As with many of the discussed diversity issues, often more than one factor comes into play. For instance, when developing applications for autistic children, it is important to keep in mind the special needs of children as well as the special needs of those with autism. Designing applications for worldwide consumption requires that social and cultural diversity issues be considered along with specific issues for design for children. The point here is that while in theory the details of specific issues are discussed in isolation, in reality, many of these issues occur together and must be considered as a network of issues that may influence each other and come into play concurrently.

4.7.2 Older Users

The realities of old age present a set of challenges to technology design and delivery that have significant impact on the outcome of access and utilization by the elderly. With children, the experience level they have is minimal and so their interactions with technology are oftentimes the formative and impressionable experiences with the tasks being dealt with and with technology itself. On the other hand, with elderly users, the amount of experience is on the other end of the spectrum: a vast set of memories from experiences in the past compose a large repertoire. This naturally influences their feelings toward technology. Older users may feel a sense of resistance to certain technologies, especially when dealing with applications for tasks that people are used to completing without technology, such as online banking systems. The feeling of being "forced" to adapt to technology during the later years of life can add to these feelings of resistance. Many applications

such as writing, shopping, and so on, have now become computer based. The Internet is now a preferred mode of communication, where messages are delivered instantly. These advances in technology have created a "keep up or be left out" paradigm, and many older users are unable to manage the emerging multitude of technological innovations. In addition to an emotional situation that may create resistance to technology, problems for older users are further compounded by various disabilities and impairments that are a common effect of the aging process (see Chapter 8 of this handbook). Memory loss, associated with aging, is often seen with the elderly. Learning and remembering instructions for technology use is further complicated by limited memory. Cataract is a common cause of vision impairment in older adults. Complications from illnesses, such as diabetic retinopathy caused by diabetes, can also contribute to vision impairments. Hearing loss and arthritis are also ailments that are commonly seen in older individuals. As one's age increases, so does the risk for developing impairments such as these. The level of computer experience and skill also varies greatly among users over the age of 65 today. Some of these individuals work in fields where the integration of technology is continual and seamless over the years. This gradual influx of technology has made it easier for these users to adapt to the growing computerization of society. In other cases, particularly in nontechnical fields and in areas of the world where technological integration has been slower to follow than the Western world, the intrusion of technology has been quick and sudden. Many manual procedures have hurriedly been replaced with computers to keep up with the rest of the world. This difference in the ways in which technology has been implemented, combined with various cultural and socioeconomic issues, has resulted in older users with varying levels of computer experience. Researchers, realizing the importance of computer experience as a variable factor in performance, have conducted studies on the utility of various combinations of multimodal feedback for older computer users (Jacko et al., 2004).

Individuals over the age of 65 will comprise 20% of the population of the United States by the year 2030 (U.S. Department of Health and Human Services Administration on Aging, 2003). The diverse landscape of America includes people from a variety of different cultures who speak different languages. The U.S. Department of Health and Human Services Administration of Aging estimates that over 16.1% of elders (those who are over the age of 65) in the United States are minority elders and that this percentage will increase by 217% by the year 2030. In comparison, the population of white elders will increase by 81%. Considering the language needs as well as understanding the cultural background of the population is important when designing technology for older users. This point is driven home when we see the statistics for the Hispanic older population in the United States. In 2004, the Hispanic older population comprised 6% of the total older adult population. By the year 2050, 17.5% of the older population will be comprised of Hispanic adults (see Figure 4.2). In addition to the growing population of elders in the United States, these numbers are increasing on the global scale as well (see Figure 4.3). It is estimated that, for the first time

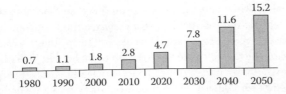

FIGURE 4.2   Population and projection of Hispanic adults over age 65. From U.S. Census Bureau. (http://www.census.gov).

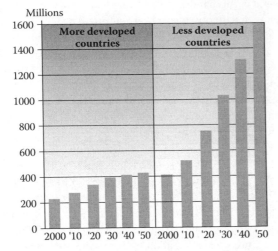

FIGURE 4.3   Aging: a global phenomenon. From U.S. Census Bureau, International Database; and the U.S. Department of Health and Human Services Administration on Aging. (http://www.aoa.gov).

in history, the population of older adults will exceed the population of children (ages 0 to 14) in the year 2050. Almost 2 billion people will be considered older adults by 2050 (U.S. Department of Health and Human Services Administration on Aging, 2003). The pervasive availability and use of technology requires that the access needs of older users be considered during design. This is particularly true when considering the enormous numbers of older individuals, because this indicates that a large percentage of technology consumers will be in this age group. In addition, it is important to be aware of the social and cultural issues pertaining to these users, since aging is a global phenomenon and not just restricted to English-speakers or those who reside in the United States. Again, interconnectedness between various diversity issues plays an important role.

Technology can be of use to older adults in a variety of ways. The first and most obvious use of technology is as a tool for communicating with friends, relatives, and colleagues. E-mail and instant messaging have provided ways for older adults and their caregivers to communicate with family even if they are physically distant. Communication can also help to create and maintain networks of friends. Maintaining communities in this way can improve feelings of well-being and help to reduce feelings of isolation. The Internet is a source of vast information and can be specifically beneficial to older adults looking for information on nursing homes, prescriptions, illnesses, alternative therapies, and so on. Many tasks, which once involved leaving one's home

and making a trip to a specific location, can now be performed online. Computers are now common in the workplace and are replacing many older manual methods of performing tasks. With more and more people aging in today's world, the importance of improving computer access for older adults is critical. The computerization of the workplace generally implied less physical activity but increased cognitive activity. This must be taken into consideration especially for older users, since there are cognitive changes as people age (Czaja and Lee, 2002).

To design technology for older users, the first step is to understand these users, their special needs and conditions. As mentioned, many physical conditions affect older users. Impairments to the sensory systems affect how users perceive and process information.

Motor skills are also affected in older adults and there are significant changes associated with response time, coordination abilities, and flexibility (Czaja and Lee, 2002). A word processing task involves typing on a keyboard, viewing this information on a computer screen, and using the mouse to save data and make changes. All these stages can be seriously affected by various impairments of old age. Decreased coordination ability can make the task of saving and editing extremely complex because of the requirement to balance mouse and keyboard activities. In addition, changes in cognitive abilities are also associated with aging. The gradual decline of cognitive capabilities with increasing age has consequences for technology use. The training process for older users is typically longer than for younger adults. Increased complexity in design can lead to decreased quality of performance from older adults due to heightened cognitive strain. While older users can use and benefit from a variety of technologies, due to the natural effects of aging, the training time, help required, and response times are all higher than for a younger population of users (Czaja and Lee, 2002). Various recommendations have been put forth in the design of technology for older adults. A summary of some of these suggestions is provided in the following (Czaja and Lee, 2002; Jacko et al., 2004):

- Special attention should be given to design and layout of information, with improved contrast to reduce screen glare, enlargement of information presented on the screen, etc.
- Careful organization of information to facilitate easy search and find tasks.
- Analysis of input devices to determine which device will be the easiest to use—the use of a mouse appears to be a source of complexity, and research into alternative input methodologies such as speech-recognition will be valuable.
- Design of software should consider potential limitations in memory and other cognitive abilities on older users, and should not rely extensively on remembering information.
- Cues and feedback are important forms of communication in which the user learns if the task has been completed as expected, the level of progress, and if errors have occurred. Offering effective feedback combinations for

older users can help to minimize cognitive burden and can assist in task completion.

- Feedback itself must not be complicated and should not require extensive cognitive deciphering.

4.8 Gender

4.8.1 Stereotypes and Societal Issues

Differences in the way we perceive things, process information, and feel toward objects and persons can be conceived of as gender-based issues. Men and women, and their child-counterparts, boys and girls, have numerous obvious differences, and the question arising is whether there is also a difference in how the two genders relate to technology. Designers keep beliefs about boys' and girls' attitudes to computers in mind while developing applications: technology for girls is created with a learning-tool model, while similar technology for boys is designed in a game-type format (Cassell, 2002). Considering gender as a marker for differences leads to the societal constructs differentiating women and men. A stereotypical conception of the masculine is to associate masculinity with tools, technology, machinery, cars, and the like. It is encouraging to note that recent studies have shown that in technology-based science classes girls show equal proficiency as boys (Mayer-Smith et al., 2000). Nowadays, many conventional ideas are being changed and challenged, with men and women working in nontraditional roles and crossing boundaries set decades, sometimes even centuries, ago. The pervasiveness of technology has contributed to this equal access, with men and women requiring, demanding, and utilizing technology in every facet of life.

The gender differences that are operative in society are sometimes reflected in the language used in technology. A language of exclusion that relies primarily on masculine pronouns such as "he," "his," and "him" can be an inhibiting factor for women accessing technology. It has been argued that this exclusionary language in technology mirrors the power imbalance in society as a whole (Wilson, 1992). Some technology applications do not reflect women's interest and roles adequately, and instead promote stereotypes. For instance, in many video games in the past, women rarely played a major role and were often relegated to playing passive roles of women who needed rescue (Cassell, 2002). This is a unique kind of access problem, where even though there is no physical or cognitive impairment that may affect usability or performance, a psychological alienation and lack of motivation can result in decreased access by women. In 2006, a BBC report concluded that women enjoy video games, and that video game makers need to recognize this (BBC News, 2004). Several games such as the *Legend of Zelda*, *The Sims*, and *The Prince of Persia: The Sands of Time* are appealing to women—the report makes a special note of *The Prince of Persia* and its very interesting storyline involving its players (BBC News, 2004). The study points out that women prefer games that take less time to learn, as opposed to games that require

extensive time commitment and complex controls. A game like *Lara Croft, Tomb Raider* that projects the role of a women as an active, strong, and aggressive leader presents an interesting case study in the area of intersection between gender and technology. On the one hand, the concept of a woman being the hero in the world of action and adventure, traditionally assigned to men, could be viewed as subverting norms of gender. Still, such a concept, despite this transgressive femininity, can distance women users by fostering the stereotype of women as thin, sensual, and physically attractive individuals. In addition to video games, e-mail provides another interesting example of gender as a diversity factor. A study of e-mail use and perception among men and women showed that while women perceived e-mail differently from men, their use of e-mail was not different (Gefern and Straub, 1997).

The issues surrounding the topic of technology and gender raise the question of how to develop technologies that are universally accessible by men and women. Cassell tackles this issue by pointing out available options (Cassell, 2002). One line of thought is that because much of technology is designed for men by men, there should be separate applications designed for women catering to their needs and interests. This includes a wide range of applications from video games for entertainment purposes, to advanced financial software for use in professional contexts. This suggestion is challenged by critics who foresee that such a trend in divisive technology would exacerbate the current dichotomy between men and women. Cassell is an advocate of a position he calls the "philosophy of undetermined design." Although Cassell speaks specifically with regard to video games, this principle can be applied to many technical applications, and encourages the design of technology, which is comprehensive in nature and allows the users to engender themselves as they choose to. With respect to video games, this engenderment would happen based on what activities the user chooses to participate in. This kind of design permits users to customize their experience and does not box them within preconceived ideas of gender.

Gender differences are an important and unique issue in user diversity. Differences caused by gender association can be subtle and difficult to quantify. Providing customizable experiences for users provides an effective way to ensure that users, irrespective of gender, can personalize their interaction with technology. While providing this level of personalization may present enormous design challenges, the resulting product would be one that enables all users, men and women, to experience the product in its maximal effectiveness.

4.9 Conclusion

In this chapter, the importance of user diversity for universal access in technology has been discussed. There are different forms of diversity, and each of these affects the user's interaction with technology in different ways. For example, certain physical disabilities can make it very difficult for a user to utilize the widely available technological tools. Diversity issues also arise out of disparities in skill level and cognitive abilities. In addition, societal factors, cultural and linguistic differences, age, and gender can all play a role in shaping a user's unique interaction with technology. In many such situations, user interaction can be improved if some changes are made during the design process, after careful consideration of the nature of diversity and the particular needs of the heterogeneous group of users.

The ideal and practice of universal access in technological design are pivotal. A comprehensive recognition and understanding of various diversity issues ranging from physical and cognitive differences to sociocultural and gender issues will lead to more participatory and inclusive technological communities. The importance of diversity issues in technology design also brings to the forefront the interdisciplinary nature of this field—technologists, doctors, psychologists, economists, and various other experts are required to provide a complete picture of the needs and requirements of users. Of course, the most important group of people is the users themselves, whose specific requirements and capabilities have the potential to give new direction to technology's march toward universal access.

References

Barry, M. and Pitt, I. (2006). Interaction design: A multidimensional approach for learners with autism, in the *Proceedings of the 2006 Conference on Interaction Design and Children*, 7–9 June 2006, Tampere, Finland, pp. 33–36. New York: ACM Press.

BBC News (2004). Women take a shine to video games. http://news.bbc.co.uk/1/hi/technology/3615278.stm.

Cassell, J. (2002). Genderizing human-computer interaction, in *The Human-Computer Interaction Handbook: Fundamentals, Evolving Technologies and Emerging Applications* (J. A. Jacko and A. Sears, eds.), pp. 401–412. Mahwah, NJ: Lawrence Erlbaum Associates.

Castells, M. (1999). *Information Technology, Globalization and Social Development*. United Nations Research Institute for Social Development (UNRISD) discussion paper. Code: DP114 (27 pages).

Czaja, S. J. and Lee, C. C. (2002). Designing computer systems for older adults, in *The Human-Computer Interaction Handbook: Fundamentals, Evolving Technologies and Emerging Applications* (J. A. Jacko and A. Sears, eds.), pp. 413–427. Mahwah, NJ: Lawrence Erlbaum Associates.

Dix, A., Finlay, J., Abowd, G., and Beale, R. (1998). *Human Computer Interaction* (2nd edition). Hertfordshire, UK: Prentice Hall Europe.

Drigas, A. S., Vrettaros, J., Stavrou, L., and Kouremenos, D. (2004). E-learning environment for deaf people in the e-commerce and new technologies sector. *WSEAS Transactions on Information Science and Applications* 5: 1189–1196.

Evans, D. G., Drew, R., and Blenkhorn, P. (2000). Controlling mouse pointer position using an infrared head-operated joystick. *IEEE Transactions on Rehabilitation Engineering* 8: 107–117.

Freitas, D. and Ferreira, H. (2003). On the application of W3C guidelines in website design from scratch, in *Universal Access in HCI: Inclusive Design in the Information Society* (Volume 4 of the *Proceedings of HCI International 2003*; C. Stephanidis, ed.), pp. 955–959. Mahwah, NJ: Lawrence Erlbaum Associates.

Gabriel, M. and Benoit, U. (2003). The web accessible for all: Guidelines for seniors, in *Universal Access in HCI: Inclusive Design in the Information Society* (Volume 4 of the *Proceedings of HCI International 2003*; C. Stephanidis, ed.), pp. 872–876. Mahwah, NJ: Lawrence Erlbaum Associates.

Gardner, H. (1993a). *Frames of Mind: The Theory of Multiple Intelligences*. New York: Basic Books.

Gardner, H. (1993b). *Multiple Intelligences*. New York: Basic Books.

Garson, D. G. (1995). *Computer Technology and Social Issues*. Hershey, PA: Idea Group Publishing.

Gefern, D. and Straub, D. W. (1997). Gender differences in the perception and use of e-mail: An extension to the technology acceptance model. *MIS Quarterly* 21: 389–400.

Gordon, W. A. (2001). *The Interface between Cognitive Impairments and Access to Information Technology*. http://www.acm.org/sigaccess/newsletter/sept05/Sept05_01.pdf.

Grammenos, D., Savidis, A., and Stephanidis, C. (2005). UA-Chess: A universally accessible board game, in *Universal Access in HCI: Exploring New Interaction Environments* (Volume 7 of the *Proceedings of HCI International 2005*; C. Stephanidis, ed.) [CD-ROM]. Mahwah, NJ: Lawrence Erlbaum Associates.

Hanna, L., Risden, K., and Alexander, K. (1997). Guidelines for usability testing with children. *Interactions* 4: 9–14.

Jacko, J. A., Barreto, A. B., Scott, I. U., Chu, J. Y. M., Vitense, H. S., Conway, F. T. et al. (2002a). Macular degeneration and visual icon use: Deriving guidelines for improved access. *Universal Access in the Information Society* 1: 197–206.

Jacko, J., Emery, V. K., Edwards, P. J., Ashok, M., Barnard, L., Kongnakorn, T., et al. (2004). The effects of multimodal feedback on older adults' task performance given varying levels of computer experience. *Behaviour & Information Technology* 23: 247–264.

Jacko, J. A., Scott, I. U., Sainfort, F., Moloney, K. P., Kongnakorn, T., Zorich, B. S., et al. (2002b). Effects of multimodal feedback on the performance of older adults with normal and impaired vision, in *Universal Access: Theoretical Perspectives, Practice, and Experience* (Proceedings of the 7th ERCIM International Workshop "User Interfaces for All"; N. Carbonell and C. Stephanidis, eds.), pp. 3–22. Berlin Heidelberg: Springer.

Jacko, J. A., Vitense, H. S., and Scott, I. U. (2002c). Perceptual impairments and computing technologies, in *The Human-Computer Interaction Handbook: Fundamentals, Evolving Technologies and Emerging Applications* (J. A. Jacko and A. Sears, eds.), pp. 504–522. Mahwah, NJ: Lawrence Erlbaum Associates.

Knepshield, S. (2001). Design for users with color-vision deficiency: Effective color combinations, in *Universal Access in HCI: Towards an Information Society for All* (Volume 3 of the *Proceedings of HCI International 2001*; C. Stephanidis, ed.), pp. 521–524. Mahwah, NJ: Lawrence Erlbaum Associates.

Larson, H. and Gips, J. (2003). A web browser for people with quadriplegia, in *Universal Access in HCI: Inclusive Design in the Information Society* (Volume 4 of the *Proceedings of HCI International 2003*; C. Stephanidis, ed.), pp. 226–230. Mahwah, NJ: Lawrence Erlbaum Associates.

Leonard, V. K., Jacko, J. A., and Pizzimenti, J. J. (2006). An investigation of handheld device use by older adults with age-related macular degeneration. *Behaviour & Information Technology* 25: 313–332.

Malkewitz, R. (1998). Head pointing and speech control as a hands-free interface to desktop computing, in the *Proceedings of the 3rd International ACM Conference on Assistive Technologies (ASSETS 98)*, 15–17 April 1998, Marina del Rey, CA, pp. 182–188. New York: ACM Press.

Marcus, A. (2001). International and intercultural user interfaces, in *User Interfaces for All: Concepts, Methods, and Tools* (C. Stephanidis, ed.), pp. 47–63. Mahwah, NJ: Lawrence Erlbaum Associates.

Marcus, A. (2002). Global and intercultural user-interface design, in *The Human-Computer Interaction Handbook: Fundamentals, Evolving Technologies and Emerging Applications* (J. A. Jacko and A. Sears, eds.), pp. 441–463. Mahwah, NJ: Lawrence Erlbaum Associates.

Mayer-Smith, J., Pedretti, E., and Woodrow, J. (2000). Closing of the gender gap in technology enriched science education: A case study. *Computers & Education* 35: 51–63.

Mirenda, P., Wilk, D., and Carson, P. (2000). A retrospective analysis of technology use patterns of students with autism over a five-year period. *Journal of Special Education Technology* 15: 5–16.

Moore, M. and Calvert, S. 2000. Brief report: Vocabulary acquisition for children with autism: Teacher or computer instruction. *Journal of Autism and Developmental Disorders* 30: 359–362.

Neale, H., Cobb, S., and Wilson, J. (2001). Involving users with learning disabilities in virtual environment design, in *Universal Access in HCI: Towards an Information Society for All* (Volume 3 of the *Proceedings of HCI International 2001*; C. Stephanidis, ed.), pp. 506–510. Mahwah, NJ: Lawrence Erlbaum Associates.

Neale, H., Cobb, S., and Wilson J. (2003). Involving users with learning disabilities in virtual environment design, in *Universal Access in HCI: Towards an Information Society for All* (Volume 3 of the *Proceedings of HCI International 2003*; C. Stephanidis, ed.), pp. 506–510. Mahwah, NJ: Lawrence Erlbaum Associates.

Newell, A. F., Carmichael, A., Gregor, P., and Alm, N. 2002. Information technology for cognitive support, in *The Human-Computer Interaction Handbook: Fundamentals,* *Evolving Technologies and Emerging Applications* (J. A. Jacko and A. Sears, eds.), pp. 464–481. Mahwah, NJ: Lawrence Erlbaum Associates.

Newell, A. F. and Gregor, P. (1997). Human computer interfaces for people with disabilities, in *Handbook of Human-Computer Interaction* (2nd edition) (M. G. Helander, T. K. Landauer, and P. V. Prabhu, eds.), pp. 813–824. Amsterdam: North-Holland Elsevier.

Piaget, J. (1970). *Science of Education and the Psychology of the Child.* New York: Orion Press.

Prevent Blindness America (2002). *Vision Problems in the U.S.: Prevalence of Adult Vision Impairment and Age-Related Eye Disease in America.* http://www.preventblindness.net/site/DocServer/VPUS_report_web.pdf?docID=1322.

Price, G. A. (2006). Creative solutions to making the technology work: Three case studies of dyslexic writers in higher education. *Research in Learning Technology* 14: 21–38.

Ramstein, C. (1996). Combining haptic and Braille technologies: Design issues and pilot study, in the *Proceedings of the Second Annual ACM Conference on Assistive Technologies*, 11–12, Vancouver, Canada, pp. 37–44. New York: ACM Press.

Roy, D. M., Panayi, M., Erenshteyn, R., Foulds, R., and Fawcus, R. (1994). Gestural human-machine interaction for people with severe speech and motor impairment due to cerebral palsy, in the *Conference Companion on Human Factors in Computing Systems (CHI 94)*, pp. 313–314. New York: ACM Press.

Sears, A., Lin, M., Jacko, J., and Xiao, Y. (2003). When computers fade: Pervasive computing and situationally induced impairments and disabilities, in *Human-Computer Interaction: Theory and Practice (Part II)* (Volume 2 of the *Proceedings of HCI International 2003*; C. Stephanidis and J. Jacko, eds.), pp. 1298–1302. Mahwah, NJ: Lawrence Erlbaum Associates.

Sears, A. and Young, M. (2002). Physical disabilities and computing technologies: An analysis of impairments, in *The Human-Computer Interaction Handbook: Fundamentals, Evolving Technologies and Emerging Applications* (J. A. Jacko and A. Sears, eds.), pp. 482–503. Mahwah, NJ: Lawrence Erlbaum Associates.

Sjöström, C. (2001). Using haptics in computer interfaces for blind people, in *CHI '01 Extended Abstracts on Human Factors in Computing Systems*, pp. 245–246. New York: ACM Press.

U.S. Department of Health and Human Services Administration on Aging (2003) (last update). *Challenges of Global Aging.* http://www.aoa.gov/press/prodsmats/fact/pdf/fs_global_aging.doc.

Wilson, F. (1992). Language, technology, gender and power. *Human Relations* 45: 883–904.

World Health Organization (2000). *International Classification of Functioning, Disability, and Health.* http://www.who.int/classifications/icf/en.

Young, M. A., Tumanon, R. C., and Sokal, J. O. (2000). Independence for people with disabilities: A physician's primer on assistive technology. *Maryland Medicine*, 1: 28–32.

Motor Impairments and Universal Access

Simeon Keates

5.1 Introduction

For people with functional impairments, access to, and independent control of, a computer can be an important part of everyday life. For example, people whose impairments prevent communication through writing or speaking can, with appropriate technology, perform these activities with computer assistance. Improved computer access has also been shown to give significant gains in educational success and employment opportunities for people with impairments. However, to be of benefit, computer systems must be accessible. That is, after all, the underlying message of this handbook.

Computer use often involves interaction with a graphical user interface (GUI), typically using a keyboard, mouse, and monitor. However, people with motor impairments often have difficulty with accurate control of standard pointing devices. Conditions such as cerebral palsy, muscular dystrophy, Parkinson's disease, and spinal injuries can give rise to symptoms that affect a user's motor capabilities. Symptoms relevant to computer operation include joint stiffness, paralysis in one or more limbs, numbness, weakness, bradykinesia (slowness of movement), rigidity, impaired balance and coordination, tremor, pain, and fatigue. These symptoms can be stable or highly variable, both within and between individuals, and can restrict the extent to which a keyboard and mouse are useful.

In a study commissioned by Microsoft (Forrester Research, Inc., 2003) it was found that one in four working-age adults have some dexterity difficulty or impairment. The most prevalent conditions that give rise to motor impairments include rheumatic diseases, stroke, Parkinson's disease, multiple sclerosis, cerebral palsy, traumatic brain injury, and spinal injuries or disorders. Cumulative trauma disorders represent a further significant category of injury that may be specifically related to computer use. While many of these conditions may seem remote for many computer users, repetitive strain injury and carpal tunnel syndrome are not. Thus, consideration of the prevalence and effects of motor impairments in the user population is an important facet of designing for universal access.

5.2 Prevalence of Motor Impairments

It is worth beginning by considering the prevalence of motor impairments in the context of universal access. There are several reasons for this.

First, much research in universal access has focused on vision impairments, blindness in particular. This focus is not the result of considering other functional or sensory impairments to be somehow less important. Instead, it is the result of a highly discernible impairment (white canes, guide dogs, etc.), an impairment that is easy to simulate (close your eyes or remove your monitor), and one that clearly affects a user's ability to interact with a GUI. However, as will be shown later in this chapter, blindness is a comparatively rare, though serious, impairment. Motor impairments and, indeed, cognitive impairments, are much more common.

Second, investigation of the prevalence of motor impairments also helps provide a framework for understanding the nature of the impairments. Such a framework is important for communicating to the designers of new computer systems and interfaces how motor impairments affect users.

Estimates of the prevalence of disability derived from any study depend on the purpose of the study and the methods used (Martin et al., 1988). Since disability has no scientific or commonly agreed upon definition (Pfeiffer, 2002), a major problem lies in the confusion over terminology. However, the ICIDH-2: International Classification of Functioning (ICF), Disability and Health (WHO, 2007) represents a rationalization of the terminology frequently used. The ICF defines disability as "any restriction or lack (resulting from an impairment) of ability to perform an activity in the manner or within the range considered normal for a human being" (WHO, 2001).

This definition has been used widely for both disability research (Martin et al., 1988; Grundy et al., 1999) and design research (Pirkl, 1994). However, such language is now generally considered too negative, and it is preferable to describe users in terms of their capabilities rather than disabilities. Thus, "capability" describes a continuum from high (i.e., "able-bodied") to low representing those that are severely "disabled." Data that describe such continua provide the means to define the populations that can use given products, thus leading to the possibility of evaluating metrics for a product's accessibility.

5.2.1 American Community Survey

The U.S. Census Bureau's 1999–2004 American Community Survey (USCB, 2004) adopted a straightforward approach to defining what constitutes a disability. Respondents were asked if they had any kind of disability, defined here as "a long-lasting sensory, physical, mental or emotional condition" (ACSO, 2007). Table 5.1 shows the prevalence of disabilities in the U.S. adult (16+) population recorded by the survey.

In Great Britain, the Office of National Statistics commissioned two surveys in the late 1980s and 1990s that attempted to describe the prevalence and severity of impairments across the entire British population (i.e., the combined populations of England, Scotland, and Wales, but not Northern Ireland) in a more rigorous manner. The surveys involved over 7500 respondents, sampled to provide representative coverage of the population.

5.2.2 The Survey of Disability in Great Britain

The Survey of Disability in Great Britain (Martin et al., 1988) was carried out between 1985 and 1988. It aimed to provide up-to-date information about the number of disabled people in Britain with different levels of severity of functional impairment and their domestic circumstances. The survey used 13 different types of disabilities ranging from locomotion to stomach issues, and gave estimates of the prevalence of each type. It showed that musculoskeletal complaints, most notably arthritis, were the most commonly cited causes of disability among adults.

An innovative feature of the survey was the construction of an overall measure of severity of disability, based on a consensus of assessments of specialists acting as judges. In essence, the severity of all 13 types of disability is established, and the 3 highest scores combined to give an overall score, from which people are allocated to 1 of 10 overall severity categories. A scale that runs from a minimum possible 0.5 to a maximum possible 13.0 represents each impairment. Note though, that not all of the scales extend across this complete range, with some having maximum values of only 9.5, for example.

TABLE 5.1   The Prevalence of Disabilities in the U.S. Adult (16+) Population

Respondents	Percentage of Total	Margin of Error
Population 16 years +	*220,073,798*	*+/− 129,242*
With any disability	16.0	+/− 0.1
With a sensory disability (i.e., with *blindness, deafness, severe vision or hearing impairment*)	4.7	+/− 0.1
With a physical disability (i.e., with a condition that substantially limits *walking, climbing stairs, reaching, lifting or carrying*)	10.6	+/− 0.1
With a mental disability (i.e., with a condition that makes it difficult to *learn, remember or concentrate*)	5.2	+/− 0.1
With a self-care disability (i.e., with a condition that makes it difficult to *dress, bathe or get around inside the home*)	3.1	+/− 0.1
With a go-outside-home disability (i.e., with a condition that makes it difficult to *go outside the home alone to a shop or doctor's office*)	4.9	+/− 0.1
With an employment disability[a] (i.e., with a condition that makes it difficult to *work at a job or business*)	5.6	+/− 0.1

Source: USCB, *American Community Survey,* http://factfinder.census.gov/jsp/saff/SAFFInfo.jsp? _pageId=sp1_acs&_submenuId=, 2004.

[a] Data for employment disability collected for ages 16–64 years only.

A person's sum physical impairment is derived using a weighted sum as shown in Equation 5.1. The weighted sum is then mapped from the resulting 0 to 18.5 (the maximum possible upper limit) scale to a 0 to 10 scale.

$$\text{Weighted sum} = \text{worst score} + 0.4 \times \text{2nd worst} + 0.3 \times \text{3rd worst} \tag{5.1}$$

5.2.3 The Disability Follow-Up Survey

The Disability Follow-Up Survey (DFS) (Grundy et al., 1999) to the 1996/97 Family Resources Survey (Semmence et al., 1998) was designed to update information collected by the earlier Survey of Disability in Great Britain (Martin et al., 1988). For the purposes of this chapter, 7 of the 13 separate capabilities proposed in the Family Resources Survey and used in the DFS are of particular relevance, specifically:

- Locomotion
- Reaching and stretching
- Dexterity
- Seeing
- Hearing
- Communication
- Intellectual functioning

These individual impairments may be grouped into three overall capabilities:

- *Motor*—locomotion, reaching and stretching, dexterity
- *Sensory*—seeing, hearing
- *Cognitive*—communication, intellectual functioning

The survey results showed that an estimated 8,582,200 adults in Great Britain (GB)—that is, 20% of the adult population—had a disability according to the definitions used. Of these 34% had mild levels of impairment (categories 1 and 2—i.e., high capability), 45% had moderate impairment (categories 3 to 6—i.e., medium capability), and 21% had severe impairment (categories 7 to 10—i.e., low capability). It was also found that 48% of the disabled population was aged 65 or older and 29% was aged 75 years or more.

5.2.4 Multiple Capability Losses

Traditionally, universal access research has tended to focus on accommodating single, primarily major, capability losses. Unfortunately, many people do not have solely single functional impairments, but several. This is especially true when considering older adults. Consequently, it is important to be aware of the prevalence of not only single, but also multiple capability losses. Therein lies a problem, as most user data focus on single impairments.

Fortunately, both the American Community Survey and the DFS provide valuable information for analyzing multiple capability losses. Tables 5.2 and 5.3 summarize the data extracted from those surveys. It is evident that in both surveys at least half of those respondents with some loss of capability have more than one loss of capability.

The comparative magnitudes of prevalence of motor and sensory impairments from the DFS (6.71 million and 3.98 million, respectively) are self-evident.

As discussed earlier, many designers and researchers automatically assume that universal access is really just enabling access for people who are blind. In practice, designing an interface that is easier to manipulate is likely to enable more people to use it than, say, supporting a Braille display. Only a comparatively small proportion of people who are blind have the skills to read Braille.

As discussed earlier, blindness is an important impairment that should not be overlooked when designing for universal access, but has comparatively low prevalence. Even within the sensory impairment category, 1.93 million people in Great Britain have vision impairments compared with 2.9 million with hearing impairments. Note that 1.93 million + 2.9 million does not equal the 3.98 million in Table 5.3 because approximately 1 million people have both some hearing and some vision impairment, especially among older adults, and so the 3.98 million figure has been corrected to remove such double-counting. Of those 1.93 million people with vision impairments, the vast majority have low vision (i.e., they can see to some extent, but have difficulty reading regular size print, even with spectacles). Only a small percentage (less than 20%) of people with a vision impairment are classified as blind. Thus people who are blind constitute approximately only 5% of the total disabled population within Great Britain.

5.2.5 Aging and Motor Impairments

In virtually every country in the developed world, the population is aging and aging rapidly. Countries such as Japan have even reached the stage of no longer being considered "aging," but "aged" with 28% of its population projected to be over 65

TABLE 5.2 Multiple Capability Losses as Reported in the 2004 American Community Survey

Respondents	Percentage of Total	Margin of Error
Population 5 years and over	*264,965,834*	*+/- 65,181*
Without any disability	85.7	+/- 0.1
With one type of disability	6.7	+/- 0.1
With two or more types of disabilities	7.6	+/- 0.1

TABLE 5.3 Multiple Capability Losses for Great Britain: Total Population ~46.9 Million Adults

Loss of Capability	Number of GB 16+ Population	Percentage of GB 16+ Population
Motor	6,710,000	14.3
Sensory	3,979,000	8.5
Cognitive	2,622,000	5.6
Motor only	2,915,000	6.2
Sensory only	771,000	1.6
Cognitive only	431,000	0.9
Motor and sensory only	1,819,000	3.9
Sensory and cognitive only	213,000	0.5
Cognitive and motor only	801,000	1.7
Motor, sensory, and cognitive	1,175,000	2.5
Motor, sensory, or cognitive	8,126,000	17.3

by 2025. A more delicate choice of phrasing would be to regard the populations as "maturing" (rather like a good wine). In a report published in 2002, the Population Division of the United Nations' Department of Economic and Social Affairs reported that:

> In 1950, there were 205 million persons aged 60 or over throughout the world … At that time, only 3 countries had more than 10 million people 60 or older: China (42 million), India (20 million), and the United States of America (20 million). Fifty years later, the number of persons aged 60 or over increased about three times to 606 million. In 2000, the number of countries with more than 10 million people aged 60 or over increased to 12, including 5 with more than 20 million older people: China (129 million), India (77 million), the United States of America (46 million), Japan (30 million) and the Russian Federation (27 million). Over the first half of the current century, the global population 60 or over is projected to expand by more than three times to reach nearly 2 billion in 2050.… By then, 33 countries are expected to have more than 10 million people 60 or over, including 5 countries with more than 50 million older people: China (437 million), India (324 million), the United States of America (107 million), Indonesia (70 million) and Brazil (58 million). (UN, 2002)

Keeping an aging workforce productively employed is a key challenge. The costs of premature medical retirement are difficult to evaluate, but the Royal Mail in the United Kingdom conducted what is believed to be the first survey of its kind to estimate the cost to its business. It found that preventable premature medical retirement was costing the company over $200 million per year, without taking into consideration the costs of recruiting replacement staff or the loss of organizational memory (the knowledge and skills accumulated over the years by experienced employees). At the time, the company was losing $800 million per year and so the costs of preventable premature medical retirement became a major target for cost reductions.

Their definition of "preventable" premature medical retirement was where an employee was deemed capable of doing a job, just not one of the jobs available within the Royal Mail's working environment (Coy, 2002).

Implicit in the argument that the aging of the population is relevant to universal access is the fact that the aging process is associated with certain decreases in user capabilities—see the DFS data discussed earlier (see also Chapter 8, "Age-Related Difference in the Interface Design Process," for further details). In many cases, these are fairly minor losses of an individual's capabilities, but the minor losses can often have a cumulative effect. Thus, someone whose eyesight is not quite as sharp as it was, whose keeps needing to turn the volume up a little bit louder on the television, and whose fingers are not quite as nimble as they once were, may find some products as difficult to use as someone with a single, but more severe, impairment. In particular, conditions such as arthritis and Parkinson's disease have strong links to decreased motor capabilities in older adults. The effects on computer access will be examined later in this chapter.

5.3. Effects of Functional Impairments on Universal Access

Having looked at the prevalence of impairments, it is helpful to think about how each of those impairment types affects someone's ability to use a computer. To illustrate how different users' capabilities can influence the difficulties that those users can expect to encounter, Table 5.4 shows the kinds of difficulties that users with specific impairments may encounter when trying to use a computer and the kinds of assistive technology that they may use.

Table 5.4 represents broad difficulties and solutions. It is also instructive to think more deeply about how users with different impairments may be affected by common graphical user interface activities, for example clicking an onscreen button.

Consider a small button or icon on a software interface. Someone who is blind would not be able to see the button, or locate it. Someone with low vision may be able to see that there

TABLE 5.4 Common Issues Facing Users with Functional Impairments When Trying to Use a Computer and Examples of the Assistive Technology That May Be Used

	Vision
Issues	• Cannot use the mouse for input • Cannot see the screen • May need magnification and color contrast
Assistive technology	• Screen readers/voice output • Braille displays • Screen magnifiers
	Hearing
Issues	• Cannot hear audio, video, system alerts, or alarms
Assistive technology	• Closed captioning • Transcripts • "Show sounds" (MS Windows)/"Flash screen" (Mac OS)
	Motor
Issues	• Limited or no use of hands • Tremor • Limited range of movement, speed, and strength
Assistive technology	• Alternate input (e.g., voice) • Access keys • Latches that are easy to reach and manipulate • Single switches as alternatives to standard point and click devices • Onscreen keyboards • Operating system (OS)-based keyboard filtering
	Cognitive/Learning
Issues	• Difficulty reading and comprehending information • Difficulty writing
Assistive technology	• Spell checkers • Word prediction aids • Reading/writing comprehension aids

Source: IBM, *Conducting User Evaluations with People with Disabilities,* http://www.ibm.com/able/resources/userevaluations.html, 2005.

is a button, but not be able to recognize the type of button or read the text on it. Someone with a motor impairment may be able to see it and recognize it, but may not be able to position the mouse pointer or keyboard focus over it. Someone with a cognitive impairment or learning disability may be able to see, recognize (note: assuming that it is not described solely by words and that the learning disability is not a literary one), and activate the button, but may not know what it does. Next, consider a small button on a piece of hardware. Many of the same issues apply. Blind users may not be able to see or locate the button; low-vision users may not be able to recognize it; motor-impaired users may not be able to activate it; and cognitive-impaired users may not know what it does.

Having examined a brief overview of how different impairments affect computer access, specific solutions for users with motor impairments will be discussed. These solutions are typically software based or hardware based, but can also be used in a complementary combination. The next section will begin by examining the software solutions first.

5.4 Common Assistive Technology Software Solutions for Motor Impairments

Please note in the following subsections, *mouse* is used as a generic term to encompass the range of cursor control input devices that consist of a two-dimensional analogue input to control the cursor position and one or more buttons that act as target activation devices.

5.4.1 Adapting the Conventional Keyboard

For relatively mild coordination difficulties or tremor, there are several options built into most modern operating systems that are designed to filter out extraneous keyboard input. One variant of this strategy is filter keys (MS Windows), where repeated key-presses within a set time less than a particular threshold value will be ignored. This filters the input for users with tremor who may depress a key more than once for a single character input. Slow keys (Mac OS) achieves a similar effect to bounce keys, in this case inserting a delay between when a key is pressed and when it is recognized by the operating system.

Another powerful option is sticky keys (both MS Windows and Mac OS), which registers the pressing of modifier keys such as Shift and Control and holds them active until another key is pressed. This removes the need for the user to press multiple keys simultaneously to activate keyboard shortcuts, hence simplifying the degree of coordination demanded for the input.

Finally, mouse keys (both MS Windows and Mac OS) allows the numeric keypad on most standard keyboards to be used to control the cursor for users with difficulty operating input devices such as mice and trackballs.

5.4.2 Onscreen Keyboard Emulators

If a user is able to control a cursor control device, such as a mouse or trackball, but does not have sufficient coordination, strength, or reach to use a keyboard, an onscreen keyboard emulator may be of use. This is a software application that displays a purely graphical version of the keyboard on the computer screen. Pressing a key (effectively a button) creates a keystroke message into the message-handling code of the operating system. This message is identical to that generated by the pressing of a key on a standard keyboard.

For users that cannot control the cursor control device well enough to hold it in place while making a selection (e.g., causing the mouse to move while trying to press the left button), many keyboard emulators support dwell and scanning inputs. These inputs use the two principal modes of mouse (and other input device) inputs—specifically cursor control (analogue) and button pressing (digital). Dwell input works by the user hovering the cursor over a specific onscreen target for a preset time duration (the threshold). If the cursor remains sufficiently steady, usually over a single onscreen key for example, then the keyboard emulator generate a corresponding key press event.

Scanning input works by highlighting successive sections of the onscreen keyboard. The user presses the mouse button (or other input device button) when the section containing the desired key is highlighted. The contents of that section of the onscreen keyboard are then scanned and so on recursively until an individual key is selected. Perhaps the most important features of scanning input that affect their overall usability are the movement and duration thresholds. Selecting the correct thresholds for a user is very important. A movement tolerance threshold that is too aggressive (i.e., small) will be difficult for a user to keep the cursor within. Meanwhile, one that is too relaxed (i.e., big) will cause false positives. Similarly, with the time duration settings, a short duration threshold will cause false positives, while a long duration threshold may be hard to maintain.

Research has been performed on automatically adjusting scanning thresholds based on user history and a degree of success has been seen (e.g., Lesher et al., 2000). Software

keyboards are addressed in more detail in Chapter 29, "Virtual Mouse and Keyboards for Text Entry," of this handbook.

5.4.3 Word Prediction

One commonly used technique for increasing the interaction rate of onscreen keyboard emulators is to couple the use of such a utility with a word prediction application. Word prediction applications can reduce the number of keystrokes required to type in a complete word by offering a list of numerically indexed completed word suggestions. The suggestions are refined based on every key that the user presses, until the desired word is suggested. At that point the user stops entering letters and usually enters the number that corresponds to the desired word.

Such systems can be effective for many keyboard emulator users, but the degree of success depends on the quality of word suggestions, especially their likelihood to include the desired word. Predictive adaptive lexicons (e.g., Swiffin et al., 1987) work by modifying the underlying lexicons based on the observed vocabulary used by the user and can increase the effectiveness of the word prediction application.

5.4.4 Cursor Control Modifications

Many users are able to control input devices such as mice to some degree, but not quite well enough to not benefit from assistance. For example, users with Parkinson's disease often can perform gross (ballistic) movements to move the cursor close to an onscreen target, but find it difficult to perform the fine homing movements required to get onto the target because of the tremor associated with their medical condition.

SteadyClicks was developed by researchers at IBM as a method of helping such users by offering a movement threshold, within which the cursor is effectively frozen at the point where the mouse button is pressed (Trewin et al., 2006). Moving outside the threshold cancels the button press. SteadyClicks has been shown to offer significant benefits to users who exhibit particular button pressing difficulties, such as moving the mouse while pressing the mouse button (slipping).

Area cursors (Worden et al., 1997) are another technique that can assist users with fine motor difficulties. These cursors work by expanding the focal point of the cursor from the typical 1 pixel to a larger area, say 12 pixels square. The increase in cursor size equates to an increase in target size, and so, by Fitts' law (Fitts, 1954), the speed of interaction increases by making the target easier to acquire. However, area cursors are of limited use when there are multiple targets that are directly adjacent to each other.

Sticky icons (Worden et al., 1997) are another technique with potential benefits for users with fine motor difficulties. Sticky icons work by automatically adjusting the mouse gain when the cursor traverses across them. In other words, the cursor slows down while crossing over them. In this case, the presence of many sticky icons on the screen can also slow down the overall interaction rate, as the cursor movement is slowed down many times en route to the target.

5.5 Common Assistive Technology Hardware Solutions

To facilitate access to computers for user with motor impairments, several input systems have been developed over the years to remove or ameliorate the problems presented by conventional keyboard and mouse input. This section summarizes the principal types of input devices currently available commercially and presents a critique of their respective performance. It will begin with the more common off-the-shelf solutions, progressing to the more innovative technological advances. The aim here is not to provide a comprehensive description of all input devices, since new technological solutions move quickly, but rather to provide an indication of the generic flavors of assistance available.

5.5.1 Trackballs

Trackballs are more straightforward to control and are often more accurate for a motor-impaired user than a mouse. One of the main issues with the mouse is its inherent degrees-of-freedom of movement and the even greater degrees-of-freedom of movement required in the user's arm to control that movement (shoulder, elbow, wrist, multiple fingers). For a trackball, though, once the hand is located on the device, thumb movement is all that is required to move the cursor to the extremities of the screen.

Trackballs, being direct mouse replacements, require no interface adaptation if used in conjunction with a keyboard. However, it is possible to use these as sole input devices if used in conjunction with onscreen keyboard emulators.

Some users can take a little time to get used to controlling the cursor position with a trackball, but that is normally the total extent of learning involved.

5.5.2 Keyguards

Keyguards are one of the earliest solutions to the problems of accessibility, but still meet a niche demand. Keyguards are simply attachments that fit over the standard keyboard with holes punched through above each of the keys. They provide a solid surface for resting hands and fingers on, making them less tiring to use than a standard keyboard where the hands are held suspended above. They also reduce the likelihood of accidental erroneous key-presses. Since they are a straightforward physical modification of the original input device, there is a minimal cognitive overhead involved in learning to use them.

However, they do still require a degree of coordination to use, the ability to move the hands over the full range of the keyboard and the finger strength to press the keys and withdraw the fingers afterwards. Some users have been known to experience difficulties

when pressing the keys and having their fingers curl over underneath the guard, preventing withdrawal of the fingers.

5.5.3 Joysticks

Replacing the mouse with a joystick picks up on many of the benefits of using a trackball—specifically a more stable base and reduced complexity of arm control for the user. For users of electric wheelchairs, the use of joysticks also takes advantage of the users' familiarity with them to control their chair. Rather than using typical games joysticks, though, many users prefer devices that have been developed explicitly as mouse-replacement devices. They are usually comparatively large devices with the joystick centrally mounted and with buttons on either side for single clicks, double clicks, and dragging. These can all be covered by a keyguard. On more expensive devices the speed of cursor response is adjustable. Again, these are physical replacements of the mouse that generate mouse inputs to the computer, so no modification of the interface is required for operation of these devices.

5.5.4 Isometric Input Devices

Isometric devices measure force input, rather than displacement. Consequently, these devices do not require any limb movement to generate the input, only muscle contractions. Perhaps the most common example of an isometric input device is the little keyboard joystick found in the middle of many modern laptop keyboards.

It has been postulated that some spasticity in particular is brought on by limb movement, and these devices offer a means of avoiding that. Studies performed using isometric joysticks (Rao et al., 1997) and an adapted Spaceball Avenger (Stapleford and Mahoney, 1997) have shown that the ability to position the cursor is improved for some users by using isometric devices.

5.5.5 Headsticks

Headsticks are another of the original methods of providing computer access, but are now increasingly rare to find, having been almost exclusively replaced by more modern technological solutions. They are included here more for historical interest. Headsticks typically consist of probosces angled in the center and mounted on a headband. The users have to move their entire upper body to operate the keyboard by pressing the keys with the end of the headstick, and generally experience great discomfort both from the physical exertions required and the rather limited aesthetic appeal. However, as the basic layout of the keyboard is not altered in any way, the learning effort is small and the software interface requires no adaptation.

5.5.6 Switches

The most physically straightforward input device to operate is a single switch. Switches can come in many different formats. For instance, they can be standard hand-operated ones, usually with large surface areas for easier targeting, or activated by foot or head movements. They can be wobble switches, which look like inverted pendulums and are activated by being knocked from side to side. There are also mouth switches, operated by tongue position or by sudden inhalation or exhalation (sip-puff switches). More modern approaches are based on touchless systems, such as the TouchFree mouse (Kjeldsen and Hartman, 2001).

The basic technical requirement for attempting to use a movement to operate a switch is a distinct, controllable motion. This can be as small as raising an eyebrow. If a user is capable of generating several of these motions independently, then it is possible to increase the number of switches to accommodate this and increase the information transfer bandwidth. For instance, it is possible for a user with some gross arm movement to operate a number of switches positioned near each other. Given the relatively low level of movement required to operate switches and their cheapness, they have become extremely popular as the preferred method of input for the more severely impaired users.

However, they do have drawbacks. The first main problem is that a single switch, or even an array of switches, cannot reproduce the full functionality of a keyboard and mouse directly. It is therefore necessary to use switches in conjunction with some kind of software adaptation to generate the extra information required. The most frequently used method for this is scanning input (see Chapter 35 of this handbook), usually in conjunction with an onscreen keyboard emulator (see Chapter 29, "Virtual Mouse and Keyboards for Text Entry").

Scanning input involves taking a regular array of onscreen buttons, be they symbols, letters, or keys, and highlighting regions of the screen in turn. The highlighting dwells over that region of the screen for a predetermined duration and then moves to another part of the screen, dwells there and so on until the user selects a particular region. This region is then highlighted in subregions and this continues until a particular button is selected. Therefore, this process can involve several periods of waiting for the appropriate sections of the screen to be highlighted, during which the user is producing no useful information.

Work has been performed to optimize this by automatically adjusting the dwell time through the use of a neural network, and the results showed that significant time gains were possible without significant degradation of input quality (Simpson, 1997). Also, using multiple switches can reduce the interaction time. However, interaction rates are still typically significantly slower than for standard keyboard use.

Positioning the switches is of prime importance for the user. If done badly, the switches can be more than just a painful annoyance. For example, in the case of seizure disorders, a badly positioned switch can bring on a seizure as the user attempts to reach it. Trial and error is often the only way of obtaining the best position for the switches.

The learning effort associated with using switch input depends upon the interface being used. A single switch, used in conjunction with a keyboard emulator using a standard layout, will involve relatively little learning. However, multiple switches with an unfamiliar symbol scheme can involve a significant amount of learning, although with time they may produce much higher information transfer rates.

5.5.7 Damping/Haptic Force Feedback Devices

One method for making a mouse more controllable for users with tremor, for example, is to increase the amount of damping available on the device itself. Earlier solutions to this issue were hardware based. The Computer MouseTRAP, for example, has viscous dampers mounted around the mouse (Neater, 2007). More recently, efforts have focused on filtering the input event stream from the mouse. The Assistive Mouse Adapter, for example, is a device that is connected between the mouse and computer and filters the input signal (Levine and Schappert, 2005).

Also recently, haptically enabled force feedback input devices, such as the Logitech Wingman Force Feedback mouse, have been studied extensively as they can provide significant improvements in interaction rates for users with motor impairments. While nominal improvements in interaction rates have been seen through the addition of damping effects, the most significant gains arise for the use of gravity wells around elements of the onscreen display (Hwang et al., 2003). Unfortunately, the Wingman Force Feedback mouse is not compatible with more recent versions of Windows, but other devices are coming onto the market.

5.5.8 Speech Recognition Systems

Speech is a potentially very good input medium for motor-impaired users. For many years it was regarded as almost the Holy Grail of computer access solutions. As Kyberd wrote, "Capturing the most adaptable communications medium, the voice, would seem to be the ideal solution" (Kyberd, 1991). Once speech became technically reliable enough, users would abandon their mice and keyboards in droves. However, there have been enough solutions that are satisfactorily good enough on the market for long enough now to know that this potential seems to have been somewhat overhyped. That is not to say that speech systems are not without their merits, nor that they are not invaluable to certain users, rather than they are not the panacea once thought.

This is especially true for users with motor impairments. Completing Kyberd's quote above, he completes the sentence, "but the variability of the quality of the human voice gives such devices limited success" (Kyberd, 1991). Speech recognition systems have improved significantly since this was written, but motor impairments can be accompanied by speech difficulties and these can still impede the recognition process.

Besides the technical difficulties of the actual recognition process, the environmental considerations also have to be addressed. Users with motor impairments are often very self-conscious and wish to avoid drawing attention to themselves. An input system that involves speaking aloud fails to facilitate this. However, there have been cases where speech recognition systems have been found to be a good solution.

Speech recognition systems are also intuitive and often require little learning once the vocabulary has been mastered. They are also often programmed to offer verbal cursor control and so can replace both the keyboard and mouse in the interaction process. Speech-based interaction is discussed in detail in Chapter 30, "Speech Input to Support Universal Access."

5.5.9 Head (or Hand) Motion Transducers

There are a number of systems available that use head position data as the input for controling the cursor. The majority of these systems used to be ultrasound based and involved triangulated microphones and speakers, often with a wire connection between. More recently, though, ultrasound has been superseded by computer vision systems, which often use a small camera, such as a webcam, pointed at the user's face or hand. Software then tracks particular facial features and turns the movement of those features, for example, nose tracking (Kjeldsen, 2006), into input event streams.

As with most of the mouse replacement systems, no software interface modifications are necessary to access most existing applications. However, some kind of switch device is needed to make selections. In early systems this was often a mouth-mounted sip-puff switch as most users of those systems do not have sufficiently good arm movement to operate a hand switch. More modern systems usually include a software utility for generating the equivalent to the mouse button clicks either through gesture recognition or switch simulation (c.f., the TouchFree mouse discussed earlier).

Learning to use head movements to control the cursor can take a little while as there is a lack of tactile feedback from the input device; but once used to it, users can control the cursor quite successfully. Vision systems in particular still have difficulties with background motion that can confuse the recognition algorithms.

5.5.10 Brain and Nerve Activity

One line of thought for generating inputs is to remove physical movement from the interaction process. The brain is essentially a collection of electrochemical transmitters and receptors, so if the internal signals could be accessed, then an entirely hands-free and intuitive input device would result.

Work is being performed at a number of universities, and some early success was achieved in using electroencephalograms (EEGs) to generate mouse movement (Pfurtscheller et al., 1993; Kostov and Polak, 1997). Similarly, work has been performed on the use of electromyograms (EMGs) that measure the nerve stimulation through skin patches (Patmore, et al., 2004). Combining EEGS with EMGs offers a more robust approach and technologies such as Brainfingers (Brain Actuated Technologies, 2005) are now available commercially.

One downside of both of these techniques is that they usually involve a hardwire connection between the user and the computer for the transfer of the data signals. Brain-computer interaction is discussed in Chapter 37 of this handbook.

5.6 Detailed Studies of the Effects of Motor Impairments

As discussed earlier, most GUI interaction consists of the user using mouse and keyboard input to interact with the interface. Motor impairments can adversely affect a user's ability to use either of these devices.

5.6.1 Pressing Keys

Motor impairments may raise error rates, reduce input speed and stamina, and give rise to pain when typing or using a pointing device. A number of specific typing errors related to disability have been noted in the literature (Cooper, 1983; Shaw et al., 1995; Brewster et al., 1996; Trewin and Pain, 1998, 1999). One study (Trewin and Pain, 1999) of the effects of physical disabilities on input accuracy for a group of 20 keyboard and mouse users with mild to moderate motor impairments found that typing took 2 to 3 times as long for this group, in comparison with a control group of people with no motor impairment.

An example of how different users behave fundamentally differently can be found in a simple set of experiments based around a basic user model, the model human processor (MHP; Card et al., 1983). The MHP states that the time to complete a task is described by Equation 5.2.

$$\text{Time to complete} = [x \times T_{(p)}] + [y \times T_{(c)}] + [z \times T_{(m)}] \qquad (5.2)$$

where:

- $T_{(p)}$ is the time for one perceptual cycle (i.e., the time required for the user's senses to recognize that they have sensed something).
- $T_{(c)}$ is the time to complete one cognitive cycle (i.e., the very simplest decision-making process in the brain). This can be thought of as being effectively akin to a "yes/no" question, such as "did I see something?" which would take one cognitive cycle to answer. A more complicated question would take several cognitive steps because it needs to be broken down into a series of either "yes/no" questions or other simple classification steps.
- $T_{(m)}$ is the time to produce a single, simple motor movement (e.g., a downward stab of a finger onto a key). Releasing the key would require a second motor movement.
- Finally, x, y, and z are all integer coefficients (1, 2, 3, etc.) that represent how many of each component cycle are required.

Card et al., (1983) describe a number of basic tests that can be performed to calibrate this model. Although the tests they describe are paper based, they are straightforward to implement within a computer program. Such a program can be used to screen potential user trial participants as the results it generates provide a useful indication of the level of severity of motor impairment of a particular user. For an able-bodied person, the typical observed times of each cycle from such a program are as follows (Keates et al., 2002b):

$$T_{(p)} = 80 \text{ ms}$$

$$T_{(c)} = 90 \text{ ms}$$

$$T_{(m)} = 70 \text{ ms}$$

For users with moderate to severe motor impairment, the observed $T_{(p)}$ is typically 100 ms—close to the 80 ms seen for the able-bodied users, but still 25% slower. The observed $T_{(c)}$ for the same users is usually 110 ms—compared with 90 ms for the able-bodied users. Note that even though the observed times are somewhat slower for the users with motor impairments, they are still close enough that the differences are not particularly important when considering the interaction as a whole.

However, the results for $T_{(m)}$ are quite complicated and provide an informative insight into how functional impairments can fundamentally change a person's behavior. The observed results for users with motor impairments typically fall into one of 3 bands: 100–110 ms, 200–210 ms, or 300–310 ms—all of which are significantly slower than the 70 ms observed for the able-bodied users.

The reason for the different bands is suggested by the 100 ms time differences between them. That difference is the same magnitude as either additional $T_{(c)}$ cycles or $T_{(m)}$ steps. If it is additional $T_{(c)}$ cycles that are present, then that implies that more cognitive effort is required of the user to perform each "simple" motor function than for able-bodied users. The likely cause of the $T_{(c)}$ cycles is the extra cognitive effort required to produce carefully controlled movements, for example, by having to try to suppress spasms or tremor. This extra cognitive effort has a knock-on implication that so-called "automatic" responses are not achievable for those particular users. Alternatively, if additional $T_{(m)}$ steps are present, then that implies any supposedly "simple" action, such as pressing down on a key, is not performed as a single movement, but actually several (usually) smaller ones.

Irrespective of the cause of the different bandings, the net effect is that some users simply cannot perform basic physical actions as quickly as other users can. The data presented here have been derived from users with motor impairments. Similar results are also seen in much of the literature on aging research, showing that, for example, increasing age is strongly correlated with slower response times on simple reaction tasks. However, in the case of aging in particular, older adults are often able to compensate to some degree for their increasingly slower response times through their increased experience and knowledge, acquired over the years. A direct consequence of this user variability in the performance of "simple" tasks is that trying to design a product to meet the requirement that all users be able to complete a task in the same time, whether nominally able-bodied, older, or functionally impaired, would be almost impossible for many products.

5.6.2 Cursor Control

There have been many studies comparing the cursor control of users with motor impairments with that for able-bodied users (e.g., Trewin and Pain, 1999; Keates et al., 2002a). For example, Trewin and Pain (1999) found that 14 of 20 participants with motor disabilities had error rates greater than 10% in a point-and-click task, mostly arising from the mouse slipping while clicking a mouse button, accidentally pressing the mouse button while en route to the target and difficulty positioning the cursor over the target.

Smith et al., (1999) examined the influence of age-related changes in the component skills required to use a mouse. They studied 60 participants in 3 age groups performing typical mouse actions, such as pointing, clicking, double clicking, and dragging. The older age group exhibited more slip errors, and these errors proved to be a major source of age-related differences in overall movement time. The sole predictor of slip errors was found to be motor coordination. After controlling for differences in this ability, age was not a significant predictor of these errors. In other words, the older adults exhibited more slip errors because of their motor impairments (which may or may not have been associated with the aging process) and not because of their age.

Looking for physiological differences between younger and older adults, Chaparro et al. (2000) examined joint wrist motions and grip strength for a sample of 147 adults aged over 60. Their results indicate that older men in particular experience significant decreases in wrist range of motion. When compared with prior studies of wrist flexion when using a mouse, their study suggests that this reduced range of motion is likely to have a significant impact on mouse use.

Hwang et al. (2004) studied the cursor trajectories of six motor-impaired computer users, compared with three users with no impairment. The users were asked to perform a point-and-click task, and the analysis of the results focused on analyzing the component submovements that comprised the overall movements from the start point to the targets. It was found that some motor-impaired users paused more often and for longer than able-bodied users, required up to five times more submovements to complete a task, and showed longer verification times (the pauses between the completion of the movement and the subsequent click).

Paradise et al., (2005) reported results from interviews with 30 individuals. Twenty-one difficulties with mouse use and 12 compensatory strategies were found. Interview participants reported difficulty in:

- Keeping the hand steady when navigating
- Slipping off menus
- Losing the cursor
- Moving in the desired direction
- Running out of room on the mouse pad
- Keeping the mouse ball from getting stuck

Keates and Trewin (2005) conducted a study comparing the cursor control behavior of four groups of users:

- Young adults (ages 20–30)
- Adults (ages 35–65)
- Older adults (ages 70 and older)
- Adults with Parkinson's disease (ages 48–63)

Participants used a program that presented a within-subjects repeated measures set of targets that participants had to click on like a button. There were three target sizes and target distances and each data collection session consisted of four targets of each size and distance combination generated randomly (see Figure 5.1).

The results from this user study indicated that older adults took longer to complete the task and paused more frequently than the other groups. Previous research suggested this was because of lower peak velocities and increased deceleration time (Ketcham and Stelmach, 2004). Keates and Trewin (2005) found in their study that the older adults did have a comparatively low peak velocity, although the Parkinson's users had the lowest peak velocities on average. Based solely on the difference in peak velocity between the older and younger adults, it would be expected that the movement time for the older adults would be 1.86 s instead of the 2.89 s actually observed. This 1.03 s shortfall implied that there was another mechanism causing the difference in movement time.

Ketcham and Stelmach (2004) also observed that older adults typically exhibit an extended deceleration time compared with younger adults. However, in the analysis offered by Keates and Trewin (2005) it was found that the results did not indicate that this deceleration time explained the time difference, either. Their preferred explanation was a significant increase in the number of pauses observed for older adults arising from a different cursor control strategy from that for the younger adults.

Instead of the expected strategy of a single large move toward the target experiments, followed by a (brief) homing phase as indicated by Fitts' experiments (Fitts, 1954), the older adults appeared to move with quite a different strategy (i.e., many smaller submovements—a kind of "carefully does it" strategy). This is in line with behavior observed by Hwang et al. (2004) and their suggestion that general movement models and assumptions may not apply universally.

The underlying causes of the pauses may have varied from user to user, but most likely derived from lack of experience and lack of confidence (i.e., wanting to be sure before doing anything). A detailed analysis of the pauses indicated that there was a strong correlation between the number of pauses observed and the number of target reentries (i.e., how many times the cursor moved onto the target per successful target selection—ideally there should be no reentries—only one single entry per target). Additionally, pauses were biased toward the end of the movement. These observations were combined with high numbers of direction changes in the cursor movement paths (ideally, if the cursor travels in a straight line to the target, there should be no direction changes). The summary of these data was that the users were having difficulty getting the cursor onto the target and keeping it there.

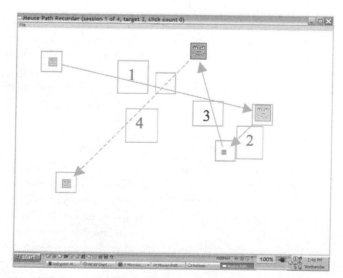

FIGURE 5.1 A screenshot of the point-and-click task showing an example sequence of the first 4 targets in a series of 37. Note the three sizes of targets and three distances to targets, along with the random angle of approach. (From Keates, S. and Trewin, S., Effects of age and Parkinson's disease on cursor positioning using a mouse, in the *Proceedings of the 7th International ACM Conference on Assistive Technologies (ASSETS '05)*, 9–12 October 2005, Baltimore, pp. 68–75, ACM Press, New York, 2005.)

Looking at the users with Parkinson's disease, the overall group times and number and distribution of pauses of the users with Parkinson's disease were consistent with the average age of the group. Their performance fell midway between that of the younger and older adults, implying an interesting question over which was more dominant—the effect of aging or the effect of the Parkinson's disease.

Keates and Trewin (2005) observed that initiating movement was sometimes difficult for the users with Parkinson's, and multiple pauses in the last half second of each movement suggested slight movement while attempting to press the mouse button.

Both the older adults and the Parkinson's users showed a small peak in the number of pauses in the first part of their movement and this occurred before the peak velocity for each movement was observed. Keates and Trewin (2005) proposed a possible explanation for this observation: both of those groups made an initial move to either locate the cursor or orient themselves with respect to the target before making the primary submovement. This reinforces the idea that both groups differ from the behavior predicted by the theoretical models developed for able-bodied users, which is in line with the observations from Hwang et al. (2004).

Finally, as a comparison, both the adult and younger adult groups exhibited movement behavior that broadly agreed with what was predicted by earlier research such as Fitts' law (Fitts, 1954). This implies that the differences observed for the older adults and Parkinson's users arose from the users and not from the experimental design of Keates and Trewin's study (2005).

5.7 Summary

This chapter has examined the issues surrounding users with motor impairments and how those impairments affect universal access in the information society. Motor impairments used to be the poor relation of universal access research, seemingly perpetually in the shadow of access for the blind. However, that situation is now changing, as more research groups and companies recognize that motor impairments are among the most prevalent conditions affecting universal access.

However, being aware that an issue needs to be addressed is only the first step in providing a solution. This chapter presented a series of examples of current and past assistive technology solutions that were designed to help provide access to computers.

Such assistance, though, is only one other part of the solution. Perhaps the most significant contribution to ensuring universal access for users with motor impairments is to ensure that the interfaces presented to them do not place undue burden on their personal capabilities.

This can only be achieved by the designers of those interfaces being aware of the needs of those users. This chapter has also shown that many of the expected user behaviors that designers have been trained on do not apply to users with motor impairments. Thus, new models of user behavior need to be developed to reflect a wider range of user capabilities.

References

Brewster, S. A., Raty, V-P., and Kortekangas, A. (1996). Enhancing scanning input with non-speech sounds, in the *Proceedings of the Second Annual ACM Conference on Assistive Technologies (ASSETS '96)*, 11–12 April 1996, Vancouver, Canada, pp. 10–14. New York: ACM Press.

Card, S. K., Moran, T. P., and Newell, A. (1983). *The Psychology of Human-Computer Interaction*. Hillsdale, NJ: Lawrence Erlbaum Associates.

Chaparro, A., Rogers, M., Fernandez, J., Bohan, M., Choi, S. D., and Stumpfhauser, L. (2000). Range of motion of the wrist: Implications for designing computer input devices for the elderly. *Disability and Rehabilitation* 22: 633–637.

Cooper, W. (1983). *Cognitive Aspects of Skilled Typewriting*. New York: Springer-Verlag.

Coy, J. (2002). Commercial perspectives on universal access and assistive technology, in *Universal Access and Assistive Technology* (S. Keates, P. Langdon, P. J. Clarkson, and P. Robinson, eds.), pp. 3–10. London: Springer-Verlag.

Fitts, P. M. (1954). The information capacity of the human motor system in controlling the amplitude of movement. *Journal of Experimental Psychology* 47: 381–391.

Forrester Research, Inc. (2003). *The Wide Range of Abilities and Its Impact on Computer Technology*. Study commissioned by Microsoft and conducted by Forrester Research, Inc.

Grundy, E., Ahlburg, D., Ali, M., Breeze, E., and Sloggett, A. (1999). *Disability in Great Britain: Results from the 1996/7 Disability Follow-up to the Family Resources Survey*. Huddersfield, UK: Charlesworth Group.

Hwang, F., Keates, S., Langdon, P., and Clarkson, P. J. (2003). Multiple haptic targets for motion-impaired users, in the *Proceedings of the SIGCHI Conference on Human Factors in Computing Systems (CHI 2003)*, 5–10 April 2003, Ft. Lauderdale, FL, pp. 41–48. New York: ACM Press.

Hwang, F., Langdon, P. M., Clarkson, P. J., and Keates, S. (2004). Mouse movements of motion-impaired users: A submovement analysis, in the *Proceedings of ACM SIGACCESS ASSETS*, 18–20 October 2004, Atlanta, pp. 102–109. New York: ACM Press.

IBM (2005). *Conducting User Evaluations with People with Disabilities*. http://www.ibm.com/able/resources/userevaluations.html.

Keates, S., Hwang, F., Langdon, P., Clarkson, P. J., and Robinson, P. (2002a). Cursor measures for motion-impaired computer users, in the *Proceedings of the 5th International ACM Conference on Assistive Technologies (ASSETS '02)*, 8–10 July 2002, Edinburgh, UK, pp. 135–142. New York: ACM Press.

Keates, S., Langdon, P., Clarkson, P. J., and Robinson, P. (2002b). User models and user physical capability. *User Modeling and User-Adapted Interaction (UMUAI)* 12: 139–169.

Keates, S. and Trewin, S. (2005). Effects of age and Parkinson's disease on cursor positioning using a mouse, in the *Proceedings of the 7th International ACM Conference on Assistive Technologies (ASSETS '05)*, 9–12 October 2005, Baltimore, pp. 68–75. New York: ACM Press.

Ketcham, C. and Stelmach, G. (2004). Movement control in the older adult, in *Technology for Adaptive Aging* (R. Pew and S. Van Hemel, eds.), pp. 64–92. Washington, DC: National Academies Press.

Kjeldsen, R. (2006). Improvements in vision-based pointer control, in the *Proceedings of the 8th International ACM SIGACCESS Conference on Computers and Accessibility*, 23–25 October 2006, Portland, OR, pp. 189–196. New York: ACM Press.

Kjeldsen, R. and Hartman, J. (2001). Design issues for vision-based computer interaction systems, in the *Proceedings of the 2001 Workshop on Perceptive User Interfaces (PUI)*, 15–16 November 2001, Orlando, FL, pp. 1–8. New York: ACM Press.

Kostov, A. and Polak, M. (1997). Brain-computer interface: Development of experimental set-up, in the *Proceedings of RESNA '97*, 20–24 June, Pittsburgh, pp. 54–56. Washington, DC: RESNA Press.

Kyberd, P. J. (1991). User interface: The vital link, in the *Proceedings of the 2nd Cambridge Workshop on Rehabilitation Robotics (CUED/B-ELECT/TR82)*, 1991 April, Cambridge, UK, pp. 169–176.

Lesher, G. W., Higginbotham, D. J., and Moulton, B. J. (2000). Techniques for automatically updating scanning delays, in the *Proceedings of RESNA 2000*, 28 June–2, July, Orlando, FL, pp. 85–87. Washington, DC: RESNA Press.

Levine, J. L. and Schappert, M. A. (2005). A mouse adapter for people with hand tremor. *IBM Systems Journal* 44: 621–628.

Martin, J., Meltzer, H., and Elliot, D. (1988). *The Prevalence of Disability among Adults*. London, UK: Her Majesty's Stationery Office.

Neater (2007). *Products: Computer MouseTRAP*. http://www.neater.co.uk/mousetrap.htm.

Paradise, J., Trewin, S., and Keates, S. (2005). Using pointing devices: Difficulties encountered and strategies employed, in *Universal Access in HCI: Exploring New Interaction Environments* (Volume 7 of the *Proceedings of the 11th International Conference on Human-Computer Interaction HCI International 2005*; C. Stephanidis, ed.). [CD-ROM]. Mahwah, NJ: Lawrence Erlbaum Associates.

Patmore, D. W., Putnam, W. L., and Knapp, R. B. (1994). Assistive cursor control for a PC window environment: Electromyogram and electroencephalogram based control, in the *Proceedings of the 2nd Annual International Conference on Virtual Reality and Persons with Disabilities*, pp. 112–114. http://www.csun.edu/cod/conf/1994/proceedings/Wec~1.htm.

Pfeiffer, D. (2002). The philosophical foundations of disability studies. *Disability Studies Quarterly* 22: 3–23.

Pfurtscheller, G., Kalcher, J., and Flotzinger, D. (1993). A new communication device for handicapped persons: The brain-computer interface, in *Rehabilitation Technology: Strategies for the European Union* (E. Ballabio, I. Placencia-Porrero, and R. Puig de la Bellacase, eds.), pp. 123–127. Amsterdam: IOS Press.

Pirkl, J. J. (1994). *Transgenerational Design: Products for an Aging Population*. New York: Van Nostrand Reinhold.

Rao, S. R., Seliktar, R., Rahman, T., and Benvenuto, P. (1997). Evaluation of an isometric joystick as an interface device for children with CP, in the *Proceedings of RESNA '97*, 20–24 June, Pittsburgh, pp. 327–329. Washington, DC: RESNA Press.

Semmence, J., Gault, S., Hussain, M., Hall, P., Stanborough, J., and Pickering, E. (1998). *Family Resources Survey: Great Britain 1996-7*. London: Department of Social Security.

Shaw, R., Loomis, A., and Crisman, E. (1995). Input and integration: Enabling technologies for disabled users, in *Extra-Ordinary Human-Computer Interaction: Interfaces for Users with Disabilities* (A. D. N. Edwards, ed.), pp. 263–278. Cambridge, UK: Cambridge University Press.

Simpson R. C. and Koester H. H. (1997). Adaptive one-switch row/column scanning for text entry, in the *Proceedings of RESNA '97*, 20–24 June 2007, Pittsburgh, pp. 318–320.

WHO (2001). *International Classification of Impairment, Disability and Health (ICF)*. Geneva: World Health Organization.

Smith, M., Sharit, J., and Czaja, S. (1999). Aging, motor control, and the performance of computer mouse tasks. *Human Factors* 40: 389–396.

Stapleford, T. A. and Mahoney, R. (1997). Improvement in computer cursor positioning performance for people with cerebral palsy, in the *Proceedings of RESNA '97*, 20–24 June, Pittsburgh, pp. 321–323. Washington, DC: RESNA Press.

Swiffin, A., Arnott, J. L., Pickering, J. A., and Newell, A. (1987). Adaptive and predictive techniques in a communication prosthesis. *Augmentative & Alternative Communication* 3: 181–191.

Trewin, S., Keates, S., and Moffatt, K. (2006). Developing steady clicks: A method of cursor assistance for people with motor impairments, in the *Proceedings of the 8th International ACM SIGACCESS Conference on Computers and Accessibility (ASSETS '06)*, 23–25 October 2006, pp. 26–33. New York: ACM Press.

Trewin, S. and Pain, H. (1998). A study of two keyboard aids to accessibility, in the *Proceedings of HCI '98 on People and Computers XIII* (H. Johnson, L. Nigay, and C. Roast, eds.), 1–4 September, Sheffield, UK, pp. 83–97. London: Springer-Verlag.

Trewin, S. and Pain, H. (1999). Keyboard and mouse errors due to motor disabilities. *International Journal of Human-Computer Studies* 50: 109–144.

UN (2002). *World Population Ageing: 1950–2050*. United Nations, Department of Economic and Social Affairs, Population Division. http://www.un.org/esa/population/publications/worldageing19502050.

USCB (2004). *American Community Survey*. http://factfinder.census.gov/jsp/saff/SAFFInfo.jsp?_pageId=sp1_acs&_submenuId=.

WHO (2007). *ICIDH-2: International Classification of Functioning, Disability and Health (ICF)*. Geneva, Switzerland: World Health Organization. http://www.who.int/classifications/icf/site/icftemplate.cfm?myurl=homepage.html&mytitle=Home%20Page.

Worden, A., Walker, N., Bharat, A., and Hudson, S. (1997). Making computers easier for older adults to use: Area cursors and sticky icons, in the *Proceedings of the SIGCHI Conference on Human Factors in Computing Systems (CHI '97)*, 22–27 March 1997, Atlanta, pp. 266–271. New York: ACM Press.

6

Sensory Impairments

Erin Kinzel and
Julie A. Jacko

6.1 Introduction

Human sensory systems, such as the eyes and ears, are central to successful interaction between users and technology. As the incidence rate of sensory impairments continues to increase, adaptations must be incorporated into computing and information technologies to accommodate the rising needs of users experiencing sensory deficits. This chapter will discuss the general structure and function of the primary human sensory systems—vision, hearing, and touch—that are vital and intrinsic to human-computer interaction (HCI), as well as some examples of common sensory-specific impairments that may affect HCI. The sensory systems reviewed in this chapter include the visual system, the auditory system, and the haptic system. Following the review of sensory systems and related impairments, some of the recent technological developments used to enhance the sensory experience of users with impairments will be discussed.

The chapter is organized into four major sections. First, visual impairments are discussed, beginning with a review of visual system function, followed by a discussion of the three most common causes of visual loss in the United States. Second, auditory impairments are discussed, beginning with a description of how the auditory system processes information. Common auditory pathologies are then discussed with the functional classifications of their effects. Third, the haptic or tactile system is described at a fundamental level, followed by a description of common pathologies or impairments. Finally, the chapter provides a brief description of some recent technological advances in HCI research, including classes of technology that have been designed and developed to aid individuals with sensory impairments, including perceptual interfaces, multimedia interfaces, multimodal interfaces, and adaptive interfaces.

6.2 Visual Impairment

6.2.1 Visual System

Figure 6.1 shows a high-level structure of the eye and some of the functional components as discussed in the following.

6.2.1.1 The Outer Eye: Cornea and Sclera

The outer surface of the eye is made up of the cornea and the sclera. The sclera is the fibrous, outer coating of the eye, which appears dense and white. It serves as the protective shell of the eye, from the cornea in the front of the eye to the outer sheath of the optic nerve in the back of the eye. The sclera is made up of three very thin layers, the outer of which contains vasculature to nourish the eye. There are holes, or canals, in the sclera that allow for the passage of blood vessels and nerves into the eye. The extraocular muscles also insert into the sclera, which control the eye's movement. The histological structures of the sclera and the cornea are very similar—the primary difference is the transparency of the cornea (versus the opaqueness of the sclera), which is attributed to the more dehydrated content of the sclera.

The cornea is a small transparent area of tissue, comparable in size and structure to the crystal of a small wristwatch. It has three main layers: the epithelium, the stroma, and the endothelium. The stroma is composed largely of collagen, which gives it a crystalline structure. The function of the epithelium and endothelium, maintaining the proper hydration balance of the stroma (which accounts for 90% of the cornea), is vital to maintaining corneal structure and ensuring its transparency. Since the cornea is avascular, it is nourished through vessels of the limbus (the border between the sclera and the cornea), the

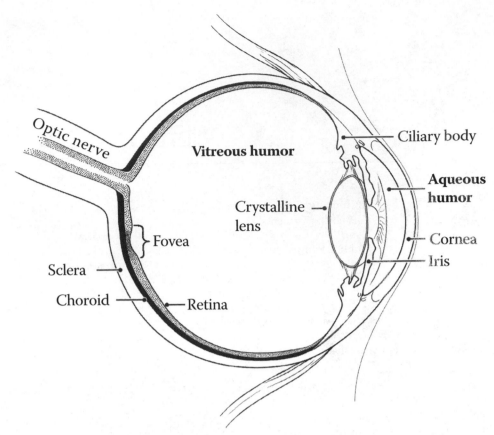

FIGURE 6.1 The structure of the eye. (Courtesy of Hoc Kho.)

aqueous (the thick, watery substance inside the eye), and the tears.

6.2.1.2 The Uveal Tract: Iris, Ciliary Body, and Choroid

The next layer of the eye, the uveal tract, is composed of the iris, the ciliary body, and the choroid. This is the main vascular layer of the eye, which provides most of the nourishment to the eye and is protected by the sclera and the cornea. The iris, which is the most visible portion of the eye, contains a layer of muscles, arranged in a circular pattern, which contracts and dilates to regulate the amount of light allowed into the eye. The empty circular area in the void of the iridial muscles is the pupil, which usually appears black when the inner eye is not illuminated by a flash of light (as with the "red eye" experienced with a camera flash). On the anterior side of this muscle is a layer of pigmented fibrovascular tissue, which appears as the colored part of the eye. Just in front of the entire iris is a chamber, which contains more aqueous fluid.

The ciliary body, which surrounds the lens, just posterior to the iris, contains muscles that attach to the lens and are responsible for shaping the lens to focus light on the retina. The lens is an avascular and almost completely transparent structure. Its main function is the refraction of light, to ensure that visual stimuli fall on the retina. Another essential role of the ciliary body is

the production of the aqueous humor, which is responsible for providing most of the nutrients for the lens and the cornea, and also washes away foreign debris. The choroid is the layer between the retina and the sclera (outer protective layer of the eye) and is composed of densely packed blood vessels. This layer nourishes the outer portion of the retina and also serves as a dark unreflective layer to prevent light from reflecting around inside the eye.

6.2.1.3 The Inner Eye: Retina

The retina, a sheet of neural tissue lining the inner side of the posterior wall of the eye, is semitransparent and multilayered. It contains photoreceptor cells, commonly called rods and cones (for their shape), which respond to photons of light and generate chemo-electric responses that create neural signals. In the center of the posterior retina is the macula, which contains more ganglion cells (neurons) than the rest of the retina. In the center of the macula is the fovea, which appears as a small dent in the center of the retina when examined with a scanning laser ophthalmoscope (SLO). The foveola is the most central portion of the fovea, in which the photoreceptors are all cones. This retinal region provides individuals with the ability to make fine visual discriminations. The signals from the photoreceptors undergo complex processing by other neurons of the retina. The retinal output takes the form of action potentials in retinal ganglion cells whose axons form the optic nerve. Several important features of visual perception can be traced to this retinal encoding and processing of light.

6.2.1.4 The Optic Nerve

The area where the optic nerve leaves the eye results in a blind spot because this break in the retinal surface contains no photoreceptors. The optic nerve, which is considered a tract of the brain, is responsible for transmitting visual information to the brain. It is estimated to contain 1.2 million nerve fibers, significantly less than the roughly 130 million receptors in the retina (Hamming and Apple, 1980). This implies that substantial preprocessing takes place in the retina before the signals are sent to the brain through the optic nerve. Damage to the optic nerve can cause irreversible blindness, since the fibers of the mammalian central nervous system are incapable of regeneration.

6.2.1.5 The Optic Tract, the Lateral Geniculate Nucleus (LGN), and the Visual Cortex

The optic nerves from both eyes meet and cross at the optic chiasm. Since each eye perceives its respective side of the visual field, as well as a portion of the other side of the visual field (via its nasal or "nose-side" retina), the information from both eyes is combined at the optic chiasm and split according to the visual field. Each half of the visual field is processed in the opposite side of the brain (i.e., right visual field in the left half of the brain, and left visual field in the right half of the brain). This split visual information feeds into the LGN, which is the location of the primary processing of the visual information. It has been shown that the LGN introduces relaying efficiencies by canceling out redundant information from the retina, but there are also other functions that are currently not completely understood. The LGN acts as the mediator between the optic tract and the visual cortex, sending projections to the visual cortex and receiving feedback. The visual cortex lies at the rear of the brain. It is the largest system in the human brain and is responsible for higher-level processing of the visual image.

6.2.2 Specific Visual Functions

6.2.2.1 Visual Field

The visual field is the total area in which objects can be seen in the peripheral vision while the eye is focused on a central point. In the clinical setting, visual field is assessed by a process called perimetry. There are two common visual field tests: static perimetry and kinetic perimetry. Static perimetry involves the presentation of stimuli to one spot of the retina in various levels of intensity to determine the threshold at which the stimulus is detected. This method is good for determining the sensitivity of the visual field in a particular area of interest on the retina. Kinetic perimetry involves the presentation of a moving stimulus of a particular size and intensity. The stimulus is moved from a nonseeing area outwardly until it is seen. Points where the object is first perceived are then connected to show the perimeter of the nonseeing area.

Deficiencies in one's visual field can largely affect interaction with technology, though the variety of effects can have different implications. Peripheral visual field loss often produces difficulty for patients in orientation and mobility functions, so activities such as driving can become extremely troublesome. Tasks such as viewing wide screens may also present difficulty, since the restricted visual field may allow for only a limited central view. On the other hand, diseases affecting the central visual field may occlude the area of the retina that is responsible for fine discrimination. Thus macular field loss often causes difficulties with activities that demand high-resolution ability such as reading (Fletcher et al., 1994). As a result, small text or icons may be impossible to discriminate. Taking into account the limitations that visual field defects can present to a user is essential in designing for the purposes of universal access.

6.2.2.2 Color Vision

Color vision is the ability to distinguish objects based on the wavelengths of the light they emit. The ability to see color can be evaluated using pseudoisochromatic color plates. Each plate is made up of circles in various colors, which result in the pattern of a number that is distinguishable from the background to a person with normal vision. Ishihara or Hardy-Rand-Rittler pseudoisochromatic color plates are designed to screen for red and green color deficiencies that generally occur congenitally. Additional tests for color vision deficits have also been developed, including the Farnsworth-Munsell 100-hue test, in which the patient must order 84 colored disks in a hue-based gradient. A less sensitive version is available in the Farnsworth Panel D-15, which is shorter and more practical (using only 15 disks) in a clinical setting.

Deficiency in color vision can have major implications for technology design, particularly in cases where color is used as the primary source for differentiation or implication. A user with a color deficiency may miss important information that could be key to proper interaction if, for example, red text is the only differentiating feature for that information. Other indications to draw the user's attention should be used to complement the color, such as underlined or bold text, to ensure that the technology is universally accessible regardless of a color deficiency.

6.2.2.3 Visual Acuity

Visual acuity refers to the smallest object or feature distinguishable by the eye at a particular distance. It is the most common measure of central visual function. Visual acuity is given in fraction form, with the denominator containing the distance at which a typical eye can distinguish an object (e.g., a letter) and the numerator designating the distance at which the patient can distinguish the object. For example, a visual acuity of 20/40 means that the patient needs an object to be twice the size or half the distance away as a typical, unimpaired eye requires. The visual acuity chart used most often in the clinical setting, the Snellen acuity chart, is made up of letters of the alphabet arranged in lines with the letters decreasing in size down the chart. The result is given with the denominator as the letter size seen by the patient and the numerator as the distance at which the test was performed.

Since visual acuity is a general measure of the ability of the eye to distinguish objects from a particular distance, many factors may contribute to this measurement. A deficiency in contrast sensitivity for example may affect the overall ability of the eye to distinguish an object. Another example is a user whose visual field is centrally occluded by macular degeneration. The need to use the peripheral visual field for fine discrimination may make this particular user's visual acuity low, since peripheral vision is not the most efficient for this type of task. Thus, if a user has a variety of factors playing into a lack of visual acuity, there may be a variety of ways for which this may be compensated when designing technology intended for universal access. For a general lack in visual acuity, high contrast as well as an increase in the size of the objects or text can be helpful for users over a variety of impairments.

6.2.2.4 Contrast Sensitivity

Contrast sensitivity refers to the ability to perceive different levels of contrast between static objects. It is defined as the reciprocal of the lowest contrast that can be detected. The minimum discernible difference in gray-scale level that the eye can detect is about 2% of full brightness. One test to measure contrast sensitivity is the Pelli-Robson chart, in which letters of a fixed size decrease in contrast. Another is the Bailey-Lovie chart, in which letters of a fixed low contrast vary in size. Contrast sensitivity is considered a more sensitive indicator of visual function and real-world visual ability than Snellen acuity, and may provide earlier detection of conditions such as retinal and optic nerve disease.

When dealing with a possible deficiency in contrast sensitivity in technology design, the key is to make the contrast as extreme as possible. Black and white is the best example of contrasting colors. If contrast cannot be increased, generally an increase in font size or icon size can help make the object more distinguishable. If dealing with severe deficiencies, a design alternative may be to use the auditory or haptic channels to convey important information.

6.2.3 Visual Impairments

6.2.3.1 Age-Related Macular Degeneration

Age-related macular degeneration (AMD) is one of the leading causes of blindness in the United States and accounts for up to 30% of all bilateral blindness among Caucasian Americans. The prevalence of severe visual loss due to AMD increases with age. It is the leading cause of irreversible visual loss in the Western world among individuals over 60 years of age. AMD is also a leading cause of low vision, broadly defined as a visual impairment interfering with an individual's ability to perform activities of daily living. In the United States, at least 10% of the population between the ages of 65 and 75 years has at least some central vision loss as a result of AMD (Jacko et al., 2003). This increases with age with 30% of individuals over the age of 75 years having vision loss due to AMD. Age is an important factor in the

progression of AMD. Other risk factors for this disease and its progression include sunlight exposure, smoking, family history, light ocular and skin pigmentation, high serum cholesterol levels, and high blood pressure (Jacko et al., 2003). The role of nutrition has not been fully identified in its impact, but a diet lacking in antioxidants and lutein may be a contributing factor as well as a diet high in fat.

The two main types of AMD are atrophic and exudative. The atrophic ("dry") form of the disease is generally a slowly progressive disease that accounts for approximately 90% of cases. It begins with characteristic yellow, acellular deposits in the macula called drusen. It is also characterized by degeneration and atrophy of the macula. Vision loss is caused through loss of photoreceptors and cells supporting the photoreceptors in the central part of the eye. Patients typically note slowly progressive central visual loss. Much less common is the exudative ("wet") form of the disease, although it is responsible for about 88% of legal blindness attributed to AMD. This form of the disease, which often occurs in association with atrophic AMD, is characterized by the growth of abnormal blood vessels beneath the macula. Bleeding, leaking, and scarring from these blood vessels eventually cause irreversible damage to the photoreceptors and supporting cells. Vision loss from wet AMD is characterized by sudden onset and rapid progression and development of visual distortions, blurry spots, or "black holes." AMD can cause profound loss of central vision, but the disease generally does not affect peripheral vision, and therefore patients typically retain their ability to move about independently.

There currently is no effective treatment to reverse the retinal damage that occurs from AMD. Different treatments of the wet form are available and may help decrease the amount of vision that is lost. In June 2006, the drug Ranibizumab was approved by the FDA for use in stopping the progression of the disease (U.S. Food and Drug Administration, 2006). This was the first therapy to show a statistically significant improvement in patient reported outcomes. Other drugs are currently under investigation, but none of these treatments can fully rectify permanent visual damage. Laser photocoagulation and photodynamic treatment of the abnormal blood vessels found in patients with the wet form of macular degeneration may help to prevent severe vision loss in some cases. Surgical rotation of the retina away from the area of abnormal blood vessels has also been effective in some cases (Machemer and Steinhorst, 1993). Other means to try to prevent further vision loss include sunglasses with ultraviolet light protection, cessation of smoking, control of serum cholesterol levels, and consumption of a diet rich in dark green leafy and orange vegetables.

6.2.3.2 Glaucoma

Glaucoma is a group of diseases usually associated with increased pressure within the eye. This increased pressure can cause damage to the cells that form the optic nerve. Primary open angle glaucoma (POAG) is the most common type of glaucoma, affecting 1.3% to 2.1% of the general population over the age of 40 years in the United States (Prevent Blindness America, 2002). It is the leading cause of irreversible blindness among African Americans and the third leading cause among Caucasians, behind AMD and diabetic retinopathy. It also accounts for between 9% and 12% of all cases of blindness. Age becomes a risk factor for Caucasians at 50 years and for African Americans at 35 years (Prevent Blindness America, 2002). People who are more likely to develop glaucoma include those who are African American, related to someone with glaucoma, very nearsighted, have had eye surgery or injury, and those who are diabetic. Extended use of steroid medications can also increase risk of glaucoma.

POAG is a persistent, gradually progressing optic neuropathy characterized by loss of peripheral vision as a result of atrophy of the optic nerve, progressing to a decline in central vision, and potentially blindness. As a result of central vision remaining relatively unaffected during early development of the disease, visual loss generally progresses without noticeable symptoms and may go undiagnosed for the beginning stages. In fact, it has been estimated that at least one half of all those who have glaucoma are unaware of it (Prevent Blindness America, 2002). POAG is typically bilateral, but in some cases it may occur asymmetrically. It is associated with increased intraocular pressure, but relatively normal eye pressure does not eliminate the possible presence of glaucoma.

Lowering intraocular pressure to a level at which optic nerve damage no longer occurs is currently the main form of glaucoma treatment. Methods of achieving this include topical or systemic medications as well as laser or conventional surgery. Medications typically work through promoting fluid excretion from the chambers of the eye, decreasing fluid production, or a combination of the two. Clinical signs of glaucoma include asymmetry or focal thinning of the neuroretinal rim, optic disc hemorrhage, and any change in the disc rim appearance or the surrounding retinal nerve fiber layer (Prevent Blindness America, 2002). Visual field testing should be performed regularly to evaluate for progressive loss of peripheral vision. Additionally, the optic nerve should be examined regularly to evaluate for evidence of progressive optic atrophy.

6.2.3.3 Diabetic Retinopathy

Diabetic retinopathy is characterized by damage to the blood vessels in the retina, resulting from complications of diabetes mellitus. Roughly 16 million Americans have diabetes mellitus, and the majority of these patients generally develop diabetic retinopathy within 20 years of their diagnosis (Prevent Blindness America, 2002). Diabetic retinopathy is the leading cause of legal blindness in Americans aged 20 to 65 years, and roughly 10,000 new cases of resulting blindness are diagnosed each year (Prevent Blindness America, 2002). One million Americans have proliferative diabetic retinopathy, and 500,000 have some form of macular edema (swelling causing visual impairment). The key risk factor for diabetic retinopathy is the extent of diabetes and the length of time a patient has had the disease. Among patients with juvenile onset diabetes, 25% will have diabetic retinopathy after 5 years. It is estimated that at 15 years, 80% of diabetics will have background retinopathy and of these, 5% to 10% will progress to proliferative changes (Prevent Blindness America, 2002). Other risk factors include smoking, high blood pressure, and

high cholesterol. Pregnancy has also been known to increase the risk for diabetic retinopathy.

Diabetic retinopathy often has no early warning signs. A person with macular edema is likely to have blurred vision, which may get better or worse throughout a given day. During the initial stages of nonproliferative diabetic retinopathy, most people do not notice any changes in their vision. As the disease progresses into proliferative diabetic retinopathy, new blood vessels form at the back of the eye and within the vitreous that fills the eye, which can bleed and blur vision. In most cases the first occurrence of this will leave just a few spots of blood in the visual field which may go away after a few hours. Without treatment, these blood vessels can bleed, cloud vision, and destroy the retina.

Early diagnosis of diabetes and the effective control of blood sugar levels and hypertension increase the ability to control the risk of developing diabetes-related retinopathy. Laser photocoagulation is used to control the leakage of fluid from retinal blood vessels and to control the growth of abnormal vessels. Panretinal laser photocoagulation treatment of eyes with high-risk proliferative diabetic retinopathy reduced the risk of severe visual loss by 50% compared with untreated control eyes (Diabetic Retinopathy Study Research Group, 1981). Surgery is often performed for nonclearing vitreous hemorrhage and for tractional retinal detachment involving or threatening the macula.

6.2.3.4 Other General Visual Impairments

While the three previously discussed visual impairments are the most commonly occurring and well-known ones in the United States, there are other eye disorders that can lead to visual impairments. Cataracts, opacity that develops in the lens, can develop for a variety of reasons, including trauma, disease, or simply advancing age. They typically progress slowly to cause vision loss but are potentially blinding if untreated. Albinism, characterized by a lack of melanin pigment, is a disorder that brings along with it several possible visual impairments. Problems particularly associated with albinism are related to the poorly developed retinal pigment epithelium (RPE) due to the lack of melanin, which tends to affect the fovea. This may result in eccentric fixation and lower visual acuity. The lack of pigment in both the RPE and the iris also causes sensitivity to light, since it prevents the blocking of stray light. Other examples of disorders include muscular problems, corneal disorders, congenital disorders, and infection. Visual impairment can also be caused by brain and nerve disorders, in which case they are usually referred to as cortical visual impairments.

All of these forms of visual impairment can have dire consequences for interacting with computers and other information technology (IT). In a traditional graphical user interface (GUI), for example, information is primarily represented through visual means, since vision is the dominant sense in humans. People with tunnel vision from glaucoma[1] or central blind spots from macular degeneration[2] may find it difficult and tiring to read an

---

[1] http://www.allaboutvision.com/conditions/glaucoma.htm.
[2] http://www.allaboutvision.com/conditions/amd.htm.

entire computer screen. Thus talking computers may be helpful in easing the strain of a primarily visual display. Some people with low vision may also find it useful to use keyboard commands instead of a mouse, because it can be easier to type a keyboard command than to use a mouse to move the cursor to a precise place on the screen. Overall, the inability of users to physically see controls, text, or visual objects on the screen presents a significant problem for use of current computing technologies. These prevalent needs for alternate means of information representation should not be ignored when designing universally accessible technologies.

6.3 Auditory Impairment

6.3.1 Auditory System

Figure 6.2 represents a high-level structure of the auditory system as well as some of the functional components discussed in the following.

6.3.1.1 Outer and Middle Ear Structure and Function

The human ear can be divided into three distinct components: the outer ear, the middle ear, and the inner ear. The outer ear begins with folds of cartilage surrounding the ear canal, which form a cup-shaped structure called the pinna. The pinna channels sound waves as they enter the ear. The reflections and directions of these waves hitting the cup structure provide information that helps the brain determine the direction in which the sound originated. The sound is amplified within the ear canal and directed toward the eardrum, or tympanic membrane, which connects to the middle ear.

The structure of the middle ear is made up of a roughly spherical air-filled cavity. Wave information travels across the air-filled middle ear cavity through a series of three delicate bones, referred to as ossicles. These bones are suspended by ligaments attached to the boney walls of the cavity. They form a bridge, with the first ossicle attaching to the eardrum while the third is attached to a small membrane called the oval window. The oval window covers the inner ear, which contains fluid rather than air. Through this bridge, the ossicles are able to act as a lever and a connection line, converting the lower-pressure eardrum sound vibrations into higher-pressure sound vibrations at the oval window. This higher pressure is necessary to move through the fluid beyond the membrane. But the sound is not uniformly amplified across this chain. There are muscles within the middle ear, attached to the ossicles that act as protection for the inner ear. If the sound entering the middle ear is above a certain threshold, the muscles tighten to reduce the movement of the ossicles and thus control the transfer of energy onto the oval window. The

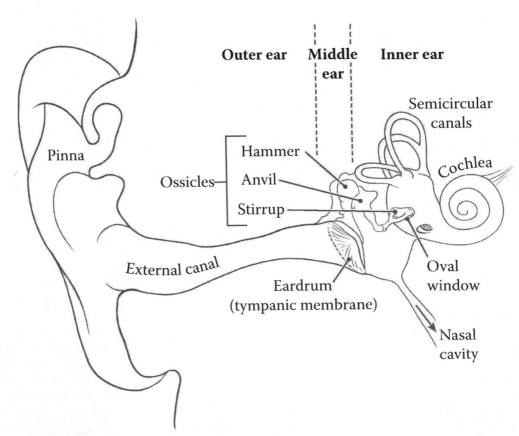

FIGURE 6.2 The structure of the auditory system. (Courtesy of Hoc Kho.)

sound information is still contained in wave form in the middle ear, and is then converted to nerve impulses in the cochlea.

6.3.1.2 Inner Ear Structure and Function

The inner ear is composed of semicircular canals and the cochlea. It is a fluid-filled chamber encased in intricate paths and passages of bones. The cochlea is shaped like a spiral snail's shell and contains the hair cells that act as auditory receptors. There are three cochlear components: the scala vestibule, scala tympani, and scala media (Levine, 2000). The scala vestibule and tympani are continuous with each other at the opening of the cochlea, while the basilar membrane separates the scala media from the scala tympani. When motion occurs in the fluid within the inner ear (generated from the ossicles of the middle ear), the basilar membrane vibrates, which causes the cilia on the hair cells to move. This bending of the hair cells initiates a neural signal (Proctor and Proctor, 1997). The neural signal then is used to form a conscious experience or perception. The cognitive processes involved with perceiving sound are complex. According to the information-processing approach, there are multiple levels of processing that occur at each stage of sensory processing. These stages range from registering the stimulus on the receptor to a final conscious representation in memory (Coren et al., 1999).

6.3.2 Elements of the Perception of Hearing

Psychologically, sounds have at least two major aspects: pitch and loudness. Complex sounds also have an important aspect called timbre, which is less fully understood. The design of computing technologies for individuals with auditory impairments must consider these fundamental elements of the perception of hearing. Pitch and timbre are primarily affected by the frequency of an auditory stimulus, which refers to how often the particles of a medium vibrate when a wave passes through it. On the other hand, loudness is typically related to the amplitude of the pressure wave, of the auditory stimulus. This relates to the intensity or the amount of energy that is transferred to the medium. For example, if energy is put into the plucking of the guitar string (that is, work is done to displace the string a certain amount from its rest position), then the string vibrates with a corresponding amplitude.

6.3.2.1 Pitch

Pitch is closely related to the frequency of a pure tone or the fundamental frequency of a sound, but is a subjective quality of sound that cannot be measured directly (Moore, 1982). Melodies found in music are attributed to the variations in pitch. In general, it is considered that the higher the frequency, the higher the pitch. But pitch is not only influenced by the fundamental frequency of a sound, but also by the pattern of harmonics (Proctor and Proctor, 1997). These and other factors are considered in the two predominant theories of auditory pitch perception.

One theory of auditory pitch perception is Weber and Bray's (1930) frequency theory. This theory suggests that the basilar membrane vibration, dependent on the frequency of the noise, is responsible for the human's fine ability to discriminate between frequencies. Another theory of auditory pitch perception, Helmholtz's place theory (1954, 1977), suggests that the location of the vibration differentiates the pitch, and that different places on the basilar membrane are affected by different frequencies (Levine, 2000). Work conducted by Von Bekesy (1960) demonstrated that waves travel down the basilar membrane from the base of the opening at a frequency corresponding to the tone, suggesting that both of these theories could be consistent.

6.3.2.2 Timbre

Timbre is related to the quality of a sound that distinguishes it from other types of sound production. Typical sounds encountered in a normal day contain multiple frequencies with relative phases, and thus the discrimination between these frequencies suggests a great deal about the quality of the sound, or timbre. Timbre is defined as the attribute of sensation in terms of which a listener can judge that two sounds of the same and pitch and loudness are different (Moore, 1982). Timbre represents the detected quality that accounts for differences in sound other than pitch (Levine, 2000). For example, timbre allows a person to distinguish between sounds generated by two different instruments, even when they are playing the same note. Different from pitch and loudness, timbre has no single scale along which timbres of different sounds can be measured, compared, or structured, thus special techniques to determine the principle dimensions of variation within a set of stimuli must be used, similar to factor analysis (Moore, 1982).

6.3.2.3 Loudness

Loudness is the quality of a sound that correlates with physical intensity. It is a subjective quality of an auditory stimulus, similar to pitch and timbre. As a subjective quality, it is very hard to measure, since true perceived loudness varies from person to person. Therefore, loudness is usually correlated with objective measures of sound intensity, generally measured in decibels (dB). Scales such as phons (equal to 1 dB SPL at a frequency of 1 kHz) and sones (equal to 40 phons, or the loudness level of a 1 kHz tone at 40 dB SPL) also allow subjective descriptions of loudness to be related to physical descriptions of frequency and intensity on standard scales of measurement.

Humans can hear sounds up to an estimated 120 dB without causing damage to the ear (Moore, 1982). The frequencies to which humans are most sensitive are in the range of 1000 to 5000 Hz. The loss of sensitivity with increasing age is much greater at high than at low frequencies. The ultimate ability to detect faint sounds is not limited by absolute sensitivity but instead it is largely affected by the level of ambient noise in the surroundings. Because of this, the environment in which systems relying on auditory cues and feedback are used must be considered during design. Auditory systems can be very beneficial modalities for visually demanding tasks, resulting from the fact that adaptation and fatigue in the auditory system are much less apparent than in the visual system (Moore, 1982). But to fully exploit this benefit, one must consider the context of use and the characteristics of the user.

6.3.3 Auditory Impairments

The degree of hearing loss can be categorized into two types, conductive or sensorineural, as seen in Table 6.1, which describes the nature and effects of some auditory impairments. Conductive hearing loss occurs when sound is not normally conducted through the outer or middle ear. Since a normally sensitive inner ear can still pick up sound, conductive hearing loss typically results in only mild to moderate impairment. These disorders can reduce an individual's sensitivity to hear sounds below 60 dB (Stein, 1988). Generally, the quality of hearing is still good, as long as the sound is amplified loud enough that it is conducted to the inner ear. On the other hand, sensorineural hearing loss stems from damage to the cochlea or auditory cranial nerve, accounting for most cases of human deafness (severe to profound hearing impairment). These disorders are more likely to be irreversible, and can result in hearing loss ranging from a drop in thresholds for a number of frequencies, to the total loss of sensation (Stein, 1988).

There is also a division made according to when the loss occurred. If the loss is present at birth or occurs within the first three years of life, it is categorized as congenital or prelingual. Prelingual means that the loss is present before the child is capable of speaking. If the loss occurs after speech development, it is termed acquired or postlingual (Stein, 1988). One of the most commonly occurring acquired hearing disorders occurs in the elderly. It is the gradual loss of high-frequency hearing, or presbycusis. An estimated 23% of the elderly between the ages of 65 to 74 years and 32% over 70 years of age are affected (National Institute on Deafness and Other Communication Disorders, 2006). Regardless of the type of hearing loss, the standard degree of impairment is defined and labeled in accordance with the ability to hear pure tones at 500, 1000, and 2000 Hz. The degree is determined by the average level of hearing intensity measured in dB. As shown in Table 6.2, a hearing impairment is considered slight if the loss is between 20 and 30 dB, mild from 30 to 45 dB, moderate from 60 to 75 dB, profound from 75 to 90 dB, and extreme from 90 to 110 dB (Cook and Hussey, 1995). Associated with these predefined hearing averages are the conjectures of sounds heard without amplification, the degree of handicap, and probable needs of an individual to resume activities of daily living, which are also shown in the Table 6.2 (Jacko et al., 2003).

The effects of hearing loss can be seen through several aspects of human functioning. Hearing disabilities could significantly affect speech and language development if acquired early in life (Rampp, 1979). In relation to this, auditory disorders may also affect reading ability. Eventually, these deficiencies in speech, language, and reading can result in learning disabilities (Rampp, 1979).

The consequences at any level of uncorrected hearing loss can be severe and the ability to hear plays an escalating role in emerging systems. For example, notification sounds are being used as prompts that must be detectible by the user. Information through sound can also be seen throughout everyday items such as phones, alarms, and doorbells. In drawing metaphors from the real world into technology, computers also make use of this form of auditory warning and notification. In addition, the auditory channel is often used to relieve visual overload, which has its advantages in application. But as more information is relayed through computing technologies, these sounds can become more specialized, thus increasing the importance of detection and discrimination. Because of the uniqueness of the growing user population, consideration in the design process must be given to making technology accessible to everyone, including individuals with hearing impairments.

6.4 Haptic Impairment

6.4.1 Haptic System

The haptic system is a perceptual channel made up of the combined input from the skin (touch and pressure) and kinesthesia (the sensation of movements of the limbs and body). The haptic system is

TABLE 6.1 Some Sources of Auditory Impairment

Location	Category	Nature	Comments
Outer ear	Conductive	Blockage	Usually wax—trivial
Outer ear	Conductive	Scarred tympanic membrane	Must be severe to affect hearing greatly
Middle ear	Conductive	Fluid	Affects movement of tympanic membrane, oval, round windows; occurs in children and affects speech production; treatment: surgical drainage
Middle ear	Conductive	Malfunctioning ossicles	Intensity loss up to 30 dB at all frequencies
Inner ear	Sensorineural	Loss of hair cells	Frequency-dependent hearing loss caused by loud sounds and certain drugs
Inner ear	Sensorineural	Other cochlea damage	Accidents, birth defects
Auditory nerve	Sensorineural	Mechanical damage	Accidents, strokes
Auditory nerve	Sensorineural	Disease damage	Poorly understood

Source: Adapted from Luce, R.D., *Sound & Hearing: A Conceptual Introduction*, Lawrence Erlbaum Associates, Hillsdale, NJ, 1993.

TABLE 6.2 Effects of Hearing Loss

Average Hearing 500–2000 Hz (ANSI)	Description of Hearing Impairment	Sounds Heard without Amplification	Degree of Handicap	Probable Needs
0–15 dB	Normal range	All speech sounds	None	None
15–15 dB	Slight hearing loss	Vowel sounds heard clearly; may miss unvoiced consonant sounds	Mild auditory dysfunction in language learning	Consideration of need for hearing aid; lip-reading; auditory training; speech therapy, preferential seating
25–40 dB	Mild hearing loss	Hears only some louder voiced speech sounds	Auditory learning dysfunction, mild language retardation, mild speech problems, inattention	Hearing aid, lip-reading, auditory training, speech therapy
40–65 dB	Moderate hearing loss	Misses most speech sounds at normal conversation level	Speech problems, language retardation, learning dysfunction, inattention	All of the above, plus consideration of special classroom situation
65–95 dB	Severe hearing loss	Hears no speech sounds of normal conversation	Severe speech problems, language retardation, learning dysfunction, inattention	All of the above, plus probable assignment to special classes
More than 95 dB	Profound hearing loss	Hears no speech or other sounds	Severe speech problems, language retardation, learning dysfunction, inattention	All of the above, plus probable assignment to special classes

Source: From Jacko, J. A., Vitense, H. S., and Scott, I. U., Perceptual impairments and computing technologies, in *The Human-Computer Interaction Handbook: Fundamentals, Evolving Technologies and Emerging Applications* (J. A. Jacko and A. Sears, eds.), pp. 504–522, Lawrence Erlbaum Associates, Mahwah, NJ, 2003. With permission.

responsible for the perception of geometric properties such as shapes, dimensions, and proportions of objects that are handled. By various manipulations such as grasping, lifting, and tracing edges, the haptic system not only extracts geometric properties, but also gives information on the weight and consistency or texture of objects.

The sensory effect of skin stimulation is termed cutaneous sensitivity. This sense of touch may simply be considered one of five human senses; however, when a person touches something or someone this gives rise to various feelings: the perception of pressure (including shape, softness, texture, vibration, etc.), relative temperature, and sometimes pain. Thus the term *touch* is actually the combined term for several senses. The skin is not uniformly sensitive to all cutaneous stimuli. Some regions, such as the fingertips, are much more sensitive than others.

There are many receptor types within the skin that are responsible for the skin's sensitivity to stimuli. The major nerve endings for hairy skin (about 95% of the skin of humans) are called basket cells. The hairless remainder of the skin, called glabrous skin, is thicker and is found on areas such as the soles of the feet, the palms of the hand, and the lips. This type of skin contains structures called encapsulated end organs, which are its primary sensory receptors. The major nerve endings in glabrous skin are Pacinian corpuscles, which are deep within the skin. The last major type, the most common, are called free nerve endings and are found in all skin regions and are not specialized to any specific skin region. All of these major types of receptors are sensitive to some sort of pressure or touch sensation. There is no conclusive evidence that stimulation of a particular type of skin receptors exclusively initiates a specific cutaneous sensation. For example, the cornea of the eye only contains free nerve endings, but it is very sensitive to pressure, temperature, and pain.

Thus, all skin receptors seem to respond somewhat to several types of sensory stimulation, and a given cutaneous sensation

may be produced by many different specialized skin receptors rather than by a single one. Neural messages sent by cutaneous receptors are transmitted to a region of the brain called the somatosensory cortex, which is represented in each hemisphere of the parietal lobe. Nerve fibers from each part of the skin surface are spatially represented in this cortex, projected and arranged so that neighboring areas of the skin are represented in neighboring regions of the cortex. Areas of the skin that are more densely populated with nerve endings are represented by larger areas of the somatosensory cortex.

Kinesthesia (also sometimes referred to as proprioception) is the sensation and perception of the position in space or movement of the body or an extremity. A key source of this stimulation is from the joints, muscles, and tendons. These areas contain a form of the Pacinian corpuscles which are stimulated by contact between joint surfaces that occurs with changes in the angles at which the bones are held. Muscles and the tendons that connect them also contain nerves that respond to changes in tension, which stimulate the perception of stretch, strain, and resistance. These stimulations may not result in a clear sensation such as seeing a particular image or hearing a sound, but the kinesthetic system is able to provide continuous information concerning spatial location and movement to the body so that it may efficiently know the position, posture, and direction of its movement without one actually needing to see it.

6.4.2 Haptic Impairments

6.4.2.1 Peripheral Neuropathy

Since a vital part of the functioning of the haptic system is the nerves within the skin, damage to the peripheral nervous system could cause serious impairment to its functioning. Peripheral neuropathy is a broad term that refers to damage to the peripheral nervous system, which results in interference in the vital connection between the central nervous system (the brain and the spinal cord) and the rest of the body. There are more than 100 identified types of neuropathy, depending on the type of nerves that are damaged (National Institute of Neurological Disorders and Stroke, 2007). Included in the wide range of symptoms that can result from this damage is the blocking of sensory information.

Among the different types of damage, sensory nerve damage causes a more complex range of symptoms, because sensory nerves have a wider, more highly specialized range of functions. Larger sensory fibers enclosed in myelin (a fatty protein that coats and insulates many nerves) register vibration, light touch, and position sense. Damage to large sensory fibers lessens the ability to feel vibrations and touch, resulting in a general sense of numbness, especially in the hands and feet. People may feel as if they are wearing gloves and stockings even when they are not. Many patients cannot recognize by touch alone the shapes of small objects or distinguish between different shapes. This damage to sensory fibers may contribute to the loss of reflexes (as can motor nerve damage). Loss of position sense often makes people unable to coordinate complex movements like walking or fastening buttons, or to maintain their balance when their eyes are shut.

Smaller sensory fibers without myelin sheaths transmit pain and temperature sensations. Damage to these fibers can interfere with the ability to feel pain or changes in temperature. People may fail to sense that they have been injured from a cut or that a wound is becoming infected. Loss of pain sensation is a particularly serious problem for people with diabetes, contributing to the high rate of lower limb amputations among this population. Pain receptors in the skin can also become oversensitized, so that people may feel severe pain (allodynia) from stimuli that are normally painless (for example, some may experience pain from bedsheets draped lightly over the body).

There is a wide variety of causes of peripheral neuropathy, which can be inherited or acquired. Forms that are inherited can be caused by genetic errors that are inborn or by new mutations in the genes. An example of this is a group of disorders called Charcot-Marie-Tooth, disease which is a result of flaws in genes responsible for manufacturing neurons or the myelin sheath. Symptoms include extreme weakening and degradation of muscles in the lower legs and feet, loss of tendon reflexes, and numbness in the lower limbs (National Institute of Neurological Disorders and Stroke, 2007). The many causes of acquired peripheral neuropathy can be organized into three major groups, including neuropathies caused by systemic disease, those caused by trauma from external agents, and those caused by infections or autoimmune disorders affecting nerve tissue.

Physical injury or trauma is the most common cause of injury to a nerve. Injury or sudden trauma caused from accidents or falls can cause nerves to be partially or completely severed, crushed, compressed, or stretched, sometimes so forcefully that they are partially or completely detached from the spinal cord. Less severe traumas, such as broken or dislocated bones, can exert damaging pressure on neighboring nerves as well as slipped disks between vertebrae compressing nerve fibers that emerge from the spinal cord. Systemic diseases that affect the entire body, such as metabolic and endocrine disorders, are another possible cause of peripheral neuropathy. Nerve tissues are highly vulnerable to damage from diseases that impair the body's ability to transform nutrients into energy, process waste products, or manufacture the substances that make up living tissue. Diabetes mellitus, characterized by chronically high blood glucose levels, is a leading cause of peripheral neuropathy in the United States. About 60% to 70% of people with diabetes have mild to severe forms of nervous system damage (National Institute of Neurological Disorders and Stroke, 2007). Infections and autoimmune disorders can also cause peripheral neuropathy. There is a wide variety of viruses and bacteria that can attack nerve tissues, some examples of which are herpes varicella-zoster (shingles), HIV, Lyme disease, diphtheria, and leprosy. Viral and bacterial infections can also cause indirect nerve damage by provoking autoimmune disorders, in which cells and antibodies of the immune system attack the body's own tissues, which typically causes destruction of the nerve's myelin sheath or axon.

Robles De-La-Torre (2006) summarizes the major effects of the loss of the haptic or tactile sense, which includes the loss of the capability to sense limb movement and position, major impairment in skilled performance, abnormal movements, and the inability to walk following the loss of somesthesis, major loss of precision and speed of movement (particularly in the hands), major difficulty performing tasks that combine significant cognitive loads and fine motor skills, major difficulty learning new or lost motor tasks, or using previous experience to guide these processes.

6.4.2.2 Kinesthetic Impairment

The sense of kinesthesia, or proprioception (the perception of the body's movement or position in space), can also suffer from impairment, although this is not as well understood. Patients who suffer from joint hypermobility or a genetic condition called Ehlers-Danlos syndrome, which results in weak connective tissue, may experience lack of bodily awareness resulting from problems in the joints, where the kinesthetic sense stems from. Kinesthesia can also be permanently impaired from viral infections such as Sacks syndrome, also referred to as proprioception deficit disorder (PDD). This rare disorder is described as a complete and total failure of the body's knowledge of itself and begins with symptoms such as a lack of coordination, ultimately culminating in full-blown proprioceptive failure and a general disembodied feeling. There is no current cure for PDD, although there have been reports of the body correcting itself with time. Most sufferers gradually replace the missing sense with those sensory abilities that remain, predominantly sight.

Haptic impairments, in addition to affecting daily functioning, can have serious effects on the ability to interact with computing technologies. Imagine, in the case of peripheral neuropathy, not being able to communicate to your limb to move to the keyboard or drag the mouse. Or once at the keyboard, imagine not being able to sense where your hand is to press a particular key, as in the case of Sack's syndrome. Actions such as gripping the mouse or clicking a button are highly integrated into the use of everyday technology. These are actions that could be extremely hard with impairments such as one of those discussed previously. Daily activities such as using an ATM could become extremely difficult without proper functioning of the haptic senses. Alternate modes of communication between the user and the technology, such as voice-activated or eye-gaze input, must be considered for these devices to be accessible to all.

6.5 Overcoming Challenges: Technology and Design for Sensory Impairments

Much research is being conducted that focuses on the design of universally accessible technology. Additionally, designers are becoming more conscious of the importance and impact of considering the various dimensions of user diversity and users' relative needs and abilities. As stated by Stephanidis et al. (1998), "universal access in the Information Society signifies the right of all citizens to obtain equitable access to, and maintain effective interaction with, a community-wide pool of information resources and artifacts." The primary idea of universal access is that all individuals should be able to access information technologies regardless of abilities, skills, requirements, and preferences (Stephanidis et al., 1999). Perceptual and adaptive interfaces are two principal examples of technological advancements representing how researchers are trying to achieve universal access.

6.5.1 Perceptual Interfaces

Research in human perception can be applied to interface design, which is the idea that inspires the design of perceptual interfaces (Reeves and Nass, 2000). The notion of perceptual design illustrates a perspective that adds human-like perceptual capabilities to the computer. This is a natural evolution of technology when considering the remarkable range of senses that humans have, and their amazing ability to integrate sensory information into perceptual models of their environments. Vision and hearing have been extensively researched, while advances in haptic research have been around since the early 1990s. The power of perceptual user interfaces comes from the combination of this understanding of natural human capabilities with computer input/output devices, and machine awareness and analysis (Turk and Robertson, 2000). Some general examples of technology combining these capabilities include speech and sound recognition and generation, computer vision, graphic animation, touch-based sensing and feedback, and user modeling (Turk and Robertson, 2000).

Multimedia and multimodal interfaces are the application of the concepts motivating perceptual interfaces. Both multimedia and multimodal interfaces also offer increased accessibility to technologies for individuals with perceptual impairments. Multimedia interfaces utilize perceptual and cognitive skills to interpret information offered to the user, whereas multimodal interfaces focus on the diversity of human communication abilities (Turk and Robertson, 2000). The power and potential of multimedia and multimodal interfaces to help individuals with perceptual impairments are described in the following two sections.

6.5.1.1 Multimedia Interfaces

Technological advancements, as well as the need for flexibility in displaying diverse forms of information, have sparked the growth of multimedia interfaces. Multimedia is the concept of using multiple forms of information and information processing to inform or entertain the user. Examples include text, audio, graphics, diagrams, animation, video, and so on. Multimedia interfaces is a broad field of research and information, but here we will only discuss multimedia through an overview of its strengths in increasing accessibility to technologies for those with sensory impairments.

The term *multimedia* is becoming increasingly interchangeable with the term *hypermedia*, which implies a degree of nonlinearity in the way that information entities are connected. Since it is becoming more and more unusual to encounter information systems that do not support some degree of nonlinearity in their interactive presentations, which may use more than one sensory modality, the boundaries between these two words are blurring (Waterworth and Chignell, 1997). The essential aspect of designing multimedia interfaces is selecting the proper forms of media and modalities and appropriately utilizing the idea of information being taken from one modality and presenting it in another (Chignell and Waterworth, 1997).

Chignell and Waterworth (1997) used three approaches to describe multimedia. First is the approach of performance, with the idea of the technology being a "play" conveyed through the media "actors." The actors are as dynamic as needed to educate and entertain given the context and the user's wants and needs. The second approach, presentation, is the idea of conveying ideas to the user through means enhanced by media such as video and animation. The last approach is a document in which ideas in text can be elaborated and enhanced through the diversity of the media used.

When designing these multimedia interfaces, consideration of the users' sensory abilities must be incorporated, as the primary concept is based on the human cognition and comprehension of the display. The strength of this technology lies in its ability to use multiple means (e.g., text, images, animation, etc.) to convey information to the user based on context and ability. Thus, a basic understanding of human sensation and perception is necessary and alternative modalities must be provided. Multimodal interfaces are a second type of perceptual interface that provide this same benefit.

6.5.1.2 Multimodal Interfaces

Multimodal interfaces utilize a wide range of perceptual capabilities to facilitate human interaction with computers. Its strengths are the increased contexts in which the technology will be used as well as the increased diversity in user accessibility. The latter of these strengths is the focus for this chapter, as it has great potential for populations with sensory impairments, allowing users the freedom to utilize their best sensory abilities or a combination of sensory modalities when interacting with technologies.

Designers are applying the concept of multimodal interfaces to make technologies accessible to users with difficulties, particularly sensory impairments. In a traditional GUI, information is primarily represented through visual means, sometimes complemented with sound. The perceptual user interface (PUI) seeks to use alternative ways of displaying information in accordance with context and user needs.

As described in previous sections, haptic senses facilitate perception of properties such as shape, texture, resistance, size, height, position, and so on. Thus, haptics can provide sensory information on various facets of interface objects. Computer scientists and engineers at the National Institute of Standards and Technology created a tactile graphic display that conveys scanned illustrations to the fingertips, such as map outlines or other graphical images. It can also translate images displayed on web pages or in electronic books using a refreshable display. Tactile means were also used by Yu et al., (2000) who used force feedback as a means of conveying numerical information. These are only some examples of tactile technologies and their application in assisting in sensory impairments; further development of this topic can be found in Chapter 33 of this handbook.

Providing additional assistance to users with sensory impairments, speech recognition technology is becoming more common with software such as Dragon Naturally Speaking (Scansoft), IBM's ViaVoice, SRI International's EduSpeak, Microsoft's Voicenet VRS, and Nuance Communications Nuance 8. These technologies allow users to interact with their computer through voice commands. Speech-driven menu navigation systems such as IN CUBE Voice Command (Command Corp.) are also available (Karshmer et al., 1994a, b). Further discussion on the topic of speech interaction can be found in Chapter 30, "Speech Input to Support Universal Access," of this handbook. However, one idea to keep in mind is that these systems need to account for the users' varying levels of hearing, speech, and cognitive abilities. A variety of injuries and impairments may cause speech degeneration, which could prevent a user from using traditional speech recognition technology (Rampp, 1979). Augmentative and alternative communication (AAC) refers to a field that studies ways in which messages and ideas can be conveyed by means other than speech (Beukelman and Mirenda, 1998). Technologies are being developed for which input can come from any number of different switches that are controlled with motions as simple as a push of a button, a puff of air, or the wrinkle of an eyebrow (Beukelman and Ansel, 1995). AAC is discussed in more detail in Chapter 47 of this handbook.

The auditory channel is also increasing in its use as an alternate means of providing users with system feedback. Some auditory technologies, such as Scansoft's Realspeak, and IBM's Websphere Voice Server, allow information to be conveyed from GUIs through electronic text-reading and speech synthesis capabilities. This IBM software also has the means to convert the text into Braille for the visually impaired (Thatcher, 1994). Audio feedback has also been used to provide a sense of depth perception to users with visual impairments by varying the location of the sound sources in a three-dimensional environment (Mereu and Kazman, 1996). More information on the use of audio technology in universal access can be found in Chapter 32, "Auditory Interfaces and Sonification," of this handbook.

Multimodal interfaces have also been found to be efficient in error avoidance and recovery, which adds to accessibility in the aspect of efficiency of use. Oviatt (1999) highlighted some of the reasons that multimodal interfaces can provide better error handling than unimodal interfaces:

- Users can choose the modality that they are most comfortable with and leverage their strengths.
- Users are free to choose the less error-prone input mode for a given context.
- When an error occurs, users have a tendency to respond by switch modes. For example, if the system is not recognizing input, the user is able to change the input mode.
- Users may feel more in control of the system with the ability to switch modes, thus become less frustrated, even at the same error rate as a unimodal system.

Overall, the variety of functionality offered to the user results in error reduction and efficiency of recover (Oviatt and Cohen, 2000). This stems from the user-centered and system-centered design ideas from which multimodal interfaces are constructed (Oviatt and Cohen, 2000; Oviatt et al., 2001). A deeper discussion of multimodal interfaces and their applications in universal access can be found in Chapter 40 of this handbook.

6.5.2 Adaptive Interfaces

Adaptive interfaces are inspired by the idea that a system could learn about the user and adapt its behavior to that individual. Adaptive interfaces have great potential for accommodating a wide range of users in a variety of work contexts. As a result, much research has been conducted on the design and implementation of adaptive interfaces.

Jameson (2003) outlined the adaptive interface's functional role of supporting system use in four categories: (1) taking over parts of routine tasks, (2) adapting the interface to fit the user, (3) giving advice about system use, and (4) controlling the dialogue. In taking over parts of routine tasks, the interface may be able to adopt the roles or activities that are difficult for the user (as a result of user's impairment), thus relieving the load and allowing the user to focus on the part of the task that the user is better equipped to complete.

In adapting the interface to fit the user, the entire interface design—menus, icons, input/output devices—can be optimized based on how the user is able to best interact with the technology. If the system is equipped to give the user advice about using it, the user can then be informed of the different options of interaction and make educated adaptations to the interface to best suit his needs. Controlling a dialogue is perhaps the most applicable form of adaptation for use with sensory impairment as the system adopts a policy for determining when and how the system should provide information, get information from the user, and perform other dialogue acts (Jameson, 2003). This capability of predicting the best method of information exchange could make a great difference in the sensory experience for users, particularly those with sensory impairments.

The concept of configurable interface designs based on user models has gotten considerable attention from researchers. Applicable user models have been defined based on visual, cognitive, motor, and other abilities that can be used to customize computer systems, with respect to both hardware and software (McMillan and Wisniewski, 1994). Approaches such as those described by pervasive accessible technology allow individuals with disabilities to use standard interface devices that adapt to the user to communicate with information technology infrastructures (Paciello, 1996). Based on an individual's disability, the implementation of a user interface management system model provides the versatility needed to adapt interfaces to fit an individual's needs. The selection of input devices, presentation of information on the screen, and choice in selection and activation method can all be adapted to fit specific user needs (Bühler et al., 1994). Adaptive applications are also available that keep track of the evolving aspects of the user, such as preferences and domain knowledge. In this case, user interface adaptation is based on a user model created by the stored information (De Bra et al., 1999).

There is great potential power in the adaptive interface in assisting a user with one or more sensory impairments. The diversity of the technology and the modes of interaction that are available provide great hope for impaired individuals in interacting with technology. Adaptive interfaces that can be customized to fit a particular individual based on his or her relative needs bring that potential power to a new level of possibilities for the user. Chapters 16 and 21 of this handbook discuss in detail the design and development of user interfaces capable of self-adaptation behavior in the context of universal access.

6.6 Conclusion

This chapter aimed to establish a basis for addressing issues of universal access from the perspective of users with sensory impairments. It examined three different types of impairments—visual, auditory, and haptic—to understand the impact that they have on human interaction with computing technologies. The functioning of these systems, as well as some specific impairments associated with each system, were described to elucidate the effects of these sensory impairments or deficits. Finally, advances in technology, including perceptual, multimodal, multimedia, and adaptive interfaces, were discussed. These emerging technologies are examples of how the (re)design of common information and computing technologies can enhance the perceptual experience of users who may have sensory impairments.

References

Beukelman, D. R. and Ansel, B. (1995). Research priorities in augmentative and alternative communication. *Augmentative and Alternative Communication* 11: 131–134.

Beukelman, D. R. and Mirenda, P. (1998). *Augmentative and Alternative Communication* (2nd ed.). Baltimore: Paul H. Brookes Publishing Co.

Bühler, C., Heck, H., and Wallbruch, R. (1994). A uniform control interface for various electronic aids, in the *Proceedings of the Fourth International Conference on Computers for Handicapped Persons* (W. L. Zagler, G. Busby, and R. R. Wagner, eds.), 14–16 September 1994, Vienna, pp. 51–58. Vienna: Springer-Verlag.

Chignell, M. and Waterworth, J. (1997). Multimedia, in *Handbook of Human Factors and Ergonomics* (2nd ed.) (G. Salvendy, ed.), pp. 1808–1861. New York: Wiley.

Cook, M. and Hussey, S. M. (1995). *Assistive Technologies: Principles and Practice*. St. Louis, MO: Mosby-Year Book.

Coren, S., Ward, L. M., and Enns, J. T. (1999). *Sensation and Perception* (5th ed.). New York: Harcourt Brace College Publishers.

De Bra, P., Houben, G., and Wu, H. (1999). AHAM: A Dexter-based reference model for adaptive hypermedia, in the *Proceedings of the Tenth ACM Conference on Hypertext and Hypermedia: Returning to Our Diverse Roots*, 21–25 February 1999, Darmstadt, Germany, pp. 147–156. New York: ACM Press.

Diabetic Retinopathy Study Research Group. (1981). Photocoagulation treatment of proliferative diabetic retinopathy: Clinical application of diabetic retinopathy study (DRS) findings. DRS Report 8. *Ophthalmology* 88: 583–600.

Fletcher, D. C., Schuchard, R. A., Livingstone, C. L., Crane, W. G., and Hu, S. (1994). Scanning laser ophthalmoscope macular perimetry and applications for low vision rehabilitation clinicians. *Ophthalmology Clinics of North America* 7: 257–265.

Hamming, A. and Apple, D. (1980). Anatomy and embryology of the eye, in *Principles and Practice of Ophthalmology* (G. A. Peyman, D. R. Sanders, and M. F. Goldberg, eds.), Vol. 1, pp. 33–68. Philadelphia: WB Saunders.

Jacko, J. A., Vitense, H. S., and Scott, I. U. (2003). Perceptual impairments and computing technologies, in *The Human-Computer Interaction Handbook: Fundamentals, Evolving Technologies and Emerging Applications* (J. A. Jacko and A. Sears, eds.), pp. 504–522. Mahwah, NJ: Lawrence Erlbaum Associates.

Jameson, A. (2003). Adaptive interfaces and agents, in *The Human-Computer Interaction Handbook: Fundamentals, Evolving Technologies and Emerging Applications* (J. A. Jacko and A. Sears, eds.), pp. 305–330. Mahwah, NJ: Lawrence Erlbaum Associates.

Karshmer, A. I., Brawner, P., and Reiswig, G. (1994a). An experimental sound-based hierarchical menu navigation system for visually handicapped use of graphical user interfaces, in the *Proceedings of the First Annual ACM Conference on Assistive Technologies (ASSETS '94)*, 31 October–1 November 1994, Marina del Ray, CA, pp. 123–128. New York: ACM Press.

Karshmer, A. I., Ogden, B., Brawner, P., Kaugars, K., and Reiswig, G. (1994b). Adapting graphical user interfaces for use by visually handicapped computer users: Current results and continuing research, in the *Proceedings of the Fourth International Conference on Computers for Handicapped Persons* (W. L. Zagler, G. Busby, and R. R. Wagner, eds.), 14–16 September 1994, Vienna, pp. 16–24. Vienna: Springer-Verlag.

Levine, M. W. (2000). *Fundamentals of Sensation and Perception* (3rd ed.). New York: Oxford University Press.

Luce, R. D. (1993). *Sound & Hearing: A Conceptual Introduction.* Hillsdale, NJ: Lawrence Erlbaum Associates.

Machemer, R. and Steinhorst, U. H. (1993). Retinal separation, retinotomy, and macular relocation II. A surgical approach for age-related macular degeneration? *Graefe's Archive for Clinical and Experimental Ophthalmology* 231: 635–641.

McMillan, W. W. and Wisniewski, L. (1994). A rule-based system that suggests computer adaptations for users with special needs, in the *Proceedings of the First Annual ACM Conference on Assistive Technologies (ASSETS '94)*, 31 October–1 November 1994, Marina del Ray, CA, pp. 129–135. New York: ACM Press.

Mereu, S. W. and Kazman, R. (1996). Audio enhanced 3D interfaces for visually impaired users, in the *Proceedings of the 1996 ACM Conference on Human Factors in Computing Systems (CHI '96)*, 14–18 April 1996, Vancouver, Canada, pp. 72–78. New York: ACM Press.

Moore, B. C. J. (1982). *An Introduction to the Psychology of Hearing* (2nd ed.). New York: Academic Press.

National Institute on Deafness and Other Communication Disorders (2006). *Presbycusis.* http://www.nidcd.nih.gov/health/hearing/presbycusis.asp.

National Institute of Neurological Disorders and Stroke (2007). *Peripheral Neuropathy Fact Sheet.* http://www.ninds.nih.gov/disorders/peripheralneuropathy/detail_peripheralneuropathy.htm.

Oviatt, S. (1999). Mutual disambiguation of recognition errors in a multimodal architecture, in the *Proceedings of the SIGCHI Conference on Human Factors in Computing Systems: The CHI Is the Limit (CHI'99)*, 15–20 May 1999, Pittsburgh, pp. 576–583. New York: ACM Press.

Oviatt, S., and Cohen, P. (2000). Multimodal interfaces that process: What comes naturally. *Communications of the ACM* 43: 45–53.

Oviatt, S., Cohen, P., Wu, L., Vergo, J., Duncan, L., Suhm, B., et al. (2001). Designing the user interface for multimodal speech and gesture applications: State-of-the-art systems and research directions, in *Human-Computer Interaction in the New Millennium* (J. Carroll, ed.), pp. 421–456. Reading, MA: Addison-Wesley Press.

Paciello, M. G. (1996). Designing for people with disabilities. *Interactions* 3: 15–16.

Prevent Blindness America (2002). *Vision Problems in the U.S.: Prevalence of Adult Vision Impairment and Age-Related Eye Disease in America.* http://preventblindness.org/vpus/2VPUS_report_web.pdf.

Proctor, R. W., and Proctor, J. D. (1997). Sensation and perception, in *Handbook of Human Factors and Ergonomics* (G. Salvendy, ed.), pp. 43–88. New York: Wiley.

Rampp, D. L. (1979). Hearing and learning disabilities, in *Hearing and Hearing Impairment* (L. J. Bradford and W. G. Hardy, eds.), pp. 381–389. New York: Grune & Stratton.

Reeves, B. and Nass, C. (2000). Perceptual bandwidth. *Communications of the ACM* 43: 65–70.

Robles De-La-Torre, G. (2006). The importance of the sense of touch in virtual and real environments. *IEEE Multimedia* 13: 24–30.

Stein, L. K. (1988). Hearing impairment, in *Handbook of Developmental and Physical Disabilities* (V. B. Van Hassek, ed.), pp. 271–294. New York: Pergamon Press.

Stephanidis, C., Salvendy, G., Akoumianakis, D., Bevan, N., Brewer, J., Emiliani, P. L., et al. (1998). Toward an information society for all: An international R&D agenda. *International Journal of Human-Computer Interaction* 10: 107–134. http://www.ics.forth.gr/hci/files/white_paper_1998.pdf.

Stephanidis, C., Salvendy, G., Akoumianakis, D., Arnold, A., Bevan, N., Dardailler, D., et al. (1999). Toward an information society for all: HCI challenges and R&D recommendations. *International Journal of Human-Computer Interaction* 11: 1–28. http://www.ics.forth.gr/hci/files/white_paper_1999.pdf.

Thatcher, J. (1994). Screen reader/2: Access to OS/2 and the graphical user interface, in the *Proceedings of the First Annual International ACM/SIGCAPH Conference on Assistive Technologies (ASSETS '94)*, 31 October–1 November 1994, Marina del Ray, CA, pp. 39–46. New York: ACM Press.

Turk, M. and Robertson, G. (2000). Perceptual user interfaces. *Communications of the ACM* 43: 32–34.

U.S. Food and Drug Administration. (2006). *FDA Approves New Biologic Treatment for Wet Age-Related Macular Degeneration.* http://www.fda.gov/bbs/topics/NEWS/2006/NEW01405.html.

Waterworth, J. A. and Chignell, M. H. (1997). Multimedia interaction, in *Handbook of Human-Computer Interaction* (M. Helander, T. K. Landauer, and P. Prabhu, eds.), pp. 915–946. New York: Elsevier Science.

Yu, W., Ramloll, R., and Brewster, S. (2000). Haptic graphs for blind computer users, in the *Proceedings of the First Workshop on Haptic Human-Computer Interaction*, 31 August–1 September 2000, Glasgow, UK, pp. 102–107. London: British Computer Society.

7

Cognitive Disabilities

Clayton Lewis

7.1 Introduction

Cognitive processes are those mental processes not involving simply sensing the environment or controlling one's movements, and cognitive disabilities are impairments in cognitive processes. They can arise in many ways, including brain injury or stroke; chromosomal abnormalities that affect the development of the brain (such as Down syndrome), producing *developmental* disabilities; severe mental illness; or effects of aging (see Chapter 8, "Age-Related Difference in the Interface Design Process"). Often the cause of a cognitive disability is unknown.

Because cognitive processes are very diverse, and complex, cognitive disabilities vary greatly in their impact. Because cognitive processes figure in many ways in the use of information technology, there are many barriers to effective use by people with cognitive disabilities.

In part because of social attitudes toward people with disabilities, many information technologists have little contact with people with cognitive disabilities (or, in some cases, have little contact that they are aware of). This chapter begins by describing several real people, as a way of illustrating some ways in which cognitive disabilities play out in people's lives, and the role information technology can play, and then presents some demographic data that place these examples in context. Against this background, the chapter then surveys the barriers to use of information technology that are associated with cognitive disabilities. It then describes the design implications of these barriers, and how design processes should be shaped to address these implications. The chapter ends with a discussion of current trends, including an increased emphasis on meeting the needs of people with cognitive disabilities in the information technology community.

7.2 Some People

7.2.1 AB

AB was "slow" in school, and was unable to obtain the postsecondary educational credentials required for some of his career aims. He reads, but not very fluently, and has difficulty understanding rapid, complex speech, or speech that includes unfamiliar words. He has trouble managing papers, and sometimes becomes flustered when plans must be changed, or a situation becomes complex. There is no "diagnosis" for AB's limitations.

AB is aware of the kinds of difficulties he encounters, and has the self-confidence to be assertive in dealing with them. For example, he will interrupt a meeting if necessary to have an unfamiliar word defined or a complex question restated. His colleagues recognize AB as a capable person with good judgment and valuable insights.

AB uses a computer for e-mail. He is helped by software that allows one of his associates to help with problems on AB's computer over the Internet. AB argues that the computer should provide a tool for looking up unfamiliar words, wherever they appear on the screen, and that, in this and other ways, the computer has great potential for improving access to information for people with cognitive disabilities.

7.2.2 Jack Horner

Jack Horner (Horner, 2004) is Regents Professor of Paleontology at Montana State University, and holder of a Macarthur Fellowship for extraordinary creativity. He has no undergraduate degree, having failed in seven attempts; he was barely able to complete secondary school. He says:

> My progress in reading, writing, and mathematics was excruciatingly slow... To this day, I struggle with the effects of dyslexia. It takes me a long time to read things, so it's an ongoing endeavor to become as well-read as I would ideally like to be. Self-paced learning is a strategy that helps me cope. Audio books are also a very helpful technology. My first publications were traumatic. I was afraid to even attempt to write something that would go to an editor. I had plenty of data, so I wasn't fearful of critical review, but I had apprehension about people seeing how little I actually knew of the English language. It's a phobia I still live with. After two junior authorships, I wrote three papers on my own, and each was published: one in the *Journal of Paleontology* and two in the British journal *Nature*. I discovered that editors would forgive my writing errors and fix them as long as the science was solid. Writing is still very difficult for me, and I would always rather a more fluent coauthor did the actual writing. I know what I can do and what I can't do, and for the things I can't do, I try to find someone to help.

7.2.3 TT

TT (Hart, 2005) is a secondary school student in a large city in the United States. With an IQ tested at about 70, and diagnosed with autism spectrum disorder, a developmental disability, TT is assigned to "life skills" classes at his school. He is not eligible for computer classes at the school, but he participated in a special tutorial program in which he learned to use the Excel spreadsheet program. During the program he was scored as attaining "independent mastery" of 98% of the Excel topics on the syllabus, which closely matched the syllabus for the mainstream course at the school. He created a database of information about video games, an interest of his. He also experimented with 3D content in a spreadsheet, and with a program that printed out messages that simulated an imminent computer crash.

7.2.4 MJ

MJ (Bergman, 1998) suffered a brain injury in a car accident. After the accident, she had problems in visual perception, memory, reasoning, reading, and writing. These difficulties led to serious financial problems, as she was unable to manage money, even with the assistance of a bookkeeper. Because of difficulty in appreciating numerical amounts, she could not control her credit card spending. She was unable to use a computer reliably, often needing help with the long sequences of steps required to turn it on and access the appropriate software. She made persistent attempts to use the software, sometimes spending several hours before giving up. Her difficulties were extremely frustrating, and, because she believed that the software she was using was the simplest available, she came to believe that she was stupid, and worthless, and that her situation was hopeless.

Customized software, carefully adapted to her needs, transformed MJ's situation. The software guides her through needed financial operations, and maintains records in a form that she can easily access. It prompts her to keep a journal, including logs of conversations, and she is able to use this material to organize her interactions with other people. Using this software, MJ has been able to manage her finances, and other aspects of her life, for more than 10 years.

7.2.5 LR

LR (Dawe, personal communication) is a 25-year-old woman with developmental disabilities. She lives in an apartment with a roommate who provides part-time care. She cannot read, but she has three part-time jobs: rolling silverware in a restaurant, providing childcare in a daycare facility, and passing out water in a retirement home. She uses the bus for some trips. She has many friends, and uses the telephone for social interaction, but sometimes calls people too often, and has trouble recognizing cues to get off the phone. She has trouble learning telephone numbers and using cell phone menus. She remembers to keep her phone charged, but does not understand why it won't work when it is not charged.

7.2.6 Abdulkader Faraax

Abdulkader Faraax (Danielsson and Jonsson, 2001) is a 10-year-old boy diagnosed with autism. His vocabulary is about 400 words, and speaking and understanding speech are difficult for him. On a talkative day he may say 30 or 40 words. Many interactions that ordinarily involve speech, such as using a taxi, are difficult for him. He uses digital photographs, 50 or 60 in a typical week, to help in communication, for example, to show his destination on a taxi ride. He and his teacher have built up a collection of 5000 photographs to represent the people and things that are important to him, such as people in his school. Besides their use in everyday communication, his teacher believes that the photographs help Abdulkader maintain awareness of people and things when they are not physically present. The teacher can also use photographs to tell Abdulkader things, such as where he can sit to work on an art project (conveyed by showing a photo of the designated table, with art supplies laid out on it).

7.3 Cognitive Disabilities

All of these people are classified as having cognitive disabilities. The examples give some sense of the diversity of these disabilities, and their impact.

As mentioned earlier, cognitive disabilities arise from impairments in cognitive processes. Because cognitive processes are complex, and have many more or less distinct aspects, cognitive disabilities can affect many aspects of cognitive function. The following list, adapted from Francik (1999), gives broad headings; each heading stands for a range of more specific functions.

- Executive functions
- Memory
- Attention
- Visual and spatial perception
- Language (including reading)
- Mathematical thinking
- Emotional control, expression, understanding
- Speed of reasoning
- Solving new problems
- Solving problems based on experience

To illustrate the finer structure of this functional classification, consider memory. Far from being a monolithic function, memory on this list has to include both *skill* learning and learning of factual (*declarative*) information, though these involve different mechanisms (Anderson, 1983; Anderson et al., 2004). Glisky (1992) reports a training study in which people with memory deficits learned to perform data entry tasks as rapidly as participants with typical memory, even though they could not remember that they had participated in the training sessions. Psychologists also distinguish *short-term* from *long-term* memory. Some people are able to retrieve information learned long ago, but cannot retain information from moment to moment in the here and now, while others can retain information only for short periods. Memory seems to act in quite different ways when material must be *recalled* and when it must only be *recognized* as familiar. In the former situation, but less so in the latter, there are powerful effects of *strategy*, so that one can *learn* to use one's memory more effectively. Any of these aspects of memory function could be impaired.

Similarly, the entry "language" in the list stands for many different functions. For example, a person may be able to process spoken language but not written language.

The entry "visual and spatial perception" refers to a crude distinction between picking up information from the environment, called *sensation* (e.g., patterns of light and dark), and interpreting the information, called *perception*. For example, people can "see" inverted faces as well as they can "see" upright ones, but most people can perform a variety of further processes, including recognition, and judging gender or emotion, much better for upright faces. Interestingly, there is good evidence that perception of faces can be impaired separately from perception of objects of other kinds (Duchaine et al., 2006). Impairment of face perception is called *prosopagnosia*; other forms of perceptual impairment, or *agnosia*, sometimes occur, such as specific impairment in the perception of text, called *alexia* (see Ghaldiali, 2004).

While specific impairment in the perception of faces is rare, there are other, much more common impairments that are also quite narrow in their impact. The conditions, called "learning disabilities" in the United States and Canada (different, more specific terms are used in the United Kingdom and Europe), can impact reading or mathematical skills, and not other mental functions. Some people with these disabilities can perform at a very high level in other areas. Jack Horner, the paleontologist, is an example. West (2000) presents brief biographical sketches of Horner and several other leaders in science, industry, and government who have learning disabilities.

While learning disabilities are narrow in their impact, it is not unusual for a person to have impairments that affect many aspects of cognitive function. MJ is an example of a person with impairments in several areas of function. The fact that a person has trouble in one area of cognitive function (e.g., reading) does not imply that she will have or will not have trouble in other areas.

The description "person with a cognitive disability" can sometimes be useful in focusing attention on difficulties people can have in performing mental operations, separately from difficulties of other kinds, for example in seeing, hearing, or moving. But it is a mistake to make too much of this separation. Many people with cognitive disabilities have other disabilities as well. Further, much of the real impact of anything commonly thought of as a disability, of whatever kind, is really better understood as a product of attitudes expressed in social and political processes (Roth, 1983; Rapley, 2004).

Thinking about people with cognitive disabilities is often shaped by implicit or explicit use of the *medical model* of disability. On this model, a person with a cognitive disability has some particular condition that can be diagnosed and perhaps cured, like a disease. The first question to ask about such a person is, What is the disease? This way of thinking is misleading in a number of ways. First, there is no diagnosis for many people with cognitive disabilities, so the model cannot usefully be applied to them. Second, the variation in cognitive functioning among people with a given diagnosis, for example Down syndrome or autism, is very large. There are people with Down syndrome who are completely nonliterate, and others who are college graduates, so the diagnosis of Down syndrome is not very helpful. Third, the focus on diagnosis, and indeed on disability itself, directs attention to the person, and away from the environment in which the person operates, while it is the environment that is often responsible for the problems people have. As designers, it is important to focus on those aspects of the environment, including the social environment, that can create or eliminate barriers for people, not on diagnoses.

7.3.1 Demographics

Because of the social consequences of disabilities, which commonly include segregation, few designers have much knowledge of people with cognitive disabilities. Designers have asked the author whether people with cognitive disabilities could use the World Wide Web, for example. The answer is that many can and do, but that the question has to be asked is an indication of how poorly people with cognitive disabilities are understood.

A survey commissioned in 2003 by Microsoft (n.d.) found that 16% of working-age computer users in the United States have some form of cognitive impairment. Employment data for the U.S. federal government show that more than 1300 people with developmental disabilities hold white-collar jobs, that is, jobs in administration or other knowledge work (United States Equal Employment Opportunity Commissions, 2005; note that these employment data are actually reported for *mental retardation*, a term, now widely deprecated, for *developmental disability*, an impairment that arises before adulthood, is not traceable to an injury or illness, and affects cognitive aspects of daily life).

These numbers make clear that there are many people with cognitive disabilities who are users of information technology, and who do the kind of work that information technology commonly supports.

While there are many people with cognitive disabilities in the workforce, many are unemployed, as is true for people with disabilities of all kinds. The employment rate for people with disabilities in the United States is less than 56%, according to the U.S. Office of Disability Employment Policy[1]; and for the European Union the rate is less than 43%, according to the European Commission Directorate General for Economic and Social Affairs.[2]

Data on the kinds of barriers related to cognitive processing that people encounter in their work, or in other aspects of their lives, are hard to obtain. Because of the prevalence of the medical model, data collection has focused on the frequency of diagnostic categories, such as developmental disability, rather than on how often people face barriers connected with memory, or attention, or other cognitive functions. The International Classification of Functioning, Disability, and Health (ICF) embodies a new approach to disability-related data (World Health Organization, n.d.). The ICF includes descriptions of activities, participation, and context, as well as descriptions of components of functioning and disabilities. In principle, the new approach can be used to tabulate barriers of different kinds, and could be useful in prioritizing research and development on eliminating or bypassing cognitive barriers.

7.4 Cognitive Barriers and Information Technology

Use of technology can involve nearly all of the cognitive functions on the previously mentioned list; hence, a wide range of cognitive barriers can arise in technology use. Functions concerning emotions may be an exception, since technology use does not normally engage the emotions. However, as MJ's situation shows, dealing with frustration can be significant. Further, some user interaction techniques are emerging that use simulated human agents in an effort to project friendliness or helpfulness (see, e.g., Brave et al., 2005). It is possible that difficulty in judging emotions could interfere with these interactions.

Here are examples of some of the cognitive barriers to successful use of computer and communication. They are organized

in the same categories as the list adapted from Francik (1999), previously shown.

7.4.1 Executive Functions

7.4.1.1 Carrying Out a Sequence of Operations

Cellular phones allow one to take pictures and send them to contacts. Doing this on one common model requires at least eight button presses, plus additional actions to scroll through the contact list. While most of these button presses are cued to some degree, some of the cues are quite cryptic, so that it is essential to keep clearly in mind what one's ultimate goal is during the entire sequence to avoid getting lost.

7.4.1.2 Managing a Stack of Goals and Subgoals

Most tasks of any complexity can be divided into parts, subtasks, which can be completed in a manner substantially independent of the main task. For some people, determining what to do next is difficult when the goal of a subtask (called a subgoal) has been completed. In some cognitive models (see, e.g., Anderson et al., 2004), this transition is managed using an information structure called a *stack*, in which the status of the main task is stored while a subtask is being worked on. For some people, this goal management is unreliable, and prompting to help them sequence the steps in a complex task is helpful (Davies et al., 2002).

A common situation in which goal management is needed in technology use is consulting help information. Here work on the main task has to be suspended, while the subtask of accessing help is carried out, and then resumed when the information sought from help has been obtained.

7.4.2 Memory

7.4.2.1 Memory in Performing a Sequence of Actions

Modern technology rarely *requires* users to carry out a procedure by simple rote memory. As in the cell phone example presented earlier, most actions have some kind of cue associated with them that can guide a user who does not know the sequence of actions required. But memory still plays a part, in many situations. Cues are often cryptic, meaning that their significance isn't completely clear when first seen. For example, in the cell phone sequence described earlier, the required action at one stage is through Option. It leads to a choice of ways to specify the recipient of the picture, but there is nothing about the cue Option that suggests that when it is first encountered. This means that memory is helpful in performing this task, in two ways. One might remember what Option means, in this situation, or one might remember that the action cued by Option is needed at this point, without regard to any interpretation of the cue.

Another consideration regarding the role of memory in using technology is the tradeoff that can be observed between reasoning and memorizing. The actions needed to perform a task usually have a logic based on their connection to underlying

---

[1] http://www.dol.gov/odep/faqs/working.htm.
[2] http://ec.europa.eu/employment_social/news/2001/dec/2666complete_en.pdf.

processes in the implementation of the task. Someone who understands this logic can often work out what to do to accomplish a task, whereas someone who does not understand it must rely on memory (or an instruction list, an external memory). Because a person with a cognitive disability may have more trouble understanding the logic of a complex system, he or she may be compelled to rely more on memory.

To illustrate these considerations, suppose you receive a document as an attachment to an e-mail message. You open the document, make some revisions, and save. At some later time you want to find the revised document, but you have difficulty, because it is among the temporary files maintained by your mail system, not among your own documents. A colleague advises you that you can avoid this problem in the future by remembering to use Save As rather than Save after revising an attachment, and to specify one of your document folders when you perform the Save As operation. If you understand the logic of Save As, and its relationship to Save, you will understand your colleague's suggestion, and will find it easy to follow in the future. If you should forget that Save As is the particular action needed, you can easily work it out. But if you do not understand the underlying logic, you will be forced to rely much more on your memory (internal or external), both for the needed action, and for what you must specify as a destination when you perform the Save As operation.

This example highlights another challenge associated with memory for actions. Both in form and in meaning, Save and Save As are quite similar, yet correct operation requires a user to discriminate between them. Technology use often requires this kind of fine discrimination: in URLs, case distinctions and the difference between .htm and .html; in programming, the distinction between X=Y and X==Y in many programming languages, and so on. As noted as long ago as 1982 in Lewis and Mack, these discriminations are a rich source of usability problems for people with typical cognitive functioning. While specific data are scant (though see the discussion in Small et al., 2005), it is likely that these problems are even more common for people with cognitive disabilities. As for other usability problems, their impact is also likely to be greater, because people with cognitive disabilities may have more trouble detecting that an error has occurred, and in recovering from it.

7.4.2.2 Procedural and Declarative Information in Technology Use

As mentioned earlier, memory for skills (procedural knowledge) involves different mechanisms from memory for facts (declarative knowledge). The example just discussed shows how these two kinds of information are intertwined in common uses of technology. One can "know what to do" as a procedure, a well-learned skill, or one can "know what to do" by deploying declarative knowledge of what particular cues mean, or how a system works. Further, as Anderson et al. (2004) argue, declarative knowledge can be proceduralized, that is, converted to skill form. A person whose declarative memory is impaired, like some of Glisky's trainees in the study cited earlier, can perform some tasks by reliance on procedural memory.

Other aspects of common technology uses make demands in which these two kinds of information are more separate. For example, keyboarding is a procedural skill called for in many technology uses. Some people with cognitive disabilities also have motor disabilities that can make keyboarding difficult. But some people have impairments in procedural memory, associated for example with severe depression, that make skill learning difficult (Budson and Proce, 2005). On the other hand, unlike declarative memory, procedural memory is often spared in patients with head injury (Milders, 1997).

Keeping track of one's files is an example of a task with heavy demands on declarative memory. Finding a file without extensive search requires recalling its name, or the names of the folders or directories within which it was placed. If recall of this information fails, one can browse, hoping to be able to recognize the name one could not recall. Such browsing may be impractically inefficient if one has many folders and cannot recall information that can be used to limit the search. Some systems allow files to be searched by contents, but even this kind of facility requires some recall of what is in a file.

Finding things can also involve spatial memory and spatial reasoning. Some people use spatial layouts rather than lists when viewing the contents of folders. Spatial memory is thought to involve different brain mechanisms from other forms of memory, and some people with cognitive disabilities have impairments in these mechanisms (Pennington et al., 2003).

Jarrold et al. (1999, 2007) have studied the relationship between verbal (linguistic) and spatial abilities, including learning, in people with different developmental disabilities. Williams syndrome is associated with relatively strong verbal abilities and weak spatial abilities, while Down syndrome shows the opposite pattern. These studies are valuable in contributing to increased understanding of the organization of cognitive functions, showing that there is a separation of mechanism between these two forms of learning. They also make a point that will be revisited in the following, that different users have different patterns of strengths and weaknesses that design needs to accommodate.

7.4.3 Attention and Perception

7.4.3.1 Interpreting a Complex Display

The home page of a typical web site has many parts, each of which contains many possible controls. The home page for the Mayo Clinic, a popular health information site in the United States, when viewed on February 25, 2007 contained 15 major, visually marked subdivisions, and a total of 107 controls, including two text entry fields, two scrolling lists, and 101 links. The two scrolling lists exposed a total of 30 additional links. A survey of web pages reported in Ivory et al. (2000) found an average of 11 distinguished text areas per page, and 40 links. Interestingly, pages in their sample ranked favorably by experts were even more complex, having an average of 18 text clusters and 60 links. Scanning a display of this complexity, looking for desired information, or for a link to follow, requires nontrivial control of visual attention (Kitajima et al.,. 2007), shifting attention from area to another, and scanning the cues associated with many controls.

Such searches may be especially difficult for some people with cognitive disabilities. Chan (2001) reviews literature on attentional deficits following brain injury, including some evidence that shifting attention is difficult for some patients. Carlin et al. (1995) present data on simple visual search by people with developmental disabilities, showing a complex pattern of impairment, with results differing according to the stimulus aspects needed to identify targets (e.g., color vs. form). Studies show superior performance by people with some cognitive disabilities in tests of visual perception using complex patterns (O'Riordan et al., 2001; Greenaway and Plaisted, 2005). As Carlin discusses, it is difficult even in a simple search task to separate the effects of attentional control and perception from other cognitive factors, such as problem solving. Problem solving, including the use of background knowledge, is certainly involved in processing web pages, as Kitajima et al. bring out. Because cues for links are nearly always textual, language processing is also involved.

7.4.3.2 Interpreting Symbols

While many cues found in interfaces, such as on web pages, are textual, many are not. Icons, small symbols, are widely used in interfaces to mark controls (usually buttons to be operated by a pointing device). The Microsoft Word screen contains 40 icons, mostly indicating operations, such as indentation, or text highlighting, but a few indicating status, such as the state of grammar checking. A further 25 icons are supplied on the same screen by the operating system. Again, most of these indicate actions, but a few indicate status, such as the charge level in the laptop's battery. One of the icons does double duty: an icon of a loudspeaker shows one the current volume level of the speaker, acting as a status indicator, but also serves as a button that exposes a volume control.

Nearly all icons embody some form that is meant to be familiar, such as the outline of a file folder or a floppy disc. This referent is not always familiar, in reality: file folders of the U.S. kind are not familiar elsewhere in the world, and hardly anyone uses floppy disks anymore. Further, the depiction is often so reduced, especially in small icons, that the form may not be easily recognized when first encountered. See Byrne (1993) for a discussion of visual search using icons.

Few icons can be interpreted without some explanation. For example, a user might encounter a small, geometric icon near the top right corner of the screen. Presumably, the user in question has seen it hundreds of times, but still may have no idea what it represents, until she selects it and finds that it reveals a menu relating to the Bluetooth wireless system. The user can now conjecture that the icon is a kind of stylized letter B, and may remember that, and interpret it in that way, when seeing the icon in the future. Like other icons, this one operates (when it works) more as a cue for remembered content than as an independent representation. The remembered content has to be obtained by exploration, as in this example, or from a short textual description that is supplied when the pointing device hovers over the icon without selecting it, or (if a more complex explanation is needed) by a help system or manual.

Research has shown that recognizing and using symbols, and even accurate pictures like photographs, is a complex skill (Stephenson and Linfoot, 1996). People with severe cognitive disabilities may require extensive training, or, in some cases, may be unable to attain mastery. Thus while some people may be helped by use of symbols rather than text in an interface, others may not be.

7.4.4 Language

7.4.4.1 Understanding Text

As discussed previously, even apparently nonlinguistic features of interfaces, like icons, usually require some interpretation of linguistic material for their use. Language in information systems is usually presented as text, requiring users to be able to carry out a chain of operations starting with visual perception of the elements of the text, proceeding through the recognition of words, taking in the assembly of words into meaningful structures, as provided for in the language, and continuing to the interpretation of these meanings in the context of the situation at hand (Friederici and Lachman, 2002).

Any of the stages in this chain can present problems. Some people with cognitive disabilities cannot carry out the early stages, and are not literate at all. Others can recognize words, but have a limited vocabulary. Some people may recognize the words in a sentence, but have trouble processing the sentence if its syntax is complex, apparently because of the demands on memory (Friederici and Lachman, 2002).

Because literacy develops over time, young children have limited vocabulary, and have trouble with complicated sentences. It has therefore become common to use stages in literacy development, usually related to grades or levels in school, to characterize the literacy of people with cognitive disabilities. But this does not work very well. The development of vocabulary is driven by experience, not just schooling, and adults with cognitive disabilities have experiences, and hence vocabulary, that do not correspond to those of schoolchildren. At the same time, range of vocabulary, and ability to process complex sentences, are not linked in a simple way (Gajar, 1989).

The practical effect of this complex situation is that people with cognitive disabilities may have trouble with the linguistic content of an interface, but there is no simple way to predict this. As Redish (2000) points out, the aspects of language assessed by readability formulas do not suffice to characterize the situation. Shorter sentences can be harder to understand than longer ones, especially if the shorter sentences are produced by cutting up longer ones, as is sometimes done to produce a better readability score for a text. Using common words may be less effective than using less common ones, depending on the context. Redish (personal communication) describes an effort to rewrite a standard lease to make it more accessible for poor readers. Replacing the term *security deposit* by a locution using more common words led to a general *decrease* in comprehension, because the adult poor readers in the study already knew what a security deposit is, from their life experience. In the face of these difficulties, rather than trying to simplify the vocabulary used in interfaces, it may be more promising to provide a means by which users can get an explanation of whatever terms are used, if they need it.

Other aspects of interfaces can have a big impact on the accessibility of linguistic material. It is obvious that a person who is blind will profit from audio presentation of text, but it is also true that this can benefit a person who has a cognitive disability that interferes with the recognition of words. While no specific data were found regarding people with cognitive disabilities, it is plausible that a *combination* of visual presentation of text with audio presentation is superior for sighted users with cognitive limitations over pure audio presentation (see Fang et al., 2006, for a review of studies of dual mode presentation). The combination may allow them to gain some of the advantages of visual presentation, such as the ability to review material by referring to the visual form, rather than having to rely on auditory memory, and the ability to link different parts of an interface via their spatial arrangement, while not having to do the work of decoding the content strictly from the text.

The informational web site thedesk.info illustrates these approaches at work. The site presents descriptions of government services of interest to people with cognitive disabilities in the United States. The creators work from descriptions that are provided by different government agencies, reorganize them under topical headings, and group together related information from different sources. They write descriptions that avoid unfamiliar and complex words, and use simple syntactic forms. They also provide audio presentation of the key texts. The site uses icons to mark major topical groupings, so as to reduce the reliance on text. The site does not have a facility for looking up unfamiliar words, though (as Redish would suggest will commonly occur) there are unfamiliar words used, even in the simplified descriptions. For example, the word *waiver* is used to describe a funded service, for historical reasons. People without experience in the support system would not recognize this unusual use, but many of the intended users of the site will understand it.

In addition to reading or otherwise comprehending text, some interfaces require users to produce it. Conducting a web search, for example, usually requires the user to type in search terms. For some users with learning disabilities, spelling is a major difficulty, and the facilities some search engines have for suggesting corrected spelling is a big help. For other users, a speech interface could be valuable.

7.4.4.2 Nonlinguistic Communication

Because some people, like Abdulkader, introduced previously, have difficulty processing language, there is interest in interaction and communication techniques that do not require it. Vanderheiden and colleagues at the Trace Center (Law and Vanderheiden, 2000) have developed EZAccess, a set of controls for common functions in information systems that can be identified using shape and color cues. Once learned, these controls make it possible to perform basic operations in a range of application systems without the need to interpret textual labels or explanations. For example, a postal kiosk patron uses these controls to navigate among choices and confirm selections (though some aspects of most transactions, such as buying postage that depends on the destination, require processing text).

Kitzing et al. (2005) review a number of projects for supporting people with language difficulties, including the one in which Abdulkader participates; see also McGrenere et al. (2003). As Danielsson and Svensk (2001) suggest, the ever-increasing ease of capturing, editing, storing, and retrieving digital images is increasing the opportunities for image-based communication and interaction.

7.4.5 Speed of Reasoning

7.4.5.1 Time-Dependent Interactions

Some interactions with information technology (IT) require responses within a certain time, and these are difficult when speed of mental operations, such as understanding language or making a choice, is reduced. For example, voice dialogues accessed by telephone commonly time out if a response is not made within an allowed interval. Some computer operations, such as logging off, may allow a certain amount of time for the user to confirm the operation before it is canceled.

Besides these cases of system-imposed speed requirements, there are situations in which operations can be too slow to be practical. According to a survey of web users seeking information for personal interest or needs (GVU, 1998), typical users give up on a search after a half hour or less. If a person's speed of reasoning is such that typical searches take an hour or more to complete, only the most pressing needs will justify persistence.

7.4.6 Solving New Problems

7.4.6.1 Using a New Tool

In addition to the problem solving that may be involved in one's own personal or professional tasks, the IT user confronts the need to solve problems that are presented by the technology itself. For example, a user recently acquired a new image-editing program, and wanted to use it to create a greeting card by combining parts of different digital photographs. The user had to give up on the new program, and do the job with an older, familiar program, because she could not figure out how to paste an image cut from one photo into another. The new program provided familiar cut-and-paste operations, but the paste operation required some follow-up action that the user could not discover in the time she was willing to invest. This kind of situation, which is not rare for a person with typical cognitive capabilities and a good deal of knowledge of computer interactions, is more common for someone whose background knowledge and problem-solving ability are limited.

7.4.6.2 Dealing with Errors

Even if one has mastered the use of a system, problems occur that require *troubleshooting*, a form of problem solving that is often even more difficult. Troubleshooting almost always requires some understanding not only of the task that one is trying to perform, but also of how the underlying system works (Kieras and Bovair, 1984). For example, a colleague recently found that he could not open a file that had been attached to an e-mail message he had received the previous week. The message referred to the file having been "shortcutted," and not available "offline." My colleague could interpret neither of these assertions, and neither was explained in the help system. He had to get technical assistance, and it emerged that "shortcutting" and "offline" referred to aspects of the implementation of his mail system of which he was completely unaware. Again, problems of this kind are common, and extremely debilitating, for users with typical cognitive capabilities and knowledge; they are more common, and more debilitating, for people with cognitive limitations.

7.4.6.3 Complex Search

Returning to a situation discussed earlier, searching for information on the web requires problem solving as well as perception of large displays (and linguistic interpretation of cues like link labels). The user has to decide which of several links looks most promising, keep track (if possible) of which links have already been explored, and the like, operations at the heart of Newell and Simon's (1972) classic account of problem solving. Unfortunately, as Pirolli (2005) shows, a relatively small decrease in the accuracy with which the user can evaluate links leads to a very large increase in the amount of searching that must be done, and hence in the time required to find something. Thus people with cognitive disabilities may find it impossible to use a site that is only frustrating for users with typical cognitive capabilities.

7.5 Implications for Design and Design Process

Given this inventory of barriers, what can be done to avoid them in designing accessible information technology? There is a parallel with design for usability, where it was recognized early on (Gould and Lewis, 1985) that design for usability has to include not only attention to particular design features, but also design processes that use actual data about effectiveness. Similarly, design for universal access has to incorporate design features that promote cognitive accessibility, but it also has to recognize that user testing is needed to ensure an adequate level of effectiveness. Because of the diversity of the audience to be addressed, there is the further need, perhaps not so prominent in design for

usability, to allow for *configurability* of interfaces so as to address different, and sometimes incompatible, needs of different users.

7.5.1 Interface Features

Many aspects of good user interface design have become generally accepted, such as using a two-stage protocol with confirmation for operations that cannot easily be reversed or providing automated or "wizard" support for complicated operations like installation. Good design for usability is important for people with cognitive disabilities as it is for other users. But there are specific interface features that have special importance for people with cognitive disabilities. These include:

- *Providing users enough time to read and use content* (see W3C, 2007, Guideline 2.2). As recognized in accessibility regulations and guidelines (for example, United States Access Board, n.d. or W3C, 2007), interfaces that require users to act or respond within a short time window are unacceptable because of the barriers they create for many people with disabilities.
- Any interface should include *a facility for looking up the meanings of unfamiliar words or phrases*, or it should be designed so as to work effectively with an assistive technology tool that provides that function (see W3C, 2007, Guideline 3.1).
- *Redundant, user-controlled modality of information* (Francik, 1999). Interfaces should support audio presentation of text, either in the interface itself or by the use of an appropriate tool. This is needed not only for users who cannot see, but also for users who can see the text, and can understand the words in the text, but cannot easily read them.
- *Spelling correction* should be provided for any spelling-sensitive interaction, for example issuing commands or entering search terms. Similarly, if an interface requires users to perform calculations, the interface should provide support (see Francik, 1999).
- *Compatibility with assistive technology.* As mentioned in connection with some of these features, users may need to use assistive technology tools, such as a screen reader. Therefore, it is very important that interfaces not include features that block the use of these tools.

Any interface should have all of the features just listed; but there are other features that are only conditionally useful, in that they are useful in some situations but are actually debilitating in other situations.

As discussed earlier, some users can process textual information relatively better than spatial information, while for others the opposite is true. Thus, for example, replacing a diagram by text in presenting a concept may help some users but harm others. No single design choice dominates for all users.

Early design guidelines for usability called for menus or screens that presented a limited number of choices, under the influence of an erroneous reading of George Miller's famous seven plus or minus two concept (LeCompte, 2000). Current design practice recognizes that wide interfaces, ones that present many choices on each screen, have substantial advantages, including requiring fewer interactions to complete a task, and lack of reliance on general descriptors for categories of choices, which are difficult to create (for early discussion, see Landauer and Nachbar, 1985). In keeping with this recognition, web pages now commonly offer tens or even hundreds of links.

While typical users are well served by this design, what about users for whom processing a complex display, or choosing among many confusing alternatives, is difficult? For them, a narrower interface that presents only a few choices at a time could be better.

As discussed in Lewis (2007), there is a tradeoff at work here. If a given number of alternatives are to be offered to users, an interface can present fewer on a given screen only by requiring users to process more screens, and hence take more steps. For some users, the difficulty of managing an interaction that requires more steps may outweigh the benefit, if any, associated with each screen being easier to process. Therefore no single design can be assumed to be superior for all users. Typical users are generally better served by wide interfaces, but some, though not all, users with cognitive limitations may benefit from narrower interfaces.

What about simply eliminating choices, so that a user confronts fewer choices *and* fewer screens? For some users, this is not helpful because they really need and use a very wide range of functions. But for other users, including users who simply do not need rarely used options, as well as users for whom processing options is difficult, this can be an excellent design direction.

Dealing with design features that are good for some users and bad for others requires configurability. Fortunately, software is sufficiently flexible that different users can have different user interfaces, chosen to meet their particular needs. Commercial software, like Microsoft Word, offers a wide range of configuration controls, including the ability to eliminate whole categories of functions from the user interface. For example, a user who never needs to include tables in documents can remove all table-related controls from the interface. The flexibility to eliminate inessential functions is a key requirement for cognitive accessibility (Francik, 1999).

While this kind of configurability offers benefits in principle, in practice problems remain. Configuring software like Word is not so easy, and few users know how to do it. Equally important, it is hard to determine what configuration would be suitable for a particular user. Making changes in the interface can make it harder to use documentation or help that refers to a standard configuration. If users have configured their personal copies of some software, this does not help them if they need to use a computer other than their own, for example in a public library. And there is the further difficulty that some important forms of configurability, for example shifting between textual and nontextual presentations, are currently beyond the reach of technology. More

work is needed on this whole complex of issues, as discussed in the following.

7.5.2 Design Process

Some important attributes of good design for people with cognitive disabilities cannot be reduced to interface features. For many interfaces, comprehensibility of text (e.g., labels on links on a web page) is crucial. But, as discussed earlier, there is no reliable way to predict that a given piece of text will or will not be comprehensible by a particular audience.

This means that the design process for comprehensible text has to include user testing for the intended audience, and it is helpful if people with cognitive disabilities are involved in creating the material, as recommended by Freyhoff et al. (1998). More generally, as has long been recognized for usability for typical users, many other aspects of an interface cannot reliably be evaluated without user testing, including whether users will notice needed controls, will interpret symbols as intended, or will realize that needed functions are possible in the interface. Design for cognitive accessibility has to include user testing.

Useful advice on user testing with people with disabilities, including people with cognitive disabilities, is becoming available (Henry, 2007). As discussed in the following, greater inclusion of people with disabilities in user studies must be a priority for the future.

While design for usability now commonly incorporates user testing, it also uses inspection methods, such as the cognitive walk-through or heuristic evaluation (Nielsen and Mack, 1994) that does not involve user testing. Inspection methods provide a structured way for designers to identify potential weak points in a design, often before the design is sufficiently realized to permit testing. Inspection methods that address accessibility are starting to become available (e.g., Paddison and Englefield, 2003), but these do not yet address cognitive accessibility. Material from Bohman (n.d. i, ii) and LD-Web (http://www.ld-web.org) could be used as the basis for an enhanced accessibility inspection that would give attention to cognitive issues.

At a higher level of process, the report by Freyhoff et al. (1998), already mentioned, describes a development process for cognitively accessible text that can be adapted as the skeleton for a design process for cognitive accessibility much more broadly. Here are the key steps, restated in terms of system design rather than document design:

- Decide on the aim of the system.
- Prepare a list of the key functions.
- Create a prototype based on the key functions.
- Check whether people with cognitive disabilities can use the prototype.
- Modify the prototype as needed and test again.

This development road map agrees well with recommended practice for design for usability, as described for example in Lewis and Rieman (1996). This convergence is not surprising, given the need for user testing in both usability and accessibility design.

7.6 Trends

Greater attention is being focused on the needs of people with cognitive disabilities. In part due to the advance of the self-advocacy movement (Dybwad and Bersani, 1996), in which people with cognitive disabilities speak out to assert their rights, people with cognitive disabilities are claiming a greater share of attention in the accessibility world. In 2007 the Web Accessibility Initiative of the World Wide Web Consortium, recognizing issues in the coverage of cognitive issues in its Web Content Accessibility Guidelines, carried out special consultation, leading to recognition of the need for future work directed at this area. Similarly, the Telecommunications, Electronic, and Information Technology Advisory Committee of the U.S. Access Board, the body responsible for accessibility regulations in the United States, has given special attention to cognitive issues. On the commercial side, Nokia included cognitive accessibility in a major review of accessibility developments.

This increased focus on cognitive accessibility will bring with it developments on a number of fronts. Because cognitive accessibility cannot be addressed only by attention to specific interface features, regulatory approaches may need to shift from a primary emphasis on checking interface features when assessing compliance to an approach that includes user testing. See Sampson-Wild (2007) for a discussion of this need in the context of web accessibility.

Technical developments to support interface configurability will accelerate. The Fluid project (http://fluidproject.org) is a major international effort to support interface configurability in community source projects in higher education. The intention is to promote the development of infrastructure for configurability, including both software structures that allow different user interface components to be selected for different users, and in allowing users to create and deploy profiles that capture their interface preferences in a portable way, available on the web, so that they can get an appropriate interface on any computer, not just their own.

Integration of accessibility, including cognitive accessibility, into design and development processes, is starting to take place. The Fluid project illustrates this trend as well. This will lead to the development of inspection methods that address cognitive accessibility. It will also lead to a very desirable increase in inclusion of people with disabilities, including cognitive disabilities, in design and development activities like focus groups for requirements development, and user tests. A positive feedback loop can accelerate these changes: as more designers and developers become more familiar with people with cognitive disabilities and their needs, barriers to inclusion are likely to diminish more rapidly.

A related trend, now in its early stages, is greater participation by self-advocates in regulatory and development processes affecting accessibility. In all spheres, technical, social, and political, participation by people with disabilities has led to great advances in accessibility, and the same can be expected from people with cognitive disabilities.

References

Anderson, J. R. (1983). *The Architecture of Cognition*. Cambridge, MA: Harvard University Press.

Anderson, J. R., Bothell, D., Byrne, M. D., Douglass, S., Lebiere, C., and Qin, Y. (2004). An integrated theory of the mind. *Psychological Review* 111: 1036–1060.

Bergman, M. M. (1998). Cognitive orthotics enhance the lives of users with cognitive deficits, in the *Proceedings of the Technology and Persons with Disabilities Conference (CSUN) 1998*, March 1998, Los Angeles. http://www.csun.edu/cod/conf/1998/proceedings/csun98_160.htm.

Bohman, P. (n.d. i). Cognitive disabilities part 1: We still know too little, and we do even less. WebAIM. http://www.webaim.org/articles/cognitive/cognitive_too_little.

Bohman, P. (n.d. ii). Cognitive disabilities part 2: Conceptualizing design considerations. WebAIM. http://www.webaim.org/articles/cognitive/conceptualize.

Brave, S., Nass, C., and Hutchinson, K. (2005). Computers that care: Investigating the effects of orientation of emotion exhibited by an embodied computer agent. *International Journal of Human-Computer Studies* 62: 161–178.

Budson, A. E. and Proce, B. H. (2005). Current concepts: Memory dysfunction. *New England Journal of Medicine* 352: 692–699.

Byrne, M. D. (1993). Using icons to find documents: Simplicity is critical, in the *Proceedings of the INTERACT '93 and CHI '93 Conference on Human Factors in Computing Systems (InterCHI-93)*, 24–29 April 1993, Amsterdam, The Netherlands, pp. 446–453. New York: ACM Press.

Carlin, M. T., Soraci, S., Goldman, A. L., and McIlvane, W. (1995). Visual search in unidimensional arrays: A comparison between subjects with and without mental retardation. *Intelligence* 21: 175–196.

Chan, R. C. K. (2001). Attentional deficits in patients with post-concussion symptoms: A componential perspective. *Brain Injury* 15: 71–94.

Danielsson, H. and Jonsson, B. (2001). Pictures as language, in the *Proceedings of the International Conference on Language and Visualization*, 8–9 November 2001, Stockholm, pp. 1–35. http://www.certec.lth.se/doc/picturesas/picturesas.pdf.

Danielsson, H. and Svensk, A. (2001). Digital pictures as cognitive assistance, in *Assistive Technology Research Series: Vol. 10, Assistive Technology: Added Value to the Equality of Life* (C. Marincek, C. Buhler, H. Knops, and R. Andrich, eds.), pp. 148–153. Amsterdam: IOS Press.

Davies, D. K., Stock, S. E., and Wehmeyer, M. L. (2002). Enhancing independent task performance for individuals with mental retardation through use of a handheld self-directed visual and audio prompting system. *Education and Training in Mental Retardation and Developmental Disabilities* 37: 209–218.

Duchaine, B. C., Yovel, G., Butterworth, E. J., and Nakayama, K. (2006). Prosopagnosia as an impairment to face-specific mechanisms: Elimination of the alternative hypotheses in a developmental case. *Cognitive Neuropsychology* 23: 714–747.

Dybwad, G. and Bersani, H. (eds.) (1996). *New Voices: Self-Advocacy by People with Disabilities*. Cambridge, MA: Brookline Books.

Fang, X., Shuang, X., Brzezinski, J., and Chan, S. (2006). A study of the feasibility and effectiveness of dual-modal information presentations. *International Journal of Human-Computer Interaction* 20: 3–17.

Francik, E. (1999). *Telecommunications Problems and Design Strategies for People with Cognitive Disabilities*. http://www.wid.org/archives/telecom.

Freyhoff, G., Hess, G., Kerr, L., Menzel, E., Tronback, B., and Van Der Veken, K. (1998). Make it simple: European guidelines for the production of easy-to-read information for people with learning disability. *Inclusion International*. http://digitalcommons.ilr.cornell.edu/gladnetcollect/270.

Friederici, A. D. and Lachman, T. (2002). From language to reading and disability: Cognitive functions and their neural basis, in *Basic Functions of Language, Reading, and Reading Disability* (E. Witruk, A. D. Friederici, and T. Lachmann, eds.), pp. 9–12. Norwell, MA: Kluwer.

Gajar, A. H. (1989). A computer analysis of written language variables and a comparison of compositions written by university students with and without learning disabilities. *Journal of Learning Disabilities* 22: 125–130.

Ghaldiali, E. (2004). Cognitive primer: Agnosia. *Advances in Clinical Neuroscience and Rehabilitation* 4: 18–20.

Glisky, E. (1992). Acquisition and transfer of declarative and procedural knowledge by memory-impaired patients: A computer data-entry task. *Neuropsychologia* 30: 899–910.

Gould, J. D. and Lewis, C. (1985). Designing for usability: Key principles and what designers think. *Communications of the ACM* 28: 300–311.

Greenaway, R. and Plaisted, K. (2005). Top-down attentional modulation in autism spectrum disorders is stimulus-specific. *Psychological Science* 16: 987–994.

GVU (1998). *9th WWW User Survey: Internet Shopping Part 2*. http://www.gvu.gatech.edu/user_surveys/survey-1998-04/reports/1998-04-Netshop2.html.

Hart, M. (2005). Autism/Excel study, in the *Proceedings of ASSETS 2005 (Seventh International ACM SIGACCESS Conference on Computers and Accessibility)*, 9–12 October 2005, Baltimore, pp. 136–141. New York: ACM Press.

Henry, S. (2007). *Just Ask: Integrating Accessibility throughout Design*. http://www.uiaccess.com/accessucd/index.html.

Horner, J. (2004). *Jack Horner: An Intellectual Autobiography*. http://mtprof.msun.edu/Spr2004/horner.html.

Ivory, M. Y., Sinha, R. R., and Hearst, M. A. (2000). Preliminary findings on quantitative measures for distinguishing highly rated information-centric Web pages, in the *Proceedings of the 6th Conference on Human Factors and the Web*, 19 June 2000, Austin, TX. http://webtango.berkeley.edu/papers/hfw00/hfw00/hfw00.html.

Jarrold, C., Baddeley, A. D., and Hewes, A. K. (1999). Genetically dissociated components of working memory: Evidence from Down's and Williams syndrome. *Neuropsychologia* 37: 637–651.

Jarrold, C., Baddeley, A. D., and Phillips, C. (2007). Long-term memory for verbal and visual information in Down syndrome and Williams syndrome: Performance on the doors and people test. *Cortex* 43: 233–247.

Kieras, D. E. and Bovair, S. (1984). The role of a mental model in learning to operate a device. *Cognitive Science* 8: 255–273.

Kitajima, M., Polson, P. G., and Blackmon, M. H. (2007). CoLiDeS and SNIF-ACT: Complementary models for searching and sensemaking on the Web. Paper presented at the *Annual Meeting of the Human Computer Interaction Consortium*, February 2007, Granby, CO. http://staff.aist.go.jp/kitajima.muneo/English/PAPERS(E)/HCIC2007.html.

Kitzing, P., Ahlsen, E., and Jonsson, B. (2005). Communication aids for people with aphasia. *Logopedics Phoniatrics Vocology* 30: 41–46.

Landauer, T. K. and Nachbar, D. W. (1985). Selection from alphabetic and numeric menu trees using a touch screen: breadth, depth, and width, in *Proceedings of the SIGCHI Conference on Human Factors in Computing Systems*, 14–18 April 1985, San Francisco, pp. 73–78. New York: ACM Press.

Law, C. and Vanderheiden, G. (2000). The development of a simple, low cost set of universal access features for electronic devices, in the *Proceedings of CUU'00*, 16–17 November 2000, Arlington, VA, pp. 118–123. New York: ACM.

LeCompte, D. C. (2000). 3.14159, 42, and 7+/–2: Three numbers that (should) have nothing to do with user interface design. *Internetworking*. http://www.internettg.org/newsletter/aug00/article_miller.html.

Lewis, C. and Mack, R. (1982). Learning to use a text processing system: Evidence from "thinking aloud" protocols, in *Proceedings of the 1982 Conference on Human Factors in Computing Systems*, 15–17 March 1982, Gaithersburg, MD, pp. 387–392. New York: ACM Press.

Lewis, C. and Rieman, J. (1996). *Task Centered User Interface Design: A Practical Guide*. http://www.hcibib.org/tcuid.

McGrenere, J., Davies, R., Findlater, L., Graf, P., Klawe, M., Moffatt, K., et al. (2003). Insights from the Aphasia Project: Designing technology for and with people who have aphasia, in the *Proceedings of the 2003 Conference on Universal Usability*, 10–11 November 2003, Vancouver, Canada, pp. 112–118. New York: ACM Press.

Microsoft (n.d.). *The Market for Accessible Technology: The Wide Range of Abilities and Its Impact on Computer Use*. http://www.microsoft.com/enable/research/phase1.aspx.

Milders, M. V. (1997). *Memory for People's Names in Closed Head Injured Patients*. Dissertation, University of Groningen. http://dissertations.ub.rug.nl/faculties/ppsw/1997/m.v.milders.

Newell, A. and Simon, H. A. (1972). *Human Problem Solving.* Englewood Cliffs, NJ: Prentice Hall.

Nielsen, J. and Mack, R. L. (eds.) (1994). *Usability Inspection Methods.* New York: John Wiley & Sons.

O'Riordan, M. A., Plaisted, K. C., Driver, J., and Baron-Cohen, S. (2001). Superior visual search in autism. *Journal of Experimental Psychology: Human Perception and Performance* 27: 719–730.

Pennington, B. F., Moon, J., Edgin, J., Stedron, J., and Nadel, L. (2003). The neuropsychology of Down syndrome: Evidence for hippocampal dysfunction. *Child Development* 74: 75–93.

Pirolli, P. (2005). Rational analyses of information foraging on the Web. *Cognitive Science* 29: 343–373.

Rapley, M. (2004). *The Social Construction of Intellectual Disability.* Cambridge, UK: Cambridge University Press.

Redish, J. (2000). Readability formulas have even more limitations than Klare discusses. *ACM Journal of Computer Documentation* 24: 132–137.

Roth, W. (1983). Handicap as a social construct. *Society* 20: 56–61.

Small, J., Schallau, P., Brown, K., and Appleyard, R. (2005). Web accessibility for people with cognitive disabilities, in *CHI '05 Extended Abstracts on Human Factors in Computing Systems*, 2–7 April 2005, Portland, Oregon, pp. 1793–1796. New York: ACM Press.

Stephenson, J. and Linfoot, K. (1996). Pictures as communication symbols for students with sever intellectual disability. *AAC Augmentative and Alternative Communication* 12: 244–255.

United States Access Board (n.d.). *Telecommunications Act Accessibility Guidelines.* http://www.access-board.gov/telecomm/rule.htm.

United States Equal Employment Opportunity Commission (2005). *Annual Report on the Federal Work Force: Fiscal Year 2004.* Table 6. http://www.eeoc.gov/federal/fsp2004/aed/table6.html.

West, T. G. (2000). The abilities of those with reading disabilities: Focusing on the talents of people with dyslexia. *LD Online.* http://www.ldonline.org/article/5867.

World Health Organization (n.d.). *International Classification of Functioning, Disability, and Health.* http://www.who.int/classifications/icf/en.

W3C (2007). *Web Content Accessibility Guidelines 2.0.* http://www.w3.org/TR/WCAG20.

8

Age-Related Differences in the Interface Design Process

Sri Kurniawan

8.1 Introduction

There are various pointers, such as government and United Nations statistics, that suggest the world is graying at a noticeable rate, especially in the developed countries. By 2020, the number of the world's older population is expected to exceed 1 billion (Spiezle, 2001). In the United Kingdom, it is predicted that by 2020, there will be 10.5 million people aged under 16 years, 27.6 million aged 16 to 49 years, and 25.2 million over 50 years compared with 11.4 million, 28.8 million, and 19.3 million in 1999, respectively (Department of Trade and Industry, 2000). About 70% of the Western world will live past 65 years and 30% to 40% past 80 years (Stuart-Hamilton, 2000). For example, in North America, at the beginning of the 20th century the average life expectancy was just 46 years, increasing to over 76 years in recent times (Spiezle, 2001). From these figures, it can clearly be evidenced that the industrialized society is becoming more of a rectangular society (meaning that nearly all people survive to old age and then die abruptly over a narrow age range at around 85 years; Fries and Crapo, 1981). Where previously the design considerations of older users have been overlooked (Tetley et al., 2000), it is now impossible for designers to continue to ignore this growing population for various reasons, including economic ones. Older persons are increasingly becoming the dominant group of customers of a variety of products and services (both in terms of number and buying power).

Although older persons should not be categorized as people with disabilities, the natural aging process carries some degenerative ability changes, which can include diminished vision, varying degrees of hearing loss, psychomotor impairments, as well as reduced attention, memory, and learning abilities. All of these changes play a role in the differences between the way older and younger persons use information and communication technology (ICT), which has to be accommodated to ensure that older persons are not disadvantaged when using ICT. This can only be realized with an understanding of the changes associated with aging.

The term *older persons* has been defined in numerous ways. A variety of research defined older people as over 40, over 50, or over 58 (Bailey, 2002). A number of gerontologists stated that 60 or 65 years of age denotes the threshold age or the onset of "old age," because at this age a number of psychological and physical changes become noticeable (Bromley, 1988). Other academics divided older persons into four groups: young old (60 to 69 years old), middle-aged old (70 to 79), old old (80 to 89) and very old old (90 years and older) (Burnside et al., 1979). Some studies referred to baby boomers as older persons, because they are entering retirement at present and as such they experience both functional decline as well as changes in lifestyle and activities (Bergel et al., 2005), which might affect their interaction with new technologies.

Older persons are also arguably the group with the widest range of characteristics. Gregor et al. (2002) captured the individual variability of older persons by categorizing this user group into three subgroups:

1. Fit older people who do not consider themselves nor appear disabled, but whose needs have changed with age.
2. Frail older people who have suffered a reduction in many of their functionalities, at least one of which may be considered a disability.
3. Disabled people who have aged, whose disabilities have affected their aging process making them dependent on other faculties that may also be declining.

One thing that is apparent, however, is that the individual variability of sensory, physical, and cognitive functioning increases with age (Myatt et al., 2000) and each of these functions will decline at different rates (Gregor et al., 2002), depending on several factors, most notably socioeconomic status and living condition. The changes associated with old age have been documented in various places, one of the recent ones being a book by Duncan et al., (2005) and the special issue of the international journal *Universal Access in the Information Society* on "Web and Aging: Challenges and Opportunities" (Zaphiris et al., 2005).

8.2 Aging and the Perceptual System

Perceptual systems consist of visual and auditory systems. Studies on age-related differences in perceptual systems found that older people experience more problems processing and learning new perceptual information than their younger counterparts do. Older persons have the ability to improve over time when learning new perceptual information, but they consistently perform well below the level of their younger counterparts (Kausler, 1994). There is some generalizable evidence that perceptual processing is also slower in older adults than it is in younger adults (Kline and Scialfa, 1996). These differences are caused by age-related changes in both the visual and the auditory system.

8.2.1 Visual System

The most common physiological changes associated with aging concern vision. There are 2 million people with poor or no vision in the United Kingdom and 90% are over 60 years of age (Royal National Institute for the Blind, 2003). Fozard (1990) suggests that problems with vision tend to appear in the early forties. At this age people have a decline in visual acuity (ability to see fine detail). One cause of this visual acuity decline is the yellowing of the lens due to discoloration of the eye's fluid. This gives the impression of looking through a yellow filter (Spector, 1982). This yellow filter may also cause higher sensitivity to glare, a condition known as night blindness (Kline and Scialfa, 1996) and reduced sensitivity to color, especially in the blue-green range (Helve and Krause, 1972). These make it difficult for older people to tell the difference between colors of a similar hue and color combination with low contrast. Older persons with color blindness normally have worsened condition with age, due to decreased blood supply to the retina (AgeLight, 2001).

At around the age of 50, older persons begin to notice difficulty in adjusting focus for near vision. This is because with age the lens becomes thicker, flatter, and less flexible. Older persons usually experience a significant decline in contrast sensitivity (the ability of individuals to detect differences in illumination levels) (Owsley et al., 1983) due to the shrinking of the pupil of the eye. The pupil is less able to change diameter, therefore letting in less light. The retina of an average 60 year old receives just 33% of the light of the retina of the average 20 year old (Armstrong et al., 1991).

At around 60 years of age, older adults may show a reduction in the width of the visual field, a condition referred to as tunnel vision at the more severe state (Cerella, 1985), a reduced ability to detect flicker, particularly in the peripheral visual field (Casson et al., 1993), and problems with persistence (the sensation of continued presence of the stimulus after presentation of the stimulus has ceased) (McFarland et al., 1958). Older persons also appear to have a decline in processing visual information (Kline and Szafran, 1975; Fozard, 1990). The ability to recognize figures that are embedded within other figures is reduced (Capitani et al., 1988); there is a decline in the ability to recognize objects that are fragmented or incomplete (although some researchers argued that this can also be caused by reduction in problem-solving ability; Salthouse and Prill, 1988; Frazier and Hoyer 1992); and there is a decline in the ability to locate a target figure in a field of distracters (Plude and Hoyer, 1986; Ellis et al., 1996; Hess et al., 1999), due to reduced ability to suppress these distracters from the focus of attention.

It should be noted that if the target location is constant there is little or no difference due to aging, as older persons would eventually learn to suppress these distracters (Farkaas and Hoyer, 1980; Carlson et al., 1995). Older people appear to benefit more than younger people when presented with advance cues indicating the future location of a visual search target (Kline and Scialfa, 1996). However, older people appear to learn visual searches at the level of the specific targets presented and unlike young people they do not show transfer of learning to new searches where the specific examples have changed but the categories have not (Fisk et al., 1997).

Aging eyes are very susceptible to fatigue and tend to be dry due to a decrease in the amount of blinking. Some older persons also experience reduced field of vision, especially in the periphery of their eyes (Hawthorn, 2000), and diminished ability to view objects clearly from a distance (Stuart-Hamilton, 2000). In addition to experiencing long and short sightedness, one-third of people 65 years of age and older will have a disease affecting their vision (Stuart-Hamilton, 2000) such as age-related macular degeneration (AMD), cataracts, glaucoma, and diabetic retinopathy.

8.2.1.1 Age-Related Macular Degeneration

AMD, sometimes known as senile maculopathy, is a genetic disease and the most common cause of severe visual impairment among older people (Ford, 1993). Macular disease refers to the breakdown or thinning of the most sensitive cells of the eye clustered in small area in the center of the retina known as the macula (Zarbin, 1998).

Macular disease affects central vision only; sufferers still can see adequately at the peripherals of their vision, a term commonly described as polo mint vision due to the hole in the center of their vision (Ford, 1993). While never resulting in total blindness, AMD is often severe enough for the sufferer to be classed as partially sighted or blind. Symptoms of macular disease usually start around the early to midfifties, typically starting in just one eye. In early stages of macular degeneration, it is difficult to read small or faint print, but as the disease worsens and spreads to both eyes, it becomes difficult even to read large print or to determine any specific details such as pictures. Fortunately, recent advances in medical treatment such as gene therapy might provide some help for AMD sufferers (Telegraph, 2002).

8.2.1.2 Cataracts

Cataract refers to the loss of transparency, or clouding, of the lens of the eye and is predominantly an age-related disease (Spector, 1982). The lens is responsible for focusing light coming into the eye onto the retina to produce clear, sharp images. However, when the lens of the eye becomes clouded, the eye is no longer able to adequately process light coming into the eye.

Cataracts are the most common cause of vision loss in people over 55 years of age (St. Lukes, 2005). Cataracts are caused by an accumulation of dead cells within the lens. As the lens is within a sealed capsule within the eye, dead cells have no way to get out and therefore accumulate over time, causing a gradual clouding of the lens.

The clouding of the lens means that less violet light enters or can reach the retina, making it harder to see colors like blue, green, and violet compared with reds, oranges, and yellows (AgeLight, 2001).

8.2.1.3 Presbyopia

Presbyopia is an age-related disorder where the eyes lose the ability to focus on objects or detail at close distances. The onset of presbyopia normally starts in the forties but is a disorder that happens to all people at some time in their lives if they live long enough (Lee and Bailey, 2005). Despite its symptoms, presbyopia is not related to nearsightedness, which is due to an abnormality in the shape of the eye. Instead, it is caused by the gradual lack of flexibility in the crystalline lens of the eye due to the natural aging process (St. Lukes, 2005). It is not a disease and cannot be avoided; however, it can easily be treated with lenses or eye surgery. People with presbyopia usually have a diminished visual field and tend to compensate for this by moving their head from side to side when reading instead of sweeping their eyes from left to right.

8.2.1.4 Glaucoma

Glaucoma is a group of diseases that can damage the optic nerve and cause blindness. While not a direct age-related disorder, it most commonly affects people over 60 or African Americans over 40 years old. Symptoms include loss of peripheral vision, starting with detail and increasing until the sufferers have a form of tunnel vision where they gradually lose all of their peripheral vision. If left untreated, this tunnel vision will continue to move inward until no vision remains.

While there are various causes of glaucoma, the most common is an open-angle glaucoma where fluid builds up in the anterior chamber of the eye causing pressure that damages the optic nerve (National Eye Institute, 2004). As with presbyopia, the sufferer has a decreased angle of vision and so must turn the head to view what a normal person could view in the peripheral vision.

8.2.2 Auditory System

Hearing also declines with age, and figures have indicated that 20% of people between 45 and 54 years of age have some hearing impairment, which rises to 75% for persons between 75 and 79 years of age (Fozard, 1990; Kline and Scialfa, 1996). In older adults there is a reduced ability to detect tones over all frequencies (Rockstein and Sussman, 1979; Schieber, 1992) but especially high-pitched sounds (Schieber, 1992). They are, therefore, often missing attention-getting sounds with peaks over 2500 Hz, making them susceptible to not noticing the sound of a fire alarm (Berkowitz and Casali, 1990; Huey et al., 1994).

Older persons show a loss in ability to localize sound, which is more apparent in persons with presbycusis (Kline and Scialfa, 1996), and a reduced ability to follow conversations in noisy surroundings, which can easily distract the older person (Hawthorn, 2000). By the age of 80 they may miss 25% of the words in a conversation (Feldman and Reger, 1967), although there is evidence that older persons can develop strategies to cope with missing words.

Older people require louder sound to hear comfortably. In a study, Coren (1994) instructed participants to listen to speech sounds and to indicate the level they preferred for listening. This experiment showed a huge difference in hearing comfort level for younger and older participants (e.g., participants of 25 years of age had median hearing comfort level at 57 decibels, whereas participants of 75 years of age had median hearing comfort level at 79 decibels).

8.3 Aging and the Cognitive System

Research related to modeling age-related slowing effects in cognitive processing has followed two streams of conflicting theories over the nature of slowing (Sliwinski and Hall, 1998). General slowing theories state that older adults' slowing search time is due to some molar process, and is independent of task-specific cognitive requirements (Groth and Allen, 2000). In other words, this stream of theories suggests that as people age, there is a general overall slowing of cognitive processing speed. The process-specific slowing theories state that age results in differential rates of slowing, depending on the task or processing domain. The larger impact seems to be with tasks that require the most cognitive processing (working memory, overall attention capacity and visual search performance). Age effects are smallest for tasks where knowledge is an important aspect of the task and largest for tasks where successful performance is primarily dependent on processing speed (Sharit and Czaja, 1994).

There is general agreement in the literature on cognitive aging that memory performance declines from early to late adulthood, and that such age-related losses in performance are much greater in relation to some tasks than in others (Grady and Craik, 2000). Studies on age differences in the cognitive system suggest that there is a decline in intellectual performance (Zajicek, 2001) and a decline in the ability to process items from long-term memory into short-term memory with aging.

Age difference in short-term memory's proficiency was thought to result from an age-related capacity change. In principle, the amount of information that can be held in short-term memory before being forgotten is less for older adults than for younger adults, although some researchers suggested that long-term memory processes might contribute to this difference (Kausler, 1994). Botwinick and Storandt (1974) found that the number of items that can be held in short-term memory averages around 6.5 items for people in their twenties to people in their fifties, but this number drops to around 5.5 for people in their sixties and seventies.

However, tests of working memory show that there is a stronger decline in the ability to process items in short-term memory as distinct from simply recalling them (Dobbs and Rule, 1990; Salthouse, 1994). Light (1990) suggested that working memory decline underlies older people's problems in text comprehension. Processing of visual information in short-term memory also slows with age (Hoyer and Rybash, 1992). It has been shown that older adults tend not to adopt strategies for organizing material to be more easily remembered unless prompted to do so (Ratner et al., 1987). Older adults appear to perform worse on spatial memory tasks (Denny et al., 1992; Cherry et al., 1993) and tend to have poorer memory for nonverbal items such as faces (Crooke and Larrabee, 1992), or map routes (Lipman and Caplan, 1992).

With long-term memory, studies have found there is a decline in episodic memory (memory for specific events) and procedural memory (memory for how we carry out tasks) (Hawthorn, 2000). Memory is particularly relevant to learning, in that to learn one must acquire the information and retain it in memory. Research shows that older adults retain skill levels in areas of expertise they have learned, although it becomes more difficult to learn a new motor skill (Cunningham and Brookbank, 1988) and more demanding to learn new complex tasks, particularly where the tasks are not meaningful to the user (Stokes, 1992). Older adults also experience a significant decline in capability on performance of memory tasks that require recall of content; however, there is little decline on memory tasks involving recognition (Rybash et al., 1995). Research also suggests that older adults tend not to adopt organizing material strategies unless informed to do so (Ratner et al., 1987), which could also suggest why older adults have poorer learning skills compared to younger adults.

The ability to form new automated responses, which is the ability to respond to stimuli automatically without conscious effort or control, particularly in visual searches, becomes more difficult (Hawthorn, 2000), and while older adults are able to learn new responses, they continue to remain attention demanding and hence contribute to cognitive load (Rogers et al., 1994). Where automated responses have been learned in older adults, these can become disruptive when learning new tasks, because it is difficult to unlearn responses where the person is unconscious of the response (Rogers et al., 1994; Hawthorn, 2000). However, some other studies pointed out that in tasks requiring complex cognitive activity (such as information search on the web), older adults performed equally well as their younger counterparts, although they took more time to finish the tasks (Kubeck et al., 1999).

Attention is the ability to focus and remember items in the face of distracting stimuli being presented, which may have to be processed simultaneously mentally (Stuart-Hamilton, 2000). Older persons experience more difficulty trying to focus and maintain attention on activities over long periods of time, or require quick and continuous scanning, which is particularly fatiguing for the user (Vercruyssen, 1996), especially during activities that require concentration on a specific task in light of distracting information (Kotary and Hoyer, 1995). Selective attention in older adults therefore becomes more difficult with age. Divided attention is the ability to attend simultaneously to and process more than one task at the same time (Hawthorn, 2000; Stuart-Hamilton, 2000). Researchers have reported that the ability to sustain divided attention in the performance of tasks declines with age, particularly in complex tasks (Salthouse and Somberg, 1982; Plude and Hoyer, 1986; McDowd and Craik, 1988; Hartley, 1992). The ability to form new automated responses, which is the ability to respond to stimuli automatically without conscious effort or control, particularly in visual searches becomes more difficult (Hawthorn, 2000), and while older adults are able to learn new responses, they continue to remain attention demanding and hence contribute to cognitive load (Rogers et al., 1994).

8.4 Aging and the Motor System

Current knowledge regarding the effect of aging on timing control in movement is largely based on research paradigms that have used reactive and/or speeded tasks conceived and analyzed within an information-processing framework (Greene and Williams, 1993). The slowing of motor skills with increasing age has been well documented. The majority of studies have used reaction time (RT; Lupinacci et al., 1993) and movement time (MT; Smith et al., 1999) as measures of performance.

The RT interval is generally regarded as a reflection of the time required for cognitive or central processing. There are conflicting conclusions from different studies on the locus of decrement of age-related slowing or impairment of RT, ranging from the response encoding processes, the central nervous system in general and perceptual-motor processes (i.e., stimulus encoding, memory set size, speed/accuracy trade-off), to external factors such as task complexity (Goggin and Stelmach, 1990).

Age-related slowing in MT was thought to be caused by task complexity and speed-accuracy trade-off. It was argued that elevated motor noise caused older adults to perform repetition to increase accuracy, which resulted in longer overall task completion time (Liao et al., 1995). However, other researchers raised the question of whether the slowing in older adults was to achieve higher accuracy or whether accuracy was the result of slower cognitive motor abilities (Goggin and Stelmach, 1990).

In studies with older adults, response times increase significantly with more complex motor tasks (Spiriduso, 1995) or in tasks with a larger number of choices (Hawthorn, 2000). Older adults perform poorly when tracking a target using a mouse (Jagacinski et al., 1995), make more submovements when using a mouse (Walker et al., 1997), and experience an increase in cursor-positioning problems if the target size is small such as the size of letters or spaces in text (Charness and Bosman, 1990). Siedler and Stelmach (1996) also reported that older adults have "less ability to control and modulate the forces they apply."

Some older people suffer from age-related diseases that affect their motor abilities, such as multiple sclerosis, arthritis, osteoporosis, stroke, and Parkinson's disease. Multiple sclerosis (MS) is a disorder of the central nervous system marked by weakness, numbness, a loss of muscle coordination, and problems with vision, speech, and bladder control. Arthritis is inflammation of joints causing pain, swelling, and stiffness. Osteoporosis is loss of normal bone density, mass, and strength, leading to increased porousness and vulnerability to fracture. Stroke refers to damage to the brain caused by interruption to its blood supply or leakage of blood outside of vessel walls. Depending upon where the brain is affected and the extent of the decreased blood supply to the brain, paralysis, weakness, a speech defect, aphasia, or death may occur. Finally, Parkinson's disease is a progressive disorder of the nervous system marked by muscle tremors, muscle rigidity, decreased mobility, stooped posture, slow voluntary movements, and a masklike facial expression. As the symptoms indicate, any of these diseases can severely affect older persons' motor abilities.

8.5 Aging and Expertise

Expertise was thought to either reduce the age differences in performance or to benefit older and younger people equally, depending on the type and difficulty of the task (Morrow et al., 1992, 1994). It was suggested that the expertise differences may reflect decreased demands on working memory capacity, and age declines may reflect reduced capacity. However, a different study suggested that although spatial visualization was a good predictor of mnemonic performance, age-related decrements

were attenuated but not eliminated by the advantages of expertise (Lindenberger et al., 1992).

There are mixed results regarding age and expertise effects on perceptual-motor skills. While Salthouse and Somberg (1982) found nearly equal improvements for young and old adults by facilitating expertise (through training), Smith et al. (as quoted in Salthouse, 1989) found that young adults acquired mnemonic skills more easily.

8.6 Behavioral Changes in Older People

There are some notable behavioral changes associated with advanced aging. The first notable change is increased cautiousness (hesitancy about making responses that may be incorrect) (Salthouse, 1991). This causes older adults to take more time to perform tasks. Speed-accuracy trade-off analysis reveals that older people do not always make more errors than younger people; they just take longer to finish a task. One most commonly cited explanation for this change is the decline in speed across a variety of situations (Birren, 1970). An older person has longer reaction times, and it has been suggested that this is caused by inefficient central nervous system (CNS) functioning. Indeed, the CNS is also at the root of sensory and perceptual changes that occur with age. To cope with these changes, older adults modify their behavior and attempt to compensate, resulting in, among others, increased cautiousness and a lack of confidence. Providing assurance to an older user that the user is in the right (or wrong) path to their information target can alleviate the lack of confidence in older persons. One piece of good news is that the influence of learned habits on behavior is unchanged with age (Grady and Craik, 2000).

8.7 Older Persons as Interactive Technology Users

People often underestimate the interest of older people on technology. It is undeniable that aging-related functional decline has some impact on their use of technology. In addition, the lack of exposure to more advanced features, even if it is merely due to their decision not to learn or adopt new technology, means that they are less up-to-date with the ever-changing modern technology (Kurniawan, 2006). However, various studies on older persons and technology show that after older persons are aware of the benefit of technology, they are keen to learn and are willing to adopt new technology. There is a long list of technology that older people around the world use on a regular basis. In this chapter, however, only two are discussed: computer and the Internet, and mobile phones, for a specific reason. These are the two everyday technologies that potentially benefit older persons the most.

8.7.1 Computer and Internet Use

Coulson (2000) asserts that computer use among older adults has been increasing by about 15% every year since 1990, and the

U.S. Department of Commerce (2000) asserts that being digitally connected has become a requirement to function in this information-rich society.

Research suggests that regular Internet and e-mail access can have a positive effect on the quality of life and well-being of older adults because computer use can provide a virtual social network for older persons with restricted mobility and thereby can potentially reduce their sense of isolation, resulting in lower negative emotion. McConatha et al. (1994) researched the effects of providing 14 Philadelphia nursing home residents between the ages of 59 and 89 years with personal computers and access to the Internet. After 6 months of using the computer and Internet facilities, the researchers found that on retesting, the group scored significantly lower on depression and significantly higher on cognitive ability with the overall mean score for cognitive ability rising by 14%. They concluded that "using the Internet allowed the elderly participants to keep their minds active and help combat depression … (and that these) … health benefits became apparent almost immediately" (p. 240).

Czaja et al. (1993) also found that many of the restrictions and social isolation problems that have been found to face elderly people either in their own homes or retirement complexes on a daily basis could be alleviated through the use of computers and online technology. The study by White et al. (1996) on the Internet and e-mail use in a retirement community in the United States also supports these findings. Lawhorn et al. (1996) suggest that e-mail and Internet use among older people encourages socialization and a sharing of experiences among their fellow peer groups. Similarly Furlong (1989), investigating whether communicating with others through SeniorNet, a computer network, could improve the lives of older persons, found that it functioned as an effective emotional support system, especially for nursing home residents.

From this research it can clearly be seen that Internet access is particularly important for older users, as it can provide social support and feelings of well-being. However, aging-related impairment can pose barriers for older persons to learn and master the use of computers and the Internet. In addition, because many of the current cohort of older persons did not grow up with computers, it is difficult for them to transfer their existing knowledge of other technology to master the Internet. Therefore, Cody et al. (1999) asserts the importance of providing computer skills to the older population. They continue to assert that providing Internet access to this group

> provides a number of significant benefits, including the ability to enroll in distance learning courses on-line for life-long education, increased knowledge of news, current events and medical/health breakthroughs, increased connectivity with family members who may live far away, increased intergenerational communication, increased perceptions of social support and the ability to feel mentally alert, challenged, useful and to feel "younger." (Furlong, 1989, pp. 269–270)

In making the Internet accessible to older users, web designers are opening the doors to provide many new opportunities to this group as well as people with disabilities. A U.K. government department has produced a report advocating that the quality of life of older people can be improved by harnessing the Internet. They state this can occur if products and services are inclusively designed to ensure they can be easily used by the increased number of older adults, many of whom may have some functional impairment. The report also suggests that by compensating for these impairments, the Internet can potentially enable older adults to maintain an independent living and social interaction (Department of Trade and Industry, 2000).

In summary, as well as the emotional and psychological support, the Internet can provide independence to this age group, which Zajicek (2004) states is "very important" for over 80% of the older adult population, along with living in their own homes. It allows users to complete activities such as shopping online, banking, and research on health matters, which would otherwise be more difficult, particularly for the less mobile users within this group.

There is another reason why facilitating the Internet for older people should be a priority for many companies. The *Wall Street Journal* reported that in 2005 baby boomers accounted for 42% of all U.S. households and controlled 50% of all consumer spending; as such they can be an important e-commerce group (Greene, 2005). A Department of Trade and Industry report supports this finding, which asserted in relation to elderly persons using information communication technology that businesses do not appear "to be aware of the potential market opportunities, or if they are aware to feel that it is worthy of attention" (Department of Trade and Industry, 2000). Keates et al. (2000) also support this claim although they commented that the majority of the industry still continues to produce products primarily aimed at the younger population. Finally, Stephanidis et al. (1999) advocate that by creating accessible web sites, organizations are promoting socially responsible behavior, which apart from providing good publicity for the organization, will also enable the organization to be perceived as an employer of choice and therefore will attract a more diverse workforce.

Contrary to popular stereotypes, research has shown that seniors are able to learn the use of computers and enjoy doing so (Furlong, 1989; Czaja et al., 1993; Lawhorn et al., 1996). Web site designers and organizations should therefore make a conscious effort not to preclude this section of the population when designing their web sites. One of the myths that must be overcome is that elderly people hold more negative attitudes toward computers and therefore are less likely to use them. In a study conducted by Czaja and Sharit (1998), consisting of 384 older users completing three real-world tasks over 3 days, it was found that there were no age-related differences in attitudes, although there were age-related effects for the dimensions of control, efficacy, and control over systems. Older people are less likely to have had long histories of computer use as their work careers would have been over or at least in the last stages when the Internet was taking off in the late 1990s (U.S. Department of Commerce, 2000).

Additionally, formal education of anyone over 65 years of age in the year 2000 would have taken place before mainframe computers were even common, and therefore it is understandable that seniors will have had less experience than younger users on the Internet (U.S. Department of Commerce, 2000). Holt and Morrell (2002) have found that older adults are increasingly surfing the web with growing confidence, which is destroying the myths about their reluctance to use computers, even though they are 50% less likely to own a computer within their own homes compared with their younger counterparts.

Czaja and Sharit's study (1998) found that the nature of computer experience has a direct impact on the attitude change for the use of a computer system. An ICM survey conducted in 2002 also revealed that all senior users in their study that gave information technology (IT) a chance became "hooked" and two-thirds of IT users in the 55 years and older age group agreed that the Internet has had a positive impact on their lives (ICM, 2002). Other studies support this and show that older people's attitude toward computers becomes more positive when they have had a positive experience of using them (Jay and Willis, 1992; Marquie et al., 1994; Baldi, 1997). One study additionally found that older people that were more familiar with computers experienced less apprehension in using computers, but also felt less "left out" in an increasingly technological world (Morris, 1994).

From the previous discussion it is clear that older persons can benefit immensely by using computers and the Internet, and, contrary to negative perceptions, older adults are receptive to computers if given the opportunity to learn and use them. This in turn will assist older persons to continue to live their lives independently and carry out activities and tasks essential for an acceptable quality of life (Czaja, 1997).

According to SeniorWatch (2003), some 49 million older people across Europe have used a computer at least once in their lives, and around 27 million currently live in a household with Internet access. Similar results have been found in surveys conducted in the United Kingdom. The so-called silver surfers now represent 12% of Internet users in the United Kingdom with 37% of 60 to 64 year olds now online at home (Guardian Newspaper, 2003). The survey also revealed a higher level of computer ownership (50%) among the 60 to 64 age group than among the 18 to 30 year olds (46%) (Guardian Newspaper, 2003).

Older persons increasingly use various new communication media such as e-mail, computer-mediated communication (CMC), and blogs, with more and more enjoying being a part of virtual communities (Kanayama, 2003). It is worth mentioning that a U.S. study in 2005 found that there is a considerable percentage of older Internet users who read blogs. Table 8.1 provides the breakdown of the statistics of these blog readers.

One older blogger who created a wave recently is Peter, a 79-year-old English pensioner living in Leicester under the user name of Geriatric1927 on the web site YouTube. After posting his first two introductory videos, Peter began a series that he called "Telling It All," which aims to provide an autobiographical account for other YouTubers covering things such as his experiences during World War II, his love for motorcycles, and family matters. At the time of writing, in total his series of autobiographical videos had been viewed more than 2 million times and he had been named by YouTube the second most subscribed channel of all time (27,624 subscribers). He received an average of 834 text responses and 16 video responses per video blog posted, except his first video (with 163 video responses). The popularity of Peter's video-blogs was soon noticed by the media. Both British and American major media networks (e.g., the *Guardian* newspaper in the United Kingdom) reported the novelty of having an older person using new media technologies and establishing bridges with the younger generations. A dedicated Wikipedia page has even been created about Geriatric1927.

8.7.2 Mobile Phone Use

Statistics show that older persons' ownership of mobile phones has increased significantly in recent years. The Office of National Statistics reported that between the years 2001 and 2003, the largest increase of the number of people owning or using a mobile phone occurred among the older age groups, with the proportion of people ages 75 and over with a mobile phone nearly doubling. It was also recorded that 53% of people ages 65 to 74 owned a mobile phone in 2003 (National Statistics, 2003). However, mobile phone users over the age of 60 years usually avoid using more complex functions and only use mobile phones for very limited purposes, such as for calling or texting in emergency situations (Coates, 2001). Users of this age group were reported to experience problems related to displays that are too small and difficult to see; buttons and characters that are too small, causing them to push wrong numbers frequently; too many functions; non-user-friendly menu arrangement; unclear instructions on how to find and use a certain function; and services that are too expensive (Kurniawan et al., 2006). The Help the Aged society identified that the "mobile phone is not too complicated to use compared to the Internet and other modern technology that the younger generation is more used to using" (BBC Wales News, 2003), so it should arguably be feasible for

TABLE 8.1   Blog Readers and Internet Users: Statistics by Age

Ages	Internet Users (% of Population)	Internet Users Who Read Blogs (% of Population)	Blog Readers (% of Population)
18–29	91	44	40
30–49	88	37	33
50–64	75	34	26
65+	33	28	9

older users to use mobile phones effectively through some sort of training intervention.

One thing to note is that older persons are well informed on advanced functions. A study on the use of mobile phones by older women shows that they are keen to understand, enthusiastic to learn, and are quite well informed about some advanced features of mobile phones such as MMS (multimedia messaging services). Terms such as *network provider, roaming,* and *satellite communication* were discussed, and the participants showed evidence of some degree of understanding of these terms, either through their interaction with the younger generation (mostly grandchildren) or their own experience (e.g., holidaying abroad). When asked to suggest features that should be implemented in an ideal phone, these older women came up with features that are quite advanced, such as a button to place phone numbers or calls into a blacklist or a hardwired panic button (Kurniawan, 2006).

Abascal and Civit (2001) stated that mobile communication technology for older persons has the main aims to support personal communication, provide a sense of security, act as a means for social integration (especially in remote areas that are unreachable by landline telephones), and enhance older persons' autonomy. However, the authors warned that this technology can also trigger social isolation (mobile communication encourages less face-to-face interaction), loss of privacy (users may feel that they can no longer have quiet time for themselves because they can be contacted anywhere and anytime), and economical problems (mobile phones and services are still costly, especially for those with small pensions).

Mikkonen et al. (2002) performed a study asking older persons to come up with innovative and out-of-the-box ideas on the use of mobile technology. Participants were very detailed and focal when talking about the characteristics of mobile technology that would benefit them. They suggested that the main functions of mobile technology are for maintaining and developing social relationships and providing health and security services. However, most participants were only willing to pay a small amount of money, around €4 to €10, for even the most innovative services (e.g., one-stop help center or security bracelet that can send calls for help) and most said that they preferred to start those innovative services "sometime in the future" rather than immediately.

Melenhorst et al. (2001) used focus group discussions to investigate perceived context-related benefits of mobile phones for older persons. The focus group stated that the three main benefits of mobile phones were to keep in touch (especially with someone emotionally close who lives more than half an hour away), to set time for a leisure activity with friends, and to share exciting good news immediately.

However, just like other studies on age-related differences in the use of interactive technology had found, there are differences in the patterns of use of mobile phones between older and younger persons. Ziefle and Bay's study (2005) shows that older adults' mental model of mobile phone menus was mostly hierarchical, although occasionally it was linear. When asked to organize existing functions, older users tended to create shallower menus and allocated fewer functions to the appropriate superordinate functions than younger users did.

8.8 Conclusion

This chapter has discussed the changes that occur with aging. Although it is apparent that most functional abilities decline with aging, not all are doom and gloom. Some abilities (e.g., those related to semantic memory) do not decline until very late in life. In addition, various studies pointed out that older persons are able to learn new skills as well as their younger counterparts and are able to perform some tasks equally well as younger persons do. In addition, the influence of learned habits on behavior is unchanged with aging.

Older persons are arguably the fastest-growing segment of potential customers of new technology and as such it would be economically wise for designers of this technology to consider the impairments that come with aging and how to facilitate effective interaction given these limitations. As noted earlier, older users may be less confident when it comes to using technology, even when they arrive from a generation that has grown up with computers, and they are likely to be more nervous about personalization that reflects their developing needs if that involves making changes themselves. The key message is that configuration needs to be simple and applied in such a way that users can see the effect of personalization immediately. Another obvious solution is to apply universal design principles to enable users with a wide range of characteristics to use the products effectively.

References

Abascal, J. and Civit, A. (2001). Universal access to mobile telephony as a way to enhance the autonomy of elderly people, in the *Proceedings of the 2001 EC/NSF Workshop on Universal Accessibility of Ubiquitous Computing: Providing for the Elderly*, 22–25 May 2001, Alcaser do Sal, Portugal, pp. 93–99. New York: ACM Press.

AgeLight LLC (2001). *Technology and Generational Marketing Strategies: Interface Design Guidelines for Users of All Ages.* http://www.agelight.com/webdocs/designguide.pdf.

Armstrong, D., Marmor, M. F., and Ordy, J. M. (1991). *The Effects of Aging and Environment on Vision.* New York: Plenum Press.

Bailey, B. (2002). Age classifications. *UI Design Newsletter.* http://www.humanfactors.com/downloads/jul02.asp.

Baldi, R. A. (1997). Training older adults to use the computer: Issues related to the workplace, attitudes and training. *Educational Gerontology* 23: 453–465.

BBC Wales News (2003). *Mobile Use Rise for Over-65s.* http://news.bbc.co.uk/1/hi/wales/2802299.stm.

Bergel, M., Chadwick-Dias, A., and Tullis, T. (2005). Leveraging universal design in a financial services company. *SIGACCESS Accessible Computing Newsletter* 82: 18–24.

Berkowitz, J. P. and Casali, S. P. (1990). Influence of age on the ability to hear telephone ringers of different spectral content, in the *Proceedings of the Human Factors and Ergonomics*

Society, 34th Annual Meeting, 8–12 October 1990, Orlando, FL, pp. 132–136. Santa Monica, CA: HFES.

Birren, J. E. (1970). Toward on experimental psychology of aging, cited in *The Psychology of Adult Development and Aging* (C. Eisdorfer and P. Lawton, eds.). Washington, DC: American Psychological Association.

Botwinick, J. and Storandt, M. (1974). *Memory Related Functions and Age*. Springfield, IL: Charles C. Thomas.

Bromley, D. B. (1988). *Human Ageing: An Introduction to Gerontology* (3rd ed.). Bungay, UK: Penguin.

Burnside, I. M., Ebersole, P., and Monea, H. E. (eds.) (1979). *Psychosocial Caring throughout the Lifespan*. New York: McGraw Hill.

Capitani, E., Della, S. S., Lucchelli, F., Soave, P., and Spinnler, H. (1988). Perceptual attention in aging and dementia measured by Gottschaldt's hidden figures text. *Journal of Gerontology: Psychological Sciences* 43: 157–163.

Carlson, M. C., Hasher, L., Connelly, S. L., and Zacks, R. T. (1995). Aging, distraction and the benefits of predictable location. *Psychology and Aging* 10: 427–436.

Casson, E. J., Johnson, C. A., and Nelson-Quigg, J. M. (1993). Temporal modulation perimetry: The effects of aging and eccentricity on sensitivity in normals. *Investigative Ophthalmology and Visual Science* 34: 3096–3102.

Cerella, J. (1985). Age-related decline in extrafoveal letter perception. *Journal of Gerontology* 40: 727–736.

Charness, N. and Bosman, E. (1990). Human factors and design, in *Handbook of the Psychology of Aging* (3rd ed.) (J. E. Birren and K. W. Schaie, eds.), pp. 446–463. San Diego, CA: Academic Press.

Cherry, K. E., Park, D. C., and Donaldson, H. (1993). Adult age differences in spatial memory: Efforts of structural content and practice. *Experimental Aging Research* 19: 333–350.

Coates, H. (2001). *Mobile Phone Users: A Small-Scale Observational Study*. http://www.aber.ac.uk/media/Students/hec9901.html.

Cody, M. J., Dunn, D., Hoppin, S., and Wendt, P. (1999). Silver surfers: Training and evaluating Internet use among older adult learners. *Communication Education* 48: 269–286.

Coren, S. (1994). Most comfortable listening level as a function of age. *Ergonomics* 37: 1269–1274.

Coulson, I. (2000). Technology challenges for gerontologists in the 21st century. *Educational Gerontology* 26: 307–315. http://www.asaging.org/at/at-226/infocus_nihtolau.html.

Crooke, T. H. and Larrabee, G. J. (1992). Changes in facial recognition memory across the adult lifespan. *Journal of Gerontology: Psychological Sciences* 47: 138–141.

Cunningham, W. R. and Brookbank, J. W. (1988). *Gerontology: The Psychology, Biology and Sociology of Ageing*. New York: Harper and Row.

Czaja, S. (1997). Computer technology and older adults, in *Handbook of Human-Computer Interaction* (2nd ed.) (M. E. Helander, T. K. Landauer, and P. Prabhu, eds.), pp. 797–812. New York: Elsevier.

Czaja, S. J., Guerrier, J. H., Nair, S. N., and Landauer, T. K. (1993). Computer communication as an aid to independence for older adults. *Behaviour and Information Technology* 12: 197–207.

Czaja, S. J. and Sharit, J. (1998). Age differences in attitudes toward computers. *Journal of Gerontology: Psychological Sciences* 53B: 329–340.

Denny, N. W., Dew, J. R., and Kihlstrom, J. F. (1992). An adult development study of encoding of spatial location. *Experimental Aging Research* 18: 25–32.

Department of Trade and Industry (2000). Ageing population panel: 5 applications of information and communications technology taskforce. *Foresight Making the Future Work for You*. http://www.foresight.gov.uk/servlet/Controller/ver=27/userid=2/Ageing_Report_5.pdf.

Dobbs, A. R. and Rule, B. G. (1990). Adult age differences in working memory. *Psychology and Aging* 4: 500–503.

Duncan, J., McLeod, P., and Phillips, L. H. (2005). *Measuring the Mind: Speed, Control and Age*. Oxford: Oxford University Press.

Ellis, R. D., Goldberg, J. H., and Detweiler, M. C. (1996). Predicting age-related differences in visual information processing using two-stage queuing model. *Journal of Gerontology: Psychological Sciences* 51B: 155–165.

Farkaas, M. S. and Hoyer, W. J. (1980). Processing consequences of perceptual grouping in selective attention. *Journal of Gerontology* 35: 207–216.

Feldman, R. M. and Reger, S. N. (1967). Relations among hearing, reaction time and age. *Journal of Speech and Hearing Research* 10: 479–495.

Fisk, A. D., Rogers, W. A., Cooper, B. P., and Gilbert, D. K. (1997). Automatic category search and its transfer: Aging, type of search and level of learning. *Journal of Gerontology: Psychological Sciences* 52B: 91–102.

Fries, J. F. and Crapo, L. M. (1981). *Vitality and Aging: Implications of the Rectangular Curve*. San Francisco: W. H. Freeman.

Ford, M. (1993). Coping again: Better sight for elderly people with central vision loss. *Broadcasting Support Services* 91: 6–28.

Fozard, J. L. (1990). Vision and hearing in aging, in *Handbook of Mental Health and Aging* (J. E. Birren, R. B. Sloane, and G. D. Cohen, eds.), pp. 150–170. San Diego, CA: Academic Press.

Frazier, L. and Hoyer, W. J. (1992). Object recognition by component features. *Experimental Aging Research* 18: 9–15.

Furlong, M. S. (1989). An electronic community for older adults: The Senior Network. *Journal of Communication* 39: 145–153.

Goggin, N. L. and Stelmach, G. E. (1990). Age-related deficits in the performance of cognitive-motor skills, in *Aging and Cognition: Mental Processes, Self Awareness and Interventions* E. A. Lovelace, ed.), pp. 135–155. Amsterdam: North-Holland Elsevier.

Grady, C. L. and Craik, F. I. M. (2000). Changes in memory processing with age. *Current Opinion in Neurobiology* 10: 224–231.

Greene, K. (2005, September 26). When we're all 64. *The Wall Street Journal*. http://esd.mit.edu/HeadLine/coughlin_wsj_article_092605.pdf.

Greene, L. S. and Williams, H. G. (1993). Age-related differences in timing control of repetitive movement: Application of the Wing-Kristofferson model. *Research Quarterly for Exercise and Sport* 64: 32–38.

Gregor, P., Newell, A. F., and Zajicek, M. (2002). Designing for dynamic diversity: Interfaces for older people, in the *Proceedings of the Fifth International ACM Conference on Assistive Technologies (ASSETS'02)*, 8–10 July 2002, Edinburgh, Scotland, pp. 151–156. New York: ACM Press.

Groth, K. E. and Allen, P. A. (2000). Visual attention and aging. *Frontiers in Bioscience* 5: 284–297.

Guardian Newspaper (2003). *Over 60s Reach for the Mouse*. http://www.guardian.co.uk/online/news/o,12597,994115,00.html.

Hartley, A. A. (1992). Attention, in *The Handbook of Aging and Cognition* (F. I. M. Craik and T. A. Salthouse, eds.), pp. 3–49. Hillsdale, NJ: Lawrence Erlbaum Associates.

Hawthorn, D. (2000). Possible implications of aging for interface designers. *Interacting with Computers* 12: 507–528.

Helve, J. and Krause, U. (1972). The influence of age on performance in the Panel D15 colour vision test. *Acta Ophthalmologica* 50: 896–901.

Hess, S., Detweiler, M. C., and Ellis, R. D. (1999). The utility of display space in keeping-track of rapidly changing information. *Human Factors* 41: 257–281.

Holt, B. J. and Morrell, R. W. (2002). Guidelines for web site design for older adults: The ultimate influence of cognitive factors, in *Older Adults, Health Information and the World Wide Web* (R. W. Morrell, ed.), pp. 109–132. Mahwah, NJ: Lawrence Erlbaum Associates.

Hoyer, W. J. and Rybash, J. M. (1992). Age and visual field differences in computing visual spatial relations. *Psychology and Aging* 7: 339–342.

Huey, R. W., Buckley, D. S., and Lerner, N. D. (1994). Audible performance of smoke alarm sounds, in the *Proceedings of the Human Factors and Ergonomics Society, 38th Annual Meeting*, 24–28 October 1994, Nashville, Tennessee, pp. 147–151. Santa Monica, CA: HFES.

ICM (2002). *IT, The Internet, and Older People*. http://www.icmresearch.co.uk/reviews/2002/it-internet-old-people.htm.

Jagacinski, R. J., Liao, M. J., and Fayyad, E. A. (1995). Generalized slowing in sinusoidal tracking in older adults. *Psychology and Aging* 9: 103–112.

Jay, G. M. and Willis, S. L. (1992). Influence of direct computer experience on older adults' attitudes toward computers. *Journal of Gerontology: Psychological Sciences* 47: 250–257.

Kanayama, T. (2003). Ethnographic research on the experience of Japanese elderly people online. *New Media and Society* 5: 267–288.

Kausler, D. H. (1994). *Learning and Memory in Normal Aging*. San Diego: Academic Press.

Keates, S., Lebbon, C., and Clarkson, P. J. (2000). Investigating industry attitudes to universal design, in the *Proceedings of the RESNA Annual Conference: Technology for the New Millennium*, 28 June–2 July 2000, Orlando, Florida, pp. 276–278. Arlington, VA: RESNA Publications.

Kline, D. W. and Scialfa, C. T. (1996). Sensory and perceptual functioning basic research and human factors implications, in *Handbook of Human Factors and the Older Adult* (A. D. Fisk and W. A. Rogers, eds.), pp. 27–54. San Diego, CA: Academic Press.

Kline, D. W. and Szafran, S. (1975). Age differences in backward monoptic masking. *Journal of Gerontology* 30: 307–311.

Kotary, L. and Hoyer, W. J. (1995). Age and the ability to inhibit distractor information in visual selective attention. *Experimental Aging Research* 21: 159–171.

Kubeck, J. E., Miller-Albrecht, S. A., and Murphy, M. D. (1999). Finding information on the World Wide Web: Exploring older adults' exploration. *Educational Gerontology* 25: 167–183.

Kurniawan, S. (2006). An exploratory study of how older women use mobile phones, in the *Proceedings of the 8th International Conference on Ubiquitous Computing (UbiComp 2006)*, 17–21 September 2006, Orange County, California, pp. 105–122. Heidelberg: Springer.

Kurniawan, S. H., Mahmud, M., and Nugroho, Y. (2006). A study of the use of mobile phones by older persons, in *Extended Abstracts of Conference on Human Factors in Computing Systems (CHI'06)*, 24–27 April 2006, Quebec, Canada, pp. 989–994. New York: ACM Press.

Lawhorn, T., Ennis, D., and Lawhorn, D. C. (1996). Senior adults and computers in the 1990's. *Educational Gerontology* 22: 193–201.

Lee, J. and Bailey, G. (2005). *Presbyopia: All about Sight*. http://www.allaboutvision.com/conditions/presbyopia.html.

Liao, M., Jagacinski, R. J., and Greenberg, N. (1995). Quantifying the performance limitations of older adults in a target acquisition task, in the *Proceedings of the Human Factors and Ergonomics Society 39th Annual Meeting*, 9–13 October 1995, San Diego, California, p. 961. Santa Monica, CA: HFES.

Light, L. L. (1990). Memory and language in old age, in *Handbook of the Psychology of Aging* (3rd ed.) (J. E. Birren and K. W. Schaie, eds.), pp. 275–290. San Diego, CA: Academic Press.

Lindenberger, U., Kliegl, R., and Baltes, P. B. (1992). Professional expertise does not eliminate age differences in imagery-based memory performance during adulthood. *Psychology and Aging* 9: 585–593.

Lipman, P. D. and Caplan, L. J. (1992). Adult age differences in memory for routes: Effects of instruction and spatial diagram. *Psychology and Aging* 7: 435–442.

Lupinacci, N. S., Rikli, R. E., and Jones, C. J. (1993). Age and physical activity effects on reaction time and digit symbol substitution performance in cognitively active adults. *Research Quarterly for Exercise and Sport* 64: 144–150.

Marquie, J. C., Thon, B., and Baracat, B. (1994). Age influence on attitudes of office workers faced with new computerised technologies. *Applied Ergonomics* 25: 130–142.

McConatha, D., McConatha, J. T., and Dermigny, R. (1994). The use of interactive computer services to enhance the quality of life for long term care residents. *The Gerontologist* 34: 553–556.

McDowd, J. M. and Craik, F. I. M. (1988). Effects of aging and task difficulty on divided attention performance. *Journal of Experimental Psychology: Human Perception and Performance* 14: 267–280.

McFarland, R. A., Warren, A. B., and Karis, C. (1958). Alteration in critical flicker frequency as a function of age and light. *Journal of Experimental Psychology* 56: 529–538.

Melenhorst, A., Rogers, W. A., and Caylor, E. C. (2001). The use of communication technologies by older adults: Exploring the benefits from the users perspective, in the *Proceedings of the HFES 45th Annual Meeting*, 8–12 October 2001, Minneapolis, pp. 221–225. Santa Monica, CA: HFES Press.

Mikkonen, M., Väyrynen, S., Ikonen, V., and Heikkila, M. O. (2002). User and concept studies as tools in developing mobile communication services for the elderly. *Personal and Ubiquitous Computing* 6: 113–124.

Morris, J. M. (1994). Computer training needs of older adults. *Educational Gerontology* 20: 541–555.

Morrow, D., Fitzimmons, C., and Von Leirer, P. A. (1994). Aging, expertise and narrative processing. *Psychology and Aging* 9: 376–388.

Morrow, D., Von Leirer, P. A., and Fitzimmons, C. (1992). When expertise reduces age differences in performance. *Psychology and Aging* 7: 134–148.

Myatt, E. D., Essa, I., and Rogers, W. (2000). Increasing the opportunities for ageing in place, in the *Proceedings of the ACM Conference on Universal Usability*, 16–17 November 2000, Arlington, Virginia, pp. 39–44. New York: ACM Press.

National Eye Institute (2004). *What Is Glaucoma?* http://www.nei.nih.gov/health/glaucoma/glaucoma_facts.asp#1.

National Statistics (2003). *Adult Mobile Phone Ownership or Use: By Age, 2001 and 2003: Social Trends.* http://www.statistics.gov.uk/STATBASE/ssdataset.asp?vlnk=7202.

Owsley, C., Sekuler, R., and Siemsen, D. (1983). Contrast sensitivity throughout adulthood. *Vision Research* 23: 689–699.

Plude, D. J. and Hoyer, W. J. (1986). Aging and the selectivity of visual information processing. *Psychology and Aging* 1: 1–9.

Ratner, H. H., Schell, D. A., Crimmins, A., Mittleman, D., and Baldinelli, L. (1987). Changes in adult prose recall: Aging or cognitive demands. *Developmental Psychology* 23: 521–525.

Rockstein, M. J. and Sussman, M. (1979). *Biology of Aging.* Belmont, CA: Wandsworth.

Rogers, W. A., Fisk, A. D., and Hertzog, C. (1994). Do ability related performance relationships differentiate age and practice effects in visual search? *Journal of Experimental Psychology, Learning, Memory and Cognition* 20: 710–738.

Royal National Institute for the Blind (2003). *Older People.* http://www.rnib.org.uk/xpedio/groups/public/documents/code/public_rnib002034.hcsp.

Rybash, J. M., Roodin, P. A., and Hoyer, W. J. (1995). *Adult Development and Aging.* Chicago: Brown and Benchmark.

Salthouse, T. A. (1989). Ageing and skilled performance, in *Acquisition and Performance of Cognitive Skills* (A. Coley and J. Beech, eds.), pp. 247–263. New York: Wiley.

Salthouse, T. A. (1991). *Theoretical Perspectives on Cognitive Aging.* Hillsdale, NJ: Lawrence Erlbaum Associates.

Salthouse, T. A. (1994). The aging of working memory. *Neuropsychology* 8: 535–543.

Salthouse, T. A. and Prill, K. A. (1988). Effects of aging on perceptual closure. *American Journal of Psychology* 101: 217–238.

Salthouse, T. A. and Somberg, B. L. (1982). Skilled performance: Effects of adult age and experience on elementary processes. *Journal of Experimental Psychology: General* 111: 176–207.

Schieber, F. (1992). Aging and the senses, in *Handbook of Mental Health and Aging* (J. E. Birren, R. B. Sloane, and G. D. Cohen, eds.), pp. 251–306. New York: Academic Press.

SeniorWatch (2003). European SeniorWatch Observatory and Inventory. http://www.seniorwatch.de.

Sharit, J. and Czaja, S. (1994). Ageing, computer-based task performance and stress: Issues and challenges. *Ergonomics* 37: 559–577.

Siedler, R. and Stelmach, G. (1996). Motor control, in *Encyclopedia of Gerontology* (J. E. Birren, ed.), pp. 177–185. San Diego, CA: Academic Press.

Sliwinski, M. J. and Hall, C. B. (1998). Constraints on general slowing: A meta-analysis using hierarchical linear models with random coefficients. *Psychology and Aging* 13: 164–175.

Smith, M. W., Sharit, J., and Czaja, S. J. (1999). Aging, motor control and the performance of computer mouse tasks. *Human Factors* 41: 389–396.

Spector, A. (1982). Aging of the lens and cataract formation, in *Aging and Human Visual Functions* (R. Sekuler, D. Kline, and K. Dismukes, eds.), pp. 27–43. New York: Alan R. Liss.

Spiezle, C. D. (2001). Interface design guidelines for users of all ages. *Agelight Technology and Generational Marketing Strategies Paper.* http://www.agelight.com/webdocs/designguide.pdf.

Spiriduso, W. W. (1995). Aging and motor control, in *Perspectives in Exercise Science and Sports Medicine: Exercise in Older Adults* (D. R. Lamb, C. V. Gisolfi, and E. Nadel, eds.), pp. 53–114. Carmel, IN: Cooper.

Stephanidis, C., Salvendy, G., Akoumianakis, D., Arnold, A., Bevan, N., Dardailler, D., et al. (1999). Toward an information society for all: HCI challenges and R&D recommendations. *International Journal of Human-Computer Interaction* 11: 1–28.

St. Lukes Eye (2005). *Cataracts.* http://www.stlukeseye.com/Conditions/Cataracts.asp.

Stokes, G. (1992). *On Being Old: The Psychology of Later Life.* London: The Falmer Press.

Stuart-Hamilton, I. (2000). *The Psychology of Ageing: An Introduction* (3rd ed.). London: Jessica Kingsley Publishers.

Telegraph (2002). *Does the Answer Lie in Our Genes?* http://www.telegraph.co.uk/health/main.jhtml?xml=/health/2002/02/26/hgene26.xml.

Tetley, J., Hanson, E., and Clarke, A. (2000). Older people, telematics and care, in *Care Services for Later Life: Transformations and Critiques* (A. Warnes, L. Warren, and M. Nolan, eds.), pp. 243–258. London: Jessica Kingsley Publications.

U.S. Department of Commerce (2000). Falling through the net: Toward digital inclusion. *A Report on American's Access to Technology Tools.* http://search.ntia.doc.gov/pdf/fttn00.pdf.

Vercruyssen, M. (1996). Movement control and the speed of behaviour, in *Handbook of Human Factors and the Older Adult* (A. D. Fisk and W. A. Rogers, eds.), pp. 55–86. San Diego, CA: Academic Press.

Walker, N., Philbin, D. A., and Fisk, A. D. (1997). Age-related differences in movement control: Adjusting submovement structure to optimize performance. *Journal of Gerontology: Psychological Sciences* 52B: 40–52.

White, H., McConnell, E., Clipp, E., et al. (1996). Surfing the net in later life: A review of the literature and pilot study of computer use and quality of life. *Journal of Applied Gerontology* 18: 358–378.

Zajicek, M. (2001). Special interface requirements for older adults, in the *Proceedings of the EC/NSF Workshop on Universal Accessibility of Ubiquitous Computing: Providing for the Elderly*, 22–25 May 2001, Alcacer do Sal, Portugal. http://virtual.inesc.pt/wuauc01/procs/pdfs/zajicek_final.pdf.

Zajicek, M. (2004). Successful and available: interface design exemplars for older users. *Interacting with Computers* 16: 411–430.

Zaphiris, P., Kurniawan, S. H., and Ellis, R. D. (2005). Web and aging: Challenges and opportunities. *Universal Access in the Information Society, Special Issue on Web and Ageing* 4: 1–2.

Zarbin, M. A. (1998). Age-related macular degeneration: review of pathogenesis. *European Journal of Ophthalmology* 8: 199–206.

Ziefle, M. and Bay, S. (2005). How older adults meet complexity: Aging effects on the usability of different mobile phones. *Behaviour and Information Technology* 24: 375–389.

9

International and Intercultural User Interfaces

Aaron Marcus and
Pei-Luen Patrick Rau

9.1 Introduction

The concept of *universal access* implies the availability of, and easy access to, computer-based products and services among all peoples in all countries worldwide. Successful computer-based products and services developed for users in different countries and among different cultures consist of partially universal, or general, solutions and partially unique, or local, solutions to the design of user interfaces. By managing the user's experience with familiar structures and processes, the user's surprise at novel approaches, as well as the user's preferences and expectations, the user-interface designer can achieve compelling forms that enable the user interface to be more usable, useful, and appealing. Globalization of product and service distribution, as with other manufacturing sectors, requires a strategy and tactics for the design process that enable efficient product development, marketing, distribution, and maintenance. Globalization of user-interface design, whose content and form are so much dependent upon visible languages and effective communication, improves the likelihood that users throughout the world will have a more successful user experience. Consequently, they are likely to be more productive and satisfied with computer-based products and services.

Demographics, experience, education, and roles in organizations of work or leisure characterize users. Their individual needs, as well as their group roles, define their tasks. User-

centered, task-oriented design methods account for these aspects and facilitate the attainment of effective user-interface design.

User interfaces conceptually consist of metaphors, mental models, navigation, appearance, and interaction. In the context of this chapter, these terms may be defined as follows (Marcus, 1995, 1998):

- *Metaphors*: Essential concepts conveyed through words and images, or through acoustic, or tactile means. Metaphors concern both overarching concepts that characterize interaction, as well as individual items, like the Trashcan standing for deletion within the desktop metaphor.
- *Mental models*: Organization of data, functions, tasks, roles, and people in groups at work or play. The term, similar to, but distinct from, cognitive models, task models, user models, and the like, is intended to convey the organization observed in the user interface itself, which is presumably learned and understood by users and which reflects the content to be conveyed as well as the available user tasks.
- *Navigation*: Movement through mental models, afforded by windows, menus, dialogue areas, control panels, and so on. The term implies process, as opposed to structure, that is, sequences of content potentially accessed by users, as opposed to the static structure of that content.

TABLE 9.1 Examples of Differing Displays for Currency, Time, and Physical Measurements

Item	U.S. Example	European Example	Asian Example
Currency	$1,234.00 (U.S. dollars)	DM1.234 (German marks)	¥1,234 (Japanese yen)
Time measures	8:00 pm, August 24, 2009 8:00 pm, 8/24/09	20:00, 24 August 2009 (England) 20:00, 24.08.09 (Germany, traditional) 20:00, 2009-08-24 (ISO 8601 Euro standard)	20:00, 2009.08.24, or Imperial Heisei 21, or H21 (Japan, based on year of Emperor's reign)
Physical measures	3 lb, 14 oz 3' 10", 3 feet and 10 inches	3.54 kg, 8.32 m (England) 3,54 kg, 8,32 m (Euro standard)	3.54 kg, 8.32 m in Roman or Katakana characters (Japan)

- *Appearance*: Verbal, visual, acoustic, and tactile perceptual characteristics of displays. The term implies all aspects of visible, acoustic, and haptic languages (e.g., typography or color; musical timbre or cultural accent within a spoken language; and surface texture or resistance to force).
- *Interaction*: The means by which users communicate input to the system and the feedback supplied by the system. The term implies all aspects of command-control devices (e.g., keyboards, mice, joysticks, microphones), as well as sensory feedback (e.g., changes of state of virtual graphical buttons, auditory displays, and tactile surfaces).

For example, an application, its data, the graphical user interface (UI) environment, and the underlying hardware, all contribute to the functional and presentational attributes of the user interface. An advanced English text editor, working within the Microsoft Vista UI environment, on a mouse- and keyboard-driven Intel processor-based PC, presents one set of characteristics. The liquid crystal displays (LCDs) and buttons on the front panel of a French paper copier, or the colorful displays and fighter-pilot-like joysticks for a children's video game on a Japanese Sony game machine, present alternative characteristics.

This chapter discusses the development of user interfaces that are intended for users in many different countries with different cultures, languages, and groups, in the context of a global communication society. The text presents a survey of important issues, as well as recommended steps in the development of user interfaces for an international and intercultural user population. With the rise of the World Wide Web and application-oriented web sites, the challenge of designing good user interfaces that are inherently accessible by users around the globe has become an immediate, practical matter, not only a theoretical issue. This topic is discussed from a user perspective, not a technology and code perspective. The chapter will: (1) introduce fundamental definitions of globalization in user-interface design; (2) review globalization in the history of computer systems; and (3) demonstrate why globalization is vital to the success of computer-based communication products.

9.2 Globalization

9.2.1 Definition of Globalization

Globalization refers to the worldwide production and consumption of products and services and includes issues at international,

intercultural, and local scales. In an information-oriented society, globalization affects most computer-mediated communication, which, in turn, affects user-interface design. The discussion that follows refers particularly to user-interface design.

Internationalization issues refer to the geographic, political, linguistic, and typographic issues of nations, or groups of nations (see some examples in Table 9.1). An example of efforts to establish international standards for some parts of user interfaces is the International Standards Organization's (ISO's) draft color legibility standards for cathode-ray tube (CRT) devices in Europe (ISO, 1989). Another example is the legal requirement for bilingual English and French displays in Canada, or the quasi-legal denominations for currency, time, and physical measurements, which differ from country to country.

Intercultural issues refer to the religious, historical, linguistic, aesthetic, and other, more humanistic, issues of particular groups or peoples, sometimes crossing national boundaries. Examples include calendars that acknowledge various religious time cycles, color/type/signs/terminology (see Table 9.2) reflecting various popular cultures, and organization of content in web search criteria reflecting cultural preferences.

Localization refers to the issues of specific, small-scale communities, often with unified language and culture, and, usually, at a scale smaller than countries, or significant cross-national ethnic regions. Examples include affinity groups (e.g., "twenty-somethings," or Saturn automobile owners), business or social organizations (e.g., German staff of DaimlerChrysler, or Japanese golf club members), and specific intranational groups (e.g., India's untouchables, or Japanese housewives). With the spread of web access, *localization* may come to refer to groups of shared interests that may also be geographically dispersed.

9.2.2 History of Globalization in Computing

In the early years of computers, 1950 to 1980, globalization was not a significant issue. IBM and other early computer and software manufacturers produced English-based, U.S.-oriented products with text-based user interfaces for national and world markets. Large, expensive computers required an elite engineering-oriented professional group to build, maintain, and use such systems. This group consisted of either native English speakers or those with some fluency in English. Personal computers eventually emerged, but with textual user interfaces. In the 1980s, workstations and eventually low-cost personal computers and

TABLE 9.2   Examples of Differing Cultural References

Item	North American/European Example	Middle-Eastern Example	Asian Example
Sacred colors	White, blue, gold, scarlet (Judeo-Christian)	Green, light blue (Islam)	Saffron yellow (Buddhism)
Reading direction	Left to right	Right to left	Top to bottom

Item	United States	France, Germany	Japan
Web search	Culture implies "high culture"; does not usually imply political discussions	Culture implies political as well as ethnographic discussions	Culture implies aesthetic discussions, e.g., tea ceremony
Sports references	Baseball, football, basketball	Soccer	Sumo wrestling, baseball

software brought dramatic increases in the worldwide population of users.

In the 1980s worldwide markets drove the development of multilingual editions of operating systems and software applications, which, by the late 1980s, had begun to acquire graphical UIs, such as those for the Apple Macintosh, Motif, OpenLOOK, and Microsoft Windows. In addition, many countries outside of the United States became software suppliers for both local and international markets.

In the mid-1990s, with the rise of the World Wide Web and nearly instant global access and interaction via the Internet, determining the target customers for global communication became immediately an issue. According to Alvarez et al. (1998), the European Information Technology Observatory's 1997 data stated that there were approximately 90.6 million Internet users globally, including 41.9 million in the United States. In addition, only 10% of people in the world were native speakers of English, yet 70% of web sites were English only.

In the 21st century, the growth of the European Union (EU), and the rise of China and India as producers and consumers of computer-based telecommunications, has changed dramatically the growth of multilingual web-based communication. In the EU, it is typical that web sites must appear in more than a dozen languages. In China, a company like TenCent has 80% of the Chinese search market, far greater than that of Google in the West. There is continuing strong interest among users outside of North America to have more content in native languages.

9.2.3 Survey of the Literature

The literature on the subject of globalization for user interfaces has continued to grow. The growth of anthropologists and ethnographers in product and service development indicate the greater attention to globalization. Three early references are the texts by Nielsen (1990), Fernandes (1995), and DelGaldo and Nielsen (1996).

The book edited by Nielsen, *Designing User Interfaces for International Use*, published in 1990, focuses on issues of language, translation, typography, and textual matters in user interfaces (including GUIs). Additionally, Ossner (in Nielsen, 1990, p. 11ff) introduces semiotics-based analysis of icons and symbols and applies that knowledge to computer-based training (CBT).

Fernandes' book, *Global Interface Design: A Guide to Designing International User Interfaces*, published in 1995, is much more oriented to GUIs and provides chapters about topics such as language, visual elements, national formats, and symbols, together with design rules. The book is notable for providing chapters about culture, cultural aesthetics, and even business justification for globalization, a topic of significant importance to the managers of research and development organizations and groups, who must argue for the merits of considering globalization in product manufacturing budgets.

The book edited by DelGaldo and Nielsen, *International User Interfaces*, published in 1996, addresses topics covered in the previous books, but adds consideration of many more cultural issues and their impact on user interface design. A notable contribution is that of Ito and Nakakoji (DelGaldo et al., 1996, pp. 105ff), which compares North American and Japanese user interfaces. They comment, for example, on Japanese users' lack of familiarity with typewriters and its effect on the use of word-processing software. This software used metaphorical references to typewriters, such as tab stops and margin settings, which were more familiar to Western users and more suited to 100-character alphanumeric keyboards and typesetting, rather than the 6000 characters of Japanese-Chinese symbols used in Japan.

A more recent, valuable reference is *Usability and Internationalization of Information Technology*, edited by Nuray Aykin (2005). The book is a fairly complete and up-to-date compendium of practical information on culture, design issues, and usability engineering. In addition, three lengthy case studies treat systems for children, user requirements in China, and a cross-cultural study of web-based travel planning.

In the last decade, especially with the rise of the web, more and more articles have appeared in conference proceedings and publications of professional organizations. Examples of these organizations in the United States include the Association for Computing Machinery's Special Interest Group on Computer-Human Interaction (ACM/SIGCHI), the International Conference in Human Computer Interaction (HCI International), the Usability Professionals Association (UPA), and the Society for Technical Communication (STC), all of which have had technical sessions, panels, and theme publications devoted to globalization and user interface design. In addition, the American Institute of Graphic Arts has created the Center for Cross-Cultural Design and maintains a web site devoted to this topic. Also of special value are the international annual conferences of the International Workshop for Internationalization of Products and Services (IWIPS), which have specialized in this topic since

1999, and the international conferences of the Localization Industries Standards Association (LISA) based in Switzerland.

Currently, a web search via Google returns 1.3 million citations that refer to user interfaces, globalization, internationalization, and/or localization. This large collection, almost four times the number 5 years ago, indicates abundant literature on these topics.

9.2.4 Advantages and Disadvantages of Globalization

The business justification for globalization of user interfaces is complex. If the content (functions and data) is likely to be of value to a target population outside of the original market, it is usually worthwhile to plan for international and intercultural factors in developing a product or service, so that it may be efficiently customized. Rarely can a product or service achieve global acceptance with a "one size fits all" solution. Developing a product or service for international, intercultural audiences usually involves more than merely a translation of verbal language. Visible (or otherwise attainable) language must also be revised, and other user-interface characteristics may need to be altered.

Developing products and services that are ready for global use, while increasing initial development costs, gives rise to potential for increased international sales. However, for some countries, monolithic domestic markets may inhibit awareness of, and incentives for, globalization. Because the United States has been such a large producer and consumer of software, it is not surprising that some U.S. manufacturers have targeted only domestic needs. However, others in the United States have understood the increasingly valuable markets overseas. To develop multiple versions efficiently, languages, icons, and other components must be easily swapped. For example, to penetrate some markets, the local language is an absolute requirement (e.g., in France).

Some software products are initiated with international versions (but usually released in sequence because of limited development resources). Other products are retrofitted to suit the needs of a particular country, language, or culture, as needs or opportunities arise. In some cases, the latter, ad hoc solution may suffer because of the lack of original planning for globalization.

Several case studies demonstrating the value of localized products appear in Fernandes (1995, pp. 161ff).

9.2.5 Globalization Design Process

The general user-interface development process, enhanced to address globalization issues, may be summarized as follows. This process is generally sequential with partially overlapping steps, some of which are, however, partially, or completely iterative. Additionally, the order in which these steps are performed within subsequent iterations may be modified to better suit the needs of the development process. For example, the evaluation step may be carried out prior to, during, or after the design step:

- *Plan*: Define the challenges or opportunities for globalization; establish objectives and tactics; determine budget, schedule, tasks, development team, and other resources. Globalization must be specifically accounted for in each item of project planning; otherwise, cost overruns, delays in schedule, and lack of resources are likely to occur.
- *Research*: Investigate dimensions of global variables and techniques for all subsequent steps (e.g., techniques for analysis, criteria for evaluation, media for documentation, etc.). In particular, identify items among data and functions that should be targets for change, and identify sources of national/cultural/local reference. User-centered design theory emphasizes gathering information from a wide variety of users; globalization refines this approach by stressing the need to research adequately users' wants and needs according to a sufficiently varied spectrum of potential users, across specific dimensions of differentiation. In current practice, this variety is often insufficiently considered.
- *Analyze*: Examine results of research (e.g., challenges or opportunities in the prospective markets), refine criteria for success in solving problems or exploiting opportunities (write marketing or technical requirements), determine key usability criteria, and define the design brief, or primary statement of the design's goals. At this stage, globalization targets should be itemized.
- *Design*: Visualize alternative ways to satisfy criteria using alternative prototypes; based on prior or current evaluations, select the design that best satisfies criteria for both general good user interface design and globalization requirements; prepare documents that enable consistent, efficient, precise, and accurate implementation.
- *Implement*: Build the design to complete the final product (e.g., write code using appropriate tools). In theory, planning and research steps will have selected appropriate tools that make implementing global variations efficient.
- *Evaluate*: At any stage, review or test results in the marketplace against defined criteria for success, for example, conduct focus groups, test usability on specific functions, gather sales and user feedback. Identify and evaluate matches and mismatches, then revise the designs to strengthen effective matches and reduce harmful mismatches. Testing prototypes or final products with international, intercultural, or specific localized user groups is crucial to achieving globalized user-interface designs.
- *Document*: Record development history, issues, and decisions in specifications, guidelines, and recommendation documents. As with other steps, specific sections or chapters of documents that treat globalization issues are required.

9.2.6 Critical Aspects for Globalization: General Guidelines

Beyond the user-interface development process steps identified in the previous section, the following guidelines can assist developers in preparing a checklist for specific tasks. Recommendations are grouped under user-interface development terms referred to earlier:

9.2.6.1 User Demographics

- Identify national and cultural target user populations and segments within those populations and then identify possible needs for differentiation of user-interface components and the probable cost of delivering them.
- Identify potential savings in development time through the reuse of user-interface components, based on common attributes among user groups. For example, certain primary (or top-level) controls in a web-based, data-retrieval application might be designed for specific user groups to aid comprehension and improve appeal. Lower-level controls, on the other hand, might be more standardized, unvarying formal elements (sometimes called "widgets").

9.2.6.2 Technology

- Determine the appropriate media for the appropriate target user categories.
- Account for international differences to support platform, population, and software needs, including languages, scripts, fonts, colors, file formats, and so on.

9.2.6.3 Metaphors

- Determine optimum minimum number of concepts, terms, and primary images to meet target user needs.
- Check for hidden miscommunication and misunderstanding.
- Adjust the appearance, orientation, and textual elements to account for national or cultural differences. For example, in relation to metaphors for operating systems, Chavan (1994) has pointed out that Indians relate more easily to the concept of bookshelf, books or notebooks, chapters or sections, and pages, rather than the desktop, file folders, and files with multiple pages.

9.2.6.4 Mental Models

- Determine optimum minimum varieties of content organization to meet target user needs.

9.2.6.5 Navigation

- Determine need for navigation variations to meet target user requirements, determine cost-benefit, and revise as feasible.

9.2.6.6 Appearance

- Determine optimum minimum variations of visual and verbal attributes. Visual attributes include layout, icons and symbols, typography, color, and general stylistic aesthetics. Verbal attributes include language, formats, and ordering sequences. For example, many Asian written languages, such as Chinese and Japanese, contain symbols with many small strokes. This factor seems to lead to an acceptance of higher visual density of marks in complex public information displays than is typical for Western countries.

9.2.6.7 Interaction

- Determine optimum minimum variations of input and feedback to meet target user requirements. For example, because of web access-speed differences for users in countries with slow access, it is usually important to provide text-only versions, without extensive graphics, as well as alternative text labels to avoid graphics that take considerable time to appear. As another example, some Japanese critics believe that office groupware applications from Northern European countries match personal communication needs of Japanese users more closely than similar applications from the United States.

9.2.7 Specific Appearance Guidelines

Because of space limitations in this chapter about a complex topic, complete, detailed guidelines cannot be provided for all of the user-interface design terms listed in the previous section. Some detailed guidelines for one important topic, visual and verbal appearance, are provided in the following. Further details can be found in works by Aykin, DelGaldo, Fernandes, and Nielsen (e.g., Nielsen, 1990; Fernandes, 1995; DelGaldo et al., 1996; and Aykin, 2005).

9.2.7.1 Layout and Orientation

- As appropriate, adjust the layout of menus, tables, dialogue boxes, and windows to account for the varying directions and size of text.
- If dialogue areas use sentence-like structure with embedded data fields or controls, these areas will need special restructuring to account for language changes that significantly alter sentence format. For example, German sentences often have verbs at the ends of sentences, while English and French place them in the middle.
- As appropriate, change layout of imagery that implies or requires a specific reading direction. Left-to-right sequencing may be inappropriate or confusing for use with right-to-left reading scripts and languages.
- Check for misleading arrangements of images that lead the viewer's eye in directions inconsistent with language reading directions.

- For references to paper and printing, use appropriate printing formats and sizes. For example, the 8.5 x 11 inch standard office letterhead paper size in the United States is not typical in many other countries that use the European A4 paper size of 210 x 297 mm with a square-root-of-two rectangular proportion.

9.2.7.2 Icons and Symbols

- Avoid the use of text elements within icons and symbols to minimize the need for different versions to account for varying languages and scripts.
- Adjust the appearance and orientation to account for national or cultural differences. For example, using a postal letterbox as an icon for e-mail may require different images for different countries.
- As a universal sign set reference, consider using as basic icon/symbol references the signs that constitute the international signage sets developed for international safety, mass transit, and communication (for examples, see American Institute of Graphic Arts, 1981; Olgyay, 1995; and Pierce, 1996).
- Avoid puns and local, unique, charming references that will not transfer well from culture to culture. Keep in mind that many universal signs are covered by international trademark and copyright use, such as Mickey Mouse and the "Smiley" smiling face. In the United States, the familiar smiling face is not a protected sign, but it is in other countries.
- Consider whether selection symbols, such as the X or checkmarks, convey the correct distinctions of selected and not-selected items. For example, some users may interpret an X as crossing out what is not desired rather than indicating what is to be selected.
- Be aware that office equipment such as telephones, mailboxes, folders, and storage devices differ significantly from nation to nation.

9.2.7.3 Typography

- Use fonts available for a wide range of languages required for the target users.
- Consider whether special font characters are required for currency, physical measurements, and so on.
- Ensure appropriate decimal, ordinal, and currency number usage. Formats and positioning of special symbols vary from language to language.
- Use appropriate typography and language for calendar, time zone, addresses, and telephone/fax references.

9.2.7.4 Color

- Follow perceptual guidelines for good color usage. For example, use warm colors for advancing elements and cool colors for receding elements; avoid requiring users to recall in short-term memory more than 5 ± 2 different coded colors.
- Respect national and cultural variations in colors, where feasible, for the target users.

9.2.7.5 Aesthetics

- Respect, where feasible, different aesthetic values among target users. For example, some cultures have significant attachment to wooded natural scenes, textures, patterns, and imagery (e.g., the Finnish and the Japanese), which might be viewed as exotic or inappropriate by other cultures.
- Consider specific culture-dependent attitudes. For example, Japanese viewers find disembodied body parts, such as eyes and mouths, unappealing in visual imagery.

9.2.7.6 Language and Verbal Style

- Consider which languages are appropriate for the target users, including the possibility of multiple languages within one country. For example, English and French within Canada.
- Consider which dialects are appropriate within language groupings and check vocabulary carefully (e.g., for British vs. American terms in English, Mexican vs. Spanish terms in Spanish, or Mainland China vs. Taiwanese terms in Mandarin Chinese).
- Consider the impact of varying languages on the length and layout of text. For example, German, French, and English versions of text generally have increasingly shorter lengths.
- Consider the different alphabetic sorting or ordering sequences for the varied languages and scripts that may be necessary and prepare variations that correspond to the alphabets. Note that different languages may place the same letters in different locations, for example, Á comes after A in French, but Å after Z in Finnish.
- Consider differences of hyphenation, insertion point location, and emphasis, that is, the use of bold, italic, quotes, double quotes, brackets, and so on.
- Use appropriate abbreviations for such typical items as dates, time, and physical measurements. Remember that different countries have different periods of time for weekends and the day on which the week begins.

9.2.8 Design for Chinese Population

Although every cultural segment deserves some attention, major segments have a greater impact on development of universal access to information systems. It is beyond the scope of this chapter to deal with even a few. Attention is given to the Chinese population as an example of increasing importance. With the rapid economic development and a widespread use of computers and the Internet in China, cultural characteristics of Chinese users have become significant issues to those companies that need to design user interfaces for Chinese users. This section discusses specific issues for this major segment of the human population, approximately 1.2 billion people, about which not much has been known and published in Western professional and business environments.

9.2.8.1 Icons and Symbols

Chinese characters are often used in the form of graphics. Chinese calligraphy uses Chinese characters to express emotions and esthetic feelings.

Chinese signs (icons and symbols) have powerful historical and cultural meaning. Some examples are the following.

The Chinese dragon, although not a real animal, nevertheless is a very important symbol for Chinese art and architecture. Chinese knots (traditional Chinese decorative knots), often used for more than fastening and wrapping, are also a special kind of symbol for the Chinese. Each knot is woven from one piece of thread and named by its shape and meaning (Li, 2004). French et al. (2002) identified language and localized graphical design as cultural issues for Taiwanese e-finance web sites. The double-fish sign conveys the message of prosperity, or "always more than enough."

Chinese architecture, like the Great Wall, is often used as the symbol of Chinese culture. Chinese palaces, temples, pagodas, and guardian lions have not changed substantially for over 1000 years.

9.2.8.2 Color

Color symbolism has special meaning within Chinese culture. Red is the symbol of happiness, luck, and joy for Chinese people. Red is used for all celebrations, including weddings and Chinese New Year, and in ancient traditional imperial buildings, such as those in the Forbidden City. Yellow, which sounds in Mandarin Chinese like the word for "royal," is used as the royal color in ancient China (Hang, 2004). White, considered the *opposite* of red, represents mourning and sorrow in Chinese culture. The ancient Chinese used green to describe the color of growing plants, but green also means cheating and shame for men. In contemporary China, joking about wearing a green cap for a Chinese man can be extremely offensive.

9.2.8.3 Language and Reading Style

In Chinese, each character is pronounced as a syllable (Hoosain, 1986). Reading and writing Chinese characters are associated strongly with processing visual codes. Cai et al. (2001) found that character style and number of strokes both have a significant impact on the legibility threshold of characters. Goonetilleke et al. (2002) found that Hong Kong Chinese use predominantly horizontal search patterns, while the Mainland Chinese change their search pattern depending on the layout presented. Nonnative Chinese readers, on the other hand, do not seem to show any preference on scanning strategy for a given layout.

9.2.8.4 Information Architecture

The mental models of user interfaces play an important role for users, and different structures or organizations within user interfaces may be needed to accommodate different styles of categorization. Choong and Salvendy (1999) and Rau et al. (2004) found a significant difference between the functional and thematic assignments for Chinese participants in relation to making errors. For Chinese participants, in Taiwan as well as in Mainland China, advantages were associated with thematic structure in terms of reduced error rate.

Two variables in Hall's model of culture (Hall, 1959, 1984, 1989, 1990; Hall and Hall, 1990)—context of communication and time orientation—have been studied by some researchers for user-interface design (Plocher and Zhao, 2002; Zhao, 2002; Zhao et al., 2002; Rau and Liang, 2003). Rau and Liang (2003) found that participants with a preference for high-context communication style were more disorientated during browsing a web-based service than those with low-context communication. *Polychronic* participants, that is, those preferring to do many things at once in parallel browsed information faster and took fewer steps than *monochronic* participants, that is, those who prefer doing things one at a time sequentially (for these terms, see Hall, 1959, pp. 177–78, or Hall, 1990, pp. 173–174). Zhao (2002) studied the effect of time orientation on browsing performance and found that polychronic participants performed browsing tasks fasters than neutral participants when participants were not familiar with the information architecture (mental model and navigation) of the browsing materials.

9.3 A Case Study

9.3.1 Planet SABRE

One early example of user-interface globalization is the design for Planet SABRE, the graphical version of the SABRE Travel Information Network (STIN), one of the world's largest private online networks. Planet SABRE was used exclusively by travel agents. The firm of one of the authors of this chapter worked closely with STIN marketing and engineering staff over a period of approximately 4 years (1996 to 2000) in developing the user interface for Planet SABRE.

The SABRE system contained approximately 42 terabytes of data about airline flights, hotels, and automobile rentals and enabled almost $2 billion of bookings annually. The system sustained up to 1 billion hits per day.

The Planet SABRE user-interface development process emphasized achieving global solutions from the beginning of the project. For example, stated requirements mentioned allowing for

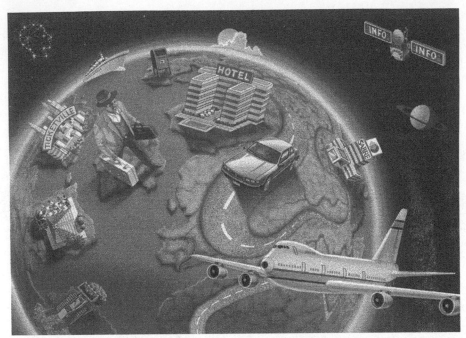

FIGURE 9.1 Example of Planet SABRE home screen showing typical icons for passenger, airline booking, hotel rental, car rental, and e-mail (post box).

the space needs of multiple languages for labels in windows and dialogue boxes.

Besides supporting multiple languages (English, Spanish, German, French, Italian, and Portuguese), the user-interface design enabled the switching of icons for primary application modules, so that they would be more gender-, culture-, and nation- appropriate (see Figures 9.1 and 9.2).

Figure 9.1 shows the initial screen of Planet SABRE, with icons representing the primary applications or modules within the system conveyed through the metaphor of objects on the surface of a planet, or in outer space. The postal box representing the e-mail functions depicts an object that users in the United States would recognize immediately. However, users in many other countries would have significant difficulty in recognizing this object because postal boxes come in very different physical forms.

Figure 9.2 shows a prototype version of a Customizer dialogue box, in which the user can change or select preferences; in particular, certain icons can be swapped, so that they appear in more recognizable images. This change could also be accomplished for other icons, such as the depiction of the passenger. Once changed in the Customizer, the icon's appearance is switched throughout the user interface wherever the icon appears.

At every major stage of prototyping, designs developed in the United States were taken to users in international locations for evaluation. The user feedback was relayed to the development team and affected later decisions about all aspects of the user-interface design.

More recent examples of the influence of culture on global corporate web site design are to be found in Marcus and

Baumgartner's (2004b) study of 12 corporate web sites for business-to-business and business-to-consumer sites (e.g., McDonalds, Coca-Cola, Hitachi, and Siemens). Their 60-page study details differences of metaphors, mental models, navigation, interaction, and appearance related to the culture dimensions of power distance, individualism versus collectivism, gender roles, uncertainty avoidance, and long-term time orientation.

9.4 Conclusions and Future Research Issues

The concept of *universal access* requires significant attention to globalization issues in the user-interface development process. Progress in technology increases the number and kinds of functions, data, platforms, and users of computer-based communication media. The challenge of enabling more people and more kinds of people to use this content and these tools effectively will depend increasingly upon global solutions. By recognizing the need for, and benefit to, users of user-interface designs intended for international and intercultural markets, developers will achieve greater success and increased profitability through the global distribution and acceptance of their products.

The recommendations provided in this chapter are an initial set of heuristics that will assist developers in achieving global solutions to their product and service development. Design methodologies must support globalization throughout the development process.

In addition, it is likely that international and intercultural references will change rapidly, requiring frequent updating of

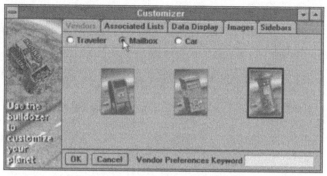

FIGURE 9.2 Example of dialogue box in the customizer application, by means of which users can change the icons to become more culturally relevant.

designs. Future work on global user interface design may address the following issues:

- How might global user interfaces be designed to account for different kinds of intelligence? Gardner (1985) has identified the following dimensions of intelligence. These dimensions of competence, comprehension, and fluency suggest users might have varying strengths of conceptual competence with regard to using user interfaces on an individual basis, but these might also vary internationally, or interculturally, due to influences of language, history, or other factors:
  - Bodily-kinesthetic: Sports, dance
  - Interpersonal: Interaction with others, empathy
  - Linguistic: Words, oration, reading, writing
  - Logical-mathematical: Logic, abstraction, induction, numerical manipulation
  - Naturalistic: Orientation to nature, animals, nurturing
  - Intrapersonal: Introspection, self-reflection
  - Spatial: Visualization, mentally manipulating objects
  - Musical: Hearing, music, rhythm,
  - Other intelligences, e.g., moral or spiritual
- How might metaphors, mental models, and navigation be designed precisely for different cultures that differ by such dimensions as age, gender, national or regional group, or profession? The author has posed this as a question to the user-interface analysis and design community (Marcus, 1993). The topic is discussed broadly in DelGaldo et al. (1996). Further, what means can be developed to enable these variations to be produced in a cost-effective manner using templates?

The taxonomic analyses of global issues for user interfaces, the theoretical basis for their component selection, the criteria for their evaluation, and their design methodology are emerging rapidly in the user-interface development field across many platforms, for example, client-server application UIs, web application UIs, mobile device UIs, and vehicle UIs. Designers should be aware of the scope of the activity, locate sources of insight, and incorporate professional techniques in their development process to improve the value and success of their international and intercultural products and services.

9.5 Resources

The following is a list of organizations that provide information about international standards and intercultural issues:

- **American National Standards Institute (ANSI)**
  This organization analyzes and publishes U.S. standards, including those for icons, color, terminology, user interfaces, and so on.
  Contact information:
  American National Standards Institute (ANSI)
  11 West 42nd Street, 13th Floor, New York, NY 10036
  Telephone: (212) 642-2000
  E-mail: info@ansi.org
  URL: http://www.ansi.org
- **China National Standards Organization**
  This organization analyzes products to be imported to China.
  Contact information:
  China Commission for Conformity of Electrical Equipment (CCEE) Secretariat
  Postal address: 2 Shoudu Tiyuguan, NanLu, 100044, P.R. China
  Office address: 1106, 11th floor, 2 Shoudu Tiyuguan, NanLu, Beijing, P.R. China
  Telephone: +86-1-832-0088, ext. 2659
  Fax: +86-1-832-0825
- **East-West Center**
  This organization, formerly funded by the U.S. Congress, is a center for technical and cultural interchange among Pacific Rim countries. Their research and publications cover culture and communication.
  Contact information:
  East-West Center
  1601 East-West Road, Honolulu, HI 96848-1601
  Telephone: (808) 944-7111
  Fax: (808) 944-7376

E-mail: ewcinfo@eastwestcenter.org
URL: http://www.ewc.hawaii.edu

- **Information Technology Standards Commission of Japan (ITSCJ)**
  This organization analyzes and publishes Japanese standards for information technology, including standards for icons, color, terminology, and user interfaces.
  Contact information:
  Information Processing Society of Japan
  Kikai Shinko Building, No. 3-5-8 Shiba-Koen,
  Minato-ku, Tokyo 105, Japan
  Telephone: +81-3-3431-2808
  Fax: +81-3-3431-6493

- **International Standards Organization**
  This organization analyzes and publishes world standards for all branches of industry and trade, including standards for icons, color, terminology, user interfaces, and so on.
  Contact information:
  International Standards Organization
  Geneva, Switzerland
  Telephone: +41-22-749-0111
  Fax: +41-22-733-3430
  E-mail: central@iso.ch
  URL: http://www.iso.ch
  Note: Of special interest is ISO 8601's international time and date standards. Information about ISO's 8601 standards, particularly about time and date standards, may be found at the following URLs:
  http://www.aegis1.demon.co.uk/y2k/y2kiso.htm
  http://www.roguewave.com/products/resources/exchange/iso8601.html

- **Japan National Standards Organizations**
  These organizations analyze and publish Japanese standards, including those for icons, color, terminology, user interfaces, and so on.
  Contact information:
  Japanese Industrial Standards Committee (JISC)
  Agency of Industrial Science and Technology
  Ministry of International Trade and Industry
  1-3-1, Kasumigaseki, Chiyoda-ku, Tokyo 100, Japan
  Telephone: +81-3-3501-9295/6
  Fax: +81-3-3580-1418

- **World Wide Web Consortium**
  This organization provides information relevant to globalization issues, including accessibility. Two URLs of interest:
  http://www.w3.org/International, for information about internalization
  http://www.w3.org/WAI (for information about accessibility)

Acknowledgments

Aaron Marcus thanks his staff at Aaron Marcus and Associates, their client SABRE Travel Information Network, and Constantine Stephanidis, the editor of this book, for their assistance in preparing this chapter. We also acknowledge the books by Aykin (2005), DelGaldo and Nielsen (1996), Fernandes (1995), and Nielsen (1990), which provided a basis for many points raised in this chapter. Finally, we acknowledge Peter Simlinger, Dipl.-Ing, Director, International Institute for Information Design, Vienna, Austria, and Andreas Schneider, President, Institute for Information Design Japan, Tokyo, Japan, for their advice about international time and space metrics.

References

Alvarez, G. M., Kasday, L. R., and Todd, S. (1998). How we made the web site international and accessible: A case study, in the *Proceedings of the Fourth Human Factors and the Web Conference*, 5 June 1998, Basking Ridge, NJ. [CD-ROM]. Holmdel, NJ.

American Institute of Graphic Arts (1981). *Symbol Signs.* Hastings House, NY: Visual Communication Books.

Aykin, N. (ed.) (2005). *Usability and Internationalization of Information Technology.* Mahwah, NJ: Lawrence Erlbaum Associates.

Cai, D. C., Chi, C. F., and You, M. L. (2001). The legibility threshold of Chinese characters in three-type styles. *International Journal of Industrial Ergonomics* 27: 9–17.

Chavan, A. L. (1994). *A Design Solution Project on Alternative Interface for MS Windows.* Masters thesis. Royal College of Art, London.

Choong, Y. Y. and Salvendy, G. (1999). Implications for design of computer interfaces for Chinese users in Mainland China. *International Journal of Human Computer Interaction* 11: 29–46.

DelGaldo, E. and Nielsen, J. (eds.) (1996). *International User Interfaces.* New York: John Wiley and Sons.

Eco, U. (1976). *A Theory of Semiotics.* Bloomington: Indiana University Press.

Fernandes, T. (1995). *Global Interface Design: A Guide to Designing International User Interfaces.* Boston: AP Professional.

French, T., Minocha, S., and Smith, A. (2002). eFinance localisation: An informal analysis of specific eCulture attractors in selected Indian and Taiwanese sites, in *Designing for Global Markets 4: Proceedings of the 4th International Workshop on Internationalization of Products and Systems* (J. Corondao, D. Day, and B. Hall, eds.), 11–13 July 2002, Austin, TX, pp. 9–21. Austin, TX: Products and Systems Internationalization.

Gardner, H. (1985). *Frames of Mind: The Theory of Multiple Intelligences.* New York: Basic Books.

Goonetilleke, R. S., Lau, W. C., and Shih, H. M. (2002). Visual search strategies and eye movements when searching Chinese character screens. *International Journal of Human-Computer Studies* 57: 447–468.

Hall, E. T. (1959). *The Silent Language.* New York: Anchor Books.

Hall, E. T. (1984). *Dance of Life: The Other Dimension of Time.* Yarmouth, ME: Intercultural Press.

Hall, E. T. (1989). *Beyond Culture.* New York: Anchor Press.

Hall, E. T. (1990). *The Hidden Dimension*. New York: Anchor Press.

Hall, E. T. and Hall, M. (1990). *Understanding Cultural Differences: Germans, French and Americans*. Yarmouth, ME: Intercultural Press.

Hang, C. (2004). *Chinese Color Theory: The Symbolism of Color in Traditional Chinese Culture*. http://home.earthlink.net/~jamesdferrell/colortheory.org/D_ColortheoryChineseColorTheory.htm.

Hoosain, R. (1986). Perceptual processes of the Chinese, in *The Psychology of the Chinese People* (M. H. Bond, ed.), pp. 38–72. Oxford, UK: Oxford University Press.

ISO (1989). *Computer Display Color*. Draft Standard Document 9241-8, Geneva.

Iwayama, M., Tokunaga, T., and Tanaka, H. (1990). A method of calculating the measure of salience in understanding metaphors, in *Proceedings of AAAI '90*, 29 July–3 August 1990, Boston, Massachusetts, pp. 298–303.

Lanham, R. A. (1991). *A Handlist of Rhetorical Terms* (2nd ed.). Berkeley: University of California Press.

Li, G. (2004). *Chinese Knots*. http://www.chinataiwan.org/web/webportal/W5088611/A5165888.html.

Marcus, A. (1992). *Graphic Design for Electronic Documents and User Interfaces*. Reading, MA: Addison-Wesley.

Marcus, A. (1993). Human communication issues in advanced UIs. *Communications of the ACM* 36: 101–109.

Marcus, A. (1995). Principles of effective visual communication for graphical user interface design, in *Readings in Human-Computer Interaction* (2nd ed.) (R. M. Baecker, J. Grudin, W. Buxton, and S. Greenberg, eds.), pp. 425–441. Palo Alto: Morgan Kaufman.

Marcus, A. (1998). Metaphor design in user interfaces. *The Journal of Computer Documentation, ACM/SIGDOC* 22: 43–57.

Marcus, A. (2006). Cross-cultural user-experience design, in *Diagrammatic Representation and Inference, Proceedings of the 4th International Conference Diagrams 2006* (D. Barker-Plummer, R. Cox, and N. Swoboda, eds.), 28–30 June 2006, Stanford, California, pp. 16–24. Berlin: Springer-Verlag.

Marcus, A. and Baumgartner, V-J. (2004a). A practical set of culture dimensions for global user-interface development, in the *Proceedings of Computer Human Interaction: 6th Asia Pacific Conference, APCHI 2004* (M. Masoodian, S. Jones, and B. Rogers, eds.), 29 June–2 July 2004, Rotorua, New Zealand, pp. 252–261. Berlin: Springer-Verlag. http://www.springerlink.com/app/home/contribution.asp?wasp=99eqac4jwr2ruwbcat33&referrer=parent&backto=issue,26,79;journal,309,1949;linkingpublicationresults,1:105633,1.

Marcus, A. and Baumgartner, V-J. (2004b). Mapping user-interface design components vs. culture dimensions in corporate websites. *Visible Language Journal* 38: 1–65.

Marcus, A., Baumgartner, V-J., and Chen, E. (2003). User-interface design vs. culture, in the *Proceedings of the International Conference on Internationalization of Products and Services (IWIPS 2003)*, 17–19 July 2003, Berlin, pp. 67–78. Austin, TX: Product and Systems Internationalization.

Marcus, A. and Gould, E. W. (2000a). Crosscurrents: Cultural dimensions and global web user-interface design. *Interactions* 7: 32–46.

Marcus, A. and Gould, E. W. (2000b). Cultural dimensions and global web user-interface design: What? So what? Now what?, in the *Proceedings of the 6th Conference on Human Factors and the Web*, 19 June 2000, Austin, Texas. http://www.amanda.com/resources/hfweb2000/hfweb00.marcus.html.

McLuhan, M. (1964). *Understanding Media: The Extensions of Man*. New York: McGraw-Hill.

Neale, D. C. and John, M. C. (1997). The role of metaphors in user interface design, in *Handbook of Human-Computer Interaction* (2nd ed.) (M. Helander, T.K. Landauer, and P. Prabhu, eds.), pp. 441–462. Amsterdam: Elsevier Science.

Nielsen, J. (ed.) (1990). Advances in human factors/ergonomics. *Designing User Interfaces for International Use* (Vol. 13). Amsterdam: Elsevier Science.

Olgyay, N. (1995). *Safety Symbols Art*. New York: Van Nostrand Reinhold [available also on diskette].

Pierce, T. (1996). *The International Pictograms Standard*. Cincinatti, OH: ST Publications.

Plocher, T. A. and Zhao, C. (2002). Photo interview approach to understanding independent living needs of elderly Chinese: a case study, in the *Proceedings of the 5th Asia-Pacific Conference on Computer Human Interaction (APCHI 2002)* (G. Dai, ed.), 1–4 November 2002, Beijing, China. Beijing: Science Press.

Rau, P.-L. P., Choong, Y-Y., and Salvendy, G. (2004). Effectiveness of icons and textural design in human computer interfaces: A cross culture study of Chinese in mainland China and Taiwan and Caucasians population. *International Journal of Industrial Ergonomics* 34: 117–129.

Rau, P.-L. P. and Liang, S.-F. M. (2003). A study of cultural effects on designing user interface for a Web-based service. *International Journal of Services Technology and Management* 4: 480–493.

Zhao, C. (2002). *Effect of Information Structure on Performance of Information Acquiring: A Study Exploring Different Time Behavior: Monochronicit/Polychronicity*. PhD dissertation. Institute of Psychology, Chinese Academy of Sciences.

Zhao, C., Plocher, T., Xu, Y., Zhou, R., Liu, X., Liang, S-F. M., and Zhang, K. (2002). Understanding the polychronicity of Chinese, in the *Proceedings of the 5th Asia-Pacific Conference on Computer Human Interaction (APCHI 2002)* (G. Dai, ed.), 1–4 November 2002, Beijing, pp. 189–195. Beijing: Science Press.

III

Technologies for Diverse Contexts of Use

10

Accessing the Web

Vicki L. Hanson,
John T. Richards,
Simon Harper, and
Shari Trewin

10.1 Introduction and Background

The World Wide Web is the single largest repository of electronic information, art, and digital culture in the world. This gives it a special significance when it comes to governmental participation, shopping, entertainment, healthcare, social networking, and economic issues such as employment and service. In short, a more independent life for many can begin with successful web access.

> For me being online is everything. It's my hi-fi, my source of income, my supermarket, my telephone. It's my way in.

This quote, taken from a blind user, sums up the sentiments experienced when talking with many disabled users, and drives home the importance of web accessibility in the context of independent living. Most services, both governmental and private, and nearly all significant information resources now have web interfaces. Access to these interfaces vastly extends the understanding and effective reach of both abled and disabled users.

Two key attributes contribute to the impact of this technology. First, the information is distributed over probably hundreds of millions of servers but accessible from any user's computer anywhere in the world. Second, the information is digital and generally represented in such a way that it can be transformed into different formats to be used more effectively by people with disabilities. These two attributes give the web its universal appeal and importance. The flattening of the global economy is also driving the web to become a universal medium for commerce. Competing effectively in most economic sectors now requires a viable web presence. With the ability to buy anything from anywhere, from the specialist baker in a foreign country, to the pizza delivery company at the end of the street, the accessibility of services has major implications for business on both a local and global scale.

The web allows companies to reach beyond their normal local orbits and be relevant to distant communities, and it is for this reason that governments and participation in political activity and government services are also increasing. Reaching distant communities, communities that may not have otherwise participated, has a knock-on effect for health, well-being, and social participation. Indeed, participation in all these areas is enhanced when the web is introduced, even when that participation is on a geographically local scale. From setting up neighborhood watch sites to local service information for healthcare, the web is becoming just another way of communicating, exchanging, and informing audiences both locally and globally. While the rapid rise of the web has left digital ethnographers and social anthropologists at something of a loss as to its exact nature, one thing is certain—the web provides a telepresence into shops, communities, and entertainment, both local and at a distance, that users would never have otherwise experienced. This is important for everybody, but especially important for communities of people who would otherwise be barred from full participation in these activities.

These communities include those whose disabilities hinder their efforts to find employment, to interact more fully with society at large, and to freely use technology without assistance. However, with the growth of the web, and a move to more thought- and communication-based activities (the emerging knowledge economy), there is the very real possibility of disabled people finding productive, fulfilling, and socially empowering lives because of web access. While the point has not yet been reached where web access is required by able-bodied people for daily living, it is probably not that far away. For disabled people, however, that threshold may have already been crossed.

This chapter discusses issues related to universal access for users with disabilities. Although the lion's share of attention in accessibility has been directed at users with vision disabilities, a number of people with other disabilities have difficulties accessing the web. This chapter considers issues related to hearing loss, motor impairment, cognitive disability, and aging as they relate to web access. This latter population, older adults, presents some very interesting issues of web access due to the fact that in many cases there are multiple sources of disability. Thus, the discussion of older adults can be viewed as a window into topics of combined disability. Although not specifically discussed in this chapter, the number of people who experience disability with web access is not limited to those with permanent disability. Many people, as a result of injury, face temporary disability that will cause them to experience one or more of the issues discussed in this chapter.

Following the presentation of needs of disabled users, this chapter describes some solutions. Guidelines to address web accessibility are described, followed by discussions of new and emerging technologies that contribute to universal access to the web.

10.2 Universal Access to the Web

Universal access within the context of web accessibility aims to help people with disabilities to perceive, understand, navigate, interact, and contribute to the web. Disabled web surfers may use intermediate technologies to gain access to web and computer functionality. In general, technology such as magnifiers and screen readers or specialized keyboards exists to provide access to any software on the computer. Such technology is not always required for the web, however, since some accessibility features are incorporated directly into web document standards and web browser features.

There are millions of people who have disabilities that affect their use of the web. Currently most web sites have accessibility barriers that make it difficult or impossible for many people with disabilities to use them (Petrie et al., 2004). Usability is considered critical to universal access. Therefore, this chapter explores technologies that not only meet the minimum requirements for web accessibility, but also those that can help provide an effective browsing experience. In this spirit, technologies that often go beyond requirements specified by web access guidelines are considered.

10.2.1 Vision

Globally, in 2002 more than 161 million people were visually impaired, of whom 124 million had low vision and 37 million were blind. However, refractive error as a cause of visual impairment was not included, which implies that the actual global magnitude of visual impairment is greater (World Health Organization, 2004). Worldwide, for each blind person, an average of 3.4 people have low vision, with country and regional variation ranging from 2.4 to 5.5. These figures—the first global estimates since the early 1990s—are the best current estimates of the global burden of visual impairment and are the result of

new studies carried out in nearly all World Health Organization[1] regions, which have substantially updated the epidemiological data. For instance, of 878 million Europeans, 3 million are blind, 13 million have low vision, and 16 million are visually impaired (World Health Organization, 2004). This represents 4% of the total European population and 7% of the global population of people with visual impairments. However, these figures appear likely to change significantly. Indeed, based on the 1990 census, both figures are rising in developed countries due to an aging population. More than 82% of all people who are blind are 50 years of age and older, although older adults represent only 19% of the world's population. Therefore, technology to afford visually impaired users access to resources is becoming increasingly important as the global population becomes proportionally older. Chapter 6, "Sensory Impairments," of this handbook offers a more detailed discussion of vision-related problems.

Visual impairment is the most addressed disability in the web accessibility spectrum. It is a challenging area because it requires the visual interaction model, and the HTML structure created to support this model, to be moved into a predominantly serial interaction paradigm. To bridge this gap, assistive technologies have been created for differing levels of visual impairment from low vision to profound blindness. These range from color manipulation software (Wakita and Shimamura, 2005),[2] through browser magnification and legibility enhancing systems (Richards and Hanson, 2004), to specialist browsing systems (Ramakrishnan et al., 2004), finishing with applications for the most extreme cases of blindness, those to audibly read the screen (JAWS, 2006). These screen readers address text-to-speech translations by audibly reading documents, top-left to bottom-right, through a text-to-speech synthesizer. This is acceptable in most instances, as the screen reader also translates the focused user event handler (such as a toolbar or menu system), and so the application's normal user event mechanisms can be used for interaction by a visually impaired user. Screen readers are discussed in detail in Chapter 28, "Screen Readers," of this handbook.

Initial attempts at screen reading, based on screen scraping, simply read the screen in a left to right, top to bottom order, often shuffling elements from semantically distinct regions that just happened to be adjacent to one another. As the visual complexity of web pages increased, these screen readers become inadequate because of the reliance of web documents on context, linking, and deeper document structure to convey information in a useful way. Because of this, web browsers and web page readers for visually impaired users have evolved to access this deeper document structure, directly examining the structured page source or the browser's internal representation of that source, the document object model (DOM). By examining the role being played

---

[1] The 10th revision of the WHO International Statistical Classification of Diseases, Injuries and Causes of Death, defines low vision as visual acuity of less than 6/18, but equal to or better than 3/60, or corresponding visual field loss to less than 20°, in the better eye with best possible correction. Blindness is defined as visual acuity of less than 3/60, or corresponding visual field loss to less than 10°, in the better eye with best possible correction. Visual impairment includes low vision as well as blindness.

[2] For color-blind users.

by portions of the text, it was hoped that more complex meanings (associated with style, color, etc.) could be derived. However, when interacting with complex web documents these readers, although better than screen scrapers, still do not capture enough of the document structure to provide the visually impaired user a browsing experience as rich and meaningful as that available to a sighted user.

To fully support these users, a better understanding of the perception of information by visually impaired users must be gained, reflecting genuine perceptual and cognitive processes rather than simply relying on a straightforward application of existing technology. This understanding may be limited to audio-only processes or may eventually extend to novel cross-modal interactions that relate to web accessibility and the display of web documents for visually impaired users, such as haptics. There is a wealth of still-to-be-discovered knowledge pertaining to web accessibility, transcoding, cross-modal interaction, and perceptual analysis, and its study may result in a set of novel tools and techniques to inform the transformation process.

In summary, many visually disabled users consider the web to be a primary source for information, employment, and entertainment (Jacobs et al., 2005). Indeed, from questioning and contact with many blind users we have discovered that the importance of the web cannot be overestimated.

10.2.2 Hearing

Significant hearing loss affects a large number of people. Demographic surveys indicate that nearly 10% of the population experiences significant hearing loss, but the numbers increase sharply with age. The number is less than 2% for children under 17 years of age, but as high as 29% for adults over the age of 65 (Holt et al., 1994; NIDCD, n.d.; RNID, n.d.). Chapter 6, "Sensory Impairments," of this handbook discusses hearing-related problems in detail.

The WAI guidelines state that alternative formats should be provided for sounds or spoken web content. The largely visual nature of the web has, in many ways, proven advantageous for people who are deaf or hard of hearing. The web provides visual access to information and services and communication applications such as instant messaging and e-mail, and video transmission of chats have enabled deaf and hard of hearing people to more easily interact with others.

As web content becomes increasingly reliant on an audio channel, however, new barriers are presented to users who have a hearing loss. Even deaf or hard of hearing individuals who are skilled lip-readers cannot understand material that is presented in the form of a voiceover or with the face of the speaker partially obscured, poorly lit, or presented at a low frame rate.

The current explosion of interest in Internet games and virtual worlds is an excellent example of environments in which increasingly sophisticated technology has the effect of increasingly disenfranchising users with disabilities. Consider that early forms of games and virtual worlds were text based. Now those have been replaced by sophisticated visual "worlds" with accompanying auditory effects. In many cases, critical information is only presented via sound. Users who do not have access to this information will be unable to participate in the game or interact fully in the virtual world. While some sounds might be considered the audio equivalent of eye candy (e.g., musical soundtracks that primarily enhance emotional impact), equal access dictates that deaf and hard of hearing users be given at least an indication of the presence of such ear candy.

Sound events can be appealingly presented in visual form in many cases. Alerting sounds and off-screen events can be visually represented by images (Kimball, 2007). Such images not only make events accessible, but can represent fun multimedia alternatives, particularly in game environments (Trewin et al., 2008). Speech, however, must be represented by other means. Captioning is most commonly used to provide a visual representation of speech.

10.2.2.1 Captioning

Most people are familiar with subtitles that are used in foreign films to provide language translations. Captioning is similar in purpose. With captioning, speech is transcribed and the text is displayed onscreen, synchronized with the video and audio (at the level of individual sentences or phrases). Ideally, sound events (such as music, alerts, or environmental sounds) are indicated in the captions so that the deaf or hard-of-hearing person is aware of these events and can respond to them as needed. Captions can be open (meaning that they are always visible) or closed (meaning that they can be turned on and off as controlled by standard web media applications).

A number of options exist for web content providers to caption material (see Core Techniques for Web Content Accessibility Guidelines 1.0, 2000; National Center for Accessible Multimedia, n.d.). An issue that often arises is whether to provide verbatim captioning or captioning at a reading level directed at the average reading level of deaf and hard of hearing users. To understand this issue, it is necessary to realize that the age at which a hearing loss is acquired has a number of important implications for a person's ultimate language skill. Many find it surprising that hearing loss at a young age often makes it difficult to acquire skilled reading (for detailed discussions of the issues involved, see Conrad, 1979). Nevertheless, the most generally accepted approach to captioning is to provide a verbatim transcript (Hanson, 2007).

10.2.2.2 Signing

Sign languages are commonly used in deaf communities worldwide (Padden and Humphries, 1988). Contrary to many expectations, there is not one common sign language throughout the world. Rather, there exist a number of sign languages, including a distinctly different British Sign Language (BSL), American Sign Language (ASL), and Australian Sign Language (Auslan), despite the commonalities of English across these geographies. For individuals who use sign language, a signed interface may be appropriate for web sites. Sign language interfaces are considered in-depth by Huenerfauth and Hanson (see Chapter 38, "Sign Language in the Interface: Access for Deaf Signers," of the current volume), so the present discussion highlights only specific issues that impact web access.

In contrast to captioning, which generally seeks to present speech and sound visually, sign language interfaces are typically considered a way of making written web content accessible to sign language users who may lack the reading proficiency needed to understand the text on a web site (see Chapter 38, "Sign Language in the Interface: Access for Deaf Signers," of the current volume; Kennaway et al., 2007). Often this is the case on web sites or applications intended for deaf children who might not yet have acquired reading skills, but it may be considered appropriate for other web sites to ensure equal access for sign language users.

Video chats, vBlogs, and video in e-mail are excellent examples of web applications that allow signers to easily communicate. These web applications have been enabled by the growing prevalence of high-speed networks combined with new video compression technologies. While natural sign language using video is ideal from the standpoint of the signing user, its use may be better suited for communication applications than for language translations of print in web sites. While video of a person signing will present the highest quality of signing, it still requires substantial local storage or high-speed links to stream remote content. Once captured, it also tends to be quite difficult to edit as the corresponding text content changes, making it costly to maintain over time. Some sites use video effectively, such as the British Deaf Association.[3] The number of such sites is limited, however.

The inflexibility of video has prompted research on the use of signing avatars. These avatars provide an animated depiction of human signers. The goal of avatar technology is to one day have avatars that can automatically translate text into a native sign language. A glimpse of the future of avatar technology can be seen in the ASL signing avatar demonstrated on DeafWorld.com.[4] For now, human intervention is required to generate such high-quality signing by avatars. Avatars that rely solely on machine translations are limited to providing sign renderings of print that follow the text word for word, rather than being translations into a native sign language having fundamentally different syntax. A current research focus is on natural language translation, with a deep understanding of linguistics providing the rules for machine translation needed for avatars to display native sign languages (see Chapter 38, "Sign Language in the Interface: Access for Deaf Signers," of the current volume; Kennaway et al., 2007).

10.2.3 Motor Skills

Many people with motor impairments rely heavily on the web to provide access to services and opportunities they would otherwise be unable to use independently. They form a very diverse group (see Chapter 5, "Motor Impairments and Universal Access," of this handbook), but share a number of core web access requirements. The most prevalent causes of motor impairment are musculoskeletal disorders, including arthritis and carpal tunnel syndrome. Worldwide, an estimated 9% to 27% of the population is affected to some degree by a musculoskeletal disorder (Huisstede et al., 2006).

The prevalence of physical impairment in the population is rising. This is partly due to demographic changes in the population—as the population ages, people are more likely to acquire some limitation in physical abilities (Lethbridge-Çejku et al., 2006). It is also partly due to improvements in healthcare. More people are able to survive traumatic events such as stroke or brain injury. These may give rise to severe physical impairments, depending on the part of the brain that has been damaged. The prevalence of Parkinson's disease is expected to double in the next 25 years, fueled by the aging of the population (Dorsey et al., 2007). Improvements in antenatal care are also leading to a rise in the prevalence of cerebral palsy (Krigger, 2006; Odding et al., 2006).

10.2.3.1 Devices

Most individuals use both a keyboard and a mouse to access the web. Some people with limited dexterity use a keyboard but no pointing device, and rely on key commands to control web browsers and navigate pages. Some use speech commands to browse the web. Others have a method of pointing and selecting a position on the screen, but no keyboard. For example, they may point by eye gaze and make selections by dwelling on a target. They enter text by clicking letters on a software keyboard on the screen (an onscreen keyboard), which are then sent to the browser by the keyboard software. Most onscreen keyboards also incorporate word prediction capabilities, where the user can select a predicted word with a single action (see also Chapter 29, "Virtual Mouse and Keyboards for Text Entry," of this handbook).

An enormous variety of inventive solutions are employed by those who find that the keyboard and mouse do not meet their needs (see examples in Alliance for Technology Access, 2004). Individuals with very limited motion can use one or two binary signals generated by a physical switch, tongue switch, EMG sensor, or other device, and can control a computer by scanning through the available options (manually or automatically) and selecting the items of interest. A switch user will typically control a web browser indirectly through keystrokes entered on a scanning onscreen keyboard (see also Chapter 35, "Automatic Hierarchical Scanning for Windows Applications," of this handbook).

Given the diversity of solutions used by this population, it is very important that web pages be designed for device independence, without the assumption that the user has both a pointing device and a keyboard. The W3C WAI guidelines provide recommendations for achieving this (Chisholm et al., 2001; Caldwell et al., 2006). For example, it is recommended that pages should not use event handlers that are triggered only by mouse events, because this makes them inaccessible to a keyboard-only or speech-only user.

---

[3] http://bda.org.uk/index.php.
[4] http://www.deafworld.com/videos/?vid=80L2Xc0K8Jg. See also http://www.vcom3d.com/vault_files/forest_asl.

Because all of these input techniques can provide text input, the standard approach to providing universal access is to ensure that browsers and web sites can be accessed and controlled via keystrokes. This provides a basic level of access, but can be very inefficient (Schrepp, 2006). For example, a switch user must repeatedly scan through the letters of her keyboard to the Tab key, and each Tab key press advances her just one step through the set of browser options. For low-bandwidth users, a specialized browser can provide far more efficient and effective key-stroke-based access (Mankoff et al., 2002). Mankoff et al. (2002) also identified several useful transcoding techniques for supporting low-bandwidth input, including the addition of links to support backwards and forwards navigation at each paragraph break, other links for skipping unwanted sections, and specialized form widgets with built-in scanning capabilities.

10.2.3.2 Input Rate

One study of mouse users with mild to moderate motor impairments (Trewin and Pain, 1999) found that selecting onscreen items typically took 2 to 3 times as long for this group as for a comparison group with no impairment, even with error correction time excluded. On the web, every link selection may require considerable time and effort. Indeed, one major study of web accessibility found that the usability problems most frequently reported by people with physical impairments were lack of clarity in site navigation mechanisms, and confusing page layout (Disability Rights Commission, 2004). This finding probably reflects the high cost of taking a wrong path and the effort required to navigate around a page.

Typing speed is also affected. While a speed of 30 to 40 words per minute is normal for average QWERTY typists (Starner, 2004), a keyboard user with a movement disorder might type at around 10 words per minute. An eye-gaze typist might produce 1 to 15 words per minute (Majaranta and Räihä, 2002). For a scanning onscreen keyboard user, 1 to 5 words per minute is not unusual (Brewster et al., 1996).

To accommodate slower typing rates, web pages can be made more usable by reducing the amount of free text entry to a minimum, providing preselected defaults wherever possible, and allowing the user to request extra time to fill in information on forms that have a time-out.

These design guidelines go beyond the provisions of the W3C's WAI and UAAG guidelines, but have already been identified as being helpful for mobile web users, who also have slower input rates, higher error rates, and a higher cost of navigation errors than desktop users (Rabin and McCathieNevile, 2006; Trewin, 2006). Many mobile web devices do not have a pointing device, and so support for keyboard navigation is also important for this community. Researchers are actively working to incorporate this core capability into Web 2.0 technologies (Schwerdtfeger and Gunderson, 2006).

10.2.3.3 Dexterity Demands

Physical impairments affect movement in many different ways. If movement is painful, tiring, or difficult to control, then the level of dexterity demanded by a web site will have a significant effect on the usability of that site. A further contributing factor is that techniques such as speech, eye-gaze pointing, or EMG can be inherently error prone or difficult to control. People with severe motor impairments are often pioneers in the use of such technologies, at a stage when they are not robust or reliable enough to become more generally popular.

The design features described previously that minimize the typing required on a web page can help to minimize the physical demands of interacting with a page. Web designers can also assist by providing error recovery mechanisms such as confirmation dialogues for irreversible actions. If a user has typed a password incorrectly several times (being unable to check what was typed), a response that does not lock the user's account is obviously preferable.

In the Disability Rights Commission study described earlier, the third most common web site problem reported by people with physical impairments was small text and graphics (Disability Rights Commission, 2004), which would be difficult to select. Although there has been research into mechanisms for making pointing and clicking on small targets easier (Worden et al., 1997; Grossman and Balakrishnan, 2005), there is little operating system or browser support currently available. Making clicking physically easier, therefore, means making the targets larger. While browsers do provide the ability to enlarge text, other nontext clickable objects such as arrows for expanding sections of tree views often cannot be made larger except by magnifying the whole page. This introduces extra navigation actions and makes it more difficult to understand the context of the web page as a whole. Cascading menus also introduce significant motor demands, because they require the user to follow a specific path through the desired sequence of submenus. The longer a continuous action sequence must be maintained, the more likely it is for movement disorders such as jerks or spasms to disrupt it.

Many users of pointing devices would benefit tremendously from having the ability to specify a minimum target size at the browser level, in a similar way to choosing a preferred font size. Web pages should therefore be designed to accommodate changes to the way that basic clickable elements are drawn, in addition to font size changes.

In summary, there will always be great diversity in the assistive solutions used by individuals with physical impairments, and the speed and accuracy with which they are able to control the browser. However, they share a need for web pages that are designed for device independence, with minimal dexterity demands, clear navigation mechanisms, and error recovery mechanisms. A good universal design should optimize both ease of keyboard access and ease of direct selection.

10.2.4 Cognition

The term *cognitive disability* covers a broad range of different issues. These can include learning disabilities (such as dyslexia), memory, attention, problem solving, mental illness, intellectual abilities, processing of sensory information, and more (Newell et al., 2007). Within each class of ability, the level of impairment will also vary, making this a very ill-defined category. Related issues are discussed in details in Chapter 7, "Cognitive Disabilities," of this handbook.

Compared with other disability areas, the topic of cognition has received little attention (Jiwnani, 2001; Bohman, 2004). Perhaps this is due to the widely differing needs of users in this population. Perhaps it is due to an assumption that the numbers of people with cognitive disabilities are few, a perception that will likely change with recent attention to autism spectrum disorder (Rice, 2007) and traumatic brain injury (Centers for Disease Control and Prevention, 2006). Perhaps it is due to assumptions that people with cognitive impairments will not use the web. Perhaps it is due to a lack of activism on the part of members of this demographic in demanding access. Whatever the reason or combination of reasons, the fact remains that many individuals with a cognitive limitation do wish to have fuller access to the web. Many even have successful, professional careers for which web access is required. Web authors should assume that their target audience *will* include people with cognitive impairments.

Extensions to current web accessibility guidelines have been suggested to improve access for some groups of users with cognitive impairments (Poulson and Nicolle, 2004). Many of the web access barriers reported by people with cognitive impairments are familiar to all users, but felt more acutely by this group. Many design guidelines aimed at improving cognitive accessibility are basic techniques for effective communication. It is not surprising, then, that sites that design for this group can be more effective for everyone (Nielsen, 2005).

A number of methods for supporting users with cognitive limitations can be implemented. Among these are navigational supports, reduced distractions, memory supports, and provision of text-to-speech. More specifics as to how web design can aid in the usability of web pages by people with cognitive disabilities are considered here:

- Web pages, and the process of navigating the web, can be complex, but there are steps that authors can take to reduce this complexity and make web sites more accessible to people who have difficulty with complex multistep tasks. According to the WAI guidelines, the most accessible sites are those that use consistent navigation mechanisms, and keep to commonly used navigation conventions. Clearly structuring the information on the page is also helpful.
- Spatial ability, situation awareness, and the ability to focus and concentrate on a specific task are all important cognitive skills for effective web navigation (Small et al., 2005). For some individuals, it can be very difficult to maintain the necessary level of awareness and focus. Animated elements on web pages are a common source of irritation to many users, but the impact on usability for an easily distractible person can be profound.
- Memory is also an issue in web navigation. It is easy for anyone to get lost in the sheer vastness of the web, unable to remember or figure out where he is, how he got there, and how to get to where he wants to be. For sites that incorporate lengthy tasks with multiple steps, context information such as "Step 2 of 4" can help to keep users oriented. In hierarchically structured sites, the use of "breadcrumbs" to show position in the site's page tree can also help.
- The Back button, although the most common means of returning to recently viewed pages, uses a stack-based model that makes entire branches of the recent history unavailable. Many users rely heavily on going back, but do not understand this model well. History and bookmark mechanisms that use text labels, and search mechanisms that require accurate spelling, can also cause navigation difficulties (Harryson et al., 2004). Supplementing the text labels with icons or page thumbnails can be very effective in promoting recognition in history and bookmark lists. Much mainstream research is devoted to investigating ways to support web navigation, with suggestions, including graphical history representations, and alternative back and history mechanisms (Ayers and Stasko, 1995; Cockburn et al., 2003).
- Many web sites require users to register, and log on with a user ID and password to access their personal account information. Remembering all of these passwords can be difficult, and so browser tools that offer to remember previously typed IDs and passwords are very useful for people with memory limitations.

Given the wide variety of issues that fall under the category of cognition, space does not permit a complete discussion of all the issues and how the types of techniques mentioned here might apply. To give some sense of how these solutions work with specific users, two populations are discussed: users with developmental disabilities and those with dyslexia.

Developmental disabilities (such as Down syndrome or autism spectrum disorders) clearly affect a person's ability to utilize the web. Clicking links and browser controls such as Page Up and Page Down appear to be of relatively little problem to this user group, while entering URLs presents great difficulty (Harryson et al., 2004). On individual pages, consistent page navigation, content supported by images, and structured browsing that guides users through a more limited number of selections are important (Brewer, 2005; Sevilla et al., 2007). Users with developmental disabilities also can benefit from text-to-speech technologies that read web content aloud and from pages that have relatively little content viewable at one time to reduce distractions (Richards and Hanson, 2004).

Dyslexia is a neurological problem that affects language. It can take many forms, of which the most commonly known is difficulty with reading and writing (Shaywitz et al., 1998; Dyslexia Research Institute, 2007; National Center for Learning Disabilities, 2007). Not surprisingly, dyslexic users can benefit from having the ability to listen to the content of web pages to be read aloud. A number of design issues also can serve to make web sites more usable for dyslexic children and adults, including providing memory supports and text options such as color contrast, font size, and spacing between words and lines of text (Elkind, 1998; Vassallo, 2003; Web Design for Dyslexic Users, 2006). The importance of providing the ability for individual users to make changes for their own needs is highlighted in research with this population, showing great individual variability, for example, in the font options when using software that allows them to adapt pages to their own needs (Gregor and Newell, 2000; Hanson et al., 2005).

10.2.5 Aging

Contrary to conventional wisdom, many older adults are eager to utilize current computer technologies, with access to the web being considered key (Fox, 2004; Czaja and Lee, 2007). Age-related issues for access to technology are discussed in Chapter 8, "Age-Related Difference in the Interface Design Process," of this handbook.

Older adults face challenges in accessing the web. These challenges come from two sources.

The first challenge for older adults is that many are novice computer users (Newell et al., 2006). With respect to the web, problems arise due to confusing terminology and a bewildering array of options on the interface of conventional browsers. Various attempts, some commercial, have been made to make web tasks easier for novice users. Some older adults, however, are retired from jobs or still hold jobs that require the use of the web; many others have simply used it for their own personal interests for many years. A number of organizations worldwide, such as SeniorNet in the United States and Old Kids in China, help novice users get started with computers and get connected online (Xie, 2006). In this same spirit, software applications geared specifically for the older adult have been devised, particularly with novice users in mind (Generations On Line, 2006; Newell et al., 2006; Caring Family, 2007; ElderMail, 2007). Although there is currently a digital divide between those over 70 and other age groups in terms of web use, demographics suggest that this divide will shrink in the coming years. Today, adults over age 70 continue to use the web less often than other age groups, but as the baby boomers become this older generation, the number of skilled web users is expected to surge, creating what has been called a *silver tsunami* (Fox, 2001).

The second challenge for older web users comes from the fact that nearly all older adults have some declines that impact computer use (if only the fact that multifocal glasses must be worn to read content on the screen and these glasses typically are not well suited to reading text on a typically positioned display). Declines of vision, hearing, dexterity, and cognition are common. Previous discussions in this chapter suggest how some of these declines will impact older users.

The complexity of web technologies, particularly some of the emerging Web 2.0 technologies, pose particular challenges for novice older users (Chadwick-Dias et al., 2007; Zajicek, 2007). While navigating the web and dealing with new social interactions and dynamically changing content appears almost effortless for younger users, the cognitive declines typical of older adults (Czaja and Lee, 2007) make the learning and use of such technologies difficult. Older adults typically read content more carefully, and read more of it before initiating an action compared to younger users (Tullis, 2007). They are also less likely to utilize new collaborative technologies such as blogs and wikis, and are less likely to download newer media types such as videos and music.

Perceptual difficulties with vision and hearing are also problematic. With sound and video becoming more prominent on the web, older users can benefit from captioning provided for users with hearing loss. Declining vision, however, is most often discussed by older users as a problem. Reduced acuity, color perception, and contrast discrimination accompany normal aging (Parker and Scharff, 1998; Faye and Stappenbeck, 2000). These declines lead to difficulties in reading small text, text that is closely surrounded by other visual elements, and text that has complex font styles or lowered contrast due to poor color choices on pages. Web design guidelines for older adults stress the use of larger fonts, high contrast (black on white or white on black) and increased spacing between text elements (NIA, 2002). Some of the necessary adaptations are built into conventional browsers (such as Internet Explorer, Firefox, Opera, and Safari), although they are typically accessed via multiple levels of menus and complex dialogue boxes, making it difficult for older adults to find and utilize these features (Sa-nga-ngam and Kurniawan, 2007). Although older adults wish to use their available vision as much as possible for browsing, speech is often helpful in augmenting their reading. Potential problems with understanding synthetic speech (Czaja and Lee, 2007) and long messages (Zajicek, 2003), exacerbated by faulty memories and erroneous models of navigation (Zajicek, 2001, 2003; Hanson et al., 2007), suggest that speech, while useful, will not ensure the needs of older users are met.

In addition to these perceptual problems, older users may have difficulty using a mouse and keyboard due to illnesses or injuries that limit dexterity. There is evidence that older adults use different movement strategies than younger adults, with lower peak velocities, and many submovements, and that fine positioning over a target is particularly difficult for this group (Keates and Trewin, 2005). Many would be well served by having larger targets on the screen that would not only make clicking easier, but also make seeing the targets easier.

As declines for older people are often in more than one area, their combination can make accessibility more challenging than for users with a single disability. Consider, for example, an older adult who has both low vision and hand tremors. Finding assistive devices for magnification, speech, and mouse correction that work well together may be difficult (as these devices are not tested together) to impossible (the devices make conflicting demands on underlying system resources). Interestingly, despite this tendency to have multiple limitations, older adults do not tend to view themselves as disabled. This fact is critical in understanding that this population is less likely than their younger counterparts to use special devices, such as assistive input technologies. Particularly among novice users, there is a tendency to view problems as due to their own lack of understanding of the technology. This rarely leads to the realization that nonstandard software and hardware are available and can make computers easier to use. As such, older users will be more likely to struggle with standard technology.

10.2.6 Web Accessibility Guidelines

Web accessibility depends on diverse aspects of web development and interaction technologies working together, including web software (tools), web developers (people), and content (e.g., font and color choices, element sizes, layout, etc.). The W3C Web Accessibility Initiative (WAI)[5] recognizes these difficulties and provides guidelines for each of these interdependent components: (1) Authoring Tool Accessibility Guidelines (ATAG), which address software used to create web sites (Treviranus et al., 2000, 2004); (2) Web Content Accessibility Guidelines (WCAG) (2006), which address the information in a web site, including text, images, forms, sounds, and so on (Chisholm et al., 2001; Caldwell, 2008); and (3) User Agent Accessibility Guidelines (UAAG), which address web browsers and media players, and their relationships to assistive technologies (Jacobs et al., 2002). There are other organizations that have also produced guidelines (e.g., IBM, RNIB, AFB, Macromedia, etc.), but the WAI guidelines integrate many of the key points of the others.

Of these issues addressed by guidelines, the ones for creating accessible web pages (WCAG) tend to be the most widely known. There is, however, no homogeneous set of guidelines that designers can easily follow. In the web accessibility field, there are also other best-practice efforts (Engelen et al., 1999; Craven and Bophy, 2003) that mainly focus on creating tools to ensure accessibility through means such as validation, transformation, and web site repair. Validation tools analyze pages against accessibility guidelines and return a report or a rating (Chisholm and Kasday, 2005; McCathieNevile and Abou-Zahra, 2005). Repair tools, in addition to validation, try to repair the identified problems (Chisholm and Kasday, 2005). Although there has been extensive work in the development of validation, repair, and transformation tools, automation is still limited. While it is likely that there are certain accessibility problems for which solutions cannot be fully automated (e.g., checking the quality of alternative text provided for images), those solutions that can be automated are still hindered by the complexity of web content analysis and the incompleteness of available transformation technologies.

10.2.7 Individual Coping Strategies

To overcome problems not yet addressed by tools and technologies, many disabled users use inventive coping strategies to enhance their performance when interacting with the web. These strategies are often aimed at dealing with complex and inappropriately marked-up pages and aim to make "bad" web pages "good" by applying previously successful strategies. Coping strategies are more successful when the user is experienced, but are often complicated and only partially successful in many cases.

These strategies can be grouped into two types: specific strategies (based on the users' experience of interacting with a particular web page) and generic strategies (applicable to most web pages). As an example of the former, some visually disabled users count the number of Tabs to skip things at the top of the Google results pages. Others use Ctrl+F to search for the text "Similar Pages," which enables them to find the first result. Examples of the latter often involve techniques for moving to the top or bottom of any page. Not surprisingly, users often indicate that they do not like pages that frequently change their structure, and prefer pages that are stable such as Google. This is mainly because users cannot apply familiar strategies when pages dynamically change structure. Both strategy types can also be used to train novice users to better cope with complex pages, which are quite common on the web. Such strategies, typically based on the design of particular pages, tend to cause users to return to just a handful of sites, however, with new sites being visited only when necessary (Yesilada et al., 2007). This means that nearly all the web, including the so-called long-tail (Brynjolfsson et al., 2003), is off-limits to the majority of visually impaired users.

10.3 Web Technology

This section presents a survey of current web technology, focusing primarily on aspects that have impacted web accessibility. Following this is a consideration of where the web appears to be going and the accessibility challenges and opportunities likely associated with emerging web technologies.

10.3.1 Existing Web Technologies and Accessibility

The web has gone through a remarkable evolution from its beginnings in 1990. The first web sites were extremely simple, consisting of little more than trivially formatted text, links, and an occasional image or two. Pages were marked up using a simple language, the hypertext markup language or HTML, and links to other pages were included as simple references according to the URL scheme. Tooling was also simple—a basic text editor and command-line FTP client were sufficient for the creation and publication of web content.

10.3.1.1 Early Evolution

The initial exponential growth of the web was powered, in part, by the fact that people could learn how to create web pages by copying the work of others. It was generally a simple matter to "View Source" and see how someone had achieved a certain layout or effect. The underlying markup could then be copied, forming the basis for new work. This had the effect of making

---

[5] Web Accessibility Initiative (WAI). http://www.w3.org/WAI.

web design patterns, both good ones and bad ones, somewhat viral, spreading through the population of content creators much like a virus spreads through a population of people (or, more recently, computers).

Unfortunately, bad design patterns, once established, had a tendency to persist. An example of a bad pattern still common today can be found in the use of tables to control page layout. The earliest web pages consisted of simple text separated by headings and further separated into paragraphs. A richer page model soon evolved that had a navigation bar (with links to other sections of the site) and an area showing the content for the current section. The navigation bar was often implemented as a column in an HTML table. Of course, some HTML tables were used (some might say *properly* used) for assembling data into semantically meaningful rows and columns. But many were used merely for layout. This is not a problem for normally sighted users, since the two uses are clearly and visibly different. But for visually impaired users it is not at all clear how a screen reader should read table content—should the columns dominate with the links inside a column being read as a related set, or should the rows dominate with column headings and associated data being read in pairs (e.g., "temperature, 70°, humidity, 80%")?

10.3.1.2 Separating Content and Presentation

Originally, HTML included presentation directives within the content itself. This conjugation of content and presentation style proved to be problematic for maintaining the device independence of the language. The desire to maintain device independence is closely linked with web pages becoming increasingly accessible, since the mechanisms allowing remapping of content to different devices go a long way toward allowing it to be mapped to different interaction modalities on any given device. This separation of content and presentation style is achieved through a confluence of two technologies: HTML and cascading style sheets (CSS) and lately their XML equivalents, XHTML (see "XHTML and XForms," later in this section) and XSL transformations (XSLT). In this model, the HTML file contains the content and the style file (CSS or XSLT) contains the presentation directives. Both files are transmitted separately to the client and assembled by the browser into the DOM tree on which the browser's final rendering is based. In this way both authors and readers can independently attach styles (e.g., fonts, colors, and spacing) to HTML documents. The CSS language is human readable and writable, and expresses style in common desktop publishing terminology. One of the fundamental features of CSS is that style sheets cascade; authors can attach a preferred style sheet, while the reader can provide a personal style sheet to further modify the final presentation.

10.3.1.3 Malformed Content

Another result of being able to easily copy and adapt others' content was that somewhat malformed HTML came to dominate the web. If a page worked well enough to be seen in the user's browser, it was good enough to copy and modify. The errors thus propagated were generally not catastrophic (if they

were, the page wouldn't have rendered well enough in the first place to warrant copying). A typical error might just be a failure to match beginning and ending tags. But the effect of these errors in the aggregate was that browsers needed to be forgiving (employing heuristics to make reasonable inferences about the authors' intentions) or most content simply could not be rendered. And the effect of the browser's being forgiving was that more and more errors were injected into the worldwide collection of web content, since they were not detected by the casual page author whose only test consisted of seeing whether their browser showed the page the way they intended. Even the *tools* that soon became available to simplify the creation of HTML were often guilty of generating malformed content. In the end, one estimate is that something in excess of 9 out of 10 existing web pages came to contain one or more HTML errors (Richards and Hanson, 2004).

While browser developers might have the resources to unscramble most malformed HTML, creators of accessibility technology may not. This generally suppresses the development of web accessibility technology. It also makes potentially attractive architectures, such as content transcoding by an intermediary server, exceedingly difficult since all the heuristics needed by a browser to cope with syntactically incorrect markup need to be replicated in the intermediary. One reaction to this problem has been the creation of stricter standards such as XHTML, which are enforced by the browser (see the following).

10.3.1.4 Rich Data Types

Early web content had only a few data types. There was, principally, text, marked up using simple HTML. There were also images, generally encoded using either the GIF or JPEG formats. Text could be transformed for accessibility purposes (barring some of the difficulties mentioned previously) rather easily. Images could be magnified as well since the formats were well known and rather straightforward. The situation now is much more complex. A wide range of objects, such as Java applets and Flash movies, can be and often are embedded in web pages. Little can be done to make most of these objects more accessible. They display their content however they do it, and no access to their underlying "model" is provided to support content transformations (although this is changing somewhat in the case of Flash, for example). Since these technologies offer a richer palette to the creative web designer, the resulting objects have typically proven to be quite complex, requiring good vision to see their parts and fine motor control to interact with them. Finally, the web user's overall task has become more complicated since the relevant run-time support for these data types must be downloaded and installed in each browser that is used. While this can be easy enough, there are cases where a skilled support staff must be engaged to find and install the necessary software. For disabled users, this can pose a formidable barrier.

10.3.1.5 Dynamic Content

Early web content was static. It was composed by an author as part of a page and rendered by a web browser. The only things a user could do was read the content, follow links to other pages, and submit simple forms back to the server. Starting with the introduction of JavaScript, however, web content started to become much more dynamic. Content in forms could be checked for correctness and completeness prior to submission. New content could be revealed on the page based on user actions without server involvement. In general, this dynamism yielded richer and more compelling user experiences. But it also made the job of creating accessibility transformations much more difficult. Consider, for example, the not uncommon case of web page content being a mix of static elements and dynamic script-driven content. Because the scripts are invoked in response to browser events (such as onLoad) or user events (such as onClick), the content is often created *in parallel* with or *after* the static content has been interpreted and represented in the browser's internal DOM. Browser-side accessibility transformations (such as that found in web page readers for the visually impaired) typically operate on the DOM. If the content is static, the code providing the transformation can be structured with the assumption that transformations can be applied to a DOM that is not changing. If the content is dynamic, the code must be structured to work on incomplete and changing DOMs. Complicating matters, the events signaled to this code by current browsers are not always complete and correct. As a result, it is not always clear when a DOM has been fully built either following initial page load or in response to a DOM-changing user event. The increasing use of Web 2.0 technologies, such as Asynchronous JavaScript and XML (AJAX; see the following), will only make these problems more vexing. Indeed, it is sometimes necessary to annotate the DOM with markers to indicate what has been transformed in a prior pass (Richards et al., 2007). This allows for touch-up of incompletely transformed DOM sections when a user action indicates that section of the document to now be of interest. Going forward, it will be important to develop a set of patterns making dynamic Web 2.0 content more accessible.

10.3.1.6 Breakdown of Browser Navigation Models

The basic mechanics of web navigation are confusing for many, especially for those with an imperfect understanding of how web technologies work or those with some cognitive disability. As discussed previously, many people have trouble comprehending the (hidden) stack model that maintains the browser's within-session page history. To make matters worse, this complex model is increasingly being violated by the implementation techniques underlying various forms of dynamic and session-dependent web content. As a result, the browser's Back button no longer reliably moves the user back through visited pages. Clicking it may now bring up a question about reposting form data. In other cases (e.g., in some web portal frameworks) clicking the Back button causes either nothing to happen or the working context to be completely abandoned (sometimes requiring the user to log back in). A similar problem has come to plague the primary means of between-session navigation—bookmarks. Bookmarks no longer reliably link to a specific

page if its content is dynamically generated based on session data. The recently articulated *REST* model (Fielding, 2000), in addition to simplifying the notion of a web service, has the desirable attribute that Back button and bookmark behaviors are more predictable.

10.3.1.7 Transcoding

Many approaches attempt to influence web content authors to create more accessible pages. As previously discussed in this chapter, these approaches include guidelines, automated validation, and best practices. They all rely, however, on authors being concerned enough about accessibility to follow them. Problems still exist with poorly built and old pages, and so solutions that aim to change bad content to good content have been developed. These solutions are known collectively as *transcoding* techniques.

As typically defined, transcoding is a technology used to adapt web content so it can be viewed on any of the increasingly diverse set of devices found in today's market. Transcoding in this context normally involves syntactic changes (Hori et al., 2000) or semantic rearrangements and chunking (Myers, 2005). However, both normally rely, to some extent, on *annotations* of the web page itself (Asakawa and Takagi, 2000). The goal of annotations for web content transcoding is to provide better support for assistive technologies, and thus for disabled users and users of small screen devices. However, annotation is expensive for content authors, because each page must be annotated with information regarding its structural context. This information must be applied to each page manually before the transcoding can work. Of course some systems exist that attempt to transcode pages based on annotations being automatically included in pages (normally from database-driven sites with templates for page creation), but these often rely on bespoke technologies, and solutions are only appropriate for one site.

Semantic transcoding aims to overcome some of the problems of using bespoke systems and technology. Using this approach, a page's encoded semantics provide the information needed to correctly infer the page's internal structure and purpose, thereby allowing the page's content to be presented in a different form that fully preserves the meaning. For the time being, however, such approaches are limited to pages built in accordance with a known template, greatly easing the analysis and transformation chore.

Each of these types of transcoding is fraught with problems. This is especially the case when sighted users and visually impaired users wish to use the same page. Automatic transcoding based on removing parts of the page results in too much information loss, and manual annotation is nearly impossible when applied to dynamic web sites (let alone the entirety of the web). Most systems use their own bespoke proxy servers or client side interfaces, and these systems require a greater setup cost in terms of user time. In practice, existing transcoding systems lean toward solving the problems of one user group and so destroy the content/structure/context for other nontarget groups.

10.3.2 Emerging Web Technologies and Accessibility

Web evolution shows no signs of slowing down. While it is very difficult to make predictions about the future of the web, a few clear trends are starting to emerge.

10.3.2.1 XHTML and XForms

The lax enforcement of HTML syntax has served as an impetus for the creation of XHTML (http://www.w3.org/TR/xhtml11). If a page is declared as containing XHTML, the syntax is checked by the browser and errors are simply signaled; no attempt is made to correct them. This makes the browser simpler and should, over time, increase the ratio of well-formed to malformed pages. Both effects would make the job of content transformation, including content transformation performed by an intermediary, much easier. Unfortunately, XHTML adoption has lagged with browsers either implementing only a portion of it or providing full error checking only if requested by the user. It remains to be seen whether the accessibility advantages associated with this technology will be realized.

XForms has arisen, in part, to provide a cleaner separation between client side data and appearance. This is a variation on the model-view separation long advocated within software engineering. In general, such a separation should make accessibility transformations more straightforward, since only the view portion of the web page needs to be modified. Indeed, one of the primary implementers of the XForms standard are mobile device manufacturers who face the challenge (related to accessibility) of providing a semantically correct re-rendering of page content originally developed for much more capable devices.

10.3.2.2 Web 2.0

A loose cluster of concepts is gaining currency under the banner of Web 2.0 (O'Reilly, 2005). Some of these concepts fall in the space of collaborative content and information development (e.g., wikis, blogs, and content tagging). Others fall in the area of content utilization (e.g., RSS and Atom, data remixing and mashups). Others represent new business models for web applications and services. Still others are best thought of as newer web programming models providing some combination of lighter-weight application development (e.g., through the use of dynamic scripting languages), more open application development (e.g., harnessing the talents of a loose confederation of developers), or richer user experiences (e.g., using AJAX for in-place data-driven updating of web pages without a page reload). It is worth noting that this latter set of concepts poses the greatest challenge to web accessibility going forward. This can be exemplified considering the potential impact of open-source development on web accessibility. Unless the governance model underlying this development has explicitly embraced web accessibility, it is unlikely that individual contributors will ensure that each new feature is accessible (although a good example of an open-source development project with sustained focus on accessibility is

provided by Mozilla, *Mozilla accessibility*[6]). Additionally, as has been mentioned, the dynamic, data-driven application model associated with AJAX makes DOM-based accessibility transformations more difficult. Time will tell which of these concepts will come to dominate the future web. It is clear, however, that web accessibility will require ongoing efforts by standards bodies, tool builders, and content developers.

10.3.2.3 Semantic Web

The vision of the Semantic Web, as articulated by Tim Berners-Lee (1999), is of a web in which resources are accessible not only to humans, but also to automated processes. The automation of tasks depends on elevating the status of the web from machine-readable to something called machine-understandable. The key idea is to have data on the web defined and linked in such a way that meaning is explicitly interpretable by software processes rather than just being implicitly interpretable by humans. To realize this vision, it will be necessary to annotate web resources with metadata (i.e., data describing the resource's content and nature). Such metadata will, however, be of limited value to automated processes, unless they share a common understanding as to their meaning. This sharing of meaning will be achieved partly through the use of ontologies. An ontology, defined using the new W3C language *OWL*, is a collection of shared terms that can be communicated between both people and applications. A reasoning engine can then make inferences about the relationships between items within an ontology and support queries over collections of items.

The Semantic Web is a vision that will take years to come to fruition. There are aspects of this vision that promise a more accessible future. First, increased machine processing may eliminate much of the tedium now associated with finding web content. For a sense of what we may expect, compare the current difficulty of maintaining a useful set of bookmarks (which tend to rapidly age into a useless jumble of often broken links) with the ease of conducting a Google search to find just what one needs in the moment of need. Of course, this barely hints at what might happen when information finding is guided not just by keywords and patterns of interpage links but by a partial *understanding* of the meaning of the distributed content (and when search occurs not just for fractions of a second but for the hours and days we might anticipate in a future world of long-running software agents). A second benefit for web accessibility may well derive from a greatly expanded use of metadata that allows for transforming content more meaningfully. Recalling the example of the use of tables to control page layout, tables annotated with their purpose (layout vs. data structuring) would allow a screen reader to behave more sensibly than is possible now. In the case of a table used for page layout, the screen reader could vocalize the navigation bar links as a coherent set (treating the column as primary). In the case of a table used for data aggregation, it could read the column names and associated data values together (treating the row as primary).

---

[6] Mozilla accessibility. http://www.mozilla.org/access.

10.3.2.4 Widgets

The last emerging web technology to be discussed points beyond the current world of generic browsers making explicit requests of discreet servers to a world of highly customized information gadgets delivering fresh information in glanceable ambient displays. Various so-called widget run-times are now becoming available to make this possible. These include operating-specific run-times (such as the desktop environments provided by Apple and Microsoft) and cross-platform run-times (such as the widget environments from Google, Yahoo, and Mozilla). Widgets are relatively easy to build, tending to be both small (from a coding perspective) and familiar (using a standard set of browser technologies such as HTML, XML, CSS, and JavaScript). While widgets can often run *in* a browser, they more typically run *free* of the browser and without any unneeded browser interface apparatus. At the time of this volume, thousands of widgets are already available for download. Of course, the proliferation of widgets may lead to a confusing set of nonstandard interface techniques and require the user to laboriously manage an ever growing widget collection. However, widgets may provide the ability to create purpose-built information environments perfectly suited for a person's preferences and abilities. For example, a not too distant future can be anticipated in which a severely motor-impaired individual using a scanning interface could efficiently control a custom-built information display rather than a complex generic browser.

10.4 Conclusions

The World Wide Web is a vital resource for people with disabilities. But even with well-motivated regulations and a growing set of relevant standards, web accessibility remains a challenge. In part this is because there are no good ways of automatically transforming all existing web content to meet the diverse needs of people with widely varying vision, hearing, motor, cognitive, and age-related disabilities. This situation will improve over time through both increased developer awareness and the utilization of new technologies. As more content becomes better structured, and as more metadata and semantic information are added, it can be expected that more people will take full advantage of the web.

There is no catch-all solution to providing accessibility to web-based resources for disabled users. Each disability requires a different set of base access technologies and the flexibility for those technologies to be personalized by the end-user.

References

Alliance for Technology Access (2004). *Computer Resources for People with Disabilities: A Guide to Assistive Technologies, Tools, and Resources for People of All Ages* (4th ed.). Alameda, CA: Hunter House Publishers.

Asakawa, C. and Takagi, H. (2000). Annotation-based transcoding for non-visual web access, in the *Proceedings of the Fourth International ACM Conference on Assistive Technologies,* 13–15 November 2000, Arlington, Virginia, pp. 172–179. New York: ACM Press.

Ayers, E. and Stasko, J. (1995). Using graphic history in browsing the World Wide Web, in the *Proceedings of the Fourth International World Wide Web Conference,* 11–14 December 1995, Boston. http://www.w3.org/Conferences/WWW4/Papers2/270.

Berners-Lee, T. (1999). *Weaving the Web.* London: Orion Business Books.

Bohman, P. (2004). *Cognitive Disabilities Part 1: We Still Know Too Little, and We Do Even Less.* http://www.webaim.org/articles/cognitive/cognitive_too_little.

Boyer, J. M. (2007). XForms. http://www.w3.org/TR/Xforms11.

Brewer, J. (2005). *How People with Disabilities Use the Web.* http://www.w3.org/WAI/EO/Drafts/PWD-Use-Web.

Brewster, S., Raty, V., and Kortekangas, A. (1996). Enhancing scanning input with non-speech sounds, in the *Proceedings of the Second Annual ACM Conference on Assistive Technologies (ASSETS '96),* 11–12 April 1996, Vancouver, British Columbia, pp. 10–14. New York: ACM Press.

Brynjolfsson, E., Hu, Y., and Smith, M. D. (2003). Consumer surplus in the digital economy: Estimating the value of increased product variety at online booksellers. *Management Science* 49. April 2003 working paper version available via the Social Science Research Network.

Caldwell, B., Chisholm, W., Slatin, J., and Vanderheiden, G. (eds.) (2006). *Web Content Accessibility Guidelines 2.0.* http://www.w3.org/TR/2004/WD-WCAG20-20041119.

Caldwell, B., Cooper, M., Reid, L. G., Vanderheiden, G., and White, J. (2008). *Web Content Accessibility Guidelines 2.0.* http://www.w3.org/TR/2008/CR-WCAG20-20080430; http://www.w3.org/TR/WCAG20.

Caring Family (2007). http://www.caringfamily.com/public/service/how_it_works.cfm.

Centers for Disease Control: Facts about Traumatic Brain Injury (TBI) (2006). http://www.cdc.gov/ncipc/tbi/FactSheets/Facts_About_TBI.pdf.

Chadwick-Dias, A., Bergel, M., and Tullis, T. (2007). Senior surfers 2.0: A re-examination of the older web user and the dynamic web, in *Universal Access in Human Computer Interaction: Coping with Diversity, Proceedings of the 4th International Conference on Universal Access in Human-Computer Interaction* (appears as Volume 5 of the combined Proceedings of HCI International 2007; C. Stephanidis, ed.), 22–27 July 2007, Beijing, pp. 868–876 (LNCS 4554). Berlin, Heidelberg: Springer.

Chisholm, W. and Kasday, L. (2005). *Evaluation, Repair, and Transformation Tools for Web Content Accessibility.* http://www.w3.org/WAI/ER/existingtools.html.

Chisholm, W., Vanderheiden, G., and Jacobs, I. (eds.). (2001). Web content accessibility guidelines 1.0. *Interactions* 8: 35–54.

Cockburn, A., Greenberg, S., Jones, S., Mckenzie, B., and Moyel, M. (2003). Improving web page revisitation: Analysis, design, and evaluation. *IT and Society* 1: 159–183.

Conrad, R. (1979). *The Deaf Schoolchild*. London: Harper & Row.

Core techniques for web content accessibility guidelines 1.0 (2000). http://www.w3.org/TR/WCAG10-CORE-TECHS/#audio-information.

Craven, J. and Bophy, P. (2003). Non-visual access to the digital library: The use of digital library interfaces by blind and visually impaired people. *Library and Information Commission Research Report* 145. Manchester Centre for Research in Library and Information Management.

Czaja, S. J. and Lee, C. C. (2007). Information technology and older adults, in *Human-Computer Interaction Handbook: Fundamentals, Evolving Technologies and Emerging Applications* (2nd ed.) (A. Sears and J. A. Jacko, eds.), pp. 777–792. Mahwah, NJ: Lawrence Erlbaum Associates.

Disability Rights Commission (2004). *The Web: Access and Inclusion for Disabled People*. Manchester: Disability Rights Commission.

Dorsey, E., Constantinescu, R., Thompson, J., Biglan, K., Holloway, R., Kieburtz, K., et al. (2007). Projected number of people with Parkinson disease in the most populous nations, 2005 through 2030. *Neurology* 68: 384–386.

Dyslexia Research Institute (2007). http://www.dyslexia-add.org.

ElderMail (2007). https://eldermail.net.

Elkind, J. (1998). Computer reading machines for poor readers. *Perspectives* 24: 4–6.

Engelen, J., Evenepoel, F., Bormans, G., Astbrink, G., Bühler, C., Ekberg, J., et al. (1999). Producing web pages that everyone can access. *TIDE-Harmony Project (DE1226)*. http://www.stakes.fi/cost219/webdesign.htm.

Faye, E. E. and Stappenbeck, W. (2000). *Changes in the Aging Eye*. New York: Lighthouse International.

Fielding, R. T. (2000). Architectural styles and the design of network-based software architectures. PhD dissertation, University of California, Irvine. http://www.ics.uci.edu/~fielding/pubs/dissertation/rest_arch_style.htm.

Fox, S. (2001). *Wired Seniors, Pew Internet and Family Life*. http://www.pewinternet.org/pdfs/PIP_Wired_Seniors_Report.pdf.

Fox, S. (2004). *Older Americans and the Internet*. http://www.pewinternet.org/pdfs/PIP_Seniors_Online_2004.pdf.

Generations On Line (2006). http://www.generationsonline.com.

Gregor, P. and Newell, A. F. (2000). An empirical investigation of ways in which some of the problems encountered by some dyslexics may be alleviated using computer techniques, in the *Proceedings of the Fourth International ACM Conference on Assistive Technologies, Assets '00*, 13–15 November 2000, Arlington, VA, pp. 85–91. New York: ACM Press.

Grossman, T. and Balakrishnan, R. (2005). The bubble cursor: Enhancing target acquisition by dynamic resizing of the cursor's activation area, in the *Proceedings of CHI 2005: ACM CHI Conference on Human Factors in Computing Systems*, 2–7 April 2005, Portland, OR, pp. 281–290. New York: ACM Press.

Hanson, V. L. (2007). Computing technologies for deaf and hard of hearing users, in *Human-Computer Interaction Handbook: Fundamentals, Evolving Technologies and Emerging Applications* (2nd ed.) (A. Sears and J. A. Jacko, eds.), pp. 885–893. Mahwah, NJ: Lawrence Erlbaum Associates.

Hanson, V. L., Brezin, J., Crayne, S., Keates, S., Kjeldsen, R., Richards, J. T., et al. (2005). Improving Web accessibility through an enhanced open-source browser. *IBM Systems Journal* 44: 573–588.

Hanson, V. L., Richards, J. T., and Lee, C. C. (2007). Web access for older adults: Voice Browsing?, in *Universal Access in Human Computer Interaction: Coping with Diversity* (*Proceedings of the 4th International Conference on Universal Access in Human-Computer Interaction*; Volume 5 of the Combined Proceedings of HCI International 2007; C. Stephanidis, ed.), 22–27 July 2007, Beijing, China, pp. 904–913 (LNCS 4554). Berlin, Heidelberg: Springer.

Harryson, B., Svensk, A., and Johansson, G. I. (2004). How people with developmental disabilities navigate the Internet. *British Journal of Special Education* 31: 138–142.

Holt, J., Hotto, S., and Cole, K. (1994). *Demographic Aspects of Hearing Impairment: Questions and Answers* (3rd ed.). http://gri.gallaudet.edu/Demographics/factsheet.html.

Hori, M., Kondoh, G., and Ono, K. (2000). Annotation-based web content transcoding, in the *Proceedings of the Ninth International World Wide Web Conference*, 15–19 May 2000, Amsterdam, The Netherlands, pp. 197–211.

Huisstede, B., Bierma-Zeinstra, S., Koes, B., and Verhaar, J. (2006). Incidence and prevalence of upper-extremity musculoskeletal disorders. A systematic appraisal of the literature. *BMC Musculoskeletal Disorders* 7. http://www.pubmedcentral.nih.gov/articlerender.fcgi?artid=1434740.

Jacobs, I., Gunderson, J., and Hansen, E. (eds.) (2002). *User Agent Accessibility Guidelines 1.0*. http://www.w3.org/TR/2002/REC-UAAG10-20021217.

Jacobs, J., Hammerman-Rozenberg, R., Maaravi, Y., Cohen, A., and Stessman, J. (2005). The impact of visual impairment on health, function and mortality. *Aging Clin Exp Res* 17: 281–286.

JAWS (2006). *JAWS® for Windows*. http://www.freedomscientific.com/fs_products/JAWS_HQ.asp.

Jiwnani, K. (2001). *Designing for Users with Cognitive Disabilities*. http://www.otal.umd.edu/uupractice/cognition.

Keates, S. and Trewin, S. (2005). Effect of age and Parkinson's disease on cursor positioning using a mouse, in the *Proceedings of the 7th International ACM SIGACCESS Conference on Computers and Accessibility (Assets '05)*, 9–12 October 2005, Baltimore, pp. 68–75. New York: ACM Press.

Kennaway, J. R., Glauert, J. R., and Zwitserlood, I. (2007). Providing signed content on the Internet by synthesized animation. *ACM Transactions on Computer-Human Interaction (TOCHI)* 14.

Kimball, R. (2007). *Games[CC]: See the Sound*. http://gamescc.rbkdesign.com.

Krigger, K. (2006). Cerebral palsy: An overview. *American Family Physician* 73: 91–104. http://www.aafp.org/afp/20060101/91.html.

Lethbridge-Cejku, M., Rose, D., and Vickerie, J. (2006). Summary health statistics for U.S. adults: National Health Interview Survey, 2004. National Center for Health Statistics. *Vital and Health Statistics* 10. http://www.cdc.gov/nchs/data/series/sr_10/sr10_228.pdf.

Majaranta, P. and Räihä, K. (2002). Twenty years of eye typing: Systems and design issues, in the *Proceedings of the 2002 Symposium on Eye Tracking Research and Applications*, 24–27 March 2002, New Orleans, pp. 15–22. New York: ACM Press.

Mankoff, J., Dey, A., Batra, U., and Moore, M. (2002). Web accessibility for low bandwidth input, in the *Proceedings of ASSETS 2002: The Fifth International ACM Conference on Assistive Technologies*, 8–10 July 2002, Edinburgh, UK pp. 17–24. New York: ACM Press.

McCathieNevile, C. and Abou-Zahra, S. (2005). *Evaluation and Report Language (EARL) 1.0.* http://www.w3.org/TR/EARL10.

Myers, W. (2005). BETSIE: BBC Education Text to Speech Internet Enhancer. *British Broadcasting Corporation (BBC) Education.* http://www.bbc.co.uk/education/betsie.

National Center for Accessible Multimedia (n.d.). http://ncam.wgbh.org/richmedia.

National Center for Learning Disabilities (2007). http://www.ncld.org/content/view/447/391.

Newell, A. F., Carmichael, A., Gregor, P., Alm, N., and Waller, A. (2007). Information technology for cognitive support, in *Human-Computer Interaction Handbook: Fundamentals, Evolving Technologies and Emerging Applications* (2nd ed.) (A. Sears and J. A. Jacko, eds.), pp. 811–828. Mahwah, NJ: Lawrence Erlbaum Associates.

Newell, A. F., Dickinson, A., Smith, M. J., and Gregor, P. (2006). Designing a portal for older users: A case study of an industrial/academic collaboration. *ACM Transactions on Computer-Human Interaction* 13: 347–375.

NIA (2002). *Making Your Web Site Senior Friendly: A Checklist.* http://www.nlm.nih.gov/pubs/checklist.pdf.

NIDCD (n.d.). *Statistics about Hearing Disorders, Ear Infections, and Deafness.* http://www.nidcd.nih.gov/health/statistics/hearing.asp.

Nielsen, J. (2005). *Lower Literacy Users.* http://www.useit.com/alertbox/20050314.html.

Odding, E., Roebroeck, M., and Stam, H. (2006). The epidemiology of cerebral palsy: Incidence, impairments and risk factors. *Disability and Rehabilitation* 28: 183–191. http://www.medscape.com/medline/abstract/16467053?src=emed_ckb_ref_0.

O'Reilly, T. (2005). *What Is Web 2.0.* http://www.oreillynet.com/pub/a/oreilly/tim/news/2005/09/30/what-is-web-20.html.

Padden, C. and Humphries, T. (1988). *Deaf in America: Voices from a Culture.* Cambridge, MA: Harvard University Press.

Parker, B. A. and Scharff, L. F. (1998). Influences of contrast sensitivity on text readability in the context of a graphical user interface. Unpublished manuscript. http://hubel.sfasu.edu/research/agecontrast.html.

Petrie, H., Hamilton, F., and King, N. (2004). Tension, what tension?: Website accessibility and visual design, in the *Proceedings of the International Cross-Disciplinary Workshop on Web Accessibility*, 7–8 May 2004, Banff, Canada, pp. 13–18.

Poulson, D. and Nicolle, C. (2004). Making the internet accessible for people with cognitive and communication impairments. *Universal Access in the Information Society* 3: 48–56.

Rabin, J. and McCathieNevile, C. (eds.) (2006). *Mobile Web Best Practices 1.0. Basic Guidelines.* W3C Recommendation. http://www.w3.org/TR/mobile-bp.

Ramakrishnan, I. V., Stent, A., and Yang, G. (2004). Hearsay: Enabling audio browsing on hypertext content, in the *Proceedings of the 13th International Conference on World Wide Web (WWW '04)*, 17–22 May 2004, New York, pp. 80–89. New York: ACM Press.

Rice, C. (2007). Prevalence of autism spectrum disorders: Autism and developmental disabilities monitoring network, 14 sites, United States, 2002. *Morbidity and Mortality Weekly Report, Surveillance Summaries* 56, pp. 12–28. http://www.cdc.gov/MMWR/preview/mmwrhtml/ss5601a2.htm.

Richards, J. T. and Hanson, V. L. (2004). Web accessibility: A broader view, in the *Proceedings of the Thirteenth International ACM World Wide Web Conference (WWW '04)*, 17–22 May 2004, New York, pp. 72–79. New York: ACM Press.

Richards, J. T, Hanson, V. L., Brezin, J., Swart, C., Crayne, S., and Laff, M. (2007). Accessibility works: Enhancing Web accessibility in Firefox, in *Universal Access in Human-Computer Interaction: Applications and Services, Proceedings of the Fourth International Conference on Universal Access in Human-Computer Interaction* (appears as Volume 7 of the combined Proceedings of HCI International 2007; C. Stephanidis, ed.), 22–27 July 2007, Beijing, pp. 133–142 (LNCS 4556). Berlin, Heidelberg: Springer.

RNID Statistics (n.d.). http://www.rnid.org.uk/information_resources/aboutdeafness/statistics.

Sa-nga-ngam, P. and Kurniawan, S. (2007). An investigation of older persons' browser usage, in *Universal Access in Human Computer Interaction: Coping with Diversity, Proceedings of the 4th International Conference on Universal Access in Human-Computer Interaction* (appears as Volume 5 of the combined Proceedings of HCI International 2007; C. Stephanidis, ed.), 22–27 July 2007, Beijing, pp. 1000–1009 (LNCS 4554). Berlin, Heidelberg: Springer.

Schrepp, M. (2006). On the efficiency of keyboard navigation in Web sites. *Universal Access in the Information Society* 5: 180–188.

Schwerdtfeger, R. and Gunderson, J. (2006). Roadmap for accessible rich internet applications. W3C Working Draft. http://www.w3.org/TR/aria-roadmap.

Sevilla, J., Herrera, G., Martínez, B., and Alcantud, F. (2007). Web accessibility for individuals with cognitive deficits: A comparative study between an existing commercial Web and its cognitively accessible equivalent. *ACM Transactions on Computer-Human Interaction (TOCHI).* http://doi.acm.org/10.1145/1279700.1279702.

Shaywitz, S. E., Shaywitz, B. A., Pugh, K. R., Fulbright, R. K., Constable, R. T., Mencl, W. E., et al. (1998). Functional disruption in the organization of the brain for reading in dyslexia. *Proceedings of the National Academy of Sciences of the United States of America* 95: 2636–2641.

Small, J., Schallau, P., Brown, K., and Appleyard, R. (2005). Web accessibility for people with cognitive disabilities, in *CHI '05 Extended Abstracts on Human Factors in Computing Systems*, pp. 1793–1796. New York: ACM Press.

Starner, T. (2004). Keyboards Redux: Fast mobile text entry. *IEEE Pervasive Computing* 3: 97–101.

Treviranus, J, McCathieNevile, C., Jacobs, I., and Richards, J. (eds.) (2000). *Authoring Tool Accessibility Guidelines 1.0.* http://www.w3.org/TR/2000/REC-ATAG10-20000203.

Treviranus, J, McCathieNevile, C., Richards, J., and May, M. (eds.) (2004). *Authoring Tool Accessibility Guidelines 2.0.* W3C Working Draft. http://www.w3.org/TR/2004/WD-ATAG20-20041122.

Trewin, S. (2006). Physical usability and the mobile Web, in *Proceedings of the WWW 2006 International Cross-Disciplinary Workshop on Web Accessibility (W4A): Building the Mobile Web: Rediscovering Accessibility?* 23–26 May 2006, Edinburgh, Scotland, pp. 109–112. New York: ACM Press.

Trewin, S., Laff, M., Cavender, A., and Hanson, V. L. (2008). Accessibility in virtual worlds, in *CHI '08 Extended Abstracts on Human Factors in Computing Systems*, pp. 2727–2732. New York: ACM Press.

Trewin, S. and Pain, H. (1999). Keyboard and mouse errors due to motor disabilities. *International Journal of Human-Computer Studies* 50: 109–144.

Tullis, T. S. (2007). Older adults and the web: Lessons learned from eye-tracking, in *Universal Access in Human Computer Interaction: Coping with Diversity, Proceedings of the 4th International Conference on Universal Access in Human-Computer Interaction* (appears as Volume 5 of the combined Proceedings of HCI International 2007; C. Stephanidis, ed.), 22–27 July 2007, Beijing, pp. 1030–1039 (LNCS 4554). Berlin Heidelberg: Springer.

Vassallo, S. (2003). *Enabling the Internet for People with Dyslexia.* http://www.e-bility.com/articles/dyslexia.php.

Wakita, K. and Shimamura, K. (2005). Smartcolor: Disambiguation framework for the colorblind, in the *Proceedings of the 7th international ACM SIGACCESS Conference on Computers and Accessibility (Assets '05)*, 9–12 October 2005, Baltimore, pp. 158–165. New York: ACM Press.

Web Content Accessibility Guidelines Working Group (WCAG WG) (2006). http://www.w3.org/WAI/GL.

Web Design for Dyslexic Users (2007). http://www.dyslexia.com/qaweb.htm.

Worden, A., Walker, N., Bharat, K., and Hudson, S. (1997). Making computers easier for older adults to use: Area cursors and sticky icons, in the *Proceedings of CHI 1997: ACM CHI Conference on Human Factors in Computing Systems*, 22–27 March 1997, Atlanta, pp. 266–271. New York: ACM Press.

World Health Organization (2004). *Magnitude and Causes of Visual Impairment.* http://www.who.int/mediacentre/factsheets/fs282/en.

Xie, B. (2006). Perceptions of computer learning among older Americans and older Chinese. *First Monday* 11. http://firstmonday.org/issues/issue11_10/xie/index.html.

Yesilada, Y., Stevens, R., Harper, S., and Goble, G. (2007). Evaluating DANTE: Semantic transcoding for visually disabled users. *ACM Transactions on Computer-Human Interaction (TOCHI)* 14.

Zajicek, M. (2001). Supporting older adults at the interface, in *Universal Access in HCI: Towards an Information Society for All* (Volume 3 of the Proceedings of the 9th International Conference on Human-Computer Interaction (HCI International 2001; C. Stephanidis, ed.), 9–12 September 2001, Singapore, pp. 454–458. Mahwah, NJ: Lawrence Erlbaum Associates.

Zajicek, M. (2003). Patterns for encapsulating speech interface design solutions for older adults, in the *Proceedings of the 2nd ACM SIGCHI and SIGCAPH Conference Computers and Universal Usability, CUU 2003*, 10–11 November 2003, Vancouver, Canada, pp. 54–60. New York: ACM Press.

Zajicek, M. (2007). Web 2.0: Hype or happiness?, in the *Proceedings of the 2007 International Cross-Disciplinary Conference on Web Accessibility (W4A '07)*, 7–8 May 2007, Banff, Canada, pp. 35–39. New York: ACM Press.

Handheld Devices and Mobile Phones

Anne Kaikkonen,
Eija Kaasinen, and
Pekka Ketola

11.1 Definition and Requirements of Mobility

11.1.1 Mobility

One way to define *mobility* is by saying that it is about going from place A to place B. It also means being away from somewhere, such as home. These are, however, simplified pictures of mobility. People have different motivations for moving around, and while on the move they also may stop and seek a quiet spot in which to do their activities. Location also has different meanings for people in different situations—and the perceptions of location and time are different in different cultures and may change over time.

Kristoffersen and Ljungberg (1999) define three types of mobility, depending on people's motivation to move around: traveling, visiting, and wandering. Traveling is going from one

place to another. A daily life example could be commuting: going to work from home. Visiting is spending time in a place temporarily before moving to another place, for example, visiting a friend or going to the movies when returning home from work. Wandering is extensive local mobility in a building or local area, like spending time in a mall.

Kakhira and Sørensen (2002) argue that mobility is not just being on the move but, far more important, is related to the interaction between mobile people—this is the way in which people interact with each other in their social lives. Therefore, they suggest expanding the mobility concept by three interrelated dimensions of human interaction: spatial, temporal, and contextual mobility. Spatial mobility means that not only the person moves, but also the objects carried by the person move, as well as symbols and spaces. Temporal mobility is related to how people perceive time and space. Contextual mobility means the change of environment, other objects, and people.

A more detailed discussion of mobility involving issues of transportation is provided in Chapter 49, "Intelligent Mobility and Transportation for All" in this handbook.

11.1.2 Contexts of Use

Mobility faces the challenge that contexts of use vary a lot and may change even in the middle of usage situations. The variable usage contexts need to be taken into account when designing mobile devices and services. The initial assumption of mobile devices and services is that they can be used "anywhere." This assumption may not always be correct; the environment and the context create challenges in use. Using mobile phones is prohibited in some environments, and in some places there may not be network coverage.

The user can be physically or temporarily *disabled*. In dark or bright environments it may be hard to see the user-interface elements. In a crowded place it may be difficult to carry a voice conversation over the phone—even more difficult than in a face-to-face situation in which you can use nonverbal cues to figure out what the other person says if you do not hear every word. In social communication the context does play an important role; people often start telephone conversations by asking about the other person's physical location (e.g., "Where are you?") to figure out whether the context of the other person allows for the phone call.

In addition to physical limitations, it may not be socially acceptable to use a mobile device in every situation. Phone usage, especially calling, may irritate bystanders (Love and Perry, 2004). There are also situations where even text messaging would not be appropriate. Having a phone on in places with microphones may disrupt the audio systems, even if the phone would be on silent. The use of cameras is prohibited in some places and hence also the use of camera phones may not be acceptable. Using a mobile phone in a public place may change the perception of the nature of the place. Puro (2002) says that people create their own private places in the middle of public spaces; they may be physically present in place but mentally and socially elsewhere. Also Cui and Roto (2008) discuss this phenomenon as a motivation to browse on mobile phones in public places. There is no doubt that this affects the way other people in the physical location approach the person using the phone, and how people tend to forget their environment while they are speaking on the phone.

11.1.3 Devices and Services

Mobile devices are especially crucial for personal communication, but they are being used for many other tasks as well. In many areas, mobile devices are replacing their fixed line (telephone) or simply bigger (television, computer) counterparts. It is increasingly important to ensure that all users have access to mobile devices and services.

Still a few years ago there was a smaller number of companies involved in mobile service development: mobile device manufacturers, mobile network operators, and a small number of companies developing hardware and software for the device manufacturers and operators. Today the situation is different, as there are many more companies involved in the entire development and service provision process. There are still mobile device manufacturers, but a great deal of device hardware and software is being developed by other companies than the manufacturers themselves. Mobile devices have turned into application platforms for which many different companies may offer applications. There are also several companies involved in developing add-on hardware such as user-interface devices.

Mobile networks already provide data rates sufficient for multimedia services, and 3G networks even for real-time video connections. Broadcast mobile TV with add-on services has been launched in many countries. Mobile devices are increasingly equipped with fast WLAN connections. Local connectivity possibilities are offered by technologies such as Bluetooth, NFC (Near Field Communication), and ULP (Ultra Low Power Bluetooth). Local connectivity facilitates mobile devices to easily access data from other devices or the environment. Different networks are suitable for different usage purposes and different contexts of use. In the future, the devices will increasingly include several network capabilities, giving flexible connectivity possibilities for the users, but also raising challenges in selecting which network to use.

When mobile applications connect to services, there is the network in-between. The network is maintained by network operator (or carrier), and contains elements manufactured by a large number of companies. The service or portal is developed by a portal/service developer that can also be a network operator. The content can be developed in practice by anyone—professional company or individual persons. As an example, Figure 11.1 shows how many different players can be involved when a user downloads a piece of music to his device. Even though there are several players involved, the user should get transparent access to the content that he is interested in.

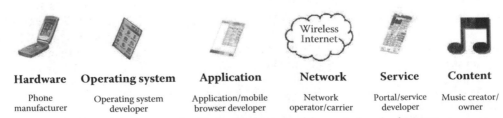

Hardware	Operating system	Application	Network	Service	Content
Phone manufacturer	Operating system developer	Application/mobile browser developer	Network operator/carrier	Portal/service developer	Music creator/owner

FIGURE 11.1 An example of different companies involved in the development of the music download process.

11.2 Mobile Users

11.2.1 For Whom Is the Technology Designed?

In principle, mobile technology is developed for everybody. But taking a look at the devices and services, one has to admit that mobile technology is developed mostly for healthy young adults in developed countries. The devices are small and have shining surfaces; they have quite small keys, and displays have text with small fonts. The trend toward small and aesthetic devices causes problems even for young people to hold the device, press the right buttons, and see the key labels and the text on the display. As the amount of applications and features in mobile devices increases, the structure of the overall system becomes increasingly complex.

Small children, elderly people, and people with functional limitations and impairments are often challenged when using mobile terminals and services, even if their requirements are increasingly taken into consideration by the industry, more commonly applying design-for-all principles in product development. Fully accessible terminals and services are still rare and difficult to achieve.

The main usability challenges with mobile devices, applications, and services are related to the widespread user population, the restrictions of the devices and technical infrastructures, as well as the varying and demanding contexts of use. Short instantaneous usage sessions are typical of mobile usage, and the usage often takes place in parallel with some other activity. Especially actions that include excess user input are prone to usability problems, due to the limited user-interface facilities on mobile devices. Transparent menus and clear feedback are especially important for younger and older users, but they will benefit all users. With mobile consumer services, the widespread population indicates a large variety of client devices, bringing in the need to adapt the services for different devices. Mobile services should be accessible with all devices, not just the most modern ones. However, limitations on device functionality may make some tasks unsuitable for some device types.

If the situation during the use cycle of terminals and services seems to be somewhat difficult, it is even more worrying during the setup and configuration phase of these products, when the user starts using a new terminal, system, or service for the very first time, or wants to replace an existing terminal with a new one. Even if it would be important, it is unusual that design-for-all efforts are taken into consideration for the steps needed to activate and configure mobile terminals and services. It is often thought that small children, elderly people, and people with functional limitations and impairments have someone to help with the device configurations.

Barriers to the use of services also include complexity of setup, unclear pricing structures, including roaming, and poor upgrade paths. The most commonly reported customer problems are setup and initial configuration topics, network failures, and wireless hardware issues. Configuring a mobile terminal to use a service is one of the most difficult tasks the user faces.

Mobile terminals are fundamentally based on communication with other devices and services. The primary target for the developers is to optimize the device and support mechanisms for the setup of the devices and clear implementation of the primary service use cases.

11.2.2 Different Special Groups and Disabilities

Depending on the individual case, users with deficiencies may require different kinds of attention. It is important to provide multimodal input and output methods to allow all users to choose the interaction patterns that suit them best.

It is important to keep in mind that in mobile devices, multimodal input and output methods help all users, not only those with permanent disabilities. Having vibration or other ways than sound to alert the user is good not only for hearing-impaired users, but for all users. In the mobile context, there is often background noise. Visually impaired users may benefit from navigation system voice prompts or initiating a voice call if those are designed properly. Voice prompting helps other users in situations when the visual focus is not on the device, such as when driving. Possibilities of having large text and good contrasts on display, and designing keyboards with tactile cues, are good for visually impaired users, but also for everybody.

In addition to the situations in which anyone can benefit on multimodality, design for all is also important as everyone grows old or can be temporarily disabled, due to illness, for example. In such situations people want to use the devices and services they are familiar with.

11.2.2.1 Guidelines

1. Provide multimodal input and output: audio, visual, sensory when possible.
2. Provide ways to modify input and output to match personal needs.
3. If possible, provide text-to-speech and speech-to-text functions.
4. Provide ways to adjust audio volume.
5. Make sure there is an option for large font and very good contrast on display.
6. Provide keys with clear tactile cues.

11.2.3 Aging Users as Users of Mobile Phones

Aging users may have both physical and cognitive deficiencies that restrict their ability to use mobile devices (see Chapter 8, "Age-Related Difference in the Interface Design Process"). Designing for very old people (fourth age) may be particularly challenging; they usually have more than one sense impaired and cannot compensate with other senses. Today's older users have very different backgrounds and experiences in relation to technology than younger generations, as presented in the special issue of *Interactions* (Livingston, 2007). These experiences affect perception of the system and its functionality.

This situation will change when the generations born after World War II become older—the generations that have higher education, are familiar with the technology, and have become accustomed to using new technology. This is not often taken into consideration in papers addressing the needs of older users in relation to technology (Zieffle and Bay, 2005; Kurniawan et al., 2006; Pattison and Stedmon, 2006). Lähteenmäki and Kaikkonen (2005) show that even among aged users, there is a big difference between people ages 52–65 and 70–81. Younger users have been better educated and have used mobile phones (and other communication technology) more frequently, and their usage patterns have been more diverse compared to older users.

Even if we manage to overcome some of the cognitive barriers with more technically experienced older users, there are natural physiological changes that happen when people grow old. The physical age affects requirements people have regarding their products, because most of them have some minor, slowly developing changes in vision, hearing, dexterity, and learning speed. However, learning speed is one of those attributes that might be a smaller problem when the education level and technology knowledge base of the older population increases over time.

According to Tuomainen and Haapainen (2003), relatively small size (e.g., size of Nokia 6210) and a stylish look turned out to be important factors for the active elderly. These were even the main drivers when choosing a phone to buy. This could be interpreted that not all aging users want to choose devices that look like ones designed for special groups, but rather devices that anyone could use and buy. The biggest problem they see in use of the mobile phones by aging users is the lack of support when learning to use the phone. There are several phone models designed for elderly users, though the one that seems to be most successful is the Japanese Raku-Raku (see, e.g., http://www.symbian.com)—the phone targeted for aged users that does not look like a device found only in stores with special aids for people with disabilities.

Ziefle and Bay (2004) studied mental models of cellular phone menus with users of different ages. Their studies demonstrated that users' mental models of how a mobile phone menu is structured significantly influenced their navigation performance. Their results indicate that older users today may need more transparent menu structures with hints that give feedback of the menu structure and where in the structure the user currently is.

Pattison and Stedmon (2006) divide the problems older users face in four different categories: vision, hearing, hand function, and cognitive process. It is clear that the first three changes will affect the use of mobile technology into the future, with many users now familiar with technology aging. It remains to be seen how well the older users of tomorrow can compensate for the changes in learning process and memory with their existing skills with technology, as they are more familiar with technology and learning to use new interaction strategies than today's older users. Leikas and Penttilä (2007) find three different perspectives to solutions to mobile technology for elderly users. The

first is improvements that compensate for decreasing senses. This offers solutions very similar to the ones Pattison and Stedmon offer in their four categories. In addition to these physiological and cognitive challenges, Leikas and Penttilä bring up two application-and service-focused approaches: mobile phones as information channels and health monitoring services. These can apply to any user groups. Key issues in the development of these services and applications are, for example, simplifying the processes and interactions, minimizing the typing, simplifying and helping with everyday life (e.g., different kinds of reminders). Leikas and Penttilä (2007) also found that elderly people often have problems with setting up the devices, for example, transferring data from their old phone to a new one. Elderly people often found user manuals too complex and were missing simplified guidance.

11.2.3.1 Guidelines

1. Design systems that are, by default, suitable for all.
2. Taking ergonomics into consideration:
   a. Vision: Good contrast, decent font size, and adequate illumination.
   b. Hearing: Different frequencies in audio, avoiding high frequencies in feedback, applying multiple channels for feedback.
   c. Dexterity (properly sized, easily distinguishable keys, not too small devices, no slippery materials) is taken into consideration.
3. Simplifying interactions whenever possible.
4. Providing clear, step-by-step instructions for device, application, or service use. Make it clear for users where the help can be found (both on-paper and on-the-device guidance may be needed).

11.2.4 Small Children as Users of Mobile Phones

Children are often assumed to be able to act naturally with technology; however, small children's cognitive development is still not yet complete. Children may not have problems with senses: most can see, hear, and feel well enough, but the complexity of the mobile phones may be too much for them cognitively and motorically.

For example, according to Piaget's theory of cognitive development, small children perceive the environment differently from adults. Until the age of 12, children process information in a different way compared to adults. Despite criticism of Piaget's theory, the different ways of explaining for the differences, the different approaches of taking cultural issues and education into consideration, no one has argued that children think in the same way as adults. Children do not have the same experience of life as adults. These differences affect the perception of the environment in different ways; for example, small children have difficulty differentiating commercials from TV programs (Bjurström, 1994).

Cognitive skills of small children make interface design challenging. Children may not read well, and their vocabulary is not as good as those of teenagers. Kano et al. (2006) argue that all the sentences used in text input tests are not suitable for children, due to the fact that they handle areas that do not belong to children's everyday life. The eye-hand coordination is not as good as with adults or even teenagers. The ability to construct a coherent perception of the system structures is different from adults. Young children also often lack the capability to perceive how their actions influence others or themselves—for example, what kinds of risks there are to giving personal contact or location information in a service. Even teenagers may not understand which information on a network game account allows another person to steal virtual property or even a whole virtual identity; they may not understand fully how to protect certain information from others.

Nevertheless, children do use mobile technology today. Even if the devices and services were not previously designed particularly for children, children should be taken into consideration if there is knowledge that they would use the mobile systems.

The ETSI guidelines for the design and deployment of information and communication technology (ICT) products and services used by children (ETSI, 2005) include several guidelines that apply for mobile device design as well. These guidelines take into consideration both cognitive capabilities of children, as well as motor skills: for example, because children have less force in muscles, and the fine motor skills of small children are not yet fully developed, children may have difficulty pressing several buttons at the same time or in a row within a short period of time.

11.2.4.1 Guidelines

1. Simple interaction makes it easy for children to use mobile phones' applications and services.
2. Visual cues are good for children who cannot read well; these are also beneficial in markets with illiterate adults.
3. Do not apply systems that require pressing multiple keys at the same time or in a row.
4. Children need to be protected, for example, systems that allow children to share their personal information, such as location, without permission of a parent, or to provide information not suitable for their age, are risky.

11.2.5 Cultural Differences, User Needs, and Technology Environments

People in different countries, regions, and cultures have different needs. The differences need to be considered if devices or services are designed to be used in different countries and cultures (see Chapter 9, "International and Intercultural User Interfaces"). Localization is not only translation, but requires more understanding about people's habits.

In the United States, the PCs seem to have a bigger role in people's lives than in Asia or even in Europe, where the mobiles are used a great deal by the majority of the population, and where many people do not narrowly use mobiles only for voice calls, but also for information searches. In some parts of Asia the mobile phone may be the only personal device used for accessing Internet services. In Ipsos Insight (2006), the percentage of households having mobile phones was reported to be higher in some Asian countries than in the United States—with Japan having a significant number of adults with experience on mobile Internet use. In some countries, like India, the phone may be often shared with friends or family members. In emerging markets, like India and Africa, basic issues like durability of the device and battery life (and ways to charge a battery) play a much bigger role than in developed countries. Also illiteracy challenges are real in emerging markets.

Public statistical information about local sales, age groups, education, and so on, give useful hints for designing products for local markets. To really tackle the challenges of accessibility and special needs, hands-on user research is always required locally with the people. Even the street noise has very different impacts on accessibility in Tokyo compared, for example, to Gränna (a small city in Sweden).

11.2.5.1 Guidelines

1. Take cultural issues into consideration, especially regarding durability, dust, and water resistance.
2. Improve ways of power supply for systems in developing markets.
3. Target for special needs of the market.
4. Consider language options and literacy versus illiteracy.

11.2.6 Tourist/Visitor Perspective

Mobile services are increasingly used abroad or in unfamiliar places. It is easy to think of use cases: navigating in nonfamiliar environments, finding out where the nearest pharmacy is, searching for what events are taking place, and so on. The list is endless and gives potential to many companies providing services.

When providing services to visitors, it is good to remember to provide language alternatives and to use simple language that can be understood by nonnative users. It is also good to remember that for nonnative users it is more difficult to understand the spoken language than the written language. As service selection may vary in different places, the visitors should easily get an overview of available services, key features of the services, and how they are taken into use.

Despite the potential, the tourist/visitor perspective in mobile environments is not very well addressed in reality. Currently, even if services exist and a visitor would be aware of them, there are problems in accessing them. Most network operators (or service providers) have roaming agreements and limitations that cover voice calls and text messaging, but Internet use, including sending pictures, is sometimes agreed upon with a limited number of operators. For example, 3G services may not be available due to roaming limitations. These practical barriers have not been studied widely. Research has mainly addressed the needs of tourists and visitors in a specific environment, like in historically significant city areas or museums (e.g., Cheverst et al., 2000, 2001). The focus of these studies has been on analyzing the information needs of visitors and suitable interactions with the technology.

A visitor may experience that in some situations a service is available, but in a different situation it is not, or the connection is dropped in the middle of the task, as the phone switched to another network operator's network. In the worst case, the user is forced to manually select the network, but that might lead to coverage problems when making voice calls.

Also roaming cost is a barrier. Even very limited use of data services may increase the phone bill more than users can accept. Data cost is often difficult to predict even in the user's own network (Roto et al., 2006). One way of solving this problem is providing WLAN (WiFi is currently based on WLAN), and easy ways of finding the hot spots. Roto and Laakso (2005) discuss problems and solutions related to network finding. The benefits are considered to be not only for tourists but for all users. The network-related problems that many users experience (difficulty of finding suitable network, cost of the use) do not usually come in research reports, as tests are often run only in one location or in one country. However, this has been a known problem, and to solve it at least in part, the European Union announced a directive in 2007 to cut the costs of roaming.

11.2.6.1 Guidelines

1. Design systems where data transfer cost is minimized and users can predict the cost of total use of the service. For example, apply ways to use free networks and provide services that can be downloaded on the device so that minimal amounts of data are transferred when traveling.
2. Provide services that directly address the needs of users, not the needs of service providers.
3. Use simple language so that nonnative users can understand easily.

11.3 Special Characteristics of Mobile Devices and Services

11.3.1 Ergonomics and Human Factors

Ergonomics is present in all uses of mobile devices. Devices are held in hands, viewed from short distances, listened to, attached to wires and other devices, kept in pockets and bags, or laid on varying surfaces to be grabbed into the hands again.

The main areas of mobile ergonomics are physical ergonomics, visual ergonomics, audio ergonomics, and ergonomics in the use of peripherals.

- *Physical ergonomics* deals with the size and shape of the device, materials, and the ergonomics of input elements such as keys. When physical ergonomics is well implemented, the device matches comfortably with different postures, and does not cause pain or fatigue during use.
- *Visual ergonomics* is related to the visibility and readability of user-interface elements. For example, keypad backlights, graphics illumination, and contrasts need to be optimized for different contexts of use. Viewing visual content needs to be considered, such as mobile TV usage, while the device is in the hand or on the table.
- *Audio ergonomics* is about the optimal hearing experience in different situations. In quiet environments more discreet device behavior is expected and needed than in noisy environments. The same applies for audio content, such as music, podcasts, and videos.
- *The ergonomics of peripherals* considers the usage of many types of wired and wireless accessories, such as headsets, loop sets, audio speakers, and external displays (TV).

Each area of ergonomics can be optimized for specific needs, such as the needs of elderly persons. For example, Kurniawan (2007) proposes six areas that need special attention: buttons, display, shape, color, size, and hardwired functions.

11.3.1.1 Guidelines

1. Design for ergonomics. Apply style guides and set ergonomics targets. Understand the possible special needs by target users.
2. Evaluate and iterate ergonomics throughout the development.
3. Verify the interplay and communication of design disciplines.
4. Consider both temporarily and permanently impaired users. They are found in all user groups. Provide visual, physical, and audio clues for device operation.
5. Provide multimodal input and output channels.

11.3.2 Miniaturization

Mobile devices tend to get smaller all the time. The challenging target of 100 g and <10-mm-thin devices has been reached long ago in all mobile device categories.

Miniaturization is enabled by advances in material technologies, user-interface technologies, and production technologies. The smaller the user interface is, the more there is need for:

- Output methods that are based on other ways than seeing, like audio and haptics.
- Input methods that are different from physical buttons, such as touch elements, voice input, and gestures.

Miniaturization affects user interface characteristics, for example, in the following ways:

- Physical input elements shrink, such as the sizes and topologies of keys. Key topology may change due to miniaturization.

- Virtual input elements are applied. For example, the physical keyboard is replaced with a virtual keyboard.
- Display becomes the dominant user-interface element.
- Sensors are used to detect user activities and the environment.
- Software-based interaction increases, such as adaptation and prediction of user intentions.

Miniaturization affects user experience, for example, in the following ways:

- The use of the device requires more attention and provides less concrete tactility.
- Structure clues vanish. The familiar physical attributes become smaller or change. For example, consider the evolution from analogue clock (clock hands) to digital clock (digital numbers). Learning a new key topology requires more cognitive effort.
- Concrete interface elements vanish or change. For example, wireless connection replaces cable, or cable is attached with a smaller plug than in the previous device.
- Interface elements merge and fit to new places, for example, touch input sensors are under physical keypad.
- Sensors are used to enhance user interface behavior.
- Interface elements are provided based on the user interface context, such as virtual keys for music playing, when appropriate, or special contextual functions available on the touch user interface.

The development becomes more demanding as the size gets smaller. The miniaturization seems to affect especially the older users of mobile technology. Small device, small keys, and small font size create problems for them (Pattison and Stedmon, 2006; Leikas and Penttilä, 2007). Often it is not possible just to optimize earlier design, but the interaction needs to be redesigned wholly or partly. More attention is needed in fine tuning and calibrating interaction elements.

11.3.2.1 Guidelines

1. Conduct proper assessment to identify the critical interface areas due to miniaturization, for example, usability context analysis (Thomas and Bevan, 1996).
2. Decide early (in the concept development) whether you will design new or optimize earlier design.
3. Pay attention to both interface consistency and interaction details, for example, the accuracy of a virtual keypad.
4. Make sure there are necessary usability testing tools to study miniaturic devices, for example, detailed physical mockups or realistic software simulations.

11.3.3 Mobile Web Services

With mobile services, the requirements for universal access arise both from the abilities of the user and from the context of use. The context may influence the mental and physical resources that the user can devote to using the mobile service. Accessibility features designed for disabled users may as well be useful for able-bodied users in demanding mobile contexts of use, for example, walking on a busy street or driving (Roto, 2006).

Mobile devices have long included dedicated applications for local personal information management. Today it is possible to fluently use remote services with the devices as wireless connections have developed. Internet services are available with specific microbrowsers or task-specific applications and mobile widgets. Services can even be available as add-ons on mobile TV. The browsers on different devices vary both by manufacturer and by the model, and they can provide web services specially designed for mobile use or ordinary web services. Accessibility issues are related to both the browser and the services.

Screen readers that convert the displayed text on the mobile handset into speech are an essential accessibility tool for blind and visually impaired people. Screen readers for mobile devices are available as commercial software, and they can also be used by sighted users to get eyes-free access to services, for instance while driving. Both the browsers and the services should be designed for screen reader use, and other accessibility tool compatibility by following the guidance described in the following.

Mobile users can access ordinary web services, especially when the services are designed for that purpose. Services designed especially for mobile users not only take into consideration limitations of a small device, but may utilize location data or other context data available on the mobile device. The data can be used as such in the mobile service or be used to adapt the selection of services, contents of the services, or appearance of the service accordingly (Kaasinen, 2005). Mobile-specific features in services or devices can also utilize the fact that mobile phones are almost always with the user and always connected to the network—this is important for security services, for example. An interesting future possibility is context tags (Kaasinen et al., 2005) that facilitate activating contextually relevant services by simply touching radiofrequency identifier (RFID) tags embedded in the environment.

The growing variety of mobile devices, networks, and other infrastructures is a major challenge for designing mobile services. Services need to be available on the different devices and usable with different connected interaction devices. Furthermore, the services may need to adapt to changes in technical infrastructures, such as the available network, even on the move. Interoperability is a critical factor for wider use of services. Interoperability problems may prevent the use of services, for example, when the user changes devices or moves to a different technical environment. The goal of universal services that would work optimally on any device and network has turned out to be extremely ambitious, as the variety of devices and technical infrastructures grows. Eventually, both devices and services need to be able to adapt.

A good starting point is a simple service, suitable for any device. In this way, no devices will be excluded. Then the usability and the attractiveness of the service can be improved by providing parallel versions that utilize the unique features of each device and browser (Kaasinen, 2005). Preferably, the content and the presentation should be adapted according to the setup of the client device, in which the input and output modalities are selected and fine-tuned according to the physical or sensory capabilities of the user and by the context of use.

In the mobile environment, the content should be provided in a form that suits the variety of client devices and different input and output devices and tools. The proposed recommendation for mobile web best practices by the World Wide Web Consortium (W3C, 2006) emphasizes keeping both the content and the presentation simple. The most relevant content should be the easiest available. The user should get a clear overview of what is available in the service. A consistent style throughout the service assists the user in creating a mental model of the service. User interface elements such as navigation bars, tables, frames, and pop-up windows should be used with care, and not in their most complex forms, as mobile devices or the interaction devices used may only support them in a limited fashion. For example, the Java and Flash capabilities of microbrowsers have great variations. The size of graphics and use of colors need to be considered to enable access also through modest devices. The service should work also in text-only mode, scripting disabled, without pointing device, and without colors.

Style sheets can be used to define how the service should look on different devices. It will still be important to include alternative content for different user interface elements and metadata describing the content to support the adaptation taking place in client browsers or proxy servers.

Tedious text input should not prevent providing users with services where they can contribute as content providers. Free text input should be accompanied with the alternative of choosing from ready-made options. The latter is easier for novice users. Although downloading may be slow, it should not be interpreted so that mobile users would be satisfied with less content. Mobile users should be provided with all the content available, but the content should be provided to them in small portions and should be organized in a way suitable for their device and their context of use (Kaasinen, 2005). Users should also be able to choose whether they want to use mobile tailored or full web content.

Chandler et al. (2007) present a toolkit with design heuristics for mobile web services. The list highlights basic usability issues, shortcomings that currently inhibit disabled users from using mobile services. In addition to user interface issues, error recovery and aid in task completion are emphasized.

Design heuristics for mobile web (Chandler et al., 2007):

- Allow flexible access to the service
- Enable user style sheets
- Optimize color and contrast
- Use text effectively to improve readability
- Provide consistent page layout
- Provide content that can easily be scanned
- Include easy-to-use navigation
- Provide meaningful links
- Provide accessible forms
- Use JavaScript carefully
- Promote accessible multimedia
- Provide informative feedback and prompts
- Aid recovery and prevent errors
- Aid task completion

Personalization and adaptivity (Section 11.5) and setting up the service (Section 11.6) are also important issues regarding the accessibility of mobile services.

11.3.3.1 Guidelines

1. Utilize available design guidelines and heuristics, such as Proposed Recommendation for Mobile Web Best Practices (W3C, 2006) and Design Heuristics for Accessibility (Chandler et al., 2007).
2. Keep content and presentation well-structured and consistent.
3. Give a clear overview of what is available in the service.
4. Do not restrict user input or access to information.
5. Start with a simple service suitable for any user device, then provide alternative content and presentation that utilizes specific features of more versatile devices and input/output (I/O) devices.

11.3.4 Services Targeted to People with Limited Abilities

Mobile devices can be used as platforms for assistive services. Recent research has looked at these devices, for example, as wayfinding aids for blind people, route guidance for people with motor disabilities, video communication for deaf people, and memory aids for older users and people with cognitive disabilities (Roe, 2007).

Sometimes a solution developed for disabled people is helpful for able-bodied people as well, as in the case of short message service (SMS) based emergency services that were introduced in Finland to serve deaf and hard of hearing people (Roe, 2007). Other examples of research efforts using mobile devices include sound transcription tools that help deaf or hard of hearing people to be aware of audio events in their environment (Matthews et al., 2006), memory aid with remote communication to enable third parties to remotely enter data into the device (Szymkowiak et al., 2005), route guidance for people with cognitive disabilities (Carmien and Fischer, 2005), and a variety of supports for tasks of daily living for people with cognitive disabilities.

Mobile devices are already being used as platforms for assistive technology products. For example, mobile devices are used as cognitive support, providing time-based audio messages and picture cues to remind of daily routines (e.g., Schedule Assistant, http://www.ablelinktech.com/_handhelds). Mobile devices are also used as platforms for communication software for speech-impaired users who communicate with symbol languages (e.g., mageTalk, http://www.imagetalk.fi/en/products_and_services). Speaking navigation systems (e.g., TalkNav, http://www.talknav.com) are available for visually impaired people.

Solutions utilizing mainstream technologies as platforms are moderate in cost compared to solutions entirely designed for disabled people (Winterberg, 2007). Mainstream applications and services can also be used for special purposes. For instance, video calls (between mobile devices, and between mobile devices and desktop computers) can be used for sign language communication. This would also facilitate remote sign language interpreting. However, the special purpose of use may set extra requirements for the services. For example, tests conducted by the Swedish Handicap Institute (Richter, 2007) showed that there are big differences between phone models regarding their suitability for sign language calls. Wide camera angle and large screen are essential, as well as sufficient visual quality (frame rate, low transmission delays, display clarity). An important issue is the possibility of signing with both hands, a parameter that is dependent on several factors. Among other things, it is important that the telephone is simple to position at a distance from the user on a table or the like, and also that the camera angle is sufficiently wide for both hands and face to be seen.

Navigation guides can be tailored for disabled and older users. A navigation guide can adapt both the presentation and the contents according to the user; for instance, it can give spoken directions to a blind user and guide routes that are accessible for a wheelchair user (Lindström, 2007). Goodman et al. (2004) found landmark-based navigation guidance beneficial especially for older users. In their system, the landmarks were presented as photos and the route guidance would be, for example, "Walk toward this tower." This kind of concrete guidance outperformed map-based guidance.

Virtanen and Koskinen (2004) emphasize providing disabled users with service entities that support them throughout their activity. Their system, called *Noppa*, utilizes commercial services and devices for improving public transport accessibility for blind users by creating access to passenger information with a personal mobile device. Their approach provides the users with an unbroken trip chain, including trip planning, finding a stop/station, finding an entrance to the station, navigating inside the station, finding the right platform and waiting place, knowing when the right vehicle arrives, finding a vehicle entrance, payment, finding a seat, departing at the right stop, navigating inside the station, finding the exit of the station, and finding the destination. *Noppa* covers different modes of traveling—buses, trains, trams, and flights—by utilizing common web services for passenger information and personal navigation. For issues related to mobility and transportation, see also Chapter 49, "Intelligent Mobility and Transportation for All."

An interesting possibility is to utilize the mobile device as a memory aid as the device is with the user almost all the time (Hayes et al., 2004). In the system demonstrated by Hayes et al., the mobile service facilitated recovering audio content from the user's recent past, and helped the user to recall, for instance, recent discussions.

Regarding different user groups, it is useful to think widely about the possibilities that mobile technologies would offer to them. For instance, Winterberg (2007) suggests the following usages for mobile technologies in supporting people with dementia: (1) structuring daily life, (2) navigation, (3) remote communication, (4) finding and tracking the user, (5) alarm when the user leaves a specified area, and (6) combinations of 1–5. It is easy to extend the list. Accessibility barriers can be lowered by providing ways for the user to learn and practice new skills. For example, a user interface can explicitly or implicitly help in learning to read or a new language, or simple games can train a person with motor disabilities or in need of rehabilitation.

Research is needed to find solutions that suit the users' needs, are ethically responsible, and have a reasonable cost in relation to alternative solutions (Winterberg, 2007). Ethical issues are crucial in solutions targeted to older and disabled people, as these solutions are often safety critical. Users may become dependent on the technology, and the technical solution may remarkably affect the user's quality of life. The usage practices that will evolve around the products and services should be anticipated in the design. Mobile technology may be used to improve care, and thus the quality of life, but it may also be used to replace human contact—a trend that will require ethical assessment.

11.3.5 Innovating with and for Elderly and Disabled Users

Universal access is not just about adapting technology to the capabilities of people, but also utilizing technical possibilities to the advantage of different user groups. The users themselves are the best experts in this, and should have an active part in the innovation and design activities. For example, codevelopment with older and disabled users is needed when innovating new technical solutions that could be utilized in their lives. Eventually, the success of a new innovation is fundamentally dependent on the users' motivation to use a specific new solution (Shin, 2007).

Leikas (2007) reports a study in which they collected over 4000 ideas of how to better utilize mobile technologies in the everyday lives of senior citizens. They organized 13 idea-generation workshops with almost 750 older citizens from different parts of Finland (see Figure 11.2). The collected service ideas were classified into three groups. The first category was *services that compensate flaws in senses*, for example, using a mobile phone as a hearing aid, using the camera of the phone as a magnifying glass, or using services that help in remembering things. The second category was *personalized information services*, such as special offers or guidance in selecting clothes according to the weather forecast. The third category of services was *health-monitoring services* that would automatically call for help when needed.

FIGURE 11.2   Senior citizens innovating novel usage possibilities for mobile technologies. (From Leikas, J. and Lehtonen, L., http://www.vtt.fi/inf/pdf/tiedotteet/2007/T2389.pdf, 2007; photo courtesy of Ari Ijäs.)

11.3.5.1 Guidelines

1. Consider how mobile technologies could be utilized to the advantage of different user groups.
2. Take the users themselves into the innovation process for new services.
3. Utilize mainstream technologies to ensure affordable solutions.
4. Design service entities to support the users throughout their activities.
5. Assess and evaluate the solutions with regard to ethical issues.
6. Consider what kinds of usage practices the products will generate, and if those are ethically acceptable.

11.3.6 Location and Positioning

Location information is becoming an integral part of mobile devices. Location information is necessary for map and way-finding services. In addition, location information can be utilized in location-aware services, providing the users with contextually relevant information and services. With satellite navigation systems such as global positioning system (GPS), the location of a user can be defined very accurately (2–20 meters). An ordinary GPS cannot be used indoors and may not work everywhere outdoors, either, if the view to the sky is limited. An important challenge for the future is to improve the reliability of navigation systems, so that people can rely on positioning devices in their everyday lives. This is especially important for blind people, who will need very accurate navigation support (Roe, 2007).

A mobile phone can be located by the telecom operator within the network. In urban areas, the accuracy can be as good as 50 meters, whereas in rural areas the accuracy may be several kilometers. The location can be detected anywhere where the phone can be used, also indoors. The accuracy does not allow for navigation tasks, but can be utilized in locating where the user approximately is. This has turned out to be useful, for example, in locating emergency calls. The approximate location information can also be used in informing the user about local facilities such as shops, bus stops, ATMs, and so on.

There are systems that combine the capabilities of GPS positioning and cellular positioning, providing more accurate and continuous information about the location both indoors and outdoors, and in different contexts, such as dense areas with high buildings.

The user can also be located and identified at a service point, utilizing, for example, WLAN, Bluetooth, infrared, near-field communication (NFC), or RFID tags. Proximity information can be utilized in the so-called hot spots that provide information or topical guidance in the hot spot area. The coverage of these systems is limited to the area where they are installed. Special software and hardware may be needed in the user's device although WLAN and Bluetooth are fairly common in current mobile devices.

11.3.6.1 Guidelines

1. Functions and services that use location information should be designed to work also in situations where location information is not available.
2. The accuracy of the location should be sufficient for the purposes that the location is used for. The user should be made aware of the restrictions.

11.3.7 Personal Computer as a Support Station

Mobile devices often require a support station for installation, operation, and maintenance. Practically all mobile devices today are designed to be supported by PCs. Whenever possible, the owner of a mobile device is expected to manage tasks such as software installations, pairing of devices, and defining how devices are expected to work together.

Typical tasks needing PC support are:

- Installation: PC software, pairing, configuration of PC and mobile device
- Content support: Data backup, synchronization, gateway to transfer content to and from the Internet, content management (albums, playlists, editing)
- User support: User guidance in electronic form (PDFs, online support), upgrades, product information
- USB charging

Users tend to perform the same tasks from different devices, such as reading e-mails from a home PC, in an Internet café, or with a mobile device. In developing markets, having a PC at home is not as common as in developed countries. Users may not have a PC to store their own content, take backups, and so on. Therefore, in developing countries, alternative ways to upload content to a device and perform other activities need to be designed. This user need is addressed, for example, in the rapidly emerging online storage, content management, and publishing services.

PCs are the most complex consumer digital terminals that are widely used in a home environment. A wide variety of customization setups can be performed. However, the flexibility and power of these terminals present particular challenges in setup activities:

- Home PCs may have many users within the family. Each user has personal configuration needs.
- Upgrading the PC requires the reinstallation and reconfiguration of software and software related to peripherals such as scanners and printers.
- New peripherals must be configured.
- Licensing agreements must be understood to avoid unnecessary expenses when upgrading or renewing PC hardware and software.
- In all countries, people may not commonly have PCs at home, but they use work PCs and devices at Internet cafes to access the Internet. In these cases they cannot install applications to the PC or take backups.

Problems in access and usability are primarily related to the installation phase. These problems are often difficult to solve, unless the user has expertise in technology. In the worst case, the installation harms the use of other PC functionalities, for example, due to the automatic setting of default functions or reconfiguring the behavior of I/O ports.

11.3.7.1 Guidelines

1. If the PC is required to support the mobile device, reflect this in all phases of design and development.
2. Design for easy installation. Conduct proper error analysis of what could go wrong.
3. Preconfigure as much as possible for device pairing, default functionality, and Internet-based functions.

4. Provide easy-to-access problem-solving support. For example, "Hints & Tricks" is always useful.
5. Test the interoperability and operation with different devices.
6. Do not change the PC behavior without user confirmation.
7. Keep in mind that there are mobile users who do not have a personal computer. Plan an alternative way for configurations and other activities.
8. Plug and play with zero manual configuration is the optimal use case for setup. In cases where this is not possible, there should be both clear software assistance and user guidance.

More detailed guidance can be found in the setup guidelines by ETSI (2006a).

11.3.8 Interoperability

Mobile services need to work on different devices, and the devices need to work within different technical infrastructures. Interoperability is about the fluent use across different device combinations, connection methods, and services. In practice, the user can send a message to a friend even if she uses different kinds of mobile devices and has a different service provider; mobile services should be used with different devices. Also, the interoperability of accessories and environments (at home, in car, etc.) must be considered. The user experiences interoperability, for example, in the following situations:

- When upgrading the device to a new one
- In the communication between mobile devices (directly or via service)
- In the use of a cellular network or Internet service
- In the use of accessories
- In downloading new software
- When roaming (when traveling in a different country)
- When using memory cards

Interoperability failures typically lead to problems that the user is not able to solve. Usually, from a user's perspective, the activity successfully performed earlier just does not work in a specific situation, and the user does not have any idea what went wrong. Some examples of interoperability problems:

- A multimedia message cannot be sent from a mobile phone because the file size is larger than the network allows.
- A long text message (>160 characters) is not sent properly to the recipient since the recipient device supports only short messages (<160 characters).
- An Internet service does not work or the user cannot send multimedia messages when traveling and the phone is roaming (e.g., packet data is not supported).
- An old memory card is not compatible with a new device.

- An Internet service does not work in the mobile device, for example, due to lack of special software or drivers needed.

Different parties in the mobile domain have responsibility in improving interoperability. Device manufacturers, network operators, and service developers have the possibility to improve the situation for the users. By testing interoperability and by maintaining consistency across product definitions, at least the basic activities can be performed regardless of device or network used.

11.3.8.1 Guidelines

1. The user should be provided information about known interoperability problems (FAQs, etc.).
2. The compatibility of old accessories should be checked during the purchase.
3. The software and hardware compatibilities should be checked during the purchase.
4. The device's user interface should guide the user in case of service/network interoperability problems.

11.3.9 Security and Trust

Mobile devices and services are part of the everyday lives of users. As users are increasingly dependent on these services in their everyday activities, trust becomes an important design goal. User trust in mobile services includes perceived reliability of the technology, information, and functions provided, reliance on the service in the planned usage situations, and the user's confidence that he can keep the service under control and that the service will not misuse personal data (Kaasinen, 2005). Users should be aware of the risks in using the product, and should be provided with information about the reliability of the service, so that they can assess whether they can rely on the service in the planned usage situations. User expectations should be in balance with the actual trustworthiness of the solutions, including technical reliability, accuracy, privacy protection, and the possibility of keeping the service under control. Product information should take care of informing the user about these issues.

Mobile devices can connect to the Internet and to other devices, and download services and applications. Mobile operating systems, such as Linux and S60, are open and anyone can develop applications for those. This openness also entails some drawbacks. Harmful functions, such as mobile viruses, may enter the devices, and users should have ways to protect their devices and recover. This includes both technical solutions and guidance on recommended practices.

Mobile services use and provide increasing amounts of different measurement data for the user, in addition to location data. The accuracy of the data needs to be analyzed in the design: Does the user have the right conception of the measurement accuracy? Is the accuracy sufficient for the kinds of tasks for which the user will be using the service? Does the user get feedback on the freshness of the data and its accuracy, especially if these vary according to the usage situation? And finally, are the data protected in case the phone is stolen or a virus enters the device?

Users must feel and really be in control of the services. This includes several aspects related to security: security and trust in using functions and services (virus protection, certificates, etc.), security in maintaining one's own data (synchronization, product upgrade), security in protecting one's own data (backup and restore, during product repair, after theft, etc.), and physical security due to mobility and contexts (theft, breaking, humidity, etc.). For example, it is very common to see expired certificates in banking and other services requiring trust—and users gracefully accepting those when prompts are given about expiration. This requires that the user has a conception of the functionality of the service, even if all the details may not be clear.

Mobile payment is gaining popularity. Several organizations already provide methods for simple mobile payment; new methods are introduced, and users are starting to adopt these methods actively. A common approach is to link the payment to a phone bill. While mobile payment is made easy, the requirements for device safety, information clarity, user control, and handling of error situations become critical.

The privacy of the user should be protected even if the user would not explicitly require it. Users tend to be trusting, and they cannot be aware of all the possible privacy threats. The user should get clear feedback on which personal data are collected, where they are stored, and who is using the data and how (Kaasinen, 2005).

11.3.9.1 Guidelines

1. It should be ensured that the user feels and is in control.
2. To provide perceived security, there should be sufficient information available about the risks and solutions.
3. There should be support for the use of security functions, such as a PIN code.
4. There should be network-based security functions.
5. In the design of security functions metaphors, interaction flows and terms should preferably be familiar from earlier products or elsewhere.
6. There should be mobile ways for users to remove/add/upgrade security on the network features.
7. The user should be informed of the service providers' name, their security and privacy certifications, and professional affiliations.
8. The service provider needs to take care that security certificates are kept valid.
9. There should be security guidance and solutions for cases when the mobile device is lost or stolen.

More detailed information about security guidelines for mobile users can be found in FICORA (2006).

11.3.10 Dealing with Errors and Problems

Complex technical and business environments are challenging for users who access services with mobile devices having limited user interface capabilities. Pressure for faster development cycles does not allow extensive usability studies, and the services may end up on the market with severe usability and technical problems, as well as inadequate content. Perhaps an even bigger problem than ease of use has been the technical reliability of the services.

Special attention needs to be put on recovering from error situations that are typical for mobile devices, such as lost connections, missing service, or broken links. Following a link can be unsatisfactory also because the page includes elements or content that cannot be optimally shown on the device. As mobile networks may be slow, the system should provide a progress indicator, so the user can follow the progress of uploads and downloads.

Preventing error situations, and appropriately supporting the user when an error occurs, are important issues. Repeated malfunctions that the user cannot understand or solve are a major source of bad usage experiences. User errors should be prevented by all means, for example, by trying to interpret, correct, or complete user input (Kaasinen, 2005).

The design should set target values for tolerable error rates. Errors cannot always be prevented, but the design should have strategies for error situations: how to recognize or forecast error situations, how to overcome these situations, and how to keep the user informed about what is happening and what actions should be taken.

As users are increasingly dependent on mobile services in their everyday lives, even battery power may become crucial. Roto (2006) suggests that the services should adapt to diminishing battery power and adapt their functionality accordingly. Kaasinen (2005) suggests that the user should be provided with estimates of battery life with selected service options.

11.3.10.1 Guidelines

1. User support should provide clear guidance about how to act in case of problems and errors.
2. The mobile system should recover fluently from error situations.
3. The system should prevent user errors, for example, by suggesting corrections to user input.

11.3.11 Product Maintenance and Upgrade

The life cycles of personal electronics are different. This is a special challenge for the maintenance of personal content and configuration. During the life cycle of a PC, a mobile device can be replaced several times. Also, during the life cycle of a mobile device, the PC may be replaced.

Product upgrade represents a critical phase in maintaining the daily work practices, functionality of online services, and preservation of data and content stored in the device. Product upgrade may also cause difficulties for the usage of existing physical peripherals. Problems are related to, for example, physical setup difficulties, combinatory problems, interoperability failures, or local wireless connection problems.

Manufacturers are developing solutions to improve the customer's capability to take products into use and upgrade. Examples of good solutions and applicable technologies in the consumer markets exist. Televisions and radios are able to scan automatically for available channels, PC MP3 players are able to search for music titles from the Internet without the user's effort, and a digital camera can be taken into use without reading a page of a manual.

The most common areas related to mobile product maintenance are:

- Maintaining personal content, including the device settings and configuration
- Maintaining functionality
- Dealing with transitions between operation systems
- Continuing to use existing software, hardware, and peripherals

The future of data-intensive mobile terminals will present yet more issues to be solved. The following list describes some technical challenges related to personal data management:

- Built-in memory: The memory capacity varies between terminals. The new terminal does not necessarily have more memory. Even if a software tool enables the transfer of data between terminals, it is not always the case that the product replacement succeeds, due to memory limitations.
- Memory extension: Several smart terminals can use memory cards. The supported sizes, technologies, and capacities vary between terminals. An old memory terminal may not physically fit into the new terminal.
- Data transfer: To transfer the data between terminals using wired or wireless connection, both terminals need to be on. Most terminals cannot be powered on without SIM. In many cases, the user has only one SIM.
- The user may not know what data can, is, or should be transferred in product upgrade.

11.3.11.1 Guidelines

1. Maintain data: The product replacement must not disrupt the use of existing data content.
2. Maintain settings and applications: The product replacement must not disrupt existing work practice.
3. Maintain accessory compatibility: The product replacement must not disrupt the use of existing accessories without obvious reason.
4. Allow several replacement methods to cope with incompatible software and hardware.
5. Prevent error possibilities and loss of data.

6. Provide immediate feedback that product replacement was successful.

7. Provide guidance regarding product upgrade.

More detailed guidance can be found in the setup guidelines by ETSI (2006a).

11.4. Mobile User Interfaces

User interfaces in dedicated mobile devices, such as music players, can be optimized for specific functions. They are easy to learn and predict, and they can provide dedicated physical controls for primary functions. User interfaces in multipurpose devices, such as advanced mobile phones, are more difficult to optimize for several purposes. Dedicated controls are typically provided for one to two primary functions, such as call keys and cameras, but otherwise the operation of the device is based on general navigation methods (HW or SW), and context-sensitive functions. The user interface of a multipurpose device needs to deal with special cases, such as interruptions due to an incoming call, multitasking, and different contexts of use (contrasts, discreet audio, etc.).

The design of computer interfaces focuses on the software user interface, and the hardware interface (QWERTY keyboard) is more or less standard across the manufacturers. Mobile user interfaces consist of careful integration of mechanics, display, software, connectors, and audiovisual elements, as can be seen in Figure 11.3. Also, services provided by PC, cellular network, and the Internet are part of the interface design.

Mobile interaction components can be divided into four different categories: user interface, external interface, service interface, and PC interface. The *user interface* deals with the user interaction. It includes input and output devices and techniques, and industrial and mechanical design and application (software) factors. The *external interface* consists of the physical and wireless connections to accessories and other devices, such as the loopset, in the vicinity. This interface helps in the use of the device but is not physically part of it. The *service interface* connects the device to online services. The use of service interface requires both software configuration and agreements with the service providers. Most mobile devices *require* the use of a PC at some phase of the life cycle (*PC interface*).

Novel user interaction tools and concepts for mobile devices are studied widely. In the future, many of those concepts can be used to improve accessibility. Interesting developments are, for example, tactons, structured tactile messages that can be used to communicate information nonvisually (Brown et al., 2006), nonspeech audio guidance (Brewster et al., 2003), gesture-based interfaces (Ailisto et al., 2006), and gestures combined with 3D audio (Marentakis and Brewster, 2004). On the other hand, new user interaction concepts may raise new accessibility challenges to be solved. For instance, touch-based interaction with tags embedded in the environment will require different physical abilities than current mobile phone usage (Välkkynen et al., 2006). Noticing the tags should not be based only on visual cues, to ensure access to visually impaired users. It is worth noting that embedded tags can in fact create new possibilities for visually impaired people as the tags can provide information about

User interface

•Keyboard, text input keys, text inputs method, graphics

•User interface-specific keys: soft & touch keys, navigational tool, power on/off key

•Specific keys: call management, camera, music, etc.

•Mechanics: handling the battery, SIM and memory cards, connectors, hinge, slide, camera

•Audio, voices, audio-visual effects, LEDs, backlights

•Ergonomics, balance, touch and feel, materials

•Touch inputs, haptics, sensors

•Software: display, visual quality, UI style interaction, icons, indicators, localization, consistency

•Applications

External interface

•Accessories (charger, etc.)

•Add-on devices (loopset, Bluetooth keypad, etc.)

•SIM card, memory card

•Carrying solutions

•Manuals

•Web support

•Online help

•Tutorials, demos

PC interface

•Bluetooth pairing, USB

•Synchronization

•Backup

•Mobile device as modem

•Etc.

Service interface

•Connectivity (Cellular, Bluetooth, WiFi), roaming

•Operator services & configurations

•Service agreements and policies

•(Mobile) Internet services

•Context-specific services

•GPS

FIGURE 11.3 Mobile user interface.

the environment through the mobile device as it interprets the information read from the tag.

11.5 Personalization and Adaptivity

11.5.1 Personalization: From Personal Ringtones to End-User Programming

Personalization of software means developing software to be more responsive to the unique and individual needs of each user. Personalization is often proposed as a solution to cope with information overflow in mobile services, by making personally relevant information or services more easily available to the user. Users may express the need for personalization and enjoy the results, but are not prepared to do much about it themselves (Nielsen, 1998). Often, personalization setup turns out to be too tedious for the users. Personalization should be voluntary, and strongly supported for novice users (Kaasinen, 2005).

_ Today, personalization with mobile phones mainly takes the form of personally selected ringtones, logos, and background images. Future visions enhance personalization toward end-user programming, where people shape their own services by utilizing available service components. The design for all approach should not be interpreted in such a way that the same solutions would fit everybody. Disabled and older people are individuals, and personalization options should be available for them. This includes requirements for accessible personalization tools and providing the personalization alternatives that disabled and elderly people would need.

Personalization may affect the means of interaction, the presentation, the content, or accessories. Personalization can be system or user initiated, the former often being described as customization. Personalization can be done manually, by utilizing user or group profiles, or the system can learn from the previous behavior of the user (Norros et al., 2003). The users can also be provided with ready-made service packages to choose from. The offerings can be based on the user profile and sent as a push service if the user accepts it. These kinds of automatic offerings would be especially useful in giving the user an overview about locally available services (Kaasinen, 2005). Pushing content to users onto the small screen is often more disturbing than in desktop environments. Although push facilitates easy access to situationally relevant content, push features should be introduced with care, and the user should be able to fine-tune or cancel the push feature easily (Kaasinen, 2005).

Personalization, service configuration, and special applications are typically lost when the product is changed to a new one. This may cause severe problems, especially when these features support the specific needs of disabled users. User profiles are usually defined separately for each individual service. As the selection of services grows, defining profiles separately for each individual service might become quite tedious for the users. There will be a need to copy the profile from one service

to another. User profile and user identity should always be kept separate to protect the privacy of the user.

11.5.1.1 Guidelines

1. Remember that disabled users are individuals: personalization should be available to all users.
2. Ensure the accessibility of personalization tools.
3. Study carefully what kind of personalization alternatives different user groups would need.
4. Ease personalization by providing ready-made service package alternatives or guided personalization services.
5. Consider supplying new service offerings to be sent automatically based on user profiles if the user accepts that.
6. Ensure that the user is able to refine the personalization with her mobile device on the fly.
7. Ease personalization by letting users copy their profiles from one service to another.
8. Consider the continuity of personalization in case the product is changed to a new one.

11.5.2 Device and Content Adaptation

Device and content adaptation is crucial in mobile services; services should adapt according to user device and available network. W3C (2006) suggests providing as reasonable an experience as possible given device limitations, and not excluding access from any particular class of devices, except when it is necessary due to device limitations. In addition, adaptation should be used to exploit the specific capabilities of the actual client device where appropriate. Adaptation can take place in the server, in the network, or in the client device.

Device adaptation can be made by providing alternative content in the service. Then the adaptation basically is about determining the device type and choosing the most appropriate set of previously prepared content to match the device characteristics (W3C, 2006). The other extreme includes dynamically formatting content and presentation, not only according to the device in question, but also according to the dynamic properties of the device and the context of use, such as with attached interaction devices (W3C, 2006).

In mobile environments, the elements of the usage context may vary. Users are different, and they use mobile devices and services for different tasks, even for tasks not anticipated in the design, and in unexpected places. The technical and service infrastructure may differ, and they may even change in the middle of a usage session; for example, the network or the positioning system may change when the user moves from one place to another. Similarly, the service infrastructure, for instance, the available services and applications, may change. The physical context may vary considerably in terms of illumination, background noise, temperature, and weather. The use of the device may affect the social situation in which the users find themselves, or the social situation may affect the way the users use the system.

Due to the varying contexts of use, mobile services cannot be designed only for specific contexts of use, but need to be adapted to different users and usage situations. Together, personalization and context awareness can greatly improve the accessibility of mobile services by providing the users with contextually and personally relevant information and functions. The services can even be invoked and the information provided to the user automatically. Context awareness can be complemented with personalization to adapt to user preferences that in different contexts vary from one individual to another. A major design challenge is to predict the personal needs in varying contexts of use without requiring constant attention from the user in the form of personalization dialogues and refining questions. For instance, the user could get different service selections for different cities that he regularly visits, and a tentative service selection for any other city.

Sometimes people are losing their abilities gradually, for example, people suffering from dementia. In such cases, it would be beneficial that their mobile services would adapt to this process by gradually refining the features of the service, as well as by introducing additional features. This would require tools for monitoring the current status of the user. Alternatively, the status could be updated by supporting persons.

11.5.2.1 Guidelines

1. Enable device adaptation with mobile services. A minimum goal is a reasonable user experience with any device and connected interaction devices.
2. Exploit the known unique features of different devices to improve accessibility and usability.
3. Utilize context awareness to ease or even to automate access to situationally relevant content and services.

11.6 Readiness for First Use

Users face practical problems when starting to use a new device, system, or service for the very first time, or when replacing an existing device with a new one. Users may even have difficulties in understanding what the product or service can or cannot do (Ikonen et al., 2002; Palen and Salzman, 2002). The root causes can be found in expanding communication features and users' limited capability to deal with the setup procedures and problem situations. Problems are often very difficult to solve and help is required from others. Impaired or elderly users especially have special requirements or limitations that may cause severe difficulty in the first use of a particular device or service.

The life cycle of a mobile device is typically short, only 2 to 3 years, and even that is decreasing. Changing the personal product or service to a new one presents a potential disruption in the life management. At the end of the life cycle, the device is normally replaced with a new one. Typical first-use cases are:

- New physical product is taken into use.
- Service or application is used for the first time.
- Product is enhanced by physical or SW extension.

- Product is taken for repair (a temporary replacement product is given).
- Product is replaced (upgraded) with a new product.

When a mobile device is taken out of the package, the user needs to:

1. Work with several items, including manuals, warranty papers, cables, detachables (battery, SIM), and accessories.
2. Start using the system through a user interface with setup tasks and basic service activation.

The user has an out-of-box experience several times during the product life cycle. The first happens when the product is initially taken into use. It is faced again when the product is replaced with a new product. In addition, expanding a system with new components or updating components or having several products in active use require similar activities as when taking the product into use, such as unpacking, installing software, and synchronizing. Hence, product design needs to consider novice users (new product), casual users (product replacement), and expert users (product updates). Design for accessibility in the first use should address all elements of the device and phases of the life cycle. Some considerations:

1. Provide clear descriptions of product features and capabilities.
2. Provide descriptions to deal with the most common first-use problems.
3. Explain the sources of assistance.
4. Consider the special needs of impaired and elderly users.

11.6.1 User Guidance

Users of mobile services experience difficulties trying to set up, configure, and access data services like e-mail, the Internet, or various messaging services (SMS, MMS, chat, and voice mail) through their mobile devices (see, e.g., Westendorp, 2007). Users lack the expertise necessary to configure and set up their terminals, services, and applications appropriately. Furthermore, even the configuration of terminal properties to the desired behavior is often beyond the users' abilities.

Understandable setup procedures and the availability of educational material become very important. Good setup guidelines can help to improve the service/terminal upgrade process, a considerable barrier to accessibility and efficient use of services. In combination with efficient user guidance, providing information on the availability and functionality of new, upgraded releases can be beneficial to the user, the terminal manufacturer, and the operator.

Problems related to insufficient user guidance with mobile devices are constantly reported, in spite of the offering of related guidelines and standards, including the following:

- The user guide is not complete (i.e., the information is not there).

- The information cannot be found (i.e., the information is there but not where the user needs it).
- The language of the user guide is inadequate (i.e., the language is too abstract, uses unknown abbreviations, uses technical and/or foreign language terms).
- The structure of the guide is inadequate (i.e., alphabetical feature list as opposed to the likely order in which users encounter or use features).
- The explanation of how to use a feature is too abstract (i.e., the subject index leads the user to a particular page on which the feature is explained, but an explanation of how to invoke the feature or of how to get to a particular branch in the menu tree is explained elsewhere).
- The information cannot be perceived adequately (i.e., older users may find it unacceptable to read print in 8-point fonts).
- The functionality/software implementation is not frozen at the time the user guide has to be completed (or sometimes the software implementation has to be changed at an even later stage); the user guide is therefore wrong and has to be corrected in later editions.
- Often, there is no obvious or useful linkage between the printed user guide, support in the web, and in-device help if these are created by different suppliers without a guiding framework.
- More detailed user education guidance and guidelines can be found in ETSI (2006b).

11.6.2 Practical Problems

Most problems and support contacts are related to the first 10 minutes of use, when the user is still unfamiliar with the product. The problems are often related to constructing the device (e.g., inserting battery, memory card, SIM card, etc.), powering on, configuring, and doing the first concrete actions.

Based on the review of some manufacturer and service provider FAQ forums, the most common problem areas are (Ketola, 2006):

1. *Power on.* Power-on topics concern:
   - Physically setting the phone ready for power on.
   - What to do if the phone does not power on.
   - Charging the battery and related problems.
   - The phone does not attach to a network (no network, calls not allowed, SIM card warnings).

   Power-on problems are common problems. A typical guidance is to start over again. The four phases in power on are the physical setup (e.g., inserting battery and charging), pressing power key, doing the initial setup activities, and letting the device complete the startup. In the collected data, problems were experienced in all phases.

2. *First use.* First use problems concern:
   - Setup problems and configuration.
   - Having practical problems in using the feature.

   - Understanding the feature and its usage in general.
   - Understanding how online elements behave.
   - Cost of using the feature.

   With a new phone the user primarily wants to continue using familiar features needed for daily life. Secondarily, the user tries to adopt and understand new functions if they are useful or interesting enough. Familiar features, such as voice mail and text messaging, are seen as straightforward configuration questions among the first-use topics (How do I set up my voice mail?). New functions are primarily seen as "what" questions, such as "What is media messaging?" "What can you transfer between devices?" and "Where do my data go?"

3. *Local connectivity.* Problems are mostly related to Bluetooth and infrared connections, but not often to cable connections. Infrared connectivity topics are primarily related to the question of using the phone as a modem. Bluetooth problems concern connectivity with accessories, other phones, and PCs. They are about troubleshooting emerging connectivity and interoperability problems and setting up for the first connection.

4. *PC connectivity.* Each manufacturer has FAQs on PC connectivity and provides several support solutions. Operators provide only little support on PC connectivity. The most common PC connectivity problems are related to:
   - Finding out if the PC and phone are compatible.
   - Establishing and troubleshooting the connection between the PC and phone with cable, Bluetooth, or infrared.
   - Changing existing connection settings in the PC software.
   - Setting up the synchronization features.
   - Using the phone as a modem for the PC.
   - Transferring documents from the phone to the PC and from the PC to the phone.

   Most mobile phones are able to connect with the PC. The usage model of content-rich smart phones is based on several PC connectivity use cases, such as backup, synchronization, and content libraries. There are technical problems and usability problems. Technical problems cover connectivity, interoperability, and combinatory issues. Usability problems are related to user's limited technical skills, insufficient guidance, difficulties in problem-solving support, and human errors.

5. *Unexpected behavior.* Several FAQs answer the questions about why the phone behaves in an unexpected way. Most of these seem to be system errors ("Why does my phone turn itself off?"), sometimes service errors ("Why can't I make calls?"), and some are related to activities in the user interface ("What are the scrolling messages on my screen?"). Unexpected behavior seems to be typical phenomena in complex consumer devices. In error situations,

the user often does not understand what caused the unexpected behavior, what it means, how it can be fixed, or if it is an error at all.

6. *Getting support.* Two common questions are often seen: "How do I contact you?" and "Where do I get user manuals?" Support functions are sometimes presented in the front pages (top level or second level) of the online services, and sometimes in the background. In some cases, the user is required to know the exact phone model before finding the correct support. The more hidden the support element is, the more questions related to support accessibility appeared. Product user guides (in PDF format) were widely available in the support pages. Some manufacturers and operators provide automated configuration support for setting the most important communication functions. A typical solution is a settings wizard that sends the configuration information to the phone.

7. *Using the services.* Several support comments discuss download problems, especially related to music. The devices' connectivity capabilities and functions, based on online services, are rapidly increasing. Also the usage of these services is steadily increasing. The same services are often accessible both via the mobile phone and via PC, such as e-mail, text messaging, or music download. Music topics deal with complex combinations and task flows, including online purchases, required software in the PC, connectivity between the phone and the PC, and finally the supported music playback formats in the phone. Several service topics are about using the device outside the home country ("What happens when I travel abroad?").

8. *Upgrade.* The problems and difficulties related to, for example, backup and restore are not yet visible in FAQs. Operators provide help in upgrading the phone and updating the service plans.

11.6.2.1 Guidelines

1. Maintain and publish information about typical problems and solutions.

11.6.3 Starting to Use the Services

Mobile devices have either one-way (e.g., GPS, FM radio) or two-way communication capabilities (cellular technologies, Bluetooth, WLAN). Devices with one-way communication can typically be taken into use with no configuration. Devices with two-way communication typically require configuration.

The biggest barriers to an increased use of services are complexity of setup, unclear pricing structures, including roaming, and poor upgrade paths. Trust also plays an important role in service uptake, especially in mobile commerce services. If services are provided by an operator, then billing and accountability

is straightforward. Services may also be offered by third parties not associated with the operator, as is the case with most Internet services. In these cases, users must build trust with a service provider who may be outside their operator's portal. To help users to build trust, services should provide enough information on security and privacy. The certificates should also be up to date to avoid intimidating error messages.

Mobile services such as tourist guides, event information, and shopping guides may be available only locally or in certain contexts. The user should be able to identify, understand, and take into use these services easily while on the move. As the selection of available services grows, it will also become increasingly important to easily get rid of unnecessary services.

Users may need help with:

- Information about the service and availability
- Getting details for configuration
- Configuring (manually vs. automatically)

More detailed guidance can be found in the setup guidelines by ETSI (2006a).

11.6.4 Purchasing Support

Mobile devices are sold practically everywhere. The level of user guidance varies from zero to extended personal support. More than half of the mobile phones globally are sold by the operators directly. These devices contain a lot of operator-specific software, configurations, and user support material. The primary access to user support is via the channel provided by the operator. Varying parts of the configuration are locked and cannot be changed by the user. The remaining parts of mobile phones are sold as operator-independent devices—for example, in department stores. The level of user support in the purchasing phase depends largely on the skills of the specific salesperson. These phones contain only minimum preconfigurations, practically as defined in the factory. The user can (and has to) tune the phone configuration herself. The primary access for support is the manufacturer's help line.

Call-center staff and point-of-sales staff are human experts operating on behalf of the manufacturer (call center) or of the vendor (point of sales). They usually receive information from the manufacturer or service provider and support the user in product choice and handling.

Visually impaired or blind users often learn about new products by sales staff, friends, or relatives explaining how a new product is handled. Audio guides are a very good way for them to learn about the product independently from others, and at a time they can choose to seek help.

11.6.4.1 Guidelines

1. Provide appropriate distribution channels for large-print versions of user guides (e.g., point-of-sales staff, mailing request via post card, or download from the manufacturer's or service provider's web site).

2. Consider briefing call center staff and possibly also point-of-sales staff on how to provide user education to deaf users via text services or signing interpreters.

11.7 Design and Evaluation Methods

11.7.1 Test and Evaluation Methods

The evaluation of the mobile devices follows the same principles as the evaluation of any system; in evaluation it is common to use the usability test protocol in a way similar to that described in several handbooks—for example, in Rubin's (1994) *Handbook of Usability Testing.*

As the mobile usage context and users differ from fixed situations, there are a few things that need to be addressed more carefully. It is important to test normal situations with disabled users, and extreme contexts with normal users (temporarily disabled users), to find out if multimodality and mobility are appropriately designed to support users in mobile situations. Mobile contexts differ from the laboratory environment and challenge the user in many ways. The ecological validity of usability studies is crucial. The effect of the context to the evaluation is the main differentiating topic in evaluation of the mobile devices and services compared to fixed system (e.g., PC software) evaluation. The main concern is if the test results from the laboratory test are valid also when the system is used in real-world contexts.

Several comparison studies during past years have explored whether laboratory tests are sufficient in any occasion (Kjeldskov et al., 2004; Baillie and Schatz, 2005; Betiol and de Abreu Cybis, 2005; Kaikkonen et al., 2005; Duh et al., 2006). These studies differ from each other, and results seem to be contradictory. The main reason for this contradiction lays in the definition of usability. The studies finding biggest differences between two test locations define usability to include a wider picture of the user experience, and not the rather narrow ISO 9241-11 standard definition of usability. The main conclusion from these studies, however, is that when usability is defined in relation to the problems in user interaction with a system, as in the ISO 9241-11 standard (1998), the same problems are found both in laboratory and in field tests. The problems that differ are more related to user experience that includes the state of the user and more contextual elements.

When testing pure user interaction, it may not be worth conducting usability tests of mobile applications or devices in the field, not because the problems would not be found in the field, but because the field testing consumes more time and resources. Kaikkonen et al. (2005) estimate that conducting usability tests in the field takes one-third more time than laboratory tests. As field testing includes more unpredictabilities, it also should be predefined in more detail. Especially if the test is run by using several moderators, in field tests there are more issues to be defined beforehand, such as moderator prompting, timing between questions, how to react to external interruptions, and to what extent the test user behavior is controlled.

The recommendation for most testing needs is to use the available resources to do several quick laboratory tests iteratively during the design process, rather than concentrate efforts on only one field test.

In some, cases laboratory testing is not enough. For example, services like navigation applications often have to be tested outside the laboratory due to technical restrictions (such as GPS not working inside), but also, more importantly because of cognitive reasons; it may be difficult to evaluate how people perceive the environment and how they can map themselves in the system and in space, if the study is done in one place. In other words, it is easier to really test if the users find their way than to simulate it in a laboratory.

When testing in real contexts, it is important to choose a location where using the phone is socially acceptable. People are usually well aware of social norms related to phone usage in public places in their own environment, and breaking norms may make the test users feel uncomfortable, as Palen et al. reported in their study (2000). If the test is not focusing on user behavior when breaking social norms, the test location for a field test should be in a place where people normally use mobile devices. Specifically, it should be socially acceptable to use a phone in the test location. Phone usage, even text messaging or keeping a phone in hand, may not be acceptable everywhere.

Places such as cafés, cinemas, transportation venues, and streets have different social codes depending on how they are built. Even if people are not consciously aware of these architectural aspects, they tend to behave in the manner that is expected by the designers (Fyfe, 1998). When testing in a location that is not familiar to the test leader, it would be good to observe people in the test location and consult the local people prior to the test. Testing in a socially unacceptable place may also create unnecessary tension for users, and they may not be able to concentrate on other issues.

Testing mobile services or devices with special user groups in a laboratory setting does not usually differ from testing any other system. However, in field testing special attention needs to be paid to the safety of the test users when they have limitations in their senses or cognitive states. Even if users are instructed to keep safety in mind during the test situation, the tested system and overall situation increase the cognitive load, and users may not be able to concentrate on the environment.

11.7.1.1 Guidelines

1. When testing the usability of the system, consider testing in a laboratory.
2. Do not test services or systems that may be socially (or otherwise) unacceptable in test location if the acceptance and reactions to it is not your focus.
3. Do not test mobile phones in environments where it is not socially acceptable to use a phone.

4. When testing in an unfamiliar location, find out if local people are using phones there.
5. Pay extra attention to users' safety in field tests, especially if users have disabilities or cognitive limitations.

11.7.2 Ethical Issues in Testing

Ethical issues are important when developing mobile services and when evaluating them. Mobile devices are considered to be personal devices, and this perception includes the need for controlling what is happening in the device. Users need to know what information the service they are using is spreading around. Information about the user cannot be forwarded without user consent; for example, location sharing needs to be active sharing, happening only with permission by the user. The user also needs to be able to choose when and to whom personal location information is shared. Location sharing in most cases should be reciprocal. The main principle for ethical design and evaluation is protecting the user from any feature or situation that might be harmful.

Use of logging software is often useful in user testing and evaluation. In this way, the use of the different features can be detected and the user does not need to remember what was used and when. When using these tools, it is absolutely crucial to inform the user about what is logged and how; user permission to log behavior should always be acquired. The information that is not needed should not be gathered; for example, when trying to find out the use of the communication tools in a mobile phone, the content of the text messages should not be gathered.

Protecting the privacy of the test user is the responsibility of each test leader. In many countries there are rules concerning how to protect the test user; for example, the number of people who know the identification of the test users should be minimal, and users should not be able to be identified from the test data. Ethical guidelines have been developed both by usability professionals and psychologists, which can be utilized in design and evaluation (e.g., APA, 2002; UPA, 2005; The British Psychological Society, 2006).

11.7.2.1 Guidelines

1. Do not design features or implementations that may be harmful for a user.
2. Protect user's privacy and control in design.
3. Protect user's privacy in test setup and in relation to private information. Do not store the test data in such a place that it would be accessible to others.

11.8 Future Visions

Certain development trends are obvious and have a major impact on mobile accessibility:

- Mobile devices, especially those developed for communication purposes, are pervading fast throughout cultures and countries. The needs and habits of existing users will evolve, and new user groups will bring new needs regarding devices and services. How people deal with daily tasks, such as simply arranging meetings and get-togethers while commuting, takes new shapes.
- Wired phones are disappearing from homes and offices. In many developed and less-developed countries, the number of mobile subscriptions already exceeds the number of fixed-line subscriptions. People rely increasingly on mobile devices and services in their daily lives.
- Technology development steadily continues enriching devices and services. Internet services will be commonly available in mobile devices. Devices keep shrinking, and user interfaces are not going toward any specific common standard—instead a multitude of different user interface solutions remains.
- Mobile devices are being equipped with more functions and applications, such as radio, music player, TV, camera, remote controller, navigation system, and so on. Integrating many different functions into a device may make the user interface complicated and the device nonintuitive (Norman, 1998). Hence, there also will be dedicated devices that provide just a few functions. On the other hand, it might be impractical to have several mobile devices for different purposes, especially if the usage is occasional. Both multipurpose and dedicated mobile devices will be available in the future.

11.8.1 Discussion

How will the *accessibility* of mobile devices develop?

Many benefits are emerging for people with disabilities. Devices and services for specific needs are easier to find due to increasing offerings. Assistive technologies, such as haptic feedback, text-to-speech solutions, and video communication, open up new possibilities. New service opportunities are emerging, such as mobile healthcare and way finding. Overall, the technological readiness to develop accessible solutions for people with disabilities is better than ever before. Norros et al. (2003) predict that mobile devices will not be just entities in their own right, but will become tools to interact with and to get information about the environment. These characteristics will become possible as the context can be measured, identified, or predicted. As a mobile device is capable of sensing the environment, it can increasingly support disabled people to compensate for flaws in their senses. For instance, the device can "see" information in the environment and report about it to a visually impaired person by speech, or it can "hear" information and indicate it to a hearing impaired user by light or vibration. For cognitively impaired people, the device could explain environmental information in the user's own language.

On the other hand, due to business and market drivers, mainstream solutions will not by default improve accessibility.

Mobile devices will have restrictions due to trade-offs between ease of use and portability. Touch and display intensive user interfaces are increasingly applied, which is a severe challenge for some disabled users, such as for blind users. Smaller products are difficult to use by people with motor limitations. Content in user interfaces is becoming richer with flashy animations, making it difficult to handle by very old and young users. In addition, the more people rely on their mobile support and service access, the more attention needs to be paid to the continuity of this support across devices, services, operator domains, and so on.

The growing variety of hardware and software capabilities gives additional possibilities for new kinds of mobile applications and services. Significant technical changes are expected to occur in equipping phones with many kinds of add-on technology to gather information and to interact with the environment. Local connectivity and positioning are the technical forerunners in this area. However, the variety of devices and infrastructures also introduces new challenges for service providers in offering their services for different client environments. The growing variety of options in individual mobile devices increases configuration challenges for end-users.

Islam and Fayad (2003) point out the necessity of developing devices in parallel with mobile infrastructures: new content types require new types of devices, new network options require devices that can choose the most suitable network in each context of use, hardware add-ons require additional battery power, more applications and personal content require more memory capacity, and so on.

To have successful mobile service business in the future, users should be able to find services easily from networks and from the surrounding environment in real time. The environment has to inform the users about the services available around them, provide easy ways to take the services into use, and adapt the services and user interface according to the usage context (Alahuhta et al., 2004). The total cost of service usage should be made very clear for users when using services in their own network as well as when roaming. In the future, reasonably priced flat fee rates for service use will become more common inside one's own network, but service roaming cost is equally important. The cost of the service usage should be in line with the value of the service—users will not be willing to pay for services they do not find useful.

The future of mobile services seems to have several alternatives that may all come true. It is therefore important to discuss the consequences of different alternatives. Mobile services will increasingly be available locally or contextually, providing users with services that optimally suit their current context of use. As service adaptation techniques develop, fixed network services will increasingly be available and automatically adapted also for mobile users. Mobile users will be provided with disposable applications that they can easily download and take into use when needed. When these are no longer needed, they will simply be thrown away. In addition to all of this, there will probably

be small dedicated information appliances that are designed for a certain purpose, including all the necessary software and hardware ready to use.

Therefore, universal access constitutes a major basic criterion for the acceptance of mobile solutions. Manufacturers, application developers, service providers, and other stakeholders are forced to consider this sooner or later. Accessibility issues need to be considered in all aspects of design, both in ensuring the accessibility of mainstream mobile technologies and in utilizing new mobile technologies for accessibility solutions. Basically, all user groups should be considered in the design, and no user group should be excluded.

References

Ailisto, H., Pohjanheimo, L., Välkkynen, P., Strömmer, E., Tuomisto, T., and Korhonen, I. (2006). Bridging the physical and virtual worlds by local connectivity-based physical selection. *Personal and Ubiquitous Computing*, 10: 333–344.

Alahuhta, P., Jurvansuu, M., and Pentikäinen, H. (2004). *Roadmap for Network Technologies and Services.* Helsinki: Tekes.

APA (2002). *Ethical Principles of Psychologists and Code of Conduct.* http://www.apa.org/ethics/code2002.html.

Baillie, L. and Schatz, R. (2005). Exploring multimodality in the laboratory and the field, in the *Proceedings of the 7th International Conference on Multimodal Interfaces (ICMI 2005)*, 3–7 October 2005, Trento, Italy, pp. 100–107. New York: ACM Press.

Betiol A. and de Abreu Cybis, W. (2005). Usability testing of mobile devices: A comparison of three approaches, in the *Proceedings of Human-Computer Interaction (INTERACT 2005)*, 12–16 September 2005, Rome, pp. 470–481. Berlin/Heidelberg: Springer-Verlag.

Brewster, S., Lumsden, J., Bell, M., Hall, M. and Tasker, S. (2003). Multimodal "eyes-free" interaction techniques for wearable devices, in the *Proceedings of the SIGCHI Conference on Human Factors in Computing Systems (CHI2003)*, 5–10 April 2003, Fort Lauderdale, FL, pp. 473–480. New York: ACM Press.

The British Psychological Society (2006). *Code of Ethics and Conduct.* http://www.bps.org.uk/the-society/ethics-rules-charter-code-of-conduct/ethics-rules-charter-code-of-conduct_home.cfm?&redirectCount=0.

Brown, L. M., Brewster, S. A., and Purchase H. C. (2006). Multidimensional tactons for non-visual information presentation in mobile devices, in the *Proceedings of the 8th Conference on Human-Computer Interaction with Mobile Devices and Services* (MobileHCI '06), 12–15 September 2006, Espoo, Finland, pp. 231–238. New York: ACM Press.

Bjurström, E. (1994). Children and television advertising: A critical study of international research concerning the effects of TV-commercials on children. *Swedish Consumer Agency Report* 8. http://www.konsumentverket.se/Documents/in_english/children_tv_ads_bjurstrom.pdf.

Carmien, S. and Fischer, G. (2005). Tools for living and tools for learning, in *Universal Access in HCI: Exploring New Dimensions of Diversity, Volume 8 of the Proceedings of the 11th International Conference on Human-Computer Interaction (HCI International 2005)* (C. Stephanidis, ed.), 22–27 July 2005, Las Vegas. [CD-ROM] Mahwah, NJ: Lawrence Erlbaum Associates.

Chandler, E., Dixon, E., and Tyler, S. (2007). Mobile phone evaluation toolkit, in *Towards an Inclusive Future: Impact and Wider Potential of Information and Communication Technologies* (P. Roe, ed.), pp. 255–262. Brussels: Cost219ter. http://www.tiresias.org/cost219ter/inclusive_future/index.htm.

Cheverst, K., Davies, N., Mitchell, K., Friday, A., and Efstratiou, C. (2000). Developing a context-aware electronic tourist guide: Some issues and experiences, in the *Proceedings of the SIGCHI Conference on Human Factors in Computing Systems (CHI 2000)*, 1–6 April 2000, The Hague, Netherlands, pp. 17–24. New York: ACM Press.

Cheverst, K., Smith, G., Mitchell, K., Friday, A., and Davies, N. (2001). The role of shared context in supporting co-operation between city visitors. *Computers & Graphics* 25: 555–562.

Cui, Y. and Roto, V. (2008). How people use web on mobile devices, in *Proceedings of World Wide Web Conference*, 21–25 April 2008, Beijing, pp. 905–914. New York: ACM Press.

Duh, H. B.-L., Tan, G. C. B., and Chen, V. H. (2006). Usability evaluation for mobile device: A comparison of laboratory and field tests, in the *Proceedings of the 8th Conference on Human-Computer Interaction with Mobile Devices and Services (MobileHCI '06)*, 12–15 September 2006, Espoo, Finland, pp. 181–186. New York: ACM Press.

ETSI (2005). ETSI EG 202 423. Guidelines for the design and deployment of ICT products and services used by children. http://portal.etsi.org/stfs/STF_HomePages/STF266/STF266.asp.

ETSI (2006a). ETSI EG 202 416. Human factors (HF); user interfaces; setup procedure design guidelines for mobile terminals and services. http://portal.etsi.org/stfs/STF_HomePages/STF285/STF285.asp.

ETSI (2006b). ETSI EG 202 417. ETSI guide human factors (HF); user education guidelines for mobile terminals and services. http://portal.etsi.org/stfs/STF_HomePages/STF285/STF285.asp.

FICORA (2006). Finnish Communications Regulatory Authority (FICORA). *Information Security Guidelines for Mobile Phone Users*. http://www.ficora.fi/mobiiliturva/english/tietoturvauhkia.html.

Fyfe, N. R. (1998). Introduction, in *Images of the Street*, pp. 1–10. New York: Routledge.

Goodman, J., Gray, P., Khammampad, K., and Brewster, S. (2004). Using landmarks to support older people in navigation, in the *Proceedings of the 6th International Symposium on Mobile Human-Computer Interaction (MobileHCI '04)* (S. Brewster and M. Dunlop, eds.), 13–16 September 2004, Glasgow, Scotland, pp. 38–48. Berlin/Heidelberg: Springer-Verlag.

Hayes, G. R., Patel, S. N., Truong, K. N., et al. (2004). The personal audio loop: Designing a ubiquitous audio-based memory aid, in the *Proceedings of the 6th International Symposium on Mobile Human-Computer Interaction (MobileHCI '04)* (S. Brewster and M. Dunlop, eds.), 13–16 September 2004, Glasgow, Scotland, pp. 168–179. Berlin/Heidelberg: Springer-Verlag.

Ikonen, V., Ahonen, A., Kulju, M., and Kaasinen, E. (2002). Trade description model: Helping users to make sense of the new information technology products, in *Electronic Commerce: Theory and Applications, Proceedings of 2nd International Interdisciplinary Conference on Electronic Commerce* (B. Wiszniewski, ed.), 13–15 November 2002, Gdansk, Poland, pp. 57–63.

Ipsos Insight (2006). *Mobile Phones Could Soon Rival the PC as World's Dominant Internet Platform: The Face of the Web.* http://www.ipsos-na.com/news/pressrelease.cfm?id=3049.

Islam, N. and Fayad, M. (2003). Thinking objectively: Toward ubiquitous acceptance of ubiquitous computing. *Communications of the ACM* 46: 89–92.

ISO 9241-11 (1998). Ergonomic requirements for office work with visual display terminals. *(VDT)s—Part 11: Guidance on Usability.* International Standard.

Kaasinen, E. (2005). User acceptance of mobile services: Value, ease of use, trust, and ease of adoption. PhD dissertation, VTT Information Technology, Espoo, Finland. 151 s. VTT Publications 566. http://www.vtt.fi/inf/pdf/publications/2005/P566.pdf.

Kaasinen, E., Tuomisto, T., and Välkkynen, P. (2005). Ambient functionality: Use cases, in the *Proceedings of Smart Objects & Ambient Intelligence Joint Conference*, 12–14 October 2005, Grenoble, France, pp. 51–56.

Kaikkonen, A., Kallio, T., Kekäläinen, A., Kankainen, A., and Cankar, M. (2005). Usability testing of mobile applications: A comparison between laboratory and field testing. *Journal of Usability Studies* 1: 4–16.

Kakhira, M. and Sørensen, C. (2002). Mobility: An extended perspective, in the *Proceedings of the 35th Annual Hawaii International Conference on System Sciences*, 7–10 January 2002, Big Island, HI, pp. 131–142 Big Island, HI: IEEE.

Kano, A., Read, J. C., and Dix, A. (2006). Children's phrase set for text input method evaluations, in the *Proceedings of the 4th Nordic Conference on Human-Computer Interaction: Changing Roles (NordicCHI 2006)*, 14–18 October 2006, Oslo, Norway, pp. 449–452. New York: ACM Press.

Ketola, P. (2006). On out-of-box experience and online support, in the *Proceedings of 20th International Symposium on Human Factors in Telecommunication*, 20–23 March 2006, Sophia-Antipolis, France. IGI Group.

Kjeldskov, J., Skov, M. B., Als, B. S., and Høegh, R. T. (2004). Is it worth the hassle? Exploring the added value of evaluating the usability of context-aware mobile systems in the field, in the *Proceedings of the 6th International Symposium on Mobile Human-Computer Interaction (MobileHCI '04)* (S.

Brewster and M. Dunlop, eds.), 13–16 September 2004, Glasgow, Scotland, pp. 61–73. Berlin/Heidelberg: Springer-Verlag (LNCS 3160).

Kristoffersen, S. and Ljungberg, F. (1999). Mobile use of IT, in the *Proceedings of the 22nd Information Systems Research Seminar in Scandinavia (IRIS 22): Enterprise Architectures for Virtual Organizations, Volume 2* (T. Käkölä, ed.), 7–10 August 1999, Keuruu, Finland, pp. 271–284. Finland: University of Jyväskylä.

Kurniawan, S. (2007). Mobile phone design for older persons. *Interactions, Special Issue: Designing for Seniors* 14: 24–25.

Kurniawan, S., Mahmud, M., and Nugroho, Y. (2006). A study of the use of mobile phones by older persons, in the *Proceedings of the CHI '06 Extended Abstracts on Human Factors in Computing Systems*, pp. 989–994. New York: ACM Press.

Lähteenmäki, M. and Kaikkonen, A. (2005). Older adults adopt new communication technology, in *Universal Access in HCI: Exploring New Dimensions of Diversity, Volume 8 of the Proceedings of the 11th International Conference on Human-Computer Interaction (HCI International 2005)* (C. Stephanidis, ed.), 22–27 July 2005, Las Vegas [CD-ROM]. Mahwah, NJ: Lawrence Erlbaum Associates.

Leikas, J. (2007). Idea movement of aging citizens: Lessons learned from innovation workshops, in the *Proceedings of the 4th International Conference on Universal Access in Human-Computer Interaction*, 22–27 July 2007, Beijing, pp. 923–931. *Lecture Notes in Computer Science* 4556.

Leikas, J. and Lehtonen, L. (2007). Ikääntyvien Ideali-ike. Käyttäjälähtöisellä innovoinnilla elämänmakuisia mobiilipalveluja. Research Notes 2389. Tampere, Finland: VTT Tiedotteita (http://www.vtt.fi/inf/pdf/tiedotteet/2007/T2389.pdf).

Leikas, J. and Penttilä, M. (2007). The needs and expectations of aging citizens: A potential for mobile terminal developers and service providers, in *Challenges for Assistive Technology* (G. Eizmendi, J. M. Azkoitia, and G. Craddock, eds.), pp. 203–207. Amsterdam: IOS Press.

Lindström, J-I. (2007). Safe navigation with wireless technology, in *Towards an Inclusive Future: Impact and Wider Potential of Information and Communication Technologies* (P. Roe, ed.), pp. 9–23. Brussels: Cost219ter. http://www.tiresias.org/cost219ter/inclusive_future/index.htm.

Livingston, J. (2007). Designing for seniors: Innovations for graying times. *Interactions, Special Issue: Designing for Seniors* 14.

Love, S. and Perry, M. (2004). Dealing with mobile conversations in public places: Some implications for the design of socially intrusive technologies, in the *CHI '04 Extended Abstracts on Human Factors in Computing Systems*, pp. 1195–1198. New York: ACM Press.

Marentakis, G. and Brewster, S. A. (2004). A study on gestural interaction with a 3D audio display, in the *Proceedings of the 6th International Symposium on Mobile Human-Computer Interaction (MobileHCI '04)* (S. Brewster and M. Dunlop,

eds.), 13–16 September 2004, Glasgow, Scotland, pp. 180–191. Berlin/Heidelberg: Springer-Verlag.

Matthews, T., Carter, S., Pai, C., Fong, J., and Mankoff, J. (2006). Scribe4Me: Evaluating a mobile sound transcription tool for the deaf, in the *Proceedings of the 8th International Conference of Ubiquitous Computing (UbiComp 2006)*, 17–21 September 2006, Orange County, CA, pp. 159–176. Berlin/Heidelberg: Springer-Verlag.

Nielsen, J. (1998, October 4). *Personalization Is Over-Rated, Jakob Nielsen's Alertbox.* http://www.useit.com/alertbox/981004.html.

Norman, D. A. (1998). *The Invisible Computer.* Cambridge, MA: MIT Press.

Norros, L., Kaasinen, E., Plomp, J., and Rämä, P. (2003). Human-technology interaction. Research and design. VTT road-map. *VTT Tiedotteita—Research Notes* 2220. http://www.vtt.fi/inf/pdf/tiedotteet/2003/T2220.pdf.

Palen, L. and Salzman, M. (2002). Beyond the handset: Designing for wireless communications usability. *ACM Transactions on Human Computer Interaction* 9: 125–151.

Palen, L., Salzman, M., and Youngs, E. (2000). Going wireless: Behavior & practice of new mobile phone users, in the *Proceedings of the 2000 ACM Conference on Computer Supported Cooperative Work (CSCW 2000)*, 2–6 December 2000, Philadelphia, pp. 201–210. New York: ACM Press.

Pattison, M. and Stedmon, A. (2006). Inclusive design and human factors: Designing mobile phones for older users. *PsychNology* 4: 267–284.

Puro, J-P. (2002). Finland: A mobile culture, in *Perpetual Contact: Mobile Communication, Private Talk, Public Performance* (J. E. Katz and M. Aakhus, eds.), pp. 19–29. New York: Cambridge University Press.

Richter, A. (2007). *Mobile Videotelephony: Test of 3G Telephones.* http://www.hi.se/global/pdf/2007/07335_pdf_MobileVideotelephony.pdf.

Roe, P. (ed.) (2007). *Towards an Inclusive Future: Impact and Wider Potential of Information and Communication Technologies.* Brussels: Cost219ter. http://www.tiresias.org/cost219ter/inclusive_future/index.htm.

Roto, V. (2006). Web browsing on mobile phones: Characteristics of user experience. PhD dissertation, Helsinki University of Technology.

Roto, V., Geisler, R., Kaikkonen, A., Popescu, A., and Vartiainen, E. (2006). Data traffic costs and mobile browsing user experience, presented at *MobEA IV Workshop on Empowering the Mobile Web, in conjunction with WWW2006 Conference.* http://public.research.att.com/~rjana/MobEA-IV/PAPERS/MobEA_IV-Paper_7.pdf.

Roto, V. and Laakso, K. (2005). Mobile guides for locating network hotspots. Presented at the *Workshop on HCI in Mobile Guides, in conjunction with MobileHCI 2005 Conference.* September 2005, Salzburg, Austria. New York: ACM Press.

Shin, D.-H. (2007). User acceptance of mobile Internet: Implication for convergence technologies. *Interacting with Computers* 19: 472–483.

Szymkowiak, A., Morrison, K., Gregor, P., Shah, P., Evans, J. J., and Wilson, B. A. (2005). A memory aid with remote communication using distributed technology. *Personal and Ubiquitous Computing* 9: 1–5.

Thomas, C. and Bevan, N. (1996). *Usability Context Analysis: A Practical Guide Version 4.04*. https://dspace.lboro.ac.uk/dspace/handle/2134/2652.

Tuomainen, K. and Haapanen, S. (2003). Needs of the active elderly for mobile phones, in *Universal Access in HCI: Inclusive Design in the Information Society, Volume 4 of the Proceedings of the 10th International Conference on Human-Computer Interaction (HCI International 2003)* (C. Stephanidis, ed.), 22–27 June 2003, Crete, Greece, pp. 494–498. Mahwah, NJ: Lawrence Erlbaum Associates.

UPA (2005). *Code of Professional Conduct for Usability Practitioners*. http://www.upassoc.org/about_upa/leadership/code_of_conduct.html.

Välkkynen, P., Niemelä, M., and Tuomisto, T. (2006). Evaluating touching and pointing with a mobile terminal for physical browsing, in the *Proceedings of the 4th Nordic Conference on Human-Computer Interaction: Changing Roles (NordiCHI 2006)*, 14–18 October 2006, Olso, Norway, pp. 28–37. Oslo, Norway: Association for Computing Machinery.

Virtanen, A. and Koskinen, S. (2004). Information server concept for special user groups, in the *Proceedings of 11th World Congress on ITS*, October 2004, Nagoya, Achi, Japan, pp. 18–24 Nagoya, Japan: ITS.

W3C (2006, November 2). *Mobile Web Best Practices 1.0*. http://www.w3.org/TR/mobile-bp.

Westendorp, P. H. (2007). *Configuring a Mobile Phone for MMS Usage: A User Experience Test*. Delft, The Netherlands: Delft University of Technology.

Winterberg, E. (2007). Ways of using mobile telephones by people with dementia, in *Towards an Inclusive Future: Impact and Wider Potential of Information and Communication Technologies* (P. Roe, ed.), pp. 75–89. Brussels: Cost219ter. http://www.tiresias.org/cost219ter/inclusive_future/index.htm.

Ziefle, M. and Bay, S. (2004). Mental models of a cellular phone menu: Comparing older and younger novice users, in the *Proceedings of the 6th International Symposium on Mobile Human-Computer Interaction (MobileHCI '04)* (S. Brewster and M. Dunlop, eds.), 13–16 September 2004, Glasgow, Scotland, pp. 25–37. Berlin/Heidelberg: Springer-Verlag.

12

Virtual Reality

Darin Hughes,
Eileen Smith,
Randall Shumaker, and
Charles Hughes

12.1 Introduction

Virtual reality (VR) refers to technologies that can permit users to experience and interact with environments that are entirely computer generated. A large part of such environments are visual, using computer screens, multiscreen projection systems, shutter glasses to provide 3D effects, and various kinds of head-mounted displays, but may also include rich audio capability, haptic (tactile) interfaces, various means for locomotion within the virtual world, and even olfaction (smell). The environments created using these techniques may represent the real world, or an entirely artificial world, or some combination of both. For the purposes of this chapter, two additional concepts will be considered within the scope of VR: mixed reality (MR) and augmented reality (AR). Milgram and Kishino (1994) defined what they call the virtuality continuum (Figure 12.1) that has at one end sensed reality in the real world, and at the other extreme VR in which the sensed world is entirely artificial. In between these is the world of MR, in which some elements of the world being sensed and with which a user is interacting are physically real while some are artificially generated.

Discussions are ongoing about the terminology and specific attributes of each. While these discussions can be energetic and entertaining, this chapter will use the terms MR and AR in a specific and more narrowly focused way. MR for the purpose of this chapter will be environments that contain both real and artificially created elements in various modalities (visual, auditory, haptic, olfactory) with the mix determined by the needs of the application. In particular, elements of the environment with which the user must interact directly will generally be real physical elements; those that the intended user must see, hear, or otherwise sense, but that they do not directly manipulate, can be entirely virtual, that is, computer generated. AR for the purposes of this chapter will be technologies that allow users to operate in the real world with additional information provided through artificial means. AR thus could include projection of artificially generated information on the visual, auditory, haptic, or other user sensory channel to provide information that would otherwise not be available to the user. This might be additional information or artificially generated substitutes for impaired senses.

One of the key features that may distinguish VR from other assistive and accommodation technologies is that a central goal of VR is to hide or help the user ignore the computer, interface technologies, and other technical elements so that the user can interact with the system in a natural way, much as one would act in the real world. This is important for therapeutic uses where the goal is to help people learn or relearn important life skills,

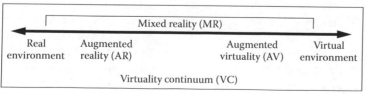

FIGURE 12.1 Milgram's virtuality continuum is the most widely cited terminology standard for VR concepts.

but is particularly critical for helping people with serious physical or emotional limitations to function more effectively in the real world, or to actively participate in such virtual worlds as Second Life. This latter application is receiving much research attention and interest, and may be a way for seriously disabled individuals to maintain rich social interactions, virtually travel, and even construct and inhabit personas that better reflect their roles in the virtual world. Virtual worlds are beginning to be used for some business meetings, and there are virtual showrooms for real products present in these environments. It seems likely that, if this trend continues, such environments could become an important vehicle for accessibility.

12.2 Technologies

12.2.1 Mobile Experiences

In its simplest form, a visually based MR or AR experience is possible with any device that has a camera, a display, and sufficient computational power and storage to merge the captured real image with virtual content. One can even provide such experiences with devices having low computing and storage capabilities, so long as fast wireless network access is available to send and process requests and data, and return the merged image for display. Even low-end versions of today's cell phones meet these minimal criteria. Higher end devices such as the Apple iPhone and similar advanced cellular telephones, and combined personal digital assistant (PDA)/phones, meet more stringent criteria.

Since MR experiences typically involve more than visual content, the built-in audio capabilities of mobile devices can provide a low-end audio augmentation to real-world sounds. Conveniently, this situation is greatly improved in modern devices that support Bluetooth wireless communication. Bluetooth uses very little power, providing communication between devices in close proximity so long as very high data bandwidth is not required. Moreover, Bluetooth simplifies the discovery and setup of services, using a protocol by which each device advertises its services, for instance, as a speaker, headset (speaker and microphone), or sensor/actuator (haptic capabilities).

To select meaningful virtual content associated with a particular real scene, an MR/AR system must know its user's location and orientation. Although mobile devices in the cell phone, PDA category do not have the capability to precisely know their position and orientation, most do know their approximate location and, with the aid of natural landmarks, can compute an estimated orientation. For mobile AR applications this makes it possible to add labels to buildings and provide guidance. For more complex MR applications, position/orientation knowledge might be used to provide visual, aural, and haptic cues as a user is walking toward an intersection. The modality can be tailored to the user's needs. So, for instance, a person with visual and aural limitations might get a tap on the shoulder as a warning of potential danger. This type of system can be used as part of a program to help a patient generalize skills developed in therapy sessions (see Figure 12.2).

FIGURE 12.2 An application example using cellular technology: identifying buildings with Nokia's mobile augmented-reality prototype.

12.2.2 Higher-End Visual Experiences

12.2.2.1 Walls, CAVES, and Flats

The next level of technology for MR/AR is to use a video wall. In this case, rear projection is commonly used. Simple versions of these just provide flat imagery. The content can be controlled by a user's gestures and movements. When three-dimensional content is displayed, users must wear specialized glasses. Some setups (passive stereo) are designed for a group that views the content as more of a show than an interactive experience; others (active stereo) allow a leader to move around with all other users having shared content from their personal perspective. Multiple walls can be used with seamless transitions.

The ultimate multiple wall environment is a CAVE, an immersive environment that surrounds the user with three (front and two sides) to six walls (four walls, ceiling, and floor) (Cruz-Neira et al., 1992). Each sidewall and the ceiling are rear-projected; the floor is down-projected (see Figure 12.3). As with active systems, the CAVE operates with a leader and all others along for the ride.

Digital flats, as used in FlatWorld (Pair et al., 2003), are rear-projected screens used in conjunction with rooms and other open areas, providing views into other rooms or exterior settings. By combining flats, real space, and immersive audio, one can create an MR experience as shown in Figure 12.4.

The approach described in this subsection constrains the virtuality to be behind the projection surface separating the virtual from the real. Here, users cannot walk into the virtual settings and so the virtual objects are adjacent rather than integrated with the real.

FIGURE 12.3 A virtual vacation from inside the CAVE at the University of Sao Paulo.

FIGURE 12.4 A scene from FlatWorld at the Institute for Creative Technologies.

12.2.2.2 Head-Mounted Displays

Head-mounted displays (HMDs) used in purely virtual settings dominate the user's visual system, blocking out the real. In contrast, MR HMDs are designed to support experiences in which the real world is blended with the virtual (Uchiyama et al., 2002). Two types of see-through techniques are used: optical and video.

In the optical HMDs, the real world is seen directly, and the virtual is overlaid using optical techniques. There is no delay in seeing the real; thus, the user's visual and vestibular systems are synchronized. A common downside is that the virtual objects generally appear in front of the real.

When video see-through is used, the HMD has a pair of cameras on the outside and a pair of displays on the inside. The cameras capture a left-eye/right-eye view of the real world. This is brought into a workstation where the real is blended with the virtual, and this mixed scene is delivered to the user's eyes, creating

a stereo view. The advantage of the video over the optical HMD is the ease with which virtual objects can be intermixed (foreground, middle ground, background) with real objects, and the virtual objects look less ghostly than with optical. The downside is that the clarity of the real world is reduced and there is latency from capture to display. This latency is somewhat offset by the fact that the virtual world latency is properly synchronized with the real-world objects and the HMD is typically open on the sides, providing a peripheral view that matches a person's vestibular system (see Figure 12.5).

12.2.2.3 Blue Screen Surround

As previously pointed out, CAVEs use rear projection to achieve an environment that surrounds participants in a virtual setting. In general, such a setup is costly, especially if stereo content is to be delivered. A less expensive alternative is to place users in an area that is enclosed by a curtain or wall that is uniformly colored. Such a surround is typically called a blue or green screen, as these are the most common colors used. To use this as a CAVE, users can wear video see-through HMDs. As a user looks around the area, the video captured is analyzed to determine which pixels are of the chosen chromatic color (called the chroma key). In a simple binary approach, only pixels close to the chroma key are replaced by virtual content. This tends to create jagged edges around objects, especially around such fuzzy areas as hair and clothes. This problem can be substantially mitigated by using partial transparency, so that unambiguous pixels are either completely transparent or completely opaque, but unclear cases are assigned transparency values based on how close they are to the chroma key as shown in Figure 12.6. This technique has been used to create context in therapy sessions, such as with brain-injured patients or with behavioral problems such as stuttering. These applications are discussed later.

The advantage with a chroma key approach is that each of any number of participants may be presented with a personalized context or point of view.

FIGURE 12.5 Altered reality as seen through a video see-through an HMD.

 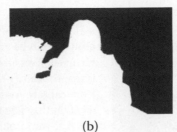 

(a)   (b)   (c)

FIGURE 12.6   (a) Original shot; (b) matte; and (c) composited result seen through an HMD.

12.2.3 Extending the Experiences with Aural Assets

12.2.3.1 Spatialized Sound

Auditory cues provide a special benefit for use in interactive system design with unique properties that other modalities do not provide, or do not provide as well (see also Chapter 32, "Auditory Interfaces and Sonification," of this handbook). In particular, audition is a 360-degree sense that functions three dimensionally. Humans can perceive sounds above and below, in front and behind the head, and to the left and right. Sounds can be heard from around corners and behind walls. Sounds provide a sense of place and help the listener with environmental recognition. In simulations, movies, and other synthetic experiences, audio has been shown to increase a sense of presence and immersion. Temporal acuity with audition is 10 times faster than its visual counterpart (Sanders and McCormick, 1993).

In the form of language, sound provides for immediate and clear communication of both complex information and direct commands. Abstract sounds such as tones, beeps, "boops," and so on, can provide for very quick encoding and transmission of certain types of information. Naturalistic sounds such as "bird chirping," "paper crumpling," and so on, can provide environmental cues but can also be encoded through the use of metaphor and analogy to represent more complex messages (McAdams, 1993).

Audio interfaces are particularly useful for the blind and have been shown to effectively assist in web navigation (Parente, 2004), computer programming (Sanchez and Aguayo, 2005), memory enhancement (Sanchez and Flores, 2004), accessibility to digital libraries (Pun et al., 1998), and education (Calder et al., 2007).

12.2.3.2 Speaker Setup

There are a variety of methods for delivering audio to users of an MR, VR, or interactive simulation. Multichannel surround sound systems can deliver dynamic audio cues across multiple locations within a 2D plane at the height of the speakers. In articulating the y axis (up/down), the surround sound can consist of multiple tiers with the potential for simultaneous above, below, and at-head-level arrangements. Traditional single-tier surround sound configurations only reproduce audio along the x and y axes (left/right and front/back). The addition of multiple tiers allows for a more three-dimensional audio experience. The

goal of sound accuracy is enhancement of listener enjoyment or creating a desired effect such as tension or alerting. Effective spatial impression creates a subjectively pleasing and situationally realistic sound environment (Soulodre et al., 2003).

Point source speakers are used when accuracy is critical and audio assets need to be delivered at a predetermined position. This typically consists of a speaker device that is not part of a surround system, but is used to fill in the soundscape by providing audio that is either inappropriate or less effective when used as part of a surround system. Point source speakers are used for a variety of applications, including wearable devices, special effects, prop-embedded sounds, and intimate audio (voices "inside" your head). Personal audio includes on-body speakers such as two-way radios, ear buds, and other such devices.

12.2.3.3 Perception of Audio Sources

As a special effect, alternative speaker locations can be used to heighten tension and mood. One such example is the addition of a speaker on the participant's body for dramatic effect. Intimate audio refers to sounds that are very close to the personal space of a listener and may convey a strong emotional impact or even a sense of violation. One example of intimate audio is an effect played through a helmet-embedded speaker behind the head. This intimate audio attempts to simulate extreme close proximity through the use of physically close speakers and creative sound design.

Directed sound is a new technology sometimes referred to by the trademark names Hypersonic Sound (http://www.atcsd.com) or Ultrasound (http://www.audiospotlight.com). Directed sound encodes audio into a high-frequency beam (well above the range of human hearing) that "distorts" in a predictable manner when it collides with a surface and reproduces the encoded sound. The effect of directed sound is that the audio is perceived to be emanating from the source at which it is pointed, not at its source of origin. There are two distinct effects that can be created with directed sound: virtual sound sources, and private audio. Virtual sound sources are sounds that appear to be transmitted by the object at which the directed sound is pointed and can produce the effect of speakers embedded anywhere in an environment. Additionally, virtual sound sources can be moved so that a sound appears to travel along a surface or set of surfaces. Private audio is sounds that are directed at a participant's head. These sounds are perceived by listeners as coming from inside their own heads. They have the effect of being dramatically louder to the individual at which they are directed than to other participants who may be in the general area. Private audio can be used for delivering soft sounds such as whispers or unique effects such as buzzing flies. Additional applications could include simulating internal thoughts in the form of "thoughts-out-loud."

Headphone-based 3D systems are effective tools for creating complete immersion, as they provide the sound designer with complete control over an auditory environment (Begault, 1999). The 3D headphone system uses an inertial tracking system to place sounds in space relative to the participant's head orientation. Headphone systems without tracking have the major problem of sounds moving as the user's head moves. In other words, if there is a sound source to the right of the listener, when she turns her head to look at it, the sound will still be to her right side. In this situation, the participant would be constantly turning to try to find the sound but would never find it in front of her. A 3D headphone system with tracking keeps the virtual sound in the appropriate location and, when the participant turns to look in the direction of a sound, the sound will be positioned in front of him. The main drawback to any headphone-based system is that real-world audio can be blocked except in specially-designed open headphones. The potential downside in open headphones is that the sounds are no longer private, in that others can hear the information being transmitted to nearby individuals.

12.2.4 Haptics

MR experiences generally involve senses beyond the visual and audio typical of VR. In particular, special effects generally include traditional scenography and show control technology, providing lights, smoke, and moving objects, such as doors that open and close. Haptic vests, which provide physical sensations of touch, are often used for increasing intensity or providing cues.

In addition to typical haptic devices such as vests, haptic audio devices are employed as well. Haptic audio refers to sound that is felt more than heard. Such effects can be achieved using subwoofers and mounted "bass shakers" that physically vibrate floors, walls, and other surfaces. They can be used to increase the sense of realism and impact, but can also be used to provide informational cues.

In the case of haptic vests, pressure points can be used to alert participants to the direction of targets or potential threats. These feedback mechanisms can be used in coordination with detection devices. As a personalized, tracked audio display within a haptic vest, it provides directional cues without cluttering up the already intense acoustic audioscape. With the use of speakers that vibrate more for feeling than for hearing, an intimate communication of stimulating points in the body provides the approximate orientation of objects that may not be heard or seen. This information can give an immediate sense of the direction of a threat and its proximity (e.g., by making the vibration's intensity vary with the distance to the object). It works in essence like a tap on the shoulder to tell the user of a direction without

adding to or distracting from the visual or acoustic noise levels. This message is transferred to an alternative sense and thus allows for these critical data to cut through the clutter of the audiovisual simulation. With objects outside of the line of sight, this approach can significantly reduce a user's response time or be an alternative for unsighted or deaf participants.

12.2.5 Tracking

As an MR experience requires that we know where a user is located and how that user's head is oriented, these systems generally need some form of tracking. This can be done with magnetic, acoustical, optical, inertial, or some combination of these devices. Magnetic systems tend to have moderate cost and easy setup, but can be rendered inaccurate by nearby metal interference. Acoustical systems are generally designed to work with inertial systems (sensors such as those found in the popular Wii game controller and the iPhone). These hybrid tracking systems are very accurate, but also quite expensive. Inertial systems by themselves are good for orientation but not position. Optical systems often operate by emitting light to reflective markers or sensing light produced by light-emitting diodes (LEDs), where the markers or LEDs are mounted on the device or person being tracked. The cost of such systems can be high when they involve highly accurate sensors mounted in large areas or moderate when lower cost cameras are employed in smaller areas. An even lower cost solution involves using commodity cameras or the cameras in a video see-through HMD with either natural or specially designed markers. These lower cost approaches are particularly appropriate with mobile MR/AR and for applications where precision is not required.

12.3 VR/MR Applications in Therapy

The basic elements that comprise MR systems are real-world capture, blending real and synthetic content, and experience delivery. A key aspect of MR is that it creates contextually meaningful, multisensory experiences. This means that MR, when done right, mimics a real environment and its multisensory cues so effectively that the user is transported viscerally to the intended destination. Moreover, MR gives an advantage over the real environment, in that it allows data capture and analysis.

MR presents unique challenges in its requirement to seamlessly integrate interacting virtual objects, audio landscapes, visual presentations, haptic feedback, and show-control devices with real-world objects such as human participants, props, and physical settings (Hughes et al., 2006). One of the greatest challenges is getting the real and virtual objects to interact visually in such a way that they are perceived as being in the same physical space. This means that virtual objects must hide and be hidden by real objects. Moreover, each must cast shadows on the other. Concurrent with this visual blending, audio must integrate with other senses; for example, if a virtual object falls, the sound of it interacting with both real and virtual objects must be heard. Where appropriate, this may also need to be integrated with haptic senses as the ground vibrates or even olfactory senses as broken branches of plants give off odor. While getting these real/virtual interactions to be physically correct is computationally demanding, to make this work, all such interactions must be carried out in real time, so that the visual, aural, haptic, and olfactory effects are properly synchronized with the user's expectations drawn from real-world experience.

In therapeutic applications of virtual environments, researchers develop experiences that are appropriate to support therapists in their programs to improve an individual's or group's performance. Therapeutic VR applications in the domain of mental health are discussed in detail in Chapter 52, "Cybertherapy, Cyberpsychology, and the Use of Virtual Reality in Mental Health," of this handbook.

An example application in this domain involved capturing a traumatic brain injury (TBI) patient's kitchen as a 3D model that we then overlaid on an accurately laid-out physical replica. The replica was constructed using plywood, and inexpensive appliances and cabinets. When viewed through a video see-through HMD, the patient had the same visual and haptic cues as in his home kitchen. Since the counters, coffeemaker, and so on were real, he could perform tasks like making coffee, toasting bagels, and preparing a bowl of cereal. The lessons learned, seen vividly in Figure 12.7, carried back to his home, where he was successful in making his own breakfast and in helping his wife find items such as coffee cups (a generalization of the activity he performed in the MR experience), whereas before the experiment he had carried out these tasks through a seemingly random search (Fidopiastis et al., 2006).

Subsequent experiments began with capturing appropriate environments (a restaurant for the stuttering and an open field for the upper body injury activities). The subjects in the restaurant experience were outfitted with wireless/wearable arousal meters (WAM) to reveal the existence of any correlation between presumed stressful events, the patient's arousal, and the onset of stuttering. Not surprisingly, the correlation was very strong and provided additional diagnostic information to the therapist. Similarly, WAM technology was used to study the correlation between physical tasks (catching bugs) motivated by the MR setting used for patients with upper body injuries. This latter experiment is very preliminary and was actually performed with uninjured subjects, so current conclusions are related to the efficacy of the system in providing motivating experiences, rather than its success in relieving the consequences of physical injuries. Subsequent studies are focusing on these latter goals and on creating a system that is easy for therapists to use both in creating therapeutic activities and in assessing their effectiveness.

12.3.1 MR for Assessment and Treatment of Posttraumatic Stress Disorder

Posttraumatic stress disorder (PTSD) is one of the most serious psychological conditions affecting active duty soldiers and veterans returning from Iraq and Afghanistan, with reports estimating that 11% to 17% of those returning from combat duty in Iraq suffer from major depression, generalized anxiety disorder, or PTSD (Hoge et al., 2004). To avoid a repeat of psychological effects of the Vietnam conflict, after which an estimated

FIGURE 12.7 A mixed-reality kitchen used for experiments in cognitive rehabilitation. From top to bottom, left panes show TBI patient at home, low-cost MR kitchen, and the view of this kitchen through HMDs (or an observer camera), respectively. The right pane provides a view of the actual environment and three screens (as projected from monitors): left monitor shows MR view; middle monitor screen shows captured data on user's path through space; and the right monitor is video of TBI patient at home.

830,000 veterans experienced chronic combat-related PTSD (Rothbaum et al., 2001), implementation of early screening and treatment is imperative.

It may be hypothesized that MR, with its ability to more effectively contextualize experiences, will achieve even better results than those already observed in VR studies such as those carried out at the Virtual Reality Medical Center (VRMC; Wiederhold and Wiederhold, 2008). As examples of this approach and how it can support rapid reuse, five existing experiences can be considered. The first is the kitchen scenario for TBI patient therapy; second, the restaurant scenario for therapy to treat stutterers; third, a square in a village scenario for a situational awareness training exercise; fourth, a forward observer experience; and fifth, a warehouse used to measure the window in which a PTSD subject might effectively operate.

The kitchen scenario was designed for a subject to carry out simple tasks such as making breakfast. No stressful events were added when this was used for the TBI patient. However, it would take only minutes to change the scenario so that a toaster could pop with a loud noise and even emit water vapor that appears to be smoke. Doors could slam and any number of other distractions could be under the control of a therapist or automatically triggered. The restaurant scenario can be pleasant or suddenly unsettled by the sound of thunder, the sound and visual appearance of breaking dishes, or a fight occurring in the street outside the window by the subject's table. The village scene, originally used to test a participant's situational awareness in a combat area, can become a vacation spot one is visiting on holiday where someone suddenly appears from out of an alley. The observer scene, in which forward observers are scanning a desert area while hidden behind a rock, can become a pleasant setting on a beach or in a forest, where such stressors as loud noises or sudden intrusions by others occur. The warehouse experience measures participant ability to perform simple tasks related to inventory management, but can include discordant sounds and visual distractions, such as a loader knocking over a palette of boxes, used to determine the range of such distractions a subject can tolerate. Interestingly, each of these also has a use in a calming scenario that might be applied after the stressful event.

The advantage of MR over VR is that it incorporates multiple senses in a naturalistic and context-specific way. In the kitchen scene, one can rest against the counter, placing a coffee cup there if desired. Equally important, one can smell and taste the coffee. While synthetic means are available for all senses, for the present taking the subject beyond the visual and auditory cues tends to be more limited, unnatural, and expensive. This may change in the future, but until a truly multimodal synthetic experience can be better synthesized and integrated, MR has much to offer.

As with VR, MR allows one to capture all motion and actions of the subject in the context in which they occurred. The captured data can be analyzed and used in a postscenario review of the experience. It could also be used to create behaviors for virtual actors as part of the production process that develops a large set of reusable behaviors. While this is a wonderful potential opportunity with an excellent library of scenes upon which to draw, there are a significant number of technical difficulties in segmenting, cataloguing, and coherently reassembling such scenes. In the long-term, it is expected that this will produce very effective and believable scenes, but much work remains to be done.

12.4 VR/MR Applications for Sensory, Physical, and Social Accessibility

While therapy addresses one aspect of accessibility, the broader contribution of this work lies in developing and assessing techniques, sensors, and processing methods that can be applied much more broadly to accessibility. Overcoming limitations in vision, audition, and haptic capability are obvious applications of the technologies described earlier, but there is growing interest in addressing mobility and social and emotional accessibility issues as well. There is considerable ongoing research in supportive technologies for vision, auditory, and haptic interfaces, all of which will be useful in VR/MR. A number of interesting and representative efforts are cited here, but there are many other excellent projects under way that investigate supportive technology for specific modalities. The individual assistive technologies that can be employed in VR/MR are not unique in themselves; more important are the fundamentally multimodal nature of VR/MR and the immersive nature of the experience. Taken together these provide a more fertile environment for sensory substitution and effective use of cross-sensory interfaces than would otherwise be possible.

In a hierarchy of information bandwidth capability, the channels of primary importance in VR/MR are visual, auditory, and haptic. While there has been some work in chemical sensing, this will probably always be secondary if only because the potential bandwidth is low and the cost relative to potential contribution is high. Mapping among the primary modalities is the obvious way to compensate for individual limitations. The literature reports many excellent examples. Obviously mapping a lower bandwidth channel to a higher one provides more flexibility, but many successful approaches go in the other direction. An excellent overview of sensory substitution and associated issues may be found in *Sensory Substitution and the Human-Machine Interface* (Bach-y-Rita and Kercel, 2003). This publication also introduces some ideas about brain plasticity and neural issues that are of value in better understanding how these sensory substitutions actually function.

Sensory augmentation and redundancy to provide enhanced capability may also be useful in VR/MR for accessibility. This was studied as a means for improving situational awareness under conditions of high cognitive workload for helicopter pilots (Raj et al., 2000). This study provided additional spatial cuing information using a haptic vest, demonstrated significant improvement in spatial awareness under high workload conditions, and may point the way for better multimodal assistive technologies. Haptic visualization may serve as an additional information channel for "visualization" of multiparameter scientific data, for augmentation as in the helicopter pilot example, and of specific interest here, as a substitution for a missing or impaired sense. An excellent overview of the equipment, issues, and growth of this area was recently published (Roberts and Paneels, 2007). More directly applicable to the issue of accessibility is a recent paper addressing haptic rendering of visual data for visually impaired users (Moustakas et al., 2007). This publication specifically addresses automation of the process of converting 2D and 3D images into haptic maps usable by visually impaired individuals, although it does not directly address how to apply this in VR/MR.

For general accessibility to virtual environments, particularly online virtual worlds, haptic interfaces will need substantially more development. For rehabilitation, entertainment, and training where equipment cost may be less critical, MR, perhaps augmented with haptic vests and other such devices, remains the best near-term option for providing haptic stimulation in virtual environments. Auditory substitution is another matter. The potential for a rich and spatially correct soundscape for VR/MR described earlier means that in dealing with virtual environments visually impaired individuals will be able to use many of the navigation skills they already possess. There is also the prospect for augmenting this by synthesizing directionally correct audio cues such as rustling leaves, footfalls, and other ambient sounds that represent key features in visual images. Totally artificial sounds may be used in this way as well. This is termed *sonification* (Mahler et al., 2006; see also Chapter 32, "Auditory Interfaces and Sonification") and is seen as useful for alerting, data understanding, and cuing mechanisms, as well as the role being considered here for sensory substitution.

In addition to dealing with physical limitations, many of these techniques are applicable to the much more general issues of accessibility in society using technology for mitigation. In particular, the development of cost-effective and flexible VR and MR equipment, successful laboratory-based VR and MR applications, and the emergence and acceptance of online virtual worlds creates substantial opportunities for enhancing accessibility to the broader world, not just assisting in the use of technology. Two online VRs, Second Life (SL) and World of Warcraft (WoW), have been cited as examples of highly successful online VR applications that may have implications well beyond entertainment and social interaction. A recent presentation by the Gartner Group, a respected industry research and analysis company, specifically addressed the emergence of virtual worlds as an important enterprise opportunity (Prentice and Jones, 2007). Among their conclusions was that "by YE 2011, 80% of active internet users (and Fortune 500 enterprises) will have a 'second life'—but not necessarily IN 'Second Life.'" Similarly, there has been considerable interest in using SL and WoW as platforms for research in such topics as economics, the social sciences, and similar areas, where experimentation in the real world or with large heterogeneous populations is normally difficult or impossible.

A recent article in the journal *Science* (Bainbridge, 2007) discusses current online virtual worlds and their technological descendents as virtual laboratories for experiments of all sorts, with a particular focus on social interaction and economic issues. Some related research has already been reported on virtual worlds' suitability for modeling epidemics (Lofgren and Fefferman, 2007), where the response of real individuals to features of the synthetic world is shown to mirror similar situations in the real world. For example, in 2005 an infectious element was introduced into the World of Warcraft that had the unexpected consequence of producing a full-blown epidemic with features that strongly reflected what had been observed in real epidemics. Given the expected widespread acceptance and use for social and commercial purposes, and even for research on social, economic, and environmental issues (Fiore et al., 2007), providing means for broad accessibility will be an important objective. This trend suggests another important application of VR/MR for accessibility where the issue is mobility, emotional, or psychological, rather than sensory.

While current online virtual-world environments do not, and in the short to middle term will not, provide as rich a visual environment or the multisensory capabilities of some of the applications mentioned earlier, they do have the virtue of becoming readily available. Interestingly, for certain populations the simplified sensory environment and even the time delays inherent in current networked "worlds" may be a useful feature (Salman, 2006). While there has been considerable interest in ways to support a variety of populations, most efforts are still in the early exploratory stages of implementation and evaluation. As a result, not much is reported in the technical literature documenting experimental results; however, the range of applications proposed, prototyped, or under way is so extensive, and the early results are promising enough, that we need to take these developments seriously. One very positive indicator is the degree to which industry, the social sciences, and various special needs communities are taking this technology into new realms that in turn will likely benefit everyone. The projects described in the following are not intended to represent a comprehensive survey of emerging efforts, but rather are a snapshot of interesting projects representing a range of physical and nonsensory accessibility issues.

Recently, *New Scientist* (http://www.newscientist.com), a weekly scientific news magazine designed for a general audience, published a series of articles on SL as a venue for removing social barriers for people with various disabilities, situations, or conditions. Two of these are notable in that they report useful results and the development of online communities that have applied these environments essentially intact as a means for improved social accessibility using VR technology. The 27 June 2007 issue included an article about individuals with Asperger's syndrome employing SL as a venue for social interaction with others on an equal footing, without the communication impediments that this syndrome can cause. The simplified body language of avatars, interaction through text, the acceptability of delays in responses, all features of existing online virtual worlds, in themselves provide accessibility for these individuals. This is also true for those with physical limitations. The 22 August 2007 issue of *New Scientist* reports on the use of virtual worlds by physically disabled people as a means for participating in many aspects of social life on an equal footing with the nondisabled.

In online virtual worlds, a user can choose an avatar arbitrarily, including gender and even species. The persona that is chosen as a person's virtual-world representation may reflect the individual's actual situation, or it may represent how such a person wishes to be seen. The same 22 August 2007 issue of *New Scientist* previously mentioned also describes a transgendered individual able to interact in the virtual world as the preferred gender, and quadriplegics are able to walk, dance, and even fly. Here again the medium itself accommodates access not just to technology, but also to the larger social world. Many such applications, ranging from accessibility for special populations to serious research in economics and social sciences, could be undertaken immediately using the virtual world's metaphor in creative ways. Various interesting and sometimes surprising applications are discussed in an article in *Newsweek* magazine (Bennett and Beith, 2007). Much of this discussion is based on anecdotes now, but serious studies are under way that are expected to validate the results reported so far.

While applications that can use existing technologies more or less intact are many, there are some conditions for which VR and MR technology must be adapted to provide access. As the highest bandwidth communication channel for humans, vision is a key element in most VR and MR applications. For web-based VR, it may be the only interface element for presentation to the user. There are available tools that allow low vision users better access to computing resources, as well as ongoing research to create methods for improved access to web sites that may contain multimedia content. For games and game-like online environments, 3D visual representations of the virtual world may be the only interface output modality, making access for the blind problematical for the moment.

12.5 Conclusions

VR and the many variants discussed here are relatively new additions to the accessibility toolkit and as such have not reached a level of maturity and stability comparable to other techniques. Even so, the therapeutic and augmentation examples cited here are just samplings of many recent and current research efforts to bring these methods to everyday use. The growth in technical capabilities that can reasonably be expected in the next 5 years means that capability that now can only be afforded by businesses and higher end-users will be available to home users within that time. The kinds of opportunities for accessibility that this represents for users of all abilities is hard to estimate, but it is clear that so far we have just touched on the most obvious uses.

References

Bach-y-Rita, P. and Kercel, S. (2003). Sensory substitution and the human-machine interface. *TRENDS in Cognitive Sciences* 7: 546–551.

Bainbridge, W. S. (2007). The scientific research potential of virtual worlds. *Science* 317: 472–476.

Begault, D. (1999). Auditory and nonauditory factors that potentially influence virtual acoustic imagery, in *AES 16th International Conference on Spatial Sound Reproduction*, 10–12 April 1999, Rovaniemi, Finland, pp. 13–26. New York: Audio Engineering Society.

Bennett, J. and Beith, M. (2007). Why millions are living virtual lives online. *Newsweek International*. http://aplink.wordpress.com/2007/07/24/newsweek-why-millions-are-living-virtual-lives-online-itsreal.

Calder, M., Cohen, R., Lanzoni, J., Landry, N., and Skaff, J. (2007). Teaching data structures to students who are blind, in the *Proceedings of the 12th Annual SIGCSE Conference on Innovation and Technology in Computer Science Education ITiCSE 2007*, 25–27 June 2005, Dundee, Scotland, pp. 87–90. New York: ACM Press.

Cruz-Neira, C., Sandin, D. J., DeFanti, T. A., Kenyon, R., and Hart, J. C. (1992). The CAVE, audio visual experience automatic virtual environment. *Communications of the ACM* 35: 64–72.

Fidopiastis, C. M., Stapleton, C. B., Whiteside, J. D., Hughes, C. E., Fiore, S. M., Martin, G. A., et al. (2006). Human experience modeler: Context driven cognitive retraining to facilitate transfer of training. *CyberPsychology and Behavior* 9: 183–187.

Fiore, S. M., Harrison, G. W., Hughes, C. E., and Rutström, E. E. (2007). Virtual experiments and environmental policy. Paper presented at *Frontiers of Environmental Economics*, 26–27 February 2007, Washington, DC.

Hoge, C. W., Castro, C. A., Messer, S. C., McGurk, D., Cotting, D. I., and Koffman, R. L. (2004). Combat duty in Iraq and Afghanistan, mental health problems, and barriers to care. *The New England Journal of Medicine* 351: 13–22.

Hughes, C. E., Stapleton, C. B., and O'Connor, M. (2006). The evolution of a framework for mixed reality experiences, in *Emerging Technologies of Augmented Reality: Interfaces and Design*, pp. 198–216. Hershey, PA: Idea Group.

Lofgren, E. and Fefferman, N. (2007). The untapped potential of virtual game worlds to shed light on real world epidemics. *The Lancet Infectious Diseases* 7: 625–629.

Mahler, T., Bayerl, P., Neumann, H., and Weber, M. (2006). Visual attention in auditory display, in *Perception and Interactive Technologies, Proceedings of the International Tutorial and Research Workshop, PIT 2006* (E. André, L. Dybkjær, W. Minker, H. Neumann, and M. Weber, eds.), 19–21 June 2006, Berlin, pp. 65–72. Berlin/Heidelberg: Springer-Verlag.

McAdams, S. (1993). Recognition of sound sources and events, in *Thinking in Sound: The Cognitive Psychology of Human Audition* (S. McAdams and E. Bigand, eds.), pp. 146–198. New York: Oxford University Press.

Milgram, P. and Kishino, A. F. (1994). Taxonomy of mixed reality visual displays. *IEICE Transactions in Information and Systems* E77-D: 1321–1329.

Moustakas, K., Nikolakis, G., Kostopoulos, K., Tzovaras, D., and Strintzis, M. G. (2007). The force field haptic rendering method: Application in haptic access to visual data for the training of the visually impaired. *IEEE MultiMedia* 14: 62–72.

Pair, J., Neumann, U., Piepol, D., and Swartout, W. (2003). FlatWorld: Combining Hollywood set-design techniques with VR. *IEEE Computer Graphics and Applications* 23: 12–15.

Parente, P. (2004). Audio enriched links: Web page previews for blind users, in the *Proceedings of the 6th International ACM SIGACCESS*, 18–20 October 2004, Atlanta, pp. 2–8. New York: ACM Press.

Prentice, P. and Jones, N. (2007). Virtual worlds: Real opportunities, paper presented at *Gartner Symposium ITxpo 2007*, 22–26 April 2007, San Francisco.

Pun, T., Roth, P., Petrucci, L., and Assimacopoulos, A. (1998). An image-capable audio Internet browser for facilitating blind user access to digital libraries, in the *Proceedings of the Third ACM Conference on Digital Libraries*, 23–26 June 1998, Pittsburgh, pp. 301–302. New York: ACM Press.

Raj, A., Kass, S., and Perry, J. (2000). Vibrotactile displays for improving spatial awareness, in the *Proceedings of the Human Factors and Ergonomics Society Annual Meeting, Volume 1, Cognitive Ergonomics*, 30 July–August 4 2000, San Diego, pp. 181–184. Santa Monica, CA: Human Factors and Ergonomics Society.

Roberts, J. and Paneels, S. (2007). Where are we with haptic visualization?, in the *Proceedings of the Second Joint Eurohaptics Conference and Symposium on Haptic Interfaces for Virtual Environment and Teleoperator Systems WHC*, 22–24 March 2007, Tsukuba, Japan, pp. 316–323. Washington, DC: IEEE Computer Society.

Rothbaum, B. O., Hodges, L. F., Ready, D., Graap, D. K., and Alarcon, R. D. (2001). Virtual reality exposure therapy for Vietnam veterans with posttraumatic stress disorder. *Journal of Clinical Psychiatry* 62: 617–622.

Salman, S. (2006). Autism community forges virtual haven. *The Guardian*, 8 March.

Sanchez, J. and Aguayo, F. (2005). Blind learners programming through audio, in the *CHI '05 Extended Abstracts on Human Factors in Computing Systems*, 2–7 April 2005, Portland, Oregon, pp. 1769–1772. New York: ACM Press.

Sanchez, J. and Flores, H. (2004). Memory enhancement through audio, in the *Proceedings of the 6th International ACM SIGACCESS Conference on Computers and Accessibility*, 18–20 October 2004, Atlanta, pp. 24–31. New York: ACM Press.

Sanders, M. S. and McCormick, E. J. (1993). *Human Factors in Engineering and Design* (7th ed.). New York: McGraw-Hill.

Soulodre, G., Lavoie, M., and Norcross, S. (2003). Objective measures of listener envelopment in multichannel surround systems. *Journal of the Audio Engineering Society* 51: 826–840.

Uchiyama, S., Takemoto, K., Satoh, K., Yamamoto, H., and Tamura, H. (2002). MR platform: A basic body on which mixed reality applications are built, in the *Proceedings of the International Symposium on Mixed and Augmented Reality*, 30 September–1 October 2002, Darmstadt, Germany, pp. 246–256. Los Alamitos, California: IEEE.

Wiederhold, B. K. and Wiederhold, M. D. (2008). Virtual reality for posttraumatic stress disorder and stress inoculation training. *Journal of CyberTherapy and Rehabilitation* 1: 23–35.

Biometrics and Universal Access

Michael C. Fairhurst

13.1 Individual Identity and the Information Society

The *information society* provides both opportunities and challenges for its citizens, and the designers of interactive systems must be increasingly aware of the responsibility to ensure that the opportunities to access, utilize, and benefit from all sources of information are made as widely available as possible, while similarly ensuring that the challenges that arise in endeavoring to achieve this are addressed in an appropriate manner. Hence, the concept of *universal access* is critical if the huge potential benefits of technological developments are to be most effectively exploited by as large a number and as wide a range of individuals as possible.

Universal access is a concept that is very broad in its compass and of immense potential importance with the increasing incorporation of technology routinely into more and more aspects of everyday life. An important element in the design of many systems, especially those that impinge on the utilization and manipulation of essentially private or particularly sensitive areas of life (for example, access to controlled areas, availability of financial or medical records, and so on), is the way in which access is controlled, security and integrity of data guaranteed, and monitoring of system use achieved (see also Chapter 57, "Security and Privacy for Universal Access," of this handbook). Thus, for many systems, and on an increasing scale, the nature of the approach adopted to the management of security is a vital issue in achieving an information society in which universal access is not just facilitated (desirable though this undoubtedly is), but also controlled and regulated in an appropriate way. This

tension between facilitation and regulation will be an important theme discussed later on in this chapter.

However, the principal focus of this chapter is the underlying issue of system access management in its broadest sense, as a supporting component in achieving the general goal of universal access. In particular, this chapter will examine a specific approach to access management in systems where the principal element in guaranteeing security of access is the exploitation of so-called biometric data, relating to the physiological or behavioral characteristics of an individual.

In fact, universal access in its most generally understood sense implies the need for an awareness of the wide range of characteristics (physical, intellectual, behavioral, and so on) of individuals, and an analysis of the extent to which these are similar or dissimilar with respect to other individuals. The system designer wishing to support universal access then generally seeks to develop information-processing strategies that will minimize these differences and their effects, especially with regard to the way in which they affect the user and the nature of the user/system interaction. Later, however, it will be shown that it is precisely this characteristic that underlies the tension referred to previously.

Establishing an individual's identity may involve one of two (related) processes:

1. Recognition ("This is John Smith"). This may be seen as a many:1 mapping problem.
2. Verification ("I confirm you are John Smith as you have claimed"). This is a 1:1 mapping problem.

In fact (2) is really a special case of (1) and is the problem that most commonly arises in the scenarios of interest that are discussed in the following. Indeed, this chapter focuses primarily on the verification problem. Traditionally, three broad approaches address the question of how to establish individual identity, and these are typically set out in the following way:

- Possession of a token
  - This corresponds to "something you have."
    - Examples might include a physical key, a swipe card, etc.
- Possession of specific knowledge
  - This corresponds to "something you know."
    - Examples might include a PIN, a password, etc.
- Possession of particular physical characteristics
  - This corresponds to "something you are."
    - Examples might include a fingerprint, a characteristic voice pattern, etc. More examples will be considered in the following.

In principle, this last option should provide the highest degree of security in establishing or confirming individual identity. This is the domain of *biometrics*, which is the principal focus of the rest of this chapter.

13.2 Introduction to Biometrics

Biometric technologies are now rapidly emerging as important components in regulating access to places, goods, and services, and significant application opportunities exist across a wide range of scenarios, including electronic commerce, security monitoring, database access, forensic investigation, and tele-medicine (Deravi and Lockie, 2000). Indeed, any situation where automated identification of individuals and identity management are required falls within the sphere of biometrics. To consider some particular examples, a computer system may be configured to offer a sign-on facility based directly on handwritten information, such as a user's individual signature, to replace (or perhaps to complement and enhance) password-based log-on control. Online monitoring of system usage could be achieved by means of a terminal-mounted camera linked to face recognition/verification software that can ensure that users are not interchanged after approved initial access. Similarly, access to a secure physical environment can be restricted by means of, for example, a fingerprint verification system or hand geometry terminal as a means of increasing security, as opposed to, for example, human visual inspection of badges (often careless and error-prone) or swipe-card control (where cards can be stolen or passed between users in collusion). It is also clear even from these simple examples that biometric identity checking offers many different options in terms of the *biometric modality* (particular source of identification measurement) adopted, and thus seems to offer some potential for exploitation where universality is sought.

The discipline of biometrics may be formally described as the measurement of specific attributes or features of an individual with the aim of being able to distinguish that individual from all others. Some examples of common biometric modalities of current interest (more details can be found, e.g., in Fairhurst, 1997; Obaidat and Sadaoun, 1997; Doddington, 1998; Jain et al., 1999; Daugman, 2004; Munro et al., 2005; Srisuk et al., 2005; Bowyer et al., 2006; Han and Bhanu, 2006; Li et al., 2006; Shen et al., 2006; Sheng et al., 2007; and Wu et al., 2006) are the following:

- Facial features
- Voice characteristics
- Fingerprints
- Handwritten signature
- Iris patterns
- Hand shape
- Hand vein patterns
- Keystroke dynamics
- Odor
- Ear shape
- Gait patterns
- Retinal blood vessel patterns

This list shows an extensive range of possible biometric modalities, illustrative rather than exhaustive, since many others could also easily be envisaged. From the point of view of this chapter, of course, the existence of so many options is especially useful, since it allows a choice to be made to support identity checking across a range of different individuals, different environments, and different application domains. Although for most biometrics, widespread practical exploitation has been a comparatively recent phenomenon (indeed, practical implementation of biometric systems for routine general use is an area that is still moving toward full maturity), research interest in all forms of biometric measurement has a long history, and experience has clearly shown that different biometric modalities offer their own particular advantages and disadvantages. Similarly, an important factor in choosing an appropriate biometric for a given task is the degree of acceptability that it is accorded within the prospective user community. Although the introduction of biometric authentication can sometimes be accompanied by skepticism or even outright opposition (e.g., because of poor reliability, concerns about privacy and civil liberties, etc.), there is general agreement that in many applications biometrics-based schemes offer advantages over other approaches to identity checking.

To pick up on some of the important issues set out previously, the chapter will explore aspects of universality that are important in relation to conventional biometrics-mediated identification schemes. First, however, the basic principles of operation of a biometric identification system will be reviewed.

13.3 Operating Characteristics of a Biometrics-Based System

The general framework for verifying identity using a biometric system is illustrated schematically in Figure 13.1. Of course, any

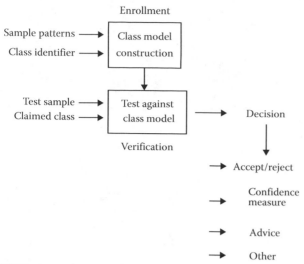

FIGURE 13.1 Schematic of a biometric identity verification system.

interaction with a biometric system requires a sensor (sometimes modality-specific, sometimes more general-purpose) to capture an appropriate sample from which an objective measurement can be made directly or indirectly. The sensor adopted will naturally depend on the biometric modality being used, but may be an entirely standard device or may require more special-purpose features. For example, a simple camera may suffice for face recognition, while fingerprint processing requires a sensor that, although standard in many ways, would not have applications beyond its immediate specific identity-checking task domain. It is also necessary to allow the user to indicate to the system a claimed identity, and again this may take a variety of forms. A simple PIN or password will often suffice, but it is equally possible to use a card-based system or any other appropriate technology or input interface.

It is then necessary to consider two distinct operational phases for system deployment (although, in principle, there may be opportunities for blurring this distinction in certain circumstances). The first phase is the *enrollment phase*. This requires a potential system user to present a sample or, more commonly, a set of samples, of his chosen biometric as the basis for constructing a reference model (often referred to as an enrollment *template*) of known provenance, against which future samples can be tested. It is clearly important to ensure that at this stage the user's identity is rigorously investigated and robustly established, since the success of future biometrics-based identity verification will depend entirely on the veracity of claimed identity at this stage. Likewise, it is important that the reference model is based on samples that reflect the true nature and natural variability of the biometric in question for that individual. After enrollment, the system will hold tagged information about the registration of a specified user, and a template/reference model that embodies and reflects the biometric data of that user with respect to the modality to be adopted.

In the second phase, the *operational phase*, the system can be used to check the identity of any individual who is prepared to donate a biometric sample. In use, an individual will declare to the system a claimed identity (by means of an appropriate mechanism such as PIN, card swipe, etc.) which must, of course, correspond to a registered system user. The user is then invited to donate a biometric sample that can subsequently be checked against the reference held for the individual whose identity is claimed. This allows the system to make a judgment as to whether the user is genuine (i.e., the biometric data can confirm the claimed identity) or an imposter (the biometric data are inconsistent with the claimed identity). This process involves simply a comparison between the measurement derived from the current biometric sample and the reference model held for the claimant. However, it is apparent that the possibility of an exact match is highly unlikely, both because the reference model will generally be a composite model derived from several samples, but principally because natural variability and prevailing operational conditions will be reflected in the characteristics of the sample used for comparison.

Thus, the basic mechanism for checking identity requires the measurement of the degree of match between sample and reference, which can be presented as either a measure of similarity, or of difference. If the latter is adopted, then the decision-making process involved can be characterized in the following way. Let the *distance* between the test sample and the reference model be d, and let T be a threshold chosen to reflect a degree of difference allowable to accept that the sample confirms true claimed identity.

Then:

- If $d < T$ the system accepts the claimed identity as the true identity.
- If $d > T$ the system will reject the claimed identity as not corresponding to the true identity.

Though the concept is very simple, it nevertheless raises an important issue, which is how to choose the value of T that should most appropriately be adopted in a given situation. Hence, the implications of this choice and the effects it is likely to have must be considered more carefully. In particular, one should be aware that absolute and unequivocal decisions may often be difficult to achieve, and this means that it is necessary to consider several factors relating to the possible occurrence of errors arising in the system.

First, it must be recognized that, even within the simple framework set out here, there is scope for the system to make incorrect decisions purely on the basis of inappropriate choice of the threshold value, T. For example, there are two obvious and fundamental scenarios that relate to the possibility of an incorrect decision being made by the system (generally expressed as a rate of occurrence across a large number of trials), designated, respectively, as the false rejection rate (FRR) and the false acceptance rate (FAR). A false rejection occurs when a genuine sample from a user claiming her true identity is determined by the system not to meet the verification criteria and is rejected by the system. On

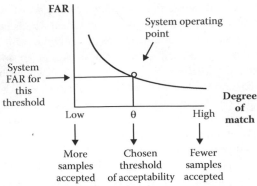

FIGURE 13.2 Variation of a system error rate with verification threshold.

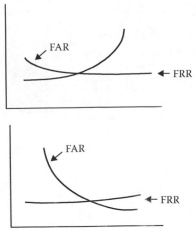

FIGURE 13.4 Observing the effect of asymmetrical FAR/FRR curves.

the other hand, a false acceptance occurs when a sample from an individual other than that whose identity is claimed meets the verification criteria and is accepted as genuine.

It is apparent that these two rates are likely to be inversely related. For example, requiring a very high degree of match between sample and stored reference before identity is verified will, of course, reduce the likelihood of an imposter meeting the criteria. However, requiring a very high degree of match might also make it more difficult for a genuine user to meet the criteria, thus increasing the likelihood of a false rejection occurring. Figure 13.2 shows the typical variations that might be expected when conducting tests to measure, for example, FAR in a system as a function of choice of T, while Figure 13.3 shows both FAR and FRR curves superimposed on the same axes. The point at which these two curves cross (i.e., the point at which FAR = FRR) is often referred to as the system equal error rate (EER), and is sometimes taken as a rough indicator of system performance. Of course, the EER conveys very little information about the shape of the two curves, which may not be symmetrical, and should therefore only be used as a broad guide to performance.

It is now clear that the requirements of the target application scenario will need to be studied carefully before a choice of T can be made. Thus, an application where high security is required will aim to keep FAR low, but perhaps with a consequential rise in FRR, while the reverse may be the case in an application where high security is not a priority, especially where large throughput of users is likely to be required. Figure 13.4 shows how asymmetrical FAR/FRR curves will impact the nature of the trade-off between these two error rates in practice.

Of course, the discussion so far has considered error rates purely in terms of how an informed choice about decision thresholds can be made. However, a typical biometric system may be susceptible to problems relating to a variety of potential sources of error. For example, other possible error sources to be considered might include:

- *Verification algorithm*: The matching algorithm could itself be weak, or susceptible to failure under specific circumstances. Such a fundamental problem at the technology level requires radical design modifications at the most basic level.
- *Data characteristics*: Unreliability may arise because the data available, either at enrollment or as captured in relation to the test sample, may not conform to the conditions required for consistency (e.g., between enrolling and subsequently using a system) or may not be representative of the variability inherent in an individual's biometric.
- *Operational environment*: The conditions under which samples are collected may have a very significant impact on system performance. For example, lighting changes during capture of a visual biometric (e.g., for face recognition purposes) can very significantly affect the quality of the captured sample and the degree of similarity measured between the test sample and the stored model.
- *System interaction*: Humans are notoriously unreliable in their ability to perform consistently in unfamiliar situations, or to observe detailed rules that might otherwise remove sources of data variability. Indeed, there may be situations in which deliberate noncooperation is to the benefit of a system user (in endeavoring to circumvent the identification process, for example), or where a simple lack of understanding can lead to poor data samples being captured.
- *Human behavior*: Humans are quite capable of exhibiting inherently variable behavioral patterns. This may not necessarily be attributable to noncooperation or a deliberate

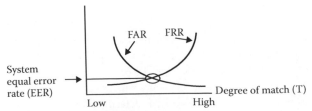

FIGURE 13.3 FAR, FRR, and EER.

attempt to undermine, but can nevertheless lead to difficulties in establishing reliable operation. Behavioral biometrics (the handwritten signature is a case in point) are especially susceptible to this problem, and this inherent unreliability can result in some potential system users being labeled as "goats"—individuals with inherently unreliable biometrics (Doddington, 1998). This point will be further discussed later.

There are, therefore, many possible factors that will determine the performance characteristics of a biometric identification system. Judicious design of a system "end-to-end," together with some understanding of how to set an appropriate balance between FAR and FRR, will be required if optimality is to be achieved in practice. This is also the principal reason why it may be necessary to think more broadly about optimizing system performance and increasing robustness, as will be discussed later.

13.4 Characterizing Biometric Processing

Perhaps a starting point for understanding the important issues in seeking to move toward universality in this area is to note that choice of a suitable modality can rarely be made without taking account of a wide variety of factors.

First, it is helpful to recognize that biometric modalities may be characterized and classified in different ways, and each categorization provides an insight into the nature and utility of the modality according to different criteria. For example, biometric measurements may be categorized as either *physiological* or *behavioral*. The first type, examples of which include iris patterns, fingerprints, and so on, relate to inherent physiological characteristics of an individual, while the second type, such as handwritten signatures and keystroke dynamics, arise from activities carried out by that individual, either those that occur spontaneously or, in some cases, those that are specifically learned. An understanding of the different ways in which biometrics can be categorized will clearly have relevance in relation to the extent to which a biometric can be universally applied. There may be two aspects to this issue, however:

1. Adoption of a particular physiological biometric will be contingent upon specific physiological characteristics of an individual that should be easy to determine. For example, a conventional voice recognition system is highly likely to be unsuitable among a population of individuals with significant speech impairments.
2. Adoption of a behavioral biometric certainly requires the user to be capable of expressing the form of behavior that underpins the modality chosen. Thus, handwritten signature verification is unlikely to be successfully implemented for an individual with severe motor impairment that affects writing skills. However, there is a more subtle factor that comes into play here, because a common

problem with behavioral biometrics is that the specific measurements they require are frequently found to be more variable than is generally the case with physiological measurements. Moreover, this variability can change very significantly among individuals. The handwritten signature is a particularly good example, because it is well known that some individuals generate (perfectly legitimate) signature samples that vary substantially at different times. As noted previously, such individuals are often referred to as "goats" in a population, and the inherent instability in their biometrics leads to problems of reliable identification, sometimes to a degree that may necessitate their exclusion from a system user population unless a satisfactory alternative can be found.

Factors such as these are clearly very relevant to a consideration of how to maximize the applicability of biometric identity checking, but this is not the only way in which biometric modalities may usefully be categorized in this respect. For example, other possibilities might include the following.

13.4.1 Visual/Nonvisual Measurements

Of particular relevance to the underlying technologies as well as, in many cases, the patterns of interaction required during use is the observation that the origin and extraction of the raw data required for the identification/verification process should be considered. More specifically, many biometric modalities are based on the extraction and processing of visual information, with image-processing tools being the primary required technology. Facial recognition is self-evidently a prime example of a visual biometric, but iris-scan technology, fingerprint processing, and hand-shape recognition are also commonly adopted modalities that rely almost entirely on visually derived information. Other modalities, however, do not use visual information in the features extracted for identification processing. Speech-based identification is a good example, while identification based on handwriting, especially the handwritten signature, often relies on an integration of visual and nonvisual information, depending on the mix of static (derived directly from the signature image itself) and dynamic (derived from the time-based pattern of signature execution) features that can be made available.

13.4.2 Contact/Noncontact Measurements

Some biometric modalities inherently require direct contact with a sensor or a specified target surface for the extraction of biometric measurements. Good examples of this type of situation are fingerprint acquisition and hand-shape recognition. Other modalities, however, do not rely on such direct contact for the acquisition of the raw biometric data for their operation, including face recognition and speech processing. This may clearly have an implication for certain potential user populations and, indeed, in relation to some cultural characteristics.

13.4.3 Overt/Covert Measurements

It is the nature of security applications that there can sometimes be a need for the acquisition of data to be covert. It can therefore be useful to explore a categorization of biometric data depending on whether the measurements involved require the direct knowledge and/or cooperation of the target subject, or whether they can be carried out without such knowledge. With current sensors, for example, it is difficult to imagine routinely collecting fingerprint images from a subject covertly, unless some easy way can be found of routinely capturing latent fingerprints. Gait analysis, and to some extent face recognition, however, can be attempted remotely without the explicit cooperation of the subject.

Such attempts to categorize biometric modalities are particularly helpful in selecting a biometric for deployment in a specific application domain, and illustrate well the difficulties of adopting a single biometric even in situations where the inherent security requirements are the highest priority consideration. More important, the discussion demonstrates well the even greater difficulty in adopting any one individual modality to achieve the degree of optimality aimed for in the context of universal access. This is exactly why it is important to consider some alternative strategies, which will be discussed later in this chapter. However, even when the main concern is only to maximize the applicability and accessibility of each modality separately, it is clear that some of the issues described previously will be very important to achieve the widest possible take-up of the security benefits afforded by biometrics-based technologies.

Of immediate interest, however, is that the discussion so far points to an apparent paradox when trying to find a single integrated solution to the aim of achieving an optimum framework for universal access in the context of biometrics processing in general. It is possible to set out this paradox in very simple terms, as follows. The problem is that two different perspectives must be held in balance, and the two competing requirements that emerge must be reconciled. Thus, on the one hand, biometric measurements seek to identify individual characteristics such that the measurements made define and emphasize the *differences* between individuals in a population. For example, it is important to locate and record minutiae in fingerprint images, because it is precisely these features that will give the best chance of distinguishing between two different individuals. On the other hand, the underlying philosophy of universal access is in many ways the opposite, because the underlying aim is not to seek out and emphasize differences, but rather to exploit characteristics that relate to the inherent *similarities* between individuals where this can be achieved, or otherwise to identify and overcome the differences that naturally exist between individuals. Although, of course, rather different strata of activity are involved in this argument, the point made is a valid one. One can imagine, for example, a scenario in which fingerprint processing is chosen as a means of optimally ensuring security in controlling access to a particular environment. This choice may have been made to guarantee high performance in relation to criteria based on finding an appropriate balance between preventing unauthorized access and not falsely rejecting authorized personnel. However, such a choice would immediately create a difficulty for any individual with inherently poor fingerprints (e.g., some manual workers whose fingerprints may be damaged or especially worn) or, perhaps more important, an individual with a disability that prevents reliable interaction with the fingerprint-sensing device.

Consequently, it is more productive, and may even be necessary, to move away from any notion of optimizing an individual modality and to contemplate a broader approach to most effectively exploit biometric identification strategies in the framework of a universal access scenario. One potentially very effective approach is to adopt a scheme that aims to make available more than one possible source of biometric information, and this leads to the principles of multimodal biometric systems.

13.5 Multimodal Biometrics

As discussed in the previous section, current approaches to personal identity authentication using a single biometric technology are limited, principally because no single biometric measurement is generally considered simultaneously sufficiently accurate, obtainable, and user-acceptable for universal application. An alternative strategy is to make multiple modalities available, with measurements being taken within some or all of these modalities concurrently (or, at least, within a short single acquisition session), with the resulting multisource data providing the basis on which identification processing is carried out. An added advantage of this approach is that techniques for sensor and classifier fusion are already very highly developed (see, e.g., Ho et al., 1994; Fairhurst and Rahman, 2000; Rahman and Fairhurst, 2003).

In principle, of course, any combination of modalities can be considered. Examples can readily be found of work that integrates modalities in different combinations (Su and Silsbee, 1996; Chen et al., 1997; Chibelushi et al., 1997; Hong and Jain, 1998; Jain and Ross, 2004; Marcialis and Roli, 2004; Wu et al., 2006). In practice, there may be an argument for seeking a combination of modalities that minimize capture effort at the point of sample acquisition (it is not difficult to imagine a sensor configuration at a user interface that allows simultaneous collection of, for example, fingerprint data and hand-shape data), but ultimately many factors will influence the choice of modalities to be made available. Irrespective of specific choice, however, it is clear that the fundamental principle of accessing and combining in some appropriate way biometric data from more than one measurement source immediately offers potential advantages with respect to the effectiveness of the identification/verification process, of which the principal elements are:

- *Greater reliability and improved accuracy*: More information on which to base an identification decision, especially when this is drawn from different and uncorrelated sources, should increase the confidence of decision making and decrease the likelihood of errors being made.

- *Greater resistance to malicious attack*: The risks of unauthorized access to a location, a service, or a source of data can be significantly reduced by decreasing the vulnerability of a system to data collected from impersonation attempts and "spoofing," particularly at the sensor level. An attacker able successfully to impersonate another user with respect to one modality is likely to find rapidly increasing degrees of difficulty if required to develop successful impersonation strategies in several very different modalities.

- *Greater flexibility for the user*: In fact, in the present context, this is perhaps the most important and beneficial feature of a multimodal approach, since the availability of more than one modality immediately raises the option of affording some degree of *choice* to a potential user. In principle, this choice could be offered purely on the basis of personal preference or cultural predisposition, but, crucially when considering issues of universality, it also prevents exclusion on the grounds of inability to provide any specific measurement required by any one of the modalities available. Thus, in a multimodal scheme, a user with severely restricted hand movement who is asked for a hand-oriented biometric sample could opt instead to rely on a facial image rather than donating, for example, a fingerprint, to establish identity.

The principle of biometric multimodality, therefore, can be seen as a possible key to resolving the tensions inherent in the differing perspectives noted previously. On the one hand, the wider multisource data provided can be exploited to increase accuracy and robustness in identity checking, while simultaneously compensating for inherent differences that will naturally occur across a population of system users.

These significant advantages cannot be achieved, however, without incurring a cost. A multimodal scheme will self-evidently add to system complexity, if only because it is necessary to incorporate either multiple sensors or to construct new or special-purpose devices. More important, however, and especially when it is the element of choice and flexibility that is the primary consideration, there are likely to be very serious implications with regard to system implementation. First, there will be more data to process, not to mention an added processing layer needed to execute the (optimized) combination of the multiple data streams or modality-specific subdecisions that a multimodal system will generate. Second, integrating biometric identity checking into a typical overall application domain can lead to inherent decision-level complexities even in straightforward application scenarios. Multimodality, and dealing with issues associated with choice and preferences and their relation to the differing characteristics and performance profiles of different individual modalities, can add significantly to the level of complexity encountered in practice.

Overall, then, the following three (related) issues can be identified as being of key importance in realizing a multimodal biometric identification system and optimally exploiting its potential in the context of universal access:

- Decision-making and data fusion
- Processing complexity
- Facilitating choice and adaptability

In fact, the first of these areas is an inherent issue in all multimodal biometric systems, and relates closely to the more generic area of multiclassifier pattern recognition. This area will not be addressed in detail here, but the fundamentals and some effective current approaches are well documented in the literature. Further details can be found, for example, in Fairhurst and Rahman (2000); Ho et al. (1994); and Rahman and Fairhurst (2003).

The second two issues are less well researched, but are particularly pertinent to a discussion focused on universal access. The following sections will explore two aspects of the design and implementation of multimodal biometric systems of direct relevance to these issues.

13.6 Managing Complexity in the Implementation of Multimodal Schemes

A first issue to consider concerns the general principles of a strategy for the effective implementation of a multimodal biometric scheme that can inherently embrace the flexibility and adaptability required for effective deployment in a situation where universality is seen as a particular target.

The scheme proposed here is based on the use of intelligent agent technology to manage the complexity arising from the use of multibiometric configurations. Intelligent autonomous agents and multiagent systems (Wooldridge and Jennings, 1995; Wooldridge, 1997) are software (sub)systems that interact with their environment and are capable of autonomous action in relation to responding to changing situations, seeking goals and interacting with other agents to meet predetermined performance criteria. They are adaptive, mobile, and "social" and are well suited to problem solving in situations where multiple perspectives exist in a task domain, of which multimodal biometrics provides an excellent example. This type of scenario often demands cooperation, coordination, and negotiation between agents (a classic example is where the requirement of a service provider for establishing an appropriate level of trust in a user's identity may need to be balanced against the confidentiality of information and/or ease of use of a system) and is thus a promising framework within which to develop a strategy for implementing multibiometric systems. At a lower level of design, the interacting agents may need to address tasks such as the location of data across multiple repositories, the reliable handling of multiple authorization levels, and (especially relevant here) user interface modifications to meet the requirements of individuals, environments, or changing circumstances; the properties of intelligent agents must be well matched to tasks such as these.

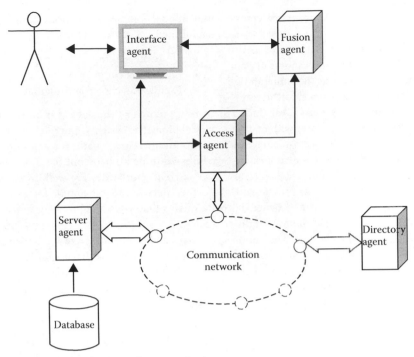

FIGURE 13.5 Schematic of a bioagent architecture for secure database access.

It is helpful to explore issues relating to system architecture and implementation in more detail in a specific illustrative application, and a typical scenario might be one where authorized access to a remote secure database of confidential information must be regulated and managed. A schematic of a possible agent-based architecture for multimodal identity checking (referred to as a *bioagent architecture*) appropriate to this type of environment is depicted in Figure 13.5, which shows how the generic framework for the biometrics system shown in Figure 13.1 would map to an integrated agent-based structure to realize the scenario imagined.

In this scheme, the processing infrastructure required to take advantage of the benefits of secure but flexible biometric access control deploys five interacting agent systems. Collectively, they perform the evaluation of the biometric measurements extracted from the system user via the modality sensors available, analyze issues relating to levels of confidence in identity, authority levels of the user, properties of the target data, and they negotiate the release or otherwise of the data item requested. The agents involved in this architecture each have a particular role to play in the overall process, which is briefly discussed in the next sections.

13.6.1 Interface Agent

The interface agent manages the pattern of interaction between the user and the system, and defines the set of biometric measurements that must be taken from the user in a particular situation. It also evaluates the corresponding confidence characteristics of the biometric recognition modules themselves. In a noisy environment (using the term in the sense of direct aural noise), for example, a voice recognition module is very likely to be associated with a low confidence in any identification decision. This agent also defines the nature of the interaction with the users according to the user characteristics.

The interface agent monitors and manages the interactions of the user during biometric capture operations. This interface may also be responsible for the capture of other important nonbiometric information such as, for the voice modality for example, a sample of the prevailing background noise. Analysis can be performed on these samples to determine the quality of the acquired data, and this can be used to help the agent to analyze possible enrollment and/or verification failures. Not only can the results from this type of analysis by the interface agent help to optimize the decision-making processes based on the measurements available, but they may also be used to provide feedback to user to improve future performance.

13.6.2 Fusion Agent

The fusion agent is responsible for the fusion of the biometric measures, relating to the different modalities in use, taken from the user. Its main role is to determine and execute an appropriate procedure for combining the information obtained from the several different biometric measures extracted in this multimodal scenario. The design of the fusion agent requires a detailed knowledge of the biometric modalities available, the characteristics of the measurements taken, and the levels of confidence they can generate.

The main goal of the fusion agent is the combination of the evidence obtained from the different biometric samples provided by the user, in response to input from the interface agent, which will have provided the biometric samples and environmental information obtained from the user during the verification phase. The global confidence score relating to the identity of the current user produced by the data fusion will be passed to the access agent for transmission to the server agent.

13.6.3 Access Agent

The access agent is responsible for negotiating access to a required data item on behalf of the user. This agent receives access information passed on from the interface agent. It locates the data, contacts the source(s) of the information item requested, and negotiates its release with the server agent.

Among the main functions of the access agent, the most important activity is the negotiation of the release of the information item requested. To achieve this goal, the access agent will need to negotiate with the server agent for the release of the information requested. In some circumstances, as part of this negotiation, a remeasurement of the biometric samples, as well as the recalculation of the combined output, may be required, This would occur, for example, where a lack of confidence has been recorded in relation to the initial biometric capture.

13.6.4 Directory Agent

The directory agent is responsible for storing and updating all relevant information about the location of information sources and/or services within the overall information network. In the search for an information item, this agent can also suggest the best way of accessing the required information (e.g., in a situation where several databases contain the information specified), based on the nature of the data, prevailing network traffic conditions, and so on.

13.6.5 Server Agent

The server agent is responsible for acting on behalf of the target database to guarantee that the information to be released remains secure, its release occurring only when a sufficient degree of confidence about the identity and access rights of the current user requesting the information has been established. As discussed previously, a negotiation process takes place between the access and server agents, taking into account the level of security of the information to be released (the higher the degree of required security associated with a piece of information, the more sure the system needs to be that the user requesting this information is genuine and authorized), the level of encryption of the data to be transmitted, and the degree of confidence that the transaction is a fraud-free process.

This agent is also responsible for detecting fraudulent access to the databases and may also maintain a log of access attempts. For example, if some unexpected pattern of data access is attempted, a security process will be activated to discover whether there is any suspicion of fraudulent system penetration, perhaps with consequences for the criteria subsequently imposed on identification confidence levels. As part of the negotiation phase, the server agent must ensure that the user wishing to access the information requested is genuine and authorized, and thus acts on behalf of the owner of the stored information.

Thus, an overall system model can be visualized that reflects the information flow associated with the processes of requesting the release of a data item, checking and confirming identity of the requester, locating the requested data item, and the negotiations that might be involved. These might include establishing an appropriate confidence in the identification decision matched to the degree of confidentiality associated with that item, and the authority levels of the individuals concerned. More important in the present context, there might also be negotiations concerning the specific modality used, the need for additional identity information (perhaps by invoking an alternative modality), and so on. A conceptual data flow model is shown in Figure 13.6, which makes clear the information pathways invoked (perhaps iteratively) in this type of processing structure.

A further discussion of this type of structure, which has been evaluated in an experimental form, and a more detailed example of how the proposed scheme can be applied in a practical

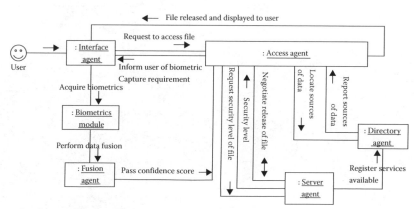

FIGURE 13.6 Information flow patterns in a secure client-server transaction.

scenario (specifically, in an e-health environment) can be found in Deravi et al. (2003).

13.7 A Framework for an Intelligent Interface for Biometric Verification

The type of analysis outlined previously in fact points to a further key advantage of the agent structure proposed here, since one of the functions of the agent cluster might be to monitor user performance and build up a model of user behavior with respect to individual devices. This in turn can stimulate the feedback of information to the user about patterns of use and areas of difficulty in interaction, an issue that clearly has a particular resonance in the context of universal access. Finally, then, as a pointer to the future, we will briefly sketch out the basic principles on which such an interface support model could be developed.

The system components required for a possible adaptive intelligent interface structure are shown in Figure 13.7. This interface structure effectively forms an intermediate layer between the biometric device interface, responsible for sample acquisition and subsequent management of the enrollment and verification procedures, and the user interface itself.

Generally, as has already been seen, a biometric system requires the specification of some form of verification threshold to determine the precise conditions under which a successful verification can be considered to have occurred. Successful verification at increasingly higher security levels, as reflected in this threshold, indicates that the system has increased confidence in the identity of the user.

A novel implementation of the general concept of *interface utility* (Brown et al., 1998) is used for user modeling in the proposed adaptive system. In this implementation, utility is characterized with respect to the levels of security that the user is able to achieve. In other words, the agent's degree of "satisfaction" that is associated with a current transaction is directly related to the confidence it has in the veracity of the claimed identity of the user.

In general terms, the goal of the overall system is to try to maximize its current utility over the period of time for which it is being used, with the value of utility directly used to determine the behavior of the agent with respect to the user. For example, an agent with a low utility score will aggressively attempt to aid the user with the donation of higher quality samples through the use of an extended user assistance protocol, while an agent with a relatively high utility score is less likely to offer the degree of assistance the low utility agent is required to adopt. However, this does not preclude invoking specific help if the agent determines that the user is experiencing difficulty in donating samples for any reason.

The behavior of the overall agent entity is based upon the current utility value for the user. After each session, the user logs are interrogated to determine the overall utility score, calculated in the following way. An overall mean utility score is calculated for the session, and this value is then scaled using a utility transformation function that generates a percentage figure for the session of interest. This figure is scaled based on the maximum figure of utility that the user can produce with respect to the security level and quality setting at which the user templates were acquired. This normalized session utility score represents the performance of the user over the most recent session. This figure is used to determine a set of "behavior bands," which determine the degree and type of user feedback that the system exhibits to attempt to increase the performance of the user.

In this way, it is possible to define the basis of a strategy whereby intelligent interaction between the required user group and a biometric-based access-regulation system can both improve the reliability of the identity-checking process itself, yet also be embedded within an effective management scheme implemented for maximal flexibility through the use of adaptive intelligent agent technology. This is an important area for future development.

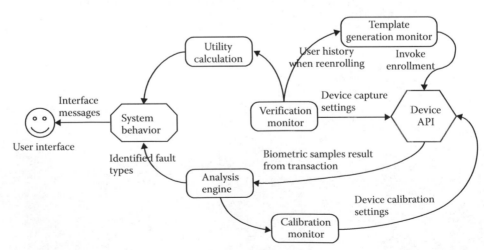

FIGURE 13.7 Adaptive interface layer.

13.8 Future Directions

It should be clear from the previous discussions that biometric technologies are now at a level of maturity that allows satisfactory implementation to be achieved in appropriate application contexts. Protection of personal computing devices using fingerprint data is perhaps one of the most generally encountered scenarios, but it is apparent that other widely accessible opportunities are increasingly being taken up. Considering some examples based on the U.K. experience, most U.K. airports now offer air travelers the option to register with the *IRIS* scheme, a "fast-track" route through passport control which checks identity using iris-scan technology, while the U.K. government is continuing to implement a program aimed at rolling out a national ID card holding biometric data about the cardholder. On the horizon are reported plans to adopt biometric identity checking during both the construction of the site of the London Olympics in 2012, and at the games themselves.

Nevertheless, a number of research issues remain unresolved, and if biometric systems are to develop and gain more widespread deployment, these remaining challenges must be addressed. To give just four examples:

- *Individual modalities*: It is undoubtedly the case that most individual modalities have limitations that restrict their application or, at least, may require constraints to be placed on operating conditions. Face recognition techniques, for example, can be severely affected by lighting conditions, or by variable subject poses, and so on (for trends toward 3D, see Bowyer et al., 2006). Facial imaging may be effective in addressing some of these areas, while iris scans at a distance or "on the move" will make the technology more flexible and attractive.
- *Multibiometric solutions*: These are increasingly seen as a way of increasing robustness in performance while also offering greater flexibility and choice for the user, as noted previously. Perhaps equally important, multibiometrics represent one possible strategy for dealing with the issue of exception handling (ensuring that maximum inclusiveness is achieved by having a mechanism for dealing with outliers in a user population), which becomes increasingly important when dealing with large-scale applications involving large populations.
- *Vulnerability and countermeasures*: As biometric systems become more widespread, there will be a correspondingly increasing incentive to attack them to access the assets they protect. Thus, it is important to give more attention both to understanding the vulnerabilities of systems (Osoka et al., 2007) and to devising appropriate countermeasures to attack. "Liveness detection" is one obvious strategy (i.e., incorporating sensing mechanisms that can distinguish, perhaps by measuring temperature, between a live finger and a dead one, or using the natural involuntary motion of the iris to guard against spoofing using artifacts), but methods based on provoking (random) spontaneous

activity and monitoring responses, or other more direct computational techniques (Kollreider et al., 2007) provide one of several alternatives.

- *Privacy and security*: Concerns remain over the problem of biometric data being compromised as a result, for example, of a template being accessed by an unauthorized user. This has resulted in a growing interest in "revocable biometrics" (Ratha et al., 2007), where the raw measurements are unidirectionally transformed in forming the usable template. In the event of compromise, an alternative transform can be substituted without rendering the raw data unusable.

These selected examples show that, although already providing real solutions to real problems, the field of biometrics is likely to remain a vibrant research field for the foreseeable future if the true potential of the technologies is to be realized.

13.9 Conclusions

This chapter has provided an introduction to the use of biometrics as a technology for supporting the introduction of individual identity verification in a range of practical scenarios relating to the management of access to physical locations, systems, and data. The chapter has set this issue explicitly in the context of interactions in the information society, and has considered in particular the exploitation of biometrics within a framework defined by the principles of universal access.

Different biometric modalities have been shown to offer fundamentally different advantages and disadvantages. This, on the one hand, means that there is no automatic choice for a single modality of universal applicability, while, on the other hand, indicates that the range available provides a basis for individual choice and flexibility. A discussion of the conditions under which universal access might be achieved has pointed to the potential value of multimodal configurations to support the principle of flexibility in harnessing the benefits that different biometric modalities can collectively offer. A further benefit of this approach, of course, is the possibility of increasing levels of performance in terms of accuracy and reliability, but there is also an additional cost arising from the increased complexity of function, which the integration of multiple biometric identity evidence sources implies.

A possible approach to addressing implementation issues that arise in a typical multimodal-processing environment has been examined, showing how a structure based on intelligent agent technology can provide a general framework within which efficiency of implementation can be coupled with other desirable features required to manage effectively a complete individual identification-processing chain.

The need for increased security, and the demands of ensuring that these needs are balanced against the need for privacy, trust, and inclusiveness among individuals, suggests that this is an area where future developments will be rapid and of increasing importance to the development of a technologically oriented society.

References

Bowyer, K. W., Chang, K., and Flynn, P. (2006). A survey of approaches and challenges in 3D and multi-modal 3D+2D face recognition. *Computer Vision and Image Understanding* 101: 1–15.

Brown, S. M., Santos, E., and Banks, S. B. (1998). Utility theory-based user models for intelligent interface agents, in *Advances in Artificial Intelligence, Proceedings of the 12th Biennial Conference of the Canadian Society for Computational Studies of Intelligence (AI'98)*, June 1998, Vancouver, Canada, pp. 378–392. Berlin/Heidelberg: Springer-Verlag.

Chen, K., Wang, L., and Chi, H. (1997). Methods of combining multiple classifiers with different features and their application to text-independent speaker recognition. *Pattern Recognition and Artificial Intelligence* 11: 417–445.

Chibelushi, C. C., Mason, J. S. D., and Deravi, F. (1997). Feature level data fusion for bimodal person recognition, in the *Proceedings of the 6th International IEE Conference on Image Processing and Its Applications*, 14–17 July 1997, Dublin, 1: 399–403.

Daugman, J. (2004). How iris recognition works. *IEEE Transactions on Circuits and Systems for Video Technology* 14: 21–30.

Deravi, F., Fairhurst, M. C., Guest, R. M., Mavity, N. J., and Canuto, A. M. D. (2003). Intelligent agents for the management of complexity in multimodal biometrics. *Universal Access in the Information Society* 2: 293–304.

Deravi, F. and Lockie, M. (2000). *Biometric Industry Report: Market and Technology Forecasts*. Oxford: Elsevier Advanced Technology.

Doddington, G. (1998). *Sheep, Goats, Lambs and Wolves: An Analysis of Individual Differences in Speaker Recognition Performance*. [CD-ROM]. National Institute of Standards and Technology.

Fairhurst, M. C. (1997). Signature verification revisited: promoting practical exploitation of biometric technology. *Electronics and Communication Engineering Journal* 9: 273–280.

Fairhurst, M. C. and Rahman, A. F. R. (2000). Enhancing consensus in multiple expert decision fusion, in the *IEE Proceedings of Vision, Image and Signal Processing* 147: 39–46.

Han, J. and Bhanu, B. (2006). Individual recognition using gait energy image. *IEEE Transactions on Pattern Analysis and Machine Intelligence* 28: 316–322.

Ho, T. K., Hull, J. J., and Srihari, S. N. (1994). Decision combination in multiple classifier systems. *IEEE Transactions on Pattern Analysis and Machine Intelligence* 16: 66–75.

Hong, L. and Jain, A. (1998). Integrating faces and fingerprints for personal identification. *IEEE Transactions on Pattern Analysis and Machine Intelligence* 20: 1295–1307.

Jain, A. K., Bolle, R., and Pankanti, S. (eds.) (1999). *Biometrics: Personal Identification in Networked Society*. Boston: Kluwer.

Jain, A. K. and Ross, A. (2004). Multibiometric systems. *Communications of the ACM* 47: 34–40.

Kollreider, K., Fronthaler, H., Faraj, M. I., and Bigun, J. (2007). Real-time face detection and motion analysis with application in liveness assessment. *IEEE Transactions on Information Forensics and Security* 2: 548–558.

Li, B., Zhang, D., and Wang, K. (2006). Online signature verification based on null component analysis and principal component analysis. *Pattern Analysis and Applications* 8: 345–356.

Marcialis, G. L. and Roli, F. (2004). Fusion of appearance-based face recognition algorithms. *Pattern Analysis and Applications* 7: 151–163.

Munro, D. M., Rakshit, S., and Zhang, D. (2005). DCT-based iris recognition. *IEEE Transactions on Pattern Analysis and Machine Intelligence* 29: 586–595.

Obaidat, M. S. and Sadaoun, B. (1997). Verification of computer users using keystroke dynamics. *IEEE Transactions on Systems, Man, and Cybernetics* 27: 261–269.

Osoka, A., Fairhurst, M. C., and Hoque, S. (2007). A novel approach to quantifying risk in biometric systems performance, in the *Proceedings of the 4th IET Visual Information Engineering 2007 Conference*, 25–27 June 2007, London.

Rahman, A. F. R. and Fairhurst, M. C. (2003). Multiple classifier decision combination strategies: A review. *International Journal on Document Analysis and Recognition* 5: 166–194.

Ratha, N. K., Chikkerur, S., Connell, J. H., and Bolle, R. M. (2007). Generating cancelable fingerprint templates. *IEEE Transactions on Pattern Analysis and Machine Intelligence* 29: 561–572.

Shen, L., Bai, L., and Fairhurst, M. C. (2006). Gabor wavelets and general discriminant analysis for face identification and verification. *Image and Vision Computing* 25: 553–563.

Sheng, W., Howells, G., Fairhurst, M. C., and Deravi, F. (2007). A mimetic fingerprint matching algorithm. *IEEE Transactions on Information Forensics and Security* 2: 402–412.

Srisuk, S., Petrou, M., Kurutach, W., and Kadyrov, A. (2005). A face authentication system using the trace transform. *Pattern Analysis and Applications* 8: 50–61.

Su, Q. and Silsbee, P. L. (1996). Robust audiovisual integration using semicontinuous hidden Markov models, in the *Proceedings of the 4th International Conference on Spoken Language Processing*, 3–6 October 1996, Philadelphia, pp. 42–45. Washington, DC: IEEE.

Wooldridge, M. (1997). Agent-based software engineering. *IEEE Transactions on Software Engineering* 144: 26–37.

Wooldridge, M. and Jennings, N. R. (1995). Intelligent agents: Theory and practice. *Knowledge Engineering Review* 10: 115–152.

Wu, X., Zhang, D., and Wang, K. (2006). Fusion of phase and orientation information for palmprint authentication. *Pattern Analysis and Applications* 9: 103–111.

Interface Agents: Potential Benefits and Challenges for Universal Access

Elisabeth André and
Matthias Rehm

14.1 Introduction

Interface agents offer new interaction styles, since they are not based on direct manipulation but instead on the indirect control of a computer system. According to Kozierok and Maes (1993), an interface agent is a "semi-intelligent, semiautonomous system which assists a user in dealing with one or more computer applications" (p. 81). In this guise, an interface agent can be seen as the user's contact person responsible for interacting with the underlying computer system. Interface agents are often represented as embodied entities that make use of gestures, facial expressions, and speech to interact with the user. Such agents are also called embodied conversational agents (ECA) (Cassell et al., 2000).

Allowing for communication styles common in human-human dialogue, embodied conversational agents can release users unaccustomed to technology from the burden of learning and familiarizing themselves with less native interaction techniques. In this way, they may greatly help people get started, especially novice users. Furthermore, they broaden the bandwidth of communication by rendering interactions multimodal. For example, agents that produce realistic visual speech have helped hearing-impaired users acquire language skills (Massaro, 2004). A personification of the interface may contribute to a feeling of trust in the system by removing the anonymity from the interaction. As a result, people will form social relationships with an interface agent, which may become a companion and even a friend to the user. If well designed, virtual agents may lead to a more enjoyable interaction and increase the user's willingness to communicate with a computer system. By serving as mediators between a user and an application, they have the potential to make information technology accessible to a larger community of users. Before being able to exploit them for universal access, however, some essential problems need to be solved. First of all, the humanization of interfaces raises users' expectations. They may perceive the embodied conversational agent as a human-like being endowed with versatile conversational skills, and thus expect social and emotional behaviors. As a consequence, it seems indispensible to equip embodied conversational agents with social and emotional behaviors—an aspect that has played a minor role in the design of traditional interfaces. Second, the users' physical and mental capabilities, as well as their preferences and needs, widely differ, which means that not all users will feel comfortable with the same agent. Rather, agents need to be distinguished from one another not only by their audio-visual appearance, but also through their habits, personalities, social roles, culture, gender, and so on. This chapter explores the potential of embodied conversational agents as mediators between heterogeneous user groups and computer systems, and investigates the technological requirements that need to be met to prepare them for universal access.

14.2 How to Design Embodied Conversational Agents for All

Multimodal corpora have proven useful at different stages of the development process of embodied conversational agents. The basic idea is to derive general behavior patterns from recordings of human beings. In the following, a corpus-based development cycle for embodied conversational agents is first described, and then the implications for the creation of universal interface agents are discussed.

14.2.1 Development Cycle for Embodied Conversational Agents

The design of an embodied conversational agent that emulates aspects of human-human face-to-face conversation is an interdisciplinary effort that may greatly benefit from models developed in the cognitive sciences, social psychology, conversational analysis, and many other related fields. Cassell's study-model-build-test development cycle (Cassell, 2007) describes a methodological approach for modeling communicative behavior for embodied conversational agents that is based on observations of humanlike behaviors. Figure 14.1 gives an overview of the different steps in this development cycle.

The first step is the collection of data from humans that are engaged in a dialogue with other humans. To acquire knowledge about human behavior, researchers rely on a large variety of resources, including recordings of users in natural or staged situations, TV interviews, Wizard of Oz studies, and motion-capturing data. Various annotation schemes have been designed to code relevant information for multimodal behaviors (for an overview, see, e.g., Dybkjær and Bernsen, 2004) ranging from gesture analysis (e.g., Frey et al., 1983; McNeill, 1992), to general movement analysis (see, e.g., Lamb and Watson, 1979) to the coding of facial expressions (e.g., Ekman and Rosenberg, 1998). An example of an annotated multimodal corpus is MUMIN (Allwood et al., 2005), which focuses on the analysis of gestures and facial displays that accompany communicative phenomena, such as feedback, turn-taking, or sequencing.

Corpus work has a long tradition in social sciences, where it is employed as a descriptive tool to gain insight into human communicative behavior. Thanks to the increased multimedia abilities of computers, a number of tools for video analysis, such as Anvil (Kipp, 2001), have been developed over the last decade, allowing for the standardized annotation of multimodal data. In addition, attempts are being made to exploit methods from signal processing to validate or replace manual annotations (see, e.g., Martin et al., 2007).

The easiest way in which to make use of the collected data would be a direct replication of human-human behaviors in an agent. This method would, however, require collecting data for

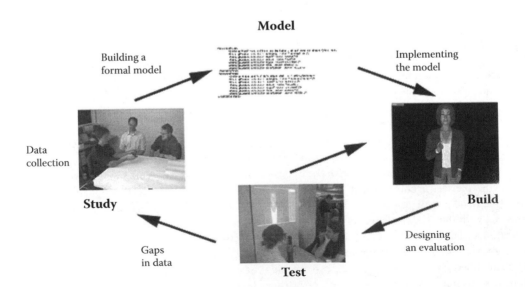

FIGURE 14.1 Development cycle for embodied conversational agents.

each possible conversational situation an agent would encounter. Since such an approach is not feasible, the generation of an embodied conversational agent should rather be driven by a formal model. Such models are built by extracting relevant parameters of the collected multimodal data, such as the frequency of certain kinds of gesture. It is important to note that in most cases formal models are not built from scratch. Rather, the data analysis serves to refine existing models. For instance, Rehm and André (2007) used Brown and Levinson's Theory of Politeness (1987), which focused on verbal means in politeness tactics, and conducted a study to identify regularities in the use of gestures.

The resulting models of human-human conversational behavior are then used as a basis for the implementation of agents that replicate the behaviors found within the models. This means that the primary interest is not in verifying or extending a theory of human behavior, but rather in exploiting the data specifically to derive control parameters/models of behavior control for agents. To accomplish this task, abstract behaviors specified in a given markup language (see Section 14.3) are usually mapped onto actual animation parameters for an agent. Basically, two types of use of corpus data should be distinguished: direct use (e.g., to generate animations that directly correspond to the behavior found in the data) and indirect use of corpus data (e.g., to extract abstract rules that govern the planning and generation process) (see Foster, 2007).

To evaluate the resulting system, experiments in which humans are confronted with embodied conversational agents following the model are set up. Depending on the outcome of such experiments, the developers might decide to acquire new data from humans, so that the existing models may be refined.

14.2.2 Implications for Universal Access

Corpus studies of human behavior provide necessary insight in behavioral patterns that can serve as parameters for the control of an embodied conversational agent. Nevertheless, corpora represent behaviors of individual people in specific situations. As a consequence, the resulting embodied conversational agent might not be the ideal communicative partner for a particular user.

Thus, the question arises of how an agent's behavior can be tailored to a user's preferences and needs. A straightforward approach would be to record a large number of subjects with different cultures, ages, personalities, and so on, and then design an embodied conversational agent that exactly replicates their behavior. Since the collection and annotation of corpora is time-consuming and cumbersome, researchers usually focus on a small set of parameters, even though it is hard to separate out the effect of the single parameters. For instance, it might be observed that the spatial extension of gestures in a corpus is unusually high. In Germany, this might easily be attributable to the recorded persons' extrovert personality trait. But what if the data were from a Southern European country like Italy? Then the observed expressive behavior might well just be the standard behavior pattern in this culture (e.g., Ting-Toomey, 1999). Rehm

et al. (2007) recorded and annotated about 20 hours of culture-specific interactions from Germany and Japan in three standardized scenarios (first meeting, negotiation, status difference). To identify the potential influence of the subjects' personalities on their behaviors, Rehm and colleagues also analyzed the personality traits of the recorded subjects, even though the focus of their research was the simulation of culture-specific differences in behavior.

A great challenge for the future is the decontextualization of multimodal corpora and their automated adaptation to a new context. The extent to which such behavior adaptation is possible depends on the generation approach employed. Approaches that heavily rely on raw data, such as recorded audio and video, are notoriously more difficult to adapt than approaches that generate behaviors from scratch based on high-level parameters (see also Section 14.7).

14.3 Markup Languages for Specifying Individualized Multimodal Behaviors

Over the past few years, a variety of markup languages and associated techniques have been developed to facilitate behavior control of embodied conversational characters. Markup languages ideally allow specifying agent behaviors on a level abstracting from the technical limitations of the underlying agent system or hardware requirements. Thus, only the multimodal communicative behavior has to be specified; the specifics of rendering this behavior is delegated to the available animation system and might be adjusted to hardware restrictions (see also Section 14.9.4).

Current examples of such markup languages are MURML (Kranstedt et al., 2002) or APML (DeCarolis et al., 2004). The Multimodal Utterance Representation Markup Language (MURML) provides a means to describe gestures and their relation to accompanying speech in a form-based way, relying on a notation borrowed from McNeill's ideas (1992) and HamNoSys, an annotation system for German sign language (Prendinger and Ishizuka, 2004). The Affective Presentation Markup Language (APML) defines an agent's communicative behavior on a more abstract level. Communicative functions corresponding to certain behavior patterns, such as "inform" or "question," are identified. Moreover, emotional facial expressions as well as coexpressive gestures can be specified by simply relating the corresponding tags to the appropriate words of the utterance.

A first attempt toward a unified language was undertaken with the Function Markup Language (FML) and the Behavior Markup Language (BML), which integrate the advantages of the aforementioned languages and are meant to be a clear interface between behavior planning and realization (Kopp et al., 2006). As the examples of MURML and APML show, an agent's behavior has to be specified on different levels of granularity. FML defines basic semantic units for a communicative event, which can be enriched with additional information on affective, discursive, or other functions. BML then specifies the coordination and synchronization of multimodal behavior, such as head movement, gestures, and speech that will later be interpreted and rendered on the fly by the underlying agent system. Ideally, such languages may be used to control a variety of rendering modules, and thus constitute an important step toward device-independent interface characters.

14.4 Equipping Conversational Agents with Perceptive Capabilities

To realize the vision of user interfaces for all, it is necessary to overcome the strong asymmetry in communication channels. In most cases, the user's means of expressing himself are restricted to typed input, which limits access especially for people with certain disabilities. In the following, the first approaches are reported seeking to analyze nonverbal communicative signals from the user with the aim of improving the robustness of multimodal analysis, recognizing the user's level of attention and regulating the flow of the interaction.

The kiosk agent Mack uses gaze as a deictic device, as well as a feedback and turn-taking mechanism (see the work by Nakano et al., 2003). Based on an analysis of human-human conversation, Sidner et al. (2005) implemented a conversational robot that is able to track the face of the conversational partner and adjust its gaze toward her. Empirical studies by Sidner and colleagues and Nakano and colleagues indicate that gaze is an excellent predictor of conversational attention in multiparty conversations. For instance, looking at the conversational partner is interpreted as a sign of engagement, while looking around the room for some time is taken as evidence of disinterest. In addition, the user's eye gaze is used as an indicator of a successful grounding process. In face-to-face communication, grounding refers to the process by which conversational partners ensure that there is an agreement about what is being said and meant. For instance, if the user's gaze does not follow the deictic gesture of an agent, the system may interpret this as evidence of a failed grounding process, and responds accordingly.

Hoekstra et al. (2007) used an eye tracker to equip two conversational agents with perceptive skills (see Figure 14.2). By analyzing and interpreting the user's eye gaze in real time, a presentation may be dynamically adapted to the user's current interest. For example, the user's attention may be directed to the window on the right in Figure 14.2. As a consequence, the agents either interrupt their current presentation and comment on the view or try to regain the user's attention. Perceptive presentation agents that are completely controlled by the user's eye gaze have a lot of potential for users with motor disabilities. A great challenge is to respond to the users' eye gaze behavior in an unobtrusive manner. For instance, frequent changes in the presentation based on the analysis and interpretation of the users' eye gaze may leave users with the feeling that they are under permanent observation, and this may irritate them.

14.5 Modeling Social Behaviors

The success of an interface character in terms of user acceptance and interface efficiency very much depends on the character's ability to socially interact with the human user. As a

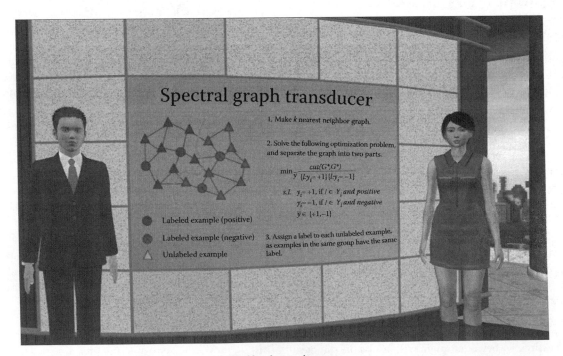

FIGURE 14.2   Attentive presentation agents that are controlled by the user's eye gaze.

consequence, characters should have a great deal of social competence that manifests itself in a number of different abilities.

14.5.1 Emotional Sensitivity

Picard (2000) created considerable awareness for the role of affect in human–computer interaction. Nevertheless, there is an ongoing debate as to whether it is necessary for a computer system to model emotions as well. Yet, when striving for believable virtual personalities, the simulation of emotional behaviors seems indispensable. For instance, Martinovsky and Traum (2002) demonstrated by means of user dialogues with a training system and a telephone-based information system that many breakdowns in man-machine communication could be avoided if the machine was able to recognize the emotional state of the user and respond to it more sensitively.

In fact, the modeling of affective characters has become a major research trend. Most of the current approaches rely on the so-called OCC model, named after its authors Ortony, Clore, and Collin (1988). This theory views emotions as arising from a valenced reaction to events and objects in light of agent goals, standards, and attitudes. For instance, an agent interacting with a user should have a different emotional response depending on whether it is able to assist the user or not. The reason why this model is attractive to computer scientists probably lies in the fact that it defines different types of emotions in terms of emotion-triggering conditions that can be easily formalized as rules in a computer-based reasoning system.

The need to equip an agent with emotional sensitivity becomes obvious in the context of computer-based learning environments. Similar to human teachers, virtual tutors should not always reveal their real emotions. Instead, they should express emotions deliberately, with the goal of increasing learning. An early example of an affective pedagogical agent is the Cosmo System (Lester et al., 2000) where the agent's pedagogical goals drive the selection and sequencing of emotional behaviors. For instance, a congratulatory act triggers a motivational goal to express admiration that is conveyed with applause. To convey appropriate emotive behaviors, agents such as Cosmo need to appraise events not only from their own perspective, but also from the perspective of others.

An important prerequisite to achieve emotional sensitivity in virtual agents is the availability of robust methods for emotion recognition. Emotional cues that have been investigated include postures and facial expressions (Kapoor and Picard, 2005), acoustic and prosodic features of speech (Ai et al., 2006) as well as physiological signals (Bosma and André, 2004). As discussed previously, there is a significant asymmetry in communication channels that also affects an agent's emotional sensitivity. While the behaviors of some agents are highly expressive, they do not have sophisticated perceptive skills to perceive the user's emotional state. One approach to mitigate this problem is to combine a recognition component with an appraisal model. Such an approach was presented by Conati (2002), who developed a probabilistic framework to derive a learner's motivational state both from his bodily reactions as well as from the state of the interaction. For instance, a learner may reveal her motivational state by her facial or vocal expression. Additional evidence for a positive or negative motivational state may be obtained by considering whether the learner has successfully completed a task or not.

14.5.2 Social Group Dynamics

Most approaches simulating social behaviors have been designed for the classical face-to-face setting in which one user interacts with a virtual character. When dealing with socially more complex settings, such as multiparty conversations, scalability becomes a considerable challenge, since one has to consider the social context of an individual as well as others.

While earlier work focused on one specific aspect of social behavior, such as the expression of socially desirable emotions, more recent research has aimed at the operationalization of complex models of social behavior between members of a group, including emotions, personality, and social roles, as well as their dynamics. For instance, Prendinger and Ishizuka (2001) investigate the relationship between an agent's social role and the associated constraints on emotion expression. They allow a human scriptwriter to specify social distance and social power relations between the characters involved in an application, such as a multiplayer game scenario. Another approach was taken by Rist and Schmitt (2002) aiming at emulating dynamic group phenomena in human-human negotiation dialogues based on sociopsychological theories of cognitive consistency dynamics (Osgood and Tannenbaum, 1955).

To simulate the social behaviors of agents as members of a group, a number of social simulation tools have been developed (Prada and Paiva, 2005; Pynadath and Marsella, 2005; Rehm et al., 2005). Individual agents are assigned a name, gender, marital status, age group, social status, sex orientation, and personality. The five-factor model of McCrae (McCrae and John, 1992), for instance, has proven to be very useful in modeling personality traits and is widely employed. The use of this technique has led to the consensus that five major factors or dimensions account for most of the variation in human personality. Although different researchers use slightly different terms to describe them, they can be summarized as open, conscientious, extravert, agreeable, and neurotic.

These frameworks not only represent characteristics of individual agents, such as personality and social status, but also take into account an explicit representation of social relationships between individual agents, relationships between agents and the groups they belong to, and the attitude of agents toward objects. Interpersonal relationships are usually described by parameters such as the degree of liking, familiarity, trust, and commitment. These values are either specified in advance by the user or derived from known properties of the agent's profile. For instance, agents with a similar social status are considered as trusting each other more than agents with a different social status.

In addition, such frameworks simulate how such relationships develop over time depending on the initial settings of the agent and group parameters and the interactions between the individual agents. For example, agents that are engaged in more positive interactions are more likely to develop a relationship of trust.

14.5.3 Cultural Interaction

Whereas a number of approaches exist to tailor an agent's behavior to emotional and/or personality parameters, few researchers have so far taken up the challenge of modeling the influences culture has on behavior, though, according to Hofstede (e.g., 2001), patterns of behavior are to a large degree dependent on one's cultural context, which in turn provides heuristics (*mental programs*) for behaving and interpreting behavior. The role of culture in universal access in general is addressed in Chapter 9, "International and Intercultural User Interfaces," of this handbook. With respect to agents more specifically, the challenge has been identified (e.g., Payr and Trappl, 2004), but realizations are often superficial, concentrating, for example, on avatar appearance for specific cultural groups (e.g., Krenn et al., 2004). De Rosis et al. (2004) illustrate this problem with their survey of the Microsoft Agents web site, which shows that the appearance, as well as the animations, of the characters are almost all based on Western cultural norms. They only found four non-Western style agents, which moreover exhibited only a reduced set of animations. Apart from imposing Western cultural standards on all users, the danger lies in a very low acceptance of such agents by users with different cultural backgrounds. It has been shown that the cultural background and behavioral consistency of an agent matter. Iacobelli and Cassell (2007) examined the influence of ethnicity on the interaction behavior of children and found that ethnicity was not only being determined by the outward appearance of the character, but also by specific verbal and nonverbal behavior patterns. Takeuchi et al. (1998) examined cultural effects on the reciprocal behavior of American and Japanese subjects and showed how subjects reacted toward a computer that helped them on a task: American subjects showed reciprocal behavior if the same computer needed their help on a later task, while Japanese subjects demonstrated reciprocal behavior if the computer was a "member" of the same group as the previous computer.

Core et al. (2006) work in the area of military training and present a comprehensive training simulation that focuses on the acquisition of social negotiation skills. They identify culture as one determinant for this social skill, especially for building or destroying trust between the agent and the user, which is a key ingredient of successful negotiation.

Rehm et al. (2007) exemplified how the user's cultural background can be inferred relying on a Bayesian network model of cultural dimensions and how this information could then be used to adapt an agent's expressive behavior. Their model so far works with synthetic cultures described in the literature but will be extended with empirical data from a large-scale corpus study (Rehm et al., 2008).

A different approach is described by Isbister et al. (2000) who realized a cross-cultural video-conferencing system (between American and Japanese students) that features a so-called Helper agent, which intervenes if communication between participants is disrupted. The agent then tries to find safe topics that allow each interlocutor to commence the conversation. To make the agent acceptable to participants from both cultures and to prevent the agent from corresponding to either an American or a Japanese stereotype, it was designed to be equally (un)familiar for both cultures by taking on the appearance of a dog.

14.6 Supporting Multiple Modalities, Multiple Conversational Styles, and Multiple Platforms

To foster the development of universal user interface agents, platforms that support multiple input and output modalities, various conversational styles, as well as heterogeneous hardware environments, are required. In addition, the platforms should take into account a large variety of application scenarios. So far, no platform has met the requirements of a universal user interface agent. Nevertheless, serious research efforts addressing one or more aspects of a universal user interface agent have been conducted.

14.6.1 Supporting Multimodal Discourse Acts

For the realization of collaborative human-computer dialogue, the Collagen framework (Rich and Sidner, 1998), which relies on a framework of collaborative discourse theory to generate multimodal discourse acts, has proven useful. The advantage of Collagen lies in the fact that it clearly separates the task model (which is application-dependent) from the discourse model, which may be used across different domains. Collagen has been used in a wide range of applications. Examples include the generation of posture shifts in the REA agent (Cassell et al., 2001a), the determination of tutorial strategies for the Steve agent (Rickel et al., 2002), and the creation of head nods for the robotic Mel agent (Sidner et al., 2005).

14.6.2 Supporting Multiple Conversational Styles

Rist et al. (2003) presented an architecture that allows for switching on-the-fly between plot-based scripting approaches and character-based improvisation approaches and that supports a clear separation between the specification of scripting knowledge (a knowledge-engineering task) and the required computational machinery for behavior generation (an implementation task). The architecture has been used to realize various agent applications, including product presentations given by a team of characters, as well as improvised role-play. In addition, the architecture supports both single user interactions and interactions with a group of humans.

14.6.3 Supporting Multiple Platforms

Barakonyi and Schmalstieg (2006) observe that most applications with humanoids today rely on a fixed set of hardware and software components. Furthermore, they usually reside on the user's desktop, are integrated in web applications, or inhabit virtual worlds. Hardly any attempts are being made to populate augmented realities (see Chapter 12, "Virtual Reality") that are partially real and partially virtual with virtual humans. To increase the flexibility of virtual humans, Barakonyi and Schmalstieg present an augmented reality framework that integrates work on autonomous agents with work on ubiquitous computing. The framework supports the creation of a large variety of agent-based applications, such as mobile agents embodied by virtual and physical objects, as well as agents that are able to migrate from one augmented reality application to another (see Figure 14.3).

14.7 How to Create Individualized Characters

A necessary step toward the creation of interface agents that are universally accessible is the realization of individual behaviors. To tailor a character to a specific user, it is necessary to move beyond the construction of generic agents. The individual profile of an agent is reflected by a large number of factors, including:

- The agent's audiovisual appearance
- The agent's role
- The agent's personality
- The agent's expressivity

14.7.1 The Agent's Audiovisual Appearance

An important component of a character's profile is its audiovisual appearance. Empirical evidence for this is, for instance, provided by Dryer (1999), who presented subjects with a set of animated characters to measure their perception of the characters' personalities. He found that characters perceived as extraverted and agreeable tended to be represented by rounder shapes, bigger faces, and happier expressions, while characters perceived as extraverted and disagreeable were typically represented through bold colors, big bodies, and erect postures.

Most projects on embodied conversational agents employ just one character over a period of several years. To allow for the efficient creation of a multitude of differently looking characters, de Heras Ciechomski et al. (2005) started from a basic repertoire of template meshes that were then adapted by using different textures, colors, and scaling factors. The objective of their work, however, was not to create interface characters for all, but to achieve the necessary randomness for believable crowd simulation. The need to adapt a character's visual appearance and personality to a particular user has also been recognized by commercial game developers. For example, games such as *The Sims,* provide the user with an interface to create characters from basic components, such as different hairstyles and outfits,

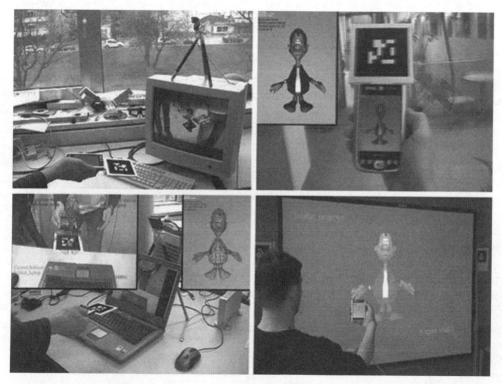

FIGURE 14.3   Agents migrating from one platform to the other. (From Barakonyi, I. and Schmalstieg, D., Ubiquitous animated agents for augmented reality, in the *Proceedings of IEEE and ACM International Symposium on Mixed and Augmented Reality 2006 (ISMAR '06)*, October 2006, Santa Barbara, CA.)

and to choose accessories, such as glasses, for them. In addition, users may assign them certain personalities.

However, having a number of characters with a different audiovisual appearance available is quite different from choosing the best character for a particular user. Studies by Gulz and Haake (2006) revealed that users seem to have a preference for stylistic agents over realistic agents. Baylor (2005) found that more realistic agent images led to a greater transfer of learning. Many more studies are necessary before general guidelines can be derived for a character's audiovisual appearance.

14.7.2 Approaches to Tailored Behavior Control

To suit different kinds of people, the agent needs to be equipped with communicative behaviors that reflect different personalities, emotions, and social roles. A straightforward approach is to equip the character with a library of manually authored scripts that determine what a particular character might do in a given situation. At run-time, the remaining task is to choose a suitable script from the library that will help accomplish a given task and at the same time meet the constraints of the current situation and the character's personality profile, role, and emotional state.

A particular problem with manually authored scripts and script libraries is that the author has to anticipate scripts for all possible situations and tasks, while allowing sufficient variation to avoid characters that behave in a monotonous and exaggeratedly predictable way. Furthermore, the manual scripting of characters can become quite complex and error-prone, since synchronization issues have to be considered. Moreover, creating scripts manually is not feasible for many applications, since it would require anticipating the needs of all potential users and preparing presentations for them. As a consequence, manually authored scripts are not an option when attempting to develop user interface agents for all.

André et al. (2000) presented a plan-based approach to behavior generation that selects plan operators representing different communication templates depending on (1) the agents' roles, (2) their attitude toward the product, (3) some personality traits (extravert vs. introvert, agreeable vs. not agreeable), and (4) their interests in certain aspects relevant for cars (e.g., the car's relation to prestige, comfort, sportiness, friendliness to the environment, cost, etc.). Based on these settings, a variety of different sales dialogues between multiple characters can be generated for the same product.

While the approach proposed by André and colleagues starts from a communicative goal that is decomposed into elementary dialogue acts, the BEAT system (Cassell et al. 2001b) starts from text that is automatically converted into embodied speech using a generate-and-filter approach. A shallow analysis of the text identifies, for instance, the theme and rheme, where *theme* denotes the topic of the utterance and *rheme* the information that is given about this topic. Based on this structure, the text is annotated with plausible gestures based on rules that are derived from studies of human-human dialogue.

Modifiable filters are then applied to trim the gestures down to a set appropriate for a particular character. Furthermore, some of the initially proposed gestures may not be suitable for co-occurring physically, so that some of them have to be filtered out. The application of various filters enables the generation of behavior variants reflecting different personalities. For instance, specific filters may be used for an introverted character resulting in a communicative behavior exhibiting fewer gestures.

Another promising approach is the use of Bayesian networks to model the relationship between emotion and its behavioral expression. Bayesian networks enable us to deal explicitly with uncertainty, which is of great advantage when modeling the connections between emotions and the resulting behaviors. An example of such an approach was presented by Ball and Breese (2000) who constructed a Bayesian network that estimated the likelihood of specific body postures and gestures for individuals with different personality types and emotions. For instance, a negative emotion increased the probability that an agent would say "Oh, you again" as opposed to "Nice to meet you!"

14.7.3 Tuning the Agent's Expressivity

In the approaches discussed previously, individual differences in behavior are mainly reflected by the propositional content of the utterances. The question arises as to how to achieve individualism in the agent's gestures, facial expressions, and speech. A key issue is defining strategies to combine signals while ensuring consistency between verbal and nonverbal communicative behaviors. An agent that portrays extroversion in its gestures but reflects introversion by its gaze behaviors may alienate the user and thus lead the interaction to fail.

One approach to make agents believable and consistent in their behaviors is to rely on high-level parameters. For instance, the EMOTE system by Chi et al. (2000) is based on dance annotation as described by Laban. The system is able to modify the execution of a given behavior by changing movement qualities, in particular the Laban principles of effort and shape. Hartmann et al. (2006) made use of six dimensions of expressivity that were derived from perceptual studies (Gallaher, 1992). Figure 14.4 depicts the same gesture with different degrees of spatial extent, which corresponds to one of the six dimensions.

Expressivity parameters are partly related to the personality of the recorded subject (Abrilian et al., 2006), but also depend on social or cultural norms (e.g., more gesticulation in Southern Europe vs. Northern Europe; Ting-Toomey, 1999). In this way, their agent is able to portray idiosyncratic behaviors that are characterized by different levels of expressivity. The settings of the parameters may be derived both from studies of the literature and corpus analyses (see Section 14.2).

The advantage of both methods is that they enable the modulation of action performance at a high level of abstraction. Furthermore, they rely on a small set of parameters that may affect different parts of the body at the same time. The hypothesis behind the approaches is that behaviors that manifest

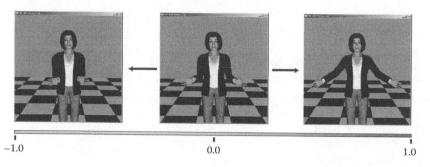

FIGURE 14.4 Variation of the spatial extent parameter. From left to right: low, medium, and high spatial extent. (From Hartmann, B., Mancini, M., and Pelachaud, C., Implementing expressive gesture synthesis for embodied conversational agents, Gesture Workshop, LNAI, Springer-Verlag, 2005.)

themselves through various channels with consistent expressivity parameters will lead to a more believable agent behavior.

14.8 Mapping Character Profiles to Users

Humans tend to associate the same stereotypical behaviors with virtual personalities as they do with real personalities. As a consequence, variables describing an agent (e.g., age, gender, ethnicity) are supposed to have an important impact on man-machine communication. A number of empirical studies were undertaken to investigate which variables influence a user's attitude toward the agent, as well as the effectiveness of the agent's presentation. A lot of attention has been paid to the question of whether users prefer characters that are similar to themselves and whether such characters are more effective, for example, as virtual trainers or consultants.

Nass et al. (2000) have shown that computer agents representing a user's ethnic group are perceived as socially more attractive and trustworthy. Bickmore and Cassell (2001) noticed that users with different personalities respond differently to social language. Introverted users perceived a virtual real estate agent as significantly more trustworthy in a pure task-based face-to-face conversation than in conversations over the phone or conversations, including social talk, while extroverts trusted the agent the least in this situation. Van Vugt et al. (2006) observed that users preferred nonideal, fatter characters as e-health advisors. Especially, users with a similar body shape trusted these agents more than they did slim characters. Baylor et al. (2006) observed that females interacting with an agent developed more positive attitudes toward math and science than females who did not interact with an agent. Their studies seem to indicate that pedagogical agents may help increase the interest of women in engineering.

These studies show that there are a number of factors influencing the user's attitude toward an agent and its effectiveness. In particular, it is difficult to generalize the results to other user groups and interaction scenarios. It is not clear to what extent the investigated effects hold when moving to different user groups and interaction scenarios.

A systematic investigation of the crucial variables that influence a user's attitude toward the agent and the effectiveness of the agent is required to help us achieve the vision of interface characters for all.

14.9 Challenges

This section reports on new research trends that aim to realize interface agents for all.

14.9.1 Cross-Cultural Issues

Recently, adapting to the user's cultural background has come into the focus of research on interface agents (see Payr and Trappl, 2004). This is an important research question, because much of one's behavior patterns are shaped by one's cultural context (Hofstede, 2001). Some preliminary approaches to this problem have already been presented in Section 14.5.3. Although the verbal and nonverbal behavior of embodied conversational agents has become increasingly sophisticated, this behavior is primarily based on a Western cultural background, due to the available agent systems and their predefined animation sequences. According to Ting-Toomey (1999), many misunderstandings in face-to-face communication arise due to misinterpretations of nonverbal cues. Thus, employing an agent that adheres to culturally determined behavior programs would enhance the efficiency of information delivery.

One of the challenges to be faced is how to represent cultural influences in such a way that they help tailor the agent's behavior to the user. For personality traits, McCrae's five-factor model (McCrae and John, 1992) has widely been employed. Schmitt et al. (2007) conducted a study across 56 nations investigating the worldwide distribution of personality traits and identified a number of interesting culture-specific patterns. Hall (1966) defines low- and high-contact cultures and associates different behavior patterns with both categories. Allwood (1985) distinguishes different levels in the interaction process at which misunderstandings in the process of intercultural communication can occur. Hofstede (2001) provides a dimensional model which promises to be well suited for integration in computational applications.

Cultures are defined along five dimensions and linked to different behavioral or mental programs, as Hofstede calls them. This information could be exploited for the behavior generation of an interface agent since, if the agent's profile is represented along Hofstede's dimensions, its behaviors can be adapted to the cultural background of the user. A similar, value-oriented approach is presented by Schwartz and Sagiv (1995). Values are defined as goals that serve as guiding principles of behavior. These values are based on three universal requirements (biological needs, coordinated social interaction, and group functioning). Cultures now differ in which values they relate to these universal needs and how they prioritize different values. It remains to be shown how these different goal structures can be reflected in specific verbal and nonverbal behaviors of agents.

14.9.2 Long-Term Adaptation

Most agent-based applications only consider short-term interactions with agents. At best, agents adapt their behavior to the situational context and the current interaction with the user. However, they do not maintain an explicit representation of interactions with users over a series of conversations. As a consequence, they do not permit long-term adaptation. As a first step toward long-term adaptation, Bickmore and Cassell (2005) presented a relational model that included three dimensions: solidarity, familiarity, and affect. During the dialogue, these dimensions are continuously updated. In particular, the system makes a controlled use of small talk to positively influence the user's trust in the agent as a function of the three dimensions.

Characters that accompany a user over a longer period of time should be able to remember earlier conversations. That is, a discourse history over a series of interactions needs to be maintained. To increase the believability of virtual agents, Ho and Watson (2006) propose to equip them with autobiographic knowledge containing the agents' individual history and have it constructed dynamically during their lifetime.

14.9.3 Agents That May Take On Different Roles

Similar to a human being, a character may encounter users in different roles. For instance, it may first assist a person as an insurance consultant and later engage with her in a conversation as a private individual. When switching from one role to the other, the character's outfit may change. Kruppa (2006) discusses the use of accessories, such as glasses, to visualize a character's role. The character's appearance is, however, only one means to indicate a role. In addition, the character's behavior should adapt to its current role. For example, users would expect a more relaxed and casual behavior from a character in the role of a sport companion than from a character serving as a finance advisor.

One of the first systems that make use of characters taking on different roles is Crosstalk, an interactive installation featuring a car sales scenario. In Crosstalk, characters may be both on-duty as well as off-duty. When a visitor approaches the Crosstalk installation, the characters signal by their body posture that they are now on-duty and start with a product presentation. If the user leaves the installation, the characters step out of their roles as professional presenters and switch to an off-duty mode where they engage in a chat. Nevertheless, the scripts for these different roles are more or less hardcoded. That is, there is no explicit representation of a character's role that guides the behavior planning processing.

14.9.4 Migrating Characters

In the future, characters will not just reside on the user's desktop, but instead need to be integrated in the user's natural environment. They will have to remain operational even when moving to a different environment. First of all, the performance of the device on which the character runs influences the choice of components that may be used to analyze user input and respond to it. For example, the reduced computing power of a mobile device compared to that of a stationary PC limits the complexity of the character model and its animation. When switching to a mobile device with a tiny display, a full body representation of the character might no longer be appropriate, and only its head and shoulders might be shown. As a consequence, the communicative skills of a character need to degrade gracefully when moving from a stationary to a mobile device. Guaranteeing a consistent appearance and behavior of the character regardless of the different technical setups provides a great challenge. After all, the user should still be able to recognize a character in a new environment. Furthermore, the use of mobile devices requires greater consideration of the context of use. The character needs to be able to detect environmental changes by means of sensors and to adapt to changing environmental setups in a very flexible manner. For example, in a noisy environment the use of speech as a means of communication is less appropriate and the character might rely on speech bubbles instead. Finally, the agent needs to account for different types of environments. For instance, the agent may switch to different means of communication depending on whether it is assisting the user at home or in public. Some users might prefer to switch to headphones when getting advice from the character in public.

André et al. (2005) describe a traversable interface that enables a character to cross the border between a purely virtual environment and a physical one. When moving from one space to another, the user has to correctly correlate different visual representations of one and the same object. Therefore, a difficult problem from the perspective of user-interface design is the question of how to make the transfer between different worlds as smooth as possible. An example of a traversable interface is shown in Figure 14.5. On the right-hand side, a character jointly explores with the user a tabletop application that combines virtual buildings of the city center of Augsburg with a real city map being laid out on a real table. On the left-hand side, they move to a virtual 3D representation of the city of Augsburg.

FIGURE 14.5 Traversable interface.

Kruppa (2006) describes a migrating character that accompanies the user in instrumented environments and is able to move from one output device to the other without appearing inconsistent in its behaviors. For instance, in the Peach project, the character acts as a virtual museum guide that first shows up on a real painting and then moves upon request to the user's mobile device. In another application, the character serves as a shopping assistant and may appear both on the user's mobile device as well as on the wall behind a shopping shelf using a steerable projector.

14.9.5 Satisfying the Needs of Heterogeneous User Groups at the Same Time

Embodied conversational agents can support individuals or groups of people, and the groups can be dynamically formed. For instance, an agent acting as a museum guide may interact with a single user, but also with groups of people. The great challenge is to generate a presentation that satisfies the individual needs of most users that watch it at the same time. A first approach was presented by Kruppa (2006) combining personal devices and stationary presentation systems to provide presentations that correspond to the interests of all users sharing a stationary information system. The basic idea is to convey the main information on the stationary presentation system and to augment this information by personalized information that is presented on the users' individual device.

14.9.6 Setting Up Guidelines for the Design and Evaluation of Embodied Conversational Agents That Support Universal Access

Traditionally, human-computer interaction has concentrated on the usability of applications. That is, the objective was to design applications that enable a user to perform a certain task as effectively and efficiently as possible, and the applications were primarily regarded as tools. Embodied conversational agents are usually not treated as pure tools, but may even form social relationships with human users. Therefore, the question arises of establishing to what extent traditional usability guidelines, such as the predictability of system behaviors, still apply. For instance, an embodied conversational agent that exhibits a monotonous behavior is likely to bore the user and decrease his willingness to interact with it. A certain degree of unpredictability might be even desired here. Not only factors that influence the usability of an application, but also factors that help to engage users and increase joy of use might be of relevance when designing and evaluating applications that make use of an embodied agent (cf., Waern and Höök, 2001). In particular, the ability of embodied agents to engage in social interactions with human users needs to be explored in the context of universal access. Valuable inspirations may come from the area of affective interfaces, focuses on new ways of capturing a user's experience (cf., Höök, 2004; Sengers, 2004).

14.10 Conclusion

Embodied conversational agents offer great promise. However, they also pose high demands on the design of appropriate interactions, since users will expect a certain degree of humanlike verbal and nonverbal conversational behaviors on the part of such agents. It is therefore necessary to enrich embodied conversational agents with social competencies. To replicate human behavior in an agent, corpus-based approaches have proven useful. Yet the decontextualization of corpus data and their automated adaptation to a new context is still an unresolved issue. As a consequence, the agents' capabilities to adapt dynamically to the preferences and needs of specific users are still strongly limited. Asymmetry in communication channels is considered a major barrier to universal access. While agents rely on gestures, facial expressions, and speech to present information, the users' input options are usually restricted to written input, which again makes the systems that employ such agents less accessible, especially to those with disabilities. Indeed most chatterbots on commercial web pages still allow for typed user input only.

Finally, it needs to be emphasized that embodied conversational agents are not considered the only possible interaction metaphor to ensure universal access. Some users clearly prefer a direct manipulation-based interface over an agent-based interface—at least for carrying out specific tasks (cf., the notion of a multiple metaphor environment that was first introduced by the FRIEND21 project; Ueda, 2001). A sizeable body of empirical work is required to shed light on the type of situation in which an embodied agent can positively contribute to the interaction and on how to design agents that best meet this requirement. In particular, longitudinal studies on the effect that agents might have on learning are needed.

Acknowledgments

Part of the work described in this chapter was funded by the German Research Foundation (DFG) under research grant RE 2619/2-1. Special thanks to Dieter Schmalstieg and Catherine Pelachaud for permission to use Figures 14.3 and 14.4 (copyright held by their organizations).

References

Abrilian, S., Martin, J. C., Buisine, S., and Devillers, L. (2006). Perception of movement expressivity in emotional TV interviews, in *HUMAINE Summerschool*, 22–28 September 2006, Genova, Italy. Genova: DIST.

Ai, H., Litman, D. J., Forbes-Riley, K., Rotaru, M., Tetreault, J., and Purandare, A. (2006). Using system and user performance features to improve emotion detection in spoken tutoring dialogs, in the *Proceedings of the Ninth International Conference on Spoken Language Processing (Interspeech 2006, ICSLP)*, 17–21 September 2006, Pittsburgh, pp. 797–800. Bonn, Germany: ISCA Archive.

Allwood, J. (1985). Intercultural communication, in *Tvärkulturell kommunikation* [*Papers in Anthropological Linguistics*]. University of Göteborg, Department of Linguistics. Gothenburg, Sweden.

Allwood, J., Cerrato, L., Jokinen, K., Navarretta, C., and Paggio, P. (2005). The mumin annotation scheme for feedback, turn management and sequencing, in *Gothenburg Papers in Theoretical Linguistics 92: Proceedings from The Second Nordic Conference on Multimodal Communication*, 7–8 April 2005, Gothenburg, Sweden, pp. 91–109. Gothenburg: Gothenburg University.

André, E., Dorfmüller-Ulhaas, K., and Rehm, M. M. (2005). Engaging in a conversation with synthetic characters along the virtuality continuum, in the *Proceedings of the 5th International Symposium on Smart Graphics*, 22–24 August 2005, Frauenworth Cloister, Germany, pp. 1–12. Berlin/Heidelburg: Springer-Verlag.

André, E., Rist, T., van Mulken, S., Klesen, M., and Baldes, S. (2000). The automated design of believable dialogues for animated presentation teams, in *Embodied Conversational Agents* (J. Cassell, S. Prevost, S., J. Sullivan, and E. Churchill, eds.), pp. 220–255. Cambridge, MA: MIT Press.

Barakonyi, I. and Schmalstieg, D. (2006). Ubiquitous animated agents for augmented reality, in *Proceedings of IEEE and ACM International Symposium on Mixed and Augmented Reality (ISMAR'06)*, 22–25 October 2006, Santa Barbara, CA. Washington, DC: IEEE Computer Society.

Baylor, A. L. (2005). Preliminary design guidelines for pedagogical agent interface image, in *IUI '05: Proceedings of the 10th International Conference on Intelligent User Interfaces*, 9–12 January 2005, San Diego, pp. 249–250. New York: ACM Press.

Baylor, A. L., Rosenberg-Kima, R. B., and Plant, E. A. (2006). Interface agents as social models: The impact of appearance on females' attitude toward engineering, in *CHI '06: CHI '06 Extended Abstracts on Human Factors in Computing Systems*, 22–27 April 2006, Quebec, Canada, pp. 526–531. New York: ACM Press.

Bickmore, T. and Cassell, J. (2001). Relational agents: a model and implementation of building user trust, in the *Proceedings of the SIGCHI Conference on Human Factors in Computing Systems (CHI '01)*, 31 March–5 April 2001, Seattle, pp. 396–403. New York: ACM Press.

Bickmore, T. and Cassell, J. (2005). Social dialogue with embodied conversational agents, in *Advances in Natural, Multimodal Dialogue Systems* (J. van Kuppevelt, L. Dybkjaer, and N. O. Bernsen, eds.), 30: 23–54. Netherlands: Springer-Verlag.

Bosma, W. and André, E. (2004). Exploiting emotions to disambiguate dialogue acts, in *IUI '04: Proceedings of the 9th International Conference on Intelligent User Interface*, 13–16 January 2004, Island of Madeira, Portugal, pp. 85–92. New York: ACM Press.

Brown, P. and Levinson, S. C. (1987). *Politeness: Some Universals in Language Usage*. Cambridge: Cambridge University Press.

Cassell, J. (2007). Body language: Lessons from the near-human, in *Genesis Redux: Essays in the History and Philosophy of Artificial Intelligence* (J. Riskin, ed.), pp. 346–374. Chicago: University of Chicago Press.

Cassell, J., Nakano, Y. I., Bickmore, T. W., Sidner, C. L., and Rich, C. (2001a). Nonverbal cues for discourse structure, in the *Proceedings of the 39th Annual Meeting of the Association of Computational Linguistics (ACL 2001)*, 6–7 July 2001, Toulouse, France, pp. 114–123. Morristown, NJ: Association for Computational Linguistics.

Cassell, J., Sullivan, J., Prevost, S., and Churchill, E. (eds.) (2000). *Embodied Conversational Agents*. Cambridge, MA: MIT Press.

Cassell, J., Vilhjálmsson, H. H., and Bickmore, T. W. (2001b). BEAT: The Behavior Expression Animation Toolkit, in the *Proceedings of SIGGRAPH 2001*, 12–17 August 2001, Los Angeles, pp. 477–486. New York: ACM Press.

Chi, D. M., Costa, M., Zhao, L., and Badler, N. I. (2000). The EMOTE model for effort and shape, in the *Proceedings of the 27th Annual Conference on Computer Graphics and Interactive Techniques*, 23–28 July 2000, New Orleans, pp. 172–182. New York: ACM Press.

Conati, C. (2002). Probabilistic assessment of user's emotions during the interaction with educational games. *Applied Artificial Intelligence* 16: 555–575.

Core, M., Traum, D., Lane, H. C., Swartout, W., Gratch, J., Lent, M. V., and Marsella, S. (2006). Teaching negotiation skills through practice and reflection with virtual humans. *SIMULATION* 82: 685–701.

DeCarolis, B., Pelachaud, C., Poggi, I., and Steedman, M. (2004). APML, a mark-up language for believable behavior generation, in *Life-Like Characters: Tools, Affective Functions and Applications* (H. Prendinger and M. Ishizuka, eds.), pp. 65–85. New York: Springer.

de Heras Ciechomski, P., Schertenleib, S., Maïm, J., Maupu, D., and Thalmann, D. (2005). Real-time shader rendering for crowds in virtual heritage, in the *Proceedings of the 6th International Symposium on Virtual Reality, Archaeology and Cultural Heritage (VAST 2005)* (M. Mudge, N. Ryan, and R. Scopigno, eds.), November 2005, Pisa, Italy, pp. 91–98.

de Rosis, F., Pelachaud, C., and Poggi, I. (2004). Transcultural believability in embodied agents: A matter of consistent adaptation, in *Agent Culture: Human-Agent Interaction in a Multicultural World* (S. Payr and R. Trappl, eds.), pp. 75–106. New York: Lawrence Erlbaum Associates.

Dryer, D. C. (1999). Getting personal with computers: How to design personalities for agents. *Applied Artificial Intelligence* 13: 273–295.

Dybkjær, L. and Bernsen, O. (2004). Recommendations for natural interactivity and multimodal annotation schemes, in *Proceedings of the LREC'2004 Workshop on Multimodal Corpora*, May 2004, Lisbon, Portugal, pp. 5–8. Paris: ELRA/ELDA.

Ekman, P. and Rosenberg, E. (eds.) (1998). *What the Face Reveals: Basic & Applied Studies of Spontaneous Expression Using the Facial Action Coding System (FACS)*. New York: Oxford University Press.

Foster, M. E. (2007). Issues for corpus-based multimodal generation, in the *Proceedings of the Workshop on Multimodal Output Generation (MOG 2007)*, 25–26 January 2007, Aberdeen, Scotland. Enschede, The Netherlands: Universiteit Twente.

Frey, S., Hirschbrunner, H. P., Florin, A., Daw, W., and Crawford, R. A. (1983). A unified approach to the investigation of nonverbal and verbal behavior in communication research, in *Current Issues in European Social Psychology* (W. Doise and S. Moscovici, eds.), pp. 143–199. New York: Cambridge University Press.

Gallaher, P. E. (1992). Individual differences in nonverbal behavior: Dimensions of style. *Journal of Personality and Social Psychology* 63: 133–145.

Gulz, A. and Haake, M. (2006). Design of animated pedagogical agents: A look at their look. *International Journal of Human Computer Studies* 64: 322–339.

Hall, E. T. (1966). *The Hidden Dimension*. New York: Doubleday.

Hartmann, B., Mancini, M., and Pelachaud, C. (2006). Implementing expressive gesture synthesis for embodied conversational agents, in the *Proceedings of the 6th International Gesture Workshop (GW 2005)*, 18–20 May 2005, Berder Island, France, pp. 188–199. Berlin: Springer-Verlag.

Ho, W. C. and Watson, S. (2006). Autobiographic knowledge for believable virtual characters, in the *Proceedings of the 6th International Conference on Intelligent Virtual Agents (IVA 2006)* (J. Gratch, M. Young, R. Aylett, D. Ballin, and P. Olivier, eds.), 21–23 August 2006, Marina del Rey, California, pp. 383–394. Berlin/Heidelberg: Springer-Verlag.

Hoekstra, A., Prendinger, H., Bee, N., Heylen, D., and Ishizuka, M. (2007). Presentation agents that adapt to users visual interest and follow their preferences, in the Proceedings of the 5th International Conference on Computer Vision Systems (ICVS 2007), 21–24 March 2007, Bielefeld, Germany. Bielefeld, Germany: Applied Computer Science Group.

Hofstede, G. (1991). *Cultures and Organizations: Intercultural Cooperation and Its Importance for Survival, Software of the Mind*. New York: McGraw Hill.

Hofstede, G. (2001). *Cultures Consequences: Comparing Values, Behaviors, Institutions, and Organizations across Nations*. Thousand Oaks, London: Sage Publications.

Höök, K. (2004). User-centered design and evaluation of affective interfaces, in *From Brows to Trust: Evaluating Embodied Conversational Agents* (Z. Ruttkay and C. Pelachaud, eds.), pp. 127–160. New York: Kluwer Academic Publishers.

Iacobelli, F. and Cassell, J. (2007). Ethnic identity and engagement in embodied conversational agents, in *Intelligent Virtual Agents (IVA)*, pp. 57–63. New York: Springer.

Isbister, K., Nakanishi, H., Ishida, T., and Nass, C. (2000). Helper agent: Designing an assistant for human-human interaction in a virtual meeting space, in *CHI '00: Proceedings of the SIGCHI Conference on Human Factors in Computing Systems*, 1–6 April 2000, The Hague, The Netherlands, pp. 57–64. New York: ACM Press.

Kapoor, A. and Picard, R. W. (2005). Multimodal affect recognition in learning environments, in *MULTIMEDIA '05: Proceedings of the 13th Annual ACM International Conference on Multimedia*, 6–11 November 2005, Singapore, pp. 677–682. New York: ACM Press.

Kipp, M. (2001). Anvil: A generic annotation tool for multimodal dialogue, in the *Proceedings of the 7th European Conference on Speech Communication and Technology*, 3–7 September 2001, Aalborg, Denmark, pp. 1367–1370. Bonn, Germany: ISCA Archive.

Kopp, S., Krenn, B., Marsella, S., Marshall, A.N., Pelachaud, C., Pirker, H., et al. (2006). Towards a common framework for multimodal generation: The behavior markup language, in the *Proceedings of the 6th International Conference on Intelligent Virtual Agents, IVA 2006* (J. Gratch, M. Young, R. Aylett, D. Ballin, and P. Olivier, eds.), 21–23 August 2006, Marina del Rey, California, pp. 205–217.

Kozierok, R. and Maes, P. (1993). A learning interface agent for scheduling meetings, in *IUI '93: Proceedings of the 1st International Conference on Intelligent User Interfaces*, 4–7 January 1993, Orlando, FL, pp. 81–88. New York: ACM Press.

Kranstedt, A., Kopp, S., and Wachsmuth, I. (2002). MURML: A Multimodal Utterance Representation Markup Language for conversational agents, in *Embodied Conversational Agents: Let's Specify and Compare Them, Proceedings of AAMAS 2002* (A. Mariott, C. Pelachaud, T. Rist, Z. Ruttkay, and H. Vilhjalmsson, eds.), 15–19 July 2002, Bologna, Italy.

Krenn, B., Neumayr, B., Gstrein, E., and Grice, M. (2004). Lifelike agents for the Internet: A cross-cultural case study, in *Agent Culture: Human-Agent Interaction in a Multicultural World* (S. Payr and R. Trappl, eds.), pp. 197–229. Mahwah, NJ: Lawrence Erlbaum Associates.

Kruppa, M. (2006). Migrating characters: Effective user guidance in instrumented environments. PhD dissertation, Saarland University, Saarbrücken.

Lamb, W. and Watson, E. (1979). *Body Code: The Meaning in Movement*. London: Routledge & Kegan Paul.

Lester, J. C., Towns, S. G., Callaway, C. B., Voerman, J. L., and FitzGerald, P. J. (2000). Deictic and emotive communication in animated pedagogical agents, in *Embodied Conversational Agents*, pp. 123–154. Cambridge, MA: MIT Press.

Martin, J. C., Caridakis, G., Devillers, L., Karpouzis, K., and Abrilian, S. (2007). Manual annotation and automatic image processing of multimodal emotional behaviours: Validating the annotation of TV interviews. *Personal and Ubiquitous Computing: Special Issue on Emerging Multimodal Interfaces.* http://www.springerlink.com/content/e760255k7068332x/fulltext.pdf.

Martinovsky, B. and Traum, D. (2003). Breakdown in humanmachine interaction: The error is the clue, in the *Proceedings of the ISCA Tutorial and Research Workshop on Error Handling in Dialogue Systems*, 28–31 August 2003, Switzerland, pp. 11–16. Bonn, Germany: ISCA Archive.

Massaro, D. W. (2004). Symbiotic value of an embodied agent in language learning, in *HICSS '04: Proceedings of the Proceedings of the 37th Annual Hawaii International Conference on System Sciences (HICSS'04) Track 5*, 5–8 January 2004, Hawaii. Washington, DC: IEEE Computer Society.

McCrae, R. R. and John, O. P. (1992). An introduction to the five factor model and its applications. *Journal of Personality* 60: 175–215.

McNeill, D. (1992). *Hand and Mind: What Gestures Reveal about Thought*. Chicago: University of Chicago Press.

Nakano, Y. I., Reinstein, G., Stocky, T., and Cassell, J. (2003). Towards a model of face-to-face grounding, in the Proceedings of the 41st Annual Meeting on Association for Computational Linguistics, Volume 1, 7–12 July 2003, Sapporo, Japan, pp. 553–561. East Stroudsburg, PA: Association for Computational Linguistics.

Nass, C., Isbister, K., and Lee, E. J. (2000). Truth is beauty: Researching embodied conversational agents, in *Embodied Conversational Agents* (J. Cassell, S. Prevost, J. Sullivan, and E. Churchill, eds.), pp. 374–402. Cambridge, MA: MIT Press.

Ortony, A., Clore, G., and Collins, A. (1988). *The Cognitive Structure of Emotions*. New York: Cambridge University Press.

Osgood, C. E. and Tannenbaum, P. H. (1955). The principle of congruity in the prediction of attitude change. *Psychological Review* 62: 42–55.

Payr, S. and Trappl, R. (eds.) (2004). *Agent Culture: Human-Agent Interaction in a Multicultural World*. London: Lawrence Erlbaum Associates.

Picard, R. (2000). *Affective Computing*. Cambridge, MA: MIT Press.

Prada, R. and Paiva, A. (2005). Intelligent virtual agents in collaborative scenarios, in the *International Conference on Intelligent Virtual Agents*, 12–14 September 2005, Kos, Greece, pp. 317–328. London: Springer-Verlag.

Prendinger, H. and Ishizuka, M. (2001). Social role awareness in animated agents, in *AGENTS '01: Proceedings of the Fifth International Conference on Autonomous Agents*, 28 May–1 June 2001, Montreal, Canada, pp. 270–277. New York: ACM Press.

Prendinger, H. and Ishizuka, M. (eds.) (2004). *Life-Like Characters: Tools, Affective Functions, and Applications*. New York: Springer-Verlag.

Pynadath, D. V. and Marsella, S. C. (2005). PsychSim: Modeling theory of mind with decision-theoretic agents, in *IJCAI-05, Proceedings of the Nineteenth International Joint Conference on Artificial Intelligence*, 30 July–5 August 2005, Edinburgh, Scotland, pp. 1181–1186. San Francisco: Morgan Kaufmann Publishers.

Rehm, M. and André, E. (2007). More than just a friendly phrase: Multimodal aspects of polite behavior in agents, in *Conversational Informatics: Engineering Approaches* (T. Nishida, ed.). New York: Wiley.

Rehm, M., André, E., and Nischt, M. (2005). Let's come together: Social navigation behaviors of virtual and real humans, in the *Proceedings of the 1st International Conference on Intelligent Technologies for Interactive Entertainment (INTETAIN 2005)*, 30 November–2 December 2005, Madonna di Campiglio, Italy, pp. 124–133. Berlin: Springer-Verlag.

Rehm, M., Bee, N., Endrass, B., Wissner, M., and André, E. (2007). Too close for comfort? Adapting to the users cultural background, in the *Proceedings of the International Workshop on Human-Centered Multimedia*, 27 October 2006, Santa Barbara, California, pp. 85–94. New York: ACM Press.

Rehm, M., Nakano, Y., Huang, H. H., Lipi, A. A., Yamaoka, Y., and Grüneberg, F. (2008). Creating a standardized corpus of multimodal interactions for enculturating conversational interfaces, in the *Proceedings of IUI-Workshop Enculturating Interfaces (ECI 2008)*. 13 January 2008, Gran Canaria.

Rich, C. and Sidner, C. (1998). Collagen: A collaboration manager for software interface agents. *User Modeling and User-Adapted Interaction* 8: 315–350.

Rickel, J., Lesh, N., Rich, C., Sidner, C. L., and Gertner, A. S. (2002). Collaborative discourse theory as a foundation for tutorial dialogue, in the *Proceedings of the 6th International Conference on Intelligent Tutoring Systems*, 2–7 June 2002, Biarritz, France, pp. 542–551. London: Springer-Verlag.

Rist, T., André, E., and Baldes, S. (2003). A flexible platform for building applications with life-like characters, in *IUI '03: Proceedings of the 8th International Conference on Intelligent User Interfaces*, 12–15 January 2005, Miami, Florida, pp. 158–165. New York: ACM Press.

Rist, T. and Schmitt, M. (2002). Applying socio-psychological concepts of cognitive consistency to negotiation dialog scenarios with embodied conversational characters, in *AISB Symposium on Animated Expressive Characters for Social Interactions*, 4–5 April, London, England London: Imperial College.

Schmitt, D. P., Allik, J., McCrae, R. R., and Benet-Martínez, V. (2007). The geographic distribution of big five personality traits: Patterns and profiles of human self-description across 56 nations. *Journal of Cross-Cultural Psychology* 38: 173–212.

Schwartz, S. H. and Sagiv, L. (1995). Identifying culture-specifics in the content and structure of values. *Journal of Cross-Cultural Psychology* 26: 92–116.

Sengers, P. (2004). The engineering of experience, in *Funology: From Usability to Enjoyment* (M. A. Blythe, K. Overbeeke, A. F. Monk, and P. C. Wright, eds.), pp. 19–29. Norwell, MA: Kluwer Academic Publishers.

Sidner, C. L., Lee, C., Kidd, C. D., Lesh, N., and Rich, C. (2005). Explorations in engagement for humans and robots. *Artificial Intelligence* 166: 140–164.

Takeuchi, Y., Katagiri, Y., Nass, C. I., and Fogg, B. J. (1998). Social response and cultural dependency in human-computer interaction, in the *Proceedings of PRICAI 1998*, 22–27 November 1998, Singapore, pp. 114–123. Berlin/Heidelberg: Springer-Verlag.

Ting-Toomey, S. (1999). *Communicating Across Cultures*. New York: The Guilford Press.

Ueda, H. (2001). The friend21 framework for human interface architectures, in *User Interfaces for All: Concepts, Methods, and Tools* (C. Stephanidis, ed.), pp. 245–270. Mahwah, NJ: Lawrence Erlbaum Associates.

van Vugt, H. C., Konijn, E. A., Hoorn, J. F., and Veldhuis, J. (2006). Why fat interface characters are better e-health advisors, in *Intelligent Virtual Agents, 6th International Conference, IVA 2006* (J. Gratch, M. Young, R. Aylett, D. Ballin, and P. Olivier, eds.), pp. 21–23 August 2006, Marina del Rey, California, pp. 1–13. Berlin/Heidelberg: Springer-Verlag.

Waern, A. and Höök, K. (2001). Interface agents: A new interaction metaphor and its applications to universal accessibility, in *User Interfaces for All: Concepts, Methods, and Tools* (C. Stephanidis, ed.), pp. 295–317. Mahwah, NJ: Lawrence Erlbaum Associates.

IV

Development Lifecycle of User Interfaces

15

User Requirements Elicitation for Universal Access

Margherita Antona,
Stavroula Ntoa,
Ilia Adami, and
Constantine Stephanidis

15.1 Introduction

User requirements elicitation is a fundamental phase of the development process of interactive product and services. This is evident in the principles of user-centered design (Norman and Draper, 1986), and has led to the development and practice of a wide variety of methods and techniques, mostly originating from the social sciences, psychology, organizational theory, creativity and arts, as well as from practical experience (Maguire and Bevan, 2002). Many of these techniques are based on the direct participation of users or user representatives in the process of formulating their own technological needs. However, the vast majority of available techniques has been developed with the "average" able-bodied user and the working environment in mind.

In universal access this precondition no longer holds, and, as pointed out in Chapter 1 of this handbook, the basic design principle of "knowing the user" becomes "knowing the diversity of users." In this context, the issue of gathering and specifying user requirements needs to be revisited. First, the consideration of accessibility becomes of foremost importance to avoid exclusion from design. This is a difficult obstacle to overcome, because of the limited current experience of designers and developers in identifying and addressing accessibility issues and applying accessibility knowledge and guidelines in a user-oriented fashion (Zimmermann and Vanderheiden, 2008). Part II of this handbook provides an overview of the main accessibility issues related to various target user groups with disabilities.

Second, the issue arises of which methods and techniques can be fruitfully employed to gather user requirements for diversity, and how such techniques need to be used, revised, and modified to optimally achieve this purpose, also taking into account the wide variety of technologies and contexts of use intrinsic in universal access.

Previous work in this area has mainly addressed the design of specific technologies for disabled users. For example, the UserFit methodology (Poulson et al., 1996) addresses the user-centered design of assistive technologies, and integrates a variety of requirements elicitation methods, identifying the problems that arise in their application. These include obtaining a representative sample of users with different types of impairments, gathering precise and comprehensive information from users who may have communication difficulties, or may be tired or confused, using combinations of techniques, obtaining specialist advice to correctly apply a method and ensure successful feedback, and following ethical procedures when the participants are not able to give their consent.

Under a universal access perspective, the collection of user requirements is further complicated by the fact that, following a design for all approach, more than one group of users with diverse characteristics and requirements need to be taken into account and involved.

This chapter aims at analyzing, based on a review of the related literature, the main issues arising in the context of user requirements elicitation for universal access, taking into account diverse target user groups, and in particular motor-impaired people, blind and visually impaired people, deaf people, people with cognitive or learning disabilities or with communication impairments, children and older users. The selection of groups is not exhaustive, and is not intended to address all the dimensions of diversity affecting the study of user requirements for universal access. For example, cultural issues are also of fundamental importance, as well as context diversity. However, it provides an overview of the variety of methods and of the complexity of the issues involved. Section 15.2 introduces the addressed user groups with respect to requirements elicitation. Section 15.3 overviews some popular and emerging user requirements elicitation methods, outlining the main issues that arise when each method is applied involving nontraditional user groups. Finally, Section 15.4 discusses the emerging need for a more systematic approach to user requirements analysis in universal access.

15.2 Target User Groups

15.2.1 Motor-Impaired People

The nature and causes of physical impairments are various; however, the most common problems faced by individuals with physical impairments include poor muscle control, weakness, and fatigue; difficulty in walking; talking, seeing, speaking, sensing or grasping (due to pain or weakness); difficulty reaching things; and difficulty doing complex or compound manipulations (push and turn) (Vanderheiden, 1997; see also Chapter 5, "Motor Impairments and Universal Access"). Individuals with severe physical impairments usually must rely on assistive devices, and the most commonly used assistive devices include mobility aids, manipulation aids, communication aids, and computer-interface aids (Vanderheiden, 1997).

As a result, the involvement of motor-impaired users in the requirements elicitation process mainly presents practical and organizational problems, which may, however, hamper the results. For example, access to buildings, as well as user transportation issues, should be taken into account. Furthermore, the physical fatigue that might be provoked in testing or discussion sessions should be also taken into account. It is imperative that hardware and software accessibility requirements are included in the investigation, especially for users who have motor impairments of upper limbs.

15.2.2 Blind and Visually Impaired People

Blindness means anatomic and functional disturbances of the sense of vision of sufficient magnitude to cause total loss of light perception, while visual impairment refers to any deviation from the generally accepted norm (see Chapter 6, "Sensory Impairments," for a thorough discussion of blindness and visual impairments). Visual acuity of an individual may vary from very poor vision to awareness of light but not of shapes, to no perception of light at all (Vanderheiden, 1997). Low vision includes problems (after correction) such as dimness of vision, haziness, film over the eye, foggy vision, extreme near- or farsightedness,

distortion of vision, spots before the eyes, color distortions, visual field defects, tunnel vision, no peripheral vision, abnormal sensitivity to light or glare, and night blindness.

Therefore, all the written material that may be used during requirements elicitation should be provided to blind and visually impaired users in the appropriate form. In more detail, blind users should be provided with such material either in Braille, if they are familiar with the Braille language, or in accessible electronic form, if they use computers and assistive hardware and software. Low-vision participants would benefit from printouts in large fonts and high text versus background contrast. When addressing color-blind participants, it should be ensured that semantic information is provided with the use of appropriate typography and not through color.

In addition, in group discussion techniques, the facilitator should make sure that all the material that is presented is explained to blind and visually impaired people orally.

15.2.3 Deaf People

Hearing impairment includes any degree and type of auditory disorder, on a scale from slight to extreme (see Chapter 6, "Sensory Impairments"). Familiar coping strategies for hearing-impaired people include the use of hearing aids, sign language, lip-reading, and telecommunication devices for the deaf (TDDs) (Vanderheiden, 1997). Furthermore, for individuals whose deafness occurred prelingually, it may also affect speech, as prelingually deaf persons can learn to speak, but their speech is usually difficult for most people to comprehend (Schein, 1981).

Consequently, to allow deaf participants to communicate effectively during a user requirements elicitation approach, sign language translators will be needed and the group participants should be instructed to speak at an appropriate pace, providing translators the needed time. In the methods in which only the deaf participant and the facilitator are involved, a sign language translator may not be necessary if the user is comfortable with lip reading. Finally, since many people with hearing disabilities from birth have problems with reading and writing (see also Chapter 38 of this handbook), any written material should be kept as simple as possible.

15.2.4 People with Cognitive or Learning Disabilities and People with Communication Impairments

Cognitive disability entails a substantial limitation in one's capacity to think, including conceptualizing, planning, and sequencing thoughts and actions, remembering, interpreting subtle social cues, and understanding numbers and symbols. Cognitive disabilities can also stem from brain injury, Alzheimer's disease and dementia, severe and persistent mental illness, and stroke (see also Chapter 7 of this handbook). Therefore, cognitive disabilities are many and diverse, individual differences are often very pronounced for this user group, and it is particularly difficult to abstract and generalize the issues involved in researching user requirements for this part of the population.

An approach that has been found to be effective for addressing this lack of generalizability problem is to work with a small number of users initially, designing a system targeted to their needs, and subsequently evaluating the system with a broader group (Moffat et al., 2006).

The user requirements collection process may be very complex, given the communication difficulties between the design team and the involved participants. Multidisciplinary teams may be required for working with these users, involving psychologist, language therapist, and other rehabilitation specialists. Very often, design methods are preferred that do not require the direct involvement of users in requirements analysis. Alternatively, relatives and caretakers are also involved as proxies for design input (e.g., Fischer and Sullivan, 2002).

Newell et al. (2002) point out the serious ethical implications related to the involvement of people with cognitive disabilities in the design process (e.g., in obtaining informed content), suggesting that the standard user-centered design methodology is not appropriate for this target user group, and proposes user-sensitive inclusive design as an approach targeted to capturing individual differences related to disability, and in particular cognitive dysfunctions.

For some particular subgroups, the direct involvement in the user requirements elicitation process has been investigated thoroughly. For example, Francis et al. (2009) discuss the issue of involving users with autism and Asperger's in the design of assistive technologies. One of the main identified problems concerns the potential for misunderstandings and the difficulties in clarifying misconceptions. Limited communication and cognitive skills may render the process complex. Fear of failure and lack of motivation may also make it particularly difficult to engage people with autism or Asperger's syndrome in the design process.

The difficulties of conducting research with adult individuals with learning disabilities are discussed in Hall and Mallalieu (2003). This user group is considered in some aspects similar to children, in particular regarding limited awareness of software application potential, communications difficulties (including low literacy levels), high willingness to agree with the analyst, and potentially limited social and interaction abilities. Finally, Astell et al. (2008) discuss particular difficulties that arise in involving people with dementia in the design process. These include obtaining informed consent, determining their requirements, eliciting their views, and evaluating prototype systems. In addition, there are difficulties related to including both family caregivers and professional care staff in the development process.

15.2.5 Children

Nowadays, children are exposed from a very early age to a wide range of technologies, including multimedia systems, electronic toys and games, and communication devices. They are immersed in media from their first steps. A survey contacted by the Henry J. Kaiser Family Foundation showed that in the United States nearly half (48%) of all children 6 years old and under have used a computer, and more than one in four (30%) have played video games. By the time they are in the 4-to-6-year-old range, 7 out of 10 have used a computer (Rideout et al., 2003). The market of children's toys includes a plethora of computer programs and electronic games with complex interfaces and interaction systems that are designed and developed specifically for children. This high level of exposure to technologies from such early ages renders the assumption that today's children will become tomorrow's power users of every technological advancement safe and logical.

The emergence of children as an important new consumer group of technology dictates the importance of supporting them in a useful, effective, and meaningful way for their needs. Designing for all should take into account that children have their own likes, dislikes, curiosities, and needs that are different from adults. Therefore, children should be regarded as a different user population with its own culture and norms (Heller, 1998).

However, gathering user requirements from this group is not an easy task, and bringing them into the designing process is even more complicated. The difficulty in the process arises from several factors. Children go to school for most of their days; there are existing power structures, biases, and assumptions between adults and children that need to be overcome, and children often have difficulty expressing their opinions and thoughts, especially when it comes to abstract concepts and actions (Druin, 2002).

Since the 1990s, there has been a growth of literature about children and human-computer interaction (HCI) issues, and the active involvement of children in the technology development process has been investigated (Druin, 1999b). Apart from the traditional means of gathering user requirements, a few other novel approaches have been developed and used in projects involving children. Section 15.3.12 of this chapter describes two case studies that used such nontraditional techniques.

15.2.6 Older People

There is overwhelming evidence that the population of the developed world is aging. The European Commission has predicted that between 1995 and 2025 the United Kingdom alone will see a 44% rise in people over 60, while in the United States the baby boomer generation, which consists of about 76 million people and is the largest group ever in the United States, is heading toward retirement (Marquis-Faulkes et al., 2005; see also Chapter 8 of this handbook). This large and diverse in its physical, sensory, and cognitive capabilities user group can benefit from technological applications that can enable them to retain their independent living, and ultimately reduce healthcare expenditure.

However, gathering requirements from this group can be a complex and difficult process. On the one hand, age-related impairments are sensitive personal matters that users are reluctant to discuss. Older adults also often exhibit a fundamental mistrust of technologies in general, which makes them reluctant to participate in design experiments (Newell et al., 2007). On the other hand, designers and developers of technology often have difficulty grasping the extent of the effects that age-related impairments have in the everyday activities of older people. To overcome these difficulties, researchers have been adapting and adjusting existing traditional design techniques so that they ease the process of involving this user group in the early stages of the design process. Section 15.3.11 describes examples of techniques that managed to successfully involve the users in the design process and proved to be a positive experience rich in lessons learned for everyone involved.

15.3 An Overview of User Requirements Elicitation Methods and Techniques

This section provides an overview of established as well as more recent approaches to user requirements elicitation. Based on a literature search, those aspects of each method and techniques that affect its application in the context of universal access are discussed.

In practice, more than one of these methods is usually employed in a design case. The choice can be dictated by several criteria, including the suitability of each method to be used with the involvement of users with specific characteristics, abilities, or limitations. Another relevant consideration is that in universal access many potential users of new technologies may be non-users of current technologies, or may not even be familiar with technology at all. In this case, it is important that the selected methods do not require from the users previous experience with similar systems.

15.3.1 Brainstorming

Brainstorming, originated from early approaches to group creativity (Osborn, 1963), is a process where participants from different stakeholder groups engage in informal discussion to rapidly generate as many ideas as possible. All ideas are recorded, and criticism of ideas forbidden. This technique is often used in the early phases of design to set the preliminary goals for a project or target system. One of the advantages of using brainstorming is that it promotes freethinking and expression, also among involved users, and allows the discovery of new and innovative solutions to existing problems. Brainstorming can be supported by various technological means, the most common of which are video (Mackay, Ratzer, and Janecek, 2000) and group support systems (Davison, 2000).

Moffat et al. (2004) discuss the use of brainstorming techniques with users with aphasia. Discussion support through the use of images can be useful in some cases, depending on the expression abilities of the involved users. Brainstorming is also often used to gather initial design goals and ideas with

the involvement of relatives and caretakers, leaving the direct involvement of users to subsequent design phases when some prototype can be shown to them.

Overall, brainstorming can be considered as appropriate when the users to be involved have good communication abilities and skills (not necessarily verbal), but can also be adapted to the needs of other groups. This may have implications in terms of the pace of the discussion and generation of ideas.

15.3.2 Direct Observation

Popular methods of exploring the user experience come from field research in anthropology, ethnography, and ethnomethodology (Beyer and Holtzblatt, 1997). Ethnographic methods are based on four basic principles (Blomberg et al., 2002):

- *Natural settings*: The foundation in ethnography is field work, where people are studied in their everyday activities.
- *Holism*: People's behaviors are understood in relation to how they are embedded in the social and historical fabric of everyday life.
- *Descriptive*: The ethnographers describe what people actually do, not what they should do. No judgment is involved.
- *Members' point of view*: The ethnographers create an understanding of the world from the point of view of those studied.

Direct observation is one of the hallmark methods of ethnographic approaches. It involves an investigator viewing users as they conduct some activity. The goal of field observation is to gain insight into the user experience as experienced and understood within the context(s) of use. Examining the users in context is claimed to produce a richer understanding of the relationships between preference, behavior, problems, and values.

Observation sessions are usually video-recorded, and the videos are subsequently analyzed. A less effective alternative is taking notes. This technique is often used in conjunction with others, such as interviews and diaries (see following subsections). Obtaining the cooperation of users is vital, so the interpersonal skills of the observer are important. The observer needs to be unobtrusive during the session, and only pose questions when clarification is necessary. The effectiveness of observation and other ethnographic techniques can vary, as users have a tendency to adjust the way they perform tasks when knowingly being watched. Davies et al. (2004) make a sound case for field studies and direct observation when designing with users with cognitive disabilities or aphasia, as this method does not rely on the participants' communication abilities. Similar observations may hold for people with autism. Ethnography, therefore, appears to be a valuable component of any user-centered or participative design activity with this group.

Shinohara (2006) reports on the use of observational studies with blind users to develop design insights for enhancing interactions between a blind person and everyday technological artifacts found in their home such as wristwatches, cell phones, or software applications. Analyzing situations where work-arounds compensate for task failures reveals important insights for future artifact design for the blind, such as, for example, tactile and audio feedback, and facilitation of user independence.

A difficulty with direct observation studies is that they may in some cases be perceived as a form of invasion of the users' space and privacy, and therefore may not be well accepted, for example, by disabled or older people who are not keen to reveal their problems in everyday activities.

15.3.3 Activity Diaries and Cultural Probes

Diary keeping is another ethnographically inspired method that provides a self-reported record of user behavior over a period of time (Whyte, 1984). The participants are required to record activities they are engaged in during a normal day. Diaries allow identifying patterns of behavior that would not be recognizable through short-term observation. However, they require careful design and prompting if they are to be employed properly by participants. Diaries can be textual, but also visual, employing pictures and videos.

Generalizing the concept of diaries, cultural probes, which originated in the traditions of artist-designers (Andreotti and Costa, 1996), are based on kits containing a camera, a voice recorder, a diary, postcards, and other items (Gaver et al., 1999). Cultural probes are claimed to allow the users great freedom and control over the self-reporting process, and have been successfully employed for user requirements elicitation in home settings with sensitive user groups, such as former psychiatric patients and the elderly (Crabtree et al., 2003).

Reading and writing a paper-based diary may be a difficult process for blind users and users with motor impairments. Therefore, diaries in electronic forms or audio-recorded diaries should be used in these cases. Lazar et al. (2007) report on a case study where electronic diaries were successfully used to investigate the causes of frustration of blind users when accessing the web through screen readers.

15.3.4 Surveys and Questionnaires

User surveys, originating from social science research (Alreck and Settle, 1995), involve administering a set of written questions to a sample population of users, and are usually targeted to obtaining statistically relevant results. Questionnaires are widely used in HCI, especially in the early design phases. Questionnaires need to be carefully designed to obtain meaningful results (Oppenheim, 2000). Questions may be closed with fixed responses and open where the respondents are free to answer as they wish. Various scales are used in questionnaires for the users to rate their responses. The simple and comprehensible formulation of questions is vital. Questions must also be focused to avoid gathering large amounts of irrelevant information.

Questionnaires can be administered in several ways, for example, by post, e-mail, or the web, and the effectiveness of mail versus web-based questionnaires has been largely discussed in the literature (Andrews et al., 2003).

Research shows that there are age differences in the way older and younger people respond to questionnaires. For example, older people tend to use the "Don't know" response more often than younger people. They also seem to use this answer when they are faced with questions that are complex in syntax. Their responses also seem to avoid the extreme ends of ranges. Researchers have found ways around this problem. For example, Eisma et al. (2004) found that having the researcher administer the questionnaire directly to the user helped to retrieve more useful and insightful information. Dickinson et al. (2002) found that in-home interviews were effective in producing a wealth of information from the user that could not have been obtained by answering a questionnaire alone.

Since questionnaires and surveys address a wide public, and it is not always possible to be aware of the exact user characteristics (i.e., if they use Braille or if they are familiar with computers and assistive hardware and software), they should be available either in alternative formats or in accessible electronic form. An example of an accessible questionnaire design process is described in a survey conducted to elicit the requirements of disabled users regarding e-Government services (Margetis et al., 2008).

15.3.5 Interviews

Interviews are another ethnographically inspired user requirements collection method (Gubrium and Holstein, 2002). In HCI, it is a commonly used technique where users, stakeholders, and domain experts are questioned to obtain information about their needs or requirements in relation to a system (Macaulay, 1996). Interviews can be unstructured (i.e., no specific sequence of questions is followed), structured (i.e., questions are prepared and ordered in advance), or semi-structured (i.e., based on a series of fixed questions with scope for the user to expand on their responses). The selection of representative users to be interviewed is important to obtain useful results. Interviews on a customer site by representatives from the system development team can be very informative. Seeing the environment also gives a vivid mental picture of how users work with existing systems and how a new system can support them (Mander and Smith, 2002).

In Smith-Jackson et al. (2003), a case study is reported involving blind and motor-impaired users in semistructured interviews. The objective of the study was to identify accessibility issues in mobile phones. In this study, the questions in the interview were associated to brief scenarios (see Section 15.3.9) illustrating features and functionalities of mobile phones to the users who were not familiar with them. Overall, the experience is reported to be positive.

With older people, interviews as a means for gathering user requirements have also proven to be an effective method, but in-house interviews can be even more productive, because they tend to lead to spontaneous excursions into users' own experiences, and demonstrations of various personal devices used (Eisma et al., 2004). Careful sequencing and delivery of simple sentence structures that avoid the use of abstract concepts is recommended. Obviously, interviews present difficulties when deaf people are involved, and sign language translation may be necessary. Interviews are often avoided when the target user group is composed of cognitively and communication-impaired people.

Recently, a trend to conduct interviews online using chat tools has been emerging. Crichton and Kinash (2003) report on the application of this method with blind users. An obvious consideration in this respect is that the used chat tool must be accessible and compatible with screen readers.

15.3.6 Group Discussions

Focus groups are inspired from market research techniques (Greenbaum, 1998). They bring together a cross-section of stakeholders in a discussion group format. The general idea is that each participant can act to stimulate ideas, and that by a process of discussion, a collective view is established (Bruseberg and McDonagh-Philp, 2001). Focus groups typically involve 6 to 12 persons, guided by a facilitator. Several discussion sessions may be organized.

The main advantage of focus groups regarding requirements elicitation from users with disabilities is that it does not discriminate against people who cannot read or write and they can encourage participation from people reluctant to be interviewed on their own or who feel they have nothing to say (Kitzinger, 1995). During focus groups, various materials can be used for review, such as, for example, storyboards (see Section 15.3.9).

Organizing a group discussion that includes participants with disabilities requires a considerable amount of preparation and is dependent on each participant's communication skills (Poulson et al., 1996). This method should not be employed for requirements elicitation if the target user group has severe communication problems. Moreover, it is important that the discussion leader manages effectively and efficiently the discussion, allowing all users to actively participate in the process regardless of their disability.

Focus groups have been used for eliciting expectations and needs from the learning disabled, as it was felt that they would result in the maximum amount of quality data. They allow a range of perspectives to be gathered in a short time period in an encouraging and enjoyable way. This is important, as typically people with learning disabilities have a low attention span. The satisfaction of the participants is also an important factor to be taken into account (Hall and Mallalieu, 2003).

Concerning older people, related research has found that it is not easy to keep a group of older people focused on the subject being discussed. Participants tend to drift their discussions off the subject matter as for them the focus group meeting is a chance to socialize. Thus, it is important to provide a social gathering as part of the experience of working with IT

researchers rather than to treat them simply as participants (Newell et al., 2007).

Kurniawan (2006) reports on a focus group study on the use of mobile phones by women aged 60 years and over. The study addresses usage patterns, problems, benefits, ideal phone design, and desired and unwanted features, as well as cooperative learning processes when encountering an unfamiliar mobile phone. The findings of the focus groups were used for elaborating a survey questionnaire to obtain quantitative data.

15.3.7 Empathic Modeling

Empathic modeling is a technique intended to help designers and developers put themselves in the position of a disabled user, usually through disability simulation. This technique has been first applied to simulate age-related visual changes in a variety of everyday environmental tasks, with a view to eliciting the design requirements of the visually impaired in different architectural environments (Pastalan, 1982). Empathic modeling can be characterized as an informal technique, and there are no specific guidelines on how to use it.

A variety of modeling techniques for specific disabilities through simple equipment are available (Poulson et al., 1996; Nicolle and Maguire, 2003; Fulton et al., 2005). Visual impairment due to cataracts can be simulated with the use of an old pair of glasses with Vaseline, while total blindness is easier to simulate using a scarf or a bandage tied over the eyes. Total hearing loss, on the other hand, can be easily simulated using earplugs. Furthermore, software application simulating visual and hearing impairments have been developed to help designers understand the interaction difficulties experienced by users (Goodman et al., 2007).

Some upper limb mobility impairments can be simulated with the use of elastic bands and splints, while others, such as lack of motor control, are quite difficult to simulate. Cognitive disabilities are also particularly difficult to simulate (Svensk, 1997). Simulators can also only communicate certain aspects of what it is like to have a disability, failing, for example, to account for context, support, and coping strategies (Goodman et al., 2006). A set of flexible and graded simulators for vision, dexterity, and reach and stretch is discussed in Cardoso and Clarkson (2006).

15.3.8 User Trials

In user trials, a product is tested by "real users" trying it out in a relatively controlled or experimental setting, following a standardized set of tasks to perform. User trials are performed for usability evaluation purposes (see Chapter 20, "The Evaluation of Accessibility, Usability, and User Experience"). However, the evaluation of existing or competitive systems, or of early designs or prototypes, is also a way to gather user requirements (Maguire and Bevan, 2002). While there are wide variations in where and how a user trial is conducted, every user trial shares some characteristics (Dumas and Redish, 1993). The primary goal is to improve the usability of a product having participants who are representative of real users to use the product carrying out real tasks while being observed, and the data that are collected are later analyzed. In field studies, the product or service is tested in a "real-life" setting.

Most of the related concerns regarding users with disabilities have already been mentioned in other requirements elicitation methods, such as user observation and focus groups. In user trials, however, an appropriately equipped room needs to be available for each session. When planning the test, it should be taken into account that trials with older users and users with disabilities may require more time than usual to complete the test without anxiety and frustration.

Research on the use of the most popular methods has indicated that modifications to well-established user trial methods are necessary when users with disabilities are involved. For example, two studies (Chandrashekar et al., 2006; Roberts and Fels, 2006) explore the required adaptations to the think-aloud protocol that have to be applied when carrying out user trials with deaf users and blind users respectively.

Furthermore, it is very important to explicitly emphasize during the instructions that it is the product that is being tested and not the user (Poulson et al., 1996), since a trial may reveal serious problems with the product, to the extent that some tasks may not be possible to carry out. Therefore, it is important that the users do not feel uncomfortable or attribute the product failure to their disability.

Finally, when the user trial participants are users with upper limb motor impairments and poor muscle control, it should be ensured that testing sessions are short, so as to prevent excessive fatigue.

15.3.9 Scenario, Storyboards, and Personas

Scenarios are widely used in requirements elicitation and, as the name suggests, are narrative descriptions of interactive processes, including user and system actions and dialogue. Scenarios give detailed realistic examples of how users may carry out their tasks in a specified context with the future system (Carroll, 1995, 2000). The primary aim of scenario building is to provide examples of future use as an aid to understanding and clarifying user requirements and to provide a basis for later usability testing. Scenarios can help identify usability targets and likely task completion times.

Storyboards are graphical depictions of scenarios, presenting sequences of images that show the relationship between user actions or inputs and system outputs. Storyboarding originated in the film, television, and animation industry (Hart, 1998). A typical storyboard contains a number of images depicting features such as menus, dialogue boxes, and windows. Storyboards may vary regarding the level of detail, the inclusion of text, the representation of people and emotions, the number of frames, and the way time-passing is indicated (Truong et al., 2006). Storyboards can be developed using various tools (e.g., Microsoft PowerPoint and similar software).

Another scenario-related method is called personas (Cooper, 1999), where a model of the user is created with a name, personality, and picture, to represent each of the most important user groups. The persona model is an archetypal representation of real or potential users. It is not a description of a real user or an average user. The persona represents patterns of users' goals and behavior, compiled in a fictional description of a single individual. Potential design solutions can then be evaluated against the needs of a particular persona and the tasks users are expected to perform.

The scenario approach was used in Antona et al. (2007) as a vehicle to discuss potential benefits and challenges of ambient intelligence (AmI) for disabled people. In this case study, generic AmI scenarios were re-elaborated by considering modifications occurring in the case that the principal character in each scenario has some disability.

Zimmermann and Vanderheiden (2008) propose a methodology based on the use of scenarios and personas to capture the accessibility requirements of older people and people with disabilities and structure accessibility design guidelines. The underlying rationale is that the use of these methods has great potential to make this type of requirement more concrete and comprehensible for designers and developers who are not familiar with accessibility issues.

According to Goodman et al. (2006), however, really reliable and representative personas can take a long time to create. Additionally, personas may not be well suited to presenting detailed technical information, for example, about disability, and their focus on representative individuals can make it more complex to capture the range of abilities in a population (see also Chapter 19, "Tools for Inclusive Design").

The use of storyboarding with disabled or older users does not appear to be common. However, it is self-evident that storyboarding is not optimal for blind users, while it requires particular care for users with limited vision or color-blindness. On the contrary, it would appear to be a promising method for deaf or hearing-impaired users.

15.3.10 Prototyping

A prototype is a concrete representation of part or all of an interactive system. It is a tangible artifact, does not require much interpretation, and can be used by end-users and other stakeholders to envision and reflect upon the final system (Beaudouin-Lafon and Mackay, 2002). Prototypes serve different purposes and thus take different forms. Off-line prototypes (also called paper prototypes) include paper sketches, illustrated storyboards, cardboard mockups, and videos. They are created quickly, usually in the early stages of design, and are usually thrown away when they have served their purpose. Online prototypes, on the other hand, include computer animations, interactive video presentations, and applications developed with interface builders. Prototypes also vary regarding their level of precision, interactivity, and evolution. With respect to the latter, rapid prototypes are created for a specific purpose and then thrown away, iterative prototypes evolve, either to work out some details (increasing their precision) or to explore various alternatives, and evolutionary prototypes are designed to become part of the final system.

Collaborating on prototype design is an effective way to involve users in design (see Section 15.3.11). Prototypes help users articulate their needs and reflect on the efficacy of design solutions proposed by designers.

Research has indicated the use of prototypes is more effective than classic methods for user requirements elicitation, such as interviews and focus groups, when designing innovative systems for people with disabilities, since potential users may have difficulty imagining how they might undertake familiar tasks in new contexts (Petrie et al., 1998). Using prototypes can be a useful starting point for speculative discussions, enabling users to provide rich information on details and preferred solutions (Engelbrektsson et al., 2004).

Prototypes are usually reviewed through user trials (see Section 15.3.8), and therefore all considerations related to user trials and evaluation are pertinent. An obvious corollary is that prototypes must be accessible to be tested with disabled people. This may be easier to achieve with online prototypes, closely resembling the final system, than with paper prototypes.

15.3.11 Cooperative and Participatory Design

Cooperative inquiry is a partnership design process that has its roots in the Scandinavian projects of the 1970s (Bodker et al., 1988). Since its introduction, cooperative inquiry as a method has been adapted and expanded, and examples of its practices can be found throughout the literature of the human-computer interaction field.

Likewise, the participatory design approach has its roots in the Scandinavian workplace democracy movement and is a set of theories, practices, and studies related to end-users as full participants in activities that lead to the development and design of technology (Muller, 2002). Participatory design may adopt a wide variety of techniques, including brainstorming, scenario building, interviews, sketching, storyboarding, and prototyping, with the full involvement of users.

Traditionally, partnership design techniques have been used for gathering user requirements from adult users. However, in the past few years a number of research projects have shown ways to adapt these techniques to benefit the design of technology process for nontraditional user groups, such as children and the elderly.

Cooperative inquiry has been widely used to enable young children to have a voice throughout the technology development process (Druin, 1999a, 1999b, 2002), based on the observation that although children are emerging as frequent and experienced users of technology, they were rarely involved in the development process. In these efforts, alterations were made to the traditional user requirement gathering techniques used in the process to meet the children's needs. For example, the adult researchers used note-taking forms, whereas the kids used drawings with small amounts of text to create cartoonlike flow charts. Overall,

involving children in the design process as equal partners was found to be a very rewarding experience and one that produced exciting results in the development of new technologies (Druin, 1999a).

Designing technology applications to support older people in their homes has also shown an increase in necessity as the developed world is aging. However, designing for this group of users is not an easy process as developers and designers often fail to fully grasp the problems that this user group faces when using technologies that affect their everyday lives. HCI research methods need to be adjusted when used on this user group. They have to take into consideration that older adults experience a wide range of age-related impairments, including loss of vision, hearing, memory, and mobility, which ultimately also contribute to a loss of confidence and difficulty in orientation and absorption of information (Zajicek, 2006; see also Chapter 8 of this handbook). Participatory design techniques can help designers reduce the intergenerational gap between them and older people, and help them better understand the needs of this group of users (Demirbileka and Demirkan, 2004). When older people participate in the design process from the start, their general fear toward using technology decreases, because they feel more in control and confident that the end result of the design process has truly taken their needs into consideration.

Wu et al. (2004) discuss the adoption of participatory design to involve users with amnesia. A set of reviewing techniques for reducing the demands made on explicit memory during the design process were developed and applied. Overall, the investigation conducted has allowed for identifying assumptions of participatory design that have been established with normal cognitively functioning populations, and to devise more flexible and supporting ways of conducting participatory design.

15.3.12 Recent and Emerging Approaches

In recent years, specific approaches and techniques targeted to nontraditional user groups have emerged, often reported through design case studies. The main goal of these techniques is to immerse the user group inside the technology design process from its early stages as an active member and not just as an informant, or a tester. This section overviews some of them.

The KidReporter method (Bekker et al., 2003) aims mainly at enabling children to contribute their opinion to a design problem through a choice of activities that finally results in a newspaper with the children's ideas about a topic. KidReporter was used in the context of a project whose aim was to design an electronic educational interactive game for a zoo. Two classes of children ages 9 and 10 participated in making a newspaper about a zoo. The researchers gathered information from children about their interests in the zoo, through activities such as making photos and descriptions of the photos, holding interviews, writing articles, and filling in questionnaires. Some teachers and parents assisted in supervising the various tasks that the children participated in.

The outcome of the implementation of the KidReporter method indicated that even though it requires a lot of planning, it brings benefits that definitely outweigh the difficulties. Its main strengths are that it is appealing and fun for children, and that the submethods for gathering information make it possible to make stronger inferences about children's opinions.

Mission from Mars (Dindler et al., 2005) is another novel approach targeted to children, which has been applied in the early phases of the development process of an electronic school bag. The method's main goal is to provide a framework for questioning specific user requirements according to elements in children's practices. The design process emphasizes fun and playfulness, facilitated by the notion that children are talking to a Martian. Seven children ages 10 and 11 and five members of the design team participated in a 3-hour session. The children were divided into three groups, which took turns presenting their material through cameras to Martians who were interested in the way schools function on earth and especially how pupils spend their time during the day. According to the researchers involved in this project, the Mission from Mars method offered an opportunity to the technical team to engage with the children, establish the necessary level of confidentiality through role-play, and get to the actual requirement for making a design that was meaningful to its users in context. The main strengths of the method are that it is another playful and motivating framework for both children and designers and the shared narrative space makes it possible to ask questions that would be impossible to raise in a conventional setting.

The use of drama has also emerged as a user requirements elicitation technique, mainly oriented to older adult users. In Marquis-Faulkes et al. (2005) the application of drama in the design of a system that provides fall detection and movement monitoring to support older people living at home is reported. To gather user requirements for the design of the system, a theater group was hired to develop and perform four scenarios based on material from focus groups and anecdotal evidence. The scenarios featured older people falling at home with different outcomes, and caretakers discussing an older person's needs. The scenarios were filmed and shown to three different groups of older people and a group of sheltered housing wardens to provoke discussions. The research team found that the outcome of discussions following the video scenarios explored effectively the user requirements early in the design cycle. They felt that drama was an extremely useful method of provoking discussion at the pre-prototyping stage and provided many insights that they believed would not have been obtained without such techniques being utilized.

Pastiche techniques (Blythe and Dearden, 2008) employ fictional characters and settings borrowed from literary and popular culture to create a space for the discussion of new technological developments and user experience. Both pastiche scenarios and pastiche personae can be used. The scenarios developed by the designers of the project are then used as discussion documents with the users. This practice creates a common and safe ground on which to discuss highly personal matters that older people may be experiencing in their everyday lives. In the case study reported in Blythe and Dearden (2008), this technique was used in a project supporting a befriending and shopping scheme for older people. The scenarios were written in the style of Dickens' *Christmas Carol* and had as their main character Ebeneezer Scrooge. Changes in how the shopping service is run are envisaged as consequences of the events that occur in the chain of scenarios. The research team found the use of pastiche techniques a low-cost, high-speed alternative to established approaches to personae and scenarios. Pastiche can provide a common ground among design stakeholders and an engaging and stimulating basis for discussion. Its main limitation lies in the difficulty of finding suitable characters that are familiar to all the members of the design team.

Finally, an emerging approach to user requirements elicitation based on arts is reported in Vickers et al. (2008). A project is described that sought to get a group of older people to think creatively about their needs and desires for technological support through the medium of paint. The approach was found to show promise, as it allowed information to be gathered in an environment that is comfortable and familiar using methods already known by the participants and that they find enjoyable. It provides a complement (or possible alternative) to standard protocols and has the potential benefit of extracting even richer information as the primary task for participants is enjoyable in its own right, and is not associated with an interrogative process.

15.4 Challenges under a Universal Access Perspective

The previous section provided an overview of established as well as emerging user requirements elicitation methods, discussing their application for the target user groups taken into consideration in Section 15.2.

Table 15.1 summarizes the results of this study, suggesting an indicative path toward method selection for the considered groups.

The following general considerations also emerge from the previous analysis:

- Practical and organizational aspects of the elicitation process play an important role when nontraditional user groups are involved, and are mentioned by many of the reviewed literature as critical to the success of the entire effort. Their importance should not be underestimated.
- Very few methods can be used as they stand when addressing diverse user groups. One of the main issues is therefore how to appropriately adapt and fine-tune methods to the characteristics of the involved people. Much of the literature reviewed in this chapter moves along these lines and proposes potential solutions for some target user groups.
- User requirements elicitation is mostly based on communication between users and other stakeholders in the design process. Therefore, the communication abilities of the involved users should be a primary concern. It is

TABLE 15.1 Summary of User Requirements Elicitation Methods

User Requirements Elicitation Methods and Techniques	Disability				Age	
	Motion	Vision	Hearing	Cognitive/ Communication	Children	Elderly
1. Brainstorming	✓	✓	■	■	■	■
2. Direct observation	✓	✓	✓	✓	✓	✓
3. Activity diaries and cultural probes	■	■	✓	■	■	✓
4. Survey and questionnaires	■	■	■	⊠	■	■
5. Interviews	✓	✓	■	⊠	■	■
6. Group discussions	✓	✓	■	⊠	■	■
7. Empathic modeling	✓	✓	✓	⊠	⊠	⊠
8. User trials	■	■	■	■	■	■
9. Scenarios and personas	✓	✓	✓	✓	✓	✓
10. Prototyping	✓	✓	✓	✓	✓	✓
11. Cooperative and participatory design	✓	✓	✓	■	■	■
12. Art-based approaches					✓	✓

✓ Appropriate.
■ Needs modifications and adjustments.
⊠ Not recommended.

important to remove potential communication obstacles and provide support to the users. Communication forms alternative to interviews and questionnaires (such as, e.g., pictures, drama, and art) can be taken into consideration when appropriate to enhance users' expressive means and encourage free expression.

- Some methods, such as, for example, empathic modeling and scenarios (see Sections 15.3.7 and 15.3.9, respectively) do not necessarily require user involvement, although they are often elaborated with the users' participation or reviewed by representative users. In any case, they should be accurate enough in rendering the users' potential problems and needs, and therefore should be based on in depth-knowledge of the users' characteristics and abilities.

- Requirements elicitation is progressively becoming an online rather than a face-to-face activity. Many of the reported methods can be supported by appropriate technology. In this case, every care should be taken that the hardware and software used are accessible to the involved users and compatible with their preferred assistive technologies. Requirements elicitation can also take place virtually through the web. This is a potentially interesting direction to investigate, as it can contribute to reducing time and costs, address users' and researchers' moving, as well as facilitate less stressful communication. On the other hand, however, face-to-face communication can allow better reciprocal understanding.

- Traditional HCI debates regarding in-laboratory versus in-field methods, group-based versus individual-based methods, short-term versus long-term, and structured versus unstructured methods, maintain and increase their timeliness in the context of universal access. There

is no available cookbook or even guide through the possible combinations of choices. Omnipresence and interweaving of technology in the social fabric of life would provide an argument in favor of longer-term, in-field, and group-based approaches. On the other hand, the difficulty of following and monitoring users in their own social environments would argue for the opposite. Along with the evolution of universal access toward AmI, large-scale research infrastructures are emerging as full-immersion environments where users can be brought into contact with new technologies, thus offering direct experience for diverse user groups (Stephanidis et al., 2007). The main advantage is offered by the possibility of combining contextual and laboratory-based methods, simulating and validating scenarios, and obtaining precious data from the continuous monitoring of user activities in various environments.

Overall, the considerations put forward in this chapter clearly point to the need for rethinking methodologically user requirements elicitation in universal access, and developing an articulated framework to orient and guide researchers and practitioners among the many dimensions of diversity that may play a role in the adoption, application, and modification of existing approaches, as well as toward the elaboration of new methods.

15.5 Summary and Conclusions

This chapter has discussed user requirements elicitation under a universal access perspective, based on the (arbitrary) selection of a number of target user groups and a literature review of existing approaches and methods that are reported to have been used for such groups. The considered user groups are motor-impaired people, people with different sensory disabilities, people with cognitive, learning, or communication disabilities, children, and older people. The addressed methods include brainstorming, direct observation, activity diaries and cultural probes, surveys and questionnaires, interviews, group discussions, emphatic modeling, user trials, scenarios, storyboards and personas, prototyping, cooperative and participative methods, as well as emerging approaches based on playing, arts, drama, and literature.

This investigation is by no means complete, both in terms of addressed groups and included methods, and does not propose any practical way out of the dilemma faced in each universal access design project: how to know users' and contexts' diversity. However, it provides a clear indication, through case review, of the complexity of the issues involved in requirements elicitation in universal access, and raises the need for further work in this area to systematically address the subject, thus resulting in additional guidelines and more concrete solutions.

References

Alreck, P. L. and Settle, R. B. (1995). *The Survey Research Handbook: Guidelines and Strategies for Conducting a Survey* (2nd ed.). Burr Ridge, IL: Irwin.

Andreotti, L. and Costa, X. (eds.) (1996). *Situationists: Art, Politics, Urbanism.* Barcelona: Museo d'Art Contemporani de Barcelona.

Andrews, D., Nonnecke, B., and Preece, J. (2003). Electronic survey methodology: A case study in reaching hard-to-involve Internet users. *International Journal of Human-Computer Interaction* 16: 185–210.

Antona, M., Burzagli, L., Emiliani, P.-L., and Stephanidis, C. (2007). The ISTAG scenarios: a case study, in *Towards an Inclusive Future: Impact and Wider Potential of Information and Communication Technologies, Section 4.1 of Chapter 4 "Ambient Intelligence and Implications for People with Disabilities"* (P. R. W. Roe, ed.), pp. 158–187. Brussels: COST219ter.

Astell, A., Alm, N., Gowans, G., Ellis, M., Dye, R., and Vaughan, P. (2008). Involving older people with dementia and their carers in designing computer based support systems: Some methodological considerations. *Universal Access in the Information Society.* http://www.springerlink.com/content/x100235u06860736.

Beaudouin-Lafon, M. and Mackay, W. E. (2002). Prototyping Development and Tools, in *The Human-Computer Interaction Handbook: Fundamentals, Evolving Technologies and Emerging Applications Book Contents* (J. A. Jacko and A. Sears, eds.), pp. 1006–1031. Mahwah, NJ: Lawrence Erlbaum Associates.

Bekker, M., Beusmans, J., Keyson, D., and Lloyd, P. (2003). KidReporter: A user requirements gathering technique for designing with children. *Interacting with Computers* 15: 187–202.

Beyer, H. and Holtzblatt, K. (1997). *Contextual Design.* San Francisco: Morgan Kaufmann.

Blomberg, J., Burrell, M., and Guest, G. (2002). An ethnographic approach to design, in *The Human-Computer Interaction Handbook: Fundamentals, Evolving Technologies and Emerging Applications* (J. A. Jacko and A. Sears, eds.), pp. 964–986. Mahwah, NJ: Lawrence Erlbaum Associates.

Blythe, M. and Dearden, A. (2008). Representing older people: Towards meaningful images of the user in design scenarios. *Universal Access in the Information Society.* http://www.springerlink.com/content/y301q8g027g0l76v/fulltext.pdf.

Bodker, S., Ehn, P., Knudsen, J. L., Kyng, M., and Madsen, K. H. (1988). Computer support for cooperative design, in the *Proceedings of the Second Conference on Computer-Supported Cooperative Work (CSCW '88)* (I. Grief and L. Suchman, eds.), pp. 377–394. New York: ACM Press.

Bruseberg, A. and McDonagh-Philp, D. (2001). New product development by eliciting user experience and aspirations. *International Journal of Human Computer Studies* 55: 435–452.

Cardoso, C. and Clarkson, P. J. (2006). Impairing designers: Using calibrated physical restrainers to empathize with users, in the *Proceedings of the 2nd International Conference for Universal Design in Kyoto,* 22–26 October 2006, Kyoto, Japan. Kyoto, Japan: International Association for Universal Design.

Carroll, J. M. (1995). *Scenario-Based Design: Envisioning Work and Technology in System Development.* Chichester: Wiley.

Carroll, J. M. (2000). *Making Use Scenario-Based Design of Human Computer Interactions.* London: MIT Press.

Chandrashekar, S., Stockman, T., Fels, D. I., and Benedyk, R. (2006). Using think aloud protocol with blind users: A case for inclusive usability evaluation methods, in the *Proceedings of the 8th International ACM SIGACCESS Conference on Computers and Accessibility (ASSETS 2006),* 23–25 October 2006, Portland, Oregon, pp. 251–252. New York: ACM Press.

Cooper, A. (1999). *The Inmates Are Running the Asylum: Why High Tech Products Drive Us Crazy and How to Restore the Sanity.* Indianapolis: Sams Publishing.

Crabtree, A., Hemmings, T., Rodden, T., Cheverst, K., Clarke, K., Dewsbury, G., et al. (2003). Designing with care: Adapting cultural probes to inform design in sensitive settings, in the *Proceedings of OzCHI 2003, New Directions in Interaction: Information Environments, Media & Technology,* 26–28 November 2003, Brisbane, Australia, pp. 4–13. Brisbane, Australia: The University of Queensland.

Crichton, S. and Kinash, K. (2003). Virtual ethnography: Interactive interviewing online as method. *Canadian Journal of Learning and Technology* 29: 101–115. http://www.cjlt.ca/content/vol29.2/cjlt29-2_art-5.html.

Davies, R., Marcella, S., McGrenere, J., and Purves, B. (2004). The ethnographically informed participatory design of a PD application to support communication, in the *Proceedings of the 6th International ACM SIGACCESS Conference on Computers and Accessibility Table of Contents,* 18–20 October 2004, Atlanta, pp. 153–160. New York: ACM Press.

Davison, R. (2000). The role of groupware in requirements specification. *Group Decision and Negotiation* 9: 149–160.

Demirbileka, O. and Demirkan, H. (2004). Universal product design involving elderly users: A participatory design model. *Applied Ergonomics* 35: 361–370.

Dickinson, A., Eisma, R., Syme, A., and Gregor, P. (2002). UTOPIA: Usable technology for older people: Inclusive and appropriate, in *A New Research Agenda for Older Adults, Proceedings of BCS HCI* (S. Brewster and M. Zajicek, eds.), 2–6 September 2002, London, pp. 38–39. London: BCS Press.

Dindler, C., Eriksson, E., Iversen, O. E., Lykke-Olesen, A., and Ludvigsen, M. (2005). Mission from Mars: A method for exploring user requirements for children in a narrative space, in the *Proceedings of the 2005 Conference on Interaction Design and Children (IDC 2005),* 2–10 June, Boulder, CO, pp. 40–47. New York: ACM Press.

Druin, A. (1999a). Cooperative inquiry: Developing new technologies for children with children, in the *Proceedings of CHI '99,* 15–20 May 1999, Pittsburgh, pp. 592–599. New York: ACM Press.

Druin, A. (ed.) (1999b). *The Design of Children's Technology.* San Francisco: Morgan Kaufmann.

Druin, A. (2002). The role of children in the design of new technology. *Behaviour and Information Technology* 21: 1–25.

Dumas, J. S. and Redish, J. C. (1993). *A Practical Guide to Usability Testing.* Westport, CT: Greenwood Publishing Group.

Eisma, R., Dickinson, A., Goodman, J., Syme, A., Tiwari, L., and Newell, A. (2004). Early user involvement in the development of Information Technology-related products for older people. *Universal Access in the Information Society* 3: 131–140.

Engelbrektsson, P., Karlsson, I. C. M., Gallagher, B., Hunter, H., Petrie, H., and O'Neill, A-M. (2004). Developing a navigation aid for the frail and visually impaired. *Universal Access in the Information Society* 3: 194–201.

Fischer, G. and Sullivan, J. (2002). Human-centered public transportation systems for persons with cognitive disabilities: Challenges and insights for participatory design. Paper presented at the *Participatory Design Conference (PDC '02),* 23–25 June 2002, Malmö University, Sweden. http://www.cs.colorado.edu/~l3d/clever/assets/pdf/gf-pdc2002-mfa.pdf.

Francis, P., Balbo, S., and Firth, L. (2009). Towards co-design with users who have autism spectrum disorders. *Universal Access in the Information Society.*

Fulton, S. J., Battarbee, K., and Koskinen, I. (2005). Designing in the dark: Empathic exercises to inspire design for our nonvisual senses, in the *Proceedings of the International Conference on Inclusive Design 2005,* 5–8 April 2005, London. http://hhrc.rca.ac.uk/archive/hhrc/programmes/include/2005/proceedings/pdf/fultonsurijane.pdf.

Gaver, B., Dunne, T., and Pacenti, E. (1999). Design: Cultural probes. *Interactions* 6: 21–29.

Goodman, J., Clarkson, J., and Langdon, P. (2006). Providing information about older and disabled users to designers, in the *Workshop on HCI, the Web and the Older Population, in the context of British HCI 2006,* 12 September 2006, London. http://www-edc.eng.cam.ac.uk/~jag76/hci_workshop06/goodman_et_al.pdf.

Goodman, J., Langdon, P., Clarkson, J., Caldwell, N. H. M., and Sarhan, A. M. (2007). Equipping designers by simulating the effects of visual and hearing impairments, in the *Proceedings of the 9th International ACM SIGACCESS Conference on Computers and Accessibility (ASSETS 2007),* 14–17 October 2007, Tempe, Arizona, pp. 241–242. New York: ACM Press.

Greenbaum, T. L. (1998). *The Handbook for Focus Group Research* (2nd ed.). London: Sage.

Gubrium, J. F. and Holstein, J. A. (eds.) (2002). *Handbook of Interview Research: Context and Method.* Thousand Oaks, CA: Sage.

Hall, L. and Mallalieu, G. (2003). Identifying the needs and expectations of users with learning disabilities, in *Universal Access in HCI: Inclusive Design in the Information Society, Volume 4 of the Proceedings of the 10th International Conference on Human-Computer Interaction (HCI International 2003)* (C. Stephanidis, ed.), 22–27 June 2003, Crete, Greece, pp. 837–841. Mahwah, NJ: Lawrence Erlbaum Associates.

Hart, J. (1998). *The Art of the Storyboard: Storyboarding for Film, TV, and Animation.* Burlington, MA: Focal Press.

Heller, S. (1998). The meaning of children in culture becomes a focal point for scholars. *The Chronicle of Higher Education* A14–A16.

Kitzinger, J. (1995). Qualitative research: Introducing focus groups. *British Medical Journal* 311: 299–302.

Kurniawan, S. (2006). An exploratory study of how older women use mobile phones, in the *Proceedings of the 8th International Conference on Ubiquitous Computing (UbiComp 2006),* 17–21 September 2006, Orange County, CA, pp. 105–122. Berlin/Heidelberg: Springer-Verlag.

Lazar, J., Allen, A., Kleinman, J., and Malarkey, C. (2007). What frustrates screen reader users on the Web: A study of 100 blind. *International Journal of Human-Computer Interaction* 22: 247–269.

Macaulay, L. A. (1996). *Requirements Engineering.* New York: Springer-Verlag.

Mackay, W. E., Ratzer, A. V., and Janecek, P. (2000). Video artifacts for design: Bridging the gap between abstraction and detail, in the *Proceedings of the 3rd Conference on Designing Interactive Systems: Processes, Practices, Methods, and Techniques,* 17–19 August 2000, New York, pp. 72–82. New York: ACM Press.

Maguire, M. and Bevan, N. (2002). User requirements analysis: A review of supporting methods, in the *Proceedings of IFIP 17th World Computer Congress on Usability: Gaining a Competitive Edge,* 25–30 August 2002, Montreal, Canada, pp. 133–148. Norwell, MA: Kluwer.

Mander, R. and Smith, B. (2002). *Web Usability for Dummies*. New York: Hungry Minds.

Margetis, G., Ntoa, S., and Stephanidis, C. (2008). Requirements of users with disabilities for e-government services in Greece, in *Computers Helping People with Special Needs, Proceedings of the 11th International Conference (ICCHP 2008)* (K. Miesenberger, J. Klaus, W. Zagler, and A. Karshmer, eds.), 9–11 July 2008, Austria, pp. 438–445. Berlin/Heidelberg: Springer-Verlag.

Marquis-Faulkes, F., McKenna J. S., Newell, F. A., and Gregor, P. (2005). Gathering the requirements for a fall monitor using drama and video with older people. *Technology and Disability* 17: 227–236.

Moffatt, K., Findlater, L., and Allen, M. (2006). Generalizability in research with cognitively impaired individuals, in *ACM CHI 2006 Workshop on Designing for People with Cognitive Impairments*, 22–27 April 2006, Montreal, Canada. http://www.cs.ubc.ca/~joanna/CHI2006Workshop_Cognitive Technologies/positionPapers/15_CHI2006_workshop_moffatt.pdf.

Moffatt, K., McGrenere, J., Purves, B., and Klawe, M. (2004). The participatory design of a sound and image enhanced daily planner for people with aphasia, in the *Proceedings of the 2004 Conference on Human Factors in Computing Systems*, 24–29 April 2004, Vienna, Austria, pp. 407–414. New York: ACM Press.

Muller, J. M. (2002). Participatory design: The third space in HCI, in *The Human-Computer Interaction Handbook: Fundamentals, Evolving Technologies and Emerging Applications Book Contents* (J. A. Jacko and A. Sears, eds.), pp. 1051–1068. Mahwah, NJ: Lawrence Erlbaum Associates.

Newell, A., Arnott, J., Carmichael, A., and Morgan, M. (2007). Methodologies for involving older adults in the design process, in *Universal Access in HCI: Coping with Diversity, Volume 5 of the Proceedings of the 12th International Conference on Human-Computer Interaction (HCI International 2007)* (C. Stephanidis, ed.), 22–27 July 2007, Beijing, pp. 982–989. Berlin/Heidelberg: Springer-Verlag.

Newell, A. F., Carmichael, A., Gregor, P., and Alm, N. (2002). Information technology for cognitive support, in *The Human-Computer Interaction Handbook: Fundamentals, Evolving Technologies and Emerging Applications* (J. A. Jacko and A. Sears, eds.), pp. 464–481. Mahwah, NJ: Lawrence Erlbaum Associates.

Nicolle, C. and Maguire, M. (2003). Empathic modelling in teaching design for all, in *Universal Access in HCI: Inclusive Design in the Information Society, Volume 4 of the Proceedings of the 10th International Conference on Human-Computer Interaction (HCI International 2003)* (C. Stephanidis, ed.), 22–27 June 2003, Crete, Greece, pp. 143–147. Mahwah, NJ: Lawrence Erlbaum Associates.

Norman, D. A. and Draper, S. W. (1986). *User Centered System Design: New Perspectives on Human-Computer Interaction*. Mahwah, NJ: Lawrence Erlbaum Associates.

Oppenheim, B. (2000). *Questionnaire Design*. New York: Continuum International Publishing Group.

Osborn, A. F. (1963). *Applied Imagination: Principles and Procedures of Creative Problem-Solving* (3rd rev. ed.). New York: Charles Scribner's Sons.

Pastalan, L. A. (1982). Environmental design and adaptation to the visual environment of the elderly, in *Aging and Human Visual Function* (R. Sekuler, D. Kline, and K. Dismukes, eds.), pp. 323–333. New York: Alan R. Liss.

Petrie, H., Johnson, V., Furner, S., and Strothotte, T. (1998). Design lifecycles and wearable computers for users with disabilities, in the *Proceedings of the First Workshop on Human Computer Interaction with Mobile Devices*, 21–23 May 1998, Glasgow, Scotland. GIST Technical Report 698-1.

Poulson, D., Ashby, M., and Richardson, S. (eds.) (1996). *USERfit: A Practical Handbook on User-Centred Design for Assistive Technology*. Brussels: ECSC–EC–EAEC.

Rideout, J. V., Vandewater, A. E., and Wartella, A. E. (2003). *Zero to Six Electronic Media in the Lives of Infants, Toddlers, and Preschoolers*. A Kaiser Family Foundation Report.

Roberts, V. L. and Fels, D. I. (2006). Methods for inclusion: Employing think aloud protocols in software usability studies with individuals who are deaf. *International Journal of Man-Machine Studies* 64: 489–501.

Schein, J. (1981). Hearing impairments and deafness, in *Handbook of Severe Disability: A Text for Rehabilitation Counsellors, Other Vocational Practitioners and Allied Health Professionals* (W. C. Stolov and M. R. Clowers, eds.), pp. 395–407. Washington, DC: United States Government Printing.

Shinohara, K. (2006). Designing assistive technology for blind users, in the *Proceedings of the 8th International ACM SIGACCESS Conference on Computers and Accessibility (Assets '06)*, 23–25 October 2006, Portland, Oregon, pp. 293–294. New York: ACM Press.

Smith-Jackson, T., Nussbaum, M., and Mooney, A. (2003). Accessible cell phone design: Development and application of a needs analysis framework. *Disability & Rehabilitation* 25: 549–560.

Stephanidis, C., Antona, M., and Grammenos, D. (2007). Universal access issues in an ambient intelligence research facility, in *Universal Access in Human-Computer Interaction: Ambient Interaction, Volume 6 of the Proceedings of the 12th International Conference on Human-Computer Interaction (HCI International 2007)* (C. Stephanidis, ed.), 22–27 July 2007, Beijing, pp. 208–217. Berlin/Heidelberg: Springer-Verlag.

Svensk, A. (1997). Empathic modelling (the sober version), in the *Proceedings of the 4th European Conference for the Advancement of Assistive Technology (AAATE'97)*, 29 September–2 October 1997, Thessaloniki, Greece. http://www.certec.lth.se/doc/empathicmodelling.

Truong, K. N., Hayes, G. R., and Abowd, G. D. (2006). Story boarding: An empirical determination of best practices and effective guidelines, in the *Proceedings of the 6th Conference on Designing Interactive Systems (DIS '06)*, 26–28 June 2006, State College, PA, pp. 12–21. New York: ACM Press.

Vanderheiden, G. C. (1997). Design for people with functional limitations resulting from disability, ageing, or circumstance, in *Handbook of Human Factors and Ergonomics* (3rd ed.) (G. Salvendy, ed.), pp. 1395–1397. New York: John Wiley & Sons.

Vickers, P., Banwell, L., Heaford, S., and Sainz de Salces, F. J. (2008). Painting the ideal home: Using art to express visions of technologically supported independent living for older people in north-east England. *Universal Access in the Information Society*. http://www.springerlink.com/content/734181t8208k713q.

Whyte, W. E. (1984). *Learning from the Field: A Guide from Experience*. Newbury Park, CA: Sage.

Wu, M., Richards, B., and Baecker, R. (2004). Participatory design with individuals who have amnesia, in the *Proceedings of the Eighth Conference on Participatory Design: Artful Integration: Interweaving Media, Materials and Practices*, 27–31 July 2004, Toronto, Canada, vol. 1, pp. 214–223. New York: ACM Press.

Zajicek, M. (2006). Aspects of HCI research for older people. *Universal Access in the Information Society* 5: 279–286.

Zimmermann, G. and Vanderheiden, G. (2008). Accessible design and testing in the application development process: Considerations for an integrated approach. *Universal Access in the Information Society* 7: 117–128.

16

Unified Design for User Interface Adaptation

Anthony Savidis and
Constantine Stephanidis

16.1 Introduction

New user-interface design methods become necessary when there is a need to capture particular properties of interactive systems, which cannot be explicitly or sufficiently represented through existing design approaches.[1] For instance, the identification of graphical constraints among interface objects is not usually carried out in the context of task analysis. Hence, if the explicit representation of graphical constraints is necessary (e.g., as an input to the implementation phase), a dedicated design process needs to be carried out for the extractioan of such necessary design information. Also, in situations where a single design approach does not fulfill the requirements of the design process, the combination of alternative design techniques is applied, leading to hybrid design methodologies (e.g., combination of task analysis and graphic design methods).

New engineering methodologies and development tools emerge in cases where existing methods and instruments are not sufficient to effectively address the development challenges for new categories of interfaces. In the context of universal access, the need for revisiting interface engineering methodologies and tools has been identified in the recent past. While various approaches and tools exist today, they mainly fall within the domain of automatically adapted interfaces. Recent engineering work in the field of universal access has identified the genuine need for design spaces (Stephanidis and Savidis, 2001) that can support user interface adaptation, evolution, and extensibility, while marrying together alternative design artifacts even for the same target design context; such design spaces may constitute the ground for advanced interface implementation processes. This chapter reports a design methodology that reflects the overall objective of universal access, and in particular the construction of evolving unified user interface design spaces. Interface engineering requirements (see Chapter 18 of this handbook) are appropriately taken into consideration in the presented design methodology.

16.1.1 The Design Problem

Design for all addresses potentially all users and usage contexts: *anyone, anyplace, anytime*. Its main objective is to ensure that each end-user is provided the most appropriate interactive experience, supporting accessible and high-quality interaction.

---

[1] This chapter is based upon Savidis, A. and Stephanidis, C. (2004). Unified user interface design: Designing universally accessible interactions. *Interacting with Computers* 16: 243–270. Reprinted here with slight modifications by concession of Elsevier.

Clearly, producing and enumerating distinct interface designs through the conduct of multiple design processes is an impractical solution, since the overall cost for managing in parallel such a large number of independent design processes, as well as for transforming each produced interface version to a target software implementation, would have unacceptable costs for both the design and the software implementation phases.

Instead, a design process is required that may lead to a single design outcome that appropriately links and organizes the differentiating aspects of the resulting interactive application around common abstract design structures and patterns, making it far easier to: (1) map to a target software system implementation; and (2) maintain, update, and extend the design itself.

The need for introducing alternative design artifacts for the same specific design context (such as a particular subtask) emerges from the fact that design for all encompasses largely varying user and usage context parameters. Consequently, when designing for any particular dialogue design context, it is likely that differentiating values of such parameters dictate the design of diverse dialogue artifacts. This issue introduces two important requirements for a suitable design method.

The first is that such a method should offer the capability to associate multiple alternative dialogue artifacts to a particular single design context, due to the varying design parameters, by enabling the unambiguous association of each alternative artifact with its corresponding values of the parameters.

The second is that the method should emphasize capturing of the more abstract structures and patterns inherent in the interface design, enabling the hierarchical incremental specialization toward the lower physical level of interaction, and making it possible to introduce alternative dialogue patterns as closely as possible to the physical design. This makes it easier for the design space to be updated and evolve, since modifications and extensions due to the consideration of additional values of the design parameters (e.g., considering new user and usage context attribute values) can be applied locally closer to the lower levels of the design, without affecting the rest of the design space.

To briefly explain the need for supporting alternative dialogue artifacts for the same design context, an example from a real-life application will be used. The AVANTI web browser (Stephanidis et al., 2000) has been developed to enable web access by supporting adaptation to the individual user, as well as to the context of use. During the interface design phase, while concentrating on the design context of the link dialogue task, alternative designs have been dictated, due to the considered user and usage context parameters, as shown in Figure 16.1.

Because the different designed artifacts have been part of the final AVANTI browser interface design, a design representation formalism was needed to enable their copresence within the resulting design space, and to clearly associate each artifact to the link-selection task and its corresponding values of the user and usage context parameters. A loose design notation is employed in Figure 16.1 to show hierarchical task analysis (subtask sequencing is omitted for clarity), as well as the need for alternative incarnations of a single task (e.g., styles S2/S3 for link selection, styles S1/*Se* for load confirmation, and styles S4/S5 for link targeting). During run-time, depending on the particular end-user and usage-context attribute values, the

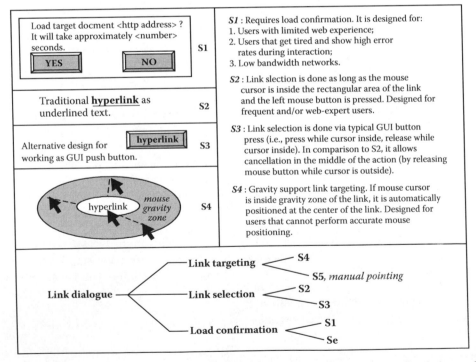

FIGURE 16.1 Designing alternative artifacts for the Link dialogue task. Se is used to indicate an "empty" style (i.e., no load confirmation dialogue supported). S5 is the typical manual link targeting GUI dialogue through the mouse. (From Savidis, A. and Stephanidis, C., *Interacting with Computers*, 16, 243–270, 2004.)

appropriate corresponding implemented artifacts are activated in the AVANTI browser.

16.1.2 The Proposed Solution

The unified interface design method proposes a specific design process to cater to the management of an evolving design space, in which alternative design artifacts can be associated with variations of the design problem parameters. The key elements of the unified user interface design process are:

- Hierarchical design discipline building upon the notion of task analysis, empowered by the introduction of *task level polymorphism*
- Iterative design process model, emphasizing *abstract task analysis* with incremental polymorphic physical specialization
- Formalization *of run-time relationships* among the alternative design artifacts, when associated with the same design context
- Documentation forms recording the *consolidated design rationale* of each alternative design artifact

16.2 Related Work

As it has been previously mentioned, in the context of design for all, given any particular design context within the overall design space, it is critical to enable copresence of multiple alternative design artifacts, each associated with well-defined values of the design parameters. Hence, the designer should be enabled to organize alternative artifacts together as part of the final interface design, by specifying their association both to the particular design context (e.g., a subtask), and to the concrete set of values of the design parameters (i.e., user and usage context attribute values). This constitutes the key required ingredient of any design process that aims to support design for diversity in the context of design for all. Additionally, the design process should provide all the necessary formal instruments for documenting the design rationale, as well as the particular run-time relationships of the various alternative artifacts, for effective design and run-time organization.

This section reviews a wide spectrum of key existing design methods, outlining their focus ranging from the low-level aspects of interaction to higher-level design artifacts, and addressing their adequacy to accommodate design diversity.

16.2.1 Task-Oriented Design Methods

Understanding user tasks is often considered one of the key aspects of the design process. Methods such as hierarchical task decomposition (Johnson et al., 1988), task-action grammars (Payne, 1984), hierarchical task analysis (Kirwan and Ainsworth, 1992), and task-based design (Wilson and Johnson, 1995) are perhaps the most representative examples of task-based design methods. The concepts of task and task model

have evolved into several lines of thought, giving rise to different ways to model tasks and structure the related knowledge (see Diaper and Stanton, 2003 for a more recent account of perspectives, methods, and techniques). Typically, approaches to tasks can vary with respect to what is intended by task, what kind of knowledge is related to tasks, the level of granularity at which tasks are to be described, in which phases of the design and development process tasks are useful, and how tasks should be represented and fed into user interface development (Bomsdorf and Szwillus, 1999).

Despite the common recognition that task modeling is a fundamental component of user interface design, such techniques do not encompass the capability to differentiate and represent design alternatives for the same task, mapping to varying design parameters. Instead, the design outcome reflects only a specific instance of the design parameters, and conveys only a single task-based structure. Consequently, following the design formalism of task-based methods, the differentiation at various levels of the task model would need to be represented via distinct alternative task-hierarchies, that is, alternative designs. Each of those distinct design instances would need to be potentially produced for each different snapshot of values of the design parameters, leading to a large number of alternative designs.

Clearly, such an approach becomes impractical for the design as well as the implementation process, due to the prohibitive cost of maintaining and engineering a large set of alternative design instances. One might observe that all plausible task design instances are likely to exhibit structural similarities, hence it would be expected that appropriate design and implementation procedures could take advantage of such commonalities by deploying reusability techniques, while organizing appropriately the varying artifacts into different levels of abstraction. However, following the definition of those methods, it is clear that there is no way to support task-structure variability and explicit association of differentiated task-patterns with varying design parameters. Moreover, there is no additional support to designers for managing abstract design, other than the inherent modeling capability offered by the hierarchical task model.

16.2.2 Asynchronous Models of User Actions

These techniques are based on the asynchronous nature of user actions, providing models for representing concurrent user-activities, mainly through operators for expressing ordering relationships and progress control. They usually rely upon the theory of reactive systems, while the most representative examples in this category are communicating sequential processes (CSP) (Hoare, 1978), UAN (Hartson et al., 1990) and PPS (Olsen, 1990). These models suffer from two main problems: (1) they are mostly suited for representing temporal action-oriented dialogue properties, rather than constructional (i.e., they cannot represent structural interface relationships); and (2) they are quite low level, by being very close to directly computable models such as an implementation language rather than to a design language, rendering them more appropriate for small design projects. Additionally, the various operators supplied for combining user actions correspond to run-time relationships (e.g., "this" should be done before "that"), while they do not provide means for representing design-oriented relationships (e.g., "this" subdialogue is defined for "those" user attribute values).

16.2.3 Interaction-Object-Oriented Design Methods

Such methods, which are less commonly used, emphasize the modeling of the constructional aspects of the user interface, aiming mainly toward a design model that can be mapped automatically or semiautomatically, via appropriate transformation tools (such as special-purpose compilers), to an implementation form. Design information such as task hierarchies may be defined as additional features within the object model. An example of such a design technique is OBSM (Kneer and Szwillus, 1995), while earlier work (e.g., Desoi et al., 1989; Manheimer, Burnett, and Wallerns, 1989) has mainly concerned specific user interface development tools, in which the object-based design specification is used to automatically generate the software implementation. Object-based methods are targeted toward the modeling of the structural interface properties, leaving no space for either behavioral or structural differentiation when the design parameters dictate diverse design decisions.

16.2.4 Visual/Graphical Design Methods

Such methods are a type of straightforward physical design, in the sense that concrete visual interface snapshots are produced, via an iterative *sketching and refinement* interface prototyping process. The design process is conducted by managing design guidelines, usability criteria, artistic design, and task-oriented knowledge, toward more effective visual communication. Such methods realize an *artifact generation* process, which can be employed at various stages of the interface design process, whenever the design needs to be instantiated into specific physical (e.g., visual) forms. Clearly, this type of technique cannot support the overall organization of the design process, starting from the early stages, when diverse design parameters impose different design decisions, but it is mostly suited for later stages when design artifacts need to be crystallized and finalized.

16.2.5 Scenario-Based Design Methods

Such methods capture key usage scenarios, annotated with design documentation that can be directly communicated to various levels of the design process, for further review and enhancement. Such design documentation constitutes valuable reference material for early prototyping and engineering. Examples of use of scenario-based design are reported in Royer (1995) and Potts (1995). The employment of scenarios is mainly a complementary design technique, and has not been applied on its own to: (1) manage the conduct of the overall design process; and (2) facilitate the construction of the finalized design space as an appropriate organization of documented dialogue patterns.

16.2.6 Design Rationale Methods

Design rationale methods support argumentation about design alternatives and record the various design suggestions, as well as their associated assessments. They are employed for the generation of design spaces that capture and integrate design information (Bellotti, 1993) from multiple sources, such as design discussions and diverse kinds of theoretical analyses. Questions, options, and criteria (QOC; McLean and McKerlie, 1995) design rationale is the most common method for constructing design spaces, capturing argumentation about multiple possible solutions to design problems. QOC can be used to characterize the nature of contributions from diverse approaches, highlighting strengths and weaknesses (Bellotti, 1993).

It is clear that the employment of methods based on design rationale for constructing design spaces leads to a comprehensive collection of alternative solutions (i.e., candidate design decisions), addressing their respective design problems, from which the most appropriate for their purpose is finally chosen (i.e., final design decisions). Hence, possible alternatives are likely to exist during the design process, but are removed from the final design, since only the best choice (for each particular design problem) should be documented as an input to the implementation phase. In comparison, this is fundamentally different from design for all, where alternative dialogue artifacts, representing final design decisions associated with particular user and usage context attribute values need to be concurrently present in the final design space.

16.2.7 Hybrid Design Methods

Hybrid design methods borrow elements mainly from task-based, scenario-based, and object-based design techniques. Usually, one of these basic techniques becomes the centerpiece of the design process, while additional information can be extracted by performing parallel design steps through the deployment of any other method. For example, TADEUS (Stary, 1996) builds upon the task model, providing also organizational and work-flow information. The technique proposed in Kandle (1995) is based on a task-model augmented with semiformalized usage scenarios. In Butler et al. (1997), a mixture of task analysis, process simulation, and object definition methods is proposed to conduct a multilevel design process in which initial workflow design information is transformed into business object component (BOC) definitions, which are then converted to the object modeling technique (OMT), being mostly appropriate for the software design phase. Hybrid models, until now, have emphasized the quick transition to an implementation phase, and have been applied to design projects in which workflow information is a necessary ingredient (i.e., business process engineering and reengineering). The known methods do not show the capability to model diversity of design parameters, and inherent polymorphism of resulting artifacts, because the design outcomes can only reflect a single snapshot of values of the design parameters.

16.3 Fundamentals of Unified User Interface Design

On the grounds of the previous discussion, in the context of a design-for-all process aiming to address multiple values of user and usage-context parameters, the *unified user-interface design method* has been defined so as to address three main objectives:

1. To enable the collection and effective organization of all potentially distinct design alternatives, as they are dictated for their particularly associated dialogue design contexts, into a single unified user interface design space.
2. To produce a design space in which, for each alternative design artifact, its particular recorded design rationale, the run-time relationships with the rest of alternative artifacts within the same design context, as well as the specific associated design parameter values, are represented in a well-defined computable form that can be directly translated by user interface developers into an implementation structure.
3. To support design evolution by enabling the effective extension of alternative dialogue artifacts at different design contexts, as new user and usage-context attribute values need to be addressed. While design for all is practically considered to be an ideal target, when intended to be addressed in a one-shot fashion during development, it becomes an arguably manageable task when tackled through development strategies that support extension and evolution.

Some of the distinctive properties of this method are elaborated in the following by addressing their links with human-computer interaction (HCI) design and by providing an overview of the outcomes and design deliverables in the context of a unified user-interface design process.

16.3.1 General Characteristics

The unified user interface design method extends the traditional design inquiry by focusing explicitly on *polymorphism* as an aid to designing and populating the final design space with alternative required dialogue artifacts, when addressing diverse user and usage contexts. In this context, design polymorphism becomes the vehicle that empowers the design method with the ability to represent in the final design space, in a well-organized way, the design pluralism that is inherent in any type of design process that aims to address a space of diverse problem-parameter values.

In terms of conduct, the method is related to hierarchical task analysis, with the difference that alternative polymorphic decomposition schemes can be employed (at any point of the hierarchical task analysis process), where each decomposition seeks to address different values of the driving design parameters. This approach leads to the notion of *polymorphic task decomposition*, which is characterized by the pluralism of plausible design options consolidated in the resulting task hierarchy. Overall, the method is based on the new concept of polymorphic task decomposition, through which any task (or subtask) may be decomposed in an arbitrary number of alternative subhierarchies. The design process realizes an exhaustive hierarchical decomposition of well-defined task categories, starting from the abstract level, by incrementally specializing in a polymorphic fashion (as different design alternatives are likely to be associated with varying user and usage context attribute values), toward the physical level of interaction. The outcomes of the method include: (1) the design space that is populated by collecting and enumerating design alternatives; (2) the polymorphic task hierarchy that comprises alternative concrete artifacts; and, (3) for each produced design artifact, the recorded design rationale that has led to its introduction, together with run-time relationships with the rest of the design alternatives for the same subtask.

16.3.2 The Design Space

Design alternatives are necessitated by the different design parameters and may be attached to various levels of the overall task hierarchy, offering rich insight into how a particular task or subtask may be accomplished by different users in different contexts of use. Since users differ with regard to their abilities, skills, requirements, and preferences, tentative designs should aim to accommodate the broadest possible range of capabilities across different contexts of use. Thus, instead of restricting the design activity to producing a single outcome, designers should strive to compile design spaces containing plausible alternatives.

16.3.3 Polymorphic Task Hierarchies

A polymorphic task hierarchy combines three fundamental properties: (1) *hierarchical decomposition*; (2) *polymorphism*; and (3) *task operators*. The hierarchical decomposition adopts the original properties of hierarchical task analysis (Johnson et al., 1988) for incremental decomposition of user tasks to lower level actions. The polymorphism property provides the design differentiation capability at any level of the task hierarchy, according to particular user and usage-context attribute values. Finally, task operators, which are based on the powerful CSP language for describing the behavior of reactive systems (Hoare, 1978), enable the expression of dialogue control flow formulae for task accomplishment. Those specific operators, taken from the domain of reactive systems and process synchronization, have been selected due to their appropriateness in expressing temporal relationships of user actions and tasks (see Figure 16.2). However, designers may freely employ additional operators as needed (i.e., the set is not closed), or may choose to document dialogue sequencing and control outside the task structure in natural language, when it engages more comprehensive algorithmic logic (e.g., consider the verbally documented precondition "*if the logged user is a guest, no sign-in is required, else, the access privileges should be checked and the sign-in dialogue is activated before chat*").

Operator	Explanation	Representation
before	*Task sequencing, documenting that task. A must be performed before task B.*	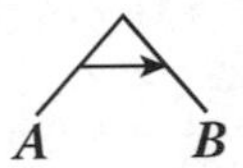 *A B*
or	*Task parallelism, documenting that task. A may be performed before, after, or in parallel to task B.*	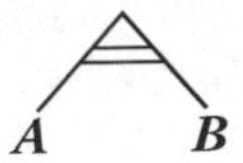 *A B*
xor	*Task exclusive completion, documenting that either A or B must be performed, but not both.*	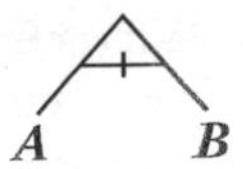 *A B*
*	*Task simple repetition, documenting that A may be performed zero or more times.*	*A**
+	*Task absolute repetition, documenting that A must be performed at least one time.*	*A+*

FIGURE 16.2 Basic task operators in the unified user interface design method. (From Savidis, A. and Stephanidis, C., *Interacting with Computers*, 16, 243–270, 2004.)

The concept of polymorphic task hierarchies is illustrated in Figure 16.3. Each alternative task decomposition is called a decomposition style, or simply a *style*, and is to be given by designers an appropriate descriptive name. Alternative task subhierarchies are attached to their respective styles. The example polymorphic task hierarchy of Figure 16.3 indicates the way two alternative dialogue styles for a Delete File task can be designed, one exhibiting direct manipulation properties with object-function syntax (i.e., the file object is selected prior to operation to be applied) with no confirmation, the other realizing modal dialogue with a function-object syntax (i.e., the delete function is selected, followed by the identification of the target file) and confirmation.

Additionally, the example demonstrates the case of physical specialization. Since selection is an abstract task, it is possible to design alternative ways for physically instantiating the selection dialogue (see Figure 16.3, lower part): via scanning techniques for motor-impaired users, via 3D hand-pointing on 3D auditory cues for blind people, via enclosing areas (e.g., irregular "rubber banding") for sighted users, and via Braille output and keyboard input for deaf-blind users. The unified user-interface design method does not require the designer to follow the polymorphic task decomposition all the way down the user-task hierarchy, until primitive actions are met. A nonpolymorphic task can be specialized at any level, following any design method chosen by the interface designer. For instance, in Figure 16.3 (lower part) graphical mockups are employed to describe each of the alternative physical instantiations of the abstract selection task. It should be noted that the interface designer is not constrained to using a particular model, such as CSP operators, for describing user actions for device-level interaction (e.g., drawing, drag-and-drop, concurrent input). Instead, an alternative may be preferred, such as an event-based representation, for example, ERL (Hill, 1986) or UAN (Hartson and Hix, 1989).

As discussed in more detail in the subsequent sections, design polymorphism entails a decision-making capability for context-sensitive selection among alternative artifacts, so as to assemble a suitable interface instance, while task operators support temporal relationships and access restrictions applied to the interactive facilities of a particular interface instance.

16.3.4 Adaptation-Oriented Design Rationale

When a particular task is subject to polymorphism, alternative subhierarchies are designed, each being associated with different user and usage context parameter values. A running interface, implementing such alternative artifacts, should encompass decision-making capability, so that, before initiating interaction with a particular end-user, the most appropriate of those artifacts are activated for all polymorphic tasks. Hence, polymorphism can be seen as a technique potentially increasing the number of alternative interface instances represented by a typical hierarchical task model. In this sense, the polymorphic task model theoretically represents the power set of all plausible design instantiations.

If polymorphism is not applied, a task model represents a singular interface design (i.e., a design instance), on which further run-time adaptation is restricted; in other words, *there is a fundamental link between adaptation capability and polymorphism on design artifacts*. This issue will be further clarified with the use of an example. Consider the case where the design process reveals the necessity of having multiple alternative subdialogues available concurrently to the user for performing a particular task. This scenario is related to the notion of multimodality,

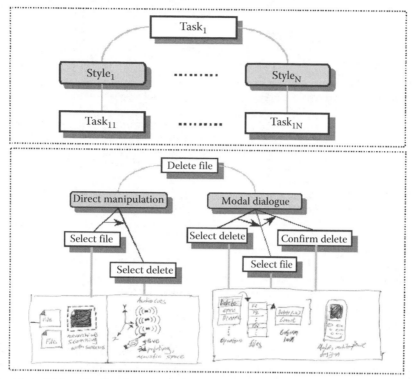

FIGURE 16.3 The polymorphic task hierarchy concept, where alternative decomposition "styles" are supported (top part), and an exemplary polymorphic decomposition, which includes physical design annotation (bottom part). (From Savidis, A. and Stephanidis, C., *Interacting with Computers*, 16, 243–270, 2004.)

which can be more specifically called *task-level multimodality*, in analogy to the notion of *multimodal input,* which emphasizes pluralism at the input-device level. The example will consider the physical design artifact of Figure 16.4, which depicts two alternative dialogue patterns for file management: one providing direct manipulation facilities, and another employing command-based dialogue. Both artifacts can be represented as part of the task-based design, in two ways:

1. Through polymorphic decomposition (see Figure 16.5, upper part), where each of the two dialogue artifacts is defined as a distinct alternative style; the two resulting styles are defined as being *compatible,* which implies that they may coexist at run-time (i.e., the end-user may freely use the command line or the interactive file manager interchangeably).

2. Via unimorphic decomposition (see Figure 16.5, lower part), where the two artifacts are defined to be concurrently available to the user, *within the same interface instance,* via the *or* operator; in this case, the interface design is hard-coded, representing a single interface instance, without needing further decision making.

The advantages of the polymorphic approach are: (1) it is possible to make only one of the two artifacts available to the user, depending on user parameters; (2) even if, initially, both artifacts are provided to end-users, when a particular preference is dynamically detected for one of those, the alternative artifact can

be dynamically disabled; and (3) if more alternative artifacts are designed for the same task, the polymorphic design is directly extensible, while the decomposition-based design would need to be turned into a polymorphic one (except for the unlikely case where it is still desirable to provide all defined subdialogues concurrently to the user).

From the implementation point of view, the polymorphic decomposition requires the explicit representation of the associated design rationale in a directly computable form, such as logic expressions, preconditions, and decision algorithms. Additionally, particular emphasis is put in the translation of the design rationale to a run-time decision-making kernel for

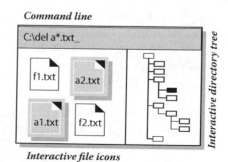

FIGURE 16.4 A design scenario for alternative concurrent subdialogues to perform a single task (i.e., task multimodality for file management). (From Savidis, A. and Stephanidis, C., *Interacting with Computers*, 16, 243–270, 2004.)

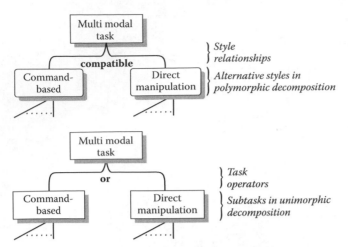

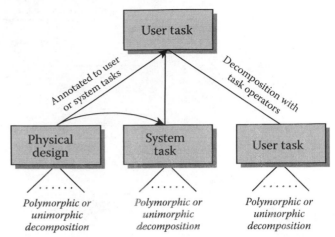

FIGURE 16.5 Two ways for representing design alternatives when designing for task-level multimodality: (1) via polymorphism, adding run-time control on pattern activation (top version); and (2) via the operator, hard-coding the two alternatives in a single-task implementation (bottom version). (From Savidis, A. and Stephanidis, C., *Interacting with Computers*, 16, 243–270, 2004.)

FIGURE 16.6 The three artifact categories in the unified user interface design method, for which polymorphism may be applied, and how they relate to each other. (From Savidis, A. and Stephanidis, C., *Interacting with Computers*, 16, 243–270, 2004.)

conditional dialogue component activation (see Chapters 21 and 22 of this handbook).

16.4 The Conduct of Unified User Interface Design

This section provides a consolidated account of how the unified interface design method can be practiced.

16.4.1 Categories of Polymorphic Artifacts

In the unified user interface design method there are three categories of design artifacts, all of which are subject to polymorphism on the basis of varying user and usage context parameter values. These three categories are (see Figure 16.6):

1. *User tasks*, relating to what the user has to do; user tasks are the center of the polymorphic task decomposition process.
2. *System tasks*, representing what the system has to do, or how it responds to particular user actions (e.g., feedback); in the polymorphic task decomposition process, system tasks are treated in the same manner as user tasks.
3. *Physical design*, which concerns the various physical interface components on which user actions corresponding to the associated user task are to be performed; the physical structure may also be subject to polymorphism.

System tasks and user tasks may be freely combined within task formulas, defining how sequences of user-initiated actions and system-driven actions interrelate. The physical design, providing the interaction context, is always associated with a particular user or system task. It provides the physical dialogue pattern associated with a task-structure definition. Hence, it plays the role of annotating the task hierarchy with physical design information. An example of such annotation is shown in Figure 16.2, where the physical designs for the Select Delete task are explicitly depicted.

In some cases, given a particular user task, there is a need for differentiated physical interaction contexts, depending on user and usage context parameter values. Hence, even though the task decomposition is not affected (i.e., the same user actions are to be performed), the physical design may have to be altered. One such representative example is relevant to changing particular graphical attributes on the basis of ethnographic user attributes. For instance, Marcus (1996) discusses the choice of different iconic representations, background patterns, visual message structure, and so on, on the basis of cultural background (see also Chapter 9, "International and Intercultural User Interfaces").

However, there are also cases in which the alternative physical designs are dictated due to alternative task structures (i.e., polymorphic tasks). In such situations, each alternative physical design is directly attached to its respective alternative *style* (i.e., subhierarchy).

In summary, the rule for identifying polymorphism for physical design artifacts is:

- If alternative physical designs are assigned to the same task, then attach a polymorphic physical design artifact to this task; the various alternative designs depict the styles of this polymorphic physical artifact (see Figure 16.7A).
- If alternative designs are needed due to alternative task structures (i.e., task-level polymorphism), then each alternative physical design should be attached to its respective style (see Figure 16.7B).

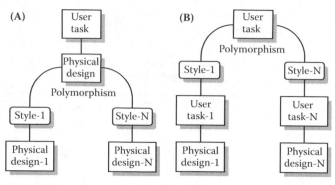

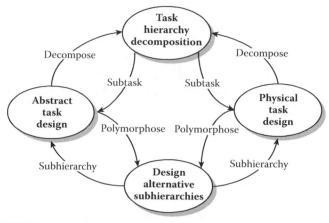

FIGURE 16.7 Representation of alternative physical artifacts: (1) in the case of the same nonpolymorphic task and (2) in the case where polymorphic task decomposition is needed. (From Savidis, A. and Stephanidis, C., *Interacting with Computers*, 16, 243–270, 2004.)

16.4.2 Design Steps in Polymorphic Task Decomposition

User tasks, and in certain cases, system tasks, need not always be related to physical interaction, but may represent *abstraction* on either user or system actions. For instance, if the user has to perform a selection task, then, clearly, the physical means of performing such a task are not explicitly defined, unless the dialogue steps to perform selection are further decomposed. This notion of continuous refinement and hierarchical analysis, starting from higher-level abstract artifacts, and incrementally specializing toward the physical level of interaction, is fundamental in the context of hierarchical behavior analysis, either regarding tasks that humans have to perform (Johnson et al., 1988), or when it concerns functional system design (Saldarini, 1989). At the core of the unified user interface design method is the *polymorphic* task decomposition process, which follows the methodology of abstract task definition and incremental specialization, where tasks may be hierarchically analyzed through various alternative schemes.

In such a recursive process, involving tasks ranging from the abstract task level to specific physical actions, decomposition is applied either in a traditional *unimorphic* fashion, or by means of alternative styles. The overall process is illustrated in Figure 16.8; the decomposition starts from abstract or physical task design, depending on whether top-level user tasks can be defined as being abstract or not. The next subsection provides a description of the various transitions (i.e., design specialization steps) from each of the four states illustrated in the process state diagram of Figure 16.8.

16.4.2.1 Transitions from the Abstract Task Design State

An abstract task can be decomposed either in a polymorphic fashion, if user and usage context attribute values pose the necessity for alternative dialogue patterns, or in a traditional manner, following a unimorphic decomposition scheme. In the case of a unimorphic decomposition scheme, the transition is realized via a *decomposition* action, leading to the *task hierarchy*

FIGURE 16.8 The polymorphic task decomposition process in the unified user interface design method. (From Savidis, A. and Stephanidis, C., *Interacting with Computers*, 16, 243–270, 2004.)

decomposition state. In the case of a polymorphic decomposition, the transition is realized via a *polymorphose* action, leading to the *alternative subhierarchies design* state.

16.4.2.2 Transitions from the Alternative Subhierarchies Design State

Reaching this state means that the required alternative dialogue styles have been identified, each initiating a distinct subhierarchy decomposition process. Hence, each such subhierarchy initiates its own instance of polymorphic task decomposition process. While initiating each distinct process, the designer may either start from the *abstract task design* state, or from the *physical task design* state. The former is pursued if the top-level task of the particular subhierarchy is an abstract one. In contrast, the latter option is relevant in case the top-level task explicitly engages physical interaction issues.

16.4.2.3 Transitions from the Task Hierarchy Decomposition State

From this state, the subtasks identified need to be further decomposed. For each subtask at the abstract level, there is a *subtask* transition to the *abstract task design* state. Otherwise, if the subtask explicitly engages physical interaction means, a *subtask* transition is taken to the *physical task design* state.

16.4.2.4 Transitions from the Physical Task Design State

Physical tasks may be further decomposed either in a unimorphic fashion, or in a polymorphic fashion. These two alternative design possibilities are indicated by the *decompose* and *polymorphose* transitions, respectively.

16.4.3 Designing Alternative Styles

The polymorphic task model provides the design structure for organizing the various alternative dialogue patterns into a unified form. Such a hierarchical structure realizes the fusion of all potential distinct designs, which may be explicitly enumerated given the relevant design parameters. Apart from the polymorphic organization model, the following primary issues need to be also addressed: (1) when polymorphism should be applied; (2) which are the user and usage context attributes that need to be considered; (3) which are the run-time relationships among alternative styles; and (4) how the adaptation-oriented design rationale, connecting the designed styles with particular user and usage context attribute values, is documented.

16.4.3.1 Identifying Levels of Potential Polymorphism

In the context of the unified user interface design method, and as part of the polymorphic task decomposition process, designers should always assert that every decomposition step (i.e., those realized either via the *polymorphose* or through the *decompose* transitions of Figure 16.8) satisfies all constraints imposed by the combination of target user and usage context attribute values. These two classes of parameters will be referred to collectively as *decision parameters/attributes*. During the design process, there may be situations where a particular task decomposition (for user or system tasks, as well as for physical design) does not address some combination(s) of the decision attribute values (i.e., those values are excluded), and so the design for this task is considered incomplete. Such an issue can be resolved during a later design iteration, by constructing the necessary alternative subhierarchy (-ies) for the subject task, appropriately addressing the excluded decision attribute values.

16.4.3.2 Constructing the Space of Decision Parameters

This section discusses the definition of user attributes, which are of primary importance, while the construction of context attributes (e.g., environment and computing-platform parameters) may follow the same representation approach. In the unified user interface design method, end-user representations may be developed using any suitable formalism encapsulating user characteristics in terms of attribute-value pairs. There is no predefined set of attribute categories. Some examples of attribute classes are: general computer use expertise, domain-specific knowledge, role in an organizational context, motor abilities, sensory abilities, mental abilities, and so on.

The value domains for each attribute class are chosen as part of the design process (e.g., by interface designers, or human factors experts), while the value sets need not be finite. The broader the set of values, the higher the differentiation capability among individual end-users. The unified user interface design method does not pose any restrictions as to the attribute categories considered relevant, or the target value domains of such attributes. Instead, it seeks to *provide only the framework in which the role of user and usage context attributes constitute an explicit part of the design process*. It is the responsibility of interface designers to choose appropriate attributes and corresponding value ranges, as well as to define appropriate design alternatives. A simple example of an individual user profile complying with the attribute/value scheme is shown in Figure 16.9. For simplicity, designers may choose to elicit only those attributes from which differentiated design decisions are likely to emerge.

16.4.3.3 Relationships among Alternative Styles

The need for alternative styles emerges during the design process, when it is identified that some particular user and/or usage-context attribute value is not addressed by the dialogue artifacts that have already been designed. Starting from this observation, one could argue that "all alternative styles, for a particular polymorphic artifact, are mutually exclusive to each other" (in this context, exclusion means that, at run-time, only one of those styles may be "active").

However, there exist cases in which it is meaningful to make artifacts belonging to alternative styles concurrently available in a single adapted-interface instance. For example, Figure 16.4 presents a case where two alternative artifacts for file management tasks, a direct-manipulation one and a command-based one, can both be present at run-time. In the unified user-interface design method, four design relationships between alternative styles are distinguished (see Figure 16.10), defining whether alternative styles may be concurrently present at run-time. These four fundamental relationships reflect pragmatic, real-world design scenarios.

Computer knowledge	expert	**frequent**	average	casual	naive
Web knowledge	very good	good	average	**some**	limited	none
Ability to use left hand	perfect	good	some	limited	**none**
Age	30				
Gender	male		**female**		

FIGURE 16.9   An example of a user profile, as a collection of values from the value domains of user attributes, extended from Savidis et al. (1997). (From Savidis, A. and Stephanidis, C., *Interacting with Computers*, 16, 243–270, 2004.)

Exclusion	*Relates many styles.* Only one from the alternative styles may be present.
Compatibility	*Relates many styles.* Any of the alternative styles may be present.
Substitution	*Relates two groups of styles together.* When the second is made "active" at run-time, the first should be "deactivated."
Augmentation	*Relates one style with a group of styles.* On the presence of any style from the group at run-time, the single style may be also "activated."

FIGURE 16.10 Design relationships among alternative styles and their run-time interpretation. (From Savidis, A. and Stephanidis, C., *Interacting with Computers*, 16, 243–270, 2004.)

16.4.3.4 Exclusion

The exclusion relationship is applied when the various alternative styles are deemed to be usable only within the space of their target user and usage context attribute values. For instance, assume that two alternative artifacts for a particular subtask are being designed, aiming to address the user expertise attribute: one targeted to users qualified as novice, and the other targeted to expert users. Then, these two are defined to be mutually exclusive to each other, since it is probably meaningless to concurrently activate both dialogue patterns. For example, at run-time a novice user might be offered a functionally simple alternative of a task, where an expert user would be provided with additional functionality and greater freedom in selecting different ways to accomplish the same task.

16.4.3.5 Compatibility

Compatibility is useful among alternative styles for which the concurrent presence during interaction allows the user to perform certain actions in alternative ways, without introducing usability problems. The most important application of compatibility is in *task multimodality*, as it has been previously discussed (see Figure 16.5 where the design artifact provides two alternative styles for interactive file management).

16.4.3.6 Substitution

Substitution has a very strong connection with adaptivity techniques. It is applied in cases where, during interaction, it is decided that some dialogue patterns need to be substituted by others. For instance, the ordering and the arrangement of certain operations may change on the basis of monitoring data collected during interaction, through which information such as frequency of use and repeated usage patterns can be extracted. Hence, particular physical design styles would need to be cancelled, while appropriate alternatives would need to be activated. This sequence of actions—cancellation followed by activation—is the realization of substitution. Thus, in the general case, substitution involves two groups of styles: some styles are cancelled and substituted by other styles that are activated afterwards.

16.4.3.7 Augmentation

Augmentation aims to enhance the interaction with a particular style that is found to be valid, but not sufficient to facilitate the user's task. To illustrate this point, let us assume that, during interaction, the user interface detects that the user is unable to perform a certain task. This would trigger an adaptation (in the form of adaptive action) aiming to provide task-sensitive guidance to the user. Such an action should not aim to invalidate the active style (by means of style substitution), but rather to augment the user's capability to accomplish the task more effectively, by providing informative feedback. Such feedback can be realized through a separate, but compatible, style. It follows, therefore, that the augmentation relationship can be assigned to two styles when one can be used to enhance the interaction while the other is active. Thus, for instance, the adaptive prompting dialogue pattern, which provides task-oriented help, may be related via an augmentation relationship with all alternative styles (of a specific task), provided that it is compatible with them.

16.4.4 Recording Design Documentation in Polymorphic Decomposition

During the polymorphic task decomposition process, there is a set of design parameters that need to be explicitly defined (i.e., given specific values) for each alternative subhierarchy defined. The aim is to capture, for each *subtask*, the design logic for deciding possible alternative styles, by directly associating user, usage context parameters, and design goals with the constructed artifacts (i.e., styles). These parameters are: (1) *users and usage contexts* (specific user and usage context attribute values addressed by a style); (2) *targets* (concrete design goals for a particular style); (3) *properties* (the specific differentiating/distinctive interaction properties of a style, potentially in comparison to other styles); and (4) *relationships* (how is a style related with other alternative styles). The values of these parameters are recorded during the decomposition process for each style in the form of a table. In Figure 16.11, an example is shown for

Task: Delete file	
Style: Direct manipulation	**Style:** Modal dialogue
Users & contexts: Expert, frequent, average	**Users & contexts:** Casual, naïve.
Targets: Speed, naturalness, flexibility	**Targets:** Safety, guided steps
Properties: Object first, function next	**Properties:** Function first, object next
Relationships: Exclusion (with all)	**Relationships:** Exclusion (with all)

FIGURE 16.11 An example of recording design documentation. (From Savidis, A. and Stephanidis, C., *Interacting with Computers*, 16, 243–270, 2004.)

the definition of these parameters regarding the two alternative styles for Delete File task (see Figure 16.3).

The purpose of design documentation is to capture design rationale associated with the various polymorphic artifacts. In this context, the notion of design rationale has a fundamentally different objective with respect to well-known design space analysis methods. In the latter case, design rationale mainly represents argumentation about design alternatives and assessments (Bellotti, 1993) before reaching final design decisions, while in the case of unified user-interface design, design rationale records the different user and usage-context attributes, as well as design objectives underpinning the already made (i.e., final) design decisions.

The set of four parameters previously defined serves mostly as an indexing method for organizing final design decisions, with primary keys the particular *subtask*, and the *users and usage contexts* parameters. The outcome of the unified user interface design approach is a single hierarchical artifact, composed of user- and usage context-oriented final design decisions, associated with directly computable parameters (i.e., task, user, and usage context attributes). This unified structure is very close to an implementation-oriented software organization model, thus potentially constituting a valuable asset for the implementation phase.

16.5 Scenarios of Unified User Interface Design

This section briefly discusses specific design scenarios that occurred in the application of the unified interface design process in the development of the adaptable and adaptive AVANTI web browser (Stephanidis et al., 2000). Some key examples of polymorphism application in the design of the adaptable and adaptive AVANTI web browser are presented. The target user audience for the AVANTI project was: able-bodied, motor-impaired, and blind users, with differing computer use expertise, supporting use in various physical environments (office, home, public terminals at stations/airports, PDAs, etc.).

16.5.1 Link Dialogue Task

In existing web browsers (e.g., Netscape Communicator, Microsoft Internet Explorer, and SUN HotJava), links in web documents are activated by pressing the left mouse button while the cursor resides within the area of a link. In Figure 16.1, the polymorphic design of the link dialogue (for textual links) has been presented. There are three steps in link selection, according to the new design: (1) targeting to the desirable link (can be done via two styles, S4 and S5); (2) actually selecting a link (which can be done in two alternative ways, S2 and S3); and (3) requiring confirmation for loading target document (also done in two ways, S1 and Se). The Se denotes an empty alternative subhierarchy (i.e., the task is considered to be directly accomplished without any user actions). The design documentation for polymorphism is provided in Figure 16.12 (only a brief summary is presented for clarity), together with the relationships between the designed styles (notice that dynamic style updates need to be explicitly mentioned).

16.5.2 Document Loading Control Task

This task concerns typical operations that browsers provide to enable users to control the web page to be loaded and displayed (e.g., forward/backward, home, reload/stop, bookmarking, load options). In Figure 16.13, two alternative style designs are shown, primarily designed for casual and naive users, which provide only a subset of the full range of operations (more advanced functions are only revealed to expert/frequent/average users). The relationships among the two styles are indicated in Figure 16.14. Also, an adaptive prompting style is designed for the Page Loading Control task, aiming to help the user in performing this task, thus being an augmentation of the previous two alternative styles.

16.5.3 Page Browsing Task

The requirement to provide accessibility of the resulting browsers by motor-impaired users necessitated the design of dialogues for enabling motor-impaired users to explore page

- *S2* mutually exclusive with *S3*
- *S1* mutually exclusive with *Se*
- *S4* compatible with *S5*

- *S1* substitutes *Se* dynamically (if high error rates are detected, or in network bandwidth reduction)

- *Se* substitutes *S1* dynamically (if in a satisfactory interaction history, link activation has been always followed by positive confirmation)

FIGURE 16.12   Relationships among alternative styles of Link Selection task. (From Savidis, A. and Stephanidis, C., *Interacting with Computers*, 16, 243–270, 2004.)

contents and activate links. One globally applied technique has been to support hierarchical scanning of all objects in the interface via binary switches (see Chapter 35 of this handbook). In this case, motor-impaired users can have access on the original visual interface designed for able users, through alternative switch-based input techniques. This approach proved to be very good for all tasks, except for the case of page browsing as users spend a lot of time for switching between the scroll-bar (of the page presentation) and the visible page contents to identify desirable information and links. This has resulted in two alternative styles, illustrated in Figure 16.15, which augment the page presentation style, and are mutually exclusive. In Figure 16.15, style S1, the summary of links is presented in a window on the left of the web page (i.e., all links collected and presented together), while in style S2, the document display on the left is automatically adjusted, so that the highlighted link is always visible. Dashed arrows indicate that document context, including the associated link, is above or below visible portion, while solid arrows are attached at visible links and point to the exact link position in the document visible area; the number displayed above or below the scroll-bar of the "links' summary listbox" indicates how many other

links are included above or below the first or last displayed link within this listbox.

In Figure 16.16, the combination of alternative styles at run-time results in different automatically produced interface versions for the AVANTI browser. Each of the different styles corresponds to specific user and usage context attribute values (i.e., the middle version, with toolbars and link-summary pane removed corresponds to kiosk installations). Finally, in Figure 16.17, augmentation styles to provide accessible alternative interaction for motor-impaired users via switch-based scanning are activated, to enable accessible web browsing. An extended discussion on the unified user interface design and implementation of the AVANTI web browser, as an in-depth case study, may be found in Stephanidis et al. (2000).

16.6 Summary and Conclusions

This chapter has presented the unified user interface design method in terms of primary objective, underlying process, representation, and design outcomes. Unified user interface design is intended to enable the "fusion" of potentially distinct design alternatives, suitable for different user groups and contexts of use,

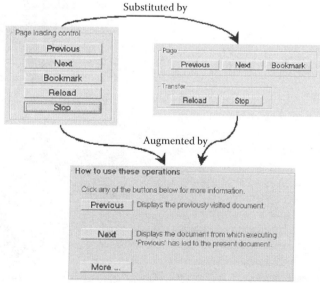

FIGURE 16.13   Designed styles for the Page Loading Control task. (From Savidis, A. and Stephanidis, C., *Interacting with Computers*, 16, 243–270, 2004.)

- The top-left style is the style to be initially active ("casual" or "naïve" values on "application expertise" user attribute).

- The top-right style is to dynamically substitute the previous style, in case that during monitoring it is observed that the user has used the operations successfully and become familiar with them. The new style logically groups operations with a title and prepares the ground for the more advanced group called "options" to be included in future interaction sessions with this user.

- The bottom style augments the particular active style, and it provides adaptive prompting/helping for carrying out the operations in cases where inability to perform the task or high error rates are dynamically detected.

FIGURE 16.14 Relationships between designed styles for Page Loading Control task. (From Savidis, A. and Stephanidis, C., *Interacting with Computers*, 16, 243–270, 2004.)

into a single unified form, as well as to provide a design structure that can be easily translated into a target implementation. By this account, the method is considered to be especially relevant for the design of systems that are required to exhibit adaptable and adaptive behavior, to support individualization to different target user groups and usage contexts. In such interactive applications, the design of alternative dialogue patterns is necessitated due to the varying requirements and characteristics of end-users.

In terms of process, the method postulates polymorphic task decomposition as an iterative engagement through which abstract design patterns become specialized to depict concrete alternatives suitable for the designated situations of use. Through polymorphic task decomposition, the unified user interface design method enables designers to investigate and encapsulate adaptation-oriented interactive behaviors into a single design construction. To this effect, polymorphic task decomposition is a prescriptive guide of what is to be attained, rather than how it is to be attained, and thus it is orthogonal to many existing design instruments.

The outcomes of unified user interface design include the polymorphic task hierarchy and a rich design space that provides the rationale underpinning the context-sensitive selection among design alternatives. A distinctive property of the polymorphic task hierarchy is that it can be mapped into a corresponding set of specifications from which interactive behaviors can be generated. This is an important contribution of the method to HCI design, since it bridges the gap between design and implementation, which has traditionally challenged user-interface engineering.

It is argued that interaction design becomes increasingly a knowledge-intensive endeavor. In this context, designers should be prepared to cope with large design spaces to accommodate design constraints posed by diversity in the target user population and the emerging contexts of use in the information society. To this end, analytical design methods, such as unified user interface design, are anticipated to become necessary tools for capturing and representing the global design context of interactive products and services. Moreover, adaptation is likely to predominate as a technique for addressing the compelling requirements for customization, accessibility, and high quality of interaction. Thus, it must be carefully planned, designed, and accommodated into the life cycle of an interactive system, from the early exploratory phases of design, through to evaluation, implementation, and deployment.

The unified user interface design method has been applied, tested, refined, formulated, and documented in its present form

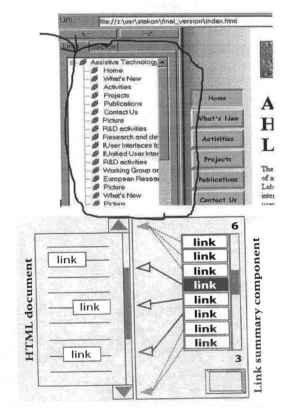

FIGURE 16.15 The Link Summary styles: S1 (top) and S2 (bottom) for Page Browsing task. (From Savidis, A. and Stephanidis, C., *Interacting with Computers*, 16, 243–270, 2004.)

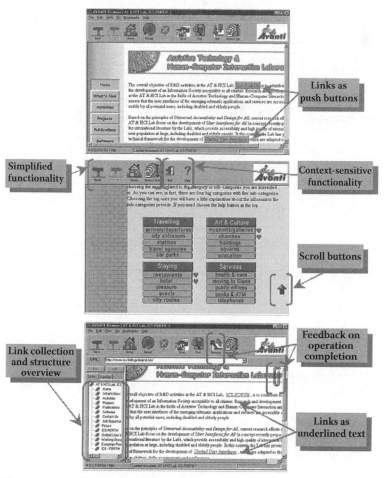

FIGURE 16.16 Three different interface versions of the AVANTI browser produced by adaptation-oriented selection of different alternative styles. (From Savidis, A. and Stephanidis, C., *Interacting with Computers*, 16, 243–270, 2004.)

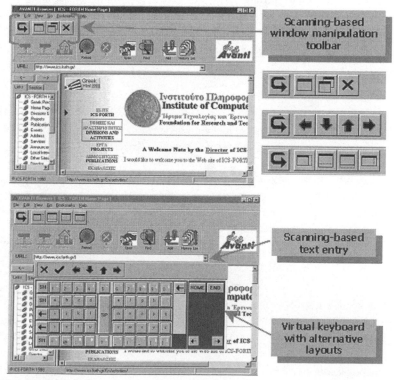

FIGURE 16.17 Alternative augmentation-oriented styles for motor-impaired user access activated at run-time in the AVANTI browser. (From Savidis, A. and Stephanidis, C., *Interacting with Computers*, 16, 243–270, 2004.)

through a number of research projects partially funded by the European Commission (TP1001 ACCESS[2]; ACTS AC042–AVANTI[3]; IST-1999-20656–PALIO[4]; IST-1999-14101 IS4ALL[5]), or national government funding (EPET-II: NAUTILUS[6]), and has been embedded in a process-oriented design support tool (Antona et al., 2006).

Also, unified user-interface design has been taught in various tutorials (Stephanidis et al., 1997; Stephanidis et al., 1999; Stephanidis et al., 2001a,b), attracting numerous participants from industry and academia, while stimulating fruitful discussions and exchange of ideas, that in many cases helped in enhancing different practical aspects of the original method.

Acknowledgments

The ACCESS TP1001 (Development Platform for Unified ACCESS to Enabling Environments) project was partially funded by the TIDE Program of the European Commission and lasted 36 months (January 1, 1994 to December 31, 1996). The partners of the ACCESS consortium are CNR-IROE (Italy), Prime Contractor; ICS-FORTH (Greece); University of Hertfordshire (United Kingdom); University of Athens (Greece); NAWH (Finland); VTT (Finland); Hereward College (United Kingdom); RNIB (United Kingdom); Seleco (Italy); MA Systems & Control (United Kingdom); and PIKOMED (Finland).

The AVANTI AC042 (Adaptable and Adaptive Interaction in Multimedia Telecommunications Applications) project was partially funded by the ACTS Program of the European Commission and lasted 36 months (September 1, 1995 to August 31, 1998). The partners of the AVANTI consortium are ALCATEL Italia, Siette Division (Italy), Prime Contractor; IROE-CNR (Italy); ICS-FORTH (Greece); GMD (Germany); VTT (Finland); University of Siena (Italy); MA Systems and Control (UK); ECG (Italy); MATHEMA (Italy); University of Linz (Austria); EUROGICIEL (France); TELECOM (Italy); TECO (Italy); and ADR Study (Italy).

References

Antona, M., Savidis, A., and Stephanidis, C. (2006). A process-oriented interactive design environment for automatic user interface adaptation. *International Journal of Human Computer Interaction* 20: 79–116.

Bellotti, V. (1993). Integrating theoreticians' and practicioners' perspectives with design rationale, in the *Proceedings of the INTERCHI'93 Conference on Human Factors in Computing Systems*, 24–29 April 1993, Amsterdam, The Netherlands, pp. 101–106. New York: ACM Press.

Bomsdorf, B. and Szwillus, G. (1998). From task to dialogue: Task-based user interface design. *SIGCHI Bulletin* 30: 40–42.

Butler, K., Espositio, C., and Klawitter, D. (1997). Designing more deeper: Integrating task analysis, process simulation & object definition, in the *Proceedings of the ACM DIS'97 Symposium on Designing Interactive Systems*, 18–20 August 1997, Amsterdam, The Netherlands, pp. 37–54. New York: ACM Press.

Desoi, J., Lively, W., and Sheppard, S. (1989). Graphical specification of user interfaces with behavior abstraction, in the *Proceedings of the CHI'89 Conference on Human Factors in Computing Systems*, 30 April–4 May 1989, Austin, Texas, pp. 139–144. New York: ACM Press.

Diaper, D. and Stanton, N. (eds.) (2003). *The Handbook of Task Analysis for Human-Computer Interaction.* London: Lawrence Erlbaum Associates.

Hartson, H. R., Siochi, A. C., and Hix, D. (1990). The UAN: A user-oriented representation for direct manipulation interface design. *ACM Transactions on Information Systems* 8: 181–203.

Hartson, R. and Hix, D. (1989). Human-computer interface development: Concepts and systems for its management. *ACM Computing Surveys* 21: 241–247.

Hill, R. (1986). Supporting concurrency, communication and synchronisation in human-computer interaction: The Sassafras UIMS. *ACM Transactions on Graphics* 5: 289–320.

Hoare, C. A. R. (1978). Communicating sequential processes. *Communications of the ACM* 21: 666–677.

Johnson, P., Johnson, H., Waddington, P., and Shouls, A. (1988). Task-related knowledge structures: Analysis, modeling, and applications, in *People and Computers: From Research to Implementation, Proceedings of HCI '88* (D. M. Jones and R. Winder, eds.), 5–9 September 1988, Manchester, U.K., pp. 35–62. Cambridge: Cambridge University Press.

Kandle, H. (1995). Integration of scenarios with their purposes in task modeling, in the *Proceedings of the ACM DIS'95 Symposium on Designing Interactive Systems*, 23–25 August 1995, Ann Arbor, MI, pp. 227–235. New York: ACM Press.

Kirwan, B. and Ainsworth, L. K. (1992). *A Guide to Task Analysis.* London: Taylor & Francis.

Kneer, B. and Szwillus, G. (1995). OBSM: A notation to integrate different levels of user interface design, in the *Proceedings of the ACM DIS'95 Symposium on Designing Interactive Systems*, 23–25 August 1995, Ann Arbor, MI, pp. 25–31. New York: ACM Press.

Manheimer, J, Burnett, R., and Wallerns, J. (1990). A case study of user interface management system development and application, in the *Proceedings of the CHI'90 Conference on Human Factors in Computing Systems*, 1–5 April 1990, Seattle, pp. 127–132. New York: ACM Press.

Marcus, A. (1996). Icon design and symbol design issues for graphical interfaces, in *International User Interfaces* (E. Del Galdo and J. Nielsen, eds.), pp. 257–270. New York: John Wiley and Sons.

---

[2] Development platform for unified access to enabling environments.
[3] Adaptive and adaptable interactions for multimedia telecommunications applications.
[4] Personalized access to local information and services for tourists.
[5] Information society for all.
[6] Unified web browser for people with disabilities.

McLean, A. and McKerlie, D. (1995). *Design Space Analysis and Use-Representations*. Technical Report EPC-1995-102, Rank Xerox.

Olsen, D. (1990). Propositional production systems for dialog description, in the *Proceedings of the ACM CHI'90 Conference on Human Factors in Computing Systems*, 1–5 April 1990, Seattle, pp. 57–63. New York: ACM Press.

Payne, S. (1984). Task-action grammars, in the *Proceedings of IFIP Conference on Human-Computer Interaction: INTERACT '84*, 4–7 September 1984, London, vol. 1, pp. 139–144. Amsterdam: Elsevier North-Holland.

Potts, C. (1995). Using schematic scenarios to understand user needs, in the *Proceedings of the ACM DIS'95 Symposium on Designing Interactive Systems*, 23–25 August 1995, Ann Arbor, MI, pp. 247–256. New York: ACM Press.

Royer, T. (1995). Using scenario-based designs to review user interface changes and enhancements, in the *Proceedings of the ACM DIS'95 Symposium on Designing Interactive Systems*, 23–25 August 1995, Ann Arbor, MI, pp. 277–246. New York: ACM Press.

Saldarini, R. (1989). Analysis and design of business information systems, in *Structured Systems Analysis*, pp. 22–23. New York: MacMillan Publishing.

Savidis, A., Stephanidis, C., and Akoumianakis, D. (1997). Unifying toolkit programming layers: A multi-purpose toolkit integration module, in the *Proceedings of the 4th Eurographics Workshop on Design, Specification and Verification of Interactive Systems (DSV-IS '97)* (M. D. Harrison and J. C. Torres, eds.), 4–6 June 1997, Granada, Spain, pp. 177–192. Berlin: Springer-Verlag.

Stary, C. (1996). Integrating workflow representations into user interface design representations. *Software Concepts and Tools* 17:173–187.

Stephanidis, C., Akoumianakis, D., and Paramythis, A. (1999). *Coping with Diversity in HCI: Techniques for Adaptable and Adaptive Interaction. Tutorial no. 11 in the 8th International Conference on Human-Computer Interaction (HCI International '99)*, 22–26 August 1999, Munich. http://www.ics.forth.gr/hci/publications/tutorials.html#HCI_International_99.

Stephanidis, C., Paramythis, A., Sfyrakis, M., and Savidis, A. (2000). A case study in unified user interface development: The AVANTI web browser, in *User Interfaces for All: Concepts, Methods, and Tools* (C. Stephanidis, ed.), pp. 525–568. Mahwah, NJ: Lawrence Erlbaum Associates.

Stephanidis, C. and Savidis, A. (2001). Universal access in the information society: Methods, tools and interaction technologies. *Universal Access in the Information Society* 1: 40–55.

Stephanidis, C., Savidis, A., and Akoumianakis, D. (1997). Unified interface development: Tools for constructing accessible and usable user interfaces. *Tutorial no. 13 in the 7th International Conference on Human-Computer Interaction (HCI International '97)*, 24–29 August 1997, San Francisco. http://www.ics.forth.gr/hci/publications/tutorials.html#HCI_International_97.

Stephanidis, C., Savidis, A., and Akoumianakis, D. (2001a). Engineering universal access: Unified user interfaces, in *Tutorial in the 1st Universal Access in Human-Computer Interaction Conference (UAHCI 2001), jointly with the 9th International Conference on Human-Computer Interaction (HCI International 2001)*, 5–10 August 2001, New Orleans. http://www.ics.forth.gr/hci/files/uahci_2001.pdf.

Stephanidis, C., Savidis, A., and Akoumianakis, D. (2001b). Universally accessible UIs: The unified user interface development, in *Tutorial in the ACM Conference on Human Factors in Computing Systems (CHI 2001)*, 31 March–5 April 2001, Seattle. http://www.ics.forth.gr/hci/files/CHI_Tutorial.pdf.

Wilson, S. and Johnson, P. (1995). Empowering users in task-based approach to design, in the *Proceedings of the ACM DIS'95 Symposium on Designing Interactive Systems*, 23–25 August 1995, Ann Arbor, MI, pp. 25–31. New York: ACM Press.

Designing Universally Accessible Games

Dimitris Grammenos,
Anthony Savidis, and
Constantine Stephanidis

17.1 Introduction

Nowadays, electronic (i.e., computer and console) games constitute a key source of entertainment. According to Entertainment Software Association (ESA) (2008), "sixty-nine percent of American heads of households play computer and video games." And, in contrast to the past, where their vast majority constituted young male gamer players, currently the gaming population comprises a wide range of people of all ages and sexes, 10% to 20% of which is disabled (Gwinn, 2007).

Unfortunately, most existing electronic games are quite demanding in terms of motor and sensor skills needed for interaction control, while they often require specific, quite complicated, input devices and techniques. This fact renders such games inaccessible to a large percentage of people with disabilities, and in particular to the blind, and those with severe motor impairments of the upper limbs. Furthermore, regarding human-computer interaction issues, computer games have fundamental differences from all the other types of software applications for which accessibility guidelines and solutions are already becoming widely available.

So far, due to the rather limited market size, little attention has been paid to the development of accessible electronic games, and even less to games that can be concurrently played by players with diverse personal characteristics, requirements, or (dis)abilities. Typically, when referring to accessible games, it is implied that such games are purposefully developed so that they can be played by a particular group of disabled people.

However, this approach has several drawbacks and is in contrast with current social, ethical, technical, and legal trends that all converge toward software applications and services that can be used by anyone, from anywhere and at anytime, or, in other words, toward universal access (Stephanidis et al., 1998). Furthermore, in a broader sense, the term *accessibility* should not be related just to the disabled, since very often user preferences (e.g., a user does not like to use a mouse) and technological (e.g., small screen) or environmental (e.g., noise, sun glare) constraints may also set barriers to access to, and use of, an application or service.

In this context, this chapter introduces the concept of universally accessible games (UA-Games) as a novel approach to creating games that are proactively designed to be concurrently accessible by people with a wide range of diverse requirements or disabilities. First, background information is provided on the various types of disabilities affecting game accessibility, the way that people with disabilities play games, and guidance and approaches for achieving game accessibility. Then, related work is presented, followed by an introduction to the concept and goals of universally accessible games. The main part of the chapter is dedicated to the step-by-step presentation of a structured method for designing universally accessible games, which is based on the unified user-interface design method (Savidis and Stephanidis, 2004; see also Chapter 16 of this handbook). Finally, some conclusions are offered along with directions for future work.

17.2 Background and Related Work

17.2.1 Disabilities Affecting Game Accessibility

There are several types of disabilities that can affect game accessibility. These disabilities may be permanent or temporary. Although there is no single universally accepted classification, an indicative list of impairments includes the following:

1. *Visual impairments*: This category mainly comprises blindness, low vision, and color-blindness (see also Chapter 6 of this handbook). Blindness implies a total (or almost total) loss of vision in both eyes, while low vision (or partial sight) implies severe visual impairment that cannot be totally corrected through the use of visual aids (e.g., glasses or lenses). People with low vision may be able to read text and distinguish forms, objects, and pictures under specific conditions (e.g., very large fonts, high contrast, particular lighting conditions) but usually also rely on other senses, such as hearing and touch. Color-blindness refers to the inability to discriminate differences in colors, mainly red and green.

2. *Motor or dexterity impairments*: Motor and dexterity impairments include the total absence of limbs or digits, and paralysis, lack of fine control, instability or pain in the use of fingers, hands, wrists, or arms (see also Chapter 5 of this handbook). Individuals with motor impairments mainly face difficulties in using standard input devices, such as the keyboard and the mouse.

3. *Hearing disabilities*: Hearing disabilities may range from total deafness (i.e., the person is not able to hear at all), to slight loss of hearing (the person can sense sounds and speech, but finds it hard to identify their content). Deaf people communicate using sign and written language, while hard of hearing individuals may rely on lip-reading and hearing aids (see Chapters 6 and 38 of this handbook).

4. *Cognitive disabilities*: This is a very broad category, which roughly includes difficulties in the performance of mental tasks (see Chapter 7, "Cognitive Disabilities"). These can range from limited and focused problems affecting a very specific cognitive function (e.g., the ability to understand math), to severe cases (e.g., brain damage) where the individual is unable to take care of daily living activities. The most common types of cognitive disabilities are: mental retardation, language and learning disabilities (e.g., dyslexia), head injury and stroke, Alzheimer's disease (i.e., memory retention problems), and dementia. Cognitive impairments are often localized (Seeman, 2002) and thus, people with cognitive impairments can be of average (or above) intelligence. Sometimes an impaired cognitive function may be combined with one or more overdeveloped cognitive functions.

5. *Speech impairments*: Speech impairments are quite rare and sometimes are combined with other disabilities, but they do not indicate limited intelligence. Individuals with speech impairments may have articulation problems (e.g., stuttering), be unable to speak loudly or clearly, or even to speak at all. Obviously, they have problems in using speech recognition systems. Depending on the severity of their case, they may use communication aids to substitute speech (see Chapter 47 of this handbook).

6. *Illiteracy*: Illiteracy is the lack of ability to read and write in any language. Although illiteracy is not a disability, it can create considerable barriers to game accessibility and thus it is treated in the overall context of game accessibility.

According to a study commissioned by Microsoft Corporation and conducted by Forrester Research (Microsoft, 2003), among adult computer users in the United States:

- One in four (25%) have a visual difficulty or impairment.
- Nearly one in four (24%) have a dexterity difficulty or impairment.
- One in five (20%) have a hearing difficulty or impairment.
- About one in seven (16%) have a cognitive difficulty or impairment.
- Very few (3%) have a speech difficulty or impairment.

Age-related disabilities are frequently referred to as a separate category (see Chapter 8 of this handbook), but all related problems fall within the categories discussed previously.

17.2.2 How People with Disabilities Play Games

Overall, game accessibility problems are usually tackled through a combination of (Figure 17.1):

- *Third-party assistive technologies*: Special hardware or software suitable, or compensating to some extent, for a specific disability (e.g., mouth-operated joystick, camera-based mouse, screen reader).
- *In-game interaction techniques*: Input and output practices integrated in the game's software that, on the one hand, are appropriate for the disabled person's interaction capabilities and needs and, on the other hand, can work with, and take advantage of, any available assistive technologies, for example, visual keyboard, scanning, speech input.
- *Game content annotation and adaptation*: Encapsulation of semantic or redundant information for all of the game's content, so that it can be delivered in a format that can be optimally perceived and used through the employed assistive technologies and interaction techniques.

17.2.3 Game Accessibility Guidance

A key tool for designing and implementing any type of accessible software application is the related consolidated know-how,

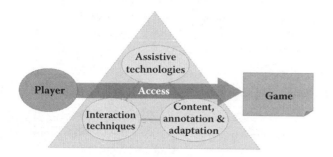

FIGURE 17.1 Approaches for tackling game accessibility problems. (From Grammenos, D., Savidis, A., and Stephanidis, C., unified design of universally accessible games, in *Universal Access in Human-Computer Interaction: Applications and Services—Volume 7 of the Proceedings of the 12th International Conference on Human-Computer Interaction (HCI International 2007)*, 607–616, Springer-Verlag, Berlin/Heidelberg, 2007.)

which usually comes in the form of collections of guidelines. Usually, when such guidelines obtain general consensus and reach a certain maturity level, they become national or international standards. For example, regarding the accessibility of general-purpose software applications and web services, both official standards, for example, Section 508 in the United States (Access Board, 2000), the international standard ISO/TS 16071 (ISO 2003), and worldwide accepted guidelines, for example, the W3C-WAI guidelines (W3C, 2004), exist.

On the contrary, currently, there are no official guidelines, nor any established worldwide initiatives in the domain of game accessibility, and evidently no related governmental or legislative actions. Still, there are two international task forces working toward shaping and consolidating game accessibility guidelines (Ossmann and Miesenberger, 2006):

1. The Games Accessibility Special Interest Group (GA-SIG) of the International Game Developers Association published in 2004 a white paper (IGDA GA-SIG, 2004) about game accessibility, which included a list of possible approaches for providing accessibility in games. Also, more recently, the SIG has created a Top Ten list of actions game developers can take to start increasing the accessibility of their games with minimal effort on their part and without greatly affecting (and perhaps even improving) general game play (IGDA GA-SIG, 2006).

2. A group led by the University of Linz and the Norwegian company MediaLT, which aims to create something similar to the W3C/WAI Web Content Accessibility Guidelines (MediaLT, 2006), based on a set of *Guidelines for the Development of Entertaining Software for People with Multiple Learning Disabilities* published by MediaLT in 2004 (MediaLT, 2004) and the guidelines published by GA-SIG.

Based on existing guidelines, the basic strategies for making games accessible can be summarized as follows:

- *Visual impairments*
  - Content annotation with semantic information
  - Provision of the content through alternative modalities, such as audio and tactile (in the form of Braille)
  - Support for content enlargement (e.g., control of font size, zooming facilities)
  - Customization of color combinations to improve contrast and simplification of visual complexity (e.g., replacing background images with solid colors) to improve legibility
  - Support of input through the keyboard, Braille devices, and speech
  - Ability to serially and hierarchically browse the content and interaction objects in a logical order

- *Motor or dexterity impairments*
  - Support/provision of alternative input devices and techniques, such as switches, specialized
  - keyboards, mice, trackballs and joysticks, scanning, visual keyboards and speech
  - Speed and timing control and adjustment to suit different response times
  - Ability to serially and hierarchically browse the content and interaction objects in a logical order

- *Hearing disabilities*
  - Visual representations of auditory information
  - Augmentation of speech with sign language
  - Sound level control

- *Cognitive disabilities*

This is probably the hardest category to address, since sometimes, depending on the type and level of disability, solutions must be provided on an individual basis. In general, all related solutions include:

- Provision of alternative (simplified, illustrated) versions of the content
- Simplification of tasks (e.g., through step-by-step procedures and wizards)
- Avoidance of blinking and flashing at particular rates that can cause photosensitive epileptic seizures in susceptible individuals

- *Speech impairments*
  - Support of alternative input/communication methods when speech is required

- *Illiteracy*
  - Content simplification
  - Provision of textual content through illustrations, audio, and video

It should be noted that each impairment can have different severity levels, possibly requiring different solutions, and that sometimes people have combinations of disabilities, thus raising compatibility issues between different approaches.

17.2.4 Technical Approaches

From a technical point of view, up to now two main approaches have been adopted to address the issue of computer game accessibility:

1. Inaccessible games become "operationally" (i.e., at runtime) accessible through the use of third-party assistive technologies, such as screen readers, mouse emulators, or special input devices. In practice, there are serious barriers and bottlenecks inherent in the absence of compatibility efforts during the development of computer games and assistive technology systems. However, when some sort of compatibility is achieved, this is typically the result of either customized low-level adaptations (i.e., hacking) or pure coincidence, rather than the outcome of appropriate design considerations.
2. Accessible games are developed from scratch, however, as targeted merely to people with a particular disability, such as audio-based games for blind people, and single-switch games for people with severe motor impairments of the upper limbs.

Following the first approach, typically a limited form of accessibility can be achieved, as well as poor interaction quality and usability. This can be attributed to several reasons:

- Longer interaction times are required, because input and output are not optimized to fit the devices and techniques used.
- Often only part of the full available functionality can be made accessible.
- Error prevention cannot be supported and interaction can be particularly error-prone; for example, when a mouse emulator is used selecting small areas can be very difficult.
- The approach is limited to reproducing the offered functionality, instead of redesigning it to suit the particular user needs.
- Extensive configuration of physical interaction parameters is required (e.g., mapping between the assistive technologies used and the supported devices) to achieve usability.
- Insurmountable implementation barriers can arise; for example, graphical images cannot be automatically reproduced in a nonvisual form.
- Design conflicts may arise, for example, having to reproduce drag & drop dialogue technique for blind players.
- Upward compatibility is not offered; when a new version of a game is developed, the accessibility adjustments will have to be reimplemented.

Although the second approach is much more promising, it still has two major drawbacks:

1. There is a significant trade-off between the cost of developing high-quality accessible games and the expected return on investment, assuming that the target user group reflects a limited market population.

2. There is an apparent hazard due to the potential segregation between able and disabled gamers, essentially leading to social exclusion.

17.2.5 Related Work

Until about 4 years ago, the domain of game accessibility has mainly been a concern of groups of disabled people (e.g., *Audyssey* online gaming magazine and *AudioGames.net* for the blind,[1] *DeafGamers*[2] for the deaf) and small-scale specialized companies producing products specifically targeted to people with a particular disability (e.g., GamesForTheBlind.com and BSC GAMES[3] for the blind, Arcess[4] and Brillsoft[5] for the motor impaired).

More recently, game accessibility started gaining some limited but increasing attention from both the scientific community and the game industry. For example, McCrindle and Symons (2000) have developed a 3D Space Invaders game that combines audio and visual interfaces with force feedback, and Westin (2004) created Terraformers, a first-person shooter game that can be played with its visual 3D graphics layer on or off, and is intended to support players with all degrees of visual ability or impairment through a sophisticated layer of sound. Bierre et al. (2005) provided an overview of game accessibility problems faced by people with disabilities and how they try to overcome them. and discuss the accessibility features included in a handful of commercial games. Atkinson et al. (2006) discuss a number of issues that are key to making first-person shooter games accessible to the blind and vision-impaired and elaborate on the approach followed in the context of the of the AGRIP project.[6]

Gamasutra,[7] which is worldwide the most visited and referenced web site for the video game industry and professionals, also featured a number of articles related to game accessibility. Bierre (2005) was the first to introduce the topic in this audience. Buscaglia (2006) was concerned with the related legal issues, Zahand (2006) approached the topic from the industry's point of view, and Folmer (2007) suggested the consolidation and use of related interaction design patterns as a means for improving game usability and accessibility.

Beyond the limited published related knowledge, the domain of game accessibility is also characterized by a serious lack of related structured methods and software tools. In 2007, the Bartiméus Accessibility Foundation has released a beta version of Audio Game Maker,[8] a free application that enables visually impaired people to develop sound-only computer games. Furthermore, in the same year, an open source project entitled

---

[1] http://www.angelfire.com/music4/duffstuff/audyssey.html.
[2] http://www.deafgamers.com.
[3] http://www.bscgames.com.
[4] http://arcess.com.
[5] http://www.brillsoft.com.
[6] http://www.agrip.org.uk.
[7] http://gamasutra.com.
[8] http://www.audiogamemaker.com.

"Game Accessibility Suite"[9] was started, aiming to create a suite of utilities to make modern Win32 games more accessible, for example, by adding subtitles, slowing them down, and adapting input controllers.

17.3 Universally Accessible Games

As mentioned earlier, existing approaches to game accessibility suffer from several limitations. Furthermore, there is a substantial lack of related structured methods and support tools for creating accessible games. In this context, universally accessible games (UA-Games; Grammenos et al., 2005) constitute a novel approach to game accessibility that aims to overcome existing limitations by allowing the creation of games that are:

- Inherently accessible to all the potential user groups, without the need for further adjustments, or third-party assistive software applications
- Concurrently accessible and playable, cooperatively or competitively, by people with diverse abilities

The concept of UA-Games is rooted in the principles of design for all (Stephanidis et al., 1998), aiming to achieve game accessibility coupled with high interaction quality, also putting forward the objective of creating games that are concurrently accessible to people with diverse abilities. The underlying vision is that through such games people will be able to have fun and compete on an equal basis, while interacting easily and effectively, irrespective of individual characteristics, deployed technologies, or location of use.

At this point, it should be clarified that when referring to games that are universally accessible, it is meant that these games can be played by all people who can potentially play them, but may currently be restrained from doing so due to game design flaws. It is obvious that there will always be games that, due to their intrinsic characteristics, cannot be made accessible to a range of people (e.g., complex strategy games for the cognitively disabled), or when made accessible may have no meaning or interest for those people (e.g., a "find the song title from listening to a melody" game for a deaf person).

The potential impact of UA-Games is threefold:

1. They open up and enhance an entertaining social experience that would otherwise be unavailable to a significant percentage of people.
2. They allow for social interaction among people who may never have (or even could have) interacted with each other.
3. They considerably expand the size and composition of the potential market of the computer games industry.

Up to now, in the context of the UA-Games research activity,[10] the following results have been achieved:

- *Unified design for UA-Games*: A systematic design approach followed and tested during the creation of four games.
- *The concept of parallel game universes*: A way for creating multiplayer games where people with diverse abilities can play cooperatively, or against each other, while at the same time experiencing the game in an optimally adapted way (Grammenos, 2006).
- *Four games* that have a twofold role, acting both as proofs of concept and as case studies:
  - UA-Chess: A universally accessible web-based chess (Grammenos et al., 2005)
  - Access Invaders: A universally accessible multiplayer and multiplatform version of Space Invaders (Grammenos et al., 2006)
  - Game Over!: A universally inaccessible game, meant to be used as a game accessibility educational tool (Grammenos, 2008)
  - Terrestrial Invaders: A game with numerous accessibility features that can address most of the accessibility guidelines that Game Over! violates (Grammenos, 2008)

This chapter focuses on the first of the above-mentioned issues, namely unified design for UA-Games.

17.4 Unified Design of Universally Accessible Games

A key prerequisite to effectively accommodate the particularly broad spectrum of diverse interaction requirements imposed by universally accessible games is to first design the interactive game space at an abstract level, in a representation-independent way, eliminating all references to the physical-level of interaction (e.g., input/output devices, rendering, low-level dialogue). Once this is accomplished, the next step is to appropriately capture the lower-level design details, incrementally specializing toward the physical level of interaction by addressing particular user characteristics. To this end, a design approach capable of representing an open set of alternative physical designs under a common abstract design umbrella is the unified user interface design method (Savidis and Stephanidis, 2004; see also Chapter 16 of this handbook). This method reflects a process-oriented discipline emphasizing abstract task definition with incremental *polymorphic* physical specialization.

Since unified design was originally targeted to the creation of accessible user interfaces, it had to be adapted to cater to the intrinsic characteristics and particular needs of game design. The basic steps in applying unified design to the development of accessible games are summarized in Figure 17.2. As shown, it is a highly participatory, user-centered, iterative process, as: (1) throughout the overall life cycle, the direct involvement of several representative end-users (gamers) with diverse characteristics, as well as domain experts (usability, accessibility, gaming, etc.) is promoted for the continuous assessment of the

---

[9] http://sourceforge.net/projects/gameaccess.
[10] http://www.ics.forth.gr/hci/ua-games.

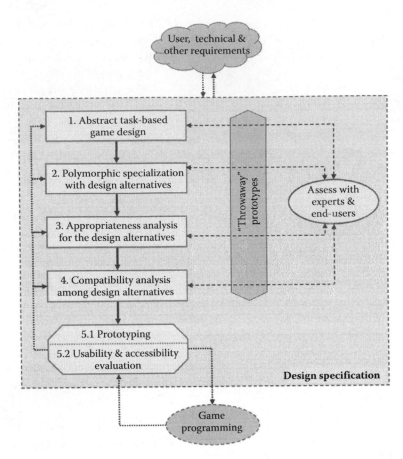

FIGURE 17.2 Applying unified design to the development of universally accessible games. (From Grammenos, D., Savidis, A., and Stephanidis, C., unified design of universally accessible games, in *Universal Access in Human-Computer Interaction: Applications and Services—Volume 7 of the Proceedings of the 12th International Conference on Human-Computer Interaction (HCI International 2007)*, 607–616, Springer-Verlag, Berlin/Heidelberg, 2007.)

design outcomes in each step; and (2) it is possible to return to a previous design step, in case, for instance, more information is required, some design artifacts have to be revisited, or the design parameters must be further specialized.

Quite often, to evaluate the decisions made at a specific step or to weigh alternatives before committing to them, it is required to quickly create small-scale temporary prototypes, known as throwaway prototypes. These may range from rough handmade sketches to simple programs. Prototyping (Floyd, 1984) is an essential part of the iterative design process, since it provides a low-cost, tangible means for gathering early and meaningful user feedback, and, at a later stage, can also serve as a common reference point, as well as a concrete, unambiguous, documentation medium for communicating design specifications to game programmers.

At this point, it should be noted that game programmers are also involved in the whole process with a twofold role: (1) they provide input about technical requirements and restrictions, as well as about the feasibility and cost of alternative design solutions; and (2) they develop and tweak the required electronic prototypes.

17.4.1 Abstract Task-Based Game Design

The goal of this first step is to break down the high-level tasks performed by people when playing the particular game—irrespective of the medium they use to play it with—as well as the actions they perform, the objects they act upon, and the knowledge they need to have. The hierarchical decomposition adopts the original properties of hierarchical task analysis (Johnson et al., 1988) for incremental decomposition of user tasks to lower level actions. In this context, it is essential to focus on the basic logical game activities and constituents, identifying their semantic attributes and relevant regulations independently of the way these can be physically instantiated to be accessible or usable to particular user groups.

As an example, the result of the decomposition for the game of chess is illustrated in Figure 17.3, where tasks are divided in two broad categories: (1) *game-play tasks*, comprising user actions directly related to the game goal and content (i.e., the board and the pieces); and (2) *game-control tasks*, which include "peripheral" user actions that affect the game state and the way that player-game interaction is performed.

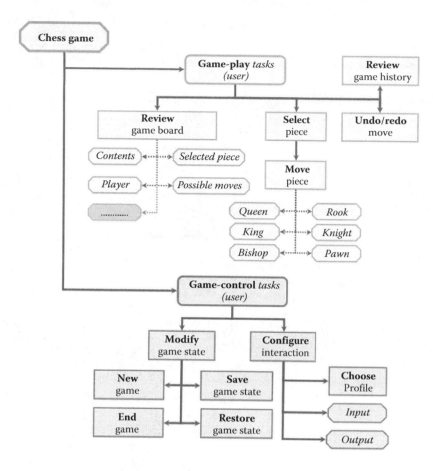

FIGURE 17.3 Abstract task decomposition for the game of chess. (From Grammenos, D., Savidis, A., and Stephanidis, C., unified design of universally accessible games, in *Universal Access in Human-Computer Interaction: Applications and Services—Volume 7 of the Proceedings of the 12th International Conference on Human-Computer Interaction (HCI International 2007)*, 607–616, Springer-Verlag, Berlin/Heidelberg, 2007.)

17.4.2 Polymorphic Specialization with Design Alternatives

In this step, the abstract tasks resulting from the previous one are mapped to multiple low-level, physical alternative interactive designs, meeting target user attributes. In this context, accessibility barriers that can possibly emerge in each task when performed by a particular user group are identified, and suitable alternative interaction methods and modalities are selected. An example of how the abstract task entitled "Select Piece" (from the example illustrated in Figure 17.3) can be mapped to alternative low-level, physical, alternative interactive designs is presented in Figure 17.4.

17.4.3 Appropriateness Analysis for the Design Alternatives

A matrix is constructed correlating the perceived appropriateness of each selected design alternative for every user attribute. The rows of the matrix represent distinct user attributes, while the columns represent alternative interactive designs (see Figure 17.5). Each cell, depending on the suitability of the particular design for the specific user attribute, is filled in with one of the symbols depicted in Table 17.1. The alternative interactive designs appropriateness matrix can be filled in by reviewing related literature, using previous know-how in the field, as well as by questioning domain experts and representatives of the target user groups.

A basic design goal is that for every abstract task, there should be at least one "ideal" or "appropriate" input and one output design alternative for each user attribute and target *user profile*. A *user profile* is a collection of user attributes (e.g., novice, sighted, hand-motor impaired gamer). The appropriateness of a design alternative for a specific user profile can be inferred by merging the corresponding rows of the matrix that contain attributes of this profile, as follows:

- If the design alternative is *inappropriate* for any of the user attributes, then it is deemed inappropriate for the entire profile.
- If the design alternative is *neutral* for a specific attribute, it means that it is not related to this attribute and thus does not affect the design's appropriateness for it.
- In all the other cases (i.e., *ideal, appropriate, could be used*), the lowest appropriateness value supersedes all the others.

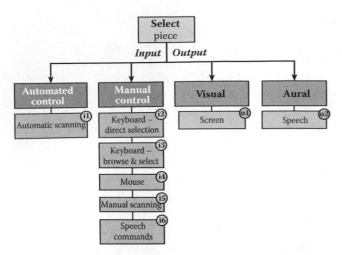

FIGURE 17.4 Example of mapping abstract tasks to alternative interactive designs. (From Grammenos, D., Savidis, A., and Stephanidis, C., unified design of universally accessible games, in *Universal Access in Human-Computer Interaction: Applications and Services—Volume 7 of the Proceedings of the 12th International Conference on Human-Computer Interaction (HCI International 2007)*, 607–616, Springer-Verlag, Berlin/Heidelberg, 2007.)

Thus, for example, Figure 17.6 illustrates the appropriateness matrix for the case of a low-vision, novice player who can use just a single switch. As can be seen, this is a particularly difficult case, since the available solutions are very limited and not optimal. Nevertheless, it can still be ensured that the game is accessible in this case.

17.4.4 Compatibility Analysis among Design Alternatives

When alternative interactive designs have been identified, it is essential that cases where two or more alternatives are mutually incompatible are pinpointed, so that they can be avoided.

TABLE 17.1 Alternative Interactive Designs Appropriateness Symbols

Symbol	Meaning
✓ (ideal)	Explicitly designed for this user attribute
⬆ (appropriate)	Suitable, but possibly not the best choice
⬇ (could be used)	If nothing else is available, it can be used, though not recommended
✗ (inappropriate)	Totally inappropriate; will result in posing an accessibility barrier
= (neutral)	Does not have any effect on the particular user attribute

Source: From Grammenos, D., Savidis, A., and Stephanidis, C., unified design of universally accessible games, in *Universal Access in Human-Computer Interaction: Applications and Services—Volume 7 of the Proceedings of the 12th International Conference on Human-Computer Interaction (HCI International 2007)*, 607–616, Springer-Verlag, Berlin/Heidelberg, 2007.

User attributes	Design alternatives							
	i1	i2	i3	i4	i5	i6	o1	o2
Full vision	=	=	=	=	=	=	✓	⬆
Low vision	⬇	⬆	⬆	⬇	⬇	✓	⬇	✓
Blind	⬇	⬆	✓	✗	⬇	✓	✗	✓
No hand-motor impairment	⬇	✓	⬆	✓	✗	✓	=	=
Uses multiple switches	⬇	✗	⬇	✗	✓	✓	=	⬇
Uses single switch	✓	✗	✗	✗	✗	✓	=	⬇
Mild cognitive impairment	⬇	⬇	⬇	⬆	⬇	⬇	⬆	⬆
Expert	⬇	✓	⬇	⬆	⬇	⬇	=	⬇
Novice	⬇	⬇	⬆	✓	⬇	⬆	✓	✓

✓: Ideal   ⬆: Appropriate   ⬇: Could be used   ✗: Inappropriate   =: Neutral

FIGURE 17.5 Example of appropriateness matrix for the design alternatives. (From Grammenos, D., Savidis, A., and Stephanidis, C., unified design of universally accessible games, in *Universal Access in Human-Computer Interaction: Applications and Services—Volume 7 of the Proceedings of the 12th International Conference on Human-Computer Interaction (HCI International 2007)*, 607–616, Springer-Verlag, Berlin/Heidelberg, 2007.)

To this purpose, several related *compatibility matrices* need to be devised. A *compatibility matrix* has as rows and columns all the alternative interactive designs that can potentially be concurrently active at a particular point in time. If two designs are compatible, then the corresponding cell is filled with a tick (✓); otherwise, the cell is filled with an **X**. Figure 17.7 presents a compatibility matrix created for the alternative input designs of the example presented in Figure 17.4.

17.4.5 Prototyping, Usability, and Accessibility Evaluation

As soon as a "stable" version of the whole (or just a discrete part of) game design is available, indicative electronic prototypes of the game can be developed, showcasing the alternative interactive properties of its user interface for the different target user groups. The usability and accessibility of these prototypes should definitely be evaluated with representative end-users. In

User attributes	Design alternatives							
	i1	i2	i3	i4	i5	i6	o1	o2
Low vision, novice, single-switch	⬇	✗	✗	✗	✗	⬆	⬇	⬇

FIGURE 17.6 Appropriateness matrix for a low-vision, novice player who can use just a single switch. (From Grammenos, D., Savidis, A., and Stephanidis, C., unified design of universally accessible games, in *Universal Access in Human-Computer Interaction: Applications and Services—Volume 7 of the Proceedings of the 12th International Conference on Human-Computer Interaction (HCI International 2007)*, 607–616, Springer-Verlag, Berlin/Heidelberg, 2007.)

	i1	i2	i3	i4	i5	i6
i1		✓	✗	✓	✗	✓
i2	✓			✓	✓	✓
i3	✗	✓		✓	✗	✓
i4	✓	✓	✓		✓	✓
i5	✗	✓	✗			✓
i6	✓	✓	✓	✓	✓	

✓: Compatible   ✗: Incompatible

FIGURE 17.7 Compatibility matrix example for the alternative input designs of Figure 17.4. (From Grammenos, D., Savidis, A., and Stephanidis, C., unified design of universally accessible games, in *Universal Access in Human-Computer Interaction: Applications and Services—Volume 7 of the Proceedings of the 12th International Conference on Human-Computer Interaction (HCI International 2007)*, 607–616, Springer-Verlag, Berlin/Heidelberg, 2007.)

this respect, a quick, handy, and very effective informal evaluation method is *thinking aloud* (Nielsen, 1993). According to this method, one (or sometimes more) evaluator observes a gamer interacting with the system, asking vocalization of thoughts, opinions, and feelings to understand the gamer's mental model—the way she thinks and performs tasks and find out any mismatches between the user's mental model and the system model. Conversations are usually recorded, so that they can be analyzed later on. Furthermore, to support the evaluation process, a list of indicative tasks is used that prompts participants to explore the full available game functionality. After playing the game, a small debriefing session can be held, where participants are asked about their overall impression of their interaction with the game, their personal preferences, likes or dislikes, as well as their suggestions regarding potential improvement and modifications.

The outcomes of this step can considerably aid in validating, correcting, and updating design decisions, as well as in developing new ideas for improving the accessibility and playability of the final game. When the design specification is considered mature, it can then be propagated for further development. Of course, as parts of the game and its functionality are being developed, it is highly desirable to regularly perform usability and accessibility testing.

17.4.6 Abstract Task-Based Game Design and Action Games

A basic design differentiation between a turn-based board game, such as chess, that was used as an example up to now, and action games is the degrees of freedom along which the game can be modified to become accessible. More specifically, in the case of board games, only the user interface (i.e., the way the game is presented and controlled) can be adapted to better match a

particular player's characteristics. The game's rules (logic) and content are fixed, and any deviations are not possible. On the other hand, action games are more flexible in this respect, since they usually have a main goal, but they do not impose restrictions on how this can be achieved, and do not have any globally established strict rules and specific content.

To provide an illustrative example, when making a chess game accessible, the possible adaptations can only affect how the board and pieces are rendered and the way the player can select and move the pieces. The type or number of the pieces, the rules they follow for moving, or what a player has to do to win cannot be changed (since this would be a different game). On the contrary, when creating an accessible version of Space Invaders, beyond changing how the player's spaceship is controlled and presented, it is possible to completely revamp the characteristics of the attacking alien ships (e.g., number, speed, firepower, size) and even the rules of the game (e.g., allow the player to destroy any alien, but only a specific alien to destroy the player, change the initial number of the player's lives, etc.).

Thus, a key difference in relation to the aforementioned design example of the chess game is that, at the stage of identifying accessibility barriers related to the game's interface, barriers that stem from the game's content and rules should also be identified, as well as possible design strategies for overcoming them. In this respect, the abstract task decomposition of a Space Invaders type of game is illustrated in Figure 17.8. This time, tasks are divided into three—instead of two—categories:

1. *Game-play tasks*, comprising user actions directly related to the game content
2. *Game-control tasks*, which include peripheral user actions that affect the game state and the way that player-game interaction is performed
3. *Game-logic tasks*, which are performed by the system to create and control the various active game elements

17.5 Conclusions and Future Work

The experience consolidated during the development of four UA-Games games has shown that the creation of universally accessible games is a demanding but manageable and achievable task that entails:

- Designing for context independence at an abstract level, without considering specific interaction modalities, metaphors, techniques, or devices, and separating the content and the related mechanics from the way that these can be accessed by, and presented to, the user
- Mapping abstract design elements to coherent, usable, and accessible interaction designs based on the users' individual characteristics
- Creating user interfaces that can support alternative interaction methods and modalities that can coexist and cooperate

- Creating user interfaces that are able to adapt to alternative user profiles (i.e., sets of preferences, requirements and needs, and contexts of use)
- Adopting inclusive and participative design approaches for instance, through considering the broadest possible population during design and having representatives from as many categories as possible participating and providing input to all the development phases
- Designing based on incomplete knowledge, as games can target highly diverse audiences
- Providing open and extensible interaction design so that, later on, it will be still possible to expand the design to cater for more user categories and contexts of use
- Designing for nontypical user groups, which may have nothing in common with the designer

To this end, analytical design methods, such as the one presented in this chapter, are anticipated to become necessary tools for capturing, representing, and managing the large and complex design spaces of universally accessible games,

addressing the compelling requirements for customization, accessibility, and high quality of interaction. The presented design method has been formulated, applied, tested, refined, and documented in its present form through the development of four computer games in the context of the UA-Games.

In this context, related identified areas for future work include:

1. *Game accessibility guidelines and consolidated knowledge*: There is a clear need for the creation of a stable and commonly agreed upon version of valid game accessibility-related guidelines, which will form a *de facto* standard for game developers worldwide. In this regard, it is crucial that specific interpretations, as well as examples of application of the guidelines, are provided for the various available distinct game genres.

2. *Game accessibility evaluation methods*: A natural consequence of the lack of guidelines is the absence of a structured method for formally assessing, either with experts or end-users, the accessibility of games. Thus, work in this

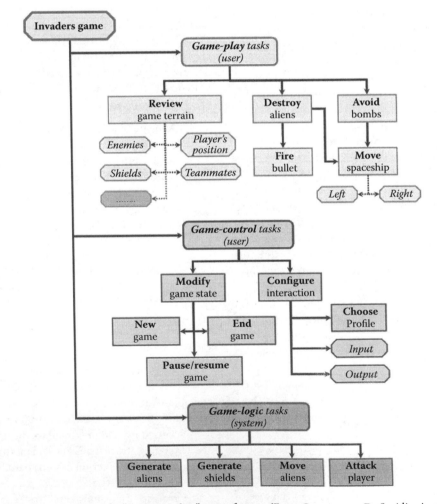

FIGURE 17.8 Abstract task decomposition of a "Space Invaders" type of game. (From Grammenos, D., Savidis, A., and Stephanidis, C., unified design of universally accessible games, in *Universal Access in Human-Computer Interaction: Applications and Services—Volume 7 of the Proceedings of the 12th International Conference on Human-Computer Interaction (HCI International 2007)*, 607–616, Springer-Verlag, Berlin/Heidelberg, 2007.)

direction would greatly aid the design process, but also the assessment of its outcomes.

3. *Design support tools*: The large number of diversification parameters that designers should take into account when designing universally accessible games can easily get out of hand. Computer-aided game design tools are definitely required to make possible the creation of large-scale games and ensure the avoidance of errors, conflicts, and omissions.

4. *Development support tools*: Two key barriers in accessible game development are the lack of related know-how and the requirement for spending additional resources. Such barriers can be easily overcome through the existence of development tools that allow easily and rapidly implementing and integrating accessibility features and techniques in games.

5. *A new game genre*: A key research challenge that has not been investigated at all until now is related to designing new types of games that have inherently the potential to be universally accessible. In other words, besides devising strategies for making accessible existing types of games, one could also consider all the currently available accessibility methods and attributes and based on them create a new type of game in which all (or at least most) of them can be successfully applied and integrated.

References

Access Board (2000). *Electronic and Information Technology Accessibility Standards.* http://www.access-board.gov/sec508/508standards.htm.

Atkinson, M. T., Gucukoglu, S., Machin, C. H. C., and Lawrence, A. E. (2006). Making the mainstream accessible: What's in a game? In the *Proceedings of ICCHP 2006*, 9–11 July 2006, Linz, Austria, pp. 380–387. Berlin: Springer-Verlag.

Bierre, K. (2005). Improving game accessibility. *Gamasutra.* http://www.gamasutra.com/features/20050706/bierre_01.shtml.

Bierre, K., Chetwynd, J., Ellis, B., Hinn, D. M., Ludi, S., and Westin, T. (2005). Game not over: Accessibility issues in video games, in *Universal Access in HCI: Exploring New Interaction Environments, Volume 7 of the Proceedings of the 11th International Conference on Human-Computer Interaction (HCI International 2005)*, 22–27 July 2005, Las Vegas [CD-ROM]. Mahwah, NJ: Lawrence Erlbaum Associates.

Buscaglia, T. (2006). Game law: Everybody conga? *Gamasutra.* http://www.gamasutra.com/features/20060428/buscaglia_01.shtml.

Entertainment Software Association (ESA) (2008). *Game Player Data.* http://www.theesa.com/facts/gamer_data.php.

Floyd, C. (1984). A systematic look at prototyping, in *Approaches to Prototyping* (R. Budde, K. Kuhlenkamp, L. Mathiassen, and H. Zullighoven, eds.), pp. 1–18. Berlin: Springer-Verlag.

Folmer, E. (2007). Designing usable and accessible games with interaction design patterns. *Gamasutra.* http://www.gamasutra.com/view/feature/1408/designing_usable_and_accessible_.php.

Grammenos, D. (2006). The theory of parallel game universes: A paradigm shift in multiplayer gaming and game accessibility. *Gamasutra.* http://www.gamasutra.com/features/20060817/grammenos_01.shtml.

Grammenos, D. (2008). Game over: Learning by dying, in the *Proceedings of the 26th Annual ACM CHI Conference on Human Factors in Computing Systems (CHI 2008)*, 5–10 April 2008, Florence, Italy, pp. 1443–1452. New York: ACM Press.

Grammenos, D., Savidis, A., Georgalis, Y., and Stephanidis, C. (2006). Access invaders: Developing a universally accessible action game, in *Computers Helping People with Special Needs, Proceedings of the 10th International Conference (ICCHP 2006)* (K. Miesenberger, J. Klaus, W. Zagler, and A. Karshmer, eds.), 12–14 July 2006, Linz, Austria, pp. 388–395. Berlin/Heidelberg: Springer-Verlag.

Grammenos, D., Savidis, A., and Stephanidis, C. (2005). UA-Chess: A universally accessible board game, in *Universal Access in HCI: Exploring New Interaction Environments, Volume 7 of the Proceedings of the 11th International Conference on Human-Computer Interaction (HCI International 2005)*, (C. Stephanidis, ed.), 22–27 July 2005, Las Vegas [CD-ROM]. Mahwah, NJ: Lawrence Erlbaum Associates.

Gwinn, E. (2007). Disabled gamers want more than 'fluffy' choices. *Chicago Tribune.* http://www.aionline.edu/about-us/news/ChicagoTribFlorio.pdf.

IGDA GA-SIG (2004). *Accessibility in Games: Motivations and Approaches.* http://www.igda.org/accessibility/IGDA_Accessibility_WhitePaper.pdf.

IGDA GA-SIG (2006). *Game Accessibility Top Ten.* http://www.igda.org/wiki/index.php/Top_Ten

ISO (2003). *ISO/TS 16071:2003: Ergonomics of Human-System Interaction: Guidance on Accessibility for Human-Computer Interfaces.* Geneva, Switzerland: International Standards Organization.

Johnson, P., Johnson, H., Waddington, P., and Shouls, A. (1988). Task-related knowledge structures: Analysis, modeling, and applications, in *People and Computers IV* (D. M. Jones and R. Winder, eds.), pp. 35–62. Cambridge: Cambridge University Press.

McCrindle, R. J. and Symons, D. (2000). Audio space invaders, in the *Proceedings of the 3rd International Conference of Disability, Virtual Reality & Associated Technologies*, pp. 59–65. Reading, UK: University of Reading.

MediaLT (2004). *Guidelines for the Development of Entertaining Software for People with Multiple Learning Disabilities.* http://www.medialt.no/rapport/entertainment_guidelines.

MediaLT (2006). *Guidelines for Developing Accessible Games.* http://gameaccess.medialt.no/guide.php

Microsoft (2003). *Accessible Technology Market Research.* http://www.microsoft.com/enable/research/default.aspx.

Nielsen, J. (1993). *Usability Engineering*. Boston: Academic Press.

Ossmann, R. and Miesenberger, K. (2006). Guidelines for the development of accessible computer games, in the *Proceedings of ICCHP 2006*, 9–11 July 2006, Linz, Austria, pp. 403–406. Berlin: Springer-Verlag.

Savidis, A. and Stephanidis, C. (2004). Unified user interface design: Designing universally accessible interactions. *International Journal of Interacting with Computers* 16: 243–270.

Seeman, L. (2002). Inclusion of cognitive disabilities in the web accessibility movement, in the *Proceedings of the Eleventh International World Wide Web Conference*, 7–11 May 2002, Honolulu, Hawaii [CD-ROM]. WWW2002, on behalf of the International World Wide Web Conference Committee. http://www2002.org/CDROM/alternate/689/index.html.

Stephanidis, C., Salvendy, G., Akoumianakis, D., Bevan, N., Brewer, J., Emiliani, P. L., et al. (1998). Toward an information society for all: An international R&D agenda. *International Journal of Human-Computer Interaction* 10: 107–134.

Westin, T. (2004). Game accessibility case study: Terraformers a real-time 3d graphic game, in the *Proceedings of the Fifth International Conference on Disability, Virtual Reality and Associated Technologies*, 20–22 September 2004, Oxford, U.K., pp. 95–100. Reading, UK: University of Reading.

World-Wide Web Consortium (W3C) (2004). *W3C-WAI Resources: Guidelines*. http://www.w3.org/WAI/Resources/#gl.

Zahand, B. (2006). Making video games accessible: Business justifications and design considerations. *Gamasutra*. http://www.gamasutra.com/features/20060920/zahand_01.shtml.

Software Requirements for Inclusive User Interfaces

Anthony Savidis and
Constantine Stephanidis

18.1 Introduction

In the context of the information society, the notion of *computing platform* concerns a wide range of devices, apart from traditional desktop computers, such as public-use terminals, phones, TVs, car consoles, and home appliances.[1] Today, such computing platforms are mainly delivered with embedded operating systems (such as Windows Mobile, Java Micro Edition, or Symbian), while the various operational capabilities (e.g., interaction) and supplied services are controlled through software. As a result, it is expected that virtually anyone may potentially use interactive software applications, from any context of use, to carry out any particular task. In this context, interactive software products should be developed to address the demands of "running anywhere" and being "used by anyone." It is argued that existing software development processes have to be revisited, as they suffer from two fundamental problems:

- They are designed for the so-called average user. However, in the context of universal access, where potentially anyone is a computer user, it is not useful to define an average user case.
- They are typically developed for desktop computer systems (high-resolution display, mouse and keyboard), usually running windowing interactive environmen ts.

Instead, in the context of universal access, the following are expected: (1) a wide range of computing platforms, with large variations of interactive peripheral equipment; and (2) new ways of interaction maximally designed for particular usage contexts and/or individual user characteristics, departing from the old traditional Windows metaphor (Canfield Smith et al., 1982).

This new situation has been already identified by the introduction of new concepts, which have been intensively supported through related research and development efforts, such as user interfaces for all (Stephanidis, 2001a), ubiquitous computing (Abowd and Mynatt, 2000), migratory interfaces (Bharat and Cardelli, 1995), wearable interfaces (Bass et al., 1997), ambient intelligence (IST Advisory Group, 2003; see also Chapter 60 of this handbook), disappearing computer (IST/FET, 2001), and plastic user interfaces (Calvary et al., 2001). These concepts share many common objectives, principles, and semantics, consolidated in a situation depicted within Figure 18.1, showing three key layers of diversity in the context of universally accessible interactions.

This chapter discusses a consolidated account of an in-depth technical analysis, originated from real-life development experience in the course of six medium-to-large scale research projects, spanning across a time frame of about 10 years, and presents a definition and classification of the key software engineering requirements and the necessary functionality to produce universally accessible interactive applications and services. Those research projects have been partially funded by the European

---

[1] This chapter is based on the article Savidis, A. and Stephanidis, C. (2006). Inclusive development: Software engineering requirements for universally accessible interactions. *Interacting with Computers* 18: 71–116. Reprinted here with slight modifications by concession of Elsevier.

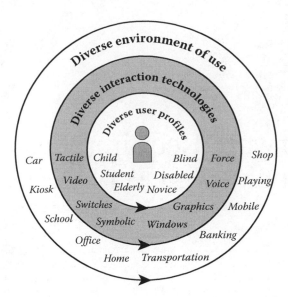

FIGURE 18.1 The three key layers of diversity in the context of universally accessible interactions. (From Savidis, A. and Stephanidis, C., *Interacting with Computers*, 18, 71–116, 2006.)

Commission (TP1001, ACCESS[2]; IST-1999-20656, PALIO[3]; ACTS AC042, AVANTI[4]; IST-1999-14101, IS4ALL[5]; IST-2000-25286, 2WEAR[6]), or by national funding agencies (EPET-II: NAUTILUS[7]).

The key objective in this context is the identification of commonly occurring development requirements, genuinely emerging from the primary need to support universally accessible interactions—anywhere, for anyone, anytime—by offering an appropriate classification scheme denoting the generic and representative categories of software engineering requirements. In this context, the focal point of the discussion is shifted from any particular development tool-type, like interface toolkits, toward a concrete set of top-level implementation issues, while outlining the profile of the necessary development instrumentation to effectively address such issues, drawing from real-life experience. Hence, some of the identified requirements concern enhancements of the development process, others relate to necessary implementation mechanisms of the host tool, while some emphasize the need for better software engineering strategies. The results of this analysis are summarized in four general categories of development requirements, each representing a particular dimension of universally accessible interactions:

- *Need for new forms and styles of interactions*: In view of the wide variety of computing platforms, situations of use, individual end-users, and tasks to be carried out through interactive software products, it is critical to provide the

most appropriate designs to ensure the highest possible interaction quality, supporting intuitive interaction, ease of use, as well as efficiency, effectiveness, and user satisfaction. The development of new metaphors, different from the desktop graphical-style traditions, may be necessary in some situations, and may constitute the foundation for experimenting with future interaction paradigms.

- *Need for effective manipulation of diverse interaction elements*: The development of universally accessible interactions encompasses diverse interaction elements, which, on the one hand, may be originated from different interaction metaphors (e.g., windowing and rooms; Savidis and Stephanidis, 1995a), while, on the other hand, can be realized into alternative physical forms (e.g., 2D/3D visual graphical, auditory, tactile). As a result, it is crucial that interface tools supply to interface developers all the necessary implementation mechanisms for the manipulation of such diverse categories of interaction elements.

- *Need to support automatic adaptation to individual users and contexts of use*: To maximize the quality of the delivered interfaces, it is imperative to support user and usage context awareness, while enabling the interface to adapt itself on-the-fly to the particular end-user and usage context. Automatic interface adaptation implies that the software encompasses and organizes appropriately alternative dialogue patterns in an implementation form, inherently requiring appropriate software engineering for the run-time activation and control of dynamic dialogue components.

- *Need to support interface mobility through ambient interactions*: To support user mobility in an open computational environment, it is necessary to address typical scenarios in which the environment is the interface. The latter requires facilities for dynamic discovery, control, and utilization of remote dynamically exposed user interface microservices embedded in environment devices.

18.2 Identification of Requirements

18.2.1 The Need for New Interaction Metaphors

The proliferation of advanced interaction technologies enables the construction of new interactive experiences in popular application domains. Thus, for example, educational software titles provide new interactive computer embodiments based on metaphoric themes that children are familiar with, such as the playground, or interactive books. The key role of metaphors is to provide users with more natural means to attain a wider range of tasks in a manner that is effective, efficient, and satisfactory, by providing a better cognitive fit between the interactive embodiment of the computer and real-world analogies with which users are already familiar. It is expected that new metaphors will depart significantly from graphical window-based interaction, which, inspired from the Star interface (Canfield Smith et al., 1982), was intended to meet the demands of able-bodied users working in an office environment.

[2] Development Platform for Unified Access to Enabling Environments (1995–1997).
[3] Personalized Access to Local Information and Services for Tourists (2000–2003).
[4] Adaptive and Adaptable Interactions for Multimedia Telecommunications Applications (1997–2000).
[5] Information Society for All (2001–2003).
[6] A Runtime for Adaptive and Extensible Wireless Wearables (2001–2003).
[7] Unified Web Browser for People with Disabilities (1999–2001).

The foreseen transition from the current massive use of WIMP interfaces to the post-WIMP era has been identified early in Van Dam (1997), with primary emphasis on new forms and metaphors of graphical interaction. The need for sensing environment events and supporting context-specific input in developing future interactive systems has been identified in the context toolkit (Salber et al., 1999). Additionally, in the white paper of the recently formed Accessibility Group of the Independent Game Developers Association (IGDA) (IGDA Accessibility, 2004), the effort toward marrying universal access and video games clearly poses new design challenges for effective and accessible metaphoric interactions (see also Chapter 17 of this handbook).

Currently, the design and realization of interaction metaphors is not supported by appropriate reusable software-libraries and toolkits. Rather, it is internally hard-coded within the user interface software of interactive applications and services. Existing multimedia-application construction libraries are usually too low-level, requiring that developers undertake the complex task of building all metaphoric interaction features from primitive interaction elements. For instance, even though various educational applications provide virtual worlds familiar to children as an interaction context, the required "world" construction kits for such specialized domains and metaphors are lacking. This is analogous to the early period of graphical user interfaces (GUIs), when developers used a basic graphics package for building window-based interactive applications. However, the evolution of GUIs into a de facto industry standard did not take place until tools for developing such user interfaces became widely available. Similarly, it is argued that the evolution of new metaphors to facilitate the commercial development of novel applications and services targeted to the population at large will require the provision of the necessary implementation support within interface tools. This chapter provides a metaphor development methodology that may be employed to pursue the design and implementation of new interaction metaphors, exposing the key demands for crafting metaphor-specific interface toolkit libraries, to practically assist in building new forms and styles of interactions. In this context, the methodology itself becomes a necessary technical ingredient that developers should possess, so as to effectively address the challenge of building new interaction metaphors. Two metaphor development cases are presented: a metaphor-flexible toolkit for nonvisual interaction, and a collection of new graphics-intensive-based interaction methods.

18.2.2 The Need for Manipulating Diverse Collections of Interaction Objects

As discussed in the previous section, the effective deployment of new interaction metaphors depends on the practical availability of the necessary tools to support the implementation of the dialogue patterns and artifacts embodied in those metaphors (e.g., visual windowing, 3D auditory, tactile, or switch-based scanning dialogues). As in the case of GUIs, the primary means to construct metaphoric interactions are likely to be in the form of implemented reusable interaction elements provided by software libraries commonly known as *toolkits* (e.g., OSF/

Motif, MFC, InterViews, Xaw/Athena widget set, JFC). Such tools provide programming facilities to mainly: (1) manipulate interaction objects and construct object hierarchies; (2) handle incoming input-device events; and (3) display lower-level graphic primitives.

Traditionally, interaction objects (e.g., windows, buttons, scrollbars, check-boxes, etc.) have been treated as the most important category of interaction elements, since the largest part of existing interface development toolkits is exclusively devoted to providing rich sets of interaction objects, appropriately linking to the underlying functionality. Moreover, interaction objects constitute a common vocabulary for both user interface designers and programmers, even though the type of knowledge typically possessed by each group is rather different. Thus, designers usually have more detailed knowledge regarding the appropriateness of different interaction objects for particular user tasks, while programmers have primarily implementation-oriented knowledge. In any case, a "button" or a "window" has the same meaning for both designers and programmers, when it comes to the specific physical entity being represented. Interface toolkits have traditionally provided a means for bridging the gap between lexical-level design and implementation. In other words, they provide a vehicle for the direct mapping of common lexical-level user tasks, such as selection, command, or text entry, revealed through a design activity, to their respective implementation constructs available in a target toolkit, like a menu, a push button, or a text field.

In the course of developing applications targeted to radically different user groups, engaging largely diverse styles of interactions, the necessity has been experienced of introducing new, functionally extended forms of interaction objects (as compared to traditional graphical toolkit elements). Those extended functional needs have emerged in the context of real-life special-purpose application developments, such as a dual hypermedia electronic book (Petrie et al., 1997), an advanced interpersonal communicator (Kouroupetroglou et al., 1996), and a full-fledged user-adapted web browser (Stephanidis et al., 2001). Such applications were targeted to diverse user groups, differing with respect to physical, sensory, and motor abilities, preferences, domain-specific knowledge, and role in organizational context. In this context, various interaction technologies and interaction metaphors had to be employed, including graphical windowing environments, auditory/tactile interaction, rooms-based interaction metaphors, and so on. In the course of these developments, the following set of commonly recurring functional requirements has been consolidated, regarding the required implementation facilities to manipulate interaction objects:

- *Integration* of third-party collections of interaction objects, the latter offered as independent toolkit libraries, while delivering novel interaction facilities possibly complying to alternative interaction metaphors
- *Augmentation*, for instance, provision of additional interaction techniques on top of existing interaction objects offered by the currently available toolkits, in cases where

the original dialogue for such objects is considered inadequate for specific combinations of target user groups and contexts of use characteristics

- *Expansion* of the original set of interaction objects by introducing new custom-made interaction objects, when particular newly designed interactive behaviors are not implemented by the utilized toolkits
- *Abstraction* applied on interaction objects even when those comply to different interaction metaphors, through the delivery of abstract interaction objects supporting the construction of logical dialogue kernels that are completely relieved from physical interaction characteristics, while appropriately supporting polymorphic physical-level binding

18.2.3 The Need for Automatic Interface Adaptation

In the context of universal access to information society technologies, due to the large diversity of potential target user groups and contexts of use, it is unrealistic to expect that a single-minded interface design instance will ensure high-quality interaction. Instead, it is most likely that alternative design decisions will have to be taken. Therefore, the ability of the interface to automatically adapt to the individual end-user, as well as to the particular context of use, is a required key property of universally accessible interactions. From a development point of view, such a facility reveals the need for explicit design and implementation of alternative interactive ways to enable different end-users to carry out the same task (i.e., alternative dialogue patterns for the same subtasks). During run-time, the software interface relying upon computable user-oriented knowledge (i.e., internally stored profiles) is responsible for assembling the eventual interface on-the-fly, by collecting and gluing together all the various end-user-specific constituent dialogues components. This type of initial best-fit automatic interface configuration, originally introduced in the context of the unified user-interfaces development framework (Stephanidis, 2001b), has been called *adaptability*. Additionally, the dynamic interface behavior to enable a continuous enhancement of the dialogue with the user by changing the original interface, based on interaction monitoring, has been commonly referred to as adaptivity or adaptive interaction. In this context, the software engineering of the user interface to accommodate such run-time behavior requires sophisticated techniques for software architecture and organization (Savidis and Stephanidis, 2001b; Stephanidis et al., 2001; see also Chapter 21 of this handbook).

18.2.4 The Need for Ambient Interactions

The concept of ambient computing reflects an infrastructure where users are typically engaged in mobile interaction sessions, within environments constituted by dynamically varying computational resources (see Chapter 60 of this handbook). In ambient interaction scenarios, the user carries a very small processing unit, for example, the size of a credit card, with an embedded operating system and wireless networking, including short-range radio networking like Bluetooth. Additionally, the user may optionally collect any number of wearable wireless gadgets. Once the On button of the processing unit is pressed, the system boots, and then seeks for in-range devices capable of hosting interaction. When such devices are detected, they are appropriately employed to support interaction. At some point, some devices get out of range (i.e., they are lost), and the system tries to use some other available devices to maintain interaction. If the available devices do not suffice for the current interaction purposes, the dialogue is considered stalled. When new devices become in-range (i.e., they are discovered), the system tries to engage those devices in interaction, either to resume dialogue from a stalled state, or to further optimize it by offering a better interaction alternative. This notion of mobile ambient interactions is illustrated in Figure 18.2, which depicts two layers of dynamically engaged interaction-capable devices: (1) inner layer wearable devices, which the user may or may not carry, depending on the situation, and which are not anticipated to vary on-the-move as frequently as environment devices; (2) outer layer ambient devices, for instance, the particular set of devices falling inside a wireless communication range with the mobile

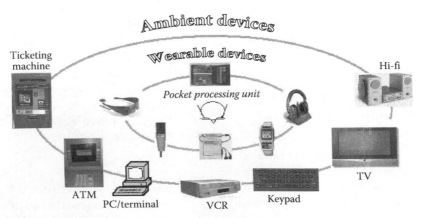

FIGURE 18.2 The two main layers of dynamically engaged ambient computing resources in the context of distributed mobile interactions. (From Savidis, A. and Stephanidis, C., *Interacting with Computers*, 18, 71–116, 2006.)

processing unit, which will normally vary according to the particular user location.

18.3 Analysis of Requirements

The identified software engineering requirements reflect two key levels of implementation functionality, that can be either supported by interface development tools, or need to be manually crafted by interface developers: (1) required functionality, being mandatory to address the identified software requirements; and (2) recommended functionality, offered to optimally address the identified requirements.

18.3.1 Metaphor Development

The metaphor development process is split in three distinct phases (see Figure 18.3): (1) *design* of the required metaphoric representation, which entails both the selection of suitable metaphoric entities and the definition of their computer equivalents in terms of presentation properties, interactive behaviors, and relationships; (2) *realization* of the interactive embodiment of the metaphor through the selection of media and modalities, interaction object classes, and associated attributes; and (3) *implementation* of a metaphor realization, through the provision of interface development software libraries, which comprise dialogue elements that comply with that particular metaphor realization.

Metaphor realization and metaphor implementation are distinguished to account for the fact that there may be many realizations of a real-world metaphor and many implementations for a particular realization of a metaphor. For instance, a room can be realized visually as a graphical entity, as in Card and Henderson (1987), but also nonvisually, as in COMONKIT (Savidis and Stephanidis, 1995a) and AudioRooms (Mynatt and Edwards, 1995). Additionally, various implementations can be built for a particular metaphor realization. This is the case with various existing window-based toolkits (e.g., OSF/Motif, MFC, InterViews, etc.), which may differ with respect

to their software implementations and programming models. However, all windowing toolkits implement a common set of dialogue techniques corresponding to the visual realization of the desktop metaphor. It follows, therefore, that such a distinction between metaphor realization and metaphor implementation is important, because it allows modifications to be introduced at a particular level without necessarily affecting the higher levels.

18.3.1.1 User-Oriented Design of Metaphoric Interaction

During the metaphor design and realization stages, specific user attribute values need to be considered. Hence, the resulting metaphor design(s) and realization(s) are directly associated to those user attribute values. One such representative example concerns the design and realization of the desktop metaphor, which is currently reflected in all windowing interactive environments. The original design had considered the needs of an average person working in an office and performing tasks primarily engaging information conveyed on paper. The resulting realization has been targeted toward sighted users, and has been based on the effective exploitation of the visual information-processing capability. It is argued that both accessibility and usability problems may arise when trying to deploy interaction metaphors across user populations other than those originally considered during the metaphor design and realization stages. The following are two examples of cases that can be characterized as less-than-perfect metaphor use:

- *Windowing interaction for blind users*: This scenario is typically reflected in existing screen readers, aiming to provide access to windowing applications by blind users (see Chapter 28 of this handbook). In this case, visual realizations of the desktop metaphor are reproduced in a nonvisual form. However, the metaphor realization is even closer to sighted user needs than the metaphor design itself, since specific visual interaction means are considered. In conclusion, fundamental entities (e.g., windows,

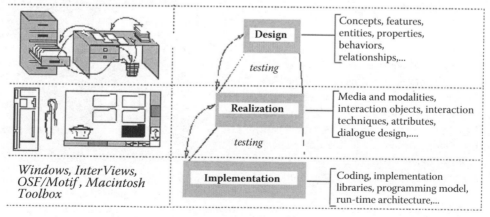

FIGURE 18.3　Metaphor development stages in the context of the unified user-interface development approach. (From Savidis, A. and Stephanidis, C., *Interacting with Computers*, 18, 71–116, 2006.)

icons, visual cues) and relationships (e.g., overlapping, spatial arrangement) in the desktop metaphor require considerable further investigation to verify whether their reproduction in a nonvisual form is meaningful.

- *Windowing interaction for children (preschool, early school)*: Various software products for educational or entertainment purposes have been produced that are targeted to children of the preschool or early school age, many of them working under popular windowing environments. Hence, the desktop metaphor is directly employed for interaction. However, some of the common properties of windowing environments, such as concurrency of input actions, multitasking (e.g., many applications), intuitive data exchange among applications (e.g., copy/paste), direct manipulation and direct activation (e.g., drag and drop), and so on, are mainly directed toward business/office tasks in a working environment. This problem has been recognized at an early point, leading to a new generation of *edutainment* software products, demonstrating a large amount of custom-made metaphoric interaction strategies like cartoons and animation, storytelling, and live characters.

18.3.1.2 The Key Role of Top-Level Containers

Containers are those classes of interaction objects that may physically enclose arbitrary instances of interaction objects. In running interactive applications, container object instances that are not enclosed within other containers are called *top-level containers*. For instance, windows providing interactive management facilities, which are not included within other windows, are called top-level windows. When designing metaphoric interaction, there can be many real-world analogies that are *transferred* in the interaction domain. Hence, practically, multiple distinct metaphors may be combined. For example, in windowing applications, the following interaction object classes are typically met, each representing a specific real-world analogy:

- Windows—*sheets of paper*
- Push buttons, sliders, potentiometers, and gauges—*electric devices*
- Check boxes—*form-filling*
- Menus—*restaurant*
- Icons—*signs*

Naturally, the original real-world physical regulations are effectively broken when containment relationships are designed (e.g., none would expect to see an electric button on a sheet of paper in the real world, while push buttons are normally embedded within windows). The effect of such containment relationships is that interaction metaphors are embedded at various levels in interactive applications. Related work in the past has investigated the design aspects of embedded interaction metaphors (Carroll et al., 1988).

In the context of universal access, the focus has been on the identification of those interactive entities that play a key role in providing the overall metaphoric nature of an interaction environment. It is likely that different interaction metaphors will have to be provided to diverse users to achieve accessible and high-quality interaction. The detection of those entities that largely affect the overall metaphoric look and feel of the interactive environment allows for providing different metaphoric representations for those entities, to derive alternative metaphoric environments. This potentially alleviates the overhead of designing from scratch alternative metaphoric artifacts. The resulting methodological framework to address this issue is based on the following principle:

The overall interaction metaphor is characterized and primarily conveyed by the metaphoric properties of top-level containers, while all embedded interaction objects are physically projected within the interaction space offered by the top-level containers.

This principle is depicted in Figure 18.4, where it clearly appears that embedded objects cannot alter the original characterization of the overall interaction metaphor.

18.3.1.3 A Metaphor Development Case for Accessibility

The practical applicability of this principle has been demonstrated within two specific research efforts: (1) the development of a nonvisual toolkit called COMONKIT (Savidis and Stephanidis, 1995b), providing a single top-level container with rooms-based interaction, and many standard interaction object classes like menu, push button, etc.; and (2) the development of a nonvisual toolkit called HAWK (Savidis et al., 1997), providing a generic container object (capable of realizing various metaphoric representations), as well as various conventional interaction objects classes (as in COMONKIT).

Testing the principles discussed previously in COMONKIT quickly led to the need for: (1) providing further variations on nonvisual presentation and feedback; and (2) supplying

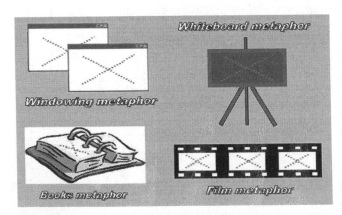

FIGURE 18.4 Some representative design scenarios of metaphoric elements, demonstrating how top-level containers largely affect the overall interactive experience of the resulting metaphoric environment. (From Savidis, A. and Stephanidis, C., *Interacting with Computers*, 18, 71–116, 2006.)

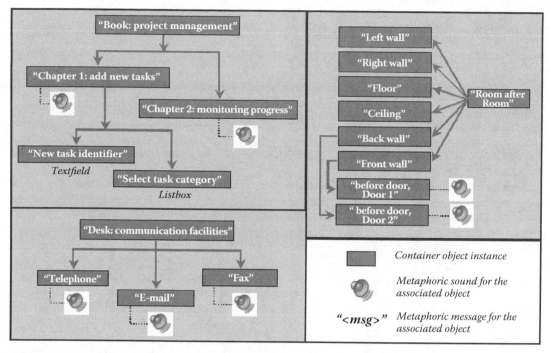

FIGURE 18.5 Conveying alternative metaphoric representations using digital audio effects and synthesized speech messages through container instances for nonvisual interaction object hierarchies using the HAWK development toolkit. (From Savidis, A. and Stephanidis, C., *Interacting with Computers*, 18, 71–116, 2006.)

alternative top-level metaphoric entities, like books, desk, and library, with genuine nonvisual realizations. This gave rise to the design and implementation of the HAWK toolkit, providing a generic container that may realize alternative metaphoric representations.

The generic container class in the HAWK toolkit provides various presentation and dialogue attributes through which alternative metaphoric nonvisual styles can be derived, by appropriately combining messages and sound feedback: (1) synthesized speech message (speech message to be given when the user focuses on the object); (2) Braille message (message displayed on Braille device when the user focuses on the object); (3) on-entry digitized audio file (to be played when the user focuses on the object); and (4) on-exit digitized audio file (to be played when the user leaves the object).

Figure 18.5 depicts the assignment of specific values to the container presentation attributes to derive alternative metaphoric representations. Three container instances, realizing books, desk-top, and rooms metaphors, respectively, are defined. In the case of nonvisual interaction, it has been relatively easy to design such parameterized metaphoric representations, due to the simplicity of the output channels (i.e., audio, speech, and Braille). This approach has been validated both with respect to its usability as an engineering method, as well as with respect to the usability of the produced interfaces (Savidis et al., 1997), while the HAWK toolkit has been used for the implementation of a nonvisual electronic book (Petrie et al., 1997), and the nonvisual component of a user-adaptable web browser (Stephanidis et al., 2001).

18.3.1.4 Advanced Graphical Metaphors

Apart from metaphors specifically designed for blind users, the introduction of radically new graphical interaction metaphors for sighted users has also been investigated. In this context, the experimentation target has been primarily twofold: (1) to study the usability of new highly interactive and visually dynamic artifacts, going beyond the traditional windowing style; and (2) to analyze the development barriers inherent in the implementation of those demanding artifacts, as well as the potential to directly combine them with existing windowing implementation toolkits. In this context, two parallel design and implementation efforts have been carried out:

- *The development of direct-manipulation immersive hierarchical information spaces*: In this context, a 3D space-efficient method to render hierarchical structures has been designed, named *inverted umbrella trees*, as opposed to typical cone trees. The resulting interaction toolkit has been employed to construct a real-time animated 3D interactive file manager (see Figure 18.6).
- *The development of dynamic real-time animation-based effects for Windows*: In this framework, specific categories of real-life phenomena, for example, fire, smoke, icing, dissolving (see Figure 18.7), have been simulated with an animation engine relying upon heuristic particle systems. Those effects have been designed to provide metaphoric feedback methods for application-specific events such as:

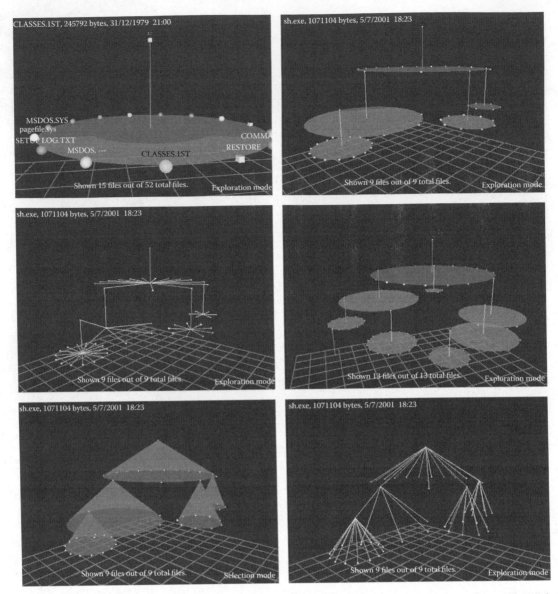

FIGURE 18.6 The implementation of a 3D direct-manipulation navigator in hierarchical information spaces, reflecting the rendering of two key metaphoric representations: (1) the newly designed inverted umbrella trees (first four snapshots) and (2) traditional cone trees (last two snapshots). (From Savidis, A. and Stephanidis, C., *Interacting with Computers*, 18, 71–116, 2006.)

- Resource demanding computation (net, processor, disc)
- Illegal data access (potential security leak)
- Failure to complete a requested operation (compile errors, failed search, could not save file)
- Application system crash
- Resource access errors (hard disc, net)
- Virus infection
- Application that could not terminate (zombies)
- Application halts (hanged) due to an error
- Application put in standby mode by user control
- Application idle for too long

From the usability point of view, early evaluation experiments indicated a high degree of acceptance by end-users, especially from young people who expressed particular satisfaction due to the presence of such appealing graphical artifacts within traditionally 2D rectangular windows. Additionally, within informal interviews, it became clear that the expectations of young people for interaction quality are today raised at a surprising level, mainly due to the tremendous progress on rendering and simulation quality of video games, which are widely popular among children, teenagers, and young adults. Moreover, most young users expressed the opinion that the static rectangular appearance of windowing applications is rather boring and obsolete.

Although the small-scale usability evaluation gave positive evidence for the introduction of more dynamic metaphoric phenomena within traditional interactive spaces, the software engineering conclusions are clearly less enthusiastic and encouraging. To implement the 3D rendering and navigation of hierarchical structures, OpenGL, a very popular and powerful

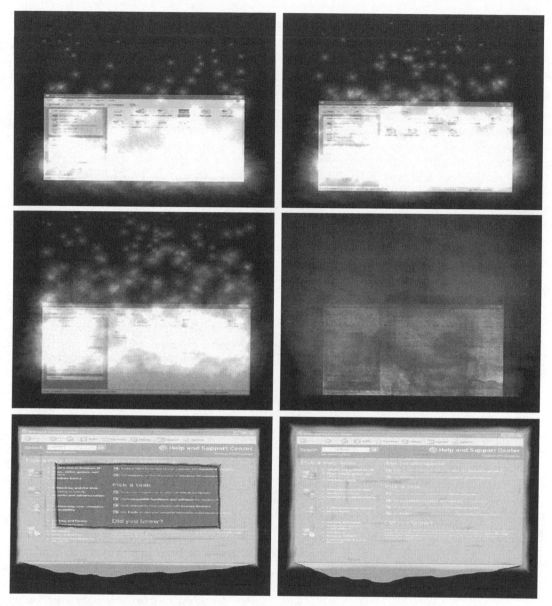

FIGURE 18.7 The implementation of real-time animation-based feedback effects for windowing applications, reflecting the dynamic rendering using heuristic particle systems of three key metaphoric events: fire, smoke, and progressive icing. (From Savidis, A. and Stephanidis, C., *Interacting with Computers*, 18, 71–116, 2006.)

portable software library for 3D graphics, has been employed. In this context, there was no way to interoperate or mix with typical windowing interaction elements to support the initial target of implementing real-time navigation in 3D, while providing details and parameter editing with form-based interfaces projected in the 2D plane. In other words, while programming 3D worlds, one should either completely forget 2D interaction elements, or pay the overhead of implementing those from scratch. Regarding the second experiment, although it concerned the implementation of 2D real-time animation effects, the display regulations and constraints of the windowing environment practically forbid the real-time implementation of dynamic effects within the normal windowing environment, especially when those effects cross the window display boundaries. In this case, it was only possible to implement dynamic animation effects with direct screen access mode, a mode that is not interoperable with normal windowing mode, treating windows merely as colored bitmaps. The overall implementation-oriented conclusion from this comprehensive experiment with radically new metaphors is that the present implementation regulations of windowing-style toolkits pose severe restrictions for the injection of graphics-intensive interactive metaphoric entities when the latter depart from the desktop metaphor. It is clear that the existing implementations are still very constraining, and do not yet consider openness and interoperability with potentially novel emerging interaction styles. Developers may only implement such artifacts in segregation, that is, as specific closed-world applications, or otherwise pay the huge overhead of implementing from scratch a complete development framework for their metaphors, for example, an alternative to the windowing toolkit.

18.3.2 Toolkit Integration

Development tools are considered to support toolkit integration if they allow importing *any* particular toolkit, so that *all* interaction elements of the imported toolkit effectively become implementationally available. For instance, if an interface builder providing interactive graphical construction techniques supports toolkit integration, then it should be possible to use the original construction techniques to also manipulate the imported object classes. Toolkit integration does not assume any particular interface building method, and may be supported by tools with various methods for interface implementation, such as programming-based, interactive constructions, state-based, event-based, demonstration-based, fourth-generation languages, and so on.

An explicit distinction needs to be made between the toolkit integration requirement and the multiplatform capability of certain toolkits. In the latter case, a single toolkit is provided with multiple (hard-coded) implementations across different operating systems, available when the toolkit product is released, that is, multiplatform toolkits such as XVT, ILOG Views, Qt, Zinc, ZooLib, Open Amulet, and JFC. In the former case, a tool is made open so that its users can take advantage of a well-documented functionality for connecting to arbitrary toolkits. The need for importing toolkits is evident in cases where the interaction elements originally supported by a particular interface development tool do not suffice. This is a possible scenario if interface development for diverse user groups needs to be addressed. For instance, in the context of dual interface development (Savidis and Stephanidis, 1995b), where interfaces concurrently accessible by sighted and blind users are built, nonvisual interaction techniques are required together with typical graphical interaction elements. Existing windowing toolkits do not supply such interaction techniques. Hence, integration of special-purpose, nonvisual interaction toolkits, such as COMONKIT (Savidis and Stephanidis, 1995a) or HAWK (Savidis et al., 1997) is necessary. In the more general case of universal access, it can be argued that scenarios necessitating toolkit integration are likely to emerge, not only as a result of user diversity, but also as a consequence of the proliferation of interaction technologies and the requirement for portable and platform-independent user interface software.

When toolkit integration is supported, interface tools supply mechanisms to developers for importing third-party development toolkits. Currently, a very small number of interface tools support this notion of *platform connectivity*, that is, the capability to implementationally connect to any particular target toolkit platform (e.g., OSF/Motif, MFC, Xt/Athena, etc.), enabling developers to manipulate its interaction elements as if they were an integral part of the interface tool. The first tool known to provide comprehensive support for toolkit integration was SERPENT (Bass et al., 1990), a user-interface management system (UIMS; Myers, 1995), where the toolkit layer was termed *lexical technology layer*. The architectural approach developed in the SERPENT UIMS revealed key issues related to the programmatic interfacing of toolkits. Toolkit integration has also been supported by the HOMER UIMS (Savidis and Stephanidis, 1998), developed to facilitate the construction of dual user interfaces. The HOMER UIMS provided a powerful integration model, which is general enough to enable the integration of nonvisual interaction libraries with traditional visual windowing toolkits. More recently, the I-GET UIMS (Savidis and Stephanidis, 2001a) provided more comprehensive toolkit integration facilities, enabling multiple toolkits to be imported and deployed concurrently through the I-GET programming language (Savidis, 2004). The MFC toolkit integration and deployment process are detailed in (Hatziantoniou et al., 2003).

18.3.2.1 Required Functionality for Toolkit Integration

The required development tool properties for toolkit integration are intended to characterize a particular tool with respect to whether it supports some degree of openness, so that interaction elements from external (to the given interface tool) toolkits can be utilized (subject to some particular implementation restrictions). For a user interface tool to support toolkit integration, the required properties are twofold:

- Ability to link or mix code at the software library level (e.g., combining object files, linking libraries together)
- Support for documented source-code hooks to support interconnections at the source code level (e.g., calling conventions, type conversions, common errors and compile conflicts, linking barriers)

If the required properties are present, it is possible for the developer to import and combine software modules that utilize interaction elements from different toolkits. In particular, the properties are satisfied in those cases where the imported toolkit provides new styles of interaction elements. This is, for instance, the case when the interface tool is a programming-based library of windowing interaction elements and the target toolkit offers audio-processing functionality for auditory interaction. In such cases, potential conflicts can be easily resolved and elements from the two toolkits can be combined. In contrast to this scenario, the required tool properties are currently not met when the imported toolkit supplies similar categories of interaction elements with the interface tool being used. For example, when trying to combine various libraries of graphical interaction elements delivered for the same programming language, various conflicts may arise, such as:

- Link conflicts at the binary library level, due to commonly named functions or global objects
- Compiling conflicts, due to commonly named data structures, classes, or name spaces

- Execution conflicts, due to system-level conflicting configurations, conflicting access to shared resources, or inability to combine in parallel the toolkit main loops

18.3.2.2 Recommended Functionality for Toolkit Integration

In this context, a development tool should offer additional means to effectively support toolkit integration. These features, constituting the comprehensive set of recommended tool properties for toolkit integration, are:

- Support for well-behaved (i.e., functionally robust and reliable) and well-documented (i.e., developers should be able to predict functional behavior from comprehensive documentation resources) compilation and linking cycles for interfaces utilizing the imported toolkits; this applies to all types of interface building methods, not only to methods supported by programming-oriented tools.
- Possibility to adopt a single implementation model for all imported toolkits. Thus, when the development tool provides visual construction methods, the same facilities should allow manipulation of interface elements from all the integrated toolkits.
- Possibility to change aspects of the programming interface (i.e., the programmable view) of any of the imported toolkits. This would minimize the effort required for programmers to become familiar with the programming style of the newly imported toolkits.
- Effective resolution of the typical conflicts, such as linking, compilation, and execution, arising from the combination of multiple style-similar toolkits.
- Importability of any type of interface toolkit, irrespective of the style of interaction supported (e.g., windowing toolkits, auditory/tactile toolkits, VR-based interaction toolkits).
- Toolkit interoperability (programmers should be enabled to combine toolkits for creating cross-toolkit object hierarchies.

Today, no single interface tool is reported in the literature that exhibits this comprehensive set of recommended properties. Regarding the notion of a single programming interface, multiplatform toolkits already provide adequate support, by means of a single predefined programming layer. Only UIMS tools like SERPENT (Bass et al., 1990), HOMER (Savidis and Stephanidis, 1995b), and I-GET (Savidis and Stephanidis, 2006; see also Chapter 27 of this handbook) supply adequate support for toolkit integration, by enabling the establishment of developer-defined programming interfaces on top of software toolkits. When a single programming interface is supported for all platforms, one important question concerns the look and feel of supported interaction objects across those platforms. There are three alternative approaches in this case:

- *Employing the native interaction controls of the platform*: For instance, making direct use of the platform's controls, along with their presentational and behavioral attributes (e.g., as in Java's early version of the AWT library, or multiplatform toolkits like XVT, Qt, Zinc, etc.)
- *Mimicking the native controls of the platform*: For instance, providing controls that are capable of altering their presentation and behavior to match the respective attributes of the target platform's native controls (e.g., as in Java's JFC Swing library)
- *Providing custom interaction elements across all platforms*: For instance, providing custom controls that look and behave the same across platforms, independently of the platforms' native controls (e.g., Tcl/Tk; Ousterhout, 1994).

Regarding toolkit interoperability and mixed toolkit hierarchies (see Figure 18.8), only the Fresco User Interface System (X Consortium, 1994) is known to support cross-toolkit hierarchies. In particular, Fresco facilitates the mixing of InterViews-originated objects with Motif-like widgets, relying upon the CORBA implementation of user interface elements as distributed replicated objects. The latter is technically distinguished from multiple look-and-feel support offered by a single toolkit,

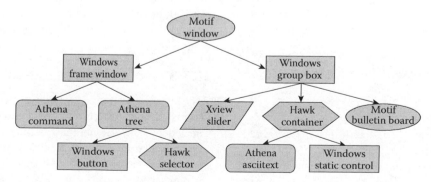

FIGURE 18.8 A hypothetical implementation scenario for cross-toolkit mixed interaction-object hierarchies, where object instances from Windows, Motif, Xt/Athena, and Xview are engaged. (From Savidis, A. and Stephanidis, C., *Interacting with Computers*, 18, 71–116, 2006.)

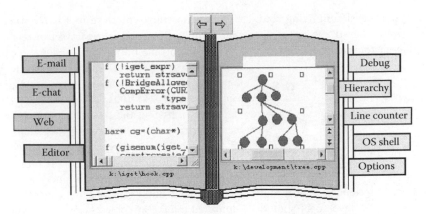

FIGURE 18.9 A hypothetical dialogue artifact for mixing container interaction objects of different software toolkits (i.e., books and windowing) complying with different interaction metaphors. (From Savidis, A. and Stephanidis, C., *Interacting with Computers*, 18, 71–116, 2006.)

like JFC (by Sun Soft Corp.) or WDS (by Infinite Monkeys Inc.), as it allows run-time look-and-feel mixing.

It should be noted that, even though elements from different toolkits may be combined, possibly employing a different look and feel, consistency is not necessarily compromised. Figure 18.9 outlines a scenario depicting the combination of a windowing toolkit with a toolkit implementing the book metaphor, exemplifying support for cross-toolkit object hierarchies at the implementation level. The advantages of mixing multiple toolkits are more evident in those cases where combined toolkits offer container objects with different metaphoric representations (like the example of Figure 18.9), thus practically leading to the fusion of alternative metaphors. In this context, toolkit interoperability plays a critical software engineering role, defined technically as follows:

> To enable the construction of interfaces by combining interaction objects from different toolkits, possibly complying with different metaphors that may impose diverse physical realizations. To enable embedded objects display and function inside the physical context of containers, through generic embedding and interoperation protocols.

Toolkit interoperability is a very demanding requirement, not yet supported by user interface toolkits, necessitating development methods and techniques for the following issues:

- *Space negotiation protocols*, to allow embedded objects to request and negotiate physical space provided by container objects, e.g.:
  ```
  "request space: 2D, graphic,
  rectangle[120x45]"
  "provide space: 2D, graphic, color[16
  bits], rectangle[12, 45, 120, 90]"
  ```
- *Space representation geometry*, to allow embedded objects to query the type of physical space and its relevant parameters offered by container objects, e.g.:
  ```
  "class: graphic, dimensions: 2, axis:
  yes[X: integer, Y: integer]"
  ```

```
"class: auditory, dimensions: 0, axis:
no, attributes: yes[volume: integer]"
```

- *Projection logic and constraints*, to allow embedded objects to define the way they can be projected in the context of container objects, e.g.:
  ```
  "transform: auditory → graphic"
  "transform: o1.id → o2.label"
  "transform: if (o1.audiofile ≠ empty)
  then [o2 = button, o2.notify =
  [play(o1.audiofile)]]"
  ```
- *Semantic annotation and mapping regulations*, enabling embedded objects to encapsulate application-oriented semantic data, so that the transformation to a projected entity may reflect the encapsulated semantics, e.g.:
  ```
  "if (o1.semantic.role = (emphatic))
  then o2.font = (typeface.Bold.12)]"
  ```

These issues are only representative, while the supplied examples outline roughly the need for protocols and microlanguages to allow for specifying and implementing on-the-fly interoperation logic, independent from the particular representation method used in the examples. It is believed that interoperability is an issue deserving further attention, as it can significantly boost the capability to implement new categories of artifacts through the open concurrent deployment of multiple toolkits.

18.3.3 Toolkit Augmentation

Augmentation is defined as the design and implementation process through which additional interaction techniques are injected into the original (native) interaction elements supplied by a particular toolkit, thus leading to improved accessibility or enhanced interaction quality for specific user categories. Newly introduced interaction techniques become an integral part of the existing interaction elements, while existing applications that make use of the toolkit inherit the extra dialogue features, without requiring revisiting of their implementation (e.g., through recompilation, or relinking, in the case of programming-based implementation approaches).

The need for toolkit augmentation arises mainly from shortcomings or design limitations regarding the supplied interaction entities of existing user interface development toolkits. Because the majority of interactive software products are built by utilizing such toolkits, these shortcomings are propagated to the resulting applications. For instance, graphical interaction libraries do not support voice-controlled interaction. Thus, nonvisual access to the interface, in a situation where direct visual attention is not possible (e.g., while driving) cannot be supported. Another typical example where augmentation is required is the case of accessibility of window-based interaction by motor-impaired users. In this case, additional interaction techniques are needed to enable a motor-impaired user to access the user interface through specialized input devices (e.g., binary switches). In both cases, augmentation implies the development of new interaction techniques, as well as the integration of (support for) special-purpose input/output (I/O) devices (e.g., voice I/O hardware, binary switches).

In Savidis et al. (1997), the augmentation of the basic windows object library to provide switch-based access through automatic/manual scanning is discussed (see also Chapter 35 of this handbook). One of the most important enhancements in this work has been the decomposition and augmentation of the user dialogue for performing window management. All top-level window interaction objects have been augmented with an additional accessible toolbar, supporting scanning interaction, thus providing all window-management operations in an accessible form (see Figure 18.10). Apart from top-level windows (e.g., FrameWindow class in the windows object library), augmented dialogues for the remaining object categories (e.g., button categories, container classes, composite objects, and text-entry objects) have been also designed and implemented.

18.3.3.1 Required Functionality for Toolkit Augmentation

The required user interface development tool properties for toolkit augmentation depict the set of functional capabilities that enable augmented object classes to be introduced into the original software libraries of a toolkit. The set of these properties is not bound to any particular programming language or category of programming languages (e.g., procedural, object-oriented, scripting). However, it is assumed that typical development functionality, such as introducing interaction object classes, accessing or modifying object attributes, and implementing callbacks or event handlers, is supported. The required facilities are:

- Provision of support for device integration. For example, when one considers input devices, this may require

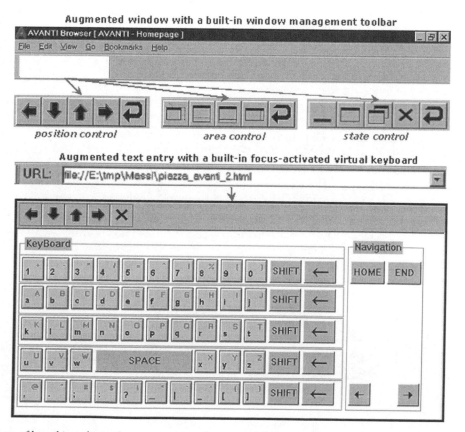

FIGURE 18.10 Snapshots of key object classes from an augmented version of the windows object library; all classes of top-level windows have been augmented with an additional toolbar, supporting scanning interaction, providing accessible window management operations. (From Savidis, A. and Stephanidis, C., *Interacting with Computers*, 18, 71–116, 2006.)

low-level software to be written and implemented either through a *polling-based* scheme (i.e., continuously checking device status), or through a *notification-based* scheme (i.e., device-level software may asynchronously send notifications when device input is detected). In either case, the newly introduced input device will have to be accompanied by the necessary application programming extensions, for instance, event class and low-level event generation.

- Provision of methods to manipulate the focus object (i.e., the object to which input from devices is redirected). During interactive episodes, different interaction objects normally gain and lose input focus via user control (e.g., through mouse or keyboard actions in windowing environments). To augment interaction, it is necessary to have programmatic control of the focus object, since any device input originating from the additional peripheral devices will need to be communicated to the current focus object.

- Programmable manipulation of the object hierarchy. When implementing augmented interaction, it is necessary to provide augmented analogies of the user's control actions, as the user must be enabled to "navigate" within the interface. Hence, the hierarchical structure of the interface objects must be accessible to the programmer when implementing the augmented navigation dialogue.

The relationship between the native object classes and the augmented object classes is a typical inheritance relationship, since augmented toolkit object classes inherit all the features of the respective original classes (see Figure 18.11). This scheme can be implemented either via subclassing, if the toolkit is provided in an object-oriented programming (OOP) framework, or via composition, in the case of non-OOP frameworks. In the latter case, composition is achieved through the definition of an augmented object structure that comprises the original object features (by directly encompassing an instance of the original object class), as well as the newly introduced augmented properties. Such an explicit instantiation is necessary to physically realize the toolkit object, which, in the case of subclassing, would be automatically carried out. Additionally, the interface tool should provide facilities for specifying the mapping between the attributes of the toolkit object instance and its corresponding features defined as part of the new object structure (e.g., procedural programming, constraints, monitors).

18.3.3.2 Recommended Functionality for Toolkit Augmentation

The comprehensive set of recommended tool properties for full support of toolkit augmentation introduces additional functional properties that need to be present in the development tool. These properties, primarily targeted toward enabling an easier and more modular implementation of the augmented object classes, are (see also Figure 18.12):

- Support for the extension of object attributes and methods on top of the original toolkit classes, that is, without affecting their original programming interface. This alleviates the problem of defining new object classes, while enabling already existing applications to directly make use of the augmented dialogue features without modifications at the source code level.

- Provision of a syntactically visible (i.e., accessible to programming entities outside the scope of the object class) extensible constructor, where additional interactive behavior can be added. This allows installing new event handlers and performing all the necessary initializations directly at the original class level. The notion of a constructor (in its object-oriented sense) may also be supported by nonprogramming-oriented interface tools, by means of user-defined initialization scripts.

- Provision of a modular device installation/integration layer, so that new device input can be attached to the toolkit input-event level. This facilitates the management of additional peripheral devices, through the original event management layer of the given interface development tool.

The most important advantage of a tool exhibiting the previously mentioned comprehensive set of recommended properties for toolkit augmentation is the elimination of the need to introduce new specialized object classes encompassing the augmented interaction features. As a result, it is not necessary to make changes to existing applications. In comparison, when only the required properties are supported, the augmented capabilities are conveyed by newly defined object classes, not originally deployed within previously developed applications. If developers desire introducing some additional dialogue control logic within existing applications, so as to take advantage of the augmented

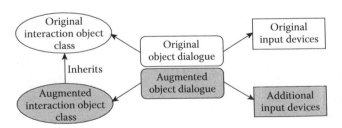

FIGURE 18.11 Relationship between original and augmented toolkit classes in the context of the required functionality to support toolkit augmentation. (From Savidis, A. and Stephanidis, C., *Interacting with Computers*, 18, 71–116, 2006.)

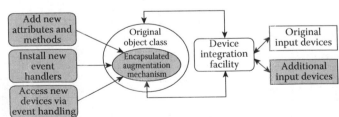

FIGURE 18.12 Toolkit augmentation to support the recommended functionality, reflecting the required capabilities for introducing dynamic object extensions. (From Savidis, A. and Stephanidis, C., *Interacting with Computers*, 18, 71–116, 2006.)

object attributes and methods, modifications, recompilation, and relinking are evidently required in all cases. Even if no changes are needed in existing applications, recompilation or relinking may still be necessary (especially in the case of purely compiled languages and static linkage to the toolkit libraries).

18.3.4 Toolkit Expansion

Expansion over a particular toolkit is defined as the process through which toolkit users (i.e., user-interface developers) introduce new interaction objects not originally supported by that toolkit. An important requirement of toolkit expansion is that all newly introduced interaction objects are made available, in terms of the manipulation facilities offered, in exactly the same manner as original interaction objects—in effect rendering add-on objects indistinguishable from original objects from the developer's point of view.

A typical case of toolkit expansion is the introduction of new interaction object classes, built on top of existing interaction facilities, to embody an alternative interactive metaphor, such as the case of a booklike metaphor in Moll-Carrillo et al. (1995). Toolkit expansion may also be necessitated by domain-specific or application-specific functionality that needs to be presented as a separate self-contained interactive entity (for example, a temperature-pressure graphical interaction object, to be employed in the implementation of a factory process control system, used for temperature and pressure visualization and control).

One of the early toolkits providing expansion support was the generic Xt toolkit, built on top of the Xlib library for the X Windowing System. The Xt mechanism provides a template widget structure, where the developer has to provide some implemented constructs. The mechanism of Xt is complex enough to turn expansion to an expert's programming task. Other approaches to expansion concern toolkit frameworks supported by OOP languages, such as C++ or JAVA. If key superclasses are distinguished, with well-documented members providing the basic interaction object functionality, then expansion becomes a straightforward subclassing task. This is the typical case with OOP toolkits like the MS Windows object library or InterViews.

Apart from user interface programming toolkits, the expansion mechanism is also supported in some higher level development tools, such as user interface management systems (UIMSs). In Peridot (Myers, 1988), the demonstration-based method for defining interactive behaviors, leads to the introduction of interaction objects that can be subsequently recalled and employed in interface construction. This capability can be viewed as expansion functionality. The Microsoft Visual Basic development environment for interface construction and behavior scripting is currently supported with various peripheral tools from third-party vendors. One such tool, called VBXpress, introduces expansion capabilities by supporting the interactive construction of new VBX interaction controls.

Finally, a more advanced approach to toolkit expansion concerns distributed object technologies for interoperability and component-based development. New interaction objects can be introduced through the utilization of a particular tool, while being employed by another. This functionality is accomplished on the basis of generic protocols for remote access to various software resources, supporting distribution, sharing, functionality exposure, embedding, and dynamic invocation. Microsoft has been the first to allow ActiveX controls to be embedded in JavaBeans containers. JavaSoft's Migration Assistant accomplishes exactly the opposite (or symmetric) link, thus enabling JavaBeans to work inside ActiveX containers. The result is that today there is software enabling the interoperation of ActiveX and JavaBeans components in both directions. For programmers using one of these component categories, this capability is an expansion of the set of available interaction controls. For instance, ActiveX programmers can use directly JavaBeans objects within ActiveX containers.

18.3.4.1 Required Tool Properties for Toolkit Expansion

The required development tool properties relate to the presence of an appropriate object expansion framework supported by the user interface development tool. The most typical forms of such an expandable object framework are:

- *Super class*: Expansion is achieved by taking advantage of the inheritance mechanism in OOP languages, while expanded objects are defined as classes derived from existing interaction object classes. Examples of such an approach are the MS Windows object library and the InterViews toolkit.

- *Template structure*: In this case, an object implementation framework is provided, requiring developers to fill in appropriate implementation gaps (i.e., supply code), mainly relevant to dialogue properties, such as visual attributes, display structure, and event handling. The most representative example of this approach is the Xlib/Xt widget expansion model of the X Windowing System. The JavaBeans approach is a more advanced version of an object implementation framework.

- *Application programming interface (API)*: In this case, resource manipulation and event propagation correspond to services and event notifications, realizing object management APIs that are built on top of standardized communication protocols. This approach is usually blended with an object-oriented implementation framework, providing a way to combine objects irrespective of their binary format, thus achieving open, component-based development. The ActiveX model is the most typical example. The OMG CORBA model, though not providing a standardized API, allows customized APIs to be built (e.g., the Fresco User Interface System; X Consortium, 1994).

- *Physical pattern*: In this case, newly introduced object classes are built via interactive construction methods. For example, construction could start from basic physical structures (e.g., rectangular regions), adding various physical attributes (e.g., textual items, colors, borders, icons), defining logical event categories (e.g., selected), and implementing behavior via event handlers (e.g., highlighting on gaining focus, returning to normal state upon losing focus). The way in which physical patterns are supported varies depending on the tool. For instance, Microsoft Visual Basic provides an exhaustive definition and scripting approach, while Peridot (Myers, 1988) offers a demonstration-based approach.

- *4GL model*: Fourth-generation interface development languages support the combination of their interaction object model with the dialogue construction methods, allowing new object classes to be built. These dialogue implementation methods are to be utilized for implementing the interactive behavior of the new objects. The I-GET UIMS is a typical example of an interface tool supporting a 4GL expansion model (Savidis et al., 2002).

18.3.4.2 Recommended Functionality for Toolkit Expansion

The comprehensive set of development tool properties for toolkit expansion includes one additional recommended functional property, namely:

- *Closure*: If an interface tool is to fully support object expansion, then it should allow developers to implement the dialogue for new interaction objects via its native dialogue construction facilities (see Figure 18.13).

In other words, developers should be allowed to define dialogues for new interaction objects via the facilities they have already been using for implementing conventional interfaces. For instance, in an interface builder, the full functionality for expansion is available only when interactive object design and implementation is facilitated.

18.3.5 Toolkit Abstraction

Toolkit abstraction is defined as the ability of the interface tool to support manipulation of interaction objects that are entirely decoupled from physical interaction properties. Abstract interaction objects are high-level interactive entities reflecting generic behavioral properties with no input syntax, interaction

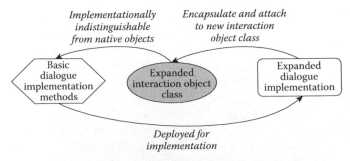

FIGURE 18.13 Closure property in maximally supported expansion: the resulting expanded objects are constructed through the original dialogue implementation facilities, made indistinguishable under a development perspective, from the native object classes. (From Savidis, A. and Stephanidis, C., *Interacting with Computers*, 18, 71–116, 2006.)

dialogue, and physical structure. However, during execution, abstract interaction objects are automatically mapped to physical interaction objects of the employed toolkit. An example of an abstract interaction object (namely a *selector*) is provided in Figure 18.14. Such an abstract interaction object has only two properties: the number of options and the selection (as an index) made by the user. Additionally, it may encompass various other programming attributes, such as a callback list (i.e., reflecting the *select* method), and a Boolean variable to distinguish among multiple-choice and single-choice logical behaviors.

As illustrated in Figure 18.14, multiple physical interaction styles, possibly corresponding to different interaction metaphors, may be defined as physical instantiations of the abstract *selector* object class. When designing and implementing interfaces for diverse user groups, even though considerable structural and behavioral differences are naturally expected, it is still possible to capture various commonalities in interaction syntax, by analyzing the structure of subdialogues at various levels of the task hierarchy. To promote effective and efficient design, implementation, and refinement cycles, it is crucial to express such shared patterns at various levels of abstraction to support modification only at a single level (i.e., the abstract level). Such a scenario requires implementation support for: (1) organizing interaction objects at various levels of abstraction; (2) enabling developers to define the way in which abstract objects may be mapped (i.e., physically instantiated) to appropriate physical artifacts; and (3) providing the means to construct dialogues composed of abstract interface objects. Abstract interaction objects can be employed for the design and implementation of generic reusable dialogue components that do not reflect physical interaction properties at development time. In this sense, such dialogue patterns are not restricted to any particular user group or interaction style. The introduction of the intermediate physical instantiation levels is also required, so that abstract forms can be mapped to concrete physical structures. By automating such an instantiation mechanism, development for diverse target user groups is facilitated at an abstract layer, while the physical realization is automated on the basis of an appropriate object instantiation mechanism. Chapter 27 of this handbook discusses in detail a user-interface programming language that employs abstract interaction objects.

The notion of abstraction has gained increasing interest in software engineering as a solution toward recurring development problems. The basic idea behind abstraction is the establishment of software frameworks that clearly separate the implementation layers relevant only to the general problem class, from the specific software engineering issues that emerge when the problem class is met with alternative instantiations. The same approach applies to the development of interactive systems to allow a dialogue structure composed of abstract objects to be retargeted to various alternative physical forms through an automation process configured and controlled by the developer.

18.3.5.1 Required Functionality for Toolkit Abstraction

The required development functionality for toolkit abstraction is targeted toward facilitating interface construction based on

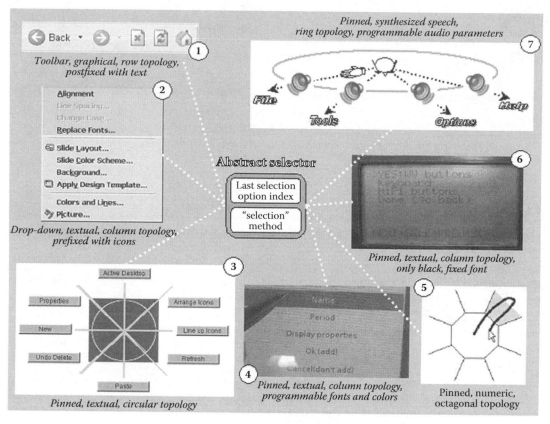

FIGURE 18.14 Alternative instantiations of an abstract selector varying with respect to topology, display medium, content of options, input devices, and appearance attributes. (1) and (2) are from WNT, (3) is from Bronevetsky (2003), (4) and (6) are from the 2WEAR Project, (5) is from McGuffin et al. (2002), and (7) is from Savidis et al. (1996). (From Savidis, A. and Stephanidis, C., *Interacting with Computers*, 18, 71–116, 2006.)

abstract objects. Additionally, some high-level implementation issues reveal the complexity of explicitly programming abstract objects if the interface development tool does not support them inherently. The required functionality for toolkit abstraction is:

- *Closed set of abstractions*: For instance, predefined collection of abstract interaction object classes is provided.
- *Bounded polymorphism*: For example, for each abstract object class C, a predefined mapping scheme S_C is supported, for the run-time binding of abstract instances to physical instances, the latter belonging to a predefined list of alternative physical object classes $P_1, \ldots, P_{n(C)}$.
- *Controllable instantiation*: For each abstract object instance I of a class C, it is possible to select at development time the specific physical class $P_j \in S_C$ to which instance I will be bind at run-time.

These properties enable the developer to instantiate abstract objects while having control over the physical mapping schemes that will be active for each abstract object instance. Mapping schemes define the candidate classes for physically realizing an abstract object class.

An approach to implement the software structure accommodating the required functionality for abstraction is provided in Figure 18.15. As shown, abstract interaction objects reflect concrete program classes, which delegate their physical instantiation as concrete physical object classes to a respective mapping scheme class. The key point to this design is the mapping scheme class, which bridges classes of abstract interface objects with classes of physical interface objects, while also preserving the independence among the abstract and physical interface object classes. Abstract objects upon instantiation never directly instantiate physical object classes, but instead request their mapping scheme instance object to perform physical instantiation (through the *Create* function). The interface programmer may extract or even modify the mapping scheme instance of an abstract object, and may alter the physical instance of its associated abstract object (i.e., by calling *Destroy* followed by a *Create* with the desirable physical object class name).

18.3.5.2 Recommended Functionality for Toolkit Abstraction

The recommended functionality introduced here can be used to judge whether interface tools provide powerful methods for manipulating abstractions, such as defining, instantiating, polymorphosing, and extending abstract interaction object classes. Support for such facilities entails the following:

- *Open abstraction set*: For instance, facilities to define new abstract interaction object classes.

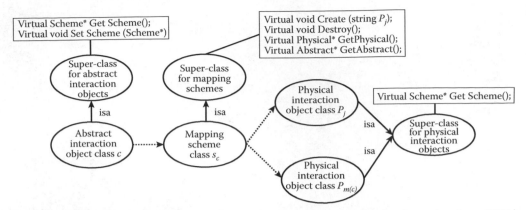

FIGURE 18.15 The software programming structure to implement the required functionality for abstraction, in an OOP language, enabling abstract objects to on-the-fly retarget to alternative mapping schemes as well as to alternative physical object instances. (From Savidis, A. and Stephanidis, C., *Interacting with Computers*, 18, 71–116, 2006.)

- *Open polymorphism*: For instance, methods to define alternative schemes for mapping abstract object classes to physical object classes, so that, for example, an abstract selector may be mapped to a visual column menu and a nonvisual list-box.
- *Physical mapping logic*: For instance, facilities for defining run-time relationships between an abstract instance and its various concurrent physical instances. This may require the definition of attribute dependencies among the physicals and abstract instances, together with propagation of callback notifications (e.g., if a logical event occurs in the context of a physical instance, its associated abstract instance must be appropriately notified).
- *Physical instance resolution*: When an abstract instance I of class C is employed in interface implementation, syntactic access to all plausible physical instances of classes $P_j \in S_C$ should be facilitated.

Currently, the recommended functionality can be normally accommodated in general-purpose object-oriented programming (OOP) language like C++ or Java, requiring demanding software patterns to be manually programmed by interface developers. Additionally, the I-GET language (Savidis, 2004; see also Chapter 27 of this handbook) provides genuine support for the specification of abstract interaction objects, with polymorphic instantiation relationships and multiple physical mapping schemes, while facilitating controllable instantiation and syntactical resolution of the physical instance.

18.3.6 Automatic Interface Adaptation

To accomplish the run-time delivery of user and usage-context adapted user interfaces, developers need to implement ways of manipulating during run-time alternative dialogue components (see Chapter 21 of this handbook). In this context, the proposed functionality is not distinguished into required or recommended, as with previously discussed software engineering requirements for handling interaction objects. Instead, a comprehensive set of functional requirements is defined to support the adaptation-oriented manipulation of dialogue components. These requirements are described in the following.

18.3.6.1 Dialogue Component Model and Dynamic Interface Assembly

This requirement reflects the necessity to provide interface developers with a genuine component mode, so as to support a straightforward mapping from the design domain of dialogue design patterns to the implementation domain of fully working dialogue components. Additionally, the effective run-time manipulation of dialogue components requires facilities for dynamic component instantiation and destruction, in an imperative or declarative manner. In this context, *imperative* means that developers add instantiation or destruction statements as part of a typical program control flow (i.e., via statements or calling conventions). *Declarative* means that the instantiation or destruction events are associated with declarative constructs, such as precondition-based activations or notification handlers. Normally, instantiation or destruction of components will be coded by developers in those points within the implementation that certain conditions dictating those events are satisfied. For this purpose, the declarative approach offers the significant advantage of relieving developers from the burden of algorithmically and continuously testing those conditions during execution, for each component class. Normally, in general-purpose programming-based toolkits the delivery of those facilities is mostly trivial; however, in specialized interface development instruments (e.g., task-based development or model-based development) the software engineering methods offered for component manipulation are less powerful.

The software organization of components should reflect the hierarchical task-oriented discipline of the interface design. This implies that some components may be dependent on the presence of other, hierarchically higher, dialogue components. This reflects the need to make the interface context for particular subtasks available (to end-users of the interface) if and only if the interface context for *ancestor* tasks is already available. For instance, the Save File As dialogue box may appear only if the Editing File interface is already available to the user.

Additionally, it is critical to support for orthogonal expansion of interface components. More specifically, when adding new dialogue components, or even interaction-monitoring components, the overall implementation structure should encapsulate the appropriate placeholders to accommodate such component extensions. Finally, the activation of components should be orchestrated to take place on the fly, reflecting the genuinely run-time decision for the end-user best-fit dialogue components. In this context, the organization structure of the user interface should not reflect a particular hard-coded interface instance, but has to effectively accommodate the dynamic process of hierarchical interface assembly and delivery from run-time-chosen components. An appropriate way to address such implementation requirements is parametric polymorphic containment hierarchies.

18.3.6.2 Parametric Polymorphic Containment Hierarchies

In the context of the AVANTI user-adapted web browser (Stephanidis et al., 2001), it has been necessary to support physical dialogue components for which the contained items could vary on-the-fly, since alternative designed and implemented versions of such embedded components had to be supported (see Figure 18.16). This functional feature required the software engineering of container components to support effectively such dynamic containment, through methods of parametrization and abstract APIs, for instance, polymorphism. Some

similarities with dynamic interface assembly can be found in typical web-based applications delivering dynamic content. The software engineering methods employed in such cases are based on the construction of application templates (technologies such as Active Server Pages by Microsoft, ASP, or Java Server Pages, JSP, by JavaSoft, are usually employed), with embedded queries for dynamic information retrieval, delivering to the user a web page assembled in that moment. In this case, there are no alternative embedded components, just content to be dynamically retrieved, while the web-page assembly technique is mandatory when HTML-based web pages are to be delivered to the end-user (in HTML, each time the content changes, a different HTML page has to be written). However, in case a full-fledged embedded component is developed (e.g., as an ActiveX object or Java Applet), no run-time assembly is required, since the embedded application internally manages content extraction and display, as a common desktop information retrieval application.

The software implementation is organized in hierarchically structured software templates, in which the key placeholders are parametrized container components. This hierarchical organization mirrors the fundamentally hierarchical constructional nature of interfaces. The ability to diversify and support alternatives in this hierarchy is due to containment parametrization, while the adapted assembly process is realized by selective activation, engaging remote decision making on the basis of end-user and usage context information.

In Figure 18.17, the concept of parametric container hierarchies is illustrated. Container classes expose their containment

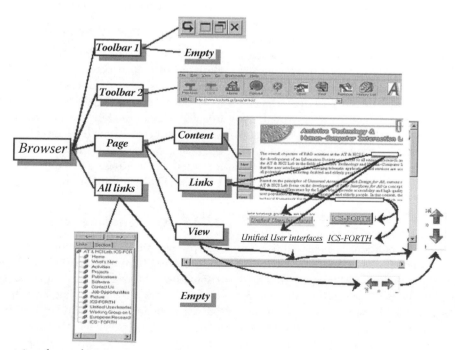

FIGURE 18.16 Parametric polymorphic containment with variant constituent components in the AVANTI browser. The indication Empty signifies components whose presence may have to be omitted upon dynamic interface delivery for certain user categories. (From Savidis, A. and Stephanidis, C., *Interacting with Computers*, 18, 71–116, 2006.)

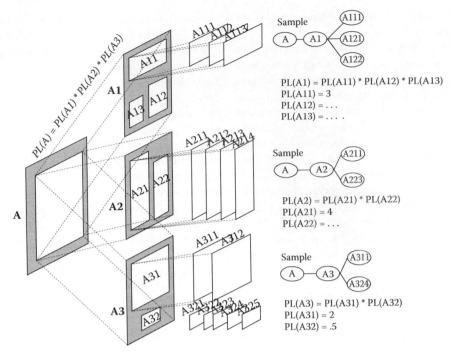

FIGURE 18.17 The notion of dynamic polymorphic hierarchical containment in automatically adapted interactions to cater for the run-time interface assembly process. (From Savidis, A. and Stephanidis, C., *Interacting with Computers*, 18, 71–116, 2006.)

capabilities and the type of supported contained objects by defining abstract interfaces (i.e., abstract OOP classes) for all the contained component classes. These interfaces, defined by container class developers, constitute the programming contract between the container and the contained classes. In this manner, alternative derived contained-component classes may be instantiated at run-time as constituent elements of a container. Following the definition of polymorphic factor *PL*, which provides a practical metric of the number of possible alternative run-time configurations of a component, the *PL* of the top-level application component gives the number of all possible alternative dynamically assembled interface instances (see also Figure 18.17). Notably, this does not reflect the total number of legal interface instances, as the combination of alternatives is not freely supported, but provides significant evidence of the potential polymorphic factor of such a hierarchical adaptation-oriented interface structure.

18.3.6.3 Dynamically Controlled Interaction Monitoring

To support adaptive interface behavior, the run-time collection and analysis of interaction monitoring information is required. This approach, which has been traditionally employed in adaptive interface research (Dieterich et al., 1993), has been also implemented in the context of the AVANTI browser. To achieve dynamically controlled interaction monitoring, all dialogue components need to expose (i.e., implement) a common programming interface (i.e., abstract class), for installing or uninstalling monitoring functionality (mainly event handlers). From a software engineering point of view, the effective management and control of interaction monitoring requires the careful design

of standard programming APIs, for all dialogue components, as well as the separation of the interaction monitoring logic, from the typical dialogue control logic of each component. This will enable the run-time orchestration of interaction monitoring, so as to collect interaction events and collate an interaction history, the latter constituting the basis on which to draw inferences regarding the particular end-user.

18.3.6.4 User Profiles and Decision Making

In automatically adapted interactions, the storage of a user profile is mandatory, necessitating the employment of appropriate user-model representation methods (see Chapter 21 of this handbook). Additionally, the run-time necessity for adaptation-oriented decision making (i.e., deciding when and how adaptation is to be performed) requires appropriate decision-logic representation methods. Various relevant technical approaches are discussed in Kobsa and Pohl (1995), Vergara (1994), and Savidis and Stephanidis (2001b).

18.3.7 Ambient Interactions

The primary motivation for ambient interactions is based on the idea that future computing platforms will not constitute monolithic "all-power-in-one" devices, but will likely support open interconnectivity enabling users to combine the facilities offered by distinct devices. Physically distributed devices may be either wearable or available within the ambient infrastructure (either stationary or mobile), and may be connected via a wireless communication link for easier deployment. Operationally, each such device will play a specific role by exposing different processing capabilities or functions, such as character display, pointing, graphics, audio playback, speech synthesis, storage, network access, and so on. From a hardware point of view, such devices may be wristwatches, earphones, public displays, home appliances, office equipment, car electronics, sunglasses, ATMs, and so on. To allow the dynamic engagement and coordination of such computing devices, a central management and control point is needed, which should be reconfigured, adapted, and fine-tuned by the end-user. Small portable processing units with embedded client applications, supporting on-the-fly device employment, are a particularly promising infrastructure for such dynamically formed ambient computing clusters.

The applications running on such portable machines should be state safe regarding failures or disconnections of externally utilized devices, while simultaneously offering comprehensive facilities to the end-user for the management of alternative device-composition configurations. The technical challenges for service-oriented composition depend on whether it concerns internal processing services or user interface elements. In this context, the reported work addresses the issue of dynamic composition from user interface microservices hosted by dynamically engaged devices. Some of the foreseen key application domains, which would largely benefit from this approach, are infomobility and navigation, intelligent office environments, smart homes, and mobile entertainment. The specific functional requirements for ambient interactions, relying upon the experience acquired in the development of the Voyager development framework for ambient interactions (Savidis and Stephanidis, 2003a, 2003b), in the context of the 2WEAR Project (see Acknowledgements), are:

- *Device discovery and wireless networking*: Even though this requirement might seem mostly related to core systems' developments, it is imperative that interface developers manage the on-the-fly detection of any in-range environment I/O devices that can be used for interaction purposes, while at the same time they should also be supplied with all the necessary instrumentation for handling wireless short-range dynamic communication links (e.g., the Bluetooth L2CAP library).
- *Device-embedded user-interface microservices*: It is necessary to implement the run-time query of interaction-specific device capabilities (e.g., text display support, supported number of text lines, presence of a software cursor, etc.). This feature implies the provision of well-documented standardized service models for dynamically available remote user interface devices, along with the definition and implementation of the concrete protocols for run-time control and coordination.
- *Automatic and on-demand dialogue reconfiguration*: To cope with the dynamic presence or disengagement of remote I/O devices, the detection of loss of connection through typical network programming libraries is needed. Additionally, it is imperative to allow end-users to dynamically reconfigure the ambient interface, offering

the on-demand deployment of alternative interaction-capable devices from the local environment infrastructure. Finally, the support for predefined reconfiguration scenarios for the automatic retargeting of the devices exploited by the interface is critical for allowing automatic dialogue reconfiguration when, during interaction, particular I/O resources get out of wireless communication range or fail.

- *State persistence and abstract interaction objects*: The key characteristic of ambient interactions is the inherent remote distribution of user interface I/O microservices within the surrounding computational environment. Such I/O resources may support a range of facilities, such as character input, text display, picture display, audio output, hardware push buttons, on/off switches, etc. Since failure and loss of connection may take place at any time, it is important to ensure that the dialogue state is centrally maintained within the mobile interface application kernel. Arguably, the most appropriate way to program such a behavior is via abstract interaction objects (Desoi et al., 1989; Duke and Harrison, 1993; Savidis and Stephanidis, 1995b; Wise and Glinert, 1995).

An in-depth technical analysis of the previously mentioned functional requirements, together with detailed design propositions for software library API and run-time architecture may be found in Savidis and Stephanidis (2002), while the software-design evaluation process and results are reported in Savidis and Stephanidis (2003b).

18.4 Discussion and Conclusions

Interaction objects play a central role in interface development. Consequently, a large part of the software for implementing commercially available interface tools is dedicated to the provision of comprehensive collections of graphical interaction objects. The basic layer providing the implementation of interaction objects is the *toolkit layer*, while interface tools typically provide additional layers on top of that. This chapter has discussed interaction-object-based interface development under the perspective of universally accessible interactions, identifying four key categories of software functionality to effectively manipulate interaction objects, namely integration, augmentation, expansion, and abstraction. Currently, there are variable degrees of support for the investigated basic mechanisms. Regarding toolkit integration, the vast majority of commercial tools offer multiplatform support in a hard-coded manner, rather than providing open mechanisms for connecting to arbitrary toolkits. Toolkit augmentation is supported in most programming-based interface tools, while higher-level development tools are very weak in this perspective. Toolkit expansion is also supported in most programming-oriented interface tools, but the considerable overhead required, as well as the inherent implementation complexity, turns the expansion task into an activity primarily targeted to expert programmers. Regarding higher-level development tools, there is an increasing number of commercially available graphical construction tools supporting expansion, while there are currently only two 4GL-based interface tools supporting expansion. Finally, toolkit abstraction, although it is the most important mechanism in terms of coping with universal access, is also the least supported mechanism in existing interface tools, requiring primarily handmade solutions with demanding software structures. The I-GET 4GL-based UIMS (Savidis and Stephanidis, 2001a) offering the I-GET pattern-reflecting language for user interface programming with explicit support for virtual interaction objects (Savidis, 2004; see also Chapter 27 of this handbook) is the only tool reported in the literature to exhibit the full set of properties required for supporting abstraction.

Software toolkits, as collections of implemented user-interface elements, typically reflect the design of particular interaction metaphors, such as the desktop/windowing metaphor. In this context, top-level container objects play a primary role since they largely affect the overall metaphoric experience. In this sense, the pursuit for new styles and forms of metaphoric interaction can be mainly focused on the design of appropriate novel container objects, as has been demonstrated in previous work such as the generic containers of the Hawk toolkit (see Figure 18.5) and the inverted umbrella trees of the immersive file manager (see Figure 18.6). The ability to hierarchically mix interaction object instances from different toolkits opens new opportunities for deployment and experimentation with mixed-toolkit artifacts. However, the technical ground for such open interoperability is not yet prepared, requiring software-intensive research efforts to design, implement, test, and deploy open toolkit interoperability infrastructures.

The need for user-interface adaptation in universally accessible interactions requires methods to accommodate dynamic user-adapted interface behavior. To implement automatic adaptation, interactive software applications should encompass the capability of appropriately delivering on-the-fly adapted interface instances, performing appropriate run-time processing that engages the selection and activation of distinct implemented dialogue components. The run-time assembly process of an interface requires an appropriate structure, which accommodates variability of contained components within container components, exhibiting the properties of dynamic polymorphic containment hierarchies. Additionally, interaction monitoring is necessary to collect and analyze interaction history so that user-oriented preferences can be drawn.

Finally, the recently introduced notion of ambient interaction reflects various alternative scenarios and technical solutions, all complementarily contributing toward the goal of smoothly and transparently integrating computing artifacts within the physical environment. Ambient interaction refers to interaction hosted by environment devices during user mobility. More specifically, in ambient interactions users carry a miniaturized pocketsize processing unit (PPU), without any I/O modules, encompassing various user-chosen embedded software applications. Those applications should be capable of dynamically detecting the presence of devices within the surrounding environment, and of using them so as to realize meaningful interaction with the user. To effectively implement such a scenario, a number of key technical prerequisites emerge: (1) presence of wireless short-range networking technology (like Bluetooth); (2) ability to discover the presence of ambient devices on the move; (3) standard communication protocols for querying and controlling the services of ambient devices; (4) programmer's APIs for dynamic user interface microservice utilization; and (5) ability for dynamic interface reconfiguration to the changing computational environment, while ensuring interface state persistence.

The development of universally accessible interactions entails software engineering requirements so as to address the grand development challenges inherent in the following four fundamental layers (see Figure 18.18), directly related to the three key layers of diversity in universal access interactions (see Figure 18.1):

- *Metaphor layer*, reflecting the primary metaphoric styles and forms of the overall interactive experience, having as key challenge the development of new metaphors.
- *Platform layer*, reflecting the delivery of metaphors as software libraries encompassing collections of implemented interaction object classes, having as key challenge the effective manipulation of interaction objects.
- *User layer*, reflecting the individual requirements, characteristics, abilities, and preferences, having as key challenge the support for user-adapted interface delivery.
- *Ambient layer*, reflecting the surrounding environment as an infrastructure of dynamically engaged computational resources, having as key challenge the provision of ambient interactions.

In this context, this chapter has presented the key software engineering requirements, reflecting hands-on development experience in the course of six projects, in a time frame of 15 years. Overall, the development of universally accessible interactions is a complex task, since it engages a highly demanding interface design process, together with a programming-intensive

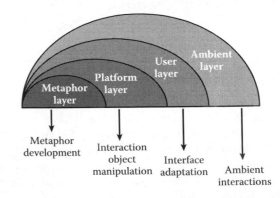

FIGURE 18.18 The four fundamental layers in universally accessible interactions, with their associated development challenges, constituting the basis of the software engineering requirements analysis. (From Savidis, A. and Stephanidis, C., *Interacting with Computers*, 18, 71–116, 2006.)

software implementation process. Surprisingly, at present all the identified software engineering requirements can be better addressed through lower-level programming instruments, since commercially available higher-level interface construction tools are mostly optimized for typical windowing user interfaces. Currently, there are various development propositions or software engineering recipes for many perspectives of ambient interactions or user-adapted interfaces, which may help in the reduction of the overall development complexity. However, to support the wide proliferation of universally accessible interactions as a common interface paradigm for future applications and services, further work is needed in the domain of user interface development tools to encompass automation mechanisms optimally suited for inclusive development.

Acknowledgments

Part of this work has been carried out in the context of the following:

- ACCESS TP1001 (Development Platform for Unified ACCESS to Enabling Environments) project, partially funded by the TIDE Program of the European Commission and lasting 36 months (January 1, 1994 to December 31, 1996). The partners of the ACCESS consortium are: CNR-IROE (Italy), Prime Contractor; ICS-FORTH (Greece); University of Hertfordshire (United Kingdom); University of Athens (Greece); NAWH (Finland); VTT (Finland); Hereward College (United Kingdom); RNIB (United Kingdom); Seleco (Italy); MA Systems & Control (United Kingdom); and PIKOMED (Finland).
- AVANTI AC042 (Adaptable and Adaptive Interaction in Multimedia Telecommunications Applications) project, partially funded by the ACTS Program of the European Commission and lasting 36 months (September 1, 1995 to August 31, 1998). The partners of the AVANTI consortium are ALCATEL Italia, Siette Division (Italy), Prime Contractor; IROE-CNR (Italy); ICS-FORTH (Greece); GMD (Germany); VTT (Finland); University of Siena (Italy); MA Systems and Control (United Kingdom); ECG (Italy); MATHEMA (Italy); University of Linz (Austria); EUROGICIEL (France); TELECOM (Italy); TECO (Italy); and ADR Study (Italy).
- 2WEAR, IST-2000-25286 (A Runtime for Adaptive and Extensible Wireless Wearables) project, partially funded by the Disappearing Computing Initiative (DCI) of the IST Program of the European Commission and lasting 36 months (January 1, 2001 to December 31, 2003). The partners of the 2WEAR consortium are Foundation for Research and Technology–Hellas, Institute of Computer Science (ICS-FORTH) (Greece), Prime Contractor; Nokia Research Centre (NRC) (Finland); Swiss Federal Institute of Technology–Zurich Institute for Computer Systems (ETHZ) (Switzerland); and MA Systems and Control Limited (United Kingdom).

References

Abowd, G. and Mynatt, E. (2000). Charting past, present, and future research in ubiquitous computing. *ACM Transactions on Computer-Human Interaction* 7: 29–58.

ACCESS Project (1996). *Development Platform for Unified Access to Enabling Environments*. London: RNIB Press.

Bass, L., Hardy, E., Little, R., and Seacord, R. (1990). Incremental development of user interfaces, in the *Proceedings of Engineering for Human-Computer Interaction* (G. Cockton, ed.), pp. 155–173. Amsterdam: Elsevier North-Holland.

Bass, L., Kasbach, C., Martin, R., Siewiorek, D., Smailagic, A., and Stivoric, J. (1997). The design of a wearable computer, in the *Proceedings of the ACM Conference on Human Factors in Computing Systems (CHI '97)*, 22–27 March 1997, Atlanta, pp. 139–146. New York: ACM Press.

Bharat, K. and Cardelli, L. (1995). Migratory applications, in *Mobile Object Systems: Towards the Programmable Internet* (J. Vitek and C. Tschudin, eds.), pp. 131–148. Heidelberg, Germany: Springer-Verlag.

Bronevetsky, G. (2003). *Circle Menus. Demo implemented in Java.* http://www.cs.cornell.edu/boom/2001sp/Bronevetsky/Circle%20Menu%20Documentation.htm.

Calvary, G., Coutaz, J., Thevenin, D., and Rey, G. (2001). Context and continuity for plastic user interfaces, in the *Proceedings of the I3 Spring Days Workshop on Continuity in Future Computing Systems*, 23–25 April 2001, Porto, Portugal, pp. 51–69. Didcot, UK: Council for the Central Laboratory of the Research Councils.

Canfield Smith, D., Irby, D., Kimball, R., Verplank, B., and Harlsem, E. (1982). Designing the star user interface. *Byte Magazine* 7: 242–282.

Card, S. K. and Henderson, D. (1987). A multiple, virtual-workspace interface to support user task switching, in the *Proceedings of the ACM Conference on Human Factors in Computing Systems and Graphics Interfaces (CHI & GI '87)* (J. Carroll and P. Tanner, eds.), 5–9 April 1981, Toronto, Canada, pp. 53–59. New York: ACM Press.

Carroll, J., Mack, R. L., and Kellogg, W. A. (1988). Interface metaphors and user interface design, in *Handbook of Human-Computer Interaction* (M. Helander, ed.), pp. 67–85. Amsterdam: Elsevier North-Holland.

Desoi, J., Lively, W., and Sheppard, S. (1989). Graphical specification of user interfaces with behavior abstraction, in the *Proceedings of the ACM Conference on Human Factors in Computing Systems (CHI '89)*, 30 April–4 May 1989, Austin, Texas, pp. 139–144. New York: ACM Press.

Dieterich, H., Malinowski, U., Kühme, T., and Schneider-Hufschmidt, M. (1993). State of the art in adaptive user interfaces, in *Adaptive User Interfaces: Principles and Practice* (M. Schneider-Hufschmidt, T. Kühme, and U. Malinowski, eds.), pp. 13–48. Amsterdam: Elsevier North-Holland.

Duke, D. and Harrison, M. (1993). Abstract interaction objects. *Computer Graphics Forum* 12: 25–36.

ERCIM News (2001, October). *ERCIM News—Special Theme: Ambient Intelligence*. http://www.ercim.org/publication/Ercim_News/enw47.

Hatziantoniou, A., Neroutsou, V., and Stephanidis, C. (2003, February). Development environments for universal access: The windows porting of the I-GET UIMS with a comprehensive test-suite of interactive applications. *Technical Report 319, ICS-FORTH*. ftp://ftp.ics.forth.gr/tech-reports/2003/2003.TR319.Dev_Env_Universal_Access.pdf.gz.

IGDA Accessibility Group (2004). *Accessibility in Games: Motivations and Approaches*. http://www.igda.org/accessibility/IGDA_Accessibility_WhitePaper.pdf.

Information Society Technologies, Future and Emerging Technologies – IST/FET (2001). *The Disappearing Computer Initiative*. http://www.disappearing-computer.org.

IST Advisory Group (2003). *Ambient Intelligence: From Vision to Reality*. ftp://ftp.cordis.lu/pub/ist/docs/istag-ist2003_consolidated_report.pdf.

Kobsa, A. and Pohl, W. (1995). The user modeling shell system BGP-MS, in *User Modeling and User-Adapted Interaction* 4: 59–106. http://www.ics.uci.edu/~kobsa/papers/1995-UMUAI-kobsa.pdf. Program code: ftp://ftp.gmd.de/gmd/bgp-ms.

Kouroupetroglou, G., Viglas, C., Anagnostopoulos, A., Stamatis, C., and Pentaris, F. (1996). A novel software architecture for computer-based interpersonal communication aids, in *Interdiscplinary Aspects on Computers Helping People with Special Needs, Proceedings of 5th International Conference on Computers Helping People with Special Needs (ICCHP '96)* (J. Klauss, E. Auff, W. Kremser, and W. Zagler, eds.), 17–19 July 1996, Linz, Austria, pp. 715–720. Vienna, Austria: Oldenberg.

McGuffin, M., Burtnyk, N., and Kurtenbach, G. (2002). FaST sliders: Integrating marking menus and the adjustment of continuous values, in the *Proceeding of the Graphics Interface 2001 Conference*, 27–29 May 2002, Calgary, Canada, pp. 35–42. http://www.graphicsinterface.org/cgi-bin/Download Paper?name=2002/174/paper174.pdf.

Moll-Carrillo, H. J., Salomon, G., March, M., Fulton Suri, J., and Spreenber, P. (1995). Articulating a metaphor through user-centered design, in the *Proceedings of the ACM Conference on Human Factors in Computing Systems (CHI '95)*, 7–11 May 1995, Denver, pp. 566–572. New York: ACM Press.

Myers, B. (1988). *Creating User Interfaces by Demonstration*. Boston: Academic Press.

Myers, B. (1995). User interfaces software tools. *ACM Transactions on Human-Computer Interaction* 12: 64–103.

Mynatt, E. and Edwards, W. (1995). Metaphors for non-visual computing, in *Extra-Ordinary Human-Computer Interaction: Interfaces for Users with Disabilities* (A. Edwards, ed.), pp. 201–220. New York: Cambridge University Press.

Ousterhout, J. (1994). *Tcl and the Tk Toolkit*. Reading, MA: Addison-Wesley Professional Computing Series.

Petrie, H., Morley, S., McNally, P., O'Neill, A-M., and Majoe, D. (1997). Initial design and evaluation of an interface to hypermedia systems for blind users, in the *Proceedings of Hypertext '97*, 6–11 April 1997, Southampton, U.K., pp. 48–56. New York: ACM Press.

Salber, D., Dey, A., and Abowd, G. (1999). The context toolkit: Aiding the development of context-enabled applications, in the *Proceedings of the 1999 Conference on Human Factors in Computing Systems (CHI '99)*, 15–20 May 1999, Pittsburgh, pp. 434–441. New York: ACM Press.

Savidis, A. (2004, January). The I-GET user interface programming language: User's guide, in *Technical Report 332, ICS-FORTH*. ftp://ftp.ics.forth.gr/tech-reports/2004/2004.TR332.I-GET_User_Interface_Programming_Language.pdf.

Savidis, A., Maou, N., Pachoulakis, I., and Stephanidis, C. (2002). Continuity of interaction in nomadic interfaces through migration and dynamic utilization of I/O resources. *Universal Access in the Information Society* 1: 274–287.

Savidis, A. and Stephanidis, C. (1995a). Building non-visual interaction through the development of the Rooms metaphor, in *Companion Proceedings of ACM Conference on Human Factors in Computing Systems (CHI '95)*, 7–11 May 1995, Denver, pp. 244–245. New York: ACM Press.

Savidis, A. and Stephanidis, C. (1995b). Developing dual interfaces for integrating blind and sighted users: The HOMER UIMS, in the *Proceedings of the ACM Conference on Human Factors in Computing Systems (CHI '95)*, 7–11 May 1995, Denver, pp. 106–113. New York: ACM Press.

Savidis, A. and Stephanidis, C. (1998). The HOMER UIMS for dual interface development: Fusing visual and non-visual interactions. *Interacting with Computers* 11: 173–209.

Savidis, A. and Stephanidis, C. (2001a). The I-GET UIMS for unified user interface implementation, in *User Interfaces for All: Concepts, Methods and Tools* (C. Stephanidis, ed.), pp. 489–523. Mahwah, NJ: Lawrence Erlbaum Associates.

Savidis, A. and Stephanidis, C. (2001b). The unified user interface software architecture, in *User Interfaces for All: Concepts, Methods, and Tools* (C. Stephanidis, ed.), pp. 389–415. Mahwah, NJ: Lawrence Erlbaum Associates.

Savidis, A. and Stephanidis, C. (2002). Interacting with the disappearing computer: Interaction style, design method, and development toolkit, in *Technical Report 317, ICS-FORTH*. ftp://ftp.ics.forth.gr/tech-reports/2002/2002.TR317.Interacting_with_the_Disappearing_Computer.pdf.gz.

Savidis, A. and Stephanidis, C. (2003a). Dynamic environment-adapted mobile interfaces: The voyager toolkit, in *Universal Access in HCI: Inclusive Design in the Information Society, Volume 4 of the Proceedings of the 10th International Conference on Human-Computer Interaction (HCI International 2003)* (C. Stephanidis, ed.), 22–27 June 2003, Crete, Greece, pp. 489–493. Mahwah, NJ: Lawrence Erlbaum Associates.

Savidis, A. and Stephanidis, C. (2003b). Interacting with the disappearing computer: evaluation of the Voyager development framework, in *Technical Report 331, ICS-FORTH*. ftp://ftp.ics.forth.gr/tech-reports/2003/2003.TR331.Interacting_Disappearing_Computer_Voyager_Development_Framework.pdf.

Savidis, A. and Stephanidis, C. (2006). Automated user interface engineering with a pattern reflecting programming language. *Journal for Automated Software Engineering* 13: 303–339.

Savidis, A., Stephanidis, C., Korte, A., Crispien, K., and Fellbaum, K. (1996). A generic direct-manipulation 3D-auditory environment for hierarchical navigation in non-visual interaction, in the *Proceedings of the ACM ASSETS'96 Conference*, 11–12 April 1996, Vancouver, Canada, pp. 117–123. New York: ACM Press.

Savidis, A., Stergiou, A., and Stephanidis, C. (1997). Generic containers for metaphor fusion in non-visual interaction: The HAWK interface toolkit, in the *Proceedings of the 6th International Conference on Man-Machine Interaction Intelligent Systems in Business (INTERFACES '97)*, 28–30 May, Montpellier, France, pp. 233–234.

Savidis, A., Vernardos, G., and Stephanidis, C. (1997). Embedding scanning techniques accessible to motor-impaired users in the WINDOWS object library, in *Design of Computing Systems: Cognitive Considerations, Proceedings of the 7th International Conference on Human-Computer Interaction,*

Volume 1 (HCI International '97) (G. Salvendy, M. J. Smith, and R. J. Koubek, eds.), 24–29 August 1997, San Francisco, pp. 429–432. Amsterdam: Elsevier, Elsevier Science.

Stephanidis, C. (ed.) (2001a). *User Interfaces for All: Concepts, Methods, and Tools*. Mahwah, NJ: Lawrence Erlbaum Associates.

Stephanidis, C. (2001b). The concept of unified user interfaces, in *User Interfaces for All: Concepts, Methods, and Tools* (C. Stephanidis, ed.), pp. 371–388. Mahwah, NJ: Lawrence Erlbaum Associates.

Stephanidis, C., Paramythis, A., Sfyrakis, M., and Savidis, A. (2001). A case study in unified user interface development: The AVANTI web browser, in *User Interfaces for All: Concepts, Methods and Tools* (C. Stephanidis, ed.), pp. 525–568. Mahwah, NJ: Lawrence Erlbaum Associates.

Van Dam, A. (1997). Post-WIMP user interfaces. *CACM* 4: 63–67.

Vergara, H. (1994). *PROTUM: A Prolog Based Tool for User Modeling*. Bericht Nr. 55/94 (WIS-Memo 10), University of Kostanz, Germany.

Weiser, M. (1991). The computer for the 21st century. *Scientific American* 265: 94–104.

Wise, G. B. and Glinert, E. P. (1995). Metawidgets for multimodal applications, in the *Proceedings of the RESNA '95 Conference*, pp. 455–457. Washington, DC: RESNA Press.

X Consortium Working Group Draft (1994). *FRESCOTM Sample Implementation Reference Manual*, Version 0.7.

19

Tools for Inclusive Design

Sam Waller and
P. John Clarkson

19.1 Introduction

The demographics of the developed world are changing; longer life expectancies and a reduced birth rate are resulting in an increased proportion of older people within the adult population. This is leading to a reduction in the potential support ratio (PSR), which is the number of people aged 15 to 64 who could support one person over 65 years of age. In 1950 the worldwide PSR was 12:1, while by 2050 it will be 4:1 globally and 2:1 in the developed world (United Nations, 1999). Almost half the U.K. adult population will be over 50 years of age by 2020 (U.S. Census Bureau, 2007). Maintaining quality of life and independent living as we grow older is becoming an imperative for sustainability of the economy, while for many organizations the aging population also results in changing market opportunities (see also Chapter 8 of this handbook).

With increasing age comes a decline in capability, yet this is often coupled with increased wealth and free time. Where previous generations accepted that capability loss and an inability to use products and services came hand in hand, the baby-boomer generation now approaching retirement is less likely to tolerate products that they cannot use, especially if due to unnecessary demands on their capabilities (Huber and Skidmore, 2003).

Typically, people are viewed as being either able-bodied or disabled, with products being designed for one category or the other. In practice, capability varies continuously, and reducing the capability demands of a product results in more people being able to use the product, as well as increased satisfaction for those who previously had difficulty.

When the capability demand of a product exceeds that of the users, they may experience frustration, difficulty, or be unable to use the products altogether (Figure 19.1). Often this is seen as the user's fault for having a poor memory, reduced strength, or imperfect vision; however, inclusive design places the responsibility with product designers to ensure that the capability levels required to use a product are as low as possible.

Frustration and difficulty with poorly designed products is something that everyone experiences at some stage. This difficulty may occur for many reasons, perhaps because of an injured arm, or long-term deterioration in eyesight, or because the user is also caring for a small child while performing the task. Indeed, research commissioned by Microsoft (2003) reported that 60% of Americans aged 18 to 64 years were likely or very likely to benefit from the use of accessible technology. Reducing design exclusion can improve the experience for a broad range of users in a wide variety of situations. Put simply, inclusive design is better design.

19.1.1 Definitions of Inclusive Design

Design refers to the structured creative process that converts an idea or perceived market need into a fully specified product, examples of which include consumer products, the provision of a service, designer graphics, a built environment, or assignment of work packages.

The term *inclusive design* and its surrounding methodology predominantly grew up in the United Kingdom to maintain quality of life and independent living for the aging population (Coleman, 2001). This had not been adequately solved through the use of assistive or medical devices that were stigmatizing, expensive, and undesirable. The British Standards Institute (2006) defines inclusive design as "the design of mainstream products and/or services that are accessible to, and usable by,

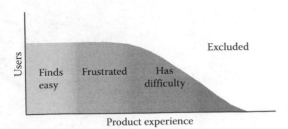

FIGURE 19.1 By meeting the needs of those who are excluded from product use, inclusive design improves product experience across a broad range of users. (From Clarkson, P. J., Coleman, R., Hosking, I., and Waller, S., *Inclusive Design Toolkit*, EDC, Cambridge, U.K., http://www.inclusivedesigntoolkit.com, 2007.)

people with the widest range of abilities within the widest range of situations without the need for special adaptation or specialized design" (p. 4).

At about the same time, design for all originated in Europe as a response to legislation such as the Disability Discrimination Act (1995), and aims to promote equality for all. It has particular prominence in fields such as web site accessibility and public transport, where access for all is a desirable goal. However, its application to mainstream product design has been limited, perhaps because the term *design for all* implies a contradiction with the commercially successful practice of targeting specific products to specific user groups and market segments.

Universal design has gained acceptance in the United States and Japan (Preiser and Ostroff, 2000), where Vanderheiden (1996) defines it as "the process of creating products (devices, environments, systems, and processes) that are usable by people with the widest possible range of abilities, operating within the widest possible range of situations (environments, conditions, and circumstances)."

The definition of inclusive design is very similar to that used for universal design; both definitions recognize that it is not always possible (or appropriate) to design one product to address the needs of the entire population, and both also focus on widening the accessibility of mainstream products. They both aim to ensure the design process delivers products that are usable, functional, and desirable. Inclusive design attaches particular importance to success for the business, which may create conflict in these three goals for the user. Nearly all decisions made throughout design activity will impact the resulting exclusion. Inclusive design provides the tools to help make these decisions and inevitable compromises in an informed manner.

Delivery of commercially successful and inclusive products requires the use of tools at three distinct levels, which will be considered in turn. At the corporate level, tools are needed to stimulate inclusive design within the organization, and manage the overall product offering to match variability in the population.

At the project level, it is essential to manage the design process such that a requirement specification is developed based on an understanding of the real user and business needs. The specification stimulates design concepts and solutions, and

provides objective criteria by which subsequent decisions can be evaluated. Users and their environments must be kept in mind throughout the design process, and this should occur through representative tools and involving real people.

Finally, evaluation tools that support decision making throughout the design process are presented, such as calculating the number of people who would be unable to use an existing product, or a proposed concept.

19.2 Corporate-Level Tools

It is unlikely that one product can fit all, but successful management of product portfolios provides an excellent means to deliver inclusive design to a variety of market segments. A desire to practice better design at a project or product level is often hampered (or stopped) by corporate-level issues, so promoting the right culture throughout the organization is the key to achieving genuine and repeatable success at a project level.

19.2.1 Product Strategy

Satisfying a range of different users or markets can often be achieved by developing a corresponding range of products in the form of platforms and portfolios (Figure 19.2). One product may not fit all, but many products can.

Successful management of portfolios provides an economic method to offer a range of products and associated feature sets that can account for the spread of capabilities and user needs within the market. The key goals of portfolio management are to maximize the value of the portfolio, ensure a balance between risk and reward, ensure strategic alignment within the company, and match the projects with available resources.

In managing a portfolio, one approach to maximizing returns and minimizing the number of projects being considered is to identify projects that have common needs and can be addressed as platform projects. This approach can effectively address a variety of user needs, with a range of related products built around a common technology platform. For example, automotive companies use a platform approach to provide a variety of performance or styling options to the customer, delivered within the budget of a single product development cycle.

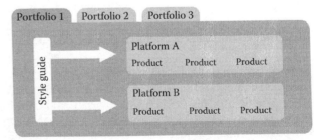

FIGURE 19.2 A portfolio contains platforms and style guides, and each platform can generate multiple products. (From Clarkson, P. J., Coleman, R., Hosking, I., and Waller, S., *Inclusive Design Toolkit*, EDC, Cambridge, U.K., http://www.inclusivedesigntoolkit.com, 2007.)

Style guides are often written to support design teams across the platforms within an organization, to help ensure effective reuse of design and understanding across the portfolio, consistency of operation (a major contributor to usability), and brand compliance. Publicly available style guides are also written to support the design process for a particular product genre, examples of which include Carmichael (1999) and Constantine and Lockwood (1999).

Market segmentation is one of the key approaches used to define and target specific markets. Dividing a market into distinct groups of buyers enables the planning of product mixes, platforms, and portfolios that offer various feature levels to suit a range of user needs and capability levels, thereby maximizing design inclusion. Platforms help resolve the inevitable compromise between being functional, usable, and desirable for users with a range of capabilities, while remaining viable for the business. See Otto and Wood (2001) for more information on developing a portfolio strategy to best exploit the market.

19.2.2 Management Support

The buy-in of senior-level management is vital to an organization's ability to deliver inclusive design, as they are well placed to implement a corporate-level strategy and culture that promote the use of appropriate tools and principles at the project level.

Management support is also required to prevent too narrow a focus on time and budget as measures of development project success. This can inhibit consideration of downstream problems and their associated costs during the earlier parts of the design process, when changes can most easily be implemented. Examples of such costs include no-fault found warranty returns, and levels of customer support. Without senior support, buy-in, and action, the implementation of inclusive design has to rely on local champions in the business, who do their best to encourage change within their sphere of influence.

The SPROC (strategy, process, resources, organization, and culture) model illustrated in Figure 19.3 is a simple way to represent the elements of a business that can have an impact on its ability to implement inclusive design.

- *Strategy*: Does the high-level business strategy support and encourage inclusive design?
- *Process*: Do existing innovation and development processes incorporate inclusive design considerations? If so, do they function well?
- *Resources*: Does the organization recognise what resources (internal or external) are required to successfully deliver inclusive design?
- *Organization*: Do organizational structures, reward systems, and metrics encourage the behaviors required for effective implementation of inclusive design?
- *Culture*: Does the combination of organizational structure, staffing, task design, and internal brand values provide for the desire and capability to deliver inclusive design?

FIGURE 19.3 To implement inclusive design, it is necessary to have a strategy that improves the design process, enables sufficient resources, motivates the organization, and promotes the right culture. (From Clarkson, P. J., Coleman, R., Hosking, I., and Waller, S., *Inclusive Design Toolkit*, EDC, Cambridge, U.K., http://www.inclusivedesigntoolkit.com, 2007.)

Management issues related to universal access are discussed in detail in Chapter 56 of this handbook. See also Keates (2007) for a five-stage plan to implement inclusive design within a business.

19.3 Project-Level Tools

Within a specific project, inclusive design tools can help manage the design process, thereby ensuring that solutions are developed in response to the real underlying needs of all the relevant stakeholders. These may include the people or organizations that use the product, which may be different from those that purchase it. Business stakeholders include those that design, manufacture, market, sell, deliver, and service the product, in addition to those that provide the finance. However, for simplicity, the *user* and the *business* will refer to all of these stakeholders throughout.

Designers are most likely to design for themselves, unless specifically directed otherwise (Cooper, 1999). Methods to keep the end-users in mind throughout the design process are therefore critical for project success, either by involving real people to support design activity, or through the development of fictitious characters that are representative of the real target users.

19.3.1 Managing the Design Process

A design challenge can arise from a variety of different contexts, such as an expansion or realignment of the overall product strategy, the availability of a new technology, or a requirement to update or repackage an existing product or service.

Good inclusive design is about making conscious and well-informed decisions throughout the design process. A great product is typically built on an understanding of the real needs of the user and the business, without any implicit bias to a particular solution. The discovery phase to develop this understanding forms the first design activity, followed by translation into a requirements specification, creation of concepts, and development of solutions.

It is often assumed that the initially perceived needs accurately represent the true problem, yet a thorough exploration of the design context is essential to ensure the real needs are identified, thus avoiding the development of a perfectly good solution to the wrong problem (Figure 19.4). Such exploration often reveals a more subtle underlying cause to the originally perceived problem.

Caswell (2006) provides an example where the client identified a need for a parking brake on wheelchairs, because several patients had been injured after falling out of them. Further exploration revealed that the real need was related to the instability of the wheelchair, which was correctly solved by adding a stabilizing outrigger. Without this exploration, a perfectly good parking brake could have been designed as a solution for the wrong problem.

19.3.1.1 Discover

The interactions between the user, business, and product form the primary focus of the discovery phase (Figure 19.5). Discovering user needs aims to uncover what attributes of the product will make it functional, usable, and desirable for the user. Market research is typically used to investigate the desires of potential users and the tasks that a product could perform to achieve these goals. Capability impairment will have a significant impact on usability, so it is essential to understand the prevalence of relevant impairments among target users.

It is also vital to discover what contexts the product could be used in, and how these different contexts affect the user's ability and willingness to interact with the product. A particular combination of a user, goal, and environment forms a scenario,

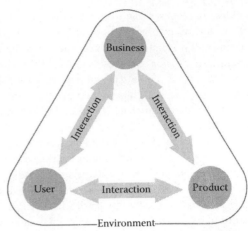

FIGURE 19.5   The interactions between the user, product, and business form the focus of the discovery phase. (From Clarkson, P. J., Coleman, R., Hosking, I., and Waller, S., *Inclusive Design Toolkit*, EDC, Cambridge, U.K., http://www.inclusivedesigntoolkit.com, 2007.)

and a representative set of scenarios provides useful design targets.

A well-articulated, prioritized, and communicated understanding of what constitutes success for the business is a prerequisite for making informed design decisions. Important business requirements can be categorized under the following headings:

- *Objectives*, such as the desired market share, margins, time to market, or return on investment
- *Resources*, such as the available budget, time scales, and personnel
- *Corporate fit*, which includes how it fits with other products, and how it supports the brand

At the end of the discovery phase, the project team should have documented answers to the questions:

- *Who* are the users and other stakeholders?
- *What* tasks will the product be used to achieve?
- *Why* does the business/user want this product?
- *When* will the product be delivered?
- *Where* will the product be used?

The project team must also understand what makes the product functional, desirable, and usable for the user, and viable for the business. These needs should be prioritized so that conflicts between them can be managed and resolved.

The project team's understanding of the real needs will have significantly improved from the initially perceived need and should continue to improve throughout the remaining stages. Subsequent outputs of the design process include the requirements specification and preliminary concepts. Knowledge gained during later design activities may lead to revisions of earlier design outputs, resulting in further cycles of iteration. Such revisions and iterations can be represented by the waterfall model of the design process, presented in Figure 19.6.

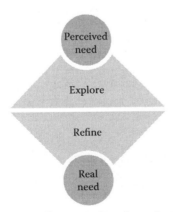

FIGURE 19.4   The initially perceived need must be explored and then refined to uncover the real needs. (From Clarkson, P. J., Coleman, R., Hosking, I., and Waller, S., *Inclusive Design Toolkit*, EDC, Cambridge, U.K., http://www.inclusivedesigntoolkit.com, 2007.)

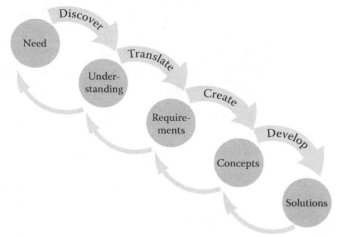

FIGURE 19.6 Waterfall model of an inclusive design process. (From Clarkson, P. J., Coleman, R., Hosking, I., and Waller, S., *Inclusive Design Toolkit*, EDC, Cambridge, U.K., http://www.inclusivedesign-toolkit.com, 2007.)

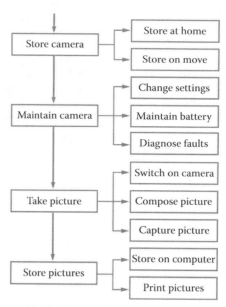

FIGURE 19.7 The beginning of a life-cycle analysis for a camera. (From Clarkson, P. J., Coleman, R., Hosking, I., and Waller, S., *Inclusive Design Toolkit*, EDC, Cambridge, U.K., http://www.inclusive designtoolkit.com, 2007.)

19.3.1.2 Translate

The knowledge gained in the discovery phase now has to be translated into a requirements specification, which structures the information in an unambiguous form that can be used to develop concepts. This translation process begins by gathering details on the functions that the product needs to perform, as a way of clarifying the interaction between the user and product (Figure 19.7). The functional description should consider the total life cycle of the product, which may include purchase, installation, use, maintenance, storage, and disposal.

A requirements matrix helps ensure that all the necessary user and business requirements have been considered. The matrix combines common requirements headings with each of the main life-cycle stages of the product (Table 19.1).

The finished requirements specification is a well-structured document that contains all the constraints that any viable design solution should satisfy. The document represents the previously discovered understanding about the user and business needs, in a structured and prioritized format suitable for generating and evaluating concepts and solutions. Items within the requirements specification should be solution independent, measurable, specific, objective, and quantified where possible.

The project team should evaluate the specification to ensure that it correctly reflects the current understanding of real needs: this understanding may have improved during the translation process. Throughout the remainder of the design process, further knowledge gained should be used to update and improve the requirements specification, although any changes made should be clearly identified, dated, and traceable.

Using the specification to generate design concepts is one way of testing both its validity and clarity. If such concepts satisfy the requirements, but not the real needs, then some aspect of the specification is likely to be flawed. See Shefelbine et al. (2002) for more detail on creating a requirements specification, or Bruce and Cooper (2000) for a broader description of requirements capture management. Klein et al. (2006) provide an example of requirements specifications for a range of inclusive digital set top box receivers, developed from a program of research to discover the user and business needs.

19.3.1.3 Create

Having identified the requirements that need to be addressed, the next phase of work requires tools and techniques to generate suitable concept solutions. Generating concepts requires stimulating and structuring creativity, then filtering and ranking the potentially large number of ideas that have been generated.

Team selection is an important part of any project, but particularly in the creative phase where a diverse range of contributors can rapidly deliver a broad range of potential ideas to address a problem. The key challenge is ensuring that this diverse team can work together in a constructive manner, which may be achieved by selecting key individuals, obtaining senior sponsorship and using effective facilitators in creative sessions. Involving users or personas with capability losses in creative sessions can also provide a stimulus for the design team to think in new directions, potentially resulting in solutions that are beneficial to all users. The Design Business Association Inclusive Design Challenge uses a design model that incorporates real users with capability loss as a stimulus for the development of mainstream design concepts with commercial potential. Further detail on the challenge can be found at http://www.hhc.rca.ac.uk/kt/challenge. The web site also describes all the inclusive design concept entries that were submitted over the past 7 years of the competition.

TABLE 19.1   Example Headings That Could Be Useful for a Requirements Specification

	Performance	Cost	Process
Design	Design performance	Design cost	Design process
Manufacturing and distribution	Manufacturing performance	Manufacturing cost	Manufacturing process
Purchasing and setup	Purchasing performance	Purchasing cost	Purchasing process
Usage	Usage performance	Usage cost	Usage process
Disposal	Disposal performance	Disposal cost	Disposal process

Source: Adapted from Shefelbine, S., Clarkson, P. J., Farmer, R., and Eason, S., *Good Design Practice for Medical Devices and Equipment: Requirements Capture*, EDC, Cambridge, U.K., 2002.

Numerous tools and techniques are available to help structure and visualize the creative process and its output. Examples include mind maps, analogy maps, life-cycle maps, and morphological grids. Without using such tools, the output of a creative session is too often a list or stack of ideas, with little awareness of what areas have been covered or not.

The creative process can produce very large numbers of ideas for consideration, the sheer volume of which can be daunting. It is therefore important to identify and agree upon a simple set of criteria to filter, rank, and evaluate the ideas. These criteria should be based on the inclusive design intent as captured by the requirements specification, and on acceptable levels of design and business risk.

The output of the creative phase is a small set of concepts. In producing a concept, the objective is to demonstrate how it can be achieved technically, the potential look and feel, the user or market need being addressed, and the value to the business. The first two objectives often involve some form of physical or virtual prototyping, allowing rapid user feedback and validation of the last two. The aim of any prototype created is to get feedback and buy-in from the key stakeholders (external and internal), so the best approach(es) to achieve this aim should be chosen.

Initially, when ideas are not well developed, it may be difficult to predict how well they could ultimately satisfy the different requirements criteria. The most pragmatic approach is to use a group of suitable individuals who can provide an educated indication of fit with all of the agreed upon criteria.

19.3.1.4 Develop

Further detailed design, manufacturing, and marketing activities are required to deliver the solution. Indeed most organizations already have robust, well-defined processes for delivering new products to market. Developing inclusively designed products should not require wholesale alteration of these existing approaches.

The key challenge is to deliver the project on time and on budget, without compromising the design intent. In reality, design teams often make numerous small decisions that individually appear to be cost-effective or pragmatic, but in combination erode the product's ability to meet the original inclusive design requirements, therefore reducing its commercial success.

When the project team develops and delivers a solution, it is imperative to evaluate it against the requirements specification (verification), and the ability to satisfy the real needs (validation). It is particularly important to evaluate the solution itself, rather than focus on the project performance as measured by development cost and time. Some useful evaluation activities include direct or indirect observation of user interaction with the product, measurement of the level of help and support required to use the product, and monitoring of the incidence of no-fault-found returns.

Some commercially successful inclusively designed products are shown in Figure 19.8, and further examples are provided at http://www.designcouncil.info/inclusivedesignresoure.

After the delivery of the solution, user-focused evaluation will help build a picture of product use that has direct relevance to the future commercial success of the product. The insights gained into how customers really use the product may also stimulate new business opportunities, identifying needs that initiate further product development. The waterfall model of the design process may be redrawn as a spiral model to reflect this iteration, presented in Figure 19.9.

See Ulrich and Eppinger (2004) or Cagan et al. (2001) for more detail on identifying real customer needs and managing the design process to convert these needs into successful products.

FIGURE 19.8 Examples of commercially successful products developed using an inclusive design approach. (Image copyright owned by Margaret Durkan, Royal College of Art Helen Hainlyn Centre.)

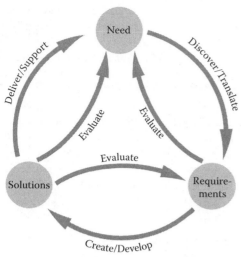

FIGURE 19.9 Spiral model of the design process. (From Clarkson, P. J., Coleman, R., Hosking, I., and Waller, S., *Inclusive Design Toolkit*, EDC, Cambridge, U.K., http://www.inclusivedesigntoolkit.com, 2007.)

19.3.2 Managing User Involvement

Users are those who will (attempt to) use the product or service once it is on the market. In practice, the type(s) of person that the product is aimed at will need to be determined, so people who represent these real end-users can be selected to engage in the design process. Involving real end-users at any or all stages of the design process can provide insight into what design solutions can and cannot be used and what goals the users want to achieve with the product.

In addition to other factors, the ages, past experiences, capabilities, and usage environments will vary widely among different users. Members of the design team are therefore often not representative users, and it is of critical importance to select an appropriate mix of users to achieve the different aims throughout the design process, as shown in Table 19.2.

Involving large numbers of users improves accuracy but can be expensive. In practice, within a single category of users, the majority of problems can be identified with 10 users (Faulkner, 2003), and at least 3 users of any one type are needed to reduce biases from user responses and observations. There are many different ways of involving users in the design process, and these can be categorized into three main types: ask, observe, or participate.

Questionnaires and other methods can be used to ask the users directly about their lives, what they want or need, or what they think of the design. However, people may struggle to correctly articulate their real needs, perhaps due to an inability to imagine what a future product could be or a poor awareness of their own habits and practices. People may also adapt their answers according to what they think the researcher wants to hear, so multiple techniques may be required to get to the heart of a user's real needs.

Controlled observation can provide deeper and unexpected insights, where users are watched and videotaped carrying out specified tasks with a product prototype. Watching and videotaping people in their daily lives also helps to understand their experiences and needs, and reveals issues related to the context of use.

Alternatively, participative methods consider users as co-designers, providing direct input into the creative process. For more information on user involvement, see Aldersley-Williams et al. (1999) and Keates (2007).

19.3.3 Using Personas

Personas bring user descriptions to life by giving them names, personalities, and lifestyles. They identify the motivations, expectations, goals, capabilities, skills, and attitudes of users, which are responsible for driving their product purchasing and usage behavior.

Although personas are fictitious, they are based on knowledge of real users. Some form of user research is necessary to generate personas that represent real end-users, rather than the opinions of the project team. It's easy to assemble a set of user characteristics and call it a persona, but it's not so easy to create personas that are truly effective tools for design and communication.

TABLE 19.2   The Types and Aims of User Involvement

	Who to Involve	**Aim**
Broad user mix	Users from a range of market segments	To understand general user requirements
Boundary users	Users on the limit of being able to use the product	To identify opportunities for design improvement
Extreme users	Users with a severe loss of capability	To inspire creativity during concept development
Mixed-experience users	Users with different levels of experience with similar products	To understand the impact of experience on use
Community groups	Groups of users who share experience of interacting with similar products	To provide a broad understanding of product use

Source: Clarkson, P. J., Coleman, R., Hosking, I., and Waller, S., *Inclusive Design Toolkit*, EDC, Cambridge, U.K., http://www.inclusivedesigntoolkit.com, 2007.

A good persona description is not a list of tasks or duties; it is a narrative that describes the flow of someone's day, as well as the person's skills, attitudes, environment, and goals. A persona answers critical questions that a task list does not, such as why the persona wants to use the product and whether he can focus on one thing at a time, or experience lots of interruptions.

It is preferable to keep the number of personas required to illustrate key goals and behavior patterns to a minimum. There is no magic number, but evidence from experienced and successful designers suggests that between four and eight personas would usually suffice to provide a focus for a single product.

When designing a persona, it is important to focus first on the information that is critical for design, such as behavior patterns, goals, environments, and attitudes. Then, the persona can be brought to life by adding a few personal details and facts, such as what the persona does after work. With a little personality, personas can become useful design targets. An example persona is illustrated in the next section, which is part of a set contained within Clarkson et al. (2007). More detail on personas can be found in Pruitt and Adlin (2006).

19.3.3.1 Example Persona

David is a 60-year-old granddad. Recently retired, David spends most days pottering around the garden and playing golf. On weekends, he enjoys walking in the countryside with his wife, Carol. He is currently searching on the Internet to arrange a walking holiday for them.

Although generally fit and healthy, David has become hard of hearing but struggles to accept it. He does have a hearing aid and, like his reading glasses, it is another thing for which Carol often has to search. David doesn't see himself as old, particularly as he helps care for his mother-in-law, Rose.

19.4 Evaluation Tools

Decisions and compromises made throughout the design process will impact the number of people able to use the finished product. This section presents tools for evaluating alternative options during the design process, so that such decisions can be made in an informed manner.

19.4.1 Expert Appraisal

Expert appraisal is the evaluation of a product or service by someone who has the professional training or experience to make an informed judgment on the design. Ideally, this person should not be biased by former involvement with the project because familiarity with any product or task makes it seem simpler and easier. Expert appraisal can identify possible causes of design exclusion, suggest improvements to reduce this exclusion, and increase user satisfaction.

Experts may include usability professionals, engineers, other designers, or those with suitable knowledge of the product type or its particular environment. It is essential that the expert has a sound knowledge of the range of users that need to be considered and the circumstances of their interaction. Achieving reliable results through this method often requires the participation of several specialists, so that different perspectives and problems can be identified.

It is desirable to have an internal person who acts as a user champion who can provide information about the user's perspective throughout the design process. Expert appraisal is usually used to detect critical problems and to provide priorities for exploration with users. However, it can also provide valuable improvement suggestions throughout the design process. See Poulson et al. (1996) for more detail on expert appraisal.

19.4.2 Capability Loss Simulators

Capability loss simulators are devices that designers can use to reduce their ability to interact with a product. Physical simulators are devices that can be worn to impair movement or vision, while software simulators modify an audio clip or photo image, so that fully able people perceive the information as though they have a capability loss.

These simulators can provide a quick and cheap method to help designers empathize with those who have capability losses, increase their understanding of the different losses, and simulate how exclusion occurs during product interaction. The cost, speed, and ease of access means that these simulators can be used both early on and repeatedly throughout the design process.

However, simulators do not replicate capability loss with perfect accuracy, and many simplistic simulators only provide a loose approximation of the effects. No simulators can replicate the experience of living with capability reduction on a daily basis, and the decline in cognitive ability cannot be meaningfully reproduced by simulation. Simulators are helpful to increase empathy with users who have reduced capability, but should not be considered as a replacement for involving real people with such losses, and should not be solely relied upon to draw conclusions regarding usability from the perspective of those who are impaired.

During physical simulation of capability loss, the person assessing the product wears items to reduce the functional ability of one or more parts of the body. Spectacles smeared with grease can also be used to simulate decreases in vision capability, as shown in Figure 19.10. Simple simulators can be created from everyday products, such as sports braces to reduce freedom of movement in key parts of the body. Cambridge University has designed a glove to simulate reduced dexterity capability, also shown in Figure 19.10. Organizations such as Visual Impairment North East (http://www.vine-simspecs.org.uk) supply glasses that aim to represent the visual effects caused by different eye conditions, while Mobilistrictor (http://www.mobilistrictor.co.uk) supplies a full bodysuit that restricts motion to simulate aging.

Software simulation of vision ability loss involves modifying a digital image to show what it might look like when viewed with a variety of different vision conditions, allowing quick comparison

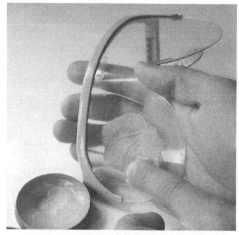 

FIGURE 19.10 Simple vision simulators can be created using glasses and grease, while Cambridge University has developed an advanced glove that reduces dexterity ability. (From Clarkson, P. J., Coleman, R., Hosking, I., and Waller, S., *Inclusive Design Toolkit*, EDC, Cambridge, U.K., http://www.inclusivedesigntoolkit.com, 2007.)

of different design concepts. Hearing simulators modify an audio file to simulate hearing conditions such as the natural deterioration caused by aging. Different sounds can be compared to see how they are affected by this loss, and the effect of the ambient background noise can also be appreciated. Example software simulators are freely available to try out on http://www.inclusivedesigntoolkit.com, which also contains details of other similar simulators.

19.4.3 Assessing Demand and Exclusion

A demand assessment measures the level of ability that a product or service demands to use it. Products with high levels of demand will exclude more people, and can also result in frustration or embarrassment. In many cases, a certain level of demand is inevitable, and a demand assessment helps to make rational decisions regarding trade-offs. However, in many other cases products demand an unnecessarily high level of ability, perhaps because the designers did not consider that other people have lower capabilities than themselves, and may need to use the products in environments that further hamper these capabilities.

A model of product interaction is presented first, followed by a subjective and simple method for assessing demands, followed by a more detailed method that uses capability data to estimate the number of people excluded. Finally, a set of graphs is presented that illustrate the levels of exclusion that result from demands on seven different capability categories that are typically used during product interaction.

19.4.3.1 A Model of Product Interaction

Any interaction with a product or service typically involves cycles where the user perceives, thinks, acts, observes the resulting change, and so on (Figure 19.11). The environmental context of the interaction may enhance or hinder these cycles. For exam-

ple, low, or indeed high, ambient light levels can compromise a user's ability to read.

Perceiving and acting both require sensory and motor capabilities. In addition, the body's sensory and motor resources are controlled by the brain and therefore require cognitive capability.

For example, perceiving text on a product can rely on the hands to move and orientate the product for visual examination; or the eyes could guide the fingers to press particular buttons. However, for the most part, perceiving requires sensory capability, thinking requires cognitive capability, and acting requires motor capability.

The following seven capability categories are identified as being the most relevant for product interaction.

- *Vision* is the ability to use the color and brightness of light to detect objects and to discriminate between different surfaces or the detail on a surface.
- *Hearing* is the ability to understand specific tones or speech in environments that may be noisy and to tell where the sounds are coming from.
- *Thinking* is the ability to process information, hold attention, store and retrieve memories, and select appropriate responses and actions.
- *Communication* is the ability to understand other people and express oneself to others (this inevitably overlaps with vision, hearing, and thinking).
- *Locomotion* is the ability to move around, bend down, climb steps, and shift the body between standing, sitting, and kneeling.
- *Reach and stretch* is the ability to put one or both arms out in front of the body, above the head, or behind the back.
- *Dexterity* is the ability of one or both hands to perform fine finger manipulation, pick up and carry objects, or grasp and squeeze objects.

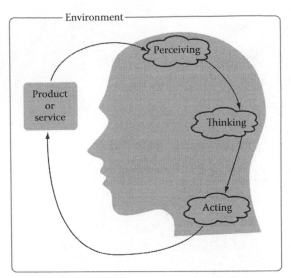

FIGURE 19.11 An interaction with a product involves cycles where the user's capabilities are used to perceive, think, and then act. (From Clarkson, P. J., Coleman, R., Hosking, I., and Waller, S., *Inclusive Design Toolkit*, EDC, Cambridge, U.K., http://www.inclusivedesigntoolkit.com, 2007.)

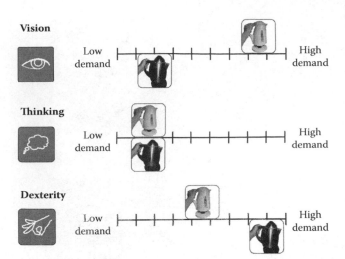

FIGURE 19.12 A simple assessment can provide an effective and visual comparison between different products or concepts according to the demand levels in different categories. The power button of the darker kettle is easier to see, but the lid is easier to open for the lighter kettle. (From Clarkson, P. J., Coleman, R., Hosking, I., and Waller, S., *Inclusive Design Toolkit*, EDC, Cambridge, U.K., http://www.inclusivedesigntoolkit.com, 2007.)

Sensory capability includes vision and hearing. Cognitive capability includes thinking and communication. Motor capability includes locomotion, reach and stretch, and dexterity.

19.4.3.2 A Simple Method for Measuring Demand

At their simplest, scales for demand in each of the seven categories range from low to high, where low and high provide a relative measure when one product or scale is compared to another (Figure 19.12). Although the scale measurements are crude, they are easy to use as an initial tool and can provide an effective visual comparison between alternative products or concepts.

Many people, especially those in older age groups, experience more than one capability loss in the form of multiple minor impairments. For this reason, estimating the number of people who would be excluded from using a product requires a single data source that covers all the capabilities required for cycles of product interaction.

19.4.3.3 The Disability Follow-Up Survey

The Disability Follow-Up to the 1996/97 Family Resources Survey (Grundy et al., 1999) remains the most recent Great British data source to consider all the relevant capabilities for product interaction. This survey was performed to help plan welfare support for disabled people.

People were selected for the Disability Follow-Up Survey if they met certain criteria, such as being in receipt of incapacity benefit. Approximately 7500 participants were asked up to 300 questions regarding whether they were able to perform certain tasks such as:

Can you see well enough to read a newspaper headline?

The results were collated to provide estimates for the national prevalence of disability. According to the definition used in the survey, 17.8% of the GB adult population have less than full ability in one or more of the seven capability categories. The GB adult population was 45.6 million people at the time of the survey.

The prevalence of disability measured within any data set depends on the threshold that is considered fully able. For each of the seven capability categories, this threshold level was set to represent an approximately equivalent reduction in quality of life, thus enabling valid comparisons between the prevalence of disability within each category. The results of this survey are first considered in detail for the sensory capabilities, followed by a comparison of exclusion levels that result from demands in each of the seven capability categories. High demand is defined such that a product with this level of demand would exclude anyone with less than full ability, so the levels of demand are also broadly comparable between the different capability categories.

A person with full vision ability can see well enough to recognize a friend across the road without difficulty, and read ordinary newsprint without difficulty. A person with full hearing ability can hear well enough to follow a conversation in a noisy environment without great difficulty, and follow a TV program at a volume that others find acceptable. The Disability Follow-Up Survey found that 8.7% of the GB adult population have less than full ability in one or both of vision and hearing.

Using a single data source for all of the different capability types allows for a detailed consideration of some interesting options. Figure 19.13 shows that 4.2% would be excluded from a product that demands full vision ability, and 6.3% excluded for full hearing ability, while 8.7% would be excluded for a product that requires both full vision and full hearing ability, and only

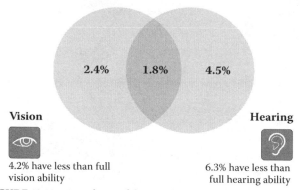

FIGURE 19.13 Prevalence of the population with less than full ability in vision and hearing, where the overlapping circles indicate the population that has capability losses in both categories. (From Clarkson, P. J., Coleman, R., Hosking, I., and Waller, S., *Inclusive Design Toolkit*, EDC, Cambridge, U.K., http://www.inclusivedesigntoolkit.com, 2007.)

1.8% would be excluded if the product could be used with either full vision or full hearing ability.

The prevalence data for vision from the Disability Follow-Up Survey are now presented in more detail. The survey results were first published in terms of the number of people with specific levels of disability, illustrated in Figure 19.14. Level V1 refers to extreme disability, V9 is mild disability, and V10 is full vision ability.

For the purpose of inclusive design, it is more useful to consider levels V1, V2, and V3, and so on, as being increasing levels of vision ability. For example, those in ability level V3 can tell by the light where the windows are (opposite of V1), and they can see the shapes of furniture in a room (opposite of V2), but they cannot recognize a friend if close to his or her face.

To estimate design exclusion, it is necessary to sum together the ability bands, to work out the total number of people who would be unable to perform a specific task. For example, supposing a product required the user to see well enough to read a newspaper headline. The number of people excluded would be the sum of people in categories V1 to V5, which is approximately 1% of the GB adult population.

However, if the required task is not specifically mentioned on the scale, then some judgment will be required to position the task between existing ability levels. The same data are now presented in a suitable format to achieve this.

19.4.3.4 Assessing Exclusion

The data from the Disability Follow-Up Survey can also be presented in a format suitable for assessing design exclusion directly, where the ability level that a product demands to use it is directly plotted against the number of people who will be excluded (Figure 19.15).

The Disability Follow-Up Survey does not make any reference to the environment, or to fatigue caused by repeated actions. Until better data are obtained, these factors can only be accounted for by using judgment to modify the demand level appropriately.

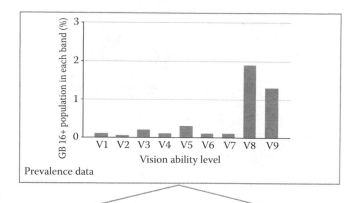

V1	Cannot tell by the light where the windows are
V2	Cannot see the shapes of furniture in a room
V3	Cannot see well enough to recognize a friend if close to his face
V4	Cannot see well enough to recognize a friend at arm's length away
V5	Cannot see well enough to read a newspaper headline
V6	Cannot see well enough to read a large print book
V7	Cannot see well enough to recognize a friend across the room
V8	Has difficulty recognizing a friend across the road
V9	Has difficulty reading ordinary newspaper prints
V10	Full vision ability
Vision ability levels	

FIGURE 19.14 Definitions of the vision ability levels from the Disability Follow-Up Survey and the corresponding number of people within each level. (From Clarkson, P. J., Coleman, R., Hosking, I., and Waller, S., *Inclusive Design Toolkit*, EDC, Cambridge, U.K., http://www.inclusivedesigntoolkit.com, 2007.)

19.4.3.5 Comparison between Demands on Different Capabilities

The demand and exclusion graphs for all seven capability categories are presented in Figures 19.16 to 19.22, to enable comparison between the exclusion levels that result from approximately equivalent demand levels in each of these categories. The complete set of statements used to describe each demand level can be found in Clarkson et al. (2007), along with the prevalence data for all seven capability categories.

A general graph of ability would be expected to follow a bell-shaped curve, in which case the number of people excluded would be expected to smoothly increase as the demand level increases. Indeed, the graphs for hearing, thinking, and reach and stretch (Figures 19.17, 19.18, and 19.21) follow this pattern, implying that losses within these capabilities occur in a gradual manner. However, the shape of the vision graph (Figure 19.16) implies that very few people are totally blind, while many more people have mild problems with vision. Reducing the vision demand made by a product from high demand to moderate demand makes a significant difference to the number of people able to use it, while further reductions to the demand level will not make as much difference.

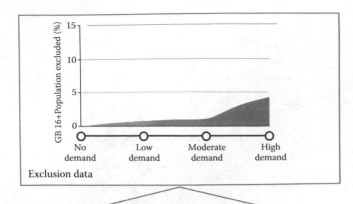

No demand

The user is not required to perceive anything by sight

Low demand

The user is required to have sufficient ability to do things like read a newspaper headline or recognize a friend at arm's length away

Moderate demand

The user is required to have sufficient ability to do things like read a large print book (approx. 16 pt text) or recognize a friend across the room

High demand

The user is required to have sufficient ability to do things like read ordinary newsprint (approx. 9 pt text) without difficulty or recognize a friend across the road without difficulty

Vision demand levels

FIGURE 19.15 The data from the Disability Follow-Up Survey is presented in a format suitable for assessing design exclusion. (From Clarkson, P. J., Coleman, R., Hosking, I., and Waller, S., *Inclusive Design Toolkit*, EDC, Cambridge, U.K., http://www.inclusivedesign toolkit.com, 2007.)

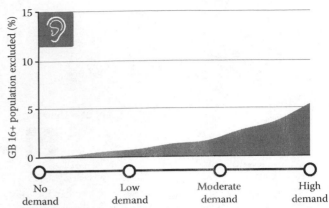

FIGURE 19.17 Demand and exclusion graph for hearing. (From Clarkson, P. J., Coleman, R., Hosking, I., and Waller, S., *Inclusive Design Toolkit*, EDC, Cambridge, U.K., http://www.inclusivedesigntoolkit. com, 2007.)

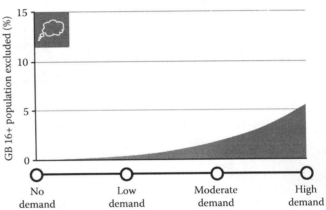

FIGURE 19.18 Demand and exclusion graph for thinking. (From Clarkson, P. J., Coleman, R., Hosking, I., and Waller, S., *Inclusive Design Toolkit*, EDC, Cambridge, U.K., http://www.inclusivedesigntoolkit. com, 2007.)

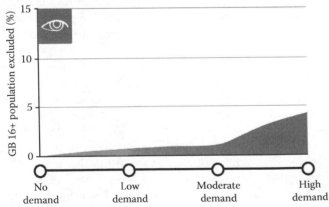

FIGURE 19.16 Demand and exclusion graph for vision. (From Clarkson, P. J., Coleman, R., Hosking, I., and Waller, S., *Inclusive Design Toolkit*, EDC, Cambridge, U.K., http://www.inclusivedesign toolkit.com, 2007.)

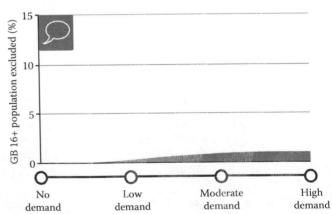

FIGURE 19.19 Demand and exclusion graph for communication. (From Clarkson, P. J., Coleman, R., Hosking, I., and Waller, S., *Inclusive Design Toolkit*, EDC, Cambridge, U.K., http://www.inclusivedesign toolkit.com, 2007.)

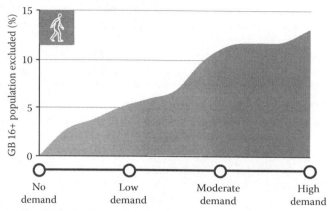

FIGURE 19.20 Demand and exclusion graph for locomotion. (From Clarkson, P. J., Coleman, R., Hosking, I., and Waller, S., *Inclusive Design Toolkit*, EDC, Cambridge, U.K., http://www.inclusivedesign toolkit.com, 2007.)

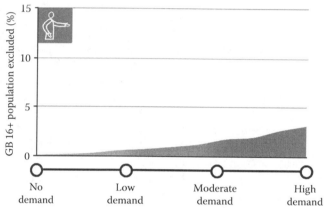

FIGURE 19.21 Demand and exclusion graph for reach and stretch. (From Clarkson, P. J., Coleman, R., Hosking, I., and Waller, S., *Inclusive Design Toolkit*, EDC, Cambridge, U.K., http://www.inclusive designtoolkit.com, 2007.)

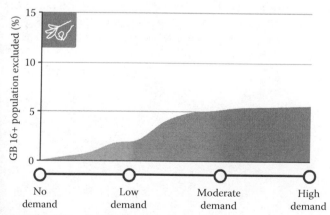

FIGURE 19.22 Demand and exclusion graph for dexterity. (From Clarkson, P. J., Coleman, R., Hosking, I., and Waller, S., *Inclusive Design Toolkit*, EDC, Cambridge, U.K., http://www.inclusivedesigntoolkit. com, 2007.)

Locomotion difficulties (Figure 19.20) are significantly more prevalent than any other category. For product design, the most relevant aspect of locomotion is the ability to bend down to perform actions at knee level or below, and large numbers of people would benefit from eliminating this bending demand. Also included within locomotion is the ability to climb steps, which should be avoided wherever possible. For dexterity demand (Figure 19.22), the majority of change occurs between moderate demand and low demand. Moderate demand corresponds to a task where the user has to use both hands, such as opening the lid of an instant coffee jar, while low demand represents a task that could be completed with either hand, such as squeezing a small sponge.

After the exclusion caused by the demands on each of the seven capabilities has been identified, the slope of each graph at the corresponding demand level provides an indication of the likely improvement available from small changes to the design. For example, given a particular product, a slight reduction in vision demand could make a significant difference to the number of people excluded, whereas a major reduction in dexterity demand might be required to achieve any real benefit. For a set of tasks, evaluating the exclusion that corresponds with each task allows identification of the pinch points for exclusion, enabling design effort to be targeted to the greatest effect.

Comparing alternative design choices according to their impact on population exclusion allows these decisions to be made in an informed and rational manner. However, obtaining meaningful numbers requires greater levels of knowledge, resources, and time than simpler methods that subjectively measure demand on scales that range from low to high.

Since any product interaction usually requires a particular sequence of tasks, and each task may require the use of multiple capabilities, a fairly complex calculation for exclusion is often required. Estimating the number of people who would be unable to use a product based on multiple different capability demands requires an exclusion calculator, which is available from http://www.inclusivedesigntoolkit.com.

19.5 Conclusions

Inclusive design is a methodology that results in better designed mainstream products that are accessible to a wider range of people. At a corporate level, the tools for effective design ensure that the best use is made of portfolios, platforms, and product derivatives to maximize market potential. The SPROC model emphasizes the importance of buy-in from senior-level management for inclusive design, and provides a simple model for the aspects of a business that will impact its successful implementation.

Within a specific design project, inclusive design tools ensure that solutions are developed to satisfy the real user and business needs. The requirements specification captures these needs, and presents them in a manner suitable for creating and evaluating concepts and solutions. Involving personas and real users enables designers to keep the variety of users and environments in mind throughout.

Decisions and compromises have to be made throughout any design process. Evaluation tools such as expert appraisal, impairment simulators, or assessment of demand and exclusion enable such decisions to be made in an informed and rational manner, with due respect for the resulting impact on exclusion.

References

Aldersey-Williams, H., Bound, J., and Coleman, R. (1999). *The Methods Lab: User Research for Design*. London, U.K.: Design for Ageing Network. http://www.education.edean.org.

British Standards Institute (2006). *BS7000-6: Managing Inclusive Design: Guide*. London: BSI British Standards.

Bruce, M. and Cooper, R. (2000). *Creative Product Design: A Practical Guide to Requirements Capture Management*. Chichester, U.K.: Wiley.

Cagan, J., Vogel, C. M., and Nussbaum, B. (2001). *Creating Breakthrough Products: Innovation from Product Planning to Programme Approval*. Upper Saddle River, NJ: Prentice-Hall.

Carmichael, A. (1999). *Style Guide for the Design of Interactive Television Services for Elderly Viewers*. Winchester, U.K.: Independent Television Commission.

Caswell, D. (2006). *Creative Problem Solving*. London, Canada: Society for Teaching and Learning in Higher Education.

Clarkson, P. J., Coleman, R., Hosking, I., and Waller, S. (2007). *Inclusive Design Toolkit*. Cambridge, U.K.: EDC. http://www.inclusivedesigntoolkit.com.

Coleman, R. (2001). *Living Longer: The New Context for Design*. London: Design Council.

Constantine, L. L. and Lockwood, L. A. D. (1999). *Software for Use: A Practical Guide to the Models and Methods of the Usage Centred Design*. Boston: Addison-Wesley Professional.

Cooper, A. (1999). *The Inmates Are Running the Asylum*. Indianapolis: Sams.

Disability Discrimination Act (1995). http://www.opsi.gov.uk.

Faulkner, L. (2003). Beyond the five-user assumption: Benefits of increased sample sizes in usability testing. *Behavior Research Methods, Instruments Computers* 35: 379–383.

Grundy, E., Ahlburg, D., Ali, M., Breeze, E., and Sloggett, A. (1999). *Research Report 94: Disability in Great Britain*. London: Corporate Document Services.

Huber, J. and Skidmore, P. (2003). *The New Old: Why Baby Boomers Won't Be Pensioned Off*. London: Demos. http://www.demos.co.uk.

Keates, S. (2007). *Designing for Accessibility: A Business Guide to Countering Design Exclusion*. London: Lawrence Erlbaum Associates.

Klein, J., Scott, N., Sinclair, K., Gale, S., and Clarkson, P. J. (2006). *The Equipment Needs of Consumers Facing Most Difficulty Switching to Digital TV*. http://www.digitaltelevision.gov.uk.

Microsoft (2003). *The Wide Range of Abilities and Its Impact on Computer Technology: A Research Study Commissioned by Microsoft Corporation and Conducted by Forrester Research Inc*. http://www.microsoft.com.

Myerson, J. and Gheerawo, R. (2004). Inclusive design in practice: Working with industry, in the *Proceedings of the Designing for the 21st Century III Conference (D21³)*, 14–16 July 2004, Manchester, U.K. http://www.designfor21st.org.

Otto, K. and Wood, K. (2001). *Product Design: Techniques in Reverse Engineering and New Product Development*. Upper Saddle River, NJ: Prentice Hall.

Poulson, D., Ashby, M., and Richardson, S. (1996). *USERfit: A Practical Handbook on User-Centred Design for Rehabilitation and Assistive Technology*. Leicestershire, U.K.: HUSAT Research Institute for the European Commission.

Preiser, W. and Ostroff, E. (2000). *Universal Design Handbook*. New York: McGraw-Hill.

Pruitt, J. and Adlin, T. (2006). *The Persona Lifecycle: Keeping Users in Mind throughout the Design Process*. San Francisco: Morgan-Kaufmann.

Shefelbine, S., Clarkson, P. J., Farmer, R., and Eason, S. (2002). *Good Design Practice for Medical Devices and Equipment: Requirements Capture*. Cambridge, U.K.: EDC.

Ulrich, K. T. and Eppinger, S. D. (2004). *Product Design and Development* (3rd ed.). New York: McGraw-Hill.

United Nations (1999). *The World at 6 Billion*. http://www.popin.org.

U.S. Census Bureau (2007). *International Data Base*. http://www.census.gov/ipc/www/idb.

Vanderheiden, G. (1996). *Universal Design: What It Is and What It Isn't*. Madison, WI: University of Wisconsin–Madison, Trace Research and Development Center. http://trace.wisc.edu/docs/whats_ud/whats_ud.htm.

20

The Evaluation of Accessibility, Usability, and User Experience

Helen Petrie and
Nigel Bevan

20.1 Introduction

This chapter introduces a range of evaluation methods that allow developers to create interactive electronic systems, products, services, and environments[1] that are both easy and pleasant to use for the target audience. The target audience may be the broadest range of people, including people with disabilities and older people, or it may be a highly specific audience, such as university students studying biology. eSystems are also specifically developed for people with particular disabilities to assist them in dealing with the problems they encounter due to their disabilities (commonly such technologies are called assistive technologies); these include screen readers for blind computer users and computer-based augmentative and alternative communication systems for people with speech and language disabilities (Cook and Polgar, 2008).

The chapter will introduce the concepts of accessibility, usability, and user experience as the criteria against which developers should be evaluating their eSystems, and the user-centered iterative design life cycle as the framework within which the development and evaluation of these eSystems can take place. Then a range of methods for assessing accessibility, usability, and user experience will be outlined, with information about their appropriate use and strengths and weaknesses.

20.2 Accessibility, Usability, and User Experience

Developers work to create eSystems that are easy and straightforward for people to use. Terms such as *user friendly* and *easy to use* often indicate these characteristics, but the overall technical term for them is *usability*. The ISO 9241 standard on *Ergonomics of Human System Interaction*[2] (Part 11, 1998) defines *usability* as:

> The extent to which a product can be used by specified users to achieve specified goals with effectiveness, efficiency and satisfaction in a specified context of use.

Effectiveness is defined as the accuracy and completeness with which users achieve specified goals; *efficiency* is defined as the resources expended in relation to the accuracy and completeness with which users achieve those goals; and *satisfaction* is defined as "freedom from discomfort, and positive attitudes towards the use of the product." Although not components of the ISO definition, many practitioners (Gould and Lewis, 1985; Shackel, 1990, 1991; Stone et al., 2005; Sharp et al., 2007) have long considered the following aspects part of *usability*:

---

[1] For ease of reading, we will use the term *eSystems* or simply *systems* to refer to the full range of interactive electronic products, services, and environments, which includes operating systems, personal computers, applications, web sites, handheld devices, and so on.

[2] This standard was originally called *Ergonomic Requirements for Office Work with Visual Display Terminals*. A program of revision and expansion of the standard is currently underway.

- *Flexibility*: The extent to which the system can accommodate changes desired by the user beyond those first specified
- *Learnability*: The time and effort required to reach a specified level of use performance with the system (also known as *ease of learning*)
- *Memorability*: The time and effort required to return to a specified level of use performance after a specified period away from the system
- *Safety*: Aspects of the system related to protecting the user from dangerous conditions and undesirable situations

ISO standards for software quality refer to this broad view of usability as *quality in use*, as it is the user's overall experience of the quality of the product (Bevan, 2001).

The previous discussion shows that usability is not given an absolute definition, but is relative to the users, goals, and contexts of use that are appropriate to the particular set of circumstances. For example, if one is developing an online airline booking system for professional travel agents to use at work, the requirements or criteria for usability components such as efficiency and learnability will undoubtedly be different than if one is developing a web site for the general public to book airline tickets. People who use an eSystem on a daily basis for their work will be prepared to put higher levels of time and effort into learning to use the system than those who are using an eSystem only occasionally; however, they may also have higher requirements for efficiency.

Like usability, *accessibility* is a term for which there is a range of definitions. It usually refers to the use of eSystems by people with special needs, particularly those with disabilities and older people. ISO 9241-171 (2008b) defines accessibility as:

> [T]he usability of a product, service, environment or facility by people with the widest range of capabilities

This definition can be thought of as conceptualizing accessibility as simply usability for the maximum possible set of *specified users* accommodated; this fits within the *universal design* or *design for all* philosophy (see Section 20.3.2; see also the chapters in Part I of this handbook). However, accessibility is also used to refer to eSystems that are specifically usable by people with disabilities. For example, the Web Accessibility Initiative (WAI),[3] founded by the World Wide Web Consortium (W3C) to promote the accessibility of the web, defines web accessibility to mean

> that people with disabilities can use the Web. More specifically, Web accessibility means that people with disabilities can perceive, understand, navigate, and interact with the Web. (WAI, 2006)

The WAI definition suggests that accessibility is a subset of usability (i.e., that accessibility is only concerned with issues for

a subset of users, being older and disabled people), whereas the ISO definition suggests that usability is a subset of accessibility (that accessibility is about issues for the largest possible range of users, including older and disabled people). This highlights the current lack of consensus about accessibility. However, for practical purposes, when discussing the development of eSystems for mainstream (i.e., nondisabled, younger) users and the problems that these users have with such systems, usability is the term used; whereas, when the development of eSystems for disabled and older users and the problems these users have with such systems, accessibility is the term used.

User experience (often abbreviated to UX) is the newest term in the set of criteria against which an eSystem should be evaluated. It has arisen from the realization that as eSystems become more and more ubiquitous in all aspects of life, users seek and expect more than just an eSystem that is easy to use. Usability emphasizes the appropriate achievement of particular tasks in particular contexts of use, but with new technologies such as the web and portable media players such as iPods, users are not necessarily seeking to achieve a task, but also to amuse and entertain themselves. Therefore the term *user experience*, initially popularized by Norman (1998), has emerged to cover the components of users' interactions with, and reactions to, eSystems that go beyond effectiveness, efficiency, and conventional interpretations of satisfaction.

Different writers have emphasized different aspects of UX: these are not necessarily contradictory to each other, but explore different aspects of and perspectives on this very complex concept. For example, Hassenzahl and Tractinsky (2006) (see also Hassenzahl, 2006; Hassenzahl et al., 2006) delineate three areas in which UX goes beyond usability:

- *Holistic*: As previously discussed, usability focuses on performance of and satisfaction with users' tasks and their achievement in defined contexts of use; UX takes a more holistic view, aiming for a balance between task-oriented aspects and other non-task-oriented aspects (often called *hedonic* aspects) of eSystem use and possession, such as beauty, challenge, stimulation, and self-expression.
- *Subjective*: Usability has emphasized objective measures of its components, such as percentage of tasks achieved for effectiveness and task completion times and error rates for efficiency; UX is more concerned with users' subjective reactions to eSystems, their perceptions of the eSystems themselves and their interaction with them.
- *Positive*: Usability has often focused on the removal of barriers or problems in eSystems as the methodology for improving them; UX is more concerned with the positive aspects of eSystem use, and how to maximize them, whether those positive aspects be joy, happiness, or engagement.

Dillon (2001), while sharing the view that a move beyond usability is needed in the design and evaluation of eSystems, suggests that an emphasis on three key issues of users' interaction with eSystems is also required:

---

[3] http://www.w3c.org/WAI.

- *Process*: What the user does (e.g., navigation through a web site, use of particular features, help, etc.). This allows the development of an understanding of users' moves, attention, and difficulties through an eSystem.
- *Outcomes*: What the user attains (e.g., what constitutes the goal and end of the interaction). This allows an understanding of what it means for the user to feel accomplishment or closure with the eSystem.
- *Affect*: What the user feels; this includes the concept of satisfaction from the definition of usability, but goes beyond that to include all emotional reactions of users, which might be empowered, annoyed, enriched, or confident. This allows the development of an understanding of users' emotional interaction with eSystems and what interaction means for users.

The new ISO standard 9241-210 (2009) defines UX as:

A person's perceptions and responses that result from the use or anticipated use of a product, system or service.

Bevan (2008) suggests that the definition of usability can be extended to encompass user experience by interpreting satisfaction as including:

- *Likability*: The extent to which the user is satisfied with the perceived achievement of pragmatic goals, including acceptable perceived results of use and consequences of use.
- *Pleasure*: The extent to which the user is satisfied with the perceived achievement of hedonic goals of stimulation, identification, and evocation (Hassenzahl, 2003) and associated emotional responses, for example, Norman's (2004) visceral category.
- *Comfort*: The extent to which the user is satisfied with physical comfort.
- *Trust*: The extent to which the user is satisfied that the product will behave as intended.

UX is still a concept that is being debated, defined, and explored by researchers and practitioners (see, e.g., Law et al., 2008). However, it is clear that this concept is already an important part of the evaluation of eSystems and will become more important in the future.

20.3 Design and Evaluation Processes: Iterative User-Centered Design and Inclusive Design

In considering when and how to conduct evaluations of eSystems, it is necessary first to situate evaluation within the overall design and development process. Software engineers have long used some form of the *waterfall* process of development (see, e.g., Sommerville, 1995) in which phases such as requirements definition, system and software design, implementation and unit testing, integration and system testing, and operation

and maintenance are temporally and organizationally distinct. When each phase is complete, a set of documentation summarizing that phase is handed to the next phase for it to start. Experts such as Sommerville acknowledge that this is a theoretical idealization, and that in practice adjustment is required between phases, captured in a *spiral* model of development. As Sommerville notes:

[T]he development stages overlap … the process is not a simple linear model but involves a sequence of iterations of the development activities. (p. 7)

However, those working on the development of highly interactive eSystems argue that the design and development process must be explicitly iterative and user-centered to address the difficulties of fully understanding user requirements, and developing eSystems that provide usable and pleasant experiences for users.

20.3.1 Iterative, User-Centered Design

A typical iterative user-centered design and development process is illustrated in Figure 20.1. The phases of the process are as follows.

20.3.1.1 Understanding Users, Tasks, Contexts

This might involve studying existing style guides, guidelines, or standards for a particular type of system; interviewing current or potential users of an eSystem about their current system, its strengths and weaknesses, and their expectations for a new or redesigned eSystem; and conducting an ethnographic (Ball and Omerod, 2000) or context-of-use (Beyer and Holtzblatt, 1997) study of a particular situation. All this contributes to an initial understanding of what the eSystem should do for users and how it should be designed. It is advisable to encapsulate the information gained at this phase in a user requirements document (complementing system requirements documents), that can then be used to track how the subsequent design and development work meets these initial requirements and can be updated to reflect changes in the understanding of the user requirements. A Common Industry Specification for Usability Requirements (CISU-R) has been proposed to provide a standard format for specifying and reporting user requirements and performance and satisfaction criteria (but not UX criteria) (NIST, 2007). This specification also proposes formats specifying the context(s) of use for an eSystem and test method and context of testing for evaluations.

20.3.1.2 Design

Initial design ideas can now be explored. It is often important to explore the design space as much as possible, to consider alternative designs and how they will meet users' needs, rather than immediately settling on one design. This will also facilitate the next stage.

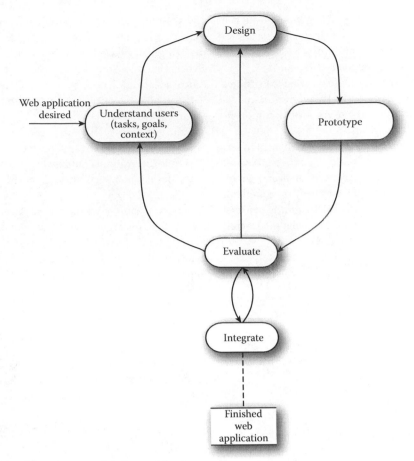

FIGURE 20.1 A typical iterative user-centered design and development process.

20.3.1.3 Prototype

Once an initial potential design, or hopefully a range of potential designs, has been developed, then prototypes can be built (Snyder, 2003). These can take many forms, from very simple to complex (often called low fidelity to high fidelity), from sketches on paper with no interactivity, to Microsoft PowerPoint or Adobe Flash animations with considerable interactivity. In fact, for the initial prototypes, it is usually better to make them obviously simple and unfinished, as that allows people involved in evaluations to realize that it is acceptable to criticize them. Prototypes might also only address part of the functionality of an eSystem, but it is important to explore particular design problems before considerable effort is put into full implementation and integration of components of an eSystem.

In producing prototypes one might realize that some design ideas are not going to be feasible, and this is the first loop of iteration, as it will feed back into the design process.

20.3.1.4 Evaluate

The heart of the process, and the figure, is evaluation. Prototypes can be evaluated by experts and particularly by potential or current users, using a variety of methods (see Section 20.4.3). A number of iterations of evaluation, designing, and prototyping may be required before acceptable levels of usability, accessibility, and user experience are reached. A document that encapsulates the target levels may also be helpful, and again this can be used to track how successive prototypes meet these levels. The evaluations can feed back to both the design phase and to the understanding of the users, their tasks and contexts. Because people are such complex entities, even an eSystem designed on the basis of a very good understanding of users from previous and current work will be unlikely to succeed on the first prototype. As Whiteside et al. (1988) commented, "users always do surprising things" (p. 799). A number of iterations of prototyping and designing are likely to be required. Nielsen and Sano (1994) reported that in designing a set of icons for the Sun Microsystems web site, 20 iterations of the icon designs proved to be necessary. This is quite a high number of iterations, but the main design and development process took only a month, with four main evaluations. Three to five iterations would seem much more typical. It should also be noted that the iterative user-centered design and development of the interactive aspects of an eSystem can usually go on in parallel with backend developments, so this iterative user-centered process should not hold up the overall development of the eSystem.

20.3.1.5 Integration and Final Implementation

Once the design of the various components of an eSystem has reached acceptable levels of usability, accessibility, and user experience, integration of components and final implementation of the interactive systems may be required. Prototypes of eSystems or components may not be implemented in the same language and/or environment as the final eSystem. Once such implementation and integration has taken place, a further evaluation may be appropriate to ensure any issues that relate to using the integrated system are addressed. Finally, once the eSystem is released to users, an evaluation of its use in real contexts may be highly beneficial. Both these final evaluations can feed back into understanding of the users, their tasks and contexts, and into the design process, if not for this version of the eSystem, then for subsequent versions.

20.3.2 Inclusive Design

In considering the iterative user-centered design process outlined in the previous section, it should be clear that including people with disabilities and older people among the evaluators can be part of this process, and that target levels for accessibility can play an important role in the overall process. This is in contrast with many writers who only include a consideration of disabled and older people at the end of the design and development process. However, it is clear that if the full range of potential users is considered from the beginning of the process, the overhead of considering the needs of disabled and older users is minimal—it is simply part of the overall design process. On the other hand, if one designs only for young, mainstream users and then attempts to expand the process for disabled and older users at a late stage, one is contradicting the user-centered design process and it is very likely that complex and expensive retrofitted solutions will be necessary for these users. In some cases, it is impossible to retrofit a solution to include the needs of particular disabled and older users, and the design process really needs to be started again. For example, in the case of producing an eSystem using Adobe Flash, if accessibility issues are considered from the beginning, there is no particular additional cost for making the system accessible to disabled users; however, experience has shown that if accessibility is only considered late in the development process, it is almost impossible to retrofit a solution for disabled users.[4]

A number of terms have been coined to cover the inclusion of disabled and older users and their needs in the design process: *universal design* (a termed coined by Ron Mace; see, e.g., Story et al., 1998), widely used in North America; *design for all*, used more commonly in Europe (see EDeAN, 2007); *barrier-free design*, and *inclusive design*. One difficulty is that all these terms suggest that *all* people should be included; for example, universal design is defined as:

> [T]he design of products and environments to be usable by all people, to the greatest extent possible, without the need for adaptation or specialized design.

While this is an honorable aim, it is an ideal for which to aim but in practice cannot be met. The literal interpretation can

rightly frighten designers and developers, who cannot see how it can be achieved, and may put them off attempting to develop accessible eSystems at all. It is important to get the problem in perspective: designers and developers need to start thinking beyond the needs of young, mainstream, and technology-literate users, and seriously consider the needs of the full range of users who might wish to use the eSystem they are developing.

It is very easy to fail to recognize the full range of users who might be interested in using or need to use a particular eSystem. For example, in developing an online payment system for a road taxing system, designers might think that only drivers, who by definition have good sight, will be interested in or need to use the system. However, a visually disabled friend of a driver may wish to pay the road tax when they are given a lift. Therefore, such an eSystem needs to be accessible to users with visual disabilities as well as fully sighted ones.

In addition, designers need to be aware that people with visual disabilities in particular will use assistive technologies to help them access many eSystems, particularly if they are accessing them in the workplace or at home (the situation for eSystems to be used in public places, such as automatic banking machines and ticket machines, is more problematic, if alternatives are not available). This includes screen readers used by blind people (see Chapter 28 of this handbook), screen magnification programs used by partially sighted people, and a variety of alternative input devices used by people with physical disabilities (Cook and Polgar, 2008). This means that the designers of a particular eSystem do not need to solve *all* the accessibility problems.

One can think of the population of users addressed by an eSystem as dividing into three groups. For users who do not use an assistive technology in the context of use, as many users as possible should be accommodated (this will include mainstream and older users); for users who do use an assistive technology, the system should work smoothly with assistive technologies (and will need evaluation with those assistive technologies to ensure that is the case); the final group—people who cannot use the system with or without an assistive technology—should ideally be an empty set.

Some universal design/design-for-all approaches propose guidelines to assist in the design of eSystems to meet the needs of disabled and older users. Such guidelines are indeed useful, and will be discussed in Section 20.4.2. However, both the use of universal design/design-for-all guidelines and evaluation with disabled and older users should be integrated into the iterative user-centered design process for the most effective development of eSystems that are usable and pleasant for the widest range of users.

20.4 Methods for Evaluation

Methods for usability, accessibility, and UX evaluation can be grouped into the following categories:

- Automated checking of conformance to guidelines and standards
- Evaluations conducted by experts

---

[4] This is not a criticism specifically made of Adobe Flash, as Adobe has worked diligently to make it accessible; this situation holds for many technologies.

- Evaluations using models and simulations
- Evaluation with users or potential users
- Evaluation of data collected during eSystem usage

Several methods are based on the use of guidelines and standards, so an outline and discussion of relevant guidelines and standards will be given first, and then outlines and discussions of the various methods will be presented.

20.4.1 Guidelines for Accessibility and Usability

20.4.1.1 Accessibility Guidelines

Guidelines on accessibility for disabled and older users are available for a number of different types of eSystems. For example, the IBM Human Ability and Accessibility Center[5] provides guidelines in the form of easy to follow checklists with hyperlinks to rationales to explain the need for the guideline, development techniques, and testing methods. There are checklists for:

- Software
- Web sites and applications
- Java applications
- Lotus notes
- Hardware
- Documentation

Three ISO standards containing accessibility guidelines were published in their final form in 2008 (see also Chapter 55 of this handbook):

- ISO/IEC 10779 (2008c): Office equipment accessibility guidelines for elderly persons and persons with disabilities
- ISO 9241-20 (2008a): Ergonomics of human-system interaction, Part 20: Accessibility guidelines for information/communication technology (ICT) equipment and services
- ISO 9241-171 (2008b): Ergonomics of human-system interaction, Part 171: Guidance on software accessibility

The key set of guidelines for assessing the accessibility of web sites is the Web Content Accessibility Guidelines developed by the WAI (see Chapter 10 of this handbook). The first version of these guidelines (WCAG1) was published in 1999 (WAI, 1999). The second version of WCAG (WCAG2; see WAI, 2008) was finalized at the end of 2008. However, it is expected that both WCAG1 and WCAG2 will be used in parallel for some time.

WCAG1 includes 14 high-level accessibility guidelines, which are broken down into 65 specific checkpoints. Each checkpoint is assigned a priority level (Priority 1, 2, and 3). A web page or document must satisfy Priority 1 (P1) checkpoints; otherwise, according to WAI "one or more groups [of disabled people] will find it *impossible* to access information in the document." If Priority 2 (P2) checkpoints are not satisfied, one or more groups of disabled people will find it *difficult* to access information in the

document. If Priority 3 (P3) checkpoints are not satisfied, one or more groups of disabled people "will find it *somewhat difficult* to access information." If a web page or site passes all the P1 checkpoints, it is said to be Level A conformant; if it passes all P1 and all P2 checkpoints, it is Level AA conformant; finally if it passes all P1, P2, and P3 checkpoints, it is Level AAA conformant.

WCAG2 carries forward many of the ideas of WCAG1, including the three levels of conformance (the complexity of priorities and levels has been removed, so only the three levels A, AA, and AAA are now used). However, rather than being organized around the 14 high-level guidelines, it is now organized around four accessibility principles:

- Content must be perceivable.
- Interface components in the content must be operable.
- Content and controls must be understandable.
- Content should be robust enough to work with current and future user agents (including assistive technologies).

Each principle is associated with a list of guidelines addressing the issues around that principle. Many of the checkpoints from WCAG1 are retained, but the organization is more logical.

Another set of guidelines often mentioned in relation to web accessibility is Section 508 of the Rehabilitation Act of the United States Federal government (see Chapter 53, "Policy and Legislation as a Framework of Accessibility"). In fact, this legislation requires federal agencies to make all their electronic and information technologies (not only web sites) accessible to people with disabilities. In practice, this means that anyone who is supplying eSystems to the U.S. federal government is obliged to make them accessible. A set of standards have been developed to specify what accessibility means for different types of eSystems,[6] and those for web sites are very similar to WCAG.

Accessibility guidelines and advice also exist for many specific technologies that are used for producing eSystems and for specific domains. For example, in relation to specific technologies:

- Adobe Flash (see Regan, 2004, and resources at the Adobe Accessibility Resource Center[7])
- Content management systems
- Joomla! (see O'Connor, 2007)
- Eclipse open source development platform (see Accessibility Features in Eclipse[8] and the Eclipse Accessibility Tools Framework Project[9])
- Java (see resources at Sun Accessibility[10])
- Microsoft products (resources at the Microsoft Accessibility Developer Center[11])

---

5 http://www-03.ibm.com/able.

6 See http://www.section508.gov.
7 http://www.adobe.com/accessibility.
8 http://help.eclipse.org/help33/index.jsp?topic= /org.eclipse.platform.doc. user/concepts/accessibility/accessmain.htm.
9 http://www.eclipse.org/actf.
10 http://www.sun.com/accessibility.
11 http://msdn.microsoft.com/en-us/accessibility/default.aspx.

TABLE 20.1 Summary of Web Content Accessibility Guidelines Version 2.0

Principle	Guidelines
1. Perceivable: Information and user interface components must be presentable to users in ways they can perceive.	1.1 Text alternatives: Provide text alternatives for any nontext content so that it can be changed into other forms people need, such as large print, Braille, speech, symbols, or simpler language. 1.2 Time-based media: Provide alternatives for time-based media. 1.3 Adaptable: Create content that can be presented in different ways (e.g., simpler layout) without losing information or structure. 1.4 Distinguishable: Make it easier for users to see and hear content, including separating foreground from background.
2. Operable: User interface components and navigation must be operable.	2.1 Keyboard-accessible: Make all functionality available from a keyboard. 2.2 Enough time: Provide users enough time to read and use content. 2.3 Seizures: Do not design content in a way that is known to cause seizures. 2.4 Navigable: Provide ways to help users navigate, find content, and determine where they are.
3. Understandable: Information and the operation of user interface must be understandable.	3.1 Readable: Make text content readable and understandable. 3.2 Predictable: Make web pages appear and operate in predictable ways. 3.3 Input assistance: Help users avoid and correct mistakes.
4. Robust: Content must be robust enough that it can be interpreted reliably by a wide variety of user agents, including assistive technologies.	4.1 Compatible: Maximize compatibility with current and future user agents, including assistive technologies.

Source: Web Accessibility Initiative (WAI), *Web Content Accessibility Guidelines 2.0*, http://www.w3.org/TR/WCAG20, 2008.

20.4.1.2 Usability Guidelines and Standards

Guidelines and standards for ensuring good usability of eSystems have been developed for many years. They range from the high-level guidelines (or heuristics) proposed by Nielsen and Molich (Molich and Nielsen, 1990; Nielsen and Molich, 1990; see also Nielsen, 1994) and Shneiderman (Shneiderman and Plaisant, 2005) (see Tables 20.2 and 20.3, respectively) to the much more detailed guidelines provided by the ISO 9241 standard. Parts 12 to 17 of the ISO 9241 series of standards contain very detailed user interface guidelines. Although these are an excellent source of reference, they are very time consuming to employ in evaluations. A further problem with the detailed ISO guidelines is that they need to be interpreted in relation to a particular interface environment (such the Microsoft Windows operating system or the web).

Detailed guidelines for web design are also available. The most comprehensive, well-researched, and easy to use set has been produced by the U.S. Government Department of Health and Human Services (HHS) (2006). This provides 207 guidelines derived from about 500 cited publications. Each guideline contains:

- A brief statement of the overarching principle that is the foundation of the guideline
- Comments that further explain the research/supporting information
- Citations to relevant web sites, technical and/or research reports supporting the guideline
- A score indicating the "Strength of Evidence" that supports the guideline
- A score indicating the "Relative Importance" of the guideline to the overall success of a web site
- One or more graphic examples of the guideline in practice

Some examples of the guidelines are presented in Table 20.4.

While no set of guidelines can be totally comprehensive, the HHS guidelines appear to be more complete and easier to use than the equivalent ISO standard 9241-151 (Bevan and Spinhof, 2007).

20.4.1.3 Problems with Guidelines

Although guidelines would appear to provide objective criteria against which to evaluate a system, they present a number of difficulties.

- The large number of guidelines require substantial effort to learn and apply appropriately.
- For a thorough evaluation, every page or screen should be evaluated against every applicable guideline, which would be very time consuming. Selecting representative screens or pages may miss some issues.
- Following guidelines usually improves an eSystem, but they are only generalizations so there may be particular circumstances where guidelines conflict or do not apply (e.g., because of the use of new features not anticipated by the guideline).
- It is difficult to apply guidelines appropriately without also having expertise in the application domain, and for accessibility guidelines, expertise in accessibility. For example, Petrie et al. (2006) reported that due to lack of experience with disabled people and their technologies, developers often do not have the conceptual framework needed to apply disability-related guidelines.

Evaluation of the characteristics of the user interface can anticipate and explain many potential usability and accessibility problems, and can be carried out before a working system is

TABLE 20.2 Neilsen's Usability Heuristics

Visibility of system status	The system should always keep users informed about what is going on through appropriate feedback within reasonable time.
Match between system and the real world	The system should speak the user's language, with words, phrases, and concepts familiar to the user, rather than system-oriented terms. Follow real-world conventions, making information appear in a natural and logical order.
User control and freedom	Users often choose system functions by mistake and will need a clearly marked "emergency exit" to leave the unwanted state without having to go through an extended dialogue. Support undo and redo.
Consistency and standards	Users should not have to wonder whether different words, situations, or actions mean the same thing. Follow platform conventions.
Error prevention	Even better than good error messages is a careful design that prevents a problem from occurring in the first place. Either eliminate error-prone conditions or check for them and present users with a confirmation option before they commit to the action.
Recognition rather than recall	Minimize the user's memory load by making objects, actions, and options visible. The user should not have to remember information from one part of the dialogue to another. Instructions for use of the system should be visible or easily retrievable whenever appropriate.
Flexibility and efficiency of use	Accelerators—unseen by the novice user—may often speed up the interaction for the expert user such that the system can cater to both inexperienced and experienced users. Allow users to tailor frequent actions.
Aesthetic and minimalist design	Dialogues should not contain information that is irrelevant or rarely needed. Every extra unit of information in a dialogue competes with the relevant units of information and diminishes their relative visibility.
Help users recognize, diagnose, and recover from errors	Error messages should be expressed in plain language (no codes), precisely indicate the problem, and constructively suggest a solution.
Help and documentation	Even though it is better if the system can be used without documentation, it may be necessary to provide help and documentation. Any such information should be easy to search, focused on the user's task, list concrete steps to be carried out, and not be too large.

Source: Nielsen, J., Heuristic evaluation, in *Usability Inspection Methods* (J. Nielsen and R. L. Mack, eds.), John Wiley & Sons, New York, 1994.

available. However, evaluation of detailed characteristics alone can never be sufficient, as this does not provide enough information to accurately predict the eventual user behavior.

To be sure of product usability/accessibility requires user testing. Although a user test is the ultimate test of usability and accessibility, it is not usually practical to evaluate all permutations of user type, task, and environmental conditions.

A number of guidelines include ratings of the importance of different guidelines. As discussed in Section 20.4.1.1, WCAG1 and WCAG2 include three levels of priority that indicate their importance in relation to the accessibility of web sites for people with disabilities. As discussed in Section 20.4.1.2, the HHS guidelines also provide a rating for the "Relative Importance" of the guideline to the overall success of the web site. Few studies

TABLE 20.3 Shneiderman's Eight Golden Principles of Good Interface Design

Strive for consistency	Consistent sequences of actions should be required in similar situations; identical terminology should be used in prompts, menus, and help screens; and consistent commands should be employed throughout.
Enable frequent users to use shortcuts	As the frequency of use increases, so do the user's desires to reduce the number of interactions and to increase the pace of interaction. Abbreviations, function keys, hidden commands, and macro facilities are very helpful to an expert user.
Offer informative feedback	For every operator action, there should be some system feedback. For frequent and minor actions, the response can be modest, while for infrequent and major actions, the response should be more substantial.
Design dialogue to yield closure	Sequences of actions should be organized into groups with a beginning, middle, and end. The informative feedback at the completion of a group of actions gives the operators the satisfaction of accomplishment, a sense of relief, the signal to drop contingency plans and options from their minds, and an indication that the way is clear to prepare for the next group of actions.
Offer simple error handling	As much as possible, design the system so the user cannot make a serious error. If an error is made, the system should be able to detect the error and offer simple, comprehensible mechanisms for handling the error.
Permit easy reversal of actions	This feature relieves anxiety, since the user knows that errors can be undone; it thus encourages exploration of unfamiliar options. The units of reversibility may be a single action, a data entry, or a complete group of actions.
Support internal locus of control	Experienced operators strongly desire the sense that they are in charge of the system and that the system responds to their actions. Design the system to make users the initiators of actions rather than the responders.
Reduce short-term memory load	The limitation of human information processing in short-term memory requires that displays be kept simple, multiple page displays be consolidated, window-motion frequency be reduced, and sufficient training time be allotted for codes, mnemonics, and sequences of actions.

Source: Shneiderman, B. and Plaisant, C., *Designing the User Interface: Strategies for Effective Human-Computer Interaction*, Pearson Education, Boston, 2005.

TABLE 20.4   Research-Based Web Design and Usability Guidelines

Category	Example
1. Design process and evaluation	Establish user requirements
2. Optimizing the user experience	Provide printing options
3. Accessibility	Do not use color alone to convey information
4. Hardware and software	Design for common browsers
5. The home page	Show all major options on the home page
6. Page layout	Set appropriate page lengths
7. Navigation	Provide feedback on user's location
8. Scrolling and paging	Eliminate horizontal scrolling
9. Headings, titles, and labels	Use unique and descriptive headings
10. Links	Provide consistent clickability cues
11. Text appearance	Use black text on plain, high-contrast backgrounds
12. Lists	Order elements to maximize user performance
13. Screen-based controls (widgets)	Distinguish required and optional data entry fields
14. Graphics, images, and multimedia	Use video, animation, and audio meaningfully
15. Writing web content	Avoid jargon
16. Content organization	Facilitate scanning
17. Search	Ensure usable search results

Source: U.S. Department of Health and Human Sciences, *Research-Based Web Design & Usability Guidelines.* http://www.usability.gov/guidelines, 2006.

have investigated the validity of these ratings, but two recent studies (Harrison and Petrie, 2006; Petrie and Kheir, 2007) have found no correlation in the ratings given by both disabled and mainstream users of actual problems that they have encountered and the ratings of those problems as given by WCAG1 and HHS. Therefore, the ratings need to be treated with a certain amount of caution and further studies of this issue are required.

20.4.2 Automated Checking of Conformance to Guidelines or Standards

20.4.2.1 When to Use Automated Checking

When initial prototypes or initial versions of full implementations are available.

20.4.2.2 Why Use Automated Checking?

To ensure that initial prototypes and initial versions of final implementations meet appropriate guidelines and standards and do not contain basic accessibility and usability problems.

20.4.2.3 Tools for Automated Accessibility Checking

The development of WCAG provided a considerable interest in creating tools to automatically check whether web sites and pages conform with the guidelines, as many of the WCAG checkpoints seemed amenable to such automated checking. One of the first such tools, and the most well-known, the Bobby Tool, is no longer available, although it is still mentioned in the literature. A comprehensive list of these tools is maintained on the WAI web site.[12] There appear to be no automatic accessibility checking tools for other types of eSystems.

Although automated accessibility checking has its role in the evaluation of web sites, its strengths and weaknesses need to be understood. Many WCAG checkpoints cannot be checked automatically, and for example only 23% of the WCAG checkpoints were checked by the Bobby tool (Cooper and Rejmer, 2001). Even a single checkpoint may require several tests to check whether it has been passed, some of which can be automated and some of which cannot.

Take, for example, WCAG1 Checkpoint 1.1: Provide a text equivalent for every nontext element. An automated tool can check whether there is an alternative description on every image, which can be a very useful function in evaluating a large web site with many images. However, no automatic tool can check whether the alternative descriptions are accurate and useful (Petrie et al., 2005). So all of these following alternative descriptions that have been found should fail to meet Checkpoint 1.1, but an automatic checking tool would pass them:

- Blah blah blah (possibly a forgotten placeholder?)
- Image of an elephant (the image was actually of a cat)
- Person at computer (gave no indication of why the image was included)

Thus, particular care needs to be taken in interpreting what it means when an automatic checking tool returns no failures for a particular WCAG checkpoint. This means that there have been no failures for the tests that the tool has been able to make on the checkpoint, but not necessarily that there are no failures at all for the checkpoint. Unfortunately, many tools fail to make clear what tests they are conducting on each checkpoint, so it is difficult for those using such tools to evaluate the accessibility of web sites to accurately assess their output. No automated tools exist for automated accessibility checking of eSystems other than web sites and pages.

---

[12] http://www.w3.org/WAI/ER/existingtools.html#General.

TABLE 20.5　Types of Expert-Based Evaluation Methods

Guidelines	Task scenarios	
	No	Yes
None	Expert review	Usability walkthrough
		Pluralistic walkthrough
		Cognitive walkthrough
General guidelines	Heuristic inspection	Heuristic walkthrough
	Preliminary accessibility review	
Detailed guidelines	Guidelines inspection	Guidelines walkthrough
	Conformance evaluation for accessibility	

20.4.2.4 Tools for Automated Usability Checking

There are some automated tools that automatically test for conformance with basic usability guidelines. A review of tools by Brajnik (2000) found that the commercial LIFT[13] tool (which also covers accessibility) made the most comprehensive range of checks, with the free Web Static Analyzer Tool (WebSAT[14]) being the next most effective. Although these tools are useful for screening for basic problems, they only test a very limited scope of usability issues (Ivory and Hearst, 2001).

20.4.3 Evaluations Conducted by Experts

20.4.3.1 When to Use Expert Evaluations

When initial prototypes are available.

20.4.3.2 Why Conduct Expert Evaluations?

- To identify as many accessibility and usability issues as possible to eliminate them before conducting user-based evaluations
- Because there are too many pages or screens to include all of them in user-based evaluations
- Because it is not possible to obtain actual or potential users for evaluations
- Because there is insufficient time for user testing
- To train developers in accessibility and usability issues

Expert-based methods ask one or more accessibility/usability or domain experts to work through an eSystem looking for accessibility or usability problems. These experts can use guidelines or they can work through task scenarios that represent what users would typically do with an eSystem (see Section 20.4.4; see Table 20.5; see Gray and Salzman, 1998). Usability methods that do not use task scenarios are referred to as reviews or inspections, while task-based evaluations are referred to as walkthroughs. For accessibility, expert-based methods have tended not to include task scenarios (although there is no reason why they should not), but are divided into preliminary accessibility reviews, which use a small number of key guidelines, and conformance evaluations for accessibility, which use the full set of WCAG1 or WCAG2.

20.4.3.3 Heuristic Evaluation

The most popular type of expert evaluation is *heuristic evaluation* (Nielsen, 1994). Heuristic evaluation originally involved a small set of evaluators examining each eSystem element to identify potential usability problems. The evaluators use a set of heuristics (such as those in Table 20.2) to guide them and rate the potential problems for how severe they are or how important they are to eliminate before an eSystem is released. Usually, a four-level rating scheme is used (1 = cosmetic problem only: need not be fixed unless extra time is available on project; 2 = minor usability problem: fixing this should be given low priority; 3 = major usability problem: important to fix, so should be given high priority; 4 = usability catastrophe: imperative to fix this before product can be released). Several evaluators are usually involved, as each individual typically only finds about one-third of the problems (Nielsen, 1994). Although heuristic evaluation can be carried out by people who are not trained in usability methods, better results are obtained by trained experts (Desurvire et al., 1992).

Since it was originally proposed by Nielsen, heuristic evaluation has been adapted in many ways. Rather than inspecting individual elements, it is often carried out by asking the evaluator to step through typical user tasks. This can combine heuristic evaluation with some of the benefits of a cognitive walkthrough. Evaluators may also work as a group, identifying potential problems together, but then rating them individually and privately, so that they do not influence each other's ratings. This is known as a Cello evaluation.[15] The relative effectiveness of different forms of heuristic evaluation has not yet been explored.

It is often difficult to get agreement between the evaluators on exactly which heuristic is associated with a particular potential usability problem. Heuristics are a useful training aid (Cockton et al., 2003), but their value to usability experts is not clear. Experienced evaluators often dispense with assessing problems against specific heuristics, preferring to rely on their understanding of the principles and their experience of observing users encountering problems.

20.4.3.4 Expert Walkthrough Evaluations

A usability walkthrough evaluation identifies problems while attempting to achieve tasks in the same way as a user, making

---

[13] http://www.nngroup.com/reports/accessibility/software.
[14] http://zing.ncsl.nist.gov/WebTools/WebSAT/overview.html.

[15] http://www.ucc.ie/hfrg/emmus/methods/cello.html.

use of the expert's knowledge and experience of potential problems. This is usually more effective than inspecting individual pages in isolation, as it takes account of the context in which the user would be using the eSystem.

Variations include:

- *Cognitive walkthrough*: For each user action, the evaluator analyzes what the user would be trying to do, whether the interface supports the user's next step to achieve the task, and whether appropriate feedback is provided (Wharton et al., 1994). Although the original method required detailed documentation, it is often used in a more lightweight fashion (e.g., Spencer, 2000).
- *Pluralistic walkthrough*: A group of users, developers, and human factors people step through a scenario, discussing each dialogue element (Bias, 1994).

20.4.3.5 Expert Evaluations for Accessibility

Two expert evaluation methods exist for assessing web sites and web-based applications for accessibility and will be outlined. No comparable methods exist for assessing the accessibility of other types of eSystems; this could be done on an ad hoc basis by an expert, or it would be preferable to employ user evaluation instead (see Section 20.4.5).

Conducting a full expert evaluation of a web site or web-based application for conformance to the WCAG1 or WCAG2 guidelines is a considerable undertaking. So an initial step is to undertake a *preliminary accessibility review*.[16] This involves:

- Selecting a representative set of pages from the web site or screen from the application
- Testing the pages/screens with a range of graphical browsers (e.g., Internet Explorer, Firefox), making the following adjustments:
  - Turning off images, and checking whether appropriate alternative text for the images is available
  - Turning off the sound, and checking whether audio content is still available through text equivalents
  - Using browser controls to vary font size: verify that the font size changes on the screen accordingly, and that the page is still usable at larger font sizes
  - Testing with different screen resolution, and/or by resizing the application window to less than maximum, to verify that horizontal scrolling is not required
  - Changing the display color to grayscale (or print out page in grayscale or black and white) and observe whether the color contrast is adequate
  - Without the mouse, using the keyboard to navigate through the links and form controls on a page (e.g., using the Tab key), making sure that all links and form controls can be accessed, and that the links clearly indicate what they lead to

A number of browser extensions and plug-in evaluation tools are available to make conducting these tests more efficient, for example the AIS Toolbar[17] for Internet Explorer (available in a wide range of languages) and Opera (currently available in English only), the Accessibar add-on,[18] and WAVE Toolbar for Firefox.[19]

- Testing the pages/screens using a specialized browser, such as a text browser (e.g., Lynx[20]), or a screen reader such as JAWS[21] or WindowEyes.[22] Screen readers are sophisticated programs with considerable functionality—an expert user, whether sighted or blind, is needed to use these programs effectively.

The results of such a preliminary accessibility review can guide the further development of a web site or web-based application. Once the web site or web-based application has been developed, it is important to undertake a full accessibility audit. The WAI has outlined the methodology for undertaking such an audit,[23] which is similar to the methodology for the preliminary accessibility review, but includes manual checking of all applicable WCAG checkpoints.

A group of European Union funded projects[24] has developed a detailed standard methodology for the expert accessibility evaluation of web sites, the Unified Web Evaluation Methodology (UWEM). Complete details on how to conduct a UWEM evaluation can be found on their web site,[25] and covers not only the evaluation procedures, but also statistical methods for sampling, critical path analysis, computer-assisted content selection, manual content selection, and interpretation and the aggregation and integration of test results.

The WAI have developed a standard reporting format for accessibility evaluation reports that is also used by the UWEM. Details of this are available on the UWEM and WAI[26] web sites.

20.4.3.6 Advantages and Disadvantages of Expert Evaluation

Expert usability evaluation is simpler and quicker to carry out than user-based evaluation and can, in principle, take account of a wider range of users and tasks than user-based evaluation, but it tends to emphasize more superficial problems (Jeffries and Desurvire, 1992) and may not scale well for complex interfaces (Slavkovic and Cross, 1999). To obtain results comparable with user-based evaluation, the assessment of several experts must be combined. The greater the difference between the knowledge and experience of the experts and the real users, the less reliable are the results.

---

[16] http://www.w3.org/WAI/eval/preliminary.html.

[17] http://www.visionaustralia.org.au/ais/toolbar.

[18] http://firefox.cita.uiuc.edu.

[19] http://wave.webaim.org/toolbar.

[20] http://lynx.browser.org.

[21] http://www.freedomscientific.com/fs_products/software_jaws.asp.

[22] http://www.gwmicro.com/Window-Eyes.

[23] http://www.w3.org/WAI/eval/conformance.html.

[24] The Web Accessibility Cluster, see www.wabcluster.org.

[25] http://www.wabcluster.org/uwem1_2.

[26] http://www.w3.org/WAI/eval/template.html.

20.4.4 Evaluations Using Models and Simulations

20.4.4.1 When to Use Evaluations Using Models and Simulations

- When models can be constructed economically
- When user testing is not practical

20.4.4.2 Why Use Evaluations Using Models and Simulations?

- If time to complete tasks is critical, for example, for economic or safety reasons

Model-based evaluation methods can predict measures such as the time to complete a task or the difficulty of learning to use an interface. Some models have the potential advantage that they can be used without the need for any prototype to be developed. Examples are the use of the keystroke level model (Mayhew, 2005), the goals, operators, methods, and selections (GOMS) model and the ACT-R model of human cognitive processes (St. Amant et al., 2007). However, setting up a model currently usually requires considerable effort, so model-based methods are cost effective in situations where other methods are impracticable, or the information provided by the model is a cost-effective means of managing particular risks. Further information on the use of models can be found in Pew and Mayor (2007).

20.4.5 Evaluations with Users

20.4.5.1 When to Use

- At all stages of development, if possible
- At the final stage of development, at least

20.4.5.2 Why Conduct User-Based Evaluations?

- To provide evidence of the accessibility and usability of an eSystem in real use by the target audience
- To provide evidence of accessibility and usability (or lack thereof) for developers or management

20.4.5.3 Types of User-Based Evaluations

In user-based methods, target users undertake realistic tasks which the eSystem is designed to support in realistic situations, or as realistic situations as possible. A minimum of assistance is given by those running the evaluation, except when participants get completely stuck or need information not readily available to them.

There are many practical details of planning and executing user evaluations, with excellent explanations in books such as Rubin and Chisnell (2008) and Dumas and Redish (1999) and the chapter by Lewis (2005). The interested reader is recommended to study one of these before undertaking user evaluations.

There are different types of user-based methods adapted specifically for formative and summative evaluations (see Table 20.6):

- Formative methods focus on understanding the user's behavior, intentions, and expectations to understand any problems encountered, and typically employ a "think-aloud" protocol.
- Summative methods measure the product usability or accessibility, and can be used to establish and test user requirements. Summative usability testing may be based on the principles of ISO 9241-11 and measure a range of usability components such as effectiveness, efficiency, and satisfaction. Each type of measure is usually regarded as a separate factor with a relative importance that depends on the context of use.

20.4.5.4 Selecting the Sample Size

While the cost benefits of usability evaluation are well established (Bias and Mayhew, 2005), there is no way to be sure that all the important usability problems have been found by an evaluation. Deciding how many participants to include in formative evaluations depends on the target percentage of problems to be identified, and the probability of finding problems (Lewis, 2006). Usability test sample size requirements for a particular desired percentage of problems can be estimated by calculating the probability of finding problems either based on previous similar usability evaluation, or from initial results of an ongoing study. A recent survey (Hwang and Salvendy, 2007) found probabilities in the range 0.08 to 0.42. This would correspond to evaluating with between 3 and 19 participants to find 80% of the problems, or between 4 and 28 participants to find 90% of the problems. Complex web sites and web-based applications in which different users may explore different aspects are likely to have lower probabilities.

Iterative testing with small numbers of participants is preferable, starting early in the design and development process. Even data from one participant may be valuable in identifying previously unnoticed flaws (Medlock et al., 2002), although the appropriateness of any changes should be confirmed by testing with subsequent participants. If carrying out a single, user-based evaluation late in the development life cycle (again, this is not the best procedure, as evaluation should be conducted on several iterations), it is typical to test with at least eight users (or more if there are several distinct target user types).

For summative evaluation, the number of participants depends on the confidence required in the results (i.e., the acceptable probability that the results were only obtained by chance). If there is little variance in the data, a sample of as few as eight participants of one type may be sufficient. If there are several types of users, other sources of variance, or if success rate is being measured (see ISO 20282-2, 2006), 30 or more users may be required.

TABLE 20.6 Purposes of User-Based Evaluation

Purpose	Description	When in Design Cycle	Typical Sample Size (per group)	Considerations
Early formative evaluations				
Exploratory	High level test of users performing tasks	Conceptual design	5–8	Simulate early concepts, for example with very low-fidelity paper prototypes
Diagnostic	Give representative users real tasks to perform	Iterative throughout the design cycle	5–8	Early designs or computer simulations; used to identify usability problems
Comparison	Identify strengths and weaknesses of an existing design	Early in design	5–8	Can be combined with benchmarking
Summative usability testing				
Benchmarking/ competitive	Real users and real tasks are tested with existing design	Prior to design	8–30	To provide a basis for setting usability criteria; can be combined with comparison with other eSystems
Final	Real users and real tasks are tested with final design	End of design cycle	8–30	To validate the design by having usability objectives as acceptance criteria; should include any training and documentation

20.4.5.5 Conducting Evaluations with Disabled and Older Users

There are a number of issues related to conducting evaluations with disabled and older users that need to be raised. It is appreciated that finding samples of disabled and older people willing and able to take part in evaluations is not easy (Petrie et al., 2006) and it may be that remote evaluations could be used to overcome this problem. Petrie et al. (2006) discuss the advantages and disadvantages of remote evaluations. Another method of overcoming this issue might seem to be using able-bodied users to simulate users with disabilities (e.g., by blindfolding people to simulate visual impairment). This is not a sensible solution to the issue at all—people who are visually impaired have developed strategies to deal with their situation, and so suddenly putting sighted people into a blindfolded situation is not at all comparable. Designers and developers may gain some useful insight into the situations of disabled and older users by experiencing simulations, but the usefulness and ethics of these are debated and highly controversial (Kiger, 1992; Burgstahler and Doe, 2004).

An important issue to consider when conducting evaluations with disabled users is whether they will use assistive technologies in using the eSystem under evaluation. If so, the correct versions and the preferred configurations of assistive technologies need to be provided for participants in the evaluation to ensure that the results of evaluations are valid. This can be an expensive and time-consuming undertaking. Again, if a suitable range of assistive technologies in not available in a testing laboratory, it may be easier to undertake the evaluation via remote testing (Petrie et al., 2006).

Finally, if evaluations are undertaken with disabled and older people, it is important that the needs of participants in the evaluation are taken carefully into consideration. Personnel running the evaluations need to be sensitive to the needs of the particular groups, such as visually disabled people, people in wheelchairs, and so on. Local organizations of disabled people can often provide training in disability awareness that can be very helpful to undertake before embarking on such evaluations. Issues to consider include:

- How will the participants come to the evaluation location (is public transport accessible, is the building easy to find for someone visually impaired)?
- Is the location itself accessible for the participants (e.g., are appropriate toilet facilities available, is there somewhere for guide dogs to exercise, etc.)?
- Are explanatory materials and consent forms available in the appropriate alternative formats (e.g., large print, Braille, Easy Read)?
- Will the pace of the evaluation be suitable for the participants (e.g., older participants may appreciate a slower pace of evaluation)?

20.4.5.6 Evaluating User Satisfaction and User Experience in User-Based Evaluations

As noted in Section 20.2, satisfaction and user experience move beyond performance-based measures that have traditionally been the focus of user-based evaluations. These aspects of the evaluation of eSystems can be assessed in a variety of ways, for example using Kansei techniques from consumer product development (Schütte et al., 2004). However, the simplest way is with rating scales and questionnaires. Psychometrically designed questionnaires (e.g., SUS, for usability; Brooke, 1996; or AttrakDiff for user experience; Hassenzahl et al., 2003) will give more reliable results than ad hoc questionnaires (Hornbaek, 2006). See Hornbaek (2006) for examples of other validated questionnaires.

20.4.6 Evaluation of Data Collected during eSystem Usage

20.4.6.1 When to Use Data during eSystem Usage

When planning to improve an existing eSystem.

20.4.6.2 Why Use Data during eSystem Usage?

Provides nonintrusive data about the use of a current eSystem.

20.4.6.3 Satisfaction Surveys

Satisfaction questionnaires distributed to a sample of existing users provide an economical way of obtaining feedback on the usability of an existing eSystem.

20.4.6.4 Web Server Log Analysis

Web-based logs contain potentially useful data that can be used to evaluate usability by providing data such as entrance and exit pages, frequency of particular paths through the site, and the extent to which search is successful (Burton and Walther, 2001). However, it is very difficult to track and interpret individual user behavior (Groves, 2007) without some form of page tagging combined with pop-up questions when the system is being used, so that the results can be related to particular user groups and tasks.

20.4.6.5 Application Instrumentation

Data points can be built into code that count when an event occurs, for example in Microsoft Office (Harris, 2005). This could be the frequency with which commands are used or the number of times a sequence results in a particular type of error. The data are sent anonymously to the development organization. These real-world data from large populations can help guide future design decisions.

20.5 Conclusion

This chapter has introduced a range of evaluation methods that assist developers in the creation of eSystems that are both accessible and usable. There are many details in the use of these techniques on which the interested reader will need to follow up; the many references provided at the end of the chapter will provide further information on particular techniques. With all these methods, the best results will be obtained by someone experienced in their use. However, anyone planning to embark on evaluation should not be put off by the complexity of the situation, but should start with some of the simpler expert evaluations and simple user-based evaluations. With practice and patience, expertise in these areas can be developed.

References

Ball, L. and Omerod, T. (2000). Putting ethnography to work: The case for a cognitive ethnography of design. *International Journal of Human-Computer Studies* 53: 147–168.

Bevan, N. (2001). Quality in use for all, in *User Interfaces for All: Methods, Concepts, and Tools* (C. Stephanidis, ed.), pp. 353–368. Mahwah, NJ: Lawrence Erlbaum Associates.

Bevan, N. (2008). Classifying and selecting UX and usability measures, in the *Proceedings of the International Workshop on Meaningful Measures: Valid Useful User Experience Measurement (VUUM)*, (E. L.-C. Law, N. Bevin, G. Christou, M. Springett, and M. Lárusdóttir, eds.), 18 June 2008, Reykjavik, Iceland, Toulouse, France: IRIT Press. http://cost294.org/vuum.

Bevan, N. and Spinhof, L. (2007). Are guidelines and standards for web usability comprehensive?, in *Human-Computer Interaction–Interaction Design and Usability (Part I), Volume 1 of the HCI International 2007 Conference Proceedings* (J. Jacko, ed.), 22–27 July 2007, Beijing, pp. 407–419. Berlin, Heidelberg: Springer-Verlag.

Beyer, H. and Hotzblatt, K. (1997). *Contextual Design: Defining Customer-Centred Systems.* San Francisco: Morgan Kaufmann.

Bias, R. G. (1994). Pluralistic usability walkthrough: Coordinated empathies, in *Usability Inspection Methods* (J. Nielsen and R. L. Mack, eds.), pp. 63–76. New York: John Wiley & Sons.

Bias, R. G. and Mayhew, D. J. (eds.) (2005). *Cost-Justifying Usability: An Update for the Internet Age.* San Francisco: Morgan Kaufmann.

Brajnik, G. (2000). Automatic web usability evaluation: What needs to be done?, in the *Proceedings of 6th Human Factors and the Web Conference 2000*, 19 June 2000, Austin, TX. http://users.dimi.uniud.it/~giorgio.brajnik/papers/hfweb00.html.

Brooke, J. (1996). SUS: A "quick and dirty" usability scale, in *Usability Evaluation in Industry* (P. Jordan, B. Thomas, and B. Weerdmeester, eds.), pp. 189–194. London: Taylor and Francis.

Burgstahler, S. and Doe, T. (2004). Disability-related simulations: If, when, and how to use them in professional development. *Review of Disability Studies* 1: 8–18.

Burton, M. and Walther, J. (2001). The value of web log data in use-based design and testing. *Journal of Computer-Mediated Communication* 6. http://jcmc.indiana.edu/vol6/issue3/burton.html.

Cockton, G., Woolrych, A., Hall, L., and Hindmarch, M. (2003). Changing analysts' tunes: The surprising impact of a new instrument for usability inspection method assessment, in the *Proceedings of HCI-2003: People and Computers XVII* (P. Palanque, P. Johnson, and E. O'Neill, eds.), 8–11 September 2003, Udine, Italy, pp. 145–161. London: Springer-Verlag.

Cook, A. and Polgar, J. M. (2008). *Assistive Technologies: Principles and Practice.* St. Louis: Mosby/Elsevier.

Cooper, M. and Rejmer, P. (2001). Case study: Localization of an accessibility evaluation, in the *CHI '01 Extended Abstracts on Human Factors in Computing Systems*, 31 March–5 April 2001, Seattle, pp. 141–142. New York: ACM Press.

Desurvire, H. W., Kondziela, J. M., and Atwood, M. E. (1992). What is gained and lost when using evaluation methods other than empirical testing, in the *Proceedings of the HCI '92 Conference on People and Computers' VII*, 15–18 September 1992, York, U.K., pp. 89–102. Cambridge, U.K.: Cambridge University Press.

Dillon, A. (2001). Beyond usability: Process, outcome and affect in human computer interactions. *Canadian Journal of Library and Information Science* 26: 57–69.

Dumas, J. S. and Redish, J. C. (1999). *Practical Guide to Usability Testing*. Bristol, U.K.: Intellect Books.

EDeAN (2007). *European Design for All eAccessibility Network*. http://www.edean.org.

Gould, J. D. and Lewis, C. (1985). Designing for usability: Key principles and what designers think. *Communications of the ACM* 28: 300–311.

Gray, W. D. and Salzman, M. C. (1998). Damaged merchandise? A review of experiments that compare usability evaluation methods. *Human-Computer Interaction* 13: 203–261.

Groves, K. (2007). *The Limitations of Server Log Files for Usability Analysis: Boxes and Arrows*. http://www.boxesandarrows.com/view/the-limitations-of.

Harris, J. (2005). *An Office User Interface Blog*. http://blogs.msdn.com/jensenh/archive/2005/10/31/487247.aspx.

Harrison, C. and Petrie, H. (2006). Severity of usability and accessibility problems in eCommerce and eGovernment websites, in *Computers and People XX: Proceedings of British Computer Society Human Computer Interaction Conference (BCS-HCI 06)*, 11–15 September 2006, London, pp. 255–262. London: British Computer Society.

Hassenzahl, M. (2003). The thing and I: Understanding the relationship between user and product, in *Funology: From Usability to Enjoyment* (M. Blythe, C. Overbeeke, A. F. Monk, and P. C. Wright, eds.), pp. 31–42. Dordrecht, The Netherlands: Kluwer.

Hassenzahl, M. (2006). Hedonic, emotional and experiential perspective on product quality, in *Encyclopedia of Human Computer Interaction* (C. Ghaoui, eds.), pp. 266–272. Hershey, PA: Idea Group.

Hassenzahl, M., Burmester, M., and Koller, F. (2003). AttrakDiff: Ein Fragebogen zur Messung wahrgenommener hedonischer und pragmatischer Qualität [AttrakDiff: A questionnaire for the measurement of perceived hedonic and pragmatic quality], in J. Ziegler and G. Szwillus (eds.), *Mensch and Computer 2003: Interaktion in Bewegung* (J. Ziegler and G. Szwillus, eds.), pp. 187–196. Stuttgart/Leipzig, Germany: B. G. Teubner

Hassenzahl, M., Law, E. L.-C., and Hvannberg, E. T. (2006). User experience: Towards a unified view, in the *Proceedings of the 2nd COST294-MAUSE International Open Workshop* (E. L.-C. Law, E. T. Hvannberg, and M. Hassenzahl, eds.), 14–18 October 2006, Oslo, Norway. http://www.cost294.org.

Hassenzahl, M. and Tractinksy, N. (2006). User experience: A research agenda. *Behaviour and Information Technology* 25: 91–97.

Hornbæk, K. (2006). Current practice in measuring usability: Challenges to usability studies and research. *International Journal of Human-Computer Studies* 64: 79–102.

Hwang, W. and Salvendy, G. (2007). What makes evaluators find more usability problems? A meta-analysis for individual detection rates. *HCI* 1: 499–507.

International Standards Organization (1998). *ISO 9241-11: Ergonomic Requirements for Office Work with Visual Display Terminals (VDTs). Part 11: Guidance on Usability*. Geneva: International Standards Organization.

International Standards Organization (2006). *ISO 20282-2: Ease of Operation of Everyday Products. Part 2: Test Method for Walk-Up-and-Use Products*. Geneva: International Standards Organization.

International Standards Organization (2008a). *ISO 9241-20: Ergonomics of Human-System Interaction. Part 20: Accessibility Guidelines for Information/Communication Technology (ICT) Equipment and Services*. Geneva: International Standards Organization.

International Standards Organization (2008b). *ISO 9241-171: Ergonomics of Human-System Interaction. Part 171: Guidance on Software Accessibility*. Geneva: International Standards Organization.

International Standards Organization (2008c). *ISO/IEC 10779: Office Equipment Accessibility Guidelines for Elderly Persons and Persons with Disabilities*. Geneva: International Standards Organization.

International Standards Organization (2009). *ISO DIS 9241-210: Ergonomics of Human-System Interaction. Part 210: Human-Centred Design Process for Interactive Systems (Formerly Known as 13407)*. Geneva: International Standards Organization.

Ivory, M. Y. and Hearst, M. A. (2001). State of the art in automating usability evaluation of user interfaces. *ACM Computing Surveys* 33: 470–516.

Jeffries, R. and Desurvire, H. (1992). Usability testing vs. heuristic evaluation: Was there a contest? *SIGCHI Bulletin* 24: 39–41.

Kiger, G. (1992). Disability simulations: Logical, methodological and ethical issues. *Disability and Society* 7: 71–78.

Law, E., Roto, V., Vermeeren, A., Kort, J., and Hassenzahl, M. (2008). Towards a shared definition of user experience, in the *CHI '08 Extended Abstracts on Human Factors in Computing Systems*, pp. 2395–2398. New York: ACM Press.

Lewis, J. R. (2005). Usability testing, in *Handbook of Human Factors and Ergonomics* (3rd ed.) (G. Salvendy, ed.). New York: John Wiley & Sons.

Lewis, J. R. (2006). Sample sizes for usability tests: Mostly math, not magic. *Interactions* 13: 29–33.

Mayhew, D. J. (2005). Keystroke level modeling as a cost justification tool, in *Cost-Justifying Usability: An Update for the Internet Age* (R. G. Bias and D. J. Mayhew, eds.). San Francisco: Morgan Kaufmann.

Medlock, M. C., Wixon D., Terrano, M., Romero, R., and Fulton, B. (2002). Using the RITE method to improve products: A definition and a case study, in the *Proceedings of Usability Professionals Association 2002*, July 2002,

Orlando, FL. http://www.microsoft.com/downloads/results. aspx?pocId=&freetext=rite%20method.

Molich, R. and Nielsen, J. (1990). Improving a human-computer dialogue. *Communications of the ACM* 33: 338–348.

National Institute of Standards and Technology (NIST) (2007). *Common Industry Specification for Usability: Requirements (IUSR: CISU-R v0.90)*. http://zing.ncsl. nist.gov/iusr.

Nielsen, J. (1994). Heuristic evaluation, in *Usability Inspection Methods* (J. Nielsen and R. L. Mack, eds.). New York: John Wiley & Sons.

Nielsen, J. and Molich, R. (1990). Heuristic evaluation of user interfaces, in the *Proceedings of CHI'90: ACM Annual Conference on Human Factors in Computing Systems*, 1–5 April 1990, Seattle, pp. 249–256. New York: ACM Press.

Nielsen, J. and Sano, D. (1994). SunWeb: User interface design for Sun Microsystem's internal web, in the *Proceedings of the 2nd World Wide Web Conference '94: Mosaic and the Web*, pp. 547–557. http://www.useit.com/papers/ sunweb.

Norman, D. A. (1998). *The Invisible Computer*. Cambridge, MA: MIT Press.

Norman, D. A. (2004). *Emotional Design: Why We Love (or Hate) Everyday Things*. New York: Basic Books.

O'Connor, J. (2007). *Joomla! Accessibility*. Birmingham, U.K.: Packt Publishing.

Petrie, H., Hamilton, F., King, N., and Pavan, P. (2006). Remote usability evaluations with disabled people, in the *Proceedings of CHI 2006: Conference on Human Factors in Computing Systems*, 24–27 April 2006, Montreal, Canada, pp. 1133–141. New York: ACM Press.

Petrie, H., Harrison, C., and Dev, S. (2005). Describing images on the Web: A survey of current practice and prospects for the future, in the *Proceedings of 3rd International Conference on Universal Access in Human-Computer Interaction, part of HCI International 2005* (C. Stephanidis, ed.). Mahwah, NJ: Lawrence Erlbaum Associates.

Petrie, H. and Kheir, O. (2007). The relationship between accessibility and usability of websites, in the *Proceedings of CHI '07: ACM Annual Conference on Human Factors in Computing Systems*, 28 April–3 May 2007, San Jose, CA, pp. 397–406. New York: ACM Press.

Pew, R. W. and Mayor, A. (eds.) (2007). *Human-System Integration in the System Development Process: A New Look*. National Academies Press. http://books.nap.edu/openbook. php?record_id=11893&page=240.

Regan, B. (2004). *Best Practices for Accessible Flash Design*. San Francisco: Macromedia. http://www.adobe.com/resources/ accessibility/best_practices/bp_fp.html.

Rubin, J. and Chisnell, D. (2008). *Handbook of Usability Testing: How to Plan, Design, and Conduct Effective Tests*. New York: John Wiley & Sons.

Schütte, S., Eklund, J., Axelsson, J. R. C., and Nagamachi, M. (2004). Concepts, methods and tools in Kansei Engineering. *Theoretical Issues in Ergonomics Science* 5: 214–232.

Shackel, B. (1990). Human factors and usability, in *Human-Computer Interaction: Selected Readings* (J. Preece and L. Keller, eds.). Hemel Hempstead, U.K.: Prentice Hall.

Shackel, B. (1991). Usability: Context, framework, definition, design and evaluation, in *Human Factors for Informatics Usability* (B. Shackel and S. Richardson, eds.), pp. 21–37. Cambridge, U.K.: Cambridge University Press.

Sharp, H., Rogers, Y., and Preece, J. (2007). *Interaction Design: Beyond Human-Computer Interaction*. London: John Wiley & Sons.

Shneiderman, B. and Plaisant, C. (2005). *Designing the User Interface: Strategies for Effective Human-Computer Interaction*. Boston: Pearson Education.

Slavkovic, A. and Cross, K. (1999). Novice heuristic evaluations of a complex interface, in the *CHI '99 Extended Abstracts on Human Factors in Computing Systems*, pp. 304–305. New York: ACM Press.

Snyder, C. (2003). *Paper Prototyping: The Fast and Easy Way to Define and Refine User Interfaces*. San Francisco: Morgan Kaufmann.

Sommerville, I. (1995). *Software Engineering* (5th ed.). Harlow, U.K.: Addison-Wesley.

Spencer, R. (2000). The streamlined cognitive walkthrough method, working around social constraints encountered in a software development company, in the *Proceedings of CHI 2000: ACM Annual Conference on Human Factors in Computing Systems*, 1–6 April 2000, The Hague, The Netherlands, pp. 353–359. New York: ACM Press.

St. Amant, R., Horton, T. E., and Ritter, F. E. (2007). Model-based evaluation of expert cell phone menu interaction. *ACM Transactions on Computer-Human Interaction (TOCHI)* 14: 1–24.

Stone, D., Jarrett, C., Woodroffe, M., and Minocha, S. (2005). *User Interface Design and Evaluation*. San Francisco: Morgan Kaufmann.

Story, M., Mueller, J., and Mace, R. (1998). *The Universal Design File: Designing for People of All Ages and Abilities*. Raleigh, NC: North Carolina State University. http://www.design. ncsu.edu/cud/pubs_p/pud.htm.

U.S. Department of Health and Human Sciences (2006). *Research-Based Web Design & Usability Guidelines*. http://www.usability.gov/guidelines.

Web Accessibility Initiative (WAI) (1999). *Web Content Accessibility Guidelines 1.0*. http://www.w3.org/TR/WCAG10.

Web Accessibility Initiative (WAI) (2006). *Introduction to Web Accessibility*. http://www.w3.org/WAI/intro/accessibility.php.

Web Accessibility Initiative (WAI) (2008). *Web Content Accessibility Guidelines 2.0*. http://www.w3.org/TR/WCAG20.

Wharton, C., Rieman, J., Lewis, C., and Polson, P. (1994). The cognitive walkthrough: A practitioner's guide, in *Usability Inspections Methods* (J. Nielsen and R. L. Mack, eds.), pp. 105–140. New York: John Wiley & Sons.

Whiteside, J., Bennett, J., and Holtzblatt, K. (1988). Usability engineering: Our experience and evolution, in *Handbook of Human-Computer Interaction* (M. Helander, ed.), pp. 791–817. Amsterdam: Elsevier North Holland.

V

User Interface Development: Architectures, Components, and Tools

21

A Unified Software Architecture for User Interface Adaptation

Anthony Savidis and
Constantine Stephanidis

21.1 Introduction

Today, software products support interactive behaviors that are biased toward the typical, or average able-bodied user, familiar with the notion of the desktop and the typical input and output peripherals of the personal computer.[1]

This has been the result of software developers' assumptions regarding the target user groups, the technological means at their disposal, and the types of tasks supported by computers. Thus, the focus has been on "knowledgeable" workers, capable and willing to use technology in the work environment, to experience productivity gains and performance improvements.

The progressive evolution of the information society has invalidated (at least some of) the assumptions in the previous scenario. The fusion between information technologies, telecommunications, and consumer electronics has introduced radical changes to traditional markets and complemented the business demand with a strong residential component. At the same time, the type and context of use of interactive applications are radically changing, due to the increasing availability of novel interaction technologies (e.g., personal digital assistants, kiosks, cellular phones, and other network-attachable equipment) that progressively enable nomadic access to information.

This paradigm shift poses several challenges: users are no longer only the traditional able-bodied, skilled, and computer-literate professionals; product developers can no longer *know* who their target users will be; information is no longer relevant only to the business environment; and artifacts are no longer bound to the technological specifications of a predefined interaction platform. In this context, users are potentially *all* citizens of an emerging information society who demand customized solutions to obtain timely access to virtually any application, from anywhere, at any time.

This chapter introduces the concept of unified user interfaces and points out some of the distinctive properties that render it an effective approach toward universal access within the information society. Subsequently, the unified user interface architecture is introduced, in the framework of an approach conveying a new perspective on the development of user interfaces, which provides a principled and systematic approach toward coping with diversity in the target users groups, tasks, and environments of use. A design methodology for unified user interfaces is reported in Chapter 16 of this handbook, while Chapter 17 presents the specific instantiation of such a design methodology in the context of designing universally accessible electronic games. Chapter 18 discusses key development requirements for building universally accessible interactions in a user interface adaptation perspective.

21.2 The Concept of Unified User Interfaces

The notion of unified user interfaces originated from research efforts aiming to address the issues of accessibility and interaction quality for people with disabilities (ACCESS Project, 1996;

---

[1] This chapter is based on Savidis A. and Stephanidis C. (2004). Unified user interface development: Software engineering of universally accessible interactions. *Universal Access in the Information Society* 3: 165–193. Reprinted here with slight modifications by concession of Springer.

see also Chapter 2, "Perspectives on Accessibility: From Assistive Technologies to Universal Access and Design for All," of this handbook). The primary intention was to articulate some of the key principles of *design for all* in a manner that would be applicable and useful to the conduct of human-computer interactions (HCI). Subsequently, these principles have been extended, appropriately adapted, compared with existing techniques, intensively tested and validated in the course of several projects, and formulated in an engineering code of practice that depicts a concrete proposition for interface development in the light of universal access.

A unified user interface is the interaction-specific software of applications or services that is capable of self-adapting to the individual end-user requirements and contexts of use. Such an adaptation may reflect varying patterns of interactive behavior, at the physical, syntactic, or semantic level of interaction, to accommodate specific user- and context-oriented parameters. Practically speaking, from the end-user point of view, a unified user interface is actually an interface that can automatically adapt to the individual user attributes (e.g., requirements, abilities, and preferences), as well as to the particular characteristics of the usage context (e.g., computing platform, peripheral devices, interaction technology, and surrounding environment). Therefore, a unified user interface realizes the combination of:

- *User-adapted behavior*: The automatic delivery of the most appropriate user interface for the particular end-user (*user awareness*)
- *Usage-context adapted behavior*: The automatic delivery of the most appropriate user interface for the particular situation of use (*usage context awareness*)

Hence, the characterization *unified* does not have any particular behavioral connotation, at least as seen from an end-user perspective. Instead, the notion of *unification* reflects the specific software engineering strategy needed to accomplish this behavior, emphasizing the proposed development-oriented perspective. More specifically, to realize this form of adapted behavior, a unified user interface reflects the following fundamental development properties:

- It encapsulates alternative dialogue patterns (i.e., implemented dialogue artifacts), for various dialogue design contexts (i.e., a subtask, a primitive user action, a visualization), appropriately associated to the different values of user- and usage-context-related attributes. The need for such alternative dialogue patterns is dictated by the design process: given any particular design context, for different user- and usage-context-attribute values, alternative design artifacts are needed to accomplish optimal interaction.
- It encapsulates representation schemes for user and usage-context parameters, internally utilizing user and usage-context information resources (e.g., repositories, servers), to extract or to update user and usage-context information.
- It encapsulates the necessary design knowledge and decision-making capability for activating, during

run-time, the most appropriate dialogue patterns (i.e., interactive software components), according to particular instances of user and usage-context attributes.

The property of unified user interfaces to encapsulate alternative, mutually exclusive, design artifacts, which can be purposefully designed and implemented for a particular design context, constitutes one of the main contributions of this research work within the interface software engineering arena. As will be discussed in subsequent sections, this notion of adaptation had not been previously supported. Previous work in adaptive interfaces had put emphasis primarily on adaptive interface updates, driven from continuous interaction monitoring and analysis, rather than on optimal design instantiation before interaction begins.

A second important contribution concerns the advanced technical properties of unified user interfaces, associated with the encapsulation of interactive behaviors targeted to the various interaction technologies and platforms. This requires, from the engineering point of view: (1) the organization of alternative dialogue components on the basis of their corresponding dialogue design context (i.e., all dialogue alternatives of a particular subtask are placed around the same run-time control unit); (2) the use of different interface toolkit libraries, for example, Java for graphical user interfaces (GUIs) and HAWK (Savidis et al., 1997) for nonvisual interaction; and (3) embedding alternative dialogue control policies, since the different toolkits may require largely different methods for interaction management (e.g., interaction object control, display facilities, direct access to I/O devices). Those issues have not been explicitly addressed by previous work on adaptive systems; none of the known systems has been targeted to supporting universally accessible interactions, since the various known demonstrators and their particular architectures were clearly focused on the GUI interface domain.

A third contribution concerns the specific architectural proposition that is part of the unified user interface development approach. This proposition presents a code of practice that provides an engineering blueprint for addressing the relevant software engineering issues: distribution of functional roles, component-based organization, and support for extension and evolution. In this context, the specific engineering issues addressed by the proposed code of practice are:

- How implemented dialogue components are organized, where they reside, and which are the algorithmic control requirements to accomplish run-time adaptation?
- How should interaction monitoring be organized, and what are the algorithmic requirements for the control of monitoring during interaction?
- Which are the steps involved in employing existing implemented interactive software, and how can it be extended or embedded in the context of a unified user interface implementation?
- Where is user-oriented information stored and in what form; in what format is user-oriented information communicated and between which components?
- What components are required to dynamically identify information regarding the end-user?

- During run-time, a unified user interface decides which dialogue components should comprise the delivered interface, given a particular end-user and usage context; where is the implementation of such adaptation-oriented decision-making logic encapsulated and in what representation or form?
- How are adaptation decisions communicated and applied, and what are the implications for the organization of implemented dialogue components?
- What are the communication requirements between the various components of the architecture to perform adaptation?
- How is a dynamic interface update performed, based on dynamically detected user attributes?
- Which of the components of the architecture are reusable?
- How do the components of the architecture evolve as the unified user interface is extended to cater for additional user- and usage-context-parameter values?
- Which implementation practices and tools are better suited for each component of the architecture?

21.3 Related Work

Previous efforts in development methods, some of which were explicitly targeted toward universally accessible interactions, fall under four general categories: (1) research work related to user interface tools, concerning prevalent architectural practices for interactive systems; (2) developments targeted to enabling the user interface to dynamically adapt to end-users by inferring particular user attribute values during interaction (e.g., preferences, application expertise); (3) special-purpose tools and architectures developed to enable the interface to cater to alternative modalities and interaction technologies; and (4) alternative access systems, concerning developments that have been targeted to manual or semiautomatic accessibility-oriented adaptations of originally nonaccessible systems or applications.

21.3.1 User Interface Software Architectures

Following the introduction of graphical user interfaces, early work in user interface software architectures had focused on window managers, event mechanisms, notification-based architectures, and toolkits of interaction objects. Those architectural models were quickly adopted by mainstream tools, thus becoming directly encapsulated within the prevailing user interface software and technology (UIST). Today, all available user interface development tools support object hierarchies, event distribution mechanisms, and callbacks at the basic implementation model. In addition to these early attempts in identifying architectural components of user interface software, there have been other architectural models, with a somewhat different focus that, however, did not gain as much acceptance in the commercial arena as was originally expected. The *Seeheim* model (Green, 1985), and its successor, the *Arch* model (UIMS Developers Workshop, 1992), have been mainly defined with the aim to

preserve the so-called principle of separation between interactive and noninteractive code of computer-based applications. These models became popular as a result of the early research work on user interface management systems (UIMS) (Myers, 1995), while in the domain of universal access systems they do not provide concrete architectural and engineering guidelines other than those related to the notion of separation.

Apart from these two architectural models, mainly referring to the *interlayer* organizational aspects of interactive applications, there have been two other implementation-oriented models with an object-oriented flavor: the model view controller (MVC) model (Goldberg, 1984; Krasner and Pope, 1988) and the presentation-abstraction-control (PAC) model (Coutaz, 1990). Those models focus on *intralayer* software organization policies, by providing logical schemes for structuring the implementation code. All four models, though typically referred to as architectural frameworks, are today considered as metamodels, following the introduction of the term for UIMS models in UIMS Developers Workshop (1992). They represent abstract families of architectures, since they do not meet the fundamental requirements of a concrete software architecture, as defined by Jacobson et al. (1997) where an architecture should provide a structure, as well as the interfaces between components, by defining the exact patterns by which information is passed back and forth through these interfaces. Additionally, the key flavor in those approaches is the separation between semantic content and representation form, by proposing engineering patterns enabling a clear-cut separation, so that flexible visualizations and multiple views can be easily accomplished.

21.3.2 Adaptive Interaction

Most previous work regarding system-driven, user-oriented adaptation concerns the capability of an interactive system to dynamically (i.e., during interaction) detect certain user properties, and accordingly decide various interface changes. This notion of adaptation falls in the *adaptivity* category (i.e., adaptation performed after the initiation of interaction), based on interaction monitoring information; this has been commonly referred to as *adaptation during use*. Although adaptivity is considered a good approach to meeting diverse user needs, there is always the risk of confusing the user with dynamic interface updates. Work on adaptive interaction has addressed various key technical issues concerning the derivation of appropriate constructs for the embodiment of adaptation capabilities and facilities in the user interface; most of those technical issues, if relevant to unified user interfaces, will be appropriately discussed in subsequent sections of this chapter.

Overall, in the domain of adaptive systems, there have been mainly two types of research work: (1) theoretic work on architectures; and (2) concrete work regarding specific systems and components demonstrating adaptive behavior. On the one hand, theoretic work addressed high-level issues, rather than specific engineering challenges, such as the ones identified earlier in this chapter. Those architectures convey useful information for the general system structure and potential control flow, mainly helping to understand the system as a whole. However, since they are mainly conceptual models, they cannot be deployed to derive the detailed software organization and the regulations for run-time coordination as is needed in the software implementation process.

On the other hand, developments in adaptive systems have mainly focused on particular parts or aspects of the adaptation process, such as user modeling or decision making, and less on usage context adaptation or interface component organization. Additionally, the interlinking of the various components, together with the corresponding communication policies employed, primarily reflected the specific needs of the particular systems developed, rather than a generalized approach facilitating reuse. Finally, while most known adaptive systems addressed the issue of dynamic interface adaptation, there has been no particular attention to the notion of individualized dynamic interface assembly before initiation of interaction. This implies that the interface can be structured on-the-fly by dialogue components maximally fitting end-user and usage context attributes.

This notion of optimal interface delivery before interaction is initiated constitutes the most important issue in universally accessible interactions; unless the initial interface delivered to the user is directly accessible, there is no point to apply adaptivity methods, since the latter assume that interaction is already taking place.

21.3.2.1 Which Architecture for Adaptivity?

Although concrete software architectures for adaptive user interfaces have not been clearly defined, there exist various proposals as to what should be incorporated in a computable form into an adaptive interactive system. In Dieterich et al. (1993), these are characterized as categories of computable artifacts, and are considered the types of models that are required within a structural model of adaptive interactive software. An appropriate reference to Hartson and Hix (1989) is also made regarding structural interface models, suggesting that structural models of human-computer interface serve as frameworks for understanding the elements of interfaces and for guiding the dialogue developers in their construction.

However, developers require concrete software architectures for structuring and engineering interactive systems, and software systems in general. From this point of view, the information provided in Dieterich et al. (1993) does not fulfill the requirements of an interface structural model, as defined in Hartson and Hix (1989), nor of a software architecture, as defined in Jacobson et al. (1997) and Mowbray and Zahavi (1995). This argument will be further elaborated in subsequent sections, as various aspects of previous work on adaptive interfaces are incrementally analyzed.

21.3.2.2 User Models versus User-Modeling Frameworks

In all types of systems aiming to support user adaptivity in performing certain tasks, both embedded user models and user-task models have played a significant role. In Kobsa and Wahlster (1989), an important distinction is made between the user-modeling component, encompassing methods to represent user-oriented information, and the particular user models as such, representing an instance of the knowledge framework for a particular user (i.e., individual user model), or user group (i.e., a stereotype model). However, this distinction explicitly associates the user model with the modeling framework, thus necessarily establishing a dependency between the adaptation-targeted decision-making software (which would need to process user models) and the overall user-modeling component. This remark reveals the potential architectural hazard of rendering an adaptive system "monolithic": since the user model is linked directly with the modeling component, and decision making is associated with user models, it may be deemed necessary or appropriate that all such knowledge categories be physically located together. More recent work has reflected the technical benefits of physically splitting the user-modeling component from the adaptive inference core, putting the responsibility of decision making directly to applications, while enabling remote sharing of user-modeling servers by multiple clients (Kobsa and Pohl, 1995).

21.3.2.3 Alternative Dialogue Patterns and the Need for Abstraction

The need for explicit design as well as run-time availability of design alternatives has been already identified in the context of interface adaptation (Browne et al., 1990). In view of the need for managing alternative interaction patterns, the importance of abstractions has been identified, starting from the observation that design alternatives constructed with an adaptation perspective are likely to exhibit some common dialogue structures. In Cockton (1987) it is claimed that flexible abstractions for executable dialogue specifications are a necessary condition for the success of adaptable human-computer interfaces. This argument implies that an important element in the success of adaptive systems is the provision of implemented mechanisms of abstraction in interactive software, allowing the flexible run-time manipulation of dialogue patterns.

21.3.2.4 Dynamic User Attribute Detection

The most common utilization of internal dialogue representation has involved the collection and processing of interaction monitoring information. Such information, gathered at run-time, is analyzed internally (through different types of knowledge processing) to derive certain user attribute values (not known prior to the initiation of interaction), which may drive appropriate interface adaptivity actions. A well-known adaptive system employing such techniques is MONITOR (Benyon, 1984). Similarly, for the purpose of dynamic detection of user attributes, a monitoring component, in conjunction with a UIMS, is employed in the AIDA system (Cote Muñoz, 1993). An important technical implication in this context is that dialogue modeling must be combined with user models. Thus, as discussed earlier, it becomes inherently associated with the user-modeling component, as well as with adaptation-targeted decision-making software. Effectively, this biases the overall adaptive system architecture toward a monolithic structure, turning the development of adaptive interface systems into a rather complicated software engineering process. It is argued that such an engineering tactic, placing all those components together within a monolithic system implementation, is a less than optimal architectural option. Moreover, considering that this approach has been adopted to address the issue of dynamic user attribute detection, it will be shown that a more flexible, distributed, and reuse-enabled engineering structure can be adopted to effectively pursue the same goal.

This argument is also supported by the fact that in most available interactive applications, internal executable dialogue models exist only in the form of programmed software modules. Higher-order executable dialogue models (which would reduce the need for low-level programming), as those previously mentioned, have been supported only by research-oriented UIMS tools. Conversely, the outcome of interface development environments, like VisualBasic, is, at present, in a form more closely related to the implementation world, rendering the extraction of any design-oriented context difficult or impossible. Hence, on the one hand, dynamic user attribute detection will necessarily have to engage dialogue-related information, while on the other hand, it is unlikely that such required design information is practically extractable from the interaction control implementation.

21.3.2.5 System-Initiated Actions to Perform Adaptivity

The final step in a run-time adaptation process is the execution of the necessary interface updates at the software level. In this context, four categories of system-initiated actions to be performed at the dialogue control level have been distinguished (Cockton, 1993) for the execution of adaptation decisions: (1) *enabling* (i.e., activation or deactivation of dialogue components); (2) *switching* (i.e., selecting one from various alternative preconfigured components); (3) *reconfiguring* (i.e., modifying dialogue by using predefined components); and (4) *editing* (i.e., no restrictions on the type of interface updates). The preceding categorization represents more a theoretical perspective, rather than reflecting an interface engineering one. Furthermore, the term component denotes mainly visual interface structures, rather than referring to implemented subdialogues, including physical structure or interaction control.

In this sense, it is argued that it suffices to define only two action classes, applicable on interface components: (1) *activate* components; and (2) *cancel* activated components (i.e., deactivate). As will be discussed, these two actions directly map to the implementation domain (i.e., activation means instantiation of software objects, while cancellation means destruction), thus considerably downsizing the problem of modeling adaptation actions.

21.3.2.6 Structuring Dialogue Implementation for Adaptivity

The notion of *interface component* refers to implemented subdialogues provided by means of prepackaged, directly deployable,

software entities. Such entities increasingly become the basic building blocks in a component-based software assembly process, highly resembling the hardware design and manufacturing process. The need for configurable dialogue components has been identified in Cockton (1993) as a general capability of interactive software to visualize some important implementation parameters, through which flexible fine-tuning of interactive behaviors may be performed at run-time.

However, the analysis in Cockton (1993) is based on a theoretical ground, and mainly identifies requirements, without proposing specific approaches to achieving this type of desirable functional behavior. For instance, the proposed distinction among scalar, structured, and higher-order objects does not map to any interface engineering practice. Moreover, the definition of adaptation policies as changes at different levels does not provide any concrete architectural model, or reveal any useful implementation patterns. The results of such theoretical studies are good for understanding the various dynamics involved in adaptive interaction; however, they do not provide added-value information for engineering adaptive interaction.

It can be concluded from the preceding discussion that the incorporation of adaptation capabilities into interactive software is far from trivial and cannot be attained through traditional approaches to software architecture. A software architecture that can accommodate the adaptation-oriented requirements of unified user interfaces is described in the next section.

21.4 Unified User Interface Development

21.4.1 The Strategy

A unified user interface consists of run-time components, each with a distinctive role in performing at run-time an interface assembly process, by selecting the most appropriate dialogue patterns from the available implemented design space (i.e., the organized collection of all dialogue artifacts produced during the design phase). A unified user interface does not constitute a single software system, but becomes a distributed architecture consisting of independent intercommunicating components, possibly implemented with different software methods/tools and residing at different physical locations. These components cooperate together to perform adaptation according to the individual end-user attributes and the particular usage context. At run-time, the overall adapted interface behavior is realized by two complementary classes of system-initiated actions:

- Adaptations driven from initial user and context information, acquired without performing interaction monitoring analysis (i.e., what is "known" before starting observing the user or the usage context)
- Adaptations decided on the basis of information inferred or extracted by performing interaction monitoring analysis (i.e., what is "learned" by observing the user or the usage context)

The former behavior is referred to as *adaptability* (i.e., initial automatic adaptation, performed before initiation of interaction) reflecting the capability of the interface to automatically tailor itself initially to the attributes of each individual end-user. The latter behavior is referred to as *adaptivity* (i.e., continuous automatic adaptation), and characterizes the capability of the interface to cope with the dynamically changing/evolving characteristics of users and usage contexts. Adaptability is crucial to ensure accessibility, since it is essential to provide, before initiation of interaction, a fully accessible interface instance to each individual end-user. Adaptivity can be applied only on accessible running interface instances (i.e., ones with which the user is capable of performing interaction), since interaction monitoring is required for the identification of changing or emerging decision parameters that may drive dynamic interface enhancements. The complementary roles of adaptability and adaptivity are depicted in Figure 21.1, while the key differences among these two adaptation methods are illustrated in Table 21.1. This

fundamental distinction is made due to the different run-time control requirements between those two key classes of adaptation behaviors, requiring different software engineering policies.

21.4.2 The Unified User Interface Software Architecture

In this section, the detailed run-time architecture for unified user interfaces is discussed, in compliance with the definitions of architecture provided by the object management group (OMG) (Mowbray and Zahavi, 1995) and Jacobson, Griss, and Johnson (1997), according to which an architecture should supply an organization of components, description of functional roles, detailed communication protocols or appropriate application programming interfaces (APIs), and key component implementation issues. The adopted architectural components are outlined by specifying: (1) their functional role; (2) their run-time behavior; (3) their encapsulated context; and (4) the proposed

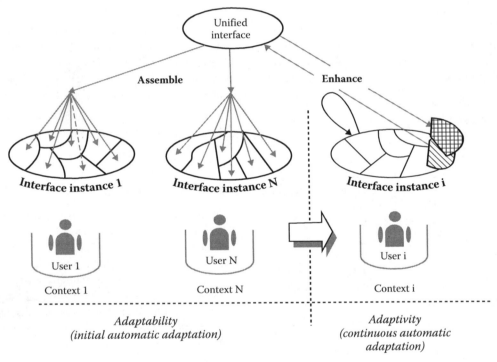

FIGURE 21.1 The complementary roles of adaptability (left) and adaptivity (right) as realized in unified user interfaces to provide user and usage-context adapted behavior. (From Savidis, A. and Stephanidis, C., *Universal Access in the Information Society*, 3, 165–193, 2004.)

TABLE 21.1   Key Differences between Adaptability and Adaptivity in the Context of Unified User Interfaces

Adaptability	Adaptivity
1. User and usage-context attributes are considered known prior to interaction.	1. User/usage-context attributes are dynamically inferred/detected.
2. "Assembles" an appropriate initial interface instance for a particular end-user and usage context.	2. Enhances the initial interface instance already "assembled" for a particular end-user and usage context.
3. Works before interaction is initiated.	3. Works after interaction is initiated.
4. Provides a user-accessible interface.	4. Requires a user-accessible interface

Source: From Savidis, A. and Stephanidis, C., *Universal Access in the Information Society*, 3, 165–193, 2004.

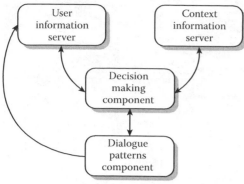

FIGURE 21.2 The four basic components of the unified user interface macro-architecture outlining run-time communication links. (From Savidis, A. and Stephanidis, C., *Universal Access in the Information Society*, 3, 165–193, 2004.)

implementation method. The components of the unified user interface architecture are (see Figure 21.2):

- Dialogue patterns component (DPC)
- Decision-making component (DMC)
- User information server (UIS)
- Context parameters server (CPS)

21.4.2.1 UIS

21.4.2.1.1 Functional Role

To supply user attribute values: (1) known off-line, without performing interaction monitoring analysis (e.g., motor/sensory abilities, age, nationality, etc.); and (2) detected online, from real-time interaction-monitoring analysis (e.g., fatigue, loss of orientation, inability to perform the task, interaction preferences, etc.).

21.4.2.1.2 Run-Time Behavior

The user information server plays a twofold role: (1) it constitutes a server that maintains and provides information regarding individual user profiles; and (2) it encompasses user representation schemes, knowledge processing components, and design information, to dynamically detect user properties or characteristics.

21.4.2.1.3 Encapsulated Content

This component may need to employ alternative ways of representing user-oriented information. In this sense, a repository of user profiles serves as a central database of individual user information (i.e., registry). In Figure 21.3, the notion of a profile structure and a profile instance, reflecting a typical list of typed attributes is shown; this model, though quite simple, is proved in real practice to be very powerful and flexible (can be stored in a database, thus turning the profile manager to a remotely accessed database). Additionally, more sophisticated user representation and modeling methods can be also employed, including support for stereotypes of particular user categories. In case dynamic user attribute detection is to be supported, the

content may include dynamically collected interaction monitoring information, design information and knowledge processing components, as it is discussed in the implementation techniques. Systems such as BGP-MS (Kobsa, 1990), PROTUM (Vergara, 1994), or USE-IT (Akoumianakis et al., 1996) encompass similar techniques for such intelligent processing. More elaborated approaches to user modeling are discussed in Chapter 24, "User Modeling: A Universal Access Perspective," of this handbook.

21.4.2.1.4 Implementation

From a knowledge representation point of view, static or preexisting user knowledge may be encoded in any appropriate form, depending on the type of information the user information server should feed to the decision-making process. Moreover, additional knowledge-based components may be employed for processing retrieved user profiles, drawing assumptions about the user, or updating the original user profiles. Figure 21.4 presents the internal architecture of the user information server employed in the AVANTI browser, a universally accessible web browser developed following the unified user interface development methodology (Stephanidis et al., 2001). It should be noted that the first version of the AVANTI browser produced in the context of the AVANTI Project employed BGP-MS (Kobsa and Pohl, 1995) for the role of the UIS. The profile manager has been implemented as a database of profiles. The two other subsystems (i.e., monitoring manager, modeling and inference) are needed only in case dynamic user attribute detection is required.

The interaction monitoring history has been implemented as a time-stamped list of monitoring events (the structure of monitoring events is described in the analysis of communication semantics) annotated with simple dialogue design context information (i.e., the subtask name). In the user models, all the types of dynamically detected user attributes have been identified (e.g., inability to perform a task, loss of orientation—those were actually the two dynamically detectable attributes required by the design in the AVANTI browser). Each such attribute is associated with its corresponding behavioral action patterns. In the specific case, the representation of the behavioral patterns has been implemented together with the pattern-matching component, by means of state automata. For instance, one heuristic pattern to detect loss of orientation has been defined as "the user moves the cursor inside the web page display area, without selecting a link, for more than N seconds." The state automaton starts recording mouse moves in the page area, increasing appropriately a weight variable and a probability value, based on incoming monitored mouse moves, while finally triggering detection when no intermediate activity is successfully performed by the user. This worked fine from an implementation point of view. However, all such heuristic assumptions had to be extensively verified with real users, so as to assert the relationship between the observable user behavior and the particular inferred user attributes. This is a common issue in all adaptive systems that employ heuristics for detecting user attributes at run-time, practically meaning that the validity of the "assumptions inferred" is dependent on the appropriateness of the specific user-action patterns chosen.

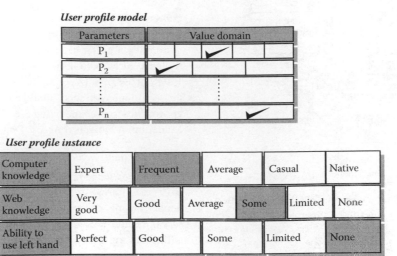

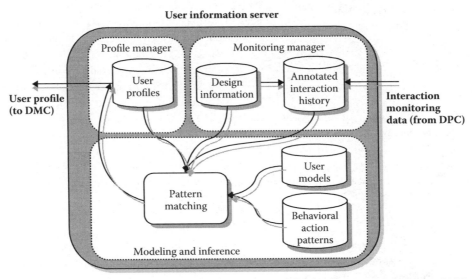

FIGURE 21.3 The typed attribute list profile model (i.e., database tuple) and sample profile instance. (From Savidis, A. and Stephanidis, C., *Universal Access in the Information Society*, 3, 165–193, 2004.)

FIGURE 21.4 Internal micro-architecture of the user information server. User profile is posted to the DMC, whereas interaction-monitoring information is received from the DPC. (From Savidis, A. and Stephanidis, C., *Universal Access in the Information Society*, 3, 165–193, 2004.)

21.4.2.2 CPS

21.4.2.2.1 Functional Role

The purpose of this component is to supply context attribute values (machine and environment) of two types: (1) (potentially) invariant, meaning unlikely to change during interaction (e.g., peripheral equipment); and (2) variant, dynamically changing during interaction (e.g., environment noise, or failure of particular equipment, etc.). This component is not intended to support device independence, but to provide *device awareness*. Its purpose is to enable the decision-making component to select those interaction patterns, which, apart from fitting the particular end-user attributes, are also appropriate for the type of equipment available on the end-user machine.

21.4.2.2.2 Run-Time Behavior

The usage-context attribute values are communicated to the decision-making component before the initiation of interaction. Additionally, during interaction, some dynamically changing usage-context parameters may also be fed to the decision-making component for decisions regarding *adaptive* behavior. For instance, the initial decision for selecting feedback may lead to the use of audio effects. Then, the dynamic detection of an increase in environmental noise may result in a run-time decision to switch to visual feedback (the underlying assumption being that such a decision does not conflict with other constraints).

21.4.2.2.3 *Encapsulated Content*

This component encompasses a listing of the various invariant properties and equipment of the target machine (e.g., handheld binary switches, speech synthesizer for English, high-resolution display mode at 16 bits, 1024 × 768, noisy environment, etc.). In this context, the more information regarding the characteristics of the target environment and machine is encapsulated, especially concerning I/O devices, the better adaptation can be achieved (information initially appearing redundant is likely to be used in future adaptation-oriented extensions).

21.4.2.2.4 *Implementation*

The registry of environment properties and available equipment can be implemented easily as a profile manager in the form of a database. Such information will be communicated to the decision-making component as attribute/value pairs. If usage-context information is to be dynamically collected, such as environment noise, or reduction of network bandwidth, the installation of proper hardware sensors or software monitors is mandatory.

21.4.2.3 Decision-Making Component

21.4.2.3.1 *Functional Role*

The role of this component is to decide, at run-time, the necessary adaptability and adaptivity actions, and to subsequently communicate those to the DPC (the latter being responsible for applying adaptation-oriented decisions).

21.4.2.3.2 *Run-Time Behavior*

To decide about adaptation, this component performs a kind of rule-based knowledge processing, so as to match end-user and usage-context attribute values to the corresponding dialogue artifacts, for all the various dialogue contexts.

21.4.2.3.3 *Encapsulated Content*

This module encompasses the logic for deciding the necessary adaptation actions, on the basis of the user and context attribute values, received from the UIS and the context parameters server, respectively. Such attribute values will be supplied to the decision-making component, prior to the initiation of interaction within different dialogue contexts (i.e., initial values, resulting in initial interface adaptation), as well as during interaction (i.e., changes in particular values, or detection of new values, resulting in dynamic interface adaptations).

In the proposed approach, the encapsulated adaptation logic should reflect predefined decisions during the design stage. In other words, the inference mechanisms employ well-defined decision patterns that have been validated during the design phase of the various alternative dialogue artifacts. In practice, this approach leads to a rule-based implementation, in which embedded knowledge reflects adaptation rules that have been

already constructed and documented as part of the design stage. This decision-making policy is motivated by the assumption that if a human designer cannot decide upon adaptation for a dialogue context, given a particular end-user and usage-context, then a valid adaptation decision cannot be taken by a knowledge-based system at run-time.

21.4.2.3.4 *Implementation*

The first remark regarding the implementation of decision making concerns the apparent "awareness" regarding: (1) the various alternative dialogue artifacts (how they are named, e.g., *virtual keyboard*; for which dialogue context they have been designed, e.g., *http address text field*); (2) user and usage-context attribute names, and their respective value domains (e.g., attribute *age*; being *integer* in *range 5...110*). The second issue concerns the input to the decision process, being individual user and usage-context attribute values. Those are received at run-time from both the UIS and the context information server, either *by request* (i.e., the DMC takes the initiative to request the end-user and usage-context profile at startup to draw adaptability decisions), or *by notification* (i.e., when the UIS draws assumptions regarding dynamic user attributes, or when the CPS identifies dynamic context attributes).

The third issue concerns the format and structure of knowledge representation. The developments carried out so far have proved that a rule-based logic implementation is practically adequate. Moreover, all interface designers engaged in the design process emphasized that this type of knowledge representation approach is far closer to their own way of rule-based thinking in deciding adaptation. This remark has led to excluding, at a very early stage, other possible approaches, such as heuristic pattern matching, weighting factor matrices, or probabilistic decision networks. A rule-based language for adaptation in unified user interfaces is discussed in Chapter 22 of this handbook.

The final issue concerned the representation of the outcomes of the decision process in a form suitable for being communicated and easily interpreted by the DPC. In this context, it has been practically proved that two categories of dialogue control actions suffice to communicate adaptation decisions: (1) *activation* of specific dialogue components; and (2) *cancellation* of previously activated dialogue components.

These two categories of adaptation actions provide the expressive power necessary for communicating the dialogue component manipulation requirements that realize both adaptability and adaptivity (see Table 21.2). Substitution is modeled by a message containing a series of *cancellation* actions (i.e., the dialogue components to be substituted), followed by the necessary number of *activation* actions (i.e., which dialogue components to activate in place of the cancelled components). Therefore, the transmission of those commands in a single message (i.e., *cancellation* actions followed by *activation* actions) is to be used for implementing a substitution action. The need to send in one message packaged information regarding the canceled component, together with the components that take its place, emerges

TABLE 21.2 The User Interface (UI) Component Manipulation Requirements (left) for Adaptability and Adaptivity and Their Expression via Cancellation/Activation Adaptation Actions

	Adaptability	Adaptivity
Initial selection of UI components	• Via **activation** actions	• *Not for adaptivity*
Dynamic selection of UI components	• *Not for adaptability*	• Via **activation** actions
Dynamic cancellation of UI components	• *Not for adaptability*	• Via **cancellation** actions
Dynamic substitution of UI components	• *Not for adaptability*	• Via a message containing **cancellation** actions, followed by the necessary number of **activation** actions

Source: From Savidis, A. and Stephanidis, C., *Universal Access in the Information Society*, 3, 165–193, 2004.

TABLE 21.3 Communicated Messages between the User Information Server and the Decision-Making Component

UIS→DMC	
Message class	Exporting an end-user profile
Content structure	Sequence of *<Attribute, Value>* pairs
When communicated	When requested by the decision-making component
Example	{ <"domain knowledge," "limited">, <"visual ability," "sighted">, <"age," "35">, <"motor ability," "fine">, ...}
Message class	Dynamic detection of user parameters
Content structure	*Parameter, Dialogue Context*
When communicated	Each time an inference is made by the user information server
Example	{ "user confused," "Link Selection Task" }

DMC→UIS	
Message class	Requesting user profile
Content structure	Message contains just request header (i.e., content is empty)
When communicated	Before decision making is initiated in the decision-making component
Example	{}
Message class	Requesting explicitly the value of a user attribute
Content structure	*Attribute*
When communicated	As needed by the decision-making component
Examples	{ "age" }, { "domain knowledge" }, { "web knowledge" }

Source: From Savidis, A. and Stephanidis, C., *Universal Access in the Information Society*, 3, 165–193, 2004.

when the implemented interface requires knowledge of all (or some) of the newly created components during interaction. For instance, if the new components include a container (e.g., a window object) with various embedded objects, and if upon the creation of the container information on the number and type of the particular contained objects is needed, it is necessary to ensure that all the relevant information (i.e., all engaged components) is received as a single message. It should be noted that, since each *activation/cancellation* command always carries its target UI component identification (see Table 21.3), it is possible to engage in substitution request components that are not necessarily part of the same physical dialogue artifact. Also, the decision to apply substitution is the responsibility of the DMC.

One issue regarding the expressive power of activation and cancellation decisions categories concerns the way dynamic interface updates (i.e., changing style or appearance, without closing or opening interface objects) can be effectively addressed. The answer to this question is related to the specific connotation attributed to the notion of a dialogue component. A dialogue component may not only implement physical dialogue context, such as a window and embedded objects, but may concern the activation of dialogue control policies, or be realized as a particular sequence of interface manipulation actions. In this sense, the interface updates are to be collected in an appropriate dialogue implementation component (e.g., a program function, an object class, a library module) to be subsequently activated (i.e., called) when a corresponding activation message is received. This is the specific approach taken in the AVANTI browser, which, from a software engineering point of view, enabled a better organization of the implementation modules around common design roles.

21.4.2.4 DPC

21.4.2.4.1 Functional Role

This component is responsible for supplying the software implementation of all the dialogue artifacts that have been identified in the design process. Such implemented components may vary

Dialogue patterns component

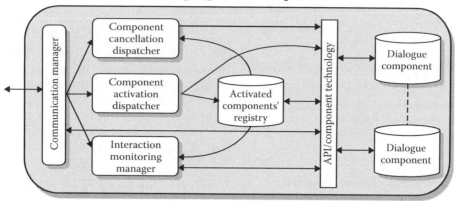

FIGURE 21.5 Internal micro-architecture of the dialogue patterns component. (From Savidis, A. and Stephanidis, C., *Universal Access in the Information Society*, 3, 165–193, 2004.)

from dialogue artifacts that are common across different user and usage-context attribute values (i.e., no adaptation needed), to dialogue artifacts that will map to individual attribute values (i.e., alternative designs have been necessitated for adapted interaction). Additionally, as it has been previously mentioned, apart from implementing physical context, various components may implement dialogue sequencing control, perform interface manipulation actions, maintain shared dialogue state logic, or apply interaction monitoring.

21.4.2.4.2 Run-Time Behavior

The DPC should be capable of applying at run-time, *activation* or *cancellation* decisions originated from the DMC. Additionally, interaction-monitoring components may need to be dynamically installed/uninstalled on particular physical dialogue components. This behavior will serve the run-time interaction monitoring control requests from the UIS, so as to provide continuous interaction monitoring information back to the UIS for further intelligent processing.

21.4.2.4.3 Encapsulated Content

The DPC either embeds the software implementation of the various dialogue components, or is aware of where those components physically reside, by employing dynamic query, retrieval, and activation methods. The former is the typical method that can be used if the software implementation of the components is provided locally, by means of software modules, libraries, or resident installed components. Usually, most of the implementation is to be carried out in a single programming language. The latter approach reflects the scenario in which distinct components are implemented on top of component-ware technologies, usually residing in local/remote component repositories (also called registries or directories), enabling reuse with dynamic deployment.

In the development of the AVANTI browser, a combination of these two approaches has been employed, by implementing most of the common dialogue components into a single language

(actually in C++, by employing all the necessary toolkit libraries), while implementing some of the alternative dialogue artifacts as independent Active X components that were located and employed on-the-fly. The experience from the software development of the AVANTI browser has proved that: (1) the single language paradigm makes it far easier to perform quick implementation and testing of interface components; (2) the component-based approach largely promotes binary format reuse of implemented dialogue components, while offering far better support for dynamic interface assembly, which is the central engineering concept of unified user interfaces (this issue we will be elaborated upon further on in this chapter). A programming language purposefully developed to facilitate user interface adaptation is presented in Chapter 27 of this handbook.

21.4.2.4.4 Implementation

The micro-architecture of the DPC internally employed in the AVANTI browser, as outlined in Figure 21.5, emphasizes internal organization to enable extensibility and evolution by adding new dialogue components. Additionally, it reflects the key role of the DPC in applying adaptation decisions. The internal components are:

- The *activation dispatcher*, which locates the source of implementation of a component (or simply uses its API, if it is a locally used library), to activate it. In this sense, activation may imply a typical instantiation in object-oriented programming (OOP) terms, calling of particular service functions, or activating a remotely located object. After a component is activated, if cancellation is to be applied to this component, it is further registered in a local registry of activated components. In this registry, the indexing parameters used are the particular dialogue context (e.g., subtask, i.e., "http address field"), and the artifact design descriptor (i.e., unique descriptive name provided during the design phase, i.e., "virtual keyboard"). For some categories of components, cancellation may not be defined during the design process, meaning there is no reason to register those at run-time for possible future cancellation (e.g., components with a temporal nature that only perform some interface update activities).

- The *cancellation dispatcher,* which locates a component based on its indexing parameters and calls for cancellation. This may imply a typical destruction in OOP terms, calling internally particular service functions that may typically perform the unobtrusive removal of the physical view of the canceled component, or the release of a remote object instance. After cancellation is performed, the component instance is removed from the local registry.

- The *monitoring manager,* which plays a twofold role: (1) it applies monitoring control requests originated from the UIS, by first locating the corresponding dialogue components, and then requesting the installation (or uninstall) of the particular monitoring policy (this requires implementation additions in dialogue components, for performing interaction monitoring and for activating or deactivating the interaction monitoring behavior); and (2) it receives interaction-monitoring notifications from dialogue components, and posts those to the UIS.

- The *communication manager,* which is responsible for dispatching incoming communication (activation, cancellation, and monitoring control) and posting outgoing communication (monitoring data, and initial adaptation requests). One might observe that there is also an explicit link between the dialogue components and the communication manager. This reflects the initiation of interaction in which the dialogue control logic (residing within dialogue components) requests iteratively the application of decision making (from the DMC). Such requests will need to be posted for all cases involving dialogue component alternatives for which adapted selection has to be appropriately performed.

- The *dialogue* components, in which the real implementation of physical dialogues, dialogue control logic, and interaction monitoring methods are typically encompassed. In practice, it is hard to accomplish isolated implementation of the dialogue artifacts as independent black boxes that can be combined and assembled on-the-fly by independent controlling software. In most designs, it is common that physical dialogue artifacts are contained inside other physical artifacts. In this case, if there are alternative versions of the embedded artifacts, it turns out that to make containers fully orthogonal and independent with respect to the contained, one has to support intensive parameterization and pay a heavier implementation overhead. However, the gains are that the implementation of contained artifacts can be independently reused across different applications, while in the more monolithic approach, reuse requires deployment of the container code (and recursively, of its container too, if it is contained as well).

21.4.2.5 Adaptability and Adaptivity Cycles

The completion of an adaptation cycle, being either adaptability or adaptivity, is realized in a number of distributed processing stages performed by the various components of the unified architecture. During these stages, the components communicate with each other, requesting or delivering specific pieces of information. Figure 21.6 outlines the processing steps for performing the initial adaptability cycle (to be executed only once), as well as the two types of adaptivity cycles (i.e., one starting from "dynamic context attribute values," and another starting from "interaction-monitoring control"). Local actions indicated within components (in each of the four columns) are either outgoing messages, shown in bold typeface, or necessary internal processing, illustrated via shaded rectangles.

21.4.3 Intercomponent Communication Semantics

This section presents the communication protocol among the various components in a form emphasizing the rules that govern the exchange of information among various communicating parties, as opposed to a strict message syntax description (see Tables 21.3 to 21.6 for a complete description of the protocol syntax). Hence, the primary focus is on the semantics of communication, regarding: (1) the type of information communicated; (2) the content it conveys; and (3) the usefulness of the communicated information at the recipient component side.

In the unified software architecture, there are four distinct bidirectional communication channels (outlined in Figure 21.7), each engaging a pair of communicating components. For instance, one such channel concerns the communication between the UIS and the DMC. Each channel practically defines two protocol categories, one for each direction of the communication link, for example, UIS→DMC (i.e., type of messages sent from UIS to DMC) and DMC→UIS (i.e., type of messages sent from DMC to UIS). The description of the protocols for each of the four communication channels follows.

21.4.3.1 Communication between the UIS and the DMC

In this communication channel, there are two communication rounds: (1) prior to initiation of interaction, where the DMC requests the user profile from the UIS, and the latter replies directly with the corresponding profile as a list of attribute/value pairs; and (2) after the initiation of the interaction, each time the UIS detects some dynamic user attribute values (on the basis of interaction monitoring), it communicates those values immediately to the DMC. In Table 21.3, the syntax of the messages communicated between the UIS and the DMC is defined and simple examples are provided.

21.4.3.2 Communication between the UIS and the DPC

The communication among these two components aims to enable the UIS to collect interaction-monitoring information, as well as to control the type of monitoring to be performed. The UIS may request monitoring at three different levels: (1) *task* (i.e., when *initiated* or *completed*); (2) *method* for interaction objects (i.e., which logical object action has been accomplished, such as

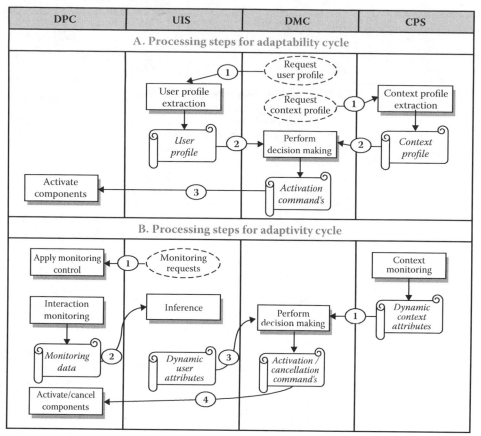

FIGURE 21.6 Processing steps, engaging communication among architectural components, to perform initial adaptability cycles (A), as well as the two types of adaptivity cycles (B). Requests originated from DPC are indicated with dashed ovals; communicated messages are shown with special banners; processing points are drawn with shaded rectangles; and logical ordering is designated with numeric oval labels. (From Savidis, A. and Stephanidis, C., *Universal Access in the Information Society*, 3, 165–193, 2004.)

"pressing" a "button" object); and (3) *input event* (i.e., specific device event, such as "mouse move" or "key press").

In response to monitoring control messages, the DPC will have to: (1) activate or cancel the appropriate interaction-monitoring software modules; and (2) continuously export monitoring data, according to the monitoring levels requested, back to the UIS (initially, no monitoring modules are activated by the DPC). In Table 21.4, the syntax of messages communicated between the UIS and the DPC is defined, and simple examples are provided.

21.4.3.4 Communication between the DMC and the DPC

As previously mentioned, inside the DPC alternative implemented dialogue artifacts are associated with their respective dialogue context (e.g., subtask). As part of the design process, dialogue artifacts had been assigned an indicative name, unique across the rest of alternatives for the same dialogue context. Each such implemented dialogue artifact, associated with particular dialogue contexts, will be referred to as a style. Styles of a given dialogue context can thus be identified by reference to their designated name.

At startup, before initiation of interaction, the initial interface assembly process proceeds as follows: the top-level dialogue component requests for its associated user-task the names of the *styles* (i.e., implemented dialogue components) that have to be activated, so as to realize the adaptability behavior. Recursively, each container dialogue component repeats the same process for every embedded component. For each received request, the DMC triggers the adaptability cycle and responds with proper activation messages. Additionally, after initiation of interaction, the DMC may communicate dynamic style activation/cancellation messages to the DPC, as a consequence of the dynamic detection of user or usage-context attribute values (originated from UIS). Table 21.5 defines the syntax of messages communicated between the DMC and the DPC, and shows simple examples.

21.4.3.5 Communication between the DMC and the CPS

The communication between these two components is very simple. The DMC requests the various context parameter values (i.e., usage-context profile), and the CPS responds accordingly. During interaction, dynamic updates on certain context property values are to be communicated to the DMC for further processing (i.e., possibly new inferences will be made). Table 21.6 defines the classes of messages communicated between the DMC and the CPS, and shows simple examples.

TABLE 21.4 Communicated Messages between the User Information Server and the Dialogue Patterns Component

UIS→DPC

Message class	Controlling the level of monitoring
Content structure	*Status, Level* *Status* = **on** or **off** *Level* = *TaskLevel* or *EventLevel*, or *MethodLevel* *EventLevel* = **event**, *ObjectName, EventCategory* *TaskLevel* = **task**, *TaskName* *MethodLevel* = **method**, *ObjectName, MethodCategory*
When communicated	As needed by the user information server inference mechanisms

Examples	`{ on, task, "link selection" }`
	`{ off, event, "HTMLPageWindow," "KeyPress" }`
	`{ on, method, "BWDPageButton," "Pressed" }`
	`{ on, method, "HTMLPageScrollbar," "Scrolled" }`

DPC→UIS

Message class	Monitoring data
Content structure	*TaskBased* or *EventBased* or *MethodBased* *TaskBased* = **task**, *TaskName, TaskAction* *TaskAction* = **initiated** or **completed** *EventBased* = **event**, *ObjectName, EventCategory, EventData* *EventData* = Sequence of <*Parameter, Value*> pairs *MethodBased* = **method**, *ObjectName, MethodCategory*
When communicated	When the corresponding user actions are performed

Examples	`{ task, "link selection," initiated }`
	`{ task, "link selection," completed }`
	`{ event, "PageLoadingControlToolbar," "KeyPress," <"key," "a"> }`
	`{ event, "BWDPageButton," "MouseButtonPress," <"button," "2"> }`
	`{ method, "StopLoadingButton," "Pressed" }`
	`{ method, "ReloadButton," "Pressed" }`

Source: From Savidis, A. and Stephanidis, C., *Universal Access in the Information Society*, 3, 165–193, 2004.

TABLE 21.5 Communicated Messages between the Decision-Making Component and the Dialogue-Patterns Component

DPC→DMC

Message class	Requesting active styles for a particular dialogue context
Content structure	*DialogueContext*
When communicated	Upon startup, to initiate the necessary dialogue patterns
Examples	`{ "link selection" }, { "page loading control" }` `{ "page display control" }, { "toolbar expert" }`

DMC→DPC

Message class	Posting decisions regarding component activation/cancellation
Content structure	*Decision* =(*DecisionType, ComponentSignature*)+ *DecisionType* = **activation** or **cancellation** *ComponentSignature* = *TaskName, ComponentName*
When communicated	• At the end of the adaptability cycle, prior to initiation of interaction • At the end of each adaptivity cycle, after initiation of interaction

Examples	`{activation, "link selection," "direct"}`
	`{activation, "stop loading," "by confirmation"}`
	`{cancellation, "link selection," "direct"}`
	`{activation, "link selection," "by confirmation"}`
	`{activation, "confirm_ok_font," "times28"}`
	`{activation "confirm_ok_color," "rgb{0,255,0}"}`
	`{activation, "confirm_cancel_color," "rgb{255,0,0}" }`

Source: From Savidis, A. and Stephanidis, C., *Universal Access in the Information Society*, 3, 165–193, 2004.

TABLE 21.6 Communicated Messages between the Decision-Making Component and the Context-Parameters Server

DMC→CPS	
Message class	Requesting context parameter values
Content structure	This message contains only the message header (i.e., empty content)
When communicated	Prior to initiation of interaction, beginning of adaptability cycle
Examples	{ }

CPS→DMC	
Message class	Responding with a list of usage-context parameter values
Content structure	Sequence of *<Attribute, Value>* pairs
When communicated	When requested by the decision-making component
Examples	{ <"screen resolution," "640, 480">, <"mouse available," "yes">, <"terminal position," "120 cm">, <"software volume ctrl," "yes">, <"speech synthesis," "yes"> }
Message class	Dynamic usage-context attribute values
Content structure	*Attribute, Value*
When communicated	Each time a usage-context attribute value is dynamically modified
Examples	{ "environment noise," "65 dB" } { "user in front of terminal," "no" } { "user in front of terminal," "yes" }

Source: From Savidis, A. and Stephanidis, C., *Universal Access in the Information Society*, 3, 165–193, 2004.

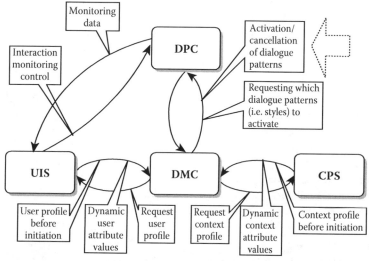

FIGURE 21.7 Type of information communicated among the components of the unified architecture, in the four basic bidirectional communication channels. The dashed arrow on the top right corner indicates the final communication step in adaptability and adaptivity cycles, which is performed by posting activation or cancellation decisions to DPC. (From Savidis, A. and Stephanidis, C., *Universal Access in the Information Society*, 3, 165–193, 2004.)

21.5 Summary, Conclusions, and Future Work

The concept of the unified user interface reflects a software engineering paradigm that addresses effectively the need for interactions automatically adapted to the individual end-user and the particular usage context. Following this engineering paradigm, interactive software applications should encompass the capability to appropriately deliver on-the-fly an adapted interface instance, performing appropriate run-time processing that engages:

- Utilization of user- and usage-context-oriented information (e.g., profiles), as well as the ability to detect dynamically user and usage-context attributes during interaction.
- Management of appropriate alternative implemented dialogue components, realizing alternative ways for physical-level interaction
- Adaptation-oriented decision making to: (1) choose, before initiation of interaction, the most appropriate dialogue components comprising the delivered interface, given any particular dialogue context, for the particular

end-user and usage-context profiles (i.e., adaptability); and (2) appropriately alter the initially delivered interface instance, according to dynamically detected user and usage-context attributes (i.e., adaptivity)

- Run-time component coordination and control to dynamically assemble or alter the target interface, the latter on-the-fly composed from the set of dynamically selected constituent dialogue components

The unified user interface development strategy provides a distributed software architecture with well-defined functional roles (i.e., which component does what), intercommunication semantics (i.e., which component requests what and from whom), control flow (i.e., when to do what), and internal decomposition (i.e., how the implementation of each component is internally structured). One of the unique features of this development paradigm is the emphasis on dynamic interface assembly for adapted interface delivery, reflecting a software engineering practice with: repository-oriented component organization, parametric containers with abstract containment APIs, and common interaction-monitoring control with abstract APIs. Although the method itself is not intended to be intensively prescriptive from the low-level implementation point of view, specific successful practices that have been technically validated in the real field, regarding decision making and dynamic user-attribute detection, are also discussed, focusing on micro-architecture details and internal functional decomposition.

The unified user interface development method constitutes a systematic and well-tested software engineering approach to practically address the issue of universal access for computer-based applications. It introduces the fundamental notion of adapted interface delivery before initiation of interaction, while proposing the technical approach to cope with the inherent run-time dynamic interface assembly process. Currently, software applications are provided with a single user interface targeted to the so-called average user, while users are typically required to learn applications, practically adapting themselves to their particular interactive characteristics. In the context of an information society, having large populations with diverse user requirements, abilities, and preferences, it is easily proved that the concept of an average user has no scientific substance (e.g., it is meaningless to define an average user for: sighted and blind users, children and elderly people, novice and expert users, etc.). Consequently, to support accessible and high-quality interaction, it is clear that alternative user interfaces need to be delivered, maximally fitting individual end-user attributes. Additionally, we are faced with an increasing production of computer-based applications that are to be used beyond the traditional desktop environment, which constantly and radically affect: working style, entertainment, transportation, communication, leisure, healthcare, information access, home activities, shopping, and generally everyday life.

In this continuous shift toward a largely computerized society, consolidated in the concept of an information society, the equal engagement of all citizens is put forth as both an ethical and a political prerequisite, but more importantly, as a legal one.

In this context, software vendors are faced with the challenge of delivering a wide range of software applications and services, for a variety of computing platforms, targeted to diverse user groups. Naturally, software firms will be activated in such new markets only if a cost-effective engagement strategy can be formulated. Consequently, to effectively support progress in this direction, it is critical that development paradigms are supplied realistically addressing the software development problem of universally accessible interactions.

The unified user interface development paradigm not only delivers an in-depth technical proposition, but it also consolidates process-oriented wisdom for building universally accessible interactions. For instance, the required technical knowledge to build user and usage-context on-the-fly adapted interfaces engages: user modeling, task design, action patterns, cognitive psychology, rule-based systems, network communication and protocols, multiplatform interfaces, component repositories, and core software engineering. Apparently, the development of system monoliths, encapsulating the necessary computable artifacts from those domains of required expertise, is not a viable development strategy. Additionally, software developers prefer incremental engagement strategies, allowing the stepwise entrance to new potential markets, by delivering successive generations of products encompassing layers of novel characteristics. Similarly, to pursue the production of software applications directly supporting universal access (i.e., by anyone, at any place, and at any time), one has to systematically set up a concrete strategy supporting evolutionary development, software reuse, incremental design, and modular construction.

Apart from reuse, as has been discussed earlier, evolution and incremental development are two fundamental features of unified user interface development. Those are reflected in the ability to progressively extend a unified user interface, by incrementally encapsulating computable content in the different parts of the architecture, to cater to additional users and usage-contexts, by designing and implementing more dialogue artifacts, and by embedding new rules for the decision-making logic. Those characteristics are particularly important because they identify a viable software engineering process to practically accomplish universal access, providing an answer to potential criticism concerning the potential feasibility of universal access.

In conclusion, unified user interface development is a paradigm that proposes a specific software engineering approach for crafting self-adapting interactive applications, according to the given individual user characteristics and contexts of use. In support of universally accessible interactions, this approach enables incremental development and facilitates the expansion and upgrade of dialogue components as an ongoing process, entailing the continuous engagement and consideration of new design parameters, and new parameter values.

Acknowledgments

The unified interface development method has been originally defined in the context of the ACCESS TP1001 (Development Platform for Unified ACCESS to Enabling Environments) project, partially funded by the TIDE Program of the European Commission and lasting 36 months (January 1, 1994 to December 31, 1996). The partners of the ACCESS consortium are CNR-IROE (Italy), Prime Contractor; ICS-FORTH (Greece); University of Hertfordshire (United Kingdom); University of Athens (Greece); NAWH (Finland); VTT (Finland); Hereward College (United Kingdom); RNIB (United Kingdom); Seleco (Italy); MA Systems &Control (United Kingdom); and PIKOMED (Finland).

The first large-scale application of the unified interface development method has been carried out in the context of the AVANTI AC042 (Adaptable and Adaptive Interaction in Multimedia Telecommunications Applications) project, partially funded by the ACTS Program of the European Commission and lasting 36 months (September 1, 1995 to August 31, 1998). The partners of the AVANTI consortium are ALCATEL Italia, Siette Division (Italy), Prime Contractor; IROE-CNR (Italy); ICS-FORTH (Greece); GMD (Germany); VTT (Finland); University of Siena (Italy); MA Systems & Control (United Kingdom); ECG (Italy); MATHEMA (Italy); University of Linz (Austria); EUROGICIEL (France); TELECOM (Italy); TECO (Italy); and ADR Study (Italy).

References

ACCESS Project (1996). *The ACCESS Project: Development Platform for Unified Access to Enabling Environments*. London: RNIB Press.

Akoumianakis, D., Savidis, A., and Stephanidis, C. (1996). An expert user interface design assistant for deriving maximally preferred lexical adaptability rules, in the *Proceedings of the 3rd World Congress on Expert Systems*, February 1996, Seoul, Korea, pp. 1298–1315. New York: Cognizant Communication Corporation.

Benyon, D. (1984). MONITOR: A self-adaptive user-interface, in the *Proceedings of IFIP Conference on Human-Computer Interaction: Volume 1, INTERACT '84*, 4–7 September 1984, London, pp. 335–341. Amsterdam, The Netherlands: Elsevier Science.

Browne, D., Norman, M., and Adhami, E. (1990). Methods for building adaptive systems, in *Adaptive User Interfaces* (D. Browne, M. Totterdell, and M. Norman, eds.), pp. 85–130. London: Academic Press.

Cockton, G. (1987). Some critical remarks on abstractions for adaptable dialogue managers, in the *Proceedings of the 3rd Conference of the British Computer Society, People & Computers III, HCI Specialist Group*, 7–11 August 1987, Devon, U.K., pp. 325–343. Cambridge, U.K.: Cambridge University Press.

Cockton, G. (1993). Spaces and distances: Software architecture and abstraction and their relation to adaptation, in *Adaptive User Interfaces: Principles and Practice* (M. Schneider-Hufschmidt, T. Kühme, and U. Malinowski, eds.), pp. 79–108. Amsterdam, The Netherlands: Elsevier Science.

Cote Muñoz, J. (1993). AIDA: An adaptive system for interactive drafting and CAD applications, in *Adaptive User Interfaces: Principles and Practice* (M. Schneider-Hufschmidt,

T. Kühme, and U. Malinowski, eds.), pp. 225–240. Amsterdam, The Netherlands: Elsevier Science.

Coutaz, J. (1990). Architecture models for interactive software: Failures and trends, in *Engineering for Human-Computer Interaction* (G. Cockton, ed.), pp. 137–151. Amsterdam, The Netherlands: Elsevier Science.

Dieterich, H., Malinowski, U., Kühme, T., and Schneider-Hufschmidt, M. (1993). State of the art in adaptive user interfaces, in *Adaptive User Interfaces: Principles and Practice* (M. Schneider-Hufschmidt, T. Kühme, and U. Malinowski, eds.), pp. 13–48. Amsterdam, The Netherlands: Elsevier Science.

Goldberg, A. (1984). *Smalltalk-80: The Interactive Programming Environment*. Reading, MA: Addison-Wesley Publishing.

Green, M. (1985). Report on dialogue specification tools, in *User Interface Management Systems* (G. Pfaff, ed.), pp. 9–20. New York: Springer-Verlag.

Hartson, R. and Hix, D. (1989). Human-computer interface development: Concepts and systems for its management. *ACM Computing Surveys* 21: 241 247.

Jacobson, I., Griss, M., and Johnson, P. (1997). Making the reuse business work. *IEEE Computer* 30: 36–42.

Kobsa, A. (1990). Modeling the user's conceptual knowledge in BGP-MS, a user modeling shell system. *Computational Intelligence* 6: 193–208.

Kobsa, A. and Pohl, W. (1995). The user modelling shell system BGP-MS. *User Modeling and User-Adapted Interaction* 4: 59–106.

Kobsa, A. and Wahlster, W. (eds.) (1989). *User Models in Dialog Systems*. Berlin: Springer-Verlag.

Krasner, G. E. and Pope, S. T. (1988). A description of the model view controller paradigm in the Smalltalk-80 system. *Journal of Object-Oriented Programming* 1: 26–49.

Mowbray, T. J. and Zahavi, R. (1995). *The Essential CORBA: Systems Integration Using Distributed Objects*. New York: John Wiley & Sons.

Myers, B. (1995). User Interfaces Software Tools. *ACM Transactions on Human-Computer Interaction* 12: 64–103.

Savidis, A., Stergiou, A., and Stephanidis, C. (1997). Generic containers for metaphor fusion in nonvisual interaction: The HAWK interface toolkit, in the *Proceedings of the 6th International Conference on Man-Machine Interaction Intelligent Systems in Business (INTERFACES '97)*, 28–30 May 1997, Montpellier, France, pp. 194–196. Paris, France: EC2 & Development.

Stephanidis, C., Paramythis, A., Sfyrakis, M., and Savidis, A. (2001). A case study in unified user interface development: The AVANTI web browser, in *User Interfaces for All: Concepts, Methods, and Tools* (C. Stephanidis, ed.), pp. 525–568. Mahwah, NJ: Lawrence Erlbaum Associates.

UIMS Developers Workshop (1992). A meta-model for the runtime architecture of an Interactive System. *SIGCHI Bulletin* 24: 32–37.

Vergara, H. (1994). *PROTUM: A Prolog Based Tool for User Modeling*. Bericht Nr. 55/94 (WIS-Memo 10), Kostanz, Germany: University of Kostanz.

22

A Decision-Making Specification Language for User Interface Adaptation

Anthony Savidis and
Constantine Stephanidis

22.1 Introduction

22.1.1 Dynamic Interface Adaptability

The notion of automatic user interface adaptation reflects the capability of interactive software to adapt during run-time to the individual end-user, as well as to the particular context of use, by delivering a most appropriate interaction experience (see Chapters 16 and 21 of this handbook).[1]

The storage location, origin, and format of user-oriented information may vary. For example, information may be stored in profiles indexed by unique user identifiers, may be extracted from user-owned cards, may be entered by the user in an initial interaction session, or may be inferred by the system through continuous interaction monitoring and analysis. Additionally, usage-context information (e.g., user location, environment noise, network bandwidth, etc.) is normally provided by special-purpose

equipment, like sensors, or system-level software. To support optimal interface delivery for individual user and usage-context attributes, it is required that for any given user task or group of user activities, the implementations of the alternative best-fit interface components are appropriately encapsulated.

Upon startup and during run-time, the software interface relies on the particular user and context profiles to assemble the eventual interface on the fly, collecting and gluing together the constituent interface components required for the particular end-user and usage context. This type of best-fit automatic interface delivery, called *interface adaptability,* has been originally introduced in the context of unified user interface development (Savidis and Stephanidis, 2001). In this context, run-time adaptation-oriented decision making is engaged, so as to select the most appropriate interface components for the particular user and context profiles, for each distinct part of the user interface. This logical distinction of the run-time processing steps to accommodate interface adaptability is effectively outlined in the run-time architecture for automatically adapted interactions discussed in Chapter 21 of this handbook.

The role of the decision-making subsystem is to effectively drive the interface assembly process by deciding which interface

---

[1] This chapter is based on Savidis, A., Antona, M., and Stephanidis, C. (2005). A decision-making specification language for verifiable user-interface adaptation logic. *International Journal of Software Engineering and Knowledge Engineering* 15: 1063-1094. Reprinted here with modifications by concession of World Scientific Publishing.

components need to be selectively activated. The interface assembly process has inherent software engineering implications on the software organization model of interface components. More specifically, as for any *task context* (i.e., part of the interface to support a user activity or task) alternative implemented incarnations may need to coexist, conditionally activated during runtime due to decision making, the need to accommodate interface context *polymorphism* arises. In other words, there is a need to organize interface components around their particular task contexts, enabling task contexts to be supported through multiple (i.e., polymorphic) deliveries. This contrasts with traditional nonadapted interfaces in which all task contexts have singular implementations. The key software requirements for user interface design and implementation have been addressed as follows:

- The *unified user interface design method* (Savidis and Stephanidis, 2004a; see also Chapter 16, "Unified Design for User Interface Adaptation") reflects the hierarchical discipline of user interface construction, emphasizing the hierarchical organization of task contexts, which may have an arbitrary number of designed alternatives called *design styles*, shortly *styles*. In this framework, the concept of polymorphic task-context hierarchies (polymorphic tasks) has been introduced. Each alternative style is explicitly annotated with its corresponding user and context attributes (i.e., its adaptation design rationale).
- *Dynamic polymorphic containment hierarchies* (Savidis and Stephanidis, 2004b) provide a software engineering method for implementing interface components exhibiting typical hierarchical containment, while enabling the dynamic establishment of containment links among contained and container components. Additionally, all

interface components reflect the organizational model of polymorphic task hierarchies, indexing uniquely components according to their particular design-time task-context and style identifiers.

The design and implementation of alternative interface components around hierarchically organized task contexts have been employed in the AVANTI Project (see acknowledgements in this chapter) for the development of the AVANTI web browser (Stephanidis et al., 2001), supporting interface adaptability. In Figure 22.1, an excerpt from the polymorphic interface component organization of the AVANTI browser is shown, to accommodate implementation of interface adaptability.

This chapter specifically focuses on effective support for decision-making implementation. In particular, it reports a method, which has been purposefully elaborated to be easier for designers to directly assimilate and deploy, in comparison to programming-based approaches using logic-based or imperative-oriented programming languages. In this context, the decision-making specification language (DMSL) has been designed and implemented. Additionally, the software engineering deployment approach of the DMSL language is presented, discussing a well-tested architecture to cope with dynamic interface assembly, further generalizing and elaborating toward accomplishing dynamic software adaptability (i.e., software adaptively assembled according to deployment requirements). In this framework, the key addressed issues are:

- *Decision-making specification language* (DMSL)
  - Supports localized decision blocks for each task context
  - Built-in user and context decision-parameters with run-time binding of values

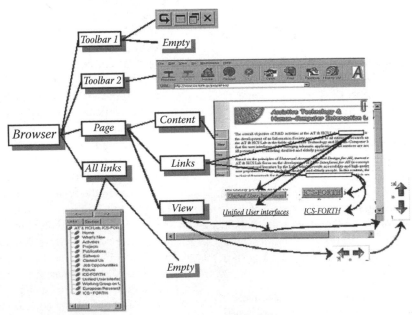

FIGURE 22.1 Parametric polymorphic containment with variant constituent components in the AVANTI browser. "Empty" indicates components whose presence may have to be omitted upon dynamic interface delivery for certain user categories. (From Savidis, A. and Stephanidis, C., *International Journal of Software Engineering and Knowledge Engineering*, 15, 1063–1094, 2005.)

- Can trigger other decision blocks, supporting modular chain evaluations
- Interface component reference made through unique design identifiers
- Supports activation and cancellation commands for interface components
- Accompanied with rule-patterns for interface component relationships
- Has been proved from real practice to be understandable and usable by designers (Stephanidis et al., 2004a; Antona et al., 2006)
- Is supported by an interactive tool for computer-assisted adaptation design
- *Dynamic software adaptability*, reflecting an architectural generalization of the largely tested and applied interface adaptability software engineering approach as follows:
  - From interface components to software components
  - From physical interface containment to architectural containment
  - From user and context parameters to software deployment parameters
  - From task contexts to architectural contexts
  - From interface-based indexing to architectural-role indexing
  - From dynamic interface assembly to dynamic software assembly

22.2 Related Work

22.2.1 Interface Adaptation

Most of the work regarding system-driven, user-oriented adaptation concerns the capability of an interactive system to dynamically (i.e., during interaction) detect certain user properties, and accordingly decide upon various interface changes. This notion of adaptation falls in the *adaptivity* category, that is, adaptation performed after the initiation of interaction, based on interaction monitoring information, and has been commonly referred to as *adaptation during use*. In adaptive interaction, the collection and processing of interaction-monitoring information is performed and analyzed through different types of knowledge processing to derive certain user attribute values (not known prior to the initiation of interaction), which may drive appropriate interface update actions. A well-known adaptive system employing such techniques is MONITOR (Benyon, 1984). Similarly, for the purpose of dynamic detection of user attributes, a monitoring component is employed in the AIDA system (Cote Muñoz, 1993), while the BGP-MS system (Kobsa and Pohl, 1995) implemented a sophisticated user-modeling server.

In such adaptive systems, decision making is characterized by two key properties: (1) it is encapsulated in system implementation, not made editable by user interface designers; and (2) its primary target is the inference of dynamic user characteristics, such as preference to particular information elements or confusion in

performing a task, instead of interface component selection as targeted in user interface adaptability. The development of decision kernels for dynamic interface component selection and delivery, upon interaction startup, has been originally introduced in the definition of unified user interface development, and later applied in the context of large-scale adaptable application developments, as such the AVANTI web browser (Stephanidis et al., 2001) the PALIO adaptable multimedia information system (Stephanidis et al., 2004b). In this context, the DMSL language has been designed to optimize the development process by offering an instrument, directly usable and deployable by designers, particularly suited for adaptation-oriented decision-making specification.

22.2.2 Software Adaptability

In the domain of user interface development, reusable software objects had been employed from the early days of graphical user interfaces (GUIs). The broad deployment of binary-reusable (i.e., compiled) components is carried in user interface software engineering from the early appearance of software libraries of interaction object classes, such as Windows Object Library, Xt/Athena, MAC Toolbox, and OSF/Motif. More recently, the development of key component-ware technologies has been carried out in parallel with advanced GUI libraries, such as COM (known formerly as ActiveX) for MFC and Java Beans for JFC. Although the notion of a software component technically denotes the packaging of software in the form of a binary-level reusable object over a component-ware technology, the term *component* is used in this chapter to generally refer to any independent part of a software system that plays a distinct functional role. In this context, interface components range from typical interaction object classes (e.g., menus, windows, buttons, check boxes, etc.), to more composite interactive artifacts serving specific roles (e.g., file menu, title bar, settings' dialogue box, print dialogue box, URL bar, etc.). Currently, primary emphasis is put on flexible component deployment during development time, rather than on automatic software assembly from constituent interface components. The need for software adaptability has been identified in Fayad and Cline (1996), mainly emphasizing static software properties such as extensibility, flexibility, and performance tunability, without negotiating the automatic and dynamic software assembly. Similarly, in Netinant et al. (2000), adaptability is also considered a key static property of software components, which can be pursued through aspectual decomposition (i.e., by employing aspect-oriented programming methods).

22.3 Decision-Making Logic Implementation

22.3.1 Activation and Cancellation Decisions for Interface Components

The outcome of a decision-making process is a sequence of *activation* and *cancellation* commands of named interface components, which are to be appropriately applied in the interface assembly process (see also Chapter 21 of this handbook). The necessity of a component coordination command-set in implementing adaptation has been identified very early in Cockton (1993), while the capability to manage dynamic interface component selection with just two fundamental commands has been introduced in the context of unified user interface development (Savidis and Stephanidis, 2001, 2004b). In this context, the functional role of those commands in dynamic interface assembly is defined as follows:

- *Activation* implies the necessity to deliver the corresponding component to the end-user. Effectively, delivery may imply instantiation (i.e., instance creation) of the respective component class, in a way dependent on the implementation form of the component (i.e., for OOP classes, dynamic instantiation suffices, for component-ware technologies, replication and object reference extraction is required).

- *Cancellation* implies that a previously activated component needs to be removed on the fly from the interface delivered to the end-user. In this case, cancellation is typically performed by destruction of the corresponding instance.

22.3.2 Outline of the Language

The decision-making logic is defined in independent decision blocks, each uniquely associated with a particular task context; at most one block per distinct task context may be supplied. The decision-making process is performed in independent sequential *decision sessions*, and each session is initiated by a request of

the interface assembly module for execution of a particular *initial decision block*. In such a decision session, the evaluation of an arbitrary decision block may be performed, while the session completes once the computation exits from the initial decision block.

The outcome of a decision session is a sequence of activation and cancellation commands, all of which are directly associated with the task context of the initial decision block. Those commands are posted back to the interface assembly module as the product of the performed decision-making session. In Figure 22.2, an example decision block is shown (an excerpt of the implementation of the decision logic AVANTI browser; see also Figure 22.1), for selecting the best alternative interface components for the "link" task context. The interface design relating to this adaptation decision logic is provided in Figure 22.3.

The primary decision parameters are end-user and the usage-context profiles, defined as two built-in objects (i.e., user and context) whose attributes are syntactically accessible in the form of named attributes. The binding of attribute names to attribute values is always performed at run-time. The encapsulation of composite attributes in user and context profiles is easily allowed due to the syntactic flexibility of attributes reference. For instance, user.abilities.vision and user.abilities.hearing are syntactic sugar for user."abilities.vision" and user."abilities.hearing", where "abilities.vision" and "abilities.hearing" are two distinct independent ordinal attributes of the user built-in object. Consequently, even though all attributes in the DMSL language are semantically scalar, the flexibility of attribute names allows syntactical simulation of aggregate structures. Additionally, in Figure 22.3, the chain evaluation of other decision blocks

```
taskcontext link [
        evaluate linktargeting;
        evaluate linkselection;
        evaluate loadconfirmation;
]

taskcontext linktargeting [
        if (user.abilities.pointing == accurate) then
                activate "manual pointing";
        else
                activate "gravity pointing";
]

taskcontext linkselection [
        if (user.webknowledge in {good, normal}) then
                activate "underlined text";
        else
                activate "push button";
]

taskcontext loadconfirmation [
        if (user.webknowledge in {low, none} or context.net == low) then
                activate "confirm dialogue";
        else
                activate "empty";
]
```

FIGURE 22.2 An example of a simple decision block to select the most appropriate delivery of web links for the individual end-user; notice that names in italics are not language keywords but are treated as string constants (i.e., user.webknowledge is syntactic sugar for user."webknowledge"). From Savidis, A. and Stephanidis, C., *International Journal of Software Engineering and Knowledge Engineering*, 15, 1063–1094, 2005.

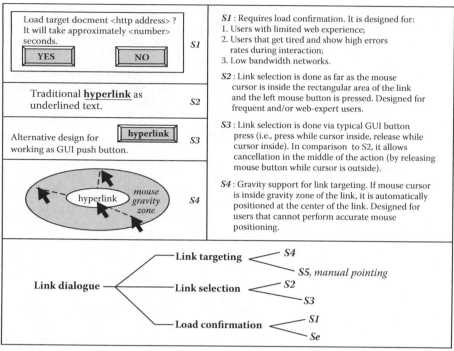

FIGURE 22.3 The link selection task context with its various subtask contexts and the associated design logic, which is encapsulated within the decision blocks of Figure 22.3. Se is used to indicate an "empty" style (i.e., no load confirmation dialogue supported). S5 is the typical direct pointing of links using the mouse. From Savidis, A. and Stephanidis, C., *International Journal of Software Engineering and Knowledge Engineering*, 15, 1063–1094, 2005.

through the `evaluate` command is illustrated. The latter can be employed when the adaptation decisions for a particular task context require decision making for particular subtask contexts.

Upon startup, the interface assembly module causes the execution of decision sessions for all polymorphic task contexts in a hierarchical manner (see Figure 22.4), so that the required alternative interface components, given the particular end-user and usage context, are effectively *marked* for interface delivery. Subsequently, the assembly process is performed, hierarchically instantiating and gluing together all marked interface components with the interface components of unimorphic task contexts.

As appears in the right part of Figure 22.4, it is possible that more than a single alternative style can be selected for a particular

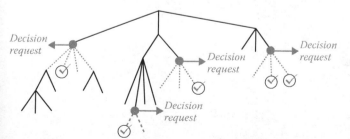

FIGURE 22.4 Illustrating the hierarchical posting of decision requests, causing decision sessions for each polymorphic task context (shown with decomposition alternatives as dashed lines), and marking of selected alternative styles (i.e., interface components) after each decision session completes. From Savidis, A. and Stephanidis, C., *International Journal of Software Engineering and Knowledge Engineering*, 15, 1063–1094, 2005.

polymorphic task context. This is dependent on the particular design rationale of the alternatives styles, while DMSL does not restrict decision blocks to output only a single activation command. Additionally, as will be explained next, the relationships among the alternative styles of a polymorphic task context are completely formalized in the DMSL language, associated with well-defined rule patterns for implementing the decision block of polymorphic task contexts. This serves two key objectives: (1) guiding designers in the organization and implementation of decision blocks; and (2) allowing developers to implement interactive design instruments that automate the generation of decision blocks from the design relationships of alternative components. In Section 22.3.4, the *Mentor* interactive adaptation-design tool will be described (Antona et al., 2006), which exploits this feature.

22.3.3 Relationships among Alternative Styles and Associated Rule Patterns

The emergence of alternative styles of polymorphic task contexts during adaptation design aims primarily to address the varying user and usage-context attribute values. For instance, as appears in the example of Figure 22.2 and Figure 22.3, the degree of the end-user web expertise leads to alternative styles for interactively supporting link selection. However, although this remark may lead to an initial assumption that all styles are mutually exclusive, there are additional design relationships among alternative styles, as demonstrated in the context of unified user interface design (Savidis and Stephanidis, 2004a; see also Chapter 16, "Unified Design for User Interface Adaptation"). Those relationships are as follows:

- *Exclusion* or *incompatibility* is applied if the various alternative styles are deemed to be usable only within the scope of their associated user and usage-context attribute values, because from the usability point of view it is inappropriate to concurrently instantiate both styles.
- *Compatibility* is applicable among alternative styles for which the concurrent presence during interaction allows the user to perform certain actions in alternative ways, without introducing usability problems.
- *Augmentation* aims to enhance the interaction with another particular style that is found to be valid, but not sufficient to facilitate the effective accomplishment of the supported user task. For instance, if during interaction it is detected that the user is unable to perform a certain task, task-sensitive guidance through a separate but compatible style could be delivered. In other words, the augmentation relationship is assigned to two styles when one can be used to enhance the interaction while the other is active (see Figure 22.5).
- *Substitution*, exhibiting a very strong link with adaptivity techniques, is applied in cases where, during interaction, it is decided that some styles need to be substituted by others. For instance, the ordering, arrangement, or availability of certain operations may change (see Figure 22.5) on the basis of interaction monitoring and extraction of information regarding frequency of use and repeating usage patterns. In this case, some styles would need to be canceled, while others would need to be activated.

In the DMSL language those relationships are not injected as a part of the semantics, but, alternatively, concrete rule patterns are delivered, effectively mapping those relationships to implementation skeletons of decision blocks. This gives adaptation designers the freedom not to necessarily adopt those particular design relationships, in case, for instance, they do not choose to employ unified design as the adaptation-design approach. In Figure 22.6, the DMSL decision-rule patterns are provided, for the previously described style relationships.

22.3.4 The Mentor Tool Supporting Interactive Adaptation Design

As previously mentioned, the design of adaptation entails the definition of polymorphic realizations of task contexts through alternative designed styles that are associated together by means of exclusion, compatibility, augmentation, or substitution relationships. To provide tool-based assistance for such a demanding design process, it is necessary to enable the effective interactive manipulation of the overall design space of task polymorphic contexts and their alternative styles.

In the mentor tool for interactive adaptation design, the polymorphic hierarchical decomposition of task contexts, either related to user tasks, system tasks (e.g., feedback), or physical design (e.g., graphical design), is genuinely supported. As depicted in Figure 22.7, the task context *FileManagement* is decomposed into alternative subtasks, like *DeleteFile*. The latter is defined to have multiple alternative decompositions, that is, it is a polymorphic subtask context, where alternative decompositions denote distinct uniquely named alternative styles, for example, the styles named *DeleteFile_DirectManipulation* and

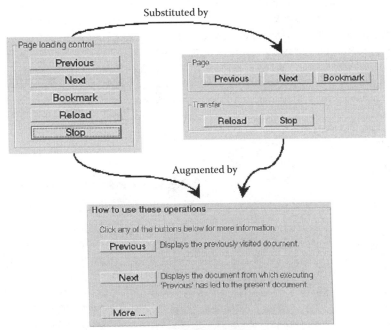

FIGURE 22.5   Alternative styles and their design relationships for the "page loading control" task context of the AVANTI browser. From Savidis, A. and Stephanidis, C., *International Journal of Software Engineering and Knowledge Engineering*, 15, 1063–1094, 2005.

Exclusion (S$_1$, S$_2$)	Compatibility (S$_1$, S$_2$)

if (S$_1$.*cond*) then
    activate S$_1$;
else
if (S$_2$. *cond*) then
    activate S$_2$;
	if (S$_1$.*cond*) then
	    activate S$_1$;
	if (S$_2$. *cond*) then
	    activate S$_2$;

Substitution (S$_1$ by S$_2$)	**Augmentation (S$_1$ by S$_2$)**

if (S$_2$. *cond and* isactive *(S$_1$)*) then [
    cancel S$_1$;
    activate S$_2$;
]
else
if (S$_1$.*cond*)
    activate S$_1$;
	if (S$_1$.*cond*)
	    if (not isactive(S$_1$)) then
	        activate S$_1$;
	    else
	    if (S$_2$. *cond*) then
	        activate S$_2$;

FIGURE 22.6 The decision rule patterns associated to the relationships among alternative styles; the style condition is the Boolean expression engaging the user and context attribute values for which the style is designed. From Savidis, A. and Stephanidis, C., *International Journal of Software Engineering and Knowledge Engineering*, 15, 1063–1094, 2005.

DeletFile_ModalDialogue of Figure 22.7. The key issue in the context of such adaptation design is the design rationale for alternative styles, such as, why a task-context should be polymorphic. As it has been discussed earlier, alternative styles are required when, given a particular task-context and diverse user or usage-context attribute values, the delivery of a single style clearly constitutes, from a usability point of view, a less than perfect design decision.

Consequently, in interactive adaptation-design support, designers should be given the capability to manipulate: (1) user and usage-context profiles; and (2) the definition of conditions for alternative styles, in the form of expressions engaging user

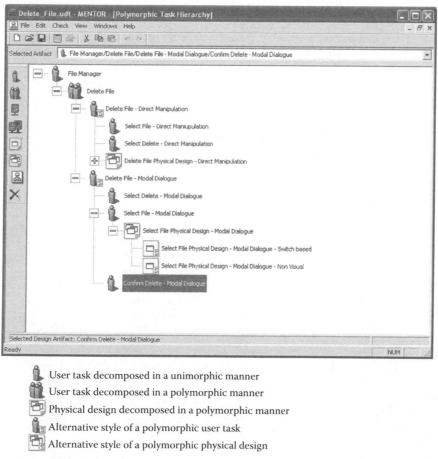

    User task decomposed in a unimorphic manner
    User task decomposed in a polymorphic manner
    Physical design decomposed in a polymorphic manner
    Alternative style of a polymorphic user task
    Alternative style of a polymorphic physical design

FIGURE 22.7 A snapshot of adaptation-design in the Mentor tool, showing the structure of a polymorphic task hierarchy for the File Manager task context and its relevant subtask contexts. From Savidis, A. and Stephanidis, C., *International Journal of Software Engineering and Knowledge Engineering*, 15, 1063–1094, 2005.

and usage-context profile parameters, such as the conditions of DMSL language rules. Both of those key features are supported in the mentor tool as illustrated in Figure 22.8. The role of stereotypes is crucial in adaptation design, as it allows user groups to be easily represented and directly referenced within style conditions. This has the following advantages:

- It relieves designers from the repetition of conditions when styles for the same user group are met in different task contexts.
- It makes design conditions self-documented, since stereotype identifiers make the design rationale far more explicit.
- It allows global changes in stereotype conditions without the need to manually change all individual style conditions.

Stereotypes are explicitly supported in the DMSL language, while the mentor tool generates DMSL-compliant stereotype definitions, according to the interactively designed stereotypes (see upper part of Figure 22.8). Stereotypes are referenced through their unique identifier while they may be also combined with other stereotypes in condition expressions. In Figure 22.9, examples of stereotype definition and deployment in the DMSL language are provided.

The consistency checking of conditions is automatically performed in the Mentor tool, using knowledge on the types and value domains of all engaged user and usage-context attributes. This allows potential errors of unsatisfactory expressions or type mismatches to be detected and reported to interface designers. Some examples showing the detection of inconsistencies for stereotype conditions are provided in Figure 22.10.

22.4 Dynamic Software Adaptability

22.4.1 Dynamic Polymorphic Containment Hierarchies

A key architectural implication due to the functional requirement of dynamic interface assembly (see Chapter 18, "Software Requirements for Inclusive User Interfaces," of this handbook) is the specific organization of implemented interface components to enable dynamically established containment hierarchies.

In nonadaptable unimorphic interactive applications, developers typically program the hierarchical structure of the user interface through hard-coded parent-child associations that are determined during development time. However, in the context of adapted interface delivery, the component containment hierarchies should support two key features: (1) parent-child associations are always decided and applied during run-time; and (2) multiple alternatively candidate contained instances are expected for container objects. The interface component organization method of dynamic polymorphic containment hierarchies is illustrated in Figure 22.11. This model has been employed for the implementation of the AVANTI browser, as shown in Figure 22.1. Following Figure 22.11, *PL* indicates the polymorphism factor, which provides the total number of all potential different run-time incarnations of an interface component, recursively defined as the product of the polymorphic factors of constituent component classes.

The dynamic interface assembly process reflects the hierarchical traversal in the polymorphic containment hierarchy, starting from the root component, to *decide, locate, instantiate,* and *initiate* appropriately every target contained component (see Figure 22.12). This process primarily concerns the interface

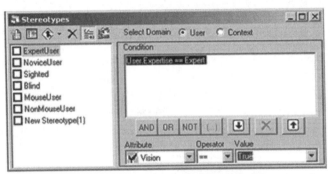

Constructing and editing a stereotype in the form of condition-expressions engaging user and usage context attribute values

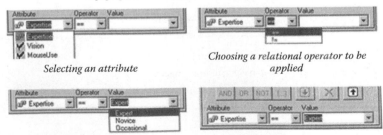

Selecting an attribute

Choosing a relational operator to be applied

Choosing an attribute value

Finalizing the sub-expression

FIGURE 22.8 Supporting user and usage-context profile manipulation, as well as condition editing (the latter for both styles and profile stereotypes). From Savidis, A. and Stephanidis, C., *International Journal of Software Engineering and Knowledge Engineering,* 15, 1063–1094, 2005.

```
stereotype user          Blind       : user.vision == false;
stereotype user          Novice      : user.webknowledge in { low, none };
stereotype context       LowNet      : context.net == low;
stereotype user          BlindNovice : Blind and Novice;

taskcontext loadconfirmation [
        if (Novice or LowNet) then
                activate "confirm dialogue;
        else
                activate "empty";
]
```

FIGURE 22.9　Definition and deployment of stereotypes; the decision block of Figure 22.3 is redefined to employ stereotypes, showing the increased readability of the adaptation rules. From Savidis, A. and Stephanidis, C., *International Journal of Software Engineering and Knowledge Engineering*, 15, 1063–1094, 2005.

components that implement polymorphic task-contexts. From the implementation point of view, the following software design decisions have been made:

- The task-context hierarchy has been implemented as a tree data structure, with polymorphic nodes triggering decision-making sessions (see also Figure 22.4).
- Interface components have been implemented as distinct independent software modules, implementing generic containment application programming interfaces (APIs),

while exposing a singleton control-API for dynamic instantiation and name-based lookup.

- The interface assembly procedure is actually carried out via two successive hierarchical passes:
  - Execution of decision sessions, to identify the specific styles for polymorphic task contexts, that will be part of the eventually delivered user interface.
  - Interface construction, through instantiation and initiation of all interface components for the decided styles.

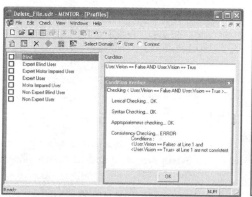

Inconsistent because of inconsistent values of the same attribute

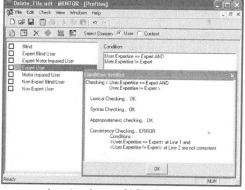

Inconsistent because of different operators for the same attribute and value

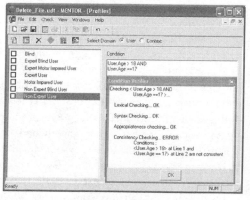

Inconsistent because of inconsistent operator and value combination for the same attribute

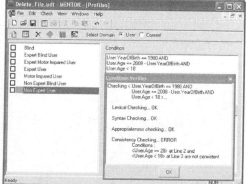

Resulting integer value incompatible with subsequent comparison operator and value

FIGURE 22.10　Examples of condition consistency checking in the construction of stereotypes. From Savidis, A. and Stephanidis, C., *International Journal of Software Engineering and Knowledge Engineering*, 15, 1063–1094, 2005.

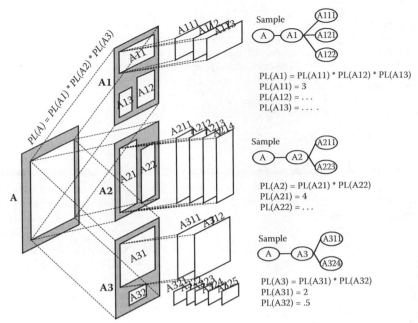

FIGURE 22.11 The notion of dynamic polymorphic hierarchical containment in interface adaptability to cater for the run-time interface assembly process. From Savidis, A. and Stephanidis, C., *International Journal of Software Engineering and Knowledge Engineering*, 15, 1063–1094, 2005.

22.4.2 Architectural Generalization from Dynamic Interface Assembly

The architectural generalization of the dynamic interface assembly method is a normal manual transformation of the basic architectural entities of interface adaptability to the general software architecture domain. As discussed in Section 22.2.2, the adopted notion of software adaptability reflects the functional properties of automatic software assembly, through decision making that relies upon run-time software adaptation parameters. It should be noted that this is a fundamentally different target from methods related

to software evolution, which focus on the automated transformation and evolution of software structures at development time, according to diverse software requirements. The key architectural generalization aspects are detailed within Figure 22.13.

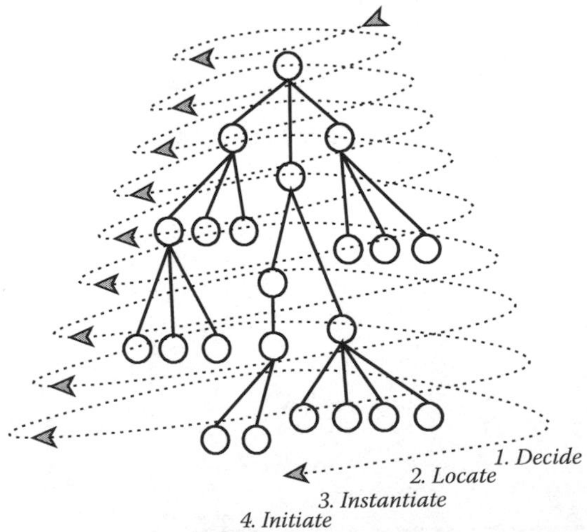

This generalization leads to an augmented vocabulary for the software architecture domain, mainly introducing the meta-elements necessary to accommodate run-time software assembly driven by decision-making for deployment adaptation, as illustrated in Figure 22.14.

Containment APIs are effectively transformed to functionality abstraction APIs, the latter normally differentiating per polymorphic architectural context and distinct application domain, which all alternative candidate components should thoroughly implement. The generalization of interface-adaptation decision parameters, that is, user and user-context profiles, concerns the *software deployment parameters* of Figures 22.13 and 22.14, conveyed as generic profile structures with a domain-specific interpretation. From the DMSL language point of view, such an extension can be trivially accommodated, by accumulating *user* and *context* built-in profiles in a generic profile-structure named *params*. For instance, the user and context profiles become syntactically visible as `params.user` and `params.context`, respectively. In the appendix at the end of this chapter the original DMSL grammar for interface adaptability is provided, together with its slightly updated version for decision making in dynamic software assembly.

Domain-specific parameters may reveal software deployment characteristics, which constitute the basis for choosing alternative best-fit software components, while stereotypes may reflect particular deployment scenarios. For example, `params.memory` and `params.storage` could be used to choose alternative implementations of algorithms with varying main and

FIGURE 22.12 Illustration of the dynamic interface assembly process as an incremental hierarchical construction procedure. From Savidis, A. and Stephanidis, C., *International Journal of Software Engineering and Knowledge Engineering*, 15, 1063–1094, 2005.

1. Decide
2. Locate
3. Instantiate
4. Initiate

Original architectural semantics	Generalized architectural semantics
Hierarchical interface structure	Hierarchical architectural structure
Task context	Architectural context
Interface component	Software component
User and usage-context attributes	Software deployment parameters
User and context stereotypes	Software deployment scenarios
Polymorphic task context	Polymorphic architectural context
Alterative styles	Alternative encapsulated components
Task decomposition	Architectural decomposition
Task-based component indexing	Architectural role component indexing
Interface component containment	Architectural containment
Generic containment APIs	Functional-role abstraction APIs

FIGURE 22.13 Generalizing the architectural semantics of dynamic interface assembly from the interface development domain to the software development domain. From Savidis, A. and Stephanidis, C., *International Journal of Software Engineering and Knowledge Engineering*, 15, 1063–1094, 2005.

secondary memory requirements. Naturally, the same behavior can be accomplished by injecting decision making within the software itself. However, through the separation and externalization of the design logic from the implementation components, better reusability is accomplished through orthogonal component combinations. Apart from interface assembly and the potential for dynamic software assembly, the same approach has been effectively employed in the context of adapted information delivery, realized through dynamic content assembly, as discussed in the next section.

22.4.3 Dynamic Query Assembly for Content Adaptability

In the context of the PALIO project (see acknowledgments), the DMSL language and the software engineering method for dynamic software assembly have been effectively employed for adaptable information delivery (Stephanidis et al., 2004b) over mobile devices to tourist users. The decision-making process was based on parameters such as nationality, age, location, interests or hobbies, time of day, visit history, and group information (i.e., family, friends, couple, colleagues, etc.). The information model reflected a typical relational database structure, while content retrieval was carried out using XML-based SQL

queries. In this context, to enable adapted information delivery instead of implementing hard-coded SQL queries, query patterns have been designed with specific polymorphic placeholders filled in by dynamically decided concrete subquery patterns. For instance, as seen in Figure 22.15, particular data categories or even query operations may be left "open," with multiple alternatives, depending on run-time content-adaptation decision making.

The implementation of dynamic query assembly has been realized through: (1) the hierarchical representation of the polymorphic query structure as a tree (i.e., in the same way as hierarchical task-contexts); (2) the easy implementation of the alternative subquery patterns as text fragments, due to the textual nature of XML-based SQL queries; and (3) the incremental assembly of the eventual query from its constituent textual elements.

22.5 Summary and Conclusions

This chapter has presented the DMSL language for adaptation-oriented decision-making specification and an appropriate software engineering approach to accommodate dynamic software adaptability. The language has been intensively applied and tested in the course of various developments targeted in supporting user and usage-context interface adaptation, like the AVANTI

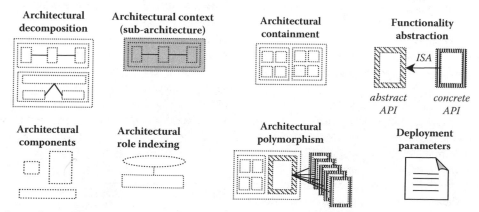

FIGURE 22.14 The key architectural meta-elements in the context of dynamic software adaptability, as transformations of the corresponding elements from the interface adaptability architectural domain. From Savidis, A. and Stephanidis, C., *International Journal of Software Engineering and Knowledge Engineering*, 15, 1063–1094, 2005.

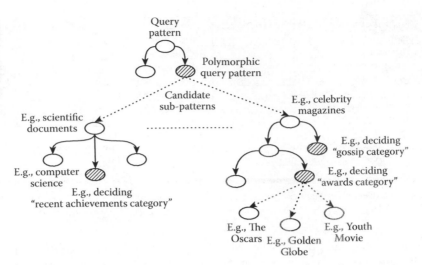

FIGURE 22.15 Query patterns with polymorphic placeholders having multiple alternative candidate subpatterns, selected through decision making. From Savidis, A. and Stephanidis, C., *International Journal of Software Engineering and Knowledge Engineering*, 15, 1063–1094, 2005.

browser (Stephanidis et al., 2001) of the AVANTI Project (see acknowledgments), and adapted information services, like the PALIO system (Stephanidis et al., 2004b) of the PALIO Project (see acknowledgments). Additionally, the DMSL language has been employed in the Mentor interactive tool (Antona et al., 2006) for adaptation-design supporting the polymorphic hierarchical decomposition of task-contexts (Savidis and Stephanidis, 2004b), generating decision rules in the DSML language.

In all these developments, the DMSL language played a crucial software engineering role, effectively enabling the separation between the decision-making logic from the repository of interface components and their run-time coordination. As a result, the decision-making logic has been made editable and extensible directly by interface designers, while interface component reuse has been largely promoted due to the orthogonal combinations inherent in the organization model of dynamic polymorphic containment hierarchies. The DMSL language reflects the polymorphic hierarchical design model, in which alternative design-parameter values (e.g., user/context/deployment attributes) require the presence of alternative design solutions (e.g., styles, software components) in different design contexts (e.g., task contexts or architectural contexts). Additionally, the DMSL language comes with a specific software meta-architecture to better encapsulate its decision-making facilities for adapted software delivery.

The binding of decision parameter attributes, decision block identifiers, and style names is always performed at run-time, enabling rules to be edited independently of the software implementation of interface components. Such loose coupling between the decision logic and the coordinated components allows decision kernels to be effectively reused across applications of the same domain, as far as the same design-time naming conventions are applicable. This accounts not only for the whole decision logic as a single reusable unit, but also for specifically selected decision blocks. For instance, the decision blocks related to the Help

task-context, and all its related subtask contexts, may be directly reused across applications offering interactive help facilities. The latter necessitates reuse of the corresponding implemented interface components, something that can be directly accommodated when those particular components implement abstract APIs for dynamic containment and help facilities, as opposed to hardwiring with the overall interface implementation.

The DMSL language and the dynamic software assembly approach have been also employed for adapted information retrieval. In this case, the implementation model has been radically different in comparison to the typically programmed interface components in adapted interface delivery, being actually realized through XML-based SQL queries. The first step was the identification of the elements that had to dynamically vary due to adaptation (i.e., query subpatterns), followed by the definition and study of the superstructure encapsulating all alternative elements. From this initial analysis, the emergence of a polymorphic hierarchy became quickly evident, while the assembly of adapted query-instances through combinations of textual patterns was easily crystallized as an appropriate implementation technique.

Overall, the described decision-making language is suited to address the development of systems that have to accommodate diversity on the fly, through dynamically decided and delivered software artifacts. In such cases, run-time decision making relying upon diversity-parameters becomes mandatory, while the software organization of the system's implementation around architectural polymorphism is considered to be a fundamental architectural property. Currently, the DMSL language is used in the context of the ASK-IT[2] and OASIS[3] projects to provide user

---

[2] ASK-IT IST-511298 (Ambient Intelligence System of Agents for Knowledge-Based and Integrated Services for Mobility Impaired Users; http://www.ask-it.org).

[3] OASIS IST-215754 (Open Architecture for Accessible Services Integration and Standardization).

interface dynamic adaptation on a range of diverse interaction devices, including PDAs and mobile phones.

22.6 Appendix

22.6.1 DMSL Grammar for Interface Adaptation

```
logic ::= { block | stereotype }
block ::= `dialogue' (string | name) compound
compound ::= `[` { stmt } `]'
stmt ::=      ifst | (`activate' | `cancel' |
`evaluate') expr `;' | compound
ifst ::= `if' `(' boolexpr `)' `then' stmt
[ `else' stmt ]
expr := primary | boolexpr | arithexpr
primary ::=  const | param | funccall | `-`
expr | `not' expr | name
param ::= (`user' | `context') `.' (string |
name)
funccall ::= libfunc `(' [ expr { `,' expr }
] `)'
const ::= `true' | `false' | number | string
| name
boolexpr ::= expr boolop expr | expr `in' set
arithexpr ::= expr arithop expr
arithop ::= `+' | `-' | `*' | `/' | `%'
boolop ::= `or' | `and' | `<' | `>' | `<=' |
`>=' | `=' | `!='
set ::= `{` [ expr {`,', expr } ] `}'
libfunc ::= `isactive' | `tonumber' |
`hasattr'
stereotype ::= `stereotype' (`user' |
`context' ) name `:' boolexpr `;'
```

22.6.2 DMSL Grammar Modifications for Software Adaptation

```
block ::= `component' (string | name)
compound
param ::= `params' `.' (string | name)
stereotype ::= `stereotype' name `:'
boolexpr `;'
```

Acknowledgments

Part of the reported work has been carried out in the context of the AVANTI AC042 (Adaptable and Adaptive Interaction in Multimedia Telecommunications Applications) project, was partially funded by the ACTS Program of the European Commission, and lasted 36 months (1995–1998). The partners of the AVANTI consortium are ALCATEL Italia, Siette Division (Italy), Prime Contractor; IROE-CNR (Italy); ICS-FORTH (Greece); GMD (Germany); VTT (Finland); University of Siena (Italy); MA Systems & Control (United Kingdom); ECG (Italy); MATHEMA (Italy); University of Linz (Austria); EUROGICIEL (France); TELECOM (Italy); TECO (Italy); and ADR Study (Italy).

Part of this work was also carried out in the context of PALIO IST-1999-20656 (Personalized Access to Local Information and Services for Tourists), was partially funded by the IST Program of the European Commission, and also lasted 36 months (2000–2003). The partners of the PALIO consortium are ASSIOMA S.p.A. (Italy), Prime Contractor; CNR-IROE (Italy); Comune di Firenze (Italy); ICS-FORTH (Greece); GMD (Germany); Telecom Italia Mobile S.p.A. (Italy); University of Sienna (Italy); Comune di Siena (Italy); MA Systems & Control (United Kingdom); and FORTHnet S.A. (Greece).

References

Antona, M., Savidis, A., and Stephanidis, C. (2006). A process-oriented interactive design environment for automatic user interface adaptation. *International Journal of Human Computer Interaction* 20: 79–116.

Benyon, D. (1984). MONITOR: A self-adaptive user-interface, in the *Proceedings of IFIP Conference on Human-Computer Interaction: Volume 1, INTERACT '84*, 4–7 September 1984, London, pp. 335–341. Amsterdam, The Netherlands: Elsevier Science.

Cockton, G. (1993). Spaces and distances: Software architecture and abstraction and their relation to adaptation, in *Adaptive User Interfaces: Principles and Practice* (M. Schneider-Hufschmidt, T. Kühme, and U. Malinowski, eds.), pp. 79–108. Amsterdam, The Netherlands: Elsevier Science.

Cote Muñoz, J. (1993). AIDA: An adaptive system for interactive drafting and CAD applications, in *Adaptive User Interfaces: Principles and Practice* (M. Schneider-Hufschmidt, T. Kühme, and U. Malinowski, eds.), pp. 225–240. Amsterdam, The Netherlands: Elsevier Science.

Fayad, M. and Cline, M. (1996). Aspects of software adaptability. *CACM* 39: 58–59.

Kobsa, A. and Pohl, W. (1995). The user modeling shell system BGP-MS, in *User Modeling and User-Adapted Interaction* 4: 5–106.http://www.ics.uci.edu/~kobsa/papers/1995-UMUAI-kobsa.pdf. Program code: ftp://ftp.gmd.de/gmd/bgp-ms.

Netinant, P., Constantinides, C. A., Elrad, T., and Fayad, M. E. (2000). Supporting aspectual decomposition in the design of adaptable operating systems using aspect-oriented frameworks, in the *Proceedings of 3rd Workshop on Object-Orientation and Operating Systems ECOOP-OOOWS*, June 2000, Sophia Antipolis, France, pp. 36–46.

Savidis, A. and Stephanidis, C. (2001). The unified user interface software architecture, in *User Interfaces for All: Concepts, Methods, and Tools* (C. Stephanidis, ed.), pp. 389–415. Mahwah, NJ: Lawrence Erlbaum Associates.

Savidis, A. and Stephanidis, C. (2004a). Unified user interface design: Designing universally accessible interactions. *International Journal of Interacting with Computers* 16: 243–270. http://authors.elsevier.com/sd/article/S0953543804000025.

Savidis, A. and Stephanidis, C. (2004b). Unified user interface development: Software engineering of universally accessible interactions. *Universal Access in the Information Society* 3: 165–193.

Stephanidis, C., Antona, M., and Savidis, A. (2004a). Design for all: Computer assisted design of user interface adaptation, in *Handbook of Human Factors and Ergonomics* (3rd ed.) (G. Salvendy, ed.), pp. 1459–1484. New York: John Wiley & Sons.

Stephanidis, C., Paramythis, A., Sfyrakis, M., and Savidis, A. (2001). A case study in unified user interface development: The AVANTI web browser, in *User Interfaces for All: Concepts, Methods, and Tools* (C. Stephanidis, ed.), pp. 525–568. Mahwah, NJ: Lawrence Erlbaum Associates.

Stephanidis, C., Paramythis, A., Zarikas, V., and Savidis, A. (2004b). The PALIO framework for adaptive information services, in *Multiple User Interfaces: Cross-Platform Applications and Context-Aware Interfaces* (A. Seffah and H. Javahery, eds.), pp. 69–92. Chichester, U.K.: John Wiley & Sons.

Methods and Tools for the Development of Unified Web-Based User Interfaces

Constantina Doulgeraki,
Nikolaos Partarakis,
Alexandros Mourouzis, and
Constantine Stephanidis

23.1 Introduction

Recently, computer-based products have become associated with a great amount of daily user activities, such as work, communication, education, entertainment, and so on. Their target population has changed dramatically. Users are no longer only traditional able-bodied, skilled, and computer-literate professionals. Instead, users are potentially all citizens. At the same time, the type and context of use of information and services is radically changing due to the increasing availability of novel access platforms, such as personal digital assistants, kiosks, cellular phones, and other network-attachable equipment, which progressively enable nomadic access to information and services (Stephanidis, 2001). These changes signal the emergence of the information society, the success of which relies to a significant degree on the opportunities offered to all citizens to obtain and maintain access to a society-wide pool of information resources and communication facilities, given the variety of context. To this end, *accessibility* in its broader sense (as the "ability to access" the functionality, and possible benefit, of some system, irrespective of where and how it runs) is considered to be of paramount importance.

Recent approaches to accessibility entail a purposeful effort to embed accessibility into products by taking appropriate actions

a priori, as early as possible throughout the design and development life cycle and, ultimately, to deliver products that can be tailored for use by the widest possible end-user population (see also Chapter 2 of this handbook).

In the case of web accessibility, the prevalent approach among modern web-based information and communication systems is to deliver a single user interface design that meets the requirements of the average user. However, the average user is a nonexisting imaginary individual, and often the profiles of a large portion of the population, especially that of people with disability, elderly people, novice users, and users on the move, differ radically. Recent efforts towards web accessibility involve the development and use of related design guidelines (see also Chapter 10 of this handbook). The use of guidelines (and evaluation tools) is today the most widely adopted approach to creating accessible web content. Yet, the overall penetration level of this approach into mainstream developments remains unsatisfying. This is mainly due to the fact that developers and industries wishing to deliver products that align with web content accessibility guidelines must first invest in special training and modify their work processes.

This chapter presents a novel approach to the development of inclusive web-based interfaces by introducing *adaption* as an inherent characteristic of web-based user interfaces.

23.1.1 Web-Based User Interfaces: Opportunities and Challenges

Web-based user interfaces (WUIs) constitute a particular type of UI that accepts input and provides output by generating web pages that are transported via the Internet and viewed by the user using a web browser program. Despite the universality of the web and the predominant role of WUIs in the evolving information society, current approaches to web design hardly embrace the principles of *design for all* and the notion of adaptation. In fact, only very recently, few web applications started providing some sort of adaptation to their users. Indicatively, the *iGoogle*[1] web site and the *Microsoft Sharepoint portal server*[2] offer to their users the ability to customize the UI, such as repositioning, minimizing and maximizing web page elements, defining the number of search results to be displayed, and personalizing the color settings. Clearly, although some of these features are well appreciated by some users, they cannot alone support the delivery of *universal access*, regardless of the user's (dis)abilities, skills, preferences, and context of use. For instance, the particular ways in which most of these features are implemented render the produced web pages completely inaccessible for blind individuals that use screen readers.

Consequently, today the main challenge for WUI designers and researchers is to come up with approaches that can meet diverse user needs and requirements in various contexts. This is a difficult and demanding task for web developers, and needs to be supported through appropriate methods and tools.

23.1.2 Toward Unified Web-Based Interfaces

This chapter proposes a novel approach, based on *web portal technologies* and UI adaptation techniques, to embed personalized accessibility features deep into web design. A portal typically includes a number of facilities for access and navigation of information, socialization, collaboration, and trade. It can be defined as a web site acting as a gateway on the way to access a multitude of online destinations that are somehow related and that can be referred to as the contents of the portal. For example: (1) web pages, web sites, and other web-based applications (including other portals); (2) people connected on the web[3]; and (3) digital resources.[4] such as documents, multimedia, software, and so on. Typically, portals include also various *portlets*, which are reusable web components that display relevant information to portal users, such as news, e-mail, weather information, and discussion forums.

An adaptation of the unified user interfaces methodology (see Chapters 16 and 21 of this handbook) is proposed here for developing web portals following a design-for-all approach and

for ensuring the delivery of *unified web-based interfaces*[5] (UWIs) that support adaptation to diverse user needs, (accessibility) requirements, and contexts of use.

To further facilitate web developers in following the proposed method in practice, an advanced developer's toolkit for UWI, called *EAGER*,[6] has been developed. By means of EAGER, a developer can produce web portals that have the ability to adapt to the interaction modalities, metaphors, and UI elements most appropriate to each individual user, according to profile information containing user and context-specific parameters.

23.2 Related Work

Previous related work mainly focuses on three complementary research directions: *web accessibility*, *web usability*, and *adaptation techniques for the web*.

23.2.1 Web Accessibility

Accessibility is traditionally associated with disabled and elderly people and reflects the efforts devoted to the task of meeting prescribed code requirements for use by people with disabilities (Bergman and Johnson, 1995; Story, 1998). Due to recent technological developments (e.g., proliferation of interaction platforms, such as wireless computing, wearable equipment, user terminals), the range of the population that may gradually be confronted with accessibility problems extends beyond the population of disabled and elderly users.

Accessibility is subject, among others, to the technological platforms through which information is accessed. In many cases, *assistive technology* (AT) products that were proved to be widely useful were also integrated into mainstream operating systems. *Alternative access-based systems* is used to define systems that support additional input and output devices and provide alternative interaction metaphors in terms of the operating system or the graphic environment. These systems are widely used, in conjunction with AT, by the majority of disabled users for accessing the web. Most modern operating systems support a number of services of this category to make possible at least the initiation of interaction with these systems. For example, Microsoft Windows Vista,[7,8] Microsoft Windows XP,[9,10] Linux (De La Rue

---

[1] http://www.google.com/ig.

[2] http://sharepoint.microsoft.com/sharepoint.

[3] In these cases we usually use the notion *community portals*.

[4] In these cases we usually use the notion *digital library portals*.

[5] The notion of unified web-based interfaces is used here to refer to web-based UIs that comply with the *unified user interfaces* (U²I) methodology (Stephanidis, 2001).

[6] EAGER: "toolkit for embedding accessibility, graceful transformation, and ease of use in web-based products."

[7] Accessibility tutorials for Windows Vista: http://www.microsoft.com/enable/products/windowsvista/default.aspx.

[8] Accessibility in Windows Vista: http://www.microsoft.com/enable/products/windowsvista/default.aspx.

[9] Windows XP accessibility resources: http://www.microsoft.com/enable/products/windowsxp/default.aspx.

[10] Windows XP accessibility tutorials: http://www.microsoft.com/enable/training/windowsxp/default.aspx.

2002) and *MacOs X*[11] include aids for visual impairments, such as screen magnifiers and screen readers-narrators (see Chapter 28 of this handbook). Additionally, some operating systems support adjusting the screen to white on black or grayscale for improving the contrast of the screen. Targeting motor impairments, modern operating systems support *onscreen keyboards* (see Chapter 29 of this handbook), *sticky keys, key commands,* and *speech recognition* (see Chapter 30 of this handbook). Finally, in the case of hearing impairments, operating systems support flashing the screen instead of playing an audio system beep, as well as quick-access key commands to adjust the system volume.

Services offered by modern operating systems enable users with a disability to enter the world of computing, by allowing initiating interaction with a computer system. However, web accessibility involves totally different issues. Often, facilities offered by operating systems and the incorporation of additional input or output devices cannot cope with the barriers set by the modern web, due to the diversity and complexity of the involved technologies and approaches. For example, some web applications use client-side scripting to such an extent that they are rendered inaccessible, even with the help of AT in conjunction with the aforementioned operating system features. It is therefore clear that the accessibility options provided by modern operating systems are valuable for enabling the initiation of interaction with a computer system, but cannot cope with the numerous barriers set for people with disabilities by modern web applications (see Chapter 10, for an overview of the main problems encountered by users with various disabilities in interacting with the web).

23.2.1.1 User Agents (Web Browsers)

Popular *user agents* (web browsers), such as Internet Explorer, Netscape Navigator, Firefox and Opera encapsulate several accessibility settings and features. These include support on zooming in on a web page to magnify text, images, and controls on the page. Additionally, users may choose colors, text size, and font style used on web pages to make web pages easier to see by changing the text, background, link, and hover colors. Advanced accessibility settings include expanding alternative text for images, moving system caret with focus/selection changes, resetting text size to medium for new windows and tabs. Users may also turn off or turn on pictures, animations, and videos, music, and other sounds included in web pages. Some web pages provide a rich interactive experience with Java applets. However, users who rely on keyboard navigation may experience problems with some Java applets that automatically set focus, and do not provide a way to "break out" of the applet and navigate to the rest of the web page. Modern browsers support particular setting to disable Java and JavaScript. Finally, some browsers include keyboard and mouse shortcuts to reduce the need for special users to reach mouse or keyboard.

However, some web sites are designed to defeat browser settings, and therefore people with disability may encounter several difficulties in accessing the web even when a web browser's accessibility settings are properly configured. As an example, a web browser's text resizing feature does not work well in web sites that do not meet accessibility guidelines and define text size using hard-coded or absolute sizes.

23.2.1.2 Web Content Accessibility: Evaluation Tools

As mentioned in the previous section, AT or other alternative access systems alone are not in the position to ensure accessibility to web-based products and services. Web content needs to be implemented in appropriate ways that allow graceful transformation of the content in alternative modalities. A number of guidelines collections for web accessibility have been developed (Gunderson, 1996; Richards, 1996; Vanderheiden et al., 1996; World Wide Web Consortium, 1999). The *Web Content Accessibility Guidelines* (WCAG) (World Wide Web Consortium, 1999) developed by the World Wide Web Consortium (W3C) provide recommendations for making web content accessible to people with disability. In general, for a web site to comply with accessibility standards, it should have at least the following characteristics:

- Valid (X)HTML
- Valid CSS for pages layout
- At least WAI-A (preferably AAA) compliance with the Web Accessibility Initiative's (WAI's) WCAG
- Compliance with Section 508 of the U.S. Rehabilitation Act (see Chapter 53 of this handbook)

The process of using, or testing conformance to, widely accepted accessibility guidelines is time consuming and requires additional developer expertise in accessibility. To address this issue, several tools have been developed enabling the semiautomatic checking of HTML documents. Such tools make the development of accessible web content easier, especially due to the fact that the checking of conformance does not rely solely on the expertise of developers.

The usage of guidelines and tools is today the most widely adopted process by web authors for creating accessible web content (see Chapter 20 of this handbook). This approach has proven valuable for bridging a number of barriers faced today by people with disabilities. Unfortunately, many limitations arise due to a number of reasons. Guidelines themselves pose several issues regarding their ease of use and adoption by developers. On the other hand, accessibility tools as application or online services interpret each and every accessibility guideline literally, without having the ability to check it as part of the page. For example,[12] one of the W3C accessibility guidelines states that a table must include a hidden summary to be read aloud to screen-reader users before reading through the table content. However, there may be a heading directly before the table to describe what the

---

[11] MacOs X user's manual: http://www.apple.com/macosx/features/universalaccess.

[12] http://www.webcredible.co.uk/user-friendly-resources/web-accessibility/automated-tools.shtml.

TABLE 23.1   Web Usability Heuristics

Web Usability Heuristics	Description
Visibility of system status	Each page has to be branded with the section it belongs to, and links to other pages should be clearly marked.
Match between system and the real world	Language used has to be simple and comprehensive to serve people from diverse backgrounds.
User control and freedom	Site has to provide several controls to users to assist customizations of the site and navigation in it.
Consistency and standards	Content and links wording has to be used consistently to avoid user confusion. Web "standards" HTML specifications, accessibility guidelines, etc. have to be followed.
Error prevention	User input has to be checked before submitting to prevent errors, but has to be double checked after submission.
Recognition rather than recall	Labels and descriptive links have to be used to inform users about where they are, by looking on current page.
Flexibility and efficiency of use	Pages have to be easily bookmarked and offer intelligent bookmarking to specific corner of the site.
Aesthetic and minimalist design	Content of web pages has to be separated in different levels of detail and provide alternative ways to access them. Content has to be also separated to relevant content chunks and provide access to them by different links.
Help users recognize, diagnose, and recover from errors	For every error message a solution (or a link to a solution) has to be provided on the error page.
Help and documentation	Help and documentation has to be integrated in the site. Help has to be context-sensitive for each page of the site.

Source: From Dougelraki, C., Partarakis, N., Mourouzis, A., and Stephanidis, C., A development toolkit for unified web-based user interfaces, in the *Proceedings of Computers Helping People with Special Needs: 11th International Conference (ICCHP 2008)*, 10–15 March 2008, Linz, Austria, pp. 346–353, Springer-Verlag, Berlin/Heidelberg, 2008.

table is about, so the hidden summary will just repeat what the previous heading said.

Accessibility tools cannot check the web site's content structure; thus, a web site may be perfectly coded and conform to the highest coding standards by an accessibility tool, but still be inaccessible for some web users if its content is poorly structured. Another example concerns the use of alternative text. When a web site includes alternative text for all the images, an accessibility tool will report a pass for this guideline, but it cannot check if the alternative texts provide an appropriate description of the images. Accessibility tools, in addition to errors, also identify warnings, which are basically guidelines that cannot be checked automatically, and have to be checked manually. As a result, automated accessibility tools can be useful and save a large amount of time, but their use needs to be carefully integrated in a wider accessibility evaluation methodology.

23.2.1.3 Development Environments and Tools

Commercially available frameworks for developing web applications so far have not paid particular attention to incorporating the aforementioned guidelines into their products. Most frameworks rely only on producing XHTML valid markup, and not on making guidelines a part of the logic used for producing the output. Additionally, the trend of making web applications more client dependent, and therefore decreasing the server side load, has produced solutions that largely rely on client-side scripting, thus worsening interaction for disabled users. The recent web developer's tool *Visual Studio 2005*[13] includes some controls that can be used for accessibility compliance, such as skip navigation functions and images that support an alternative text and an extended description. Additionally, this developer's tool encapsulates an accessibility validation program that checks if the code generated by the pages is valid XHTML and adheres to the WAI accessibility standards of W3C.

23.2.2 Web Usability

Usability is now recognized as an important software quality attribute, and plays a key role to make the web friendly to all target users. General usability principles are applied to the web by means of usability heuristics proposed by experts.

Generally, there are several resources that include usability principles and guidelines in detail, but are interspersed in books, on the web, and in papers. Additionally, usability is a subjective issue; therefore, to ensure a web site's usability, after applying several usability principles, a usability testing process has to be applied too with real users. Web usability evaluation is an essential element of quality assurance, and a true test of how people actually use a web site. An example of the most established usability heuristics for the web can be found in Instone (2000), building on 10 general usability heuristics defined in Nielsen (1993), and are presented in Table 23.1.

23.2.3 Adaptation Techniques for the Web

23.2.3.1 Alternative User Agents

To ensure seamless access to the World Wide Web, several approaches involved the development of special-purpose user agents (web browsers). The *AVANTI Web Browser* (Stephanidis et al., 2001; see also Chapter 2, "Perspectives on Accessibility: From Assistive Technologies to Universal Access and Design for All") facilitates static and dynamic adaptations to adapt to the skills, desires, and needs of each user, including people with visual and motor disabilities. *WebAdapter* (Hermsdorf, 1998) is a web agent that provides accessibility functionalities for blind, visual, and physically impaired people. *pwWebSpeak32*[14] is a commercially available web agent designed and developed by SoundsLink for users who wish to access the Internet in a nonvisual or combined auditory and visual way. Another special-purpose agent is proposed in Henricksen and Indulska (2001), including

---

[13] http://www.it-analysis.com/content.php?articleid=13021.

[14] http://www.soundlinks.com/pwgen.htm.

sophisticated adaptation mechanisms to provide context-aware behavior and user interfaces.

In general, special-purpose agents offer a very promising approach to web accessibility and usability. Special-purpose agents can deliver a number of facilities that include alternative interaction modalities and UI elements, support for a number of assistive technologies and input/output devices, as well as text-to-speech and speech-recognition facilities. These agents offer the advantages of desktop-based processing together with the positive features of intermediary frameworks, acting themselves as a proxy between the web and the user. However, these approaches are limited by the fact that the user must have the actual product installed on the computer used to gain access to the web. Therefore, these facilities must either be presented as commercial products or be embedded in existing mainstream web browsers for the aforementioned benefits to reach their actual beneficiaries.

23.2.3.2 Intermediary Agents

Intermediary agents can be considered as filtering and transformation tools that build alternative versions of web pages based on disability category, user preferences, or heuristic rules. *WebFace* (Alexandraki et al., 2004) constitutes a representative intermediate agent supporting accessibility with respect to physical or perceptual disabilities, or combinations thereof. *Web Adaption Technology*[15] developed by IBM Research proposes a method of making web pages accessible without requiring the use of assistive technologies. Several other intermediary frameworks are specifically designed for people with vision impairments, and focus on removing sticky user interface elements or on transforming them to exploitable elements by assistive technologies. More specifically, these frameworks transform a web page from a graphics-heavy and inaccessible version to a text-only version that is easily accessible by visually impaired users. Some of the most well-known systems in this category are the *Personalizable Accessible Navigation* (Iaccarino et al., 2006), the *Access Gateway* system (Brown and Robinson, 2001), the *Textualise* system,[16] the *Accessibility Transformation Gateway* (Parmanto et al., 2005), the *IBM system* described in Han et al. (1998), *BETSIE*,[17] *Crunch* (Gupta and Kaiser, 2005; Gupta et al., 2005), *Muffin*[18] and *RabbIT*[19]. *Web-Based Intermediaries* (WBI) (Barrett and Maglio, 1998, 1999a, and 1999b) is a special dynamic framework that includes a text-to-speech service (Barra et al., 2003) supporting the speaking of the text of HTML pages during their displaying to end-users. *Web Page Transformation* (Chen et al., 2005) is another framework that offers a web page transformation algorithm to browse web pages on small devices.

In conclusion, the concept of intermediary agents is considered a very promising approach for enabling universal accessibility on the web. However, practical experience has highlighted

a number of issues that tend to reduce the universality of the approach. Many web sites do not produce HTML code for each element appearing in a web page. There are examples of web sites where the rendered code is entirely in a client scripting language. Additionally, malformed HTML code tends to make these web sites unreadable by proxies (due to the issues arising in the process of parsing malformed documents). An ad hoc solution to this issue was the development of specialized filters for each web site or portal parsed by an intermediary agent. However, the current situation of the web is far from allowing a general solution to this problem. Finally, the diversity of the target user population cannot be addressed by just "fixing" the rendered HTML output. Sometimes there is a need to perform additional operations, such as replacement of interaction elements and modalities, to cope with a diverse target user population. This is clearly not supported in the intermediary agents approach.

23.2.3.3 Self-Adapting Web-Based Systems

In addition to intermediary agents, there are web systems that encapsulate adaptation. *E-Victor* (Graef, 2000) is an eCommerce system that has been developed as a component-based application. Components are grouped into services that offer functionality to application users (user services) or to other services (internal service). The E-Victor system supports alternative navigation, layout, and user interaction components using the WebComposition Markup Language. Although this approach has important results, the use of a specific markup language reduces the advantage to be used as a general solution for web development. Additionally, this approach focuses on a specific system, and does not present a generic framework to be potentially applied for the majority of web sites or sites developed with a specific technology. Finally, E-Victor does not support accessibility adaptations.

In Taib and Ruiz (2006), a finite state machine algorithm was proposed, which gathers information on the clicking styles of the user and categorizes them into predefined profiles during progress through the task. A default profile gets weighted by the modality used at each step of the navigation. Finally, the system applies predefined presentation templates for every step of the process, progressively adapting to the interaction style of the user. A visual profile (using mainly images) would receive a shorter description with many images, while a text profile would get only one image and a longer text. A multimedia profile would get a video description, a speech profile, a spoken description, and an iconic profile of a digest style, including bullet points and iconic information. This approach uses statistical values to choose among predefined patterns of interaction styles that fit to specific user interaction styles. Although this algorithm can derive conclusions about user presentation preferences (e.g., image style, text style) it cannot infer interaction preferences. Additionally, this method supports only alternative information representation, and does not offer the means to be extended to support alternative UI elements, dialogue controls, and so on.

---

[15] http://www.webadapt.org.

[16] http://aquinas.venus.co.uk.

[17] http://www.bbc.co.uk/education/betsie.

[18] http://muffin.doit.org.

[19] http://rabbit-proxy.sourceforge.net.

23.3 A Framework for Embedding Adaptation in Web User Interfaces

Until recently, adaptive techniques have had limited impact on the issue of universal access. The unified user interfaces development methodology (Savidis and Stephanidis, 2004a; Stephanidis, 2001; see also Chapter 16, "Unified Design for User Interface Adaptation," and Chapter 21, "A Unified Software Architecture for User Interface Adaptation") has been proposed as a complete technological solution for supporting universal access of interactive applications and services. A unified user interface has been defined as comprising a single (unified) interface specification that exhibits the following properties (Savidis and Stephanidis, 2004b):

1. It embeds representation schemes for user and usage-context parameters and accesses user and usage-context information resources (e.g., repositories, servers) to extract or update such information.

2. It is equipped with alternative implemented dialogue patterns (i.e., implemented dialogue artifacts) appropriately associated with different combinations of values for user- and usage-context-related parameters. The need for such alternative dialogue patterns is identified during the design process, when, given a particular design context, for differing user and usage-context attribute values, alternative design artifacts are deemed necessary to accomplish optimal interaction.

3. It embeds design logic and decision-making capabilities that support activating, at run-time, the most appropriate dialogue patterns according to particular instances of user and usage-context parameters, and is capable of interaction monitoring to detect changes in parameters.

As a consequence, a unified interface realizes:

- User-adapted behavior (user awareness), for instance, the interface is capable of automatically selecting interaction patterns appropriate to the particular user.
- Usage context-adapted behavior (usage context awareness), for instance, the interface is capable of automatically selecting interaction patterns appropriate to the particular physical and technological environment

From a user perspective, a unified user interface can be considered as an interface tailored to personal attributes and to the particular context of use, while from the designer perspective it can be seen as an interface design populated with alternative designs, each alternative addressing specific user and usage-context parameter values. Finally, in an engineering perspective, a unified user interface is a repository of implemented dialogue artifacts, from which the most appropriate according to the specific task context are selected at run-time by means of an adaptation logic supporting decision making (Savidis and Stephanidis, 2004b).

At run-time, the adaptations may be of two types: (1) *adaptability* (i.e., initial automatic adaptation driven by user and context information known prior to the initiation of interaction); and (2) *adaptivity* (i.e., continuous automatic adaptation driven by information acquired through interaction monitoring).

Adaptability is crucial to ensure accessibility, since it is essential to provide, before initiation of interaction, a fully accessible interface instance to each individual end-user and context of use (Stephanidis, 2001a). Adaptivity can be applied only on accessible running interface instances (i.e., ones with which the user is capable of performing interaction), since interaction monitoring is required for the identification of changing/emerging decision parameters that may drive dynamic interface enhancements, and reflects the interface's capability to cope with the dynamically changing or evolving user and context characteristics.

As discussed in Chapter 21, "A Unified Software Architecture for User Interface Adaptation," the concept of unified user interfaces is supported by a specific architecture, depicted in Figure 23.1.

In this chapter, the unified user interfaces method and architecture are appropriately adapted for the web. This adapted approach is referred to as UWI.

23.3.1 Unified Web-Based Interfaces

23.3.1.1 Adaptive and Adaptable Behavior

WUI adaptations must take into account a wider collection of parameters, such as context and user-specific attributes (e.g., input/output devices, disabilities, user attitude toward technology, etc.). The adaptation of an application can occur in different

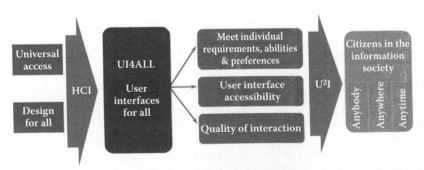

FIGURE 23.1 Overview of the unified user interfaces architecture. (From Doulgeraki, C., Partarakis, N., Mourouzis, A., and Stephanidis, C., A development toolkit for unified web-based user interfaces, in the *Proceedings of Computers Helping People with Special Needs: 11th International Conference (ICCHP 2008)*, 10–15 March 2008, Linz, Austria, pp. 346–353, Springer-Verlag, Berlin/Heidelberg, 2008.)

ways and can cover a number of aspects of the application or its environment

The adaptation of an application can be classified according to which aspects of the application are adapted. Generally, the web application model described in Gaedke et al. (1999) distinguishes five orthogonal aspects of an application:

1. *Content*: Adaptation of content affects the content such as text, graphics, or any other media type or data used or displayed by the application. This type of adaptation is most common on the web. UWI supports adaptations that automatically modify the presentation and conceived behavioral attributes of interactive elements. As an example, images can be transformed as normal images, as simple text containing the image's alternative text, and as a hyperlink that downloads the image and has as text the image's alternative text.
2. *Navigation*: Adaptation of navigation adapts the navigational structure of a web application hiding or modifying links. UWI supports navigation adaptations. Some examples include the linearization of the whole navigation of the portal in a top navigation bar to facilitate blind users, and step-by-step navigation that reduces the number of links that a motor-impaired user has to scan.
3. *Layout*: Adaptation of layout changes the way information is presented to a user visually. This can be done to accommodate different types of displays or to satisfy preferences of aesthetic, cultural, and so on, a user may have. The proposed framework supports layout adaptations, as it offers alternative template layouts depending on screen resolution, devices, disability, and so on.
4. *User interaction*: Adaptation of user interaction changes the way the user interacts with the application. An application might adapt, for example, offering a wizard-based interface to less experienced users and a single page form to other users. The UWI framework supports conditional activation and deactivation of multiple interaction modalities based on the user profile, including alternative task structures, alternative syntactic paradigms, task simplification and adaptable and adaptive help facilities run-time task guidance.
5. *Processing*: Adaptation of processing changes the way user input is processed. For example, a product request of a person that has placed many large orders in the past might be processed differently from that of a previously unknown person. A UWI framework doesn't support such adaptations. These kinds of adaptations cannot be addressed by a generic framework, and rely solely on the implementation of each web application.

On the other hand, the adaptation of an application can also be classified, according to how it takes place, into the following three distinct categories:

- Manual adaptation, which can take place during the development cycle of an application where a developer can modify the application to meet different user requirements or run on a different platform. It can also take place through manual reconfiguration of an application. The UWI framework supports this kind of adaptation by providing users with a mechanism to extend their generic profile by selecting among a number of specific interaction and accessibility settings. Therefore, each user selects an individual adjustment of the web application.
- Automated adaptation, which does not require any effort on the part of a user, administrator, or developer. Instead, it is performed automatically by the application or the framework the application runs in, and can take place at regular time intervals or can be event driven. The proposed framework supports this kind of adaptation. For example, based on statistic values, the system can identify situations where a user cannot complete a task, and activates a help provision component.
- Semiautomated adaptation, which requires the user or another person to provide certain information through questionnaires, feedback buttons, and so on, for the application to be adapted. The information obtained does not directly describe how the application should be adapted, but is used by the system to determine a suitable application configuration. UWI supports this kind of adaptation by providing users with a generic profile where general information such as disability and language is stored. It also supports decision logic to decide on adjustments of the web application based on the general information.

The next sections provide a detailed discussion of the proposed architecture for designing and implementing UWI.

23.3.1.2 The UWI Methodology

The UWI methodology is derived from the aforementioned architectural structure for enabling the development of unified user interfaces. Figure 23.2 presents a general overview of the UWI architecture and the engaged communication channels. The basic components involved are as follows:

- The *user information component*, responsible for collecting and propagating user specific attributes.
- The *context information component*, responsible for collecting and propagating attributes varying by the context of use.
- The *decision-making component*, in charge of the overall decision making regarding the conditional activation/deactivation of interaction elements.
- The *designs repository*, a repository of the alternative interaction styles design to be implemented in the Dialogue Controls Component.
- The *dialogue controls component*, a repository of alternative interaction styles to be conditionally activated or deactivated to form the final user interface.

The engaged communications channels are used for propagating user- and context-specific parameters from the user and context information component to the decision-making component. On the other hand, the decision-making component is

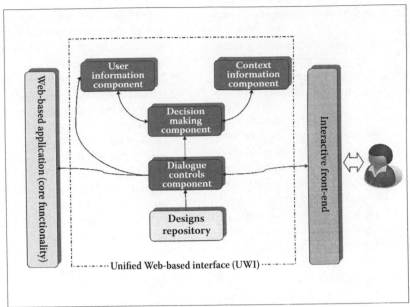

FIGURE 23.2 Architecture of UWIs. (From Doulgeraki, C., Partarakis, N., Mourouzis, A., and Stephanidis, C., A development toolkit for unified web-based user interfaces, in the *Proceedings of Computers Helping People with Special Needs: 11th International Conference (ICCHP 2008)*, 10–15 March 2008, Linz, Austria, pp. 346–353, Springer-Verlag, Berlin/Heidelberg, 2008.)

responsible for propagating its decisions to the dialogue controls component to inform the activation/deactivation of interaction elements. The aforementioned communication channels are bidirectional for the decision-making component to be able to propagate decisions that may result in changes on the user- or context-specific parameters. The overall process results in the activation of the appropriate interaction elements to be used for rendering the final interactive front-end.

Each part of the presented architecture is described in detail in the following sections to clarify its role in the adaptation process, as well as its internal behavior.

23.3.1.2.1 User Information Component

The scope of the user information component (UIC) is to provide information regarding user profiles. A user profile initially contains attributes specified by the user prior to the initiation of interaction or during interaction, and may be modified during interaction if relevant changes are monitored.

Figure 23.3 presents the detailed UIC internal architecture. The user profiles repository is a database containing all the user-specific parameters. On the other hand, the user components repository collects information regarding the conditional activation/deactivation of interactive elements per user as propagated by the decision-making component. The interaction-monitoring module provides appropriate mechanisms to monitor and record user interactions and propagate the related data to the user profiles repository. Such data include the successful or unsuccessful completion of actions, the popularity of navigation options, and so on.

The core element of the UIC is the profiling module, which is responsible for propagating user profile information to the

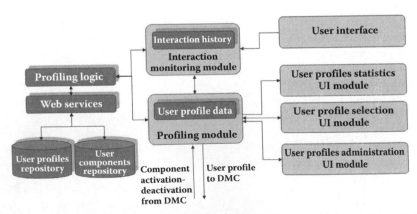

FIGURE 23.3 User information component architectural model. (From Doulgeraki, C., Partarakis, N., Mourouzis, A., and Stephanidis, C., A development toolkit for unified web-based user interfaces, in the *Proceedings of Computers Helping People with Special Needs: 11th International Conference (ICCHP 2008)*, 10–15 March 2008, Linz, Austria, pp. 346–353, Springer-Verlag, Berlin/Heidelberg, 2008.)

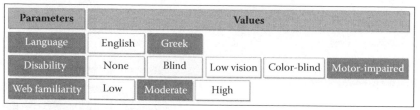

FIGURE 23.4 Example of user profile instance. (From Doulgeraki, C., Partarakis, N., Mourouzis, A., and Stephanidis, C., A development toolkit for unified web-based user interfaces, in the *Proceedings of Computers Helping People with Special Needs: 11th International Conference (ICCHP 2008)*, 10–15 March 2008, Linz, Austria, pp. 346–353, Springer-Verlag, Berlin/Heidelberg, 2008.)

decision-making component (DMC), and additionally acts as a front end providing all the required functionality to be used by the specialized profiling UI modules.

These modules include:

- The *user profiles statistics UI module*, which presents statistics based on the data collected from user selections regarding profiles, accessibility, and interaction preferences.
- The *user profile selection UI module*, which supports selection among predefined user profiles and manual configuration of the available settings.
- The *user profiles administration UI module* allows the administration of the predefined profiles.

Figure 23.4 depicts an example of the attribute-value-based user profile model. Similar considerations hold for the context information component presented in the next section.

23.3.1.2.2 Context Information Component

The context information server of the derived architecture was initially intended to collect and propagate context attribute values (machine and environment) of two types:

1. (Potentially) invariant, meaning unlikely to change during interaction (e.g., peripheral equipment)
2. Variant, dynamically changing during interaction (e.g., due to environment noise, the failure of particular equipment, etc.)

This component was therefore intended to support device awareness. In the context of web application, the attributes to be supported dynamically by this component are radically decreased, due to the lack of methods for capturing and propagating such information from the client side.

Despite the aforementioned limitations, the role of this module in the extended architecture remains significant, and its internal architecture is presented in Figure 23.5. As shown in this figure, the CIC shares functionality with the UIC. The profiling module is also used to collect and propagate context-specific parameters to different subsystems. The context-monitoring module has the responsibility to monitor context changes and propagate this information to the user-profiling module. This module in turn enriches the user-context profile repository with these context-specific attributes to be used in the process of decision making.

In case dynamic user attribute detection is desired, the content may include dynamically collected interaction-monitoring information, design information, and knowledge-processing components.

23.3.1.2.3 Decision-Making Component

The DMC (Figure 23.6) makes decisions based on the conditional activation and deactivation of UI components and on the user attributes provided by the UIC and the repository of alternative dialogue patterns. The core of this component consists of a number of if-then-else rules representing the design space of the user interface. More specifically, this module performs the overall decision making of when, why, and how adaptation

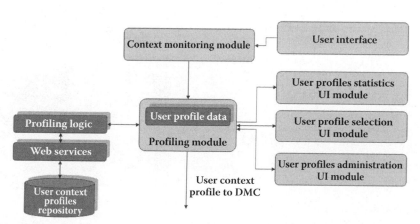

FIGURE 23.5 Context information component architectural model. (From Doulgeraki, C., Partarakis, N., Mourouzis, A., and Stephanidis, C., A development toolkit for unified web-based user interfaces, in the *Proceedings of Computers Helping People with Special Needs: 11th International Conference (ICCHP 2008)*, 10–15 March 2008, Linz, Austria, pp. 346–353, Springer-Verlag, Berlin/Heidelberg, 2008.)

will occur, based on the mapping of user attributes to selections among alternative dialogue patterns.

For example, the decision logic for presenting links can be as follows:

- "web knowledge" in {very good, good} → "underlined text"
- "web knowledge" in {normal, low} → "push buttons"

An example decision rule is presented below.

Decision-Making Logic: An Excerpt

```
switch(disability)
{
 case (long) Disablity.Blind:
 {
 return DisplayStatistics.Table;
 }
 case (long) Disablity
      ColorBlindProtanope:
 {
 return DisplayStatistics.
      GraphicalProtanope;
 }
 case (long) Disablity.
      ColorBlindDeuteranope:
 {
 return DisplayStatistics.
      GraphicalDeuteranope;
 }
 case (long) Disablity.
           ColorBlindTritanope:
 {
      return DisplayStatistics.
           GraphicalTritanope;
 }
 case (long) Disablity.MotorImpaired:
 {
      return DisplayStatistics.
           Graphical;
 }
 case (long) Disablity.LowVision:
 {
      return DisplayStatistics.Table;
 }
 case (long) Disablity.None:
 default:
 {
 return DisplayStatistics.Graphical;
 }
}
```

23.3.1.2.4 Dialogue Controls Component

The role of the dialogue controls component (DCC) (Figure 23.7) is to apply interface adaptation decisions propagated by the DMC, and additionally to structure the final interface using

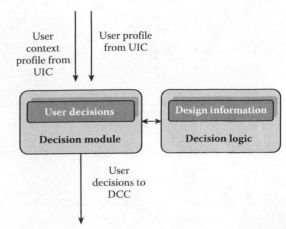

FIGURE 23.6 Decision-making component architectural model. (From Doulgeraki, C., Partarakis, N., Mourouzis, A., and Stephanidis, C., A development toolkit for unified web-based user interfaces, in the *Proceedings of Computers Helping People with Special Needs: 11th International Conference (ICCHP 2008)*, 10–15 March 2008, Linz, Austria, pp. 346–353, Springer-Verlag, Berlin/Heidelberg, 2008.)

the selected dialogue components. More specifically, this component (1) provides the implementation of the alternative dialogue components of a self-adapting interface in the form of dynamic libraries; (2) moderates and administrates the alternative dialogue components; and (3) maintains a record of user interaction with alternative dialogue components.

23.3.1.2.5 Designs Repository

The design repository component is developed during the design cycle of a web application, and provides the DCC with the designs of the alternative dialogues controls in a form of abstract design and polymorphism. Polymorphic decomposition leads from abstract design pattern to a concrete artifact (see also Chapter 16, "Unified Design for User Interface Adaptation").

Figure 23.8 depicts an example of polymorphic task hierarchy, illustrating how two alternative dialogue styles for an Upload File task may be designed. Alternative decomposition styles are depicted in the upper part of the figure, and an exemplary

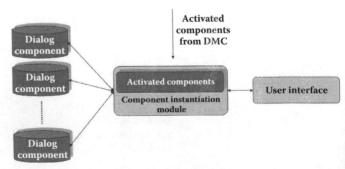

FIGURE 23.7 Dialogue controls component architectural model. (From Doulgeraki, C., Partarakis, N., Mourouzis, A., and Stephanidis, C., A development toolkit for unified web-based user interfaces, in the *Proceedings of Computers Helping People with Special Needs: 11th International Conference (ICCHP 2008)*, 10–15 March 2008, Linz, Austria, pp. 346–353, Springer-Verlag, Berlin/Heidelberg, 2008.)

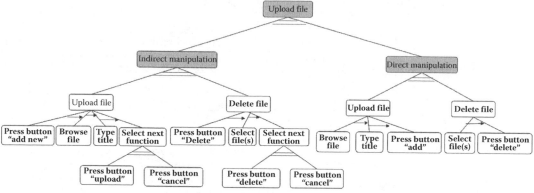

FIGURE 23.8 The polymorphic task hierarchy concept. (From Doulgeraki, C., Partarakis, N., Mourouzis, A., and Stephanidis, C., A development toolkit for unified web-based user interfaces, in the *Proceedings of Computers Helping People with Special Needs: 11th International Conference (ICCHP 2008),* 10–15 March 2008, Linz, Austria, pp. 346–353, Springer-Verlag, Berlin/Heidelberg, 2008.)

polymorphic decomposition in the lower part. Figure 23.9 includes physical design annotations corresponding to the alternative styles.

In the design process, the following primary decisions need to be taken:

- At which points of a task hierarchy polymorphism should be applied, based on the considered (combinations of) user and usage-context attributes.
- How different styles behave at run-time. This is performed by assigning to pair(s) of style (groups) design relationships.

These decisions need to be documented in a design rationale recorded by capturing, for each subtask in a polymorphic hierarchy, the underlying design logic, which directly associates user usage-context parameter values with the designed artifacts.

As a minimum requirement, such a rationale should document (Savidis and Stephanidis, 2004a) the related task, the design targets leading to the introduction of the style, the supported execution context, the style properties, and the design relationships with alternative styles. In Table 23.2, an instance of such documentation record for the task Upload Files is depicted, adopting a tabular notation.

23.3.2 Extensibility

One of the most prominent features of a UWI framework is its ability to extend, thus promoting code reusability and development efficiency. New interaction elements can be easily incorporated in the framework by adding the required decision logic for supporting their conditional activation and deactivation. Additionally, the framework can be extended to support new interaction modalities and global application settings. For example, to produce high-quality interfaces on a PDA, new styles can be produced to facilitate the interaction and display requirement of a PDA. Additionally, whenever new functionality is required, the process for extending the framework to support new interaction elements must be applied. The overall process of decision making does not change.

23.4 EAGER: A Toolkit for Web Developers

The *EAGER toolkit,* targeted to facilitate the development of UWIs, has been developed in Microsoft Visual C# .NET. For each of the UWI components described in the previous section, the corresponding elements of EAGER are presented in the following subsections.

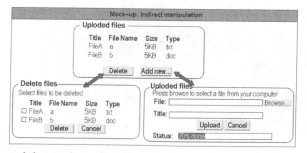

FIGURE 23.9 Physical design alternatives (for File uploader). (From Doulgeraki, C., Partarakis, N., Mourouzis, A., and Stephanidis, C., A development toolkit for unified web-based user interfaces, in the *Proceedings of Computers Helping People with Special Needs: 11th International Conference (ICCHP 2008),* 10–15 March 2008, Linz, Austria, pp. 346–353, Springer-Verlag, Berlin/Heidelberg, 2008.)

TABLE 23.2 An Instance of Design Rationale Documentation

	Task: Upload Files	
Style	Indirect manipulation	Direct manipulation
Targets	Simplicity, guided steps	Speed, effectiveness
Parameters	User (novice)	User (expert)

Properties	Upload file:
	Press button Add New first, browse file next,
	  type title next, press button Upload
	Delete file:
	Press button Delete first, select file(s) next,
	  press button Delete
		Upload file:
		Browse file first, type title
		  next,
		  press button Add
		Delete file:
		Select file(s) first, press button
		  Delete next

Relationships	Exclusive	Exclusive

23.4.1 UIC Implementation

As described in the previous section, UIC consists of the following modules: user profile repository, user components repository, profiling module, interaction-monitoring module, user profiles statistics UI module, user profile selection UI module, and user profiles administration UI module (see Figure 23.3).

In EAGER, the user profiles repository stores information about basic user attributes as presented in Figure 23.10. These attributes (language, disability, web familiarity) can be thought of as the minimum level of information to be used for performing decision making by the EAGER toolkit.

As already discussed, the DMC propagates the decisions made regarding the conditional activation/deactivation of interaction elements to the UIC. These decisions are mapped to a more detailed scheme and are stored by the UIC. Additionally, users are provided with the ability to manually override the default decision making, and alter their personal setup for enriching their individual profile based on personal interaction and accessibility preferences. An excerpt of the extended user attributes is presented in Figure 23.11.

On the other hand, the profiling module is used for propagating user attributes that expose functionality for enabling the retrieval of basic and extended user characteristics. The interaction-monitoring module enables the collection and propagation of attributes regarding user actions. The module responsible for performing these operations exposes functionality for monitoring the outcome of user actions.

The user profile statistics UI module is responsible for generating statistics based on the data collected from user profiles and personal user setups. The information provided by this module can be used to understand the behavior and preferences of actual users to get feedback and enrich the decision logic of the framework.

The user profile selection UI module acts as a front end to the final users of applications developed using the EAGER toolkit. The UI module enables end-users to enter their profile information for the decision making to take place. Additionally, specific UI interfaces are provided for manually overriding the system's decisions. Using these interfaces, all the user-specific settings regarding activation or deactivation of interactive elements can be set. In this way, the EAGER framework provides not only the underlying infrastructure, but also specific user-based or system-based administrative facilities and in-depth presentation of the user profile selection UI module in terms of functionality, user interface features, and its incorporation into an actual portal implementation.

The user profiles administration UI module is an administrative facility that enables the generation of predefined user profiles. These profiles can in turn be published and used by the application end-users to perform a quick setup of their interface. A novice user, for example, can select among a descriptive predefined user profile instead of using the user profile selection UI module to edit specific user profile parameters.

23.4.2 Context Information Component Implementation

As described in the previous section, CIC consists of the user context profiles repository and the context-monitoring module (see Figure 23.5). Their implementation using EAGER is presented in detail in the following subsections.

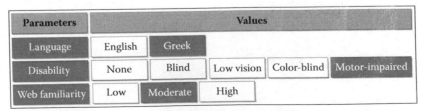

FIGURE 23.10 User profile basic attributes specified prior to initiation of interaction. (From Doulgeraki, C., Partarakis, N., Mourouzis, A., and Stephanidis, C., A development toolkit for unified web-based user interfaces, in the *Proceedings of Computers Helping People with Special Needs: 11th International Conference (ICCHP 2008)*, 10–15 March 2008, Linz, Austria, pp. 346–353, Springer-Verlag, Berlin/Heidelberg, 2008.)

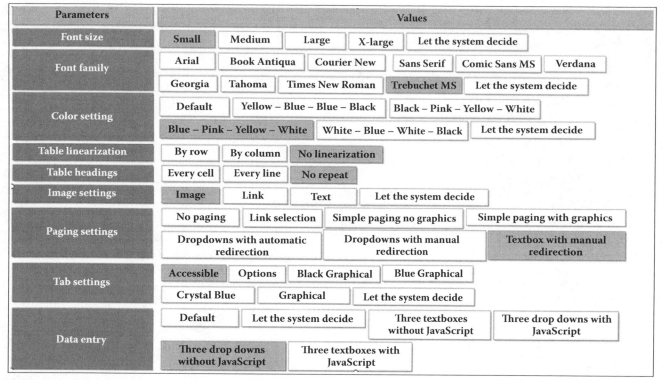

FIGURE 23.11 Excerpt of user profile with extended attributes specified during interaction. (From Doulgeraki, C., Partarakis, N., Mourouzis, A., and Stephanidis, C., A development toolkit for unified web-based user interfaces, in the *Proceedings of Computers Helping People with Special Needs: 11th International Conference (ICCHP 2008),* 10–15 March 2008, Linz, Austria, pp. 346–353, Springer-Verlag, Berlin/Heidelberg, 2008.)

23.4.2.1 User Context Profiles Repository

The user context profiles repository stores information such as the technical equipment used for accessing a web site (desktop, palmtop, mobile phone, etc.) or the assistive technologies that are used (screen reader, screen magnifier, etc.), as presented in Figure 23.12. This kind of information is of particular importance for enabling the decision-making component to infer the presentation characteristics and limitations of each individual user request.

23.4.2.2 Context-Monitoring Module

The context-monitoring module plays an important role for identifying the presentation characteristics of each client in terms of the device and display resolution used. The implemented architecture incorporates the appropriate functions for requesting information by clients accessing the application and extracting data such as the browser used (e.g., Firefox, mobile Internet Explorer, etc.) and its capabilities (e.g., JavaScript support, CSS support, etc.).

23.4.3 DMC

The repository of mappings contains design information regarding the user interface for ensuring the usefulness of occurring adaptations. More specifically, the DMC has the following functionality. Initially, if only the basic user and context attributes are available, it performs decision making based on these attributes and the incorporated repository of mappings. The process of decision making is more complex when specific user interaction and accessibility settings are present. In this case, extended settings are overridden only if not explicitly selected by the user. For the decisions to be propagated by the DMC component after each decision-making session is completed, a new repository of final settings is created for each user. This component

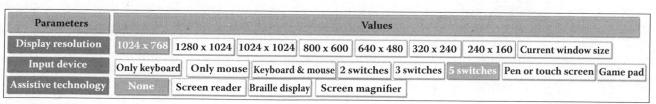

FIGURE 23.12 Context-specific attributes. (From Doulgeraki, C., Partarakis, N., Mourouzis, A., and Stephanidis, C., A development toolkit for unified web-based user interfaces, in the *Proceedings of Computers Helping People with Special Needs: 11th International Conference (ICCHP 2008),* 10–15 March 2008, Linz, Austria, pp. 346–353, Springer-Verlag, Berlin/Heidelberg, 2008.)

is responsible for generating the final set of user characteristics and for propagating the decisions made. The decision logic for performing decision making and generating the final set of user characteristics is encapsulated.

23.4.4 DCC Implementation

23.4.4.1 Dialogue Controls Hierarchy

This subsection presents the dialogue controls hierarchy. More specifically, the namespace ics.Adaptation.DialogueControls contains all the designer-enabled controls to be used by application developers for building adaptive applications. Each control contained in this namespace represents an interaction element that can be instantiated at run-time to a number of different styles. These styles are invisible to the end-users of the adaptation library, and are contained in a different namespace.

23.4.4.2 Examples of Dialogue Controls Implementation

23.4.4.2.1 Content Adaptation

An interesting example of content adaptation concerns images. The way that a web application presents images can be altered according to the specific user characteristics and presentation requirements. Different image-displaying schemes are facilitated

by the framework through the image UI components hierarchy presented in Figure 23.13.

23.4.4.2.2 Layout Adaptation

Templates affect the basic characteristics of each page controlling the positioning to be applied on page content, the available containers of page content, and the presentation characteristics used for rendering page content. To achieve this, a hierarchy of different template variations has been developed, as presented in Figure 23.14.

23.4.4.2.3 Navigation Adaptation

In many occasions, the existence of several navigation schemes in different screen positions may influence the usability of a web application, especially when the scanning of a complete page for each selection is required. In this case, an alternative way to navigate through pages is required. The full navigation scheme was implemented to provide a navigation alternative that would be available from a specific screen location. The full navigation hierarchy is presented in Figure 23.15.

23.4.4.2.4 Interaction Adaptation

An example of interaction adaptation concerns date entry, a frequent task for users of web applications. A number of variations

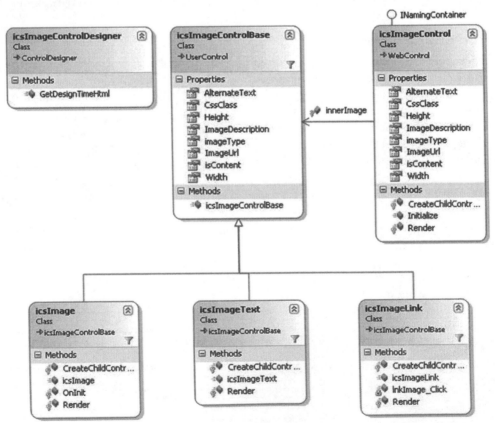

FIGURE 23.13 Images hierarchy. (From Doulgeraki, C., Partarakis, N., Mourouzis, A., and Stephanidis, C., A development toolkit for unified web-based user interfaces, in the *Proceedings of Computers Helping People with Special Needs: 11th International Conference (ICCHP 2008)*, 10–15 March 2008, Linz, Austria, pp. 346–353, Springer-Verlag, Berlin/Heidelberg, 2008.)

FIGURE 23.14   The hierarchy of the potential template variations. (From Doulgeraki, C., Partarakis, N., Mourouzis, A., and Stephanidis, C., A development toolkit for unified web-based user interfaces, in the *Proceedings of Computers Helping People with Special Needs: 11th International Conference (ICCHP 2008),* 10–15 March 2008, Linz, Austria, pp. 346–353, Springer-Verlag, Berlin/Heidelberg, 2008.)

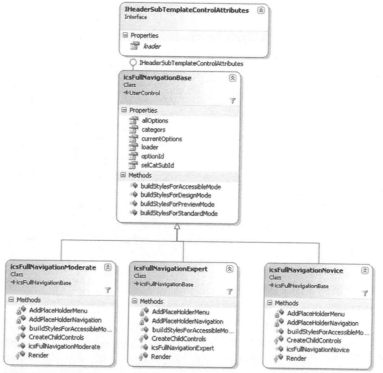

FIGURE 23.15   Full navigation subtemplate control attributes. (From Doulgeraki, C., Partarakis, N., Mourouzis, A., and Stephanidis, C., A development toolkit for unified web-based user interfaces, in the *Proceedings of Computers Helping People with Special Needs: 11th International Conference (ICCHP 2008),* 10–15 March 2008, Linz, Austria, pp. 346–353, Springer-Verlag, Berlin/Heidelberg, 2008.)

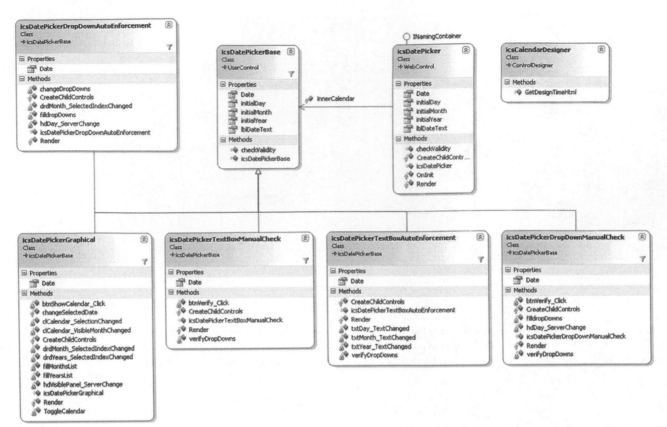

FIGURE 23.16 Date entry diagram. (From Doulgeraki, C., Partarakis, N., Mourouzis, A., and Stephanidis, C., A development toolkit for unified web-based user interfaces, in the *Proceedings of Computers Helping People with Special Needs: 11th International Conference (ICCHP 2008)*, 10–15 March 2008, Linz, Austria, pp. 346–353, Springer-Verlag, Berlin/Heidelberg, 2008.)

in style is facilitated in EAGER according to specific user characteristics (Figure 23.16).

23.4.4.3 Accessibility Evaluation of UI Elements

To evaluate all the outcomes of EAGER and to prove that offer accessibility, the following evaluations were performed:

- All the autonomous user interface elements were evaluated in all the alternative styles externally from a portal using at least one evaluation accessibility tool checking for conformance with W3C *Web Content Accessibility Guidelines* (WCAG) and U.S. Section 508. Evaluating these elements in a portal's environment is particularly difficult due to the wide range of alternative UI elements combinations.
- All the user interface elements that are related to color were evaluated to assess color effectiveness.

An exhaustive evaluation of all the UI elements was carried out using the Watchfire Bobby[20] tool that checks conformance with W3C WCAG and U.S. Section 508 in parallel. A report with the errors and warnings found was produced. For the errors that were identified, it was initially explored if it was possible to correct them. The errors that were correctable were corrected at

---

[20] Watchfire Bobby Version 5.10.2.3.

once. For the rest of the errors, the accessibility level that the specific controls conform was noted. For the warnings that were identified, the same procedure was carried out.

For the user interface elements that are related to colors, W3C guidelines were followed. To conform to WCAG, foreground and background color combinations should provide sufficient contrast when viewed by someone with low vision or color-blindness, or when viewed on a black-and-white screen. The formula suggested by W3C to determine the brightness of a color is:

$$((Red\ value \times 299) + (Green\ value \times 587) + (Blue\ value \times 114)) / 1000$$

Two colors provide good color visibility if the brightness difference is greater than 125 and the color difference is greater than 500.

There are several accessibility tools that support contrast checking, but they examine only the difference between foreground and background colors for text elements without taking into account the absolute position of the HTML elements, and it is possible to have visually a different background on the site compared with the one from the tool. Additionally, dynamic changes and mouse over effects (such as hover) are ignored by the tool. For the UI elements that provide image output, the

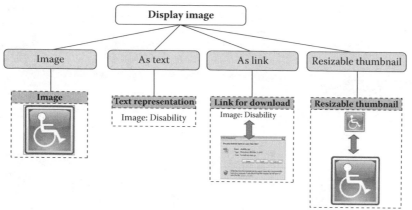

FIGURE 23.17 Image alternative representations. (From Doulgeraki, C., Partarakis, N., Mourouzis, A., and Stephanidis, C., A development tool-kit for unified web-based user interfaces, in the *Proceedings of Computers Helping People with Special Needs: 11th International Conference (ICCHP 2008)*, 10–15 March 2008, Linz, Austria, pp. 346–353, Springer-Verlag, Berlin/Heidelberg, 2008.)

online tool *Vischeck*[21] was used to transform output images to images that people with color-blindness see to evaluate them.

23.4.5 Designs Repository

23.4.5.1 Content-Related Alternative Designs

A good example of content-related design alternatives concerns images. Images appear usually as content of web portals. Blind or low-vision users are not interested in viewing images but only in reading the alternative text of them that describes the image. To facilitate blind and low-vision users, two design alternatives were produced, which are presented in Figure 23.17.

The text representation of the image simply does not present the image, but only a label with the prefix "Image:" followed by the alternative text of the image. The second representation, targeted to users with visual impairments, is the same as the first with the difference that, instead of a label, a link is included that leads to the specific image giving the ability of saving the image. In particular, a blind user may not wish to view an image but may wish to save it to a disk and use it properly.

---

[21] http://www.vischeck.com/vischeck/vischeckImage.php.

In addition to this method, another design was produced that can be selected as a preference by web portal users in which the images are represented as a thumbnail bounding the size that holds on the web page. A user who wishes to view the image in normal size may click on it.

In Table 23.3, the design rationale of the alternative images design is presented.

23.4.5.2 Layout-Related Alternative Designs

A good example of layout-related alternative designs concerns templates. A portal template generally maps to the generic scheme that incorporates the containers hosting contents. As presented in Figure 23.18, two generic template styles were designed. The linearized template style contains all the containers (top navigation, content, bottom navigation) in a linear form. On the other hand, the columns template style has three alternative styles where top and bottom navigation are placed on the top and bottom positions, and the middle container is split in two, three, or four columns, respectively, for the two, three, four columns template.

According to the design rationale presented in Table 23.4, the linearized template supports speed, naturalness, and flexibility

TABLE 23.3 Design Rationale of the Images Alternatives

		Task: Display Image		
Style	Image	As Text	As Link	Resizable Thumbnail
Targets		Facilitate screen reader and low-vision users in order not to be in difficulties with image viewing	Facilitate screen reader and low-vision users in order not to be in difficulties with image viewing but with the capability to save or view an image	Speed, bandwidth saving, better utilization of screen size, compact view
Parameters	User (default)	User (blind or low vision)	User (blind or low vision) and user preference	Platform (mobile) (all-all, directly selected)
Properties	View image	Read image alternative text	Read image alternative text or select link named as the image alternative text to save or view the image	View image thumbnail and select it to view it in normal size
Relationship	Exclusive	Exclusive	Exclusive	Exclusive

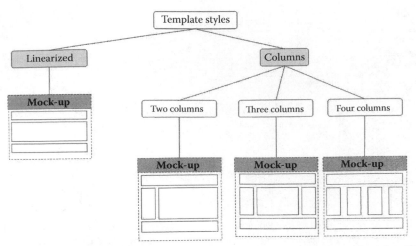

FIGURE 23.18   Template representation styles. (From Doulgeraki, C., Partarakis, N., Mourouzis, A., and Stephanidis, C., A development toolkit for unified web-based user interfaces, in the *Proceedings of Computers Helping People with Special Needs: 11th International Conference (ICCHP 2008)*, 10–15 March 2008, Linz, Austria, pp. 346–353, Springer-Verlag, Berlin/Heidelberg, 2008.)

TABLE 23.4   Design Rationale of the Template Styles

	Task: Template Styles	
Style	Linearized	Columns
Targets	Accessibility, speed, naturalness, flexibility, minimize communication with server	Speed, flexibility, cover optimum screen size
Parameters	User (blind, low vision), platform (mobile)	User (no visual impairments)
Properties		
Relationships	Exclusive	Exclusive

for blind or low-vision users, whereas the columns templates sustain speed, flexibility, and optimum screen size for users with no visual impairments. The alternative columns templates are intended to be used to support content flexibility.

Template size constitutes another significant aspect that is associated with the screen resolution in which the portal will be presented. As presented in Figure 23.19 and Table 23.5, the template size may be resized according the device screen resolution to cover the optimum screen size.

When the resolution is 800×600, the template covers all the surface of the screen, whereas for 1024×768 resolutions the template has on its left and right sides a small unexploited area to maximize readability of the contents. For resolutions greater than 1024×768, the width of the empty area left and right of the template is increased according to screen resolution.

23.4.5.3 Navigation-Related Alternative Designs

Navigation constitutes one of the main mechanisms that a web portal user uses. Multiple alternatives of the navigation mechanism were designed to support individual user abilities and preferences. These are presented in Figure 23.20.

The linearized navigation for novice users (see Figure 23.21) offers a linear form for all the navigation links of the portal, and in parallel a step-by-step navigation is supported. Initially, the user has to select among navigation hierarchies, next

among entire navigation elements, and finally among entire navigation subelements. In each step, the previous hierarchy is available to navigate back to another navigation hierarchy or navigation element. This step-by-step navigation mechanism offers a guided navigation to novice users with vision impairments to enhance accessibility, flexibility, and usability of the portal.

Linear navigation is targeted to moderate with visual impairments. Initially, the user selects among navigation hierarchies and then the available navigation elements for the selected navigation hierarchy are presented, along with a navigation path through which the user may navigate back to the navigation hierarchy. Through this procedure the user has to scan limited navigation options using the screen reader, know each time which pages are browsed, and always have an efficient way to navigate back to the navigation hierarchies thanks to the path mechanism (see Figure 23.22).

Linear navigation for expert users with visual impairments resembles linear navigation for moderate users, but without the path mechanism. In this way, the expert has the ability to navigate back to the navigation hierarchy, but is not notified about the web page browsed each time (see Figure 23.23). In this way, the expert can browse through the navigation mechanism quickly, without having the screen reader always reading the path of the entire page.

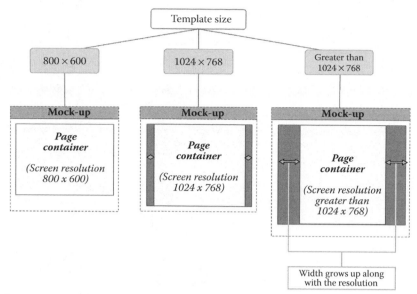

FIGURE 23.19 Template alternatives according to device resolution. (From Doulgeraki, C., Partarakis, N., Mourouzis, A., and Stephanidis, C., A development toolkit for unified web-based user interfaces, in the *Proceedings of Computers Helping People with Special Needs: 11th International Conference (ICCHP 2008)*, 10–15 March 2008, Linz, Austria, pp. 346–353, Springer-Verlag, Berlin/Heidelberg, 2008.)

TABLE 23.5 Design Rationale of the Templates Alternatives According to Device Resolution

	Task: Template Styles		
Style	800×600	1024×768	Greater than 1024×768
Targets	Cover optimum screen size	Cover optimum screen size	Cover optimum screen size
Parameters	Device resolution: 800×600	Device resolution: 1024×768	Device resolution: greater than 1024×768
Properties			
Relationships	Exclusive	Exclusive	Exclusive

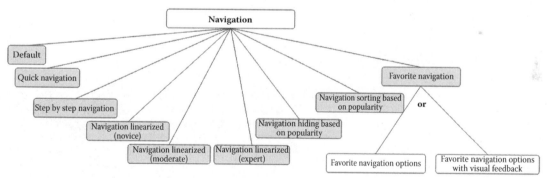

FIGURE 23.20 Navigation alternatives. (From Doulgeraki, C., Partarakis, N., Mourouzis, A., and Stephanidis, C., A development toolkit for unified web-based user interfaces, in the *Proceedings of Computers Helping People with Special Needs: 11th International Conference (ICCHP 2008)*, 10–15 March 2008, Linz, Austria, pp. 346–353, Springer-Verlag, Berlin/Heidelberg, 2008.)

23.4.5.4 Interaction-Related Alternative Designs

A good example of design alternatives related to interaction is that of date entry. Five alternatives were designed for date entry, and are presented in Figure 23.24. As recorded in Table 23.7, the first design includes three dropdowns where the user has to select three items (year, month, day) using dropdowns. These dropdowns are constructed in such a way that when a user selects a specific year and month, the appropriate calculations are made based on leap years and number of days that each month includes so that only the valid days are placed in the days' dropdown. This option is targeted to novice users, because of the simplicity and error prevention that it offers.

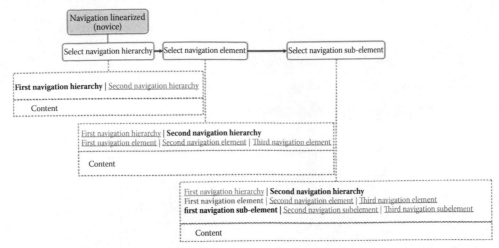

FIGURE 23.21 Navigation linearized (novice). (From Doulgeraki, C., Partarakis, N., Mourouzis, A., and Stephanidis, C., A development toolkit for unified web-based user interfaces, in the *Proceedings of Computers Helping People with Special Needs: 11th International Conference (ICCHP 2008),* 10–15 March 2008, Linz, Austria, pp. 346–353, Springer-Verlag, Berlin/Heidelberg, 2008.)

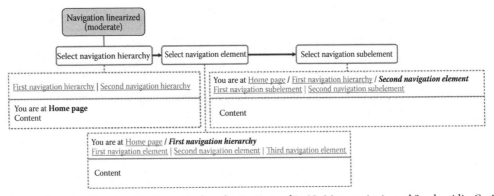

FIGURE 23.22 Navigation linearized (moderate). (From Doulgeraki, C., Partarakis, N., Mourouzis, A., and Stephanidis, C., A development toolkit for unified web-based user interfaces, in the *Proceedings of Computers Helping People with Special Needs: 11th International Conference (ICCHP 2008),* 10–15 March 2008, Linz, Austria, pp. 346–353, Springer-Verlag, Berlin/Heidelberg, 2008.)

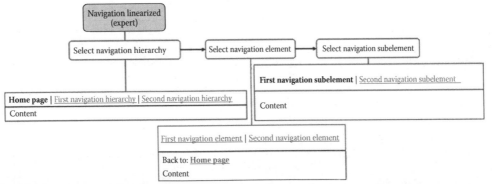

FIGURE 23.23 Navigation linearized (expert). (From Doulgeraki, C., Partarakis, N., Mourouzis, A., and Stephanidis, C., A development toolkit for unified web-based user interfaces, in the *Proceedings of Computers Helping People with Special Needs: 11th International Conference (ICCHP 2008),* 10–15 March 2008, Linz, Austria, pp. 346–353, Springer-Verlag, Berlin/Heidelberg, 2008.)

The second design looks like the first design, with the difference that the validation of the user input is made manually. After the selection of the appropriate values, the user has to push the button Ok to validate the selected date. In case of a mistake, an error in red appears guiding the user to correct the mistake.

This design artifact is targeted to users with visual impairments, since it does not necessitate extended screen reading and is simple in use.

As shown in Figures 23.25 to 23.29, in textbox design with automated date enforcement design the user has to type year,

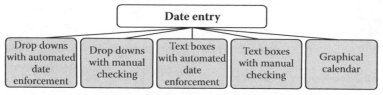

FIGURE 23.24 Date entry alternatives. (From Doulgeraki, C., Partarakis, N., Mourouzis, A., and Stephanidis, C., A development toolkit for unified web-based user interfaces, in the *Proceedings of Computers Helping People with Special Needs: 11th International Conference (ICCHP 2008)*, 10–15 March 2008, Linz, Austria, pp. 346–353, Springer-Verlag, Berlin/Heidelberg, 2008.)

TABLE 23.6   Design Rationale of the Step-by-Step Navigation (Novice, Moderate, Expert)

	Task: Navigation		
Style	Navigation linearized (novice)	Navigation linearized (moderate)	Navigation linearized (expert)
Targets	Accessibility, flexibility, usability	Accessibility, flexibility, usability, limited reading by the screen reader	Accessibility, speed, flexibility, usability, limited reading by the screen reader
Parameters	User (blind or low vision and novice web expertise), platform (mobile)	User (blind or low vision and moderate web expertise), platform (mobile)	User (blind or low vision and expert web expertise), platform (mobile)
Properties	Navigation hierarchy first, navigation element next (for desired navigation hierarchy), navigation subelement next (for desired navigation element)	Navigation hierarchy first, navigation element next (for desired navigation hierarchy), navigation subelement next (for desired navigation element)	Navigation hierarchy first, navigation element next (for desired navigation hierarchy), navigation subelement next (for desired navigation element)
Relationships	Exclusive	Exclusive	Exclusive

TABLE 23.7   Design Rationale of Date Input Alternatives

	Task: Date Entry				
Style	Dropdowns with automated date enforcement	Dropdown with manual checking	Text boxes with automated date enforcement	Text boxes with manual checking	Graphical calendar
Targets	Simplicity, error prevention easiness	Facilitate screen reader, effectiveness	Speed, effectiveness	Facilitate screen reader, effectiveness	Limited necessity of user input, simplicity
Parameters	User (novice), platform (mobile)	User (blind or low vision)	User (expert)	User (blind or low vision) and expert)	User (moderate, motor impaired), platform (mobile)
Properties	Select year first, month next, and day at last	Select year first, month next, day next, and press button Ok at last	Type year first, month next, and day at last	Type year first, month next, day next, and press button Ok at last	Type day first, '/' next, month next, '/' next, year at last or click on Calendar icon and select date on the virtual calendar
Relationships	Exclusive	Exclusive	Exclusive	Exclusive	Exclusive

month, and day in the textboxes. The validation of the date would take place when the user would try to complete the particular action. This date input alternative is designed for expert web users to improve the speed and effectiveness of use. A different design is provided for users who have visual impairments and are web experts, too. Simply, in this design the user has to push the button Ok to validate the date inserted. For moderate and motor-impaired users, another design was produced that offers a virtual calendar from where the user has monthly previews, may navigate through years and months using two dropdowns, and can select a date by clinking on it when the appropriate date is met. This design is characterized as suitable for motor-impaired users because it does not necessitate much user input, and is suitable for moderate users too because it offers a calendar-like graphical representation of dates that is more familiar to the users.

23.4.6 Benefits for Developers

Using the EAGER toolkit instead of the Microsoft's Visual Studio alone presents a number of advantages in terms of the developer's performance and efficiency, and in particular regarding time and development efforts.

The complexity of the UI design effort is radically reduced due to the flexibility provided by the EAGER toolkit for designing interfaces at an abstract task-oriented level. Using EAGER, designers are not required to be aware of the low-level details

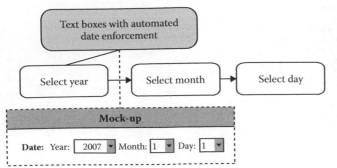

FIGURE 23.25 Date entry (dropdowns, automated date enforcement). (From Doulgeraki, C., Partarakis, N., Mourouzis, A., and Stephanidis, C., A development toolkit for unified web-based user interfaces, in the *Proceedings of Computers Helping People with Special Needs: 11th International Conference (ICCHP 2008)*, 10–15 March 2008, Linz, Austria, pp. 346–353, Springer-Verlag, Berlin/Heidelberg, 2008.)

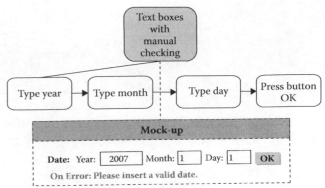

FIGURE 23.28 Date entry (text boxes, manual checking). (From Doulgeraki, C., Partarakis, N., Mourouzis, A., and Stephanidis, C., A development toolkit for unified web-based user interfaces, in the *Proceedings of Computers Helping People with Special Needs: 11th International Conference (ICCHP 2008)*, 10–15 March 2008, Linz, Austria, pp. 346–353, Springer-Verlag, Berlin/Heidelberg, 2008.)

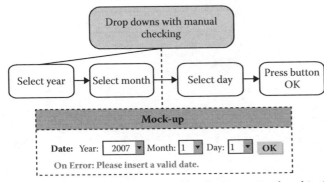

FIGURE 23.26 Date entry (dropdowns). (From Doulgeraki, C., Partarakis, N., Mourouzis, A., and Stephanidis, C., A development toolkit for unified web-based user interfaces, in the *Proceedings of Computers Helping People with Special Needs: 11th International Conference (ICCHP 2008)*, 10–15 March 2008, Linz, Austria, pp. 346–353, Springer-Verlag, Berlin/Heidelberg, 2008.)

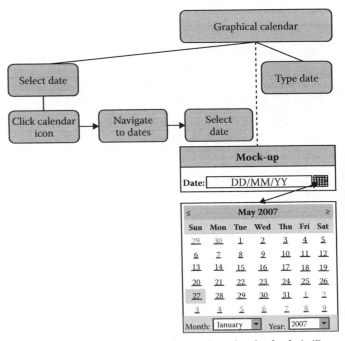

FIGURE 23.29 Date entry (text box and graphical calendar). (From Doulgeraki, C., Partarakis, N., Mourouzis, A., and Stephanidis, C., A development toolkit for unified web-based user interfaces, in the *Proceedings of Computers Helping People with Special Needs: 11th International Conference (ICCHP 2008)*, 10–15 March 2008, Linz, Austria, pp. 346–353, Springer-Verlag, Berlin/Heidelberg, 2008.)

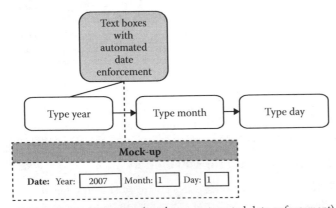

FIGURE 23.27 Date entry (text boxes, automated date enforcement). (From Doulgeraki, C., Partarakis, N., Mourouzis, A., and Stephanidis, C., A development toolkit for unified web-based user interfaces, in the *Proceedings of Computers Helping People with Special Needs: 11th International Conference (ICCHP 2008)*, 10–15 March 2008, Linz, Austria, pp. 346–353, Springer-Verlag, Berlin/Heidelberg, 2008.)

introduced in representing interaction elements, but only of the high-level structural representation of a task and its appropriate decomposition into subtasks, each of which represents a basic UI and system function.

On the other hand, the process of designing the actual front end of the application using a markup language is radically decreased in terms of time, due to the fact that developers initially have to select a radically increased number of interface

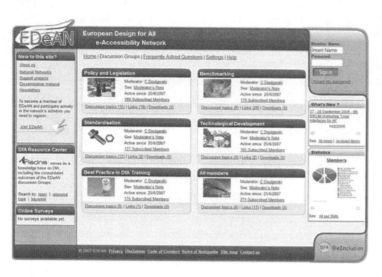

(Crystal-blue)

(Silver-black)   (Cyan-blue)

FIGURE 23.30   The public area of the EDeAN portal (various skins). (From Doulgeraki, C., Partarakis, N., Mourouzis, A., and Stephanidis, C., A development toolkit for unified web-based user interfaces, in the *Proceedings of Computers Helping People with Special Needs: 11th International Conference (ICCHP 2008)*, 10–15 March 2008, Linz, Austria, pp. 346–353, Springer-Verlag, Berlin/Heidelberg, 2008.)

components each of which represents a far more complex facility. Additionally, developers do not have to spend time editing the presentation characteristics of the high-level interaction element, due to the internal styling behavior.

The actual process of transforming the initial design into the final web application using traditional UI controls introduces a lot of coding in conjunction the higher level elements offered by the EAGER toolkit. Here it is worthy to notice that the EAGER toolkit currently consists of a total number of 55K pure code lines at the availability of web portal developers. Furthermore the incorporation of EAGER's higher level elements make a portal's code more usable, more readable, and especially safe, due to the fact that each interaction component introduced by this framework is designed separately, developed and tested introducing a high level of code reuse, efficiency, and safety.

Additionally, UIs developed with the EAGER toolkit can adapt according to specific user and context parameters, and therefore are rendered in a number of variations. It is therefore clear that when using a standard UI toolkit a monolithic interface is created, whereas when using the EAGER toolkit dynamically adaptable interfaces are generated.

23.4.7 The EDeAN Portal Case Study

A prototype portal was developed, as proof-of-concept, following the UWI methodology by means of the EAGER toolkit. To elucidate the benefits of EAGER, an already existing portal was selected and redeveloped from scratch to identify and compare the advantages of using EAGER, both at the developer's site, in terms of developer's performance, as well as at the end-user site,

in terms of the user-experience improvement. In particular, the original portal of the European Design for All e-Accessibility Network (EDeAN), namely Hermes,[22] was redesigned and implemented using the EAGER development framework.[23]

The new version of the Hermes portal (see Figures 23.30 and 23.31) introduces a number of major improvements in terms of

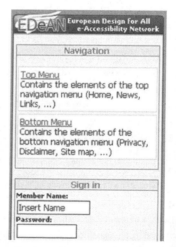

FIGURE 23.31   The public area in the mobile EDeAN portal. (From Doulgeraki, C., Partarakis, N., Mourouzis, A., and Stephanidis, C., A development toolkit for unified web-based user interfaces, in the *Proceedings of Computers Helping People with Special Needs: 11th International Conference (ICCHP 2008)*, 10–15 March 2008, Linz, Austria, pp. 346–353, Springer-Verlag, Berlin/Heidelberg, 2008.)

---

[22] Old version of the EDeAN portal: http://www.edean.org.

[23] The new EDeAN portal was designed and developed in the context of the Dfa@eInclusionProject (http://www.dfaei.org).

functionality, and is the first portal developed using the unified web interfaces framework presented in the previous sections of this chapter. All the modules already existing in the previous version and the newly added modules were altered to use the interaction artifacts offered by the EAGER development framework. Additionally, several different modules were implemented, such as digital library, training courses, and a complete user-profiling mechanism. This effort provided valuable feedback in a number of directions. Initially, the development of such a large-scale application proved the viability and consistency of the toolkit and elucidated its ability to stand as a horizontal and efficient development tool.

23.5 Conclusions and Future Work

This chapter proposes a novel approach to the development of web user interfaces that is based on the unified user interfaces methodology. The proposed framework, intended as an alternative to traditional design for the average user, guides the development of UWI and aims to ensure accessibility and usability for users with diverse characteristics. The EAGER toolkit further facilitates web developers in effectively following the proposed approach in practice. In this context, a number of UI elements were designed in various forms (polymorphic task hierarchies) according to specific user and context parameter values. This phase provided feedback to the actual development process of EAGER, which involved the development of the alternative interaction elements and the mechanisms for facilitating the dynamic activation/deactivation of interaction elements and modalities based on individual user interaction and accessibility preferences.

Another key feature of the EAGER toolkit is its ability to be extended and include an unlimited number of alternative interaction modalities and elements. This process mainly entails the design and coding of the alternative interactions styles. Then, they can be easily incorporated in the existing toolkit, simply by modifying the decision logic for supporting their conditional activation and deactivation. Additionally, existing web applications or parts of applications implemented with .NET can be easily altered to encapsulate the EAGER toolkit attributes and, thereby, rendered accessible and usable for various user categories, including novice users, users of assistive technologies or portable devices, and so on. Notably, it has been estimated that applications implemented from scratch using the EAGER toolkit may contain up to 50% less lines of pure code in total when compared to traditionally developed applications, showing that, thanks to abstraction, EAGER generates shorter, more robust and comprehensive source codes.

In the EDeAN portal case study, the correlation of the various alternative designs of UI elements to user and context parameters (i.e., the automatic adaptation according to generic, predefined profiles) has been made on a normative basis. Therefore, it cannot currently be claimed to be optimal, and needs to be further verified in the future, for example, through feedback from user trials in real contexts. However, this work has made clear that the developed framework allows embedding in web-based applications such decision-making logics and automatic adaptation facilities for the benefit of accessibility and better user experience.

Along the same line, the produced UI element designs are not claimed to be optimal, for example, in terms of aesthetics. However, it was made clear that it is feasible to develop web-based interfaces that can support and import alternative designs for fully accessible and personalized ways of interactions, without payoffs in terms of aesthetics or inclusiveness.

Concerning additional enhancements of the EAGER toolkit, several advanced and intelligent techniques have been identified that can improve its effectiveness and efficiency. These include advanced built-in mechanisms to support:

- Rapid configuration of the layout and behavior of individual interfaces according to the selections of users with similar profiles
- Selection of a particular conformance level to W3C-WAI guidelines

Other potential future directions concern the integration of EAGER with a design support tool for the unified user interface design method (e.g., Antona, Savidis, and Stephanidis, 2006), as well as the development of a web-based version of the EAGER toolkit to support the development of UWIs using technologies other than .NET.

Overall, UWIs and the EAGER toolkit are considered a significant contribution toward embedding accessibility, graceful transformation, and ease of use for all in future and existing web-based applications, and, ultimately, toward universal access.

References

Alexandraki, C., Paramythis, A., Maou, N., and Stephanidis, C. (2004). Web accessibility through adaptation, in the *Proceedings of Computers Helping People with Special Needs: 9th International Conference (ICCHP 2004)* (K. Miesenberger, J. Klaus, W. Zagler, and D. Burger, eds.), 7–9 July 2004, Paris, pp. 302–309. Berlin/Heidelberg: Springer-Verlag.

Antona, M., Savidis, A., and Stephanidis, C. (2006). A process-oriented interactive design environment for automatic user interface adaptation. *International Journal of Human Computer Interaction* 20: 79–116.

Barra, M., Grieco, R., Malandrino, D., Negro, A., and Scarano, V. (2003). TextToSpeech: A heavy-weight Edge computing service, in the *Poster Proceedings of 12th International World Wide Web Conference*, 20–24 May 2003, Budapest, Hungary. New York: ACM Press.

Barrett, R. and Maglio, P. (1998). Adaptive communities and web places, in the *Proceedings of the 2nd Workshop on Adaptive Hypertext and Hypermedia, HYPERTEXT 98*, 20–24 June 1998, Pittsburgh. New York: ACM Press.

Barrett, R. and Maglio, P. (1999a). Intermediaries: An approach to manipulating information streams. *IBM Systems Journal* 38: 629–641.

Barrett, R. and Maglio, P. (1999b). WebPlaces: Adding people to the Web, in the *Proceedings of 8th International World Wide*

Web Conference, 11–14 May 1999, Toronto, Canada. http://www.almaden.ibm.com/u/pmaglio/pubs.html. New York: ACM Press.

Bergman, E. and Johnson, E. (1995). Towards accessible human-computer interaction, in *Advances in Human-Computer Interaction* (J. Nielsen, ed.) 5: 87–113. Norwood, NJ: Ablex Publishing Corporation.

Brown, S. S. and Robinson P. (2001). A World Wide Web mediator for users with low vision, in the *Proceedings of CHI 2001 Workshop No. 14 Universal Design: Towards Universal Access in the Information Society*, 31 March–5 April 2001, Seattle. http://www.ics.forth.gr/proj/at-hci/chi2001/files/brown.pdf.

Chen, Y., Xie, X., Ma, W.-Y., and Zhang, H.-J. (2005). Adapting web pages for small-screen devices. *IEEE Internet Computing* 9: 50–56.

De La Rue, M. (2002). *Linux Accessibility: HOWTO v3.1*. http://tldp.org/HOWTO/Accessibility-HOWTO.

Doulgeraki, C., Partarakis, N., Mourouzis, A., and Stephanidis, C. (2008). A development toolkit for unified web-based user interfaces, in the *Proceedings of Computers Helping People with Special Needs: 11th International Conference (ICCHP 2008)*, 10–15 March 2008, Linz, Austria, pp. 346–353. Berlin/Heidelberg: Springer-Verlag.

Gaedke, M., Schempf, D., and Gellersen, H. (1999). WCML: An enabling technology for the reuse in object-oriented Web engineering, in the *Poster-Proceedings of the 8th International World Wide Web Conference (WWW8)*, 11–14 May 1999, Toronto. http://www.teco.edu/~gaedke/paper/1999-www8-poster.pdf.

Graef, G. (2000). Adaptation of web-applications based on automated user behaviour analysis, in the *Proceedings of the Second Annual Conference on World Wide Web Applications (WWW 2000)*, 6–9 September 2000, Johannesburg, South Africa, pp. 72–75. Cape Town, South Africa: Cape Peninsula University of Technology.

Gunderson, J. (1996). *World Wide Web Browser Access Recommendations*. Champaign: University of Illinois at Urbana/Champaign.

Gupta, S. and Kaiser, G. (2005). Extracting content from accessible web pages, in *W4A '05: Proceedings of the 2005 International Cross-Disciplinary Workshop on Web Accessibility (W4A)*, 10 May 2005, Chiba, Japan, pp. 26–30. New York: ACM Press.

Gupta, S., Kaiser, G. E., Grimm, P., Chiang, M. F., and Starren, J. (2005). Automating content extraction of HTML documents. *World Wide Web* 8: 179–224.

Han, R., Bhagwat, P., Lamaire, R., Mummert, T., Perret, V., and Rubas, J. (1998). Dynamic adaptation in an image transcoding proxy for mobile web browsing. *IEEE Personal Communications* 5: 8–17.

Henricksen, K. and Indulska, J. (2001). Adapting the web interface: An adaptive web browser, in the *Proceedings of the Australasian User Interface Conference 2001, Volume 23*, 29 January–1 February 2001, Queensland, Australia, pp. 21–28. IEEE.

Hermsdorf, D. (1998). WebAdapter: A prototype of a WWW browser with new special needs adaptations, in the *Proceedings of the XV IFIP World Computer Congress, Computers and Assistive Technology (ICCHP 98)* (A. Edwards, A. Arato, and W. Zagler, eds.), 31 August–4 September 1998, Vienna, Austria, pp. 151–160. Vienna, Austria: Austrian Computer Society.

Iaccarino, G., Malandrino, D., and Scarano, V. (2006). Personalizable edge services for web accessibility, in the *Proceedings of the 2006 International Cross-Disciplinary Workshop on Web Accessibility (W4a): Building the Mobile Web: Rediscovering Accessibility?*, Volume 134, 22–23 May 2006, Edinburgh, Scotland, pp. 23–32. New York: ACM Press.

Instone, K. (2000). *Usability Heuristics for the Web*. http://web.archive.org/web/19990429162604/webreview.com/wr/pub/97/10/10/usability/sidebar.html.

Nielsen, J. (1993). *Usability Engineering*. New York: Academic Press Limited.

Parmanto, B., Ferydiansyah, R., Zeng, X., Saptono, A., and Sugiantara, I. W. (2005). Accessibility transformation gateway, in the *Proceedings of 38th Hawaii International Conference on System Sciences*, 3–6 January 2005 Waikoloa, HI, p. 183a. http://ieeexplore.ieee.org/stamp/stamp.jsp?arnumber=1385605&isnumber=30166.

Richards, J. (1996). *Guide to Writing Accessible HTML*. Toronto: University of Toronto.

Savidis, A. and Stephanidis, C. (2004a). Unified user interface design: Designing universally accessible interactions. *International Journal of Interacting with Computers* 16: 243–270.

Savidis, A. and Stephanidis, C. (2004b). Unified user interface development: Software engineering of universally accessible interactions. *Universal Access in the Information Society* 3: 165–193.

Stephanidis, C. (2001). The concept of unified user interfaces, in *User Interfaces for All: Concepts, Methods, and Tools* (C. Stephanidis, ed.), pp. 371–388. Mahwah, NJ: Lawrence Erlbaum Associates.

Stephanidis, C., Paramythis, A., Sfyrakis, M., and Savidis, A. (2001). A case study in unified user interface development: The AVANTI web browser, in *User Interfaces for All: Concepts, Methods, and Tools* (C. Stephanidis, ed.), pp. 525–568. Mahwah, NJ: Lawrence Erlbaum Associates.

Story, M. F. (1998). Maximising usability: The principles of universal design. *Assistive Technology* 10: 4–12.

Taib, R. and Ruiz, N. (2006). Multimodal interaction styles for hypermedia adaptation, in the *Proceedings of the International Conference on Intelligent User Interfaces (IUI'06)*, 29 January–1 February 2006, Sydney, Australia, pp. 351–353. New York: ACM Press.

Vanderheiden, G., Chisholm, W., and Ewers, N. (1996). *Design of HTML Pages to Increase Their Accessibility to Users with Disabilities, Strategies for Today and Tomorrow*. Technical Report. Trace R&D Centre: University of Wisconsin–Madison.

World Wide Web Consortium (1999). *Web Content Accessibility Guidelines 1.0. W3C Recommendation*. http://www.w3.org/TR/WCAG10.

User Modeling: A Universal Access Perspective

Ray Adams

24.1 Introduction

The focus of this chapter is the fascinating subject of user modeling, a theme that has a long track record (e.g., Whitefield, 1989; Dix et al., 2004). In particular, the spotlight is on the necessary knowledge, skills, and understanding required to appreciate, develop, and apply user models to the evaluation and solution of interactive system design requirements. This focus will need to draw upon both cognitive science and computer science, from both theoretical and practical perspectives.

All the chapters in this handbook are interrelated; for example, Chapters 25 and 61 of this handbook are particularly relevant to this chapter. Chapters in Part I provide a valuable introduction and overview of universal access that constitutes the context for the present chapter. Chapters in Part II address many of the issues of user diversity that create the need for effective user models. However, all the chapters are relevant, since they contribute to a better understanding of the context of universal access, theory, and practice, in the present and future.

24.2 User Modeling in the Information Society

This section describes the information society as the context for user modeling. Today, many modern cultures can be

characterized as information societies in which information and the consequent knowledge are seen as forming their core products and services. Therefore, access to information becomes crucial to the effective functioning of these modern cultures, and exclusion from it has substantial costs in terms of injustice, inequity, unprofitability, and inefficiency. The driving force behind the development of the World Wide Web itself has been to create accessible electronic information, resources, and communications (Berners-Lee, 2005). Status as a citizen in the information society is predicated on the possession of suitable access to information society technologies.

If this scenario of the inclusive information society has any validity, then accessible information society technology (IST) may have profound influences not only on the ways in which people access and interact with technologies, but also on how they relate to each other and on the social structures and rules that are generated in so doing. In the near future, everyday life may be both empowered and governed by IST and related social structures. Clearly, all citizens of the information society will need access to both social technologies and technology-supported social structures.

While intended users vary in psychomotor, perceptual, cognitive, and social dimensions (Maybury, 2001, 2003), a focus on the information society inevitably raises the importance of social considerations such as social intelligence, not only because everybody differs in this respect, but also because specific disabilities impact social intelligence in different ways (Adams and Gill, 2007a, 2007b).

User modeling is a powerful way to capture the demand-characteristics, diversity, and distinctiveness of the intended populations of users of a system or artifact. However, this concept is made even more powerful by inclusion in the context of universal access and accessibility (Stephanidis, 2001a). The drive toward universal access (i.e., pervasive, anytime, anywhere computing power) places user modeling at the center of universal design, programming, and accessibility evaluation for interactive artifacts and systems. Universal access itself places a number of useful demands on artifacts design.

It is proposed here that there are multiple types of accessibility that are required for a user to function properly. At least five levels of accessibility can be defined and loosely correspond to the Internet-layered model (Moseley, 2007), as shown in Table 24.1.

They are as follows. First, universal access requires the possession of suitable and affordable hardware platforms through joint or personal ownership (1: hardware access), as well as, second, apposite connectivity to networks and centers of communication (2: connectivity access). Third, it is also predicated upon the possession of an accessible interface that matches the needs and preferences of the intended users (3: interface access). Fourth, the system design and structure should support navigability and access to the contents of a system (4: cognitive access; Adams, 2007; Keates et al., 2007). Finally, fifth, functionality is not usually an end in itself, but a means to the end of using a system to access our objectives (5: goal access). Goal access is probably the most important type of accessibility, although access will fail without all five levels being addressed adequately.

In addition to these factors, there are at least three additional contributory factors to respect when considering accessibility. First, consider the types of responses required by a system (physical or mental) that may create additional access problems. For example, a system design may assume a high level of manual dexterity but an individual user may lack the manual dexterity to make the required hand movements, lack the reaction time to respond quickly enough, or may work in a confined space. Second, consider the notion of usability, defined as the ease of use. This also should be considered as part of the acceptability of a system (Norman, 1986). For example, consider an application that supports a function requested by the users (such as conference calls) but makes the process of making those calls too difficult to be acceptable to the intended users. Third, consider the design of the underlying markup languages of web sites (e.g., HTML, XHTML, etc.) that can of themselves create either accessibility problems or solutions and so need very careful consideration. For example, an otherwise well-presented web site may not support screen readers very well due to an excessive use of frames. Even so, well-designed code is necessary but not sufficient to guarantee that all users will find a web site to be acceptable. The Web Accessibility Initiative is a major related project to define web site design standards for accessibility (Brewer, 2004).

User modeling relates well to interface, cognitive, and goal accessibility, by capturing vital information about interface, cognitive, and goal requirements and preferences. The growing demand for universal access to information in the evolving information society produces an inexorable move toward more and more powerful and interlinked technological solutions. In this context, user requirements must be captured by more powerful user models, based upon more advanced user-sensitive methods. In addition, current work is looking more seriously at artifacts, systems, and services that are not only functional and accessible, but more attractive aesthetically and can provide smarter responses to the demands placed upon them (Lawrence and Tavakol, 2007). This theme will be taken up again in the final section of this chapter.

24.3 User Diversity

24.3.1 The Typical User

User-centered system design (UCSD) (Norman, 1986) has made a significant contribution to interactive system design by drawing attention to the importance of including the intended users in the design and evaluation processes. Even so, there are still numerous treatments of software engineering where consideration of the users is at best implicit. Equally, there are applications of UCSD that take account only of the typical user, but not of the full diversity of users. The requirements of a range of users cannot easily be captured by consideration of an average user alone. Nevertheless, user-centered system design can be seen as a significant advance over previous egocentric design methods, where the designers essentially designed for themselves.

TABLE 24.1 Comparing Types of Accessibility and the Internet Layered Model

Comparing	Accessibility Types	Internet Layered Model
1	Hardware access	Physical
2	Connectivity access	Data link
3	Interface access	Network
4	Cognitive access	Transport
7	Goal access	Application

24.3.2 Beyond the Typical User

User profiling provides a way to move away from one-size-fits-all thinking, particularly when focused on atypical users, and constitutes the starting point for effective user modeling. Schofield et al. (2004) have argued that virtual learning environments do not automatically benefit all students with disabilities, but that the introduction of student profiling is much more likely to be associated with improved, cost-effective virtual learning.

What are the relevant dimensions of user diversity within the context of universal access?

Two sides of such diversity are disability and capability. Considering disability, one approach is based on the concept of the syndrome, in which commonly associated symptoms are used to classify people by syndromes. While this was once a widely accepted approach in cognitive neurology, it is now generally agreed upon that it is insufficiently accurate to be satisfactory (Ellis and Young, 1996). As it is no longer enough to use broadly defined syndromes, current research has taken two directions: first, the construction of more finely tuned categories; and second, a focus on individual profiles. Since there is no evidence of the emergence of an acceptable, new categorization system, the field has moved instead toward a focus on individual profiles based on case studies and clarified by generic, experimental research. If so, then the user model or profile approach can be considered more effective than any revised, syndrome-based approach.

24.3.3 User Modeling Approaches to Address User Diversity

The inclusion cube (Keates and Clarkson, 2003), which draws explicitly on the model human processor (MHP) of Card et al. (1983), provides a useful way to describe and quantify the dimensions of inclusion. The cube identifies three dimensions of disability/capability, namely, sensory, cognitive, and psychomotor, and these provide an excellent starting point for the development of a generic model of the intended user that reflects both disabilities and abilities in three dimensions, supported by the operationalization of a set of cognitive capability scales for user modeling (Langdon et al., 2004). Related to the inclusion cube, but derived from Broadbent's (1984) Maltese cross theory, Simplex Two (see Figure 24.2) provides nine dimensions of human performance: input system, feedback reception, executive function, working memory, emotional evaluation, mental models, long-term memory, simple response systems, and complex, learned responses.

The MHP is based on a simple and elegant architecture, including short-term memory, long-term memory, and intervening cognitive processes. Information flow is unidirectional. A simple depiction is shown in Figure 24.1

The MHP theory was never intended as a full theory of human cognition, but as a tractable guide for system designers and evaluators, providing a simple but coherent model of

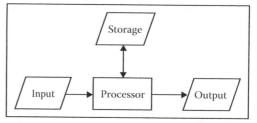

General model of computer information processing system

FIGURE 24.1 Model human processor.

the intended users based on valid psychological axioms. While this approach has been criticized for its incompleteness, it was never intended to be complete, merely to capture and apply the main relevant points of current theories in cognition (Dix et al., 2004). MHP is, in fact, an engineering model of human performance supported by GOMS (goals, operators, methods, and selection rules) as a method of task analysis. The concept of the production rule is at the center of the instantiation of GOMS, a concept picked up by more recent theories like ACT (Anderson, 1983, 1990; Anderson and Lebiere, 1998), Soar (Newell, 1990) and EPIC (Kieras and Meyer, 1997, 2000). MHP and GOMS are clearly useful for PC-based interfaces, but may not be so applicable to the emerging, new types of technology (Byrne, 2001).

MHP is an alternative to complex cognitive models and can be seen as a stripped-down psychological theory, one that can be used by designers and other computer scientists, but without the irrelevant complexities of a fully fledged theory. The MHP (Card et al., 1983) is one outstanding theory that deliberately takes this tactic. MHP has the dual advantages of being simple enough to be applicable and also being developed directly to solve IT design problems. Over the years, the status of MHP as a psychological process theory has been strongly criticized. Such criticisms have been due, in part, to a failure to appreciate that MHP was never intended as a full psychological theory, but rather an application of cognitive psychology to act as a guide to system design and evaluation. Some subsequent research has attempted, with some success, to develop MHP further (Liu et al., 2006). They developed the queuing network-model human processor (QN-MHP) in an attempt to integrate the queuing network approach with the symbolic approach as a basis for cognitive modeling. The overall aim was to unify theoretical and methodological aspects of cognitive modeling and the development of usable, human-computer interaction (HCI) simulation methods.

However, a number of major paradigm shifts have occurred in system design and in the issues considered important in this field. So MHP perhaps is best seen as a brilliant trailblazer that can encourage us to develop similar new ways of dealing with interactive technology in all its forms, with user-supportive architectures and with adaptable and self-adaptive new systems. However, it has considerable value as both a teaching aid and to inspire new approaches to accessibility. For example, the inclusive design approach (Keates and Clarkson, 2003) draws explicitly on MHP to construct the inclusion cube. This is a major component of their inclusive design theory. This is an important, new approach that encourages designers to quantify the proportion of all intended users who are included or excluded. MHP provides a simple architecture of human cognition, including perceptual and response components, short-term and long-term memory, plus cognitive processing. It was intended to capture the modal model of the then current understanding of human cognition, but to be sufficiently simplified for use in practice by designers and students. It did a good job at the time, and is still useful to teach applicable psychological principles and to inspire new approaches.

GOMS (goals, operators, methods, and selection rules; see the following for an explanation) started firmly on the foundation of MHP. While MHP is still an important part of a teacher's, researcher's, and designer's repertoire, it may be overlooked by those whose interest in user interface design is from a software engineering perspective (e.g., Lauesen, 2005). More surprisingly, MHP, the conceptual foundation, is often overshadowed by its more pragmatic spinoff, GOMS. For example, Benyon et al.'s (2005) excellent book on designing interactive systems focuses usefully on GOMS, but mentions MHP only in passing (p. 100), without an index entry for MHP. Sharp et al. (2007), with an apparent greater emphasis on social and cultural factors than on basic cognitive concerns, do not refer to MHP by name. Even Dix et al. (2004) in their comprehensive text on human-computer interaction give more space to GOMS than MHP.

There are many different strains of GOMS. First, the original version produced by Card et al. (1983) is commonly referred to as CMN-GOMS. The second main version of GOMS is the keystroke level model (KLM). This second version of GOMS is often referred to as KLM-GOMS (John and Kieras, 1996). CMN-GOMS deals with inferable cognitive processes and observable, physical actions. KLM-GOMS focuses, as the name implies, on observable, physical actions. As such, it is conceptually simpler than CMN-GOMS. Both approaches are relatively behavioristic in their approach, but no more than, say, Turing's test. Both are based on the assumption that an overall task can be decomposed effectively and meaningfully into subtasks. The KLM-GOMS technique takes several simplifying assumptions that make for a slightly simpler and slightly restricted version of GOMS. A third version of GOMS is the natural GOMS language (NGOMSL) (Kieras, 1997). This version adds natural language as a basis with which to build GOMS models and a procedure with which to build GOMS models. NGOMSL methods are represented by an underlying cognitive theory known as cognitive complexity theory (CCT). CCT provides NGOMSL with the ability to incorporate internal operators such as using information in working memory or establishing new subgoals. NGOMSL draws upon cognitive psychology to estimate the time required to learn how to achieve tasks. A fourth version of GOMS is CPM-GOMS (CPM standards for critical path method) (John et al., 2002). It supports parallel information processing by the user; however, it

is more complex than other versions of GOMS and so requires a steeper learning curve for the practitioner to acquire.

GOMS seems to be able to serve any theory that makes the assumption of the decomposition of tasks into discrete psychological processes and physical movements. While GOMS comes directly out of MHP, potentially it can rest on any well-founded theory of the users' perceptual, cognitive, and psychomotor processes. It is worth defining the previous four constructs of GOMS, though they may be readily transparent to most, if not all, readers. Goals indicate that the users want to achieve certain results through the use of an interactive system (e.g., write a book). Operators are the basic responses that a user must make (e.g., keystroke, mouse click, hand gesture, etc.). Methods refer to the achievement of subgoals through the carrying out of sequences of actions (e.g., use of menus vs. use of special key movements). Thus methods are pitched at a more abstract level than are operators. Selection rules are the basis upon which users select the most appropriate methods with which to achieve their goals (e.g., "use special key movements for very familiar tasks, but use menu options for less familiar tasks"). GOMS has both strengths and weaknesses (Olson and Olson, 1995). Benefits of GOMS models include: the easy comparison of two or more systems in terms of execution time and learning time; and the ease to modify and measure the consistency of systems. Limitations of GOMS include: it is aimed generally at expert users, not novices or intermediates; it assumes errorless performance; it cannot predict errors or error rate; sessional factors such as fatigue are not accommodated; individual differences are ignored; and the focus is on usability but not on functionality or accessibility. As shown in this chapter, new approaches are aiming to overcome these limitations.

However, the further elaboration of this level of detail may be of more interest to the researcher and of less interest to the practitioner (e.g., the designer, consultant, personal advisor, etc.) who requires a well-justified list of dimensions with which to capture crucial user characteristics. It may be useful to show these dimensions upon which user-specific requirements can be based, and so Table 24.2 brings together not only the intrinsic dimensions of human cognitive performance, but also a set of additional, extrinsic dimensions that relate to them: age, culture, language/s, demographics, IT training, and income level, producing fifteen measurements in all. The former group of factors is likely to be reflected directly in the specific user models, while the latter may influence generic aspects of the system, such as language of use, or call upon user stereotypes such as those related to age. This list is not intended to be exhaustive. For example, some practitioners may wish to add gender, numeracy, literacy, or personality (Adams, 2004). The previous factors are depicted by Table 24.2, the completion of which could provide a suitable basis for a specific user model.

At the moment, the psychological dimensions of user diversity supplied by Simplex Two (Adams, 2007) are treated holistically. Provision has been made for the development of subsystems that

TABLE 24.2   Dimensions of User Diversity

Psychological Factors: Structure for a Cognitive User Profile			
Inclusion cube	Simplex 2	Subsystems	Measure
Input	Input system		
	Feedback reception		
Cognition	Executive function		
	Working memory		
	Emotional evaluation		
	Mental models		
	Long-term memory		
Response	Simple response systems		
	Complex, learned responses		
Additional factors			
Age			
Culture			
Languages			
Demographics			
IT training level			
Income level			

may be important for research purposes. But there may also be the risk of adding unwanted complexity to the user-monitoring (evaluation) and user-modeling processes. There is a trade-off between power/complexity on the one hand and simplicity/ease-of-use on the other hand.

24.4   Role and Importance of User Modeling in Universal Access

The importance of user modeling in universal access rests on its contributions to:

- Capturing the central aspects of user requirements and preferences
- Achieving the seemingly impossible task of capturing critical user diversity
- Portraying the specific needs of atypical users (Newell and Gregor, 2000)
- Developing powerful new application development methodologies for emerging platform innovations
- Achieving inclusive design
- Attaining e-accessibility
- Developing new assessment methods

These are all the essential ingredients that support the aspiration of universal access, as expressed in the development of the inclusive information society. User modeling is therefore an important component of universal access methods for interactive system design.

User modeling can also be complemented by user monitoring, which is vital if user models are to be based upon accurate and up-to-date evidence. User monitoring supports a range of valuable functions. Adequate user models must be based on a broad range of user characteristics and their dimensions of diversity, which are intrinsic to the inclusive information society. In turn, they allow for user-specific recommendations, for the creation of alternative new and emerging interaction techniques and devices to support diversity in user interaction, working with the various dimensions of diversity in the user population, including, but not limited to, various forms of disability. User monitoring, mediated by effective user models, can support system design, personalization, and adaptation, as well as supporting smarter interactive systems that are capable of automatically adapting to diverse user needs and requirements.

User monitoring is critical since user characteristics and requirements are rarely static. Some skills, interests, capabilities, and disabilities are illness related; others reflect age-related changes (positive or negative) and systematic changes associated with education, culture, IST experience, and vocational changes; while still others relate to session-based changes such as fatigue, loss of concentration, and so on, and interactions between sessional changes and longer term factors such as head injury. The aspiration is that smarter systems can both capture initial (pre-interaction) characteristics of the user (e.g., preferences and skill levels) and also monitor ongoing changes in user characteristics, requirements, and psychophysiology, and adapt accordingly in appropriate and acceptable ways during interaction.

User monitoring captures user data of different types in a range of different ways. For example, one can simply ask the users for their views, get them to try out prototypes, record online performance, or even conduct rigorous laboratory experiments or prolonged field trials. Each method has its own strengths and weaknesses. Advocates of one approach alone, as shown in the following, are rarely correct. Methods can be seen as falling along a dimension from informal to formal. Relatively informal methods often have the advantages of immediacy, timeliness, breadth, and obvious relevance. But, their data may also prove to be too subjective, unreliable, and unreplicable. The data from well-designed, laboratory-based experiments may prove to be precise, well focused, reliable, and replicable. Conversely, their data may be too narrowly focused, of limited applicability, and of limited relevance.

Recent work on developing user profiles has found that user-related problems of insight and awareness can actually limit the value of certain approaches, particularly when these methods are used in isolation. In this context, insight is defined as the user having an overall, long-term understanding of their strengths and weaknesses. Awareness is defined as a shorter-term appreciation that their current or recent performance has been good or bad, perhaps even being able to comment on their working standards. Adams and Langdon (2004) explored the relationship between performance awareness and overall insight and concluded that it is more complex than originally predicted. Insight and awareness are not necessarily synonymous and can occur independently of each other. In particular, they found that insight problems can occur without awareness problems, but not the reverse, in a series of case studies of individuals with various disabilities, including, but not limited to, head-injury-related problems. The original paper gives the fuller details of each case (Adams and Langdon, 2004). Table 24.3 shows the results.

Clearly, insight and awareness do not necessarily share the same underlying psychological mechanisms. Many clients with acquired disabilities gradually develop a degree of insight into their difficulties, but it is clear that this development can be slow and painful. If an individual can possess some awareness of current performance, he should be able to use that awareness to build up insight into strengths and weaknesses. Surprisingly, this does not always happen. In four of the five in-depth case studies, poor insight accompanied good performance awareness. The link from awareness to insight itself may be damaged. If so, considerable effort may be required to build an adequate level of insight. There may be a good case for the inclusion of checks of insight and awareness within a user-centered design methodology. Clearly, measurement of user characteristics may rely on assumptions about their levels of insight or awareness, particularly when there is some doubt about user insight or awareness.

24.5 Inclusiveness, Adaptability, Adaptivity, and Augmentation

User modeling provides an effective basis for inclusiveness, adaptability, adaptivity, and augmentation (Adams, 2006b). Inclusiveness requires a focus not only on technological concepts combined with innovative solutions, but also on generic and specific user requirements captured through user-specific models. An adaptable system refers to a system that has the capability to be personalized or adapted to match user-specific requirements by entering user-specific model data before run-time. It can also be used to capture user-specific feedback after system use. An adaptive system is one that has the facility to update during run-time such a user-specific model, based on monitored user performance (see also Chapters 16 and 21 of this handbook).

Augmentation refers to the provision of additional, system-based cognitive resources to compensate an individual for relatively excessive demands from the environment of use (see also Chapter 7 of this handbook). In each case, a user-specific model provides the necessary data for the necessary adaptation function to be carried out successfully. Another way of looking at this use of the user model is through the concept of distributed cognition (Sharp et al., 2007, pp. 129–131), where the possibility of user overload is compensated for by the provision of externally distributed cognitive resources in the environment. For example, a personal digital assistant (PDA) prompting device can augment a person's own faulty memory processes to provide navigational or decision support. The key concept is to employ the user model to identify relevant strengths and also important areas where augmentation or cognitive support would be welcome.

TABLE 24.3 User Cases with Insight and Awareness Problems Based on the Concept of Universal Access

Case Study	Insight	Awareness of Performance
A1.1	* Underestimates abilities	* Slightly underestimates
A1.2	* Underestimates abilities	* Slightly underestimates
A1.3	* No insight into problems	OK
A1.4	* Insight problems, incorrect estimation	OK
A1.5	* Only minor insight problems	OK
A1.6	* Only minor insight problems	OK
A1.7	* No insight into problems	* Low awareness
A1.8	OK	OK
A1.9	* No insight into problems	* Low awareness

An * indicates a problem.

24.6 User Modeling and Unified User Interface Design

The user modeling and unified user interface design (UUID) methodology (Savidis et al., 2001; see also the relevant chapters of this handbook) is motivated by the perceived need for "a proactive approach to cater to the requirements of the broadest possible end-user population" (Stephanidis, 2001a). The aim is to meet the challenge of accessibility for all and of the quality of the user–system interaction (Adams and Russell, 2006). UUID methodology defines a single interface specification that addresses all intended user groups. Increasingly, intended users of accessible systems are seen as not just people with disabilities or older adults, but the widest population at large. UUID is intended to remove the need for post-hoc adaptations or add-ons, which are often purely reactive in design. Two key themes of the UUID methodology are adaptability and adaptivity. The former (adaptability) is defined as the propensity of an interactive system to allow for change with which to meet user requirements on the basis of information about the user prior to the start of a session of interaction. The latter (adaptivity) reflects the ability of a system to alter its manner of interaction from information gleaned while interacting with the user. Both adaptability and adaptivity require the measurement of accessibility and user models to enhance accessibility and improve the available user-specific models.

UUID provides a structured context within which these two themes can be deployed effectively, that is, on a rational and systematic consideration of the design space as opposed to an arbitrary, idiosyncratic, or subjective approach. Thus UUID allows for a systematic consideration of the numerous design decisions that must or could be made. This is particularly important as the number of possible and available technological solutions are increasing dramatically, fueled by the development of new platforms and new modes of interaction. Savidis and Stephanidis (2001) point out that software adaptation can be made dynamically, keeping pace with concurrent functional enhancements. Moseley (2005) shows that hardware can be adapted too and that dynamic and reconfigurable hardware systems are available that support adaptation and adaptivity based on a special kind of semiconductor chips known as field-programmable gate arrays (FPGAs). Thus, the potential for design overload grows further, and the deployment of methodologies such as UUID, to cope with it, becomes even more important.

UUID provides a substantial repertoire of conceptual building blocks, toolkits, and a grammar for toolkit expansion based on the requisite tool properties (Savidis and Stephanidis, 2001; see also Chapter 18, "Software Requirements for Inclusive User Interfaces"). The UUID paradigm also has a well-developed software architecture (see Chapter 21 of this handbook), including the following orthogonal components:

- Dialogue patterns component (DPC)
- Context parameters server (CPS)
- User information server (UIS)
- Decision-making component (DMC) in conjunction with the run-time model

To enable system-driven adaptation to take place effectively and efficiently, all the necessary parameters, decision-making logic rules, and interface artifacts are encapsulated in the software in a computational form, thus creating a viable run-time system environment. The four components are related to each other by the communication routes shown in Figure 24.4 (Section 24.7).

The UIS is the primary concern in the context of this chapter. This is because the role of this server is to record, maintain, and update the user models of the intended users of the system. The UIS starts with some form of user identification such as a password, bio-sign, or smart card. This allows the UIS to identify and select the relevant specific user models and to draw any relevant, consequent implications for adaptation. Preexisting user knowledge can be encoded in a variety of appropriate forms, reflecting the type of information the UIS needs to send to the DMC. Additional knowledge-based components can be used to process the retrieved user profiles, applying assumptions about the user and updating the original user profiles. Systems such as USE-IT (Stephanidis, 2001b) are used for this task.

The UIS can perform both an initial manipulation of user profiles prior to the start of interaction and a collection and processing of data about run-time interaction events to make additional inferences about the users, such as changes in performance levels, preferences, task orientation, fatigue, task failure, and so on. In combination with context-of-use data from the CPS, information is fed through to the DMC that makes the decisions about the adaptation options, qualitative or quantitative, based on a predefined set of rules. This information consists of messages conveying user attribute values.

24.7 A Framework for User Modeling

This section introduces a brief framework as a useful way to view user modeling, based upon the key concepts of theory, generic models, and specific models. Later on, the concept of simplistic theory is also introduced and defined. One first step in this approach to user modeling is to construct a theory of the relevant psychological processes of the intended users. For example, as shown in Figure 24.2, Simplex Two (Adams, 2006a) is a theory of human cognition that has been developed to underpin user modeling and system accessibility assessment.

Simplex has been validated by two large-sample, qualitative meta-analyses (Adams, 2007). As with many theories of cognitive psychology, it is intended to supply a cognitive, multiprocessor architecture that can be applied to each and every individual. Unlike most mainstream psychological theories, it also focuses upon individual differences in capabilities and disabilities, by providing a basis for building individual profiles. Another powerful example of an underpinning theory would be

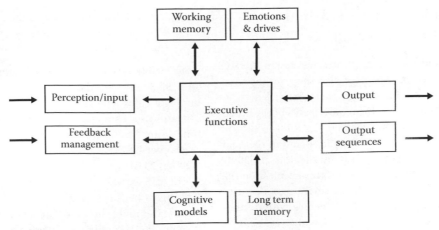

FIGURE 24.2 Cognitive architecture of a user based on Simplex Two.

the triarchic theory of human intelligence of Sternberg, which has the potential to be applied to cognitive user modeling, both overall and at the individual level, but it has not yet been applied in this way (Sternberg, 2000). A theory is defined here as "a set of statements or principles devised to explain a group of facts or phenomena, especially one that has been repeatedly tested or is widely accepted and can be used to make predictions about natural phenomena." In principle, one could consider any suitable underpinning theory (e.g., Simplex Two or Sternberg's triarchic theory). In this chapter, Simplex Two is used as an example, not because it is considered more suitable than other theories, but because it provides a typical view.

The second step is to create a generic user model derived from the adopted theory, in the case of this chapter that is the Simplex theory, to capture the essential features of that theory for a given design purpose. A model is defined in this chapter as "a depiction or simulation of a theory that assumes the theory to be true and captures aspects of the theory that are relevant to the task at hand." In particular, a generic model is defined as that type of model "that can be used to define a set of specific models that share a common architecture but are distinguished by the possession of different qualitative or quantitative features."

Thus, a generic model can be used to depict the nature of a theory and so subject that theory to tests of its sufficiency, consistency, or predictability, the results of which can then be fed back into the design of the theory. The term generic or (general) can be used in at least two ways. In this chapter, a generic model is intended as a model that captures the essence of all possible users, but does not distinguish between different users. For the purpose of focusing on a specific user, a specific user model is needed (see Figure 24.3). However, a user model might be applicable only to the requirements of a specific application. As defined in Kass and Finin (1991), the concept of a general user model can be used to derive user models (for the same users) across different applications. In this chapter, the term generic model is used to indicate generalization across users, and the term general model to be generalizable across different systems or applications.

As an example, a theory of human memory can be considered that specifies two distinct types of memory store: memory A and memory B. A generic model to depict that theory might consist of a system of two box files and a set of reference cards. It could then be proposed that each incoming item to be remembered is written on a card and entered into one of the boxes. When a memory is full, new items can only be added by displacing existing items. The creation of such a model might be used to help to understand where the theory needed further specification (e.g., how is the memory store selected in each case).

The next step is the generation of a specific user model to capture the skills and requirements of a specific user or set of users. Such a specific model is a member of the class of all specific models defined by the generic model. In the previous example, a specific model might specify the capacity of each memory type (A or B); so that the capacity of memory A is seven items and that of memory B is 100 items. These would be average values and reflect the average user. In practice, intended users may vary in their values. For example, one user may have a capacity of five units for A and 120 items for B, while a second user's values may be different (A = 12; B = 80). Thus, at this point after the first three steps have been carried out, a theory of user psychology, a generic model of that theory, and a specific model about a subset of users or even a single user are available for practical application. As shown previously, the move from a generic model to a specific model is through the assignment of specific parametric values for each relevant aspect of the generic model, within a stated level of uncertainty (Kay, 2001; Adams and Langdon, 2003).

The subsequent fourth step is the specification of a way to encapsulate user information to be applied within an existing adaptable or adaptive system or for an inclusive design

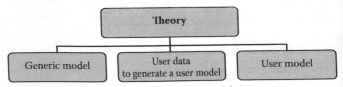

FIGURE 24.3 Relating theory and user models.

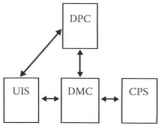

FIGURE 24.4 Architecture of a UUID-based, adapting system.

methodology such as UUID (Stephanidis, 2001a, 2001b; see also Chapter 16, "Unified Design for User Interface Adaptation") to create a new system. This is shown in Figure 24.4.

As introduced previously, Simplex Two is a cognitive psychology theory of human performance. It specifies nine components of human cognition or mental activities related to each other by a defined architecture that centers on an executive function. This theory can be depicted by a generic model in diagrammatic form, consisting of simple boxes and arrows, in which the nine components are linked by a set of arrows to indicate the flow of information within the system (see Figure 24.2). A specific model (of a specific user or set of users) can be produced by evaluating the user (or set of users) on each of the nine components, and assigning a descriptive or numerical (nominal, ordinal, or interval scales) value to each component, or even subcomponents, indicating the user's capabilities or disabilities.

The specific (i.e., calibrated) models are then used to evaluate the match of intended users and existing systems and to develop new systems. They can also constitute a design component of inclusive system design methodologies. They are used by adaptable and adaptive systems, the former to adapt to a known user profile as captured by the user model, the latter as a starting point for updating that profile based on user performance.

24.8 The Contribution of Computing Science

Computing science contributes to user modeling in a number of emerging ways. These ways include methods to capture and hold user models in the system so that systems designs and system interfaces can be customized to meet the needs of an individual user. Second, computing science approaches to user modeling also allow for the design and creation of adaptable systems such that the system design and interface can adapt to each different user as their individual profiles are entered into the system before run-time. Third, computing science approaches to user modeling also allows for the design and creation of adaptive systems such that a user model can be adapted during run-time to allow the system to learn about changes in the user's requirements and to adapt accordingly.

Emerging methods (such as UUID) can provide a conceptual and practical basis for such adapted, adaptable, and adaptive designs for families of systems. An early example of an adaptive system is the Friends21 framework (Ueda, 2001), a system that

employs a learning process to build user and task models. These models are maintained by the system, holding information in a user information manager, covering user interests, skills, and preferences. An example of a universal access methodology is the unified user interface architecture (Savidis and Stephanidis, 2001; see also Chapter 21, "A Unified Software Architecture for User Interface Adaptation"), which places user models at the center of the design process by means of the UIS, as discussed in this chapter. Such powerful approaches generate both system implementations that reflect the properties of known user models and have the capability to learn and adapt to new or emerging user models (Kay, 2001).

The concept of the stereotype has been employed to provide systems for user modeling, including the UM toolkit (Kay, 2001) and BGP-MS (belief, goal, and plan-maintenance system) developed by Kobsa (1995, 2001). For example, someone may be designated a novice, while someone else might be nominated as an intermediate or an expert, with all the qualities associated with that designation. Stereotypes often identify the most plausible user characteristics and are often used as default profiles. One advantage of such an approach is that further user attributes are inferred from the initial information, but it is worth noting that such inferred attributes are not always correct—and so have to be checked and utilized carefully. Stereotypes may also be used to decide to which communities a person belongs, such that the intersection of communities can capture the diversity of different people. A stereotype can be seen as at least a first approximation in the development of a more detailed model.

Computing science provides several representation languages to portray the details of user models. These include, inter alia, XML, UML (see also Chapter 26 of this handbook), and Java. For example, the primary use for XML is recognized as the description of the meaning of data (Kraynak, 2005; Moseley, 2005). For example, the profile of an individual user, using Simplex Two, might be captured by XML (extensible markup language) as follows:

```xml
<?xml version="1.0" encoding="ISO-8859-1"?>
<user>
<ID>Smith</ID>
<perception>weak</perception>
<feedback>medium</feedback>
<working_memory>strong</ working_memory>
<LTM>weak</LTM>
<mental_modeling>medium</mental_modeling>
<executive>strong</executive>
<emotions>weak</emotions>
<output>medium</output>
<complex_output>strong</complex_output>
</user>
```

This XML profile assigns a value (in this case: low, medium, or high) to each of the nine cognitive components identified by Simplex Two. "<user>" is the root element and each subsequent line captures both the data (low, medium, or high) and the meaning of the data (e.g., output). Two qualities of XML are important to note. First, the simple syntax of XML must be implemented more carefully than, for example, that of HTML to be parsed. Second, it is clear enough to be read by both users and computers.

UML (unified modeling language) is another powerful way to use modeling as a basis for designing good systems and to take a user-centered approach to capture the perspectives of different users (Stevens and Pooley, 2000). UML is a diagrammatic language defined by both its syntax and semantics. The syntactic rules define which diagrams are legitimate and which are not. The semantic rules determine the significance of those acceptable diagrams. UML provides the system designer with a family of diagram types that support the visualization of the design issues and solutions that make for high-quality designs.

Java can also provide a powerful, conceptual basis for dynamic user modeling, in which key concepts are depicted by objects or classes that send messages or methods call to each other to exchange and process information. For example, Ye and Herbert (2004) developed a framework for manual or automatic adaptation of user interface design to match current and emerging user characteristics. They produced a Java-based framework that supported the introduction or removal of adaptation mechanisms using an XML compliant, generic user interface markup language (XUL). Java can be used to implement artificial neural networks for user modeling (Watson, 1997), though some authors have expressed strong reservations about the validity of neural networks for psychological, functional, and user modeling (Coltheart, 2004).

The development of user models should not be confined to providing specific user models. Thus, there is a need for general user model systems (see previous discussion) that can provide and adapt user models across different systems or applications, since user requirements are likely to vary systematically across them (e.g., different levels of expertise). Kass and Finin (1991) presented GUMS, a general user-modeling shell that provides a platform for the creation, maintenance, and updating of user models. It employs three techniques to represent user knowledge: stereotypes (see previous discussion), explicit default rules, and failure as negation. The acquisition of new models is handled by GUMAC, the general user model acquisition component. New models can be acquired explicitly (by the feeding in of new information) or implicitly (based on user interactions with the system). The latter is guided by a set of user model acquisition rules. This work has shown that general user modeling across different systems is not only beneficial, but is also a practical proposition (Kass and Finin, 1991; Neal and Shapiro, 1991). A further consideration of universal access methodologies and associated software architectures (UUID; Savidis and Stephanidis, 2001) is provided in the following section.

24.9 The Contribution of Cognitive Science

A concern for the requirements and preferences of specific users or groups of user when using information-processing technology

leads almost inevitably to the consideration of human information processing, cognitive psychology of the user, the overall architecture of human psychology, individual psychological differences, and the aetiology of specific cognitive disabilities (Ellis and Young, 1996). There is substantial evidence that user modeling can exert powerful influences on the effective utilization of computer technology, for example, effective web navigation (Juvina and van Oostendorp, 2004), message handling (Jokinen et al., 2004), adapted speech support for disabled users of web browsers (Heim et al., 2006), and agent-based interactions (Benyon et al., 2005). Cognitive science provides a number of powerful, user-relevant theories. A consideration of the full range of cognitive theories is beyond the scope of this chapter, but a limited set of theories will be discussed as they are very relevant to user modeling.

Atkinson and Shiffrin (1968) produced a pioneering theory to human cognition, focusing on short-term memory and on human information processing as a way to conceptualize human cognition. They influenced the MHP (Card et al., 1983), which is one of the first and best-known approaches to both human cognition and interactive system design. Broadbent's Maltese cross provided a further, conceptual advance, affording a well-defined executive function and supporting the two-way flow of information. Broadbent's memory-store model has been based on a considerable volume of high-quality research and was presented in a seminal paper (Broadbent, 1984). It includes four memory stores and a well-delineated executive function, that is, input memory, output memory, working memory, long-term memory, and a linking executive function, and was presented as a simplistic model (see the following) of human memory. However, it is clear that it has the potential to become a cohesive architecture of human cognition. It and MHP have been compared with a powerful von Neumann machine.

The work on the Simplex (Adams, 2007) approach to user modeling began with the Maltese cross as a conceptual framework. This provided a structured basis for selecting appropriate user assessment methods more systematically. However, our need to give emphasis to cognition and performance rather than memory led to the development of Simplex One, in which each of the five components were given modular status (Pinker, 2002), possessing both memory and processing capability. This created a more powerful explanatory model that is more like a parallel processor than a serial processing system or von Neumann machine. Further work indicated that nine dimensions were needed rather than the five of Simplex One (Adams, 2007). This led to the formulation of Simplex Two. This new model provides nine dimensions, each of which is treated holistically to evaluate a specific user's requirements. It has proved to be very useful in practical terms and has been used in a large number of individual assessments and case studies. Equally important, it has proved to be of theoretical value and has been validated by two large-sample, qualitative meta-analyses, confirming the nine components, which were, in fact, the only factors that were found in both analyses (Adams, 2007). Further work will look at other candidate factors such as episodic memory and social intelligence, which were identified in one (out of the two) study only. However, Baddeley (2000) suggests that episodic memory should be viewed as a component of working memory, so perhaps it should not be seen as a separate module. Work on social intelligence is reported by Adams and Gill (2007a, 2007b).

Double dissociation (defined in the following) is a vitally important principle when distinguishing between different requirements for different users. Traditionally, in neuropsychology, people were simply grouped together by common symptoms (e.g., poor ability to process pictures). However, the concept of double dissociation is a more powerful way of distinguishing between subgroups of users (Ellis and Young, 1996). For example, an individual may perform badly on a memory task, but do well on a task requiring sustained attention. In this case, a single dissociation has been found between the two tasks. So a group of such users who show the same profile could be distinguished from a group who did well on both tasks.

However, this contrast could simply be reflecting the possibility that memory task was more sensitive because it was more difficult than the attention task. The difference between the two groups may have nothing to do with a specific cognitive function such as memory. If, on the other hand, a second group did well on the memory task and badly on the attention task, then this difference cannot be explained by task difficulty and is more likely to be related to selective impairment of the cognitive functions involved in the two tasks. While the example given here uses cognitive functions, the same argument applies to most aspects of ability or disability. Adams and Langdon (2004) provide case studies in which double dissociation is demonstrated between two individuals with difference profiles, as shown in Figure 24.5. Only a full interaction, such as this example, can rule out practice effects as an explanation. In summary, then, a double dissociation occurs when a user or group of users performs better on one task rather than another task, while a second user or group of users shows the opposite by being better at the second task.

Now we consider the concept of the simplistic theory and contrast it with other types of theory to explore the practical implications of adopting different types of theory. Broadbent referred to his Maltese cross approach as *simplistic*, a term that has been taken further and formally defined by Adams and Langdon (2005). They defined a simplistic theory as one that both captures the key results and concepts of research, but also is sufficiently clear and simple to guide relevant practitioners. In these terms, Atkinson and Shiffrin's (1968) was not a simplistic theory but had

the potential to be so. The MHP was intended to be a simplistic theory, since it aimed to capture the then-current state of the art in cognitive psychology and aimed to solve problems of interactive system design in HCI. Broadbent was not only an accomplished theoretician, but he was also a famous applied psychologist. So, while he did not focus on more specific areas of application such as HCI, he strongly expected good theories to be applied to the solution of real-world problems such as HCI. Finally, Simplex One and Simplex Two were both intended to be simplistic theories, particularly to provide a research-grounded guide to user modeling and the design of accessible, interactive systems.

The MHP drew upon extant information-processing theories of human cognition such as that of Atkinson and Shiffrin (1968), and included core functions such as input, short-term memory, long-term memory, output, and cognitive processes. However, it is concise enough to be a workable guide for practitioners such as designers, consultants, rehabilitation workers, teachers, and so on. The key features of MHP include: information processing, unidirectional flow of information, short-term memory but no explicit executive function, all representative of the then extant conceptual consensus.

In addition, there are some theories of intermediate complexity that are relevant to user modeling and interaction design. Two examples are taken from Norman (1986, 1998, 2004) and show the increasing complexity of such models. First, Norman (1986) presented the theory of action that captures the activities of an interactive system user by seven types of action:

- Goal establishment
- Intention formation
- Action specification
- Action execution
- Perception of state
- Interpretation of state
- Comparison of state to goals and intensions

To relate this theory to system design, two types of gap were postulated that had to be overcome by good design: an execution gap and an evaluation gap. An execution gap is a gap between a user's goal for action and the means to execute that goal. An evaluation gap is the gap between the way a system is understood and the way it is intended to be understood. Closing these gaps is said to reduce the potential for cognitive overload of a given design. Critics of this approach claim that human actions are not as structured as this theory supposes (Sharp et al., 2007).

Second, Norman (2004) has proposed the emotional design model that focuses on the importance of human emotional factors in human activities. This approach postulates three different levels of the human brain that support qualitatively different types of emotional expression: visceral (automatic), behavioral (usability), or reflective (significance and meaning). Again, critics of this approach claim that human emotions are much more complex and variable than this theory supposes (Sharp et al., 2007). However, such criticisms can, arguably, be leveled at all simplistic models and may not diminish their practical usefulness.

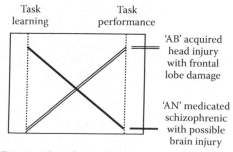

FIGURE 24.5 The relationship between task learning and task performance for two individuals tested with the Stroop assessment test.

Finally, there are a number of powerful, cognitive frameworks that could form the basis for cognitive user models. They include ACT-R (adaptive control of thought–rational; Anderson, 1983; Anderson and Lebiere, 1998), SOAR (Newell, 1990), COGENT (Fox, 1980; Fox and Cooper, 1997), and ICS (interacting cognitive subsystems; Barnard, 1999; Barnard et al., 2000). A full review of such models is beyond the scope of this chapter, but these models all appear to share a number of important qualities and they deserve attention within the context of cognitive user modeling for universal access.

ACT-R is intended to provide the architecture of human cognition, both for purposes of simulating and understanding intelligent human behavior. It is intended to develop into a system that can perform the full range of human cognitive activities. Further information is available from the ACT-R web site (http://www.act-r.psy.cmu.edu.com). ACT-R is focused on creating both a coherent theory of cognitive processes and the application of such a theory to the solution of problems of practical importance. As such, it continues the efforts of great psychologists such as Donald Broadbent (e.g., Broadbent, 1957, 1971) and Gordon Bower (e.g., Hilgard and Bower, 1966) in attempting to combine psychological theory and practical application. Anderson has a deliberately equal concentration on applying theory to specific problems and developing an explanatory theory of cognition.

The structure of ACT-R aims to capture key findings from cognitive psychology research, though it also resembles a programming language or development environment. Thus, it is possible to write programs (or models) in ACT-R that combine ACT-R's axioms of cognition with predictions about specific task requirements. The next step in this methodology is to compare the performance of the model with that of human participants in the same task, and so test these predictions. This closely parallels the previous treatment of theories, generic and specific models for user modeling.

ACT-R is made up of at least three components: (1) modules (or programs), (2) buffers, and (3) a pattern matcher. There are two types of module: perceptual-motor and memory. Memory is further divided into declarative memory (knowledge for facts) and procedural memory (knowledge that enables us to carry out skilled actions). The latter is modeled by production systems in ACT-R. Access to the modules is via ACT-R's buffers. The pattern matcher searches for the required productions that are implemented one at a time (i.e., in a serial manner). Thus, the production system that produces responses, both internal and external, is a serial system. In addition, ACT-R possesses powerful subsystems that represent underlying cognitive processes such as perceptual analyses of the surrounding that are essentially parallel in nature. ACT-R does not appear to have an overall executive function, but conflict resolution mechanisms deal with dissonant outputs from the parallel processes.

ACT-R research has generated over 500 publications (as shown in their web site), and this count does not include publications about earlier versions of ACT-R and predecessor models. Anderson et al. (2004) provide a very readable account of ACT-R. Byrne (2001) is a good example of work that applies the ACT-R approach to cognitive user modeling in an HCI context, exploring menu design and menu selection. In this context, a well-designed interactive design is considered to be based upon models of the cognitive, perceptual, and motor capabilities of the user, the task, and the technology used (Adams, 2005a; Adams and Langdon, 2003). Byrne used a variant of ACT-R called ACT-R/PM, where PM stands for perceptual motor. He used two earlier models of a menu selection task and some data from an eye-tracking task to generate a new series of models based on ACT-R/PM, such that the final version presented a satisfactory if not perfect simulation of the data.

Besides its applications to cognitive psychology, ACT-R has been applied to the following practical problems: the development of user models to assess different computer-human interfaces, intelligent e-learning systems, and computer-based cognitive agent systems for e-training. ACT-R combines specific axioms about human cognition with the demands of each task domain to produce task-specific solutions. Thus, ACT-R is a very good candidate for a cognitive user model. Anderson et al. (1997) discuss the ACT-R approach to user modeling. They report how ACT-R has been extended with a theory about the interaction between higher-level processes and a visual interface. This theory takes account of the way in which visual attention can be distributed across a screen and how information can be encoded in a way that ACT-R can use. ACT-R was used to simulate several classic results that relate speed and selectivity to ways in which visual attention can move across a visual display. Also, ACT-R makes clear predictions about the time taken to search and select from menus. These predictions were confirmed. It was concluded that ACT-R could interact with visual computer displays with the same performance as that of typical human subjects. On this basis, it is argued that ACT-R is well suited to provide a model for tasks involving computer-human interactions.

Clearly, ACT-R does a very good job of modeling human cognition. It also captures the performance of typical subjects on well-defined tasks using visual computer displays. The next steps might be to develop the capacity to model the cognitive profiles of: (1) different subgroups of intended system users, (2) users with specific skills impairments, or (3) users with special needs. This would provide a very robust foundation for building powerful cognitive user models in the future. Thus, ACT-R has important implications for both HCI and for universal access.

Cogent (Fox, 1980; Fox and Cooper, 1997) is a modeling environment for cognitive theories and models in general and functionally modular box and arrow depictions in particular. It builds upon the concepts of functional modularity (from cognitive psychology) and object-oriented design (from computer science). Models can be produced in terms of interacting processes and buffers. For example, a model can be produced for medical diagnostic tasks where working memory is set out as a distinct information store (Fox, 1980; Fox and Cooper, 1997). In these and other publications, these authors developed a range of cognitive models within the Cogent framework. They conclude that different assumptions about the specific knowledge representation are not critical to the predictive accuracy of Cogent. They also reported that manipulation of the retrieval functions supported the conclusion that the fit between model and subject performance disappears when alternate retrieval assumptions are considered (e.g., when more important symptoms are accessed before less important symptoms in a medical decision-making task). Cogent models also showed that the same behavior could be obtained from a system that has a simple propositional rule system (Fox, 1980), a first-order diagnosis system, or a generic decision procedure. Overall, Fox and Cooper (1997) focus on human learning and the computational models produced within Cogent and are able to generate performance data on specific tasks that are very similar to the data obtained from human participants. The authors distinguish between cognitive and algebraic approaches to human decision-making. They also argue that algebraic techniques do not take account of cognitive factors (e.g., memory limitations and specific subject strategies), individual subject differences (knowledge and experience) and situational factors (e.g., specific task requirements) and so are too limited to model the decision-making process. Fox and Cooper (1997), therefore, support a cognitive approach and their work is very relevant to a cognitive approach to user modeling.

The Cogent modeling environment provides a number of important features (as described in-depth on the Cogent web site http://cogent.psyc.bbk.ac.uk), including, but not limited to, a visual programming environment, a range of standard cognitive modules (memory buffers, structured knowledge bases, rule-based processes, connectionist networks, intermodule communication, interprocess communication, input/output (I/O) sources and sinks), and a rule-based modeling language and an implementation system.

Cogent acts as a programming language and as a framework with which new models of human cognition can be developed. On this basis, it should be possible to build variants of user models that distinguish between different groups of intended users of systems. So far, there appear to be no studies with Cogent that address the issues of user diversity, but Cogent is a powerful language and modeling system with a range of tools to support development of models and it has substantial potential for user modeling in universal access.

ICS was developed by Philip Barnard at Cambridge, United Kingdom, and is intended as a broadly based constituent analysis of human cognitive resources (e.g., Barnard, 1999; Barnard et al., 2000). It is a highly modular approach such that its architecture is based on a set of functionally independent modules communicating over a data network. Each module has the same internal functionality (i.e., the representation, storage, and transformation of information) but operates in different domains of processing. The modules are governed by a common set of processing principles. Barnard (1999) used ICS to predict and simulate a number of robust findings from short-term memory research, including influences of modality of presentation, word length, articulatory suppression, phonological similarity, suffixes, grouping, and very rapid presentation.

Barnard et al. (2000) has also applied the ICS mental architecture to account for the core signs and symptoms of mental disabilities, and in particular schizophrenia. Four sources of variation are posited to underpin information-processing activities: variation in self-representation, variation in modes of processing, variation in rates of change in the content of mental images, and variation in synchronization of exchanges of information between subsystems. Barnard concluded that in schizophrenia, the two subsystems specialized to handle (1) propositional and (2) implicational meaning become out-of-step with each other, resulting in cognitive dysfunctions and, often, confusion. At short feedback delays, the constituents of implicational representations become intermingled, providing conditions for thought disorder. At longer feedback delays, stream separation leads to conditions allowing the formation of abnormal implicational models.

Teasdale and Barnard (1993) employed ICS as a basis for understanding normal and dysfunctional cognitive-affective relationships and how they could be modified. Nine interacting cognitive subsystems were proposed, each specialized for handling a specific type of information. The emphasis in ICS is on the importance, as part of the total cognitive configuration-producing emotion, of a schematic synthetic level of processing to integrate both propositional meaning and direct sensory contributions. Processing at this overall level corresponds subjectively to a holistic sense or feeling rather than to thoughts or images. The influence of depression and dysphoric mood states was also investigated, as they are often associated with quantitative or qualitative shifts in performance across a range of memory tasks. They investigated the recollection of autobiographical events and word lists in dysphoric states (an emotional state characterized by anxiety, depression, or unease). In both autobiographical recall and in recognition memory, recollection in the dysphoric group was lower than recollection in matched controls. They found that recollection performance was influenced by two factors: (1) the degree of differentiation of schematic mental models and (2) the executive mode that predominates when memory tasks are carried out, within the context of the ICS model.

Teasdale and Barnard (1993) also explored ICS and emotions; in particular, they looked at depression and bipolar disorders, distinguishing two sorts of necessary representation. First, theories in the science base need to be expressed clearly. Second, representations need to be bridged from real-world behavior to basic theory and from theory back to the real world. It is suggested that macrotheories of the "normal" human mental architecture such as ICS could help synthesize basic theoretical accounts of diverse psychopathologies, without recourse to special-purpose clinical-cognitive theories of those particular psychopathologies or even specific symptoms. This suggestion is illustrated through concrete examples of how the essence of quite complex basic theory can be translated into a simpler representational format to help clinicians conceptualize a psychopathological state and pinpoint relevant variables that might be changed by therapeutic interventions. It is also pointed out that the inevitable problems of complexity in theories such as ICS can be overcome.

ICS researchers are active in the field of HCI, and their research is of potentially great relevance to universal access. For example, ICS has been used to explore the use of cinematographic techniques for computer interface design (Barnard et al., 2000). The conclusion was that computer interfaces could be made more usable if their designers made use of cinematography techniques, which have evolved to guide the viewer through a narrative despite frequent discontinuities in the presented scene (i.e., cuts between shots). They argued that the ICS psychological model of watching film could support such transference, and presented an extended account. This model allows the identification of colocation of objects of interest in the same screen position before and after a cut. Eye movements were measured while participants watched a commercially released motion picture in its entirety, in its original theatrical format. Predictions were made about the use of colocation. Results were broadly in line with collocation predictions. It was concluded that ICS does afford valuable insights into the human cognitive processes of the comprehension of film (Barnard et al., 2000).

Barnard (1999) and colleagues explored cognitive modeling as a basis for modeling user requirements. They provided the ICS architecture for human perception, cognition, and action and showed how it can be used to approximate models of human cognitive activity that can support practical interactive system design. ICS can also be used to model affective-cognitive interactions and the implications of mental illness for cognition, namely human cognitive dysfunctions. Clearly, it has some current capacity for modeling individual variations. If this is so, then ICS can capture important aspects of human psychology to provide a basis for user modeling. In addition, ICS explicitly deals with human cognitive dysfunctions and so can capture some of the human factors that are involved in human error, long-term or short-term. This capability to deal with human problems can be applied to those human problems associated with usability and accessibility in user modeling and universal access.

SOAR is another cognitive architecture introduced in the 1980s by Allen Newell and colleagues. It is intended to deal with both AI and psychological aspects of cognition; for example, it can be used for creating AI entities and for modeling human cognition. At the onset, the SOAR architecture (Newell, 1990) was focused on symbolic reasoning. Recently, extra features have been added, for example, episodic memory (Nuxoll and Laird, 2004) and the activation of the contents of working memory to support forgetting in working memory (Chong, 2003), as pioneered by the ACT-R family of models (Anderson and Lebiere, 1998). SOAR has a reputation for being difficult to learn and to use, but, nevertheless, has significant potential.

SOAR is essentially a production rule-based cognitive model with two types of knowledge: (1) working memory (short-term, declarative) and (2) production rules (long-term, procedural). Production rules consist of actions and conditions. If the conditions of a production match the contents of working memory, then that production fires, which creates (or removes) one or more elements in working memory. Chunking is the main mechanism for learning in SOAR. Its architecture has several

unique characteristics, including that simultaneous production firing is allowed. Many production systems permit one production to fire at a time. SOAR allows only those productions whose conditions match working memory and thus are not in conflict with each other to fire in parallel, avoiding possible conflicts that might arise from allowing the simultaneous firing of any productions.

SOAR creates a special type of knowledge structure called an operator. A production can propose an operator by creating it in the working memory. Multiple operators can be proposed in a given situation, but only one can be selected. Selection is controlled by the creation of preferences. SOAR operators extend the traditional match-fire cycle to a three-phase decision cycle: propose, select, and apply. SOAR distinguishes between two types of persistence in working memory. Working memory elements that are created as part of an application of an operator persist indefinitely; they remain in working memory until they are explicitly removed. Second, there are those working memory elements that are removed as soon as they are no longer needed. SOAR has a perceptual subsystem of independent modules for each input channel. The provided data go straight into working memory. The response subsystem also consists of independent modules for each output channel. They accept commands from working memory and execute them. Their progress can then be monitored through sensors that are fed back into the system via the perception subsystem. All perceptual and motor behavior is mediated through working memory. SOAR is a complex, powerful, if difficult to use, cognitive framework. It can generate a family of models. The power of SOAR also suggests that it could be developed to capture the essentials of intended users of systems and even to capture user diversity as required by user modeling.

The four complex cognitive theories considered previously seem to have much of value in common. Each does not, in fact, generate a single model, but produces a whole family of models, which can include versions that make crucially different predictions for critical results or that capture different phenomena or tasks or different types of user. If so, they are best seen as powerful frameworks for the production of a significant number of cognitive models with varying predictions and properties. These frameworks may be powerful enough to potentially deal with all known findings about human cognition.

However, to make useful predictions about human cognitive performance or to tackle a new problem domain, additional specifications may often (if not always) be needed for each cognitive framework. In addition, a further level of detail is often required to simulate specific user cognition problems. At the moment, it is still difficult to identify diagnostic tests that would distinguish between these theories and determine which are best. If they all have the full power of modeling languages, as seems likely, then they may always stand as optional alternatives to be chosen by convenience of application, design criteria, familiarity or user preferences, and so on.

More intriguing is the possibility that each of these frameworks will generate only a finite set of cognitive models. Such a finite set can be tested exhaustively, in principle, by repeated specification of necessary details. In practice, however, this may not yet be feasible. Nevertheless, it may yet prove possible eventually to distinguish between them. At the present time, despite the current level of complexity achieved to date, such frameworks or model families are in their growth and development phases. This is a testament to the complexity of human cognition. Perhaps these frameworks will move into a discriminatory phase, but, for now, they offer an intriguing basis for the creation of complex, cognitive user models. In the future, it will be found out if they may or may not be distinguishable from each other in principle or in practice. If the latter, then they may all provide a valuable basis for cognitive user models as judged on their scientific merits. They may or may not be distinguishable from each other in practice.

A critical aspect of these models is that they all capture a considerable amount of current knowledge about intelligent human behavior and cognition. Equally interesting, they all manage to capture cognitive phenomena in strikingly different ways, with different architectures, though there is a move toward greater synergy and comparison between them. Their capture of the essentials of human cognition is perhaps their strongest achievement. If so, they are well placed to meet the challenge of people such as Broadbent and Bower (see previous discussion), combining theoretical insights with practical problem solving. This is not to say that they have not had success in this domain, but clearly they still have unmet potential.

In the context of universal access, user models based on any one of the cognitive models should be able to reflect individual differences in sensory impairments, psychomotor impairment, cognitive deficits, special skills, and extreme environments (Adams, 2005b). To date, the four models vary significantly in the extent to which they have approached both theoretical and practical problems.

Because of their complexity and power, all four complex frameworks considered here will require a substantial learning investment from the researcher or practitioner. This is not necessarily a bad point, since one would expect to make such an investment to achieve the power and resources on offer. However, this point does underline the attractiveness of simplistic theories (such as MHP, GOMS, Simplex, etc.) that are more accessible to their users. SOAR, in particular, has been said to be very hard to learn and equally hard to use. ACT-R provides tutorials, and Cogent provides tutorial help and substantial tool support. If so, user-sensitive system designers may be reluctant to become serious users themselves. However, at a more practical level, a simple awareness with such complex theories may, at least, make designers more aware of some of the psychological and pedagogic factors that they need to consider during the design of new systems and their evaluation. The complexity of these frameworks is perhaps the reason why most researchers use only one of these cognitive families, but there has always been a subtheme of comparison or combination. This is particularly relevant to the U.S.-based systems, which all acknowledge the influence of MHP. The potential applicability of these cognitive theories (or families of theories) to universal access is well

founded upon their power and complexity; the evidence for it is yet to be established.

24.10 Multidisciplinary Issues: The Value for Different Readers

The concept of the user model is of value to a number of areas, including cognitive, clinical, occupational psychology and neuropsychology, rehabilitation, assistive technology, HCI, interactive system design, and related practitioner areas. While the focus of the present chapter has been on an appreciation of the role of user modeling in the context of universal access, it is clear from the issues discussed here and in other chapters of this handbook, that user modeling has a much wider scope for application. For example, the construction of user models not only provides guidance to support practitioner skills in learning methods, human factors, diagnosis, treatment, vocational assessment, team building, analysis, and rehabilitation, but also offers a testing ground for the theories that they purport to depict. While the design of a model assumes the theory to be true, nevertheless the attempts to model a theory test its sufficiency, consistency, and predictability, as mentioned previously.

Vocational rehabilitation relies on the practitioner possessing sufficient information and insight into the strengths and weaknesses of an individual. User modeling, particularly when based on well-established theories, can capture the complex details of an individual's requirements and so act as an effective guide to the rehabilitation expert. Similarly, a user model can provide a systematic basis for the identification of the most appropriate assistive technology solutions. In this case, such models can result from a variety of sources, from a simple conversation with an individual to an in-depth assessment lasting two to three days.

Turning to HCI and interactive system design, the user model can be used to generate relevant design heuristics (Adams and Langdon, 2003), to capture user requirements in the face of user diversity and complexity, and to explore the adequacy of the design of an interactive system by complementing the use of volunteers. The main point is that high-quality user modeling can be used both to develop germane theory and also to guide best practice by practitioners in related fields. This conclusion underlines the importance of simplistic theories, which, as discussed previously, both (1) aim to capture the state of the art in research and (2) guide best practice.

24.11 Confidentiality, Privacy, and User Modeling

The associated issues of accessibility and user modeling raise cautionary thoughts about the invasion of privacy or the exploitation of personal information. People are not always consistent in the information they divulge. For example, it is commonplace for web sites that offer free products or services to require the completion of an onscreen form giving personal information often accompanied by a request to use that information to distribute marketing information. It seems counterintuitive that one may be willing to provide the information online, but not to provide the same information to a stranger in the street who was offering an identical free product! What information would people want to provide for a user model for a system? Would that include, for example, preferences for movies, cognitive skills or disabilities, sexual orientation, or age? How would a user consider a system that could infer some of individual key characteristics from performance? What about the use of radiofrequency identification (RFID) technology, in which information from tags can be read automatically? When would this be seen as excessively intrusive? Perhaps "preference level for privacy" should be included as part of any user model in such circumstances.

24.12 The Future of User Modeling and Monitoring

The future of user modeling and monitoring is certainly exciting. Practitioners will increasingly have access to efficient and effective ways to capture key characteristics and requirements without having to commit to the excessively high workloads currently required by some approaches. Powerful and accessible theories of user psychology will be required to generate effective user models that capture the changing needs of diverse individuals, working at the design stage through emerging methodologies such as UUID and being applicable to a diverse range of systems through adaptation and adaptivity. These systems will be increasingly smart and responsive to user needs, and will often be invisible in the environment (Newman and Lemming, 1995; Stephanidis, 2001c). They may also be embedded in physical environments, acting invisibly, seamlessly, pervasively, and accessibly (Streitz, 2007). Aesthetics is an emerging aspect of acceptable design that could be conducive to universal accessibility, while catering to user diversity. Lawrence and Tavakol (2007) have presented their approach to effective web site design that they call balanced web site design (BWD), which attempts to build on the potential, but often missed, synergies between accessibility, usability, and aesthetics. Significant progress in all these areas will contribute to the emergence of the inclusive information society.

References

Adams, R. (2004). Universal access through client-centered cognitive assessment and personality, in *User-Centered Interaction Paradigms for Universal Access in the Information Society, Proceedings of the 8th ERCIM International Workshop "User Interfaces for All"* (C. Stary and C. Stephanidis, eds.), 28–29 June 2004, Vienna, Austria, pp. 3–15. Heidelberg: Springer-Verlag.

Adams, R. (2005a). Natural computing, in *Natural Computing and Interactive System Design* (R. Adams, ed.), pp. 5–34. Harlow, U.K.: Pearson.

Adams, R. (2005b). *Natural Computing and Interactive System Design.* Harlow, U.K.: Pearson.

Adams, R. (2006a). E-accessibility: The challenges of measurement. Presented at the *International Design for All Conference (DfA 2006)*, 13–15 September 2006, Rovaniemi, Finland. http://dfasuomi.stakes.fi/EN/dfa2006/rovaniemi/programme/parallel/eaccAdams.htm.

Adams, R. (2006b). Applying advanced concepts of cognitive overload and augmentation in practice: The future of overload, in *Foundations of Augmented Cognition* (2nd ed.) (D. Schmorrow, K. M. Stanney, and L. Reeves, eds.). Arlington, VA: Strategic Analysis.

Adams, R. (2007). Decision and stress: Cognition and e-accessibility in the information workplace. *Universal Access in the Information Society* 5: 363–379.

Adams, R. and Gill, S. (2007a). User modeling and social intelligence, in *Universal Access in Human Computer Interaction: Coping with Diversity, Volume 5 of the Proceedings of the 12th International Conference on Human-Computer Interaction (HCI International 2007)* (C. Stephanidis, ed.), 22–27 July 2007, Beijing, pp. 584–592. Heidelberg: Springer-Verlag.

Adams, R. and Gill, S. P. (2007b). Augmented cognition, universal access, and social intelligence in the information society, in *Foundations of Augmented Cognition, Volume 16 of the Proceedings of the 12th International Conference on Human-Computer Interaction (HCI International 2007)* (D. Schmorrow and L. Reeves, eds.), 22–27 July 2007, Beijing, pp. 231–240. Heidelberg: Springer-Verlag.

Adams, R. and Langdon, P. (2003). SIMPLEX: A simple user check-model for inclusive design, in *Universal Access in HCI: Inclusive Design in the Information Society, Volume 4 of the Proceedings of the 10th International Conference on Human-Computer Interaction (HCI International 2003)* (C. Stephanidis, ed.), 22–27 June 2003, Crete, Greece, pp. 13–17. Mahwah, NJ: Lawrence Erlbaum Associates.

Adams, R. and Langdon, P. M. (2004). Assessment, insight and awareness in design for users with special needs, in *Designing for a More Inclusive World* (S. Keates, P. Langdon, P. J. Clarkson, and P. Robinson, eds.), pp. 49–58. London: Springer-Verlag.

Adams, R. and Langdon, P. (2005). Exploring cognitive factors for universal access, in *Universal Access in HCI: Exploring New Dimensions of Diversity, Volume 8 of the Proceedings of the 11th International Conference on Human-Computer Interaction (HCI International 2005)* (C. Stephanidis, ed.), 22–27 June 2005, Las Vegas [CD-ROM]. Mahwah, NJ: Lawrence Erlbaum Associates.

Adams, R. and Russell, C. (2006). Lessons from ambient intelligence prototypes for universal access and the user experience, in *Universal Access in Ambient Intelligence Environments, Proceedings of the 9th ERCIM Workshop "User Interfaces for All"* (C. Stephanidis and M. Pieper, eds.), 27–28 September 2006, Bonn, Germany, pp. 229–243. Heidelberg: Springer-Verlag.

Anderson, J. R. (1983). *The Architecture of Cognition*. Cambridge, MA: Harvard University Press.

Anderson, J. R. (1990). *The Adaptive Character of Thought*. Hillsdale, NJ: Lawrence Erlbaum Associates.

Anderson, J. R., Bothell, D., Byrne, M. D., Douglass, S., Lebiere, C., and Qin, Y. (2004). An integrated theory of the mind. *Psychological Review* 111: 1036–1060.

Anderson, J. R. and Lebiere, C. (1998). *The Atomic Components of Thought*. Mahwah, NJ: Lawrence Erlbaum Associates.

Anderson, J. R., Matessa, M., and Lebiere, C. (1997). ACT-R: A theory of higher level cognition and its relation to visual attention. *Human-Computer Interaction* 12: 439–462.

Atkinson, R. C. and Shiffrin, R. M. (1968). Human memory: A proposed system and its control processes, in *The Psychology of Learning and Motivation: Advances in Research and Theory* (K. W. Spence, ed.), Vol. 2, pp. 89–195. New York: Academic Press.

Baddeley, A. D. (2000). The episodic buffer: A new component of working memory. *Trends in Cognitive Sciences* 4: 417–423.

Barnard, P. J. (1999). Interacting cognitive subsystems: Modeling working memory phenomena with a multi-processor architecture, in *Models of Working Memory* (A. Miyake and P. Shah, eds.), pp. 298–339. Cambridge, U.K.: Cambridge University Press.

Barnard, P. J., May, J., Duke, D., and Duce, D. (2000). Systems interactions and macrotheory. *Transactions on Computer Human interface* 7: 222–262.

Benyon, D., Turner, P., and Turner, S. (2005). *Designing Interactive Systems: People, Activities, Contexts and Technologies*. Harlow, U.K.: Pearson.

Berners-Lee, T. (2005). WWW at 15 years: Looking forward, in the *Proceedings of the 14th International Conference on World Wide Web*, 10–14 May 2005, Chiba, Japan, p. 1. New York: ACM Press.

Brewer, J. (2004). Web accessibility highlight and trends, in *ACM International Conference Proceeding Series, Volume 63, Proceedings of the 2004 International Cross-Disciplinary Workshop on Web Accessibility (W4A)*, 17–22 May 2004, New York, pp. 51–55. New York: ACM Press.

Broadbent, D. E. (1957). A mechanical model for human attention and immediate memory, *Psychological Review* 64: 205–215.

Broadbent, D. E. (1961). *Behaviour*. London: Eyre & Spottiswoode.

Broadbent, D. E. (1971). *Decision and Stress*. London: Academic Press.

Broadbent, D. E. (1984). The Maltese cross: A new simplistic model for memory. *The Behavioral and Brain Sciences* 7: 55–94.

Byrne, M. D. (2001). ACT-R/PM and menu selection: Applying a cognitive architecture to HCI. *International Journal of Human-Computer Studies* 55: 41–84.

Card, S. K., Moran, T. P. and Newell, A. (1983). *The Psychology of Human-Computer Interaction*. Hillsdale, NJ: Lawrence Erlbaum Associates.

Chong, R. (2003). The addition of an activation and decay mechanism to the Soar architecture, in the *Proceedings of the 5th International Conference on Cognitive Modeling*, 9–12 April 2003, Bamberg, Germany. Bamberg, Germany: Univereristats-Verlag.

Coltheart, M. (2004). Brain imaging, connectionism and cognitive neuropsychology. *Cognitive Neuropsychology* 21: 21–25.

Dix, A., Finlay, J., Abowd, G., and Beale, R. (2004). *Human-Computer Interaction* (3rd ed.). London: Prentice Hall.

Ellis, A. W. and Young, A. W. (1996). *Human Cognitive Neuropsychology: A Textbook with Readings*. Hove, U.K.: Psychology Press.

Fox, J. (1980). Making decisions under the influence of memory. *Psychological Review* 87: 190–211.

Fox, J. and Cooper, R. (1997). Cognitive processing and knowledge representation in decision making under uncertainty, in *Qualitative Theories of Decision Making* (R. W. Scholz and A. C. Zimmer, eds.), pp. 83–106. Lengerich, Germany: Pabst.

Heim, J., Nilsson, E. G., and Skjetne, J. H. (2006). User profiles for adapting speech support in the Opera Web Browser to disabled users, in *Universal Access in Ambient Intelligence Environments, Proceedings of the 9th ERCIM Workshop "User Interfaces for All"* (C. Stephanidis and M. Pieper, eds.), 27–28 September 2006, Bonn, Germany, pp. 154–172. Heidelberg: Springer-Verlag.

Hilgard, E. R. and Bower, G. H. (1966). *Theories of Learning*. New York: Appleton Century Crofts.

John, B., Vera, A., Matessa, M., Freed, M., and Remington, R. (2002). Automating CPM-GOMS, in the *Proceedings of the SIGCHI Conference on Human Factors in Computing Systems: Changing Our World, Changing Ourselves* (CHI 2002), 20–25 April 2002, Minneapolis. New York: ACM Press.

John, B. E. and Kieras, D. E. (1996). Using GOMS for user interface design and evaluation: Which technique? *ACM Transactions on Computer-Human Interaction* 3: 287–319.

Jokinen, K., Kanto, K., and Rissanen, J. (2004). Adaptive user modelling in AthosMail, in *User-Centered Interaction Paradigms for Universal Access in the Information Society, Proceedings of the 8th ERCIM International Workshop "User Interfaces for All"* (C. Stary and C. Stephanidis, eds.), 28–29 June 2004, Vienna, Austria, pp. 149–158. Heidelberg: Springer-Verlag.

Juvina, I. and van Oostendorp, H. (2004). Individual differences and behavioral aspects involved in modeling web navigation, in *User-Centered Interaction Paradigms for Universal Access in the Information Society, Proceedings of the 8th ERCIM International Workshop "User Interfaces for All"* (C. Stary and C. Stephanidis, eds.), 28–29 June 2004, Vienna, Austria, pp. 77–95. Heidelberg: Springer-Verlag.

Kass, R. and Finin, T. (1991). General use modeling: A facility to support intelligent interaction, in *Intelligent User Interfaces* (J. W. Sullivan and S. W. Tyler, eds.), pp. 111–128. New York: Addison-Wesley.

Kay, J. (2001). User modeling for adaptation, in *User Interfaces for All: Concepts, Methods, and Tools* (C. Stephanidis, ed.), pp. 271–294. Mahwah, NJ: Lawrence Erlbaum Associates.

Keates, S., Adams, R., Bodine, C., Czaja, S., Gordon, W., Gregor, P., et al. (2007). Cognitive and learning difficulties and how they affect access to IT systems. *Universal Access in the Information Society* 5: 329–339.

Keates, S. and Clarkson, J. (2003). *Countering Design Exclusion: An Introduction to Inclusive Design.* London: Springer-Verlag.

Kieras, D. E. (1997). A *Guide to GOMS Model Usability Evaluation using NGOMSL*, in *The Handbook of Human-Computer Interaction* (2nd ed.). New York: Elsevier.

Kieras, D. E. and Meyer, D. E. (1997). An overview of the EPIC architecture for cognition and performance with application to human-computer interaction. *Human-Computer Interaction* 12: 391–438.

Kieras, D. E. and Meyer, D. E. (2000). The role of cognitive task analysis in the application of predictive models of human performance, in *Cognitive Task Analysis* (J. M. Schraagen and S. F. Chipman, eds.), pp. 237–260. Mahwah, NJ: Lawrence Erlbaum Associates.

Kobsa, A. (1995). Supporting user interfaces for all through user modeling, in the *Proceedings of 6th International Conference on Human-Computer Interaction (HCI International 1995)*, 9–14 July 1995, Yokohama, Japan, pp. 155–157. http://www.ics.uci.edu/~kobsa/papers/1995-HCI95-kobsa.pdf.

Kobsa, A. (2001). Generic user modeling systems. *User Modeling and User-Adapted Interaction* 11: 49–63.

Kraynak, J. (2005). *Master Visually: Creating Web Pages.* Hoboken, NJ: John Wiley & Sons.

Langdon, P., Keates, S., and Clarkson, P. J. (2004). New cognitive capability scales for inclusive design, in *Designing for a More Inclusive World* (S. Keates, P. J. Clarkson, P. Langdon, and P. Robinson, eds.), pp. 59–69. London: Springer-Verlag.

Lauesen, S. (2005). *User Interface Design: A Software Engineering Perspective.* Harlow, U.K.: Pearson Education.

Lawrence, D. and Tavakol, S. (2007). *Balanced Website Design: Optimising Aesthetics, Usability and Purpose.* London: Springer-Verlag.

Liu, Y., Feyen, R., and Tsimhoni, O. (2006). Queuing network-model human processor (QN-MHP): A computational architecture for multitask performance in human-machine systems. *ACM Transactions on Computer-Human Interaction* 13: 37–70.

Maybury, M. T. (2001). Intelligent user interfaces for all, in *User Interfaces for All: Concepts, Methods, and Tools* (C. Stephanidis, ed.), pp. 65–80. Mahwah, NJ: Lawrence Erlbaum Associates.

Maybury, M. T. (2003). *Intelligent Interfaces for Universal Access: Challenges and Promise.* http://www.mitre.org/work/tech_papers/tech_papers_01/maybury_intelligent.

Moseley, R. (2005). Self-adaptation and reflection in dynamic reconfigurable systems, in *Natural Computing and Interactive System Design* (R. Adams, ed.), pp. 84–112. Harlow, U.K.: Pearson.

Moseley, R. (2007). *Developing Web Applications.* West Sussex, U.K.: John Wiley & Sons.

Neal, J. G. and Shapiro, S. C. (1991). Intelligent multi-media interface technology, in *Intelligent User Interfaces* (J. W. Sullivan and S. W. Tyler, eds.), pp. 11–43. New York: Addison-Wesley.

Newell, A. (1990). *Unified Theories of Cognition.* Cambridge, MA: Harvard University Press.

Newell, A. F. and Gregor, P. (2000). User sensitive inclusive design: In search of a new paradigm, in the *ACM Conference on Universal Usability, Proceedings of the 2000 Conference on Universal Usability*, 16–17 November 2000, Arlington, VA, pp. 39–44. New York: ACM Press.

Newman, W. and Lemming, M. (1995). *Interactive System Design.* New York: Addison Wesley.

Norman, D. (1986). Cognitive engineering, in *User Centered System Design* (D. Norman and S. W. Draper, eds.). Hillsdale, NJ: Lawrence Erlbaum Associates.

Norman, D. (1998). *The Invisible Computer.* Boston, MA: MIT Press.

Norman, D. (2004). *Emotional Design: Why We Love (or Hate) Everyday Things.* New York: Basic Books.

Nuxoll, A. and Laird, J. (2004). A cognitive model of episodic memory integrated with a general cognitive architecture, in the *Proceedings of the International Conference on Cognitive Modeling*, 30 July–1 August 2004, Pittsburgh. Mahwah, NJ: Nuxoll and Laird LEA.

Olson, J. R. and Olson G. M. (1995). The growth of cognitive modeling in human-computer interaction since GOMS, in *Readings in Human-Computer Interaction: Toward the Year 2000* (R. M. Baecker, J. Grudin, W. Buxton, and S. Greenberg, eds.), pp. 603–625. San Francisco: Morgan Kaufmann.

Pinker, S. (2002). *The Blank Slate: The Modern Denial of Human Nature.* London: Allen Lane.

Savidis, A., Akoumianakis, D., and Stephanidis, C. (2001). The unified user interface design method, in *User Interfaces for All: Concepts, Methods and Tools* (C. Stephanidis, ed.), pp. 417–440. Mahwah, NJ: Lawrence Erlbaum Associates.

Savidis, A. and Stephanidis, C. (2001). Development requirements for implementing unified user interfaces, in *User Interfaces for All: Concepts, Methods and Tools* (C. Stephanidis, ed.), pp. 441–468. Mahwah, NJ: Lawrence Erlbaum Associates.

Schofield, S., Hine, N., Arnott, J., Joel, S., Judson A., and Rentoul, R. (2004). Virtual learning environments: Improving accessibility using profiling, in *Designing for a More Inclusive World* (S. Keates, P. Langdon, P. J. Clarkson, and P. Robinson, eds.), pp. 41–48. London: Springer-Verlag.

Sharp, H., Rogers, Y., and Preece, J. (2007). *Interaction Design: Beyond Human-Computer Interaction* (2nd ed.). Chichester, U.K.: John Wiley & Sons.

Stephanidis, C. (2001a). User interfaces for all: New perspectives into human-computer interaction, in *User Interfaces for All: Concepts, Methods and Tools* (C. Stephanidis, ed.), pp. 3–17. Mahwah, NJ: Lawrence Erlbaum Associates.

Stephanidis, C. (ed.) (2001b). *User Interfaces for All: Concepts, Methods and Tools.* Mahwah, NJ: Lawrence Erlbaum Associates.

Stephanidis, C. (ed.) (2001c). Universal access in HCI: Towards an information society for all, in the *Proceedings of*

HCI International 2001, Volume 3 of (the 9th International Conference on Human-Computer Interaction), 5–10 August 2001, New Orleans. Mahwah, NJ: Lawrence Erlbaum Associates.

Sternberg, R. J. (ed.) (2000). *Handbook of Intelligence* (2nd edition). Cambridge, U.K.: Cambridge University Press.

Stevens, P. and Pooley, R. (2000). *Using UML: Software Engineering with Objects and Components.* New York: Addison Wesley.

Streitz, N. A. (2007). From human-computer interaction to human-environment interaction: Ambient intelligence and the disappearing computer, in *Universal Access in Ambient Intelligence Environments, Proceedings of the 9th ERCIM Workshop "User Interfaces for All"* (C. Stephanidis and M. Pieper, eds.), 27–28 September 2007, Bonn, Germany, pp. 3–13. Heidelberg: Springer-Verlag.

Teasdale, J. D. and Barnard, P. J. (1993). *Affect, Cognition and Change: Remodelling Depressive Thought.* Hove, U.K.: Lawrence Erlbaum Associates.

Ueda, H. (2001). The Friends21 framework for human interface architectures, in *User Interfaces for All: Concepts, Methods, and Tools* (C. Stephanidis, ed.), pp. 245–270. Mahwah, NJ: Lawrence Erlbaum Associates.

Watson, M. (1997). *Intelligent Java Applications.* San Francisco: Morgan Kaufman.

Whitefield, A. (1989). Constructing appropriate models of computer users: The case of engineering designers, in *Cognitive Ergonomics and the Human-Computer Interface* (J. Long and A. Whitefield, eds.). Cambridge, U.K.: Cambridge University Press.

Ye, J. and Herbert, J. (2004). Framework for user interface adaptation, in *User-Centered Interaction Paradigms for Universal Access in the Information Society, Proceedings of the 8th ERCIM International Workshop "User Interfaces for All"* (C. Stary and C. Stephanidis, eds.), 28–29 June 2004, Vienna, Austria, pp. 167–174. Heidelberg: Springer-Verlag.

Model-Based Tools: A User-Centered Design for All Approach

Christian Stary

25.1 Introduction

The underlying vision of user interfaces for all is to offer an approach for developing computational environments that cater to the broadest possible range of human abilities, skills, requirements, and preferences. Consequently, user interfaces for all should not be conceived as an effort to advance a single solution for everybody, but rather, as a new perspective on HCI that alleviates the obstacles pertaining to *universal access* in the information society. (Stephanidis, 2001, p. 7)

Recognizing the diversity of humans or users, *design for all* (also termed *universal design*) advocates a design perspective that eliminates the need for "special features" and fosters individualization and end-user acceptability (Stephanidis, 2001). To construct systems for individual support and user adaptation, developers need to acquire and represent user requirements. They also need to know how to deal with variable user needs along the different phases of development. Design tools have to support the management of various user needs, including their successive transformation to interaction technologies (including platforms). Resulting user interfaces should be adaptable and flexible with respect to devices, user roles, and tasks (cf. Baresi et al., 2006). Since the advent of mobile interaction devices and web-based hypermedia, the development of adaptable user interfaces has gained new momentum (Brusilovsky, 2002). The design space of hypermedia comprises the context of use, user characteristics, the content to be displayed and/or manipulated, and interaction styles.

Model-based design approaches are expected to incorporate context information, including task and user characteristics, while taking an implementation-oriented engineering perspective on development. They are oriented toward multiple perspectives on represented design information and modular development components, the so-called models. Each model, such as the task, user, or interaction model, is intended to support the representation and construction of interactive systems from a component and behavior perspective (cf. Paternò, 1999).

The interplay of models should reflect the methodological approach that the designers need to follow for development. As such, development techniques should not only capture the structure and behavior of an interactive application, but also the capability to adapt to different situations of use. Corresponding tools are targeted to facilitate user interface (UI) design through the provision of design assistance and execution of design specifications (automation) (Myers et al., 2000). The latter contain the context of UIs (Calvary et al., 2003).

However, most of the current model-based approaches are oriented toward a straightforward engineering perspective dedicated to structured refinement rather than exploring design inputs for user-centered individualization. The traditional model-based approach is as follows (Paternò and Santoro, 2003): a task model is located at the top level of a model hierarchy. It comprises all activities that the users need to perform to reach specified goals. Attributes describe these activities along a hierarchy following temporal and/or causal relationships. The task model also comprises (business-) objects that users have to manipulate for task accomplishment. The abstract UI model is located below the task model in the model hierarchy. It covers the logical structure for interactive task accomplishment in terms of structured interaction styles, such as graphical user interfaces (GUIs). The relationship to tasks and user goals is not represented in that model, which is, however, of crucial importance to develop design alternatives (Wood, 1998). Tasks do not propagate to the concrete UI model located at the bottom of the hierarchy in a transparent way. Rather, this model represents the way information is presented to users, in terms of visual, textual, or acoustic information, and implementation details, such as media properties and platform parameters. This model forms the basis for the actual, physical implementation of a UI. As development proceeds, information is added or inserted into models. These additions may alter the task model (Caffiau et al., 2007), and, as such, the design space.

Alternative approaches, such as K-MAD (Baron et al., 2006), keep the various models networked rather than hierarchically structured (for top-down development). They focus on the exploitation of task model elements, allowing for different paths to accomplish tasks and representations of actions. They might be refined to system actions at a later stage or remain human ones—an issue of high relevance for divergent user group support. The models' relationships are kept for both design and implementation purposes. It is the designer's task (based on a given model-based method) to capture this relevant information, and the model-based tool's task to support design *and* implementation.

For instance, once the dialogue model changes, its effects on the structure of the tasks, including their sequence of use, should be revised. A typical case concerns office tools. Once drag-and-drop can be used for attaching files to e-mails, the sequence of subtasks to complete a mail might change and should be revised. As soon as such effects are evaluated from a user perspective prior to implementation, the development effort can be reduced.

This chapter takes a closer look at the design task, based on collected empirical data that provide a proper basis for functional requirements of design-support tools. Such requirements also enable a structured comparison of existing model-based design for all (DfA) tools. The remainder of the chapter is structured accordingly. In Section 25.2, the empirical data and corresponding DfA concepts are reviewed. Section 25.3 promotes the capabilities of advanced model-based DfA tools based on the requirements gathered in Section 25.2. Section 25.4 concludes the chapter.

25.2 The Design Task Leads to Tool Requirements

In this section, empirical findings emerging from a conducted survey (Stary, 2002; Hailpern and Tarr, 2006) and relevant to DfA are reviewed to develop an understanding of the design task (Section 25.2.1). Based on these findings, the role of design representations and design support can be discussed from the perspective of tool users (Sections 25.2.2 and 25.2.3).

25.2.1 DfA Is Based on User Task Accomplishment

Although UI design is often a trade-off (Norman, 1988), it targets a specification of the structure and behavior of an application under development (Traetteberg, 2002). In the course of design, results from analysis are progressively refined to develop an architecture, scenarios of interaction, and a flow of control for implementation (Mirel, 2004). Prior to implementation, a comprehensive conceptual model should be specified, taking into account the structure of an application and all possible behaviors (Wood, 1998). Design support techniques, such as the unified user interface design method (Stephanidis, 2001; see also Chapter 16 of this handbook), attempt to relate various design alternatives to operational task descriptions at the level of conceptually modeled interaction styles. User interaction should be tuned to individual work or task flows rather than having to follow precoded sequences of interaction elements.

For interactive-application development not only functional issues are considered to be crucial, but also how the users organize and accomplish their work tasks, and how this information propagates to the implementation phase of an application (Sebilotte, 1994; Wilson and Johnson, 1996; Clerckx et al., 2004; Günter and Grote, 2006). Taking into account the work tasks of users requires the use of methods from occupational psychology, work or design sciences (Hackos and Redish, 1998; Diaper and Stanton, 2004; Mirel, 2004). Nevertheless, task-oriented user interface development should be a seamless while structured procedure (cf. Beyer and Holtzblatt, 1998, p. 389f.; Bodart et al., 1994).

Interaction design practices (cf. Hammond et al., 1983; Gould and Lewis, 1985; Rosson et al., 1987; Grudin and Poltrock, 1989; Johnson and Johnson, 1998), including model-driven software development (Stahl and Völter, 2006), focus on programming (of widget hierarchies). They are neither intertwined with representing or communicating design alternatives, nor driven by customization capabilities (Czarnecki et al., 2006). However, design representations should serve as containers for ideas, carrying their own context, and evolve iteratively with continuous improvements (Floyd, 1987; Bodker, 1998). A survey study reveals that the designers' background seems to influence their behavior, as several divergent practices could be identified (Stary, 2002): loosely versus tightly structured procedures, formal versus informal modeling, user versus technology orientation.

Most of the interviewed designers for the survey shared a fundamental aspect of design. User information resulting in perspectives on tasks and technology has to be mapped to a software architecture and a corresponding behavior specification. Some of the designers stick to technolgical units, such as class specification, and capture whatever is technologically feasible (bottom up). Others try to map complex organizational behavior patterns to discrete units of operation (top down), regardless of technological constraints. In the first case, context (i.e., users and task characteristics), the capabilities of technology triggers the development, while in the latter case, technology is adapted to the utmost extent according to the context of use.

Design refers to both a (nonlinear) process and a result (Furnas, 2000). In particular, "design processes are themselves nontrivial and so at any point in time must be, recursively, one design result of earlier design processes" (p. 218f.). Design is likely to be iterative. It involves a variety of activities, in particular the mapping of user tasks and characteristics, including domain- or user-specific ontologies to system structures and operations (Ulich et al., 1991). As a result, it serves as reference to all results from decision making and provides programmers with relevant implementation details (Howard, 1997). As Darcher recently formulated, "the discipline of design therefore provides the interface between understanding and creation" (2006, p. 254). Targeting toward user support "design is about making *choices* concerning *which* sequences of action the user should be able to perform, *which* design elements are used and *how* they composed to support this behavior." Design choices address "how tasks are mapped to dialog structure, how formal specifications are derived from design sketches and how dialog structures are decomposed" (Traetteberg, 2002, p. 139).

Mapping these findings to DfA, the DfA design task can be considered the set of the following activities:

- Representing various user needs in terms of their abilities, tasks, and organization of work
- Specifying technical system capabilities
- (Re-)arranging acquired information according to requirements and constraints (making design choices)
- Transforming them to implementation details to provide device/platform details and code (specifications)

Accomplishing DfA should lead to structure and behavior specifications of technical systems enabling universal access. As the empirical data show, DfA is understood quite differently by individual designers, leading to various work styles. These work styles could be supported by tools enabling flexible arrangement and transformation of information based on flexible representation schemes.

25.2.2 Design Representations Serve as Focal Point

From a development perspective, the initial challenge in the course of DfA is to learn about users and situations of use, and to capture tasks and domain semantics intelligible to users and developers. Such a representation needs to be comprehensive with respect to an application's scope; otherwise additional development cycles are likely to be necessary (Züllighoven, 2005). When designers refine user needs to (executable) models of task accomplishment, design representations serve a variety of purposes. They allow explaining the mapping of work processes and situations to (predefined) program functions and interaction facilities. In this way, various ways can be revealed for users to control an application. This information can be utilized to discuss the effectiveness and efficiency of work procedures, which in turn

might enrich the design space for interaction paths. Besides providing recovery strategies from errors, design representations can cover a wide range of feedback on the effects of users' action. The language for representation and the notational process support should comprise highly expressive and semantically rich constructs. The latter should empower the involved developers with a tool for communication and provide a comprehensive basis for further (functional) refinement and implementation (Stary, 2002).

However, software and user interface developers, domain-expert users, and usability specialists (especially contextually oriented ones) are likely to speak different languages when they describe and explain work, work flows, goals, decision points, and relevant strategies and knowledge (Bodker and Gronbaek, 1991; Beyer and Holtzblatt, 1998). Any mismatch should be made transparent, explaining different notations, domain, and/ or design ontologies. Ontologies facilitate the integrity of design ideas and reflect user values, priorities, goals, and constraints (Van Welie et al., 1998; Clerckx et al., 2004).

As different techniques for task analysis are used that usually feature nontechnical aspects (Keates, 2006), design representations should allow a variety of perspectives rather than narrow the design space to a particular viewpoint (e.g., programming widget hierarchies). In addition, notations used for specification should accurately reflect an application's context and its semantics. Otherwise, communication problems might be caused among participating users, analysts, designers, programmers, and evaluators in the course of development. They should rather share the same context and ontology when tasks are acquired, or design inputs are reviewed. Only then are users likely to accept certain technological support for accomplishing their tasks. To that respect, semiformal approaches, open to novel semantic constructs and structures, seem to be effective for transferring task and design information (Herrmann et al., 2000; Oppl and Stary, 2005; Hemmecke and Stary, 2007).

Unifying notations or specification languages have the capability to support the successive transformation of tasks to executable items at the user interface. However, even highly developed representation techniques, such as unified modeling language (UML), do not focus on the identification and mapping of design-relevant information to computable elements and relationships (Scogings and Phillips, 2004; Sottet et al., 2006). Besides extensions to standard modeling languages, such as to UML proposed by da Silva and Paton (2000), Obrenovic and Starcevic (2004), and others, viewpoints have been provided to support the design procedure and to reduce the complexity for development. According to Nuseibeh et al., (1994, p. 76), viewpoints comprise:

- *Style*: Notation required to represent knowledge, such as UML
- *Work plan*: Development actions, strategy, and process to address development process knowledge
- *Domain*: Area of concern addressing specification knowledge

- *Specification*: Partial system description, capturing specification knowledge
- *Work record*: Development history capturing specification knowledge

Such contextual schemes should comprise structural and process items at different layers of abstraction (Czarnecki et al., 2006). The designer's work space is composed of the following set of activities on representations, as D'Souza and Wills (1998) have observed from object-oriented developments:

- Refinement and abstraction
- Encapsulation of structure and behavior
- Mapping from one notation to another
  - Coherence checking
  - Accuracy and traceability assurance

Hence, representations play a crucial role in the course of design, both from a process and from a result perspective (Van Harmelen, 2001). As a means for quality assurance and communication, they have to keep design elements and results from various stages in development intelligible for all involved parties. Their notational capacity has to capture information according to the scope of an application in a form semantically correct and sufficiently detailed to implement the specified structure and behaviors. As Traetteberg puts it, more effort should be put into understanding:

- How (design representation) language is used throughout the design process
- The need for integration of modeling languages across perspectives
- How design knowledge can be integrated into a model-based approach to UI design (2002, p. 141)

As model-based approaches allow representing various perspectives, tools should provide the capability to integrate the different viewpoints:

If task-based user interface design is to be supported, the different models must be *integrated* within a tool. The models must be specified, together with links between models from the same design phase, as well as links between models from different design phases. (Bomsdorf and Szwillus, 1999, p. 294)

25.2.3 Tools Needed to Meet Design-Task Requirements

Representation schemes could dominate the design process, as notations and design-language ontologies might emphasize certain aspects (e.g., GUI design), if not trigger the entire design process itself. This might facilitate design procedures; however, we have to keep in mind that "design is a complex process that requires both creativity and a firmly grounded understanding of users, tasks and environments" (Hackos and Redish, 1998, p. 296). A typical design path starts with a list of tasks to be supported. This list is then expanded to task scenarios that themselves are propagated to scenarios of use. Required are both high- and low-level task descriptions (i.e., how tasks are performed), which are then put into the context of user profiles (p. 336f). Consequently, design-support tools have to incorporate results from user, work, and task analysis for refinement and transformation to executable structures and functions (Limbourg and Vanderdonckt, 2004b).

Taking into account that understanding of the scope of design, features of effective support tools should enable the activities listed in Table 25.1. Features are abstract descriptions of design activities that might be supported by different software-system functions. The list of activities in Table 25.1 summarizes empirical findings of model-driven software engineering and usability engineering. The activities to be featured have been structured along the order in which designers specify and modify models. In this way, a development procedure can be established for

TABLE 25.1 Development Activities

(Model-Based) Development Phases	Design and Implementation Tasks
Work, user, and task analysis	Ontology building
Business/task modeling	Refine/abstract
User's conceptual model analysis	Relate
Systems analysis	Encapsulate
• Data flow	Check coherence
• Objects	
User modeling	Refine/abstract
Data modeling	Relate (associate)
Interaction modeling	Encapsulate
Application model modeling	Check coherence
	Check completeness
	Check conformance
Initial prototyping (on paper)	Sketch
Usability testing 1	Check accuracy
Prototyping with dataflow and interface	Instantiate specification
Usability testing 2	Check accuracy
Implementation according to design	Instantiate specification

Source: Adapted from Stary, C., *Behavior and Information Technology*, 21, 425–440, 2002; enriched with data provided by Hailpern and Tarr (2006).

both approaches, for sequential (or hierarchical) and nonlinear design.

The presented analysis leads to a pool of items that is refined and transformed to (executable) elements. Although each item is assigned to a particular perspective, its processing occurs in accordance with others as represented in the models. Design mainly focuses on refinements within the scope of analysis, ensuring consistency and completeness of information. Evaluation is required to adjust design proposals to user needs and task requirements in terms of accurately mapping the latter to interactive system elements (Ulich et al., 1991; Luyten et al., 2003; Limbourg and Vanderdonck, 2004b). The transition from analysis to design corresponds to mapping complex, contingent human behaviors of work to rule-bound events and properties of task-oriented interface elements. A smooth transition should not be hindered through problems with skills, lack of structured procedures, and incompatibilities of method, representation, and notation. Software-engineering principles should not dominate design, and thus bias refinement and transformation toward implementation.

At the representational side, ontologies are not only used for analysis and design, but also to handle the transition from analysis to design (Calvary et al., 2003). Both analysis and design have to be understood in terms of procedural and structural knowledge (process and results). Analysis mostly focuses on participative handling of organizational and user (task) knowledge (O'Neill and Johnson, 2004), whereas design on software architecture (Clerckx et al., 2004). This gap can be bridged through step-by-by-step refinements (in terms of tasks, roles, and data) as well as enrichments (in terms of adding artifact details, such as interaction modalities), and the integration of initially separated knowledge categories.

Documentation plays a crucial role for process and result specifications, in particular in checking coherence and conformance to acquired requirements. The data provided by Stary (2002) reveal that documentation is still a controversial issue among developers, due to a variety of compatibility problems. On one hand, developers use a mix of methods in the course of analysis; on the other hand, they use different notations for design. Finally, they are not used to document transitions between phases, in particular between analysis and design. The latter is required not only to communicate development knowledge to users and developers, but also to trace development activities and understand their rationale, at least from one of the addressed perspectives. This property requires structured openness when modeling user's work. Any documentation should capture the user's language of problem solving (Mirel, 2004).

25.3 Model-Based Tool Support

Taking into account the empirical results given in Section 25.2 requires models and model handling for task-oriented user interaction (modeling for interactivity, cf. Mirel, 2004, 96f.) in addition to detailing information for (device) implementation. Such a procedure fills the (indefinite) design space "where UI developers and application developers may collaborate in terms

of a UI design contract elaboration" (Da Silva and Paton, 2000, p. 120). A proper tool serves as a repository for acquired, manipulated, and processed information in the course of development. Hence, its representation scheme, and the corresponding activity and communication support for developers and users play a crucial role (Johnson and Johnson, 1998; O'Neill and Johnson, 2004). Schemes representing models are expected to provide an accurate level of abstraction and to cover multiple perspectives. They should reduce complexity without neglecting context. Section 25.3.1 focuses on the identification of models, Section 25.3.2 on tuning models, and Section 25.3.3 on prototyping and adaptation support. The perspective taken for review strives for the Gestalt design *and* technical feasibility.

25.3.1 Model Identification

As already sketched in the introduction, the identification of models has been dominated by a top-down approach to the design and implementation of UIs (Paternò, 1999):

- *Task and domain (object) model*: Located at the top layer of abstraction, this model is expected to capture all activities that need to be performed to reach user goals. They might be detailed through properties (attributes), and arranged along a hierarchy representing causal or temporal mutual relationships. In addition, this model is expected to capture (business) objects that have to be manipulated to perform tasks.
- *Abstract user interface model*: Located below the top layer, it contains the logical structure required for task accomplishment at the UI. Abstract elements of interaction styles are captured in their mutual context (e.g., an area for data manipulation in a browser-like way). The relationships to user tasks or goals are encoded (e.g., to indicate that at a certain point in time options for navigating in a task hierarchy have to be offered). The final codality of information for presentation (text, graphics a.t.l.) and user control (modality) are not specified at that layer of abstraction.
- *Concrete user interface model*: All abstract interaction elements are refined to more concrete ones. The initial abstract specification of style elements is replaced with one containing concrete attributes and relationships between interaction elements, including the modality of interaction. At that point, platform- and media-specific information is attached to enable physical access to an application.

The instantiation of the concrete model leads to an actual user interface, utilizing (standard) programming techniques, such as, for example, Java.

The envisioned interplay of the models in the course of task-based design can be exemplified as follows. Consider booking a course at the university. It requires several subtasks, such as finding proper options, selecting class data, and so on. This information is kept on the task (model) level, in addition to information for selection and manipulation, such as the available courses.

The subsequent abstract UI model captures all (categories of) interaction elements that are supposed to support each subtask. For instance, to identify course options an entry field for search has to be provided. Finally, the concrete UI model captures all specific interaction elements. For instance, in a web interface, search entry fields are supported by text bars after selecting a search engine (and site). The actual user interface can be constructed based on the specification of the concrete UI model.

Such a top-down approach focusing on layered design perspectives facilitates the use of multiple platforms and various arrangements of interaction elements. It encodes rather than making transparent Gestalt design knowledge—knowledge that is mainly created in the course of mapping the results of user and work analysis (located in the top-level task/domain model) to abstract interaction elements. The initially acquired (and represented) context of an interactive application does not guide directly concrete UI designs. It is rather assumed that the abstract UI model provides a sufficient design space for DfA.

In addition, it is assumed that user goals can be represented in terms of tasks or domain objects. Users or their mental models are not specified explicitly (e.g., in a user model). User modeling targets the acquisition and representation of relevant design information about users (McTear, 2000). Its content bridges the "gap between what users know and what they need to know" for successful interaction (Shneiderman, 2000, p. 90). As such, it should be used as a primary source of information for meeting individual user needs through adaptation. The latter is directed by user goals, user experiences, device diversity, and UI behavior. Issues related to user modeling are discussed in detail in Chapter 24. of this handbook

The advent of AI (artificial intelligence) techniques for user modeling has led to a shift from more or less implicit specifications of user properties in traditional model-based design toward explicit representations. They can be updated automatically by applications (Pohl and Nick, 1999). They represent the system's current beliefs about the knowledge and preferences users have. As a consequence, the functionality as well as the interface behavior of a software system may change significantly over time. The drivers for those changes correspond to those already indicated previously (Brusilovsky, 2002): user task experience in terms of qualifications and skills required for task accomplishment, user motivation to accomplish tasks through attitude data, and UI experience addressing the level of interactive-media competence.

DfA as a contextualized process throughout development requires a shift in model-based design (support) towards:

- A more detailed and flexible representation of users, work tasks, and domain elements, at least to the same extent as UI representations
- Explicit conceptual relationships among design elements (inter- and intramodel-specific), considering universal access from a usability and software-engineering perspective in a more integrative but self-contained way

Consequently, a DfA-oriented approach has to take a more networked rather than hierarchic perspective on model-based development. It requires a set of models that still enables the specification of an application model (in the sense sketched previously), but details the user- and use-relevant elements, for example, in terms of a task, user, domain, and an interaction model. The traditional engineering perspective has to be enriched with design variants oriented toward interactivity rather than implementation. The following set of models lays the ground for model-based DfA tools that recognize the need for supporting diverse user communities and situations of use:

- The *task model* contains details of the organization at hand, the tasks, and the users' perception of work. Task modeling includes modeling the objectives users want or have to meet using an interactive application, probably in response to particular situations or events, as well as the different activities that users have to perform to accomplish their tasks, as, for example, given by global business processes.
- The *user model* contains a role model by reflecting specific views on tasks and data (according to the functional roles of users in organizations). It also captures individual user characteristics, capabilities, or specific needs that developers have to take into account for adaptation.
- The *domain (data) model* addresses all data, material, and resources required for task accomplishment in the application domain.
- The *interaction model* is composed of device and style specifications that designers need to construct a UI or to generate a UI prototype. Hence, the structure of modalities and its related behavior are captured in that model, both at an abstract and a concrete level. For the developers' convenience, platform-specific specifications can become part of the interaction model. In case behavior specifications are missing, the preprogrammed, platform-inherent behavior might restrict the solution space for DfA.
- The *adaptation model* contains all rules that trigger a particular structure and behavior of an interactive application. As such, it enables interaction adapted to user-model elements, such as left-hand assignments to interaction media (e.g., mouse buttons) based on consistent representations. It also enables the multiple presentation of task-information for navigation (e.g., tree views and acoustic trees). Because its scope includes all the previously listed models and their relationships, adaptation may concern task, data, interaction elements, and their intertwining.
- The *application model* integrates all five models from a static and dynamic perspective. In this way, it defines the entire structure and behavior of an interactive application. For interaction the users' privileges, preferences, special needs, and tasks to be performed are taken into account when mapped to a concrete device or widgets of the interaction model.

Of particular interest is the set of relationships that constitute design context. For instance, elements of the problem domain

TABLE 25.2 Sample Inter- and Intramodel Conceptual Relationships in K-MAD

	Task Model	Role Model	Object Model	Interaction Model
Task Model	Is-part-of	Is-performed-by	Is-related-to	Is-linked-to
Role Model	Performs		Handles	
Object Model				Is-presented
Adaptation Model	<condition>		<condition>	Dialogue control

(data) model can be derived from tasks, since each task requires certain data for accomplishment. In this way, the problem domain model is related to the task model by an "is-based-on" relationship in TADEUS (Stary, 2000). Table 25.2 shows some extracted relationships between the different models in K-MAD (Baron et al., 2006; Caffiau et al.. 2007). The user model is named role model, and the problem domain data model is named object model. In that approach the task model is closely linked to the interaction model. In this way, any change in the task or interaction model (e.g., introducing drag and drop for setting up a mail) can be reflected from the interaction and the task perspective, switching from an implementation perspective to design. The <condition> indicates a variable element.

Similar concepts can be found in other approaches, for example, ProcessLens (Dittmar et al., 2004; Stary, 2006). There, for example, the task model provides a declarative "before" relationship to enrich task hierarchies, and trigger behavior specifications. The domain data model is derived from the task model using the "is-based-on" relationship. Hereby, the content required for task accomplishment can be assigned explicitly to elements from the task model. To model task-based navigation, the task model "is-presented" by means of the interaction model. User model elements are conceptually linked to the task, user, and domain model, since users are likely to handle tasks in a particular way or role. They also might inform other users, control task accomplishment, and create data.

In the context of DfA, data have to be linked to presentation and manipulation media. In this way, the data model is coupled with the interaction model. The adaptation model is tied in for several reasons:

- Tasks can be accomplished in various ways.
- Users may act in different organizational roles.
- Domain data might serve different purposes according to their context of use.
- Information might be presented at the UI in various ways, involving task elements for navigation, domain data for content manipulation, and device properties for presentation.

25.3.2 Tuned Representation and Manipulation of Models

Given the set of reference models in Section 25.3.1, traditional model-based approaches can be described, as shown in Figure 25.1. Although user tasks and application domain data are captured, design is restricted to refining an abstract UI specification

up to the point platforms can be used for implementation. The relationships are named according to the rationale of the primarily layered engineering approach: intertwining associates elements in a variety of ways, whereas sublayers restrict associations to hierarchical relationships. User modeling and adaptation are not considered.

As the empirical investigations have shown, the different modeling activities described so far (task modeling, interaction modeling, etc.) might not be performed in a linear way, except starting with task or user modeling (top-down design followed by middle-out). In this way, a focal point is set up for refinement and prototyping. Modifications might not only affect a single model, but also relationships between model elements. When context should be kept in the course of DfA activities, continuous tuning of representations is required, according to the various ontologies and their relationships, involving the adaptation model (see Section 25.3.3).

ProcessLens includes user and adaptation modeling for DfA, similar to Teallach (Griffiths et al., 1991) or Dygimes (Clerckx et al., 2004), but in a context-sensitive way. It bridges the gap between structured software development and usability engineering through supporting:

- Acquisition and representation of tasks and organizational information by providing an ontology for business intelligence
- Mapping of acquired information to interaction features by providing explicit design relationships

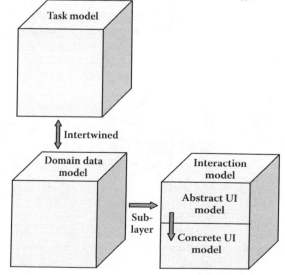

FIGURE 25.1 Traditional model-based design.

- Intertwined specification of structure and behavior of an interactive application—a feature that enables the automated generation of prototypes, ensuring early feedback and involvement of users.

As such, ProcessLens does not only show "what will take place" and "how this is to take place" (Bodker, 1998, p. 120), but *why this is to take place* from a user task perspective. It implements both the quest for capturing nonfunctional requirements and constraints and the quest for understandable traceability (e.g., Jarke, 1998; Sutcliffe et al., 1999). Hence, (re)presentations and prototypes are not only containers for design ideas, but also serve as a basis for collaborative reflection on artifacts and for active involvement of task performers.

25.3.2.1 Interfacing Analysis: Ontology Building

Because it attempts to support analysis, design, and implementation, ProcessLens has to provide a representation technique for user and work analysis. The results from analysis serve as input to subsequent design activities, starting with task modeling and leading to an application model.

Because the transformation from analysis to design should be as seamless as possible (minimal reduction of semantics, minimal loss of context for further developments), an ontology has been developed featuring the most important elements from work environments for model-based design. It comprises the *organization* and/or *organizational units* that represent the structure of the addressed organization. It captures *roles*, together with the *activities* and the *materials* that are processed in the course of task accomplishment. User task accomplishment might be based on temporal and causal *constraints* between activities. Materials may either be data or raw materials being processed to accomplish tasks. They are assigned to activities and, finally, represent the results of work/business processes. Besides the common object-oriented relationships has-subclass, has, and has-part, a variety of *conceptual relationships* are provided to network organizational units, roles, tools, materials, and activities. For instance, Employs relates organizational units to roles, Handles denotes the responsibility of a role for a particular activity, and Creates denotes the creation of material through an activity (see Stary, 2000).

Seamless development is enabled through direct mapping of the entities and relationships to UML representations in terms of the task, user, and problem domain data model, and, eventually the interaction model (in case of existing interactive computer support). Most of the conceptual entities (organizational unit, task activity, material, tool) represent nodes (class names), whereas structural, temporal, and causal relationships correspond to semantic links between classes of objects in diagrams.

25.3.2.2 Design

The ontology used for specification empowers the analyst and the designer with a semantically rich representation of the universe of discourse, and finally, of the application. The initial designer's task is to set up a task model (see also Figures 25.2 and 25.3). This model contains all those relevant activities that are considered to be relevant by the task performers. Then, refinements and supplementary specifications according to the users' roles, abilities, and preferences, to domain data-processing requirements, to interface styles and architectures have to be performed, until the structure and behavior of the application model enables

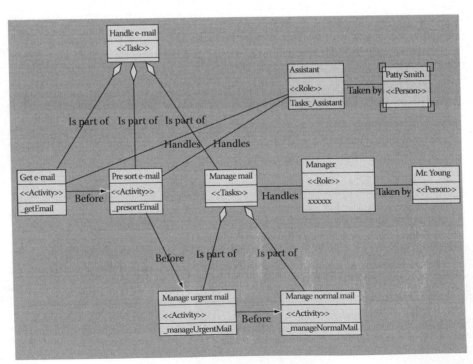

FIGURE 25.2 ProcessLens DfA-approach.

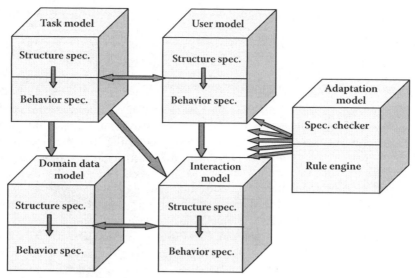

FIGURE 25.3 Sample integrated task/role representation in ProcessLens.

prototyping. A complete application model requires at least the specification of a task, its underlying data, and the assignment of interaction features to the task and the data for presentation and manipulation. Once this information has been provided, the application model can be directly processed by the ProcessLens prototyping engine. Table 25.3 summarizes the designer's tasks and the conceptual relationships stemming from the model-driven ontology.

25.3.2.3 Task Modeling: Structure

Based on the acquired user and task data, those activities that are expected to be supported through the interactive software are selected. They might be decomposed into further activities, and related to each other according to the required sequence of accomplishment (applying the "before" relationship). This information is captured in a class diagram. The activities are represented as classes (containing identifiers, but no methods and attributes), and the structural and dynamic relationship are represented as links between the classes (see also Figure 25.3).

25.3.2.4 Task Modeling: Behavior

This part of the task model contains all procedures to be followed for task accomplishment, including the input/output behavior of the application from the task perspective. As a consequence, each class diagram of the structure specification of the task model located at the bottom of the hierarchy is related to a set of activity diagrams. It represents the accomplishment of

a subtask (activity). For instance, to update mail data, the mail record has to be scanned, loaded, and then modified according to the required changes. Input data might be typed in, and the output might be the display of the entire record on the screen. In this way, the first refinement of behavior is specified in a procedural way through state transitions of task objects. Before that integration, the process of task accomplishment is captured through declarative descriptions in the structure specification of the task model (e.g., using "before" relationships between subtasks in ProcessLens).

25.3.2.5 User Modeling: Structure

The structure of the user model does not only comprise functional user group definitions as the organization of tasks and roles requires, but also the individual characteristics of users in terms of personal profiles. There are two ways to define functional user groups. One way is based on the structure of the organization and their units of operation. For instance, each staff member of a certain department has a particular set of privileges, such as the accounting staff having the right to access payroll data. The other way to identify user profiles is based on professional skills, such as recruiting, which staff members have regardless of their organizational assignment and position. The first option facilitates the identification of access permits to data according to the structure of the organization, whereas the latter facilitates the definition of professional profiles, as well as the assignment of staff members to more than one organizational unit, according to their skills and qualifications. Actually, both views have to be integrated to gain sufficient insight into the organization

TABLE 25.3 Conceptual Relationships and Design Activities in ProcessLens

Set Up	Refine and Abstract	Relate
Employs	Is a	Handles, creates, concerns, informs, controls, requires,
Has	Has part	before, is based on, corresponds to, taken by

and the user population that are going to be supported. A thorough work analysis should provide data from both perspectives. In ProcessLens they are captured through constructing a class diagram accounting for both perspectives.

The specification of the user model has to be performed in line with the other models. Hence, one of the primary results of this step is a diagram in which classes are named according to the functional roles required for task accomplishment. Figure 25.3 shows part of a structure specification of a user model.

In addition, as already indicated, the structural user model might comprise information required for interaction, namely individual profiles with respect to abilities and needs (e.g., for left-handed users). Since the profile data are specified from a user perspective, but concern interaction, they require assignment to elements of the interaction model (e.g., mouse settings for left-handed users). This issue is addressed by the adaptation model in the same way; each user can only access those domain data that she is enabled to according to her role for accomplishing a certain task. In this way, the domain model, the user, and the task model are related.

25.3.2.6 User Modeling: Behavior

The behavior specification of the user model describes the work process from the perspective of a particular functional role. For instance, it might capture a high-level description of a workday in terms of activities as perceived by a user. These activities might then be directly linked to activities of the task model (behavior specification; Figure 25.4).

25.3.2.7 Domain (Data) Modeling: Structure

The designer has to define the classes of data required for task accomplishment. For instance, e-mailing requires the creation of the class mail. For context-sensitive specification and seamless development, the "is-based-on" relationship is applied between activities and data. An identifier, attributes, and operations have to be specified for each class. Setting up a data model is also required to provide information for the integration of the

data-related functionality with the interaction facilities later on in the development process (such as assigning text fields to mail data that are expected to be entered when a certain manipulation task has to be performed).

25.3.2.8 Domain (Data) Modeling: Behavior

For each data class derived from a task activity, a life cycle has to be provided. For instance, the life cycle of mail has to be defined, according to the attributes and methods specified in the class mail. In case of multiple involvement of a class in several tasks, such as Person in several subtasks, the dynamic specification integrates different behavior specifications into a single activity diagram. The model captures the entire life cycle of data, while the privileges for access according to the tasks and the roles of users are captured in the adaptation model.

Such a self-contained domain data model allows checking whether each class is actually required for task accomplishment, and, vice versa, whether data are missing for task accomplishment (checking for task-completeness with respect to work objects).

Since many model-based approaches (e.g., Clerckx et al., 2004; Wolff et al., 2005) use concurrent task trees to represent tasks and generate UIs, a direct comparison to those approaches reveals some differences. Table 25.4, an excerpt from Dittmar et al. (2004), shows the difference with respect to homogeneity and integrity of specifications as well as the provision of conceptual relationships.

25.3.2.9 Interaction Modeling: Structure

The initial step in interaction modeling in ProcessLens concerns the setup of generic interaction features. In case of platform-specific solutions, such as for post-WIMP GUIs, the structure of the elements, and styles have to be loaded from resource scripts. In assigning tasks and user actions to presentation elements of a platform-dependent design saves time and effort for specification. When using generic structures, further structural refinement and adjustment are required.

TABLE 25.4 Contextual Task Modeling Support in ProcessLens and CTTE

	Task	Task Domain	Actor	Task ⇔ Domain	Task ⇔ Actor
ProcessLens	- Task hierarchy - Sequential temporal relations between tasks - Activities with behavior	Data objects comprising attributes and behavior specification	Organization units, roles, persons, including behavior specification	- Predefined static relations (e.g., creates) - Synchronization of corresponding behavior	Predefined static relations (e.g., handles)
	Class and activity diagrams to describe structure and behavior of model elements (conform to UML)				
CTTE	- Cooperation tree to control task trees - Explicit temporal relations between sibling tasks	Informal description	Roles	None	- One task tree for each role - Simple concept of coordination

Source: Excerpted from Dittmar, A., Forbrig, P., Heftberger, S., and Stary, C., Tool support for task modeling: A "constructive" exploration, in the *Proceedings of the 11th International Conference on "Design, Specification, and Verification of Interactive Systems" (EHCI-DSVIS'04)*, 11–13 July 2004, Hamburg, Germany, pp. 59–74, Springer-Verlag, Berlin/Heidelberg, 2004.

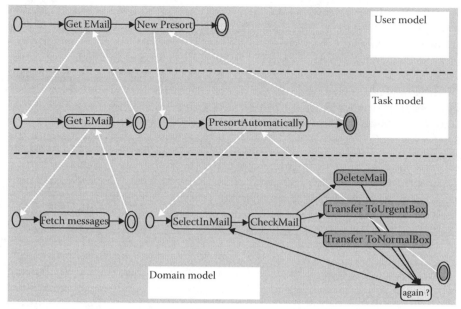

FIGURE 25.4 Activity diagrams of the role element *Assistant* (above the dotted line) and of the activity elements *Get EMail* and *Presort EMail* of the task model (left and right bottom), including their synchronization (the white directed links denote synchronization transitions).

In particular, platform-specific structures have to be adjusted to particular styles, since they provide a variety of arrangements (e.g., through recursive structures, such as container objects). Before the tasks and the problem domain data are assigned to the selection and grouping of the interaction elements and styles, in this step the fundamental look and feel of the interactive application at hand can be specified.

The next step in interaction modeling can be considered the first move toward application-specific design. The selected (and eventually prearranged) interaction classes are further refined and tuned with other model elements to achieve fully customized interaction features—the type of relationship to be set between the generic interaction elements and the data depends on the used platform.

25.3.2.10 Interaction Modeling: Behavior

As for the previous models, for some of the classes of the generic structural interaction model an activity diagram has to be specified. These diagrams have to be related to activity diagrams of the task, user, and domain model to specify task and user conforming interaction, following the same principle as shown in Figure 25.4.

25.3.2.11 Application Modeling

The specification of an application requires the synchronization of the specifications achieved so far. Special interaction symbols used in the coordination diagram (see Figure 25.4) enable the visualization of the global behavior of the application. Checking for completeness is based on (1) tracing the key events for task accomplishment; (2) checking for the completeness of links between interaction-relevant tasks, operations on data, data and interaction elements; and (3) checking for possible side effects, such as deadlocks.

25.3.2.12 Functional Specification

Once the overall behavior has been specified, each of the methods of the class/object specifications has to be detailed in the course of functional specification. The operations can be specified through selection of methods provided in platforms or using formal languages.

25.3.2.13 Prototyping

Prototyping is enabled in ProcessLens based on the application model. It does not require a functional specification at the level of data manipulation, but on the interaction level (e.g., window management). In the course of prototyping instances of interaction model classes are generated and their life cycles are instantiated as specified in the synchronized activity diagrams.

Other model-based approaches advise design processes similar to ProcessLens, such as Dygimes (Clerckx et al., 2004, p. 285f.):

1. Create a context-sensitive task specification with the ConcurTaskTrees notation.
2. Create UI building blocks for the separate tasks.
3. Relate the UI building blocks with the tasks in the task specification.
4. Define the layout using constraints.
5. Define custom properties for the UI appearance (e.g., preferred colors, concrete interactors, etc.).
6. Generate a prototype and evaluate it (the dialogue model and presentation model are calculated automatically).
7. Change the task specification and customization until satisfied.

In this approach, UI building blocks (abstract UI model, being part of the interaction model) are assigned to CTTE (2006) elements (part of the task model). According to defined constraints (part of the adaptation model), a context-specific task specification is constructed, leading to so-called annotated task sets. These are refined to a dialogue and presentation model (as part of the interaction model). Both models are fed to a run-time environment that is triggered by individual layout constraints (stemming from a user model), device profiles, and available widgets (stemming from an interaction model).

The early addressing of UI components is due to the assignment of UI building blocks (i.e., structure specifications) to each task following task modeling. It triggers a linear sequence that has to be iterated for each modification of interaction designs. It also determines the interaction space for each iteration. In approaches of this kind, the task flexibility depends on the variety of work paths in the task tree specification.

For interaction modeling besides generic UML structures, for example, for modalities (Obrenovic and Starcevic, 2004), XML derivates, such as XIML (Puerta and Eisenstein, 2002), XUL, or USIXML (Limbourg and Venderdonckt, 2004a), are widely used (see Chapter 26 of this handbook). They facilitate transformations as well as the dynamic creation of models. For instance, XIML as UI description language supports the specification of various (linked) interaction models simultaneously. Its definition allows capturing task, user, presentation, and dialogue models as well as platform details. The presentation and dialogue model cover the structure and behavior part of an interaction model. However, there is no executable relationship between the various interaction model parts.

Using XUL and CTTE Wolff et al. (2005) transform models on the basis of a task model. Tasks are assigned to elements of the abstract UI model (i.e., interaction model) as views. Transitions between these views and tasks are also defined in XUL. The device dependence for actual UI generation is achieved through a mapping-rule base for each device. In this way, design information is kept self-contained in terms of the addressed models as well as interconnected to develop design alternatives (Gestalt) and implementation variants. The view concept instantiates a particular conceptual relationship (e.g., the relationship "is-presented" is used when interaction model elements are assigned to elements of a task model). The mapping mechanism can be considered part of an adaptation model.

25.3.3 Adaptation Support

Capturing variability requires a proactive design approach (Marakas, 2001). Proactivity means to ensure at design time the utmost flexibility of the UI. It is supported by adaptability (Baresi et al., 2006), meaning the capability of an interactive system to support their users according to their functional role, individual characteristics, tasks to be accomplished, and situation of use, even at the device level (cf. Ye and Herbert, 2004). A unifying notation and specification scheme ranging from

analysis via design to implementation support is of great benefit, since it allows addressing the entire design space in a unifying way. Otherwise, notational mappings are required. For adaptation the set of representations have to be consistent, in particular when specific concepts and relationships are used throughout development. Consistency checkers, e.g., being part of the adaptation model in ProcessLens (i.e., the specification checker in Figure 25.2), lay the ground for the adaptability of UI components. Their algorithms process conceptual relationships and the concerned models.

25.3.3.1 Ensuring Consistency

Conceptual model constructs beyond object-oriented ones, such as the handles relationship between elements of the user model and the task model in ProcessLens, requires special treatment of relationships. Every time a relationship between class specifications is set, the correctness of its use has to be checked. It requires a procedural specification of the semantics of each relationship. If relationships were to be used with various meanings, the resulting specification would be ambiguous. For instance, the relationship handles can only be used between functional user roles (specified in the user model) and task specifications of the task model.

For a complete check, structure and behavior specifications have to be processed. In ProcessLens, the structure specification is checked with respect to whether a relationship (1) is used correctly (i.e., whether the entities selected for applying the relationship correspond to the defined syntax), and (2) is used completely (i.e., all the necessary information is provided to set the relationship, such as checking the existence of activity diagrams corresponding to abstract task specifications). For behavior specification, it is checked whether the objects and/or classes connected by the relationship behave according to the semantics of this relationship (e.g., the behavior of data objects corresponds to the sequence implied by "before").

The algorithms used in ProcessLens are designed in such a way that they scan design representations. Basically, the meaning of conceptual relationships is mapped to constraints concerning class and activity diagrams of particular models. The checker indicates an exception, in case the specifications do not comply with the semantically grounded syntax. For instance, the correct use of the "before" relationship in the task model requires meeting the following constraints in the structure part ("before" can only be put between task class specifications) and in the behavior part. In the diagram, it is checked whether all activities of task 1 (assuming task 1 "before" task 2) have been completed before the first activity in a behavior specification of task 2 is started.

25.3.3.2 Ensuring Adaptability

Based on consistent specifications, the mental load caused by UI components can be checked, as well as the suitability for individualization, both considered to be crucial for a human-centered DfA approach. To achieve minimal mental load when accomplishing work tasks, the number of user inputs for navigation and manipulation of data has to be minimized, and the software output has to be adjusted to the task and user characteristics. Such a check concerns each of the models. The task-specific path following synchronized activity diagrams should not contain any procedure or step that is not required for task accomplishment. It has to be minimal, in the sense that only those navigation and manipulation tasks are activated that directly contribute to the accomplishment of the addressed user task.

The ProcessLens engine is able to check the minimal length of paths provided for interactive task accomplishment. Initially, it ensures that the task is actually an interactive task (meaning that there exists a presentation of the task at the user interface) as well as there exists an interaction modality to perform the required interactive data manipulation to accomplish the addressed task. The rule engine (see Figure 25.2) determines whether the task in the task-model activity diagram is linked to an element of a class diagram of the interaction model to present the task at the UI (through the relationship "is-presented"). Then it checks links throughout the entire specification (application model) whether the task is linked to data that have to be manipulated interactively to accomplish a task—an "is-based-on" relationship has to be found. Each interactive task also requires a link to interaction-model elements, which present the task data for interactive manipulation to the user. Finally, each task requires a link to a user role (expressed through "handles").

Once the completeness of a task specification has been checked, each path related to that task in an activity diagram is checked. The engine tries to find a path that is not related to the data manipulation of that task. If such a path cannot be found, users are provided only with those interaction features required to complete the addressed task through the assigned role. There might be different paths for the same task when handled through different roles or instances of roles. ProcessLens can handle multiple task interaction procedures checking the adaptation model. In this way suitability of individualization according to ISO 9241, Part 10, can be implemented. Interactive systems support suitability for individualization if they allow for adaptation to the user's individual needs and skills for a given task. Following this understanding, DfA requires interactive applications that can be adapted (1) to the individual organization of tasks; (2) to individual user roles and profiles; and (3) to various interaction styles and devices.

Adaptability with respect to the organization of tasks means that a particular task might be accomplished in several ways. Hence, a specification enabling flexibility with respect to tasks contains more than a single path in an activity or class diagram within a task (implemented through "before" relationships) to deliver a certain output given a certain input. Adaptability with respect to user roles means that a user might have several roles and even switch between them, eventually leading to different perspectives on the same task and data. Adaptability with respect to interaction devices and styles (i.e., the major concern for polymorph multimedia application development) does not only require the provision of several devices or styles based on a common set of interaction elements, it also requires the availability of more than a single way to accomplish interaction tasks at the UI, for instance, direct manipulation (drag and drop) and menu-based window management. The latter can be checked again at the activity diagram level, namely through checking whether a particular operation, such as closing a window, can be performed along different state transitions. For instance, closing a window can be enabled through a menu entry or a button located in the window bar.

In case a user is assigned to different tasks in a flexible way or changes roles (to accomplish a certain task), assignments of and between tasks and roles have to be flexible. A user might also want to switch between modalities or change interaction styles (e.g., when leaving the office to switch to a mobile device, the assignment of interaction elements and procedures to data and tasks have to be modified accordingly). Changing assignments requires links between the different entities that are activated or deactivated at a certain point in time. It requires the existence of conceptual relationships, as well as their dynamic manipulation at run-time.

To ensure adaptability (1) the designer has to set up the proper design space, and (2) modifications have to occur in line with the semantics and derived syntax of the design space. The first objective can be met through providing relationships between design items, both within a model (e.g., "before"), and between models (e.g., "handles"). The first relationship enables flexibility with each of the perspectives mentioned previously (e.g., the organization of tasks), whereas the second relationship allows for flexible tuning of components, in particular the assignment of different styles of interaction to tasks, roles, and domain data.

The second objective can be met allowing for manipulating relationships according to the syntax and semantics of the representation scheme (i.e., specification language), and providing algorithms to check the correct use of the language as well as the changes related to the manipulation of relationships. For instance, the "before" relationship can only be set between subtasks in the task model, since the "before" algorithm processes the design representation. In case the relationship is modified (e.g., "before") between subtasks, the restrictions to the concerned activity diagrams have to be enforced accordingly.

25.4 Conclusion

Today's model-based design tools exploit task models to interaction and application models with respect to function *and* Gestalt. In this way, user-centered DfA takes into account users, tasks, and software-engineering principles (for interface and functional development). An equilibrium between the functional and the nonfunctional world requires different perspectives or viewpoints on design information. In the course of development these perspectives represent self-containing parts of a DfA space. Establishing such a space requires dedicated information structures as well as mechanisms for mutually tuning its elements. DfA is supported through context-sensitive procedures for transformation and refinement to tuned functional components implementing universal access.

Establishing a framework of models enables tool development that captures design inputs and supports the development of design alternatives as well as the transformation to the implementation level. The models have to capture adaptation mechanisms, since they have to be considered the keys for universal accessibility. Adaptability has to be established at design time and become an inherent Gestalt element for the support of various users and their needs.

References

Baresi, L., Di Nitto, E., and Ghezzi, C. (2006). Toward open-world software: Issues and challenges. *IEEE Computer* 39: 36–43.

Baron, M., Lucquiaud, V., Autard, D., and Scapin, D. (2006). K-MADe: Un environment pour le noyau du modèle de description de l'activité (K-MADe: An environment for model-based activity specifications) in the *Proceedings of the 18th French-Speaking Conference on Human-Computer Interaction (IHM'06)*, 18–21 April 2006, Montreal, Canada. http://kmade.sourceforge.net/data/2006-ihm06-mbvldads.pdf.

Beyer, H. and Holtzblatt, K. (1998). *Contextual Design: Defining Customer-Centered Systems.* San Francisco: Morgan Kaufmann.

Bodart, F., Hennerbert, A., Leheureux, J., Provot, I., and Vanderdonckt, J. (1994). A model-based approach to presentation: A continuum from task analysis to prototype, in the *Proceedings of the 1st Eurographics Workshop on Design, Specification, Verification of Interactive Systems (DSV-IS'94)* (F. Paternò, ed.), 8–10 June 1994, Bocca di Magra, Italy, pp. 25–39. Berlin: Springer-Verlag.

Bodker, S. (1998). Understanding representation in design. *Human-Computer Interaction* 13: 107–125.

Bodker, S. and Gronbaek, K. (1991). Cooperative prototyping: Users and designers in mutual activity. *Man-Machine Studies* 34: 453–478.

Bomsdorf, B. and Szwillus, G. (1999). CMF: A coherent modeling framework for task-based user interface design, in the *Proceedings of the 3rd International Workshop on Computer-Aided Design of User Interfaces (CADUI'99)*, 21–23 October 1999, Louvain-la-Neuve, Belgium, pp. 293–304. Dordrecht: Kluwer.

Brusilovsky, P. (2002). Domain, task, and user models for an adaptive hypermedia performance support system, in the *Proceedings of the International Conference on Intelligent User Interfaces 2002 (IUI'02)*, 13–16 January 2002, San Francisco, pp. 23–30. New York: ACM Press.

Caffiau, P., Girare, P., Scapin, D., and Guittet, L. (2007). Generating interactive applications from task models: A hard challenge, in the *Proceedings of the 6th Workshop on Task Models and Diagrams for User Interface Design (TAMODIA 2007)*, 7–9 November 2007, Toulouse, France, pp. 267–272. Berlin/Heidelberg: Springer-Verlag.

Calvary, G., Coutaz, J., Thevenin, D., Limbourg, Q., Bouillon, L., and Vanderdonckt, J. (2003). A unifying reference framework for multi-target user interfaces. *Interacting with Computers* 15: 289–308.

Clerckx, T., Luyten, K., and Coninx, K. (2004). Generating context-sensitive multiples device interfaces from design, in the *Proceedings of the 5th International Conference on Computer-Aided Design of User Interfaces (CADUI'04)*, 14–16 January 2004, Madeira Island, Portugal, pp. 283–296. Dordrecht: Kluwer.

CTTE (2006). *The ConcurTaskTrees Environment.* http://giove.cnuce.cnr.it/ctte.html.

Czarnecki, K., Antkiewicz, M., Hwan, Ch., and Kim, P. (2006). Multi-level customization in application engineering: Developing mechanisms for mapping features to analysis models. *Communications of the ACM* 49: 61–65.

Da Silva, P. P. and Paton, N. (2000). UMLi: The unified modeling language for interactive applications, in the *Proceedings of the 3rd International Conference "The Unified Modeling Language, Advancing the Standard" (UML'00)*, 2–6 October 2000, York, U.K., pp. 117–132. Berlin/Heidelberg: Springer-Verlag.

Darcher, D. (2006). Consilience for universal design: The emergence of a third culture. *Universal Access in the Information Society* 5: 253–268.

Diaper, D. and Stanton, N. (2004). *The Handbook of Task Analysis for Human-Computer Interaction.* London: Lawrence Erlbaum Associates.

Dittmar, A., Forbrig, P., Heftberger, S., and Stary, C. (2004). Tool support for task modeling: A "constructive" exploration, in the *Proceedings of the 11th International Conference on "Design, Specification, and Verification of Interactive Systems" (EHCI-DSVIS'04)*, 11–13 July 2004, Hamburg, Germany, pp. 59–74. Berlin/Heidelberg: Springer-Verlag.

D'Souza, D. and Wills, A. C. (1998). *Objects, Components and Frameworks in UML: The Catalysis Approach.* Reading, U.K.: Addison-Wesley.

Floyd, Ch. (1987). Outline of a paradigm change in software engineering, in *Computers and Democracy* (G. Bjerknes, P. Ehn, and M. Kyng, eds.), pp. 191–211. Aldershot, U.K.: Avebury.

Furnas, G. W. (2000). Future design mindful of moras. *Human-Computer Interaction* 15: 205–261.

Gould, J. D. and Lewis, C. (1985). Designing for usability: Key principles and what designers think. *Communications of the ACM* 28: 300–311.

Griffiths, T., Barclay, P. J., McKirdy, J., Paton, N. W., Gray, P. D., Kennedy, J., et al. (1999). Teallach: A model-based user interface development environment for object databases, in the *Proceedings of User Interfaces to Data Intensive Systems (UIDIS)*, 5–6 September 1999, Edinburgh, Scotland, pp. 86–96. New York: IEEE Press.

Grudin, J. and Poltrock, S. (1989). User interface design in large corporations: Communication and coordination across disciplines, in the *Proceedings of the SIGCHI Conference on Human Factors in Computing Systems: Wings for the Mind (CHI'89)*, Austin, TX, pp. 197–203. New York: ACM Press.

Günter, H. and Grote, G. (2006). Redefining task interdependence in the context of supply networks, in the *Proceedings of the 13th European Conference on Cognitive Ergonomics: Trust and Control in Complex Socio-technical Systems (ECCE'06)*, 20–22 September 2006, Zurich, Switzerland, pp. 55–63. New York: ACM Press.

Hackos, J. T. and Redish, J. C. (1998). *User and Task Analysis for Interface Design*. New York: John Wiley & Sons.

Hailpern, B. and Tarr, P. (2006). Model-driven development: The good, the bad, and the ugly. *IBM Systems Journal* 45: 451–461.

Hammond, N., Jorgensen, A., MacLean, A., Barnard, P., and Long, J. (1983). Design practice and interface usability: Evidence from interviews with designers, in the *Proceedings of the SIGCHI Conference on Human Factors in Computing Systems (CHI'83)*, 12–15 December 1983, Boston, Massachusetts, pp. 40–44. New York: ACM Press.

Hemmecke, J. and Stary, C. (2007). The tacit dimension of user tasks: Elicitation and contextual representation, in the *Proceedings of the 5th International Workshop on Task Models and Diagrams for User Interface Design (TAMODIA 2006)*, 23–24 October 2007, Hasselt, Belgium, pp. 308–323. Berlin/Heidelberg: Springer-Verlag.

Herrmann, T., Hoffmann, M., Loser, K.-U., and Moysich, K. (2000). Semistructured models are surprisingly useful for user-centered design, in the *Proceedings of the Fourth International Conference on the Design of Cooperative Systems (COOP'2000)* (R. Dieng, A. Giboin, L. Karsenty, and G. DeMichelis, eds.), 23–26 May 2000, Mediathel of Sofia Antipolis, France, pp. 159–174. Amsterdam: IOC Press.

Howard, S. (1997). Trade-off decision making in user interface design. *Behaviour & Information Technology* 16: 98–109.

Jarke, M. (1998). Requirements tracing. *Communications of the ACM* 41: 32–36.

Johnson, H. and Johnson, P. (1998). Integrating task analysis into system design: Surveying designer needs. *Ergonomics* 32: 1451–1467.

Keates, S. (2006). Pragmatic research issues confronting HCI practitioners when designing for universal access. *Universal Access in the Information Society* 5: 269–278.

Limbourg, Q. and Vanderdonckt, J. (2004a). Addressing the mapping problem in user interface design with UsiXML, in the *Proceedings of the 3rd Annual Conference on Task Models and Diagrams (TAMODIA 2004)*, 15–16 November 2004, Prague, Czech Republic, pp. 155–163. New York: ACM Press.

Limbourg, Q. and Vanderdonckt, J. (2004b). Comparing task models for user interface design, in *The Handbook of Task Analysis for Human-Computer Interaction* (D. Diaper and N. Stanton, eds.), pp. 135–154. London: Lawrence Erlbaum Associates.

Luyten, K., Clerckxs, T., Conin, K., and Vanderdonckt, J. (2003). Derivation of a dialog model from a task model by activity chain extraction, in the *Proceedings of the 10th International Workshop on Interactive Systems. Design, Specification, and Verification (DSV-IS'03)*, 11–13 June 2003, Madeira Island, Portugal, pp. 203–217. Berlin/Heidelberg: Springer-Verlag.

Marakas, G. M. (2001). *Systems Analysis and Design: An Active Approach*. Upper Saddle River, NJ: Prentice Hall.

McTear, M. F. (2000). Intelligent interface technology: From theory to reality. *Interacting with Computers* 12: 323–336.

Mirel, B. (2004). *Interaction Design for Complex Problem Solving: Developing Useful and Usable Software*. San Francisco: Morgan Kaufman.

Myers, B. A., Hudson, S. E., and Pausch, R. F. (2000). Past, present, and future of user interface software tools. *ACM TOCHI* 7: 3–28.

Norman, D. A. (1988). *The Psychology of Everyday Things*. New York: Basic Books.

Nuseibeh, B., Kramer, J., and Finkelstein, A. (1994). Framework for expressing the relationships between multiple views in requirements specification. *IEEE Transactions on Software Engineering* 20: 760–773.

Obrenovic, Z. and Starcevic, D. (2004). Modeling multimodal human-computer interaction. *IEEE Computer* 37: 65–72.

O'Neill, E. and Johnson, P. (2004). Participatory task modeling: Users and developers modelling users' tasks and domains, in the *Proceedings of the 3rd Annual Conference on Task Models and Diagrams (TAMODIA 2004)*, 15–16 November 2004, Prague, Czech Republic, pp. 67–74. New York: ACM Press.

Oppl, S. and Stary, C. (2005). Towards human-centered design of diagrammatic representation schemes, in the *Proceedings of the Fourth International Workshop on Task Models and Diagrams for User Interface Design (TAMODIA 2005)*, 26–27 September 2005, Gdansk, Poland, pp. 55–62. New York: ACM Press.

Paternò, F. (1999). *Model-Based Design and Evaluation of Interactive Applications*. Berlin/Heidelberg: Springer-Verlag.

Paternò, F. and Santoro, C. A. (2003). Unified method for designing interactive systems adaptable to mobile and stationary platforms. *Interacting with Computers* 15: 347–364.

Pohl, W. and Nick, A. (1999). Machine learning and knowledge representation in the LabUr approach to user modeling, in the *Proceedings of the 7th International Conference on User Modeling (UM 1999)*, 20–24 June 1999, Banff, Canada, pp. 179–188. New York: Springer-Verlag.

Puerta A. and Eisenstein J. (2002). XIML: A common representation for interaction data, in the *Proceedings of the 7th International Conference on Intelligent User Interfaces (IUI 2002)*, 13–16 January 2002, Miami, pp. 214–215. New York: ACM Press.

Rosson, M. B., Maass, S., and Kellog, W. A. (1987). Designing for designers: An analysis of design practice in the real world, in the *Proceedings of the SIGCHI/GI Conference on Human Factors in Computing Systems and Graphics Interface*

(CHI'87), 5–9 April 1987, Toronto, Canada, pp. 137–142. New York: ACM Press.

Scogings, C. and Phillips, C. (2004). Linking task and dialog modeling: Toward an integrated software engineering method, in *The Handbook of Task Analysis for Human-Computer Interaction* (D. Diaper and N. Stanton, eds.), pp. 551–566. London: Lawrence Erlbaum Associates.

Sebilotte, S. (1994). From users' task knowledge to high-level interface specification. *Human-Computer Interaction* 6: 1–15.

Shneiderman, B. (2000). Universal usability: Pushing human-computer interaction research to empower every citizen. *Communications of the ACM* 43: 85–91.

Sottet, J. S., Calvary, G., Favre, J. M., Demeure, A., and Coutaz, J. (2006). Towards mappings and model transformations for consistency of plastic user interfaces, in the *Workshop "The Many Faces of Consistency in Cross-Platform Design," CHI 2006*, 22–27 April 2006, Quebec, Canada. New York: ACM Press.

Stahl, T. and Völter, M. (2006). *Model-Driven Software Development: Technology, Engineering, Management*. Chicester: John Wiley & Sons.

Stary, C. (2000). TADEUS: Seamless development of task-based and user-oriented interfaces. *IEEE Transactions on Systems, Man, and Cybernetics* 30: 509–525.

Stary, C. (2002). Shifting knowledge from analysis to design: Requirements for contextual user interface development. *Behavior and Information Technology* 21: 425–440.

Stary, C. (2006). Ensuring task conformance and adaptability of polymorph multimedia applications, in *Handbook of Research on Mobile Multimedia, Vol. I* (I. K. Ibrahim, ed.), pp. 291–310. Hershey, PA: Idea Publishing.

Stephanidis, C. (ed.) (2001). *User Interfaces for All: Concepts, Methods, and Tools*. Mahwah, NJ: Lawrence Erlbaum Associates.

Sutcliffe, A. G., Economou, A., and Markis, P. (1999). Tracing requirements errors to problems in the requirements engineering process. *Requirements Engineering* 4: 134–151.

Traetteberg, H. (2002). Using user interface models in design, in the *Proceedings of the Fourth International Conference on Computer-Aided Design of User Interfaces (CADUI'02)*, 15–17 May 2002, Valenciennes, France, pp. 131–142.

Dordrecht, Germany: Kluwer.

Ulich, E., Rauterberg, M., Moll, T., Greutmann, T., and Strohm, O. (1991). Task orientation and user-oriented dialog design. *International Journal of Human-Computer Interaction* 3: 117–144.

Van Harmelen, M. (ed.) (2001). *Object Modeling and User Interface Design*. New York: Addison Wesley.

Van Welie, M., van der Veer, G. C., and Eliëns, A. (1998). An ontology for task world models, in the *Proceedings of the Fifth International Eurographics Workshop on Design, Specification and Verification of Interactive Systems (DSV-IS'98)*, 3–5 June 1998, Abingdon, U.K., pp. 57–70. Berlin/Heidelberg: Springer-Verlag.

Wilson, S. and Johnson, P. (1996). Bridging the generation gap: From work tasks to user interface design, in the *Proceedings of the Second International Workshop on Computer-Aided Design of User Interfaces (CADUI'96)*, 5–7 June 1996, Namur, Belgium, pp. 77–94. Namur, Belgium: Presses Universitaires de Namur.

Wolff, A., Forbrig, P., Dittmar, A., and Reichart, G. (2005). Linking GUI elements to tasks: Supporting an evolutionary design process, in the *Proceedings of the Fourth International Workshop on Task Models and Diagrams for User Interface Design (TAMODIA 2005)*, 26–27 September 2005, Gdansk, Poland, pp. 27–34. New York: ACM Press.

Wood, L. E. (ed.) (1998). *User Interface Design: Bridging the Gap from User Requirements to Design*. Boca Raton, FL: CRC Press.

XUL (n.d.). *XML User Interface Language*. http://www.mozilla.org/projects/xul.

Ye, J.-H. and Herbert, J. (2004). Framework for user interface adaptation, in *User-Centered Interaction Paradigms for Universal Access in the Information Society, Proceedings of the 8th ERCIM International Workshop "User Interface for All"* (C. Stary and C. Stephanidis, eds.), 28–29 June 2004, Vienna, pp. 167–174. Berlin/Heidelberg: Springer-Verlag.

Züllighoven, H. (2005). *Object-Oriented Construction Handbook: Developing Application-Oriented Software with the Tools and Material Approach*. San Francisco: Morgan Kaufmann.

26

Markup Languages in Human-Computer Interaction

Fabio Paternò and
Carmen Santoro

26.1 Introduction

Markup languages are increasingly used in human-computer interaction (HCI) to represent relevant information and process it. Interest in them has been stimulated, in particular, by the growing availability of many potential interaction devices and the need for supporting a wide variety of users, including the disabled. The XML meta-language has become the *de facto* standard for creating markup languages, and a variety of XML-based languages have been proposed to address various aspects relevant to HCI. A markup language is a set of words and symbols useful for identifying and describing the different parts of a document. It combines text and related extra information, expressed using markup symbols intermingled with the primary text in a hierarchical structure of elements and attributes.

Some reasons can be identified at the basis of XML's popularity. First, different from some markup languages that are purpose-specific (e.g., HTML for describing document appearance) and that cannot be reused for a different goal, XML is a self-describing format in which the markup elements represent the information content. Thus, XML completely leaves the interpretation of such data to the application that reads them, and information content is separated from information rendering, making it easy to provide multiple views of the same data. By leaving the names, hierarchy, and meanings of elements/attributes open and definable, XML lays the foundation for creating custom and modular (new formats can be defined by combining and reusing other formats) XML-based markup languages. Also, XML has a plain text format, which means that it is both human and machine-readable. The wide availability of tools for text file authoring software facilitates rapid XML document authoring and maintenance,

and cross-platform interoperability. This was not so easy before XML's advent when most data interchange formats were proprietary "binary" formats, and therefore not easily shared by different software applications or across different computing platforms. Moreover, the strict syntax and parsing requirements allow the appropriate parsing algorithms to remain simple, efficient, and consistent. XML is a robust, logically verifiable format based on international standards and is unencumbered by licenses or restrictions. Lastly, it is well supported. Thus, by choosing XML, it is possible to access a large and growing community of tools, services, and technologies based on it (XLink, XPointer, XSLT, but also RDF and the semantic web).

XML-based languages have been considered to address various aspects relevant to HCI. For example, as pervasive computing evolves, interactive application developers should cope with the problem of providing solutions for simultaneous deployment on a growing number of platforms for disparate users in a wide variety of contexts. Because developing ad hoc solutions might result in a significant overhead, one feasible solution is abstracting from the presentation details of the specific medium used for the interaction, and focusing more properly on aspects related to the semantic of the interaction, namely what is the expected result that an interaction should reach. Thus, capturing and describing the essence of what a user interface should be can be obtained by identifying logical descriptions that contain semantic information and are able to highlight the main aspects to consider.

In addition, applications for mobile users force designers to take into account a number of related aspects that could affect the way in which interactive applications might change not only their rendering but also the functionality provided to the users, depending on the context in which the interaction takes place. Such aspects might include the different categories of users representing the expected target population, the environmental conditions in which interaction might occur, the particular device that might be used, and so on. Therefore, such different aspects should be described by appropriate specifications. Such logical specifications are usually described using XML-based languages.

The first issue connected with the adoption of XML-based languages for specifying user interface-related aspects at different abstraction levels is providing automatic tools able to understand such specifications, and consequently supporting the designers in deriving effective user interfaces. Another issue is semantically correlating such specifications to each other, so as to provide designers with different views of the same interactive system, along with automatic tools able to transform such different specifications.

This chapter is intended to extensively address and discuss the aforementioned topics. After a brief introduction about the usefulness of markup languages in HCI, the chapter discusses what aspects relevant to HCI can be addressed through the XML-based languages that have been proposed (abstract/concrete user interface descriptions, tasks, users, devices, environments, etc.). Next, it is discussed how guidelines can be formalized in markup languages and how such information can be exploited to support user interface designers and evaluators, in particular in the accessibility area. Moreover, two subsections are devoted to describing how markup languages can be used to support, respectively, the implementation of interactive applications and the transformations of UI models. A section is dedicated to markup languages for multidevice interfaces, which is a particularly important and timely topic. This section includes reviewing and discussing existing approaches that exploit XML-based languages for describing information useful for designing interactive systems. Furthermore, a subsequent section is dedicated to the motivations and usefulness for transforming the different specifications (i.e., for forward/reverse engineering, redesign for different platforms, multidevice support, etc.). Last, some conclusions are drawn and indications are provided of current challenges in the area of markup languages specifically used for supporting the development of future pervasive interactive systems exploiting ambient intelligence.

26.2 What Can Be Specified with Markup Languages

Markup languages are a very general tool to represent any type of information; they have been found particularly useful to represent several types of aspects in HCI:

- *Abstract descriptions of interactive systems*: There are several possible views on an interactive system that differ depending on the abstraction level considered. The purpose of such levels is to highlight the semantic aspects, removing low-level details that are not particularly important.

- *Context of use*: People can interact with systems in a wide variety of contexts of use, which differ in terms of the user characteristics, the technology available (in terms of interaction resources and modalities, connectivity, etc.), the surrounding environment (level of noise, luminosity, etc.), and the social context.

- *Guidelines*: An increasing number of guidelines have been proposed to help in accessibility and usability evaluation.

- *Implementation languages*: In particular for web environments, many types of implementation languages based on XML have been proposed, which also support different modalities: form-based interfaces are described by XHTML and XUL, while vectorial graphics is supported through SVG (scalable vector graphics, see http://www.w3.org/Graphics/SVG), multimodal and multimedia applications can be obtained by X+V (http://www.voicexml.org/specs/multimodal/x+v/12) and SMIL (http://www.w3.org/AudioVideo).

- *Transformations*: With so many XML-based languages used for describing both models and implementation languages, and used in an interoperable way among different tools, the problem emerges to define transformations able to translate such languages from one to another.

Figure 26.1 shows the different ways in which XML-based descriptions have been used in HCI (see boxes with bold labels). One of the most common uses is identifying XML-based languages for specifying different descriptions that are relevant for context-dependent interactive applications. This includes not only different descriptions of a user interface according to various abstraction levels (task and object level, abstract level, and the concrete one), but also conditions regarding the current context that are deemed relevant for the design and evaluation of the interactive application (since they can have an impact on both of them). Another goal for which XML-based languages are more and more used is the specification of final implementation languages for user interfaces.

Also, the mechanism underlying XML, which allows tagging the different parts of a description of a user interface (at various levels, from the most abstract one to the final level) provides twofold benefits. On the one hand, XML tagging-based languages allow for providing a logical meaning to the different parts of the description. On the other hand, the use of such XML-based languages is a stimulus for developing XML-based languages to specify the transformations that can relate such various descriptions with each other. Last, XML-based descriptions have also been used to describe guidelines, which are meant to be provided as input to tools for evaluating interactive applications.

26.2.1 Abstract Descriptions

In the research community in model-based design of user interfaces there is a general consensus on what useful logical descriptions are (Szekely, 1996; Paternò, 1999; Calvary et al., 2003; see also Chapter 25 of this handbook):

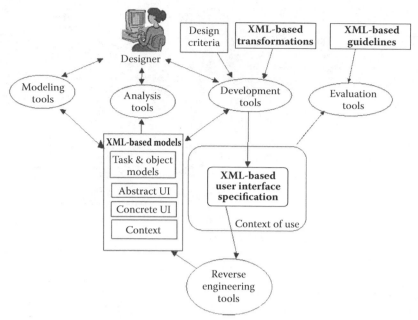

FIGURE 26.1 Using XML-based specifications in HCI.

- The task and object level, which reflects the user view of the interactive system in terms of logical activities and objects manipulated to accomplish them
- The abstract user interface, which provides a modality-independent description of the user interface
- The concrete user interface, which provides a modality-dependent but implementation-language-independent description of the user interface
- The final implementation, in an implementation language for user interfaces

Thus, for example, the task "select vegetarian menu" implies the need for a selection object at the abstract level, which indicates nothing regarding the platform and modality in which the selection will be performed (it could be through a gesture or a vocal command or a graphical interaction). When moving to the concrete description, a specific platform has to be assumed, for example, the graphical PDA, and a specific modality-dependent interaction technique needs to be indicated to support the interaction in question (e.g., selection could be through a radio-button or a dropdown menu), but nothing is indicated in terms of a specific implementation language. After choosing an implementation language, the last transformation from the concrete description into the syntax of a specific user interface implementation language is performed. The advantage of this type of approach is that it allows designers to focus on logical aspects and take into account the user's view right from the earliest stages of the design process.

Figure 26.1 shows how various XML-based descriptions can be used at different stages in the life cycle of the interactive software application. Figure 26.2 shows how data contained in different XML-based descriptions can be used both at run-time and design time in the process of generating user interfaces.

At design time, the idea is to precompute once the different versions of a UI for the various platforms the designers want to address. To this aim, various levels of abstraction are considered to obtain a refinement mechanism able to transform high-level abstraction descriptions to more concrete specifications, up to the final implementation. Such transformations are carried out by taking into account information provided in various relevant models (user, platform, environment, etc.) and should also be specified to transform concrete user interface descriptions to abstract ones to maintain model consistency.

Furthermore, some data contained in such models can be more or less relevant, depending on the transformation (between two different abstraction levels) considered. For instance, depending on the characteristics specified in the model of the platform currently considered, an appropriate transformation will be supported for obtaining a concrete user interface from an abstract one, so as to take into account the features and resources of the device at hand.

The data contained in such models are also useful at run-time, since they are considered when a dynamic reconfiguration is needed as a consequence of some occurring events. For example, if the user changes a device, the reconfiguration phase has to calculate again the effects of the changes occurred, and then trigger a dynamic generation of the user interface so as to adapt the UI to the new device. Section 26.3 provides a description of different markup languages for multidevice user interfaces.

26.2.2 Context of Use

The context of use is defined by a number of aspects (user, device, environment, social context), which can be formalized at various extents. In particular, the user and the device have been the subject of various proposals aiming to represent the associated

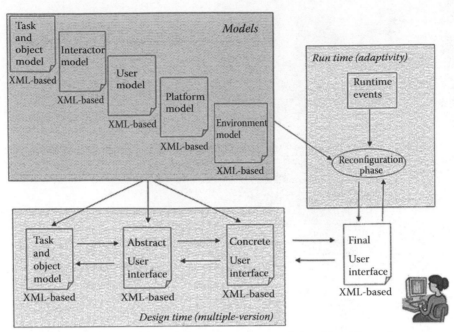

FIGURE 26.2 Generating user interfaces by using XML-based descriptions at design time and run-time.

important information. In general, in user modeling the goal is to represent the level of knowledge or interest of a user regarding a number of topics (see also Chapter 24 of this handbook). This information can be represented at different levels of detail: binary (known/unknown), qualitative (good/medium/poor), and quantitative (i.e., probability that the user knows a piece of information). The user model can be used to adapt the user interface. Several aspects of the user interface can adapt depending on the user: presentation (e.g., in the choice of layout, colors, fonts, etc.), content (depending on the level of knowledge of the user), and navigation (links can be enabled or disabled depending on user interests). Ubiquitous computing offers new chances and challenges to the field of user modeling. With the markup language UserML, Heckmann and Krueger (2003) propose a language for combining partial user models in a ubiquitous and comprehensive computing environment, where all different kinds of systems work together to satisfy the user's needs.

The main idea of UserML is to enable communication about partial user models via the Internet. Therefore, one goal of UserML is the representation of partial user models. Using XML as a knowledge representation language has the advantage that it can be used directly in the Internet environment. One disadvantage is that the nested structures of the XML tags only represent a tree, while often the structure of a graph is needed. The approach used for UserML is, therefore, a modularized approach (see Figure 26.3) in which several categories will be connected via identifiers (IDs) and references to identifiers (ID-Refs). The content of a UserML document is divided into MetaData, UserModel, InferenceExplanations, ContextModel, DomainModel and DeviceModel. In this way, the tree structure of XML can be extended to represent graphs. Therefore, a two-level approach is proposed. At the first level, a simple

XML structure is offered for all entries of the partial user model. At the second level, there is the ontology that defines the categories. The advantage of this approach is that different ontologies can be used with the same UserML tools, thus different user modeling applications can use the same framework and keep their individual user model elements. In the following excerpt, an example is presented of a partial user model that uses categories from the ontology "UserOL"

The UserModel consists of an unbounded list of UserData entries. Each one defines the category, the range, and the value.

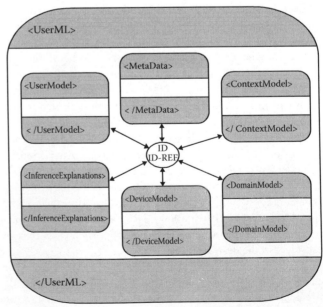

FIGURE 26.3 UserML containers, connected via IDs and ID-REFs.

```
<UserModel>
  <UserData id="231">
    <category>userproperty.timepressure</category>
    <range>low-medium-high</range>
    <value>high</value>
    <ontology>"http://www.u2m.org/UserOL/"</ontology>
  </UserData>
  <UserData id="224">
    <category>userproperty.walkingspeed</category>
    <range>slow-normal-fast</range>
    <value>fast</value>
    <ontology>"http://www.u2m.org/UserOL/"</ontology>
  </UserData>
  <UserData id="122">
    <category>usercontext.location</category>
    <range>airport.location</range>
    <value>X35Y12</value>
    <ontology>"http://www.u2m.org/UserOL/"</ontology>
  </UserData>
</UserModel>
```

The reference to the ontology can also be set by default in the UserModel element. An alternative approach would have been to encode the user modeling knowledge into the XML elements <timepressure>, <psychological-state>, <typing-behavior>.

An implementation architecture has also been elaborated of a general user model editor, which is based on UserML. The current implementation of this editing tool transforms UserML into XForms with XSLT.

Regarding the devices, the need for describing their features derives from the increasing availability of various types of interactive devices. Thus, to have applications able to adapt to their features, there should be a way to represent them. This is particularly important in the area of mobile phones, in which the possible characteristics are the most variable ones. The generic Composite Capabilities/Preference Profiles (CC/PP) framework (http://www.w3.org/2001/di) provides a mechanism through which a mobile user agent—a client, such as a browser, that performs rendering within a mobile device—can transmit information about the mobile device. The user agent profile (UAProf, 2001; http://www.openmobilealliance.org/release_program/uap_v20.html) is based on the CC/PP framework. It includes device hardware and software characteristics, information about the network the device is connected to, and other attributes. It is possible to identify a device through the header of HTTP requests. All the devices complying with UAProf have a CC/PP description of their characteristics in a repository server, which can be queried for obtaining the related information. The description of the devices is in the *Resource Description Framework* (see http://www.w3.org/RDF), another XML-based language. When a mobile device sends a request, it also informs about the URL where its profile is located through a specific field in the request called *X-Wap-Profile*. For example, the *x_wap_profile* for a Nokia N9500 is:

x_wap_profile:http://nds1.nds.nokia.com/uaprof/N9500r100.xml

The textbox on the following page shows an excerpt of a profile for a mobile device.

Another proposal in this area is WURFL (wireless universal resource file) (Passani, 2005), which is an XML configuration file that contains information about capabilities and features of several wireless devices. The main objective is to collect as much information as possible about all the existing wireless devices that access WAP pages, so that developers will be able to build better applications and better services for the users. This proposal aims to support web applications for mobile devices. The goal is to programmatically abstract away devices' differences, avoid the current situation wherein we need to modify applications whenever a new device ships, and avoid the current situation that we need to track new devices that ship (particularly those in uninteresting markets). The basic idea is to provide a global database of all devices and their capabilities. The underlying assumption is that browsers are different, but they also have many features in common, and that browsers/devices coming from the same manufacturer are most often an evolution of the same hardware/software. In other words, differences between, for example, a Nokia 7110 and a Nokia 6210 are minimal; devices from different manufacturers may run the same software. For example, Openwave Systems provides the browser to Siemens, Motorola, Alcatel, Mitsubishi, Samsung, Panasonic, Telit, and Sagem. WURFL has created a compact, small, and easy way to update a matrix. WURFL is based on the concept of a family of devices. All devices are descendents of a generic device, but they may also descend from more specialized families (such as those who use the same browser). Its goal is to overcome some limitations of the UAProf standard developed in the OMA. UAProf

```xml
<?xml version="1.0"?>

<rdf:RDF
    xmlns:rdf= "http://www.w3.org/..."
    xmlns:prf="http://www.openmobilealliance.org/..."
    xmlns:mms="http://www.wapforum.org/..."
    xmlns:pss5="http://www.3gpp.org/...">
    <rdf:Description rdf:ID="Profile">
      ......

    <prf:component>
      <rdf:Description rdf:ID="HardwarePlatform">
        ......

        <prf:PixelAspectRatio>1x1</prf:PixelAspectRatio>
        <prf:PointingResolution>Pixel</prf:PointingResolution>
        <prf:ScreenSize>640x200</prf:ScreenSize>
        <prf:ScreenSizeChar>29x5</prf:ScreenSizeChar>
    <prf:StandardFontProportional>Yes</prf:StandardFontProportional>
        <prf:SoundOutputCapable>Yes</prf:SoundOutputCapable>
        <prf:TextInputCapable>Yes</prf:TextInputCapable>
        <prf:Vendor>Nokia</prf:Vendor>
        <prf:VoiceInputCapable>No</prf:VoiceInputCapable>
      ......

      </rdf:Description>
    </prf:component>
    ......

  </rdf:Description>
</rdf:RDF>
```

seems to rely too much on someone else setting up the infrastructure to request profiles. There are cases of manufacturers just associating the profile of different phones into a new one. WURFL can be installed in any site and does not need to access device profiles from a repository on the Internet.

26.2.3 Guidelines

To assure a certain level of accessibility and usability of user interfaces (UIs), several design criteria and guidelines have been proposed in the literature.

Guidelines are intended for developers and designers: they are general principles that can be followed to improve application accessibility and usability. Many countries have adopted legislation that imposes some level of compliance to accessibility guidelines. This has raised interest in tool support. The goal of tool support is not to completely replace human evaluators and designers, but to help them manage the complexity of many existing web sites, apply evaluation criteria in a consistent manner, and thereby make their work more efficient. Some tools for this purpose already exist, such as Bobby, LIFT (http://www.usablenet.com), and A-Prompt (http://aprompt.snow.utoronto.ca), but the increasing request for support poses new issues that deserve better answers. In particular, three important areas have

been identified for accessibility tools: support for checking multiple sets of guidelines, code correction, and reporting.

The issue of multiple guidelines involves many factors. One is that there are various organizations that issue standard guidelines; some are international standards (such as the W3C/WAI, or the ISO FDIS 9241-171 and TS 16071), and others are national (such as Section 508). They often share the same body of knowledge but also have slight differences that designers have to consider in some contexts. Another reason is that, while accessibility guidelines usually provide an important step forward to make sites accessible to everyone, including the disabled, often their mere application does not provide usable results for some classes of users. Indeed, being able to access information is not enough. In fact, although a service may be accessible to certain types of users (such as the blind), it still may not be sufficiently usable to such users. While accessibility and usability are closely related to each other, they address slightly different issues. Accessibility aims to increase the number of users who can access a system, this means to remove any potential technical barrier that does not allow the user to access the information. Usability aims to make the user interaction more effective, efficient, and satisfactory. Thus, it is possible to have systems that are usable but not accessible, which means that some users cannot access the information, but those who can access it are supported in such a way as to find it easily. It is also possible to have systems accessible but not usable, which means that all users can access the desired information, but this can only occur after long and tedious interactions. Consequently, there is also a need for integrating these two aspects to obtain interactive services for a wide variety of users, including people who are disabled.

This section mainly focuses on guidelines for web applications, because the web is currently the most common UI environment. Each guideline can include one or more checkpoints. Checkpoints are technical solutions that support the application/evaluation of the criteria and usually correspond to specific implementation constructs that guarantee the satisfaction of the associated guideline. For example, the guideline "Logical partition of interface elements" expresses the concept of well-structuring and organizing the page content. So, it provides a general principle that should be taken into account by developers during web site design. Then, developers can decide how to apply this criterion. Usually, several solutions can be adopted. For example, the web page content could be structured by using frames, or blocks <div> customizable by cascading style sheet (CSS) properties. Alternatively, the content within the page could be visualized by embedding it in the layout or data tables. Moreover, long page content could be partitioned through heading levels, paragraphs, or specific page parts could be marked with "hidden labels." So, all these cases apply the same general guideline (i.e., partitioning the content), but use different technical solutions.

The analysis of web site accessibility and usability by means of guidelines, similar to other inspection methods used in usability/accessibility assessment, requires observing, analyzing, and interpreting the web site characteristics. Since these activities require high costs in terms of time and effort, there is great interest in developing tools that automate them in various phases. However, even the use of automatic tools has problems, such as the length and detailed nature of reports that make them difficult to interpret, accessibility guidelines require developers to fully understand their requirements, and often they still need some manual inspection. Besides, another issue has to be considered when automatic support is used is the "repair process." In fact, even if accessibility and usability problems are detected automatically, repairing them can require a lot of effort. A completely automatic repair process is not easy to implement; a semiautomatic support for this important phase is advisable. However, several common evaluation tools do not provide any support for repair functionality, and those tools that provide some support require working with the underlying HTML code, which can be tedious and difficult to do because of the many low-level details to handle. In this perspective, MAGENTA (Multi-Analysis of Guidelines by an ENhanced Tool for Accessibility) (Leporini et al., 2006) has been developed to assist developers in handling web pages and guidelines.

Many automatic analysis tools were developed to assist evaluators by automatically detecting and reporting guideline violations and in some cases making suggestions for fixing them. EvalIris (Abascal et al., 2004) is an example of a tool that provides designers and evaluators with a good support to easily incorporate new additional accessibility guidelines. This is obtained through a language for guidelines represented by an XML schema, which has been used to specify WAI guidelines. The result report of the single page evaluation is provided through a representation defined in XML as well. MAGENTA is another tool that supports the verification of guidelines expressed by an XML-based language (GAL: guidelines abstraction language). One advantage of this choice is that the guidelines are specified externally to the tool. Thus, if a new set of guidelines has to be checked, the tool can still be used without modifying its implementation, as long as the new guidelines are specified through such abstraction language. MAGENTA also aims at addressing other ways to support designers, such as providing interactive support for correcting the web pages violating the guidelines considered. Another contribution in this area is DESTINE (Beirekdar et al., 2005), which supports W3C and Section 508 guidelines specified through another language for guidelines and it provides reports that include statistics at different levels (site, page, guideline). However, even this tool does not address the possibility of supporting the repair of web pages. Imergo (Mohamad et al., 2004) is another interesting contribution in the area of tools for accessibility, which aims to provide integrated support for content management systems and engineered support for multiple guidelines. MAGENTA provides a different solution to such an issue, based on the definition of an abstract language for guidelines and support for automatic repair. It allows developers to visualize the reports without performing again the evaluation process because the reports are stored in external XML files.

26.2.4 Markup Languages for Implementation of Interactive Applications

XUL is an example of a markup language for creating UIs. It is a part of the Mozilla browser and related applications, and is available as part of Gecko (http://developer.mozilla.org/en/docs/Gecko). With XUL and other Gecko components, developers can create sophisticated applications without special tools. XUL was designed for creating the UI of the Mozilla application, including the web browser, mail client, and page editor. XUL may be used to create these types of applications. However, it may also be used for any web application, for instance, when developers need to be able to retrieve resources from the network and require a sophisticated UI. Like HTML, in XUL developers can create an interface using a markup language, use CSS style sheets to define appearance, and use JavaScript for behavior. Programming interfaces for reading and writing remote content over the network and for calling web services are also available. Unlike HTML, however, XUL provides a powerful set of UI widgets for creating menus, toolbars, tabbed panels, and hierarchical trees to give a few examples. This means that developers do not have to look for third-party code or include a large block of JavaScript in their application just to handle a popup menu. XUL has all of these elements built-in. In addition, the elements are designed to look and feel just like those on the user's native platform, or designers can use standard CSS to create their own look.

A markup language for vectorial graphics is scalable vector graphics (SVG), which, together with JavaScripts, supports interactive direct manipulation graphical interfaces. The vocal modality is supported through VoiceXML (http://www.w3.org/TR/voicexml20). X+V supports multimodal interfaces by combining XHTML and VoiceXML. SMIL supports multimedia presentations, including video or audio streams combined with texts and images.

26.2.5 Transformations

As previously discussed, XML can be used to formalize various aspects, concepts, and models relevant in HCI. This knowledge often needs to be transformed in design, development, and evaluation. For example, one model can be converted into another model of the same system. A transformation is a set of rules and techniques used for this modification. A description of a transformation may be a rule specified in natural language, an algorithm in an action language, or in a model mapping language. Therefore, there is the problem of transforming different XML-based specifications one into another. There are various types of transformations; for example, abstraction moves toward more abstract representations, and refinement moves toward more concrete representations. Mappings are a specific type of transformation that associates one element of a representation with one or multiple elements of another representation. Section 26.4 will provide a more detailed description of different markup-based approaches for supporting transformations among different XML-based descriptions.

26.3 Existing Markup Languages for Multidevice UIs

Looking at the historical evolution of development software languages (represented in Figure 26.4), one can notice that there is

a tendency to increase the abstraction levels to make them more manageable and expressive. In particular, in recent years there has been a focus on device-independent markup languages to address the issues raised by multidevice environments and models (see, e.g., the evolution toward model-driven architectures of UML and related languages).

Souchon and Vanderdonckt (2003) provided an early discussion of various model-based approaches, mainly for multidevice UIs, considering the models supported, the method applied, the tools associated, the implementation languages and the platforms supported, along with the number of tags, the expressivity, and the openness of the environments. Since then many other proposals have been put forward in this area. For example, in TADEUS (Müller et al., 2001) authors present a concept of device-independent UI design based on the XML-based technology. Specifically, it is an approach for integrating task and object knowledge into the development process and its underlying representations, with a special focus on mobile devices. The model-based approach of TADEUS is based on user, task, and business-object models. It allows the computer-based development of interactive systems, with the three models at the basis for the development of the application logic and the UI design.

A UI is considered a simplified version of the MVC model (Burbeck, 1987) and is separated into a model component and a presentation component. The model component (also called abstract interaction model) describes the feature of the UI on an abstract level, while the UI objects with their representation are specified in the presentation component. During the development, a mapping transforming the abstract interaction model to the specific interaction model is necessary. TADEUS uses a representation of both models. The transformation process is described by attribute grammars: for different platforms different grammars are necessary. The XML-based technology is used in this approach for specifying the description of the abstract interaction model, the description of specific characteristics of different devices, and the specification of the transformation process.

In this section it is not possible to mention all the XML-based languages that have been proposed to address the issues raised

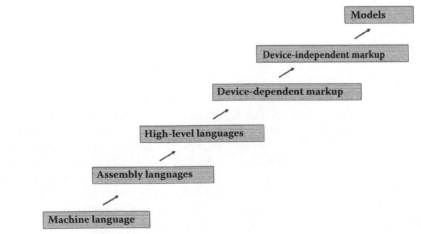

FIGURE 26.4 Historical evolution of software languages.

by multidevice interfaces. Thus, a subset is discussed that should be sufficient to illustrate the relevant issues and the possible solutions.

26.3.1 XIML

The extensible interface markup language (XIML) (http://www.ximl.org) is an extensible XML-based specification language for multiple facets of multiple models in a model-based approach, developed by a forum headed by RedWhale software. It was introduced as a solution that enables a framework for the definition and interrelation of interaction data items. As such, XIML can provide a standard mechanism for applications and tools to interchange interaction data and to interoperate within integrated UI engineering processes, from design, to operation, to evaluation.

The XIML language draws mainly from two foundations. One is the study of ontologies and their representation, and the other is the work on interface models (Szekely et al., 1995; Puerta, 1997; Puerta and Eisenstein, 2001). From the former, XIML draws the representation principles it uses; from the latter, it derives the types and nature of interaction data. The basic structure of XIML consists of components, relations, and attributes:

- *Components*: XIML is an organized collection of interface elements that are categorized into one or more major interface components. The language does not limit the number and types of components that can be defined or the number and types of elements under each component. However, it is expected that an XIML specification would support a relatively small number of components with one major type of element defined per component. In its first version (1.0), XIML predefines five basic interface components, namely task, domain, user, dialogue, and presentation. The first three of these can be characterized as contextual and abstract, while the last two can be described as implementational and concrete:

  - *Task*: The task component captures the user tasks that the interface supports. The component defines a hierarchical decomposition of tasks and subtasks, along with the expected flow among those tasks and the attributes of those tasks. The granularity of tasks is not set by XIML, so valid tasks can range from simple one-step actions (e.g., Enter Date, View Map) to complicated multistep processes (e.g., Perform Contract Analysis). An XML excerpt of a task specified in XIML is shown below:

```
<TASK_MODEL ID='tm1'>
  <TASK_ELEMENT ID='t1' name='Make
  annotation'>
  <TASK_ELEMENT ID='t1.1'
  name='Select location'/>
  <TASK_ELEMENT ID='t1.2'
  name='Enter note'/>
  <TASK_ELEMENT ID='t1.3'
  name='Confirm Annotation'/>
  </TASK_ELEMENT>
</TASK_MODEL>
```

- *Domain*: The domain component is an organized collection of data objects and classes of objects that is structured into a hierarchy. This hierarchy is similar in nature to that of an ontology, but at a very basic level. Objects are defined via attribute-value pairings. Objects to be included in this component are restricted to those that are viewed or manipulated by a user and can be either simple or complex types. For example, "Date," "Map," and "Contract" can all be domain objects. An excerpt of the domain component is shown here:

```
<DOMAIN_MODEL ID='dm1'>
  <DOMAIN_ELEMENT ID='d1.1'
  name='map annotation'>
  <DOMAIN_ELEMENT ID='d1.1.1'
  name='location'/>
  <DOMAIN_ELEMENT ID='d1.1.2'
  name='note'/>
  <DOMAIN_ELEMENT ID='d1.1.3'
  name='entered_by'/>
  <DOMAIN_ELEMENT ID='d1.1.4'
  name='timestamp'/>
  </DOMAIN_ELEMENT>
</DOMAIN_MODEL>
```

- *User*: The user component defines a hierarchy—a tree—of users. A user in the hierarchy can represent a user group or an individual user. Therefore, an element of this component can be "Doctor" or can be "Doctor John Smith." Attribute-value pairs define the characteristics of these users. As defined today, the user component of XIML is expected to capture data and features that are relevant in the functions of design, operation, and evaluation. An XML excerpt for the user component is shown below:

```
<USER_MODEL ID='umodel'>
  <USER_ELEMENT ID='u1.1'
  NAME='field researcher'>
  <USER_ELEMENT ID='u1.1.1'
  NAME='field
  supervisor'/>
  <USER_ELEMENT ID='u1.1.2'
  NAME='field
  geologist'/>
  </USER_ELEMENT>
  <USER_ELEMENT ID='u1.2'
  NAME='analyst'/>
</USER_MODEL>
```

- *Presentation*: The presentation component defines a hierarchy of interaction elements that comprise the concrete objects that communicate with users in an interface. Examples of these are a window, a push button, a slider, or a complex widget such as an ActiveX control to visualize stock data. It is generally intended that the granularity of the elements in the presentation component will be relatively high, so that the logic and operation of an interaction element are separated from its definition. In this manner, the rendering of a specific interaction element can be left entirely to the corresponding target display system.
- *Dialogue*: The dialogue component defines a structured collection of elements that determine the interaction actions that are available to the users of an interface. For example, a "Click," a "Voice response," and a "Gesture" are all types of interaction actions. The dialogue component also specifies the flow among the interaction actions that constitute the allowable navigation of the UI. This component is similar in nature to the Task component, but it operates at the concrete levels, as opposed to the abstract level of the Task component.

- *Relations*: The interaction data elements captured by the various XIML components constitute a body of knowledge that can support organization and knowledge-management functions for UIs. A relation in XIML is a definition or a statement that links any two or more XIML elements either within one component or across components. By capturing relations in an explicit manner, XIML creates a body of knowledge that can support design, operation, and evaluation functions for UIs. The run-time manipulation of those relations constitutes the operation of the UI. However, XIML does not specify the semantics of those relations, which is left up to the specific applications that utilize XIML.
- *Attributes*: In XIML, attributes are features or properties of elements that can be assigned a value. The value of an attribute can be one of a basic set of data types, or it can be an instance of another existing element. Multiple values are allowed, as well as enumerations and ranges. The basic mechanism in XIML for defining the properties of an element is to create a number of attribute-value pairs for that element. In addition, relations among elements can be expressed at the attribute level or at the element level. As in the case of relations, XIML supports definitions and statements for attributes.

One of the main disadvantages of XIML is that a rather simple notion of task models is supported by this approach, for which tool support is not currently available. Another drawback connected to XIML is the fact that, different from the majority of the other UI description languages, it is developed within a software company, and therefore its use is protected by copyright.

26.3.2 UIML

An example of a language that has addressed the multidevice interface issue is the UI Markup Language (UIML) (http://www.uiml.org; Abrams et al., 1999; VT Course, n.d.), an XML-compliant language that supports a declarative description of a UI in a device-independent manner. This has been developed mainly by Harmonia.

In UIML, a UI is simply a set of interface elements with which the user interacts. These elements may be organized differently for different categories of users and families of appliances. Each interface element has data (e.g., text, sounds, images) used to communicate information to the user. Interface elements can also receive information from the user using interface artifacts (e.g., a scrollable selection list) from the underlying application. Since the artifacts vary from appliance to appliance, the actual mapping (rendering) between an interface element and the associated artifact (widget) is done using a style sheet. Run-time interaction is done using events. Events can be local (between interface elements) or global (between interface elements and objects that represent an application's internal program logic; i.e., the backend). Since the interface typically communicates with a backend to perform its work, a run-time engine provides for that communication. The run-time engine also facilitates a clean separation between the interface and the backend.

UIML describes a UI in the following sections (see Figure 26.5): *structure, style, content, behavior* (which compose the interface section) and *presentation* and *logic sections*.

```
<uiml>
    <interface>
        <structure> ...</structure>
        <style> ...</style>
        <content> ...</content>
        <behavior> ...</behavior>
    </interface>
    <peers>
        <logic> ...</logic>
        <presentation>...</presentation>
    </peers>
</uiml>
```

- *Structure*: The <structure> element contains a list of <part> elements. Each <part> element describes some abstract part of a UI and it has at least a class name. Here is an example:

```
<structure>
    <part id="TopLevel" class="JFrame">
        <part id="Label" class="JLabel"/>
        <part id="Button" class="JButton"/>
    </part>
</structure>
```

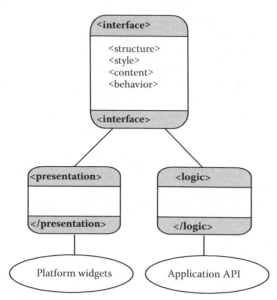

FIGURE 26.5 UIML structure.

The UI excerpt shown contains three parts: one named *TopLevel,* which is an instance of class *JFrame,* one named Label, which is an instance of class *JLabel,* and one named Button, which is an instance of class *JButton.* Note that Label and Button are nested inside *TopLevel,* which implies a relationship between the three parts. If two versions of a UI for two very different devices are considered (such as a mobile phone and a PC), two different <structure> elements will be used. That is because a mobile phone display can only show a few lines of text (e.g., one menu), while a desktop PC could display complex UIs (e.g., containing many pull-down menus).

- *Style:* The <style> element contains a list of <property> elements, giving presentation properties of the parts. Here's an example:

```
<style>
  <property part-name="TopLevel"
  name="background">blue</property>
</style>
```

This excerpt sets property *background* to the value *blue* in the part named *TopLevel.* It is worth noting that UIML itself does not define property names like *background* or values like *blue*—these are part of the vocabulary, whose name is given later in the <presentation> element. When designing a UI for multiple devices, multiple <style> elements can be included. One will represent <style> that is common to all devices; others will represent <style> information that is specific to certain devices. UIML uses a cascade mechanism similar to CSS to combine different <style> elements for a specific device.

- *Content:* The <content> element contains the text, images, and other content that goes into the UI. The <content> element contains one or more scalar values (e.g., a string) or data models (e.g., a list of strings to populate a menu),

each identified by a name and referenced in other parts of the UIML document by the <reference> element. One benefit to separating content from structure is that this facilitates internationalization. Another benefit is that different devices need different words in their messages. For instance, a phone with a small display may need a shorter message than a PC, while a voice interface might sound stilted if the same text as a PC's textual message is used. A UIML document could contain different <content> elements for the phone, PC, and voice UIs.

- *Behavior:* The <behavior> element contains a set of rules to define how the UI reacts to a stimulus, either from a user or from the external programming environment. For instance, a user might click a button. The external programming environment might send an asynchronous message that should be displayed in the UI. The following is a simple example:

```
<behavior>
  <rule>
    <condition>
      <event class="actionPerformed"
      part-name="Button"/>
    </condition>
    <action>
      <property part-name="Label"
      name="text">Button pressed<
      </property>
    </action>
  </rule>
</behavior>
```

When the <condition> is satisfied, the <action> is executed. Specifically, the <condition> is satisfied whenever the Java *actionPerformed* listener receives an *actionEvent* on a part whose name suggests that it is a button (*Button*). The previous example uses the Java AWT/Swing vocabulary for UIML. The <action> sets a property named *text* for the part *Label* and changes the text in label *Label* to a say "Button pressed" when the UI part named "Button" is clicked.

- *Logic:* The <logic> element defines the application programming interface (API) of business logic with which the UI interacts. Here is an example of the *<logic>* element:

```
<logic>
  <d-component id="counter" maps-
  to="org.something.simplecounter">
    <d-method id="increment" return-
    type="int" maps-to="count"/>
  </d-component>
</logic>
```

This UIML allows the other parts of UIML to call method *increment()* in a component named *Counter.*

- *Presentation*: The final part is the <presentation> element. In virtually all UIML documents, this simply names a vocabulary file.

```
<presentation base="Java_1.3_
Harmonia_1.0"/>
```

A UIML implementation reads the UIML vocabulary file, and then interprets all the part class names (e.g., *JLabel*, *JButton*, *JFrame*) and property names (e.g., *text*) according to the mappings in the vocabulary file.

One of the main disadvantages of UIML is the fact that it does not support the task level. Some research on how to integrate task models with UIML started some years ago at Virginia Tech (Alí et al., 2003), but the results have not been incorporated in the Liquid environment supporting UIML. Another shortcoming of UIML is that it provides a single language to define different types of UIs; therefore, there is the need to design separate UIs for each device.

26.3.3 TERESA XML

TERESA XML supports the various possible abstraction levels. The task level, which describes the activities that should be supported by the system, is specified in a hierarchical manner. The temporal relationships occurring between the different tasks are expressed using the CTT notation (Paternò, 1999; Berti et al., 2004). Then, at the abstract level, an abstract UI is composed of a number of presentations and connections among them. While each presentation defines a set of interaction techniques perceivable by the user at a given time, the connections define the dynamic behavior of the UI, by indicating what interactions trigger a change of presentation and what the next presentation is.

There are *abstract* interactors indicating the possible interactions in a platform-independent manner: for instance, the type of interaction to be performed (e.g., selection, editing, etc.) can be indicated without any reference to concrete ways to support such a performance (e.g., selecting an object through a radio

button or a pulldown menu, etc.). This level also describes how to compose such basic elements through some composition operators. Such operators can involve one or two expressions; each of them can be composed of one or several interactors or, in turn, compositions of interactors. In particular, the composition operators have been defined taking into account the type of communication effects that designers aim to achieve when they create a presentation (Mullet and Sano, 1995). They are:

- *Grouping*: Indicates a set of interface elements logically connected to each other
- *Relation*: Highlights a one-to-many relation among some elements; one element has some effects on a set of elements
- *Ordering*: Some kind of ordering among a set of elements can be highlighted
- *Hierarchy*: Different levels of importance can be defined among a set of elements

It is also worth noting that, at the abstract level, some work has also been conducted regarding the issue of how to connect the interactive part of a software application with the functional core, namely the set of application functionalities independent of the media and the interaction techniques used to interact with the user. This aspect is very relevant in addressing the problem of generating dynamic pages with TERESA.

The *concrete* level is a refinement of the abstract UI: depending on the type of platform considered, there are different ways to render the various interactors and composition operators of the abstract UI. The concrete elements are obtained as a refinement of the abstract ones. For example, a navigator (an abstract interactor) can be implemented through a textlink, an imagelink, or a simple button, and in the same way, a single choice object can be implemented using a radio button, a list box, or a dropdown menu. The same holds for the operators; for example, the grouping operator in a desktop platform can be refined at the concrete level by a number of techniques, including both unordered lists by row and unordered lists by column (apart from classical grouping techniques such as fieldsets, bullets, and colors). The

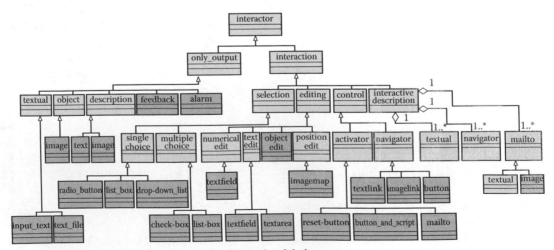

FIGURE 26.6   An excerpt of the concrete user interface for the graphical desktop.

small capability of a mobile phone does not allow implementing the grouping operator by using an unordered list of elements by column; therefore, this technique is not available on this platform. In a vocal device, a grouping effect can be achieved through inserting specific sounds or pauses or using a specific volume or keywords.

In the following, excerpts are reported of the concrete UI (CUI) languages specified in TERESAXML considering two platforms: desktop and mobile. They are both obtained as a refinement of the abstract language. Such refinements are added at the lower levels of the hierarchical structure of the languages. As it is possible to see from the following excerpts, the desktop CUI is defined by a number of presentations and settings, which are used for the page generation. The mobile devices need additional information concerning their capabilities.

Each presentation contains interactors and interactor compositions, the connections to other presentations, and provides a description of specific properties, such as title, header, background, and so on.

Considering more in detail the interactors specification, one can see how it is possible to specify a "single selection" interactor in the CUI desktop.

The main differences in the concrete languages are in the concrete interactors associated with an abstract one. For example, considering a "single selection," it can be observed that in a mobile device the "list boxes" are not suitable.

Each concrete language takes into account the interaction resources available. Thus, for example, the grouping composition can be supported in several ways in the desktop, but only a subset is supported in mobile devices.

One advantage of this approach based on multiple levels of abstraction is that all the concrete interface languages share the same structure and add concrete platform-dependent details to the abstract language regarding the possible attributes for implementing the logical interaction objects and the ways to compose them. All languages in this approach, for any abstraction level, are defined in terms of XML to make them more easily manageable and to allow for their export/import in different tools.

Another advantage of this approach is that maintaining links among the elements in the various abstraction levels allows for the possibility of linking semantic information (such as the activity that users intend to do) and implementation levels, which can be exploited in many ways. A further advantage is that designers of multidevice interfaces do not have to learn all the details of the many possible implementation languages, because the environment allows them to have full control over the design through the logical descriptions and leave the implementation to an automatic transformation from the concrete level to the target implementation language. In addition, if a new implementation language needs to be addressed, the entire structure of the environment does not change, but only the transformation from the associated concrete level to the new language has to be

```
<!ELEMENT concrete_desktop_interface    (default_settings, presentation+)>

<!ELEMENT default_settings     (background, font_settings, operators_settings)

<!ELEMENT concrete_mobile_interface    (device_type, default_settings, presentation+)>

<!ELEMENT device_type (big | medium | small)>

<!ELEMENT big EMPTY>

<!ATTLIST big

    graphic_support (%option;) #REQUIRED>

<!ELEMENT medium EMPTY>

<!ATTLIST medium

    graphic_support (%option;) #REQUIRED>

<!ELEMENT small EMPTY>

<!ATTLIST small

    graphic_support CDATA #FIXED "no">
```

```
<!ELEMENT presentation       (presentation_properties, connection*, (interactor |
                             interactor_composition))>

<!ELEMENT presentation_properties     (title, background, font_settings, top)>
```

```
<!ELEMENT interaction    (selection | editing | control | interactive_description)>
<!ELEMENT selection    (single | multiple)>
<!ELEMENT single    (radio_button | list_box | drop_down_list)>
<!ATTLIST single
    cardinality (%cardinality_value;) #REQUIRED>
<!ELEMENT radio_button (choice_element+)>
<!ATTLIST radio_button
    label CDATA #REQUIRED
    alignment (%element_selection_alignment;) #REQUIRED>
<!ELEMENT choice_element EMPTY>
<!ATTLIST choice_element
    label CDATA #REQUIRED
    value CDATA #REQUIRED>
```

```
<!ELEMENT single (radio_button | drop_down_list)>
```

```
CUI-desktop
<!ELEMENT grouping (fieldset?, bullet?, background_color?,    position)>
```

```
CUI-mobile
<!ELEMENT grouping ((fieldset | bullet), position)>
```

added. This is not difficult because the concrete level is already a detailed description of how the interface should be structured.

26.3.4 UsiXML

The semantics of the UsiXML models (Limbourg et al., 2004; Vanderdonckt, 2005) are based on meta-models expressed in terms of UML class diagrams, from which the XML schema definitions are derived. Currently, there is no automation between the initial definition of the semantics and their derivation into XML schemas. Only a systematic method is used for each new release.

All model-to-model transformations are themselves specified in UsiXML to keep only one UIDL throughout the development life cycle. Model-to-code transformation is ensured by appropriate tools that produce code for the target context of use or platform. For reverse engineering, code-to-model transformations are mainly achieved by derivation rules that are based on the mapping between the meta-model of the source language (e.g., HTML) and the meta-model of the target language (i.e., UsiXML).

In UsiXML, a concrete UI model consists of a hierarchical decomposition of concrete interaction objects. A CIO is defined as any UI entity that users can perceive, such as text, image, or animation, and/or manipulate, such as a push button, a list box, or a check box. A CIO is characterized by various attributes such as, but not limited to: ID, name, icon, content, defaultContent, defaultValue. Each CIO can be subtyped into sub-CIOs depending on the interaction modality chosen: graphicalCIO for GUIs, auditoryCIO for vocal interfaces, 3DCIO for 3D UIs, and so on. Each graphicalCIO inherits properties from its parent CIO and has specific attributes such as: isVisible, isEnabled, fgColor, and bgColor to depict foreground and background colors, and so on. Each graphicalCIO is then subtyped into one of the two possible categories: graphicalContainer, for all widgets containing other widgets such as page, window, frame, dialogue box, table, box and their related decomposition, or graphicalIndividualComponent, for all other traditional widgets that are typically found in such containers. A graphical IndividualComponent cannot be further decomposed. The model supports a series of widgets defined as graphicalIndividualComponents such as: textComponent, videoComponent, imageComponent, imageZone, radioButton, toggleButton, icon, checkbox, item, comboBox, button, tree, menu, menuItem, drawingCanvas, colorPicker, hourPicker, datePicker, filePicker, progressionBar, slider, and cursor. Thanks to this progressive inheritance mechanism, each final element of the concrete UI inherits properties from these levels, depending on the category it belongs to. The properties populating the concrete level of UsiXML have been chosen because they belong to the intersection of the property sets of major UI toolkits, such as Windows GDI, Java AWT and Swing, HTML. In this way, a CIO can be specified independently from the fact that it will be further rendered in HTML, VRML, or Java (this quality is often referred to as the property of *implementation language independence*).

UsiXML aims to address the development of UIs for multiple contexts of use, and has the advantage of providing a graphical syntax for a majority of constituent models. Also, unlike XIML, UsiXML's language specifications are freely available and not protected by copyright. However, UsiXML renderers are still at a development stage, and cannot compete with the numerous implementations of UIML.

26.3.5 Pebbles

In the Pebbles project at Carnegie Mellon University in Pittsburgh, a personal universal controller (PUC) environment has been developed, which supports the downloading of logical descriptions of appliances and the automatic generation of the corresponding UIs (Nichols et al., 2002).

A PUC engages in two-way communication with everyday appliances, first downloading a specification of the appliance's functions, and then automatically creating an interface for controlling them. The specification of each appliance includes a high-level description of every function, a hierarchical grouping of those functions, and a description of the availability of each function relative to the appliance's state. The PUC architecture has four parts: appliance adaptors, a specification language, a communication protocol, and interface generators.

The assumption is that each appliance has its own facility for accepting connections from a PUC. The peer-to-peer characteristic of the PUC architecture allows a more scalable solution than other systems, such as ICrafter (Ponnekanti et al., 2001) and UIA (Eustice et al., 1999), which rely on a central server to manage connections between interfaces and appliances. A PUC could discover appliances by intermittently sending out broadcast requests. To connect the PUC to any real appliance,

an appliance adaptor has to be built. An adaptor should be built for each proprietary appliance protocol that the PUC communicates with. To make the construction of new appliance adaptors easy, an adaptor development framework has been created. It is implemented entirely in Java, and manages all PUC communication for the adaptor, allowing the programmer to concentrate on communicating with the appliance.

A description of an appliance's functions is available, so that the PUC can automatically generate an interface. This description does not contain any information about look or feel: decisions about look and feel are left up to each interface generator. The PUC specification language is XML based. Some of the key information contained are:

- *State variables and commands*: Interface designers must know what can be manipulated on an appliance before they can build an interface for it. Therefore, the manipulable elements are represented as state variables. Each state variable has a given type that tells the interface generator how it can be manipulated. Commands are also useful when an appliance is unable to provide feedback about state changes back to the controller, either by manufacturer choice or a hardware limitation of the appliance.

- *Type information*: Each state variable must be specified with a type so that the interface generator can understand how it may be manipulated. In PUC, seven generic types may be associated with a state variable: Boolean, integer, fixed point, floating point, enumerated, string, custom.

- *Label information*: The interface generator must also have information about how to label the interface components that represent state variables and commands. Providing this information might be difficult, because different interface modalities require different kinds of label information. In PUC this information is provided with a generic structure called the label dictionary, in which every label contained, whether it is phonetic information or plain text, will have approximately the same meaning. Thus the interface generator can use any label within a label dictionary interchangeably.

- *Dependency information*: The two-way communication feature of the PUC allows it to know when a particular state variable or command is unavailable. This can make interfaces easier to use, because the components representing those elements can be disabled. The specification contains formulas that specify when a state or command will be disabled depending on the values of other state variables, currently specified with three types of dependencies: equal-to, greater-than, and less-than. Each state or command may have multiple dependencies associated with it, combined with the logical operations AND and OR. These formulas can be processed by the PUC to determine whether a component should be enabled when the appliance state changes.

Following is an excerpt of specification with PUC for a to-do list application (see PUC Documentation):

```
<spec>
  <groupings>
    <group name="Commands" is-a="list-
    commands" priority="10">
    <command name="Add" is-a="list-add"
    priority="8">
      <labels>
        <label>Add To-Do Item</label>
        <label>Add To-Do</label>
        <label>Add</label>
      </labels>
    </command>

<command name="Delete" is-a="list-remove"
priority="7">
  <labels>
      <label>Delete To-Do Item</label>
      <label>Delete To-Do</label>
      <label>Delete</label>
  </labels>
  <active-if>
    <greaterthan state="ToDo.List.
    Length"><constantvalue="0"/>
    </greaterthan>
    <defined state="ToDo.List.Selection"/>
  </active-if>
</command>
<state name="SortBy" priority="5">
  <type>
  <enumerated>
    <item-count>3</item-count>
  </enumerated>
  <value-labels>
    <map index="1">
      <labels>
        <label>Category</label>
      </labels>
    </map>
    <map index="2">
      <labels>
        <label>Completion Date</label>
        <label>Date</label>
      </labels>
    </map>
    <map index="3">
      <labels>
        <label>Completed</label>
      </labels>
    </map>
  </value-labels>
  </type>
  <labels>
    <label>Sort By</label>
          <label>Sort</label>
        </labels>
```

```
      <active-if>
        <greaterthan state="ToDo.List.
        Length">
          <constant value="0"/>
        </greaterthan>
      </active-if>
    </state>
  </group>
  <group>
  ...
  </group>
  </groupings>
</spec>
```

In this XML excerpt there is a grouped expression with two commands (Add and Delete) that can be used on the manipulable elements (actions in the to-do list) and also a state for sorting them, depending on different criteria: category, completion date, completed. The specification contains also formulas that specify when a state or command will be disabled depending on the values of other state variables: for instance, as it appears in the previous specification, the Delete command is available only if the length of the to-do list is greater than 0.

The communication protocol enables appliances and PUC adaptors to exchange information bidirectionally and asynchronously. The protocol is XML-based and defines six messages, two sent by the appliance and four sent by the PUC. The PUC can send messages requesting the specification, the value of every state, a change to a particular state, or the invocation of a command. The appliance can send the specification or the current value of a particular state. When responding to a PUC request for the value of every state, the appliance will send a current value message for each of its states.

The PUC architecture has been designed to be independent of the type of interface presented to the user. Generators have been developed for two different types of interfaces: graphical and speech. The graphical interface generator takes an arbitrary description written in the PUC specification language and makes use of the group tree (specifying the states variables and commands to be included in the UI) and dependency information (containing formulas specifying when a state or a command will be disabled depending on the values of other state variables) to create the related UI. The actual UI components that represent each state variable and command are chosen using a decision tree; the components are then placed into panels according to the inferred structure, and laid out using the group tree; the final step of the generation process instantiates the components and places them on the screen.

The speech interface generator creates an interface from a PUC specification using USI (universal speech interface) interaction techniques. The speech-interface generator differs from the graphical interface in that it connects to multiple PUC adaptors (if multiple adaptors can be found on the network), requests the device specifications, and then uses those specifications collectively to configure itself so that it can control all devices with the same speech grammar. This includes building a grammar for the parser, and a language model and pronunciation dictionary for the recognizer.

One of the main disadvantages of this approach is that its application area is limited to home appliances that require similar interfaces.

26.3.6 XForms

XForms is an XML application that represents the next generation of forms for the web, and has introduced the use of abstractions to address new heterogeneous environments. The primary difference when comparing XForms with HTML Forms, apart from XForms being in XML, is the separation of the data being collected from the markup of the controls collecting the individual values. This not only makes XForms more tractable, by making it clear what is being submitted and where, but it also eases reuse of forms, since the underlying essential part of a form is no longer bound to the page it is used in. A second major difference is that XForms, while designed to be integrated into XHTML, is no longer restricted only to be a part of that language, but may be integrated into any suitable markup language. In the XForms approach, forms are composed of a section that describes what the form does, called the XForms model, and another section that describes how the form is to be presented.

By splitting traditional XHTML forms into three parts—XForms model, instance data, and UI—it separates presentation from content, allows reuse, gives strong typing, reduces the number of roundtrips to the server, as well as offers device independence and a reduced need for scripting.

- *Model*: The data layer that describes the form's data and logic. In other words, it is the nonvisible definition of an XML form as specified by XForms.
- *Instance data*: An internal tree representation of the values and state of all the instance data nodes associated with a particular form.
- *UI items*: The presentation layer that lets the user input data that will be stored in the data layer. XForms UI items are used to display data from the XForms model.

XForms creates a separate data layer inside the form, allowing for collecting data from the form and copying the data into a separate block of XML that can be formatted as preferred. Separating the data layer from the presentation layer makes XForms device independent. The data model can be used for all devices. The presentation can be customized for different UIs, like mobile phones, handheld devices, and Braille readers for the blind. Since XForms is device independent and based on XML, it is also possible to add XForms elements directly into other XML applications, such as VoiceXML (speaking web data) and SVG. UI controls encapsulate all relevant metadata such as labels, thereby enhancing accessibility of the application when using different modalities. XForms UI controls are generic and suited for device independence. Therefore, the high-level nature of the UI controls, and the consequent intent-based authoring of

the UI, make it possible to retarget the user interaction to different devices.

In particular, XForms aims to separate presentation from content through the definition of a set of platform-independent, general-purpose controls, and focus on the goal (or intent) behind each form control. Indeed, the list of XForms controls includes objects such as select (choice of one or more items from a list), trigger (activating a defined process), output (display-only of form data), secret (entry of sensitive information), and so on, rather than refer to concrete examples like radio buttons, checkboxes, and so forth, which are hardwired to specific representations of such controls. This kind of logical description locates the types of abstractions supported by XForms at the abstract UI level. However, the task level is not explicitly addressed. In addition, XForms is basically aimed at expressing form-based UIs and less for supporting other modalities (e.g., vocal).

26.4 Transforming Markup Language-Based Specifications

There are various solutions for transforming different XML-based specifications one into another. One solution is represented by XSLT (http://www.w3.org/TR/xslt), which is an XML-based specification allowing the transformation of a XML document into another format. This format can be XHTML, XML, or anything else. Consequently, XSLT could be used to transform models. Nevertheless, it appears difficult to directly use XSLT on a real system for some reasons:

- Writing an XSLT program is long and tedious.
- One important problem with XSLT is its poor readability and the high cost of maintenance for associated programs.
- Executing an XSLT program is not user-friendly for models transformation. There are no error messages that depend on the application domain. For example, the XSLT processor does not inform the user about nonexisting concepts.

An example of an approach using XSLT transformations is TADEUS (Müller et al., 2001). The creation of the XSL-based model description is based on the knowledge of available UIOs for specific representations. It is necessary to know which values of properties of a UIO are available in the given context. The XML-based device-dependent abstract interaction model (skeleton) and available values of properties are used to create an XSL-based model description specifying a complete UI. In addition, through XSL transformation a file describing a specific UI will be generated. The XSL transformation process consists of two subprocesses: (1) creation of a specific file representing the UI (WML, VoiceXML, HTML, Java, etc.) and (2) integration of content (database, application) into the UI.

Another approach is based on graph transformations (GT) techniques that has been developed in UsiXML (Limbourg et al., 2004) to formalize explicit transformations between any pair of models (except for the implementation level). The reasons for this choice were that it is (1) visual: every element within a GT-based language has a graphical syntax; (2) formal: GT is based on a sound mathematical formalism (algebraic definition of graphs and category theory); and (3) it enables verifying formal properties on represented artifacts. Furthermore, the formalism applies equally to all levels of abstraction of UIs. The implementation level is the only model that cannot be supported by graph transformations, because it would have supposed that any markup or programming language to be supported should have been expressed in a meta-model to support transformations between meta-models: the one of the initial language to the one of the desired specification language. It was observed that to address this problem, the powerfulness of GT techniques was not needed and surpassed by far other experienced techniques, such as derivation rules.

To support the manipulation of models in UsiXML, two software tools were developed: TransformiXML (Limbourg et al., 2004) and IdealXML (Montero et al., 2005). TransformiXML consists of a Java application that triggers transformations of models expressed by graph grammars. However, TransformiXML takes a long time for its processing and the performance is slow. IdealXML is a Java-based application containing the graphical editor for the task model, the domain model, and the abstract model. It can also establish any mapping between these models either manually (by direct manipulation) or semiautomatically (by calling TransformiXML).

Another approach is followed in the TERESA tool (Mori et al., 2004), which supports various transformations: from task-model-related information to abstract UI, from abstract UI to concrete interface for the specific platform, and, finally, an automatic UI generation.

In the first transformation, the goal is transforming the task-based specification of the system into an interactor-based description of the corresponding abstract UI. It is worth pointing out that within TERESA it is also possible to access the inverse transformation, since for each interactor the tool is able to automatically identify and highlight the related task, so that designers can immediately spot the relation. Both the static arrangement of interactions in the same presentation and the dynamic behavior of the abstract UI are derived by analyzing the semantic of the temporal operators included in the task model specification. For instance, a sequential operator between two tasks implies that the related presentations will be sequentially triggered: this will be rendered, at the abstract level, by associating a connection between two different abstract presentations (with each presentation supporting the performance of just one task), so that the performance of the first task will trigger the activation of the second presentation and render the sequential ordering. On the contrary, a concurrency operator between two tasks implies that the associated interactors is presented at the same time, so as to support concurrency between the connected activities; therefore, the abstract objects supporting their performance will be included in the same presentation.

The transformation from abstract UI to concrete UI represents an example of a model-to-model mapping, as it allows for mapping one abstract UI model to a number of related concrete UI models. Each concrete UI specification is associated with a specific interaction platform (e.g., the desktop, the mobile platform, the vocal platform, the digital TV, etc.). The generated concrete UIs can be further customized according to a number of parameters made available to the designer. For instance, one can consider an abstract interactor supporting the task of selecting an item for a set of possible elements: this abstract interactor can be mapped onto a radio button, or a pulldown menu, or a vocal selection (on a vocal platform).

The last transformation from concrete UI to implementation is a model-to-text mapping connecting a concrete UI for a specific platform to one or multiple types of final UIs, depending on a particular selected implementation language. Currently, transformations for implementations in a number of languages (XHTML, XHTML MP, VoiceXML, X+V, Java for Digital TV, JSP) are available, and others are under development.

However, the problem of such transformations is that they are hardcoded within the TERESA tool. A first solution to overcome this problem is to introduce a declarative mechanism to specify mappings between models that can be dynamically used. Indeed, in this approach, XSLT processing can be used to support model-to-model transformation. This solution implies writing a single XSLT transformation that will accept (1) a generic transformation and (2) configuration parameters (both in XML) as source trees and produce a specialized transformation as the result tree. With this approach it is possible to define mappings between abstract and concrete UI specifications (model-to-model transformation) and between concrete and implementation UI models (model-to-text transformations). The objective of this approach is to be the most general possible.

Another type of transformation supports the inverse process—abstraction. It uses reverse engineering techniques able to take the UI of existing applications for any platform and then build the corresponding logical descriptions. Early work in reverse engineering for UIs was motivated by the need to support maintenance activities aiming to reengineer legacy systems for new versions using different UI toolkits (Moore, 1996), in some cases even supporting migration from character-oriented UIs to graphical UIs. More recently, interest in UI reverse engineering has received strong impetus from the advent of mobile technologies and the need to support multidevice applications. To this end, a good deal of work has been dedicated to UIs reverse engineering to identify corresponding meaningful abstractions. Other studies have investigated how to derive the task model of an interactive application starting with the logs generated during user sessions (Hudson et al., 1999). However, this approach is limited to building descriptions of how the UI was actually used, which is described by the logs, but is not able to provide a general description of the tasks supported, which includes even those not considered in the logs. Previous work in reverse engineering has addressed only one level or platform at a time. For example, ReversiXML (Bouillon et al., 2004) has focused on creating a concrete/abstract UI from HTML pages for desktop systems. WebRevenge (Paganelli and Paternò, 2002) has addressed the same types of applications to build only the corresponding task models.

In general, there is a lack of approaches able to address different platforms, especially involving different interaction modalities, and to build the corresponding logical descriptions at different abstraction levels. ReverseAllUIs (Bandelloni et al., 2007) aims to overcome this limitation.

26.5 Conclusions

This chapter has discussed the many aspects that can be formalized through markup languages (based on XML) in HCI. The current trend is to further exploit such tools to obtain more intelligent environments and better support interoperability. One important aspect in the design of a markup language is to identify the relevant information, avoiding a plethora of low-level details that complicate its processing. Another important issue is the availability of tools able to support the editing, application, and transformation of markup languages. Tools need to be implemented in such a way that if modifications to the language are made, then the tools that exploit their information do not require making profound changes in the implementation.

Lastly, the challenges raised by ambient intelligence call for solutions able to model not only the interdependencies and configuration options of ambient technologies, but also their behavior, privacy/visibility effects, and reliability in order to provide a rich description of such environments.

References

Abascal, J., Arrue, M., Fajardo, I., Garay, N., and Tomás, J. (2004). Use of guidelines to automatically verify Web accessibility. *Universal Access in the Information Society* 3: 71–79.

Abrams, M., Phanouriou, C., Batongbacal, A., Williams, S., and Shuster, J. (1999). UIML: An appliance-independent XML user interface language, in the *Proceedings of the 8th International World Wide Web Conference*, 11–14 May 1999, Toronto, Canada. http://www8.org/w8-papers/5b-hypertext-media/uiml/uiml.html.

Alí, F., Perez-Qinones, M., and Abrams, M. (2003). Building MultiPlatform user interfaces with UIML, in *Multiple User Interfaces: Cross-Platform Applications and Context-Aware Interfaces* (A. Seffah and H. Javahery, eds.), pp. 95–118. New York: John Wiley & Sons.

Bandelloni, R., Paternò, F., and Santoro, C. (2007). Reverse engineering cross-modal user interfaces for ubiquitous environments, in the *Proceedings of the Engineering Interactive Systems 2007 (EIS 2007)*, 22–24 March 2007, Salamanca, Spain. Berlin/Heidelberg: Springer-Verlag.

Beirekdar, A., Keita, M., Noirhomme, M., Randolet, F., Vanderdonckt, J., and Mariage, C. (2005). Flexible reporting for automated usability and accessibility evaluation of web sites, in the *Proceedings of the IFIP TC13 International Conference on Human-Computer Interaction 2005 (INTERACT 2005)*, pp. 281–294. Berlin/Heidelberg: Springer-Verlag.

Berti, S., Correani, F., Paternò, F., and Santoro, C. (2004). The TERESA XML language for the description of interactive systems at multiple abstraction levels, in the *Proceedings of the Workshop on Developing User Interfaces with XML: Advances on User Interface Description Languages 2004*, May 2004, Gallipoli, Italy, pp. 103–110. http://giove.isti.cnr.it/cameleon/cp25.html.

Bouillon, L., Vanderdonckt, J., and Chow, K. C. (2004). Flexible re-engineering of Web sites, in the *Proceedings of the 8th ACM International Conference on Intelligent User Interfaces (IUI'2004)*, 13–16 January 2004, Madeira Island, Portugal, pp. 132–139. New York: ACM Press. http://st-www.cs.uiuc.edu/users/smarch/st-docs/mvc.rtf.

Burbeck, S. (1987). *Applications Programming in Smalltalk-80: How to Use Model-View-Controller (MVC)*. Softsmarts.

Calvary, G., Coutaz, J., Thevenin, D., Limbourg, Q., Bouillon, L., and Vanderdonckt, J. (2003). A unifying reference framework for multi-target user interfaces. *Interacting with Computers* 15: 289–308.

Eustice, K. F., Lehman, T. J., Morales, A., Munson, M. C., Edlund, S., and Guillen, M. (1999). A universal information appliance. *IBM Systems Journal* 38: 575–601.

Heckmann, D. and Krueger, A. (2003). A user modeling markup language (UserML) for ubiquitous computing, in the *Proceedings of the 9th International Conference on User Modeling 2003 (UM 2003)*, 22–26 June 2003, Johnstown, PA, pp. 393–397. Berlin/Heidelberg: Springer-Verlag.

Hudson, S., John, B., Knudsen, K., and Byrne, M. (1999). A tool for creating predictive performance models from user interface demonstrations, in the *Proceedings of the 12th Annual ACM Symposium on User Interface Software and Technology (UIST'99)*, 7–10 November 1999, Asheville, NC, pp. 93–102. New York: ACM Press.

Leporini, B., Paternò, F., and Scorcia, A. (2006). Flexible tool support for accessibility evaluation. *Interacting with Computers* 18: 869–890.

Limbourg, Q., Vanderdonckt, J., Michotte, B., Bouillon, L., and Lopez, V. (2004). UsiXML: A language supporting multi-path development of user interfaces, in the *Proceedings of the 9th IFIP Working Conference on Engineering for Human-Computer Interaction Jointly with 11th International Workshop on Design, Specification, and Verification of Interactive Systems (EHCI-DSVIS'2004)*, 11–13 July 2004, Hamburg, Germany, pp. 200–220. Berlin/Heidelberg: Springer-Verlag.

Mohamad, Y., Stegemann, D., Koch, J., and Velasco, C. A. (2004). Imergo: Supporting accessibility and Web standards to meet the needs of the industry via process-oriented software tools, in the *Proceedings of the 9th International Conference on Computers Helping People with Special Needs (ICCHP 2004)* (K. Miesenberger, J. Klaus, W. Zagler, and D. Burger, eds.), 7–9 July 2004, Paris, France, pp. 310–316. Berlin/Heidelberg: Springer-Verlag.

Montero, F., Jaquero, V. L., Vanderdonckt, J., Gonzalez, P., Lozano, M. D., and Limbourg, Q. (2005). Solving the mapping problem in user interface design by seamless integration in IdealXML, in the *Proceedings of the 12th International Workshop on Design, Specification, and Verification of Interactive Systems (DSV-IS'2005)* (M. Harrison, ed.), 13–15 July 2005, Newcastle, U.K., pp. 161–172. Berlin/Heidelberg: Springer-Verlag.

Moore, M. M. (1996). Representation issues for reengineering interactive systems. *ACM Computing Surveys* 28.

Mori, G., Paternò, F., and Santoro, C. (2004). Design and development of multi-device user interfaces through multiple logical descriptions. *IEEE Transactions on Software Engineering* 30: 507–520.

Müller, A., Forbrig, P., and Cap, C. (2001). Model-based user interface design using markup concepts, in the *Proceedings of the 8th International Workshop on Interactive Systems: Design, Specification, and Verification (DSV-IS 2001)*, 13–15 June, Glasgow, Scotland, pp. 16–27. Berlin/Heidelberg: Springer-Verlag.

Mullet, K. and Sano, D. (1995). *Designing Visual Interfaces*. Upper Saddle River, NJ: Prentice Hall.

Nichols, J., Myers, B. A., Higgins, M., Hughes, J., Harris, T. K., Rosenfeld, R., et al. (2002). Generating remote control interfaces for complex appliances, in the *Proceedings of the ACM Symposium on User Interface Software and Technology (UIST 2002)*, 27–30 October 2002, Paris, France, pp. 161–170. New York: ACM Press.

Paganelli, L. and Paternò, F. (2002). Automatic reconstruction of the underlying interaction design of Web applications, in the *Proceedings of the 14th International Conference on Software Engineering and Knowledge Engineering (SEKE 2002)*, 15–19 July 2002, Ischia, Italy, pp. 439–445. New York: ACM Press.

Passani, L. (2005). *Welcome to the WURFL the Wireless Universal Resource File*. http://wurfl.sourceforge.net.

Paternò, F. (1999). *Model-Based Design and Evaluation of Interactive Application*. Berlin/ Heidelberg: Springer-Verlag.

Ponnekanti, S. R., Lee, B., Fox, A., Hanrahan, P., and Winograd, T. (2001). ICrafter: A service framework for ubiquitous computing environments, in the *Proceedings of the International Conference on Ubiquitous Computing (UBICOMP 2001)*, 30 September–2 October 2001, Atlanta, pp. 56–75. Berlin/ Heidelberg: Springer-Verlag.

PUC Documentation (n.d.). http://www.pebbles.hcii.cmu.edu/puc/specification.html#examplespec.

Puerta, A. (1997). A model-based interface development environment. *IEEE Software* 14: 40–47.

Puerta, A. and Eisenstein, V. (2001). XIML: A common representation for interaction data, in the *Proceedings of the 7th International Conference on Intelligent User Interfaces (IUI 2001)*, 13–16 January 2001, Madeira Island, Portugal, pp. 214–215. New York: ACM Press. http://people.csail.mit.edu/jacobe/papers/mui_chapter.pdf.

Souchon, N. and Vanderdonckt, J. (2003). A review of XML-compliant user interface description languages, in the *Proceedings of 10th International Conference on Design, Specification, and Verification of Interactive Systems (DSV-IS 2003)*, 11–13 June 2003, Madeira Island, Portugal, pp. 377–391. http://www.isys.ucl.ac.be/bchi/publications/2003/Souchon-DSVIS2003.pdf.

Szekely, P. (1996). Retrospective and challenges for model-based interface development, in the *Proceedings of the 3rd International Workshop on Design, Specification, and Verification of Interactive Systems (DSV-IS'96)*, pp. 1–27. Vienna: Springer-Verlag.

Szekely, P., Sukaviriya, P., Castells, P., Muthukumarasamy, J., and Salcher, E. (1995). Declarative interface models for user interface construction tools: The MASTERMIND approach, in *Engineering for Human-Computer Interaction* (L. J. Bass and C. Unger, eds.), pp. 120–150. London: Chapman & Hall.

UAProf (2001). *Wireless Application Protocol*. http://www.openmobilealliance.org/tech/affiliates/wap/wap-248-uaprof-20011020-a.pdf.

Vanderdonckt, J. (2005). A MDA-compliant environment for developing user interfaces of information systems, in the *Proceedings of the 16th Conference on Advanced Information Systems Engineering (CAiSE 2005)*, pp. 16–31. Berlin/ Heidelberg: Springer-Verlag.

VT Course (n.d.). *Virginia Tech Master Course on Internet Software: UIML Lesson*. http://mit.iddl.vt.edu/courses/cs5244/coursecontent/mod5/lesson2/uiml5.html.

27

Abstract Interaction Objects in User Interface Programming Languages

Anthony Savidis

27.1 Introduction

The notion of abstraction has gained much attention in software engineering as a solution for recurring development problems.[1] The basic idea has been the establishment of software frameworks clearly separating those implementation layers relevant only to the nature of the problem, from the engineering issues, which emerge when the problem class is instantiated in practice in various different forms. The same philosophy, when applied to developing interactive systems, means employing abstractions for building dialogues so that a dialogue structure composed of abstract objects can be retargeted to various alternative physical forms, through an automatic process controlled by the developer. In the context of user interface development, interaction objects play a key role in implementing the constructional and behavioral aspects of interaction. In this context, numerous software libraries exist, such as MFC, JFC, GTK+, and so on, offering comprehensive collections of object classes whose instantiation by the running program effectively results in the interactive delivery of graphical interaction elements; such libraries are commonly known as interface toolkits. Currently, there are no similar libraries for abstract interaction objects, practically implying that their implementation and programming linkage to concrete interface toolkits has to be manually crafted by client programmers.

However, there are a few design models, in certain cases accompanied with incomplete suggested design patterns, on what actually constitutes abstract interaction objects and their particular software properties. Past work in the context of abstract interaction objects (Blattner et al., 1992; Duke and Harrison, 1993; Duke et al., 1994; Foley and Van Dam, 1983; Savidis and Stephanidis, 1998) reflects the need to define appropriate programming versions relieved from physical interaction properties such as color, font size, border, or audio feedback, and only reflecting an abstract behavioral role (i.e., why an object is needed). This definition makes a clear distinction of abstract interaction objects from multiplatform interaction objects, the latter merely forming generalizations of similar graphical interaction objects met in different toolkits, through standardized application programming interfaces (APIs). The software API of generalized objects is easily designed in a way offering fine-grained control to the physical aspects of interaction objects, since the target object classes exhibit very similar graphical properties. For instance, a multiplatform "push button" offers attributes like position, color, label, border width, and so on, since all graphical interface toolkits normally offer programming control of such push-button object attributes.

In Figure 27.1, the large diversity of interaction objects supporting "selection from an explicit list of options" is demonstrated. Such alternative instantiations differ so radically with respect to physical appearance and behavior, practically turning the quest for a common multiplatform generalized API, offering fine-grained programming control over the physical characteristics, to a technically unachievable task. However, as is also depicted in Figure 27.1, from a programming point of view there is a common denominator among these different physical forms,

---

[1] This chapter is based on Savidis, A. (2005). Supporting virtual interaction objects with polymorphic platform bindings in a user interface programming language, in *Rapid Integration of Software Engineering Techniques, Proceedings of the International Workshop RISE 2004* (N. Guelfi, ed.), 26 November 2004, Luxembourg, pp. 11–23. Berlin/Heidelberg: Springer. Reprinted here with slight modifications by concession of Springer.

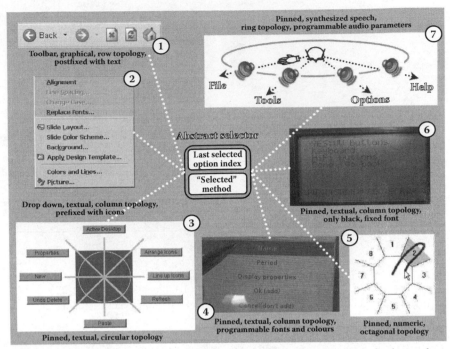

FIGURE 27.1 Alternative incarnations of an abstract Selector varying with respect to topology, display medium, content of options, input devices, and appearance attributes. (1) and (2) are from WNT, (3) is from Bronevetsky (2003), (4) and (6) are from the 2WEAR Project,2, (5) is from McGuffin et al. (2001), and (7) is from Savidis et al. (1996). (From Savidis, A., Supporting virtual interaction objects with polymorphic platform bindings in a user interface programming language, *Rapid Integration of Software Engineering Techniques, Proceedings of the International Workshop Rise 2004* (N. Guelfi, ed.), pp. 11–23, Springer LCNS: 3475, 2005. With permission.)

which can constitute an initial design toward an abstract selector object. This abstract entity is concerned only with: (1) the index of the last option selected by the user; and (2) the logical notification that an option has been selected (i.e., "Selected" method). Since this entity has no physical attributes, there is still an open issue regarding the way fine-grained programming control of physical views is to be allowed when coding with abstract selectors.

Abstraction for interaction objects gains more practical value in domains where the implemented interfaces should have dynamically varying physical forms, depending on the interaction facilities offered by the particular host environment. Such an issue becomes of primary importance in domains like ubiquitous computing and universal access. Today there are no actual recipes for implementing and linking abstract interaction objects to different interface toolkits. In this context, this chapter presents a user interface programming language (i.e., a domain-specific language), which encapsulates and implements a specific software programming pattern for abstract interaction objects, while also offering to the programmer declarative methods for the easy definition and deployment of abstract interaction objects. In particular, I-GET supports the development of unified user interfaces (see Chapters 16 and 21 of this handbook) of this handbook. The next sections provide an overview of the programming facilities offered in the context of the I-GET user interface programming language for

the definition and deployment of abstract interaction objects (Savidis, 2004).

27.2 Definition of Virtual Interaction Object Classes

Abstract interaction objects are explicitly supported in the I-GET language through the keyword virtual, subject to specific deployment regulations. Therefore, abstract interaction objects are another category of domain-specific classes supporting compile-time type-safety. Definitions of key virtual object classes are provided in Figure 27.2. The source code for the complete definition of typical virtual object classes is surprisingly very small, while some of the supported features, such as the provision of constructor and destructor blocks, are not practically needed. At the header of each virtual object class, the identifiers of the imported toolkits for which it actually constitutes an abstraction need to be explicitly enumerated, separated with commas. In Figure 27.2, the defined virtual classes are applicable to four toolkits: Xaw, MFC, Hawk (Savidis et al., 1997), and JFC.

In Figure 27.3, the physical mapping schemes of the state virtual object class are defined for the MFC and Hawk imported toolkits. The keyword instantiation issues the beginning of a specification block encompassing the logic to physically instantiate a virtual class to concrete toolkit interaction objects, through alternative

```
#define ALL MFC, Xaw, Hawk, JFC          virtual Button (ALL) [
virtual Selector (ALL) [                      public:
    public:                                   method Pressed;
    method        Selected;                   constructor [ ]
    word          UserChoice = 0;             destructor [ ]
    constructor [ ]                       ]
    destructor [ ]
]                                         virtual Message (ALL) [
                                              public:
virtual Container (ALL) [                      string label = " ";
    public:                                   constructor [ ]
    constructor [ ]                           destructor [ ]
    destructor [ ]                        ]
]
                                          virtual Textfield (ALL) [
virtual State (ALL) [                         public:
    public:                                   string Text = " ";
    bool      State = true;                   method Changed;
    method Changed;                           constructor [ ]
    constructor [ ]                           destructor [ ]
    destructor [ ]                        ]
]
```

FIGURE 27.2 The complete definition of the most representative virtual interaction object classes (no code has been omitted). (From Savidis, A., Supporting virtual interaction objects with polymorphic platform bindings in a user interface programming language, *Rapid Integration of Software Engineering Techniques, Proceedings of the International Workshop Rise 2004* (N. Guelfi, ed.), pp. 11–23, Springer LCNS: 3475, 2005. With permission.)

mapping schemes. For each distinct toolkit, a separate instantiation definition has to be provided. At the top of Figure 27.3 two macros are defined. The first, named EQUALITY, employs monitors to establish nonexclusive additive equality constraints between two variables (see Savidis, 2004; Savidis and Stephanidis, 2006). This is a more general approach than built-in constraints (see Savidis, 2004; Savidis and Stephanidis, 2006), which upon activation supersede the previously active constraint on a variable. The second, named TERNARY, is used for syntactic convenience to simulate the ternary operator, not supported in the I-GET language. In each instantiation definition, there are arbitrary mapping schemes as subsequent distinct blocks (labels 1 in Figure 27.3), where each such block starts with a header engaging two identifiers separated by a colon (shaded lines in Figure 27.3).

The first identifier is a programmer-decided descriptive scheme name, for example, ToggleButton, which has to be unique inside the context of the container instantiation definition, while the second identifier is the name of the toolkit class, for example, ToggleButton, to which the virtual class is mapped (this need not be unique inside an instantiation). Even though the scheme name and the toolkit class name need not be the same, it was chosen to follow a naming policy in which the scheme identifier is the same as its associated toolkit class.

In each scheme, the programmer supplies the code to maintain a consistent state mapping between the virtual instance, syntactically accessible through {me}, and the particular physical instance, syntactically accessible through {me}Toolkit, for example, {me}MFC or {me}Hawk. In this context, state consistency is implemented by: (1) the equality of the virtual instance attributes with the corresponding physical instance attributes, which is implemented through the monitor-embedding EQUALITY

macro (see label 2 in Figure 27.3), or through explicit monitors when no direct type conversions are possible (see labels 3 in Figure 27.3); and (2) the artificial method notification for the virtual instance, when the corresponding method of the physical instance is triggered (see labels 4 in Figure 27.3). After all scheme blocks are supplied, a default scheme name is assigned, for example, default ToggleButton or default RadioButton. Such a scheme is activated automatically by default upon virtual class instantiation, if no other scheme name is explicitly chosen.

27.3 Declaration and Deployment of Virtual Object Instances

The instantiation definitions for each different imported toolkit can be provided in separately compiled files, while being optionally linked as distinct software libraries, called instantiation libraries. During user interface development, programmers have to link the necessary instantiation libraries with the overall user interface compiled code. At run-time, when a virtual instance is created, it requests the realization of its physical instantiation from every linked instantiation library (see Figure 27.4, step 1).

As a result, each instantiation class, for instance, a singleton class, will create an instance of the appropriate mapping scheme (see Figure 27.4, step 2), that is, either the default, or a scheme explicitly chosen upon virtual instance declaration. Then, the newly created scheme instance automatically produces an instance of its associated toolkit class (see Figure 27.4, step 3). Next, the scheme activates any locally defined monitors, constraints, or method implementations (see Figure 27.4, step 4), which actually establish the run-time state mapping between the virtual instance and the newly created toolkit instance.

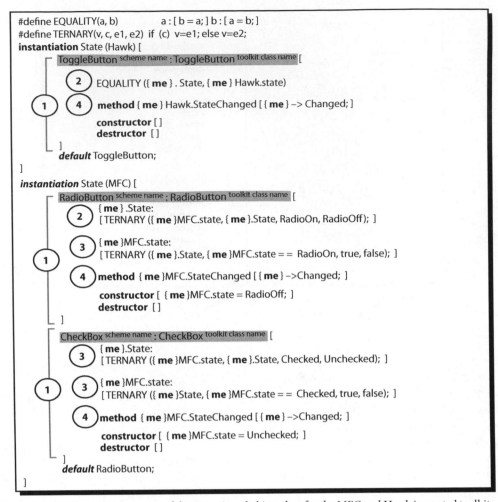

FIGURE 27.3 The logic for polymorphic mapping of the state virtual object class for the MFC and Hawk imported toolkits. The numbers include directives for the virtual class definition file, and the toolkit interface specification files are omitted for clarity. (From Savidis, A., Supporting virtual interaction objects with polymorphic platform bindings in a user interface programming language, *Rapid Integration of Software Engineering Techniques, Proceedings of the International Workshop Rise 2004* (N. Guelfi, ed.), pp. 11–23, Springer LCNS: 3475, 2005. With permission.)

The resolution of the {me} and {me}Toolkit language constructs, in the context of scheme classes generated by the compiler, is also illustrated in Figure 27.4: {me} is mapped to the X virtual instance, while {me}Toolkit is mapped to the Z toolkit instance. As can be observed from the previous description of physical instantiation, a single virtual instance is always mapped to a number of concurrently available toolkit instances, this number being equal to the total instantiation libraries actually linked. In practice, this implies that, during run-time, virtual instances can be delivered with plural physical instantiations. For instance, if one links together the MFC and Xaw instantiation libraries, all virtual instances have dual physical instantiations for both Xaw and MFC. If the running user interface is connected upon startup with the respective toolkit servers of those imported toolkits, then the user interface will be consistently replicated in two forms, at each toolkit server machine. A scenario of such run-time interface replication with two concurrent instances for windowing toolkit servers may not be considered particularly beneficial for end-users. However, if the interface is replicated

for toolkit servers of toolkits offering complementary modalities to interactions objects, such as, for instance, the MFC–Hawk (Savidis et al., 1997) or the Xaw–Hawk toolkit pairs, then the resulting interface provides an augmented physical realization of the application dialogue in complementary interoperable physical forms (i.e., it is unified user interface). Such interfaces can be effectively targeted to a broad audience, including user groups with different interaction requirements, offering concurrently various user interface instances (e.g., graphical, auditory, and tactile).

In Figure 27.5 an example is provided showing the implementation of a simple confirmation dialogue through virtual object instances. When declaring virtual object instances, programmers may optionally choose for each toolkit any of the named schemes supplied in the corresponding instantiation definition.

Additionally, since virtual instances are delivered with multiple physical instantiations, the corresponding physical parent instances have to be explicitly supplied per toolkit. For instance, the definition ok: parent(MFC) = {cont}MFC denotes that the

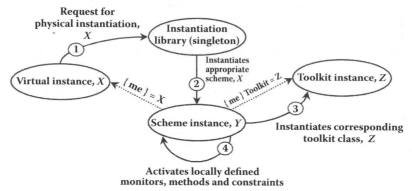

FIGURE 27.4 The automatic run-time steps to physically instantiate a virtual class for each target toolkit. (From Savidis, A., Supporting virtual interaction objects with polymorphic platform bindings in a user interface programming language, *Rapid Integration of Software Engineering Techniques, Proceedings of the International Workshop Rise 2004* (N. Guelfi, ed.), pp. 11–23, Springer LCNS: 3475, 2005. With permission.)

physical parent for the MFC physical instantiation for the ok virtual instance is the MFC physical instantiation of the cont virtual instance. The I-GET language provides syntactic access to each of the alternative physical instantiations of a virtual instance through explicit toolkit qualification. For example, the expression {ok}MFC provides syntactic visibility to the MFC instantiation of the ok virtual instance, which, as it is reflected in the default scheme of the MFC instantiation, is actually an MFC Button instance. As depicted in Figure 27.5, virtual objects enable singular object declarations and method implementations, while also providing fine-grained control to physical aspects through physical scheme selection, physical-instance hierarchy control, and syntactic visibility of toolkit-specific instances. The latter, apart from appearance control, allows specialized toolkit-

specific method implementations to be supplied. Figure 27.5 depicts the implementation of methods {yes}Hawk.Pressed and {no}MFC.Pressed.

27.4 Code Generation

The support of virtual object classes having polymorphic plural instantiations (see Chapter 16 of this handbook), through facilitating scheme selection upon virtual instance declarations, is one of the most demanding and complex code generation patterns.

The run-time organization to accomplish this functional behavior has been illustrated earlier in Figure 27.4, introducing also the logical differentiation of classes among virtual objects,

```
#define PARENT(o) \
    parent(MFC)={o}MFC :parent(Xaw) = {o}Xaw :parent(Hawk) = {o}Hawk

agent ConfirmQuit (string text) [
    virtual Container      cont  : scheme (MFC) = FrameWindow
                                 : scheme (Xaw) = PupupWindow;
    virtual Message        msg   : PARENT(quit);
    virtual Button         yes   : PARENT(quit);
    virtual Button         nol   : PARENT(quit);


    method {yes}.Pressed [ terminate; ]
    method {no}.Pressed [ destroy {myagent}; ]

    method {yes}Hawk.Pressed  [ printstr("Hawk YES."); ]
    method {no}MFC.Pressed    [ printstr("MFC NO."); ]

constructor [
        {msg}.label = text;
        {yes}Xaw.label = {yes}MFC.text = {yes}Hawk.msg = "Yes";
        {no}Xaw.label = {no}MFC.text = {no}Hawk.msg = "No";
        {msg}Xaw.borderWidth = 4;
        {msg}Xaw.bgColor = "green";
        . . .
    ]
    destructor [...]
]
```

FIGURE 27.5 An example of a unified implementation of a confirmation dialogue, engaging virtual object instances, retargeted automatically to the MFC, Xaw, and Hawk toolkits. (From Savidis, A., Supporting virtual interaction objects with polymorphic platform bindings in a user interface programming language, *Rapid Integration of Software Engineering Techniques, Proceedings of the International Workshop Rise 2004* (N. Guelfi, ed.), pp. 11–23, Springer LCNS: 3475, 2005. With permission.)

```
class _INSTStateMFC;
class _INSTStateHawk;

#define INST_ARGS    Agent*, _VICState*, unsigned, LexicalClass*

class _VICState : public VirtualObject {
        friend class State_Initializer;
        friend class _INSTStateMFC;
        friend class _INSTStateHawk;
private:
        static _INSTStateMFC*              (*instantiateMFC) (INST_ARGS);
        static void                        (*deleteMFC) (INSTStateMFC*);         ①
        static _INSTStateHawk*             (*instantiateHawk) (INST_ARGS);
        static void                        (*deleteHawk) (INSTStateHawk*);
        _INSTStateMFC*         _instMFC;           ②
        _INSTStateHawk*        _instHawk;
        MethodList             Changed_methods;
        SmartType<bool>        _VARState;          ③
public:
        void AddMethod_Changed (MethodFunc f, void* owner)
                { Changed_methods.Add(f, owner); }
        _VICState (
                Agent*          owner,
                unsigned        MFCscheme,
                unsigned        Hawkscheme,
                LexicalClass*   MFCparent,
                LexicalClass*   Hawkparent,
        ) {
④    [ _instMFC = (*instantiateMFC) (owner, this, MFCscheme, MFCparent) ;
      [ _instHawk = (*instantiateHawk) (owner, this, Hawkscheme, Hawkparent) ;
        }
        ~_VICState ( ) {
                (*deleteMFC) ( _instMFC);          ⑤
                (*deleteHawk) (_instHawk);
        }
};

class State_Initializer {
        public :
        static unsigned flagMFC, flagHawk;
        State_Initializer (void);
};
static State_Initializer State_initializer;
```

FIGURE 27.6   The code-generated class definition for the state virtual object class; for simplicity, only the items corresponding to the MFC and Hawk target toolkits are shown, since for the Xaw and JFC toolkits, the resulting code generation is similar. (From Savidis, A., Supporting virtual interaction objects with polymorphic platform bindings in a user interface programming language, *Rapid Integration of Software Engineering Techniques, Proceedings of the International Workshop Rise 2004* (N. Guelfi, ed.), pp. 11–23, Springer LCNS: 3475, 2005. With permission.)

instantiation definitions, and mapping schemes. Such logical distinction of roles is also reflected in code production, leading to the generation of appropriate classes from those three key categories. Figures 27.6 and 27.7 present the code generated for the State virtual object class of Figure 27.2, and its respective instantiation definitions of Figure 27.3. Overall, the code generation for virtual classes provides the ground for multiple concurrent physical instantiations, by delivering the placeholder as well as the activation mechanism for the toolkit-specific instantiation definitions. This capability to automatically activate any instantiation definitions of virtual classes, once their respective instantiation library is linked with the user interface code, is the most demanding feature.

Following Figure 27.6, the produced header file encompasses first the forward declarations of the instantiation classes for each target toolkit (e.g., class _INSTStateHawk for the Hawk toolkit).

The generated virtual class (e.g., _VICState) encapsulates pointer variable declarations for all potential instantiation classes (see label 2), like _INSTStateMFC* _instMFC for MFC. Additionally, all instantiation classes are defined as friends for the generated virtual class. The code generation for any local definitions made inside virtual classes is collected in one fragment (label 3).

The adopted approach toward automatic support for multiple instantiation, according to the particular instantiation libraries linked, is based on explicit instantiation and destruction functions pairs, for example, instantiateMFC and deleteMFC (see label 1), one for each target toolkit (e.g., MFC). The implementation idea is that, initially, those functions will be supplied with a default empty implementation by the generated virtual class (see labels 6 and 7). Then, during run-time, each linked instantiation class sets, upon global data initialization, its specific pair of fully implemented instantiation and destruction functions. To ensure

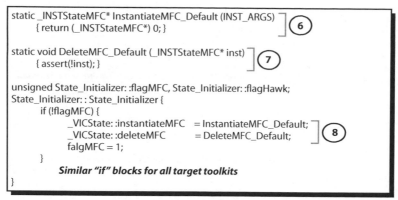

```
static _INSTStateMFC* InstantiateMFC_Default (INST_ARGS)
        { return (_INSTStateMFC*) 0; }                              ⑥

static void DeleteMFC_Default (_INSTStateMFC* inst)
        { assert(!inst); }                                         ⑦

unsigned State_Initializer: :flagMFC, State_Initializer: :flagHawk;
State_Initializer: : State_Initializer {
        if (!flagMFC) {
                _VICState: :instantiateMFC    = InstantiateMFC_Default;
                _VICState: :deleteMFC          = DeleteMFC_Default;      ⑧
                falgMFC = 1;
        }
        Similar "if" blocks for all target toolkits

}
```

FIGURE 27.7   The code-generated implementation file (excerpts) for the state virtual object class. (From Savidis, A., Supporting virtual interaction objects with polymorphic platform bindings in a user interface programming language, *Rapid Integration of Software Engineering Techniques, Proceedings of the International Workshop Rise 2004* (N. Guelfi, ed.), pp. 11–23, Springer LCNS: 3475, 2005. With permission.)

that the pair set by such instantiation classes cannot be overwritten during initialization by the default pair of the virtual class, a technique introduced in Schwarz (1996) for safely initializing static variables in C++ libraries is used. This technique uses a static flag per toolkit (e.g., flagMFC for MFC), which is to be unconditionally set by the respective instantiation classes after its specific pair of instantiation and destruction functions of the virtual class is set (this will be discussed later in the context of code generation of instantiation classes).

At the virtual class side, the default functions per toolkit are set only if the corresponding flag is not set (see label 8); consequently, irrespective of the order of initialization, the initializations made by instantiation classes can never be overwritten. Within the produced virtual class, those functions are called to actually perform the physical instantiation (in the constructor) or destruction (in the destructor) for each target toolkit (see label 4 for instantiation, and label 5 for destruction).

As depicted in Figure 27.8, instantiation definitions are generated as instantiation classes (e.g., _INSTStateMFC), while each embedded mapping scheme is produced as a distinct scheme class (e.g., _SCHStateMFCRadioButton). Scheme classes encompass two key member pointers, holding their run-time associated virtual instance and physical instance respectively (see

```
class _SCHStateMFCRadioButton : public Scheme {
        public :
        _VICState*                    myvirtual;
        _PICRadioButtonMFC*           myphysical;            ①
        Code generation here for constructs defined in the scheme block
        _SCHStateMFCRadioButton (Agent*, _State*, LexicalClass*);
        ~_SCHStateMFCRadioButton( );

};

class _SCHStateMFCCheckBox : public Scheme { ... };

class _INSTStateMFC {
        public :                         ②
        Scheme* scheme;
        _INSTStateMFC (Agent*, _VICState*, unsigned, LexicalClass*);
        ~_INSTStateMFC ( ) { assert (myphysical); delete myphysical; }
};

class StateMFC_Initializer {
        public :
        static unsigned flag;
        stateMFC_Initializer (void);                       ③
};
static StateMFC_Initializer stateMFC_Initializer;
```

FIGURE 27.8   The code generation of the header file for the instantiation definition of state virtual class for the MFC toolkit. (From Savidis, A., Supporting virtual interaction objects with polymorphic platform bindings in a user interface programming language, *Rapid Integration of Software Engineering Techniques, Proceedings of the International Workshop Rise 2004* (N. Guelfi, ed.), pp. 11–23, Springer LCNS: 3475, 2005. With permission.)

```
_SCHStateMFCRadioButton::_SCHStateMFCRadioButton (
        Agent* a, _VICState* v, LexicalClass* p
) {
        myvirtual = v;
        myphysical = new _PICRadioButtonMFC (p);          ──┐  ①
        Code generation here for initializations of constructs defined in the scheme block
}

_INSTStateMFC: :_INSTStateMFC (
        Agent* a, _VICState* v, unsigned i, LexicalClass* p
) {
   ┌─ switch (i)  {
②      case 1: scheme = new _SCHStateMFCRadioButton(a, v, p); break;
        case 2: scheme = new _SCHStateMFCCheckBox(a, v, p); break;
   └─ }
}

static _INSTStateMFC* InstantiateMFC (
        Agent* a, _VICState* v, unsigned i, LexicalClass* p          ──┐  ③
) { return new _INSTStateMFC (a, v, i, p); }

static void DeleteMFC (_INSTStateMFC* inst)          ──┐  ④
        ( assert (inst); delete inst;}

unsigned StateMFC_Initializer: : flag;
StateMFC_Initializer: : StateMFC_Initializer (void) {
        if (!flag) {
            _VICState: :instantiateMFC = InstantiateMFC;          ──┐  ⑤
            _VICState: :destroyMFC      = DestroyMFC;
            flag = State_Initializer: :flagMFC = 1;
        }
}
```

FIGURE 27.9 Key fragments of the generated implementation file for the instantiation definition of state virtual class for the MFC toolkit. (From Savidis, A., Supporting virtual interaction objects with polymorphic platform bindings in a user interface programming language, *Rapid Integration of Software Engineering Techniques, Proceedings of the International Workshop Rise 2004* (N. Guelfi, ed.), pp. 11–23, Springer LCNS: 3475, 2005. With permission.)

label 1) (e.g., myvirtual of type _VICState* and myphysical of type _PICRadioButtonMFC*). In the code generation of the scheme blocks, the compiler always resolves the {me} expression as myvirtual, and the {me}Toolkit expression as myphysical. As can be observed from Figure 27.8, instantiation classes like _INSTStateMFC are very simple upon code generation, encompassing a superclass Scheme* scheme pointer (see label 2), which holds the particular run-time active scheme. Finally, the generated header file employs the technique for static data initialization (see label 3), so as to ensure that schemes are safely initialized prior to any run-time use (see also Figure 27.9). In Figure 27.9, key fragments of the generated implementation file for the instantiation definition of the State virtual class are supplied. As previously discussed, scheme classes are responsible for the automatic creation of physical instances for their associated lexical class. Following this need, as reflected in Figure 27.9, the constructor of the _SCHStateMFCRadioButton scheme class (see label 1) first stores in the myvirtual member the caller virtual instance, and then performs the creation of a _PICRadioButtonMFC lexical object instance, stored in myphysical. The instantiation of the appropriate scheme classes is performed inside the constructor of the container instantiation class, through a switch statement (see label 2) over a parameter (e.g., the i) that provides the order of appearance of the desirable scheme within the instantiation definition. This number is easily defined by the compiler upon virtual instance declaration, being the order of either the programmer-supplied scheme or the default scheme.

Additionally, the compiler produces the key pair of functions for the instantiation (see label 3) and destruction (see label 4) of the generated instantiation class. Those functions are appropriately assigned upon initialization to the corresponding members of the virtual class (see label 5). After the assignment is performed, the corresponding flag is set (e.g., flagMFC), thus disabling overwriting of those functions with the default implementations due to virtual class initialization.

27.5 Discussion and Conclusions

Through the presented language constructs, the developer is enabled to define and instantiate abstract object classes, while having control on the physical mapping schemes that will be active for each abstract object instance at run-time; mapping schemes explicitly define the alternative candidate physical classes to physically realize abstract object class.

The need for having multiple physical instances active, all attached to the same abstract object instance (i.e., plural instantiation), can be exploited in case the alternative physical forms are compatible, while their copresence results in added-value

Polymorphism in virtual interaction objects	Polymorphism in typical OOP languages
• Aims to support alternative morphological realisations (i.e., polymorphism with its direct physical meaning).	• Aims to support reuse and implementation independence from different toolkits (i.e., polymorphism with its metaphoric meaning).
• Instantiation is applied directly on abstract object classes.	• Instantiation is always applied on derived non-abstract classes.
• Multiple physical instances, manipulated via the same abstract object instance, may be active in parallel at a time.	• References to an abstrat always refer to a single derived object instance at a time.

FIGURE 27.10 Key differences with respect to polymorphism and abstract objects between general-purpose OOP languages and polymorphism as supported through virtual object classes. (From Savidis, A., Supporting virtual interaction objects with polymorphic platform bindings in a user interface programming language, *Rapid Integration of Software Engineering Techniques, Proceedings of the International Workshop Rise 2004* (N. Guelfi, ed.), pp. 11–23, Springer LCNS: 3475, 2005. With permission.)

interactions. For instance, in the context of unified user interface development (see Chapter 21, "A Unified Software Architecture for User Interface Adaptation"), two or more concurrently active instances are required (e.g., a visual and a nonvisual) for each abstract object instance. The notions of polymorphic physical mapping and plural instantiation have fundamentally different functional requirements in this context (see also Chapter 16, "Unified Design for User Interface Adaptation") with respect to polymorphism of superclasses in object-oriented program (OOP) languages. The key differences are outlined in Figure 27.10.

Clearly, the traditional schema of abstract/physical class separation in OOP languages by means of class hierarchies and ISA relationships cannot be directly applied for implementing the abstract/physical class schema as needed in interface development. An explicit run-time architecture is required, where connections among abstract and physical instances are explicit programming references, beyond the typical instance of run-time links from ISA hierarchies.

References

Blattner, M. M., Glinert, J. A., and Ormsby, G. R. (1992). Metawidgets: Towards a theory of multimodal interface design, in the *Proceedings of COMPSAC '92*, pp. 115–120. Los Alamitos, CA: IEEE Computer Society Press.

Bronevetsky, G. (2003). *Circle Menus*. http://www.cs.cornell.edu/boom/2001sp/Bronevetsky/Circle%20Menu%20Documentation.htm.

Duke, D. J., Faconti, G. P., Harrison, M. D., and Paternò, F. (1994). Unifying views of interactors, in the *Proceedings of the Workshop on Advanced Visual Interfaces*, 1–4 June 1994, Bari, Italy, pp. 143–152. New York: ACM Press.

Duke, D. and Harrison, M. (1993). Abstract interaction objects. *Computer Graphics Forum* 12: 25–36.

Foley, J. and Van Dam, A. (1983). *Fundamentals for Interactive Computer Graphics* (1st ed.). Boston: Addison-Wesley.

McGuffin, M., Burtnyk, N., and Kurtenbach, G. (2001). FaST sliders: Integrating marking menus and the adjustment of continuous values. *Graphics Interface 2001*. http://www.graphicsinterface.org/cgi-bin/DownloadPaper?name=2002/174/paper174.pdf.

Savidis, A. (2004). *The I-GET User Interface Programming Language: User's Guide*. Technical Report 332, ICS-FORTH. ftp://ftp.ics.forth.gr/tech-reports/2004/2004.TR332.I-GET_User_Interface_Programming_Language.pdf.

Savidis, A. and Stephanidis, C. (2006). Automated user interface engineering with a pattern reflecting programming language. *Journal for Automated Software Engineering* 13: 303–339.

Savidis, A., Stephanidis, C., Korte, A., Crispien, K., and Fellbaum, K. (1996). A generic direct-manipulation 3D-auditory environment for hierarchical navigation in non-visual interaction, in the *Proceedings of the ACM ASSETS'96 Conference*, 11–12 April 1996, Vancouver, Canada, pp. 117–123. Paris: EC2 & Developpement.

Savidis, A., Stergiou, A., and Stephanidis, C. (1997). Generic containers for metaphor fusion in non-visual interaction: The HAWK interface toolkit, in the *Proceedings of the 6th International Conference on Man-Machine Interaction Intelligent Systems in Business (INTERFACES '97)*, 28–30 May 1997, Montpellier, France, pp. 194–196.

Schwarz, J. (1996). Initializing static variables in C++ libraries, in *C++ Gems* (S. Lippman, ed.), pp. 237–241. New York: SIGS Books.

VI

Interaction Techniques and Devices

Screen Readers

Chieko Asakawa and
Barbara Leporini

28.1 Introduction

A screen reader is a kind of assistive technology software that assists people with little or no functional vision in interacting with computers. There are various types of assistive technologies to support various types of disabilities. Some are purely software applications, while others require devices or provide options to use devices in addition to the software. A screen reader basically works as a software application using text-to-speech software, but a Braille pin-display[1] can be added to generate real-time and refreshable Braille output. Screen readers were mainly developed for blind people, while screen magnifiers are for people with weak eyesight. The applicability of screen readers has been expanded, and they are often used in combination with screen magnifying technology to assist people with weak eyesight (Paul and Dillin, 1998) or for people with cognitive and learning impairments.

This chapter first describes the basic concept of a screen reader. Then, it discusses design principles and interaction technologies of a screen reader by considering some frequently used examples, such as standard graphical user interfaces (GUIs) and the web. The survey includes the technical challenges in developing an application such as a screen reader from a developer's perspective and the problems in interacting nonvisually with visual interfaces from a user's perspective.

---

[1] BAUM Retec AG (http://www.baum.de/en/products/brailledisplay/index.php).

28.1.1 Screen Reader: Function and Main Characteristics

A screen reader is essential software for blind people to interact with computers, as it allows users to access computers effectively and intuitively using nonvisual means. The primary and most critical goal of a screen reader is to capture the information being displayed on the screen, perhaps with keyboard assistance, and represent the information as speech or as Braille. The screen reader does this by interpreting and optimizing the captured visual information to be suitable and understandable as nonvisual information.

Capturing the displayed information and scanning the keyboard input heavily depend on the operating system, so various screen readers have been developed to support various operating systems such as Microsoft Windows, Linux, Apple Macintosh OS X, and so on.

In other words, the screen reader is a piece of software that mediates between a user and an operating system (and its applications), thus assisting blind users by representing the standard visual user interface.

Blind people need to explore not only the content, but also software interfaces. To interact with an application, it is critical to obtain information about the interfaces, such as the control elements, the icons, the menus and menu items, and other GUI controls. To explore the content effectively and meaningfully, it is often helpful to obtain rich text information such as font sizes, font styles and colors, and the logical structure of the content such as the headings, paragraphs, tables, lists, and so on. Thus, a screen reader needs to analyze the content structure and rich text information, as well as to interpret the standard visual interfaces.

All of the captured and interpreted information needs to be presented to a user who can only use keyboard operations. Therefore, each screen reader provides a set of navigational commands to help users interact with applications effectively and obtain information quickly and accurately.

There are various methods for capturing the displayed information and for analyzing and interpreting it even with the same operating system. As a result, there are various types of screen readers, for example, to support Windows XP, currently available on the market. Each screen reader follows its own approach and has its own pros and cons.

A screen reader is generally capable of providing both speech and Braille output. A text-to-speech (TTS; Schroeter, 2005) system is usually packaged with a screen reader as its default TTS system. Therefore, users do not need to purchase a TTS separately. However, if users have a preferred TTS, the default TTS system can be changed from a selection menu.

A TTS system is a speech synthesis system to convert text to speech. They are used for various purposes, not only for screen readers, but also with other applications that provide speech output. There are various TTS systems, and support for various languages and screen readers can use these TTS systems to support multiple languages. Braille output may also be enabled for users who have a Braille pin display connected to their computers.

28.1.2 Screen Readers' Classifications

A first classification of screen reader types is based on the type of operating system with which the screen reader has to interact. The two main classes can be considered the command line interface screen readers, which interact with text-based interfaces, and the GUI screen readers, which interact with a graphical interface. The main differences are due to the operating system structure. In a text-based operating system, the screen reader has to interpret simpler content than in a graphical structure. In fact, a textual interface is basically made of text characters. In contrast, in a graphical operating system, the screen reader's role is more demanding. A GUI is more complex, and the object structure is rather difficult to interpret. The content displayed on the screen includes text, icons, graphics, menus, and control elements. Several applications can be in use at the same time, and there are other complications. Since GUIs have already become standard interfaces for personal computers, current screen reader products are capable of dealing with GUIs. More details on the basic technologies appear in Section 28.6.

28.1.3 Commercially Available Screen Readers

Many screen readers have been developed and are available in the market. Popular screen readers vary in each country, according to language differences, cultural differences, application differences, and other differences in the local requirements. There are some globally popular screen readers such as JAWS,[2] Window-Eyes,[3] and Hall for Windows.[4] There are also simple versions of screen readers built into various operating systems. These screen readers are used by more than 95% of blind users of the web (Thatcher, 2006).

28.1.3.1 JAWS (Freedom Scientific)

JAWS for Windows is one of the most popular screen readers in the world. JAWS, an acronym for Jobs Access With Speech, was developed by Freedom Scientific. It is characterized by its customization functions with a scripting system (called JAWS Script[5]; Lazzaro, 2003), which allows it to deal with a wide range of GUI applications. It is available in standard or professional editions. A timed demonstration version can be downloaded from the Freedom Scientific web site (http://www.freedomscientific.com). The demo version is fully functional.

28.1.3.2 Window-Eyes (GW Micro)

Window-Eyes is another popular screen reader in the global market. A timed demonstration is available from GW Micro. The demo version is fully functional as well (http://www.gwmicro.com/Window-Eyes/Demo).

28.1.3.3 Hal

Hal is a screen reader developed in the United Kingdom by Dolphin Computer Access (http://www.dolphincomputeraccess.com). This screen reader is available in two versions: Hal Standard or Hal Professional. The timed demonstration version of Hal is fully functional as well.

28.1.3.4 Narrator (Microsoft)

Narrator[6] is built-in screen reader software bundled with Microsoft Windows XP and Vista. Any Windows user can use it simply by activating the program, but the functionality is not sophisticated enough for typical usage. Users who want to use computers for their education or work generally need more practical and powerful screen readers.

28.1.3.5 VoiceOver

VoiceOver[7] is a screen reader for Macintosh OS X, developed by Apple. It has been built-in and bundled with the operating

---

[2] Freedom Scientific. Low Vision, Blindness and Learning Disability Adaptive and Assistive Software and Hardware Technology (http://www.freedomscientific.com).

[3] Windows Eyes (http://www.gwmicro.com).

[4] Hal for Windows (http://www.dolphincomputeraccess.com/products/hal.htm).

[5] Basics of Scripting, Freedom Scientific (http://www.freedomscientific.com/fs_support/doc_screenreaders.asp).

[6] Hear text read aloud with Narrator, Microsoft (http://www.microsoft.com/enable/training/windowsvista/narrator.aspx).

[7] Apple–Accessibility–VoiceOver (http://www.apple.com/accessibility/voiceover).

system since version 10.4 (Tiger). This screen reader is the only choice for the Macintosh OS X series, and still has some limitations, but it is maturing and shows promise.

28.1.3.6 Open Source Screen Readers

For UNIX systems, there are no commercial screen readers for X-Windows, the standard windowing system of UNIX, but there are shareware and open source projects to make X-Windows accessible for people who are blind. In particular, the open source software GNOME (GUI shell for UNIX and Linux) project has a screen reader project called Gnopernicus (http://www.baum. ro/gnopernicus.html). There is at least one sophisticated access system for text-based UNIX, Emacspeak (http://emacspeak. sourceforge.net/), but the Emacs editing environment is more often accessed with terminal software under Windows, using Windows screen readers. Recently, a new and promising screen reader named ORCA (http://live.gnome.org/Orca) has become available. Orca is a free and open source assistive technology for the Linux environment. Orca provides access to applications and toolkits that support the accessibility application programming interface (API) for Linux platform called AT-SPI (assistive technology service provider interface), which is also used by the GNOME desktop. Orca is thus a part of the GNOME platform. As a result, Orca is already included by default in a number of operating systems and distributions, including Open Solaris and Ubuntu. An open source screen reader for Windows is now available, the nonvisual desktop access (NVDA) through the NonVisual Desktop Access project (http://www.nvda-project. org). This project began in 2006, and though development progress has been rapid and shows high potential, it should still be regarded as alpha software as of early 2008.

28.1.3.7 Self-Voicing Programs

A screen reader is an approach to create a read-aloud interface for nonsighted users that deal with the same displays as seen by sighted users. It provides more or less full access to existing applications, but sometimes there are limitations on the usability compared to the degree of control enjoyed by sighted users. The self-talking approach tries to provide an ideal interface for some program that is specially optimized for visually impaired users. A self-voicing application is an application that provides an aural interface without requiring a separate screen reader. One of its limitations is that the user can only interact with that specific application and not with others. A prominent group of self-voicing applications are talking web browsers. Examples of these kinds of applications are Home Page Reader (Asakawa, 1998) and Connect Out-loud.[8]

A more recent approach uses self-voicing capabilities added to browsers or other applications. For example, Firefox has the FireVox[9] (Thiessen and Chen, 2007) plug-in for voice access and Acrobat Reader has a basic read-aloud function without any

screen reader. Such specific applications tend to have limited utility for blind people, because their features are limited and they lack navigation reading commands. Typically, the user can only read the content sequentially. In contrast, a screen reader allows users to navigate more freely.

28.1.3.8 Server-Side Screen Readers

The word *screen reader* usually refers to client-side software to access visual information nonvisually. However, server-side voicing technologies also have a long history, and they can allow blind users to access a wide variety of information. The "1-800-hypertext" research project was one of the earliest systems that allowed users to access web information through a telephone. Since then, various voice portals have been developed and applied in specific domains. Recently, server-side TTS methods are becoming popular, supported by increasing network bandwidth. The Readspeaker[10] is a service company to provide reading sound files to various web sites. When a user accesses a web page that is enabled by the service, a server on the Readspeaker site converts the page into a sound file by using a server-side TTS engine, and embeds a link to the sound file in the page.

28.1.3.9 Screen Readers for Mobile Platforms

Finally, screen readers for mobile devices have been developed, such as for smart phones or for personal digital assistants (PDAs). Examples are Mobile Speak and Mobile Speak Pocket (http://www.codefactory.es/index2.htm) and Talks (http://www. nuance.com/talks).

28.1.4 Brief Historical Background

As described previously, a screen reader is defined as a software component that outputs the information on the screen to a speech synthesis and/or a Braille display as controlled by keyboard operations. It has been a key technology to make computers accessible to blind people. Even before computer displays appeared, the Optacon (Linvill and Bliss, 1966), which translated print into tactile vibrating images, could be used by the first blind programmers. The Optacon was used until screen readers were developed, even after computer displays were in use.

Before screen readers became available, there was some previous work in the early 1980s on alternative approaches.[11] For example, SAID became an IBM product as the 3270 Talking Terminal in 1981 (Keates, 2006), but only as a remote terminal for certain IBM mainframe computers. The basic SAID design concept was transferred to the screen readers for DOS (Thatcher, 1994). Several screen readers for DOS became commercially available, such as JAWS in 1989, and they helped blind people around the world use computers.

However, the DOS era ended and the GUI became widespread. This created various problems for screen reader developers and users, because GUIs were originally inaccessible without vision.

---

[8] Connect Outloud, Freedom Scientific (http://www.freedomscientific. com/fs_products/software_connect.asp).

[9] FireVox (http://firevox.clcworld.net).

[10] Readspeaker (http://www.readspeaker.com).

[11] Microvox (http://members.tripod.com/werdav/t2smicrv.html).

In 1989, Berkeley Systems (BSI) developed OutSpoken for the Macintosh GUI. This was the first screen reader product for GUI applications, and influenced screen readers for Microsoft Windows. Later on, the IBM Screen Reader/2 was developed (Schwerdtfeger, 1991; Thatcher, 1994), and the first version of JAWS for Windows appeared in 1995. The developers of these screen readers invented various methods to read aloud the text on a screen. The off-screen model (see Section 28.5) is one of the most important methods, which contributed toward making GUIs accessible in the mid-1990s. The scripting capability was also an important invention to make complicated GUIs accessible nonvisually (see Section 28.3.1). Some screen readers handle multiple languages and multiple operating systems. Others were developed in certain countries for local languages (e.g., Japanese, Chinese, Korean, etc.) as well for certain operating systems (e.g., UNIX, Mac OS).

28.2 Screen Reader Functioning

28.2.1 Basic Concepts

A screen-reading program interprets the content shown on the screen to provide a blind person with a certain accessible modality, such as speech (using a speech synthesizer) or a written form (using a refreshable Braille display). Regardless of the output modality (voice or Braille format), the screen reading software attempts to recognize the displayed content and interpret it to extract correct and useful information. To provide the content in an adequate manner that is useful for a blind person, it is necessary to give the user a lot of complex information, not only simple labels and text, but also the types of objects. For instance, a screen reader has to tell the user about the interactive elements (such as dialogue boxes, buttons, edit fields, radio buttons, or checkboxes). In practice, it has to recognize if a text string is just content (text), or is part of a much more complex object (such as part of a menu, within a set of buttons, or linked to a checkbox). A screen-reading program seeks to "interpret" the user interface to provide as much information as possible. Once the content has been "identified," it can be presented to the user through the speech synthesizer or refreshable Braille display.

A speech synthesizer, which "speaks" the text sent to it from the screen-reading program, is usually software that works with a sound card. A refreshable Braille display or Braille terminal is an electromechanical device for displaying Braille characters, usually by means of raising bumps through holes in a flat surface. Braille is also a generic term for such a code that maps the characters of a writing system to a group of six, or in some cases eight, raised dots, which can be read with the fingers by blind people (http://www.rnib.org.uk/xpedio/-groups/public/documents/publicwebsite/public_everydaybraille.hcsp). A Braille display is not like a full computer screen. Because these displays are expensive, usually only 40 or 80 Braille cells are available. This means that to explore or read some information, the user needs to navigate using a limited number of characters at a time. Even if special shortcuts are available for the Braille display, this navigation tends to be rather slow.

However, because Braille literacy rates are low, speech output is the most common assistive technology for users with visual impairments (Zhao et al., 2006). Nevertheless, various activities can be performed better by reading through a Braille display. Examples include reading program code or musical notation, or correcting errors. For this reason, a blind user may switch between the speech synthesizer and Braille display, or use both at the same time depending on circumstances.

28.2.2 General Features

As mentioned in the previous section, when a user interacts with a computer via a screen reader, the rendered content must be gathered, interpreted, and then converted into spoken or haptic output. That is, the screen reader recites using the voice synthesizer modality the content of what is displayed on the screen, perhaps word by word, line by line, or link by link, or it can write using the tangible Braille modality, so that the users can directly read it by themselves.

Essentially, a screen reader allows a person to control a computer GUI using a keyboard rather than a mouse. Many of the keyboard commands that a blind person uses are the same alternative keyboard commands a sighted person can use on the PC. For example, to open the File menu the Alt-F shortcut can be used. To close the File menu, it is possible to press the Alt key again or alternatively use the Escape key. In general, almost all of the commands can be invoked via the keyboard. In addition, the screen reader often offers special commands to emulate the mouse pointer and its behavior. A blind person can use these special keyboard commands provided by the screen reader to read information or to obtain system information. For example, most screen readers offer keystroke commands to read the title bar, the status bar; the current line, phrase, or word; and so on. In short, common screen-reading features include capabilities needed to handle and navigate in the user interface:

- The full screen, usually an application window or the desktop
- A user-defined area of the screen, such as a customized frame or a dialogue window
- Control elements (buttons, radio buttons, or checkboxes), including their status (e.g., selected or checked), menus and items, and so on
- An entire document
- A paragraph, sentence, line, word, individual letters, or the phonetic equivalent of a letter
- Properties and features, such as capital letters, punctuation, symbols, font size, font type, and so on
- System messages, including system-tray notifications

To provide all of these features, the screen-reading software needs to interact with and to analyze the user interface. To do this, it is necessary that the user interface (UI) be well designed and developed with consideration of specific accessibility principles and techniques. Otherwise, the screen reader will not be able to meaningfully interpret the content and the types of the elements.

An application can be navigated by a blind user via a screen reader if the UI has been designed well and in an accessible manner. In addition, if more specific usability principles are applied, the application will also be simple to support for people with other interaction requirements. These conditions constitute the basis for an accessible and usable application.

28.2.3 Customization

Screen readers are optimized to read major application software (e.g., Microsoft Word) at the time of their development. It means that screen readers are not good at reading newly developed or updated applications. Also, screen readers are not good at reading local or specialized (but important) applications. For example, support for a program for call center operation may be required for users to get a job working there.

Support for user customization is a key feature to give flexibility to a screen reader program, so that it can cope with newly developed, updated, or local applications without modifying the core of the screen reader program. Most screen readers have functions to reassign shortcut keys to improve usability. For example, a user can unify the keyboard shortcuts among frequently used application programs. A labeling function allows users to add readable text to inaccessible buttons and images. Once a user figures out the meaning of an "inaccessible image," it can become accessible by registering the label for the image.

The functionality of customization differs for each screen reader program. Therefore, this function is one of the main criteria when selecting an appropriate screen reader for a user. If a user will only use a few major application programs, then there are many choices. Some other user might need to access some specialized application, and thus may require a screen reader with sophisticated customization capabilities.

Usually a set of utilities and functions are available for most screen readers to allow finding appropriate settings for various customization features. For example, many screen readers have utilities such as a dictionary manager, a graphics labeler, and a keyboard configuration utility. Following is the list of management utilities to control user-level configurations in JAWS version 8.0. This list is based on the online manual for JAWS 8.0.

- *The Configuration Manager* allows a user to modify many settings that determine what information JAWS reads, and how much of it is read.
- *The Dictionary Manager* lets a user change the way JAWS speaks words, phrases, abbreviations, or symbols. A user can also assign a sound to play, speak the word in a different language, or any combination of the various options.
- *The Graphics Labeler* allows users to label icons, toolbar buttons, and pictures in an application or on the Internet.
- *The Keyboard Manager* controls the assignment of keystrokes to JAWS activities.
- *The Prompt Creator* allows a user to customize prompts (descriptions) for controls in dialogue boxes.

- *The Frame Viewer* allows a user to manage frames (rectangular areas of a window or the screen, which JAWS will then monitor for any activity that might take place).
- *The Customize List View* is a function to allow a user to optimize how JAWS provides information about columns and items in list views. List views are areas in a window or dialogue box that contain one or more items, such as files, folders, records, and so on.
- *Script Manager* is a script-editing program with all the features necessary for a user to create scripts, that is, a small computer program that controls how JAWS reacts, and what is heard (see Section 28.3.2). All JAWS activities assigned to keystrokes are scripts.

28.2.4 Learning to Use a Screen Reader

For a blind person, learning to interact with a Windows or web application through a screen reader requires a lot of effort. Several aspects and activities must be understood and many commands must be memorized. A blind person beginning to work with a computer has to understand several aspects of the goals and context. Certain specific mental models must be built in a different manner with respect to sighted users to perceive the environment and the spatiality in which the goals will be pursued. Usually, the operating systems and applications have been developed so that they are visually intuitive and easy to perceive. Unfortunately, the approaches generally used exploit visual principles and graphical features. This means that for blind users those principles are not helpful, so different strategies must be used when teaching people with vision impairments how to use such systems.

The greatest problem encountered when explaining how to use an application and a screen reader is due to the lack of a global overview. Therefore, a blind user has to mentally track and memorize more information than a sighted user.

Screen readers offer many functions that can be activated through shortcuts. A blind user must learn the complex structure of a GUI and at the same time acquire various screen reader commands. This requires great effort. For this reason, it is important that usability principles are applied not only to operating systems and applications, but also to screen-reading programs. Kurniawan et al. (2003) investigated the usability of a screen reader and the mental models of blind users in the Windows environment. In particular, the authors explored blind users' mental models and strategies in coping with the Windows environment. The study found that blind users possess a functional or structural mental model (or a combination) and they have a rich and highly procedural strategy for coping with a new Windows environment and application. Barnicle (2000) investigated and discussed a usability testing technique used to identify and examine obstacles that individuals who are blind and who use screen reading technology encountered while attempting to complete a series of tasks using Windows and associated software applications.

28.3 Screen-Reading Technologies

28.3.1 Introduction

To support screen reading, it is necessary to obtain the information that is on the screen, in synchronization with the user's keyboard operations. Over the years, screen readers have used various methods to obtain visual information.

As mentioned in Section 28.1.2, screen readers can be classified into two main categories: textual or graphical environments. For example, screen reader products in the DOS days literally read the 80x25 display buffer to say what was on the screen. The ASCII codes stored in the display buffer were sent directly to a hardware device to produce synthesized speech. With the advent of the GUI, screen readers had to become much smarter. Now, they can capture text as it is written to the display through the display driver.

28.3.2 Basic Technologies

Figure 28.1 shows an abstract architecture of a typical screen reader. These functions are invoked as the basic steps of screen reading. The capabilities of a screen reader can be measured by comparing these features with other screen readers.

28.3.2.1 Basic Steps of Screen Reading

Step 1

Detecting a user's operations by capturing keyboard events and invoking internal commands based on keyboard configuration files

When a user presses a key on the keyboard, the screen reader captures it by referring to the keyboard configuration. A user can reassign keyboard shortcuts by modifying the configuration. Keyboard events can be assigned to built-in commands or to commands written in scripts. For example, when a user presses the Down cursor key in a word processor program, a command to read the next line is invoked. For JAWS the invoked command's name is "SayNextLine."

Step 2

Obtaining textual information on screen through an off-screen model, accessibility API, and/or an internal model based on configuration files

Each screen reader has various methods to obtain textual information from the visual screen. In Figure 28.1, these methods are classified into three approaches, which will be explained later in this section.

Step 3

Constructing readable information based on configuration files.

The quantity and quality of the readable text are the keys to improving the usability of screen readers and applications. Therefore, each screen reader tries to optimize the readable text by precisely modifying it through various techniques. Some screen readers have functions to control the readable text by using configuration files, such as dictionaries, specific texts for labels, and scripts. (See the list of configuration managers in Section 28.2.3.)

Step 4

Presenting information through TTS engines and/or Braille pin displays

The final step is the presentation phase. The constructed texts are presented through speech or Braille channels, which are the two major methods for practical nonvisual presentations. Most screen readers can handle these two channels, but differ in their coverage of specific TTS engines and Braille pin displays. It is important to check the quality of the default (bundled) speech engine and the coverage of pin displays before purchasing a screen reader.

28.3.2.2 Step Discussion and Analysis

Of these steps, Steps 2 and 3 are the most characteristic phases for screen reading. In the remainder of this section, these two steps are described in more detail to give an idea of basic screen reader technology.

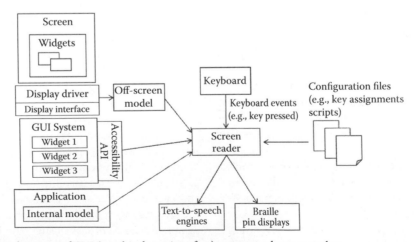

FIGURE 28.1 The typical architecture of GUI (graphical user interface) systems and screen readers.

For the second step (obtaining textual information), various methods have been elaborated. One of the oldest methods is a way to intercept the display driver function calls or to inspect the video memory, and then create a text-based data model for screen reading. These sophisticated technologies create what is called an off-screen model (OSM), which is, in effect, a database of all the text displayed on the graphical screen. When a screen reader user requests information, it is read from this database, the OSM, rather than from the screen itself.

This was the key technology to obtain text information from GUI components, especially in the early days of GUI accessibility efforts. However, with this approach the screen reader sometimes needs to use unsupported functions of the operating systems. That is why this method is not recommended by operating system vendors. These technical conflicts between accessibility enablement and the use of unsupported functions led to fully supported API-based methods.

An accessibility API is an interface by which an application (the server) exposes its GUI and content to another application (the client). Through the accessibility API, the client discovers, represents, and modifies the server's UI and content.

Thus, an accessibility API is an official API for screen reading supported by a GUI operating system vendor. Each modern GUI system provides an API for screen readers. Examples include Microsoft Active Accessibility (MSAA) for Windows (XP and earlier), UI Automation (UIA) for Windows Vista, the Java Accessibility API for Java, the Mac Accessibility API under Mac OS X, the Gnome Accessibility API for GNOME systems, ATK under Linux, and IAccessible2 for various platforms.

Clients, such as a screen-reading program, use the accessible object interfaces as the basic means to retrieve information about the server's UI and content. An accessible object that represents an interactive UI component supports interfaces to simulate the user's input. It is the server's responsibility to implement and expose the accessible object interfaces for every UI component and content fragment. Accessible objects can represent a UI component such as a button or a menu item. In Gonzalez and Guarino Reid (2005) useful references on APIs can be found. These APIs share the following sets of functions:

- *Event mechanisms*: Mechanisms to notify the system of events calling for verbalizations, such as when a button was clicked, the focus changed on a widget, or a key was pressed.
  - *Widgets role information*: Information about the types of GUI widgets, such as buttons, checkboxes, slide-bars, and pull-down menus. The roles are used to notify the user of the types of operations available for an active widget.
  - *State of widgets*: Information about the status of each GUI widget. Each widget has a predefined set of available "states," such as checked or unchecked for checkboxes and open or closed for nodes in tree views.

Screen readers can obtain this information through the supported API channel. One of the problems with this approach is that each application should behave properly to expose the information. Each application vendor is responsible for appropriately implementing the API capabilities in their applications, but they are not accessibility professionals. Screen reader vendors have expertise, but they cannot change the nonvisual interfaces implemented by application vendors. Also, it is hard to improve nonvisual usability independently of the application release cycle when using this accessibility API approach. That is why the off-screen-model approach is still widely used by screen readers.

Another problem is the capabilities of APIs. For example, the current Flash player (version 9) supports the MSAA to make Flash content accessible. However, MSAA does not have enough coverage of roles and states to present highly visual Flash UIs, such as the "accordion widget," which did not exist when MSAA was developed. This is why IAccessible2 was defined by extending MSAA's coverage, but is not widely used at this time.

The issue of bridges is another problem area. In modern operating systems, various GUI libraries and virtual machines (VMs) are running inside each application. Java applications are running on Java VM software, .Net applications are running on top of a .Net virtual machine called CLI (common language infrastructure), and Flash content is running on top of the Flash Virtual Machine. The problem is that each GUI library and each virtual machine has its own accessibility API, and does not use the operating system's API. For example, Java (with the Swing GUI library) on Windows only provides Java Accessibility API functions, and therefore the API should be dynamically translated into MSAA to make Java applications accessible. The translation component is called a bridge. In the case of Java-MSAA, the bridge is called the Java Access Bridge, and each user needs to install this bridge to access Java applications with the Windows screen readers. A bridge often slows down screen-reading speed, since it translates much information in real-time. In addition, API translation sometimes fails because of the gap between API functions. The virtual machine approach is the current trend of computing, so it is expected that this problem will loom larger in the future.

The internal model approach is another method used to obtain information. The internal model is an abstract data structure underlying the graphical representation. This is useful to create a semantically meaningful nonvisual interface, since the internal model is a structured and semantically rich internal representation of the information on the screen. This approach is widely used for reading content, such as web pages and word processor documents. Usually, the internal models are not visible outside of the applications that create them, but some applications provide access to internal models to allow application developers to build higher-level applications. One example is Internet Explorer (IE). IE provides access to the internal content model, the HTML document object model (DOM), to allow custom applications (e.g., Visual Basic applications) to use the IE components as general web containers as a part of their UIs. That is how screen readers can obtain live HTML DOM, and provide nonvisual access based on it. Before this method became available, each screen reader could only track the web browsers by using the off-screen model. It was impossible to provide usable interfaces based on tag structures as described in Section 28.4.

Modern screen readers use these three methods (off-screen modeling, using the accessibility API, and internal modeling) simultaneously to provide maximum access to visual information. Because of progress in the visual techniques of GUIs, the off-screen model approach is becoming less and less successful in obtaining meaningful information. Therefore, it is expected that future screen readers will rely only on the internal models and the accessibility APIs.

In Step 3, a screen reader generates readable text to be sent to a speech engine by assembling the readable text from Step 2. The readable text is the key to defining the usability of the screen reader, so each screen reader tries to optimize the readable text to be competitive with other screen readers. Most screen reader vendors implement reading logic in their core component code, and this means they cannot be customized by users, and it is also difficult for the vendors themselves to improve the readable text. Notwithstanding, a few high-end screen readers provide highly customizable functions by using the scripting capability.

Scripts are small programs used to provide an optimized UI for users. This is the most powerful approach to precisely control readable reading text in accordance with the users' intentions. Figure 28.2 shows an excerpt from a script file for Microsoft Word as a simple example of script in JAWS (Lazzaro, 2003). This script allows users to redefine the shortcut key combination for justifying a paragraph to something other than the default Control + J. The first line contains the name of the script. In this case, the name is "JustifyText." The second line throws a keyboard event (Control + J) to the application. This means that a user can freely define a shortcut key combination, and then this script sends the appropriate keyboard event (Control + J) to the application. Line 3 is a pause to wait for the justification to be completed. Line 4 is an "if" statement to determine whether or not the justification command succeeded. If the command was successfully applied to the paragraph, then Line 5 provides a "formatted" message to notify the user of success.

28.3.3 Technologies for the Web

With the advent of the web, the way in which screen readers work has changed yet again. Now screen readers look at the DOM of the page to know what the browser would display. For the web, as well as for some applications, screen readers are not reading the

```
1: Script JustifyText()
2:    TypeKey (ksJustifyText) ; CtrlJ in English
3:    Pause ()
4:    if paragraphIsJustified() then
5:    SayFormattedMessageWithVoice(vctx_message,
      ot_status,msgJustified,cMsgSilent)
6:    endIf
7: EndScript
```

FIGURE 28.2 Example of JAWS' Script (simplified from a JAWS script for Microsoft Word).

screen anymore; instead, the screen reader is using the DOM to provide a speech rendering of data that makes up a web page.

It is very important to recognize that this screen-reading process is converting a two-dimensional page to a one-dimensional text string, whether spoken or displayed in Braille. This is aptly called linearization of the page. Related to HTML code, the simplest way of picturing linearization is by imagining an HTML document stripped of all its tags, leaving just the text together with the textual values of some attributes such as alt and title. The resulting text file is the linearized version of the page. Different parts of that text document will be presented to a screen reader user through a synthesized speech or Braille modality.

Another way to picture the linearization process is to read the page from left to right and from top to bottom. Tables are read left to right and top to bottom before continuing, and each cell is read completely before continuing to the next cell. Of course, there is more to this linear view than just characters. It must also include form elements and links to capture the function of the page.

28.4 Interacting with a GUI

28.4.1 General Concepts

When personal computers appeared in the market, they only had text-based (or character-based) UIs. Users could interact with computers with textual commands by using keyboards, and this interaction process could easily be verbalized for screen readers. In contrast, systems using a GUI require users to "see" the screen and to "point" at targets. This interface is very intuitive for the sighted majority, but it was barely accessible for screen reader users. In the late 1990s, the developers of screen readers faced challenges in dealing with GUIs, and sought methods and architectures to control GUIs using TTS and a keyboard instead of the screen and a pointing device.

The basic concept of an alternative to vision is a keyboard-based exploration method. Sighted users move their eyes to scan a GUI and find the necessary information for their operations. Blind users move the focal position on the screen by using the keyboard, and the screen reader must read aloud information about the cursor position.

The focus is an operating system property, provided the application is well-designed. The focus indicates the component of the GUI currently selected. For example, text entered at the keyboard or pasted from the clipboard is sent to the component that currently has the focus. When the user moves by using the Tab key, the focus moves among interactive components, also called controls. More precisely, the focus only moves among the interactive interface elements, such as edit fields, buttons, combo-boxes, checkboxes, and so on. Not all of the content shown on the screen is interactive, and noninteractive content does not receive the focus. Content that is only displayed (without interaction) is only for reading (and no action from users is required or expected). Therefore, when moving with the Tab key, the focus skips directly to interactive elements, avoiding the noninteractive parts (such as text). As a result, a blind user may receive only a partial perception of the UI.

For frequently used commands, applications usually provide shortcut keys. A shortcut key is a combination of keys to invoke a command without moving the cursor to the specific GUI component assigned to invoke that command. For Microsoft Windows, lists of keyboard shortcuts are available in places such as the Windows help pages and on the Microsoft web site (http://support.microsoft.com/kb/126449). Shortcut keys can drastically reduce the required time for operations, but users need to learn how the keyboard shortcuts work and memorize the frequently used ones to use them effectively with GUIs.

If all applications were developed using only the Windows standard controls and also provided shortcut keys, then the focus movement and shortcut keys would provide a minimal level of access to applications. However, because the Windows standard controls are not visually rich enough and focus on usability for sighted users, modern applications and popular applications are often designed to be visually usable with their own nonstandard UI components. This trend requires screen readers to be continually updated to access certain applications.

To make nonstandard visual interfaces accessible, various techniques are used. Additional exploratory methodologies have been developed to allow blind users to get information about the current window. Screen readers usually offer special commands to freely explore the content visualized on the screen. For example, JAWS has two cursor modes: JAWS cursor and invisible cursor. The JAWS cursor mode corresponds to the system mouse cursor. The invisible cursor mode is somewhat similar to the JAWS cursor, but is not directly linked to the mouse cursor. These cursors are used to read static text to which the system focus does not have access, or to access other parts of windows where the system mouse cursor cannot be moved. Hence, besides moving the mouse cursor the JAWS cursor is also used to provide a quick overview of the information currently available on the screen. Usually that cursor is limited to the current window, the main application, or dialogue window that has the focus, to avoid confusion due to mixed partial contents.

These exploration operations are extremely tedious and time consuming, so each screen reader provides functions to create scripts to automate them. Scripts are powerful tools for constructing appropriate readable text from the available information in applications (see Section 28.5). These scripts can be assigned not only to standard operations, but also to special shortcut keys. Screen reader vendors continue to update these scripts for major applications, and to package them with their screen reader products.

Users need to learn how the keyboard shortcuts work and they also need to memorize at least the frequently used ones to effectively work with GUIs. This is the most significant difference between the sighted and the blind in using GUIs. This difference is extremely critical, especially for beginners. GUIs were originally developed for sighted users to help them use the OS intuitively while minimizing the learning needed to start using a computer. These helpful visual effects in the GUIs cannot be conveyed easily to blind users. Therefore, they need to learn how to operate the GUIs by spending more time when they first start using the computer, while sighted beginners can generally start using it very quickly. The learning curves for sighted and blind users are completely different.

28.4.2 Example 1: Basic Operations

A navigation method for basic Windows operations while using a screen reader is outlined in the following. Note: This is not the only method for nonvisual operations with a screen reader.

The Start menu can be opened with Ctrl+Esc or by pressing the Windows key. Screen reader users often register the frequently used applications on the Start menu so they can quickly find the application to be opened. After the Start menu is opened, the Up and Down cursor keys can be used for moving between items. To make this faster, alphabetic shortcut keys can be used. In Figure 28.3, starting on the third-level menu, if M is pressed once, the focus moves to Microsoft Office Access, and the screen reader will say "Microsoft Office Access 2003." With the next press of M, it moves to Microsoft Office Excel. The screen reader reads the text information in focus, synchronized with the keyboard movements. When the Enter key is pressed on Microsoft Word, that program is opened.

On the initial screen of Word, the focus starts in the editing area, and the screen reader can read the focal information along with the keyboard movements. The Alt key is used to move to the menu, and Esc is used to go back to the editing area from the menu area. If the focus is deeper in the menus, for example in a second-level menu below the top level, the Esc key must be pressed twice to go back to the editing area. The Esc key will go back to a previous level until the editing area is reached. The cursor keys also move the focus item in these menus.

Dialogue box navigation is different from the cursor key operations in menus. It basically uses the Tab key. Figure 28.4 is the dialogue to open a file in Microsoft Word. The initial position is in the File name field. To move to the next field, the Tab is used. To go back to the previous field, Shift+Tab is used. In a dialogue box there can be various types of controls, such as another dialogue box, a list box, a combo box, an edit box, buttons, or trees. The cursor Up and Down keys are used to move between items in a combo box or list box. Following all keyboard movements, the screen reader outputs the information at the focal position. A dialogue box can be closed by pressing Enter or an OK button (retaining any selections) or Esc (ignoring any selections).

The basic operations can mainly be performed by using the cursor keys, Tab key, Enter, Esc, and Alt. For more advanced and effective operations, more keyboard shortcuts should be used and for more effective and even faster operations, screen reader commands can be combined.

28.4.3 Example 2: Using an E-Mail Program

This section discusses the example of a common e-mail application, Eudora, to better illustrate how a blind user interacts with a GUI.

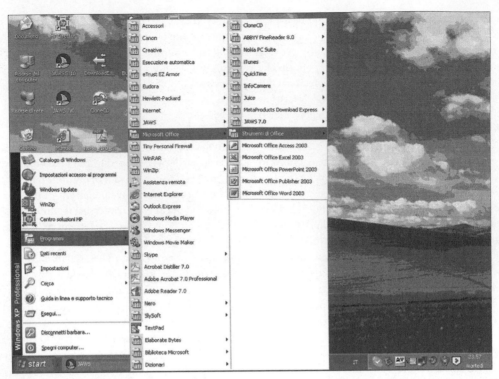

FIGURE 28.3   Start menu: the Microsoft Office 2003 menu.

The next figure shows a composition window where an outgoing message is being edited. The screenshot shows the Eudora window where various elements and components are opened and displayed. First, the figure shows the main window of the Eudora application. That window contains the standard components of a window, such as the application title, the menu bar, a toolbar, and so on. In the main application window a child window (a message composition window) is opened and rendered. Since the subwindow is smaller than its parent window, some components of the main window are still visible. A tree view of the available mailboxes is open and located on the left behind the child window. In addition, the subwindow for composing mail itself contains other information and control elements: all of the message components, such as headers, message body, and a set of icons related to possible actions on the message (such as a priority-setting combo box, a return receipt option, and so on).

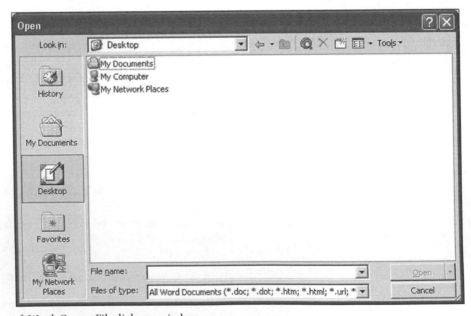

FIGURE 28.4   Microsoft Word: Open a File dialogue window.

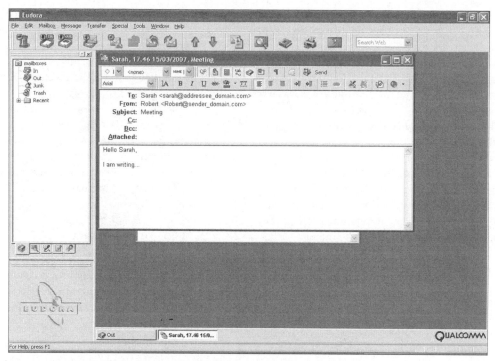

FIGURE 28.5 Eudora e-mail application with a composition window open to edit a new outgoing message.

Several of the elements are standard control components (i.e., elements that can receive the focus), while others are just small clickable icons. As a result of using these windows, the UI is quite complex and information rich. In such cases, the interaction with the UI will not be intuitive or simple to explore without vision.

Let us see how a blind user interacts with and explores such a complicated window. First, when a child window (i.e., a subwindow) opens, if the user moves using the Tab key, the focus shifts among the edit fields: "To:", "Subject", "CC:", "BCC:", and message body. Basically, this means that the Tab key is now moving only within the child window. To move the focus out of the child window, other commands have to be used. For instance, to activate the main menu bar the Alt key can be pressed, and then the user can move among the menus and submenus.

Nevertheless, in such a situation, the graphical and more structured elements, such as the mailboxes shown on the left, or the graphical toolbar in the child window, cannot be reached by the system focus. There is no nonvisual way for those elements to get the system focus, because those functions have been designed solely for mouse interactions without keyboard navigation.

Therefore, to allow a blind user to interact with and explore those components, a different modality is needed. Usually, such an interaction modality is offered by the screen reader commands. For example, to explore those UI areas that cannot receive the focus, an "exploratory modality" is available. For example, the screen reader JAWS offers a mouse pointer simulation, which allows the user to freely move among all graphical and textual components. Thus, by using the JAWS cursor or the invisible cursor, the blind user can hover the cursor over the toolbar icons or the mailboxes. Because the JAWS pointer emulation is directly

linked to the real mouse pointer, by clicking with the left or right mouse buttons the user can then interact with these graphical elements. To allow the screen reader to recognize the graphical elements (e.g., icons), all of these images must be labeled with some text. This can be done with a special automatic function of the screen reader, which can use the tool tip values as possible labels. If this is not possible, because the application developer has not defined any tool tip value for the graphical components, then the user can manually label the graphical elements (e.g., Ctrl+Insert+G in JAWS) with short descriptions, but this may require human assistance from a sighted person.

As a specific example, when the JAWS cursor or invisible cursor passes over the menu bar, caused by moving line by line (with the Up and Down arrows), then the entire line is read. Moving word by word, each component is read separately. Therefore, when the menu bar is read using the Up arrow key, the entire line is read, so the user perceives the menu elements as one message: "file, edit, mailbox, message, transfer, ... help." The entire line is read together. In contrast, when moving through the focus, each menu item is read one at a time. To read the next item, the user must move the focus or the mouse pointer.

Let us consider another example, still using Figure 28.5. When in the exploratory mode, the JAWS cursor moves over the two lines corresponding to the addressee and sender fields, the "To:" and "From:" lines. These lines are read by the screen reader as "Trash To: Sarah <sarah@addressee_domain.com>" and "Recent From: Robert <robert@sender_domain.com." The two words *trash* and *recent* are read as the first words of the lines, because they appear on the left in the vertical list of mailboxes. Effectively, what happens is that the screen reader captures the

entire line, including what is displayed outside of the open child window, thus including the text appearing inside the main Eudora window. In brief, the JAWS screen reader is not able to understand and arrange information in a logical way when exploring line by line using the JAWS cursor. The screen reader takes all of the information displayed in a line and reads it as it is visualized. The user needs to interpret and arrange the information read by the screen reader. For this reason, this kind of exploration is more difficult, and when possible a blind person will find it easier to move using the Tab key to explore by listening to how the focus changes. Unfortunately, this is not always possible, because some applications are not designed and developed in a standard and accessible manner. This means that the developers have to apply specific rules when developing a GUI if they want the application to be accessible and easy to interact with.

In conclusion, when exploring a page through focus movement, the user is in a more functional modality; when using an exploratory mode, the user can learn how the content is placed as well as the noninteracting parts (i.e., elements that do not receive the focus). To stay orientated in a UI, both modalities are useful and necessary. At first, when the interface is not well understood, the user may explore using the JAWS cursor or the invisible cursor. Later, following the changing focus (while using the Tab key) will become more functional and effective. The exploratory method can also be used to obtain information on some of the nonfunctional informational elements (such as the status bar).

28.5 Interacting with a Web Interface

28.5.1 General Concepts

All screen readers have some kind of specialized web mode for reading web pages, because there is usually no focus point on a web page, such as a text cursor, selector, or focus rectangle. For some screen readers such as JAWS or Hal, the web mode is called Virtual Focus, because the screen reader must create its own focus point. Virtual Focus deals with the page content as a document. In practice, the content is provided to the user as a single page. This means that the content needs to be linearized by the screen reader (see Section 28.3.3).

When a blind user navigates in a web page, basically there are three main modalities to explore the content:

1. *Using the arrow keys*: This option allows users to read sequentially all of the content, line by line on the screen. The screen reader announces the text, form fields, links, buttons, heading levels, and so on, along with the content.
2. *Using the Tab key*: This way, the focus is moved only among the control elements. Therefore, the user moves on links, edit fields, buttons, checkboxes, and so on. In this mode, the blind person is only able to read the text of links, labels of buttons, and so on, but the text and information written elsewhere is not read. It is clear that with this mode the user will not have a global view of the page content.

Therefore, it is very important that the links or control elements have labels that are self-explanatory.
3. *Using special commands*: This allows quickly reaching specific parts or components.

In practice, blind users rarely listen to a page in full. Often, they quickly navigate to the controls of the page (i.e., the interaction elements). More experienced users navigate within the page content by using special commands offered by the screen reader. For example, to navigate inside tables' content, special JAWS commands are available, such as the Ctrl+Alt+Arrow keys to move among the cells or the "l" command to jump to the next list.

28.5.2 Web Accessibility

In recent years, the use of web sites has been expanding, and the number of users accessing them is steadily increasing. For this reason, it is important that the information be easily reachable by all, including people with disabilities. A web site can be said to be accessible if it can be used by everyone, including people with disabilities. In addition to people who explore web pages by using screen readers (with voice synthesizers or Braille displays) or other assistive technologies, people using low-bandwidth technologies such as cellular phones, black and white screens, talking browsers used with telephone, and so on, should also be included. For this reason, accessibility is complicated by the fact that a web site is not a published piece of work so much as a living document that can be interpreted in different ways by different browsers and on different platforms.

In the context of web site design, accessibility is a measure of how easy it is to access, read, and understand the content of a web site. Web applications have significant differences compared to classical GUIs. Basically, the structure of the web site (e.g., number of pages, links between pages, etc.) is directly related to and may be affected by the information that is available on the web site (amount, type, etc.), while the structure of GUIs is usually static. Therefore, the user interactions and the screen reader interpretations differ from stand-alone applications.

To make a web site more accessible, the web designer should follow some simple criteria and guidelines that do not constrain the graphical features (such as images and icons). To improve accessibility it would be better to provide textually equivalent information with each multimedia element. There are various organizations that issue standard guidelines. Some are international (such as the W3C/WAI), and others develop national standards (such as Section 508[12] in the United States). Often, collections of guidelines share the same body of knowledge, but are slightly different, so designers have to consider the contexts.

The Web Accessibility Initiative Interest Group[13] (WAI-IG) of the World Wide Web Consortium investigates problems in

---

[12] Section 508: The Road to Accessibility (http://www.section508.gov).
[13] The recommendations developed by the WAI-IG and the new work of the W3C group are available at http://www.w3.org/TR/WCAG10 (5 May 1999) and http://www.w3.org/TR/2004/WD-WCAG20-20040730, respectively.

accessing web resources and produces guidelines for web content, authoring tools, and user agent accessibility. A set of 14 main guidelines, called the Web Content Accessibility Guidelines 1.0 (WCAG 1.0, 1999), was created by this group, which is now working on version 2.0 of the recommendations (WCAG 2.0, 2007). In the United States other accessibility guidelines partially overlapping WCAG 1.0 have been defined by the federal government[14] and other guidelines are specified by several national or international organizations.

Generally speaking, making a web site more accessible and usable requires considerable effort by developers in handling the web page code and many specific design guidelines. The developers have to decide which principles to use for the specific cases, how to apply them, and when. Evaluating web pages also requires a lot of effort. From this perspective, several tools have been proposed in the literature to support designers and evaluators in applying the guidelines to web sites. Web accessibility is discussed in detail in Chapter 10, "Accessing the Web," of this handbook.

28.5.3 Interfaces of Voice Web Browsers

A screen reader deals with web page content in a manner very different from visual rendering. The interactions require a certain expertise in advanced screen reader and browser commands, and users must often make considerable effort to orient themselves within the page content. However, user interactions are greatly improved and simplified if the content of web interfaces is appropriately arranged and structured according to a logical flow.

Screen readers are capable of various navigation commands to provide full access to accessible web content. The framework of web accessibility helps support nonvisual access by providing three main types of information about web pages. The first type is text equivalents for visual objects, such as alternative texts for images. The second type is landmarks to support nonvisual navigation, such as heading tags. The third type is semantic structures, such as table headers for a table. Any screen reader is capable of verbalizing such information, thus allowing users to navigate the page by using landmarks and semantic structures.

Concerning web pages, the new generation of screen readers interprets the HTML code as it is structured, thus considering different tags, while the first generation could not, and generated a continuous line of information without any kind of separation. In spite of the possibilities for recognizing different tags in a page (such as tables, headings, lists, etc.) certain problems are still common.

The JAWS screen reader takes the page's tag structure and serializes the content (text, links, edit fields, buttons, cells, and so on). Also, frames or blocks <div> are lined up, without taking into account any specific positions assigned by cascading style sheet (CSS) properties. The web page is dealt with by the screen reader as a document with a virtual cursor that allows moving line by line or word by word using the arrows keys, or link by link using the Tab key.

Screen readers such as JAWS interpret the code as it was written, and line up the page content in the form of a single column. Thus, the order in which the <div> blocks, the frames, and the paragraphs are written is very important. The content flow interpreted by the screen reader can be very different from its rendering on the screen. A left-side menu, for example, may be read after the content on the right if its code is written afterward so as to correspond to the content on the right side. Its rendering on the left can be controlled by CSS properties (e.g., flow: left). Also, tables must be linearized. When a table is linearized, the contents of the cells become a series of paragraphs in sequence. The cells should make sense when read in row order and should include structural elements (creating paragraphs, headings, lists, etc.), so that the page makes sense after linearization.

Usually a web page is explored by using the Tab key (link by link), by using arrow keys (line by line or word by word), or through special commands (to jump to certain elements). However, it is important to remember that a blind user typically prefers to handle the page link by link (with the Tab key) or using arrow keys to read sequentially as in a document. Any special commands are mostly used by the more experienced users to move quickly around the pages. Furthermore, many special screen reader commands operate well only if the developer has used specific tags or attributes, or if appropriate criteria have been followed. Hence, it is important to favor navigation via the keyboard by assigning a scale of importance to the links, by applying shortcuts to main elements, by using specific tags such as <H1>, and so on.

Also, for web pages, what is offered in a visual layout differs from the layout provided for listening. Often when developers design a web page, they provide some useful information by means of visual features, such as position, color, separating blank spaces, formatting features, and so on. For example, some secondary information is put on the side, so that users can recognize it immediately. It is important to provide the same meaning to blind users by other means (e.g., using a table, a heading, a hidden label, etc.).

In conclusion, the screen reader basically announces every word on a page, line by line, word by word, link by link, in a sequential way. Due to the drawbacks described in Section 28.2, reading a web page can be somewhat laborious. By following specific design criteria, as well as by applying appropriate principles, web navigation can be made more efficient and satisfactory.

28.5.4 An Example of Web Interaction

To understand how appropriate tags can positively affect the web interaction, this chapter can be used as an example, assuming that this chapter were available in a poorly structured HTML format (e.g., with no heading styles). To find the main sections of the document, such as 28.1 and 28.2, a blind user could try to look for numbered sections by searching for ".1," ".2," ".3"; if the paragraphs were numbered, the reader could get an idea of the document structure by searching for progressive paragraph numbers. Although this approach would allow users to find the

---

[14] Further information is available at http://www.section508.gov.

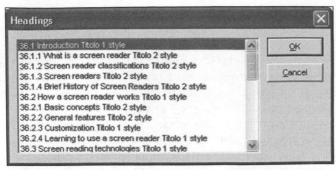

FIGURE 28.6 Chapter structure generated through a JAWS command.

main sections of the page, it is not at all suitable. However, if the sections and pages were not even numbered, then all of the drawbacks described previously would have a negative impact on both web page navigation and reading the web content. To get an overview of the page, the user would have to read it sequentially to get information about its structure. Thus, even if a page is available in an HTML format that seems to be accessible, it may still be difficult to read efficiently. When reading a document sequentially (line by line), the user may only develop an idea of the document structure after reading the entire thing, but that would require a considerable effort.

In contrast, the inclusion of appropriate features, such as heading styles, when the document is being prepared would allow the reader to get an overview of the content. Figure 28.6 shows the structure of this chapter. This list of paragraphs and subparagraphs was generated automatically with a JAWS command (Insert+F6). The user can understand the chapter structure by reading the list of paragraphs with the arrow keys. By pressing the Enter key on a title the focus skips to the corresponding content, so the navigation through the chapter is also easier and faster.

28.6 Designing a UI

28.6.1 Introduction to UIs

Usually the analysis of digital obstacles for the disabled only addresses accessibility, although usability is also fundamental for simplifying both navigation and interaction for users using assistive devices or users with disabilities. The difficulties in providing universal access can be addressed through the application of the principles of accessibility and usability. Usability is a multidimensional concept, since it can refer to several aspects whose importance depends on the application domain. Although accessibility and usability are closely related, accessibility is aimed more specifically at making a web site (or application) available to a wider user population, while the goal of usability is to make users' experiences using the web site more efficient and satisfying.

Therefore, for a blind person it is very important that an application be not only accessible, but also usable. An assistive technology, such as a screen reader, gets information from the UI of the application or web page. Therefore, it is necessary that a UI is first of all accessible to allow the screen reader to obtain the information needed. In addition, to facilitate the interactions of a blind user, it is also important that the interface is usable. Thus, an appropriate design for the UI is crucial for improving the accessibility and the usability of a system.

To make graphical interfaces and web sites accessible and usable, various guidelines and criteria have been proposed. However, when designing and developing an application, various aspects should be taken into account. First, designers should be aware which UI features are affected by accessibility and usability issues. In addition, developers should be aware of how assistive technologies (such as a screen reader) work and how they interpret the UI.

28.6.2 Auditory Perception versus Visual Layout

For users interacting by means of assistive technologies, the UI layout and structure are crucial. Traveling through the web or operating a stand-alone application is difficult for users with visual impairments since the web pages or program windows are designed for visual interactions (Goble et al., 2000). In contrast to sighted users, screen reader users cannot see the implicit structural and navigational knowledge encoded within the visual presentation of web pages (Yesilada et al., 2004). Therefore, when navigating by screen reader, the user may perceive windows or page content in a different manner from what is actually shown on the screen. For example, when navigating a web page, the content perceived through a screen reader is different from its visual structure. The differences can cause frustration for blind people. Page layout may result in confusing screen reader feedback, poorly designed or unlabeled forms can make navigation very difficult, and the lack of alternative text for pictures can be mystifying (Lazar et al., 2007). To solve or reduce these inconveniences, developers should be aware of how the content is perceived, so that they can design the UI appropriately.

This section discusses the main differences between visual rendering and aural perception. More details on how a blind user perceives web content and the information displayed on the screen were presented in Sections 28.4 and 28.5.

To understand the differences between visual rendering (basically relying on mouse interactions) and aural perception (usually based on keyboard interactions), it is fundamental to know how the screen reader deals with web content or displayed information. Additionally, the way a blind user works with the application affects the perception of the UI.

When designing a UI, developers pay attention to structuring the content in such a way that the users easily identify the main parts and functions. Additionally, UIs are becoming increasingly complex and rich in information to make the content more attractive and full of functions. If no appropriate techniques are used to position and highlight the main functions and content, it becomes difficult to accomplish tasks. Therefore, features such as position, form, font and color, component arrangement, and styles (bold, underlining, etc.) are used to visualize information. If such elements are appropriately used, the users are able to easily identify the crucial text and desired functions.

The focus of design is on how to place the content and components in the UI. Design principles are usually based on a visual rendering, so that the UI is pleasant to look at and work with. Frequently there is little consideration of the logical flow among elements and content. However, the structure of a UI can heavily affect interactions by users who are constrained to use assistive technologies, such as a screen reader.

In software applications, generally speaking the screen reader will follow the system focus. That means the focal point skips from component to component in a limited area. When jumping element by element (usually with the Tab key) a blind user perceives each single component without context. If the logical flow is not appropriate, the user may have to skip many useless elements before finding the needed components. At the same time, when working in this way, the user has to gradually build a mental map to get an overview of the overall UI. Such exploration of a UI may require a long time. In this context, specific visual features used to focus attention, such as position, styles, and so on, are not effective for blind users. For easier identification of the main components by users who are interacting through a keyboard, alternative strategies and features should be used. For example, the functions in toolbars shown with intuitive icons should be replicated in menus with textual labels and shortcuts.

For web interfaces the screen reader rearranges the content to obtain a logical flow. Visual structures such as frames or columns are lost. This means that the content should be logically structured so that blind users can use other techniques to grasp the structures. For example, when a result is shown at a specific point within a page, specific visual properties are often used to immediately emphasize that result. Other alternative modalities can be used to provide the same emphasis. Elements like hidden labels or shortcuts can improve web navigation and make the structure of content more obvious.

28.6.3 Aspects Affecting Perception

As already described, a screen reader is an application that attempts to identify and interpret what is being displayed on the screen. Then the content from the screen can be presented to a blind user as speech through a voice synthesizer or as a Braille display. People with limited vision often use screen magnifiers, which face some of the same problems addressed by screen readers. Some of the problems and difficulties encountered when interacting through a screen reader can also occur when interacting with a screen magnifier.

To simplify user interactions, the developers who design a product should be aware of the main needs of users, as well as of their interaction modalities with the system (Leporini and Paternò, 2004). Knowledge of how screen readers work can make the design process easier. In Sections 28.4 and 28.5, the way a blind user works with a GUI or web page was described

with interaction examples. In this section, the main difficulties encountered by a visually impaired user are summarized. These problems are mainly due to differences between visual perception and aural or haptic perceptions.

The main problems for a blind person interacting through a screen reader are:

- *Lack of context*: Using a screen reader (or magnifier) the user accesses only small portions of the content and may lose the overall context.
- *Information overload:* The unchanging, low-information portions of the site (menus, frames with banners, etc.) may overload the reading memory because the user has to listen to all of the items frequently, thus slowing down navigation. Also, in software application UIs there are numerous icons, symbols, toolbars, child windows, and so on. The arrangement of these elements also affects the complexity of the UI.
- *Excessive sequencing in reading the information*: The commands for navigating and reading can force users to follow the page content sequentially, causing frustration.
- *Keyboard navigation*: Blind users cannot use the mouse (to point, scroll, select) when moving around the page, but rely on keyboard commands, such as the Tab key and arrow keys. This makes navigation within a page slow.
- *Screen reader interpretation*: The screen reader deals with web page content in a manner that differs greatly from visual rendering. CSS properties permit arranging the content (e.g., text blocks, tables, columns, etc.) in several ways independently of its position in the HTML code. In a different way, a screen reader reads the content in the same order as it appears in the source code file by ignoring the positioning CSS properties. The behavior when highlighting important UI elements (e.g., through color or font size) is similar. Although the screen reader is able to detect the rendered properties, the results are not equivalent to the visual results.

In conclusion, there are many differences between visual and aural perceptions exposed by screen readers. Designers should apply specific guidelines and appropriate designs based on the interface's purposes and the assistive technologies required. They should concentrate on good designs to improve the interface usability for both visual and aural navigation.

28.6.4 GUI Design

When designing a GUI interface, the developer should keep in mind how a blind user usually works with the interface. In Section 28.3, interaction with a GUI was described. Starting from that overview, when designing a GUI, important aspects to consider are:

- *Focus*: Which component should have the default focus, and how should the focus move between components? These are difficult but important questions in UI design.

Giving the focus to a wrong object means that the user has to waste time moving the focus. Giving the focus to the appropriate object at each point can significantly enhance the users' experiences. When defining the logical flow assigned to interactive components (the Tab visiting order) the designer should think about the most appropriate order. The Tab order of the controls in a UI determines the sequence in which the focus will change when the user uses the Tab key. The Tab order is usually right to left within each row of controls, thus assigning a logical order to interactive elements. When designing a UI, the developer should think about the most logical orders for the tasks that the users wish to accomplish.

- *Selection of functions and commands*: A nonvisually impaired user typically performs a simple function by clicking on a graphical icon. A graphical toolbar composed of several intuitive icons is probably the selection modality preferred by the sighted user. Since selecting modalities makes it easier to select functions, UIs typically offer several toolbars with various graphical icons for related functions. Though selecting a function with a click is generally easy, reaching a toolbar icon is not simple for a blind person, even if the screen reader simulates a mouse pointer. Sometimes the screen reader is not even able to correctly recognize the graphical icons, calling for an alternative modality. A first solution is to make sure that explicit menu items for all functions are available in the UI. Also, shortcuts should be available for important functions. Then users who interact through the keyboard can select functions or commands by exploring the menus and submenus or by learning a key combination (such as Ctrl+s to save a file).

- *Additional multimedia features*: Visual representations can communicate certain kinds of information much more rapidly and effectively than nonvisual methods. For this reason, the interface design should try to maintain a comparable degree of expressiveness in both the visual and aural versions. For example, to inform the user about a successful or failed event, in addition to a visual icon or textual message, a short sound can be played. If different sounds are used consistently for various kinds of events, a blind user can easily understand the status of a process. In practice, it is suggested to communicate information and event status through various modalities: textual, graphical, and aural.

28.6.5 Web UI Design

The developer should be aware of how the screen reader handles the web page layout, and how blind or visually impaired users perceive page content and interact with the interface. To make web pages and web sites accessible, developers have to apply the relevant accessibility guidelines. For the web, various accessibility guidelines and recommendations are available.

The most important are the WCAG 1.0 and 2.0 (draft) already introduced in Section 28.5.1. Usability guidelines for the web aimed at improving navigation for people with vision impairments have been proposed (Theofanos and Redish, 2003; Leporini, 2004; Leporini and Paternó, 2008). Such usability guidelines are based on blind user interactions with the web through a screen reader. By observing blind people interacting with web pages, guidelines have been identified and tested to evaluate if those users really gain benefits from the proposed criteria.

To make web sites and services more accessible and usable, some additional aspects should be considered. To this end, the main issues in UI design to take into account for web navigation are:

- *Page content structure*: The page content should be structured in sections and subsections by using heading levels. In some cases, hidden labels, which have no visual impact but which are used by the screen reader, can help.

- *Quick navigation*: Importance should be assigned (using the Tab index attribute) so users can reach the most important elements quickly. Lower values should be assigned to secondary links. Also, shortcuts should be associated with the main elements (e.g., navigation bar, login edit fields, etc.) and links to the pages for results.

- *Additional multimedia features*: What is offered as a visual layout differs from the presentation for hearing. When developers design a web page, they provide useful information using visual features, such as position, color, separating blank spaces, formatting features, and similar features. For example, secondary information may appear at the side so that users can recognize the relationship immediately. It is important to provide the same message for blind users by other means (e.g., using a table, a heading, or a hidden label, etc.).

28.7 Future Directions

Screen readers have been evolving to provide better UIs for visually impaired users. There have been great successes in allowing blind users to access and control visual interfaces by using nonvisual methods. Basic access is here, and various efforts are continuing to maintain access, even though visual interfaces are becoming even more complicated and rich. However, usability is still limited because of the large number of necessary key combinations and the complexity of the operations.

The primary future challenge for screen readers is obviously the improvement of usability. For sighted users, GUIs can be operated by applying the simple principle of "see and click." This simplicity drastically expanded the usage of computers from engineers and scientists to almost anyone in society. At the same time, the nonvisual access provided by screen readers is too complex and there are no general principles for nonvisual operations. It is clear that next-generation screen readers should provide simpler and more usable interfaces for a wider variety of visually impaired people.

New possibilities for improving blind users' interaction are already emerging. The framework of web accessibility is one of the avenues leading to these advances. In this framework, voice browsers obtain information from the active HTML structure, and they will provide logical navigation interfaces for users. The interface will be totally different from the "see and click" world, but optimized for nonvisual use with a limited number of key combinations. Simple UIs will contribute to shortening the learning curves for new applications. These advances will contribute to improve education and job opportunities for visually impaired people. We hope that screen readers will evolve in these directions, and finally implement simple principles for nonvisual access in the near future.

References

Asakawa, C. (1998). User interface of a home page reader, in the *Proceedings of ASSETS '98*, pp. 149–156. New York: ACM Press.

Barnicle, K. (2000). Usability testing with screen reading technology in a windows environment, in the *Proceedings of the 2000 Conference on Universal Usability (CUU-00)*, 22–23 November 2000, Arlington, VA, pp. 102–109. New York: ACM Press.

Goble, C., Harper, S., and Stevens, R. (2000). The travails of visually impaired web travellers, in the *Proceedings of the Eleventh ACM on Hypertext and Hypermedia*, 30 May–3 June 2000, San Antonio, TX, pp. 1–10.

Gonzalez, A. and Guarino Reid, L. (2005). Platform-independent accessibility API: Accessible document object model, in the *Proceedings of the 2005 International Cross-Disciplinary Workshop on Web Accessibility (W4A '05)*, 10–14 May 2005, Chiba, Japan, pp. 63–71. New York: ACM Press.

Keates, S. (2006). SIGACCESS member profile: Jim Thatcher. *ACM SIGACCESS Accessible Computing* 85: 56–56. http://portal.acm.org/citation.cfm?id=1166118.1166132&coll-GUIDE&dl-GUIDE&CFID=17289008&CFTOKEN=71669297.

Kurniawan, S. H., Sutcliffe, A. G., Blenkhorn, P. L., and Shin, J. (2003). Investigating the usability of a screen reader and mental models of blind users in the Windows environment. *International Journal of Rehabilitation Research* 26: 145–147.

Lazar, J., Allen, A., Kleinman, J., and Malarkey, C. (2007). What frustrates screen reader users on the web: A study of 100 blind users. *International Journal of Human-Computer Interaction* 22: 247–269.

Lazzaro, J. (2003). An introduction to JAWS scripting. *AFB AccessWorld* 4.

Leporini, B. and Paternò, F. (2004). Increasing usability when interacting through screen readers. *Universal Access in the Information Society* 3: 57–70.

Leporini, B. and Paternò, F. (2008). Applying web usability criteria for vision-impaired users: Does it really improve task performance? *International Journal of Human-Computer Interaction* 24: 17–47.

Linvill, J. G. and Bliss, J. C. (1966). A direct translation reading aid for the blind, in the *Proceedings of the Institute of Electrical and Electronics Engineers* 54: 40–51.

Paul, W. F. and Dillin, D. (1998). Low vision combined screen-reading and screen enlargement: A necessity in the workplace, in the *Proceedings of CSUN 1998*, 17–23 March 1998, Los Angeles. http://www.csun.edu/cod/conf/1998/proceedings/csun98_106.htm.

Schroeter, J. (2005). Text-to-speech synthesis, in *Electrical Engineering Handbook* (3rd ed.). http://www.research.att.com/~ttsweb/tts/papers/2005_EEHandbook/tts.pdf.

Schwerdtfeger, R. S. (1991). Making the GUI talk. *BYTE* 16: 118–128.

Thatcher, J. (1994). Screen reader/2: Access to OS/2 and the graphical user interface, in the *Proceedings of the First Annual ACM Conference on Assistive Technologies (Assets '94)*, 31 October–1 November 1994, Marina del Rey, CA, pp. 39–46. New York: ACM Press.

Thatcher, J. (2006). Assistive technology: Screen readers and browsers, in *Web Accessibility*, pp. 103–124. Berkeley, CA: Apress.

Theofanos, M. F. and Redish, J. (2003). Guidelines for accessible and usable Web sites: Observing users who work with screen readers. *Interactions* X: 38–51.

Thiessen, P. and Chen, C. (2007). Ajax live regions: ReefChat using the fire vox screen reader as a case example, in the *Proceedings of the 2007 International Cross-Disciplinary Conference on Web Accessibility (W4A '07)*, 7–8 May 2007, Banff, Canada, pp. 136–137. New York: ACM Press.

WCAG 1.0 (1999). *Web Content Accessibility Guidelines 1.0*. Web Accessibility Initiative (WAI), World Wide Web Consortium. http://www.w3.org/wai.

WCAG 2.0 (2007). *Web Content Accessibility Guidelines 2.0*. W3C Working Draft. http://www.w3.org/WAI/GL/WCAG20.

Yesilada, Y., Harper, S., Goble, C., and Stevens, R. (2004). Screen readers cannot see, in *Web Engineering*, pp. 445–458. New York: Springer-Verlag.

Zhao, H., Plaisant, C., Shneiderman, B., and Lazar, J. (2006). *A Framework for Auditory Data Exploration and Evaluation with Geo-Referenced Data Sonification*. HCIL Technical Report (HCIL-2005-28), Human-Computer Interaction Lab, University of Maryland, College Park, MD.

Virtual Mouse and Keyboards for Text Entry

Grigori Evreinov

29.1 Introduction

Textual communication has a significant impact on cognitive development, social inclusion, and integration of all people of any age. While people have equal rights, they have different skills and abilities. At any moment an accident or disease can lead to a number of disorders and even to the loss of perceptual, motor, or cognitive ability. Being bereft of any way to communicate their needs, people can be considered "communication disabled." In such a case, it is difficult to provide appropriate assistance (Boissiere, 2003).

Fortunately, alternative human-computer interaction techniques allow fitting almost any human ability to support and even increase bandwidth for communicating alphanumeric data with computer support in comparison with traditional writing methods (Hashimoto et al., 1996; Paradiso et al., 2000; Lukaszewicz, 2003; Lyons et al., 2004). Both the nondisabled and people with speech disorders often choose simple gestures during a dialogue to confirm or reject any statement of another person or to express some intention (Toyama, 1998; Kawato and Ohya, 2000; Kjeldsen, 2001; Hanheide et al., 2005). Gestures can be produced by hand, head, or through facial expression. In the case of a disabled person, the ability of caregivers to make a right interpretation of gestures depends on many factors such as the situation, message content, and the previous communication experience with the person. Automatic gesture recognition is a promising technique but often operates with a limited vocabulary. Different methods vary in accuracy and reliability (see Chapter 34 of this handbook).

Speech input (see Chapter 30 of this handbook) yet is limited in robustness and flexibility and is inappropriate for certain tasks and environments. When a disabled person has to communicate private information he cannot rely on speech input or other people. Banking terminals (such as an ATM, see Chapter 48 of this handbook) are still not equipped with biometric data readers, and the alphanumeric codes provide universal and ubiquitous authorized access to confidential information. Therefore, text entry skills and assistive tools are considered a great challenge, as well as a necessity, in providing equal access and an appropriate communication level for all.

Key entry techniques were initially designed for able-bodied persons having average fingers dexterity and normal vision. However, for a very broad range of motor disorders any key layouts can be completely inaccessible. Additionally, the lack of typing skills is usually diminished due to visual feedback. Because physical keys are accessible for touch without activation of their function, blind people can successfully utilize discrete and assembled keys such as keypads and keyboards.

Whether the habitual layout of physical keys being utilized in other text entry techniques can benefit blind users, relying on their previous experience, is still an open question. Congenitally blind people who learned Braille at an early age and still regularly use this technique as adults (Braille display and Braille keyboard) do not desire to acquire typing skills on a regular keyboard (Arato et al., 2004; Juhasz et al., 2006). In contrast, for elderly people or people who acquired blindness as a result of disease or an accident, and who were computer users before that, it might be difficult to learn Braille because of low tactile sensitivity or due to other problems. These people can utilize a regular computer keyboard (Keyboard Layouts, 2007) augmented with a screen reader (speech synthesis), nonspeech sounds (earcons), and haptic feedback cues (tactons) when appropriate. However, hearing loss, noisy environments, or silent conditions might also restrict the use of screen readers.

With the advent of touch screens discrete and continuous gestures became a popular input technique (Goldberg, 1997; Nesbat, 2003; Zhai and Kristensson, 2003). A vast body of empirical studies on alternative text entry systems have been carried out to propose reliable methods that would be easy to learn and universally accessible (Hansen et al., 2003b; Harbush and Kühn, 2003; Miniotas et al., 2003; Beck et al., 2004; Evreinov and Raisamo, 2004). In fact, touch screen applications and onscreen keyboards are not accessible for the blind as-is. Onscreen keyboards are often considered to exclude visually impaired users who cannot benefit directly from the techniques based on the visual feedback. The same situation is observed with virtual mouse and video-based input techniques. Still, this is not the full truth. Even touch input devices and video-based input techniques can be accessible without visual feedback (Evreinov and Raisamo, 2002b; Arato et al., 2004; Wobbrock et al., 2004; Yfantidis and Evreinov, 2004).

In many cases, developers and researchers do not make a difference between virtual, software, and onscreen keyboards. The terms are often used as synonyms to describe, as a rule, a software simulation of the physical keyboard. However, these terms have emerged to discriminate three different categories of the human-computer interaction techniques that have been designed as alternatives to a regular physical keyboard.

A virtual keyboard (VKbd) is an interaction technique for input binary codes that does not require a pressure sensor being physically presented within a touch area, while the technique can be used for inputting the text, computer commands, and simulation musical instruments (e.g., piano keyboard). In a wide sense, a virtual keyboard is a metaphor to facilitate the user in shaping the cognitive model of the interaction space and technique. Herewith, a visualization of virtual keys can be optional (see more in Section 29.3).

The onscreen keyboard (OSK) is an interaction technique for input binary codes. OSK presents a layout of software buttons simulating a function of physical keys, which are being projected onto the flat surface of the screen and can be activated with the use of any other input device (mouse, joystick, touch screen, etc.). The OSK can be used to extend the functioning of the regular keyboard to input specific commands or symbols (see Section 29.4). Herewith, visualization is a key feature of the OSKs.

Alternative methods of human-computer interaction have been developed as an integration of different techniques and algorithms to facilitate communication between people and machines and to support user performance at an acceptable level when the conventional methods have failed (Mehta, 2009). Interaction techniques should be strictly adaptive to personal cognitive and sensory-motor abilities of the user. There are many constraints in text entry as a part of human-human-computer interaction (MacKenzie et al., 2009). People want to communicate using a natural language and skills they already have, in a real time, in any conditions, as fast as possible.

Nevertheless, multifunctionality of a single key can easily be realized using an OSK (Evreinov and Raisamo, 2002a, 2004; Mehta, 2003; Seveke, 2006). The universality of the OSK as an imaging tool and an external memory aid is obvious when people with diverse needs can use alternative input techniques to produce commands or textual messages.

No layout is perfect for a virtual keyboard or software buttons of the OSK, and no communication techniques exist that can fully satisfy the requirements of users with diverse characteristics and abilities. Perceptual (touch, kinesthetic, proprioception, vision, and hearing), cognitive (attention, memory, speech perception, and decision making), and motor skills (movement coordination and behavioral stereotypes) can essentially be worsened because of a disease's progress (e.g., agnosia, aphasia, apraxia). Therefore, research has striven to design a wide spectrum of text entry systems and to increase their efficiency.

29.2 Text Entry: Universality and Diversity

Irrespective of the technique used, text entry can be considered a universal paradigm for human-computer interaction, data input, and presentation. Typing is a way to create data that encode semantic information using a discrete (binary) input of symbols and the commonly accepted array of the keys. Being visualized in a readable form, the encoded information can immediately be perceived and comprehended by the typist and other readers. Different typing systems have been designed to encode the knowledge-specific content to process the textual information, music notations, and mathematical and chemical expressions.

Being developed in Europe, the earlier prototypes of the typewriter keyboard and, later, computer control consoles, were oriented to the alphabetic writing systems. Alphabets have different length, the frequency of the letters' use, and special (national) characters, for example, diacritics and accents in Spanish, French, Danish, Turkish, and other languages. As computers spread around the globe (Yfantidis and Evreinov, 2004), information has to be accessible and manageable in other languages using also nonalphabetic writing systems such as Chinese, Korean, Japanese, Hindi, Arabic, and others. Language-specific features regarding text entry of Hebrew, Asian, and Arabic scripts have recently been reviewed and presented in MacKenzie and Tanaka-Ishii (2007, Part 3). This chapter focuses on text entry for the alphabetic writing system.

Full-size keyboards normally provide a physical space for each letter commensurable with an average fingertip size of about 1.75 cm^2. Taking into account an error of the finger positioning, physical buttons are arranged with gaps of about 3 mm around each touch surface. Thus, even a regular keyboard having 52 keys occupies a significant part of a desktop. Conventional computer keyboards usually have 83 to 101 or more keys, to enter alphanumeric characters and punctuation marks, and to make text editing as convenient as possible.

Typing should allow creating equations and performing file operations with a document. The integrated modifier keys such as Shift, Control, Alt, and Alt-Gr provide multifunctionality of the basic keyset. Thus, the degree of freedom (DOF) for a regular keyboard is extended to 1.5 (Kolsch and Turk, 2002). Text entry systems should provide efficient pointing and selection of any symbol from the alphabet set. Still, mapping of the multiple characters to a physical key may cause visual overload and ambiguity of the primary feedback, and cognitive and motor problems for novices.

Text entry is not only a sequence of clicks; it also involves a pretyping phase, the duration of which depends on the task, language features, and typing experience. Usually, the typist has to simultaneously coordinate different activities. Language and speech comprehension depend on cognitive skills and abilities such as memory, analysis, and synthesis. To imagine a word (through activation of a cognitive model), to divide it into syllables or directly into a sequence of characters, a basic knowledge of grammar is required. To encode the word, typing and grammar rules have to be appropriately integrated. The detection of the keys sequentially or by chord requires sensory-motor experience, finger-wrist (hand-eye) coordination, and knowledge of the keyboard layout. To perform key(s) selection, fine motor skills and selective attention are needed. Finally, to compare the result of typing to the cognitive model of the word using secondary feedback (visual or speech signals) requires perceptual-cognitive abilities. Thus, there are many different reasons that can hinder or delay text entry.

29.3 Virtual Keyboards

A virtual keyboard is an interaction technique for input binary codes that does not require a pressure sensor being physically presented within a touch area, while the technique can be used for input of text, computer commands, and simulation of musical instruments (e.g., piano keyboard). The sensing space can be controlled using remote sensors (electrical, optical, ultrasonic) and different tracking techniques (Paradiso et al., 2000; Matsui and Yamamoto, 2001; Kolsch and Turk, 2002). Virtual keyboards can include discrete areas for each symbol or a reduced number of the sensitive areas (SA) activated sequentially (multitap methods) or by chord (Lightglove). Virtual keys present sensitive areas of the virtual working space, and they can be activated immediately when the input device detects any changes within the SA or after a short delay (dwell time) to discriminate between intentional and accidental actions (Hansen et al., 2003a). Herewith, the visualization of the virtual keys can be implemented in different ways, but is not obligatory. A sequential activation of spatial positions forms particular motions (gestures; see Chapter 34,

"Vision-Based Hand Gesture Recognition for Human-Computer Interaction"). The possibilities of such a technique for text entry are discussed in Zhai and Kristensson (2003).

Virtual keyboards differ in how the primary feedback is presented for pointing and selection. A confirmation action for the virtual key selection can physically be separated from pointing. The particular sensor can detect any local action, for example, an electrical muscle activity or teeth chattering (Hashimoto, et al., 1996), breath pressure, eye blinks, or frowning (Surakka, et al., 2004), which can be used to confirm a selection of the virtual key. Pointing can be performed through gestures produced by hand, head, finger, face, nose (Gorodnichy and Roth, 2004), mouth (de Silva et al., 2004), tongue (Salem and Zhai, 1997; Lukaszewicz, 2003), lips, eyes, brows, or the whole body movement, for example, as a displacement of the center of mass (Evreinov et al., 1999).

Various devices can be used to monitor typing on VKbd, for example, wristwatches and electronic pens equipped with a small liquid crystal display (LCD), cell phones, pocket PCs, as well as peripheral displays or virtual displays based on the IO2 technology (a.k.a. free space or midair video display) (BeamOne HoloTouch Evaluation System, 2006; IO2 Technology, 2006). Visualization can facilitate user input showing the entire keyboard, a symbol selected, or just an indication of the markers of the particular spatial positions that have to be activated. When discrete areas of any keyboard are distributed in a two-dimensional (or 3D) input space, it is obvious that the typist will be supported with the primary kinesthetic and proprioceptive feedback accompanying a relocation of hands and fingers. However, haptic signals provide low spatial accuracy, which is normally compensated by visual feedback when the VKbd is presented as a visual image projected on any hard surface (VKey, 2006) or gaseous one (BeamOne HoloTouch Evaluation System, 2006; Heliodisplay, IO2 Technology, 2006). The visual manifestation of the layout works as an external memory aid, and some people can benefit from monitoring their input.

Only a small number of techniques support two-hand text entry for 10-finger typing (non-chord based) on a full-size virtual keyboard; for example, VKey and VKPC are equipped with the Lumio's sensing module (Du et al., 2005; VKey, 2006). Other solutions allow the sequential input of the alphanumeric characters using different combinations of signals from a limited number of sensors, interpreters, and predictors (MacKenzie et al., 2001; Miniotas et al., 2003; Evreinova et al., 2004). When many discrete spatial positions have to be selected sequentially by direct pointing, then spatial factors, such as distance and size, of the virtual key limit text entry speed (see Fitts law—Fitts, 1954; see also MacKenzie et al., 1999; MacKenzie and Soukoreff, 2002; MacKenzie, 2003). Therefore, different optimization concepts, which had been designed for OSKs and mobile text entry, can be appropriate for VKbds (see also Sections 29.4 and 29.5).

Different approaches exist facilitating pointing and selection of any virtual key using appropriate primary and secondary feedback cues to decrease cognitive load. Symbols are commonly grouped according to their function: letters, ciphers, punctuation marks, signs, and so on. Linguistic features allow dividing the functional groups into frequently and less frequently used symbols. The functional groups of the virtual keys can be reassembled in accordance with their priority and accessibility into the virtual layers. That is, one layer (a functional group of the keys) is directly visible, while other layers can be visible on demand. The layers can be toggled using auxiliary keys (modifiers; Yfantidis and Evreinov, 2005), gestures, dwell time, speech cues, and additional sensors. Finally, groups of symbols can be rearranged according to their proximity. While on a physical keyboard, to support coordinated two-hand typing, frequent adjacent letter pairs should be located at opposite sides; for sequentially entered digraphs on the virtual keyboard their letters should almost be equally accessible perceptually, cognitively, and with a similar amount of motion. That is, these letters have to be close to each other in space, time, and within the mental model of the key-to-character mapping (Zhai et al., 2000).

Primary feedback before entering the symbol is very important for both novices and expert typists. When the number of functional groups or virtual keys is less than 10, it is possible to apply nonspeech, speech-similar sounds, or vibro-tactile cues as landmarks or navigational beacons (Rinott, 2005). Navigation within the virtual layout can also be augmented by a natural position of the fingers. As discussed in Evreinov and Raisamo (2002b), when the third finger can detect a raised tactile landmark (pin) on the key "5" of the physical keypad, on the key "F" and "J" of the regular keyboard, adjacent keys can easily be located with the second and fourth fingers, as well as the thumb, and be recognized due to layout knowledge and finger memory. Hence, it is not necessary to sonify each position of the virtual key or the entire user input (motions) by a continuous sound or vibration when the difference in spatial locations can be presented through kinesthetic and proprioceptive cues of the related fingers. For example, for blind navigation within 12 virtual keys having the layout of 3 by 4 sensitive areas in a working space, only the left or middle keys within each row (or rows) could have a different frequency of the MIDI sounds. The left, middle, and right virtual keys can easily be detected according to a colocation of three fingers. The stereo balance of the sound volume can be used. However, when pointing is performed by head movements without visual feedback, people can easily discriminate the lateral head displacements concerning anatomical landmarks (a relative position of neck, shoulders, or thorax). In such a case, it is not necessary to add any modulation of audio signals. Only the fact of an acquisition of the virtual key has to be confirmed by the secondary feedback signal (to finalize input and make virtual interaction perceivable). Optimizing feedback can significantly decrease the perceptual and cognitive load.

In addition to knowledge about landmarks, which have to facilitate pointing and selection of the virtual keys, the user of the VKbd has to learn all mapping features on how to type any symbol. Because it is often impossible to recognize or predict an unfinished letter (e.g., a Braille symbol), a secondary feedback about the character to be entered cannot be given in time. Sometimes, learning a new text entry technique is similar to the acquisition of a new language. Therefore, skills and knowledge transfer are very important usability factors, which have to be taken into account in designing text entry techniques. Any mnemonic rules, hints, and feedback cues can facilitate training and memorizing the way of encoding the letters and mapping the virtual keys (Yfantidis and Evreinov, 2004). Elderly people who are already familiar with Braille can experience stress during training to learn an alternative text entry technique. Still, they can benefit from the habitual encoding of the letters using a spatial layout of the virtual keys, which resembles a regular Braille cell, even when a sequential one-handed Braille input slows down typing to about 6 wpm (Arato et al., 2004; Juhasz et al., 2006).

29.3.1 The Reduced Key Input

Decreasing the number of keys to 3 to 12 requires a particular key-to-symbol mapping (MacKenzie et al., 1999; MacKenzie, 2002a, 2002b; Sandnes et al., 2003). A reduced keyboard with high multifunctionality of the virtual keys increases cognitive load during text entry. The layout of spatial positions and the user activity have to be optimized for learning and typing. Empirical studies have proven that when the typist can easily memorize access to each symbol (encoding), then skill acquisition requires less effort and time, higher typing speed is achieved, and the technique is less error prone (Wobbrock et al., 2004; Yfantidis and Evreinov, 2004; Evreinova and Evreinov, 2005; Wobbrock et al., 2006).

Different optimization methods for text entry with a reduced number of physical keys have been proposed, which can also be applicable for the VKbds. Being used by mobile phone manufacturers, T9 is the most famous disambiguating technology. When some letters of words are sequentially entered with the use of different keys of the ambiguous keyboard, T9 allows predicting words and phrases using a linguistic database (dictionary). However, the prediction worsens by any error, and additional keystrokes are needed to select the desired word, to edit, or even to retype the phrase (MacKenzie and Soukoreff, 2002). Being continuously supervised to validate both the characters entered and a prediction list of words to make a decision and confirm a right selection, the method increases perceptual and cognitive load.

Figure 29.1 illustrates that, depending on the number of keys, some alphabet characters can be entered using a single keystroke, while other characters have to be entered using a sequence of two keystrokes. A regular keyboard (QWERTY in Figure 29.1) allows entering each character using a single keystroke, for instance, with one degree of freedom (DOF = 1) when no modifiers are used. The combinatorial methods (DOF = 1.5) require additional user perceptual, motor and cognitive activity.

Supposing 12 SAs on the VKbd should encode 27 alphabet characters to be entered (26 letters and the space character), 9 SAs can encode any symbol, and 3 SAs can be used as modifiers.

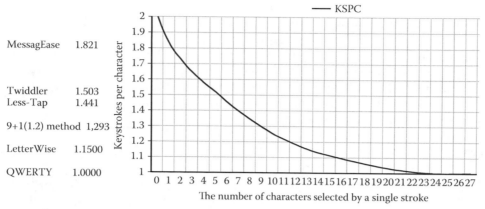

FIGURE 29.1 Typing optimization on the reduced 12-key telephone keypad when frequently used alphabet characters in English are accessible by a single keystroke while other characters are entered using two keystrokes. Several methods on the left illustrate typing speed in KSPC. (From Mackenzie, I. S., *Proceedings of the Fourth International Symposium on Human Computer Interaction with Mobile Devices*, Berlin/Heidelberg: Springer-Verlag, 2002, pp. 195–210.)

However, 9 letters, such as frequently used characters, can immediately be entered by a single stroke without any additional activity. Hence, the space character having the highest rate of use (Table 29.1) can be assigned to one of the SAs being reserved as modifier. Thus, 10 letters (see "9+1(1.2) method" in Figure 29.1) could be entered by a single stroke (gesture or any other way), and 17 characters could be entered using two sequential keystrokes. The number of the SAs as well as the number of modifiers (additional sensors) can vary (Figure 29.1), and predicted performance can be estimated as the measure of the user motor activity in keystrokes per character (KSPC). Herewith, the SAs can be implemented as the finger movement sensors, and the modifiers could detect, for example, the hands or head position (slope, rotation).

According to the VKbd definition, the sensors have to be placed outside of the fingertip (touch area). Examples of different technologies are: Senseboard (Sweden), which records the specific movements (a local pressure; Du et al., 2005) of metacarpals; Lightglove (United States), based on the optical reflectance matrix detecting fingertip movements; Scurry (Samsung), including two angular-velocity sensors (gyroscopes) and two axis acceleration sensors on four fingers and others (Fukumoto and Tonomura, 1997; Pratt, 1998; Rosenberg, 1998). Using the video-based input system it would be possible to detect any body movements remotely. However, for real-time detection of the multiple features, there is no agreed-upon recognition approach achieving appropriate balance between the computational costs and reasonable reliability.

29.3.2 Toward a Universal Paradigm

Figure 29.2 illustrates the method of text entry "9+1(1.2)" where 9 letters (F) and the space character can be selected by a single stroke or gesture. Other characters can be entered by two strokes (composite gestures) using the modifiers Left (L) and Right (R). Such a technique requires fewer keystrokes (1.293) without any additional input processing than MultiTap (2.034) and many

other typing techniques. For example, LetterWise, which is based on prefix disambiguation algorithm (i.e., the probability of the letter sequences in a language), requires 1.150 KSPC (MacKenzie et al., 2001). When it is possible to use both hands, the "9+1(1.2)" method can outperform many other techniques, because the modifiers can be activated simultaneously with the nondominant hand, taking no extra time during sequential typing.

By virtue of the definition, virtual keyboards should be hidden or transparent to the user and the application. For example, Isokoski has proposed the use of three or four off-screen targets (the sensitive areas) and the modifier target to encode alphabet characters making sequential pointing and selection of these targets by a video-based eye tracker (Isokoski, 2000; Isokoski and Raisamo, 2000). The author utilized combinatorial methods, such as the Morse code and the MDITIM (minimal device independent text input method) technique, and also placed outside the monitor screen and tested other layouts of the targets, such as the QuikWriting (Perlin, 1998) and Cirrin (Mankoff and Abowd, 1998) keyboards. However, the encoding systems required learning and practice to memorize the sequence of abstract actions to input each symbol, while fine motor skills were needed to manage the full-featured keyboards. Thus, the number of the sensitive areas (off-screen targets) of the VKbd was the main constraint concerning the abilities of the potential users.

L F R	L F R	L F R
L F R	L F R	L F R
L F R	L F R	L F R
L	Space	R

FIGURE 29.2 An illustration of the "9+1(1.2)" text entry method. Symbols in L-position (of the L-F-R sensitive area) are selected with a prior activation of the L sensitive area; R-position is selected with a prior activation of the R sensitive area. Symbols in F-position and the Space character are directly entered by a single stroke (gesture) of the corresponding sensitive area.

Yfantidis and Evreinov have developed the Gesture Driven Software Button (GDSB) text entry technique (Yfantidis and Evreinov, 2004), and used for the VKbd the character layout (ERATIONS_), which is similar to the example discussed in Section 29.5. The alphabet characters were split into three groups (layers), and each letter was associated with one of eight directional gestures. Access to the layers was performed in a cyclic manner using dwell time at the starting point of any gesture. Since directional gestures provided a location-independent selection for 24 characters, other functionality was implemented through the substitution of the characters by the needed symbol, letter, or command using also dwell time. For instance, the character "S" was changed to "space," "C" to "Z," and "G" to "J," "N" to "Next line," "D" to "Backspace," and so on. As reported in Yfantidis and Evreinov (2004), after 100 words have been entered (training) the typing speed was about 12 wpm. The text entry speed increased to 18 wpm when modifier keys were used instead of the dwell time to change the layers (groups of the characters) with the nondominant hand (Yfantidis and Evreinov, 2004). No visual feedback was provided in two modes of the proposed text entry technique. The monitoring of the characters entered was accompanied by recorded speech sounds.

Rinott proposed to sonify gesture-based text entry, and introduced a haptic-auditory text entry technique called SonicTexting (Rinott, 2005) for continuous rounded thumb gestures using the Keybong joystick input device. Two sound modes have been implemented for beginners and experts. In the beginner mode, the typist received a primary speechlike feedback as in a loop manner phoneme (A, E, I, V, R, N in Figure 29.3) referred to the main direction (gesture), or as a mix of two phonemes in the case of the intermediate direction (e.g., "QS, QS...", "UT, UT... UT", "JK" etc.). The distance to each letter was a function of the sound volume within associated characters, and the eight main directions had different pitch of sounds. In the expert mode, percussive sounds were used for navigation, different pitch of sounds was assigned to eight directions, and the sound attack

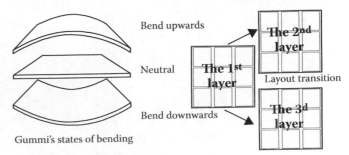

FIGURE 29.4 Text entry for a bendable computer. (Adapted from Schwesig, C., Poupyrev, I., and Mori, E., Gummi: A bendable computer, in the *Proceedings of the SIGCHI Conference on Human Factors in Computing Systems (CHI 2004)*, 24–29 April 2004, Vienna, Austria, pp. 263–270. ACM Press, New York, 2004.)

was proportional to the velocity of the movement. For instruction purposes, the subjects used the sounds map, but expert users were able to write the SMS using only the sound feedback. The author mentioned that the Keybong joystick could provide the vibration cues to signal the difference between the groups of characters. The published results were obtained using only the perceptual haptic and sound feedback.

Schwesig et al. (2004) presented another text entry system transparent for application (which can be transparent as well for expert users) that used a clock face topology of the characters arranged in three layers that have been switched by bendable computer screen ("Gummi"), as shown in Figure 29.4. The pointing within each layer was implemented through directional bending, and a slight bending in the opposite direction was used to make a selection of the characters.

The dynamic range of efforts ±100N was needed to manipulate the entire device (bendable PDA) that cannot be applicable for continuous typing due to ergonomic constraints. The interaction technique resembles shaping or "sculpting" the characters. However, when muscle efforts could be decreased with a three-dimensional position sensor, which looks reasonable for manipulating on the VKbd in the absence of visual feedback, as such a way provides enhanced haptic information. A perceivable shape of the bendable computer surface is unique regarding each virtual character's location, which can facilitate learning and make the technique less error prone.

29.4 OSK Concepts

OSK is an interaction technique for input binary codes. OSK presents a layout of software buttons simulating a function of physical keys, which are being projected onto a flat surface of the screen, and can be activated with the use of any other input device (mouse, joystick, touch screen etc.). As follows from the definition, onscreen visual image (a software label/button) is the principal feedback in the OSK.

To provide full control over the computer by extending the input functionality of the physical keys, a regular keyboard could be presented visually. In 1978, Rudi Grimm, Max Syrbe,

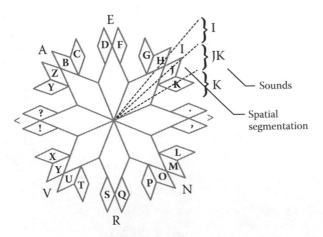

FIGURE 29.3 The layout of the characters in SonicTexting, Version 5. (Adapted from Rinott, M., SonicTexting, in the *CHI '05 Extended Abstracts on Human Factors in Computing Systems*, 2–7 April 2005, Portland, OR, pp. 1144–1145, ACM Press, New York, 2005.)

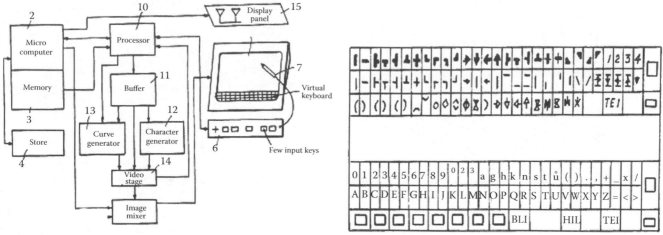

FIGURE 29.5 The input/output-color-screen system and the layout of the first onscreen keyboard. Figures 1 and 4 in the original document were adapted from Grimm et al. (1978).

and Herbert Truck from Fraunhofer-Gesellschaft Zur Förderung der angewandten Forschung e.V. (Munich, Germany) received the U.S. Patent 4,079,450 for "input-output-color-screen system" (Grimm et al., 1978). The system comprised a color display, a microcomputer, and a light pen as a direct input device for screen coordinates. The OSK was composed of symbols arranged in a section of the screen. The keyboard could be activated manually through a light pen to select the desired symbols (Figure 29.5).

As can be seen in Figure 29.5, the layout of the OSK ("virtual" in the original; Grimm, Syrbe, and Truck, 1978) looks like the input console of early computers. An interactive onscreen image has brought obvious benefits. Using a character generator has allowed creating any layout of any symbols in any language and changing them in real time. Extending functionality of the graphic controls was a great milestone for the development of the graphical user interface (GUI). As the OSK could be activated with almost any input device, the problem of computer accessibility for people who would significantly benefit from augmented communication was raised.

In 1986, Carol M. Auer with colleagues from AT&T Bell Laboratories (Murray Hill, NJ) introduced an application for a "computer interface device," which included a flat screen display and a touch-sensitive element (Auer et al., 1988). They also described simulated keyboards that can be displayed on the monitor. Simulated keyboards could generate appropriate control signals in response to activation of the simulated keys (Auer et al., 1988). Figure 29.6 depicts one of the first images of the QWERTY OSK designed for a touch screen.

At that time, the layout of physical keys had been changed several times to increase the functionality of the keyboard and to improve the touch-typing performance for functional keys (edit, navigation, modifiers). To attract potential users, manufacturers and designers used a familiar regular QWERTY layout. In contrast to the dissemination of the myth about the motor skill transfer (Mobience, 2004; Sandnes and Jian, 2004; Ahn and

Kim, 2005; see also discussion in Section 29.5), other researchers believed that an optimized layout could increase pointing accuracy and speed up typing (Montgomery, 1980).

The progress in microminiaturization of computers and the emergence of handheld and pocket computers, touch-sensitive input devices, and graphical applications were powerful stimuli for the development of alternative text input methods. Due to reduced size, typing was hard and slow on small-screen devices. Short travel distances of the finger or stylus decreased a kinesthetic feedback to a minimum. Typing with two hands was almost impossible, except when the physical key could modify the OSK layout (Yfantidis and Evreinov, 2004).

Designers of onscreen and reduced keyboards often omit and ignore the cognitive cost and perceptual processes related with sequential text entry. The movements of fingers, hand, and eyes play an important role in visual analysis, whereas kinesthetic signals from those directly follow to the motor cortex (the main integrative structure of the brain) and provide both feed-forward and feedback components during text entry (Grush, 2004). In the optimization models the pointing and selection task to attain the needed key (the sensitive area) was often considered a single-target task. Nevertheless, the cognitive load is enormously increased in the presence of other letters acting as distractors. The OSK significantly distracts the typist, grabs the attention, and impedes the ability to do something different from typing. The visual search with divided attention during a sequential input demands a strong concentration and cross-modal coordination to keep pace with internal speech.

The statistical optimization of the interface for text input (the keyboard layout) can be based on the language modeling and the linguistic features, as discussed in Section 29.5. Many publications and patents are available concerning optimization algorithms for text entry based on touch input techniques. An example is USPTO U.S. Class 345/173. The performance and predictive modeling for text entry have been widely discussed in the literature (MacKenzie, et al. 1999; MacKenzie et al., 2001;

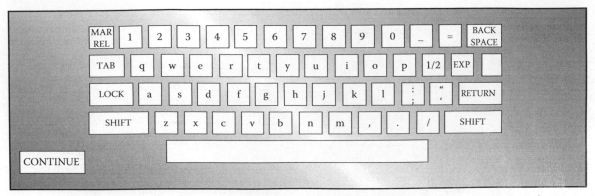

FIGURE 29.6 The QWERTY onscreen keyboard designed for touch screen. (Adapted from Auer, C. M., Castagno, D. L., Haley, Jr., A. W., Moore, IV, H. H., O'Leary, S. E., Paley, S. J., et al., *Computer Interface Device*, U.S. Patent 4,725,694, 1988.)

MacKenzie and Soukoreff, 2002; MacKenzie and Tanaka-Ishii, 2007). However, a statistically optimal behavior to point and select the letter sequentially by the finger or stylus movement for an individual user may be inconvenient. Making a specific movement with an input device is determined to a high degree by individual capabilities and disabilities of the user. Even linear motion for some people can be easy in one direction, while another person would never utilize this direction to avoid strain, pain, or spastic involuntary movements or tremor. Diverse optimizations of the OSKs, which were primarily intended for stylus input, have also been tested with alternative input devices and techniques. While it is obvious that eye and head movements are very different from hand gestures, continuous text entry was also adapted for eye typing (see Chapter 36 of this handbook).

The visual field is overloaded with graphical elements, which can dynamically change (Raynal, 2006). The dynamic optimization of the letter layout contradicts basic psychological principles and diminishes the meaning of the OSK as an external memory aid. To decrease the cognitive load in the systematic visual search of a letter, the OSK layout and its changes should be predictable for the typist. In such a case, the text input technique will form visual routines, primitive visual operations, to identify the target property (a symbol, its location and surrounding) available to the user for selecting and manipulating (Toyama, 1998; Keyboard Layouts, 2007).

Can the user benefit from the OSK when all characters are almost equally accessible? Can the user benefit from the dynamically changing layout of symbols? Is it possible to find the perfect and universal trade-off between typing speed, user skills, and abilities? These are still open questions.

29.4.1 Reduced Key Input

Many attempts have been made to adapt different versions of onscreen keyboards for physically disabled people. In some cases, disabled people can only use a single button to make a selection from alternatives that are being presented sequentially (Evreinov and Raisamo, 2002a, 2004; Mehta, 2003; Seveke, 2006). The OSKs were integrated with a scanning method to navigate and dwell time to select any software control such as a text and list box, hyperlink and icon, a menu item or command button, by using a single switch input.

There are different constraints in the application of the scanning technique (see Chapter 35, "Automatic Hierarchical Scanning for Windows Applications"), for instance, cyclicity and hierarchy. Cyclicity requires sequential access to the onscreen items. Herewith, a huge number of the alternatives (27 to 106 and more) make difficult not only navigation itself, but also waiting for the short period of time until the necessary item becomes accessible (Simpson and Koester, 1999; Lesher et al., 2000; Evreinov and Raisamo, 2004). The related problems can easily be illustrated with the examples presented in Figure 29.7. Fortunately, groupings of symbols and special functions (drawing, editing), which can be accessible on demand, have been developed in the graphical interfaces. However, phrase-based techniques use a complicated hierarchical structure with a huge number of branches to specify a phrase category. To type the sentence "what is the date today" using eLocutor (Mehta, 2009), it is necessary to make 6 clicks that take about 13 seconds after the first word is entered. Before the first word is entered, it is necessary to attain the appropriate group of words and perform 6 clicks without any mistake (Figure 29.7B). This process can take about 14 seconds or more, depending on the number of alternatives (groups, branches, words).

Scanning methods are often augmented by prediction techniques, which also can be trained to a user-generated language model (dictionary) (Hansen et al., 2003b; Harbush and Kühn, 2003). Different solutions have been proposed to improve hierarchical single-switch navigation based on prediction algorithms concerning the task, the situation, and the context, enhanced by individual topic-oriented vocabularies. Some methods are effective, for example, the DynaVox Systems (DynaVox, 2008), when a phrase kit (dictionary) is small enough.

However, direct text input without hierarchical groups and prediction could also work well when the number of keystrokes per alphabet would be minimized (Miniotas et al., 2003; Poirier and Belatar, 2006). The OSKs, which are based on special techniques for people with disabilities, are not intended to speed up

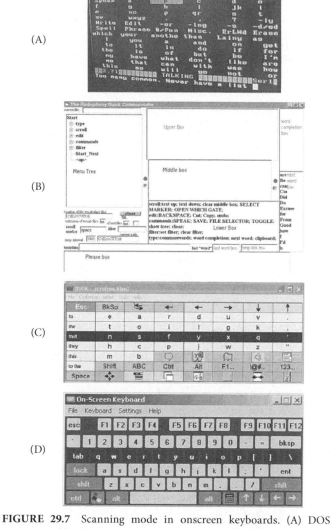

(A)

(B)

(C)

(D)

FIGURE 29.7 Scanning mode in onscreen keyboards. (A) DOS version used by Stephen Hawking for a long time. (From Mehta, A., *e Locutor version 2*, 2009.) (B) The menu groups of the Radiophony quick communicator (From Mehta, A., *e Locutor version 2*, 2009.) (C) WiViK onscreen keyboard 2006. (D) Microsoft OSK in Windows 2000.

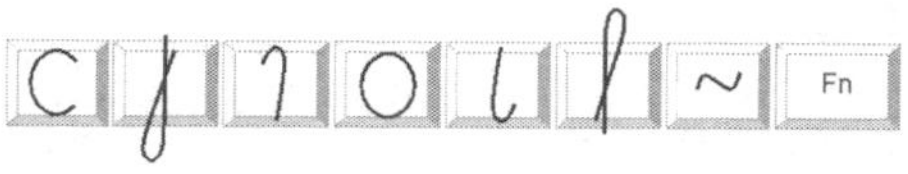

FIGURE 29.8 The OSK of the symbol creator adapted for eye-gaze typing. (Adapted from Miniotas, D., Spakov, O., and Evreinov, G., Symbol Creator: An alternative eye-based text entry technique with low demand for screen space, in the *Proceedings of the 9th IFIP TC13 International Conference on Human-Computer Interaction (INTERACT'03)* (M. Rauterberg, M. Menozzi, and J. Wesson, eds.), 1–5 September 2003, Zurich, Switzerland, pp. 137–143, IOS Press, Amsterdam, The Netherlands, 2003.)

typing, and some experience is needed to achieve a reasonable typing speed and accuracy depending on the individual user abilities.

Other methods aim at reducing the number of software controls (buttons, labels, icons) of the OSK, which have to be pointed out or scanned and selected using single-switch and video-based input techniques (see Chapter 35, "Automatic Hierarchical Scanning for Windows Applications"). For instance, the Symbol Creator technique (Figures 29.8 and 29.10) relies to a great extent on the experience in school handwriting (Miniotas, Spakov, and Evreinov, 2003), but it could also be easily learned. Western scripts have several handwriting styles for elementary school-aged children. The loops and other shapes provide systematic steps for letter analysis and efficient motor and memory cues for children. These basic elements were used in a systematic way to yield a relatively small set of segments to be used for assembling letters and symbols for text input. Similar decomposition of the letters is suitable for East Asian and other languages. The waiting time in the scanning mode could significantly be decreased with a small number of segments to encode the letters. The OSK of the Symbol Creator for a mobile device has a layout of 3 by 3 software buttons, as shown in Figure 29.11 (bottom right). Such an approach compensates for an increased number of keystrokes. Fortunately, a reasonable KSPC index can be found for the particular character-based alphabets when cognitive load can be kept to a minimum. For the English alphabet KSPC was equal to two (2.0342) using multitap mode on a mobile phone keypad (MacKenzie, 2002a; Tian-Jian Jiang et al., 2007).

29.4.2 Abbreviated Text Input

The number of keystrokes is a critical parameter for motion-impaired users. Various methods based on spelling models and language models have recently been developed (Shieber and Baker, 2003; Shieber and Nelken, 2006; SIBYLLE, 2006; Wandmacher and Antoine, 2006) to decrease the number of keystrokes per word. These methods can be combined with other optimization techniques.

For instance, abbreviated text input (Shieber and Baker, 2003) allows entering text in compressed form using a simple stipulated abbreviation method that reduces the number of characters to be entered by about three times. As stated by the authors, the use of statistical language-processing techniques can decode (expansion) the abbreviated text with a residual word error rate of about 3% or less. Because text processing is independent from the user, the cognitive overload from switching attention to the system's actions is eliminated. During the reported tests, the keystroke saving rate varied from 30% for French to 50% for Swedish and German (Beck et al., 2004).

Prediction accuracy depends on the specific language and on a statistical language model (e.g., based on n-grams). Accuracy could be increased when the typist and system could utilize the same dictionary of abbreviations. However, for some users it is not obvious that abbreviated typing is a convenient way of entering text. Abbreviated text input breaks habitual routines in segmentation of words (1-, 2-, or 3-phoneme syllabic input). It demands a new mental model and a variable pace of typing, learning the dictionary of abbreviations and, consequently, increasing the cognitive cost of text entry.

TABLE 29.1 The Reduction of the English Alphabet

English letter frequencies (the numbers under the letters) according to WiViK on-screen keyboard (2006)															
a	e	f	g	h	i	k	l	m	n	o	p	r	s	t	u
66	104	19	16	45	60	5	34	21	58	63	17	51	53	76	22
The letter candidates to be removed															
		v	J(z)	ch	y	q(c/x)					b		c(z/x)	d	w(v)
		8	1(1)	5.6	14	1(26/2)					12		26(1/2)	33	16(8)
The reduced English alphabet and the possible frequencies of the letters															
a	e	f	g	h	i	k	l	m	n	o	p	r	s	t	u
66	104	27	~17	50	74	~31	34	21	58	63	29	51	~79	109	~38

29.4.3 The Reduced Alphabet

Many people intuitively use phonetic typing to encode messages in the native language using the regular English keyboard. In most European alphabets, the number of letters could be reduced to increase text entry rate in unofficial communication. For instance, using substitutions of the phonetically similar consonants in the English alphabet can reduce it to 16 letters without loss of readability (Table 29.1). Similar to abbreviation-expansion, the spell checker could retrieve automatically the correct form of the words printed with a modified spelling. For instance, using the Microsoft Word spell checker the words *question, circular, conventional, dynamic, zooming,* and *discussion* can invariantly be retrieved when they were entered as *kuestion, sirkular, konfentional, tunamik, sooming,* and *tiskussion.* Problems were found in retrieving about 3.5% to 5% of the words, such as *device (tefise), windows (uintous),* and *drawing (trauing).*

The test words were taken from Words Frequencies (2004) according to the recommendations in MacKenzie (2004). Figure 29.9 illustrates the brief instruction (A) and the OSK layout (B) used during the test of the reduced English alphabet. The average text entry speed in experiments with novices varied from 18 to 25 wpm after 1 hour of practice. The subjects were surprised at how easily they were able to type modified ("wrong") words.

Finally, it has to be noted that the usability of the OSK is greatly affected by the personal skills and abilities (perceptual, motor, and cognitive) required to manage the physical input device (touch screen and stylus, mouse, joystick or a single switch) used to activate onscreen software controls. Limited sensory or motor abilities can completely prevent the use of the OSK.

29.5 Design Principles for the Reduced Keyboard

Character mapping (encoding) can be different for various text entry techniques and devices. To minimize skills and resources when the mental model of the letters pointing and selection is shaped, the user activity such as perceptual, motor, and cognitive, has to be well coordinated (see more details in Zhai et al., 2000). The characters' colocation, grouping, and mnemonics, as well as hints and feedback cues, have an impact on typing performance (Wobbrock and Myers, 2006; MacKenzie and Tanaka-Ishii, 2007). This section discusses the basic principles that can be used for designing a reduced keyboard.

The novices easily can memorize six or more letters in a sequence when their combination resembles the word (e.g., QWERTY, FITALY). When such a sequence has a high probability of occurrence of the letters and groups of characters in the language corpus, their close location could benefit the typist by decreasing the amount of motion during text input.

For instance, TION is a unique combination of four characters (quadgram), which does not contain a space (before and inside) and has the probability of occurrence of about 5.2 in 1000 English words. More accurately (Table 29.2, left) the occurrence of this quadgram is 1,211.366 in 232,907.808 possible four-letter combinations (MacKenzie, 2003). The quadgram TION contains the most frequently used characters (Table 29.2, right) and other digrams of these letters (IT, TI, ON, TO, NT), which are also frequently used in the English language. Thus, it would be suitable to make such letters equally accessible with a similar amount of motion to select them.

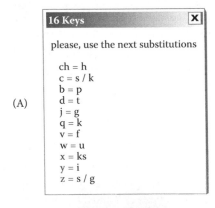

FIGURE 29.9 The rules (A) and the 16-key layout of the OSK (B) used during the test text on iPAQ pocket PC.

The letters E, R, A, and S have also a high probability of occurrence. The sequential access with a single keystroke (gesture) to eight characters and the letter space can provide entering about 70.7% of English text. One more frequently used letter (in English), H, would increase the directly entered number of characters up to 75.2% (Sokey, 2006). Similar considerations have been applied in designing different techniques. However, the issue remains open of how to arrange the frequently and less frequently used characters so that the user would easily reach them.

The virtual keyboard has form-factor restrictions, that is, a space where sensors can detect user activity. The SAs can be arranged as a grid (Figure 29.2) or as a one-row matrix (Figure 29.10). A linear layout cannot contain many sensitive areas, because it would slow down typing speed due to the physical distance between areas (see Fitts law; Fitts, 1954). Nine characters, the spacebar, and the modifier should directly be accessible (Figure 29.1) to provide the means for entering other characters and symbols.

Primary feedback can give a hint to the user to facilitate pointing. Grouping the sensitive areas by three can decrease the number of primary cues. Even for experienced typists, the significance of primary cues is not decreased. Usually, they unconsciously utilize the natural cues provided by tactile, kinesthetic, and proprioceptive afferents. Audio and haptic signals can be utilized as the external primary (navigational) cues (landmarks and beacons). MIDI can be used to easily coordinate sound parameters and user input. For instance, three groups of the SAs can be marked with three musical notes (e.g., do, fa, and la). Each time the typing system detects changes within the specific group of the SAs (Figure 29.3, groups 1 to 3), the typist can hear a corresponding sound. The MIDI notes can also change audio properties, such as balance and timbre, when the typist navigates within each group. Any suitable earcon can be assigned to the modifier location (4) in Figure 29.10. Speech cues are usually utilized instead of visual feedback as secondary signals to confirm the execution of an operation such as backspace, next line,

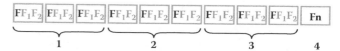

FIGURE 29.10 The linear arrangement of the reduced number of sensitive areas. Symbols F (in a bold) are directly entered by a single stroke (gesture). Symbols F_1 and F_2 can be selected in combination with or prior activation of the Fn sensitive area (modifier), for example, using a single stroke and double stroke (opposite gestures) accordingly.

upper or lower case, or access to a specific group of the symbols. Each character entered can also be sonified.

Thus, on the reduced keyboard direct access can be appropriate for the characters E, R, A, T, I, O, N, S, and the letter space. The location of the character S is quite frequently at the end of English words before the letter space ("_"). The probability of occurrence of the digram S_ is 10,860.471 of 505,863.847 words (Table 29.3, left). Other characters have less probability of being at the end of words after the character N (Table 29.3, right). Therefore, TIONS_ is a very promising combination for closely located SAs (virtual keys on the VKbd or software buttons on the OSK).

By taking into consideration the data of Table 29.4, the most favorable layouts are ERATIONS_ and REATIONS_, which can facilitate sequential typing on the reduced VKbd, OSK, or physical keyboard. The utterance of the first combination of the letters is phonetically similar to the word *aerations*, the second combination resembles the word *rea(c)tions*. In both cases, the user can easily memorize at once the layout and access to nine letters (SAs) using a single stroke. However, this is not claimed to be an ideal and universal solution for text entry, and it only illustrated an approach where linguistic features and dictionary-based statistics were used to optimize typing on the reduced keyboard in a way that could speed up learning the novel characters arrangement.

To accomplish the layout of the reduced keyboard, the other 18 characters have to be distributed within the same SAs. The

TABLE 29.2 Quadgrams

_the	7027.378	space	997.552	176
the_	5777.107	E	586.316	104
of	2789.403	T	432.382	76
and_	2594.123	A	376.459	66
_and	2429.793	O	355.384	63
in	1695.860	I	341.068	60
to	1468.146	N	330.665	58
ing_	1399.790	S	301.210	53
tion	**1211.366**	R	286.713	51
ion_	1057.976	H	254.365	45
…	…	…	…	…
Combinations in total	232907.808	in words	5666.538	1000

Quadgrams with occurrence greater than 1,000.000 within the word frequencies database (Word Frequencies, 2004). The space character is marked by "_" (the part on the left). The first ten frequently used characters in the English language having the highest occurrences are shown on the right.

TABLE 29.3   Relative Frequency of Occurrence of Last Letters in Word Endings

e_	**18403.847**	ns_	**464.686**
s_	**10860.471**	ne_	445.872
d_	9443.914	nr_	0
t_	9143.009	na_	20.938
r_	**5309.971**	…	…
a_	**2386.285**		
…	…		
Combinations in total	505863.847	Combinations in total	300869.920

The relative frequency of occurrence of the last letters in word endings is shown in the part on the left. The relative frequency of occurrence of digrams "ns," "ne," "nr," and "na" in word endings is shown in the part on the right. The space character is marked by "_".

same goal should be pursued: the user should easily memorize the second function of each sensitive area and easily recall them when required.

There are different ways to combine less frequently used characters and provide appropriate access to activate them. For example, those characters can be split into two groups. Herewith, one of these groups of the less frequently used characters, together with the first group (ERATIONS_) composed of the frequently used characters, should support entering not less than 90% of the text. The next two groups of the characters could be: CLPDFGMHZ (21.2%) and QUBKVWYXJ (8.1%) as shown in Figure 29.11 (top left). In such a case, an appropriate modifier (stroke, gesture) or dwell time can be used to encode entering the second and the third groups of characters associated with the same sensitive areas. As the first group of characters was composed of vowels, it would be difficult to combine consonants into the patterns that resemble the word. Instead of memorizing the secondary layout of letters, it would be easy to imagine the transformation of each character of the first group into two alternative letters associated with the same sensitive area. The phonetic connection and graphic similarity of the symbols can be utilized. Furthermore, as shown in Figure 29.11, acronyms and homonyms, as well as mute vowels (e.g., CLiP, QUBe, mute vowels are shown in lowercase), can simplify shaping the mental model of the local rules: mapping the characters across virtual layers (groups).

The third group (QUBKVWYXJ) contains only rarely used characters. These characters could be entered with a similar amount of motion as the second group (CLPDFGMHZ) by making keystrokes in conjunction with a modifier key (additional sensitive areas or physical buttons) used to enable/activate the specific layer, as shown in Figure 29.2. Therefore, the position of the letter U does not seem to be a bad choice when thinking about the use of homonym QUB or QUBiK to memorize the subgroup of sensitive areas. Graphic similarity (Figure 29.11, right) is often used in designing alternative text entry to rely on previous user experience, for example, in handwriting (Evreinov and Raisamo, 2002a,b; Jannotti, 2002; Miniotas et al., 2003; Wobbrock, Myers, and Aung, 2004; Belatar and Poirier, 2005; Sandnes and Huang, 2006; Wobbrock et al., 2006).

Supposing a possibility to transfer previous user experience of typing on a regular keyboard, the authors of the Mobience layout (Mobience, 2004; Figure 29.11, bottom left) have attempted to apply projective mapping (homography) in designing an easily learnable reduced keyboard for mobile devices with a layout of 3×3 keys or sensitive areas. Unfortunately, it is difficult or even impossible to transfer typing skills acquired with a regular keyboard (two-hands or one-hand typing experience) to interact with any different input device, as some authors suppose. A behavioral stereotype is an elaborate dynamical virtual structure of neuronal excitation, which involves and integrates perceptual, cognitive, and motor activity. When the character layouts are rearranged to be utilized for entering text on a mobile device or with the VKbd, a different mental model of user activity is required (Mobience, 2004; Ahn and Kim, 2005). Even changing

TABLE 29.4   Trigram Occurrences of the Combinations of Letters

ERA	**190.643**	**ERAT**	**72.023**
REA	**466.206**	**REAT**	**132.296**
AER	0	AERT	
RAE	4.118	RAET	0
EAR	412.032	EART	0
ARE	665.663	ARET	7.237
…	…	…	…
Combinations in total	300869.920	Combinations in total	232907.808

Trigram occurrences of the combinations of the letters "E," "R," and "A" (the part on the left) and quadgrams of their combinations with "T" within the word frequencies database (Word Frequencies, 2004) (the part on the right).

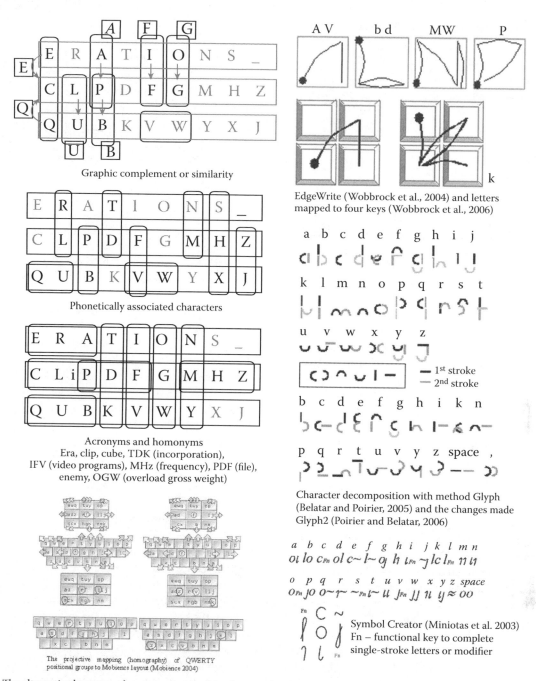

FIGURE 29.11 The alternative layouts and mnemonic rules based on previous typing and writing experience.

two keys on the QWERTY (U.S./U.K.) layout can brake typing on the QWERTZ (German, Austria) keyboard. The rhythm of splitting the words into syllables or characters (segmental-syllabic model) has a significant impact on typing rhythm (Keller and Zellner, 1996; Sandnes and Jian, 2004). In turn, the rhythm of entering text can modulate internal diction.

Letter decomposition into a set of the segments (Figure 29.11, right) allows creating symbols through a sequence of actions instead of typing (entering the character by a single action).

Table 29.5 illustrates the encoding systems that have been implemented and tested with the OSKs. The six-dots contracted Braille was included for comparison, as it was also implemented

in a version of the OSK on the iPAQ pocket PC. The Braille OSK was comprised of nine sensitive areas: six SAs for entering Braille dots, the SA for the letter space, the SA for a backspace operation, and the SA acting as a delimiter to validate a completion of the character entry. The concept of creating symbols is universal on a physical level. Composition (encoding) of symbols can be implemented with the help of any discrete user actions: by physical keys, using a stylus tap or stroke techniques, by directional gestures, using eye and head tracking techniques, scrolling or scanning and a single key.

When symbols were composed of a variable number of segments (one to three primitives), for example, Braille (Arato et al.,

TABLE 29.5   Reduced Text Input Systems Using Letter (De)composition

Encoding System (OSK Version)	Intuitively Encoded Symbols	Number of Segments	KSPC	KSPA	wpm
Braille (Arato et al., 2004)	N/A	6+(space) +1(delimiter)	2,5238+1	85+27	4,5-6 (finger)
Glyph (Belatar and Poirier, 2005)	74%	6+1(delimiter)	2.065+1	61+27	4-14 (stylus)
Glyph2 (Poirier and Belatar, 2006)	33%	6+1(delimiter)	2	54	7 (stylus)
Symbol Creator (Miniotas et al., 2003)	70%	7+1(delimiter)	2	54	12 (eye)-20 (stylus)

Reduced text input systems using the letter (de)composition from a set of primitives (segments) and which have been implemented as the OSKs. KSPA, key strokes per alphabet; wpm, words per minute.

2004) and Glyph (Belatar and Poirier, 2005) techniques, a delimiter was needed, but it significantly increased the perceptual-motor and cognitive cost of text entry. When symbols were created using only one or two KSPC (Miniotas, Spakov, and Evreinov, 2003; Poirier and Belatar, 2006), the delimiter was needed to complete the symbols comprised of a single segment. Still, the delimiter activated beforehand can be useful as a modifier key. The perceptual model of such a technique is quite simple, and involves just the knowledge of cursive handwriting. It allows the user to easily memorize the OSK layout. For instance, in Symbol Creator and Glyph, 70% and 74%, respectively, of alphabet characters were intuitively encoded, while other letters and symbols were encoded conditionally. A blind user typing with the help of Symbol Creator on the keypad is about three times faster than using Braille code. The minimum number of sensitive areas could significantly improve scanning techniques with a single-switch typing.

The concept of the keyboard is universal in a wide sense. The keyboard presents an array of the physical, virtual, or software entities that can be activated by physical action or signals produced directly or indirectly (through a special device) by the user. The keyboard elements can produce binary signals, which can be integrated into a command or message, or be converted into another signal. To efficiently interact with a keyboard, the user has to learn how to activate keyboard elements and the meaning of the binary signals being produced. Multifunctionality of keyboard elements causes cognitive overload of the user. Being presented simultaneously, the functions of each element can distract and cause perceptual overload. The size of the keyboards and access to their elements require particular motor skills and experience to avoid selection errors. Still, to adapt the keyboard to the particular user needs the meanings of the signals and commands can be assigned to keyboard elements in a way that facilitates learning and interaction.

The universality of the virtual and onscreen keyboards for text entry is still limited by the language the typist can use. However, the development of symbolic languages (see Chapter 39, "Visible Language for Global Mobile Communication: A Case Study of a Design Project in Progress") will promote an emergence of advanced intellectual keyboards as universal tools for human-device integration and human-human communication. As an example, the universal remote control can recognize the type of the system and change the protocol of the control

signals accordingly, while the layout of the keyboard elements remains the same for the user (Zimmermann et al., 2002; Saito et al., 2003).

Besides the designing of the text entry techniques as the specific integration of the user activity (perceptual, motor, and cognitive) and significant progress in developing the devices that can record and process user input (e.g., see Chapter 36, "Eye Tracking," and Chapter 37 "Brain-Body Interfaces"), there is a great challenge in the methods and models for evaluating the usability and accessibility of the proposed solutions, algorithms, and technology. Jacob Wobbrock has recently presented a comprehensive survey on measures of text entry performance (MacKenzie and Tanaka-Ishii, 2007, pp. 47–74). Wobbrock and Myers (2006) also present a detailed analysis of the input stream for character-level errors.

29.6 Virtual Mouse

Graphical interfaces allow the user to communicate with computer systems using discrete or/and continuous n-dimensional input, different physical movements, and micromovements of skin, eyes, and metacarpals. A relative displacement of the object(s) can be recorded using diverse physical principles, which allow detecting differences between signals corresponding to the particular position of the object(s). Herewith, one of the object(s) acts as the reference point, the reference pattern (texture) or coordinate system. Another object is being affected by human action. Thus, human actions can be transformed into the output signals of the motion detector and then be used to control computer functions, for example, displacements or movements of the pointer on the computer monitor, or to execute a specific command to control the system state.

In earlier input techniques, interaction with images such as setting the markers, selecting part of an image, and creating graphic primitives was based on an optical sensor (e.g., a lighting pen) using detection and modulation of the beam of the cathode ray tube display screen (Gindrum, 1970; Clark, 1971; Pieters, 1974; Iwamura, Hamada, and Hayashi, 1976). Most of the features and tricks, which nowadays seem to be inherent in advanced graphical input devices, were designed in the beginning of the 1980s with the help of analogue circuits and discrete electronic components, for example, Sketchpad with a lighting pen (Sutherland, 1980). Because of limited computational capacity, the earlier graphical applications such as computer-aided design (CAD) systems did not support the multitasking mode and could not simultaneously process analogue input during run-time. CAD systems were mostly oriented to the use of discrete (binary) keyboard input. Designing the mouse and graphical input stimulated the development of a new class of computer peripheral components—the human interface devices (HID), which were developed to standardize processing the human input and computer output signals (Human Interface Device, 2008).

On the other hand, the quasi-graphical user interface in DOS already contained primitive objects, which can be considered ancestors of the widgets nowadays used as the fundamental bricks of GUIs. For these objects, controls have been assigned particular properties, events, and procedures, which were intended to make interaction with computers more user-friendly. Thus, the specifications were given for both software and physical objects sensitive to continuous user activity. By mediating the user actions, the HID driver or specific software module could immediately translate the signals of the input device into the commands or the values of the variables in the program. In particular, a pointing device, such as a mouse or joystick, is sensitive to hand movements of the user. The driver of the pointing device translates these data into coordinates of the screen pointer. In addition, a pointing device can be equipped with physical buttons (single or multiple) to support system-dependent operations (command input). Essentially, physical mice were designed to be in physical contact with the hand of the user and to detect movements in relation to its supporting surface (desktop).

As analogue input devices were not intended for blind users, there was no need to make linear the transformation function of their relative displacement into the coordinates of the pointer. Strongly supported by visual feedback, the user had to operate the input device doing any hand movements to better fit the pointer to the desired location. Sensitivity of the mouse (the coordinates transfer function) is usually adjusted with indexes of speed and acceleration. Due to recording the relative displacements, the mouse can be elevated and placed in another desktop location without changing the coordinates of the screen pointer. This mode allows avoiding a cumulative error of displacement caused by friction and slippage in moving parts of earlier prototypes and, later, the optic-related problems with low-resolution CMOS sensors. Relativity of the mouse coordinates is often referred to as mouse mode, in contrast to joystick mode, which means absolute positioning regarding the absolute reference point of the coordinates (0,0) usually associated with the upper left corner of the computer screen (system function) or the center of the screen in some applications.

The virtual mouse is an input device that simulates and supports the basic mouse functions such as pointing, selection, and access to contextual menus. It detects and transforms into a position of the screen pointer the relative displacements of any objects being affected by the human activity. Essentially, virtual mice can have or not have a physical contact with the user or with the object the displacement of which is being detected and recorded. If appropriate, the virtual mouse can also support

joystick mode. Technically, the virtual mouse is a motion detector that can use different physical principles to record the spatial displacement of interaction objects (Paradiso et al., 2000).

29.6.1 Basic Principles of the Techniques

Figure 29.12 illustrates different arrangements of the sensor, interaction objects, display device, and the features that are recorded to detect changes produced by the user activity. In earlier prototypes, user activity such as head, hand, and finger movements was detected by an array of photoelements (Kumar and Segen, 2001), through capacitive and inductive sensors (Gard, 1999). However, removing the hand from the detection area did not allow storing the last position of the pointer, the coordinates of which were controlled by hand. Instead of direct input, human gestures can be detected as a result of actions applied to a physical object, such as a box, or a stick, the spatial position of which can be recorded (Figure 29.12d). This feature was taken into account in designing the regular mouse and, later, the virtual mouse as presented in Erdem et al. (2002). The authors recorded movements of the passive object within a detection area (viewing area) of the sensor using the video-based tracking technique. The mouse buttons were defined as the particular (color) regions where the user could make a mouse click when covering the button region with a finger. Of course, such a technique is only an extension of the inverse paradigm of the optical mouse.

Figure 29.12b represents an embedded (CMOS) sensor of the interaction object1, which can track the reference texture (R), conditionally shown on a side edge, of the supporting surface doing successive frames (snapshots), which are compared by autocorrelation to ascertain the direction and amount of movement (Blalock et al., 1998). The texture could have the particular pattern appearing in Figure 29.13 to mark off spatial positions as it was implemented in the Anoto Pen (Saiseau, 2005).

Virtual mice use the active mode—radar-like techniques based on emitting acoustic, infrared, and short waves and synchronous detection of reflected signals, for example, their attenuation or phase modulation due to the displacement of the target object. In passive detection mode natural lighting (daylight or fluorescent lamps) is usually used. The object can also act as an active source of the directed signals (acoustic, electromagnet, and light/laser emission) (Miyaoku et al., 2004; Mühlehner and Miesenberger, 2004; SIM, 2006) or landmarks (Figure 29.12c). Cantzler and Hoile used screen corners of the monitor as the landmarks to determine the position and orientation of cameras regarding the monitor (Cantzler and Hoile, 2003). The infrared emission of human skin/body was used as well. In the enhanced mode, the features of the detection area could be provided through the particular gradient of an external radiation, such as magnitude and frequency modulation or polarization of the magnetic, electrostatic, and acoustic (ultrasound) fields (Kumar and Segen, 2001; Matsui and Yamamoto, 2001; Soundbeam, 2006; Stanley, 2006).

The markers array can enhance motion detection techniques. This method can be used to detect and delimit multiple features during object tracking to add functionality to the virtual pointer

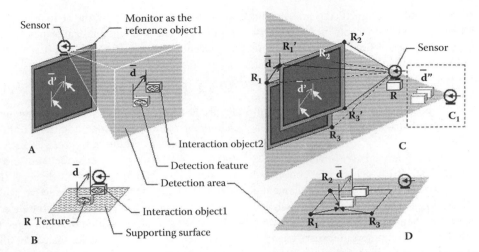

FIGURE 29.12 Recording the relative displacement ($\bar{d}$ – displacement vector) of the interaction object(s), one of which acts as the reference. Displacement ($\bar{d}$) is transformed into ($\bar{d}'$) of the pointer on the monitor screen.

The case A: The sensor associated with the reference object1 tracks object2.
The case B: The sensor associated with the interaction object1 tracks the reference pattern (R texture).
The case C: The sensor associated with the reference object (R) tracks the monitor displacement ($\bar{d}$) using the object features (R_1, R_2, R_3).
The case C_1: The sensor tracks the monitor displacement ($\bar{d}$) and deviation ($\bar{d}''$) of the object R.
The case D: The sensor tracks the displacement ($\bar{d}$) of the object within the detection area by triangulation (R_1, R_2, R_3).

(SIM, 2006). The markers can improve the signal-noise ratio and robustness of their detection. Active markers can be produced using different technology that is more or less expensive. But the markers being attached to the user can also worsen usability of the virtual human-computer interaction (on TrackIR, 2006).

As the input device, a motion detector can track the external object or the particular features of the object using video-based tracking techniques. In the last decade, new algorithms for real-time object recognition and tracking have been developed demanding less computational complexity and being appropriate for a large variety of objects with different colors and textures (Comaniciu et al., 2000; Kamenick et al., 2005). Video-based tracking possesses greater features and functionality. For example, instead of tracking the particular object2 (Figure 29.12A), a camera can detect its own displacement or the displacement of the object mechanically associated with the camera. Such a technique is especially appropriate for handheld devices equipped

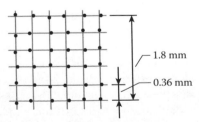

FIGURE 29.13 The micropattern structure of the Anoto patterned paper. (Adapted from Saiseau, A., *Le papier numérique, une opportunité pour les imprimeurs*, http://cerig.efpg.inpg.fr/Note/2005/anoto-papier.htm and http://www.anotogroup.com/cldoc/aog36.htm, 2005.)

with cameras (personal digital assistants, smart phones, ultra-mobile PC).

29.6.2 Review of Case Studies

Ballagas et al. (2005) demonstrated that the mobile phone could be used as a pointing device for public terminals and large displays. The successive background images captured by the camera were processed to determine the relative motion of the mobile phone in a vertical plane (x, y) with a possibility of rotating the phone (α) in the same plane.

Gai et al. (2005) have evaluated a similar tracking method using a NOKIA 6630 camera phone. Their algorithm was based on the detection of a feature (600 candidate points) such as an edge or a corner and tracking this feature within the scene. Shifting the image sample in various directions the algorithm calculates the vector of maximum change within a restricted search area (15 by 15 pixels). However, as stated by the authors, the algorithm occupies a considerable percentage of the CPU resources of the mobile phone. Mouse pointing was considered slow and tracking unreliable. The camera key OK was used as the left mouse key.

Zhang (2003) showed that fingertip tracking on an arbitrary quadrangle-shaped plane object could be used to simulate mouse and joystick input. The projective mapping (homography) between the image plane and the screen allows generating various system events. Toggling and activation of the commands was implemented through dwell time. The author demonstrated that the algorithms and techniques were flexible, robust, and accurate for interaction with Windows applications. In particular, the functionality of vision-based interaction with fingers and paper was illustrated through the use of a calculator, painting, entering text, and manipulating 3D objects. Still, a preliminary calibration of the neutral position of the plane object is an essential procedure before fingertip detection and tracking.

Tsang and Pun studied the possibility of controlling a computer by free finger movements in the air (Tsang and Pun, 2005). They developed a method for skin-color detection by color segmentation, based on the chrominance component of the video input stream and density regularization to reinforce the regions of skin-color pixels. To minimize computation, the authors used an effective search region restricted to the area of 60 by 60 pixels. The algorithm continuously shifts the search region in such a way that the center of the search coincides with the last fingertip location. However, the restricted detection window limits the speed of finger motion that should be no faster than 68 pixels per second, at a frame rate of CMOS sensor of 24 fps and resolution of 320 by 240 pixels. Any faster movement will fail to be detected. A typical resolution of monitors is about 75 pixels per inch (i.e., 17 inches at 1024 by 768 pixels) or higher, meaning that the speed of finger motion should be less than 1 inch per second. A click function was considered a specific motion that was recorded during the last 20 positions of the fingertip.

A vision-based simulation of the computer mouse demonstrated by Argyros and Lourakis (2006) was implemented using tracking and recognition of hand gestures combined with fingertip detection and interpretable vocabularies. Herewith, one hand was assigned to control pointer movements and the other hand or both hands were used to simulate regular commands as well as activation/deactivation of the virtual mouse.

Changbo et al. have designed a system for hand tracking and gesture recognition—Virtual Mouse. The tracking technique was based on an improved version of the Condensation algorithm, and active shape model (51 points). The recognition was based on Hidden Markov Models. The system works in nearly real time; the mouse's action was determined after about a two-frame delay (Hu et al., 2000).

Gorodnichy and Roth (2004) used a convex-shaped nose feature for face tracking implemented with two cameras (stereo tracking). The described technique was robust and accurate regardless of the orientation of the head and the camera. However, to accomplish a command mode the authors additionally used voluntary eye blinks, which many people consider an inconvenient method for making a selection. The authors also did not consider the computational capacity of the method.

The combination of a laser beam and video input has potential in the design of virtual mice or joysticks. However, a conventional approach such as when a motion sensor detects a position of the laser spot would greatly depend on human factors, such as personal perceptual and motor skills. Hand jitter and dexterity, deficiency of visual-motor coordination, and the haptic (kinesthetic) sense have an impact on user performance when a laser pointer is used. The authors who studied laser pointers reported that pointing and selection are slow, error-prone, and inaccurate due to hand unsteadiness (Myers et al., 2002). Still,

a laser is attractive to test new ideas, algorithms, and concepts such as optical (beam color and size) and spatial (beam direction) modulation. Mühlehner and Miesenberger have recently discussed an alternative mouse system called the HeadControl+ (2004). The prototype consists of a laser pointer headset and web camera to track a position of the laser beam spot. As stated by the authors, the virtual keyboard can be projected onto the navigation area within the field of view of the camera, for example, "in front of, next to or over the computer display" (Mühlehner and Miesenberger, 2004, p. 775). Among other things, inappropriate projective mapping between head gestures and movements of the laser spot could cause additional problems. Still, it is difficult to estimate the approach presented conceptually without thorough empirical studies.

Most of the authors and inventors aim at designing a virtual mouse using video-based tracking techniques. However, the target groups of end-users of the virtual mouse are very different in their abilities, from nondisabled people to those who can control only head or eye movements. Virtual mice have to be adapted to various settings and conditions that usually impose many constraints, such as lighting and occlusion, the number of dimensions and a range of movements to be recorded, distance to sensor, tracking latency, robustness, pointing speed and accuracy, delimiters of continuous tracking, simulation of mouse buttons functionality, and so on. The performance of the motion detector is a crucial parameter of virtual mice. Therefore, the adopted techniques and algorithms are different even within the group of video-based trackers (Betke etal, 2002; Evreinova, Evreinov, and Raisamo, 2006; EyeTwig.com, 2006; Face MOUSE, 2006). When a virtual mouse is used only in Windows applications, the technique should mimic, at least, the click features of a regular mouse, such as single click, double click, and right click, and common operations, such as select, launch, copy, paste, and, optionally, drag and drop. When the virtual mouse and virtual keyboard are used together, the interaction techniques must be coordinated to benefit the user from such integration.

29.7 Conclusion

Different solutions and principles, which have been implemented to improve the usability of text entry, were collected to illustrate progress and indicate possible problems and alternatives. Both the basic and advanced human-computer interactions are still oriented on a point-and-click paradigm. Even the stylus-based input on mobile devices often mimics the mouse interaction model developed for desktop interfaces.

There are many text entry features and the problems of user interface for text entry that can be done more efficiently. The number of tasks needed for everyday work of the regular user is quite restricted. Usually, people are concerned with a remote communication service and network, the applications to process office documents (text, tables, and presentations) and entertainment activities (games, sport, music, movies). Direct/physical manipulation cannot be an efficient style of interaction for numerous tasks and applications when trial-and-error is a sole way to find the best solution (game play, web search, text writing). Herewith we have to note that most of the applications can easily be connected to the web and so be considered part of a global database. There is already about 2×10^9 textual documents in different formats (txt, doc, rtf, pdf, docx) and about 0.5×10^9 images indexed by Google (retrieved on May 10, 2008), which could be used to augment communication abilities for elderly people and users with special needs.

But a rich interface design having a multi or multiple extension puts high demands on user abilities and discourages people from using smart tools and applications. Any further design improvements extending mouse tricks and touch screen interaction (gestures, hovering) demand more cognitive effort and motor skills.

The user intuitively prefers to perform only the minimum of habitual actions to interact with the minimum set of interface components and subtasks. Personal preferences are conservative, and often an innovation (application or technique) suffers from neglecting the cognitive factors in the interaction model. To optimize human-computer communication, the interaction techniques should be strictly adaptive to personal cognitive and sensory-motor abilities.

Multifunctionality and accessibility of any key can easily be realized using OSKs. The universality of the OSK as an external memory aid is obvious when people with diverse needs use any suitable alternative input techniques to communicate information and to produce commands.

Virtual input tools could augment interaction and make it more natural and intuitive. Still, virtual techniques were mostly developed as case studies. However, technology progress will promote ubiquitous video input as a way for user-device integration with intelligent systems. Being applied in perceptual, attentive, and affective interfaces, video-based interaction demonstrates a wideband input. Rather than primitive tracking (delivering coordinates of any feature), video input aims to provide a system with information about the user attention and intent.

Probably in the near future, through games and entertainment computing virtual interaction techniques can become a conventional and universal input. But, is it possible to find the perfect and universal trade-off between typing speed, user skills, and abilities? This is still an open question.

References

Ahn, J. and Kim, M. H. (2005). General-purpose text entry rules for devices with 4x3 configurations of buttons, in the *Proceedings of the International Conference on Computational Science and its Appearance (ICCSA 2005)*, 9–12 May 2005, Singapore, pp. 223–231. Berlin/Heidelberg: Springer-Verlag.

Arato, A., Juhasz, Z., Blenkhorn, P., Evans, G., and Evreinov, G. (2004). Java-powered Braille Slate Talker, in the *Proceedings of the 9th International Conference on Computers Helping People with Special Needs (ICCHP 2004)*, 7–9 July 2004, Paris, pp. 506–513. Berlin/Heidelberg: Springer-Verlag.

Argyros, A. A. and Lourakis, M. I. A. (2006). Vision-based interpretation of hand gestures for remote control of a computer mouse, in the *Proceedings of the HCI'06 Workshop in conjunction with ECCV'06*, 7–13 May 2006, Graz, Austria, pp. 40–51. Berlin/Heidelberg: Springer-Verlag.

Auer, C. M., Castagno, D. L., Haley, Jr., A. W., Moore, IV, H. H., O'Leary, S. E., Paley, S. J., et al. (1988). *Computer Interface Device*. U.S. Patent 4,725,694.

Ballagas, R., Rohs, M., and Sheridan, J. G. (2005). Mobile phones as pointing devices, in the *Proceedings of the Workshop on Pervasive Mobile Interaction Devices (PERMID) at PERVASIVE 2005*, May 2005, Munich, Germany. http://media.informatik.rwth-aachen.de/materials/publications/ballagas2005b.pdf.

BeamOne HoloTouch Evaluation System (2006). http://www.holodemo.com/Datasheets/BeamOneDatasheet.pdf.

Beck, C., Seisenbacher, G., Edelmayer, G., and Zagler, W. L. (2004). First user test results with the predictive typing system FASTY, in the *Proceedings of the 9th International Conference on Computers Helping People with Special Needs (ICCHP 2004)*, 7–9 July 2004, Paris, pp. 813–819. Berlin/Heidelberg: Springer.

Belatar, M. and Poirier, F. (2005). Entrée de données pour les systèmes interactifs nomades. Glyph: écrire avec 7 touches en un seul geste. Rapport de stage Master2 Recherche IHM. 2005. (Data input for interactive systems. Glyph: Writing with 7 keys in a single gesture). M.S. thesis, Université de Bretagne Sud, Campus de Tohannic, Vannes, France.) http://www-valoria.univ-ubs.fr/Mohammed.Belatar/docs/RAPPORT_DEA.pdf.

Betke, M., Gips, J., and Fleming, P. (2002). The camera mouse: Visual tracking of body features to provide computer access for people with severe disabilities. *IEEE Transactions on Neural Systems and Rehabilitation Engineering* 10: 1–10.

Blalock, T. N., Baumgartner, R. A., Hornak, T., and Smith, M. T. (1998). *Method and Device for Tracking Relative Movement by Correlating Signals from an Array of Photoelements*. U.S. Patent 5729008. http://www.patentstorm.us/patents/5729008.html.

Boissiere, P. (2003). An overview of existing writing assistance systems, in the *Proceedings of the IFRATH Workshop 2003*, 16–17 October 2003, Paris, pp. 158–165. New York: ACM Press.

Cantzler, H. and Hoile, C. (2003). A novel form of a pointing device, in the *Proceedings of Vision, Video, and Graphics 2003*, 10–11 July 2003, Bath, U.K., pp. 1–6. Geneva: EUROGRAPHICS Association.

Clark, R. J. (1971). *Light Probe Circuit for Persistent Screen Display System*. U.S. patent 3,579,225.

Comaniciu, D., Ramesh, V., and Meer, P. (2000). Real-time tracking of non-rigid objects using mean shift, in the *Proceedings of the IEEE Conference of Computer Vision and Pattern Recognition (CVPR'00)*, Volume 2, 13–15 June 2000, Hilton Head Island, SC, pp. 142–149. Los Alamitos, CA: IEEE Computer Society.

de Silva, G. C., Lyons, M. J., and Tetsutani, N. (2004). Vision based acquisition of mouth actions for human-computer interaction, in the *Proceedings of the 8th Pacific Rim International Conference on Artificial Intelligence (PRICAI 2004)*, 9–13 August 2004, Auckland, New Zealand, pp. 959–960. Berlin/Heidelberg: Springer-Verlag.

Du, H., Oggier, T., Lustenberger, F., and Charbon, E. (2005). A virtual keyboard based on true-3D optical ranging, in the *Proceedings of the British Machine Vision Conference (BMVC 2005), Volume 1*, 5–8 September 2005, Oxford, U.K., pp. 220–229. Oxford, U.K.: Alden Group Ltd.

DynaVox (2008). *DynaVox Technologies Product Information.* http://www.dynavoxtech.com.

Erdem, A., Erdem, E., Yardimci, Y., Atalay, V., and Çetýn, A. E. (2002). Computer vision based mouse, in the *Proceedings of the IEEE International Conference on Acoustics, Speech, and Signal Processing (ICASSP '02), Volume 4*, 13–17 May 2002, Orlando, FL, pp. 4178–4181. Los Alamitos, CA: IEEE Computer Society.

Evreinov, G., Agranovski, A., Yashkin, A., and Evreinova, T. (1999). PadGraph, in *Human-Computer Interaction: Communication, Cooperation, and Application Design, Volume 2 of the Proceedings of HCI International'99 (the 8th International Conference on Human-Computer Interaction)* (H.-J. Bullinger and J. Ziegler, eds.), 22–27 August 1999, Munich, Germany, pp. 985–989. London: Lawrence Erlbaum Associates.

Evreinov, G. and Raisamo, R. (2002a). Cyclic input of characters through a single button manipulation, in the *Proceedings of the 8th International Conference on Computers Helping People with Special Needs (ICCHP 2002)*, 15–20 July 2002, Linz, Austria, pp. 259–266. Berlin/Heidelberg: Springer-Verlag.

Evreinov, G. and Raisamo, R. (2002b). Tactile pointer for touch screen manipulations, in the *Proceedings of HANDICAP 2002 2ème Conférence "pour l'essor des technologies d'assistance."* 13–14 June 2002, Paris, pp. 115–120. L'Institut Federativf de Recherche sur les Aides Techniques pour Personnes Handicapees. Paris: IFRATH.

Evreinov, G. and Raisamo, R. (2004). Optimizing menu selection process for single-switch manipulation, in the *Proceedings of the 9th International Conference on Computers Helping People with Special Needs (ICCHP 2004)*, 7–9 July 2004, Paris, pp. 836–844. Berlin/Heidelberg: Springer-Verlag.

Evreinova, T. and Evreinov, G. (2005). Four-key text entry augmented with color blinking feedback for print-handicapped people with ocular pathology, in the *Proceedings of the 1st International Workshop on IT Design for All (ITdesign4all 2005) in Conjunction with DEXA 2005*, 22–26 August 2005, Copenhagen, Denmark, pp. 891–895. Los Alamitos: IEEE Press.

Evreinova, T., Evreinov, G., and Raisamo, R. (2004). Four-key text entry for physically challenged people, in the *Adjunct Proceedings of the 8th ERCIM Workshop "User Interfaces for All,"* 28–29 June 2004, Vienna, Austria. http://www.ui4all.

gr/workshop2004/files/ui4all_proceedings/adjunct/techniques_devices_metaphors/16.pdf.

Evreinova, T., Evreinov, G., and Raisamo, R. (2006). Video as input: Spiral search with the sparse angular sampling, in the *Proceedings of the 21st International Symposium on Computer and Information Sciences (ISCIS 2006)* (A. Levi, E. Savas, H. Yenigün, S. Balcisoy, and Y. Saygin, eds.), 1–3 November 2006, Istanbul, Turkey, pp. 542–552. Berlin/Heidelberg: Springer-Verlag.

EyeTwig.com (2006). http://www.eyetwig.com.

FaceMOUSE (2006). http://www.aidalabs.com.

Fitts, P. M. (1954). The information capacity of the human motor system in controlling the amplitude of movement. *Journal of Experimental Psychology* 47: 381–391.

Fukumoto, M. and Tonomura, Y. (1997). Body coupled FingeRing: Wireless wearable keyboard, in the *Proceedings of the Conference on Human Factors in Computer Systems (ACM CHI 1997)*, 22–27 March 1997, Atlanta. http://www.sigchi.org/chi97/proceedings/paper/fkm.htm.

Gai, Y-B., Wang, H., and Wang, K-Q. (2005). A virtual mouse system for mobile device, in the *Proceedings of the 4th International Conference on Mobile and Ubiquitous Multimedia (MUM 2005), Volume 154*, 8–10 December 2005, Christchurch, New Zealand, pp. 127–131. New York: ACM Press.

Gard, M. D. (1999). *Computer Interface Device.* U.S. Patent 5,990,865. International class G06F 3/033.

Gindrum, R. J. (1970). *Light Pen Detection Verification System.* U.S. Patent 3,509,350.

Goldberg, D. (1997). *Unistrokes for Computerized Interpretation of Handwriting.* U.S. Patent 5,596,656.

Gorodnichy, D. O. and Roth, G. (2004). Nouse "use your nose as a mouse" perceptual vision technology for hands-free games and interfaces. *Image and Vision Computing* 22: 931–942.

Grimm, R., Syrbe, M., and Truck, H. (1978). *Input-Output-Color-Screen System.* U.S. Patent 4,079,450.

Grush, R. (2004). The emulation theory of representation: motor control, imagery, and perception. *Behavioral and Brain Sciences* 27: 377–442.

Hanheide, M., Bauckhage, C., and Sagerer, G. (2005). Combining environmental cues and head gestures to interact with wearable devices, in the *Proceedings of the 7th International Conference on Multimodal Interfaces*, 4–6 October 2005, Torento, Italy, pp. 25–31. New York: ACM Press.

Hansen, J. P., Johansen, A. S., Hansen, D. W., Itoh, K., and Mashino, S. (2003a). Command without a click: Dwell time typing by mouse and gaze selections, in the *Proceedings of the 9th IFIP TC13 International Conference on Human-Computer Interaction (INTERACT'03)* (M. Rauterberg, M. Menozzi, and J. Wesson, eds.), 1–5 September 2003, Zurich, Switzerland, pp. 121–128. Amsterdam, The Netherlands: IOS Press.

Hansen, J. P., Johansen, A. S., Hansen, D. W., Itoh, K., and Mashino, S. (2003b). Language technology in a predictive, restricted on-screen keyboard with ambiguous layout for severely disabled people, paper presented at the *EACL 2003 Workshop on Language Modeling for Text Entry Methods.* http://www.it-c.dk/research/EyeGazeInteraction/Papers/Hansen_et_al_2003a.pdf.

Harbush, K. and Kühn, M. (2003). Towards an adaptive communication aid with text input for ambiguous keyboards, in the *Proceedings of the 10th Conference of the European Chapter of the Association for Computational Linguistics (EACL'03),* 12–17 April 2003, Budapest, Hungary, pp. 207–210. Morristown, NJ: Association for Computational Linguistics.

Hashimoto, M., Yonezawa, Y., and Itoh, K. (1996). New mouse-function using teeth-chattering and potential around eyes for the physically challenged, in *Interdisciplinary Aspects in Computers Helping People with Special Needs, Proceedings of the 5th International Conference on Computers Helping People with Special Needs (ICCHP'96)* (J. Klaus, E. Auff, W. Kremser, and W. L. Zagler, eds.), 17–19 July 1996, Linz, Austria, pp. 93–98. Munich, Germany: R. Oldenbourg Verlag GmbH.

Hu, C., Liang, L., Ma, S., and Lu, H. (2000). Virtual mouse: Inputting device by hand gesture tracking and recognition, in the *Proceedings of the Third International Conference on Advances in Multimodal Interfaces (ICMI 2000),* 14–16 October 2000, Beijing, China, pp. 88–95. London: Springer-Verlag.

Human Interface Device (2008). http://en.wikipedia.org/wiki/Human_interface_device.

IO2 Technology (2006). http://www.io2technology.com.

Isokoski, P. (2000). Text input methods for eye trackers using off-screen targets, in the *Proceedings of Eye Tracking Research & Applications Symposium 2000,* 6–8 November 2000, Palm Beach Gardens, FL, pp. 15–22. New York: ACM Press.

Isokoski, P. and Raisamo, R. (2000). Device independent text input: A rationale and an example, in the *Proceedings of the International Working Conference on Advanced Visual Interfaces (AVI 2000)* (V. Di Gesù, S. Levialdi, and L. Tarantino, eds.), 23–26 May 2000, Palermo, Italy, pp. 76–83. New York: ACM Press.

Iwamura, M., Hamada, N., and Hayashi, Y. (1976). *Light Pen Detection System.* U.S. Patent 3,997,891.

Jannotti, J. (2002). *Iconic Text Entry Using a Numeric Keypad.* http://www.jannotti.com/papers/iconic-uist02.pdf.

Juhasz, Z., Arato, A., Bognar, G., Buday, L., Eberhardt, G., Markus, N. et al. (2006). Usability evaluation of the MOST mobile assistant (SlatTalker), in the *Proceedings of the 10th International Conference on Computers Helping People with Special Needs (ICCHP 2006),* 9–11 July 2006, Linz, Austria, pp. 1055–1062. Berlin/Heidelberg: Springer-Verlag.

Kamenick, T., Koenadi, A., Qiu, Z. J., and Wong, S. Y. (2005). *Webcam Face Tracking.* CS540 Project Report. http://pages.cs.wisc.edu/~jerryzhu/cs540/project/report/webcam/final_report.html.

Kawato, S. and Ohya, J. (2000). Real-time detection of nodding and head-shaking by directly detecting and tracking the between-eyes, in the *Proceedings of the Fourth IEEE International Conference on Automatic Face and Gesture Recognition 2000,* 28–30 March 2000, Grenoble, France, pp. 40–45. Los Alamitos, CA: IEEE Computer Society.

Keller, E. and Zellner, B. (1996). A timing model for fast French. *York Papers in Linguistics* 17: 53–75. http://cogprints.org/885/00/KellerZellnerTimingModel.pdf.

Keyboard Layouts (2007). http://en.wikipedia.org/wiki/Keyboard_layout.

Kjeldsen, R. (2001). Head gestures for computer control, in the *Proceedings of the IEEE ICCV Workshop on Recognition, Analysis, and Tracking of Faces and Gestures in Real-Time Systems,* July 2001, Vancouver, Canada, pp. 61–68. Los Alamitos, CA: IEEE Computer Society.

Kolsch, M. and Turk, M. (2002). *Keyboards without Keyboards: A Survey of Virtual Keyboards.* http://www.cs.ucsb.edu/research/trcs/docs/2002-21.pdf.

Kumar, S. and Segen, J. (2001). *Gesture-Based Computer Interface.* U.S. Patent 6,222,465. International class G06F 3/00.

Lesher, G. W., Higginbotham, D. J., and Moulton, B. J. (2000). Techniques for automatically updating scanning delays, in the *Proceedings of the RESNA 2000 Annual Conference: Technology for the New Millennium,* pp. 85–87. Arlington, VA: RESNA Press.

Lukaszewicz, K. (2003). The ultrasound image of the tongue surface as input for man/machine interface, in the *Proceedings of the 9th IFIP TC13 International Conference on Human-Computer Interaction (INTERACT'03)* (M. Rauterberg, M. Menozzi, and J. Wesson, eds.), 1–5 September 2003, Zurich, Switzerland, pp. 825–828. Amsterdam, The Netherlands: IOS Press.

Lyons, M. J., Chan, C., and Tetsutani, N. (2004). MouthType: Text entry by hand and mouth, in the *CHI '04 Extended Abstracts on Human Factors in Computing Systems,* pp. 1383–1386. New York: ACM Press.

MacKenzie, I. S. (2002a). KSPC (keystrokes per character) as a characteristic of text entry techniques, in the *Proceedings of the Fourth International Symposium on Human Computer Interaction with Mobile Devices,* pp. 195–210. Berlin/Heidelberg: Springer-Verlag.

MacKenzie, I. S. (2002b). Mobile text entry using three keys, in the *Proceedings of the Second Nordic Conference on Human-Computer Interaction (NordiCHI 2002),* 19–23 October 2002, Aarhus, Denmark, pp. 27–34. New York: ACM Press.

MacKenzie, I. S. (2003). *Research in Advanced User Interfaces: Models, Methods, and Measures.* http://www.cs.uta.fi/~scott/mmm.

MacKenzie, I. S. (2004). *A Note on Phrase Sets for Evaluating Text Entry Techniques.* http://www.yorku.ca/mack/RN-PhraseSet.html.

MacKenzie, I. S., Kober, H., Smith, D., Jones, T., and Skepner, E. (2001). LetterWise: Prefix-based disambiguation for mobile text input, in the *Proceedings of the ACM Symposium on User Interface Software and Technology (UIST 2001),* 11–14 November 2001, Orlando, FL, pp. 111–120. New York: ACM Press.

MacKenzie, I. S. and Soukoreff, R. W. (2002). Text entry for mobile computing: Models and methods, theory and practice. *Human-Computer Interaction* 17: 147–198.

MacKenzie, I. S. and Tanaka-Ishii, K. (eds.) (2007). *Text Entry Systems: Mobility, Accessibility, Universality*. San Francisco: Morgan Kaufmann.

MacKenzie, I. S., Zhang, S. X., and Soukoreff, R. W. (1999). Text entry using soft keyboards. *Behaviour & Information Technology* 18: 235–244.

Mankoff, J. and Abowd, G. D. (1998). Cirrin: A word-level unistroke keyboard for pen input, in the *Proceedings of UIST'98 ACM Symposium on User Interface Software and Technology*, 1–4 November 1998, San Francisco, pp. 213–214. New York: ACM Press.

Matsui, N. and Yamamoto, Y. (2001). A new input method of computers with one CCD camera: Virtual keyboard, in the *Proceedings of the 8th TC13 IFIP International Conference on Human-Computer Interaction (INTERACT'01)*, 9–13 July 2001, Tokyo, Japan, pp. 678–679. Amsterdam, The Netherlands: IOS Press.

Mehta, A. (2009). *eLocutor version 2*. http://www.radiophony.com/html_files/download2.html.

Miniotas, D., Spakov, O., and Evreinov, G. (2003). Symbol creator: An alternative eye-based text entry technique with low demand for screen space, in the *Proceedings of the 9th IFIP TC13 International Conference on Human-Computer Interaction (INTERACT'03)* (M. Rauterberg, M. Menozzi, and J. Wesson, eds.), 1–5 September 2003, Zurich, Switzerland, pp. 137–143. Amsterdam, The Netherlands: IOS Press.

Miyaoku, K., Higashino, S., and Tonomura, Y. (2004). C-blink: A hue difference-based light signal marker for large screen interaction via any mobile terminal, in the *Proceedings of the 17th Annual ACM Symposium on User Interface Software and Technology (UIST 2004)*, 24–27 October 2004, Santa Fe, NM, pp. 147–156. New York: ACM Press.

Mobience (2004). *Character Arrangements, Input Methods and Input Device*. International application #PCT/KR2004/000577. http://www.mobience.com/solution_layout_alphabet2.html.

Montgomery, E. H. (1980). *Data Input System*. U.S. Patent 4,211,497.

Mühlehner, M. and Miesenberger, K. (2004). HeadControl+: A multi-modal input device, in the *Proceedings of the 9th International Conference on Computers Helping People with Special Needs (ICCHP 2004)*, 7–9 July 2004, Paris, pp. 774–781. Berlin/Heidelberg: Springer-Verlag .

Myers, B. A., Bhatnagar, R., Nichols, J., Peck, Ch. H., Kong, D., Miller, R., and Long, A. Ch. (2002). Interacting at a distance: Measuring the performance of laser pointers and other devices, in the *Proceedings CHI'2002: Human Factors in Computing Systems*, 20–25 April 2002, Minneapolis, MN, pp. 33–40.

Nesbat, S. B. (2003). A system for fast, full text entry for small electronic devices, in the *Proceedings of the 5th International Conference on Multimedia Interfaces (ICMI'03)*, 5–7 November 2003, Vancouver, Canada, pp. 4–11. New York: ACM Press.

on TrackIR (2006). http://www.naturalpoint.com.

Paradiso, J., Hsiao, K., Strickon, J., Lifton, J., and Adler, A. (2000). Sensor systems for interactive surfaces. *IBM Systems Journal* 39: 892–914.

Perlin, K. (1998). Quikwriting: Continuous stylus-based text entry, in the *Proceedings of the 11th Annual ACM Symposium on User Interface Software and Technology (UIST'98)*, 1–4 November 1998, San Francisco, pp. 215–216. New York: ACM Press.

Pieters, L. A. (1974). *Information Selection in Image Analysis System Employing Light Scanning*. U.S. Patent 3,832,485.

Poirier, F. and Belatar, M. (2006). Glyph 2: une saisie de texte avec deux appuis de touche par caractère, principes et comparaisons, (Glyph 2: Text input using two keystrokes per character – principles and comparisons) in the *Proceedings of the 18th International Conference of the Association Francophone d'Interaction Homme-Machine, ACM International Conference Proceeding Series, Volume 133*, 18–21 April 2006, Montreal, pp. 159–162. New York: ACM Press.

Pratt, V. R. (1998). *Thumbcode: A Device-Independent Digital Sign Language*. http://boole.stanford.edu/thumbcode.

Raynal, M. (2006). Le système KeyGlass: aide à la saisie de caractères pour personnes handicapées moteur. Retours sur une première expérimentation, (The system KeyGlass: Assistance with the input of characters for motor-handicapped people. The results of the first experimentation) in the *Proceedings of Handicap 2006*, 7–9 June 2006, Paris, pp. 155–160.

Rinott, M. (2005). SonicTexting, in the *CHI '05 Extended Abstracts on Human Factors in Computing Systems*, 2–7 April 2005, Portland, OR, pp. 1144–1145. New York: ACM Press.

Rosenberg, R. (1998). *Computing without Mice and Keyboards: Text and Graphic Input Devices for Mobile Computing*. PhD thesis, University College, London.

Saiseau, A. (2005). *Le papier numérique, une opportunité pour les imprimeurs*. (The digital paper, an opportunity for the printers) http://cerig.efpg.inpg.fr/Note/2005/anoto-papier.htm; http://www.anotogroup.com/cldoc/aog36.htm.

Saito, A., Minami, M., Kawahara, Y., Morikawa, H., and Aoyama, T. (2003). Smart baton system: A universal remote control system in ubiquitous computing environment, in the *Proceedings of the ICCE '03*, 2–5 December 2003, Hong Kong, pp. 308–309.

Salem, C. and Zhai, S. (1997). An isometric tongue pointing device, in the *Proceedings of the Conference on Human Factors in Computer Systems (ACM CHI 1997)*, 22–27 March 1997, Atlanta. http://www.sigchi.org/chi97/proceedings/technote/cs.htm.

Sandnes, F. E., Arvei, A., Thorkildssen, H. W., and Bruverud, J. O. (2003). *TriKey: Mobile Text-Entry Techniques for Three Keys*. http://www.iu.hio.no/~frodes/trikey.

Sandnes, F. E. and Huang, Y.-P. (2006). Chording with spatial mnemonics: Automatic error correction for eyes-free text entry. *Information Science and Engineering* 22: 1015–1031.

Sandnes, F. E. and Jian, H.-L. (2004). Pair-wise variability index: Evaluating the cognitive difficulty of using mobile text entry systems, in the *Proceedings of the 6th International Symposium on Mobile Human-Computer Interaction (MobileHCI 2004)*, 13–16 September 2004, Glasgow, Scotland, pp. 347–350. Berlin/Heidelberg: Springer-Verlag.

Schwesig, C., Poupyrev, I., and Mori, E. (2004). Gummi: A bendable computer, in the *Proceedings of the SIGCHI Conference on Human Factors in Computing Systems (CHI 2004)*, 24–29 April 2004, Vienna, Austria, pp. 263–270. New York: ACM Press.

Seveke, E. (2006). *MauSi Scan.* http://www.computer-fuer-behinderte.de.

Shieber, S. M. and Baker, E. (2003). Abbreviated text input, in the *Proceedings of the 8th International Conference on Intelligent User Interfaces (IUI 2003)*, 12–15 January 2003, Miami, pp. 293–296. New York: ACM Press.

Shieber, S. M. and Nelken, R. (2006). Abbreviated text input using language modeling. *Natural Language Engineering* 1: 1–19.

SIBYLLE (2006). *SIBYLLE: An Adaptive Communication Aid.* http://www.sir.blois.univ-tours.fr/~antoine/SIBYLLE/Sibylle_en/Sibylle_en.html.

SIM (2006). *Self-Identified Markers for Motion Capture Systems.* http://www.ptiphoenix.com/products/visualeyez/simarker.php.

Simpson, R. C. and Koester, H. H. (1999). Adaptive one-switch row-column scanning. *IEEE Transactions on Rehabilitation Engineering* 7: 464–473.

Sokey (2006). http://www.mobience.com.

Soundbeam (2006). http://www.soundbeam.co.uk.

Stanley, M. (2006). *Non Contact Human-Computer Interface.* U.S. Patent 20060238490. International class G09G 5/00.

Surakka, V., Illi, M., and Isokoski, P. (2004). Gazing and frowning as a new technique for human-computer interaction. *ACM Transactions on Applied Perception* 1: 40–56.

Sutherland, I. E. (1980). *Sketchpad: A Man-Machine Graphical Communication System.* New York: Garland Publishers.

Tian-Jian Jiang, M., Zhan, J., Lin, J., Lin, J., Hsu, W. (2007). An automated evaluation metric for Chinese text entry. *Journal of Computation and Language (CoRR).* http://arxiv.org/abs/0704.3662.

Toyama, K. (1998). Look, Ma—No hands! Hands-free cursor control with real-time 3D face tracking, in the *Proceedings of the 1998 Workshop on Perceptual User Interfaces (PUI '98)*, 4–6 November 1998, San Francisco. http://www.cs.ucsb.edu/conferences/PUI/PUIWorkshop98/Papers/Toyama.pdf.

Tsang, W-W. M. and Pun, K-P. (2005). A finger-tracking virtual mouse realized in an embedded system, in the *Proceedings of 2005 International Symposium on Intelligent Signal Processing and Communication Systems (ISPACS 2005)*, December 2005, Hong Kong, pp. 781–784. Los Alamitos, CA: IEEE Computer Society.

VKey (2006). http://www.lumio.com.

Wandmacher, T. and Antoine, J-Y. (2006). Training language models without appropriate language resources: Experiments with an AAC system for disabled people, in the *Proceedings of the 5th European conference on Language Resources and Evaluation (LREC 2005)*, 24–26 May 2006, Genoa, Italy. http://www.sir.blois.univ-tours.fr/~antoine/articles/06_LREC_Wandmacher_Antoine.pdf.

WiViK on-screen keyboard (2006). http://www.wivik.com.

Wobbrock, J. O. and Myers, B. A. (2006). Analyzing the input stream for character-level errors in unconstrained text entry evaluations. *ACM Transactions on Computer-Human Interaction (TOCHI)* 13: 458–489.

Wobbrock, J. O., Myers, B. A., and Aung, H. H. (2004). Writing with a joystick: A comparison of date stamp, selection keyboard and EdgeWrite, in the *Proceedings of Graphics Interface 2004 Conference (GI '04)*, 17–19 May 2004, London, Canada, pp. 1–8. Waterloo, Canada: Canadian Human-Computer Communications Society.

Wobbrock, J. O., Myers, B. A., and Rothrock, B. (2006). Few-key text entry revisited: Mnemonic gestures on four keys, in the *Proceedings of the ACM Conference on Human Factors in Computing Systems 2006 (CHI '06)*, 22–27 April 2006, Quebec, Canada, pp. 489–492. New York: ACM Press.

Word Frequencies (2004). University of Tampere. http://www.cs.uta.fi/%7Escott/mmm/WordFrequencies.xls.

Yfantidis, G. and Evreinov, G. (2004). Adaptive blind interaction technique for touchscreens. *Universal Access in Information Society* 4: 344–353.

Yfantidis, G. and Evreinov, G. (2005). Blind text entry for mobile devices, in *Assistive Technology: From Virtuality to Reality, Proceedings of AAATE 2005* (A. Pruski and H. Knops, eds.), 6–9 September 2005, Lille, France, pp. 246–250. Amsterdam, The Netherlands: IOS Press.

Zhai, S., Hunter, M., and Smith, B. A. (2000). The Metropolis keyboard: An exploration of quantitative techniques for virtual keyboard design, in the *Proceeding of the 13th Annual ACM Symposium on User Interface Software and Technology (UIST 2000)*, 5–8 November 2000, San Diego, pp. 119–128. New York: ACM Press.

Zhai, S. and Kristensson, P-O. (2003). Shorthand writing on stylus keyboard, in the *Proceedings of the SIGCHI Conference on Human Factors in Computing Systems 2003 (CHI 2003)*, 5–10 April 2003, Fort Lauderdale, FL, pp. 97–104. New York: ACM Press.

Zhang, Z. (2003). Vision-based interaction with fingers and papers, in the *Proceedings of the International Symposium on the CREST Digital Archiving Project (CREST03)*, 23–24 May 2003, Tokyo, Japan, pp. 83–106.

Zimmermann, G., Vanderheiden, G., and Gilman, A. (2002). Prototype implementations for a universal remote console speci.cation, in the *Proceedings of the CHI '02*, 20–25 April 2002, Minneapolis, pp. 510–511. New York: ACM Press.

30

Speech Input to Support Universal Access

Jinjuan Feng and
Andrew Sears

30.1 Introduction

Speech-based interactions allow users to communicate with computers or computer-related devices without the use of a keyboard, mouse, buttons, or any other physical interaction device. By leveraging a skill that is mastered early in life, speech-based interactions have the potential to be more natural than interactions using other technologies such as the keyboard. Based on the input and output channels being employed, speech interactions can be categorized into three groups: spoken dialogue systems, speech output systems, and speech recognition systems. Spoken dialogue systems include applications that utilize speech for both input and output, such as telephony systems and speech-based environment control systems with voice feedback. Speech output systems include applications that only utilize speech for output while leveraging other technologies, such as the keyboard and mouse, for input. Screen access software, which is often used by individuals with visual impairments, is an example of speech output. Speech recognition systems include applications that utilize speech for input and other modalities for output, such as speech-based cursor control in a GUI (graphical user interface) and speech-based dictation systems (see Table 30.1). The focus of this chapter is on those interactions where speech is used to provide input to some kind of computing technology. When discussing speech-based input, potential applications can be divided into three major categories, which are most easily distinguished based on the size of the vocabulary that the system recognizes, but the reality is that there is no clear dividing line that separates these categories. Vocabulary size is a continuous variable with systems recognizing as few as two or as many as tens of thousands of words. In the subsequent discussion, both speech and nonspeech output are considered. Typical applications include:

- Telephony systems, which tend to use small input vocabularies as well as speech output, environmental control applications with small input vocabularies that may support speech or nonspeech output
- Speech-based interactions with GUIs can support navigation, window manipulations, and various other command-based interactions with widely varying input vocabularies ranging from just a few words to several hundred
- Dictation applications, which support users as they compose e-mails, letters, and reports as well as smaller tasks such as filling in portions of forms where free-form input is allowed

From the perspective of universal access (UA), speech-based interactions should be considered one of a set of tools available to help address the goal of ensuring that information technologies are accessible by all citizens as they address a variety of tasks in diverse contexts. While UA is concerned with addressing the needs of all possible users, three populations are of particular interest: children, older adults, and individuals with disabilities. For older users, who may be experiencing age-related visual and physical impairments, the graphical interface with a keyboard and mouse for input can present a variety of challenges, while speech can offer a natural style of interaction and reduce the need for physical interactions. Educational software, toys, and various web sites use speech to provide information to children, which serves as a natural and potentially easy-to-learn input solution. Perhaps the most obvious population that could benefit from speech-based interactions are individuals with physical impairments that hinder their use of more traditional input devices, such as the keyboard and mouse. For these users, speech can provide effective, inexpensive interaction solutions.

By taking a broad view of UA, we need to address the needs of a diverse population of users as well as a variety of tasks and varied environments in which these tasks are accomplished. As technology advances, the notion of computing devices has evolved from the traditional desktop computers to include an endless array of mobile devices and the extensive use of computing devices that are embedded in larger systems. Possibilities include entering data using a mobile device while walking (Price et al., 2006), in-vehicle reminder or navigation systems (Zajicek and Jonsson, 2006), and communication devices for firefighters (Jiang et al., 2004). In each of these scenarios, the user's hands may be busy with another task (e.g., driving a car, conducting medical procedures) and the traditional keyboard and mouse may be inaccessible or inappropriate, making speech a compelling input alternative.

While speech has been highlighted as a promising technology for decades, and there are clear opportunities to leverage the capabilities of this technology, adoption rates for speech-based applications are largely disappointing (Gotte, 2000) and user-based evaluations are often discouraging (e.g., Koester, 2004). Frequently, this lack of success has been attributed to the substantial challenges associated with the nature of human speech and limitations in existing design. Perhaps the most critical obstacle to date has been recognition errors and the cumbersome recovery process associated with these unavoidable errors (e.g., Karat et al., 1999; Sears et al., 2001). For speech-based applications to be successful, designers must understand these challenges and design systems to leverage the strengths and minimize the impact of the weaknesses of the technology.

This chapter provides an overview of the application of speech recognition in the context of UA. It begins by describing several common applications of speech recognition, highlighting both the current and potential users of these systems. Next, research on speech-based solutions to UA challenges is examined, with a focus on those activities that address speech-based input. Based on existing research, a range of issues are discussed to consider when designing, evaluating, and disseminating speech-based systems, before concluding with a discussion of future trends and unaddressed issues.

There are several related bodies of research that are beyond the scope of the chapter. First, systems that employ speech output, such as screen access software for individuals with visual impairments, are not discussed. For more information regarding this topic, the readers are referred to Chapter 28 of this handbook, as well as to the proceedings of the ASSETS conference (e.g., ASSETS 2006).

Second, this chapter does not discuss typical spoken dialogue systems/conversational user interface since research in this field is rarely motivated by the goal of UA. For more information about spoken dialogue systems, the reader is referred to Chapter 31, "Natural Language and Dialogue Interfaces," as well as to Lai et al. (2007) and the proceedings of the CHI conference (e.g., CHI 2007). There are also numerous books published on this topic such as *Voice User Interface Design* by Cohen et al. (2004) and *Voice Interaction Design* by Harris (2005). Finally, speech is often used in multimodal systems. For more information on multimodal interactions, including the use of speech in such interactions, the reader is referred to Chapter 40 of this handbook and to Oviatt (2007).

30.2 Speech-Based Applications

As highlighted previously, there are a wide variety of speech-based applications, which can be categorized in several ways. One critical criterion to consider is the size of the input vocabulary, as this influences the technology that is needed, the range of possibilities that the application can support, and the likelihood of recognition errors. The remainder of this section discusses several important categories of speech-based applications, the size of the input vocabulary supported by these applications, the underlying technology required to support the application, and the intended user population. For a more comprehensive review of the underlying technologies, as well as additional discussion of design principles for speech-based applications, see Lai et al. (2007).

30.2.1 Small Vocabulary Solutions

Many applications of speech recognition use small vocabularies, which can significantly reduce recognition error rates. One common example is speech-based command and control systems that allow users to interact with computers or computer-related devices by speaking predefined commands. Such systems often translate the user's utterances into system commands by matching the acoustic signal with models associated with each of the currently available commands. In some systems, these models are predefined for all available commands, while other systems allow users to customize these models. For example, voice dialing on a mobile phone allows users to record the name to be associated with a phone number. As long as the names remain distinct, accuracy can be maintained. However, similar sounding names can lead to recognition errors.

Telephony applications are another common example of a small vocabulary speech-based interaction. These systems are widely used to handle information inquiries, order processing, and a variety of other tasks. Some systems use a simple menu system, allowing users to select the appropriate option at each stage using simple spoken commands. Other systems try to establish a dialogue with the user in an attempt to gather the necessary information and complete the transaction as quickly as possible. In both cases, the speech recognition engine has a relatively small set of words that it expects at any given time. When designed effectively, such systems can be easy to learn to use, while allowing many transactions to be completed quickly.

Feedback to the user can be delivered through visual or auditory cues, depending on the nature of the system. With voice dialing, the system may play back the recorded name that matches the spoken command while simultaneously displaying the individual's name and phone number. With environmental control systems, the user may see a light, television, or other device turn on or off. If a visual display is used, the spoken command may be displayed, so the user can confirm that it is accurate, or it may simply be executed requiring the user to undo the resulting action in the case of a recognition error.

Unlike dictation systems, the speech recognition engines used to support command and control systems are often optimized for use with a small vocabulary, ranging from just a few to several hundred words. A small vocabulary tends to increase recognition accuracy, but a variety of factors can increase the number of recognition errors. Poorly designed command sets, which result in multiple commands that sound alike, will produce more recognition errors than well-designed distinct command sets. This can be a serious challenge when users are allowed to define the commands, as is the case with voice dialing.

Command and control systems have been used for a variety of tasks. For instance, multiple speech-based command and control solutions exist that emulate the conventional keyboard and mouse, allowing users to manipulate traditional GUIs via speech. These solutions are used by some individuals with physical disabilities who have difficulty using more traditional interaction solutions. Speech-based command and control systems are used for environmental control applications, allowing users to manipulate thermostats, lights, televisions, and any number of other devices. Such applications can be more than convenient, providing increased independence for many individuals.

As technology evolves, people are using mobile devices to access information from less predictable and less stable environments. Using devices in a noisy environment, or an environment where the background noise changes, can have a significant impact on the accuracy of the speech recognition process. This is effectively illustrated by voice dialing, which can be used under widely varying conditions. As many readers may have experienced, voice dialing is more likely to produce erroneous results when used in a noisy environment. At times this could be as simple as failing to recognize any input, but a noisy environment can also cause input to be recognized incorrectly which, in this case, can result in a call being placed to the wrong

individual. Creating custom models for the spoken commands can help reduce errors, but requires more time and introduces the additional challenge of knowing who the user is before processing any input.

30.2.2 Large Vocabulary: Dictation Systems

Speech-based dictation systems can allow users to generate large quantities of text via speech. Like the applications discussed previously, these systems use a speech recognizer to translate an audio signal into text. Perhaps the key difference is that dictation systems use a much larger vocabulary, which often exceeds 20,000 words. As a consequence of the larger vocabulary, dictation applications tend to be less accurate than command and control systems, and it is normally recommended that users create a personal speech profile to improve recognition accuracy. Vendors have claimed recognition accuracy rates as high as 99% for a number of years (e.g., Dragon NaturallySpeaking, 2006). However, accuracy rates observed when users completed composition-oriented tasks representative of typical usage scenarios have been substantially lower than those reported by vendors (e.g., Sears et al., 2001). Research has also confirmed that speech-based solutions for correcting recognition errors have been slow, error prone, and frustrating. Multiple studies concluded that users spent more time correcting errors when using hands-free solutions than they did dictating (Karat et al., 1999; Sears et al., 2001). When feasible for the user, environment, and task, effective multimodal error correction solutions can provide an efficient alternative.

Numerous injuries or diseases can impair motor function in the hands or arms without affecting speech. Examples include amputation, high-level spinal cord injuries (SCI),[1] amyotrophic lateral sclerosis (ALS)[2] at specific stages, and repetitive strain injuries (RSI).[3] For individuals with these or comparable injuries or diseases, large vocabulary dictation systems can provide a powerful alternative to the traditional keyboard and mouse, allowing them to generate a variety of documents such as e-mails, papers, and business reports. Importantly, the ability to produce such documents can significantly increase both educational and career opportunities. For the general public, such systems may serve as a useful alternative reducing the risk of keyboard or mouse-based RSI.

Table 30.1 summarizes the typical speech-based applications. As illustrated, speech-based command and control systems have been employed to address varying needs of individual users, challenges introduced by the environment in which the interactions occur, as well as the need to support a variety of tasks.

---

[1] Injuries that occur when the spinal cord (a collection of nerves extending from the base of the brain through the spinal column) is compressed, cut, damaged, or affected by disease.

[2] Also known as "Lou Gehrig's disease," ALS is a progressive neurodegenerative disease that affects nerve cells in the brain and the spinal cord.

[3] Injuries that occur from repeated physical movements, damaging tendons, nerves, muscles, and other soft body tissues, especially those in hands and arms.

TABLE 30.1 Summary of Speech Interaction Applications Employed in the Field of Universal Access

| | Diversified Population | | | | | Diversified Tasks | Diversified Contexts of Use |
| | Various Impairments | | | Age-Related Diversity | | | |
	Physical Impairments	Perceptual Impairments	Cognitive Impairments	Older Users	Younger Users		
Small vocabulary command and control systems	Interaction with GUI Environment control	Not explored	Not explored	Environment control	Toys Educational software	In vehicle driving aids (to be fully explored)	Home environment In vehicle driving aids (to be fully explored)
Large vocabulary Dictation systems (voice recognition systems)	Communication tasks Generating lecture notes	Hearing aids Generating lecture notes	Generating lecture notes	Not explored	Not explored	Healthcare field for generating reports	Not explored
Spoken dialogue systems	Telephony systems	Telephony systems	Not explored	Telephony systems	Not explored	Eyes busy tasks Hands busy tasks	Mobile environment
Speech output systems	Screen access for users with severe disabilities	Screen access software	Text to speech to facilitate reading	Reminders Driving assistant	Educational software	Eyes busy tasks Hands busy tasks	Mobile environment

Dictation systems provide critical support for communications activities for individuals with upper body physical impairments. One area that has received little attention is the use of speech-based solutions by individuals with cognitive impairments.

30.3 Research Issues in Speech Interaction

Extensive research has investigated the design, implementation, and evaluation of speech interfaces, especially speech-based command and control solutions. Since a comprehensive review of research on speech-based interactions is beyond the scope of this chapter, the primary goal here is to provide a summary of the current state of knowledge with respect to designing speech applications for diverse users and contexts of use. A secondary goal is to highlight unsolved problems in need of additional research. This section begins by discussing how speech interactions have been used by individuals with physical, hearing, visual, and cognitive impairments. Next, it discusses the use of speech applications by older users and children. It concludes with an examination of the relationship between speech applications and the environments in which they are used. The focus is on articles that explicitly address the needs of specific user populations or the use of speech recognition technologies in more nontraditional environments such as a moving vehicle. As a result, much of the research investigating the use of speech recognition for telephony applications is not reviewed.

30.3.1 Speech-Based Interactions for Users with Physical Impairments

Speech-based interactions may be most useful when a disease or injury results in impaired motor function in the hands or arms without affecting the individual's ability to speak. In these situations, the traditional keyboard and mouse can be difficult or impossible to use, limiting an individual's access to both education and employment (see Chapter 5 of this handbook). Currently, employment rates for individuals with disabilities are low. For example, only approximately 25% of working age individuals with high-level SCI are employed (SCI Information Network, 2006). For this group of individuals who cannot use traditional input techniques effectively, speech recognition provides a viable alternative for interacting with computing devices.

Three domains appear to offer significant promise with respect to individuals with physical impairments: (1) environmental control systems that allow for increased independence, (2) command and control systems that support interactions with standard GUIs available on many computing systems, and (3) dictation systems that allow individuals to generate e-mails, reports, papers, and other written materials.

30.3.1.1 Environmental Control Systems

Environmental control systems employ various techniques, such as speech recognition, to provide access to equipment in home or office environments. Early speech-enabled environmental control systems supported a very limited vocabulary and strictly defined dialogues, allowing users to interact with a limited number of devices. As natural language processing and speech recognition become more robust, environmental control systems are supporting increasingly more flexible dialogues and larger vocabularies. This, in turn, allows users to control more devices. There has been extensive research on environmental control systems, including research that produced the Ubiquitous Talker (Nagao and Rekimoto, 1995), context-sensitive natural-language-based interactions (Coen et al., 1999), an agent-based architecture for speech dialogues (Quesada et al., 2001), and the speech graffiti technique (Tomko and Rosenfeld, 2004). While somewhat relevant, these papers focus on technical and implementation-oriented concerns, and do not directly address the design of effective solutions for individuals with physical impairments. As a result, they are not reviewed in detail within this chapter.

With the increasing number of digital devices present in the home or office environment, correctly identifying the target device becomes more challenging. One alternative is to move from a speech-only solution to multimodal interactions. For example, Wilson and Shafer (2003) developed a system, XWand, that allows users to identify the target device by pointing a wand. The device is then controlled using either speech or gestures. For example, a user could turn on a light by pointing the wand and saying "Turn on." While an experiment was conducted to evaluate the accuracy of the pointing solution, information was not provided regarding the accuracy of the speech recognizer, user preferences, or the overall effectiveness of the system.

More recently, Davis et al. (2003) proposed a framework for capturing context-aware user profiles to improve speech systems for severely disabled users. Leong et al. (2005) went a step further, developing and evaluating CASIS, a context-aware speech system for controlling devices in either a home or office environment. Special attention was given to reducing recognition errors, as well as the impact of those errors that did occur. More specifically, items in the n-best list were reordered using additional information on the physical context, such as the user's location and room temperature. A small user study, which did not include individuals with disabilities, suggested that additional information about the context of use reduced the error rate by 41%.

Most articles that discuss speech-enabled environmental controls focus on theory, systems architecture, and implementation. Unfortunately, even when systems are implemented, they are rarely subjected to a comprehensive user-based evaluation. Further, when user studies are conducted, some of the decisions regarding the designs of these studies raise questions leaving the validity of the results uncertain. For example, both Wilson and Shafer (2003) and Leong et al. (2005) used participants with technical backgrounds that may make generalizing the results to the public difficult. It is even more difficult to determine how their results may relate to the experiences of individuals with physical disabilities. Studies including participants that are more representative of the intended target users would provide more informative results regarding the efficacy of the proposed solutions.

30.3.1.2 Command and Control Systems for GUIs

To interact with information technologies that use a GUI, users need an effective method for controlling the cursor location with the most common solution being the traditional mouse. However, users with physical impairments that affect their hands or arms can experience significant difficulty using a mouse. Older users with decreased fine motor control, including tremors, may also find it challenging to use a mouse, especially when the object they are selecting is small (e.g., Keates and Trewin, 2005). Numerous alternatives have been explored, such as a mouth stick, head-tracking devices (e.g., LoPresti and Brienza, 2004), eye-controlled interactions (e.g., Lankford, 2000), speech-based solutions (e.g., Sears et al., 2003a), and electrophysiological technologies (e.g., Birch et al., 2002). Among these alternatives, speech is particularly promising, since the underlying technology is sufficiently mature to support commercial applications, it is relatively inexpensive, the necessary hardware is readily available, and the setup process is simple.

Manaris and Harkreader (1998) explored the use of speech recognition to generate keystrokes and mouse events with the goal of providing individuals with upper-body motor-control impairments more effective access to traditional GUIs. Two navigation techniques were available: (1) direction-based navigation that resulted in continuous cursor movements (e.g., "Move left" followed by "Stop") and (2) target-based navigation that caused the cursor to jump to one of five predefined regions of the screen. Manaris et al. (2002) described two user-based studies of their solution, with the first including three users with upper-body impairments that interfered with the use of a keyboard and mouse. Participants completed a series of tasks using their preferred input method, as well as the new speech-based solution. All participants were comfortable with the new interaction solution, and data entry rates were comparable between their new speech-based solution and the participants' preferred input methods. The second study compared the new interface to a handstick with 43 able-bodied users transcribing one paragraph using a handstick and a second paragraph using the speech interface. The speech interface allowed for a 37% decrease in task completion times, a 74% increase in typing rates, and a 63% decrease in error rates.

Several papers investigated a variety of alternatives for improving the direction-based navigation solution employed by Manaris and Harkreader (1998). For example, Karimullah and Sears (2002) evaluated the efficacy of a predictive cursor designed to help users compensate for the processing delays associated with speech recognition. A total of 28 able-bodied participants completed a series of target selection tasks, half using the standard direction-based solution and half using a solution that was enhanced with a second cursor designed to predict where the actual cursor would stop if a command were issued at that moment. Interestingly, the predictive cursor did not result in improvements, suggesting that users were reasonably adept at compensating for the speech recognition processing delays. In a follow-up study, Sears et al. (2002) allowed users to vary the speed of the cursor, confirming that a variable speed cursor can allow for improved performance, especially when the object being selected is small.

Grid-based solutions position the cursor using recursive grids (most often 3x3 grids), which allow users to drill down until the cursor is in the desired location on the screen (Kamel and

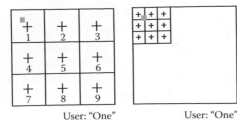

FIGURE 30.1 Demonstration of grid-based technique. (From Dai, L., Goldman, R., Sears, A., and Lozier, J., *Behaviour and Information Technology*, 24, 219–230, 2005.)

Landay, 2002). As illustrated in Figure 30.1, to select the gray square the user must start by saying "one" (i.e., selecting the top left cell). At this point, they would likely say "one" again, but they could also say "two" since the target spans the border of these two cells. Two variations of the grid-based solution were compared by Dai et al. (2005). The first provides nine cursors, one in the center of each cell. The second solution provides a single cursor in the middle cell. While the nine-cursor solution allowed users to select targets more quickly, the single-cursor solution resulted in lower error rates.

Researchers have also investigated the use of nonverbal vocalizations to control the cursor. De Mauro et al. (2001) proposed a system that generated continuous cursor movements in response to simple vocalizations. Unlike most implementations where words are mapped to commands, de Mauro et al. mapped the sounds of individual vowels to commands. Unfortunately, no data were provided regarding the efficacy of their navigation mechanism. Similarly, Harada et al. (2006) developed a vocal joystick that allowed users to control the cursor by varying vocal parameters such as vowel quality, loudness, and pitch. Their system processes vocal characteristics in every audio frame, transforming the input into various control parameters to manipulate the cursor. As a result, the system is highly responsive, with changes in vocal characteristics being mapped rapidly to changes in the system state. A small study with four expert users without physical impairments, who were part of the research team, confirmed that Fitts' law could be used to describe expert performance with the vocal joystick. A second study with nine novices compared the vocal joystick with a grid-based solution and the traditional speech-based cursor control technique explored by Manaris, Sears, and others. The vocal joystick and grid-based solution allowed for comparable target selection times, while the more traditional speech-based cursor technique was significantly slower.

Mihara et al. (2005) presented a system in which multiple ghost cursors are aligned vertically or horizontally with the actual cursor. The system supports rapid discrete cursor movements (e.g., saying "seven" causes the cursor to jump to the seventh ghost cursor) and slower continuous cursor movements using nonverbal vocalizations to fine tune the cursor location (e.g., "Ahhhh..." causes the cursor to move slowly in a predefined direction). Five individuals with no physical impairments, all of whom were members of the research laboratory,

completed a series of tasks using the new cursor control solution. Target selection times were promising when compared to previously reported results for other speech-based solutions, but participants did report some fatigue when using nonverbal vocalizations to control the cursor.

Sporka et al. (2006) explored the control of computer games using speech and nonverbal vocalizations. Two versions of an arcade game of Tetris were implemented. One version used speech commands (e.g., "left," "right") to rotate objects, while the other used humming. Twelve able-bodied participants used both solutions. The humming solution was approximately 2.5 times faster than the speech-controlled solution while simultaneously reducing errors. The humming solution was even more beneficial as the distance the object had to move or the difficulty of the game increased.

While each of these techniques allows the cursor to be positioned anywhere on the screen, they do not take advantage of contextual information that may be available. For example, icon names can allow for direct selection, and knowledge that an individual is editing a text document can allow for word-by-word navigation. Target-based navigation solutions exploit contextual information in an effort to make speech-based navigation more efficient. For instance, Christian et al. (2000) explored speech-based navigation for the web. Users could read the words that served as links to navigate within or between documents. Their system automatically associated numbers with links, providing a second method of navigating that can be particularly useful if multiple links share the same labels (e.g., multiple links to "More" information). While errors were minimal with the speech-based solution, tasks could be completed significantly faster with a mouse. More important, the tasks were constrained such that no other commands were active and users could not dictate text. As a result, the reported error rates are not representative of what would be expected in the context of a dictation-oriented application. Table 30.2 summarizes the speech or nonverbal cursor control techniques available for interacting with GUI as well as text editing.

30.3.1.3 Dictation Solutions

Dictation systems can allow users to generate a large quantity of text without typing. Combined with the speech-based command and control solutions, these systems enable users to complete a full range of computer-related activities essential for employment or education purposes. As a result, dictation systems can be powerful tools for individuals with physical impairments that make it difficult or slow to use other interaction solutions.

Pieper and Kobsa (1999) developed a speech-based system that allowed severely impaired users who are bedridden to interact with a computer. Starting with a commercially available system, the authors adapted the system to more effectively match the environment in which it would be used. For example, instead of using a standard screen for output, information was displayed on the ceiling so that users could read it while lying in bed. The speech recognition engine was adjusted to filter out the background noise produced by respiratory devices. One patient

TABLE 30.2   Overview of Cursor Control Techniques Using Speech or Nonverbal Sounds

Technique	Description	Example	Appropriate Application Domain	Example References
Direction-based (continuous)	Move the cursor continuously by specifying movement direction and the position to stop	Move up…stop	GUI	Manaris and Harkreader (1998); Karimullah and Sears (2002)
Direction-based (discrete)	Move the cursor by specifying movement direction	Move up, move up four lines	Text editing	Christian et al. (2000); Sears et al. (2003a)
Target-based	Move the cursor by specifying the destination of movement	Select 'reader'; start 'WORD'	Text editing, GUI	Sears et al. (2003a)
Anchor-based	A variation of direction-based technique with carefully selected words and targets as navigation anchors to speed up navigation process	Next, previous, move up	Text editing	Feng and Sears (2004)
Grid-based	Select the target using recursive grids	One, one, six	GUI	Kamel and Landay (2002); Dai et al. (2005)
Nonverbal control	Move the cursor continuously by issuing vocal sounds	Ahhhh….	GUI	de Mauro et al. (2001); Harada et al. (2006)
Ghost cursor	Select the target by specifying the number of the ghost cursor closest to the target (may be combined with continuous movement techniques for final selection)	Seven, thirteen	GUI Text editing	Mihara et al. (2005)

with ALS used the system for 7 months to write poetry, diaries, and letters. Significant training was required for the user to use the system effectively. The researchers also confirmed that commands could not be more than three syllables due to the respiratory cycle of the breathing device. Overall, the user was satisfied with the system.

Manasse et al. (2000) examined how a 19-year-old female survivor of a traumatic brain injury (TBI) interacted with a dictation system to generate text. The participant had both cognitive challenges (e.g., difficulty in reading, dictation, and writing) and motor challenges (e.g., ataxia and tremors in arms and legs). Using a standard keyboard, the participant could type 10 words per minute. While the participant experienced a recognition error rate of approximately 20%, she quickly mastered the navigation and error correction commands. Interestingly, the participant generated more text using the standard keyboard, but produced more complex sentences when using the speech recognition system.

While recognition accuracy has increased, wide adoption of speech recognition systems continues to be hindered by the error-prone nature of the technology and the cumbersome error recovery process. While dictating text should be the primary focus, several studies report that experienced users still spent a vast majority of their time correcting errors (Karat et al., 1999; Sears et al., 2001). Sears et al. (2001) investigated how individuals

with high-level SCI employed voice recognition systems to compose text documents. Seven individuals with high-level SCI and seven individuals with no physical impairments completed four dictation tasks each. No differences were identified between SCI participants and traditional users in terms of recognition error rates, navigation commands failure rates, and overall productivity. However, participants with SCI tended to interrupt their dictation to correct errors more often, and were more satisfied with their interactions (Oseitutu et al., 2001; Sears et al., 2001).

Sears et al. (2003a) explored the challenges encountered with speech-based navigation when using speech recognition to produce text documents. Both target- and direction-based navigation commands failed frequently, with error rates ranging between 15% and 19%. The vast majority of failed navigation commands were associated with recognition errors, invalid commands (e.g., say "choose day" instead of "select day" or "correct day"), and long pauses in the middle of issuing a command. Failed commands resulted in a variety of consequences, but cascading errors were arguably the most significant challenge. Cascading errors refer to situations when new errors occur while the user is correcting an earlier error (Karat et al., 1999). In these situations, users could spend significant time correcting errors and become quite frustrated before they could continue dictating new text. Several modifications were proposed for the navigation commands with a follow-up study confirming that the proposed revisions reduced failure rates as well as the severity of the consequences experienced when commands did fail.

Feng and Sears (2004) proposed a new navigation technique that employs selected words or targets in text documents as navigation anchors to speed up the target selection process. The system offers six simple, short navigation commands: Next, Previous, Move up, Move down, Move right, and Move left. The Next and Previous commands move the cursor between navigation anchors. The four Move commands move the cursor to the specified direction by one line or one word at a time. Since all six commands are short, easy to construct, and phonetically distinctive, the failure rates of those commands are significantly lower compared to target-based technique. The navigation anchors allow the user to skip multiple words and, in many cases, land the cursor directly onto the desired target by a single Next command. A preliminary user study suggested that the anchor-based technique achieves satisfactory efficiency at very low error rates (i.e., approximately 5%).

While making individual commands more effective can allow for more efficient interactions, Halverson et al. (1999) confirmed that user experience must also be considered. More specifically, Halverson et al. found that novice able-bodied users tended to fixate on a single error correction strategy and that this often led to frustrating and inefficient experiences. While more experienced users changed strategies more readily, the tendency of novices to fixate on a single solution has important implications. Since an unsatisfactory initial experience reduces the likelihood of continued usage, it may be useful to ensure that novices are directed toward reliable interaction mechanisms even if these are not necessarily the most powerful solutions available. While speech is a more natural solution for generating text, and speech-based spatial navigation solutions continue to be less efficient than the mouse, researchers continue to develop new alternatives, and refine existing, speech-based alternatives for pointing tasks.

30.3.1.4 Programming Solutions

In addition to producing generic text documents, researchers have investigated the use of speech recognition to complete programming tasks. Unlike standard dictation applications, programming normally involves the use of numerous user-defined labels that may not be part of any predefined vocabulary (e.g., countEvent, myPointer). Therefore, effective solutions for spelling words that are not recognized using the predefined vocabulary are necessary. Arnold et al. (2000) proposed a VocalProgramming system that integrated speech recognition and a predefined syntax for the programming language. Hubbell et al. (2006) developed a syntax-directed graphical editor for programmers with physical impairments. A preliminary usability test with a small number of participants, including one individual with severe physical impairments, suggested that the performance of the system was acceptable. Similar to speech-based word-processing systems, recognition error rates were high and recognition errors had a significant impact on the effectiveness of the system.

30.3.2 Speech Interaction for Users with Cognitive, Hearing, or Language Impairments

Speech recognition has been applied in clinical language remediation to help individuals with cognitive, speech, language, or hearing impairments develop language and communication skills. At the same time, few studies have investigated the use of speech for people with hearing and language impairments from the human-computer interaction perspective.

Lehman (1998) described a system called Simone Says that provided an interactive environment for language remediation for individuals, especially children, with pervasive developmental disorders. Pervasive development disorders are also known as autistic spectrum disorders (ASD), and usually result in impairments in reciprocal social interaction as well as communication skills. Simone Says led children through the normal developmental sequence to achieve a set of specific linguistic goals (e.g., vocabulary, basic syntax, conversational turn taking). A graphical interface supported by speech recognition and natural language processing provided a simple and natural communication environment. A Wizard-of-Oz experiment was proposed to validate several design issues, such as tolerable recognition error rate, but empirical results were not reported.

Wade et al. (2001) conducted a preliminary study that investigated the use of voice recognition by individuals with aphasia. Since the standard training process of speech recognition software is highly demanding for users with aphasia, a new method of training is explored, aiming at bypassing the standard training process. Six individuals with aphasia together with five nonimpaired users participated in the study. Different from the standard training process, participants used a vocabulary of 50 words and 24 phrases to train the speech software over a maximum of five sessions. The result suggested that there was no significant difference in word level accuracy between participants with aphasia and nonimpaired participants. However, participants with aphasia had lower phrase level accuracy as compared to nonimpaired participants. This study demonstrated the potential of the new training method in facilitating individuals with aphasia to access speech technology.

Cox et al. (2002) developed TESSA, a prototype system designed to facilitate communications between a deaf person and a post office clerk. The clerk's speech was processed using speech recognition with the resulting text separated into one or more phrases. The phrases were transformed into British Sign Language (BSL) and presented using an avatar. Ambiguous tasks, combined with several sources of errors, introduced significant challenges. Environmental noise resulted in recognition errors, recognized text was occasionally divided into phrases incorrectly, and transforming phrases into BSL sometimes introduced additional errors. An empirical evaluation involving seven prelingually deaf users and three post office clerks evaluated the third step: transforming phrases into signs. Participants were quite positive, providing an average rating of 2.2 out of 3 when assessing the acceptability of the signs produced by the system. The accuracy of the speech recognition process, and the impact of speech recognition errors on overall system performance, were not discussed.

30.3.3 Speech Interaction for Users with Impairments in an Educational Setting

Information technologies are playing an increasingly important role in education. While speech recognition (SR) has proven useful in a variety of educational applications, including educational software (e.g., Russell et al., 1996; Shi et al., 2003), speech rehabilitation (Lehman, 1998), and special education (Wade et al., 2001), this section focuses on the use of speech recognition to facilitate learning and note taking. Speech recognition has been used to transform lectures into text in real time, allowing students to read and annotate lecture notes during class. While this could prove beneficial for all students, the primary focus has been on students with impairments. For instance, students with impaired hearing usually require interpreters, note takers, or some form of assistive listening device to capture important information from lectures. Students with physical impairments that slow their typing or writing could also use a speech recognition system to produce or supplement their notes. Similarly, students with learning disabilities that may hinder their ability to process information presented orally, or their ability to produce written notes, could also benefit.

Bain et al. (2002) reported an ongoing project investigating the use of SR-based lecture transcription as a learning tool in classroom settings, including the impact of recognition errors on performance and user acceptance. While recognition accuracy rates of 98% have been reported, results under realistic usage conditions tend to be lower and transcribing lectures is even more challenging as lectures are characterized by extemporaneously generated speech. Bain et al. (2002) noted that recognition error rates are insufficient if the goal is to understand how errors impact transcription comprehension as some errors impact comprehension more than others. They also noted that the time and effort involved in correcting recognition errors was problematic.

Hewitt et al. (2005) developed SpeakView, a system that produced live captioning for lectures in a classroom setting. Working in parallel with Microsoft PowerPoint, the system displayed the captions in conjunction with the instructor's materials. Unlike most existing commercial speech recognition systems, which require users to dictate punctuation, SpeakView automatically adds punctuation to the text. Sixty-nine students evaluated the system through a study in which an instructor presented two London tourist attractions alongside PowerPoint slides. One talk was given with captions generated by SpeakView at the bottom of the screen, the other without any captions. No information was provided regarding the recognition accuracy of the automatically generated captions. The majority of the participants expressed a preference for presentations with captions, indicating that talks with captions were easier to understand, and that they felt they were able to answer more questions correctly with

the captions. While these systems are promising, many questions remain unanswered, including student tolerance for errors, the impact of environment noise, whether students should be allowed to annotate or edit notes during class, and how such a system alters the learning environment, including student participation in the classroom.

30.3.4 Speech Interaction for Children

Researchers and practitioners have long acknowledged the potential of speech interactions for younger users, including infants, children, and teenagers. For instance, the Foundation for Blind Children (FBC) specifically lists speech recognition as one interaction solution for blind children who have not learned typing (FBC, 2007). At the same time, children present unique challenges, since existing models that drive speech recognition engines were developed using adult speech, which differs significantly from that of children (Aist et al., 1998). As a result, the recognition accuracy for children tends to be lower and more variable when compared to adults (Giuliani and Gerosa, 2003). Fortunately, significant research has focused on constructing better acoustic and linguistic models for children to improve recognition accuracy (e.g., Aist et al., 1998; Gustafson and Sjolander, 2002). Given the algorithmic focus of these efforts, they are not reviewed here as they are beyond the scope of this chapter.

Nix et al. (1998) developed a speech recognition system, Watch Me! Read, to teach reading skills to 5- to 7-year-old emerging readers. Two specific challenges were explored: (1) creating a child-appropriate acoustic model to improve recognition accuracy, and (2) designing an effective system in the presence of significant system errors. An initial test with an adult acoustic model yielded a recognition accuracy rate of 75% for children. A new acoustic model was developed using 48,000 instances of spoken words from 5- to 7-year-old boys and girls from dialectally diverse regions of the United States, yielding a recognition accuracy rate of 95%. With the new acoustic model, the system achieved a false negative rate of 5% (the system judged the child's speech as incorrect when the word was pronounced correctly) and a false positive rate of 4% (the system judged the child's speech as correct when the word was pronounced incorrectly). To boost the child's tolerance of these system errors, it was suggested that the system evaluation of the child's speech be conveyed more subtly. Instead of indicating that the child's pronunciation was incorrect, the system presented a simpler task when detecting an incorrect response.

More recently, Oviatt (2000) explored the design of appropriate and robust conversational interfaces for children by comparing the speech of children and adults when interacting with an animated character. Ten children between 6 and 10 years of age participated in the study, with results suggesting that children had significantly higher disfluency rates than adults. More specifically, 25% of the children's utterances contained disfluencies or idiosyncratic lexical contents, including invented words, concatenated words, or mispronounced words. These findings provided insights regarding the development of more accurate acoustic and linguistic models as well as how speech interfaces could be designed to be more tolerant of errors.

Arunachalam et al. (2001) adopted a different perspective, investigating both politeness and frustration in the language used by children when interacting with computer systems via speech with the goal of assessing the user experience. Speech data were collected from 160 children 6 to 14 years of age. Using a Wizard-of-Oz experiment allowed the children to interact with an error-free speech-based system. A discourse analysis found that younger children (8 to 10 years) were less likely to use overt politeness markers (e.g., Thank you, Please) or more polite information requests (e.g., Can you, Will you, Could you) as compared to older children (11 to 14 years). In addition, younger children expressed frustration verbally (e.g., using words such as "shut up," "heck," "damn") more than older children. The difference identified between the two age groups confirmed that diversity exists even though ages only differed by a few years, suggesting different interface solutions may be needed.

Researchers also explored the possibility of using speech interactions to detect speech or hearing impairments in infants. Fell et al. (1996) developed EVA, an early vocalization analyzer, to help predict articulation and language difficulties based on early babbling activities. The system consisted of a segmenter that provided start and stop times for each utterance and a post-processor that counted the utterances. EVA was tested using four infants 4 to 6 months of age. An audio recording approximately 60 minutes in duration was collected while each infant was playing with a quiet toy. The audio file was subsequently analyzed by both EVA and a human judge. The results suggested that EVA and the human judge had 92.8% agreement on the number of utterances, 79.8% agreement on duration of utterances, and 87.3% agreement on frequency of utterances. The system could be used as a standard for the analysis and evaluation of utterances of both normal and at-risk infants to predict later speech or language disorders.

30.3.5 Speech Interactions in Dynamic Environments

Context-aware computing attempts to leverage information about the context in which the interactions are occurring to make those interactions more efficient (Moran and Dourish, 2001). Sears et al. (2003b) introduced the phrase "situationally induced impairments" to describe the temporary challenges users experience due to the environment in which they are working or the activities they are engaged in. For example, the vibration experienced in a moving vehicle may cause difficulty typing. Dim, variable, or excessively bright lighting, such as that experienced at night, outdoors, or walking outside on a sunny day, may cause difficulty reading from a PDA screen. Similarly, physical exertion or environmental noise may complicate speech-based interactions.

While the most common data entry solutions for mobile devices rely on small keyboards or stylus-based interactions, these solutions tend to require both visual attention and the use of both hands. As a result, existing input techniques are not necessarily appropriate for use while in motion. Speech is a promising alternative for mobile data entry, since it allows the user to move freely, can be used while one's hands are busy with another activity, and does not require the user to look at the device to enter text. Unfortunately, the effectiveness of speech interaction can be reduced by motion as well as environmental factors, including noise. Several studies have investigated how these factors affect the efficacy of speech-based interactions in the context of mobile interactions. For instance, Rollins (1985) explored changes in speaking patterns associated with noisy environments and how these changes affect speech recognition. Holzman (2001) examined the impact of noise and lighting on the use of a portable medical documentation system that used speaker-independent speech recognition for data entry.

More recently, Entwistle (2003) studied the effect of physical exertion on speech recognition accuracy. Three conditions were investigated: dictation while rested, dictation after light physical exercise, and dictation after harder physical exercise. Under the second and the third conditions, participants exercised for a specific amount of time and then read a newspaper article to a speech recognition system. The results suggested that recognition accuracy decreased with increased physical exertion. Since data were collected after participants stopped exercising, the results were attributed to changes in speaking patterns following physical exertion, and not with ongoing physical exertion or motion itself.

Building on Entwistle's observations, Price et al. (2006) conducted a study investigating the impact of motion on speech recognition performance. Participants completed speech-based data entry tasks while seated or walking. Error rates increased significantly when participants were walking while the impact of the conditions experienced while completing the standard enrollment process varied. When users enrolled while seated, but completed tasks while walking, recognition errors increased. In contrast, if users enrolled while walking and completed tasks while seated, recognition errors decreased. The authors concluded that additional studies were necessary to accurately interpret these results.

Others have proposed or developed mobile systems with speech-based input. For instance, Manaris et al. (1999) proposed a user interface model for UA to mobile computing devices using speech recognition to provide access to a virtual keyboard and mouse. Sawhney and Schmandt (2000) developed the Nomadic Radio, a wearable computing device for managing voice and text-based messages. Users used speech to navigate among messages and compose new messages with incoming text messages being presented using speech synthesis. A preliminary evaluation was conducted with two novice users for a period of 3 days. The results suggest that the conceptual model for browsing messages nonvisually in an audio interface might be challenging for users. Increasingly, speech-based interactions are being employed in commercial products designed for use in mobile environments. For example, GPS systems that accept input via speech are currently being developed (Peterson, 2006).

The studies and systems discussed in this section confirm the potential of speech interactions for diverse user populations and usage scenarios. At the same time, several challenges were highlighted, including:

1. Recognition errors continue to be a significant problem that must be addressed in all applications.
2. Speech-based navigation poses significant challenges, and navigation in nonvisual environments introduces additional challenges.
3. Speech-based interfaces can introduce additional cognitive challenges as users may be required to remember commands instead of recognizing the desired alternative as in menu-based systems.
4. Environmental factors, including both noise and movement, can affect recognition accuracy. At this point, the relationship between enrollment condition, usage condition, and recognition errors needs to be studied in more detail.
5. Speech-based interactions can lead to both privacy and security concerns.

30.4 Lessons Learned for Design, Evaluation, and Dissemination of Speech Applications

Speech-based input can be useful for individuals with impairments that hinder their use of traditional input devices. At the same time, recognition errors introduce significant challenges, vocabularies vary among user groups, and speech-based interfaces can introduce additional cognitive demands. While visual feedback may reduce the benefits of eyes-free interactions, it can also help users understand how well their speech is being recognized and what commands may currently be valid. While many issues are in need of additional research, existing research provides several important insights regarding the use of speech as an input solution. These lessons are derived both from experiences studying and implementing speech applications for users with physical impairments and from related research. Many of the challenges involved in conducting user-based studies of speech-based systems are similar to those encountered when studying any other interactive information technology. For a discussion of additional issues that should be considered when testing speech-based applications, the reader is referred to Lai et al. (2007).

30.4.1 Get Target Users Involved

As with any other application, it is important to get representatives from the target user group involved as early as possible when designing speech-based interfaces. While this is a common recommendation within the human-computer interaction (HCI) community, involving representative users may be even more important for speech applications. First, the success of a speech application depends on understanding the vocabulary, speech patterns, and interaction preferences of the target users. If the dialogue effectively leverages the users' vocabulary, the interaction will be more natural, out-of-vocabulary problems will be reduced, and commands will be easier for users to learn, remember, and retrieve. Second, different user communities may adopt different interaction strategies when completing the same tasks. For example, it was found that users with high-level spinal cord injuries adopted different interaction strategies as compared to users without impairments (Oseitutu et al., 2001). While it is common to use participants without disabilities in early proof-of-concept studies, results confirm that this must be done with caution.

Getting target users involved can be a challenge, especially when the intended users are individuals with impairments, older users, or children. Finding and recruiting a sufficient number of participants with impairments can often be a challenge. Even when sufficient participants are recruited, conducting user studies with individuals with impairments can be more complicated and time consuming when compared to studies involving computer users without disabilities. Studies including individuals with disabilities may need to be conducted at the participants' home, workplace, or a medical facility, increasing the time and expense, as well as the challenges associated with having limited control over the experimental conditions (e.g., Feng et al., 2005). Others have reported that running focus groups involving older users was challenging as the older users' attention often drifted away from the current topic (Zajicek, 2006).

30.4.2 Design for Dynamic Capabilities

Ideally, solutions will be designed to accommodate physical, perceptual, and cognitive capabilities that may change with time. Examples include the gradual changes often associated with getting older, changes associated with progressive diseases such as ALS, or conditions where capabilities vary from day to day such as multiple sclerosis, or the evolving skills and knowledge of children as they get older. Situationally induced impairments associated with the environment in which users are interacting or the activities they are engaged in introduce additional challenges, as these difficulties tend to be both short-term and highly dynamic. As a result, individuals experiencing situationally induced impairments are less likely to develop strategies to overcome these challenges. Speech-based interactions may be one piece of the solution, but multimodal interfaces also offer significant promise, since appropriate interaction techniques could be selected based on the current circumstances.

30.4.3 Dialogue Design Is the Key

In essence, a speech interface is a dialogue between user and computer. An effective dialogue will leverage the users' language, vocabulary, and communication conventions. When designing speech-based dialogues, one must carefully consider the trade-off between efficiency and reliability. For speech applications, larger vocabularies are associated with more recognition errors. As a result, some commands that offer significant functionality may also be more error prone. For example, simple navigation commands such as Move up or Move down can be highly reliable, but may only move the cursor up or down one line within a text document. In contrast, Select target can allow the user to move the cursor directly to any word on the screen, but is also more likely to result in recognition errors, since Target can be replaced by any word in the current document. Research confirms that carefully manipulating this trade-off between efficiency and reliability can improve overall system performance.

30.4.4 Design for Error

Due to the nature of human speech, which is ambiguous, inconsistent, and highly context dependent, recognition errors are largely unavoidable in speech-based applications. Since new speech is recognized in the context of previously recognized words, recognition errors often occur in clusters with one error being followed immediately by another. A similar pattern has been observed for spoken commands with additional errors occurring as a user works to correct the consequences of an earlier error. These cascading errors can result in lengthy, frustrating delays. When designing speech-based applications, it is critical to provide acceptable, if not ideal, solutions for correcting the consequences of recognition errors. Several issues that are worth consideration include:

- *Multimodal solutions for error correction*: While speech-based navigation solutions exist, other interaction modalities tend to provide more efficient alternatives. When feasible, designers should consider offering multimodal error correction solutions that allow users to correct errors using a variety of approaches such as speech, keyboard, mouse, stylus, or a touch screen.
- *Reduce the consequences of failed commands*: Traditionally, researchers and developers have invested significant time in the effort to reduce error rates. In contrast, the consequences of these errors have been understudied. With traditional interaction technologies, such as the keyboard and mouse, this approach often provides acceptable solutions since system errors are quite rare. In contrast, system errors (i.e., recognition errors) are quite common in speech applications and the consequences of these errors can vary dramatically. Previous research has confirmed that system usability can be improved significantly by reducing the consequences of failed commands when those failures cannot be eliminated (Sears et al., 2003b).
- *Encourage effective strategy development*: When users can choose among a set of alternatives, different interaction strategies will be developed. The adoption of an ineffective

strategy can significantly degrade the interaction experience and, in many cases, can lead to abandonment of the technology. Therefore, when designing speech-based interactions, the designers should consider the various strategies users may develop and encourage or guide users toward more effective strategies.

30.4.5 Learning Effects

Speech-based interactions differ from those that utilize the traditional keyboard and mouse. While speaking is quite natural, users must learn how to dictate, memorize built-in commands, adapt to the presence of recognition errors, and master error recovery solutions. At the same time, users must also develop interaction strategies. Novices tend to fixate on a single interaction strategy. With experience users are more likely to change their strategies depending on the current situation. Unfortunately, if novices adopt inefficient strategies, they can become frustrated, increasing the likelihood they will abandon the technology.

While a number of studies investigated the performance of novice users or users with extended use of speech systems, little has been learned about how interaction strategies are developed or how they evolve as users gain experience. Feng et al. (2005) reported a longitudinal study that explicitly investigated how users learned to use a dictation application. Participants completed 6 to 8 hours of dictation tasks over a span of 8 days. Productivity, measured in corrected words per minute, increased from 11 wpm on day one to 18 wpm by the end of the study. Given appropriate guidance, participants were able to develop efficient speech-based navigation strategies. One set of commands, which reliably moved the cursor short distances, was used when the desired word was close to the cursor and a fundamentally different approach, which was more error prone but could move the cursor long distances with a single command, was employed when the desired word was far from the cursor. This study suggests that when multiple alternatives are provided, each should have clear strengths and weaknesses such that users can learn when each approach should be used. While this study provided valuable insights, significant research is needed to better understand how strategies are developed over longer periods, with more diverse applications, and more diverse users.

30.4.6 Dissemination

Dissemination is a critical but largely underinvestigated step in the life cycle of assistive technology. It has been reported that about one-third of all assistive technologies are abandoned (Scherer, 1996), but the abandonment rate for speech technologies may be even higher. For example, Goette (2000) interviewed individuals with physical or perceptual impairments and reported that 42.5% had abandoned speech recognition systems. Goette reported that adoption was more likely if user expectations were realistic. Users who successfully adopted speech systems expected to devote more time and effort than stated in the manual before becoming proficient. Unfortunately, many users have unrealistic views regarding speech recognition, expecting their initial interactions to be as smooth as a conversation with another person. Given such expectations, it is not surprising that those users were disappointed after interacting with an error-prone speech recognition system. In fact, many users are frustrated during their initial interactions with speech applications due to unnatural dialogues, frequent recognition errors, and difficulty in navigation. Therefore, it is critical that users be educated regarding the challenges associated with speech-based interactions such that they can develop more realistic expectations toward the technology. An overstated and unrealistic description of the technology may attract users to try the system, but will quickly lead to frustration and will likely turn them away from the technology.

For the same reason, sufficient training should be provided on important issues such as strategy development and error recovery. A well-designed help function that supports both user-initiated queries and system-initiated guidance may help users tackle the challenges of speech interactions.

30.5 The Future of Speech Interaction

There are many important unsolved questions in the realm of speech recognition. Among them, four areas have been understudied and have the potential to lead to far-reaching results. First, the field of speech interaction is in crucial need of more effective user and task models. Most earlier findings regarding speech interactions are highly case specific and difficult to be generalized to other contexts. Development of systematic user and task models will provide valuable guidance to the design and evaluation of speech systems, and assist in the generalization of research findings.

Second, generalizable guidelines and standards would be of great value as developers seek to implement more effective speech-based solutions. The lack of effective guidelines and standards makes the design, development, and implementation of speech applications both case dependent and time consuming. While there has been significant progress, due primarily to the effort of the World Wide Web Consortium (W3C 2007), more work is needed. Useful resources and examples include the Voice Browser project, VoiceXML documentation, and Speech Recognition Grammar Specifications for web access applications involving speech technology.

Third, more long-term user studies of speech interactions in realistic conditions are needed. Currently, little is known about expert performance with speech applications, and even less is known about how this expertise is developed. Most extant studies examine novice users, users with a few hours' experience, and in some instances users who had spent several dozen hours interacting with speech applications. In each study, predefined, highly focused tasks were completed in a controlled laboratory setting. Long-term studies that last several months or even years are needed to examine important questions, including:

- For which tasks are speech-based applications used when these applications are made available in work or home environments?
- What recognition error rates do users experience when completing realistic tasks at home and at work?
- What are the adoption/abandonment rates of various speech applications?
- How long do users invest before deciding whether they will adopt or abandon speech technology?
- What are the major factors that influence the adoption/ abandonment decision?

Finally, it is important to explore the potential of speech interactions in a variety of additional domains. Within the field of UA, the majority of previous speech interaction research has focused on the use of speech as an alternative input solution for users with motor impairments. In recent years, some researchers have investigated issues relevant for older users, children, and individuals with hearing, cognitive, or communication impairments. However, this research is still quite preliminary, and any findings that have been reported are still in need of additional validation. There is also a disconnection between research in the area of UA and other related domains such as rehabilitation, speech pathology, and special education. Effective collaboration across domains could produce significant advances.

Acknowledgments

This material is based upon work supported by the National Science Foundation (NSF) under Grant No. IIS-0328391 and the U.S. Department of Education under Grant No. H133G050354. Any opinions, findings, and conclusions or recommendations expressed in this material are those of the authors and do not necessarily reflect the views of NSF or the U.S. Department of Education.

References

Aist, G., Chan, P., Huang, X., Jiang, L., Kennedy, R., Latimer, D., et al. (1998). How effective is unsupervised data collection for children's speech recognition?, in the *Proceedings of the 5th International Conference on Spoken Language Processing (ICSLP 1998)* (R. H. Mannell and J. R. Ribes, eds.), December 1998, Sydney, Australia, pp. 3171–3174. Canberra City, Australia: Australian Speech Science and Technology Association, Inc.

Arnold, S., Mark, L., and Goldthwaite, J. (2000). Programming by voice, VocalProgramming, in the *Proceedings of the Fourth International ACM Conference on Assistive Technologies (ASSETS 2000)*, 13–15 November 2000, Arlington, VA, pp. 149–155. New York: ACM Press.

Arunachalam, S., Gould, D., Andersen, E., Byrd, D., and Narayanan, S. (2001). Politeness and frustration language in child-machine interaction, in the *Proceedings of EUROSPEECH 2001 Scandinavia, 7th European*

Conference on Speech Communication and Technology, 2nd INTERSPEECH Event (EUROSPEECH-2001) (P. Dalsgaard, B. Lindberg, H. Benner, and Z. Tan, eds.), 3–7 September 2001, Aalborg, Denmark, pp. 2675–2678. http://www.isca-speech.org/archive/eurospeech_2001/e01_2675.html.

ASSETS (2006). *8th International ACM SIGACCESS Conference on Computers and Accessibility*, 23–25 October 2006, Portland, OR. http://www.acm.org/sigaccess/assets06.

Bain, K., Basson, S., and Wald, M. (2002). Speech recognition in university classrooms: Liberated learning project, in the *Proceedings of the Fifth International ACM Conference on Assistive Technologies (ASSETS 2002)*, 8–10 July 2002, Edinburgh, Scotland, pp. 192–196. New York: ACM Press.

Birch, G., Bozorgzadeh, Z., and Mason, S. (2002). Initial on-line evaluations of the LF-ASD brain-computer interface with able-bodied and spinal-cord subjects using imagined voluntary motor potentials. *IEEE Transactions on Neural Systems and Rehabilitation Engineering* 10: 219–24.

CHI (2007). *ACM SIGCHI Conference on Human Factors in Computing Systems 2007*, 28 April–3 May 2007, San Jose, CA. http://www.chi2007.org.

Christian, K., Kules, B., Shneiderman, B., and Youssef, A. (2000). A comparison of voice controlled and mouse controlled web browsing, in the *Proceedings of the Fourth International ACM Conference on Assistive Technologies*, 13–15 November 2000, Arlington, VA, pp. 72–79. New York: ACM Press.

Coen, M., Weisman, L., Thomas, K., and Groh, M. (1999). A context sensitive natural language modality for an intelligent room, in the *Proceedings of the 1st International Workshop on Managing Interactions in Smart Environments (MANSE'99)*, December 1999, Dublin, Ireland, pp. 68–79.

Cohen, M., Giangola, J., and Balogp, J. (2004). *Voice User Interface Design*. Boston: Addison-Wesley.

Cox, S., Lincoln, M., Tryggvason, J., et al. (2002). TESSA, a system to aid communication with deaf people, in the *Proceedings of the Fifth International ACM Conference on Assistive Technologies (ASSETS 2002)*, 8–10 July 2002, Edinburgh, Scotland, pp. 205–212. New York: ACM Press.

Dai, L., Goldman, R., Sears, A., and Lozier, J. (2005). Speech-based cursor control using grids: Modeling performance and comparisons with other solutions. *Behaviour and Information Technology* 24: 219–230.

Davis, A., Moore, M., and Storey, V. (2003). Context-aware communication for severely disabled users, in the *Proceedings of the 2003 Conference on Universal Usability (CUU 2003)*, 16–17 November 2003, Vancouver, Canada, pp. 106–111. New York: ACM Press.

de Mauro, C., Gori, M., Maggini, M., and Martinelli, E. (2001). *Easy Access to Graphical Interfaces by Voice Muse*. Technical report, Università di Siena.

Dragon NaturallySpeaking (2006). *Dragon NaturallySpeaking Professional 9*. http://talktoyourcomputer.com/products_dnspro90.html.

Entwistle, M. S. (2003). The performance of automated speech recognition systems under adverse conditions of human exertion. *International Journal of Human-Computer Interaction* 16: 127–140.

FBC (2007). *Foundation for Blind Children*. http://www.the-fbc.org/productsAndServices/speechreco.html.

Fell, H., Ferrier, L., Mooraj, Z., Benson, E., and Schneider, D. (1996). EVA, an early vocalization analyzer, an empirical validity study of computer categorization, in the *Proceedings of the Second Annual ACM Conference on Assistive Technologies (ASSETS 1996)*, 11–12 April 1996, Vancouver, Canada, pp. 57–63. New York: ACM Press.

Feng, J., Karat, C-M., and Sears, A. (2005a). How productivity improves in hands-free continuous dictation tasks: Lessons learned from a longitudinal study. *Interacting with Computers* 17: 265–289.

Feng, J. and Sears, A. (2004). Using confidence scores to improve hands-free speech-based navigation in continuous dictation systems. *ACM Transactions on Computer-Human Interaction* 11: 329–356.

Feng, J., Sears, A., and Law, C. M. (2005b). Conducting empirical experiments involving users with spinal cord injuries, in *Universal Access in HCI: Exploring New Dimensions of Diversity, Volume 8 of the Proceedings of the 11th International Conference on Human-Computer Interaction (HCI International 2005)* (C. Stephanidis, ed.), 22–27 July 2005, Las Vegas [CD-ROM]. Mahwah, NJ: Lawrence Erlbaum Associates.

Giuliani, D. and Gerosa, M. (2003). Investigating recognition of children's speech, in the *Proceedings of the IEEE International Conference on Acoustics, Speech, and Signal Processing, Volume 2 (ICASSP 2003)*, April 2003, Hong Kong, pp. 137–140. Piscataway, NJ: IEEE.

Goette, T. (2000). Keys to the adoption and use of voice recognition technology in organizations. *Information Technology & People* 13: 67–80.

Gustafson, J. and Sjolander, K. (2002). Voice transformations for improving children's speech recognition in a publicly available dialogue system, in the *Proceedings of the International Conference on Spoken Language Processing (ICSLP 2002)* (J. H. L. Hansen and B. Pellom, eds.), 16–20 September 2002, Denver, pp. 297–300. http://www.niceproject.com/publications/VoiceTransformations.pdf.

Halverson, C., Horn, D., Karat, C-M., and Karat, J. (1999). The beauty of errors: Patterns of error correction in desktop speech systems, in the *Proceedings of the 7th IFIP Conference on Human-Computer Interaction (INTERACT'99)*, 30 August–3 September 1999, Edinburgh, Scotland, pp. 133–140. Amsterdam, The Netherlands: IOS Press.

Harada, S., Landay, J., Malkin, J., Li, X., and Bilmes, J. (2006). The vocal joystick: Evaluation of voice-based cursor control techniques, in the *Proceedings of the 8th International ACM SIGACCESS Conference on Computers and Accessibility (ASSETS 2006)*, 23–25 October 2006, Portland, OR, pp. 197–204. New York: ACM Press.

Harris, R. (2005). *Voice Interaction Design, Crafting the New Conversational Speech Systems*. San Francisco: Morgan Kaufmann Publishers.

Hewitt, J., Lyon, C., Britton, C., and Mellow, B. (2005). SpeakView: Live captioning of lectures, in *Universal Access in HCI: Exploring New Dimensions of Diversity, Volume 8 of the Proceedings of the 11th International Conference on Human-Computer Interaction (HCI International 2005)* (C. Stephanidis, ed.), 22–27 July 2005, Las Vegas, Nevada [CD-ROM]. Mahwah, NJ: Lawrence Erlbaum Associates.

Holzman, T. G. (2001). Speech-audio interface for medical information management in field environments. *International Journal of Speech Technology* 4: 209–226.

Hubbell, T., Langan, D., and Hain, T. (2006). A voice activated syntax-directed editor for manually disabled programmers, in the *Proceedings of the 8th International ACM SIGACCESS Conference on Computers and Accessibility (ASSETS 2006)*, 25–26 October 2006, Portland, OR, pp. 205–212. New York: ACM Press.

Jiang, X., Hong, J., Takayama, L., and Landay, J. (2004). Ubiquitous computing for firefighters: Field studies and prototype of large displays for incident command, in the *Proceedings of the SIGCHI Conference on Human Factors in Computing Systems (CHI 2004)*, 24–29 April 2004, Vienna, Austria, pp. 679–686. New York: ACM Press.

Kamel, H. and Landay, J. (2002). Sketching images eyes-free: A grid-based dynamic drawing tool for the blind, in the *Proceedings of the Fifth International ACM Conference on Assistive Technologies (ASSETS 2002)*, 8–10 July 2002, Edinburgh, Scotland, pp. 33–40. New York: ACM Press.

Karat, C. M., Halverson, C., Karat, J., and Horn, D. (1999). Patterns of entry and correction in large vocabulary continuous speech recognition systems, in the *Proceedings of the SIGCHI Conference on Human Factors in Computing Systems: The CHI Is the Limit (CHI '99)*, 15–20 May 1999, Pittsburgh, pp. 568–575. New York: ACM Press.

Karimullah, A. and Sears, A. (2002). Speech-based cursor control, in the *Proceedings of the Fifth International ACM Conference on Assistive Technologies (ASSETS 2002)*, 8–10 July 2002, Edinburgh, Scotland, pp. 178–185. New York: ACM Press.

Keates, S. and Trewin, S. (2005). Effect of age and Parkinson's disease on cursor positioning using a mouse, in the *Proceedings of the 7th International ACM SIGACCESS Conference on Computers and Accessibility (ASSETS 2005)*, 9–12 October 2005, Baltimore, pp. 68–75. New York: ACM Press.

Koester, H. H. (2004). Usage, performance, and satisfaction outcomes for experienced users of automatic speech recognition. *Journal of Rehabilitation Research and Development* 41: 739–754.

Lai, J., Karat, C-M., and Yankelovich, N. (2007). Conversational speech interfaces and technologies, in *The Human-Computer Interaction Handbook* (J. Jacko and A. Sears, eds.), pp. 381–390. Mahwah, NJ: Lawrence Erlbaum Associates.

Lankford, C. (2000). Effective eye-gaze input into Windows, in the *Proceedings of the 2000 Symposium on Eye Tracking Research*

& Applications, 6–8 November 2000, Palm Beach Gardens, Florida, pp. 23–27. New York: ACM Press.

Lehman, J. F. (1998). Using speech and natural language technology in language intervention, in the *Proceedings of the 3rd International ACM Conference on Assistive Technologies (ASSETS '98)*, 15–17 April 1998, Marina del Rey, CA, pp. 19–26. New York: ACM Press.

Leong, L. H., Kobayashi, S., Koshizuka, N., and Sakamura, K. (2005). CASIS: A context-aware speech interface system, in the *Proceedings of the 10th International Conference on Intelligent User Interfaces (IUI 2005)*, 9–12 January 2005, San Diego, pp. 231–238. New York: ACM Press.

LoPresti, E. and Brienza, D. (2004). Adaptive software for head-operated computer controls. *IEEE Transactions: On Neural Systems and Rehabilitation Engineering* 12: 102–111.

Manaris, B. and Harkreader, A. (1998). SUITEKeys: A speech understanding interface for the motor-control challenged, in the *Proceedings of the 3rd International ACM Conference on Assistive Technologies (ASSETS '98)*, 15–17 April 1998, Marina del Rey, CA, pp. 108–115. New York: ACM Press.

Manaris, B., Macgyvers, V., and Lagoudakis, M. (1999). Universal access to mobile computing devices through speech input, in the *Proceedings of the 12th International Florida Artificial Intelligence Research Society Conference (FLAIRS-99)*, 3–5 May 1999, Orlando, FL, pp. 286–292. Menlo Park, CA: AAAI Press.

Manaris, B., Macgyvers, V., and Lagoudakis, M. (2002). A listening keyboard for users with motor impairments: A usability study. *International Journal of Speech Technology* 5: 371–388.

Manasse, B., Hux, K., and Rankin-Erickson, J. (2000). Speech recognition training for enhancing written language generation by a traumatic brain injury survivor. *Brain Injury* 14: 1015–1034.

Mihara, Y., Shibayama, E., and Takahashi, S. (2005). The migratory cursor: Accurate speech-based cursor movement by moving multiple ghost cursors using non-verbal vocalizations, in the *Proceedings of the 7th International ACM SIGACCESS Conference on Computers and Accessibility (ASSETS 2005)*, 9–12 October 2005, Baltimore, pp. 76–83. New York: ACM Press.

Moran, T. and Dourish, P. (eds.) (2001). Introduction to the special issue on "Context-Aware Computing." *Human-Computer Interaction* 16: 1–8.

Nagao, K. and Rekimoto, J. (1995). Ubiquitous talker: Spoken language interaction with real world objects, in the *Proceedings of the International Joint Conference on Artificial Intelligence (IJCAI 1995)*, 20–25 August 1995, Montreal, pp. 1284–1291. San Marco, CA: Morgan Kaufmann.

Nix, D., Fairweather, P., and Adams, B. (1998). Speech recognition, children, and reading, in the *CHI 98 Conference Summary on Human Factors in Computing Systems*, 18–23 April 1998, Los Angeles, pp. 245–246. New York: ACM Press.

Oseitutu, K., Feng, J., Sears, A., and Karat, C-M. (2001). Speech recognition for data entry by individuals with spinal cord injuries, in *Universal Access in HCI: Towards an Information Society for All, Volume 3 of the Proceedings of the 9th International Conference on Human-Computer Interaction (HCI International 2001)* (C. Stephanidis, ed.), 5–10 August 2001, New Orleans, pp. 402–406. Mahwah, NJ: Lawrence Erlbaum Associates.

Oviatt, S. (2000). Talking to thimble jellies: Children's conversational speech with animated characters, in the *Proceedings of the Sixth International Conference on Spoken Language Processing (ICSLP 2000), Volume 3*, 16–20 October 2000, Beijing, pp. 877–880.

Peterson, R. (2006). IBM strives for superhuman speech tech. *PC Magazine*. http://www.pcmag.com/article2/0,1895, 1915071,00.asp.

Pieper, M. and Kobsa, A. (1999). Talking to the ceiling: An interface for bed-ridden manually impaired users, in the *CHI '99 Extended Abstracts on Human Factors in Computing Systems*, pp. 9–10 (extended abstract). New York: ACM Press.

Price, K. J., Lin, M., Feng, J., Goldman, R., Sears, A., and Jacko, J. A. (2006). Motion does matter: An examination of speech-based text entry on the move. *Universal Access in the Information Society* 4: 246–258.

Quesada, J. F., Garcia, F., Sena, E., Bernal, J. A., and Amores, G. (2001). Dialogue managements in a home machine environment: Linguistic components over an agent architecture. *SEPLN* 27: 89–98.

Rollins, A. (1985). Speech recognition and manner of speaking in noise and in quiet, in the *Proceedings of the SIGCHI Conference on Human Factors in Computing Systems (CHI'85)*, pp. 197–199. New York: ACM Press.

Russell, M., Brown, C., Skilling, A., Series, R., Wallace, J., Bonham, B., et al. (1996). Applications of automatic speech recognition to speech and language development in young children, in the *Proceedings of the Fourth International Conference on Spoken Language (ICSLP '96), Volume 1*, 3–6 October 1996, Philadelphia, pp. 176–179. Piscataway, NJ: IEEE.

Sawhney, N. and Schmandt, C. (2000). Nomadic radio: Speech and audio interaction for contextual messaging in nomadic environments. *ACM Transactions on Computer-Human Interaction* 7: 353–383.

Scherer, M. J. (1996). Dilemmas, challenges, and opportunities, in *Living in the State of Struck* (M. Scherer, ed.). Cambridge, MA: Brookline Books.

SCI Information Network (2006). *Spinal Cord Injury: Facts and Figures at a Glance*. http://www.spinalcord.uab.edu/show.asp?durki=21446.

Sears, A., Feng, J., Oseitutu, K., and Karat, C-M. (2003a). Speech-based navigation during dictation: Difficulties, conse-

quences, and solutions. *Human Computer Interaction* 18: 229–257.

Sears, A., Karat, C-M., Oseitutu, K., Karimullah, A., and Feng, J. (2001). Productivity, satisfaction, and interaction strategies of individual with spinal cord injuries and traditional users interacting with speech recognition software. *Universal Access in the Information Society* 1: 4–15.

Sears, A., Lin, M., Jacko, J., and Xiao, Y. (2003b). When computers fade pervasive computing and situationally induced impairments and disabilities, in *Human-Computer Interaction: Theory and Practice (Part II), Volume 2 of the Proceedings of HCI International 2003 (10th International Conference on Human-Computer Interaction)* (C. Stephanidis and J. Jacko, eds.), pp. 1298–1302. Mahwah, NJ: Lawrence Erlbaum Associates.

Sears, A., Lin, M., and Karimullah, A. S. (2002). Speech-based cursor control: Understanding the effects of target size, cursor speed, and command selection. *Universal Access in the Information Society* 2: 30–43.

Shi, Y., Xie, W., Xu, G., et al. (2003). The smart classroom: Merging technologies for seamless tele-education. *IEEE Pervasive Computing* 2: 47–55.

Sporka, A., Kurniawan, S., Mahmud, M., and Slavik, P. (2006). Non-speech input and speech recognition for real-time control of computer games, in the *Proceedings of the 8th International ACM SIGACCESS Conference on Computers and Accessibility (ASSETS 2006)*, 22–25 October 2006, Portland, OR, pp. 213–220. New York: ACM Press.

Tomko, S. and Rosenfeld, R. (2004). Speech graffiti vs. natural language: Assessing the user experience, in the *Proceedings of HLT/NAACL 2004*, 2–7 May 2004, Boston. http://www.cs.cmu.edu/~usi/papers/HLT04.pdf.

Wade, J., Petheram, B., and Cain, R. (2001). Voice recognition and aphasia: Can computers understand aphasic speech? *Disability & Rehabilitation* 23: 604–613.

Wilson, A. and Shafer, S. (2003). XWand: UI for intelligent spaces, in the *Proceedings of the SIGCHI Conference on Human Factors in Computing Systems (CHI 2003)*, 5–10 April 2003, Fort Lauderdale, FL, pp. 545–552. New York: ACM Press.

World Wide Web Consortium (W3C) (2007). *Voice Brower Activity*. http://www.w3.org/Voice.

Zajicek, M. (2006). Aspects of HCI research for older people. *Universal Access in the Information Society* 5: 279–386.

Zajicek, M. and Jonsson, I. (2006). In-car speech systems for older adults: Can they help and does the voice matter? *International Journal of Technology Knowledge and Society* 2: 55–64.

31

Natural Language and Dialogue Interfaces

Kristiina Jokinen

31.1 Introduction

Dialogue management technology has already for several years enjoyed a level of maturity where speech-based interactive systems are technologically possible and spoken dialogue systems have become commercially viable. Various systems that use speech interfaces exist and range from call routing (Chu-Carroll and Carpenter, 1999) to information-providing systems (Aust et al., 1995; Zue, 1997; Raux et al., 2005; Sadek, 2005) and speech translation (Wahlster, 2000), not to mention various VoiceXML-type applications that enable speech interfaces on the web (VoiceXML Forum).[1] The common technology is based on recognizing keywords in the user utterance and then linking these to appropriate user goals and further to system actions. This allows regulated interaction on a particular topic, using a limited set of words and utterance types (see also Chapter 30, "Speech Input to Support Universal Access," of this handbook). While spoken dialogue systems deploy flexibility and modularity of agent-based architectures (such as DARPA Communicator; Rudnicky et al., 1999; Seneff et al., 1999), the applications are usually designed to follow a stepwise interaction script and to direct the user to produce utterances that fit into predefined utterance types, by designing clear and unambiguous system prompts. Spoken dialogue technology also uses statistical classifiers that are trained to classify the user's natural language utterances into a set of predefined classes that concern possible topics or problem classes that the user may want to get help about. In particular, they are deployed in the so-called How May I Help You (HMIHY) technology (Gorin et al., 1997), which also includes context-free grammars to guarantee a high degree of accuracy

of speech recognition and word-spotting techniques to pick the requested concepts from the user input. The HMIHY-type interfaces have been influential in the development of speech-based interactive systems, although typically they do not include deep language understanding components.

In technology and commercial application development, the focus has been on relatively simple tasks that would not need sophisticated interaction capabilities beyond individual word and phrase recognition. Some evidence has been found for supporting the view that the users actually prefer a command-based interface to one with human-like communication capabilities: the users want to get the task done, efficiently, and with as little trouble as possible, and it does not matter whether they can express themselves using natural language with the same freedom and fluency as in human-human communication. It must be noted, however, that in similar situations with a human operator, the users do not expect fancy conversation capabilities either but decent task completion: the users would reward quick and accurate service and not necessarily chatty but socially nice conversations. It may thus be that the alleged preference of command interfaces to more natural communication is due to the tasks that do not support natural language conversation rather than to the user's dislike of this mode of human-computer interaction, if it were available and working properly for them. Consequently, in research communities, interaction models strive for more humanlike spoken language capabilities, to understand spontaneous speech, and to model reasoning processes. The focus has especially been on error management and handling misconceptions, on conversational features of speech as well as on multimodal aspects of communication. This research helps in understanding human communicative behavior, but it also serves as an inspiration for advanced spoken

---

[1] VoiceXML Forum (http://www.voicexml.org).

language technology that aims at developing adaptive interfaces and handling spontaneous natural language utterances.

However, both practical and research prototypes often fail to reach the level of user satisfaction that would allow the users to enjoy the interaction. The reasons have been pinned down to bad speech recognizer performance, as well as to straightforward dialogues that are directed toward task completion. In system initiative dialogues, the users are also required to follow the system designers' conceptual models, instead of being able to express their goals in their own way. It is thus assumed that solutions, where natural language communication is supported both on the technical and conceptual level, are likely to succeed better. There are several reasons for this. First, speech creates an illusion of real interaction capabilities, and the users apply their intuitive human language strategies to interact with the system, even though they may be well aware of their partner as a computer (cf. studies by Reeves and Nass, 1999). Although the users can learn to speak to the computer (i.e., learn the "computerese" language), and may prefer a straightforward interface that is not too humanlike, it is the human manner of interaction that is used as the standard for comparing the system performance; in other words, the viewpoint for evaluating human-computer interaction is the human-human interaction. Second, applications in which spoken interaction would show its power are characterized by complex domains that are difficult to model in the detailed and exhaustive way required by the state-based technology. For instance, trip planning and assistance in machine maintenance fall in this category: the speakers may not know what the different options and alternatives are concerning their preferences for a planned trip, or how to classify maintenance problems using technical terminology, but can reach satisfactory solutions in these tasks through natural language communication. However, planning requires separate knowledge of the possible plans and their prerequisites, and a number of technical maintenance operations require specialized knowledge of the relevant terminology and functioning of the machine. The HMIHY-type modeling of interaction by combining domain information and dialogue acts in few dialogue states quickly becomes infeasible: the number of possible dialogue states is huge and the listing of all possible paths through the state space (i.e., all possible ways to conduct a dialogue with the user) simply becomes impossible. In these applications, separate models for reasoning about dialogue strategies and task planning would be necessary, and a natural language interface would allow flexibility both for the user to express requests and statements, and for the system to present complex information. Third, in interface design, a new metaphor has also appeared to replace the traditional "computer as a tool" metaphor, namely that of "computer as an agent" (Jokinen, 2003, 2009). The new metaphor regards the human-computer interaction as a cooperative activity between two agents: the system and the user. The system mediates between the user and the backend application, and assists the users in achieving their goals. The system is expected to act in a manner of an intelligent rational agent, so the metaphor presupposes the system's capability to communicate through natural language.

The challenge that speech and language technology faces in this context is not so much in producing systems that enable interaction in the first place, but to design and build systems that allow interaction in a natural way: to provide models and concepts that enable experimentation with complex natural language interactive systems, and to test hypotheses for flexible human-computer interaction. The non-natural aspects of interfaces are usually traced back to the lack of robustness in speech recognition, deficient natural language understanding, and inflexible interaction strategies, but there are also other aspects that are required before interactive applications can reach the same level as their human counterparts in robustness and versatile performance. In addition to improved interaction strategies, natural language interfaces are also to be extended in their knowledge management and reasoning capabilities, so as to support inferences concerning the user's intentions and beliefs behind the observed utterances. Since interaction can take place via speech, text, and signing, special attention is also to be paid to multimodality (see Chapter 40 of this handbook for more detail about multimodal user interfaces).

The goal of building natural interactive systems thus becomes close to studying intelligent interaction in general. The design and development of more natural dialogue systems require that the aspects of natural communication that facilitate intelligent and intuitive interaction are taken into account, and this, on its part, demands better understanding of how people ground their intentions in the environment they live in, and how they communicate their intentions to other agents, which may well be automatic devices and interactive systems besides human beings. In the ubiquitous technological context (Weiser, 1991), where the users are envisaged to interact with various kinds of intelligent devices which populate the environment, it may seem less bizarre to think that the design of natural language interactive systems is, in fact, analogous to the construction of intelligent systems in the artificial intelligence research. The use of natural language consists of planning and performing complex actions on the basis of observations of the communicative context, and this accommodates with the notion of intelligent agents in artificial intelligence as defined in the authoritative textbook by Russel and Norvig (2003, p. vii): "The main unifying theme is the idea of an intelligent agent. We define AI as the study of agents that receive percepts from the environment and perform actions. Each such agent implements a function that maps percept sequences to actions...." Interactive natural language systems can basically be defined in a similar way as implementing a function that maps the computer agent's perceptions of the communicative context to communicative actions.

Without going deeper into the philosophical aspects of intelligent interaction with machines, it should be pointed out that naturalness is a problematic concept even when consideration is limited to the techniques and models that enable implementation of intelligent interaction. Often, it simply refers to the use of natural language as a mode of interaction, but can also be understood in a wider sense that focuses on the natural and intuitive aspects of the interaction in general. It can be attached to modeling human-human communication and building systems that try to mimic human communication, and it can also refer to applications that try to take advantage of the users' natural ways of giving and receiving information. The wider perspective of natural interaction, however, presupposes understanding of what it actually means to communicate in natural ways in different situations. For instance, natural language is by far the best means of explaining and negotiating complex abstract ideas, while gesturing is more natural when pointing and giving spatial information. On the other hand, spoken language is not suitable if one is concerned about disturbing others or privacy issues. Much of the speakers' attitudes and emotions are also expressed by facial expressions and body posture, which are natural means for conveying nonverbal information. In human-computer interaction, quite the opposite is the case: graphical interfaces with touch screen, mouse and keyboard are more natural than the abstract ways of expressing oneself in natural language: they seem to provide the user with more concrete control over the task than language-based conversations. When considering universal access to digital information in general, the natural means of human-computer interaction include possibilities for interacting with both spoken and written text, as well as with various other interactive modes such as tactile or gaze-based interaction. It must be emphasized that novel interface types are not only introduced as alternatives for users with disabilities, but as supporting natural interaction in general. For instance, Milekic (2002) talks in favor of tangible interfaces that allow grabbing, and emphasizes that they are easier, more natural, and more effective to use, since they apply biological knowledge of how it feels to touch and grab things: this enables manipulation of objects without explicit formalization of what one is doing.

To develop natural language dialogue interfaces, it is thus necessary to consider the communicative situation as a whole. Jokinen (2009) discusses the system's communicative competence, which includes the following: physical feasibility of the interface, efficiency of reasoning components, natural language robustness, and conversational adequacy. The first aspect refers to the enablements for communication such as the user having access to digital information and being able to use the system in a natural way, while the second one refers to algorithms and architectures that enable the system to conduct robust and natural interaction with the user. The two last aspects take the system's language-processing and interaction capabilities into account: analysis and generation of utterances, managing dialogue strategies, and adaptation to the user. According to this view, it is important to investigate the coordination of different input and output modes (speech, text, pen, touch, eye movement, etc.), so as to utilize appropriate modalities for natural exchange of information between different users in a wide variety of communicative situations. Consequently, natural interaction can be considered an approach to interface design that attempts to empower different users in various everyday situations to exploit the strategies they have learned in human-human communication, with an ultimate aim of constructing intelligent and intuitive interfaces that are aware of the context and the user's individual needs. The

notion of natural interaction thus refers to the system's ability to support functionality that the user finds intuitive and easy (i.e., the interactive system should *afford* natural interaction; cf. Norman, 2004). Affordance contains natural language as an essential mode for the user to interact with the system.

This chapter discusses natural language dialogue interfaces in detail. The chapter is structured as follows. Section 31.2 provides an overview of natural language interaction and natural interfaces. Section 31.3 develops the view of complex interactive systems as communicating agents, and it also provides a short overview of the constructive dialogue model, and discusses contextual understanding of dialogues. Section 31.4 presents a route navigation system as an example of natural language applications, and discusses its evaluation from the point of view of the user's expectations and experience. Finally, Section 31.5 draws conclusions concerning system evaluation and the system's communicative capability.

31.2 Natural Language Interaction

31.2.1 Linguistic Approach

Natural language is both a cognitive and cultural phenomenon: it is purposeful activity by individual speakers, but it is also a speaker-independent entity regulated by the norms of the language community. It is used to communicate one's ideas and thoughts, and it is also a means to build social reality through interaction. The focus of linguistic research has thus been on the structure of language on one hand, and on the use of language on the other hand. The former includes research on the rules and regularities of morphology (word formation) and of syntax (phrase and sentence parsing), while the latter concerns semantic and pragmatic inferencing as a function of the interaction and communicative context.

An important aspect of language use is to understand the propositional meaning that the observed linguistic entities (written sentences, spoken utterances, signed expressions) convey. The meaning can be constructed on the basis of the order of the elements (*John ate a fish* vs. *A fish ate John*) but is also dependent on the context in which the elements occur. For example, the word *table* may have different interpretations in the request *You should move that table somewhere else* (a piece of furniture or a figure in a document), and the phrase *thank you* may have different functions (a sign of gratitude or a sign of wanting to finish a conversation). Hence, natural language communication does not only consist of attaching words together according to certain grammar rules, but of interpreting messages and relating their meaning to the existing world. It is an activity through which individual speakers create cognitive representations of the world.

Linguistic meanings exhibit both individual and social dimensions: they are simultaneously represented in the cognitive processes of the speakers as well as in the cultural schemes that the speakers have created, and continuously create, through interaction and communication. On the other hand, in linguistic anthropology and conversation analysis, Gumperz and Hymes

(1972) have suggested that language exists only in the interaction with other members of the language community. Linguistic meaning is the shared understanding that emerges among the speakers in the communicative situations, and thus it is learned with respect to the speakers and situations. However, this view does not explain why language is also independent from the individuals and their interactions: it is not the case that meanings need to be created and learned separately in each speech event, but the speakers can, in fact, rely on some shared code that already exists within the language community. In other words, there is a link between the speaker's individual knowledge of the world and the natural language used in the language community.

In the classic work by Grice (1957, 1989), two types of meaning are postulated: the *linguistic meaning* refers to the semantic interpretation of the utterance, and the *speaker meaning* refers to the pragmatic interpretation whereby the intentions of the speaker are spelled out. Successful communication requires that the speaker has an intention to deliver a message, and that the listener recognizers the speaker's intention to communicate something. The signals sent by the speaker are thus meant to be understood as symbols with certain meanings, and the listener is expected to react to them in a relevant manner.

The intentions of the speaker are encoded in the discourse functions of the contributions. Following the speech act theory of Austin (1962) and Searle (1979), utterances function as actions that have certain prerequisites and that are aimed at fulfilling the speaker's intentions. Utterances may contain performative verbs that explicitly indicate the act performed (such as promise, baptize, etc.), but usually the act is to be conventionally inferred on the basis of the utterance content and its context (requests, statements, acknowledgments). In dialogue research, the term *dialogue act* is used to emphasize the fact that the original notion of speech acts is extended and modified to cover various dialogue properties (see, e.g., Bunt and Girard, 2005, for a taxonomy and recognition of dialogue acts). Dialogue acts have been influential in the development of plan-based dialogue systems (e.g., TRAINS, Allen et al., 1996) and, although they may not constitute an explanation of the dialogue as such, they can be used as convenient labels and abstractions of the speaker's mental state to structure dialogues and learn dialogue strategies.

Through dialogues the speakers exchange meaningful information for the purpose of achieving certain (linguistic or nonlinguistic) goals. Consequently, dialogue contributions are linked together: the speakers act in a rational and consistent manner to reach their goals. The dialogues show global coherence in the overall dialogue structure, and local coherence in the sequences of individual utterances (Grosz and Sidner, 1986). The former captures a general task structure and is often modeled with the help of a task hierarchy, a dialogue topic, or the speakers' intentions while the latter realizes the speakers' attentional state, in a focus stack and builds upon the cohesive means between consecutive utterances like the use of pronouns (*A: Take the bus number 73. B: Does* **it** *stop by the hospital?*) or ellipsis (*A: yes, it stops by the hospital. B: and [does it stop] by the station?*).

The dialogue is also teamwork, and the partners collaborate on the underlying task (see articles by Cohen and Levesque and Grosz and Sidner in Cohen et al., 1990, as well as recent work on the collaborative interface agent in Collagen, Rich et al., 2005). The participants also cooperate on building shared knowledge via grounding and feedback (Clark and Schaefer, 1989), as well as on coproducing dialogue contributions (Fais, 1994), and being engaged in the communicative activity in the first place (Ideal Cooperation; see Section 31.3.1).

The dynamic and changing nature of natural language makes the modeling of natural language dialogues difficult. The semantics of words evolves, new concepts develop, and also different dialogue strategies are learned so as to cope with novel circumstances. It is impossible simply to itemize the facts necessary for modeling relevant actions and events, so learning and adaptation are necessary. As argued in Jokinen (2000), interactive systems need to be equipped with a dynamic update procedure that emulates learning through interaction. Learning to associate concepts with situations in which the concepts are used enables adaptation to different circumstances. Adaptation and learning are pertinent features especially for applications that aim at coping with complex tasks, such as negotiations and planning, but they can improve robustness even in practical information-providing systems like a train timetable or restaurant guides. For instance, a user who is engaged in a dialogue concerning good restaurants may suddenly ask about train timetables and the leaving of the last train. Usually, dialogue systems treat such questions as off-domain interruptions, and the user is requested either to continue on the original topic, or to confirm that she wants to quit the dialogue. However, the question may also be motivated by the user's need to catch the last train after an evening out, and it indicates indirectly the user's preferences for the location of the restaurant. Thus, the user may intend to link it to the topic of restaurant choices through a world knowledge association. These kinds of associations are of course impossible to anticipate or list exhaustively in advance, but the associations between the domain and world knowledge can be learned from interactive situations: the bond between a concept and its associtited content can be activated and reinforced if the user happens to appear in a similar situation again, but if the co-occurrence appears to be an unexpected one-time event, the bond between the two items is decreased. Similar kinds of learning experiments have been conducted within dialogue research, especially on optimizing dialogue management strategies such as confirming or providing helpful information, by dialogue simulation using reinforcement learning (e.g., Walker et al., 1998; Levin et al., 2000; Scheffler and Young, 2002; Williams and Young, 2006).

31.2.2 Natural Language Interfaces

The extensive use of natural language in interactive systems suggests that the role of a natural language front-end is to be redefined. It is not simply an interface that connects user commands to a set of possible automatic reactions, but a special software component that initiates, maintains, and records interaction between human users and computational applications. Dialogue systems thus contain a particular dialogue management component, which initiates interaction and creates abstract representations that are further manipulated and processed in the reasoning components of the system's architecture.

As a consequence, the interface has to be distinguished from the system's language capabilities. The interface refers to different physical devices through which human-computer interaction is enabled, such as the computer screen, mouse, keyboard, and microphones. Natural language, however, should not be mixed with the medium, since it brings in an extra dimension compared with the command and menu-based systems: that of symbolic communication (cf., discussion on the medium, code, and modality in Jokinen, 2009). Natural language expressions form a particular code for interaction, analogous to a set of gestures or color signs, except that its expressive capacity is richer and more varied. Besides offering a wide range of conventional symbols for presenting information, a natural language interface also presupposes minimum conversational capability (i.e., an ability to interpret utterances within the dialogue context and to hypothesize possible intentions behind the utterance usage).

A top-level view of a dialogue system is presented in Figure 31.1. The natural language front-end consists of an input analyzer and an output generator. The former includes language-understanding components such as a morphological and syntactic analyzer, and a topic-spotting and semantic-interpretation module, while the latter includes components dealing with the planning and generation of system utterances. Spoken dialogue systems also include a speech recognizer, which often integrates language-understanding components, and a speech synthesizer, which speaks out the system utterances. The dialogue manager is the component that manages interaction: it deals with decisions on how to react to the user input and how to consult the backend database, and it also updates the dialogue history. The task manager is a special reasoning component that takes care of the reasoning and access to the backend application. It may be part of the dialogue manager, especially if the application-related reasoning is included in the dialogue strategy design, but to emphasize the need for intelligent reasoning and inference, it is depicted as a component of its own in Figure 31.1. Separate information storages are usually maintained to keep track of the

task- and dialogue-related information dynamically created and exchanged in the course of the dialogue, (shared context info storage), as well as to encode user preferences (user models).

Figure 31.1 also depicts interface languages between the components. The user and the system interact with natural language, which is parsed and semantically analyzed in the natural language front-end. Semantic interpretation is often a keyword or pattern-based "shallow" interpretation process, but can also deploy sophisticated parsing techniques to uncover syntactic-semantic relations between the constituents. If the input is spoken language, it would first go through a speech recognizer, which produces an n-best list of possible recognized utterances with a confidence score or a network of words from which the correct interpretation can be picked. The interpreted result, the semantic representation of the utterance, contains semantic predicates and arguments representing the utterance meaning as well as a dialogue act label representing the speaker's intention. For instance, if the user has informed the system *I'd like to go to the station*, the semantic representation may look as a conjunction of predicates as follows:

$$[\text{inform}(x,y),\ \text{user}(x)\ \&\ \text{system}(y)\ \&\ \text{travel}(t,x)\ \&\ \text{destination}(t,s)\ \&\ \text{station}(s)\ \&\ \text{def}(s)]$$

The predicates may also be linked to ontological concepts, in which case the representation should be called *conceptual representation*. For instance, in the above representation, the predicate "travel" may represent a general meaning of the verb "go", but it can also stand for a generic concept of traveling to somewhere; from a linguistic point of view, the two representations are at different descriptive levels, but in practical work, the fine distinction is usually not necessary.

The semantic (or conceptual) representation can further instantiate a frame (i.e., a representation of the task or application knowledge that can be used as a basis for various reasoning processes). For instance, a travel frame may contain slots for a departure and a destination place, a departure time and an arrival time, as well as for a route between the two places, and look like the following flat structure:

$$[\text{depPlace}=P;\ \text{depTime}=T;\ \text{destPlace}=D;\ \text{destTime}=S;\ \text{route}=R]$$

The semantic representation may also trigger a plan recognition process and instantiate a plan for the purpose of reaching a certain goal. A plan consists of a sequence of actions that are executed in a certain order if their preconditions hold (cf. dialogue acts and plan-based dialogue systems mentioned in the previous section). The dialogue manager may also maintain an agenda, which is simply a list of actions that it needs to perform, including dialogue acts with the user, system acts to update the information storages, and database commands to access the backend database.

The actual implementation of the different representations is commonly performed with the help of XML, which provides a standard interface language that is easy to process and integrate with other system modules such as the speech and multimodal components (cf. W3C standards of SRGS, SISR, and EMMA).

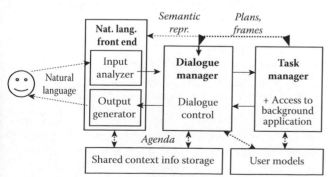

FIGURE 31.1 A top-level view of a dialogue system. (From Jokinen, K., *Constructive Dialogue Management: Speech Interaction and Rational Agents.* New York: John Wiley & Sons, 2009.)

This chapter does not go into detail about the different dialogue management techniques (scripts, frames, and agent-based approaches), but the reader is referred to the overview in McTear (2004). As for references and implementation of dialogue systems, see also Jokinen et al. (2002), and Jokinen (2003, 2008).

31.3 Complex Systems as Communicating Agents

Interactive systems are mostly task-based applications that model straightforward question-answer interactions, and may thus seem rather simple from the point of view of human conversations. The HMIHY-type applications inherently expect quick and clear interaction, and the efficiency of the interface (i.e., the efficiency of the natural language dialogue) is measured with regard to the length of the dialogue. As mentioned previously, expectations in these tasks concern efficient and clear interaction rather than socially adapted conversations, and even if the partner is a human service agent, efficiency counts as a major factor for the satisfactory service. It is thus plausible that the reason for natural language interfaces not being as popular as one might expect on the basis of the fundamental status of natural language in human communication is related to the fact that the underlying tasks are too simple and familiar to support spontaneous use of the whole range of human conversational capabilities. Moreover, human-computer situations are single and serial: the users interact with the system one at the time, and perform one task at a time. Natural language, however, is better suited for a different type of communicative behavior: it shows its power in complicated multiparty multitask situations where various issues need to be negotiated and also problems can occur. Natural language is used to clarify misunderstandings and to correct wrong information, as well as to reason on the situation itself. It is important to note that to solve problems, the partners should be able to discuss the information that has been exchanged in the dialogue, and for this, it is essential to reason on what was said, why it was said, and what could have been said instead (i.e., to master meta-level communication that concerns the language and communication itself, not just objects and actions in the world). Natural language is also essential for the coordination and communication of social interaction, to understand subtle signals related to acceptable and polite interaction, and to avoid imposing or embarrassing the partner.

Returning to ubiquitous and context-aware communication, natural language communication seems to provide a realistic interface for the intelligent environment. The ubiquitous computing paradigm envisages that pervasive and mobile communication technologies will be deployed through a digital environment that is aware of the presence of mobile and wireless appliances (see also Chapter 60 of this handbook). The environment is adaptive and responsive to the users' needs, habits, and emotions, and ubiquitously accessible via natural interaction. The use of text, speech, graphics, touch, and gaze allows natural input/output modalities for the user to interact with the backend application, and the users can choose the modality that best suits the particular circumstance and the users' preferences. In this kind of context, the intelligence of the environment is perceived through a two-way communication between the user and the environment, whereby the user can express various preferences and intentions in a natural manner, and the system is capable of appropriately responding to them. As argued above, this type of communication is best conducted through natural language, since it offers the most versatile and flexible means for conveying the participants' intended meanings.

An example of the intelligent interaction management is organization of offline data (texts, music, photos, videos, e-mails) so that it supports fast information retrieval according to some topical principles that relate to the conversational topic that the user is talking about. Below are two relevant aspects of conversation in this respect: cooperation and context management, both of which affect the smoothness of communication in ubiquitous contexts.

31.3.1 Cooperation

As already mentioned, language is purposeful behavior by rational agents. For instance, Allwood (1976) defines communication as activity by rational agents bound by social obligations. A dialogue system's desired behavior can also be grounded on the notion of cooperative and appropriate communication (Allwood et al., 2000; Jokinen, 1996, 2009). Rational cooperation can be seen as emerging from the partners' communicative capabilities that maintain interaction on the basis of relevant and truthful information. It is based on the speakers' observations about the world and on their reasoning, within the dialogue context, about the new information being exchanged in the dialogue contributions. In human-computer interaction, cooperation manifests itself in the system properties that enable the user to interact with the system: robustness of data processing and appropriate presentation of the information to the user (Jokinen, 2009). Robustness is thus not only a quantitative measure of the system's response capabilities, it also subsumes qualitative evaluation of the system's communicative competence.

In human-human conversations, cooperation refers to a participant's overt behavior that seems to convey the participant's willingness, benevolence, and ability to provide relevant responses that address what the partner has questioned. According to Allwood (1976), the speakers are engaged in ideal cooperation if they:

- Have the same goal
- Consider each other cognitively
- Consider each other ethically
- Have trust that the partner behaves according to these terms

The first requirement means that the agents must cooperate *on* something, that is, they must share intentions to achieve a certain goal. Cognitive consideration refers to the agent's deliberation on the fulfillment of the goal in the most reasonable way. An important dimension in ideal cooperation is ethical consideration, which obliges the agents to treat their partners as rational motivated agents as well. This means that rational agents should not only attempt to fulfill their own goals, but they should not prevent other agents from fulfilling their goals either. Besides bringing aspects of politeness, indirectness, voluntary help, and so on, into rational action, ethical considerations also account for the agents' seemingly irrational behavior, such as volunteering for a tedious job to save someone else from doing it (increase own pain instead of pleasure), or choosing an inefficient method that would allow easier interaction with others (act in an incompetent way). Ethical consideration thus functions as a counterforce to cognitive reasoning. Finally, the mutual trust binds the agents' acts together and provides a basis for understanding the speaker's meanings: by assuming that the partner communicates according to principles of ideal cooperation, the agent can recognize intentions behind the communicative acts, deliberate on the possible alternatives, and decide on the appropriate response.

In interactive systems, cooperation and communicative obligations are usually hardcoded in the control structure, and the system's reactions are aimed at producing the most straightforward response. The system cannot reason on the meta-level about different cooperation strategies which, as mentioned previously, would be necessary in negotiations and resolving misunderstandings and other problematic situations. In the context of obiquitous information technology; however, models of rationality and cooperation can improve the quality of practical systems and services: by implementing these principles in the reasoning process, interactive systems can be made more flexible and reactive.

Constructive dialogue modeling (CDM) (Jokinen, 2009) implements ideal cooperation as part of the construction of a shared model of two communication-related tasks: evaluating the partner's goal, and planning an appropriate response to it. The former concerns how the partner's goal can be accommodated in a given context, and results in strategic decisions about appropriate next goals. The latter concerns how the goal can be realized in different contexts. From the agent's point of view, communication is a cycle that consists of an analysis of the partner's contribution, the evaluation of the new information in regard to the agent's own knowledge and intentions, and reporting the evaluation results back to the partner as the agent's (communicative) reaction. While the evaluation of new information is motivated by the agent's cognitive consideration within the changed context, the reporting of the result is influenced by ethical considerations: the agent needs to inform the partner of the new situation, so as to allow the partner to work toward the shared goal with all the relevant information.

The agents respond to the changes in the context in which the interaction takes place, but they are also capable of planning and taking initiatives to fulfill their own goals. Since the evaluation of the exchanged information takes place in the context of the agent's plans and goals, the agent's reaction may not always be as expected by the partner. In this case, ideal cooperation obliges the agent to provide a reason for the failure to act according to the cooperation expectations, possibly together with an apology, so as to allow the partner to reevaluate the feasibility of her original plan. The reevaluation may then lead to further negotiations to reconcile the participants' contradictory intentions. It is important to notice that mere refusal to act without any apparent reason would usually be interpreted as a sign of unwillingness to cooperate, and would result in an open conflict.

The main features of the CDM agents are as follows:

- The agents are rational and cooperative.
- The agents exchange new information.
- The agents construct mutual knowledge through interaction.
- The agents plan and realize utterances locally.
- The agents use general conversational principles (ideal cooperation).

In a ubiquitous context, the cooperation between human and the intelligent environment can be modeled based on the same principles of ideal cooperation and the CDM agents. For instance, a request to switch on the light or a question about the last bus is usually responded to by complying with the request: switching the light on or providing the relevant information. In case the agent wishes to refuse the request, it is necessary to produce a relevant reason (e.g., the fuse is blown/Sorry, I haven't got the timetables/I haven't got a permission to give the information/I don't think taking the bus is a good idea/It would be more economical and ecological not to switch on the lights, etc.).

31.3.2 Context Understanding

Human-human communication involves smooth coordination of a number of knowledge sources: characteristics of the speakers, topic, and focus of the conversation, meaning and frequency of lexical items, communicative context, physical environment, world knowledge, and so on. The following human-human dialogue between a service agent and a customer (Interact corpus, Jokinen et al., 2002) exemplifies these aspects: the overall dialogue structure is nondeterministic, and the agent's guidance shows flexible and considerate interaction strategy.

A: I'd like to ask about bus connection to Malmi hospital from Herttoniemi metro station—so is there any possibility there to get a bus?

L: Well, there's no direct connection—there's the number 79 that goes to Malmi but it doesn't go to the hospital, it goes to Malmi station.

A: Malmi station? Oh, yes—we've tried that last time and it was awfully difficult.

L: Well, how about taking the metro and changing at Sörnäinen, or Hakaniemi if that's a more familiar place?

A: Well, Hakaniemi is more familiar, yes.

L: Ok, from there you can take the bus 73.

A: 73?

L: Yes, it leaves Hakaniemi just there where you exit from the metro to the bus stop, next to the marketplace.

A: So, it's by the marketplace that 73 leaves from?

L: Yes.

A: And it's not there where the other buses leave from in front of Metallitalo?

L: No, it's there right when you come out from the metro.

A: And it goes to the hospital?

L: Yes, it has a stop just by the hospital.

A: Ok, it must be a better alternative than the bus we took to the station, we didn't know which way to continue and nobody knew anything and we had to take the taxi....

L: What a pity—there would have been the number 69, though. It leaves close to the terminal stop of number 79 and goes to the Malmi hospital.

A: I see, so 79 to the station and then 69?

L: Yes.

A: Are they on the same stop?

L: Well, not on the same stop but very close to each other, anyway.

A: Close to each other? Ok, well thank you for your help.

L: Thank you, good-bye.

A: Good-bye.

This dialogue shows how the speakers cooperate with each other on building a shared understanding of what is the best bus route, and also how they pay attention to the partner's emotional state and needs. For instance, the agent senses the customer's frustration and introduces a simpler route via Hakaniemi using metro and bus, but as the customer returns to her earlier frustrating experience, the agent provides information of this option, too. Language is also related to the context in which the dialogue takes place. The speakers make frequent references to the physical environment (*change in X*, *close to each other*, *Hakaniemi*, *Malmi station*), and the spatial and visual environment directs their interpretation and generation of linguistic utterances (*it's there right when you come out from the metro*). In other words, language is grounded in the communicative context. Grounding is a part of natural communication, and in linguistics, it also refers to the building of the shared understanding of the dialogue goal (Clark and Schaefer, 1989). This takes place by the partner giving feedback to the speaker about how the presented new information is being understood (as in the above dialogue by using the feedback particles *ok*, *I see*), on the basis of which the speaker can then clarify the content as necessary.

In interactive systems, the context is included in the general usage scenario. The assumed context may not be the one the user is in, however. Especially in mobile situations, the systems should interact by making dynamic references to the physical environment, and also allow pointing, gestures, and gazing as input modes, since purely verbal expressions may become rather clumsy. Moreover, information should be presented in accordance with the communicative context, since the user's attention is not necessarily directed toward the system and the service that it provides, but is often divided between the service and some primary task such as talking to people (cf. the example previously of intelligent data organization). This requires awareness of the context and the user's intentions, as well as an ability to interpret the user's verbal utterances with respect to the possible goals as are apparent in the context.

31.4 Example of an Interactive Navigation System

This section briefly presents one location-based service, the multimodal map navigation system (MUMS), and its evaluation concerning cooperation in mobile context. The MUMS system (Hurtig and Jokinen, 2006; Jokinen and Hurtig, 2006) is a multimodal route navigation system that aims at providing the user with real-time travel information and mobile navigation in Helsinki. It was developed in a technological cooperation project among Finnish universities, supported by the Finnish Technology Agency TEKES and several IT companies. The main goal was to build a robust practical application that would allow the users to use both spoken language commands and pen-pointing gestures as input modalities, and also output information in both speech and graphical modes. The system is based on the Interact system (Jokinen et al., 2002), which aimed at studying methods and techniques for modeling rich natural language interaction in situations where the interaction had not been functional or robust enough. Some of the design principles were to allow spoken language errors and hesitations in user input, and to support intuitive interaction modes for information presentation.

Figure 31.2 presents the general architecture of the MUMS system. The PDA client only has a lightweight speech synthesizer, while the server handles all processing of the information. The touch screen map can be scrolled and zoomed, and the inputs are recorded simultaneously and time-stamped for later processing. The architecture is described in more detail in Hurtig and Jokinen (2006). A screenshot of the system output is presented in Figure 31.3. It corresponds to a situation where the system answers a user question in spoken and graphical form: *The tram 7 leaves from the Opera stop at 13:46. There are no changes. The arrival time at the Railway Station is at 13:56.*

The evaluation of the system aimed at establishing the difference between the users' expectations and experience with the system, as well as studying how the background information affects the users' views about the system properties, especially its cooperation with the user. It was hypothesized that the system's communicative capability would have a positive effect on the perceived cooperation, and consequently, on the user's experience of the system. The users had varying levels of experience with speech and tactile interfaces, but none had used the MUMS or a similar system before. They were given a short demonstration of the system use before the actual evaluation sessions, and were divided into two groups depending on the background information given to them about the system they were about to use: one group was instructed to interact with a speech interface that also had a tactile input option, while the other group was instructed to interact with a tactile system with spoken dialogue capabilities. Both groups were then given the same scenario-based tasks

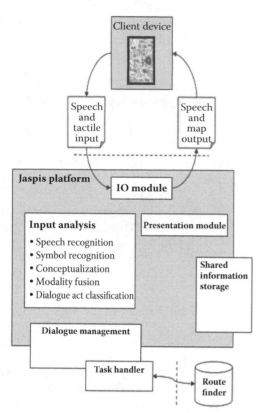

FIGURE 31.2 General architecture of MUMS (From Hurtig, T. and Jokinen, K., Modality fusion in a route navigation system, in the *Proceedings of the IUI 2006 Workshop on Effective Multimodal Dialogue Interfaces*, 29 January 2006, Sydney, Australia, pp. 19–24, 2006.)

FIGURE 31.3 A screenshot in MUMS.

(Kanto et al., 2003) that dealt with using the system to find means of public transportation to reach certain places. The expected and perceived system performance was measured by asking the users to fill in the same evaluation form twice: first to estimate their expectations of the system performance before the actual tasks, and again after completing the tasks. The questions were organized into six groups concerning the user's perception of the system's speech and graphical interface; the user's perception of the system's functionality and consistence; the user's perception of the system's responses and their appropriateness; the user's perception of the system taking the user into account and the easiness of completing the task; the user's eagerness in future use of the system; and the overall assessment of the system.

The evaluation results are reported in detail in Jokinen and Hurtig (2006), and show that user expectations were fulfilled and the system seems to have offered a genuinely positive experience. Here the differences in the perceived performance measures and the overall change between the user expectations and the actual use of the system are briefly discussed.

As expected, the users' prior knowledge of the system influenced their evaluations. Both groups gave very positive reviews for the system's multimodal aspects, and felt that both touch and speech are important in the interaction. The speech group, however, seemed to emphasize the contribution of the system's speech and graphical representation to the intelligibility of the

system's output, and they were willing to use a tactile interface unimodally (i.e., they believed that a tactile interface was more usable even if not combined with speech). The tactile group, on the other hand, seemed to think that several modalities make interaction flexible and the system easy to use, but they also believed that a unimodal tactile interface would not be as usable as the multimodal one. There was also some evidence that the tactile group was more willing to use a unimodal speech system than the speech group, but the differences were not significant. The speech group also felt, more than the tactile group, that the system was slow, even though the response time of the system was not affected by the mode of user input. This supports the claim that speech presupposes rapid and understandable communication. Moreover, it is interesting that the tactile group was slightly more positive at the use of speech input and output than what they had expected, whereas the speech group was disappointed with the use of the unimodal speech system. The speed of the system is thus important, especially if the users expect that the system is meant for spoken interaction. Analogously, the tactile group was more critical toward the map qualities, and in fact, the difference between the user's expectations and perceived system qualities was in absolute terms negative. The speech group, on the other hand, perceived the use of the map quite positively.

The results show that the priming effect comes to play a role in system evaluation. The tasks given to the two groups and the system that the groups used were exactly the same, but the expectations and the experience varied significantly between the groups. In general, system properties like understandability and pleasantness of the speech output are highly evaluated, and the users were unanimous that the system with both speech and tactile information is preferable to a unimodal one. However, compared with the speech group, the tactile group was positively surprised at the system capabilities and the system's functionality: the system was reported to be helpful, considerate, and cooperative, functioning in a consistent way and capable of recognizing the spoken input usually at the first try. In the conducted evaluations, speech recognition worked in a similar fashion for both groups, so the differences cannot be associated solely to speech recognition problems. Rather, the answer seems to lie again in the predisposition of the users and their expectations concerning whether they interact with a speech interface that uses tactile input, or with a tactile interface that can speak as well. As mentioned previously, the users automatically adjust their assessment of the system's communicative capabilities with respect to what can be expected from a system, and the use of spoken language seems to bring in tacit assumptions of fluent human communication and expectations of the system possessing similar communication capabilities.

31.5 Conclusions

This chapter has discussed natural language communication as an important aspect of an interactive system's functionality and usability. When dealing with present-day interactive services, the users are often forced to adapt their human communication methods to the requirements of the technology. Simultaneously, with the development of a ubiquitous environment, a growing number of interactive applications will appear, and expectations for fluent and intelligent communication become higher. It is thus argued that in the future intelligent environments, the system's communicative capability becomes important, and it is necessary to support the system's communicative capability to understand and provide natural language expressions, recognize new and old information, reason about the topic of interaction, and adapt to the user's different dialogue strategies. Also, intelligent software is needed to take into account requirements for complex interaction: the users' knowledge and intentions, variation in their viewpoints and interests, and the context in which interaction takes place. The design of intelligent dialogue interfaces is thus compared to the design and development of intelligent AI agents that would be able to map their observations of the environment onto appropriate communicative and domain actions.

However, the more complex systems are constructed, the more complicated it is to evaluate the system and try to determine appropriate features that guide user assessments. Especially for systems that exploit natural language understanding and reasoning, multimodality, and complex task domains, evaluation is not a straightforward matter of getting a task done, but also involves the user's experience of the system and interaction itself. For instance, the evaluation of the MUMS system reported in this chapter provides some evidence that the user's perception of the system depends on the system's communicative capabilities related to the underlying task as well as on the predisposition that the user has towards the system in the first place (e.g., if the system is a speech interface which would support natural interaction, or a tactile-based tool that need not address requirements of natural language interaction). It was further assumed that the perceived cooperation, as manifested through natural language interaction, would affect the user's evaluation most. This was confirmed on the level of appropriate responses and the easiness to complete the task, but it was difficult to pin down cooperation in the overall assessment of the MUMS system, since other factors also influenced the evaluation (e.g., slowness of the system was considered one of the most disappointing features).

The standard evaluation criteria are usefulness (what the system is for), efficiency (how well the system performs), and usability (is the user satisfied). The system's objective performance and the user's subjective view of the usage of the system are taken into account, with the goal to maximize the objective function of user satisfaction. However, users also assess the system's quality with respect to the service that the system provides: they look at the system from the point of view of its usefulness in helping them to achieve certain goals or offering them some extra value, even if the application is not optimal in its objective performance.

Quality evaluation can be operationalized by determining those properties and functions of the system that contribute to the user's positive perception of the system and the service it provides. Once the features have been determined, their impact on the system performance can be quantified and compared as described previously: by calculating the difference between the user's expectations and experience before and after the use of the system, and checking how the different features have contributed to the overall change. When considering natural language interfaces, an important aspect in their quality evaluation is the system's communicative competence: success of interaction measured in terms of dialogue and cooperation principles.

This chapter has argued in favor of natural language interfaces as a means to support an intuitive approach for designing and interacting with computer applications. Interactive systems should afford natural interaction, and thus be able to support services that the users find intuitive to use. It is emphasized that in intelligent ubiquitous environments, natural language offers the most versatile possibilities to conduct interactions (from simple command prompts to complicated natural conversations) and, in fact, also affords the most natural interface mode to interact with intelligent agents. It is assumed that the system's communicative competence measures the system's capability to cooperate with the user on a shared task, be it to find information, to negotiate some planning options, or to entertain, and thus affects the user's perception of how user-friendly, flexible, and competent the interface agent is.

References

Allen, J. F., Miller, B. W., Ringger, E. K., and Sikorski, T. (1996). A robust system for natural spoken dialogue, in the *Proceedings of the 1996 Annual Meeting of the Association for Computational Linguistics (ACL'96)*, 24–27 June 1995, Santa Cruz, CA, pp. 62–70.

Allwood, J. (1976). Linguistic communication as action and cooperation, in *Gothenburg Monographs in Linguistics* 2. Göteborg University, Department of Linguistics, Göteborg, Sweden.

Allwood, J., Traum, D., and Jokinen, K. (2000). Cooperation, dialogue, and ethics. *International Journal for Human-Computer Studies* 53: 871–914.

Aust, H., Oerder, M., Seide, F., and Steinbiss, V. (1995). The Philips automatic train timetable information system. *Speech Communication* 17: 249–262.

Austin, J. L. (1962). *How to Do Things with Words*. Cambridge, MA: Harvard University Press.

Bunt, H. and Girard, Y. (2005). Designing an open, multi-dimensional dialogue act taxonomy, in the *Proceedings of DIALOR'05*, 9–11 June 2005, Loria, France, pp. 37–44.

Chu-Carroll, J. and Carpenter, B. (1999). Vector-based natural language call routing. *Computational Linguistics* 2: 361–388.

Clark, H. and Schaefer, E. (1989). Contributing to discourse. *Cognitive Science* 13: 259–294

Cohen, P. R., Morgan, J., and Pollack, M. E. (eds.) (1990). *Intentions in Communication*. Cambridge, MA: MIT Press.

Fais, L. (1994). Conversation as collaboration: Some syntactic evidence. *Speech Communication* 15: 231–242.

Gorin, A. L., Riccardi, G., and Wright, J. H. (1997). How May I Help You? *Speech Communication* 23: 113–127.

Grice, H. P. (1957). Meaning. *The Philosophical Review* 66: 377–388.

Grice, H. P. (1989). *Studies in the Way of Words*. Cambridge, MA: Harvard University Press.

Grosz, B. J., Joshi, A. K., and Weinstein, S. (1995). Centering: A framework for modeling the local coherence of discourse. *Computational Linguistics* 21: 203–225.

Grosz, B. and Sidner, C. (1986). Attention, intentions, and the structure of discourse. *Computational Linguistics* 12: 175–203.

Gumperz, J. and Hymes, D. (ed.) (1972). *Directions in sociolinguistics: The ethnography of communication*. New York: Holt, Rinehart and Winston.

Hurtig, T. and Jokinen, K. (2006). Modality fusion in a route navigation system, in the *Proceedings of the IUI 2006 Workshop on Effective Multimodal Dialogue Interfaces*, 29 January 2006, Sydney, Australia, pp. 19–24.

Jokinen, K. (1996). Cooperative response planning in CDM: Reasoning about communicative strategies, in *TWLT 11. Dialogue Management in Natural Language Systems* (S. LuperFoy, A. Nijholt, and G. Veldhuijzen van Zanten, eds.), pp. 159–168. Enschede: Universiteit Twente.

Jokinen, K. (2000). Learning dialogue systems, in *From Spoken Dialogue to Full Natural Interactive Dialogue: Theory, Empirical Analysis and Evaluation*, in the *Proceedings of Workshop at the 2nd International Conference on Language Resources and Evaluation* (LREC), Athens, Greece. (L. Dybkjaer, ed.), pp. 13–17.

Jokinen, K. (2003). Natural interaction in spoken dialogue systems, in the *Proceedings of the HCI International 2003 Conference, volume 4*, 22–27 June 2003, Crete, Greece, pp. 730–734. Mahwah, NJ: Lawrence Erlbaum Associates.

Jokinen, K. (2008). User interaction in mobile navigation applications, in *Map-Based Mobile Services: Design, Interaction and Usability* (L. Meng, A. Zipf, and S. Winter, eds.),

pp. 168–197. Springer Series on Geoinformatics and Cartography. Heidelberg/Berlin: Springer-Verlag.

Jokinen, K. (2009). *Constructive Dialogue Management: Speech Interaction and Rational Agents*. New York: John Wiley & Sons.

Jokinen, K. and Hurtig, T. (2006). User expectations and real experience on a multimodal interactive system, in the *Proceedings of Interspeech Conference*, 17–21 September 2006, Pittsburgh. http://www.isca-speech.org/archive/interspeech_2006/i06_1815.html.

Jokinen, K., Kerminen, A., Kaipainen, M., Jauhiainen, T., Wilcock, G., Turunen, M., et al. (2002). Adaptive dialogue systems: Interaction with Interact, in the *Proceedings of the 3rd SIGdial Workshop on Discourse and Dialogue* (K. Jokinen and S. McRoy, eds.), 11–12 July 2002, Philadelphia, pp. 64–73.

Kanto, K., Cheadle, M., Gambäck, B., Hansen, P., Jokinen, K., Keränen, H., and Rissanen, J. (2003). Multi-session group scenarios for speech interface design, in *Human-Computer Interaction: Theory and Practice (Part II)* (C. Stephanidis and J. Jacko, eds.), volume 2, pp. 676–680. Mahwah, NJ: Lawrence Erlbaum Associates.

Levin, E., Pieraccini, R., and Eckert, W. (2000). A stochastic model of human-machine interaction for learning dialog strategies. *IEEE Transactions on Speech and Audio Processing* 8: 11–23.

McTear, M. (2004). *Spoken Dialogue Technology: Toward the Conversational User Interface*. London: Springer-Verlag.

Milekic, S. (2002). Towards tangible virtualities: Tangialities, in *Museums and the Web 2002: Selected Papers from an International Conference* (D. Bearman and J. Trant, eds.). http://www.archimuse.com/mw2002/papers/milekic/milekic.html.

Norman, D. A. (2004). *Emotional Design: Why We Love (or Hate) Everyday Things*. Cambridge, MA: Basic Books.

Raux, A., Langner, B., Black, A., and Eskenazi, M. (2005). Let's go public! Taking a spoken dialog system to the real world, in the *Proceedings of Interspeech 2005*, 4–8 September 2005, Lisbon, Portugal. http://www.cs.cmu.edu/~awb/papers/is2005/IS051938.PDF.

Reeves, N. and Nass C. (1996). *The Media Equation: How People Treat Computers, Television, and New Media Like Real People and Places*. New York: Cambridge University Press.

Rich, C. C., Sidner, C. L., Lesh, N., Garland, A., Booth, S., and Chimani, M. (2005). DiamondHelp: A collaborative task guidance framework for complex devices, in the *Proceedings of the 20th AAAI Conference and the 17th Innovative Applications of Artificial Intelligence Conference*, 9–13 July 2005, Pittsburgh, pp. 1700–1701. Pittsburgh: AAAI Press/The MIT Press.

Rudnicky, A., Thayer, E., Constantinides, P., Tchou, C., Shern, R., Lenzo, K. et al. (1999). Creating natural dialogs in the Carnegie Mellon Communicator System, in the *Proceedings of the 6th European Conference on Speech Communication and Technology (Eurospeech-99)*, September 1999, Genoa, Italy, pp. 1531–1534.

Russel, S. and Norvig, P. (2003). *Artificial Intelligence: A Modern Approach* (2nd ed.). Upper Saddle River, NJ: Prentice Hall.

Sadek, D. (2005). ARTIMIS rational dialogue agent technology: An overview, in *Multi-Agent Programming: Languages, Platforms and Applications, Volume 15*, pp. 217–243. New York: Springer-Verlag.

Scheffler, K. and Young, S. (2002). Automatic learning of dialogue strategy using dialogue simulation and reinforcement learning, in the *Proceedings of Human Language Technology*, 2–7 May 2002, Boston, pp. 12–18.

Searle, J. R. (1979). *Expression and Meaning: Studies in the Theory of Speech Acts*. Cambridge, U.K.: Cambridge University Press.

Seneff, S., Lau, R., and Polifroni, J. (1999). Organization, communication, and control in the GALAXY-II conversational system. In the *Proceedings of the 6th European Conference on Speech Communication and Technology (Eurospeech 99)*, September 1999, Genoa, Italy, pp. 1271–1274.

Wahlster, W. (ed.) (2000). *Verbmobil: Foundations of Speech-to-Speech Translation*. Berlin/Heidelberg: Springer-Verlag.

Walker, M. A., Fromer, J. C., and Narayanan, S. (1998). Learning optimal dialogue strategies: A case study of a spoken dialogue agent for email, in the *Proceedings of the 36th Annual Meeting of the Association for Computational Linguistics and 17th International Conf. on Computational Linguistics*, 10–14 August 1998, Montreal, Canada, pp. 1345–1352.

Weiser, M. (1991). The computer for the twenty-first century. *Scientific American* 265: 94–104.

Williams, J. D. and Young, S. (2006). Partially observable Markov decision processes for spoken dialog systems. *Computer Speech and Language* 21: 393–422.

Zue, V. (1997). Conversational interfaces: Advances and challenges, in the *Proceedings of Eurospeech 97*, 22–25 September 1997, Rhodes, Greece, pp. KN 9–18.

32

Auditory Interfaces and Sonification

Michael A. Nees and
Bruce N. Walker

32.1 Introduction

Auditory interfaces and sonification—information display by means of nonspeech audio (Kramer et al., 1999)—have been the subject of increasing interest in recent decades (for reviews, see Kramer et al., 1999; Frysinger, 2005). With the advent of ubiquitous digital technologies, high-fidelity sound samples have become increasingly easy and inexpensive to produce and implement (Hereford and Winn, 1994; Flowers et al., 2005). Perhaps more important, however, an increasing awareness of the shortcomings and limitations of traditional visual interfaces has spurred research on sound as a viable mode of information display. Nonspeech audio cues have been implemented to varying degrees in interface design, ranging from nonspeech audio as a complement or supplement to existing visual displays (e.g., Brown et al., 1989; Brewster, 1997), to hybrid systems that integrate nonspeech audio with other audio technologies (e.g., screen readers; see Morley et al., 1999; Stockman et al., 2005). Attempts have even been made to develop interfaces (usually for the visually impaired) where feedback and interaction are driven primarily by sounds (e.g., Edwards, 1989a, 1989b; Mynatt, 1997; Bonebright and Nees, in press).

Despite the potential utility of sound in interface design, a recent survey of experts in human-computer interaction (HCI) and usability (Frauenberger et al., 2007a) reported that only about 58% of respondents had designed with audio in any form.

Nonspeech audio and sonification represent an important tool for universally accessible interface design, yet most interface designers consider speech audio first (and perhaps exclusively) when implementing audio in a system. Perhaps as a relic of the limited sound production capabilities of early personal computers (see Flowers et al., 2005), perceptions (and in some cases legitimate concerns) linger that sounds in interfaces are a minimally informative annoyance to the user.

This chapter argues that appropriately chosen and implemented nonspeech sounds can be a pleasant, informative, and integral part of interface design, and interfaces with nonspeech audio can promote adherence to at least five of the seven principles of universal design (Connell et al., 1997; McGuire et al., 2006), including (1) equitable use; (2) flexibility in use; (3) simple and intuitive use; (4) perceptible information; and (5) tolerance for error.

The current chapter seeks to provide an introduction to nonspeech auditory information display and an overview of the relevant issues and critical decision points regarding the use of nonspeech audio in interfaces. The discussion is guided by the theme that nonspeech auditory displays can universally enhance the human operator's experience with human-machine systems. As this chapter focuses on the potential benefits of nonspeech audio, the interested reader is referred to other chapters in this volume (e.g., Chapters 28, 30, and 40 of this handbook) for a complete discussion of the range of interface options available to the auditory or multimodal display engineer.

32.2 Appropriate Uses of Nonspeech Auditory Display

The best-practice use of nonspeech audio in interfaces requires a careful consideration of the types of users, tasks, and environments where the system will be implemented (for more detailed discussions, see Kramer, 1994; Barrass, 1997; Nees and Walker, 2007). To the extent that nonspeech audio is able to effectively convey the intended message, obvious accessibility benefits are incurred by certain types of system users (i.e., equitable use, see Connell et al., 1997; McGuire et al., 2006), particularly the 161 million people worldwide who are blind or visually impaired (Resnikoff et al., 2004). Screen readers (see Chapter 28 of this handbook) have been quite effective at making text (and other verbal information) accessible for blind and visually impaired people across a wide variety of digital systems (Tobias, 2003). Other aspects of the interface (e.g., spatial, pictorial, or iconic information, etc.), however, cannot be easily represented with a simple text translation, and the inherent limitations introduced by a text-to-speech display system may introduce new navigation and usability difficulties, especially when the original materials (e.g., web pages, etc.) were not developed with a consideration of screen reader accessibility (Mankoff et al., 2005).

While accessibility for special populations has been one driving force in auditory display research, certain task dependencies and environmental conditions may render the affordances of nonspeech audio beneficial for most users of a system. For example, recent advances in technology have paradoxically expanded the realm of visual information display toward opposite extremes in physical size. Portable devices like cell phones, mp3 players, and even laptop computers continue the trend toward smaller physical dimensions, thereby leaving appreciably less space (or perhaps even no space) for a visual display to occupy (see, e.g., Brewster, 2002). Fixed workstations, on the other hand, have become characterized by multiple visual displays with increasingly large physical sizes, due in part to increases not only in the affordability of displays but also in the expanded computing power to support multiple concurrent displays. As a result, visually intensive workstations and other multitasking situations may overburden the visual modality (Grudin, 2001). System limitations from both small and large visual displays are universally applicable and not unique to any particular type of user, and the inclusion of nonspeech audio in some interfaces can promote universal design principles such as flexibility in use and perceptible information (Connell et al., 1997; McGuire et al., 2006).

In addition to these display-related interface design challenges, environmental conditions external to the system may impose further obstacles for the use of traditional, visual-only displays. Line of sight with a visual display may be obscured (e.g., a firefighter in a smoke-filled room) or unstable (e.g., a jogger viewing an mp3 player's display). Other task dependencies may introduce additional demands on the human visual system that prevent the concurrent use of a visual display (e.g., when navigating or using mobile devices while walking, driving, or performing any

other visually demanding task). Audition requires no physical or stable line-of-sight with a display device (Kramer, 1994), which again allows for equitable use, flexibility in use, and perceptible information, and the inclusion of audio cues may even introduce more tolerance for error (Connell et al., 1997; McGuire et al., 2006) into the system than visual displays alone.

Another notable property of the human auditory system is its sensitivity to the temporal aspects of sound (Bregman, 1990; Kramer, 1994; Flowers and Hauer, 1995; Flowers et al., 1997; Kramer et al., 1999). In many instances, response times for auditory stimuli are faster than those for visual stimuli (Kramer, 1994; Spence and Driver, 1997). Furthermore, people can resolve subtle temporal dynamics in sounds more readily than in visual stimuli; thus, the rendering of data into sound may manifest periodic or other temporal information that is not easily perceivable in visualizations (Flowers et al., 2005). Audition, then, may be the most appropriate modality for simple and intuitive (Connell et al., 1997; McGuire et al., 2006) information display when data have complex patterns, express meaningful changes in time, or require immediate action.

32.3 A Brief Taxonomy of Nonspeech Audio and Sonification

While nonspeech audio has an important role to play in interface design, the specific types of nonspeech sounds that could be used to solve a given interface design challenge are numerous and diverse. Proposed categorical descriptions of nonspeech sounds generally have been arranged according to form (i.e., according to the parameters of the sound) or function (i.e., with respect to the role of the sound within a system) with some convergence between these approaches. A brief description of the types of nonspeech sounds used in interface design is offered here; for a summary, see Table 32.1. The current discussion is organized roughly according to the functions of sounds in interfaces, but in reality the definitional boundaries for nonspeech audio sounds tend be vague and overlapping. For more discussion on taxonomic descriptions of nonspeech auditory displays, the interested reader is referred to Kramer (1994), Walker and Kramer (2004, 2006a, 2006b), and de Campo (2007), whose sonification design map organized the relationships between nonspeech auditory displays along several quantitative continua.

32.3.1 Alarms, Alerts, and Warnings

Alarms, alerts, and warnings are generally brief, infrequent, unsubtle sounds designed to capture a person's attention. Traditionally alerts and warnings convey binary status information about an event's onset or offset (Edworthy and Hellier, 2006). For example, a doorbell informs a dwelling's occupants that someone is at the door (i.e., the alert indicates the onset of an event, the arrival of a visitor); this alert does not indicate who is outside, or what they might want. Alerts and warnings usually convey that immediate (or at least temporally proximal) action

TABLE 32.1 Common Classes of Auditory Displays with Their Typical Forms and Functions in Systems

Common Auditory Display Classes	Typical Forms or Characteristic Sound Manipulations	Common Functions in Systems
Alarms	Brief, simple sounds that capture attention	Alerting, warning
Auditory icons	Environmental sounds; ecologically relevant sounds	Object, status, and process indicators; auditory menus
Earcons	Brief, abstract motifs with rule-based iterations	Object, status, and process indicators; auditory menus
Spearcons	Brief, accelerated speech	Object, status, and process indicators; auditory menus
Auditory graphs	Data mapped to frequency	Data exploration aids
Audification	Periodic data sampled within audible range drive frequency	Data exploration aids
Model-based sonifications	Various	Data exploration aids
3D audio displays	Virtual spatial audio via HRTFs	Spatial-orienting cues; navigation aids
Soundscapes	Various, often naturalistic	Ongoing status indicators; monitoring aids
Audio in arts and entertainment	Various	Sonification as art; aids for enhanced and accessible experiences of exhibitions, games, etc.

is required, and Haas and Edworthy (1996) found that higher frequency, rate, and intensity all contribute to more perceived urgency in an auditory alarm signal.

32.3.2 Object, Item, and Status Indicators and Auditory Menus

Sounds such as earcons (e.g., Blattner et al., 1989; Brewster et al., 1993; McGookin and Brewster, 2004; Bonebright and Nees, 2007a), auditory icons (e.g., Gaver, 1989; Keller and Stevens, 2004; Bonebright and Nees, 2007), and spearcons (Walker et al., 2006; Palladino and Walker, 2007) are examples of status and process indicators. Like alerts and warnings, these sounds tend to be brief, but they provide informative cues about the nature of the underlying action or event. These sounds are often used to facilitate tasks such as scrolling (Brewster et al., 1994), pointing, clicking, and dragging with the mouse (Winberg and Hellstrom, 2003), or moving files, and so on, in the interface. *Earcons* are abstract, artificial sounds that bear no ecological relationship to the represented process or event (e.g., beeps, chimes, abstract sound motives, etc.; see Blattner et al., 1989). *Auditory icons* are more natural sounds that have some real-world relationship with their referent process or event (Gaver, 1989), although the degree of ecological relatedness may vary (see Keller and Stevens, 2004). The abstract nature of earcons allows for flexibility in representation, as such abstract sounds can be assigned to most any object, item, or process in an interface. A trade-off exists, however, in that the user is required to learn the association between sounds and their referents; for large catalogues of abstract sounds, users may be unwilling or unable to learn the meaning of the sounds (Watson and Kidd, 1994). Research has shown that auditory icons are generally easier to learn and remember than earcons (Bonebright and Nees, 2007; for a review, also see Edworthy and Hellier, 2006), but auditory icons are less flexible in that some objects, items, and processes have no inherent, ecological sound association (e.g., What sound should represent a Save command?).

Recently, an alternative to earcons and auditory icons has emerged that may be able to ameliorate some of the flexibility-learnability trade-off in interface sounds. *Spearcons* use temporally compressed speech to represent objects, items, or processes with sound (Walker, Nance, and Lindsay, 2006; Palladino and Walker, 2007).[1] Spearcons have been shown to outperform both earcons and auditory icons (Walker et al., 2006) and may be especially useful in the design of flexible auditory menus (Palladino and Walker, 2007) or for representing a large number of items.

32.3.3 Data Representation and Exploration

Rather than offering a brief indication of a transitory system state, auditory displays for data exploration use sound to represent information from an entire (usually quantitative) data set. *Auditory graphs* (for representative work, see Flowers and Hauer, 1992, 1993, 1995; Brown and Brewster, 2003; Smith and Walker, 2005; Nees and Walker, 2007) are typical examples of sonifications designed for data exploration purposes. Auditory graphs most commonly use changes in auditory frequency to correspond to changes in data values along the visual y axis, while time corresponds to the visual x axis. Nees and Walker (2007) recently proposed a conceptual psychological model of auditory graph comprehension. They argued that the advantages of visual graphs, namely the emergence of otherwise unnoticed patterns and data features in plots of data, can be preserved in auditory representations of quantitative data. In much the same way as individual data points combine to form cohesive patterns in a visual graph, sequences of notes in auditory graphs are grouped according to Gestalt principles and can convey equivalent information (Nees and Walker, 2008).

Exploratory work has also examined auditory versions of numerous traditional display formats, including auditory scatterplots (e.g., Flowers et al., 1997; Bonebright et al., 2001), box-whisker plots (Flowers and Hauer, 1992; Peres and Lane, 2003, 2005), histograms (Flowers and Hauer, 1993), multidimensional data sets (Hermann and Hunt, 2005), and tabular data

---

[1] Whether or not spearcons are recognized by listeners as speech may depend upon the listener's abilities and experience, as well as the word or phrase that is accelerated. As the name implies, spearcons can be viewed as a hybrid of speech and nonspeech auditory displays.

(Stockman et al., 2005). These efforts have commonly relied on variations of the pitch-time display format described previously, and the variety of displays that have been developed suggest auditory analogues or alternatives for many visual graphical displays. *Audification*, for example, shifts the waveforms of periodic data into the audible range of frequencies for data exploration (e.g., seismographs; see Dombois, 2001), while *model-based sonifications* represent multidimensional datasets as virtual, interactive objects that systematically drive sound via user input (see Section 32.4.4; Hermann and Hunt, 2005).

As an alternative to traditional visualizations, auditory displays of quantitative information may: (1) make data accessible for visually impaired students and scientists, thereby promoting collaborative efforts; (2) provide an immersive, multimodal, and more effective educational experience for students of math and science; (3) allow for the detection of otherwise unnoticed patterns and anomalies in data; and (4) offer an equivalent, alternative mode of information display in circumstances where visual information display is inadequate (see, e.g., Kramer, 1994; Kramer et al., 1999; Nees and Walker, 2007). These advantages epitomize the spirit and principles of universal design (Connell et al., 1997; McGuire et al., 2006).

32.3.4 3D Audio Displays and the Symbolic Representation of Spatial Relationships in Graphical User Interfaces

A number of studies have confirmed that auditory signals can direct visual attention to a spatial location (e.g., Mondor and Amirault, 1998; McDonald et al., 2000; Eimer, 2001; Brock et al., 2002; see also Schmitt et al., 2000; Spence et al., 2004), and spatial manipulations of audio have been shown to facilitate a three-dimensional visual search (Bolia et al., 1999). Thus, spatial audio has been recognized as an important means of capturing, orienting, or guiding attention (Kramer, 1994). Current technology allows for the delivery of *3D* or *virtual spatial audio*: a two-point sound source (e.g., headphones) in conjunction with head-related transfer functions (HRTFs) can induce the perception that a sound originated from an external environmental source (Wightman and Kistler, 1983, 1989; Walker and Lindsay, 2005; Folds, 2006).

In addition to orienting applications, virtual spatial audio cues have been successfully implemented as audio-only navigational aids, where the virtual spatial location of an audio beacon guides the user along a specified path to a destination.[2] Examples of this approach include the system for wearable audio navigation (SWAN; Walker and Lindsay, 2005; Wilson et al., 2007), and the personal guidance system (PGS; Golledge et al., 1991;

Loomis et al., 1993; Loomis et al., 2005). The SWAN system generally employs spatialized nonspeech sounds, whereas the PGS has usually used spatialized speech.

Walker and Lindsay tested a number of different types of audio beacons, including pink noise bursts (i.e., broad-spectrum noise bursts with equal power per octave), a sonar-like ping, and pure tones, and the broad-spectrum pink noise cue was found to be particularly effective for guiding navigation (Walker and Lindsay, 2006a, 2006b). While a wealth of data support the feasibility of nonspeech audio as a navigation aid, it should be noted that performance outcomes for navigation were negatively impacted by the introduction of a (particularly difficult) concurrent speech discrimination secondary task (Walker and Lindsay, 2006a). The practical costs of these laboratory-induced performance decrements for the dual-task are unclear, and more research is needed to clarify how competing auditory signals may or may not result in interference for navigation systems and indeed all auditory displays (see Section 32.4.7).

While spatial audio has been shown to effectively direct attention and guide navigation through physical space on a gross, or macro level (e.g., from upwards of several inches), much research has been directed at the representation of spatial relationships with sound for smaller physical spaces, such as the dimensions (i.e., the screen size) of traditional visual displays. For example, lateralized audio (e.g., left-right stereo panning) has been used in conjunction with frequency cues (with higher frequency corresponding to higher spatial position) to provide auditory representations of the spatial relationships between objects on a computer screen (Winberg and Hellstrom, 2003). Other interfaces have used increasing pitch to represent movement from left to right and up and down on the screen (Edwards, 1989b), while yet other approaches have used combinations of pitch manipulations and the number of sounds presented to indicate position within a grid of rows and columns on a computer display (Bonebright and Nees, in press). Some of these projects have been targeted at visually impaired users; some have specifically targeted sighted users. Nevertheless, the approaches are inherently universal in that they promote alternative and flexible means of interaction with interfaces for many users.

Despite the insights gained from such studies, there remains no inherent, standard, or even clearly best way to use sound to convey the spatial relationships between objects in user interfaces. A major design dilemma, then, involves the extent to which audio interfaces should maintain the conventions of visual interfaces (Mynatt and Edwards, 1992), and indeed most attempts at auditory display seek to emulate or translate elements of visual interfaces to the auditory modality. While retrofitting visual interfaces with sound can offer some consistencies across modalities, the constraints of this approach may hinder the design of auditory interfaces, and native auditory interfaces would likely sound much different from interfaces designed with a relative visual counterpart in mind. While visual objects exist primarily in space, auditory stimuli occur in time. A more appropriate approach to auditory interface design, therefore, may require designers to focus more strictly on auditory capabilities. Such

---

[2] Virtual spatial audio cues seem particularly suited to indicate where an operator should look or move in physical space. It should be noted, however, that attempts to map virtual audio-spatial location to nonspatial data (e.g., using stereo panning and higher or lower virtual spatial elevation to represent quantities for conceptual dimensions, etc.; see Roth et al., 2002) have been less successful, perhaps owing to systematic misperceptions of virtual elevation (see Folds, 2006).

interfaces may present the items and objects of the interface in a fast, linear fashion over *time* (see, e.g., Eiriksdottir et al., 2006) rather than attempting to provide auditory versions of the spatial relationships found in visual interfaces. This approach can often lead to the deployment of enhanced auditory menus with a mix of speech and nonspeech components. Such advanced interfaces are relatively novel (compared to simpler text-to-speech menus). Ongoing research in advanced auditory menu-based interfaces looks promising, and will generally provide better interfaces for most users (Yalla and Walker, 2007).

32.3.5 Soundscapes and Background Auditory Displays

Many continuous auditory stimuli can be allowed to fade to the extreme periphery of conscious awareness, yet meaningful changes in such ongoing sounds are still noticed (Kramer, 1994). Designers have taken advantage of this auditory capability with *soundscapes*—ambient, continuous sonifications—to facilitate a human operator's awareness of dynamic scenarios (e.g., a bottling plant, Gaver et al., 1991; financial data, Mauney and Walker, 2004; a crystal factory, Walker and Kramer, 2005). Soundscapes often have been designed to mimic natural, ongoing auditory stimuli (e.g., a thunderstorm with rain), and parameters of the soundscape are mapped to particular variables in a multidimensional data set (e.g., Mauney and Walker, 2004). While the listener may not necessarily act upon every change in the soundscape, the display allows for ongoing monitoring and awareness of a changing situation.

32.3.6 Arts and Entertainment

Researchers and musicians have long recognized the potentially unique aesthetic or entertainment value of data-driven (i.e., sonified) music[3] (see, e.g., Quinn and Meeker, 2001), and the International Conference on Auditory Display has regularly featured a concert performance (e.g., International Conference on Auditory Display, 2004, 2006). A recent push in research, however, has taken the notion of sonification as entertainment a step further by advocating for enhanced and accessible exhibitions (e.g., museums, aquaria, zoos, etc.). People with disabilities, particularly the visually impaired, have been shut out of many of the educational and entertainment ("edutainment") experiences offered at traditional exhibitions. While virtual, online-accessible museums are one possible solution to the problem (see Anable and Alonzo, 2001), a remote virtual experience lacks many important aspects (including the novelty and excitement) of a live visit to the actual sites of educational and culturally meaningful exhibitions. While recommendations for real museum accessibility are available (Salmen, 1998), the audio component of accessibility has primarily involved text-to-speech conversions of plaques and verbal materials—a practice that does not capture the most interesting aspects of dynamic exhibitions.

Walker and colleagues (Walker et al., 2006, 2007) have recently begun developing a system for sonifying the real-time dynamics of an aquarium. The movements of the fish are tracked (e.g., with computer vision) and translated to continuous, non-speech (and often musical) auditory representations. The result is a soundscape whereby categorical information about the types of fish can, for example, be represented by instruments of different timbre, while movements of the fish can be conveyed by other dimensions of sound such as pitch, tempo, loudness, or spatial location. Similar innovative approaches may enhance the experience of both static and dynamic exhibitions for many users, as supplementary audio may provide for a more immersive environment in museums, zoos, and aquaria where line-of-sight contact with the exhibit may be obscured by crowds or by perceptual or mobility impairments.

Another important development in accessible entertainment has been an increased interest in auditory games (also see Chapter 17 of this handbook). Audio-only interfaces have been developed for traditionally visual games such as the "Towers of Hanoi" (Winberg and Hellstrom, 2001) and "Tic-Tac-Toe" (Targett and Fernstrom, 2003). More elaborate attempts at audio-only gaming have also begun to appear, including an auditory role-playing game based on the Beowulf story (Liljedahl et al., 2007). Liljedahl et al. argue that audio-only gaming offers players the opportunity to construct rich, unconstrained internal images of the game's landscape from the suggestive nature of the sounds. Interestingly, a recent prototype for an audio-only computer soccer game may actually be able to offer constructive insights for both blind and sighted players on the real soccer field (Stockman et al., 2007).

32.4 Design Considerations for Auditory Interfaces

Theoretical accounts of human interactions with sonification and other nonspeech auditory display design have been slow to develop, in part due to the highly interdisciplinary nature of the field (Nees and Walker, 2007). Recently, however, a number of authors have taken steps toward elaborating sonification theory and organizing the extant knowledge base, including de Campo's sonification design space map (de Campo, 2007), Frauenberger, Stockman, and Bourguet's audio design survey (2007a) and framework (2007b), and Nees and Walker's model of auditory graph comprehension (2007). Despite these recent advances in the field, concrete and specific sonification design guidelines that are grounded in literature and theory are still not generally available. While researchers have described guidelines for nonspeech auditory displays (Hereford and Winn, 1994; Watson and Kidd, 1994; Brown et al., 2003; Flowers, 2005; Edworthy and Hellier, 2006), these attempts have generally provided advice for particular instantiations of auditory displays as opposed to generalized recommendations or comprehensive descriptions for the entire scope of nonspeech audio. Furthermore, in at least one case it has been shown that adherence to published standards for auditory

---

[3] Also see http://www.tomdukich.com/weather%20songs.html.

displays *did not* even ensure the identifiability of sounds (see Lacherez et al., 2007). Rather than articulating what would necessarily be an incomplete list of rules or guidelines here, this chapter offers a broader discussion of the critical issues for implementing nonspeech audio in interface design. Careful consideration of these topics will help to ensure the appropriate deployment of sound in a system and offer a universally accessible and enhanced interface experience for many populations of users.

32.4.1 Detectability and Discriminability

An auditory display is useless if the listener cannot hear the sounds in the system's environment of operation. Research in psychoacoustics has provided ample descriptions of minimum thresholds for detection of sounds along a number of relevant auditory dimensions (e.g., Hartmann, 1997), while masking theories have made valuable predictions about the human listener's ability to hear a sound signal against noise (for a discussion, see Watson and Kidd, 1994). The highly controlled testing conditions for such stimuli, however, can be drastically different from the environments where auditory displays will actually be used by listeners. Accordingly, ecologically plausible testing conditions for applications of auditory displays have been recommended (Watson and Kidd, 1994; Brewster, 2002; see also Walker and Kramer, 2004). Another concern is central or informational masking, whereby sounds are masked at higher levels beyond the cochlea in the auditory system. This variety of masking is not well understood, nor can it readily be predicted by extant models of the acoustic periphery (see Durlach et al., 2003). While the requirement of detectability for auditory information may seem straightforward, the interface designer may encounter problems if simple detection is not given due consideration during the design process.

Given that a sound can be heard by the human listener in the system's environment of operation, a second basic consideration is the discriminability of sounds with distinct meanings in the interface. Like detection, researchers have studied the discriminability of sounds along a wealth of dimensions such as pitch (e.g., Stevens et al., 1937; Turnbull, 1944), loudness (e.g., Stevens, 1936), tempo (e.g., Boltz, 1998), and duration (e.g., Jeon and Fricke, 1997), to name but a few. Again, however, the stimuli and controlled conditions for data collection in such studies may not precisely translate to the real-world scenarios where auditory interfaces will be used, and the designer is cautioned to proceed with an awareness of both the psychoacoustic discriminability of manipulated dimensions of sounds as well as the further additional constraints imposed by the tasks and environments for which the system is designed. Two sounds that carry different pieces of information must be distinguished to ensure that the operator will perceive the intended message.

32.4.2 Annoyance

The potential for sounds to annoy the user is a concern for auditory interface design (Kramer, 1994; Frauenberger et al.,

2007b). Edworthy (1998) described the independent nature of sound aesthetics and performance outcomes. Sounds that annoy the user may be ignored or turned off, even when the presence of auditory cues enhances user performance with the system. Likewise, sounds may enhance the aesthetic experience of an interface without improving performance with the system. Some have suggested that musical nonspeech sounds (e.g., sounds from the MIDI instrument base) with their richer harmonic and acoustic features, are easier to perceive than pure tones and simple waveform sounds (Ramloll et al., 2001; Brown et al., 2003; Childs, 2005). Simply using musical sounds, however, will not guarantee a pleasant experience of the auditory interface for all users, tasks, and environments. Bonebright and Nees (in press) recently found that four different types of earcons (including both pitched musical instruments and pure-tone-based variations) as well as a speech condition all led to auditory displays that were rated as "neutral" to somewhat "annoying" in the context of the study task, which was a dual-task listening and orienting paradigm. Another study found that high-pitched interface sounds can be particularly annoying (Bonebright and Nees, 2007). This makes it clear that developing an auditory interface is, in all regards, a design task, with all the inherent difficulties associated with design. It is encouraging, however, that other research has shown that users can be very satisfied with abstract, nonspeech sounds similar to those used by Bonebright and Nees (e.g., Morley et al., 1999).

In general, very little research has addressed the role of aesthetics in auditory display design and many questions remain regarding how to make aesthetically pleasing interface sounds. It remains advisable to pilot sounds with a representative sample of the target user group to eliminate particularly annoying and displeasing sounds, unless such sounds are invoked with a specific intent (e.g., as an alarm tied to a critical, rare event, etc.). Another possible solution involves customizability, where users are given a choice of instruments or sound types, all of which can convey equivalent information. Regardless of the approach, evaluation of aesthetics needs to be longitudinal, since preferences can evolve, and acceptance can increase or decrease as the user becomes more familiar with the interface.

32.4.3 Mappings, Scalings, Polarities

Mapping refers to the dimension of sound that is employed to vary with and thus represent changes in data. For example, an auditory display of temperature could map changes in data to changes in a number of acoustic parameters, such as pitch, loudness, or tempo. In general, groups of listeners have shown some concurrence about which aspects of sound are good for portraying certain conceptual dimensions of data. Nees and Walker (2007) give a detailed discussion and justification of the convention of mapping pitch to y-axis spatial location in auditory graphs, and pitch generally offers a robust mapping dimension for quantities (Brown et al., 2003; Flowers, 2005). Some sound dimensions (e.g., loudness) are often not very effective representations of data for both perceptual and practical reasons (Neuhoff et al., 2002; Walker and Kramer, 2004). Walker has attempted to determine the appropriate acoustic dimension for a given type of data by examining mappings between numerous conceptual data dimensions (e.g., temperature, pressure, danger) and three acoustic dimensions (pitch, tempo, and spectral brightness; Walker, 2002, 2007). Pitch, for example, generally maps well to changes in temperature, but tempo is not particularly effective for this conceptual dimension. Future research should extend and expand upon this approach to guide interface designers toward best-practice mapping choices. Currently, designers should be warned that not all acoustic mappings are equally effective for representing a given conceptual data dimension, and best-practice design decisions for interfaces will arise from an awareness of empirical data and usability pilot testing. As auditory display design requires explicit decisions regarding mapping, a variety of sources should be consulted to attain an awareness of the varieties of mappings available for nonspeech auditory display designers (e.g., Bonebright et al., 2001; Neuhoff et al., 2002; Walker, 2002, 2007; Brown et al., 2003; Edworthy et al., 2004; Flowers, 2005). Redundant or dual mappings (i.e., mapping more than one acoustic dimension to changes in data) may further facilitate comprehension of the display (Kramer, 1994; Bonebright and Nees, 2007).

Following the selection of an acoustic mapping for data, the polarity of the data-to-display relationship must be considered. Increases in a given acoustic dimension (e.g., pitch, tempo, etc.) are most often mapped to increases in the data represented (a positive mapping polarity; Walker, 2002, 2007), but listeners agree that some conceptual data dimensions are better represented with a negative polarity mapping. For example, listeners might agree that increasing pitch suggests increasing temperature, yet the same group of listeners may feel that decreasing pitch offers a more intuitive representation of increasing size. Walker and Lane (2001) showed that some polarity mappings were reversed for visually impaired as compared to sighted listeners. While positive polarities may generally capture listener intuitions (Brown et al., 2003), interface designers should be mindful of user populations and conceptual data dimensions for which this convention is violated. Walker (2002, 2007) provided data for the preferred polarities for many conceptual data dimensions, and usability testing is advisable when evidence regarding a specific polarity relationship is not available.

Along with polarity, the auditory display designer must also consider the amount of change in an acoustic dimension that will be used to represent a unit of change in the data. Magnitude estimation has been employed to describe the intuitive slopes for scaling frequency to a number of conceptual data dimensions (Walker, 2002, 2007), and the conceptual data dimension being represented impacts the choice of scaling factor in the display. For example, equal quantitative changes (e.g., a one-unit increase) in different conceptual data dimensions (e.g., temperature and size) are not necessarily best represented by the same

change in the acoustic display dimension. A match between the listener's preferred or intuitive internal scaling function and the display's scaling function may facilitate comprehension of the information presented, particularly when judgments of absolute or exact values are required. Where feasible, scaling factors should be chosen to match the intuitive user preferences for representing change in a given conceptual dimension (for a number of empirically determined scaling slopes, see Walker, 2002, 2007). Brown et al. (2003) have further suggested minimum (MIDI note 35, ~61.7 Hz) and maximum (MIDI note 100, ~2637 Hz) scaling anchors. Again, the interface designer is encouraged to consult available empirical data as guidance, but ultimately empirical findings, design experience and expertise, and usability pilot testing will converge to determine the best-practice for a given application.

32.4.4 Interactivity

Interfaces for different scenarios may vary considerably in the degree to which interactivity is allowed or encouraged. Some auditory interfaces, such as alarms, may simply be activated by a particular system condition and occur without any opportunity for the user to actively adjust or manipulate the display; non-interactive sounds in interfaces have been referred to as tour based (Franklin and Roberts, 2004) or concert mode (Walker and Kramer, 1996). Other auditory components of an interface may allow for a particular sound message to be replayed, which may often be appropriate given the transient nature of sound. Displays at the extreme end of the interactivity spectrum may allow for elaborate user control of and immersion in the display, including pausing, scanning, scrubbing, skipping backward and forward, and zooming in and out of display dimensions. Such interactivity, called query based (Franklin and Roberts, 2004) or conversation mode (Walker and Kramer, 1996), may be especially helpful for tasks involving data exploration and analysis (see Brown et al., 2002). For model-based sonifications, user control is imperative and drives the presentation of sounds in an entirely active data exploration process (see Hermann and Hunt, 2005). The inclusion of interactive control over the auditory components of a system warrants a consideration of the role of audio in the system and the extent to which such features aid the user in the task at hand versus the cost and potential negative effects of building interactive control into the interface. Of course, the controls that enable this interactivity should also be designed to support universal access (see, e.g., Chapters 29 and 33 of this handbook).

32.4.5 Individual Differences

Important individual differences may influence the interpretation of auditory displays such that different users may interpret the same sounds to have different meanings. If a technology strives toward universal access, the interface designer should be aware of the range and variety of individual differences that must be accommodated (e.g., Meyer and Rose, 2000). For universal design, then, individual differences represent not only a crucially important design challenge, but also an opportunity to meet the needs of diverse populations of users. Individual difference variables that may be relevant to the interpretation of auditory displays include cognitive abilities (e.g., memory and attention), musical ability, listening skills, learning styles, and perceptual abilities. As discussed in the following, researchers have only just begun to examine the role of these individual difference variables in the comprehension of auditory displays. It is also important to note that researchers have yet to consider potential cultural influences on the interpretation of auditory displays or sounds in general, and these types of studies could provide valuable and heretofore lacking insight regarding cross-cultural differences or similarities in meaning-making for sounds.

Walker and Lane (2001) found differences between groups of visually impaired and sighted listeners in magnitude estimation tasks. As mentioned previously, this study indicated that in some situations visually impaired and sighted listeners intuit the same polarities for data-to-display mappings, but in other cases different polarities result. Sighted individuals, for example, preferred a positive polarity when mapping frequency to the conceptual dimension "number of dollars," whereas visually impaired individuals preferred a negative polarity. Auditory interface designers must empirically examine and anticipate these potential conflicting intuitions across user groups and take caution against unknowingly creating a display that is biased against universal access.

Researchers have further suggested that the transient nature of auditory displays may impose inordinate burdens on memory (Morley et al., 1999; Frauenberger et al., 2007b), a concern that warrants a consideration of the impact of cognitive abilities (e.g., memory, attention, etc.) on auditory display performance. Walker and Mauney (2004) studied the impact of individual differences in cognitive abilities on auditory magnitude estimation tasks. They found some evidence that cognitive abilities affected the interpretation of auditory displays. Listeners with better scores on working memory capacity (WMC) and nonverbal reasoning measures performed better on the magnitude estimation task than those listeners who had lower scores on WMC and nonverbal reasoning tests; however, the scaling slope of the data-to-display mappings did not seem to be affected by cognitive abilities, musical experience, or demographic variables (Walker and Mauney, 2004). Mauney (2006) investigated cognitive abilities and musical experience as predictors of frequency and tempo discrimination. Participants completed the Operation span (O-span) task as a measure of working memory capacity and the Raven's progressive matrices task as a measure of nonverbal reasoning. Results showed that performance on the Raven's and O-span tests seemed to predict some, but not all, tested frequency and tempo discrimination thresholds, with better cognitive abilities associated with lower thresholds. As this pattern of results suggested, the role of cognitive abilities in the comprehension of auditory displays is not well understood, although there is reason to believe that further research will yield stable relationships between certain cognitive abilities

and performance with auditory stimuli. The generally transient nature of auditory displays may impose memory demands that could exacerbate individual differences in cognitive variables, so good auditory interface design will require the intuitive use of audio that does not require memorization of large catalogues of sounds.

Researchers have long predicted that the special training and listening abilities of musicians would translate to superior performance with auditory displays as compared to nonmusicians, and a few studies have found such a relationship (e.g., Neuhoff et al., 2002; Sandor and Lane, 2003; Lacherez et al., 2007). In general, however, many researchers have reported weak to nonexistent relationships between musical experience and performance with auditory displays (see Watson and Kidd, 1994; Bonebright et al., 2001; Walker, 2002; Nees and Walker, in press). Watson and Kidd (1994) suggested that the comprehension of auditory displays may simply require perceptual acuity (as opposed to musical ability per se), which is likely a variable that is distributed homogeneously across musicians and nonmusicians. Furthermore, while nonmusicians may not be formally trained in music theory, most adult listeners have at least acquired a wealth of *implicit* knowledge about the rules, structures, and relationships between sounds in music (Bigand, 1993). This implicit knowledge may be enough to perform tasks with auditory interfaces, which generally require no responses related to explicit musical knowledge. Finally, brief, valid tools exist for measuring musical ability, and the use of surrogate measures (e.g., self-reported years of musical experience) may not be capturing enough of the variance in actual musical ability to detect meaningful relationships (Nees and Walker, 2007).

32.4.6 Training and Skill Acquisition

While accessibility often implies that a system should be intuitive and easily understood even by novice users, novel interfaces such as nonspeech auditory displays may require at least some minimal explanation or instruction for the user. Watson and Kidd (1994) accurately pointed out that many people will be unwilling to commit to extensive training in the meaning of sounds in an interface, yet brief training (i.e., under 30 minutes) has been shown to positively impact performance with auditory displays. Smith and Walker (2005) showed that brief training for a point estimation task resulted in better performance than no training. Walker and Nees (2005b) also demonstrated that a brief training period reduced performance error by 50% on a point estimation sonification task. Although to date, little attention has been paid to the issue of training sonification users, recent and ongoing work is examining exactly what types of training methods are most effective for different classes of sonifications (e.g., Walker and Nees, 2005a). Studies that have explicitly analyzed performance data over time (i.e., across trials or blocks of trials) have suggested that performance improves with experience with the novel displays (Walker and Lindsay, 2006b; Nees and Walker, 2008; Bonebright and Nees, in press), but the upper limits of performance with auditory displays remain unknown (Walker and Nees, 2005a; Nees and Walker, 2007). Longitudinal studies of skill acquisition with auditory interfaces are needed.

32.4.7 Concurrent Sounds

Numerous studies have shown that the discriminability and identifiability of sounds decrease as the number of concurrently presented sounds increase (Bonebright et al., 2001; Ericson et al., 2003; McGookin and Brewster, 2004; Walker and Lindsay, 2006a; Lacherez et al., 2007). Theory and research alike suggest, however, that such problems can be somewhat ameliorated to the extent that acoustic cues allow for concurrently presented sounds to be parsed into separate streams (Bregman, 1990). To this end, researchers have suggested that spatial separation of different data (e.g., presenting different data series to left and right headphone channels; see Bonebright et al., 2001; Brown et al., 2003), the use of distinct timbres for different data series (Bonebright et al., 2001; McGookin and Brewster, 2004), and staggering the onsets of concurrent messages (McGookin and Brewster, 2004) may all facilitate the segregation of concurrent audio information. While pitch is also an effective cue for parsing concurrent auditory streams, it is often used to represent dynamic, noncategorical information in auditory displays and may not be an appropriate dimension for promoting the separation of different data series. It should further be noted that the concurrent presentation of distinct channels of auditory information probably has a limit, beyond which the distinct streams of information will become impractical to parse and perceive (Flowers, 2005). This theoretical limit is likely dependent upon not only the number of concurrently presented sounds, but also their qualitative characteristics. Bonebright and Nees (in press), for example, found little to no interference for the comprehension of speech passages in the presence of a concurrent orienting task with earcons. Care should be taken when interfaces or environmental circumstances allow for overlapping sounds, as more research is needed to clarify the limits of perception for simultaneous auditory input.

32.4.8 Delivery of Audio: Hardware

A growing majority of digital devices come equipped with high-fidelity sound production capabilities off-the-shelf. The auditory component of many interfaces may require little or no modification to hardware, but rather a design philosophy that takes better advantage of the existing capability to improve system accessibility with audio. The hardware considerations for the delivery of an audio interface may vary across different use scenarios, however, and Walker and Lindsay (2005) described a number of the challenges encountered when designing their system for wearable audio navigation (SWAN). The SWAN project encountered logistical constraints beyond the auditory interface itself, including technical limitations such as unreliability in sensors (e.g., the fallibility of GPS

and other technologies that attempt to precisely determine a mobile user's location) as well as practical limitations in battery power, size, and durability of a wearable, mobile computer (also see Chapter 11 of this handbook).

Similarly, many attempts at auditory interfaces have been coupled with custom input devices (e.g., Morley et al., 1999; Winberg and Hellstrom, 2003), but such improvisations may not be necessary to the success of an auditory interface. While novel or emerging hardware technologies may eventually transform many of the ways in which people interact with a system (see Chapters 29 and 33 of this handbook), existing, off-the-shelf capabilities of most hardware already allow for the implementation of auditory interfaces that could offer enhanced accessibility for many users.

For delivering sound, audio-capable systems have traditionally relied upon speakers or headphones, both of which are inexpensive options for producing audio of sufficient fidelity for most applications of auditory displays. Privacy and the potentially intrusive nature of delivering sounds through speakers are interrelated, basic concerns. When used in the presence of other people, speakers not only may compromise a user's privacy, but also can interfere with the activities of those nearby or cause annoyance. Headphones may circumvent these problems, but having one's ears covered by headphones introduces new difficulties for interacting with and maintaining awareness of one's surroundings. Blind users, for instance, gather a majority of their environmental information from sound, and they will generally be unwilling to cover their ears, even to use a potentially beneficial system.

One potential solution that is actively being researched is *bonephones*—bone-conduction headphones. Small transducers sit on the mastoid behind the ear and vibrate the skull, effectively stimulating the cochlea directly and bypassing the outer and middle ear. The ears remain uncovered, but the delivery of private audio messages is still possible. With minimal equalization, bonephones have been shown to have similar psychoacoustic signatures as headphones with regard to thresholds (Walker and Stanley, 2005), and early research suggests that virtual spatialized audio is possible with bonephones (Stanley and Walker, 2006). The devices, however, are currently not widely available to consumers, and more research is needed to clarify the potential role for interference between audio delivered via bonephones and concurrent stimulation from environmental sound sources.

32.4.9 Delivery of Audio: Software

As described throughout this chapter, most digital devices have off-the-shelf hardware and software capabilities for sound production, and the success of auditory interfaces will primarily be a function of empirically based design philosophies that embrace the use of sound. No standard add-on software packages exist for the general production of custom nonspeech audio for use in interfaces. Many laboratories involved in research on auditory displays, however, have developed purpose-specific sonification software packages that are often available as free, open-source downloads. Applications for representing data with sound include NASA's Mathtraxx,[4] the Oregon State University Science Access Project's Accessible Graphing Calculator,[5] the Georgia Tech Sonification Lab's Sonification Sandbox[6] (see Walker and Cothran, 2003; Davison and Walker, 2007) and Auditory Abacus[7] (Walker et al., 2004). Stockman et al. (2005) are working on a prototype for a software package that works with Microsoft Excel and CSound to allow sonification of cells of spreadsheets, while Hetzler and Tardiff (2006) have also developed an Excel plug-in for data sonification. Cook (2007) recently described a number of software development projects aimed at analyzing and synthesizing environmental sounds. Other resources of interest for the auditory interface designer include the web site of the International Community for Auditory Display[8] as well as the AUDITORY electronic mail list,[9] both of which offer access to experts with years of collective experience in implementing sounds for research and application.

32.5 Conclusions

Sonification and auditory interfaces can enhance and improve the universal accessibility of a system for a number of users, tasks, and environments. The thoughtful and informed addition of nonspeech audio to an interface, especially as one important element of a holistic approach to universal design, can enhance and improve the accessibility and usability of a system. Nonspeech audio is uniquely suited to convey particular types of information and to ameliorate some of the limitations imposed by traditional visual interfaces. For the visually impaired, computers and other digital technologies have dramatically impacted and will continue to improve access to education, employment, and an overall higher quality of life (Gerber, 2003; Tobias, 2003), and nonspeech auditory displays can fill gaps in accessibility related to alerting or warning functions, status or process updates, ongoing monitoring tasks, and even data exploration. The relevance of nonspeech audio to interface design extends well beyond affordances for the visually impaired (e.g., Griffith, 1990). The benefits of universally usable interfaces should extend system capabilities for many users during visually intensive tasks or in environments where vision is not the ideal modality for information display. Auditory interfaces and sonification can be major contributors to compliance with at least five of the seven principles of universal design (Connell et al., 1997; McGuire et al., 2006), including (1) equitable use; (2) flexibility in use; (3) simple and intuitive use; (4) perceptible information; and (5) tolerance for error.

---

[4] http://prime.jsc.nasa.gov/mathtrax.

[5] http://dots.physics.orst.edu/calculator.html.

[6] http://sonify.psych.gatech.edu/research/sonification_sandbox/index.html.

[7] http://sonify.psych.gatech.edu/research/audio_abacus/index.html.

[8] http://www.icad.org.

[9] http://www.auditory.org.

Flowers (2005) asked whether sound should be a standard component of desktop interfaces. This chapter has suggested that nonspeech sound is an underused and underinvestigated tool for the development of universally accessible interfaces. It further suggested that audio can be implemented immediately and cheaply in most existing interfaces, with little or no modifications to existing software and hardware. Ultimately, the potential of sound to benefit many users of an interface will only be unlocked when researchers commit to explore the best-practice role of sound in interfaces and when designers actively implement audio in interfaces.

References

Anable, S. and Alonzo, A. (2001). Accessibility techniques for museum web sites, in the *Proceedings of Museums and the Web (MW 2001)*, 14–17 March 2001, Seattle. http://www.archimuse.com/mw2001/papers/anable/anable.html.

Barrass, S. (1997). *Auditory Information Design*. PhD dissertation, Australian National University, Canberra, Australia.

Bigand, E. (1993). Contributions of music to research on human auditory cognition, in *Thinking in Sound: The Cognitive Psychology of Human Audition* (S. McAdams and E. Bigand, eds.), pp. 231–377. New York: Oxford University Press.

Blattner, M. M., Sumikawa, D. A., and Greenberg, R. M. (1989). Earcons and icons: Their structure and common design principles. *Human-Computer Interaction* 4: 11–44.

Bolia, R. S., D'Angelo, W. R., and McKinley, R. L. (1999). Aurally aided visual search in three-dimensional space. *Human Factors* 41: 664–669.

Boltz, M. G. (1998). Tempo discrimination of musical patterns: Effects due to pitch and rhythmic structure. *Perception and Psychophysics* 60: 1357–1373.

Bonebright, T. L. and Nees, M. A. (2007). Memory for auditory icons and earcons with localization cues, in the *Proceedings of the International Conference on Auditory Display (ICAD2007)*, 26–29 June 2007, Montreal, Canada, pp. 419–422. http://sonify.psych.gatech.edu/~mike/Nees%20articles/BonebrightNeesetal2001ICAD.pdf.

Bonebright, T. L. and Nees, M. A. (in press). Using an orienting auditory display during a concurrent listening task. *Applied Cognitive Psychology*.

Bonebright, T. L., Nees, M. A., Connerley, T. T., and McCain, G. R. (2001). Testing the effectiveness of sonified graphs for education: A programmatic research project, in the *Proceedings of the International Conference on Auditory Display (ICAD2001)*, 29 July–1 August 2001, Helsinki, Finland, pp. 62–66. Espoo, Finland: Laboratory of Acoustics and Audio Signal Processing and the Telecommunications Software and Multimedia Laboratory.

Bregman, A. S. (1990). *Auditory Scene Analysis: The Perceptual Organization of Sound*. Cambridge, MA: MIT Press.

Brewster, S. (1997). Using non-speech sound to overcome information overload. *Displays* 17: 179–189.

Brewster, S. (2002). Overcoming the lack of screen space on mobile computers. *Personal and Ubiquitous Computing* 6: 188–205.

Brewster, S., Wright, P. C., and Edward, D. E. (1994). The design and evaluation of an auditory enhanced scrollbar, in the *Proceedings of the SIGCHI Conference on Human Factors in Computing Systems (CHI 94)*, 24–28 April 1994, Boston, pp. 173–179. New York: ACM Press.

Brewster, S., Wright, P. C., and Edwards, A. D. N. (1993). An evaluation of earcons for use in auditory human-computer interfaces, in the *Proceedings of the SIGCHI Conference on Human Factors in Computing Systems (CHI 93)*, 24–29 April 1993, Amsterdam, The Netherlands, pp. 222–227. New York: ACM Press.

Brock, D., Stroup, J. L., and Ballas, J. A. (2002). Using an auditory display to manage attention in a dual task, multi-screen environment, in the *Proceedings of the International Conference on Auditory Display (ICAD2002)*, 2–5 July 2002, Kyoto, Japan. http://www.cogsci.rpi.edu/cogworks/down/ BrockEtAl_ICAD02.pdf.

Brown, L. M. and Brewster, S. (2003). Drawing by ear: Interpreting sonified line graphs, in the *Proceedings of the International Conference on Auditory Display (ICAD2003)*, 6–9 July 2003, Boston, pp. 152–156. Boston: Boston University Publications Production Department.

Brown, L. M., Brewster, S., Ramloll, R., Burton, M., and Riedel, B. (2003). Design guidelines for audio presentation of graphs and tables, in the *Proceedings of the International Conference on Auditory Display (ICAD2003)*, 6–9 July 2003, pp. 284–287. Boston: Boston University Publications Production Department.

Brown, L. M., Brewster, S., and Riedel, B. (2002). Browsing modes for exploring sonified line graphs, in the *Proceedings of the 16th British HCI Conference*, 2–6 September 2002, London, pp. 2–5. http://www.dcs.gla.ac.uk/~stephen/papers/HCI2002-brown.pdf.

Brown, M. L., Newsome, S. L., and Glinert, E. P. (1989). An experiment into the use of auditory cues to reduce visual workload, in the *Proceedings of the ACM CHI 89 Human Factors in Computing Systems Conference (CHI 89)*, 30 April–4 May 1989, Austin, TX, pp. 339–346. New York: ACM Press.

Childs, E. (2005). Auditory graphs of real-time data, in the *Proceedings of the International Conference on Auditory Display (ICAD2005)*, 6–9 July 2005, Limerick, Ireland. http://sonify.psych.gatech.edu/ags2005/pdf/AGS05_Childs.pdf.

Connell, B. R., Jones, M., Mace, R., Mueller, J., Mullick, A., Ostroff, E., et al. (1997). *The Principles of Universal Design, Version 2.0*. Raleigh, NC: The Center for Universal Design.

Cook, P. R. (2007). Din of an "iquity": Analysis and synthesis of environmental sounds, in the *Proceedings of the International Conference on Auditory Display (ICAD2007)*, 26–29 June 2007, Montreal, Canada, pp. 167–172. Montreal, Canada: Schulich School of Music, McGill University.

Davison, B. K. and Walker, B. N. (2007). Sonification Sandbox reconstruction: Software standard for auditory graphs, in the *Proceedings of the International Conference on Auditory Display (ICAD2007)*, 26–29 June 2007, Montreal, Canada pp. 509–512. Montreal, Canada: Schulich School of Music, McGill University.

de Campo, A. (2007). Toward a data sonification design space map, in the *Proceedings of the International Conference on Auditory Display (ICAD2007)*, 26–29 June 2007, Montreal, Canada, pp. 342–347. Montreal, Canada: Schulich School of Music, McGill University.

Dombois, F. (2001). Using audification in planetary seismology, in the *Proceedings of the International Conference on Auditory Display (ICAD2001)*, 29 July–1 August 2001, Helsinki, Finland, pp. 227–230. Espoo, Finland: Laboratory of Acoustics and Audio Signal Processing and the Telecommunications Software and Multimedia Laboratory.

Durlach, N. I., Mason, C. R., Kidd, G., Arbogast, T. L., Colburn, H. S., and Shinn-Cunningham, B. (2003). Note on informational masking. *Journal of the Acoustical Society of America* 113: 2984–2987.

Edwards, A. D. N. (1989a). Modelling blind users' interactions with an auditory computer interface. *International Journal of Man-Machine Studies* 30: 575–589.

Edwards, A. D. N. (1989b). Soundtrack: An auditory interface for blind users. *Human-Computer Interaction* 4: 45–66.

Edworthy, J. (1998). Does sound help us to work better with machines? A commentary on Rautenberg's paper "About the importance of auditory alarms during the operation of a plant simulator." *Interacting with Computers* 10: 401–409.

Edworthy, J. and Hellier, E. (2006). Complex nonverbal auditory signals and speech warnings, in *Handbook of Warnings* (M. S. Wogalter, ed.), pp. 199–220. Mahwah, NJ: Lawrence Erlbaum Associates.

Edworthy, J., Hellier, E. J., Aldrich, K., and Loxley, S. (2004). Designing trend-monitoring sounds for helicopters: Methodological issues and an application. *Journal of Experimental Psychology: Applied* 10: 203–218.

Eimer, M. (2001). Crossmodal links in spatial attention between vision, audition and touch: Evidence from event-related brain potentials. *Neuropsychologia* 39: 1292–1303.

Eiriksdottir, E., Nees, M. A., Lindsay, J., and Stanley, R. (2006). User preferences for auditory device-driven menu navigation, in the *Proceedings of the Human Factors and Ergonomics Society 50th Annual Meeting*, 16–20 October 2006, San Francisco, pp. 2076–2078. Santa Monica, CA: Human Factors and Ergonomics Society.

Ericson, M. A., Brungart, D. S., and Simpson, B. D. (2003). Factors that influence intelligibility in multitalker speech displays. *International Journal of Aviation Psychology* 14: 313–334.

Flowers, J. H. (2005). Thirteen years of reflection on auditory graphing: Promises, pitfalls and potential new directions, in the *Proceedings of the International Conference on Auditory Display (ICAD2005)*, 6–9 July 2005, Limerick, Ireland. http://sonify.psych.gatech.edu/ags2005/pdf/AGS05_Flowers.pdf.

Flowers, J. H., Buhman, D. C., and Turnage, K. D. (1997). Crossmodal equivalence of visual and auditory scatterplots for exploring bivariate data samples. *Human Factors* 39: 341–351.

Flowers, J. H., Buhman, D. C., and Turnage, K. D. (2005). Data sonification from the desktop: Should sound be part of standard data analysis software? *ACM Transactions on Applied Perception* 2: 467–472.

Flowers, J. H. and Hauer, T. A. (1992). The ear's versus the eye's potential to assess characteristics of numeric data: Are we too visuocentric? *Behavior Research Methods, Instruments and Computers* 24: 258–264.

Flowers, J. H. and Hauer, T. A. (1993). "Sound" alternatives to visual graphics for exploratory data analysis. *Behavior Research Methods, Instruments and Computers* 25: 242–249.

Flowers, J. H. and Hauer, T. A. (1995). Musical versus visual graphs: Cross-modal equivalence in perception of time series data. *Human Factors* 37: 553–569.

Folds, D. J. (2006). The elevation illusion in virtual audio, in the *Proceedings of the Human Factors and Ergonomics Society 50th Annual Meeting*, 16–20 October 2006, San Francisco, pp. 1576–1580. Santa Monica, CA: Human Factors and Ergonomics Society.

Franklin, K. M. and Roberts, J. C. (2004). A path based model for sonification, in the *Proceedings of the Eighth International Conference on Information Visualization (IV '04)*, 14–16 July 2004, London, pp. 865–870. Washington, DC: IEEE.

Frauenberger, C., Stockman, T., and Bourguet, M.-L. (2007a). A survey on common practice in designing audio user interface, in the *Proceedings of the 21st British HCI Group Annual Conference (HCI 2007), Volume 1*, 22–27 July 2007, Beijing, pp. 187–195. Montreal, Canada: Schulich School of Music, McGill University.

Frauenberger, C., Stockman, T., and Bourguet, M.-L. (2007b). Pattern design in the context space: A methodological framework for auditory display design, in the *Proceedings of the International Conference on Auditory Display (ICAD2007)*, 26–29 June 2007, Montreal, Canada, pp. 513–518. Montreal, Canada: Schulich School of Music, McGill University.

Frysinger, S. P. (2005). A brief history of auditory data representation to the 1980s, in the *Proceedings of the International Conference on Auditory Display (ICAD2005)*, 6–9 July 2005, Limerick, Ireland. http://sonify.psych.gatech.edu/ags2005/pdf/AGS05_Frysinger.pdf.

Gaver, W. W. (1989). The SonicFinder: An interface that uses auditory icons. *Human-Computer Interaction* 4: 67–94.

Gaver, W. W., Smith, R. B., and O'Shea, T. (1991). Effective sounds in complex systems: The ARKola simulation, in the *Proceedings of the ACM Conference on Human Factors in Computing Systems (CHI 91)*, 28 April–5 May 1991, New Orleans, pp. 85–90. New York: ACM Press.

Gerber, E. (2003). The benefits of and barriers to computer use for individuals who are visually impaired. *Journal of Visual Impairment and Blindness* 97: 536–550.

Golledge, R. G., Loomis, J. M., Klatzky, R. L., Flury, A., and Yang, X. L. (1991). Designing a personal guidance system to aid navigation without sight: Progress on the GIS component. *International Journal of Geographical Information Systems* 5: 373–395.

Griffith, D. (1990). Computer access for persons who are blind or visually impaired: Human factors issues. *Human Factors* 32: 467–475.

Grudin, J. (2001). Partitioning digital worlds: Focal and peripheral awareness in multiple monitor use, in the *Proceedings of the 2001 SIGCHI Conference on Human Factors in Computing Systems (CHI 01)*, 31 March–5 April 2001, Seattle, pp. 458–465. New York: ACM Press.

Haas, E. C. and Edworthy, J. (1996). Designing urgency into auditory warnings using pitch, speed and loudness. *Computing and Control Engineering Journal* 7: 193–198.

Hartmann, W. M. (1997). *Sounds, Signals and Sensation: Modern Acoustics and Signal Processing*. New York: Springer-Verlag.

Hereford, J. and Winn, W. (1994). Non-speech sound in human-computer interaction: A review and design guidelines. *Journal of Educational Computer Research* 11: 211–233.

Hermann, T. and Hunt, A. (2005). An introduction to interactive sonification. *IEEE Multimedia* 12: 20–24.

Hetzler, S. M. and Tardiff, R. M. (2006). Two tools for integrating sonification into calculus instruction, in the *Proceedings of the International Conference on Auditory Display (ICAD2006)*, 19–24 June 2006, London, pp. 281–284. London: Department of Computer Science, Queen Mary, University of London.

International Conference on Auditory Display (2004). Listening to the mind listening: Concert of sonifications at the Sydney Opera House, from the *Proceedings of the International Conference on Auditory Display (ICAD2004)*, 6–9 July 2004, Sydney, Australia [Concert]. http://www.icad.org/websiteV2.0/Conferences/ICAD2004/concert.htm.

International Conference on Auditory Display (2006). Global music: The world by ear, from the *Proceedings of the International Conference on Auditory Display (ICAD2006)*, 19–24 June 2006, London [Concert]. London: Department of Computer Science, Queen Mary, University of London.

Jeon, J. Y. and Fricke, F. R. (1997). Duration of perceived and performed sounds. *Psychology of Music* 25: 70–83.

Keller, P. and Stevens, C. (2004). Meaning from environmental sounds: Types of signal-referent relations and their effect on recognizing auditory icons. *Journal of Experimental Psychology: Applied* 10: 3–12.

Kramer, G. (1994). An introduction to auditory display, in *Auditory Display: Sonification, Audification and Auditory Interfaces* (G. Kramer, ed.), pp. 1–78. Reading, MA: Addison-Wesley.

Kramer, G., Walker, B. N., Bonebright, T., Cook, P., Flowers, J., Miner, N., et al. (1999). *The Sonification Report: Status of the Field and Research Agenda. Report Prepared for the National Science Foundation by Members of the International Community for Auditory Display*. Santa Fe, NM: International Community for Auditory Display (ICAD).

Lacherez, P., Seah, E. L., and Sanderson, P. M. (2007). Overlapping melodic alarms are almost indiscriminable. *Human Factors* 49: 637–645.

Liljedahl, M., Papworth, N., and Lindberg, S. (2007). Beowulf: A game experience built on sound effects, in the *Proceedings of the International Conference on Auditory Display (ICAD2007)*, 26–29 June 2007, Montreal, Canada, pp. 102–106. Montreal, Canada: Schulich School of Music, McGill University.

Loomis, J. M., Golledge, R. G., and Klatzky, R. L. (1993). Personal guidance system for the visually impaired using GPS, GIS and VR technologies, in the *Proceedings of the First Annual International Conference, Virtual Reality and Persons with Disabilities*, 17–18 June 1993, Millbrae, CA, pp. 71–74.

Loomis, J. M., Marston, J. R., Golledge, R. G., and Klatzky, R. L. (2005). Personal guidance system for people with visual impairment: A comparison of spatial displays for route guidance. *Journal of Visual Impairment and Blindness* 99: 219–232.

Mankoff, J., Fait, H., and Tran, T. (2005). Is your web page accessible? A comparative study of methods for assessing web page accessibility for the blind, in the *Proceedings of the ACM CHI Conference on Human Factors Computing Systems (CHI 05)*, 2–7 April 2005, Portland, OR, pp. 41–50. New York: ACM Press.

Mauney, B. S. and Walker, B. N. (2004). Creating functional and livable soundscapes for peripheral monitoring of dynamic data, in the *Proceedings of the International Conference on Auditory Display (ICAD2004)*, 6–10 July 2004, Sydney, Australia. http://www.icad.org/websiteV2.0/Conferences/ICAD2004/papers/mauney_walker.pdf.

Mauney, L. M. (2006). *Individual Differences in Cognitive, Musical and Perceptual Abilities*. Master's thesis, Georgia Institute of Technology, Atlanta. http://smartech.gatech.edu/bitstream/1853/13972/1/mauney_lisa_m_200612_mast.pdf.

McDonald, J. J., Teder-Salejarvi, W. A., and Hillyard, S. A. (2000). Involuntary orienting to sound improves visual perception. *Nature* 407: 906–908.

McGookin, D. K. and Brewster, S. (2004). Understanding concurrent earcons: Applying auditory scene analysis principles to concurrent earcon recognition. *ACM Transactions on Applied Perception* 1: 130–150.

McGuire, J. M., Scott, S. S., and Shaw, S. F. (2006). Universal design and its applications in educational environments. *Remedial and Special Education* 27: 166–175.

Meyer, A. and Rose, D. H. (2000). Universal design for individual differences. *Educational Leadership* 58: 39–43.

Mondor, T. A. and Amirault, K. (1998). Effect of same- and different-modality spatial cues on auditory and visual target identification. *Journal of Experimental Psychology: Human Perception and Performance* 24: 745–755.

Morley, S., Petrie, H., O'Neill, A.-M., and McNally, P. (1999). Auditory navigation in hyperspace: Design and evaluation of a non-visual hypermedia system for blind users. *Behaviour and Information Technology* 18: 18–26.

Mynatt, E. D. (1997). Transforming graphical interfaces into auditory interfaces for blind users. *Human-Computer Interaction* 12: 7–45.

Mynatt, E. D. and Edwards, W. K. (1992). Mapping GUIs to auditory interfaces, in the *Proceedings of the 5th Annual ACM Symposium on User Interface Software and Technology*, 15–18 November 1992, Monterrey, CA, pp. 61–79. New York: ACM Press.

Nees, M. A. and Walker, B. N. (2007). Listener, task and auditory graph: Toward a conceptual model of auditory graph comprehension, in the *Proceedings of the International Conference on Auditory Display (ICAD2007)*, 26–29 June 2007, Montreal, Canada, pp. 266–273. http://sonify.psych.gatech.edu/~walkerb/publications/pdfs/2007ICAD-NeesWalker.pdf.

Nees, M. A. and Walker, B. N. (2008). Data density and trend reversals in auditory graphs: Effects on point estimation and trend identification tasks. *ACM Transactions on Applied Perception* 5: Article 13.

Neuhoff, J. G., Kramer, G., and Wayand, J. (2002). Pitch and loudness interact in auditory displays: Can the data get lost in the map? *Journal of Experimental Psychology: Applied* 8: 17–25.

Palladino, D. and Walker, B. N. (2007). Learning rates for auditory menus enhanced with spearcons versus earcons, in the *Proceedings of the International Conference on Auditory Display (ICAD2007)*, 26–29 June 2007, Montreal, Canada, pp. 274–279. http://sonify.psych.gatech.edu/~walkerb/publications/pdfs/2007ICAD-PalladinoWalker.pdf.

Peres, S. C. and Lane, D. M. (2003). Sonification of statistical graphs, in the *Proceedings of the International Conference on Auditory Display (ICAD2003)*, 6–9 July 2003, Boston, pp. 157–160. Boston: Boston University Publications Production Department.

Peres, S. C. and Lane, D. M. (2005). Auditory graphs: The effects of redundant dimensions and divided attention, in the *Proceedings of the International Conference on Auditory Display (ICAD2005)*, 6–9 July 2005, Limerick, Ireland, pp. 169–174.

Quinn, M. and Meeker, L. D. (2001). Research set to music: The climate symphony and other sonifications of ice core, radar, DNA, seismic and solar wind data, in the *Proceedings of the International Conference on Auditory Display (ICAD2001)*, 29 July–1 August 2001, Helsinki, Finland. http://www.acoustics.hut.fi/icad2001/proceedings/papers/quinn.pdf.

Ramloll, R., Brewster, S., Yu, W., and Riedel, B. (2001). Using non-speech audio sounds to improve access to 2D tabular numerical information for visually impaired uses, in the *Proceedings of the IHM-HCI 2001*, 10–14 September 2001, Lille, France, pp. 515–530.

Resnikoff, S., Pascolini, D., Etya'ale, D., Kocur, I., Pararajasegaram, R., Pokharel, G. P., et al. (2004). Global data on visual impairment in the year 2002. *Bulletin of the World Health Organization* 82: 844–851.

Roth, P., Kamel, H., Petrucci, L., and Pun, T. (2002). A comparison of three nonvisual methods for presenting scientific graphs. *Journal of Visual Impairment and Blindness* 96: 420–428.

Salmen, J. P. (1998). *Everyone's Welcome: The Americans with Disabilities Act and Museums*. Washington, DC: American Association of Museums.

Sandor, A. and Lane, D. M. (2003). Sonification of absolute values with single and multiple dimensions, in the *Proceedings of the International Conference on Auditory Display (ICAD2003)*, 6–9 July 2003, Boston, pp. 243–246. Boston: Boston University Publications Production Department.

Schmitt, M., Postma, A., and De Haan, E. (2000). Interactions between exogenous auditory and visual spatial attention. *Quarterly Journal of Experimental Psychology* 53A: 105–130.

Smith, D. R. and Walker, B. N. (2005). Effects of auditory context cues and training on performance of a point estimation sonification task. *Applied Cognitive Psychology* 19: 1065–1087.

Spence, C. and Driver, J. (1997). Audiovisual links in attention: Implications for interface design, in *Engineering Psychology and Cognitive Ergonomics Vol. 2: Job Design and Product Design* (D. Harris, ed.), pp. 185–192. London, U.K.: Ashgate Publishing.

Spence, C., McDonald, J. J., and Driver, J. (2004). Exogenous spatial-cuing studies of human cross-modal attention and multisensory integration, in *Cross-Modal Space and Cross-Modal Attention* (C. Spence and J. Driver, eds.), pp. 277–320. Oxford, U.K.: Oxford University Press.

Stanley, R. and Walker, B. N. (2006). Lateralization of sounds using bone-conduction headsets, in the *Proceedings of the Annual Meeting of the Human Factors and Ergonomics Society (HFES 2006)*, 16–20 October 2006, San Francisco, pp. 1571–1575. http://sonify.psych.gatech.edu/~walkerb/publications/pdfs/2006HFES-StanleyWalker.pdf.

Stevens, S. S. (1936). A scale for the measurement of a psychological magnitude: Loudness. *Psychological Review* 43: 405–416.

Stevens, S. S., Volkmann, J., and Newman, E. B. (1937). A scale for the measurement of the psychological magnitude pitch. *The Journal of the Acoustical Society of America* 8: 185.

Stockman, T., Hind, G., and Frauenberger, C. (2005). Interactive sonification of spreadsheets, in the *Proceedings of the International Conference on Auditory Display (ICAD2005)*, 6–9 July 2005, Limerick, Ireland, pp. 134–139. www.icad .org.

Stockman, T., Rajgor, N., Metatla, O., and Harrar, L. (2007). The design of interactive audio soccer, in the *Proceedings of the International Conference on Auditory Display (ICAD2007)*, 26–29 June 2007, Montreal, Canada, pp. 526–529.

Targett, S. and Fernstrom, M. (2003). Audio games: Fun for all? All for fun? in the *Proceedings of the International Conference on Auditory Display (ICAD2003)*, 6–9 July 2003, Boston, pp. 216–219. Boston: Boston University Publications Production Department.

Tobias, J. (2003). Information technology and universal design: An agenda for accessible technology. *Journal of Visual Impairment and Blindness* 97: 592–601.

Turnbull, W. W. (1944). Pitch discrimination as a function of tonal duration. *Journal of Experimental Psychology* 34: 302–316.

Walker, B. N. (2002). Magnitude estimation of conceptual data dimensions for use in sonification. *Journal of Experimental Psychology: Applied* 8: 211–221.

Walker, B. N. (2007). Consistency of magnitude estimations with conceptual data dimensions used for sonification. *Applied Cognitive Psychology* 21: 579–599.

Walker, B. N. and Cothran, J. T. (2003). Sonification Sandbox: A graphical toolkit for auditory graphs, in the *Proceedings of the International Conference on Auditory Display (ICAD2003)*, 6–9 July 2003, Boston, pp. 161–163. http://sonify.psych.gatech.edu/~walkerb/publications/pdfs/2003ICAD-WalkerCothran-Sandbox.pdf.

Walker, B. N., Godfrey, M. T., Orlosky, J. E., Bruce, C., and Sanford, J. (2006). Aquarium sonification: Soundscapes for accessible dynamic informal learning environments, in the *Proceedings of the International Conference on Auditory Display (ICAD2006)*, 15–20 July 2006, Madrid, pp. 238–241. http://sonify.psych.gatech.edu/~walkerb/publications/pdfs/2006ICAD-WalkerGodfreyOrloskyBruceSanford.pdf.

Walker, B. N., Kim, J., and Pendse, A. (2007). Musical soundscapes for an accessible aquarium: Bringing dynamic exhibits to the visually impaired, in the *Proceedings of the International Computer Music Conference (ICMC 2007)*, 27–31 August 2007, Copenhagen, Denmark, pp. 268–275. http://sonify.psych.gatech.edu/~walkerb/publications/pdfs/2007ICMC-WalkerKimPendse.pdf.

Walker, B. N. and Kramer, G. (1996). Human factors and the acoustic ecology: Considerations for multimedia audio design, in the *Proceedings of the Audio Engineering Society 101st Convention*, 8–11 November 1996, Los Angeles. http://sonify.psych.gatech.edu/~walkerb/publications/pdfs/1996AES-HF_ecology.pdf.

Walker, B. N. and Kramer, G. (2004). Ecological psychoacoustics and auditory displays: Hearing, grouping and meaning making, in *Ecological Psychoacoustics* (J. Neuhoff, ed.), pp. 150–175. New York: Academic Press.

Walker, B. N. and Kramer, G. (2005). Mappings and metaphors in auditory displays: An experimental assessment. *ACM Transactions on Applied Perception* 2: 407–412.

Walker, B. N. and Kramer, G. (2006a). Auditory displays, alarms and auditory interfaces, in *International Encyclopedia of Ergonomics and Human Factors* (2nd ed.) (W. Karwowski, ed.), pp. 1021–1025. New York: CRC Press.

Walker, B. N. and Kramer, G. (2006b). Sonification, in *International Encyclopedia of Ergonomics and Human Factors* (2nd ed.) (W. Karwowski, ed.), pp. 1254–1256. New York: CRC Press.

Walker, B. N. and Lane, D. M. (2001). Psychophysical scaling of sonification mappings: A comparison of visually impaired and sighted listeners, in the *Proceedings of the International Conference on Auditory Display (ICAD2001)*, 29 July–1 August 2001, Helsinki, Finland, pp. 90–94. http://sonify.

psych.gatech.edu/~walkerb/publications/pdfs/2001ICAD-WalkerLane-Paper.pdf.

Walker, B. N. and Lindsay, J. (2005). Using virtual environments to prototype auditory navigation displays. *Assistive Technology* 17: 72–81.

Walker, B. N. and Lindsay, J. (2006a). The effect of a speech discrimination task on navigation in a virtual environment, in the *Proceedings of the Human Factors and Ergonomics Society 50th Annual Meeting*, 16–20 October 2006, Santa Monica, CA, pp. 1538–1541. http://sonify.psych.gatech.edu/~walkerb/publications/pdfs/2006HFES-WalkerLindsay.pdf.

Walker, B. N. and Lindsay, J. (2006b). Navigation performance with a virtual auditory display: Effects of beacon sound, capture radius and practice. *Human Factors* 48: 265–278.

Walker, B. N., Lindsay, J., and Godfrey, J. (2004). The Audio abacus: Representing a wide range of values with accuracy and precision, in the *Proceedings of the International Conference on Auditory Display (ICAD2004)*, 6–9 July 2004, Sydney, Australia. http://www.icad.org/websiteV2.0/Conferences/ICAD2004/papers/walker_lindsay_godfrey.pdf.

Walker, B. N. and Mauney, L. M. (2004). Individual differences, cognitive abilities and the interpretation of auditory graphs, in the *Proceedings of the International Conference on Auditory Display (ICAD2004)*, 6–9 July 2004, Sydney, Australia. http://sonify.psych.gatech.edu/~walkerb/publications/pdfs/2004ICAD-WalkerLMauney.pdf.

Walker, B. N., Nance, A., and Lindsay, J. (2006). Spearcons: Speech-based earcons improve navigation performance in auditory menus, in the *Proceedings of the International Conference on Auditory Display (ICAD2006)*, 15–20 July 2006, Madrid, pp. 95–98. http://sonify.psych.gatech.edu/~walkerb/publications/pdfs/2006ICAD-WalkerNanceLindsay.pdf.

Walker, B. N. and Nees, M. A. (2005a). An agenda for research and development of multimodal graphs, in the *Proceedings of the International Conference on Auditory Display (ICAD2005)*, 6–9 July 2005, Limerick, Ireland, pp. 428–432. http://sonify.psych.gatech.edu/~walkerb/publications/pdfs/2005ICADAGS-WalkerNees.pdf.

Walker, B. N. and Nees, M. A. (2005b). Brief training for performance of a point estimation sonification task, in the *Proceedings of the International Conference on Auditory Display (ICAD2005)*, 6–9 July 2005, Limerick, Ireland, pp. 276–279. http://sonify.psych.gatech.edu/~walkerb/publications/pdfs/2005ICAD-WalkerNees-training.pdf.

Walker, B. N. and Stanley, R. (2005). Thresholds of audibility for bone-conduction headsets, in the *Proceedings of the International Conference on Auditory Display (ICAD2005)*, 6–9 July 2005, Limerick, Ireland, pp. 218–222. www.icad.org

Watson, C. S. and Kidd, G. R. (1994). Factors in the design of effective auditory displays, in the *Proceedings of the International Conference on Auditory Display (ICAD1994)*, 7–9 November 1994, Santa Fe, NM, pp. 293–303. Santa Fe, NM: Santa Fe Institute.

Wightman, F. L. and Kistler, D. J. (1983). Headphone simulation of free-field listening I: Stimulus synthesis. *Journal of the Acoustical Society of America* 85: 858–867.

Wightman, F. L. and Kistler, D. J. (1989). Headphone simulation of free-field listening II: Psychophysical validation. *Journal of the Acoustical Society of America* 85: 868–878.

Wilson, J., Walker, B. N., Lindsay, J., Cambias, C., and Dellaert, F. (2007). SWAN: System for wearable audio navigation, in the *Proceedings of the 11th International Symposium on Wearable Computers (ISWC 2007)*, 11–13 October 2007, Boston, MA. IEEE.

Winberg, F. and Hellstrom, S. O. (2001). Qualitative aspects of auditory direct manipulation: A case study of the Towers of Hanoi, in the *Proceedings of the International Conference on Auditory Display (ICAD2001)*, 29 July–1 August 2001, Helsinki, Finland, pp. 16–20. Espoo, Finland: Laboratory of Acoustics and Audio Signal Processing and the Telecommunications Software and Multimedia Laboratory.

Winberg, F. and Hellstrom, S. O. (2003). Designing accessible auditory drag and drop, in the *Proceedings of the 2003 Conference on Universal Usability (CUU2003)*, 10–11 November 2003, Vancouver, Canada, pp. 152–153. New York: ACM Press.

Yalla, P. and Walker, B. N. (2007). *Advanced Auditory Menus*. Georgia Institute of Technology GVU Center Technical Report # GIT-GVU-07-12. Atlanta: Georgia Institute of Technology.

33

Haptic Interaction

Gunnar Jansson and
Roope Raisamo

33.1 Introduction[1]

Hands have an impressive ability to perform different kinds of manipulation tasks, from working with miniature details to lifting heavy objects. The excellent performance of the hand in natural contexts is based on the efficiency of the interaction between sensors and muscles in the hand and the environment, based on the biological development of humans over time. This is a fact robot developers have had to be quite aware of. Even if vision is important for information about manual activities, the most important information is obtained from the hand itself as a highly competent sense organ.

Even if the feet have less importance than the hands, the feet are important sense organs, too. They can detect information about the ground, its slant, material, and hardness. Sighted people are to a large extent informed about these features via their eyes, but blind people have mainly to rely on their feet. The capacities of the feet are also demonstrated by the fact that people without hands can train their feet to be astonishingly efficient substitutes for their hands.

The sensory capacities of the hands and the feet have often been subsumed under the skin senses. Tactual information obtained via the skin has a basic importance for performance, but that is not the whole story. When the hands are functioning, they are often performing very precise movements, meaning that there is a close cooperation between the sensors in the skin and sensors in the muscles, tendons, and joints (the kinesthetic sense) in the guidance of the movements. The neural system coordinates efficiently the different kinds of sensory information with the muscles performing the movements (Cholewiak and Collins, 1991; Wing et al., 1996; Johnson, 2002).

The aim of this chapter is to provide an overview of basic issues within the haptic interaction area, as well as the use of haptics in present-day technical applications. Haptics has clear benefits in universal access interfaces, as it supports both vision and hearing, and offers additional possibilities to present or manipulate information. The same interfaces may be used by all people in eyes-busy situations, or by visually impaired people in all of their computer use. In addition, haptics also has potential to help people with motor problems, as force feedback can be used to add strength or stability to motor actions.

33.2 An Overview of Haptics

This section introduces the basic terminology and concepts related to the haptic sense in interactive systems.

[1] For a related overview of a partially overlapping area for this chapter, see Jansson (2008).

33.2.1 Touch, Haptics, and Activity

When touch is described as a skin sense, it is often considered a passive receiver of stimulation from the environment. In opposition to this view, pioneers such as Katz in his seminal work (Katz, 1989) and Gibson (1962) emphasized the observers as active explorers of the environment. A hand is regarded as a perceptual system that relies on exploration to collect information, and active touch is in many contexts considered superior to passive touch.[2] To stress the importance of activity, the sense is often called *active touch* or *haptics*. The latter term will most often be used here, as well as the adjective *haptic* sometimes alternating with *touch* and *tactile* or *tactual* in contexts where these terms are traditional. The use of the latter two terms is not consistent in the literature, but tactile is often used for the physical stimulus (i.e., in tactile pictures) and tactual for perceptual aspects.

33.2.2 Exploratory Procedures

The movements performed to collect information via the hands are usually not random, but specific for obtaining particular kinds of information. Lederman and Klatzky (1987) suggested a number of basic exploratory procedures, among others, *lateral motion* for perceiving texture, *pressure* for perceiving hardness, *static contact* for perceiving temperature, *unsupported holding* for perceiving weight, *enclosure* (enclosing the object in a hand or both hands) for perceiving global shape and volume, and *contour following* for perceiving global shape and exact shape. Other exploratory procedures suggested are wielding to get information about several properties of an object (Turvey and Carello, 1995) and shaking, for example, a container with liquid, to be informed about the amount it contains (Jansson et al., 2006b). One of the problems with haptic displays is that they sometimes require non-natural exploratory procedures that, at least initially, decrease their potential usefulness.

33.2.3 Haptics Compared with Vision

In the real world, vision and haptics can provide both common and different kinds of information about the environment. Sometimes haptics provides more advanced information than vision, for instance concerning the weight of objects and the

texture and hardness of surfaces. The most important aspect where haptics lags behind vision is in providing an overview of a scene. While vision gives a practically immediate overview, there are often severe problems in getting an overview of a scene haptically. It may be a laborious and time-consuming task requiring several explorations of the scene before an overview is obtained.[3] Also when used in other ways, haptics often works slower than vision. Such differences are important to consider when dividing the tasks between visual and haptic displays.

The space covered by the two senses is also a very important difference. Vision allows the sensing of information kilometers away, while haptics is mainly restricted to the space within arm's reach. There are possibilities of extending the haptically reachable space with tools, for instance, the long cane for the visually impaired and, to some extent, via information projected onto the skin. Both of these mechanisms are discussed in the following. Thus, haptics can provide spatial information, but it offers a more limited capacity than vision for overview and 3D space.

33.2.4 The Capacity of Bare Fingers in Real Environments

When investigating the usefulness of technical devices utilizing haptics, it is instructive to take into account the natural functioning of haptics. An example is the situation when someone is searching for an object in a pocket or in a bag without the help of vision. With haptics, it is possible to identify the object, grasp it with suitable force, and take it out for the desired use. The potential to identify common objects by haptics is close to perfect within a few seconds (Klatzky et al., 1985).

Katz (1989) gives many examples of the capacity of haptics in such situations. For instance, it is remarkable that haptics can demonstrate transparency capabilities, as when physicians by palpation of the surface of a body can obtain information about the conditions of an organ under the skin and fat layers. A related property of haptics is remote touching, that is, the experience of a distant object via some medium. Physicians can perceive properties of the inner parts of the body also via instruments. Visually impaired persons with a long cane can perceive the properties of the ground at the end of the tip of the cane when touching it with the cane. The hand can thus pick up information via a tool and locate the information to the end of the tool. More details about visually impaired people's use of haptics can be found in books edited by Ballesteros Jiménez and Heller (2004), Heller (2000), Heller and Schiff (1991), and Schiff and Foulke (1982).

33.2.5 Limitations of Haptics

Even if haptics has an often underestimated capacity, it has also limitations that should be considered in planning for applications. A great number of receptors in the skin, muscles, tendons,

---

[2] There is some controversy about the superiority of active touch. There is experimental evidence that active and passive touch give equivalent results for the perception of texture (Lederman, 1981) and small patterns (Vega-Bermudez et al., 1991). Magee and Kennedy (1980) found even better results for passive than for active touch in identifying objects in raised line drawings, interpreting the result to depend upon favorable conditions for attention to the perceptual task in the passive case. Further experimental analysis was reported by Richardson et al. (2004), and a theoretical discussion was presented by Hughes and Jansson (1994). The outcome of this discussion was that active exploration is favorable for the efficiency of haptics for most tasks, but there may be tasks where it is not necessary. Symmons et al. (2004) found the results to a large extent to be task-dependent. Johnson (2002) suggested that passive touch requires more concentration and that the difference between the two kinds of touch is similar to the difference between situations with dim and bright light in vision.

---

[3] However, there are also sometimes possibilities to identify objects at a "haptic glance," that is, a short contact with the object, especially when the observer has hypotheses about what object to expect, and the identification is based on local properties such as texture (Klatzky and Lederman, 1995).

and joints are involved and have to be provided with a minimum stimulation to be experienced by the user (see Johnson, 2002, for details). There are two kinds of such limitations: absolute thresholds, which are necessary to get any experience at all, and difference thresholds, also called just noticeable differences (JNDs), that allow discriminating between different amounts of a property. The thresholds are different depending upon stimulus variable and body site, as well as the observer's age. Below, some examples of difference thresholds will be discussed.

An overview of difference thresholds relevant for haptic interfaces was provided by Tan (2007). For example, when the task is to compare the hardness of two objects by squeezing them, the resistance force of the harder object has to be about 7% larger than the resistance force of the less hard object (Pang et al., 1991). To discriminate between force directions, the difference needs to be 33° irrespective of reference direction (Tan et al., 2006). The threshold may vary if visual information is simultaneously available, in different ways when the two kinds of information are congruent and incongruent, respectively (Barbagli et al., 2006). To estimate the resolution of fingertip positions during active motion, basic data about discrimination and identification of finger joint-angle are needed. Such data were provided by Tan et al. (2007).

33.2.6 Natural Capacity of Haptics Should Be Considered When Evaluating Technical Aids

It is important to remember the potentials of haptics in real environments when the efficiency of technical aids is investigated. Likewise, it is important to consider under which conditions the performance can be expected to be successful; especially, which stimulus properties are available and which exploratory procedures are possible to perform. When a device fails to provide the information wanted, important reasons are probably that it is not successfully utilizing the capabilities of haptics, that is, it does not provide the stimulus properties and exploration options that are essential for the functioning of haptics. As will be discussed below, there are usually many constraints on these basic requirements for the working of present-day haptic displays in comparison with what is naturally available.

33.2.7 Haptics Research and Universal Access

Haptics is a research area that covers several disciplines from physiology and psychology to computer science, human-computer interaction, robotics, and electrical engineering. Typically, productive research in this emerging area requires multidisciplinary collaboration combining the latest hardware solutions, software, and understanding of human functioning. To start with, controlled experimental studies are needed to understand the use of new haptic technology. Based on the results of this basic research, novel hardware and software prototypes can be constructed and tested with actual users. Only after iterative research and development can end-user applications be successfully created and distributed to enhance universal access in the information society.

33.3 Haptic Low-Tech Aids for the Visually Impaired

Low-tech aids such as the long cane, the guide dog, Braille, and embossed pictures are often used by the visually impaired. In all these cases, interaction between the haptic sense and the environment is important.

The *cane* as an aid for walking without guidance of vision has been used for centuries. It allows users to extend their arms to reach the ground and objects in front of them. In spite of its technical simplicity, a present-day long cane can inform its user about many features of the environment, information that is helpful in the guidance of walking (Guth and Rieser, 1997; Jansson, 2000b). The haptic stimulus properties providing this information are vibrations in the cane and force feedback during the handling of the cane.

It may seem strange to consider the guide dog for the visually impaired pedestrian as a technical aid, but it may be thought of in that way. The dog and the person form a unit together with the uniting harness (a girdle encircling the dog's chest with a handle for the user). With a hand on the handle, a visually impaired pedestrian interacts with the dog for both controlling the guide dog and receiving information about its activities.

A visually impaired person can read text by using the hands to obtain a tactile equivalent of visual letters and other symbols, the most common version being *Braille,* coded within a six-point (or sometimes eight-point) rectangular matrix that is presented in embossed form and read by the fingers (Foulke, 1991). There is also a mechanical Braille version consisting of matrices of pins to be discussed in the following.

Also pictorial information can be presented as *embossed pictures.* Visually impaired people can explore embossed lines and patterns of embossed units, and much effort has been made to make them useful (e.g., see Edman, 1992). Some kinds of pictures can be interpreted without much difficulty (Kennedy et al., 1991), and it has also been reported that blind people can draw pictures (in two dimensions) of three-dimensional objects and scenes (Kennedy, 1993). However, more complex embossed pictures are more difficult to perceive haptically, the main problem being getting an overview of the picture. The necessary exploratory movements can take considerable time and they sometimes do not result in an understanding of the picture. Further, tactile pictures are typically two-dimensional, and it is usually difficult to perceive the three-dimensional properties of the scene haptically, even if some possibilities have been reported. Kennedy (1993) described the understanding of perspective by blind people, while Jansson and Holmes (2003) reported on the haptic perception of depth in pictures via texture gradients.

Tactile maps are embossed pictures depicting geographical conditions. They are useful in many contexts, but they suffer from the difficulty of providing an easy overview. There have been many efforts to make them as useful as possible; for overviews, see Bentzen (1997), Edman (1992; Chapter 6) and Jansson (2003).

The main lesson from the success of several low-tech aids is that the usefulness of a technical aid is *not* a function of its technical complexity. Technically very simple aids, such as the long cane and the guide dog, can be very efficient, if they make available the proper information and utilize the natural capacities of the haptic sense. The interaction of the user with a long cane and the guide dog is very natural. It is a kind of tool use similar to many other activities. The user moves the hand with the tool, thereby performing acts in the environment and getting information back as a result of the activity. In the case of the guide dog, the activities of the dog are added, but interactions with moving objects are also natural. In the reading of embossed pictures the natural capacity of haptics to pick up properties of objects via edges, as well as textures and other characteristics of surfaces is utilized. However, the common restriction to 2D information is a deviation from what is most common in the real world, and this is one of the problems of embossed pictures. Braille is a bit different, as matrices of this kind are not often naturally appearing in the environment. That is probably the reason why it usually takes considerable time to learn how to read Braille. However, the task seems to be sufficiently similar to natural activities to make efficient reading possible after suitable training.

33.4 Matrices of Point Stimuli

Matrices of point stimuli can be seen as representations of physical surfaces with low spatial resolution, and they are used because of technical and economical problems with producing haptic devices with high spatial resolution. Such matrices have been constructed in efforts to replace or supplement low-tech aids with technology providing more advanced information. One option is to present an extended matrix in contact with the skin. Such a matrix makes it possible to form a pattern within the matrix by dynamically elevating some of the pins above the rest of the matrix. The pins can be either static or vibrating. Many such devices have been built; an overview is provided by Kaczmarek and Bach-y-Rita (1995). A device with electrical stimulation to the tongue has also been developed (Bach-y-Rita et al., 2003). Matrices of point stimuli of different kinds have been developed and will be discussed in the following.

33.4.1 Matrices for Large Skin Areas

Even if the long cane and the guide dog are very important, these low-tech aids have constraints. The long cane does not cover the space beyond about 1 meter in front of the pedestrian and no space above waist level (when used normally). The guide dog cannot be made responsible for orientation in the larger environment. These constraints have motivated many efforts to construct mobility devices that compensate for the constraints, called ETAs (electronic travel aids). Compensating senses may be hearing or haptics, sometimes in combination.

An early device intended for the guidance of locomotion was the Electrophthalm (Starkiewicz and Kuliszewski, 1963), which reproduced visual information from a camera onto a matrix of pins fastened on the forehead; a later version with a larger matrix (Palacz and Kurcz, 1978) is shown in Figure 33.1. A related device is the tactile vision substitution system (TVSS), some stationary versions presenting information via vibrating pins to the back or, concerning mobile versions, via electrodes to the chest (Bach-y-Rita, 1972); a version of the latter kind is also shown in Figure 33.1.

In these devices, the camera on the head picks up environmental visual information that is transformed to a tactile matrix attached to the forehead, chest, or back. The user interacts with the environment by moving the camera when turning the head, not with the skin surface getting the tactile information. The pattern changes produced by motions in the environment and movements of the user are picked up by the video camera. One task important for orientation and mobility is to localize and identify objects at a distance in the three-dimensional space, another to guide the user to reach goal objects and to avoid obstacles during walking. These are other tasks than those usually performed by haptics. The skin can be seen to work in a way analogous to the retina of the eye receiving changing stimulations caused by the observer's eye and body movements (Bach-y-Rita, 1972). However, for the tactile displays there are apparent limitations in spatial resolution, which reduce the possibilities of object identification, as well as collecting distance information. In spite of this, it has been reported that users can perceive objects external to the body as localized in the space in front of them (White et al., 1970). An important prerequisite for this effect is reported to be that the user is active in obtaining the information, thus by moving the camera. The reports are anecdotal, however, and there are also conflicting reports of not attaining external localization in spite of long training (Guarniero, 1974, 1977).

Tactile information provided by a device such as the Electrophthalm and a user active in getting the information can guide walking (Jansson, 1983). In an experiment the walking person had to make a slalom walk around two poles and toward a goal pole guided only by tactile information from a 12 × 18 matrix of

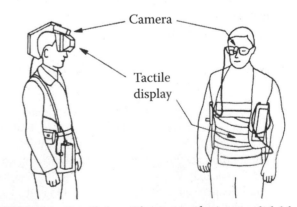

FIGURE 33.1 Two devices with a matrix of point stimuli fed from a camera, the Electrophthalm (left) and a version of the Tactile Visual Substitution System (right). (Based on photographs taken by Gunnar Jansson.)

vibrators on the forehead. Even if the tactually guided walk was not as smooth as the one that is visually guided, it was possible.

A related experiment was made with a stationary TVSS with 20×20 pins onto the back of the user (Jansson and Brabyn, 1981), where the task was to bat a ball while running in front of the seated player. In this case the camera was stationary, as well as the location of the matrix of pins. The changes on the matrix were caused by the motions of the ball and the user's movements of the bat. The total time of a game was about 3.5 sec, during which time the player had to pick up the tactile information, organize the response, and perform appropriate movements. Trained players could perform this task quite well.

Both these experiments demonstrate that tactile information obtained via a camera can guide movements in tasks of this kind. However, the visual information transformed to tactual information in these cases is very restricted compared with the information in real environments, black and white representations of a few simple objects. A tactile display of this kind cannot sufficiently forward complex information necessary for guiding movement in ordinary environments. The Electrophthalm and the TVSS are interesting examples of the capacity of haptic information, but they have not been further developed to practically usable travel aids, probably because of these restricting features.

33.4.2 Small Matrices for Reading Text

There are two main ways of reading text with the aid of haptic displays, either by a matrix presenting a symbol system such as Braille, or by a matrix producing copies of visual letters. In the former case, the haptic display presents protruding pins in the traditional Braille arrangement. There are commercially available devices presenting a row or rows of refreshable Braille cells consisting of pins as an extension of the ordinary computer keyboard (i.e., http://www.papenmeier.de/reha/rehae.htm).

In the latter case, a reading aid makes direct reading of visual text possible by transforming the visual pattern into a corresponding tactual one. Such an aid is the Optacon (Optical to Tactile Converter). It consists of a camera that the user moves with one hand over a text, whereby it is transformed into copies of the letters within a matrix of 5×20 vibrators (its last version, Optacon II). This changing pattern is picked up by a finger pad on the other hand (Bliss, 1978). The Optacon was a success at first, but a drawback was that reading the tactile patterns was a slow process for most people and required considerable training to reach a reasonable speed. With a new technology based on optical reading of a text and transforming it into speech it was much easier to get access to the text, which decreased the demand for a device such as the Optacon, and its production was terminated. However, it has also after that been used in research on different aspects of haptic perception.

33.4.3 Larger Matrices Presented to Finger Pads

There are matrices with many more pins than those just discussed. A matrix such as the Optacon requires that the camera is moved over the different parts of a visual figure and that the successive tactual information is perceptually integrated into a total scene. An alternative to a moving camera like the Optacon is to have a larger pin matrix that is explored by moving fingers. As the spatial resolution with present technology is usually a few millimeters, quite a number of pins would be needed to simultaneously cover a larger scene.

The largest matrix built so far is the Dot Matrix Display DMD 120060, which has a total size of 18×36 cm containing 60×120 pins with a center distance between the pins of 3 mm (e.g., http://www.metec-ag.de/company.html). Such a distance is much larger than the spatial resolution of touch on the finger pad, and the display thus does not utilize the full capacity of touch. This display was intended for the visually impaired, but it is quite expensive and only a few devices have been built. Further development of the device is in progress.

A stationary haptic display near the capacity of touch is a display with 20×20 vibrators with a center distance of 0.4 mm and a total size of 8×8 mm (Pawlik et al., 1998). This is a display constructed for basic research on touch and exists so far only in one copy. However, it can give information about the capacity of touch at maximum spatial resolution. A display closer to application with lower spatial resolution (3 mm) but allowing also individual height variation (in 0.1 mm steps to a maximum of 10 mm) of the pins has also been developed (Shinohara et al., 1998).

33.4.4 Audio-Tactile Information with the Aid of a Touch Tablet

Computers can of course be used for production of tactile maps. GISs (geographic information systems) may be a suitable starting point for the content of the map, but as they often contain too much information to be readable haptically, the information has to be reduced. Michel (1999) demonstrated that there are differences in the suitability of the formats of the systems for reduced information presentation.

The efficiency of tactile maps can be increased by combining tactile and coordinated auditory information, by putting a tactile map onto a touch tablet connected to a computer. When the map-reader presses specific points on the tactile map, auditory information is obtained, for instance the name of a geographic feature. Pioneering work was made by Parkes (1988) with the NOMAD device. Related work is the interactive auditory learning tool TACTISON (Burger et al., 1993) and the dialogue system AUDIO-TOUCH (Lötzsch, 1995). Increased efficiency in reading tactile maps when they are enhanced with auditory information in this way has been experimentally demonstrated (Holmes et al., 1995). A system for the production of audio-tactile graphic information, including an embosser, was developed for the reading of virtual maps by Viewplus (http://www.viewplus.com/products/touch-audio-learning). With this system, the interaction with the map is similar to that with traditional tactile maps on a touch tablet, but the starting point was a virtual map. An evaluation of this system suggested that it may be useful, but that improved tactile information is required (Jansson and Juhasz, 2007).

33.4.5 Tactile Computer Mouse

An ordinary computer mouse is a well-known example of interaction between a computer and its user. Visual feedback is usually provided via a computer screen and gives the user information about the location and movements of the mouse. A tactile mouse has an added function by providing information via a pattern of vibrating or stationary protruding pins within a matrix on the top of a mouse. The dynamic pattern within the matrix is usually the result of the user's hand movements, but it can of course also be rendered through a computer program. Auditory information may be provided to improve the efficiency. One example is the VTPlayer (http://www.virtouch2.com) that is a mouse with two 4 × 4 arrays of pins, each intended for one finger. In addition to kinesthetic information, it provides simultaneous information about features of, for instance, a geographical map. It has been shown that such a map has potential for teaching geographical facts to visually impaired readers, but revisions can be expected to improve its functioning (Jansson et al., 2006a). The tactile mouse is meant to be used by visually impaired people, but as they do not get the visual feedback, they have problems in staying informed about the location and motions of the cursor on the screen. They get kinesthetic information, but that may be in conflict with the tactual information. For instance, moving the tactile mouse horizontally with its body oriented nonperpendicularly to the direction of the motion path results in an oblique motion path of the cursor on the screen. There may also be problems in knowing the location when the mouse is lifted and replaced in a new position, as well as differences in the final location of the cursor appearing when the mouse has been moved at different speeds. In any case, there have been efforts to investigate the usefulness of haptic mice for visually impaired people in several contexts, such as reading graphs presented both tactually and auditorily (Wall and Brewster, 2006), giving directional information during navigation in a virtual environment (Pietrzak et al., 2006), and teaching science (Wies et al., 2001).

33.5 Haptic Displays

Haptic displays are for haptics what computer screens are for vision and loudspeakers for hearing. They are robotic devices allowing haptic manipulation and information about virtual objects and scenes via a computer. They are of interest both for people with vision as complements to visual and auditory information, and for people without vision as a substitute for this sense.

The term *haptic display* has in practice mainly been used for advanced robotic gadgets for haptic exploration of virtual objects. Other devices, such as tactile pictures and mice, could be said to belong to this category, but traditionally they are not classified in this way. So far, more than 30 different haptic displays have been constructed (http://haptic.mech. northwestern. edu/intro/gallery). For general overviews, see Burdea (1996), Burdea and Coiffet (2003), and McLaughlin et al. (2002). Two examples of haptic displays studied in the context of universal access are shown in Figure 33.2.

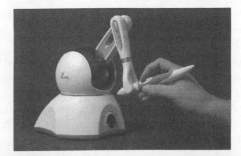

FIGURE 33.2 Two examples of haptics displays studied in the context of universal access: the one-point display Phantom OMNI (top). (See http://www.sensable.com; printed with permission by SensAble Technologies, Inc.); the two-point display GRAB (bottom). (See http://www.grab-eu.com; printed with permission by M. Bergamasco and C. A. Avizzano, PERCRO laboratory, Scuola Superiore Sant'Anna, Pisa, Italy.)

Haptic displays provide force feedback by a "collision" between the virtual representation of the endpoint of a stylus, a thimble, or similar tool in a user's hand, and the surface of a virtual object. In addition to the shape of objects, surface properties such as hardness/softness, texture, and friction can be rendered in 3D for haptic exploration.

The most interesting property of the haptic displays as a potential aid for the visually impaired is that they make available direct haptic information about the 3D aspects of an object or a scene, information that is difficult to obtain haptically from 2D depictions of 3D objects and scenes. A basic problem for the usefulness of the haptic displays concerns whether this information is sufficient and can have a form making it useful. For visually impaired people, it is especially important to consider its usefulness for haptics alone, without being combined with vision, and also the possibilities of enhancing haptics with auditory information.

33.5.1 Information Available via a Haptic Display

The enormous potential of haptics when functioning in natural contexts is only partly utilized by the haptic displays developed so far. The most important constraints concern *number and size of contact surfaces*. When the bare hand is used naturally, there are several contact surfaces and each of them has an extension of roughly at least a finger pad. In present-day haptic displays, the number of contacts is quite low, in most cases just one. The contact surface is also, except in a few devices, only a tiny point.

These are drastic differences from natural haptics, with important effects for the efficiency of haptic displays.

There are several drawbacks with having only one contact area available, as no simultaneous information from several contact areas can be obtained; only successive information is available. The use of one finger only may be natural in some contexts, as Symmons and Richardson (2000) found for exploration of a 2D tactile picture. However, the situation is different when a 3D object is explored. The use of only one finger means, among other things, that the number of available exploratory procedures decreases. For instance, enclosure, an important exploratory procedure for perceiving global shape where several fingers are grasping an object, cannot be used. Klatzky et al. (1993) found a deterioration of performance between five fingers and one finger, and Lederman and Klatzky (2004a) investigated other constraints and provided a theoretical context for the effects of such constraints. Jansson and Monaci (2006) obtained related results (Figure 33.3, bare fingers) with the difference between one finger and two fingers being the largest one. However, the main difference in the latter experiment concerned the size of the contact area that has a huge effect on the availability of spatially distributed information.

33.5.1.1 The Importance of Spatially Distributed Stimulation with High Spatial Resolution

In real situations haptics has a high spatial resolution, especially at the exploring fingertips (around 1 mm). This spatially distributed stimulation is missing in a one-point haptic display.

Lederman and Klatzky (1999) simulated the restricted information condition at a virtual contact with such a display by putting rigid sheaths on the fingertips in a real situation. Most of the perceptual measures, including tactile spatial acuity and 2D edge detection, showed substantial impairment. In a related experiment, Jansson and Monaci (2006) demonstrated that if the fingertips are equipped with such rigid sheaths during identification of real objects, the performance is drastically impaired (Figure 33.3, fingers with sheaths). The difference between one finger and two fingers was still there, but it was reduced. The large differences were those between bare fingers and fingers with sheaths. The amount of information at the contact areas is thus of central importance, more important than the number of contacts. The situations in presently available haptic displays were simulated in these experiments by the situations with sheaths, and the results demonstrate that the effects of the reduced stimulation available via these haptic displays are very large. For the development of haptic displays there is no considerable advantage, at least concerning identification of virtual objects, by increasing the number of contacts, if each of them does not contain spatially distributed information. A related result, showing no improvement with up to three contact points, was obtained in an experiment with a haptic display (Frisoli et al., 2005a).

An effort to increase the spatial distribution in the context of rendering virtual textiles was made in the EU project HAPTEX (http://haptex.miralab.ch). Textiles are presently rendered visually, and the goal of HAPTEX was to add to this a corresponding

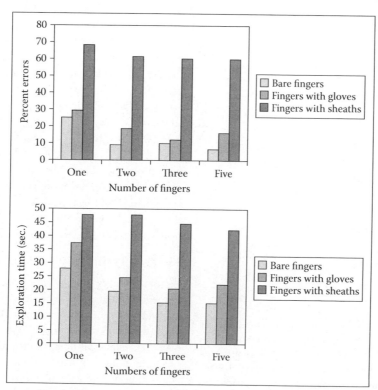

FIGURE 33.3 Proportion of errors and exploration time as functions of number of fingers and spatially distributed information. (From Jansson, G. and Monaci, L., *Virtual Reality*, 9, 243–249, 2006. With permission.)

rendering of haptic information allowing real-time haptic interaction with textiles (Magnenat-Thalman et al., 2007). Among the efforts within this project was to develop hardware and software able to simulate the complex viscoelastic properties of textiles. The stimulator array for a fingertip consisted of 24 moving contactors with 2-mm interdistances, thus a relatively high spatial resolution. Together with specially developed software the system made possible, in some cases, for users to match real and virtual textiles. If more technical developments are made to create spatially distributed information with high spatial resolution, it is reasonable to expect huge leaps in the usefulness of haptic displays.

33.5.2 Reduced Information Presentation

The present display types without spatially distributed information at the contact points will probably be dominant for several years, because of the technical complexity of providing useful extended information. However, there is potential with constrained information, and some issues that are relevant for applications to visually impaired people will be discussed in the following.

33.5.2.1 Identification of Shape

Even with information reduced to one contact area with small extension it is possible to identify simple regular shapes with the help of only haptic information, according to an experimental study (Jansson et al., 1999) using a display allowing exploration of virtual objects via a handheld stylus or a fingertip in a thimble attached to a robot arm, a PHANToM 1.5 (http://www.sensable.com). However, efficiency was much smaller than with

bare fingers (Figure 33.4). The proportion of correct identifications was perfect in the natural conditions, but only around 0.7 in the two haptic display conditions. The exploration time was much longer in these conditions (around 30 sec) than in the natural conditions (about 2 sec). It is also interesting that the two haptic display conditions had very similar results. Whether the participants used the stylus or the thimble version of the PHANToM did not matter. This suggests that the important property when using such a display is the constraint to *one* contact point, not the way of handling it.

Both proportion correct and exploration time in identification of object shapes improve with the size of the object, at least up to 100 mm (Jansson, 2000a), but especially exploration time can be expected to increase with larger object shapes. The performance deteriorates when the shapes are more complex (Jansson and Larsson, 2002).

33.5.2.2 Judgment of Surface Properties

Textures are important properties for the identification of objects. Hollins et al. (1993) made multidimensional scaling of textures and found two main dimensions—soft-hard and smooth-rough—and one weaker dimension—sticky-slippery. In contrast with shape, virtual texture seems potentially easy to judge via present-day haptic displays. In an experiment where judgments of the roughness of real and virtual sandpapers were both explored with a stylus, the judgments were very similar (Jansson et al., 1999). Heller (1989) found that textures can be judged by visually impaired people about as well as by sighted people, which suggests that vision is not necessary for texture perception. That texture gradients can provide 3D information similar to visual gradients was demonstrated by Holmes et al. (1998). Several aspects of perceiving textures via a probe were investigated by Lederman et al. (1999) and Klatzky et al. (2003), and an analysis of many properties of texture perception can be found in Lederman and Klatzky (2004b). The problems in rendering realistic haptic textures were discussed by Choi and Tan (2004).

In spite of the problems in providing realistic material properties to the virtual objects, it is important to try to make such properties easily identifiable, as they are salient for recognizing the objects easily (Klatzky et al., 1993; Klatzky and Lederman, 2007). An inexpensive way to render some textures, at least roughness, is to use a force-feedback mouse (Klatzky and Lederman, 2006; Lederman et al., 2006).

33.5.2.3 Training in Exploratory Procedures

Virtual textures may be more successfully perceived than the shape of virtual objects, because the exploratory procedure for judging texture is much simpler than those for judging shape. For judging texture it is sufficient just to make any movement over the surface, while it is much more complicated when shape has to be judged. Practice in judging shape can be expected to improve performance, which was shown in experiments (Jansson and Ivås, 2001). Already a few days of one-hour practice sessions gave large improvements for a majority of the participants. An

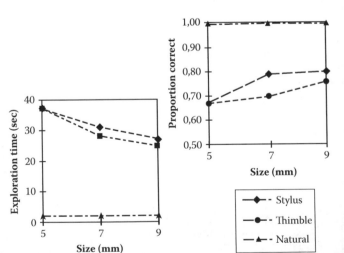

FIGURE 33.4 Proportion correct and exploration time for identification of simple shapes in three sizes during three conditions. (From Jansson, G. and Billberger, K., The PHANToM used without visual guidance, in the *First PHANToM Users Research Symposium (PURS99)*, 21–22 May 1999, Deutsches Krebsforschungszentrum, Heidelberg, Germany, http://mbi.dkfz-heidelberg.de/purs99/proceedings/Jansson.pdf, 1999.)

important practical conclusion of this result is that there is a great risk of underestimating the usefulness of a haptic display, if the participants have not sufficient practice in exploration with the specific display.

33.5.2.4 Colocation of Visual and Haptic Space

Objects are usually perceived to occupy the same location visually and tactually in natural contexts. In virtual worlds they are often not colocated, that is, the visual presentation is made on a screen and the tactual presentation on another location, which may be a drawback not only for sighted people, but also for visually impaired people with remaining vision. Advantages with colocation may be expected, as that is the natural condition. Some such effects were shown for a targeting task (Wall et al., 2002) and for perception of form (Jansson and Öström, 2004). Under stereo conditions in the latter study there was a significant effect in judging the depth dimension of a distorted object. Informally it was also found that performing tasks such as finding knobs and regaining contact with lost virtual objects were facilitated under colocation conditions.

33.5.3 A General Problem with Complex Haptic Devices Used in Practice

A general problem concerning technically complicated devices such as those presented previously is that they may have shown their potential in laboratory contexts, but their usefulness in real situations may be more problematic. The intended users may find them difficult to use, possibly because the devices are not sufficiently adapted to the functioning of haptics (Hatwell, 2006). Setting up good psychophysical studies is difficult because of the need to choose good methods, find exact definitions of proximal stimuli, and handle new hardware. Therefore, close cooperation between several disciplines, including psychology and engineering, is important (Tan, 2006). Guidelines for tactile and haptic interaction between computer users and relevant devices are under development by ISO (International Standards Organization; van Erp et al., 2006).

33.6 Haptic Displays as Aids for the Visually Impaired

Haptic displays are often used together with visual and sometimes also with auditory displays, for instance, in medical training and for computer games. When they are used by people with severe visual impairment, these people have to work without the extra information obtained from vision, which makes the task more difficult, especially concerning getting an overview and not losing contact with an object explored, as well as finding interesting parts of the object to explore. In spite of this, haptic displays may be useful for visually impaired people, as it is the only way in some contexts for getting direct information (Jansson et al., 1999; Brewster and Murray-Smith, 2001; Yu et al., 2001; Sjöström, 2002).

One context where haptic displays is potentially useful for visually handicapped people is in the experience of art at museums, as it is usually not allowed to touch statues at museums, for instance. The EU project PURE-FORM (http://www.pureform.org) aimed to develop a haptic display for exploration of virtual copies of statues (Bergamasco and Prisco, 1998; Bergamasco et al., 2001; Frisoli et al., 2002). The display with the corresponding virtual visual copy is presented in Figure 33.5. This display is intended for users in general, but it has, of course, a special interest for visually impaired people as a unique possibility of a direct experience of statues on their own (Jansson et al., 2003). A prototype was evaluated at four European museums (Frisoli et al., 2005b). Sighted participants found in general the haptic experience amusing and instructive, and so did the visually impaired participants, but to a somewhat lower degree. A problem for the visually impaired was probably that the haptic display, in spite of the increased 3D information it provided, did not solve the general problem of haptic perception without vision, namely that of obtaining an overview of an object or scene. Some help with the overview is needed, for instance, verbal introductory explanation and/or guidance for exploration of the object or scene.

In another EU project called GRAB (http://www.grab-eu.com), a new haptic and audio virtual environment was developed (Bergamasco et al., 2006). In the GRAB haptic display, which has two arms with thimbles, the user puts one finger in each thimble. For the GRAB system three applications were developed and evaluated: a game, a chart data explorer, and a city-map explorer. The evaluations indicated its usefulness for visually impaired people concerning these applications, and provided suggestions for improvements. An example is the evaluation of the maps application (GRAB, 2004).

Sjöström (2002) implemented several games for blind children. For example, a haptic battleship game, a haptic mathematics program, and a painting application in which textures correspond to different colors were developed. Patomäki et al. (2004) constructed several simple games for visually impaired 4- to 6-year-old children making use of dynamic haptic effects with the PHANTOM device. The study showed that the PHANTOM device had potential even for small children, but the results suggested that the focus group should consist of a bit older children. In a later project (Saarinen et al., 2006), multimodal applications for explorative learning were constructed for 6- to 10-year-old visually impaired children (Figure 33.6). The results were promising, as the children could easily use the system making use of haptic, auditory, and visual modalities, and they also learned new facts about space, the solar system, and the Earth. A special usability testing process was developed to allow testing the systems with visually impaired children (Raisamo et al., 2006a).

Recently, the EU-funded project MICOLE (IST-2003-511592 STP, http://micole.cs.uta.fi/) studied a collaborative learning environment for visually impaired children that makes use of cross-modal presentation of information and provides haptic and auditory support for navigation in the interface. The inclusive interface architecture makes it easier to implement similar

FIGURE 33.5 PURE-FORM display with a user holding a finger in one of the two thimbles exploring to experience the virtual haptic copy of a statue together with corresponding virtual visual copy. (Printed with permission by Antonio Frisoli and Massimo Bergamasco, PERCRO, Scuola Superiore Sant'Anna, Pisa, Italy.)

FIGURE 33.6 A visually impaired girl being shown the multimodal learning environment before starting to use it. (From Raisamo, R., Hippula, A., Patomäki, S., Tuominen, E., Pasto, V., and Hasu, M., *IEEE Multimedia*, 13, 70–76, 2006.)

applications in the future. It has been published on the web under the LGPL license.

In addition to making use of high-quality haptic displays, affordable game controllers supporting force feedback or vibro-tactile feedback have been used to create software that is usable for visually impaired children. Raisamo et al. (2006b) presented a tactile memory game that was usable and interesting to them (Figure 33.7). This shows that the devices don't need to be the

FIGURE 33.7 A visually impaired boy is playing a haptic memory game with an inexpensive vibro-tactile gamepad. (From Raisamo, R., Patomäki, S., Hasu, M., and Pasto, V., *Interacting with Computers* 19, 196–205, 2006.)

most sophisticated ones, as long as the design of the application is such that it makes the best use of the devices available.

33.7 Conclusions

Regarding universal access, it is fundamental to provide people with reduced functioning of senses such as vision and hearing with alternative options. Haptics has a great potential to provide such options. It does not have the capacity of vision to provide immediate overview and to cover a distance in space, but it has its strength in providing information about objects and events in near space. The bare hands in real environments can very competently perform many tasks, including identification of objects and judging surface properties. The success of many low-tech haptic aids for the visually impaired, for instance the long cane, the guide dog, and Braille, demonstrates that the competence of haptics can be used to open up new possibilities of traveling and access to written information. It should be noted that the success is not based on advanced technology, but on the adaptation of the substituting devices to the functioning of the substituting sense. For instance, the long cane is an excellent example of the use of tools, where haptics has been useful for thousands of years.

The main challenge for modern technically advanced aids is to develop devices that are well adapted to the natural functioning of haptics. This is not a simple task. Haptics has not been as intensely studied scientifically as vision and hearing, even if

haptics research has increased much during the last decades. Much basic perception research remains to be done to acquire a better understanding about the functioning of haptics, which is important for the development of both hardware and software. Even if humans have a great capacity to learn how to use new gadgets, the best results are obtained if the natural capacities are utilized.

An illustrative example of the needs of adapting hardware to the functioning of haptics is the present situation concerning haptic displays. The hand is an enormously versatile perceptual system with a large amount of sensors distributed over a skin area covering a highly movable "mechanical device" that can be used for exploration. In this activity another large number of sensors in the muscles, tendons, and joints are involved. Compared to this impressive perceptual system for information, present-day haptic displays are very simple. Most of them have one contact point that is used to poke the virtual world. There have been efforts to increase the number of contact points as well as the distribution of sensors over a larger area, but the results have been far from approaching the complexity of the human hand. There are devices with more than one contact point and some with extended contact surfaces, but nothing very similar to the hand. The greatest technical challenge is to provide a contact surface that simulates the perceptual aspects of the human skin.

Even if the technical development has a long way to go to provide an ideal haptic device, the devices developed so far can improve universal access, as has been shown previously. However, because of the potentials of haptics, much more useful tools for this purpose can be expected in the future.

Acknowledgments

The authors' research reported here was funded by the Swedish Council for Research in Humanities and Social Sciences, the Bank of Sweden Tercentenary Fund, the Swedish Transport and Communication Research Board, the Academy of Finland, and the European Union projects TIDE-BP-1148 – MOBIC, IST-2000-29580-PURE-FORM, IST-2003-511592 STP MICOLE, and IST-2004-002114-ENACTIVE.

References

Bach-y-Rita, P. (1972). *Brain Mechanisms in Sensory Substitution.* New York: Academic Press.

Bach-y-Rita, P., Kaczmarek, K. A., and Tyler, M. E. (2003). A tongue-based tactile display for portrayal of environmental characteristics, in *Virtual and Adaptive Environments: Applications, Implications, and Human Performance* (J. Hettinger and M. W. Haas, eds.). Mahwah, NJ: Lawrence Erlbaum Associates.

Ballesteros Jiménez, S. and Heller, M. A. (eds.) (2004). *Touch, Blindness, and Neuroscience.* Madrid: UNED Press.

Barbagli, F., Salisbury, K., Ho, C., Spence, C., and Tan, H. Z. (2006). Haptic discrimination of force direction and the influence of visual information. *ACM Transactions on Applied Perception* 3: 125–135.

Bentzen, B. L. (1997). Orientation aids, in *Foundations of Orientation and Mobility* (2nd ed.) (B. B. Blash, W. R. Wiener, and R. L. Welsh, eds.), pp. 284–316. New York: AFB Press.

Bergamasco, M., Avizzano, C., Di Petri, G., Barbagli, F., and Frisoli, A. (2001). The museum of pure form: System architecture, in the *Proceedings of the 10th IEEE International Workshop on Robot and Human Interactive Communication*, 18–21 September 2001, Paris, pp. 112–117. Piscataway, NY: IEEE Press.

Bergamasco, M., Avizzano, C. A., Frisoli, A., Ruffaldi, E., and Marcheschi, S. (2006). Design and validation of a complete haptic system for manipulative tasks. *Advanced Robotics* 20: 367–389.

Bergamasco, M. and Prisco, G. (1998). Design of an anthropomorphic haptic interface for the human arm, in *Robotic Research, the Eight International Symposium* (Y. Shirai and S. Hircse, eds.), pp. 278–289. London: Springer-Verlag.

Bliss, J. C. (1978). Reading machines for the blind, in *Active Touch: The Mechanism of Recognition of Objects by Manipulation, a Multidisciplinary Approach* (G. Gordon, ed.). Oxford, U.K.: Pergamon Press.

Brewster, S. and Murray-Smith, R. (eds.) (2001). *Haptic Human-Computer Interaction.* Berlin: Springer-Verlag.

Burdea, G. (1996). *Force and Touch Feedback for Virtual Reality.* New York: John Wiley & Sons.

Burdea, G. and Coiffet, P. (2003). *Virtual Reality Technology* (2nd ed.). New York: John Wiley & Sons.

Burger, D., Mazurier, C., Cesarano, S., and Sagot, J. (1993). *Non-Visual Human-Computer Interactions, Prospects for the Visually Handicapped, Colloque INSERM* (D. Burger and J.-C. Sperandio, eds.), volume 228, pp. 97–114. Montrouge, France: John Libbey Eurotext.

Choi, S. and Tan, H. Z. (2004). Toward realistic haptic rendering of surface textures. *IEEE Computer Graphics and Applications* 24: 40–47.

Cholewiak, R. and Collins, A. (1991). Sensory and physiological basis of touch, in *The Psychology of Touch* (M. A. Heller and W. Schiff, eds.). Hillsdale, NJ: Lawrence Erlbaum Associates.

Edman, P. K. (1992). *Tactile Graphics.* New York: American Foundation for the Blind.

Foulke, E. (1991). Braille, in *The Psychology of Touch* (M. A. Heller and W. Schiff, eds.), pp. 219–233. Hillsdale, NJ: Lawrence Erlbaum Associates.

Frisoli, A., Jansson, G., Bergamasco, M., and Loscos, C. (2005b). Evaluation of the pure-form haptic displays used for exploration of works of art at museums, in the *Proceedings of Worldhaptics 2005*, 18–20 March 2005, Pisa, Italy [CD-ROM]. Pisa, Italy: IEEE Computer Society Press.

Frisoli, A., Simoncini, F., and Bergamasco, M. (2002). Mechanical design of a haptic interface for the hand, in the *Proceedings of 2002 ASME DETC 27th Biennial Mechanisms and Robotics Conference*, 29 September–2 October 2002, Montreal, Canada, pp. 25–32. Montreal: ASME Press.

Frisoli, A., Wu, S. L., Ruffaldi, E., and Bergamasco, M. (2005a). Evaluation of multipoint contact interfaces in haptic perception of shapes, in *Multi-Point Interaction with Real and Virtual Objects* (F. Barbagli, D. Prattichizzo, and K. Salisbury, eds.), volume 18, pp. 177–188. Berlin/Heidelberg: Springer-Verlag.

Gibson, J. J. (1962). Observations on active touch. *Psychological Review* 69: 477–491.

GRAB (2004). *Results of the Validation of the Maps Application.* GRAB IST-2000-26151, Deliverable 16/3. Bilbao, Spain: Foundación LABEIN.

Guarniero, G. (1974). Experience of tactile vision. *Perception* 3: 101–104.

Guarniero, G. (1977). Tactile vision: A personal view. *Journal of Visual Impairment and Blindness* 71: 125–130.

Guth, D. A. and Rieser, J. J. (1997). Perception and the control of locomotion by blind and visually impaired pedestrians, in *Foundations of Orientation and Mobility* (2nd ed.) (B. B. Blash, W. R. Wiener, and R. L. Welsh, eds.), pp. 9–38. New York: AFB Press.

Hatwell, Y. (2006). A survey of some contemporary findings on haptic perception. *Plenary Talk at the Eurohaptics International Conference (EH 2006).* Abstract available in the Conference Proceedings, p. 3. Paris: Eurohaptics 2006.

Heller, M. A. (1989). Texture perception in sighted and blind observers. *Perception and Psychophysics* 45: 49–54.

Heller, M. A. (ed.) (2000). *Touch, Representation and Blindness.* Oxford, U.K.: Oxford University Press.

Heller, M. A. and Schiff, W. (eds.) (1991). *The Psychology of Touch.* Hillsdale, NJ: Lawrence Erlbaum Associates.

Hollins, M., Faldowski, R., Rao, S., and Young, F. (1993). Perceptual dimensions of tactile surface texture: A multidimensional scaling analysis. *Perception and Psychophysics* 54: 697–705.

Holmes, E., Hughes, B., and Jansson, G. (1998). Haptic perception of texture gradients. *Perception* 27: 993–1008.

Holmes, E., Michel, R., and Raab, A. (1995). Computerunstützte Erkundung digitale Karten dursch Sehbehinderte [Computer supported exploration of digital maps for visually impaired people], in *Taktile Medien. Kolloquium über tastbare Abbildungen für Blinde* [Tactile media: Symposium on touchable pictures for the blind] (W. Laufenberg and J. Lötzsch, eds.), pp. 81–87. Marburg, Germany: Deutsche Blindenstudienanstalt e. V. Carl-Strehl-Schule.

Hughes, B. and Jansson, G. (1994). Texture perception via active touch. *Human Movement Science* 13: 301–333.

Jansson, G. (1983). Tactile guidance of movement. *International Journal of Neuroscience* 19: 37–46.

Jansson, G. (2000a). Basic issues concerning visually impaired people's use of haptic displays, in the *Proceedings of the 3rd International Conference on Disability, Virtual Reality and Associated Technologies* (P. Sharkey, A. Cesarani, L. Pugnatti, and A. Rizzo, eds.) 23–25 September 2000, Sardinia, Italy, pp. 33–38. Reading, U.K.: University of Reading. http://www.icdvrat.reading.ac.uk/2000/papers/2000_05.pdf.

Jansson, G. (2000b). Spatial orientation and mobility of people with vision impairment, in *The Lighthouse Handbook on Visual Impairment and Vision Rehabilitation* (B. Silverstone, M. A. Lang, B. P. Rosenthal, and E. E. Faye, eds.), pp. 359–375. New York: Oxford University Press.

Jansson, G. (2003). Tactile maps: Overview of research and development, in *Tactile Maps, Guidance in Map Production* (Y. Eriksson, G. Jansson, and M. Strucel, eds.), pp. 45–78. Stockholm: The Swedish Braille Authority.

Jansson, G. (2008). Haptics as a substitute for vision, in *Assistive Technology for Visually Impaired and Blind People* (M. A. Hersh and M. A. Johnson, eds.). London: Springer-Verlag.

Jansson, G., Bergamasco, M., and Frisoli, A. (2003). A new option for the visually impaired to experience 3D art at museums: Manual exploration of virtual copies. *Visual Impairment Research* 5: 1–12.

Jansson, G. and Billberger, K. (1999). The PHANToM used without visual guidance, in the *First PHANToM Users Research Symposium (PURS99)*, 21–22 May 1999, Deutsches Krebsforschungszentrum, Heidelberg, Germany. http://mbi.dkfz-heidelberg.de/purs99/proceedings/Jansson.pdf.

Jansson, G., Billberger, K., Petrie, H., Colwell, C., Kornbrot, D., Fänger, J., et al. (1999). Haptic virtual environments for blind people: Exploratory experiments with two devices. *International Journal of Virtual Reality* 4: 10–20.

Jansson, G. and Brabyn, L. (1981). Tactually guided batting. *Uppsala Psychological Reports*, No. 304, Department of Psychology, Uppsala University, Uppsala, Sweden.

Jansson, G. and Holmes, E. (2003). Can we read depth in tactile pictures? Potentials suggested by research in tactile perception, in *Art beyond Vision: A Resource Guide to Art, Creativity, and Visual Impairment* (E. Axel and N. Levant, eds.), pp. 146–156. New York: Art Education for the Blind and American Foundation for the Blind.

Jansson, G. and Ivås, A. (2001). Can the efficiency of a haptic display be increased by short-time practice in exploration?, in *Haptic Human-Computer Interaction* (S. Brewster and R. Murray-Smith, eds.), pp. 85–91. Heidelberg: Springer-Verlag.

Jansson, G. and Juhasz, I. (2007). The reading of virtual maps without vision. Paper presented at *XXIII International Cartographic Conference (ICC 2007)*, 4–10 August 2007, Moscow.

Jansson, G., Juhasz, I., and Cammilton, A. (2006a). Reading virtual maps with a haptic mouse: Effects of some modifications of the tactile and audio-tactile information. *British Journal of Visual Impairment* 24: 60–66.

Jansson, G., Juslin, P., and Poom, L. (2006b). Liquid-specific properties can be utilized for haptic perception of amount of liquid in a vessel put in motion. *Perception* 35: 1421–1432.

Jansson, G. and Larsson, K. (2002). Identification of haptic virtual objects with different degrees of complexity, in the *Proceedings of Eurohaptics 2002* (S. A. Wall, B. Riedel, A. Crossan and M. R. McGee, ed.), 8–10 July 2002, Edinburgh, Scotland, pp. 57–60. Edinburgh, U.K.: Edinburgh University.

Jansson, G. and Monaci, L. (2006). Identification of real objects under conditions similar to those in haptic displays: Providing spatially distributed information at the contact areas is more important than increasing the number of areas. *Virtual Reality* 9: 243–249.

Jansson, G. and Öström, M. (2004). The effects of co-location of visual and haptic space on judgments of forms, in the *Proceedings of the 4th International Conference on Eurohaptics 2004* (M. Buss and M. Fritschi, eds.), 5–7 June 2004, Munich, pp. 516–519. Munich: Technische Universität München.

Johnson, K. (2002). Neural basis of haptic perception, in *Stevens' Handbook of Experimental Psychology, Vol. 1. Sensation and Perception* (3rd ed.) (H. Pashler and S. Yantis, eds.), pp. 537–583. New York: John Wiley & Sons.

Kaczmarek, K. A. and Bach-y-Rita, P. (1995). Tactile displays, in *Virtual Environments and Advanced Interface Design* (W. Barfied and T. Furness III, eds.), pp. 349–414. New York: Oxford University Press.

Katz, D. (1989). *The World of Touch* (translated by L. E. Kreuger; original work published in 1925). Hillsdale, NJ: Lawrence Erlbaum Associates.

Kennedy, J. M. (1993). *Drawing and the Blind: Pictures to Touch.* New Haven, CT: Yale University Press.

Kennedy, J. M., Gabias, P., and Nicholls, A. (1991). Tactile pictures, in *The Psychology of Touch* (M. A. Heller and W. Schiff, eds.), pp. 263–299. Hillsdale, NJ: Lawrence Erlbaum Associates.

Klatzky, R. L. and Lederman, S. J. (1995). Identifying objects from a haptic glance. *Perception and Psychophysics* 57: 1111–1123.

Klatzky, R. L. and Lederman, S. J. (2006). The perceived roughness of resistive virtual textures: I. Rendering by a force-feedback mouse. *ACM Transactions on Applied Perception* 3: 1–14.

Klatzky, R. L. and Lederman, S. J. (2007). Object recognition by touch, in *Blindness, Brain Plasticity, and Spatial Functioning* (J. Rieser, D. Ashmeed, F. Ebner, and A. Com, eds.). Mahwah, NJ: Lawrence Erlbaum Associates.

Klatzky, R. L., Lederman, S. J., Hamilton, C., Grindley, M., and Swendsen, R. H. (2003). Feeling textures through a probe: Effects of probe and surface geometry and exploratory factors. *Perception and Psychophysics* 65: 613–631.

Klatzky, R. L., Lederman, S. J., and Metzger, V. A. (1985). Identifying objects by touch: An "expert system." *Perception and Psychophysics* 37: 299–302.

Klatzky, R. L., Loomis, J. M., Lederman, S. J., Wake, H., and Fujita, N. (1993). Haptic identification of objects and their depictions. *Perception and Psychophysics* 54: 170–178.

Lederman, S. J. (1981). The perception of surface texture by active and passive touch. *Bulletin of the Psychonomic Society* 18: 253–255.

Lederman, S. J. and Klatzky, R. L. (1987). Hand movements: A window into haptic object recognition. *Cognitive Psychology* 19: 342–368.

Lederman, S. J. and Klatzky, R. L. (1999). Sensing and displaying spatially distributed fingertip forces in haptic interfaces for teleoperators and virtual environment systems. *Presence* 8: 86–103.

Lederman, S. J. and Klatzky, R. L. (2004a). Haptic identification of common objects: Effects of constraining the manual exploration process. *Perception and Psychophysics* 66: 618–628.

Lederman, S. J. and Klatzky, R. L. (2004b). Multisensory texture perception, in *Handbook of Multisensory Processes* (G. A. Calvert, C. Spence, and B. E. Stein, eds.), pp. 107–122. Cambridge, MA: MIT Press.

Lederman, S. J., Klatzky, R. L., Hamilton, C. L., and Ramsey, G. I. (1999). Perceiving roughness via a rigid probe: Psychophysical effects of exploration speed and mode of touch. *Haptics-e* 1: 1–20.

Lederman, S. J., Klatzky, R. L., Tong, C., and Hamilton, C. (2006). The perceived roughness of resistive virtual textures: Effects of varying viscosity with a force-feedback device. *ACM Transactions on Applied Perception* 3: 15–30.

Lötzsch, J. (1995). Von audio-taktilen Grafiken zu interaktiven 3D-Modellen [From audio-tactile Graphics to interactive 3D models], in *Taktile Medien. Kolloquium über tastbare Abbildungen für Blinde* (Tactile media: Symposium on touchable pictures for the blind) (W. Laufenberg and J. Lötzsch, eds.), pp. 130–136. Marburg, Germany: Deutsche Blindenstudienanstalt e. V. Carl-Strehl-Schule.

Magee, L. E. and Kennedy, J. M. (1980). Exploring pictures tactually. *Nature* 283: 287–288.

Magnenat-Thalman, N., Volino, P., Bonanni, U., Summers, I. R., Bergamasco, M., Salsedo, F., and Wolter, F.-E. (2007). From physics-based simulation to the touch of textiles: The HAPTEX project. *International Journal of Virtual Reality* 6: 35–44.

McLaughlin, M. L., Hespanha, J. P., and Sukhatme, G. S. (2002). *Touch in Virtual Environments: Haptics and the Design of Interactive Systems*. Upper Saddle River, NJ: Prentice Hall.

Michel, R. (1999). *Interaktiver Layoutentwurf für individuelle taktile Karten* [Interactive Layout Plan for Individual Tactile Maps]. PhD dissertation, Der Fakultät für Informatik, Otto-von-Guericke Universität, Magdeburg, Germany.

Palacz, O. and Kurcz, E. (1978). Przydatnos´c´ zmodyfikowanego elektroftalmu EWL-300 wg Starkiewicza dla niewidomych [The usefulness of modified Electrophthalm designed by Starkiewicz for the blind]. *Klinika Oczna* 48: 61–63.

Pang, X. D., Tan, H. Z., and Durlach, N. I. (1991). Manual discrimination of force using active finger motion. *Perception & Psychophysics* 49: 531–540.

Parkes, D. (1988). "Nomad": An audio-tactile tool for the acquisition, use and management of spatially distributed information by visually impaired people, in the *Proceedings of the 2nd International Symposium on Maps and Graphics for Visually Impaired People*, 20–22 April 1988, London (A. F. Tatham and A. G. Dodds, eds.) pp. 24–29. London: King's College.

Patomäki, S., Raisamo, R., Salo, J., Pasto, V., and Hippula, A. (2004). Experiences on haptic interfaces for visually impaired young children, in the *Proceedings of the 6th International Conference on Multimodal Interfaces (ICMI'04)*, 13–15 October 2004, State College, PA, pp. 281–288. New York: ACM Press.

Pawlik, D. T. V., van Buskirk, C. P., Killebrew, J. H., Hsiao, S. S., and Johnson, K. O. (1998). Control and pattern specification for a high density tactile array, in the *Proceedings of ASME Dynamic Systems and Control Division, Vol. DSC-64*, 15–20 November 1998, Anaheim, CA. American Society of Mechanical Engineers (http://www.asme.org).

Pietrzak, T., Pecci, I., and Martin, B. (2006). Static and dynamic tactile cues experiments with VTPlayer nouse, in the *Proceedings of the Eurohaptics International Coference (EH 2006)*, 3–6 July 2006, Paris, pp. 63–69.

Raisamo, R., Hippula, A., Patomäki, S., Tuominen, E., Pasto, V., and Hasu, M. (2006a). Testing usability of multimodal applications with visually impaired children. *IEEE Multimedia* 13: 70–76.

Raisamo, R., Patomäki, S., Hasu, M., and Pasto, V. (2006b). Design and evaluation of a tactile memory game for visually impaired children. *Interacting with Computers* 19: 196–205.

Richardson, B. L., Symmons, M. A., and Wuillemin, D. B. (2004). The relative importance of cutaneous and kinesthetic cues in raised line drawings identification, in *Touch, Blindness, and Neuroscience* (S. Ballesteros Jiménez and M. A. Heller, eds.), pp. 247–250. Madrid: UNED Press.

Saarinen, R., Järvi, J., Raisamo, R., Tuominen, E., Kangassalo, M., Peltola, K., and Salo, J. (2006). Supporting visually impaired children with software agents in a multimodal learning environment. *Virtual Reality* 9: 108–117.

Schiff, W. and Foulke, E. (1982). *Tactual Perception: A Sourcebook.* Cambridge, U.K.: Cambridge University Press.

Shinohara, M., Shimizu, Y., and Mochizuki, A. (1998). Three-dimensional tactile display for the blind. *IEEE Transactions on Rehabilitation Engineering* 6: 249–256.

Sjöström, C. (2002). *Non-visual Haptic Interaction Design: Guidelines and Applications.* PhD dissertation, Division of Rehabilitation Engineering Research, Department of Design Sciences (CERTEC), Lund Institute of Technology, Lund, Sweden.

Starkiewicz, W. and Kuliszewski, Y. (1963). Active energy radiating system: The 80-channel elektroftalm, in the *Proceedings of the International Congress on Technology and Blindness.* New York: American Foundation for the Blind.

Symmons, M. and Richardson, B. (2000). Raised line drawings are spontaneously explored with a single finger. *Perception* 29: 621–626.

Symmons, M. A., Richardson, B. L., and Wuillemin, D. B. (2004). Active versus passive touch: Superiority depends more on the task than the mode, in *Touch, Blindness, and Neuroscience* (S. Ballesteros Jiménez and M. A. Heller, eds.), pp. 179–185. Madrid: UNED Press.

Tan, H. Z. (2006). The role of psychophysics in haptic research: An engineer's perspective. *Plenary Talk at Eurohaptics International Conference (EH 2006)*, 3–6 July 2006, Paris, France. Abstract available in the Proceedings, p. 5.

Tan, H. Z. (2007). All about thresholds: An overview of human haptic perception of mechanical properties, in the *Proceedings of the 4th International Conference on Enactive Interfaces*, 20–22 November 2007, Grenoble, France, p. 19. Grenoble, France: Association ACROE INPG.

Tan, H. Z., Barbagli, F., Salisbury, K., Ho, C., and Spence, C. (2006). Force-direction discrimination is not influenced by reference force direction. *Haptics-e* 4: 1–6.

Tan, H. Z., Srinivasan, M. A., Reed, C. M., and Durlach, N. I. (2007). Discrimination and identification of finger joint-angle position using active motion. *ACM Transactions on Applied Perception* 4: 1–14.

Turvey, M. T. and Carello, C. (1995). Dynamic touch, in *Perception of Space and Motion* (W. Epstein and S. Rogers, eds.), pp. 401–490. San Diego: Academic Press.

van Erp, J. B. F., Carter, J., and Andrew, I. (2006). ISO's work on tactile and haptic interaction guidelines, in the *Proceedings of the Eurohaptics International Conference (EH 2006)*, July 2006, Paris, pp. 467–470. Paris: Eurohaptics 2006.

Vega-Bermudez, F., Johnson, K., and Hsiao, S. S. (1991). Human tactile pattern recognition: Active versus passive touch, velocity effects, and pattern of confusion. *Journal of Neurophysiology* 65: 531–546.

Wall, S. A. and Brewster, S. (2006). Feeling what you hear: Tactile feedback for navigation of audio graphs, in the *Proceedings of the SIGCHI Conference on Human Factors in Computing Systems (CHI 2006)*, 22–27 April 2006, Quebec, Canada, pp. 1123–1132. New York: ACM Press.

Wall, S. A., Paynter, K., Shillito, A. M., Wright, M., and Scali, S. (2002). The effect of haptic feedback and stereo graphics in a 3D target acquisition task, in the *Proceedings of Eurohaptics 2002*, 8–10 July 2002, Edinburgh, Scotland, pp. 23–29. Edinburgh, U.K.: University of Edinburgh.

White, B. W., Saunders, F. A., Scadden, L., Bach-y-Rita, P., and Collins, C. C. (1970). Seeing with the skin. *Perception and Psychophysics* 7: 23–27.

Wies, E. F., Gardner, J. A., O'Modhrain, S., and Bulatov, V. L. (2001). Web-based touch display for accessible science education, in *Haptic Human-Computer Interaction* (S. A. Brewster and R. Murray-Smith, eds.), pp. 52–60. Berlin: Springer-Verlag.

Wing, A. M., Haggard, P., and Flanagan, J. R. (eds.) (1996). *Hand and Brain: The Neurophysiology and Psychology of Hand Movements.* San Diego: Academic Press.

Yu, W., Ramloll, R., and Brewster, S. (2001). Haptic graphs for blind computer users, in *Haptic Human-Computer Interaction* (S. Brewster and R. Murray-Smith, eds.), pp. 41–51. Berlin/Heidelberg: Springer-Verlag.

Vision-Based Hand Gesture Recognition for Human-Computer Interaction

Xenophon Zabulis,
Haris Baltzakis, and
Antonis Argyros

34.1 Introduction

In recent years, research efforts seeking to provide more natural, human-centered means of interacting with computers have gained growing interest, particularly in the context of universal access. A particularly important direction is that of perceptive user interfaces, where the computer is endowed with perceptive capabilities that allow it to acquire both implicit and explicit information about the user and the environment. Vision has the potential of carrying a wealth of information in a nonintrusive manner and at a low cost; therefore, it constitutes a very attractive sensing modality for developing perceptive user interfaces. Many approaches for vision-driven interactive user interfaces resort to technologies such as head tracking, face and facial expression recognition, eye tracking, and gesture recognition.

This chapter focuses on vision-based recognition of hand gestures. The first part of the chapter provides an overview of the current state of the art regarding the recognition of hand gestures as these are observed and recorded by typical video cameras. A complete review of computer vision-based technology for hand gesture recognition is a very challenging task. Despite the fact that the provided review is not complete, an effort was made in this chapter to report research results pertaining to the full cycle of visual processing toward gesture recognition, covering issues from low-level image analysis and feature extraction to higher level interpretation techniques. Nevertheless, this chapter does not discuss several interesting techniques that require special

cameras or setups. In that sense, this chapter does not report on methods based on cameras operating beyond the visible spectrum (e.g., thermal cameras, etc.), active techniques (e.g., those that require the projection of some form of structured light), and invasive techniques that require modifications of the environment (e.g., that the user wears gloves of a particular color distribution or with particular markers).

The second part of the chapter presents a specific approach taken to gesture recognition, intended to support natural interaction with autonomous robots that guide visitors in museums and exhibition centers. The proposed gesture recognition system builds on a probabilistic framework that allows the utilization of multiple information cues to efficiently detect image regions that belong to human hands. Tracking over time is achieved by a technique that can simultaneously handle multiple hands that may move in complex trajectories, occlude each other in the field of view of the robot's camera, and vary in number over time. Dependable hand-tracking, combined with fingertip detection, facilitate the definition of a small, simple, intuitive hand gestures vocabulary that can be used to support robust human-robot interaction. Sample experimental results presented in this chapter confirm the effectiveness and the efficiency of the proposed approach, meeting the robustness and performance requirements of this particular form of interaction.

The chapter concludes with a brief description of the challenges that still need to be addressed toward achieving vision-based hand gesture recognition for human computer interaction and universal access.

34.2 Computer Vision Techniques for Hand Gesture Recognition

Most of the complete hand interactive systems can be considered to be composed of three layers: detection, tracking, and recognition. The detection layer is responsible for defining and extracting visual features that can be attributed to the presence of hands in the field of view of the camera(s). The tracking layer is responsible for establishing temporal data association between successive image frames so that, at each moment in time, the system may be aware of what is where. Moreover, in model-based methods, tracking also provides a way to maintain estimates of model parameters, variables, and features that are not directly observable at a certain moment in time. Last, the recognition layer is responsible for grouping the spatiotemporal data extracted in the previous layers and assigning the resulting groups with labels that are associated with particular classes of gestures. In this section, research on these three identified subproblems of vision-based gesture recognition is reviewed.

34.2.1 Detection

The primary step in gesture recognition systems is the detection of hands and the segmentation of the corresponding image regions. This segmentation is crucial because it isolates the task-relevant data from the image background, before passing them to the subsequent tracking and recognition stages. A large number of methods have been proposed in the literature that utilize several types of visual features and, in many cases, their combination. Such features are skin color, shape, motion, and anatomical models of hands. A comparative study on the performance of some hand segmentation techniques can be found in Cote et al. (2006).

34.2.1.1 Color

Skin color segmentation has been utilized by several approaches for hand detection. A major decision toward providing a model of skin color is the selection of the color space to be employed. Several color spaces have been proposed, including RGB, normalized RGB, HSV, YCrCb, YUV, and so on. Color spaces efficiently separating the chromaticity from the luminance components of color are typically considered preferable. This is due to the fact that by employing chromaticity-dependent components of color only, some degree of robustness to illumination changes can be achieved. Terrillon et al. (2000) review different skin chromaticity models and evaluate their performance.

To increase invariance against illumination variability, some methods (Martin and Crowley, 1997; Bradski, 1998; Kampmann, 1998; Francois and Medioni, 1999; Herpers et al., 1999a; Kurata et al., 2001) encode pixel colors in color spaces like HSV (Saxe and Foulds, 1996), YCrCb (Chai and Ngan, 1998), and YUV (Argyros and Lourakis, 2004b; Yang et al., 1998). In contrast to the RGB color space, these color spaces encode the pixel luminance independently of its chrominance. The histogram of the 2D chrominance component is expected to exhibit a strong peak for image regions where human skin appears. This facilitates invariance to shadows, illumination changes, as well as the orientation of the surface (skin) relative to the light source(s). Having selected a suitable color space, the simplest approach for defining what constitutes skin color is to employ bounds on the coordinates of the selected space (Chai and Ngan, 1998). These bounds are typically selected empirically (i.e., by examining the distribution of skin colors in a preselected set of images). Another approach is to assume that the probabilities of skin colors follow a distribution that can be learned either offline or by employing an online iterative method (Saxe and Foulds, 1996).

Several methods (Kjeldsen and Kender, 1996; Saxe and Foulds, 1996; Starner et al., 1998; Dominguez et al., 2001; Jones and Rehg, 2002; Argyros and Lourakis, 2004b; Sigal et al., 2004) utilize precomputed color distributions extracted from statistical analysis of large datasets. For example, in Jones and Rehg (2002), a statistical model of skin color was obtained from the analysis of thousands of photos on the web. In contrast, methods such as those described in Kurata et al. (2001) and Zhu et al. (2000) build a color model based on collected samples of skin color during system initialization.

When using a histogram to represent a color distribution (e.g., Kjeldsen and Kender, 1996; Wu et al., 2000; Jones and Rehg, 2002), the color space is quantized and, thus, the level of quantization affects the shape of the histogram. Parametric models of the color distribution have been used in the form of a single Gaussian distribution (Kim et al., 1998; Yang et al., 1998; Cai and Goshtasby, 1999) or a mixture of Gaussians (Jebara and Pentland, 1997; Jebara et al., 1997; Raja and Gong, 1998; Raja et al., 1998; McKenna et al., 1999). Maximum-likelihood estimation techniques can be thereafter utilized to infer the parameters of the probability density functions. In another parametric approach (Wu et al., 2000), an unsupervised clustering algorithm to approximate color distribution is based on a self-organizing map.

The color of human skin varies greatly across human races or even between individuals of the same race. Additional variability may be introduced due to changing illumination conditions and/or camera characteristics. Therefore, color-based approaches to hand detection need to employ some means for compensating for this variability. In Sigal et al. (2004) and Yang and Ahuja (1998), an invariant representation of skin color against changes in illumination has been pursued, but not with conclusive results. In Yang et al. (1998), an adaptation technique estimates the new parameters for the mean and covariance of the multivariate Gaussian skin color distribution, based on a linear combination of previous parameters. However, most of these methods are still sensitive to quickly changing or mixed lighting conditions. A simple color comparison scheme is employed in Dominguez et al. (2001), where the dominant color of a homogeneous region is tested as if occurring within a color range that corresponds to skin color variability. Other approaches (Martin and Crowley, 1997; Bradski, 1998; Kurata et al., 2001) consider skin color to be uniform across image space and extract the pursued regions through typical region-growing and pixel-grouping techniques. More advanced color segmentation techniques rely on histogram matching (Ahmad, 1994), or employ a simple look-up table approach (Quek et al., 1995; Kjeldsen and Kender, 1996), based on the training data for the skin and possibly the pixel's surrounding areas. In Francois and Medioni (1999) and Herpers et al. (1999a), the skin-colored blobs are detected by a method that employs scan lines and a Bayesian estimation approach.

In general, color segmentation can be confused by background objects that have a color distribution similar to human skin. A way to cope with this problem is based on background subtraction (Rehg and Kanade, 1994; Gavrila and Davis, 1996). However, background subtraction is typically based on the assumption that the camera system does not move with respect to a static background. To solve this problem, the dynamic correction of background models and/or background compensation was investigated in Blake et al. (1999) and Utsumi and Ohya (1998b).

In another approach (Ahmad, 1994), the two image blobs at which the hand appears in a stereo pair are detected based on skin color. The hands are approximated by an ellipse in each image and the axes of the ellipses are calculated. The orientation of the hand in 3D is computed by corresponding the two pairs of axes in the two images. The method in Herpers et al. (1999b) and MacLean et al. (2001) also uses a stereoscopic pair to estimate the position of hands in 3D space. The binocular pair could pan and tilt and, in addition, the zoom and fixation distance of the cameras was software controlled. The estimated distance and position of the hands were utilized, so that the system could focus attention on the hands of the user, by rotating, zooming, and fixating accordingly.

Skin color is only one of many cues to be used for hand detection. For example, in cases where the faces also appear in the camera's field of view, further processing is required to distinguish hands from faces (Wren et al., 1997; Yoon and Kim, 2004; Zhou and Hoang, 2005). Thus, skin color has been utilized in combination with other cues to obtain better performance. Stereoscopic information has been utilized mainly in conjunction with skin color cue to enhance the accuracy of hand localization. In Triesch and von der Malsburg (1998), stereo is combined with skin color to optimize the robustness of tracking and in Etoh et al. (1991) to cope with occlusions. In Yuan et al. (1995), skin detection is combined with nonrigid motion detection and in Derpanis et al. (2004) skin color was used to restrict the region where motion features are to be tracked. An important research direction is, therefore, the combination of multiple cues. Two such approaches are described in Azoz et al. (1998) and Shimada et al. (1998).

34.2.1.2 Shape

The characteristic shape of hands has been utilized to detect them in images, in multiple ways. Much information can be obtained by just extracting the contours of objects in the image. If correctly detected, the contour represents the shape of the hand and is therefore not directly dependent on viewpoint, skin color, and illumination. On the other hand, the expressive power of 2D contours can be hindered by occlusions or degenerate viewpoints. In the general case, contour extraction based on edge detection results in a large number of edges that belong to the hands but also to irrelevant background objects. Therefore, sophisticated postprocessing approaches are required to increase the reliability of such an approach. In this spirit, edges are often combined with (skin-) color and background subtraction/motion cues.

In the 2D/3D drawing systems of Krueger (1991, 1993) and Utsumi and Ohya (1998a, 1998b), the user's hand is directly extracted as a contour by assuming a uniform background and performing real-time edge detection in this image. Examples of the use of contours as features are found in both model (Kuch and Huang, 1995)- and appearance-based techniques (Gavrila and Davis, 1996; Pavlovic et al., 1996). In Downton and Drouet (1991), finger and arm candidates are selected from images, through the clustering of the sets of parallel edges. In a more global approach (Gavrila and Davis, 1995), hypotheses of hand 3D models are evaluated by first synthesizing the edge image of a 3D model and comparing it against the acquired edge image.

Local topological descriptors have been used to match a model with the edges in the image. In Belongie et al. (2002), the *shape context* descriptor is proposed, which characterizes a particular point location on the shape. This descriptor is the histogram of the relative polar coordinates of all the other points of the shape. Detection is based on the assumption that corresponding points on two different shapes will ideally have a similar shape context. The descriptor has been applied to a variety of object recognition problems (Belongie et al., 2002; Mori and Malik, 2002), with limited background clutter. In Sullivan and Carlsson (2002), all topological combinations of four points are considered in a voting matrix, and one-to-one correspondences are established using a greedy algorithm.

Background clutter is effectively dealt with in Isard and Blake (1996, 1998a), where particle filtering is employed to learn which curves belong to a tracked contour. This technique makes shape models more robust to background noise, but shape-based methods are better suited for tracking an object once it has been acquired. The approach in Segen and Kumar (1999) utilizes as input hand images against a homogeneous and planar background. In this method, the illumination is such that the hand's shadow is cast on the background plane. By corresponding high-curvature features across the silhouettes of the hand and its shadow, depth cues such as vanishing points are extracted and the hand's pose is estimated. In Wang et al. (2008), a *variable shape structure* descriptor is proposed to detect hands (and other shapes in which some shape parts can be repeated) from image edges in the presence of heavy clutter. The approach is based on hidden Markov models (HMMs) (see Section 34.2.3.6) and finds the globally optimal registration of a model with the image features, as quantified by the chamfer distance metric.

Certain methods focus on the specific morphology of hands and attempt to detect them based on characteristic hand shape features such as fingertips. The approaches in Argyros and Lourakis (2006b), Maggioni (1995), and Vaillant and Darmon (1995) utilize curvature as a cue to fingertip detection. Another technique that has been employed in fingertip detection is template matching. Templates can be images of fingertips (Crowley et al., 1995) or fingers (Rehg and Kanade, 1995), or generic 3D cylindrical models (Davis and Shah, 1994b). Such pattern-matching techniques can be enhanced with additional image features, such as contours (Rehg and Kanade, 1994). The template-matching technique was utilized also in Crowley et al. (1995) and O'Hagan and Zelinsky (1997), with images of the top view of fingertips as prototypes. The pixel resulting in the highest correlation is selected as the position of the target object. Apart from being very computationally expensive, template matching cannot cope either with scaling or rotation of the target object. This problem was addressed in Crowley et al. (1995) by continuously updating the template.

In Song and Takatsuka (2005), the fingertip of the user was detected in both images of a calibrated stereo pair. In these images, the two points at which this tip appears establish a stereo correspondence, which is utilized to estimate the fingertip's position in 3D space. In turn, this position is utilized by the system to estimate the distance of the finger from the desk and, therefore, determine if the user is touching it. In Jennings (1999), a system is described for tracking the 3D position and orientation of a finger using several cameras. Tracking is based on the combination of multiple sources of information, including stereo (range) images, color segmentation, and shape information. The hand detectors in Azarbayejani and Pentland (1996) and Brand et al. (1997) utilize nonlinear modeling and a combination of iterative and recursive estimation methods to recover 3D geometry from blob correspondences across multiple images. These correspondences were thereafter utilized to estimate the translation and orientation of blobs in world coordinates. In Argyros and Lourakis (2006a), stereoscopic information is used to provide 3D positions of hand centroids and fingertips, but also to reconstruct the 3D contour of detected and tracked hands in real time. In Yin and Xie (2003), stereo correspondences of multiple fingertips have been utilized to calibrate a stereo pair. In the context of fingertip detection, several heuristics have also been employed. For example, for deictic gestures it can be assumed that the finger represents the foremost point of the hand (Maggioni, 1995; Quek et al., 1995). Many other indirect approaches for the detection of fingertips have been employed, such as image analysis using specially tuned Gabor kernels (Meyering and Ritter, 1992). The main disadvantage of fingertips as features is that they can be occluded by the rest of the hand. A solution to this occlusion problem involves the use of multiple cameras (Rehg and Kanade, 1994; Lee and Kunii, 1995). Other solutions are based on the estimation of the occluded fingertip positions, based on the knowledge of the 3D model of the gesture in question (Rehg and Kanade, 1995; Shimada et al., 1998; Wu and Huang, 1999; Wu et al., 2001).

34.2.1.3 Learning Detectors from Pixel Values

Significant work has been carried out on finding hands in gray level images based on their appearance and texture. In Wu and Huang (2000), the suitability of a number of classification methods for the purpose of view-independent hand posture recognition was investigated. Several methods (Cui et al., 1995; Cui and Weng, 1996b; Quek and Zhao, 1996; Triesch and von der Malsburg, 1996, 1998) attempt to detect hands based on hand appearances, by training classifiers on a set of image samples. Appearance-based methods assume that the intraclass variability of hand gestures performed by different people is lower compared to the variability of a specific gesture performed by different people. To reduce misclassifications, members of the gesture vocabulary are required to exhibit a characteristic and distinct visual appearance. Still, automatic feature selection constitutes a major difficulty. Several papers consider the problem of feature extraction (Quek and Zhao, 1996; Triesch and von der Malsburg, 1996, 1998; Nolker and Ritter, 1998) and selection (Cui et al., 1995; Cui and Weng, 1996b), with limited results regarding hand detection. The work in Cui and Weng (1996b) investigates the difference between the most discriminating features (MDFs) and the most expressive features (MEFs) in the classification of motion clips that contain gestures. It is argued that MEFs may not be the best for classification, because the features that describe some major variations in the class are, typically, irrelevant to how the subclasses are divided. MDFs are selected by multiclass, multivariate discriminate analysis and have a significantly higher capability to catch major differences between classes. The pertinent experiments indicate that MDFs are superior to the MEFs in automatic feature selection for classification.

More recently, methods based on a machine learning approach called boosting have demonstrated very robust results in face and hand detection. Due to these results, they are reviewed in more detail in the following. Boosting is a general method that can be used for improving the accuracy of a given learning algorithm (Schapire, 2003). It is based on the principle that a highly accurate or "strong" classifier can be derived through the linear combination of many relatively inaccurate or "weak" classifiers. In general, an individual weak classifier is required to perform only slightly better than random. As proposed in Viola and Jones (2001) for the problem of hand detection, a weak classifier might be a simple detector based on basic image block differences efficiently calculated using an integral image.

The AdaBoost algorithm (Freund and Schapire, 1997) provides a learning method for finding suitable collections of weak classifiers. For training, it employs an exponential loss function that models the upper bound of the training error. The method utilizes a training set of images that consists of positive and negative examples (hands and nonhands, in this case), which are associated with corresponding labels. Weak classifiers are added sequentially into an existing set of already selected weak classifiers to decrease the upper bound of the training error. It is shown that this is possible if weak classifiers are of a particular form (Schapire and Singer, 1999; Friedman et al., 2000). AdaBoost was applied to the area of face and pedestrian detection (Viola and Jones, 2001; Viola, Jones, and Snow, 2003) with impressive results. However, this method may result in an excessive number of weak classifiers. The problem is that AdaBoost does not consider the removal of selected weak classifiers that no longer contribute to the detection process. The FloatBoost algorithm proposed in Li and Zhang (2004) extends the original AdaBoost algorithm in that it removes an existing weak classifier from a strong classifier if it no longer contributes to the decrease of the training error. This results in a more general and, therefore, more efficient set of weak classifiers.

In the same context, the final detector can be divided into a cascade of strong classifier layers (Viola and Jones, 2001). This hierarchical structure is composed of a general detector at the root, with branch node being increasingly more appearance-specific as the depth of the tree increases. In this approach, the larger the depth of a node, the more specific the training set becomes. To create a labeled database of training images for the tree structure, an automatic method (Ong and Bowden, 2004) for performing grouping of images of hands at the same posture is proposed, based on an unsupervised clustering technique.

34.2.1.4 3D Model-Based Detection

A category of approaches utilize 3D hand models for the detection of hands in images. One of the advantages of these methods is that they can achieve view-independent detection. The employed 3D models should have enough degrees of freedom to adapt to the dimensions of the hand(s) present in an image.

Different models require different image features to construct feature-model correspondences. Point and line features are employed in kinematic hand models to recover angles formed at the joints of the hand (Rehg and Kanade, 1995; Shimada et al., 1998; Wu and Huang, 1999; Wu, Liu, and Huang, 2001). Hand postures are then estimated, provided that the correspondences between the 3D model and the observed image features are well established. Various 3D hand models have been proposed in the literature. In Rehg and Kanade (1994) and Stenger et al. (2002), a full hand model is proposed that has 27 degrees of freedom (DOF) (6 DOF for 3D location/orientation and 21 DOF for articulation). In Lin et al. (2002) and Wu et al. (2005), a *cardboard model* is utilized, where each finger is represented by a set of three connected planar patches. In Goncalves et al. (1995), a 3D model of the arm with 7 parameters is utilized. In Gavrila and Davis (1996), a 3D model with 22 DOF for the whole body with 4 DOF for each arm is proposed. In McCormick and Isard (2000), the user's hand is modeled much more simply, as an articulated rigid object with three joints comprised by the first index finger and thumb.

In Rehg and Kanade (1994), edge features in the two images of a stereoscopic pair are matched to extract the orientation of in-between joints of fingers. These are subsequently utilized

for model-based tracking of the hands. In Nolker and Ritter (1998), artificial neural networks that are trained with body landmarks are utilized for the detection of hands in images. Some approaches (Lee and Kunii, 1995; Heap and Hogg, 1996a, 1996b) utilize a deformable model framework to fit a 3D model of the hand to image data. The fitting is guided by forces that attract the model to the image edges, balanced by other forces that tend to preserve continuity and evenness among surface points (Heap and Hogg, 1996a, 1996b). In Lee and Kunii (1995), the process is enhanced with anatomical data of the human hand that are incorporated into the model. Also, to fit the hand model to an image of a real hand, characteristic points on the hand are identified in the images, and virtual springs are implied that pull these characteristic points to goal positions on the hand model.

34.2.1.5 Motion

Motion is a cue utilized by a few approaches to hand detection. The reason is that motion-based hand detection demands a very controlled setup, since it assumes that the only motion in the image is due to hand movement. Indeed, early works (e.g., Freeman and Weissman, 1995; Quek, 1995; Cui and Weng, 1996b) assumed that hand motion is the only motion occurring in the imaged environment. In more recent approaches, motion information is combined with additional visual cues. In the case of static cameras, the problem of motion estimation reduces to that of background maintenance and subsequent subtraction. For example, in Cutler and Turk (1998) and Martin et al. (1998) such information is utilized to distinguish hands from other skin-colored objects and cope with lighting conditions imposed by colored lights. The difference in luminance of pixels from two successive images is close to zero for pixels of the background. By choosing and maintaining an appropriate threshold, moving objects are detected within a static scene.

In Yuan et al. (1995), a novel feature, based on motion residue, is proposed. Hands typically undergo nonrigid motion, because they are articulated objects. Consequently, hand detection capitalizes on the observation that for hands interframe appearance changes are more frequent than for other objects such as clothes, face, and background.

34.2.2 Tracking

Tracking, or the frame-to-frame correspondence of the segmented hand regions or features, is the second step in the process toward understanding the observed hand movements. The importance of robust tracking is twofold. First, it provides the interframe linking of hand/finger appearances, giving rise to trajectories of features in time. These trajectories convey essential information regarding the gesture and might be used either in a raw form (e.g., in certain control applications such as virtual drawing, the tracked hand trajectory directly guides the drawing operation) or after further analysis (e.g., recognition of a certain type of hand gesture). Second, in model-based methods, tracking also provides a way to maintain estimates of model parameter variables and features that are not directly observable at a certain moment in time.

34.2.2.1 Template-Based Tracking

This class of methods exhibits great similarity to methods for hand detection. Members of this class invoke the hand detector at the spatial vicinity that the hand was detected in the previous frame, so as to drastically restrict the image search space. The implicit assumption for this method to succeed is that images are acquired frequently enough.

Correlation-based feature tracking is directly derived from the previous approach. In Crowley et al. (1995) and O'Hagan and Zelinsky (1997), correlation-based template matching is utilized to track hand features across frames. Once a hand has been detected in a frame, the image region in which it appears is utilized as the prototype to detect the hand in the next frame. This technique is employed for a static camera in Darrell et al. (1996) to obtain characteristic patterns (or "signatures") of gestures, as seen from a particular view. The work in Hager and Belhumeur (1996) deals also with variable illumination. A target is viewed under various lighting conditions. Then, a set of basis images that can be used to approximate the appearance of the object viewed under various illumination conditions is constructed. Tracking simultaneously solves for the affine motion of the object and the illumination. Real-time performance is achieved by precomputing *motion templates*, which are the product of the spatial derivatives of the reference image to be tracked and a set of motion fields.

Some approaches detect hands as image blobs in each frame and temporally match blobs that occur in proximate locations across frames. Approaches that utilize this type of blob tracking are mainly the ones that detect hands based on skin color, the blob being the correspondingly segmented image region (e.g., Argyros and Lourakis, 2004b; Birk et al., 1997). Blob-based approaches are able to retain tracking of hands even when there are great variations from frame to frame.

Extending this approach, deformable contours, or *snakes*, have been utilized to track hand regions in successive image frames (Cootes and Taylor, 1992). Typically, the boundary of this region is determined by intensity or color gradient. Nevertheless, other types of image features (e.g., texture) can be considered. The technique is initialized by placing a contour near the region of interest. The contour is then iteratively deformed toward nearby edges, to better fit the actual hand region. This deformation is performed through the optimization of an energy functional that sums up the gradient at the locations of the snake, while, at the same time, favors the smoothness of the contour. When snakes are used for tracking, an active shape model is applied to each frame. The convergence of the snake in that frame is used as a starting point for the next frame. Snakes allow for real-time tracking and can handle multiple targets and complex hand postures. They exhibit better performance when there is sufficient contrast between the background and the object (Cootes et al., 1995). On the contrary, their performance is compromised in cluttered backgrounds. The reason

is that the snake algorithm is sensitive to local optima of the energy function, often due to ill foreground/background separation or large object displacements and/or shape deformations between successive images.

Tracking local hand features has been employed in specific contexts only. The methods in Baumberg and Hogg (1994) and Martin et al. (1998) track hands in image sequences by combining two motion estimation processes, both based on image differencing. The first process computes differences between successive images. The second computes differences from a background image that was previously acquired. The purpose of this combination is to achieve increased robustness near shadows.

34.2.2.2 Optimal Estimation Techniques

Feature tracking has been extensively studied in computer vision. In this context, the optimal estimation framework provided by the Kalman filter (Kalman, 1960) has been widely employed in turning observations (feature detection) into estimations (extracted trajectory). The reasons for its popularity are real-time performance, treatment of uncertainty, and the provision of predictions for the successive frames.

In Argyros and Lourakis (2004b), the target is retained against cases where hands occlude each other, or appear as a single blob in the image, based on a hypothesis formulation and validation/rejection scheme. The problem of multiple blob tracking was investigated in Argyros and Lourakis (2004a), where blob tracking is performed in both images of a stereo pair and blobs are corresponded, not only across frames, but also across cameras. The obtained stereo information not only provides the 3D locations of the hands, but also facilitates the potential motion of the observing stereo pair, which could be thus mounted on a robot that follows the user. In Breig and Kohler (1998) and Kohler (1997), the orientation of the user's hand was continuously estimated with the Kalman filter, to localize the point in space that the user indicates by extending the arm and pointing with the index finger. In Utsumi and Ohya (1999), hands are tracked from multiple cameras, with a Kalman filter in each image, to estimate the 3D hand postures. Snakes integrated with the Kalman filtering framework (see the following) have been used for tracking hands (Terzopoulos and Szeliski, 1992). Robustness against background clutter is achieved in Peterfreund (1999), where the conventional image gradient is combined with optical flow to separate the foreground from the background. Optical flow is also used in Kim and Lee (2001), to estimate the direction and magnitude of the target's motion along the length of the snake. The resulting motion estimates are used to initialize the snake just before each iteration, thus improving accuracy and convergence. The success of combining optical flow is based on the accuracy of its computation and, thus, the approach is best suited for the case of static cameras.

Treating the tracking of image features within a Bayesian framework has been long known to provide improved estimation results. The works in Fablet and Black (2002), Hue et al.

(2002), Isard and Blake (1998b), Isard and McCormick (2001), Koller-Meier and Ade (2001), and Vermaak et al. (2002) investigate the topic within the context of hand and body motion. In Wren et al. (1997), a system tracks a single person by color segmentation of the image into blobs; it then uses prior information about skin color and topology of a person's body to interpret the set of blobs as a human figure. In Bregler (1997), a method is proposed for tracking human motion by grouping pixels into blobs based on coherent motion, color, and temporal support using an expectation-maximization (EM) algorithm. Each blob is subsequently tracked using a Kalman filter. Finally, in McCormick and Blake (1999) and McCormick and Isard (2000), the contours of blobs are tracked across frames by a combination of the iterative closed point (ICP) algorithm and a factorization method to determine global hand pose.

The approach in Black and Jepson (1998c) reformulates the eigenspace reconstruction problem (reviewed in Section 34.2.3.2) as a problem of robust estimation. The goal is to utilize the previous framework to track the gestures of a moving hand. To account for large affine transformations between the eigenspace and the image, a multiscale eigenspace representation is defined and a coarse-to-fine matching strategy is adopted. In Leonardis and Bischof (1996), a similar approach was proposed that uses a hypothesize-and-test approach, instead of a continuous formulation. Although this approach does not address parameterized transformations and tracking, it exhibits robustness against occlusions. In Gupta et al. (2002), a real-time extension of the work in Black and Jepson (1998c), based on eigen-tracking (Isard and Blake, 1998a), is proposed. Eigenspace representations have been utilized in a different way in Baumberg and Hogg (1994) to track articulated objects by tracking a silhouette of the object, which was obtained via image differencing. A spline was fit to the object's outline and the knot points of the spline form the representation of the current view. Tracking an object amounts to projecting the knot points of a particular view onto the eigenspace. Thus, this eigenspace-based method uses the shape (silhouette) information instead of the photometric one (image intensity values).

In Utsumi and Ohya (1999), the 3D positions and postures of both hands are tracked using multiple cameras. Each hand position is tracked with a Kalman filter and 3D hand postures are estimated using image features. This work deals with the mutual hand-to-hand occlusion inherent in tracking both hands, by selecting the views in which there are no such occlusions.

34.2.2.3 Tracking Based on the Mean Shift Algorithm

The mean shift algorithm (Cheng, 1995) is an iterative procedure that detects local maxima of a density function by shifting a kernel toward the average of data points in its neighborhood. The algorithm is significantly faster than exhaustive search, but requires appropriate initialization.

The mean shift algorithm has been utilized in the tracking of moving objects in image sequences. The work in Comaniciu et al. (2000, 2003) is not restricted to hand-tracking, but can be used to track any moving object. It characterizes the object

of interest through its color distribution as this appears in the acquired image sequence, and utilizes the spatial gradient of the statistical measurement toward the most similar (in terms of color distribution similarity) image region. An improvement of the previous approach is described in Chen and Liu (2001), where the mean shift kernel is generalized with the notion of the trust region. Contrary to mean shift, which directly adopts the direction toward the mean, trust regions attempt to approximate the objective function and, thus, exhibit increased robustness toward being trapped in spurious local optima. In Bradski (1998), a version of the mean shift algorithm is utilized to track the skin-colored blob of a human hand. For increased robustness, the method tracks the centroid of the blob and also continuously adapts the representation of the tracked color distribution. Similar is also the method proposed in Kurata et al. (2001), except the fact that it utilizes a Gaussian mixture model to approximate the color histogram and the EM algorithm to classify skin pixels based on the Bayesian decision theory.

Mean shift tracking is robust and versatile for a modest computational cost. It is well suited for tracking tasks where the spatial structure of the tracked objects exhibits such a great variability that trackers based on a space-dependent appearance reference would break down very fast. On the other hand, highly cluttered background and occlusions may distract the mean shift trackers from the object of interest. The reason appears to be its local scope in combination with the single-state appearance description of the target.

34.2.2.4 Particle Filtering

Particle filters have been utilized to track the position of hands and the configuration of fingers in dense visual clutter. In this approach, the belief of the system regarding the location of a hand is modeled with a set of particles. The approach exhibits advantages over Kalman filtering, because it is not limited by the unimodal nature of Gaussian densities that cannot represent simultaneous alternative hypotheses. A disadvantage of particle filters is that for complex models (such as the human hand) many particles are required, a fact that makes the problem intractable especially for high-dimensional models. Therefore, other assumptions are often utilized to reduce the number of particles. For example, in Isard and Blake (1998a), dimensionality is reduced by modeling commonly known constraints due to the anatomy of the hand. Additionally, motion capture data are integrated in the model. In McCormick and Blake (1999) a simplified and application-specific model of the human hand is utilized.

The CONDENSATION algorithm (Isard and Blake, 1998a), which has been used to learn to track curves against cluttered backgrounds, exhibits better performance than Kalman filters, and operates in real-time. It uses factored sampling, previously applied to the interpretation of static images, in which the probability distribution of possible interpretations is represented by a randomly generated set. Condensation uses learned dynamic models, together with visual observations, to propagate this random set over time. The result is highly robust tracking of agile motion. In McCormick and Isard (2000), the "partitioned sampling" technique is employed to avoid the high computational cost that particle filters exhibit when tracking more than one object. In Laptev and Lindeberg (2001), the state space is limited to 2D translation, planar rotation, scaling, and the number of outstretched fingers.

Extending the CONDENSATION algorithm, the work in Mammen et al. (2001) detects occlusions with some uncertainty. In Perez et al. (2002), the same algorithm is integrated with color information; the approach is based on the principle of color histogram distance but, within a probabilistic framework, the work introduces a new Monte Carlo tracking technique. In general, contour tracking techniques, typically, allow only a small subset of possible movements to maintain continuous deformation of contours. This limitation was overcome to some extent in Heap and Hogg (1996b), who describe an adaptation of the CONDENSATION algorithm for tracking across discontinuities in contour shapes.

34.2.3 Recognition

The overall goal of hand gesture recognition is the interpretation of the semantics that the hand(s) location, posture, or gesture convey. Basically, there have been two types of interaction in which hands are employed in the user's communication with a computer. The first is the control applications such as drawing, where the user sketches a curve while the computer renders this curve on a 2D canvas (Wu et al., 2001; Lin et al., 2002; Wu et al., 2005). Methods that relate to hand-driven control focus on the detection and tracking of some feature (e.g., the fingertip, the centroid of the hand in the image, etc.) and can be handled with the information extracted through the tracking of these features. The second type of interaction involves the recognition of hand postures, or signs, and gestures. Naturally, the vocabulary of signs or gestures is largely application dependent. Typically, the larger the vocabulary is, the hardest the recognition task becomes. Two early systems indicate the difference between recognition (Birk et al., 1997) and control (Mapes and Moshell, 1995). The first recognizes 25 postures from the International Hand Alphabet, while the second was used to support interaction in a virtual work space.

The recognition of postures is a topic of great interest on its own, because of sign language communication (see Chapter 38, "Sign Language in the Interface: Access for Deaf Signers," of this handbook). Moreover, it forms the basis of numerous gesture-recognition methods that treat gestures as a series of hand postures. Besides the recognition of hand postures from images, recognition of gestures includes an additional level of complexity, which involves the parsing, or segmentation, of the continuous signal into constituent elements (see Ong and Ranganath, 2005, for a survey on automatic sign language analysis). In a wide variety of methods (e.g., Triesch and von der Malsburg, 1998), the temporal instances at which hand velocity (or optical flow) is minimized are considered observed

postures, while video frames that portray a hand in motion are sometimes disregarded (e.g., Birk et al., 1997). However, the problem of simultaneous segmentation and recognition of gestures without being confused with intergesture hand motions remains a rather challenging one. Another requirement for this segmentation process is to cope with the shape and time variability that the same gesture may exhibit (e.g., when performed by different persons or by the same person at different speeds).

The fact that even hand posture recognition exhibits considerable levels of uncertainty casts the previous as processing computationally complex and error-prone. Several of the reviewed works indicate that lack of robustness in gesture recognition can be compensated by addressing the temporal context of detected gestures. This can be established by letting the gesture detector know of the grammatical or physical rules that the observed gestures are supposed to express. Based on these rules, certain candidate gestures may be improbable. In turn, this information may disambiguate candidate gestures, by selecting to recognize the most likely candidate. The framework of HMMs, which is discussed later in this section provides a suitable framework for modeling the context-dependent reasoning of the observed gestures.

34.2.3.1 Template Matching

Template matching, a fundamental pattern-recognition technique, has been utilized in the context of both posture and gesture recognition. In the context of images, template matching is performed by the pixel-by-pixel comparison of a prototype and a candidate image. The similarity of the candidate to the prototype is proportional to the total score on a preselected similarity measure. For the recognition of hand postures, the image of a detected hand forms the candidate image, which is directly compared with prototype images of hand postures. The best matching prototype (if any) is considered the matching posture. Clearly, because of the pixel-by-pixel image comparison, template matching is not invariant to scaling and rotation.

Template matching was one of the first methods employed to detect hands in images (Freeman and Weissman, 1995). To cope with the variability due to scale and rotation, some authors have proposed scale and rotational normalization methods (e.g., Birk, Moeslund, and Madsen, 1997), while others equip the set of prototypes with images from multiple views (e.g., Darrell and Pentland, 1993). In Birk et al. (1997), the image of the hand is normalized for rotation based on the detection of the hand's main axis, and then, scaled with respect to hand dimensions in the image. Therefore, in this method, the hand is constrained to move on a planar surface that is frontoparallel to the camera. To cope with the increased computational cost when comparing with multiple views of the same prototype, these views were annotated with the orientation parameters (Fillbrandt et al., 2003). Searching for the matching prototype was accelerated by searching only in relevant postures with respect to the one detected in the previous frame. A template composed of edge directions was utilized in Freeman and Roth (1995). Edge detection is performed on the image of the isolated hand and edge orientations are computed. The histogram of these orientations is used as the feature vector. The evaluation of this approach showed that edge orientation histograms are not very discriminative, because several semantically different gestures exhibit similar histograms.

A direct approach of including the temporal component into the template-matching techniques has been proposed in Darrell et al. (1996), and Darrell and Pentland (1993, 1995). For each input frame, the (normalized) hand image region is compared to different views of the same posture and a 1D function of responses for each posture is obtained; due to the dense posture parameterization this function exhibits some continuity. By stacking the 1D functions resulting from a series of input frames, a 2D pattern is obtained and utilized as a template.

Another approach to creating gesture patterns that can be matched by templates is to accumulate the motion over time within a *motion* or *history* image. The input images are processed frame-by-frame and some motion-related feature is detected at each frame. The detected features, from all frames, are accumulated in a 2D buffer at the location of their detection. The obtained image is utilized as a representation of the gesture and serves as a recognition pattern. By doing so, the motion (or trail) of characteristic image points over the sequence is captured. The approach is suited for a static camera observing a single user in front of a static background. Several variations of this basic idea have been proposed. In Bobick and Davis (1996, 2001) the results of a background subtraction process (human silhouettes) are accumulated in a single frame and the result is utilized as the feature vector. In Bradski and Davis (2000, 2002) and Davis (2001), an extension of the previous idea encodes temporal order in the feature vector, by creating a *history gradient*. In the accumulator image, older images are associated with a smaller accumulation value and form a fading-out pattern. Similar is the approach in Cutler and Turk (1998), where the accumulation pattern is composed of optical flow vectors. The obtained pattern is rather coarse and even with the use of a user-defined rule-based technique the system can distinguish only among a very small vocabulary of coarse body gestures. In Yang et al. (2002), an artificial neural network is trained to learn motion patterns similar to the previous method. In Ionescu et al. (2005), a single hand was imaged in a control environment that featured no depth and its (image) skeleton was computed; the accumulation of such skeletons along time, in a single image, was used as the feature vector.

34.2.3.2 Methods Based on Principal Component Analysis

Principal component analysis (PCA) methods have been directly utilized mainly in posture recognition. However, this analysis facilitates several gesture recognition systems by providing the tokens to be used as input to recognition.

PCA methods require an initial training stage, in which a set of images of similar content is processed. Typically, the intensity values of each image are considered as values of a 1D vector, whose dimensionality is equal to the number of pixels in the image; it is assumed or enforced that all images are of equal size. For each such set, some basis vectors are constructed that can be used to approximate any of the (training) images in the set. In the case of gesture recognition, the training set contains images of hands in certain postures. This process is performed for each posture in the vocabulary, which the system should later be able to recognize. In PCA-based gesture recognition, the matching combination of principal components indicates the matching gesture as well. This is because the matching combination is one of the representatives of the set of gestures that were clustered together in training, as expressions of the same gesture. A problem of eigenspace reconstruction methods is that they are not invariant to image transformations such as translation, scaling, and rotation.

PCA was first applied to recognition in Sirovich and Kirby (1987) and later extended in Turk and Pentland (1991) and Murase and Nayar (1995). A simple system is presented in Ranganath and Arun (1997), where the whole image of a person gesturing is processed, assuming that the main component of motion is the gesture. View dependence is compensated by creating multiple prototypes, one for each view. As in detection, the matching view indicates also the relative pose to the camera. To reduce this complexity in recognition, the system in Birk et al. (1997) rotationally aligns the acquired image with the template, based on the arm's orientation. It, therefore, stores each gesture prototype in the same orientation.

PCA systems exhibit the potential capability of compressing the knowledge of the system by keeping only the principal components with the *n* highest eigenvalues. However, in Birk et al. (1997), it was shown that this is not effective if only a small number of principal components are to be kept. The works in Colmenarez and Huang (1996) and Moghaddam and Pentland (1995) attempt to select the features that best represent the pattern class, using an entropy-based analysis. In a similar spirit, in Cui et al. (1995) and Cui and Weng (1996b) features that better represent a class (expressive) are compared to features that maximize the dissimilarity across classes (discriminative), to suggest that the latter give rise to more accurate recognition results. A notable extension of this work is the utilization of the recognition procedure as feedback to the hand segmentation process (Cui and Weng, 1996a). In that respect, authors utilize the classification procedure in combination with hand detection to eliminate unlikely segmentations.

34.2.3.3 Boosting

The learning methods reviewed in Section 34.2.1.3 exhibit remarkable performance in hand detection and hand posture recognition, but limited application in hand gesture recognition. Here, characteristic examples of the use of these methods for posture recognition are reviewed.

In Lockton and Fitzgibbon (2002), a real-time gesture recognition system is presented. Their method is based on skin-color segmentation and is facilitated by a boosting algorithm (Freund and Schapire, 1997) for fast classification. To normalize for orientation, the user is required to wear a wristband so that the hand shape can be easily mapped to a canonical frame. In Tomasi et al. (2003), a classification approach was proposed, together with parameter interpolation to track hand motion. Image intensity data were used to train a hierarchical nearest neighbor classifier, classifying each frame as one of 360 views, to cope with viewpoint variability. This method can handle fast hand motion, but it relies on clear skin color segmentation and controlled lighting conditions. In Wachs et al. (2002), the hand is detected and the corresponding image segment is subsampled to a very low resolution. The pixels of the resulting patterns are then treated as n-dimensional vectors. Learning in this case is based on a C-means clustering of the training parameter space.

34.2.3.4 Contour and Silhouette Matching

This class of methods mainly refers to posture recognition and is conceptually related to template matching in that it compares prototype images with the hand image that was acquired, to obtain a match. The defined feature space is the edges of the previous images. The fact that a spatially sparser feature is utilized (edges instead of intensities) gives rise to the employment of slightly different similarity metrics in the comparison of acquired and prototype images. In addition, continuity is favored to avoid the consideration of spurious edges that may belong, for example, to background clutter.

In Borgefors (1988) and Gavrila and Davis (1996), chamfer matching (Barrow et al., 1977) is utilized as the similarity metric. In Stenger et al. (2002), matching is based on an unscented Kalman filter, which minimizes the geometric error between the profiles and edges extracted from the images. The same edge image features are utilized in Deutscher et al. (2000), and recognition is completed after a likelihood analysis. The work in Borgefors (1988) applies a coarse-to-fine search, based on a resolution pyramid of the image, to accelerate the process. In an image-generative approach (Gavrila and Davis, 1996), the edges of idealized models of body postures are projected onto images acquired from multiple views and compared with the true edges using chamfer matching while a template hierarchy is also used to handle shape variation. In Olson and Huttenlocher (1997), a template hierarchy is also utilized to recognize 3D objects from different views and the Hausdorff distance (Huttenlocher et al., 1993) is utilized as the similarity metric. In a more recent approach, the work in Cheung et al. (2000) utilizes the robust *shape context* (Belongie et al., 2002) matching operator.

Research by Athitsos and Sclaroff (2001, 2002, 2003) and Rosales et al. (2001) utilizes chamfer matching between input and model edge images. The model images are *a priori* synthesized with the use of a data glove. The number of model images is very high ($\approx 10^5$) to capture even minute differences in postures. To cope with this amount of data, the retrieval is performed hierarchically, by first rejecting the greatest proportion of all database views, and then ranking the remaining candidates in order of similarity to the input. In Athitsos and Sclaroff (2002), the chamfer-matching technique was evaluated against edge

orientation histograms, shape moments, and detected finger positions.

The use of silhouettes in gesture recognition has not been extensive, probably because different hand poses can give rise to the same or similar silhouette. Another reason is that silhouette matching requires alignment (or else, point-to-point correspondence establishment across the contour of the silhouette), which is not always a trivial task. Also, matching of silhouettes using their conventional arc-length descriptions (or *signatures*) is very sensitive to deformations and noise. Due to the local nature of edges, perceptually small dissimilarities of the acquired silhouette with the prototype may cause large metric dissimilarity. Thus, depending on the metric, the overall process can be sensitive even to small shape variations, which are due to hand articulation in-between stored poses of the hand. To provide some flexibility against such variations, the work in Stenger et al. (2006) aligns the contours to be matched using the ICP algorithm (Besl and McKay, 1992). A more effective approach for dealing with this variability is presented in Segen and Kumar (1999), where the intrusions and protrusions of the hand's silhouette are utilized as classification features.

In Lanitis et al. (1995), a simple contour-matching technique was proposed that targeted posture recognition. In Kuch and Huang (1995), contour matching is enabled mainly for control and coarse hand modeling. The approach in Heap and Samaria (1995) employs a silhouette alignment and matching technique to recognize a prototype hand-silhouette in an image and subsequently track it. Finally, polar-coordinate descriptions of the contours points, or *signatures* (Brockl-Fox, 1995) and *size functions* (Uras and Verri, 1995) have been used. Also similar is the approach in Shimada et al. (2001), which, after extracting the silhouette of a hand, computes a silhouette-based descriptor that the recognition will be based upon. Because this descriptor is a function of the contour's arc length, it is very sensitive to deformations that alter the circumference of the contour and, thus, the authors propose a compensation technique. In addition, to reduce the search space of each recognition query, an adjacency map indexes the database of models. In each frame, the search space is limited to the adjacent views of the one estimated in the previous frame.

34.2.3.5 Model-Based Recognition Methods

Most of the model-based gesture recognition approaches employ successive approximation methods for the estimation of their parameters. Since gesture recognition is required to be invariant of relative rotation, intrinsic parameters such as joint angles are widely utilized. The strategy of most methods in this category is to estimate the model parameters, e.g. by inference or optimization, so that the extracted features match a model.

In an early approach (Quek, 1995), the 3D trajectory of hands was estimated in the image, based on optical flow. The extremal points of the trajectory were detected and used as gesture classification features. In Campbell et al. (1996), the 3D trajectories of hands are acquired by stereo vision and utilized for HMM-based learning and recognition of gestures. Different feature vectors were evaluated as to their efficacy in gesture recognition. The results indicated that choosing the right set of features is crucial to the obtained performance. In particular, it was observed that velocity features are superior to positional features, while partial rotational invariance is also a discriminative feature.

In Davis and Shah (1994a), a small vocabulary of gestures is recognized through the projection of fingertips on the image plane. Although the detection is based on markers, a framework is offered that uses only the fingertips as input data and permits a model that represents each fingertip trajectory through space as a simple vector. The model is simplified in that it assumes that most finger movements are linear and exhibit minute rotational motion. Also in Kang and Ikeuchi (1993), grasps are recognized after estimating finger trajectories from both passive and active vision techniques (Kang and Ikeuchi, 1991). The authors formulate the grasp-gesture detector in the domain of 3D trajectories, offering, at the same time, a detailed modeling of grasp kinematics (see Kang and Ikeuchi, 1991, for a review on this topic).

The approach in Bobick and Wilson (1997) uses a time-collapsing technique for computing a prototype trajectory of an ensemble of trajectories to extract prototypes and recognize gestures from an unsegmented, continuous stream of sensor data. The prototype offers a convenient arc-length parameterization of the data points, which is then used to calculate a sequence of states along the prototype. A gesture is defined as an ordered sequence of states along the prototype, and the feature space is divided into a relatively small number of finite states. A particular gesture is recognized as a sequence of transitions through a series of such states, thus casting a relationship to HMM-based approaches (see Section 34.2.3.6). In Kawashima and Matsuyama (2003), continuous states are utilized for gesture recognition in a multiview context. In Campbell et al. (1996), the 3D trajectories of hands when gesturing are estimated based on stereoscopic information and, in turn, features of these trajectories, such as orientation, velocity, and so on, are estimated. Similarly, for the purpose of studying two-handed movements, the method in Shamaie and Sutherland (2005) estimates features of 3D gesture trajectories.

In Wren and Pentland (1997), properties such as blob trajectories are encoded in 1D functions of time and then matched with gesture patterns using dynamic temporal warping (DTW). In Eisenstein et al. (2003), a framework is presented for the definition of templates encoding the motion and posture of hands using predicates that describe the postures of fingers at a semantic level. Such data structures are considered to be semantic representations of gestures and are recognized, via template matching, as certain gesture prototypes.

The approach presented in Black and Jepson (1998a, 1998b) achieves the recognition of gestures given some estimated representation of the hand motion. For each pair of frames in a video sequence, a set of parameters that describe the motion are computed, such as velocity or optical flow. These parameter vectors form temporal trajectories that characterize the gesture. For a

new image sequence, recognition is performed by incrementally matching the estimated trajectory to the prototype ones. Robust tracking of the parameters is based on the CONDENSATION tracking algorithm (Isard and Blake, 1996). The work in Isard and Blake (1998c) is also similar to the previous method, showing that the CONDENSATION algorithm is compatible with simple dynamical models of gestures to simultaneously perform tracking and recognition. The work in Gong et al. (1999) extends the previous approach by including HMMs to increase recognition accuracy.

In Lin et al. (2002) and Wu et al. (2005), the hand gesture is estimated by matching the 3D model projections with observed image features, so that the problem becomes a search problem in a high-dimensional space. In such approaches, tracking and recognition are tightly coupled, since by detecting or tracking the hand the gesture is already recognized. In Rosales et al. (2001), the low-level visual features of hand joint configuration were mapped with a supervised learning framework for training the mapping function. In Wu and Huang (2000), the supervised and the unsupervised learning framework were combined and, thus, incorporated a large set of unlabeled training data. The major advantage of using appearance-based methods is the simplicity of their parameter computation. However, the mapping may not be one-to-one, and the loss of precise spatial information makes them less suited for hand position reconstruction.

34.2.3.6 HMMs

A hidden Markov model (HMM) is a statistical model in which a set of hidden parameters is determined from a set of related, observable parameters. In an HMM, the state is not directly observable, but instead, variables influenced by the state are. Each state has a probability distribution over the possible output tokens. Therefore, the sequence of tokens generated by an HMM provides information about the sequence of states. In the context of gesture recognition, the observable parameters are estimated by recognizing postures (tokens) in images. For this reason and because gestures can be recognized as a sequence of postures, HMMs have been widely utilized for gesture recognition. In this context, it is typical that each gesture is handled by a different HMM. The recognition problem is transformed to the problem of selecting the HMM that matches best the observed data, given the possibility of a state being observed with respect to context. This context may be spelling or grammar rules, the previous gestures, cross-modal information (e.g., audio), and others. An excellent introduction and further analysis on the approach, for the case of gesture recognition, can be found in Wilson and Bobick (1995).

Early versions of this approach can be found in Yamato et al. (1992), Schlenzig et al. (1994a), and Rigoll et al. (1996). There, the HMMs were performing directly on the intensity values of the images acquired by a static camera. In Morguet and Lang (1997), the edge image combined with intensity information is used to create a static posture representation or a search pattern. The work in Rigoll et al. (1998) includes the temporal component, in an approach similar to that of Bobick and Davis (1996), and HMMs are trained on a 2D motion image. The method operates on coarse body motions and visually distinct gestures, executed on a plane that is frontoparallel to the camera. Images are acquired in a controlled setting, where image differencing is utilized to construct the required motion image. Incremental improvements of this work have been reported in Eickeler et al. (1998).

The work in Vogler and Metaxas (1998) proposes a posture recognition system whose inputs are 3D reconstructions of the hand (and body) articulation. In this work, HMMs are coupled with 3D reconstruction methods to increase robustness. In particular, moving limbs are extracted from images using the segmentation of Kakadiaris et al. (1994) and, subsequently, joint locations are recovered by inferring the articulated motion from the silhouettes of segments. The process is performed simultaneously from multiple views and the stereo combination of these segmentations provides the 3D models of these limbs which are, in turn, utilized for recognition.

In Starner et al. (1998), the utilized features are the moments of skin-color-based blob extraction for two observed hands. Grammar rules are integrated in the HMM to increase robustness in the comprehension of gestures. This way, posture combinations can be characterized as erroneous or improbable depending on previous gestures. In turn, this information can be utilized as feedback to increase the robustness of the posture recognition task and, thus, produce overall more accurate recognition results. The approach in Lee and Kim (1999) introduces the concept of a threshold model that calculates the likelihood threshold of an input (moments of blob detection). The threshold model is a weak model for the superset of all gestures in the vocabulary and its likelihood is smaller than that of the correct gesture model for a given gesture, but larger than for a nongesture motion. This can be utilized to detect if some motion is part of a gesture or not. To reduce the states model, states with similar probability distributions are merged, based on a relative entropy measure. In Wren and Pentland (1997), the 3D locations that result from stereo multiple-blob tracking are input to an HMM that integrates a skeletal model of the human body. Based on the 3D observations, the approach attempts to infer the posture of the body.

Conceptually similar to conditional-based reasoning is the *causal analysis approach*. This approach stems from work in scene analysis (Brand et al., 1993), which was developed for rigid objects of simple shape (blocks, cubes, etc.). The approach uses knowledge about body kinematics and dynamics to identify gestures through human motor plans, based on measurements of shoulder, elbow, and wrist joint positions in the image plane. From these positions, the system extracts a feature set that includes wrist acceleration and deceleration, effort to lift the hand against gravity, size of gesture, area between arms, angle between forearms, nearness to body, and so on. Gesture filters use this information, along with causal knowledge on human interaction with objects in the physical world, to recognize gestures such as opening, lifting, patting, pushing, stopping, and clutching.

34.2.4 Complete Gesture Recognition Systems

Systems that employ hand-driven human-computer communication, interpret the actions of hands in different modes of interaction depending on the application domain. In some applications the hand or finger motion is tracked to be replicated in some kind of 2D or 3D manipulation activity. For example, in a painting application the finger may sketch a figure in thin air, which, however, is to be replicated as a drawing on the computer's screen. In other cases, the posture, motion, and/or gesture of the user must be interpreted as a specific command to be executed or a message to be communicated. Such a specific application domain is sign language understanding for the hearing impaired. Most of the systems presented in this subsection fall in these categories; however, there are some that combine the previous two modes of interaction. Finally, a few other applications focus on gesture recognition for understanding and annotating human behavior, while others attempt to model hand and body motion for physical training.

The use of a pointing finger instead of the mouse cursor appears to be an intuitive choice in hand-driver interaction, as it has been adopted by a number of systems—possibly due to the cross-culture nature of the gesture as well as its straightforward detection in the image. In Fukumoto et al. (1994), a generic interface that estimates the location and orientation of the pointing finger was introduced. In Crowley et al. (1995), the motion of the user's pointing finger indicates the line of drawing in a FingerPaint application. In Quek (1996), 2D finger movements are interpreted as computer mouse motion in a FingerMouse application. According to Ahmad (1994), the 3D position and planar orientation of the hand are tracked to provide an interface for navigation around virtual worlds. In Wu et al. (2004), tracking of a human finger from a monocular sequence of images is performed to implement a 3D blackboard application; to recover the third dimension from the two-dimensional images the fact that the motion of the human arm is highly constrained is utilized.

The Digital Desk Calculator application (Wellner, 1993) tracked the user's pointing finger to recognize numbers on physical documents on a desk and to perform calculations. The system in Siegl et al. (2007) utilizes the direction of the pointing gesture of the user to infer the object that the user is pointing at on a desk. In Krueger and Froehlich (1994), a responsive workbench allows the user to manipulate objects in a virtual environment for industrial training via tracking of the user's hands. More recently the very interesting system in Bauckhage et al. (2004, 2005) attempts to recognize actions performed by the user's hands in an unconstrained office environment. The system applies attention mechanisms, visual learning, and contextual and probabilistic reasoning to fuse individual results and verify their consistency, and also attempts to learn novel actions performed by the user.

In Argyros and Lourakis (2006b), a vision-based interface for controlling a computer mouse via 2D and 3D hand gestures is presented. Two vocabularies are defined: the first depends only on 2D hand-tracking, while the second makes use of 3D information and requires a second camera. The second condition of operation is of particular importance, because it allows the gesture observer (a robot) to move along with the user. In another robotic application (Kortenkamp et al., 1996), the user points the finger and extends the arm to indicate locations on the floor to instruct a robot to move to the indicated location.

Applications where hand interaction facilitates the communication of a command or message from the user to the system require that the posture and motion of hands are recognized and interpreted. Early gesture recognition applications supported just a few gestures that signified some basic commands or concepts to the computer system. For example, in Darrell and Pentland (1993), a monocular vision system supported the recognition of a wide variety of yes/no hand gestures. In Schlenzig et al. (1994b), a rotation-invariant image representation was utilized to recognize a few hand gestures such as hello and good-bye in a controlled setup. The system in Cohen et al. (1996) recognized simple natural gestures, such as hand trajectories that comprised circles and lines.

Some systems combine the recognition of simple gestures with manipulative hand interaction. For example, in Wilson and Oliver (2003) stereo-vision facilitates hand-tracking and gesture-recognition in a graphical user interface (GUI) that permits the user to perform window-management tasks without the use of the mouse or keyboard. The system in Berry et al. (1998) integrated navigation control gestures into the BattleView virtual environment. The integrated gestures were utilized in navigating as well as moving objects in the virtual environment. Hand-driven 3D manipulation and editing of virtual objects is employed in Pavlovic et al. (1996) and Zeller et al. (1997) in the context of a virtual environment for molecular biologists. In Segen and Kumar (1998), a hand-gesture interface is proposed that allows the manipulation of objects in a virtual 3D environment by recognizing a few simple gestures and tracking hand motion. The system in Hopf et al. (2006) tracks hand motion to rotate the 3D content that is displayed in an autostereoscopic display. In the system of Hoch (1998) the user interacts in front of a projection screen, and where interaction in physical space and pointing gestures are used to direct the scene for filmmaking.

In terms of communicative gestures, the sign language for the hearing impaired has received significant attention (Cui et al., 1995; Starner and Pentland, 1995; Waldron, 1995; Grobel and Assan, 1997; Starner et al., 1998; Vogler and Metaxas, 1999; Bauer and Hienz, 2000; Vogler and Metaxas, 2001; Martinez et al., 2002; Tanibata et al., 2002; Yang et al., 2002). Besides providing a constrained and meaningful data set, it exhibits significant potential impact in society, since it can facilitate the communication of the hearing impaired with machines through a natural modality for the user. In Imagawa et al.

(1998), a bidirectional translation system between Japanese Sign Language and Japanese was implemented to help the hearing impaired communicate with normal speaking people through sign language. Among the earliest systems is the one in Starner and Pentland (1995), which recognized about 40 signs of American Sign Language, and was later extended (Starner et al., 1998) to observe the user's hands from a camera mounted on a cap worn by the user. Besides the recognition of individual hand postures, the system in Martinez et al. (2002) recognized motion primitives and full sentences, accounting for the fact that the same sign may have different meanings depending on context. The main difference of the system in Yang et al. (2002) is that it extracts motion trajectories from an image sequence and uses these trajectories as features in gesture recognition in combination with recognized hand postures. Other examples of sign language recognition systems are discussed in Chapter 38 of this handbook.

Gestures have been utilized in the remote control of a television set via hand gestures in Freeman (1999), where an interface for video games is also considered. In Kohler (1997), a more general system for the control of home appliances was introduced. In Lee and Kim (1999), a gesture recognition method was developed to spot and recognize about 10 hand gestures for a human-computer interface, instantiated to control a slide presentation. The systems in Cairnie et al. (2000), McAllister, et al. (2000) and Zobl et al. (2004) recognize a few hand postures for the control of in-car devices and nonsafety systems, such as radio/CD, AC, telephone, and navigation systems, with hand postures and dynamic hand gestures, in an approach to simplify the interaction with these devices while driving. Relevant to control of electronic devices, in MacLean et al. (2001) a system is presented for controlling a video camera via hand gestures with commands such as zoom, pan, and tilt. In Triesch and von der Malsburg (1996), a person-independent gesture interface was developed on a real robot; the user is able to issue commands such as how to grasp an object and where to put it. The application of gesture recognition in teleoperation systems has been investigated in Sheridan (1993), to pinpoint the challenges that arise when controlling remote mechanisms in such large distances (earth to satellite), where the roundtrip time delay for visual feedback is several tenths of a second.

Tracking and recognizing body and hand motion has also been employed in personal training. The system in Brand et al. (1997) infers the posture of the whole body by observing the trajectories of hands and the head, in constrained setups. In Davis and Bobick (1998), a prototype system for a virtual personal aerobics trainer was implemented that recognizes stretching and aerobic movements and guides the user into a training program. Similarly, in Becker (1997) a virtual t'ai chi trainer is presented. Recently, Sony (Fox, 2005) introduced a system that tracks body motion against a uniform background and features a wide variety of gaming and personal training capabilities. The ALIVE II system (Maes et al., 1995) identifies full body gestures to control artificial life creatures, such as virtual pets and companions that, sometimes, mimic the body gestures of the user. Gestures such as pointing the arm are interpreted by the simulated characters as a command to move to the indicated location. In addition, the user can issue gesture-driven commands to manipulate virtual objects. In Cutler and Turk (1998), the authors present a hand and body gesture-driven interactive virtual environment for children.

The system in Quek (2000) attempts to recognize free-form hand gestures that accompany speech in natural conversations and that provide a complementary modality to speech for communication. A gesture-recognition application is presented in Ju et al. (1997), where an automatic system for analyzing and annotating video sequences of technical presentations was developed. In this case, the system passively extracts information about the presenter of the talk. Gestures such as pointing or writing are recognized and utilized in the annotation of the video sequence. Similarly, in Black and Jepson (1998a), a system that tracks the actions of the user on a blackboard was implemented. The system can recognize gestures that command the system to "print," "save," "cut," and so on, the contents of the blackboard.

34.3 An Approach to Human-Robot Interaction Based on Hand Gestures

This section presents the development of a prototype gesture recognition system intended for human-robot interaction. The application at hand involves natural interaction with autonomous robots installed in public places such as museums and exhibition centers. The operational requirements of such an application challenge existing approaches, in that the visual perception system should operate efficiently under totally unconstrained conditions regarding occlusions, variable illumination, moving cameras, and varying background. Moreover, since no training of users can take place (users are assumed to be visitors of museums/exhibitions), the gesture vocabulary needs to be limited to a small number of natural, generic, and intuitive

gestures that humans use in their everyday human-to-human interactions.

The proposed gesture recognition system builds upon a probabilistic framework that allows the utilization of multiple information cues to efficiently detect regions belonging to human hands (Baltzakis et al., 2008). The utilized information cues include color information, motion information through a background subtraction technique (Grimson and Stauffer, 1999; Stauffer et al., 1999), expected spatial location of hands within the image, as well as velocity and shape of the detected hand segments. Tracking over time is achieved by a technique that can handle hands that may move in complex trajectories, occlude each other in the field of view of the robot's camera, and vary in number over time (Argyros and Lourakis, 2004b). Finally, a simple set of hand gestures is defined based on the number of extended fingers and their spatial configuration.

34.3.1 The Proposed Approach in Detail

A block diagram of the proposed gesture recognition system is illustrated in Figure 34.1. The first two processing layers of the diagram (i.e., processing layers 1 and 2) perform the detection task (in the sense described in Section 34.2), while processing layers 3 and 4 correspond to the tracking and recognition tasks, respectively. In the following sections, details on the implementation of the individual system components are provided.

34.3.1.1 Processing Layer 1: Estimating the Probability of Observing a Hand at the Pixel Level

Within the first layer, the input image is processed to identify pixels that depict human hands. Let U be the set of all pixels of an image. Let M be the subset of U corresponding to foreground pixels (i.e., a human body) and S be the subset of U-containing pixels that are skin colored. Accordingly, let H stand for the sets of pixels that depict human hands. The relationships between these mentioned sets are illustrated in the Venn diagram shown in Figure 34.2. The implicit assumption in the

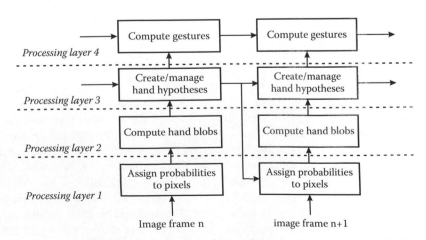

FIGURE 34.1   Block diagram of the proposed approach for hand-tracking and gesture recognition. Processing is organized into four layers.

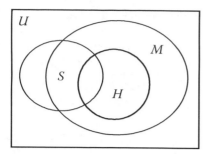

FIGURE 34.2 The Venn diagram representing the relationship between the pixel sets U, M, S, and H.

previous formulation is that H is a subset of M, which means that hands always belong to the foreground. It is also important that, according to this model, all pixels belonging to hands are not necessarily assumed to be skin-colored.

Accordingly, let S and H be binary random variables (i.e., taking values in $\{0, 1\}$), indicating whether a pixel belongs to S and H, respectively. Also, let M be a binary variable (determined by the employed foreground subtraction algorithm) that indicates whether a pixel belongs to M. Let L be the 2D location vector containing the pixel image coordinates, and let T be a variable that encodes a set of features regarding the currently tracked hypotheses (the contents of T will be explained later in this section). Given all of these conditions, the goal of this processing layer is to compute whether a pixel belongs to a hand, given (a) the color c of a single pixel, (b) the information m on whether this pixel belongs to the background (i.e., $M = m$), and (c) the values l and t of L and T, respectively. More specifically, the conditional probability $P_h = P(H = 1 | C = c, T = t, L = l, M = m)$ needs to be estimated.[1]

To perform this estimation, the Bayesian network shown in Figure 34.3 is assumed. The nodes in the graph of this figure correspond to random variables that represent degrees of belief on particular aspects of the problem. The edges in the graph are parameterized by conditional probability distributions that represent causal dependencies between the involved variables. It is known that

$$P(H=1|c,t,l,m)=\frac{P(H=1,c,t,l,m)}{P(c,t,l,m)} \quad (34.1)$$

By marginalizing the numerator over both possible values of S and the denominator over all four possible combinations of S and H (the values of S and H are expressed by the summation indices s and h, respectively), P_h can be expanded as:

$$P_h=\frac{\displaystyle\sum_{s\in\{0,1\}}P(H=1,s,c,t,l,m)}{\displaystyle\sum_{s\in\{0,1\}}\sum_{h\in\{0,1\}}P(h,s,c,t,l,m)} \quad (34.2)$$

By applying the chain rule of probability and by taking advantage of the variable (in-)dependencies implied by the graph of Figure 34.3b, one obtains:

$$P(h,s,c,t,l,m)=P(m)P(l)P(t|h)P(c|s)P(s|l,m)P(h|l,s,m) \quad (34.3)$$

Finally, by substituting to Equation 34.1, one obtains:

$$P_h=\frac{P(t|H=1)\displaystyle\sum_{s\in\{0,1\}}P(c|s)P(s|l,m)P(H=1|l,s,m)}{\displaystyle\sum_{h\in\{0,1\}}P(t|h)\displaystyle\sum_{h\in\{0,1\}}P(c|s)P(s|l,m)P(h|l,s,m)} \quad (34.4)$$

Details regarding the estimation of the individual probabilities that appear in Equation 34.4 are provided in the following sections.

34.3.1.1.1 Foreground Segmentation

It can be easily verified that when $M = 0$ (i.e., a pixel belongs to the background), the numerator of Equation 34.4 becomes zero as well. This is because, as already mentioned, hands have been assumed to always belong to the foreground. This assumption simplifies computations because Equation 34.4 should only be evaluated for foreground pixels.

To compute M, the foreground/background segmentation technique proposed by Stauffer and Grimson (Grimson and Stauffer, 1999; Stauffer et al., 1999) is employed, which uses an adaptive Gaussian mixture model on the background color of each image pixel. The number of Gaussians, their parameters, and their weights in the mixture are computed online.

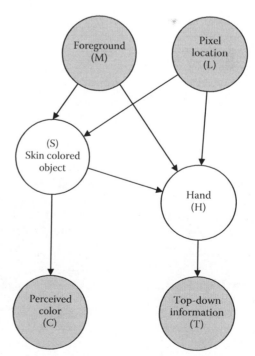

FIGURE 34.3 The proposed Bayes net.

---

[1] Capital letters are used to indicate variables, and small letters are used to indicate specific values for these variables. For brevity, the notation $P(x)$ is also used to refer to probability $P(X = x)$, where X is any of the previously defined variables and x a specific value of this variable.

34.3.1.1.2 The Color Model

$P(c|s)$ is the probability of a pixel being perceived with color c given the information on whether it belongs to skin or not. To increase robustness against lighting variability, colors are transformed to the YUV color space. Following the same approach as in Yang et al. (1998) and Argyros and Lourakis (2004b), the Y (luminance) component is completely eliminated. This makes C a two-dimensional variable encoding the U and V (chrominance) components of the YUV color space.

$P(c|s)$ is obtained offline through a separate training phase with the procedure described in Argyros and Lourakis (2004b). Assuming that C is discrete (i.e., taking values in $[0.255]^2$) the result can be encoded in the form of two, 2D look-up tables; one table for skin-colored objects ($s = 1$) and one table for all other objects ($s = 0$). The rows and the columns of both look-up tables correspond to the U and V dimensions of the YUV color space.

34.3.1.1.3 The Spatial Distribution Model

A spatial distribution model for skin and hands is needed to evaluate $P(s|l,m)$ and $P(h|l, s,m)$. These two probabilities express prior probabilities that can be obtained during training and are stored explicitly for each location l (i.e., for each image pixel). To estimate these probabilities, a set of four different quantities are computed offline during training.

These quantities are depicted in Table 34.1 and indicate the number of foreground pixels found in the training sequence for every possible combination of s and h. As already discussed, only computations for foreground pixels are necessary. Hence, all training data correspond to $M = 1$. Therefore, $P(s|l,M = 1)$ and $P(h|l, s,M = 1)$ can be easily expressed in terms of s_{00}, s_{01}, s_{10}, and s_{11} as:

$$P(s|l,M=1)=\frac{P(s,M=1,l)}{P(M=1,l)}=\frac{s_{0s}+s_{1s}}{s_{00}+s_{01}+s_{10}+s_{11}} \quad (34.5)$$

Similarly:

$$P(h|l,s,M=1)=\frac{P(h,s,M=1,l)}{P(s,M=1,l)}=\frac{s_{hs}}{s_{0s}+s_{1s}} \quad (34.6)$$

34.3.1.1.4 Top-Down Information Regarding Hand Features

Within the second and third processing layers, pixel probabilities are converted to blobs (second layer) and hand hypotheses, which are tracked over time (third layer). These processes are described later in Sections 34.3.1.2 and 34.3.1.3, respectively. Nevertheless, as Figure 34.1 shows, the third processing layer

TABLE 34.1 Quantities Estimated during Training for the Spatial Distribution Model

$h = 0$		$h = 1$	
$s = 0$	$s = 1$	$s = 0$	$s = 1$
s_{00}	s_{01}	s_{10}	s_{11}

of image n provides top-down information exploited during the processing of image $n+1$ at layer 1. For this reason, the description of the methods employed to compute the probabilities $P(t|h)$ that are further required to estimate P_h is deferred to Section 34.3.1.3.

34.3.1.2 Processing Layer 2: From Pixels to Blobs

This layer applies hysteresis thresholding on the probabilities determined at layer 1. These probabilities are initially thresholded by a strong threshold T_{max} to select all pixels with $P_h > T_{max}$. This yields high-confidence hand pixels that constitute the seeds of potential hand blobs.

A second thresholding step, this time with a "weak" threshold T_{min}, is used to recursively characterize pixels with probability $T_{min} < P_h < T_{max}$. These pixels are characterized as hand pixels, if they are adjacent to pixels already characterized as hand pixels.

A connected components labeling algorithm is then used to assign different labels to pixels that belong to different blobs. Size filtering on the derived connected components is also performed to eliminate small, isolated blobs that are attributed to noise and do not correspond to meaningful hand regions.

Finally, a feature vector for each blob is computed. This feature vector contains statistical properties regarding the spatial distribution of pixels within the blob and will be used within the next processing layer for data association.

34.3.1.3 Processing Layer 3: From Blobs to Object Hypotheses

Within the third processing layer, blobs are assigned to hand hypotheses, which are tracked over time. Tracking over time is realized through a scheme that can handle multiple objects that may move in complex trajectories, occlude each other in the field of view of a possibly moving camera, and whose number may vary over time. For the purposes of this chapter,[2] it suffices to mention that a hand hypothesis h_i is essentially represented as an ellipse $h_i = h_i (cx_i, cy_i, \alpha_i, \beta_i, \theta_i)$ where (cx_i, cy_i) is the ellipse centroid, α_i and β_i are, respectively, the lengths of the major and minor axis of the ellipse, and θ_i is its orientation on the image plane. The parameters of each ellipse are determined by the covariance matrix of the locations of blob pixels that are assigned to a certain hypothesis. The assignment of blob pixels to hypotheses ensures (1) the generation of new hypotheses in cases of unmatched evidence (unmatched blobs), (2) the propagation and tracking of existing hypotheses in the presence of multiple, potential occluding objects, and (3) the elimination of invalid hypotheses (i.e., when tracked objects disappear from the scene of view).

34.3.1.3.1 Top-Down Information Regarding Hand Features Revisited

For each tracked hand hypothesis, a feature vector T is generated, which is propagated in a top-down direction to further assist the

---

[2] For the details of this tracking process, the interested reader is referred to Argyros and Lourakis (2004b).

assignment of hand probabilities to pixels at processing layer 1. The feature vector T consists of two different features:

1. The average vertical speed v of a hand, computed as the vertical speed of the centroid of the ellipse modeling the hand. The rationale behind the selection of this feature is that hands are expected to exhibit considerable average speed v compared to other skin-colored regions such as heads.
2. The ratio r of the perimeter of the hand contour over the circumference of a hypothetical circle having the same area as the area of the hand. The rationale behind the selection of this feature is that hands are expected to exhibit high r compared to other objects. That is, $r = \frac{1}{2}\rho / \sqrt{\pi\alpha}$, where ρ and α are the hand circumference and area, respectively.

Given v and r, $P(t|h)$ is approximated as:

$$P(t|h) \approx P(v|h)P(r|h) \tag{34.7}$$

$P(t|h)$ is the probability of measuring a specific value t for the feature vector T, given the information of whether a pixel belongs to a hand or not. A pixel is said to belong to a hand depending on whether its image location lies within the ellipse modeling the hand hypothesis. That is, the feature vector T encodes a set of features related to existing (tracked) hands that overlap with the pixel under consideration.

In the actual implementation, both $P(v|h)$ and $P(r|h)$ are given by means of one-dimensional look-up tables that are computed offline during training. If there is more than one hypothesis overlapping with the specific pixel under consideration, the hypothesis that yields maximal results is chosen for $P(t|h)$. Moreover, if there is no overlapping hypothesis at all, all of the conditional probabilities of Equation 34.7 are substituted by the maximum values of their corresponding look-up tables.

34.3.1.4 Processing Layer 4: Recognizing Hand Gestures

The application considered in this chapter involves natural interaction with autonomous mobile robots installed in public places, such as museums and exhibition centers. Since the actual users of the system will be untrained visitors of a museum/exhibition, gestures should be as intuitive and natural as possible. Moreover, the challenging operational requirements of the application at hand impose the absolute need for gestures to be simple and robustly interpretable. Four simple gestures have been chosen to comprise the proposed gesture vocabulary, which is illustrated in Figure 34.4.

All four employed gestures are static gestures, i.e., gestures in which the information to be communicated lies in the hand and finger posture at a certain moment in time. More specifically, the employed gestures are as follows:

- *The Stop gesture*: The user extends a hand with all five fingers stretched to stop the robot from its current action.
- *The Thumbs-Up gesture*: The user performs a thumbs-up sign to approve or answer yes to a question by the robot.

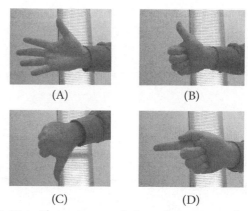

FIGURE 34.4 The gesture vocabulary of the proposed approach: (A) the Stop gesture, (B) the Thumbs-Up gesture, (C) the Thumbs-Down gesture, (D) the Point gesture.

- *The Thumbs-Down gesture*: The user expresses disapproval or answers no to a question by doing the thumbs-down gesture.
- *The Point gesture*: The user points to a specific exhibit or point of interest to ask the robot to guide the user there.

It is also important that because of the generic nature of the employed gestures, their actual meaning can be interpreted by the robot based on specific, contextual information related to the scenario of use.

To robustly recognize the gestures constituting the gesture vocabulary, a rule-based technique is employed that relies on the number and the posture of the distinguishable fingers (i.e., the number of detected fingertips corresponding to each tracked hand hypothesis and their relative location with respect to the centroid of the hypothesis).

34.3.1.4.1 Finger Detection

Fingertip detection is performed by evaluating a curvature measure of the contour of the blobs that correspond to each hand hypothesis as in Argyros and Lourakis (2006b). The employed curvature measure assumes values in the range [0.0, 1.0] and is defined as:

$$K_l(P) = \frac{1}{2}\left(1 + \frac{\overrightarrow{P_1 P} \cdot \overrightarrow{P_2 P}}{\|\overrightarrow{P_1 P}\|\|\overrightarrow{P_2 P}\|}\right) \tag{34.8}$$

where P_1, P, and P_2 are successive points on the contour, P being separated from P_1 and P_2 by the same number of contour points. The symbol (·) denotes the vector dot product. The algorithm for finger detection computes $K_l(P)$ for all contour points of a hand and at various scales (i.e., for various values of the parameter l). A contour point P is then characterized as the location of a fingertip if both of the following conditions are met:

- $K_l(P)$ exceeds a certain threshold for at least one of the examined scales.
- $K_l(P)$ is a local maximum in its (scale-dependent) neighborhood of the contour.

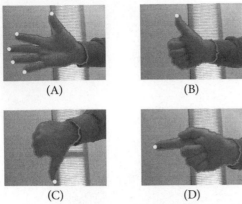

FIGURE 34.5 Fingertip detection. Fingers are denoted as white circles.

TABLE 34.2 Rules Used to Recognize the Four Gestures of the Selected Vocabulary

Gesture	Visible Fingertips	Orientation ϕ (in degrees)
Stop	5	Irrelevant
Thumbs-Up	1	$\varphi \in [60; 120]$
Thumbs-Down	1	$\varphi \in [240; 300]$
Point	1	$\varphi \in [0, 60] \cup [120, 240] \cup [300, 360]$

Evaluation of curvature information on blob contour points has been demonstrated in the past (Argyros and Lourakis, 2006b) to be a robust way to detect fingertips.

A significant advantage of contour features like fingertips is that in most cases they can be robustly extracted regardless of the size of the blob (i.e., distance of the observer), lighting conditions, and other parameters that usually affect color- and appearance-based features. Figure 34.5 shows some examples from a fingertip detection experiment. The extracted contour points of each hand are depicted in blue. The detected fingertips are marked with white circles. In the reported experiments, the curvature threshold of the first criterion was set to 0.7.

34.3.1.4.2 Recognizing a Gesture

As already mentioned, all employed gestures are static, that is, gestures in which the information to be communicated lie in features obtained at a specific moment in time. The employed features consist of the number of distinguishable fingers (i.e., fingers with distinguishable fingertips) and their orientation φ with respect to the horizontal image axis. To compute the orientation φ of a particular finger, the vector determined by the hand's centroid and the corresponding fingertip is assumed.

To recognize the four employed gestures a rule-based approach is used. Table 34.2 summarizes the rules that need to be met for each of the four gestures in the vocabulary. Moreover, to determine the specific point in time when a gesture takes place, three additional criteria have to be satisfied.

- *Criterion 1*: The hand posture has to last for at least a fixed amount of time t_g. In the actual implementation of the system, a minimum duration of half a second is employed (i.e., $t_g = 0.5$ sec). Assuming a frame rate of 30 Hz, this means that to recognize a certain posture, the hand posture has to be maintained for a minimum of 15 consecutive image frames.
- *Criterion 2*: The hand that performs the gesture has to be (almost) still. This is determined by applying the requirement that the hand centroid remains within a specific threshold radius r_g for at least t_g seconds. In the conducted

experiments an r_g value of about 30 pixels has been proven sufficient to ensure that the hand is almost standstill.

- *Criterion 3*: The speed of the hand has to be at its minimum with respect to time. To determine whether the hand speed has reached its minimum, a time lag t_l is assumed (fixed to about 0.3 sec in our experiments).

34.3.2 Experimental Results

The proposed approach has been assessed using several video sequences containing people performing various gestures in indoor environments. Several videos of example runs are available on the web.[3]

This section presents results obtained from a sequence depicting a man performing a variety of hand gestures in a setup that is typical for human-robot interaction applications (i.e., the subject is standing at a typical distance of about 1 m from the robot looking toward the robot). The robot's camera is installed at a distance of approximately 1.2 m from the floor. The resolution of the sequence is 640 × 480 and it was obtained with a standard, low-end web camera at 30 frames per second. Figure 34.6 depicts various intermediate results obtained at different stages of the proposed approach. A frame of the test sequence is shown in Figure 34.6A. Figure 34.6B depicts the result of the background subtraction algorithm, that is, $P(M)$. To achieve real-time performance, the background subtraction algorithm operates at down-sampled images of dimensions 160 × 120. Figure 34.6C depicts P_h, that is, the result of the first processing layer of the proposed approach. The contour of the blob and the detected fingertip that correspond to the only present hand hypothesis is shown in Figure 34.6D. As can be verified, the algorithm manages to correctly identify the hand of the depicted man. Also, in contrast to what would happen if only color information were utilized, neither skin-colored objects in the background nor the subject's face is falsely recognized as a hand.

Figure 34.7 shows six more frames out of the same sequence. In all cases, the proposed approach has been successful in correctly identifying the hands of the person and in correctly recognizing the performed gesture. The presented results were obtained at a standard 3 GHz personal computer, which was able to process images of size 640 × 480 at 30 Hz.

---

[3] http://www.ics.forth.gr/xmpalt/research/gestures/index.html.

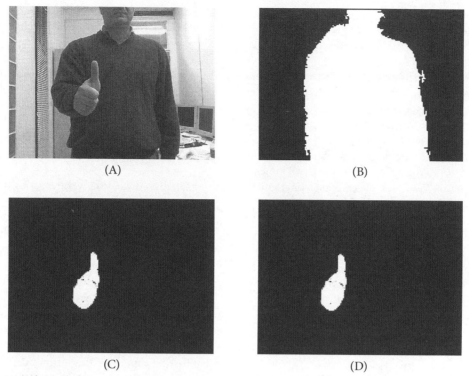

FIGURE 34.6 The proposed approach in operation: (A) original frame, (B) background subtraction result, (C) pixel probabilities for hands, and (D) contour and fingertip detection.

34.4 Open Challenges for Vision-Based Hand Gesture Recognition

Although many techniques for hand gesture recognition have been proposed, robust hand gesture recognition is still considered an unsolved problem. Despite the huge progress in the field over the last years, the development of machines with perceptual capabilities comparable to those of humans is still a distant goal that requires major improvements over a large number of research fields. There are still many major challenges to be addressed:

- *Detection and tracking under realistic conditions*: Although current hand detection and tracking techniques work well in relatively constrained environments, their performance under realistic conditions still remains an unsolved problem. Varying illumination conditions, image noise, cluttered backgrounds, occlusions, and camera motions are some of the parameters that may deteriorate the performance of most of the state-of-the-art approaches. Providing robust features is required to arrive at systems able to robustly perform in more general settings.
- *Combination of multiple visual cues*: Different approaches to hand detection exhibit different weaknesses, depending on the visual cue that they are based upon. For example, skin color detection provides a valuable cue for identifying image regions within which hands may appear; however, it is of limited use because of the possible occurrence of background with similar color. Shape-based techniques suffer in the presence of cluttered backgrounds and complex illumination. An open challenge is the appropriate exploitation of several such local visual cues to facilitate the robust detection of hands.

- *View-independent recognition*: Most current systems are viewpoint dependent (i.e., they require that the subject stands with a prespecified pose with respect to the camera). Pose invariant methods for gesture recognition are required to make such systems effective in a wider set of situations.

- *Handling of the inherent gesture intraclass variability*: There are no clear boundaries among the numerous classes of different gestures. People use different gestures to communicate the same thing or use the same gesture to communicate different things. Even the same person rarely performs the same gesture two times in exactly the same way. Performance may depend on psychological factors, fatigue, stress, environmental conditions, age, gender, ethnic background, and so on. Further investigation of these issues is expected to shed more light on the development of computer vision algorithms for gesture recognition.

- *Continuous gesture recognition*: Identifying when a gesture starts and when a gesture ends in a continuous sequence of hand motions remains a significant challenge that vastly affects the performance of current systems. Even for posture recognition systems (i.e., systems that only utilize information contained in a single frame) spotting the right frame out of continuous videostream without confusion with intergesture hand motions remains a challenging problem.

- *Recognition of hand gestures in context*: Hand gestures are usually coupled with other human activities such as speech or face gestures and body postures. A natural way to improve performance on gesture recognition is by combining information in multimodal systems.

- *Gesture recognition and learning*: Undoubtedly, machine learning is to play a central role in future developments in the area of hand-based human-computer interaction. The large variability of application domains and corresponding vocabularies demand increased automation in the customization of such systems. Learning and tutoring of a system seems a proper approach to build systems with robust performance and with adaptability to the user, the task, and the context in a certain application domain.

- *Recognition of two-handed gestures*: Most of the current approaches treat the case on single-handed gestures. Dealing with two-handed gestures introduces several challenges to the detection and tracking of hands and the recognition of their gesture.

- *Creation of more standard data sets and benchmarks*: As the field of hand gesture recognition grows and becomes more mature, the need for standard data sets and benchmarks is becoming more important. Unfortunately, neither large public-domain datasets, nor standard benchmarking procedures currently exist, making the comparison of individual approaches very difficult. Of particular importance is the availability of spontaneous data acquired from subjects that are not aware of the recording. It is expected that the creation of such datasets and benchmarks will have a significant impact on the effectiveness of research efforts and the rate of progress in the field.

34.5 Summary

This chapter has reviewed several existing methods for supporting vision-based human-computer interaction based on the recognition of hand gestures. The provided review covers research work related to all three individual subproblems of the full problem, namely, detection, tracking, and recognition. Moreover, an overview is provided of some integrated gesture recognition systems.

Additionally, a novel gesture recognition system has been presented intended for natural interaction with autonomous robots that guide visitors in museums and exhibition centers. The proposed gesture recognition system builds on a probabilistic framework that allows the utilization of multiple information cues to efficiently detect image regions that belong to human hands. Tracking over time is achieved by a technique that can simultaneously handle multiple hands that may move in complex trajectories, occlude each other in the field of view of the robot's camera, and vary in number over

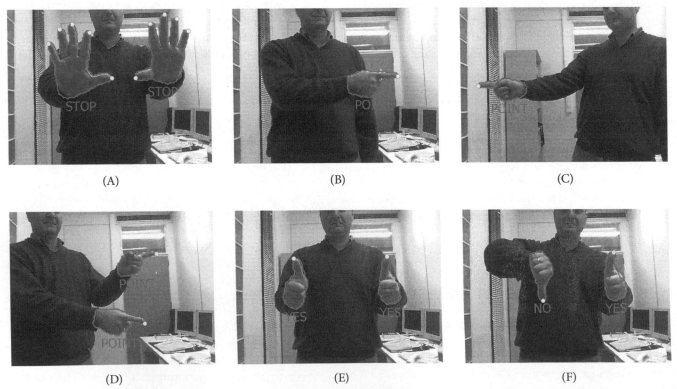

(A)   (B)   (C)

(D)   (E)   (F)

FIGURE 34.7 Six frames of a sequence depicting a man performing gestures in an office environment.

time. Dependable hand-tracking, combined with fingertip detection, facilitate the definition of a small and simple hand gesture vocabulary that is both robustly interpretable and intuitive to humans interacting with robots. Experimental results presented in this chapter confirm the effectiveness and the efficiency of the proposed approach, meeting the run-time requirements of the task at hand. Nevertheless, and despite the vast amount of relevant research efforts, the problem of efficient and robust vision-based recognition of natural hand gestures in unprepared environments still remains open and challenging, and is expected to remain of central importance to the computer vision community in the forthcoming years.

Acknowledgments

This work was partially supported by EU-IST NoE MUSCLE (FP6-507752), the Greek national GSRT project XENIOS, and the EU-IST project INDIGO (FP6-045388).

References

Ahmad, S. (1994). A usable real-time 3D hand tracker, in *Asilomar Conference on Signals, Systems and Computers*, 31 Oct–2 Nov, Asilomar, CA, pp. 1257–126. Washington, DC: IEEE.

Argyros, A. A. and Lourakis, M. I. A. (2004a). 3D tracking of skin-colored regions by a moving stereoscopic observer. *Applied Optics* 43: 366–378. London: Springer-Verlag.

Argyros, A. A. and Lourakis, M. I. A. (2004b). Real-time tracking of multiple skin-colored objects with a possibly moving camera, in the *Proceedings of the European Conference on Computer Vision 2004*, 11–14 May 2004, Prague, pp. 368–379. Washington, DC: IEEE.

Argyros, A. A. and Lourakis, M. I. A. (2006a). Binocular hand tracking and reconstruction based on 2D shape matching, in the *Proceedings of the International Conference on Pattern Recognition*, 20–24 August 2006, Hong Kong, pp. 207–210. Washington, DC: IEEE.

Argyros, A. A. and Lourakis, M. I. A. (2006b). Vision-based interpretation of hand gestures for remote control of a computer mouse, in *Proceedings of European Conference on Computer Vision Workshop on HCI*, 13 May 2006, Graz, Austria, pp. 40–51.

Athitsos, V. and Sclaroff, S. (2001). 3D hand pose estimation by finding appearance-based matches in a large database of training views, in *IEEE Workshop on Cues in Communication*, 9 December 2001, Kauai, HI, pp. 100–106. Washington, DC: IEEE.

Athitsos, V. and Sclaroff, S. (2002). An appearance-based framework for 3D hand shape classification and camera viewpoint estimation, in *IEEE Conference on Face and Gesture Recognition*, 10–12 April 2002, Southampton, U.K., pp. 45–50. Washington, DC: IEEE.

Athitsos, V. and Sclaroff, S. (2003). Estimating 3D hand pose from a cluttered image, in the *Proceedings of IEEE Computer Vision and Pattern Recognition, Volume 2*, 16–22 June 2003, Milwaukee, Wisconsin, 432–439. Washington, DC: IEEE.

Azarbayejani, A. and Pentland, A. (1996). Real-time self-calibrating stereo person-tracker using 3-d shape estimation from blob features, in the *Proceedings of the International Conference on Pattern Recognition (ICPR 1996)*, August 1996, Vienna, pp. 99–108. Washington, DC: IEEE.

Azoz, Y., Devi, L., and Sharma, R. (1998). Reliable tracking of human arm dynamics by multiple cue integration and constraint fusion, in the *Proceedings of the IEEE Computer Vision and Pattern Recognition (CVPR 1998)*, 23–25 June 1998, Santa Barbara, CA, pp. 905–910. Washington, DC: IEEE.

Baltzakis, H., Argyros, A., Lourakis, M., and Trahanias, P. (2008). Tracking of human hands and faces through probabilistic fusion of multiple visual cues, in the *Proceedings of the International Conference on Computer Vision Systems (ICVS 2008)*, 12–15 May 2008, Santorini, Greece, pp. 33–42. London: Springer-Verlag.

Barrow, H., Tenenbaum, R., Bolles, J., and Wolf, H. (1977). Parametric correspondence and chamfer matching: Two new techniques for image matching, in *International Joint Conference in Artificial Intelligence*, 22–25 August 1977, Cambridge, MA, pp. 659–663.

Bauckhage, C., Hanheide, M., Wrede, S., Kaster, T., Pfeiffer, M., and Sagerer, G. (2005). Vision systems with the human in the loop. *EURASIP Journal on Applied Signal Processing* 14: 2375–2390.

Bauckhage, C., Hanheide, M., Wrede, S., and Sagerer, G. (2004). A cognitive vision system for action recognition in office environments, in the *Proceedings of the IEEE Computer Vision and Pattern Recognition (CVPR 2004), Volume 2*, 27 June–2 July 2004, Washington, DC, pp. 827–833. Washington, DC: IEEE.

Bauer, B. and Hienz, H. (2000). Relevant features for video-based continuous sign language recognition, in *IEEE International Conference on Automatic Face and Gesture Recognition*, 28–30 March 2000, Grenoble, France, pp. 440–445. Washington, DC: IEEE.

Baumberg, A. and Hogg, D. (1994). Learning flexible models from image sequences, in the *Proceedings of the European Conference on Computer Vision 1994, Volume 1*, 2–6 May 1994, Stockholm, pp. 299–308. London: Springer-Verlag.

Becker, D. A. (1997). *Sensei: A real-time recognition, feedback, and training system for T'ai chi gestures*. MS thesis, Media Lab, MIT, Cambridge, MA.

Belongie, S., Malik, J., and Puzicha, J. (2002). Shape matching and object recognition using shape contexts. *IEEE Transactions on Pattern Analysis and Machine Intelligence* 24: 509–522.

Berry, G., Pavlovic, V., and Huang, T. (1998). Battleview: A multimodal HCI research application, in *Workshop on Perceptual User Interfaces*, 5–6 November 1998, San Francisco, pp. 67–70. New York: ACM Press.

Besl, P. J. and McKay, N. D. (1992). A method for registration of 3d shapes. *IEEE Transactions on Pattern Analysis and Machine Intelligence* 14: 239–256.

Birk, H., Moeslund, T. B., and Madsen, C. B. (1997). Real-time recognition of hand alphabet gestures using principal component analysis, in the *Proceedings of the Scandinavian Conference on Image Analysis 1997*, 29 June–2 July 1997, Halmstad, Sweden, pp. 261–268. Pattern Recognition Society.

Black, M. and Jepson, A. (1998a). A probabilistic framework for matching temporal trajectories: Condensation-based recognition of gesture and expression, in the *Proceedings of the European Conference on Computer Vision 1998, Volume 2*, 2–6 June 1998, Freiburg, Germany, pp. 909–924. London: Springer-Verlang.

Black, M. and Jepson, A. (1998b). Recognizing temporal trajectories using the condensation algorithm, in *IEEE International Conference on Automatic Face and Gesture Recognition*, 10–12 April 1998, Southampton, U.K., pp. 16–21. Washington, DC: IEEE.

Black, M. and Jepson, A. (1998c). Eigentracking: Robust matching and tracking of articulated objects using a view-based representation. *International Journal of Computer Vision* 26: 63–84.

Blake, A., North, B., and Isard, M. (1999). Learning multi-class dynamics, in the *Proceedings of Advances in Neural Information Processing Systems, Volume 11*, 29 November–4 December 1999, Denver, pp. 389–395. Cambridge, MA: MIT Press.

Bobick, A. and Davis, J. (1996). Real-time recognition of activity using temporal templates, in *IEEE Workshop on Applications of Computer Vision*, 21–24 February 1996, Austin, TX, pp. 39–42. Washington, DC: IEEE.

Bobick, A. and Davis, J. (2001). The representation and recognition of action using temporal templates. *IEEE Transactions on Pattern Analysis and Machine Intelligence* 3: 257–267.

Bobick, A. and Wilson, A. D. (1997). A state-based approach to the representation and recognition of gesture. *IEEE Transactions on Pattern Analysis and Machine Intelligence* 19: 1325–1337.

Borgefors, G. (1988). Hierarchical chamfer matching: A parametric edge matching algorithm. *IEEE Transactions on Pattern Analysis and Machine Intelligence* 10: 849–865.

Bradski, G. (1998). Real time face and object tracking as a component of a perceptual user interface, in *IEEE Workshop on Applications of Computer Vision*, 21–24 February 1998, Austin, TX, pp. 214–219. Washington, DC: IEEE.

Bradski, G. and Davis, J. (2000). Motion segmentation and pose recognition with motion history gradients, in *IEEE Workshop on Applications of Computer Vision*, 4–6 December 2000, Palm Springs, FL, pp. 238–244. London: Springer-Verlag.

Bradski, G. and Davis, J. (2002). Motion segmentation and pose recognition with motion history gradients. *Machine Vision and Applications* 13: 174–184.

Brand, M., Birnbaum, L., and Cooper, P. (1993). Sensible scenes: Visual understanding of complex structures through causal analysis, in *AAAI Conference on Artificial Intelligence*, 11–15 July 1993, Washington, DC, pp. 45–56. Cambridge, MA: AAAI Press/MIT Press.

Brand, M., Oliver, N., and Pentland, A. (1997). Coupled hidden Markov models for complex action recognition, in the *Proceedings of IEEE Computer Vision and Pattern Recognition*, 17–19 June 1997, San Juan, Puerto Rico, pp. 994–999. Washington, DC: IEEE.

Bregler, C. (1997). Learning and recognizing human dynamics in video sequences, in the *Proceedings of IEEE Computer Vision and Pattern Recognition*, 17–19 June 1997, San Juan, Puerto Rico, pp. 568–574. Washington, DC: IEEE.

Breig, M. and Kohler, M. (1998). *Motion Detection and Tracking under Constraint of Pantilt Cameras for Vision-Based Human Computer Interaction.* Technical Report 689, Informatik VII, University of Dortmund/Germany.

Brockl-Fox, U. (1995). Real-time 3D interaction with up to 16 degrees of freedom from monocular image flows, in *International Workshop on Automatic Face and Gesture Recognition*, 26–28 June 1995, Zurich, Switzerland, pp. 172–178. Washington, DC: IEEE.

Cai, J. and Goshtasby, A. (1999). Detecting human faces in color images. *Image and Vision Computing* 18: 63–75.

Cairnie, N., Ricketts, I., McKenna, S., and McAllister, G. (2000). Using finger-pointing to operate secondary controls in an automobile, in *Intelligent Vehicles Symposium 2000, Volume 4*, 4–5 October 2000, Dearborn, MI, pp. 550–555. Amsterdam, The Netherlands: IOS Press.

Campbell, L., Becker, D., Azarbayejani, A., Bobick, A., and Pentland, A. (1996). Invariant features for 3-d gesture recognition, in *IEEE International Conference Automatic Face and Gesture Recognition*, 13–16 October 1996, Killington, VT, pp. 157–162.

Chai, D. and Ngan, K. (1998). Locating the facial region of a head and shoulders color image, in *IEEE International Conference on Automatic Face and Gesture Recognition 1998*, 14–16 April 1998, Nara, Japan, pp. 124–129. Washington, DC: IEEE.

Chen, H. and Liu, T. (2001). Trust-region methods for real-time tracking, in the *Proceedings of the International Conference on Computer Vision, Volume 2*, 7–14 July 2001, Vancouver, Canada, pp. 717–722. Washington, DC: IEEE.

Cheng, Y. (1995). Mean shift, mode seeking, and clustering. *IEEE Transactions on Pattern Analysis and Machine Intelligence* 17: 790–799.

Cheung, G., Kanade, T., Bouguet, J., and Holler, M. (2000). A real time system for robust 3D voxel reconstruction of human motions, in the *Proceedings of IEEE Computer Vision and Pattern Recognition, Volume 2*, 13–15 June 2000, Hilton Head, SC, pp. 714–720. Washington, DC: IEEE.

Cohen, C., Conway, L., and Koditschek, D. (1996). Dynamical system representation, generation, and recognition of basic oscillatory motion gestures, in the *International Conference on Automatic Face and Gesture Recognition*, 14–16 October 1996, Killington, VT, p. 60. Washington, DC: IEEE.

Colmenarez, A. and Huang, T. (1996). Maximum likelihood face detection, in *International Conference on Automatic Face and Gesture Recognition*, 14–16 October 1996, Killington, VT, pp. 307–311.

Comaniciu, D., Ramesh, V., and Meer, P. (2000). Real-time tracking of non-rigid objects using mean shift, in the *Proceedings of IEEE Computer Vision and Pattern Recognition*, 13–15 June 2000, Hilton Head, SC, pp. 142–149.

Comaniciu, D., Ramesh, V., and Meer, P. (2003). Kernel-based object tracking. *IEEE Transactions on Pattern Analysis and Machine Intelligence* 25: 564–577. Washington, DC: IEEE.

Cootes, T. F. and Taylor, C. J. (1992). Active shape models: Smart snakes, in *British Machine Vision Conference*, 21–24 September 1992, Leeds, U.K., pp. 266–275. London: Springer-Verlag.

Cootes, T. F., Taylor, C. J., Cooper, D. H., and Graham, J. (1995). Active shape models: Their training and applications. *Computer Vision and Image Understanding* 61: 38–59.

Cote, M., Payeur, P., and Comeau, G. (2006). Comparative study of adaptive segmentation techniques for gesture analysis in unconstrained environments, in *IEEE International Workshop on Imagining Systems and Techniques 2006*, 29 April 2006, Minori, Italy, pp. 28–33. Washington, DC: IEEE.

Crowley, J., Berard, F., and Coutaz, J. (1995). Finger tracking as an input device for augmented reality, in *International Workshop on Gesture and Face Recognition*, 26–28 June 1995, Zurich, Switzerland, pp. 195–200. Washington, DC: IEEE.

Cui, Y., Swets, D., and Weng, J. (1995). Learning-based hand sign recognition using SHOSLF-M, in *International Workshop on Automatic Face and Gesture Recognition 1995*, Zurich, Switzerland, pp. 201–206. Washington, DC: IEEE.

Cui, Y. and Weng, J. (1996a). Hand segmentation using learning-based prediction and verification for hand sign recognition, in the *Proceedings of IEEE Computer Vision and Pattern Recognition*, 18–20 June 1996, San Francisco, pp. 88–93. Washington, DC: IEEE.

Cui, Y. and Weng, J. (1996b). Hand sign recognition from intensity image sequences with complex background, in the *Proceedings IEEE Computer Vision and Pattern Recognition*, 18–20 June 1996, San Francisco, pp. 88–93. Washington, DC: IEEE.

Cutler, R. and Turk, M. (1998). View-based interpretation of real-time optical flow for gesture recognition, in *Proceedings of International Conference on Face and Gesture Recognition 1998*, 14–16 April 1998, Nara, Japan, pp. 416–421. IEEE Computer Society.

Darrell, T., Essa, I., and Pentland, A. (1996). Task-specific gesture analysis in real-time using interpolated views. *IEEE Transactions on Pattern Analysis and Machine Intelligence* 18: 1236–1242.

Darrell, T. and Pentland, A. (1993). Space-time gestures, in *Proceedings of IEEE Computer Vision and Pattern Recognition*, 15–17 June 1993, New York, pp. 335–340. Washington, DC: IEEE.

Darrell, T. and Pentland, A. (1995). Attention driven expression and gesture analysis in an interactive environment, in *Proceedings of International Conference on Automatic Face and Gesture Recognition*, September 1995, Zurich, Switzerland, pp. 135–140. Washington, DC: IEEE.

Davis, J. (2001). Hierarchical motion history images for recognizing human motion, in *IEEE Workshop on Detection and Recognition of Events in Video 2001*, 8 July 2001, Vancouver, Canada, pp. 39–46.

Davis, J. and Bobick, A. (1998). Virtual pat: A virtual personal aerobic trainer, in *Workshop on Perceptual User Interfaces 1998*, 5–6 November 1998, San Francisco, pp. 13–18. New York: ACM Press.

Davis, J. and Shah, M. (1994a). Recognizing hand gestures, in *Proceedings of European Conference on Computer Vision 1994*, 2–6 May 1994, Stockholm, pp. 331–340. London: Springer-Verlag.

Davis, J. and Shah, M. (1994b). Visual gesture recognition. *Vision, Image, and Signal Processing* 141: 101–106. London: Spring-Verlag.

Derpanis, K., Wildes, R., and Tsotsos, J. (2004). Hand Gesture Recognition within a Linguistics-Based Framework, in *Proceedings of 8th European Conference on Computer Vision*, 11–14 May 2004, Prague, pp. 282–296. Berlin/Heidelberg: Springer-Verlag.

Deutscher, J., Blake, A., and Reid, I. (2000). Articulated body motion capture by annealed particle filtering, in *Proceedings of IEEE Computer Vision and Pattern Recognition, Volume 2*, 13–15 June 2000, Hilton Head, SC, pp. 126–133. Washington, DC: IEEE.

Dominguez, S. M., Keaton, T., and Sayed, A. H. (2001). A robust finger tracking method for wearable computer interfacing. *IEEE Transactions on Multimedia* 8: 956–972.

Downton, A. and Drouet, H. (1991). Image analysis for model-based sign language coding, in *International Conference of Image Analysis and Processing 1991*, 17–19 September 1991, Dresden, Germany, pp. 637–644. World Scientific.

Eickeler, S., Kosmala, A., Rigoll, G., Jain, A., Venkatesh, S., and Lovell, B. (1998). Hidden Markov model based continuous online gesture recognition, in *International Conference on Pattern Recognition 1998, Volume 2*, 16–20 August 1998, Brisbane, Australia, pp. 1206–1208. Washington, DC: IEEE.

Eisenstein, J., Ghandeharizadeh, S., Golubchik, L., Shahabi, C., Donghui, Y., and Zimmermann, R. (2003). Device independence and extensibility in gesture recognition, in *Proceedings of IEEE Virtual Reality 2003*, 22–26 March 2003, Los Angeles, pp. 207–214. Washington, DC: IEEE.

Etoh, M., Tomono, A., and Kishino, F. (1991). Stereo-based description by generalized cylinder complexes from occluding contours. *Systems and Computers in Japan* 22: 79–89.

Fablet, R. and Black, M. (2002). Automatic detection and tracking of human motion with a view-based representation, in *Proceedings of European Conference on Computer Vision 2002*, 26 June–2 July 2002, Dublin, Ireland, pp. 476–491. Washington, DC: IEEE.

Fillbrandt, H., Akyol, S., and Kraiss, K. F. (2003). Extraction of 3D hand shape and posture from images sequences from sign language recognition, in *Proceedings of European Conference on Computer Vision 2002*, May 2003, Copenhagen, Denmark, pp. 181–186. London: Springer-Verlag.

Fox, B. (2005, August 23). Invention: Magic wand for gamers. *New Scientist*. http://www.newscientist.com/article/dn7890.

Francois, R. and Medioni, G. (1999). Adaptive color background modeling for real-time segmentation of video streams, in *International Conference on Imaging Science, Systems, and Technology 1999*, 28 June–1 July 1991, Las Vegas, pp. 227–232. Washington, DC: IEEE.

Freeman, W. (1999). Computer vision for television and games, in *International Workshop on Recognition, Analysis and Tracking of Faces and Gestures in Real-Time Systems 1999*, 26–27 September 1999, Corfu, Greece, p. 118. Washington, DC: IEEE.

Freeman, W. and Roth, M. (1995). Orientation histograms for hand gesture recognition, in *Proceedings of International Conference on Automatic Face and Gesture Recognition*, September 1995, Zurich, Switzerland, pp. 296–301. Washington, DC: IEEE.

Freeman, W. and Weissman, C. (1995). Television control by hand gestures, in *International Workshop on Automatic Face and Gesture Recognition 1995*, 26–28 June 1995, Zurich, Switzerland, pp. 179–183. Washington, DC: IEEE.

Freund, Y. and Schapire, R. (1997). A decision-theoretic generalization of on-line learning and an application to boosting. *Journal of Computer and System Sciences* 55: 119–139.

Friedman, J., Hastie, T., and Tibshiranim, R. (2000). Additive logistic regression: A statistical view of boosting. *Annals of Statistics* 28: 337–374.

Fukumoto, M., Suenaga, Y., and Mase, K. (1994). "Finger-pointer": Pointing interface by image processing. *Computers and Graphics* 18: 633–642.

Gavrila, D. and Davis, L. (1995). Towards 3D model-based tracking and recognition of human movement: A multi-view approach, in *International Workshop on Automatic Face and Gesture Recognition 1995*, 26–28 June 1995, Zurich, Switzerland, 272–277. Washington, DC: IEEE.

Gavrila, D. and Davis, L. (1996). 3-D model-based tracking of humans in action: A multi-view approach, in *Proceedings of IEEE Computer Vision and Pattern Recognition*, 18–20 June 1996, San Francisco, pp. 73–80. Washington, DC: IEEE.

Goncalves, L., di Bernardo, E., Ursella, E., and Perona, P. (1995). Monocular tracking of the human arm in 3D, in *Proceedings of International Conference on Computer Vision*, 20–23 June 1995, Boston, pp. 764–770. Washington, DC: IEEE.

Gong, S., Walter, M., and Psarrou, A. (1999). Recognition of temporal structures: Learning prior and propagating observation augmented densities via hidden markov states, in *Proceedings of International Conference on Computer Vision*, 20–25 September 1999, Corfu, Greece, pp. 157–162. Washington, DC: IEEE.

Grimson, W. E. L. and Stauffer, C. (1999). Adaptive background mixture models for real time tracking, in *Proceedings of IEEE Computer Vision and Pattern Recognition*, 23–25 June 1999, Ft. Collins, CO, pp. 246–252. Washington, DC: IEEE.

Grobel, K. and Assan, M. (1997). Isolated sign language recognition using hidden Markov models, in *IEEE International Conference on Systems, Man, and Cybernetics 1997*, 12–15 October 1997, Orlando, FL, pp. 162–167. Washington, DC: IEEE.

Gupta, N., Mittal, P., Roy, S., Chaudhury, S., and Banerjee, S. (2002). Condensation-based predictive eigentracking, in *Indian Conference on Computer Vision, Graphics and Image Processing 2002*, 16–18 December 2002, Ahmadabad, India, pp. 49–54. Mumbai, India: Allied Publishers.

Hager, G. and Belhumeur, P. (1996). Real-time tracking of image regions with changes in geometry and illumination, in *Proceedings of IEEE Computer Vision and Pattern Recognition*, 18–20 June 1996, San Francisco, pp. 403–410. Washington, DC: IEEE.

Heap, A. and Samaria, F. (1995) Real-time hand tracking and gesture recognition using smart snakes in the *Proceedings of Interface to Real and Virtual Worlds*, 20 June 1995, Montpellier, France. Washington, DC: IEEE.

Heap, T. and Hogg, D. (1996a). 3D deformable hand models, in *Gesture Workshop on Progress in Gestural Interaction 1996*, 19 March 1996, York, U.K., pp. 131–139. London: Springer-Verlag.

Heap, T. and Hogg, D. (1996b). Towards 3D hand tracking using a de-formable model, in *IEEE International Conference Automatic Face and Gesture Recognition 1996*, 14–16 October 1996, Killington, VT, pp. 140–145. Washington, DC: IEEE.

Herpers, R., Verghese, G., Darcourt, K., Derpanis, K., Enenkel, R., Kaufman, J., et al. (1999a). An active stereo vision system for recognition of faces and related hand gestures, in *International Conference on Audio- and Video-Based Biometric Person Authentication 1999*, 22–23 March 1999, Washington, DC, pp. 217–223. London: Springer-Verlag.

Herpers, R., Verghese, G., Derpanis, K., McCready, R., MacLean, J., Levin, A., et al. (1999b). Workshop on recognition, analysis, and tracking of faces and gestures in real-time systems, in *International Workshop on Recognition, Analysis, and Tracking of Faces and Gestures in Real-Time Systems 1999*, 26–27 September 1999, Corfu, Greece, pp. 96–104. Washington, DC: IEEE.

Hoch, M. (1998). A prototype system for intuitive Film planning, in *Automatic Face and Gesture Recognition 1998*, 14–16 April 1998, Nara, Japan, pp. 504–509. Washington, DC: IEEE.

Hopf, K., Chojecki, P., Neumann, F., and Przewozny, D. (2006). Novel autostereoscopic single-user displays with user interaction, in *Proceedings of SPIE 2006, Volume 6392*, 15–19 January 2006, San Jose, CA. Society of Photo-Optical Instrumentation.

Hue, C., Le Cadre, J., and Perez, P. (2002). Sequential Monte Carlo methods for multiple target tracking and data fusion. *IEEE Transactions on Signal Processing* 50: 309–325.

Huttenlocher, D., Klanderman, G., and Rucklidge, W. (1993). Comparing images using the Hausdorff distance. *IEEE Transactions on Pattern Analysis and Machine Intelligence* 15: 850–863.

Imagawa, K., Lu, S., and Igi, S. (1998). Color-based hands tracking system for sign language recognition, in *International Conference on Face and Gesture Recognition 1998*, 14–16 April 1998, Nara, Japan, pp. 462–467. Washington, DC: IEEE.

Ionescu, B., Coquin, D., Lambert, P., and Buzuloiu, V. (2005). Dynamic hand gesture recognition using the skeleton of the hand. *EURASIP Journal on Applied Signal Processing* 2005: 2101–2109.

Isard, M. and Blake, A. (1996). Contour tracking by stochastic propagation of conditional density, in *Proceedings of European Conference on Computer Vision 1996*, 14–18 April 1996, Cambridge, U.K., pp. 343–356. London: Springer-Verlag.

Isard, M. and Blake, A. (1998a). CONDENSATION: Conditional density propagation for visual tracking. *International Journal of Computer Vision* 29: 5–28.

Isard, M. and Blake, A. (1998b). ICONDENSATION: Unifying low-level and high-level tracking in a stochastic framework, in *Proceedings of European Conference on Computer Vision 1998*, 2–6 June 1998, Freiburg, Germany, pp. 893–908. London: Springer-Verlag.

Isard, M. and Blake, A. (1998c). A mixed-state condensation tracker with automatic model-switching, in *Proceedings of International Conference on Computer Vision*, 4–7 January 1998, Bombay, India, pp. 107–112. Washington, DC: IEEE.

Isard, M. and McCormick, J. (2001). Bramble: A Bayesian multiple-blob tracker, in *Proceedings of International Conference on Computer Vision, Volume 2*, pp. 34–41. Washington, DC: IEEE.

Jebara, T. and Pentland, A. (1997). Parameterized structure from motion for 3D adaptive feedback tracking of faces, in *Proceedings of IEEE Computer Vision and Pattern Recognition*, 17–19 June 1997, San Juan, Puerto Rico, pp. 144–150. Washington, DC: IEEE.

Jebara, T., Russel, K., and Pentland, A. (1997). Mixture of eigenfeatures for real-time structure from texture, in *Proceedings of International Conference on Computer Vision*, 2 January 1998, Bombay, India, pp. 128–135. Washington, DC: IEEE.

Jennings, C. (1999). Robust finger tracking with multiple cameras, in *IEEE Workshop on Recognition, Analysis and Tracking of Faces and Gestures in Real-Time Systems 1999*, 26–27 September 1999, Corfu, Greece, pp. 152–160. Washington, DC: IEEE.

Jones, M. J. and Rehg, J. M. (2002). Statistical color models with application to skin detection. *International Journal of Computer Vision* 46: 81–96.

Ju, S., Black, M., Minneman, S., and Kimber, D. (1997). Analysis of gesture and action in technical talks for video indexing, in *Proceedings of IEEE Computer Vision and Pattern Recognition*, 17–19 June 1997, San Juan, Puerto Rico, pp. 595–601. Washington, DC: IEEE.

Kakadiaris, I., Metaxas, D., and Bajcsy, R. (1994). Active part decomposition, shape and motion estimation of articulated objects: A physics-based approach, in *Proceedings of IEEE Computer Vision and Pattern Recognition*, 21–23 June 1994, Seattle, pp. 980–984. Washington, DC IEEE.

Kalman, R. E. (1960). A new approach to linear filtering and prediction problems. *Transactions of the ASME—Journal of Basic Engineering* 82:35–42.

Kampmann, M. (1998). Segmentation of a head into face, ears, neck and hair for knowledge-based analysis-synthesis coding of video-phone sequences, in *Proceedings of International Conference on Image Processing, Volume 2*, 4–7 October 1998, Chicago, pp. 876–880. Washington, DC: IEEE.

Kang, S. and Ikeuchi, K. (1991). *A Framework for Recognizing Grasps*. Technical Report CMU-RI-TR-91-24, Robotics Institute, Carnegie Mellon University, Pittsburgh.

Kang, S. and Ikeuchi, K. (1993). Toward automatic robot instruction for perception: Recognizing a grasp from observation. *IEEE Transactions on Robotics and Automation* 9: 432–443.

Kawashima, H. and Matsuyama, T. (2003). Multi-viewpoint gesture recognition by an integrated continuous state machine. *Systems and Computers in Japan* 34: 1–12.

Kim, S., Kim, N., Ahn, S., and Kim, H. (1998). Object oriented face detection using range and color information, in *IEEE International Conference on Automatic Face and Gesture Recognition 1998*, 14–16 April 1998, Nara, Japan, pp. 76–81. Washington, DC: IEEE.

Kim, W. and Lee, J. (2001). Visual tracking using snake for object's discrete motion, in *IEEE International Conference on Robotics and Automation 2001, Volume 3*, 21–26 May 2001, Seoul, Korea, pp. 2608–2613. Washington, DC: IEEE.

Kjeldsen, R. and Kender, J. (1996). Finding skin in color images, in *IEEE International Conference Automatic Face and Gesture Recognition 1996*, 14–16 October 1996, Killington, VT, pp. 312–317. Washington, DC: IEEE.

Kohler, M. (1997). System architecture and techniques for gesture recognition in unconstrained environments, in *International Conference on Virtual Systems and MultiMedia 1997, Volume 10–12*, 10–12 September 1997, Geneva, Switzerland, pp. 137–146. Washington, DC: IEEE.

Koller-Meier, E. and Ade, F. (2001). Tracking multiple objects using the condensation algorithm. *Journal of Robotics and Autonomous Systems* 34: 93–105.

Kortenkamp, D., Huber, E., and Bonasso, R. (1996). Recognizing and interpreting gestures on a mobile robot, in *National Conference on Artificial Intelligence 1996*, 4–8 August 1996, Portland, OR, pp. 915–921. Cambridge, MA: AAAI Press/MIT Press.

Krueger, M. (1991). *Artificial Reality II*. Reading, MA: Addison-Wesley.

Krueger, M. (1993). Environmental technology: Making the real world virtual. *Communications of the ACM* 36: 36–37.

Krueger, W. and Froehlich, B. (1994). The responsive workbench. *IEEE Computer Graphics and Applications* 14: 12–15.

Kuch, J. and Huang, T. (1995). Vision based hand modeling and tracking for virtual teleconferencing and telecollaboration, in *Proceedings of International Conference on Computer Vision)*, 20–23 June 1995, Cambridge, MA, pp. 666–671. Washington, DC: IEEE.

Kurata, T., Okuma, T., Kourogi, M., and Sakaue, K. (2001). The hand mouse: GMM hand-color classification and mean shift tracking, in *International Workshop on Recognition, Analysis and Tracking of Faces and Gestures in Real-time Systems 2001*, 13 July 2001, Vancouver, Canada, pp. 119–124. Washington, DC: IEEE.

Lanitis, A., Taylor, T., Cootes, C., and Ahmed, T. (1995). Automatic interpretation of human faces and hand gestures using flexible models, in *Proceedings of International Conference on Automatic Face and Gesture Recognition*, September 1995, Zurich, Switzerland, pp. 98–103. Washington, DC: IEEE.

Laptev, I. and Lindeberg, T. (2001). Tracking of multi-state hand models using particle filtering and a hierarchy of multi-scale image features, in *Proceedings of Third International Conference on Scale-Space and Morphology in Computer Vision*, 7–8 July 2001, Vancouver, Canada, pp. 63–74. London: Springer-Verlag.

Lee, H.-K. and Kim, J. H. (1999). An HMM-based threshold model approach for gesture recognition. *IEEE Transactions on Pattern Analysis and Machine Intelligence* 21: 961–973.

Lee, J. and Kunii, T. L. (1995). Model-based analysis of hand posture. *IEEE Computer Graphics and Applications* 15: 77–86.

Leonardis, A. and Bischof, H. (1996). Dealing with occlusions in the eigenspace approach, in *Proceedings of IEEE Computer Vision and Pattern Recognition*, 18–20 June 1996, San Francisco, pp. 453–458. Washington, DC: IEEE.

Li, S. and Zhang, H. (2004). Multi-view face detection with float-boost. *IEEE Transactions on Pattern Analysis and Machine Intelligence* 26: 1112–1123.

Lin, J., Wu, Y., and Huang, T. S. (2002). Capturing human hand motion in image sequences, in *Proceedings of IEEE Workshop on Motion and Video Computing 2002*, 5–6 December 2002, Washington, DC, pp. 99–104. Washington, DC: IEEE.

Lockton, R. and Fitzgibbon, R. (2002). Real-time gesture recognition using deterministic boosting, in *Proceedings of British Machine Vision Conference*, September 2002, Cardiff, U.K., pp. 817–826. London: British Machine Vision Association.

MacLean, W., Herpers, R., Pantofaru, C., Wood, C., Derpanis, K., Topalovic, D., and Tsotsos, J. (2001). Fast hand gesture recognition for real-time teleconferencing applications, in *International Workshop on Recognition, Analysis and Tracking of Faces and Gestures in Real-time Systems 2001*, 13 July 2001, Vancouver, Canada, pp. 133–140. Washington, DC: IEEE.

Maes, P., Darrell, T., Blumberg, B., and Pentland, A. (1995). The alive system: Full-body interaction with autonomous agents, in *Computer Animation Conference 1995*, 19–21 April 1995, Geneva, Switzerland, pp. 11–18. Washington, DC: IEEE.

Maggioni, C. (1995). GestureComputer: New ways of operating a computer, in *International Workshop on Automatic Face and Gesture Recognition*, September 1995, Zurich, Switzerland, pp. 166–171. Washington, DC: IEEE.

Mammen, J. P., Chaudhuri, S., and Agrawal, T. (2001). Simultaneous tracking of both hands by estimation of erroneous observations, in *Proceedings of British Machine Vision Conference*,

10–13 September 2001, Manchester, U.K., pp. 83–92. London: British Machine Vision Association.

Mapes, D. J. and Moshell, M. J. (1995). A two-handed interface for object manipulation in virtual environments. *PRESENSE: Tele-operators and Virtual Environments* 4: 403–416.

Martin, J. and Crowley, J. (1997). An appearance-based approach to gesture-recognition, in *International Conference on Image Analysis and Processing 1997*, 17–19 September 1997, Florence, Italy, pp. 340–347. London: British Machine Vision Association.

Martin, J., Devin, V., and Crowley, J. (1998). Active hand tracking, in *IEEE Conference on Automatic Face and Gesture Recognition 1998*, 14–16 April, Nara, Japan, pp. 573–578. Washington, DC: IEEE.

Martinez, A., Wilbur, B., Shay, R., and Kak, A. (2002). Purdue RVL-SLLL ASL database for automatic recognition of American sign language, in *International Conference on Multimodal Interfaces 2002*, 14–16 October 2002, Pittsburgh, pp. 167–172. Washington, DC: IEEE.

McAllister, G., McKenna, S., and Ricketts, I. (2000). Towards a non-contact driver-vehicle interface, in *Proceedings of IEEE Intelligent Transportation Systems 2000*, 1–3 October 2000, Dearborn, MN, pp. 58–63. Washington, DC: IEEE.

McCormick, J. and Blake, A. (1999). A probabilistic exclusion principle for tracking multiple objects, in *Proceedings of International Conference on Computer Vision*, 20–23 June 1999, Cambridge, MA, pp. 572–578. Washington, DC: IEEE.

McCormick, J. and Isard, M. (2000). Partitioned sampling, articulated objects, and interface-quality hand tracking, in *Proceedings of European Conference on Computer Vision 2000*, 26 June–1 July 2000, Dublin, Ireland, pp. 3–19. Washington, DC: IEEE.

McKenna, S., Raja, Y., and Gong, S. (1999). Tracking color objects using adaptive mixture models. *Image and Vision Computing* 17: 225–231.

Meyering, A. and Ritter, H. (1992). Learning to recognize 3D-hand postures from perspective pixel images, in *Artificial Neural Networks II*, 7-11 June 1992, Baltimore, MD, pp. 821–824. Amsterdam, The Netherlands: Elsevier Science Publishers.

Moghaddam, B. and Pentland, A. (1995). Maximum likelihood detection of faces and hands, in *International Conference on Automatic Face and Gesture Recognition*, September 1995, Zurich, Switzerland, pp. 122–128. Washington, DC: IEEE.

Morguet, P. and Lang, M. K. (1997). A universal HMM-based approach to image sequence classification, in *Proceedings of International Conference on Image Processing*, 26–29 October 1997, Washington, DC, pp. 146–149. Washington, DC: IEEE.

Mori, G. and Malik, J. (2002). Estimating human body configurations using shape context matching, in *Proceedings of European Conference on Computer Vision 2002, Volume 3*, May 2002, Copenhagen, Denmark, pp. 666–680. Washington, DC: IEEE.

Murase, H. and Nayar, S. (1995). Visual learning and recognition of 3d objects from appearance. *International Journal of Computer Vision* 14: 5–24.

Nolker, C. and Ritter, H. (1998). Illumination independent recognition of deictic arm postures, in *Annual Conference of the IEEE Industrial Electronics Society 1998*, 31 August–4 September 1998, Aachen, Germany, pp. 2006–2011. Washington, DC: IEEE.

O'Hagan, R. and Zelinsky, A. (1997). Finger Track: A robust and real-time gesture interface, in *Australian Joint Conference on Artificial Intelligence 1997*, 23–29 August 1997, Aichi, Japan, pp. 475–484. London: Springer-Verlag.

Olson, C. and Huttenlocher, D. (1997). Automatic target recognition by matching oriented edge pixels. *IEEE Transactions on Image Processing* 6: 103–113.

Ong, E. and Bowden, R. (2004). A boosted classifier tree for hand shape detection, in *Automatic Face and Gesture Recognition*, 17–19 May 2004, Seoul, Korea, pp. 889–894. Washington, DC: IEEE.

Ong, S. C. W. and Ranganath, S. (2005). Automatic sign language analysis: A survey and the future beyond lexical meaning. *IEEE Transactions on Pattern Analysis and Machine Intelligence* 27: 873–891.

Pavlovic, V., Sharma, R., and Huang, T. (1996). Gestural interface to a visual computing environment for molecular biologists, in *International Conference Automatic Face and Gesture Recognition (FG 1996)*, 14–16 October 1996, Killington, VT, pp. 30–35. Washington, DC: IEEE.

Perez, P., Hue, C., Vermaak, J., and Gangnet, M. (2002). Color-based probabilistic tracking, in *Proceedings of the European Conference on Computer Vision 2002*, May 2002, Copenhagen, Denmark, pp. 661–675. London: Springer-Verlag.

Peterfreund, N. (1999). Robust tracking of position and velocity with Kalman snakes. *IEEE Transactions on Pattern Analysis and Machine Intelligence* 10: 564–569.

Quek, F. (1995). Eyes in the interface. *Image and Vision Computing* 13: 511–525.

Quek, F. (1996). Unencumbered gesture interaction. *IEEE Multimedia* 3: 36–47.

Quek, F. (2000). Gesture, speech, and gaze cues for discourse segmentation, in *Proceedings of IEEE Computer Vision and Pattern Recognition*, 13–15 June 2000, Hilton Head, SC, pp. 247–254. Washington, DC: IEEE.

Quek, F., Mysliwiec, T., and Zhao, M. (1995). Finger mouse: A freehand pointing interface, in *IEEE International Workshop on Automatic Face and Gesture Recognition 1995*, 17–19 May 1995, Seoul, Korea, pp. 372–377. Washington, DC: IEEE.

Quek, F. and Zhao, M. (1996). Inductive learning in hand pose recognition, in *IEEE Automatic Face and Gesture Recognition 1996*, 14–16 October 1996, Killington, VT, pp. 78–83. Washington, DC: IEEE.

Raja, S. and Gong, S. (1998). Tracking and segmenting people in varying lighting conditions using colour, in *International Conference on Automatic Face and Gesture Recognition 1998*, 14–16 April 1998, Nara, Japan, pp. 228–233. Washington, DC: IEEE.

Raja, Y., McKenna, S., and Gong, S. (1998). Colour model selection and adaptation in dynamic scenes, in *Proceedings of European Conference on Computer Vision 1998*, 2–6 June 1998, Freiburg, Germany, pp. 460–475. London: Springer-Verlag.

Ranganath, S. and Arun, K. (1997). Face recognition using transform features and neural networks. *Pattern Recognition* 30: 1615–1622.

Rehg, J. and Kanade, T. (1994). Digiteyes: Vision-based hand tracking for human-computer interaction, in *Workshop on Motion of Non-Rigid and Articulated Bodies 1994*, 11–12 November 1994, Austin, TX, pp. 16–24. Washington, DC: IEEE.

Rehg, J. and Kanade, T. (1995). Model-based tracking of self-occluding articulated objects, in *Proceedings of the International Conference on Computer Vision*, 20–23 June 1995, Cambridge, MA, pp. 612–617. Washington, DC: IEEE.

Rigoll, G., Kosmala, A., and Eickeler, S. (1998). High performance real-time gesture recognition using hidden Markov models, in *Proceedings of the International Gesture Workshop on Gesture and Sign Language in Human-Computer Interaction 1998*, 17–19 September 1998, Bielefeld, Germany, pp. 69–80. London: Springer-Verlag.

Rigoll, G., Kosmala, A., and Schusterm, M. (1996). A new approach to video sequence recognition based on statistical methods, in *Proceedings of the International Conference on Image Processing*, *Volume 3*, September 1996, Lausanne, Switzerland, pp. 839–842. Washington, DC: IEEE.

Rosales, R., Athitsos, V., Sigal, L., and Sclaroff, S. (2001). 3D hand pose reconstruction using specialized mappings, in *Proceedings of the International Conference on Computer Vision*, 7–14 July 2001, Vancouver, Canada, pp. 378–385. Washington, DC: IEEE.

Saxe, D. and Foulds, R. (1996). Toward robust skin identification in video images, in *IEEE International Conference on Automatic Face and Gesture Recognition 1996*, 14–16 October 1996, Killington, VT, pp. 379–384. Washington, DC: IEEE.

Schapire, R. (2003). The boosting approach to machine learning: An overview, in *MSRI Workshop on Nonlinear Estimation and Classification 2002*, *Volume 171*, 19–29 March 2003, Berkeley, CA, pp. 149–172. Springer-Verlag.

Schapire, R. and Singer, Y. (1998). Improved boosting algorithms using confidence-rated predictions. *Machine Learning* 37:297–336.

Schlenzig, J., Hunter, E., and Jain, R. (1994a). Recursive identification of gesture inputs using hidden Markov models, in *IEEE Workshop on Applications of Computer Vision 1994*, 5–7 December 1994, Sarasota, FL, pp. 187–194. Washington, DC: IEEE.

Schlenzig, J., Hunter, E., and Jain, R. (1994b). Vision based hand gesture interpretation using recursive estimation, in *Asilomar Conference Signals, Systems, and Computers 1994*,

Volume 2, 31 October–2 November 1994, Pacific Grove, CA, pp. 1267–1271. Washington, DC: IEEE.

Segen, J. and Kumar, S. (1998). Fast and accurate 3D gesture recognition interface, in *Proceedings of International Conference on Pattern Recognition,* 17–20 August 1998, Queensland, Australia, pp. 86–91. Washington, DC: IEEE.

Segen, J. and Kumar, S. S. (1999). Shadow gestures: 3D hand pose estimation using a single camera, in *Proceedings of IEEE Computer Vision and Pattern Recognition,* 23–25 June 1999, Ft. Collins, CO, pp. 479–485. Washington, DC: IEEE.

Shamaie, A. and Sutherland, A. (2005). Hand tracking in bimanual movements. *Image and Vision Computing* 23: 1131–1149.

Sheridan, T. (1993). Space teleoperation through time delay: Review and prognosis. *IEEE Transactions on Robotics and Automation* 9: 592–606.

Shimada, N., Kimura, K., and Shirai, Y. (2001). Real-time 3-d hand posture estimation based on 2-d appearance retrieval using monocular camera, in *International Workshop on Recognition, Analysis and Tracking of Faces and Gestures in Real-time Systems 2001,* 13 July 2001, Vancouver, Canada, pp. 23–30. Washington, DC: IEEE.

Shimada, N., Shirai, Y., Kuno, Y., and Miura, J. (1998). Hand gesture estimation and model refinement using monocular camera: Ambiguity limitation by inequality constraints, in *IEEE International Conference on Face and Gesture Recognition 1998,* 14–16 April 1998, Nara, Japan, pp. 268–273.

Siegl, H., Hanheide, M., Wrede, S., and Pinz, A. (2007). An augmented reality human-computer interface for object localization in a cognitive vision system. *Image and Vision Computing* 25:1895–1903.

Sigal, L., Sclaroff, S., and Athitsos, V. (2004). Skin color-based video segmentation under time-varying illumination. *IEEE Transaction Pattern Analysis and Machine Intelligence* 26: 862–877.

Sirovich, L. and Kirby, M. (1987). Low-dimensional procedure for the characterization of human faces. *Journal of the Optical Society of America* 4: 519–524.

Song, L. and Takatsuka, M. (2005). Real-time 3D finger pointing for an augmented desk, in *Australasian Conference on User Interface, Volume 40,* 16–19 January 2005, Tasmania, Australia, pp. 99–108. Darlinghurst, Australia: Australian Computer Society.

Starner, T. and Pentland, A. (1995). Visual recognition of American sign language using hidden Markov models, in *IEEE International Symposium on Computer Vision,* 19–21 November 1995, Coral Gables, FL, pp. 265–270. Washington, DC: IEEE.

Starner, T., Weaver, J., and Pentland, A. (1998). Real-time American sign language recognition using desk and wearable computer-based video. *IEEE Transactions on Pattern Analysis and Machine Intelligence* 20: 1371–1375.

Stauffer, C., Eric, W., and Grimson, L. (1999). Adaptive background mixture models for real-time tracking, in *Proc. IEEE Computer Vision and Pattern Recognition (CVPR 1999),* 23–25 June 1999, Ft. Collins, CO, pp. 2246–2252. Washington, DC: IEEE.

Stenger, B., Mendonca, R., and Cippola, R. (2004). Model-based 3D tracking of an articulated hand, in *Proceedings of IEEE Computer Vision and Pattern Recognition (CVPR 2004),* 27 June–2 July 2004, Washington, DC, pp. 126–133.

Stenger, B., Thayananthan, A., Torr, P., and Cipolla, R. (2006). Model-based hand tracking using a hierarchical Bayesian filter. *IEEE Transactions on Pattern Analysis and Machine Intelligence* 28: 1372–1384.

Sullivan, J. and Carlsson, S. (2002). Recognizing and tracking human action, in *Proceedings of European Conference on Computer Vision 2002, Volume 1,* May 2002, Copenhagen, Denmark, pp. 629–644. London: Springer-Verlag.

Tanibata, N., Shimada, N., and Shirai, Y. (2002). Extraction of hand features for recognition of sign language words, in *International Conference on Vision Interface 2002,* 27–29 May 2002, Calgary, Canada, pp. 391–398.

Terrillon, J., Shirazi, M., Fukamachi, H., and Akamatsu, S. (2000). Comparative performance of different skin chrominance models and chrominance spaces for the automatic detection of human faces in color images, in *Proceedings of International Conference on Automatic Face and Gesture Recognition,* 28–30 March 2000, Grenoble, France, pp. 54–61. Washington, DC: IEEE.

Terzopoulos, D. and Szeliski, R. (1992). Tracking with Kalman snakes, in *Active Vision* (A. Blake and A. Yuille, eds.), pp. 3–20. Cambridge MA: MIT Press.

Tomasi, C., Petrov, S., and Sastry, A. (2003). 3D tracking = classification + interpolation, in *Proceedings of International Conference on Computer Vision, Volume 2,* 13–16 October 2003, Nice, France, pp. 1441–1448. Washington, DC: IEEE.

Triesch, J. and von der Malsburg, C. (1996). Robust classification of hand postures against complex background, in *IEEE Automatic Face and Gesture Recognition 1996,* 14–16 October 1996, Killington, VT, pp. 170–175. Washington, DC: IEEE.

Triesch, J. and von der Malsburg, C. (1998). A gesture interface for human-robot-interaction, in *Proceedings of International Conference on Automatic Face and Gesture Recognition,* April 1998, Nara, Japan, pp. 546–551.

Turk, M. and Pentland, A. (1991). Eigenfaces for recognition. *Journal of Neuroscience* 3: 71–86.

Uras, C. and Verri, A. (1995). Hand gesture recognition from edge maps, in *International Workshop on Automatic Face and Gesture Recognition 1995,* 26–28 June 1995, Zurich, Switzerland, pp. 116–121. Washington, DC: IEEE.

Utsumi, A. and Ohya, J. (1998a). Direct manipulation interface using multiple cameras for hand gesture recognition, in *Proceedings of the IEEE International Conference on Multimedia Computing and Systems,* 28 Jun–1 Jul 1998, Austin, Texas, pp. 264–267. Washington DC: IEEE.

Utsumi, A. and Ohya, J. (1998b). Image segmentation for human tracking using sequential-image-based hierarchical adaptation, in *Proceedings of IEEE Computer Vision and Pattern*

Recognition, 23–25 June 1998, Santa Barbara, CA, pp. 911–916.

Utsumi, A. and Ohya, J. (1999). Multiple-hand-gesture tracking using multiple cameras, in *Proceedings of IEEE Computer Vision and Pattern Recognition*, 23–25 June 1999, Ft. Collins, CO, pp. 473–478. Washington, DC: IEEE.

Vaillant, R. and Darmon, D. (1995). Vision-based hand pose estimation, in *International Workshop on Automatic Face and Gesture Recognition 1995*, 26–28 June 1995, Zurich, Switzerland, pp. 356–361. Washington, DC: IEEE.

Vermaak, J., Perez, P., Gangnet, M., and Blake, A. (2002). Towards improved observation models for visual tracking: Selective adaptation, in *Proceedings of European Conference on Computer Vision 2002*, May 2002, Copenhagen, Denmark, pp. 645–660. London: Springer-Verlag.

Viola, P. and Jones, M. (2001). Robust real-time object detection, in *IEEE Workshop on Statistical and Computational Theories of Vision 2001*, July 2001, Vancouver, Canada, pp. 1–25. Washington, DC: IEEE.

Viola, P., Jones, M., and Snow, D. (2003). Detecting pedestrians using patterns of motion and appearance, in *Proceedings of International Conference on Computer Vision*, 13–16 October 2003, Nice, France, pp. 734–741. Washington, DC: IEEE.

Vogler, C. and Metaxas, D. (1998). ASL recognition based on a coupling between HMMs and 3D motion analysis, in *Proceedings of International Conference on Computer Vision*, 4–7 January 1998, Bombay, India, pp. 363–369. Washington, DC: IEEE.

Vogler, C. and Metaxas, D. (1999). Toward scalability in ASL recognition: Breaking down signs into phonemes, in *International Gesture Workshop on Gesture-Based Communication in Human-Computer Interaction 1999*, 17–19 March 1999, Gif-Sur-Yvette, France, pp. 211–224. London: Springer-Verlag.

Vogler, C. and Metaxas, D. (2001). A framework for recognizing the simultaneous aspects of American sign language. *Computer Vision and Image Understanding* 81: 358–384.

Wachs, J., Kartoun, U., Stern, H., and Edan, Y. (2002). Real-time hand gesture telerobotic system, in *World Automation Congress 2002, Volume 13*, 9–13 June 2002, Orlando, FL, pp. 403–409. Washington, DC: IEEE.

Waldron, M. (1995). Isolated ASL sign recognition system for deaf persons. *IEEE Transactions on Rehabilitation Engineering* 3: 261–271.

Wang, J., Athitsos, V., Sclaroff, S., and Betke, M. (2008). Detecting objects of variable shape structure with hidden state shape models. *IEEE Transactions on Pattern Analysis and Machine Intelligence* 30: 477–492.

Wellner, P. (1993). The DigitalDesk calculator: Tangible manipulation on a desk top display, in *ACM Symposium on User Interface Software and Technology 1993*, 3–5 November 1993, Atlanta, pp. 27–33. New York: ACM Press.

Wilson, A. and Bobick, A. (1995). Learning visual behavior for gesture analysis, in *IEEE Symposium on Computer Vision 1995*, 19–21 November 1995, Coral Gables, FL, pp. 229–234. Washington, DC: IEEE.

Wilson, A. and Oliver, N. (2003). Gwindows: Robust stereo vision for gesture-based control of windows, in *International Conference on Multimodal Interfaces 2003*, 5–7 November 2003, Vancouver, Canada, pp. 211–218. New York: ACM Press.

Wren, C. R., Azarbayejani, A., Darrell, T., and Pentland, A. (1997). PFinder: Real-time tracking of the human body. *IEEE Transactions on Pattern Analysis and Machine Intelligence* 19: 780–785.

Wren, C. and Pentland, A. (1997). Dynamic models of human motion, in *IEEE International Conference of Automatic Face and Gesture Recognition 1997*, 14–16 April 1997, Nara, Japan, pp. 22–27. Washington, DC: IEEE.

Wu, A., Hassan-Shafique, K., Shah, M., and da Vitoria Lobo, N. (2004). Virtual three-dimensional blackboard: Three-dimensional finger tracking with a single camera. *Applied Optics* 43: 379–390.

Wu, Y. and Huang, T. T. (1999). Capturing human hand motion: A divide-and-conquer approach, in *Proceedings of International Conference on Computer Vision*, 20–25 September 1999, Corfu, Greece, pp. 606–611. Washington, DC: IEEE.

Wu, Y. and Huang, T. S. (2000). View-independent recognition of hand postures, in *Proceedings of IEEE Computer Vision and Pattern Recognition, Volume 2*, 13–15 June 2000, Hilton Head, SC, pp. 84–94. Washington, DC: IEEE.

Wu, Y., Lin, J., and Huang, T. S. (2005). Analyzing and capturing articulated hand motion in image sequences. *IEEE Transactions on Pattern Analysis and Machine Intelligence* 27: 1910–1922.

Wu, Y., Liu, Q., and Huang, T. (2000). An adaptive self-organizing color segmentation algorithm with application to robust real-time human hand localization, in *ACCV 2000*, 8–11 January 2000, Taipei, pp. 1106–1111. London: Springer-Verlag.

Wu, Y., Liu, Q., and Huang, T. (2001). Capturing natural hand articulation, in *Proceedings International Conference on Computer Vision*, 7–14 July 2001, Vancouver, Canada, pp. 426–432. Washington, DC: IEEE.

Yamato, J., Ohya, J., and Ishii, K. (1992). Recognizing human action in time-sequential images using hidden Markov model, in *Proceedings of IEEE Computer Vision and Pattern Recognition*, 15–18 June 1992, Urbana-Champaign, IL, pp. 379–385. Washington, DC: IEEE.

Yang, J., Lu, W., and Waibel, A. (1998). Skin-color modeling and adaptation, in *ACCV 1998*, 8–10 January 1998, Hong Kong, pp. 687–694. Washington, DC: IEEE.

Yang, M. and Ahuja, N. (1998). Detecting human faces in color images, in *Proceedings of International Conference on Image Processing*, 4–7 October 1998, Chicago, pp. 127–130. Washington, DC: IEEE.

Yang, M. H., Ahuja, N., and Tabb, M. (2002). Extraction of 2D motion trajectories and its application to hand gesture recognition. *IEEE Transactions on Pattern Analysis and Machine Intelligence* 24: 1061–1074.

Yin, X. and Xie, M. (2003). Estimation of the fundamental matrix from uncalibrated stereo hand images for 3D hand gesture recognition. *Pattern Recognition* 36: 567–584.

Yoon, S. M. and Kim, H. (2004). Real-time multiple people detection using skin color, motion and appearance information, in *Proceedings of IEEE International Workshop on Robot and Human Interactive Communication*, September 2004, Kurashiki, Japan, pp. 331–334. Washington, DC: IEEE.

Yuan, Q., Sclaroff, S., and Athitsos, V. (1995). Automatic 2D hand tracking in video sequences, in *IEEE Workshop on Applications of Computer Vision 1995*, February 1995, Austin, TX, pp. 250–256. Washington, DC: IEEE.

Zeller, M., Phillips, C., Dalke, A., Humphrey, W., Schulten, K., Huang, S., et al. (1997). A visual computing environment for very large scale biomolecular modeling, in *IEEE International Conference Application-Specific Systems, Architectures and Processors*, 23–25 July 1994, Samos, Greece, pp. 3–12. Washington, DC: IEEE.

Zhou, J. P. and Hoang, J. (2005). Real time robust human detection and tracking system, in *Proceedings of IEEE Computer Vision and Pattern Recognition, Volume III*, 20–25 June 2005, San Diego, pp. 149–149. Washington, DC: IEEE.

Zhu, X., Yang, J., and Waibel, A. (2000). Segmenting hands of arbitrary color, in *Proceedings of International Conference on Automatic Face and Gesture Recognition*, 28–30 March 2000, Grenoble, France, pp. 446–455. Washington, DC: IEEE.

Zobl, M., Nieschulz, R., Geiger, M., Lang, M., and Rigoll, G. (2004). Gesture components for natural interaction with in-car devices, in *International Gesture Workshop in Gesture-Based Communication in Human-Computer Interaction*, 15–17 April 2004, Genoa, Italy, pp. 448–459. Berlin/Heidelberg: Springer-Verlag.

Automatic Hierarchical Scanning for Windows Applications

Stavroula Ntoa,
Anthony Savidis, and
Constantine Stephanidis

35.1 Introduction

Computer users with motor impairments of upper limbs often face difficulties in accessing interactive applications and services using standard input devices. Several technological approaches attempt to deal with the issues raised by motor impaired users' interaction with graphical environments, each addressing a different user category according to their individual characteristics and abilities. One of these methods is scanning, which is mainly targeted to users with poor muscle control, weakness, fatigue, or severe mobility problems in general.

This chapter discusses the automatic hierarchical scanning technique for Windows applications, a method that enables motor-impaired users to work with any application running in Microsoft Windows, by using only a binary switch as an input device, thus overcoming problems related to the inability to use traditional input devices, such as the keyboard or the mouse. The sections that follow describe the scanning technique in detail, present a review of scanning approaches in research and industry, and introduce the automatic hierarchical scanning technique through the case study of a scanning tool named FastScanner.

35.2 The Scanning Technique

Scanning is an interaction method addressing the needs of users with hand motor impairments. The main concept behind this technique is to eliminate the need for interacting with a computer application using traditional input devices, such as with a mouse and keyboard. Therefore, all the interactive objects composing a graphical user interface are being sequentially focused and highlighted (e.g., by a colored marker), while users can select to interact with the object currently having the focus by activating a switch. To eliminate the need for using a keyboard to type in text, an onscreen keyboard is usually provided.

During scanning, the focus marker scans the interface and highlights interactive objects sequentially, in a predefined order (e.g., from top to bottom and from left to right). Furthermore, scanning can be either automatic or manual. In the first case, the marker automatically moves from one interface element to the next after a predefined time interval of user inactivity (i.e., not pressing the activation switch), while the time interval can usually be customized according to user needs. In manual scanning, the user moves the focus marker to the next interface element whenever desired with the use of a switch.

Activation switches can vary from hand, finger, foot, tongue, or head switches to breath-controlled switches or eye-tracking switches (ABLEDATA, 2008; Enabling Devices, 2008). Figure 35.1 presents some indicative switches of the aforementioned types. Furthermore, any keyboard key (e.g., the space key) or mouse click can be used as a switch.

35.3 Review of Scanning Approaches

Scanning approaches can be classified in two main categories:

- Approaches targeted at embedding scanning in the development of software applications
- Approaches targeted at providing scanning techniques compatible for use with existing software applications

The first category follows a proactive philosophy (Stephanidis and Emiliani, 1999) targeted to application developers to create

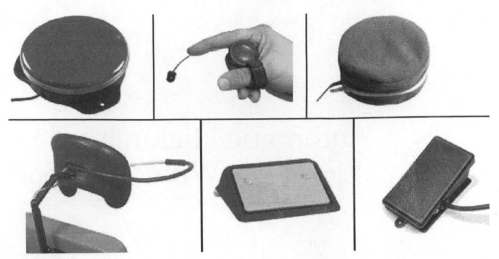

FIGURE 35.1 Types of switches.

accessible products and services without the need for *a posteriori* adaptation. This category can be further classified to:

1. Augmented libraries, enhanced with interaction objects inherently supporting scanning (Savidis et al., 1997). An example of an application based on this approach is discussed in Chapter 44, "Developing Inclusive e-Training," of this handbook. The main drawback of this approach is that application and services developed with such libraries face the problem of becoming obsolete once the next generation of the operating system they have been developed for is introduced.
2. Policy initiatives by mainstream actors, such as Microsoft Accessibility Developer Center (Microsoft Corporation, 2008a) and Sun Microsystems Accessibility Program (Sun Microsystems, 2008a), which facilitate software developers in creating more accessible products.

On the other hand, approaches addressing directly the needs of motor-impaired users themselves can be categorized according to the following criteria:

1. Accessibility features of operating systems
2. Applications with embedded scanning
3. Scanning tools, aiming to offer more generic solutions to users with hand motor impairments

Accessibility features of operating systems aim to make computer use easier for users with disabilities. Since most of them are not directly related to scanning, they are only shortly presented in this chapter for completeness purposes. The features that address the interaction requirements of motor-impaired users include (Linux Online, 2008; Microsoft Corporation, 2008b; Sun Microsystems, 2008b) but are not limited to: key combinations (e.g., Ctrl+Shift) that can be provided by pressing one key at a time rather than simultaneously, mouse cursor move and mouse button operations that can be carried out with the use of the keyboard, brief or repeated keystrokes that can be ignored and the repeat rate adjusted, onscreen keyboard with scanning facilities, options to adjust how fast a mouse button should be clicked for a selection to be made, and more.

Applications with embedded scanning are developed so as to support scanning in the first place and are accessible to people with motor impairments. However, they only partially address the interaction requirements of users. For example, a user needs more than one application to carry out a variety of everyday tasks (e.g., web browser, e-mail client, entertainment software, educational software, document authoring software, etc.). Therefore, users with motor impairments should employ various applications with embedded scanning techniques, possibly facing interoperability issues, and should often update to the latest version of each such application. Furthermore, a major drawback of these approaches is their increased cost of development and maintenance.

Finally, scanning tools enable users to operate the graphical environment of the operating system, eliminating the need for using various specialized applications for carrying out everyday tasks. Keyboard and mouse emulation programs with embedded scanning are popular among scanning tools, since they intend to ensure user interaction without the use of the traditional keyboard and mouse. Several research and industrial efforts have been targeted toward keyboard emulation, suggesting a variety of approaches, such as alphanumeric keyboards (Gnanayutham et al., 2004), chorded keyboards (Lin et al., 2006), keyboards enhanced with mouse emulation options (EnableMart, 2008) or with words and phrases abbreviation (Words+, 2008) and word prediction (WiViK, 2008) features. On the other hand, mouse emulation with scanning support software allows users to control the mouse pointer using one or two switches (Applied Human Factors, 2008; Gus Communications, 2008). Chapter 29 of this handbook discusses keyboard and mouse emulation in detail. Finally, tools that provide scanning of interface elements based on their screen location (Cooper and Associates, 2008) or their place in the hierarchy of objects comprising the active window (Ntoa et al., 2004), provide switch access to motor impaired users.

35.4 Automatic Hierarchical Scanning

The hierarchical scanning method provides access to all the interactive interface elements of the currently active window, based on their place in the hierarchy, by dynamically retrieving the window hierarchical structure. The main advantage of this technique in comparison to other scanning approaches is that it ensures rapid interaction, avoiding time-consuming sequential access to all interactive interface elements.

Issues related to hierarchical scanning are illustrated in this chapter through the case study of the FastScanner tool (Ntoa et al., 2004), which provides access to Microsoft Windows applications without recourse to any subsequent modification, through the use of binary switches as an alternative to traditional input devices. In the following subsections, the FastScanner architecture is presented, the hierarchy retrieval and filtering process is analyzed, the classification of interactive objects into categories of scanning objects is explained, and an example of scanning dialogue is provided. Finally, issues related to the tool's graphical user interfaces as well as evaluation issues are discussed.

35.4.1 FastScanner Architecture

FastScanner retrieves the initial hierarchy of the objects comprising the active application window through the Microsoft Active Accessibility platform, as presented in Figure 35.2. However, since the initial hierarchy retrieved is radically different from the actual hierarchy of interface elements that are visible to the user, a filtering stage is required before establishing the final objects' hierarchy. Once the final hierarchy is determined, the scanning dialogue is initiated, having the window itself as the currently focused and highlighted object.

As shown in Figure 35.2, in the context of the FastScanner tool, all the interactive objects that may compose an application have been classified into categories, and their behavior during

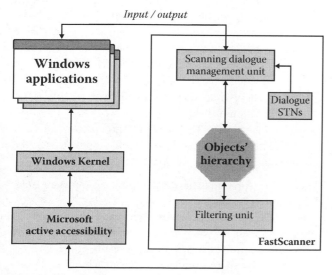

FIGURE 35.2 FastScanner architecture.

a scanning dialogue has been modeled using State Transition Networks (Parnas, 1969; Wasserman, 1985). Furthermore, user input is interpreted by the scanning dialogue management unit according to the type of the object currently in focus, and the corresponding output is produced based on the state transition networks. Finally, when user interaction causes a transformation in the current application hierarchy, the application itself notifies the Active Accessibility platform, through the Windows kernel. The Microsoft Active Accessibility platform in turn notifies the FastScanner application, which reconstructs appropriately the final objects' hierarchy.

35.4.2 Hierarchy Retrieval and Filtering

The objects' tree-structured hierarchy in FastScanner is retrieved through Microsoft Active Accessibility, having as tree root the active application window itself. During this process, once an object is added to the hierarchy, all of its children are examined and if necessary added to the hierarchy. To add an element to the hierarchy, the following properties are retrieved and stored:

- *Name*: It should be noted that an object's name cannot effectively identify it, since there are several nameless window objects. For example:
  - The title bar of a window is a nameless object.
  - The minimize button of the title bar has the name Minimize.
- *Description*: It is a short text providing a brief description of the objects' visual appearance. For example:
  - The description text for a window title bar is as follows: "Displays the name of the window and contains controls to manipulate it."

It should be mentioned that this attribute is empty for many objects as well.

- *Role*: It is a property of all windows objects, explaining their interactive role. For example:
  - The role property for a window title bar is "title bar."
  - The role property for a button is "push button."
- *Value*: This property is used only in some objects, when there is a need for value specification. For example:
  - The value attribute of a scroll bar represents the bar place as a percentage, from 0 (top) to 100 (bottom).
- *Action*: This attribute is empty for noninteractive objects, while for interactive objects it describes the action that a user can perform on them. For example:
  - The action attribute for a button is "Press."
  - The action attribute for a menu is "Open."
- *State*: There are several values for this attribute, describing objects' state. Examples of state values are as follows:
  - "Normal," when no special state should be indicated.
  - "Unavailable," for objects that are not interactive.
  - "Invisible," for objects that are not visible.

- "Selected," for checkboxes that are selected.
- *Coordinates*: Refers to information regarding the screen location of an object by recording the top left corner coordinates as well as the object width and height.
- *Number of children.*
- *Window handle*: This attribute is provided by the operating system and can uniquely identify a window.

It should be noted that the objects comprising the final hierarchical structure are not only the interactive window objects, but container objects as well, whose role is to group the objects contained. Examples of such container objects are windows, group boxes, and frames. Although in graphical user interfaces (GUIs) no user interaction is supported for container objects, in scanning dialogue they are of great importance, as they accelerate interaction by allowing users to directly skip the scanning of large groups of objects.

Additionally, before actually adding an object to the hierarchy, and once its properties have been retrieved, they are examined to determine whether the object should be actually added to the hierarchy. In more detail, the object attributes that are relevant are:

- *Role*: There are several objects that can be directly excluded from the final objects' hierarchy due to their role, which reveals that user interaction is not possible with these objects. Such object roles are: alert, animation, border, caret, character, chart, clock, cursor, diagram, dial, grip, equation, graphic, helpballoon, hotkeyfield, indicator, progressbar, rowheader, separator, slider, sound, statusbar, tooltip, and whitespace. Furthermore, static text is also excluded from the hierarchy, since no user interaction is involved.
- *State*: By assessing an object's state it can be determined whether it is visible and available for interaction. If these two conditions are not satisfied, the object is not added to the final hierarchy.
- *Coordinates*: Several objects, although having a valid role and state, are actually not visible to the user, due to their placement. By inspecting their screen location, all these objects are eliminated from the hierarchy.

Finally, if during the aforementioned checks, it is determined that an object should not be included in the hierarchy, its parent's "number of children" attribute is reduced by one. In the case where the parent node is left without any children at all and it is not an interactive object, then the parent node is also removed from the hierarchy and its parent is recursively updated. In the case where the parent node is not interactive and left with one child it remains in the hierarchy. However, in this case, an extra object attribute is added, indicating that the scanning dialogue should not affect this object, but directly its child.

As a result, after the tree enhancement process, where all the unnecessary nodes are eliminated, the final object hierarchy includes only the necessary interactive elements. To better illustrate the tree enhancement process, an example is presented.

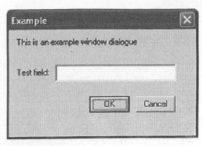

FIGURE 35.3 Example window for hierarchy retrieval.

The initial objects' hierarchy for a window consisting of a text field and two buttons, as presented in Figure 35.3, consists of 142 objects. On the other hand, the final enhanced objects' hierarchy consists of just 10 objects. The initial hierarchy retrieved and the enhanced one are represented in Figures 35.4 and 35.5, respectively.

35.4.3 Interactive Objects' Categories

Interactive objects' classification into categories according to their behavior in scanning dialogue leads to the design and development of state transition networks, modeling objects' interactive behavior. This is a significant unit of the FastScanner architecture, since it allows the effective and efficient management of the scanning dialogue. The objects' categories as defined for the FastScanner application are:

- *Text input objects*: They represent interface elements used to provide text to an application, such as text fields. To address the needs of motor-impaired users, a virtual keyboard supporting scanning has been designed and embedded in FastScanner.
- *Simple objects*: They represent interface elements that are directly associated with a single user action. When using the traditional input devices, interaction with these objects is achieved with a single mouse click. Well-known examples of such elements are buttons.
- *Selection objects*: These objects usually belong to a list or a tree structure, and users need to select the objects before interacting with them. When using the traditional input devices, selection of these objects is achieved with a single mouse click, while interaction with them is achieved with a double mouse click.
- *Container objects*: Their role is to group other interactive elements. When using the traditional input devices, these objects do not entail any user interaction at all.

Table 35.1 represents the most typical examples of objects belonging to each category.

Since each one of the previous object categories is characterized by a different interactive behavior, and to better address the needs of scanning users, a different scanning dialogue technique was designed and implemented for each object category. Nevertheless, all the interaction techniques are based on the

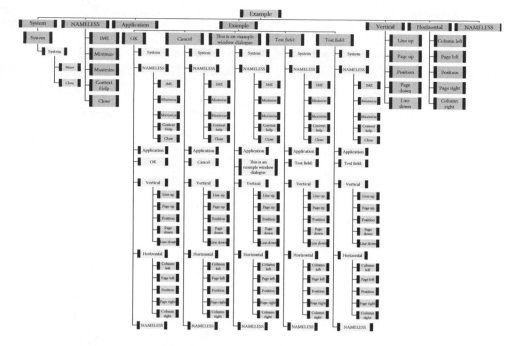

FIGURE 35.4 Initial hierarchical tree of the example window.

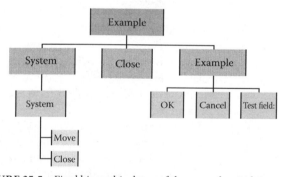

FIGURE 35.5 Final hierarchical tree of the example window.

same interaction principles concerning dialogue states and user actions. In more detail, there are three types of dialogue states that are supported in FastScanner and that are explicitly indicated to the users by the color of the focus marker:

- *Entry state*: It is represented by the green color of the marker and indicates that the object is active and ready to be used.
- *Exit state*: It is represented by the red color of the marker and indicates that the dialogue is ready to move to the next interactive interface element.
- *Selection state*: It is used only in the case of selection objects, such as the items of a list; it is represented by the orange color of the marker and indicates that the enclosed object is ready to be selected.

On the other hand, user input is provided through the use of switches. FastScanner has been designed to support three types of switches, and therefore three types of input commands:

- *Select*: Initiation of the selection action indicates that the user wishes to proceed with the action that the dialogue state indicates. Therefore, if the scanning dialogue state is *entry*, the user will interact with the current object. Otherwise, if the scanning dialogue state is *exit*, the dialogue will move to the next interactive interface element.
- *Next*: Initiation of the next action indicates that the user wishes to change the dialogue state to the next available one.
- *Reverse*: Initiation of the reverse action indicates that the user wishes to reverse scanning order from top-down to bottom-up and vice versa.

Furthermore, the possibility for time scanning has been added to FastScanner to help users interact with a Microsoft Windows application more efficiently. When working in time-scanning mode, users have to initiate only the *select* actions. The *next* action is automatically initiated by the system when a specific time interval elapses without a user action.

Based on these scanning dialogue principles and the interactive properties of each object category, as well as model users' interaction with an application using FastScanner, a state transition network (STN) diagram has been designed and implemented for each object category. All the STN diagrams are represented in Figures 35.6 through 35.9. In general, for all object categories, the dialogue is initiated in *exit* state and the following general principles are followed:

- If the dialogue is in *exit* state and the user initiates the *select* action, then scanning focus moves to the next object in the scanning hierarchy.

TABLE 35.1   Classification of Windows Object Types

Text Input Objects	Simple Objects	Selection Objects	Container Objects
Text field	Drop-down menu buttons	List items	Title bars
Text area	Menu buttons	Combination list items	Menu bars
	Buttons		Toolbars
	Menu items		Scrollbars
	Option buttons		Group boxes
	Checkboxes		List boxes
	Spin buttons		Drop-down list boxes
	Links		Drop-down menus
			Page tab control
			Tables
			Windows
			Dialogue boxes
			Frames

- If the dialogue is in *exit* state and the user initiates the *next* action, then scanning focus remains on the same object and the dialogue state is turned to *entry*. There is only one exception to this principle, in the case of selection objects.
- If the dialogue is in *entry* state and the user initiates the *next* action, then scanning focus remains on the same object and the dialogue state is turned to *exit*.
- If at any time the user initiates the *reverse* command, both the currently focused object and the dialogue state remain unaffected; however, the order in which objects will be scanned is reversed.

The STN dialogue diagrams that follow describe the scanning behavior for each object category. In all cases, the aforementioned principles are followed.

When an *entry* dialogue state occurs at a text entry object, a *select* action sets off the virtual keyboard to allow the user to input the desired text. Once the text entry task is completed, the user closes the virtual keyboard and the focus is set again to the text entry object in *entry* state.

When the scanning dialogue is focused on a simple object at *entry* state, a *select* action activates the simple object, for example, if the object is a button then it is pressed.

Scanning dialogue with selection objects is initiated in the *exit* state and entails one additional state, the *selection* state. If the user triggers the *select* action, then the dialogue moves to the next object in the scanning hierarchy. However, if the user triggers the *next* action, then dialogue state is changed to *selection*. From this state, when the *select* command is issued, then the active object is selected, otherwise if the *next* command is issued the dialogue state changes to *entry*. For example, in the case of the Microsoft Windows Explorer, if a user wishes to select a folder to copy it, then the user should move the scanning focus to the list item representing the folder, set the status to *selection*, and initiate the *select* action. This would correspond to a single mouse click on the folder item. On the other hand, if the user wishes to open the folder to view the contained files, then once the folder list item is in focus, the scanning state should be set to *entry* and the *select* action should be initiated. This would correspond to a double mouse click on the folder item.

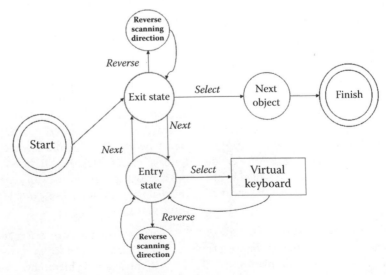

FIGURE 35.6   Dialogue STN for text entry objects.

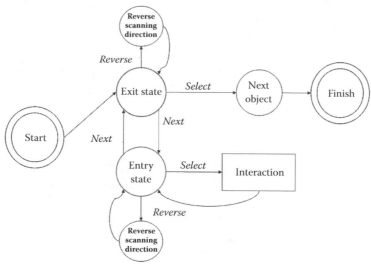

FIGURE 35.7 Dialogue STN for simple objects.

In container objects, if a *select* action is initiated when the dialogue state is *entry*, then the scanning focus moves to the first contained object.

Besides the previously mentioned STNs, some additional ones implementing scanning dialogue for more experienced users were designed as well, aiming to accelerate user interaction. The quick dialogue STNs are presented in Figures 35.10 through 35.13.

The main difference is that quick scanning dialogue does not include the *exit* state at all. In more detail, when scanning focuses on an object, the dialogue is initiated at *entry* state. If the user presses the switch corresponding to the *select* command, then the appropriate interaction according to the object type will be initiated. In the case of text entry objects, interaction entails displaying the virtual keyboard, while in the case of container objects interaction means moving the scanning focus to the first contained object. Since there is no *exit* state, when

the *next* command is provided in *entry* state, the dialogue moves to the next object in the case of text entry, simple and container objects. In the case of selection objects, the dialogue moves to the next available state (i.e., the *selection* state).

In conclusion, the tool supports two types of user input, namely, time-based and manual, and two modes of function, namely, standard and quick scanning. The summary of these operation modes is provided in Table 35.2.

35.4.4 Scanning Dialogue Example

After having discussed the FastScanner object types and how user interaction has been modeled according to the object type, and to better illustrate the functionality of FastScanner, a step-by-step example is presented in this section. The example refers to the scenario of opening the folder "My Documents" using the quick dialogue mode.

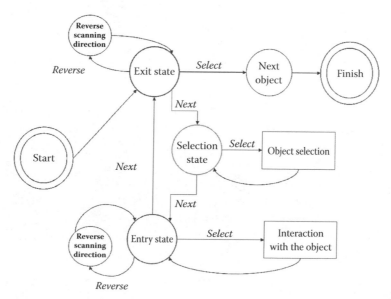

FIGURE 35.8 Dialogue STN for selection objects.

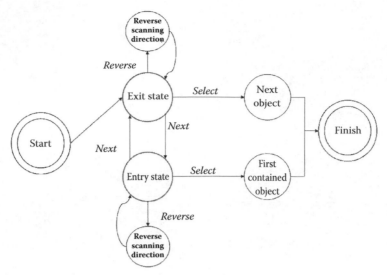

FIGURE 35.9 Dialogue STN for container objects.

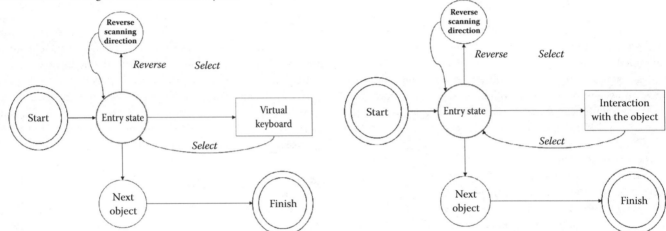

FIGURE 35.10 Quick dialogue STN for text entry objects.

FIGURE 35.11 Quick dialogue STN for simple objects.

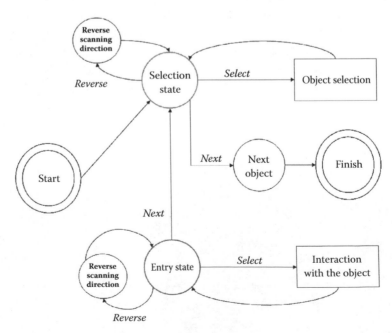

FIGURE 35.12 Quick dialogue STN for selection objects.

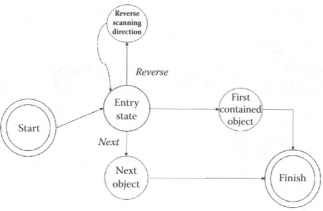

FIGURE 35.13   Quick dialogue STNs for container objects.

TABLE 35.2   Summary of FastScanner Operation Modes

Standard Mode	
Timed Input: Number of Switches	Manual Input: Number of Switches
1. Select (mandatory). Confirms that the user wishes to continue with the selected dialogue state. 2. Reverse (optional). Reverses scanning order.	1. Select (mandatory). Confirms that the user wishes to continue with the selected dialogue state. 2. Next (mandatory). Changes the dialogue state. 3. Reverse (optional). Reverses scanning order.

Quick Mode	
Timed Input: Number of Switches	Manual Input: Number of Switches
1. Select (mandatory). Interaction with the currently focused object. 2. Reverse (optional). Reverses scanning order.	1. Select (mandatory). Interaction with the currently focused object. 2. Next (mandatory). Moves scanning dialogue to the next available object (or state in the case of selection objects only). 3. Reverse (optional). Reverses scanning order.

1. Scanning dialogue initially focuses on the window itself (i.e., the Microsoft Windows Explorer window).

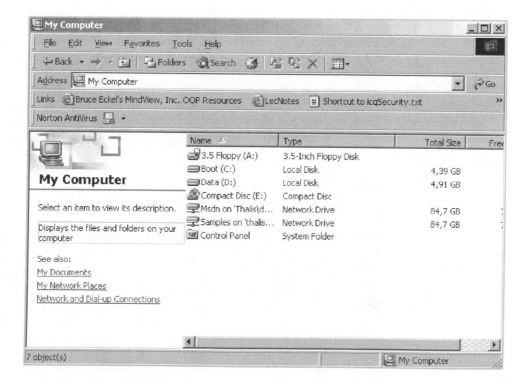

2. Following a *select* command, scanning focuses on the first contained object.

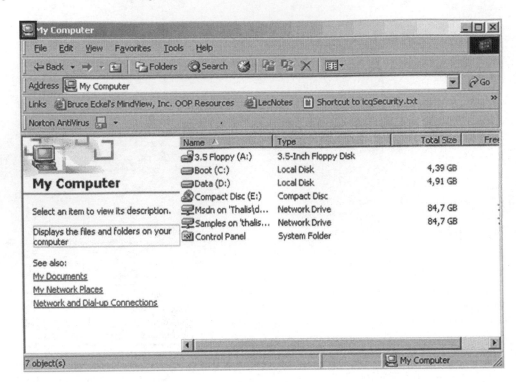

3. The *next* action moves the focus marker to the next element, for instance, the title bar, in the *entry* state.

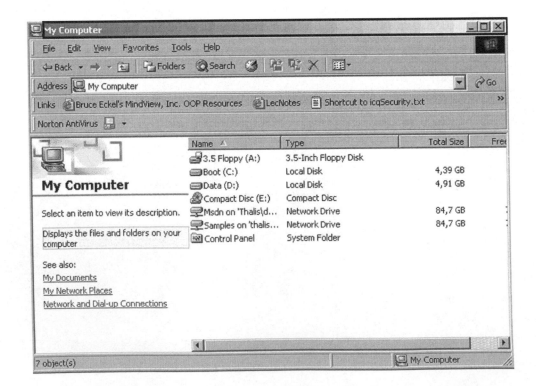

4. Since there is no need to interact with one of the objects contained in the title bar, the *next* action is initiated and the dialogue focuses on the next object (the toolbars container), in the *entry* state.

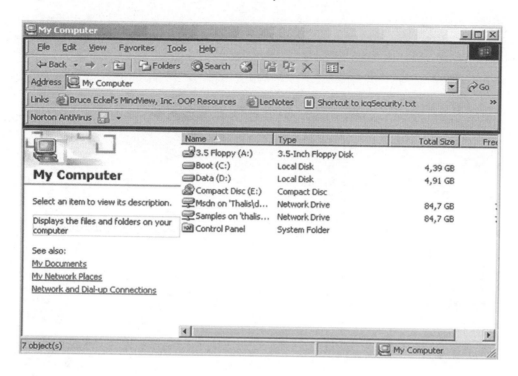

5. Since there is no need to interact with one of the objects contained in the toolbars container, the *next* action is initiated and the scanning dialogue focuses on the next object (the main window area), in the *entry* state.

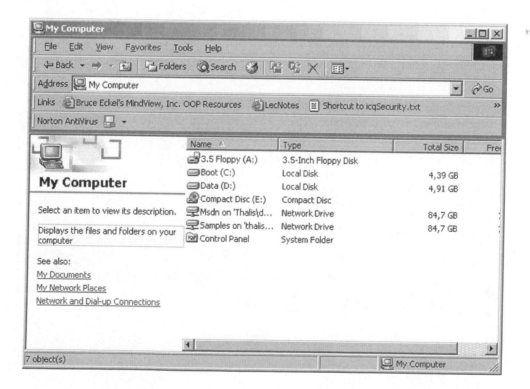

6. Since the target object is included in this container, the *select* action is provided and the dialogue moves to the first contained element.

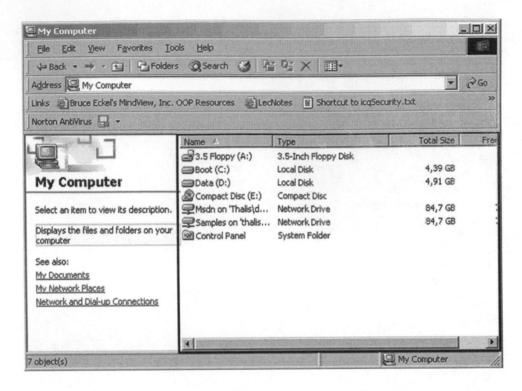

7. Once again, since there is no need to interact with one of the objects contained in the highlighted area, the *next* action is initiated and scanning focuses on the next object, which is the target object, i.e., a link to the folder "My Documents."

8. Following a *select* command, scanning focuses on the next object, which is the target object (i.e., a link to the folder named "My Documents").

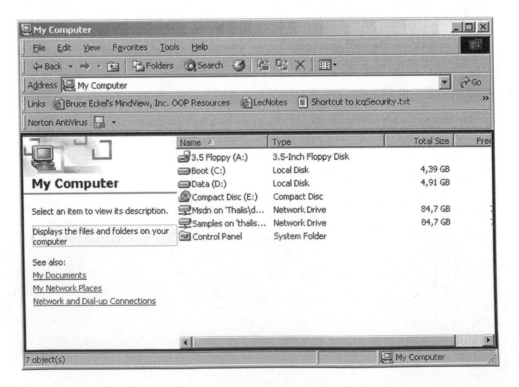

9. By triggering the *select* command, the currently active object will be activated and the folder named "My Documents" will be opened.

35.4.5 Graphical User Interfaces

The graphical user interfaces of the tool aim to allow users to perform actions that are usually carried out with the use of a pointing device, such as moving or resizing a window, as well as to personalize their interaction with the tool.

In more detail, the tool should provide users with a way to select which one of the currently opened windows is the one they wish to interact with. Therefore, when the user initiates the FastScanner tool, a window appears, presenting all the application windows that are currently open, and asks the user to select the desired application. This dialogue is depicted in Figure 35.14a. Interaction with the initial dialogue of the tool takes place through scanning, and with the use of the binary switches, as described earlier. Once the user selects the Scan button, the requested application comes to the foreground and the scanning dialogue is transferred to it.

Furthermore, one additional dialogue box is used for window moving and resizing actions, as well as for activating the scanning focus marker and scanning input settings dialogues. The main settings dialogue box is presented in Figure 35.14b and can be triggered by pressing the *select* and *next* switch at the same time in manual scanning mode, or by activating the *next* switch in time-scanning mode. In more detail, the user can move the

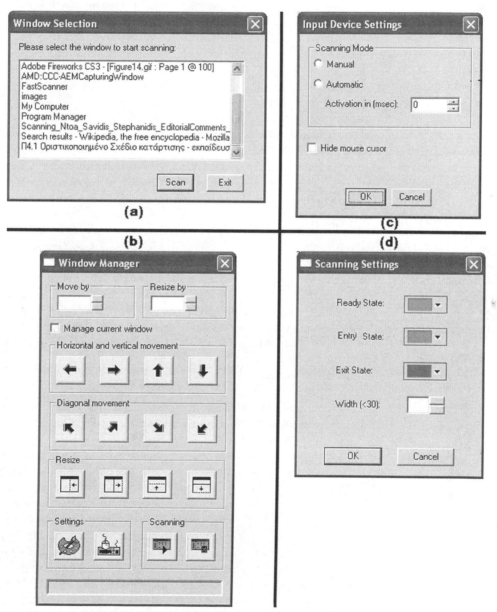

FIGURE 35.14 GUI elements of the tool: (1) initial window, (2) settings dialogue, (3) border settings (color and width), and (4) input device settings.

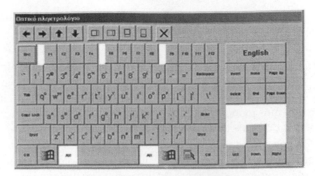

FIGURE 35.15 Virtual keyboard of the FastScanner tool.

window of the application by selecting one of the eight buttons of the first two rows, or resize the window by selecting one of the four buttons of the third row of the main settings dialogue. The button with the palette initiates the border settings window, which is displayed in Figure 35.14d and allows the user to set the color for each one of the dialogue states (entry, exit, and selection), as well as the width of the focus marker. The button with the input devices in the initial settings window initiates the input device settings window, which is displayed in Figure 35.14c and is used for enabling or disabling the time-scanning mode and specifying the time interval for the automatic triggering of the "next" action. The last two buttons of the initial settings window can be used for pausing and restarting FastScanner and for terminating the operation of the tool.

Finally, one additional window that was designed and is embedded in the FastScanner tool is that of the virtual keyboard, which is presented in Figure 35.15. The virtual keyboard was created using the QWERTY layout, by omitting the numeric pad keys and enhancing it with an additional window functionalities toolbar. The functionalities toolbar provides controls for the management of the keyboard window itself. In more detail, the first four buttons move the keyboard left, right, up, and down, the next four buttons resize the window and the last button closes the virtual keyboard window. Finally, the button named English above the Insert, Home, and Page Up buttons is used to change the keyboard language from English to Greek and vice versa.

User tests carried out for the keyboard indicated that users were rather slow in typing text. Therefore, buttons were grouped into clusters, as shown in Figure 35.16. User tests of the enhanced virtual keyboard with groups provided better results, which were further optimized for experienced users who would work effectively with the *reverse* switch as well.

35.4.6 Events Handling

As windows change their structure and their properties according to user actions, it was important to monitor user actions and interpret their results on the currently active window to update the scanning hierarchy whenever necessary. More specifically, whenever a user acts on a windows object, the operating system sends a message to all the applications that are hooked to the

events of a specific window. This message consists of certain parameters, defining the event, the object that generated it (by providing its window handle), as well as the object type.

FastScanner hooks the event messages posted for the currently scanned window and acts accordingly when necessary. For example, if the message indicates that the window has been moved, all the object properties of the scanning hierarchy, referring to the objects' location on the screen, are updated to the latest information. As a result, when necessary, the objects' hierarchy is being traversed and the necessary objects are added, updated, or deleted.

It should be noted, however, that the operating system posts several messages for one user action performed on a windows object, and therefore these messages have to be filtered and interpreted according to the current context. For example, when a window comes to the foreground, four messages are posted: EVENT_SYSTEM_FOREGROUND, EVENT_OBJECT_LOCATIONCHANGE, EVENT_OBJECT_REORDER, and EVENT_OBJECT_FOCUS. However, the hierarchy should be constructed only once to achieve a fast system response time. Events handling in FastScanner is a complex process and required prioritizing the importance of the posted messages based on their significance, their uniqueness, and their consistency to avoid unnecessary hierarchy reconstruction.

35.4.7 Evaluation

To assess the usability of the FastScanner tool, a user-based laboratory usability evaluation has been performed, involving ten participants, five with motor disabilities of upper limbs and five able-bodied users simulating temporary inability to use their hands (Ntoa et al., 2004). The evaluation was based on an appropriate scenario of common tasks performed by a typical computer user, such as writing documents, reading and sending an e-mail, and navigating in the web. An analysis of the performance results indicated that participants were in general effective in carrying out the tasks. In particular, they achieved the majority of the requested task goals, and the number of errors was small in average. Additionally, the majority of errors were corrected, while rarely the users asked the observer for help. Furthermore, participants' interaction with the FastScanner application was considered rather efficient, since they did not need a very long

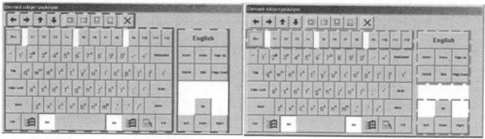

FIGURE 35.16 Button groups as defined in virtual keyboard.

time to carry out the requested tasks, and they rarely needed to consult the instruction booklet. Finally, the analysis of the users' responses to the satisfaction questionnaires suggested that they were overall satisfied by the tool. Participants were on average less satisfied during the execution of the first scenario, entailing text typing. A detailed inspection of their answers to the questionnaires revealed that they met difficulties in using the virtual keyboard to enter text. However, it was noticed that able-bodied users were more dissatisfied than disabled users, probably due to the fact that they are acquainted with more rapid interaction for entering text.

35.5 Summary and Conclusions

This chapter discussed current approaches to scanning and has thoroughly presented an approach to automatic hierarchical scanning in a windows environment through the case study of a tool developed according to this method, the FastScanner tool. FastScanner allows user interaction through the use of binary switches, and by providing the appropriate graphical user interfaces users can work efficiently with any application they wish.

To achieve fast user interaction, FastScanner adopts the hierarchical scanning technique and provides access to the objects of the selected window according to their place in the window hierarchy and not based on their screen location as in row-column scanning. User interaction is also accelerated by the use of container objects, allowing users to skip scanning large groups of objects. Furthermore, the tool supports two types of user input, namely time-based and manual, and two modes of function, namely standard and quick scanning. In time-based input, users are able to interact with the use of just one switch. Standard scanning mode addresses first-time users of the tool, while quick scanning mode serves the needs of expert users.

Evaluation of the tool indicated that participants were in general effective and efficient in carrying out the tasks. In particular, they achieved the majority of the requested task goals, the number of errors was small in average, and they completed the tasks in reasonable time without the need for any help. Finally, the analysis of the users' responses to the satisfaction questionnaires suggested that they were overall satisfied by the tool, although they were first-time users.

References

ABLEDATA (2008). *Electro-mechanical Switches.* http://www.abledata.com/abledata.cfm?pageid=19327&top=11324&deep=2&trail=22,11316.

Applied Human Factors (2008). *ScanBuddy Assistive Software.* http://www.ahf-net.com/Scanbuddy.htm.

Cooper, R. J. and Associates (2008). *CrossScanner Switch Method.* http://rjcooper.com/cross-scanner/index.html.

EnableMart (2008). *REACH On-Screen Keyboard Suite.* http://www.enablemart.com/Catalog/Word-Prediction/REACH-On-Screen-Keyboard-Suite.

Enabling Devices (2008). *Capability Switches.* http://enablingdevices.com/catalog/capability-switches.

Gnanayutham, P., Bloor, C., and Cockton, G. (2004). Soft keyboard for the disabled, in the *Proceedings of the 9th International Conference on Computers Helping People with Special Needs (ICCHP 2004)* (J. Klaus, K. Miesenberger, D. Burger, and W. Zagler, eds.), 7–9 July 2004, Paris, pp. 999–1002. Berlin/Heidelberg: Springer-Verlag.

Gus Communications, Inc. (2008). *Gus! Scanning Cursor.* http://www.gusinc.com/scancur.html.

Lin, Y. L., Chen, M. C., Yeh, Y. M., Tzeng, W. J., and Yeh, C. C. (2006). Design and implementation of a chorded on-screen keyboard for people with physical impairments, in the *Proceedings of the 10th International Conference on Computers Helping People with Special Needs (ICCHP 2006)* (K. Miesenberger, J. Klaus, W. Zagler, and A. Karshmer, eds.), 9–11 July 2006, Linz, Austria, pp. 981–988. Berlin/Heidelberg: Springer-Verlag.

Linux Online (2008). *Linux Accessibility HOWTO.* http://www.linux.org/docs/ldp/howto/Accessibility-HOWTO/index.html.

Microsoft Corporation (2008a). *Microsoft Accessibility Developer Center.* http://msdn2.microsoft.com/en-us/accessibility/default.aspx.

Microsoft Corporation (2008b). *Accessibility in Microsoft Products.* http://www.microsoft.com/enable/products/default.aspx.

Ntoa, S., Savidis, A., and Stephanidis, C. (2004). FastScanner: An accessibility tool for motor impaired users, in the *Proceedings of the 9th International Conference on Computers Helping People with Special Needs (ICCHP 2004)* (J. Klaus, K. Miesenberger, D. Burger, and W. Zagler, eds.), 7–9 July

2004, Paris, pp. 796–804. Berlin/Heidelberg: Springer-Verlag.

Parnas, D. L. (1969). On the use of transition diagrams in the design of a user interface for an interactive computer system, in *Proceedings of the 1969 ACM National Conference,* 26–28 August 1969, pp. 379–385. New York: ACM Press.

Savidis, A., Vernardos, G., and Stephanidis, C. (1997). Embedding scanning techniques accessible to motor impaired users in the WINDOWS object library, in *Design of Computing Systems: Cognitive Considerations, Proceedings of the 7th International Conference on Human-Computer Interaction (HCI International '97), Volume 1* (G. Salvendy, M. J. Smith, and R. J. Koubek, eds.), 24–29 August 1997, San Francisco, pp. 429–432. Amsterdam, The Netherlands: Elsevier Science.

Stephanidis, C. and Emiliani, P. L. (1999). Connecting to the information society: A European perspective. *Technology and Disability Journal* 10: 21–44.

Sun Microsystems, Inc. (2008a). *Sun Microsystems Accessibility Program: Developer Information.* http://www.sun.com/access/developers/index.html#app.

Sun Microsystems, Inc. (2008b). *Solaris Accessibility Quick View.* http://www.sun.com/software/star/gnome/accessibility/quickview.xml.

Wasserman, A. I. (1985). Extending state transition diagrams for the specification of human-computer interaction. *IEEE Transactions on Software Engineering* SE-11: 699–713.

WiViK (2008). *WiViK On-Screen Keyboard (Virtual Keyboard) Software.* http://www.wivik.com.

Words+ (2008). *EZ Keys Version 2.50a.* http://www.words-plus.com/website/products/soft/ezkeys.htm.

36

Eye Tracking

Päivi Majaranta,
Richard Bates, and
Michael Donegan

36.1 Introduction

People use their eyes mainly for observation, but people also use gaze to enhance communication; for example, staring at somebody soon causes a reaction: "What? Do you want something?" Similarly, an intense look at a water jug may be enough to motivate someone at a dinner party to pour more water for you. Thus the direction of a person's gaze not only allows observation by that person of the world around them, but also reveals and communicates their focus of visual attention to the wider world (Just and Carpenter, 1976). It is this ability to observe the visual attention of a person, by human or machine, that allows eye gaze tracking for communication.

In some cases, eye gaze may be the only communication option available for a person. For example, after a severe accident a person may not be able to speak, in which case, a doctor may ask the person to "look up" or "look down" as an indication of understanding and agreement. This method of communication can be expanded from a simple yes or no command to a full communication system by adding meaningful objects in the view of a user. An example of this approach is the gaze communication board (Figure 36.1). Here a board has pictures, commands, or letters attached to it, with the user selecting items on the board by looking at them. The person, or interpreter, on the other side of the transparent board interprets the message by following the eye movements of the user onto the differing targets. Such a system illustrates the simple communication power of eye gaze tracking.

Manual eye gaze tracking systems such as the E-Tran frame are not always convenient, private, practical, or possess all of the communication functions a user may wish to use. Hence, computer-based gaze communication systems have been developed where an eye tracking device and a computer replace the manual communication board. In these eye tracking systems letters (or any other symbols, images, or objects) are shown on a computer screen placed in front of the user. The user simply points and selects these items by looking at them, with an eye tracking device recording their eye movements and a computer program analyzing and interpreting their eye movements in place of the human operator of the E-Tran frame. Such a system forms a basic gaze communication system.

This chapter will briefly summarize the history of these eye tracking systems, and how the technology has developed and improved during the years, making eye control a real choice for people with disabilities. Thus, the focus of this chapter is on interactive use of eye tracking in real time as a form of assistive technology. Data collection and offline analysis of eye movements, and its diverse application to eye movement research in physiology, psychology, marketing, usability, and so on, are out of the scope of this chapter. For an overview of the various

FIGURE 36.1 A gaze communication board ("E-Tran frame"). The person on the other side of the board acts as a human eye tracker and interprets the direction of gaze through the transparent board.

applications of eye tracking, see, for example, Duchowski (2002) or Jacob and Karn (2003).

Various gaze communication systems have been developed since the late 1970s (Majaranta and Räihä, 2002), and pointing with the eyes has been found to be fast and easy for the user (Stampe and Reingold, 1995). However, interpreting a person's intentions from their eye movements is not a trivial task. The eye is primarily a perceptual organ, not normally used for control, so the question arises how casual viewing can be separated from intended gaze-driven commands. If all objects on the computer screen would react to the user's gaze, it would cause a so-called Midas touch (or perhaps Midas gaze) problem: "everywhere you look something gets activated" (Jacob, 1991). This problem can be avoided by *dwell time*, where objects are only activated with an intentionally prolonged gaze. Dwell time and other solutions are discussed in the following sections that introduce the basics of gaze input and eye control.

The obvious users of eye control technology are those for whom it is a necessity, for example, people who have lost most motor control of their body and only have control over eye movements. This may seem like a rare disability, but with modern medicine enabling longer-term survival from often severe injury or ongoing disability, such conditions are becoming increasingly common. Table 36.1 shows an estimate of the numbers of people with disabilities who might benefit from gaze-based communication in the European Union (EU) (Jordansen et al., 2005).

For example, a brain stem stroke may leave the person in a "locked-in" state: fully aware and awake but almost totally paralyzed. In this locked-in syndrome eye movements are often the only voluntary movement not affected by the paralysis. Another example is amyotrophic lateral sclerosis (ALS, or motor neuron disease, MND), which is a progressive, incurable disease in which the person gradually loses control over all voluntary muscle movement as the motor neuron system degenerates. In the late stages of ALS, eye movement may well be the only controllable movement left. In both examples, personality, memories, and intelligence may be left intact, and, although they may not be able to speak, this does not mean sufferers of these conditions cannot communicate. In fact, communication via eye movements is still possible. Note that in some cases all voluntary muscle movement is lost, including eye movements, and even more advanced techniques are needed, such as brain-computer interfaces (BCI; see Chapter 37 of this handbook; see also Wolpaw et al., 2002).

People who are unable to move but have good control over eye movements have traditionally been the best candidates for eye-tracking systems. As they do not move, this immobility can make tracking easier in contrast to people who can move (voluntarily or involuntarily, e.g., in the case of cerebral palsy) and are much harder to track due to their movements. However, eye control may still be a genuine *choice* for both types of users, as eye control can be faster and less tiring than, for example, a manual switch-based system or a head-pointing-based system. Eye movements are extremely fast, and gaze pointing locates and points at a target long before a manually controlled mouse cursor may reach it. Since people look at things before they act on them, this speed gives the impression that eye movements anticipate their action (Land and Furneaux, 1997), greatly enhancing interaction and communication.

Recent advances in technology have considerably improved the quality of eye-tracking systems, such that a far broader group of people may now benefit from eye control. This is illustrated

TABLE 36.1 Total Population in the EU That Might Benefit from Gaze-Based Communication

Condition	Total Population
ALS/MND	27,000
Multiple sclerosis	135,000
Cerebral palsy	900,000
Quadriplegia (spinal cord injury)	36,000
Spinal muscular atrophy	54,000
Muscular dystrophy	126,000
Brainstem stroke	688,000
Traumatic brain injury	675,000
* Total	2,641,000

* Of overall total of 450 million people in EU.

later in the chapter, where a series of case studies from user trials are reported showing the potential benefits of eye control technology.

36.2 History, Trends, and Current Technologies

36.2.1 The History of Eye Tracking

Early in the field of eye gaze tracking, eye movements were studied mainly to observe the nature of human eye movements, rather than to use these movements for communication. The first *eye-tracking devices* that produced objective and accurate data were highly invasive and uncomfortable. For example, the system developed by Delabarre in the late 1800s used an eyecup with a lever extending to draw the eye movements on a smoked drum. The eyecup was attached directly to the eye surface (which required anesthetization with cocaine) and had a hole in it through which the test subject could see (Wade and Tatler, 2005). A breakthrough in eye movement research was the later development of the first "noninvasive" eye-tracking apparatus by Dodge and Cline in the early 1900s (Wade and Tatler, 2005). This was based on photography and light reflected from the cornea of the eye (the shiny reflective surface of the eye). Many basic properties and types of eye movements were categorized using Dodge and Cline's camera-based device or its later improved versions. The "Dodge Photochronograph" is considered the inspirer and first ancestor of the current video-based, corneal reflection eye-tracking systems discussed later.

The development of computing power enabled gathering of eye-tracking data in real time, as well as the development of assistive technology systems aimed directly at people with disabilities (e.g., Ten Kate et al., 1979; Levine, 1981; Friedman et al., 1982; Yamada and Fukuda, 1987; and Hutchinson et al., 1989, all of whom focused primarily on users with disabilities). These first systems were typically based on eye typing, where the user could produce text by using the focus of gaze as a means of input. One of the earliest eye-typing systems, the Eye-Letter-Selector (Ten Kate et al., 1979), is shown in Figure 36.2. Here eye movements were detected by two phototransistors attached to the spectacles' frames (the frames are located on top of the device in Figure 36.2).

The Eye-Letter-Selector could not track the direction of the gaze to allow direct selection of individual characters on the keyboard. Instead, it detected eye movements to the left or right and used these as a single or double eye-controlled switch system (Ten Kate et al., 1979). To enable typing the system adopted a column-row scanning procedure. Columns of letters were highlighted automatically one by one in sequence across the keyboard. When the scanning reached the column where the desired letter was, the user-activated eye-switch by looking right. The system then scanned the rows in sequence down the selected (highlighted) column, and again, when the desired letter was reached, the user could select it by looking right. This enabled slow but effective eye typing.

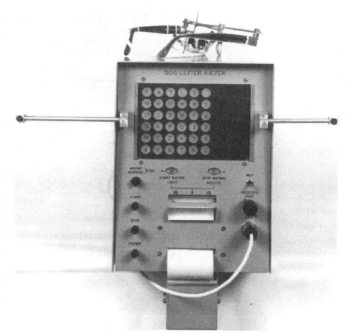

FIGURE 36.2 The Eye-Letter-Selector detected rough horizontal eye movements.[1]

36.2.2 Contemporary Technologies

Current eye-tracking technologies have evolved from the early systems such as the Eye-Letter-Selector into a range of technologies: electro-oculography (EOG) where the user wears small electrodes around the eye to detect the eye position (Figure 36.3), scleral contact lens/search coil where the user wears a contact lens with a magnetic coil on the eye that is tracked by an external magnetic system, video-oculography (VOG) or photo-oculography (POG) where still or moving images are taken of the eye to determine the position of the eye, and finally video-based combined pupil/corneal reflection techniques that extend VOG by artificially illuminating both the pupil and cornea of the eye for increased tracking accuracy (Figure 36.4; Duchowski, 2003).

Examining each of these approaches in turn, EOG-based systems may be seen as impractical for everyday use, because they require electrodes to be placed around the eye to measure the skin's electric potential differences. There are, however, EOG systems that are successfully used for augmentative and alternative communication (see, e.g., Gips et al., 1993; Hori et al., 2006). The EagleEyes system developed by Gips et al. (1993) has improved the quality of life of numerous users (Figure 36.3). There are some drawbacks, however, as some people may not wish to have electrodes placed on their face, and the electrodes can also fall off if the user perspires (Betke et al., 2002). EOG-based systems are not, however, sensitive to changes in lighting conditions (especially outdoor lighting), which is a considerable problem with video-based systems. As the EOG potential is

---

[1] A detailed, illustrated description of the system and its later variations is available at http://www.ph.tn.tudelft.nl/~ed/ELS-Handi.html.

FIGURE 36.3 Eye painting with EagleEyes (http://www.eagleeyes.org). (Gips, J. and Olivieri, P., EagleEyes: An eye control system for persons with disabilities, in the *Eleventh International Conference on Technology and Persons with Disabilities*, Los Angeles, March 1996, http://www.cs.bc.edu/~eagleeye/papers/paper1/paper1.html.)

proportional to the angle of the eye in the head, an EOG-based mouse pointer is moved by changing the angle of the eyes in the head (EagleEyes, 2000). The user can move the mouse cursor by moving the eyes, the head, or both. More information about the EOG-based EagleEyes system is available in DiMattia et al. (2001) or at http://www.eagleeyes.org.

Systems that use contact lenses or in-eye magnetic coils are mainly used for psychological or physiological studies that require high accuracy (these systems can be very accurate to a fraction of a degree). Here gaze tracking is used to provide an objective and quantitative method of recording the viewer's point of regard. Such information can be used for medical and psychological research to gain insight into human behavior and perception (see, e.g., Rayner, 1995).

VOG and POG camera (video or still) based systems are considered to be the least obtrusive, and thus are best suited for interactive applications that react to the user's gaze at some level (Morimoto and Mimica, 2005). These systems tend to be inaccurate, and so are enhanced using pupil detection combined with corneal reflection to provide a point of regard

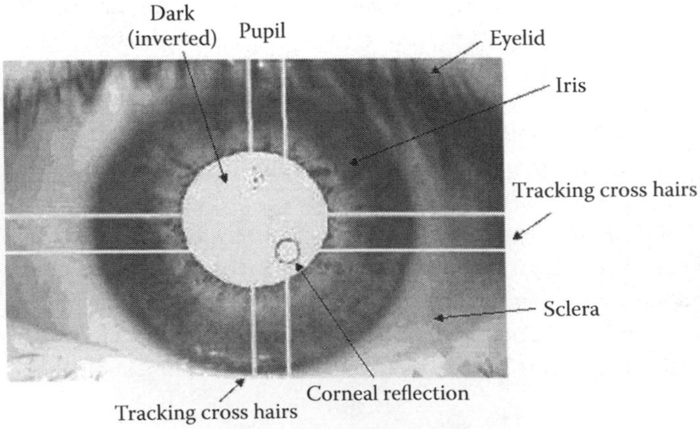

(POR) measurement, which means that the system can calculate the direction of gaze (Duchowski, 2003). In practice, at least two reference points are required for the gaze point calculation. By measuring the corneal reflection(s) (from an infrared artificial light source aimed off-axis at the eye) relative to the center of the pupil, the system can compensate for inaccuracies and also for a limited degree of head movement. Gaze direction in these systems is calculated by measuring the changing relationship between the moving dark pupil of the eye and the essentially static reflection of the infrared light source back from the surface of the cornea. This approach relies on shining infrared light (to avoid the tracked subject squinting) at an angle onto the cornea of the eye, with the cornea producing a reflection of the illumination source (Figure 36.4).

In operation the corneal reflection remains approximately constant in position during eye movement; hence, the reflection will remain static during rotation of the eye and changes in gaze direction, thus giving a basic eye and head position reference. This reflection also provides a simple reference point to compare with the moving pupil, and so enables calculation of the gaze direction vector of the eye (for a more detailed explanation see Duchowski and Vertegaal, 2000).

Most of the currently available eye control systems are video-based (VOG) with corneal reflection, and hence this chapter concentrates mostly on these video-based systems.

36.2.3 Currently Available Eye Control Systems

Only a minority of the tens of currently available eye-tracking systems[2] are targeted at people with disabilities. Most of the systems use the same basic technical principles of operation, but

FIGURE 36.4 Video frame from a VOG system showing eye corneal reflection and pupil detection.

Labels in figure: Dark (inverted), Pupil, Eyelid, Iris, Tracking cross hairs, Sclera, Corneal reflection, Tracking cross hairs

[2] For a list of eye trackers for eye movement research, analysis, and evaluation, see http://www.cogain.org/eyetrackers/eyetrackers-for-eye-movement-research.

TABLE 36.2 Commercially Available (Video-Based) Eye-Control Systems[3]

Eye Response Technologies: ERICA
(http://www.eyeresponse.com)
Mouse emulation enables full control of Windows (typing, e-mail, web browsing, games etc.). Portable, flexible mounting options. Environmental control and remote IR control available as accessories. Touch screen and head control also possible. Comes with tablet/laptop PC, available for both Windows and Macintosh.

LC Technologies: Eyegaze
(http://www.eyegaze.com)
Dedicated, eye-controlled keyboard and phrases, which allow quick communication, synthesized speech. Access to Internet and e-mail. Play eye-controlled games (included), run computer software, operate a computer mouse. Includes also support for book reading, and control of lights and appliances (environmental control).

Tobii Technology: MyTobii
(http://www.tobii.com)
Dedicated eye typing, e-mail and gaze-controlled games included. Includes mouse emulation that can be used to control Windows. Dwell time, a switch, or an eye blink can be used to click. Tracks both eyes. Good tolerance to head movements. Long-lasting calibration with minor drifting. Accessories include a mounting arm. Available in several languages.

EyeTech Digital Systems: Quick Glance
(http://www.eyetechds.com)
Mouse emulation enables full control of Windows (typing, e-mail, web browsing, games, etc.). A switch or an eye blink can be used to click, in addition to dwell time selection. Several models available, with varying properties. Allows moderate head movements. Portable, comes with a tablet PC. Tracking module can also be purchased separately. Available in several languages.

Metrovision: VISIOBOARD
(http://www.metrovision.fr)
Mouse emulation enables full control of Windows (typing, e-mail, web browsing, games, etc.). Clicking can be done by dwell time (staring), eye blinks, or an external switch. Allows moderate head movements. Mounting arm for people in seated or lying position (cannot be attached to a wheelchair).

H.K. EyeCan: VisionKey
(http://www.eyecan.ca)
Head-mounted, lightweight, fully portable eye communication system. Comes with a stand-alone, separate control unit with a display (attached to eye glass frames) and a voice synthesizer, so no computer is needed when on the move. A standard USB keyboard interface is provided for computer control. Compatible with Windows or Macintosh. Provides scanning options for people with limited eye control. Independent of (nonviolent) head/body movements.

what makes certain systems suitable for people with disabilities are the applications (software) that are supported or come with the system and the (technical) support and accessories provided by the manufacturers and retailers. For a disabled person an eye control system is a way of communicating and interacting with the world and may be used extensively, daily, and in varying conditions by people with varying needs and (dis)abilities. Thus, reliability, robustness, safety, and mounting issues must be carefully taken into account, in addition to ease of use and general usability.

Currently available commercial eye control systems targeted at people with disabilities are listed in Table 36.2. The list only includes video-based systems that are operated by eye movements. Systems that use eyes as simple switches, such as systems based solely on eye blinks, are not listed, as these are not regarded as full eye-tracking systems and tend to offer more limited communication. The EOG-based EagleEyes is also excluded, as it is not sold, even though it is available (for free) to qualified users from a foundation in the United States (for more information, see http://www.eagleeyes.org). All the systems listed in Table 36.2 are in daily use by people with disabilities.

36.3 Basics of Gaze Input

36.3.1 The Nature of Eye Movements

Briefly addressing the nature of eye movements, we know that we look at things by holding our gaze relatively still[4] on an object for a short while, long enough for the human brain to perceive the nature of the object. Such a *fixation* typically lasts approximately 200 to 600 ms. Between fixations, gaze jumps rapidly from one object to another, with these *saccades* typically lasting approximately 30 to 120 ms. Saccades are ballistic movements; once a saccadic jump has been started, it cannot be interrupted nor can its direction be changed. In addition to saccadic eye movements, eyes can also smoothly follow a moving target, known as *(smooth) pursuit*. Normal eye movement is thus made from fixations on objects of interest joined by rapid saccades between those objects, with occasional smooth pursuit of moving objects.

---

[3] For an up-to-date list, see http://www.cogain.org/eyetrackers.
[4] The eyes make very small, rapid movements even during fixations to keep the nerve cells in the retina active and to correct small drifting in focus. These "tremors" and "microsaccades" are so small that they are of little importance for practical applications of eye tracking.

Examining the retina within the eye, the size of the high-acuity field of vision, the *fovea*, gives accurate vision that subtends an angle of about 1 degree from the eye. To illustrate this approximately, this angle from the eye corresponds to an area about the size of a thumbnail when looking at it with the arm straightened. Everything inside this foveal area is seen in detail, with everything outside this narrow field seen indistinctly. Thus, people only see a small portion of any full scene in front of them accurately at a time—it is this narrow vision that generates the need to move the eyes rapidly around to form a full view of the world. The farther away from the fovea an object is, the less detailed it appears to the human eye. The remaining peripheral vision provides cues about where to look next, and also gives information on movement or changes in the scene in front of the viewer (for more information about eye movements and visual perception, see, e.g., Haber and Hershenson, 1973).

Since the foveal area of acute vision is fairly small, and because people actually need to direct their gaze nearly directly toward an object of interest to get an acute view of the object (within 1 degree or so), tracking the gaze direction is possible—if the eye is pointing at an object, the user is probably looking at and perceiving that object.

36.3.2 Calibration

Before a VOG eye-tracking system can calculate the direction of gaze, it must be calibrated for each user. This is usually done by showing a few (e.g., nine equally spaced) points on the screen and asking the user to gaze at the points, one at a time (Figure 36.5). The images of the eye are analyzed by the computer and each image is associated with a corresponding screen coordinate. These main points are used to calculate any other point on screen via interpolation of the data. The accuracy of such systems is very much dependent on a successful calibration.

Most current eye-tracking devices achieve an accuracy of 0.5° visual angle from the user (this is the equivalent of a region of approximately 15 pixels on a 17" display with resolution of 1024 × 768 pixels viewed from a distance of 70 cm). The practical accuracy of the system may be less because of "drifting," where over time the measured point of gaze drifts away from the actual point of gaze. This drift is caused by the changes in the characteristics of the eyes, and is mainly due to changes in pupil size, and excessive movement of the head resulting in the eye moving away from the clear view of the camera and the original calibration position. The effects of drifting can be taken into account and be dynamically corrected to some extent (Stampe and Reingold, 1995). Inaccuracy in pointing is corrected by realigning the possibly inaccurate measured gaze position onto the center of any object selected. It is (often correctly) assumed that the user is looking at the center of the object he wishes to select. Thus, if the measured point of gaze does not match the coordinates on the center of the object, it is possible to correct the drift by realigning the measured gaze position onto the center of the object—where the user is most probably looking at—thus correcting the drift.

Some eye-tracking systems have additional techniques for preventing drifting. Using data from both eyes may help, as the system may continue with data from one eye if the other is lost. For example, the Tobii tracker uses averaged data from both eyes to minimize the drifting effects (Tobii, 2006). This binocular averaging enables a long-lasting calibration with very little drifting and saves the user from continuous recalibrations.

A VOG eye tracker must have an unobstructed view of the eye and pupil to be able to track the eye. Eyelids or lashes may partially cover the pupil and ambient light or reflections from the environment may cause problems. Spectacle lenses or frames may cause extra reflections, and when contact lenses are used, the reflection is obtained from the surface of the contact lens instead of the cornea. This can cause problems if the lenses are displaced over time, causing degradation in tracking accuracy. Problems may also be prevented or minimized by careful setup, for example, by minimizing changes in the lighting conditions and positioning the camera so that it has a clear view of the user's eye (Goldberg and Wichansky, 2003). Finally, most

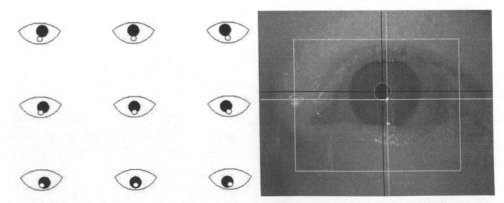

FIGURE 36.5 An illustration of pupil movements (black circles) and near stationary corneal reflections (smaller white circles) as seen by eye tracker's camera at each of the nine calibration points (left) and an image taken by an eye tracker's video camera (right). Note how the relationship between pupil and corneal reflection changes as the eye gaze direction changes.

eye-tracking systems have problems with severe involuntary head or eye movements. Certain medical conditions[5] may also prevent a successful calibration (Donegan et al., 2005). In some cases, calibration may be totally impossible or very inaccurate. If the calibration fails, some systems can be used with a default calibration, and special filtering of eye movements can be applied if the user has eye movement disorders (Charlier et al., 1997).

36.3.3 Gaze Pointing

Eye movements are so rapid that it is not always easy to realize how much and how often the eye moves. Gaze is easily attracted (or distracted) by movement in the peripheral vision, resulting in unwanted flicks away from objects of interest. Eye movements are also largely unconscious and automatic; people do not normally need to think where to look. When needed, however, one can control gaze at will, which makes eye control possible.

Gaze pointing, or placing the computer mouse cursor where the user is looking on the computer screen, is an intuitive method that requires little training (Stampe and Reingold, 1995), as it mimics the operation of a normal desktop mouse. However, it should be noted that for a profoundly disabled person who does not have prior experience on any method of computer control, it may take time to master a gaze-pointing eye control system (Donegan et al., 2006b; Gips et al., 1996).

36.3.4 Midas Touch and Selection Techniques

As the same communication modality, gaze, is used for both perception (viewing the information and objects on a computer screen) and control (manipulating those objects by gaze), the system should be able to distinguish casual viewing from the desire to produce intentional commands. This way the system can avoid the Midas touch problem, where all objects viewed are unintentionally selected (as introduced earlier; Jacob, 1991). The obvious solution is to combine gaze pointing with some other modality for selection. If the person is able to produce a separate "click," then this click can be used to select the focused item. This can be a separate switch, a blink, a wink, a wrinkle on the forehead, or even smiling or any other muscle activity available to that person (Ware and Mikaelian, 1987; Surakka et al., 2003, 2004; Fono and Vertegaal, 2005; Monden et al., 2005). In addition, blinks and winks can be detected from the same video signal used to analyze eye movements, removing the need for additional switch equipment. As muscle activity may be extremely faint or weak, it is typically measured by electromyography (EMG) of any available working muscles. Some systems are based solely on blinks or winks (using the eyes as kind of switches), without tracking gaze direction. For more information about such systems, see, for example, Barreto et al. (2000) or Grauman et al. (2003).

If a user is only capable of moving their eyes, separate switches are not an option, and the system must be able to separate casual viewing from intentional eye control. The most common solution is to use dwell time, a prolonged gaze, with a duration longer than a typical fixation (typically, 500 to 1000 ms; see, e.g., Hansen et al., 1995, 2003a; Istance et al., 1996; Velichkovsky et al., 1997; Majaranta and Räihä, 2002). Most current eye control systems provide adjustable dwell time as one of the selection methods. Requiring the user to fixate for a long time does reduce false selections, but it is uncomfortable for the user, as fixations longer than 800 ms are often broken by blinks or saccades (Stampe and Reingold, 1995). A long dwell time may also be tiring to the eyes and hinder concentration on the task (Majaranta et al., 2006).

Another solution for the Midas touch problem is to use a special selection area (Yamada and Fukuda, 1987) or an onscreen button (Ohno, 1998). For example, in the "quick glance" method developed by Ohno (1998), each object that could be selected was divided into two areas: command name and selection area. Selection was made by first fixating briefly on the command (to determine the name or type of the command) and then confirming that a selection was required by fixating briefly on the selection area. Alternatively, if the user is experienced and knows the locations of commands, she needs to only glance directly at the selection area associated with that command.

Sometimes gaze is unfocused or undirected when the attention of a user is not directed at interaction with the screen. For example, a user may be concentrating or thinking about something, and while thinking his eyes may wander about the screen and accidentally or unconsciously point at an object. Since the eyes are always "on" (they are always pointing unless closed), there is always a risk (and an annoyance and fear) of accidentally staring at an object that is then unintentionally selected. This results in the user feeling that she may not fully relax. Thus, in addition to a long enough dwell time, it is beneficial to the user if eye control can be paused with, for example, an onscreen Pause command to allow free viewing of the screen without the fear of Midas touch (Donegan et al., 2006a).

36.3.5 Accuracy Limitations

The accuracy of the measured point of gaze is a problem if a user wishes to use gaze as her main method to control a standard graphical computer interface. Many of the target objects on typical graphical user interfaces are smaller than the area of high acuity vision (subtending less than 1 degree across at a normal viewing distance from the screen). Even if eye trackers were perfectly accurate, the size of the fovea restricts the practical accuracy of systems to about 0.5° to 1°. Everything inside the foveal region is seen in detail, and the eye may be up to 1 degree away from the object and still perceive it. Also, attention can be retargeted within the foveal region at will without actually moving the eyes. Thus, gaze is not as accurate an input device in comparison to other devices such as a desktop hand mouse, but it can be much faster at pointing due to the speed of the eye if the target objects on screen are large enough (Ware and Mikaelian, 1987; Sibert and Jacob, 2000).

---

[5] For more information about medical conditions that may allow or prevent using an eye-gaze system, see, for example, http://www.eyegaze.com/content/who-able-use-eyegaze-edge.

Thus increasing the size of the targets on the screen makes them easier to "hit" and improves the performance of eye gaze input, and this results in objects designed for eye gaze control often being quite large on screen. However, having only a few, large buttons on screen at a time prevents the use of full-size keyboards such as a full QWERTY keyboard. Instead, keys and controls can be organized hierarchically in menus and submenus, and special techniques such as automatic word prediction can be used to speed up the text entry process, with ambiguous or constantly changing and adapting keyboard layouts (see, e.g., Frey et al., 1990; Hansen et al., 2003b). As an example, GazeTalk (Figure 36.9) has large buttons that support users who have difficulty getting or maintaining a good calibration, or it may be used to enable the use of an eye tracker with a low spatial resolution. It also aids typing speed by changing the keyboard layout using a language model to predict the next most probable letters.

It is important to acknowledge that eye control can be an option to consider even with a very poor calibration. Making onscreen objects much larger can make a difference between a user being able to use an eye-tracking device or not being able to use it at all (Donegan et al., 2005).

36.4 Assistive Applications of Eye Tracking

When an eye tracker is used as an assistive device, it provides a way of communication for a person who cannot talk, and a way of interacting with the world for a person whose mobility is restricted. This section discusses ways of implementing the most common functions of eye control, and provides a few examples of gaze-controlled applications, discussing special design issues that arise from using gaze input.

36.4.1 Mouse Emulation

As introduced earlier, a common way of implementing eye control is to use eye movements to control the mouse cursor. Binding eye movements directly to mouse movements to create an eye mouse may seem an easy solution; however, there are several issues that have to be taken into account.

Eyes move constantly, and eyes make small corrective movements even when fixating. If the cursor of such an eye mouse followed eye movements faithfully without any *smoothing*, the cursor movement would appear very jerky and it would be difficult to concentrate on pointing, as the cursor itself would attract attention (Jacob, 1993). Applying proper smoothing (by averaging data from several gaze points) "dampens" down the jitter, making visual feedback more comfortable and less disturbing (Lankford, 2000). Smoothing the cursor may also assist in maintaining the pointer on target long enough for it to be selected. On the other hand, smoothing slows down the cursor movement. Some applications, such as action games or the Dasher text entry system (Ward and MacKay, 2002), benefit from faster response. Thus, it should be possible to adjust the amount of smoothing (Donegan et al., 2005).

If calibration is poor, the cursor may not be located exactly where the user looks, but a few pixels offset. This may cause users to try and look at the cursor that is displaced away from the actual point of gaze: as the user moves his gaze to look at the cursor, being a few pixels offset, the cursor again moves away from the gaze point. This causes users to chase the cursor that is always a few pixels away from the point the user looks at (Jacob, 1995). Experienced users may learn either to ignore the cursor or to take advantage of the visual feedback provided by the cursor to compensate for any slight calibration errors by adjusting their own gaze point accordingly to bring the cursor onto an object (Donegan et al., 2006b).

The accuracy of eye pointing is restricted to about 1 degree or less, depending on the success of the calibration. If screen resolution is set low and large icons are used, people with good, stable eye control may be able to use standard graphical user interfaces (such as Windows) directly by eye gaze (see, e.g., Donegan et al., 2006b). Special techniques, such as zooming or temporarily enlarging an object on the screen (Bates and Istance, 2002) or a fisheye lens (Ashmore et al., 2005) help in selecting tiny objects such as menu items or shortcut buttons in a typical Windows environment. Figure 36.6 shows an example of using the Zoom tool included in the Quick Glance's Eye Tools menu to magnify a portion of an image.

A mouse click can be executed by dwell time, blink, switch, or any other selection method described previously. In addition to single (left) mouse click, right click, double click, and dragging are also needed if full mouse emulation is desired. These functions are typically provided in a separate (Mouse Click) menu, such as the Quick Glance's Eye Tools menu (EyeTech, 2005) shown in Figure 36.6. An alternative solution is, for example, to use a short dwell time for a single click and longer dwell time for a double click (Lankford, 2000). Feedback on the different stages of the dwell time progress can be shown in the cursor itself by changing its appearance. It may, however, be difficult for some people to understand the different stages.

The main benefit of using mouse emulation is that it enables access to any windows, icons, menus, pointer based (WIMP) graphical user interface. In addition, and as important, it enables the use of any existing access software, such as environmental control applications or dwell click tools. There are also applications that allow the user (or a helper) to design her own special access keyboards with varying target sizes and layouts (for more information of the possibilities and examples of applications, see Donegan et al., 2005).

36.4.1.1 Eye Mouse versus Head Mouse

Perhaps the closest alternative to eye pointing for the users detailed in Table 36.1 is using the head to point if they retain some head control. Obviously, the main difference between eye pointing and head pointing is that, when pointing with a head mouse, the eyes are free for viewing. If the user has good head control, a head mouse can also be quite accurate. Considering the

FIGURE 36.6 Quick Glance's Eye Tools menu provides a zoom tool for target magnification in addition to mouse actions (double click, right click, dragging, etc.), quick calibration correction, and other options.

availability and price of head mice[6] (even a mid-price-range eye control system would be far more expensive than a head mouse), a head mouse could be a better choice than an eye mouse for those who can use it, although anecdotal evidence suggests concerns over prolonged exposure of the neck to repetitive pointing tasks.

Bates and Istance (2002, 2003) compared eye mouse and head mouse in a real-world test that consisted of various simple tasks using a word processor and an Internet browser. Overall, head mouse performance and user satisfaction for the head mouse was higher than for eye mouse. However, the results suggest that an eye mouse could exceed the performance of a head mouse and approach that of a hand mouse if target sizes were large. Performance increased with increased practice; with experienced eye mouse users reaching head mouse performance, though it seems to require more training to master an eye mouse than a head (or hand) mouse.

Hansen et al. (2004) obtained similar results when comparing eye typing with input by head or hand. They tested eye performance using the onscreen keyboards Dasher (Ward and MacKay, 2002) and GazeTalk (Hansen et al., 2001) in Danish and in Japanese. Gaze interaction was found to be just as fast as head interaction, but more erroneous than using head or hand mouse.

The user should have a choice of choosing eye or head mouse, depending on the task and the physical condition of the user, as suggested by a user who tried eye control and was impressed by it (Donegan et al., 2005). For her, eye control felt more natural and requires less effort than either the mouthstick (her main interaction method) or head mouse.

36.4.2 Typing by the Eye

Communication is a fundamental human need, and difficulties in communication may lead to loneliness and social isolation. Developers of eye control systems are well aware of this and so eye typing is typically the first application implemented and tried out by users with an eye control system, with such eye-typing systems being available since the late 1970s (Majaranta and Räihä, 2002).

In a typical eye-typing system, there is an onscreen keyboard (there is no need to adhere to a QWERTY layout), an eye-tracking device that tracks the user's gaze, and a computer that analyzes the user's gaze behavior. To type by gaze, the user focuses on the desired letter by looking at one of the keys on the onscreen keyboard. Selection is typically made by dwell time, or alternatively, using any of the selection methods discussed earlier. Typed text appears in the input field, often located above the keyboard (Figure 36.7).

Eye typing can be slow, typically below 10 words per minute (wpm), due to dwell time durations setting a limit on the maximum typing speed. So, for example, with a 500-ms dwell time and a 40-ms saccade from one key to another, the maximum speed would be 22 wpm. In practice, entry rates are far below that (Majaranta and Räihä, 2007). People need time for cognitive processing, to think what to type next, to search for the next key on the keyboard, to correct errors, and so on.

Ward and MacKay (2002) developed a writing method, Dasher, that enables efficient text entry by gaze using continuous (gaze) pointing gestures. Dasher requires no selection or dwell time at all. Everything is done by using the direction of gaze. The user navigates by pointing at the letter(s) on the right side of the screen. As soon as the cursor points at a letter, the letter starts to fly or expand and move left toward the center of the window. The letter is selected (written) when it crosses the vertical line in the center. For example, in Figure 36.8, the letter *t* has just been selected. This movement and expansion results in the letter

---

[6] For a comparison of head pointers, see "Communication aid survey" at http://www.ace-centre.org.uk.

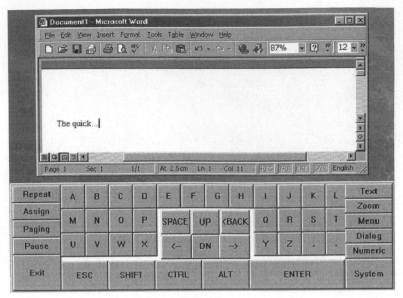

FIGURE 36.7 EC Key (Istance et al., 1996), a typical gaze-driven keyboard.

becoming large and easy to point at just before it crosses the vertical line, enabling accurate selection of the letter. Canceling is done simply by looking left of the center line, which causes the flying to reverse and the typing to be undone. Dasher embeds the prediction of next letters into the writing process itself. Letters that are more probable get more space and appear larger on the screen, and therefore are easier to locate with the cursor. In Figure 36.8, the user is in the middle of entering gaze writing, with letters *writ* already entered and pointing at *ing*. Writing is efficient and easy because several letters can be selected with one gesture. Learning to write with Dasher takes more time than learning to eye type with an onscreen keyboard, but it is much faster once mastered. Dasher is the fastest gaze-writing system at the moment (25 to 34 wpm). Further information about different approaches to text entry by gaze is available in Majaranta and Räihä (2007).

Communication is more than just producing text, and eye typing feels especially slow in face-to-face communication situations, as its entry rate is far below that of human speech (150 to 250 wpm). Ready-made greetings and phrases can speed up everyday communication, and a speech synthesizer can give the user a voice and the ability to speak aloud. However, the synthesized voice may not feel right if it is not age and gender specific (Friedman et al., 1982). Since the user must look at a computer monitor to communicate, this greatly alters the communication between the user and other people. The normal eye-to-eye contact of typical communication is broken, and facial expressions cannot be so easily viewed by the user whose attention is focused on the screen. Due to this loss of facial communication, a see-through communication board may even feel more natural for everyday communication because people maintain a face-to-face connection (see Figure 36.1). The communication board can also be used everywhere—it is always reliable and does not crash. Besides, an experienced

human interpreter[7] is a far more effective word predictor than any of the computer-based systems that do not understand conversational context and situation and are not able to understand humor, and so on. There is still a need for more effective textual gaze communication.

36.4.3 Drawing by the Eye

It is simple to bind the mouse movement to the eye movement and then just select the paint tool, as is done in Figure 36.3, where a thick line with varying colors follows the user's eye movements. However, drawing recognizable objects (houses, trees, people) is not easy with free-eye drawing (Tchalenko, 2001). The characteristics of eye movements prevent using the eyes as a pencil to sketch a finely defined shape. For example, since eye movements are ballistic, it is easy to draw a fairly straight direct line from point A to B just by glancing at the starting and ending point. However, trying to draw slowly or trying to draw a curved line is hard, since the eye does not easily fixate on empty white space and does not move smoothly but in saccadic jumps. Thus, an eye-drawn circle would not be a smoothly curved circle, but would be more like a multi-angled polygon with many small jumps. The human eye needs a moving target to follow to initiate smooth (pursuit) movement.

The Midas touch problem is also strongly present: is the user moving his eyes to draw, or is he looking at the drawing? A method for distinguishing between drawing and looking is needed. Hornof et al. (2004) developed EyeDraw, which implements a set of eye-controlled tools for drawing and painting.

---

[7] After years of practice, one may learn the locations of letters and no longer need the board; each letter has its position in thin air (http://www.cogain.org/media/visiting_kati).

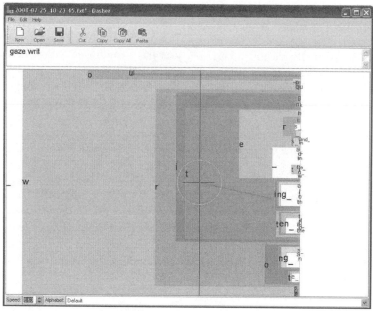

FIGURE 36.8 Dasher predicts the next probable, consecutive letters.

Using the tools, the user manages the drawing process assisted by the tools, rather than attempting free-eye drawing. For example, to draw a line, the user first looks at the Draw Line button in the tools menu. The button is highlighted after dwelling on it to show that the Draw Line tool is active. To draw the line, the user then moves the cursor to the location where the drawing should start and dwells on it. The cursor color changes to show that the starting point has been defined. From now on, a straight line, with its other end fixed on the starting point, starts to follow the user's gaze. Again, the user has to dwell on the selected location to define an ending point for the line. By changing the color of the cursor, the program provides feedback on the current state, which can be either looking or drawing.

Defining the starting point by staring at a blank drawing surface is somewhat difficult, as the eye does not easily fixate without a target object of interest. Also, if calibration is inaccurate and the cursor offset from the actual gaze position, then looking at the eye-driven cursor for a visual anchor point does not work, as fixating at the cursor would only move it away from focus due to calibration inaccuracy, causing the user to chase the cursor across the screen. Therefore, EyeDraw provides a grid of dots that act as visual anchors and help in placing the starting and ending points, literally, on the dot. EyeDraw has been successfully used by people with disabilities to produce drawings, although younger children may get frustrated as they may not have the patience to learn the tools and different states (Hornof and Cavender, 2005). For them, free-eye drawing provides an easy start with immediate positive feedback.

In addition to eye typing and eye drawing, there are several (dedicated) eye-controlled applications, such as eye music (Hornof and Sato, 2004), Internet browsing, e-mail, games, and so on. Such applications are included in many of the commercial eye-control systems targeted at people with disabilities (see Table 36.2).

36.5 Visual and Auditory Feedback

Appropriate feedback is important when the same modality is used for both control and perception. When using gaze to control an application and select objects on screen, gaze is engaged in the input process: the user needs to look at an object to select it. This means that the user cannot simultaneously control an object and view the effects of the action, unless the effect appears on the object itself. For example, if the user is entering text by gaze, she cannot see the text appear in the text input field at the same time she selects a letter by "eye pressing" a key on an onscreen keyboard. To review the text written so far, the user needs to move her gaze from the onscreen keyboard to the typed text field. This looking back and forth can become excessive, especially as novices often shift their gaze between the keyboard and the text input field to review the text written so far (Bates, 2002). This shifting can be reduced by adding auditory feedback, for example, an audible click, or by speaking out each letter as they are written. Appropriate feedback also increases performance and improves accuracy. Even a slight improvement in performance makes a difference in a repetitive task such as text entry, in which the effect accumulates character by character (Majaranta et al., 2006).

Experienced users learn to cope with having to use the same modality for input (control) and output (feedback); they complete the task (e.g., scrolling) before gazing at the results (Bates, 2002). Providing appropriate feedback on the dwell time progress (see Figure 36.9) and the selection process may significantly improve performance and make eye control more pleasant for the user (Majaranta et al., 2006). There is a fundamental difference in using dwell time as an activation command compared to, for example, a button click. When manually clicking a button, the user makes the selection and defines the exact moment when the selection is made. Using dwell time, the user only initiates the action; the system makes the actual selection after a predefined interval. When physically clicking a button, the user also feels and hears the button click. Such extra confirming (auditory or tactile) feedback is missing when an eye press is used to click, and so must be provided.

Many eye tracking manufacturers provide applications that are specifically developed for eye control. The main advantage of such dedicated applications is that the many special requirements of gaze interaction already introduced in this chapter can better be taken into account. So, for example, the layout and the structure of the application can be designed so that items are large enough to be easily accessible by gaze. The feedback provided by a dedicated application can be implemented so that it supports the process of gaze pointing and dwell time selection. As introduced previously, the constant movement of the eye cursor may disturb some users. This may be avoided by hiding the cursor completely, but then another kind of (visual) feedback is needed. This feedback can be shown directly on the item being gazed at by highlighting the button, for example, by bordering the button or changing its background color.

Showing the feedback on the center of the focused item, rather than the actual (potentially slightly inaccurate) position of the gaze, seems to be especially useful for some users (Donegan et al., 2006b). If the feedback is animated inward or toward the center of the button, it may help the user to maintain his gaze focused in the center for the full duration of the dwell time (Majaranta et al., 2006). This, in return, may be helpful if calibration is poor, as when the feedback is shown at the center of a gaze-responsive button, the calibration appears to be perfect to the user, encouraging the user to feel confident when using gaze.

36.6 Results from User Trials

This section summarizes some of the key findings from the user trials conducted within COGAIN, a European Network of Excellence on Communication by Gaze Interaction (http://www.cogain.org). The network combines the efforts of researchers, manufacturers, and user organizations for the benefit of people with disabilities. COGAIN works to spread information about eye-control systems, to identify users' real needs, to improve the quality of eye-control systems, and to develop new hardware and software solutions (Bates et al., 2006). The authors of this chapter are members of the network.

An extensive study on user requirements by Donegan et al. (2005) shows that, to date, eye control can effectively meet only a limited range of user requirements, and that it can only be used effectively by a limited number of people with disabilities. Furthermore, the range of applications that are suitable for easy and effortless control by the eye is limited. COGAIN works to make eye control accessible to as many people as possible.

The results reported here summarize some of the key findings from COGAIN user trials and usability studies (mainly Donegan et al., 2006b, but also Donegan et al., 2005, 2006a). The trials have involved many users from across Europe, with varying disabilities, such as ALS/MND, locked-in syndrome, multiple sclerosis, (athetoid or dyskinetic) cerebral palsy, severe brain damage, and spinal muscular atrophy. Several eye-control systems from different manufacturers have been used in the trials, including both brief one-time trials as well as a series of long-term trials.

36.6.1 Potential Users with a Wide Range of Abilities and Disabilities

When listing potential user groups of eye-control technology, people with ALS are usually among the highest priority as people who most need and benefit from eye control. In the late stages of ALS, control over all other body movements may be lost, but the person can still move his eyes. COGAIN user

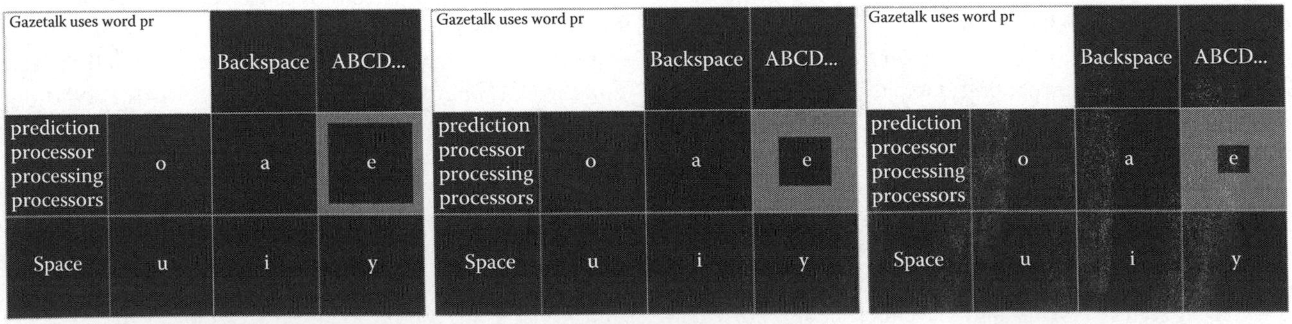

FIGURE 36.9   GazeTalk[8] provides visual feedback on the dwell time progress on letter *e*.

---

[8] GazeTalk (Hansen et al., 2001) is freely available at http://www.cogain.org/downloads.

trials, conducted by Politecnico di Torino and the Torino ALS Centre, have confirmed that people with ALS may greatly benefit from eye control. The participants especially appreciated the ability to use applications independently. Such independence is not possible with their other methods of communication, such as the E-Tran frame (shown in Figure 36.1), which relies on a communication partner. Eye control also facilitated more complex communication. The participants felt that they were able to express more than only their primary needs (for detailed results, see Donegan et al., 2006b). It is important to spread information about eye control and its potential. This lack of information was highlighted as before participating in COGAIN user trials the majority of the participants were not aware that it is possible to write a letter, play chess, send an e-mail, or communicate needs, emotions, and problems just by eye gaze alone.

Eye control suits people with ALS well, because they have good visual, cognitive, and literacy skills. They also do not have involuntary movement, so their eyes are fairly easy to track. However, there are people with a wide range of complex disabilities who might also benefit greatly from eye control, but who find it difficult because of involuntary head movement, visual difficulties, or learning difficulties. The ACE Centre in the United Kingdom and DART, the regional center for augmentation and alternative communication (AAC) and computer access in western Sweden, deliberately chose to work with users with complex disabilities who still might benefit from eye control. They found strong indications that eye-control technology can have a positive impact on communication ability and, subsequently, quality of life. Results also showed that there are many issues that may lead to failure or success in trialing gaze control with users with complex disabilities. Examples of such key issues are discussed in the following, and suggestions for a successful assessment are made.

36.6.2 Initial Setup and Hardware Challenges

36.6.2.1 Positioning

The quality and accuracy of gaze control are directly related to the success of the calibration. Having a flexible mounting system was found to be necessary in all cases of the user trials, and this enabled the eye-control system to be safely positioned for optimal comfort, function, and visibility. Most eye trackers have an optimal distance and optimal angle in relation to the user's eyes, and it was found that poor positioning may lead to poor or totally failed calibration.

For example, one participant (Bjorn in Figure 36.10) used the system in a side-lying position where he maintained a very stable head position. He had excellent eye control and was able to have full control over all Windows applications by controlling the cursor with his eyes, combined with an onscreen keyboard. However, those participants who had either involuntary head movement or visual difficulties (e.g., nystagmus) were found to be less accurate, though in most cases this could be compensated

FIGURE 36.10 A side-lying position enables maintaining a very stable head position.

for by personalized grid-based interface software such as The Grid, Rolltalk, SAW, and so on.[9]

36.6.2.2 Environmental and Situational Factors

It was found that a range of environmental (e.g., lighting conditions) and situational factors (e.g., anxiety) may contribute to the success—or failure—of the calibration. As eye control is a new technology, there is great interest in attending eye-control user trials. However, having too many people around may disturb the user and make concentration difficult. During the trials, it also became apparent that eye control is often tried when all other options have failed, and so expectations are high. If the trial is a failure, it should be made clear that it was not the fault of the user. It is important that the user (and anybody involved) retains a positive attitude for any future trials.

36.6.2.3 Finding the Best System for Each User

Some of the participants tried out several systems before an appropriate system was found. Finding a system that accommodates the physical and perceptual needs of the user may be essential for the success of the trial. For example, many of the users had spasm or involuntary head (or eye) movement that required a tracker that was able to accommodate to them.

For example, Michael has difficulty with head control. In addition, he has involuntary side-to-side eye movement (nystagmus), which becomes more severe when he is tired. The nystagmus made calibration very difficult or impossible, depending on the eye-tracking device. When the nystagmus was at its most severe level, none of the currently available trackers were able to cope with it. When Michael did achieve a successful calibration, the

---

[9] The Grid is available from Smartbox (http://www.smartboxat.com); Rolltalk is available from Igel (http://www.rolltalk.com); and SAW is available from the ACE Centre (http://www.ace-centre.org.uk).

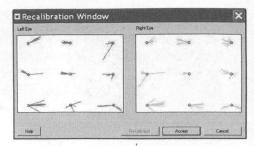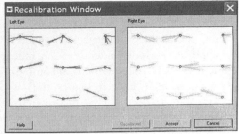

FIGURE 36.11 Success in calibration varied depending on the severity of nystagmus. Calibration on the left enables using a grid with smaller cells compared to the calibration on the right that shows large involuntary movement around the nine target calibration points.

increased speed, comfort, and satisfaction (in comparison with switches) made him extremely enthusiastic about eye control. When the nystagmus was not severe (see Figure 36.11), he was able to use a 2 × 4 grid. However, he preferred to write using a less efficient 2 × 2 grid because it required less effort from him.

36.6.2.4 Customized Calibration

It was found useful to be able to manually adjust the calibration process. Some of the participants (especially those with learning disabilities or brain damage) had trouble following the automatic calibration procedure. Some of the users also had trouble in visually accessing the full screen, and one participant was only able to move his eyes horizontally. It was found to be useful if the assistant could manually adjust the size and appearance of the calibration target as well as its location and timing.

For users who have no experience with eye control it may be hard for them to understand what they are expected to do during calibration. Calibration can be a tedious process. For example, Fredric (in Figure 36.12) has a squint, which made it difficult for the eye tracker to recognize his eyes. A manual calibration was eventually achieved, with a lot of prompting and support from the staff at DART. At the second trial, he was unable to complete the calibration. It is assumed that the calibration process is not motivating enough—there should be pictures and sounds to look at to make it more interesting. Luckily, in his case, the default calibration worked reasonably well.

36.6.3 User's Abilities and Customized Software

36.6.3.1 Visual Representation and Feedback

Many of the participants had limitations in their visual and/or cognitive abilities. For them, it was essential that the visual representation and organization of the onscreen objects was adjustable to make them clearly visible and comprehensible (pictures, symbols, text, size, colors, and so on).

Proper visual and auditory feedback was found to be essential. For example, some people were not able to use dwell time to select objects if a visible cursor was shown, as the movement of the eye-driven cursor constantly distracted their eyes away from the target—but when feedback was given on the button itself and the cursor was hidden, they were able to dwell on it.

36.6.3.2 Progressive Trials

It was found to be helpful if gaze-driven activities were graded (easy/simple to difficult/complex) and carried out in many sessions over time. Some of the participants had not been able to access technology independently before, so any computer-based access method would require time for learning. Moving a cursor with eyes or using dwell time to select something (and avoiding a Midas touch at the same time) are new skills, even to people with perfect cognitive, literacy, and visual skills.

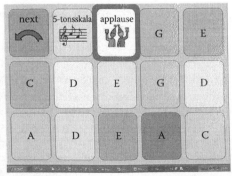

FIGURE 36.12 Fredrik (on the left), concentrating hard on playing guitar with the computer. A screen shot of a simple music playing application (on the right).

36.6.3.3 Introductory Activities with Immediate Positive Feedback

For some users with complex disabilities it was especially useful to try out eye control with very simple tasks, such as tasks where they could not fail. The easy start and immediate positive feedback gave a good basis for further trials with more demanding tasks. Playing music with the eyes was one such application that could be used in the initial trials: whenever the user was able to successfully dwell on a button, a musical sound was played. Many users found this activity rewarding and fun, and the success motivated them to try out more demanding tasks.

For example, Fredric, who suffered major brain damage when he was partially drowned, had not shown much interest or concentration on anything after the accident. Thus, it was not known how severe his brain damage was. He has no voluntary motor control, apart from his eyes. He was able to play music with his eyes and seemed to enjoy it a lot (Figure 36.12). It was the first time (after the accident) that any of his carers had seen him concentrate as hard and as long on anything.

36.6.3.4 User's Engagement and Motivation

The level of satisfaction and engagement gained from eye control appears to be relative to the level of disability. In particular, the adult participants with ALS, who were unable to speak or move any of their limbs, were particularly highly motivated to learn a new method of communication, and felt that eye control gave them hope when all other methods had failed.

Obviously, the tasks and the applications should be user focused; applications should be interesting and motivating to the specific user (considering their age, gender, interests, and so on). Being offered a possibility to independently access something that really interested the user may be the turning point in progress, as was true in Helen's case. By the age of nine, she had never been able to access any technology independently. All deliberate actions caused violent involuntary movement, even the effort of trying to speak causes spasm. However, it was found that her involuntary movement was not induced by eye control. Using eye control Helen is now able to write e-mails independently, in a relaxed, easy way. However, reaching the point where she could type by eye was not an easy or a quick process. Even though both her reading and spelling abilities were within her age range, the first trials with eye typing failed. She only selected random letters and showed no interest in it. She was more interested in stories that had been prepared for her, and she learned the basics of eye control by using her gaze to turn the pages of an onscreen book (Figure 36.13).

Nearly a year after the first trials, a personalized, eye-controlled e-mail utility was introduced to her. The possibility of independently writing e-mails was the turning point and made her really motivated to learn to eye type. Helen's eye-typing interface is illustrated in Figure 36.14. Because of involuntary head movement, the onscreen targets need to be fairly large. It takes two steps for her to type a letter, first selecting a group, and then the individual letter.

36.6.4 The KEE Approach: The Key for Successful Access to Eye Control Technology

Summarizing the experiences from the user trials, Donegan and Oosthuizen (2006) formulated the KEE concept for eye control and complex disabilities. It summarizes three key issues that enhance even the most complex users' chances of successful access to eye-control technology.

The KEE approach to trialing and implementing eye-control technology is:

- **K**nowledge-based, that is, founded on what is known of the user's physical and cognitive abilities
- **E**nd-user focused, that is, designed to meet the end-users' interests and needs
- **E**volutionary, that is, ready to change in relation to the end-user's response to eye-control technology and software provided

KEE is a tool for a successful assessment of whether or not eye control is—or can be made—accessible for a certain user. Following the KEE approach, "a different way of unlocking the door to eye control technology can be found for each user" (Donegan and Oosthuizen, 2006).

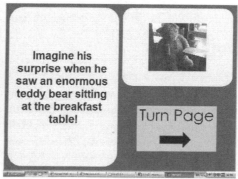

FIGURE 36.13 Helen learns to use her gaze to turn the page of an electronic storybook.

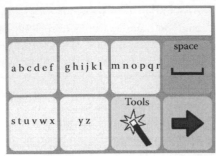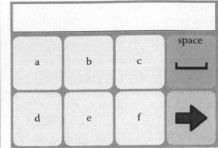

FIGURE 36.14 Helen's eye typing interface.

36.7 Conclusions

In Europe alone, the number of potential beneficiaries of eye-tracking technology amounts to several hundreds of thousands of people but, as yet, only a small number of these people are actually using eye control (Jordansen et al., 2005). For many of them, eye control is potentially the quickest, least tiring, and most reliable form of access to technology—by far.

The COGAIN user trials reported previously show that even if the first trial with eye control fails, it does not necessarily mean that eye-control technology is not suited for a certain user. With appropriate (or properly adjusted) hardware and software, eye-control technology can become accessible even to people with the most complex of disabilities.

Many of the potential users are already using special software for communication. If their condition deteriorates or if they for any other reason move to eye control, they should be able to continue using familiar programs. Existing applications should be made as eye-friendly as possible. There are a variety of applications especially directed at users with disabilities, such as environmental control applications, that would be highly beneficial for eye-control users if they were made eye-friendly.

The prices of the commercially available systems are still fairly high (thousands or tens of thousands of Euros). Even though current eye-tracking devices are reasonably accurate and fairly easy to use, they are not yet affordable to everybody. Several lower cost eye trackers are being developed (for examples of such low-cost systems, see, e.g., Corno et al., 2002; Fejtová et al., 2004; Li et al., 2006). However, to reduce the prices of commercial eye trackers, it is necessary to mainstream gaze interaction for a mass market (Hansen et al., 2005).

36.8 Future Directions

Using the eye as an input device is essential for disabled people who have no other means of controlling their environment. However, as has been discussed and demonstrated in this chapter, the eye is primarily a perceptual organ, and there are many problems in using the eye as a control medium. Therefore, one should not expect eye control to become an appealing option in mainstreaming applications, such as applications that are widely used by the general public. Instead of using the eyes as an explicit control medium, an application can make use of the information of the user's eye movements subtly in the background (being eye-aware) without disturbing the user's normal, natural viewing (Jacob, 1993).

Extending this concept, some of the most promising areas for eye-tracking applications are so-called attentive interfaces. Attentive user interfaces (Vertegaal, 2003) benefit from the information of the user's area of interest (AOI), and change the way information is displayed or the way the application behaves depending on the assumed attentive state of the user. By monitoring the user's eye movements the application knows more about the user's state and intentions, and is able to react in a more natural way, thus helping the user to work on the task instead of interacting with the computer (Nielsen, 1993).

Simply detecting the presence of eyes or recognizing eye contact can enhance the interaction substantially. EyePliances (Shell et al., 2003) are appliances and devices that respond to human eye contact. For example, an Attentive TV will automatically pause a movie when eye contact is lost, deducing nobody is watching the video, and "Look-To-Talk" devices (Shell et al., 2004) know when they are spoken to, for example, a lamp that reacts to the turn on/off command only when a person is looking at it. For a review of attentive applications, see, for example, Hyrskykari et al. (2005).

Gaming is another example of promising new areas for eye tracking. Adding gaze input can enhance the gaming experience and increase the immersion of a video game (Smith and Graham, 2006). Furthermore, eye interaction in games is fun and easy to learn, and perceived as natural and relaxed by the players (Jönsson, 2005). Gaze can be used in various ways, for example, to control the movement of the character or to aim in a first-person shooting game. Even though using gaze does not seem to be any faster or more efficient than the traditional controlling methods, it is not much worse, either (Isokoski and Martin, 2006). Results from user trials are promising, but more research is needed.

If eye trackers become more common, gaze input could also enhance interaction in standard desktop environments. For example, gaze could indicate the currently focused window in a desktop environment (Fono and Vertegaal, 2005), manual mouse pointing could be enhanced by automatically warping the cursor to or near the object the user is looking at (Zhai et al., 1999), or a web page could automatically scroll down (or up) following the gaze of the user reading it or searching information from it (Nakano et al., 2004). Gaze pointing combined with keyboard click for selection can be more efficient or at least comparable to using conventional mouse click—and is strongly preferred by most users (Kumar et al., 2007).

Eye control can offer great possibilities, but it is important to understand its limitations. As a mother of a young boy noted after an eye-control trial: "it was ironic that the more fun he had, the more he laughed and the more his eyes closed, making it impossible for the tracker to pick up his eyes."

Acknowledgments

This work was supported by the COGAIN Network of Excellence on Communication by Gaze Interaction. We would like to thank all COGAIN partners, users, and their helpers who participated in the user studies or otherwise contributed to this work.

References

Ashmore, M., Duchowski, A. T., and Shoemaker, G. (2005). Efficient eye pointing with a fisheye lens, in the *Proceedings of Graphics Interface 2005 (GI '05)*, 9–11 May 2005, Victoria, Canada, pp. 203–210. Ontario, Canada: Canadian Human-Computer Communications Society.

Barreto, A. B, Scargle, S. D., and Adjouadi, M. (2000). A practical EMG-based human-computer interface for users with motor disabilities. *Journal of Rehabilitation Research and Development* 37: 53–64.

Bates, R. (2002). Have patience with your eye mouse! Eye-gaze interaction with computers can work, in the *Proceedings of the 1st Cambridge Workshop on Universal Access and Assistive Technology, CWUAAT*, 25–27 March 2002, Cambridge, U.K., pp. 33–38. http://rehab-www.eng.cam.ac.uk/cwuaat/02/7.pdf.

Bates, R., Donegan, M., Istance, H. O., Hansen, J. P., and Räihä, K-J. (2006). Introducing COGAIN: Communication by gaze interaction, in *Designing Accessible Technology, Part II "Enabling Computer Access and the Development of New Technologies"* (J. Clarkson, P. Langdon, and P. Robinson, eds.), pp. 77–84. London: Springer-Verlag.

Bates, R. and Istance, H. O. (2002). Zooming interfaces! Enhancing the performance of eye controlled pointing devices, in the *Proceedings of the 5th International ACM Conference on Assistive Technologies (ASSETS 2002)*, 8–10 July 2002, Edinburgh, Scotland, pp. 119–126. New York: ACM Press.

Bates, R. and Istance, H. O. (2003). Why are eye mice unpopular? A detailed comparison of head and eye controlled assistive technology pointing devices. *Universal Access in the Information Society* 2: 280–290.

Betke, M., Gips, J., and Fleming, P. (2002). The camera mouse: Visual tracking of body features to provide computer access for people with severe disabilities. *IEEE Transactions on Neural Systems and Rehabilitation Engineering* 10: 1–10.

Charlier, J., Buquet, C., Dubus, F., Hugeux, J. P., and Degroc, B. (1997). VISIOBOARD: A new gaze command system for handicapped subjects. *Medical and Biological Engineering and Computing*, 35, 416, supplement D90.OS1.03. http://perso.wanadoo.fr/metrovision/pdf/1997_charlier.pdf.

Corno, F., Farinetti, L., and Signorile, I. (2002). A cost-effective solution for eye-gaze assistive technology, in the *Proceedings of IEEE International Conference on Multimedia and Expo (ICME2002), Volume 2*, 26–29 August 2002, Lausanne, Switzerland, pp. 433–436. http://www.cad.polito.it/pap/db/icme2002.pdf.

DiMattia, P., Curran, F. X., and Gips, J. (2001). *An Eye Control Teaching Device for Students without Language Expressive Capacity: EagleEyes*. Lampeter, Wales: Edwin Mellen.

Donegan, M. and Oosthuizen, L. (2006). The KEE concept for eye-control and complex disabilities: Knowledge-based, end-user focused and evolutionary, in the *Proceedings of the 2nd Conference on Communication by Gaze Interaction (COGAIN 2006)*, 4–5 September 2006, Turin, Italy, pp. 83–87. http://www.cogain.org/cogain2006/COGAIN2006_Proceedings.pdf.

Donegan, M., Oosthuizen, L., Bates, R., Daunys, G., Hansen, J. P., Joos, M., et al. (2005). D3.1 User requirements report with observations of difficulties users are experiencing. *Communication by Gaze Interaction (COGAIN)*. IST-2003-511598: Deliverable 3.1. http://www.cogain.org/results/reports/COGAIN-D3.1.pdf.

Donegan, M., Oosthuizen, L., Bates, R., Istance, H., Holmqvist, E., Lundälv, M., et al. (2006b). D3.3 report of user trials and usability studies. *Communication by Gaze Interaction (COGAIN)*. IST-2003-511598: Deliverable 3.3. http://www.cogain.org/results/reports/COGAIN-D3.3.pdf.

Donegan, M., Oosthuizen, L., Daunys, G., Istance, H., Bates, R., Signorile, I., et al. (2006a). D3.2 Report on features of the different systems and development needs. *Communication by Gaze Interaction (COGAIN)*. IST-2003-511598: Deliverable 3.2. http://www.cogain.org/results/reports/COGAIN-D3.2.pdf.

Duchowski, A. T. (2002). A breadth-first survey of eye tracking applications. *Behavior Research Methods, Instruments and Computers* 34: 455–470.

Duchowski, A. T. (2003). *Eye Tracking Methodology: Theory and Practice*. London: Springer-Verlag.

Duchowski, A. T. and Vertegaal, R. (2000). *Eye-Based Interaction in Graphical Systems: Theory and Practice*. Course 05, SIGGRAPH 2000. New York: ACM Press.

EagleEyes (2000). *EagleEyes for Windows: User Manual*. Boston: Boston College. http://www.cs.bc.edu/~eagleeye/manuals.html.

EyeTech (2005). *Quick Glance 2 User's Guide*. EyeTech Digital Systems.

Fejtová, M., Fejt, J., and Lhotská, L. (2004). Controlling a PC by eye movements: The MEMREC project, in the *Proceedings of the 9th International Conference on Computers Helping People with Special Needs (ICCHP 2004)*, 7–9 July 2004, Paris, pp. 770–773. Berlin/Heidelberg: Springer-Verlag.

Fono, D. and Vertegaal, R. (2005). EyeWindows: Evaluation of eye-controlled zooming windows for focus selection, in the *Proceedings of the SIGCHI Conference on Human Factors in Computing Systems (CHI '05)*, 2–7 April 2005, Portland, OR, pp. 151–160. New York: ACM Press.

Frey, L. A., White, K. P., Jr. and Hutchinson, T. E. (1990). Eye-gaze word processing. *IEEE Transactions on Systems, Man, and Cybernetics* 20: 944–950.

Friedman, M. B., Kiliany, G., Dzmura, M., and Anderson, D. (1982). The eyetracker communication system. *Johns Hopkins APL Technical Digest* 3: 250–252.

Gips, J., DiMattia, P., Curran, F. X., and Olivieri, P. (1996). Using EagleEyes—An electrodes based device for controlling the computer with your eyes—to help people with special needs, in *Interdisciplinary Aspects on Computers Helping People with Special Needs, Proceedings of the 5th International Conference on Computers Helping People with Special Needs (ICCHP '96)* (J. Klaus, E. Auff, W. Kremser, and W. Zagler, eds.), 17–19 July 1996, Linz, Austria, pp. 630–635. Vienna: R. Oldenburg.

Gips, J., Olivieri, C. P., and Tecce, J. J. (1993). Direct control of the computer through electrodes placed around the eyes, in *Human-Computer Interaction: Applications and Case Studies (Proceedings of HCI International 1993)* (M. J. Smith and G. Salvendy, eds.), 8–13 August 1993, Orlando, FL, pp. 630–635. Amsterdam, The Netherlands: Elsevier Science.

Goldberg, J. H. and Wichansky, A. M. (2003). Eye tracking in usability evaluation: A practitioner's guide, in *The Mind's Eye: Cognitive and Applied Aspects of Eye Movement Research* (J. Hyönä, R. Radach, and H. Deubel, eds.), pp. 493–516. Amsterdam, The Netherlands: North Holland.

Grauman, K., Betke, M., Lombardi, J., Gips, J., and Bradski, G. R. (2003). Communication via eye blinks and eyebrow raises: Video-based human-computer interfaces. *Universal Access in the Information Society* 2: 359–373.

Haber, R. N. and Hershenson, M. (1973). *The Psychology of Visual Perception*. London: Holt, Rinehart and Winston.

Hansen, J. P., Andersen, A. W., and Roed, P. (1995). Eye-gaze control of multimedia systems, in *Symbiosis of Human and Artifact, Proceedings of the 6th International Conference on Human Computer Interaction (HCI International 1995)* (Y. Anzai, K. Ogawa, and H. Mori, eds.), 9–14 July 1995, Tokyo, pp. 37–42. Amsterdam, The Netherlands: Elsevier Science.

Hansen, J. P., Hansen, D. W., and Johansen, A. S. (2001). Bringing gaze-based interaction back to basics, in *Universal Access in HCI: Towards an Information Society for All, Volume 3 of the Proceedings of the 9th International Conference on Human-Computer Interaction (HCI International 2001)* (C. Stephanidis, ed.), 5–10 August 2001, New Orleans, pp. 325–328. Mahwah, NJ: Lawrence Erlbaum Associates.

Hansen, J. P., Hansen, D. W., Johansen, A. S., and Elvesjö, J. (2005). Mainstreaming gaze interaction towards a mass market for the benefit of all, in *Universal Access in HCI: Exploring New Interaction Environments, Volume 7 of the Proceedings of the 11th International Conference on Human-Computer Interaction (HCI International 2005)* (C. Stephanidis, ed.) 9–14 July 1995, Tokyo [CD-ROM]. Mahwah, NJ: Lawrence Erlbaum Associates.

Hansen, J. P., Johansen, A. S., Hansen, D. W., Itoh, K., and Mashino, S. (2003a). Command without a click: Dwell time typing by mouse and gaze selections, in the *Proceedings of the 9th IFIP TC13 International Conference on Human-Computer Interaction (INTERACT '03)* (M. Rauterberg, M. Menozzi, and J. Wesson, eds.), 1–5 September 2003, Zurich, Switzerland, pp. 121–128. Amsterdam, The Netherlands: IOS Press.

Hansen, J. P., Johansen, A. S., Hansen, D. W., Itoh, K., and Mashino, S. (2003b). Language technology in a predictive, restricted onscreen keyboard with ambiguous layout for severely disabled people. *EACL 2003 Workshop on Language Modeling for Text Entry Methods*, 13 April 2003, Budapest, Hungary. http://www.it-c.dk/research/EyeGazeInteraction/Papers/Hansen_et_al_2003a.pdf.

Hansen, J. P., Tørning, K., Johansen, A. S., Itoh, K., and Aoki, H. (2004). Gaze typing compared with input by head and hand, in the *Proceedings of the 2004 Symposium on Eye Tracking Research and Applications (ETRA 2004)*, 22–24 March 2004, San Antonio, TX, pp. 131–138. New York: ACM Press.

Hori, J., Sakano, K., and Saitoh, Y. (2006). Development of a communication support device controlled by eye movements and voluntary eye blink. *IEICE Transactions on Information and Systems* 89: 1790–1797.

Hornof, A. J. and Cavender, A. (2005). EyeDraw: Enabling children with severe motor impairments to draw with their eyes, in the *Proceedings of the ACM Conference on Human Factors in Computing Systems (CHI 2005)*, 2–7 April 2005, Portland, OR, pp. 161–170. New York: ACM Press.

Hornof, A., Cavender, A., and Hoselton, R. (2004). EyeDraw: A system for drawing pictures with eye movements, in the *Proceedings of the 6th International ACM SIGACCESS Conference on Computers and Accessibility (ASSETS 2004)*, 18–20 October 2004, Atlanta, pp. 86–93. New York: ACM Press.

Hornof, A. and Sato, L. (2004). EyeMusic: Making music with the eyes, in the *Proceedings of the Conference on New Interfaces for Musical Expression (NIME04)* (M. J. Lyons, ed.), 3–5 June 2004, Hamamatsu, Japan, pp. 185–188. Singapore: National University of Singapore.

Hutchinson, T. E., White, K. P., Martin, W. N., Reichert, K. C., and Frey, L. A. (1989). Human–computer interaction using eye-gaze input. *IEEE Transactions on Systems, Man, and Cybernetics* 19: 1527–1534.

Hyrskykari, A., Majaranta, P., and Räihä, K.-J. (2005). From gaze control to attentive interfaces, in *Universal Access in HCI: Exploring New Interaction Environments, Volume 7 of the Proceedings of the 11th International Conference on Human-Computer Interaction (HCI International 2005)* (C. Stephanidis, ed.) 22–27 July 2005, Las Vegas [CD-ROM]. Mahwah, NJ: Lawrence Erlbaum Associates.

Isokoski, P. and Martin, B. (2006). Eye tracker input in first person shooter games, in the *Proceedings of the 2nd Conference on Communication by Gaze Interaction (COGAIN 2006)*, 4–5 September 2006, Turin, Italy, pp. 76–79. http://www.cogain.org/cogain2006/COGAIN2006_Proceedings.pdf.

Istance, H. O., Spinner, C., and Howarth, P. A. (1996). Providing motor impaired users with access to standard Graphical User Interface (GUI) software via eye-based interaction, in the *Proceedings of the 1st European Conference on Disability, Virtual Reality and Associated Technologies (ECDVRAT 1996)*, 8–10 July 1996, Maidenhead, U.K., pp. 109–116. http://www.icdvrat.reading.ac.uk/1996/papers/1996_13.pdf.

Jacob, R. J. K. (1991). The use of eye movements in human-computer interaction techniques: What you look at is what you get. *ACM Transactions on Information Systems* 9: 152–169.

Jacob, R. J. K. (1993). Eye movement-based human-computer interaction techniques: Toward non-command interfaces, in *Advances in Human-Computer Interaction* (H. R. Hartson and D. Hix, eds.), Volume 4, pp. 151–190. Norwood, NJ: Ablex Publishing Co.

Jacob, R. J. K. (1995). Eye tracking in advanced interface design, in *Virtual Environments and Advanced Interface Design* (W. Barfield and T. A. Furness, eds.), pp. 258–288. New York: Oxford University Press.

Jacob, R. J. K. and Karn, K. S. (2003). Eye tracking in human-computer interaction and usability research: Ready to deliver the promises (section commentary), in *The Mind's Eye: Cognitive and Applied Aspects of Eye Movement Research* (J. Hyönä, R. Radach, and H. Deubel, eds.), pp. 573–605. Amsterdam, The Netherlands: Elsevier Science.

Jordansen, I. K., Boedeker, S., Donegan, M., Oosthuizen, L., di Girolamo, M., and Hansen, J. P. (2005). D7.2 Report on a market study and demographics of user population. *Communication by Gaze Interaction (COGAIN)*. IST-2003-511598: Deliverable 7.2. http://www.cogain.org/results/reports/COGAIN-D7.2.pdf.

Jönsson, E. (2005). *If Looks Could Kill: An Evaluation of Eye Tracking in Computer Games*. Master's thesis, Royal Institute of Technology, Stockholm, Sweden.

Just, M. A. and Carpenter, P. A. (1976). Eye fixations and cognitive processes. *Cognitive Psychology* 8: 441–480.

Kumar, M., Paepcke, A., and Winograd, T. (2007). EyePoint: Practical pointing and selection using gaze and keyboard, in the *Proceedings of the ACM Conference on Human Factors in Computing Systems (CHI 2007)*, 28 April–3 May 2007, San Jose, CA, pp. 421–430. New York: ACM Press.

Land, M. F. and Furneaux, S. (1997). The knowledge base of the oculomotor system. *Philosophical Transactions: Biological Sciences* 352: 1231–1239.

Lankford, C. (2000). Effective eye-gaze input into Windows, in *Proceedings of the Eye Tracking Research and Applications Symposium 2000 (ETRA'00)*, 6–8 November 2000, Palm Beach Gardens, FL, pp. 23–27. New York: ACM Press.

Levine, J. L. (1981). *An Eye-Controlled Computer*. Research report RC-8857, IBM Thomas J. Watson Research Center, Yorktown Heights, NY.

Li, D., Babcock, J., and Parkhurst, D. J. (2006). openEyes: A low-cost head-mounted eye-tracking solution, in the *Proceedings of the ACM Eye Tracking Research and Applications Symposium 2006 (ETRA 2006)*, pp. 95–100. New York: ACM Press.

Majaranta, P., MacKenzie, I. S., Aula, A., and Räihä, K.-J. (2006). Effects of feedback and dwell time on eye typing speed and accuracy. *Universal Access in the Information Society* 5: 199–208.

Majaranta, P. and Räihä, K.-J. (2002). Twenty years of eye typing: Systems and design issues, in the *Proceedings of the Symposium on Eye Tracking Research and Applications 2002 (ETRA 2002)*, 25–27 March 2002, New Orleans, pp. 15–22. New York: ACM.

Majaranta, P. and Räihä, K.-J. (2007). Text entry by gaze: Utilizing eye-tracking, in *Text Entry Systems: Mobility, Accessibility, Universality* (I. S. MacKenzie and K. Tanaka-Ishii, eds.), pp. 175–187. San Francisco: Morgan Kaufmann.

Monden, A., Matsumoto, K., and Yamato, M. (2005). Evaluation of gaze-added target selection methods suitable for general GUIs. *International Journal of Computer Applications in Technology* 24: 17–24.

Morimoto, C. H. and Mimica, M. R. M. (2005). Eye gaze tracking techniques for interactive applications. *Computer Vision and Image Understanding* 98: 4–24.

Nakano, Y., Nakamura, A., and Kuno, Y. (2004). Web browser controlled by eye movements, in the *Proceedings of the IASTED International Conference on Advances in Computer Science and Technology 2004 (ACST 2004)*, 22–24 November 2004, St. Thomas, US Virgin Islands, pp. 93–98. Calgary: ACTA Press.

Nielsen, J. (1993). Noncommand user interfaces. *Communication ACM* 36: 82–99.

Ohno, T. (1998). Features of eye gaze interface for selection tasks, in the *Proceedings of the 3rd Asia Pacific Computer-Human Interaction (APCHI' 98)*, 15–17 July 1998, Kanagawa, Japan, pp. 176–182. Washington, DC: IEEE Computer Society.

Rayner, K. (1995). Eye movements and cognitive processes in reading, visual search, and scene perception, in *Eye Movement Research: Mechanisms, Processes and Applications* (J. M. Findlay, R. Walker, and R. W. Kentridge, eds.), pp. 3–22. Amsterdam, The Netherlands: North Holland.

Shell, J. S., Vertegaal, R., Cheng, D., Skaburskis, A. W., Sohn, C., Stewart, A. J., et al. (2004). ECSGlasses and EyePliances: Using attention to open sociable windows of interaction, in the *Proceedings of the Symposium on Eye Tracking Research and Applications 2004 (ETRA 2004)*, 22–24 March 2004, San Antonio, TX, pp. 93–100. New York: ACM Press.

Shell, J. S., Vertegaal, R., and Skaburskis, A. W. (2003). EyePliances: Attention-seeking devices that respond to visual attention, in the *Extended Abstracts of Human Factors in Computing Systems (CHI 2003)*, 5–10 April 2003, Fort Lauderdale, FL, pp. 770–771. New York: ACM Press.

Sibert, L. E. and Jacob, R. J. K. (2000). Evaluation of eye gaze interaction, in the *Proceedings of the ACM Human Factors in Computing Systems Conference (CHI 2000)*, 1–6 April 2000, The Hague, The Netherlands, pp. 281–288. New York: ACM Press.

Smith, J. D. and Graham, T. C. (2006). Use of eye movements for video game control, in the *Proceedings of the 2006 ACM SIGCHI international Conference on Advances in Computer Entertainment Technology (ACE '06)*, 14–16 June 2006, Hollywood, CA. New York: ACM Press.

Stampe, D. M. and Reingold, E. M. (1995). Selection by looking: A novel computer interface and its application to psychological research, in *Eye Movement Research: Mechanisms, Processes and Applications* (J. M. Findlay, R. Walker, and R. W. Kentridge, eds.), pp. 467–478. Amsterdam, The Netherlands: Elsevier Science.

Surakka, V., Illi, M., and Isokoski, P. (2003). Voluntary eye movements in human–computer interaction, in *The Mind's Eye: Cognitive and Applied Aspects of Eye Movement Research* (J. Hyönä, R. Radach, and H. Deubel, eds.), pp. 73–491. Amsterdam, The Netherlands: Elsevier Science.

Surakka, V., Illi, M., and Isokoski, P. (2004). Gazing and frowning as a new technique for human–computer interaction. *ACM Transactions on Applied Perception* 1: 40–56.

Tchalenko, J. (2001). Free-eye drawing. *Point: Art and Design Research Journal* 11: 36–41.

Ten Kate, J. H., Frietman, E. E. E., Willems, W., Ter Haar Romeny, B. M., and Tenkink, E. (1979). Eye-switch controlled communication aids, in the *Proceedings of the 12th International Conference on Medical and Biological Engineering*, August 1979, Jerusalem, Israel, pp. 19–20.

Tobii (2006). *User Manual: Tobii Eye Tracker and ClearView Analysis Software*. Tobii Technology AB.

Velichkovsky, B., Sprenger, A., and Unema, P. (1997). Towards gaze-mediated interaction: Collecting solutions of the "Midas touch problem," in the *Proceedings of the IFIP TC13 International Conference on Human-Computer Interaction (INTERACT 1997)*, 14–18 July 1997, Sydney, Australia, pp. 509–516. London: Chapman and Hall.

Vertegaal, R. (2003). Attentive user interfaces. *Communications of the ACM* 46: 30–33.

Wade, N. J. and Tatler, B. W. (2005). *The Moving Tablet of the Eye: The Origins of Modern Eye Movement Research*. Oxford, U.K.: Oxford University Press.

Ward, D. J. and MacKay, D. J. C. (2002). Fast hands-free writing by gaze direction. *Nature* 418: 838.

Ware, C. and Mikaelian, H. H. (1987). An evaluation of an eye tracker as a device for computer input, in the *Proceedings of the SIGCHI/GI Conference on Human Factors in Computing Systems and Graphics Interface (CHI and GI '87)*, 5–9 April 1987, Toronto, Canada, pp. 183–188. New York: ACM Press.

Wolpaw, J. R., Birbaumer, N., McFarland, D. J., Pfurtscheller, G., and Vaughan, T. M. (2002). Brain-computer interfaces for communication and control. *Clinical Neurophysiology* 113: 767–791.

Yamada, M. and Fukuda T. (1987). Eye word processor (EWP) and peripheral controller for the ALS patient. *IEEE Proceedings Physical Science, Measurement and Instrumentation, Management and Education* 134: 328–330.

Zhai, S., Morimoto, C., and Ihde, S. (1999). Manual and Gaze Input Cascaded (MAGIC) pointing, in the *Proceedings of the SIGCHI Conference on Human Factors in Computing Systems: The CHI Is the Limit (CHI'99)*, 15–20 May 1999, Pittsburgh, pp. 246–253. New York: ACM Press.

37

Brain-Body Interfaces

Paul Gnanayutham
and Jennifer George

37.1 Introduction

As medical technology not only extends the human natural life span, but also leads to increased survival from illness and accidents, the number of people with disabilities is constantly increasing. In comparison to all types of disability, those to the brain are among the most likely to result in death or permanent disability. In the European Union, brain injury accounts for 1 million hospital admissions per year (Council of Europe, 2002). Injury is the leading cause of death for children in Europe. For every child that dies from injuries, another 160 children are admitted to a hospital for a severe traumatic brain injury (Vincenten, 2001). Each year in the United States, an estimated 1.4 million people sustain a brain injury (Langlois et al., 2004; North American Brain Injury Society, 2005). Studies have reported personality changes attributed to brain damage, which contribute to the perception of those with brain injury as social misfits. As a result of this, individuals with brain damage often face difficulty in adjusting to their injuries, causing extreme isolation and loneliness (DeHope and Finegan, 1999). Brain-damaged patients typically exhibit deficiency in memory, attention, concentration, analyzing information, perception, language abilities, and emotional and behavioral areas (Serra and Muzio, 2002). In the United Kingdom, out of every 100,000 of the population, between 100 and 150 people suffer a brain damage (Tyrer, 2005). Some cannot communicate, recreate, or control their environment due to severe motor impairment.

Brain-body interfaces have been defined as a real-time communication system designed to allow a user to voluntarily send messages without sending them through the brain's normal output pathways such as speech, gestures, or other motor functions, but only using bio-signals from the brain (Wolpaw et al., 1997, 2000). This type of communication system is needed by disabled individuals who have parts of their brain active, but have no means of communicating with the outside world (Wolpaw et al., 2003). There are two types of brain-body interfaces, namely *invasive* (signals obtained by surgically inserting probes inside the brain) and *noninvasive* (electrodes placed externally on part of the body). Existing assistive technology is of limited if any use for many people with this disability. Over the last four decades, advances have been made in the design and construction of aids for the brain injured and motor impaired. These fall into two categories. The first requires the use of some movement, speech, or breath to operate a computer input device. Examples include:

- *HeadMouse*: Using wireless optical sensors that transform head movements into cursor movements on the screen (Origin Instruments, 1998)
- *Tonguepoint*: A system mounted on a mouthpiece (Salem and Zhai, 1997)
- *Eye-tracking*: A system follows eye movements (Senso-Motoric Instruments, 2007)

Many disabled people have impairments that make it impossible for them to use any devices in this category. Research has been carried out on the brain's electrical activities since 1925 (Kozelka and Pedley, 1990). Brain-body interfaces, also called brain-computer interfaces or brain-machine interfaces, provide new augmentative communications channels for those with severe motor impairments.

The rest of this chapter covers: the brain and the type of brain injuries that could be sustained, bio-potentials that could be obtained from the brain, usage of bio-potentials to develop a brain-body interface, applications of brain-body interfaces, and the challenges and the future of brain-body interfaces.

37.2 Brain

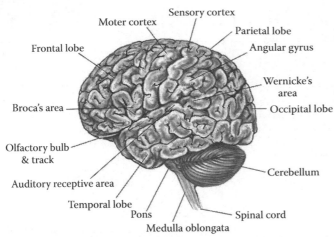

FIGURE 37.1 Brain map. (Courtesy of http://www.headinjury.com.)

The three main components of the brain are the cerebellum, cerebrum, and brain stem (pons and medulla oblongata). The cerebellum is located between the brain stem and the cerebrum. The cerebellum controls facial muscle coordination, and damage to this area affects the ability to control facial muscles, thus affecting signals (eye movements and muscle movements) needed by brain-body interfaces (Figure 37.1).

The cerebrum is the largest part of the brain and sits on top of the cerebellum and contains large folds of brain matter in grooves (Kalat, 1995). The cerebrum is divided into two hemispheres, the right and the left. The dividing point is a deep groove called the longitudinal cerebral fissure. The left hemisphere controls the right side of the body, while the right hemisphere controls the left side of the body. The cerebrum is the section where thoughts are created and memory is stored. The associated brain waves may be used in brain-body interfaces. The cerebrum also has five lobes, which are the frontal lobe, occipital lobe, temporal lobe, parietal lobe, and insular lobe. Injury to the cerebrum can leave a person fully aware of her surroundings but unable to react to any events happening in the surroundings (Berkow et al., 1997). The frontal lobe contains the motor cortex, which creates alpha brain waves. The occipital lobe contains the visual cortex. The visual cortex affects visual perception, which creates brain waves (Schmolesky, 2006). The temporal lobe contains the cranial nerve and auditory cortex (Berkow et al., 1997). Damage to this region may affect a person's hearing. The parietal lobe contains the primary somatosensory cortex. Damage to this area of the brain affects the ability to use bio-potentials to manipulate a brain-body interface. The insular lobe affects emotion, and damage

to this region may affect a person's ability to relax when using a brain-body interface.

The brain stem controls basic functions such as eating, respiration, and heart rate (Fridlund, 1994) and also controls cognition (Berkow et al., 1997). It is connected to the spinal cord and covered by a small flap of brain tissue known as the dura. The cranial nerves that carry the signals to control facial movements also originate in the brain stem, hence the brain stem is of interest when using brain-body interfaces.

There are two main stages in traumatic brain injury, the primary and the secondary. The secondary brain injury occurs as a response to the primary injury. In other words, primary brain injury is caused initially by trauma, amyotrophic lateral sclerosis, brain stem stroke, and so on, but includes the complications that can follow, such as damage caused by lack of oxygen, and rising pressure and swelling in the brain. A brain injury can be seen as a chain of events beginning with the first injury, which occurs in seconds after the accident, and being worsened by a second injury, which happens in minutes and hours after this, depending on when skilled medical intervention occurs (Kalat, 1995; Berkow et al., 1997; Schmolesky, 2006). There are three types of primary brain injury: closed, open, and crush. Closed head injuries are the most common type, and are so called because no break of the skin or open wound is visible. Open head injuries are not so common. In this type of injury, the skull is opened and the brain exposed and damaged. In crush injuries the head might be caught between two hard objects. This is the least common type of injury, and often damages the base of the skull and nerves of the brain stem rather than the brain itself. Individuals with brain injury require frequent assessments and diagnostic tests (Sears and Young, 2003). Most hospitals use either the Glasgow Coma Scale for predicting early outcome from a head injury (e.g., whether the person will survive) or Rancho Levels of Cognitive Functioning for predicting later outcomes of head injuries (Roy, 2004).

37.3 Bio-Potentials for Brain-Body Interfaces

This section describes the bio-potentials that can be used in brain-body interfaces. Bio-potentials are electrical signals from the brain that can be obtained from the skull, forehead, or other parts of the body (the skull and forehead are predominantly used because of the richness of bio-potentials in these areas). Each bio-potential has its own unique characteristics, such as amplitude, frequency, method of extraction, and time of occurrence. Each brain-injured patient (apart from persistent vegetative state patients) can produce one or more of these bio-potentials with differing degrees of consistency. Brain-injured patients can operate brain-body interfaces depending on the reliability of the bio-potential that they can produce. Current brain-body interfaces can transfer data up to 68 bits/second (Gao et al., 2003).

37.3.1 Electroencephalography

Electroencephalography (EEG) measures electrical brain activity that results from thoughts or imagined movements (Kalcher et al., 1994). Electroencephalographic signals can be collected by electrodes placed on the scalp or forehead (Berkow et al., 1997). The amplitude can vary between 10 and 100 μV when measured on the scalp or forehead. Electroencephalography covers a frequency spectrum of 1 to 30 Hz, and is divided into five classes. Authorities on electroencephalography dispute the exact frequency demarcation points of the five classes (Berg et al., 1998). Robinson sampled electroencephalographic signals from 93 participants and classified them as delta, theta, alpha, beta, and high beta (Robinson, 1999). Robinson's classification will be used throughout this chapter. Delta and theta waves cannot be used in brain-body interfaces because they indicate unconsciousness, deep sleep, or stress in an individual. Alpha, beta, and high beta electroencephalographic signals can be used as bio-potentials for brain-body interfaces.

37.3.2 Alpha Waves (Mu Waves)

Alpha waves have a frequency range of 8 to 12 Hz. The alpha wave is collected through electrodes placed over a large fold in the brain known as the central sulcus (Kozelka and Pedley, 1990) or on the forehead (Berg et al., 1998). Eye closures often produce strong signals that also affect electrical activity in the alpha range. Kalcher et al. (1994) say that movement of a limb or imagined movement of a limb also affects alpha waves. Alpha waves can be used for controlling brain-body interfaces.

37.3.3 Beta Waves

Beta waves have a frequency range of 12 to 20 Hz. It has been observed (Berg et al., 1998) that people with brain lesions have diminished capabilities to manipulate beta waves. In Berg's work, military pilots used a brain-body interface with beta settings to control one axis of the cursor in a flight simulator thus creating a brain-body interface.

37.3.4 High Beta Waves

High beta waves have a frequency range of 20 to 30 Hz. Facial movements often produce strong signals at approximately 45 Hz that also affect electrical activity in both the beta and high beta ranges (Berg et al., 1998). High beta waves can be used for controlling brain-body interfaces.

37.3.5 Electromyography

Electromyography (EMG) measures an electrical signal resulting from a contracted muscle (Berkow et al., 1997). The moving of an eyebrow, for example, is a muscle contraction that produces waves at 18 Hz, but resonates throughout the electroencephalographic spectrum (Berg et al., 1998). Electromyographic signals can be collected on the arms, legs, or face, because muscle contractions may occur there. Electromyographic signals have an amplitude range of 0.2 to 2000 μV.

37.3.6 Electro-oculography

Electro-oculographic (EOG) signals are low-frequency signals derived from the resting potential (corneal-retinal potential) by ocular or eyeball movements (Tortora and Derrickson, 2006). Eyeball movements affect the electroencephalographic spectrum in the delta and theta regions between 1.1 and 6.25 Hz (Roy, 2004). Electro-oculographic signals have an amplitude range of 1 to 4 mV. Electro-oculographic signals can be used for navigating a cursor to left or right on a computer screen using a brain-body interface.

37.3.7 Slow Cortical Potentials

Slow cortical potentials (SCPs) are signals of the cerebral cortex that can be collected from the scalp surface. They are electroencephalographic oscillations in the frequency range 1 to 2 Hz (Kotchoubey et al., 1997) and can be positive or negative. The signals can be 5 to 8 μV, and a person may be trained to change the amplitude of slow potential signals to indicate a selection such as for a spelling device. Birbaumer and his team developed a spelling device that had letters appearing on a screen at random. The user could choose letters using slow cortical potentials, thus constructing words (Birbaumer et al., 1999; Hinterberger et al., 2003).

37.3.8 Evoked Potential

Another signal detected in the electroencephalographic range is the evoked potential (EP), also known as an event-related brain potential (ERP). Evoked potential can be a positive or negative signal and can occur at various times after visual or auditory stimuli. Evoked potentials occur when a person concentrates on an object. Evoked potentials are of relatively low amplitude signals with a range of 1 to 10 μV in comparison with electroencephalographic signals (10 to 100 μV). When someone sees or hears anything that is especially meaningful to him then a special response is produced such as steady-state visual evoked potential, P300 and N400. Electroencephalography measures all brain activity at any point in time, while the evoked potential is that part of the activity associated with the processing of a specific event (post stimuli). One application could be dialing a telephone number pad using evoked potentials.

37.3.9 Steady-State Visual Evoked Potential/ Steady-State Visual Evoked Responses

Steady-state visual evoked potentials (SSVEPs), also known as steady-state visual evoked responses (SSVERs), are obtained when users can indicate their interest in specific stimuli by choosing to attend or ignore it (Cheng et al., 2002; Gao et al., 2003). This allows a user to send information by voluntarily modulating her attention, through SSVEP (e.g., choosing buttons flashing at different rates on a virtual telephone keypad to make a phone call). SSVEP uses the 4- to 35-Hz frequency range. SSVEPs transfer data at high data transfer rates (68 bits/s) at time intervals of 100 to 1000 ms after the stimuli.

37.3.9 P300

The P300 (also called P3) is a component of the evoked potential range of brain waves. P300 displays a brain wave with positive amplitude, peaking at around 300 ms after task-relevant stimuli. This signal occurs in the delta (0.5 to 4 Hz) and theta (4 to 7 Hz) frequency range. Kotchoubey and his team (2001, 2002) investigated bio-potentials in patients with severe brain damage. They used oddball tasks (two stimuli with different probabilities, e.g., 80/20) using signals such as sine tones, complex tones, or vowel sounds *o* and *i*, to elicit P300 waves from 25 out of 33 patients. The P300 is perhaps the most studied evoked potentials component in investigations of selective attention and information processing (e.g., for choosing letters on a computer screen to communicate) in comparison to the other components of the evoked potentials (Farwell and Donchin, 1988; Patel and Azzam, 2005). The key stroke level model gives an average of 200 to 280 ms for an average typist to type a character or press a key on a keyboard (Card et al., 1983; Kieras, 2005). The times given by key stroke level model compare favorably with the P300 task-relevant stimuli, but the participants using the P300 will have problems processing the letters on screen at this slow speed since our brain processes information in chunks (Hinterberger et al., 2005).

37.3.10 N400

The N400 is a component of the evoked potential range of brain waves. N400 displays a brain wave with negative amplitude, peaking at around 400 ms triggered by unexpected linguistic stimuli. The N400 is most pronounced over centroparietal regions of the scalp and tends to be larger over the right hemisphere. This brain wave is used for applications similar to that of P300 (Spencer et al., 2004).

37.3.11 Electrocochleography

Electrocorticographic (ECoG) signals are obtained by recording brain surface signals with electrodes located on the surface of the cortex (invasive method). It is an alternative to data taken noninvasively by electrodes outside the brain on the skull such as in electroencephalography, electromyography, and evoked potential. The bio-potentials obtained by electrocochleography are applied in a brain-body interface in similar ways to electroencephalographic signals but electrocochleography records at 300 to 1000 μV amplitude and has a frequency of 40 Hz (Tran et al., 1997).

37.3.12 Low-Frequency Asynchronous Switch Design

The low-frequency asynchronous switch design (LF-ASD) was introduced as an invasive brain-body interface technology for asynchronous control applications. The low-frequency asynchronous switch design operates as an asynchronous brain switch (ABS) that is activated only when a user intends to control. The switch is placed on the scalp. It maintains an inactive state output when the user is not meaning to control the device (i.e., he may be idle, thinking about a problem, or performing some other action). The LF-ASD is based on electroencephalographic signals in the 1- to 4-Hz frequency range (Borisoff et al., 2004) with an amplitude of 10 to 100 μV.

37.3.13 Local Field Potential

Signals can be recorded in the human frontal cortex using implanted microwires in the sensorimotor regions of the neocortex, which exhibit synchronous oscillations in the 15- to 30-Hz frequency range and have an amplitude of 6 μV. These signals are also prominent in the cerebellum and brain stem sensorimotor regions. These signals are called local field potentials (LFPs). Multiple electrodes can be used to record these local field potentials, which can be synchronized with the execution of trained and untrained movements of limbs. Local field potentials provide an excellent source of information about the cognitive state of the subject and can be used for neural prosthetic applications (Harrison et al., 2004; Kennedy et al., 2004).

37.3.14 Role of Bio-Potentials in Brain-Body Interfaces

Electroencephalography gives access to one bio-potential (brain waves) that can be found in every brain-injured patient, but the amplitude of this signal is rather small (10 to 100 μV). However, in the absence of any other signal, electroencephalography can be used in brain-body interfaces (BBIs). Electromyographic signals (muscle movements) and electrooculographic signals (eye movements) are two bio-potentials with high amplitude (1 to 4 mV) that can be used in brain-body interfaces, but users must be able to move their muscles and eyes in a controlled manner to apply these two bio-potentials. The latter could also be used to operate other assistive devices, such as an eye tracker or switch. There are other bio-potentials, positive and negative, which occur after a period of a stimulus to indicate selection, such as slow cortical potentials, steady-state visual evoked potential, P300, and N400. Researchers have tried to use these bio-potentials for spelling devices and other information-processing brain-body interfaces, with limited success.

It can be concluded that electromyographic and electrooculographic signals are the two front-runners for the most suitable bio-potentials for noninvasive brain-body interfaces because they are high-amplitude bio-potentials, which can be easily produced by a patient in comparison to other bio-potentials. Experiments with bio-potentials obtained by invasive means are limited in comparison to noninvasive bio-potentials, due to the medical intervention needed to access the neurons and the risks involved in opening the skull. The signals obtained are noise free in comparison with the noninvasive bio-potentials. Electroencephalographic signals, electromyographic signals, and electrocochleographic signals are three examples of bio-potentials obtained by invasive technology. From these three bio-potentials, electrocochleographic signals offer the highest amplitude (300 to 1000 μV) and are the strongest contenders to using invasive technology. The risks, difficulties, and requirements involved in setting up an invasive brain-body interface system make noninvasive brain-body interfaces the preferred choice for an assistive technology device.

37.4 Components of Brain-Body Interfaces

Noninvasive technology involves the collection of control signals for the brain-body interface without the use of any surgical techniques, with electrodes placed on the face, skull, or other parts of their body. The signals obtained are first amplified, then filtered and thereafter converted from an analogue to a digital signal (Figure 37.2). Various electrode positions are chosen by the developers, such as electrode caps, electrode headbands with different positions and number of electrodes or the International 10-20 system (Pregenzer et al., 1994). Authorities dispute the number of electrodes needed for collection of usable bio-potentials (Berg et al., 1998). Junker recommends using three electrodes for collecting signals (Junker, 1997) while Keirn and Aunon (1990) recommend using six electrodes. Chatrian and his team claim that at least 20 electrodes are needed (Chatrian et al., 1996). The caps may contain as many as 256 electrodes, though typical caps use 16, 32, 64, or 128 positions, and each cap has its own potential sources of error. High-density caps can yield more information, but in practice they are hard to utilize for

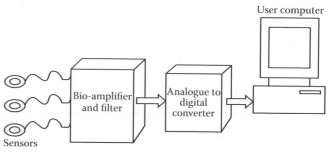

FIGURE 37.2 Noninvasive brain-body interface.

real-time communications (Nunez et al., 1999). This is due to the fact that the bio-potentials obtained from these large numbers of electrodes could possibly need extensive offline processing to make any sense of what the user is trying to express. There is only one agreed upon standard for the positions and number of electrodes, which is the International 10-20 system of electrodes (Jasper, 1958).

Invasive electrodes can give better noise to signal ratio and obtain signals from a single or small number of neurons (Figure 37.3). However, signals collected from the brain require expensive and dangerous surgical measures. There are two types of electrodes used for invasive brain-body interfaces. If signals need to be obtained with the least noise and from one or few neurons, neurotrophic electrodes are used (Kennedy, 1999; Siuru, 1999; Kennedy et al., 2000). The other choice is the Utah Intracranial Electrode Array (UIEA), which contains 100 penetrating silicon electrodes, placed on the surface of the cortex with needles penetrating into the brain, which can be used for recording and simulating neurons (Maynard et al., 1997; Spiers et al., 2005). Neuron discrimination (choice of single or a group of neurons) does not play any part in the processing of signals in brain-body interfaces (Sanchez et al., 2005).

Brain activity produces electrical signals that can be read by electrodes placed on the skull, forehead, or other part of the body (the skull and forehead are predominantly used because

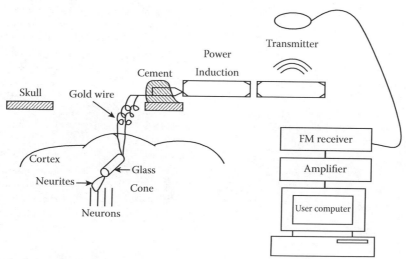

FIGURE 37.3 Invasive brain-body interface.

of the richness of bio-potentials in these areas). Algorithms then translate these bio-potentials into instructions to direct the computer, so that people with brain injury have a channel to communicate without using the normal channels. Various research groups have developed brain-body interfaces using electrocochleography, electroencephalography, magnetic resonance imaging, and slow cortical potentials (Table 37.2). Invasive electrodes can give better signal to noise ratio and obtain signals from a single or small number of neurons. Invasive brain-body interfaces have been developed using electroencephalography,

electromyography, electrocochleography, magnetic resonance imaging, slow cortical potentials, neuroprosthetic signals, and low-frequency asynchronous switch.

The last 10 years have seen many research groups developing brain-body interfaces, either noninvasive (Table 37.1) or invasive (Table 37.2) types. Tables 37.1 and 37.2 show a summary of research conducted during this period.

Most of the brain-body interfaces listed in Tables 37.1 and 37.2 were used for basic communications, such as research completed by authors of this chapter, who developed an assistive

TABLE 37.1 Summary of Noninvasive Brain-Body Interfaces

Dates	Researcher/Research Group	Brain-Body Interfaces	Participants	Outcome
1997–1999	Craig et al.	Alpha wave-based	21 able-bodied and 16 disabled	95% able and 93% disabled; used eye closure to switch devices (Craig et al., 1997, 1999)
1997–2000	Kostov and Polak	EEG-based	1 able-bodied and 1 disabled	70%–85% success in moving a cursor in real-time (Kostov and Polak, 1997a,b, 2000)
1991–1998	Wolpaw et al.	EEG-based	5 able-bodied	41%–90% success in moving a cursor around a screen (Wolpaw and McFarland, 1994; Wolpaw et al., 1997; Miner et al., 1998)
1990	Keirn and Aunon	EEG-based	5 able-bodied	90% success in choosing one of six letters on a screen (Keirn and Aunon, 1990)
1990	Barreto et al.	EEG- and EMG-based	6 able-bodied	Moving a cursor around a screen and also mouse clicks; success rate not given (Barreto et al., 1999)
1990	Knapp and Lusted	EEG-, EMG-, and EOG-based	1 disabled	Move a cursor; no other data available (Lusted and Knapp, 1996)
1996	Knapp and Lusted	EEG-, EMG-, and EOG-based	6 able-bodied	65% success in hitting a target on screen (Barea et al., 2000)
1999–2002	Doherty et al.	EEG-, EMG-, and EOG-based	3 disabled	60% success in hitting a target on screen (Doherty et al., 2001)
1994	Pfurtscheller et al.	EEG-, EMG-, and EOG-based	4 able-bodied	50% success in Extend a bar on screen (Kalcher et al., 1994)
1999	Pfurtscheller et al.	EEG-, EMG-, and EOG-based	4 able-bodied	87% success in Extend a bar on screen (Neuper et al., 1999)
2001	Pfurtscheller et al.	EEG-, EMG-, and EOG-based	3 able-bodied	70%–95% success in Extend a bar on screen (Pfurtscheller and Neuper, 2001)
1988–2000	Donchin et al.	P300-based	10 able-bodied and 4 disabled	Able-bodied selected six to eight letters per minute while disabled selected three per minute (Donchin et al., 2000)
1998–2003	Bayliss and Bollard	P300-based	5 able-bodied	50%–90% success in completing virtual driving (Bayliss, 2003; Hinterberger et al., 2005)
1999–2003	Birbaumer et al.	SCP-based	5 disabled	75% success in using the developed spelling device (Birbaumer, 2003; Schalk et al., 2004)
2003–2007	Birbaumer et al.	EEG-, fMRI-, and SCP-based	5 able-bodied and 6 disabled	Average one letter per minute (Birbaumer et al., 2003)
2002–2007	Birch and Mason	LF-ASD-based	5 able-bodied and 2 disabled	78% able and 50% disabled; success in producing signals (Borisoff et al., 2004)
2002–2007	Cheng et al.	SSVER-based	13 able-bodied	62% success in sending information to a computer (Beverina et al., 2003)
2003–2007	Weiskopf et al.	fMRI-based	No data available	No data available (Weiskopf et al., 2003)
2000–2008	Akoumianakis et al.	Adaptive user interfaces- and interface agents-based	No data available	No data available (Akoumianakis et al., 2000; Friedman et al., 2007)
2004–2008	Navarro	Bluetooth-based	No data available	No data available (Navarro, 2004)

TABLE 37.2 Summary of Invasive Brain-Body Interfaces

Dates	Researcher/Research Group	Brain-Body Interfaces	Participant	Outcome
1999–2000	Kennedy et al.	EEG- and EMG-based	5 able-bodied and 2 disabled	78% able and 50% disabled; success in producing signals (Kennedy et al., 1999, 2000)
1999–2007	Levine et al.	ECoG-based	13 able-bodied	62% success in sending information to a computer (Levine et al., 1999)
2005–2007	Birbaumer et al.	ECoG-based	No data available	Not data available (Lal et al., 2005)
2004–2007	Stanford University	Neuroprosthetic-based	No data available	No data available (Harrison et al., 2004)
2002–2007	Tsinghua University	Motor functions-based	No data available	No data available (Musallam et al., 2004)

device called the Cyberlink (Junker, 1997; Metz and Hoffman, 1997; Berg et al., 1998; Cyberlink–Brainfinger, 2007), that combined eye movement, facial muscle, and brain wave bio-potentials detected at the user's forehead to navigate a cursor to hit the targets shown below thereby communicating using their bio-potentials (Gnanayutham et al., 2003, 2005; Gnanayutham, 2004a,b, 2005).

Vision-impaired participants and comatose participants were the two groups of nonverbal quadriplegic brain-injured people who could not be included in the previous study. Further studies showed how the vision impaired could also be included in using brain-body interfaces to communicate (Gnanayutham and George, 2007, 2008). The efforts appearing in Tables 37.1 and 37.2 have carried out extensive work and created many applications. For example, the P300 has been used for choosing letters on a computer screen to communicate (Farwell and Donchin, 1988; Patel and Azzam, 2005); SSVER has been used to send information by voluntarily modulating the attention (e.g., choosing buttons flashing at different rates, on a virtual telephone keypad to make a phone call). One of the well-known applications for electromyography as a BBI is HaWCoS: the hands-free wheelchair control system developed at the University of Siegen in Germany (Felzer and Freisleben, 2002). Hence, brain-body interfaces can be used to communicate by using targets that contain words or phrases, switch devices for on/off, launch computer applications, spelling by choosing individual words, dialing a telephone,

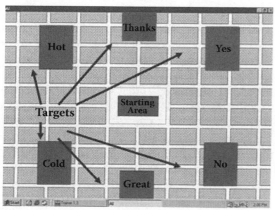

FIGURE 37.4 Interface used by authors' participants to communicate.

surfing the net, operating rehabilitation robots, controlling wheelchairs, and so on. The challenges that need to be dealt with by brain-body interfaces consist of choosing the appropriate bio-potentials for the individual users, making provisions for individuals to use the brain-body interface in a controlled and consistent way with minimum training and frustration.

37.5 Conclusions

Diagnostics and measurements of brain injuries have progressed, but accessible and reliable brain-body interfaces are necessary to make progress in rehabilitating brain-injured patients, as well as to improve the quality of life of patients. Brain-body interfaces have not been shown to be dependable enough for main software manufacturers to integrate them into mainstream operating systems and applications. This trend is likely to continue, unless computer manufacturers see a need to invest in this area to help this group of computer users who need to operate a computer using bio-potentials.

The pace of research is increasing, and good progress is being made in the area of assistive technology. The last 10 years have seen more than 30 research groups working on developing brain-body interfaces, both invasive and noninvasive types. Extensive work has been carried out, and many applications have been created, for example, for spelling, surfing the net, operating robots, and controlling wheelchairs, and for real-time manipulation of bio-potentials obtained from the brain. Many brain-body interface research applications are laboratory implementations, with limited test results obtained from the brain-injured community as shown in Tables 37.1 and 37.2. This technology is promising but more evaluation has to be done with the disabled participants to show the capabilities of brain-body interfaces in the field. The evaluations have to be real-time with portable equipment that can be used in the field. Despite the potential shown by many of the brain-body interface devices, limited use is made by the disabled community. Hence, there is a clear need to take this technology outside the laboratory and into the field to nursing homes and hospitals. Brain-body interfaces are used in laboratories, and encouraging results are obtained, but hospitals and nursing homes (especially government-run ones) are very reluctant to try this research due to ethical reasons, and are worried about being sued if things go wrong. Permissions have to be obtained from disabled participants, their guardians, and also the hospital or nursing home in which they are located before any evaluation can be carried out using a brain-body interface. The hospitals and nursing homes the researchers worked with were either private or located abroad.

Research is being done in positron emission tomography and functional magnetic resonance imaging in the identification of residual cognitive function in persistent vegetative state patients to investigate the possible use as a brain-body interface. Research is also being done in wearable wireless brain-body interfaces where technology such as Bluetooth is proposed for transmitting and receiving signals from the participant. More than 30 centers all over the world have been set up during the last 10 years in the area of brain-body interfaces, and extensive research is being done in using bio-potentials to communicate, recreate, and control the environment. Conferences and workshops are being used by researchers to collaborate in this area. Standards such as BCI2000 have been agreed upon to standardize the results. The authors of this chapter believe it will only be a matter of time before this technology filters into the daily use of the disabled community.

Acknowledgments

Thanks to Dr. Chris Bloor, Prof. Gilbert Cockton, and Dr. Eamon Doherty and to the following institutes: Vimhans Delhi, Mother Teresa's Mission of Charity Delhi, Holy Cross Hospital Surrey, Castel Froma, Leamington Spa, and the Low Vision Unit of the National Eye Hospital Colombo.

References

Akoumianakis, D., Savidis, A., and Stephanidis, C. (2000). Encapsulating intelligent interactive behaviour in unified user interface artefacts. *Interacting with Computers* 12: 383–408.

Barea, R., Boquete, L., Mazo, M., López, E., and Bergasa, L. M. (2000). E.O.G. guidance of a wheel chair using neural networks. *IEEE Computer Society* 4: 668–671.

Barreto, A. B., Scargle, S. D., and Adjouadi, M. (1999). A real-time assistive computer interface for users with motor disabilities. *SIGCAP Newsletter* 64: 6–16.

Bayliss, J. D. (2003). The use of the P3 evoked potential component for control in a virtual apartment. *IEEE Transactions on Neural Systems and Rehabilitation Engineering* 11: 113–116.

Berg, C., Junker, A., Rothman, A., and Leininger, R. (1998). *The Cyberlink Interface: Development of A Hands-Free Continuous/Discrete Multi-Channel Computer Input Device.* Small Business Innovation Research Program (SBIR) Phase II Final Report, Brain Actuated Technologies, Yellow Springs, OH.

Berkow, R., Beers, M., Bogin, R., and Fletcher, A. (1997). *The Merck Manual of Medical Information.* Whitehouse Station, NJ: Merck Research Laboratories.

Beverina, F., Palmas, G., Silvoni, S., Piccione, F., and Glove, S. (2003). User adaptive BCIs: SSVEP and P300 based interfaces. *PsychNology Journal* 1: 331–354.

Birbaumer, N. (2003a). The thought translation device: A brain-computer interface for paralyzed, in *ISPRM, 2nd World Congress*, May 2003, Prague, pp. 117–121. Israel: Monduzzi Editore.

Birbaumer, N., Ghanayim, N., Hinterberger, T., Iversen, I., Kotchoubey, B., Kübler, A., Perelmounter, J., Taub§, E., and Flors, H. (1999). A spelling device for the paralysed. *Nature* 398: 297–298.

Birbaumer, N., Hinterberger, T., Kübler, A., and Neumann, N. (2003b). The Thought-Translation Device (TTD): Neurobehavioral mechanisms and clinical outcome. *IEEE Transactions on Neural Systems and Rehabilitation Engineering* 11: 120–123.

Borisoff, J. F., Mason, S. G., Bashashati, A., and Birch, G. E. (2004). Brain-computer interface design for asynchronous control applications: Improvements to the LF-ASD asynchronous brain switch. *IEEE Transactions on Biomedical Engineering* 51: 985–992.

Card, S., Moran, T., and Newell, A. (1983). *The Psychology of Human-Computer Interaction.* Hillsdale, NJ: Lawrence Erlbaum Associates.

Chatrian, G., Bergamasco, B., Bricolo, B., Frost, J., and Prior, P. (1996). IFCN recommended standards for electrophysiological monitoring in comatose and other unresponsive states. Report on IFCN committee. *Electroencephalography and Clinical Neurophysiology* 99: 103–122.

Cheng, M., Gao, X., and Gao, S. (2002). Design and implementation of a brain-computer interface with high transfer rates. *IEEE Transactions on Biomedical* 49: 1181–1186.

Council of Europe (2002). *Towards Full Social Inclusion of Persons with Disabilities.* Social, Health and Family Affairs Committee, Document 9632, Communication Unit of the Assembly.

Craig, A., Kirkup, L., McIsaac, P., and Searle, A. (1997). The mind as a reliable switch: Challenges of rapidly controlling devices without prior learning, in *Interact 97* (S. Howard, J. Hammond, G. Lindgaard, eds.), July 1997, Sydney, Australia, pp. 4–10. London: Chapman & Hall.

Craig, A., McIsaac, P., Tran, Y., Kirkup, L., and Searle, A. (1999). Alpha wave reactivity following eye closure: A potential method of remote hands free control for the disabled. *Technology and Disability* 10: 187–194.

Cyberlink–Brainfinger (2007). http://www.brainfingers.com.

DeHope, E. and Finegan, J. (1999). Self determination model: An approach to develop awareness for survivors of traumatic brain injury. *NeuroRehabilitation* 13: 3–12.

Doherty, E. P., Cockton, G., Bloor, C., and Benigno, D. (2001). Improving the performance of the cyberlink mental interface with the yes/no program, in *CHI 2001*, 20 March 2001, Seattle, pp. 69–76. New York: ACM Press.

Donchin, E., Spencer, K. M., and Wijesinghe, R. (2000). The mental prosthesis: Assessing the speed of a P300-based brain-computer interface. *IEEE Transactions on Rehabilitation Engineering* 8: 174–179.

Farwell, L. and Donchin, E. (1988). Talking off the top of your head: Toward a mental prosthesis utilizing event-related brain potentials. *Electroencephalography and Clinical Neurophysiology* 70: 510–523.

Felzer, T. and Freisleben, B. (2002). HaWCoS: The "hands-free" wheelchair control system, in the *Proceedings of ASSETS 2002*, 8–10 July 2002, Edinburgh, Scotland, pp. 127–134. New York: ACM Press.

Fridlund, A. (1994). *Human Facial Expression: An Evolutionary View*. San Diego: Academic Press.

Friedman, D., Leeb, R., Guger, C., Steed, A., Pfurtscheller, G., and Slater, M. (2007). Navigating virtual reality by thought: What is it like? *PRESENCE* 16: 100–110.

Gao, X., Xu, D., Cheng, M., and Gao, S. (2003). A BCI-based environmental controller for the motion-disabled. *IEEE Transactions on Neural Systems and Rehabilitation Engineering* 11: 137–140.

Gnanayutham, P. (2004a). Assistive technologies for traumatic brain injury. *ACM SIGACCESS Newsletter* 80: 18–21.

Gnanayutham, P. (2004b). The state of brain body interface devices. *UsabilityNews*. http://www.usabilitynews.com.

Gnanayutham, P. (2005). Personalised tiling paradigm for motor impaired users, in the *Proceedings of HCI International 2005 Conference*, 22–27 July 2005, Las Vegas [CD-ROM]. Mahwah, NJ: Lawrence Erlbaum Associates.

Gnanayutham, P., Bloor, C., and Cockton, G. (2003). AI to enhance a brain computer interface, in *Universal Access in HCI: Inclusive Design in the Information Society, Volume 4 of the Proceedings of the 10th International Conference on Human-Computer Interaction (HCI International 2003)* (C. Stephanidis, ed.), 22–27 June 2003, Crete, Greece, pp. 1397–1401. Mahwah, NJ: Lawrence Erlbaum Associates.

Gnanayutham, P., Bloor, C., and Cockton, G. (2005). Discrete acceleration and personalised tiling as brain body interface paradigms for neurorehabilitation, in the *Proceedings of ACM CHI 2005*, 2–7 April 2005, Portland, pp. 261–270. New York: ACM Press.

Gnanayutham, P. and George, J. (2007). Inclusive design for brain body interfaces, in *Foundations of Augmented Cognition 3rd Edition, Proceedings of the 3rd International Conference on Augmented Cognition in the Context of HCI International 2007 Conference* (D. Schmorrow and L. Reeves, eds.), 22–27 July 2007, Beijing, pp. 102–111. Berlin: Springer-Verlag.

Gnanayutham, P. and George, J. (2008). Developing brain body interfaces for the visually impaired in *Current Advances in Computing, Engineering and Information* (P. Petratos and P. Dandapani, eds.), pp. 41–54, Athens, Greece: ATINER.

Harrison, R. R., Santhanam, G., Krishna, V., and Shenoy, K. (2004). Local field potential measurement with low-power analogue integrated circuit, *International Conference of the IEEE EMBS*, September 2004, San Francisco, pp. 4067–4070. San Francisco: IEEE EMBS.

Hinterberger, T., Schmidt, S., Neumann, N., Mellinger, J., Blankertz, B., Gabriel, C., and Birbaumer, N. (2003). *Brain-Computer Communication and Slow Cortical Potentials*. http://ida.first.fhg.de/publications/HinSchNeuMelBlaCurBir04.pdf.

Hinterberger, T., Wilhelm, B., Mellinger, J., Kotchoubey, B., and Birbaumer, N. (2005). A device for the detection of cognitive brain functions in completely paralyzed or unresponsive patients. *Biomedical Engineering, IEEE Transactions* 52: 211–220.

Jasper, H. (1958). The ten-twenty electrode system of the International Federation in Electroencephalography and Clinical Neurophysiology. *Electroencephalographic Clinical Neurophysiology* 10: 371–375.

Junker, A. (1997, December 2). U.S. Patent 5,692,517.

Kalat, J. (1995). *Biological Psychology* (5th ed.). Pacific Grove, CA: Brooks Cole Publishing Company.

Kalcher, J., Flotzinger, D., Gölly, S., Neuper, G., and Pfurtscheler, G. (1994). Brain-computer interface (BCI) II, in *Proceedings of the 4th International Conference ICCHP 94*, Vienna, Austria, pp. 171–176. Berlin/Heidelberg: Springer-Verlag.

Keirn, A. and Aunon, J. (1990). Man-machine communication through brainwave processing. *IEEE Engineering in Medicine and Biology Magazine* 9: 55–57.

Kennedy, J. (1999). *Preparing Your Laboratory Report*. http://www.psywww.com/tipsheet/labrep.htm.

Kennedy, P., Adams, K., Bakay, R., Goldwaithe, J., Montgomery, G., and Moore, M. (1999). Direct control of a computer from the human central nervous system, in *Brain-Computer Interface Technology Theory and Practice First International Meeting Program and Papers*, 16–20 June 1999, Rensselaerville, NY, pp. 65–70.

Kennedy, P., Andreasen, D., Ehirim, P., King, B., Kirby, T., Mao, H., and Moore, M. (2004). Using human extra-cortical local field potentials to con troll a switch. *Journal of Neural Engineering* 1: 72–77.

Kennedy, P. R., Adams, K., Bakay, R. A. E., Goldwaithe, J., and Moore, M. (2000). Direct control of a computer from the human central nervous system. *IEEE Transactions on Rehabilitation Engineering* 8: 198–202.

Kostov, A. and Polak, M. (1997a). Brain-computer interface: Development of experimental setup, in *Resna 97 Proceedings*, June 1997, Pittsburgh, pp. 54–56. Arlington, VA: RESNA Press.

Kostov, A. and Polak, M. (1997b). Prospects of computer access using voluntary modulated EEG signal, in *Brain & Consciousness Proceedings EPCD Symposium*, 22–23 September 1997, Belgrade, Yugoslavia, pp. 233–236. Belgrade, Yugoslavia: ECPD of the United Nations University for Peace.

Kostov, A. and Polak, M. (2000). Parallel man-machine training in development of EEG-based cursor control. *IEEE Transactions on Rehabilitation Engineering* 8: 203–205.

Kotchoubey, B., Lang, S., Baales, R., Herb, E., Maurer, P., Merger, G., Schmalohr, D., Bostanov, V., and Birbaumer, N. (2001). Brain potentials in human patients with extremely severe diffuse brain damage. *Neuroscience Letters* 301: 37–40.

Kotchoubey, B., Lang, S., Bostanov, V., and Birbaumer, N. (2002). Is there a mind? Electrophysiology of unconscious patients. *News in Physiology Sciences* 17: 38–42.

Kotchoubey, B., Schleichert, H., Lutzenberger, W., and Birbaumer, N. (1997). A new method for self-regulation of slow cortical potentials in a timed paradigm. *Applied Psychophysiology and Biofeedback* 22: 77–93.

Kozelka, J. and Pedley, T. (1990). Beta and MU rhythms. *Journal of Clinical Neurophysiology* 7: 191–207.

Lal, T. N., Hinterberger, T., Widman, G., Schröder, M., Hill, N. J., Rosenstiel, W., Elger, C. E., Schölkopf, B., and Birbaumer, N. (2005). Methods towards invasive human brain computer interfaces. *Advances in Neural Information Processing Systems* (L. K. Saul, Y. Weiss, L. Bottou, eds.), pp. 737–744. Cambridge, MA: MIT Press.

Langlois, J. A., Ruthland-Brown, W., and Tomas, K. E. (2004). *Traumatic Brain Injury in the United States: Emergency Department Visits, Hospitalizations, and Deaths*. Centers for Disease Control and Prevention, National Center of Injury Prevention and Control. http://www.cdc.gov/ncipc/pub-res/pubs.htm.

Levine, S. P., Huggins, J. E., BeMent, S. L., Rohde, M. M., Passaro, E. A., Kushwaha, R. K., Schuh, L. A., Ross, D. A., Elisevich, K. V., and Smith, B. J. (1999). A direct brain interface based on event related potential. *IEEE Transaction Rehabilitation Engineering* 8: 180–185.

Lusted, H. and Knapp, R. (1996). Controlling computers with neural signals. *Scientific American* 5: 58–63.

Maynard, E. M., Nordhausen, C. T., and Normann, R. A. (1997). The Utah Intracranial Electrode Array: A recoding structure for potential brain-computer interfaces. *Electroencephalography Clinical Neurophysiology* 102: 228–239.

Metz, S. and Hoffman, B. (1997). Mind operated devices. *Cognitive Technology* 2: 69–74.

Miner, L. A., McFarland, D. J., and Wolpaw, J. R. (1998). Answering questions with an electroencephalogram-based brain-computer interface. *Archives of Physical Medical and Rehabilitation* 79: 1029–1033.

Musallam, S., Corneil, B. D., Greger, B., Scherberger, H., and Anderson, R. A. (2004). Cognitive control signals for neural prosthetics. *Science* 305: 258–262.

Navarro, K. F. (2004). Wearable, wireless brain computer interfaces in augmented reality environments, in *ITCC 2004*, April 2004, Las Vegas, pp. 643–647. Las Vegas: IEEE Computer Society.

Neuper, C., Schlögl, A., and Pfurtscheller, G. (1999). Enhancement of left-right sensorimotor EEG differences during feedback-regulated motor imagery. *Journal of Clinical Neurophysiology* 16: 373–382.

North American Brain Injury Society (2005). *Brain Injury Facts*. http://www.nabis.org/public/bfacts.shtml.

Nunez, P. L., Sibertstein, R. B., Shi, Z., Carpenter, M. R., Srinivasan, R., Tucker, D. M., Doran, S. M., Cadusch, P. J., and Wijesinghe, R. S. (1999). EEG coherency II: Experimental comparisons of multiple measures. *Clinical Neurophysiology* 110: 469–486.

Origin Instruments (1998). *HeadMouse: Head-ControlledPointing for Computer Access*. http//www.orin.com/access/head mouse/index.tm.

Patel, S. H. and Azzam, P. N. (2005). Characterization of N200 and P300: Selected studies of the event-related potential. *International Journal of Medical Sciences* 2: 147–154.

Pfurtscheller, G. and Neuper, C. (2001). Motor imagery and direct brain-computer communication. *Proceedings of the IEEE* 89: 1123–1134.

Pregenzer, M., Pfurtscheller, G., and Flotzinger, D. (1994). Selection of electrode positions for an EEG-based brain computer interface (BCI). *Biomedizinische Technik* 39: 264–269.

Robinson, D. (1999). The technical, neurological and psychological significance of "alpha," "delta" and "theta" waves confounded in EEG evoked potentials: A study of peak latencies. *Clinical Neurophysiology* 110: 1427–1434.

Roy, E. A. (2004). *The Anatomy of a Head Injury*. http://www.ahs.uwaterloo.ca/~cahr/headfall.html.

Salem, C. and Zhai, S. (1997). An isometric tongue pointing device, in the *Proceedings of CHI 1997*, 22–27 March 1997, Atlanta, pp. 22–27. New York: ACM Press.

Sanchez, J. C., Principe, J. C., and Carne, P. R. (2005). Is neuron discrimination preprocessing necessary for linear and non-linear brain machine interface models?, in the *Proceedings of HCI International 2005 Conference*, 22–27 July 2005, Las Vegas [CD-ROM]. Mahwah, NJ: Lawrence Erlbaum Associates.

Schalk, G., McFarland, D. J., Hinterberger, T., Birbaumer, N., and Wolpaw, J. R. (2004). BCI2000: A general-purpose brain-computer interface (BCI) System. *IEEE Transactions on Biomedical Engineering* 51: 1034–1043.

Schmolesky, M. (2006). *The Primary Visual Cortex*. http://web-vision.med.utah.edu/VisualCortex.html.

Sears, A. and Young, M. (2003). Physical disabilities and computing technologies: An analysis of impairments, in *The Human-Computer Interaction Handbook* (J. A. Jacko and A. Sears, eds.), pp. 482–503. Mahwah, NJ: Lawrence Erlbaum Associates.

SensoMotoric Instruments (2007). http://www.smi.de.

Serra, M. and Muzio, J. (2002). The IT support for acquired brain injury patients the design and evaluation of a new software package, in the *Proceedings of the 35th Annual Hawaii International Conference on System Sciences (HICSS 2002)*, 7–10 January 2002, Big Island, HI, pp. 1814–1821. Washington, DC: IEEE Computer Society.

Siuru, B. (1999). A brain/computer interface. *Electronics Now* 70: 55–56.

Spencer, D. K., Kravitz, C., and Hopkins, M. (2004). Neural correlates of bimodal speech and gesture comprehension. *Brain and Language* 89: 253–260.

Spiers, A., Warwick, K., and Gasson, M. (2005). Assessment of invasive neural implant technology, in the *Proceedings of HCI International 2005 Conference*, 22–27 July 2005, Las Vegas [CD-ROM]. Mahwah, NJ: Lawrence Erlbaum Associates.

Tortora, G. J. and Derrickson, B. (2006). *Principles of Anatomy and Physiology* (11th ed.). Hoboken, NJ: John Wiley & Sons.

Tran, T. A., Spencer, S. S., Javidan, M., Pacia, S., Marks, D., and Spencer, D. D. (1997). Significance of spikes recorded on intraoperative electrocorticography in patients with brain tumor and epilepsy. *Epilepsia* 38: 1132–1139.

Tyrer, S. (2005). *UK Acquired Brain Injury Forum*. http://www.ukabif.org.uk/index.htm.

Vincenten, J. (2001). *Priorities for Child Safety in the European Union: Agenda for Action*. Amsterdam, The Netherlands: ECOSA.

Weiskopf, N., Veit, R., Mathiak, K., Grodd, W., Goebel, R., and Birbaumer, N. (2003). Physiological self-regulation of regional brain activity using real-time functional magnetic resonance imaging (fMRI): Methodology and exemplary data. *NeuroImage* 19: 577–586.

Wolpaw, J., Birbaumer, N., Heetderks, W. J., McFarland, D. J., Peckham, P. H., Schalk, G., Donchin, E., Quatrano, L. A., and Robinson, C. J. (2000). Brain-computer interface technology: A review of the First International Meeting (J. Wolpaw and T. Vaughan, eds.). *IEEE Transactions on Rehabilitation Engineering* 8: 164–173.

Wolpaw, J. R., Flotzinger, D., Pfurtscheller, G., and McFarland, D. (1997). Timing of EEG based cursor control. *Journal of Clinical Neurophysiology* 14: 529–538.

Wolpaw, J. R. and McFarland, D. J. (1994). Multi-channel EEG-based brain-computer communication. *Electroencephalography and Clinical Neurophysiology* 90: 444–449.

Wolpaw, J. R., McFarland, D. J., Vaughan, T. M., and Schalk, G. (2003). The Wadsworth Centre Brain-Computer Interface (BCI) research and development program. *IEEE Transactions on Neural Systems and Rehabilitation* 11: 204–207.

World Health Organization (2005). *Disability, Including Prevention, Management and Rehabilitation*. Report by the Secretariat A58/17, April 2005, pp. 373–392. Washington, DC: World Health Organization Publications.

Sign Language in the Interface: Access for Deaf Signers

Matt Huenerfauth and
Vicki L. Hanson

38.1 Introduction and Background

This chapter introduces the reader to several important and inter-related issues in deafness, language literacy, and computer accessibility. Throughout this discussion, the chapter will motivate the use of sign language technologies in the interface of computing systems to improve their accessibility for deaf signers.[1] The chapter discusses the advantages of sign language interfaces for deaf signers (for a discussion of other technologies for deaf and hard of hearing users, see, e.g., Hanson, 2007), and highlights important applications of this technology, as well as the challenges that arise when integrating sign language technologies into real applications. Many of the sign language generation and understanding technologies discussed in this chapter are still in the research stage of development, and so another focus of this chapter is to explain what makes this technology challenging to build, survey the progress of the field, and discuss current issues that researchers are tackling. This chapter will help the reader understand what technologies are currently available and what direction the field is expected to take in the coming decades.

A question that might first arise is why such interfaces are needed. By and large, interactions with computers involve reading or writing text. Deaf signers, at first blush, would not appear to be disadvantaged in their ability to read and write. However, interfaces requiring reading and writing also have the potential to disenfranchise many deaf users (see also Chapter 6 of this handbook). As this may seem counterintuitive, the chapter begins with a discussion of deafness, sign language, and literacy to indicate why this is the case.

38.1.1 Deafness, Sign Language, and Literacy

Millions of deaf and hard of hearing people worldwide use a sign language to communicate. Sign languages are naturally occurring languages with linguistic structures (e.g., grammars, vocabularies, word order, etc.) distinct from spoken languages. For instance, American Sign Language (ASL) is the primary means of communication for an estimated 500,000 people in the United States (Mitchell et al., 2006). ASL is a full natural language that includes various linguistic phenomena that make it distinct from English (Lane et al., 1996; Neidle et al., 2000; Liddell, 2003).

There are a number of factors that determine whether an individual with hearing loss will use a sign language, including family circumstances, educational experiences, age of onset of hearing loss, and degree of hearing loss. Signers comprise a deaf community, whose membership is determined more by a shared language than by degree of hearing loss (Padden and Humphries, 1988, 2005). In fact, people who experience hearing loss as adults tend not to become signers or members of this

[1] In this chapter, individuals with hearing loss are considered as being deaf or hard of hearing, the terminology designated by the World Federation of the Deaf (*What Is Wrong with the Use of These Terms: "Deaf-Mute," "Deaf and Dumb," or "Hearing-Impaired"?* National Association of the Deaf, http://www.nad.org/site/pp.asp?c=foINKQMBF&b=103786).

community. Contrary to popular expectation, sign languages are not universal (Klima and Bellugi, 1979); countries and locales around the world have their own native sign languages shared by members of deaf communities in those areas.

Important for the present discussion is the fact that these signed languages are not based on the spoken languages of the region. People are often surprised to learn, for example, that ASL is more similar to French Sign Language, from which it originated, than it is to British Sign Language (BSL). Thus, despite the common written language shared by the deaf communities in America and Great Britain, the sign languages of these two communities are not similar (Lane, 1976). Deaf individuals often acquire a sign language as their first language and are most fluent and comfortable in this first language. For these individuals, sign language interfaces are highly desirable.

Sign language interfaces are a necessity for that subset of the deaf population with difficulty in reading and writing. Despite the fact that many deaf individuals are skilled readers, not all deaf signers develop this level of proficiency. The reasons may be varied, but this phenomenon is replicated worldwide, regardless of sign language or written language of the country. For example, studies have shown that the majority of deaf high school graduates in the United States have only a fourth-grade English reading level (Holt, 1993)—this means that deaf students around age 18 have a reading level more typical of 10-year-old hearing students. This literacy issue has become more significant in recent decades, as new information and communications technologies have arisen that place an even greater premium on written language literacy in modern society.

To focus the discussion in this chapter and to keep it more concrete, examples from ASL will be primarily used. There are many other sign languages used around the world that are also the subject of computational research: British Sign Language, Japanese Sign Language, Polish Sign Language, and others. These languages are linguistically distinct from ASL—and from each other. While each language has its own distinct grammar and vocabularies, they share many features with ASL: the use of multiple parts of the signer's body in parallel, the use of locations in space around the signer to represent entities under discussion, and the modification of individual signs to indicate subtleties of meaning. Thus, the majority of technologies developed for one of these languages can therefore be adapted (by changing the vocabulary of signs and some of the grammar rules) for use with other sign languages. Some of this work for other sign languages will be discussed later in this chapter.

38.1.1.1 Deaf Accessibility Tools and English Literacy

Many accessibility responses for deaf users simply ignore part of the problem—often designers make the assumption that the deaf users of their tools have strong English reading skills. For example, television closed captioning converts an audio English signal into visually presented English text on the screen; however, the reading level of this text may be too high for many deaf viewers. While captioning makes programming accessible to a large number of hearing, deaf, and hard of hearing users, a number of deaf users may be cut off from important information contained in news broadcasts, educational programming, political debates, and other broadcasts that have a more sophisticated level of English language. Communications technologies like teletype telephones (sometimes referred to as telecommunications devices for the deaf or TDDs) similarly assume the user has English literacy. The user is expected to both read and write English text to have a conversation. Many software designers incorrectly assume that written English text in a user interface is always accessible to deaf users. Few software companies have addressed the connection between deafness and literacy, and so few computer user interfaces make sufficient accommodation for deaf users.

A machine translation system from English text into ASL animations could increase the accessibility of all of these technologies for signers. Instead of presenting written text on a television screen, telephone display, or computer monitor, each could instead display ASL signing. An automated English-to-ASL machine translation (MT) system could make information and services accessible when English text captioning is too complex, or when an English-based user interface is too difficult to navigate.

In addition, technologies for *recognizing* sign language could also benefit deaf signers. The ability to input commands to a computing system using ASL would make the interaction more natural for deaf signers, and the ability of the system to translate sign language input into English text or speech could open additional avenues of communication for deaf signers with low levels of English literacy.

The ultimate sign language interface tool would be one that could recognize sign language input while also having the ability to output sign language from spoken utterances or text. Such a tool would allow easy interaction between deaf signers and hearing speakers. It would also allow deaf signers natural and easy access to computers and other devices. However, as will be discussed later in this chapter, a great deal of research remains to be done to make this tool a reality. Today, both production and, even more so, recognition systems are in relatively early stages of development.

38.1.2 Sign Languages

There is a common misconception that the reason why many deaf people have difficulty reading text is that it is presented in the form of letters/characters in a writing system. It would follow that if every word of a written language sentence was replaced with a corresponding sign (the assumption is also made that such a correspondence always exists), then deaf signers would be able to understand the text. This is generally not true. In fact, deaf individuals can be observed signing word for word as they read text. Thus, they have no trouble with print, *per se*. After signing each word in an English sentence, however, they may not have understood the meaning of the sentence because of the grammatical differences between the languages (Hanson and Padden, 1990). By and large, presentation of ASL signs in

English word order (and without the accompanying ASL linguistic information contained in facial expressions, eye gaze, etc.) would not be understandable to a deaf user. The differences between English and ASL are significant enough that fluency in one language does not imply fluency in the other. For an ASL signer, reading English is analogous to an English speaker reading a foreign language.

38.1.2.1 Other Forms of Signing Communication

There are many forms of signing communication that are not full languages; for instance, Signed English (SE) is a form of manual signing that is distinct from ASL but is not a full natural language. There are several different styles of SE communication, but all of them encode an English sentence into a set of signs performed by the signer's hands. SE uses many of the same signs as ASL (and some additional signs of its own). SE retains English sentence structure and word order, and it is most commonly used in educational settings. Fingerspelling is another method of signing communication in which the letters of the English alphabet are conveyed using special handshapes to spell words during signing. Signers typically reserve fingerspelling for titles, proper names, and other specific situations. The extensive use of English-like structure leads to a nonfluent ASL communication that is difficult to understand.

38.1.2.2 Some Sign Language Linguistic Issues

To illustrate the complexity of sign languages, and why this presents such a technical challenge for machine translation, this section discusses some interesting phenomena that occur in ASL. Many of the issues discussed in the following have parallels in other sign languages used around the world.

ASL is a visual language in which the signer's facial expression, eye gaze, head movement, shoulder tilt, arm movements, and handshapes convey linguistic information; however, it is not enough to know how a signer's body moves to understand an ASL sentence. It is also necessary to remember how the "signing space" around the body has been filled with imaginary placeholders that represent entities under discussion (Meier, 1990; Neidle et al., 2000). These locations are a *conversational state* that signers must remember. During a conversation, when a new entity is mentioned, a signer can use various ASL constructions to associate that entity with a 3D location in the signing space:

- Determiners and certain post-noun-phrase adverbs point out a 3D location for an entity (Neidle et al., 2000).
- Some nouns can be signed outside their standard location to associate the entity to which they refer with a 3D location in the signing space.

After establishing placeholders in space (Neidle et al., 2000), the movements of many other ASL constructions are spatially parameterized on these locations:

- Personal, possessive, and reflexive pronouns involve pointing movements toward the placeholder location of the entity being referred to.

- Some ASL verbs change their movement path, hand orientation, or other features to indicate the 3D placeholder location of their subject, object, or both. What features are modified and whether this modification is optional depends on the verb (Padden, 1988; Liddell, 2003).
- While signing ASL verb phrases, signers can use combinations of head-tilt and eye-gaze to indicate a verb's subject and object (Neidle et al., 2000).
- While signing possessive pronouns or noun phrases, signers can use their head-tilt to indicate the possessor and their eye-gaze to indicate the possessed entity (Neidle et al., 2000).
- Signers can tilt their torso toward locations in the signing space on opposite sides of their body when conveying a contrastive relationship between two entities (this is often called "contrastive role shift").

By changing a verb's movement path or by performing head-tilt and eye-gaze during verb phrases signing, the identity of the subject/object of the sentence can be expressed without performing a noun phrase for each. If their identity is conveyed this way, the signer may optionally drop these noun phrases from the sentence (Neidle et al., 2000). Signers also often topicalize one of the noun phrases in a sentence—establishing that entity as an important focus of discussion. To topicalize a noun phrase, it is performed at the start of the sentence (instead of at its original place in the sentence) with the signer's eyebrows raised. Because a topicalized noun phrase is no longer performed at the subject or object position in the sentence, the role it fulfills in the sentence could be ambiguous. Verbs with movement modifications or head-tilt/eye-gaze indicate the identity of their subject and object and can disambiguate sentences that have undergone topicalization movement.

Generally, the locations chosen for this pronominal use of the signing space are not topologically meaningful; that is, one imaginary entity being positioned to the left of another in the signing space doesn't necessarily indicate that the two entities are located the first at the left of the second in the real world. Other ASL expressions are more complex in their use of space, and position invisible objects around the signer to topologically indicate the arrangement of entities in a 3D scene being discussed. ASL constructions called *classifier predicates* allow signers to use their hands to represent an entity in the space in front of them and to position, move, trace, or re-orient this imaginary object to indicate the location, movement, shape, or other properties of some corresponding real-world entity under discussion. A classifier predicate consists of the hand in one of a set of semantically meaningful shapes as it moves in a 3D path through space in front of the signer. For example, to convey the sentence "the car drove up to the house and parked next to it," signers use two classifier predicates. Using a handshape for bulky objects, they move one hand to a location in front of their torso to represent the house. Next, using a moving vehicle handshape, their other hand traces a 3D path for the car that stops next to the house. To produce these two classifier predicates, there must be a spatial model of how the house and the car in the scene are arranged (Huenerfauth, 2004).

The three-dimensional nature of classifier predicates makes them particularly difficult to generate using traditional computational linguistic methods designed for written languages. Instead of producing a string of signs that convey the information about the scene (as a string of words might be arranged to produce a written language sentence), during classifier predicates, signers actually convey 3D information directly to their audience using the space in front of their bodies. During other spatial constructions, the signer's own body is used to represent a character in the narrative (often called *narrative role shift* or *body classifiers*); signers show comments and actions from the perspective of people under discussion. Signers can also use the signing space in these different ways simultaneously to convey meaning (Liddell, 2003).

Written/spoken languages typically lengthen a sentence by appending morphemes or adding words to incorporate additional information. Sign languages, however, make use of their many channels to incorporate additional information by modifying the performance of a sign, performing a meaningful facial expression during a sentence, or making use of the space around the signer. This leads to the interesting finding that the rate of proposition production is similar in signed and spoken languages, despite the fact that a sign takes longer to produce than does a word (Klima and Bellugi, 1979). A sign language sentence consists of several simultaneous independently articulated parts of the body: the eye gaze; the head tilt; the shoulder tilt; the facial expression; and the location, palm orientation, and handshape of the signer's two hands. Temporally coordinating these channels over time is an important part of a correct ASL utterance—if the timing relationships between movements of different parts of the body are not correct, then the meaning of the ASL sentence is typically affected. For instance, the direction in which the eyes gaze during the performance of an ASL verb sign can be used to indicate the object of that verb, but the eye gaze must occur during the verb sign itself to convey this meaning.

38.2 Producing Sign Language Output

For spoken languages, a computer system can display written text onto the screen for the user. For sign languages, this approach is generally not possible. The coordinated use of multiple parts of a signer's body during a sign language performance and the use of 3D space around the signer (especially during classifier predicates) can be challenging to encode in a written representation. While several sign language writing systems have been proposed, most are difficult to use for people who are not linguists (Newkirk, 1987) and some involve drawing symbolic two-dimensional diagrams of sign movements (Sutton, 1998). Thus, none have gained significant popularity among the deaf community.

Computer-friendly notation schemes have also been developed (Prillwitz et al., 1989; Kuroda et al., 2001) and have been used by some sign language researchers (Bangham et al., 2000;

Marshall and Sáfár, 2004); however, as with sign language writing systems, they have not been adopted by signers for writing. Without a community of users that accept and have developed literacy skills in one of these writing systems, none can be used as output on a sign language interface. Therefore, the output must be displayed in the form of video or animation of a humanlike character signing.

38.2.1 Sign Language Video

To address these literacy issues, a number of applications have been developed that display videos of humans performing sign language. These interfaces have been employed not only for making audio and speech materials accessible to signers (e.g., Petrie et al., 2004; Efthimiou and Fotinea, 2007; Kennaway et al., 2007; Link-it, 2007; RNID, 2007), but also for teaching reading and writing to deaf signers (e.g., Padden and Hanson, 2000; AILB, 2007). For example, Petrie et al. (2004) created a sign interface that used signed videos to present tooltip information in an application. Deaf signers were found to overwhelmingly prefer the sign video over spoken (video), graphical, and text information. For instructional purposes, Hanson and Padden (1990) displayed stories in ASL and print (English), allowing young deaf signers to compare the two and learn about correspondences between ASL and English. Research showed that these children were able to use their first language (ASL) to improve comprehension of written sentences and to help in the writing of English.

While using videos of human sign language performances can be appropriate when there are a finite set of sentences that a system must ever convey to the user (or when there is a single message that it needs to convey that is known ahead of time), it is difficult to use videos as the basis for a computer system that must generate/assemble novel signed sentences.

One might imagine that sign language generation system could be created by recording a dictionary of videos containing a large number of signs performed by the same human signer (standing in approximately the same location in the camera frame). To build novel sentences, it might be expected that it is sufficient to simply concatenate together a set of videos—one for each sign in the sentence—to produce the output video. Unfortunately, there are three major challenges with this approach: (1) smoothing the transitions between the signs on each video so that the assembled output video does not appear jumpy, (2) handling signs whose movement paths are calculated based on a complex set of 3D factors (and are not known until performance time), and (3) handling the many possible combinations of movements on the different parts of the signer's body (a different version of each sign would need to be recorded for every possible combination of facial expression, head tilt, eye gaze, shoulder tilt, etc.).

It is quite difficult to reassemble samples of video of individual signs into a coherent-looking presentation of a sign language message. Most successful sign language generation systems have instead chosen to create animations of a 3D humanlike character that moves to perform a sign language message. This approach has the advantage of allowing the system to more easily blend together individual signs into a smooth-looking sign language sentence, as well as to generate sentences that involve more complex modifications/inflections to the standard dictionary form of a sign to accommodate grammatical requirements of the language. For instance, many sign language verb forms can modify their movement path to indicate the subject/object in the signing space. It would be difficult to pre-record a different version of each verb sign for the motion path between every possible starting/ending location in the signing space. Using an animation-based approach, the system can instead synthesize a particular version of a performance that may never have been recorded from a human signer.

38.2.2 Animations of Sign Language

A major area of Natural Language Processing (NLP) research is the design of software that can translate a sentence from one language into another automatically. The process of automatically translating from a sentence in a *source* language into a sentence in a *target* language is generally referred to as *machine translation* or simply *translation*. Broadly speaking, machine translation software can be thought of as operating in several stages:

1. If the input to the software is an audio recording of someone speaking, then speech recognition software is used to convert the sounds into a text string for each sentence.
2. This written text sentence (in the source language) is analyzed by linguistic software to identify its syntactic structure.
3. A semantic representation of the sentence is produced that represents its meaning.
4. If multiple sentences are being translated as a group (i.e., a paragraph or document), then the software may need to reorganize the way in which the sentences are sequenced or how concepts are introduced (according to the linguistics of the target language).
5. For each output sentence (which will be produced in the target language), a semantic representation is produced for what information content that sentence should convey.
6. The syntactic structure of each sentence is determined (e.g., the software plans the verb, subject, and object of each sentence).
7. A written language text string is produced for each sentence (from its syntactic representation).
8. If the desired output of the software is an audio presentation of the target language sentence spoken aloud, then speech synthesis software is used to convert from the written language text string into an audio file.

NLP researchers also study how to build software that focuses on a subset of the processing steps required for machine translation software. For example, in some applications, the goal is not to translate between languages, but instead the task is to convey some information using sentences in a human language. This task is known as *natural language generation* or simply *generation*, and it loosely corresponds to the processing steps 5 to 7. Thus, generation can be thought of as a subset of the translation process—specifically, it is the final half of the process during which a sentence is constructed from semantic information.

The lack of a written form for most sign languages means that these traditional processing stages work somewhat differently. Instead of a written string, many sign language systems will create some type of script (generally in a proprietary format for each system) that specifies the movements for an animated character. Instead of speech output, sign language systems produce an animation of a humanlike character performing the sentence (based on the information encoded in the script). In the field of sign language computational linguistics, the focus has generally been on building software to translate from written language into sign language (rather than on sign to text translation)—partially motivated by several accessibility applications that would benefit from technology that translates in this direction.

The remainder of this section will describe several sign language animation systems; these systems vary in the extent to which they implement the NLP processing steps listed at the beginning of this section. For example, sign language synthesis systems convert from a script of a sign language performance into an animation; the task of such systems is analogous to that of the text-to-speech synthesis software (step 7 in the previous numbered list). The sign language translation systems face a greater technical challenge; they must address all of the complex steps in a machine translation process from written text into sign language animation.

38.2.2.1 Virtual Signing Characters

Research into virtual reality human modeling and animation has reached a point of sophistication where it is now possible to construct a model of the human form that is articulate and responsive enough to perform sign languages. The level of quality of such human avatar animations has increased such that human signers can now view the onscreen animations and successfully interpret the movements of the avatar to understand its meaning (Wideman and Sims, 1998). However, just because graphics researchers know how to move the character, this doesn't mean that sign language generation software is available. Given a written language text or some other semantic input representation, a computational linguistic component would need to tell the animated character what to do (assuming the correct instruction set for the interface between the linguistics and the animation components has been determined).

Graphics researchers have built many types of animated human characters for use in simulations or games; some have developed characters specifically to perform sign language movements (Wideman and Sims, 1998; Elliot et al., 2005, 2008; Huenerfauth, 2006b). A major challenge in creating an animated signing character is ensuring that the character is sufficiently articulate to perform the necessary hand, arm, face, head, and eye movements that are required to perform a particular sign language. Graphics researchers have incorporated anatomical information to produce more accurate models of the human

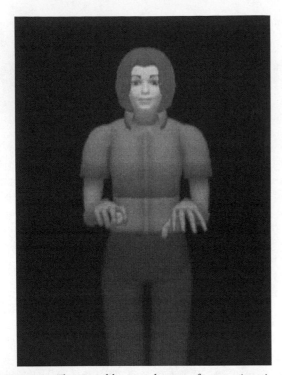

FIGURE 38.1 The virtual human character from an American Sign Language generation system. (From Huenerfauth, M., *Generating American Sign Language Classifier Predicates for English-To-ASL Machine Translation*. PhD dissertation, Computer and Information Science, University of Pennsylvania, Philadelphia, 2006. Copyright held by author.)

body; attempts to animate sign language have motivated developments in representing facial expression (Craft et al., 2000), thumb articulation (McDonald et al., 2001), and joint movement (Tolani et al., 2000; Zhao et al., 2000).

38.2.2.2 Sign Language Animation Scripts

A sign language animation system must decide what message it plans to convey and then produce a sign language output script for an animated character to follow; the format of this linguistics-animation interface specification is an open area of research. Some sign language animation scripts represent the movements of individual signs in a lexicon, and others are used to encode the syntactic structure of entire sentences. Various sign language generation projects have invented their own script format, and each approach has advantages and disadvantages that affect the system's output quality.

A common approach for representing a sign language sentence performance is to use a string of *glosses*, words in the local written language that loosely correspond to the meaning of particular signs. For instance, there is a somewhat conventional set of English words that are used to identify each sign in ASL. When an animation system uses a string-of-glosses representation to specify the sign language sentence prior to animation, they typically augment the representation with limited facial expression

and body movement information (Zhao et al., 2000; Marshall and Sáfár, 2004). Unfortunately, this approach encodes the non-manual portions of the performance in a very limited manner, and it does not handle ASL phenomena in which the movement paths of the hands are determined by the way the signer associates objects with locations in space around the body (i.e., there is not a gloss for every variation of those signs).

Other approaches to symbolically specifying a sign language performance for an animated character encode more detail about the movements that compose individual signs. These approaches allow the system to specify a sign language performance that includes phenomena that are more complex than a simple concatenation of a string of manual signs, and they also allow the system to encode the way in which the movements of the signer should blend from one sign to the next (Speers, 2001). Because they encode information at a subsign level, some of these representations are not specific to encoding a single sign language—for example, the Signing Gesture Markup Language (SiGML) was designed to encode British, Dutch, and German Sign Language—and potentially many others (Bangham et al., 2000). Newer sign language specification scripts have included better support for encoding the simultaneous movement of multiple parts of the signer's body, which must be temporally coordinated (Huenerfauth, 2006a). This particular representation can also use information about how the locations in space around the signer have been associated with objects under discussion to calculate novel motion paths for the hands—thus enabling it to more easily encode complex sign language phenomena (Huenerfauth, 2006b).

38.2.2.3 Sign Language Synthesis Systems

Some research projects have focused primarily on the final portion of the sign language production process: the synthesis step from a sign language specification to an animation of a human-like character. These systems are generally not linguistic in nature; they often employ sophisticated models of human figures, restrictions on the articulatory capabilities of the human form, graphic animation technology, and databases of stored ASL sign animations to produce a smooth and understandable presentation of the sign language message. These systems can be loosely categorized along a concatenative versus articulatory axis (Grieve-Smith, 1999, 2001). Concatenative systems assemble an animation by pasting together animations for the individual signs that compose the message; generally, these systems apply some form of smoothing operation so that the resulting output does not appear jerky (Ohki et al., 1994; Bangham et al., 2000; Petrie and Engelen, 2001; Petrie et al., 2005; Vcom3D, 2007). Articulatory systems derive an animation at run-time from a motion-script by reading the movement instructions from this symbolic script and then manipulating a model of a human figure (Messing and Stern, 1997; Grieve-Smith, 1999, 2001; Lebourque and Gibet, 1999; Bangham et al., 2000; Saksiri et al., 2006). Articulatory systems have greater potential for producing the variety of movements that compose a fluent sign language performance and for adapting to new advances in the understanding of sign language linguistics (see Figure 38.2).

If a designer wants to incorporate sign language animations into an application, then one way to do this is to use one of the sign language synthesis systems mentioned previously. The designer must be proficient in sign language, and he must be able to anticipate the entire set of sign language messages that the application would need to display. The designer would then hand-code each sign language performance for the animated character (using the script format), and then this script can be passed off to the synthesis system to produce the animation output (possibly at run-time). This design approach can work for applications in which a small number of user interface prompts must be expressed in sign language, but it would not work well for a system that needed to deliver media content to users (since this content would not be known ahead of time). The requirement that the programmer be proficient in sign language can also make this design approach impractical. For these reasons, some natural language processing researchers are studying ways to generate sign language output from a written language input specification or other data source. This way, developers without expertise in sign language could incorporate sign language animation technology into their applications that use changing content.

38.2.3 Generation and Translation Systems

This section describes several systems that produce sign language animation output from an input that is an English (or other written language) sentence to be translated or from some other input data source of content to be conveyed to the user. There is a spectrum of system designs: from some systems that merely produce a transliteration of a written language sentence to other systems that produce a grammatically correct sign language

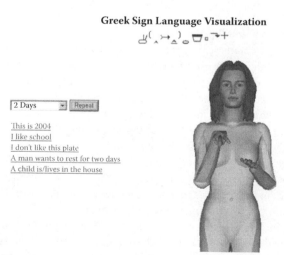

FIGURE 38.2 Screenshot from an application that synthesizes animations of Greek Sign Language. (From Karpouzis, K., Caridakis, G., Fotinea, S-E. and Efthimiou, E., *Computers & Education*, 49, 54–74, Elsevier, 2007.)

performance. Those systems that are transliterations of a written language provide for automatic sign generation (with almost no human intervention required), but they provide less of an accessibility advantage for deaf signers. The systems with correct sign language output are more complex to implement (and until machine translation technology further improves they will still require a human intermediary to check the correctness of their output), but they have the potential to make more information and services accessible to deaf signers by translating information from written language text into sign language animation.

38.2.3.1 Fingerspelling Systems

While some fingerspelling animation systems simply concatenate pictures or animations of sign language fingerspelled letters, more sophisticated systems manipulate a human hand model to create a natural flow and timing between the letters (Davidson et al., 2000, 2001). Much of this software has been designed not for communicative purposes but rather for sign language education—people learning to interpret fingerspelling at more fluent speeds can use these educational systems to automatically produce an animation to practice with. An animation of a human character fingerspelling every word of an English sentence would generally not be understandable to a deaf user with low levels of written language literacy (since it is merely an encoding of the original text). However, as discussed previously, sign languages sometimes use fingerspelling for proper names, titles, and other specific words; therefore, fingerspelling software is an important subcomponent of a full written language-to-sign language translation system.

Researchers are also studying how to build other important subcomponents of a complete written language-to-sign language translation system. Some have explored the design of sign animation databases (Crasborn et al., 1998; Furst et al., 2000) and software tools for linguistic informants to help build these databases (Wolfe et al., 1999; Toro et al., 2001). Some researchers have used motion-capture data glove technology to collect 3D coordinates of a signing performance (Ohki et al., 1994; Lu et al., 1997; Bangham et al., 2000; Verlinden et al., 2001), and others have generated sign animations from some style of symbolic encoding of each sign (Lu et al., 1997; Grieve-Smith 1999, 2001).

38.2.3.2 Transliteration Systems

Transliteration systems convert from text to sign, retaining the word order and some grammatical structures from the text in the sign output. There have been several previous research systems designed to convert from English into Signed English or into an English-like form of pseudo-ASL signing (Grieve-Smith, 1999, 2001; Bangham et al., 2000). With little structural divergence between English text and Signed English animation, the architecture of these systems is typically a simple dictionary look-up process. For each word of the English input text string, the system will look up the correct sign in the Signed English dictionary (containing animations of each sign), and create an output animation that concatenates together all of the signs in the sentence into a complete animation. Children and deaf adults with low English literacy skills, who would be the target users of English-to-ASL machine translation software, would not generally find this form of English-like signing understandable. While Signed English output may be more understandable for deaf signers than a fingerspelling system, actual sign language animation of ASL would be more useful.

38.2.3.3 Sign Language Translation Systems

There have been several research projects that have focused on generating animations of sign language using a written language input string. These systems analyze the linguistic structure of the input text. The grammatical structure, word order, and vocabulary of the text are translated into the appropriate sign language grammatical structure, word order, and vocabulary. Such systems produce a script that specifies the sign language performance—generally using sign language synthesis software to produce an actual animation output in which a humanlike character performs the sign language sentence.

There are several dimensions along which the quality of these translation systems can be measured: the variety of grammatical structures they can understand or generate; the subtlety of variations they can generate for individual signs; the vocabulary size (of written language words or sign language signs); the degree to which they correctly use the face, eyes, and body of the signer; whether or not they can use the space around the signer to position objects under discussion; whether or not they can generate complex spatial phenomena in sign language (such as classifier predicates); whether they produce smoothly moving and realistic animations of a signer; and whether the sign language translation they select for the written language input sentence is accurate/understandable.

Early demonstration systems were capable of producing only a small number of sentences with a limited variety of grammar structures. While those systems did not have the robustness needed for full translation, they helped identify the complexities involved in building higher-quality translation systems. These short-lived projects also played an important role in developing animation dictionaries of specific signs for various sign languages. Translation systems have been created for ASL (Zhao et al., 2000; Speers, 2001), Chinese Sign Language (Xu and Gao, 2000), German Sign Language (Bungeroth and Ney, 2004), Irish Sign Language (Veale et al., 1998), Japanese Sign Language (Ohki et al., 1994; Lu et al., 1998; Tokuda and Okumura, 1998; Adachi et al., 2001), Polish Sign Language (Suszczańska et al., 2002), Sign Language of the Netherlands (Verlinden et al., 2001), and others.

More recent sign language translation systems have been designed that can handle richer linguistic phenomena, be more easily scaled up to handle large vocabularies, are actually deployed in sample applications, or that can handle complex linguistic phenomena in sign language. Recent projects have focused on a number of sign languages, including, for example, ASL (Davidson et al., 2001; Huenerfauth, 2006a,b), British Sign Language (Marshall and Sáfár, 2004, 2005; Sharoff et al., 2004), Greek Sign Language (Karpouzis et al., 2007), Irish Sign Language (Morrissey and Way, 2005), Japanese Sign Language (Shionome et al., 2005), and South African Sign Language (van Zijl and Barker, 2003).

Current sign language systems still require the intervention of a human to ensure that the sign language output produced is accurate. As the quality and linguistic coverage of these systems improve over time, it eventually may be possible for computer interface designers to request automatic translation of written language text into sign language animations. The long-term goal of research on sign language generation and translation is broad-coverage systems that can handle a wide variety of written language input sentences and successfully translate them into sign-language animations. The sign language sentences that are created should be fluent translations of the original written language text, and they should be able to incorporate the full range of linguistic phenomena of that particular sign language. In the coming years, sign language translation systems should be able to more accurately translate sentences with a broader vocabulary and structure, successfully use the space around the signer's body to represent objects under discussion, and generate more complex linguistic phenomena, such as spatial verbs (Marshall and Sáfár, 2005) or classifier predicates (Huenerfauth, 2006b).

38.3 Understanding Sign Language Input

Deaf signers may benefit not only from interfaces that change text into sign language, but also from interfaces that can recognize signing. Sign recognition has the goal of automatically converting the sign language performance of a human user into a computational representation of the performance—that allows the computer to identify the meaning of the user's signing and possibly to later translate it into text or speech. These technologies have been investigated for a number of years, primarily to address the need to facilitate communication between signers and nonsigners. They also have the potential to provide an alternative means of natural language input to computers. While individual projects have focused on particular sign languages, the statistical nature of most research in this area makes this technology easily adaptable to a variety of sign languages.

A key difference between sign language recognition and the more general problem of gesture recognition is that the linguistic structure of sign language input can be used by some machine learning techniques to help determine the likelihood of predicting the next sign a human will perform based on the frequency of some signs following others or the syntactic structure of the sentence. Thus, despite the more complex body movements of sign language, the fact that the performance has a linguistic structure can help guide the recognition process.

Some attempts at recognizing and understanding sign language performed by humans have focused solely on the recognition of fingerspelling (Takahashi and Kishino, 1991; Bowden and Sarhadi, 2002). Fingerspelling recognition alone, however, will not be greatly beneficial to deaf signers, given the same literacy issues discussed previously. However, because fingerspelling is sometimes used during sign language, a fingerspelling recognizer would be a necessary subcomponent of a full sign language recognition system.

Sign recognition systems have used two types of technology: camera-based systems (Bauer and Heinz, 2000; Xu et al., 2000; Vogler and Metaxas, 2001; Kapuscinski and Wysocki, 2003; Starner et al., 2004; Zhang et al., 2004; Yang et al., 2005) and motion-capture-based systems (Waldron and Kim, 1995; Braffort, 1996; Vamplew, 1996; Liang and Ouhyoung, 1998; Cox et al., 2002; Yuan et al., 2002; Brashear et al., 2003; Vogler and Metaxas, 2004). The camera-based systems capture video of a human sign language performance and use machine vision software to locate the parts of the signer's body in the image, identify the 3D location of each part of the body, identify the configuration (pose or handshape) of that part of the body, and then attempt to identify the sign (or other aspect of the sentence) that the person is performing. The motion-capture systems use a variety of sensors (e.g., infrared light beacons, radiofrequency sensors, or gyroscopes) to identify the location or angle of parts of the signer's body. The signer will wear a set of motion-capture cybergloves and sometimes other sensors on the body as part of a motion-capture suit. These systems use the data from the sensors to calculate the location and configuration of parts of the signer's body, and then they identify what subportion of a sign language sentence is currently being performed.

Just as the task of producing a sign language animation could be broken into several stages (generation, synthesis, etc.), the work of a sign language input system can be divided into two important subtasks: recognition and understanding.

38.3.1 Recognition

In the recognition phase, the individual elements of the performance must be identified. While some researchers have attempted to recognize the linguistic elements of facial expressions used during a sign language performance (Vogler and Goldenstein, 2005), most systems focus on the linguistic elements of the signer's hands and attempt to identify only the individual signs in the sentence. Various factors can make the recognition of parts of the signing performance a difficult task: variations in lighting conditions, differences in appearance between signers, changing handshapes or facial expressions/appearance, occlusion of parts of the body, blending of signs into one another during a performance, grammatical modifications to signs that cause them to vary from their standard dictionary form, and other variations in the way that sentences are performed by different signers (intersigner variation) or the same signer on different occasions (intrasigner variation).

Depending on the application, it is possible to make the task of a sign language recognizer easier by asking the signer to perform sign language at a slower speed, asking the signer to pause briefly between each sign that is performed (Waldron and Kim, 1995; Grobel and Assam, 1997), limiting the vocabulary size supported, restricting the variety of sentences that can be performed (e.g., fixed list of sentences that can be identified, small set of sentence templates, a limited grammar, a grammar without some complex sign language features, etc.), allowing the recognizer to process data offline (instead of real-time), training the system to identify the signing of a single person (instead of a variety of signers), and so on. Most work on sign language recognition is based on statistical learning approaches such as neural networks (Murakami and Taguchi, 1991; Vamplew, 1996), independent component analysis (Windridge and Bodden, 2004), or hidden Markov models (Bauer and Kraiss, 2001; Brashear et al., 2003; Vogler and Metaxas, 2004; Yang et al., 2005). Research on recognizing continuous signing (full sentences without pauses added between signs) for a variety of sign languages is surveyed in Loeding et al. (2004).

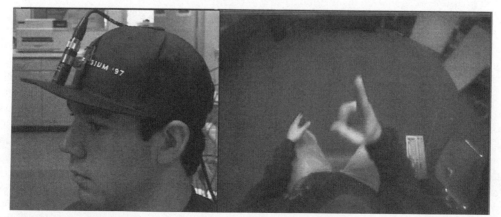

FIGURE 38.3 A sign recognition system using a hat-mounted camera that obtains an overhead image of the signer's hands. (From Brashear, H., Starner, T., Lukowicz, P., and Junker, H., Using multiple sensors for mobile sign language recognition, in the *Proceedings of the 7th IEEE International Symposium on Wearable Computers*, 21–23 October 2003, White Plains, NY, pp. 45–52, Elsevier, IEEE, 2003.)

38.3.2 Understanding

During the second phase of a sign language input system (the understanding phase), the system will attempt to determine the meaning of a sentence that the person is signing. The challenges that arise during this phase are primarily linguistic in nature and are similar to those encountered by researchers attempting to build natural language understanding software for other languages. If the sign language input is being used as part of an application that only expects to receive a small set of possible commands (or receive input sentences that discuss only a limited set of topics), then the task of the understanding system is somewhat easier. Because most research on sign recognition is still in the early phases of development, there has been little computational linguistic research on the understanding of sign language input (since without high-quality sign recognition systems, there is less current need for software to perform the understanding phase). There has been some work on translating from sign language input into written language output (Wu et al., 2004), but since the accuracy level (and vocabulary size) is often quite limited for most recognition systems, work on this understanding phase is still preliminary.

38.4 Sign Language in Applications

With the proliferation of computers and Internet technology in modern society, it is important for someone to be able to access computers to fully participate in society, culture, government, and communication. Because of the dominance of print on the web, there is a danger that a digital divide will disenfranchise those deaf users who are not skilled readers. In addition, the increased use of broadband communications, media, and Internet technology is making the transmission of audio and video more common, creating other potential sources of difficulty for deaf users. The information on the auditory channel of this media is inaccessible to deaf users without special captioning, and text captioning alone may not be readable by all deaf users. Technology presented earlier in this chapter that can generate animations of sign language or understand sign language input can be used to increase the accessibility of computers and other forms of information technology for these users.

38.4.1 Sign Language on a User Interface

Computer interfaces typically contain large amounts of information presented in the form of written language text. This written text can take many forms: elements of the interface (such as the labels on buttons or menus), the content on the computer (names of files, textual information inside documents), or media displayed on the computer (web pages from the Internet, captioning text on streaming media). It is commonplace for software developers to create user interfaces that can be localized/internationalized into different languages. The major difference between localizing software for a sign language is that since there is generally no written form, the sign language translation cannot be written on the buttons, menus, or application windows of the computer screen.

More research is needed to determine how to best display sign language on an interface. For instance, for users with partial literacy skills, it may be desirable to simultaneously display the text and corresponding sign on an interface, rather than removing the text when signing is displayed. However, without more user-based studies, it is unclear what design is best for deaf users with various levels of written language literacy. Human-Computer interaction (HCI) research is needed to determine how to design user interfaces that allow deaf users to direct their attention on an interface containing a mix of onscreen written language text and an animated signing character in one portion of the interface.

Earlier in this chapter, a linguistic phenomenon present in many sign languages, namely classifier predicates, was discussed. During these constructions, signers draw a miniature version of a scene that they are discussing in the air in front of their bodies using linguistically determined handshapes and movements to show the arrangement or motion of objects. Because of the complex and unusual nature of this phenomenon, most sign language animation systems are not designed to generate them. This is unfortunate because it would be extremely useful for an onscreen sign language character to be able to perform classifier predicates when it is part of the interface of a computer system (Huenerfauth, 2007). Since the sign language cannot be statically written on elements of the interface, the animated character will frequently need to refer to and describe elements of the surrounding screen when giving instructions about how to use the system. When discussing a computer screen, a human signer will typically draw an invisible version of the screen in the air with her hand and use classifier predicates to describe the layout of its components and explain how to interact with them. After the signer has drawn the screen in this fashion, she can refer to individual elements by pointing to their corresponding location in the signing space. Making reference to the onscreen interface is especially important when a computer application must communicate step-by-step instructions or help-file text. Written language-illiterate users may also have limited computer experience; so, conveying this type of content may be especially important.

For computer software developers who wish to make their programs accessible to deaf signers, using a sign language generation system to produce animations in help systems may be more practical than videotaping a human signer. There would be a significant investment of resources needed to record and update such videos, and there is another challenge: variations in screen size, operating system, or user-configured options may cause the icons, frames, buttons, and menus on an interface to be arranged differently. An animation-based sign language system may be able to produce variations of the instructions for each of these screen configurations dynamically—producing a video of a human signer for each would be impractical.

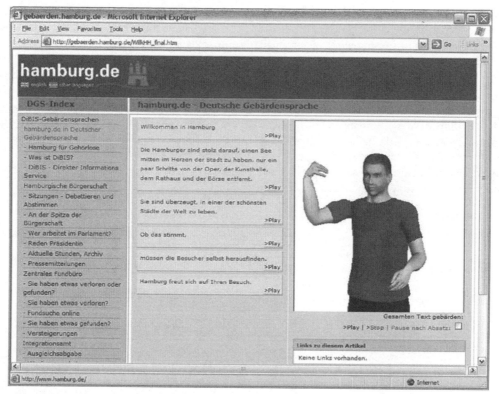

FIGURE 38.4 Screenshot of a web site augmented with a sign language animated character from the European Union's eSIGN project. (From Elliott, R., Glauert, J., Kennaway, J., Marshall, I., and Safar, E. (2008), *Universal Access in the Information Society*, 6, 375–391.)

38.4.2 Sign Language in Communication Systems

There are some forms of communication technology that allow deaf signers to interact with sign language; for instance, video-phone services, video relay services, and Internet webcam systems. In all of these systems, signers who are in two geographically remote locations can communicate using a video camera and a display screen that shows the other signer. When two signers are interacting with such a system, the communication is generally accessible (aside from occasional video quality or time delay issues with these technologies that may impair communication). However, when video relay systems involve a human sign language interpreter who is facilitating a meeting with both deaf and hearing participants, it can be difficult to maintain a smooth conversational flow due to the inability of users to make eye contact correctly via videophone or to see the body movements of remote conversational participants. Researchers are investigating various technologies to improve the turn-taking interactions in video conference interpreted settings (Konstantinidis and Fels, 2006).

While these video-based services offer an exciting new communication option for deaf signers, they are still only deployed in a limited fashion. There is significant expense in the video conferencing equipment and in hiring a sign-language interpreter when both hearing and deaf participants wish to communicate. The necessary video quality needed for sign language is also not yet possible to transmit over most wireless phone networks; so,

these applications are generally limited to nonmobile settings. In lieu of video-based technologies, many deaf people use teletype telephones (TTYs), TDDs, two-way text pages, mobile phone text-messaging, or Internet-based instant-message programs for day-to-day communication. Unfortunately, all of these means of communication still require strong written language reading and writing skills to interact with hearing persons. Software to automatically translate from written language text into sign language animation (to be displayed on the screen of these communication devices) could be used to make these technologies more accessible to deaf signers with low levels of written language literacy.

38.4.2.1 Future Potential of Speech-to-Sign Devices

Translation technology also has exciting future applications when combined with speech recognition software. By incorporating a microphone, speech-to-text software, and a written language-to-sign-language translation component into a handheld computer, one could produce a conversational interpreting tool to provide real-time interpreting services for deaf signers in contexts where hiring a live interpreter would be impossible. It is important to note that all of these technologies (speech recognition, language understanding, machine translation, sign language generation, sign language synthesis) have the potential to make errors during their processing. In fact, none of these technologies are currently able to perform anywhere near the level of humans at these activities. When embedded into an application in which one technology provides the input for the next one in a pipelined architecture, these errors can aggregate. Prototype systems like this have been built for a limited set of sentences (Sagawa et al., 1997) and if the domain of language use is thus limited, then the accuracy rate of speech and translation technologies increases—making their immediate use more practical. For instance, the TESSA project was a prototype system developed to facilitate interactions between hearing and deaf people during transactions in a post office (Cox et al., 2002). It will take many more years of development of each of these technologies, however, for handheld speech-to-sign devices to be able to reach a usable level of accuracy for everyday interactions.

38.4.2.2 Translation Quality and Ethical Considerations

This issue of software errors applies to any setting in which translation software will be used to translate written language text into sign language (or vice versa). No machine translation system between any pair of languages is perfect—even those developed for languages that have been computationally studied for several decades. Sign languages have been the focus of such research for a much shorter time, and translation technology for them will continue to develop for many years to come. In applications in which users understand that the translations may not be perfect and that they cannot fully trust the translation provided, many machine translation technologies for written languages have been successfully deployed. For instance, people browsing a web site in a language they do not speak can use online machine translation software to produce a version of the page in their local language. While the translation may not be perfect, users can get the basic meaning of the page (and perhaps request a full translation of the page by a human translator at a later time). Similarly, in the near future, deaf signers may be able to benefit from machine sign language translation technologies in various applications as long as they are aware that signed translations may not be as accurate as those provided by human interpreters.

It will be a challenge for future accessibility designers to determine when machine sign language translation technologies reach a "good enough" level of linguistic coverage and accuracy such that they feel their users will begin to benefit from the sign language animations they provide. (Technology for understanding sign language input and translating it into written language output will develop later over time.) These developers must weigh the implications of providing imperfect sign language translations of written language material versus providing no translation (and potentially leaving the material inaccessible for many users). Sign language computational linguists should make the limitations of their translation software clear so that potential users (such as software developers) can use the technology appropriately.

It is essential for developers of accessibility software not to overestimate the accuracy of sign language animation technologies, and it is equally important for them to understand the limited benefits that animations of signed transliteration systems (such as Signed English) have for users with limited written language literacy. Given the relatively young state-of-the-art of sign language technologies, service providers (e.g., governments, companies, media outlets, etc.) must be careful not to prematurely deploy these technologies in the name of accessibility. The goal of sign language translation systems has generally been to provide additional assistance to deaf signers in settings in which human interpreters are not possible, not to replace interpreters in settings in which they are currently deployed. Sign language interpreters perform valuable services in many situations, allowing for fluent two-way communication between deaf and hearing people. No computer system is capable of providing the same level of sophisticated and subtle translation that a qualified professional interpreter can. The development of sign language translation technology will take several more decades, and the burden falls to technologists to inform potential users of the current state-of-the-art and its limitations.

38.4.3 Sign Language Captioning for Media

Many deaf or hard of hearing people with literacy challenges can have difficulty understanding the written language text on the captioning provided with television programs, Internet media content, or at some live events. There exists some controversy about the language level to be used for captioning. Simply put, the issue revolves around the question of whether captions should be verbatim transcripts or whether simplified captioning should be provided. Verbatim captioning is generally preferred by users themselves, even though it may not be accessible to some (National Institutes of Health, 2002). Instead of providing simplified written language text, automatic translation software could be used to convert this text into a sign language animation that could be displayed for these users. Of course, many of the translation quality and ethical issues discussed in the communications section previously also apply when translating captions into sign language animation for deaf signers.

There are several options as to how television captioning text could be made more accessible for deaf signers in the future. Broadcasters could run translation software on their written language closed captioning text (to produce a sign language animation script). If the transmission is not of a live event (and the broadcasters have some linguistic expertise in sign language), then they could manually fix any translation errors in the sign language script. Finally, they could transmit the script over the network, and sign language synthesis software on the user's receiving device (e.g., television, computer) would generate the animation of a humanlike character performing the sign language sentences specified by the script. Various approaches for the transmission of this alternative captioning would need to be explored, and standards for the transmission of the script would need to be established. Alternatively, the sign language animation could be synthesized prior to transmission and sent as a secondary video feed with the original signal. If this video-feed approach is used, then more bandwidth may be required in the transmission but potentially fewer standards would need to be established.

Similar English literacy issues arise when deaf signers need to access nonelectronic written materials. Deaf students who need to read classroom textbooks or deaf adults who wish to access magazines, printed government publications, and other documents could benefit from a tool that incorporates a text scanner, optical character recognition software, and a written language-to-sign language translation component to make this information more accessible. Just like the speech-to-sign handheld system described previously, systems that rely on multiple stages of processing with potentially imperfect technologies can aggregate the errors made at one stage (and pass them on to the next). However, in this case, optical character recognition software has reached a sufficient level of accuracy that a scanner-based system could become practical as the quality of sign language translation technology continues to improve.

38.4.4 Applications Focused on Sign Language

The previous sections have discussed ways in which standard computer applications, communications tools, and media could be made more accessible for deaf signers. However, the development of sign language generation and recognition technology actually makes several new types of computer applications possible that may benefit deaf signers. Instead of translating content that was originally created in a written language, these applications would include original content in the form of sign language animation. Educational software can be created to help users learn sign language literacy skills (by watching sign language animations or performing sign language that is recognized by the system) or to help users learn other academic content (through explanation in the form of sign language animation). Sign language scripting software can also be created to allow users to create and edit sign language animations much like word-processing software allows editing of written language content.

38.4.4.1 Educational Tools to Learn Sign Language

Software to teach users sign language could also benefit from sign language generation and recognition software. Animation technology could be used to produce demonstrations of sign language using virtual human characters, which the user could view from any angle, zoom in or out, and slow down (more easily than a videotape). Recognition technology is being used to allow users to practice performing sign language sentences, and the system could provide the user with feedback about his performance (Brashear et al., 2006). Written language to sign language translation technology could also be incorporated into educational software—it could translate novel sentences into sign language and demonstrate them for the user. This would allow the user to request particular sentences that she is interested in learning, and the animation technology could make the educational software more engaging and interactive.

Like all computerized language learning software, a sign language learning program may help users feel more comfortable learning a language than they would with classroom instruction

or in-person conversational practice. While live interaction is extremely important for language learning, sometimes students are too intimidated to enroll in classes. These issues are particularly important when considering the case of hearing parents who should learn sign language because they have learned their child is deaf (or has special communication needs that make the use of sign language appropriate). The complex emotions that these parents sometimes experience when learning about their child's status can make them uncomfortable or resistant to enrolling in formal sign language classes. A computer program that could teach these parents basic sign language could be an extremely useful way to get them interested in the language and build their confidence to enroll in sign language learning activities.

38.4.4.2 Educational Tools to Learn through Sign Language

Sign language generation technology also has important applications in educational computer programs for deaf and hard of hearing students. Just like the "Sign Language on a User Interface" section discussed previously, educational programs could have a user interface with a sign language character on the screen to help students better understand the information presented and use the software. The user interface of any standard children's educational program could be extended with an onscreen character in this way. It would also be possible to build educational software especially for deaf students with original content in the form of sign language animations (instead of translating content originally designed in a written language and for hearing students). Particularly effective in this context would be content addressing issues of particular interest to deaf students or using culturally appropriate storylines or scenarios (Hanson and Padden, 1990); it could also take advantage of the rich capabilities of sign languages to convey spatial concepts or narratives to teach important content.

38.4.4.3 Sign Language Word Processing or Scripting Software

Because nearly all sign languages lack a standard orthography and writing system that is widely accepted by their users, it is currently very difficult to store, process, and transmit sign language via computer. Video is currently the best means for capturing information in sign language, and while digital video editing software and faster Internet connections are making video easier to work with on a computer, sign language still remains a much less computer-friendly language than written languages. As sign language generation and translation technologies are being developed, researchers are designing new computer encodings of sign language that should be easier for users to store, edit, process, and transmit with computer technology. These representations are much smaller in size than video, and they can be modified prior to being turned into animation by sign language synthesis software. In this way, it may be possible to produce sign language word-processing (or *sign-processing*) software that could allow people to write, edit, and replay sign language information on their computer. Aside from increasing

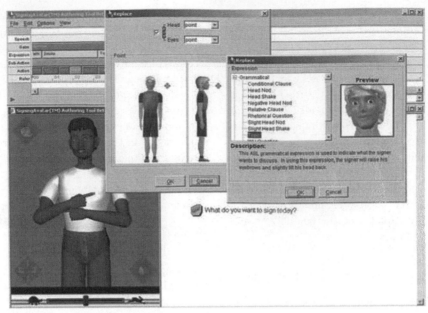

FIGURE 38.5 Screenshot from the user's guide for Sign Smith Studio, a sign language authoring tool software that allows users to script animations of signed English or American Sign Language. (Image courtesy of VCom3D, Inc. All rights reserved.)

computer accessibility for deaf signers, the development of a standard method of writing more sign languages on a computer could have implications on sign language literature and deaf education.

There have been some systems that allow users to script a sign language performance that can be replayed as animation (Elliott et al., 2005; Vcom3D, 2007). The Vcom3D commercial software automatically produces a Signed English animation when given an English text string; however, the content developer needs to carefully script the signing character's movements to produce a more ASL-like form of output using a graphical user interface. Currently, the animated character in systems like this cannot be controlled at a sufficient level of detail to produce a fully fluent sign language performance, and significant expertise is still needed by a content developer who wants to produce an ASL animation. Automatic English-to-ASL translation software could enhance systems like this to help ASL-naïve users produce an ASL animation from an English input string.

38.5 Conclusion

At present, there is no easy means for providing sign language to facilitate access for deaf signers. Automatic translation systems exist for providing fingerspelling or creating sign renderings for each written word in a text. Both of these, as discussed, do little to make text more accessible to a deaf signer. Similarly, the recognition of fingerspelling or individual signs does little to improve interactions for deaf signers. For this reason, in the area of sign language and interfaces, attention is currently focusing on natural language processing and machine translation technologies, with the goal of providing signers a means of interacting using their native sign language skills. While realization

of the ultimate goal of natural sign language interaction is many years away, the encouraging news is that progress is being made. Linguistic advances in understanding natural sign languages, combined with advances in computer technology, may one day allow deaf signers to have computer applications available in sign language.

Acknowledgments

Portions of this chapter were written as part of a visiting professorship for Vicki L. Hanson at the University of Dundee, made possible through an award from the Leverhulme Trust.

References

Adachi, H., Yosizawa, S., Fujita, M., Matsumoto, T., and Kamata, K. (2001). *Analysis of News Sentences with Sign Language and Sign Translation Processing*. IPSJ SIGNotes Natural Language Abstract No. 091-003.

AILB (2007). *E-Learning Environments for Deaf Adults*. http://www.fit.fraunhofer.de/projects/mobiles-wissen/ailb_en.html.

Bangham, J. A., Cox, S., Elliot, J. R., Glauert, J. R. W., Marshall, I., Rankov, S., and Wells, M. (2000). Virtual signing: Capture, animation, storage and transmission: An overview of the ViSiCAST project. *IEEE Seminar on Speech and Language Processing for Disabled and Elderly People*, 13 April 2000, London, UK. Piscataway, NJ: IEEE.

Bauer, B. and Heinz, H. (2000). Relevant features for video-based continuous sign language recognition, in the *Proceedings of the 2000 International Conference on Automatic Face and Gesture Recognition*, 28–30 March 2000, Grenoble, France, pp. 440–445. Piscataway, NJ: IEEE.

Bauer, B. and Kraiss, K.-F. (2001). Towards an automatic sign language recognition system using subunits, in *Revised Papers from the International Gesture Workshop on Gesture and Sign Languages in Human-Computer Interaction*, 18–20 April 2001, London, pp. 64–75. London: Springer-Verlag.

Bowden, R. and Sarhadi, M. (2002). A non-linear model of shape and motion for tracking finger spelt American Sign Language. *Image and Vision Computing* 20: 597–607.

Braffort, A. (1996). ARGo: An architecture for sign language recognition and interpretation, in the *Proceedings of the Gesture Workshop on Progress in Gestural Interaction*, 19 March 1996, York, U.K., pp. 17–30. New York: ACM Press.

Brashear, H., Henderson, V., Park, K., Hamilton, H., Lee, S., and Starner, T. (2006). American Sign Language recognition in game development for deaf children, in the *Proceedings of the 8th International ACM SIGACCESS Conference on Computers and Accessibility (ASSETS '06)*, 23–25 October 2006, Portland, OR, pp. 79–86. New York: ACM Press.

Brashear, H., Starner, T., Lukowicz, P., and Junker, H. (2003). Using multiple sensors for mobile sign language recognition, in the *Proceedings of the 7th IEEE International Symposium on Wearable Computers*, 21–23 October 2003, White Plains, NY, pp. 45–52. Piscataway, NJ: IEEE.

Bungeroth, J. and Ney, H. (2004). Statistical sign language translation, in the *Proceedings of the Workshop on Representation and Processing of Signed Languages in the Context of the 4th International Conference on Language Resources and Evaluation (LREC'04)*, May 2004, Lisbon, Portugal, pp. 105–108. http://www-i6.informatik.rwth-aachen.de/~bungeroth/lrec04.pdf.

Cox, S., Lincoln, M., Tryggvason, J., Nakisa, M., Wells, M., Tutt, M., Abbott, S. (2002). TESSA, a system to aid communication with deaf people, in the *Proceedings of the 5th International ACM Conference on Assistive Technologies (ASSETS 2002)*, 8–10 July 2002, Edinburgh, Scotland, pp. 205–212. New York: ACM Press.

Craft, B., Hinkle, D., Sedgwick, E., Alkoby, K., Davidson, M. J., Carter, R., et al. (2000). An approach to modeling facial expressions used in American Sign Language. Presented at the 2000 DePaul CTI Research Conference, November 2000, Chicago.

Crasborn, O., van der Hulst, H., and van der Kooij, E. (1998). SignPhon: A database tool for crosslinguistic phonological analysis of sign languages. Presented at the 6th International Conference on Theoretical Issues in Sign Language Research, 12–15 November 1998, Washington, DC.

Davidson, M. J., Alkoby, K., Sedgwick, E., Berthiaume, A., Carter, R., Christopher, J., et al. (2000). Usability testing of computer animation of fingerspelling for American Sign Language. Presented at the 2000 DePaul CTI Research Conference, 4 November 2000, Chicago.

Davidson, M. J., Alkoby, K., Sedgwick, E., Carter, R., Christopher, J., Craft, B., et al. (2001). Improved hand animation for American Sign Language, in the *Proceedings of the Technology and Persons with Disabilities Conference 2001 (CSUN 2001)*, 19–24 March 2001, Los Angeles. http://www.csun.edu/cod/conf/2001/proceedings/0092davidson.htm.

Efthimiou, E. and Fotinea, S-E. (2007). An environment for deaf accessibility to educational content, in the *Proceedings of the 1st International Conference on Information and Communication Technology and Accessibility (ICTA 2007)*, 12–14 April 2007, Hammamet, Tunisia, pp. 125–130. http://www.ilsp.gr/docs/amea/ICTA07.pdf.

Elliott, R., Glauert, J. R. W., and Kennaway, J. R. (2005). Developing techniques to support scripted sign language performance by a virtual human, in *Universal Access in HCI: Exploring New Dimensions of Diversity, Volume 8 of the Proceedings of the 11th International Conference on Human-Computer Interaction (HCI International 2005)* (C. Stephanidis, ed.) 22–27 July 2005, Las Vegas [CD-ROM]. Mahwah, NJ: Lawrence Erlbaum Associates.

Elliott, R., Glauert, J., Kennaway, J., Marshall, I., and Safar, E. (2008). Linguistic modeling and language-processing technologies for avatar-based sign language presentation. *Universal Access in the Information Society* 6: 375–391.

Furst, J., Alkoby, K., Berthiaume, A., Chomwong, P., Davidson, M. J., Konie, B., et al. (2000). Database design for American Sign Language, in the *Proceedings of the ISCA 15th International Conference on Computers and Their Applications (CATA-2000)*, 29–31 March 2000, New Orleans, pp. 427–430. Cary, NC: International Society for Computers and Their Applications.

Grieve-Smith, A. B. (1999). English to American Sign Language machine translation of weather reports, in the *Proceedings of the 2nd High Desert Student Conference in Linguistics* (D. Nordquist, ed.), 26–28 March 1999, Albuquerque, NM. Albuquerque, NM: High Desert Linguistics Society. http://www.unm.edu/~grvsmth/portfolio/mt-weath.html.

Grieve-Smith, A. B. (2001). SignSynth: A sign language synthesis application using Web3D and Perl, in the *Proceedings of the 4th International Workshop on Gesture and Sign Language Based Human-Computer Interaction* (I. Wachsmuth and T. Sowa, eds.), 18–20 April 2001, London, pp. 134–145. London: Springer-Verlag.

Grobel, K. and Assam, M. (1997). Isolated sign language recognition using hidden Markov models. *Proceedings of the 1997 IEEE International Conference on Systems, Man, and Cybernetics* 1: 162–167.

Hanson, V. L. (2007). Computing technologies for deaf and hard of hearing users, in *Human-Computer Interaction Handbook: Fundamentals, Evolving Technologies and Emerging Applications* (2nd ed.) (A. Sears and J. A. Jacko, eds.), pp. 885–893. Mahwah, NJ: Lawrence Erlbaum Associates.

Hanson, V. L. and Padden, C. (1990). Bilingual ASL/English instruction of deaf children, in *Cognition, Education, and Multimedia: Exploring Ideas in High Technology* (D. Nix and R. Spiro, eds.), pp. 49–63. Hillsdale, NJ: Lawrence Erlbaum Associates.

Holt, J. (1993). Demographic, Stanford Achievement Test—8th edition for deaf and hard of hearing students: Reading comprehension subgroup results. *American Annals of the Deaf* 138: 172–175.

Huenerfauth, M. (2004). Spatial representation of classifier predicates for machine translation into American Sign Language, in the *Proceedings of the Workshop on Representation and Processing of Signed Languages in the Context of the 4th International Conference on Language Resources and Evaluation (LREC'04)*, 26–28 May 2004, Lisbon, Portugal. http://eniac.cs.qc.edu/matt/pubs/huenerfauth-2004-lrec-classifier-predicate-representations.pdf.

Huenerfauth, M. (2006a). Representing coordination and non-coordination in American Sign Language animations. *Behaviour & Information Technology* 25: 285–295.

Huenerfauth, M. (2006b). *Generating American Sign Language Classifier Predicates For English-To-ASL Machine Translation*. PhD dissertation, Computer and Information Science, University of Pennsylvania, Philadelphia.

Huenerfauth, M. (2007). Misconceptions, technical challenges, and new technologies for generating American Sign Language animation. *Universal Access in the Information Society* 6: 419–434.

Kapuscinski, T. and Wysocki, M. (2003). Vision-based recognition of Polish Sign Language, in the *Proceedings of the Symposium on Methods of Artificial Intelligence (AI-METH 2003)*, 5–7 November 2003, Gliwice, Poland, pp. 67–68. Gliwice, Poland: Polish Association for Computational Mechanics.

Karpouzis, K., Caridakis, G., Fotinea, S-E., and Efthimiou, E. (2007). Educational resources and implementation of a Greek Sign Language synthesis architecture. *Computers & Education* 49: 54–74.

Kennaway, J. R., Glauert, J. R. W., and Zwitserlood, I. (2007). Providing signed content on the Internet by synthesized animation. *ACM Transactions on Computer Human Interaction (TOCHI)* 14: 1–29.

Klima, E. S. and Bellugi, U. (1979). *The Signs of Language*. Cambridge, MA: Harvard University Press.

Konstantinidis, B. and Fels, D. (2006). Hand waving apparatus for effective turn-taking (HWAET) using video conferencing for deaf people, in the *Proceedings of the 3rd Cambridge Workshop on Universal Access and Assistive Technology (CWUAAT 2006)*, 10–12 April 2006, Cambridge, U.K., pp. 77–82. Cambridge, U.K.: Cambridge University Press.

Kuroda, T., Tabata, Y., Murakami, M., Manabe, Y., and Chihara, K. (2001). Sign language digitization and animation, in *Universal Access in HCI: Towards an Information Society for All, Volume 3 of the Proceedings of the 9th International Conference on Human-Computer Interaction (HCI International 2001)* (C. Stephanidis, ed.), 5–10 August 2001, New Orleans, pp. 363–367. Mahwah, NJ: Lawrence Erlbaum Associates.

Lane, H. (1976). *The Wild Boy of Aveyron: A History of the Education of Retarded, Deaf and Hearing Children*. Cambridge, MA: Harvard University Press.

Lane, H., Hoffmeister, R., and Bahan, B. (1996). *A Journey into the Deaf World*. San Diego: DawnSign Press.

Lebourque, T. and Gibet, S. (1999). A complete system for the specification and generation of sign language gestures,

in the *Proceedings of the International Gesture Workshop on Gesture-Based Communication in Human-Computer Interaction*, 17–19 March 1999, Gif-sur-Yvette, France, pp. 227–238. London: Springer-Verlag.

Liang, R.-H. and Ouhyoung, M. (1998). A real-time continuous gesture recognition system for sign language, in the *Proceedings of the 3rd International Conference on Automatic Face and Gesture Recognition*, Spring 1998, Nara, Japan, pp. 558–565. Piscataway, NJ: IEEE Press.

Liddell, S. (2003). *Grammar, Gesture, and Meaning in American Sign Language*. Cambridge U.K.: Cambridge University Press.

Link-it (2007). http://www.sit.se/net/Specialpedagogik/In+English/Educational+materials/Deaf+and+Hard+of+Hearing/Products/Link-it.

Loeding, B. L., Sarkar, S., Parashar, A., and Karshmer, A. I. (2004). Progress in automated computer recognition of sign language, in the *Proceedings of the 9th International Conference on Computer Helping People with Special Needs (ICCHP 2004)* (K. Miesenberger, ed.), 7–9 July 2004, Paris, pp. 1079–1087. Berlin/Heidelberg: Springer-Verlag.

Lu, S., Seiji, I., Matsuo, H., and Nagashima, Y. (1997). Towards a dialogue system based on recognition and synthesis of Japan Sign Language, in the *Proceedings of the Gesture and Sign Language in Human-Computer Interaction: International Gesture Workshop*, September 1997, Biefeld, Germany, p. 259. Berlin/Heidelberg: Springer-Verlag.

Marshall, I. and Sáfár, E. (2004). Sign language generation in an ALE HPSG 2004, in the *Proceedings of the 11th International Conference on Head-Driven Phrase Structure Grammar Center for Computational Linguistics* (S. Müller, ed.), 3–6 August 2004, Belgium, pp. 189–201. Stanford, CA: CSLI Publications.

Marshall, I. and Sáfár, E. (2005). Grammar development for sign language avatar-based synthesis, in *Universal Access in HCI: Exploring New Dimensions of Diversity, Volume 8 of the Proceedings of the 11th International Conference on Human-Computer Interaction (HCI International 2005)* (C. Stephanidis, ed.), 22–27 July 2005, Las Vegas [CD-ROM]. Mahwah, NJ: Lawrence Erlbaum Associates.

McDonald, J., Toro, J., Alkoby, K., Berthiaume, A., Carter, R., Chomwong, P., et al. (2001). An improved articulated model of the human hand. *The Visual Computer* 17: 158–166.

Meier, R. P. (1990). Person deixis in American Sign Language, in *Theoretical Issues in Sign Language Research* (S. Fischer and P. Siple, eds.), Volume 1, pp. 175–190. Chicago: University of Chicago Press.

Messing, L. and Stern, G. (1997). *Sister Mary*. Unpublished manuscript.

Mitchell, R., Young, T. A., Bachleda, B., and Karchmer, M. A. (2006). How many people use ASL in the United States? Why estimates need updating. *Sign Language Studies* 6: 306–335.

Morrissey, S. and Way, A. (2005). An example-based approach to translating sign language, in the *Proceedings of the Workshop on Example-Based Machine Translation*, 12–16 September 2005, Phuket, Thailand, pp. 109–116. Kyoto, Japan: Asia-Pacific Association for Machine Translation.

Murakami, K. and Taguchi, H. (1991). Gesture recognition using recurrent neural networks, in the *Proceedings of the SIGCHI Conference on Human Factors in Computing Systems: Reaching through Technology (CHI '91)*, 18–20 May 1991, Pittsburgh, pp. 237–242. New York: ACM Press.

National Institutes of Health (2002). *Captions for Deaf and Hard of Hearing Viewers*. NIH Publication No. 00-4834. http://www.nidcd.nih.gov/health/hearing/caption.asp#edit.

Neidle, C., Kegl, J., MacLaughlin, D., Bahan, B., and Lee, R. G. (2000). *The Syntax of American Sign Language: Functional Categories and Hierarchical Structure*. Cambridge, MA: The MIT Press.

Newkirk, D. (1987). *SignFont Handbook*. San Diego: Emerson and Associates.

Ohki, M., Sagawa, H., Sakiyama, T., Oohira, E., Ikeda, H., and Fujisawa, H. (1994). Pattern recognition and synthesis for sign language translation system, in the *Proceedings of the 1st International ACM Conference on Assistive Technologies (ASSETS 1994)*, 31 October–3 November 1994, Marina del Rey, CA, pp. 1–8. New York: ACM Press.

Padden, C. (1988). *Interaction of Morphology and Syntax in American Sign Language. Outstanding Dissertations in Linguistics*. Series IV. New York: Garland Press.

Padden, C. and Hanson, V. L. (2000). Search for the missing link: The development of skilled reading in deaf children, in *The Signs of Language Revisited: An Anthology to Honor Ursula Bellugi and Edward Klima* (K. Emmorey and H. Lane, eds.), pp. 435–447. Mahwah, NJ: Lawrence Erlbaum Associates.

Padden, C. and Humphries, T. (1988). *Deaf in America: Voices from a Culture*. Cambridge, MA: Harvard University Press.

Padden, C. and Humphries, T. (2005). *Inside Deaf Culture*. Cambridge, MA: Harvard University Press.

Petrie, H. and Engelen, J. (2001). MultiReader: A multimodal, multimedia reading system for all readers, including print disabled readers, in *Assistive Technology: Added Value to the Quality of Life* (C. Marincek et al., eds.), pp. 61–69. Amsterdam, The Netherlands: IOS Press.

Petrie, H., Fisher, W., Weimann, K., and Weber, G. (2004). Augmenting icons for deaf computer users, in *CHI '04 Extended Abstracts on Human Factors in Computing Systems*, pp. 1131–1134. New York: ACM Press.

Petrie, H., Weber, G., and Fisher, W. (2005). Personalisation, interaction and navigation in rich multimedia documents for print-disabled users. *IBM Systems Journal* 44: 629–636.

Prillwitz, S., Leven, R., Zienert, H., Hanke, T., and Henning, J. (1989). *HamNoSys. Version 2.0; Hamburg Notation System for Sign Languages. An Introductory Guide*. Volume 5 of the International Studies on Sign Language and Communication of the Deaf. Hamburg, Germany: Signum Press.

RNID (2007). *BSL: Index of Sign Language Clips*. http://www.rnid.org.uk/bsl/bsl_video_clips_index.

Sagawa, H., Takeuchi, M., and Ohki, M. (1997). Description and recognition methods for sign language based on gesture components, in the *Proceedings of the 2nd International Conference on Intelligent User Interfaces (IUI 97)*, 6–9 January 1997, Orlando, FL, pp. 97–104. New York: ACM Press.

Saksiri, B., Ferrell, W. G., and Ruenwongsa, P. (2006). Virtual sign animated pedagogic agents to support computer education for deaf learners. *ACM SIGACCESS Accessibility and Computing* 86: 40–44. New York: ACM Press.

Sharoff, S., Hartley, A., and Llewellyn-Jones, P. (2004). Sentence generation in British Sign Language. Paper presented at the 3rd International Conference on Natural Language Generation (INLG04), 14–16 July 2004, Brockenhurst, U.K.

Shionome, T., Kamata, K., Yamamoto, H., and Fischer, S. (2005). Effects of display size on perception of Japanese Sign Language: Mobile access in signed language, in *Universal Access in HCI: Exploring New Dimensions of Diversity, Volume 8 of the Proceedings of the 11th International Conference on Human-Computer Interaction (HCI International 2005)* (C. Stephanidis, ed.), 22–27 July 2005, Las Vegas [CD-ROM]. Mahwah, NJ: Lawrence Erlbaum Associates.

Speers, d'A. L. (2001). *Representation of American Sign Language for Machine Translation*. PhD dissertation, Department of Linguistics, Georgetown University, Washington, DC.

Starner, T., Weaver, J., and Pentland, A. (2004). Real-time American Sign Language recognition using desk and wearable computer based video. *IEEE Transactions on Pattern Analysis and Machine Intelligence* 20: 1371–1375.

Suszczańska, N., Szmal, P., and Francik, J. (2002). Translating Polish texts into sign language in the TGT system, in the *Proceedings of the 20th IASTED International Multi-Conference, Applied Informatics (AI 2002)*, 18–21 February 2002, Innsbruck, Austria, pp. 282–287. Calgary, Canada: ACTA Press.

Sutton, V. (1998). *The Signwriting Literacy Project*. Paper presented at the Impact of Deafness on Cognition AERA Conference, 13–14 April 1998, San Diego.

Takahashi, T. and Kishino, F. (1991). Gesture coding based in experiments with a hand gesture interface device. *SIGCHI Bulletin* 23: 67–73.

Tokuda, M. and Okumara, M. (1998). Towards automatic translation from Japanese into Japanese Sign Language, in *Assistive Technology and Artificial Intelligence, Applications in Robotics, User Interfaces and Natural Language Processing* (V. O. Mittal, H. A. Yanco, J. M. Aronis, and R. C. Simpson, eds.), pp. 97–108. London: Springer-Verlag.

Tolani, D., Goswami, A., and Badler, N. (2000). Real-time inverse kinematics techniques for anthropomorphic limbs. *Graphical Models and Image Processing* 62: 353–388.

Toro, J., Furst, J., Alkoby, K., Carter, R., Christopher, J., Craft, B., et al. (2001). An improved graphical environment for transcription and display of American Sign Language. *Information* 4: 533–539.

Vamplew, P. (1996). *Recognition of Sign Language Using Neural Networks*. PhD dissertation, University of Tasmania, Hobart, Australia.

van Zijl, L. and Barker, D. (2003). South African Sign Language machine translation system, in the *Proceedings of the 2nd International Conference on Computer Graphics, Virtual Reality, Visualisation and Interaction in Africa (Afrigraph 2003)*, 3–5 February 2003, Cape Town, South Africa, pp. 49–52. New York: ACM Press.

Vcom3D (2007). *Sign Language Software—Studio*. http://www.vcom3d.com/Studio.htm.

Veale, T., Conway, A., and Collins, B. (1998). The challenges of cross-modal translation: English to sign language translation in the Zardoz system. *Machine Translation* 13: 81–106.

Verlinden, M., Tijsseling, C., and Frowein, H. (2001). A signing avatar on the WWW. Paper presented at the International Gesture Workshop 2001, 18–20 April 2001, London. http://www.visicast.sys.uea.ac.uk/Papers/IvDGestureWorkshop2000.pdf.

Vogler, C. and Goldenstein, S. (2005). Analysis of facial expressions in American Sign Language, in *Universal Access in HCI: Exploring New Dimensions of Diversity, Volume 8 of the Proceedings of the 11th International Conference on Human-Computer Interaction (HCI International 2005)* (C. Stephanidis, ed.), 22–27 July 2005, Las Vegas [CD-ROM]. Mahwah, NJ: Lawrence Erlbaum Associates.

Vogler, C. and Metaxas, D. (2001). A framework for recognizing the simultaneous aspects of American Sign Language. *Computer Vision and Image Understanding* 81: 358–384.

Vogler, C. and Metaxas, D. (2004). Handshapes and movements: Multiple-channel ASL recognition, in the *Proceedings of the 5th International Gesture Workshop on Gesture-Based Communication in Human-Computer Interaction (GW 2003)*, 15–17 April 2003, Genoa, Italy, pp. 247–258. Berlin/Heidelberg: Springer-Verlag.

Waldron, M. B. and Kim, S. (1995). Isolated ASL sign recognition system for deaf persons. *IEEE Transactions on Rehabilitation Engineering* 3: 261–271.

Wideman, C. J. and Sims, E. M. (1998). Signing avatars, in the *Proceedings of the Technology and Persons with Disabilities Conference 1998 (CSUN 1998)*, 17–23 March 1998, Los Angeles. http://www.csun.edu/cod/conf/1998/proceedings/csun98_027.htm.

Windridge, D. and Bowden, R. (2004). Induced decision fusion in automatic sign language interpretation: Using ICA to isolate the underlying components of sign, in the *Proceedings of the 5th International Workshop on Multiple Classifier Systems (MCS 2004)*, 9–11 June 2004, Cagliari, Italy, pp. 303–313. Berlin/Heidelberg: Springer-Verlag.

Wolfe, R., Alkoby, K., Barnett, J., Chomwong, P., Furst, J., Honda, G., et al. (1999). An interface for transcribing American Sign Language, in *ACM SIGGRAPH 99 Conference Abstracts and Applications, International Conference on Computer Graphics and Interactive Techniques*, p. 229. New York: ACM Press.

Wu, C. H., Chiu, Y. H., and Guo, C. S. (2004). Text generation from Taiwanese Sign Language using a PST-based language model for augmentative communication. *IEEE Transactions on Neural Systems and Rehabilitation Engineering* 12: 441–454.

Xu, L. and Gao, W. (2000). Study on translating Chinese into Chinese Sign Language. *Journal of Computer Science and Technology* 15: 485–490.

Xu, M., Raytchev, B., Sakaue, K., Hasegawa, O., Koizumi, A., Takeuchi, M., and Sagawa, H. (2000). A vision-based method for recognizing non-manual information in Japanese Sign Language, in the *Proceedings of the 3rd International Conference on Advances in Multimodal Interfaces (ICMI 2000)*, 14–16 October 2000, Beijing, pp. 572–581. London: Springer-Verlag

Yang, X., Jiang, F., Liu, H., Yao, H., Gao, W., and Wang, C. (2005). Visual sign language recognition based on HMMs and autoregressive HMMs, in the *Proceedings of the 6th International Gesture Workshop on Gesture in Human-Computer Interaction and Simulation (GW 2005)*, 18–20 May 2005, Berder Island, France, pp. 80–83. Berlin/Heidelberg: Springer-Verlag.

Yuan, Q., Gao, W., Yang, H., and Wang, C. (2002). Recognition of strong and weak connection models in continuous sign language, in the *Proceedings of the 16th International Conference on Pattern Recognition, Volume 1 (ICPR'02)*, 11–15 August 2002, Quebec, Canada, pp. 75–78. Washington, DC: IEEE Computer Society.

Zhang, L., Fang, G., Gao, W., Chen, X., and Chen, Y. (2004). Vision-based sign language recognition using sign-wise tied mixture HMM, in the *Proceedings of the 5th Pacific Rim Conference on Advances in Multimedia Information Processing (PCM 2004)*, 30 November–4 December 2004, Tokyo, pp. 1035–1042. Berlin/Heidelberg: Springer-Verlag.

Zhao, L., Kipper, K., Schuler, W., Vogler, C., Badler, N., and Palmer, M. (2000). A machine translation system from English to American Sign Language, in the *Proceedings of the 4th Conference of the Association for Machine Translation in the Americas on Envisioning Machine Translation in the Information Future*, 10–14 October 2000, Cuernavaca, Mexico, pp. 54–67. London: Springer-Verlag.

Visible Language for Global Mobile Communication: A Case Study of a Design Project in Progress

Aaron Marcus

39.1 Introduction

39.1.1 Universal Visible Languages

To the extent that many or most of the earth's inhabitants use language, one might say that language itself is a universal means to access the past experience of human civilization. The widespread use of specific languages in the past, such as the Roman Empire's and later the Catholic Church's use of Latin, the extensive use of French in European diplomacy up to the 19th century, the Chinese Empire's use of Mandarin, or the global use of English in commerce today serve as examples. Over the centuries, many different theorists and practitioners have attempted to augment natural languages (those emerging from the idiosyncratic flow of technological, economic, social, military, and other factors) by attempting to design more rational artificial, universal spoken and written languages. Such universal written languages are called technically *pasigraphies* and are described, among other places, at the LangMaker web site (2007). The most famous pasigraphy is perhaps Gottfried Wilhelm Leibniz' *Characteristica Universalis* (Cohen, 1954; Leibniz web site, 2007) begun in about 1677. Others are George Dalgarno's Universal Language begun in 1661 (Dalgarno web site, 2007) and John Wilkins' Analytical Language begun in 1668 (Wilkins web site, 2007).

Some authors have proposed systems inspired by contact with alien, other-worldly experiences (Weilgart, 1974); others are based on internal, rational desires for a more orderly, *a priori* constructed system than natural languages produce. These rational languages, both verbal and visual, are ones in which the formal structure of the signs reveals the semantics of the term or symbol.

Today, perhaps the best-known universal verbal language is Esperanto, which means "one who hopes." (Esperanto web site, 2007). Ludwig Zamenhof, under the pseudonym Doktoro Esperanto, developed the language in 1887, with the specific objective of creating an international second language. In 1905, Zamenhof published his *Fundamento de Esperanto*, which set forth the basic principles of the language's structure and formation. Esperanto is probably the most successful of these artificial international languages, but the number of speakers is estimated to be somewhere between 100,000 and 2 million people (Encyclopedia Britannica web site, 2007). An organization founded in 1908, the Universala Esperanto-Asocio has members in approximately 83 countries, and there are approximately 50 national Esperanto associations that use Esperanto (see, e.g., http://www.esperanto-usa.org). There is an annual World Esperanto Congress; over 100 periodicals are published in the language; and the *Encyclopedia Britannica* estimates that more than 30,000 books have been published in Esperanto. Such universal verbal languages are beyond the scope of this chapter.

An offshoot of these systems are those that focus on nontextual signs, or visible languages, intended for easy learning and use by people all over the world, a kind of visual Esperanto. In the last century, C. K. Bliss in Australia invented Blissymbolics in 1949, a language of visual, nonverbal signs, which is described in his fundamental work *Semantography* (Bliss, 1965). Bliss attempted to convince the United Nations to declare Blissymbolics a world auxiliary visible language. Blissymbolics has been used to assist people who are suffering from aphasia, a medical condition that hinders them in using, reading, or understanding written verbal languages. However, such people are nevertheless able to read, write, and use the visual symbols in the 3,000 signs of Blissymbolics because the damaged verbal processing center is circumvented (Bliss web site, 2007).

In September 1982, Bliss granted an exclusive, noncancellable and perpetual, worldwide license to the Blissymbolics Communication Institute (1985, now Blissymbolics Communication International) based in Toronto, Ontario, Canada, for the application of Blissymbols, for use by disabled persons and persons having communication, language, and learning difficulties. Although the author knew of Bliss' system and communicated with him in the 1970s, and was aware of the Toronto Bliss center, because of copyright, license, and legal factors, the concept design described in this chapter did not use his system, but another similar one is described in the following, which is legally and commercially more available.

A related system of signs is known as Picture Communication Symbols (PCS), originally developed by Mayer-Johnson, LLC, for use in augmentative and alternative communication (AAC) systems. PCS consists of between 3,000 and 7,000 color and black-and-white signs. The PCS comprises a central set of approximately 5,000 symbols, supplemented by general-purpose and country-specific libraries, for a total of 12,000 signs. The AAC systems may use computer graphics systems (Dynamyte) or low-technology media, such as a communication board. PCS signs have been translated into 40 different languages. People can also develop their own PCS signs for specific communication needs with people who have limited speech capabilities.

Several studies have found PCS to be more self-evident than other graphic sign systems such as Blisssymbols (Mizuko, 1987). A graphic symbol is self-evident or transparent if "the shape, motion, or function of the referent is depicted to such an extent that meaning of the symbol can be readily guessed in the absence of the referent" (Fuller and Lloyd, 1991, p. 217). Because of their high self-evident quality, PCS signs are easy to learn by children with little or no speech. Several studies report that children with cognitive disabilities learn PCS easily. This system, although reportedly successful for a distinct set of users, was not chosen for the studies reported in this chapter because of the copyright, licensing, and legal factors surrounding their use.

Other systems exist for specific groups of people with disabilities. A system called Lingraphica has been developed (Lingraphica web site, 2007) by Dr. Richard D. Steele (Steele, 2005) with others and deployed as a speech-generating device designed specifically for adults with aphasia. The system combines images, animation, text, and spoken words to provide communication for those who have lost their speech, enabling users to express their needs, desires, and wishes. The Lingraphica system also provides unlimited speech practice to maximize the return of natural language skills. Dr. Steele (with two others)

has a patent on this technique (Steele et al., 1999), in which language-oriented information is communicated via a graphical user interface, including the ability to vary the playback speed of the audiovisual aspect of an icon, moving among different layers of a hierarchical database, and text searching and matching with icons. This system, while interesting and evidently valuable, is beyond the scope of the concerns of this chapter, because of the use of illustrations, text, and so on, as well as the copyright, patent, licensing, and legal factors.

The previously described recent systems show potentially universal visible language systems applied practically to a specific set of users with disabilities of language use. Their commercial success suggests the viability of universal visible language systems for larger, more general groups of people.

The graphic designer and sign designer Yukio Ota introduced in 1964 his own version of a universal sign language called LoCoS (Ota, 1973a, 1973b), which stands for Lovers Communication System. Ota invented the LoCoS language while he was in Italy in 1964, and he eventually published a Japanese LoCoS reference book in 1973 (Ota, 1973b). This URL shows that reference book: http://locosworld.net. An English translation is currently in preparation. Ota has presented lectures about LoCoS around the world since he designed the signs, and he published several articles in English explaining his system (e.g., Ota 1973a). He has promoted the use of the language worldwide for the past 40 years. The author has written about Ota's work (Marcus, 2003), and the author's firm maintains an extranet about LoCoS at this URL: http://clients.amanda.com/locos.

One of the significant features of LoCoS is that it can be learned in 1 day. Participants at Ota's lectures have been able to write him messages after hearing about the system and learning the basics of its vocabulary and grammar. Mr. Ota has not patented or protected this system and actively encourages people around the world to experiment with it. In addition, together with colleagues in Japan and elsewhere, he has conducted continual investigation and design of enhancements of the system. Some research articles about the language have appeared (e.g., Vanhauer, 2006).

39.1.2 Universal Visible Languages and Mobile Devices

With the rise of mobile device usage worldwide (e.g., more than 500 million users in China alone), it seems reasonable to consider how universal visible languages might augment communication, not only for people with disabilities, but for the general public. The author believes circumstances of global communication, user's increasing orientation to visual communication as well as verbal communication (e.g., interest in and use of video, cinema, graphic signs for mobile communication, etc.), and new technology might avoid the fate of Esperanto, which has had only limited success.

Machine-to-machine translation also poses an alternative to universal visible languages that might be used in mobile devices. Systems such as the Vermobil system (Wahlster, 2000; Eichler,

2004) achieve adequate quality of speech-to-speech translation using so-called corpus-centered computation (Sumita, 2002), in which translation knowledge is extracted from large document collections called corpora, or by more traditional approaches using machine translation for written languages. In one report (Sumita, 2002), high-quality translation was demonstrated in the domain of travel conversation (a scenario mentioned in the following). However, the author believes that many or most of these translation systems are still in developmental stages, much like some computer-based visible language systems, and that visible language systems offer the advantages of people-centered design, comparable or perhaps lower development costs, and aspects of appeal that seem to justify further research and development. The chapter describes one such investigation.

Based on this background, the author's firm worked with Mr. Ota over a period of several months in 2005, and in the ensuing months since then in an unfunded, limited, informal user-centered design project centering on the use of LoCoS with mobile devices. The initial objective was to study how LoCoS might be used in the context of a mobile device, such as a mobile phone, and to design concept prototypes that would enable others to understand and evaluate the merits of further research and development. This chapter presents an introduction to LoCoS, the design issues presented by trying to adapt LoCoS to a mobile device (called m-LoCoS), an initial set of prototype screens, and future design challenges.

39.1.3 Basics of LoCoS

As mentioned, LoCoS is an artificial, nonverbal, generally nonspoken, visible language system designed for use by any human being to communicate with others who may not share spoken or written natural languages. Individual signs may be combined to form expressions and sentences in somewhat linear arrangements, as shown in Figure 39.1.

The signs may be combined into complete LoCoS expressions or sentences, formed by three horizontal rows of square area

typically reading from left to right. Note this culture/localization issue: many, but not all symbols could be flipped left to right for readers/writers accustomed to right-to-left verbal languages. The main contents of a sentence are placed in the center row. Signs in the top and bottom rows act as adverbs and adjectives, respectively. Looking ahead to the possible use of LoCoS in mobile devices with limited space for sign display, a mobile-oriented version of LoCoS can use only one line. The grammar of the signs is similar to English (subject-verb-object). This aspect of the language, also, is an issue for those users used to other paradigms from natural verbal languages.

LoCoS differs from alphabetic natural languages in that the semantic reference (sometimes called *meaning*) and the visual form are closely related. LoCoS differs from some other visible languages; e.g., Bliss symbols, mentioned previously, use more abstract symbols, while LoCoS signs are more iconic. LoCoS is similar to, but different from, Chinese ideograms, like those incorporated into Japanese Kanji signs. LoCoS is less abstract in that symbols of concrete objects like a road sign show pictures of those objects. Like Chinese signs or Kanji, one sign refers to one concept, although there are compound concepts. According to Ota, LoCoS reuses signs more efficiently than traditional Chinese signs. Note that the rules of LoCoS did not result from careful analysis across major world languages for phonetic efficiency. Although LoCoS does have rules for pronunciation (see Figure 39.2), it is rarely used; audio input/output was not explored.

LoCoS has several benefits that would make it potentially usable, useful, and appealing as a sign language displayable on mobile devices. First, it is easy to learn in a progressive manner, starting with just a few basics. The learning curve is not steep, and users can guess correctly at new signs. Second, it is easy to display; the signs are relatively simple. Third, it is robust. People can understand the sense of the language without knowing all signs. Fourth, the language is suitable for mass media and the general public. People may find it challenging, appealing, mysterious, and fun.

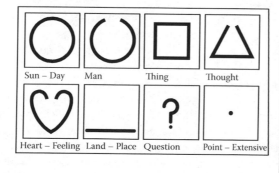

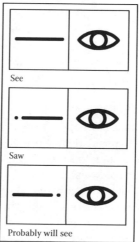

FIGURE 39.1   Individual and combined signs.

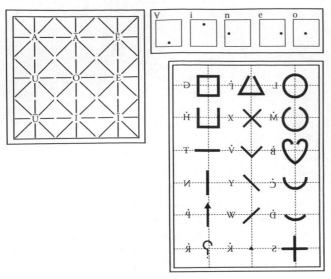

FIGURE 39.2 Pronunciation of LoCoS.

39.2 Initial Design of m-LoCoS

39.2.1 Universal Visible Messaging

LoCoS can be used in a universal visual messaging application, as opposed to text messaging; people who do not speak the same language can communicate with each other. People who need to interact via a user interface that has not been localized to their own language normally would find the experience daunting. People who speak the same language but want to communicate in a fresh, new medium may find LoCoS especially appealing (e.g., teenagers and children). People who may have some speech or accessibility issues, as noted previously, may find m-LoCoS useful.

Currently, initial prototype screens showing how LoCoS could be used in mobile devices have been developed. The HTML prototype screens have been developed using a Motorola V505 and a Nokia 7610 mobile phone. A LoCoS-English dictionary was begun and is in progress. Future needs include expanding LoCoS, exploring new, different visual attributes for the signs of LoCoS, including color, animation, and nonlinear arrangements (which Yukio Ota calls LoCoS 2.0), and further developing the prototype so that it is more complete and interactive.

The assumptions and objectives for m-LoCoS include the following.

For the developing world, there is remarkable growth in the use of mobile phones. China has approximately 500 million phones, larger than the U.S. population, and India is growing rapidly. People seem to be willing to spend up to 10% of their income for phones and service, which is often their only link to the world at large. For many users, the mobile phone is the first one that they have ever used. In addition, in developing countries literacy levels may be low, especially familiarity with computer user interfaces. Thus, if mobile voice communication is expensive and unreliable, mobile messaging may be slower but

cheaper, and more reliable. Texting may be preferred to voice communication in some social settings. M-LoCoS may make it easier for people in developing countries to communicate with each other and with those abroad. The fact that LoCoS can be learned in 1 day makes it an appealing choice.

In the industrialized world, young people (i.e., ages 2 to 25) have a high aptitude for learning new languages and user interface paradigms. It is a much-published phenomenon that young people like to text message, in addition, and sometimes in preference, to talking on their mobile phones. In Japan, additional signs, called emoticons, have been popular for years. In fact, newspaper accounts chronicle the rise of *gyaru-moji* ("girl-signs") (Chari, 2004; *Japan Times*, 2005), a "secret" texting language of symbols improvised by Japanese teenage girls. They are a mixture of Japanese syllables, numbers, mathematical symbols, and Greek characters. Even though *gyaru-moji* take twice as long for input as standard Japanese, they are still popular. This phenomenon suggests that young people might enjoy sign-messaging using LoCoS, even if input were slightly more difficult. The signs might be unlike anything they have used before, they would be easy to learn, they would be expressive, and they would be aesthetically pleasing. A mobile-device-enabled LoCoS might offer a fresh new way to send messages.

39.2.2 User Profiles and Use Scenarios

Regarding users and their use-context, although 1 billion people use mobile phones now, there are another 1 billion people, many in developing countries, who have never used *any* phone before. Keep in mind that a mobile phone's entire user interface could be displayed in LoCoS, not only the messaging applications, but all applications, including voice conversations. For younger users interested in a "cool" or "secret" (from their parents) form of communication in the industrialized world, they would be veteran mobile phone users. LoCoS could be an add-on application for their existing phones, and the success of *gyaru-moji* in Japan, as well as emoticon use (Niedermeier, 2001), suggests that a mobile LoCoS could be successful. Finally, one could consider the case of travelers in countries that do not speak the traveler's language. In this situation, LoCoS could be helpful for any kind of mobile device.

Bearing in mind these circumstances three representative user profiles and use scenarios were developed for future design of m-LoCoS applications. Use Scenario 1 concerns the micro-office in a less-developed country: Srini is a man in a small town in India. User Scenario 2 concerns young lovers in a developed country: Jack and Jill, boyfriend and girlfriend, in the United States. Use Scenario 3 concerns a traveler in a foreign country: Jaako is a Finnish tourist in a restaurant in France. Each of these use scenarios is described in the following.

Use Scenario 1: Micro-office in a less-developed country. Srini in India lives in a remote village that does not have running water, but just started having access to a new wireless network. The network is not reliable or affordable enough for long voice conversations, but is adequate for text messaging. Srini's mobile phone is the only means for non-face-to-face communication with his business partners. Srini's typical communication topic is this: should he go to the next village to sell his products, or wait for the prices to rise?

Use Scenario 2: Young lovers in the United States. Jack and Jill, boyfriend and girlfriend, text message each other frequently, using 5 to 10 words per message, and 2 to 3 messages per conversation thread. They think text messaging is cool (i.e., highly desirable). They think it would be even cooler to send text messages in a private, personal, or secret language not familiar to most people looking over their shoulders or somehow intercepting their messages.

Use Scenario 3: Tourist in a foreign country. Jaako, a Finnish tourist in a restaurant in Paris, France, is attempting to communicate with the waiter; however, he and the waiter do not speak a common language. A typical restaurant dialogue would be: "May I sit here?" "Would you like to start with an appetizer?" "I'm sorry; we ran out of that." "Do you have lamb?" All communication takes place via a single LoCoS-enabled device. Jaako and the waiter take turns reading and replying, using LoCoS.

These use scenarios have not yet been incorporated into user tests with the designs concepts currently, but they are ready to serve at this future step.

39.2.3 Design Implications and Design Challenges

The design implications for developing an m-LoCoS are that the language must be simple and unambiguous, input must occur quickly and reliably, and several dozen LoCoS signs must fit onto one mobile-device screen. Another challenge is that LoCoS as a system of signs must be extended for everyday use. Currently, there are about 1,000 signs, as noted in the LoCoS guidebook published in Japanese (Ota, 1987). However, these signs are not sufficient for many common scenarios. It has been estimated that about 3,000 signs are required. This number of signs is approximately equal to what comprises Basic Chinese, a limited set of signs for essential communication in that language. The new signs cannot be arbitrary, but should follow the current patterns of LoCoS and be appropriate for contexts a half-century after its invention. Even supposedly universal, timeless

sign systems like Isotype from Otto Neurath's group (Ota, 1973; Neurath, 1980) featured some signs that, almost a century later, are hard to interpret. One example is a small pyramidal shape representing sugar, based on a familiar commercial packaging of individual sugar portions in Europe in the early part of the 20th century.

Another design challenge for m-LoCoS is that the mobile phone user interface itself should utilize LoCoS (optionally, like language switching). For the user in developing countries, it might be the case that telecom manufacturers and service providers might not have localized, or localized well, the user interface to the specific users' preferred language. This would enable the user to more comfortably rely on a single language for the controls and for help, as well as within applications. For users in more developed countries, the cool factor or the devotee's interest in LoCoS would make a LoCoS-based user interface desirable. Figure 39.3 shows an initial sketch of such signs.

Not only must the repertoire of the current LoCoS signs be extended, but also the existing signs must be revised to update them, as mentioned earlier for Isotype's signs. Despite Ota's best efforts, some of the signs are culturally or religiously biased. It is difficult to make signs that are clear and pleasing to everyone in the world. What is needed is a practical compromise that achieves tested success with the major cultures of the target users. Examples of current challenges are shown in Figure 39.4. The current LoCoS sign for *restaurant* might often be mistaken for a *bar* because of the wineglass sign inside the building sign. The cross as a sign for *religion* might not be understood correctly, thought appropriate, or even be welcome in Moslem countries.

Another challenge would be to enable and encourage users to try LoCoS. Target users must be convinced to try to learn the visible language in 1 day. Non-English speakers might need to accommodate themselves to the English subject-verb-object structure. In contrast, in Japanese, the verb comes last, as it does in German dependent phrases. Despite Ota's best efforts, some expressions can be ambiguous. Therefore, there seems to be a need for dictionary support, preferably on the mobile device itself. Users should be able to ask, "What is the LoCoS sign for X, if any?" or "What does this LoCoS sign mean?"

Displaying LoCoS on small screens is a fundamental design challenge. There are design trade-offs among the

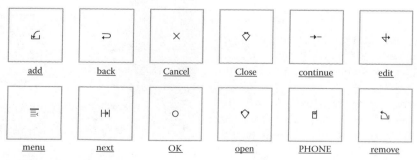

FIGURE 39.3 Sketch by the author's firm of some user interface control signs based on LoCoS.

FIGURE 39.4 LoCoS signs for priest and restaurant.

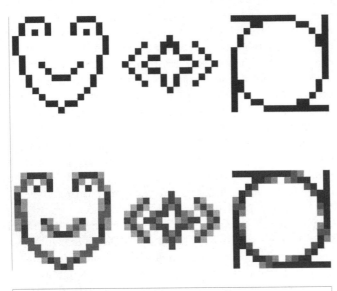

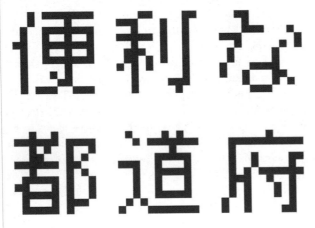

FIGURE 39.5 Examples of signs drawn with and without anti-aliasing.

dimensions of legibility, readability, and density of signs. The first question that arises concerns the dimensions of a sign in pixels. Figure 39.5 shows some comparative sketches of small signs.

Japanese phones and web sites often seem to use 13 × 13 pixels. In m-LoCoS, it was decided to use 15 × 15 pixels for the signs. This density is the same as smaller, more numerous English signs. Within the design team, there was some discussion about whether signs should be anti-aliased; unfortunately, not enough

was known about support of mobile devices with grayscale pixels to know what to recommend. Are signs easier to recognize and understand if anti-aliased? This issue is a topic for future user research.

39.2.4 Classifying, Selecting, and Entering Signs

There are several issues related to how users can enter LoCoS signs quickly and reliably. Users may not know for sure what the signs look like. What the user has in mind might not be in the vocabulary yet, or might not ever become a convention.

One solution to inputting signs is to select a sign from a list (menu), the technique used in millions of Japanese mobile phones. In these circumstances, an issue is how to locate one of 3,000 signs by means of a matrix of 36 signs that may be displayed in a typical 128 × 128 pixel screen (or a larger number of signs in the larger displays of many current high-end phones). The current prototype uses a two-level hierarchy to organize the signs. Each sign is of 18 domains of subject matter. Each domain's list of signs is accessible with 2 to 3 keystrokes; 3,000 signs divided into 18 domains would yield approximately 170 signs per domain, which could be shown in 5 screens of 36 signs each. A three-level hierarchy might also be considered. As with many issues, these would have to be user-tested carefully to determine optimum design trade-offs. Figure 39.6 shows a sample display.

To navigate among a screen-full of signs to a desired one, numerical keys can be used for eight-direction movement from a central position at the 5-key, which also acts as a Select key. For cases in which signs do not fit onto one screen (i.e., more than 36 signs), the 0-key might be used to scroll upward or downward with one or two taps. There are challenges with strict hierarchical navigation. It seems very difficult to make intuitive the taxonomy of all concepts in a language. Users would have to learn

FIGURE 39.6 Sample prototype display of a symbol menu for a dictionary.

which concept is in which category. Shortcuts would be needed for frequently used signs.

In addition, there are different (complementary) taxonomies. Form taxonomies could group signs that look similar (e.g., those containing a circle). Properties taxonomies could group signs that are concrete versus abstract, artificial versus natural, microscaled versus macroscaled, and so on. Schemas (domains in the current prototype) would group *apple* and *frying pan* in the same domain because both are in the food/eating schema.

Most objects and concepts belong to several independent (orthogonal) hierarchies. Might it not be better to be able to select from several? This challenge is similar to multifaceted navigation in mobile phones. It is also similar to the "20 Questions" game, but would require fewer questions because users can choose from up to one dozen answers each, not just two choices. Software should sort hierarchies presented to users by most granular to more general chunking. It is also possible to navigate two hierarchies with just one key press.

A realistic, practical solution would incorporate context-sensitive guessing of what sign the user is likely to use next. The algorithm could be based on the context of a sentence or phrase the user is assembling, or it could be based on what signs/patterns the user frequently selects. Figure 39.7 illustrates multiple categories selection scheme.

Of course, if the phone has a camera, as most recent phones do, the user could always write signs on paper and send that image-capture to a distant person or show the paper to a person nearby. However, the user might still require and benefit from a dictionary (in both directions of translation) to assist in assembling the correct signs for a message.

There are other alternatives to navigate-and-select paradigms. For example, the user could actually draw the signs, much like Palm Graffiti, but this would require a mobile device with a touch screen (as earlier PDAs and the Apple iPhone and its competitors provide). One could construct each sign by combining, rotating, and resizing approximately 16 basic shapes. Ota has also suggested another, more traditional approach, the LoCoS

	Concrete	Abstract	Don't know
Man-made	1	2	3
Naturally occurring	4	5	6
Both	7	8	9
Don't know	*	0	#

FIGURE 39.7 Possible combinations of schema choices for signs.

keyboard, but this direction was not pursued. The keyboard is illustrated in Figure 39.8.

Still another alternative is the Motorola iTAP technique, which uses stroke-order sequential selection. In recent years, there were approximately 320 million Chinese phones in use, with 90 million using text messaging in 2003, using sign input via either Pinyin or iTAP. m-LoCoS might be able to use sequential selection, or a mixed stroke/semantic method. Figure 39.9 shows examples of stroke-order sign usage for Chinese input.

39.2.5 Future Challenges

Beyond the challenges and issues described previously, there are other more general ones to secure a successful design and implementation of m-LoCoS on mobile devices that would enable visible language communication among disparate, geographically distant users.

For example, the infrastructure challenges are daunting but seem surmountable. One would need to establish protocols for encoding and transmitting LoCoS over wireless networks. In conjunction, one would need to secure interest and support from telecom hardware manufacturers and mobile communication services.

As a user-centered design project, working prototypes must be developed and deployed among a wide variety of users who can test them in many different languages and use contexts. Using feedback from such tests is a standard method of gaining insight into improvements in hardware, software, user interface, and design method.

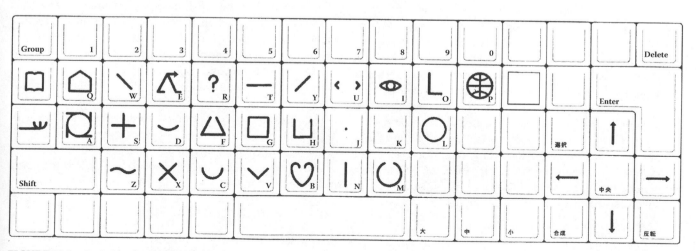

FIGURE 39.8 LoCoS keyboard designed by Yukio Ota.

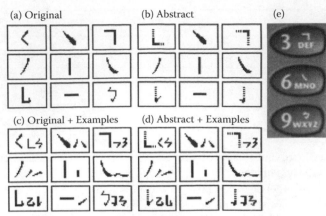

(a) Original (b) Abstract (e)

(c) Original + Examples (d) Abstract + Examples

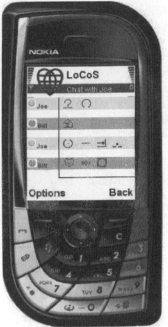

FIGURE 39.9 Examples of stroke-order sequential selection. (From Lin, M. and Sears, A., Graphics matters: A case study of mobile phone keypad design for Chinese input, in the *CHI '05 Extended Abstracts on Human Factors in Computing Systems*, pp. 1593–1596, ACM Press, New York, 2005. With permission.)

Literal translation of the "chat"
Joe: where? you
Bill: Restaurant
Joe: I will go there
Bill: happy new year

FIGURE 39.10 Example of a prototype chat depicted with LoCoS on a mobile phone.

Such future development depends upon funded tasks. Means of further development are currently being discussed with several major phone handset manufacturers and mobile phone service companies in several countries. It seems too early to evaluate the overall merits of the language or the approach.

39.3 Conclusion: Current and Future Prototypes

The author's firm, with the assistance and cooperation of Yukio Ota, investigated the design issues and designed prototype screens for m-LoCoS in early 2005, with subsequent adjustments since that time. About 1,000 signs were assumed for LoCoS, which is not quite sufficient to converse in modern, urban, and technical situations. As noted earlier, Basic Chinese and several modern visible language systems consist of 3,000 to 5,000 signs.

There is a need for a larger community of users and contributors of new signs. The current prototype is a set of designed screens that have been transmitted as images and show the commercial viability of LoCoS. Figure 39.10 shows a sample screen.

Among the next steps contemplated for the development of m-LoCoS are how to develop an online community for interested students, teachers, and users of LoCoS. For this reason, an extranet about LoCoS at the URL indicated earlier was designed and implemented. In addition, new sign designs to extend the sign set and to update the existing one, ideal taxonomies of the language, working interactive implementations on mobile devices from multiple manufactures, and the resolution of technical and business issues mentioned previously lie ahead. Of special interest to the design community is research into LoCoS 2.0, which is currently underway through Yukio Ota and his colleagues in Japan. The following objectives have also been discussed:

- Alternative two-dimensional layouts
- Enhanced graphics
- Color of strokes, including solid colors and gradients

- Font-like characteristics (e.g., thick-thins, serifs, cursives, italics, etc.)
- Backgrounds of signs (solid colors, patterns, photos, etc.)
- Animation of signs
- Additional signs from other international sets (e.g., vehicle transportation, operating systems, etc.)

m-LoCoS, when implemented in an interactive prototype on a commercial mobile device, would be ready for a large deployment experiment. This would provide a context to study its use and suitability for typical work and leisure environments. In addition, the deployment would provide a study situation for trying out LoCoS 2.0 enhancements.

Although only a small, initial step, this investigation demonstrates the viability of visible language systems in mobile phone usage context. As mentioned earlier, other commercial systems exist, such as Blissymbolics, PCS, and Lingraphics. Together with the open system of LoCoS (i.e., usage not covered by licensing and other legal constraints), many opportunities for planning, analysis, design, and evaluation lie ahead.

This initial case study shows the steps and some of the issues of a user-centered design process for making a visible language system accessible to users of mobile devices. Such developments may lead to progress in systems for specific groups with disabilities or may benefit the general population and lead to some form of alternative or augmented visible language system for global mobile communication.

Acknowledgments

The author acknowledges with thanks the anonymous reviewers of the draft version of this chapter who suggested many improvements. He also acknowledges the inspiration and assistance of Yukio Ota, president, Sign Center, and professor, Tama Art University, Tokyo, Japan. In addition, the author thanks designer/analyst Dmitry Kantorovich of the author's firm for his extensive assistance in preparing earlier versions of this text and for the figures used in it. This chapter is based on an article by the author in the *Proceedings of Human-Computer Interface International 2007* (Marcus, 2007).

Further Reading

American Institute of Graphic Arts (1981). *Symbol Signs.* New York: Hastings House Publishers.

Arnstein, J. (1983). *The International Dictionary of Graphic Symbols.* London: Kogan Page Ltd.

Chevalier, J. and Gheerbrant, A. (1996). *The Penguin Dictionary of Symbols.* London: Penguin Group.

Dreyfuss, H. (1984). *Symbol Source Book.* New York: John Wiley & Sons.

Green, P. and Burgess, W. T. (1980). *Debugging a Symbol Set for Identifying Displays: Production and Screening Studies.* Ann Arbor: University of Michigan UM-HSRI-80-64.

Hahner, W. and Heidrich, W. (1990). *A Study of Icons from a Graphic Design Perspective: Parts 1, 2, and 3. Introduction to Human-Computer Interaction.* London: The London Institute Central, St. Martins College of Art and Design.

Haller, R. and Kinross, R. (1991). *Otto Neurath Band 3: Gesammelte Bildpädagogische Schriften* [Otto Neurath, Volume 3: Collected Picture-Pedagogical Writings]. Vienna, Austria: Holder-Pichler-Tempsky.

ISO 7001 (1990). *International Standard: Public Information Symbols.* Geneva, Switzerland: International Organization for Standardization.

Johansen, J. D. and Larsen, S. E. (2002). *Signs in Use: An Introduction to Semiotics.* New York: Routledge.

Lionni, P. (2001). *Facts of Life–2.* Mainz, Germany: Hermann Schmidt.

Liungman, C. G. (1994). *Dictionary of Symbols.* New York and London: W. W. Norton and Company.

Marcus, A. (2003). Universal, ubiquitous, user-interface design for the disabled and elderly. Fast Forward column. *Interactions* 10: 23–27.

Modley, R. (1976). *Handbook of Pictorial Symbols: 3,250 Examples from International Sources.* New York: Dover Publications.

Olgyay, N. (1995). *Safety Symbols Art.* New York: Van Nostrand Reinhold.

Pierce, T. (1996). *The International Pictograms Standard.* Cincinnati: ST Publications.

Sayer, J. R. and Green, P. (1988). *Current ISO Automotive Symbols versus Alternatives: A Preference Study.* Warrendale, PA: Society of Automotive Engineers.

Shepherd, W. (1971). *Shepherd's Glossary of Graphic Signs and Symbols.* New York: Dover Publications.

University of Reading (1975). *Graphic Communication through ISOTYPE.* Reading, U.K.: University of Reading.

Zavalani, T. (1972). *Outlines from Jet Era Glyphs: A Utilitarian Graphic Sign System.* Sherman Oaks, CA: Self-Published.

References

Bliss, C. K. (1965). *Semantography (Blissymbolics)* (2nd ed.). Sydney, Australia: Semantography Publications.

Blissymbolics Communication Institute (1985). *1985 Annual Report.* Toronto: Blissymbolics Communication Institute.

Bliss web site (2007). *Blisssymbolics Communication International.* http://www.blissymbolics.org.

Chari, N. (2004). *Gyari-Moji.* http://www.venchar.com/2004/02/gyarumoji.html.

Cohen, J. (1954). On the project of a universal character. *Mind* 63: 49–63.

Dalgamo web site (2007). *Ars Signorum.* http://www.langmaker.com/db/Dalgarno%27s_Universal_Language.

Eichler, F. (2004). *AI Heute* (AI Today), p. 7. http://www.logic.at/lvas/185170/03-Eichler.pdf.

Encyclopedia Britannica web site (2007). *Esperanto.* http://www.britannica.com/ebc/article-9363908.

Esperanto web site (2007). http://www.esperanto.com.

Fuller, D. and Lloyd, L. (1991). Toward a common usage of iconicity terminology. *Augmentative and Alternative Communication* 7: 215–220.

Japan Times (2005, April 14). Japanese girls devise their own written language. http://www.kanjiclinic.com/kc69final.htm.

LangMaker web site (2007). *Pasigraphy.* http://www.langmaker.com/db/Pasigraphy.

Leibniz web site (2007). *Characteristica Universalis.* http://en.wikipedia.org/wiki/Characteristica_universalis.

Lin, M. and Sears, A. (2005). Graphics matters: A case study of mobile phone keypad design for Chinese input, in the *CHI '05 Extended Abstracts on Human Factors in Computing Systems*, pp. 1593–1596. New York: ACM Press.

Lingraphica web site (2007). *Lingraphica.* http://www.aphasia.com and http://www.lingraphicare.com.

Marcus, A. (2003). Icons, symbols, and more. Fast Forward column. *Interactions* 10: 28–34.

Marcus, A. (2007). LoCoS Mobile UI: Universal language for global communication, in *Human-Computer Interaction, HCI Intelligent Multimodal Interaction Environments (Part III), Volume 3 of the Proceedings of the 12th International Conference on Human-Computer Interaction (HCI International 2007)* (J. A. Jacko, ed.), 22–27 July 2007, Beijing, pp. 144–153. Berlin/Heidelberg: Springer-Verlag.

Mizuko, M. (1987). Transparency and ease of learning of symbols represented by Blisssymbols, PCS, and Picsyms. *Augmentative and Alternative Communication* 3: 129–136.

Neurath, O. (1980). *International Picture Language.* Reading, U.K.: University of Reading.

Niedermeier, K. (2001). *Emoticons.* Mainz, Germany: Hermann Schimidt.

Ota, Y. (1973a). LoCoS: An experimental pictorial language. *Icographic* 6: 15–19. Published by ICOGRADA, the International Council of Graphic Design Associations, London.

Ota, Y. (1973b). *LoCoS: Lovers Communications System* (in Japanese). Tokyo: Pictorial Institute.

Ota, Y. (1987). *Pictogram Design.* Tokyo: Kashiwashobo.

PCS web sites (2007). http://www.cricksoft.com/uk/products/symbols/pcs.aspx and http://www.mayer-johnson.com.

Steele, R. D. (2005). *Introduction to Lingraphica.* http://www.convention.asha.org/2005/handouts/293_Steele_Richard_074693_110205121900.pdf.

Steele, R. D., Gonsalves, R. F., and Leifer, L. J. (1999). *Method of Communication Using Sized Icons, Text, and Audio.* U.S. Patent 5973694. http://www.patentstorm.us/patents/5973694.html and http://www.freepatentsonline.com/5973694.html.

Sumita, E. (2002). Corpus-centered computation, in the *Proceedings of the ACL-2002 Workshop on Speech-to-Speech Translation: Algorithms and Systems, Volume 7,* July 2002, Philadelphia, pp. 1–8. Morristown, NJ: Association for Computational Linguistics.

Vanhauer, M. (2006). *Overcoming the Language Barrier: Das Potential der internationalen Symbolsprache LoCoS in der Mensch-Maschine-Kommunikation.* Hochsc1hule Offenburg, Germany: Hochschule Offenburg, Fachbereich Medien und Informations-wesen, Thesis No. 163704, Registered 6 December 2006.

Wahlster, W., (ed.) (2000). *Vermobil: Foundations of Speech-to-Speech Translation.* Berlin/Heidelberg: Springer-Verlag.

Weilgart, J. W. (1974). *A UI: The Language of Space.* Decorah, IA: Cosmic Communication.

Wilkins web site (2007). *Wilkins' Philosophical Language.* http://www.langmaker.com/db/Wilkins%27_Analytical_Language.

40

Contributions of "Ambient" Multimodality to Universal Access

Noëlle Carbonell

40.1 Introduction

For over 20 years, research advances on modalities and multimodality in human-computer interaction have had a major beneficial effect on the implementation of universal access in the coming information society. Increase in the range of available modalities and styles of multimodal interaction makes it possible to compensate for a growing diversity of physical disabilities, and thus to provide a larger community of disabled users with easy computer access and appropriate facilities for browsing and processing digital information. Research on the implementation of emerging concepts subsumed by the coined phrase *ambient intelligence* and the concomitant development of new contexts of use, such as user mobility, provide new opportunities for further developing the implementation of universal access, and for significantly extending the actual population of physically disabled people who will benefit from improved access to digital information.

The first part of the chapter demonstrates how the development of research on ambient intelligence has the potential to boost advances toward the generalization of effective universal access. For comprehensive definitions of concepts, such as universal access and accessibility (meaning accessible computing), see the first three chapters in this book.

In the second and third parts of this chapter this claim is further supported. A few recent scientific results are presented,

with a view to illustrating the possible fruitful contributions of research on the design and implementation of novel modalities and ambient intelligence environments, to the spreading of universal access into society. Major software implementation issues are discussed in the fourth part, and a software architecture is proposed for the design and implementation of new multimodal user interfaces. This software architecture takes into account the diversification of interaction modalities and contexts of use resulting from, and concomitant with, recent advances in research on ambient intelligence. More precisely, the organization and content of the chapter can be summed up as follows:

- Section 40.2 starts with some definitions of the main original concepts that characterize ambient intelligence (Section 40.2.1); then it shows how research focused on the implementation of these concepts has stimulated scientific efforts on the design and implementation of new interaction modalities (Section 40.2.2) and opened up new promising computer application areas; it closes by demonstrating, through short examples, how ambient intelligence may contribute to advancing the implementation of universal access, thanks to the diversification of interaction modalities and contexts of use (Section 40.2.3).
- Section 40.3, after a few definitions, describes recent scientific advances in input modality and multimodality

processing, and suggests their potential for improving computer accessibility.

- Section 40.4 briefly surveys recent large-scale research projects aimed at developing ambient intelligence environments, and hints at the benefits that could be derived from these attempts for implementing assisted living facilities that may significantly improve everyday life for a majority of people with physical disabilities.
- Section 40.5 proposes and describes a novel software architecture model meant to provide effective assistance in the design of multimodal user interfaces appropriate for interacting with ambient intelligence environments; a scenario of use is proposed that demonstrates the feasibility of implementing pervasive computing and context-awareness concepts, and suggests the potential of assisted living for people with disabilities.
- Section 40.6 wraps up the chapter; it briefly surveys a few research projects that contribute to implementing ambient intelligence concepts with a view to advancing assisted living applications and universal access to ambient intelligence emerging environments. Some research directions are proposed on issues that slow down progress toward effective assisted living.

40.2 Ambient Intelligence, a Spur to Advancing Universal Access

Ubiquitous or pervasive computing, and the disappearing computer,[1] are concepts that, during the 5th European Framework Programme (FP5: 1998–2002), motivated and inspired numerous research initiatives and projects, such as, for instance, Ambient Agoras.[2] This project aimed at:

> providing situated services, place-relevant information, and feeling of the place ("genius loci") to users, so that they feel at home in the office… via ambient displays embedded in the environment and mobile devices that are used in a combined way… to communicate for help, guidance, work, or fun.

During the 6th Framework Programme (2002–2006), emphasis shifted to more ambitious objectives in an area referred to by the umbrella phrase *ambient intelligence*.

This section first summarizes and comments on available definitions of these concepts. Then, it shows how attempts at designing and experimenting with ambient intelligence environments stimulated research on, and investigation of, the potential of alternative interaction paradigms and modalities to substitute for direct manipulation in such environments. The third subsection illustrates the benefits that can be derived from both scientific results in this emerging research area and future

applications of ambient intelligence concepts toward advancing the implementation of universal access.

40.2.1 Ambient Intelligence: Concepts and Objectives

Ambient intelligence refers to a broad research area. Pioneer researchers coined a few phrases to characterize major scientific objectives and challenges in this domain. These definitions are reported and commented upon in the following.

40.2.1.1 Pervasive and Ubiquitous Computing

According to Want et al. (2005, p. 14):

> *pervasive computing* aims to integrate computation into our daily work practice to enhance our activities without being noticed. In other words, computing becomes truly invisible.

In 1996, Mark Weiser had already used similar terms to characterize ubiquitous computing[3]:

> it is invisible, everywhere computing that does not live on a personal device of any sort, but is in the woodwork everywhere.

Advances in computing and electronics are opening up new computer interaction possibilities. Faster networks and wireless technologies make it possible to access interconnected smart artifacts that behave in a coordinated way, using other devices than the standard peripherals of desktop PCs (i.e., mouse, keyboard, screen and speakers). These trends introduce new off-the-desktop application scenarios involving mobile users, mobile devices, and richer forms of interaction with cooperative artificial agents accessible from everywhere.

To sum up, the long-term objective of ubiquitous computing and ambient intelligence is to enable users to interact with a dynamic suite of devices embedded in their everyday environment, so that they may be unaware of many of them. Phrases like *disappearing computer* or *context-aware systems/environments* refer to major implications of these concepts.

At present, the prevalent vision of research on ambient intelligence is, according to Philips, which has been involved in the implementation of this concept since 1999:

> people living easily in digital environments in which the electronics are sensitive to people's needs, personalized to their requirements, anticipatory of their behavior and responsive to their presence.[4]

---

[1] http://www.disappearing-computer.net.

[2] http://www.ambient-agoras.org.

[3] See http://www.ubiq.com/hypertext/weiser/UbiHome.html.

[4] http://www.research.philips.com/technologies/syst_softw/ami/index.html.

To transform this vision into reality, current research focuses on the design of context-aware and proactive integrated services meant to improve daily life. These services will be accessible through interconnected hardware and middleware embedded in everyday objects at home or while on the move. For instance, AMIGO,[5] an integrated project in the 6th European Framework Programme, aims at improving people's lives by "realizing the full potential of home networking" by developing interoperable middleware and attractive user services. Future home networked environments will address all vital user aspects:

home care and safety, home information and entertainment, and ambience sharing for advanced personal communication.

To achieve such a goal, collaborations between specialists (researchers and engineers) in mobile and home networking, software development, consumer electronics, and domestic appliances are needed. Emphasis on system awareness to context, proactive user support, and integration of services distinguishes ambient intelligence from previous research on pervasive computing and projects within the framework of the Disappearing Computer European Initiative[6] (2001–2003). For short introductions to pioneer and recent research on ambient intelligence, see Shadbolt (2003). Context awareness, that is perception of, and sensitivity to, the environment, and proactiveness are major components of human intelligence, while home networking (see, e.g., Georgia Tech's Aware Home[7]) is necessary for implementing integrated services using hardware and middleware embedded in the daily environment. See Section 40.4 for short descriptions of major research projects on ambient intelligence. A detailed discussion of ambient intelligence and its current evolution is provided in Chapter 60 of this handbook.

40.2.1.2 Context Awareness

Context awareness or contextual awareness is a key feature of ubiquitous computing systems. Context here means situational information relevant to the interaction between a user and an application:

A context-aware intelligent environment is a space in which a ubiquitous computing system has contextual awareness of its users and the ability to maintain consistent, coherent interaction across a number of heterogeneous smart devices. (Shafer et al., 2001)

According to Starner (1999, pp. 25–26), a context-aware interactive system must be capable of:

- Perceiving the users and their interaction environment through various sensors, some of which may be wearable

- Interpreting and modeling sensor data on the users (e.g., their current location, activity, physiological, and emotional state), their physical environment, the interaction, and task progress
- Interacting with the user through a contextually driven user interface or distributed interaction devices embedded in everyday objects

The concept of system context awareness can now be easily implemented, thanks to the availability of sensors capable of providing useful information on the user, the system, and the interaction environment. Development tools are proposed for implementing context awareness in Dey et al. (2001).

To sum up, context aware systems or software agents are to be capable of perceiving features in the interaction environment; these features vary according to the application domain and system role (e.g., virtual tour guide, smart assistant to home activities). Context awareness is discussed in detail in Chapter 59 of this handbook.

40.2.1.3 Mobile Computing

Ubiquitous computing and ambient intelligence imply mobile computing. Therefore, user mobility appears as a major design requirement for applications that aim at implementing these concepts appropriately. Obviously, monolithic standard graphical user interfaces intended for stand-alone PCs cannot meet this requirement. It is then necessary to break with traditional styles of human-computer interaction. In particular, user interface functionalities have to be replicated in peripheral devices distributed in objects accessible to the user during daily activities everywhere and at any time. These devices will be embedded either in objects in the user's everyday physical environment, or in wearable objects, according to the activity the user is engaged in and the service or assistance he is meant to receive from the smart distributed system.

40.2.1.4 Interface Plasticity

Interaction with such distributed user interfaces has to be flexible, as it takes place in various physical settings and, most often, while users are engaged in some daily routine activity (e.g., driving or cooking). These new contexts of use, which restrict the availability of users' perceptual and motor resources, induce constraints upon input and output medium choices. In particular, the volume and content of information exchanges between user and system have to be tailored to fit the specific throughput and expressive power of available interaction devices (e.g., PDAs) and modalities. The implementation of such constraints raises new research issues that constitute a fast developing research area focused on achieving user interface plasticity (Calvary et al., 2001), especially through what may be called *adaptive multimodality*.

In addition, to be truly flexible, context-aware interactive systems for mobile users have to be capable of adaptive interaction and user support, that is, dynamic adaptation to the current user's profile and its evolution in the course of interaction. In particular, adaptivity is needed in order to take account of evolutionary perceptual and motor impairments (Stephanidis, 2001).

---

[5] http://www.research.philips.com/technologies/syst_softw/ami/index. html.
[6] http://www.disappearing-computer.net.
[7] http://www.cc.gatech.edu/fce/ahri.

Research on adaptive user interfaces has recently developed into an established and flourishing research area. User modeling in the context of universal access is discussed in detail in Chapter 24 of this handbook. Readers may also refer to Jameson (2003) for a comprehensive survey of adaptive user models, or consult the proceedings of the biannual User Modeling Conference and the UMAI journal[8] for information on current research in this area.

To conclude, the feasibility of user interfaces enabling interaction with smart artifacts embedded in everyday physical environments or in wearable objects offers the opportunity to design new styles of multimodal interaction. It also makes it possible to consider the design and implementation of novel, sophisticated forms of assistance to everyday life activities, thus extending the range of present computer applications significantly. The next section describes the recent diversification of interaction modalities. Section 40.2.3 discusses the potential benefits that can be derived from the development of ambient intelligence research and applications in order to advance the implementation of universal access.

40.2.2 Ambient Intelligence: An Incentive to the Diversification of Interaction Modalities

Enhancing daily activities through smart artifacts and agents creates new contexts of use where direct manipulation may prove to be inappropriate as an interaction paradigm. Enriching and diversifying input/output modalities appears to be a major requisite for successful implementation of ambient intelligence concepts.

Switching from one daily routine activity to another often implies switching interaction modalities. In addition, enabling ubiquitous computing implies the capability to accommodate user mobility, and therefore to invent new interaction devices that can be embedded in ordinary objects belonging to the user's everyday environment, or in wearable objects. Significant research advances in the design and implementation of new modalities and combinations of modalities are needed to provide users of future ambient intelligence environments with easy and natural access to smart agents embedded in everyday objects, while on the move or engaged in other activities.

40.2.2.1 Speech Supplemented with Pointing Gestures

Speech, which is said to be the most natural human communication means, appears as a particularly well-suited interaction modality in any situation where the use of mouse and keyboard is impossible or awkward, for instance, while the user is on the move or engaged in another manual activity (see also Chapter 30

of this handbook). However, speech alone is not an appropriate substitute for mouse and keyboard when used for interacting with visual interfaces. Complex linguistic phrases are often needed to express spatial references precisely; such phrases are costly to elaborate and significantly increase cognitive workload, so that users may be reluctant to use speech for interacting with graphical interfaces. In addition, reliable understanding of such phrases is still beyond the capabilities of state-of-the-art systems.

Speech combined with 2D or 3D gestures makes it possible to overcome this limitation of expressiveness. Using 2D pointing gestures simultaneously with deictic phrases is an easy and efficient way of designating objects and positions on the screen; this form of multimodal interaction, first proposed in Bolt (1980), may be used as an appropriate substitute for keyboard and mouse in many contexts of use. For instance, a finger (Carbonell and Dauchy, 1999; Robbe et al., 2000) or a pen (Oviatt, 1997) can be used to designate positions on a touch screen or a graphics pad/tablet, respectively. At present, 3D hand and arm movements are most often considered for pointing at large screens, especially immersive virtual reality environments (e.g., reality centers or caves); cameras are currently able to track and interpret pointing movements with fair accuracy (Carbini et al., 2004; Demirdjian et al., 2005); sensors or specific pointing equipments (e.g., data gloves, magic wands) can also be used for this purpose.

40.2.2.2 Gaze as an Alternative Pointing Modality

Gaze plays an important role in human face-to-face communication. Its use as a substitute modality for pointing gestures can now be considered, due to recent, significant advances in gaze-tracking technology (see also Chapter 36 of this handbook). Eye trackers have become much more accurate and reliable, imprecision being currently inferior to half a degree of visual angle. Head-mounted eye trackers, which enable eye movement capture while moving, have become sensibly less intrusive. In particular, their size and weight have decreased drastically so that they can be fixed to spectacles. Besides, the precision of nonintrusive, vision-based, tracking techniques is improving steadily. This technological evolution may explain the recent growth of research on the analysis and modeling of users' spontaneous or controlled eye movements, and the elicitation of their gaze strategies in realistic human-computer interaction environments. For instance, eye trackers are currently used to evaluate graphical user interfaces (Goldberg et al., 2002), for example, to elicit users' gaze patterns on multimedia documents (Lohse, 1997), or to assess the influence of display spatial layout on users' efficiency and visual comfort during search tasks (Pearson and van Schaik, 2003; Simonin, Kieffer, and Carbonell, 2005). They are also used within research projects focused on the implementation of gaze as an alternative pointing modality that can appropriately replace mouse selection or even hand gesture pointing in some interaction environments and applications. For interacting with large digital displays, such as walls or immersive reality visual environments, hand gestures present the same limited pointing

---

[8] Proceedings of User Modeling Conferences are published by Springer-Verlag in Lecture Notes on Computer Science volumes (Lecture Notes in Artificial Intelligence subseries); see http://www.springer.com/west/home/computer/lncs. Issues of the *User Modeling and User-Adapted Interaction Journal of Personalization Research* are available at http://www.springerlink.com.

capabilities as gaze. In particular, both modalities cannot express the third dimension (depth) of 3D virtual scenes; magic wands (Ciger et al., 2003), which are meant to overcome this limitation, are not intuitive to use. Speech augmented with pointing gestures seems to be more appropriate than gestures alone for selecting and manipulating virtual objects in 3D environments. In such contexts of use, gaze is more suitable *a priori* than hand or arm gestures as a pointing modality, as designation of virtual objects or positions in a 3D virtual scene by gaze is more accurate, faster, and less tiring than using pointing gestures. In addition, hands are free and can be used to carry out other tasks. Gaze, thanks to these features, may prove to be very useful in ambient intelligence environments, especially for pointing at remote augmented everyday objects and smart artifacts while carrying out other tasks in the environments. Gaze coupled with speech may also prove to be useful for interacting with information systems and applications in parallel with domestic activities or while on the move.

40.2.2.3 Haptics: A Promising Output Modality?

Haptics may be viewed as an input/output modality, due to its privileged role in the sensory-motor loop (Rasmussen, 1986); tactile and force feedback are so closely intertwined with gestures that it is difficult to separate haptic perception from physical action. According to Gibson, haptic perception is "the sensibility of the individual to the world adjacent to his body by use of his body" (from Kennedy, 1993, p. 8). It is closely related to the concept of active touch, which stipulates that more information is gathered when a motor plan (movement) is associated with the sensory system, and to kinesthesis, which refers to the ability to feel movements of the limbs and body. In human-computer interaction (HCI) contexts, tactile and force feedback has to be coupled to create true haptic sensations; as force feedback devices are mostly used for controlling software applications, haptics may be viewed, in HCI contexts, as an input and output modality.

At present, the appropriateness of this modality for interacting with large displays and immersive virtual reality environments seems questionable: exoskeletons are still cumbersome and impede movements. On the other hand, the use of haptics for handling virtual or augmented objects and smart artifacts is promising. However, present force feedback devices seldom provide tactile sensations, and tactile devices provide poor quality tactile sensations and no force feedback. Further research and technological advances are therefore needed before haptics can be implemented into user interfaces convincingly (see also Chapter 33 of this handbook).

Some multiuser prototype applications have already been proposed using, for instance, augmented tabletops (Hinrichs et al., 2006), interactive surfaces (Schmidt et al., 2003; Wigdor et al., 2006), or smart clothing.[9] However, interaction with most of these prototypes is limited: contact with the interactive

surface or material activates mere visual feedback, to the exclusion of tactile and kinesthetic responses.

The next section lists the main likely effects of scientific and technological efforts to develop effective ambient intelligence environments and applications on the implementation of universal access in the information society.

40.2.3 Ambient Intelligence: An Opportunity for Advancing Universal Access

40.2.3.1 Diversification of Interaction Modalities

Most research bent on implementing ambient intelligence aims at assisting users in their daily activities. This implies a novel outlook on interaction, as workstations will be replaced by smart agents embedded in everyday objects and computing tasks will be viewed, in many contexts of use, as secondary tasks that only contribute to the completion of primary daily activities. In addition, these secondary tasks will often have to be carried out in parallel with primary activities. For instance, users of future ambient intelligence environments will be given the means to look up a street name on a digital city map while walking through the city, or to participate in a remote audio meeting while doing chores. Therefore, they will be placed in situational contexts where they will be confronted with similar difficulties to those encountered by physically disabled users, inasmuch as they will have to distribute their perceptual and motor capabilities between two tasks at least, a current daily activity and a computing task. As a consequence, they will need new forms of multimodality to cope with these parallel activities, and new modalities to interact with software agents embedded in everyday life objects. This explains why most current research projects on ambient intelligence lay emphasis on the design of new human-computer interaction modalities and flexible multimodal interaction. Some major issues on multimodal interaction are then common to both ambient intelligence and universal access research areas, which may prove to be most beneficial, especially for advancing the implementation of universal access.

Many solutions initially designed for enabling users of ambient intelligence applications to carry out computing tasks while performing some daily activities will prove to be appropriate for providing physically disabled users with improved computer access. Future commercialized products meant for ambient intelligence users, such as new interaction devices with sophisticated software for processing new modalities and combinations of modalities, will also satisfy the needs and requirements of disabled users at reasonable expense, since ambient intelligence applications are meant for very large user communities in the general public.

First, the great diversity of available new modalities will make it possible to provide physically disabled users with the adaptable or adaptive user interfaces necessary for addressing their specific disabilities, which may evolve in the course of time. Taking into account the possible evolutions of disabilities through personalized interaction is indeed crucial for an increasing number of

---

[9] For smart clothing or "interactive wear," see the Infineon company web site at http://www.infineon.com.

physically disabled users and seniors. These user communities need personalized multimodal interaction. A wide range of input modalities is necessary for accommodating the large spectrum of motor disabilities and exploiting at best actual individual motor capabilities. Synergistic use of multiple output modalities is also required for providing users with reduced multisensory acuity (e.g., senior users) with sufficient multimodal redundancy.

Second, flexible synergistic multimodality will be useful for ensuring satisfactory computer access for users with complex disabilities (e.g., some kinds of palsies or visual deficiencies). Thus, significant advances in the implementation of universal access will be made rapidly and at reasonable cost.

For instance, successful implementation of exclusive multimodality for ensuring user interface plasticity (Thevenin and Coutaz, 1999) may prove to be very useful for providing perceptual or motor-disabled users with easy access to electronic information services. Solutions have already been proposed, and products will soon appear on the market for adapting information presentation and input formats to interaction devices endowed with diverse information exchange capabilities and supporting varied media and modalities. The requirements and needs of mobile and nomadic users being roughly similar to those of a wide range of physically impaired users, applying these solutions to the latter category of users will significantly improve their access to computers. For example, automatic word completion is currently an active research topic in the mobile computing scientific community. The aim is to improve text entry facilities on mobile devices such as PDAs and smart phones. Enhanced word completion would greatly improve the usability and efficiency of gaze-activated virtual keyboards used by severe motor-disabled users; it would also facilitate text input for users who have difficulty using a keyboard, due to physical, cognitive, or cultural limitations.

Translation of information expressed in one medium into another medium, which is required for achieving user interface plasticity or automating search through multimedia databases, is also necessary for providing users with some specific perceptual disability with access to data stored on a medium they cannot access due to their functional limitations. Some of these translations/conversions are difficult to automate and costly to perform manually. For example, visual information has to be indexed to be accessible to blind users. Such indexing is also required for automating search in large collections of video recordings of meetings that will be collected in future smart rooms (see Section 40.4 for brief descriptions of some major, recent or still running, research projects on the design and implementation of smart rooms). In the same way, converting audio information into another modality (e.g., text) may prove to be useful for improving both multimedia data processing, and access to digital audio for deaf users. Thus, convergence between the needs and expectations of both the general public and users with some perceptual impairment is an invaluable asset for speeding up both the design of solutions that will meet the requirements of users with perceptual disabilities and the implementation of possible solutions into appropriate commercial products.

On the whole, research efforts on the implementation of ambient intelligence concepts are likely to significantly improve the accuracy of speech recognition, natural language understanding, and nonintrusive gesture and gaze tracking. They will also improve the usability and utility of these interaction modalities. In addition, the spreading of such intelligent environments into the general public will entail major reductions of equipment costs. This scientific and technological evolution will greatly benefit users with motor, auditory, or visual impairments, who will be offered meaningful, reliable, and easy-to-use interaction facilities, especially cheap and flexible mono- or multimodal command languages based on a great diversity of input modalities.

40.2.3.2 Diversification of Contexts of Use

The emphasis laid by ambient intelligence on supporting users in daily activities, both at home and on the move, does not only imply embedding software assistance agents and user interfaces in ordinary objects in the user's everyday environment. To assist users efficiently in their daily tasks, it is also necessary that agents have perceptual capabilities. For instance, a virtual guide has to be aware of the position of the user who is visiting an outdoor archaeological or historical site in order to be able to provide her with appropriate information about the ruins she is getting near to, through her PDA or any other digital channel and equipment (Pitarello, 2003). It is necessary to equip digital agents with sensors (e.g., a GPS in this example) so as to endow them with sufficient awareness to the user and the environment for being capable of helping the user effectively and efficiently.

Context-aware digital assistants have a great potential for promoting social integration of users with physical disabilities, beyond improving their access to computing and electronic information resources. The main contribution of system context awareness to helping these users overcome their motor or perceptual limitations lies in the new possibilities that the implementation of this feature opens up for increasing their autonomy and their effective participation in collaborative and social activities.

For instance, awareness to the user location makes it possible to guide visually impaired people through places unknown to them. Context-aware agents could be used for monitoring seniors during their daily home activities; in particular, electronic surveillance may be useful for warning mobile emergency medical services or other concerned services of incidents or accidents (e.g., falls, faints, etc.) that may befall older people living alone; it will help these people continue living at home autonomously. Context-aware agents could also help perceptual and motor-disabled people to complete daily tasks, thus contributing to their autonomy.

Some research projects on ambient intelligence, such as Oxygen and augmented multiparty interaction (AMI) (see Section 40.4), focus on supporting mobile computing and remote meetings; other initiatives, such as the Computers in the Human Interaction Loop (CHIL) project (see Section 40.4), aim at augmenting face-to-face communication. Outcomes of these projects will prove to be useful for enabling users with motor or perceptual disabilities to participate in face-to-face remote meetings or virtual sightseeing tours.

Context awareness together with user interface distribution all over the user environment may significantly contribute to improving the daily life and social integration of people with physical disabilities. They may play an important role in the implementation of assisted living, one of the main research themes that the 7th European Framework Programme (FP7, 2007–2013) is promoting and supporting.

The next section presents a few recent scientific advances in multimodal interaction that may significantly improve computer access for users with perceptual or motor impairments.

40.3 Benefits of Advances in Multimodal Interaction for Universal Access

A few definitions are first presented and explained. Then, recent results on emerging interaction modalities and the processing of multimodal inputs are presented, and their potential contributions to improving computer access are briefly discussed.

40.3.1 Definitions

Only basic concepts are defined in this section, that is, modality versus medium and multimodality versus multimedia.

40.3.1.1 Modality versus Medium

The term *modality* refers to the use of a medium, or channel of communication, as a means to express and convey information (Coutaz and Caelen, 1991; Maybury, 1993). Within the framework of a communication exchange, the sender translates concepts (symbolic information) into physical events that are conveyed to the recipient through an appropriate medium, and the recipient interprets the incoming signal in terms of abstract symbols. These processes involve the user's senses and motor skills and, symmetrically, the system input/output devices. According to the taxonomy presented in Bernsen (1994), which focuses on output modalities exclusively, several modalities or modes may be supported by the same medium (e.g., text and graphics used as output modalities). The definitions of medium and modality proposed by Maybury summarize the distinction currently made between their meanings as follows:

> by media we mean the carrier of information such as text, graphics, audio, or video. Broadly, we include any necessary physical interactive device (e.g., keyboard, mouse, microphone, speaker, screen). In contrast, by mode or modality we refer to the human senses (more generally agent senses) employed to process incoming information, e.g., vision, audition, and haptics. (Maybury, 2001, p. 382)

In other words, *modality* refers to the perception or production of meaningful information, that is, the translation or interpretation of physical events into symbols (perception) and, symmetrically, the coding of symbols into physical events (production). The set of devices necessary for acquiring, storing, and transmitting one type of physical representation is called a medium.

40.3.1.2 Multimodality versus Multimedia

Multimodality refers to the simultaneous or alternate use of several modalities. Integrating new modalities into user interfaces appears as a useful interaction paradigm for advancing the implementation of universal access.

The difference in meaning between multimedia and multimodality stems directly from the distinction between medium and modality. Multimedia systems have the capability of storing data conveyed through several input media, and, symmetrically, to convey data to the user through a set of appropriate output media. In addition, multimodal user interfaces are capable of communicating information to the user through several modalities. They have also the capability to understand multimodal commands from the user.

To characterize the various possible styles of multimodality (i.e., combinations of modalities), a taxonomy composed of four classes has been proposed (Coutaz and Caelen, 1991). As this chapter focuses on issues relating to usage rather than implementation, only three classes in this taxonomy need to be considered:

- Exclusive multimodality, which characterizes multimodal sequences of unimodal messages
- Synergistic multimodality, which refers to multimodal commands or, in other words, to the simultaneous use of several modalities to formulate a single information message
- Concurrent multimodality, which is the simultaneous use of several modalities for expressing different, unrelated information messages, that is, the use of one modality per independent information message

Another taxonomy that is useful for designing user-centered multimodal interaction and for assessing the usability of modality selection and assignment characterizes multimodality as follows. Two or more different modalities can be used complementarily or redundantly in a single information message. Each modality can be assigned to a specific type of information (e.g., one modality for system warnings, another for help messages, etc.). Two modalities are considered equivalent if they have the same power of expression (i.e., one modality can be used as a substitute for the other). Complementarity, Assignment, Redundancy, and Equivalence constitute the CARE properties (Coutaz et al., 1995).

To illustrate these definitions, the main benefits that interface designers who aim at improving computer accessibility can draw from the implementation of multimodality in current desktop computer environments are briefly summarized. These benefits are threefold. First, exclusive multimodality can advance access to computing facilities, as it provides users with the means to choose among available media and modalities according to the specific usability constraints induced by the current context of use (Oviatt, 2000, 2003) or their actual motor and perceptual capabilities. Second, alternative media, modalities, and styles of multimodality are necessary for providing physically disabled users with satisfactory computer access (Carbonell, 1999). In particular, a significant advance in the implementation of universal computer access may be achieved, thanks to the integration of speech into standard user interfaces as an alternative or supplementary input and output modality. Last, synergistic multimodality, which represents an indisputable improvement over unimodal interaction in terms of expressive power, usability, and efficiency, has also the potential to improve computer accessibility: intermodality redundancy may facilitate system message perception and understanding to users with reduced perceptual capabilities (e.g., seniors), while complementarity may be used to provide users with severe motor disabilities with more flexible and efficient input facilities.

40.3.2 Recent Advances in Multimodal Interaction

Efforts to implement ambient intelligence concepts, such as the design of smart homes and meeting rooms (see Section 40.4), have stimulated research on interaction modalities and their combined use. Significant scientific advances have been achieved in various directions:

- Accuracy and robustness of speech recognition, especially for speech recorded in adverse environments, for instance, on the street, in cars, or in meeting rooms
- Accuracy and reliability of nonintrusive video tracking of hand gestures, head and body movements, with a view to implementing remote gesture interaction with large screens (boards or walls) in virtual, augmented, or mixed reality settings, and to analyzing the behaviors of participants in smart meeting rooms (e.g., relative positions and groupings of participants, gaze/head direction)
- Design and implementation of new modalities (e.g., gaze and, to a lesser degree, haptics) and new styles of multimodal interaction
- Accuracy and flexibility of multimodal processing, especially fusion of input modalities

These scientific advances, which concern input modalities and multimodality, are detailed in the remainder of the section.

40.3.2.1 Speech Processing

Performances of speech recognizers have greatly improved during the last few years, due to the development of research on the design of smart meeting rooms and applications for mobile users. A significant increase in the robustness of speech recognition to noise (Misra et al., 2005) has been obtained by partners in the AMI European project on augmented multiparty interaction during business meetings (see Section 40.4). Regarding mobility, speech recognition in vehicles has motivated a growing research interest (Ahn and Ko, 2005; Park and Ko, 2006).

These advances make it possible to increase the range and diversity of human-computer interaction situations where speech can be considered and used as an appropriate input modality. They have the potential to enhance the usability and efficiency of user interfaces meant for the general public, and to improve computer accessibility for users with motor or hearing disabilities. Issues related to speech input are discussed in detail in Chapter 30, "Speech Input to Support Universal Access," of this handbook.

Integrating speech into standard user interfaces as an alternative, equivalent, input modality provides users on the move with easy computer access (Oviatt, 2000). Synergistic multimodal commands combining speech with deictic gestures and/or body movements have been shown to be more efficient than direct manipulation in some application contexts. For instance, the case study reported in Cohen et al. (1998) suggests that synergistic multimodal interaction involving pen and speech used complementarily may prove more efficient, in terms of speed mainly, than direct manipulation, at least for map-based tasks. According to the authors, "this advantage holds in spite of a 68% multimodal success rate, including the required error correction." In addition, speech used synergistically with pointing gestures has greater power of expression and flexibility than direct manipulation. The usability of this style of multimodality is also higher, since speech is currently viewed as the most natural human means of expression and communication.

Enhanced speech recognition will obviously improve computer accessibility for users with severe motor disabilities that often distort speech production and reduce physical mobility. In addition, it will make it possible to assist people with varied motor disabilities in their daily activities at home, in meetings, on trains, and so on, that is, in noisy environments where robust speech recognition is needed for interacting with smart assistance agents. Advanced speech recognizers capable of processing spontaneous speech from various speakers in varied ambient noise conditions may also contribute to the social inclusion of hearing-impaired people by providing them with real-time accurate orthographic transcripts of television debates, meetings, informal face-to-face discussions, and chats.

In addition, numerous coming applications meant for the general public are stimulating research focused on enhancing text-to-speech algorithms, that is, automatic translation of texts into speech synthesis utterances. Enhanced text-to-speech facilities are useful for offsetting display size limitations of PDAs, smart phones, and pocket PCs in mobile computing applications. In addition, high-quality voice synthesis of text messages is required for endowing embodied conversational agents (ECAs), which are swarming over the web with intelligible and humanlike speech capabilities; for an introduction to research on ECAs, see Catrambrone et al. (2004). Significant advances in text-to-speech software have also the potential to highly improve computer accessibility for partially sighted and blind users.

40.3.2.2 Vision-Based Detection and Tracking of Persons

Recent efforts on nonintrusive video detection and tracking of body motion have resulted in significant improvements in time processing and robustness (i.e., Garcia and Delakis, 2003; Viola and Jones, 2004), for rapid and reliable face detection (Argyros and Lourakis, 2004; Viola and Jones, 2004), for real-time tracking of skin-colored body parts, and for recovering 3D human body pose from image sequences (Agarwal and Triggs, 2006). Main efforts have been aimed at improving robustness and reducing video equipment complexity and cost; for instance, Agarwal and Triggs (2006) focus on estimating human pose from cluttered monocular images. This progress opens up numerous prospective application domains:

- Remote gesture interaction with large screens and immersive virtual reality environments, especially hand and arm pointing at virtual 3D objects in multimodal (speech and gesture) interaction environments (Nickel and Stiefelhagen, 2003; Carbini et al., 2004), or manipulation of virtual or augmented objects (Demirdjian et al., 2005). Issues related to hand gesture recognition are addressed in Chapter 34, "Vision-Based Hand Gesture Recognition for Human-Computer Interaction," of this handbook.
- Participant and group tracking in smart meeting rooms (McKenna et al., 2000; Nakanishi et al., 2002; Terrillon et al., 2004; Carbini et al., 2005).
- Animation of avatars representing participants in remote meetings, and avatar animation steered by user movements (Schreer et al., 2005), an emerging research area with many promising prospective applications in game, remote collaborative activity, and groupware settings.

Higher accuracy in detection and tracking of human body parts using cheaper video equipment opens up new research directions for implementing assisted living. In particular, it is possible now to consider the design and development of software agents capable of monitoring people's activities in their home environments. Awareness of a person's current pose, gestures and movements will enable these agents to determine the person's current activity and to detect possible critical situations (e.g., domestic incidents/accidents, health problems). Such knowledge is necessary for assisting people in their daily activities appropriately, especially by anticipating their needs and expectations; it is also essential for ensuring coarse (but continuous and nonintrusive) monitoring of people's physical condition, activities, and mental states (e.g., through appropriate biological sensors and/or facial expression interpretation software). Therefore, smart agents endowed with visual context awareness capabilities have the potential to help many people with disabilities to achieve autonomous living. Coupled with robotic aids, they will enable seniors and people with motor disabilities, vision impairment, or incapacitating chronic diseases, to continue living in their homes. People with cognitive, mental, and psychological problems may also benefit from system awareness of their current mental and emotional state.

In addition, accurate vision-based gesture and body part tracking will make it possible to interpret sign languages reliably and, using text-to-speech software, to enable deaf users to actively participate in meetings and informal discussions (Starner et al., 1998). This will contribute to their social integration. Sign language understanding is discussed in detail in Chapter 38, "Sign Language in the Interface: Access for Deaf Signers."

40.3.2.3 Emerging New Interaction Modalities: Gaze and Haptics

Regarding computer accessibility, gaze and haptics appear to be two promising complementary modalities, as they contribute to making up for two different disabilities. Controlled gaze may be used by severe motor-disabled users for interacting with standard applications, while haptics may help blind or ill-sighted users to get, explore, or interact with spatial information representations, such as maps, graphics, or musical scores. However, their utility and usability are different, as well as the maturity of their implementation in human-computer interaction.

First, eye tracking technology is more mature, flexible, and usable than haptic input/output technology (see Chapter 36, "Eye Tracking"). Accuracy of gaze point detection is in the order of 0.5 degree of visual angle. Equipments can now be embedded in computer screens, and head movements of some amplitude are allowed. Lightweight head-mounted eye trackers allow freer head and body movements within a radius of a few meters without loss of accuracy. Due to recent progress in size miniaturization and weight reduction, such eye trackers can even be attached to spectacles, which results in increased accuracy and reduced intrusiveness. Contrastingly, haptic input/output devices are rather cumbersome and intrusive. Many of them provide only one type of feedback, tactile (e.g., Braille cells) or kinesthetic. In addition, the quality of haptic feedback is still rather poor; for instance, gloves provide more realistic sensations than styles, since the contact surface of a stylus with virtual objects in the haptic space is reduced to a point. To obtain realistic tactile and kinesthetic feedback, cumbersome intrusive devices are needed, for instance a tactile glove together with a powered exoskeleton around the user's hand and forearm. Major enhancements are still needed for haptics technology to become actually usable.

Second, gaze used as an input modality is much more flexible and usable than haptic input and output modalities, at least for the time being. Controlled gaze appears to be an appropriate substitute for mouse pointing and selection. For instance, gaze-controlled virtual keyboards have been currently used by motor-disabled people for over 20 years (Marajanta and Räihä, 2002). This implies that gaze can provide motor-impaired users with easy access to current user interfaces based on icon manipulation and selection of items in menu systems and, in the long-term, to graphic editors (see Hornof et al., 2004). Moreover, the speed of gaze-controlled text entry is bound to improve in the near future as automatic word completion techniques are one of the major issues on which ubiquitous computing research is currently focusing with a view to endowing mobile devices (e.g., cell phones or PDAs) with easy rapid text entry (MacKenzie et al., 2006; Wobbrock et al., 2006). Besides, gaze-contingent displays,[10] which are now operational (Duchowski et al., 2004), offer attractive gaze-controlled zooming facilities for multi-scale visual exploration of complex visual scenes (Pomplum et al., 2001). Useful applications for vision impaired users are also emerging; see, for instance, the SOLAIRE project meant to facilitate reading for users with reduced visual field (Tlapale et al., 2006).

On the other hand, the versatility and complexity of human haptic capabilities have not yet been thoroughly investigated. It is difficult, at least for the time being, to define the potential contribution of haptics, a twofold substitute modality for visual interaction, to improving computer accessibility for vision-impaired users, although research on haptics has been active for about 10 years in various areas: manipulation of virtual reality objects (Gregory et al., 2000; Baxter et al., 2001), ambient intelligence (Basdogan et al., 2000), and blind users (Colwell et al., 1998; Grabowski and Barner, 1998; Eriksson, 1999; Fritz and Barner, 1999; Sjöström et al., 2003). Numerous empirical or experimental studies involving blind users have been conducted on the efficiency of haptic feedback for exploring graphical data (Tornil and Baptiste-Jessel, 2004; Brewster and King, 2005; Brown et al., 2005; Klok et al., 2005) and, to a lesser extent, music scores (Williams et al., 2004). Issues concerning the effectiveness of multimodal (haptic and audio) rendering of graphical displays (e.g., diagrams, drawings) have been given much attention. Nevertheless, definite results and conclusions have yet to be reached (Dufresne et al., 1995; Yu et al., 2002; Yu and Brewster, 2003). This may be due, at least partly, to the present limitations of haptic devices. However, the main difficulty is by far the currently insufficient basic knowledge of kinesthetic perception, compared to vision, and the absence of a meaningful theory of haptic perceptual processes. Further research, especially studies of the actual expressive power of this complex input/output modality, is needed; otherwise, progress in the effective implementation of haptics in human-computer interfaces is liable to be difficult and slow.

40.3.2.4 Multimodal Input Processing

Tailor-made multimodal input languages are necessary for providing some disabled users with acceptable computer access. This is the case of people with severe motor impairments and senior users with multiple perceptual and motor disabilities. In addition, as discussed in Section 40.2.2, novel modalities and styles of multimodality are to be integrated into user interfaces for future applications meant to implement ubiquitous computing or ambient intelligence concepts and, therefore, assisted living, which will prove to be beneficial to people with disabilities. In this context, synergistic multimodality based on emerging modalities and new combinations of modalities appears as a promising approach for succeeding in creating user interfaces that will come up to the needs and expectations of user communities suffering from severe motor disabilities or moderate perceptual and motor impairments.

Semantic interpretation, or fusion, of synergistic multimodal commands is a complex process. Input data are most often noisy, and the signal-to-noise ratio varies from one medium or input channel to another. In addition, the main characteristic of human expression is its great inter-individual variability, whatever medium and modality are considered. Statistical approaches are currently used to achieve robust recognition of monomodal input data, at the expense of collecting large training corpora for each modality, while knowledge-based approaches have been used to perform multimodal fusion until recently; see, for instance, the event-driven algorithm proposed in Nigay and Coutaz (1993), or the semantic approach presented in Johnston (1998), which operates on representations of monomodal chunks using first-order logic. Knowledge-based fusion algorithms can process speech- and gesture-based commands appropriately in standard contexts of use, that is, interaction with desktop computers equipped with a touch screen or a graphical tablet (Oviatt, 1997). However, their robustness to noise is insufficient for accurately processing multimodal inputs recorded in ambient intelligence environments that aim at assisting users in their everyday life activities at home, or while on the move or participating in social events. "Cocktail party noise" sensibly degrades the performance of present speech recognizers; remote visual interpretation of 3D hand or body gestures is much more imprecise than virtual object designation (i.e., selection) on a graphical tablet or a touch screen, using a pen or a finger respectively. The recent diversification of interaction environments and the resultant appearance of new, less precise, input modalities for coping with these new interaction situations may explain recent renewed interest in research on modality fusion; it may explain, in particular, the emergence of statistical approaches for performing modality fusion. To illustrate this evolution, the Smartkom prototype (Wahlster, 2002; Pfleger, 2004; Engel and Pfleger, 2006), which implements an original multilayered knowledge-based approach for performing modality fusion and fission, can be compared with the sophisticated statistical fusion method proposed in Christoudias et al. (2006).

The next section illustrates how recent research achievements in ambient intelligence may contribute to advancing universal access through the development of assisted living for people with physical disabilities. A few meaningful examples selected from recent or running research projects are briefly described.

40.4 Potential Benefits of Advances in Ambient Intelligence to Assisted Living

This section briefly presents a few research projects that contribute to the implementation of ubiquitous computing and ambient intelligence, especially by proposing realistic application scenarios and appropriate distributed computer environments

---

[10] A gaze-contingent display is a display that responds to where an observer looks within it.

and architectures, or by developing new technologies for implementing context awareness and flexible multimodal interaction. Projects have been selected that have the potential to inspire innovative applications for assisting users with physical disabilities in their daily activities and thus to advance the implementation of assisted living.

40.4.1 EasyLiving

EasyLiving is one of the first major projects[11] that attempted to implement intelligent everyday environments. Several experimentations were performed using a dedicated test laboratory equipped with distributed sensors and input/output devices embedded in the daily physical environment of the user. The main software tools developed include:

- A geometric model of the physical environment, for achieving context awareness; situational information relevant to the interaction is captured using computer vision techniques (for tracking people) and sensors (for tracking events, changes in the physical environment, etc.)
- Dynamic user interface adaptation (i.e., adaptivity), multiple sensor/actuator modalities and modality fusion capabilities, for providing flexible multimodal interaction
- Automatic calibration of sensors and semiautomatic building of a geometric model of the user physical environment, for facilitating the development of applications
- Device-independent communication and data protocols, for ensuring system extensibility

In Shafer et al. (2001), a realistic scenario of interaction with EasyLiving is described to illustrate the new interaction facilities that a context-aware intelligent environment can offer to users in their daily activities at home. For instance, lights in a room can be remotely turned on, using a speech command. Also, sitting in the living room on a sofa that faces the TV set automatically activates the display of a menu with various options (e.g., "Watch a movie," "Play music," etc.). Modalities used for information exchanges between the user and the distributed system can change according to the situational context (exclusive multimodality). For instance, a switch—instead of a speech command—may be used to turn on the lights in a room, so as to avoid disturbing a light sleeper in an adjoining room. Similarly, the system may resort to speech synthesis to convey an urgent warning to the user when asleep, instead of displaying it.

40.4.2 The Portable Help Desk: An Application of the Activity/Attention Framework

In Smailagic et al. (2001), a design framework for context-aware applications is proposed. It is based on the assumption that:

context aware applications are built upon at least two fundamental services, spatial awareness and temporal awareness. (p. 38)

Temporal awareness amounts to the schedule of public and private events. Mobile computing is often performed simultaneously with other attention-consuming primary tasks (e.g., walking, driving, etc.), so that users may neglect either the current primary activity or the less important activity (i.e., computing). The aim of the activity/attention framework is to reduce user distraction and optimize the simultaneous execution of a primary task and a mobile computing task, using a *distraction matrix*, which characterizes user activities by the amount of attention the user requires. Individual computer activities[12] are categorized by the amount of distraction the user introduces in the primary activity.

Smailagic et al. (2001) also present some interactive applications that were developed to validate this design framework, demonstrate how to use it, and illustrate the benefits that can be gained using it. For example, the portable help desk (PHD) enables a mobile user to get maps of the immediate vicinity, including static and dynamic information. Thus, while walking (or moving in a wheelchair) on a campus, the user can obtain information on the location, contact information, and availability of members of the academic staff, using the PHD.

40.4.3 Oxygen, Ozone, CHIL, and AMI

More recent or running research projects are focused on further developing and implementing core concepts of pervasive computing and ambient intelligence.

40.4.3.1 Oxygen

The Oxygen project at MIT[13] aimed at supporting highly dynamic and varied human activities at home, at work, and on the move. This ambitious project laid emphasis on enabling users to work together with other people through space and time.

Since the end of Oxygen, research on ambient intelligence at MIT Media Laboratory has diversified. The main current research direction is twofold. One research group is engaged in designing interfaces that are more responsive to human needs and actions in everyday life contexts of use, thanks to greater intelligence, interactivity, and immersive capacity. For instance, Laibowitz et al. (2006) present a network of wireless sensors and wearable displays designed to facilitate group interaction in large meetings. Research efforts of another group are centered on the implementation of "smart cities." In Mitchell (2005) a presentation and discussion of this concept is provided. Current scientific activities of this group are focused on the design of innovative props to individual and social activities in cities, for instance,

---

[11] The project started in 1997 and seems to have been active until 2001. For further information, see the Microsoft web site at http://research.microsoft.com/easyliving.

[12] Authors consider three classes of computer activities: information, communication, and creation tasks.

[13] This project was active until the end of 2004 at least. See the Oxygen web site at http://oxygen.lcs.mit.edu.

customizable environments (e.g., cars, apartments) or props to enhance children's activities on playgrounds.

40.4.3.2 Ozone

Ozone,[14] an EU-funded project (FP5, 2001–2004), pursued objectives roughly similar to those of Oxygen, since it aimed to:

> investigate, define and implement/integrate a generic framework to enable consumer-oriented ambient intelligence applications.

The main difference between the two projects lies in the user tasks and activities to be supported. While Oxygen was focused on supporting collective work, the main utility and usability objectives of Ozone were to:

> enhance the quality of life by offering relevant information and services to the individual, anywhere and at any time.

A major outcome of this European project is the implementation of research results in two demonstrators, one for home activities, the "home demonstrator," and one for nomadic use, the "away demonstrator." See Ozone (2005, Final Report, Section 5.5); see also Diederiks et al. (2002) for applications involving home activities.

40.4.3.3 CHIL and AMI

The CHIL and AMI European projects (FP6) focused on enhancing and augmenting communication between users; see the web sites of these projects.[15]

40.4.3.3.1 CHIL—Computers in the Human Interaction Loop (2004–2006)

This project aimed at creating:

> environments in which computers serve humans who focus on interacting with other humans.

Partners intended to achieve this goal thanks to the design and implementation of computer services that would be capable, from the observation of people interacting face-to-face with each other, to infer the state of their activities and their intentions. To be well accepted, such services would have to be endowed with perceptual context awareness, so as to provide users with implicit assistance whenever possible. This anticipation-based form of assistance is currently assumed to raise higher user satisfaction than responding to explicit help requests, since it interferes less with the user's current activity and requires less effort on the user's part than explicit requests. The main research issues addressed include:

- Multimodal recognition and interpretation of all available perceptual cues in the environment, with a view to

explaining human collaborative activities and anticipating participants' intentions
- Quantifiable cognitive and social models of interactions between humans

Smart rooms, namely office and lecture rooms, are the privileged applications that CHIL used for demonstration and evaluation purposes.

40.4.3.3.2 AMI—Augmented Multiparty Interaction (2004–2006)

This project focused on the design and implementation of new multimodal technologies capable of supporting human interaction in the context of smart meeting rooms and remote meeting assistance agents. It aimed to make human interaction more effective in real time, by developing new tools for supporting cooperative work and assisting navigation in multimodal meeting recordings. The project addressed several difficult research issues in two main areas:

- Multimodal input processing, namely multilingual speech recognition and speaker tracking, gesture interpretation, handwriting recognition and shape tracking, integration of modalities and multimodal dialogue modeling
- Content abstraction, including multimodal information indexing, summarizing, and retrieval

Outcomes of the project include several demonstrators (e.g., an offline meeting browser and an online remote meeting assistant), and dissemination of large collections of annotated multimodal meeting recordings.

40.4.3.3.3 AMIDA—Augmented Multiparty Interaction with Distance Access (2006–2009)[16]

This European project in progress is continuing the research undertaken in AMI. However, the emphasis has been slightly shifted, and is now on supporting live meetings with remote participants, using affordable sensors (e.g., webcams and microphones). The main targeted application domain is advanced videoconferencing systems with new functionalities, such as remote monitoring, interactive accelerated playback, shared context and presence.

The next section proposes and describes a software architecture designed for the development of advanced user interfaces that implement the new concepts presented and illustrated in this section and the preceding ones.

40.5 Software Architecture for Advanced Multimodal User Interfaces

Specific on-the-shelf components are now available that can process and interpret data from a wide range of input devices reliably. However, these components are monomodal, in the sense that they are dedicated to a specific medium and modality. There is not yet, outside research prototypes, any software platform capable of interpreting multimodal input data accurately.

---

[14] http://www.extra.research.philips.com/euprojects/ozone.

[15] http://chil.server.de/servlet/is/101/ and http://hmi.ewi.utwente.nl/Projects/ami.html.

[16] http://hmi.ewi.utwente.nl/project/AMIDA.

Symmetrically, software on the market is available for the generation of monomodal output messages conveyed through various media, whereas the generation of multimodal presentations is not yet supported. This section presents a design approach and software architecture that makes it possible to:

- Interpret users' multimodal commands or manipulations/ actions using partial monomodal interpretations, each interpretation being elaborated by a dedicated component that processes the specific input data stream transmitted through one of the available media (i.e., fusion process of events or data). Here, *multimodal* refers to synergistic multimodality, which raises complex implementation issues (e.g., modality fusion).
- Match these global interpretations with appropriate functions in the current application software, or translate them into appropriate commands (i.e., execution calls of the appropriate functions in the kernel of the considered software application).
- As regards system multimodal outputs, break up the information content of system messages into chunks, then assign the resulting data chunks to appropriate modalities, and input each of them into the relevant monomodal generation/presentation component. This treatment is viewed as a data *fission* process in contrast with multimodal input fusion.

In most current applications, stereotyped system messages only need to be implemented in order to achieve efficient interaction; therefore, simple techniques can be applied to generate appropriate multimodal system messages. Modality selection can be easily performed using available ergonomic criteria. As regards accessibility, the World Wide Web Consortium's (W3C) Web Access Initiative (WAI) has designed accessibility guidelines[17] that have been implemented in software tools such as Bobby.[18] Efficient generation software components also exist for many output modalities (except for haptics, a modality which still needs to motivate further research studies), namely, speech, sound, graphics, and text.

On the other hand, the fusion of multimodal inputs and the resulting global interpretation in terms of executable function calls are much more complex, due mainly to the imprecision of monomodal recognizers and interpreters, the errors of which add up. The generic software architecture presented in the remainder of this section is meant to overcome these difficulties in contexts where the information exchanges between user and software do not exceed the expressive power of direct manipulation (Shneiderman, 1993). In these contexts, the semantic interpretation of users' multimodal utterances amounts to matching utterances with appropriate functions in a specific software application; this simplicity ensures robust human-computer interaction while limiting development complexity and cost.

The proposed architecture is based on a five-layer user interface model, each layer being designed according to component programming principles. Intercomponent information exchanges implement the W3C SOAP message exchange protocol.[19] Generic interpreters and generators on the market can be used for processing monomodal inputs and outputs, as the dialogue between user and system has the same expressive power as direct manipulation, which limits both the volume and complexity of information exchanges.

Software development issues are not considered here, since they have motivated the publication of numerous manuals and best-practice case studies; see, for instance, Bass et al. (1998) for design implementation issues, and Clements et al. (2001) for testing and evaluation methods. The same observation holds for component-based programming; readers may refer to Szyperski et al. (2002) among others for detailed information.

40.5.1 Overall Software Architecture

A multimodal user interface is considered generic if:

- Additional modalities (i.e., monomodal interpreters and/ or generators) can be easily plugged in, and modalities implemented in the user interface plugged out.
- It can be easily interfaced with the functional kernel of any standard application software.

To fulfill these requirements a multilayer architecture is needed. The five-layer ARCH metamodel[20] appears to be a better candidate than the three-layer Seeheim architecture,[21] due to its finer granularity, hence, its greater flexibility. A slightly modified version of the original ARCH model is represented in Figure 40.1. This new version takes into account the W3C recommendations for the design of multimodal interactive software. In particular, it complies with the W3C requirement that advocates the "separation of data model, presentation layer and application logic."[22]

The role of the five components represented in Figure 40.1 is briefly described.

40.5.1.1 Functional Core

This layer includes all functions in the software application that operate on objects in the application domain; it constitutes the kernel of the software application.

---

[17] http://www.w3.org/WAI.

[18] Bobby (http://bobby.watchfire.com), a Watchfire Corporation product, is meant to help web application designers comply with available accessibility guidelines by exposing accessibility problems in web pages and suggesting solutions to repair them.

[19] See W3C (2003b) for an overview of this message exchange framework.

[20] See UIMS Tool Developers Workshop (1992) and, for an application of this slinky metamodel to object-oriented design, Carrière and Kazman (1997).

[21] cf. Pfaff and Hagen (1985).

[22] See W3C (2003a).

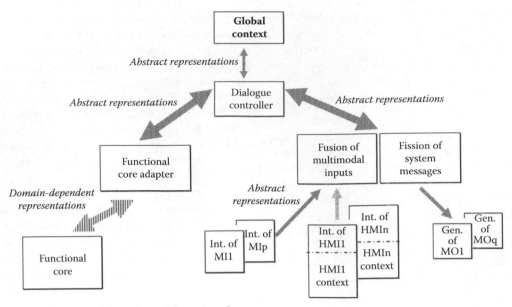

FIGURE 40.1 Overall architecture of the multimodal user interface.

- "Fusion of …" and "Fission of …": These components represent the multimodal interpreter and generator.
- "Int." and "Gen." designate interpreters and generators.
- "MI1,"…, "MIp," "MO1," …, and "MOq" designate monomodal input/output data.
- HMI1…HMIn designate home monitoring devices; their private contexts store data from sensors.

(From Carbonell, N., Multimodal interfaces: A generic design approach, in *Universal Access in Health Telematics: A Design Code of Practice* (C. Stephanidis, ed.), Figure 2, p. 221, Springer-Verlag, Berlin/Heidelberg, 2005.)

40.5.1.2 Functional Core Adapter

The functional core adapter serves as an interface between the functional core and the dialogue controller. In particular, it is in charge of translating:

- Abstract meaningful representations of the user's multimodal commands (coming from the dialogue controller) into appropriate executable requests to the functional core
- Symmetrically, results of the functional core activities into abstract representations intended for the dialogue controller

These representations of the meanings of user inputs and system outputs are independent of the input/output modalities[23] used.

The functional core adapter also sorts out conflicting interpretations of multimodal inputs provided by the dialogue controller, using knowledge of the application functionalities and contextual information (on the system and interaction current states). It may select one of these functionalities, and then notify the dialogue controller of its choice, so that the latter can update the dialogue history appropriately and, possibly, inform the user accordingly. It may also reject all interpretations, and request the dialogue controller to initiate a clarification subdialogue with the user.

40.5.1.3 Dialogue Controller

The dialogue controller or manager is the key component in the multi-thread interactive system. It is responsible for task-level synchronization and scheduling. In other words, it controls both the flow of commands and events from the user toward the functional core and, symmetrically, the flow of results from the functional core and events from the application domain, toward the user. It manipulates abstract representations of input and output information.

To synchronize input, output, and external events, it orders them relatively to each other from the time stamps attached to them, in accordance with the W3C requirements.[24] For managing and controlling the dialogue with the user (initiative of exchanges, clarification subdialogues, etc.), it may use information from the dialogue context. Typically, the dialogue between the user and the system may be user driven or application driven; mixed initiative may also be considered.

The dialogue controller uses an abstract representation format for exchanging information with the functional core adapter and the multimodal interpreter-generator (see the previous discussion about the functionalities of the functional core adapter).

---

[23] In compliance with, and as a generalization of, the recommendation of the W3C on the semantics of input generated by UI components (W3C, 2003a, Section 40.2.6, "Semantics of input generated by UI components").

[24] In particular, see the MMI-A17 requirement in W3C (2003a).

40.5.1.4 Multimodal Interpreter and Generator

The multimodal interpreter performs the fusion of all partial interpretations produced by the various monomodal input components, with a view to building valid commands. Fusion algorithms operate on time-stamped monomodal inputs from the user. This component also ranks competing global interpretations of monomodal or multimodal chunks according to the confidence scores computed by monomodal interpreters.

As for the multimodal generator, its main function is to split up into chunks the information content of the current message generated by the functional core. These information chunks are processed by the appropriate monomodal generators, which build the monomodal constituents of the multimodal presentation of the overall message. The assignment of information chunks to medias follows general ergonomic and semantic criteria. The multimodal generator also defines the temporal and spatial organizations of the message presentation to the user. Before passing the abstract representation of the information content of a message to the multimodal generator, the dialogue controller may modify and/or annotate it to take into account the current user profile and the current interaction context/environment (using the global context described in the following).

In the case of exclusive multimodality, the fusion (or fission) processor is not activated. The abstract representation generated by the monomodal interpreter (or the dialogue controller) is simply forwarded to the dialogue controller (or to the appropriate monomodal generator).

40.5.1.5 Monomodal Interpreters and Generators

These components include, for each modality, the software[25] necessary for:

- Interpreting input events or signals from the user, then creating abstract representations of monomodal interpretations (monomodal interpreters)
- Generating outputs from the abstract representations of information chunks delivered by the multimodal generator (monomodal generators)

Any plug-in software on the market can be used as a monomodal component, provided that it is encapsulated appropriately. In particular, interpreters should be capable of translating monomodal recognition results into "phrases" in the common abstract representation language, while generators should have the necessary knowledge for translating abstract representations into appropriate internal specific representations. In addition, each generator should include a styling and a rendering module according to W3C (2003c).

40.5.1.6 Communication between Components

Information exchanges between components have been limited to pairs of components as shown in Figure 40.1, in order to simplify message management. They are implemented using message-passing techniques in order to facilitate the use/reuse of available components or plug-ins.

Message content and structure are defined according to the XML SOAP framework proposed by the W3C,[26] thus making distributed implementations of the proposed architecture over the Internet easier to develop. SOAP provides a framework for the exchange of structured and typed application-specific information between peers in a decentralized distributed environment. It is a one-way message exchange protocol that can be used to create interaction patterns, for instance request-response. A SOAP message is an XML Infoset[27] that includes a description of its information content. SOAP also provides the means to specify the actions to be taken by a SOAP node on reception of a message.

40.5.1.7 Context

For applications involving sophisticated interaction techniques, such as dynamic adaptation or natural language interaction, it is possible to add a global context component to the architecture (see Figure 40.1). This component, which is consulted and updated by the dialogue controller exclusively, may be viewed as a repository of dynamic knowledge of the overall context of the user-application interaction. It may include a history of interactions, a description of the current user's profile, information on the current state of the application and display, a set of constraints implied by the current context of use (e.g., mobile device, PC), and so on.

Some components may need specific contextual information. For instance, dynamic adaptation of speech recognition algorithms to the type and intensity of background noise results in a significant improvement of speech recognition accuracy. The functional core adapter also needs specific contextual information. It should keep some trace of the interaction semantics or, at least, be aware of the last activated application function and of the current state of the application software. This information is needed, for instance, to process conflicting or incomplete interpretations and commands referring to functions that have not been implemented in the functional core.

It is possible to take such requirements into account by integrating specific context components into the proposed architecture (see instances in Figure 40.1), and restricting access rights[28] (to the information stored in such repositories) to the components that need them. However, as soon as a piece of contextual information proves to be useful for several components, it will be stored in the global context rather than in a private repository, so as to retain the benefits of the simple message management strategy that has been chosen. Distribution of information

---

[25] Especially monomodal device drivers and signal handlers. See W3C (2003a).

[26] See W3C (2003b).

[27] XML Information Set (Infoset) is a W3C specification describing an abstract data model of an XML document in terms of a set of information items. The definitions in the XML Information Set specification are meant to be used in other specifications that need to refer to the information in a well-formed XML document.

[28] For instance, consultation and updating rights.

in the global context according to components' needs will be assigned to the dialogue controller, which will use a push strategy. This choice, which favors component interconnection simplicity over execution speed, facilitates component adjunction and withdrawal, a major advantage of component-based software architectures.

40.5.2 Scenario of Use: Tele-Care Application

To illustrate how the proposed generic user interface architecture can be tailored to meet the specific requirements and constraints of real applications, this section briefly describes how this architecture can be used for implementing an interaction scenario in the area of Tele-Care: tele-home monitoring of sick or physically disabled people. This scenario is a short summary of a case study developed in Carbonell (2005).

40.5.2.1 Scenario Description

The objective is to control a set of monitoring devices connected to a distant medical center. These devices, which have been installed in the home of a senior patient suffering from a severe renal disease,[29] are capable of transmitting (through the Internet) measurements to the medical center in charge of the patient where they are stored in the patient's health records. Monitoring devices are connected to a PC in the patient's home; this PC manages remote data exchanges with the medical center and interactions between the patient and the distant medical center or the local monitoring devices.[30] Most measurement procedures involve interactions with the patient. This patient is also suffering from the usual chronic pathologies related to aging, which progressively reduce older people's mobility, dexterity, and visual and auditory acuity. The patient is at an early stage of this aging process, and can live at home autonomously.

Providing such patients with adaptable or adaptive multimodal facilities for (1) dialoguing with the distant medical staff in charge of home care, (2) accessing their health records, and (3) interacting easily with home monitoring devices, is likely to improve their daily life significantly.

Thanks to recent advances and achievements in ambient intelligence research, patients will have the possibility to choose among a wide range of available input/output devices and modalities according to their actual physical capabilities and the home activity they are currently involved in. Accessibility and mobility requirements of patients with perceptual and motor impairments who need tele-home monitoring can be appropriately met with present available media and modalities.

The next section exemplifies how the generic software architecture model proposed in Section 40.5.1 can be tailored to meet these specific user interface design requirements.

---

[29] For example, end-stage renal failure treated by continuous ambulatory peritoneal dialysis.

[30] For example, warnings and instructions on the operation of the monitoring devices in addition to measurements.

40.5.2.2 Multimodal User Interface Architecture

The application of the software architecture model in Figure 40.1 to the design of the patient's user interface is explained in this section.

40.5.2.2.1 Overall Software Architecture

The user interface software architecture should be distributed, since it is meant to run on a networked hardware configuration including several home monitoring device processors and a PC supervisor. In this context, it is advisable to use the SOAP message exchange framework for implementing information exchanges between software components in the interface; see (W3C, 2003b) for details.

Information exchanges between the patient and the application are standard, in the sense that they can be expressed adequately using direct manipulation. A simple command language is sufficient for interacting effectively with monitoring devices and setting up communication with the medical center. Substituting flexible multimodal interaction for direct manipulation is easy to implement. See the design recommendations in W3C (2006b).

40.5.2.2.2 Global Context

The global context component will mainly include the current user's profile, which the dialogue controller needs to be aware of in order to implement supplementary multimodality as a means for enhancing accessibility. According to W3C (2003a):

> An application makes supplementary use of multimodality if it allows to carry every interaction (input or output) through to completion in each modality as if it was the only modality.

This profile may be established and updated by the user, possibly with the help of the system; alternatively, it may be defined automatically by the system from the data in the patient's electronic health record, which the controller will obtain from the application functional core via the functional core adapter.

Adaptability will be sufficient for achieving accessibility in standard contexts of use where multimodality aims at compensating for the patient's perceptual and motor deficiencies, which develop very slowly in the course of time. As for adaptivity, it will prove to be useful for dynamically updating input or output modalities according to the patient's current location and activity. However, extensive research efforts are still needed in order to be capable of identifying and tracing users' daily home activities accurately (Oliver and Horvitz, 2003; Yu and Balland, 2003). At present, implementation of dynamic adaptation to the user's current environment and activity should be viewed as a research objective rather than as an available mature technology.

To implement adaptability (or adaptivity), the global context has to include device profiles for specifying features such as media type, processing requirements, performances, location, and so on (W3C, 2003c). This information may be of use to the dialogue controller for (1) selecting output modalities according to user profiles and/or contexts of use, (2) advising users in their choices of input modalities (in cases when user profile definition or updating facilities are available), and (3) taking the initiative for changing the currently active modality or combination of modalities whenever its efficiency deteriorates due to external events or changes in the interaction environment.

40.5.2.2.3 Monomodal Interpreters and Generators

Some interpreters and generators may need specific contextual information, which will be stored in private context repositories. For instance, sophisticated speech recognizers which model background noise and use the resulting models to improve recognition accuracy need private context repositories for storing incoming data generated by appropriate sensors in the user's environment.

40.5.2.2.4 Integration of Home Monitoring Devices as Components of the User Interface

Integration of home monitoring devices into the patient's user interface (see Figure 40.1) will greatly facilitate the implementation of effective and easy patient-device interactions. Thus, the user interface will be able to inform patients of daily results, explain measurements, and compare them with results of the previous day, or with weekly and monthly averages. The dialogue controller could generate this information and tailor its content and presentation to the patient's profile. It could also assist patients in carrying out measurement procedures by providing them with timely instructions during procedure execution.

To implement these extensions and integrate home monitoring devices in the patient's user interface, the most simple and effective approach is to consider them as specific monomodal input (and/or output) devices (see Figure 40.1), since the main processing to be carried out on the raw data generated by monitoring devices is to translate them into abstract representations "understandable" by the dialogue controller. Thus, the dialogue component will be able to design suitable output messages (regarding both information content and presentation) from these representations; this feedback will be conveyed to the patient through the user interface output channels. If need be, reversed information flows may be considered (i.e., from the dialogue controller toward home monitoring devices) for processing information that the patient has to deliver to home monitoring devices in the course of measurement procedures.

Using the latter extension, the user interface could help patients to carry out interactive measurement procedures by providing them with appropriate input facilities, namely: effective input error correction and prevention, efficient guidance and assistance in procedure execution, and appropriate personalized multimodal feedback. The second information flow is not represented in Figure 40.1, hence the absence of arrows directed toward the set of HMIs.

40.5.3 Conclusions

The proposed architecture is generic (i.e., application independent), provided that application-specific information is inputted

into the components as initialization parameters. This information should be specified in text form, using XML files with appropriate metadata and structures, in order to facilitate the implementation of components as plug-ins. Components will have to ensure translation of this information into proprietary internal formats, directly, or, most often, through encapsulated software. For any given application, this approach may be used to feed in specific static information to components in the user interface, for instance, input and output speech vocabularies to the speech interpreter and generator, respectively, or dialogue strategies (e.g., initiative of exchanges) to the dialogue controller.

The implementation of sophisticated functionalities in a multimodal user interface implies annotating abstract representations of the user's inputs and system outputs with meta-information, such as recognition scores, time stamps, and so on. Such meta-information is necessary for translating user inputs into valid commands/requests to the application core, and system outputs into calls to appropriate presentation software tools. EMMA,[31] the Extensible MultiModal Annotation language proposed by the W3C, may provide a satisfactory solution to this issue in the short-term. Although EMMA is intended as a data format for the interface between input processors and interaction management systems, and is currently viewed as a flexible mark-up language for representing human input to a multimodal application, its usage may be (and, hopefully, will be) extended to annotate system output and information exchanges between system components, since its focus is on meta-information.

Further information on the W3C activities related to the architecture and development of multimodal interactive software can be found in W3C (2006a,b). Further technical information on EMMA can be found in W3C (2005). As for SOAP, the home page of the XML Protocol Working Group[32] includes links and pointers to current work on this protocol.

40.6 General Conclusions and Perspectives

40.6.1 Potential Contributions of Ambient Intelligence Research to Universal Access Progress

This chapter aimed at demonstrating and illustrating the potential benefits that can be derived from research advances in the implementation of ambient intelligence for promoting universal access in the fast-developing information society. Ambient intelligence, a concept that subsumes ubiquitous (or pervasive) and mobile computing, refers to an emerging promising research area, the aim of which is to enhance our current activities at home or on the move by assisting them through invisible software agents embedded in our daily life environment. Advancing universal access implies (1) enabling everybody, including people

---

[31] See W3C (2003d).

[32] http://www.w3.org/2000/xp/Group.

with physical disabilities, to have easy access to all computing resources and sophisticated information services that will soon be made available to the general public and expert users all over the world, and (2) providing people with disabilities with advanced assisted living.

Section 40.2 has shown that advances in the implementation of ambient intelligence have the potential to contribute to the progress of universal access by extending the range of interaction modalities available to users with physical disabilities, and by providing the necessary technology for implementing enhanced assisted living. The claim put forward and discussed is that advances in research on ambient intelligence and the development of applications implementing this concept for both the general public and expert users may contribute to the progress of universal access significantly.

This claim is supported by the following main argument. As targeted ambient intelligence applications aim at assisting users in a wide range of daily activities, these activities will have to be performed simultaneously with appropriate interactive computer support. Therefore, users of such applications will be placed in situational contexts where they will be confronted with similar difficulties to those encountered by physically disabled users, inasmuch as they will have to share their perceptual and motor capabilities between several tasks, daily activities, and computing task(s). The convergence between the needs and requirements of small user communities made up of people suffering from similar disabilities, and those of a future, very large community, including both the general public and expert users, is a great opportunity for these smaller user communities. They will indirectly benefit from the outcomes of research and development efforts focused on satisfying the needs and expectations of the larger user community. Thus, the likely spread, in the near future, of ambient intelligence applications into the general public will significantly contribute to advancing the implementation of universal access and its growth throughout society.

Section 40.3 has focused on recent advances in the processing of some input modalities, such as speech, gestures, gaze, or haptics and synergistic combinations of modalities, all of which are currently viewed as appropriate substitutes for direct manipulation when the use of keyboard, mouse, and standard screen is awkward or impossible. The benefits of these advances for users engaged in other manual activities or on the move, or for users with motor or perceptual disabilities, are suggested. Recent advances in the processing of output modalities, and in particular speech synthesis (Van Santen et al., 2005) and text-to-speech translation (Naranayan and Alwan, 2005) have not been discussed, because they improve the effectiveness of existing applications (e.g., screen readers), but do not open up new classes of applications for people with disabilities. As for research on talking heads, embodied conversational agents and affective computing (Beskow et al., 2004; Schröder, 2006), it has progressed spectacularly in recent years, and its most promising applications will benefit people with mental disorders or social integration problems (Bosseler and Massaro, 2003), an area outside the scope of this chapter.

Section 40.4 has surveyed some major recent or running projects in the area of ambient intelligence, which illustrate the possible contribution of recent technological advances, such as wireless sensor networks, smart rooms, and live meetings with remote participants, toward the implementation of enhanced assisted living for people with physical disabilities.

Section 40.5 has proposed and described a software architecture intended for truly multimodal user interfaces (i.e., implementing synergistic multimodality). This architecture takes into account the recent diversification of modalities and the emergence of context-aware systems distributed in the user's life environment.

40.6.2 Major Research Issues in Ambient Multimodality, Assisted Living, and Accessibility

Research efforts have already been undertaken toward taking advantage of scientific and technological advances in ambient intelligence with a view to improving computer accessibility and developing universal access in the information society.

Most current efforts are bent toward benefiting from recent advances in modality and multimodality processing, which research on ambient intelligence has contributed to stimulate (see Sections 40.2.3.1 and 40.2.3.2). For instance, Akram et al. (2006) propose a nonintrusive camera-based system for tracking user movements (face or body) and translating them to pointer movement on the screen; this system enables desktop users with motor disabilities to navigate to all areas of the screen even with very limited physical movement. The study reported in Heim et al. (2006) is focused on the definition of appropriate user profiles for adapting speech assistance in web browsing to users with reading and writing disorders; future work will take into account users with motor impairments. An accessible multimodal prototype of the Pong game is presented in Savidis et al. (2006); it enables blind and visually impaired users to play this classic computer game using force feedback and an auditory display, namely, a grid of nine (soon 24) loudspeakers.

Some recent research on accessible computing has also addressed issues relating to the accessibility of ambient intelligence environments, especially those created for supporting user mobility. Thus, Arning and Ziefle (2006) report a series of studies meant to assess the usability of PDAs for senior users, based on comparisons between performances obtained by younger and older adults in personal data management tasks. Ziefle and Bay (2005) analyze the effects of aging on the usability of mobile phones. Sanchez and Aguayo (2006) describe the design, implementation, and evaluation of an instant messaging environment for blind users; the prototype that has been developed on a PocketPC uses a nine-key virtual keyboard and text-to-speech technology as input and output modalities, respectively. Arrue et al. (2006) propose a flexible software tool (EvalAccess MOBILE) for assessing the global accessibility of web-based applications implemented on mobile devices.

A few research studies focus on accessibility issues raised by other ambient intelligence concepts and features. For instance, Ortiz et al. (2006) presents an empirical study of the possible effects of the presence of an ECA on the subjective judgments and behaviors of elderly users; however, the topic of this study is only indirectly related to ambient intelligence, despite the title of the chapter, as participants interact with classic desktop applications through a standard PC workstation. Adams and Russell (2006), on the other hand, address a major issue raised by the emergence and development of ambient intelligent environments and applications: the accessibility of such environments. This pioneer study attempts to define a methodology for assessing the usability of ambient intelligence applications for users with disabilities, based on the evaluation of several existing prototypes using a simple cognitive model: analysis of older adults' interactions with two different prototypes and a case study centered on user-participative design of a PDA application called *ambient friend*.

Permeation of ambient intelligence concepts and scientific advances through research on assisted living has not yet reached the same maturity. Few accomplished research work has yet been published in major international journals or conferences (i.e., see Nehmer et al., 2006). However, numerous initiatives for promoting research on ambient assisted living (AAL) are fast developing in the European scientific community. Thus, the joint research project called BelAmI[33] includes AAL among its research themes. The European Union has also been supporting AAL applied research projects since 2003; two calls for research proposals have been issued during the Sixth Framework Programme (Information Society Technologies, 2002–2006); other calls for stimulating research on AAL will be launched during the Seventh Framework Programme.

This brief survey of research activities that attempt to bridge the gap between ambient intelligence and universal access suggests that further scientific efforts are needed in order to achieve fruitful interpenetration between these two areas. In particular, efforts should focus on the following objectives and challenges:

- Make the most of the new interaction modalities that ambient intelligence research has made available to user interface designers, for improving computer accessibility and universal access in the information society.
- Take advantage of intelligent networked devices embedded in everyday life objects and new multimodal interaction facilities developed within the framework of ambient intelligence research, in order to implement flexible AAL environments for people with physical disabilities and, later on, for people with cognitive impairments. For a comprehensive review of research issues concerning universal access to ambient intelligence environments, see Emiliani and Stephanidis (2005).

- Address accessibility issues relating to the usability of ambient intelligence interactive environments and applications. Until now, these issues have been addressed by a few research studies only, due, maybe, to the lack of operational prototypes. Meaningful research results relating to these issues are essential for advancing the implementation of accessibility and universal access in the coming information society. They could also significantly contribute to the ergonomic assessment of these environments, especially by providing data on the actual usability of the new interaction media and modalities proposed for overcoming interaction constraints stemming from the new contexts of use developed within the framework of ambient intelligence.

References

Adams, R. and Russell, C. (2006). Lessons from ambient intelligence prototypes for universal access and the user experience, in *Universal Access in Ambient Intelligence Environments, Proceedings of the 9th ERCIM Workshop "User Interfaces for All"* (C. Stephanidis and M. Pieper, eds.), 27–28 September 2006, Bonn, Germany, pp. 229–243. Berlin/Heidelberg: Springer-Verlag.

Agarwal, A. and Triggs, B. (2006). Recovering 3D human pose from monocular images. *IEEE Transactions on Pattern Analysis and Machine Intelligence* 28: 44–58.

Ahn, S. and Ko, H. (2005). Background noise reduction via dual-channel scheme for speech recognition in vehicular environment. *IEEE Transactions on Consumer Electronics* 51: 22–27.

Akram, W., Tiberii, L., and Betke, L. (2006). A customizable camera-based human computer interaction system allowing people with disabilities autonomous hands-free navigation of multiple computing tasks, in *Universal Access in Ambient Intelligence Environments, Proceedings of the 9th ERCIM Workshop "User Interfaces for All"* (C. Stephanidis and M. Pieper, eds.), 27–28 September 2006, Bonn, Germany, pp. 28–42. Berlin/Heidelberg: Springer-Verlag.

Argyros, A. A. and Lourakis, M. I. (2004). Real time tracking of multiple skin-coloured objects with a possibly moving camera, in the *Proceedings of the 8th European Conference on Computer Vision (ECCV 2004)*, 11–14 May 2004, Prague, pp. 368–379. Berlin/Heidelberg: Springer-Verlag.

Arning, K. and Ziefle, M. (2006). Barriers of information access in small screen device applications: The relevance of user characteristics for a transgenerational design, in *Universal Access in Ambient Intelligence Environments, Proceedings of the 9th ERCIM Workshop "User Interfaces for All"* (C. Stephanidis and M. Pieper, eds.), 27–28 September 2006, Bonn, Germany, pp. 117–136. Berlin/Heidelberg: Springer-Verlag.

Arrue, M., Vigo, M., and Abascal, J. (2006). Automatic evaluation of mobile web accessibility, in *Universal Access in Ambient Intelligence Environments, Proceedings of the 9th ERCIM*

---

[33] BelAmI partners include Fraunhofer IESE, several universities (Kaiserslautern, Budapest, and Szeged), and the Bay Zoltan Research Foundation in Budapest.

Workshop "User Interfaces for All" (C. Stephanidis and M. Pieper, eds.), 27–28 September 2006, Bonn, Germany, pp. 229–243. Berlin/Heidelberg: Springer-Verlag.

Basdogan, C., Ho, C.-H., and Srinivasan, M. A. (2000). An experimental study on the role of touch in shared virtual environments. *ACM Transactions on Computer-Human Interaction* 7: 443–460.

Bass, L., Clements, P., and Kazman, R. (1998). *Software Architecture in Practice* (2nd ed.). Boston, MA: Addison-Wesley.

Baxter, W. V., Scheib, V., Lin, M. C., and Manocha, D. (2001). dAb: Interactive haptic painting with 3D virtual brushes, in the *Proceedings of the 28th Annual Conference on Computer Graphics and Interactive Techniques*, pp. 461–468. New York: ACM Press.

Bernsen, N.-O. (1994). Foundations of multimodal representations: A taxonomy of representational modalities. *Interacting with Computers* 6: 347–371.

Beskow, J., Cerrato, L., Granström, B., House, D., Nordenberg, M., Nordstrand, M., and Svanfeldt, G. (2004). Expressive animated agents for affective dialogue systems, in *Affective Dialogue Systems* (E. André, L. Dybkjaer, W. Minker, and P. Heisterkamp, eds.), pp. 240–243. Berlin/Heidelberg: Springer-Verlag.

Bolt, R. A. (1980). Put-that-there: Voice and gesture at the graphics interface, in the *Proceedings of the 7th International Conference on Computer Graphics and Interactive Techniques*, 14–18 July 1980, Seattle, pp. 262–270. New York: ACM Press.

Bosseler, A. and Massaro, D. W. (2003). Development and evaluation of a computer-animated tutor for vocabulary and language learning for children with autism. *Journal of Autism and Developmental Disorders* 33: 653–672.

Brewster, S. A. and King, A. (2005). The design and evaluation of a vibrotactile progress bar, in the *Proceedings of World Haptics 2005*, 18–20 March 2005, Pisa, Italy, pp. 499–500. Washington, DC: IEEE Press.

Brown, L. M., Brewster, S. A., and Purchase, H. C. (2005). A first investigation into the effectiveness of tactons, in the *Proceedings of World Haptics 2005*, 18–20 March 2005, Pisa, Italy, pp. 167–176. Washington, DC: IEEE Press.

Calvary, G., Coutaz, J., and Thévenin, D. (2001). A unifying reference framework for the development of plastic user interfaces, in the *Proceedings IFIP TC13.2-WG2.7 Working Conference EHCI'01* (M. Reed Little and L. Nigay, eds.), May 2001, Toronto, Canada, pp. 173–192. Berlin/Heidelberg: Springer-Verlag.

Carbini, S., Viallet, J.-E., and Bernier, O. (2004). Pointing gesture visual recognition by body feature detection and tracking, in the *Proceedings of the International Conference on Computer Vision and Graphics (ICCVG 2004)*, 22–24 September 2004, Warsaw, Poland, pp. 203–208. Berlin/Heidelberg: Springer-Verlag.

Carbini, S., Viallet, J.-E., Bernier, O., and Bascle, B. (2005). Tracking body parts of multiple persons for multiperson multimodal interface, in the *Proceedings of IEEE International Workshop on Human-Computer Interaction (ICCV-HCI), at the 10th IEEE International Conference on Computer Vision (ICCV 2005)*, 17–20 October 2005, Beijing, pp. 16–25. Berlin/Heidelberg: Springer-Verlag.

Carbonell, N. (1999). Multimodality: A primary requisite for achieving an information society for all, in the *Proceedings of HCI International 1999, Volume 2* (H.-J. Bullinger and J. Ziegler, eds.), 22–26 August 1999, Munich, Germany, pp. 898–902. Mahwah, NJ: Lawrence Erlbaum Associates.

Carbonell, N. (2005). Multimodal interfaces: A generic design approach, in *Universal Access in Health Telematics: A Design Code of Practice* (C. Stephanidis, ed.), pp. 209–223. Berlin/Heidelberg: Springer-Verlag.

Carbonell, N. and Dauchy, P. (1999). Empirical data on the use of speech and gestures in a multimodal human-computer environment, in the *Proceedings of HCI International 1999, Volume 1* (H.-J. Bullinger and J. Ziegler, eds.), 22–26 August 1999, Munich, Germany, pp. 446–450. Mahwah, NJ: Lawrence Erlbaum Associates.

Carrière, J. and Kazman, R. (1997). *Assessing Design Quality from a Software Architectural Perspective*. Technical Report, Software Engineering Institute, Carnegie Mellon University, Pittsburgh.

Catrambrone, R., Stasko, J., and Xiao, J. (2004). ECA as user interface paradigm, in *From Brows to Trust: Evaluating Embodied Conversational Agents* (Z. Ruttkay and C. Pelachaud, eds.), pp. 239–267 (Human-Computer Interaction Series, Vol. 7). Norwell, MA: Kluwer Academic Publishers.

Christoudias, C. M., Saenko, K., Morency, L.-P., and Darrell, T. (2006). Co-adaptation of audio-visual speech and gesture classifiers, in the *Proceedings of the 8th ACM International Conference on Multimodal User Interfaces (ICMI'06)*, 2–4 November 2006, Banff, Canada, pp. 84–91. New York: ACM Press.

Ciger, J., Gutierrez, M., Vexo, F., and Thalmann, D. (2003). The magic wand, in the *Proceedings of the 19th Spring Conference on Computer Graphics*, 24–26 April 2003, Budmerice, Slovakia, pp. 119–124. New York: ACM Press.

Clements, P., Kazman, R., and Klein, M. (2001). *Evaluating Software Architecture: Methods and Case Studies*. Boston, MA: Addison-Wesley.

Cohen, P. R., Johnston, M., McGee, D., Oviatt, S. L., Clow, J., and Smith, I. (1998). The efficiency of multimodal interaction: A case study, in the *Proceedings of the 5th International Conference on Spoken Language Processing (ICSLP'98), Volume 2* (R. H. Mannell and J. Robert-Ribes, eds.), 30 November–4 December 1998, Sydney, Australia, pp. 249–252. Sydney, Australia: ASSTA.

Colwell, C., Petrie, H., Kornbrot, D., Hardwick, A., and Furner, S. (1998). Haptic virtual reality for blind computer users, in the *Proceedings of the 3rd International ACM Conference on Assistive Technologies (ASSETS'98)*, 15–17 April 1998, Marina del Rey, CA, pp. 92–99. New York: ACM Press.

Coutaz, J. and Caelen, J. (1991). A taxonomy for multimedia and multimodal user interfaces, in the *Proceedings of the*

1st ERCIM Workshop on Multimodal HCI, November 1991, Lisbon, Portugal, pp. 143–148. http://iihm.imag.fr/publs/1991/ERCIM91_PoleIHMM.ps.gz.

Coutaz, J., Nigay, L., Salber, D., Blanford, A., May, J., and Young, R. M. (1995). Four easy pieces for assessing the usability of multimodal interaction: The CARE properties, in the *Proceedings of the 5th IFIP International Conference on Human-Computer Interaction, Bringing People Together (INTERACT'95)* (K. Nordby, P. Helmersen, D. Gilmore, and S. Amesen, eds.), 27–29 June 1995, Lillehammer, Norway, pp. 115–120. London: Chapman & Hall.

Demirdjian, D., Ko, T., and Darrell, T. (2005). Untethered gesture acquisition and recognition for virtual world manipulation. *Virtual Reality* 8: 222–230.

Dey, A. K., Salber, D., and Abowd, G. D. (2001). A conceptual framework and a toolkit for supporting the rapid prototyping of context-aware applications. *Human-Computer Interaction* 16: 97–166.

Diederiks, E., van de Sluis, R., and van de Ven, R. (2002). Sociability and mobility concepts for the connected home, in *Universal Access: Theoretical Perspectives, Practice and Experience (Proceedings of the 7th ERCIM International Workshop on User Interfaces for All)* (N. Carbonell and C. Stephanidis, eds.), 23–25 October 2002, Paris, pp. 442–457. Berlin/Heidelberg: Springer-Verlag.

Duchowski, A. T., Cournia, N., and Murphy, H. (2004). Gaze-contingent displays: A review. *Cyber Psychology and Behavior* 7: 621–634.

Dufresne, A., Martial, O., and Ramstein, C. (1995). Mutimodal user interface system for blind and 'visually occupied' users: Ergonomic evaluation of the haptic and auditive dimensions, in the *Proceedings of the 5th IFIP International Conference on Human-Computer Interaction, Bringing People Together (INTERACT'95)* (K. Nordby, P. Helmersen, D. Gilmore, and S. Amesen, eds.), 27–29 June 1995, Lillehammer, Norway, pp. 163–168. London: Chapman & Hall.

Emiliani, P.-L. and Stephanidis, C. (2005). Universal access to ambient intelligence environments: Opportunities and challenges for people with disabilities. *IBM Systems Journal* 44: 605–619. http://www.research.ibm.com/journal/sj/443/emiliani.html.

Engel, R. and Pfleger, N. (2006). Multimodal fusion, in *SmartKom: Foundations of Multimodal Dialogue Systems* (W. Wahlster, ed.), pp. 223–235 (Cognitive Technologies Series, Part III). Berlin/Heidelberg: Springer-Verlag.

Eriksson, Y. (1999). How to make tactile pictures understandable to the blind reader, in the *Proceedings of the 65th International Federation of Library Associations and Institutions Council and General Conference*, 20–28 August 1999, Bangkok, Thailand. http://www.ifla.org/IV/ifla65/65ye-e.htm.

Fritz, J. P. and Barner, K. E. (1999). Design of a haptic data visualisation system for people with visual impairments. *IEEE Transactions on Rehabilitation Engineering* 7: 372–384.

Garcia, C. and Delakis, M. (2003). Convolutional face finder: A neural architecture for fast and robust detection. *IEEE Transactions on Pattern Analysis and Machine Intelligence* 26: 1408–1423.

Goldberg, J. H., Stimson, M. J., Lewenstein, M., Scott, N., and Wichansky, A. M. (2002). Eye tracking in web search tasks: Design implications, in the *Proceedings of the 2000 Symposium on Eye Tracking Research and Applications (ETRA 2002)*, 25–27 March 2002, New Orleans, pp. 51–58. New York: ACM Press.

Grabowski, N. A. and Barner, K. E. (1998). Data visualisation methods for the blind using force feedback and sonification, in the *Proceedings of the Annual Conference of the International Society for Optical Engineering–SPIE, Volume 3524*, 27–28 January 1998, San Jose, CA, pp. 131–139. Bellingham, WA: SPIE.

Gregory, A., Ehmann, S., and Lin, M. C. (2000). inTouch: Interactive multiresolution modeling and 3D painting with a haptic interface, in the *Proceedings of the IEEE Virtual Reality 2000 Conference (VR 2000)*, 18–22 March 2000, New Brunswick, NJ, pp. 45–52. IEEE Computer Society. http://computer.org/proceedings/vr/0478/0478toc.

Heim, J., Nilsson, E. G., and Skjetne, J. H. (2006). User profiles for adapting speech support in the Opera web browser to disabled users, in *Universal Access in Ambient Intelligence Environments, Proceedings of the 9th ERCIM Workshop "User Interfaces for All"* (C. Stephanidis and M. Pieper, eds.), 27–28 September 2006, Bonn, Germany, pp. 154–172. Berlin/Heidelberg: Springer-Verlag.

Hinrichs, U., Carpendale, S., and Scott, S. D. (2006). Evaluating the effects of fluid interface components on table top collaboration, in the *Proceedings of the Working Conference on Advanced Visual Interfaces (AVI 2006)*, 23–26 May 2006, Venice, Italy, pp. 27–34. New York: ACM Press.

Hornof, A. J., Cavender, A., and Hoselton, R. (2004). EyeDraw: A system for drawing pictures with eye movements, in the *Proceedings of the 6th International ACM SIGACCESS Conference on Computers and Accessibility (ASSETS 2004)*, 18–20 October 2004, Atlanta, pp. 86–93. New York: ACM Press.

Jameson, A. (2003). Adaptive interfaces and agents, in *The Human-Computer Interaction Handbook: Fundamentals, Evolving Technologies and Emerging Applications* (J. Jacko and A. Sears, eds.), pp. 305–330. Mahwah, NJ: Lawrence Erlbaum Associates. Revised edition in press. Revised chapter for the 2nd edition at http://dfki.de/~jameson/abs/Jameson06Handbook.html.

Johnston, M. (1998). Unification-based multimodal parsing, in the *Proceedings of the 27th International Conference on Computational Linguistics and the 36th Annual Meeting of the Association for Computational Linguistics (COLING'98)*, 10–14 August 1998, Montreal, pp. 624–630. Orlando, FL: Morgan Kaufmann Publishers.

Kennedy, J. (1993). *Drawing and the Blind: Pictures to Touch*. New Haven, CT: Yale University Press.

Klok, M., Uzan, G., Chêne, D., and Zijp, S. (2005). Blind gestural interaction: An experimental approach, in the *Proceedings of*

6th International Workshop on Gesture in Human-Computer Interaction and Simulation, Ile de Berder, France, 18–20 May 2005, pp. 18–20.

Laibowitz, M., Gips, J., Aylward, R., Pentland, A., and Paradiso, J. A. (2006). Information processing in sensor networks, in the *Proceedings of the 5th International Conference on Information Processing in Sensor Networks (IPSN'06)*, 19–21 April 2006, Nashville, TN, pp. 483–491. New York: ACM Press.

Lohse, G. L. (1997). Consumer eye movement patterns on yellow pages advertising. *Journal of Advertising* 26: 61–73.

MacKenzie, I. S., Chen, J., and Oniszczak, A. (2006). Unipad: Single stroke text entry with language-based acceleration, in the *Proceedings of the ACM 4th Nordic Conference on Human-Computer Interaction: Changing Roles (NordiCHI'06)*, Oslo, Norway, 14–18 October 2006, pp. 78–85. New York: ACM Press.

Marajanta, P. and Räihä, K.-J. (2002). Twenty years of eye typing: Systems and design issues, in the *Proceedings of the 2002 Symposium on Eye Tracking Research and Applications (ETRA 2002)*, 25–27 March 2002, New Orleans, pp. 15–22. New York: ACM Press.

Maybury, M. T. (ed.) (1993). *Intelligent Multimedia Interfaces.* Menlo Park, CA: AAAI/MIT Press.

Maybury, M. T. (2001). Universal multimedia information access, in *Universal Access in HCI: Towards an Information Society for All (Volume 3 of the Proceedings of HCI International 2001)* (C. Stephanidis, ed.), pp. 382–386. Mahwah, NJ: Lawrence Erlbaum Associates.

McKenna, S. J., Jabri, S., Duric, Z., Rosenfeld, A., and Wechsler, H. (2000). Tracking groups of people. *Computer Vision and Image Understanding* 80: 42–56.

Misra, H., Ikbal, S., Sivadas, S., Bourlard, H. and Hermansky, H. (2005). Multi-resolution spectral entropy feature for robust ASR, in the *Proceedings of the IEEE International Conference on Acoustics, Speech, and Signal Processing (ICASSP 2005), Volume 1*, 18–23 March 2005, Philadelphia, pp. 253–256. IEEE Computer Society. http://ieeexplore.ieee.org/servlet/opac?punumber=805.

Mitchell, W. J. (2005). *Placing Words: Symbols, Space, and the City.* Boston, MA: MIT Press.

Nakanishi, Y., Fujii, T., Kiatjima, K., Sato, Y., and Koike, H. (2002). Vision-based tracking system for large displays, in the *Proceedings of the 4th International Conference on Ubiquitous Computing (UBICOMP 2002)*, 29 September–1 October 2002, Goteborg, Sweden, pp. 152–159. Berlin/Heidelberg: Springer-Verlag.

Naranayan, S. and Alwan, A. (eds.) (2005). *Text to Speech Synthesis: New Paradigms and Advances.* Upper Saddle River, NJ: Prentice Hall.

Nehmer, J., Karshmer, A., Becker, M., and Lamm, R. (2006). Living assistance systems: An ambient intelligence approach, in the *Proceedings of the 28th International Conference on Software Engineering (ICSE 2006)*, Orlando, FL, 19–25 May 2006, pp. 43–50. New York: ACM Press.

Nickel, K. and Stiefelhagen, R. (2003). Pointing gesture recognition based on 3D tracking of face, hands and head orientation, in the *Proceedings of the 5th ACM International Conference on Multimodal Interfaces (ICMI 2003)*, 5–7 November 2003, Vancouver, Canada, pp. 140–146. New York: ACM Press.

Nigay, L. and Coutaz, J. (1993). A design space for multimodal systems: Concurrent processing and data fusion, in the *Proceedings of the Joint International Conference on Human Factors in Computing Systems (CHI'93 & INTERACT'93)* (S. Ashlund, K. Mullet, A. Henderson, E. Hollnagel, and T. White, eds.), 24–29 April 1993, Amsterdam, The Netherlands, pp. 172–178. New York: ACM Press.

Oliver, N. and Horvitz, E. (2003). Selective perception policies for guiding sensing and computation in multimodal systems: A comparative analysis, in the *Proceedings of the 5th International Conference on Multimodal Interfaces (ICMI'03)*, 5–7 November 2003, Vancouver, Canada, pp. 36–43. New York: ACM Press.

Ortiz, A., del Puy Carretero, M., Oyarzun, D. Yanguas, J. J., Buiza, C., Gonzalez, M. F. and Etxberria, I. (2006). Elderly users in ambient intelligence: Does an avatar improve the interaction?, in *Universal Access in Ambient Intelligence Environments, Proceedings of the 9th ERCIM Workshop "User Interfaces for All"* (C. Stephanidis and M. Pieper, eds.), July 2006, Beijing, pp. 99–114. Berlin/Heidelberg: Springer-Verlag.

Oviatt, S. L. (1997). Multimodal interactive maps: Designing for performance. *Human-Computer Interaction* 12: 93–129.

Oviatt, S. L. (2000). Multimodal system processing in mobile environments, in the *Proceedings of the 13th Annual ACM Symposium on User Interface Software and Technology (UIST'2000)*, 5–8 November 2000, San Diego, pp. 21–30. New York: ACM Press.

Oviatt, S. L. (2003). Multimodal interfaces, in *The Human-Computer Interaction Handbook: Fundamentals, Evolving Technologies and Emerging Applications* (J. Jacko and A. Sears, eds.), pp. 286–304. Mahwah, NJ: Lawrence Erlbaum Associates.

Ozone (2005). *New Technologies and Services for Emerging Nomadic Societies.* http://www.extra.research.philips.com/euprojects/ozone/public_public_documents.htm#TOP.

Park, J. and Ko, H. (2006). Achieving a reliable compact acoustic model for embedded speech recognition system with high confusion frequency model handling. *Speech Communication* 48: 737–745.

Pearson, R. S. and van Schaik, P. (2003). The effect of spatial layout of and link colour in web pages on performance in a visual search task and an interactive search task. *International Journal of Human Computer Studies* 59: 327–353.

Pfaff, G. and Hagen, P. J. W. (eds.) (1985). *Seeheim Workshop on User Interface Management Systems.* Berlin/Heidelberg: Springer-Verlag.

Pfleger, N. (2004). Context-based multimodal fusion, in the *Proceedings of the 6th ACM International Conference on*

Multimodal User Interfaces (ICMI'04), 13–15 October 2004, State College, PA, pp. 265–272. New York: ACM Press.

Pittarello, F. (2003). Accessing information through multimodal 3D environments: Towards universal access. *International Journal on Universal Access in the Information Society* 2: 189–204.

Pomplum, M., Ivanovic, N., Reingold, E. M., and Shen, J. (2001). Empirical evaluation of a novel gaze-controlled zooming interface, in the *Proceedings of the 9th International Conference on Human-Computer Interaction (HCI International 2001)*, 5–10 August 2001, New Orleans, Louisiana, pp. 1333–1337. Mahwah, NJ: Lawrence Erlbaum Associates.

Rasmussen, J. (1986). *Information Processing and Human-Machine Interaction: An Approach to Cognitive Engineering.* Amsterdam, The Netherlands: System Science and Engineering.

Robbe, S., Carbonell, N., and Dauchy, P. (2000). Expression constraints in multimodal human-computer interaction, in the *Proceedings of the ACM International Conference on Intelligent User Interfaces (IUI'2000)* (H. Lieberman, ed.), 9–12 January 2000, New Orleans, pp. 225–229. New York: ACM Press

Sanchez, J. and Aguayo, F. (2006). Mobile messenger for the blind, in *Universal Access in Ambient Intelligence Environments, Proceedings of the 9th ERCIM Workshop "User Interfaces for All"* (C. Stephanidis and M. Pieper, eds.), 27–28 September 2006, Bonn, Germany, pp. 369–385. Berlin/Heidelberg: Springer-Verlag.

Savidis, A., Stamou, A., and Stephanidis, C. (2006). An accessible multimodal Pong Game Space, in *Universal Access in Ambient Intelligence Environments, Proceedings of the 9th ERCIM Workshop "User Interfaces for All"* (C. Stephanidis and M. Pieper, eds.), 27–28 September 2006, Bonn, Germany, pp. 405–418. Berlin/Heidelberg: Springer-Verlag.

Schmidt, A., Strohbach, M., van Laerhoven, K., and Gellersen, W. K. (2003). Ubiquitous interaction: Using surfaces in everyday environments as pointing devices, in *Universal Access, Theoretical Perspectives, Practice and Experience (Proceedings of the 7th ERCIM International Workshop on User Interfaces for All)* (N. Carbonell and C. Stephanidis, eds.), October 2003, Paris, pp. 263–279. Berlin/Heidelberg: Springer-Verlag.

Schreer, O., Tanger, R., Eisert, P., Kauff, P., Kaspar, B., and Englert, R. (2005). Real-time avatar animation steered by live body motion, in the *Proceedings of the 13th International Conference on Image Analysis and Processing (ICIAP 2005)*, 6–8 September 2005, Cagliari, Italy, pp. 147–154. Berlin/Heidelberg: Springer-Verlag.

Schröder, M. (2006). Expressing degree of activation in synthetic speech. *IEEE Transactions on Audio, Speech and Language Processing* 14: 1128–1136.

Shadbolt, N. (2003). Ambient intelligence. *IEEE Intelligent Systems* 18: 2–3.

Shafer, S. A., Brumitt, B., and Cadiz, J. J. (2001). Interaction issues in context-aware intelligent environments. *Human-Computer Interaction* 16: 363–378.

Shneiderman, B. (1993). Direct manipulation: A step beyond programming languages, in *Sparks of Innovation in Human-Computer Interaction* (B. Shneiderman, ed.), pp. 13–37. Norwood, NJ: Ablex Publishing Corporation.

Simonin, J., Kieffer, S., and Carbonell, N. (2005). Effects of display layout on gaze activity during visual search, in the *Proceedings of the IFIP TC13 International Conference on Human-Computer Interaction (INTERACT'05)* (M.-F. Costabile and F. Paterno, eds.), 12–16 September 2005, Rome, Italy, pp. 1054–1058. Berlin/Heidelberg: Springer-Verlag.

Sjöström, C., Danielsson, H., Magnusson, C., and Rassmus-Gröhn, K. (2003). Phantom-based haptic line graphics for blind persons. *Visual Impairment Research* 5: 13–32.

Smailagic, A., Siewiorek, D., Anhalt, J., Gemperle, F., Salber, D., and Weber, S. (2001). Towards context aware computing: Experiences and lessons. *IEEE Intelligent Systems* 16: 38–46.

Starner, T. (1999). *Wearable Computing and Contextual Awareness.* PhD dissertation, MIT Media Laboratory, Cambridge, MA.

Starner, T., Weaver, J., and Pentland, A. (1998). Real-time American sign language recognition using desk and wearable computer-based video. *IEEE Transactions on Pattern Analysis and Machine Intelligence* 20: 1371–1375.

Stephanidis, C. (2001). Adaptive techniques for universal access. *User Modeling and User Adapted Interaction* 11: 159–179.

Szyperski, C., Gruntz, D., and Murer, S. (2002). *Component Software: Beyond Object-Oriented Programming* (2nd ed.). New York: Addison-Wesley and ACM Press.

Terrillon, J.-C., Pilpre, A., Niwa, Y., and Yamamoto, K. (2004). DRUIDE: A real time system for robust multiple face detection tracking and hand posture recognition in colour video sequences, in the *Proceedings of the 17th International Conference on Pattern Recognition (ICPR 2004), Volume 3*, 23–26 August 2004, Cambridge, U.K., pp. 302–305.

Thevenin, D. and Coutaz, J. (1999). Plasticity of user interfaces: Framework and research agenda, in the *Proceedings of the IFIP TC.13 Conference on Human-Computer Interaction (INTERACT'99)* (A. Sasse and C. Johnson, eds.), 30 August–3 September 1999, Edinburgh, Scotland, pp. 110–117. Amsterdam, The Netherlands: IOS Press.

Tlapale, E., Bernard, J.-B., Castet, E., and Kornprobst, P. (2006). *The SOLAIRE Project: A Gaze-Contingent System to Facilitate Reading for Patients with Scotomatas.* Technical Report n°0326, pp. 1–16. Rocquencourt, France: INRIA.

Tornil, B. and Baptiste-Jessel, N. (2004). Use of force feedback pointing devices for blind users, in *User-Centered Interaction Paradigms for Universal Access in the Information Society, Proceedings of the 8th ERCIM Workshop on User Interfaces for All* (C. Stary and C. Stephanidis, eds.), pp. 479–486. Berlin/Heidelberg: Springer-Verlag.

UIMS Tool Developers Workshop (1992). A metamodel for the runtime architecture of an interactive system. *SIGCHI Bulletin* 24: 32–37.

Van Santen, J., Kain, A., Klabbers, E. and Mishra, T. (2005). Synthesis of prosody using multi-level unit sequences. *Speech Communication* 46: 365–375.

Viola, P. and Jones, M. J. (2004). Robust real time face detection. *International Journal of Computer Vision* 57: 137–154.

W3C (2003a). *Multimodal Interaction Requirements, W3C Note*. http://www.w3.org/TR/mmi-reqs.

W3C (2003b). *SOAP Version 1.2 – W3C Recommendation*. http://www.w3.org/TR/soap.

W3C (2003c). *Multimodal Interaction Framework – W3C Note*. http://www.w3.org/TR/mmi-framework.

W3C (2003d). *Requirements for EMMA – W3C Note*. http://www.w3.org/TR/EMMAreqs.

W3C (2005). *EMMA: Extensible Multimodal Annotation Markup Language*. Working draft. http://www.w3.org/TR/emma.

W3C (2006a). *Third Working Draft for the Multimodal Architecture and Interfaces*. http://www.w3.org/TR/2006/WD-mmi-arch-20061211.

W3C (2006b). *Common Sense Suggestions for Developing Multimodal User Interfaces*. http://www.w3.org/TR/2006/NOTE-mmi-suggestions-20060911.

Wahlster, W. (2002). SmartKom: Fusion and fission of speech, gestures and facial expressions, in the *Proceedings of the 1st International Workshop on Man-Machine Symbiotic Systems*, 25–26 November 2002, Kyoto, Japan, pp. 213–225. Kyoto, Japan: University of Kyoto, Matsuyama Laboratory.

Want, R., Farkas, K. I., and Narayanaswami, C. (2005). Guest editors' introduction: Energy harvesting and conservation. *IEEE Pervasive Computing* 4: 14–17.

Wigdor, D., Shen, C., and Balakrishnan, R. (2006). Table-centric interactive spaces for real-time collaboration, in the *Proceedings of the Working Conference on Advanced Visual Interfaces (AVI 2006)*, 23–26 May 2006, Venezia, Italy, pp. 103–107. New York: ACM Press.

Williams, R. L., Srivastava, M., Conaster, R., and Howell, J. N. (2004). Implementation and evaluation of a haptic playback system. *Haptics-e: The Electronic Journal of Haptics Research* 3. http://www.haptics-e.org/Vol_03/he-v3n3.pdf.

Wobbrock J. O., Myers, B. A., and Chau, D. H. (2006). In-stroke word completion, in the *Proceedings of the 19th ACM Symposium on User Interface Software and Technology (UIST'06)*, 15–18 October 2006, Montreux, Switzerland, pp. 333–336. New York: ACM Press.

Yu, C. and Balland, D. H. (2003). A multimodal learning interface for grounding spoken language in sensory perceptions, in the *Proceedings of the 5th International Conference on Multimodal Interfaces (ICMI'03)*, 5–7 November 2003, Vancouver, Canada, pp. 164–171. New York: ACM Press.

Yu, W. and Brewster, S. (2003). Evaluation of multimodal graphs for blind people. *Universal Access in the Information Society* 2: 105–127.

Yu, W., Reid, D., and Brewster, S. A. (2002). Web-based multimodal graphs for visually impaired people, in *Universal Access and Assistive Technology, Proceedings of the 1st Cambridge Workshop on Universal Access and Assistive Technology (CWUAAT'02)* (S. Keates, P. Langdon, P. J. Clarkson, and P. Robinson, eds.), 25–27 March 2002, Cambridge, U.K., pp. 97–108. London: Springer-Verlag.

Ziefle, M. and Bay, S. (2005). How older adults meet complexity: Aging effects on the usability of different mobile phones. *Behavior and Information Technology* 24: 375–389.

VII

Application Domains

Vocal Interfaces in Supporting and Enhancing Accessibility in Digital Libraries

Tiziana Catarci,
Stephen Kimani,
Yael Dubinsky, and
Silvia Gabrielli

41.1 Introduction

The importance of libraries cannot be overstated. Libraries do facilitate the sharing of expensive resources, the preservation and organization of artifacts and ideas, and the bridging of the gap between people and ideas. Advances in technology have enabled the digitization of documents, text, sounds, and images in such a way that information, including library artifacts, is preserved and made available virtually anytime, anywhere via Internet-connected devices. The digital library enables its users to leverage this information by bridging the gap between the context of information and the application of knowledge. A key consideration in this respect is the distinction between the digital collection and the true digital library (Eisen et al., 2003). A digital collection is the mere gathering of information, whereas the digital library brings the digital collection alive by providing accessible usability of the collection via mechanisms used to search, browse, accumulate, synthesize, and correlate information into knowledge. A general mistake is to rely solely on technology in transforming information into knowledge, as can be discerned in the quote, "Digital library in a broader context is nothing but a database. The objective of the digitization process should be 'empowerment of the people'" (The Energy and Resources Institute, 2004, p. 8).

The involvement or incorporation of the user or customer in the systems development life cycle can help in gaining a deeper understanding of the user requirements. This is because requirements do evolve from the user's context and from the direct use of any system, including digital libraries. User participation or incorporation is therefore instrumental in guiding the development and evaluation process of any digital library. This is no less true when it comes to the requirements of users with disabilities. There may be many legitimate and reasonable motivations for supporting accessibility. Some have often viewed accessibility as an issue related to government agencies and the agencies' suppliers. However, supporting accessibility is mandatory for many organizations. For instance, on June 25, 2001, the U.S. federal government put into effect Section 508 of the Rehabilitation Act, which mandates that IT products/tools purchased or developed by federal agencies must provide equal accessibility to people with disabilities. Some establishments adopt accessibility for various other reasons, such as: it has the potential to create a market opportunity, it tends to be ultimately beneficial to all users, and it involves innovative technology. Interesting and inspiring as such reasons may be, it is important to ensure that the motivation for supporting accessibility stems from a commitment to the provision of equal opportunities for accessing resources for people with disabilities by the meeting of user needs (Mirabella et al., 2004). It has been previously observed that as long as companies and government agencies view accessibility primarily in terms of compliance with regulations and technical specifications, rather than as a way of supporting the work practices and requirements of people with disabilities, equal opportunity will remain a disguise (Nielsen, 2001). Supporting accessibility is a step of commitment in enabling equal access to resources for users with disabilities. It is worth pointing out that, although there exist various guidelines, standards, and techniques for facilitating accessibility, most of them have been developed by adopting a domain-independent perspective, which is sometimes insufficient to achieve accessibility goals in specific application areas, such as in the case of e-learning and the creation of eContent for people with disabilities. Previous studies have shown that W3C guidelines are often more suited to ensure technical aspects of accessibility (i.e., that a visual content is readable by a blind person through the support of assistive technologies, such as screen readers) rather than conceptual ones, which are more related to the usability and quality of experience the disabled user is provided with when accessing that specific content (Di Blas et al., 2004; Gabrielli et al., 2006).[1]

It is worth noting that speech interfaces have been shown to be especially useful as an input/output medium for the visually impaired, blind, and physically impaired users (see also Chapter 30 of this handbook). Furthermore, provisions and developments targeting users with disabilities usually end up being of benefit in the long run also to the population at large. Vocal interfaces tend to be useful in hands-busy, eyes-busy, mobile, and hostile/difficult settings to virtually all types of users. Vocal interfaces do pose some challenges to effective access and interaction. However, they have a great role to play in supporting and enhancing accessibility. This is true as well when it comes to accessibility to digital libraries (DLs). This chapter describes how vocal interfaces can be used to support and enhance accessibility in digital libraries, and how the user/customer can be involved in the development process through an appropriate integration of user-centered methodology and agile methods.

This work has been carried in the context of the DELOS Network of Excellence on Digital Libraries (http://www.delos.info). The DELOS Network of Excellence on Digital Libraries conducted a joint program of activities aimed at integrating and coordinating the ongoing research activities of the major European teams working in Digital Library-related areas with the goal of developing the next-generation digital library technologies. In the domain of user interfaces and visualization, the ultimate goal was to develop methodologies, techniques, and tools to establish a theoretically motivated and empirically supported frame of reference for designers and researchers in the field of user interfaces and visualization techniques for digital libraries, so as to enable future DL designers and developers to meet not only the technological, but also the user-oriented requirements in a balanced way.

This chapter first provides some background information pertaining to digital libraries and accessibility, followed by a discussion of the role and place of vocal interfaces in supporting and enhancing accessibility. Then efforts in appropriating vocal interfaces to support and enhance accessibility in digital libraries are described. After that, the evaluations carried out on the proposed vocal interface are discussed.

41.2 Background Knowledge

41.2.1 Digital Libraries

One particularly worthwhile definition of digital libraries is given by Borgman (1999) while talking about competing visions of DLs:

> Digital libraries are a set of electronic resources and associated technical capabilities for creating, searching and using information. In this sense they are an extension and enhancement of information storage and retrieval systems that manipulate digital data in any medium (text, images, sounds; static or dynamic images) and exist in distributed networks. The content of digital libraries includes data, metadata that describe various aspects of the data (e.g., representation, creator, owner, reproduction rights), and metadata that consist of links or relationships to other data or metadata, whether internal or external to the digital library. (p. 234)

According to the Digital Library Federation, "Digital libraries are organizations that provide the resources, including the specialized staff, to select, structure, offer intellectual access to, interpret, distribute, preserve the integrity of, and ensure the persistence over time of collections of digital works so that they are readily and economically available for use by a defined community or set of communities."[2]

Many other definitions can be found. However, some common points may be identified across different definitions:

- DLs can comprise digital as well as nondigital entities.
- The realm of libraries is constituted not only of library objects but also of associated processes, actors, and communities.
- The content of DLs can be extremely heterogeneous.

It appears that the bottom-line DL issue in this matter is to provide a coherent view of the (possibly) large collection of the available materials (Lynch and Garcia-Molina, 1995). With the foregoing statement in mind DLs can be regarded, to a large extent, as an information environment. In this environment, there are producers and consumers of information and knowledge (and collaborators); producers may also be consumers at the same time, and vice versa.

41.2.2 Accessibility

While accessibility has often been narrowly associated with disabilities, in the context of universal access, accessibility "implies the global requirement for access to information by individuals with different abilities, requirements and preferences, in a variety of contexts of use; the meaning of the term is intentionally broad to encompass accessibility challenges as posed by diversity in: (1) the target user population profile (including people

---

[1] More material relevant to EU initiatives related to eAccessibility and DL can be found at http://europa.eu.int/information_society/activities/digital_libraries/what_is_dli/index_en.htm and http://europa.eu.int/information_society/policy/accessibility/rtd/index_en.htm.

[2] http://www.diglib.org/about/dldefinition.htm.

with special needs) and their individual and cultural differences; (2) the scope and nature of tasks (especially as related to the shift from business tasks to communication and collaboration-intensive computer-mediated human activities); (3) the technological platforms and associated devices through which information is accessed" (Stephanidis et al., 1998, p. 3). There exist various guidelines, standards, and techniques (Caldwell et al., 2005; Theofanos and Redish, 2003) for facilitating accessibility with respect to aspects such as hardware, software, and the web.

41.3 Vocal Interfaces and Accessibility

Auditory interfaces may be speech-based or non-speech-based or a combination of both. Non-speech voice may take the form of musical sounds and natural sounds (e.g., alarms and beeps).

Speech as an interaction medium is slow for presenting information. It is also transient and thus difficult to review or edit. It also can easily cause cognitive overload and thus interfere with tasks that require significant cognitive resources (Shneiderman, 2000). Speech input, especially in an open or public setting can exhibit reduced privacy, making users feel less secure in the process (Maguire, 1999). There are also challenges concerning possible (or vulnerability to) distractions and disruptions (such as in the case of noise pollution).

As indicated by Shneiderman (2000), speech interfaces are sometimes helpful for hands-busy, eyes-busy, mobility-required, or hostile environments. Moreover, speech has been shown to be useful for store-and-forward messages, alerts in busy settings, and input/output for visually impaired, blind, and physically impaired users. According to Maguire (1999), speech user interfaces can offer important benefits to visually impaired users. Maguire also indicates that a commercial study that the author conducted found that a small sample of visually impaired users were very positive about the concept of a speech-based bank machine. It is worth noting that the accuracy of dictation input has been on the increase. However, adoption outside the disabled-user community has been slow compared to visual interfaces (Shneiderman, 2000).

However, for instance, Shneiderman notes: "I expect speech messaging, alerts, and input-output for blind or motor-impaired users to grow in popularity" (Shneiderman, 2000, p. 65). In fact, other user groups and applications are most likely to benefit from vocal interfaces:

Dictation designers will find useful niches, especially for routine tasks. There will be happy speech-recognition users, such as those who wish to quickly record some ideas for later review and keyboard refinement (e.g., content authors like tourist guides that need to be inspired by the surrounding place/location to create meaningful contents and experiences for tourists). Telephone-based speech-recognition applications, such as voice dialing, directory search, banking, and airline reservations, may be useful complements to graphical user interfaces." (Shneiderman, 2000, p. 65)

It is interesting to observe that products and devices such as human digital assistants based on text-to-speech voice and avatars are raising much interest nowadays, since they offer benefits and interaction support in web information systems (Aberg and Shahmehri, 2001, 2003) to many different user groups.

In the general research area of speech user interfaces (SUI), the work by Oviatt et al. (1994) focused on constrained (i.e., guided) versus unconstrained (i.e., unguided) speech interfaces. Oviatt et al. found that a more structured speech interface reduced the number of words, the length of utterances, and the amount of information included in a single utterance. Furthermore, they found that users preferred the constrained interfaces to the unconstrained ones by a factor of two to one. These findings have direct implications on the user interface design of applications for users with disabilities, because a highly structured interface can reduce the computational requirements of the speech-recognition engine and can result in a generally more pleasant user experience.

The need to make computers more accessible to people with disabilities has been a hot topic in the last years, with some high-profile accessibility initiatives such as Web Content Accessibility Guidelines (WCAG), Section 508 of the Rehabilitation Act in the United States, and the EU's eAccessibility initiative (see Chapter 53 of this handbook). The increasing participation of the community of computer users with disabilities consequently brings about an increased demand for the involvement of this community in the area of computer-supported entertainment, including computer games (Sporka et al., 2006c). This brings about the need for understanding the benefits and drawbacks of different assistive input devices and techniques that may be used for playing computer games. In arcade games, people with motor impairments are usually at a disadvantage when playing against people without motor impairments due to their disabilities. Although such games require a limited number of commands, the commands have to be given rapidly—a challenging problem many assistive devices are not able to cope with. There exist various aids for helping people with motor impairments to use interactive systems more effectively, such as, for instance: Sticky Keys; voice recognition systems; and pointers controlled by eye, mouth, or head movements. While most of these aids have been helpful, some of them incur high cost, thereby making acoustic input, such as speech-recognition systems, a popular choice (Dai et al., 2004). Non-speech input has been gaining popularity within the human-computer interaction (HCI) community. Non-speech input has been reported and evaluated in many different contexts (Sporka et al., 2006a) such as: the entry of values in a voice-operated remote control of a TV set (Igarashi and Hughes, 2001), an application for launching predefined UNIX commands upon reception of corresponding melodies (Watts and Robinson, 1999), emulating the keyboard (Sporka et al., 2006a), and emulating the mouse (Sporka et al., 2004, 2006b; Bilmes et al., 2005).

When considering specifically the domain of digital libraries, there are several efforts that are worthy of consideration (i.e., Hauptmann, 1995; Bolchini and Paolini, 2006; Catarci, 2006), which are discussed in the following.

In discussing the importance of aural digital libraries, Bolchini (in Catarci, 2006) reports several scenarios in which a user cannot rely on the visual channel to use a web site (or an interactive application in general):

- *Visual impairment*: The visual channel is completely ruled out (blind users) or severely affected (severe sight problems).
- *Small devices*: Users of palm-held devices, smart phones, and mobile devices in general can only partially rely on the screen.
- *Context-related scenarios*: These are situations (e.g., while driving a car, walking in a city or in a museum) in which looking at the screen is not desirable or possible, since it is dangerous, or it is preferable to look elsewhere (e.g., to the countryside, the buildings, the monuments, the paintings, etc.). The application should convey messages to the users and still let them focus also on the surrounding environment.

In the research efforts mentioned previously, a set of requirements for effective aural interaction with DLs has been defined, also taking into account existing assistive technologies, such as screen readers (Bolchini and Paolini, 2006). Some new requirements that must be taken into account are summarized in Table 41.1.

This chapter discusses an interaction paradigm for DLs that also exploits the user's physical context/environment, and is based on a concept that is familiar in the field of libraries, for example, the library catalogue. It should, however, be stated that guidelines such as the ones previously mentioned could be considered for refining the proposed design or development process.

Hauptmann (1995) describes the place of speech recognition in a digital video library, outlining how it can be used for transcript creation from video, alignment with manually generated transcripts, query interface, and audio paragraph segmentation. Hauptmann in the reported work finds out that speech-recognition accuracy varies a lot based on the quality and type of target data, and that reasonable recall and precision can be obtained with moderate speech-recognition accuracy. These findings are interesting for the purposes of this chapter because they concern speech interfaces and a digital video library. However, this chapter proposes a more natural and seamless link between the physical and digital realms while interacting with library artifacts, also paying close attention to users with disabilities.

41.4 Vocal Interfaces in Enhancing and Supporting Accessibility in Digital Libraries

As part of the NoE DELOS user interfaces and visualization cluster activities, various nonconventional paradigms for accessing and interacting with digital libraries besides the common search and search refinement mechanisms have been investigated. One of the nonconventional access and interaction paradigms is catalogue browsing (Catarci, 2006). It is within this paradigm that speech-based interaction has been incorporated to support and enhance accessibility of the underlying digital library. The first step of this research effort was a survey of existing literature/works pertaining to physical and online catalogues, followed by the tool development process for this access and interaction paradigm.

The initial survey of existing literature/works on catalogues would contribute to the elaboration of user requirements especially at the outset. To elaborate, refine, and situate/customize the user requirements for catalogue browsing, the users or customers were incorporated in the process through their physical presence and through relevant data collection methods. Lubetzky (Wilson, 1989) suggested that a library catalogue has the following two functions:

- *The finding function (documents function)*: Based on the author, the title, or the subject of the book, on the catalogue, the patron can determine whether the library has the book.
- *The colocation function (works function)*: The catalogue should help to determine what the library has by a particular author or on a particular subject, and the catalogue should show the various editions or translations of a given work.

TABLE 41.1 Aural Requirements: Synoptic Table

Concerns	Requirements
Info architecture	R1: Provide the user with an aural quick glance of the application content, whenever necessary.
	R2: Provide the user with an aural "semantic map" of the whole application.
	R3: Provide an executive summary of any list of items (especially the long ones).
	R4: Define strategies to partition long lists in smaller, meaningful chunks.
	R5: Provide "semantic back" mechanisms emphasizing the history of visited pieces of content, rather than the sequence of physical pages.
	R6: Provide a semantic navigation mechanism to go "up" to the last visited list of items.
Page navigation	R7: Keep consistency across pages by creating aural page templates.
	R8: Minimize the number of templates.
	R9: Allow the user quickly grasping how the page is organized by communicating its structure.
	R10: Read first the key message of the page (e.g., the content), and then the other sections.
	R11: Allow the user to access directly a section of interest at any time.

As a basic requirement, an online catalogue should support the primary functions of a card catalogue: finding and colocation functions. It is also important to take into consideration the observation that the online catalogue can help to better identify and characterize the nature, scope, and orientation of library entities (Fattahi, 1997). However, the design of online catalogues should not be too much constrained or limited to the type of functions supported by card catalogues, but the design should explore the new possibilities offered by technology to better match emerging needs and requirements of online behavior. In fact, it has been argued that online catalogues are still hard to use because they are often designed without sufficient understanding of searching behavior (Borgman, 1996). An assessment of the effectiveness of online catalogue design should not be based on its success in matching queries, but rather by its success in answering questions. All these processes may help to disambiguate or take into account the context of the users' information need, thus enabling them to find appropriate answers and acquire a better understanding of knowledge structures in a certain domain. It should also be pointed out that the design of online catalogue systems has often failed to consider or exploit the social collaborative dimension of searching behavior that can be easily observed in physical libraries. An online catalogue system should never permit a user's search attempt to fail to retrieve one or more bibliographic records for review and action, and provide assistance to the searcher where necessary; and never assume the display of a bibliographic record is the end of a search, as bibliographic records can be generative, or serve as information seeds to fertilize subsequent searching (Hildreth, 1995). In view of this, some of the basic initial user requirements that should be addressed include:

- Complexity of searching material that has been classified according to librarians' taxonomies
- Dynamic evolving information needs
- Time constraints, channel or device constraints
- Digital convergence

The developed tool for catalogue browsing and access paradigm (CBAP) is geared toward demonstrating and validating the catalogue browsing concept. To that end, a high-level DL solution was proposed based on the catalogue paradigm to provide library access and seamless interaction with physical and digital entities (Bertini et al., 2004). Such a solution relies on mobile computing (Bertini et al., 2005; Kimani and Bertini, 2005) and the catalogue paradigm to facilitate the library artifact access and seamless interaction, since it proposes a more natural link between the card catalogue artifact (as used by visitors of a physical library) and online catalogue interfaces (as part of the services provided by DLs to actors).

The types of transitions that have been previously considered in other design domains (e.g., Rogers et al., 2002) and that should be supported by the proposed solution are:

- *Physical to digital*: The user could be provided with system functionality for reading the barcode of a physical information resource (e.g., a book or article) to digitally collect further details related to it. Interesting benefits of this facility are envisaged for both end-users of physical/digital libraries (faster access to the information looked for, possibly through handheld devices) and for librarians (e.g., information updating/maintenance for both card and online catalogues).
- *Physical to physical (through digital) or digital to physical*: Further services to support users' activities within the physical library could be provided by the same barcode reading facility. Examples could be the activation of visual signals to help the user locate the physical book (item) on shelves or even the instantiation of an automatic delivery service within the library.
- *Digital to digital*: This refers to a set of functionalities that could be designed to better link information provided by online catalogues to other information services available through the DL. It also includes the possibility for users to customize or personalize their search, according to their needs, devices, and contexts.

It is worth noting that in the tool development process for the proposed DL solution, an appropriate integration of the user-centered methodology (Norman and Draper, 1986) and agile methods (Beck and Andres, 2005) have been adopted. The agile approach to software development emphasizes individuals and interactions[3] in the process of software development. It provides a framework for constructing software products through an iterated and incremental approach, where each iteration produces a working code and other artifacts valuable to customers and the project, in a highly collaborative fashion, to produce a quality product in a cost-effective and timely manner, to meet the changing requirements of its stakeholders, and to minimize the risk of failure of the product. In the integrated development process, the incorporation/involvement of the customer or the user is ensured through his physical presence and through relevant requirements gathering techniques. As indicated earlier, this would help in elaborating, refining, and situating/customizing the user requirements for catalogue browsing.

In the initial prototype, the developed user interface is a speech-based mobile interface to a DL (Dubinsky et al., 2007). Beyond the search activity, two additional features were defined. The first feature is enabling vocal commands for artifact searching and localization. The second feature is enabling artifact localization in the physical library using a digital positioning system. Speech input is enabled for navigating the application, whereas speech output is used for providing the user with information about position to support artifact localization. Table 41.2 summarizes the set of detailed requirements for the application as they were presented and prioritized by the customer at the beginning of the first release.

The customer further emphasized that nowadays libraries involve several kinds of artifacts such as DVDs, audio and video CDs, and so on. This note is significant when analyzed using the

---

[3] See also the agile manifesto at http://agilemanifesto.org.

TABLE 41.2 Customer Stories for CBAP Prototype

Customer Story	Priority
1. Mobility:	
a) The application is web-based and able to run on mobile devices.	I
b) The user can move while using the application (a librarian with a bunch of books in one hand and a PDA in the other).	II
2. CBAP: The user interface should be inspired by the typical catalogue cards box.	I
3. Navigating between physical and digital realms—location and speech:	
a) The user should be able to search to find a book of interest. The information can be filtered by queries on topics and/or authors. It should be useful to readers and librarians. [digital→digital]	I
b) The user should be able to receive instructions about the physical location of the book that he/she searches for (as a reader) or wants to put back on the shelf (as a librarian). That is, the application has to provide information about the path to follow inside the library to be able to physically hold the book or put it back on the correct shelf. [digital→physical]	II
c) The application has to show the path (see 3.b) with an output speech interface. In general, the system has to provide a speech I/O infrastructure. Example: the librarian is walking or has a bunch of books on his/her hands, so he/she prefers voice over looking at the PDA screen. Real-world example: GPS. [digital→physical]	I
d) The application must be able to understand and execute commands given by a vocal input. Example: the librarian has a bunch of books on his/her hands so cannot easily press any PDA button. [physical→digital]	II
4. Artifacts for navigating between physical and digital realms—barcodes:	
a) The application should provide a barcode reading infrastructure. [physical→digital]	I
b) The application should demonstrate the identification of physical items and related digital data by reading their barcodes. [physical→digital]	II
5. Related search: The application should suggest, given an item, similar-related ones. Example: once he/she reads a book, the reader is particularly satisfied and curious, so wants to know if there is something else written by the same author, or other same arguments, etc. Real-world example: web interfaces that present "Who bought this, also bought…."	II
6. Clustering: The application should be able to cluster related items when showing them. Example: the librarian wants to retrieve information about everything produced by the same author that the library has.	II
7. New items disposal: If new products arrive, the application should be able to suggest to the librarian where to put them, considering places appropriate to the library's shelves management. A rules interface is to be provided by the system for properly disposing into the library.	II
8. Top list: The application should be able to highlight the "top" products as the "most popular" or "latest" charts.	II
9. Advertisement: In case of new artifacts, the application should send an e-mail to inform users that new products are available.	II
10. Pictures: The application should show pictures of the items, to let a user recognize them through a visual approach.	II
11. Digital libraries: The application should be based on a new digital library to be developed for the project's sake.	II

library and catalogue metaphor, since, as perceived by the users, the traditional setting of the library includes mainly books.

The first release of CBAP was performed by two developers during 4 months (from the middle of May until the beginning of September 2006) and was composed of four iterations. Customer collaboration and evaluation by users were emphasized during the process. Measures were taken to control the progress.

The implementation of the CBAP prototype is performed using the Opera 8.5 browser (W3C-compliant) that supports VoiceXML and XHTML for Microsoft Windows XP systems. In addition, it provides a small-screen view that enables development for mobile applications. Figure 41.1 presents CBAP screenshots of the guided search interface (A), search result interface (B) and the book localization interface (C).

41.5 Evaluating the Vocal Interface

The evaluation process of CBAP is composed of evaluation iterations that each examine the artifacts of the previous development iteration and result in design changes to be considered during the current or next development iteration. The first development iteration (di1) provides its artifacts. During the second development iteration (di2), the first evaluation iteration (ei1) took place to evaluate and reflect on the artifacts produced in

the first development iteration (di1) and further to decide upon changes that should be introduced. During the third development iteration (di3), the second evaluation iteration (ei2) took place to evaluate and reflect on the artifacts produced in the second development iteration (di2), and so on. Therefore, the development and evaluation evolved side by side while the evaluation starts one iteration after the development starts: di1, di2+ei1, di3+ei2, and so on. Each iteration took a period of 3 to 5 weeks, and, as previously mentioned, the first CBAP release was composed of four such iterations. The process of combining the agility concepts with ongoing user evaluation that contributes to the design and is performed by the team members follows recent trends to embrace rapid and iterative methods and be part of the team for the sake of continuous HCI design (Norman, 2006).

In the first two evaluation iterations the user groups were identified to include librarians and readers, and questionnaires and semistructured interviews were prepared to better understand user needs. In the third evaluation iteration, a cooperative evaluation was performed with two users to learn about users' behavior with the system and identify major problems. After the fourth iteration ended, and the first release was over, a controlled experiment was planned and conducted for the purpose of the evaluation of the speech interface. In what follows, the performed experiment is described to illustrate the notion of

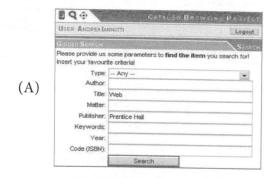

(A)

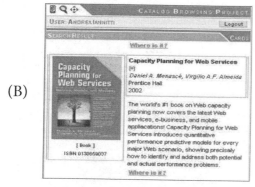

(B)

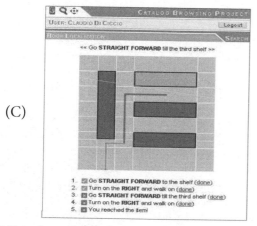

(C)

FIGURE 41.1 Screenshots from the CBAP prototype.

evaluation that is side by side with development. The experiment is small in scale so no statistics are used. Still, quantitative data are provided that enable decision making to ensure better design of the user interfaces.

A within-subjects experiment was conducted with six non-disabled participants who are computer science students in different levels, three male and three female. The experimental task included login to the system, search activities, and book localization activity. The experiment was conducted in a lab environment, and the localization part was mocked as in a physical library. The digital library for the experimental purpose was simulated by an internal data structure. The task can be performed using speech (S) or without speech (non-S). Each of the participants performed the task in both modes S and non-S, while three participants followed S and then non-S, and three followed non-S and then S. Before starting the experiment, each participant filled out an

attitude questionnaire and received 10-minute training on how to use CBAP. After the experiment each of the participants filled out a questionnaire to reflect on the activities.

The remainder of this section presents the experimental qualitative and quantitative data. Table 41.3 presents the answers of the participants to some initial attitude questions with respect to speech, where SD means that the participant strongly disagree with the statement, D means disagree, A means agree, and SA means strongly agree. When for a specific statement the sum of answers is less than six, this means that some participants did not answer on this question.

As can be observed from Table 41.3, the attitudes of the participants with respect to speech interfaces are mixed and do not follow a consistent approach. According to participants' answers, though speech interfaces are fun, they are also annoying, and though they like them, they do not always prefer them. It is, however, worth noting that in other experiments (e.g., Smith et al., 2000) users significantly preferred a speech-augmented user interface.

The aforementioned questionnaire also had some open questions asking the participants to provide features that they considered important to be included in speech interfaces, advantages and disadvantages of speech interfaces, and a personal scenario that happened to them when using such interfaces. Following are some of the expressions of participants as they answered those questions:

- "[consider important] using realistic voices"
- "...I had to provide some information to an automatic operator—it was boring waiting for its answers."
- "[disadvantage] It can take several minutes to interact with speech interfaces."
- "[advantage] They can become friendly."

The answers given by the participants bring out two main aspects about speech interfaces, which are user friendliness and user interaction. The participants consider speech interfaces as both friendly and not friendly, or as both fast and slow. For example, one participant answered the following in two consecutive rows, "[advantage] faster than normal interfaces"; "[disadvantage] a user may wait too long before achieving [his/her] purpose."

After filling out the questionnaire, the participants received a one-page users' guide and when they completed reading it and had no more questions, they received the task page according to

TABLE 41.3 Participants' Attitude to Speech Interfaces

Statement	SD	D	A	SA
I like interfaces with speech features.	1	1	3	1
I have experience with speech interfaces.	1		5	
I use speech interface when I can.	1	3	2	
People I know do not like speech interfaces.		4	1	
Speech interfaces are slow.	1	2	1	1
I feel uncomfortable with speech interfaces.		2	4	
Speech interfaces are fun.	1		4	1
Speech interfaces are annoying.		4	2	
I expect to use more speech interfaces in the future.			3	3
I prefer interfaces that do not include speech.		3	3	

TABLE 41.4 Averaged Search Time (in Minutes)

Group	Averaged Search Duration	Averaged Non-S Search Duration	Averaged S Search Duration
Non-S → S	54.66	28	81.33
S → non-S	26.58	14	39.16

their appropriate experimental order of S and non-S. An automatic time measure, which was developed as part of the system, provides the time stamps of the login/logout and the time stamps of each search start/end. Table 41.4 presents the averaged time in minutes that was invested on the two search activities by both experimental groups.

As can be observed, the S→non-S group performed the entire task almost twice as fast as the non-S→S group. When looking into the data of speech and nonspeech per each group, it can be seen that the participants in both groups performed the speech task slower then the nonspeech task. This implies that the speech interface is harder to use and should be improved. Related work has also found that experienced users can provide music input faster with speech user interface (Smith et al., 2000). Another observation is that although the speech task required more time from the participants, they better learned the system when first using it with the speech option.

After completing the CBAP task, participants were asked to fill out a questionnaire to reflect on their own activities. Table 41.5 presents their level of agreement to some statements.

As can be observed, most participants find it hard to use CBAP in its current stage, though it was fun and they expect to use such interfaces in the future.

The same questionnaire had some open questions asking the participants to describe what they liked about CBAP, what are the problems they have encountered, their severity ranking between 1 (not so important) and 5 (very important), and their recommendations on how to deal with the specific problem. Following are some of the expressions of participants answering those questions:

- "I expect the system to vocally recognize also the value I want to search."
- "It was easy to use; funny to use."
- "[rank 4] Instructions too fast."
- "It's been a new experience to me."
- "[rank 3] too sensitive to pronunciation."

TABLE 41.5 Participants Reflect after Using CBAP

Statement	SD	D	A	SA
I like searching using speech commands.			4	2
I have experience with vocal GPS.	1	2	3	
I prefer to work with the silence mode.			2	4
People will feel that the speech mode is too slow.			2	4
I feel uncomfortable with the system I use.	1	1	3	1
It was fun.			4	2
It was annoying.	1	4	1	
I expect I will see such systems in the future.			6	

- "[rank 5] unstable."
- "[I like] the GPS system."
- "I like activating commands by voice."
- "[rank 5] sometimes it doesn't understand what I say."
- "[rank 5] I have to repeat."
- "[recommendation] try to translate to Italian; it should be more flexible with pronunciation."
- "[I like] moving the cursor by speaking."

Examining the answers, it was concluded that subsequent work should focus on implementing speech for all interface features and improving the online usage information. This is based on the observation that when users are introduced to a speech-based interface, they expect it to be fully speech-based; that is, no use of the keyboard is required at all. Further, they expect to receive vocal online help to assist them in the process of using the application.

As the application becomes more and more stable, users with disabilities will also gradually be directly involved. This will include evaluations with blind users or visually impaired users. Considering the close relationship between libraries and learning, and thus between digital libraries and e-learning, the design of the experiment involving blind or visually impaired users could appropriately borrow from a similar experiment carried out in the VICE e-learning project (Catarci et al., 2008).

The participants and materials for the experiment are summarized in Table 41.6. Each user is given the CBAP application and a description of tasks to perform with the application. The participants are provided with an informed consent form and a demographics questionnaire. The participants are expected to interact with the CBAP application to accomplish the indicated tasks. The aim is to test accessibility directly with the users and to obtain other suggestions by this experience, such as possible difficulties faced while trying to perform the tasks and any additional information on how to improve the application to make it more accessible. A questionnaire will be used to collect the foregoing information.

At the end of the evaluation, the evaluator holds a short interview to clarify or probe further the user's experience while interacting with the CBAP application. The interview also provides an opportunity for the user to raise some issues or questions. Some of the main lessons learned from the experiments in VICE involving blind users or visually impaired users will be taken into consideration. These include:

- While transforming a user interface element for blind users, one of the main difficulties is that designers often do not have sufficient knowledge on how speech technologies work and how blind users interact.
- One of the greatest problems in designing an accessible user interface stems from the fact that designers can hardly foresee which assistive technology will be used by the impaired user (e.g., speech-based tools such as screen readers often do not allow users to distinguish links already visited from those not yet visited). It is probably for this reason that some users prefer sequential navigation to hypertextual navigation).

TABLE 41.6 Evaluation with Blind or Visually Impaired Users

Participants	10 visually impaired or blind usersar
Materials for each participant	1. Informed consent form
	2. Demographics questionnaire
	3. CBAP application
	4. Description of the tasks to be carried with the CBAP application (and corresponding instructions)
	5. Questionnaire (pertaining to the participant's experience while interacting with the CBAP application)

41.6 Conclusions

This chapter has described the importance of assuring accessibility in digital libraries, focusing on how vocal interfaces can be appropriately used to support and enhance accessibility in digital libraries. We have discussed one of our efforts in this direction, the CBAP. An appropriate integration of the agile software development and the user-centered methodology has been adopted, thereby providing rapid and iterative development (i.e., the design progresses/evolves while being refined/improved based on user evaluation and customer feedback).

With respect to the methodologies used, the conducted experiment led to the conclusion that the user evaluation is fostered by the process agility, and that product development benefits from keeping the design updated according to the evaluation outcomes. In addition, a set of evaluation tools including automated features is built and refined during the process and can be further used.

Concerning catalogue browsing, it was found that users accept this access paradigm in a natural manner as if they were used to browsing over a catalogue. They enjoy giving vocal instructions to the system and receiving vocal localization guidance from the system. Still, they expect the system to understand as much as possible natural language utterances and be more sensitive to languages and pronunciation.

References

Aberg, J. and Shahmehri, N. (2001). An empirical study of human Web assistants: Implications for user support in Web information systems, in the *Proceedings of the SIGCHI Conference on Human Factors in Computing Systems (CHI 2001)*, 31 March–5 April 2001, Seattle, Washington, pp. 404–411. New York: ACM Press.

Aberg, J. and Shahmehri, N. (2003). Live help systems, in *Human Factors and Web Development* (J. Ratner, ed.), pp. 287–309. Mahwah, NJ: Laurence Erlbaum Associates.

Beck, K. and Andres, C. (2005). *Extreme Programming Explained* (2nd ed.). La Porte, IN: Addison-Wesley.

Bertini, E., Calì, A., Cartaci, T., Gabrielli, S., and Kimani, S. (2005). Interaction-based adaptation for small screen devices, in the *Proceedings of the 10th International Conference on User Modeling (UM 2005)*, 24–29 July 2005, Edinburgh, Scotland, pp. 277–281. Berlin/Heidelberg: Springer-Verlag.

Bertini, E., Catarci, T., and Kimani, S. (2004). Library and Ubicomp: Supporting seamless interaction with physical and digital entities. Paper presented in the context of the Workshop on Interaction Design for CSCL in Ubiquitous Computing at Mobile HCI'04, 13–16 September 2004, Glasgow, Scotland.

Bilmes, J. A., Li, X., Malkin, J., Kilanski, K., Wright, R., Kirchhoff, K., et al. (2005). The vocal joystick: A voice-based human-computer interface for individuals with motor impairments, in the *Proceedings of Human Language Technology Conference and Conference on Empirical Methods in Natural Language Processing (HLT/EMNLP 2005)*, 6–8 October 2005, Vancouver, Canada, pp. 995–1002. ACL Anthology.

Bolchini, D. and Paolini, P. (2006). Interactive dialogue model: A design technique for multi-channel applications. *IEEE Transactions on Multimedia* 8: 529–541.

Borgman, C. L. (1996). Why are online catalogs still hard to use? *Journal of the American Society of Information Science* 47: 493–503.

Borgman, C. L. (1999). What are digital libraries? Competing visions. *Information Processing and Management* 35: 227–243.

Caldwell, B., Chisholm, W., White, J., and Vanderheim, G. (2005). *Web Content Accessibility Guidelines 2.0*, Working Draft 23. http://www.w3.org/TR/WCAG20.

Catarci, T. (ed.) (2006). *Deliverable D4.0.1: WP4 Progress Report on Tasks 4.5-4.10.* http://delos.dis.uniroma1.it/C2/Deliverables/Document%20Library.

Catarci, T., De Giovanni, L., Gabrielli, S., Kimani, S., and Mirabella, V. (2008). Scaffolding the design of accessible eLearning content: A user-centered approach and cognitive perspective. *Cognitive Processing* 9. Berlin/Heidelberg: Springer-Verlag.

Dai, L., Goldman, R., Sears, A., and Lozier, J. (2004). Speech-based cursor control: A study of grid-based solutions, in the *Proceedings of the 6th International ACM SIGACCESS Conference on Computers and Accessibility (ASSETS 2004)*, 18–20 October 2004, Atlanta, pp. 94–101. New York: ACM Press.

Di Blas, N., Paolini, P., and Speroni, M. (2004). "Usable accessibility" to the web for blind users, in the *Adjunct Proceedings of the 8th ERCIM Workshop on User Interfaces for All (UI4ALL 2004)*, 28–29 June 2004, Vienna, Austria. Vienna: Springer-Verlag.

Dubinsky, Y., Catarci, T., and Kimani, S. (2007). Using catalogue browsing for speech-based interface to a digital library, in the *Proceedings of the IASTED International Conference on Human Computer Interaction (IASTED-HCI 2007)*, 14–16 March 2007, Chamonix, France. Calgary, Canada: ACTA Press.

Eisen, M., Greenstein, D., Martin, R. S., McRobbie, M. A., Pitts, L. H., Stanton, R., et al. (2003). *Proceedings of the 2003 Symposium "Emerging Visions for Access in the Twenty-first Century Library,"* 21–22 April 2003, University of California. Vienna: Springer-Verlag.

The Energy and Resources Institute (2004). *Report on the International Conference on Digital Libraries.* Organized by The Energy and Resources Institute in partnership with Department of Culture, Ministry of Tourism and Culture, Government of India, New Delhi, 24–27 February 2004.

Fattahi, R. (1997). *The Relevance of Cataloguing Principles to the Online Environment: A Historical and Analytical Study.* PhD dissertation, University of New South Wales, Sydney, Australia.

Gabrielli, S., Mirabella, V., Kimani, S., and Catarci, T. (2006). A boosting approach to the development of eContent for learners with special needs. *Journal of Educational Technologies & Society* 9: 17–26.

Hauptmann, A. G. (1995). Speech recognition in the Informedia Digital Video Library: Uses and limitations, in the *Proceedings of the 7th IEEE International Conference on Tools with AI (ICTAI-95)*, November 1995, Washington, DC, pp. 288–294. Washington, DC: IEEE.

Hildreth, C. R. (1995). The GUI OPAC: Approach with caution. *The Public-Access Computer Systems Review* 6: 6–18.

Igarashi, T. and Hughes, J. F. (2001). Voice as sound: Using non-verbal voice input for interactive control, in the *Proceedings of the 14th Annual ACM Symposium on User Interface Software and Technology (UIST 2001)*, 11–14 November 2001, Orlando, FL, pp. 155–156. New York: ACM Press.

The Energy and Resources Institute (2004). *Report on the International Conference on Digital Libraries.* Organized by The Energy and Resources Institute in partnership with Department of Culture, Ministry of Tourism and Culture, Government of India, New Delhi, 24–27 February 2004.

Kimani, S. and Bertini, E. (2005). Exploiting ubiquitous computing to support digital library tasks, in *Emergent Application Domains in HCI, Volume 5 of the Proceedings of HCI International 2005* (J. A. Jacko and V. K. Leonard, eds.), 22–27 July 2005, Las Vegas [CD-ROM]. Mahwah, NJ: Laurence Erlbaum Associates.

Lynch, C. and Garcia-Molina, H. (1995). *Interoperability, Scaling, and the Digital Libraries Research Agenda.* IITA Digital Libraries Workshop. http://diglib.stanford.edu/diglib/pub/reports/iita-dlw/main.html#2.

Maguire, M. (1999). A review of user-interface design guidelines for public information kiosk systems. *International Journal of Human-Computer Studies* 50: 263–286.

Mirabella, V., Kimani, S., Gabrielli, S., and Catarci, T. (2004). Accessible e-Learning material: A no-frills avenue for didactical experts. *New Review of Hypermedia and Multimedia* 10: 165–180.

Nielsen, J. (2001). Beyond accessibility: Treating users with disabilities as people, in *Alertbox.* http://www.useit.com/alertbox/20011111.html.

Norman, D. (2006). Why doing user observations first is wrong. *ACM Interactions* 13: 50, 63.

Norman, D. and Draper, S. (1986). *User Centered System Design: New Perspectives on Human-Computer Interaction.* Mahwah, NJ: Lawrence Erlbaum Associates.

Oviatt, S. L., Cohen, P. R., and Wang, M. (1994). Toward interface design for human language technology: Modality and structure as determinants of linguistic complexity. *Speech Communication* 15: 283–300.

Rogers, Y., Scaife, M., Gabrielli, S., Smith, H., and Harris, E. (2002). A conceptual framework for mixed reality environments: Designing novel learning activities for young children. *Presence* 11: 677–686.

Shneiderman, B. (2000). The limits of speech recognition. *Communications of the ACM* 43: 63–65.

Smith, L. A., Chiu, E. F., and Scott, B. L. (2000). A speech interface for building musical score collections, in the *Proceedings of the 5th ACM Conference on Digital Libraries (DL 2000)*, 2–7 June 2000, San Antonio, TX, pp. 165–173. New York: ACM Press.

Sporka, A. J., Kurniawan, S. H., Mahmud, M., and Slavik, P. (2006). Non-speech input and speech recognition for real-time control of computer games, in the *Proceedings of the 8th International ACM SIGACCESS Conference on Computers and Accessibility (ASSETS 2006)*, 23–25 October 2006, Portland, OR, pp. 213–220. New York: ACM Press.

Sporka, A. J., Kurniawan, S. H., and Slavik, P. (2004). Whistling User Interface (U³I), in *User-Centered Interaction Paradigms for Universal Access in the Information Society, Proceedings of the 8th ERCIM International Workshop "User Interfaces for All" (UI4ALL 2004)* (C. Stary and C. Stephanidis, eds.), 28–29 June 2004, Vienna, Austria, pp. 472–478. Berlin/Heidelberg: Springer-Verlag.

Sporka, A. J., Kurniawan, S. H., and Slavik, P. (2006a). Non-speech operated emulation of keyboard, in *Designing Accessible Technology, Part III "Assistive Technology and Rehabilitation Robotics," Proceedings of the Cambridge Workshop on Universal Access and Assistive Technology (CWUAAT 2006)*, April 2006, Cambridge, U.K., pp. 145–154. London: Springer-Verlag.

Sporka, A. J., Kurniawan, S. H., and Slavik, P. (2006b). Acoustic control of mouse pointer. *Universal Access in the Information Society* 4: 237–245.

Stephanidis, C., Akoumianakis, D., Sfyrakis, M., and Paramythis, A. (1998). Universal accessibility in HCI: Process-oriented design guidelines and tool requirements, in the *Proceedings of the 4th ERCIM Workshop "User Interfaces for All"* (C. Stephanidis and A. Waern, eds.), 19–21 October 1998, Stockholm, Sweden. Crete, Greece: ICS-FORTH.

Theofanos, M. F. and Redish, J. (2003). Guidelines for accessible and usable web sites: Observing users who work with screenreaders. *Interactions* 10: 36–51.

Watts, R. and Robinson, P. (1999). Controlling computers by whistling, in the *Proceedings of Eurographics 1999*, 8–9

August 1999, Cambridge, UK. Oxford: Oxford University Computing Services.

Wilson, P. (1989). The second objective, in *The Conceptual Foundations of Descriptive Cataloging* (E. Svenonius, ed.), pp. 5–16. San Diego: Academic Press.

Theories and Methods for Studying Online Communities for People with Disabilities and Older People

Ulrike Pfeil and
Panayiotis Zaphiris

42.1 Introduction

In recent years, the Internet is increasingly used for social interactions and collaborations. The emergence of social networking web sites (e.g., http://www.myspace.com, http://www.facebook.com, and http://www.linkedin.com) and virtual worlds (e.g., http://www.secondlife.com and http://www.activeworlds.com) show the growing tendency to use the Internet to interact with others online (Lipsman, 2007). Also, the popularity of traditional ways of online communication (e.g., chats and discussion boards) is constantly increasing. In a survey, Horrigan et al. (2001) found that 84% of Internet users had used an online community. In addition to retrieving information from the Internet, people are now starting to use online settings to meet other people, develop friendships, play, and exchange experiences and support (Preece and Ghozati, 2001; Verhaagen, 2005; Bausch and Han, 2006; Lipsman, 2007). What does this new trend mean for the research area of inclusive design? Can online communities be facilitated to support people with disabilities and older people in their daily life?

Research in the area of computer-mediated communication (CMC) has already started to investigate the impact of CMC characteristics on the social interactions between people in online environments. Distinct characteristics of CMC are, for example, the anonymity in online communities, the constant availability and access to some forms of online communities (e.g., bulletin boards), the role of deception and honesty in online communities, and also the degree to which the technology of the online community allows for the exchange of nonverbal cues (e.g., facial expressions). While the impact of these characteristics on communication has been researched intensively in the context of communication among able-bodied people, little research has been done to investigate the benefits and challenges of online communities for people with disabilities and older people.

The distinct characteristics of online communities are of particular importance for people with disabilities and the elderly. Exploratory research showed that online communities offer an opportunity for people with disabilities to meet other people in an environment where their disability is not visible, allowing them to communicate free from stigmatization (Bowker and Tuffin, 2002, 2003). Furthermore, meeting others with similar experiences offers people with disabilities and older people the opportunity to exchange factual and emotional support in a group of likeminded and understanding people. Often, these online communities are characterized by a high level of empathy and social support (Pfeil and Zaphiris, 2007). Online communities offer older people and people with disabilities the opportunity to interact with others without leaving the house, which can prevent isolation and decrease loneliness (Bradley and Poppen, 2003). However, research has found that there are also special challenges when it comes to online communities for people with disabilities, as deception and misbehavior in online communities for people with disabilities can have harmful consequences, especially for a vulnerable user group (Bowker and Tuffin, 2003). Further research is needed to investigate the benefits and challenges of online communities for people with disabilities and the elderly, and to fully understand how online communities can be utilized to support these user groups in their daily life.

The chapter begins with an elaboration on the characteristics of online communities and a discussion of studies that have investigated online communities for people with disabilities and older people. Theories and methodologies in the area of CMC that can be used to study online communities for people with disabilities and older people are then described. Finally, an overview of the current status of this research area and its implication for practitioners is provided.

42.2 What Are Online Communities?

Generally, online communities are referred to as settings, where people can meet and communicate with each other online (Preece et al., 2003). However, a lot of discussions arise when it comes to finding a common definition for the term *online community*. Some researchers consider online communities as an environment where support, empathy, and friendships develop (e.g., Rheingold, 1993); others are rather interested in the analysis, design, and evaluation of different technologies that support online communication and group building (Preece, 2000; De Souza and Preece, 2004; Maloney-Krichmar and Preece, 2005). According to Preece and Maloney-Krichmar (2005), an online community consists of "people who come together for a particular purpose, and who are guided by policies […], and supported by software." Rheingold (1993) uses the term *virtual community* and describes it as "social aggregations that emerge from the Net when enough people carry on those public discussions long enough, with sufficient human feeling, to form webs of personal relationships in cyberspace" (p. 5).

The Internet offers many opportunities to support online communication. Different technologies such as bulletin boards, usenet groups, listservs, newsgroups, video conferencing, chats, MUDs (multi-user dungeon, dimension, or sometimes domain), 3D virtual worlds (e.g., SecondLife and ActiveWorlds) and wikis each have their own attributes and support a specific type of communication. To categorize different forms of technology that provide the opportunity to communicate online, researchers commonly divide them into asynchronous and synchronous communication technologies and distinguish them according to different media that they support (e.g., text, graphic, audio, video). Asynchronous and synchronous communication technologies differ in the requirement for people to be online at the same time. Chats, for example, can only be used by people who are online and access the chat at the same time, and are therefore synchronous. Bulletin boards, in contrast, do not necessarily require people to be online at the same time, and are therefore an example for asynchronous technologies. Another characteristic is the media that the respective technology supports. Whereas most newsgroups, for example, support mainly text, other technologies, like video conferencing and online virtual game environments, support text, graphics, video, and audio (Zaphiris et al., 2006). The asynchronous/synchronous dimension and the different media that these technologies support both influence the nature of communication within the respective community.

Online communities can emerge in different forms. Some are similar to public places like a park or a street, where members do not feel a strong personal commitment to the online community. In contrast, other online communities are closely knit and participants of these communities feel a strong bond between each other and therefore return regularly. Personal commitments and a feeling of belonging are more common in the latter kind of communities. Also, the size of online communities can have implications on the structure of the community. It can vary extensively between only a few members up to hundreds and thousands. Larger communities have to be more formal and include roles and conventions, whereas smaller online communities might be more flexible and informal (Butler et al., 2002).

Rather than physical proximity, researchers use the nature and strength of relationships among members of a community to determine the characteristics of that community (Wellman and Gulia, 1999). Jones (1997) states that communities online are formed around similar interests of the participants. Similar needs and/or experiences of the participants, the provision of resources valuable to the participants, the engagement in social support, and the development of friendship may also be characteristics of an online community (Wellman, 2000).

Butler et al. (2002) investigated the reason why people invest effort to maintain the vitality of their online communities. They stated that people can gain different kinds of benefits out of the participation in an online community: informational (get valuable information on a topic of interest), social (develop social ties, e.g., friendships), visibility (make yourself known beyond the borders of your current location) and altruistic (find benefit in helping others). However, Chang et al. (1999) state that the content of the communication, the dynamics of the interactions, and the benefits of participation vary from different online communities. It is therefore necessary to study online communities from the user's perspective.

42.2.1 Studying Online Communities

A plethora of methods exist that can enable us to get a good understanding of the social activities that take place in online communities. The benefits of some of the more classic techniques are briefly described:

- *Interviews*: Interviews can be used to gain insights about general characteristics of the participants of an online community and their motivation for participating in the community under investigation. The data collected come straight from the participants of the online communities, whereby they are able to provide feedback based on their own personal experiences, activities, thoughts, and suggestions.

- *Questionnaires*: For online communities, a questionnaire can be used to elicit facts about the participants, their behavior and their beliefs/attitudes. Like interviews, questionnaires are an important technique for collecting user opinions and experiences they have had through the use of CMC and their overall existence in online communities.
- *Personas*: For online communities, personas can be used to better understand the participants of the community and their background (Cooper, 1999). Personas can also be used as a supplement to Social network Analysis (described later in this chapter) to get a greater overview of the characteristics of key participants of a community. Using personas, web developers gain a more complete picture of their prospective and/or current users and are able to design the interfaces and functionality of their systems, to be more personalized and suited for the communication of the members of their online communities.
- *Social network analysis (SNA)*: Social network analysis (Wasserman and Faust, 1994) is a very valuable technique when it comes to analyzing online communities, as it can provide a visual presentation of the community and, more important, it can provide qualitative and quantitative measures of the dynamics of the community.

42.3 Online Communities and People with Disabilities

In the following, studies that investigated online communities for people with disabilities and older people are discussed. Particular focus is placed on the topics of empowerment through participation in online communities, opportunities and challenges due to the anonymity and the masked identity in online communities, and the exchange of social support in online communities for people with disabilities and older people.

42.3.1 Empowerment through Participation in Online Communities

Some characteristics of online communities can be especially valuable for people with disabilities and older people. One advantage of online communities, for example, is that people can meet and interact with others without having to leave their homes. People who experience mobility impairments find it sometimes difficult to meet others face-to-face, because transportation is complicated or exhausting for them. As online communities offer social interactions without having to move outside the home, this barrier can be overcome.

Also, for people who have difficulties in communicating orally, text-based online communities offer an alternative way to communicate with others. Asynchronous communication especially allows people to construct messages at their own pace and time, which is particularly beneficial for people who have difficulties engaging in oral face-to-face conversations or who have a slower typing speed (Bowker and Tuffin, 2003). Most asynchronous online communities allow people to view previous messages, which might be a helpful support for people with memory impairments (Lewin, 2001).

Furthermore, online communities offer people the possibility to get in touch with others who experience a similar life situation. Especially for people with disabilities who would like to get to know others that have the same or a similar disability, online communities offer the possibility to get access to a lot of people. Friendships between people are possible despite large geographic distances and people can exchange experiences and support. Meeting others who share common life experiences can empower people and create a sense of community. Bakardjieva and Smith (2001) found that people who are in a similar situation built online communities that are characterized by a high level of empathy and trust in which people exchange their knowledge and experiences.

Tilley et al. (2006) focused on the use of online communities by people with severe physical disabilities. Based on interview findings, they constructed the virtual community model, a theory that describes the main characters of online communities for physically impaired people. At the center of their model, the authors state that information and communication technology (ICT) can help people with physical disabilities to gain a sense of control over their lives, moving toward a more independent way of living. The greater the severity of the disability, the more independence people gain from participating in online environments, often using assistive technologies to do so. Furthermore, they found that participation in online community reduces isolation and shifts the work paradigm for people with physical disabilities. This can lead to an empowerment of people with disabilities resulting in increased inclusion and participation in societal activities. This finding is similar to Roulstone's (1998) research, which concluded that online communication allows people with disabilities to present themselves competently and show their abilities and skills without being prejudiced according to their impairment. Participants in his research stated that this is especially engaging in the work environment, as online communication offers them a way to communicate their work and skills on an equal level as able-bodied people (Roulstone, 1998).

42.3.2 The Role of Anonymity and Identity

Online communities offer people with disabilities the possibility to socially interact with others and present themselves outside their identity as a disabled person. Due to the lack of visual cues in many of the online communities, impairments are not visible and can be masked (Cromby and Standon, 1999). Concerning the benefits and challenges of anonymity in online communities, Bargh (2002) mentioned that this characteristic can be both beneficial and harmful. The advantages and disadvantages seem to be exaggerated for people with disabilities, as this user group is believed to benefit most from anonymous communication, but at the same time also presents a group vulnerable toward deceptive and rude behavior.

Bowker and Tuffin (2003) interviewed 21 people with physical and sensory disabilities and asked them about their engagement in online communities and their strategies and thoughts about anonymity and deception. Findings illustrated that users develop strategies to keep themselves safe from possible harmful behaviors of others. Participants mentioned that they are cautious when interacting with strangers in online communities, try to separate their online interactions from their offline lives, and take time to carefully evaluate the communication partners based on their messages to make a judgment about their trustworthiness. Also, participants mentioned that they do not feel especially vulnerable toward misbehavior in online communities in comparison to able-bodied people, as the danger to suffer from dishonest behavior is the same for everybody participating in online communities. They felt that their disability does not pose an additional danger in online community participation.

Also, participants reported engaging in small deceptive acts themselves to experience an identity that does not include their disability. Participants mentioned that they withhold information about being disabled, because they fear that people will react in an intolerant way. Based on experiences in offline conversations, participants said that a lot of people have stereotypes and react in a negative way to them. Not revealing their disability protects them from such reactions in online communities. The deception in these situations is justified, as it is seen as a protection from possible harmful treatments from others (Bowker and Tuffin, 2003).

Bowker and Tuffin (2002) also investigated how people with disabilities choose to disclose information about themselves and their disability. Their interview of participants reported that they choose to disclose information depending on the topic and purpose of the online communication that they engage in; for example, they consider it to be appropriate to talk about their disability in online communities for people with disabilities, but they do not consider it to be necessary to reveal that information in an online community that is not connected to the disability. The anonymity of people in online communities was found to offer people a choice and control about what information they would like to disclose about themselves. Participants mentioned that they liked being able to participate on an equal level as nondisabled people in online communities. Participants reported that they enjoy being judged according to the content of their messages in the online community instead of their physical appearance or impairments.

In Bowker and Tuffin's (2002) study, participants also mentioned that when they disclose their disability, this disclosure has much less effect on the communication compared to conversations in offline settings. Even if the communicators are informed about the disability, they are not permanently confronted with it because it is not visible in the exchanged texts. Not only for other people, but also for people with disabilities themselves, social interaction in online communities offers a mental break from their condition. Having control over the degree, time, and pace in which they choose to inform other members of the online community about their disability offers a choice to people with disabilities that they often do not have in offline settings (Bowker and Tuffin, 2002). The equalizing nature of online communications and the opportunity to alter the characteristics of one's online identity (e.g., Turkle 1995) seem to be of special value for people with disabilities. Often, asynchronous online communities are favored, because they eliminate possible differences in the speed with which people with disabilities can write and read messages.

42.3.3 The Exchange of Social Support

Investigating the impact of online communication on people's lives, Bradley and Poppen (2003) focused explicitly on the question of whether using the Internet and engaging in online social interactions increase or decrease the loneliness of people with disabilities and older people. Providing equipment, Internet access, and if necessary training to housebound older and disabled people and caregivers, they investigated the impact that Internet usage had on their perceived level of loneliness. Their findings showed that participants engaged in social interactions with each other via the Internet, and developed friendships in online environments. A follow-up questionnaire a year later illustrated that the satisfaction with social interactions increased significantly during that year, suggesting that participation in online communities does indeed decrease the level of loneliness for housebound older people and people with disabilities.

Concerning online communication by older people, research showed that older people develop friendships in online communities (Xie, 2005; Pfeil and Zaphiris, 2007) and that the social networks emerging in online communities help them to cope with stressful life situations (Wright, 1999). In a comparison between the social networks of newsgroups for younger and older people, Zaphiris and Sarwar (2006) found that the newsgroup for older people had more consistencies and stability in activity and behaviors of its participants. Although the teenage newsgroup had a higher number of visitors, messages per person, and on average longer messages, the newsgroup for older people had higher numbers of replies to messages and therefore showed a higher degree of interactivity, responsiveness, and reciprocity. Overall, the study concluded that older people can be as active as younger users of CMC and can form more stable and interactive groups with emerging natural leaders and influencers through CMC.

Being of a similar age and experiencing similar life situations are believed to increase the level of support and understanding in online communities (Bakardjieva and Smith, 2001). This indicates that online communities where older people or people with disabilities meet others who experience a similar situation are very supportive and characterized by a friendly and understanding atmosphere. Pfeil and Zaphiris (2007) conducted a qualitative content analysis of 400 messages of a discussion board for older people. The results from this study showed that older people used the discussion board most frequently to talk about their own situation and problems. Secondly, activities and messages that nurture the community feeling within an online community were also very common. The fact that others within the online community had experienced similar situations led to a high level of understanding and support within the online community. Both general uplifting support as well as serious personalized support were identified in the exchanged messages. Messages were often very emotional and personal. Older people developed a sense of community and were frequently checking on other member's well-being, which showed that they were concerned about each other. Within the online community, older people could both seek and provide support, which was found to be empowering for this target population. This study showed that older people do indeed engage in supportive activities in online communities and that online communities provide an empathic space for older people to engage in social interactions (Pfeil and Zaphiris, 2007).

Investigating 1,472 messages of an online community for people with disabilities, Braithwaite et al. (1999) sorted the messages into a category system investigating characteristics of support. Their findings show that emotional support was most common within the messages (40%), followed by informational support (31.7%), esteem support (18.6%), and network support (7.1%). Tangible support that involves practical support was least frequent (2.7%). Braithwaite et al. (1999) conclude that online communities can provide an environment where people can actively help one another to manage some of the physical and social limitations imposed by disability.

42.4 Theories for Studying Online Communities

In the following section, different CMC theories will be discussed, starting with the presentation of key characteristics of the social presence theory (Short et al., 1976), which discusses the extent to which communication media support the awareness of the copresence of the communication partners. Then, the social information processing (SIP) theory (Walther, 1992, 1996) will be discussed. SIP theory proposes that the restrictions imposed by limited opportunities to convey nonverbal cues in CMC can be overcome, and, given sufficient time, CMC can be as intimate as face-to-face communication. Lastly, social identity/deindividuation (SIDE) theory (Spears and Lea, 1992) is discussed. It deals with CMC in group settings and concludes that in online communities, members are more likely to act according to group norms opposed to individual norms. Each theory focuses on a specific aspect of social interactions in online communities. The impact and use of the theories to research about online communities for people with disabilities and older people are investigated.

42.4.1 Social Presence Theory and the Cues-Filtered Out Approach

Social presence theory (Short et al., 1976) deals with the ability of the mediating technology to create a sense of copresence during a communication process. The underlying concept is that the degree to which mediated communication can convey social presence influences the way people interact with each other. Social presence theory helps to explain how different kinds of mediating technologies influence social behavior and communication. Social presence is a quality of the communication medium and describes the degree to which a communicator is aware of the presence of his communication partner. The higher the social presence, the more accurate is the perception of the communication partner. Thus, social presence not only includes the spoken word, but is also conveyed via nonverbal cues, like body language, facial expressions, and the tone of the voice.

As mediating technologies are often limited in the ability and variety of conveying nonverbal or voice-cues, this restriction is believed to have an influence on the characteristics of the communication, as it makes it more difficult for the communicators to get a sense of the social presence of her communication partner (Culnan and Markus, 1987; Walther, 1993). For example, text-based communication environments do not utilize the exchange of facial expressions, and communicators cannot hear each other's voices. This reduces the number of clues about the communicators' characteristics and current inner states (Short et al., 1976). Especially the lack of seeing the other person's physical appearance is believed to strongly influence the way people interact in text-based online environments, as it has a significant effect on the overall impression of this person. In cases where people communicate with each other only via text, communicators use the content of exchanged messages to develop an overall impression of the other person. Wallace (1999) concluded that once the communicators have developed an impression of the communication partner in an online environment, people resist changing their impression, even if they are confronted with evidence that their current impression is not completely correct. This can have both positive and negative consequences for communication.

As people have a lower awareness of their communication partners, and less possibilities to convey their personal characteristics and current emotional state, it is common that misunderstandings occur in these communications. Once such misunderstandings occur, it is often difficult to solve them successfully, as they can easily escalate. As people are less aware of the presence of others, they become less inhibited and less afraid of possible negative consequences of their actions, and sometimes engage in more extreme behavior than they would in an offline setting (Kiesler et al., 1984). This can lead to violent attacks against other members of the online community. Especially in online communities that deal with emotional and sensitive topics, misunderstandings can easily hurt members. The danger of serious consequences based on misunderstandings need to be especially considered in online communities for people with disabilities and older people. As such online communities are often dealing with sensitive topics, misunderstandings might result in serious consequences (e.g., people might get hurt and drop out of the online community).

Flaming can also prevent the development of close relationships and trust in online environments. Studies have indicated that in some online communities, the feeling of togetherness and the warm and supportive atmosphere is constantly stressed and mentioned within the messages exchanged to prevent people from misusing the trust and supportive atmosphere within the online community (Pfeil and Zaphiris, 2007). People seem to be especially careful about what they say in online communities and often stress that it is only from their viewpoint and mention that others might think differently about this. Abbreviations like IMHO (In My Humble Opinion) and emoticons (e.g., ☺ and ☹) are being used to make messages sound less aggressive (Wallace, 1999). Studies that investigated the degree of flaming in online communities reported that misbehavior is least common in online support communities in which people talk about emotional and personal situations (Preece and Ghozati, 2001).

Some scholars believe that CMC cannot achieve the same intensity and intimacy as face-to-face communication. The lack of nonverbal cues, such as eye contact and body lean, hinders communication from becoming personal. CMC is described as "impersonal, unsociable, cold, and insensitive" (Lea and Spears, 1995, p. 214). According to this opinion, online communication should focus on distant, task-oriented, and informational message exchange, as it is not appropriate for the exchange of emotions and social interactions. However, empirical research suggests that it is indeed possible to engage in personal and emotional communication in online communities and people even develop friendships and relationships online (Rheingold, 1993; Preece, 1998, 1999; Pfeil and Zaphiris, 2007).

As mentioned previously, a reduced awareness of the social presence of the other members within an online community reduces inhibitions. This can also increase the courage to disclose personal information (Walther, 1996). This phenomenon explains the high level of emotional self-disclosure, as especially found in health communities (Preece, 1998, 1999) or in communities for older people (Pfeil and Zaphiris, 2007). Once people start to disclose information about themselves, they also encourage others to do the same (Wallace, 1999; Pfeil and Zaphiris, 2007). This is especially the case for online communities in which people share a common illness or experience a similar difficult situation, as it is the case for online communities for people with disabilities and older people. Knowing that other people are in a similar situation, exchanging and sharing experiences often lead to a higher level of trust in online communities, which in turn encourages people to disclose more information about themselves. Wallace (1999) referred to this phenomenon as reciprocity in self-disclosure, and considered it to play a very important role in online communities. The exchange of emotions and experiences made in relation to a person's disability and life situation can increase the feeling of togetherness within an online community and lead to strong bonds and friendships between its members.

Also, the lack of visual and audible cues makes identity deception easier. As Turkle (1995) noted, people often take on different identities in online communities to act out different aspects of themselves. Again, identity deception can be seen as positive, as it allows people to take on a different identity from their offline selves and try out new and different aspects of their personality.

This can be particularly liberating for people with disabilities, as they can appear to be free of the disability in an online community. Not being stigmatized by the visible aspects of disabilities gives people the opportunity to act independent of their physical conditions, and therefore alter parts of their identity. In online games and online 3D environments like SecondLife, members can often create their online identity in the form of avatars, and therefore determine the physical appearance that they will make for others.

On the other hand, the opportunity to hide and alter one's identity can also be misused and people might appear as somebody they are not. One example of this is reported in Bowker and Tuffin (2003), and describes an incident when an able-bodied person appeared in an online community as a severely disabled girl. After a while, the same person also posted as the boyfriend of this girl and both triggered a lot of concern and emotional support from other members of the online community. Once the lie had been detected, it had a severe impact on the atmosphere within the online community, as other members felt betrayed and hurt by the deception (Van Gelder, 1991, cited in Bowker and Tuffin, 2003).

42.4.2 Social Information Processing Theory

The social presence theory concentrates on the lack of nonverbal cues in CMC and argues that communication lacking physical presence and visual and audible cues does not facilitate the development of friendships and close relationships. To explain the findings of empirical research that it is indeed possible to establish close friendships in online communities, Walther (1992) developed the SIP theory. The SIP theory states that CMC is as useful as face-to-face conversation for establishing closeness and friendships between communicators. According to the SIP theory, relationships between communicators grow when people reveal information about themselves and use that information to build an impression of the other person. Similar to the social presence theory, Walther (1992) mentioned that the lack of nonverbal cues restricts communication, but he also states that the communicators can make up for this disadvantage by describing impressions of oneself in verbal cues. Thus, although CMC is restricted by a lack of nonverbal cues, it benefits from verbal and textual cues. When communicators have learned to adapt to the situation, relationships and friendships maintained by CMC can be as strong as their offline counterparts (Walther, 1992).

Walther (1996) goes even further by introducing the *hyperpersonal perspective*, which proposes that some relationships are more close and intimate online than they would be offline, due to selective self-presentation and overemphasis on similarities between the communicators. He states that due to the absence of nonverbal cues, people tend to overemphasize the attributes that they receive via CMC, and therefore construct an idealized impression of their communication partner. The asynchronous nature of online communities and the possibility to edit messages before sending them is believed to strengthen this phenomenon, as people have more time to read and reflect on a message, think about and construct a new message, carefully censor the self-presentation that they want to make toward the other person (Walther, 1996). Kanayama (2003) stated that the opportunity to take time to understand and reflect on each other's messages is beneficial especially for older people, as it gives them the freedom to write and decode messages in their own pace to develop satisfactory social relationships. The absence of physical presence prevents judgment of appearances like age, gender, attractiveness, and disability on first sight. Instead, this judgment is based on the information that the communicators reveal about themselves (Walther, 1992).

Also, online communicators are often given the opportunity to change their appearance and attractiveness in online communication, be it through posting a very positive picture of oneself (or even a picture of somebody else) or describing oneself in a very positive and attractive light. This leads to the fact that others believe that she or he is more attractive. Walther named this phenomenon self-fulfilling prophecy. Walther (1996) himself explains that hyperpersonal CMC occurs

> when users experience commonality and are self-aware, physically separated, and communicating via a limited-cues channel that allows them to selectively self-present and edit; to construct and reciprocate representations of their partners and relations without the interferences of environmental reality. (p. 33)

Critics of the hyperpersonal perspective argue that it lacks providing explanations for negative relational outcomes in CMC, as it concentrates mainly on accounting for positive developments of relationships.

On the other hand, Sourkup (2000) criticizes that the SIP theory does not account for the current technical multimedia applications of CMC. He argues that CMC can indeed convey nonverbal cues, and that a multimedia approach to CMC theory needs to be established that will account for these possibilities. Furthermore, Sourkup (2000) criticizes Walther's assumption that a CMC message contains less information than face-to-face communication. He argues that

> If an e-mail message is in a three-dimensional format and contains video, audio and animation, that message certainly contains as much (if not more) social information as many face-to-face comments. Rather than viewing CMC as "limited" or "purely verbal" communication, multimedia CMC should be viewed as a unique context with many complex communicative qualities. (p. 414)

42.4.3 SIDE Theory

Spears and Lea (1992) explore the social and psychological dimensions of CMC and conclude that CMC does not per definition restrict social activities. They state that CMC and group

forming in CMC settings might be highly influenced by social group norms that are established in the setting. In contrast to the beliefs that anonymity frees people's behavior and leads to more equality among group members, Postmes et al. (1998) state that anonymity in online communities shifts the focus from the individual identity to social identity. Individuation is established through visual contact, profile pictures, and proximity, whereas de-individuation occurs when communications lack these cues.

Postmes et al. (1998) state that social identity develops in an online community when its members are in a state of de-individuation. This means that people do not see themselves as individuals in a group where "the I" interacts with "others," but they establish a "we" feeling in which the awareness of individuals is switched into awareness of a group identity. Instead of acting due to general norms driven by the thought of the consequences of misbehavior, people act according to the social identity as established in the respective group or community. The shift from individual to social identity allows people in online communities to take on a new aspect of their identity. Depending on the kind of online community, people with disabilities and the elderly might want to stress their age and disability or might want to ignore it.

Similarly, Spears and Lea (1992) believe that the lack of visual cues and the greater distance in CMC communication leads to de-individuation. But unlike previous studies, they do not conclude that this de-individuation results in misbehavior or rude activities (e.g., flaming). Instead, they reason that members behave according to the group identity. In cases where flaming is within the group norm, this might lead to more rude behavior, but in cases where the group norm is otherwise, flaming is not prevalent. This is also in line with findings from Preece and Ghozati (2001) that different levels of flaming and empathic content can be found in online communities with different topics. For example, flaming was more prevalent in online communities that discussed sport than in those that discussed health issues. In contrast, health communities had more empathic content than online communities around a sport. According to the SIDE theory this is due to the different group norms in these online communities. Consequently, people might engage in different behavior, depending on the online community that a person is participating in. People with disabilities might want to choose to place emphasis on their disability in certain online communities, but not in others (Bowker and Tuffin, 2002).

Group norms are established through group members' behavior, which leads to a distinctive group identity and group boundaries. Members of an online community often meet around a commonality, and this sense of similarity is exaggerated due to the lack of contrasting cues. This leads to overemphasized group solidarity. Walther (1996) states that "when participants are led to perceive that they are in a group relationship, each tends to hold a 'social self-categorization' rather than an 'individual self-categorization,' [which leads to the] attributions of greater similarity and liking with one's partners" (p. 18). In cases where members were seeing themselves as a part of a community, they established in-group favorism, which clearly distinguishes group members of the in-group from out-group people (Postmes et al., 1998). Behavior and norms of the in-group are accepted and preferred to out-group behavior.

Critiques of the SIDE theory mention that its conclusions are mainly based on experiments in laboratory settings, in which communication was controlled. Complete anonymity and only text-based communication was investigated, which is rarely the case in CMC. Postmes et al. (1998) themselves state that the types of anonymity created in experimental settings are not generally found in "real life" on the Internet. Additionally, with the help of pictures, avatars, user names, and so on, current multimedia applications allow for more than only text-based communication to construct an identity (Sourkup, 2000).

42.5 Conclusion

This chapter has provided an introduction to the topic of CMC and how online communities are being used by people with disabilities and the elderly. It is evident from the literature review provided that an established body of literature is available, which can provide a strong theoretical basis for the study of such communities. Moreover, there exists an established set of methodologies that have been successfully used in the study of online communities in general. In the authors' view, what is missing is a research agenda on (1) identifying whether these theories and methods are applicable to the study of online communities for people with disabilities and (2) applying them to the study of online communities for people with disabilities. New forms of CMC (e.g., Second Life, 3D virtual worlds) have the potential of providing valuable virtual social spaces for people with special needs. Research in this area can reveal valuable insights in the way these virtual spaces could be utilized for the benefit of these groups.

References

Bakardjieva, M. and Smith, R. (2001). The Internet in everyday life: Computer networking from the standpoint of the domestic user. *New Media and Society* 3: 67–83.

Bargh, J. A. (2002). Beyond simple truths: The human-internet interaction. *Journal of Social Issues* 58: 1–8.

Bausch, S. and Han, L. (2006). *Social Networking Sites Grow 47 Percent, Year Over Year, Reaching 45 Percent of Web Users.* http://www.nielsen-netratings.com/pr/pr_060511.pdf.

Bowker, N. and Tuffin, K. (2002). Disability discourses for online identities. *Disability and Society* 17: 327–344.

Bowker, N. and Tuffin, K. (2003). Dicing with deception: People with disabilities' strategies for managing safety and identity. *Journal of Computer-Mediated Communication* 8. http://jcmc.indiana.edu/vol8/issue2/bowker.html.

Bradley, N. and Poppen, W. (2003). Assistive technology, computers, and Internet may decrease sense of isolation for homebound elderly and disabled persons. *Technology and Disability* 15: 19–25.

Braithwaite, D. O., Waldron, V. R., and Finn, J. (1999). Communication of social support in computer-mediated groups for people with disabilities. *Health Communication* 11: 123–151.

Butler, B., Sproull, L., Kiesler, S., and Kraut, R. (2002). Community effort in online groups: Who does the work and why, in *Leadership at a Distance* (S. Weisband and L. Atwater, ed.). Mahwah, NJ: Lawrence Erlbaum Associates.

Chang, A., Kannan, P. K., and Whinston, A. B. (1999). Electronic communities as intermediaries: The issues and economics, in the *Proceedings of the 32nd Annual Hawaii International Conference on System Sciences, Volume 5,* 5–8 January 1999, Maui, HI, p. 5041. Washington, DC: IEEE.

Cooper, A. (1999). *The Inmates Are Running the Asylum: Why High-Tech Products Drive Us Crazy and How to Restore the Sanity.* Indianapolis: Sams.

Cromby, J. and Standon, P. (1999). Cyborgs and stigma: Technology, disability, subjectivity, in *Cyberpsychology* (A. J. Gordo-Lopez and I. Parker, eds.), pp. 95–112. New York: Routledge.

Culnan, M. J. and Markus, M. L. (1987). Information technologies, in *Handbook of Organizational Communication* (F. Jablin, L. L. Putnam, K. Roberts, and L. Porter, eds.), pp. 420–443. Newbury Park, CA: Sage.

De Souza, C. S. and Preece, J. (2004). A framework for analyzing and understanding online communities. *Interacting with Computers* 16: 579–610.

Horrigan, J. B., Rainie, L., and Fox, S. (2001). *Online Communities: Networks That Nurture Long-Distance Relationships and Local Ties.* Washington, DC: Pew Internet and American Life Project.

Jones, Q. (1997). Virtual-communities, virtual-settlements and cyber-archaeology: A theoretical outline. *Journal of Computer-Mediated Communications* 3. http://jcmc.indiana.edu/vol3/issue3/jones.html.

Kanayama, T. (2003). Ethnographic research on the experience of Japanese elderly people online. *New Media and Society* 5: 267–288.

Kiesler, S., Siegel, J., and McGuire (1984). Social psychological aspects of computer-mediated communications. *American Psychologist* 39: 1123–1134.

Lea, M. and Spears, R. (1995). Love at first byte? Building personal relationships over computer networks, in *Understudied Relationships: Off the Beaten Track* (J. Wood and S. Duck, eds.), pp. 197–237. Newbury Park, CA: Sage.

Lewin, M. (2001). *Equal with Anybody, Computers in the Lives of Older People.* PhD disseration, University of Stirling, Stirling, Scotland. http://www.odeluce.stir.ac.uk/mlewin/Dissertation.htm.

Lipsman, A. (2007). *Social Networking Goes Global.* Reston, VA: ComScore World Metrix. http://www.comscore.com/press/release.asp?press=1555.

Maloney-Krichmar, D. and Preece, J. (2005). A multilevel analysis of sociability, usability and community dynamics in an online health community. *Transactions on Computer-Human Interactions (TOCHI)* 12: 201–232.

Pfeil, U. and Zaphiris, P. (2007). Patterns of empathy in online communication, in the *Proceedings of the SIGCHI Conference on*

Human Factors in Computing Systems (CHI '07), 28 April–3 May 2007, San Jose, CA, pp. 919–928. New York: ACM Press.

Postmes, T., Spears, R., and Lea, M. (1998). Breaching or building social boundaries? SIDE-effects of computer-mediated communication. *Communication Research* 25: 689–715.

Preece, J. (1998). Empathic communities: Reaching out across the Web. *Interactions* 5: 32–43.

Preece, J. (1999). Empathic communities: Balancing emotional and factual communication. *Interacting with Computers* 12: 63–77.

Preece, J. (2000). *Online Communities: Designing Usability, Supporting Sociability*. Chichester, U.K.: John Wiley & Sons.

Preece, J. and Ghozati, K. (2001). Observations and explorations of empathy online, in *The Internet and Health Communication: Experience and Expectations* (R. R. Rice and J. E. Katz, eds.), pp. 237–260. Thousand Oaks, CA: Sage Publications.

Preece, J., and Maloney-Krichmar, D. (2005). Online communities: Design, theory, and practice. *Journal of Computer-Mediated Communication* 10. http://jcmc.indiana.edu/vol10/issue4/preece.html.

Preece, J., Maloney-Krichmar, D., and Abras, C. (2003). History and emergence of online communities, in *Encyclopedia of Community* (B. Wellman, ed.). Great Barrington, MA: Berkshire Publishing Group.

Rheingold, H. (1993). *The Virtual Community: Homesteading on the Electronic Frontier*. Reading, MA: MIT Press.

Roulstone, A. (1998). Researching a disabling society: The case of employment and new technology, in *The Disability Reader: Social Sciences Perspectives* (T. Shakespeare, ed.), pp. 110–128. London: Cassel Educational Ltd.

Short, J., Williams, E., and Christie, B. (1976). *The Social Psychology of Telecommunications*. London: John Wiley & Sons.

Soukup, C. (2000). Building a theory of multimedia CMC: An analysis, critique, and integration of computer-mediated communication theory and research. *New Media and Society* 2: 407–425.

Spears, R., and Lea, M. (1992). Social influence and the influence of the "social" in computer-mediated communication, in *Contexts of Computer-Mediated Communication* (M. Lea, ed.), pp. 30–65. London: Harvester-Wheatsheaf.

Tilley, C. M., Bruce, C. S., Hallam, G., and Hills, A. P. (2006). A model for a virtual community for people with long-term, severe physical disabilities. *Information Research*, 11. http://informationr.net/ir/11-3/paper253.html.

Turkle, S. (1995). *Life on the Screen: Identity in the Age of the Internet*. New York: Simon and Schuster.

Van Gelder, L. (1991). The strange case of the electronic lover, in *Computerization and Controversy: Value Conflicts and Social Choices* (C. Dunlop and R. Kling, eds.), pp. 364–375. Boston: Academic Press.

Verhaagen, D. (2005). *Parenting the Millennium Generation: Guiding Our Children Born between 1982 and 2000*. Westport, CT: Praeger Publishers.

Wallace, P. (1999). *The Psychology of the Internet*. Cambridge, U.K.: Cambridge University Press.

Walther, J. (1992). Interpersonal effects in computer-mediated interaction: A relational perspective. *Communication Research* 19: 52–90.

Walther, J. B. (1993). Impression development in computer-mediated interaction. *Western Journal of Communication* 57: 381–398.

Walther, J. (1996). Computer-mediated communication: Impersonal, interpersonal, and hyperpersonal interaction. *Communication Research* 23: 3–43.

Wasserman, S. and Faust, K. (1994). *Social Network Analysis: Methods and Applications*. Cambridge, U.K.: Cambridge University Press.

Wellman, B. (2000). Changing connectivity: A future history of Y2.03K. *Sociological Research Online*, 4. http://www.socresonline.org.uk/4/4/wellman.html.

Wellman, B. and Gulia, M. (1999). Net surfers don't ride alone: Virtual communities as communities, in *Networks in the Global Village* (B. Wellman, ed.), pp. 331–366. Boulder, CO: Westview Press.

Wright, K. B. (1999). Computer-mediated support groups: An examination of relationships among social support, perceived stress, and coping strategies. *Communication Quarterly* 47: 402–414.

Xie, B. (2005). Getting older adults online: The experiences of SeniorNet (USA) and OldKids (China), in *Young Technologies in Old Hands: An International View on Senior Citizen's Utilization of ICT* (B. Jaeger, ed.), pp. 175–204. Copenhagen, Denmark: DJOF Publishing.

Zaphiris, P., Ang, J., and Laghos, A. (2006). Online communities, in *The Human-Computer Interaction Handbook* (J. A. Jacko and A. Sears, eds.). Mahwah, NJ: Lawrence Erlbaum and Associates.

Zaphiris, P. and Sarwar, R. (2006). Trends, similarities, and differences in the usage of teen and senior public online newsgroups. *ACM Transactions on Computer-Human Interaction (TOCHI)* 13: 403–422.

Computer-Supported Cooperative Work

Tom Gross and
Mirko Fetter

43.1 Introduction

Computer-supported cooperative work develops novel concepts and systems with a special focus on the social interaction in groups and communities of users. Universal access develops novel concepts and systems with a special focus on the individual cognitive, perceptual, and physical strengths and weaknesses of users. Bringing these domains together entails novel interesting challenges for cooperative universal access.

Computer-supported cooperative work as a *research field* has several predecessors, which will be described in the next section. The first workshop gathered 20 participants interested in technical support for social interaction in 1984. Another conference took place in 1986. Since 1988, the ACM organizes an official Computer-Supported Cooperative Work conference series biannually in North America, and since 1989 the European Computer-Supported Cooperative Work conference series takes place biannually in Europe.

The term *computer-supported cooperative work (CSCW)* was coined by Irene Greif who organized the first workshop in 1984. Irene Greif (1988) writes that CSCW is "computer-assisted coordinated activity such as communication and problem solving carried out by a group of collaborating individuals" (p. xi). Wilson (1991) defines CSCW similarly and also points out that it combines a basic understanding of how users interact with each other and the development of technologies for networks, hardware, and software. Other authors have slightly broader perspectives. For instance,

Bowers and Benford (1991) point out that "in its most general form, CSCW examines the possibilities and effects of technological support for humans involved in collaborative group communication and work processes" (p. v). Ellis et al. (1991) write that CSCW

> looks at how groups work and seeks to discover how technology (especially computers) can help them work.…Even systems designed for multi-user applications, such as office information systems, provide minimal support for user-to-user interaction. This type of support is clearly needed, since a significant portion of a person's activities occur in a group, rather than an individual, context. (p. 39f)

The related term *groupware* was defined by Johnson-Lentz and Johnson-Lentz (1982) as computer-based technology that supports social group processes and by Ellis et al. (1991) as "computer-based systems that support groups of people engaged in a common task (or goal) and that provide an interface to a shared environment" (p. 40).

There has been some debate on the four characters of CSCW, and of groupware (Bannon and Schmidt, 1989) as well as of social software. However, today all terms—CSCW, groupware, and social software—can be seen from a rather *broad perspective*. Like Marca and Bock (1992) who write:

> The construction of tools to foster group intellectual work should be no surprise. We have had such tools

for a long time, and yet have tended to take them for granted. Isn't a slate or blackboard a piece of groupware? How about an overhead or slide projector? A megaphone or loudspeaker? The ordinary mail system? Babbage contributed a number of crucial ideas to our mail system, such as the postage stamps. What about a telegraph or telephone? Bell's work on the telephone started from his interest in developing a tool to help deaf students work with others. That he failed in his original goal does not diminish the magnitude of the telephone system as a groupware tool. (p. v)

Similarly, Lynch et al. (1990) point out that CSCW and groupware can be seen as a paradigm rather than as a narrow area of research or category of systems and prototypes. They write:

> Groupware is distinguished from normal software by the basic assumption it makes: groupware makes the user aware that he is part of a group, while most other software seeks to hide and protect users from each other … software that accentuates the multiple user environment, co-ordinating and orchestrating things so that users can see each other, yet do not conflict with each other. (p. 160)

Marca and Bock (1992) also write that:

> Groupware is a conceptual shift; a shift in our understanding. The traditional computing paradigm sees the computer as a tool for manipulating and exchanging data. The groupware paradigm, on the other hand, views the computer as a shared space in which people collaborate; a clear shift in the relationship between people and information. (p. 60)

This chapter applies this broad perspective and defines CSCW to be a research field of computer science and related areas that have the primary focus on understanding and supporting groups of users, their social interaction, and their interaction with technology; groupware is defined as the technological concepts, systems, and prototypes supporting social interaction among groups of users.

Universal access moves from a perspective that "recognises, respects, values, and attempts to accommodate a very wide range of human abilities, skills, requirements, and preferences in the design of computer-based products and operating environments" and it tries to cope with a diversity in the target user group, the users' tasks, and the contexts of use. To provide universal access the products and services are often designed to change their look-and-feel according to the respective user and the respective context of use (Stephanidis and Savidis, 2001).

Bringing these domains—CSCW and universal access— together entails novel interesting challenges with respect to universal access for groups and communities, particularly for those bringing together users with heterogeneous cognitive, perceptual, and physical skills. This chapter provides a contribution from the perspective of CSCW. The remainder of this chapter

glances at the emergence of CSCW, introduces basic social and technical concepts, provides an overview of systems and prototypes, and discusses the convergence of CSCW and universal access.

43.2 Emergence of CSCW and Social Software

This section describes the emergence of CSCW starting with early visions of technical support for social interaction to CSCW and to social software and ambient intelligence.

43.2.1 Early Visions on Technical Support for Social Interaction

Some early visionaries had innovative ideas and developed novel concepts, systems, and prototypes for the technical support of networked social interaction.

Vannevar Bush developed ideas for a concept and system to support elegant storing of data and retrieval of data in groups. He wrote that:

> When data of any sort are placed in storage, they are filed alphabetically or numerically, and information is found (when it is) by tracing it down from subclass to subclass.…The human mind does not work that way. It operates by association. With one item in its grasp, it snaps instantly to the next that is suggested by the association of thoughts. (1945, Section 6)

The *Memex* system that his group developed facilitated the scanning of paper documents and allowed the storing of large amounts of data. It was based on the concept of an associative memory and provided trails through the data to provide users with guidance when navigating through the massive amount of data. While the system was based on microfilms and is technologically outdated, the basic concepts of Memex are still relevant for today's hypertext systems.

Douglas Engelbart was inspired by Bush's concepts and suggested innovative concepts and tools for augmenting humans' intellect (1968). Engelbart suggested that users and computers should evolve together, and that computers should be seen as tools to help users comprehending and solving complex problems that they would not be able to solve without technology. Engelbart and his group developed the *oN-Line System (NLS)* that had many concepts from modern desktop publishing and cooperation support such as text processing, spreadsheets, spontaneous audio and desktop video conferencing among coauthors, and so forth. At the IFIP Fall Joint Computer Conference in San Francisco in 1968, he demonstrated the NLS system, including a novel keyboard and a pointing device that he called a mouse and that has been in use ever since (Engelbart and English, 1968).

Finally, Joe Licklider (1960, 1968) developed great ideas and concepts for technical support of online communities that are still being developed and used today. Licklider (1968) writes:

We want to emphasise ... the increasing significance of the jointly constructive, the mutually reinforcing aspect of communication—the part that transcends "now we both know a fact that only one of us knew before." When minds interact, new ideas emerge. We want to talk about the creative aspect of communication.... Let us differentiate this from another class which we will call informational housekeeping. The latter is what computers today are used for in the main; they process payroll checks, keep track of bank balances, calculate orbits of space vehicles, control repetitive machine processes, and maintain varieties of debit and credit lists. Mostly they have not been used to make coherent pictures of not well understood situations. (p. 21)

From these early visions, ideas, concepts, prototypes, and systems it took some time until in the 1980s CSCW emerged as an official research area.

43.2.2 CSCW

CSCW can be seen on the one hand as a follow-up of the previously mentioned early visions, and on the other hand as a result of social needs (e.g., increasing globalization, cooperation, and division of labor that required technological support) and technological opportunities (e.g., improvements in basic technology of computer software and hardware as well as networking infrastructures).

Computing systems evolved from mainframe systems to networked personal computers. The first computing systems were mainframe systems, where remote terminals with no processors and with no hard disks were connected to a central host, which did all the computation. At that time, the means of communication via computers were rather poor. The vendor with by far the largest market share, IBM, at that time shipped their mainframes with a system called PROFS. This system offered some functionality for communication—users could send messages to their associates working at the same host and could schedule meetings with a shared calendaring tool. The invention of personal computers (PCs) in the early 1980s brought a dissemination of computing power. The need to share peripheral hardware, like printers and scanners, led to the connection of the PCs. More and more local area networks (LAN) and wide area networks (WAN) arose, which were the technological basis for the first groupware systems.

Organizational work styles changed from a hierarchical, monolithic, and rigid form of cooperation to flatter organizations and increased division of labor within and between companies. The division of labor within but also between different companies, global markets, and globally acting enterprises increasingly required remote collaboration.

The simultaneous diffusion of PCs and the increasing electronic data processing stimulated the request for the integration of computers and collaboration, as well as the use of computers as a means of communication and collaboration. The view of computers shifted from personal computers to *inter*personal computers; at the same time, the importance of the research on office automation declined, the term *computer-supported cooperative work* (CSCW) was created, and research on CSCW increased in importance. These trends have been reflected in a changing focus of many computer scientists from single-user applications to office automation systems and, later on, to CSCW and groupware systems. Research into CSCW brought with it investigations into the nature of interactions of individuals in groups. Concepts of situated action, shared workspaces, and the need for information about the presence and activities of others—group awareness—have emphasized the importance of spontaneous, improvised human behavior. They have considerably influenced groupware designers and contributed a lot to the efficiency of the systems developed.

CSCW as a research field was only constituted two decades ago. The first workshop in this field was held at the Massachusetts Institute of Technology (MIT) in 1984, attracted only 20 people, and was organized by Irene Greif of MIT and Paul Cashman of the Digital Equipment Corporation, who both coined the term. Since then, interest in CSCW has been growing tremendously. The first international conference with the same name took place in 1986 and already attracted 300 participants. Conferences followed in biannual intervals. The growing interest became a biannual European conference called European Conference on Computer-Supported Cooperative Work (ECSCW). Starting in 1989, conferences were held. Over the years several subdisciplines emerged, like workflow management systems and business process reengineering.

43.2.3 Social Software: Back to the Roots

Social software refers to technology that typically offers flexible support for the activities and types of social interaction performed in diverse social settings ranging from small groups to large communities. Tepper (2004) writes:

Many forms of social software are already old news for experienced technology users; bulletin boards, instant messaging, online role-playing games, and even the collaborative editing tools built into most word processing software all qualify. But there are a whole host of new tools for discussion and collaboration, many of them in some way tied to the rise of the WebLogs (or "blog"). New content syndication and aggregation tools, collaborative virtual workspaces, and collaborative editing tools, among others, are becoming popular, and social software is maturing so quickly that keeping up with it could be a full-time job in itself. (p. 19f)

Social software aims to scale up for any social setting; from this perspective it is different from CSCW and groupware, but very similar to the concepts and systems of the early visionaries.

Although the term *social software* has been used since the end of the 1980s, it has become widely known only recently with the advent of the Web 2.0. The Web 2.0 has experienced a huge hype during the last few years. It combines several important concepts (O'Reilly, 2005), such as: a clear focus on the web as a platform; collective intelligence and wisdom of crowds through peer-production of contents and social recommendations of contents; cooperative software development; and a rich user experience through novel user interface concepts and base technology.

Overall, it can be said that—besides the huge hype surrounding social software and the Web 2.0—most of the concepts have been around since the early visions, yet a slight progress in base technology and a growing recognition of the potential influence of novel technical support for social interaction can be identified.

43.2.4 Ambient Intelligence: Cooperation beyond the Desktop and the Web

Ambient intelligence provides users support for social interaction beyond the traditional graphical user interface and WIMP paradigm. Ambient intelligence is often based on ubiquitous computing technology, and aims to improve users' work and private life by analyzing and adapting to the current situation with a special focus on the presence and activities of users. In particular, it uses a broad range of opportunities for presenting information and functionality in the users' physical environment, and at the same time offers users a broad range of implicit input, allowing users to naturally interact with the environment through their presence, movements, body expressions, and language (Greenfield, 2006). Interaction-related aspects of ambient intelligence are discussed in Chapter 60 of this handbook.

Ambient intelligence has been boosted by constant improvements in sensing technology and support for precisely mechanical and electronic capturing of information about present users and their actions and social interactions (Gross et al., 2006, 2007). At the same time, the base technology such as disks for storing data, processors, and networks for extracting and inferring on the data, and novel algorithms for identifying patterns in the data have increased the means for processing the captured data (Baeza-Yates and Ribeiro-Neto, 1999). Finally, novel actuators and displays support the multimodal adaptation of the environment and presentation of information and functionality in the physical environment.

43.3 Social Concepts as Point of Departure

Since the goal of CSCW and also of related areas such as social software and ambient intelligence is to adequately support users in their social interaction and their interaction with the system, it is vital to understand some basic concepts of social science.

43.3.1 Social Entities

In social science literature, many different types of social entities can be distinguished. This section only focuses on some that are highly relevant for CSCW.

A *group* according to the *Oxford English Dictionary* is "a number of people or things that are located close together or are considered or classed together … a number of people who work together or share certain beliefs…." Typically, the number of people who form a group is limited. And groups have roles and norms that are basically expectations of the group members on how other group members should think and act in specific situations (Newcomb, 1943). Clear roles and norms reduce the coordination effort in groups. Groups also have cohesion—that is, an attraction of the group and of being a member (Festinger, 1950; Goodman et al., 1987). A *team* is often seen as a group with a shared goal. Sundstrom et al. (1990), for instance, define teams as "small groups of interdependent individuals who share responsibility for outcomes of their organizations."

A *community* according to the *Merriam-Webster* (2006) can be any of the following: "the people living in a particular area"; "a group of people with a common interest living in one place"; "a group of people sharing a common interest and relating together socially"; "the body of people in a profession or field of activity"; or "the quality or state of having many qualities in common." According to Mynatt et al. (1997) a community is a "social grouping that includes, in varying degrees: shared spatial relations, social conventions, a sense of membership and boundaries, and an ongoing rhythm of social interaction" (p. 211). A *network community* is typically technologically mediated and persistent, and supports multiple interaction styles and real-time interaction for multiusers. Often it is emphasized that communities—as compared to groups and teams—are bigger in membership, more diverse in interests, and do not necessarily have direct interaction among each individual member.

A *social network* is a "set of people or groups of people with some pattern of contacts or interactions between them" (Wasserman and Faust, 1995; Newman, 2003). The patterns, as well as the reasons for the contacts, can be very diverse, such as friendship, mutual interests, business, and so forth.

43.3.2 Social Interaction

In any of the previous social entities a variety of social interactions can take place; this section only focuses on five basic types of social interaction that are relevant for CSCW and that can take place in several types of social entities.

- *Coexistence* of multiple users means that in a specific setting there are two or more persons that are mutually aware of each other—either because they are at the same place and can perceive each other's presence, or because they have some technological means providing them with mutual information. Coexistence allows for social encounters and for spontaneous contacts, it is a prerequisite for the other types of social interaction, especially for communication.

- *Communication* is explicit exchange of symbolic messages, and implicit propagation of a user's changes to the environment or system to other users. It is important in social entities to make arrangements, to exchange ideas, and to transmit knowledge. Communication is, therefore, also an important basis for the other types of social interaction.

- *Coordination* is the management of interdependencies of activities, actors, or resources. Activities can have some procedural constraints and need to run in a specific sequence; actors can have interpersonal or intrapersonal conflicts in their roles; and resources can need to be used or shared by multiple parties (Malone and Crowston, 1992).
- *Consensus* refers to the decision making in social settings. Typically, social entities collect issues that are to be discussed and where a compromise has to be found, and positions that reflect the opinion of an individual or a subset of the social entity, and arguments that support the positions (Conklin, 1988). In the process of decision making, typically issues, positions, and arguments are collected and structured, evaluated, and votes are taken (Ellis et al., 1991).
- *Collaboration* refers to the close cooperation in a social entity, such as sharing and manipulating data together. Since collaboration often requires quite an effort for communication and coordination, it is important to structure the social entity and the group process adequately.

For the purpose of explaining the individual types of social interaction, they have been described individually. However, in practice, very often two or more types of social interaction occur together. Also, often different types of social interaction can follow each other, and interaction can occur in various modes: obtrusive or unobtrusive, embedded or symbolic, ephemeral or persistent, ad hoc or stipulated (Simone and Schmidt, 1993).

In addition to the already described types of social interaction, successful CSCW often requires the consideration of other aspects of social interaction.

- *Articulation* refers to the fact that successful social interaction often requires the people involved to discuss and negotiate the structure of the social entity and the process of social interaction. Typically, articulation is carried out continually, since frequently reactions to changing conditions and unanticipated contingencies require adapting the plan (Gerson and Star, 1986). In the articulation process, questions like who, what, where, when, how, and so forth are clarified, and participants, responsibilities, tasks, activities, and resources are aligned (Simone and Schmidt, 1993, p. 38; Strauss, 1985).
- *Situated action* reflects the fact that the users' actions and interactions are often open and cannot be anticipated completely. Coherence between intentions, actions, and effect can hardly be predicted and can often only be analyzed ex post (Suchman, 1987).

With these social concepts in mind, technical concepts will be introduced in the next section.

43.4 Technical Concepts: A CSCW View on Distributed Systems

The development of minicomputers and high-speed computer networks in the early 1970s made it feasible to move away from previous centralized systems and to build up networks of distributed systems. This shift brought up a number of advantages. Many of those advantages are of a technical nature, like for instance the bundling of computing power of multiple CPUs to process one single task or increased reliability when one CPU fails. Some authors refer to systems that primarily target these advantages as *parallel systems*. Furthermore, there are systems that are inherently distributed, as their individual parts and machines are spread over multiple places. This distribution is either due to the fact that data that accumulate locally is also locally collected but processed collectively (e.g., the sale figures of a company that result from different partners all over the world), or due to the fact that locally distributed users need technological support for working together.

Underneath many CSCW systems are clusters of *networks* and computers. Distributed systems and the software and protocols that tie these clusters together are the basics of most systems that support cooperative work. In this sense, a broad and deep understanding of the possible architectures, technologies, protocols, and so forth is crucial to develop usable CSCW systems.

With a perspective on CSCW and its understanding of networks and distributed computers as an enabler to support distributed collaboration between multiple users over networks connecting multiple computers, *distributed systems* can be defined as a "collection of independent computers that for its users feels like a single seamless system" (cf. also Tanenbaum and Van Steen, 2006, p. 2). This definition is derived from the assumption that in CSCW remote users have a deeper understanding of the fact that the distributed system is only a link or a channel between them and the other spatial distributed users. So, from the CSCW perspective, Tanenbaum's idea that it appears as a single coherent computer seems to be countervailing against this specific understanding of distributed systems. A group of people, physically colocated in a meeting room, using their laptops to share data via a wireless LAN, are relying on distributed systems just as the participants taking part in the same meeting via their computers' webcams from another country. In this sense, it is not the connection of computers but of users that is the CSCW perspective on distributed systems.

43.4.1 CSCW Classification of Distributed Systems

With this definition, a classification scheme for distributed systems from a CSCW perspective is needed to distinguish different systems on a technological level. The user-centered perspective of CSCW on distributed systems leads to a classification scheme focusing on the possibilities of interactions and the multimodality at the interface level, as well as the underlying technological concepts, for the presentation to the user. Therefore, three classes of systems are distinguished, namely, the *browser-based applications*, *desktop applications*, and *ubiquitous environments*, as shown in Figure 43.1.

The literature on distributed systems offers many other schemes for classification. Some are reported in the following for a sense of completeness.

Based on the previous convention, a distinction arises between parallel systems (multiprocessor and multicomputer) and distributed systems (multicomputer and multiuser), and only distributed systems are considered for CSCW.

Distributed systems can be characterized and distinguished according to their underlying network structures. The classic literature (Tanenbaum, 2002; Coulouris et al., 2005) differentiates with respect to network structures:

- According to types of transmission (e.g., broadcast vs. point-to-point)
- According to the geographical distance (e.g., personal area networks, local area networks, wide area networks)
- According to the type of physical connection (e.g., coaxial cables, fiber cables, microwave transmission)
- According to even coarser physical levels (e.g., between wireless and wired)

At a higher level, different types of architectures provide a scheme for the classification of distributed systems. Beside the basic and common client-server architecture, where a client node sends a request to a server node to get a response, other architectures supplement this classical two-tier architecture. For instance, the three-tier architecture moves the client intelligence to a middle tier, allowing for simpler, stateless clients, succeeded by the n-tier or multitier architecture, where the request is forwarded on the server side to other services to compute a response, which is then redirected to the client. Another common architecture is the peer-to-peer architecture, where each node—called a peer—has equal capabilities and privileges, and therefore can act as a server or a client.

These architectural principles are mostly reflected in the software, and therefore codetermine the degree of coupling of a distributed system. This leads to a last distinctive feature differentiating tightly coupled from loosely coupled software,

referring to the level of mutual dependencies between multiple systems to perform a single task from a software perspective.

43.4.1.1 Browser-Based Applications

The term *browser-based applications* summarizes all applications that are delivered to the users from a server via a network—such as the World Wide Web or an intranet—and that are presented inside the window of a web browser. The requirements to run such an application can therefore be reduced to a connection to the network and a web browser that for some purposes needs to be extended with some plug-ins.

By 1993, when the first browsers made it possible to view static HTML pages, the common gateway interface (CGI) defined a first standard that enabled web servers to run small programs and generate dynamic content as a response to client requests. Since then, numerous protocols and languages were developed to offer a variety of possibilities for the dynamic generation of content through server-side scripting, including Java server pages (JSP), PHP hypertext preprocessor, active server pages (ASP), or Java servlets. On the browser side, dynamic HTML (DHTML) technologies like JavaScript and cascading style sheets (CSS) improved the possibilities to design the user interface and interaction of web applications. The wireless application protocol (WAP) allows browsers on mobile phones or handheld computers to present information to the user in the form of the wireless mark-up language (WML).

The combination of dynamic elements on the client and server side allows developers to create *rich and highly responsive* browser-based applications referred to as rich Internet applications (RIAs). Technologies like Adobe Flash or Java Applets enable developers to embed small programs with even more interactive elements into a web page, allowing the easy integration of user interface behaviors like drag-and-drop or sliders. One recent result of the efforts to reduce the necessity to completely reload a browser page from the server as a reaction to user input is asynchronous JavaScript and XML (Ajax) (Garrett, 2005). Just like DHTML, Ajax is not a specific technology, but a new approach to combine a number of known technologies, including JavaScript and XML at the forefront.

The *interface* of browser-based applications has changed over time from simple clients that allow the presentation and

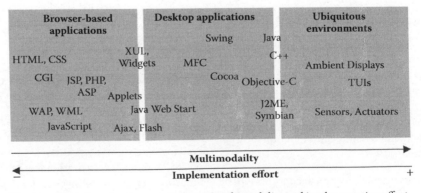

FIGURE 43.1 CSCW classification of distributed systems according to multimodality and implementation effort.

manipulation of data via forms and tables to more intuitive useable RIAs. The flexibility to run anywhere without prior installation and increased compatibility over different platforms and browsers reduce the implementation effort of browser-based applications, and make them a very common approach for deploying CSCW applications. On the downside, the browser acts as a sandbox, isolating the application from most system resources, and therefore makes it almost impossible to open browser-based applications to more channels than just visual and auditory outputs and inputs. Unfortunately, applications implemented in technologies that allow a richer experience like Flash or Ajax are often harder to interpret. Furthermore, the look-and-feel of the interactive elements created with those technologies often differs from the operating systems' look-and-feel, forcing the user to repeatedly learn how to interact with the new widgets.

There are multiple ways and a wide set of technologies to implement browser-based applications that support users to work together. Although this approach has many drawbacks, the proportion between the low implementation effort compared to the size of the reachable audience often outweighs user interface concerns.

43.4.1.2 Desktop Applications

Desktop applications contain all systems that are locally installed on a device and that present themselves to the user through a local graphical user interface (GUI) that permits a style of interaction with the system often denoted as a WIMP paradigm (i.e., Windows, Icons, Menus, and Pointing device). These systems do not necessarily depend on a network connection—as they are not delivered via such an infrastructure—but can use them to exchange data to facilitate distributed interaction.

The possibilities to *implement* such applications are manifold and reflected in the variety of programming languages and their combination possibilities. Two very basic categories can be distinguished: native applications, which are programmed in languages that are more specific to one operating system, and platform-independent applications, which run on multiple platforms but often need a virtual machine. Both types come with a set of different application programming interfaces (APIs) and libraries that facilitate the development of GUIs for the underlying program logic. Prominent combinations are Objective-C and Cocoa for Mac OS X, C++ and Microsoft Foundation Classes (MFC) on Windows, and Java and Swing for cross-platform applications.

The implementation effort is generally bigger for a desktop application than for browser-based applications, but entails a number of advantages. First of all, the familiarity with the interaction paradigms lowers the barrier to use those systems. Furthermore, systems that are written in machine-oriented languages have a better performance and additionally are able to access more system resources than browser-based applications. As a last criterion, the offline capabilities of distributed desktop applications often allow users to access offline data without the need for a network connection, in opposition to browser-based applications where data can only be accessed online.

The current orientation toward component-based software engineering will more and more enable us to put systems together from different building blocks that can be combined with each other. Under the phrase *contextual collaboration* (Hupfer et al., 2004) these ideas come to life when components with collaborative capabilities are integrated in core applications like word processors or CAD tools.

Technologies like Java Web Start, which can deploy Java desktop applications over a web browser via a network, contributed to a convergence between desktop and browser-based applications. At the same time, technologies like the Java Platform, Micro Edition (Java ME), or Symbian OS bring desktop applications to mobile devices and so pave the way to ubiquitous computing.

43.4.1.3 Ubiquitous Environments

Ubiquitous environments are defined as the accumulation of hardware and software that forms a framework of tools to support users with a specific task, embedded in daily activities and objects and so remaining widely invisible to the users.

On the one side, the post-desktop vision of human-computer interaction called ubiquitous computing (Weiser, 1991) is still a very young field in computer science. On the other hand, the research community already defined a number of concepts and paradigms that can serve as a fundament for implementation work in this area. From a technology perspective, sensors and actuators form the interface between the user and the ubiquitous environment. Sensors like accelerometers, thermometers, GPS receivers, or RFID readers allow the environment to capture information about the users and their current situation. Actuators, for example, lights or ventilators, help to channel the output of information by directly manipulating the physical environment of the user. Just as modern APIs for GUIs permit designing software that decouples the model from the view based on the model-view-controller (MVC) pattern, in the future new APIs and platforms for ubiquitous environments will allow developing *tangible user interfaces (TUI)* (Ishii and Ullmer, 1997) and *ambient displays* (Wisneski et al., 1998) in much the same way. When implemented in programming languages that can deal with the constraints of limited resources on small devices, for example, Java ME, those concepts can be integrated in everyday objects.

Ubiquitous environments will herald a new area of human-computer interaction, able to involve all modalities and therefore allow for a new naturalness in the way users interact with distributed systems.

43.4.2 Technical Interaction: Communication in CSCW Applications

The distributed nature of CSCW systems requires mechanisms that support the communication between multiple processes. Unlike single processor systems, processes in distributed systems do not rely on a shared memory. This complicates the synchronization and information exchange between processes, especially when they run on different machines. Therefore, all interprocess communication (IPC) in distributed systems is based on the messages passing. However, for the smooth flow of messages between different network types, machine architectures, operating systems, and programming languages, a set of rules and standards is needed that developers can rely on. With the ISO-specified Open System Interconnection Reference Model (ISO OSI) (Day and Zimmermann, 1983) a set of layers, rules, and interfaces are given and formalized in a number of protocols on which modern distributed systems can be found. Those protocols define a basic set of agreements on how the communication is handled between communicating nodes.

43.4.2.1 Message Formats

For the developers of distributed systems modern programming languages provide a number of libraries that encapsulate all necessary methods for communication with other platforms over those protocols. Still, there are a variety of different approaches and practices that can be applied to reach the goal of facilitating communication between the processes of a distributed system.

In the following, some standards and practices are introduced that are relevant for implementing CSCW systems and are discussed regarding their pros and cons. As a rule of thumb, the compactness of messages and the self-descriptiveness of the message formats are inversely proportional (cf. Figure 43.2). This leads to the general conclusion that choosing the right technology is often a trade-off between higher traffic volumes versus improved interoperability. In fact, low-level protocols are often used in a proprietary manner; therefore, to maintain or extend applications developed with these protocols, the developer needs a very detailed understanding of the whole system. Standardized, higher-level mechanisms often need a deeper knowledge of the technologies and methodologies of the standard, and the development produces a lot of implementation overhead that makes the employment of specialized tools often a necessity.

Besides those considerations, the applicability of the technology to communicate between heterogeneous platforms and programming languages, the general footprint in terms of processing costs, the firewall friendliness, or the likeliness of future changes and extensions to the technology should be considered as criteria for the selection of base technology.

Two models of IPC can be distinguished: the request-reply, and the stream-based model (Farley, 1998). Approaches and technologies are described in the following for both models,

with a focus on the request-reply model, which offers more areas of application and has more nuances in the field of CSCW. The stream-based approach in CSCW is often used for the distribution of continuous media—for example, in video-conferencing software.

43.4.2.2 Socket Communication and Custom Message Formats

Both approaches are low-level and often deliver a quick but proprietary, nonstandard solution for a small, straightforward problem.

Socket communication sends the data for the interprocess communication via sockets. Sockets are unique identifiers for endpoints provided by the transport service of a network. In IP-based networks two types of sockets can be distinguished: datagram sockets that send each packet individually addressed via User Datagram Protocol (UDP), and stream sockets that use the connection-oriented protocol TCP that opens a permanent connection to send its data. When a stream socket is used, a bidirectional connection for sending byte-streams is opened between two nodes. To exchange data, the receiving side needs to know the data and the sequence of the data it will receive. This works if both sides have a small, well-defined set of data to exchange that do not undergo frequent changes.

Custom message formats typically structure the data in a human-readable manner (e.g., comma-separated lists, XML-structured key-value pairs). Messages in this case are often sent via Hypertext Transfer Protocol (HTTP) request to a server-side application (e.g., realized as a servlet or CGI script) that generates the response and sends it back to the client in the same format structure. These protocols improve the extensibility and maintenance of small distributed applications while keeping the message overhead at an acceptable rate.

43.4.2.3 Special-Purpose Protocol

To increase the compatibility to third-party applications, services, or special servers, it is necessary to conform to standardized protocols. As mentioned before a protocol is an agreement for two computers to communicate with each other defining a number of rules and formats for the exchange of data. Some of these protocols (e.g., HTTP) are broadly used for a variety of purposes, sometimes diverted from their intended use. Other protocols are highly specialized for a very specific scope. An

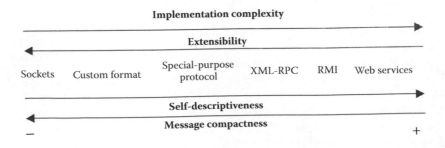

FIGURE 43.2 Message formats.

example of such a protocol is the eXtensible messaging and presence protocol (XMPP), an open protocol for instant messaging. The real-time transport protocol (RTP) defines a format for transmitting real-time data, such as audio or video. As a last example, Juxtapose (JXTA) defines an open set of XML protocols for implementing peer-to-peer infrastructures.

Special-purpose protocols are most of the time highly effective for implementing distributed systems with a special task. These protocols often define well-balanced message formats that allow for effective communication between systems, thus guaranteeing a high level of interoperability. However, the appliance of those protocols only makes sense in, and is often limited to, a very specific task.

43.4.2.4 Remote Procedure Calls and Web Services

Remote procedure calls (RPCs) (Birrell and Nelson, 1984) are a higher-level concept with the idea to transfer the mechanisms of calling a local procedure on a local computer to a remote computer to facilitate the call of functions via a network. The aim is from the perspective of the programmer that the passing of messages between the local and the remote host happens invisibly for the programmer, and that it consequently does not make a difference where the message is called. To transfer the input parameters and return values to the remote function, the methods' signature must conform to the specified set of types defined by the specification of the numerous implementations of RPC that is used. Some RPC implementations serialize—a process often referred to as *marshalling*—complex objects to send them via a network.

A widespread variant is the *XML-RPC* (Winer, 1999) specification. With its small footprint and reference implementations in many programming languages for a variety of platforms, it is an excellent tool for the implementation of highly interoperable systems. The lean design of the protocol makes XML-RPC a good choice for systems with little processing power, such as, for example, embedded systems or mobile devices. Another RPC variant, the Java API for remote method invocation (RMI), is tailored for use in Java environments and therefore allows for an easy invocation of remote Java objects.

Web services can be seen as one of the latest inventions in standardizing interprocess communications. They are designed to support interoperable machine-to-machine interaction over the web or other HTTP networks by utilizing a set of XML-based protocols and standards. A central element is the simple object access protocol (SOAP), which defines a message format that provides standards to encode instances of application-defined data types as well as for the definition of remote procedure calls and responses (Gudgin et al., 2007). Web services implementations, such as, for example, Apache Axis (Apache Software Foundation, 2006), are therefore often used to implement RPC-like calls—though this is not their primary intended purpose. However, the high payload of a SOAP message in combination with the comparable high demand for processing power that is needed for parsing the XML data outweighs the high flexibility of the protocol, especially for the application in

ubiquitous environments. Additionally, the complexity of the format makes the employment of specialized tools for the development almost indispensable.

43.5 The Design Space of CSCW

This section introduces the design space of CSCW, consisting of three core dimensions (cf. Figure 43.3):

- The user interaction with the system (i.e., classical groupware, social software, ambient intelligence)
- The social interaction among the users (i.e., coexistence, communication, coordination, consensus, collaboration)
- The system interaction among the different components of the cooperative system (sockets, custom formats, special-purpose protocols, XML-RPC, RMI, web services)

Subsequently, prominent examples of systems and prototypes for user interaction in classical groupware, social software, and ambient intelligence are provided.

43.5.1 Classical Groupware

Classical groupware systems typically support groups of users consisting of 7 to 12 persons. Traditionally, they were based on desktop applications, but recently web-based groupware has emerged. The following types of support can be distinguished: awareness support, communication support, coordination support, and collaboration support. Many systems have overlapping functionality, but from an analysis perspective the concepts become clearer with this distinction.

43.5.1.1 Awareness Support for Coexistence

Awareness support provides users with adequate mutual information to facilitate communication, coordination, and collaboration in groups. Gross et al. (2005) write:

> Awareness of the factors that define a group of items and users and tasks associated with the group helps to implement and manage services as well as devices for

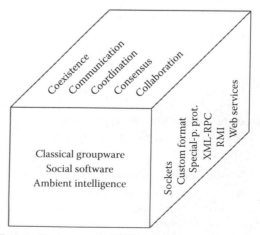

FIGURE 43.3 Design space of CSCW.

a group and its members. It should help to increase the orientation of individuals within a group and task space. (p. 325)

The general definition of awareness is "1: archaic: watchful, wary; 2: having or showing realization, perception, or knowledge" (Merriam-Webster, 2006). In the CSCW literature many definitions of awareness can be found. For instance, Dourish and Bellotti (1992) define awareness as "an understanding of the activities of others, which provides a context for your own activity" (p. 107). Gutwin and Greenberg (1995) emphasize the requirement for rapid updates and write that awareness information should be up-to-the-minute.

Four *types of awareness information* that are relevant in groupware can be distinguished: informal awareness (i.e., information on the users' presence and availability); social awareness (i.e., information on the users' attention and interest); group-structural information (i.e., information on the users' positions and roles in the group); and workspace awareness (i.e., information on the shared workspace and the artifacts contained).

Many *factors* influence the type of information, the frequency of updates, and the granularity of the information the users need; among them are: the synchronicity of the group process (i.e., do the users cooperate at the same time or sequentially); the degree of engagement (i.e., do the users cooperate intensely or is the cooperation only a background process); the similarity of tasks and views (i.e., do the users work on the same parts of a project, and do they look at the same parts); as well as privacy and disruption preferences (i.e., how much information are users willing to share, and how much information are users willing to attend to) (Gaver et al., 1992).

There are many examples of systems providing awareness support. Asynchronous awareness support systems provide asynchronous information about past user actions:

- The *session capture and replay* system exemplifies the general concept: It captures information on users' interaction with the system, and stores this information into session objects. Users can then annotate, modify, and exchange session objects, as well as replay past interaction by running session objects. The principle here is what-you-see-now-is-what-I-saw-then (WYSNIWIST) (Manohar and Prakash, 1995).
- Tracking changes in documents of Microsoft Word or related text-processing applications also allows viewing and going through past action (in this case manipulations) in the respective document.

Most synchronous awareness support systems combine functionality for capturing and presenting awareness information in real-time communication:

- *Media spaces* aim to provide users with permanent information and open communication channels between two locations. They primarily focus on rooms, rather than individual communication partners. For instance, the media space at Xerox connected offices in Palo Alto and in Portland (Bly et al., 1993). It allowed users to have chance encounters and serendipitous conversations with colleagues at the other site.
- *Traditional instant text-messaging* systems provide awareness and still protect privacy through online states, where users can explicitly indicate their state and availability (e.g., available, away, invisible') on a global level. Single-protocol clients such as ICQ, AOL Instant Messenger, Yahoo! Messenger, and Skype, as well as multiprotocol clients such as Gaim, Trillian, and Adium, are in general available for free and are widespread. Enterprise instant-messaging platforms and message-oriented middleware infrastructures (e.g., IBM Lotus Instant Messaging and Web Conferencing, Microsoft Office Live Communications, Novel GroupWise) provide technically more advanced concepts with respect to reliability, scalability, security, integration of user directories, and so on, but are often commercial.
- *Personality-based instant text-messaging* systems support selective information disclosure by allowing users to create multiple online personalities and selectively specify each personality's contacts, information sources to share, and online state to show. So, for instance, a user can be available for partners in project A with an upcoming deadline, but not for partners of projects B and C. The PRIMI platform, for instance, supports personalities through an implementation of Goffman's concept of faces and fronts (Goffman, 1959).

43.5.1.2 Communication Support

Communication support provides users with means of expressing themselves either by transmitting symbolic messages, or by propagating changes to other users. For any type of communication to work efficiently it is important that the communication partners have a common ground—that is, some common basis of language and understanding (Clark and Brennan, 1991).

There are many communication support systems. Basically, there are asynchronous communication support systems with a delay between the creation of a message and the creation of a reply, and synchronous communication support systems with instantaneous replies to messages. Depending on the format of the message, we can distinguish text-based messaging systems from audio- and video-conferencing systems.

- *E-mail* systems typically support the asynchronous exchange of text messages. Some systems support rich formats in e-mail messages; for instance, in the Andrew Message system (Borenstein and Thyberg, 1991), and the Slate system (Thomas et al., 1985) messages can include rich text, graphics, and spreadsheets. While at the beginning e-mail communication took place between one sender and one recipient, nowadays mailing lists allow group discussions.
- *Extended messaging* systems provide extended functionality to e-mail. Semistructured message systems such as the

Information Lens system (Malone et al., 1986) facilitate coordination via e-mail through semiformal messages, and interest-based filtering. The system parses the semiformal messages and reacts automatically (e.g., by forwarding relevant e-mails to interested users or by providing support for meeting arrangements such as automatic reservations for meeting rooms). Structured conversation systems such as the Cosmos system (Araujo et al., 1988) build on the concept of illocutionary acts and structure the message exchange according to typical conversation patterns (e.g., after a question, an answer is likely to follow). In Cosmos a specific language to define the structure is available. Similarly, in the COORDINATOR systems, the sender can define the message type of the expected answer (e.g., background information, commitment) (Winograd and Flores, 1987).

- *Instant text-messaging* systems are newer than their e-mail counterparts—ICQ was the first general instant messenger and has spread widely since 1996 (ICQ, 2007). They support the asynchronous exchange of text messages, but since the synchronous exchange is more prominently used, instant messaging will be described in the following.

Synchronous communication support allows users to have real-time conversations based on text, audio, and video contents:

- *Unix talk* systems allow users to have online real-time text conversations with other users on a Unix terminal; they send messages character by character from one terminal to another.
- *Internet relay chat (IRC)* systems support online real-time text conversations in multiple channels. Users can log into an IRC server and go to a virtual place called a *channel*. Channels have modes and can be public, which is the default mode, where anyone can join and online users can see each other and each other's channels; private, where online users can see others online without their channels; and secret, where no online users are shown. Messages are transmitted line by line. Users with special privileges are called *channel operators*; they can set the mode of a channel and kick out users who misbehave (Pioch et al., 1997).
- *Multiuser dungeons* (MUDs) are textual virtual realities, where the users log into a server and get a text description of the virtual reality and can navigate through the systems via text commands. MUDs allow multiple users to enter the same area and to have text-based real-time conversations. While the first MUDs focused on adventures and role-playing (e.g., AberMUD, LPMUD; Bartle, 1990), later MUDs support social encounters and online communication (e.g., TinyMUD, LamdadMOO; Curtis, 1992, 1996). Users with special privileges in MUDs are called *wizards*.
- *Instant text-messaging* systems do not only provide awareness support and support for asynchronous communication, but also for real-time text communication among online users through the line-oriented transmission of text messages.

- *Instant audio-messaging* systems support real-time speech. Examples are telephones, but also more recently voice-over-Internet-protocol (VoIP) systems such as Skype (Skype, 2007). Some operating systems have integrated audio-conferencing support (e.g., iChat in Mac OS X; Apple Computer, 2007). Audio conferencing in groups sometimes entails new challenges, such as a lack of mimics and gestures and lack of track of listeners and speakers (Nietzer, 1991).
- *Instant video-messaging* systems provide both real-time audio and video connection. Early systems extended the telephone with video transmission (e.g., PICTUREPHONE; Dorros et al., 1969). Later, many research prototypes supporting desktop video conferencing were developed (e.g., multimedia conferencing (MMConf) provides desktop video conferencing and application sharing; Crowley et al., 1990); Rapport is another early desktop video conferencing system (Ahuja et al., 1988). Examples of more recent desktop video-conferencing systems are iChat, iVisit (2007), and NetMeeting (Microsoft, 2006). The most advanced instant video-messaging systems feature multiple video cameras per participant capturing the person, the person's desk, and so forth (e.g., MERMAID; Watabe et al., 1990).
- *Media spaces* do not only provide mutual awareness of the inhabitants of remote rooms, but also permanent audio and video connections between them. So, users can have ad hoc conversations without the need to technically initiate the conversation. Meetings in media spaces having random participants and an unarranged agenda, and highly interactive and informal (Fish et al., 1990). Examples of media space systems are Polyscope (Borning and Travers, 1991), Vrooms (Borning and Travers, 1991), and Portholes (Dourish and Bly, 1992).

43.5.1.3 Coordination Support

Coordination systems arrange aims and goals, activities, actors, and interdependencies among different activities of geographically dispersed groups. For this purpose, they monitor and visualize the activities of all users. Some systems trigger users upon certain actions or results of other users. The model underlying the coordination system can be form-oriented (routing of documents through organizations), procedure-oriented (coordination of activities seen as procedures), conversation-oriented (based on the theory of articulation), or communication-structure-oriented (based on roles in conversations).

Workflow management systems are prominent examples of coordination systems (WfMC, 2007). They allow administrators to analyze cooperative processes and to define these processes with process definition tools. They have workflow enactment components that trigger processes. They also have process monitoring tools and process redefinition tools for rearranging processes in case of unforeseen contingencies. Other examples of coordination systems are resource planning systems (Egger and Wagner, 1993). Coordination support in these types of systems is addressed in many special conferences on the border of CSCW, and therefore it is not discussed in further detail in this section.

43.5.1.4 Consensus Support

Consensus support provides users with means to generate ideas, discuss ideas, and select ideas and make decisions.

Asynchronous group decision support systems (GDSS) facilitate the preparation and taking of group decisions. They help structuring, recording, and making group decisions more transparent both during the procedure of arriving at a decision, but also and very important after the decision has been taken. The latter can be important for users who join the group at a later stage, so they can view the records and see how the decision process evolved and why the decisions were made (Teufel et al., 1995).

Asynchronous GDSS are often based on the issue-based information system method (IBIS). The IBIS method structures argumentation processes into issues, positions, and arguments. Issues are the basic topics that are to be discussed. Positions are basic opinions that individual users or subgroups of users might have. Arguments are used to support one's own position or to weaken other positions. In IBIS issues, positions, and arguments are structured in a tree shape (Rittel and Kunz, 1970).

The gIBIS (graphical IBIS) system is a hypertext system based on the IBIS approach. It basically provides a graphical user interface to the issues, positions, and arguments stored in a relational database. Distributed users can access the gIBIS database via the network and create a semantic network of issues, positions, and arguments (Yakemovic and Conklin, 1990; Conklin, 1988).

Synchronous group decision support systems facilitate the finding of compromise in various ways. Compared to noncomputerized brainstorming meetings, group decision support systems offer several advantages: users can enter their opinions and contributions to the brainstorming in parallel, which can lead to a faster and broader input; the system automatically logs results and the process and provides vast support for creating the minutes; and so on. There are two basic types of group decision support systems: simple software applications, and more complex systems based on software and hardware:

- *rIBIS* (real-time issue-based information system) is a simple synchronous extension of the gIBIS system described previously. It is a hypertext-based graphical user interface that allows users to manage issues, positions, and arguments. All data are stored in a database. Users can synchronously edit and manage the IBIS-based argumentation tree. Like GROVE (see Section 43.5.1.5), rIBIS allows users to create their private views containing entries that are hidden from the other users (Rein and Ellis, 1991).

- *EMS* (electronic meeting rooms systems) are more complex and typically consist of software and hardware. They make the decision-making process more efficient by allowing multiple copresent users to prepare, carry out, and document decisions via networked computers. The computers

are typically positioned on a ring-shaped arrangement of a table with an open end to fit in a projection screen. The software of EMS typically supports meeting preparation, brainstorming, idea organization, voting, and so on.

- *GroupSystems* was among the first EMS (Applegate et al., 1986; Nunamaker et al., 1991). The CoLab system was another early, yet advanced, EMS. It was composed of six networked workstations and a huge central touch-sensitive screen with a stand-up keyboard (Stefik et al., 1987). The Dolphin systems use a Xerox Liveboard for the projection, which allows advanced pen-based input for editing hypermedia-based documents and informal structures, such as scribbles or sketches (Streitz et al., 1994; Mark et al., 1995).

43.5.1.5 Collaboration Support

Collaboration support provides users with means for effectively and efficiently working together. The organization of teamwork is more nuanced. Sharples (1993) distinguishes several forms of teamwork in collaborative writing tasks that provide some hints for the complexity of teamwork on a whole: in parallel work the team members divide the overall task in subtasks and perform them individually; in sequential work the team members also divide the overall task in subtasks, but the individual tasks are typically split chronologically and build on the results of the previous subtasks; and in reciprocal work the team members, the whole process, and its subtasks are strongly interwoven. Schmidt and Rodden (1996) point out that everyday work practice often encompasses "fluent transitions between formal and informal interaction" and "an inextricable interweaving of individual and cooperative work" (p. 157). They also write that cooperative systems should address the following issues: cooperative ensembles can become quite large; cooperative ensembles can be transient; membership can be unstable; patterns of interaction can change over time; and the processes can be distributed in time, space, and control. While all these distinctions are important, this section will—for analytical and pedagogical reasons—simply distinguish asynchronous from synchronous teamwork.

This section presents systems and prototypes supporting cooperation in teams (and some systems that support both cooperation and coordination in teams such as group decision support systems). A distinction is made between asynchronous support for users cooperating at different times and synchronous support for users cooperating in real time. Furthermore, different application scenarios such as group editing, group decision support, and shared workspaces are distinguished.

Asynchronous team support systems provide functionality for a group of users who cooperate at different times.

Asynchronous group editors in general allow users to produce and edit documents together and to observe each other's changes on the documents. Group editors either support asynchronous or synchronous cooperation. In asynchronous group editors, the team members cooperate on shared documents at different times; in synchronous group editors, the team members cooperate at the same time. In asynchronous cooperation, group members work independently of other users on their individual task. Often annotations for reviews or comments can be added after the draft is completed.

As Sharples (1993, p. 13) points out, no clear definition of the term *coauthoring* or *collaborative writing* is possible:

- An episode of collaborative writing may range from a few minutes taken to plan a joint memo, to many years for writing a coauthored book. Nor is there a clear distinction between writing and nonwriting; ideas gained from browsing through a library or talking to colleagues over lunch may be incorporated in the document. (p. 13)

Group editors typically have a functionality that is similar to single user editors, such as functions for creating new documents, opening existing documents, editing documents, and saving documents. Additionally, they provide specific group support, such as collaboration awareness (information on other users' actions that provide a context for editing activities), fault tolerance and good response time (fast and reliable run-time behavior, independent of distribution of users and data), concurrency control (consistency, even if two or more users are editing the same parts of a shared document), multiuser undo and redo (user actions should be reversible and repeatable), usable as a single-user editor (usage of editor on a single machine and without other users), and rich document structure (document templates for various types of documents, including publications and mind maps) (Prakash, 1999).

Quilt is a typical representative of asynchronous group editors. It supports asynchronous document production in a group of users using hypertext technology. Users can create base articles, and annotate them by means of linked subdocuments. Quilt enforces social roles such as coauthor, commenter, editor, or reader. Various document styles modulate these roles. For instance, in the document style exclusive, only the creator and author of a section is allowed to modify it (Fish et al., 1988; Leland et al., 1988).

Shared workspaces allow groups of users to share artifacts. They typically provide functionality for adding, changing, and deleting artifacts, as well as notifications about other users' activities in the shared workspace. The BSCW (basic support for cooperative work) system, the SharePoint Team Services, the Groove system, and the Lotus Notes systems are examples of shared workspace systems:

- The *BSCW* system (Bentley et al., 1997) allows users to create and maintain shared workspaces and documents of various types. It is based on standard web technology and can therefore be used with standard web browsers without any extensions. It has particular strengths concerning the easy and straightforward management of access rights among users, and elegant and informative awareness mechanisms such as a daily activity report informing users about activities in the workspaces in which they participate. The BSCW system has application programming

interfaces featuring WebDAV (Whitehead, 2006) and XML-RPC (UserLand Software, 2005).

- The *SharePoint Team Services* system (Whitehead, 2006) from Microsoft provides interactive web pages containing information on upcoming meetings, to-do lists, documents, and bulletin boards. Its particular strengths are customizable views by filtering, and subscriptions to notifications (Microsoft, 2007). The Groove system is very similar (Microsoft, 2005).
- The *Lotus Notes* system (Monson et al., 2006) is a product of the Lotus Development Corporation, which is part of the IBM Corporation. It provides a shared workspace based on documents and databases, and it allows customization through views and forms. A particular strength is its efficient replication of databases—that is, users can replicate the database on their own computer, then work offline, and easily synchronize the databases later on.

Synchronous team support supports cooperation among users who work together in real-time—either at the same place or at different places. Synchronous group editors allow users to produce and edit documents together and to observe each other's changes on the documents. In addition to many features from single user editors, they provide special support for collaboration awareness, fault tolerance and good response time, concurrency control, multiuser undo and redo, and so on. Some examples of group editors are as follows:

- *ShrEdit* (McGuffin and Olson, 1992; Neuwirth, 1995) is an example of a group editor with pessimistic concurrency control. ShrEdit aims to support the cooperative editing of users in face-to-face real-time design meetings. It provides concurrency control via locks of insertion points—that is, each user can lock the area of the insert and other users cannot change the locked area. However, ShrEdit does not have roles. ShrEdit provides immediate updates of the changes of all users on every user's monitor. Users can use find and track functions to request information on others' positions and activities. The design of ShrEdit deliberately leaves out communication tools and features for asynchronous editing, such as versioning of documents (McGuffin and Olson, 1992; Neuwirth, 1995).
- *GROVE* (group outline and view editor) is a representative of group editors with optimistic concurrency control. Departing from an optimistic perspective the design imposes hardly any system constraints: it does not support roles, and per default all users can read and edit all parts of the document synchronously and concurrently. Users can change this and can add locking if needed. GROVE provides elegant functionality for managing private and shared views of document parts. In a private view users can hide areas of the document from other users. Users can then change, update, and polish these areas and publish them only when they think that the respective part is adequate for the public. The public parts are updated

immediately on any change. In the shared documents private areas are shown as cloudbursts that disappear and are exchanged for the new contents only when the user discloses the respective private view. The GROVE designers did not include explicit communication support—they point out that any of the broad range of existing tools could be used (e.g., audio conferencing, telephones) (Ellis et al., 1991).

- *SubEthaEdit* is an outstanding group text editor; it received several awards (e.g., the 2003 Apple Design Award for Best Mac OS X Student Project) (TheCodingMonkeys, 2007). Using SubEthaEdit is straightforward: users simply launch the application, create a new text document or open an existing one, and start editing it. By pressing the announce button, users can publish their document and others can then easily join the editing process.

Teamrooms provide special support for cooperation: since they are based on persistent virtual rooms that provide asynchronous and synchronous support for cooperative work:

- *DIVA* is a virtual office environment based on virtual rooms, which can contain people, documents, and desks. People are represented as small avatars with a text label indicating the name of the respective user; they can freely move in the environment. Their current position as well as movements are shown by corresponding changes to the avatars. Documents are visualized by icons that are similar to document icons in traditional file systems. The copresent users for cooperation can use desks. The rooms support the interaction among copresent users and between users and documents. Rooms are equipped with audio and video tools, and audio and video connections are automatically established among the copresent users. Rooms can have purposes: either general purpose, or special purpose such as private offices or meeting rooms (Sohlenkamp and Chwelos, 1994).
- *TeamWave* (formerly known as TeamRooms) environment is a very similar system (Roseman, 1996).
- *wOrlds* environment goes beyond other team room systems as far as the overall concept is concerned (it is based on a sophisticated locales theory for the structuring in of the environment into rooms; Fitzpatrick, 1998), and as far as the underlying event notification is concerned (all actions in the environment trigger events, to which users can subscribe; Fitzpatrick et al., 2002). The wOrlds environment is based on the locales framework. Social worlds are seen as the structure and dynamics of interaction in groups, and can have various types of memberships and durations or existence. Locales are the sites and means— that is the "collections of conditions which both enabled and constrained action possibilities" (Fitzpatrick, 1998, p. 45). The locales framework is based on the work of Anselm Strauss (1993). Although it was not described by Strauss and cannot directly be derived from his work, his work was still a source of inspiration for the framework.

43.5.2 Social Software

Social software refers to technology that typically offers flexible support for multiple activities and types of social interaction for diverse social settings, ranging from small groups to large communities. Basically, in social software the same types of social interaction as in classical groupware systems can be supported— that is, coexistence, communication, coordination, consensus, and collaboration. Many of the previously described classical groupware systems can be used for communities and are not discussed again; therefore, the next subsections only focus on some special social software systems that are mainly used for communities and not as groupware.

43.5.2.1 Awareness Support

Awareness support in social software can be found in many early extensions of the World Wide Web. Examples are the sociable web, the awareness protocol, and CSCW3.

Sociable web consists of a modified web browser and a modified web server. The web browser—an extended Mosaic browser for XWindow—has similar functionality as standard web browsers plus an extra window for communication. The extra window contains a text entry field for typing messages, a listbox for displaying received messages, and a listbox for displaying icons and names of other users. Web servers are used as chat channels; users who are on the same server are displayed with icons and names in the respective listbox (Donath and Robertson, 1995).

Awareness protocol takes an approach that is different from the systems described so far; it is an open awareness protocol for the web that provides users with information about the presence of other users on a web server. The authors argue that they focus on an open, simple, and extensible protocol and introduce an auxiliary protocol. The protocol basically captures the presence of users on a web page. When a user accesses a page, the awareness client signs onto the respective page and queries the awareness server for information about other users who are currently accessing this page. When a user leaves a page, the awareness client signs off (Palfreyman and Rodden, 1996).

CSCW3 is a cooperative web browser that is based on a room metaphor where every web page is treated as a room and all present users have mutual awareness (Gross, 1997).

43.5.2.2 Communication Support

Communication support in social software is mainly based on concepts and systems from classical groupware. For communities, e-mail systems, including mailing lists, are used, as well as internet relay chat (IRC), MUDs, and instant messaging.

WebLogs are web pages that users create to communicate time-based information in text, pictures, or videos to other users (Kumar et al., 2004). WebLogs are typically presented to readers in reverse chronological order—so, readers when navigating to a WebLog can immediately see the most recent entries on top of the page. Users have diverse reasons for creating and maintaining WebLogs (Nardi et al., 2004): they do it to document their life and communicate information they consider important to family, friends, and colleagues; to comment on current developments in technology, politics, or economy; to note their thoughts and feelings as relief and catharsis; to write in front of an imaginary audience; and to contribute to a community forum.

Flash forums are systems to rapidly disseminate topic-centered information, mainly about technology news (Dave et al., 2004). Slashdot.org is a very prominent example with a huge readership (Open Source Technology Group, 2007).

43.5.2.3 Coordination Support

Coordination support in social software is also mainly based on concepts and systems from classical groupware such as resource planning systems.

Social navigation systems analyze the users' navigation behavior and present it to other users to align the online movement of users. They depart from Bush's idea of trails through the data space. The Footprints system is a prominent example (Wexelblat and Maes, 1999).

43.5.2.4 Consensus Support

Consensus support in social software and especially the IBIS approach are for communities similar to classical groupware.

Recommender systems aim to support consensus among users with similar opinions and taste. They act as social filters and analyze the behavior of users to generate user profiles, and then compare the profiles with the aim of giving users good recommendations. There are some variations of recommender systems: they can provide annotations in context (e.g., GroupLens; Konstan et al., 1997); they can help find good items (e.g., Herlocker et al., 2004; movielens, 2007).

43.5.2.5 Collaboration Support

Collaboration support in social software supports tight cooperation in communities. In social software, collaboration support often means support for sharing and exchanging information, and for creating and extending a community memory.

Wikis are a prominent example of systems supporting collaboration in communities; they allow users to easily set up and maintain web pages (wiki.org, 2002; Schwall, 2003; Cunningham, 2007). In wikis, web pages can easily be created and edited in the web browser. They are based on a simple text syntax for contents and cross-links. They are open and allow many readers to also be editors; they are incremental and build on existing contents; and they are observable since everybody can see the history of changes to all pages. The Wikipedia is a very prominent example of a wiki featuring many thousands of articles in many languages and providing them for free (Wikipedia, 2007).

43.5.3 Ambient Intelligence

Ambient intelligence (AmI) aims to improve users' work and private life by analyzing and adapting to the current situation with a special focus on the presence and activities of users. AmI is often based on ubiquitous computing technology and goes beyond the traditional WIMP paradigm of today's computers (see also Chapter 60, "Ambient Intelligence"). In particular, it uses a broad range of opportunities for presenting information and functionality in the users' physical environment, and at the same time offers users a broad range of implicit input where the system automatically captures the presence and actions of users in the room, allowing users to naturally interact with the environment through their presence, movements, body expressions, and language.

43.5.3.1 Awareness Support

Awareness support in AmI is straightforward: it takes the users' physical environment to display digital information in periphery of the users' attention, thereby informing users without disruption.

ambientROOM consists of a small room with special equipment to present information. On the ceiling "water ripples" show the speed of activities at a remote site—the smaller and faster the ripples, the more action at the other site. A spotlight is projected onto the wall to show the results of electronic field sensors in a work area as active wallpaper. The ambient sound is a soundscape that maps activity on a digital whiteboard. And the audible soundtracks use the sounds of birds and rainfall with changing sound volume and density to show the number of unread e-mails or the level of a stock portfolio (Wisneski et al., 1998).

AROMA is a system for abstract representation of presence supporting mutual awareness and provides a generic architecture for capturing, abstracting, synthesizing, and displaying presence awareness. It provides multisensory information presentation such as an audio soundscape, an electromechanical merry-go-round, an electromechanical vibrator in a chair, thermo-electric devices to control the temperature of the hand-rest, and so forth (Pedersen and Sokoler, 1997).

43.5.3.2 Communication Support

Communication support in AmI mainly presents the contents of a conversational message in the ambience of a room.

Information percolator is an ambient display medium and can present information sent by colleagues by means of 32 transparent tubes filled with water (covering an area of 1.4 m by 1.2 m) that release small bursts of air bubbles (Heiner et al., 1999). The sound of the bubbles in the tubes was deliberately kept—as the volume of the sound produced by the device roughly corresponds to the number of active tubes, the users can acoustically capture the activity level without constantly watching the information percolator.

Audio aura is a system providing serendipitous information as auditory cues in the background of the users' attention. The cues are tied to the users' actions in the physical workspace. Audio aura is based on sonic ecologies—that is, sounds topics (e.g., sound related to a beach such as seagull cries, waves, and wind) that represent a set of related events (e.g., information related to the state of personal e-mail). The system was frequently used to provide sonic ecologies related to the e-mail a user receives (Mynatt et al., 1998).

43.5.3.3 Coordination Support

Coordination support for AmI is often based on large displays in rooms that provide workers with vital information on the group process and allow them to better coordinate with each other; so, they are closely related to awareness support. For instance, large displays have been developed, used, and tested in hospitals (Bardram et al., 2006).

43.5.3.4 Consensus Support

Consensus support for AmI is not widespread; some examples are social TV applications where distributed users are supported in finding a TV program they want to watch together (Coppens et al., 2005; Oehlberg et al., 2006).

43.5.3.5 Collaboration Support

Collaboration support for AmI provides support for close cooperation beyond keyboard and mouse, GUI and WIMP.

Theater of work enabling relationships (TOWER) provides awareness information and collaboration support based on an awareness information environment in which information is captured in the electronic world and in the physical world, and is presented in a 3D multiuser virtual environment and in ambient interfaces (Prinz, 1999; Gross, 2000). A 3D multiuser virtual environment that is projected onto the wall of the respective users' office provides the members of a workgroup with a detailed representation of the shared artifacts and the actions of the group members and uses strengths of the PC such as fast processors and big monitors. The ambient interfaces complement it by presenting the awareness information in the physical environment of the group members.

Cooperative media space (CMS) virtually connects the two rooms of the authors' lab, which are in two different buildings with a 5 minute walk. It is based on Sens-ation (Gross et al., 2006, 2007). The concept should allow for seamless awareness, communication, and cooperative group interaction among the users of each room and also between the users of the two rooms. The CMS offers:

- *Pervasive presence and communication* (i.e., to facilitate chance encounters and ad hoc conversations among users in either room, a concept for identifying users when they enter the room was developed; identified users are automatically registered).
- *Smart roomstates* (i.e., adequate information on the current use of the room, in front of the door; derived from inference of sensor data such as running applications and open documents as well as noise and movement).
- *Seamless group interaction* (i.e., facilitate cooperation across all computers and displays of either room with a concept to move the cursor of their current display onto any other display and to share this other display with other users).

The CMS can be used for work-related settings, as well as for social settings such as group TV watching (Gross, 2006; Gross et al., 2008).

43.5.4 Trends in CSCW

There are several trends combining classical groupware systems, social software, and AmI. For instance, FrameDrops supports users in creating and maintaining shared web-based archives of video, picture, and text, as well as in later browsing through the archives (Gross and Kleppe, 2005). Users can capture videos and pictures on the go, annotate them, and send them to a shared FrameDrops server. The server processes the incoming data and automatically produces a new entry in the shared FrameDrops weblog. All entries can later be retrieved and viewed with a standard web browser. Subsequently, the approach of FrameDrops is presented.

FrameDrops allows users to create and insert entries in a simple manner and without knowledge of the underlying technology and transfer mechanisms. Being a mobile weblog, it can be used anyplace and anytime. Additionally, FrameDrops' web interface facilitates later changes—this is particularly convenient for entering or altering large amounts of text.

FrameDrops can automatically position entries via GPS, and put entries in a geographical as well as chronological context. This coupling of the users' coordinates in the real world fosters a stronger connection between the online world and the real world, and between online and real communities. For instance, FrameDrops stimulates real-world encounters of users who have so far only met online by providing real-time information on their current positions. This is a novel way of bridging the gap between communities in the real and in the virtual world, which has been emphasized by several authors (e.g., Rheingold, 1993).

Mobile phones of the third generation feature all means that are needed to create and publish moving images. FrameDrops uses the following: an integrated camera can be used to capture videos and pictures; an e-mail application can be used to exchange the captured data; a web browser can be used to retrieve entries later; and an integrated GPS receiver can be used to detect the user's current position. So far, the production of media was possible, but required a lot of effort: the users manually had to capture pictures and video, transfer them to a computer, edit the data on the computer, and upload the data to a server.

FrameDrops provides an easy and intuitive web interface to the information base and its individual entries. The interface is based on a Flash client that seamlessly integrates the individual entries into overview pages showing comments and static picture frames of the videos. The Flash client is lightweight and provides simple interaction for viewing the video entries.

More than 20 years ago CSCW emerged as a research area on its own—entailing a clear shift from single-user orientation toward group orientation. Grudin (1990) described this phenomenon already in 1990 when he argued that the focus of research moves outward from (1) computer hardware, to (2) software, to (3) perceptual aspects of input and display devices, to (4) cognitive aspects of gradually higher level, and finally to (5) the group-level interactions. More recently, there have been some more movements such as social software and AmI.

43.6 CSCW and Universal Access: Toward a Convergence

The basic concepts of the two domains of CSCW and universal access already have many similarities and a convergence. Extrapolating this convergence leads to interesting challenges of cooperative universal access.

As mentioned in the introduction, CSCW is a research field of computer science and related areas that have the primary focus on understanding and supporting groups of users, their social interaction, and their interaction with technology; and groupware to be the technological concepts, systems, and prototypes supporting social interaction among groups of users. At the same time, CSCW and groupware are also a conceptual shift with a view of technology as enablers for social interaction. Universal access, on the other hand, has a special focus on diverse human characteristics and individual strengths and weaknesses and aims to extend the strengths and compensate for the weaknesses.

CSCW and universal access have several existing similarities and convergences. From a CSCW perspective, many aspects of universal access can be identified. The following are some examples referring to the previous sections:

- The early visions on technical support for social interaction already had the motivation to extend the strengths and compensate for the weaknesses of users: for instance, Memex explicitly aims to extend human memory and NLS explicitly aims to augment human intellect.
- Classical groupware helps users in groups: for instance, awareness support for coexistence aims to provide users with information and reminders about the activities of the other group members and the overall state of the group endeavor.
- Social software is based on many concepts, systems, and tools for accessing the web that are described in other parts of this book, and that can be combined with any type of social software.
- Am I analyzes and adapts the current situation for users, often by means of ubiquitous computing technology. Multimodal adaptation of the environment and presentation of information and functionality in the physical environment is basically possible in any type of AmI environment.

The convergence of CSCW and universal access can entail a range of novel interesting challenges with respect to universal access for groups and communities. This chapter mentioned just a few; others are also addressed in other parts of the book. Exploring different modalities is an important prerequisite (e.g., tactile and haptic as well as olfactory can be relevant besides the dominantly supported audible and visual). Providing fast and efficient cross-modal translations (e.g., tactile and haptic for some group members, visual for others) is vital for bringing together users with heterogeneous cognitive, perceptual, and physical abilities.

Acknowledgments

We would like to thank the members of the Cooperative Media Lab—especially Tareg Egla, Christoph Oemig, and Thilo Paul-Stueve.

References

Ahuja, S. R., Ensor, J. R., and Horn, D. N. (1988). The Rapport multimedia conferencing system, in the *Proceedings of the Conference on Office Information Systems*, March 1988, Palo Alto, CA, pp. 1–9. New York: ACM Press.

Apache Software Foundation (2006). *Apache Axis 1.4.* The Apache Software Foundation. http://ws.apache.org/axis.

Apple Computer, I. (2007). *Apple—Mac OS X—iChat AV.* http://www.apple.com/macosx/features/ichat.

Applegate, L. M., Konsynski, B. R., and Nunamaker, J. F. (1986). A group decision support system for idea generation and issue analysis in organisation planning, in the *Proceedings of the Conference on Computer-Supported Cooperative Work (CSCW'86)*, 3–5 December 1986, Austin, TX, pp. 16–34. New York: ACM Press.

Araujo, R. B., Coulouris, G. F., Onions, J. P., and Smith, H. T. (1988). The architecture of the prototype COSMOS messaging system, in the *Proceedings of Research into Networks and Distributed Applications (EUTECO'88)*, Vienna, Austria, pp. 157–170. Amsterdam, The Netherlands: Elsevier.

Baeza-Yates, R. and Ribeiro-Neto, B. (1999). *Modern Information Retrieval.* Reading, MA: Addison-Wesley.

Bannon, L. J. and Schmidt, K. (1989). CSCW: Four characters in search of a context, in the *Proceedings of the First European Conference on Computer-Supported Cooperative Work (ECSCW'89)*, 13–15 September 1989, London, pp. 358–372. Dordrecht, The Netherlands: Elsevier.

Bardram, J. E., Hansen, T. R., and Soegaard, M. (2006). Large interactive displays in hospitals: Motivation, examples, and challenges, paper presented at the *Workshop on Information Visualisation and Interaction Techniques for Collaboration across Multiple Displays at the Conference on Human Factors in Computing Systems (CHI 2006)* (L. Terrenghi, R. May, and P. Baudisch, eds.), 22–27 April 2006, Quebec, Canada.

Bartle, R. (1990). *Interactive Multi-User Computer Games.* MUSE Ltd. Research Report. ftp://ftp.lambda.moo.mud.org/pub/MOO/papers/mudreport.txt.

Bentley, R., Appelt, W., Busbach, U., Hinrichs, E., Kerr, D., Sikkel, K., et al. (1997). Basic support for cooperative work on the World-Wide Web. *International Journal of Human Computer Studies* 46: 827–846.

Birrell, A. D. and Nelson, B. J. (1984). Implementing remote procedure calls. *ACM Transactions on Computer Systems (TOCS)* 2: 39–59.

Bly, S. A., Harrison, S. R., and Irvin, S. (1993). Media spaces: Bringing people together in a video, audio, and computing environment. *Communications of the ACM* 36: 28–47.

Borenstein, N. S. and Thyberg, C. A. (1991). Power, ease of use, and cooperative work in a practical multimedia message system. *International Journal of Man-Machine Studies* 34: 229–259.

Borning, A. and Travers, M. (1991). Two approaches to casual interaction over computer and video networks, in the *Proceedings of the Conference on Human Factors in Computing Systems (CHI'91)*, 27 April–2 May 1991, New Orleans, pp. 13–20. New York: ACM Press.

Bowers, J. and Benford, S.D. (eds.) (1991). *Studies in Computer-Supported Cooperative Work: Theory, Practice, and Design.* Human Factors in Information Technology Series, Vol. 8. Amsterdam, The Netherlands: North Holland.

Bush, V. (1945). As we may think. *The Atlantic Monthly* 176: 101–108.

Clark, H. H. and Brennan, S. E. (1991). Grounding in communication, in *Perspectives on Socially Shared Cognition* (L. B. Resnick, J. M. Levine, and S. D. Teasley, eds.), pp. 127–149. Washington, DC: American Psychological Association.

Conklin, B. (1988). gIBIS: A hypertext tool for exploratory policy discussion. *ACM Transactions on Office Information Systems* 6: 303–331.

Coppens, T., Vanparijs, F., and Handelkyn, K. (2005, February 4). *AmigoTV: A Social TV Experience through Triple-Play Convergence.* Compagnie Financiere Alcatel, Paris.

Coulouris, G., Dollimore, J., and Kindberg, T. (2005). *Distributed Systems: Concepts and Design* (4th ed.). Boston, MA: Addison-Wesley/Pearson Education.

Crowley, T., Milazzo, P., Baker, E., Forsdick, H., and Tomlinson, R. (1990). MMConf: An infrastructure for building shared multimedia applications, in the *Proceedings of the Conference on Computer-Supported Cooperative Work (CSCW'90)*, 7–10 October 1990, Los Angeles, pp. 329–342. New York: ACM Press.

Cunningham, W. (2007). *Wiki Design Principles.* http://c2.com/cgi/wiki?WikiDesignPrinciples.

Curtis, P. (1992). Mudding: Social phenomena in text-based virtual realities, in the *Proceedings of the Conference on the Directions and Implications of Advanced Computing.* Palo Alto, CA: Computer Professionals for Social Responsibility.

Curtis, P. (1996). *LambdaMOO Programmer's Manual: For LambdaMOO Version 1.8.0p5.* Xerox PARC, Palo Alto, CA.

Dave, K., Wattenberg, M., and Muller, M. J. (2004). Flash forums and ForumReader: Navigating a new kind of large-scale online discussion, in the *Proceedings of the ACM 2004 Conference on Computer-Supported Cooperative Work (CSCW 2004)*, 6–10 November 1983, Chicago, pp. 232–241. New York: ACM Press.

Day, J. D. and Zimmermann, H. (1983). The OSI reference model, in the *Proceedings of the IEEE* 71: 1334–1340.

Donath, J. S. and Robertson, N. (1995). The sociable web. Paper presented at *the Workshop on WWW and Collaboration at the Fourth International WWW Conference (WWW'95)*, 11–14 December 1995, Boston.

Dorros, I., Davis, C. G., Harris, J. R., Williams, R. D., Korn, F. A., Ritchie, et al. (1969). PICTUREPHONE. *Bell Laboratories Record* 47: 131–193.

Dourish, P. and Bellotti, V. (1992). Awareness and coordination in shared workspaces, in the *Proceedings of the Conference on Computer-Supported Cooperative Work (CSCW'92)*, 31 October–4 November 1992, Toronto, Canada, pp. 107–114. New York: ACM Press.

Dourish, P. and Bly, S. (1992). Portholes: Supporting awareness in a distributed work group, in the *Proceedings of the Conference on Human Factors in Computing Systems (CHI'92)*, 3–7 May 1992, Monterey, CA, pp. 541–547. New York: ACM Press.

Egger, E. and Wagner, I. (1993). Negotiating temporal orders: The case of collaborative time management in a surgery clinic. *Computer Supported Cooperative Work (CSCW)* 1: 255–277.

Ellis, C. A., Gibbs, S. J., and Rein, G. L. (1991). Groupware: Some issues and experiences. *Communications of the ACM* 34: 38–58.

Engelbart, D. and English, W. K. (1968). A research centre for augmenting human intellect, in the *Proceedings of the Fall Joint Computing Conference (FJCC'68)*, Montvale, NY, pp. 395–410. Washington, DC: AFIPS Press.

Farley, J. (1998). *Java Distributed Computing.* Sebastopol, CA: O'Reilly & Associates.

Festinger, L. (1950). Informal social communication. *Psychological Review* 57: 271–282.

Fish, R. S., Kraut, R. E., and Chalfonte, B. L. (1990). The VideoWindow system in informal communications, in the *Proceedings of the Conference on Computer-Supported Cooperative Work (CSCW'90)*, 7–10 October 1990, Los Angeles, pp. 1–11. New York: ACM Press.

Fish, R. S., Kraut, R. E., Leland, M. D. P., and Cohen, M. (1988). Quilt: A collaborative tool for collaborative writing, in the *Proceedings of the Conference on Office Information Systems*, pp. 30–37. New York: ACM Press.

Fitzpatrick, G. (1998). *The Locales Framework: Understanding and Designing for Cooperative Work.* PhD dissertation, Department of Computer Science and Electrical Engineering, University of Queensland, Brisbane, Australia.

Fitzpatrick, G., Kaplan, S., Mansfield, T., Arnold, D., and Segall, B. (2002). Supporting public availability and accessibility with elvin: Experiences and reflections. *Computer Supported Cooperative Work: The Journal of Collaborative Computing* 11: 447–474.

Garrett, J. J. (2005). *Ajax: A New Approach to Web Applications.* Adaptive Path. http://www.adaptivepath.com/publications/essays/archives/000385.php.

Gaver, W. W., Moran, T., MacLean, A., Lövstrand, L., Dourish, P., Carter, K. A., and Buxton, W. (1992). Realising a Video Environment: EUROPARC's RAVE System, in the *Proceedings of the Conference on Human Factors in Computing Systems (CHI'92)*, 3–7 May 1992, Monterey, CA, pp. 27–35. New York: ACM Press.

Gerson, E. M. and Star, S. L. (1986). Analysing due process in the workplace. *ACM Transactions on Office Information Systems* 4: 257–270.

Goffman, E. (1959). *The Presentation of Self in Everyday Life*. New York: Doubleday Anchor Books.

Goodman, P. S., Ravlin, E., and Schminke, M. (1987). Understanding groups in organisations. *Research in Organisational Behaviour* 9: 121–173.

Greenfield, A. (2006). *The Dawning Age of Ubiquitous Computing*. Indianapolis: New Riders Publishing.

Greif, I. (ed.) (1988). *Computer-Supported Cooperative Work: A Book of Readings*. Computer and People Series. San Mateo, CA: Morgan Kaufmann Publishers.

Gross, T. (1997). The CSCW3 prototype—supporting collaboration in global information systems, in *Conference Supplement of the Fifth European Conference on Computer-Supported Cooperative Work (ECSCW'97)*, 7–11 September 1997, Lancaster, U.K., pp. 43–44.

Gross, T. (2000). Towards ubiquitous cooperation: From CSCW to cooperative web computing to ubiquitous computing, in the *Proceedings of the Eighth International Information Management Talks (IDIMT 2000)*, 20–22 September 2000, Zadov, Czech Republic, pp. 145–162. Linz, Austria: Universitaetsverlag Rudolf Trauner.

Gross, T. (2006). Towards cooperative media spaces, in the *Proceedings of the Fourteenth International Information Management Talks (IDIMT 2006)*, 13–15 September 2006, Budweis, Czech Republic, pp. 141–150. Linz: Universitaetsverlag Rudolf Trauner.

Gross, T., Egla, T., and Marquardt, N. (2006). Sens-ation: A service-oriented platform for developing sensor-based infrastructures. *International Journal of Internet Protocol Technology (IJIPT)* 1:159–167.

Gross, T., Fetter, M., and Paul-Stueve, T. (2008). Toward advanced social TV in a cooperative media space. *International Journal of Human-Computer Interaction* 24: 155–173.

Gross, T. and Kleppe, M. (2005). FrameDrops: A mobile videoblog for workgroups and virtual communities, in the *Proceedings of the International ACM SIGGROUP Conference on Supporting Group Work (Group 2005)*, 6–9 November 2005, Sanibel Island, FL, pp. 128–131. New York: ACM Press.

Gross, T., Paul-Stueve, T., and Palakarska, T. (2007). SensBution: A rule-based peer-to-peer approach for sensor-based infrastructures, in the *Proceedings of the 33rd EUROMICRO Conference on Software Engineering and Advanced Applications (SEAA 2007)*, 28–31 August 2007, Lubeck, Germany, pp. 333–340. Los Alamitos, CA: IEEE.

Gross, T., Stary, C., and Totter, A. (2005). User-centered awareness in computer-supported cooperative work-systems: Structured embedding of findings from social sciences. *International Journal of Human-Computer Interaction* 18: 323–360.

Grudin, J. (1990). The computer reaches out: The historical continuity of interface design, in the *Proceedings of the Conference on Human Factors in Computing Systems (CHI'90)*, 1–5 April 1990, Seattle, pp. 261–268. New York: ACM Press.

Gudgin, M., Hadley, M., Mendelsohn, N., Moreau, J., and Nielsen, H. F. (2007). *SOAP Version 1.2 Part 1: Messaging Framework* (2nd edition). http://www.w3.org/TR/soap12.

Gutwin, C. and Greenberg, S. (1995). Support for group awareness in real-time desktop conferences, in the *Proceedings of the Second New Zealand Computer Science Research Students' Conference*, 18–21 April 1995, Hamilton, New Zealand, pp 18–21. Hamilton, New Zealand: University of Waikato.

Heiner, J. M., Hudson, S. E., and Tanaka, K. (1999). The information percolator: Ambient information display in a decorative object, in the *Proceedings of the ACM Symposium on User Interface Software and Technology (UIST'99)*, 7–10 November 1999, Asheville, NC, pp. 141–148. New York: ACM Press.

Herlocker, J. L., Konstan, J. A., Terveen, L., and Riedl, J. (2004). Evaluating collaborative filtering recommender systems. *ACM Transactions on Information Systems* 22: 5–53.

Hupfer, S., Ross, S., and Patterson, J. (2004). Introducing collaboration into an application development Environment, in the *Proceedings of the 2004 ACM Conference on Computer Supported Cooperative Work*, 6–10 November 2004, Chicago, pp. 21–24. New York: ACM Press.

ICQ Inc. (2007). *ICQ.com: Community, People Search, and Messaging Service!* http://www.icq.com.

Ishii, H. and Ullmer, B. (1997). Tangible bits: Towards seamless interfaces between people, bits and atoms, in the *Proceedings of the 1997 SIGCHI Conference on Human Factors in Computing Systems*, 18–23 April 1997, Los Angeles, pp. 234–241. New York: ACM Press.

iVisit LLC (2007). *Best Video Conferencing and Videochat Software for PC and Mac*. http://www.ivisit.com.

Johnson-Lentz, P. and Johnson-Lentz, T. 1982. Groupware: The process and impacts of design choices, in *Computer-Mediated Communication Systems: Status and Evaluation* (E. B. Kerr and S. R. Hiltz, eds.), pp. 42–55. New York: Academic Press.

Konstan, J. A., Miller, B. N., Maltz, D., Herlocker, J. L., Gordon, L. R., and Riedl, J. (1997). GroupLens: Applying collaborative filtering to usenet news. *Communications of the ACM* 40: 77–87.

Kumar, R., Novak, J., Raghavan, P., and Tomkins, A. (2004). Structure and evolution of the blogspace. *Communications of the ACM* 47: 35–39.

Leland, M., Fish, R., and Kraut, R. (1988). Collaborative document production using quilt, in the *Proceedings of the Conference on Computer-Supported Cooperative Work (CSCW'88)* 26–28, Portland, OR, pp. 206–215. New York: ACM Press.

Licklider, J. C. R. (1960). Man-computer symbiosis, in *IRE Transactions on Human Factors in Electronics* HFE-1: pp. 4–11.

Licklider, J. C. R. (1968). The computer as a communicationdevice. *Science and Technology* April: 21–31.

Lynch, K. J., Snyder, J. M., Vogel, D. M., and McHenry, W. K. (1990). The Arizona analyst information system: Supporting collaborative research on international technological trends, in *Multi-User Interfaces and Applications* (S. Gibbs and A. A. Verrijn-Stuart, eds.), pp. 159–174. Amsterdam, The Netherlands: Elsevier.

Malone, T. and Crowston, K. (1992). *Towards an Interdisciplinary Theory of Coordination.* Report Number: CCS TR# 120, Centre for Coordination Science, Sloan School of Management, Massachusetts Institute of Technology, Cambridge, MA.

Malone, T. W., Grant, K. R., and Turbak, F. A. (1986). The information lens: An intelligent system for information sharing in organisations, in the *Proceedings of the Conference on Human Factors in Computing Systems (CHI'86)*, 13–17 April 1986, Boston, pp. 1–8. New York: ACM Press.

Manohar, N. and Prakash, A. (1995). The session capture and replay paradigm for asynchronous collaboration, in the *Proceedings of the Fourth European Conference on Computer-Supported Cooperative Work (ECSCW'95)*, 11–15 September 1995, Stockholm, Sweden, pp. 149–164. Dordrecht, The Netherlands: Kluwer Academic Publishers.

Marca, D. and Bock, G. (eds.) (1992). *Groupware: Software for Computer-Supported Cooperative Work.* Los Alamitos, CA: IEEE Computer Society Press.

Mark, G., Haake, J. M., and Streitz, N. (1995). The use of hypermedia in group problem solving: An evaluation of the DOLPHIN electronic meeting room environment, in the *Proceedings of the Fourth European Conference on Computer-Supported Cooperative Work (ECSCW'95)*, 11–15 September 1995, Stockholm, Sweden, pp. 197–214. Dortrecht, The Netherlands: Kluwer Academic Publishers.

McGuffin, L. and Olson, G. (1992). *ShrEdit: A Shared Electronic Workspace.* Report Number: Technical Report 45, Cognitive Science and Machine Intelligence Laboratory, University of Michigan, Ann Arbor, MI.

Merriam-Webster (2006). *Merriam-Webster Online.* http://www.m-w.com.

Microsoft (2005). *Groove Virtual Office: Virtual Office Software for Sharing Files, Projects, and Data.* Microsoft Corporation. http://groove.net.

Microsoft (2006). *NetMeeting Home.* Microsoft Corporation. http://www.microsoft.com/windows/netmeeting.

Microsoft (2007). *Microsoft SharePoint Products and Technologies.* Microsoft Corporation. http://www.microsoft.com/sharepoint/default.mspx.

Monson, P., Alexander, M., Hartwell, S., and Sosnowski, V. (2006). *IBM Lotus Workspace: Team Collaboration 2.0.1.* http://www.redbooks.ibm.com/abstracts/redp3929.html?Open.

movielens (2007). *movielens: Movie Recommendations.* http://movielens.umn.edu/login.

Mynatt, E. D., Adler, A., Ito, M., and O'Day, V. L. (1997). Design for network communities, in the *Proceedings of the Conference on Human Factors in Computing Systems (CHI'97)*, 22–27 March 1997, Atlanta, pp. 210–217. New York: ACM Press.

Mynatt, E. D., Back, M., and Want, R. (1998). Designing audio aura, in the *Proceedings of the Conference on Human Factors*

in Computing Systems (CHI'98), 18–23 April 1997, Los Angeles, pp. 566–573. New York: ACM Press.

Nardi, B. A., Schiano, D. J., Gumbrecht, M., and Swartz, L. (2004). Why we blog. *Communications of the ACM* 47: 41–46.

Neuwirth, C. (1995). Collaborative writing: Practical problems and prospective solutions. Presented as Tutorial #11 at the *Fourth European Conference on Computer-Supported Cooperative Work (ECSCW'95)*, 11–15 September 1995, Stockholm, Sweden.

Newcomb, T. M. (1943). *Personality and Social Change.* New York: Dryden Press.

Newman, M. E. J. (2003). The structure and function of complex networks. *SAM Review* 45: 167–256.

Nietzer, P. (1991). Telekonferenzsystem mit graphischer Dialogsteuerung [Telephone Conferencing System with Graphical Dialog Control] in *Computergestuetzte Gruppenarbeit (CSCW)* [Computer-Supported Cooperative Work (CSCW)] (J. Friedrich and K.-H. Roediger, eds.), pp. 197–206. Stuttgart, Germany: Teubner.

Nunamaker, J. F., Dennis, A. R., Valacich, J. S., Vogel, D. R., and George, J. R. (1991). Electronic meeting systems to support group work. *Communications of the ACM* 34: 40–61.

Oehlberg, L., Ducheneaut, N., Thomton, J. D., Moore, G. E., and Nickell, E. (2006). Social TV: Designing for distributed, sociable television viewing, in the *Proceedings of the European Conference on Interactive Television (EuroITV 2006)*, 25–26 May 2006, Athens, Greece, pp. 251–262. New York: ACM Press.

Open Source Technology Group (2007). *Slashdot: News for Nerds, Stuff That Matters.* http://slashdot.org.

O'Reilly, T. (2005). *O'Reilly Network: What Is Web 2.0?* http://www.oreillynet.com/lpt/a/6228.

Palfreyman, K. and Rodden, K. (1996). A protocol for user awareness on the World-Wide Web, in the *Proceedings of the ACM 1996 Conference on Computer-Supported Cooperative Work (CSCW'96)*, 16–20 November 1996, Boston, MA, pp. 130–139. New York: ACM Press.

Pedersen, E. R. and Sokoler, T. (1997). AROMA: Abstract representation of presence supporting mutual awareness, in the *Proceedings of the Conference on Human Factors in Computing Systems (CHI'97)*, 22–27 March 1997, Atlanta, pp. 51–58. New York: ACM Press.

Pioch, N., Rasmussen, O., Hoyle, M. A., and Lo, J. (1997). *A Short IRC Primer.* http://www.irchelp.org/irchelp/ircprimer.html.

Prakash, A. (1999). Group editors, in *Computer-Supported Cooperative Work* (M. Beaudouin-Lafon, ed.), pp. 103–133. Chichester, U.K.: John Wiley & Sons.

Prinz, W. (1999). NESSIE: An awareness environment for cooperative settings, in the *Proceedings of the Sixth European Conference on Computer-Supported Cooperative Work (ECSCW'99)*, 12–16 September 1999, Copenhagen, Denmark, pp. 391–410. Dortrecht, The Netherlands: Kluwer Academic Publishers.

Rein, G. L. and Ellis, C. A. (1991). rIBIS: A real-time group hypertext system. *International Journal of Man Machine Studies* 34: 349–368.

Rheingold, H. (1993). *The Virtual Community*. Reading, MA: Addison-Wesley.

Rittel, H. and Kunz, W. (1970). *Issues as Elements of Information Systems*. Report Number: #131, Insitut fuer Grundlagen der Planung, University of Stuttgart, Stuttgart, Germany.

Roseman, M. (1996). TeamRooms: Network places for collaboration, in the *Proceedings of the ACM 1996 Conference on Computer-Supported Cooperative Work (CSCW'96)* 16–20 November 1996, Boston, MA, pp. 325–333. New York: ACM Press.

Schmidt, K. and Rodden, T. (1996). Putting it all together: Requirements for a CSCW platform, in *The Design of Computer-Supported Cooperative Work and Groupware Systems* (D. Shapiro, M. Tauber, and R. Traunmueller, eds.), pp. 157–176. Amsterdam, The Netherlands: Elsevier.

Schwall, J. (2003). *The Wiki Phenomenon*. http://www.schwall.de/dl/20030828_the_wiki_way.pdf.

Sharples, M. (ed.) (1993). Computer-supported collaborative writing, in *Computer-Supported Cooperative Work* (D. Diaper and C. Sanger, eds.). Heidelberg: Springer-Verlag.

Sharples, M., Goodlet, J. S., Beck, E. E., Wood, C. C., Easterbrook, S. M., and Plowman, L. (1993). Research issues in the study of computer-supported collaborative writing, in *Computer-Supported Collaborative Writing* (M. Sharples, ed.), pp. 9–28. Heidelberg: Springer-Verlag.

Simone, C. and Schmidt, K. (1993). *The COMIC Project: Computational Mechanisms of Interaction for CSCW*. Report Number: Esprit Basic Research Project 6225—Deliverable D3.1, Department of Information Science, University of Milan and Cognitive Systems Group, Risoe National Laboratory, Milan, Italy.

Skype (2007). *Skype. Take a Deep Breath*. http://www.skype.com.

Sohlenkamp, M. and Chwelos, G. (1994). Integrating communication, cooperation, and awareness: The DIVA virtual office environment, in the *Proceedings of the Conference on Computer-Supported Cooperative Work (CSCW'94)*, 13–17 September 1994, Milan, Italy, pp. 331–343. New York: ACM Press.

Stefik, M., Foster, G., Bobrow, D., Kahn, K., Lanning, S., and Suchman, L. (1987). Beyond the chalkboard: Computer support for collaboration and problem solving in meetings. *Communications of the ACM* 30: 32–47.

Stephanidis, C. and Savidis, A. (2001). Universal access in the information society: Methods, tools, and interaction technologies. *International Journal on Universal Access in the Information Society* 1: 40–55.

Strauss, A. (1985). Work and the division of labour. *The Sociological Quarterly* 26: 1–19.

Strauss, A. (1993). *Continual Permutations of Action*. New York: Aldine de Gruyter.

Streitz, N., Geissler, J., Haake, J. M., and Hol, J. (1994). DOLPHIN: Integrated meeting support across local and remote desktop environments and LiveBoards, in the *Proceedings of the Conference on Computer-Supported Cooperative Work (CSCW'94)*, 22–26 October 1994, Chapel Hill, NC, pp. 345–358. New York: ACM Press.

Suchman, L. (1987). *Plans and Situated Actions: The Problem of Human-Machine Communication*. Cambridge, MA: Cambridge University Press.

Sundstrom, E., De Meuse, K. P., and Futrell, D. (1990). Work teams: Applications and effectiveness. *American Psychologist* 45: 120–133.

Tanenbaum, A. S. (2002). *Computer Networks* (4th ed.). Upper Saddle River, NJ: Prentice Hall.

Tanenbaum, A. S. and Van Steen, M. (2006). *Distributed Systems: Principles and Paradigms* (2nd ed.). Upper Saddle River, NJ: Prentice Hall.

Tepper, M. (2004). The rise of social software. *ACM netWorker: The Craft of Network Computing* 7: 18–23.

Teufel, S., Sauter, C., Muehlherr, T., and Bauknecht, K. (1995). *Computerunterstuetzung fuer die Gruppenarbeit (Computer Support for Groupwork)*. Bonn: Addison-Wesley.

The CodingMonkeys (2007). *SubEthaEdit*. http://www.codingmonkeys.de/subethaedit.

Thomas, R. H., Forsdick, H. C., Crowley, T. R., Schaaf, R. W., Tomlinson, R. S., Travers, V. M., and Robertson, G. G. (1985). Diamond: A multimedia message system built on a distributed architecture. *IEEE Computer* 18: 65–78.

UserLand Software (2005). *XML-RPC Home Page*. http://www.xmlrpc.com.

Wasserman, S. and Faust, K. (1995). *Social Network Analysis*. Cambridge, MA: Cambridge University Press.

Watabe, K., Sakata, S., Maeno, K., Fukuoka, H., and Ohmori, T. (1990). Distributed multiparty desktop conferencing system: MERMAID, in the *Proceedings of the Conference on Computer-Supported Cooperative Work (CSCW'90)*, 7–10 October 1990, Los Angeles, pp. 27–38. New York: ACM Press.

Weiser, M. (1991). The computer for the 21st century. *Scientific American* 265: 94–100.

Wexelblat, A. and Maes, P. (1999). Footprints: History-rich tools for information foraging, in the *Proceedings of the Conference on Human Factors in Computing Systems (CHI'99)*, pp. 270–277. New York: ACM Press.

WfMC (2007). *The Workflow Management Coalition: Standards*. Workflow Management Coalition Specification, Brussels, Belgium. http://www.wfmc.org/standards/standards.htm.

Whitehead, J. (2006). *WebDAV Resources*. Microsoft Corporation. http://www.webdav.org.

wiki.org (2002). *Wiki: What Is Wiki?* http://wiki.org/wiki.cgi?WhatIsWiki.

Wikipedia (2007). *Wikipedia, the Free Encyclopedia*. http://en.wikipedia.org.

Wilson, P. (1991). *Computer-Supported Cooperative Work*. Oxford, U.K.: Intellect Books.

Winer, D. (1999). *XML-RPC Specification*. UserLand Software. http://www.xmlrpc.com/spec.

Winograd, T. and Flores, F. (1987). *Understanding Computers and Cognition: A New Foundation of Design*. New York: Addison-Wesley.

Wisneski, C., Ishii, H., Dahley, A., Gorbet, M., Brave, S., Ullmer, B., and Yarin, P. (1998). Ambient displays: Turning architectural space into an interface between people and digital information, in the *Proceedings of the First International Workshop on Cooperative Buildings: Integrating Information, Organisation, and Architecture (CoBuild'98)*, 25–26 February 1998, Darmstadt, Germany. Heidelberg: Springer-Verlag.

Yakemovic, C. K. B. and Conklin, E. J. (1990). Report on a development project use of an issue-based information system, in the *Proceedings of the Conference on Computer-Supported Cooperative Work (CSCW'90)*, 7–10 October 1990, Los Angeles, pp. 105–118. New York: ACM Press.

44

Developing Inclusive e-Training

Anthony Savidis and
Constantine Stephanidis

44.1 Introduction

The development of e-learning systems is targeted toward supporting the overall learning process with software instruments enabling: (1) learners to easily and effectively assimilate the learning material; (2) tutors to set up more productive and effective learning processes; and (3) supervisors to organize, execute, monitor, and evaluate the online learning process.[1] In this context, e-learning systems are basically information systems, however, not always structured and developed following the particular software engineering methods of the information systems field. For instance, edutainment software employs multimedia interaction methods to deliver cartoonlike, mostly drill-and-practice, sessions in a *learning by playing* fashion. In such systems, the learning content is mostly fused with the multimedia presentation logic, usually not being modeled and stored as structured content in a database.

Additionally, training systems targeted to putting trainees very close to the real working tasks and activities put primary emphasis in the implementation of *learning by doing* methods. However, there are still situations where the delivery of traditional electronic courseware is still needed, as a means of supporting distributed online courses, in processes that could be characterized as mainly involving *learning by studying* interactions. In the latter case, the learner is usually faced with interactive material that has to be explicitly studied, while the e-learning system emphasizes content delivery rather than the delivery of a new teaching style.

While e-learning systems exist for numerous thematic topics, employing teaching approaches such as those mentioned previously, there is relatively limited reported experience in applying design for all and universal access methods in the context of e-learning system development. Recently, the need to produce software applications and services accessible and optimally usable by potentially everyone has been emphasized (Stephanidis, 2001). In some development cases it is practically unavoidable to design and deliver customized system versions for specific target user groups, while in other cases the pursuit of systems capable of adapting to individual end-users is proving to be a more cost-effective goal. This chapter reports the design, implementation, evaluation, and deployment experience regarding inclusive e-training systems for *vocational training applications for people with disabilities*. In this context, two applications are presented, adopting a teaching style that borrows elements from both learning by doing and learning by studying approaches:

- An accessible canteen manager application, training hand-motor and cognitively disabled people for the cashier management of a typical canteen. In this case, the training application was also the real-life application system, while the training process has been carried out with a supervisor either present in the field or indirectly monitoring through a camera.
- A multimedia sew tutorial application, training people with cognitive disabilities in typical sewing tasks. In this case, the trainees are put in front of the real sewing devices, while the training process is based on two strategies: (1) giving students their own personal computer, requiring that they manually operate the multimedia tutorial application; and (2) using a single large projected display and loudspeakers, as a coordinated training session guided and operated by a tutor-supervisor.

---

[1] This chapter is based on Savidis, A., Grammenos, D., and Stephanidis, C. (2006). Developing inclusive e-learning systems. *Universal Access in the Information Society* 5: 51–72. Reprinted here with modifications by concession of Springer.

44.2 Training Applications

44.2.1 Canteen Manager for Motor-Impaired People

The canteen manager application has been developed with a twofold role: (1) serving as a training tool; and (2) being the real canteen-cashier application. The target groups for this training application were: (1) users with severe disabilities in the upper limbs; (2) people with nonsevere cognitive impairments; and (3) able-bodied users. Additionally, it has been necessary to support accessibility for people with both cognitive and motor impairments (i.e., combined disabilities).

In Figure 44.1, the overall system infrastructure is illustrated, mainly indicating the special switch input devices employed for scanning interaction (see Chapter 35 of this handbook for a detailed discussion of the scanning interaction technique). The type of switches shown is only indicative as the switch adapter box allows attaching various types of switches typically connected through a serial port. The discussion follows by a brief description of the key user interface design decisions, continuing with a presentation of specific prominent system development issues; then, the findings of the usability evaluation process are discussed.

44.2.1.1 Design Objectives and Requirements

The two prominent design requirements of the cashier training application have been: (1) the need to support accessibility for people with severe disabilities on the upper limbs as well as cognitively disabled users; and (2) the necessity to serve both as the initial training instrument and as the target application to be eventually deployed in real life. The foreseen real-life setup has been as follows:

- The cashier application operator (the disabled user) interacts with the customer to receive the orders and then issues the receipt that is given to the customer.
- Another member of the staff (an able user) gets the printed receipt from the customer, together with the payment, deposits the cash, and then supplies the corresponding products.

44.2.1.2 Interface Design

Following the key design requirements, the training session for the deployment phase has been eventually organized in two different ways (see Figure 44.2): (1) in direct supervised mode (initially, upon the early training sessions); and (2) in indirect supervised mode (after the user has gained adequate experience of use). The user interface had to be easy to use, comprehensible, and accessible. In this context, it has been decided to minimize the use of verbal explanations (text) as much as possible, by providing iconic representations, in most cases exact photographs of the canteen products, together with digitized speech-audio.

This design style is different from the one deployed in common cashier applications used in restaurants, cafes, or canteens, all of which largely employ textual descriptions. The initial screen of the cashier training application is provided in Figure 44.3. The product categories are provided with iconic representations (not photographs), which have been carefully chosen among alternatives after an initial interview and evaluation round with end-users, scoring the appropriateness of each alternative icon regarding comprehensibility. After a specific product category is chosen, a menu consisting of the product photographs is displayed (see Figure 44.4). The trainee may either select a product, in which case the focus moves automatically on the dialogue box to provide the number of purchased items (this is the scenario shown in Figure 44.4), or the trainee returns back to the product categories menu. When the customer request for the number of items to be ordered is verified (i.e., the dialogue box of Figure 44.4), the order is automatically added on the total sum (see upper left part of Figure 44.5).

The design decision to emphasize the use of iconic menus through exact photographic representations of products on the application menus has been proved to be suitable for the specific type of training topic and categories of trainees, as it allowed for quickly shifting to unsupervised sessions (actually indirectly supervised as the sessions were always monitored through a camera). It has been observed that trainees were capable of easily handling the sessions without any external support, a fact that allowed them to quickly increase their knowledge regarding the cashier system, while in the meantime developing their confidence and sense of independence in being able to effectively accomplish the assigned task.

The design of the user interface to support accessibility by hand-motor-impaired users has employed switch-based hierarchical scanning techniques, in particular the method developed

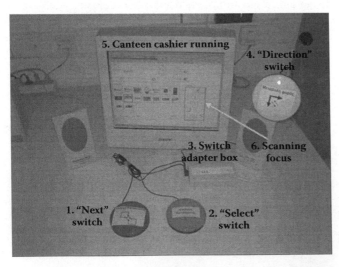

FIGURE 44.1 A snapshot of the canteen manager cashier application running, illustrating the peripheral devices and their interaction-specific role; direct input is allowed through the use of a touch screen, while the keyboard is not used. (From Savidis, A., Grammenos, D., and Stephanidis, C., *Universal Access in the Information Society*, 5, 51–72, 2006.)

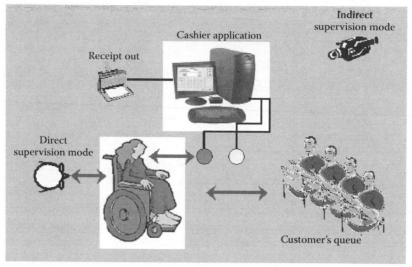

FIGURE 44.2 The organized training sessions with the cashier application in supervised and nonsupervised mode; multiple real-life scenarios for customer arrival pace were defined and applied. (From Savidis, A., Grammenos, D., and Stephanidis, C., *Universal Access in the Information Society*, 5, 51–72, 2006.)

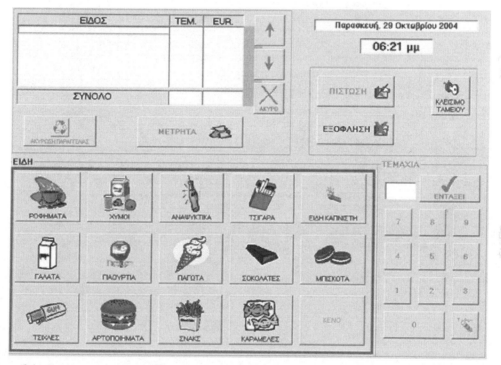

FIGURE 44.3 Snapshot of the main screen of the cashier application running (in Greek). The large rectangle plays two roles: First, it shows the only group of options (product categories) currently selectable; second, it is used for hierarchical switch-based scanning of options. (From Savidis, A., Grammenos, D., and Stephanidis, C., *Universal Access in the Information Society*, 5, 51–72, 2006.)

for augmenting the basic windows object library, originally reported in Savidis et al. (1997). Following this technique, the graphical direct-manipulation dialogue is decomposed in a dialogue structure involving sequences of only two basic actions: *next* and *select*. Typically, each of those is associated with a particular binary switch, thus enabling the thorough use of a graphical application without using the inaccessible mouse point

device and the keyboard. To accomplish this goal, it is imperative to appropriately reproduce the dialogue for every type of user interface object into subdialogues of *next* and *select* actions. The way this has been achieved for the cashier application is shown in Figure 44.6, providing the dialogue automata for the key graphical user interface objects (i.e., container objects, composite objects, and iconic buttons).

FIGURE 44.4   Product selection menu (photographs) and dialogue box to provide the number of purchased items. (From Savidis, A., Grammenos, D., and Stephanidis, C., *Universal Access in the Information Society*, 5, 51–72, 2006.)

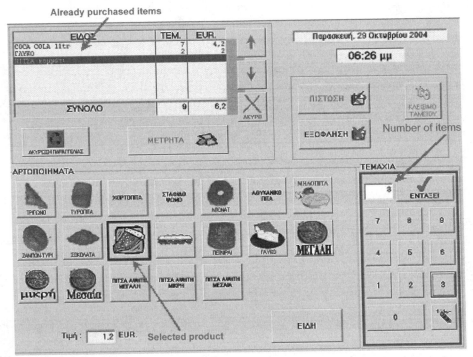

FIGURE 44.5   The view of the cashier application once the trainer has entered two items while in the process of entering a customer's order for the number of the currently selected product (which is a piece of pizza). (From Savidis, A., Grammenos, D., and Stephanidis, C., *Universal Access in the Information Society*, 5, 51–72, 2006.)

During interaction, the particular interaction object owning the scanning dialogue focus is indicated with a green bounding rectangle (the latter called the *scanning highlighter*). As shown in Figure 44.6, the dialogue with a button object is decomposed as follows:

- First, the scanning highlighter is drawn on the graphical content of the button object to indicate it has the focus.
- If the switch being associated with the *next* action is pressed, the scanning dialogue will move to the next interface object (at the same level of the object hierarchy),

meaning the current button object effectively loses dialogue focus.

- If the switch being associated with the *select* action is pressed, the button is considered to be pressed, in which case all internal (to the button object) actions are performed, so that a "button press" notification is eventually posted to the owner application (i.e., the cashier).

In case of container or composite objects, the dialogue is more complicated to allow the user control if the dialogue with such an object is to be initiated (the entry mode; see Figure 44.6,

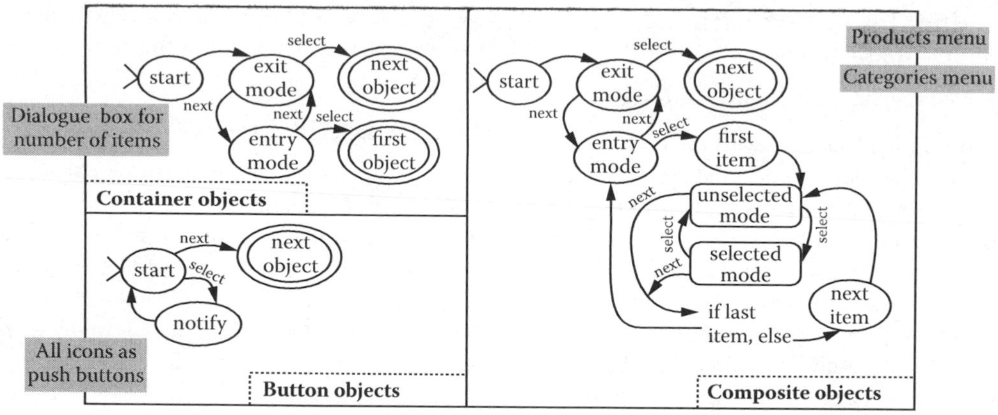

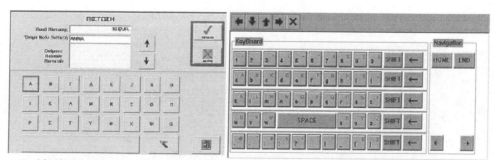

FIGURE 44.6 The dialogue automata for the three types of interface objects in the cashier application, showing the dialogue decomposition into two fundamental input actions: next (first switch) and select (second switch). (From Savidis, A., Grammenos, D., and Stephanidis, C., *Universal Access in the Information Society*, 5, 51–72, 2006.)

indicated with a green highlighter), or if the dialogue is to be skipped (including all contained/constituent objects), enabling one to speed up the scanning dialogue (the exit mode; see Figure 44.6 and left part of Figure 44.7, indicated with a red highlighter).

44.2.1.3 System Development

The cashier trainer application has been developed using Microsoft Foundation Classes (MFC), running under Windows. The scanning dialogue capabilities have been implemented as derivations of the original MFC classes, incorporating the hierarchical scanning capability by implementing entirely the dialogue automata illustrated under Figure 44.6; this constituted an implementation of the interaction-object augmentation software pattern according to the technique discussed in Savidis and Stephanidis (2001; see also Chapter 16 of this handbook). The switches have been driven (only for input) at the level of the serial port interface. The menus of product categories and products, including the photographs, graphical illustrations,

and prices, were all defined in a configuration file (in extensible mark-up language, XML). Initially raw configuration data in an INI file were used; subsequently, it was decided to employ more structured content descriptions with XML definitions (XML is largely employed for structured configuration data). An alternative solution would be to store all product-related information directly in a relational database.

Finally, it has been necessary to allow the focus object to be also changed using direct pointing (the cashier application uses a touch screen), even if the scanning dialogue is currently used. This was necessary in all cases where direct intervention from an able person was required. For instance, during the training sessions with direct supervision mode, the supervisor could intervene and select the correct items from the screen using touch pointing. Clearly, since such an action effectively modifies the focus object of the scanning dialogue (i.e., where the scanning highlighter is), it had to be internally mapped to the augmented dialogue implementation, so that the graphically selected focus object would also become the scanning focus.

FIGURE 44.7 The full-screen virtual keyboard supporting switch-based scanning for the cashier training application (left) and a full-function onscreen keyboard that has been proven to be excessively detailed for the cashier application. (From Savidis, A., Grammenos, D., and Stephanidis, C., *Universal Access in the Information Society*, 5, 51–72, 2006.)

44.2.2 Sew Trainer for People with Cognitive Disabilities

44.2.2.1 Design Objectives and Requirements

This application had a similar purpose with the cashier trainer application, in the sense that, on the one hand it is focused on a very specific job, while, on the other hand, the training application is intended for direct deployment during the field trials. The only difference was related to the fact that while the cashier trainer had a dual role, being both the training application and the real application in the field, the sew trainer played primarily the role of a multimedia tutor. However, the deployment approach for the sew trainer application for real-life training sessions had to also support different tutoring scenarios, as it was derived from the requirements of the particular target user group.

During the initial design-discussion sessions with experts in special education, coming mainly from the specific organization where the sew trainer would be eventually deployed, it came out that it was critical to support both (see also Figure 44.8): (1) individualized and group training (i.e., each trainee or group of trainees sitting in front of a PC running the sew trainer application); and (2) tutoring classroom sessions, where the system is in use by a trainer, projecting into a large screen, while all trainees sit in front of the actual sewing machines. The following sections briefly introduce the key aspects of the sew trainer application.

44.2.2.2 Interface Design

The user interface design had to be particularly simple, delivering minimal functionality while emphasizing iconic illustrations and pictures. Initially, as the target user group consisted of language-literate adults, a satisfactory reading capability was assumed. However, it was quickly understood that this was not necessarily the general case, as the involved users had varying reading capabilities or deficiencies. Hence, even for textual/verbal illustrations, it was mandated to introduce digitized speech of a relatively slow speech rate.

Another small design detail concerned the rendering of the selectable objects in the sew trainer user interface; the default behavior of the software library that has been employed for interface implementation was to draw unselectable icons or options in the so-called gray style. Since this was not well understood by users with cognitive disabilities, as was concluded during the initial evaluation sessions and interviews, the technique finally implemented was to simply draw selectable items (or groups of items) with green bounding rectangles and nonselectable items with red bounding rectangles (see Figure 44.9).

FIGURE 44.8 The different types of training sessions foreseen for the sew trainer application. (From Savidis, A., Grammenos, D., and Stephanidis, C., *Universal Access in the Information Society*, 5, 51–72, 2006.)

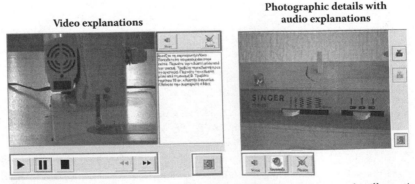

FIGURE 44.9 Snapshot of the sew trainer application showing video explanations (left) and photographic illustrations of sewing machine details (right). (From Savidis, A., Grammenos, D., and Stephanidis, C., *Universal Access in the Information Society*, 5, 51–72, 2006.)

Additionally, the first prototype included audio feedback for all buttons, in case the mouse focus entered an option (e.g., audio messages for "stop sound," "exit," etc.). However, if such a mouse focus event occurs while audio explanations are provided, the playback needs to be temporarily suspended to play the button title. It was proved that this was more confusing than helpful, so in the final design the audio playback of button titles was dropped.

In Figure 44.10, one of the main interface screens is shown, through which trainees may select the parts of the sewing machine for which they need explanation/training material. In this case, typical active areas over the graphical image were defined, providing audio feedback regarding the title of the particular associated machine accessory or sewing procedure. Selection can be carried out by either clicking on the active area or pressing the associated push button (with the photographic or video camera icon).

44.2.2.3 System Development

The sew trainer application has been developed using MFC, including the Media Player control for video and audio playback. The video has been stored in MPEG format (including audio), while the separate audio explanations in WAV files. Finally, the location of files for graphical icons, photographs, video and audio explanations were defined within a Windows INI configuration file.

44.3 Usability Evaluation

44.3.1 Approach

The approach employed for the valuation of the training applications has mainly emphasized the extraction of information concerning the overall interface quality. Subjective evaluation aims to provide tools and techniques for collecting, analyzing, and measuring the subjective opinion of users when using an interactive software system. Subjective measures are quantified and standardized indicators of psychological processes and states as perceived by the individual user. Typically, subjective evaluation, if it is to deliver maximum benefits, requires a fairly stable prototype of the interactive system. It is a type of usability evaluation, which tries to capture the satisfaction of end-users when using a particular system. It therefore does not rely upon expert's opinion or the opinion of any other intermediary evaluation actor. Some of the benefits of subjective evaluation versus other engineering techniques for usability evaluation include the following:

- Subjective evaluation measures the end-user opinion that is frequently dismissed by other engineering approaches to usability evaluation.
- Subjective evaluation techniques are very reliable, valid as well as efficient, and effective in comparison with alternatives available.

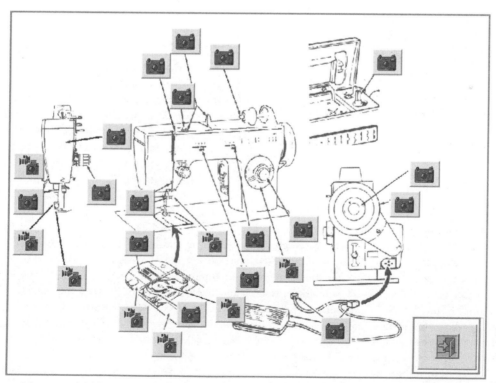

FIGURE 44.10 One of the main screens of the sew trainer application showing the details of the sewing machine for which information is provided, together with the form of the explanation (i.e., pictures or video). (From Savidis, A., Grammenos, D., and Stephanidis, C., *Universal Access in the Information Society*, 5, 51–72, 2006.)

When conducting subjective usability evaluation, the analyst should carefully select the sample users to be representative of the target user group. There are several techniques available for subjective evaluation. These include interviews, which may be structured or unstructured, the use of diary studies, as well as talk-about methods. Various questionnaire techniques have been successfully introduced and widely used in subjective evaluations. Questionnaires may be factual, attitude-based, or of a survey type. Among the most popular questionnaires are IBM Computer Usability Satisfaction Questionnaires (Lewis, 1995), which have been adopted as the instrument of this study, and the SUMI questionnaire (Kirakowski, 1994).

44.3.2 About the Questionnaires

The IBM questionnaires constitute an instrument for measuring the users' subjective opinion in a scenario-based situation. Two types of questionnaires are typically used; the first, namely ASQ (After Scenario Questionnaire), is filled in by each participant at the end of each scenario (so it may be used several times during an evaluation session), while the other one, namely CSUQ (Computer System Usability Questionnaire) is filled in at the end of the evaluation (one questionnaire per participant). Regarding metrics, it should be noted that the questionnaires adopt a seven-point scaling system where low scores are better than high scores. The result of the subjective evaluation with the IBM Computer Usability Satisfaction Questionnaires is a set of psychometrics, which can be summarized as follows:

- *ASQ* score, for participants' satisfaction with the system for a given scenario
- *OVERALL* metric, providing an indication of the overall satisfaction score
- *SYSUSE* metric, providing an indication of the system's usefulness
- *INFOQUAL* metric, providing the score for information quality
- *INTERQUAL* metric, providing the score for interface quality

44.3.3 Process

It is important to mention that subjective usability evaluation using the IBM questionnaires requires a scenario-based procedure. To this effect, a comprehensive scenario has been developed per training application to facilitate the evaluation process. After performing the scenario, the end-users were asked to fill in the ASQ questionnaire and the CSUQ questionnaire; for our specific target user groups, this process has been carried out with close cooperation with the therapists that had to question end-users and then fill in the questionnaires. As the training applications were targeted to supporting specific real-life tasks, the scenarios included real situations that end-users had to cope with. For instance, as part of one scenario for the cashier trainer application, the end-user was asked to handle "an order of two pieces of pizza and an orange juice, charging those via credit" while in other cases there were "customers" issuing such requests directly.

44.3.4 Subjects

Two independent groups were set up, one per training application. The cashier application user group consisted of 12 people as follows:

- Five hand-motor-impaired users
- Six users with cognitive disabilities
- One motor-impaired person having also cognitive disabilities (not having hand-motor impairments)

The user group for the sew trainer application consisted of seven people as follows:

- Three female adults (below 30 years old)
- One male teenager (15 years old)
- Three male adults (below 30 years old)

Before conducting the study, a couple of days were dedicated to introduce the participants (including their therapists) through online demonstrations of the training applications, in which they were allowed to comment in a thinking-aloud process, being given all the necessary explanations and clarifications. Then, they were introduced to the usage scenario, spending some time to give the appropriate guidance as to what exactly should be done.

44.3.5 Results

In the scoring of the IBM questionnaire, therapists were allowed to provide scores with fractional parts, in case they felt the discrete values could not convey accurately the desirable scoring. This has resulted in an excessive use of scores with fractional parts, which had an inherent effect on the evaluation results. In Figure 44.11 are the results for the cashier training.

From the results of Figure 44.11, the overall conclusion is that the subjective opinion of users regarding the cashier training application is good, as indicated by the fact that the overall score (OVERALL, 3.07) passes very successfully the acceptability test by taking scores far less than four (4). The best scores are observed in the interface quality metric (INTERQUAL); this was more or less anticipated, mainly due to the largely tested and verified quality of the scanning interaction techniques that have been injected in the trainer application (the five motor-impaired users gave the highest INTERQUAL score). The results of Figure 44.12 revealed also some interesting aspects about the sew trainer application. While the overall scenario-based experience with the system has been considered satisfactory (ASQ values 3.26 and 3.28, respectively) for the two basic scenarios, there has been a small deviation among the scores on information quality (INFOQUAL, 2.94) and interface quality (INTERQUAL,

Participants	Scenario A	Scenario B	Scenario C	Overall	Sysuse	Infoqual	Interqual
DK	3,1	2,9	3,43	3,1	3,24	2,9	2,6
JK	3,54	2,87	2,65	3,45	3,6	2,5	3,05
DA	3,02	3,91	3,42	2,91	3,5	2,65	2,76
MA	2,8	2,67	2,88	2,76	3,12	3,07	2,98
KP	2,99	3,03	3,12	3,1	3,45	2,65	2,93
PM	3,31	3,19	3,12	3,45	3,53	3,12	3,09
GE	2,76	2,81	2,92	2,92	3,01	2,93	3,12
YA	2,1	3,01	3,14	3,13	2,9	3,19	2,45
YP	3,23	2,78	2,82	2,9	3,31	2,92	2,67
LK	3,42	2,9	3,76	3,75	3,12	3,14	3,29
KK	2,71	3,65	3,81	2,98	2,97	3,67	2,75
MA	2,98	3,65	2,42	2,41	3,05	2,89	3,18
Partial	2,99666667	3,11416667	3,12416667	3,0716667	3,233333	2,96916667	2,90583333
ASQ avg	3,078333333						

FIGURE 44.11 The spreadsheet file with the summary of the usability evaluation results for the cashier trainer application; the averages are shown in shaded mode. (From Savidis, A., Grammenos, D., and Stephanidis, C., *Universal Access in the Information Society*, 5, 51–72, 2006.)

Participants	Scenario A	Scenario B	Overall	Sysuse	Infoqual	Interqual
JG	3,56	3,02	3,12	3,31	3,45	3,12
JL	3,76	3,62	3,45	3,44	3,12	3,19
MN	3,12	3,12	3,56	3,75	2,98	2,87
MI	3,17	3,77	3,25	3,78	2,81	3,42
SA	2,76	2,81	3,01	3,82	2,94	3,21
HO	3,41	2,92	3,32	3,32	2,51	2,99
SA	3,1	3,76	2,99	3,91	2,77	3,65
Partial	3,268571429	3,288571429	3,242857143	3,6185714	2,94	3,20714286
ASQ avg	3,278571429					

FIGURE 44.12 The spreadsheet file with the summary of the usability evaluation results for the sew trainer application. (From Savidis, A., Grammenos, D., and Stephanidis, C., *Universal Access in the Information Society*, 5, 51–72, 2006.)

3.2). The latter indicates that although trainees found effectively all the information they needed for the purposes of the scenarios, they mainly considered that the delivery could be further enhanced.

44.4 Conclusions and Future Work

The e-learning field encompasses a wide variety of e-learning systems, with different pedagogical approaches, semantic content, and interaction characteristics. The work reported in this chapter has put primary emphasis on the inclusive characteristics of e-learning systems, putting accessibility forward. The reported experience has concerned two training applications for people with cognitive disabilities and hand-motor impairments. Those systems have been entirely evaluated and are practically used in concrete settings. From the evaluation and deployment experience of the training applications, the following design-oriented conclusions have been drawn:

- A large number of early evaluation cycles needs to be organized and conducted, always in close cooperation with therapists.
- In many cases, it is necessary to implement varying design features, since there are tasks where the need for optimal design mandates design variations for some users, by enabling individualized deployment with key features defined in configuration files.

- In numerous dialogue scenarios the default graphical feedback methods do not suffice, and more graphically emphatic ways are needed.
- The usability evaluation process for work training requires intensive testing with multiple real-life scenarios.
- Accessible interaction methods need to support all together: speed of interaction, accuracy of actions, and physical stamina, which may mandate the design of an application-specific accessible dialogue (like switch-based hierarchical scanning).
- The graphical interface should be kept very simple, pop-ups are to be avoided, while the overall dialogue design should reflect a "shallow" hierarchy of alternative "screens."
- Software or hardware speech synthesis quality should be intensively tested with different users, exploiting all the applications programming interface (API) capabilities to control the speed and voice styles; in many cases the recoding of familiar voices may be the only optimal solution (for people with cognitive disabilities).

Each of the two training systems reviewed in this chapter had its distinctive design, implementation, and deployment characteristics. It is argued that because the e-learning field accounts for a very large number of information systems, any consolidated wisdom toward making a particular category of e-learning systems user-inclusive is bound to have limited chances for open deployment and reuse. In this context, the reported experience is not put forward as a recipe for all classes of inclusive e-learning systems, but primarily aims to deposit a specific consolidated know-how, while also raising awareness on the key design and implementation challenges in the pursuit of inclusive e-learning technology.

Acknowledgments

The two training applications reported in this chapter have been developed under a subcontract (2001) from the Centre of Special Training, Therapeutic Clinic of Lifelong Diseases, located in Agios Nikolaos, Crete.

References

Kirakowski, J. (1994). *Subjective Usability Measurement Inventory.* MUSIC Project. http://www.hcirn.com/ref/refk/kira94.php.

Lewis, R. J. (1995). IBM Computer Usability Satisfaction Questionnaires: Psychometric evaluation and instructions for use. *International Journal of Human-Computer Interaction* 7: 57–78.

Savidis, A. and Stephanidis, C. (2001). Development requirements for implementing unified user interfaces, in *User Interfaces for All: Concepts, Methods, and Tools* (C. Stephanidis, ed.), pp. 441–468. Mahwah, NJ: Lawrence Erlbaum Associates.

Savidis, A., Vernardos, G., and Stephanidis, C. (1997). Embedding scanning techniques accessible to motor-impaired users in the WINDOWS Object Library, in *Design of Computing Systems: Cognitive Considerations, Proceedings of the 7th International Conference on Human-Computer Interaction (HCI International '97), Volume 1* (G. Salvendy, M. J. Smith, and R. J. Koubek, eds.), 24–29 August 1997, San Francisco, pp. 429–432. Amsterdam, The Netherlands: Elsevier Science.

Stephanidis, C. (2001). *User Interfaces for All: Concepts, Methods, and Tools.* Mahwah, NJ: Lawrence Erlbaum Associates.

Stephanidis, C., Paramythis, A., Sfyrakis, M., and Savidis, A. (2001). A case study in unified user interface development: The AVANTI web browser, in *User Interfaces for All: Concepts, Methods, and Tools* (C. Stephanidis, ed.), pp. 525–568. Mahwah, NJ: Lawrence Erlbaum Associates.

Training through Entertainment for Learning Difficulties

Anthony Savidis,
Dimitris Grammenos, and
Constantine Stephanidis

45.1 Introduction

45.1.1 Motivation and Background

The overall objective of the work presented in this chapter is to provide a systematic basis for work-oriented training of people with learning difficulties, employing techniques that on the one hand better motivate trainees to be involved in the overall process, while on the other hand has the potential to provide more effective and flexible training instruments.[1]

Traditional approaches to the provision of interactive technology support for learning difficulties mainly focus on the use of various technologies, such as word processors, word predictors, spelling checkers, calculators, and the like, to compensate for the lack of particular abilities (Day and Edwards, 1996).

The focal point of the work reported in this chapter concerns the learning of real-life tasks to be carried out using mission-specific software applications, or, in other words, training for work tasks involving the deployment of software applications.

Toward this purpose, a suite of four applications, comprising two training applications and two games, has been developed following an inclusive design approach, and evaluated with users with learning difficulties. The underlying rationale of the performed work was that the combination of educational and game applications with an inclusive design approach has the potential

to offer new opportunities toward more effective, but also more rewarding training of young adults with learning difficulties.

The chapter is organized as follows. The remainder of this introductory section briefly discusses the concepts of inclusive e-Learning and e-Entertainment, and introduces the target user groups of the developed applications and the adopted approach. Section 45.2 reports the development of the two games, and Section 44.3 their evaluation. Finally, Section 45.3 summarizes and concludes the chapter.

45.1.1.1 Inclusive e-Entertainment

The role of entertainment for pedagogical purposes, supporting or even complementing the learning process, is well known in the education domain, being universally acknowledged. Today, electronic entertainment is very popular, attributing to a huge number of commercial game titles. One out of two Americans play video games, and in the United States the sales of games outnumber the sales of books (Digiplay, 2004), while the world market for video games grows at much faster rates than the cinema box office, VHS/DVD rental, and music retail (Screendigest, 2004). Additionally, the distribution of home consoles has today a total number far higher than the amount of personal computers for home use.

While there has been a lot of research and development toward the universal access (Stephanidis et al., 1998) of computer applications and services, it has only recently been addressed that there is a need for accessible entertainment through the formation of the Game Accessibility Special Interest Group of the International Game Developers Association (IGDA). This group defines *game accessibility* (GA-SIG, 2004) as "the ability to play a game even when functioning under limiting conditions. Limiting conditions can be functional limitations, or

---

[1] This chapter is based on Savidis, A., Grammenos, D., and Stephanidis, C. (2007). Developing inclusive e-learning and e-entertainment to effectively accommodate learning difficulties. *Universal Access in the Information Society* 5: 401–419. Reprinted here with modifications by concession of Springer.

disabilities—such as blindness, deafness, or mobility limitations." Also recently, there is a serious trend for introducing games for training and learning (known as *game-based learning*; Prensky, 2000) to take advantage of the unparalleled motivation and engagement that computer games can offer learners of all ages. University departments are gradually introducing computer games in their curricula to support alternative learning styles, attract student interest, and reinforce learning objectives (e.g., Giguette, 2003). Games are also promoted as policy education, exploration, and management tools (e.g., the Serious Games Initiative[2]).

45.1.2 Target User Groups

A learning disability (see also Chapter 7 of this handbook) is a neurological disorder that affects one or more of the basic psychological processes involved in understanding or in using spoken or written language.[3] Every individual with a learning disability is unique and shows a different combination and degree of difficulties. A common characteristic among people with learning disabilities is uneven areas of ability. For instance, a child with dyslexia who struggles with reading, writing, and spelling may be very capable in math and science.

People with learning disabilities are usually of average or above average intelligence. In contrast to other disabilities, such as blindness, vision, and motor impairments, a learning disability does not become immediately obvious in daily life (Pieper et al., 2003).

Learning disabilities is an umbrella term describing a number of specific learning disabilities, such as:

- *Dyslexia*: Language and reading disability
- *Dyscalculia*: Problems with arithmetic and math concepts
- *Dysgraphia*: A writing disorder resulting in illegibility
- *Dyspraxia (sensory integration disorder)*: Problems with motor coordination
- *Central auditory processing disorder*: Difficulty processing and remembering language-related tasks
- *Nonverbal learning disorders*: Trouble with nonverbal cues (e.g., body language) poor coordination, clumsy
- *Visual perceptual/visual motor deficit*: Reverses letters; cannot copy accurately; eyes hurt and itch; loses place; struggles with cutting
- *Language disorders (aphasia/dysphasia)*: Trouble understanding spoken language; poor reading comprehension

The International Classification of Functioning, Disability and Health (ICF)[4] describes how people live with their health condition (see also Chapter 4 of this handbook). ICF is a classification of health and health-related domains that describe body functions and structures, activities, and participation. In ICF, functions related to learning are classified into two main categories:

basic learning and applying knowledge. Basic learning includes copying, rehearsing, learning to read, learning to write, learning to calculate, and acquiring skills. Applying knowledge includes focusing attention, thinking, reading, writing, solving problems, and making decisions. In particular, the following definitions are provided:

- *Focusing attention*: Intentionally focusing on specific stimuli, such as by filtering out distracting noises.
- *Thinking*: Formulating and manipulating ideas, concepts, and images, whether goal-oriented or not, either alone or with others, such as creating fiction, proving a theorem, playing with ideas, brainstorming, meditating, pondering, speculating, or reflecting.
- *Reading*: Performing activities involved in the comprehension and interpretation of written language (e.g., books, instructions, or newspapers in text or Braille), for the purpose of obtaining general knowledge or specific information.
- *Calculating*: Performing computations by applying mathematical principles to solve problems that are described in words and producing or displaying the results, such as computing the sum of three numbers or finding the result of dividing one number by another.
- *Making decisions*: Making a choice among options, implementing the choice, and evaluating the effects of the choice, such as selecting and purchasing a specific item, or deciding to undertake and undertaking one task from among several tasks that need to be done.

Learning disabilities may be associated with attention-deficit/hyperactivity disorder (ADHD), which is difficulty in regulating attention (i.e., paying attention as needed and shifting attention to another task when required).[5] Individuals with ADHD may be inattentive, hyperactive, or impulsive. They may have significant difficulty in learning because they don't have ready access to the persistent mental effort required for the rote learning of facts, figures, and procedural rules.

Finally, some similarity has been observed between some forms of learning disabilities, and in particular nonverbal learning disorder and Asperger syndrome, a mild form of autism, which does not affect intelligence (Klin et al., 1995).

The work reported in this chapter is mainly targeted to young adults with learning difficulties due to learning disabilities, attention deficit disorder, and Asperger syndrome, and addresses software applications for training in specific work tasks. The underlying idea is that software educational applications, but also games, have the potential to help acquire skills that support improving the functions listed previously, namely focusing attention, thinking, reading, calculating, and making decisions, in the context of a specific work task.

Two accessible and highly configurable action games are discussed that address combined disabilities like cognitive,

---

2 http://www.seriousgames.org.

3 http://www.ldanatl.org/aboutld/parents/ld_basics/ld.asp.

4 http://www3.who.int/icf/icftemplate.cfm.

5 http://www.ldanatl.org/aboutld/resources/frames.asp?top.asp+http://www.smartkidswithld.org.

hearing, motor, and vision impairments. In particular, the two action games are accessible and configurable remakes of the classic Space Invaders and the Pong arcade games, which from early experiments proved to play a significant role in challenging and sharpening basic kinesthetic skills, orientation capabilities, short-term strategic thinking, and decision making, while highly enhancing the esteem of involved disabled players/learners. The two games where deployed concurrently with the two training applications reported in Chapter 44 of this handbook with the objective to have an amplifying effect and support faster learning cycles.

The adopted approach toward this objective was driven by the main design principle that the training process should be augmented with targeted game-based informal learning, focusing on challenging, testing, sharpening, and training particular intellectual skills and capabilities, in an overall highly stimulating, motivating, joyful, and scenario-driven metaphoric game atmosphere.

45.2 Inclusive Entertainment Applications

45.2.1 Approach

Following the successful development of two inclusive training applications, a canteen manager and a sewing tutorial (see Chapter 44, "Developing Inclusive e-Training"), it was decided to enhance the suite with entertainment applications targeted *to support implicit highly motivated and stimulating training for specific skills.* In particular, computer game genres were chosen through which some key basic skills, which were deemed as particularly important for the type of developed training applications, could be potentially sharpened. Play is a fundamental activity of children that allows them to acquire the foundations of self-reflection and abstract thinking, develop complex communication and meta-communication skills, learn to manage their emotions, and explore the roles and rules of functioning in adult society (Verenikina et al., 2003). The development of the entertainment applications was carried out from scratch, as it was necessary to introduce: (1) a number of purposeful gameplay characteristics; (2) accessibility support to accommodate sensory and motor disabilities in accordance with learning difficulties; and (3) to encapsulate extensive configuration facilities both at the level of the user interface as well as at the basic game logic.

45.2.2 King Pong

King Pong is a remake of the original Pong arcade game, one of the first computer games. It has been selected due to its simplicity, as well as because it offers a platform to sharpen various kinesthetic and intellectual capabilities in a straightforward manner, enabling the addition of different scenarios and conditional progress within the very simple game plot. Regarding

game play, King Pong resembles the classic Pong game, simulating table tennis (ping pong), where a small ball travels across the screen. Additionally, there are two paddles that are controlled by one player each. The players control their respective paddle using keyboard or joystick and only in the vertical direction. When the ball hits the borders of the playing field or one of the paddles, it bounces and its vector velocity is changed according to the angle of the impact. In contrast, if one player's paddle misses the ball, the other player scores a point. King Pong can be played either by a single player pitted against a computerized opponent or by two players each controlling a paddle. The two-player games are feasible not only by sharing the same computer, but also over a network.

The basic game has been developed to be accessible by both hand-motor-impaired and blind players. This has been accomplished with the development of an auditory media space (approximately 2W × 1.5H meters), supporting spatial audio, consisting in the final version of 24 loudspeakers (4 rows, 6 loudspeakers each, with 24 separate audio channels). The prototype of the auditory lattice is illustrated in Figure 45.1—a special-purpose wood-made setup is under construction. Game snapshots are provided in Figure 45.2.

The configuration capabilities of the game provide sufficient flexibility to therapists/tutors in fine-tuning game and interaction aspects to individual end-user requirements. For instance, ball speed, force-feedback support, and degree of automatic player-paddle positioning, scoring increase/decrease factors, paddle speed, computer opponent accuracy, and so on, can be easily configured. Additionally, the configurability of graphics allows the total alteration of the appearance of the game, as shown in Figure 45.3.

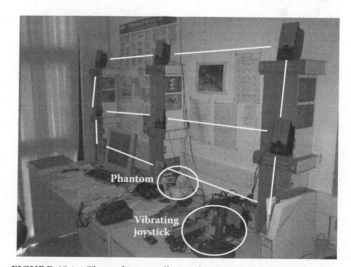

FIGURE 45.1 The auditory wall, as it has been prototypically set up for the Pong media-space game; a projection screen will be put in front of the wall so that spatial audio can augment the large visual display. (From Savidis, A., Grammenos, D., and Stephanidis, C., *Universal Access in the Information Society*, 5, 401–419, 2007.)

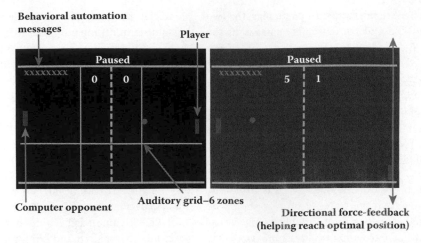

FIGURE 45.2 Snapshots of the game's basic screen (Retro profile), with some explanatory annotations; the graphics regarding user/computer paddle, ball, scene boundaries, and background are configurable. (From Savidis, A., Grammenos, D., and Stephanidis, C., *Universal Access in the Information Society*, 5, 401–419, 2007.)

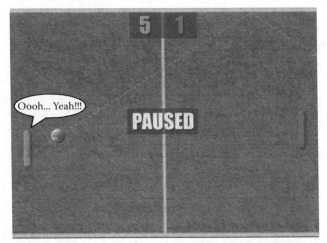

FIGURE 45.3 A radically different scene produced by selecting alternative graphics for the game objects and background; the direction of the ball is also illustrated as a means for providing help to cognitively impaired players. (From Savidis, A., Grammenos, D., and Stephanidis, C., *Universal Access in the Information Society*, 5, 401–419, 2007.)

Representative categories of training-related tasks in King Pong are defined in the following (some of those are reusable across different game types):

- Reach score N, then lose.
- Play until you hit the ball at your paddle M times.
- Make the ball hit the left/right/up/down wall.
- Play until the opponent score becomes K.
- Pause the game.
- Where is the direction of the ball after we restart?
- Lose now.
- Produce an explosion (happens after certain score increment).

- Make the computer "laugh" (happens in a certain profile when losing with 0 score).
- Try to make a score K less/more than M.

45.2.3 Access Invaders

Access Invaders is another remake of a classic game, which delivers a radically different class of action and plot. In short, the basic game play can be described as follows:

- The player controls a spaceship whose goal is to prevent a group of attacking aliens from landing on the ground by shooting them down.
- The game is played in a single fixed screen, so there is no scrolling of the game terrain.
- The player's spaceship can only move from left to right, and vice versa.
- The attacking aliens move from left to right and vice versa, but they also gradually move from the top of the screen toward the bottom, occasionally firing bombs.
- Periodically, an alien mothership crosses the top of the screen, traveling from left to right. It does not fire on the player, but if destroyed, additional bonus points are gained.
- A range of protective shields exist between the aliens and the player's spaceship, which are gradually worn out by the player's and the aliens' shots.
- The game ends when the player's spaceship has been destroyed for a fixed number of times, which is usually referred to as "lives."

Kinesthetically, it involves various categories of skills, while perceptually it requires recognition of more complicated phenomena, relevant causes, and associated effects. Due to its more rich graphical appearance, it usually becomes more attractive and appealing to end-users than the Pong game itself. In the meantime, due to the fact that it offers a larger repertoire of game-oriented events and user activities, it enables the definition of numerous alternative types of training tasks.

The availability of different profiles of the Access Invaders game is illustrated in Figure 45.4, reflecting the numerous configuration capabilities of the game.

Some of the representative training tasks in Access Invaders are:

- Destroy *N* aliens.
- Destroy the leftmost/rightmost platform.
- Destroy the *Nth* platform from the left/right.
- Set your ship at a position where there is no alien ship above.
- Set your ship at a position where there is a platform on top.
- Fire *N* bullets.
- Move left, right, and left again to return to the original position.
- How many aliens are there at the bottom row?
- Destroy a <*specify color*> alien.

45.2.4 Development Issues

The development of the inclusive games discussed in this section has been carried out using the Accessory Game Engine (AGE; Grammenos et al., 2005a). This game engine has been purposely implemented to facilitate the programming of universally accessible games, supporting:

- Multiple players, playing sequentially in a turn-taking style; the profiles of individual players vary, indicating a

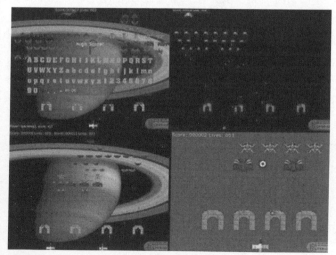

FIGURE 45.4 Four snapshots of the Access Invaders game, each depicting a different game profile (the bottom right is intended for people with low vision). (From Savidis, A., Grammenos, D., and Stephanidis, C., *Universal Access in the Information Society*, 5, 401–419, 2007.)

different type of disability (e.g., blindness, motor impairments, cognitive impairments)
- Configuration of game/interaction characteristics to individual profiles, facilitating variations of the terrain game structure and the basic game regulations
- Logical input channels, enabling multimodal physical input from multiple devices or input control units
- Multiple players, playing concurrently in a competition, or cooperation style over the network; players may have radically different user profiles, each of which may employ alternative ways for rendering the game "board," as well as interaction techniques and input/output modalities

The software architecture of the AGE game engine is outlined in Figure 45.5. The key issue is the separation of the logical structure of the active game entities (usually called *sprites*) from their particular display method, through independent renderers that comprise a particular terrain rendering policy. This enables the plugging of alternative independent renderers to support multiple game views accessible by different end-users. A similar architecture split is also adopted for input control and management, enabling effectively different physical input channels to be adopted, or even plugged-in, without the need to modify the core game implementation.

45.2.5 Integration with Inclusive Training Applications and Experimental Testing

In the integration of the inclusive training applications (see Chapter 44, "Developing Inclusive e-Training") with the games presented in this chapter, it was chosen to avoid the delivery of integrated systems in the form of typical edutainment software, since, for job-targeted training systems, any severe deviation from the original tasks to be carried out by end-users in the real field is inappropriate. Instead, a fused approach was formulated, emphasizing entertainment as a *metaphoric training process*, supported through supervised sessions, while heavily relying on metaphoric scenarios and tasks, and intensive trainee-trainer interaction. This integrated view of the training process with distinct and separate roles of the deployed software instruments, is illustrated in Figure 45.6.

In this context, four users were involved in an experiment regarding the use of the games for training purposes, and their use in combination with the training applications, as follows:

- Male, age 18, with dyspraxia due to a mild cerebral palsy combined with a learning disability (dyslexia). In daily life, problems occur in activities requiring adequate fine-motor skills, eye-hand coordination, orientation in space, and use of symbols for reading, writing, and understanding written text.
- Male, age 21, with severe attention deficit disorder in childhood, which has improved following rehabilitation and medical treatment. He still has difficulty in concentration

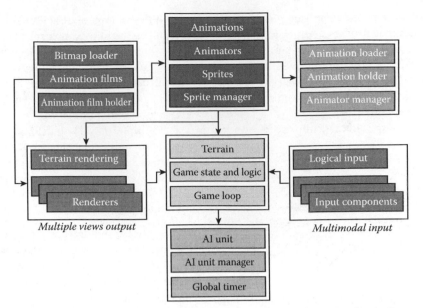

FIGURE 45.5 Overview of the key components of the AGE and their main call dependencies. (From Savidis, A., Grammenos, D., and Stephanidis, C., *Universal Access in the Information Society*, 5, 401–419, 2007.)

and in assimilating difficulties that are obvious in attempts to remember written and spoken instructions and to execute complex tasks.

- Female, age 17, with dyslexia and dyscalculia, resulting in difficulty reading, writing, and understanding written text, and inability to execute arithmetic calculations with two digit numbers.
- Male, age 19, with mild Asperger syndrome causing poor language skills, poor social behavior, and obsessive behavior.

The overall objective of the experiment was to investigate the potential of the game to enhance users' basic skills necessary for obtaining optimum benefit from the learning applications. First, the users were allowed to explore and familiarize themselves with the games. It clearly appeared that none of them had serious problems playing the games successfully. An appropriate configuration was selected for each user and each game with the help of the instructors.

Then, three combined sessions were conducted for each user. In each session, the user was asked to play one of the two games in direct supervised mode, executing specific tasks established by the instructors. The tasks were given either before startup or while the user was playing. More specifically:

- *Participant 1*: The selected gaming tasks were related to enhancing the eye-hand coordination, visual perception, dexterity, and orientation skills necessary for mouse pointing operations.
- *Participant 2*: The selected tasks were related to enhancing the width of visual and acoustic attention, concentration on tasks, and orientation within the user interface.
- *Participant 3*: The selected tasks were related to enhancing instruction understanding and counting abilities.
- *Participant 4*: The assigned tasks were mainly targeted to ensure a pleasant and rewarding gaming experience that would motivate the user to also perform some training tasks to be allowed to play again after training.

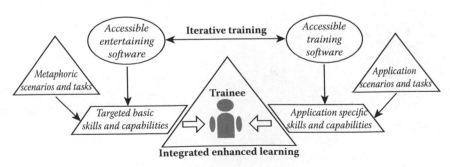

FIGURE 45.6 Integrated training combining learning and entertainment software while supporting accessibility for people having learning difficulties, which may be combined with sensory or motor disabilities. (From Savidis, A., Grammenos, D., and Stephanidis, C., *Universal Access in the Information Society*, 5, 401–419, 2007.)

The initial playing phase lasted 15 minutes. Following the games, the users were asked to practice with the training application they were already familiar with, performing a new set of tasks, with no time limits. A final phase of free game playing was allowed as a reward to the users.

During the entire duration of the sessions, the users were observed by the evaluators under laboratory conditions, and the sessions were recorded. After each session, a brief interview followed, during which the users and their instructors were asked to comment on the experience. During the playing/training sessions, it quickly became evident that the extent to which explicit job-related tasks could be given to end-users always had to be compromised by setting up an overall atmosphere of a genuine playful experience. In other words, users did not appreciate the inexcusable transformation of what was originally presented to them as primarily a game to a work activity. So, it was necessary to improvise and produce scenarios that would facilitate the smooth integration and delivery of the training tasks as purely encapsulated events within the overall gaming activity. In the interviews, all users appeared to have enjoyed the experience, and felt positive about the idea of playing while learning.

Overall the results concerning playing games are very good, also due to the fact that both games are very simple. Furthermore, after the interviews with the participants and their teachers, it came out that the social value of the concept of universally accessible games created a positive attitude. The main complaints reported were related to the speech and sound quality, as well as to the game difficulty, either because it was lower or higher than desired. This problem could have probably been alleviated by conducting user profiling sessions prior to the evaluation process, and then adequately adapting all difficulty-related game parameters.

Concerning the training aspects of the experiment, the instructors, who prior to the experiment had expressed their concern about the potential distraction that playing the game could cause with respect to the training tasks, formulated the opinion that, although this might indeed happen, playing intermixed with learning appeared to have potentially positive effects for the acquisition of specific cognitive skills. Although more training would be necessary to evaluate the users' improvements in the necessary skills for each case, the instructors could observe over the three sessions that all participants had increased motivation and interest. They also observed some unanticipated benefits of playing in various areas that were not in focus for a specific user. For example, they observed that the possibility of unlimitedly repeating tasks while playing was a considerable advantage, in particular for users with attention deficit disorder or Asperger syndrome.

45.2.6 Lessons Learned

This concurrent deployment of purposefully developed accessible games, along with the targeted mission-specific training applications, proved in the course of the conducted experiments to accommodate learning difficulties effectively. More specifically, some of the representative basic and nonbasic skills that are exercised and sharpened with the two accessible video games developed are:

- Orientation
  - Recognizing direction of a moving object
  - Predicting direction of a moving object
  - Predicting collision of moving objects
- Hand-eye/hand-ear coordination
  - Controlling motion with hand (mouse, keyboard, switches, joystick)
  - Moving objects in different directions, with different granularities of speed and displacement
  - Following a predetermined path
  - Reaching a desirable target position
  - Following the movement of another object
- Cause and effect
  - Assimilating cause and effect
  - Reproducing cause to accomplish an effect
  - Recognizing effect and understanding its cause
- Counting (with visual/auditory feedback)
  - Recognizing the total number objects in a set
  - Calculating when a desirable number of objects in a set is reached
  - Recognizing increment/decrement effect
  - Associating numeric values with explicitly enumerated objects
- Comparison
  - Numeric values (higher/lower)
  - Enumerated objects (more/less)
  - Volume and size (higher-bigger/lower-smaller)
  - Distances (farther/closer)
- Elementary arithmetic
  - Comparison (as previously)
  - Counting (as previously)
  - The notion of zero (nothing/empty)
  - Addition with objects and numeric values
  - Subtraction with objects and numeric values

45.3 Conclusions

The work reported in this chapter is mainly targeted to young adults with learning difficulties due to learning disabilities, attention deficit disorder, and Asperger syndrome. However, in principle the approach can be applied also to other target user groups, for example, more severely limiting cognitive conditions such as Down syndrome. It is also believed that at certain stages of education and training, the combination of learning and playing can be useful also for individuals with average cognitive abilities, depending on age and domain subject.

The development of two game applications has been reported. The developed applications are targeted to the effective accommodation of learning problems, also in combination with perception, fine-motor, or dexterity problems, through the combination of training and playing, with the objective of making the overall user experience more effective, pleasant, and rewarding. The obtained results are positive concerning both the game applications alone, as well as their integration into a more articulated training process. This confirms that further research on inclusive learning and edutainment is necessary to identify new, more engaging training techniques, as well as game genres suitable for enhancing the training process.

Acknowledgments

The King Pong and Access Invaders entertainment applications have been mainly developed in the context of the MICOLE project (IST-2003-511592 STP), partly funded by the Information Society Technologies Programme of the European Commission DG Information Society. The partners in the MICOLE consortium are University of Tampere (Finland); University of Glasgow (United Kingdom); University of Metz (France); Uppsala University (Sweden); Lund University (Sweden); Royal Institute of Technology KTH (Sweden); Siauliai University (Lithuania); Institute of Computer Science FORTH (Greece); University of Linz (Austria); University of Pierre and Marie Curie (France); France Telecom (France); and Reachin Technologies AB (Sweden).

The cashier and sew trainer training applications have been developed and are deployed under a subcontract (2001) from the Centre of Special Training, Therapeutic Clinic of Lifelong Diseases, Agios Nikolaos, Crete.

References

Day, S. L. and Edwards, B. J. (1996). Assistive technology for post-secondary students with learning disabilities. *Journal of Learning Disabilities* 29: 486–492.

Digiplay (2004). *Some Key Gaming Facts.* http://www.digiplay.org.uk/facts.php.

Game Accessibility Special Interest Group (GA-SIG) (2004). *Accessibility in Games: Motivations and Approaches.* http://www.igda.org/accessibility/IGDA_Accessibility_WhitePaper.pdf.

Giguette, R. (2003). Pre-games: Games designed to introduce CS1 and CS2 programming assignments. *ACM SIGCSE* 35: 288–292.

Grammenos, D., Savidis, A., and Stephanidis, C. (2005). The development of a sensory system for intelligent game creatures, in *Universal Access in HCI: Exploring New Interaction Environments, Volume 7 of the Proceedings of the 11th International Conference on Human-Computer Interaction (HCI International 2005)* (C. Stephanidis, ed.), 22–27 July 2005, Las Vegas [CD-ROM]. Mahwah, NJ: Lawrence Erlbaum Associates.

Klin, A., Volkmar, F. R., Sparrow, S. S., Cicchetti, D. V., and Rourke B. P. (1995). Validity and neuropsychological characterization of Asperger syndrome: Convergence with nonverbal learning disabilities syndrome. *The Journal of Child Psychology and Psychiatry and Allied Disciplines* 36: 1127–1140.

Pieper, M., Morasch, H., and Piéla, G. (2003). Bridging the educational divide. *Universal Access in the Information Society* 2: 243–254.

Prensky, M. (2000). *Digital Game-Based Learning.* New York: McGraw-Hill.

Screendigest (2004). *Press Release: Video Games Market Demonstrates over 100 Percent Growth in Six Years, London.* http://www.screendigest.com/reports/ils04/press_releases_31_08_2004n/view.

Stephanidis, C., Salvendy, G., Akoumianakis, D., Bevan, N., Brewer, J., Emiliani, P. L., et al. (1998). Towards an information society for all: An international R&D agenda. *International Journal of Human-Computer Interaction* 10: 107–134.

Verenikina, I., Harris, P., and Lysaght, P. (2003). Child's play: Computer games, theories of play and children's development, in the *Proceedings of Young Children and Learning Technologies.* Selected papers from the International Federation for Information Processing Working Group 3.5 Open Conference, Melbourne, Australia. Conferences in Research and Practice in Information Technology, 34 (J. Wright, A. McDougall, J. Murnane, and J. Lowe, eds.), July 2003, pp. 99–106. New York: ACM Press.

46

Universal Access to Multimedia Documents

Helen Petrie,
Gerhard Weber, and
Thorsten Völkel

46.1 Introduction

The World Wide Web was invented for reading and browsing text with a few supporting graphics. Recent developments have dramatically increased the number of web pages and have led to the inclusion of more media in these pages. Production of web sites by content management systems (CMSs) and the ease of adding information from digital cameras have helped to extend the web, encouraging participation by all kinds of authors and allowing rapid updating. CMSs have enabled authors to be more productive in generating content by formatting presentations almost automatically without the help of a designer. Blogs use a simple type of CMS, and editing tools for editing blogs focus on writing and are powerful enough for those people whose interest is in sharing their personal views and not on editing a newspaper or magazine. Multimedia content such as radio, music, speech, audio books, television, recorded lectures, and so on, are technically more challenging to produce but nevertheless attract a large audience. Web users can also participate by sharing personal photos and videos since media formats have been standardized and browsers now allow interactive access to multimedia. As a consequence, many traditional business processes have been modified; for example, paper versions of pictures have become less common and music is now sold via web shops and then distributed via high bandwidth Internet connections.

Despite the success of the web, printed books are still common and the preferred format for most readers. However, in parallel to publishing printed books (pBooks), publishers are starting to issue licenses for the production of electronic books (eBooks). However, in the eBook market, fiction and bestsellers have minor market shares compared to specialized literature, such as academic monographs. While already a niche market in the print market, printed encyclopedias are now also facing the challenge of easy navigable electronic versions that can be extended and updated much more easily than traditional multivolume printed books. For example, updating political maps can hardly keep up with the changes induced by wars or resulting from negotiations by the European Union with its neighbors. Dictionaries, another niche market, are now popular on portable devices such as pocket PCs and PDAs using software products such as the Mobipocket Reader. None of these devices, reading systems, or formats has lived up to the success predicted at the turn of the millennium, and many dedicated devices seem to have dropped out of production.

New technologies such as ePaper, of which iLiad[1] is one example, make reading and writing similar to using a notepad. When laptops eventually become really light and portable and have sufficient battery power—and assuming they have stable and user-friendly software—then electronic books will have a massive and widespread hardware platform on which to base their market. Other advantages of eBooks that will be important once the hardware problem is overcome, are the searching capabilities and the ability to store a complete library on a single device.

---

[1] http://www.irextechnologies.com/products/iliad.

Thus, eBooks will be an important information medium in the near to medium future. One interesting aspect of the recent evaluations of eBook devices and reading systems (e.g., see Press, 2000; Malama et al., 2004; Marshall and Bly, 2005) is that none of them mentioned the requirements or problems of "print-disabled" readers (Bauwens et al., 1994). Therefore this chapter focuses on access to multimedia by print-disabled users who have specific needs when reading and navigating though eBooks and web documents. Such needs were largely of a technical nature in the days of limited bandwidth. When 2400-Baud modems prevented the downloading of images, an alternative description helped save bandwidth for everyone, and at the same time enabled blind people to understand materials such as web pages. Other requirements for print-disabled people include recordings of the spoken word for many user groups, including blind, visually impaired, and dyslexic readers, and transcriptions of recorded voices for hearing-impaired and deaf people.

Print disability does not only affect access to books, newspapers, and magazines. In other domains, electronic displays have become ubiquitous. For example, electronic versions of timetables and announcements help to reduce maintenance costs for public transportation authorities. For the same reason, loudspeaker announcements on train platforms now come from a supervisor located in a central train station. To reach all passengers, each medium has its advantages and disadvantages. Visually impaired people prefer spoken announcements, while deaf and hearing-impaired people prefer visual presentation on electronic displays. An overriding problem is that each medium and technology is not always used for the same information and even may contradict each other. It is left to the reader (or listener) to apply common sense and trust one of these technologies.

The term *multimedia barrier* is used in this chapter to summarize these issues. Multimedia barriers are created by the potential lack of access by print-disabled people to particular content in particular media eBooks, by the lack of quality features in the publishing process ensuring that all readers can read and handle eBooks, and, finally, by the failure of authors to understand the requirements of all readers, including those with print disabilities.

46.2 The Multimedia Barrier

User agents with their graphical user interfaces often create multimedia barriers in eBooks, since multimedia eBooks often introduce novel control features. Lack of standardization of interaction techniques for controlling media prevent assistive technologies, such as screen readers and screen magnifiers, from working properly. An off-screen model (OSM) of the application would be required before synthetic speech or refreshable Braille could be generated (Mynatt and Weber, 1994). Web technologies have introduced an abstract layer of this type as the basic approach for all hypertext reading software though the document object model (DOM). An OSM or DOM is both of a hierarchical nature and separate content from presentation. Therefore, they can both be serialized and presented in tactile or auditory media in a structured way while preserving a navigable layout and providing sufficient overview when needed by the reader. The following sections discuss the requirements of eBook contents that enable a user agent to support reading for a variety of readers with print disabilities.

Figure 46.1 shows as an example a multimedia Spanish-English dictionary. Readers find icons being displayed to indicate audio contents. However, appropriate interaction techniques for nonvisual access to these media assets are missing. This includes both an appropriate verbal cue and handling of keyboard input to activate playback. Similar examples can be shown in other book reading software and titles such as Microsoft's Encarta. Figure 46.1 is also an example of book reading software aimed at mobile reading devices, especially mobile phones. But the same restrictions apply: screen readers for mobile phones may provide access to text, but interaction for browsing and controlling other media is very limited.

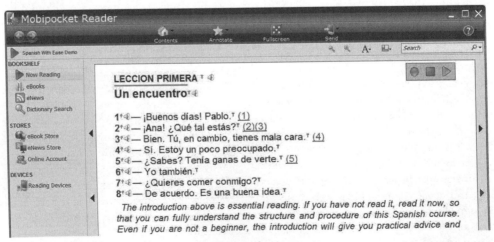

FIGURE 46.1   eBook reader indicating multimedia contents by loudspeaker icons.

Multimedia barriers are created when publishing eBooks. Publishing an eBook is simply a process of transformation if text and other media such as graphics are available in appropriate digital data formats. A multimedia barrier is created by a lack of knowledge of the requirements of alternative transcriptions of some media for some readers or the lack of availability of such transcriptions (which may have arisen for a variety of reasons, lack of time to produce them, lack of funding, etc.). It can be argued that not all graphics, such as decorative elements, can be or need to have alternative descriptions. Some special notations, such as phonemes related to foreign language terms, as shown in Figure 46.1, require standardized codes. Graphical notations for writing mathematical terms, musical scores, architectural diagrams, software engineering diagrams, or chemical formulas, to name but a few, cannot be automatically transformed because of proprietary data formats. Transcription of these media is therefore limited, and often carried out by conversion into graphics, which then loses semantic information.

Transcription usually requires human intervention to meet specific user requirements. A universal approach to the needs of disabled readers is required to understand all such user requirements, to integrate transcriptions, and to apply automatic transformations. User requirements have to be gathered for each user group.

To distinguish the needs of the different heterogeneous user groups on one hand, and on the other hand the individual reader who belongs to one of these user groups, the term *user group* as used within this chapter is defined as follows:

> A *user group* as addressed by universal usability principles is a group of print-disabled users sharing the same type of limited access for achieving a particular reading task with an eBook.

Within a particular *user group*, such as blind users, individual users may have different needs; for example, some blind users prefer to read texts in Braille whereas others prefer to read them in synthetic speech. An individual user may also have different needs at different times and in different circumstances. For example, a particular blind person may wish to read text in Braille if he is studying the text for an exam, but in synthetic speech if the text is for leisure.

There are also social factors that affect the type of access to books. There are libraries for the blind collecting and delivering Braille books and audio books. These libraries are often also publishers, as the pBooks transcribed into Braille or audio constitute new pieces of work in their own right. Borrowing such books is often limited to registered print-disabled readers. There are several eBook distribution organizations and publishers, such as BookShare,[2] that accept only disabled readers to meet restrictions required by copyright legislation or directives and the relevant access to copyright information options that have been adopted by the different countries. Universal usability

principles for these user groups can be applied to eBooks for developing universal access.

Publishers of titles containing sign language work in a similar way. There are also eBook collections available containing sign language (e.g., from the Internet Public Library[3]). However, there is little overlap in the workflow among these specialized publishers and distributors, which might yield an economy of scale and a competitive advantage. Many print-disabled readers use more than one supplier, getting audio, Braille, or large print from different sources.

46.2.1 User Requirements

The MultiReader Project (Petrie et al., 2005b) was set up to address the user requirements of readers in a more coherent manner. The project addressed the needs of five user groups: mainstream readers and four groups of print-disabled readers: blind readers, partially sighted readers, dyslexic readers, and deaf readers. Issues related to blind, partially sighted, and deaf computer users are discussed in more detail in Chapter 6, "Sensory Impairments," of this handbook. Although it is self-evident that the blind and partially sighted people will have difficulty reading print, and the reading difficulties of people with dyslexia are well known, people who are deaf may also have considerable reading problems. For example, they may have a sign language as their native language, for example, American Sign Language (ASL) or British Sign Language (BSL). In spite of their names, these languages are not the sign equivalents of spoken/written English, but are very different languages in their own right, with their own grammars and vocabularies (see also Chapter 38 of this handbook). So English is a second language for native users of ASL or BSL. In addition, because of their difficulties in learning spoken languages, the reading ages of deaf children are usually substantially behind their hearing peers (Campbell and Burden, 1995). Deaf readers appreciate graphics, video clips, and translations into a sign language. Partially sighted people need enlarged views, short lines of text with increased interline spacing, and individual color settings. A navigable reading system for both mainstream readers and those with print disabilities needs to be able to apply different navigation and reading strategies appropriate for different user groups and individuals. Although there has been considerable work on the navigational strategies of mainstream users of hypermedia systems (see Vora and Helander, 1997, for a review), there has been little work on the strategies of print-disabled users of such systems. To gather information about how people interact with and navigate through a variety of different types of reading material, a number of methods for user needs elicitation were used in the MultiReader Project: focus groups, questionnaires, and interviews with users and experts in the field.

In creating multimedia books for all users, the expectations of mainstream readers as well as the requirements of print-disabled users need to be fulfilled. The following needs have

---

[2] http://www.bookshare.org.

[3] http://www.ipl.org.

been identified within the different groups of print-disabled readers (Petrie et al., 2002). Blind readers prefer text as Braille output and/or audio, whereas partially sighted readers require appropriate colors and contrast (often higher contrast than is comfortable for fully sighted readers). Dyslexic readers also need altered contrast, although in their case, often lowered contrast is appropriate. Short line lengths and increased interline spacing is also helpful for partially sighted and dyslexic readers, as is highlighting of text phrases, sometimes in combination with audio output.

Table 46.1 is based on the user requirements elicitation exercises conducted in the MultiReader Project. Filled cells in Table 46.1 include enrichments required by particular user groups that go beyond common features provided for mainstream readers. Enrichment means the provision of content in a different, additional medium to the original. This includes text description of images, and subtitling of videos. Enrichments leave the original material unchanged but aim to increase information accessibility by providing additional material in a different medium to overcome the limitation experienced by the user group. All additional media support the user in understanding the original material. In the table, a letter *p* indicates that the enrichment is of primary interest, a letter *m* indicates if this enrichment occurs always with other media, and letter *s* indicates a feature that is of secondary importance.

Extensive use of pictorial, graphical, and video material is the kind of general enrichment increasingly found in leisure reading material. Other formats may not easily fulfill this requirement, due to restrictions of a particular domain or format.

The last requirement for presentation of information in short and simple bite-sized chunks, for ease of reading and comprehension, is difficult to measure and research is ongoing. Besides counting syllables and other lexical features, linguistic analysis is required. Jenge et al. (2006) propose analyzing morphology, lexicology, syntax, semantics, and discourse while comparing these features with text used in normalized tests of reading competence.

All of the first 10 requirements listed in Table 46.1 are items that need to be added to the content of a multimedia document. They can be referred to as *media enrichments*, as they use the very multimedia nature of a multimedia document to turn it into a universally readable document. The design of enrichments is described in more detail in the following section, where literature provides good evidence of how to do this.

46.2.1.1 Speech Output

Synthesized speech and recorded human speech are both acceptable media for allowing people (both mainstream and print-disabled) to read a book. Recorded human speech produces the highest quality speech. However, synthesized speech can be more easily controlled by the listener. Blenkhorn et al. (1993) lists a number of criteria for synthesized speech specifically for blind people:

- Speech synthesis has to have the capability to be presented fast; blind users can understand synthetic voices speaking at up to 500 words/minute.

TABLE 46.1 Requirements from Different Print-Disabled User Groups

Presentation	User Groups			
	Blind	Partially Sighted	Deaf	Dyslexic
1. Speech output	p	m		s
2. Good descriptions of images and graphics	p			
3. Audio descriptions in videos	p			
4. Vary font style and size		p	s	p
5. Vary text and background color		p	p	m
6. Enlargement of images, graphics, and video		p		
7. Text or graphic output for speech and auditory signals			p	
8. Sign language translations			p	
9. Increase line spacing, line length			m	p
10. Word-by-word or sentence-by-sentence highlighting of text			m	p
11. Extensive use of pictorial, graphic, and video material			p	
12. Presentation of information in short and simple "bite-sized" chunks for ease of reading and comprehension			p	m

The letter *p* indicates that the enrichment is of primary interest; *m* indicates if this enrichment occurs always with other media; and *s* indicates a feature that is of secondary importance.

Source: Adapted from Langer, I., Adaptation of multimedia eBooks: Universal access in HCI, in *Universal Access in HCI: Inclusive Design in the Information* Society, Volume 4 of the *Proceedings of the 10th International Conference on Human-Computer Interaction (HCI International 2003)* (C. Stephanidis, ed.), 22–27 June 2003, Crete, Greece, pp. 1442–1446, Lawrence Erlbaum Associates, Mahwah, NJ, 2003.

- The process of synthesis should be adaptable for interactive use: a user may want to stop speech at any time immediately.
- Speech synthesis should have little or no latency to synchronize output with user's input.
- Insertion of pauses should be possible to read unknown text more slowly or to change prosody.
- Spelling of words is a necessary option.

Presenting graphical notations using speech synthesis is possible for mathematical terms, but not for other domains such as chemistry and architecture.

46.2.1.2 Descriptions of Images and Graphics

Decorative images may need only briefly be described or may be left without description if they are only decorative. Petrie et al. (2005a) report on a small study among visually impaired people and their interest in images as found in web pages. Interviewees indicated that they would like the following information about images:

- Objects, buildings, people in the image
- What is happening in the image
- Purpose of the image
- Colors in the image
- Emotion, atmosphere of the image
- Location depicted in the image

Not all images have to be described. Interviewees identified a number of classes of images they were not interested in. These include uninformative or decorative images, bullets or spacers, logos (as the relevant information should be apparent elsewhere), and images that are explicitly described in text.

The order of words in a description was also seen as important, with general agreement from the interviewees that descriptions should have the most important information at the beginning of the description and be in simple language. There appears to be no optimum length for a description. However, the recommendation that two or three words were sufficient for a description was not supported, with most interviewees preferring more information rather than less, particularly if it is ordered by importance.

For other classes of graphics, for example, maps, architectural diagrams, electrical circuit diagrams, labeled diagrams, or exploded drawings, no general guidelines are available. However, guidelines do exist for describing works of art (Alonzo, 2001).

46.2.1.3 Audio Descriptions in Videos

Audio description enriches any event with complex visuals such as movies, television programs, sports events, theater performance, or visual art by additional short image descriptions during pauses in the speech, sound, and music stream. These brief explanations describe a scene to a blind listener to augment her understanding of the action. Typically, the characters' actions, gestures, emotional expressions (except invisible mental states) as well as their appearance and clothing may be described (see Fix, 2005; Office of Communications, 2007). Atmosphere and

setting should also be taken into account. Audio descriptions may also be provided for live theater performance and follow the same principles. A describer writes or generates the description for the narrator.

Selection is essential when developing an audio description, as the pauses available for delivering the description are often very limited. However, the describer should not censor any content that he does not like. It is not necessary to fill every pause with description. If a description is too long, some extension over the next scene is acceptable, especially if the following scene starts quietly or with nonverbal audio. The cocktail party effect will allow the audience to distinguish the narrator's voice even if a character speaks.

Different principles have been developed for audio description in museums and exhibitions, as well as for touch-sensitive surfaces, which are beyond the scope of this chapter.

46.2.1.4 Variations of Font Style and Size, Color, Line Spacing, Line Length

There are a large number of different user capabilities in visual acuity, contrast sensitivity, visual field, and color perception to be considered when designing textual layout (Jacko and Sears, 1998). Sets of guidelines are often aimed at particular user groups (Evett and Brown, 2005). Readers should be able to change the presentation mode whenever they need to. Some visual impairments change, for example, during the course of the day or with fatigue, requiring flexibility in the form of presentation of the text.

Alternative color schemes are required by many people with visual impairment to reduce glare, increase contrast, or to account for color vision deficiencies. High contrast can be useful for people with color vision deficiencies and elderly people with cataracts. Presenting text in negative contrast reduces screen glare and can be useful for elderly people and people with age-related macular degeneration. Negative results in white text on a black background, while another common option is yellow text on a blue background.

46.2.1.5 Enlargement of Images, Graphics, and Video

While background and foreground color can usually be very easily changed for text (although not for text that is actually graphics, which is why that option should be avoided), increasing contrast is not easily achieved for images, icons, and so on. Simple enlargement is the most common approach. Blenkhorn et al. (2003) report on magnification factors of up to 32. As the proportion of the original image that can be seen is affected by $1/n^2$, users want to control magnification factors in small steps. Even small factors of 1.5 with several difference levels of multiples of 0.1 are used by elderly people.

With magnification, a pixel becomes a block of pixels. In other words, resampling cannot introduce more information. A regular pattern is sampled by another regular pattern at a different frequency and hence artifacts are introduced. Anti-aliasing can produce smoothed shapes by computing the average of the light pattern that is represented by each pixel. A full-color image requires proper averaging of the three color components, not just the overall brightness levels. Temporal anti-aliasing can be applied in rendering animations and spans multiple frames of the presentation.

46.2.1.6 Text or Graphic Output for Speech and Auditory Signals

Transcribing speech produces captions for hearing impaired, deaf, and deafened people (National Institutes of Health, 2002). For the purposes of this chapter, subtitles as used for translating foreign language material are excluded. Most of the following guidelines are listed in the standard book by Ivarsson (1992):

- Use simple sentence structure (subject-verb-object).
- Each line and each caption should be understandable.
- Semantic completeness of each caption.
- No word separation between captions.
- Written language should correspond to spoken language.
- Profanities appear in writing stronger than spoken language and have to be softened.
- All text appearing in signs, newspapers, and so on should be transcribed.
- Voice-over should be written in italics.
- Songs, if relevant, should be written in italics.
- Lead-in and lead-out should match the rhythm of spoken language.
- Maximum two lines per caption; bottom is longer than top line.
- Match lead-in and lead-out with cuts.
- Dialogue length should match the duration of the caption.
- Color may help to identify characters.
- Reference to characters in discussions is possible by positioning captions or by verbal reference.
- Verbalization of sounds should appear synchronized.

Prillwitz (2001) reports on a study of 167 subjects using captions on German television: 92% requested more broadcasts with captions and 44% requested technical improvements; 11% wanted no simplification of language, 6% wanted simplification, and 30% wanted a one-to-one transcription of all information. In another study reported by Prillwitz (2001), among 1000 television viewers, 40% indicated they felt confused by captions.

46.2.1.7 Sign Language Translations

Chapter 38, "Sign Language in the Interface: Access for Deaf Signers," provides background information on sign language. Sign language is the primary language for many deaf people, but not necessarily for hearing impaired and people acquiring deafness later in life. Sign languages are very different from spoken languages in lexicon, grammar, and their use of space. For example, pronouns may be signed by referring with deictic gestures to one and the same position in front of the signing person.

Synchronization of multimedia documents with sign language cannot be achieved on a word level. Phrases and sentences need to be identified and matched with signs; however, the use of pronouns may be difficult to visualize in highlighting (see also Figure 46.5).

46.2.1.8 Highlighting of Text

Several assistive tools that highlight text, either sentence by sentence or word by word, can improve reading. Elkind et al. (1996) describe a study among adult dyslexic readers where speech synthesis was synchronized with text highlighting. Users selected a reading speed, selected the colors for display of text and for highlighting, and decided whether to have the synthesizer read continuously or pause after each sentence. Reading rate was clearly a major problem for the participants. Their unaided rate was very slow compared to that of a control group. Their mean rate was only 156 words per minute (wpm). This compares to a mean of 255 wpm for high school seniors and 300 to 310 wpm for college students. Through text highlighting their mean reading rate increased by 25 wpm, or 16%, to 180 wpm. However, the study also makes it clear that not all dyslexic readers benefit from text highlighting and speech output.

46.3 Other Work

Previous work on the accessibility of multimedia documents has discussed the one-document-for-all approach (Hillesund, 2002; Langer, 2003), which encapsulates all media within the eBook or document being read by the user where required assistive devices may be used, but the eBook contains all the necessary media and structural information and the reader must select or read those. For example, hypertext markup language (HTML) has been developed into a markup language for a general audience, as well as for the inclusion of text to support blind readers. Images and active objects such as applets can be described verbally by a screen reader and replace the visualization. According to the Web Content Accessibility Guidelines (WCAG; see Chisholm et al., 1999), any accessible HTML page also contains, together with the images, alternative descriptions of the images. A web page becomes even more accessible if it contains a sign language video, but few web sites currently make use of this enrichment.

Another example of a one-document-for-all approach is that of multimedia documents written in the synchronized multimedia integration language (W3C SMIL, 2005), which has been recommended by the World Wide Web Consortium (W3C) for markup of multimedia documents. SMIL synchronizes multiple media by marking them up for serial and parallel presentation, each being applied multiple times with the possibility of being nested. This markup language separates the actual media from the structural document to be transferred over the distribution channel. Thus it is possible to avoid transmission of sign language videos to users who are not deaf. Nonetheless, to be able to meet all user requirements listed in Table 46.1, the document available for transmission must be complete.

If the publisher follows a one-document-for-all approach, all media and their transcriptions have to be prepared. In addition to the author, transcribers and multiple proofreaders are involved, as few writers can address the needs of all readers and ensure quality (see Figure 46.2).

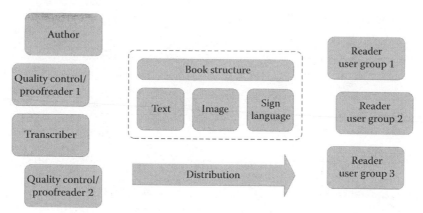

FIGURE 46.2 One-document-for-all approach to multimedia documents.

The one-document-for-all approach is applied in television broadcasting as well as in talking books. While watching television is not the typical reading scenario, this approach can combine the presentation of text with movies and sound. An accessible movie or television show contains captions (Prillwitz, 2001; National Institutes of Health, 2002) and audio description for blind television viewers. There is no structural document containing markup being distributed and hence all media are transmitted to the television. The audience cannot always select the media most appropriate for them. Depending on the broadcast system, captions or audio descriptions may be selectable as a stereo channel. However, this approach is not very flexible for all stakeholders involved. It may reduce aesthetic quality for some viewers and may even reduce performance.

The digital accessible information system (DAISY) standard addresses blind, visually impaired, and dyslexic readers through a kind of multimedia file format for audio books. The DAISY Consortium was formed in May 1996 by talking-book libraries to lead the worldwide transition from analogue to digital talking books.

There are different classes of DAISY audio books starting from audio only or text only up to hybrid books that contain both the full audio and synchronized text. All of these classes of audio books offer improved access and digitized human speech through links between the digital audio sound files and the marked up text files. It is these links that give the talking book reader access to the structure of the book, including chapters, pages, or even paragraphs. The collection of these navigation points in the book is called the navigation control center (NCC). The NCC makes it possible to announce structural information such as the chapters, headings, and page numbers (in the corresponding print book). It integrates a text file (written in a subset of HTML), multiple audio files, and a file for synchronizing text and audio (by applying a subset of SMIL). The NCC introduces internal labels for structural elements, which are referenced by SMIL files (e.g., temporal markup indicating parallel presentation of text and audio) as well as text files (containing, for example, markup for headings). Through markup of the structural elements of a book and including aforementioned labels, the NCC allows an abstract description of the interaction techniques offered by a DAISY player as part of its user interface. Portable reading devices for playing DAISY books may render the audio components and allow navigation via a small keyboard. In addition, PC-based reading software can render the text contents on Braille displays.

Originally the DAISY standard was based on XHTML. In 2005, the DAISY Consortium published an upgraded version of the standard that relied on XML. This standard was later accepted by NISO, the National Information Standards Organization for the United States, a nonprofit association accredited by the American National Standards Institute (ANSI). This ANSI/NISO Z39.86 specification has been constructed in a way to allow for modular extensions with additional functionalities. The first and, until now, only such extension is the MathML in DAISY extension, which was approved in February 2007. The ANSI/NISO Z39.86 specification has recently begun to be accepted by organizations and companies not involved with literature for the blind, but those addressing the demands of the mass market.

46.3.1 Personalization and Multimedia

User modeling has been applied in various domains, and basic terms such as adaptation and adaptivity for these advanced user interfaces are well accepted. Both concepts can be applied to either navigation or presentation while affecting the document in various ways (Brusilovsky, 2001). Adaptivity has been used for educational media, online information systems such as museums, encyclopedias, and so on. However, user models of people with such contradictory special reading needs as shown by blind and deaf users have not yet been taken into account. User models for print-disabled readers and their application to multimedia documents have previously been avoided due to a lack of understanding of the granularity[4] needed for the structure describing such a document.

---

[4] Granularity refers to the size or duration of media for which adaptation may occur.

TABLE 46.2 SMIL Features for Testing User Profile Attributes

Markup Element	Purpose
<audio systemOverdubOrSubtitle = "overdub">	Identifies a medium to be used for overdubbing
<textStream systemOverdubOrSubtitle = "subtitle">	Identifies a medium to be used for subtitling
<par systemCaptions = "on">	Tests for enabled captions in user profile setting
<audio systemAudioDesc = "on">	Identifies content suitable as audio description for blind users

Adaptation in distributed client/server applications using web technologies has been widely developed, especially for adaptation of navigation in eShops. Adaptation of presentation for multimedia content is different, as temporal features have to be taken into account. The following definition of multimedia documents is applied in this chapter (based on Steinmetz, 1993):

Multimedia documents are integrating time-independent media and at least one time-dependent medium.

Examples of time-independent media are text and graphics. Many graphical notations such as those for mathematical terms and for project management are time-independent as well, even if they represent temporal information. Some types of non-photorealistic graphics may be changed considerably, and the application may produce new views within a short time. As each rendering needs to be perceived separately, such a medium is also referred to as time-independent (Strothotte and Schlechtweg, 2002). Time-dependent media are typically video, audio, and animation. Since digitization and sampling serialize any information into time-dependent media for transfer and information exchange, time-dependent media are referred to as those whose presentation is perceived and interpreted as changing. Even a slide show in a digital picture frame with a refresh rate of an hour may be considered time-dependent.

Previous research has developed examples of adaptation of time-independent media by time-dependent media and vice versa, such as text by audio, or audio by text. HYPERAUDIO is an adaptive hypermedia museum guide that takes context, namely position, into account. Leaving an object in the middle of an audio narration can be considered a sign of low interest in that object, and causes some trigger to stop narration while updating the user model. In contrast, walking near an object that could be of interest to the user can trigger a relevant narration.

From the user's point of view a presentation is extended. This is caused either by the system (playing some media) or by the user herself exploring the newly rendered static medium. The objective of these systems is to avoid any gaps (or silences) as well as unwanted extensions as a result of adaptation. In contrast, print-disabled readers may accept such a result of adaptation to perceive the original medium and the result of any adaptation.

Adaptation of time-dependent media by time-dependent media has been developed in the MultiReader project through the use of SMIL and additional XML markup. Videos embedded in eBook pages are an example of material that requires a variety of adaptations, each having its own temporal requirements.

SMIL provides the basic markup structures for separating content from presentation even in videos. It uses content mark-up in conjunction with a set of attributes to select or control alternative presentations. For example, attributes describe the bandwidth available and allow selection among different presentation media with different resolutions. Attributes of user models used in SMIL focus on session concepts such as type of language or screen size of the viewing device, and provide only very limited information about users with special needs. Table 46.2 lists all markup elements available in SMIL 2.0 related specifically to the needs of blind and deaf users.

SMIL does not ensure consistency between the different media. It is left to the designer of an eBook to make sure, for example, that an audio description is actually in place and that captions meet the requirements of deaf viewers. Figure 46.3 provides an example of an interval diagram that visualizes the underlying temporal model of a typical SMIL document containing the title animation of a Star Wars movie, a movie scene showing the arrival of Luke Skywalker who says "Hello, Obi-Wan," as well as audio description and captions.

No limitations of this interval model arise if the designer applies the one-document-for-all approach when creating the multimedia document and takes all user requirements into account. For example, audio description might take longer to speak than the period for which the title is visible. A SMIL document can be composed such that the following scene starts after the last medium of the previous segment has ended (denoted by curly braces in Figure 46.3 and by diagonal lines during which the last image of the title animation is frozen).

If *scene 1* and the following *scene 2* adhere to this scheme, the sighted viewer will notice a still image showing the last picture of *scene 1*, while audio description goes on. For presentation at cinemas this appears to be unacceptable, unlike an individual presentation by a web browser.

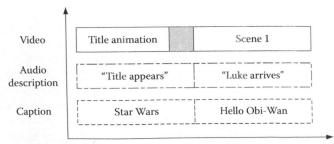

FIGURE 46.3 Interval model for a video with audio description and captions.

The drawback of the one-document-for-all approach arises if a designer of an SMIL document cannot introduce additional media without being aware of the level of granularity needed for meeting the temporal requirements of audio description and captions. If scenes themselves have to be divided into smaller temporal units, such cutting is prepared while the document is edited. Afterwards, the meta-information to describe, for example, the duration of the complete video scene, is lost in SMIL. Similar examples can be developed for sign language productions and audio books that are synchronized with words of a textual medium for highlighting. For word level synchronization the level of granularity becomes too detailed for handling by one and the same author, and experts in sign language and audio description have to support the designer of the SMIL document. Beyond SMIL's markup features, mark-up elements are needed to identify media snippets small enough for gapless composition, but also for creating a particular rendering suitable for each group of readers.

In conclusion, SMIL is only an intermediate format for a universal accessible multimedia document to be used in a web-based distribution channel. Authors are to be supported by more explicit markup elements as part of the production, generation, and distribution process to achieve better adaptation.

46.3.2 MultiReader Documents

A MultiReader document supports personalization by transformation of the contents through preprocessing contents (sign language video, inverse colors) based on a property editor for the user profile, adaptable views (font size, color, speed variations) of enriched contents, and navigation techniques based on nested formatting of media objects through appropriate markup. Markup of content is based on a combination of HTML and SMIL (XHTML+TIME) and an additional mark-up language to separate navigation techniques from content. Several enriched media documents are linked together by a main file to form the basis of a MultiReader eBook. A content preprocessor generates the final eBook based on user profiles by a twofold XSL transformation step. The first step selects a subset of all pages (waypoint) to support even very large documents. The second step of the preprocessing interprets the type of enrichment and identifies matches with the user profile. The resulting document is valid HTML+TIME (integrated with CSS files), which can be displayed by an industry standard browser. To improve accessibility of this browser, several novel user interface elements have been added, for example, icons showing sign language or a pause button.

46.3.2.1 Microformats

The type of each enrichment and medium is identified by MultiReader microformats, called *containers*. These names are valid class names of CSS but are used only during preprocessing. Hierarchical nesting of MultiReader containers is possible through nested <span>-tags, each attributed by a microformat. An overview of the microformats, including the media containers (e.g., cvsign, cnarration, cmusic, csound, canimation, cpic, cmap), is given in Table 46.3. The table also specifies their appropriateness to particular user groups (where M = mainstream, D = deaf, H = hearing impaired, B = blind, P = partially sighted, and Y = dyslexic). For better readability, containers related to audio have been removed from the table.

46.3.2.2 Profile Editor

Reader profiles are maintained in a two-level process. Level 1 matches a file containing a user profile with user groups who participated in the evaluation of the MultiReader books: mainstream reader, blind reader, readers with low vision, deaf and dyslexic readers. Level 2 is more detailed, and consists of an editor for choosing all available media as well as an accessible CSS editor to redefine each visual attribute resulting from preprocessing (see Figure 46.4).

A user's profile as maintained by the MultiReader reading program is also affected by the GUI's profile mechanisms, as well the browser's profile mechanism. In addition, an assistive device such as a screen reader controls the speech synthesizer's speed of speaking. From the user's perspective it appears to be possible

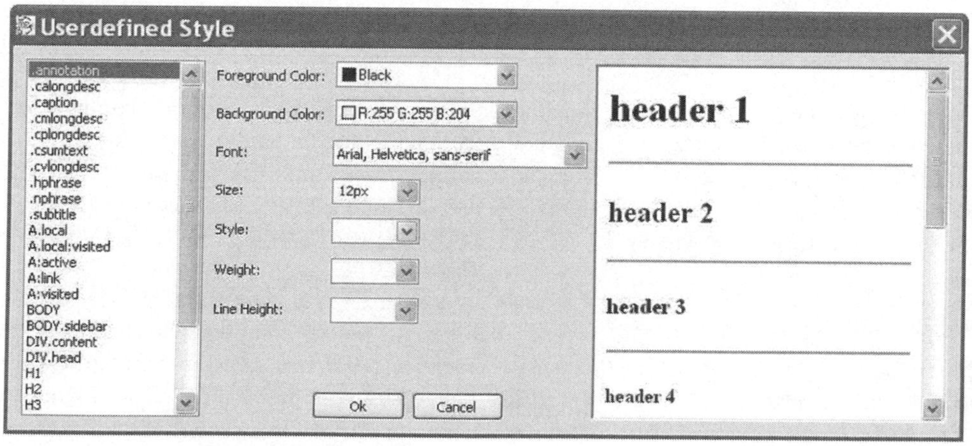

FIGURE 46.4 Microformats are integrated with CSS formatting.

TABLE 46.3   Microformats Used by MultiReader

	Profile Option	User Groups						Container Name	Content/Meaning
Video	Video	M	D	H	B	P	Y	cvideo	Video (pictures) and related media
	Narration	M	D	H	B	P	Y	cvnarration	Spoken words (audio file), original language
	Music	M	D	H	B	P	Y	cvmusic	Music
	Background sound	M	D	H	B	P	Y	cvbgsound	Nonspeech audio (e.g., birds singing)
	Subtitles		D	H				cvsubtitle	Textual representation of narration
	Text description		D	H				cvtxtdesc	Textual description of other sounds
	Audio description				B	P		cvaudiodesc	Auditory description of visible things
	Sign language		D	H				cvsign	Translation of narration
Pictoial Elements	Animation	M	D	H	B	P	Y	canimation	Slide show, but also animated GIF
	Static representation				B	P	Y	castatic	Slide show will be displayed as a number of pictures next to each other
	Audio description				B	P		caadesc	Prerecorded audio describing the animation
	Pictures	M	D	H	B	P	Y	cpic	Photographs, maps, etc.
	Long description				B	P		cplongdesc	Long description of the picture
	Maps	M	D	H	B	P	Y	cmap	SVG maps
	Long description				B	P		cmlongdesc	Textual description of the map
	Audio description				B	P		cmadesc	Prerecorded speech describing the map
Text	Sign language for text		D					ctsign	Sign language representation of text
	Synthetic speech for text	m			B	P	Y		Textual content will be read out
	Text		D	H			Y		Text (h*x*, p) will be highlighted
	by sentence		D	H			Y	chigh	
	by word		D	H			Y		

M, mainstream; D, deaf; H, hearing impaired; B, blind; P, partially sighted; Y, dyslexic.

to create contradictory settings, for example, for the volume of spoken utterances.

To reduce this complexity, a model for the validity of each attribute was developed (Weimann, 2003) and integrated with the user profiling mechanism of the operating system. Settings are distinguished for the operating system, the user profile, the session, and for a page. Page settings, such as magnification, affect a complete session but not the user profile. Next time the reading program is started, the original zoom factor is used and not a temporary value used during the last session. An example of short-term attributes is color. Changes in colors affect a single page only; the following page will use the colors as defined by the user profile.

46.3.2.3 Temporal Layout

Enrichments may lead to parallel presentation of media such as captions shown together with a video. A simple but powerful approach to support readers is to allow them to pause, repeat, or continue some time-dependent medium. If a book page contains multiple time-dependent and possibly compound media, it is not always obvious to the reader what the effect of each control operation is.

Authors of multimedia documents create parallel presentations of media for many reasons. Consider the following scenario: a cookbook shows the preparation of a dish step by step by a video and a narrator explains verbally the handling of ingredients. As some recipes require parallel operation of the oven, pan, and some manual preparation, a multiwindow metaphor will allow viewing of all activities. To turn all of these parallel presentations into an accessible document requires a structured model of the temporal layout through microformats as listed in Table 46.3.

Several groups of readers require serialization of parallel media to reduce complexity and support understanding. For example, a narrator should not be heard while text is presented by speech synthesis to a blind reader. Figure 46.5 shows a similar situation for deaf readers: (1) a sign language video explaining the title of a book page should precede (2) a video that is part of

FIGURE 46.5 Sample page of the *London Tour Guide* for a deaf reader.

a London Tour Guide and (3) may be followed by sign language transcribing some text in the book page.

Nested containers with different microformats allow inclusion or exclusion of enrichments as part of preprocessing. In general, the following guidelines have been applied for MultiReader books:

(C1) A media container with multiple audio or video elements is serialized.

(C2) The duration of a container is the sum of the duration of its media elements as selected according to the user profile making up the container.

(C3) Synchronization among container elements follows a serialized presentation.

(C4) Time-dependent containers such as audio and video can be started at the beginning, paused at any time, or skipped.

For serialization of two parallel presentations, p1 and p2, the following three cases have to be considered:

1. *Extension*: For better accessibility the duration of playback of the medium can be extended; extension may also take place beforehand.

2. *Linking*: For highlighting, captions, and so on, serialization can make use of the human reader's capability to integrate more accessible media with the original medium.

3. *Decomposition*: Long passages are decomposed into smaller passages to make the structure of the information more accessible. For example, decomposition can be achieved by sequences of nonverbal and verbal information.

Temporal layout benefits from many disciplines such as cutting videos, arranging content by a program guide for entertainment, or developing a sequence order for educational purposes.

46.4 Developing Universally Accessible Media: The Process

Current models implementing the design philosophy of user-centered design (ISO 13407, 1999) focus on computers and single users or a homogeneous user group. According to Norman (1988), user-centered design requires a good understanding of the mapping of tasks to the needs of the user by taking into account what can be made observable, the constraints and the possibility of errors. To understand the constraints and improve the system for error tolerance, an iterative design is essential for a user-centered approach.

Universal access to multimedia documents adds another dimension to the complexity of this process. While ISO 13407 requires a specification of the user, a universal accessible multimedia document is read by different user groups, each having its specific requirements as discussed previously. Figure 46.6 describes the modified process resulting from the need to address multiple user groups who share the same tasks.

The number of iterations in a user-centered design can be constrained by the quality assurance methods applied to each intermediate step. During the MultiReader project, evaluations of each user group were applied as described in Figure 46.6. The following section discusses the three iterations applied for the developing of the MultiReader books listed in Table 46.4.

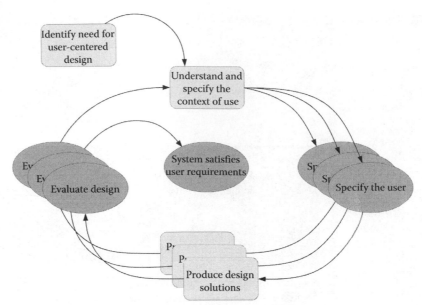

FIGURE 46.6 User-centered design for ensuring universal access to multimedia documents.

Table 46.4 does not indicate the approach to define the type of enrichments, which will be detailed in the following sections. During the first iteration, navigation techniques were explored while supporting multiple views on multimedia presentations. During a second iteration, synchronization of enrichments with content was analyzed, and in a third iteration multiple, heterogeneous reader groups participated in the evaluations conducted in London and Amsterdam.

46.4.1 Iteration 1: Navigation through Multiple Views of the eBook

A tourist guide for the German town of Wernigerode was developed and enriched by sign language video, as well as by English and German voice recorded from a native speaker. While the medieval town is small, maps are needed by most tourists in the irregular shaped lanes. Maps were coded in scalable vector graphics (SVG) to support magnification and color mapping. Besides text and images, videos show the major sightseeing points, such as town hall and castle. The enrichments were verbal descriptions of the added images, text annotations on the maps, and captions of the videos. In the reading program, readers could choose different colors for text and background, enable highlighting of sentences, and control magnification of text and maps.

The book was evaluated by 19 partially sighted, deaf, and sighted users for general accessibility and some usability problems, especially:

- Personalization of intra-document navigation structures (e.g., table of contents, indices)
- Personalization of intra-page navigation structures (e.g., to top of page, location of navigation links)

The personalization of navigation structures was positively received, but more enhancements were required. Suggestions for improvements of the user interface were to be incorporated in the next iteration and led to the redevelopment of the user interface of the browser. More consistent navigation techniques were required.

46.4.2 Iteration 2: Synchronization of Media

The Wernigerode prototype was extended with more pages; each with additional enrichments for the second iteration. These techniques were applied in a new book, a tourist guide to parts of London, which appeared to be more easily read by English-speaking participants.

Several sign language interpreters signed all the text and were videotaped. The user interface of the reading program ensuring

TABLE 46.4 Overview on MultiReader Books

Book Title	Number of Text Pages	Number of Videos	Number of Images
London Tourist Guide	110	66	44
Wernigerode Tourist Guide	60	7	60
European Cookery Book	90 recipes	4	20
Hamlet	Full text of the play and footnotes	2	10
New Perspectives on Art	60 long text pages	5	30

sentence-by-sentence highlighting was synchronized with the sign language videos. The toolbar of the program was prepared for personalization in several ways:

- Sign language video to replace each icon
- Human head video (for lip reading)
- Different icon sets supporting two levels of magnification and two color schemes
- Animated icons for dyslexic readers

User evaluation with 12 deaf, dyslexic, and mainstream users (as control group) was carried out mainly testing synchronization issues. Deaf readers appreciated sign language videos in conjunction with text, synchronized text highlighting, and toolbar icons containing sign language (Petrie et al., 2005b). Dyslexic readers did not react very positively to animated icons. Readers expressed a need for training to make best use of the detailed personalization features. While all found this aspect of the reading program very useful, the developers fine-tuned the profile attributes per user group.

46.4.3 Iteration 3: Integration

The third version of the multimedia document was evaluated with all target user groups. It incorporates the full range of features specified in the user requirements and improvements based upon the previous two iterations of evaluations. A method of synchronizing personal profiles of the reading program with the underlying GUI for each reader (Weimann, 2003) was incorporated. Readers are able to set their own personal profiles while using the system: changing speed of speech output, background and text colors, magnification of text and images, highlighting and video presentation. Further materials have been added with more multimedia enrichments, for example, text captions for videos, audio description of images, and text description of audio elements.

Content enrichments provided in this iteration included audio output for the text, improved, richer descriptions of images for blind users, and signed descriptions of images for deaf users. Adaptations of the interface included improved navigation structures, such as thematic indexes.

An evaluation with 70 print-disabled users from all the target user groups investigated all the features developed within the MultiReader system. In particular, the use of sign language videos synchronized with text highlighting for deaf readers, as well as text highlighting synchronized with simultaneous speech output for dyslexic readers, were both very successful. This evaluation also showed the importance of providing reader control for each time-dependent medium and all time-dependent enrichments. For example, dyslexic readers need to be able to control the speed of highlighting the text and blind readers need to be able to start and stop videos (particularly so that they do not interfere with the speech from a screen reader).

46.4.4 Context of Reading

The context of all these evaluations were desktop computers equipped with standard multimedia features. The *European Cookbook* was read outside of the kitchen and thus many aspects of navigation in the recipes belonging to a complete menu could not be convincingly recreated in a laboratory setting. An initial attempt toward use in the kitchen, when it is difficult to operate a keyboard, is the MultiReader book-reading software's ability to separate all interaction techniques used for reading from content and presentation. This allowed integration of a speech input facility for speaking readers. For example, browsing page by page is done through spoken commands. The limitations of this approach become apparent when gathering appropriate annotations and making them suitable for personalization.

While the tourist books addressed a mobile scenario, the reading devices were not evaluated in this context. The books were read for preparation of a tourist journey and tasks aimed at information to be gathered by the readers for a better understanding of the places to be visited.

For a true tourist book a mobile scenario should be studied. Enrichments in this respect include many services and physical features such as public transportation and accessible routes for pedestrians who are blind. Moreover the user groups to consider have to include people in wheelchairs. In the following, an application of MultiReader personalization is discussed in a follow-up study using mobile phones, which already now can be used for reading eBooks.

46.5 Mobile Scenarios

Mobile telephony has been adopted by large parts of the population. Even though mobile phones have been used mainly for telephony since only a few years ago, they have emerged as multifunctional multimedia devices usable for a multitude of purposes. Mobile phones allow synchronous communication from anywhere at anytime without being dependent on a stationary telephone. Due to the wide range of possible applications developed recently, such as asynchronous information services based on multimedia messaging services (Völkel and Weber, 2006), disabled and elderly people can benefit especially and have greater independence (Abascal and Civit, 2001).

Elderly, visually impaired, and hearing-impaired users impose additional requirements regarding the usage of mobile phones. For example, font size on the screen and keypad, as well as luminosity and contrast, are important properties (Plos and Buisine, 2006). Some users prefer black-and-white contrast, while others have difficulties when reading a very bright screen. In addition, blind people encounter problems with hard-to-feel buttons. Buttons with 0.3 mm height are a possible solution to this problem, as has been reported by Tomioka (2004). Requirements for hearing impaired people include high-contrast presentation of textual information. However, an interference with the mobile phone is likely to arise from the use of auditory prostheses, limiting the overall use of mobile phones for this user group.

Currently available mobile phones present information to the user through a great variety of different media such as text, images, sound, and video. Consequently, elderly, disabled, and in particular print-disabled people can benefit from personalized presentations of information rendered in different media types (Petrie et al., 2005b). This section therefore focuses on multimedia messaging services (MMS) as one possible solution for the implementation of presentations that are accessible for these user groups. To create an accessible presentation using multimedia messaging services, messages must be transformed. Afterwards, they consist of appropriate media for each specific user. Consequently, the structure of a message relies on a user profile that includes all specific requirements for the media arrangement. The profiling of multimedia messages is discussed in detail in the next section, followed by the discussion of a public transportation scenario.

46.5.1 Profiling of Multimedia Messaging Services

The MMS standard is a subset of the original SMIL standard that takes into account the output and input limitations of mobile phones. MMS SMIL is thus intended to meet the "minimum set of requirements and guidelines for end-to-end interoperability" (Open Mobile Alliance, 2005). MMS documents are organized into a slide show where the individual slides contain at most two regions. The top region typically contains an image or a small video clip, and the lower region is used for displaying text content. Figure 46.7 shows the scheme of the structure of a multimedia message.

The use of a subset of the SMIL language for the definition of multimedia messages allows the use of a broad range of different media types and formats. To guarantee interoperability between sender and recipient, Open Mobile Alliance (2005) published the MMS Conformance Document defining various content classes. These classes include message types allowing exclusively plain text (Text), and classes allowing an extensive use of multimedia elements such as baseline JPEG, GIF, and WMPD images, video in H.263 and MPEG4 codec, and sound in AMR-NB and SP-MIDI format (Video Basic and Video Rich). The latter classes support messages up to 300 kb. Most of the currently available mobile phones support the Video Rich content class, which allows the usage of all multimedia elements necessary for the implementation of accessible presentations.

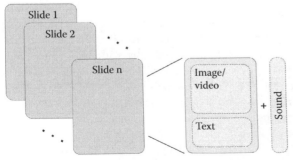

FIGURE 46.7 Basic structure of a multimedia message.

Some important differences must be considered when using multimedia messaging services for the realization of accessibility. For example, authors of web content can normally assume that users with low vision use assistive technologies such as screen magnifiers or screen readers. This assumption does not hold for mobile scenarios, as assistive technologies for mobile phones are very rare. Although screen readers such as Nuance Talks (2006) are available for mobile phones and offer hands-free access to most information and features, no support is offered for multimedia messaging services. Consequently, the presentation of multimedia messages must be personalized with respect to the individual requirements of the user. Two possible strategies are applicable for personalization, namely, client-side adaptation and server-side adaptation of the message content and its corresponding presentation.

For the client-side adaptation of multimedia messaging services, additional software in conjunction with a user profile must be deployed for each user and each mobile phone respectively. In addition, a broad range of software versions is necessary, as each mobile phone type uses proprietary software for rendering the content of multimedia messages. The use of client-side adaptation requires the delivery of all potentially necessary media as the profile of the user is not known in advance. This leads to messages that are likely to exceed the maximum size of 300 kb. Compared to client-side generation, server-side generation of personalized multimedia messaging services has some advantages that are discussed in detail in the following.

Basically, two solutions are possible for server-side generation of accessible multimedia messaging services. One possibility is the authoring of different versions of the message for each estimated user group. Although this approach certainly works, it has some important shortcomings, as the workload for authors and their necessary knowledge increase significantly. Furthermore, the integration of new user groups demands higher investment, as a specific version of all messages must be created for the new user group. When modifying existing messages, all changes have to be repeated for all versions.

The solution finally implemented follows a different approach by using user profiles stored on the server, which allows a dynamic adaptation of messages based on media transformation during run-time. In addition, user profiles might also be transmitted by the user when dealing with pull-based services permitting an increased data privacy as the profiles are stored on the client device. However, this approach does not permit the implementation of push-based services as have been described in Völkel and Weber (2005). Since every user group has specific requirements, appropriate media arrangements have been adapted (see Table 46.1) and were applied to MMSs (see Table 46.5).

Regarding the presented media arrangements, synthetic speech is used to present visual information for blind users and users with low vision. Additionally, as a conversion of information presented by pictorial elements cannot be performed automatically, authors of multimedia messages must provide an alternative textual description. This description is then taken as the input for the generation of synthetic speech.

TABLE 46.5 Media Arrangements for Different User Profiles

	Information	Blind	Partially	Deaf	Dyslexic	Meaning
Text	Text		X	X	X	Textual content
	Synthetic Speech	X	X		X	Textual content as synthesized speech
	Appearance		X	X	X	Background contrast
Pictorial Elements	Picture		X	X	X	Photos, drawings, maps, etc.
	Alt Description		X	X	X	Description of the picture
	Audio Description	X	X			Description as synthesized speech
Audio	Audio	X	X		X	Music, sound, narration, etc.
	Alt Description			X		Textual or graphical description of the audio
Video	Video	X	X	X	X	Video content
	Alt Description			X		Textual or graphical description of the video
	Audio Description	X	X			Description as synthesized speech

Considering the profile for deaf users, textual information should be presented using big fonts and high contrast. People who are deaf from birth mostly use sign language as their mother tongue and have consequently difficulties reading written language. Another option is the use of small videos of sign interpreters presenting the content of the multimedia message. However, the inclusion of sign interpreters within the production process of multimedia messages imposes additional costs. A suitable alternative is to use avatars that can produce high-quality sign language videos (Elliot et al., 2000).

The implemented solution for the generation of accessible MMS is based on the system architecture presented in Figure 46.8.

The system consists of a document management system managing all media elements such as texts, images, audio clips, and video clips. The central MMS profiling engine retrieves necessary media elements identified by parsing the basic MMS description available in SMIL format from the MMS document management component. Based on an additional user profile retrieved from the server by a user profile management component or provided directly by the client, content of nonappropriate media elements is substituted by appropriate media elements. The resulting multimedia message is then sent to an MMS gateway that handles the process of delivering the message via standard GSM/UMTS networks to the recipient.

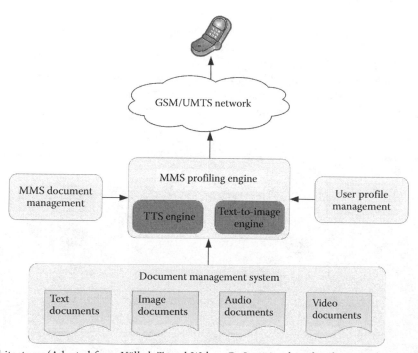

FIGURE 46.8 System architecture. (Adapted from Völkel, T. and Weber, G., Location-based and personalized information services for spas. *Workshop Space, Place & Experience in Human-Computer Interaction (INTERACT'05)*, 12–16 September 2005, Rome, Italy, http://infosci.cornell. edu/place, 2005.)

The process of automatically substituting media elements still imposes one serious problem. For example, for blind users text content will be replaced by synthetic speech. If an additional pictorial element is also present on the same slide, the alternative description of this element will again be transformed into synthetic speech. Consequently, two audio streams must be synchronized, that is, rendered in a specific order that is to be determined. The problem of determining the rendering order is one of linearization, which mainly occurs in cases where information available via different media must be presented by the same time-dependent media. This problem is referred to as the problem of coherence of time-dependent media, and is defined as the temporal coherence that is a quality mark applicable to interactive systems for their ability to replace multiple time-dependent or time-independent media with other time-dependent or time-independent media at different levels of temporal granularity (Petrie et al. 2005b).

One possible solution to ensure temporal coherence is the provision of additional information by authors of multimedia messages indicating the rendering order. Consequently, authors are required to provide more information than would be necessary for ordinary multimedia messages. However, this effort is minimal compared to an alternative authoring of a multitude of different messages for each user group. Additionally, new user profiles can be included easily as only the transformation rules must be provided.

46.5.2 Public Transport Scenario

Many mobility-impaired people such as visually impaired or blind people rely on public transport for covering distances greater than those that are easily walked. Timetables play an important role regarding ordinary public transport within city areas and its surrounding areas. Choosing appropriate bus routes for itineraries that are not known in advance relies strongly on the knowledge of corresponding departure times. This knowledge is particularly important very early in the morning and late in the evening when the frequency of transport decreases.

Timetables for public transport are mostly provided in printed form mounted within a display cabinet. The information presented is static, that is, only expected target departure times of transportation means are provided. An adaptation to short-term changes as a consequence of differing traffic is not possible. Recently, a new class of timetables has emerged and is widely deployed in many cities. Electronic boards are used to display up-to-date timetable data that are instantly updated when changes occur.

Figure 46.9 presents an example for both a paper-based timetable and an electronic timetable. The printed timetable is organized into five columns, the first showing the full hour followed by the minutes for days from Monday to Friday (second column), for Saturday (third column), and for Sunday (fourth column). The following stops and the time to reach them are presented in the fifth column. The digital display uses only three columns to display up-to-date timetable data. The number of the bus line is presented in the first column followed by the final destination of the route and either a relative time (i.e., 5 min) until the arrival of the bus, or an absolute time (i.e., 14:00).

For visually impaired people, the readability of information presented by printed timetables as well as electronic timetables imposes some serious problems. Visually impaired people might not be able to approach the display as closely as necessary, as the electronic timetable is presented well above eye level to prevent vandalism. In addition, fonts and contrast are not sufficient for many visually impaired and elderly people even using visual aids. Blind people have no possibility of accessing the information nonvisually.

To overcome the barriers imposed by printed and digital timetables for visually impaired people, the mAIS (mobile automatic information assistance system) project intents to provide an accessible information service based on standard mobile phones and multimedia messaging services. The project is carried out with academic, public, and industrial partners from northern Germany, including the cities of Flensburg, Neumünster, and Kiel, as well as local engineering companies.

Basically, two scenarios are covered by the system, the first being a request within the range of a Bluetooth connection

FIGURE 46.9 Paper-based timetable (a) and electronic timetable (b).

around the bus stop, and the second being a remote request. When the passenger reaches the bus stop, timetable data can be retrieved by establishing a Bluetooth connection to a beacon that is mounted on the electronic display. The mobile phone then queries for a personalized and accessible multimedia message containing the desired timetable data by transmitting the profile of the user. Possible bus lines of the resulting timetable data can easily be selected as this service is only available within a short range of the corresponding bus stop.

The system also offers remote access to timetable data, which is particularly useful outside the range of the deployed beacons and in case a passenger wants to pre-plan a bus ride. A standard GPRS/UMTS connection must be established by the passenger for remote requests. The personal profile in conjunction with the desired departure bus stop is transmitted to the server, which in response transmits a multimedia message containing the desired data. As in the previous scenario, the multimedia message presents the information in an accessible form.

The scenario for reading on mobile phones has been implemented as part of the mAIS project. Beacons have been deployed at specific bus stops in the cities of Flensburg, Neumünster, and Kiel. Tests were being carried out with three different user groups. The first group includes users with low vision (especially elderly people), and the second group includes blind users. A special version of the mobile client compatible with the Nuance Talks screen reader has been developed enabling speech-based access to the system. The third group serves as the control group and includes only users with no visual impairment.

Results of a field study with 55 mobility-impaired passengers (blind visually impaired, hearing impaired, deaf, wheelchair users, and a control group) indicate that users read passenger displays often well ahead of the time of departure, easily up to half an hour from a remote location before reaching the platform or bus stop (Weber et al., 2008). To ensure data protection, they want to store their user profile data on their mobile phone and release it only for processing geographic information and transforming information in a single transaction. The field study points out that server-based multimedia content preparation as described previously can be utilized not only by print-disabled but also by mobility-impaired people.

46.6 Outlook

Both MultiReader and mAIS have confirmed the heterogeneous needs of print-disabled users and the advantages of personalized multimedia information. The process of authoring this information has been identified and requirements for coherence of media have been presented.

Future work may be able to develop automatic tools for making sure that coherent media are indeed coherent and follow these requirements. However, a more promising approach will be to use either heuristic methods or tools to conclude from only a few users' feedback about the quality of universally accessible eBooks.

References

Abascal, J. and Civit, A. (2001). Universal access to mobile telephony as a way to enhance the autonomy of elderly people, in *Proceedings of the 2001 EC/NSF Workshop on Universal Accessibility of Ubiquitous Computing*, 22–25 May 2001, Alcacer do Sal, Portugal. New York: ACM Press.

Alonzo, A. (2001). A picture is worth 300 words: Writing visual descriptions for an art museum web site, in *the Proceedings of CSUN 2001: Technology and Persons with Disabilities Conference*, 19–24 March 2001, Los Angeles. Northridge, CA: California State University Northridge. http://www.csun.edu/cod/conf/2001/proceedings/0031alonzo.htm.

Bauwens, B., Engelen, J., and Evenepoel, F. (1994). Increasing access to information for the print disabled through electronic documents in SGML, in the *Proceedings of ASSETS 98*, 15–17 April 1994, Marina del Rey, CA, pp. 55–61. New York: ACM Press.

Blenkhorn, P., Evans, G., King, A., Kurniawan, S., and Sutcliffe, A. (2003). Screen magnifiers: Evolution and evaluation. *Computer Graphics and Applications IEEE* 23: 54–61.

Brusilovsky, P. (2001). Adaptive hypermedia. *User Modeling and User-Adapted Interaction* 11: 87–110.

Campbell, R. and Burden, V. (1995). Pre-lingual deafness and literacy: A new look at old ideas, in *Speech and Reading: A Comparative Approach* (B. de Gelder and J. Morais, eds.). London: Psychology Press.

Chisholm, W., Vanderheiden, G., and Jacobs, I. (eds.) (1999). *Web Content Accessibility Guidelines 1.0, W3C Recommendation 5-May-1999*. World Wide Web Consortium. http://www.w3.org/TR/WCAG10.

Elkind, J., Black, M. S., and Murray, C. (1996). Computer-based compensation of adult reading disabilities. *Annals of Dyslexia* 46: 159–186.

Elliot, R., Glauert, J. R. W., Kennaway, J. R., and Marshall, I. (2000). The development of language processing support for the ViSiCAST project, in the *Proceedings of the Fourth International ACM Conference on Assistive Technologies*, 13–15 November 2000, Washington, DC, pp. 101–108. New York: ACM Press.

Evett, L. and Brown, D. (2005). Text formats and web design for visually impaired and dyslexic readers: Clear text for all. *Interacting with Computers* 17: 453–472.

Fix, U. (2005). *Hörfilm* [Audio Description]. Berlin: Schmid.

Hillesund, T. (2002). Many outputs: Many inputs: XML for publishers and e-book designers. *Journal of Digital Information* 3. http://journals.tdl.org/jodi/article/view/jodi-62/75.

ISO 13407 Standard (1999). *Human-Centered Design Processes for Interactive Systems*. Geneva: International Organization for Standardization.

Ivarsson, J. (1992). *Subtitling for the Media: A Handbook of an Art*. Stockholm: Transedit.

Jacko, J. and Sears, A. (1998). Designing interfaces for an overlooked user group: Considering the visual profiles of partially sighted users, in the *Proceedings of the Third*

International ACM Conference on Assistive Technologies, 15–17 April 1998, Marina del Rey, CA, pp. 75–77. New York: ACM Press.

Jenge, C., Hartrumpf, S., Helbig, H., and Osswald, R. (2006). Automatic control of simple language in web pages, in the *Proceedings of the 10th International Conference on Computers Helping People with Special Needs (ICCHP 2006),* 11–13 July 2006, Linz, Austria, pp. 207–214. Heidelberg: Springer-Verlag.

Langer, I. (2003). Adaptation of multimedia eBooks: Universal access in HCI, in *Universal Access in HCI: Inclusive Design in the Information Society, Volume 4 of the Proceedings of the 10th International Conference on Human-Computer Interaction (HCI International 2003)* (C. Stephanidis, ed.), 22–27 June 2003, Crete, Greece, pp. 1442–1446. Mahwah, NJ: Lawrence Erlbaum Associates.

Malama, C., Landoni, M., and Wilson, R. (2004). Fiction electronic books: A usability study, in the *Proceedings of ECDL 04,* 12–17 September 2004, Bath, U.K., pp. 69–79. Heidelberg: Springer-Verlag.

Marshall, C. C. and Bly, S. (2005). Turning the page on navigation, in the *Proceedings of JCDL'05,* 7–11 June 2005, Denver, pp. 225–234. New York: ACM Press.

Mynatt, B. and Weber, G. (1994). Nonvisual presentation of graphical user interfaces: Contrasting two approaches, in the *Proceedings of CHI 1994,* 24–28 April 1994, Boston, pp. 166–172. New York: ACM Press.

National Institutes of Health (2002). *Captions for Deaf and Hard-of-Hearing Viewers.* NIH Publication No. 00-4834. http://www.nidcd.nih.gov/health/hearing/caption.asp#net.

Norman, D. (1988). *The Psychology of Everyday Things.* New York: Basic Books.

Nuance Talks (2006). *Convenient Audio Access to Mobile Phones.* http://www.nuance.com/talks.

Office of Communications (2007). *Guidelines on the Provision of Television Access Services.* http://www.ofcom.org.uk/tv/ifi/guidance/tv_access_serv/guidelines.

Open Mobile Alliance (2005). *MMS Conformance Document 1.2.* http://www.openmobilealliance.org/release_program/mms_v1_2.html.

Petrie, H., Fisher, W., Weber, G., Langer, I., Gladstone, K., Rundle, C., et al. (2002). Universal interfaces to multimedia documents, in the *Proceedings of 4th IEEE International Conference on Multimodal User Interfaces, IEEE,* 14–16 October 2002, Pittsburgh, pp. 319–324. Piscataway, NJ: IEEE.

Petrie, H., Harrison, C., and Dev, S. (2005a). Describing images on the Web: A survey of current practice and prospects for the future, in the *Proceedings of the HCI International 2005,* 22–27 July 2005, Las Vegas [CD-ROM]. Mahwah, NJ: Lawrence Erlbaum Associates.

Petrie, H., Weber, G., and Fisher, W. (2005b). Personalization, interaction and navigation in rich multimedia documents for print-disabled users. *IBM Systems Journal* 44: 629–636.

Plos, O. and Buisine, S. (2006). Universal design for mobile phones: A case study, in the *Proceedings of the Conference on Human Factors in Computing Systems,* 22–27 April 2006, Montreal, pp. 1229–1234. New York: ACM Press.

Press, L. (2000). From p-books to e-books. *Communications of the ACM* 43: 17–21.

Prillwitz, S. (2001). *Angebot für Gehörlose im Fernsehen und ihre Rezeption* [Services for Deaf People in TV and Their Reception]. Kiel: Unabhängige Landesanstalt für das Rundfunkwesen.

Steinmetz, R. (1993). *Multimedia Technology: Introduction and Foundations.* Berlin: Springer-Verlag.

Strothotte, T. and Schlechtweg, S. (2002). *Non-Photorealistic Computer Graphics: Modeling, Rendering, and Animation.* San Francisco: Morgan Kaufmann.

Tomioka, K. (2004). Universal design practices: Development of accessible cellular phones, in the *Proceedings of the International Conference on Universal Design "Designing for the 21st Century III."* http://www.designfor21st.org/proceedings.

Völkel, T. and Weber, G. (2005). Location-based and personalized information services for spas. *Workshop Space, Place & Experience in Human-Computer Interaction (INTERACT'05),* 12–16 September 2005, Rome, Italy. http://infosci.cornell.edu/place.

Völkel, T. and Weber, G. (2006). Scenarios for accessible multimedia messaging services, in *Universal Access in Ambient Intelligence Environments: Proceedings of the 9th ERCIM Workshop "User Interfaces for All"* (C. Stephanidis and M. Pieper, eds.), 27–28 September 2006, Bonn, Germany, pp. 211–226. Berlin/Heidelberg: Springer-Verlag.

Vora, P. R. and Helander, M. G. (1997). Hypertext and its implications for the Internet, in *Handbook of Human-Computer Interaction* (2nd ed.) (M. Helander, T. K. Landauer, and P. Prabhu, eds.). Amsterdam: Elsevier.

W3C SMIL (2005). *Synchronized Multimedia Integration Language (SMIL 2.1).* http://www.w3.org/TR/2005/REC-SMIL2-20051213.

Weber, G. (2008). Access to electronic time table displays, in *Mensch und Computer 2008* (J. Herczeg, ed.), pp. 17–26. Munich: Oldenbourg.

Weimann, K. (2003). Modeling users with special reading needs, in *Universal Access in HCI: Inclusive Design in the Information Society, Volume 4 of the Proceedings of the 10th International Conference on Human-Computer Interaction (HCI International 2003)* (C. Stephanidis, eds.), 22–27 June 2003, Crete, Greece, pp. 1487–1491. Mahwah, NJ: Lawrence Erlbaum Associates.

Interpersonal Communication

Annalu Waller

47.1 Introduction

Communication is the essence of life. Simple, everyday communication is taken for granted by most people. People communicate for a range of reasons: from expressing their needs and wants to engaging in social closeness. Communication can be achieved a myriad of ways, but spoken communication remains the natural medium for interactive communication. Interactive communication using speech also relies on the ability to hear—any restriction of either channel will have a negative impact on an individual's ability to interact.

Various techniques are used to augment or to provide an alternative to impaired speech and/or hearing. Sign languages and lip-reading are used by deaf people (see also Chapter 38 of this handbook), while technologies such as electronic voice aids (artificial larynges) function as speech prostheses when vocal cords are damaged due to cancer and other trauma. This chapter is concerned with people who have complex communication needs.

Individuals with complex communication needs (CCN) require additional support for communication due to physical and/or cognitive impairments. CCN can have devastating consequences, as is testified by the following quote by Rick Creech, a man with severe cerebral palsy:

If you want to know what it is like to be unable to speak, there is a way. Go to a party and don't talk. Play mute. Use your hands if you wish but don't use paper and pencil. Paper and pencil are not always handy for a mute person. Here is what you will find: people talking; talking behind, beside, around, over under and through, and even for you. But never with you. You are ignored until finally you feel like a piece of furniture. (Musselwhite and St. Louis, 1988, p. 104)

The inability to speak intelligibly can be due to a variety of congenital or acquired impairments. Congenital impairments, for example, cerebral palsy and Down syndrome, can impact both the physical ability to speak and the cognitive development of language. Acquired communication disorders can result from traumatic injury, for example, as a result of a spinal cord injury, head injury, or stroke, or from degenerative diseases, such as multiple sclerosis or motor neurone disease. Communication can be compromised still further by additional visual and hearing impairments.

The nature of complex communication impairments demands a multimodal approach to providing communication support. This support is referred to as augmentative and alternative communication (AAC) and is defined by the American Speech-Language-Hearing Association (ASHA) as follows:

AAC is, foremost, a set of procedures and processes by which an individual's communication skills (i.e., production as well as comprehension) can be maximized for functional and effective communication. It involves supplementing or replacing natural speech and/or writing with aided (e.g., picture communication symbols, line drawings, Blissymbols, and tangible objects) and/or unaided symbols (e.g., manual signs, gestures, and finger spelling). (American Speech-Language-Hearing Association, 2002, p. 2)

An AAC system is "an integrated group of components, including the symbols, aids, strategies, and techniques used by individuals with severe speech and language disabilities to enhance communication" (Castrogiovanni, 2008, p. 1). An AAC system can be unaided (e.g., sign language, gesture) or aided (use of a physical device or object, e.g., a computer or symbol chart). Effective AAC will employ a range of aided and unaided systems to ensure appropriate support in different situations, for example, in the classroom or work situation, at play, and at mealtimes.

AAC emerged in the late 1950s and early 1960s (Zangari et al., 1994). The advent of microprocessor technology led to the development of early speech-generation devices. Although unaided and light-tech-aided AAC, for example, the use of letter/word/symbol boards, had been in use before, electronics provide a unique means by which people with no speech can access synthetic speech output.

Speech-generation devices, or SGDs, are composed of three main components: the input system, processing, and the output system. The input system refers to how the user interacts with the device in terms of the interface vocabulary (i.e., literacy or symbol-based) and the physical access (e.g., keyboard or switch). The processing system refers to the computational techniques used to translate the user input into the system output. The output system refers to how the device outputs speech (e.g., synthesized or digitized speech, and printed output).

47.2 The Input System

47.2.1 Interface Vocabulary

SGDs can be divided into two broad areas: literacy-based and symbol-based devices. Literacy-based devices allow users to generate novel vocabulary using traditional orthography. Most interfaces use traditional orthography (the written word), although other orthographic symbol systems such as Morse code, Braille, and phonemic symbols can also be used (Beukelman and Mirenda, 2005). However, cognitive limitations have a major impact on the ability to encode and decode traditional orthography. Many individuals with congenital CCN struggle to acquire functional literacy and are thus severely restricted in harnessing the potential freedom afforded by technology (Smith, 2005). People with acquired cognitive impairments, such as, aphasia resulting from a stroke (cerebral vascular accident), also experience varying degrees of difficulty with both expressive and receptive language.

Visual images, photographs, and drawings provide alternatives to literacy-based interfaces. Words, phrases, and complete messages can be stored under pictures. In addition, published AAC symbol sets (e.g., Picture Communication System,[1] Rebus[2]) and semantic-based writing systems (e.g., Blissymbolics[3]) can be used to represent concepts (words) (Wilson, 2003). The type of picture/symbol/graphic used will depend on the iconicity (ease of recognition), transparency (guessability), opaqueness (logic organization), and learnability of the image. For instance, a photograph of a house may be transparent (i.e., it is easily recognizable), while a Blissymbol representing the emotion of happy needs logical processing (heart = feeling; up arrow = up) and is

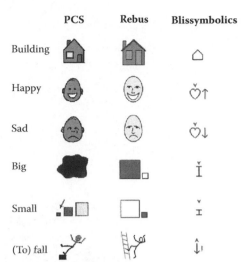

FIGURE 47.1 Examples of three symbol sets that have been developed for augmentative and alternative communication.

thus more opaque, but can be learned and extended to generate other concepts (see Figure 47.1). The more concrete a representation, the more recognizable it will be. However, representations of abstract meanings (e.g., emotions) will involve less transparent images, necessitating a longer learning curve.

47.2.2 Physical Access to SGDs

People who use aided AAC can access devices using direct access or scanning. Direct access using a conventional or adapted keyboard, mouse, or touch screen is faster and easier than scanning.

Direct access depends on the user's ability to access discrete keys or locations on a physical input device. The way in which the input device is calibrated can support or hinder access (e.g., the user may not be able to retract easily from selecting, thus necessitating the need for nonrepeating keys).

Scanning is employed when the physical impairments of the user restrict access (i.e., the input bandwidth is reduced, and is implemented using a single or multiple switches) (see Chapter 35 of this handbook). Target items are highlighted in groups or individually until they are selected by switch activation. Items are organized for linear or group scanning and are organized to reduce the number of switch activations/rate of selection. Row/column scanning is most common, where each column is scanned until one is selected. Items within the selected column are scanned. Figure 47.2 shows a row/column letter matrix organized according to letter frequency in English (e.g., the letters *e* and *a* are the most common letters in English and are thus located in the top left corner of the matrix, requiring three scans each).

Auditory scanning provides auditory feedback for each scan and is used primarily with visually impaired users, but can also benefit others. The highlighting of choices can be automated, or directed by the user (called step scanning). Care must be taken when choosing a scanning method or layout for individual

---

[1] http://www.mayer-johnson.com.
[2] http://www.widgit.com.
[3] http://www.blissymbolics.org.

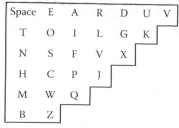

Space	E	A	R	D	U	V
	T	O	I	L	G	K
	N	S	F	V	X	
	H	C	P	J		
	M	W	Q			
	B	Z				

FIGURE 47.2 A letter matrix organized by English letter frequency.

users—a user with increased muscle tone might tense up in anticipation of activating a switch and may benefit from a directed scanning method whereby the user activates a switch to move the scan along while doing nothing to indicate a selection.

The provision of feedback is a crucial aspect of any input system. Feedback can be tactile or auditory (e.g., a tactile depression when activating a key on a keyboard or an auditory beep when scanning target items). The feedback helps users to monitor and prepare their actions. The tactile and auditory click when using a keyboard helps a typist detect when a key has been pressed, while a scanning beep helps a user anticipate when to select a switch and will confirm that a switch activation has occurred. User fatigue can vary the ability to activate switches or a keyboard. Systems should offer easy modification of scan speeds, and can also be programmed to adapt to the user's performance automatically (e.g., by monitoring deletions).

Physical access to AAC devices depends on the individual characteristics of the user and can utilize a range of physical attributes. Higginbotham et al. (2007) discuss the importance of seating and positioning in achieving the most effective and efficient access to assistive technology, and provide an overview of how technological advances in automatic speech recognition, eye tracking, motion and gesture recognition, and brain interfaces are expanding the access opportunities for people with CCN. Notwithstanding the wide range of access opportunities, the way in which user input is processed is crucial in supporting the formulation of communication.

47.3 Processing System

Although text and speech output can be accessed through different physical access systems and through a variety of vocabularies, communication rates are extremely slow—8 to 10 words per minute compared to typical conversation rates of 150 to 250 words per minute (Goldman-Eisler, 1968).

SGDs can be broadly divided into language retrieval devices and text-to-speech systems. In practice, text-to-speech systems use language retrieval techniques to improve the rate of conversation and reduce the effort required for message generation. Language retrieval devices provide access to spoken messages without the need for users to be literate, but vocabulary is restricted to what has been prestored by others.

The role of the processing system is to support the generation of communication messages by assisting in the retrieval of prestored language and by accelerating the communication rate. The generation of communication requires the user to know what he wishes to say and how he should say it. This may be difficult for users with developmental, intellectual, and linguistic difficulties. The processing system must therefore also support the pragmatics (use of language) for users.

47.3.1 Storage and Retrieval of Messages

The most basic SGDs provide a means by which users can retrieve prestored language items by selecting a button. Dedicated devices with static interfaces can range from one-button devices, such as the BIGmack, to several keys, such as the Pocket GoTalk and SuperTalker, and provide simple message retrieval (Figure 47.3).

(a)

(b)

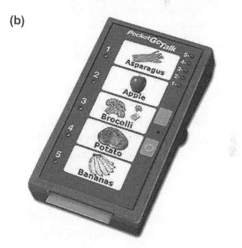

(c)

FIGURE 47.3 (a) BIGmack, (b) Pocket GoTalk, and (c) SuperTalker—simple message retrieval devices.

FIGURE 47.4 The Dynamo from Dynavox.

The amount of vocabulary on message retrieval devices can be extended by introducing levels. Different overlays can be used with devices to remind users of the different vocabulary stored on each level.

Devices that use dynamic displays, for example, the Dynavox range of devices (Figure 47.4), provide multilevel message retrieval by changing the computer display until a target utterance is selected. Although similar to level message retrieval, dynamic displays can also be viewed as encoding devices, since messages are stored under a sequence of selections. Dynamic display devices mirror the use of communication books, which are organized using topics indexed on the top sheet, or topic boards, which provide vocabulary in topic groups. As with low-technology alternatives, there is potential for vocabulary to be accessed via various routes.

AAC devices with programmable displays are also able to use graphic scenes or photographs in which hot spots access embedded messages (Dye et al., 1998). Referred to as visual scene displays (Blackstone, 2004), these devices are different from traditional dynamic display devices in that the displays are personalized and use highly contextual visual representations such as photographs and drawings. Devices such as the Dynavox range can be adapted to function as a visual scene display (VSD). VSDs have been successfully applied for people with aphasia (Dietz, McKelvey, and Beukelman, 2006) and children with CCN (Drager and Light, 2006).

Encoding describes ways in which linguistic items can be stored under code sequences, usually in devices with static input keyboards. The Minspeak system integrated a technique called Semantic Compaction (Baker, 1982, 1987), which uses a set of iconic codes as a mnemonic system to encode words, phrases, and sentences. The message, "I like hamburgers," could be encoded under "apple" (mnemonic for food) + "lightning bolt" (mnemonic for fast). Semantic compaction is used on Prentke Romich devices such as the Pathfinder Plus (Figure 47.5), but this type of message encoding is also found on a variety of other devices in different forms, such as dynamic screens.

47.3.2 Acceleration Techniques

The acceleration of slow communication rates has always been a focus of AAC research (Fould, 1980, 1987; Swiffin et al.,

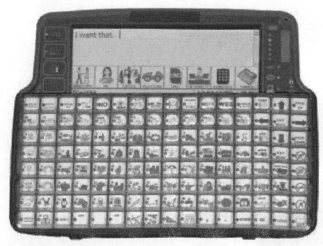

FIGURE 47.5 Pathfinder Plus from Prentke Romich.

1987; Vanderheiden and Kelso, 1987; Higginbotham, 1992). Acceleration techniques can be divided into two broad types: expansion and prediction.

Expansion describes techniques that expand abbreviation codes into fuller text. Examples of encoding techniques can be found in traditional word processors, which offer users the facility to store words or phrases under abbreviation sequences (e.g., *ph* could expand into *telephone*). Dedicated SGDs such as the LightWriter (Figure 47.6) allow users to store phrases such as "My name is Anna" under a user-defined sequence of keys (e.g., MN).

Some research has investigated automated message expansion. The Compansion project (Demasco and McCoy, 1992; McCoy and Demasco, 1995) uses natural language generation techniques to expand telegraphic sentences into full grammatical sentences (e.g., APPLE EAT JOHN is automatically generated into THE APPLE IS EATEN BY JOHN). Such software requires in-depth knowledge of syntax, morphology, and semantics to know that the apple does not eat John.

Message prediction is achieved by using statistical and syntactic knowledge of written language. Single letter prediction was thoroughly researched in the 1980s (Heckathorne et al., 1987), but the constant rearrangement of the letter display was unusable. Instead, static grids of letters (cf. Figure 47.1), which use letter frequencies, are widely used in scanning onscreen keyboards.

Letter frequencies have been utilized with more success in generating words typed on reduced keyboards (Arnott and Javed,

FIGURE 47.6 The LightWriter SL35.

1992) and character disambiguation is now part of mobile tele-communications and is available on some AAC devices.[4] Using statistical knowledge and dictionaries of words, disambiguation software can generate words typed on a restricted keyboard (e.g., nine keys on a mobile phone). Selecting the following three keys on a phone, ABC + ABC + TUV could generate CAT or BAT, depending on the frequencies of these words.

Word prediction (Swiffin et al., 1987; Higginbotham, 1992) was initially developed to assist typists with physical disabilities. Using statistical and syntactic information, it is possible to predict probable words from what has already been typed. Although the increase in communication rate is at most 50% (Higginbotham, 1992), word prediction is of benefit to children learning to read and write; and dyslexic individuals who are able to recognize the target words they wish to type (Newell et al., 1992).

Word prediction techniques have also been applied to Blissymbolics. Hunnicutt (1986) produced software that translates Blissymbol sentences into spoken Swedish sentences.

A unique application of letter and word knowledge is found in the software program Dasher (Ward and MacKay, 2002). The user floats through letters and words that appear from the right. Experiments report a communication rate of 25 words per minute. The system uses the same underlying language model as other word prediction systems, but the user interface is very different, providing a dynamic interface that does not seem to confuse users as was the case in early letter prediction systems. Dasher is available in numerous languages, including Blissymbolics.[5]

Whole message retrieval has also benefitted from linguistic prediction. Langer and Hickey (1998) developed WordKeys, a research system that allows the user to retrieve prestored messages by typing or selecting a key word that can be associated with any word in the target message. For instance, the message "I enjoy watching tennis" could be retrieve by entering the word ball, even though ball does not appear in the sentence. The associations (e.g., a ball is used in tennis) are automatically generated by a large semantic lexicon, derived from the WordNet database (Miller et al., 1993).

Research by Todman (in press) suggests that well-designed utterance-based retrieval systems that link pragmatic features and user goals can lead to faster communication without losing coherence. Experiments comparing transcriptions with and without the use of an utterance-based device showed that conversational rate and perceived communication competence was higher when the utterance-based device was used. The device, a modified version of the TALK system (Todman et al., 1995), is based on a pragmatic model that sees the progression of a conversation as a series of gradual shifts of perspectives relating to the speaker, time (past, present, future), and event-related information (what, where, who, how, and why). By changing perspective, the system can predict possible utterances.

---

4 http://www.dynavoxtech.com/products/palmtop3.aspx.
5 http://www.inference.phy.cam.ac.uk/dasher.

47.3.3 Pragmatic Support

As stated before, effective conversation is not solely dependent on communication rate, but on the way language is acquired and used. Conversation is complex and exhibits different characteristics depending on the goal, or purpose, of the conversation. Research into the purposes of interactive communication identified four main areas: (1) communication of needs and wants; (2) information transfer; (3) social closeness; and (4) social etiquette (Light, 1988).

Aided AAC has traditionally focused on providing fast access to needs-based communication (Murphy et al., 1994). Factors that influence the effectiveness of an interaction include: increased turn taking so that control of the conversation is shared; increased agreement when changing topic; and decreased disruption to conversation flow during and after a conversation breakdown (Cheepen, 1988). One factor that seems to elude nonspeaking people most is the difficulty in taking, and maintaining, control of a conversation.

Everyday communication can be divided into three main areas (Cheepen, 1988): speech-in-action; phatic communication; and free narrative. Speech-in-action refers to what Bowden and Beukelman (1988) have defined as the communication of needs and wants (i.e., goal-driven communication). Phatic communication refers to the type of conversation used to express information that has no real significance other than social contact and etiquette (e.g., greetings and farewells). Information transfer and real social closeness is communicated through free narrative.

Research into conversation modeling has led to the development of a communication system model that addresses different types of interaction (Newell, 1991): formulaic conversation, reusable conversation, and unique conversation (Figure 47.7).

Formulaic conversation includes generic speech acts such as openings, small talk, feedback, and closings (Alm, 1988), whereas unique (word-based) conversation is very particular to the current situation. Between these two ends of the conversation spectrum lies another type of conversation—reusable conversation. This includes a large amount of conversation that is neither formulaic nor unique, but is reused often on many different occasions. Such reusable conversation can be sentence/phrase-based

Type of conversation		Examples
Formulaic conversation		Hello How are you Bye bye
Reusable conversation	Needs/wants	I am hungry. I need help. Please call my assistant.
	Personal stories	I went skiing in Canada. It was very cold. I stayed with cousins... I have lived in Dundee for nine years. I used to live in Cape Town...
Unique conversation		We saw a film called ...

FIGURE 47.7   Different types of verbal conversation.

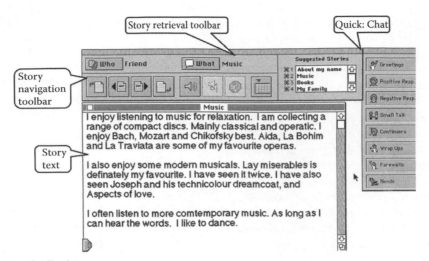

FIGURE 47.8 Screenshot of Talk:About.

(i.e., goal-driven conversation) or narrative-based (i.e., retelling anecdotes, experiences, etc.; Waller, 1992).

Goal-driven and phatic communication can be anticipated to a high degree. Most augmentative communication systems offer facilities to retrieve frequently used sentences for goal-driven communication. The acceleration techniques mentioned previously are used to aid the recall of this type of information.

Work on using automatic retrieval of phatic communication (Brophy-Arnott et al., 1992) has been shown to be an effective framework for a conversation. Implemented in Talk:About (Don Johnston Inc.[6]) as Quick:Chat, it allows users to produce greetings, feedback remarks, and so on, quickly and effectively.

The third type of communication, free narrative, depends on word-by-word communication, which is slow and frustrating. Individuals with CCN do not tend to engage in storytelling (Beukelman, 1991; Waller, 1992; Murphy et al., 1994) and yet, "the major bulk of interaction takes place through the medium of Story" (Cheepen, 1988).

47.3.4 Story-Based Communication

Communication of social closeness and expression of personality is not restricted to isolated sentences. Clarke and Clarke (1977, p. 35) comment that

hardly any of our day-to-day use of language stops after one sentence. People engage in conversations, stories, gossip, and jokes that consist of a succession of sentences in a highly organized social activity.

Stories consist of anecdotes, jokes, and experiences, and are used to promote social acceptance, social closeness, and personality projection. Stories and anecdotes are particularly

important, as past experience and the ability to relate events is an essential part of a person's makeup. Conversational narratives provide a way of forming experience and relating past experience (Quasthoff and Nikolaus, 1982) and play an important part in an individual's social and educational development. In addition, telling and retelling personal stories helps individuals to reflect on how experiences form who they are— stories provide form and meaning to life and are used to create one's self-concept (Polkinghorne, 1995).

Research into conversational narrative revealed that the basic information within a "story" remains relatively constant, although the length increases over time (Waller, 1992). The narration of story texts proved important, as the amount of text used depends on factors such as conversational partner and time constraints. This led to the concept of storing larger chunks of text for reuse instead of resorting to word-by-word text generation.

The storage of reusable conversation raises issues about the retrieval of information. Several techniques used in the fields of information retrieval and artificial intelligence assist the user in locating target texts within large databases. These approaches include the use of semantic networks (Waller, 1992), fuzzy logic (Alm et al., 1993) and story scripts (Alm et al., 1995).

The narration of stories has also received attention. It is important that users are able to control the delivery of the text, both by selecting parts to be spoken and by pausing for comments from the listener (Waller and Newell, 1997). The concept of story-based communication forms the basis of an integrated communication system called Talk:About (Waller et al., 1996). Figure 47.8 shows a screen of the system with Quick:Chat down the right-hand side, a story retrieval toolbar along the top, and a story navigation toolbar below.

The Talk:About software included features that supported characteristics essential to a narrative development in conversation, which would be critical to include in the design of future communication devices. These are the need to develop stories over time, the need to retrieve and move between appropriate

---

[6] http://www.donjohnston.com.

stories within conversation, and the need to narrate the story piece by piece to facilitate turn taking between conversational partners.

47.3.5 Accessing New Vocabulary

Developing conversational skills depends on opportunities to experience successful interactions and to acquire new vocabulary. To generate novel vocabulary, individuals require the ability to read and write. Preliterate individuals are dependent on others to provide them with new vocabulary. With this new vocabulary comes an additional learning demand in that the user has to master the retrieval of new items. Vocabulary acquisition is difficult and unnatural. Children with CCN have the additional disadvantage of not being able to play with language.

Two projects are currently underway to investigate ways in which individuals with CCN can access and experiment with novel vocabulary. The ongoing BlissWord project (Andreasen et al., 1998; Waller, 1998; Arnott et al., 1999; Waller et al., 2000; Waller and Jack, 2002) is investigating ways in which users can explore new vocabulary using Blissymbolics in a way that reduces the time required for entering new words letter by letter. The STANDUP project (Ritchie et al., 2006) has developed software that allows children with CCN to play with language and word meanings by generating new jokes.

Although Bliss lexical items (called Bliss words) can be spelled using a sequence of one or more Bliss characters, there is as yet no commercially available software specifically designed for people with CCN who use Blissymbolics to write Bliss. Existing software manipulates Bliss words as picture items, but the items cannot be broken down into characters. The current stage in the ongoing BlissWord project is the development of a Blissymbolic font that will provide a resource for future Blissymbolic applications.

An example of one such application is the implementation of a Bliss word processor that will allow users to generate novel vocabulary independently. This will allow nonorthographic users to move beyond prestored narrative to elaboration and coconstruction. Because of the generative characteristics of Blissymbolics, predictive algorithms can be applied to it to assist users in the retrieval of words based on concepts they symbolize, again possibly shortening retrieval time. Bliss words are sequenced beginning with a classifier (e.g., all emotions begin with a heart). The interface shown in Figure 47.9 illustrates the way in which Bliss characters and words can be predicted once a shape has been selected from the Bliss keyboard—the interface produces a list of Bliss words that begin with classifiers using that shape. Frequency and word lists can be used to further refine the Bliss words that are displayed. Users do not need to be literate to explore language and vocabulary. It is envisaged that video clips and spoken explanation could further augment learning through exploration.

People who use Blissymbolics can generate new words and concepts using low-tech Blissymbol boards (McNaughton, 1993). This use of Blissymbolics is mediated by another person. If, for

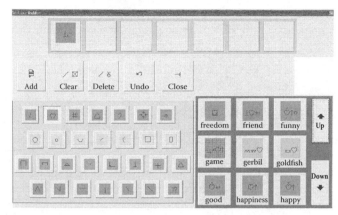

FIGURE 47.9 BlissWord interface showing the prediction of Blisswords containing a heart.

example, a person using Blissymbolics pointed to the Bliss character for *building*, followed by the Bliss character for *book*, the listener would interpret the Bliss word as *library*. The ability to use Blissymbolics as a writing medium allows users to construct new concepts and words independently.

The second work in progress is the STANDUP project (O'Mara and Waller, 2003; O'Mara, 2004; O'Mara et al., 2004; Ritchie et al., 2006; Manurung et al., 2008). It encourages play with words, which is a critical part of language development in children. Typically developing children enjoy jokes and riddles, which offer an opportunity to practice language, conversation, and social interaction skills. In particular, jokes are a type of conversational narrative and, as such, play an important role in the development of storytelling skills.

The STANDUP project has involved the development of interactive software that allows children with CCN to engage in building simple punning riddles. Puns, punning riddles, and jokes (verbal wordplay) form a natural part of children's discourse. They provide a structure within which words and sounds can be experienced and within which the normal rules of language can be manipulated.

However, children with CCN do not always have language play opportunities. Although some clinicians (e.g., King-DeBaun, 1997; Musselwhite and Burkhart, 2002) have reported on their use of verbal humor as a support for communication skills, little research has been reported on the role of humor in AAC or the role it plays in developing storytelling skills.

Most nontext AAC devices are based on the retrieval of prestored linguistic items, such as words, phrases, and sentences. Even when question type jokes are made available on a device (e.g., "What do you call a judge without fingers? … Justice thumbs"), the focus is on the order of retrieval and pragmatic use of the joke, rather than on generating novel humor. The three-year STANDUP project was designed to address this gap, and has been designed for children with CCN to build their own novel jokes.

The STANDUP program is based on the research system JAPE (Binsted et al., 1997), which generated a variety of types of novel punning riddles. In these humorous texts in a question and answer form, humor arises from some form of linguistic ambiguity within the text. Examples of such jokes produced by JAPE are:

> "What do you call a murderer with fibre? A cereal killer."

> "What's the difference between leaves and a car? One you brush and rake, the other you rush and brake."

> "What do you get when you cross a monkey and a peach? An ape-ricot."

The STANDUP project has taken the JAPE concept and designed a software program to generate jokes and an interface suitable for children with CCN. The goal was to provide users with the means to construct jokes on specific topics, using familiar vocabulary, enabling them to experiment with different forms of jokes. The resulting software is interactive, dynamic, flexible, and accessible, providing a source of language development and social interaction possibilities that enable the user to go beyond the needs and wants of assisted communication. STANDUP was evaluated with nine children with CCN (Waller et al., in press). Initial results show that the children are able to use the software to generate puns. The children are exploring new vocabulary and are eager to entertain others, which is not available to them without STANDUP.

Unlike current use of humor in AAC, where jokes are pre-stored in communication devices, STANDUP helps users generate novel puns using a computer algorithm. One of the goals for this research has been to develop a system that will allow users to create novel conversational items so that they can have control of what they use and what they discard. When typically developing youngsters engage in early humor, they construct their own jokes, which are seldom funny. It is through experimentation that children learn to manipulate the semantic simultaneously with the pragmatic use of language, and it is hoped that by providing access to novel puns, children with CCN may engage in similar learning experiences.

47.4 Output System

Output from AAC devices is either transient or permanent. Transient output includes speech and visual output while permanent output can be printed or stored electronically.

47.4.1 Speech Output

Speech-generating devices produce speech using either digitized or synthesized speech (Lloyd et al., 1997). The quality of speech output has been assessed in terms of intelligibility and comprehension (Schlosser et al., 2003). Intelligibility refers to the listener's ability to recognize phonemes and words in isolation, while comprehension brings meaning to information in the context of current or past knowledge (Kintsch and Van Dijk, 1978). Although research (Mirenda and Beukelman, 1987) reported intelligibility of single word speech synthesis as 81.7% versus natural speech as 97.2%,

intelligibility can be improved when target items are within a closed-response format (Greene et al., 1984)—96.7% speech synthesis compared to 99% with natural speech. Meaningful sentences can further improve accuracy. Intelligibility and comprehension of speech output is dependent on the quality of the speech synthesizer (Mirenda and Beukelman, 1990) and can be improved by training and practice (Schlosser et al., 2003). Although research into speech synthesis and AAC has focused on synthetic speech (Schlosser et al., 2003), it is generally believed that digitized speech is superior because of its inherent intelligibility. This is reflected in the availability of digitized speech on devices that primarily use synthesized speech.

Devices that use digitized speech allow natural speech to be recorded for later playback. The quality of digitized speech is better than synthesized speech because it is a time-sampled replication of natural speech. Devices using digitized speech are referred to as closed or whole message devices as the output vocabulary is restricted to the words, phrases, or messages that have been recorded into the device. Closed or whole message devices tend to be small and simple to operate, e.g., the Pocket GoTalk (Figure 47.3). These devices are often used to introduce potential users to speech-generation devices and are of particular use when intelligibility is paramount, the user is ambulant, the user requires minimal support, or the user is unable to formulate messages independently. The total recording time varies between devices.

Devices that use synthesized speech use text-to-speech to generate spoken output. Text-to-speech systems are classified into the categories: formant synthesis where the main acoustic features of the speech signal is modeled, concatenative synthesis where units of recorded speech (e.g., diphones, syllables, words) are concatenated using signal processing algorithms to produce smooth transitions, and articulatory synthesis where physical models of the vocal system are used to generate speech (Tasic et al., 2003).

Advances in speech synthesis have resulted in significant improvements in the quality and intelligibility, and speech synthesis systems from developers such as Nuance,[7] AT&T,[8] and Fonix[9] are utilized within many of the commercial AAC devices. Research in emotional speech synthesis offers users the ability to modify the emotion of speech output (Murray et al., 1996; Shroder, 2001). Although speech utterances (words and phrases) can be tagged to reflect different voices in some AAC devices, the ability to manipulate the emotional stress easily within interactive generated communicative messages is not yet generally available to people who use AAC. A company that has identified emotion as an important feature is BlinkTwice.[10] Their device, the Tango (Figure 47.10), uses a combination of digitized

---

[7] http://212.8.184.250/tts.

[8] http://www.naturalvoices.att.com.

[9] http://www.fonixspeech.com.

[10] http://www.blink-twice.com.

FIGURE 47.10 The Tango.

and synthetic speech from the Acapela Group.[11] Users are able to reuse prestored words/phrases/messages and generate novel vocabulary, which can be transformed into different emotions by being morphed into a range of voices from whispering to shouting by choosing a mood.

Speech output is a vital component for AAC devices: it allows users to be independent communicators and reduces the need for third-party interpretation; it enables nonspeaking individuals to communicate over the telephone; it provides a common and less threatening communication mode between nonliterate users and communication partners who may be visually impaired or may not understand the user's graphic system; it allows a user to attract attention and to communicate at a distance (Beukelman and Mirenda, 2005).

Visual output formed the primary output mode prior to the availability of speech output. However, visual output on computer screens can provide clarification of the message generated by the user of an AAC device (Higginbotham, 1989). This is often useful when speech output is compromised (Fucci et al., 1995), for example, in loud environments, but is essential if the listener has a hearing impairment. Visual output need not be restricted to text, and some AAC devices display graphic symbols with text labels. However, many devices only produce the output message using text.

47.4.1 Printed and Stored Output

Printed output remains an important output medium for AAC devices, despite the emergence of good quality speech output. Printed output can provide a permanent copy of conversation that can be read at a later date. It can also provide proof of a user's communication production; for example, a child can take evidence of work home. Participants in the STANDUP project (Black et al., 2007) were able to use a printed copy of their favorite jokes as a way to tell their jokes without access to an AAC device.

Communication can be stored on electronic media for later use. This allows users to compose longer or more permanent communications over a prolonged period of time. Experiences, which can be written down and edited over time, provide a source of prestored conversational material for interactive conversation. Printed output is no longer restricted to text, but may contain any form of graphic material, from photographs to AAC graphic representations systems.

Electronic communication provides people who use AAC with the ability to use their devices to communicate both asynchronously and synchronously. People who use e-mail are increasingly utilizing the Internet for communication. The asynchronous nature of e-mail reduces the disadvantage of the slow communication rate characteristic of AAC interaction and levels the playing field for people with CCN (Cohen and Light, 2000). An interview with Professor Stephen Hawking by John Humphries[12] provides an example of how asynchronous communication can support real-time communication. Questions were sent by the interviewer and answered by the interviewee using e-mail over an extended period of time resulting in a final interview in which the conversation flowed at a normal speaking rate.

People using speech-generation devices are able to use a combination of written and spoken communication for synchronous electronic communication. The author, a speaking colleague, and a nonspeaking colleague have successfully conducted meetings using Skype[13] software, which provides free audio, video, and text communication. The nonspeaking participant is able to use his speech-generation device as text input to the instant message facility, which complements the audio transmission of the electronic speech output.

Graphic-based e-mail communication such as BlissInternet (Lindsay et al., 1998) is becoming increasingly available. Support for graphic representation systems within the World Wide Web is being encouraged (Judson et al., 2005) and symbol-based web sites are being supported.[14]

47.5 Conclusion

SGDs can be literacy based or symbol based, although many devices are multimodal. Literacy-based systems are mainly software based (e.g., EZ-Keys, Winbag, etc.), while symbol-based systems (e.g., Pathfinder, Dynavox) have traditionally been marketed on dedicated platforms. The increasing availability of robust portable platforms such as tablets and handheld devices has seen the development of more mixed

[11] http://www.acapela-group.com/index.html.

[12] http://www.bbc.co.uk/radio4/today/interview/octnovdec_2006.shtml.

[13] http://www.skype.com/intl/en-gb/useskype.

[14] http://www.symbolworld.org/eLive.

communication software. This is welcome, as dedicated devices can cost in the region of three to four times more than other systems. However, purchasing dedicated devices includes tailored training programs and ongoing technical support.

Pragmatic approaches to augmentative system design have provided insights into ways in which such a system can be used to interact at different levels. This approach has used conversational modeling and information retrieval techniques to: (1) increase the speed of communication; and (2) allow the users to express themselves in more detail without having to concentrate on re-creating text. The ability to communicate more easily allows users to have more control over an interaction, leading to more natural conversation.

The advances in portability, physical input, and speech synthesis provide individuals with CCN with physical access to spoken and written output. However, it is in the continued development of linguistic and pragmatic support that there will be a reduction in the physical and cognitive load on the operational use of SGDs, releasing individuals with CCN to develop language and conversational skills.

References

Alm, N. (1988). *Towards a Conversation Aid for Severely Physically Disabled Non-Speaking People*. PhD dissertation, University of Dundee, Nethergate, Dundee, Scotland.

Alm, N., Morrison, A., and Arnott, J. L. (1995). A communication system based on scripts, plans and goals for enabling non-speaking people to conduct telephone conversations, in the *IEEE Conference on Systems, Man & Cybernetics*, 22–25 October 1995, Vancouver, Canada, pp. 2408–2412. Los Alamitos, CA: IEEE.

Alm, N., Nicol, M., and Arnott, J. L. (1993). The application of fuzzy set theory to the storage and retrieval of conversational texts in an augmentative communication system, in the *Proceedings of RESNA '93*, June 1993, Las Vegas, pp. 127–129. Washington, DC: The RESNA Press.

American Speech-Language-Hearing Association (2002). *Augmentative and Alternative Communication: Knowledge and Skills for Service Delivery [Knowledge and Skills]*. http://www.asha.org/policy.

Andreasen, P. N., Waller, A., and Gregor, P. (1998). BlissWord: Full access to Blissymbols for all users, in the *Proceedings of the 8th Biennial Conference of the International Society for Augmentative and Alternative Communication*, August 1998, Dublin, Ireland, pp. 167–168. Toronto: ISAAC.

Arnott, J. L., Alm, N., and Waller, A. (1999). Cognitive prostheses: Communication, rehabilitation and beyond, in the *Proceedings of the IEEE International Conference on Systems, Man and Cybernetics*, 12–15 October 1999, Tokyo, pp. 346–351. Los Alamitos, CA: IEEE.

Arnott, J. L. and Javed, M. (1992). Probabilistic character disambiguation for reduced keyboards using small text samples. *Augmentative & Alternative Communication* 8: 215–223.

Baker, B. (1982). Minspeak: A semantic compaction system that makes self expression easier for communicatively disabled individuals. *Byte* 7: 186–202.

Baker, B. (1987). Semantic compaction for sub-sentence vocabulary units compared to other encoding and prediction systems, in *RESNA Tenth Annual Conference Proceedings* (R. D. Steele and W. Gerry, eds.), 19–23 June 1987, San Jose, CA, pp. 118–120. Washington, DC: The RESNA Press.

Beukelman, D. R. (1991). Magic and cost of communicative competence. *Augmentative and Alternative Communication* 7: 2–10.

Beukelman, D. R. and Mirenda, P. (2005). *Augmentative and Alternative Communication: Management of Severe Communication Disorders in Children and Adults*. Baltimore: Paul H. Brookes Publishing Co.

Binsted, K., Pain, H., and Ritchie, G. (1997). Children's evaluation of computer-generated punning riddles. *Pragmatics and Cognition* 5: 309–358.

Black, R., Waller, A., Ritchie, G., Pain, H., and Manurung, R. (2007). Evaluation of joke-creation software with children with complex communication needs. *Communication Matters* 21: 23–27.

Blackstone, S. (2004). Clinical news: Visual scene displays. *Augmentative Communication News* 16: 1–8.

Bowden, P. and Beukelman, D. (1988). Rate, accuracy, and message flexibility: Case studies in communication augmentation strategies, in *The Vocally Impaired: Clinical Practice and Research* (L. Bernstein, ed.), pp. 295–311. Philadelphia: Grune & Stratton.

Castrogiovanni, A. (2008). *Communication Facts: Special Populations: Augmentative and Alternative Communication*. Rockville, MD: American Speech and Hearing Association.

Cheepen, C. (1988). *The Predictability of Informal Conversation*. Oxford, U.K.: Printer Publishers Ltd.

Clarke, H. H. and Clarke, E. V. (1977). *Psychology and Language*. New York: Harcourt Brace Jovanovich.

Cohen, K. and Light, J. (2000). Use of electronic communication to develop Mentor-Protégé relationships between adolescent and adult AAC users: Pilot study. *Augmentative & Alternative Communication* 16: 227–238.

Demasco, P. W. and McCoy, K. F. (1992). Generating text from compressed input: An intelligent interface for people with severe motor impairments. *Communications of the ACM* 35: 68–78.

Dietz, A., McKelvey, M., and Beukelman, D. (2006). Visual display scene (VSD): New AAC interface for persons with aphasia. *Perspectives in Augmentative and Alternative Communication* 15: 13–17.

Drager, K. and Light, J. (2006). Designing dynamic display AAC systems for young children with complex communication needs. *Augmentative & Alternative Communication* 15: 3–7.

Dye, R., Alm, A., Arnott, J. L., Harper, G., and Morrison, A. I. (1998). A script-based AAC system for transactional interaction. *Natural Language Engineering* 4: 57–71.

Foulds, R. (1980). Communication rates of nonspeech expression as a function in manual tasks and linguistic constraints, in the *Proceedings of the International Conference on Rehabilitation Engineering*, June 1980, Toronto, Canada, pp. 83–87. Washington, DC: The RESNA Press.

Foulds, R. (1987). Guest editorial. *Augmentative & Alternative Communication* 3: 169.

Fucci, D., Reynolds, M. E., Bettagere, R., and Gonzales, M. D. (1995). Synthetic speech intelligibility under several experimental conditions. *Augmentative & Alternative Communication* 11: 113–116.

Goldman-Eisler, F. (1968). *Psycholinguistics Experiments in Spontaneous Speech*. New York: Academic Press.

Greene, B. G., Manous, L. M., and Pisoni, D. B. (1984). *Perceptual Evaluation of DECtalk: A Final Report on Version 1.8. Research on Speech Perception Progress Report No. 10.* Bloomington, IN: Speech Research Laboratory, Psychology Department, Indiana University.

Heckathorne, C., Voda, J., and Leibowitz, L. (1987). Design rationale and evaluation of the Portable Anticipatory Communication Aid: PACA. *Augmentative & Alternative Communication* 3: 170–180.

Higginbotham, D. J. (1989). The interplay of communication device output mode and interaction style between non-speaking persons and their speaking partners. *Journal of Speech and Hearing Disorders* 54: 320–333.

Higginbotham, D. J. (1992). Evaluation of keystroke savings across five assistive communication technologies. *Augmentative & Alternative Communication* 8: 258–272.

Higginbotham, D. J., Shane, H., Russell, S., and Caves, K. (2007). Access to AAC: Present, past, and future. *Augmentative & Alternative Communication* 23: 243–257.

Hunnicutt, S. (1986). Bliss symbol-to-speech conversion: "Blisstalk." *Journal of the American Voice I/O Society* 3: 19–38.

Judson, A., Hine, N., Lundalv, M., and Farre, B. (2005). Empowering disabled users through the semantic web, in *First International Conference on Web Information Systems and Technologies*, 26–28 May 2005, Miami, FL. INSTICC. http://www.webist.org/2005.

King-DeBaun, P. (1997). Computer fun and adapted play: Strategies for cognitively or chronologically young children with disabilities part 1 and 2, in the Proceedings of Technology and Persons with Disabilities Conference, 18–22 March 1997, Los Angeles. Northridge, CA: California State University. http://www.dinf.ne.jp/doc/english/Us_Eu/conf/csun_97/csun97_066.html.

Kintsch, W. and Van Dijk, T. A. (1978). Towards a model of text comprehension and production. *Psychological Review* 85: 363–394.

Langer, S. and Hickey, M. (1998). Using semantic lexicons for full text message retrieval in a communication aid. *Natural Language Engineering* 4: 41–55.

Light, J. (1988). Interaction involving individuals using augmentative and alternative communication systems: State of the art and future directions. *Augmentative and Alternative Communication* 4: 66–82.

Lindsay, P., McNaughton, S., Marshall, P., Baird, E., and Guy, T. (1998). BlissInternet: The new highway for Bliss. *8th Biennial Conference of ISAAC*, 24–27 August 1998, Dublin, Ireland.

Lloyd, L. L., Fuller, D. R., and Arvidson, H. (1997). *Augmentative and Alternative Communication: A Handbook of Principles and Practices*. Needham Heights, MA: Allyn and Bacon.

Manurung, R., Ritchie, G., Pain, H., Waller, A., O'Mara, D., and Black, R. (2008). The construction of a pun generator for language skills development. *Applied Artificial Intelligence* 22: 841–869.

McCoy, K. F. and Demasco, P. W. (1995). Some applications of natural language processing to the field of augmentative and alternative communication, in the *Proceedings of the IJCAI '95 Workshop on Developing AI Applications for Disabled People*, June 1995, Montreal, Canada, pp. 97–112.

McNaughton, S. (1993). Graphic representation systems and literacy learning. *Topics in Language Disorders* 13: 58–76.

Miller, G. A., Beckwith, R., Fellbaum, C., Gross, D., Miller, K., and Tengi, R. (1993). Five papers on WordNet. International *Journal of Lexicography* 3. ftp://ftp.cogsci.princeton.edu/pub/wordnet/5papers.ps

Mirenda, P. and Beukelman, D. (1990). A comparison of intelligibility among natural speech and seven speech synthesizers with listeners from three age groups. *Augmentative & Alternative Communication* 6: 61–68.

Mirenda, P. and Beukelman, D. R. (1987). A comparison of speech synthesis intelligibility with listeners from three age groups. *Augmentative & Alternative Communication* 3: 120–128.

Murphy, J., Collins, S., and Moodie, E. (1994). The limited use of AAC systems. *Communication Matters* 8: 9–12.

Murray, I. R., Arnott, J. L., and Rohwer, E. A. (1996). Emotional stress in synthetic speech: Progress and future directions. *Speech Communication* 20: 85–91.

Musselwhite, C. and Burkhart, J. (2002). Social scripts: Co-planned sequenced scripts for AAC users, in the *Proceedings of Technology and Persons with Disabilities Conference*, 18–23 March 2002, Los Angeles.

Musselwhite, C. and St. Louis, K. (1988). Communication Programming for Persons with Severe Handicaps. Austin, TX: Pro Ed. Northridge, CA: California State University. http://www.csun.edu/cod/conf/2002/proceedings/209.htm

Newell, A. F. (1991). Assisting interaction with technology: Research and practice. *British Journal of Disorders of Communication* 26: 1–10.

Newell, A. F., Arnott, J. L., Booth, L., and Beattie, W. (1992). Effect of the PAL word prediction system on the quality and quantity of text generation. *Augmentative and Alternative Communication* 8: 304–311.

O'Mara, D. A. (2004). *Providing Access to Verbal Humour Play for Children with Severe Communication Impairment*. PhD dissertation, University of Dundee, Nethergate, Dundee, Scotland.

O'Mara, D. and Waller, A. (2003). What do you get when you cross a communication aid with a riddle? *The Psychologist* 16: 78–80.

O'Mara, D., Waller, A., Ritchie, G., Pain, H., and Manurung, H. M. (2004). The role of assisted communicators as domain experts in early software design, in the *Proceedings of the 11th Biennial Conference of the International Society for Augmentative and Alternative Communication*, 6–10 October 2004, Natal, Brazil. [CD-ROM]. Toronto: ISAAC.

Polkinghorne, D. E. (1995). Narrative configuration in qualitative analysis, in *Life History and Narrative* (J. A. Hatch and R. Wisniewski, eds.), pp. 5–24. London: Routledge.

Quasthoff, U. M. and Nikolaus, K. (1982). What makes a good story? Towards the production of conversational narratives, in *Discourse Processing* (A. Flammer and W. Kintsch, eds.). Oxford, U.K.: North-Holland Publishing.

Ritchie, G., Manurung, R., Pain, H., Waller, A., and O'Mara, D. (2006). The STANDUP interactive riddle-builder. *IEEE Intelligent Systems* March/April: 67–69.

Schlosser, R. W., Blischak, D. M., and Koul, R. K. (2003). Roles of speech output in AAC, in *The Efficacy of Augmentative and Alternative Communication* (R. W. Schlosser, ed.), pp. 471–532. San Diego: Elsevier Science.

Shroder, M. (2001). Emotional speech synthesis: A review. *EUROSPEECH-2001*, 3–7 September 2001, Aalborg, Denmark, pp. 561–564. http://www.isca-speech.org.

Smith, M. (2005). *Literacy and Augmentative and Alternative Communication*. London: Elsevier Academic Press.

Swiffin, A., Arnott, J. L., Pickering, J., and Newell, A. F. (1987). Adaptive and predictive techniques in a communication prosthesis. *Augmentative & Alternative Communication* 3: 181–191.

Tasic, J. F., Najim, M., and Ansorge, M. (2003). *Intelligent Integrated Media Communication Techniques: Cost 254 & Cost 276*. London: Kluwer Academic Publishers.

Todman, J. (in press). Whole utterance approaches in AAC. *Augmentative & Alternative Communication*.

Todman, J., Elder, L., and Alm, A. (1995). Evaluation of the content of computer-aided conversations. *Augmentative & Alternative Communication* 11: 229–234.

Vanderheiden, G. and Kelso, D. (1987). Comparative analysis of fixed-vocabulary communication acceleration techniques. *Augmentative & Alternative Communication* 3: 196–206.

Waller, A. (1992). *Providing Narratives in an Augmentative Communication System*. PhD disseration, University of Dundee, Nethergate, Dundee, Scotland.

Waller, A. (1998). Pragmatic approaches to the design of augmentative communication systems, in the *Proceedings of the IEEE International Workshop on Robotic and Human Communication Conference*, 16–20 May 1998, Leuven, Belgium.

Waller, A., Alm, N., and Don Johnston Inc. (1996). From lab to laptop: An example of technology transfer, in the *7th Biennial Conference of the International Society for Augmentative and Alternative Communication (ISAAC)*, 7–10 August 1996, Vancouver, Canada.

Waller, A., Black, R., O'Mara, D., Pain, H., Ritchie, G., and Manurung, H. (in press). Evaluating the STANDUP pun generating software with children with cerebral palsy. *ACM Transactions on Accessible Computing*.

Waller, A. and Jack, K. (2002). A predictive Blissymbolic to English translation system, *in the Fifth International ACM Conference on Assistive Technologies (ASSETS 2002)*, 8–10 July 2002, Edinburgh, Scotland.

Waller, A. and Newell, A. F. (1997). Towards a narrative based communication system. *European Journal of Disorders of Communication* 32: 289–306.

Waller, A., Oosterhoorn, E., and Andreasen, P. N. (2000). A language independent Bliss to sentence translation system. *Communication Matters* 14: 9–10.

Ward, D. J. and MacKay, D. J. C. (2002). Fast hands-free writing by gaze direction. *Nature* 418: 838.

Wilson, A. D. (2003). *Communicating with Pictures and Symbols*. Edinburgh: CALL Centre, University of Edinburgh.

Zangari, C., Lloyd, L. L., and Vicker, B. (1994). Augmentative and alternative communication: An historic perspective. *Augmentative & Alternative Communication* 10: 27–59.

48

Universal Access in Public Terminals: Information Kiosks and ATMs

Georgios Kouroupetroglou

48.1 Introduction

A transition process is taking place from a service-based community to a self-service community. In the not-too-distant past, the idea of using a kiosk to place an order in a store or restaurant, or to complete a transaction at the point of sale or check in at an airport was more a vision than a reality. Today, public access terminals can provide a self-serve method for customers or citizens to purchase items in a supermarket, obtain services or information in a movie theater, and accomplish transactions with a civil service or a hospital. Stephanidis et al. (1998) have predicted that public information systems and terminals will be increasingly used in a variety of domains. The automated teller machine (ATM), the most common public terminal, is 40 years old. In 1967, the very first ATM was installed; today the estimated number of ATMs globally is over 1.5 million, with an increasing rate of 192.32 ATMs per day (R.B.R., 2006). In the United Kingdom the number of ATM cash transactions per inhabitant was 42.3 in 2004.[1] The retail kiosks installed worldwide had reached in 2006 the number of 630,000 units (S.R.A., 2006). From the practical and technological perspective, there are three main factors that drive the adoption of public access terminals: (1) the consumer demand to save time (e.g., they don't want to wait in line), (2) the

requirement for a 24/7 service provision, and (3) the need for reducing the overall cost of products and services.

Universal access is crucial to ensure that anyone, anywhere, and at anytime is enabled to use public access terminals at one's own convenience and in an unconstrained way. The range of users requiring accessibility includes, but is not limited to, novice- to experienced-users, the elderly and the disabled, foreigners, the nonliterate, and the occasionally and temporarily disabled. Universal access is very important, taking into account the considerable diversity in users and their abilities along with the variability of usage contexts of public access terminals and the range of their types.

In the following sections, first a classification model of public access terminals is proposed, followed by a structured presentation of their user interface characteristics. Next, the relative barriers of use along with user requirements are examined. Based on existing research, a range of accessibility strategies and prototype accessible public terminals is presented. Then an overview of the available specific accessibility guidelines and standards is provided.

48.2 Classification of Public Access Terminals

A public access terminal (PAT), or in short a public terminal, can be defined as an ICT (information and telecommunication

---

[1] Interact Inc. (http://www.interac.org/en_n3_31_abmstats.html).

technologies)-based interactive system, located in a public area, indoors or outdoors, self-operated by the user, providing access to information or sales of services and products. PATs are also referred to as information transaction machines (ITMs) and point-of-decision (PoD) kiosks. Public terminals are computerized machines that may store data locally or retrieve them through a computer network. Although there are a wide variety of different types of PATs, they can be classified in two broad categories according to their informational and transactional dimensions:

- *Information PAT*: An interactive public terminal that accepts input from any user, without any kind of identification or authentication, and responds by providing free-of-charge information. They are known as public information kiosks, public kiosks, information kiosks, info-kiosks, or kiosks. Typical examples of information PATs include: web kiosks, health information kiosks (allowing the public to look up health-related information surrounding diseases, treatments, provider choices as well as drug and drug interactions), tourist or museum kiosks, guide kiosks at train and bus terminals, building directories and guides.
- *Transaction PAT*: An interactive self-service public terminal for the accomplishment of a purchase of goods or carrying out a service without the need or the assistance of a clerk. They can be further subclassified according to two independent main characteristics:
  - Financial or nonfinancial transaction PATs
  - User authentication transaction PATs

Thus, the following four types of transaction PATs can be distinguished:

- *Financial transaction PATs requiring user authentication*: e.g., ATMs
- *Financial transaction PATs without user authentication (cash/direct-only pay)*: e.g., retail or vending kiosks, photo kiosks (which allow customers to view, edit, personalize, and print digital photographs), parking kiosks, fare machines (they dispense tickets/fares for transportation)
- *Nonfinancial transaction PATs requiring user authentication*: e.g., electronic voting machines, e-government kiosks or civil kiosks (that can issue diverse civil-related documents)
- *Nonfinancial transaction PATs without user authentication*: e.g., job applications kiosk

The most widespread PAT is the ATM:

A public wall-mounted or standalone banking terminal that is customer-activated to make bank transactions (such as cash withdrawal and balance inquiry, check deposit, bill payment, and transfer between accounts) without a human teller.

An ATM allows a bank customer to conduct banking transactions from almost every other ATM machine in the world. They are also called: automatic teller machine, bank machine, cash machine, automated banking machine, money machine, hole-in-the-wall, or cashpoint.

Common are the hybrid transaction PATs, for example, accepting both credit cards and cash for bill payments. As the informational and transactional dimensions of PATs get continuous values from low to high, hybrid PATs can behave both as information and transactional public terminals. Tung and Tan had followed a classification scheme for the case of PATs in Singapore (Tung, 1999; Tung and Tan, 1998). A typical example of a hybrid PAT is a tourism kiosk that accepts input from a user and displays information (e.g., maps or available hotels), but one can also book a hotel room using a credit card or purchase bus tickets and pay using coins (Proll and Retschtzegger, 2000).

The following three emerging usage cases can be considered for PATs:

- With air travel check-in transaction kiosks customers are able to: (1) check in for international flights in a faster way; (2) view/change their seating options; (3) view the layout of the aircraft; (4) check bags; (5) update their frequent flyer number; and (6) transact in their own language by selecting from tens of available languages.
- Check-in health kiosks offer patients the option to notify the provider that they have arrived for their appointments, review and verify the time and place of their appointments that day, review documents, sign consent forms, complete patient records and arrange to pay any treatment copayments (Lovelock, 2006). Patient identification is typically handled by swiping a driver's license, credit card, or membership card.
- Integrated download kiosks allow users of mobile phones, PDAs, personal computers, or MP3 players to do the following via Bluetooth, infrared, or USB connection or a flash card reader:
  - Browse and purchase ringtones, real tones, wallpaper, and MP3 music files
  - Download games
  - Print digital photos
  - Pay cellular phone bills
  - Browse and print out mobile coupons
  - Customize and download animated greeting cards

Multifunction PATs that combine in a single kiosk solution a full ATM functionality with an integrated download kiosk, as described previously, have been launched in 2006.

48.3 User Interface Characteristics of PATs

Nowadays, PATs use a multimedia PC, which is housed in a close-fitting casing and uses a common operating system. From a technological point of view, a typical PAT configuration includes a stand-alone or networked interactive self-service device whose primary interface with the customer is through a touch screen or programmable buttons. Its diagonal screen size is between five and nine inches, with resolution being no less than quarter VGA and it possesses multimedia capabilities including moving pictures and polyphonic sounds. (Frost and Sullivan, 2004)

Many PATs connect to the Internet or an intranet, either with a landline or with wireless access, allowing users to perform structured or semistructured queries to obtain real-time information and perform transactions.

To be effective, kiosk applications often require more than a touch screen and a PC. As a result, certain manufacturers offer a number of kiosk peripherals that enhance the functionality of PATs. Some peripherals are standard features on selected models, while others are optional add-ons.

The topics related to user interaction with a PAT can be grouped in the following four issues:

1. Physical/environmental
2. User authentication
3. Operation
4. Cash/money handling

The physical/environmental issue includes:

- The type of enclosure: wall-mount, standalone, or drive-in solution
- The size and the positioning height of the terminal
- The size and the positioning of the peripherals for the operation and the money handling
- The proximity sensors
- Accessway (continuous accessible path of travel) and circulation space in front of the PAT

The proximity sensors detect (within a typical configurable range from 1.5 to 5 feet) the presence of users, and can thus serve two important purposes when integrated into a PAT. First, they can detect an approaching user and launch an audio/video greeting. Second, and more important, proximity sensors can detect when a user walks a near distance from a PAT, which is useful for automatically logging users out of secure applications. When a user walks away, a proximity sensor triggers commands to close the PAT's application.

On most modern PATs, customer authentication is accomplished by inserting into a card reader a plastic card with a magnetic stripe or a plastic smartcard with a chip that contains the card number and some security information of the cardholder, such as an expiration date or CVC (card validation code)/CVV (card verification value). The customer then verifies her identity by entering through the encrypting PIN pad (EPP) a password code, often referred to as a personal identification number (PIN). Thus, the typical peripherals of a PAT requiring user authentication are a card reader (with manual or motorized insert called a dip reader and swipe reader, respectively) and an EPP.

Recent years have seen financial institutions worldwide turn to biometric technology to deal with the pressing issue of security of PATs (Mäkinen et al., 2002; Sanderson, 2003; Asawa et al., 2005). Biometric technology is based on the scanning of a customer's fingerprint, palm, iris, face, speech, and so on. With the development of biometric solutions, there is no need to remember PIN numbers. The mass retail segment, however, has been slower than other segments to embrace biometrics. Barriers to adoption have included cost, a lack of application standards and interoperability, and customer resistance to the adoption of the new technology. First introduced in October 2004, biometric bank cards have been announced by 15 financial institutions in Japan as of December 2005. Some Indian banks have started implementing biometric applications in retail branch applications for officer authentication (Jena, 2007).

The main operation of a PAT is performed through a number of input and/or output devices:

- *Input devices*: Touch screen, number pad, function-key pushbuttons (usually close to the display), trackball, keyboard, barcode reader, bill scanner (with OCR)
- *Output devices*: Display, receipt/coupon/ printer (impact or thermal), loudspeaker, passbook/bankbook printer (dot matrix or inkjet), full page printer, jack for headphones, photo printer
- *Input/output devices*: Telephone handset, flash memory card reader, USB port, and wireless bidirectional communication via Wi-Fi, infrared, or Bluetooth

Cash and money handling is performed by bill dispensers, coin dispensers, cash/check deposits, bank-note acceptors (note trays), and multicoin selectors/acceptors.

Manufacturers have demonstrated and deployed coordination of ATMs with mobile phones[2] (see Perton, 2006), but those solutions have not yet reached the market worldwide.

48.4 Barriers of Use and User Requirements

Despite the growing penetration of PATs, there remain a significant proportion of citizens and customers who either cannot, or will not, use these machines. This happens for various reasons, including disabilities, aging, language differences, acceptability, and satisfaction. Thus, the greatest challenge for any public technology is how to accommodate the vast range of potential users. In addressing this topic it is important to determine the reasons for nonuse of PATs and consider ways of improving their design for all users. Usability issues are often interrelated with those of accessibility. Usability of PATs must be inclusive, ensuring that the needs of the elderly and of people with disabilities are addressed. But accessibility does not guarantee usability. On the other hand, there is always a need for balancing between accessibility and usability.

---

[2] NRT Technology Corp: "Gaming and Casino Solutions: QuickJack." http://www.nrtpos.com/html/gamingquickjack.shtml.

Keats and Clarkson (2004), based on detailed demographic data, provide an in-depth analysis of user capabilities in a structured way, usable for design purposes. They cover for all ages the major single and multiple functional impairments, such as motion (locomotion, dexterity, reach and stretch), sensory (vision and hearing), and cognitive (communication and intellectual). Furthermore, they have developed a model for the quantitative determination of the overall capability of a person. This approach can be applied to the four user interface topics described in Section 48.3 to quantify design exclusion of a particular PAT.

Roe et al. (1995) have cross-related the barriers of various user tasks related to terminals with various impairments. The identified barriers are graduated according to the severity of the problem they pose, ranging from "usual, no problem" to "impossible." The Australian Bankers' Association (ABA),[3] the Irish National Disability Authority,[4] and the Scientific Research Unit of the RNIB (Royal National Institute of Blind People)[5] have developed detailed user needs analysis for PATs.

Analytical (Connell et al., 2004) and empirical (Blignaut, 2004) usability evaluation methodologies have been introduced to assess the usability of PATs in real situations. The component "easy to operate," was selected in a users' survey as the most critical design parameter for a PAT. This finding is in line with users' suggestion that they would like to see more user-friendly public terminals (Tung, 2001).

Elderly users, regardless of their country of origin, complained about the complexity of the language of ATM interfaces (professional jargon) and found it difficult to cope with (Tarakanov-Plax, 2005). Other parameters relating to the usage of PATs by older adults include difficulty of learning, limited technology experience (Darch and Caltabiano, 2004), and low perceived user comfort and control (Adams and Thieben, 1991; Gilbert and Fraser, 1997; Marcellini et al., 2000).

Some studies have indicated that PATs for voting (e.g., those including a touch screen and telephone-like keypads) impose a number of difficulties for a significant minority of voters (Bederson et al., 2003; Gaston, 2005; Devies, 2006).

The barriers that illiterate or semiliterate people face when using PATs have been analyzed based on alternative interfaces, such as icon-based and speech-based interfaces (Gilbert and Fraser, 1997; Thatcher et al., 2005, 2006). The overall results indicate that users show a tendency to prefer the icon-based alternative interface to the speech-based alternative interface and the traditional text-based ATM interface.

---

[3] "Industry Standard on Automated Teller Machines (ATMs)," The Australian Bankers' Association (ABA), 2002, http://www.bankers.asn.au.

[4] "Accessibility Guidelines for Public Access Terminals," Irish National Disability Authority IT Accessibility Guidelines, Version 1.1. http://accessit.nda.ie.

[5] "Accessibility Guidelines for Public Access Terminals," http://www.tiresias.org.

48.5 Accessibility Strategies for Public Terminals

During the last few decades, extensive research has investigated accessibility options for public terminals. This section provides an overview of the current state of knowledge with respect to developing PATs for diverse users and context of use.

48.5.1 Software Tools

An architecture of a universally accessible ATM prototype was designed by Sun Microsystems for the Java Platform (Korn, 1999). Following this approach, and using the relative Java interface enhancements provided, one can implement kiosks that are able to interact with anyone who approaches them—where "anyone" includes people who speak different languages, people who may have one or more disabilities, and people who have different levels of expertise and comfort with information technology. The universally accessible ATM prototype is a proof-of-concept demonstration that illustrates how the technologies available in the Java platform can be used to create an ATM machine that automatically reconfigures its user interface to match the needs of the user. The universally accessible ATM is based on a four-tier, thin-client model. The client is written using the Java foundation classes, and utilizes the pluggable look and feel architecture, the Java accessibility support, and the Java2D functionality to render fonts in any language character set. In the middle tier, servers use the remote method invocation (RMI) protocol to supply the client with the main user interface, the language specific strings, the Swing pluggable look and feel classes and themes, and any assistive technologies the user might need. At the back end, a SQL database on a mainframe contains banking records that are queried via JDBC (Java database connectivity). Finally, users carry with them a Java card that states their language and locale, and provides URLs to the Swing pluggable look and feel, theme, and assistive technologies they require. These URLs could potentially extend to servers outside the bank's firewall, using signed applets and the JDK (Java development kit) security model. User preference information is stored on the user's ATM card. This can be a Java card or a magnetic strip card. User preference information includes the pluggable look and feel classes (Java Look and Feel, Windows Look and Feel, Audio Look and Feel); look and feel theme (low vision theme of the Java Look and Feel), and the user's country and language information. The universally accessible ATM client is configured to switch look and feels based on information located on the Java card. Because of this, users can specify alternate look and feels, such as a high-contrast or low-vision look and feel (or the equivalent theme of an existing look and feel), and so can use the ATM even though they may be color-blind or their eyesight is failing.

Stephanidis (2001) has introduced the concept of unified user interfaces (UUI), that is, interfaces self-adapting to user and usage context, using either of-line knowledge (adaptability) or real-time adaptivity. The theoretical approach (Savidis and

Stephanidis, 2001c, 2004b), the design methodology (Savidis et al., 2001; Savidis and Stephanidis, 2004a) along with an appropriate architectural framework (Savidis and Stephanidis, 2001a) for the development of UUI, as well as tools that facilitate and support the design and implementation phases (Akoumianakis and Stephanidis, 2001; Savidis and Stephanidis, 2001b) are well documented in the literature. Four levels of adaptation were identified in UUI:

- Semantic (internal functionality, information represented)
- Syntactic (dialogue sequencing, syntactic rules, user tasks)
- Constructional
- Physical (devices, object attributes, interaction techniques)

According to this approach, automatic adaptation at any level implies polymorphism, that is, for a given task, alternative interactive artifacts are designed according to different attribute values of user and usage context (Stephanidis et al., 2001b). The UUI development methodology is described in detail in Chapter 16, "Unified Design for User Interface Adaptation" and Chapter 21, "A Unified Software Architecture for User Interface Adaptation" of this handbook.

Based on the UUI paradigm, a number of accessible prototype PATs were developed and evaluated, including an instance of the web browser deployed as an example of accessible kiosk metaphor (Stephanidis et al., 2001a), and other information kiosks (Stephanidis et al., 2001b, 2004; Zarikas et al., 2001; Ntoa and Stephanidis, 2005). In these efforts, visual and nonvisual interaction are supported to satisfy the requirements of blind users, users with vision problems, and users with mobility impairments of the upper limbs. To facilitate blind users in locating the system and learning how to use it without assistance, a camera can be attached to the system, and when motion is detected, instructions of use are announced.

EZAccess (Law and Vanderheiden, 2000; Vanderheiden et al., 1999) consists of a set of interface enhancements developed by TRACE Center,[6] which can be applied to electronic products and devices so that they can be used by more people including those with disabilities. Among the main EZ features are:

- *Voice + 4 button navigation*: Gives complete access to any onscreen controls and content. This feature also provides feedback and information in a logical way such that it can be used by both sighted and nonsighted users. Typical items include onscreen text, images, and controls.
- *Touch talk*: Lets users touch onscreen text (and graphics), to hear them read (or described) aloud.
- *Button help*: Provides a way for users to instantly identify any button on the device. At any time, a person can see or hear any button's name and status. They can also get more information about what the specific button can be used for.

- *Layered help*: Provides context-sensitive information about using the device. If a person needs more help, he can press the help button repeatedly, receiving more information each time.
- *ShowCaptions*: Provides a visual presentation of any text or sounds created by the device that are not already visually displayed.

The EZ Access methodology can be applied to a wide range of interactive electronic systems, from public information and transaction machines, such as kiosks, to personal handheld devices, like remote controls, mobile phones, and PDAs (Vanderheiden, 2001). By using EZ Access, developers can design PATs that are usable not only by more people, but also in a wider range of environments and contexts. EZ Access principles have been deployed in a number of prototype PATs and electronic voting machines (Law and Vanderheiden, 1998; Vanderheiden, 1997).

By applying a seven-level approach, Keats et al. (2004) presented two examples of prototype kiosks designed to help principally older adults access online governmental information sources. The seven-level approach addresses each of the system acceptability goals identified by Nielsen (1993), namely social and practical acceptability. Practical acceptability consists of three components: utility (functionality), usability, and accessibility. Social acceptability includes attributes such as desirability.

48.5.2 Speech Technology

Among the first domains investigated as a means to incorporate accessibility in PATs were speech technologies (Fellbaum and Kouroupetraglou, 2008). Indeed, speech and voice technologies offer accessibility solutions for a wide range of disabilities (Kouroupetroglou and Nemeth, 1995; Freitas and Kouroupetroglou, 2008). The ESPRIT MASK project (Gauvain et al., 1996) involved in the design of a speech-driven interface for a public information kiosk intended to provide passengers with timetables and other information at railway stations in Paris. The project identified many challenges associated with designing a speech system for use by the public in an open space, such as ensuring robust recognition across all potential users, and capturing adequately clear audio in noisy open areas. Hone et al. (1998) investigated user attitudes toward, and performance of, speech technology in ATMs. They found that groups of impaired users, and particularly the visually impaired, were more positive about the concept of speech, citing various difficulties with current visual-manual interactions. Most nonusers, however, would not be encouraged to use ATMs with the addition of speech. Nevertheless, Manzke et al. (1998) conclude that speech output is essential for blind users to operate an ATM successfully. Additionally, acoustic feedback (e.g., beeps) as well as the natural machine noise, are very important for orientation. But sighted users should not be forced to listen to "too much" acoustic information. The MINNELLI interactive ATM (Steiger and Suter, 1994) facilitated interactions with bank customers primarily

---

[6] Trace Research and Development Center, College of Engineering, University of Wisconsin–Madison, http://trace.wisc.edu/ez/.

by the use of short animated cartoons to present information on bank services. However, the MINNELLI system required basic user training, which reduces its applicability in most public sites. Other research that has explored the use of animated "talking heads" and avatars to interact with end-users include the Digital Smart Kiosk project (Christian and Avery, 1998), the Smart Kiosk project (Christian and Avery, 2000), as well as the work presented in Cassell (2000) and Mäkinen et al. (2002). The aim of the AUGUST animated talking agent system (Gustafson et al., 1999a, 1999b) was to be able to analyze how novice users interact with a multimodal information kiosk covering several domains, placed without supervision in a public location. The animated agent had a distinctive personality, which, as it turned out, invited users from the public to try the system and also socialize, rather than just choosing straightforward information-seeking tasks. Another successful kiosk, with a broader scope than the MINELLI, was the MACK system (Stocky and Cassell, 2002; Cassell et al., 2002) based on embodied conversational agents and on information displays in mixed-reality kiosk format to display spatial intelligence. MACK provides information on residents and directions to locations at a research site. It integrates multiple input sources that include speech, gesture, and pressure. The system also exhibits a degree of spatial intelligence by utilizing awareness of its location and the layout of the building to reference physical locations. MIKI (McCauley and D'Mello, 2006) has been introduced as a three-dimensional, directory assistance-type digital persona. MIKI shares several similarities to the MACK system in that both systems use multiple input channels and provide information on people, groups, and directions to locations at a research site. However, the MACK system relies on rule-based grammars alone for speech input. This greatly restricts the scope of the questions with which a user can query the system. MIKI differs from other intelligent kiosk systems because of its advanced natural language-understanding capabilities that provide it with the ability to answer informal verbal queries without the need for rigorous phraseology.

48.5.3 Multimodal Interfaces

Others researchers have examined multimodal display of information and multimodal input in PATs. Looking at the combination of text and graphics in information display, Kerbedjiev (1998) proposed a methodology for realizing communicative goals in graphics. He suggests that more appealing multimedia presentations take advantage of both natural language and graphics. Such findings have paved the way for embodied conversational agents (ECAs), capable of natural language and its associated nonverbal behaviors. An ECA kiosk has a further advantage over graphics, as it can reference actual physical objects in the real world. Feiner and McKeown (1991) make a similar distinction between the functions of pictures and words. Pictures describe physical objects more clearly, while language is more adept in conveying information about abstract objects and relations. This research led to their COMET system, which generates text and

3D graphics on the fly (Feiner and McKeown, 1991). Similarly, Maybury's TEXTPLAN generates multimedia explanations, tailoring these explanations based on the type of communicative act required (Maybury, 1998). Taking these principles to the next level means replacing graphical representations with the physical objects themselves. To do so involves immersing the interface in a shared reality, such as a kiosk, and using an ECA to reference objects in the real world through natural language and gesture. The EMBASSI multimodal framework provides means for users with special needs to interact with otherwise not accessible PATs (Richter and Enge, 2003). On terminals extended by this framework, users can interact with the PAT through their own mobile device, thus allowing customized means for interaction. Additionally, personal information stored on the mobile device can be used to provide further assistance to the user. Based on a component-based flexible architecture, Reithinger and Herzog (2006) have recently introduced the advanced multimodal SmartKom system. Their prototype public access system combines seven different applications, which can be accessed through coherent seamless interaction: cinema program, city information, seat reservation, pedestrian navigation, document scanning, address book, and e-mail. This kind of advanced multimodal dialogue functionality requires underlying methods (Herzog and Reithinger, 2006) and development techniques (Herzog et al., 2004; Herzog and Ndiaye et al., 2006).

Interaction through pointing gesture recognition for public terminals has been investigated with positive results (Malerczyk, 2004; Malerczyk et al., 2005). Also, video-based input modalities like region tracking, gesture recognition, and pattern recognition have been implemented in a prototype mixed-reality kiosk system (Malerczyk et al., 2003). A gesture frame based on quasi-electrostatic field sensing has been recently demonstrated (Li et al., 2004). The system captures the arm gestures of the user and translates pointing gestures into screen coordinates and selection command, which become a gesture-based, hands-free interface for browsing and searching multimedia archives of an information kiosk in public spaces. Sagawa and Takeuchi (2000) developed an information kiosk with a sign language recognition system. Icon-based user interfaces for ATMs have been also introduced (Gilbert and Fraser, 1997; Thatcher et al., 2006).

48.5.4 Screen Layout Design

A clear and appealing screen layout is crucial to the success of PATs. Borchers et al. (1995) address the problem of developing such a layout, and provide several guidelines, drawn from traditional typography and Gestalt psychology, as well as from hypertext authoring and human-computer interaction. Hartson (2003) provides guidelines to help designers think about how cognitive, physical, sensory, and functional affordance can work together naturally in contextualized human-computer interaction (HCI) design or evaluation. A set of design principles and guidelines for ensuring the immediate usability of PATs can be found in Kules et al. (2003). These principles and guidelines were formulated while developing PhotoFinder Kiosk, a community photo library. Attendees of CHI 2001 successfully used this kiosk to browse and annotate collections of photographs spanning 20 years of CHI and related conferences, producing a richly annotated photo history of the field of human-computer interaction. They have used observations and log data to evaluate the PhotoFinder Kiosk and refine the guidelines.

48.5.5 Accessibility Assessment

An accessibility assessment model was developed (Lee and Cho, 2004) for calculating an accessibility index for public information kiosks. The model consists of eight terms to be considered in kiosk design and its operation. Taking into account the level of importance among the terms included in the model and the possibility of realization with current levels of technology for the terms, an index was organized to quantitatively assess the accessibility of public kiosks. Furthermore, a structured technique (Accessibility through Adaptability, ActA) for checking user interface accessibility has been designed (Stary and Totter, 2003). ActA enables the measurement in terms of a system's capability to provide accurate interaction features for individual users and their needs.

48.6 Accessibility Guidelines and Standards for Public Terminals

This section is intended to give a brief overview of the available guidelines and standards that are directly related to the accessibility domain of PAT.

The Irish National Disability Authority has developed the "Accessibility Guidelines for Public Access Terminals" as a part of its general IT Accessibility Guidelines.[4] The guidelines are grouped in two priorities: (1) Priority 1 ensures that the terminal can be used by most people with impaired mobility, vision, hearing, cognition, and language understanding; (2) priority 2 addresses ease of use, as well as additional user groups (people with cognitive impairments or multiple disabilities). Table 48.1 summarizes these guidelines. In their full version[4] each guideline is further analyzed in three parts:

1. Rationale
2. Specific directions and techniques
3. Instructions for checking

As an example, the appendix to this chapter presents the three parts for the case of Guideline 1.7.

TABLE 48.1 Accessibility Guidelines for Public Access Terminals—Irish National Disability Authority

	Priority 1
1.1	Ensure that all operable parts are reachable by people of all heights and people sitting in a wheelchair or buggy.
1.2	Ensure that displays are within sight of people of all heights and people sitting in a wheelchair or buggy.
1.3	Ensure that controls are adequately sized and sufficiently spaced to be operated by people with limited dexterity.
1.4	Ensure that operation requires minimal strength, grip, and wrist twisting.
1.5	Ensure that the terminal can be operated using only one hand.
1.6	If using a touch screen or contact-sensitive controls, do not require that it is touched by a body part.
1.7	Ensure that users with restricted or no vision can use all functions of the terminal.
1.8	Ensure that all outputs can be perceived by users with restricted or no vision.
1.9	Ensure that all outputs can be perceived by users with restricted or no hearing.
1.10	Use the simplest language possible for instructions, prompts, and outputs and, where possible, supplement it with pictorial information or spoken language.
1.11	If using cards, ensure that the card can be inserted into the card reader in its correct orientation without requiring vision.
1.12	If using biometric identification, provide an alternative access security mechanism for users who do not possess the required biological characteristic.
1.13	Do not cause the screen to flash at a frequency above 2 Hz.
1.14	When installing the terminal, ensure that users can get to it along an unobstructed path and operate it from a stable position.
1.15	Ensure that an equivalent service is available through an accessible channel for users who cannot use the terminal.
	Priority 2
2.1	Allow sufficient time to accommodate the slowest users.
2.2	Provide a way for users to cancel the whole transaction at any point and retrieve any items they have inserted.
2.3	Ensure that the user interface and task flow are similar across different functions and remains the same across repeated visits.
2.4	When deploying more than one version of a terminal, ensure that the user interfaces are similar.
2.5	Do not require users to remember a fixed supplied PIN.
2.6	Provide for users with multiple impairments.
2.7	Provide training or assistance for new users.
2.8	Ensure privacy and security during use.

Source: "Accessibility Guidelines for Public Access Terminals," Irish National Disability Authority IT Accessibility Guidelines, Version 1.1. http://accessit.nda.ie.

The Scientific Research Unit of the RNIB provides through TIRESIAS[5] a rather detailed set of accessibility guidelines for PATs as a part of its main section "Guidelines for the Design of Accessible Information and Communication Technology Systems." A short checklist is also available (see Table 48.2).

On the Section 508 home page of the Civil Rights Division, U.S. Department of Justice, a checklist is provided (see Table 48.3) as a tool for evaluating the extent to which information transaction machines are accessible to and usable by most people with disabilities.[7] This checklist is partially based on work of the Trace Research and Development Center of the University of Wisconsin–Madison.[6]

The ABA and the Accessible e-Commerce Forum, worked with representatives from member banks, other financial institutions, community groups, suppliers, and retailers to develop in 2002 voluntary industry standards that aim to improve the accessibility of electronic banking. The "Standard on Automated Teller Machines"[3] describes in detail the requirements and the corresponding specifications for the design, deployment, and operation of accessible ATMs. Such requirements cover the following eight topics:

1. *Access and location*: Exterior route, interior route, site, doors and entrances to interior ATMs, signage (location signs, door signs/room identification, Braille-tactile and visual signs, interface component signs), lighting (ATM approach lighting, ATM task area lighting, ATM display lighting), user operating space (card access slot, floor surfaces, ambient noise)
2. *ATM operation*: Input controls, cash dispenser, speech input, user identification/verification, audio (audio hardware, audio scripts, initiating audio, audio operation), operating instructions, Auslan, transaction time, supporting documentation
3. *Card swiping, insertion and withdrawal*: Card reader/ATM activation (card reader signage, card slot orientation, card slot identification, card removal)
4. *ATM display*: Color and contrast, text (background, font, and case), advertising, user display selection
5. *Keypad*: Characteristics of keys, keypad mapping, keypad layout, function keys, function display keys
6. *Outputs*: Beep feedback, printed text
7. *Security* and *privacy*
8. *Installation, maintenance and operating instructions*: Installation height and knee clearance zone, height and reach for user interface components, display, grab bar, parcel shelf, walking stick notch, waste receptacle

The ABA standards are ranked in three priority levels: Level 1 corresponds to "shall"; level 2 corresponds to "should"; and level 3 corresponds to "may." For each case there is an indication of whether the requirement/specification applies to all equipment, services, or installations or only to new programs or substantially upgraded equipment, services, or installations.

Furthermore, the following national standards address the accessibility of ATMs:

1. The Australian AS 3769-1990 "Automatic teller machines—User access"[8] includes recommendations for their design and performance to facilitate unobstructed access to a level, adequately sized, well-lit area in front of an ATM, and the provision of features of the user interface of the ATM which are within reach and operable by the greatest possible number of users.
2. The Canadian CAN/CSA-B651.1-01 (R2006)-2001 standard "Barrier-Free Design for Automated Banking Machines" (CSA 2001) covers enhanced accessibility features for the visually impaired.

"Ergonomics of human-system interaction: Guidance on accessibility for human-computer interfaces" ISO/TS 16071 (Gulliksen and Harker, 2004; ISO/TS 16071, 2002) is the first major software accessibility standard applicable to PATs. A universal access model has been proposed recently to identify areas belonging in ISO/TS 16071 that require further accessibility guidance (Carter and Fourney, 2004). In case a PAT has been designed to deliver web-based services or HTML content, it should also follow the Web Accessibility Initiative (WAI) Guidelines.[9]

The ITU E135[10] concerns the human factors aspects of public terminals for people with vision, hearing, and motion impairments. It provides a generic framework of recommendations for public terminal operating procedures that facilitates ease of use for the six basic user actions defined in ITU E134.[11] These actions include initialization, means of payment, identification, communication, next (optional), and end.

An approach allowing an individual to locate and access public information terminals would be to include a low-cost infrared link as a part of the devices' design. Users could then interface with the PAT using their own interface, such as a mobile phone, PDA, or with mobile assistive technology, such as a mobile Braille keypad (Shroff and Winters, 2006). Toward this target, research has been conducted on the development of a universal infrared link protocol standard (Novak et al., 1996) and the universal remote console communication (URCC) standard (Vanderheiden et al., 1998; Shroff and Winters, 2006).

An important issue to accomplish accessibility in PATs is related to automatic user interface generation (Krzysztof and Weld, 2004; Shroff, 2005) combined with a protocol for remote interaction with other devices. ANSI/INCITS 389, 390, 391, 392, 393-2005 is a family of standards that:

---

[7] "Information Transaction Machines Accessibility Checklist" available on the Section 508 home page of the Civil Rights Division, U.S. Department of Justice: http://www.usdoj.gov/crt/508.

[8] AS 3769-1990 "Automatic teller machines—User access."

[9] Web Accessibility Initiative Guidelines and Techniques, http://www.w3.org/WAI/guid-tech.html.

[10] ITU E135, "Human factors aspects of public telecommunications terminals for people with disabilities."

[11] ITU E134, "Human factors aspects of public terminals: Generic operating procedures."

TABLE 48.2   Short Checklist for the Accessibility of Public Access Terminals

1	Locating and accessing a terminal
1.a	Are the location signs easy to read?
1.b	Are there adequate lighting levels?
1.c	Are there queuing arrangements?
1.d	Is there a clear path for wheelchairs?
1.e	Is there a level surface?
1.f	Is there a location system for blind users?
2	Card systems
2.a	Can it be easily used by someone with poor manual dexterity?
2.b	Does the card contain user requirements?
2.c	Is there a notch on the card for orientation?
2.d	Is there embossing for card identification?
2.e	Does it use a contact card system?
3	External features, labels, and instructions
3.a	Are labels positioned to be easy to read?
3.b	Are the labels legible for someone with low vision?
3.c	Are the instructions numbered?
3.d	Are the controls reachable from a wheelchair?
3.e	Is there a funneled card entry slot?
4	Screens and interaction
4.a	Is the screen shielded from sunlight?
4.b	Is the touch screen reachable from a wheelchair?
4.c	Have you minimized parallax problems?
4.d	Are there foreign languages for screen instructions?
5	Operating instructions
5.a	Have you used simple vocabulary?
5.b	Is there an inductive loop facility?
5.c	Is there an audio jack socket?
5.d	Is there audible feedback of key input?
5.e	Is there speech output?
5.f	Have you provided a video link?
6	Keypads
6.a	Have you used the telephone layout for numeric keys?
6.b	Is there a raised dot on number 5?
6.c	Are there clear visual markings on keys?
6.d	Have you used raised or recessed keys?
6.e	Are the keys well spaced?
6.f	Are the keys internally illuminated?
6.g	Is there tactile feedback on keys?
6.h	Is generous time allowed for key input?
7	Touch screens
7.a	Is there an option to increase character size?
7.b	Are there large key fields?
7.c	Does text accompany graphical symbols?
7.d	Is there a speech output option?

(Continued)

TABLE 48.2   (*Continued*)

8	Retrieving money, cards, and receipts
8.a	Is security adequate?
8.b	Do documents protrude at least 3 cm?
8.c	Do cards protrude at least 2 cm?
8.d	Is minimum force needed to withdraw card?
9	Typefaces and legibility
9.a	Are instructions at least 16 point type size?
9.b	Have you used good contrast text?
9.c	Are there background patterns?
9.d	Have you used an easy to read typeface?
9.e	Have you used short line length?
9.f	Is the receipt readable by someone with low vision?
10	Training
10.a	Do you provide instruction booklets in clear print?
10.b	Are there instructions on audiotape?
10.c	Is assistance provided for first-time users?

Details and explanations are given in the Accessibility Guidelines for Public Access Terminals.[5]

TABLE 48.3   Information Transaction Machines Accessibility Checklist

1	Can the user change sound settings, such as volume?
2	For all visual information and cues, are there simultaneous corresponding audible information and cues?
3	Is there sufficient contrast between foreground and background colors or tones so that a person with low vision can use the technology, or is it possible for the user to select foreground and background colors?
4	Is all text information displayed large enough that it can be read by someone with low vision, or is it possible for the user to select an enlarged display?
5	Can users select speech input?
6	If speech input is used, is an alternative method available for inputting information, such as typing on a keyboard or scanning printed material, so that someone who cannot speak can use the technology?
7	For all sound cues and audible information, such as "beeps," are there simultaneous corresponding visual cues and information?
8	Is there a headphone jack to enable the user to use an assistive listening system to access audible information?
9	Can users simultaneously change the visual display settings and the sound settings?
10	Can the user read displayed output with a tactile display such as Braille?
11	Does the technology allow the user to use scanning input?
12	Is the technology manufactured such that it allows a person using a wheelchair to approach the technology, including all controls, dispensers, receptacles, and other operable equipment, with either a forward or parallel approach?
13	Is the technology manufactured so that, if the equipment is properly placed, the highest operable part of controls, dispensers, receptacles, and other operable parts fall within at least one of the following reach ranges? If a forward approach is required, the maximum high forward reach is 48 inches. If a side approach is allowed, and the reach is not over an obstruction, the maximum high side reach is 54 inches. If it is over an obstruction which is no more than 24 inches wide and 34 inches high, the maximum high side reach is 46 inches.
14	If electrical and communication system receptacles are provided, are they mounted no less than 15 inches above the floor?
15	Are all controls and operating mechanisms operable with one hand and operable without tight grasping, pinching, or twisting of the wrist?
16	Is the force required to operate or activate the controls no greater than 5 lbf?
17	Are instructions and all information for use accessible to and independently usable by persons with vision impairments?
18	Is the technology manufactured in such a way that it can be made detectable to persons with visual impairments who use canes to detect objects in their path? Note: Objects projecting from walls with their leading edges between 27 and 80 inches above the finished floor should protrude no more than 4 inches into walks, halls, corridors, passageways, or aisles. Objects mounted with their leading edges at or below 27 inches above the finished floor may protrude any amount. Freestanding objects mounted on posts or pylons may overhang 12 inches maximum from 27 to 80 inches above the ground or finished floor.
19	After you have evaluated this ITM using the checklist, have users with a wide variety of disabilities test it for accessibility. Describe the accessibility successes and problems they encountered during these exercises, including any suggestions for improvement: [space provided for answer]

Source: "Information Transaction Machines Accessibility Checklist," U.S. Department of Justice, Civil Rights Division. http://www.usdoj.gov/crt/508.

TABLE 48.4 Family of ANSI INCITS Standards to Facilitate Operation of Information and Electronic Products through Remote and Alternative Interfaces and Intelligent Agents

Standard	Short Title	Short Description
ANSI INCITS 389-2005	Universal Remote Console	Provides a framework of components that combine to enable remote user interfaces and remote control of network accessible electronic devices and services through a Universal Remote Console (URC)
ANSI INCITS 390-2005	User Interface Socket Description	Defines an extensible markup language (XML)-based language for describing a user interface socket, the purpose of which is to expose the relevant information about a target so that a user can perceive its state and operate it
ANSI INCITS 391-2005	Presentation Template	Defines a language (presentation template markup language) for describing modality-independent user interface specifications, or presentation templates associated with a user interface socket description
ANSI INCITS 392-2005	Target Description	Defines an extensible markup language (XML)-based language for the description of targets and their sockets, as used within the URC framework for discovery purposes
ANSI INCITS 393-2005	Resource Description	Defines a syntax for describing atomic resources

TABLE 48.5 Biometric Data Interchange ANSI INCITS Standards

Standard	Short Title	Short Description
ANSI INCITS 394-2004	Information Technology –Application Profile for Interoperability, Data Interchange and Data Integrity of Biometric-Based Personal Identification for Border Management	Specifies the application profile to be used when incorporating biometrics-based identification and verification into border management applications and systems; border management includes pre-arrival, arrival, stay management, departure, and database reconciliation/management
ANSI INCITS 395-2005	Information Technology—Biometric Data Interchange Formats—Signature/Sign Data	Specifies a data interchange format for representation of digitized sign or signature data, for the purposes of biometric enrollment, verification, or identification through the use of Raw Signature/Sign Sample Data or Common Feature Data
ANSI INCITS 398-2005	Information Technology—Common Biometric Exchange Formats Framework (CBEFF)	Describes a set of data elements necessary to support biometric technologies in a common way; these data elements can be placed in a single file used to exchange biometric information between different system components or between systems; the result promotes interoperability of biometric-based application programs and systems developed by different vendors by allowing biometric data interchange

- Provides a framework and protocols to create a true universal remote console
- Provides a framework for automatic UI generation
- Is device-independent
- Is platform- and OS-independent
- Supports extensible markup language (XML)
- Consists of a framework designed for multiple communication methodologies
- Allows for multimodal input/output
- Allows for integration of accessibility user devices
- Allows for complex target devices to be controlled

In Table 48.4 a short description of the family of these standards is provided.[12] Another set of related standards are those allowing biometric data interchange[13] (see Table 48.5). Based on this family of standards, one can design and develop flexible and robust PATs that support complex, multimodal, accessible, personalized, and universally usable user interfaces.

The standards referring to the accessibility of plastic cards used in PATs are presented in Table 48.6.[14]

The ITU-T P370 standard[15] deals with the magnetic field strength around the ear-cap of telephone handsets that provide for coupling to hearing aids.

Finally, the Japan Vending Machine Manufacturers Association, under its design guidelines for ATMS for the visually disabled (Hosono and Miki, 2004), has adopted a set of 15 tactile symbols (see Figure 48.1) to help users identify functions on the ATM by touch. However, these symbols have not been widely adopted outside Japan.

---

[12] ANSI/INCITS 389-2005, ANSI/INCITS 390-2005, ANSI/INCITS 391-2005, ANSI/INCITS 392-2005, ANSI/INCITS 393-2005.

[13] ANSI/INCITS 394-2005, ANSI/INCITS 395-2005, ANSI/INCITS 398-2005.

[14] EN 1332, Part 2; EN 1332, Part 4; ETR 165 (1995); ETSI ETS 300 767 (1997); ISO/IEC 10536 (2000), Part 1; ISO/IEC 10536 (2000), Part 2; ISO/IEC 10536 (2000), Part 3; ISO/IEC 10536 (2000), Part 4; ITU E136 "Tactile identifier on pre-paid telephone cards."

[15] ITU-T P370 "Magnetic field strength around the ear-cap of telephone handsets which provide for coupling to hearing aids."

TABLE 48.6 Standards Referring to the Accessibility of Plastic Cards

		Title
EN 1332		Machine-readable cards, related device interfaces, and operations
	Part 2	Dimension and location of tactile identifier for ID-1 cards
	Part 4	Coding of user requirements for people with special needs
ETR 165 (1995)		Recommendations for a tactile identifier on machine-readable cards for telecommunications terminals
ETSI ETS 300 767 (1997)		Telephone prepayment cards: tactile identifier
ISO/IEC 10536 (2000)		Identification cards: contactless integrated circuit cards
	Part 1	Design principles and symbols for the user interface
	Part 2	Dimension and location of tactile identifier for ID-1 cards
	Part 3	Keypads
	Part 4	Coding of user requirements for people with special needs
ITU E136		Tactile identifier on prepaid telephone cards

48.7 Conclusions

The rapid penetration of PATs for a wide and diverse range of application domains and key societal functions requires that people can effectively interact with them to participate equally in society. For this reason, PATs must be designed taking into consideration human diversity and the variety of user capabilities and preferences. Although most of the major manufacturers of PATs adopted the principles of universal design, there is still a long way to go before existing technology and research outcomes described in this chapter will be incorporated at a satisfactory level in practice. Also, there is a need for the recent related standards to be acknowledged by the industry and to be expanded and strengthened, along with appropriate legislation and regulation measures. Finally, there is a challenge for the research community to study innovative ways for enhancing universal access to PATs. Besides the progresses described in this chapter, the research agenda is still open for the investigation of the migration of existing accessibility knowledge in the user interface of PATs. For example, one can use in speech synthesis augmented auditory representation of texts (Xydas and Kouroupetroglou, 2001; Xydas et al., 2003) or auditory description of tabular data (Spiliotopoulos et al., 2005) and advanced multilingual personalized information objects (Calder et al., 2005) in spoken interaction, or allow symbol or sign language users to interact with PATs throughout advanced augmentative and alternative communication systems (Antona et al., 1999; Kouroupetroglou and Pino, 2001; Viglas and Kouroupetroglou, 2003) or mobile video relay services (Bystedt, 2007). Finally voice-based (Dai et al., 2005; Harada et al., 2006) or brain-computer interface (Pino et al., 2003) cursor control techniques can be explored.

48.8 Appendix

Accessibility Guideline 1.7: ***"Ensure that users with restricted or no vision can use all functions of the terminal,"*** from Accessibility Guidelines for Public Access Terminals, the Irish National Disability Authority.[4]

Users will normally access the functions of the terminal through controls such as buttons, keys, and knobs. These may or may not be visible to users who are blind, partially sighted, or color blind. However, all users with restricted vision must still be able to use all the functions. Where possible, the controls should be designed so that at least users who have partial vision

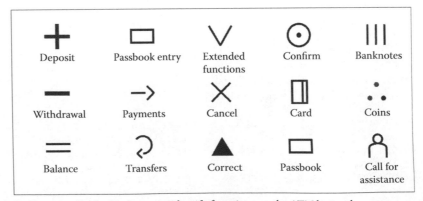

FIGURE 48.1 Tactile symbols allow visually disabled users to identify functions on the ATM by touch.

or color-blindness are able to perceive them, understand what each is for, and know how to operate them. It may also be possible to design in such a way that users who are completely blind can still perceive, distinguish, and operate the same controls. If this is not possible or extremely difficult, an alternative control method should be made available that these users can perceive and that can be used to access the full functionality.

48.8.1 Rationale

At the bank, when I'm paying a bill and I have to enter my PIN number on the little keypad they have on the counter. The keys are grey and the numbers on the keys are black, so it's hard to read. Also, the keys are very close together so it's difficult to see one from the other.

–Partially sighted bank customer

Control labels, prompts, and delivered information are usually provided as text but presented visually. Any user who cannot see to read the text will not be able to perceive the information it contains. The controls themselves have first to be perceived by the user before they can be operated. Again, this often relies on sight, so that people with restricted sight may be unable to use the terminal. A particular problem occurs with terminals that use unlabeled buttons for input, changing the prompt next to each button on successive screens. Knowing which button to press relies on visible correlations that are difficult to learn.

Because the keys have different meanings during the transaction, it's very difficult to use. It would be nice if it would quietly confirm each key press, so I know what's going on.

–Partially sighted bank customer

Light and reflections are always a problem due to the viewing angle. If you are looking down, your shadow covers the screen. Imagine you are sitting and have to stretch your head to see the screen. Once I withdrew too much money because I couldn't see the screen properly.

–Partially sighted bank customer

48.8.2 Directions and Techniques

48.8.2.1 Add Voice Output

Add voice output to speak the instructions. This can be achieved using either prerecorded audio or speech synthesis. Speech synthesis, while more flexible, is often of much poorer quality and may be difficult to understand for some users and in noisy environments. If voice output is likely to be intrusive or if the instructions give away sensitive personal information, allow the audio to be turned off during a user session and provide a standard jack socket for connecting an earphone. Inserting a jack plug should switch off the output to the external loudspeakers.

48.8.2.2 Consider Developing a Separate Audio Menu

If the terminal relies on visible correlations between changing prompts and unlabeled buttons, users may still not be able to know which button is associated with each prompt. In this case, it may be best to develop a separate audio menu that prompts the user to press a number on the keypad for each choice. This can be done along the lines of telephone interactive voice response (IVR) systems that ask the user to "press 1 for this option, press 2 for that option" etc.

48.8.2.3 Add Tactile Indicators to Buttons and Keys

It is standard practice to put a single raised dot on the 5 key to help users orientate their fingers on the keypad by touch. It is also possible to emboss Braille on keys and buttons, although this is not as widely effective as it may seem, since less than 2% of visually impaired people can read Braille. Also, Braille has less value in outdoor situations during cold weather, because tactual sensitivity is dramatically reduced at lower temperatures.

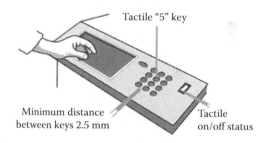

Tactile "5" key

Minimum distance between keys 2.5 mm

Tactile on/off status

KEY AND BUTTON DESIGN
Make buttons large and tactilely discernable.

48.8.2.4 Raise or Recess Buttons and Keys

Raise or recess the buttons and keys by at least 2 mm over the surrounding area.

48.8.2.5 Provide Tactile and Audio Feedback

Provide tactile and audio feedback to indicate the operation of controls. Tactile indication can be provided by requiring a gradual increase in the force to activate a control, followed by a sharp decrease as it is activated. Audio feedback can be given using a beep or click. For multiposition controls, feedback should be used to indicate the current position or status.

48.8.2.6 Label Text, Color, and Contrast

For label text, ensure that characters are at least 4 mm high but avoid using all upper case, which is more difficult to read than mixed case. For good contrast, use light-colored characters on a dark background (e.g., white or yellow on matt black or a dark color). Avoid using pale colors or patterned backgrounds for text. Also avoid red on green or yellow on blue since these combinations may cause problems for people who are color-blind. Use a typeface designed for display, such as Tiresias,[16] which has numerals with open shapes that are easier to distinguish for people with low vision.

WELL-DESIGNED LABEL TEXT

For good contrast, use light-colored characters on a plain dark background.

48.8.2.7 Do Not Rely on Color for Meaning

While color coding can be useful as an aid to recognition, it should not be relied on entirely, since over 8% of Irish males and some females have difficulty distinguishing between red and green (other forms of color-blindness are relatively uncommon).

48.8.2.8 Raise the Edges of Input Slots

Design a raised ridge around input slots such as those used for entering a card or plugging in a headphone jack. This will make them easier to locate by touch.

48.8.2.9 Allow User-Selectable Settings

Applying the previous techniques should result in a terminal that suits all users. However, in some cases, what is best for one group of users is not necessarily best for all. If this is the case, it may help if the user interface can be adapted by the user, or automatically for the user, to fit individual capabilities. For example, users who are visually impaired could choose voice output and large type, while users with good vision may prefer to have more detail and no sound. The choice could be made by the user selecting from a number of displayed options. Alternatively,

information required for the terminal to switch automatically could be encoded on a user's smart card per request.

48.8.2.10 Use the Telephone Layout for Keypads rather than the Calculator Layout

The telephone layout is recommended as the standard for public access terminals. Using this layout will ensure the most consistency with other terminals.

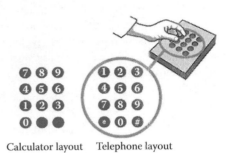

Calculator layout   Telephone layout

KEYPAD LAYOUTS

Use telephone layout for keypads.

How you could check for this …

48.8.2.11 Self-Test Early Prototypes

Designers can run simple sight tests themselves on an early prototype, by simulating various types of vision loss. Complete loss of sight can be simulated either by wearing a blindfold, turning off the lights, or putting the terminal in a black bag. To simulate partial sight, a test user who normally wears glasses could take them off. It is also possible to buy low-vision simulation glasses that simulate various types of visual impairments. In all cases, extreme care should be taken to avoid injury through loss of balance or collision with unseen objects. This may require that the test user remains seated or, if one must move around, obstacles such as floor cabling are removed in advance. Although this type of ad hoc testing will not replace proper testing with real users, it will give some insight into what it is like to be operating with reduced vision.

48.8.2.12 Test with Real Users

During development, you should test the prototype in a realistic situation with real people who have various forms of visual impairment. In particular, you should include people who are recently impaired and have not yet developed enhanced perception or coping methods.

---

[16] Tiresias typefaces, http://www.tiresias.org/fonts/index.htm.

References

Adams, A. S. and Thieben, K. A. (1991). Automatic teller machines and the older population. *Applied Ergonomics* 22: 85–90.

Akoumianakis, D. and Stephanidis, C. (2001). USE-IT: A tool for lexical design assistance, in *User Interfaces for All: Concepts, Methods, and Tools* (C. Stephanidis, ed.), pp. 469–487. Mahwah, NJ: Lawrence Erlbaum Associates.

Antona, M., Stephanidis, C., and Kouroupetroglou, G. (1999). Access to lexical knowledge in interpersonal communication aids. *Journal of Augmentative and Alternative Communication* 15: 269–279.

Asawa, T., Ohta, A., and Ando, T. (2005). Promoting universal design of automated teller machines (ATMs). *Fujitsu Scientific & Technological Journal* 41: 86–96.

Bederson, B. B., Lee, B., Sherman, R. M., Herrnson, P. S., and Niemi, R. G. (2003). Electronic voting system usability issues, in the *Proceedings of the SIGCHI Conference on Human Factors in Computing Systems*, 5–10 April 2003, Ft. Lauderdale, FL, pp.145–152. New York: ACM Press.

Blignaut, P. (2004). An empirical methodology for usability analysis of a touchscreen-based information kiosk system for African users with low levels of computer literacy, in *User-Centered Interaction Paradigms for Universal Access in the Information Society, Proceedings of the 8th ERCIM Workshop "User Interfaces for All"* (C. Stary and C. Stephanidis, eds.), 28–29 June 2004, Vienna, Austria, pp. 203–218. Berlin/Heidelberg: Springer-Verlag.

Borchers, J., Deussen, O., and Knörzer, C. (1995). Getting it across: Layout issues for kiosk systems. *ACM SIGCHI Bulletin* 27: 68–74.

Bystedt, P. (2007). Access to video relay services through the pocket interpreter (3G) and Internet (IP), in *Towards an Inclusive Future: Impact and Wider Potential of Information and Communication Technologies* (P. Roe, ed.), pp. 60–64. Brussels, Belgium: COST219ter.

Calder, J., Melengoglou, A. C., Callaway, C., Not, E., Pianesi, F., Androutsopoulos, I., et al. (2005). Multilingual personalized information objects, in *Multimodal Intelligent Information Presentation* (O. Stock and M. Zancanaro, eds.), Volume 27, pp. 177–201 (Series: Test, Speech and Language Technology). Berlin/Heidelberg: Springer-Verlag.

Carter, J. and Fourney, D. (2004). Using a universal access reference model to identify further guidance that belongs in ISO 16071. *Universal Access in the Information Society* 3: 17–29.

Cassell, J. (2000). Embodied conversational interface agents. *Communications of the ACM* 43: 70–78.

Cassell, J., Stocky, T., Bickmore, T., Gao, Y., Nakano, Y., Ryokai, K., et al. (2002). MACK: Media Lab Autonomous Conversational Kiosk. Presented at *Imagina '02*, 12–15 February 2002, Monte Carlo, Monaco.

Christian, A. D. and Avery, B. L. (1998). Digital smart kiosk project, in the *Proceedings of the CHI '98 Conference on Human Factors in Computing Systems*, pp. 155–162. New York: ACM Press.

Christian, A. D. and Avery, B. L. (2000). Speak out and annoy someone: Experience with intelligent kiosks, in the *Proceedings of the SIGCHI Conference on Human Factors in Computing Systems (CHI 2000)*, 1–6 April 2000, The Hague, The Netherlands, pp. 313–320. New York: ACM Press.

Connell, I., Blanderford, A., and Green, T. (2004). CASSM and cognitive walkthrough: Usability issues with ticket vending machines. *Behaviour and Information Technology* 23: 307–320.

CSA (2001). CAN/CSA-B651.1-01 (R2006) *Barrier-Free Design for Automated Banking Machines*. Mississauga, Canada: Canadian Standards Association.

Dai, L., Goldman, R., Sears, A., and Lozier, J. (2005). Speech-based cursor control using grids: modeling performance and comparisons with other solutions. *Behaviour and Information Technology* 24: 219–230.

Darch, U. and Caltabiano, N. J. (2004). Investigation of automatic teller machine banking in a sample of older adults. *Australasian Journal on Ageing* 23: 100–103.

Devies, A. J. (2006). *Touch Screen Not Best Choice for Disabled Voters*. http://www.votetrustusa.org/index.php?option=com_content&task=view&id=1595&Itemid=807.

Fellbaum, K. and Kouroupetroglou, G. (2008). Principles of electronic speech processing with applications for people with disabilities. *Technology and Disability* 20: 55–85.

Freitas, D. and Kouroupetroglou, G. (2008). Speech technologies for blind and low vision persons. *Technology and Disability* 20: 135–156.

Frost & Sullivan (2004). *Cost and Design of Point of Decision (PoD) Kiosks Meet Pressing In-Store Marketing Needs of Retailers*. http://www.frost.com/prod/servlet/cpo/25969614.

Gaston, C. A. (2005). A better way to vote, in the *Proceedings of the 38th Annual Hawaii International Conference on System Sciences (HICSS '05)*, 3–6 January 2005, Big Island, HI, pp. 117. Washington, DC: IEEE.

Gauvain, J., Gangolf, J., and Lamel, L. (1996). Speech recognition for an information kiosk, in the *Proceedings of the International Conference on Spoken Language Processing 1996 (ICSLP '96)*, 3–6 October 1996, Philadelphia, pp. 849–852. Washington, DC: IEEE.

Gilbert, D. K. and Fraser, E. (1997). An analysis of automatic teller machine usage by older adults: A structured interview approach. *Applied Ergonomics* 28: 173–180.

Gulliksen, J. and Harker, S. (2004). The software accessibility of human-computer interfaces-ISO Technical Specification 16071. *Universal Access in the Information Society* 3: 6–16.

Gustafson, J., Lindberg, N., and Lundeberg, M. (1999). The August spoken dialogue system, in the *Proceedings of Eurospeech '99*, 5–9 September 1999, Budapest, Hungary, pp. 1151–1154. http://www.speech.kth.se/august/eur99_augsys.html.

Gustafson, J., Lundeberg, M., and Liljencrants, J. (1999). Experiences from the development of August: A multimodal spoken dialogue system, in the *Proceedings of IDS'99*, pp. 61–64. http://www.speech.kth.se/prod/publications/files/1242.pdf.

Harada, S., Landay, J., Malkin, J., Li, X., and Bilmes, J. (2006). The vocal joystick: Evaluation of voice-based cursor control techniques, in the *Proceedings of the 8th International ACM SIGACCESS Conference on Computers and Accessibility (ASSETS 2006)*, 23–25 October 2006, Portland, OR, pp. 197–204. New York: ACM Press.

Hartson, H. R. (2003). Cognitive, physical, sensory, and functional affordances in interaction design. *Behaviour & Information Technology* 22: 315–338.

Herzog, G. and Ndiaye, A. (2006). Building multimodal dialogue applications: System integration in SmartKom, in *SmartKom: Foundations of Multimodal Dialogue Systems* (W. Wahlster, ed.), pp. 439–452. Berlin/Heidelberg: Springer-Verlag.

Herzog, G., Ndiaye, A., Merten, S., Kirchmann, H., Becker, T., and Poller, P. (2004). Large-scale software integration for spoken language and multimodal dialog systems. *Natural Language Engineering* 10: 283–305.

Herzog, G. and Reithinger, N. (2006). The SmartKom architecture: A framework for multimodal dialogue systems, in *SmartKom: Foundations of Multimodal Dialogue Systems* (W. Wahlster, ed.), pp. 55–70. Berlin/Heidelberg: Springer-Verlag.

Hone, K. S., Graham, R. M., Maguire, C., Baber, C., and Johnson, G. I. (1998). Speech technology for automatic teller machines: An investigation of user attitude and performance. *Ergonomics* 41: 962–981.

Hosono, H. and Miki, H. (2004). Universal design in practice at Oki Electric, in the *Proceedings of the Designing for the 21st Century III: An International Conference on Universal Design (D21³)*. http://www.designfor21st.org/proceedings/proceedings/plenary_iaud_oki.html.

ISO/TS 16071 (2002). *Ergonomics of Human-System Interaction: Guidance on Accessibility for Human-Computer Interfaces.* http://www.iso.org/iso/catalogue_detail?csnumber=30858.

Jena, C. (2007) Biometric ATMs for rural India. *Computer Express.* http://www.expresscomputeronline.com/20070312/technology01.shtml.

Keates, S. and Clarkson, J. (2004). *Countering Design Exclusion: An Introduction to Inclusive Design.* Berlin/Heidelberg: Springer-Verlag.

Keates, S., Clarkson, P. J., and Robinson, P. (2004). Design for participation: Providing access to e-information for older adults. *Universal Access in Information Society* 3: 149–163.

Korn, P. (1999). *Architecture of an Automatic Teller Machine Usable by All Using the Java Platform.* Sun Microsystems, Inc. http://www.sun.com/access/articles/wp-aatm.

Kouroupetroglou, G. and Nemeth, G. (1995). Speech technology for disabled and elderly people, in *Telecommunications for All* (P. Roe, ed.), pp. 186–195. Brussels, Belgium: ECSC-EC-EAEC.

Kouroupetroglou, G. and Pino, A. (2001). ULYSSES: A framework for incorporating multi-vendor components in interpersonal communication applications, in *Assistive Technology, Proceedings of the 6th European Conference for the Advancement of Assistive Technology (AAATE 2001)* (C. Marincek, ed.), 3–6 September 2001, Ljubljana, Slovenia, pp. 55–59. Amsterdam, The Netherlands: IOS Press.

Krzysztof, G. and Weld, D. (2004). SUPPLE: Automatically generating user interfaces, in the *Proceedings of the 9th International Conference on Intelligent User Interfaces (IUI'04)*, 13–16 January 2004, Island of Madeira, Portugal, pp. 93–100. New York: ACM Press.

Kules, B., Kang, H., Plaisant, C., Rose, A., and Shneiderman, B. (2003). *Immediate Usability: Kiosk Design Principles from the CHI 2001 Photo Library.* Technical Report, 21 January 2003, University of Maryland, Computer Science Department; CS-TR-4293, UMIACS; UMIACS-TR-2001-71, HCIL-TR-2001-23.

Law, C. M. and Vanderheiden, G. C. (1998). EZ access strategies for cross-disability access to kiosks, telephones and VCRs, in the *Proceedings of the 1998 International Conference on Technology and Persons with Disabilities (CSUN'98)*, March 1998, Los Angeles. http://www.dinf.ne.jp/doc/english/Us_Eu/conf/csun_98/csun98_074.htm.

Law, C. M. and Vanderheiden, G. C. (2000). The development of a simple, low cost set of universal access features for electronic devices, in the *Proceedings of the ACM Conference on Universal Usability (CUU 2000)*, 16–17 November 2000, Arlington, VA, pp. 118–123. http://www.acm.org/pubs/articles/proceedings/chi/355460/p118-law/p118-law.pdf.

Lee, S. and Cho, J. E. (2004). Evaluating accessibility of public information kiosks, in the *Proceedings of the 9th International Conference on Computers Helping People with Special Needs (ICCHP 2004)* (K. Miesenberger, D. Burger, and W. Zagler, eds.), 7–9 July 2004, Paris, pp. 76–79. Berlin/Heidelberg: Springer-Verlag.

Li, Y., Groenegress, C., Strauss, W., and Fleischmann, M. (2004). Gesture Frame: A screen navigation system for interactive multimedia kiosks, in *Gesture-Based Communication in Human-Computer Interaction, Proceedings of the 5th International Gesture Workshop (GW 2003)* (A. Camurri and G. Volpe, eds.), 15–17 April 2004, Genoa, Italy, pp. 380–385. Berlin/Heidelberg: Springer-Verlag.

Lovelock, J. (2006). Extending care to the community. *Kiosk Business Magazine, 6th Annual Kiosk Benchmark Study*, p. 14.

Mäkinen, E., Patomäki, S., and Raisamo, R. (2002). Experiences on a multimodal information kiosk with an interactive agent, in the *Proceedings of the 2nd Nordic Conference on Human-Computer Interaction (NordiCHI 2002)*, 19–23 October 2002, Aarhus, Denmark, pp. 273–276. New York: ACM Press.

Malerczyk, C. (2004). Interactive museum exhibit using pointing gesture recognition, in the *Proceedings of the 12th International Conference in Central Europe on Computer Graphics, Visualization and Computer Vision 2004 (WSGC 2004)*, 2–6 February 2004, Bory, Czech Republic, pp. 165–172. Plzen, Czech Republic: UNION Agency-Science Press.

Malerczyk, C., Dähne, P., and Schnaider, M. (2005). Pointing gesture-based interaction for museum exhibits, in *Universal Access in HCI: Exploring New Interaction Environments, Volume 7 of the Proceedings of the 11th International Conference on Human-Computer Interaction (HCI International 2005)* (C. Stephanidis, ed.), 22–27 July 2005, Las Vegas [CD-ROM]. Mahwah, NJ: Lawrence Erlbaum Associates.

Malerczyk, C., Schnaider, M., and Gleue, T. (2003). Video based interaction for a mixed reality kiosk system, in *Universal Access in HCI: Inclusive Design in the Information Society, Volume 4 of the Proceedings of the 10th International Conference on Human-Computer Interaction (HCI International 2003)* (C. Stephanidis, ed.), 22–27 June 2003, Crete, Greece, pp. 1148–1152. Mahwah, NJ: Lawrence Erlbaum Associates.

Manzke, J. M., Egan, D. H., Felix, D., and Krueger, H. (1998). What makes an automated teller machine usable by blind users? *Ergonomics* 41: 982–999.

Marcellini, F., Mollenkopf, H., Spazzafumo, L., and Ruoppila, I. (2000). Acceptance and use of technological solutions by the elderly in the outdoor environment: Findings from a European survey. *Zeitschrift fur Gerontologie und Geriatrie* 33: 169–177.

Maybury, M. (1998). Planning multimedia explanations using communicative acts, in *Readings in Intelligent User Interfaces* (M. Maybury and W. Wahlster, eds.), pp. 99–106. San Francisco: Morgan Kaufman.

McCauley, L. and D'Mello, S. (2006). MIKI: A speech enabled intelligent kiosk, in the *Proceedings of the 6th International Conference on Intelligent Virtual Agents (IVA 2006)* (J. Gratch et al., eds.), 21–23 August 2006, Marina del Rey, CA, pp. 132–144. Berlin/Heidelberg: Springer-Verlag.

Nielsen, J. (1993). *Usability Engineering*. San Francisco: Morgan Kaufmann Publishing.

Novak, M., Kaine-Krolak, M., and Vanderheiden, G. C. (1996). Development of a universal disability infrared link protocol standard, in the *Proceedings of RESNA '96 Pioneering the 21st Century*, pp. 458–460. Salt Lake City: RESNA Press. http://trace.wisc.edu/docs/irdoc_resna96/irrsna96.htm.

Ntoa, S. and Stephanidis, C. (2005). ARGO: A system for accessible navigation in the World Wide Web. *ERCIM News* 61: 53.

Perton, M. (2006). Japanese bank to allow cellphone ATM access. *Engadget*. http://www.engadget.com/2006/01/27/japanese-bank-to-allow-cellphone-atm-access.

Pino, A., Kalogeros, E., Salemis, I., and Kouroupetroglou, G. (2003). Brain computer interface cursor measures for motion-impaired and able-bodied users, in *Universal Access in HCI: Inclusive Design in the Information Society, Volume 4 of the Proceedings of HCI International 2003* (C. Stephanidis, ed.), 22–27 June 2003, Crete, Greece, pp. 1462–1466. Mahwah, NJ: Lawrence Erlbaum Associates.

Proll, B. and Retschtzegger, W. (2000). Discovering next generation tourism information systems: A tour on TIScover. *Journal of Travel Research* 39: 182–191.

R.B.R. (2006). *Global ATM Market and Forecasts to 2011*. Report of Retail Banking Research Ltd. http://www.rbrlondon.com.

Reithinger, N. and Herzog, G. (2006). An exemplary interaction with SmartKom, in *SmartKom: Foundations of Multimodal Dialogue Systems* (W. Wahlster, ed.), pp. 41–52. Berlin/Heidelberg: Springer-Verlag.

Richter, K. and Enge, M. (2003). Multi-modal framework to support users with special needs, in the *Proceedings of the 5th International Symposium on Human-Computer Interaction with Mobile Devices and Services (MobileHCI 2003)* (L. Chittaro, ed.), 8–11 September 2003, Udine, Italy, pp. 286–301. Berlin/Heidelberg: Springer-Verlag.

Roe, P., Sandhu, J., Delaney, L., Gill, J., and Narcinelly, M. (1995). Consumer overview, in *Telecommunications for All* (P. Roe, ed.). Brussels, Belgium: ECSC-EC-EAEC.

Sagawa, H. and Takeuchi, M. (2000). Development of an information kiosk with a sign language recognition system, in the *Proceedings on the 2000 Conference on Universal Usability (CUU 2000)*, 16–17 November 2000, Arlington, VA, pp. 149–150. New York: ACM Press.

Sanderson, C. (2003). *Automatic Person Verification Using Speech and Face Information*. PhD dissertation, University of Griffith, Queensland, Australia.

Savidis, A., Akoumianakis, D., and Stephanidis, C. (2001). The unified user interface design method, in *User Interfaces for All: Concepts, Methods, and Tools* (C. Stephanidis, ed.), pp. 417–440. Mahwah, NJ: Lawrence Erlbaum Associates.

Savidis, A. and Stephanidis, C. (2001a). Development requirements for implementing unified user interfaces, in *User Interfaces for All: Concepts, Methods, and Tools* (C. Stephanidis, ed.), pp. 441–468. Mahwah, NJ: Lawrence Erlbaum Associates.

Savidis, A. and Stephanidis, C. (2001b). The I-GET UIMS for unified user interface implementation, in *User Interfaces for All: Concepts, Methods, and Tools* (C. Stephanidis, ed.), pp. 489–523. Mahwah, NJ: Lawrence Erlbaum Associates.

Savidis, A. and Stephanidis, C. (2001c). The unified user interface software architecture, in *User Interfaces for All: Concepts, Methods, and Tools* (C. Stephanidis, ed.), pp. 389–415. Mahwah, NJ: Lawrence Erlbaum Associates.

Savidis, A. and Stephanidis, C. (2004a). Unified user interface design: Designing universally accessible interactions. *International Journal of Interacting with Computers* 16: 243–270.

Savidis, A. and Stephanidis, C. (2004b). Unified user interface development: Software engineering of universally accessible interactions. *Universal Access in the Information Society* 3: 165–193.

Shroff, P. (2005). *Algorithm to Automatically Generate Multi Modal User Interfaces Based on User Preferences and Abilities*. MS thesis, Marquette University, Milwaukee, WI.

Shroff, P. and Winters, J. M. (2006). Generation of multi-modal interfaces for hand-held devices based on user preferences and abilities, in the *Proceedings of the 1st*

Distributed Diagnosis and Home Healthcare (D2H2) Conference, 2–4 April 2006, Seattle, pp. 124–128. Washington, DC: IEEE.

Spiliotopoulos, D., Xydas, G., Kouroupetroglou, G., and Argyropoulos, V. (2005). Experimentation on spoken format of tables in auditory user interfaces, in the *Proceedings of the 3rd International Conference on Universal Access in Human-Computer Interaction, jointly with HCI International 2005*, 22–27 July 2005, Las Vegas [CD-ROM]. Mahwah, NJ: Lawrence Erlbaum Associates.

S.R.A. (2006). *Kiosks and Interactive Technology.* Report of Summit Research Associates, Inc. http://www.summit-res.com.

Stary, C. and Totter, A. (2003). Measuring the adaptability of universal accessible systems. *Behaviour & Information Technology* 22: 101–116.

Steiger, P. and Suter, B. A. (1994). MINELLI: Experiences with an interactive information kiosk for casual users, in the *Proceedings of the UBILAB Conference 1994, Computer Science Research at UBILAB*, September 1994, Zurich, Switzerland, pp. 124–133. Konstanz, Germany: Universitätsverlag Konstanz.

Stephanidis, C. (2001). The concept of unified user interfaces, in *User Interfaces for All: Concepts, Methods, and Tools* (C. Stephanidis, ed.), pp. 371–388. Mahwah, NJ: Lawrence Erlbaum Associates.

Stephanidis, C., Paramythis, A., Sfyrakis, M., and Savidis, A. (2001a). A case study in unified user interface development: The AVANTI web browser, in *User Interfaces for All: Concepts, Methods, and Tools* (C. Stephanidis, ed.), pp. 525–568. Mahwah, NJ: Lawrence Erlbaum Associates.

Stephanidis, C., Paramythis, A., Zarikas, V., and Savidis, A. (2004). The PALIO framework for adaptive information services, in *Multiple User Interfaces: Cross-Platform Applications and Context-Aware Interfaces* (A. Seffah and H. Javahery, eds.), pp. 69–92. Chichester, U.K.: John Wiley & Sons.

Stephanidis, C., Salvendy, G., Akoumianakis, D., Bevan, N., Brewer, J., Emiliani, P. L., et al. (1998). Toward an information society for all: An international R&D agenda. *International Journal of Human-Computer Interaction* 10: 107–134.

Stephanidis, C., Savidis, A., and Akoumianakis, D. (2001b). Tutorial on "Engineering Universal Access: Unified User Interfaces," in the *1st Universal Access in Human-Computer Interaction Conference (UAHCI 2001), Jointly with the 9th International Conference on Human-Computer Interaction (HCI International 2001)*, 5–10 August 2001, New Orleans. http://www.ics.forth.gr/hci/files/uahci_2001.pdf.

Stocky, T. and Cassell, J. (2002). Shared reality: Spatial intelligence in intuitive user interfaces, in the *Proceedings of the 7th International Conference on Intelligent User Interfaces (IUI'02)*, 12–15 January 2002, Miami, FL, pp. 224–225. New York: ACM Press.

Tarakanov-Plax, A. (2005). Design concept for ATM Machine, accessible for the elderly users in Israel, in the *Proceedings of the International Conference on Inclusive Design (INCLUDE 2005)*, 5–8 April 2005, London. http://www.hhc.rca.ac.uk/archive/hhrc/programmes/include/2005/proceedings/pdf/tarakanova.pdf.

Thatcher, A., Mahlangu, S., and Zimmerman, C. (2006). Accessibility of ATMS for the functionally illiterate through icon-based interfaces. *Behaviour & Information Technology* 25: 65–81.

Thatcher, A. Shaik, F., and Zimmerman, C. (2005). Attitudes of semi-literate and literate bank account holders to the use of automatic teller machines (ATMs). *International Journal of Industrial Ergonomics* 35: 115–130.

Tung, L. L. (1999). The implementation of information kiosks in Singapore: An exploratory study. *International Journal of Information Management* 19: 237–252.

Tung, L. L. (2001). Information kiosk for use in electronic commerce: Factors affecting its ease of use and usefulness, in the *Proceedings of the 14th Bled Electronic Commerce Conference*, 25–26 June 2001, Bled, Slovenia, pp. 329–350. Kranj, Slovenia: University of Maribor Press.

Tung, L. L. and Tan, J. H. (1998). A model for the classification of information kiosks in Singapore. *International Journal of Information Management* 18: 255–264.

Vanderheiden, G. C. (1997). Cross disability access to touch screen kiosks and ATMs, in *Design of Computing Systems, Proceedings of the 7th International Conference on Human-Computer Interaction (HCI International 1997)* (M. J. Smith, G. Salvendy, and R. J. Koubek, eds.), 24–29 April 1997, San Francisco, pp. 417–420. New York: Elsevier.

Vanderheiden, G. C. (2001). Everyone interfaces, in *User Interfaces for All: Concepts, Methods, and Tools* (C. Stephanidis, ed.), pp. 441–468. Mahwah, NJ: Lawrence Erlbaum Associates.

Vanderheiden, G. C., Law, C., and Kelso, D. (1998). Universal Remote Console Communication Protocol (URCC), in the *Proceedings of CSUN'98*, March 1998, Los Angeles. http://www.dinf.ne.jp/doc/english/Us_Eu/conf/csun_98/csun98_096.htm.

Vanderheiden, G. C., Law, C. M., and Kelso, D. (1999). EZ Access interface techniques for anytime anywhere anyone interfaces, in *CHI '99 Extended Abstracts on Human Factors in Computing Systems*, pp. 3–4. New York: ACM Press.

Viglas, C. and Kouroupetroglou, G. (2003). e-AAC: Making Internet-based interpersonal communication and WWW content accessible for AAC symbol users, in *Universal Access in HCI: Inclusive Design in the Information Society, Volume 4 of the Proceedings of HCI International 2003* (C. Stephanidis, ed.), 22–27 June 2003, Crete, Greece, pp. 276–280. Mahwah, NJ: Lawrence Erlbaum Associates.

Xydas, G. and Kouroupetroglou, G. (2001). Augmented auditory representation of e-Texts for text-to-speech systems, in the *Proceedings of the 4th International Conference on Text, Speech and Dialogue (TSD 2001)*, 11–13 September 2001, Zelezna Ruda, Czech Republic, pp. 134–141. Berlin/Heidelberg: Springer-Verlag.

Xydas, G., Spiliotopoulos, D., and Kouroupetroglou, G. (2003). Modelling emphatic events from non-speech aware documents in speech based user interfaces, in *Human-Computer Interaction, Theory and Practice, Volume 2 of the Proceedings of HCI International 2003* (C. Stephanidis and J. Jacko, eds.), 22–27 June 2003, Crete, Greece, pp. 806–810. Mahwah, NJ: Lawrence Erlbaum Associates.

Zarikas, V., Papatzanis, G., and Stephanidis, C. (2001). An architecture for a self-adapting information system for tourists, in the Proceedings of the Workshop on Multiple User Interfaces over the Internet: Engineering and Applications Trends (in Conjunction with HCI-IHM'2001), September 2001, Lille, France. http://users.encs.concordia.ca/~seffah/ihm2001/papers/zarikas.pdf.

49

Intelligent Mobility and Transportation for All

Evangelos Bekiaris,
Maria Panou,
Evangelia Gaitanidou,
Alexandros Mourouzis, and
Brigitte Ringbauer

49.1 Introduction

Humans, throughout our history, have continuously searched for new ways to facilitate movement, travel, and transportation to survive, but also to improve quality of life. Nowadays, as distance becomes less critical thanks to continuous advances in communications and transport technologies, more and more people proceed into urban and interurban traveling for business, leisure, socialization, entertainment, and education. For example, we move from one location to another to get to work, to attend a business meeting, to visit a museum or a friend in another city or country, for shopping, and so on. Clearly, mobility, the ability and willingness to move by means of self-powered motion (e.g., walking) or transportation, is a basic facilitator toward autonomous living and social inclusion.

49.1.1 Basic Concepts in Travel and Every Transport

An individual, to get from one location to another, is faced with land characteristics and infrastructures (roads, railways, airports, signage, etc.), travel and transport means (bicycle, automobiles, buses, ships, etc.), and support services and operations

(e.g., gas stations, tolls, and regulations). In this context, the process of moving from one location to another involves four basic concepts: *orientation, roaming, wayfinding,* and *travel*. The process of orientation is about determining one's location and direction with reference to one's surroundings. Roaming refers to wandering or traveling freely and with no specific destination, whereas wayfinding refers to the process of finding and choosing an appropriate path to reach a destination. Orientation and wayfinding are mainly supported by orientation, position, and indication signs, and by social navigation (Munro et al., 1999). *Social navigation* is the concept that when people are looking for information (e.g., directions, recommendations) they will turn to other people rather than use formalized information artifacts. Finally, the process of travel is the actual process of moving by any means within the physical space. Clearly, mobility highly depends on the individual's ability, supported or not, to perform all these tasks in any given context.

49.1.2 Definition of Mobility Impairment

In this chapter, the following definition is adopted. Any individual that experiences limitations in self-powered motion or in using common transport means, in terms of orientation, wayfinding, and travel, is defined as a *mobility-impaired person*. Such limitations may include total lack of ability, or the involvement of high risks, in terms of effectiveness, efficiency, safety, and security, with respect to the act of moving from one location to another, including roaming. Following this definition, the notion of mobility-impaired (MI) people includes, but is not limited to, two major populations: people with disability and the elderly. In particular, the following population groups can be considered as MI (adapted from Simões et al., 2006):

- Blind and vision-impaired people
- Deaf and people with hearing impairments
- Motor-impaired people (e.g., wheelchair or walker users)
- People with balance or stamina limitations
- People with cognitive impairments, including limitation in terms of memory, understanding, and learning
- People with psychological disorders, such as agoraphobia
- Illiterate people and inexperienced individuals, such as novice drivers and first-time travelers

Finally, the elderly can also be considered as MI people, since, due to aging, they often embody mild forms of the characteristics of the aforementioned groups. In these terms, *MI users* is a superset of elderly and disabled users that face not only traditional accessibility issues, such as these related to user aging or disability, but also issues related to user mobility.

49.1.3 Indicative Barriers to Autonomous Mobility

As a whole, transport services and facilities, daily or not, allow the active individual to perform important functions, including moving about at will, engaging in social and recreational activities when desired, and reaching business and social services when needed. However, currently major deficiencies exist in assistance to MI people in the transportation modes, which are hampering their social and working activities. In addition, stereotypical views about several groups of MI people (e.g., that it is safer for them to be accompanied while on the move or that they should not be supported to have a fast and efficient transportation as time is not critical to them), which are either no longer valid or totally false, interfere with the society's ability to increase access for all to community activities and support them in living an independent and quality life. Whether MI persons provide their own transportation in private vehicles, by walking and cycling, or in taxis and on transit using public transport, a number of barriers are omnipresent. Some are indicatively mentioned in the following list:

- In many European countries, despite considerable efforts from their governments, a significant portion of public transport infrastructure and means remains largely inaccessible (blocked pavements, steps too high, no seats close to the entrance, etc.) and constrained to the needs of the majority, rather than the specific needs of individuals.

- Adding to that, navigation information, such as direction, position, or indication signs, and operational information, such as departure announcements, train timetables, bus itineraries, and ticket fares, are often delivered in inaccessible or unusable ways for MI people.
- If a multimodal transport trajectory has to be undertaken, MI persons often become frustrated and have difficulty organizing and coping with such rather complicated and time-dependent travel itineraries.
- In developing areas, land characteristics and infrastructures change dramatically over time (e.g., changes in street plans, directions, and buildings often used as personal landmarks) and often disorient MI people and lead them into extremely confusing, if not risky, situations.
- MI people are often unhappy with traveling simply because they feel uncomfortable with leaving their houses (including in-house electrical appliances) unattended.
- During peak hours, MI people are often confronted with an overcrowded public transport system that causes them problems in terms of comfort and safety.
- Traffic jams further add to the increasing anxiety of MI drivers, who are often being subjected to aggressive behavior from impatient drivers, hence adding more stress and anxiety.
- Finding parking space at busy areas can be frustrating for all drivers, yet it can become particularly cumbersome for many MI people.
- Crime rates in public areas and transport often discourage MI people, being relatively more vulnerable, from traveling or taking advantage of social navigation.

49.2 ICT Contribution: Opportunities and Challenges

With the emergence of *information and communication technologies* (ICT), including mobile computing and ambient intelligence, new opportunities and challenges are arising with regard to the aforementioned barriers in autonomous mobility and transportation for all, especially for people at risk of exclusion. ICT, in combination with other emerging technologies, can lead to the development and intelligent integration of different data sources and services (*infomobility services*) as a means to encourage and enhance everyday mobility and travel (Naniopoulos et al., 2004). Today, ICT-based potential support to MI people on the move ranges from geographical information systems, multimedia databases, Internet-based route information, reservation systems, and planners to traffic information kiosks, and then on to pedestrian and in-car navigation services, including location-aware virtual guides (e.g., Abowd et al., 1997; Not et al., 1997; Cheverst et al., 2000; Malaka and Zipf, 2000; Poslad et al., 2001; Fink and Kobsa, 2002; Kruger et al., 2004; Panou et al., 2005).

In this area, ICTs are exponentially expanding and beginning to pervade into those categories of citizens who are largely technically illiterate or technophobic. In general, traffic information services are free of charge, provided through financing by local authorities (i.e., governments). Such services, however, are usually static and primarily offered through stationary means, such as web sites, infokiosks, variable message signs (VMSs), and variable direction signs (VDSs) (Naniopoulos et al., 2004). Mobile IT alternatives are growing even faster. Yet, in a very competitive market, focused on numbers and quick profits through trend marketing, ICT development has aimed mainly at young and agile users (e.g., *automotive navigation systems*, ANS[1]), inducing a high risk for social exclusion of a large and more quickly growing market, that of Europe's senior and disabled population. Only as a result of *a posteriori* knowledge, the corresponding research agenda was revised to address the issue of users' diversity, including research on personalization (Abowd et al., 1997; Malaka and Zipf, 2000; Poslad et al., 2001; Fink and Kobsa, 2002), developments for specific user groups, such as blind and elderly (e.g., Green and Harper, 2000; Coroama, 2003; Hub, Diepstraten, and Ertl, 2004; EasyWalk, 2008) or cognitively impaired people (Carmien and Gorman, 2003), and special developments for specific domain applications, such as social navigation support systems (Bilandzic et al., 2008) or special navigation aids for pedestrians[2] (Feiner et al., 1997; Ross and Blasch, 2000; Kolbe, 2004), wheelchair users (Gunderson, et al., 1996; Levine et al., 1999; Lankenau et al., 2003; Uchiyama et al., 2005), blind users (Loomis et al., 1998; Mori and Kotani, 1998), museum visitors (Fleck et al., 2002; Raptis et al., 2005), and for public transport disabled users (Sasaki et al., 2002; Banâtre et al., 2004).

In general, with respect to the user and the travel life cycle phase (orientation and destination decision, wayfinding and travel), the potential ICT-based support can be divided into three main areas: *pretrip support, on-trip support,* and *emergency and risk management support.*

49.2.1 Pretrip Support

Pretrip support is aimed at, or eventually results in, helping individuals in their travel preparations, such as creating a new itinerary or collecting information and guidance regarding a trip. There are many electronic sources through which travelers can find helpful pretrip information, such as timetables, flights, tickets, hotel offers, and information on the accessibility of various sites. Such sources include governmental and corporate web sites of various kinds, infokiosks, electronic message signs located at the stops, and so on. These information sources follow a generic design, which is not, however, in most cases, accessible for various user categories.

For instance, in TELSCAN's[3] collaborative testing of the EUROSCOPE-ROMANSE Triplanner infokiosk (MK I touch screen), 9 out of 13 wheelchair users could not get close enough to operate the terminal properly (Barham and Alexander, 1998). More specifically, the lowest interface elements (icon buttons) were located at a height of 1.219 m. This proved to be a particular problem throughout the evaluation for people using wheelchairs. In further tests in TELSCAN's collaborative testing of the ROMANSE Triplanner MK I touch screen, 10 out of 13 wheelchair users could not see the screen properly, because it was too high. In addition, 4 wheelchair users could not reach the touch screen, and a further 10 could only do so with difficulty. Also, within the same project, 12 out of 53 elderly or disabled users experienced problems due to parallax. This was a problem not only for those who had to look up at the screen from below, but also for one subject who was very tall (Barham and Alexander, 1998). Parallax indicates an apparent displacement of an object, caused by an actual change of point of observation. Figure 49.1 illustrates the importance of a foot recess in the kiosk to enable a person in a wheelchair to get close enough to reach the control buttons or touch screen (Nicolle and Burnett, 1999).

49.2.2 On-Trip Support

Once a trip is initiated and the user gets on the move, further support can be provided, for example, by means of information about the bus station that he needs to get to, possible traffic problems along the way, and, generally, further information

FIGURE 49.1 A wheelchair user's difficulty in reaching the ROMANSE Triplanner info-kiosk. (From Nicolle, C. and Burnett, G., eds., *TELSCAN Code of Good Practice and Handbook of Design Guidelines for Usability of Systems by Elderly and Disabled Travellers,* TELSCAN Project Deliverable 5.2, Commission of the European Communities, 1999.)

---

[1] There exist a plethora of commercial navigation systems for car navigation, many of which feature also a pedestrian mode, such as TomTom Navigator, Falk Mobile Navigator, or Destinator. For a brief overview, refer to http://en.wikipedia.org/wiki/Automotive_navigation_system.

[2] Known as *pedestrian or personal navigation assistants* (PNA).

[3] Project home page: http://hermes.civil.auth.gr/telscan/telsc.html.

about the rest of the trip. Such kinds of information can be delivered through personal mobile devices, in-vehicle systems, roadside VMS/VDSs, and message displays installed in public transportation means. The readability/legibility of the information delivered and generally the usability of the delivery platform for various MI user groups are controversial, at best.

As an example, there are many driver support systems nowadays available in cars (i.e., lane deviation warning systems, anticollision systems, driver monitoring systems, etc.). Within AIDE,[4] a review of existing user interfaces of various systems developed in eight research projects was performed based on a specific analysis template, providing the assessment results from the user's point of view. In parallel, the opinions of 16 experts on user needs for ADAS/IVIS user interfaces were captured and analyzed through a questionnaire-based survey. The results highlight that existing ADAS/IVIS are not appropriate for older drivers, or for drivers with problems in the upper extremities. The problems arise due to the small control buttons and the visual display's positioning, which are impossible to be used by disabled drivers, or the small displays and letters, which are difficult to read not only by elderly drivers, but sometimes also by typical drivers, requiring important attention time, which leads to increased risk while driving. Also, inappropriate visual display contrast and acoustic signal frequency cause problems for elderly drivers. Both elderly and disabled users need adaptation in the warning thresholds of longitudinal and lateral control systems (Panou et al., 2005).

49.2.3 Emergency Handling

Emergencies might occur at any time, and might require important reactions by drivers, transportation means users, or pedestrians. A typical example is when driving a car, where a siren of an ambulance or a police car indicates to the driver to move the car out of the way. If the driver is a person with hearing disabilities, she needs other ways of being warned (e.g., optical signal or tactile warning). If the deaf user is a pedestrian, then alerting mobile equipment is needed, else he might be in danger, if for example he attempts to cross the street at the specific time when the emergency vehicle passes by. Another risky situation is when there is an emergency in a transportation means (fire, etc.) and the passengers need to evacuate it immediately. The passengers with hearing problems might not be able to react on time, as they might not detect the situation immediately. Also, a visually impaired passenger will not be able to escape safely without the assistance of another person, putting her life at risk. These examples highlight the need for accessible, multimedia emergency information provision and evacuation plans support for all transportation means and hubs.

---

[4] Project home page: http://www.aide-eu.org.

49.2.4 A Critical Overview

Overall, the current situation regarding ICT-based research and developments for mobility and transport is characterized by:

- *Fragmentation in terms of information and service provision*: Due to development scattering, the available sources of support (e.g., infomobility services) are often single-purposed (leaving significant gaps with regard to individual user needs and user goal coverage) and spread in (hyper-) space. Integrated approaches (e.g., merging indoor user localization techniques (Bahl and Padmanabhan, 2000) with tools for outdoor environments) in the provision of support to users on the move will increase the findability, perceived usefulness, and ultimately, the user acceptance of the delivered services and products.

- *Context, network, and device dependencies*: Due to significant diversity in terms of device/platform capabilities and limitations, especially of mobile devices and displays, and due to a continuous, mainly user-driven, shifting of user location and of the *conditions of use* (Antona et al., 2007), current design and development solutions are highly challenged. To this end, the use of most services and products currently remains highly context, network and device dependent.

- *Insufficient service personalization and user interface adaptation*: Travel, especially for entertainment and tourism, is highly related to personal interests, preferences, and habits. Therefore, it is critical to support *personalization* (Searby, 2003), that is, to deliver systems that can be modified appropriately in their configuration or behavior by information about the user. Clearly, all types of personalization, *adaptability* (customization), *adaptivity*, and *intermediate cases* (Jameson, 2003a, 2003b) need to be implemented effectively in future ICT-based developments for MI users, both in terms of systems and their interfaces (Akoumianakis and Stephanidis, 1997), but of information (content) itself.

- *Lack of inclusive and "design for all" approaches*: Most developments in the area remain bounded to the original designs for the "average" user in a desktop environment. Yet, the success of new systems for MI users is subject to their ability to cope with diversity in the end-user population, the access media and devices, and the dynamic nature of today's contexts of use. Although various design approaches have been proposed to accommodate the requirements of diverse population groups into interactive interfaces (Stephanidis, 2001), including universal design principles (Story, 1998), limited effort has been put worldwide into securing universal access to mobility and (everyday) transport-related systems.

- *Inconsistency in terms of interaction quality and delivered user experience*: Design disparity, lack of conformance to established heuristics (Nielsen, 1994) for usability and accessibility, and limited attention to user-based evaluation, often lead to increased inaccessibility and complexity, rather than to reducing the already existing barriers in physical transport infrastructure and means.

- *Insufficient software and services interoperability:* Due to lack of (compliance with) technical, interoperability specifications, and due to market competitiveness and emerging strategies, most research and end-user products fail to facilitate the exchange of information with other existing systems or services.
- *Limited multilingualism:* Since the field of ICT in mobility and transport is still in a state of flux, limited attention has been paid to the development and delivery of multilingual interfaces and information (content). Clearly, this is a significant burden toward universal access.

In light of the previous list, there is an arising need for fostering the conduct of human-centered research and development to ensure ambient, intuitive, and easy access to human-sensitive travel-related information and services for all citizens, and especially for mobility impaired people.

49.3 Related Research and Development Work and Current Gaps

Table 49.1 lists some of the previous EC-funded research initiatives relevant to the target population of MI people. Through these initiatives, useful reference groundwork is available; yet significant integration advances are needed to bring all the emerging results into the realization of the intelligent mobility

and transport for all vision (see Section 49.4). The state of the art in the major areas of research on mobility and transportation is briefly highlighted in the following subsections.

49.3.1 Navigation, Localization, and Route Guidance

Today, mainly three different navigation methods are available: off-board navigation with all necessary data calculated on a central server and sent via GPRS/GSM to the end device (e.g., PDA or SmartPhone); on-board navigation with the navigation data on a CD and an on-board unit, and hybrid navigation with mixed solutions. Navigation systems for vehicles and pedestrians exist as state of the art separately. Their fusion and integration into a seamless navigation product constitute the new frontier, addressed by the 6th Framework STREP IM@GINE IT.[5]

To bring such a product to the market for MI people, however, the following key elements are still required:

- Precise and seamless user localization with such an accuracy that it can provide real-time guidance to people with limited perception of the environment, such as blind, partially sighted, and deaf users.
- Accessible route functionality of route guidance systems, according to the specific residual abilities of each MI person.

---

[5] Project home page: http://www.imagineit-eu.com.

TABLE 49.1 Indicative EC-Funded Research and Development (R&D) Projects of Interest to the Mobility and Transportation Sector

Areas of Interest	Relevant Projects
Mobile travel information	PROMISE, HANNIBAL, INFOTEN, ITSWAP, PROMISE, WH@M, WITTY, TURTLE, IMAGE
Fixed traveler information	AUSIAS, EUROSCOPE, QUARTET+, TABASCO, DISTINCT, EUROSPIN, EU-SPIRIT, TourIST
Pedestrians and passengers navigation and route guidance	PEPTRAN, IAIM, TURTLE
Intermodal information and cooperation of transport operators	EUROSPIN, EU-SPIRIT, TRIDENT, ISCOM, TRAVELGUIDE, SAMPO, SAMPLUS, INVETE
User interfaces for travel information	INFOPOLIS, TELSCAN, INVETE, TURTLE, EQUALITY, MECCS
Multimodal interfaces	PHYSTA, ERMIS, HUMAINE
System architecture and open platform design	EUROSPIN, EU-SPIRIT, PROMISE, KAREN, TITOS
Tourism information and/or cross-sector information	CRUMPET, E-TOUR, GUIDEFREE, PALIO, TOURSERVE, ESTIA, INTOURISME, DISTINCT, MONADS, TURTLE
Electronic payment systems for transport and their integration with information networks	ADEPTII, ADEPTIII, CALYPSO, DISTINCT, SIROCCO
Multimedia information environments	DISTINCT, AMIDE, FAETHON
Open geographic information systems	GEOMED, MORE
E-commerce/M-commerce/L-commerce	DISTINCT, WITTY, ELSME, PROTONET, Cash
Intelligent agent-based services	IMAGE, CRUMPET, ADAMANT, AGENT CITIES, ORESTEIA, IM@GINE IT
Specific services for elderly and disabled (E&D) users	CONFIDENT, SPORT4ALL, RACE-TUDOR, NEWT, TURTLE, DISTINCT, INDIES, EQUALITY
Specific mobile equipment for E&D users	MORE
Semantic web services	SWWS, SWWC, ACEMEDIA, KNOWLEDGEWEB

- Intermodal and personalized accessibility information (content) along each route, for the functionality to operate (referring to accessibility of pavement, transportation means, bus stops or parking lots, point of origins and destination).

Specifically on localization, in the near future several global navigation satellite systems (GNSS), such as the global positioning system (GPS) of the United States, the global navigation satellite systems (GLONASS) of Russia, and the European satellite navigation systems (GALILEO) will offer much improved accuracy, improving it from 10 to 1 meters or even better. Nevertheless, in some urban areas (where MI people need the most help and guidance) there will still be poor guidance from GNSS due to noncoverage from a sufficient number of satellites. Other techniques will be used in combination for better localization, such as GSM triangulation or "logical" localization. This is of special importance in urban areas, where buildings create the so-called urban canyon phenomenon.

In several European cities, key landmarks (intersections, traffic lights, bus stops, etc.) are being equipped with RF transmitters (i.e., in Newcastle and currently under trial in the university's campus) or even with WiFi emitters (i.e., in Sunderland through the Tyne & Wear e-Kiosk Consortium), that may offer precise localization of persons carrying passive or active tags (wearable or even integrated at their mobile) (Edwards and Blythe, 2003). The approach of using smartsigns to augment more conventional navigation and location technologies has been proven to be a low-cost and reliable approach that supports localization where due to location or the (urban) environment more conventional systems cannot operate correctly or with the degree of accuracy desired (Blythe, 1999). Yet there is a need to integrate different localization techniques and route accessibility content to offer seamless and precise navigation and route guidance services to MI people everywhere in Europe (Blythe, 2003).

49.3.2 Web and Mobile Services for Users on the Move

Web services are the next-generation application integration technology. The technology is being standardized under the auspices of the World Wide Web consortium (W3C), and numerous implementations are already available. Web services exchange data using extensible markup language (XML). Access is provided by the simple object access protocol (SOAP) running over HTTP or other transport protocol. Interfaces to web services are defined in a web service description language (WSDL) that enables the design of compatible clients. Finally, universal description, discovery and integration (UDDI) provides a universal registry, where companies can publish web services to be accessed by others. Work is in progress on a comprehensive web services security (WSS) specification that will provide a framework for ensuring that access to web services is limited to authorized clients. For the time being, the development and execution platforms for web services include Java 2, Enterprise Edition (J2EE) and Microsoft. NET application servers, offered by many middleware vendors (i.e. IBM, BEA, SUN, Oracle, Microsoft), based on proprietary s/w solutions.

While mature technology for web services is several years away, the demand for web services is rapidly growing, especially in the travel and tourism sector. According to a study carried out by the magazine *ONLINE TODAY* (2002), 69% of Internet users use the Internet to compare hotel prices, 28% for booking flight or rail tickets, and 25% for booking hotels. This vast market is gradually getting mobile. Figures 49.2 and 49.3 display this boom in the worldwide location-based services (WLS), as well as mobile consumer service revenues in Europe. And yet, the forecasted rapid gains seem to be more and more delayed and even endangered by strong competition from other regions, such as iMode, the hyper service from Japan.

The real revolution in such services is anticipated from the introduction of semantic web-enabled web services (SWWS),

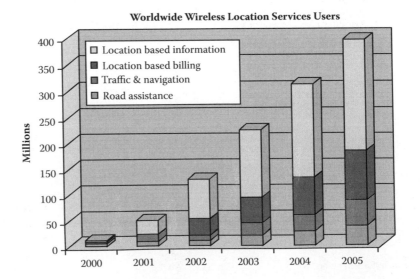

FIGURE 49.2 WLS users worldwide.

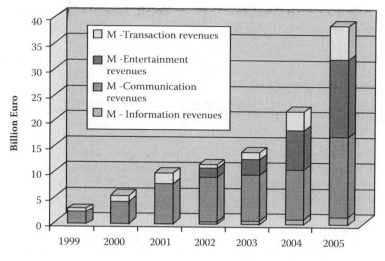

FIGURE 49.3   Mobile consumer services revenues by m-Commerce sector.

that will transform web services from a static collection of information into a distributed system of computation, making content within the web machine-processable and machine-interpretable. The application of ontologies is increasingly seen as key to enable semantics-driven data access and processing (Bussler et al., 2002).The key terms used are briefly defined here:

- A web service is an application function that can be assessed by other applications using Internet technologies.
- An ontology is a shared and machine-executable conceptual model in a specific domain of interest (Ankolenkar et al., 2001). They interweave human understanding of symbols with their machine processability.
- Semantics refers to the correct population of attributes with correct domain values. Within web services, semantics is seen as a vertical layer that may be exploited by the horizontal layers: services description, publishing, discovery, trading partners as well as service flow and composition.

Web services are defined by the web service modeling framework (WSMF) that consists of four main different elements: *ontologies* that provide the terminology used by other elements; *goal repositories* that define the problems that should be solved by web services; *web services descriptions* that define various aspects of a web service; and *mediators,* which bypass interoperability problems. However, to bring these technologies to the benefit of MI people, there are several barriers, the most prominent of which are as follows:

- While the user perceives the transport network as seamless and unlimited, neither the supply of transport services nor the provision of travel information services is in reality seamless or unlimited. If any of the nodes within the transport chain are not accessible, the overall trip is cancelled or even worse, which results in frustration for the user. Thus, reliable and seamless content on accessible

facilities needs to exist on the web, in the first place. This is not limited to the transport network, but extends to most infomobility services.

- Such services need to be interoperable and "talk to each other" or to a centralized system. Thus, common ontologies need to be defined, gradually working toward a semantic web for accessible infomobility services.
- Appropriate search engines that can deal with numerous and heterogeneous data formats, business logics, and user needs need to be developed. These need to be intelligent and adaptive, so as to cover the varying needs of different types of MI users as well as to work in a still nonstandardized web environment.

49.3.3 Intelligent Agents Technology

Agent technology has been applied to the infomobility services sector in recent years. The work conducted in a series of projects, such as IMAGE,[6] CRUMPET,[7] ADAMANT,[8] and the most recent IM@GINE IT, have addressed the state of the art, and in some cases extended it. Recently, agent technology was also applied for supporting virtual elderly assistance communities, in the context of the TeleCARE[9] project. The overall goal in TeleCARE is the design and development of a configurable framework and new technological solutions for tele-supervision and tele-assistance, based on the integration of a multiagent and a federated information management approach, including both stationary and mobile intelligent agents, combined with the services likely to be offered by the emerging ubiquitous computing and intelligent home appliances, and applied to assist elderly people. An

---

[6] See footnote 5.
[7] Project home page: http://www.ist-crumpet.org.
[8] Project home page: http://adamant.elec.qmul.ac.uk.
[9] Project home page: http://www.uninova.pt/~telecare.

agent-based approach was selected, since the following requirements that can be dealt with nicely by agent technology (Weiss, 1999) are inherent to infomobility services provisioning:

- *Distribution*: Timely and geographical distribution of users and services have to be taken into account.
- *Heterogeneity*: User support, services, devices, and networks are provided by different sources.
- *Coordination and communication*: Optimal user satisfaction can only be achieved if coordination and communication between users and service providers are provided.
- *Mobility*: Users and user devices may change their physical and logical position over time.

Moreover, recent research has proposed that elderly and disabled people compose a segment of the population that would profit very much from ambient intelligence if it is accessible (Abascal, 2004). Furthermore, O'Hare et al. (2005) advocate the use of agents as a key enabler in the delivery of ambient intelligence (AmI). Thus, in an AmI framework for servicing elderly and disabled (mobility-impaired) people, the role of agent technology is crucial.

The overall aim of CRUMPET was to implement, validate, and trial tourism-related value-added services for nomadic users (across mobile and fixed networks). The services provided by CRUMPET took advantage of integrating four key emerging technologies and applying them to the tourist domain (Poslad et al., 2001): location-aware services, personalized interaction, seamlessly accessible multimedia mobile communication, and smart component-based middleware that uses multiagent technology.

IMAGE proposed a very different business logic (use local content and service providers), but also proposed a technical solution with some advantages in comparison to CRUMPET (Moraitis et al., 2003). First was the ability to access the IMAGE services through various devices (PDAs, mobile phones, PCs) due to the fact that user profiles are stored at the IMAGE server and not at the user's device. Second, IMAGE services can be offered by any network provider who can opt to use an entirely new user interface. The IMAGE system adapts the service to user habits, and each IMAGE server has the ability to seamlessly interoperate with other servers (i.e., IMAGE servers). Crucial to supporting this ubiquitous/pervasive computing, networks are necessary to deliver a range of communications carrier services that may be available to the user though the establishment of a low-level mobile ad hoc network (MANET).

ADAMANT exploited the use of agent technology for offering infomobility services in a local area (i.e., within an airport).

The most recently proposed work in this context is the IM@GINE IT project in the e-Safety area of IST. It offers a new package of services that employs intercity travel and keeps the personalization and adaptation feature of IMAGE along with the flexibility of service provider choice. Intercity travel will be supported with more than one transport type (airplane, car, boat, rail, bus, etc.). Moreover the reservation feature and timetable management during multimodal travel will enable the user to plan a safe, comfortable, carefree journey. In IM@GINE IT, possibly heterogeneous agents will be used for the first time for infomobility services provisioning. Thus, an agent communication language (ACL), based on an ontology that is fully independent of an implementation framework, will support the communication of agents built on different platforms and frameworks (that exist in different car hardware/nomad devices). Another innovation that this project proposes is the development of middle agents that will be applied for the first time in this application area (transport and location-based services). Using ontologies and ACL this agent type is the semantic gateway for the device-resident agents to the IM@GINE IT services. It can identify collaborators of the service-requesting agents depending on location and service. TeleCARE used the AGLETS framework (AGLETS, as a mobile agents framework, provides basic inter-platform mobility and communication mechanisms, see Aglets API Documentation, v. 2.0.2; Java Aglet Community at http://aglets.sourceforge.net) to develop agents that create and manage virtual communities of elderly people. They offer services like chat and e-mail to support the virtual community. Moreover, they monitor the living status of elderly persons and are able to undertake some action (social alarm, inform a relative, etc.) when the need arises. Finally, they offer an agenda reminder service, a time bank management system (a mechanism for collaborative community building), and entertainment in the form of games, music, and education programs.

However, the approaches mentioned previously scarcely address the case of an elderly or disabled person requesting infomobility services. This is quite a different case:

- Personalization does not only refer to learning user habits. It encapsulates the need for knowledge regarding the situation that this person is into. Therefore, the personal assistant agent must employ powerful knowledge regarding the person's type of disability. For example, a route request and presentation for a wheelchair user poses different requirements than for other user groups. The accessibility features of crossroads, busses, and so on, must be taken into account. In the case of an elderly or sick person even the weather conditions need to be considered. Thus, agents with knowledge sufficient to serve each disability type must be developed. Furthermore, in the case that a person has more than one type of impairment, these agents must be able to cooperate to serve that person.
- Service discovery and provisioning (the middle-agents task) must allow for requesting services according to accessibility characteristics. More middle-agent types need to be developed, in comparison to IM@GINE IT.
- The immediate ambient plays a vital role for servicing the user (domotic services, ticketing services, etc.), and all these services must be accessible through the user's device. Here, the user agent does not only use a local service; it either needs to select and get the local service and then adapt it for the user, or it is the actual user of the service

(e.g., switch on the air conditioning) and must have the relevant knowledge and profile of the user.

The TeleCARE project offered a starting point for elderly care services specification and design. Technologically, the ASK-IT project[10] moved further since it developed agents that have reasoning capabilities and that can profile their user (intelligent and proactive agents). Agents in TeleCARE functioned using the simple stimulus-action paradigm. Moreover, ASK-IT addressed a wide range of disabled people, not just the elderly.

49.3.4 User Interface Technology

User interface research and development is a rapidly evolving field. New interaction approaches, methods, and techniques are continually being produced and modified, with important consequences on users' view of, navigation in, and interaction with computers. In parallel, a transition from desktop computers to mobile computing devices with mobile phones and palmtops is obtaining a significantly greater market penetration than conventional PCs. Taking also under consideration the recent efforts to provide computer-based interactive applications and services accessible by the broadest possible end-user population, including people with disabilities, one can conclude that the design and development of user interfaces require the revision of currently prevailing human-computer interactions (HCI) assumptions, such as that of designing for the average user in a desktop environment. This assumption needs to be incrementally replaced by the more demanding and challenging objective of designing to cope with diversity in the end-user population, the access media and devices, the contexts of use, and so on (Stephanidis, 2001). In this handbook the reader may find various chapters that present the current advances in this field. In these terms, the services for users on the move, including pretrip services, need high-quality user interfaces that address effectively diversity in the following three dimensions: (1) target user population and contexts of use; (2) categories of effectively delivered services and applications; and (3) deployment computing-platforms (i.e., PDAs, mobile, desktops, laptops).

49.4 Toward Accessible Transport: An Integrated Approach

The ICT capabilities have seemingly infinite potential usefulness to MI users, given their relatively limited mobility and specific requirements for assistive services. Indeed, the real need for such well-designed ICT is much more clear-cut than in other sectors of the EU's citizenship. This population requires and deserves design for all consideration to access easily both the Internet and mobile-based services; to do so they require (Simões et al., 2006):

- Easily used one-stop-shop service and information delivery sites, offering relevant and integrated content as needed; for instance, age-concerned information on travel, accessible transport and accommodation, events and sites of interest, and how-to advice on getting there
- Data delivery alternative methods that will adapt the information context and the user interface to individual abilities, interest, and preferences, whether derived from the user profile or implied by the user's habits
- Relevant and reliable services that are available on call by the user throughout a journey or service request, providing guidance for coping with regional variations that may affect service
- Geo-referenced services to allow the user to request info and services nearby
- Integrated general-market services that enable access to any information, with full choice, and ability to purchase the relevant service in a convenient and worry-free way
- Integration of personal and mobile computing technologies, wireless communications, context awareness, and adaptation techniques

In light of this list, the key research and development objectives include, but are not limited to:

- *Mediation of services and content*: In a pervasive, translucent, understandable (by ontologies) and managed (by web semantics) way, supporting seamless and efficient supply-demand matching (service negotiation, brokerage, etc.).
- *Seamless environment management*: Service provision everywhere, anytime, and by many mobile and/or fixed means, using alternative business models.
- *User preference and context-related driven processes*: Automatic adaptation of service content and layout (user interface) to user explicit preferences (based on user profile) and implicit preferences (based on history of use of service) as well as to the context of use (user location, traveling mode, scope of travel, such as tourist, commuter, resident, etc.). In terms of user interface, hardware optimized and innovative devices for both stationary and mobile communication give access to the services and tools by involving the appropriate modalities.
- *Flexible geo-referenced services*: Combining multimodal travel information provision with pedestrian navigation on accessible routes, both outdoors and indoors, at the required level of accuracy for each user (e.g., higher accuracy required for blind people for obstacle avoidance) and the context of use (e.g., more precision is required on the lane position while driving a car than being in the bus).
- *All within a user-confidence-based environment*: Able to handle issues of safety, reliability, security, privacy, and usability.

This approach emanates from the ASK-IT Integrated Project[11] (cofinanced by the EC within the 6th FP under the eInclusion

---

[10] http://www.ask-it.org.

[11] http://www.ask-it.org.

initiative) and views such services in a holistic way, covering integrated transportation chains (i.e., accessible use of a car, finding and using accessible parking, accessible ways to the nearest station entrance, accessible public transport, etc.) in an AmI space for the integration of functions and services for MI people across various environments, enabling the provision of personalized, self-configurable, intuitive, and context-related applications and services and facilitating knowledge and content organization and processing. The main aim is to develop AmI in semantic web-enabled services to support and promote the mobility of MI people, facilitating the provision of personalized, self-configurable, intuitive, and context-related applications and services, as well as facilitating the management and processing knowledge and information content. The developed services are available through the Internet to all kinds of devices (laptops, PCs, mobile phones, PDAs, automotive screens, etc.). The project focuses in many application areas, among which a prominent one is transportation. The targeted outcome is a service that any user could have access to through his mobile device anywhere, anytime, and adapted to his own needs and preferences.

49.5 Structuring User Needs into a Common Ontological Framework

Within the framework of the ASK-IT project, an ontological model has been built to gather all relevant data, structured in a way to effectively provide the appropriate information, according to the request the user poses each time. The information contained in the model is arranged, so as to cover the transportation needs of various user groups and different transportation modes.

49.5.1 Defining User Groups and Selecting Transport Modes

The work has been arranged in a series of steps. First of all, the user groups of interest had to be identified. An additional "All" user group was added (11 user groups in total), to include the content affecting all user groups, independently of the particular impairment, as well as the nonimpaired users. Second, the means of transportation, for which the relevant survey would be conducted, had to be selected. The aim was to include all possible ways by which a person could move, either on foot, as a passenger, or as a driver, and include the total trip from origin to destination. In this concept, eight transportation type categories were included: Airplane, Train/Metro, Bus/Tram, Ship, Car–Driver, Car–Passenger, Pedestrian, Stations. For each of these categories, content survey was undertaken, relative to each of the user groups, to identify and collect the appropriate data that would be included in the ontological model, thus forming 88 categories in total.

49.5.2 Content Survey

The next step was to identify and collect all the necessary information that would constitute the model's content. To collect the appropriate data for all of the 88 categories, an extended survey was undertaken. Information was collected from literature, previous and current research initiatives, transportation institutions and operators, as well as organizations and societies of people with disability. A great amount of information was gathered and assessed, extracting the valuable conclusions for the needs of the model. Focus was on the information related to the use of transportation systems by specific user groups, attributes that could cause problems to MI people and their appropriate values for the needs of the users, as well as the ways on which these problems could be tackled and the use of transportation modes that could be facilitated and made more user-friendly for all user groups.

49.5.3 Organization of Data

Following the target to provide the information that would be useful by practical means to the users, the data collected during the survey was organized in three groups of tables. The first group of tables contains the type of information that is essential for the user before a trip, the second includes the information that is useful during the trip, and, finally, in the third group of tables, all attributes related to both pretrip and on-trip procedures are listed together, with their suggested values, where relevant. The content organization and the model structure follow the theoretical approach of the Action and Activity theories (Wiethoff and Sommer, 2005). These theories aim to describe the behavior of a person during task completion. In brief, the Action theory sets the framework to break down the different activities into goals, subgoals, and functional units, while the dimensions of the Activity theory's minimal meaningful context define the information that is required for each goal, subgoal, and functional unit.

49.5.3.1 Pretrip Information

In the first group of tables, pretrip information is included. More specifically, pretrip information has been organized, following the general terms of the Action and Activity theory as mentioned previously, as shown in Figure 49.4. This is the basic information organization, which of course varies according to the mode of transport and its particularities.

49.5.3.2 On-Trip Information

In the second group of tables, the on-trip information needs of the predefined user groups are listed. A similar decomposition is followed here as well, as illustrated in Figure 49.5. In this case as well, the figure illustrates the basic information organization, which of course varies according to the mode of transport and its particularities.

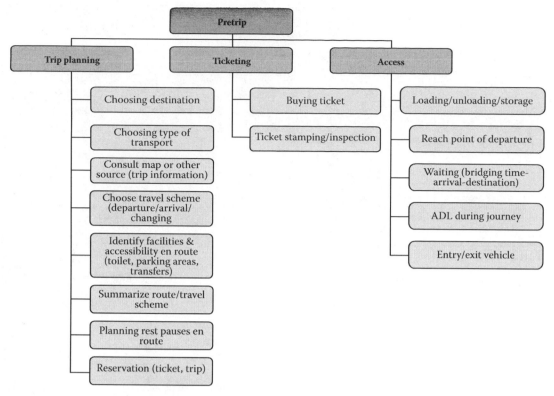

FIGURE 49.4 Organization of pretrip information needs.

49.5.3.3 Attributes

In the third group of tables, the specific attributes that determine whether a transport mode is accessible or not, as well as the corresponding recommended values (where relevant), are included. These attributes resulted from the analysis of the information needs of the two previous groups of tables and constitute the content of the web-based information service. The corresponding basic structure is illustrated in Figure 49.6. A representative extract from the BUS/TRAM table is presented in Table 49.2.

In this case as well, the decomposition varies according to the transport mode. Especially for stations (transport hubs), the structure presented in Figure 49.7 has been followed. As an example, an extract from the AIRPORT-STATIONS table can be seen in Table 49.3.

49.5.4 Prioritization Procedure

After completing level 3 tables (Attributes) for all transport modes and all user groups, the content was prioritized in terms of importance and usefulness to the users. The prioritization was implemented in a 0 to 3 scale: 0–not relevant; 1–nice to have; 2–important; and 3–essential. Prioritization was first performed by experts who assessed the attributes, based on their experience and their scientific know-how. In a second phase, interviews were conducted with real users, representing all the identified user groups, who assessed the attributes, according to their personal

needs and preferences. After comparing the two assessments, a final prioritization number was assigned to each attribute. The attributes that were finally included in the ontologies and, thus, provided to the users through the web service, are the ones with a grading equal to or greater than 2.

49.5.5 Discussion

The development of this ontology is an effort to provide to all MI persons the possibility to move with their preferred transport mode in the most possible effective way, making them aware of which of the available transport modes have the facilities to enable their easy access and use. People are thus able not only to plan their trip according to their special needs, but also to check at any time what are their options and provided facilities in any transport mode or station. In a broader context, apart from transportation, mobility of people affects other areas of human activity as well, such as tourism and leisure, social services, personal support groups, e-working, e-learning, and so on. So, the next step would be to combine ontologies relative to several kinds of activities, defining a mechanism and a suite of ontologies that interconnects diverse knowledge (structured as ontologies) from different heterogeneous frameworks of various application domains of relevance to the disabled. This kind of hyperontology could act as a middleware using a common language format, regardless of each module's structure. Thus, the development of increasingly sophisticated and effective spatial

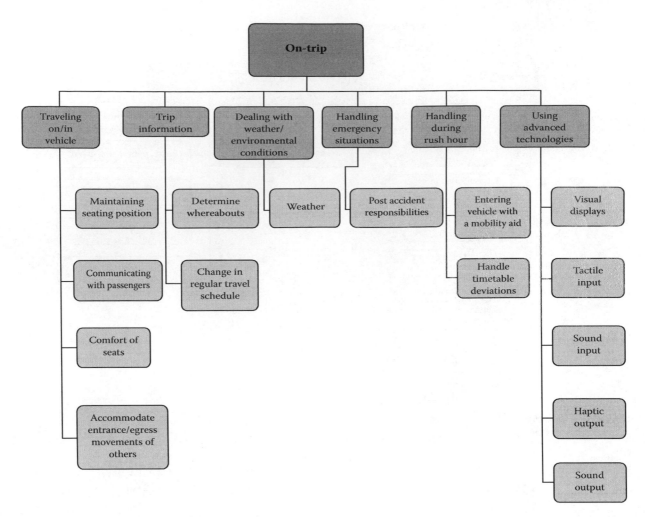

FIGURE 49.5 Organization of on-trip information needs.

assistance systems and location-based services for the disabled would be enabled. The final target is to enhance mobility for all societal groups, thus minimizing the phenomena of frustration and exclusion that are currently met in many of the user groups that are in focus.

49.6 Best Practices in User Interfaces and Services Provision

Various research groups and projects have worked toward the development of accessible user interfaces and services provision. Such best-practice examples can be found in past and present projects, for various transportation modes and infomobility services. In this section, four characteristic examples are provided, which, combined, aim at serving in a holistic way different transportation means and user groups. The information provided addresses the physical level, as well as the content and interface level.

49.6.1 Infokiosk Proper Design

This section contains key guidelines coming from the TELAID project (Nicolle and Burnett, 1999) regarding the design of public access terminals for use by elderly and disabled people (e.g., locating kiosks, display/control design, information requirements). The TELAID challenge was to ensure that IT solutions in transport consider the needs of the elderly and disabled, not only in early stages of development, but also during their use in practice, so that they can be usable by this part of the population.

49.6.1.1 Where to Locate Kiosks

Systems should be positioned at different height levels to cater to users of wheelchairs and short people (Jönsson and Fasen, 1995). Also, a sufficient turning radius should be allowed for people in wheelchairs to get to and away from the kiosk. The length of a wheelchair is usually less than 1.25 m, including the footboard, and its width is in most cases less than 0.75 m. This gives a necessary turning radius of 1.4 to 1.5 m (see Figure 49.8). If the

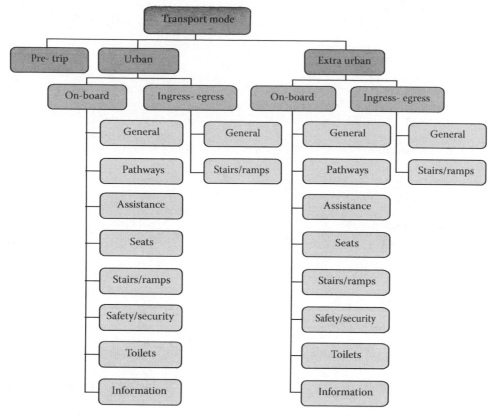

FIGURE 49.6   General organization of attributes.

environment is very bright (natural or artificial light), the kiosk should have a sun-shield, to minimize reflections. Furthermore, a knee/foot recess should be foreseen below the user interface. A recess will enable people who use wheelchairs to get close enough to view the screen and minimize the problem of reach when entering instructions.

49.6.1.2 Where to Locate the Display and Controls

It is difficult to find a compromise of fixed position for a terminal that can be used by both people in wheelchairs and standing users. Persons seated in wheelchairs generally require a low placement and a relatively wide display angle, while standing persons desire a higher placement and smaller display angle. The display should be viewable from the eye level of a person sitting in a wheelchair. Depending on the angle of the screen, the maximum height for the screen will vary (see Figure 49.9).

People with low vision should be able to get close to the screen to see it more clearly. Attention should be paid so that the terminal's outer casing does not obscure the view of the lower row of buttons/icons. If a person in a wheelchair can only approach the kiosk from the front, the maximum height of any interactive part of the terminal should not be higher than 1.2 m (Gill, 1997). The lowest height of an operable part should not be less than 0.7 m. To accommodate the needs of all travelers, an effective solution will be to provide two terminals at different heights, as shown by Figure 49.10 (Gill, 1997).

Alternatively (Figure 49.11), TELSCAN's collaborative testing of booking terminals with the SAMPLUS project found that a terminal height of 90 cm would be acceptable for more people, as long as the angle of the terminal display was adjustable (about 45 and 30°).

49.6.1.3 How to Design the Display and Controls

Wherever possible, at least 16-point type should be used, as many people with visual impairments are still able to clearly read this size. The minimum for special headings should be larger, for instance, 18- or 20-point font and larger, depending on the level of heading. The height of capital letters (not including spaces between rows) should be at least 4 mm. When using a touch screen, or when lining up function keys with the display, the problem of parallax should be considered. Parallax indicates an apparent displacement of an object, caused by an actual change of point of observation. A tall person or someone in a wheelchair may especially find this is a problem. Because of the lack of tactile feedback from a touch screen interface, users may then be unaware they have not hit the correct part of the screen. They can therefore become confused and disillusioned when the system does not obey their instructions. Using lines on the user interface leading from the key to the surface of the display will help to alleviate this problem. Type across photos or illustrations is harder to read, lacks contrast, and can confuse the eye. If maps are used, they should be simple, with good contrast and definition of important locations.

TABLE 49.2 Extract from BUS/TRAM Attributes Table

User Group			Wheelchair Users		
Information Need	Information Element	Conditions/Attribute	Value	Value Limit	Priority
Pre-trip	Availability of reduced fares for disabled or accompanying people	Yes/no	Text		3
Urban bus-on board-stairs	Width of the lift	cm	Number	min. 110 cm	3
	Depth of the lift	cm	Number	min. 140 cm	3
	Height of lift controls from the floor	cm	Number	90–110 cm	3
Urban bus-on board-doors	Width of outer door	cm		min. 90 cm	3
	Free space before and after the door	cm × cm	Number	min. 150 cm × 150 cm	3
	Door clear open width	mm	Number	min. 900 mm; preferred 1200 mm	3
	Needed force to open doors	Nt	Number	max. 10 Nt	3
	Doors possible to open with one hand	Yes/no	Text	Yes	2
Urban bus-on board-pathways	Minimum passage width	mm	Number	900 mm	3
	Maneuvering space	mm	Number	1600 mm × 2000 mm for wheelchair 180° turn; 1200 mm × 1200 mm for wheelchair 90° turn	3
Urban bus-ingress/ egress	Manually operated ramp (availability, width, gradient, working load)	Yes/no	Text	Width: 800 mm	3
		mm	Number		
		%	Number	Gradient: <12% if the ramp is unfolded onto a curb 150 mm high	
		kg	Number	Working load: 300 kg	
	Level access gap between the curb and the doorstep of the bus or tram	mm	Number	<50 mm	3
	Width of ramp	mm	Number	min. 1200 mm; min. between handrails 1000 mm (preferred 2000 mm)	3
	Gradient of ramp	%	Number	max. 1:12; preferred 1:20	3
	Length of ramp	m	Number	max. 10 m; preferred max. 50 m	3
Extra urban-on board-seats	Length of seating	mm	Number	Conventional: 1250 mm When legs outstretched: 1500 mm For wheelchair and assistant: 1750 mm For adult and assistance dog: 1500 mm	3
	Free space for mobility aids	mm	Number	Length of powered scooter/electric pavement vehicle 1500 mm Width of wheelchair (with elbows) 900 mm Width of 95th percentile manual wheelchair (excluding elbows) 695 mm Width of electric pavement vehicle or scooter 800 mm	3

Buttons should be at least 2 to 2.5 cm in width, while square keys are recommended, rather than rectangles (see Figure 49.12).

49.6.2 VMS and Public Transport Display Signs Proper Design

49.6.2.1 How to Present Messages

Advanced roadside traffic information systems like VMS and VDS play a major role in strategies trying to influence individual and collective traffic behavior. Roadside, generally location-specific infrastructure-based traffic information systems aim at efficient and safe use of the capacities of the traffic network. Traffic flows can be influenced by VMS systems, by provision of dynamic on-trip traffic information or traffic regulation messages adapted to real-time traffic conditions. VMSs need to be carefully designed in terms of information content and information presentation to meet the special needs of elderly and visually impaired drivers. Adequate font size and color, and the introduction of various pictograms, representing well-known traffic

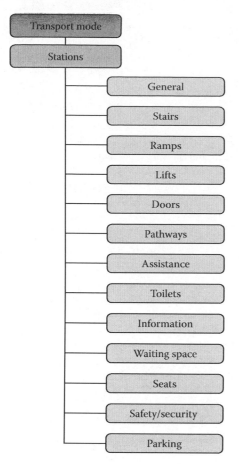

FIGURE 49.7 Organization of stations attributes.

signs, are of great importance in the given messages. Within the TRAVELGUIDE EC cofunded project, a VMS sign HMI was developed, that contained information on traffic and parking availability at the drivers' route. This VMS was optimized and a second version was finally established, where less text was used and symbols were added for easier and faster comprehension by drivers (see Figure 49.13).

An evaluation of the optimized design of the VMS that took place in Greece with four user groups, among which included disabled (six wheelchair users) and elderly (six drivers above 65 years of age) drivers. Every screen stayed on the VMS for around 5 seconds. Results regarding workload showed that there was no, or low, negative effect on workload, with different results for each user group: 1.71 for typical users, 1.32 for elderly users, and 1.78 for foreign users (there were no results reported for the disabled users) (Andreone et al., 2002). As far as transport display signs are concerned, letters and numbers should be at least 10 mm in height for every meter of viewing distance, with no lettering smaller than 22 mm in height. The size and contrast of text should be sufficient for travelers with impaired vision to read it. The precise size of text used will depend on a number of factors, including proximity of the screen to the user, ambient light levels, and clarity of the image. Therefore, the usability of the screen needs to be tested

in situ, with a range of users. Fiber-optic signs are preferred to LEDs in terms of legibility, though both are acceptable and are better than flip-over signs. Where messages are scrolled on the screen, the time of display should be sufficient for travelers to read and understand the message. A period of 10 seconds seems suitable for most. Symbols can be more easily and swiftly comprehended, provided they are clear, unambiguous, and well tested with users, including elderly people and people with disabilities. For people who have no or low vision, audible information at key points is particularly important during the course of their journey. Computerized speech information systems can be used to give the name of the next stop or station, the side of the train or metro, which will be against the platform, and other safety-related information.

49.6.2.2 Where to Locate Displays and How to Present Information

There are two possibilities for locating displays:

- *On board (i.e., on buses/trams/trains/metros, etc.):* On-board displays should be located at least 1.5 m above floor level. Route numbers should be at least 250 mm high, while the destination name should be in letters at least 125 mm high. White or bright yellow lettering on a black background is the most easily visible.
- *Off board (i.e,. at bus/tram stops/train/metro stations, etc.):* Displays with texts and symbols/icons should be placed so that they are easy to find. Displays should ordinarily be located 1.4 to 1.6 m above the ground. As found from testing (Nicolle and Burnett, 1999), this height is suitable for reading by eye and also facilitates reading by touch, if any tactile marking is included. Displays intended to be read from a distance, which are at risk of being obscured, should be placed at least 2.1 m above floor level. Character size on displays should be at least 12 mm and the relief (i.e., raising of text) at least 1 mm, if any tactile marking is included. Important displays that are meant to be seen from a distance of some meters should have a character size of 24 to 40 mm.

49.6.2.3 Required Level of Illumination

In terms of legibility, it is important that the light does not dazzle. A person with a visual impairment may need to view the screen from a very short distance or may need to use visual aids. Older eyes are also more susceptible to glare. The light should be adjustable within the area of 1000 to 5000 lux, depending on environmental conditions. The light should be free of flickering.

49.6.3 In-Vehicle Driver Assistance Systems Interface for All Users

This section contains specific guidelines to aid in the design of in-vehicle advanced driver assistance systems (ADAS) for elderly

TABLE 49.3   Extract from "AIRPORT-STATIONS" Attributes Table

User Group	Lower Limb Impairment				
Information Need	Information Element	Conditions/Attribute	Value	Value Limit	Priority
Airport-stairs	Number of steps in a flight	Number	Number	min. 3; max. 12	3
	Step height	mm	Number	min. 100 mm; max. 170 mm; preferred 150 mm	3
	Step depth	mm	Number	min. 250 mm; preferred 300 mm	3
Airport-ramps	Width of ramp	mm	Number	min. 1200 mm; min. between handrails 1000 mm; preferred 2000 mm	3
	Gradient of ramp	%	Number	max. 1:12; preferred 1:20	3
	Length of ramp	m	Number	max. 10 m; preferred max. 50 m	3
Airport-lifts	Width of lift door	mm	Number	min. 900 mm; preferred 1200 mm	3
	Width of lift cabin	mm	Number	1100 mm	3
Airport-information	Height of signs	mm	Number	1400–1600 mm from the ground	2
Airports-seats	Location of accessible seats	Text	Text		3
	Neighboring seats for personal assistants	Yes/no	Text		2
Airport-parking	Availability and location of parking spaces reserved for disabled travellers	Yes/no	Text		3
	Number of parking spaces reserved for disabled passengers	Number	Number		3
	Distance between parking spaces reserved for disabled passengers and the entrance	m	Number		3
	Accessible route from parking spaces reserved for disabled customers to the entrance	Yes/no	Text		3

Priority scale: 3, essential; 2, important; 1, nice to have; 0, not relevant.

and disabled drivers, as emanated from work performed in the context of several EC-cofunded projects over several years, such as TELAID, TELSCAN, TRANSWHEEL, TRAVELGUIDE, AGILE, AIDE, and ASK-IT.

ADAS and in-vehicle information systems (IVIS) are very important, as they assist the driver and increase safety by providing warning and information about potential hazards on the road. Some systems intervene to control the vehicle control, when a necessary action by the driver is not taken. Thus, such systems are used more and more nowadays, by all driver cohorts. Especially for MI drivers, these systems are even more important, because they assist them in tasks that they have difficulties with or that they cannot perform, endangering this way their own safety as well as the safety of others nearby. An example of a common weakness of elderly drivers and drivers with neck problems is the difficulty they may have in turning their heads right or left to check for other vehicles passing by, and specifically within the blind-spot zone, while changing lanes or overtaking. The blind-spot monitoring system provides great help for this problem, as it detects any vehicle in the blind spot of the car.

But ADAS/IVICS cannot be used optimally by several MI users, as they may not be able to use their particular modalities (i.e., deaf drivers will not hear audio warnings of ADAS or speech guidance of IVIS; color-blind drivers can't discern certain colors on the navigation map or the automotive display, etc.); they need a different intensity (i.e., louder, brighter, etc.) or may have slower reaction times due to physical or cognitive limitations and require thus adapted information/warning timing.

ADAS/IVIS rules may thus be adapted in terms of:

- Timing of warning
- Modality of warning
- Intensity of warning

In the following paragraphs, three ADAS functionalities are briefly explained, for which the relevant adaptations to HMI are proposed in Table 49.4.

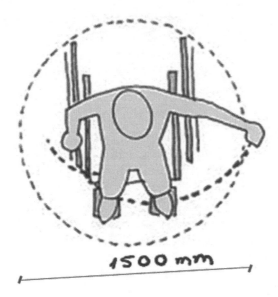

FIGURE 49.8 Necessary turning radius of a wheelchair. (From Nicolle, C. and Burnett, G., eds., *TELSCAN Code of Good Practice and Handbook of Design Guidelines for Usability of Systems by Elderly and Disabled Travellers*, TELSCAN Project Deliverable 5.2, Commission of the European Communities, 1999.)

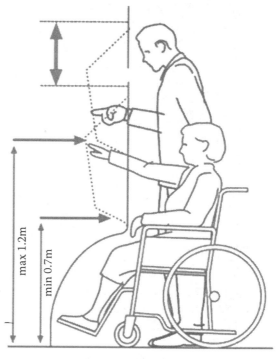

FIGURE 49.10 Placement of two terminals at different heights to accommodate the needs of diverse travelers. (From Gill, J., *Access Prohibited? Information for Designers of Public Access Terminals*, London, Royal National Institute for the Blind/CEC INCLUDE, 1997.)

Adaptive cruise control (ACC) systems keep the car at a steady speed (speed control), while maintaining a constant time gap to the car ahead (distance control). It accelerates or decelerates the car to keep the gap with the vehicle ahead, or maintains a fixed speed if the road ahead is clear. The system can be overridden by the driver or ignored at any time. Also, collision avoidance systems (CAS) or anti-collision assist (ACA) help the driver to avoid collisions by detecting other vehicles or obstacles (e.g., using radar).

Lane deviation warning (LDW) systems detect the host vehicle's position inside the lane (by measuring the distance from the lane boundaries) and warn the driver when the vehicle trajectory indicates potential hazard of exiting the lane.

Driver-monitoring systems detect the driver's physiological status, for instance: drowsiness, reduced attention, eye blinking rate, eye movements, hand force on the steering wheel, heart rate variability, and so on.

The adaptation of the relevant rules is based upon AIDE, extended also to seven MI driver groups of ASK-IT (Simões et al., 2006), as shown in Table 49.4.

The rules are very important, as, for example, older drivers tend to have longer reaction times than younger people, especially when tasks become more complex. Thus, the activation criteria for the ACC/CAS and LDW may need to be adjusted to accommodate individual differences in reaction times. Finally, a driver should be able to control the intensity, or brightness, of a visual signal, to meet individual requirements (but not below

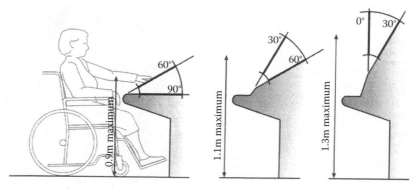

FIGURE 49.9 Terminal combination of proper height with angle of the display. (From Gill, J., *Access Prohibited? Information for Designers of Public Access Terminals*, London, Royal National Institute for the Blind/CEC INCLUDE, 1997.)

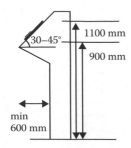

FIGURE 49.11 Terminal of 90 cm height and adjustable display, illustrating adjustable demand responsive transport booking terminal for standing and sitting users. (From Westerlund, Y. and Stahl, A., *Demand Responsive Transport (DRT) Using Advanced Telematics*, Internal Report, TELSCAN Project TR1108, Commission of the European Communities, 1999; Design solution: LO Design in Tibro with support from Västtrafik AB in Skövde.)

a set minimum). With aging, there is a progressive decrease in contrast sensitivity, or a person's ability to distinguish between light and dark. This is due to less light coming through the lens onto the retina.

In addition, some general guidelines, formulated within the TELAID project (Nicolle and Burnett, 1999) regarding positioning of specific controls and visual display elements to be accessible for use by elderly and disabled drivers, are briefly surveyed in the following.

It is suggested that touch screens are only used when the vehicle is stationary, and that a secondary keyboard designed for people with upper limb motor impairments can be attached, if required. The reason for this is that touch screens place a high visual workload on the driver and could interfere with the primary driving task (ICE Ergonomics, 1994), especially for elderly or disabled drivers, who may be functioning to the limits of their abilities. In an evaluation of a route guidance system, 30% of the elderly subjects found it difficult to enter a destination using a touch screen (Oxley et al., 1994).

Larger buttons (around 24 mm) are preferred to smaller buttons (around 4 mm). This is particularly the case for tasks of

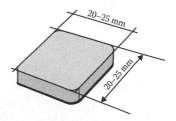

FIGURE 49.12 Proposed dimensions of control buttons for infokiosks. (From Poulson, D., Allison, G., Ashby, M., and Maguire, M., *ATTACH Guidelines for Developing Public Information Kiosks*, Loughborough, HUSAT Research Institute, 1996.)

higher priority. This is crucial, especially for drivers with dexterity problems and also because larger buttons require less attention at the button before it is activated.

The size of an in-vehicle visual display must be large enough and the contrast high enough so that the driver does not need to bend toward the display to read the information. A Zoom In icon or screen enlargement feature should be available wherever possible. However, while driving, the driver should not be required to manually zoom in and out to different scale levels. Instead, the system should automatically present the optimum amount of usable information.

In-vehicle display tasks should not require a large amount of a driver's attention. The visual distraction of an in-vehicle display has a strong link to system safety. This is particularly the case for the elderly and the disabled, who may already be reaching the limits of their attentional resources (due to either perceptual/cognitive limitations or from using car adaptations). Unfortunately, there are no established criteria for what constitutes an unacceptable level of distraction. The following cut-off criteria have been proposed in the literature:

- No in-vehicle display task should require more than 5 seconds total viewing time.
- No in-vehicle display task should require average glance duration of more than 2 seconds or need more than 4 separate glances. Drawing on the results of a series of

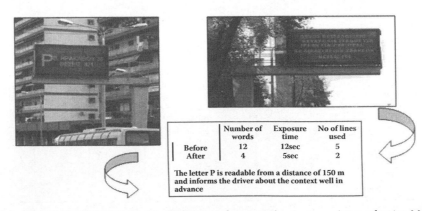

FIGURE 49.13 Old (right) and optimized VMS (left). (From Naniopoulos, A., Bekiaris, E., and Dangelmaier, M., TRAVELGUIDE: Towards integrated user friendly systems for drivers traffic information and management provision, in *International Conference on Machine Automation (ICMA 2000)*, 27–29 September 2000, Osaka, Japan, 2000.)

TABLE 49.4   ADAS Adaptation Algorithms for Elderly and Disabled Drivers: Examples

System	MI User Group	Adaptation of Timing	Adaptation of Modality	Adaptation of Intensity
ACC/ CAS	1. Hearing		Haptic, with visual as secondary (instead of vocal) warning. Alternatives: seatbelt vibration, break pulses, acceleration cut-off.	
	2. Color blindness		Acoustic, with visual as secondary. Color of visual LEDs to be adapted.	
	3. Illiterate		Simplification of message. Hearing warning with sound and vocal message "Brake!"	
	4. Elderly	Adaptation of warning timing based upon personal reaction time; or adaptation of warning timing based upon personal time-to-collision (TTC). Of course, the safety thresholds are always satisfied.		Enhance intensity and duration
	5. Upper limb		Alternative system operation without hands by: -Foot knob -Voice control -Head control	
	6. Neck problems			
LDW	1. Hearing		Haptic, with visual as secondary (instead of vocal). Alternatives: rumble strips simulation by seat movements, seatbelt vibration, visual warning by LED at external mirror or at the inside pillar of the car.	
	2. Color blindness		Adaptation of secondary visual warning (LEDs) color.	
	3. Illiterate		Simplified vocal output: sound, followed by message "Don't change lane."	
	4. Elderly	Adaptation of warning distance from the lane border according to personal time-to-line-crossing (TLC). Of course, the safety thresholds are always satisfied.		Enhanced intensity and duration
	5. Upper limb		Alternative system operation without hands by: -Foot knob -Voice control -Head control	
	6. Neck problems		Visual feedback not provided at peripheral but at central visual system. Additional visual feedback at automotive screen.	
Driver monitoring	1. Hearing		Haptic, with visual as secondary (instead of vocal). Alternatives: seatbelt vibration, seat vibration, brake pulses, etc.	
	2. Color blindness		Adaptation of colors/ frequency at the secondary warning and status LEDs at central mirror.	
	3. Illiterate		Simplified vocal outputs: sound and message, "Stop driving, you are too tired!"	
	4. Elderly	Adaptation of sensors and algorithms to focus upon elderly drivers' hypovigilance peaks (at midday instead of midnight). Eyelid camera with shutter function and monitoring algorithm adaptation.	Vocal and haptic feedback lights and LEDs are not visible, even as secondary feedback channels at midday.	Mild intensity, longer duration (not to startle the elderly driver)
	5. Upper limb		Alternative system on/off operation, without hands.	
	6. Neck problems			

Source: From (1) Panou, M. and Bekiaris, E., Towards the development of an adaptation strategy for safe and integrated infomobility services use by mobility impaired drivers, in the *Proceedings of ITS 2006,* 26–30 June 2006, Jhongli, Taiwan [CD], 2006 and (2) Visintainer, F., Panou, M., and Pagle, K., ADAS and IVICS for all, in the *Proceedings of the ASK-IT International Conference 2006,* October 2006, Nice, France [CD], 2006.

experimental trials, Zwahlen et al. (1988) proposed a design guide based on the probabilities of a vehicle deviating out of a lane while the driver glanced toward an in-vehicle display.

For in-vehicle displays, the height of text characters (letters and numbers) should be:

(distance of viewer from the display) × (tangent – visual angle)

Recommended visual angles, based on the function of the text are given (Campbell et al., 1998):

- Titles and other key elements = 0.50°, minimum
- Dynamic or critical elements = 0.33°, minimum
- Static or noncritical elements = 0.266°, minimum

Figure 49.14 highlights the relationship between eye distance, text character height, and visual angle.

In general, warnings should be available in alternative forms (i.e., visual, auditory, or tactile). This could allow both visual and hearing impaired persons to adapt the signal to their perceptual characteristics (Campbell et al., 1998). Older drivers may benefit more from an auditory alert. Tonal warning signals should have frequencies between about 500 and 3000 Hz. Hearing sensitivity at high frequencies tends to decrease with age, particularly for men. So, high frequencies above 2000 Hz for warning signals should be avoided if the older population is considered.

49.6.4 Adjustable Wheelchair for All Transportation Means

The TRANSWHEEL project (DE3013) developed a crash-proofed, high comfort, and maneuverability wheelchair and seat (see Figure 49.15), to solve in an integrated way the safety, comfort, and maneuverability problems of wheelchair users in all transportation means (Naniopoulos, 2002).

It is a crash-proofed wheelchair and tie-down system, which can be homologated as a vehicle seat, since it can withstand a 48 km/h frontal collision against deformable barrier and a 35 km/h rearward collision against a car, with dummy Hybrid III 50%. The wheelchair has a seat height of about 53 cm, which is appropriate for everyday use (sitting in, leaving), adjustable from 33 cm (the standard height of a Scudo seat) to 83 cm (to reach the same things as the average standing person), thus being able to reach a book at a high shelf of a library, open a train window, reach the fire extinguisher and an emergency phone in a building, but also appropriate to be used in all transportation means and have the same view field as any other car passenger/driver (Bekiaris et al., 2000).

This wheelchair seat, through individual pressure regulation at five or more different areas of its back and seat, can be manually or even automatically adapted to the particular requirements of the user and can change form with time, during transportation, thus offering a comfort improvement of 30% to 60%, especially during long-term transport (over 2 hours). The seating pressure is especially reduced at the ischiums, which is important for decubitus prophylaxis.

The TRANSWHEEL wheelchair user interface incorporated the project's innovative functionalities, namely comfort and elevation. Two UI prototypes were developed: a low-end and a high-end (Nicolle and Burnett, 1999). The low-end prototype is very close to the OPTIMUS wheelchair user interface, having as additional functionalities the control of the wheelchair's comfort and elevation features (see Figure 49.16). The settings of this user interface can be adjusted by either the left/right buttons or by moving the joystick accordingly, depending on the action. The settings of the low-end-user interface can be assessed by pressing the MODE button located at the middle-end of the display panel. Finally, the user can turn on/off the wheelchair's lights.

The high-end prototype is innovative in relation to current wheelchairs' user interfaces, promoting a new, easier, and friendlier way of use, by using an LCD display with touch screen and more intuitive interactive elements, so that the user can easily navigate through the user interface by just touching buttons (see Figure 49.17). When the concept of a touch screen was introduced at the initial verification user tests in Greece, users responded very positively, underlining the problems they were facing when trying to enter the wheelchair's mode and change the user interface settings using the joystick.

The high-end TRANSWHEEL wheelchair user interface has implemented many features of the initial ergonomic interaction design. Although the LCD touch screen is monochrome (blue and yellow) and thus lacks colorful representations, the overall layout has been designed to be pleasant and easy to use. To

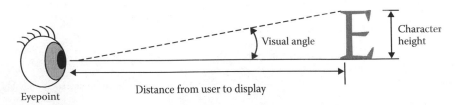

FIGURE 49.14 Relation of viewing distance to symbol height and visual angle. (From Nicolle, C. and Burnett, G., eds., *TELSCAN Code of Good Practice and Handbook of Design Guidelines for Usability of Systems by Elderly and Disabled Travellers*, TELSCAN Project Deliverable 5.2, Commission of the European Communities, 1999.)

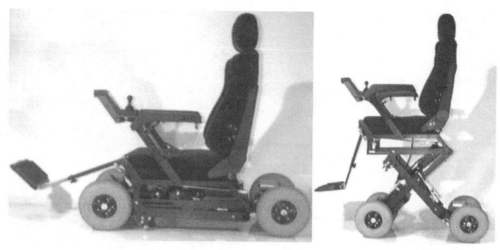

FIGURE 49.15   TRANSWHEEL wheelchair.

enhance interactivity and the system's overall high performance, the use of graphics is rather limited.

The main screen is divided in two parts: the interactive panel and the display. The display is highlighted with a double-line box and includes all the indications of the wheelchair's driving mode. The display consists of two subparts, the upper and the lower, separated with a line. The upper part is used for displaying lights, flashes, alarm, and driving locked indications. The lower part includes the following:

- Battery status located at the lower left of the box
- Wheelchair's velocity (electronic counter and percentage bar) located at the middle and lower right of the box
- Message display located below the electronic velocity counter and used to inform the driver on relevant actions (i.e., "Lights on," "Driving locked," for wheelchair's error messages, etc.)

The active part of the main screen consists of the following nine buttons: Left flash; Right flash; Alarm; Driving locked; Driving unlocked; Horn; Lights on; Lights off; and Mode button. When the user presses the Mode button, located at the UI's main screen, the settings screen appears. The settings screen has four buttons: Elevation, Comfort, Help, and Exit. The Elevation button navigates to the Elevation screen where the user can adjust the position of the wheelchair. The Comfort button guides the user to the comfort screen where the aircushions' pressure can be set. By pressing the Help button, the user enters the help screen, where help on the wheelchair's indications and driving mode buttons is available. Finally, the Exit button exits to the driving mode (main) screen.

In Figure 49.18, the right flash indication is displayed.

To adjust the pressure of an aircushion, the user presses the button that corresponds to it. In Figure 49.19, the user has selected the aircushion 5 and the system corresponds with a message guiding the user to select the pressure of the relevant aircushion. After selecting the preferred pressure, the user presses the Save button to save it and the system corresponds

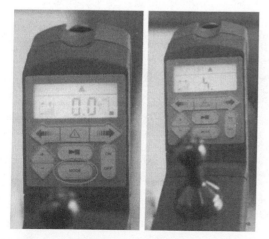

FIGURE 49.16   Low-end UI: elevation screen (left) and driving mode screen (right). (From Bekiaris, E., Nikolaou, S., Luecking, H., and Papaioannou, G., TRANSWHEEL Deliverable 5.2. *TRANSWHEEL Prototype Ergonomic Design and User Interface*, 2000.)

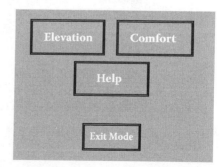

FIGURE 49.17   High-end UI settings screen. (From Bekiaris, E., Nikolaou, S., Luecking, H., and Papaioannou, G., TRANSWHEEL Deliverable 5.2. *TRANSWHEEL Prototype Ergonomic Design and User Interface*, 2000.)

with a message informing the user that the selected aircushion pressure has been saved.

After the realization of both TRANSWHEEL user interfaces, a user survey was conducted to get users' acceptance and opinions regarding its potential improvement. The high-end UI was preferred by the users in terms of ease of use and pleasantness against the low-end one. This represents just one good example of a multifunctional UI, that takes into account both autonomous operation and operation within various transportation means.

49.6.5 Trip Planning

49.6.5.1 Intermodal Travel Planning for Wheelchair Users

The prototype developed in the ASK-IT project defines in detail intermodal travel planning for wheelchair users. Wheelchair users face a number of problems concerning their mobile life. For wheelchair users, access barriers play the most important role. They struggle with architectural barriers as well as with fairly accessible, but not usefully equipped facilities. However, many wheelchair users live a very mobile life and have learned to find their way—even if they have to ask a stranger for help, or if they arrive half an hour late because the exit with the elevator led to a pathway that was too steep for them. For them, trip planning provides the opportunity to get one-stop information concerning access and barriers, which otherwise they would have collected through several web sites and lots of phone calls and even that would not be complete (e.g., nearest accessible toilet).

The prototype is designed for a normal screen (desktop or tablet PC), because the task of (more complex) travel planning is most commonly done at home by wheelchair users, as it is more comfortable that way and many of them also have upper limb impairments. So for that prototype it is assumed that the travel planning is done at the big screen and the result is synchronized then to the mobile device.[12] The prototype represents a module-

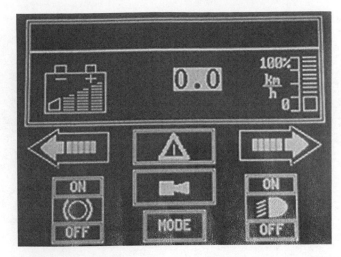

FIGURE 49.18 High-end UI main screen: flash indication. (From Bekiaris, E., Nikolaou, S., Luecking, H., and Papaioannou, G., TRANSWHEEL Deliverable 5.2. *TRANSWHEEL Prototype Ergonomic Design and User Interface*, 2000.)

spanning ASK-IT user interface concept and defines the general information architecture, the basic task flow, the wording, main graphical elements, and the overall concept. The starting screen is shown in Figure 49.20.

The prototype shows a general overview of ASK-IT services ranging from route planning, leisure activities, including leisure-related POIs, domotic services and personal assistance services to car-related services and access to e-learning services.[13] A concept for the presentation of favorite services on the basis of shortcuts as well as for the display of status information (e.g., actual position) was worked out. Main functional areas (e.g., application navigation, main content frame, etc.) were defined. The information architecture is shown in Figure 49.21.

The focus on the prototype lies on travel planning and POIs search. The tasks targeted are (selection):

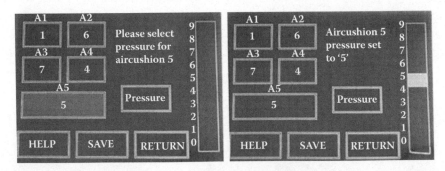

FIGURE 49.19 High-end UI comfort screen: aircushion selection and storing. (From Bekiaris, E., Nikolaou, S., Luecking, H., and Papaioannou, G., TRANSWHEEL Deliverable 5.2. *TRANSWHEEL Prototype Ergonomic Design and User Interface*, 2000.)

---

[12] This does not mean that we assume that all travel planning activities are done on a big screen. But for that user group this was confirmed in the interviews to be more comfortable for complex journeys, like holiday trips.

[13] Please note that the contents and services the prototype uses are preliminary. The actual UI implementation and content selection is realized by the individual module developers within ASK-IT and the test sites and according to the actual service specification.

FIGURE 49.20 ASK-IT travel planning for wheelchair users (home screen).

- Plan an intermodal trip (with decision aid of accessibility levels, comfort options)
- Plan a trip on the basis of a trip already done
- Plan an event (e.g., musical according to interests)
- Get current position on an active trip
- Get assistance
- Use favorite service (e.g., order a taxi to current position)
- Find an accessible POI (e.g., toilet, fully accessible restaurant) nearby
- Bookmark POIs and use them for travel planning

The design was aimed at being friendly and not overly computer-prone, so as not to scare off people with low computer experience. The design corresponds to a thick client, but it can easily be adapted to a web-based client.

In the usability test performed with 10 wheelchair users covering a variety of impairments, the overall user interface concept was confirmed and the acceptance was very high. Especially the availability of information that is hard to get nowadays was highly appreciated. An interesting result was that wheelchair users were not very sensitive about wheelchair-specific wording used in an application (e.g., "walk for 100 meters" instead of "roll for 100 meters"). It is more important that the information is correct and specific to their needs. Therefore, they appreciated a wheelchair icon in the travel mean overview instead of a pedestrian icon (see Figure 49.22), as it makes clear that the information the application gives is based on their accessibility needs.

49.6.5.2 Pedestrian Navigation on Mobile Devices

The prototype developed satisfies requirements for mobile pedestrian navigation systems considering location-based services and shows respective user interface solutions. These user interface principles are the basis of an understandable and accessible user interface. It has been iteratively optimized with expert reviews and user tests in the natural setting in the Stuttgart region (outdoor and indoor).

How people find their way in the urban environment and how they orient in the city has been well investigated by architects and psychologists. For example, Kevin Lynch (1968) identified paths, edges, districts, nodes, and landmarks as being important elements in the city enabling pedestrian navigation. To understand the layout of a city, people need to create a mental map using these elements.

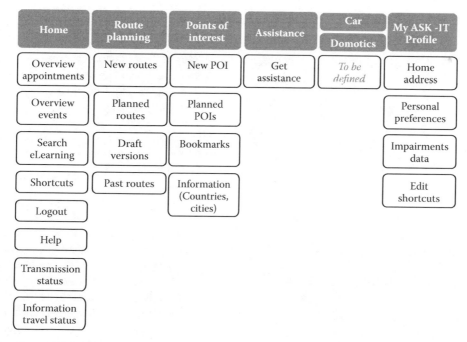

FIGURE 49.21 Information architecture user interface concept travel planning for wheelchair users.

Besides, in the past years mobile guides have been investigated and are still an important issue. There were research projects, for example, Cyberguide (Long et al., 1996), REAL (Baus et al., 2002), SAiMotion (Heidmann and Hermann, 2003), and Deep Map (Malaka and Zipf, 2000), which in part considered psychological fundamentals of spatial orientation and wayfinding (Passini, 1984). However, many have focused mainly on the technical implementation (Baus et al., 2005). In contrast, the ASK-IT prototype tries to find the optimum assistance for pedestrian navigation by defining information requirements for mobile guides that are essential. The prototype will be designed for a PDA as it is most commonly used for navigation applications.

In spite of past research findings, the design and functionality of existing navigation systems for pedestrians are still based upon navigation systems used in cars. They operate like simple route planners. After entering the start and the destination, a route is calculated. There are several characteristics that differentiate car navigation from pedestrian navigation and make pedestrian navigation much more complex, including the following:

- In a car, the driver has a fixed seating position and is looking in the driving direction. A pedestrian moves freely and the looking direction changes dynamically. Therefore, map alignment is much more complicated for the pedestrian.
- Cars stay on roads whereas the possible walking context of a pedestrian in a pedestrian zone is much wider. Therefore, guiding has to be more detailed without being annoying.
- Whereas cars operate only outdoors, pedestrian navigation in ASK-IT deals with indoor and outdoor navigation. Especially a seamless transition between these two is a challenge (Baus et al., 2002).

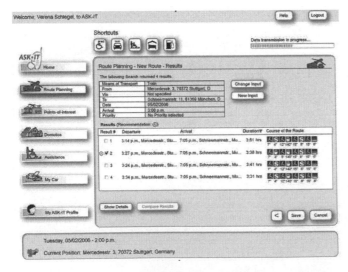

FIGURE 49.22 Results on route planning with recommendation on accessible route and wheelchair icons in course of the route, to indicate that transfer times are calculated for wheelchair users.

- For cars, meaningful decision points for navigation are given in form of simple sequential commands to change roads (e.g., "turn left in 400 m; turn left now"). For a pedestrian, orientation in enriched environments has more to do with landmarks and nodes. Sightseeing or trekking guides use descriptions like "follow the path until you reach a semi-derelict house…." So meaningful decision points for pedestrians have to be worked out systematically.
- Pedestrian wayfinding is often less target-oriented than with a car. Pottering while heading toward an (accessible) point of interest is not the situation often encountered using the car navigation system in contrast to the pedestrian use.
- Whereas a car driver experiences a more or less private context, the pedestrian is very close to other pedestrians. So speech output with verbal cues would probably be irritating to the other pedestrians and perhaps intimidating for the pedestrian to be obviously controlled by a device. Therefore a way needs to be found that allows unobtrusive guiding without forcing the pedestrian to observe the device every second and miss the scene.
- The acceptance of strict guidance is supposed to be lower for a pedestrian than for car use (i.e., the message "please turn around" repeated when a command is not followed is presumed to be more irritating for a pedestrian walking in the city center). It has also to be carefully chosen what a "wrong path" is. If the user decides to pop in a shop or get some coffee on the way, it would be annoying to be bothered with new-way suggestions by the system.
- For the car, the driving speed is more constant than for a pedestrian. So for the pedestrian the relation of actual position to destination is probably more important than for a car driver. Therefore, switching between different scales has to be supported (e.g., wayfinding scale, with direct surroundings and whole distance with indication of actual position).
- Because the pedestrian has many more options to choose from, more paths to take than a car driver, an easy mapping of digital map, cognitive map, and the real world has to be supported. Pedestrians have more decision time, so the display could be enriched with pictures of the surroundings to support mapping.

As mobile use of mobile devices often requires imprecise input, the prototype was developed to incorporate finger-based interaction as much as possible.

- *Navigation modes' guidance versus orientation*: Basically a user-centered portable navigational aid has to support three wayfinding situations (Passini, 1984): the normal navigation mode (targeted navigation), the tourist mode (aimless navigation, e.g., roaming) and the emergency mode (extremely conditional navigation, in terms of time and individual conditions/skills). The more a person is guided, the faster the person can reach a desired destination and vice versa. In case of an emergency, the user is guided most, while in the tourist mode the user is guided less. The normal navigation mode and tourist mode were worked out for ASK-IT. The system is designed for a PDA in landscape format, following the format of the visual field.

- *Overview on the interaction concept*: All highly important functions regarding wayfinding are provided at the first navigation level. As a pedestrian has to concentrate on the road traffic and other pedestrians, the interaction concept is as easy as possible and thus contributes to security. It also allows a quick use of the system while walking through the city.

The three different map views (wayfinding map, survey map, remaining route map) can be switched by pushing a button. This is similar to the functionality of a cyclometer that often provides only one button for switching between different functions. The wayfinding map provides guidance and the two other maps provide orientation. Information for guidance and for orientation is equally accessible in this concept. Also the user can change the zoom level of the wayfinding map according to her needs. Detailed maps and photos are provided at particular places only. The textual version of the route instructions can be switched on by pushing the indicator that displays the graphical route instructions. The screen layout and the described interaction elements can be seen in Figure 49.23.

49.6.5.3 Pedestrian Navigation for Blind Users

The wayfinding requirements for blind people differ significantly from the ones using their visual senses to orientate. Blind pedestrians cannot see obstacles on the pavement or spot the church in the distance for orientation. For micronavigation, that is, navigation through the immediate environment, the long cane or a guiding dog serve as mobility aids. Additional electronic mobility aids include ultrasonic, laser, and infrared technology on the market. For macronavigation, for instance, navigation through the more distant environment like the church example, mobility support is much more complicated (Petrie et al., 1997).

There have been several projects that tried to meet this challenge. The MoBIC project (Strothotte et al., 1996; Petrie et al., 1997) used a preparation system to plan a walk in an urban area combining touch tablet maps with speech output for guiding exploration. The selected route can be transferred to the MoBIC outdoor system (MoODS) that guides the user outdoors on his way. The MoODS takes the journey plan and assists travelers finding their way by matching geographic information system (GIS) information against travelers' positions (GPS). Directions, warnings, and other information are given in speech or Braille. Loomis et al. (1994) aimed at developing a portable, self-contained system that allows visually impaired individuals to travel through familiar and unfamiliar environments without the assistance of guides. The system used GPS, a geographic information system comprising a database of the environment and functions for automatic route planning and for selecting the database information. The system provides a virtual acoustic environment for travelers, guiding them to buildings and other objects in the environment by sound signals that appear to come

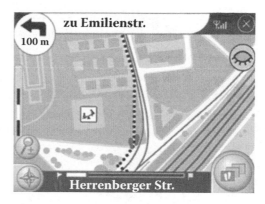

FIGURE 49.23 Wayfinding map pedestrian navigation.

from the direction of the respective object. There are also several navigation aid systems on the market.[14] An interesting recent product is Trekker.[15] It is a navigation aid for blind and visually impaired people and uses GPS and digital maps to help blind persons find their way in urban and rural areas. It complements existing aids (white canes and guide dogs) and provides information through synthesized speech output.

The ASK-IT prototype developed in this context supports blind people in wayfinding based on acoustic interface elements. Sound offers broad possibilities for unobtrusive information provision. This might help blind people to know that they are on the right way, avoiding annoying spoken information of secondary importance. Also privacy plays a major role here. Additionally, the use of speech is investigated, e.g., to clarify ambiguous situations in wayfinding. In December 2006 interviews with six severely visually impaired and blind people were performed to learn about essential information to be presented by an electronic travel aid and nice-to-have information and events. The acceptance of the sound-based concept was very high, although concerns on the challenges of sound design and concept development were raised as well. The sounds should be pleasant and not annoying. The concept can be divided in provision of orientation-related information and navigation-related information. The first class of information has the aim to orient the person in the environment without a planned route—similar to the visual scanning of the physical surroundings of an able-bodied person. The second class of information deals with navigation-related information (i.e., guide a person on a preplanned route).

The following three categories of information need to be considered for orientation:

1. *Information for orientation*: Essential information like street names or directions, detailed information like public buildings, shopping malls, and transport-related information like stations, bus stops, etc.
2. *Safety-related information*: Critical situations like street crossings and stairs and information on safe pathways.
3. *Temporary barriers*: Such as construction sites.

---

[14] For an overview, see http://www.tiresias.org/equipment/eb24.htm.
[15] http://www.humanware.ca/web/en/p_DA_Trekker.asp.

For navigation support, information such as "You are on the right/wrong way," "You have to turn right/left soon," and "Now you have to turn right/left" need to be given. Sounds that are easy to learn and to distinguish are being designed. Additionally, the concept for sound families is being investigated toward indicating information categories and to allow chunking of information. The developed sounds for orienting and guiding blind people will be also used to enhance graphical user interfaces for navigation.

49.7 Key Challenges for the Future

Several decades of accessibility research have resulted in the buildup of accessibility islands. Thus, accessible buses operate through inaccessible bus stops, accessibility information on travel routes and times is offered in an inaccessible way, and in-vehicle drivers' assistance and information systems may even endanger the drivers' safety by inappropriate timing or intensity. This situation appears like a failure to recognize the obvious: accessible transport requires developing a reliable, robust, and seamlessly accessible transportation chain, both in terms of information (pretrip, on-trip), physical accessibility attributes at the transportation means and their stations/hubs, and accompanying services. Lack of such a systematic approach results in MI users being reluctant to leave their homes (thus, being of reduced mobility) or requiring assistance to travel due to the many unforeseen barriers they may meet (thus, being nonautonomous).

Current research addresses these needs (e.g., ASK-IT), and AmI frameworks are applied to support the user in any environment and under any context of use. Still, to be widespread, reliable, and meet user confidence, research needs to intensify regarding:

- Seamless and standardized provision of dynamic, real-time, and reliable information content; through common ontologies that allow not only transport-related but also added-value services from many sectors to "talk to each other," offering to the MI user one-stop-shop information services, with content provision that is self-adapted to his needs and wants.
- Seamless, cost-effective, and reliable (of high quality of service) telecommunication services, capable of roaming from one telecom provider to another, from a country to another, and from one device to another; operable with high accuracy (for user and POI localization) both indoors and outdoors.
- Design for all-based user interfaces for all types of devices (infokiosks, PCs, mobile phones, in-vehicle platforms, etc.), capable of being self-adapted modularly to the user's residual abilities, the context of use, key environmental factors, and even the users' preferences and dynamic behavior (i.e., type of transportation means preferred, maximum distance capable of/willing to walk, if traveling as a commuter or a tourist, etc.).

- Personalized services, self-adapting their UI and the mediated content and technology used per user profile, using the previously stated content, telecommunication, and UI concepts.

These requirements call for technological, but also for commercial, political, and legal actions, involving all key players and especially "pushing" the industry (i.e., through incentives, appealing business plans for public-private partnerships in the sector, accessible public procurement schemes or even mandates) to adopt standardized, open, secure, and shared platforms for the development and provision of such services; starting with the obvious: low-cost, secure, and dynamic infomobility services and data roaming across Telcos Europewide.

The enhanced mobility of MI users is not just a politically correct, social equity target; it is a high potential new market and a high-capacity workforce, the further neglect of which is not an option in the European economy and industry if it is to remain viable.

References

Abascal, J. (2004). Ambient intelligence for people with disabilities and elderly people. Presented in the *ACM's Special Interest Group on Computer-Human Interaction (SIGCHI), Ambient Intelligence for Scientific Discovery (AISD) Workshop*, 25 April 2004, Vienna, Austria. http://www.andrew.cmu.edu/course/60-427/aisd/elderly.pdf.

Abowd, G. D., Atkeson, C. G., Hong, J., Long, S., Kooper, R., and Pinkerton, M. (1997). Cyberguide: A mobile context-aware tour guide. *Wireless Networks* 3: 421–433.

Akoumianakis, D. and Stephanidis, C. (1997). Supporting user-adapted interface design: The USE-IT system. *International Journal of Interacting with Computers* 9: 73–104.

Andreone, L., Damiani, S., Deregibus, E., Lilli, F., Morreale, D., Navone, P. , et al. (2002). Evaluation of pilots and testing. Deliverable 5 of the TRAVELGUIDE project (GRD-1999-10041). Commission of the European Communities.

Ankolenkar, A., Burstein, M., Son, T. C., Hobbs, J., Lassila, O., Martin, D., et al. (2001). *DAML-S:Semantic Markup For Web Services*. http://www.daml.org/services.

Antona, M., Mourouzis, A., and Stephanidis, C. (2007). Towards a walkthrough method for universal access evaluation, in *Universal Access in HCI, Part I of the Proceedings of the 4th International Conference on Universal Access in Human-Computer Interaction (UAHCI 2007)*, 22–27 July 2007, Beijing, pp. 325–334. Berlin/Heidelberg: Springer-Verlag.

Bahl, P. and Padmanabhan, V. (2000). Radar: An in-building RF-based user location and tracking system, in the *Proceedings of INFOCOM 2000, 19th Annual Joint Conference of the IEEE Computer and Communications Societies, Volume 2*, 26–30 March 2000, Tel Aviv, Israel, pp. 775–784. Piscataway, NJ: IEEE.

Banâtre, M., Couderc, P., Pauty, J., and Becus, M. (2004). Ubibus: Ubiquitous computing to help blind people in public transport, in the *Proceedings of Mobile HCI 2004*, 13–16 September 2004, Glasgow, Scotland, pp. 310–314. Berlin/Heidelberg: Springer-Verlag.

Barham, P. and Alexander, J. (1998). *Evaluation of Interactive Information Terminals (TRIPlanner Mk1) with Respect to Their Use by the Elderly and People with Disabilities.* A report prepared by the TELSCAN Project (TR1108) on behalf of EUROSCOPE—ROMANSE II. Commission of the European Communities.

Baus, J., Cheverest, K., and Kray, C. (2005). A survey of map-based mobile guides, in *Map-Based Mobile Services: Theories, Methods and Implementations* (L. Meng, A. Zipf, and T. Reichenbacher, eds.), pp. 197–213. Berlin: Springer-Verlag.

Baus, J., Krüger, A., and Wahlster, W. (2002). A resource-adaptive mobile navigation system, in the *Proceedings of the 7th International Conference on Intelligent User Interfaces*, 13–16 January 2002, San Francisco, pp. 15–22. New York: ACM Press.

Bekiaris, E., Nikolaou, S., Luecking, H., and Papaioannou, G. (2000). TRANSWHEEL Deliverable 5.2. *TRANSWHEEL Prototype Ergonomic Design and User Interface.* Restricted deliverable 5.2 of the TRANSWHEEL Project (DE3013). Commission of the European Communities.

Bilandzic, M., Foth, M., and De Luca, A. (2008). CityFlocks: Designing social navigation for urban mobile information systems, in the *Proceedings of the ACM Designing Interactive Systems (DIS)* (J. van der Schijff, G. Marsden, and P. Kotze, eds.), 25–27 February 2008, Cape Town, South Africa. New York: ACM Press.

Blythe, P. (1999). RFID for road tolling, road-use pricing and vehicle access control. RFID Technology (Ref. No. 1999/123). *IEE Colloquium*, 25 October 1999, pp. 811–816.

Blythe, P. (2003). Access restrictions as a move towards the sustainable city: History, evolution and future trends, in the *Proceedings of CIVITAS Trendsetter Workshop on the Accessible City*, September 2003, Prague, Czech Republic. Commission of the European Communities.

Bussler, C., Fensel, D., and Maedche, A. (2002). A conceptual architecture for semantic web enabled web services. *ACM SIGMOD Record* 31: 24–29.

Campbell, J. L., Carney, C., and Kantowitz, B. H. (1998). *Human Factors Design Guidelines for Advanced Traveler Information Systems (ATIS) and Commercial Vehicle Operations (CVO). Report No. FHWA-RD-98-057.* Seattle: Battelle Human Factors Transportation Center/Federal Highway Administration.

Carmien, S. and Gorman, A. (2003). Creating distributed support systems to enhance the quality of life for people with cognitive disabilities. Presented at the *2nd International Workshop on Ubiquitous Computing for Pervasive Healthcare Applications (UbiHealth 2003)*, October 2003, Seattle. http://l3d.cs.colorado.edu/clever/assets/pdf/ubihealth03.pdf.

Cheverst, K., Davies, N., Mitchell, K., Friday, A., and Efstratiou, C. (2000). Developing a context-aware electronic tourist guide: Some issues and experiences, in the *Proceedings of CHI'00*, 1–6 April 2000, The Hague, The Netherlands, pp. 17–24. New York: ACM Press.

Coroama, V. (2003). The chatty environment: A world explorer for the visually impaired, in the *Adjunct Proceedings of UbiComp 2003* (J. McCarthy and J. Scott, eds.), 12–15 October 2003, Seattle, pp. 221–222.

EasyWalk (2008). *Promo Ez Walk by II Village.* http://es.youtube.com/watch?v=Kuzzggqnggg; http://www.gpsgadgets.net/2008/02/17/easy-walk-and-vodafone;http://news.bbc.co.uk/2/hi/technology/6458005.stm.

Edwards, S. J., Blythe, P. T., Hamilton, N., Russell, P., and Soutter, J. (2003). SmartSign: A navigation aid for wheelchair users, in the *Proceedings of the 9th World Congress on Intelligent Transportation Systems*, 14–17 October 2003, Chicago. September Enterprise Management Associates.

Feiner, S., MacIntyre, B., Hollerer, T., and Webster, T. (1997). A touring machine: Prototyping 3d mobile augmented reality systems for exploring the urban environment, in the *Proceedings of the 1st IEEE International Symposium on Wearable Computers (ISWC '97)*, October 1997, Cambridge, MA, pp. 208–217. London: Springer-Verlag.

Fink, J. and Kobsa, A. (2002). User modeling for personalized city tours. *Artificial Intelligence Review* 18: 33–74.

Fleck, M., Frid, M., Kindberg, T., O'Brien-Strain, E., Rajani, R., and Spasojevic, M. (2002). From informing to remembering: Ubiquitous systems in interactive museums. *IEEE Pervasive Computing* 1: 13–21.

Gill, J. (1997). *Access Prohibited? Information for Designers of Public Access Terminals.* London: Royal National Institute for the Blind/CEC INCLUDE.

Green, P. and Harper, S. (2000). An integrating framework for electronic aids to support journeys by visually impaired people, in the *Proceedings of the International Conference on Computers Helping People with Special Needs*, July 2000, Karlsruhe, Germany, pp. 281–288. Wein, Austria: OCG Press.

Gunderson, R. W., Smith, S. J., and Abbott, B. A. (1996). Applications of virtual reality technology to wheelchair remote steering systems, in the *Proceedings of the 1st Euro Conference of Disability, Virtual Reality & Associated Technology*, July 1996, Maidenhead, U.K., pp. 47–56. ECDVRAT and University of Reading.

Heidmann, F. and Hermann, F. (2003). Benutzerzentrierte Visualisierung raunbezogener Informationen für ultraportable mobile Systeme (User-centered visualization of user-related information for ultraportable mobile systems) in *Visualisierung und Erschließung von Geodaten. Karthografische Schriften Band 7* (Visualization and development of spatial data) (D. Drasch and M. Sester, eds.), pp. 121–132. Bonn: Kirschbaum Verlag.

Hub, A., Diepstraten, J., and Ertl, T. (2004). Design and development of an indoor navigation and object identification system for the blind, in the *Proceedings of ACM ASSETS 2004*, 18–20 October 2004, Atlanta, pp. 147–152. New York: ACM Press.

Jameson, A. (2003a). Adaptive interfaces and agents, in *The Human-Computer Interaction Handbook: Fundamentals, Evolving Technologies and Emerging Applications* (J. Jacko and A. Sears, eds.), pp. 305–330. Mahwah, NJ: Lawrence Erlbaum Associates.

Jameson, A. (2003b). Systems that adapt to their users: An integrative overview. Tutorial presented at the *9th International Conference on User Modelling*, 22–26 June 2003, Johnstown, PA.

Kolbe, T. H. (2004). Augmented videos and panoramas for pedestrian navigation, in the *Proceedings of the 2nd Symposium on Location Based Services and TeleCartography 2004* (G. Gartner, ed.), 28–29 January 2004, Vienna, Austria. *Geowissenschaftliche Mitteilungen* 66: 45–52.

Kruger, A., Butz, A., Muller, A., Stahl, C., Wasinger, R., Steinberg, K.-E., and Dirschl, A. (2004). The connected user interface: Realizing a personal situated navigation service, in the *Proceedings of IUI 04, 2004 International Conference on Intelligent User Interfaces*, 4–7 January 2004, Orlando, FL, pp. 161–168. New York: ACM Press.

Lankenau, A., Röfer, T., and Krieg-Bruckner, B. (2003). Self-localization in large-scale environments for the Bremen Autonomous Wheelchair, in *Spatial Cognition III* (Freksa et al., eds.), pp. 34–61. Berlin: Springer-Verlag.

Levine, S. P., Bell, D., Jaros, L., Simpson, R., Koren, Y., and Borenstein, J. (1999). The NavChair assistive wheelchair navigation system. *IEEE Transactions on Rehabilitation Engineering* 7: 443–451.

Long, S., Kopper, K., Abowed, G. D., and Atkeson, C. G. (1996). Rapid prototyping of mobile context-aware applications: The Cyberguide case study, in the *Proceedings of the 2nd ACM International Conference on Mobile Computing and Networking*, 10–12 November 1996, New York, pp. 97–107. New York: ACM Press.

Loomis, J. M., Golledge, R. G., and Klatzky, R. L. (1998). Navigation system for the blind: Auditory display modes and guidance. *Presence* 7: 192–203.

Malaka, R. and Zipf, A. (2000). Deep map: Challenging IT research in the framework of a tourist information system, in *Information and Communication Technologies in Tourism 2000, Proceedings of ENTER 2000, 7th International Congress on Tourism and Communications Technologies in Tourism* (D. Fesenmaier, S. Klein, and D. Buhalis, eds.), 22–24 January 2000, Barcelona, Spain, pp. 15–27. Wien: Springer-Verlag.

Moraitis, P., Petraki, E., and Spanoudakis, N. (2003). Providing advanced, personalised infomobility services using agent technology, in *Twenty-Third SGAI International Conference on Innovative Techniques and Applications of Artificial Intelligence (AI2003)*, 15–17 December 2003, Cambridge, U.K. pp. 35–48. Cambridge, U.K.: Peterhouse College.

Mori, H. and Kotani, S. (1998). Robotic travel aid for the blind: HARUNOBU-6, in the *Proceedings of the 2nd European Conference on Disability, Virtual Reality and Assistive Technology*, 17–19 September 1998, Skovde, Sweden, pp. 193–202. ECDVRAT and University of Reading.

Munro, A. J., Hook, K., and Benyon, D. R. (1999). Footprints in the snow, in *Social Navigation of Information Space* (A. J. Munro, K. Hook, and D. R. Benyon, eds.). London: Springer-Verlag. http://www.sics.se/~kia/papers/IntroFINALform.pdf.

Naniopoulos, A. (2002). Final Report. Report of the TRANSWHEEL Project (DE3013). Commission of the European Communities.

Naniopoulos, A., Bekiaris, E., and Dangelmaier, M. (2000). TRAVELGUIDE: Towards integrated user friendly systems for drivers traffic information and management provision, in *International Conference on Machine Automation (ICMA2000)*, 27–29 September 2000, Osaka, Japan. pp. 585–590.

Naniopoulos, A., Bekiaris, E., and Panou, M. (2004). Costs and benefits of information technology systems and their application in the infomobility services: The TRAVEL-GUIDE approach, in *Economic Impacts of Intelligent Transportation Systems. Innovations and Case Studies* (E. Bekiaris and Y. Nakanishi, eds.), pp. 463–480. Oxford, U.K.: Elsevier.

National Swedish Board for Consumer Policies and the Swedish Handicap Institute (1995). *Automatic Service Machines, Service for Everybody?* Stockholm: The Swedish Handicap Institute.

Nicolle, C. and Burnett, G. (eds.) (1999). *TELSCAN Code of Good Practice and Handbook of Design Guidelines for Usability of Systems by Elderly and Disabled Travellers*. TELSCAN Project Deliverable 5.2. Commission of the European Communities.

Nielsen, J. (1994). Heuristic evaluation, in *Usability Inspection Methods* (J. Nielsen and R. Mack, eds.), pp. 25–62. New York: John Wiley & Sons.

Not, E., Petrelli, D., Stock, O., Strapparava, C., and Zancanaro, M. (1997). Person-oriented guided visits in a physical museum, in *Museums Interactive Multimedia 1997: Cultural Heritage Systems. Design and Interfaces (Selected Papers from ICHIM97), Archives and Museum Informatics* (D. Bearman and J. Trant, eds.), 1–5 September 1997, Paris, pp. 69–79. Pittsburgh: Archives & Museum Informatics.

O'Hare, G. M. P., O'Grady, M. J., Keegan, S., O'Kane, D., Tynan, R., and Marsh, D. (2004). Intelligent agile agents: Active enablers for ambient intelligence, presented at the *ACM's Special Interest Group on Computer-Human Interaction (SIGCHI), Ambient Intelligence for Scientific Discovery (AISD) Workshop*, 25 April 2004, Vienna, Austria. Springer-Verlag.

Oxley, P., Ayala, B., Alexander, J., and Barham, P. (1994). *Evaluation of Route Guidance Systems*. Deliverable No. 20 of the CEC DRIVE II EDDIT Project (V2031). Commission of the European Communities.

Panou, M. and Bekiaris, E. (2006). Towards the development of an adaptation strategy for safe and integrated infomobility services use by mobility impaired drivers, in the *Proceedings of ITS 2006*, 26–30 June 2006, Jhongli, Taiwan [CD].

Panou, M., Bekiaris, E., and Gaitanidou, E. (2005). *A Holistic Approach on In-Car HMI Elements to Formulate a Strategy for Integrated and Personalised ADAS/IVIS HMI.* Hannover, Germany: ITS Europe.

Passini, R. (1984). *Wayfinding in Architecture.* New York: Van Nostrand Reinhold Company.

Petrie, H., Johnson, V., Strothotte, T., Raab, A., Michel, R., Reichert, L., and Schalt, A. (1997). MoBIC: An aid to increase the independent mobility of blind travellers. *The British Journal of Visual Impairment* 15: 63–66.

Poslad, S., Laamanen, H., Malaka, R., Nick, A., Buckle, P., and Zipf, A. (2001). Crumpet: Creation of user-friendly mobile services personalised for tourism, in the *Proceedings of the 2nd International Conference on 3G Mobile Communication Technologies (3G 2001)*, 26–29 March 2001, London, pp. 28–32. IEE Conference Publication 477.

Poulson, D., Allison, G., Ashby, M., and Maguire, M. (1996). *ATTACH Guidelines for Developing Public Information Kiosks.* Loughborough: HUSAT Research Institute.

Raptis, D., Tselios, N., and Avouris, N. (2005). Context-based design of mobile applications for museums: A survey of existing practices, in the *Proceedings of the 7th International Conference on Human Computer Interaction with Mobile Devices and Services (MobileHCI'05), Volume 111,* 19–22 September 205, Salzburg, Austria, pp. 153–160. New York: ACM Press.

Ross, D. A. and Blasch, B. B. (2000). Wearable interfaces for orientation and wayfinding, in the *Proceedings of the ACM ASSETS'00,* 13–15 November 2000, Arlington, VA, pp. 193–200. New York: ACM Press.

Sasaki, H., Tateishi, T., Kuroda, T., Manabe, Y., and Chihara, K. (2002). Wearable computer for the blind aiming at a pedestrians' intelligent transport systems, in the *Proceedings of the 3rd International Conference on Disability, Virtual Reality and Associated Technologies,* 23–25 September 2000, Alghero, Italy, pp. 235-241. ICDVRAT/University of Reading, U.K.

Searby, S. (2003). Personalisation: An overview of its use and potential. *BT Technology Journal* 21: 13–19.

Simões, A., Gomes, A., and Bekiaris, E. (2006). *Use Cases.* Deliverable 1.1.2 of the ASK-IT project (IST-2003-511298), Commission of the European Communities.

Stephanidis, C. (ed.) (2001). *User Interfaces for All: Concepts, Methods and Tools.* Mahwah, NJ: Lawrence Erlbaum Associates.

Story, M. F. (1998). Maximising usability: The principles of universal design. *Assistive Technology* 10: 4–12.

Strothotte, T., Fritz, S., Michel, R., Raab, A., Petrie, H., Johnson, V., et al. (1996). Development of dialogue systems for a mobility aid for blind people: Initial design and usability testing, in the *Proceedings of the Second Annual ACM Conference on Assistive Technologies, Assets '96,* 11–12 April 1996, Vancouver, Canada, pp. 139–144. New York: ACM Press.

Uchiyama, H., Deligiannidis, L., Potter, W. D., Wimpey, B. J., Barnhard, D., Deng, R., and Radhakrishnan, S. (2005). A semi-autonomous wheelchair with helpstar, in the *Proceedings of the 18th International Conference on Innovations in Applied Artificial Intelligence* (M. Ali and F. Esposito, eds.), June 2005, Bari, Italy, pp. 809–818. London: Springer-Verlag.

Visintainer, F., Panou, M., and Pagle, K. (2006). ADAS and IVICS for all, in the *Proceedings of the ASK-IT International Conference 2006,* October 2006, Nice, France [CD].

Weiss, G. (1999). *Multi-Agent Systems: A Modern Approach to Distributed AI.* Cambridge, MA: MIT Press.

Westerlund, Y. and Stahl, A. (1999). *Demand Responsive Transport (DRT) Using Advanced Telematics.* Internal Report, TELSCAN Project TR1108. Commission of the European Communities.

Wiethoff, M. and Sommer, S. (2005). *User Behaviour Modelling as Framework for the Specification of User Requirements.* ASK-IT Internal Report.

Zwalen, H. T., Adams, C. C. Jr., and Debald, D. P. (1988). Safety aspects of CRT touch panel controls in automobiles, in *Vision in Vehicles II* (A. G. Gale et al., eds.), pp. 335–344. Amsterdam, The Netherlands: Elsevier North-Holland.

50

Electronic Educational Books for Blind Students

Dimitris Grammenos,
Anthony Savidis,
Yannis Georgalis,
Themistoklis Bourdenas, and
Constantine Stephanidis

50.1 Introduction

In the past, students with disabilities used to be segregated into special schools or classrooms, populated only by children sharing the same disability. This practice had a negative impact on the ability and opportunities of these students for inclusion in society. As a result, presently many countries follow a more open approach called inclusive education (Clough and Corbett, 2000), where students with disabilities participate in the mainstream classrooms. This approach provides better opportunities for inclusion to those students, but also raises new challenges to the educational system, as students with disabilities need to access the same educational material as the rest of the students.

Access to educational material, as to any other type of printed material, constitutes a major challenge for blind people. The traditional method employed to overcome this problem is to transcribe school textbooks in Braille format, using Braille typewriters and embossers (i.e., printers), or to record them in the form of audiotapes and CDs. Both of these approaches suffer from numerous drawbacks (Petrie et al., 1997), which are mainly due to their physical instantiation, as well as to the limited interaction capabilities that they can offer. Furthermore, the two approaches work complementarily, since depending on the abilities, knowledge, and preferences of the learner, as well as the current context of use, only one or both of them should be employed. Thus, the availability of content in just one form may once more result in lack of accessibility.

Nowadays, the existence of educational content in electronic format has the potential, on the one hand, to overcome the aforementioned limitations, while, on the other hand, to take advantage of the best qualities of both approaches, since electronic documents can be rendered in both audio and tactile form, but, most important, they can be augmented with numerous, added-value interaction capabilities.

In this context, this chapter introduces the concept of electronic books and the way that they can be used to provide accessible educational textbooks to blind students. Furthermore, it presents a novel software platform for developing and interacting with multimodal interactive electronic textbooks that provide a dual user interface (Savidis and Stephanidis, 1995), that is, an interface concurrently accessible by visually impaired and sighted persons. The platform, named Starlight, comprises two subsystems: (1) the Writer, facilitating the authoring of electronic textbooks, encompassing various categories of interactive exercises (Q&A, multiple choice, fill in the blanks, etc.); and (2) the Reader, enabling multimodal interaction with the created electronic textbooks, supporting various features like searching, bookmarking, replay of sentences or paragraphs, user annotations and comments, activity recording, and context-sensitive help. The chapter further discusses the competitive features of the dual user interface and of supplied functionality compared to existing electronic books. It also consolidates the key design findings, elaborating on prominent design issues, design rationale, and respective solutions, highlighting strengths and weaknesses, and outlining directions for future work.

50.2 Background and Related Work

50.2.1 How Do Blind People Use Computers?

In order to access computers, blind people employ two basic technologies: (1) text-to-speech, and (2) Braille displays (see also

Chapter 6 of this handbook). Text-to-speech technology allows the dynamic reproduction of any text using a humanlike voice. Braille is a writing system that uses six to eight raised dots in various patterns to represent letters and numbers. Due to its tactile nature, Braille can be read through touch. A Braille display is a hardware device that provides tactile output, mimicking the way blind people read Braille text on paper. A Braille display is composed of numerous "cells" (usually 40, 65, or 80) each of which contains 8 rounded plastic or metal pins that can be mechanically lifted, thus displaying a single ASCII character. Braille displays work complementary to text-to-speech. It is generally considered that speech is for speed and Braille is for accuracy. For example, a spelling mistake can be more easily detected on a Braille display than through speech. The most prominent input device for the blind is the keyboard, but in some cases the mouse is also employed.

The aforementioned technologies are used in combination with a screen reader, a software tool that can interpret what is displayed on the screen and convert it either to speech or Braille (see also Chapter 28 of this handbook). The most popular screen readers are JAWS by Freedom Scientific, Hal by Dolphin, and Window-Eyes by GW Micro. Since screen readers only present one word at a time (either through speech or Braille), it is very difficult for a blind person to get an overview of the current screen's layout and content. In fact, it is said that it is the equivalent of a sighted person trying to look at the screen through a straw.

50.2.2 Electronic Books

Electronic books are software applications that adopt the book metaphor in order to render multimedia content (text, images, audio, etc.) on a computing device, while providing related functionality, such as page browsing, table of contents, and bookmarking (see Chapter 46 of this handbook). The electronic files that contain the educational content are usually referred to as e-books (or eBooks), while the terms *reader* or *player* usually denote the interactive software applications employed for reading e-books. Currently, few such players exist that are accessible to visually impaired people, while offering high quality of interaction for content presentation and navigation. The key advantages of e-books compared to their printed counterparts with regard to accessibility by the blind are the following:

- They can be automatically read using a speech synthesizer, thus allowing eyes-free access.
- They can be rendered on a Braille display, thus allowing tactile access.
- They do not have to be physically held or flipped, thus allowing hands-free access.

Currently, there several different file formats in which an e-book may be created. Very few of them are compatible with each other, but most them are not. In most cases a different reader

application is required in order to render each format. Based on their intended use, available formats can be broadly classified in three categories:

1. *General-purpose formats*: Formats that are not only used for creating e-books but also for any kind of text. The most representative examples include the following:
   (a) *Plain text*: ASCII characters with no formatting of any type. It has a very small size and can be accessed by any application. Its main disadvantage is that, since it does not include any kind of semantic information (e.g., headings, sections), the rendering, navigation, and interaction options that can be supported are very limited.
   (b) *Hypertext markup language (HTML)*: The language used for creating web pages. E-books created in HTML can be read using any web browser (e.g., Mozilla Firefox, Microsoft Internet Explorer, Apple Safari). E-books created in HTML require more space than those in plain text, and their creation requires some programming skills or the use of related interactive tools.
   (c) *Portable document format (PDF)*: A format invented by Adobe Systems mainly to create documents the visual properties of which remain the same, independently of the software tool or platform used to read them. Since this is a proprietary format, specific software tools by Adobe are required in order both to create and read such documents. Currently it is considered the most popular e-book format.
2. *General-purpose e-book formats*: Formats that can only be used for creating e-books, such as:
   (a) *Microsoft LIT*: A proprietary format created by Microsoft. Books in this format can only be read by using the Microsoft Reader program.
   (b) *IDPF epub*: An open standard created by the International Digital Publishing Forum.
   (c) *Mobipocket*: A format based on the Open eBook standard, which is quite popular in portable devices since readers exist for most Windows, Symbian, BlackBerry, and Palm operating systems.
3. *Accessible e-book formats*: Formats used for creating accessible e-books (mainly by the blind), which are also referred to as digital talking books (DTBs):
   (a) *Digital Accessible Information System (DAISY)*: A standard aiming to make print material accessible and navigable for print-disabled persons, more formally known as ANSI/NISO z39.86 (ANSI/NISO, 2005). Most digital talking books for the blind are based on this standard. It defines the format and content of the files comprising a DTB and establishes a set of requirements for DTB playback software and devices. DAISY e-books can contain text, audio, or both.
   (b) *NIMAS*: A new standard format recently established by the National File Format (NFF) Technical Panel to

facilitate the provision of accessible versions of print textbooks to PreK–12 students with disabilities (NFF, 2005). NIMAS is actually a subset of DAISY aiming to simplify the markup complexity of the original standard by defining a minimum set of requirements for compliance. This format currently can only be used for literary-based textbooks, since mathematics and science textbooks are not supported.

50.2.3 Accessible e-Book Reading Applications

Available solutions can be broadly classified in two categories:

- Partially accessible e-books that integrate reading capabilities or are compatible with popular screen reading software
- Digital talking books developed specifically for use by blind people

The most widely used applications in the first category are the Microsoft Reader and the Adobe Reader. The Microsoft Reader (Microsoft, 2006) is equipped with quite elaborate navigation control and has a text reading feature that, however, does not work for copy-protected e-books. In addition to that, a Verbosity feature is supported, through which the current interaction focus is announced by tracking the movement of the mouse and keystrokes. This feature tries to reproduce part of the functionality offered by screen readers, since the Reader's compatibility with such applications is not guaranteed (Microsoft, 2002), but user keystrokes are not spoken, and thus user input is not accessible. Users can read an e-book from the last point they left it, or from the farthest page they have read and also get brief information about their current location in the e-book. The Adobe Reader (Adobe, 2004; WebAim, 2006) also has a Read Out Loud option, which, using an embedded speech synthesizer, can recite the contents of a PDF document but does not provide any navigation controls. Users can affect, to a limited extent, the reading order strategy and can also select to open a document to the last section where they left off. In order to achieve a higher level of accessibility, Acrobat Reader is made compatible with screen readers. Recently, Adobe has released Adobe Digital Editions,[1] a free e-book program designed especially for PDF-based e-books. Unfortunately, unlike the Acrobat Reader, this application does not presently include any accessibility features.

In short, the major drawbacks of software belonging in this category can be summarized as follows:

- Unless a screen reader is used (when and if possible), only a very limited part of the available functionality is accessible to the visually impaired users. But even then, there are several usability problems associated with screen readers (Barnicle, 2000; Parente, 2006), which also have a high cost, and, as a result, are used only by a small percentage of blind people.

---

[1] http://www.adobe.com/products/digitaleditions.

- These programs have a graphical interface that is optimized for visual use. Thus, in order for users to effectively use them, they must possess an accurate mental model and a good understanding of their visual structure and layout. These facts impose an unnecessary high mental workload, and result in erroneous, painstaking interaction.
- Feedback to most user and system events is provided solely through visual cues, thus leaving the nonsighted users wondering about the effect of their actions, or the current state of the interaction process.
- The overall effectiveness, efficiency, and quality of nonvisual interaction are quite poor.

Regarding digital talking books for the blind, the National Information Standards Organization has suggested specific guidelines for hardware/software platforms (i.e., players) that render the contents of a DTB (NISO, 1999). Several related software applications that use the DAISY format (e.g., FSReader by Freedom Scientific, gh PLAYER by gh, Victor Reader Soft by Humanware, EaseReader by Dolphin, eClipseReader by irti) but also hardware devices (e.g., PlexTalk by Plextor, ezDaisy by Telex, Victor Reader by Humanware) are available. All of them mainly provide speech output, and only one supports Braille displays.

As DAISY books have an explicit and well-defined structure, related players usually offer several alternative ways for navigating in the book's content (e.g., table of contents, browse headings, move directly to a specific location). Furthermore, typical features supported include bookmarks, information about the current position and the total length of the book, remembering the user's last position, note-taking, and highlighting.

50.2.4 Research Applications Accessible to the Blind

In addition to commercial software, a few research systems have also been developed that aimed to improve the access of people with visual impairments—and especially for the blind—to printed material.

DAHNI was an early hypermedia system with a nonvisual interface (Petrie et al., 1997). The system could be used with three input devices: keyboard, joystick, and a custom-made touch-tablet. All available commands were logically arranged on two-dimensional space, in a sideways H shape. The user would always start from a central point and then navigate through the available commands using the various input devices. Shortcut keys were also supported for expert users.

VoxBoox (Jain and Gupta, 2006) is a prototype system that automatically translates books published in HTML to VoiceXML that is enhanced with additional code in order to allow speech-based user interaction through a small set of commands (e.g., go to the beginning or end of the document, start/stop reading). Additionally, speech bookmarks can be inserted on a page or paragraph, so that, later on, users can return to it by uttering the bookmark's name.

Some more loosely related work that deals with nonvisual access to information, but not in the form of an e-book, includes:

- Research efforts that address the problem of accessing mathematics nonvisually, for example, Mathtalk (Stevens et al., 1997), AudioMath (Ferreira and Freitas, 2004), Math Genie (Karshmer et al., 2004); see also Chapter 51 of this handbook.
- The Hyperbolic Browser, a tool for navigating hierarchical structures to aid nonsighted programmers (Smith et al., 2004)
- The AVANTI web browser (Stephanidis et al., 2001) that can adapt itself for nonvisual access to the web
- HOMER (Savidis and Stephanidis, 1995), a user interface management system, which facilitates the development of dual user interfaces

50.3 The Starlight Platform

A key limitation of existing e-books that are accessible to the blind is that they are mostly designed to support the reading of novel-like books by a single end-user. Critical features that typical educational textbooks support, like recapitulation questions, exercises, exams, and personal annotations, are totally missing. Additionally, they lack functionality that is necessary for deployment in an educational setting, such as:

- Logging a student's progress (e.g., which content has been read and for how long)
- Provision of related statistics
- Full content editing by the educator
- Configuration of functionality to the student's characteristics, abilities, and requirements
- Concurrent student-educator use
- Support for multiple registered users

In this context, the rest of this chapter presents the development of an educational textbook platform, namely the Starlight platform (Grammenos et al., 2007), supporting all the various features previously mentioned, as they are not offered by available accessible e-books through a dual user interface, an interface that is manifested through alternative modalities and is concurrently appropriate for both visual and nonvisual use.

The platform comprises two subsystems:

- The Writer, facilitating the authoring of electronic textbooks, encompassing various categories of interactive exercises (Q&A, multiple choice, fill in the blanks, etc.)
- The Reader, enabling multimodal interaction with the created electronic textbooks, supporting various features like searching, bookmarking, replay of sentences/paragraphs, user annotations/comments, activity recording, and context-sensitive help

50.3.1 Development Process

The development process adhered to the basic principles of human-centered design (ISO, 1999) with the direct involvement (in different parts of the process) of two usability experts, two accessibility specialists, three specialized educators, four blind students of varying ages and two of their parents, in total. During the very early design phase, it soon became apparent that the goal of delivering a dual user interface for such comprehensive e-book functionality was highly demanding, requiring far more intensive iterative design than typically required. In this context, very quick iterations were conducted involving many design, prototyping, and evaluation sessions. For this purpose, a discount usability method was adopted (Nielsen, 1993, 1994; Wharton et al., 1994). The details of the design process are presented in Figure 50.1.

First, a requirements and task analysis phase took place, collecting data from:

- Relevant bibliography (e.g., Graziani and Arato, 1998; Morley, 1998; Smith et al., 2004)
- Commercial products
- Accessibility experts
- Target users
- Assistive technology experts
- Software engineers

The analysis of the collected information has led to the identification of the primary functionality and the preliminary interaction design, including several annotated paper sketches and digital mockups. These mockups were reviewed by a team comprising a usability expert, an accessibility specialist, an educator of students with visual impairments, and a lead software engineer. Following their comments, the initial version of the design specification was created (see Figure 50.1).

Next, the initial implementation of the system, encompassing the functionality and the user interface, was carried out. From this point on, expert evaluation (based on heuristic evaluation; Nielsen, 1994) was repeated after every user interface feature becoming available or modified; the latter was practically required many times per day. Every week, the overall platform was tested with end-users, adopting a combination of the Thinking Aloud (Nielsen, 1993) and Cognitive Walkthrough (Wharton et al., 1994) methods. The results of both processes were compiled in a prioritized list of suggested corrections and improvements. Then, this list was discussed, filtered, and finalized in a plenary meeting, based on criteria of criticality, feasibility, and development cost.

50.3.2 The Starlight Reader

This section elaborates on the key interactive properties of the Starlight Reader, presenting the corresponding design rationale, and discussing how the final design decisions were shaped by the feedback from the iterative evaluation sessions.

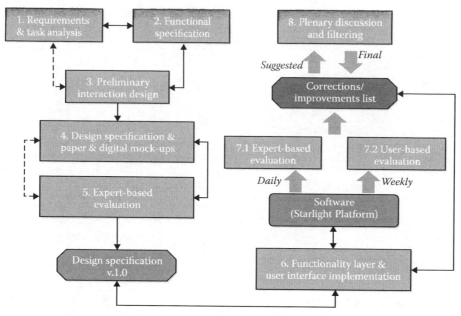

FIGURE 50.1 Outline of the development process.

50.3.2.1 Multimodality

As previously mentioned, the Starlight Reader has a dual interface comprising a visual and a nonvisual manifestation. The visual interface has similar function and use as any other Windows application. The nonvisual interface is rendered through synthetic speech, via two different voice types interchangeably—male and female—as well as through Braille displays. During the requirements analysis phase, it became apparent that the concurrent use of speech and tactile modalities was prominent, as these two modalities had to be deployed for different aspects of interaction. Speech is faster and easier to use, especially for younger students who are in the process of learning Braille, but on the other hand, accuracy and text comprehension is significantly better with Braille (García, 2004). User input is supported through the keyboard, the Braille display's hardware keys, and any type of joystick that has at least two buttons. In particular, joystick support can considerably aid the use of the system by younger children, since it is very easy to master and requires virtually no training.

50.3.2.2 Interaction Metaphor

Starlight adopts the metaphor of an electronic book, but with a variation; its content is still decomposed into chapters, sections, subsections, paragraphs, and sentences, as any typical book, but not in pages. This decision was deliberately taken due to the fact that e-books do not have to be confined by a visual restriction inherent in the physical medium used to instantiate printed books in the past. A paper page simply represents an arbitrary quantity of text, with no particular semantic connotation. In the digital world, especially when text is rendered nonvisually, this concept is meaningless and mostly misleading. In fact, it comes as no surprise that document authoring tools support automatic paging, since a page never constitutes a semantic element of the document structure, like sections or headings. The conducted expert evaluations and user tests validated this decision.

In general (see Figure 50.2), a book can contain an arbitrary number of chapters, each of which can contain text, organized in subsections of any depth, notes, exercises, and exams. Four

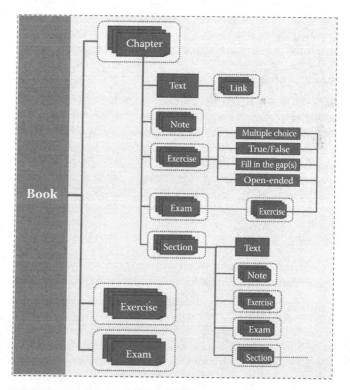

FIGURE 50.2 Structure of a Starlight book.

different types of exercises are supported: (1) multiple choice; (2) true/false; (3) fill in the blank; and (4) open-ended questions. Additionally, a sentence may include an internal link to any other position in the same e-book. The difference among recapitulation exercises that can be added at the end of each section and exams is the following:

- Recapitulation exercises can be accessed by the student as part of reading the book and while taking them. The student:
  - Is allowed to access any other part of the book
  - Directly gets feedback about correctness after providing an answer and not an overall score for all of them
  - Can attempt to answer a single exercise multiple times
- Exams can only be triggered by the educator and during them the student:
  - Is not allowed to access any part of the book
  - Can move freely among the exercises and answer them multiple times, but gets no feedback about their correctness
  - Has to explicitly state when she has completed the exam in order to get a report that contains a score and summary of the results, as well as a detailed list of all the exercises, of the answers, and whether these were correct or not

In order to be able to support the novel functionality that was required by Starlight, a new, custom-made, e-book format was used, instead of one of the currently available standards such as DAISY (ANSI/NISO, 2005) and NIMAS (NFF, 2005). The present version of this format does not support text formatting or images, but it is foreseen that this will be included in a subsequent version.

50.3.2.3 Sharing

In a typical school setting, a single computer is usually shared among several students, each of whom has individual interaction preferences, bookmarks, and notes, as well as data related to examination and test results. Unlike other e-book rendering applications, Starlight e-books support multiple students, automatically keeping track of personal preferences (e.g., voice volume and speed), content viewed, progress, exercises completed, and exam results, as well as where the student left the book.

50.3.2.4 Dual Interaction

A problem may arise in an inclusive educational setting in the case where a blind student collaborates with a sighted peer (student, teacher, parent, or friend), or the opposite. If the e-book is rendered exclusively nonvisually, such collaboration and social interaction is not possible. If the e-book has a Windows-based visual interface and is also compatible with a screen reader, then it can be presented nonvisually, but as mentioned earlier in Section 50.2.3 the interaction quality of the nonvisual interface can be considerably, or even unacceptably, compromised.

To overcome such problems Starlight implements two distinct user interfaces through nonoverlapping input/output (I/O) modalities that can work in parallel without conflicts. Thus, nonvisual use is achieved through the keyboard, speech synthesis, and Braille, while visual use is achieved through a graphical interface and the mouse.

50.3.2.5 Navigation

Navigation in electronic texts constitutes a key challenge for most users, and several solutions for improving its usability have been suggested (e.g., Shubin and Meehan, 1997). Typically, most approaches aim to effectively address context control, usually referred to as the "Where am I?" issue. In a visual setting, there are various ways to support navigation (e.g., using different font size and colors, layout) relying on immediate visual perception. However, such techniques do not apply to speech or Braille modalities, due to the sequential delivery of content, and to the fact that auditory information, in contrast to visual, is transient (Morley et al., 1999). In this context, existing navigation techniques have been integrated or adapted (e.g., Graziani and Arato, 1998; Morley, 1998; Morley et al., 1999; Smith et al., 2004), and new ones have been designed:

1. *Control over the auditory presentation of information*

   - *Play and stop*: A single key (ESCAPE), which due to its placement on the keyboard, is the most easily, rapidly, and unmistakably accessed key by a blind person.
   - *Repeat*: The last sentence heard.
   - *Navigate in the book's structure*: To the table of contents, to the start of the book, to the next/previous sentence, paragraph, section, or chapter, to the exercises or notes of the current section. From the table of contents, the user can also access the time spent in each of the chapters, sections, subsections, and so on.
   - *Navigate inside a sentence* (i.e., spelling mode): To read the current sentence or the current word, go to the start or end of the sentence, move to the next/previous word or letter. When single letters are read, information is also provided such as whether the letter is capital, if it has an accent, and if it belongs to a language other than the text's "native" one.
   - *Control speech volume and rate*.
   - *Single-key navigation*: Special care is taken so that a novice user can interact without having to rely on multiple keys and shortcuts. Thus, it is possible to go through all the initial dialogues (book selection, user login, start position) and browse all of the contents from the very first sentence till the end, using just a single key (which by default is set to be the right arrow).
   - *Start-up options*: When an e-book is opened, the user has the options to go to the last position where she left off, to the table of contents, or to the book's start.

2. Support of hypertext controls and tools

- *Move back and forth in the navigation history*: This feature is very useful in nonvisual applications, because it helps users return to a prior location when they are accidentally transferred to another one (e.g., due to the press of a wrong shortcut).

- *Bookmarking at sentence level*: Existing applications support bookmarking a page but not a specific point in it. This is impractical, as a blind person has to read through the entire text to locate the actual point of interest that was the reason of the bookmarking. In Starlight, bookmarks can be added to sentences. Furthermore, when a bookmark is read, its context (chapter, section, paragraph, etc.) is also described, so that users can easily identify the target of the bookmark.

- *Search*: The results are presented in a list, each entry carrying context information (chapter, section, etc.). When a list item is selected, the user is transferred to that point, with the option to jump to the position of the next or previous search result without returning to the results list.

3. Provision of context information and interaction safeguards

- *Where am I?* A precise description of the current context is provided. The description is created bottom-up, starting from the relative position of the current paragraph in the current section (e.g., "paragraph 3 out of 5"), all the way up to the top level (e.g., "section 2 out of 4 Section Title, chapter 4 out of 10 Chapter Title"). If the user is currently using a support facility (e.g., bookmarks, notebook, search), the name of the facility is also read.

- *Return to the last position in the book's content*: Users may directly return to their last position in the content from anywhere. This works as a safeguard, since practice has shown that users are often lost while using the book's support tools or interface and need a meaningful place (or interaction state) to return.

- *Different voice for rendering interface and content*: In a speech-based application, it is often difficult to distinguish between content and interface, since they both are rendered as words, which are often interweaved. In order to differentiate among the two, Starlight employs two distinctively different voices (i.e., male and female). As was revealed during evaluation, this feature had a significantly positive impact on the overall usability of the Reader.

- *Automatic section numbering*: Hierarchical section numbering (e.g., 3.7.1) can provide valuable and intuitive context and navigation clues. To ensure the consistency and accuracy of section numbering, it is automatically generated by the Reader.

4. Nonspeech auditory feedback: In auditory interfaces, a convenient way to provide feedback while avoiding speech overuse is nonspeech audio in the form of either structured combinations of musical sounds (or earcons; Stevens et al., 1995) or natural, iconic sounds that work just like metaphors. In Starlight, sounds are deployed to denote the initiation or completion of specific actions and events, adopting distinct everyday sound effects (e.g., Mynatt 1997) for the following actions:

- Add a bookmark.
- Denote that the current text contains a link.
- Move to next or previous section/chapter.
- Focus is transferred to another part of the book through the table of contents or by following a hyperlink, a bookmark, or a search result.
- Notebook open/close.
- Speech volume is changed.
- Spelling mode is switched on/off.

50.3.2.6 Context Sensitivity

Task-based context-sensitive menu: At any time a nonvisual menu can be activated with the commands supported in the current context. The menu is structured at two levels: on the higher level are user tasks, and within each task, the supported functions.

Task-based context-sensitive help: Help is provided for the available functionality for the active user tasks. Help content is structured similarly to the nonvisual menu.

50.3.2.7 Adaptive Prompting

Whenever the system waits for user input, or it is anticipated that the user intends to perform a certain action, the system prompts these actions together with the corresponding shortcuts. Typical cases concern yes/no dialogues, exercises, exams, the table of contents, and so on. This information is always provided at the end of the dialogue, so that the user may quickly skip it. Additionally, there is a shortcut exploration mode, where, whenever a shortcut is pressed, the system automatically announces the associated function. This feature was suggested during early formative evaluation sessions by a blind student.

50.3.2.8 Configurability

Since nonvisual applications unavoidably rely on the use of shortcuts, it is crucial that users may redefine them according to individual preferences and needs. The latter is fully supported, detecting and resolving also potential conflicts among shortcuts.

50.3.2.9 Integrated Text Editing

Starlight includes a nonvisual text editor that is used whenever text input is needed, for example, in the notebook, open-ended questions, and notes. This editor supports bilingual editing (Greek and English) and works as follows:

- *Cursor*: It is implicit, representing a character position inside the text. Insertion is performed after the current position, moving the cursor to the newly inserted character.
- *Key press*: It is read. If it is a letter, it is stated if it is capital, accented, or in English. If no key is pressed for a while, the entire current word is read.
- *Navigation*: Move the cursor a character/word to the left/right, to the start/end of the current sentence or whole text. When the cursor is moved, the word in that position is read, followed by the character at the cursor position.
- *Copy/paste*: From user tests it was found that using the typical keyboard-based MS Windows approach for selecting a piece of text (i.e., keeping the Shift key pressed while also using the arrow keys to select text) had several usability problems. For example, according to this technique, if the Shift key is pressed, then the selection is lost and the user has to start over again. Also, it requires two hands, thus prohibiting blind users from concurrently using the Braille display to locate precisely the text to select. To overcome these problems the user presses a shortcut (or selects the related option from the nonvisual menu) to anchor the start point of the selection, and then using the arrow keys can freely move to the end of the selection. Then, via the menu or through shortcuts, the selected text can be copied or cut.
- *Reading*: The user may listen to the current word or sentence, the selected text, or the entire text.

50.3.2.10 Embedded Notebook

Users may swap at any time between the e-book and the notebook in order to keep notes of what they read, write down questions, and so on. A feature that was appreciated during user testing is that the currently spoken sentence can be directly copied to the notebook, making very easy the creation of outlines and summaries of the text as it is read. Furthermore, more advanced users can insert custom anchor points (i.e., quick access bookmarks)

in their texts. Notes can also be added to each section, either by book authors at creation time (e.g., references, explanations, comments) or by the students during reading.

50.3.2.11 Adaptable Levels of Functionality

The goal of Starlight was to create a tool addressing a wide range of student ages, from the first grades to university levels. The Starlight platform offers comprehensive functionality that may not be needed for particular target users. Additionally, educators may wish to introduce all available functionality incrementally, starting with the very basics, and moving toward more elaborate options. For this purpose, every function offered by the Starlight Reader can be ranked using a number, and then the e-book author/editor can tune set the ranks of functionality actually exposed to the end-users (e.g., the students).

50.3.2.12 Privacy Considerations

As a blind user has noticed during evaluation, in contrast to a sighted person, a blind reader may not be aware of whether someone else is staring at the screen. For example, in a school setting, there might be other students copying answers, or the student may wish to keep writing/notes private from the tutor until done. To proactively avoid such situations, Starlight offers the option of blanking the screen, and having feedback of this state, in order to protect privacy.

50.3.2.13 Accessible Utilities

Every part of the Reader is accessible, including utilities such as the installation tools, book selection, and user management. Furthermore, the full user manual is provided with the Reader as an e-book itself.

50.3.2.14 Visual Interface

The visual interface of the Reader supports sighted users and users with deteriorated vision, and is targeted to peers with a supervisory role (teachers, parents), friends, or other students.

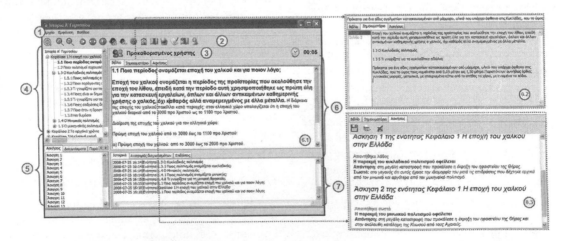

FIGURE 50.3 The visual interface of the Starlight reader.

Besides speech and Braille output, unlimited font and interaction object magnification is offered. The main interface (Figure 50.3) consists of the following parts:

- *Main menu*: Contains some general functions, such as alternative help (shortcuts, context-sensitive, user manual) and visual layout options (which components to show/hide).
- *Commands toolbar*: Provides direct one-click access to the most frequently used and needed functionality. The toolbar is context sensitive, as it is dynamically populated with functions that are relevant to the task at hand.
- *User information*: The user's name and the time that has passed since she started reading the book.
- *Book structure navigator*: Presents the book's structure. Users can reveal or hide a node's contents and can jump to a specific section by double-clicking on it. This function is particularly useful when an educator wants to quickly transfer a student to a particular position in the book. The user's current position is noted with bold letters, while unread sections are marked with red.
- *Section's contents*: Comprises four tabs, each one respectively containing a list with all the available exercises, exams, paragraphs and notes of the section that is currently selected at the book structure navigator. Users can jump to any one of them by double-clicking.
- *Main part*: May contain the following tabs:
  - *Content*: This is where the book's contents are graphically rendered. The sentence that is currently being read is presented using blue color and a much larger font.
  - *Notebook*: Here the contents of the nonvisual notebook are presented (see Section 50.3.2.9) and edited.
  - *Exercises and exams*: If, during the current session, the student has taken any exercises, an extra tab is added, which contains detailed information about the exercises' content, the student's answers, as well as whether these were correct. The educator has the option to save or print the contents of this tab. Similarly, additional tabs are added whenever the student takes any exams.
- *Logging and statistics*: This section has three tabs containing the student's: (1) navigation history (i.e., which sections she has visited and for how long; (2) a list of past exams' reports (that then can be opened to the "main part"); and (3) summarizing statistics about all the exercises that she has answered since she has started using the book.

50.3.3 The Starlight Writer

The Writer facilitates the authoring of electronic textbooks that can be accessed through the Reader. Currently, the Writer has only a visual interface and thus is not accessible to blind people. The Writer supports versioning (i.e., keeps track of the book's edition number) so that updates to existing books can be created.

In order to enhance the efficiency and usability of the Writer the following global interactive features are supported:

- *In-place context-sensitive menus*: The user can access related functionality for any part of the user interface through a menu that appears by pressing the right mouse button.
- *Undo/redo*: In order to avoid accidental loss of the user's work, an unlimited number of undo and redo actions is supported.
- *Cut, copy, and paste at any level*: Information of any type can be cut, copied, and pasted in any relevant place in the book. Thus, for example, a whole chapter with all its content can be duplicated, a subsection from a specific depth can be moved to another depth, multiple exercises can be exchanged among sections, and so on.
- *Context-sensitive help*: Through the respective menu option or by pressing F1, users can get help that is relevant to their task at hand.

The basic interface of the Writer (see Figure 50.4) consists of the following parts:

- *Main menu*: All the currently available functions can be accessed through this menu (e.g., open, save and close a book, search/replace, change interaction options, get help). Depending on the current context of use, some additional menus may appear.
- *Toolbar*: Supports quick and direct access to the most commonly used functions.
- *Book contents tree*: Provides an overview of the book's contents and also works as an easy and efficient navigation mechanism.
- *Current path (Where Am I?)*: Illustrates the path (i.e., the place in the book's hierarchy) of the piece of content that is currently being modified. For efficient use of space, only the numbering of the headings is included in the path and not their full titles. The current level is presented using a black, bold font. Previous levels are rendered in blue color and also work as hyperlinks, allowing users to easily move to higher levels.
- *Section title*: The title of the section being edited.
- *Sequential navigation buttons*: Three buttons used for browsing the book's contents in the same order that they are read by the Reader. One of them is used moving one level up in the book hierarchy, while the other two lead to the previous and next section respectively.
- *Content editor*: This is where the actual editing takes place. Its content and function changes dynamically depending on the type of the content (e.g., the book's table of contents, a specific section's contents). The content editor has a separate toolbar, which also adapts to the current context of use. When editing a section of the book, this area comprises the following components:
  - *Text editor*: Contains the actual text of the current section. The active sentence (i.e., based on where the

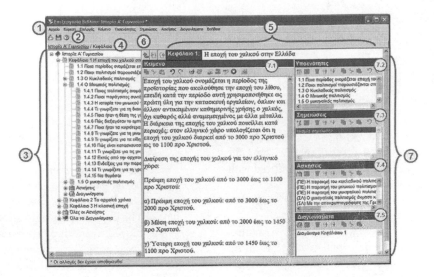

FIGURE 50.4 The user interface of the Starlight writer.

cursor is located) is highlighted. If a sentence contains a link, it is underlined. By moving the mouse over such a sentence, a tooltip appears describing the link's target. The text editor provides facilities for reading the current sentence, a piece of selected text, or the whole text, using the same voices and attributes as the Reader. This is a very useful feature, since very often alternative speech synthesizers render text in a different way, a fact that may result in changes in the written text. In addition to that, the Writer supports the use of alternative configuration files for the available speech synthesizers, thus allowing for respective speech adjustments.

- *Subsections*: A list containing all the subsections of the current section.
- *Notes*: A list of notes related to the current section.
- *Exercises*: Exercises for the current section. A single exercise may be shared among different sections and exams, thus allowing for reusability and easier maintenance.
- *Exams*: Available exams for the current section.

50.4 Evaluation

As mentioned earlier, usability and accessibility evaluation constituted an integral, inseparable part of the iterative development procedure. In addition to formative evaluation iterations, testing in real practice was performed. Some of the evaluation findings have already been mentioned in the previous section, in order to justify, explain, or support some of the presented interaction design decisions. This section refers to additional findings, some of which constitute pointers for future work and software extensions.

The Achilles' heel of Starlight, as of any other system that uses synthesized speech, is the quality of the synthetic voices. Although

all of the users that owned a screen reader admitted that the quality of the Greek speech synthesizer used in Starlight ("Ekfonitis+" by ILSP[2]) was much better than theirs, there still is a great distance to cover among synthetic speech and digitized recordings by professional narrators. The key related problem is the lack of intonation that has a negative impact on comprehension and also tires listeners after a certain period of time. Thus, when users were asked about what they thought needed to be improved, they almost unanimously responded "the voices."

Another problem of speech synthesizers relates to content creation. Each speech synthesizer reads the same text in a different way. Also, some are capable of using the context of a word to infer its correct pronunciation, while others do not. This practically means that, to create content that can be read consistently and correctly by all synthesizers, extensive manual processing of the material is required.

Another problem was related to the overall interaction, with an emphasis on user assistance. Nonvisual interaction can be quite challenging for novice users, mainly because, explicitly or implicitly, it relies on different states (or contexts). This is unavoidable, since sound (but also the Braille display) is a sequential medium, and thus only one thing can be presented at a time, unlike, for example, GUIs, where help and guidance messages can be presented in parallel with the main interface. Consequently, the user has to mentally keep track of the current context, and the user may get lost if a single change of state is not noticed. Starlight includes several methods to ensure that content changes are sufficiently highlighted, and that the user can easily retrieve the current context. Still, the main drawback of these solutions is that they are reactive, that is, the user has to do something (e.g., press the Help button, use the menu) to get some guidance, but in some cases the user may not be in a position to do that. Thus, what is additionally needed is to provide

---

[2] http://www.ilsp.gr/ekfonitis_plus_eng.html.

proactive methods (e.g., an intelligent agent) that by monitoring user activity (or inactivity) will be able to infer such problematic situations and take the initiative to remedy it.

In the same context, a popular user request was the provision of an interactive tutorial that would substantially facilitate learning how to use Starlight.

Finally, users suggested that the Starlight Reader should be also able to present:

- *Images*: Providing magnification and change of contrast options for those with deteriorated vision and hierarchical navigatable descriptions for the blind.
- *In-line sounds*: Inserted throughout the text to annotate or make it more interesting and vivid.
- *HTML pages*: A lot of educational texts are already available on the web, and users said that they would prefer to use Starlight to read them instead of a web browser with a screen reader; educators also stated that they would like to import them to existing e-books or to create new e-books.
- *Mathematics*: As they constitute a significant part of educational material for many lessons. A challenge here is how to render mathematics on a Braille display, since most of the time their presentation requires multiple Braille lines (see Chapter 51, "Mathematics and Accessibility: A Survey").

Overall, the opinion of all stakeholders (potential users, educators, specialists) about Starlight was very positive, as it is designed to support teachers and students by directly addressing existing, real needs and effectively addresses a substantial lack that schools with blind students faced up to now. Of course, there is still much room for improvements and extensions, but currently—at least for Greek schools—Starlight constitutes the only available fit solution, thus opening a new dimension to the educational process.

50.5 Conclusions and Future Work

This chapter has discussed the concept of electronic books and how they can be used to provide accessible educational textbooks to blind students. In this context, the Starlight platform for creating and reading such books was presented. More specifically, the user-centered development process followed was highlighted, as well as the interaction design characteristics and rationale of the Starlight Reader, as these were shaped after numerous iterations of evaluations both with experts and potential end-users.

In summary, the Starlight Reader builds upon the characteristics, findings, and recommendations of past related work, and extends it with novel features that on the one hand improve the usability and accessibility of electronic books, while on the other hand can considerably support the educational process by directly addressing specific related needs. Some of these features are:

- Dual user interface
- Sharing with registered users and individual configurations

- Single-key interaction
- Dual voice for separating the content from the user interface
- Bookmarking down to the sentence level
- Automatic section numbering
- Embedded on-demand notebook
- Supporting interactive exercises and examinations
- Adaptable functionality level tuned by educators
- Logging and statistics per student
- Integrated authoring support

The Starlight platform has been used for the development of eight related products. The educational content for these products was provided by Savalas Publications,[3] a leading publishing company for Greek educational books, and was adapted in order to be suitable for use by blind students, in cooperation with the Panhellenic Association of the Blind and specialized educators. This content was transformed into e-books using the Starlight Writer. Starlight e-books can work with any speech synthesis engine that supports Microsoft's Speech Application Programming Interface (SAPI). The aforementioned Starlight products also include the "Ekfonitis+" text-to-speech system developed by the Institute for Language and Speech Processing – ATHENA.

Future work includes developments to support:

- Text formatting, in-line images and sounds
- Importing e-books written in HTML, DAISY, and NIMAS
- Rendering mathematics visually on the screen, as well as through speech and Braille
- Nonvisual presentation and navigation in information-augmented images
- Deployment of the Starlight Reader on mobile devices such as 3G phones and personal digital assistants (PDAs)
- Creation of a dual user interface for the Writer, so that blind students can also edit the content of the books, or even create their own

References

Adobe (2004). *Adobe® Acrobat® 7.0 and Accessibility*. http://www.adobe.com/enterprise/accessibility/pdfs/acrobat7_accessibility_faq.pdf.

ANSI/NISO (2005). *Z39.86-2005 Specifications for the Digital Talking Book*. Bethesda, MD: NISO Press. http://www.daisy.org/z3986/2005/z3986-2005.html.

Barnicle, K. (2000). Usability testing with screen reading technology in a Windows environment, in the *Proceedings of the 2000 Conference on Universal Usability*, 16–17 November 2000, Arlington, VA, pp. 102–109. New York: ACM Press.

---

3 http://www.savalas.gr.

Clough, P. and Corbett, J. (eds.) (2000). *Theories of Inclusive Education: A Student's Guide*. London: Paul Chapman Publishing.

Ferreira, H. and Freitas, D. (2004). Enhancing the accessibility of mathematics for blind people: The AudioMath Project, in the *Proceedings of ICCHP 2004* (K. Miesenberger et al., eds.), 7–9 July 2004, Paris, pp. 678–685. Heidelberg: Springer-Verlag.

García, L. G. (2004). Text comprehension by blind people using speech synthesis systems, in the *Proceedings of ICCHP 2004* (K. Miesenberger et al., eds.), 7–9 July 2004, Paris, pp. 538–544. Heidelberg: Springer-Verlag.

Grammenos, D., Savidis, A., Georgalis, Y., Bourdenas, T., and Stephanidis, C. (2007). Dual educational electronic textbooks: The Starlight Platform, in the *Proceedings of the 9th International ACM SIGACCESS Conference on Computers and Accessibility (ASSETS 2007)*, 14–17 October 2007, Tempe, AZ, pp. 107–114. New York: ACM Press.

Graziani, P. and Arato, A. (1998). Hybrid book: A new perspective in the education and the leisure for the blind, in the *Proceedings of NTEVH 98: Telematics in the Education of the Visually Handicapped*, 2–3 June 1998, Paris. http://www.snv.jussieu.fr/inova/publi/ntevh/hybook.htm.

ISO (1999). *ISO 13407: Human-Centered Design Processes for Interactive Systems*. International Organization for Standardization.

Jain, A. and Gupta, G. (2006). VoxBoox: A system for automatic generation of interactive talking books, in the *Proceedings of the 8th International ACM SIGACCESS Conference on Computers and Accessibility (ASSETS '06)*, 23–25 October 2006, Portland, OR, pp. 275–276. New York: ACM Press.

Karshmer, A. I., Bledsoe, C., and Stanley, P. B. (2004). The architecture of a comprehensive equation for the print impaired, in the *Proceedings of ICCHP 2004* (K. Miesenberger et al., eds.), 7–9 July 2004, Paris, pp. 614–619. Heidelberg: Springer-Verlag.

Microsoft (2002). *Accessibility Tips*. http://www.microsoft.com/reader/info/support/pc/accessibility.asp.

Microsoft (2006). *Microsoft Reader Accessibility*. http://www.microsoft.com/reader/info/support/faq/accessibility.asp.

Morley, S. (1998). Digital talking books on a PC: A usability evaluation of the prototype DAISY playback software, in the *Proceedings of the 3rd International ACM Conference on Assistive Technologies (Assets '98)*, 15–17 April 1998, Marina del Rey, CA, pp. 157–164. New York: ACM Press.

Morley, S., Petrie, H., O'Neill A.-M., and McNally, P. (1999). Auditory navigation in hyperspace: Design and evaluation of a non-visual hypermedia system for blind users. *Behaviour & Information Technology* 18: 18–26.

Mynatt, E. D. (1997). Transforming graphical interfaces into auditory interfaces for blind users. *Human-Computer Interaction* 12: 7–45.

NFF (2005). *NIMAS Technical Specification, Version 1.0*. http://nimas.cast.org/about/report/index.html.

Nielsen, J. (1993). *Usability Engineering*. Boston: Academic Press.

Nielsen, J. (1994). Heuristic evaluation, in *Usability Inspection Methods* (J. Nielsen and R. Mack, eds.), pp. 25–62. New York: John Wiley & Sons.

NISO (1999). *Playback Device Guidelines: Prioritized List of Features for Digital Talking Book Playback Devices*. http://www.loc.gov/nls/z3986/background/features.htm.

Parente, P. (2006). Clique: A conversant, task-based audio display for GUI applications. *ACM SIGACCESS Accessibility and Computing*, 34–37.

Petrie, H., Morley, S., McNally, P., O'Neill, A-M., and Majoe, D. (1997). Initial design and evaluation of an interface to hypermedia systems for blind users, in the *Proceedings of Hypertext '97*, 6–11 April 1997, Southampton, U.K., pp. 48–56. New York: ACM Press.

Savidis, A. and Stephanidis, C. (1995). Developing dual user interfaces for integrating blind and sighted users: The HOMER UIMS, in the *Proceedings of the SIGCHI Conference on Human Factors in Computing Systems* (I. R. Katz, R. Mack, L. Marks, M. B. Rosson, and J. Nielsen, eds.), 7–11 May 1995, Denver, pp. 106–113. New York: ACM Press/Addison-Wesley.

Shubin, H. and Meehan, M. M. (1997). Navigation in web applications. *Interactions* 4: 13–17.

Smith, A., Cook, J., Francioni, J., Hossain, A., Anwar, M., and Rahman, M. (2004). Nonvisual tool for navigating hierarchical structures, in the *Proceedings of ASSETS 2004*, 18–20 October 2004, Atlanta, pp. 133–139. New York: ACM Press.

Stephanidis, C., Paramythis, A., Sfyrakis, M., and Savidis, A. (2001). A case study in unified user interface development: The AVANTI web browser, in *User Interfaces for All: Concepts, Methods, and Tools* (C. Stephanidis, ed.), pp. 525–568. Mahwah, NJ: Lawrence Erlbaum Associates.

Stevens, R. D., Edwards, A. D. N., and Harling, P. A. (1997). Access to mathematics for visually disabled students through multi-modal interaction. *Human-Computer Interaction* 12: 47–92.

WebAim (2006). *Accessibility Features in Acrobat Reader 7*. http://www.webaim.org/techniques/acrobat/reader.

Wharton, C., Rieman, J., Lewis, C., and Polson, P. (1994). The cognitive walkthrough method: A practitioner's guide, in *Usability Inspection Methods* (J. Nielsen and R. Mack, eds.), pp. 105–140. New York: John Wiley & Sons.

51

Mathematics and Accessibility: A Survey

Enrico Pontelli,
Arthur I. Karshmer,
and Gopal Gupta

51.1 Introduction

The study of mathematics is crucial in the preparation of students to enter careers in science, technology, engineering, and related disciplines such as the social and behavioral sciences. For many sighted students, math education poses a serious roadblock in entering technical disciplines, which has a serious impact on economic competitiveness and science-related capabilities. For visually impaired students, the roadblock is even higher, due to the additional difficulties they have to face in accessing mathematics.

Presentation of written information to blind individuals has traditionally been accomplished through the use of Braille. For presenting text, while Braille may not have been an ideal solution, it has certainly been a satisfactory one. Traditional Braille utilizes a raised character set composed of six dots per character, which limits the character set to 64 possible characters. Even for simple text, this does not represent an adequate alphabet. To solve the problem, most Braille notations use multicharacter representation. For example, "A" (capital a) is represented in American standard Braille by a sequence of two characters: ",a" (or ",a"), that is, the letter A preceded by a dot 6 (represented by a comma in North American ASCII Braille).

In recent years, there have been proponents of an eight-dot system, which could then allow an alphabet of 256 unique characters (Schweikjardt, 1998; Gardner, 2005a). For a variety of reasons that resemble the QWERTY vs. the Dvorak keyboard debate the eight-dot systems have not gained much popularity. While the text Braille issue is complex, it pales in comparison with the difficulties in the representation of math in Braille. While text is linear in nature, anything but the simplest math is not normally represented linearly. Note that Braille itself is also linear. Thus, the multitude of mathematical operators and special symbols has to be simulated by various sequences of Braille characters. Additionally, the spatially arranged structures have to be linearized so that they can be represented in Braille (e.g., the square root operation). All these requirements make math learning and teaching very complex.

In 1990 the United States passed landmark legislation, the Americans with Disabilities Act, to directly address a broad range of problems. Similar legislations have been passed by many other countries. These laws have significantly improved accessibility for people with disabilities but have had marginal success with math and the visually impaired. Other legislation such as The Education Act of 1973, The Telecommunications Act of 1996, and the aggressive implementation of Section 508 of The Education Act of 1973, have resulted in many positive changes. Unfortunately, the math education of the visually impaired has not seen any noticeable improvement.

In response, numerous projects have surfaced to bring closure to the problem. A simple Google search on the words "math blind," currently returns 1.4 million hits. The problem, however, is complex and dual-pronged: to make meaningful headway, the needs of both the students *and* their teachers have to be taken into account. Thus, any solution to the problem must make it easy for blind students to internalize ("visualize") mathematical expressions. Likewise, a good solution should not place undue burden on the teachers, in terms of preparing the material for the student, or having to learn substantial new material (e.g., learning a Braille math notation).

There have been several interesting projects over the years that have addressed these issues, and they will be discussed later in this chapter. While a generic solution has yet to be developed, progress is being made. With current interdisciplinary research projects, the underlying issues are being exposed by teaming mathematicians, cognitive psychologists, human computer interaction experts, and the user community.

51.2 A Classification of Math Accessibility Approaches

The goal in making mathematics accessible to a blind student is to ensure that an entire mathematical document is accessible. This means that not only individual mathematical expressions are accessible, but sequences of expressions that may arise, for example, in a proof, as well as text embedded between mathematical expressions, are also accessible. The approaches to making mathematics accessible can be classified into two types: *static* and *dynamic*.

- *Static approaches*: The mathematical content is statically converted into a format that is reproducible using assistive devices (such as Braille refreshable displays and other tactile devices) or that can be printed on Braille paper. In these approaches, the document is mostly viewed as a passive entity (akin to a printed document presented to a sighted user), while the active component is represented by the user, who uses an assistive device (e.g., a refreshable Braille display) to move around the document, reading parts, skipping other parts, backtracking through it, and so on.
- *Dynamic approaches*: The mathematical content is presented in a dynamic, interactive fashion. These approaches require a conversion process that allows the user to navigate the mathematical content in accordance with its mathematical structure. In this case, the document itself becomes an active component; by performing intelligent transformation of the document, its semantic structure is exposed and information overload on the user reduced.

Note that static approaches render mathematical expressions in Braille, while dynamic approaches render it using audio—alone or along with other traditional techniques, such as refreshable Braille. It should be noted that the two are complementary to each other. Experience indicates that just as the ability to read and write is important for sighted individuals (in addition to being able to hear things), the ability to write using Braille-based codes is likewise important for blind individuals as audio hearing is not enough, however interactive it may be (National Library of Service, 2000).

51.3 Static Approaches

51.3.1 Introduction

In the static approaches, sophisticated, special Braille notations have been developed for mathematics. Virtually every national Braille notation also has at least one special version for math. These math codes attempt to present complex mathematical expressions in a way that blind individuals can follow. For example, these notations include:

- The Nemeth Math code (Nemeth, 1972) is used in the United States, Canada, New Zealand, Greece, and India.
- The mathematical code in use in Germany and Austria, currently called the Marburg code (Epheser et al., 1992) was derived from the original invention of Scheid et al. (1930); see also Schweikhardt (2000) for further details.
- The French Math code (Commission Evolution du Braille Français, 2001).
- ItalBra (Biblioteca Italiana per Ciechi "Regina Margherita" O.N.L.U.S., 1998), the Italian Math code in use in Italy.
- The British Math Braille code (Braille Authority of the United Kingdom, 1987).

For a thorough survey of the geographical distribution of different Braille notations, the interested reader is referred to UNESCO (1990).

Unfortunately, these notations introduce their own problems. The Braille-based mathematical notations devised in various countries constitute a new language, and consequently must be learned by the students. Of course, this learning has to be supervised by teachers, who themselves need to be proficient in these notations. In the United States and many other countries, there is a paucity of math and science teachers (Lee, 2005) and especially trained special education teachers capable of teaching both math and complex Braille notations (Gill, 2006). With the broad acceptance of mainstreaming, and the resulting demise of special education centers for the visually impaired, the problem has become even more difficult. This is because a math teacher at a school, community college, or university may come across one or two blind students over a span of a few years, and thus does not have a very strong incentive to learn the math codes. The situation is somewhat alleviated through engagement of special education teachers who know the Braille math code and help teachers with blind students in their classes on a demand basis.

There are several national codes that have been designed for encoding mathematics in six-dot Braille. An overview of these codes is provided in the following. An overview of assistive technology tools that have been developed to make the task of learning and teaching mathematics for blind students and their teachers, respectively, is also provided.

51.3.2 Six-Dot Braille

The 6-dot Braille was invented by Louis Braille in the 1820s. It uses 6 dots arranged in 3 rows of 2 dots each, which are raised depending on the letter or symbol to be represented. There are 64 possible letters and symbols that can be represented by the 6-dot system (no raised dots represents a blank character, thus there are 63 letters/symbols where at least 1 dot is raised). The invention of Braille allowed blind individuals to read and write for the first time at speeds equal to or surpassing those of the sighted individuals in writing print text (Braille, 1829). What Braille did was to provide an alphabet to allow blind individuals to read and write. This leap opened doors to other possibilities, such as Braille-based code for mathematics, science, music, and so on. Braille-based codes for mathematics started appearing in the mid to late 1900s: for example, the Nemeth code, designed by blind mathematician Abraham Nemeth, was published in 1951, while the Spanish Math Braille code was developed in 1987.

51.3.2.1 Overview of Braille Mathematical Formats

As mentioned earlier, there are several standardized math codes developed by different countries (United States, United Kingdom, France, Italy, Spain, etc.). Most of these codes attempt to cover all of the mathematics that can be expressed in print math. These codes linearize mathematics. Note that any two-dimensional mathematical expression can be encoded in a linear sentence through the use of parentheses. Thus, the fraction $\frac{x+1}{x-1}$ can be represented as $(x + 1)/(x - 1)$. However, this process of linearization introduces too many parentheses, thus most codes will provide a wide variety of grouping symbols for indicating the beginning of a group as well as its end; for example, in Nemeth code the previous fraction is represented as ?x+1/x−1#, where "?" is the begin fraction indicator and "#" is the end fraction indicator. Using different grouping symbols for different categories of expressions is important for a linear notation that will be read left to right. If regular parentheses were used instead of "?" and "#," then the blind reader would know that he is reading a fraction only upon encountering the symbol "/." Using special symbols to indicate to the reader what he should anticipate is an important design feature while developing Braille math codes. An equally important feature is keeping the user aware of the context he is in at all times. This is because, at any given time, the user is focused on one symbol (the one on which his finger is placed) and the part to the right of the finger is not known to him at all (in contrast to a sighted user who can infer a lot of structural information in one glance), while the text to the left is in the reader's memory. Thus, in Nemeth code, for example, while writing expressions involving exponents, the level of the exponent has to be indicated explicitly each time. For example, the expression $x^{(z+2)^c} + y$ is coded as {x^{{z+2}^{c}+3}+y in LaTeX (Lamport, 1985), a popular package for math print typesetting, since to a sighted reader (or a computer) the braces indicate the scope of exponents. However, this does not work well in a Braille setting, as the reader will quickly forget the exponent level as he

moves left to right in the formula, and a considerable number of backtrackings of the finger will be needed to understand it. To make the context explicit, in Nemeth code this formula is coded as: x^z+2^^c^+3"+y. Note that the number of ^ indicates the exponent level of the expression that is to follow. The context awareness that is built into Braille math codes makes them extremely difficult to parse (Karshmer et al., 1998), however.

The various math Braille codes devised differ in

- The mapping of the Braille alphabet to ASCII as different countries use different standards; for example, unlike the American standard, there is no direct encoding of the + symbol in British math Braille code.
- The sequence of Braille characters used to denote various mathematical operators and special symbols; for example, the begin fraction and end fraction indicators.

To illustrate these differences, consider the expression

$$\frac{x+1}{x-1}$$

This expression has different encodings in different Braille national standards. The following (Archambault et al., 2005) are the encodings in some popular formats:

(Nemeth)

(French 1)

(French 2)

(ItalBra)

(Marburg)

(British)

Observe that French has two encodings: the first is the historical format, while the second makes use of the recent French Braille format, which was introduced so as to simplify the language to make it computer-processable. Markers are used to denote numerators and denominators. In all formats except for Nemeth code and ItalBra, the numerator and denominator blocks are the same; thus, a reader is aware of the fraction only when the fraction symbol is reached. Nemeth and ItalBra avoid the use of a fraction symbol by using different block markers. Not all math codes follow the design features mentioned earlier to the same degree.

Some Braille math codes also allow for spatially arranged structures, such as matrices, determinants, continued fractions, and grade school level arithmetic sum, multiplication, and division problems. Two examples of spatial arrangements (polynomial addition and a 3×2 matrix) for Nemeth code are shown in the following. Note that the sequence of "3"s is used in Nemeth code to draw a line needed in representing arithmetic sums, multiplication, long division, and so on.

	ASCII Nemeth	Nemeth Braille code
Polynomial Addition	$3x^2"+2x-25$ $+6x^2-5x$ 3333333333333	⠿⠿ ⠿⠿⠿⠿⠿⠿⠿ ⠿⠿⠿ ⠿⠿⠿⠿⠿⠿⠿⠿ ⠿ ⠿⠿⠿⠿⠿⠿⠿⠿⠿⠿⠿⠿⠿
3x2 Matrix	,{,a 0,) ,{0 ?1/,asin^2 .b#,) ,{0 ?1/,c#+?cos .b/2#,)	⠿⠿⠿ ⠿⠿⠿ ⠿⠿⠿ ⠿⠿⠿⠿⠿⠿⠿⠿⠿ ⠿⠿⠿⠿⠿ ⠿⠿⠿ ⠿⠿⠿⠿⠿⠿⠿⠿⠿⠿⠿ ⠿⠿⠿⠿⠿⠿

51.3.2.2 Translation and Back-Translation of Math Codes

Given the various math codes, one obvious problem faced is to convert print math into a specific math code (translation). Thus, tools need to be developed that achieve precisely this. Any embedded text should also be translated to text Braille. Typically, print math documents are prepared in LaTeX. Thus, the conversion task involves translating a LaTeX document to a Braille document where LaTeX math expressions are coded in math Braille code, while the embedded text is coded in text Braille. Translators have been built for translating LaTeX to various national codes. The most mature of these is the Scientific Notebook system that converts LaTeX to Nemeth code (Duxbury Systems, 2000). Other systems include those for converting to Marburg, to the French code, and to ItalBra.

Once print math and text has been translated into Braille for a blind student to read, another problem arises, namely, how does the sighted teacher read the answers prepared by a blind student, say, for homework or an exam? The blind student will typically write the answers in Braille. To facilitate this communication from student to teacher, tools are needed that will back-translate the math code and Braille text to LaTeX, so that it can be read and graded by a sighted instructor. The task of building a back-translator is significantly harder than the task of building a translator from LaTeX to Braille math, because typically the Braille math codes were designed to be back-translated by human translators, and thus have a large number of context-sensitive features that make computer processing and parsing extremely difficult. Consider the example of exponents again in Nemeth code. For instance, the expression $x^{(z+2)^c+3}$ will be coded as x^z+2^^c^+3 in Nemeth code. The exponent level of the expression c, denoted by the number of ^ symbols, is exactly one more than that of $(z+2)$. This exact relationship between the number of ^ symbols is hard to express in formal *context-free* grammars, which are traditionally used for building parsers. Due to the presence of context sensitivity, the problem of back-translation was widely considered unsolvable (Scadden, 1996), until Gupta, Karshmer, and Guo proposed advanced techniques based on *programming language semantics* and *logic programming* for achieving this back-translation (Karshmer et al., 1998; Annamalai et al., 2003). In fact, the French Braille math code was revised and changed to make it more computer processable and parsable (Moço and Archambault, 2003). A number of projects have attempted to build translators and back-translators for various math codes:

- *The Labradoor project* (Miesenberger et al., 1998) was started for converting LaTeX to Marburg, and subsequently back-translating Marburg to LaTeX (Batušić et al., 2003).
- *MAVIS* (Karshmer et al., 1998) was the pioneering project that showed how to solve the back-translation problem using language semantics and logic programming, and developed the first Nemeth Braille code to LaTeX back-translator.
- The *Insight* project (Karshmer et al., 1998; Annamalai et al., 2003; Gopal et al., 2007) further improved upon the MAVIS project to develop a complete system for back-translating math documents in Nemeth code with embedded text (in Grade II Braille) to LaTeX. The system takes, as its input, a scan of a Braille sheet (in JPG format), performs image processing to recognize the Braille dots, and produces the corresponding ASCII Braille file. The Nemeth code and Level II Braille are automatically identified and separated, and then each translated separately and merged to produce a single LaTeX file that a sighted person can then view.
- *Multi-Language Mathematical Braille Translator* (Moço and Archambault, 2003), uses the approach developed in the UMA (Universal Mathematics Access) project (Karshmer et al., 2004) to provide multilingual translation between the two French notations, the Marburg notation, and the Nemeth code. The approach is based on developing a common intermediate format in which the various notations can be translated, back and forth.
- *BraMaNet* (Schwebel, 2004) is a system that uses an XSL Style Sheet to translate MathML into French Math Braille code. It contains a user-friendly interface that can be used, in conjunction with MathType, to translate Microsoft Word documents into Braille for printing.
- *Math2Braille* (Crombie et al., 2004) is an open-source module to convert MathML 2.0 to the Braille standard used in the Netherlands.

Along with the context-sensitive nature of the specific math Braille codes (as mentioned, for example, in the discussion of the Nemeth code), the process of interconversion has to deal with the natural issues of ambiguity present in the syntactic representation of mathematics. For example, a formula like $\sum_i f(i)$ could be interpreted either as the product between two quantities ($\sum_i$ and $f(i)$) or as a summation; these two interpretations would lead to distinct Nemeth encodings, respectively:

$$⠠⠮ ⠋⠦⠊⠴ \quad \text{and} \quad ⠨⠠⠮⠋⠦⠊⠴$$

Further examples of ambiguity have been discussed in the literature. For example, the *dotless* Braille resources web site[1] analyzes the complexity of properly interpreting occurrences of the symbol x:

---

[1] http://www.dotlessbraille.org.

- x used to represent dimensions (e.g., "2×2 cm"), typically encoded in Braille using the word *by*
- x used as a multiplication symbol, encoded in Nemeth by the two-cell (dot 4 followed by dots 1–6)
- x used to denote degree of magnification (e.g., "$2\times$ zoom"), that should be transcribed as the letter indicator followed by the x symbol (i.e., in ASCII "#2|$\times$ zoom")

51.3.2.3 Universal Libraries

The previously described approaches highlight the inherent difficulty arising from the great variety existing in the digital formats (e.g., MathML, OpenMath), typographical formats (e.g., LaTeX) and Braille formats (e.g., Nemeth, Marburg) adopted to describe mathematical content. Furthermore, in many settings (e.g., educational applications) it is interesting to allow translation between arbitrary pairs of formats, as in the following examples:

- Translation from MathML to Nemeth Braille is required for students accessing mathematical content deployed on the web.
- Translation from LaTeX or MathML to Nemeth Braille is required to allow an instructor to distribute homework and notes to visually impaired students.
- Translation from Nemeth Braille to LaTeX is required to allow a student, who is typesetting his solutions using a Braille typewriter or a Braille embosser, to submit the material to the instructor.
- Translation of a mathematical document written in one Braille-based math code (say Marburg) to another Braille-based math code (say Nemeth) is necessary to facilitate communication between blind scholars, mathematicians, and engineers.

The research conducted by the iGroup UMA[2] directly addressed this problem. This work hinges on designing a common intermediate format to which various notations (for sighted as well as nonsighted) can be translated to, back and forth, as the basis of translating one notation to another. As part of this project, Palmer and Pontelli (2003) proposed OpenMath as a common intermediate format and as a basis to develop two-directional translation tools between OpenMath and the other relevant formats (e.g., Nemeth Braille, Marburg, LaTeX). The work of iGroup UMA successively evolved in a full-blown and open source *Universal Math Conversion library* (Archambault et al., 2004, 2005). The emphasis in this project is to develop an open library, easily expandable, that supports multimodal presentation of mathematical content and that can be transparently used by nonexpert users. The library provides a single API for developers of applications necessitating conversions between different mathematical formats. The library relies on a central format, and the format chosen is MathML Presentation. For each additional format, the library provides two modules (the *input* module and the *output*

module) aimed at converting to and from MathML. The library maintains, in particular, internal Braille tables describing each national Braille format. The library is highly portable and accessible by both Windows and Linux applications.

51.3.3 Mathematics-Specific Braille Extensions

The use of 6-dot Braille for the encoding of mathematical content has been widely criticized. The 6-dot format can only produce 63 different Braille cells (observe that in 6-dot Braille an unused cell or blank cell is implicitly considered a space). As a consequence, multiple sequences of Braille cells have to be used to encode distinct symbols, and many Braille cells have different meanings depending on the context in which they appear in. These types of designs have an unfortunate consequence—formats like Nemeth Braille become *context-sensitive languages* (Hopcroft et al., 2000), which are inherently hard and expensive to parse and translate.

For this reason, there have been many attempts to expand Braille codes to 8-dot formats, which allow a larger number of Braille cells (256), simplifying the encoding of a larger set of symbols and creating a better consistency with the 8-bit standard ASCII character set used on most computers. Transcription systems for publisher-provided electronic mathematics books into an 8-dot format were reported as early as 1985 (i.e., the work of W. Schweikhardt to produce mathematical content in Stuttgart Math Notation; Schweikhardt, 1985).

A number of official and semiofficial general 8-dot Braille codes have been developed, mostly in Europe, but relatively little literature has been reproduced in any 8-dot Braille code. In the context of representation of mathematical content, two relevant proposals targeting 8-dot representation of mathematical content are DotsPlus (Gardner, 2003) and LAMBDA (Edwards et al., 2006).

DotsPlus Braille gracefully extends 6-dot grade 1 standard Braille, with the exception that most double cell characters of the 6-dot font are single cells in the 8-dot font. Capital letters, for example, are encoded with an extra dot (dot 7 position) on the left side of the row below the bottom of the standard 6-dot lowercase letter cell. The novelties of DotsPlus include:

- Most punctuation marks are not Braille, but small graphic symbols; similarly, most of the characters from complex literature (e.g., mathematical symbols) are encoded as graphic symbols shaped similarly to the corresponding print symbols.
- Numbers are encoded in a single cell; the digits from 1 to 9 are encoded as in the literary Braille number mode with an additional dot-6 (the digit 0 has a distinct encoding to avoid conflict with the letter "w").

The outcome is a notation that is easier to learn and remember for the representation of mathematical content. Unusual symbols are encoded in a shape analogous to the print format,

---

[2] http://karshmer.lklnd.usf.edu/~igroupuma/index.html.

$$ax^2 + bx + c = 0$$

FIGURE 51.1 A simple equation in print format and DotsPlus format.

allowing visually impaired individuals to follow the same learning/remembering process as a sighted reader, and avoid having to memorize complex sequences of Braille cells. Furthermore, DotsPlus Braille reduces the context dependence of individual symbols in a Braille line, leading to the ability to print math in standard print format (including positional placements of symbols). Figure 51.1, drawn from Gardner (2002), shows the print format and the DotsPlus format of a simple equation. The success of DotsPlus is also related to the ability to use the Tiger Tactile Graphics and Braille Embosser (Gardner et al., 2006), which includes printer drivers for DotsPlus fonts.

The LAMBDA project introduced a linear math notation, the LAMBDA code (Edwards et al., 2006; Fogarolo, 2006; Schweikhardt et al., 2006), largely inspired by the way mathematical content is encoded in MathML. Thus, the LAMBDA code plays the role of a markup language—it provides markers to denote special types of expressions and representation of common symbols—expressed in an 8-dot Braille notation, to provide linear encoding of mathematical content. The LAMBDA code is meant to be an internal source code for the representation of mathematics, and tools have been investigated to translate LAMBDA code to national Braille formats (e.g., Italian Braille) for output purposes.

For example, the representation of a fraction requires three markers: one denoting the beginning of the fraction, one representing the end of the numerator, and one representing the end of the fraction. For example, the fraction

$$\frac{a+1}{b+1}$$

would be represented as <start fraction> a+1 <fraction symbol> b+1 <end fraction>. The three markers have different 8-dot Braille representations depending on the specific country; the previous fraction, in U.K. 8-dot LAMBDA code, would appear as (Fogarolo et al., 2005):

where ▓▓ represents the <start fraction>, ▓▓ represents the <end fraction>, and ▒▒ represents the fraction symbol.

The use of 8-dot Braille code, with its 256 possible combinations of dots, implies that many markers and symbols require multiple Braille cells. The format structure adopted in the LAMBDA code tries to appeal to logical constructions to facilitate the interpretation and recollection; in particular, many symbols are constructed using prefixes, which identify the class the symbol belongs to. For example, Greek symbols are constructed by a fixed prefix followed by the Braille code for the corresponding Latin letter; the symbol ζ is represented as:

As another example, the symbol $\cup$ (set union) is encoded as

where the first cell is a prefix denoting set operations, and the second cell is the traditional encoding of +.

The LAMBDA code is supported by a sophisticated LAMBDA editor, which recognizes and handles the hierarchical structure of a mathematical expression, automatically manages the different blocks composing it, and allows for different forms of visualization (e.g., it allows hiding the content of the blocks).

51.3.4 Other Tactile Approaches

- ViewPlus has developed a variety of tools to enhance the mathematical experience of visually impaired individuals. One of the notable products is the Accessible Graphing Calculator (AGC) (Walsh et al., 2001; Gardner et al., 2006), which offers the traditional features of a graphing calculator but with the ability to produce both aural presentation (as a varying frequency sound) and tactile presentation (using Tiger embosser) of two-dimensional graphs.

- A recent improved method for determining significant boundaries in images encoded in scalable vector graphics (SVG) has been investigated (Krufka and Barner, 2005) with the goal of producing more effective outputs on the Tiger embosser; this can be effectively used for presentation of mathematical content, thanks to the recent studies of encoding of mathematical formulae using SVG (Stanley et al., 2004).

51.4 Dynamic Approaches

51.4.1 Foundations of Math Presentation

51.4.1.1 Cognitive Foundations

Relatively little work has been done in investigating the cognitive aspects of human interaction with mathematical content. The most relevant effort in this area has been reported in Barraza et al. (2004). This work studies the perceptual and cognitive processes used by sighted individuals during equations reading; the ultimate goal is to understand these processes to the extent of being able to develop aural presentation mechanisms that provide visually impaired individuals with equivalent process capabilities. The investigation in Barraza et al. (2004) was conducted using think-aloud protocols and eye-tracking devices, and it involved separate experiments aimed at measuring

- *The capability of recalling equations*: The experiment consisted of exposure to different equations for varying periods of times, followed by presentation of distracting screens.
- *The impact of knowledge of the structure of the equation*: The experiment exposed the subjects to a preview of the structure of the equation, and measured whether this affected the time to solve the equation.
- *The extent to which subjects use a chunking approach in reading an equation*: They decompose the equation into chunks, defined by expressions within parentheses, solve such subexpressions, and store their outcome in memory.

The outcome of this study can be summarized as follows:

- The reading process is mostly a left-to-right process, moving one element at a time (similarly to the way standard text is read).
- An initial scan is typically performed to acquire the structure of the equation before proceeding to its detailed understanding.
- Readers backtrack very frequently when understanding/ solving an equation.
- Readers tend to process operators and numbers more deeply than parentheses.
- Readers chunk together parts of an equation (especially the content of a subexpression within parentheses) and process the chunk modularly; as a consequence of chunking, the readers tend to solve the equation hierarchically.

These observations are also consistent with the cognitive and user studies conducted to investigate user errors in accessing mathematics using Braille, as reported in Cahill et al. (1996).

51.4.1.2 Prosody and Speaking of Mathematics

Significant research efforts have been invested in understanding the effectiveness of different approaches to the actual speaking of mathematics.

The majority of the approaches to aural presentation of mathematical content rely on two key schemes for denoting the structure within a mathematical expression: lexical and prosody cues/indicators.

Use of lexical indicators explicitly denotes structural information through additional spoken components. For example, an expression of the type $\sqrt{x+1}$ would require the use of lexical indicators of the type "begin square root" and "end square root," leading to a possible utterance of the form "begin square root x plus one end square root." Seminal work in the investigation of lexical indicators has been conducted by Chang (1983). This approach has been widely adopted in many of the tools for aural presentation of mathematics, and it has been recognized as particularly effective when dealing with the major structure of the expression. On the other hand, various studies (e.g., Baddeley, 1992; Stevens et al., 1997) have highlighted negative aspects of lexical indicators, mostly associated with the overload imposed on the reader's working memory.

Use of prosodic indicators are modifications of features of the spoken output to capture changes in the presented structure. These may include pauses, modifications of pitch and tempo, rhythm and tone. Seminal work on the use of prosodic cues in the presentation of algebraic content has appeared in O'Malley et al. (1973), which discovered the correlation between pauses and syntactic boundaries at operators, fractions, parentheses, and so on. This investigation was refined to consider pitch, duration, and amplitude by Streeter (1978). A richer set of rules for the use of prosody cues in algebraic presentations has been presented in Stevens (1996). The various studies have also drawn a distinction between speech prosody cues and nonspeech ones. The studies of Blattner et al. (1989); Brewster et al. (1992); and Stevens (1996) suggest the use of *algebraic earcons*, which associate nonspeech sounds to different constructs—for example, a complex fraction is associated with two long notes with constant pitch separated by two silent beats (Stevens et al., 1997).

The more recent work of Fitzpatrick (2002) and Fitzpatrick et al. (2006) argues for the effectiveness of using only speech prosody; the proposal draws a parallel between the structure of mathematical expressions and the composition of English sentences, and analogous use of pausing and speaking rate to aurally capture the nesting structure of the expression. Fitzpatrick's effort aims at the design of standardized prosodic effects, instead of ad hoc approaches used by individual human readers. Fitzpatrick also advocates the combined use of prosody and lexical clues to better deal with complex expressions (Fitzpatrick et al., 2006), and to overcome limitations of existing speech synthesizers.

Another interesting overview of some of the issues connected to speaking mathematics has been presented in Fateman (2006), with considerations relative to the transition between speech synthesis and speech recognition of mathematics.

51.4.2 Presentation and Navigation Tools

51.4.2.1 AsTeR

Audio system for technical readings (AsTeR) (Raman, 1994) is a system to reformat electronic documents, typeset in LaTeX, to produce audio documents.

The first step in the AsTeR design originates from the development of a document representation model, aimed at making explicit the logical structure of the document; the model relies on attributed trees. Mathematical content is encoded in a *quasi-prefix* form, where the prefix form of the expression is enriched by *visual* attributes (e.g., left superscript, left subscript, accent, underbar); the objective is to delay the assignment of a semantics to the mathematical content, by preserving those visual components that make possible its (possibly ambiguous) presentation. AsTeR contains a recursive descent parser to extract quasi-prefix forms from the mathematical content of a LaTeX document, enriched with a complete precedence table of mathematical operators and heuristic rules to handle ambiguous notations (e.g., to ensure that an expression like $\sin 2n\pi$ is correctly interpreted as

$\sin(2n\pi)$ instead of $\sin(2)\cdot n\cdot\pi$, as the strong precedence of function application would suggest).

AsTeR introduces a rule-based language, called audio formatting language (AFL), used to map the internal document representation to audio. Rules in AFL perform transformations of the audio state, which is composed of a speech state, a sound state, and a pronunciation state. The language allows blocks, multithreads, and synchronization between modifications to different components of the state. Statements from AFL are grouped into *rendering rules* for each object of the document tree, and groups of rules can be themselves grouped into a *rendering style* (akin to the notion of style sheet). Multimodality is achieved by activating and deactivating different rendering styles.

The rendering style provided for mathematical content is built on the principle of minimizing verbosity, mostly through the use of fleeting and persistent cues, and modifications of intonation and voice inflection. In particular:

- Nesting in the expression is reflected by a change in voice that feels like falling off into the distance, along with pauses around the subexpression (whose length is dependent on the complexity of the subexpression).
- Higher and lower pitch voices are used to reflect superscripts and subscripts.
- In absence of additional information, the visual attributes are presented in the order subscript, superscript, underbar, accent, left-subscript, left-superscript.

Parenthesized expressions are conveyed by combining the change in tone with a persistent sound cue.

AsTeR also relies on the principle (described in Section 51.4.1.1) of allowing readers to acquire the top-level structure of a mathematical formula before accessing the subexpressions (chunking). This process is realized in AsTeR via *variable substitution*; this is automatically applied when the complexity of the subexpressions is sufficiently large. For example, given an expression $\frac{e_1}{e_2}$ where e_1 and e_2 are complex expressions, this would be rendered as "fraction x over y, where x is ... and y is ...."

AsTeR provides active browsing capabilities. Mathematical content can be navigated as a tree structure, with the ability of moving across siblings, as well as marking nodes and returning to marked nodes. Since the mathematical content is enriched by the visual attributes, the navigation includes specific commands for accessing such attributes. Upon entering a node, an immediate summarization (which includes a summary of the content—typically the main operator) and a contextual description (which recalls the siblings) is presented.

51.4.2.2 MathGenie

MathGenie is a comprehensive equation reading program that presents the visually impaired student with verbal renderings of the equation under study along with refreshable Nemeth Braille code output (Stanley and Karshmer, 2006). By design, the system offers a number of different techniques for browsing equations, each of which presents the structure as well as the content of the mathematics. The system was designed to run on virtually any computer running Windows 2000 or later without the aid of any other software package.

For the teacher, no knowledge of Braille is required to generate materials for visually impaired students. Through the use of any equation editor that generates MathML, the teacher can prepare materials for both sighted and visually impaired students. The MathML output of the equation editor is the input to the MathGenie equation browser. Figure 51.2 shows a snapshot of the MathGenie interface.

51.4.2.3 Design Principles and Navigation Strategies

The design of MathGenie involved research in domains, beyond the obvious computing and human-computer interaction (HCI) issues. For the first time, cognitive psychological research was employed to understand the key issues in reading and understanding mathematical equations (Karshmer and Gillan, 2003; Gillan et al., 2004). The results of the studies indicated several important issues associated with reading equations to a visually impaired student. Among the findings were: (1) casual reading of equations is highly prone to error on the part of the reader and therefore the listener, (2) structural components such as fractions, radicals, summations, parenthetical expressions, integrations, and so on are critical components and should be the foci in equation reading, (3) the preliminary glimpse of an equation seems to offer little in the process of equation cognition, and (4) the student must have the ability to navigate equations in a variety of ways. These results became the foundation of the design of MathGenie.

MathGenie relies on the use of lexical clues to represent the structure of the mathematical content. The browsing process provides the following key features (Karshmer et al., 2002):

- The default reading proceeds left to right along the expression; the access to the components of the expression can be realized either at the level of symbols or at the level of words.
- The expression can be presented in an abstract way, to highlight the hierarchical structure and abstract away small subexpressions. For example, an expression of the type $a_i = \int \frac{\sqrt{a+b}}{a-b} x^2 dx$ could be abstractly presented as
  - A subscript i equals to
  - Limit integral with
    - Lower limit square root of something, and
    - Upper limit square root of something
  - Of something dx

The depth of navigation to provide an abstract presentation is dependent on the specific symbols used in the equation.

- Users are allowed to evaluate subexpressions and replace them with the corresponding value.

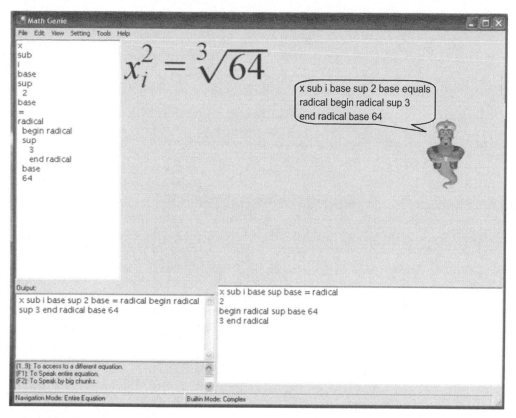

FIGURE 51.2 Snapshot of the MathGenie user interface.

- The parse tree of the expression is visible, can be traversed in various directions, including the ability to repeatedly navigate the same subexpressions.

Another important feature of the navigation process supported by MathGenie is the ability to voicemark expressions. A voicemark, similar to a voice anchor (Reddy et al., 2004), is an aural bookmarking of subexpressions, where the user can vocally assign a bookmark to any subexpression during the navigation process. Bookmarking subexpressions can be useful in understanding and internalizing expressions, allowing a step-by-step navigation, and allowing users to jump at will between voicemarked expressions—effectively allowing them to implement their own navigation strategy and customize the abstraction of expressions. Work is in progress to introduce voicemarking in the MathGenie infrastructure.

51.4.2.4 MathGenie's Architecture

Figure 51.3 shows the overall architecture of MathGenie (Karshmer et al., 2004). Input is provided in the form of (Presentation) MathML, and it offers synchronized display and aural presentation; MathGenie includes the possibility of providing speech input to control navigation, and the ability to control a refreshable Braille display.

To increase the usability of MathGenie, the browser uses a table-driven speech technique. This means that before any component is spoken, it is first found in the table for the local national language, and its correct pronunciation is then sent to the speech engine. This again permits the use of inexpensive hardware and software tools, since the least expensive speech engine can be employed and no special voice engines are required. To date, MathGenie supports English, French, and Spanish. The system includes editing tools to allow local language support to be added.

MathGenie offers synchronized aural and graphical presentation of the equation components being accessed. The graphical rendering is aimed at enhancing accessibility for individuals with low vision (by providing magnification of the equation being presented) and dyslexia (by providing color-contrasted highlighting). The visual rendering is achieved by mapping MathML to SVG.

Accessibility is further enhanced by connection to an online dictionary of mathematical terms, accessible through a simple keyboard shortcut during navigation of a mathematical expression.

51.4.2.5 Browsing Mathematics via VoiceXML

Another approach to interactively listening to and navigating mathematics has been proposed by Reddy et al. (2005).

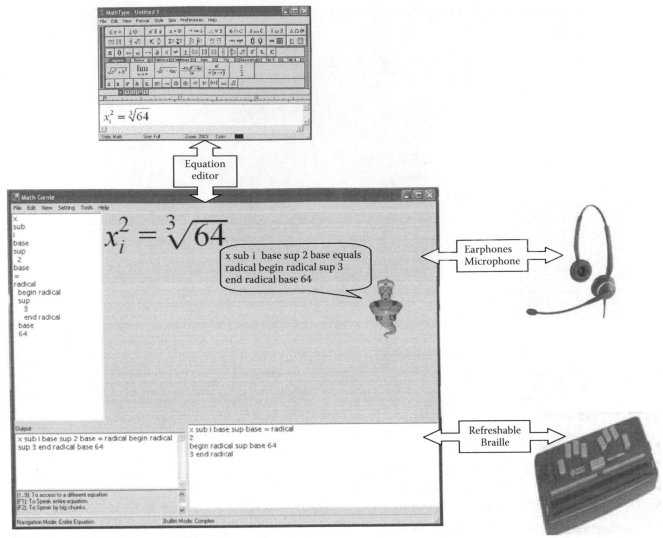

FIGURE 51.3 External architecture of the MathGenie.

This approach is based on translating an expression coded in MathML to VoiceXML (Oshry et al., 2006). They extend VoiceXML (Reddy, Annamalai, and Gupta, 2004) to make it dynamically navigable via audio. A user can give navigational commands through voice, thus no keyboard is involved in the interaction.

VoiceXML is an XML-based markup language designed for creating audio/voice-based documents. Just as a web browser renders HTML documents visually, a VoiceXML interpreter renders VoiceXML documents aurally. The VoiceXML technology has been widely adopted in the telecom industry for automated handling of calls.

In Gupta et al.'s approach, mathematical expressions coded in MathML are mechanically translated to VoiceXML via XSLT, a transformation language for XML. The VoiceXML document can then be heard with a voice browser by a visually impaired individual. However, merely hearing the VoiceXML document is not enough, since VoiceXML provides very little navigational control to a user while listening to a document. In fact,

navigational control in present versions of VoiceXML is completely controlled by the author of the VoiceXML page (Reddy et al., 2004). To make navigation more interactive and under the listener's control, the VoiceXML document is further enhanced with voice anchors. VoiceXML documents enhanced with voice anchors allow listeners to place speech labels on subexpressions of a mathematical expression. These speech labels can be used by listeners to move around the various subexpressions via simple voice utterances and voice commands. Thus, the mathematical formula can be browsed much more effectively. Note that in Reddy et al.'s approach, in contrast to other browsers for math formulas (such as MathGenie), the listener interacts with the documents and aurally browses math expressions through voice input rather than through keyboard input.

51.4.2.6 MathPlayer

MathPlayer (Soiffer, 2005) has been proposed by Design Science Inc. MathPlayer originates as a browser plug-in for the visualization of MathML content; in its successive versions, MathPlayer includes speech rules based on lexical cues (making use of explicit start/stop markers of complex operators), as well as interfaces with screen readers (e.g., JAWS). In a style similar to MathGenie, MathPlayer allows the user to navigate the more complex mathematical expressions. MathPlayer supports two types of navigation:

- *Text-based navigation*: It allows the user to move linearly through the textual representation of the expression, by navigating words that are spoken.
- *Tree-based navigation*: It allows the navigation of the tree structure of the expression, in a fashion analogous to MathGenie.

A rule library is provided to minimize the number of words used to produce an unambiguous presentation of an expression; for example, an operator applied to *single token* operands is spoken without the need of markers denoting the start and end of the expression. An expression like a/b is spoken simply as "a over b" instead of "begin fraction a fraction symbol b end fraction."

The navigation is synchronized with highlighting of the parts of the expression being accessed—a feature that is important for students with certain types of learning disabilities.

51.4.2.7 Other Systems

The MATHS (Mathematical Access for Technology and Science for Visually Disabled Users) project (Weber and Stevens, 1996) provided the foundation for many of the proposals discussed in this section. The project aimed at the development of interactive workstations (the MATHS workstations) providing speech and nonspeech audio access to mathematical formulae as well as Braille representation. The speech presentation of formulae includes the use of the correct prosody (for English and Flemish) and the ability to obtain *audio glances* of complex formulae, achieved by associating different sound instruments to different classes of terms. Braille output can be produced in Flemish Braille and Stuttgart Math notation. The MATHS workstations include also an integrated editor (the M3 Editor) for the typesetting of mathematical content.

The AudioMath project (Ferreira and Freitas, 2004) proposes a tool to provide aural presentation of mathematical content encoded in MathML Presentation. The AudioMath prototype relies on a database of rules for conversion of MathML to text (to be spoken) and the use of prosodic marks to provide pauses and modulations required to reduce ambiguity.

REMathEx (Gaura, 2002) is another sophisticated math reader based on MathML Presentation. The program views expressions according to their tree structure and the user can navigate the tree structure. The navigation process relies on the alternation of different types of activities.

- *Global expression preview*: This activity leads to the complete aural presentation of an expression up to a user-determined depth of its expression tree.
- *Selection of current node*: The user can position himself on any node of the frontier of the expression tree visited during an expression preview; this will cause the presentation of the subexpression rooted at such node.
- *Local expression preview*: This is analogous to the global expression preview, except that it is applied to the subexpression represented by the node the user is currently positioned at.

The aural presentation presents an expression up to a given depth; the elements below the set depth limit are presented using a subexpression substitution mechanism (e.g., if the subexpression below the depth limit is a/b then the system will replace it with the word *fraction*). The substitution is avoided if the complexity of the underlying expression is very small. Time delays are also used in the aural presentation to enhance the separation between components of the expression.

MathTalk (McClellan, 2007) is another tool that employs voice to facilitate learning of mathematics. It uses sophisticated voice recognition software to recognize mathematical expressions spoken by a blind user. In conjunction with Scientific Notebook, these expressions can be recorded in LaTeX, and then converted to Nemeth code.

The MathSpeak initiative (gh LLC, 2006) provides a grammar and lexicon for the preparation of mathematical material for digital books, along with a software tool, the gh PLAYER, which offers aural and visual presentation.

51.5 Beyond Reading

51.5.1 Creating and Editing Accessible Mathematics

51.5.1.1 The Infty Project

The Infty project is mainly aimed at enhancing the accessibility of mathematical content existing in printed form. The Infty system is composed of different tools, interacting via exchange of content in XML format. The key components are the following:

- *InftyReader* (Fukuda, Ohtake, and Suzuki, 2000): This tool makes use of advanced OCR technology to recognize the structure of mathematical formulae from printed text; the recognition process proceeds by first creating a network linking the symbols in the expression by labeled edges (where each label denotes the type of relation between the symbols, e.g., subscript, superscript, etc.), and then using an algorithm to detect the spanning tree of the network representing the structure of the expression.
- *InftyEditor* (Suzuki et al., 2004): This tool is a typesetting tool for scientific documents, typically used to edit the output of InftyReader. Indeed, the tool maintains in a window the image of the original printed document for comparison purposes. Mathematical content can be edited by reasoning about it in terms of its LaTeX representation. The editor includes the facility to produce an aural presentation of the expression (based on simple insertion of markers to unambiguously represent the structure of the expression).

- *Infty converters* (Suzuki et al., 2004): The output of InftyEditor is stored in an XML format (called IML). The Infty project has led to the development of converters from IML to various output formats, such as LaTeX, MathML, Unified Braille Codes, and Japanese Braille codes.
- *ChattyInfty* (Suzuki, 2005; Komada et al., 2006): This tool is an advanced component for aural presentation, along the lines of the MathGenie System (Stanley and Karshmer, 2006).

51.5.1.2 WinTriangle

WinTriangle (Gardner, 2005b) is an RTF scientific word processor, usable by both sighted and visually impaired individuals, capable of both displaying and voicing text and symbols, and providing access to a collection of markup symbols (Triangle fonts) that can be used to linearly encode virtually any scientific expression; the notation makes use of markers for the nonlinear components of the expression, for example,

$$\frac{a^2}{b}$$ is encoded as the sequence of symbols xxax(2)xxxb x

WinTriangle allows the editing and aural presentation of documents. In addition, the WinTriangle project has been enriched with an additional component, called LateX2Tri (Thompson, 2005), which converts LaTeX to RTF with Triangle fonts.

51.5.1.3 Other Proposals

- REMathEx (Gaura, 2002) is a system developed to provide the ability to read and edit mathematical expressions, encoded using MathML Presentation format. Expressions are viewed according to their parse-tree structure. The user can access the expression in its integral format or in *preview* mode, where the tree is navigated only up to a certain depth. The editing capabilities are limited to the substitution of the current subexpression (i.e., the node of the parse tree the user is currently positioned at) with a new expression. The editing is guided by an interactive dialogue, which provides aural menus for the choice of operators to insert in the expression.
- The BlindMath project (Pepino et al., 2006) integrates Scientific Work Place[3] and JAWS[4] to provide accessibility of document preparation using LaTeX to express mathematical content.

51.5.2 Working with Mathematics

The use of mathematical assistive technologies in the educational context raises the problem of moving from simple access to mathematical content to the actual *manipulation* of mathematics (i.e.,

doing mathematics, carrying out calculations, and solving exercises). This prompted researchers to explore the development of assistive environments for the manipulation of mathematics. The most advanced proposal in this area is the work of Stöger et al. (Stöger et al., 2004, 2006). Manipulation of mathematics involves tasks well beyond the simple understanding of a mathematical formula; it requires the ability to segment the formula, copy and transform parts of it, and maintain referential access to distinct parts of a formula. The Mathematical Working Environment (MaWEn) is designed to enhance formula navigation with the following features:

- Ability to identify subexpressions according to the semantic structure of the mathematical content (e.g., given the beginning of a parenthesized expression, automatically detect the complete subexpression enclosed by parentheses)
- Ability to mark and recall arbitrary subformulae as well as to conceal/reveal their presence on request
- Ability to replace subformulae with new expressions (e.g., simplify a subexpression to its value)
- Support for scratch paper (where the user can temporarily copy parts of expressions to be manipulated) and ability to call a calculator
- Libraries to apply global transformations to complete expressions (e.g., allow changing a sign globally in a parentheses expression instead of manually copying and negating each individual term)
- Intelligent replacement strategies (e.g., when evaluating an equation, each unknown has to be consistently substituted with the corresponding value)

51.6 Open Problems and Perspectives for the Future

51.6.1 Localization Problems

The previous efforts on translation between mathematical formats for mathematics have already addressed the presence of significant differences in notations across national boundaries.

The differences go beyond the specific notations, as distinct approaches, traditions, and methodologies affect the way mathematics is seen and understood. This implies that approaches to presentation of mathematics should include localization components and should customize the delivery to the specific national standards.

Relatively limited work has been conducted to enable presentation methodologies to apply localization to its output. The Universal Math Conversion library (Archambault et al., 2004) has been developed with the specific intent of facilitating the interoperation between Braille mathematical documents expressed using different national math Braille notations, including the development of a canonical subset of MathML specifically designed to facilitate the interconversion process (Archambault and Moço, 2006). The LAMBDA project (Edwards

---

[3] http://www.mackichan.com.

[4] http://www.freedomscientific.com/fs_products/software_jaws.asp.

et al., 2006) lays the foundations of the design of its eight-dot code on understanding the peculiarities of the different national codes used in different European countries. LAMBDA defines an open-separator-close tag structure to linearly represent mathematical structures. To be actually usable, the tag structures have to be concretized according to the national conventions (this was done for each country participating in the LAMBDA consortium). The instantiation process requires the following components:

- The specification of all dot configurations for the symbols, which are represented on one Braille cell (e.g., lowercase/ uppercase Latin letters, etc.).
- The definition of the notations to be described. All the mathematical notations used in an educational curriculum have to be linearly described.
- The assignment to each tag of a name to be displayed in the list of tags in the mathematical editor. The name depends on the national languages.
- The assignment to each tag of at least two names to be used in the preparation of the speech output. They are necessary to process the string to be read by the speech synthesizer.

Each localization of LAMBDA is encoded in an XML structure, to enable the LAMBDA-related tools (e.g., the mathematical editor) to easily retrieve information and set the local working environment.

MathGenie (Karshmer et al., 2004) has also been designed with language localization capabilities—its design isolates the vocabulary as a user-accessible table, and the current distribution includes both English and Spanish tables.

Nevertheless, true accessibility requires adaptation of presentation based not only on language and notation, but also on learning styles and teaching methodologies. A few studies have been conducted concerning country-specific issues in math accessibility (Kobolkova and Lecky, 2002; Meyer and Jung, 2002), but clearly more work is required to fill this gap.

51.6.2 The Importance of Context

The majority of the research on math accessibility conducted so far concentrates on enabling accessibility of math at the *formula* level. This implies that the presentation is based on analyzing exclusively the individual formula currently under consideration. This approach has inherent limitations; most formulae can only be properly interpreted and presented according to the context in which they appear. For example, a formula like $\pi(x + 1)$ could be interpreted as

- A permutation of the numbers from 0 to x+1
- The product of π (=3.14159...) and x+1
- The application of the function π to the value x+1

Relatively limited work has been conducted to incorporate *context* in the interpretation and presentation of mathematics. A first step in this direction has been discussed in Palmer and

Pontelli (2003), which addresses the accessibility of OpenMath (Caprotti et al., 2000). Contrary to other digital formats for the representation of mathematical content (e.g., MathML and LaTeX), OpenMath draws the operators used in the construction of mathematical formulae from content dictionaries (CDs), which are topic-specific collections of operator definitions—and this allows the analysis and presentation of formulae to have a clear reference to the defining CDs, which provide the appropriate semantics of each operator. The use of CDs introduces also a property of openness and extensibility in the representation of mathematical content, providing the ability to introduce operators and semantics as needed.

Additional precision in the translation process, and a better connection between semantics and presentation, can be gained by taking advantage of the context in which the formula is used (Pontelli and Palmer, 2004). This opens the problem of:

- Determining ways to represent relevant components of the document containing the formulae; for example, formulae appear as part of definitions, theorems, proofs, examples, etc.
- Determining ways to relate relevant document components to the corresponding formulae.
- Determining ways to relate documents to classes of documents. For example, documents should be related to classes of documents discussing the same topic (e.g., chapters of a book should be related to the topic of the book, such as a book on foundations of statistics).

An approach to handling context in mathematical content has been proposed by De Carvalho (2005). The author investigates a multipurpose mathematical notation, based on attributed grammars, to enable the dynamic creation and modification of the mapping between syntax and semantics. This framework could provide an adequate framework for handling flexible context-dependent interpretation of mathematical formulae.

Another natural answer to part of the previously mentioned problem comes from the recent work on OMDoc (Kohlhase, 2001; Kohlhase and Franke, 2001), which provides a natural markup framework for the encoding of mathematical concepts (e.g., lemmas, propositions, statements, proofs). The currently investigated directions are:

- Identification of the relevant document components within presentation formats (e.g., a Braille+Nemeth document) and their explicit representation within OMDoc
- Use of the knowledge encoded in the OMDoc documents to facilitate the translation process—the OMDoc information can be used to provide missing components of the semantics of the formula (e.g., the context provided by OMDoc might determine that the formula is a mathematical logic statement, thus clearly identifying the set of CDs to be used to generate an OpenMath version of the formula).

The use of OMDoc allows also for drawing knowledge from the existing rich online formalizations of mathematics (e.g., as in MBase; Kohlhase and Franke, 2001). On the other hand, the capabilities of OMDoc are limited to the representation of a single document or part of a document. It lacks the capability to describe global properties of a document (e.g., the topic of a document) or global relationships between components of the document. Even worse, the format does not address the problem of relating between and across different documents. Approaches like those presented in MBase, aimed at the creation of a repository of mathematical facts, are not adequate to the needs of accessibility of mathematics, since they are mostly meant to help automated theorem proving systems and not to directly assist human documents. This state of things suggests not only the need for better analysis algorithms, but also the need for a high-level markup language that builds on OMDoc and extends its expressive capabilities to the encoding of whole mathematical documents and organized collections of documents.

51.6.3 Integration of Components

The previous discussion highlights the complexity of providing nonvisual access to mathematical content. In particular, it is clear that a number of dependent issues have to be effectively addressed, ranging from interoperation between representation formats, adoption of different assistive devices, and modalities of presentation/navigation. These issues are strongly dependent on each other, forcing researchers to address the whole spectrum of options and issues. This has led in recent years to the development of broad projects that span a variety of aspects of accessibility of mathematical content in an integrated fashion. Two notable efforts in this direction are represented by the LAMBDA Project[5] and the International Universal Mathematics Accessibility Group (iGroup UMA).[6]

The LAMBDA project was funded by the European Union with the objective of creating an integrated system for both writing and reading mathematical content for the benefit of blind students. The strategic structure of the project relies on the development of an editor to write mathematical expressions in a linear way, capable of interacting with different assistive devices, and of a linear mathematical code, designed to be interoperable with various existing formats for mathematics (including formats used by advanced systems for the manipulation of mathematics, e.g., Mathematica). The international collaboration involves teams from the United Kingdom, Italy, Germany, Portugal, Greece, Spain, and France.

The iGroup UMA (Karshmer et al., 2004) is an international cooperation involving researchers from Ireland, Austria, France, Japan, and the United States. The goal of the cooperation is to develop tools to enhance accessibility of mathematical content across different representation formats (e.g., Braille and digital formats), across national styles and conventions, and across different levels of visual capabilities. Figure 51.4 shows the overall organization of the project. The project provides a uniform

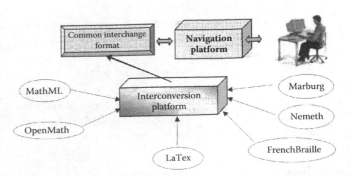

FIGURE 51.4 System Organization, UMA Project.

interconversion platform that allows two-directional translation between a wide range of formats for the representation of mathematics. The interconversion platform relies on the use of a common interchange format (MathML in the case of the current version of the project), which is used as a bridge between any pair of formats considered. The project provides also tools for the rendering and navigation of mathematical content; currently this component is realized within the previously described MathGenie.

51.7 Conclusion

The problems associated with teaching mathematics to visually impaired students are longstanding and difficult. With the advent of the reasonably inexpensive computer, many partial solutions have been put forward, as described in the body of this work. Each has added knowledge required to make incremental advances in the state of the art.

Additionally, the problems associated with math and blindness are now being studied from an interdisciplinary perspective, adding needed fundamental understanding to the design and development process. The data generated by these efforts have been invaluable to the understanding of the problem's nature and potential solution.

Finally, and most important, is the advent of real international cooperation in understanding the dimensions of the problem and their potential solution. As most governmental units do not view the problem as one of great import, and the obvious lack of potential wealth to be derived from solving the problem, the international research community has banded together to make essential breakthroughs in the past decade. This effort is to be applauded and encouraged.

The future looks brighter than ever before. With new research in basic life sciences and understanding of the brain and nervous system, we should see dramatic advances in the near future. The authors look to a bright future in which their colleagues in scientific disciplines will be playing on a level field regardless of disabilities.

Acknowledgments

The authors wish to acknowledge the collaboration and the suggestions of A. Annamalai, D. Archambault, M. Batušić,

---

[5] http://www.lambdaproject.org.

[6] http://karshmer.lklnd.usf.edu/~igroupuma/index.html.

C. Bledsoe, D. Fitzpatrick, D. Gillan, D. Gopal, H. Guo, S. Jolly, K. Miesenberger, V. Moço, B. Palmer, H. Reddy, P. Stanley, and B. Stöger. This research was partially supported by National Science Foundation Grants 0220590 and 0420407.

References

Annamalai, A., Gopal, D., Gupta, G., Guo, H., and Karshmer, A. I. (2003). INSIGHT: A comprehensive system for converting Braille based mathematical documents to LaTeX, in *Universal Access in HCI: Inclusive Design in the Information Society, Volume 4 of the Proceedings of the 10th International Conference on Human-Computer Interaction (HCI International 2003)* (C. Stephanidis, ed.), 22–27 June 2003, Crete, Greece, pp. 1245–1249. Mahwah, NJ: Lawrence Erlbaum Associates.

Archambault, D., Batušić, M., Berger, F., et al. (2005). The universal maths conversion library: An attempt to build an open software library to convert mathematical contents in various formats, in *Human-Computer Interfaces: Concepts, New Ideas, Better Usability, and Applications, Volume 3 of the Proceedings of the 11th International Conference on Human-Computer Interaction (HCI International 2005)* (M. J. Smith and G. Salvendy, eds.), 22–27 July 2005, Las Vegas [CD-ROM]. Mahwah, NJ: Lawrence Erlbaum Associates.

Archambault, D., Fitzpatrick, D., Gupta, G., Karshmer, A. I., Miesenberger, K., and Pontelli, E. (2004). Towards a universal maths conversion library, in the *Proceedings of the 9th International Conference on Computers Helping People with Special Needs (ICCHP 2004)*, 7–9 July 2004, Paris, France, pp. 664–669. Berlin/Heidelberg: Springer-Verlag.

Archambault, D. and Moço, V. (2006). Canonical MathML to simplify conversion of MathML to Braille mathematical notations, in the *Proceedings of the 10th International Conference on Computers Helping People with Special Needs (ICCHP 2006)*, 11–13 July 2006, Linz, Austria, pp. 1191–1198. Berlin/Heidelberg: Springer-Verlag.

Baddeley, A. D. (1992). *Your Memory: A User's Guide*. London: Penguin Books.

Barraza, P., Gillan, D. J., Karshmer, A. I., and Pazuchanics, S. (2004). A cognitive analysis of equation reading applied to the development of assistive technology for visually-impaired students, in the *Proceedings of the Human Factors and Ergonomics Society 48th Annual Meeting (HFES 2004)*, 20–24 September 2004, New Orleans, pp. 922–926. Santa Monica, CA: Human Factors and Ergonomics Society.

Batušić, M., Miesenberger, K., and Stöger, B. (2003). Parser for the Marburg Mathematical Braille Notation NIDRR Project: Universal math converter, in *Universal Access in HCI: Inclusive Design in the Information Society, Volume 4 of the Proceedings of the 10th International Conference on Human-Computer Interaction (HCI International 2003)* (C. Stephanidis, ed.), 22–27 June 2003, Crete, Greece, pp. 1260–1264. Mahwah, NJ: Lawrence Erlbaum Associates.

Biblioteca Italiana per Ciechi "Regina Margherita" O.N.L.U.S. (1998). *Codice Braille Italiano*. [Italian Braille Code]. Monza, Italy.

Blattner, M., Surnikawa, D., and Greenberg, R. (1989). Earcons and icons: Their structure and common design principles. *Human Computer Interaction* 4: 11–44.

Braille Authority of the United Kingdom (1987). *Braille Mathematics Notation*. Mathematics Committee. http://www.bauk.org/docs/bmn.pdf.

Braille, L. (1829). *Method of Writing Words, Music, and Plain Songs by Means of Dots, for Use by the Blind and Arranged for Them* (in French).

Brewster, S. A., Wright, P., and Edwards, A. D. N. (1992). A detailed investigation into the effectiveness of earcons, in *Auditory Display, Sonification, Audification and Auditory Interfaces, Proceedings of the First International Conference on Auditory Display* (G. Kramer, ed.), 11–12 April 1992, Vancouver, Canada, pp. 471–498. New York: Addison Wesley.

Cahill, H., Linehan, C., and McCarthy, J. (1996). Blind and partially sighted students' access to mathematics and computer technology in Ireland and Belgium. *Journal of Visual Impairment and Blindness* 90: 105–114.

Caprotti, O., Carlisle, D. P., and Cohen, A. M. (2000). *The OpenMath Standard*. The OpenMath Esprit Consortium.

Chang, L. A. (1983). *Handbook for Spoken Mathematics (Larry's Speakeasy)*. Berkeley, CA: Lawrence Livermore Laboratory, The Regents of the University of California.

Commission Evolution du Braille Français (2001). *Notation Mathematique Braille*, mise a jour de la notation mathematique de 1971. [Commission for evolution of French Braille (2001), Braille mathematical notation, update of the mathematical notation for Braille 1971]. Paris: Department of Transcription and Publishing, National Institute for Blind Youth. http://handy.univ-lyon1.fr/MH/bramanet/download/nouvelleNormeMath.pdf.

Crombie, D., Lenoir, R., McKenzie, N., and Barker, A. (2004). math2braille: Opening access to mathematics, in the *Proceedings of the 9th International Conference on Computers Helping People with Special Needs (ICCHP 2004)*, 7–9 July 2004, Paris, pp. 670–677. Berlin/Heidelberg: Springer-Verlag.

De Carvalho, J. W. (2005). *Mathematics as Game of Types*. PhD dissertation, University of Western Ontario, London, Canada.

Duxbury Systems (2000). *MegaMath Translator for MegaDots*.

Edwards, A. D. N., McCartney, H., and Fogarolo, F. (2006). Lambda: A multimodal approach to making mathematics accessible to blind students, in the *Proceedings of the 8th International ACM SIGACCESS Conference on Computers and Accessibility (ASSETS 2006)*, 23–25 October 2006, Portland, OR, pp. 48–54. New York: ACM Press.

Epheser, H., Pograniczna, D., and Britz, K. (1992). *Internationale Mathematikschrift für Blinde* [International Mathematics Notation for Blind], in Marburger Systematiken der Blindenschrift (Teil 6) (J. Hertlein and R. F. V. Witwe, eds.). Marburg, Germany: Deutsche Blindenstudienanstalt.

Fateman, R. (2006). *How Can We Speak Math? The Evolution of Mathematical Communication in the Age of Digital Libraries.* Minneapolis: University of Minnesota.

Ferreira, H. and Freitas, D. (2004). Enhancing the accessibility of mathematics for blind people: The AudioMath Project, in the *Proceedings of the 9th International Conference on Computers Helping People with Special Needs (ICCHP 2004)*, 7–9 July 2004, Paris, pp. 678–685. Berlin/Heidelberg: Springer-Verlag.

Fitzpatrick, D. (2002). Speaking technical documents: Using prosody to convey textual and mathematical material, in the *Proceedings of the 8th International Conference on Computers Helping People with Special Needs (ICCHP 2002)*, 15–20 July 2002, Linz, Austria, pp. 494–501. Berlin/Heidelberg: Springer-Verlag.

Fitzpatrick, D. (2006). Mathematics: How and what to speak, in the *Proceedings of the 10th International Conference on Computers Helping People with Special Needs (ICCHP 2006)*, 9–11 July 2006, Linz, Austria, pp. 1199–1206. Berlin/Heidelberg: Springer-Verlag.

Fogarolo, F. (2006). Maths and blind students: The LAMBDA project, CSA Vicenza, in *Handbook for the LAMBDA Maths Editor vs. 3.35* (F. Fogarolo, C. Bernareggi, F. Coppi, and E. Bortolazzi, eds.).

Fogarolo, F., Bernareggi, C., Coppi, F., and Bortolazzi, E. (2005). *Handbook for the LAMBDA Maths Editor vs. 3.35* (3rd ed., IST-2001-37139). http://www.lambdaproject.org.

Fukuda, R., Ohtake, N., and Suzuki, M. (2000). Optical recognition and Braille transcription of mathematical documents, in the *Proceedings of the 7th International Conference on Computers Helping People with Special Needs (ICCHP 2000)*, 17–21 July 2000, Karlsruhe, Germany, pp. 711–718. Wien: Österreichische Computer Gesellschaft.

Gardner, J. (2002). Access by blind students and professionals to mainstream math and science, in the *Proceedings of the 2002 International Conference on Computers Helping People with Special Needs (ICCHP'02)*, 15–20 July 2002, Linz, Austria, pp. 207–212. Berlin/Heidelberg: Springer-Verlag.

Gardner, J. (2003). DotsPlus Braille tutorial: Simplifying communication between sighted and blind people, in the *Proceedings of the 2003 CSUN International Conference on Technology and Persons with Disabilities (CSUN 2003)*, 17–22 March 2003, Los Angeles. http://www.csun.edu/cod/conf/2003/proceedings/284.htm.

Gardner, J. (2005a). *Introduction to DotsPlus® Braille.* http://dots.physics.orst.edu/dotsplus.html.

Gardner, J. (2005b). *WinTriangle: A Scientific Word Processor for the Blind.* http://www.wintriangle.com.

Gardner, J., Ungier, L., and Boyer, J. J. (2006). Braille math made easy with the Tiger Formatter, in the *Proceedings of the 10th International Conference on Computers Helping People with Special Needs (ICCHP 2006)*, 9–11 July 2006, Linz, Austria, pp. 1215–1222. Berlin/Heidelberg: Springer-Verlag.

Gaura, P. (2002). REMathEx: Reader and editor of the mathematical expressions for blind students, in the *Proceedings of the 8th International Conference on Computers Helping People with Special Needs (ICCHP 2002)*, 15–20 July 2002, Linz, Austria, pp. 486–493. Berlin/Heidelberg: Springer-Verlag.

gh LLC (2006). *Welcome to the MathSpeak Initiative.* http://www.ghmathspeak.com.

Gill, J. (2006). *Information Resources for People Working in the Field of Visual Disabilities.* London: Royal National Institute for the Blind.

Gillan, D. J., Barraza, P., Karshmer, A. I., and Pazuchanics, S. (2004). Cognitive analysis of equation readings: Application to the development of the MathGenie, in the *Proceedings of the 9th International Conference on Computers Helping People with Special Needs (ICCHP 2004)*, 7–9 July 2004, Paris, pp. 630–637. Berlin/Heidelberg: Springer-Verlag.

Gopal, D., Wang, Q., Gupta, G., Cnitnis, S., Guo, H., and Karshmer, A. I. (2007). Towards completely automatic backtranslation of Nemeth Braille Code, in *Universal Access in Human-Computer Interaction. Applications and Services, Proceedings of the 4th International Conference on Universal Access in Human-Computer Interaction, Volume 7 of the Combined Proceedings of HCI International 2007* (C. Stephanidis, ed.), 22–27 July 2007, Beijing, pp. 309–318. Berlin/Heidelberg: Springer-Verlag.

Hopcroft, J. E., Motwani, R., and Ullman, J. D. (2000). *Introduction to Automata Theory, Languages, and Computation.* Boston: Addison Wesley.

Karshmer, A. I., Bledsoe, C., and Stanley, P. (2004). The architecture of a comprehensive equation browser for the print impaired, in the *Proceedings of the 9th International Conference on Computers Helping People with Special Needs (ICCHP 2004)*, 7–9 July 2004, Paris, pp. 614–619. Berlin/Heidelberg: Springer-Verlag.

Karshmer, A. I. and Gillan, D. (2003). How well can we read equations to blind mathematics students: Some answers from psychology, in *Universal Access in HCI: Inclusive Design in the Information Society, Volume 4 of the Proceedings of the 10th International Conference on Human-Computer Interaction (HCI International 2003)* (C. Stephanidis, ed.), 22–27 June 2003, Crete, Greece, pp. 1290–1294. Mahwah, NJ: Lawrence Erlbaum Associates.

Karshmer, A. I., Gupta, G., Geiger, S., and Weaver, C. (1998). Reading and writing mathematics: The MAVIS project, in the *Proceedings of the Third International ACM Conference on Assistive Technologies (ASSETS 1998)*, 15–17 April 1998, Marina del Rey, CA, pp. 136–143. New York: ACM Press.

Karshmer, A. I., Gupta, G., and Gillan, D. J. (2002). Architecting an auditory browser for navigating mathematical expressions, in the *Proceedings of the 8th International Conference on Computers Helping People with Special Needs (ICCHP 2002)*, 15–20 July 2002, Linz, Austria, pp. 477–485. Berlin/Heidelberg: Springer-Verlag.

Karshmer, A. I., Gupta, G., Pontelli, E., et al. (2004). UMA: A system for universal mathematics accessibility, in the *Proceedings* of the 6th International ACM SIGACCESS Conference on Computers and Accessibility (ASSETS 2004), 18–20 October 2004, Atlanta, pp. 55–62. New York: ACM Press.

Kobolkova, M. and Lecky, P. (2002). Experience with access to mathematics for the blind students in Slovakia, in the *Proceedings of the 8th International Conference on Computers Helping People with Special Needs (ICCHP 2002)*, 15–20 July 2002, Linz, Austria, pp. 510–511. Berlin/Heidelberg: Springer-Verlag.

Kohlhase, M. (2001). OMDOC: Towards an Internet standard for the administration, distribution and teaching of mathematical knowledge, in the *Proceedings (revised papers) of the International Conference on Artificial Intelligence and Symbolic Computation (AISC 2001)*, pp. 32–52. London: Springer-Verlag.

Kohlhase, M. and Franke, A. (2001). MBase: Representing knowledge and context for the integration of mathematical software systems. *Journal of Symbolic Computation* 32: 365–402.

Komada, T., Yamaguchi, K., Kawane, F., and Suzuki, M. (2006). New environment for visually disabled students to access scientific information by combining speech interface and tactile graphics, in the *Proceedings of the 10th International Conference on Computers Helping People with Special Needs (ICCHP 2006)*, 9–11 July 2006, Linz, Austria, pp. 1183–1190. Berlin/Heidelberg: Springer-Verlag.

Krufka, S. E. and Barner, K. E. (2005). Automatic production of tactile graphics from scalable vector graphics, in the *Proceedings of the 7th International ACM SIGACCESS Conference on Computers and Accessibility (ASSETS 2005)*, 9–12 October 2005, Baltimore, pp. 166–172. New York: ACM Press.

Lamport, L. (1985). *LaTeX: A Document Preparation System*. Boston: Addison Wesley.

Lee, C. (2005). Filling the math/science teacher void. *UCLA Today* 26. http://www.today.ucla.edu/2005/050816news_mathscience.html.

McClellan, N. (2007). *The MathTalk System*. http://www.metroplexvoice.com/tech_notes.htm.

Meyer, E. and Jung, M. (2002). LaTeX at the University of Applied Sciences Giessen-Friedberg: Experiences at the Institute for Visually Impaired Students, in the *Proceedings of the 8th International Conference on Computers Helping People with Special Needs (ICCHP 2002)*, 15–20 July 2002, Linz, Austria, pp. 508–509. Berlin/Heidelberg: Springer-Verlag.

Miesenberger, K., Batušić, M., and Stöger, B. (1998). Labradoor: A contribution to making mathematics accessible for the blind, in the *Proceedings of the 6th International Conference on Computers Helping People with Special Needs (ICCHP 1998)*, 31 August–4 September 1998, Vienna, Austria, pp. 307–315. Wien: Österreichische Computer Gesellschaft.

Moço, V. and Archambault, D. (2003). Automatic translator for mathematical Braille, in *Universal Access in HCI: Inclusive Design in the Information Society, Volume 4 of the Proceedings of the 10th International Conference on Human-Computer Interaction (HCI International 2003)* (C. Stephanidis, ed.), 22–27 June 2003, Crete, Greece, pp. 1335–1339. Mahwah, NJ: Lawrence Erlbaum Associates.

National Library of Service (2000). *Braille: Into the Next Millennium*. Washington, DC: Library of Congress.

Nemeth, A. (1972). *The Nemeth Braille Code for Mathematics and Science Notation*. Louisville, KY: American Printing House for the Blind.

O'Malley, M. M., Kloker, D., and Dara-Abrams, B. (1973). Recovering parentheses from spoken algebraic expressions. *IEEE Transactions on Audio and Electroacoustics* 21: 217–220.

Oshry, M., Auburn, R. J., Baggia, P., Bodell, M., Burke, D., and Burnett, D. C. (2006). *Voice Extensible Markup Language*. W3C Working Draft. Cambridge, MA: W3 Consortium.

Palmer, B. and Pontelli, E. (2003). Experiments in translating and navigating digital formats for mathematics (a progress report), in *Universal Access in HCI: Inclusive Design in the Information Society, Volume 4 of the Proceedings of the 10th International Conference on Human-Computer Interaction (HCI International 2003)* (C. Stephanidis, ed.), 22–27 June 2003, Crete, Greece, pp. 1320–1324. Mahwah, NJ: Lawrence Erlbaum Associates.

Pepino, A., Freda, C., Schettino, A., Ferraro, F., Pagliara, S., and Zanfardino, F. (2005). Lecture notes in CS, in *"BlindMath": A New Scientific Editor for Blind Students*, pp. 1171–1174. Berlin/Heidelberg: Springer-Verlag.

Pontelli, E. and Palmer, B. (2004). Translating between formats for mathematics: Current approach and an agenda for future developments, in the *Proceedings of the 9th International Conference on Computers Helping People with Special Needs (ICCHP 2004)*, 7–9 July 2004, Paris, pp. 620–625. Berlin/Heidelberg: Springer-Verlag.

Raman, T. V. (1994). *Audio Systems for Technical Reading*. PhD dissertation, Cornell University, Ithaca, NY.

Reddy, H., Annamalai, N., and Gupta, G. (2004). Listener-controlled dynamic navigation of VoiceXML documents, in the *Proceedings of the 9th International Conference on Computers Helping People with Special Needs (ICCHP 2004)*, 7–9 July 2004, Paris, France, pp. 347–354. Berlin/Heidelberg: Springer-Verlag.

Reddy, H., Gupta, G., and Karshmer, A. I. (2005). Dynamic aural browsing of MathML documents with VoiceXML, in *The Management of Information: E-Business, the Web, and Mobile Computing, Volume 2 of the Proceedings of the 11th International Conference on Human-Computer Interaction (HCI International 2005)* (M. J. Smith and G. Salvendy, eds.), 22–27 July 2005, Las Vegas [CD-ROM]. Mahwah, NJ: Lawrence Erlbaum Associates.

Scadden, L. (1996). Making mathematics and science accessible to blind students through technology, in the *Proceedings of RESNA '96 Conference*, January 1996, Salt Lake City, UT. http://www.dinf.ne.jp/doc/english/Us_Eu/conf/resna96/page51.htm.

Scheid, F. M., Windau, W., Zehme, G. (1930). *System der Mathematik- und Chemieschrift für Blinde* [Mathematics and chemical notation for the blind]. Marburg, Germany: Deutsche Blindenstudienanstalt.

Schwebel, F. (2004). *BraMaNet: Logiciel de traduction des mathématiques en Braille.* [BraMaNet: Software translations of mathematics in Braille]. http://handy.univ-lyon1.fr/projects/bramanet.

Schweikhardt, W. (1985). Rechnerunterstützte übertragung einer mathematischen formelsammlung in die Stuttgarter Mathematikschrift für Blind [Computer-based transcription of a mathematical collection of formula into Stuttgart Maths Notation ("e")], in the *Proceedings of the 5th International Workshop on Computerised Braille Production* (J. M. Ebersold, T. Schwyter, and W. A. Slaby, eds.), 30 October–1 November, 1985, Winterthur, Switzerland. Eichstatt, Germany: Katholische Universitat.

Schweikhardt, W. (1998). *Stuttgarter mathematikschrift für blinde* [Stuttgart mathmatics notation for the blind]. Universität Stuttgart, Institut für Informatik http://www.vis.uni-stuttgart.de/~schweikh/webseiteSMSB.

Schweikhardt, W. (2000). Requirements on a mathematical notation for the blind, in the *Proceedings of the 7th International Conference on Computers Helping People with Special Needs (ICCHP 2000)*, 17–21 July 2000, Karlsruhe, Germany, pp. 661–670. Vienna: Österreichische Computer Gesellschaft.

Schweikhardt, W., Bernareggi, C., Jessel, N., Encelle, B., and Gut, M. (2006). LAMBDA: A European system to access mathematics with Braille and audio synthesis, in the *Proceedings of the 10th International Conference on Computers Helping People with Special Needs (ICCHP 2006)*, 11–13 July 2006, Linz, Austria, pp. 1223–1230. Berlin/Heidelberg: Springer-Verlag.

Soiffer, N. (2005). Advances in accessible web-based mathematics, in the *Proceedings of the 2005 CSUN International Conference on Technology and Persons with Disabilities (CSUN 2005)*, 14–19 March 2005, Los Angeles. http://www.csun.edu/cod/conf/2005/proceedings/2407.htm.

Stanley, P. and Karshmer, A. I. (2006). Translating MathML into Nemeth Braille code, in the *Proceedings of the 10th International Conference on Computers Helping People with Special Needs (ICCHP 2006)*, 11–13 July 2006, Linz, Austria, pp. 1175–1182. Berlin/Heidelberg: Springer-Verlag.

Stanley, P. B., Bledsoe, C., and Karshmer, A. I. (2004). Utilizing scalable vector graphics in the instruction of mathematics to the print impaired student, in the *Proceedings of the 9th International Conference on Computers Helping People with Special Needs (ICCHP 2004)*, 7–9 July 2004, Paris, pp. 626–629. Berlin/Heidelberg: Springer-Verlag.

Stevens, R. D. (1996). *Principles for the Design of Auditory Interfaces to Present Complex Information to Blind People.* PhD dissertation, University of York.

Stevens, R. D., Edwards, A. D. N., and Harling, P. A. (1997). Access to mathematics for visually disabled students through multi-modal interaction. *Human-Computer Interaction* 12: 47–92.

Stöger, B., Batušić, M., Miesenberger, K., and Haindl, P. (2006). Supporting blind students in navigation and manipulation of mathematical expressions: Basic requirements and strategies, in the *Proceedings of the 10th International Conference on Computers Helping People with Special Needs (ICCHP 2006)*, 11–13 July 2006, Linz, Austria, pp. 1235–1242. Berlin/Heidelberg: Springer-Verlag.

Stöger, B., Miesenberger, K., and Batušić, M. (2004). Mathematical working environment for the blind: Motivation and basic ideas, in the *Proceedings of the 9th International Conference on Computers Helping People with Special Needs (ICCHP 2004)*, 7–9 July 2004, Paris, pp. 656–663. Berlin/Heidelberg: Springer-Verlag.

Streeter, L. A. (1978). Acoustic determinants of phrase boundary representation. *Journal of the Acoustical Society of America* 64: 1582–1592.

Suzuki, M. (2005). *Infty Project: About ChattyInfty.* http://www.inftyproject.org/download/AboutChattyInftyE.txt.

Suzuki, M., Kanahori, T., Ohtake, N., and Yamaguchi, K. (2004). An integrated OCR software for mathematical document and its output with accessibility, in the *Proceedings of the 9th International Conference on Computers Helping People with Special Needs (ICCHP 2004)*, 7–9 July 2004, Paris, pp. 648–655. Berlin/Heidelberg: Springer-Verlag.

Thompson, D. M. (2005). LaTeX2Tri: Physics and mathematics for the blind or visually impaired, in the *Proceedings of the 2005 CSUN International Conference on Technology and Persons with Disabilities (CSUN 2005)*, 14–19 March 2005, Los Angeles. http://www.csun.edu/cod/conf/2005/proceedings/2552.htm.

UNESCO (1990). *World Braille Usage, National Library Service for the Blind and Physically Handicapped.* Washington, DC: Library of Congress.

Walsh, P., Lundquist, R., and Gardner, J. A. (2001). The audio-accessible graphing calculator, in the *Proceedings of the 2001 CSUN International Conference on Technology and Persons with Disabilities (CSUN 2001)*, 19–24 March 2001, Los Angeles. http://www.csun.edu/cod/conf/2001/proceedings/0129walsh.htm.

Weber, G. and Stevens, R. D. (1996). Integration of speech and Braille in the MATHS workstation, in the *Proceedings of the 5th International Conference on Computers Helping People with Special Needs (ICCHP 1996)*, 17–19 July 1996, Linz, Austria, pp. 617–625. Munich: R. Oldenbourg Verlag GmbH.

52

Cybertherapy, Cyberpsychology, and the Use of Virtual Reality in Mental Health

Patrice Renaud,
Stéphane Bouchard,
Sylvain Chartier, and
Marie-Pierre Bonin

52.1 Cybertherapy and Cyberpsychology

The term *cybertherapy* is a compound of two Greek roots: *kybernetes* (control) and *therapeia* (care). Cybertherapy may be defined as the field where communication and information technology is employed to support and improve health care services for various clinical populations (Satava, 1995; Riva et al., 2006). The origin of cybertherapy, both historically and technologically speaking, plunges into cybernetics—a science developed by a generation of accomplished scientists, including Norbert Wiener, who was considered the leading cyberneticist during the 1940s and 1950s. The world of cybernetics had a strong influence on the development of the first functional computer, invented in large part by John von Neumann, who worked actively with cybernetic circles. Studies were conducted on communication methods, as well as control and regulatory feedback mechanisms in machines, animals, and sociotechnical systems. Cybernetics became the catalyst of many scientific disciplines (e.g., neuroscience, computer science, artificial intelligence, complexity, and dynamic systems theories), while also contributing to the development of new communication and information-processing technologies to influence the lives of every individual.

In this sense, the overall project of cybertherapy is in straight line with the philosophy behind universal design and access (Stephanidis et al., 1998, 1999). It aims at providing technological access to health care services to as large a population of users as possible. Cybertherapy's research and development (R&D) contributes to the general endeavors toward helping populations with special needs all around the world enter the information society.

52.1.1 Cyberpsychology

Cyberpsychology lies at the heart of cybertherapy. It plays a specific role in studies pertaining to psychological phenomena coupled with new communication and information technologies, as well as those specific to the beneficial or harmful effects of such innovations on mental health. From this viewpoint, the use of new communication and information technologies for curative purposes in the broad field of psychopathology is of particular interest to cyberpsychology. This emerging scientific discipline aims to better understand and employ technologies such as personal computers, palm pilots, the Internet, cellular phones, and video games (Butcher et al., 2004; Cavanagh and Shapiro, 2004; Riva et al., 2006). Throughout its short history, however, cyberpsychology has paid a rather unique attention to virtual reality (VR). It focused on input/output technologies, hypermedia, interactivity, and the design of specialized virtual spaces. Section 52.2 looks further into these simulation and immersion technologies to better understand their clinical uses in psychology.

52.2 Virtual Reality

Since the first prototypes proposed by Morton Heilig, Myron Krueger, and especially Ivan Sutherland in the 1950s and 1960s, the essentials of understanding technological assembly required by VR have hardly changed (Ellis, 1995; Stanney and Zyda, 2002). Starting from to the simulator machine, VR's technical assembly can be arbitrarily identified according to both the *inputs* transmitted to the computer through reactions recorded from the human operator, and the *outputs* (which the computer produces via reactions with the inputs) transmitted to different sensory canals of the human operator.

Inputs are produced via a series of sensors and transducers that transform behavioral and physiological variables into physical ones, which are in turn stored in the computer's register. Motor displacements in particular are recorded by a tracking system (generally magnetic, using infrared and/or ultrasound) that isolates the coordinates specific to where the sensors are found on the human operator's body (Ellis, 1995; Foxlin, 2002). The operator is localized within a defined sensory space, and his movements are registered across a series of orientational and positional changes. As the operator moves and orients himself in a simulated area, he can perform hand, head, eye, or full-body movements. Physiological measures that characterize the state of the human operator in virtual immersion can also be transmitted to the computer (Palsson and Pope, 2002; Wiederhold et al., 2002; Wiederhold and Rizzo, 2005). In most cases, the inputs' main function is to vary the parameters that control the state of the virtual environment (i.e., the multimedia arrangement of stimuli oriented toward the human subject[1]). Furthermore, they can help analyze the behavioral dynamics that contribute to interactions with simulated objects in virtual reality for fundamental or clinical purposes (Renaud et al., 2001; Foxlin, 2002; Renaud et al., 2002; Renaud et al., 2008).

VR's technical assembly enables a thorough analysis of motor sequences, as with chronophotography—a technique developed by Étienne-Jules Marey (1830–1904) during the 19th century and described in Paul Virilio's *The Aesthetics of Disappearance* (1980). This assembly allows for an unusual incursion into the motor activities that support kinematic variations of the subjective viewpoint in virtual immersion (i.e., an analysis of the human subject's first-person experience while interacting with the simulated content). In VR, the field of vision that is developed by the subject varies simultaneously in terms of displacement and orientation following the movements that are recorded directly or indirectly using the head-mounted display. Variations in Cartesian coordinates (x, y, and z) and in Eulerian coordinates (yaw, pitch, and roll) modify in a coherent way the subject's visual experience. Registering these coordinates allows for establishing an index concerning the spatial relationship between this viewpoint and the geometric properties of virtual objects. As a result, different contexts of approach and avoidance motor actions can be used to assess subjects moving from one place to another in virtual immersion (Renaud et al., 2001, 2002a,b). Accounting for the oculomotor activity observed in immersion (which will be discussed later) enables completion of this analysis by accurately determining the portion of the VE (virtual environment) to which the subject pays overt visual attention (Duchowski et al., 2002; Renuad, 2006; Renaud et al., 2008).

As for the outputs, the computer produces a number of stimulations for different sensory segments that make up the human subject's sensorium. This particular event occurs following an analysis of the inputs, which are transmitted by the transduction of the human operator's voluntary and involuntary behaviors. The head-mounted display is composed of two small screens that allow the subject to experience stereoscopic immersion.[2] Sound, touch, olfactory, and proprioceptive stimuli may also join visual stimuli to reinforce the effect of realism in virtual immersion.

The human-computer interface that is unique to the assembly used in VR favors a continuous and coherent perceptual-motor loop in the human subject (Biocca and Levy, 1995; Ellis, 1995). Through a feedback mechanism, the outputs become a source of information for the human operator regarding her spatial position in the virtual environment. These outputs then drive the motor behaviors to adjust the virtual environment's resulting state. Through this perceptual-motor relay, immersive VR becomes possible and the illusion effect (i.e., the feeling of presence) occurs (Slater et al., 1998; Renaud et al., 2002a, 2007).

52.3 Immersion and Presence

What is VR's immersive potential? Slater and Wilbur (1997) define immersion as the measure in which a computer system can offer illusions of reality that are: inclusive (eliminating the inputs outside of the VE); demanding (mobilizing sensory modalities); panoramic (covering the visual field); and vivid (offering a good resolution of the image). Immersive potential is measured by "the feeling of presence"—a theoretical concept developed in virtual immersion and characterized first and foremost as a psychological state or subjective

---

[1] Recent studies aiming to use electroencephalographic signals for biofeedback purposes in VR have been reported. This biofeedback, mediated by the VE's intermediate, is also called neurofeedback (Allanson and Mariani, 1999). We have developed the first biofeedback prototype of male and female sexual responses mediated by VR (Renaud et al., 2006).

[2] The CAVE (Cave Automatic Virtual Environment) represents another type of visualization system used in VR (Cruz-Neira et al., 1992) and will be explained further in this chapter.

perception that causes an individual to surrender to the illusion created by an immersive technical assembly. This illusion consists in forgetting both the exterior environment and immersive technology in favor of the VE (Witmer and Singer, 1998; International Society for Presence Research, 2007). The feeling of presence is considered a product of many factors, mainly the level of immersion, the subject's level of attention, and the degree of interaction (Slater et al., 1998; Schubert et al., 2001; Renaud et al., 2007). Acting as a kind of perception, presence must have perceptual-motor determinants that tie the subjective perspective to a limited set of possible viewpoints. These perceptual processes that create the illusion of presence are most likely mediated by oculomotor behaviors, since they form the main entry to visual perception (Renaud et al., 2007, 2008).

52.4 VR and Clinical Psychology

According to the results of an investigation led in 2002 by Norcross and colleagues, the use of VR in treating psychological disorders and psychopathology will increase in the years to come (Linstone and Turoff, 1975; Norcross et al., 2002; Riva, 2006). Thanks to their research study based on the Delphi method—a systematic survey that obtains forecasts from a large panel of experts—four factors were found to be involved in a successful psychotherapeutic approach. They are: (1) efficiency—having access to methods that are used for brief and economical treatments; (2) evidence—the scientific validation of therapeutic benefits by means of a given method; (3) evolution—introducing a method to the growing list of various psychotherapeutic approaches; (4) integration—allowing a method to combine important elements of different approaches.

The costs of employing VR were very high at the beginning of the 1990s but have considerably declined since then. For instance, certain VR applications that needed expensive Silicon Graphics computer platforms some years ago are now running on simple personal computers. Similarly, the price of the first commercial head-mounted displays (HMDs) 10 years ago was outrageous. Today, however, it is possible to purchase one for a very affordable price. In other respects, over the past 15 years, the use of VR has proven to be efficient in treating many psychopathologies, and has also enabled the integration of clinical principles coming from different theoretical approaches in psychotherapy (i.e., approaches that fit in virtual reality). As a result, VR in psychology is turning out to be quite promising.

52.5 The Scientific Advantages of VR in Psychology

A literature review performed by Riva in 2005 deals with the use of VR in psychology and identifies 996 published scientific articles dealing with this topic. The author reveals that more than one-third of these articles (371) were written over the last 3 years, suggesting a strong growth as well as a growing interest for VR research, which began in 1992.[3]

According to Riva (2006), most controlled scientific studies having clinical samples of more than 10 subjects and showing efficiency in VR systems come from the integration of cognitive-behavioral or strictly cognitive approaches. These treatment methods are known in psychology for their adherence to principles of the scientific method.[4]

In fact, VR favors the scientific method in clinical research because of multiple reasons, one of them being that VR noticeably improves the external validity (also called ecological validity) of the results that it generates. When compared to over-simplified stimuli found in certain laboratory studies, those that come from VR are much closer to reality and lead to a better generalization of the results.

Contrary to what usually occurs in scientific research, this gain in external validity is not achieved to the detriment of a rigorous control of the experimental variables. Even though stimuli presentation in a VE may be closer to exterior reality, it is rigorously and faithfully the same experimental condition that occurs in each trial. As a result, the valid causative inferences based on the variables involved, and more specifically the effects of simulated conditions on behavior and the subjective reactions of human subjects, are favored (Brewer, 2000).

VR may even strengthen the internal validity of collected laboratory measures by resorting, as was explained earlier, to the recording and quantitative control of the first-person visual content experienced by a human subject in virtual immersion (Duchowski et al., 2002; Renaud et al., 2008). By analyzing in more detail the contingencies that unite subjective experience in virtual immersion with the human subject's responses (i.e., responses obtained by means of a questionnaire, as well as behavioral and physiological responses), less ambiguous causative links can be established between psychological phenomenology and its quantifiable manifestations. For example, getting no erectile response[5] from a pedophile patient placed in immersion with a virtual child victim may be more easily explained by resorting to an analysis of the first-person attentional content (Renaud et al., 2002b, 2005; Renaud, 2004; Chartier et al., 2006). Researchers or clinicians may, for instance, conclude that the aforementioned patient avoided looking at virtual sexual characters in the hopes of not feeling sexually aroused (see Figure 52.4 for a description of a system that monitors first-person experience in VR immersion). The possibility of examining subjective experience and its attentional content via

---

[3] VR in clinical psychology dates back to 1992. It was introduced by the Human-Computer Interaction Group at Clark University (North et al., 2002).

[4] Based on studies conducted by Norcross and colleagues (2002), these approaches are among some of the most promising ones.

[5] As measured with a penile plethysmograph.

FIGURE 52.1 Image of one of the virtual spiders in Côté and Bouchard's study. (From Côté, S. and Bouchard, S., *Applied Psychophysiology and Biofeedback*, 30, 217–232, 2005.)

VR's intermediate may significantly improve the value of other measures that are obtained simultaneously.

In general, when coupled with VR, the use of psychophysiological measurement techniques, such as electrocardiography and electroencephalography, greatly gains in validity (Bullinger et al., 2001; Mager et al., 2001).

52.6 Clinical Advantages of VR in Psychology

From a clinical viewpoint, the use of VR in mental health leads to a number of therapeutic benefits. (1) VR can simulate treatment contexts that are not easily accessible or practically impossible to reproduce in reality (e.g., simulating an airplane take-off, locomotion in high-altitude places, or being a potential victim of a sexual aggressor). (2) VR provides the possibility of repeating on demand a given context in virtual immersion. This controlled repetition allows researchers to better target a debilitating symptom and to accurately treat it in a patient. (3) The implementation of a clinical treatment protocol, which is both automated and controlled, ensures a better adherence to its various procedures. This benefit becomes extremely useful when struggling with dimensions of noncompliance or even malingering in some patients. (4) The recording and storage of immersive sessions facilitate records management and clinical follow-up. They also help bridge the gap between clinical practice and scientific research in psychology. (5) Patients seem to build more self-motivation through VR compared with the use of more standard methods (Rothbaum et al., 2000; Garcia-Palacios et al., 2001).

52.7 Treating Phobic Disorders through VR

Multiple clinical problems were tackled through the use of VR in clinical psychology. However, those that were subjected to the most research were associated with anxiety disorders. Studies on posttraumatic stress disorder, for instance, demonstrated the efficiency of VR-based therapy (Rothbaum et al., 1999, 2001; Basoglu et al., 2003). Phobic disorders, however, were the most thoroughly studied psychopathologies. These are characterized by extreme and irrational fear resulting from the presence or anticipation of an object or a particular situation. This fear is also accompanied by an active avoidance of the object or situation (e.g., fear of animals, heights, and blood; American Psychiatric Association, 1994). Phobic disorders constitute the most prevalent form of anxious disorders according to the National Institute of Health. In fact, between 8.7% and 18.1% of Americans suffered from a phobia. Over the last few decades, the cognitive-behavioral approach has been viewed as the preferred treatment for this psychopathology. Thanks to exposure therapy, the cognitive-behavioral approach can obtain meaningful results with phobic disorders (Foa and Kozak, 1986; Foa and McNally, 1986). Virtual reality exposure therapy (VRE), which is an extension of the cognitive-behavioral approach, uses the same treatment logic.

VRE provides environment exposure to allow the subject to experiment and ease aversive emotions associated with pathological avoidance as a result of virtual immersion, the feeling of presence and attentional engagement. Gradual control of these emotions of fear and distress, which are accompanied by subjective, physiological, and behavioral responses, is the central factor for VRE. The VE that simulates the object or the issue in question prompts a series of symptoms so that, with the help of a therapist, the patient can cope with what is virtually there. Once these symptoms are controlled *in virtuo*, the beneficial effects may be generalized in non-VR (i.e., outside of the VE, *in vivo*).

VRE was shown to be as efficient as cognitive-behavioral therapy when treating numerous phobic disorders (Anderson et al., 2004; Riva, 2006). In fact, its efficacy was demonstrated in patients with specific phobias,[6] including the fear of heights (Emmelkamp et al., 2002), the fear of flying in an airplane (North, North, and Coble, 1997; Rothbaum et al., 2000), the fear of confined spaces, such as claustrophobia (Botella et al., 1998; Bullinger et al., 1998), and the fear of spiders, or arachnophobia (Carlin et al., 1997; Hoffman et al., 2003; Côté and Bouchard, 2005).

Arachnophobia is a specific animal-type phobia that is classified among the most common phobias today. According to a study conducted in England, 32% of women and 18% of men showed anxiety or great fear in the presence of a spider (Davey, 1994). Although it can appear insignificant at first blush, this fear gives rise to major debilitating effects in people who are affected by its pathological form.

52.8 VRE's Efficacy in Treating Arachnophobia

As part of Côté and Bouchard's study (2005), the efficiency of VRE was evaluated using questionnaires (Fear of Spider Questionnaire, Spider Beliefs Questionnaire, Self-Efficacy Questionnaire, STAI-Trait Anxiety Questionnaire), along with a behavioral avoidance test (BAT). These outcome measures were obtained both before (pretreatment) and after VRE treatment (post-treatment) to assess the efficacy of the latter.

As for the BAT, a live tarantula (*Grammostola rosea*, 14 cm long) was put in a closed transparent box on a motorized platform placed on a table. This spider was located 173 cm away from the participant and was completely hidden within a cardboard box. Participants were asked to complete each step to the best of their abilities until their anxiety level was too high. The BAT performance was evaluated on a scale of 0 to 10 by giving a score to those participants who could successfully complete their last task. The BAT is composed of 10 steps, which are described as follows: participants had to sit on a bench at the end of the motorized platform. The researcher lifted the cardboard box (Step 1) and removed its lid (Step 2). After looking at the spider for 1 minute, the participants were required to move the platform closer to them by holding a button (moving the platform closer by 25 cm every time constituted Steps 3 to 9). Once the minimum distance between the participants and the platform was reached (23 cm to the chest), the participants had

to bend forward and place their face above the opening of the box and look at the spider for 1 minute (Step 10). During the BAT, they were allowed to take a short break and keep the platform still, but any pause longer than 25 seconds was considered a complete stop. For this reason, the researcher would instruct the participants to keep the platform still and look at the spider for an additional 35 seconds. The instructions that describe these steps were given to the participants prior to the start of the test.

52.8.1 Subjects

The sample consisted mostly of women (27 out of 28). Participants were between 21 and 53 years of age (mean of 34, SD = 10.3) and were all diagnosed with arachnophobia according to the Diagnostic and Statistical Manual of Mental Disorders criteria (American Psychiatric Association, 1994).

52.8.2 Treatment

52.8.2.1 Equipment

The VR environments were displayed using a computer equipped with Windows 2000 (Pentium III, 4.2 GHz, 1 Go of RAM, with an nVidia GeForce4 Ti 4200 128 MB), an Intertrax2 motion tracker from InterSense (USB model, 3 dof, update rate 256 Hz), an I-Glass SVGA head-mounted display by IO-Display (800 X 600, 26 degrees FoV diagonal) and a Gyration wireless mouse. The VR environments were created using a 3D game editor (see Figure 52.1).

The VR environments were made up of two apartments, each having many rooms that were used in three difficulty levels. The first one included pictures of spiders hanging on walls, as well as a few small live spiders that were generally still. In the second level, the sizes of these spiders ranged from 15 virtual centimeters to 2 virtual feet. They were now making more unexpected moves, but they were usually away or around the participant. Finally, in the third level, the spiders appeared in all sizes and were numerous. They were often seen moving toward the participants, some in a rather aggressive manner (e.g., precipitating toward the participants' feet whenever they entered the room). In levels 2 and 3, the participants could pick up a magazine and kill the spiders by hitting them (see Figure 52.2).

52.8.2.2 Exposure

VRE was divided into five standardized individual weekly sessions (60 minutes long). Using a predetermined hierarchy, participants would gradually approach virtual spiders until their anxiety level decreased. The last 15 minutes of the final session were devoted to a review of the participants' improvements during therapy, along with the planning of *in vivo* exposure exercises that participants could do at home to maintain their gains and prevent relapse. Explanations were provided about relapse prevention. Participants were asked to act as normally as possible at home to prevent intentional *in vivo* exposure between

---

[6] Social phobia, that is, the experience of fear in social situations and of being evaluated by others, seems to be treatable via VRE (Anderson et al., 2003).

FIGURE 52.2 Image of what the patients were seeing while killing the spiders with a magazine.

sessions that could potentially blur the results. However, they were not told to actively avoid spiders at home.

During exposure sessions, a subjective measure was administered verbally by the therapist as a way to measure anxiety. Every 5 minutes, she would ask the participants to rate their anxiety on a scale of 0 to 100 and record their answers to adjust the exposure's intensity. Following each session, the participants would stay in the waiting room for 15 minutes before leaving. This procedure was done to avoid any negative side effects of VRE *a posteriori*.

52.8.3 Results

Descriptive statistics as well as the results from repeated measure ANOVAs are reported in Table 52.1. These results show that the

treatment did in fact significantly improve all measures. None of the participants were able to reach the last two steps of the BAT before treatment. However, 60.7% succeeded in reaching step 9 after treatment and 46.4% were able to go through all 10 steps of the BAT.

52.8.4 Discussion

The results in Section 52.8.3 show that arachnophobic patients undergoing VRE may succeed in significantly reducing their subjective anxiety, in both a statistical and a clinical sense. Exposure to a virtual spider allowed these subjects to feel and control disturbing reactions that they had experienced before with real spiders. VRE may thus serve as a valid clinical tool to help phobic people conquer specific fears, such as arachnophobia.

TABLE 52.1 Results on the Outcome Measures ($N = 28$)

	Pretreatment	Post-treatment	$F(1, 27)$
Behavior avoidance test	4.25 (3.25)	8.39 (2.24)	66.4**
Fear of spider questionnaire	99.71 (22.11)	48.86 (15.16)	67.39**
Spider beliefs questionnaire—beliefs	98.79 (25.75)	62.93 (12.45)	57.61**
Spider beliefs questionnaire—self	74.54 (18.61)	47.21 (10.58)	60.68**
Self-efficacy questionnaire	34.16 (20.11)	72.13 (16.71)	70.8**
STAI—trait anxiety	34.11 (8.35)	32.18 (8.97)	4.38*

* $p < 0.05$; ** $p < 0.001$

As mentioned earlier, VRE's curative effect may be attributed to the immersion of the senses in a VE and help provoke symptoms that are in need of treatment. By triggering an information-processing system that is similar to that observed in reality when faced with phobogenic situations, VRE allows for correcting the pathological cognitive biases that are unique to phobias. In fact, these biases are associated with typical motor and physiological reactions in patients who exhibit defense and flight behaviors (Anderson et al., 2004; Côté and Bouchard, 2005). The use of VRE for phobic disorders seems to considerably weaken the links that unite problematic cognitive biases with noticeable manifestations of fear and anxiety.

Attentional bias (a specific form of cognitive bias) plays a key role in organizing phobic disorders (Foa and Kozak, 1986; Lang et al., 2000; Öhman et al., 2001; Côté and Bouchard, 2005). It is characterized by a lower detection limit of situations or objects perceived as threatening. Moreover, it orients and concentrates attention toward these situations or objects by mobilizing the individual's psychophysiological resources. In the case of arachnophobia, this attentional bias was demonstrated by using oculomotor movements as physiological markers (Lange et al., 2004; Miltner et al., 2004; Rinck and Becker, 2006; Pflugshaupt et al., 2007). Many studies reveal typical oculomotor patterns seen in arachnophobic patients who would find themselves scanning the visual information relative to their phobia more rapidly and intensely.[1] The previous explanations regarding the organization of visual attention in phobic patients are important when dealing with VRE, since this particular treatment focuses first and foremost on the need for patients to be aware of the critical simulated visual information (Wilhelm, 2005). If patients are not aware of the latter, the therapeutic experience cannot occur.

52.9 Developing a Procedure That Can Evaluate Avoidance and Phobic Attentional Processes in Virtual Immersion: Preliminary Clinical Results

All clinical processes unique to mental health require an appropriate treatment that can correct a given pathological state, as well as a diagnostic evaluation. This evaluation process is generally applied before and after treatment to verify the efficiency of the therapeutic procedure. Research studies targeted to use VR for diagnostic evaluation purposes are less numerous than those focusing on treatment, but studies on attention deficit and hyperactivity disorder (Wann et al., 1997; Rizzo et al., 1999), anxiety disorders (Renaud, Bouchard, and Proulx, 2002; Wiederhold and Wiederhold, 2004), autistic disorders (Trepagnier, Sevrechts, and Peterson, 2002), addictive behaviors (Baumann and Sayette, 2006), and deviant sexual preferences (Renaud et al., 2002; Renaud, 2004; Renaud et al., 2005) were conducted so far.

To expand knowledge of the mechanisms at work in treating phobic disorders with VRE, a diagnostic evaluation procedure was developed regarding motor behaviors that are necessary for movement and for the orientation of overt visual attention in arachnophobic patients. Furthermore, these patients were placed in immersion and were exposed to phobogenic stimuli to better understand the dynamics of behavioral avoidance and the information processing associated with it (Renaud et al., 2002a). This diagnostic evaluation procedure regarding phobic avoidance behavior represents not only the first computerized BAT, but also a valuable tool that can capture the nature of a patient's experience by witnessing events through his visual perspective.

52.9.1 Subjects

The conducted experiment involved a small sample of five women between 24 and 49 years of age (mean of 35.6, SD = 11.5) who were all diagnosed with arachnophobia according to the Diagnostic and Statistical Manual of Mental Disorders criteria (American Psychiatric Association, 1994).

52.9.2 Experimental Task and Protocol

The subjects were standing at the center of an immersive vault and were asked to move as much as possible toward a virtual tarantula (condition 1; Figure 52.3a) or a virtual sphere acting as a neutral stimulus (condition 2; Figure 52.3b). The following instructions were given to the participants: "Try not to lose sight of the spider (or sphere) while moving as much as possible toward it. You can also move backwards if fear overcomes you, and then proceed forward shortly after by approaching the spider (or sphere) as close as possible until the end of the session. Although this exercise will last 3 minutes, do not preoccupy yourself with the time. We will notify you when the session is over." The exercise was held in a virtual room that simulated a kitchen with a counter on which the spider or sphere was moving. The targets (spider or sphere) in both experimental conditions shared exactly the same kinematic properties, moving according to variable speeds and trajectories that were similar to those that a real spider could trace.

52.9.3 Equipment

A CAVE-type immersive vault was used in this study by the Université du Québec en Outaouais (Cruz-Neira et al., 1992), consisting of a cluster of four computers that generate the VE and one computer that records ocular measures. One of the four computer clusters acts as the master, while the other three are the slaves connected to a projector displayed on one of the walls of the immersive vault. All computers communicate through a network using a CISCO 100 Mbps switch. The master computer gathers the inputs provided by a human subject (keyboard, mouse, motion sensors, ocular measures) and

 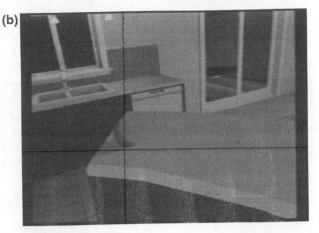

FIGURE 52.3 (a) The virtual spider and (b) the neutral target; crosshairs depict the immersed subject's momentary point of regard in the VE.

distributes them to the slaves, so that all cluster machines can calculate the changes made in the VE and also generate a report. The master and the slaves have nearly the same configuration. In fact, the only difference lies in the mother card—the master computer uses Intel D925XECV2, while the slaves use Intel D925XEBC2.

The projectors used (Electrohome Marquee 8500) are modified by Hi Rez Projections Inc. They have a resolution of 1280x1024 with a frame rate of 85 Hz. The human subject wears active Nuvision 60GX stereoscopic glasses coupled with an oculomotor tracking system (ASL model H6) (see Figure 52.4).

The graphic cards are interconnected, allowing them to be frame locked. The positional and orientational coordinates are provided by an IS-900 motion tracker from InterSense Inc. The Virtools 3.5 middleware is responsible for creating the appropriate environment and ensuring communication between the computer clusters. Finally, OpenGL 2.0 plays a role in the rasterization process to benefit from active stereoscopy.

52.9.4 Measurement Techniques

52.9.4.1 Immersive Video-Oculography

The adopted method performs gaze analysis by way of virtual measurement points (VMPs) placed over virtual objects (see Figure 52.4). The gaze radial angular deviation (GRAD) from VMPs is obtained by combining the 6 degrees-of-freedom (DOF) resulting from head movements and the 2 DOF (x and y coordinates) resulting from the eye-tracking system (Duchowski et al., 2002; Renaud, 2006; Renaud et al., 2008). While variations in the 6 DOF developed by head movements define momentary

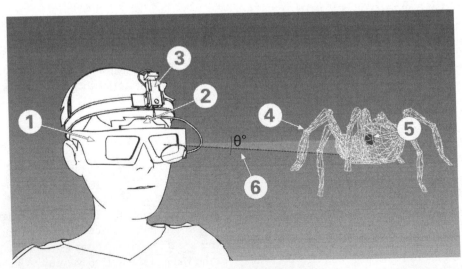

FIGURE 52.4 Subject wearing stereoscopic glasses coupled with an oculomotor tracking device: (1) IS-900 motion tracker from InterSense; (2) active Nuvision 60GX stereoscopic glasses; (3) oculomotor tracking system (ASL model H6); (4) a virtual spider in wired frame; (5) a virtual measurement point (VMP); and (6) a gaze radial angular deviation (GRAD) from the VMP.

changes in the global scene experienced in the immersive vault, the 2 DOF generated by the eye-tracking device allow line-of-sight computation relative to VMPs. The more this measure approaches zero, the closer the gaze dwells in the immediate vicinity of the selected VMP. Moreover, VMPs are locked onto and therefore move jointly with virtual objects, making it possible to examine the visual pursuit of dynamic virtual objects. Therefore, this method allows for measuring the visual response, which includes the scanning time in the vicinity of selected VMPs as well as GRAD patterns relative to VMPs (see Figure 52.5).

52.9.4.2 Computerized BAT Applied in Virtual Immersion

Avoidance behavior is measured by calculating the distance separating the patient from the VMP placed on virtual objects that will be approached by the patient (Renaud et al., 2002a). The coordinates obtained at a frequency of 60 Hz through the motion tracker are fed into a trigonometric function that calculates the distance between the patient and the virtual object. From this calculation, an accurate picture can be obtained of the temporal evolution associated with phobic avoidance (see Figure 52.5).

52.9.5 Results

Raw data are displayed in Table 52.2. Repeated analyses of variance were done to compare the subjects' responses to the neutral target (sphere) and to the phobogenic target (spider). Consequently, it was observed that the subjects were on average farther away from the phobogenic stimulus ($F(1,4) = 8.344$, $p<0.05$). This result fits usual observations during BATs involving phobic patients. As seen in Figures 52.5 and 52.6, however,

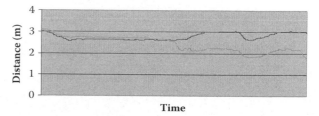

FIGURE 52.5 A three-minute sample showing distance fluctuations from the spider (in black) and the neutral object (in gray) for one representative subject. This person was getting closer to the target in the presence of the neutral object than when she was exposed to a phobogenic stimulus (in black).

the computerized BAT applied in immersion provides more detailed information on the temporal evolution of avoidance in phobic patients, which depends on where they are spatially located with respect to the phobogenic stimulus.

Next, the subjects were shown to have their attention more precisely centered on the phobogenic target when compared to the neutral one, ($F(1,4) = 9.134$, $p<0.05$). As measured by the GRAD, the phobic subjects seemed to be more easily attentive to the stimulus associated with their fears. Finally, the patients' lability of visual pursuit behavior, measured by the GRAD's standard deviation, was significantly lower when tracking the phobogenic stimulus than it was when tracking the neutral one ($F(1,4) = 8.475$, $p<0.05$). The greatest mobilization of attention toward the phobogenic stimulus is therefore characterized by a tighter control of the motor processes that maintain critical information processing.

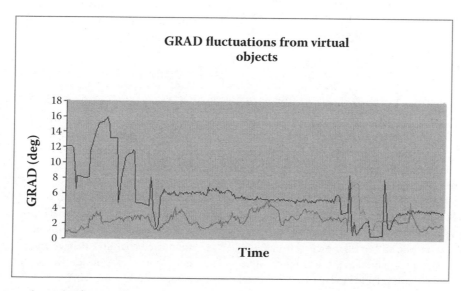

FIGURE 52.6 A five-second sample of GRAD fluctuations from the spider (gray) and the neutral object (black) for one representative subject. The closer the data approach zero, the closer the gaze is to the center of the moving target.

TABLE 52.2 Case Summaries

Subjects	Average GRAD with Neutral Target (deg)	Average GRAD with Phobogenic Target (deg)	GRAD SD with Neutral Target	GRAD SD with Phobogenic Target	Average Distance from Neutral Target (m)	Average Distance from Phobogenic Target (m)
1	8.79	8.93	6.11	4.87	0.66	0.77
2	11.69	4.37	6.69	2.62	1.30	2.24
3	9.12	5.35	7.96	2.87	1.84	2.67
4	9.74	3.44	12.22	2.29	2.42	2.78
5	9.25	6.49	6.61	4.64	1.06	1.24
Mean	9.72	5.72	7.92	3.46	1.45	1.94

52.9.6 Discussion

Although this first study contained preliminary data from a very small clinical sample, it was able to demonstrate VR's potential in the detailed analysis of motor behavior when coupled with the monitoring of attentional activity. VR systems allow measuring patients' attentional engagement in immersion, both qualitatively and quantitatively. In the qualitative sense, researchers or clinicians[1] can witness events through the subject's visual first-person perspective in immersion. They can also better understand how the patients occupy the VE and manage their visual attention in relation to critical simulated zones. As for the quantitative aspect, statistical data that are similar to those obtained in the previous section may accurately illustrate the perceptual-motor parameters reflecting the phobic patients' attentional organization, while also monitoring their progress during VRE. By resorting to this important clinical advantage, diagnostic evaluation is no longer performed exclusively before and after treatment, but over the course of the entire VRE. As a result, the patients' clinical evolution may be described more accurately, thus showing more easily the individual differences that lie in the expression of a given pathology, particularly patterns of motor avoidance and attentional bias. Adjusting the application of clinical protocols to the subjects' idiosyncrasies may therefore be greatly facilitated.

52.10 Conclusion

This chapter has observed that VR has gained scientific credibility and popularity with mental health practitioners. This is largely so because more and more credible studies can establish its curative efficiency on mental health. VR is considered an important therapeutic adjuvant because of its scientific and clinical advantages. This technology may not be able to fully replace the therapist human input in the treatment process (which is its purpose, anyways), but it can definitely allow its users to expand and develop the span of important experiences that can lead them to a more adaptive lifestyle and to their own well-being.

In addition, VR has a strong potential to attract many people. Its popularity will probably increase from year to year, as younger generations who are familiar with video games and digital communication will have to gain access to mental health care. Not only are these new generations more open to and accepting of new technologies, but they also integrate them more willingly into their lifestyle. In fact, the use of technology brings them a sense of belonging and self-recognition. It is therefore natural to see these people in a technologically oriented environment, allowing them to change their way of being, their behavior, or simply to get to know themselves better. This will without any doubt contribute in a major way to speed up the universal access general project (Stephanidis et al., 1998, 1999).

Looking at technology more closely, it can be noticed that its evolution helps to objectify subjectivity, by increasing the ways in which people can learn more about themselves and change. This *virtualization* process from the viewpoint of the self is a product of technology's speeded-up advancements. This accelerated technological development, which mirrors the accelerated transmission of information itself, contributes to reducing the historical gap between the arrival of technological innovations and the integration of the latter into the market and then into our habits and customs (Stephanidis et al., 1998, 1999; Goldstone, 2002; Dinopoulos and Sener, 2007). This Schumpeterian growth of technologies at the heart of cybertherapy finds echo in the increasing population of users, including both practitioners and beneficiaries. This observation indicates that our notion of health and of our relation to it may change for good as we enter the information society.

References

Allanson, J. and Mariani, J. (1999). Mind over virtual matter: Using virtual environments for neurofeedback training, in *the Proceedings of IEEE Virtual Reality Conference 1999 (VR'99)*, 13–17 March 1999, Houston, TX, pp. 270–273. Los Alamitos, CA: IEEE Computer Society Press.

American Psychiatric Association (1994). *Diagnostic and Statistical Manual of Mental Disorders* (4th ed.). Washington, DC: American Psychiatric Association.

Anderson, P., Jacobs, C., and Rothbaum, B. O. (2004). Computer-supported cognitive behavioral treatment of anxiety disorders. *Journal of Clinical Psychology* 60: 253–267.

Anderson, P., Rothbaum, B. O., and Hodges, L. F. (2003). Virtual reality in the treatment of social anxiety: Two case reports. *Cognitive and Behavioral Practice* 10: 240–247.

Basoglu, M., Livanou, M., and Salcioglu, E. (2003). A single session with an earthquake simulator for traumatic stress in earthquake survivors. *American Journal of Psychiatry* 160: 788–790.

Baumann, S. B. and Sayette, S. A. (2006). Smoking cues in a virtual world provoke craving in cigarette smokers. *Psychology of Addictive Behaviors* 20: 484–489.

Biocca, F. and Levy, M. R. (1995). *Communication in the Age of Virtual Reality*. Hillsdale, NJ: Lawrence Erlbaum Associates.

Botella, C., Baños, R. M., Perpiña, C., Villa, H., Alcañiz, M., and Rey, A. (1998). Virtual reality treatment of claustrophobia: A case report. *Behaviour Research & Therapy* 36: 239–246.

Brewer, M. (2000). Research design and issues of validity, in *Handbook of Research Methods in Social and Personality Psychology* (H. Reis and C. Judd, eds.), pp. 3–16. Cambridge, U.K.: Cambridge University Press.

Bullinger, A. H., Müller-Spahn, F., Mager, R., Stoermer, R., Kuntze, M. F., and Wurst, F. (2001). Virtual reality in future diagnoses and treatment. *World Journal of Biological Psychiatry* Suppl. 1: 194S.

Bullinger, A. H., Roessler, A., Hofmann, M., Kuntze, M., Hemmeter, U., and Müller-Spahn, F. (1998). Cognitive behavioral therapy of claustrophobia using a 3D-VR-system, in *Proceedings of the "Medicine Meets Virtual Reality: 6" Conference*, 28–31 January 1998, San Diego. Amsterdam, The Netherlands: IOS Press.

Butcher, J. N., Perry, J., and Hahn, J. (2004). Computers in clinical assessment: Historical developments, present status, and future challenges. *Journal of Clinical Psychology* 60: 331–345.

Carlin, A. S., Hoffman, H. G., and Weghorst, S. (1997). Virtual reality and tactile augmentation in the treatment of spider phobia: A case study. *Behaviour Research and Therapy* 35: 153–158.

Cavanagh, K. and Shapiro, D. A. (2004). Computer treatment for common mental health problems. *Journal of Clinical Psychology* 60: 239–251.

Chartier, S., Renaud, P., Bouchard, S., Proulx, J., Rouleau, J. L., Fedoroff, P., and Bradford, J. (2006). Sexual preference classification from gaze behavior data using a multilayer perceptron. *Annual Review of CyberTherapy and Telemedicine* 4: 149–157.

Côté, S. and Bouchard, S. (2005). Documenting the efficacy of virtual reality exposure with psychophysiological and information processing measures. *Applied Psychophysiology and Biofeedback* 30: 217–232.

Cruz-Neira, C., Sandin, D. J., DeFanti, T. A., Kenyon, R., and Hart, J. C. (1992). The CAVE: Audio visual experience automatic virtual environment. *Communications of the ACM* 35: 64–73.

Davey, G. C. L. (1994). Self-reported fears to common indigenous animals in an adult U.K. population: The role of disgust sensitivity. *British Journal of Psychology* 85: 541–554.

Dinopoulos, E. and Sener, F. (2007). New directions in Schumpeterian growth theory, in *Edgar Companion to Neo-Schumpeterian Economics* (H. Hanusch and A. Pyka, eds.), pp. 688–704. Cheltenham, U.K.: Edward Elgar. http://bear.cba.ufl.edu/dinopoulos/PDF/schumpeterian growth.pdf.

Duchowski, A., Medlin, E., and Cournia, N. (2002). 3-D eye movement analysis. *Behavior Research Methods, Instrument & Computers* 34: 573–591.

Ellis, S. (1995). Origins and elements of virtual environments, in *Virtual Environments and Advanced Interface Design* (T. A. Furness III and W. Barfield, eds.), pp. 14–62. New York: Oxford University Press.

Emmelkamp, P. M. G., Krijn, M., Hulsbosch, A. M., de Vries, S., Schuemie, M. J., and Van der Mast, C. A. P. G. (2002). Virtual reality treatment versus exposure in vivo: A comparative evaluation in acrophobia. *Behaviour Research & Therapy* 40: 509–516.

Foa, E. B. and Kozak, M. J. (1986). Emotional processing of fear: Exposure to corrective information. *Psychological Bulletin* 99: 20–35.

Foa, E. B. and McNally, R. J. (1986). Mechanisms of change in exposure therapy, in *Current Controversies in the Anxiety Disorders* (R. M. Rapee, ed.), pp. 329–343. New York: Guilford Press.

Foxlin, E. (2002). Motion tracking requirements and technologies, in *Handbook of Virtual Environments: Design, Implementation, and Applications* (K. Stanney, ed.), pp. 163–210. Mahwah, NJ: Lawrence Erlbaum Associates.

Garcia-Palacios, A., Hoffman, H. G., and See, S. K. (2001). Redefining therapeutic success with virtual reality exposure therapy. *CyberPsychology & Behavior* 4: 341–348.

Goldstone, G. A. (2002). Efflorescences and economic growth in world history: Rethinking the rise of the West and the industrial revolution. *Journal of World History* 13: 323–389.

Hoffman, H., Garcia-Palacios, A., Carlin, A., Furness III, T. A., and Botella, C. (2003). Interfaces that heal: Coupling real and virtual objects to treat spider phobia. *International Journal of Human-Computer Interaction* 16: 283–300.

International Society for Presence Research (2007). http://www.temple.edu/ispr.

Lang, P. J., Davis, M., and Öhman, A. (2000). Fear and anxiety: Animal models and human cognitive psychophysiology. *Journal of Affective Disorders* 61: 137–159.

Lange, W. G. T., Tierney, K. J., Reinhardt-Rutland, A. H., and Vivekananda-Smith, P. (2004). Viewing behaviour of spider phobics and non-phobics in the presence of threat and safety stimuli. *British Journal of Clinical Psychology* 43: 235–243.

Linstone, H. A. and Turoff, M. (1975). *The Delphi Method: Techniques and Applications*. Reading, MA: Addison Wesley.

Mager, R., Bullinger, A. H., Müller-Spahn, F., Kuntze, M. F., and Stoermer, R. (2001). Real-time monitoring of brain activity in patients with specific phobia during exposure therapy, employing a stereoscopic virtual environment. *CyberPsychology and Behavior* 4: 465–469.

Miltner, W. H. R., Krieschel, S., Hecht, H., Trippe, R., and Weiss, T. (2004). Eye-movements and behavioral responses to threatening and non-threatening stimuli during visual search in phobic and non-phobic subjects. *Emotion* 4: 323–339.

Norcross, J. C., Hedges, M., and Prochaska, J. O. (2002). The face of 2010: A Delphi poll on the future of psychotherapy. *Professional Psychology: Research and Practice* 33: 316–322.

North, M. M., North, S. M., and Coble, J. R. (2002). Virtual reality therapy: An effective treatment for psychological disorders, in *Handbook of Virtual Environments: Design, Implementation, and Applications* (K. M. Stanney, ed.), pp. 1065–1078. Mahwah, NJ: Lawrence Erlbaum Associates.

Öhman, A., Flykt, A., and Esteves, F. (2001). Emotion drives attention: Detecting the snake in the grass. *Journal of Experimental Psychology: General* 130: 466–478.

Palsson, O. S. and Pope, A. T. (2002). Morphing beyond recognition: The future of biofeedback technologies. *Biofeedback* 30: 14–18.

Pflugshaupt, T., Mosimann, U. P., Schmitt, W. J., et al. (2007). To look or not to look at threat? Scanpath differences within a group of spider phobics. *Journal of Anxiety Disorders* 21: 353–366.

Renaud, P. (2004). Moving assessment of sexual interest into the 21st century: The potential of new information technology. Invited speech at the *23rd Annual Research and Treatment Conference (ATSA)*, 30 October 2004, Albuquerque, NM.

Renaud, P. (2006). *Method for Providing Data to Be Used by a Therapist for Analyzing Patient Behavior in a Virtual Environment.* US Patent 7128577.

Renaud, P., Bouchard, S., and Proulx, R. (2002a). Behavioral avoidance dynamics in the presence of a virtual spider. *IEEE Transactions in Information Technology and Biomedicine* 6: 235–243.

Renaud, P., Chartier, S., and Albert, G. (2008). Embodied and embedded: The dynamics of extracting perceptual visual invariants, in *Chaos and Complexity: Recent Advances and Future Directions in the Theory of Nonlinear Dynamical Systems Psychology* (S. Guastello, D. Pincus, and M. Koopmans, eds.). New York: Cambridge University Press.

Renaud, P., Chartier, S., Albert, G., Cournoyer, L.-G., Proulx, J., and Bouchard, S. (2006). Sexual presence as determined by fractal oculomotor dynamics. *Annual Review of Cybertherapy and Telemedecine* 4: 87–94.

Renaud, P., Chartier, S., Albert, G., Décarie, J., Cournoyer, L.-G., and Bouchard, S. (2007). Presence as determined by fractal perceptual-motor dynamics. *Cyberpsychology and Behavior* 10: 122–130.

Renaud, P., Proulx, J., Rouleau, J.-L., et al. (2005). The recording of observational behaviors in virtual immersion: A new clinical tool to address the problem of sexual preferences with paraphiliacs. *Annual Review of Cybertherapy and Telemedecine* 3: 85–92.

Renaud, P., Proulx, J., Rouleau, J.-L., et al. (2006). Sexual and oculomotor biofeedback mediated by sexual stimuli presented in virtual reality. Paper presented at the *49th Annual Meeting of the Society for the Scientific Study of Sexuality*, 9–12 November 2006, Las Vegas.

Renaud, P., Rouleau, J.-L., Granger, L., Barsetti, I., and Bouchard, S. (2002b). Measuring sexual preferences in virtual reality: A pilot study. *Cyberpsychology and Behavior* 5: 1–10.

Renaud, P., Singer, G., and Proulx, R. (2001). Head-tracking fractal dynamics in visually pursuing virtual objects, in *Nonlinear Dynamics in Life and Social Sciences* (W. Sulis and I. Trofimova, eds.), pp. 333–346. Amsterdam, The Netherlands: IOS Press NATO Science Series.

Rinck, M. and Becker, E. S. (2006). Spider fearful individuals attend to threat, then quickly avoid it: Evidence from eye movements. *Journal of Abnormal Psychology* 115: 231–238.

Riva, G. (2006). Virtual reality in psychotherapy: Review. *Cyberpsychology and Behavior* 8: 220–230.

Riva, G., Botella, C., Légeron, P., and Optale, G. (eds.) (2006). *Internet and Virtual Reality as Assessment and Rehabilitation Tools for Clinical Psychology and Neuroscience.* Amsterdam, The Netherlands: IOS Press.

Rizzo, A., Buckwalter, J. G., Neumann, U., et al. (1999). Virtual environments for targeting cognitive processes: An overview of projects at the University of Southern California Integrated Media Systems. *Cyberpsychology and Behaviour* 2: 89–100.

Rizzo, A. A., Wiederhold, M. D., and Buckwalter, J. G. (1998). Basic issues in the use of virtual environments for mental health applications, in *Virtual Environments in Clinical Psychology and Neuroscience: Methods and Techniques in Advanced Patient-Therapist Interaction* (G. Riva, B. K. Wiederhold, and E. Molinari, eds.), pp. 21–42. Amsterdam, The Netherlands: IOS Press.

Rothbaum, B. O., Hodges, L., Alarcon, R., et al. (1999). Virtual reality exposure therapy for Vietnam veterans with posttraumatic stress disorder. *Journal of Traumatic Stress* 12: 263–271.

Rothbaum, B. O., Hodges, L. F., Ready, D., Graap, K., and Alarcon, R. D. (2001). Virtual reality exposure therapy for Vietnam veterans with posttraumatic stress disorder. *Journal of Clinical Psychiatry* 62: 617–622.

Rothbaum, B. O., Hodges, L. F., Smith, S., Lee, J. H., and Price, L. (2000). A controlled study of virtual reality exposure therapy for the fear of flying. *Journal of Consulting and Clinical Psychology* 68: 1020–1026.

Satava, R. M. (1995). Medical applications of virtual reality. *Journal of Medical Systems* 19: 275–280.

Schubert, T., Friedmann, F., and Regenbrecht, H. (2001). The experience of presence: Factor analytic insights. *Presence: Teleoperators and Virtual Environments* 10: 266–281.

Slater, M., Steed, A., and McCarthy, J. (1998). The influence of body movement on subjective presence in virtual environments. *Human Factors* 40: 469–477.

Slater, M. and Wilbur, S. (1997). A framework for immersive virtual environments (FIVE): Speculations on the role of presence in virtual environments. *Presence: Teleoperators and Virtual Environments* 6: 603–616.

Stanney, K. M. and Zyda, M. (2002). Virtual environments in the 21st century, in *Handbook of Virtual Environments: Design, Implementation, and Applications* (K. M. Stanney, ed.), pp. 1–14. Mahwah, NJ: Lawrence Erlbaum Associates.

Stephanidis, C., Salvendy, G., Akoumianakis, D., Bevan, N., Brewer, J., Emiliani, P. L., et al. (1998). Toward an information society for all: An international R&D agenda. *International Journal of Human-Computer Interaction* 10: 107–134.

Stephanidis, C., Salvendy, G., Akoumianakis, D., et al. (1999). Toward an information society for all: HCI challenges and R&D recommendations. *International Journal of Human-Computer Interaction* 11: 1–28.

Trepagnier, S., Sebrechts, M. M., and Peterson, R. (2002). Atypical face gaze in autism. *Cyberpsychology and Behaviour* 5: 213–217.

Virilio, P. (1980). *Esthétique de la Disparition* [The aesthetics of disappearance]. Paris: Balland.

Wann, J. P., Rushton, S. K., Smyth, M., and Jones, D. (1997). Virtual environments for the rehabilitation of disorders of attention and movement. *Studies in Health Technology and Informatics* 44: 157–164.

Wiederhold, B. K., Jang, D. P., Kim, S. I., and Wiederhold, M. D. (2002). Physiological monitoring as an objective tool in virtual reality therapy. *CyberPsychology & Behavior* 5: 77–82.

Wiederhold, B. K. and Rizzo, A. (2005). Virtual reality and applied psychophysiology. *Applied Psychophysiology and Biofeedback* 30: 183–187.

Wiederhold, B. K. and Wiederhold, M. D. (2004). *Virtual Reality Therapy for Anxiety Disorders*. Washington, DC: American Psychological Association Press.

Wilhelm, F. H., Pfaltz, M. C., Gross, J. J., Mauss, I. B., Kim, S. I., and Wiederhold, B. K. (2005). Mechanisms of virtual reality exposure therapy: The role of the behavioral activation and behavioral inhibition systems. *Applied Psychophysiology & Biofeedback* 30: 271–284.

Witmer, B. G. and Singer, M. J. (1998). Measuring presence in virtual environments: A presence questionnaire. *Presence: Teleoperators and Virtual Environments* 7: 225–240.

VIII

Nontechnological Issues

53

Policy and Legislation as a Framework of Accessibility

Erkki Kemppainen,
John D. Kemp, and
Hajime Yamada

53.1 Introduction

The Convention on the Rights of Persons with Disabilities, adopted 13 December 2006 by the General Assembly of the United Nations and ratified on 3 May 2008 by the twentieth State Party, covers many areas of life and indicates from a disability point of view important policy and legislative areas that have been agreed upon globally. The convention provides that states that ratify it should enact laws and other measures to improve disability rights, and also to abolish legislation, customs, and practices that discriminate against persons with disabilities.

When looking specifically at issues related to interaction with technology in different regions, many relevant policy areas can be identified. They have a certain identity in terms of issues, values, social institutions, and legislative systems. In the following, the situation in Europe, Japan, and the United States is discussed, highlighting their characteristic features.

A law should be understood in relation to legal principles, other laws, cases, and other legally relevant sources. Legislation changes in the course of time and partial amendments of legal acts are common. The style of legal writing may vary in different parts of the world, but accuracy is a common requirement. To have a picture of the situation in a certain country, one should use national legal data banks and other relevant sources.

53.2 The Convention on the Rights of Persons with Disabilities (2008)

It can be said that the United Nations Convention on the Rights of Persons with Disabilities and Optional Protocol is based on the idea of the diversity of people. The preamble states that disability is an evolving concept and that it results from the interaction between persons with impairments and attitudinal and environmental barriers that hinder their full and effective participation in society on an equal basis with others. The preamble recognizes the importance of accessibility to the physical, social, economic, and cultural environment; to health and education; and to information and communication, in enabling persons with disabilities to fully enjoy all human rights and fundamental freedoms.

From the functioning and activity, or the disability, point of view, many important areas of life are targeted in the convention. The titles indicate what is relevant for disability policy anywhere. The substantial articles, including an article on accessibility, are the following:

- Article 1: Purpose
- Article 2: Definitions
- Article 3: General principles
- Article 4: General obligations
- Article 5: Equality and nondiscrimination
- Article 6: Women with disabilities
- Article 7: Children with disabilities
- Article 8: Awareness-raising
- Article 9: Accessibility
- Article 10: Right to life
- Article 11: Situations of risk and humanitarian emergencies
- Article 12: Equal recognition before the law
- Article 13: Access to justice
- Article 14: Liberty and security of the person
- Article 15: Freedom from torture or cruel, inhuman, or degrading treatment or punishment
- Article 16: Freedom from exploitation, violence, and abuse
- Article 17: Protecting the integrity of the person
- Article 18: Liberty of movement and nationality
- Article 19: Living independently and being included in the community
- Article 20: Personal mobility
- Article 21: Freedom of expression and opinion, and access to information
- Article 22: Respect for privacy
- Article 23: Respect for home and the family
- Article 24: Education
- Article 25: Health
- Article 26: Habilitation and rehabilitation
- Article 27: Work and employment
- Article 28: Adequate standard of living and social protection
- Article 29: Participation in political and public life
- Article 30: Participation in cultural life, recreation, leisure, and sport
- Article 31: Statistics and data collection

In addition, international cooperation and monitoring are defined in the convention.

The following definitions, which are central to accessibility, are presented in Article 2:

"Reasonable Accommodation" means necessary and appropriate modification and adjustments not imposing a disproportionate or undue burden, where needed in a particular case, to ensure to persons with disabilities the enjoyment or exercise on an equal basis with others of all human rights and fundamental freedoms; "Universal design" means the design of products, environments, programmes and services to be usable by all people, to the greatest extent possible, without the need for adaptation or specialized design.

"Universal design" shall not exclude assistive devices for particular groups of persons with disabilities where this is needed.

With regard to design, several concepts have been used in different contexts. Universal design and design for all mean roughly the same. Translations in different languages may have different connotations.

Accessibility is one of the general principles presented in Article 3. Accessibility is about physical environment, transportation, information and communications, including information and communications technologies and systems, and facilities and services. Article 9 has been devoted to accessibility:

53.2.1 Article 9: Accessibility

1. *To enable persons with disabilities to live independently and participate fully in all aspects of life, States Parties shall take appropriate measures to ensure to persons with disabilities access, on an equal basis with others, to the physical environment, to transportation, to information and communications, including information and communications technologies and systems, and to other facilities and services open or provided to the public, both in urban and in rural areas. These measures, which shall include the identification and elimination of obstacles and barriers to accessibility, shall apply to, inter alia:*

   (a) *Buildings, roads, transportation and other indoor and outdoor facilities, including schools, housing, medical facilities and workplaces;*

   (b) *Information, communications and other services, including electronic services and emergency services.*

2. *States Parties shall also take appropriate measures to:*

   (a) *Develop, promulgate and monitor the implementation of minimum standards and guidelines for the accessibility of facilities and services open or provided to the public;*

   (b) *Ensure that private entities that offer facilities and services which are open or provided to the public take into account all aspects of accessibility for persons with disabilities;*

   (c) *Provide training for stakeholders on accessibility issues facing persons with disabilities;*

   (d) *Provide in buildings and other facilities open to the public signage in Braille and in easy to read and understand forms;*

   (e) *Provide forms of live assistance and intermediaries, including guides, readers and professional sign language interpreters, to facilitate accessibility to buildings and other facilities open to the public;*

   (f) *Promote other appropriate forms of assistance and support to persons with disabilities to ensure their access to information;*

   (g) *Promote access for persons with disabilities to new information and communications technologies and systems, including the Internet;*

(h) Promote the design, development, production and distribution of accessible information and communications technologies and systems at an early stage, so that these technologies and systems become accessible at minimum cost.

Accessibility is something general; assistive technologies are somehow tailored to correspond to individual needs. Both accessibility and assistive technologies are explicitly mentioned in the contexts of mobility and communications:

53.2.2 Article 20: Personal Mobility

States Parties shall take effective measures to ensure personal mobility with the greatest possible independence for persons with disabilities, including by:

(a) Facilitating the personal mobility of persons with disabilities in the manner and at the time of their choice, and at affordable cost;

(b) Facilitating access by persons with disabilities to quality mobility aids, devices, assistive technologies and forms of live assistance and intermediaries, including by making them available at affordable cost;

(c) Providing training in mobility skills to persons with disabilities and to specialist staff working with persons with disabilities;

(d) Encouraging entities that produce mobility aids, devices and assistive technologies to take into account all aspects of mobility for persons with disabilities.

In addition to mobility, communication is another aspect of life where accessibility is important. In the first place, accessibility is related to freedom of expression and opinion. In Article 21, which is about freedom of expression and opinion and access to information, the importance of the possibility of the choice of the form of communication is recognized.

53.2.3 Article 21: Freedom of Expression and Opinion, and Access to Information

States Parties shall take all appropriate measures to ensure that persons with disabilities can exercise the right to freedom of expression and opinion, including the freedom to seek, receive and impart information and ideas on an equal basis with others and through all forms of communication of their choice, as defined in Article 2 of the present Convention, including by:

(a) Providing information intended for the general public to persons with disabilities in accessible formats and technologies appropriate to different kinds of disabilities in a timely manner and without additional cost;

(b) Accepting and facilitating the use of sign languages, Braille, augmentative and alternative communication, and all other accessible means, modes and formats of communication of their choice by persons with disabilities in official interactions;

(c) Urging private entities that provide services to the general public, including through the Internet, to provide information and services in accessible and usable formats for persons with disabilities;

(d) Encouraging the mass media, including providers of information through the Internet, to make their services accessible to persons with disabilities;

(e) Recognizing and promoting the use of sign languages.

Reasonable accommodation and accessibility are essential also for education as defined in Article 21. It makes inclusive education a stated expectation for the state parties. In an accessible environment proper communication forms are needed.

The following measures are described in Article 24(3): (a) facilitating the learning of Braille, alternative script, augmentative and alternative modes, means and formats of communication and orientation and mobility skills, and facilitating peer support and mentoring; (b) facilitating the learning of sign language and the promotion of the linguistic identity of the deaf community; (c) ensuring that the education of persons, and in particular children, who are blind, deaf, or both deaf and blind, is delivered in the most appropriate languages and modes and means of communication for the individual, and in environments that maximize academic and social development.

Health is a dimension of life. In Article 25 concerning health it is required that persons with disabilities have the right to the enjoyment of the highest attainable standard of health without discrimination on the basis of disability. States parties shall take all appropriate measures to ensure access for persons with disabilities to health services that are gender-sensitive, including health-related rehabilitation.

In addition to the wording of Article 25 concerning health, a separate article has been devoted to rehabilitation, namely Article 26. The title "Habilitation and rehabilitation" indicates how rehabilitation should be understood. The goal of rehabilitation is maximum independence and full physical, mental, social, and vocational ability, and full inclusion and participation in all aspects of life. Rehabilitation is multidisciplinary. Also, assistive devices belong also to the context of habilitation and rehabilitation:

53.2.4 Article 26: Habilitation and Rehabilitation

1. States Parties shall take effective and appropriate measures, including through peer support, to enable persons with disabilities to attain and maintain maximum independence, full physical, mental, social and vocational ability, and full inclusion and participation in all aspects of life. To that end, States Parties shall organize, strengthen and extend comprehensive habilitation and rehabilitation services and programs, particularly in the areas of health, employment, education and social services, in such a way that these services and programs:

    (a) Begin at the earliest possible stage, and are based on the multidisciplinary assessment of individual needs and strengths;

    (b) Support participation and inclusion in the community and all aspects of society, are voluntary, and are available to persons with disabilities as close as possible to their own communities, including in rural areas.

2. *States Parties shall promote the development of initial and continuing training for professionals and staff working in habilitation and rehabilitation services.*

3. *States Parties shall promote the availability, knowledge and use of assistive devices and technologies, designed for persons with disabilities, as they relate to habilitation and rehabilitation.*

Participation in society means also participation in working life. Nondiscrimination and reasonable accommodation are important for employment and working life. According to Article 27(1), States parties recognize the right of persons with disabilities to work on an equal basis with others; this includes the right to the opportunity to gain a living by work freely chosen or accepted in a labor market and work environment that is open, inclusive, and accessible to persons with disabilities. Reasonable accommodation is essential in the context of work.

But accessibility alone is not enough to cope with challenges of life. Sufficient resources are needed. Social protection can give a safety net and measures to act in a society. Social systems have histories as well as natural, social, and cultural conditions and vary to a large extent.

As many obligations in various articles have been qualified in many ways, so is done with regard to social protection in Article 28(1): states parties recognize the right of persons with disabilities to an adequate standard of living for themselves and their families, including adequate food, clothing, and housing, and to the continuous improvement of living conditions, and shall take appropriate steps to safeguard and promote the realization of this right without discrimination on the basis of disability. Services and devices are a part of social protection: appropriate steps are required to ensure equal access by persons with disabilities to clean water services, and to ensure access to appropriate and affordable services, devices, and other assistance for disability-related needs.

Accessibility is relevant also for participation in political and public life as described in Article 29. It is required that voting procedures, facilities, and materials are appropriate, accessible, and easy to understand and use.

Access to culture is also important and technology plays a significant role, but culture still is basically about human beings. Accessibility is about places of culture and other cultural products, as defined in Article 30, which is about participation in cultural life, recreation, leisure, and sport.

There are many steps before the convention enters into force in a state. As mentioned, the convention was adopted on 13 December 2006 and was opened for signature on 30 March 2007 at the United Nations. The Convention on the Rights of Persons with Disabilities and Optional Protocol entered into force on 3 May 2008, after the convention received its 20th ratification, and the Optional Protocol 10 ratifications. But it needs to be implemented to become reality. States that ratify the convention should enact laws and other measures to improve disability rights, and also abolish legislation, customs, and practices that discriminate against persons with disabilities.

The convention prescribes a monitoring system at an international and national level. There will be international cooperation, a Committee on the Rights of Persons with Disabilities, and reports by states' parties. National implementation and monitoring includes one or more focal points with government for matters relating to the implementation of the convention.

Drafting the convention has been a long social process and so will be its implementation. Nongovernmental organizations have played an important role in the process, and they will be important partners also in the future. Information on the convention and international norms and policy guidelines can be found at http://www.un.org/esa/socdev/enable.

53.3 Europe and eAccessibility

53.3.1 Policy Framework

In a 2005 communication (Communication 2005) on "Electronic Accessibility" (eAccessibility), the European Commission calls upon its member states to do more to promote EU e-Accessibility initiatives in a concerted approach and to encourage uptake by industry. According to a political message, which is expressed in a press release, the commission is determined to improve eAccessibility, which is a prerequisite for everyone's ability to participate fully in a knowledge-based society. This holds true in particular for older persons and also for people with disabilities. Rather than proposing new legislation immediately, the commission has decided first to fully explore the possibilities available with stakeholders, including users, industry, and standardization bodies. While continuing to support ongoing measures such as standardization, design for all (DfA), web accessibility and research and technology development, the commission proposes the use of three policy levers available to member states:

- To improve the consistency of accessibility requirements in public procurement contracts in the information and communications technology (ICT) domain
- To explore the possible benefits of certification schemes for accessible products and services
- To make better use of the "e-Accessibility potential" of existing legislation (Press release, 2005)

This communication on eAccessibility contributes to the implementation of a wider initiative, i2010: A European Information Society for Growth and Employment, which presents a new strategic framework and broad policy orientations to promote an open and competitive digital economy, emphasizing ICT as a driver of inclusion and quality of life. The commission has the ambitious objective of achieving an information society for all, promoting an inclusive digital society that provides opportunities for all and minimizes the risk of exclusion.

Also the Riga Declaration (11 June 2006) confirms the EU's commitment to eAccessibility. The conference held in Riga, Latvia, included an informal meeting of ministers, where Ministers of the European Union, member states, and accession and candidate countries, European Free Trade Area (EFTA) countries, and other countries adopted a Declaration on eInclusion, which provides political guidance for future action (Ministerial Declaration, 2006).

These documents represent what are important policy objectives and measures at the time of their publication. In fact, many policy and legislative areas are potentially or actually relevant to eAccessibility.

For example, within the European Design for All Network (EDeAN) some important policy and legislative areas for Europe have been discussed, and the main topics were structured as follows: nondiscrimination, ICT, employment, public procurement, copyright, eServices, assistive technology, other areas (www.edean.org).

In the European Union, the interplay between European and national legislation is essential. This chapter refers mainly to legislation at a European level, but one should be aware that there is effective relevant legislation in member states, for example, in the field of nondiscrimination, telecommunications, or data protection. European legislation can be found at http://www.eur-lex.europa.eu. Information society issues at a European level are described in a thematic portal at http://ec.europa.eu/information_society/index_en.htm. Information on eAccessibility and eInclusion legislation and policies at a national level can be found, for example, at http://countryprofiles.wikispaces.com.

53.3.2 Nondiscrimination

Accessibility can be seen as an implication of nondiscrimination. Equality and nondiscrimination are cornerstones in many constitutions and laws in Europe. The European Union as well its member states have implemented various measures to promote equal opportunities for people with disabilities.

Article 13 (ex Article 6a) of the treaty establishing the European Community[1] states that

> without prejudice to the other provisions of this Treaty and within the limits of the powers conferred by it upon the Community, the Council, acting unanimously on a proposal from the Commission and after consulting the European Parliament, may take appropriate action to combat discrimination based on sex, racial or ethnic origin, religion or belief, disability, age or sexual orientation.

With regard to specific legal measures at a European level, a directive establishing a general framework for equal treatment in employment and occupation (Directive 2000/78/EC) lays

down a general framework for combating discrimination, for example, on the grounds of disability. The directive includes also articles about reasonable accommodation for disabled persons, and about positive action.

The directive outlaws direct discrimination as well as indirect discrimination. The latter is defined as an apparently neutral provision or practice that puts people with disabilities at a particular disadvantage compared with other persons, unless that provision or practice is objectively justified. Similarly, the directive makes employers responsible for providing reasonable accommodation for disabled people to enable them to access, participate in, and advance in employment. To make a reasonable accommodation, employers may need to respond to the specific functional abilities of a disabled person by, for example, providing or modifying equipment or facilities, or by changing practices or procedures. One kind of reasonable accommodation involves the provision of adaptive technologies or equipment for the worker who needs them.

53.3.3 ICT and Access to Terminal Equipment and Services

Accessibility with regard to ICT can be called eAccessibility. It is about electronic communications and services. In the first place access is access to equipments. But with the help of equipment and technologies, it can mean access to society. Telecommunications services, both common services, such as directory services or mobile communications, or more refined services, enable social activities that otherwise would not be possible.

Central to the legal framework are the telecommunications directives, especially Directive 1999/5/EC on radio equipment, telecommunications terminal equipment, and the mutual recognition of their conformity (RTTE). This directive recognizes the needs of disabled users by mentioning in the preamble that it should be possible to identify and add features for users with a disability and in Article 3 gives powers to the commission to decide that a certain equipment class shall be so constructed that it supports certain features to facilitate its use by users with a disability, but does not specify accessibility requirements.

Another important piece of legislation is Directive 2002/22/EC on universal service and users' rights to electronic communications networks and services (Universal Service Directive), which states in relation to the universal service in Article 3 in Chapter II that member states shall ensure that a specific set of services are made available at the quality specified to all end-users in their territory, independent of geographical location, and, in the light of specific national conditions, at an affordable price. The chapter entitled "Universal Service Obligations Including Social Obligations," covers the provision of access at a fixed location, directory enquiry services and directories, public pay telephones and special measures for disabled users. Special measures to be used by member states are to be targeted to ensure access to, and affordability of, publicly available telephone services, including access to emergency services, directory enquiry services, and directories.

---

[1] Treaty establishing the European Community (consolidated version). http://www.eur-lex.europa.eu.

Directive 2002/21/EC on a common regulatory framework for electronic communications networks and services (Framework Directive) gives tasks to member states. Article 8 states that the national regulatory authorities shall promote competition in the provision of electronic communications networks, electronic communications services, and associated facilities and services by, among other things, ensuring that users, including disabled users, derive maximum benefit in terms of choice, price, and quality. The national regulatory authorities shall also promote the interests of citizens of the European Union by, inter alia, addressing the needs of specific social groups, in particular disabled users.

User involvement is observed. Article 16(2) of Directive 2000/31/EC on certain legal aspects of information society services, in particular electronic commerce, in the internal market ("Directive on Electronic Commerce") states that member states and the commission shall encourage the involvement of associations or organizations representing consumers in the drafting and implementation of codes of conduct affecting their interests. Where appropriate, associations representing the visually impaired and disabled should be consulted so that their specific needs are taken into account.

The regulatory framework is reviewed periodically. It should be seen in the context of other relevant legislation and specific instruments, for example, public procurement.

53.3.4 Privacy and Transparency

Privacy is a fundamental human issue. Both the Council of Europe and the European Union have legislation in this field. The Council of Europe has created legislation on human rights. Article 8 of the Convention for the Protection of Human Rights and Fundamental Freedoms (Convention 1950) states that everyone has the right to respect for private and family life, and home and personal correspondence. There shall be no interference by a public authority with the exercise of this right, except such as in accordance with the law and as is necessary in a democratic society in the interests of national security, public safety, or the economic well-being of the country; for the prevention of disorder or crime; for the protection of health or morals; or for the protection of the rights and freedoms of others.

In the field of information society, European Union legislation includes the Directive 95/46/EC on the processing of personal data and Directive 2002/58/EC concerning the processing of personal data and the protection of privacy in the electronic communications sector. In addition, there is detailed national legislation in the member states.

Privacy protection is well justified. Health is typically considered a private affair. But there are also other needs. People also want to know about their government. This is a basis of democracy. Administrations should be transparent; one should have access to public documents. In Nordic countries public access is a traditional principle and regulated in detail by law (Kemppainen, 1996). The citizen has a good reason to trust a transparent government. Hence, any government also has an interest to be open and transparent.

But governments may also have matters that for good reasons should not be transparent, for example, for the sake of public security or business reasons. Furthermore, newspapers and other media may want to know more than they are told. There is also a legitimate reason for knowledge gathering in a democracy, where citizens have the task and often the duty to elect their leaders.

Hence, many, sometimes conflicting, interests are involved in the transfer of information. It is the task of legislation to balance these interests. The complexity of interests leads to the fact that the legislation is also complicated. An organized society requires trust between persons. Trust requires privacy protection and confidentiality. On the other hand, trust requires also transparency to a certain degree. Privacy issues are addressed in more detail in Chapter 57, "Security and Privacy for Universal Access," of this handbook.

53.3.5 Product Safety

Safety supports usability and accessibility. Product safety is import for consumers (Kemppainen et al., 2007). Article 1 of the Directive 85/374/EC on liability for defective products expresses the main principle that the producer shall be liable for damage caused by a defect in his product. According to Directive 2001/95/EC on general product safety, a product is deemed safe once it conforms to specific community provisions, national regulations, or certain principles. The general safety requirement is imposed by the directive on general product safety on any product put on the market for consumers or likely to be used by them, including all products that provide a service.

A product is deemed safe once it conforms to the specific community provisions governing its safety. In the absence of such provisions, the product's compliance is determined in certain other ways.

Obligations have been laid on manufacturers and distributors. The manufacturers must put on the market products that comply with the general safety requirements. They must also provide consumers with the necessary information. Distributors are obliged to supply products that comply with general safety requirements, to monitor the safety of products on the market, and to provide the necessary documents to ensure that the products can be traced. If the manufacturers or distributors discover that a product is dangerous, they must notify the competent authorities, and if necessary, cooperate with them.

The member states should put in place structures that are responsible for monitoring a product's compliance with safety requirements and taking the necessary measures in this regard, for example, prohibiting products that fail to comply from being marketed. The member states can take restrictive measures by informing the commission, which communicates the information to the other member states.

53.3.6 Public Procurement

Public procurement is a remarkable market force. Public procurement is regulated by directives. The primary purpose of these directives is to ensure that there is a properly functioning internal market, so suppliers from any member state can have equal access to the public procurement markets in any member state.

Attention has been given to how these directives can be used to further also other objectives of public policy, including environmental and social objectives. Public procurement policy and practice for accessible ICT has the potential to play a vital role in removing barriers to participation in the information society by disabled or older people.

Directives encourage the inclusion of accessibility criteria in public procurement. The preambles to the Directives (paragraph 29 of Directive 2004/18/EC and paragraph 42 of Directive 2004/17/EC) state that: "Contracting authorities should, whenever possible, lay down technical specifications so as to take into account accessibility criteria for people with disabilities or design for all users." In addition, the specific articles on technical specifications (Article 23, Paragraph 1 of Directive 2004/18/EC and Article 34, Paragraph 1 of Directive 2004/17/EC) state that: "Whenever possible [these] technical specifications should be defined so as to take into account accessibility criteria for people with disabilities or design for all users" (International Workshop on Accessibility Requirements for Public Procurement in the ICT Domain[2]).

However, it can be difficult to specify what should be the needed accessibility requirements. Standards can help. The European Commission has proposed to the member states the development of a European standard for a toolkit for eAccessibility requirements to be used in public procurement. Mandate 376 (M-376) by the European Commission has been given to the European Standardization Organizations (ESOs) to come with a solution for common requirements and conformance assessment. The main objectives of the mandate M-376 are to harmonize and facilitate the public procurement of accessible ICT products and services and to provide a mechanism through which the public procurers have access to an electronic toolkit, enabling them to make use of these harmonized requirements in the procurement process. The results might be useful also for procurement in the private sector. The work will be done by two combined project teams, an ETSI Specialist Task Force (STF) and a CEN Project Team (PT) (Placencia, 2007).

53.3.7 Assistive Technology

Assistive technology refers to products, devices, or equipment that are used to maintain, increase, or improve the functional capabilities of people with disabilities. There are wide variations in the delivery systems for assistive technology in the member

states. The way in which the provision of assistive technology is regulated and organized reflects differences in the overall social protection (European Commission, 2003).

For example, the provision of assistive technology has been an essential part of policy tradition in Nordic countries (Denmark, Finland, Iceland, Norway, and Sweden). All Nordic countries have a national system for allocation of assistive technology based on discretionary evaluation of the individual, but some systems are centralized or standardized, while others are based on local discretion and administrative control (Halvorsen and Hvinden, 2007).

53.4 Japan and Standardization

53.4.1 Legislation and Policy Measures

The Strategic Headquarters for the Promotion of an Advanced Information and Telecommunications Network Society (IT Strategic Headquarters) of Japan was established within the cabinet in 2001. The establishment was based on a law called the "Basic Law on the Formation of an Advanced Information and Telecommunications Network Society." The law provides the definition of an "advanced information and telecommunications network society," a society in which people can develop themselves creatively and vigorously in all fields of activities by acquiring, sharing, and transmitting a variety of information or knowledge on a global scale freely and safely through the Internet and other advanced information and telecommunications networks.

In January 2001, the IT Strategic Headquarters published a program, the so-called "e-Japan Strategy," which thereafter has been revised annually. The Ministry of Internal Affairs and Communications launched a research and development program called "u-Japan," which complements the e-Japan Strategy. The "u" in u-Japan stands for ubiquitous, universal, user-oriented, and unique. Ubiquitous indicates a permanent easy connection to networks anytime, anywhere by anything and anyone. Universal is a concept whereby all, including older persons and those with disabilities, can actively participate in society without thinking of an equipment and network. User-oriented is a declaration reflecting the fact that the technologies must be developed based on a user perspective; and finally, unique means creative and vigorous.

This movement was accelerated by the amendment of the Fundamental Law for the People with Disabilities, which was introduced in May 2004. Article 19, "Realization of barrier free information," was added to the law and requests the following:

> States and local authorities shall undertake the necessary measures to spread electronic computers and their related devices and other information and communications equipment that is user-friendly for people with disabilities, to enhance the convenience of the latter in their use of telecommunications and broadcasting services, and to equip facilities which provide information for people with

---

[2] http://europa.eu.int/information_society/policy/accessibility/deploy/pubproc/ws-2004-10/index_en.htm.

disabilities, allowing them to make use of information in an efficient manner and to express their own will.

Under this provision, consideration is stipulated for "promoting the computerization of administration and the use of information and communications technologies in public services." To promote these policy measures, it is important to implement accessibility functions into mainstream products.

53.4.2 Difficulty in Implementing Accessibility Functions in Mainstream Products

Chances of interacting with ICT products and services are increasing day by day. An individual cannot succeed in society without access to the ICT products and services. The issues of accessibility by people with disabilities must be urgently solved, so that they can also succeed and contribute to the society. In the countries where the population is aging, it is also necessary to include older persons in the information society.

Companies generally agree with the overarching desire to implement accessibility functions in mainstream products, but they face difficulties during product design. The greatest difficulty is that it is not easy to fully understand the needs of people with disabilities.

People in different categories have different needs. However, because of the small number of people in each category, it is not feasible for a company, even if the company tries its best, to gather needs directly from people in various categories. Thus, it is difficult for the company to design products taking into account the widest range of needs that people with disabilities have.

The ISO (International Organization for Standardization) and IEC (International Electrotechnical Commission) jointly form a committee called Joint Technical Committee (JTC1). The JTC1 established the Special Working Group on Accessibility (SWG-A) in 2004 considering the necessity of addressing ICT accessibility in global standardization activities. The SWG-A gathered the needs of people with disabilities and published a *User Needs Summary* (2008).

The summary can also be used by a company to understand the needs of people with disabilities and to analyze whether its mainstream products take into account these needs. The summary contributes to solve the first difficulty a company faces when designing products in which accessibility functions are implemented.

However, there exists a second difficulty: companies must decide which needs should be fulfilled while others are not. Usually the decision is made based on the following logic. A company evaluates individual needs from the viewpoints of the market size and the resources necessary to fulfill the need. If the market size is greater than the resource, the company decides to fulfill the need. Otherwise, the company cannot implement an accessibility function corresponding to the need. In short, economical justification is necessary.

Companies are profit-seeking entities, and profit seeking is the origin of a healthy market economy. Therefore, this behavior of companies has to be accepted. Of course, in some cases

companies recognize their social responsibility and take into account some needs that are not economically justifiable. But it is impossible for companies to respond to all needs because of their social responsibility.

The needs that are not covered by mainstream products are supported by assistive technologies. It is common that governments subsidize development and procurement of assistive technologies from a social welfare policy perspective: governments compensate a failure in the market. But this is also the very reason why it is rare to find an assistive technology solution that is used worldwide. In other words, the assistive technology market is segmented. The European Commission conducted research on this subject in 2003. A new strategy is needed to facilitate the implementation of accessibility functions in mainstream products. Otherwise, society will continue to depend on assistive technologies that are economically inefficient.

53.4.3 Voluntary Standardization: A Tool to Facilitate ICT Accessibility

In every region, development of ICT accessibility standards is one of the priority topics. In Japan, for example, a set of standards has been developed and approved since 2004. These standards are used compulsorily or voluntarily and form a tool to facilitate ICT accessibility. Accessibility standards are discussed in detail in Chapter 55, "eAccessibility Standardization," of this handbook, while the standardization process is discussed in Chapter 54, "Standards and Guidelines."

During ICT accessibility standard development, the drafting committee carefully selects needs. The first criterion to select needs is matching categories of needs and kinds of disabilities. If the standard is to determine the interface between hearing aids and telephone sets, needs that directly relate to the coupling of hearing aids and telephone sets are chosen. If the standard is for web contents, needs that relate to mobility disabilities, e.g., needs of people using wheelchairs, are not taken into account.

The second criterion is technical feasibility because standards are valuable only when they are used in the market. Needs that are not technically achievable are usually not covered by standards. The drafting committee is very careful to write clauses. If a standard requests a product to "be operable without vision" and "be operable without hearing" simultaneously, the technology hurdle to achieve both requirements is very high. If the standard requests the product to comply with at least one of these requirements, the hurdle is low. But if a product complies with only "be operable without vision," people with hearing disability cannot or may not operate the product.

In the case where the drafting committee consists of users, providers, and a neutral party, users tend to insist on higher hurdles, while providers try to lower technology hurdles. The role of a neutral party, in most cases people from academia, is important to find solutions or compromises. Many standardization organizations claim that they are industry driven, but it is important to invite third parties to the drafting committees for accessibility standards development.

A committee usually avoids hurdles that are too high to jump over by the current technology. But it is recommended not to lower the hurdle too much: even though many manufacturers can easily ship products complying with the ICT accessibility standards, many users face a psychological difficulty in using products that are labeled "accessible."

A hurdle that is too high will reduce manufacturers' incentive to design products that comply with the standards. A standard that everyone ignores is not an influential one. It must be understood that usually standards do not cover all needs. One exception is ISO 9241-20 that will recommend accessibility requirements of all ICT products and services. It is a political process among drafting committee members to select needs that must be met in standard development process.

Publication of standards is important for both users and manufacturers. Manufacturers welcome the publication of ICT accessibility standards. It is easier for manufacturers to prioritize needs by referring to standards than by selecting needs internally.

However, what is more influential and useful is that the standards can be used to inform designers. ICT products and services are in most cases designed by young designers who are situated far from people with disabilities and older persons. ICT accessibility standards are educational material for these young designers.

Users can acquire mainstream products that implement some accessibility functions from the market. Users also can understand some needs are not satisfied by the mainstream products. Users can make their own decision between buying mainstream products or relying on assistive technologies.

Provisioning of accessibility functions in mainstream products results in an important side effect. A dramatic decrease of unit production cost of ICT hardware can be expected: ICT hardware uses a huge number of semiconductor chips, the cost of which decreases by accumulating production. Because of the reduction of unit cost, for example, a variety of speaking ICT products that install voice synthesizer chips can now be found. Assistive technology developers can also use the voice synthesizer function in their product without caring about the cost. In this way, mainstream products and assistive technologies can build a collaborative relation. In the case of software as well, if an embedded software module can be used in a variety of ICT products, it also reduces the unit cost of ICT products.

The market welcomes accessible products. The success of an accessible product influences other manufacturers and facilitates further deployment of accessible products. In the best case, this chain reaction creates a trend of deeper thinking of featuring accessibility functions in the mainstream products. This is the final effect of the voluntary use of ICT accessibility standards.

Transfer of knowledge from assistive technology developers to mainstream product designers is inevitable to implement accessibility functions into mainstream products. Assistive technology developers receive rewards of the knowledge transfer that occurs usually in the form of patent licensing.

Designing of assistive technologies becomes easier when standards define the interface between assistive technologies and mainstream products explicitly. In addition, as previously mentioned, assistive technology developers can use cheaper components that were already installed in mainstream products. These development and production cost reductions positively affect the assistive technology market and facilitate the dissemination of assistive technologies. Development of voluntary standards is a business risk for assistive technology developers.

53.4.4 Use of Standards as a Mandatory Requirement for Public Procurement

In 1995, the Japanese government announced guidelines for the criteria to be used in the general evaluation of contracts and tenders for the supply of computers and services to the government as an agreement among agencies and bureaus. A statement in the announcement reads: "Items to be evaluated shall be established in conformity with international and national standards." Therefore, products and services to be supplied to the government must be designed with consideration for ICT accessibility determined in the series of national standards.

One example is the Ministry of Economy, Trade, and Industry's public procurement announcement about information systems. The announcement requests tenders to explain how the national accessibility standards are met.

In Japan, there is a public procurement system for accessible web sites for the central government. The two basic laws explained previously request the necessity of accessibility in governmental web sites. Under the two basic laws, the government developed basic plans and annual implementation plans in which target dates are determined. One example was that e-Government systems that were under development before 2006 would increase information provisioning in an accessible format by fiscal year 2006.

Web site development of these central and local governments is usually contracted to system integrators. In order to get contracts, system integrators develop and publish free of charge web accessibility checking tools. In addition, system integrators develop helper tools, such as, voice synthesizer software and/or software to change color schemes, installed in governmental web sites.

One interesting point is that the laws do not impose punishment. It is believed that the graying of Japanese society triggered the movements. According to the 2005 edition of the *White Paper on the Aged Society*, Japanese people ages 65 years or older accounted for about 20% of the national population, and this figure is expected to top 26% (approximately 33 million people) by 2015. Since technologies are rapidly evolving in the ICT area, technical specifications that are used in public procurement must be revised periodically.

53.4.5 Necessity of Conformance Assessment and Its Difficulty

Most of the accessibility requirements in voluntary standards and mandatory technical specifications are not quantitative but qualitative. This qualitative nature makes the product conformance judgment very difficult.

If a requirement is "the character size printed on operable controls shall be 5 mm or larger and the contrast ratio shall be 4:1 or higher," it is easy to judge. However, if a requirement is "a screen display device shall support advanced functions such as character enlargement and contrast adjustment," no one can determine the appropriate sizes and the contrast of characters on a screen.

Because of this ambiguity, regions are now struggling to establish accessibility conformity assessment systems. Especially for public procurement it is very necessary to establish such a system.

Telecommunications equipment that complies with the national standard JIS X8341-4 "Telecommunications equipment" carries a symbol called a *U mark* shown in Figure 53.1.

The U mark is a labeling system maintained by the Info-Communication Access Council. The council developed a checklist based on JIS X8341-4. A manufacturer uses the checklist and if it thinks a product complies with the checklist, the manufacturer sends the list to the council. The manufacturer attaches the mark to the product, and the council discloses the received checklist to the public. In short, the U mark is a self-declaration system. By 20 October 2006, 20 products carried the mark.

In a country like Japan where implementation of accessibility functions is not mandatory, it is difficult to establish a reliable conformance assessment system. There are a variety of alternatives of conformance assessment systems, namely self-declaration, self-policing mechanism, best-practitioner methods, top-runner systems, third-party testing, accessibility management system, and so on.

FIGURE 53.1 U mark attached to accessible telecommunications equipment. (Reproduction from Info-Communication Access Council.)

It is necessary to consider the pros and cons of alternatives. But what is important is to start such considerations as soon as possible, so that one can meet the necessity of developing conformance assessment systems for public procurement of accessible ICT products and services.

53.5 A U.S. Disability Civil Rights Perspective

The Americans with Disabilities Act (ADA), signed into law by President George H. W. Bush on 26 July 26 1990, requires that covered entities' programs, services, or activities afford equal access to qualified individuals with disabilities unless doing so would fundamentally alter the nature of their programs, services, or activities or would impose an undue burden. ADA Title I requires that covered employers provide all aspects of their employment opportunities to otherwise qualified individuals with disabilities in a nondiscriminatory manner. Since its passage, the ADA has withstood constitutional challenges though, in recent years, the U.S. Supreme Court has narrowed its applicability and availability to fewer persons with disabilities by various statutory interpretations. Nonetheless, it has been a model used by approximately 40 countries worldwide that have adopted a legislative framework of civil rights for their citizens with disabilities.

As a result of U.S. Supreme Court rulings over the past 10 years that significantly narrowed the definition of disability and other provisions, the ADA Amendments Act was signed into law on September 25, 2008. President Bush has indicated that he will sign such a House-passed legislation into law if no other provisions are included by the Senate. In essence, the scope of the amendments has been limited to restoring the rights thought previously granted in the 1990 ADA but narrowed by Supreme Court rulings and nothing else. Therefore, while the existence of an Internet or World Wide Web (WWW) was not even contemplated in 1990, there was great reluctance by U.S. lawmakers today to include in these 2008 ADA amendments any clear statement of public policy that, as a civil right, access to the Internet or the WWW must be required of covered entities.

A second relevant piece of U.S. disability civil rights legislation is the Rehabilitation Act of 1973, as amended, and specifically Title V thereof. Section 504 requires nondiscrimination toward people with disabilities in the provision of programs and services as well as employment opportunities by federal financial recipients, such as hospitals and health care facilities, colleges and universities, local school districts, state and local governments and any nongovernmental organization or other entity that receives federal financial assistance. Section 508 of the Rehabilitation Act and Section 255 of the Communications Act of 1996 help to form the legal backbone of accessibility in the American ICT environment.[1] In broad terms, Section 508 requires federal agencies to use accessibility as a selection criterion when procuring ICT, while Sections 255 and 251(a)(2) of the

Communications Act of 1996 require certain telecommunications-related equipment and services to be designed, developed, and fabricated to be accessible to and usable by people with disabilities, if readily achievable.

These provisions ensure that people with disabilities will have access to a broad range of products and services such as telephones, cell phones, pagers, call-waiting, and operator services, which were often inaccessible to many users with disabilities. The U.S. Federal Communications Commission enforces these requirements, in conjunction with the U.S. Department of Justice.

Despite the ADA's mandate of equal treatment and accessibility that includes information and communication technologies, people with disabilities experience discrimination in a variety of manners. The remainder of this section examines health care settings to gain some insight into how the ADA is being implemented and enforced by the U.S. Department of Justice (DOJ).

The DOJ recognizes that enhanced and emerging technologies may allow health care providers to obtain qualified interpreters more quickly, economically, and efficiently 24 hours a day. For instance, providers may utilize video interpreting services whereby a qualified sign language interpreter appears via video from a remote location on a television-like screen. A health care provider opting for this approach, however, must take the necessary steps to ensure that the appropriate hardware and software are in place to support the system, and that staff understand how to operate and maintain the equipment. This is especially critical in the hospital setting. Where a hospital purports to utilize video interpreting services but does not provide the necessary administrative and operational support to ensure the system works, a patient with a hearing disability is denied the right to fully participate in health care decisions, and family members are shut out from communicating with the hospital about their loved one.

For example, in 2006, the DOJ intervened in a private lawsuit, *Gillespie v. Dimensions Health Corp., d/b/a Laurel Regional Hospital*,[3] brought by seven deaf individuals against Laurel Regional Hospital alleging a failure to provide appropriate auxiliary aids and services, including qualified interpreters, necessary to ensure effective communication for deaf patients or deaf family members, either in the emergency department or during hospitalizations. In that case, the Maryland hospital had an older system of video interpreting services available, but hospital staff allegedly had difficulty setting up and operating the system. The picture allegedly was, at times, too blurry for a patient to clearly distinguish the arms and hands of the video interpreters, and the video camera could not be adjusted for prone patients so that the interpreter and the patient could clearly see each other's hands, arms, and heads.

In *Laurel Hospital*, the complainants alleged that the hospital failed to provide an interpreter for a deaf patient during hospitalization. The hospital allegedly did not attempt to communicate with the deaf patient in any way, but rather forced her hearing mother to function as a relay person, consecutively exchanging simplistic messages between her adult daughter and the hospital regarding her daughter's condition and treatment. The patient complained to the DOJ that her mother was often unable to communicate to her what hospital personnel had said, and that because her mother does not know sign language, the patient was forced, often unsuccessfully, to try to read her mother's lips.

In 2006, the DOJ resolved the Laurel Hospital case through a consent decree,[4] which included detailed provisions for the implementation and administration of a program to ensure effective communication with persons with hearing disabilities. The consent decree required the hospital, among other measures, to continue to provide both onsite interpreters and interpreters appearing through video interpreting services where necessary for effective communication, and to train hospital personnel to accommodate the communication needs and preferences of deaf or hard-of-hearing patients and family members. In addition, the DOJ required Laurel Hospital to satisfy specified performance standards for its video interpreting services regarding the quality and clarity of the televised video and audio, regardless of the body position of the patient, and to train hospital staff to quickly and easily set up and operate the system.

53.5.1 Virtual Public Accommodation under the ADA

Another area where people with disabilities may encounter difficulty with accessibility is accessing the Internet. One way to help meet the ADA's requirements is to ensure that web sites have accessible features for people with disabilities. An entity with an inaccessible web site may also meet its legal obligations by providing an alternative accessible way for people with disabilities to use its programs or services, such as a staffed telephone information line. These alternatives, however, are unlikely to provide an equal degree of access in terms of hours of operation and the range of options and programs available. The degree to which such alternate access is "unequal" and possibly discriminatory is the basis of recent federal litigation under The Americans with Disabilities Act and California's state disability civil rights law.[5]

There have been a few cases that directly dealt with the issue of ADA applicability to the web, and even fewer that provide guidance. A few cases settled before a court ruling, such as *National Federation of the Blind, Inc. v. AOL Time Warner, Inc.* (Civil Action No 99-12303 D Mass), where AOL agreed to publish an accessibility policy and take other steps to improve accessibility.

---

[3] *Gillespie v. Dimensions Health Corp., d/b/a Laurel Reg. Hosp.*, No. 05-73 (D. Md.). http://www.ada.gov/laurelco.htm.

[4] *Gillespie and United States v. Dimension Health Corp. d/b/a Laurel Reg. Hosp.*, No. 05-73 (D. Md. July 12, 2006). http://www.ada.gov/laurelco.htm.

[5] See *National Federation of the Blind v. Target Corp.*, discussed *infra*. This case is discussed in Section 53.5.2.

In another case, *Rendon v. Valleycrest Prods., Ltd.*,[6] plaintiffs with hearing and upper-body mobility impairments sued the producers of the television show "Who Wants to Be a Millionaire," claiming that the use of the "fastest finger" telephone selection process violated the ADA. The *Rendon* case stands for the idea that Title III discrimination can be done by off-site, nonphysical actions and procedures. The court held that there must be a nexus to a physical place.

In *Access Now, Inc. v. Southwest Airlines Co.*,[7] the plaintiffs alleged that Southwest Airline's web site was inaccessible to blind people because it did not provide the user with textual information. The court dismissed the case, holding that the ADA does not apply to private web sites. The court applied a narrow reading of the statute and held that "to fall within the scope of the ADA as presently drafted, a public accommodation must be a physical, concrete structure."[8] The court understood Title III of the ADA to govern only access to physical structures. By doing this, the court rejected the holding in *Carparts Distribution Center, Inc. v. Automotive Wholesaler's Ass'n of New England, Inc.*[9] that Title III of the ADA applies to discrimination in the access to employee medical benefit plans, even if the plans themselves are not purchased or sold on an in-person basis at a physical location. The court also found that the plaintiffs failed to state a claim upon an alternative nexus theory in that they failed to "demonstrate that Southwest's website impedes their access to a specific, physical, concrete space such as a particular airline ticket counter or travel agency."[10]

On appeal, the plaintiffs did not argue that the web site is a place of public accommodation, so the Circuit Court did not rule on this question. The plaintiffs argued that Southwest Airlines as a whole is a place of public accommodation in that it offers a travel service. The Circuit Court refrained from ruling on this theory because the District Court had not had the opportunity to rule on it previously. The Circuit Court said it does not have to deal with *Carparts* because the District Court had dealt with the plain meaning of the statute. The court distinguished itself from the *Rendon* case by saying that, in *Rendon*, the plaintiffs were denied access to a physical place, unlike in this case.

53.5.2 Current Leading Federal Case: *National Federation of the Blind v. Target Corp.*

The *National Federation of the Blind v. Target Corp.* case is currently underway in the federal district court for the Northern District of California.[11] The *National Federation of the Blind* (NFB) case is the latest in a series of ADA cases that involve arguments about whether Title III of the ADA is applicable to

conduct at nonphysical sites, including Internet sites. Although the case has not yet gone to trial, its potential implications for other companies that maintain Internet sites may be inferred from the arguments that the parties have advanced and the comments that the court has made thus far in pretrial motions.

According to the NFB, it notified Target Corporation (Target) of alleged problems that blind individuals have when trying to use its web site, www.target.com, in May 2005, and then entered into negotiations with the retailer to find a mutually acceptable solution to those problems.[12] When those negotiations broke down, the NFB initiated its class action lawsuit against Target seeking to compel Target to alter its web site so that blind individuals can use the site independently. The advocacy group also seeks unspecified damages and attorneys' fees.

Over the past several months, the parties have filed various motions, responses, and replies with the district court. The NFB amended its initial complaint to refine its allegations. Target then tried to have the court dismiss the case, and the NFB fought the attempted dismissal. In the end, the court generally sided with the NFB regarding their motion to dismiss and allowed the case to move forward.[13]

The case now proceeds on the basis of the allegations contained in the amended complaint. Those allegations include assertions that Target is violating not only Title III of the ADA, but also the California Unruh Civil Rights Act (Unruh Act) and the California Disabled Persons Act (Disabled Persons Act). It is important to note that a violation of the ADA constitutes a violation of both the Unruh and Disabled Persons Acts. Therefore, as long as the NFB can show that a violation of the ADA has occurred, the group also will have established violations of the two California statutes.

One major issue in this case is whether or not Title III of the ADA applies to web sites. In part, Title III prohibits discrimination against any person "on the basis of disability in the full and equal enjoyment of the *goods, services, facilities, privileges, advantages, or accommodations of any place of public accommodation.*"[14] In the past, plaintiffs have relied on at least two different theories when asserting a Title III violation. First, they have claimed that discrimination in the granting of full and equal enjoyment of goods, services, facilities, privileges, advantages, or accommodations occurred at a place of public accommodation. Second, they have asserted that such discrimination occurred at an off-site, sometimes nonphysical, location that has a nexus to a place of public accommodation.

The first theory—holding that an alleged violation occurred at a place of public accommodation—is simpler and perhaps more straightforward than the second. The problem with relying on the first theory in cases like the one that the NFB brought is that

---

[6] 294 F.3d 1279 (11th Cir. 2002).

[7] 227 F.Supp.2d 1312 (S.D. Fla. 2002), appeal dismissed, 385 F.3d 1324 (11th Cir. 2004).

[8] *Id.*, 1318.

[9] 37 F.3d 12 (1st Cir. 1994).

[10] 227 F. Supp. 2d at 1321.

[11] *National Federation of the Blind v. Target Corp.*, No. 3:06-cv-01802 (N.D. Cal. removed to federal court March 8, 2006).

[12] *See* NFB's amended complaint, ¶ 37.

[13] *See* Memorandum and Order regarding Motion to Dismiss (denying the bulk of the Motion to Dismiss but also denying the NFB's request for a preliminary injunction that would have required Target to immediately alter its web site).

[14] 42 U.S.C. §§ 12101 *et seq.* (emphasis added).

courts in the United States are divided over whether a nonphysical site, like a web site, constitutes a place of public accommodation under the ADA. Some courts have held that the phrase refers only to physical locations.[15] As a result, defendants, like Target, often claim that Title III of the ADA is inapplicable to web sites and that complaints about a web site's accessibility cannot form the basis of a Title III lawsuit. Probably due to this line of court decisions, the NFB has chosen not to rely on an argument that http://www.target.com constitutes a place of public accommodation.

Instead of claiming that www.target.com is a place of public accommodation, the NFB relies on the second theory—the nexus theory of Title III liability. In its Amended Complaint and Opposition to the Motion to Dismiss, the NFB asserts that the relevant public accommodation in the present case is the brick-and-mortar Target stores, but that www.target.com constitutes "a service that is by and integrated with these stores."[16] As a result, the NFB claims that the court need not find that an online web site constitutes a place of public accommodation in order to find that Target has violated the ADA. Instead, the plaintiff is making an argument that a violation of Title III of the ADA exists if the web site has a connection to traditional Target stores such that the administration of the web site constitutes a barrier to the full enjoyment of the stores, which constitute places of public accommodation.[17] Or, as the court has enunciated the NFB's claim, it is an allegation that "the inaccessibility of Target.com denies the blind the ability to enjoy the services of Target stores."[18]

The nexus theory proved most successful in the *Rendon v. Valleycrest Prods., Ltd.* case concerning the popular television game show "Who Wants to be a Millionaire?" which relied on an automated, telephone process to screen out applicants for the show.[19] In order to become eligible to appear on the game show, would-be contestants had to call a hotline and compete against other individuals to answer a series of questions by typing on a telephone keypad. No provisions were made for people with hearing or mobility impairments. The district court in the case dismissed the plaintiffs' claims that the screening process discriminated against individuals with disabilities because, the court asserted, the screening-out process was not carried out at a physical location and thereby did not constitute a place of public

accommodation.[20] The plaintiff appealed to the circuit court, however, and that court reached a different conclusion. The circuit court stated:

> [Th]ere is nothing in the text of the [ADA] to suggest that discrimination via an imposition of screening or eligibility requirements must occur on site to offend the ADA....[21] [T]he fact that the plaintiffs in this suit were screened out by an automated telephone system, rather than by an admission policy administered at the studio door, is of no consequence under the statute.[22]

While the "Who Wants to be a Millionaire?" case did not involve a web site, it is important to the analysis of the NFB case because it clearly establishes that discrimination does not have to take place at an actual place of public accommodation in order to violate Title III of the ADA.[23] Instead, actionable discrimination occurs, according to the court, if an individual with a disability faces a barrier to access to a place of public accommodation via a remote, nonphysical action or procedure.

Despite the ruling in the "Who Wants to be a Millionaire?" case, there has been no clear elaboration of what kind of nexus has to exist between an off-site action and a place of public accommodation for there to be a Title III violation. For example, it is unclear from prior case law whether the fact that activities on www.target.com do not create a physical barrier to entry for blind individuals at brick-and-mortar Target stores undermines the NFB's claims. Target argues that, in order for the nexus theory to work, www.target.com would have to be "offered in, or preclude access to, a place of public accommodation," that is, a Target store, and cites the "Who Wants to be a Millionaire?" case as support for this argument.[24] Addressing Target's argument, the court in the NFB case stated that "no court has held that under the nexus theory a plaintiff has a cognizable claim only if the challenged service prevents physical access to a public accommodation."[25] Furthermore, the court stated that:

> The case law does not support defendant's attempt to draw a false dichotomy between those services which impede physical access to a public accommodation and those merely offered by the facility. Such an interpretation would effectively limit the scope of Title III to the provision of ramps, elevators and other aids that operate to remove physical barriers to entry.[26]

[15] See, e.g., *Weyer v. Twentieth Century Fox Film Corp.*, 198 F.3d 1104, 1114 (9th Cir. 2000); and *Access Now v. Southwest Airlines, Co.*, 227 F.Supp.2d 1312, 1318 (S.D. Fla. 2002); but see *Doe v. Mutual of Omaha Ins. Co.*, 179 F.3d 557, 559 (7th Cir. 1999); and *Carparts Distribution Ctr., Inc. v. Automotive Wholesalers Assoc. of New England, Inc.*, 37 F.3d 12, 19-20 (1st Cir. 1994).

[16] Amended complaint ¶ 56. See also Opposition to Defendant Target Corporation's Motion to Dismiss, pp. 5–6.

[17] For a further discussion of this theory, see National Council on Disability, When the Americans With Disabilities Act Goes Online: Application of the ADA to the Internet and the Worldwide Web, p. 10 (July 10, 2003). http://www.ncd.gov/newsroom/publications/2003/adaInternet.htm.

[18] Memorandum and Order, p. 9.

[19] See *Rendon v. Valleycrest Productions Ltd.*, 119 F.Supp.2d 1344 (S.D. Fla. 2000).

[20] See *Id.*

[21] *Rendon v. Valleycrest Productions Ltd*, 294 F.3d 1279, 1283-84 (11th Cir. 2002).

[22] *Id.*, 1286.

[23] *Stoutenborough v. National Football League*, 59 F.3d 580, 583 (6th Cir. 1995) (noting that Title III of the ADA protects access to all of the services that a public accommodation offers without distinguishing between physical and nonphysical sites or services).

[24] See Reply in Support of the Motion to Dismiss, pp. 2–3.

[25] Memorandum and Order, p. 7.

[26] *Id.*, p. 9.

These types of arguments designed to flesh out the scope and nature of the required nexus likely will be a primary focus of the current litigation between the NFB and Target. As a result, this case may create important precedent for future analyses of Title III of the ADA, as it applies to Internet sites for retailers and other entities that maintain web sites associated with physical locations or other places of public accommodation.

Another important issue in the NFB case is whether states have the right to regulate the Internet. In its motion to dismiss, Target argued that applying the Unruh Act and the Disabled Persons Act to the case at hand would constitute state regulation of the Internet, because those acts are California state statutes.[27] Target argues that such regulation would violate the Commerce Clause of the United States Constitution, which reserves the ability to govern interstate commerce to the federal government, and that it would do so for two reasons.[28] First, according to Target, application of the state statutes to www.target.com would amount to California's governing conduct beyond the state's borders. Second, Target argues that the Internet, like railroads, is a component of national commerce that requires uniform regulation at the national level.

The NFB generally countered Target's contentions by arguing that the state statutes do not directly regulate commerce or discriminate against interstate commerce.[29] The group also asserts that states have the right to regulate activity on the Internet.[30] While Target appears to have the more solid constitutional arguments on its side, the court seems to side more with the NFB on these issues. The court's reasoning is based, at least in part, on the notion that Target could develop a separate web site just for California consumers that would comply with the state statutes. The logical extension of this argument is that all nationwide retailers would be forced to create separate web sites for each state and ensure that consumers in each state have access to the appropriate, state-specific web sites. Furthermore, the court asserts that having Target alter its entire web site to comply with California state laws would "not mean that California is regulating out-of-state conduct,"[31] although the support for this argument is unclear at best. On the other hand, the court acknowledged that there is strong precedent for the notion that only the federal government can regulate the Internet.[32] To date, the court declined to rule on the Commerce Clause arguments.[33] Presumably, these arguments will arise throughout the remainder of the litigation.

While the implications of the Commerce Clause arguments are significant, they may have little impact on the practical resolution of the NFB case. Even if Target wins these arguments, it may lose the case on ADA grounds. The ADA is a federal statute, so its potential applicability to the Internet in this case may

very well require Target and other entities that own or operate web sites as well as traditional places of public accommodation to ensure that their web sites are accessible to blind individuals and persons with other disabilities throughout the country. Why then is Target willing to challenge a state's right to govern the Internet when doing so creates significant risk not only for Target, but for other companies operating Internet sites as well? One possible reason for Target's decision is that it believes it can win these arguments on appeal in the Ninth Circuit Court of Appeals and thereby create precedent that is helpful to the retailer. Alternatively, now that Target has received the district court's order indicating that that court is at least sympathetic to the NFB's arguments on the Commerce Clause, the retailer may choose to negotiate a settlement with the NFB in order to avoid establishing negative precedent in the district court.

53.5.3 Settlement in *NFB v. Target* Case

On August 27, 2008, the National Federation for the Blind (NFB) and Target Corporation announced that they had resolved their longstanding differences regarding the accessibility of the Target website, by way of a class settlement agreement in the United States District Court, District of California.

The lawsuit began in September 2006, when NFB alleged that Target's website was inaccessible to the blind and therefore violated the Americans with Disabilities Act (ADA), the California Civil Rights Act, and the California Disabled Persons Act. In early motion to dismiss, Target requested that the court terminate the suit on the grounds that it was not legally required to make its website accessible. The Court denied the motion and found that federal and state civil rights laws do indeed apply to the Target website. The court noted that, under Title III of the ADA, individuals with disabilities are protected against discrimination in "places of public accommodation." The court in *Target* noted, however, that—under Ninth Circuit law—a "place of public accommodation" is limited to physical locations and places, and would not, therefore, encompass Internet websites. The court thus stated that "[this] Circuit has declined to join those circuits which have suggested that a "place of public accommodation" may have a more expansive meaning."

This observation, however, did not bring an end to the court's inquiry, as NFB's legal theory was not that Target.com was itself a place of public accommodation, but rather that unequal access to the website denied blind patrons the full enjoyment of the goods and services of Target stores, which themselves are places of public accommodation. NFB thus advanced the argument that Target's website violated the mandates of the ADA under the "nexus theory," which involves a claim of unequal access to a *service* of a place of public accommodation, where there is a "nexus" between the challenged service and the place of public accommodation. The court agreed with NFB and concluded that the plain language of the ADA encompasses functions of places of public accommodation that go beyond the mere "brick and mortar" location and that the ADA prohibits discrimination in the full enjoyment of the "goods, services, facilities, privileges,

---

[27] Motion to Dismiss, pp. 21–24.

[28] See *id.*

[29] See Opposition to Motion to Dismiss, p. 19.

[30] See *id.*, pp. 20-24 (citing, for example, cases upholding state child protection laws).

[31] See Memorandum and Order, p. 18.

[32] See *id.*, p. 21.

[33] *See id.*, p. 22.

advantages, or accommodations" *of* any place of public accommodation—not *in* a place of public accommodation.

The court's denial of Target's motion to dismiss was the beginning of a series of rulings that did not bode well for the retail giant. On October 2, 2007, the court certified a nationwide class of all legally blind individuals in the United States who had attempted to access Target.com, as well as a California subclass of blind individuals who had attempted to access the website.

Perhaps seeing the handwriting on the wall, Target has agreed to settle the matter. Under the terms of the settlement, among other things, Target will ensure that its website meets the requirements of the Target Online Assistive Technology Guidelines and that "blind guests using screen-reader software may acquire the same information and engage in the same transactions as are available to sighted guests with substantially equivalent ease of use." Target will also take steps to implement changes identified by NFB and Target technical personnel. Once these changes are made, NFB will certify the website through the NFB Nonvisual Accessibility Certification program. Additionally, NFM will monitor the Target website, performing quarterly testing and annual user testing. NFB will also provide periodic, one-day training sessions regarding website accessibility to the retailer's employees who are responsible for the Target website.

The settlement also has a monetary component: in order to settle all claims for damages on behalf of the California class members, Target will pay six million dollars to be allocated among the class.

The Target settlement is considered a bit of a mixed blessing by some but is overall a very welcome development. On the downside, the fact that the case has settled means that there will be no forthcoming ruling that further affirms the right to accessible websites and the internet. On the other hand, the terms of the settlement make clear that the retail giant has been "persuaded" to make its website accessible and has acquiesced to the momentum of website accessibility. In the long run, this outcome may be nearly as persuasive in the overall case for website accessibility as would another judicial opinion.

For the full text of the agreement, go to http://www.nfbtarget lawsuit.com/final_settlement.htm.

53.6 Concluding Remarks

Use of technology takes place always in some context. Policies and legislation constitute one important context. Policies and legislation change over time and place. That is why it is important to identify more fundamental policy and legislative issues relevant to universal access. In this chapter, several frameworks for analyzing and assessing policies and legislation have been presented.

In different regions, different measures are most important. Legislation, regulation, legal cases, standardization exist everywhere, but there is variation with regard to which measures have been paid the most attention. It depends on legal, social, and historical context and tradition.

The frameworks that have been discussed include a United Nations Convention on the Rights of Persons with Disabilities, which puts accessibility issues in a wider context of different areas of life.

In Europe, eAccessibility is a clear policy objective. The policy framework is presented in eAccessibility Communication and in other European initiatives. Legislative areas that can promote equal opportunities and eAccessibility include: nondiscrimination, ICT, privacy and transparency, product safety, public procurement, and assistive technology. There is relevant legislation both at a European and a national level.

In Japan, after the publication of a set of ICT accessibility standards, products that comply with standards have come into the market. But this happened not because it was forced by legislation, but voluntarily. For governmental web sites, two national laws request mandatory compliance with a standard. The Japanese position is a mixture of mandatory and voluntary use of national ICT accessibility standards.

The core legislative framework in the United States consists of the Americans with Disabilities Act, the Rehabilitation Act, and the Telecommunications Act. In federal and state courts, there is an emerging body of law that informs public policy that supports the rights of people with a variety of disabilities to claim a civil right to accessible technologies and digital information. Design for all and assistive technologies help to promote access to facilities and services.

It is important to keep global harmonization in standardization. One reason is to reduce technical barriers to trade. If technical specifications differ region by region, manufacturers need to adjust their products so that they meet technical specifications. On the other hand, if the specifications are harmonized globally, manufacturers benefit from the economy of scale. The economy of scale also benefits users of ICT products who can purchase products at lower prices.

Another reason for harmonization in standardization so that it takes accessibility needs into account is an ethical perspective, for human dignity and respect for human diversity. It is important to identify accessibility requirements that serve as many people as possible. Global harmonization encourages the movement of people, including people with disabilities and older persons, around the globe. However, for example, it is hard work for people to access a public terminal abroad, if the user interface is completely different from what they have in their homelands. But, if global harmonization is achieved, people can move more freely and frequently. Global harmonization consequently facilitates mutual understanding of people in different countries via movement of people.

References

Communication (2005). *Communication to the Council, the European Parliament, the Economic and Social Committee, and the Committee of Regions on eAccessibility.* The Commission of the European Communities. Brussels, 13.09.2005. COM(2005) 425.

Convention (1950). *Convention for the Protection of Human Rights and Fundamental Freedoms.* Council of Europe. http://www.coe.int.

Convention (2008). *Convention on the Rights of Persons with Disabilities and Optional Protocol. United Nations.* http://www.un.org/disabilities/default.asp?navid=12&pid=150.

Directive 85/374/EC on liability for defective products. http://www.eur-lex.europa.eu.

Directive 95/46/EC on the processing of personal data. http://www.eur-lex.europa.eu.

Directive 1999/5/EC on radio equipment, telecommunications terminal equipment and the mutual recognition of their conformity (RTTE). http://www.eur-lex.europa.eu.

Directive 2000/31/EC on certain legal aspects of information society services, in particular electronic commerce, in the Internal Market (Directive on Electronic Commerce). http://www.eur-lex.europa.eu.

Directive 2000/78/EC establishing a general framework for equal treatment in employment and occupation. http://www.eur-lex.europa.eu.

Directive 2001/95/EC on general product safety. http://www.eur-lex.europa.eu.

Directive 2002/21/EC on a common regulatory framework for electronic communications networks and services (Framework Directive). http://www.eur-lex.europa.eu.

Directive 2002/22/EC on universal service and user's rights to electronic communications networks and services (Universal Service Directive). http://www.eur-lex.europa.eu.

Directive 2002/58/EC concerning the processing of personal data and the protection of privacy in the electronic communications sector. http://www.eur-lex.europa.eu.

Directive 2004/17/EC of the European Parliament and of the Council of 31 March 2004 coordinating the procurement procedures of entities operating in the water, energy, transport and postal services sectors. http://www.eur-lex.europa.eu.

Directive 2004/18/EC of the European Parliament and of the Council of 31 March 2004 on the coordination of procedures for the award of public works contracts, public supply contracts and public service contracts. http://www.eur-lex.europa.eu.

European Commission (2003). *Access to Assistive Technology in the European Union.* http://ec.europa.eu/employment_social/disability/assitive_technology_study_en.pdf.

Government of Japan (2005). White Paper on the Aged Society [in Japanese]. http://www8.cao.go.jp/kourei/english/annualreport/2005/05wp-e.html.

Halvorsen, R. and Hvinden, B. (2007). Accessibility and participation for people with disabilities: Unsolved issues in the Nordic welfare states, in *Provision of Assistive Technology in the Nordic Countries* (2nd ed.). Helsinki: NUH, Nordic Centre for Rehabilitation Technology.

Kemppainen, E. (1996). Public access and privacy in an information society, in *Dreams and Realities: Information Technology in the Human Services, HUSITA 4 Conference* (B. Glastonbury, ed.), June 1996, Helsinki, Finland. Helsinki, Finland: Stakes National Research and Development Centre for Welfare and Health.

Kemppainen, E., Giovannini, C., Allen, B., Abascal, J., Soede, T., and Delaitre, S. (2007). Moral and legislative issues with regard to ambient intelligence, in *Towards an Inclusive Future: Impact and Wider Potential of Information and Communication Technologies* (P. Roe, ed.), pp. 188–205. Brussels, Belgium: COST. COST219ter. http://www.tiresias.org/cost219ter/inclusive_future/index.htm.

Ministerial Declaration (2006). *Declaration on eInclusion,* 11 June 2006, Riga, Latvia. http://ec.europa.eu/information_society/events/ict_riga_2006/index_en.htm.

Placencia, I. (2007). eAccessibility: European developments and targets, in *Towards an Inclusive Future: Impact and Wider Potential of Information and Communication Technologies* (P. Roe, ed.), pp. 234–246. Brussels, Belgium: COST. COST219ter. http://www.tiresias.org/cost219ter/inclusive_future/index.htm.

Press release (2005, September 15). *Commission Calls for Coordinated Action to Make Information and Communication Technologies More Accessible for Citizens.* http://europa.eu/rapid/pressReleasesAction.do?reference=IP/05/1144&format=HTML&aged=0&language=EN&guiLanguage=en.

User Needs Summary (2006). JTC1 SWG-A. This document will be published as TR 29138-1 from ISO/IEC JTC1 in 2008.

54

Standards and Guidelines

Gregg C. Vanderheiden

54.1 Introduction

Standards and guidelines take a wide variety of forms and serve diverse functions. Some are simply advisory and informative. Others are voluntary but have specific requirements that must be met to claim conformance. Compliance with some standards may be required as part of a purchasing contract or by regulation.

Standards and guidelines go by a wide variety of titles that do not necessarily indicate their application. Documents that fulfill the roles for standards and guidelines may be called "standards," "guidelines," "recommendations," or even "bulletins."

This chapter provides an introduction to the world of standards and guidelines, overviews some of the principal roles they play, and describes the different processes used to create them. The objective is to provide a short introduction to this area, as well as basic information regarding what it takes to participate in these processes.

Standards are very important to universal access for a number of reasons. Some standards deal with built-in accessibility features. If the same functionality varies widely from one device to another it may not be possible for a user to figure out how to activate or use the features he needs on products he encounters. Other standards deal with interconnectivity and allow users to connect their assistive technologies in standard and predictable ways to mainstream products to make them usable. Finally, performance standards ensure that the features operate in a manner that is safe and sufficient to meet users' needs.

54.2 The Role of Standards and Guidelines

Standards and guidelines are developed for a number of purposes, including:

- *Compatibility*: To allow products to work together
- *Performance*: To ensure a certain level of performance
- *Design*: To ensure that something is done in a particular way

54.2.1 Compatibility Standards

When two or more devices need to work together and interoperate, the point at which they meet must often be standardized.

This is especially necessary if the products or systems have been or will be created by different companies. Interoperability or compatibility standards must be followed precisely by both parties so that their devices will work together.

Usually, interoperability standards are voluntary but have clear conformance requirements. That is, companies are free to use the standard or not, but if they say that they follow the standard, there are very clear specifications that must be met. Unless both products follow them precisely, interoperability is compromised or fails.

The RS232 serial standard and the USB standard are examples of *interoperability* or *compatibility standards*. If products on both sides of the connection follow the standard, then plugging the two devices together should result in a successful mechanical and electrical connection.

Standards that are referred to as voluntary may be required to meet product certification (e.g., to qualify for a "Windows-ready" logo) or to meet a contract requirement (e.g., "all computers provided under these contracts shall have at least two USB 2.0 ports"). Compatibility standards can also be required by law (e.g., oxygen tank fittings).

54.2.2 Performance Standards

Performance standards are so named because they require a particular level of performance. This might be strength, load, impact, or many other types of performance. Again, the standards may be voluntary (e.g., companies may choose to meet particular standards to enhance effectiveness of their products) or implementation may be mandated in a purchase contract or by law (e.g., all wiring in the house must meet or exceed a specified level of flame resistance).

54.2.3 Design Standards

Whereas performance standards state what should be accomplished (that is, the end state), a design standard prescribes particular ways that something should be done. Where a performance standard might specify that a particular level of measurement should be reached, the design standard would establish the particular technique that should be used.

For example, a performance standard for a grab bar might state:

Instructions and all information for use shall be made accessible to and independently usable by persons with vision impairments.

A design standard related to flashing lights might state:

Visual alarm signals shall have the following minimum photometric and location features:

1. The intensity shall be a minimum of 75 candela.
2. The flash rate shall be a minimum of 1 Hz and a maximum of 3 Hz.

3. The maximum pulse duration shall be two-tenths of 1 second (0.2 sec) with a maximum duty cycle of 40%.

Standards are rarely purely design or purely performance, but are usually a combination.

54.2.4 Safety Standards

Safety standards are not really a type of standard, but are instead an application area. Safety standards include performance, design, or compatibility provisions and are often a mixture of these. The previously mentioned fire rating on household wiring is one example of a performance provision. The oxygen tank fitting standard would be compatibility, and the color coding of oxygen valves to prevent medical mishaps would be a design standard. Safety standards can be voluntary, but compliance is usually mandated by regulation.

54.2.5 Accessibility Standards

Accessibility standards are another application area. They generally include a mixture of compatibility (interoperability), performance, and design standards. Safety aspects may address prevention of acoustic injury, triggering of seizures, and other health-related issues. Interoperability provisions may ensure that products will be compatible with adaptive assistive technologies (e.g., screen readers and alternate keyboards) and personal assistive technologies (e.g., hearing aids, power wheelchairs, and prosthetics) that individuals with disabilities use. Interoperability provisions may also ensure, for example, that a communication device can communicate with a broad range of mainstream communication devices, using a wide variety of media, so that the user can select which media to use according to abilities and preferences. Design guidelines are provided in areas where specific strategies have been found to be effective and are expected to be stable over a long period of time. Performance guidelines are used where additional design flexibility is desired. Performance guidelines provide more flexibility but less clarity, and are harder to implement for designers who are not as familiar with leads and constraints of individuals with disabilities. Accessibility standards are addressed in more detail in Chapter 55, "eAccessibility Standardization," of this handbook.

54.3 A Standard by Any Other Name

A standard can go by many different names. To confuse things further, people will dispute whether a particular document is a standard or not depending on how it is was created—not just how it is used. For example, a standard that has been developed in an open process with public review is often called an "industry voluntary standard." However, some will argue that unless it is created solely by industry, it is not an industry standard. Others will argue that standards that are not developed in an open process are not real standards. Standards are developed in both open and closed processes, and by most every sector from private to public to government. Regardless of origin, process, or name, they are referred to here as standards when they serve one or more of the roles previously cited for standards. The World Wide Web Consortium (W3C) calls its standards *recommendations* (e.g., HTML,[1] CSS[2]).

The Internet Engineering Task Force that creates the standards for the Internet calls its standards RFCs (where RFC stands for request for comment). This refers to a description of a standard for a new or modified Internet or networking protocol. When standards are proposed, they are made available for public comment, so that they can be refined and agreed upon. The document that details the proposed standards is called an *Internet draft* document or I-D. When the documents are finalized, they get the RFC name. Some RFCs of wide importance get their status raised, after many years of use, to an *Internet standard*, still keeping their RFC name (e.g., RFC-1122).[3]

The standard in the United Kingdom with regard to flashing that could cause seizures in individuals with epilepsy (which the broadcast industry must meet) is called a *guidance note* and is in fact an annex. Yet even as an annex to a note, it functions as the standard for the industry and for enforcement (Ofcom, 2005).[4]

In fact, any document that serves in one of the roles mentioned previously can be thought of as a standard. Companies have all sorts of standards they adhere to, and groups of companies or whole industries may have both formal and informal standards that they follow.

54.4 Formal versus Informal Standards

One would think that formal standards that were developed through a careful process with widespread input would be more valuable or stable than informal standards, or standards that were developed quickly or by a small closed group. In fact, a large percentage of the formal standards are not used, and many have never been used. A formal public standard is not more or less likely to be followed. Its use depends on whether the standard will meet some need of industry better than other standards. Some of the most successful standards are ones that were actually practices that were used first and established as standards later. So, excellent standards can be developed in a wide variety of fashions. And although proper process can increase the quality of the standard, it cannot ensure its usefulness.

54.4.1 Need for Fairness and a Level Playing Field

The adoption of a standard by industry can provide economic advantage to those who are in a better position to use the standard (or who have intellectual property rights to some aspects of the standard). Also, standards that are enforced as regulations by governments can have significant effects on the companies or organizations being regulated. In both these cases, there is a desire for fairness and a level playing field. In addition, if the standard is to be enforced as a regulation, there is a need for all stakeholders to be able to provide input. For these reasons, it is often desirable to have an open standards process that involves all of the stakeholders (including the public sector) and which allows for review and comment as part of the development process of the standard.

Standards can be created in a wide variety of ways, including:

- Single company creates a standard and it is later adopted as an industry standard
  - (e.g., VCR and DVD standards)
- A consortium of companies creates a standard in a closed fashion
  - (e.g., standards developed by industry associations such as CEA, TIA, or alliances such as DLNA—the Digital Living Network Alliance)
- Voluntary standards organizations
  - (e.g., American National Standards Institute, ANSI; the International Organization for Standardization, ISO; European Telecommunications Standards Institute, ETSI; IEEE, originally Institute of Electrical and Electronics Engineers)
- Governmental standards
  - (e.g., standards developed by a government or by an advisory committee created by the government and then adopted in regulation)

54.5 Shall, Must, Should, Recommended, May, Optional

In reviewing standards it is important to understand the meaning of certain key words, such as *shall*, *should*, and *may*. The following is an excerpt from RFC-2119[5] that defines the use of these terms.

1. MUST: This word, or the terms "REQUIRED" or "SHALL," means that the definition is an absolute requirement of the specification.
2. MUST NOT: This phrase, or the phrase "SHALL NOT," means that the definition is an absolute prohibition of the specification.

---

[1] HTML 4.01 Specification, W3C Recommendation (1999, December 24). http://www.w3.org/TR/1999/REC-html401-19991224.

[2] Cascading Style Sheets, level 2, CSS2 Specification. http://www.w3.org/TR/1998/REC-CSS2-19980512.

[3] Requirements for Internet Hosts: Communication Layers, Internet Engineering Task Force (1989 October). http://www.ietf.org/rfc/rfc1122.txt.

[4] Ofcom Guidance Note on Flashing Images and Regular Patterns in Television (Re-issued as Ofcom Notes 25 July 2005). http://www.ofcom.org.uk/tv/ifi/guidance/bguidance/guidance2.pdf.

[5] Key words for use in RFCs to Indicate Requirement Levels, Internet Engineering Task Force (1989 October). http://www.ietf.org/rfc/rfc2119.txt.

3. SHOULD: This word, or the adjective "RECOM-MENDED," means that there may exist valid reasons in particular circumstances to ignore a particular item, but the full implications must be understood and carefully weighed before choosing a different course.

4. SHOULD NOT: This phrase, or the phrase "NOT RECOM-MENDED," means that there may exist valid reasons in particular circumstances when the particular behavior is acceptable or even useful, but the full implications should be understood and the case carefully weighed before implementing any behavior described with this label.

5. MAY: This word, or the adjective "OPTIONAL," means that an item is truly optional. One vendor may choose to include the item because a particular marketplace requires it, or because the vendor feels that it enhances the product, while another vendor may omit the same item. An implementation which does not include a particular option MUST be prepared to interoperate with another implementation which does include the option, though perhaps with reduced functionality. In the same vein, an implementation which does include a particular option MUST be prepared to interoperate with another implementation which does not include the option (except, of course, for the feature the option provides).

ISO Directives Part 2 Annex G[6] provides similar descriptions but slightly different directions. They state that:

- SHALL and SHALL NOT are to "be used to indicate requirements strictly to be followed to conform to the document and from which no deviation is permitted."
- MUST, MUST NOT, and MAY NOT are not to be used for requirements.
- SHOULD and SHOULD NOT are to "be used to indicate that, among several possibilities, one is recommended as particularly suitable, without mentioning or excluding others, or that a certain course of action is preferred but not necessarily required, or that (in the negative form) a certain possibility or course of action is deprecated but not prohibited."
- MAY and MAY NOT are to "be used to indicate a course of action permissible within the limits of the document."
- POSSIBLE, IMPOSSIBLE, and CAN are not to be used to indicate permissible courses of action.
- CAN and CANNOT are to "be used for statements of possibility and capability, whether material, physical or causal."
- MAY signifies permission expressed by the document, whereas CAN refers to the ability of a user of the document or to a possibility open to him/her.

---

[6] International Organization for Standards (ISO) (2001). ISO/IEC Directives, Part 2 Rules for the Structure and Drafting of International Standards (4th ed.).

54.6 Normative versus Informative Elements

In reading standards the terms NORMATIVE and INFOR-MATIVE are often found.

- *Normative* elements are defined in ISO Directives Part 2[6] as "elements that describe the scope of the document, and which set out provisions."
- *Informative* elements are those that "identify the document, introduce its content and explain its background, its development and its relationship with other documents" or "that provide additional information intended to assist the understanding or use of the document."

In short, normative parts of the standard are the ones that must be followed. Informative parts are just that, informative, and do not need to be followed. Note that normative portions of the standard may appear anywhere, including the appendices or annexes. For example, in ISO Directives Part 2 Annexes A, B, and I are informative but Annexes C, D, E, F, G, and H are normative.

54.7 Do You Have to Follow Standards?

The simple answer is yes, no, and sometimes.

Most standards are industry voluntary standards. They are created so that technologies can work together or work predictably. Companies are free to use the standards or not. Of course, companies may be forced to use standards by market pressures. For example, if customers all want to have their VCR play a VHS tape, then the company would be forced by market pressures to use the VHS standard. When people buy a computer, they expect it to have USB ports, and today everybody wants to buy computers that have USB 2.0 ports. It would be hard to sell computers without them.

Once a company decides to use the standard, it will want to be able to assure people that it has followed the standard. This may involve testing labs or other certifying mechanisms that will evaluate products to determine whether products that say they have used a particular standard in fact have met the mandatory parts of the standard.

In addition, conformance to some standards may be required by government regulation. Sometimes a voluntary standard or parts of one, are made mandatory through regulation. Often standards are developed specifically to meet regulatory needs. For example, when the Americans with Disabilities Act was passed, the law specifically required the U.S. Access Board to create accessibility standards. The same is true with the Section 508 standards, whose development was mandated by legislation.

It should be noted that particular care must be taken when creating voluntary standards that will likely later be mandated. Although the standardization body does not have the ability to require that the standard be followed, the language and provisions should be written very carefully so that a later mandate does not create unwanted side effects. When standards are voluntary, they can be applied when they make sense and simply bypassed for those products or situations where they do not. However, if they are not applied as per their conformance criteria, while a product can be developed and sold, it cannot claim compliance to the relevant standard. If they are mandated, however, this is no longer possible. Thus, provisions of a standard that may be mandated should be written to include exceptions where the provisions would not apply. A scoping of the applicability by the regulatory agencies might be used to determine when the entire set of standards should be applied or not, but it is difficult (and rare) to provide scoping for individual provisions within the standard.

54.8 Stages in the Development of a Standard

As noted previously, standards are developed by a variety of organizations that use different specific procedures. In general, however, most of the major standards organizations follow a process that is fairly similar.

The following illustration describes the process used to develop ISO standards.

- *Potential work item (PWI)*: This is prior to the first official stage, where a group informally gathers ideas and prepares a proposal to the standardization body. In some organizations, this might be done by a special-interest or working group prior to the formation of the technical group tasked with the development of the standard. In other cases, a standing technical group may carry out this role. In ISO, this activity can be initiated by a proposal for a potential work item that is presented to the responsible subcommittee (SC) or working group (WG) for review and processing as an NWI.

- *New work item (NWI)*: At this stage, a proposal is circulated by a subcommittee or working group, often with a rough draft of the proposed standard. In ISO, it would be circulated to all of the participating and observing member countries. Once an NWI is approved, ISO has a clock that begins ticking, and the final standard is expected to be completed within 36 to 40 months.

- *Committee draft (CD)*: Once the NWI is approved, the subcommittee assigns it to a working group to begin serious drafting of the standard. Committee drafts are prepared and, once satisfactory to the WG and SC, are released for review and comments by the national bodies. Depending upon the feedback and comments (and the complexity of the standard), multiple CDs may be created before a draft wins approval needed to move forward to the next stage.

- *Draft international standard (DIS)*: When a CD wins approval, it is usually accompanied by comments and recommended changes that need to be resolved by the working group. The working group takes each of these comments and either accepts the comment, making

appropriate changes to the draft, or must provide a rationale for not accepting it. In the end, they create a proposed DIS. At this point, the standard again goes out for a vote. In the case of some standards, there may be a delay here as it is translated into other official languages, before being sent out for review. (Standards in the ISO 9241 series may be translated into French and/or German, because they are dual processed as CEN standards for the European Community.) It is believed that at this point the standard is getting close to completion so the voting is very formal and much more attention is given to the standard. This is the first truly public point, at which ISO will sell copies of the draft to the general public. Unfortunately, in some cases, this would be the first time that some reviewing bodies may pay close attention. Depending on the results of these reviews, it is still subject to major changes. Thus manufacturers are at double jeopardy whether or not they start to use it at this time.

- *Final draft international standard (FDIS)*: Taking the comments from the DIS review, the committee now creates a final document that they believe will pass without any substantial change. Once the document has successfully passed a DIS vote and the final changes have all been made, it becomes an FDIS. At this point, it is put out for the final vote. Within FDIS, however, the vote is simply a yes or a no as comments are not solicited, although they are often made. If the vote is successful, only editorial comments will be addressed. Technical comments may be retained for consideration in future revisions of the standard.

- *International standard (IS)*: Once a standard has passed the FDIS process, it moves on to become an IS. In ISO, approval must still be granted by an internal ISO committee. The document is formatted and then posted. This final process usually adds some months to the standard process. In ISO, it can add up to 6 months due to the need for translation.

Although different standards groups have names for their working groups (working group, technical committee, subcommittee, etc.) and procedures that vary somewhat from this, they all tend to follow the same basic stages from initial proposal for a work item, through multiple drafts with reviews, to the final draft, the clearing of comments and approval.

54.9 Getting Involved in Standards

Standards are often considered a mystery and something that is carried out behind closed doors by a select group of people. The process is also often seen as arcane and hard to understand.

While creating a standard does require participation by people who are familiar with the process and the potential sinkholes, most standards working groups are a mixture of people well-versed in creating standards, people well-versed in the topic area, and people representing the various stakeholders. Many of the individuals may be new to standards development.

Participation in the standards process can also take place at a number of different levels. Participation does not always have to involve travel. Although working groups usually meet face-to-face, many groups also conduct business by telephone. In addition, being part of a group that reviews standards can give one the opportunity to review and contribute to the standards during the various review stages. The ideal accessibility standards participant (or participant who will contribute accessibility considerations for a mainstream standard) would be one who:

- Understands the technology (or technologies) under discussion intimately.
- Understands all of the disabilities that would be involved (all degrees and combinations and needs, positions, and constraints of them cumulatively).
- Understands the standards process (what it can and cannot do and how each standards group operates).
- Is motivated to do standards. This can be one of the most challenging, mind-numbing, grueling, and frustrating activities, but it is also very rewarding.
- Is someone who reaches out (no one can rely on themselves and their own knowledge and do well for all stakeholders).

However, no one comes to the standards process with all of these skills. Often effective standards working groups are composed of people who, together, bring these qualities to the table.

54.9.1 Which Standards Are Important to Be Involved In?

With regard to those interested in standards and disability, there are two major categories of standards to consider.

First, there are standards that are written specifically for the benefit of people who have disabilities. Examples would be wheelchair standards, accessibility standards, and performance standards for various types of prosthetics, orthotics, or assistive technologies.

The second category would be mainstream standards, which need to include some considerations for people with disabilities. Unfortunately, this can include a tremendously wide range of standards.

Appendix A is an example of just one group ANSI standards that relate to wheelchairs (one assistive technology). Appendix B provides a list of standards that relate to information and communication technologies (ICT) that have disability impact. While some of these relate specifically to disability, many are general mainstream standards that have disability implications or components. As can be seen, they cover a wide range of different standards organizations. It is often the case that a single product may involve standards from a variety of different standards organizations. All of these must work together to provide an effective product, and all must support disability access so that an end product is usable by people with disabilities.

54.10 Challenges in Next-Generation Standards

The rapid advances being made in technologies and materials are posing tremendous challenges for both disability-related and mainstream standards. When new technologies or new materials are introduced, the current set of standards and testing methods may no longer make sense or be effective. In the area of information and communication technologies, this should be very clear, but it can also apply to wheelchairs and prosthetics, orthotics, and assistive technologies in general. As a result, the development of effective standards that have a meaningful shelf life is becoming increasingly difficult. Recent advances that point to the merging of technologies and the human body may complicate matters even more.

In fact, with advancing technologies, industry identified two requirements for disability-related standards (especially requirements) that they say are critical:

1. That the provisions be general so that they allow for industry innovation and changing technologies.
2. That the provisions be specific so that it is clear when they have been met. It is unfair to require companies to meet standards where it is unclear exactly what is needed to meet the standard or to determine when the standard provisions have been met.

These are both real needs of industry, especially companies involved with information and communication technologies (ICT), and creating standards that can meet both is a challenge.

To address this, new strategies are going to be required. One strategy used by the W3C and its Web Content Accessibility Guidelines, is to move to provisions that are more performance based (e.g., they say what should be accomplished) while not specifying the exact method to achieve this. Two problems must be overcome, however, to use this approach. First, for people to figure out if they have met the standard or not, the provisions must be objective. That is, they must be testable enough that developers and evaluators can reliably determine whether or not the provision has been met. Making the provisions both general performance based and objective/testable is difficult.

The second problem is the need to provide developers with specific guidance and ideas for meeting the provisions. This is particularly true for developers who may not themselves be experts in cross-disability access. Notes and examples are sometimes used. It is then extremely important that they are not misunderstood to be requirements in and of themselves and so that they do not constrain the developers using alternate approaches that may better meet the desired outcome of the actual guidance. There is usually a limit on the number and types of examples that can be provided. Furthermore, examples cannot be included for all the technologies that come out after the standard is adopted and released.

To address the need for specific guidance for a wide range of technologies, including technologies that are released after the standard is adopted, the Web Content Accessibility Guidelines 2.0 (WCAG 2.0) provides an accompanying document that lists specific techniques that would be sufficient to meet the provisions in the standard. Thus, individuals who are not experts in accessibility can consult the accompanying documents (titled "Quick Reference" and "Understanding WCAG 2.0"), and simply choose any of the techniques listed as sufficient and implement them. Or, if they are more knowledgeable, they can use a different or new technique that would meet the provisions of the standard.

This construction (a set of testable guidelines accompanied by a support document listing sufficient techniques) provides both the general (yet testable) guidelines and the specific (but not required) sufficient techniques. Individuals or companies who are knowledgeable are still free to innovate and meet the requirements of the standard in other ways, thus allowing for the introduction of new techniques and techniques for future technologies.

54.11 Conclusion

Standards are an invaluable but often misunderstood element in the overall *assistive technology* and *design for all* environment. Although they take time and effort to develop, they can have profound effects on the design of both special and mainstream technologies. Participation is open to all, from industry, to academia, to the public. And participation by all of the sectors is important to the process of ensuring that the standards by which our world is designed and built include the accommodations needed to allow full inclusion. Recognition of the importance of this work both by academic institutions and by funding agencies is important if people from academia and from the consumer organizations are to be able to participate in this important activity. It is as important as many aspects of research and as difficult or more difficult. Yet, it is a critical component in creating effective assistive technologies, and compatible mainstream technologies.

54.12 Appendix A: Listing of ISO/TC 173 Wheelchair Standards (10-07)

ISO/TC 173 Assistive Products for Persons with Disability		(as of 2007-10-18)
Standard	Title	Stage
ISO/TC 173/SC 1/WG 1		
13570-1	Guidelines for the application of the ISO 7176 series on wheelchairs	TR
13570-2	Technical Report: Wheelchairs—Typical values and recommended limits or dimensions, mass, and maneuvering space as determined in ISO 7176-5	TR/NWI
7176-1	Wheelchairs—Part 1: Determination of static stability	IS
7176-2	Wheelchairs—Part 2: Determination of dynamic stability of electric wheelchairs	IS
7176-3	Wheelchairs—Part 3: Determination of effectiveness of brakes	IS 2ed
7176-4	Wheelchairs—Part 4: Energy consumption of electric wheelchairs and scooters for determination of theoretical distance range	IS
7176-4	Wheelchairs—Part 4: Energy consumption of electric wheelchairs and scooters for determination of theoretical distance range	DIS
7176-5	Wheelchairs—Part 5: Determination of overall dimensions, mass, and turning space	IS
7176-5	Wheelchairs—Part 5: Determination of dimensions, mass, and maneuvering space	DIS
7176-6	Wheelchairs—Part 6: Determination of maximum speed, acceleration, and deceleration of electric wheelchairs	IS
7176-7	Wheelchairs—Part 7: Measurement of seating and wheel dimensions	IS
7176-8	Wheelchairs—Part 8: Requirements and test methods for static, impact, and fatigue strengths	IS
7176-9	Wheelchairs—Part 9: Climatic tests for electric wheelchairs	IS
7176-10	Wheelchairs—Part 10: Determination of obstacle-climbing ability of electric wheelchairs	IS
7176-10	Wheelchairs—Part 10: Determination of obstacle-climbing ability of electric wheelchairs	DIS
7176-11	Wheelchairs—Part 11: Test dummies	IS
7176-11	Wheelchairs—Part 11: Test dummies	NWI
7176-13	Wheelchairs—Part 13: Determination of coefficient of friction of test surfaces	IS
7176-15	Wheelchairs—Part 15: Requirements for information disclosure, documentation, and labeling	IS
7176-20	Wheelchairs—Part 20: Determination of the performance of stand-up wheelchairs	CD
7176-20	Wheelchairs—Part 20: Determination of the performance of stand-up wheelchairs	NWI
7176-22	Wheelchairs—Part 22: Setup procedures	IS
7176-26	Wheelchairs—Part 26: Vocabulary	IS
7176-27	Requirements and test methods for pushrim-activated power-assisted wheelchairs (PAPAW)	NWI
7193	Wheelchairs—Maximum overall dimensions	IS

(Continued)

Standard	Title	Stage
ISO/TC 173/SC 1/WG 6		
10542-1	Technical systems and aids for disabled or handicapped persons—Wheelchair tiedown and occupant-restraint systems—Part 1: Requirements and test methods for all systems	IS
10542-2	Technical systems and aids for disabled or handicapped persons—Wheelchair tiedown and occupant-restraint systems—Part 2: Four-point strap-type tiedown systems	IS
10542-3	Wheelchair tiedown and occupant-restraint systems—Part 3: Docking-type tiedown systems	DIS
10542-4	Wheelchair tiedown and occupant-restraint systems—Part 4: Clamp-type tiedown systems	IS
10542-5	Wheelchair tiedown and occupant-restraint systems—Part 5: Systems for specific wheelchairs	DIS
10542-A	Wheelchair tiedown and occupant-restraint systems for forward facing wheelchair-seated passengers—Part A: Requirements and test methods—Frontal impact	NWI CD
10865-1	Wheelchair tiedown and occupant-restraint systems for rearward facing wheelchair-seated passengers—Part 1: Systems for accessible transport vehicles designed for use by both seated and standing passengers	CD
16840-4	Wheelchair seating—Part 4: Seating systems for use in motor vehicles	DIS
7176-19	Title Wheelchairs—Part 19: Wheeled mobility devices for use in motor vehicles	NWI
7176-19	Wheelchairs—Part 19: Wheeled mobility devices for use in motor vehicles	DIS
ISO/TC 173/SC 1/WG 8		
7176-23	Wheelchairs—Part 23: Requirements and test methods for attendant-operated stair-climbing devices	IS
7176-24	Wheelchairs—Part 24: Requirements and test methods for user-operated stair-climbing devices	IS
7176-28	Wheelchairs—Part 28: Requirements and test methods for stair-climbing devices	CD
ISO/TC 173/SC 1/WG 10		
7176-14	Wheelchairs—Part 14: Power and control systems for electric wheelchairs—Requirements and test methods	IS
7176-14	Wheelchairs—Part 14: Power and control systems for electrically powered wheelchairs and scooters—Requirements and test methods	DIS
7176-21	Wheelchairs—Part 21: Requirements and test methods for electromagnetic compatibility of electrically powered wheelchairs and motorized scooters	IS
7176-21	Wheelchairs—Part 21: Requirements and test methods for electromagnetic compatibility of electrically powered wheelchairs and scooters, and battery chargers	CD
7176-25	Wheelchairs—Part 25: Requirements and test methods for batteries and their chargers for electrically powered wheelchairs and motorized scooters	DIS
7176-25	Wheelchairs—Part 25: Batteries and chargers for powered wheelchairs and motorized scooters—Requirements and test methods	NWI
ISO/TC 173/SC 1/WG 11		
16840-1	Wheelchair seating—Part 1: Vocabulary, reference axis convention and measures for body segments, posture, and postural support surfaces	IS
16840-2	Wheelchair seating—Part 2: Determination of physical and mechanical characteristics of devices intended to manage tissue integrity—Seat cushions	IS
16840-3	Wheelchair seating—Part 3: Determination of static, impact, and repetitive load strengths for postural support devices	IS
16840-5	Wheelchair seating—Test methods for determining the pressure relief characteristics of devices intended to manage tissue integrity—Seat cushions	NWI
7176-16	Wheelchairs—Part 16: Resistance to ignition of upholstered parts—Requirements and test methods	IS

Source: Based on listing provided by Dennis Axelson, Secretariat, RESNA—Technical Standards Board, from Cooper, Ohnabe, and Hobson (2006).[7]

---

[7] Hobson, D., Cooper, R., and Ferguson-Pell, M. (2006). Standards for assistive technology, in *An Introduction to Rehabilitation Engineering* (R. Cooper, H. Ohnabe, and D. Hobson, eds.), Taylor & Francis, New York.

54.13 Appendix B: ICT Accessibility Standards

The following list is compiled from several sources, including:

- Standards relating to the accessibility of ICT systems, by Richard Hodgkinson FISTC (January 7, 2008)[8]
- Accessible real-time conversational services: Summary of overviews, specifications and standards, by Gunnar Hellström[9]
- ISO/IEC TR 29138-2, Information Technology—Accessibility Considerations for People with Disabilities—Part 2: Standards inventory (PDTR 2007)[10]
- The author's own knowledge of standards in this area

Overviews, Articles, Policy Statements

Document	Full Name	Explanation	Type
COCOM 04-08	INCOM Report	Report on eAccessibility to EU regulatory group COCOM	Priority action requirements
ITU-T web pages on accessibility	ITU-T SG 16 Work on Accessibility	Summary of work in ITU-T multimedia group and related	Web site
Accessibility presentation in IMTC-NGN workshop	Accessibility in New Emerging Networks and Services	Presentation made for joint ITU-T / IMTC workshop 2006	PDF of Powerpoint

High-Level Accessibility Standards: International and Local/Regional

Document	Full Name	Explanation	Type	Status
ISO/IEC Guide 71:2001	Guidelines to address the needs of older persons and people with disabilities when developing standards.	Guidance to writers of relevant international standards on how to take into account the needs of older persons and persons with disabilities.		Approved
BS 7000-6	Design management systems. Managing inclusive design. Guide.	UK BSI		Published
CWA 14661	Guidelines to standardizers of ICT products and services in the CEN ICT domain.	EU CEN		Published
DIN TR 124	Barrierefreie Gebrauchsgüte.	Germany DIN	—	Published
JIS X 8341-1	Guidelines for older persons and persons with disabilities—information and communications equipment, software and services—Part 1: common guidelines	Japan JIS		Published
JIS Z 8071	Guidelines for standards developers to address the needs of older persons and persons with disabilities	Japan JIS		Published
Stanca Act	Provisions to support the access to information technologies for the disabled	Italy Federal Law		Published
Korea Guidelines	Recommendation guidelines to improve accessibility for the disabled and the elderly to the IT services/IT products	Korea MIC		Published
Nordic Guidelines	Nordic guidelines for computer accessibility	Nordic Council of Ministers		Published
US 508	Section 508 of the U.S. Rehabilitation Act	USA		Published

---

[8] Hodgkinson, R. (2008). Report on International ICT Accessibility Standards Proposed, Being Developed and Recently Published. London: RNIB Digital Accessibility Team. http://www.tiresias.org/research/standards/index.htm.

[9] Accessible Real Time Conversational Services: Summary of Overviews, Specifications and Standards, by Gunnar Hellström. http://www.realtimetext.org/index.php?pagina=28.

[10] ISO/IEC TR 29138-2 (2007). Information Technology—Accessibility Considerations for People with Disabilities—Part 2: Standards inventory.

EIA/TIA Accessibility-Specific Documents

Standard	Full Name of Standard	Explanation	Type	Status
TIA/EIA 825a	A Frequency Shift Keyed Modem for Use on the Public Switched Telephone Network	The standard for TTY signals, which permitted mainstream industry to design for compatibility with TTY as technologies moved to digital.	Modem transport	Approved
TIA 1001	Standards for Text over IP (TIA 1001) Transit Gateway	U.S. effort to develop standard methods for carrying Baudot over IP networks, using voice band data and gateway approaches.	Gateway	Approved
TIA- 504-A	Telecommunications-Telephone Terminal Equipment-Magnetic Field and Acoustic Gain Requirements for Headset Telephones Intended for Use by the Hard of Hearing		Requirement	Approved
ANSI/ TIA-968-A	Telecommunications - Telephone Terminal Equipment—Technical Requirements for Connection of Terminal Equipment to the Telephone Network		Requirement	Approved
TR 45 TSB-121	2.5 mm Audio Interface For Mobile Wireless Handsets—Text Telephones (TTY)	Connector standard for wireless telephones and TTYs.	Physical	Approved
C.63/ANSI ANSI C.63.19	American National Standard for Methods of Measurement of Compatibility between Wireless Communication Devices and Hearing Aids	Measurements of wireless telephone emissions and hearing aid immunity, with predicted performance based on measures (now in use in an FCC order).	Interference	Approved
TIA IS-823	TTY/TDD Extension to TIA/EIA 136-410 Enhanced Full Rate Speech Codec		Transport	Approved
TIA IS-840	Minimum Performance Standards for Text Telephone Signal Detector and Text Telephone Signal Regenerator		Requirements	Approved
TIA TSB-121	2.5 mm Audio Interface for Mobile Wireless Handsets—Text Telephones (TTY)		Physical	Approved

EIA/TIA General Documents with Accessibility Interest

Standard	Full Name of Standard	Explanation	Type	Status
EIA 608	Recommended Practice for Line 21 Data Service (Analog Television Closed Captioning)	USA EIA		Published 1999
EIA 708 B	Digital Television (DTV) Closed Captioning	USA EIA		Published 1999
TIA IS-127-2	Enhanced Variable Rate Codec, Speech Service Option 3 for Wideband Spread Spectrum Digital Systems—Addendum 2		Speech codec	Approved
TIA IS-707-A-2	Data Services Options for Spread Spectrum Systems—Radio Link Protocol Type 3—Addendum No. 2			Approved
TIA IS-733-1	High Rate Speech Service Option 17 for Wideband Spread Spectrum Communications Systems			Approved
TIA IS-789A	Electrical Specification for the Portable Phone to Vehicle Interface		Physical	Approved
TIA/EIA-688	DTE/DCE Interface for Digital Cellular Equipment		Physical	Approved
T1.209-2003	American National Standard for Operations Administration and Maintenance and Provisioning (OAM&P)—Network Tones and Announcements	Provides an industry standard way for network routing messages to be conveyed in TTY in addition to voice.		Approved

ETSI Accessibility Standards Currently Being Developed

Standard	Full Name	Explanation	Type	Status
ETSI EG 202 487	Human Factors: User Experience Guidelines for Telecare Solutions (e-Health)	Detailed user experience and user interface guidelines, applicable to a wide range of telecare solution elements.		WD
ETSI TS 102 577	Human Factors: Public Internet Access Points (PIAPs)	Guidance for providers and operators of Public Internet Access Points (PIAPs). A design for all approach will be followed.		WD
ETSI TS 102 511	Human Factors: AT Commands for Assistive Mobile Device Interfaces	AT command protocol stacks that can be used to enable assistive devices to interwork satisfactorily with mobile terminals.		TS
ETSI ES 202 076	Human Factors: Generic Spoken Command Vocabulary for ICT Devices and Services	Minimum set of spoken commands required to control the generic and most common functions of ICT devices.	Developed in a design for all approach	
	ETSI —Human Factors: Harmonized Relay Services	Requirements for the provision of all kinds of relay services in all networks.		
ETSI TR 102 612	Human Factors: EC Mandate 376—European Accessibility Requirements for Public Procurement of Products and Services in the ICT Domain—Phase 1			WD

ETSI and 3GPP Accessibility-Specific Documents

Standard	Full Name	Explanation	Type	Status
ETSI ETR 333	Text Telephony, User Requirements And Recommendations	Describes user needs for text conversation and guidelines for standardization.	Requirement	Approved
ETSI TR 101 806	Guidelines for Telecommunication Relay Services for Text and Video	Guidelines for telecommunication relay services for text and video.	Requirement	Approved
ETSI SR 001 996	An Annotated Bibliography of Documents Dealing with Human Factors and Disability	EU ETSI		Approved
ETSI TS 122 226 3GPP TS 22.226	Global Text Telephony Service Stage 1	Global text and total conversation (GTT) service description.	Service	Approved
ETSI TS 123 226 3GPP TS 23.226	Global Text Telephony Architecture Stage 2	GTT-specific description of network architecture for text conversation and especially CTM text telephony.	Architecture	Approved
ETSI TS 126 226 3GPP TS 26.226	CTM Modem, General description	Robust and error-tolerant modem for text telephony specified for mobile networks.	Modem transport	Approved
ETSI TS 126 235 3GPP TS 26.235	Packet Switched Conversational Multimedia, Default Codecs	Includes text conversation using T.140 in RTP as RFC 4103.	Profile	Approved
ETSI TS 126 236 3GPP TS 26.236	Packet-Switched Conversational Multimedia, Transport	Includes text conversation transport RFC 4103.	Profile	Approved
ETSI EG 202 320	Duplex Universal Speech and Text Communication	Guide for text as a mainstream call component.	Requirements and implementation framework	Approved
ETSI TS 102 511	AT Commands for Assistive Mobile Device Interfaces	External interface for assistive devices.	External interface specification	Draft
ETSI TS 122 101 3GPP TS 22.101	Service Principles	Brief description of mobile text conversation.	Service	Approved
ESI TS 126 110 3GPP TS 26.110	Circuit-Switched Multimedia Telephony (3G.324)	Includes text conversation using T.140 in AL1 channel.	Multimedia call control	Approved
ETSI TS 124 008 3GPP TS 24.008	Terminal Indicators	CTM indicator from terminal is defined here.	Call control	Approved
3GPP TS 26.114	IMS Multimedia Telephony Codec Considerations	Real-time text, video, and audio are included.	Profile and codec details	Approved for 3GPP release 7

IETF Accessibility-Specific Documents

Standard	Full Name of Standard	Explanation	Type	Status
IETF RFC 4102	Registration of Text/Red	MIME registration of text with redundancy.	MIME registration	Approved
IETF – avtrfc 2793bis	RTP Payload for Text Conversation	RTP Payload for T.140 text conversation {replaced by RFC 4103}.		Approved
IETF RFC 4103	RTP Payload for Text Conversation	RTP Payload for T.140 text conversation. MIME registered as "text/t140," used in H.323 and SIP and 3GPP.	Transport	Approved replaces RFC 2793
draft-ietf-sipping-toip-08	Framework of Requirements for Real-Time Text Conversation Using SIP	Requirements and implementation guidelines for real-time text in the SIP environment.	Requirements	Approved, awaiting publication
IETF RFC 3351	User Requirements for the Session Initiation Protocol (SIP) in Support of Deaf, Hard of Hearing and Speech-Impaired Individuals	Handles transcoding and other value-added services invoked through SIP.	Requirements	Approved
IETF RFC 4351	RTP Payload for Text Conversation Interleaved in an Audio Stream	RTP payload for text, intended for transit gateways where number of ports may be an issue.	Transport	Approved
draft-hellstrom-textpreview	Presentation of Text Conversation in Real Time and en-bloc Form	Presentation of real-time text conversations.	Presentation procedures	Draft
draft-hellstrom-textgwy	Real-Time Text Interworking between PSTN and IP Networks	Interworking procedures between PSTN textphones and SIP with real-time text.	Call control and media handling	Draft
draft-hellstrom-text-turntaking	Registration of the Real-Time-Text Media Feature Tag	Registration of a capability for real-time text.	SIP session level capability	Draft
draft-hellstrom-simple-text-transmission	Coding and Transmission of Text in Real-Time and en-bloc Mode based on MSRP	Procedures for using the IM protocol MSRP for real-time text.	Transport protocol details	Draft
draft-hellstrom-text-conference	Text Media Handling in RTP based Real-Time and Message Conferences	Procedures for multiparty sessions with RTP based real-time text.	RTP session details	Draft

IETF General Documents of Specific Interest for Accessibility

Specification	Title	Explanation	Type	Status
IETF RFC 2833	RTP Payload for DTMF Digits, Telephony Tones and Telephony Signals	Encoding and transport of tones over IP.	Transport	Approved, obsoleted by RFC 4733
IETF RFC 4566	Session Description Protocol	Contains "text" as an allowable media type in multimedia calls.	Call control	Approved
IETF RFC 2198	Redundancy for RTP Payloads	Used in RFC4103 for reliability of text traffic.	Transport	Approved
IETF RFC 2805	Media Gateway Control Protocol Architecture and Requirements	Contains text gateway requirements.	Procedure requirements	Approved
IETF RFC 3840	Indicating User Agent Capabilities in the Session Initiation Protocol (SIP)	Can be useful for preference and capability indication for service invocation.	Call control	Approved
IETF RFC 3841	Caller Preferences for the Session Initiation Protocol	Can be useful for preference and capability indication for service invocation.	Call control	Approved
IETF RFC 4733	Definition of Events for Telephony Tones	Text transport in RFC 4103 mentioned.	Transport	Approved, obsoletes RFC 2833

(Continued)

IETF General Documents of Specific Interest for Accessibility (Continued)

Specification	Title	Explanation	Type	Status
IETF RFC 4734	Definition of Events For Modem, FAX, and Text Telephony Signals	Text telephone signals expanded, refers to transport of text in RFC 4103 .	Transport	Approved
IETF RFC 4504	SIP Telephony Device Requirements and Configuration	Text requirements included, referring to RFC 4103.	Device requirements	Approved
IETF RFC 4117	Transcoding Services Invocation in the Session Initiation Protocol (SIP) Using Third Party Call Control (3pcc)	Most examples valid for text relay service and text gateway invocation.	Call control procedures	Approved
IETF RFC 4597	Conference Scenarios	Real-time text included referring to RFC 4103.	Service description	Approved
draft-ietf-sipping-transc-conf	The Session Initiation Protocol Conference Bridge Transcoding Model	Mentioned that it can be used for invocation of relay services.	Procedure specification	Approved, to be published
RFC 5012	Requirements for Emergency Context Resolution with Internet Technologies	Requirements for emergency services in IP, including real-time text, referencing RFC 4103.	Service requirements	Approved
draft-ietf-ecrit-phonebcp	Best Current Practices for Communications Services in Support of Emergency Calling	Refers to real-time text for emergency calls; Refers to RFC 4504.	Service and terminal requirements	Draft
draft-ietf-ecrit-framework	Framework for Emergency Calling in Internet Multimedia	Structure for emergency services in IP; Refers to RFC 4103 for text.	Service requirements	Draft

ITU Accessibility-Specific Documents

Document	Title	Explanation	Type	Status
ITU-T Rec. V.18	Operational and Interworking Requirements for DCE:s Operating in the Text Telephone Mode	Includes automatic interworking with most legacy text telephones.	Modem transport	Approved
ITU-T Rec. V.151	Procedures for End-to-End Connection of Analogue PSTN Text Telephones over an IP Network Utilizing Text Relay	Text telephony transit through IP.	Gateway	Approved
ITU-T Rec. F.703	Multimedia Conversational Services	Defines text telephony and total conversation services.	Service description	Approved
ITU-T Rec. H.224	A Real Time Control Protocol for Simplex Applications Using the H.221 LSD/HSD/HLP Channel	Addition of client id=2 for T.140 text transport.	Transport	Approved
ITU-T Rec. H.323 Annex G	Text Conversation and Text SET	Defines T.140 text inclusion in H.323 IP multimedia.	Call control	Approved
ITU-T Rec.T.134	Text Chat Application Entity	Application for text conversation in the T.120 data conferencing concept.	Transport	Approved
ITU-T Rec. T.140	Protocol for Multimedia Application Text Conversation	Text conversation protocol for multimedia application. With amendment 1 (2000).	Presentation level	Approved
ITU-T Rec.T.140 - Addendum	Marking of Missing Characters	Replacement for characters missing after transmission.	Presentation	Approved
ITU-T H Series Supplement 1	Video Quality for Sign Language and Lip Reading	Quality characteristics of video transmission of importance for sign language and lip-reading use.	Requirement	Approved
ITU-T Rec. H.248.2v2	Packages for Fax, Text and Call Discrimination	Text telephony to text over IP gateway. Approved revision.	Gateway	Approved
ITU-T FSTP.TACL	Technical Paper: Accessibility Checklist	General accessibility checklist for standardizers.	Guideline	Approved
ITU-T F.790	Telecommunications Accessibility Guidelines for Older Persons and Persons with Disabilities	General accessibility guidelines.	Guideline	Approved

(Continued)

ITU Accessibility-Specific Documents (Continued)

Document	Title	Explanation	Type	Status
ITU-T Rec. F.724	Service Description and Requirements for Videotelephony Services over IP Networks	ITU-T		Published 2005
ITU-T Rec. F.733	Service Description and Requirements for Multimedia Conference Services over IP Networks	ITU-T		Published 2005
ITU-T Rec. F.742	Service Description and Requirements for Distance Learning Services	ITU-T		Published 2005
ITU-T Rec. H.245	Control Protocol for Multimedia Communication	ITU-T		Published 2006
ITU-T Rec. H.320	Narrow-Band Visual Telephone Systems and Terminal Equipment	ITU-T		Published 2004
ITU-T draft Annex L/H.324	Updated Draft for H.324 Annex on Text Conversation	Addition or real-time text to the H.324 multimedia environment more reliable than earlier method.	Transport	Consent

ITU General Documents of Accessibility Interest

Standard	Full Name of Standard	Explanation	Type	Status
ITU-T Recommendation H.248.21	Controlled Multimedia Conferences	Includes text	Conference control protocol	Approved
ITU-T Rec. E.135	Human Factors Aspects of Public Telecommunication Terminals for People with Disabilities	ITU-T		Published 1995
ITU-T Rec. F.700	Framework Recommendation for Multimedia Services, Annex A.3	Multimedia framework, including real-time text.	Service description	Approved
ITU-T Rec. V.250	Serial Asynchronous Automatic Dialling and Control	Contains modem control for V.18.	Call control	Approved
ITU-T Recommendation V.8	Procedures for Starting Sessions of Data Transmission over the Public Switched Telephone Network	Contains call function = text telephony.	Call control	Approved
ITU-T Recommendation V.8 bis	Procedures for the Identification and Selection of Common Modes of Operation between Data Circuit-Terminating Equipments (DCEs) and between Data Terminal Equipments (DTEs) over the Public Switched Telephone Network	Contains procedures for starting simultaneous voice and text in V.18.	Call control	Approved
ITU-T Rec. V.16	A Simultaneous Voice Plus Data Modem, Operating at a Voice Plus Data Signaling Rate of 4800 bit/s, with Optional Automatic Switching to Data-Only Signaling Rates of Up to 14 400 bit/s, for Use on the General Switched Telephone Network and on Leased Point-to-Point 2-Wire Telephone Type Circuits	ITU-T		Published 1996
ITU-T V.150.1	Modem Transit over IP	International recommendation for transport of modem over IP.	Gateway	Approved
ITU-T Rec. V.152	Procedures for Supporting Voice Band Data over IP networks	Usable for text telephony transit through IP networks.	Gateway	Approved
ITU-T Rec. H.324	Terminal for Low Bit-Rate Multimedia Communication	Addition of data channel for T.140 text.	Multimedia system	Approved

(Continued)

ITU General Documents of Accessibility Interest (Continued)

Standard	Full Name of Standard	Explanation	Type	Status
ITU-T Rec. V.61	Analog Simultaneous Voice and Data (Permits Voice Carry Over with ASCII Modems	Possible to use under V.18.	Modem transport	Approved
ITU-T Rec. H.320	ISDN Visual Telephone Systems and Terminal Equipment	Text conversation added.	Multimedia system	Approved
ITU-T Rec. H.245	Control Protocol for Multimedia Communication	Multimedia control protocol; includes setup of text channels.	Call control	Approved
ITU-T Y.1541	Network Performance Objectives for IP-Based Services		Requirement	Approved
ITU-T Y.2000-series Supplement 1	Scope for Next Generation Network, R1	Real-time text and services for people with disabilities included.	Requirements	Approved
ITU-T Y.2201	Requirements for Next Generation Networks, R1	Accessibility features and services included.	Service requirements	Approved
ITU-T F.724	IP Videophone Service Description	Real-time text included.	Service description	Approved
ITU-T F.733	Multimedia Conference Services	Real-time text included.	Service description	Approved
ITU-T F.742	Service Description and Requirements for Distance Learning Services	Real-time text included.	Service description	Approved
ITU-T J.161	Audio and Video Codec Requirements and Usage for the Provision of Bidirectional Services over Cable Television Networks Using Cable Modems	Cable TV network IP communication; text telephony and IP real-time text included.	Technical requirements	Approved June 2007
ITU-T Y.2000 SerSup1	NGN Release 1 Scope	ITU-T		Published 2006
ITU-T Y.2211	IMS-Based Real-Time Conversational Multimedia Services over NGN	Real-time text included.	Requirements specification	Consent
ITU-T Y.2012	NGN Release 1 Requirements	ITU-T		Published 2006
ITU-T Y.NGN-et-tech	Next Generation Networks—Emergency Telecommunications—Technical Considerations	Real-time text included.	Functional specification	Draft

ISO Accessibility-Related Documents

Document	Title	Explanation	Type	Status
ISO TS 14415	Ergonomics of the Thermal Environment: Application of International Standards to People with Special Requirements	ISO TC159/SC5		IS published 2005
ISO/IEC 18019	Software and System Engineering: Guidelines for the Design and Preparation of User Documentation for Application Software	ISO/IEC JTC1/SC7		IS published 2004
ISO 9241-20	Accessibility Guidelines for Information Communication/Technology (ICT) Equipment and Services	Guidelines for planning, designing, and developing evaluating and purchasing hardware and software ICT for people with disabilities.	Based upon Japanese national standard, JIS 8341-1:2004	FDIS Ballot
ISO 9241-129	Guidance on Software Individualization	Ergonomic requirements and recommendations for individualization of human-computer interactions.		NP WD under prep
ISO 9241-135	Natural Language Dialogues	User-centered design of software user interfaces for voice-controlled dialogue systems to increase usability.		PWI

(Continued)

ISO Accessibility-Related Documents (*Continued*)

Document	Title	Explanation	Type	Status
ISO 9241-136	Voice/Auditory Interaction	User-centered design of software user interfaces for voice interaction systems to increase usability.		PWI
ISO 9241-151	Ergonomic Design of World Wide Web Interfaces	User-centered design of World Wide Web user interfaces.		FDIS Ballot
ISO 9241-152	Interpersonal Communication: Usability and Accessibility of Computer-Based Data and Voice Communication			PWI
	Interoperability of Assistive Technologies and Information Technologies			PWI
ISO 9241-171	Guidance on Software Accessibility	Design of accessible software for use at work, in the home, in education, and in public places.	Replaces ISO 16071	FDIS Ballot
ISO 9241-910	Framework for Tactile and Haptic Interactions	Tactile and haptic interfaces and interactions.		WD
ISO 9241-920	Guidance on Tactile and Haptic Interactions	Haptic and tactile hardware and software interactions.		DIS Ballot in translation
ISO TR 22411	Ergonomic Data and Ergonomic Guidelines for the Application of ISO/IEC Guide 71 to Products and Services to Address the Needs of Older Persons and Persons with Disabilities	Human ability data and design considerations that are useful for standards developers to implement the principles of ISO/IEC Guide 71 into individual standard.	Technical Report	DTR
ISO 24500	Guidelines for All People, including Elderly Persons and Persons with Disabilities—Auditory Signals on Consumer Products			WD
ISO 24501	Guidelines for All People, including Elderly Persons and Persons with Disabilities—Auditory Signals on Consumer Products—Sound Pressure Levels of Signals for the Elderly and in Noisy Conditions			NP
ISO 24502	Guidelines for All People, including Elderly Persons and Persons with Disabilities—Visual Signs and Displays—Specification of Age-Related Relative Luminance and Its Use in the Assessment of Light	Based upon Japanese standard JIS S 0031:2004.		NP
ISO 24503	Guidelines for All People, including Elderly Persons and Persons with Disabilities—Marking Tactile Dots on Consumer Products	Marking tactile dots to be put on the operating parts of various consumer products that have electrically operated switches.	Based upon Japanese standard JIS S 0011-2000	WD
ISO 28803	Ergonomics of the Physical Environment—Application of International Standards to People with Special Requirements			WD
ISO/IEC 10779	Accessibility Guidelines for Office Equipment	For planning, developing, and designing office equipment to improve information accessibility required when primarily elderly persons, persons with disabilities, and persons with temporary disabilities use office equipment.	Developed from Japanese standard JIS X8341-5	DIS Ballot
ISO/IEC 24751-1	Information Technology—Individualized Adaptability and Accessibility in e-Learning, Education and Training—Part 1: Framework and Reference Model	ISO/IEC JTC1/SC36		Under development FDIS published 2006

(Continued)

ISO Accessibility-Related Documents (***Continued***)

Document	Title	Explanation	Type	Status
ISO/IEC 24751-2	Information Technology—Individualized Adaptability and Accessibility in e-Learning, Education and Training—Part 2: "Access for All" Personal Needs and Preferences for Digital Delivery	ISO/IEC JTC1/SC36		Under development; FCD published 2007
ISO/IEC 24751-3	Information Technology—Individualized Adaptability and Accessibility in e-Learning, Education and Training—Part 3: "Access for All" Digital Resource Description	ISO/IEC JTC1/SC36		Under development; FCD published 2007
ISO/IEC 24751-4	Information Technology—Individualized Adaptability and Accessibility in e-Learning, Education and Training—Part 4: Access for All Non-digital Resource Description	ISO/IEC JTC1/SC36		Under development; new work item 2007
ISO/IEC 24751-5	Information Technology—Individualized Adaptability and Accessibility in e-Learning, Education and Training—Part 5: Personal Needs and Preferences for Non-digital Resources	ISO/IEC JTC1/SC36		Under development; new work item 2007
ISO/IEC 24751-6	Information Technology—Individualized Adaptability and Accessibility in e-Learning, Education and Training—Part 6: Personal Needs and Preferences for Description of Events and Places	ISO/IEC JTC1/SC36		Under development; new work item 2007
ISO/IEC 24751-7	Information Technology—Individualized Adaptability and Accessibility in e-Learning, Education and Training—Part 7: Description of Events and Places	ISO/IEC JTC1/SC36		Under development; new work item 2007
ISO/IEC 24751-8	Information Technology—Individualized Adaptability and Accessibility in e-Learning, Education and Training—Part 8: Language Accessibility and Human Interface Equivalencies (HIEs) in e-Learning Applications	ISO/IEC JTC1/SC36		Under development; new work item 2007
ISO/IEC 24752	Universal Remote Console	Part 1: Framework. Part 2: User Interface Socket Description. Part 3: Presentation Template. Part 4: Target Description. Part 5: Resource Description.	Based on U.S. ANSI/INCITS 2005-389-393	IS in publication
SO/IEC 24756	Framework for Specifying a Common Access Profile (CAP) of Needs and Capabilities of Users, Systems and Their Environments	Framework for selecting and supporting computer-related accessibility, including accessibility supported by assistive technologies.		FCD
ISO/IEC 247861	Information Technology—User Interfaces—Accessible User Interface for Accessibility Setting on Information Devices—Part 1: General and Methods to Start	ISO/IEC JTC1/SC35		CD
ISO/IEC 24786	Accessible User Interface for Accessibility Setting on Information Devices	Requirements to make the user interface of the accessibility setting accessible.		CD
ISO/IEC 26511	Software and Systems Engineering—Requirements for Managers of User Documentation	Consistent, complete, accurate, and usable documentation.		WD
ISO/IEC 26512	Software and Systems Engineering—Requirements for Acquirers and Suppliers of User Documentation	Acquirers and suppliers of software user documentation.		NP
ISO/IEC 26513	Software and Systems Engineering—Requirements for Testers and Assessors of User Documentation	Testers and assessors of software user documentation.		WD

(*Continued*)

ISO Accessibility-Related Documents (*Continued*)

Document	Title	Explanation	Type	Status
ISO/IEC 26514	Software and Systems Engineering—Requirements for Designers and Developers of User Documentation	Designers and developers of software user documentation.		CD
ISO/IEC TR 19765	Information Technology—Survey of Icons and Symbols That Provide Access to Functions and Facilities to Improve the Use of Information Technology Products by the Elderly and Persons with Disabilities	ISO/IEC JTC1/SC35		TR published 2007
ISO/IEC TR 19766	Information Technology—Guidelines for the Design of Icons and Symbols Accessible to All Users, including the Elderly and Persons with Disabilities	ISO/IEC JTC1/SC35		TR published 2007 under revision as 1158110
ISO/IEC TR XXXXX	Information Technology—Accessibility Considerations for People with Disabilities	Part 1: User Needs Summary. Part 2: Standards inventory. Part 3: Guidance on User Needs Mapping.	Development by ISO/IEC JTC 1 - SWG-A	WD
ISO/IEC XXXXX	Accessibility Functions for Personal Computers/Information Processing	Accessibility functions regarding PCs.	Based upon JIS X8341-2	WD

Standards from Other International Bodies

Document	Title	Explanation	Type	Status
WCAG	Web Content Accessibility Guidelines 1.0	Guidelines for web content.		1.0 Approved 2.0 Last Call Draft
UAAG	User Agent Accessibility Guidelines	Guidelines for browsers and other user agents.		1.0 Approved 2.0 Working Draft
ATAG	Authoring Tools Accessibility Guidelines	Guidelines for web authoring tools.		1.0 Approved 2.0 Working Draft
HFES 200.2	Human Factors Engineering of Software User Interfaces: Part 2: Accessibility	Part 2 of HFES 200– which focuses on accessibility.		Accepted In publication
HFES 200.4	Human Factors Engineering of Software User Interfaces: Part 4: Voice Input/Output and Telephony	Part 4 of HFES 200– which focuses on voice and IVR and includes accessibility.		Accepted In publication
IMS	IMS AccessForAll Meta-data Overview	IMS		Published 2004
CEN/BT WG 185	Report on conformance schemes (second report as requested by M/376—Phase 1)	Analysis on testing and conformity schemes of products and services meeting accessibility requirements.		
BS PAS 124	Guidance on web standards	Best practice for "defining, implementing and managing organizational web standards".		WD
DSL Forum TR-122v1.01	Base requirements for consumer-oriented analog terminal adapter functionality	DSP forum ATA functionality, mentioning textphone passthrough. by V.152 and V.151.	Product profiling	Approved

Japanese Standards

Document	Title	Explanation	Type	Status
JIS X 8341-1	Guidelines for older persons and persons with disabilities—Information and communications equipment, software and services—Part 1: Common guidelines	Japan JIS		Published
JIS Z 8071	Guidelines for standards developers to address the needs of older persons and persons with disabilities	Japan JIS		Published
JBMS-71	Auditory signal	Japan JBMIA		Published 2006
JIS S0011	Guidelines for all people, including elderly and people with disabilities—Marking of tactile dots on consumer products	Japan JIS		Published 2000
JIS S 0012	Guidelines for all people, including elderly and people with disabilities—Usability of consumer products	Japan JIS		Published 2000

(*Continued*)

Japanese Standards (Continued)

Document	Title	Explanation	Type	Status
JIS S 0013	Guidelines for the elderly and people with disabilities—Auditory signals on consumer products	Japan JIS		Published 2002
JIS S 0014	Guidelines for the elderly and people with disabilities—Auditory signals on consumer products—Sound pressure levels of signals for the elderly and in noisy conditions	Japan JIS		Published 2003
JIS S 0021	Guidelines for all people, including elderly and people with disabilities—Packaging and receptances	Japan JIS		Published 2000
JIS S 0022	Guidelines for all people, including elderly and people with disabilities—Packaging and receptacles—Test methods for opening	Japan JIS		Published 2001
JISS 022-3	Guidelines for all people, including elderly and people with disabilities—Packaging and receptacles—Tactile indication for identification	Japan JIS		Published 2007
JISS 022-4	Guidelines for all people, including elderly and people with disabilities—Packaging and receptacles—Evaluation method by user	Japan JIS		Published 2007
JIS S 0025	Guidelines for all people, including elderly and people with disabilities—Packaging and receptacles—Tactile warnings	Japan		Published 2004
JIS X 8341-2	Guidelines for older persons and persons with disabilities—Information and communications equipment, software and services—Part 2: Information processing equipment	Japan JIS		Published 2004
JIS X 8341-4	Guidelines for older persons and persons with disabilities—Information and communications equipment, software and services—Part 4: Telecommunications equipment	Japan JIS		Published 2005
JIS X 8341-5	Guidelines for older persons and persons with disabilities—Information and communications equipment, software and services—Part 5: Office equipment	Japan JIS		Published 2006
JIS X 8341-2	Guidelines for older persons and persons with disabilities—Information and communications equipment, software and services—Part 2: Information processing equipment	Japan JIS		Published 2004
JIS X 8341-3	Guidelines for older persons and persons with disabilities—Information and communications equipment, software and services—Part 3: Web content	Japan JIS		Published 2004
JIS S 0011	Guidelines for all people, including elderly and people with disabilities—Marking of tactile dots on consumer products	Japan JIS		Published 2000
JIS S 0012	Guidelines for all people, including elderly and people with disabilities—Usability of consumer products	Japan JIS		Published 2000
JIS S 0013	Guidelines for the elderly and people with disabilities—Auditory signals on consumer products	Japan JIS		Published 2002
JIS S 0031	Guidelines for the elderly and people with disabilities—Visual signs and displays—Specification of age-related relative luminance and its use in assessment of light	Japan JIS		Published 2004
JIS S 0032	Guidelines for the elderly and people with disabilities—Visual signs and display—Estimation of minimum legible size for a Japanese single character	Japan JIS		Published 2003
JIS S 0033	Guidelines for the elderly and people with disabilities—Visual signs and displays—A method for color combinations based on categories of fundamental colors as a function of age	Japan JIS		Published 2006
JIS S0014	Guidelines for the elderly and people with disabilities—Auditory signals on consumer products—Sound pressure levels of signals for the elderly and in noisy conditions	Japan JIS		Published 2003
JIS T 0901	Guidelines of electronic guide system using audible signage for visually impaired persons	Japan JIS		Published 2005

Australian Guidelines

Document	Title	Explanation	Type	Status
AS 3769	Automatic teller machines: User access	Australia SA		Published 1990 Amended 1990
AS/ACIF S040	Requirements for customer equipment for use with the standard telephone service—features for special needs	Australia Australian Communications Authority		Published 2001
AS/NZS 4277	Text telecommunications—User interface requirements—For deaf people and people with hearing and speech disabilities	Australia SA		Published 1995 Amendment 1996

Canadian Accessibility Standards

Document	Title	Explanation	Type	Status
CSA T510-95	Performance and Compatibility Requirements for Telephone Sets With Loop Signaling	Canada CSA		Published 2003
CSA T516	Telecommunications—Telephone Terminal Equipment Requirements for Pay Telephone Keypads and Function Keys with Particular Regard to Use by Persons with Disabilities	Canada CSA		Published 2002
CSA B651	Accessible Design for the Built Environment	Canada CSA		Published 2004
CSA B651.1	Barrier-Free Design for Automated Banking Machines	Canada CSA		Published 2001
CSA B651.2	Accessible Design for Self-Service Interactive Devices	Canada CSA		Published 2007
CSA B659-01	Design for Aging	Canada CSA		Published 2001
CWA 14835	Guidelines for Making Information Accessible through Sign Language on the Web	EU CEN		Published

Korean Accessibility Standards

Document	Title	Explanation	Type	Status
KICS.OT10.0003	Internet Web Contents Accessibility Guideline	Korea MIC		Published 2005
TTAS.KO10.0213	Software Accessibility Guidelines 1.0	Korea TTA		Published 2006
TTAS.OT10.0073	Korean User Agent Accessibility Guidelines 1.0	Korea TTA		Published 2006
TTAS.OT10.0074	Korean Authoring Tools Accessibility Guidelines 1.0	Korea TTA		Published 2006
TTAS.KO-09.0040	Automatic Teller Machine's Accessibility Guidelines 1.0	Korea TTA		Published 2006
TTAS.OT 09.0001	Digital Talking Book Guidelines 1.0	Korea TTA		Published 2006
TTAS.KO-07.0050	The Standard for DTV Closed Caption System	Korea TTA		Published 2007

Spanish Accessibility Standards

Document	Title	Explanation	Type	Status
UNE 139801	Computer applications for people with disabilities. Computer accessibility requirements. Hardware.	Spain AENOR		Published 2003
UNE 139802	Computer applications for people with disabilities. Computer accessibility requirements. Software.	Spain AENOR		Published 2003
UNE 153010	Subtitling for deaf and hard-of-hearing people. Subtitling by teletext.	Spain AENOR		Published 2003
UNE 153020	Audio description for visually impaired people. Guidelines for audio description procedures and for the preparation of audio guides.	Spain AENOR		Published 2005

Swedish Accessibility Standards

Document	Title	Explanation	Type	Status
24 hour agency web guidelines	Guidelines for an accessible public administration	Swedish government		Published 2006

U.S. Accessibility Standards

Document	Title	Explanation	Type	Status
ADAAG	ADA Accessibility Guidelines	Guidelines for conformance with Americans with Disabilities Act		Approved Revision in Draft
36 CFR Part 1194 (Code of Federal Regulations)	Section 508 Accessibility Standards	Standards for development, purchase or use of electronic and information technology by U.S. government		Approved
47 CFR Part 7 (Code of Federal Regulations)	Section 255 Accessibility Standards (relating to: Telecommunications Act, Section 255)	Standards for manufactures and providers of telecommunication equipment and services		Approved
47 CFR Part 79.1	47 C.F.R. § 79.1 Closed Captioning of Video Programming	U.S. Federal Regulation		2004
47 CFR Part 79.1	47 C.F.R. § 79.2 Accessibility of Programming Providing Emergency Information (relating to: Telecommunications Act, Section 713)	U.S. Federal Regulation		2004

(Continued)

U.S. Accessibility Standards (*Continued*)

Document	Title	Explanation	Type	Status
TEITAC Report	Telecommunication and Electronic and Information Technology Advisory Committee Report	Federal Advisory Committee report on unifying and updating the 508 and 255 accessibility guidelines		Working Draft
VVVSG	Voluntary Voting System Guidelines	Voting standard		Accepted
ANSI/INCITS 389-2005 through 393-2005	Information Technology—Protocol to Facilitate Operation of Information and Electronic Products through Remote and Alternative Interfaces and Intelligent Agents: 389—Universal Remote Console 390—User Interface Socket Description 391—Presentation Template 392—Target Description 393—Resource Description	Universal Remote Control and Pluggable User Interfaces		Accepted
ANSI C.63.19	American National Standard for Methods of Measurement of Compatibility between Wireless Communication Devices and Hearing Aids	USA ANSI/IEEE		Published 2007

Acknowledgments

This chapter was developed with support from the National Institute on Disability and Rehabilitation Research (NIDRR), U.S. Department of Education, under grant H133E030012 Rehabilitation Engineering Research Center on Universal Interface and Information Technology Access and grant H133E040013 Telecommunication Access Rehabilitation Engineering Research Center, a collaboration of the University of Wisconsin–Trace Center, Gallaudet University, and Omnitor. Any opinions expressed herein are those of the author and not necessarily those of the funding agency.

Special thanks to Professor Jim A. Carter Jr., Department of Computer Science, University of Saskatchewan, Canada, and Gunnar Hellström, Omnitor, Sweden, for their review and comments on the chapter.

References

Cascading Style Sheets, level 2, CSS2 Specification. http://www .w3.org/TR/1998/REC-CSS2-19980512.

Hellström, G. (2005). *Accessible Real Time Conversational Services: Summary of Overviews, Specifications and Standards.* Washington, DC: ITI/INCITS. http://www.jtc1access.org/ documents/swga_108.htm.

Hobson, D., Cooper, R., and Ferguson-Pell, M. (2006). Standards for assistive technology, in *An Introduction to Rehabilitation Engineering* (R. Cooper, H. Ohnabe, and D. Hobson, eds.). Abingdon, U.K.: Taylor and Francis.

Hodgkinson, R. (2008, January 7). Report on International ICT Accessibility Standards Proposed, Being Developed and Recently Published. London: RNIB Digital Accessibility Team. http://www.tiresias.org/research/standards/index. htm.

HTML 4.01 Specification: W3C Recommendation (1999, December 24). http://www.w3.org/TR/1999/REC-html401-19991224.

International Organization for Standards (ISO) (2001). *ISO/IEC Directives: Part 2. Rules for the Structure and Drafting of International Standards* (4th ed.). Geneva: ISO/IEC.

Internet Engineering Task Force (1989). *Key Words for Use in RFCs to Indicate Requirement Levels.* http://www.ietf.org/ rfc/rfc2119.txt.

ISO/IEC TR 29138-2 (2007). *Information Technology—Accessibility Considerations for People with Disabilities—Part 2: Standards Inventory.* http://www.iso.org/iso/iso_catalogue .htm.

Ofcom (2005, July 25). Guidance Notes, Section 2: Harm and Offence Annex 1, "Ofcom Guidance Note on Flashing Images and Regular Patterns in Television (Re-issued as Ofcom Notes 25 July 2005)." http://www.ofcom.org.uk/tv/ ifi/guidance.

Requirements for Internet Hosts (1989, October). Communication Layers, Internet Engineering Task Force. http://www.ietf .org/rfc/rfc1122.txt.

eAccessibility Standardization

Jan Engelen

55.1 Standards: Brief Overview

In very general terms, producing a *standard* (*fr*: norme, standard; *de*: Norme; *es*: norma) is a voluntary action set up, almost uniquely, by commercial partners who believe that the standardization will permit easier exchanges of products and goods (see Chapter 54, "Standards and Guidelines," of this handbook for an in-depth discussion of the process of producing standards and guidelines). This implies very often that the acceptance of the standards is also voluntary and triggered by expected commercial benefits.

On the other hand, laws in many countries are referring more and more to the required acceptance of several standards (e.g., on safety or on ecological aspects). The net result of this need for standards is that nowadays many standardization initiatives are stimulated (= subsidized) by public bodies or, in Europe, directly and indirectly by the European Commission. Many guidelines have also been created by stakeholder groups.[1]

55.2 New Developments in Design for All-Related Standardization

As design for all (DfA) standardization was explicitly mentioned in the eEurope2002 plan, several new actions were established over the last years in the European context (Engelen, 2003a). Four major recent changes can be distinguished: the setup of coordinating working groups and organizations; the democratization of the standardization processes themselves; the increasing impact of nonformal standardization bodies; and the establishment of standardization-related discussion in open arenas and

---

[1] More information about the standardization processes can be found in Chapter 54, "Standards and Guidelines."

for nonspecialists. Each of those aspects will be briefly explained in the remainder of this contribution.

55.2.1 Formal versus Informal Standardization Activities

Over the last years, several standardization bodies have set up standardization-related initiatives that are easier to manage and can produce outcomes much faster. They all have special designations so that they cannot be confused with formal standards. For example, the International Organization for Standardization (ISO) has developed a new range of *deliverables*, or different categories of specifications, allowing publication at an intermediate stage of development before full consensus is reached.

Some of these standardization-related activities are summarized in Table 55.1.

55.2.2 European Initiatives

55.2.2.1 Initial Steps

In the middle of the 1990s, when information and communications technologies (ICT) systems started booming, ETSI, one of the three European standardization organizations, organized, in collaboration with the Danish Centre for Technical Aids and the European Commission, the 1996 "European Policy Workshop ICT Standardization and Disability in Europe." The three major outcomes (Brandt, 1996) retain their importance, even nowadays, although the third one (legislation) has since then been taken on board in several EU countries:

- Industry is not sufficiently aware of the market potential for accessible products.

TABLE 55.1   Examples of Standardization-Related Activities

Name	Web Site	Names of Related Activities
ISO	http://www.iso.ch	• Publicly Available Specification (PAS) • Technical Specification (TS) • Technical Report (TR) • International Workshop Agreement (IWA)
ITU-T	http://www.itu.int/ITU-T	• Recommendations
CEN	http://www.cenorm.org	• CEN Workshop Agreement (CWA)
ETSI	http://www.etsi.org	• Specialist Task Forces (STF)
CENELEC	http://www.cenelec.org	• CENELEC Workshop Agreement (CWA) • CENELEC Guides
CEN/CENELEC		• CEN/CENELEC Guides

- Standardization processes should take into account the requirements of people with disabilities, and these users should be more involved in standardization work.
- There is a need for legislation in the accessibility domain.

55.2.2.2 Coordination Initiatives

55.2.2.2.1 ICT Standards Board

The ICT Standards Board (ICTSB)[2] is an initiative from the three recognized European standards organizations, with the participation of specification providers as partners to coordinate standardization activities in the field of Information and Communications Technologies (ICT).

The ICTSB listens to requirements for standards and specifications that are based on concrete market needs and expressed by any competent source. The board then considers what standards or specifications need to be created, and how the task will be carried out (and by whom).

One of its major recent research-oriented activities was the participation in the COPRAS project (see the following).

55.2.2.2.2 Design for All and Assistive Technology Standardization Coordination Group

The Design for All and Assistive Technology Standardization Coordination Group (DATSCG)[3] was created within ICTSB as a direct response to the eEurope2002 plan. It has the following objectives:

- To ensure coordination of the ICT-related standardization work in the DfA and assistive technology (AT) fields

- To act as an overall focal point on design for all and assistive technology standardization
- To assist in organizing promotional activities on design for all and assistive technologies standardization requirements in ICT
- To promote knowledge and awareness of existing guidelines and tools by the market players

Although membership is by invitation, DATSCG tries to involve as many organizations as possible in their activities, including organizations of, or for, persons with a disability. Especially the contribution of the European Disability Forum (EDF) as representative of the final users is very important. Another important player is ANEC, "the European consumer voice in standardization," representing the consumer organizations from the European Union member states and the EFTA countries. Also the Association for the Advancement of Assistive Technology in Europe (AAATE) has an observer status in DATSCG.

DATSCG has proven to be an important channel for information exchange on standardization issues as it groups the main players in this field.

55.2.2.2.3 eAccessibility Expert Group

Mainly as a consequence of the eEurope actions for the promotion of ICT use in Europe, the European Commission created several working groups to keep an eye on the actions promised by the different EU countries and by the Commission itself.

The high-level group on the Employment and Social Dimension of the Information Society (ESDIS) was established in 1999 to support the European Commission in the analysis of the impact of the information society on employment and on social cohesion. The eEurope topics related to persons with disability and elderly persons were delegated by ESDIS to the eAccessibility expert group.[4] With respect to standardization, the eAccessibility group produced an overview document[5] by the end of 2002 (Engelen, 2003b).

After a short period of inactivity, the eAccessibility group was recreated as an expert support group for the eInclusion activities of the European Commission's Directorate General on Information Society and Media (EC-DG INFSO-H3). In 2006, it changed names and constitution into eInclusion Expert Group. Its representatives are officially delegated by the governments of all old and new EU member countries.

55.2.2.2.4 Cooperation Platform for Research and Standards

Cooperation Platform for Research and Standards (COPRAS)[6] was a support action project in the EU's 6th Framework Program, aiming to improve the interfacing, cooperation, and exchange between IST (information society technologies) research projects and ICT standardization. It was initiated by several European

---

[2] http://www.ictsb.org/ and http://www.ict.etsi.org.

[3] http://www.ictsb.org/DATSCG_home.htm and http://www.ict.etsi.org/DATSCG_home.htm.

[4] http://europa.eu.int/information_society/policy/accessibility/index_en.htm.

[5] http://ec.europa.eu/employment_social/knowledge_society/docs/eacc_dfastd.pdf.

[6] http://www.w3.org/2004/copras.

standards organizations in cooperation with the ICTSB, the coordinating forum for ICT standardization in Europe.

COPRAS addressed the challenge of better synchronizing the continuous technological development in ICT with standardization processes, thus making the benefits of these technological developments better and more easily and quickly accessible to industry and society. Its mission therefore was to stimulate, facilitate, and ease cooperation and exchange between current as well as future IST research projects and ICT standards organizations. Its activities and deliverables supported projects in finding the relevant standards organizations to signal their output to, enabling them to upgrade their results through standardization, and hence stimulate their dissemination and usage.

As one of its deliverables, COPRAS has developed a set of generic guidelines facilitating interfacing between research projects and ICT standards organizations.[7] Its ultimate goal was to bring IST research and standardization closer together, and to provide research projects as well as other stakeholders in government, industry, and society with a platform facilitating exchange between research and standardization, and furthering Europe's leading position in ICT development.

55.2.2.2.5 USEM

The USEM project[8] aims to facilitate, enhance, and increase qualification and participation of disabled or elderly users and their respective organizations in the European standardization process in IST. Furthermore, a user network will be set up to promote the exchange of experiences between user organizations involved in standardization work.

USEM supports a number of important objectives:

- To design a core curriculum for the training of end-users wanting to play a key role in standardization work
- To make sure that more users with disabilities acquire the skills needed for participation in the design and assessment of European standardization
- To improve the exchange of experiences between different user groups on a European level by user information networking
- To disseminate information and encourage the uptake of new standards
- To actively involve disabled and elderly people in the process of IST standardization
- To get users involved in European standardization actions

55.2.2.3 Standardization Efforts through Workshops and Other Collaborative Schemes

As stated in Section 55.1.3, a democratization process is taking place in the standardization arena. Besides formally established committees for creating formal standards, all standardizing bodies now have working groups and task forces where all

interested people are welcome, minimally as observers but often as contributors, too.

55.2.2.3.1 CEN Workshop Agreements in the DfA Field

Examples of the previously mentioned efforts are the establishment of a CEN Workshop on Design for All in ICT, CWA14661 "Guidelines to Standardisers of ICT Products and Services in the CEN ICT Domain"[9] and, more recently, the creation of the CEN Workshop on website certification, the Specifications for a Complete European Certification Scheme Concerning the Delivery of a Quality Mark for Web Content Accessibility—WS/WAC,[10] and the Workshop on Document Processing for Accessibility (CWA-DPA).

The WAC workshop was established to obtain a first-level European agreement on a European certification scheme concerning the delivery of a quality mark for web accessibility. Such a scheme had previously been investigated by an EU IST project (Support-EAM), part of a cluster of projects (WABcluster) that are defining an overall European methodology for assessing web accessibility in conformance with W3C WAI content guidelines.[11] This European quality mark is based on the use of a methodology for assessing web accessibility within a European certification scheme. The final agreement (June 2006) is freely available.[12]

The CEN/ISSS WS DPA workshop[13] had three key objectives, namely:

- To bring together a very large group of players in the information provision and e-publishing chain to achieve a substantial increase of accessible information at a European level
- To provide guidelines on integrating accessibility components within the document management and publishing process rather than as just a specialized, additional service
- To raise awareness and stimulate the adoption at local, regional, national, and European levels of the emerging formats and standards for the provision of accessible information and to find ways of ensuring that technological protection measures do not inadvertently impede legitimate access to information by people with print impairments

55.2.2.3.2 ETSI Specialist Task Forces

STFs are typical for ETSI. An STF is a team of highly skilled experts working together over a predefined period to draft an

---

[7] http://www.w3.org/2004/copras/docu/faq/Overview.html.

[8] http://www.usem-net.eu.

[9] www2.nen.nl/getfile?docName=198267.

[10] http://www.cenorm.be/cenorm/businessdomains/businessdomains/isss/activity/ws-wac.asp and http://www.support-eam.org.

[11] http://www.wabcluster.org.

[12] ftp://ftp.cenorm.be/PUBLIC/CWAs/e-Europe/WAC/CWA15554-00-2006-Jun.pdf.

[13] http://www.cen.eu/cenorm/businessdomains/businessdomains/isss/activity/ws-dpa.asp.

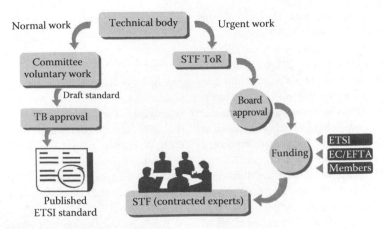

FIGURE 55.1   How STFs (right) relate to "normal standardization" work (left).

ETSI standard under the technical guidance of an ETSI technical body (TB) and with the support of the ETSI secretariat.[14]

Some of the AT and DfA-related STFs have been focusing[15] on:

- Design for all: Guidelines for ICT products and services (STF 184)
- Duplex universal speech and text (DUST) communication [e-Inclusion] (STF 267)
- Human-related technical guidelines for real-time person-to-person communication services (STF 284)
- Enabling and improving the use of mobile e-services (STF 285)
- Accessing symbols for use with video content and ICT devices (STF 286)
- User-oriented handling of multicultural issues in broadband and narrowband multimedia telecommunications (STF 287)
- AT commands for assistive mobile device interfaces (STF 304)
- Extending e-inclusion for public Internet access points (PIAPs) (STF 342)

55.2.2.3.3 COST219ter

The main objective of this collaborative European action, involving also members from the United States, Australia, and Japan, is to increase the accessibility of next-generation telecommunication network services and equipments to elderly people and people with disabilities, by design or, alternatively, by adaptation when required.[16] Several major actions toward standardization have been undertaken over the past years.

COST219ter members have been collaborating with the ITU-T work on "Total Conversation: Increased Usability of Conversational Services in Mobile and Fixed Networks" (convener: Gunnar Hellstrom, cf. Section 55.2.3.2).[17]

On March 8, 2005, COST219ter organized in Florence a specialized workshop on "eAccessibility Legislation and Policy: The Role of Standardization." Specialists of W3C, ISO, ETSI, ITU-T, the D4ALL.net project, EDeAN, and TEDICORE (Australia) presented ongoing work within their organizations or countries.[18] In 2007, the COST219ter action also published a brochure aiming at improving user involvement in standardization (Gill, 2007).

Within the COST219ter action, a special working group was set up to harmonize testing for accessibility. Especially the usability and accessibility of mobile telephones was worked out thoroughly, was tested internationally, and is available as a toolset (Chandler et al., 2007).

55.2.2.4 Public Discussions

One of the unique and recent developments in the standardization field, especially in relation to design for all, is the potentially large involvement of specialists, users, and user representatives in the discussions.

55.2.2.4.1 *European Design for All Network Standardization SIG*

The European Design for All Network (EDeAN) was established in 2002 as a response to the European eEurope2002 program for stimulation of IST use (see also Chapter 58 of this handbook). One of the action lines was the "creation of a network of major expert centers in Design for All." Another was the "Publication of Design for All standards for accessibility of information technology products, in particular to improve the employability and social inclusion of people with special needs" (as mentioned in Section 55.2.1.2.3). Although initial discussions on the latter

---

[14] http://portal.etsi.org/stfs/process/home.asp.

[15] More info about STFs: http://portal.etsi.org/Portal_STF/FullSearch. asp?Param=.

[16] http://www.tiresias.org/cost219ter.

[17] http://www.tiresias.org/cost219ter/florence/hellstrom.htm.

[18] http://www.tiresias.org/cost219ter/florence/index.htm.

topic also took place in the eAccessibility working group, a more open approach was established through the creation of a public discussion forum.

Electronic information exchange within the EDeAN network was set up by the D4ALLnet project (IST-2001-38833, Design for All Network of Excellence) that created the HERMES collaborative web-based platform,[19] developed by FORTH-ICS (Crete). D4ALLnet was a thematic network funded by the European Commission that supported the operation of EDeAN by providing an accessible web-based platform to enable virtual networking and cooperation, as well as information and knowledge exchange between EDeAN network members (Bühler and Stephanidis, 2004). Members of the EDeAN network are exchanging information within so-called special-interest groups, including one on *standardization.*

The SIG standardization group has over 100 members. Most of them are from Europe (but new EU member states are underrepresented), a few are from the United States, Australia, and Hong Kong. Also several observers of the European Commission are taking part in the discussions.

This discussion platform has been shown to provide a unique means of bringing information on ongoing standardization activities directly to persons interested in this subject but not members of formally established standardization task forces or working groups.

55.2.3 Some National Initiatives

55.2.3.1 United States

In the United States, due to its large concentration of huge software enterprises, several official and de facto organizations are active. In a recent contribution,[20] Gregg Vanderheiden, director of the Trace R&D Center at the University of Wisconsin–Madison, enumerated over 40 of these standardization groups. The number of guidelines is also growing rapidly.

In the United States, several legislative actions have been undertaken, and as could be expected, they often do not refer to

RERC work on Access Guidelines
- Guidelines for Consumer Products (1992)
- IBM Accessibility Guidelines (1993)
- Software Accessibility Guidelines for ITF (1994)
- First Web Guidelines (HTML) (1995)
- Microsoft Accessibility Guidelines (1999)
- TAAC work
  - Compiled over 1000 guidelines for TAAC Committee
- On-Line Design Tool - 255
- EITAAC work – and support
- Universal Design Principles
- Accessibility Essentials – Design Tool 2
  - User requirements
- WCAG (W3C Web Content Accessibility Guidelines)

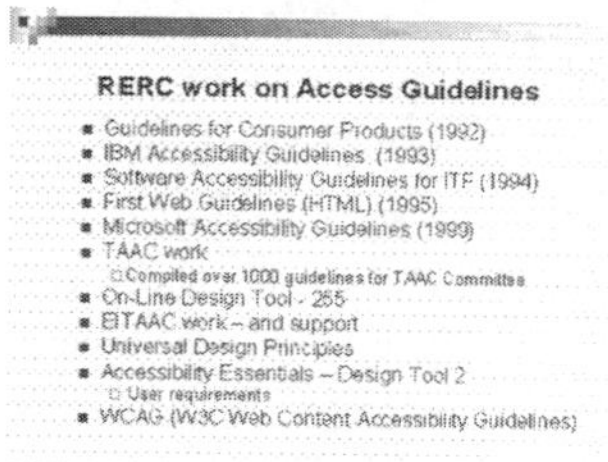

FIGURE 55.2 A small excerpt of guideline work in the United States (cf. footnote 23).

official formal standards (as there are still very few) but to guidelines made to specify the details of the laws.

Two of them are famous:

- American with Disabilities Act (ADA)
- Federal Rehabilitation Act (Section 508)

Although discussing the laws themselves falls outside the scope of this chapter (see Chapter 53, "Policy and Legislation as a Framework of Accessibility," for a discussion of legislation issues in relation to accessibility), especially Section 508 work is highly important for standardization.[21] Outside observers (e.g., from the European Union) were welcomed to participate in the 2007 revision of the 508 Guidelines.

The revision work itself appears to have become a gargantuan task. Emerging technologies have made current 508 standards obsolete. Bluetooth and wireless mobile devices, streaming web video and asynchronous Java and XML-enabled web sites have become common since the original standards were set in place.

Also Section 508 has separate standards for software applications and web applications, although nowadays many applications are running via the web. According to the original 508 rules, software applications must be accessible through keyboard shortcuts and hotkeys. On the contrary, web 508 rules do not have this requirement.

55.2.3.2 United Kingdom

In 1995, the Disability Discrimination Act (DDA) was passed to introduce new measures aimed at ending the discrimination that many disabled people face. It protects disabled people in the areas of:

- Employment
- Access to goods, facilities, and services
- The management, buying, or renting of land or property
- Education

The act is based on the principle that disabled people should not be discriminated against by service providers or those involved in the disposal or management of premises. Although the use of standards is stressed, no specific guidelines are given.

55.2.3.3 Italy

In Italy the law imposing accessibility measures to ICT systems was accepted in 2004 under the name *Stanca act* (after its main promoter, the Minister of Innovation and Technologies). It refers extensively to ISO standardization work. The law aims at drawing up a set of rules governing the criteria and requirements for guaranteeing accessibility. The guidelines intend to regulate both the operational and the organizational issues related to accessibility, as well as introduce the usability principle, defined in a similar way as ISO 9126-1 and ISO 9241-11 rules.

---

[19] http://www.edean.org.

[20] Vanderheiden, G. (2007). New, more robust models for access to mainstream technologies. Presented at COST219ter conference "Extending Horizons: Accessibility to Next Generation Networks Conference," London, BTCentre, January 2007. http://www.tiresias.org/cost219ter/extending_horizons/vanderheiden.htm.

[21] http://www.section508.gov.

55.2.3.4 Germany

On the background of the EU recommendation to adopt the WAI accessibility guidelines, part of the eEurope2002 plan, Germany has taken several legislative initiatives, including social book IX (SGB IX) and the law on the equalization of opportunities for people with disabilities (Bundesbehindertengleichstellungsgesetz, BGG). Through these actions the issue of barrier-free access at the workplace and to public infrastructure has received a new emphasis in Germany. The legislation process was performed for the first time ever with participation of organizations of end-users. The definition of barrier-free access for people with disabilities to human-made infrastructure highlights three characteristics: taking the usual way, without extra effort, and basically without assistance. For the first time, access to information technology, particularly barrier-free access to the Internet, was explicitly taken up in the BGG.

On July 24, 2002 the decree on barrier-free information technology (Barrierefreie Informationstechnik Verordnung, BITV) according to BGG § 11 was officially published by the German federal government and entered into force.[22]

55.2.3.5 Japan

Article 13-2 of the Industrial Standardization Law of Japan states that relevant ministers must enact any drafts proposed by JISC (Japanese Industrial Standards Committee) as industrial standards.

ICT accessibility is being promoted through the accessible design forum. This forum ensures that committee members are aware of aging and disability issues, and that users themselves are represented (and trained if necessary). Their activities are heavily based on ISO/IEC Guide 71 (JIS Z8071). The forum promotes the accumulation of information and know-how,

information sharing and its effective use. It also contributes to raising awareness through the organization of symposiums and participation in exhibitions.

The development of the actual accessibility standards is performed through a hierarchical approach. JISX8341-1 contains the overall framework and the common guidelines. Standards JISX8341-2 and higher then specify the accessibility requirements for the different application domains.

The forum contributes also to global harmonization efforts:

- Cooperation among Asian countries
- 2003 establishment of Japan-Korea-China Accessibility Design Committee
- 2005 explanation to relevant organizations in Singapore, Malaysia, and Thailand
- Contribution to international standardization bodies. The Japanese standards have been forwarded to the following organizations:

  - JIS X8341-1   ISO 9241-20
  - JIS X8341-2   ISO JTC1/SC35
  - JIS X8341-3   WAI's WCAG2.0
  - JIS X8341-4   ITU-T SG16
  - JIS X8341-5   ISO JTC1/SC28

The forum's impact on the Japanese society consists of the following actions:

- Procurement processes must consider conformity with international and national standards.
- There has been an amendment to the fundamental law for the people with disabilities.
- High priority for the realization of information barrier-free society.
- Accessibility considerations now frequently taken into account for web site design.

55.2.4 International Initiatives

55.2.4.1 International Organization for Standardization

The ISO is the world's largest developer of standards. Although ISO's principal activity is the development of technical standards, ISO standards also have important economic and social repercussions. ISO standards make a positive difference, not just to engineers and manufacturers for whom they solve basic problems in production and distribution, but to society as a whole. ISO has been very active, amidst a huge range of other topics, in computer usability and accessibility. A few examples of their recent standardization work related to e-accessibility are:

- ISO DIS 9241-20, "Ergonomics of human-system interaction—Part 20; Accessibility guidelines for information/communication technology (ICT) equipment and services"

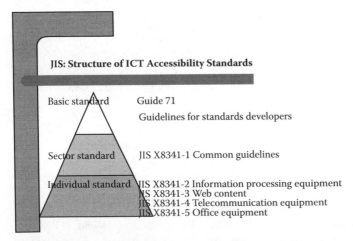

FIGURE 55.3 The Japanese scheme for building up accessibility standards in a hierarchical way. (Courtesy of Hajime Yamada.)

---

[22] Presentation by Professor Buehler at CEN-ETSI CCE conference, Nice, 2003. Proceedings available only on CD.

- ISO TS 16071, "Ergonomics of human-system interaction -- Guidance on accessibility for human-computer interfaces" (Gulliksen and Harker, 2004)

Other important ISO standardization work is referenced in the documents of JTC 1-SWGA.

55.2.4.1.1 Establishment of a Special Working Group on Accessibility

One of the major ISO initiatives in this field is the creation (2004) of an SWG on accessibility within the existing Joint Technical Committee 1 (JTC 1).

JTC 1 believes that the work in the area of information communication and technology standardization for accessibility is a major undertaking, encompassing many international, regional, and local interests. Additionally, there are significant standards efforts taking place in ISO, IEC, ITU, and the national and regional standards bodies, as well as various consortia/fora and user groups. As identified in its long-term business plan, and to be responsive to international, regional, national, and end-user requirements in the area of accessibility, JTC 1 established an SWG on accessibility to:

- Gather user requirements, being mindful of the varied and unique opportunities (direct participation of user organizations, workshops, liaisons)
- Make an inventory of all known accessibility standards efforts
- Identify areas/technologies where voluntary standards are not being addressed and suggest an appropriate body to consider the new work
- Track public laws, policies/measures, and guidelines to ensure the necessary standards are available
- Through wide dissemination of the SWG materials, encourage the use of globally relevant voluntary standards
- Assist consortia/fora, if desired, in submitting their specifications to the formal standards process

To reach these goals, the membership was kept very much open to all individuals and organizations involved in related activities. Also all documents are made public on the SWG's website.[23]

Currently, the work is organized in two task groups:

- Task Group 1 on User Requirements
- Task Group 2 on Accessibility Standards Inventory and Gap Analysis

The standards inventory is considered almost complete. As stated previously, it can be freely downloaded from the SWG's documents area.

55.2.4.2 ITU

ITU, headquartered in Geneva, Switzerland, is an international organization within the United Nations System where governments and the private sector coordinate global telecom networks and services. Telecom standardization falls under subgroup ITU-T. Within ITU-T, Study Group 16 (ITU-T-SG16) is responsible for studies relating to multimedia service capabilities, and application capabilities (including those supported for next-generation networks). This encompasses multimedia terminals, systems (e.g., network signal processing equipment, multipoint conference units, gateways, gatekeepers, modems, and facsimile), protocols, and signal processing (media coding).

Study Group 16 has established a subgroup on "Accessibility and Standardization."[24]

This group has published an accessibility checklist[25] for the makers of standards to ensure that they are taking into account the needs of those to whom accessibility to ICTs is restricted, the deaf or hard-of-hearing, for example. Such a list will help to ensure that accessibility needs are taken into account at an early stage, rather than having to retrofit existing standards.

Another important issue for SG 16 is total conversation (TC). A TC service is an audiovisual conversation service providing bidirectional, symmetric real-time transfer of motion video, text, and voice between users in two or more locations. This real-time text differs from instant messaging systems (e.g., SMS) as a TC system provides the bidirectional transfer of one character at a time. This gives the user the feel of real-time communication, just like voice or video systems that transport streaming media over IP. The concept is aimed at providing rich media real-time conversation for all people and for varying situations. This includes, but is not limited to, people that are disabled in some way (e.g., the deaf or hard-of-hearing, blind, etc.), but also people who find themselves in a situation where the complementing media—video and real-time text—together with voice fulfill the conversation needs much better than only voice. ITU-SG 16 made sure that sections on accessibility were properly integrated in at least 20 standardization documents.[26]

Recently, SG 16 started work on Recommendation F.790 for Telecommunications Accessibility Guidelines for the elderly and people with disabilities. Currently, it is still an internal working document.

55.2.5 Guidelines, Task Force Reports, Working Groups (Informal or de Facto Standards)

55.2.5.1 W3C Guidelines

In relation to design for all or universal design, there is one very well known example of activity: the Web Access Initiative[27] of the World Wide Web Consortium, which produced several guidelines on web accessibility.

---

[23] http://www.jtc1access.org.

[24] http://www.itu.int/ITU-T/studygroups/com16/accessibility.

[25] http://www.itu.int/dms_pub/itu-t/opb/tut/T-TUT-FSTP-2006-TACL-MSW-E.doc.

[26] http://www.itu.int/ITU-T/studygroups/com16/accessibility/docs/apflyer.pdf.

[27] http://www.w3.org/WAI.

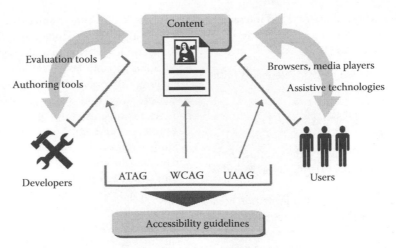

FIGURE 55.4 Essential components of web accessibility according to W3C-WAI.

Although these guidelines are almost universally accepted as the primary reference point (de facto standard) for web accessibility matters, many countries establishing legislative actions for imposing web accessibility were not able to refer to the WAI guidelines, as the W3C cannot be considered a standardization body in the proper sense of the word. Unfortunately, this has already led to several national variants of web accessibility guidelines.

55.2.5.2 ICF: International Classification of Functioning, Disability and Health

As a new member of the World Health Organization (WHO) family of international classifications, ICF[28] describes how people live with their health condition. ICF is a classification of health and health-related domains that describe body functions and structures, activities, and participation. The domains are classified from body, individual, and societal perspectives. Since an individual's functioning and disability occurs in a context, ICF also includes a list of environmental factors.

ICF is useful to understand and measure health outcomes. Strictly speaking, ICF is not a standard, but it is frequently used by funding organizations to quantify problems caused by impairments.

55.3 Activities Bypassing Formal Standardization

It has been mentioned several times that legislative processes in most countries have a strict need for referral to formal national or international standards. Especially in the domain of e-accessibility, these standards tend to be nonformal such as workshop agreements, technical specifications, guidelines, and so on. This has lead to several alternative approaches.

55.3.1 National Laws with Own Guidelines

In January 2005, a very important European colloquium was held in Paris on "Policies and Legislations in Favor of eAccessibility in Europe." Delegates from many countries have explained their national situation. The simple conclusion of the workshop is that almost no harmonization seems to exist (yet). The details can be found in the colloquium's archives.[29]

55.3.2 Procurement Rules as an Alternative to Standardization

As previously stated, the European Union faces a serious problem with supranational legislation: it is almost impossible to achieve in the accessibility domain. On the other hand, long-term experience in the United States shows that the buying power of governments and authorities can be used to impose accessibility requirements. The principle is quite simple: authorities add, in their calls for tender, special clauses on accessibility features that the products or services they want to buy will have to meet. This forces manufacturers to pay attention to the accessibility of the equipment they develop and sell. As a consequence, many, also in Europe, see procurement strategies as a way to improve the accessibility of goods and services, too.

55.3.2.1 International Workshop on Accessibility Requirements for Public Procurement (Brussels 2004)

A workshop held in October 2004 in Brussels addressed the harmonization of eAccessibility requirements to be used in the public procurement of ICT products and services and the requirements for policy implementation in this field. It was organized by the European Commission, the U.S. Access Board, and the European ICT Standards Board (ICTSB). It was supported by the European Disability Forum (EDF) with support of the eInclusion@EU project. The workshop also contributed to the USA-EU "Exchange

---

[28] http://www.who.int/classifications/icf/en.

[29] http://www.braillenet.org/colloques/policies/program.html.

of Information Regarding the Planned Use of ICT Standards in Support of Regulations and Other Public Policies" (in the field of eAccessibility policies).

55.3.2.1.1 Background

The 2005 EC Communication on Accessibility[30] stressed again that public procurements in the ICT domain are an important lever for the deployment of eAccessibility, as they have the potential to play a vital role in removing barriers to participation in the information society by disabled or older people.

A mandate (M-376) had been given by the European Commission to the European Standardization Organizations (ESOs) at the end of 2005 to come up with a solution for common requirements and conformance assessment.

The mandate (M-376 on European Accessibility Requirements for Public Procurement of Products and Services in the ICT Domain[31]) requested the ESOs to come up with, in phase I, an inventory of ICT products, existing accessibility requirements and current gaps, existing standards related to accessibility and their related testing and certification schemes. In phase II (probably starting late 2009), real standardization work will be done, including the establishment of a European Standard (EN) and a technical report (TR). Also a toolkit, that is, a collection of guidance and support material, will be made available freely.

55.3.2.1.2 Status

The start of this mandate was plagued with several problems. Mainstream industry has expressed concerns with regard to some of the certification schemes that could be considered. Several ESOs have objected for procedural reasons (the free availability of the toolkit, nonstandard working procedures, and the requirement that the work must be followed up and judged by groups with limited experience and no formal responsibility in standardization work).

As a consequence, the work has had a slow start since autumn 2007. Within CEN, the work is attributed to a special committee, whereas within ETSI an STF (cf. Section 55.2.1.3.2) was set up.[32] In June 2008 both groups organized a workshop to finalize

phase 1 of the mandate work.[33] Members of the DATSCG group (cf. Section 55.2.1.2.2) were invited to keep an eye on the integration of the work.

55.4 Conclusion

This contribution has provided a general overview of recent progress in standardization in relation to DfA and AT. This appears to be a very large field with many ongoing activities. The very important changes that are currently taking place in the standardization field have also been addressed, such as more informal work, more open discussions, public discussion fora, and standardization of procurement procedures requiring DfA solutions.

References

Brandt, A. (ed.) (1996). ICT standardization and disability in Europe, in the *Proceedings of the European Policy Workshop*, April 1996, Amsterdam, The Netherlands. Sophia Antipolis: ETSI.

Bühler, C. and Stephanidis, C. (2004). European co-operation activities promoting design for all in information society technologies, in the *Proceedings of the 9th International Conference on Computers Helping People with Special Needs (ICCHP 2004)*, 7–9 July 2004, Paris, pp. 80–87. Berlin/Heidelberg: Springer-Verlag.

Chandler, E., Dixon, E., and Tyler, S. (2007). Mobile phone evaluation toolkit, in *Towards an Inclusive Future, Impact and Wider Potential of Information and Communication Technologies* (P. Roe, ed.), p. 255. Brussels: COST219ter.

Engelen, J. (2003a). The next hot item for Assistive Technology and Design for All: Standardization, in *Assistive Technology: Shaping the Future* (G. Craddock, L. McCormack, et al., eds.), pp. 34–42. Amsterdam/Berlin: IOS Press.

Engelen, J. (2003b). The work of the eAccessibility expert group, presented at the *International Congress "Accessibility for All,"* 24–26 April 2003, Nice, France. http://web.archive.org/web/20031005201238/www.etsi.org/cce/proceedings/6_1.htm.

Gill, J. (2007). *Involving People with Disabilities in the Standardization Process*. Brussels: COST219ter. http://www.tiresias.org/about/publications/disabilities_standardisation/Standards_2007.pdf.

Gulliksen, J. and Harker, S. (2004). The software accessibility of human-computer interfaces—ISO Technical Specification 16071. *Universal Access in the Information Society* 3: 6–16.

---

[30] The Communication from the Commission to the Council, the European Parliament, the Economic and Social Committee, and the Committee of Regions, regarding eAccessibility (adopted on September 13, 2005). See http://ec.europa.eu/information_society/activities/einclusion/policy/accessibility/com_ea_2005/index_en.htm.

[31] http://www.etsi.org/WebSite/document/aboutETSI/EC_Mandates/m376en.pdf.

[32] CEN BT Working Group 185, Cenelec BT Working Group 101-5 and ETSI TC/HF (Technical Committee/Human Factors) that in practice acts via STF333.

---

[33] http://www.ictsb.org/DATSCG_conference_public_procurement.htm.

56

Management of Design for All

Christian Bühler

56.1 Introduction

This chapter looks at design for all (DfA) from a management perspective. It considers DfA as a process, which requires appropriate management at different levels.

The need to address barrier-free accessibility pushed by organizations of people with disability as a pending problem has created various activities and substantial literature around *accessibility* and the related potential design philosophies. Specifically, terms such as barrier-free design, accessible design, inclusive design, universal design, DfA, and Kyoyohin have all come about to provide a means for understanding the various aspects of accessibility. However, accessibility or universal accessibility based on the traditional understanding of disability is only the starting point for DfA. Instead of disability and age, DfA fosters diversity, and instead of special social action it fosters mainstreaming. It is the intention of DfA to reflect a new concept for design that attempts to accommodate the broadest possible range of human abilities, requirements, and preferences in the design of all products and environments. Thus, it promotes a design perspective that eliminates the need for special features and fosters individualization and end-user acceptability (Bühler and Stephanidis, 2004). It comprises a proactive strategy postulating accessibility and quality embedded in a product or service at design time. The goal is to consider the requirements of all users in the design process and to develop a solution that is suitable for all users. Of course, this would be the most inclusive design approach possible. At the same time, one would fulfill the requirements of each individual and reach the largest market

possible. DfA, therefore, includes both a social and a market component (Bühler, 1999, 2001; Ellis, 2003).

The previous definitions of DfA are oriented toward user diversity, and do not explicitly acknowledge the interdependence of user, environment, and time. The International Classification of Functioning Disability and Health, ICF (WHO, 2001) describes a bio-psycho-social model in relation to functional health. Through the introduction of context factors, such models make it very clear that functional health depends not only on static individual predispositions, but also very much on situations, circumstances, and conditions. This important issue is particularly relevant from the perspective of universal design (Vanderheiden, 1996). Besides the diversity of the users, DfA needs to reflect the broadest range of situations and circumstances (i.e., environments and temporarily changing conditions of use).

In this chapter, a broad understanding of the term is adopted. It is based on the definition in the context of the eEurope initiative, which puts focus on the integration of older people and people with disabilities in the information society. This approach to DfA consists of three principal strategies (EDeAN, 2004):

- Design of products, services, and applications, that are demonstrably suitable for most of the potential users without any modifications
- Design of products that are easily adaptable to different users (e.g., by incorporating adaptable or customizable user interfaces)
- Design of products that have standardized interfaces, capable of being accessed by specialized user interaction devices

DfA is seen as an important complement to assistive technology. The basic underlying idea is to produce products and services in a way that as many people as possible can use them directly or with the help of assistive technology. This concept contains already an important management issue. In a quickly changing world, with a very high innovation speed of mainstream products, product development time and product life cycles can become very short. Therefore, products are replaced very quickly in the market. Adapting all products and services for those who cannot operate them is time consuming, difficult to achieve in the short time cycles, costly, and may lead to discrimination. Designing and producing for as many as possible is the more economic strategy (first strategy). The second strategy is to create easily adaptable user interfaces, if the first strategy seems inappropriate. As a minimum requirement, the existence of standardized interfaces to other products (third strategy) tries to support the compatibility with special products (assistive technology). The management problem is to decide what is appropriate in each specific case and which of the three strategies should be followed. Obviously there exists no simple, general answer.

DfA strategies are of particular value for information society technologies, which change at very high speeds. This area is of course an important and interesting field for DfA. Almost all citizens are among the users, and these products and services can help to overcome traditional barriers of everyday life. Although it originated from concepts elaborated in the field of accessibility of the built environment, and research actions in assistive technology (Mace, 1998; Mueller, 1998; Bühler and Knops, 1999; Preiser and Ostroff, 2001), DfA has meanwhile been adopted as a concept for the information societies around the globe. Stakeholders in the field are encouraged by administrations and governments at a national and international level to develop, disseminate, and employ the concept of DfA. The idea is taken up by politicians, research and development (R&D) programs, and NGOs to develop an information society for all in Europe and beyond. While in Europe the term *design for all* dominates (CEU, 2001–2005), in the United States mostly *universal design* (Center of Universal Design, 1997) is used, and in Japan the concepts of *Kyoyohin* and of *universal design* are combined (Kyoyohin Foundation, 2001).

Unfortunately, a number of misconceptions are connected with DfA. These are based on the different starting points of the stakeholders, the different motivations and professional backgrounds, language problems, and confusion between motivations, strategies, approaches, and objectives. One fundamental misconception results from an error of translation but also semantics; it is connected to the meaning of the word *design* itself. Many people translate or understand the word *design* in only a pure aesthetical context and as a result of a development, rather than design, of function and usability as a process. They restrict design to the task of aesthetics designers and do not consider the task of development and construction (engineering). Designers often argue that DfA goes along with a restriction in creativity that limits the artistic aspects of design. Other misconceptions are connected with the term *all* (or *universal*).

In both cases *all* and *universal* do not address specific groups of users, but strive for covering diversity. And, in turn, if one single user is not addressed in a DfA process, this does not mean that it can't be DfA or universal design. Some people are very much impressed by the demographic shift, and consider DfA as addressing only solutions for older people and their functional restrictions. Some confuse DfA with special products for health care and rehabilitation. Others believe that DfA is just another label for accessibility. When using the term *inclusive*, some people tend to consider mainly those who are excluded. Marketing people do not see a DfA market for their specialized products. Politicians hope that mainstreaming through DfA will solve various social problems. Company officials expect small markets and costly development procedures. Users expect stigmatizing design, making one a visible old or disabled person. And the ongoing discussion provides impressions of all these misconceptions. There is obviously a strong need for education to overcome this confusion. Education needs to be part of the management strategy.

Reflecting on DfA, different management aspects on different levels for various purposes in diverse milieus and contexts are to be considered. Also the range of potential managers of DfA varies widely: politicians, administrators, company leaders, developers, engineers, designers, educators, and association leaders constitute only some examples of stakeholders potentially involved. Therefore, a very open perception of the term *management* is taken in this chapter, instead of a restricting definition or a specific management approach. Management comprises all activities needed for planning, organizing, leading, motivating, coordinating, and controlling the processes connected to DfA.

Following the introduction, the chapter discusses important levels of DfA management. The next section underpins the process character of DfA, followed by a reflection on approaches to DfA. Next, three important management issues are presented, namely general awareness, training, and cooperation with users. Finally, the chapter concludes with a summary of the current situation and recommendations.

56.2 Levels of DfA Management

It is important to note that DfA has been available for a notable period of time, but the market penetration is still rather low. Most design considerations are still based on a standard user and a business concept, targeting specific user groups. There is low awareness about the business potential of DFA products. Further, a number of misconceptions are connected with the low take-up (see Section 56.1). If a clearer picture of DfA is achieved, DfA is accepted as a good idea at a general level, but it often fails to step over into the strategies and design considerations leading to products and services. Obviously, a goal-oriented management is required from the conceptualization of DfA to its implementation in products and services in different areas. In Figure 56.1 different management levels for DfA are presented in a cascaded structure. Of course, it is possible to start with DfA considerations on a product level and implement the management and process just for one product to begin; more products may follow in case of success. Following this direction, the management and procedures are extended to the product family, brand, and company, and might be taken up by an industrial or service branch in a typical bottom-up approach (Figure 56.1). In an early stage of the concept, this is a reasonable approach. The DfA concept, however, has already gone through this phase, supported by research and development in many countries and companies (Preiser and Ostroff, 2001; European Commission, 2002; Klironomos et al., 2005). In the current situation, DfA is stimulated the other way round, in a top-down approach. The international research and development community, NGOs, international bodies like the UN, standardization organizations, EU, and national governments sponsor actions, cooperation, development of guidelines, examples of best practice, benchmarking, regulation, and legislation to support the take-up and implementation (Figure 56.1). Issues related to accessibility in legislation are discussed in Chapter 53 of this handbook).

56.2.1 Policy Level (International and National)

In the past decade, the aging of societies has particularly stimulated national governments in several countries to take action for more accessible products. Strategies combine voluntary and mandatory actions. Governments have funded research projects (CEU, 2002), support actions (CEU, 2004), elaboration of standards (EDeAN, 2006; eEurope Standards, 2007), award schemes, monitoring procedures, benchmarking, and enacting of legislation (EDeAN, 2005). These have led to action plans, the creation of national and international networks (Stephanidis et al., 1998; Universal Design Network, 1999; COST219bis, 2001; UDA, 2003; EDeAN, 2006; etc.), and arrangement of symposia and conferences. National strategies are particularly helpful to reach company CEOs and stimulate the top-down support for DfA on a branch level and within companies. Considering the list of options for national action, the discussion is ongoing whether regulative mandatory requirements or stimulative positive action yields better outcomes (CEU, 2005). Good experiences have been made by positive action with broad public dissemination like publishing lists of good products (Kyoyohin Foundation, 2001; Top of the Web, 2001; BIK 95+, 2006) or award schemes (e.g., Breaking barriers award, Design-for-all-AT Awards, 2004; BIENE, 2006).

The regulation of public procurement with respect to accessibility requirements has been a measure, for example, in the United States (U.S. Access Board, 1998). In the EU a procurement toolkit (CEU, 2005) is under development, which will involve criteria on accessibility in the application and selection process. Another measure is the stimulation of standardization. The EU has launched a number of mandates to the European standardization organizations CEN, CENELEC, and ETSI with respect to accessibility requirements and DfA (eEurope Standards, 2007). Standards can give guidance to the developers and a baseline for future reference in legislation. On the other hand, CSR (corporate social responsibility) programs and awareness campaigns try to support voluntary action (CSR, 2007). While users tend to prefer mandatory action, including third-party certification, industries very much prefer voluntary action and self-declaration. Both directions carry good potential. As DfA is seen more as a business case, mandatory action to enforce DfA seems inappropriate. However, mandatory action on accessibility requirements (as a social action) can enforce DfA solutions as a market reaction and business strategy (e.g., U.S. Access Board, 1998; BBGG, 2002). In this context, the users as a market force have a particular role with their own purchasing power. It is expected that in the aging markets (accessible) DfA products will have a competitive advantage. It is therefore most convincing if user organizations like ANEC (European Association for the Co-ordination of Consumer Representation in Standardization), organizations of older people, and organizations of people with disabilities campaign for DfA.

56.2.2 Branch Level

Umbrella organizations of branches of industries or service providers support to their members in different common affairs. One option is to identify trends and to give a higher profile to the branch as a whole. DfA is a very suitable subject in this respect. It is future oriented in terms of the aging societies, business oriented in terms of competitiveness and broadening markets, it takes up government stimulus, and it underpins corporate social responsibility. Many organizations have started activities for their members in this direction, creating working groups and workshops, providing guidelines, reacting on governmental initiatives, and performing education and training of staff.

An example of a commercially oriented initiative of companies in Japan started in 1999 to support the production and marketing of Kyoyohin products. Kyoyohin and Kyoyo services are designed to be used by as many people as possible, including the elderly and those with disabilities. The initiative has created the Kyoyohin Foundation to manage Kyoyohin products in the

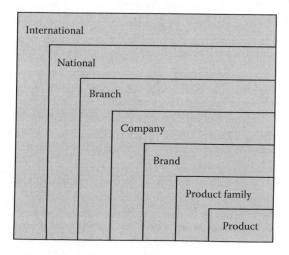

FIGURE 56.1 Levels of design-for-all management.

Japanese market. Several branches are specifically addressed. In comparison with universal design and DfA, Kyoyohin takes a downstream perspective and a down-to-earth approach (Kyoyohin Foundation, 2001). It comes up with a concept of general awareness of diversity and practical product-related guidelines. Kyoyohin targets five main principles:

1. Products/services that are accessible for persons with and without disabilities alike
2. Products/services that are not targeted specifically for special users
3. Products/services that are manufactured and used widely
4. Products/services that are reasonable in price compared to other similar products and services
5. Products/services that are consistently in the market

It is strongly market oriented and considers the benefits for the companies in a growing market of Kyoyohin products.

Another strategy at a branch level is to share resources for better accessibility of products. This could be a suitable solution in case of mandatory requirements. A group of companies identifies product characteristics and services that can be better supported jointly rather than on a company level. This approach enables companies to provide the effort needed at a comparably low cost. In this context, cooperation in standardization needs also to be mentioned. If companies agree on standards, this can help to create a common platform for reaching a better level of compatibility, usability, and accessibility in a fair way beyond competition (W3C, 1999).

56.2.3 Brand/Company Level

At a company level, several companies have already decided to adopt DfA (or universal design, including accessibility) as part of the company policy. Typically, such adoption starts with a general high-level decision leading to a statement in the overall strategy of the company, to develop products for as many people as possible, including the idea of diversity and more specifically people with disabilities and of older age (e.g., Pacific Bell, 1994; SUN, 2007). For actual implementation, DfA needs to be enforced within the company through education, design guidelines, change of procedures, and change of quality definitions and control. It is very important to establish a DfA management to give DfA requirements serious weight among other lines of the company policy. Many global companies in the information and communication technologies (ICT) field have indeed installed a policy with regard to accessibility, mainly as a result of U.S. legislation (U.S. Access Board, 1998). They have created accessibility centers, with a team cooperating with the profit centers of the company and in relevant standardization or government-related working groups. Others have placed accessibility experts into the teams of the profit centers. For many companies one can find an accessibility web site as part of the company web site.

Another approach to support all people at a brand level is to put different options into different devices or services. If, for example, a company produces cell phones, not necessarily each

and every product needs to have a functional range suitable for all. Instead, at least one phone out of a product family should cover a specific functional range, where all phones together cover the functional range for all. This strategy is not DfA in a strict sense, but a feasible solution if considered seriously. It is of course not sufficient, if the pricing for some of the devices goes beyond a certain level compared to the rest of the product family (i.e., a device matching the needs of one group would not be affordable to the group).

56.2.4 Product Level

It is indeed rather complicated to transfer the DfA philosophy into concrete characteristics at a technical level: which concrete size of a button, which weight, sound level, font size, brightness, and so on, are suitable for all? Users are different from each other, and the variety of situations and conditions of use are broad, while technology is changing quickly. Therefore, at the product level itself, DfA management needs very much to focus on the diversity of user needs in a rapidly changing technical environment. In the following, the focus on users is described as a continuous process in the product life cycle.

56.3 DfA as a Process

A product is commonly considered as the result of a development and design process, produced and sold as visualized in Figure 56.2 from the product intention to the product on the market. In this kind of process, functional specifications, technical production conditions, safety requirements, competitive aspects, and so on, are combined and lead finally to the product.

Management of DfA at a product level follows very much the same procedure. However, this forward control scheme represents a management perspective in a rather simplified manner neglecting important feedback loops. In the case of DfA, user participation plays a key role (Bühler, 2002). Considering the development process for DfA more closely, it is obvious that a kind of iteration or feedback is required at the time of development (Figure 56.3). In this way, the product characteristics are elaborated in a continuously controlled comparison of product intention and user requirements. Typical approaches to the measurement of user requirements are questionnaires, focus groups based on drafts of nonfunctional mockups, user tests with prototypes in laboratory environments, and so on.

The results of these measurements need to be transferred to the product intention level, which is symbolized in the block FBC (feedback control) and the diamond as a symbol for comparison.

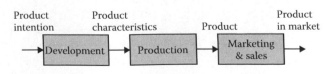

FIGURE 56.2 Product to market.

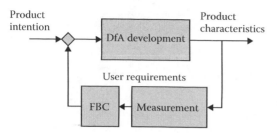

FIGURE 56.3 DfA development loop.

The difference between the intention and the measurement triggers the change of the DfA development.

A similar consideration can be made for production, marketing, and sales, which are grouped in a second feedback loop (Figure 56.4). Again, user feedback is the main focus of the second loop. This time the fully functional product is already existing and available. It is now possible to compare the satisfaction of users with the product characteristics based on the use of the real product in a continuously controlled way. The product on the market can be improved with each new production run, but basically without changing the product intention and characteristics. The measurement in this loop has left the laboratory and considers real-life experiences with the real product. The measured user satisfaction needs to be transferred at the product characteristics level, which is symbolized in the block FBC (feedback control) and the diamond as symbol for comparison. Differences between the product characteristics and user satisfaction triggers the change of the DfA production, marketing, and sales.

The product intention itself is also a result of user feedback. The loop from market research to support the creation of new products is well known. Figure 56.5 visualizes the complete process in three major feedback loops. The measurements in the third loop are taken in the actual market environment on a larger scale. The use of the product, comparison with alternative products, product usability, and accessibility are aspects of this measurement. The outcome of this research needs to be transferred at the product intention level, which is symbolized by the box FBC and the diamond as a symbol for comparison. The difference between the intention and the usability/accessibility measure triggers the change in the DfA development in an overall loop.

56.3.1 DfA User Requirements

56.3.1.1 Variety of User Requirements

In the three loops depicted in Figure 56.5, the feedback of user experience and user opinion is very important. For DfA management, it is necessary to manage the diversity of users and the differences of their real-life environments:

Diversity of Users

- Sex, age, size, weight
- Intelligence, education
- Mental and body fitness
- Perception range
- Specific skills
- Acute or chronic disease
- Disability, etc.

Different Real-Life Environments

- Home (living room, bathroom), city, rural area, shop, outdoors
- Bus, tram, tube, car, ship, plane
- School, workplace
- Cinema, theater, sports arena
- Summer, winter, rain, snow, ice
- Bright sunshine, artificial light, darkness, etc.

The high number of different potential users and scenarios of use force the restriction of the selection of users in the feedback loop for practical reasons. However, instead of concentrating on one type of user for feedback, in DfA it is necessary to involve the largest possible diversity. It is mandatory to consider different abilities of the users as proposed in the product design ideas browser (http://trace.wisc.edu/docs/browser). Lists of criteria for different disabilities and application domains can help to get an understanding of potential problems. Experts in usability, psychology, and disability might help to identify requirements. However, interaction with the users themselves is necessary. The immediate contact of users and staff in design, development, and marketing provides deeper insight and is much more authentic than the statements of experts. The choice of environment scenarios is also a crucial task. Instead of concentrating on a fixed scenario like in a laboratory, variations of the conditions of use are required.

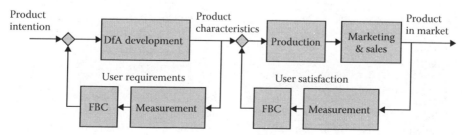

FIGURE 56.4 DfA development and market loops.

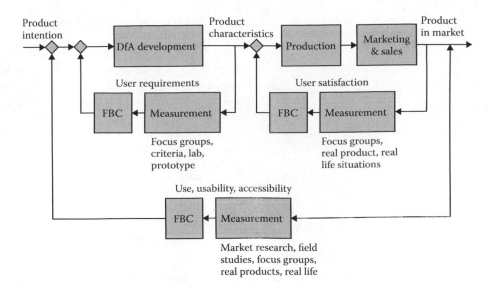

FIGURE 56.5 Management of DfA on product level.

56.3.2 Intended Result (Quality of the Product)

The quality of the DfA product comprises the quality of use of different users in a large variety of situations. It is important to note that even if one product could be used by everyone, it might not be accepted by anyone. This contradiction results from a potentially low usability of the product for all: an expert user might be bored by too many explanations, or too many time-consuming steps of low complexity, while for other users the explanations and complexity might be still rather challenging. Further, there might be characteristics supporting using conditions that are not optimal for other conditions. Obviously, the danger is to arrive at a compromise that is not acceptable by all. Instead of enlarging the market share, this would lead to a loss. A consideration closely connected is the question of a marketable price. Of course, a DfA solution, which is more expensive compared to the different solutions targeted to various user groups, might not be purchased. A DfA product or service needs to be competitive in terms of quality of the solution and market price. This consideration is taken up in the market orientation of universal design, and especially in the European three strategies approach (see Section 56.4.1).

56.4 Approaches to DfA

56.4.1 Direct Access

In the three strategies approach of DfA, as formulated in Europe (EDeAN 2004), the first strategy comprises:

> The design of products, services and applications, which are demonstrably suitable for most of the potential users without any modifications.

The result of applying this strategy would be one product for (most of) all, thus looking like the ideal case. Of course, direct access wherever possible is the first choice from a user perspective and from a business perspective. Consequently, this general design goal has the highest priority. The inherent management aspect is twofold: arriving at solutions that fulfill the requirement and assessing the feasibility (technical, production, sales and service, cost) of these solutions. The easy and positive end of the task is a cheap, smart solution for all. In reality, it becomes more complicated, as the questions emerge of who is "all," or more strictly, who is not included, and what is "demonstrably suitable," or more strictly, which solution is not satisfactory for a particular user group. Both questions have no general answer. Basically, they refer to a classical optimization problem, where a complicated cost function needs to be minimized. Elements of such a cost function comprise, for example:

- The potential diversity and number of people who can use the product directly
- Obligations through legislation and regulation
- Overall company positioning
- CSR value
- Usability and pricing compared to competitive products and hence potential number of buyers
- Cost of development, production, sales, and service
- Innovation-value
- Potential profit

These elements range from strict DfA considerations to traditional business criteria. Obviously, the outcome of the optimization depends very much on the setup of the cost function and the weight given to the different elements.

With the help of this kind of consideration, one can identify the "most of all" for whom the product can be directly used without modification. On the other hand, all others not included create a potential market for another solution or product. This leads to the second strategy of DfA.

56.4.2 Adaptation and Adaptability

The second strategy comprises the design of products which are easily adaptable to different users (e.g., by incorporating adaptable or customisable user interfaces). In the first instance this sounds more costly and complicated. However, in the ICT domain and many other fields, this strategy represents a well-known and feasible way of providing flexibility. Decomposing products into modules and offering a selection package is a very common approach for complex products or services like computers, kitchens, automobiles, assisted living services, and so on. Each user can compose a product best fitting his/her needs and wishes. While in former times kitchen tables have been available in only one height, today one can choose from different heights without extra cost. In software-based applications, users can individualize the interface within a given framework of options and settings and the introduction of personal information. Storage of user settings and profiles can be local (e.g., browser settings), but also related to a login for a service in the Internet ("my ..."). Modern approaches can measure or monitor user behavior (even mood) and propose and perform automatic self-adaptation. However, a similar management issue arises as for strategy 1: one needs to decide which "different users" and which "required performance" of the product receive attention. This can be reflected in two new elements of the cost function:

- The potential diversity and number of different users that can use the product through (interface) adaptation, customization, or self-adaptation
- The variety of adaptation, customization, or self-adaptation mechanisms to serve the purpose

The optimization of the cost function by consideration of the different elements again identifies the user group that is served by this approach and those still left out, and it leads further down to the third strategy.

56.4.3 Compatibility with Assistive Devices

The third strategy tries to take care of the rest of all users, who are not reached by the two other strategies:

The design of products, which have standardized interfaces, capable of being accessed by specialized user interaction devices.

Actually, this strategy connects to the traditional approach to provide special solutions to people with special requirements. This is not really a DfA strategy. However, if a process has started with the two other strategies and comes to the conclusion that the product will not be accessible for all users, it at least provides an option for others to add on what is needed. Classically, those are the providers of assistive technology products. A wide range of assistive technology products exists in the market to support people with disabilities. Mostly, small and medium-size companies have specialized around impairments and technologies. More than 20,000 assistive technology products are listed in databases (e.g., REHADAT, 2007). Very different needs can be supported, and the customization level goes as far as to individual solutions for the single user. A significant difference compared to DfA products lies in the financing of assistive technology products. In many countries, the public solidarity systems (tax based, insurance based, welfare based) pay for the assistive technology products according to national traditions and rules (HEART, 1994). Hence, if the effort for reaching a DfA solution according to strategy 1 or 2 is too high for a company, or would raise the product price to a noncompetitive level, the functionality can and needs to be transferred into the assistive technology range.

To add on assistive technology, developers need to interface between the mainstream product and the assistive technology product and between the assistive technology product and the user. As long as assistive technology developers need to guess about interfaces to mainstream products, maybe to create individual hardware gateways or to reinvent interfaces or use product functions that have been intended for other purposes, this may lead to inefficient and expensive solutions. A particular problem arises for mainstream products with short life cycles: once an adaptation is available for a product, the next release of this product has replaced the product and the adaptation also needs a new release. This is ineffective and costly and hardly achievable by small assistive technology companies. The availability of standardized interfaces, either public or in partner agreement, supports the developers of assistive technology and reaches out to the next group of users.

As already mentioned, if this strategy is the first and only strategy employed, it is not really DfA. Even this third strategy will probably not enable all users to use the product or service, as amending strategies, education, training, therapy, and personal assistance are needed.

56.5 DfA Awareness and Training

56.5.1 Policy Level

Technology has a great potential to support people with disabilities (Council of Europe, 2003). In this respect, it means more than just to carry out daily tasks supported by technology without human assistance. Technology can empower self-determination, decision making, and can also facilitate participation in society and improve quality of life. Technology in the information society gains the special connotation of information, knowledge transfer, communication, and remote action in eGovernment, eBusiness, and so on. The relevance of technology for the employability of people is particularly high. Without access to modern information society technologies, people become excluded from job opportunities. On the other hand, the use of technology provides new job opportunities for people with disabilities (CEU, 2001; European Commission, 2002). DfA provides solutions for direct access in the mainstream. Assistive technology as the modern concept of rehabilitation technology deals with products and services to enable people to use general premises and mainstream technology. DfA and assistive technology are complementary elements in a continuum of solutions. Users need to select an appropriate combination of DfA and special solutions to match their needs. The overall management issue of the policy level is to make sure that this continuum of solutions—either in the mainstream or as a special solution—is provided for the users at a good quality level and in an economical way. In the international and national contexts it is important to raise awareness about this potential of technology:

- Available options need to be made available in all countries.
- Public awareness for the social and general benefit need to be increased.
- (Scarce) resources need to be spent optimally.
- The mainstream and special sectors shall be stimulated to more innovation in the field of DfA and assistive technology.
- The academic education sector shall provide awareness and knowledge about the concept to students as the future key players.

An example of an awareness action is the Design for All/Assistive Technology Awards Scheme (Design-for-all-AT Awards, 2004), which was launched by the European Commission in 2004. Awards have been sponsored for products in DfA and assistive technology in the categories of daily living, mobility and transportation, communication and information, lifelong learning, employment, culture/leisure/sports. The winners were presented at the large exhibition, REHACARE International in 2004. This joint action of the public sector, different associations, and industry underpinned successfully the importance of disseminating the options and benefits of this concept for people.

Besides public awareness, education and training play an important role. Education can indeed make the difference. This has been acknowledged for a long time. Nowadays, with long and changing employment careers of people in a rapidly changing environment, the concept of lifelong learning has achieved much attention. This has clear implications for individuals as learners, the companies with specific requirements regarding the workforce, and also the providers of education. The responsibility of the public sector is very much related to education in public schools and universities. It is crucial to integrate DfA in the curricula of economics, engineering, computer science, architecture, industrial design, rehabilitation science, social science, and so on. Students in all these disciplines need to be educated for future joint efforts toward DfA. A clear understanding of the idea, the methodology, and the benefits of DfA already at high

school and university level, will favor knowledge transfer toward the next work generation. However, schools and universities are today already resources for lifelong learning. Strategically opening up for further education and new forms of teaching and learning, they can significantly contribute beyond the first education phase. It is up to the policy level to encourage schools, universities, and other educational settings to take up DfA in their courses. The support of networking, research, development of teaching material and its provision free of charge is a typical portfolio for public action.

56.5.2 Companies and Professionals

Inside the companies the level of education on DfA is a central factor for successful implementation. Without basic education on DfA, prejudices and misconceptions as described earlier will hinder DfA implementation. Awareness raising, further education and training of existing staff, and selection of educated employees are the main roads to follow. Many companies run further education schemes inside the company. The education and training on DfA should be integrated in such schemes. It is also possible to source out and cooperate with external educational experts in DfA. In this context, besides the development teams, decision makers and marketing and sales professionals should be included in the training. It is of strategic importance to include these key players who represent the major driving force in the company development.

Many professionals are organized in associations around their specific disciplines or field of work. Many professional organizations understand further education of their members is a core objective and field of activity. In this case, the professionals self-organize their subject of interest, for example, in conferences, symposia, courses, field visits, and so on.

For support of the education process, resources have been made available not only in books, journals, and brochures, but also through web sites. Typically, the resources try to provide information at different levels:

- Introduction to DfA
- Benefits of DfA
- Demographic change and disability
- Principles of DfA
- Approaches to DfA
- Methods and tools of DfA
- Specific information related to technologies, applications and services
- Case studies (mostly best practice)
- Experiences of DfA
- Links to resources
- Links to related fields (accessibility, standards, legislation)

The web site, "Universal Design Education Online" (http://www.udeducation.org), sponsored by the National Institute on Disability and Rehabilitation Research (NIDRR) of the U.S. Department of Education provides one example of such a web site.

It offers additionally the choice to visit and use the site from the point of view of the teacher or the learner. It provides also the option to submit material for publication on the site. The European network EDeAN has created the DfA resource center ARIADNE (http://dfarc.edean.org) and the site Design for All Education and Training (http://www.education.edean.org/index.php). ARIADNE offers a broad range of information, such as case studies, information about national or international projects related to DfA, best-practice reports, design tools, benchmarking tools and results, product assessment reports, and various publications. Users can choose different options to retrieve the information, and registered users are allowed to comment or add new information. Another educational site labels under inclusive design (www.inclusivedesigntoolkit.com). It contains, among other subjects, a section on corporate implications and product strategies.

It is important to note that companies often provide good public information relevant to DfA in support of their products. However, they mostly address it from the perspective of accessibility. Particularly software providers inform about options and support to create accessible formats using their products. As this is a clear consequence of legislation on accessibility, again the important role of policy and legislation needs to be acknowledged (see also Chapter 53, "Policy and Legislation as a Framework of Accessibility").

Of course the kind of education and training needs to be dedicated to the targeted group of players. It varies from short courses with aggregated knowledge to curricula on methodologies or to hands-on training.

56.6 Cooperation with User Groups

In DfA and universal design, the objective is to create products (devices, environments, systems, and services) that are usable by people with the widest possible range of capabilities, operating within the widest possible range of situations (environments, conditions, and circumstances). The definition of universal design highlights three aspects:

- Universal design is not a product but a process.
- Universal design is not for all, but for the widest possible range of abilities.
- Universal design is not for all situations, but for the widest possible range of circumstances.

Thus, the DfA approach requires at least a consideration of the needs of the broadest user group possible and measures to prepare interfaces for those excluded from consideration in the final design. An important mechanism of DfA is the participation of users in the process. To reach the broadest user group and a variety of situations, people with various abilities, including people with disabilities should be more actively involved in the process. In this context, users with different disabilities can represent diversity. To some extent, their experiences with vision, hearing, manipulation, dexterity, concentration, memory, and so on, highlight many of the problematic issues for other users

and in various contexts. Market research methods and studies, which keep the users and the developers separate, are helpful, but not sufficient. Many developers are not accustomed to discussing their work with nonspecialists directly, and vice versa. Therefore, many players in industry, research, and development have little experience in this respect.

A good starting point for dealing with the diversity of users is contact with organizations of users with disabilities. Usually, many users with disabilities join organizations focusing on their particular impairment or disability. These user organizations carry out very good field work on the basis of a democratic structure. There have not been many meeting points between user organizations of people with disabilities and industries because they have very different agendas. Therefore, contact between developers and such organizations of end users has remained generally weak. Industries, research, and development communities are in principle prepared to interact with users. On the other hand, knowledge of end-users about the new opportunities offered by technological solutions is often limited. In practice, user participation is often reduced to being the subject of study, or being involved late in the project life cycle in connection with testing of prototypes or products. A change of attitudes and approaches to user participation has been started to manage this situation. An interesting framework has been elaborated and promoted in the European context (Fortune, 2000; Bühler, 2001) and will be introduced briefly.

56.6.1 The FORTUNE Concept of User Participation

Following the description of the process of DfA development the participation of end users is obviously an important component. It needs serious implementation to receive full benefit of the users' skills and knowledge. It encompasses both user representation from the user organizations on behalf of the whole user group, and participation of the single individual reflecting his/her own experiences and needs. Careful planning for all feedback loops (as described previously) is required. Particular focus needs to be put on resource planning in terms of budget, time, and accessible premises and material. Cooperation should be guided by a partnership spirit, in which the partners (developers and end-users) work toward common goals on an equal basis. The FORTUNE concept describes a benchmarking model in this respect for user participation in projects. It provides a vision of how user participation in a project should take place. Although it was developed with a particular focus on participation of people with disabilities, it can serve as a model for a user participation in DfA. Especially management questions and attitude aspects are comparable and can be applied. Seven principles describe the ideal model (Table 56.1; FORTUNE, 2000). The first two principles deal with attitude and management. It is important to meet all users in a spirit of partnership on an equal level and with mutual respect. Further, it is preferable to involve user organizations rather than individuals only. It is indeed a weakness to select single users on an individual basis, for example, users who

TABLE 56.1 The Seven Principles of User Participation in Projects

<number of the principle, MEMO TITEL, key idea, short explanation>

❶ PARTNERSHIP

- **Cooperation is based on the idea of partnership**

  Partnership on equal level, with mutual respect. Cooperative attitude. Sharing a common affair, responsibility and influence, risk and benefit. Looking for solutions, not for problems.

❷ USER ORGANIZATION-BASED

- **Users are members or representatives of an organization of end-users**

  Representing more than the individual (own) case. Having the support of the organization for practical matters and for getting feedback from other members.

❸ EQUAL PAYMENT

- **Users receive payments on the same basis as all other partners**

  All partners in the project receive appropriate payments for their contribution. The contribution of users is not handled as a volunteering activity, but as a fully valuable contribution to the project.

❹ ACCESSIBILITY

- **All project materials, communications, and premises are made accessible to the users**

  Alternative formats for print material, appropriate communication media, accessible meeting sites, rooms and hotel accommodation, personal assistance.

❺ QUALIFIED STAFF

- **Every partner has to provide qualified staff members to the project**

  Staff members provide the right attitude, respect, expertise and skills for the project. They accept project rules and constraints like timing, budgets, confidentiality, etc.

❻ SOUND PLAN

- **The project plan contains appropriate work packages and tasks of user participation**

  User participation is planned and described with the same detail as all other items of the project plan, including responsibilities, methods, timing, and budgets.

❼ EARLY INVOLVEMENT

- **Users are partners from the very beginning of a project**

  Users are already involved in the first steps of creating a project, identifying the project idea, creation of the consortium, project planning.

are engineers or programmers. They tend to act as engineers or programmers rather than users, and often limit their view on the very own case. Representatives of user organizations, which means in this case individual users who are selected by organizations, act as user experts from an ability perspective, and on the background of a group of users. The user organization can give practical support and general backup to the selected individuals in the course of the process.

Principles 3 to 5 consider the spending of resources. The third principle is concerned with payments. It is a false assumption that users with disabilities receive social benefits, are not involved in a work process, and should provide their contribution for free. The fourth principle is related to the accessibility of the infrastructure. Although accessibility requirements seem to be pretty obvious, experiences show that unsatisfactory accessibility considerations can lead to a drawback in the process or a complete failure. According to principle 5, all partners involved need to select staff qualified with the right attitude and expertise. The roles of the different players shall be clear and accepted by all. The last two principles focus on planning. Principle 6 requires sound planning of user participation to lead to useful results. It is one option to ask one of the user organizations to play a leading role in the planning of user participation. Finally, principle 7 makes it clear that early involvement is essential for success. This refers directly to the first of the three feedback loops.

56.6.2 User Involvement in the DfA Process

As described in the process-oriented description, users need to take part in different stages. User participation can take place throughout the total value chain, from the first idea or problem identification to the release of a product on the market or during evaluation. Different methods can be applied for the involvement in different stages of the process. In Table 56.2 (FORTUNE, 2000) the selection of methods is presented and linked to different phases. This relates to what is also known as user-centered design (Norman and Draper, 1986; Schuler and Namioka, 1993). Much valuable information about user-centered design is described in the book *User-Fit* (Poulsen et al., 1996).

Many of the methods listed in Table 56.2 have a user focus and involve users in carrying out relevant investigations. However, in several instances users are involved as informants, rather than partners. From the perspective of the FORTUNE model, it is much better to involve users early in the planning of the investigations, in arriving at specifications and frameworks for carrying out investigations, the planning of criteria and specifications for usability studies, the planning of interviews and questionnaires, and other planning and frameworks for user trials and future trials.

56.7 Conclusion

DfA provides high potential for company economics as well as for social economics. The current slow take-up of the concept can be sped up by targeted management activities on different levels. The endorsement, stimulation, and incentives through policy and legislation play a key role. Many of the current activities in DfA are actually a result of legislation and regulation related to barrier-free accessibility, a social concept in relation to the inclusion of people with disabilities. The obligation of the public administration to become accessible for all has resulted in buying (accessible) for all. The request for accessible products through public procurement has stimulated a response of producers and suppliers, providing more accessible products and services. A second public driving force is awareness about the demographic change of many societies. It is connected with a change of the future markets and societal needs. The focus needs to be channeled from the average to variety and diversity.

TABLE 56.2 Overview of the Methods of User Participation and the Value Chain

USER PARTICIPATION METHODS	Problem definition	Functional specification	Prototype development	Preliminary testing	Final design + production	Final testing	Marketing and sales
Brainstorming *(Creativity processes)*	●	○	○		◗		
Creative Problem Solving (CPS) *(Creativity processes)*	●	○	○		◗		
Wizard of Oz *(Creativity processes)*	●	○	◗		◗		
Task Analysis	●	○		○	◗		
Direct Observations	○	○		●		○	
Activity Diary	●	●		◗		●	
Expert Opinion	◗	◗	◗	◗	◗	◗	◗
Questionnaires	●						●
Interviews	●	◗		◗	◗	○	○
Group Discussions	●	◗	◗		◗		○
Focus Groups	◗	●	○		●		
User Panels	◗	◗		●	●		
User Trials	◗	◗	◗	●	○	●	○
Field Trials				○		○	○
Simulation Work	○	○	●	●	○	●	
Usability Testing					●	●	

The Value Chain

Explanation:

○ No/small relevance	○ Possibly relevant	◗ Some relevance	● Very relevant

REMARK: The value of user participation in the planning process is irrespective of the method chosen!

A third important public responsibility is DfA education at schools, high schools, and universities. Activities like DfA networking, DfA information platforms, DfA research engineering centers, DfA award schemes, DfA research, public procurement guidelines, and so on, with public support provide a platform for further take-up.

At the commercial and industrial level, strategies for accessibility connected with DfA have led to significant improvements in some sectors and companies. Best practice starts with the take-up of accessibility and DfA in the company mission profile. The actual implementation follows through a detailed planning and introduction of DfA processes at the different levels of the company. Education in DfA comprises a core element of successful DfA management. Company education schemes need to include DfA in the curriculum. Public education schemes at schools, high schools, and universities can make a difference by integrating DfA in all curricula.

The requirement to deal with variety and diversity leads to user-oriented design and development concepts. Early involvement of representatives of different user groups is one strategy. Users with different (dis-)abilities representing diversity can help to foster a user-centered approach. As a benchmark for the management of user participation the FORTUNE concept provides a helpful reference.

References

BBGG (2002). *Gesetz zur Gleichstellung behinderter Menschen.* [Act on equal opportunities for disabled persons]. http://bundesrecht.juris.de/bgg/index.html.

BIENE (2006). *BIENE Award.* http://www.biene-award.de.

BIK 95+ (2006). *BIK–95+ Liste.* http://www.bik-online.info/test/95plus/index.php.

Bühler, C. (1999). *Ensuring Access for All, the Role of Telecommunications Systems for Elderly and Those with Special Needs.* EC contract 48422, European Commission. Brussels- Luxembourg: ESC-ECEAEC.

Bühler, C. (2001). Empowered participation of users with disabilities in universal design. *Universal Access in the Information Society* 1: 85–90.

Bühler, C. (2002). eEurope—Accessibility—User participation; Participation of people with disabilities and older people in the information society, in the *Proceedings of the 8th International Conference on Computers Helping People with Special Needs (ICCHP 2002)*, 15–20 July 2002, Linz, Austria, pp. 81–90. Berlin/Heidelberg: Springer-Verlag.

Bühler, C. and Knops, H. (eds.) (1999). *Assistive Technology at the Threshold of the New Millennium.* AT Research Series 6. Amsterdam, The Netherlands: IOS Press.

Bühler, C. and Stephanidis, C. (2004). European co-operation activities promoting design for all in information society technologies, in the *Proceedings of the 9th International Conference on Computers Helping People with Special Needs (ICCHP 2004)*, 7–9 July 2004, Paris, pp. 80–87. Berlin/Heidelberg: Springer-Verlag.

Center of Universal Design (1997). *Principles of Universal Design.* Version 2.0. NC State University. http://www.design.ncsu.edu/cud/about_ud/udprinciples.htm.

CEU (2001). *COM (2001) 529 Final, eEurope 2002: Accessibility of Public Web Sites and Their Content, Communication from the Commission to the Council, the European Parliament, the European Economic and Social Committee and the Committee of the Regions.* Brussels.

CEU (2002a). (European Commission) *IST Work Programme and Cross Programme Themes, Design for All.* http://cordis.europa.eu/ist/cpt/designforall.htm.

CEU (2002b). *COM (2002) 62 Final, eEurope Benchmarking Report, Communication from the Commission to the Council, the European Parliament, the European Economic and Social Committee and the Committee of the Regions.* Brussels.

CEU (2004). (European Commission) *Applications Relating to Persons with Special Needs Including the Disabled and Elderly.* http://cordis.europa.eu/ist/ka1/special_needs/projects/projects_domain.htm.

CEU (2005a). *COM (2005) 425 Final, eAccessibility, Communication from the Commission to the Council, the European Parliament, the European Economic and Social Committee and the Committee of the Regions.* Brussels.

CEU (2005b). *COM (2005) 229 Final, i2010: A European Information Society for Growth and Employment, Communication from the Commission to the Council, the European Parliament, the European Economic and Social Committee and the Committee of the Regions.* Brussels.

COST219bis (2001). *Bridging the Gap? Access to Telecommunications for All People* (P. R. W. Roe (ed.). Commission of European Communities.

Council of Europe (2003). *e-Accessibility: Improving the Access of People with Disabilities to the Knowledge Based Society,* Resolution 5165/03, OJ 14 January 2003.

CSR (2007). *CSR-Europe.* http://www.csreurope.org.

Design-for-all-AT Awards (2004). http://www.ftb-volmarstein.de/aktuell/ataward.html.

EDeAN (2004). *About.* http://www.edean.org.

EDeAN (2005). *White Paper: Promoting Design for All and e-Accessibility in Europe.* http://www.edean.org.

EDeAN (2006). *EDeAN Standardisation Documents Area.* http://www.edean.org.

eEurope Standards (2007). *eAccessibility Standardisation Activities.* http://eeurope-standards.org/activities_e-accessibility.htm.

Ellis, G. (2003). From Information Age to Inclusive Age: The economic concept of inclusion, in *AT Research Series 11,* pp. 19–24. Amsterdam, The Netherlands: IOS Press.

European Commission (2002). *eEurope 2005 Action Plan.* http://europa.eu.int/information_society/eeurope/2002/news_library/documents/eeurope2005/eeurope2005_en.pdf.

Fortune (2000). *Fortune Guide, Empowered Participation of Users with Disabilities in Projects, a Summary with the Main Results of the FORTUNE Project* (DE9231). Wetter: FTB-Verlag.

HEART (1994). *European Service Delivery Systems in Rehabilitation Technology.* iRv, Hoensbroek, The Netherlands: Horizontal European Activities in Assistive and Rehabilitation Technologies.

Klironomos, I., Kempainen, E., Aalykke, S., Rodriguez, C., and Whitney, G. (2005). *EDeAN Policy and Legislation Documents Area.* http://www.edean.org.

Kyoyohin Foundation (2001). *Kyoyo-Hin White Paper 2001.* Tokyo: The Kyoyo-Hin Foundation. http://www.kyoyohin.org/eng.

Mace, R. (1998). Universal design in housing. *Assistive Technology* 10: 21–28.

Mueller, J. (1998). Assistive technology and universal design in the workplace. *Assistive Technology* 10: 37–43.

Norman, D. A. and Draper, S. W. (1986). *User-Centred System Design.* Hillsdale, NY: Lawrence Erlbaum Associates.

Pacific Bell (1994). *Universal Design Policy: The Pacific Bell's Advisory Group for People with Disabilities (AGPD).* http://trace.wisc.edu/docs/pacbell_ud/agpd.htm.

Poulsen, D., Ashby, M., and Richardson, S. (1996). *Userfit: A Practical Handbook on User-Centred Design for Assistive Technology.* Publication of the European Commission, TIDE/User project.

Preiser, W. F. E. and Ostroff, E. (eds.) (2001). *Universal Design Handbook.* New York: McGraw-Hill.

REHADAT (2007). *Information System for Vocational Rehabilitation.* http://rehadat.de.

Schuler, D. and Namioka, A. (1993). *Participatory Design: Principles and Practices.* Hillsdale, NY: Lawrence Erlbaum Associates.

Stephanidis, C., Salvendy, G., Akoumianakis, D., Bevan, N., Brewer, J., Emiliani, P. L., et al. (1998). Toward an information society for all: An international R&D agenda. *International Journal of Human-Computer Interaction* 10: 107–134. http://www.ics.forth.gr/hci/files/white_paper_1998.pdf.

SUN (2007). http://www.sun.com/access/developers/updt.HCI.advance.htm.

Top of the Web (2001). http://www.bedstpaanettet.dk.

UDA (2003). Universal Design Alliance. http://www.universaldesign.org.

Universal Design Network (1999). http://www.universaldesign.net.

U.S. Access Board (1998). *The Rehabilitation Act Amendments (Section 508).* http://www.access-board.gov/sec508/guide/act.htm.

Vanderheiden, G. C. (1996). *Universal Design.* Trace R&D Center, University of Wisconsin–Madison. http://trace.wisc.edu/world/gen_ud.html.

W3C (1999). *World Wide Web Consortium: Web Accessibility Initiative (W3C-WAI), Web Content Accessibility Guidelines 1.0.* http://www.w3c.org/TR/WCAG10.

WHO (2001). *The International Classification of Functioning Disability and Health– ICF.* http://www3.who.int/icf/icftemplate.cfm.

57

Security and Privacy for Universal Access

Mark T. Maybury

57.1 Introduction

This chapter provides an overview of security and privacy requirements for universal access by addressing the following topics:

- *Security*: The need for confidentiality, integrity, and availability of information.
- *Privacy*: The need for individuals to control the information that pertains to them as an identifiable individual (e.g., identity, personal information, activities).[1]
- *Legal environment*: In the United States, Canada, and Europe, related to security and privacy and how this influences universal access.
- *Technology for privacy*: Technologies like encryption and anonymization that can help ensure confidentiality.
- *Personalization*: The need to tailor interaction to individual needs and situations driving a need for individualized user and discourse models.
- *Usability, security, and privacy*: How to achieve an effective balance among these seemingly inconsistent objectives.

57.2 Security

Security computing requirements include confidentiality, integrity, and availability (FIPS 199, 2004). *Confidentiality* is the control of information so that it is only shared among users with

appropriate access. *Integrity* is ensuring the information is accurate and has not been altered or manipulated in some way. Finally, *availability* means making sure that users have access where, when, and how needed.

Since the creation of information systems, engineers have been concerned with the establishment of secure computer systems (Bell and LaPadula, 1976). Hoffman (1969) surveyed social and economic benefits and risks of digital information, summarizing countermeasures to various accidental and deliberate information privacy threats (e.g., masquerading, piggyback entry, trap doors). In the interim, many new technologies have been created to facilitate data and processing security such as encryption, digital certificates (for identity and encryption), multilevel operating systems, secure databases, and anonymization. Modern network-centric computing has resulted in the need for secure networking, distributed computing and middleware (Anderson, 2008). Identification and remediation of infrastructure vulnerabilities (e.g., cve.mitre.org) and weaknesses (e.g., cwe.mitre.org) remain a continual and community-based process involving security analysts, software and hardware developers, and end-users. Malware (e.g., cme.mitre.org) is another major threat to privacy in which software often surreptitiously captures and transmits personal information without the knowledge or consent of the user.

57.3 Privacy

There have been many definitions of privacy, such as the 1890 paper by Warren and Brandeis that defined privacy as "the right

---

[1] Privacy: http://en.wikipedia.org/wiki/Privacy.

to be let alone" (p. 193). According to the *Merriam-Webster's Dictionary, privacy* is "the freedom from unauthorized intrusion." Privacy can relate to nondisclosure of one's identity, individual properties (e.g., age, telephone, property, health) or activities (e.g., investments, political voting). Inadequate privacy can increase the risk of inappropriate disclosure (e.g., of personal medical information and/or disabilities that might be in a detailed user model), identity theft, or, even more serious risks such as child abduction. The 1948 United Nations Universal Declaration of Human Rights Article 12 states:

> No one shall be subjected to arbitrary interference with his privacy, family, home or correspondence, nor to attacks upon his honour and reputation.

In 1966, John McCarthy proposed a "computer bill of rights," including items such as:

- No organization, governmental or private, is allowed to maintain files that cover large numbers of people outside of the general system.
- The rules governing access to the files are definite and well publicized, and the programs that would enforce these rules are open to any interested party, including, for example, the American Civil Liberties Union.
- An individual has the right to read his own file, to challenge certain kinds of entries in his file and to impose certain restrictions on access to his file.
- Every time someone consults an individual's file this event is recorded, together with the authorization for the access.
- If an organization or an individual obtains access to certain information in a file by deceit, this is a crime and a civil wrong. The injured individual may sue for invasion of privacy and be awarded damages.

While not necessarily of this origin, some but not all of these ideas appear in subsequent privacy principles.

Some national constitutions protect privacy explicitly (e.g., France's Declaration of the Rights of Man and of the Citizen), whereas other constitutions have been interpreted as protecting privacy. For example, via cases such as *Griswold v. Connecticut* (1965), the Supreme Court of the United States has determined that the U.S. Constitution contains "penumbras" that implicitly grant a right to privacy against government intrusion. Other countries without constitutional privacy protections have laws protecting privacy, such as the United Kingdom's Data Protection Act 1998 or Australia's Privacy Act 1988.

The U.S. Privacy Act of 1974 implemented the original set of fair information practices. The Privacy Act states:

> No agency shall disclose any record which is contained in a system of records by any means of communication to any person, or to another agency, except pursuant to a written request by, or with the prior written consent of, the individual to whom the record pertains.

In 1988 the Privacy Act was amended by the Computer Matching and Privacy Act, which requires:

- Procedural uniformity in carrying out matching programs
- Due process for subjects in order to protect their rights
- Establishment of Data Integrity Boards at each agency performing matching to ensure oversight of matching programs

Also in the United States, the Common Rule for the Protection of Human Subjects[2] mandates review and approval of research protocols involving human subjects and it explicitly addresses the need for informed consent, confidentiality, and data protection. Also in the United States, under the implementation guidance to Section 208 of the E-Government Act of 2002 (Bolten, 2003), agencies are required to:

- Conduct privacy impact assessments (and make these publicly available)
- Post privacy policies on public web sites
- Translate privacy policies into standardized machine-readable format
- Report annually to the U.S. Office of Management and Budget (OMB) on compliance

Finally, in the United States, if identifiable information was collected as part of medical treatment or monitoring, the data might fall under the HIPAA security and privacy rules. The Administrative Simplification provisions of the Health Insurance Portability and Accountability Act of 1996 (HIPAA, Title II) required the Department of Health and Human Services (HHS) to establish national standards for the security of electronic health care information. These include administrative, technical, and physical security procedures for covered entities (health care providers, health care clearinghouses, and health plans) to use to assure the confidentiality of electronic protected health information.

57.4 OECD Privacy Guidelines

While there are no binding international agreements, the Organization for Economic Co-operation and Development (OECD) Guidelines on the Protection of Privacy and Transborder Flows of Personal Data (OECD, 1980) provide a foundation for privacy protection. They aim to avoid "unlawful storage of personal data, the storage of inaccurate personal data, or the abuse or unauthorised disclosure of such" but also to ensure effective data flow to enable the "development of economic and social relations among Member countries." Thus there is a tension between protecting personal data and its capture and sharing to enhance citizen, national, or international well-being.

---

[2] The Common Rule for the Protection of Human Subjects. Subpart A of National Science Foundation 45 CFR Part 690: Federal Policy for the Protection of Human Subjects (same as 45 CFR Part 46, which pertains to HHS).

The OECD guidelines introduce the following key principles (directly quoted):

Collection Limitation Principle

There should be limits to the collection of personal data and any such data should be obtained by lawful and fair means and, where appropriate, with the knowledge or consent of the data subject.

Data Quality Principle

Personal data should be relevant to the purposes for which they are to be used, and, to the extent necessary for those purposes, should be accurate, complete and kept up-to-date.

Purpose Specification Principle

The purposes for which personal data are collected should be specified not later than at the time of data collection and the subsequent use limited to the fulfillment of those purposes or such others as are not incompatible with those purposes and as are specified on each occasion of change of purpose.

Use Limitation Principle

Personal data should not be disclosed, made available or otherwise used for purposes other than those specified in accordance with Paragraph 9 except:

a. with the consent of the data subject; or
b. by the authority of law.

Security Safeguards Principle

Personal data should be protected by reasonable security safeguards against such risks as loss or unauthorised access, destruction, use, modification or disclosure of data.

Openness Principle

There should be a general policy of openness about developments, practices and policies with respect to personal data. Means should be readily available of establishing the existence and nature of personal data, and the main purposes of their use, as well as the identity and usual residence of the data controller.

Individual Participation Principle

An individual should have the right:

a. to obtain from a data controller, or otherwise, confirmation of whether or not the data controller has data relating to him;
b. to have communicated to him, data relating to him

- within a reasonable time;
- at a charge, if any, that is not excessive;
- in a reasonable manner; and
- in a form that is readily intelligible to him;

c. to be given reasons if a request made under subparagraphs (a) and (b) is denied, and to be able to challenge such denial; and
d. to challenge data relating to him and, if the challenge is successful to have the data erased, rectified, completed or amended.

Accountability Principle

A data controller should be accountable for complying with measures which give effect to the principles stated above

Table 57.1 lists these OECD principles and relates them to privacy principles from Canada, Europe, and Asia, which are described in subsequent sections.

TABLE 57.1 Privacy Requirements across the World

OECD Principles	European Union Data Protection Directive	Canadian PIPEDA	APEC Privacy Framework
Collection Limitation	"Data must be adequate, relevant and not excessive" (28)	Limiting Collection (4) "limited to that which is necessary"	Collection Limitation (III) "relevant," lawfully/fairly obtained, "notice or consent"
Data Quality	Chapter II, Section I—Data Quality Principles	Accuracy (6) "accurate, complete, and up-to-date"	Integrity (VI) "accurate, complete … up-to-date"
Purpose Specification	(28) "purposes must be explicit and legitimate and must be determined at the time of collection of the data; whereas the purposes of processing further to collection shall not be incompatible with the purposes as they were originally specified"	Identifying Purposes (2) "identified by the organization at or before the time the information is collected"	Uses (IV) "personal information collected should only be used to fulfil the purposes of the collection"
Use Limitation	"for historical, statistical or scientific purposes" (29); restrict access to protect rights (e.g., "access to medical data may be obtained only through a health professional") (42)	Limiting Use, Disclosure, and Retention (5) "not be used or disclosed for purposes other than those for which it was collected, except with … consent or as required by law" "retained only as long as necessary"	Uses (IV) Should only be used to fulfill the purposes of the collection "except … with consent … when necessary to provide a service or product requested by the individual …by law"

(Continued)

TABLE 57.1　Privacy Requirements across the World (*Continued*)

OECD Principles	European Union Data Protection Directive	Canadian PIPEDA	APEC Privacy Framework
Safeguards	(29) "Member States furnish suitable safeguards; whereas these safeguards must in particular rule out the use of the data in support of measures or decisions regarding any particular individual"	Safeguards (7) "protected by security safeguards appropriate to the sensitivity of the information"	Security Safeguards (VII) "appropriate safeguards against risks such as loss or unauthorized access … destruction, use, modification, or disclosure
Openness	The data subject can learn of processing and "must be given accurate and full information" (38) and be informed when data are recorded or disclosed (39) and "in order to verify in particular the accuracy of the data and the lawfulness of the processing; whereas, for the same reasons, every data subject must also have the right to know the logic involved in the automatic processing of data concerning him, at least in the case of the automated decision" (41)	Openness (8) "readily available" … "policies and practices"	
Individual Participation	The data subject "must be able to exercise the right of access to data relating to him which are being processed in order to verify in particular the accuracy of the data and the lawfulness of the processing" (41)	Individual Access (9) "individual shall be informed of the existence, use, and disclosure." Can amend inaccurate or incomplete information. Must provide alternative formats for sensory disabled.	Access and correction (VIII) Reasonable access to "personal information about them" and have inaccurate information "rectified, completed, amended, or deleted"
Accountability		Accountability (1) Challenging Compliance (10) "individuals … accountable for … compliance"	Accountability (IX) "personal information controller held accountable"
	Consent (30) or necessary	Consent (3) "knowledge and consent of the individual are required for the collection, use, or disclosure"	Choice (V) "clear, prominent, easily understandable, accessible and affordable mechanisms to exercise choice in relation to collection, use and disclosure"
	The data subject must "be informed when data are recorded or disclosed" (38)	Notice	Notice (II) "personal information controller should provide clear and easily accessible statements about practices and policies with respect to personal information" "before or at the time of collection [or] as soon after as is practicable"
	Protect subject's life (31) or important public interest (e.g., public health and social protection (34))	Preventing Harm (I)	Preventing Harm (I) design "to prevent the misuse" of personal information. "Remedial measures should be proportionate to the likelihood and severity of the harm threatened by the collection use and transfer of personal information."

57.5 European Union Data Protection Directive

Unlike the United States, both the European Union (EU) and Canada have broad privacy protections based on fair information practices covering the private sector. The European Union Data Protection Directive (EC, 1995) went into effect in 1998 and consists of the following key elements (directly quoted):

27. protection of individuals must apply as much to automatic processing of data as to manual processing

28. the data must be adequate, relevant and not excessive in relation to the purposes for which they are

processed; whereas such purposes must be explicit and legitimate and must be determined at the time of collection of the data; whereas the purposes of processing further to collection shall not be incompatible with the purposes as they were originally specified

29. further processing of personal data for historical, statistical or scientific purposes is not generally to be considered incompatible with the purposes for which the data have previously been collected provided that Member States furnish suitable safeguards; whereas these safeguards must in particular rule out the use of the data in support of measures or decisions regarding any particular individual

30. the processing of personal data must in addition be carried out with the consent of the data subject or be necessary for the conclusion or performance of a contract binding on the data subject, or as a legal requirement, or for the performance of a task carried out in the public interest or in the exercise of official authority, or in the legitimate interests of a natural or legal person, provided that the interests or the rights and freedoms of the data subject are not overriding; whereas, in particular, in order to maintain a balance between the interests involved while guaranteeing effective competition, Member States may determine the circumstances in which personal data may be used or disclosed to a third party in the context of the legitimate ordinary business activities of companies and other bodies; whereas Member States may similarly specify the conditions under which personal data may be disclosed to a third party for the purposes of marketing whether carried out commercially or by a charitable organization or by any other association or foundation, of a political nature for example, subject to the provisions allowing a data subject to object to the processing of data regarding him, at no cost and without having to state his reasons

31. Whereas the processing of personal data must equally be regarded as lawful where it is carried out in order to protect an interest which is essential for the data subject's life

34. Whereas Member States must also be authorized, when justified by grounds of important public interest, to derogate from the prohibition on processing sensitive categories of data where important reasons of public interest so justify in areas such as public health and social protection

38. Whereas, if the processing of data is to be fair, the data subject must be in a position to learn of the existence of a processing operation and, where data are collected from him, must be given accurate and full information, bearing in mind the circumstances of the collection;

39. the data subject should be informed when the data are recorded or at the latest when the data are first disclosed to a third party;

41. Whereas any person must be able to exercise the right of access to data relating to him which are being processed, in order to verify in particular the accuracy of the data and the lawfulness of the processing; whereas, for the same reasons, every data subject must also have the right to know the logic involved in the automatic processing of data concerning him, at least in the case of the automated decisions referred to in Article 15 (1); whereas this right must not adversely affect trade secrets or intellectual property and in particular the copyright protecting the software; whereas these considerations must not, however, result in the data subject being refused all information;

42. Whereas Member States may, in the interest of the data subject or so as to protect the rights and freedoms of others, restrict rights of access and information; whereas they may, for example, specify that access to medical data may be obtained only through a health professional;

58. Whereas provisions should be made for exemptions from this prohibition in certain circumstances where the data subject has given his consent

The EU Data Protection Directive further defines personal data (EC, 2002) and consent as follows:

"personal data" is information relating to an identified or identifiable natural person ('data subject'); … who can be identified, directly or indirectly, … wrt to one or more factors specific to his physical, physiological, mental, economic, cultural or social identity; "the data subject's consent" shall mean any freely given specific and informed indication of his wishes by which the data subject signifies his agreement to personal data relating to him being processed.

57.6 Canadian Privacy Protection

Canadians have the "right to the protection of personal information collected, used or disclosed in the course of commercial activities, in connection with the operation of a federal work, undertaking or business or interprovincially or internationally." This is institutionalized in the Personal Information Protection and Electronic Documents Act (PIPEDA). The PIPEDA (2000) establishes principles to govern the collection, use and disclosure of personal information. The principles encompass

accountability, identifying the purposes for the collection of personal information, obtaining consent, limiting collection; limiting use, disclosure and retention; ensuring accuracy, providing adequate security, making information management policies readily available, providing individuals with access to information about themselves, and giving individuals a right to challenge an organization's compliance with these principles.

Based on the Canadian Standards Association's *Model Code for the Protection of Personal Information* (http://www.csa.ca/standards/privacy), recognized as a national standard in 1996, Canada defines 10 privacy principles[3]:

Principle 1—Accountability

An organization is responsible for personal information under its control and shall designate an individual or individuals who are accountable for the organization's compliance with the following principles.

Principle 2—Identifying Purposes

The purposes for which personal information is collected shall be identified by the organization at or before the time the information is collected.

Principle 3—Consent

The knowledge and consent of the individual are required for the collection, use, or disclosure of personal information, except where inappropriate.

Principle 4—Limiting Collection

The collection of personal information shall be limited to that which is necessary for the purposes identified by the organization. Information shall be collected by fair and lawful means.

Principle 5—Limiting Use, Disclosure, and Retention

Personal information shall not be used or disclosed for purposes other than those for which it was collected, except with the consent of the individual or as required by law. Personal information shall be retained only as long as necessary for the fulfillment of those purposes.

Principle 6—Accuracy

Personal information shall be as accurate, complete, and up-to-date as is necessary for the purposes for which it is to be used.

Principle 7—Safeguards

Personal information shall be protected by security safeguards appropriate to the sensitivity of the information.

Principle 8—Openness

An organization shall make readily available to individuals specific information about its policies and practices relating to the management of personal information.

Principle 9—Individual Access

Upon request, an individual shall be informed of the existence, use, and disclosure of his or her personal information and shall be given access to that information. An individual shall be able to challenge the accuracy and completeness of the information and have it amended as appropriate.

Principle 10—Challenging Compliance

An individual shall be able to address a challenge concerning compliance with the above principles to the designated individual or individuals accountable for the organization's compliance.

PIPEDA defines personal information as "information about an identifiable individual, but does not include the name, title or business address or telephone number of an employee of an organization." The Act establishes a privacy commissioner who conducts investigations to resolve any contraventions of the principles; otherwise disputes can be resolved in federal court. Notably, Canada also requires that an organization provide access to personal information "in an alternative format to an individual with a sensory disability who has a right of access to personal information" if a version of the information exists in that format or if "its conversion into that format is reasonable and necessary in order for the individual to be able to exercise rights."

Table 57.2 lists the various definitions of personal information across various countries. While there is much similarity, it can be noted, for example, that some definitions, such as the Canadian one, specifically exclude some information from being private such as employee names, titles, and business address/telephone.

57.7 Asian-Pacific Privacy Framework

The Asia-Pacific Economic Cooperation (APEC) Privacy Framework[4] recognizes "the enormous potential of electronic commerce to expand business opportunities, reduce costs, increase efficiency, improve the quality of life, and facilitate the greater participation of small business in global commerce."

APEC defines *personal information* as "any information about an identified or identifiable individual" and a *personal information controller* as "a person or organization who controls the collection, holding, processing or use of personal information." They further define *publicly available information* as:

> personal information about an individual that the individual knowingly makes or permits to be made available to the public, or is legally obtained and accessed from: a) government records that are available to the public; b) journalistic reports; or c) information required by law to be made available to the public.

[3] Canadian Privacy Principles: http://canada.justice.gc.ca/en/news/nr/1998/attback2.html.

[4] Asia-Pacific Economic Cooperation (APEC) Privacy Framework (2005): http://www.apec.org/etc/medialib/apec_media_library/downloads/taskforce/ecsg/pubs/2005.Par.0001.File.v1.1.

TABLE 57.2 Personal Information Definitions

Area	Act	Definition of Personal Information
Global	OECD	Information relating to an identified or identifiable individual (a natural person).
Europe	EU Directive	Information relating to an identified or identifiable natural person ("data subject"); … who can be identified, directly or indirectly, … with respect to one or more factors specific to his physical, physiological, mental, economic, cultural or social identity.
Canada	PIPEDA	Information about an identifiable individual, but does not include the name, title, or business address or telephone number of an employee of an organization.
Asia-Pacific	APEC	Any information about an identified or identifiable individual.

APEC is founded on a set of privacy principles analogous to those in OECD. These principles include (quoted directly):

I. Preventing Harm:

14. Recognizing the interests of the individual to legitimate expectations of privacy, personal information protection should be designed to prevent the misuse of such information. Further, acknowledging the risk that harm may result from such misuse of personal information, specific obligations should take account of such risk, and remedial measures should be proportionate to the likelihood and severity of the harm threatened by the collection, use and transfer of personal information.

II. Notice

15. Personal information controllers should provide clear and easily accessible statements about their practices and policies with respect to personal information that should include:

a. the fact that personal information is being collected;

b. the purposes for which personal information is collected;

c. the types of persons or organizations to whom personal information might be disclosed;

d. the identity and location of the personal information controller, including information on how to contact them about their practices and handling of personal information;

e. the choices and means the personal information controller offers individuals for limiting the use and disclosure of, and for accessing and correcting, their personal information.

16. All reasonably practicable steps shall be taken to ensure that such notice is provided either before or at the time of collection of personal information. Otherwise, such notice should be provided as soon after as is practicable.

17. It may not be appropriate for personal information controllers to provide notice regarding the collection and use of publicly available information.

III. Collection Limitation

18. The collection of personal information should be limited to information that is relevant to the purposes of collection and any such information should be obtained by lawful and fair means, and where appropriate, with notice to, or consent of, the individual concerned.

IV. Uses of Personal Information

19. Personal information collected should be used only to fulfill the purposes of collection and other compatible or related purposes except:

a. with the consent of the individual whose personal information is collected;

b. when necessary to provide a service or product requested by the individual; or,

c. by the authority of law and other legal instruments, proclamations and pronouncements of legal effect.

V. Choice

20. Where appropriate, individuals should be provided with clear, prominent, easily understandable, accessible and affordable mechanisms to exercise choice in relation to the collection, use and disclosure of their personal information. It may not be appropriate for personal information controllers to provide these mechanisms when collecting publicly available information.

VI. Integrity of Personal Information

21. Personal information should be accurate, complete and kept up-to-date to the extent necessary for the purposes of use.

VII. Security Safeguards

22. Personal information controllers should protect personal information that they hold with appropriate safeguards against risks, such as loss or unauthorized access to personal information, or unauthorized destruction, use, modification or disclosure of information or other misuses. Such safeguards should be proportional to the likelihood and severity of the harm threatened, the sensitivity of the information and the context in which it is held, and should be subject to periodic review and reassessment.

VIII. Access and Correction

23. Individuals should be able to:
    a. obtain from the personal information controller confirmation of whether or not the personal information controller holds personal information about them;
    b. have communicated to them, after having provided sufficient proof of their identity, personal information about them;
        i. within a reasonable time;
        ii. at a charge, if any, that is not excessive;
        iii. in a reasonable manner;
        iv. in a form that is generally understandable; and,
    c. challenge the accuracy of information relating to them and, if possible and as appropriate, have the information rectified, completed, amended or deleted.

24. Such access and opportunity for correction should be provided except where:
    i. the burden or expense of doing so would be unreasonable or disproportionate to the risks to the individual's privacy in the case in question;
    ii. the information should not be disclosed due to legal or security reasons or to protect confidential commercial information; or
    iii. the information privacy of persons other than the individual would be violated.

25. If a request under (a) or (b) or a challenge under (c) is denied, the individual should be provided with reasons why and be able to challenge such denial.

IX. Accountability

26. A personal information controller should be accountable for complying with measures that give effect to the Principles stated above. When personal information is to be transferred to another person or organization, whether domestically or internationally, the personal information controller should obtain the consent of the individual or exercise due diligence and take reasonable steps to ensure that the recipient person or organization will protect the information consistently with these Principles.

The APEC further provides guidelines for implementation by nation states, addressing such topics as how to:

- Maximize the benefits of privacy protections and information flows
- Give effect to the APEC Privacy Framework via a variety of means (e.g., legislative, administrative, industry, self-regulatory) in a nondiscriminatory manner
- Educate and publicize domestic privacy protections
- Ensure cooperation between the public and private sectors

- Provide for appropriate remedies in situations where privacy protections are violated
- Ensure mechanism for reporting domestic implementation of the APEC Privacy Framework

57.8 The Need for Personalization

Ensuring information access for all implies the need to capture, process, and apply individual information. Users value the ability to have personalized news delivery as well as skill and domain customized application interfaces (Light and Maybury, 2002; Maybury, 2005). This often requires capturing, representing, updating, and reasoning about explicit models of the user to include but not be limited to the user's:

- *Properties*: Age, gender, race, nationality
- *Physical and perceptual skills*: Function or dysfunction of movement, sight, smell, hearing, touch, taste
- *Preferences*: Visual and acoustic elements (e.g., volume, font size), interactive features (e.g., speed, degree of machine vs. user control and/or mixed initiative)
- *Beliefs*: User assumptions and beliefs
- *Interests*: The topics and content that are the focus of the user
- *Expertise*: Education level, subject matter expertise, linguistic skills (e.g., reading, writing, listening, speaking, foreign language)
- *Role*: System administrator, expert user, guide
- *Goals/tasks*: The current activity of the user to achieve particular objectives
- *Workload*: The physical and cognitive complexity of the user's tasks (e.g., required kinesthetic skill, multitasking, attention distracters)
- *Social network*: The roles and relationships of the user with others
- *Environment*: Location of user, environment such as ambient noise, lighting, number of other humans nearby

Even something as personal as religious beliefs or gender preference could impact the kind of art a user might be interested in when browsing a virtual museum, the kinds of products to display during an electronic commerce session, or the customization of packages to support online vacation planning. Not only might personal information about users enhance interaction, it might be absolutely necessary to ensure effective interaction. For example, reaction times differ across ages or with differing natural perceptual and motor skills, so the timeout of a dialogue box in an interface might need to be customized to the user or situation. Moreover, legislation increasingly requires accessibility for audio, visual, or motor challenged users, and treating these users as a generic class as opposed to individuals may or may not be appropriate in all cases. Unfortunately, the ubiquitous collection and processing of personal data for authentication or personalization reduces the practicality of privacy (Sparck Jones, 2003).

Table 57.3 outlines a range of universal access challenges associated with various privacy requirements. For example,

TABLE 57.3 Privacy Requirements and Implications for Universal Access

Privacy Requirement	Universal Access Challenge (User Model, Adaptation)
Collection Limitation	Limited user data limits value of tailored interaction. Making user aware of collection can be intrusive.
Data Quality	Accurate, complete, and up-to-date models are preferred but often user data are incomplete or uncertain.
Purpose Specification	Need to anticipate all potential uses.
Use Limitation	Desire to use models across sessions and application to tailor interaction without bothering user.
Safeguards	Need to encrypt models at rest and in transmission (e.g., from one platform to another) and provide for confidentiality, integrity, and availability.
Individual Protection	Need to make user models and adaptation strategies so user can inspect, understand, delete, or correct.
Openness	Some algorithms not easily explained (e.g., how do you describe hidden nodes in a neural network?).
Accountability	Need to audit access, update, and use of user model.
Anonymization	Cross-session user modeling.

collection limitations restrict the ability to capture personalized information that could help personalize interaction. Also, quality data about users are needed in order to ensure effective access or interaction, and yet data are often incomplete or uncertain. While ideally a system should identify the purpose to which collected personal information will be applied, anticipating all uses is often difficult. Moreover, better models of user attributes and interests are developed across multiple interactive sessions, and yet if private information is limited to a single session the value of more personalized interaction may not be possible. The use of encryption or anonymization to protect confidentiality and integrity may be warranted. More challenging is that algorithms may need to be explained to a user, which could be difficult when complex sets of rules and/or statistical models are applied to make decisions. Also, logging of user model creation, modification, and access may be required to ensure accountability.

Anonymization may be necessary to protect sources and identities, and yet personal information may be necessary not only for convenience but for life critical situations, such as in medicine. For example, to address the need to balance privacy and necessary access to personal information, the American Medical Informatics Association (AMIA) De-identification Challenge aims to take patient medical records and detect and redact Protected Health Information (PHI). This de-identified information can then be used freely in mining for epidemiological analyses or automated classification of diagnoses to avoid false billing or misdiagnoses. Wellner et al. (2007) have demonstrated 97% de-identification performance on a test corpus of 910 medical discharge summaries using rapid retargeting of existing language processing toolkits.

Automated methods for privacy enhancement have received a lot of attention in recent years. Active areas of research include privacy preserving, data mining (Oliveira and Zaïane, 2004), collaborative filtering with privacy, and privacy-enhanced data management.

Kobsa (2002) highlights that personalized hypermedia conflict with user privacy concerns and national privacy laws, which typically require "parsimony, purpose-specificity, and user awareness or even user consent in the collection and processing of personal data" (p. 62). He notes this is at odds with practical considerations such as the need to log and identify individual users, keep and analyze logs across sessions and users, and to make automated decisions (e.g., scoring). Compounding this, he details that many countries restrict the flow of personal information across national boarders, including the European Union, Argentina, Australia, Hong Kong, Hungary, Lithuania, New Zealand, and Taiwan.

Kobsa notes that if users can interact with anonymity (e.g., Kobsa and Schreck, 2003), privacy laws generally do not apply and users' privacy concerns disappear. Moreover, users that remain anonymous are more likely to provide data about themselves, which will improve personalization. If anonymity is not possible, Kobsa recommends a set of guidelines that address key issues in most national privacy laws and consider privacy preferences of many users (quoted):

- Make (long-term) personalization a clear purpose of your service. This may improve the accuracy of the data needed for personalization purposes, and will also facilitate the application of the next guideline.
- Provide comprehensive and intelligible advance notice to users about all data that is to be collected, processed and transferred, and indicate the purposes for which this is being done. This is likely to increase users' trust in the application and is mandated by virtually all privacy laws.
- Obtain users' informed and voluntary consent to processing their data for personalization purposes.
- Provide organizational and technical means to allow users to inspect, block, rectify, and erase both the data they provided, and specifically the assumptions the system inferred about them.
- Provide security mechanisms that are commensurate with the technical state of the art and the sensitivity of the stored user data.

Wang and Kobsa (2007) argue for the need to tailor privacy to the constraints of each individual user. They report a dynamic privacy-enhancing user modeling system that selects personalization methods during run-time that respect users' current privacy concerns as well as the privacy laws and regulations.

57.9 Universal Access Implications

Ensuring privacy and security for physically, emotionally, and cognitively challenged individuals is difficult. For example, accessibility options must be available but their use must be kept private. Those with perceptual or cognitive (e.g., memory, reasoning) challenges must be able to understand and manipulate information regarding the collection, use, and safeguards of information and yet they may be ill equipped to do so. Should the aim be to offer appropriate mechanisms to users to help decide how much information they wish to make an informed choice about disclosing? An interesting extension of this issue would then be how support is offered for someone with a cognitive impairment to make an "informed" choice. How do you ensure effective management of privacy with children or elderly or foreign users who may have limited understanding of law and society, policies, or technologies or simply a limited ability to communicate?

In addition to inadequate understanding of privacy implications, universal access places new burdens on the kind of security mechanisms that might be necessary. This is particularly challenging for confidentiality and availability. For example, how can certain security features such as authentication or identity management be provided when an individual may have short-term or long-term memory loss? What if the user can only use one medium, such as voice? Consider ensuring information security of a physically and visually impaired user interacting with an ATM or in a public kiosk. How can it be ensured that information and services are available where, when, and how users need them?

57.10 Conclusion: Secure Universal Access

Security and privacy are civil rights defended by governments. At the same time, the ability to customize interaction to users may not only be desirable but even necessary for disadvantaged users. How can it be ensured that a system provides personalized access to enhance its utility but also security and privacy? Typically, authentication to control access implies additional time and effort on the user. For example, a user would need to enter a password into a smartphone (such as a Blackberry or TREO) while walking instead of just hitting speed dial. In summary, it is necessary to balance disclosure and privacy, for example, promoting electronic commerce by protecting personal information that is collected, used, or disclosed in specific circumstances.

Acknowledgments

Special appreciation to Professor Alfred Kobsa for his insights into privacy and user modeling and to Stuart Shapiro for sharing his extensive experience with respect to privacy and security. This chapter also benefited from anonymous reviewer feedback.

References

Anderson, R. (2008). *Security Engineering: A Guide to Building Dependable Distributed Systems* (2nd ed.). New York: John Wiley & Sons.

Bell, D. and LaPadula, L. (1976). *Secure Computer Systems: Unified Exposition and Multics Interpretation.* MITRE Technical Report 2997. Bedford, MA: MITRE Corporation.

Bolten, J. (2003). *OMB Guidance for Implementing Privacy Provision of (Section 208 of) the E-Government Act of 2002.* http://www.whitehouse.gov/omb/memoranda/m03-22.html.

EC (1995). Directive 95/46/EC of the European Parliament and of the Council of 24 October 1995 on the Protection of Individuals with Regard to the Processing of Personal Data and on the Free Movement of Such Data. *Official Journal of the European Communities,* No L. 281. http://www.cdt.org/privacy/eudirective/EU_Directive_.html#HD_NM_6; http://eur-lex.europa.eu/LexUriServ/LexUriServ.do?uri=CELEX:31995L0046:EN:NOT.

EC (2002). Directive 2002/58/EC of the European Parliament and of the Council of 24 October 1995 Concerning the Processing of Personal Data and the Protection of Privacy in the Electronic Communications Sector. *Official Journal of the European Communities,* No L 201/37. http://www.spam-laws.com/docs/2002-58-ec.pdf.

FIPS 199 (2004). *Federal Information Processing Standard (FIPS) 199, Standards for Security Categorization of Federal Information and Information Systems.* Computer Security Division, National Institute of Standards and Technology. http://csrc.nist.gov/publications/fips/fips199/FIPS-PUB-199-final.pdf.

Hoffman, L. (1969). Computers and privacy: A survey. *ACM Computing Surveys* 1: 85–103.

Kobsa, A. (2002). Personalized hypermedia and international privacy. *Communications of the ACM* 45: 64–67.

Kobsa, A. and Schreck, J. (2003). Privacy through pseudonymity in user-adaptive systems. *ACM Transactions on Internet Technology (TOIT)* 3: 149–183. http://portal.acm.org/citation.cfm?id=767196&dl=acm&coll=&CFID=15151515&CFTOKEN=6184618.

Light, M. and Maybury, M. (2002). Personalized multimedia information access: Ask questions, get personalized answers. *Communications of the ACM* 45: 54–59. http://portal.acm.org/citation.cfm?id=506218.506246.

Maybury, M. (2005). Intelligent information access. Tutorial presented in the context of the 2005 *International Conference on Intelligence Analysis (IA '05),* 3–5 May 2005, McLean, VA. https://analysis.mitre.org/proceedings_agenda.htm#t6.

McCarthy, J. (1966). Information. *Scientific American* 215: 64–73.

OECD (1980). *The Organisation for Economic Co-operation and Development (OECD) Guidelines on the Protection of Privacy and Transborder Flows of Personal Data.* http://www.oecd.org/document/18/0,2340,en_2649_34255_1815186_1_1_1_1,00.html.

Oliveira, S. and Zaïane, O. (2004). Toward standardization in privacy-preserving data mining, in the *Proceedings of the 3rd Workshop on Data Mining Standards (DM-SSP*

2004), in Conjunction with KDD 2004, August 2004, Seattle, pp. 7–17. http://citeseerx.ist.psu.edu/viewdoc/summary?doi=10.1.1.58.4417.

PIPEDA (2000). *Canada's Personal Information Protection and Electronic Documents Act* (updated in 2006). http://www.privcom.gc.ca/legislation/02_06_01_e.asp.

Sparck Jones, K. (2003). Privacy: What's different now? *Interdisciplinary Science Reviews* 28: 287–292.

Wang, Y. and Kobsa, A. (2007). Respecting users' individual privacy constraints in web personalization, in the *Proceedings of the 11th International Conference on User Modeling 2007 (UM 2007)*, 25–29 June 2007, Corfu, Greece, pp. 157–166. Berlin/Heidelberg: Springer-Verlag. http://www.springer link.com/content/f6585435g5618340.

Warren, S. D. and Brandeis, L. D. (1890). The right to privacy. *Harvard Law Review* 4: 193–220.

Wellner, B., Huyck, M., Mardis, S., Aberdeen, J., Morgan, A., Peshkin, L., et al. (2007). Rapidly retargetable approaches to de-identification in medical records. *Journal of the American Medical Informatics Association* 14: 564–573.

58

Best Practice in Design for All

Klaus Miesenberger

58.1 Introduction

Outlining examples of best practice always risks not taking into account other examples that one might value being much better suited to demonstrate the issue than those selected. As up to now no extended surveys or comparative studies on best practice have been conducted, and only very general quality indicators and criteria do exist. Awards or competitions, most of the time, do not clearly outline how they value best practice in design for all, as the selection is based on the opinion and closed discussions of experts.

Therefore, this chapter is not intended to argue for best-practice examples in a competition-like way. The selection of examples in this chapter should not be seen as a valuation of single activities over others. The discussion of examples should demonstrate the difficulty of selecting best-practice cases for one's own practice and motivate us to look for other examples in similar or different domains, demonstrating the qualities of and motivations for design for all. If readers should be able to find examples that are better than those presented here and become motivated to demonstrate their qualities, the goal of this chapter has been reached. The discussion should also show the need for more detailed empirical and comparative studies that might outline indicators and reference frameworks for best practice in design for all.

The selection used in this chapter is driven by the author's context and by a European context of organizations and initiatives like the European Design for All and eAccessibility Network (EDeAN),[1] the Inclusive Design Curriculum Network (IDCNet),[2] the Design for All Network of Excellence (D4AllNet),[3] and the Association for the Advancement of Assistive Technology in Europe (AAATE).[4] It should also be mentioned that other chapters might cover aspects discussed here in more detail.

This chapter outlines best practice in design for all starting from the process-oriented nature of design for all, and thereby the fact that its qualities are inherent to final products; they are often invisible at first glance and do not advertise for design for all, as this might tend to stigmatize the design as of limited use for people with specific needs or problems. Based on this discussion, this chapter presents examples of, and references to, indicators and criteria for best practice in design for all. This enters into a demonstration of examples of best practice related to (1) awareness raising, (2) political and legal development, (3) eAccessibility, (4) social economic and business impact, (5) web accessibility, (6) software accessibility, (7) game accessibility, and (8) teaching design for all.

---

[1] EDeAN: http://www.edean.org.
[2] IDCNet: http://www.idcnet.info.
[3] D4AllNet: http://www.d4allnet.gr.
[4] AAATE: https://www.aaate.org.

58.2 The Process-Oriented and Often Invisible Nature of Design for All

Research and development over the last decades have established a profound body of knowledge in the domain of design for all. Good theories and concepts have been put forward regarding how to drive a design process toward design for all. These concepts have been proven in prototypes and demonstrators (e.g., The Center for Universal Design, 1997; Stephanidis, 2001; Darzentas and Miesenberger, 2005). There are also some first collections of examples demonstrating the qualities of design for all (e.g., IDeA[5]; The Universal Design Toolkit[6]; The Universal Design Net[7]; see also The Center for Universal Design, 2001). It is concluded that design for all offers advantages for design in ethical, political, legal, social, and also economical terms (Stephanidis, 2001; Darzentas and Miesenberger, 2005).

If the benefits and advantages are obvious, why are the principles of design for all often not taken into account? Why are the qualities of design for all often not recognized? These qualities seem to be absorbed by other, general qualities of products and services. People interested, invited, or forced to design for all, as well as experts advocating for design for all, therefore express a demand for convincing examples of best practice advertising for its qualities. Is this possible or is it a too optimistic expectation?

Design for all first of all asks for changing attitudes and for increasing awareness for the diversity of users leading to a change in the design process. Sets of criteria and principles for design for all (The Center for Universal Design, 2001; Stephanidis, 2001) outline this process-oriented nature of design for all (see Section 58.3). Design for all intends to influence and guide the general design process from its early stages until the final evaluation of products and services due to its user-centered approach and the demand for user involvement. This conceptual and process-oriented level is the core area of influence of design for all. Design for all qualities of the final products/devices/services might be recognized in two different ways:

- As special features for specific target groups, which are usually seen more as assistive features (e.g., special cars for the aging population) than as mainstream design for all
- As general improvement of the usability, which most of the time is not recognized as design for all, but as general design of high quality (e.g., good design of controls in cars)

The second option where design for all is an integral part of the general design seems to better meet with the concept of design for all. Due to this reason, the qualities that design for all brings into the process have to become inherent to the process of design, and might not be seen at a first glance in the final products or

services. They support accessibility and general usability, but they do not demonstrate or advertise for it. Design for all is a basic requirement, a prerequisite contributing to better systems and services. It is the general purpose of design which the final products advertise for, and not singular aspects like accessibility and usability for specific groups. Design is a holistic process and design for all can (only) be successful in the context of such holistic purposes and intents.

This often invisible characteristic of design for all seems to be a reason for the lack of convincing and therefore self-marketing examples. Additionally, an explicit visibility of design for all in terms of design for people with specific needs, in particular for those with disabilities, is in danger of being recognized as stigmatizing (e.g., examples of mobile phones advertising their special features for the aging population that did not meet with acceptance, both by mainstream and specific target groups). Explicit design for specific groups often might be more of an obstacle than a support to successful mainstream design.

There is a considerable number of examples of devices and features that everybody uses on a daily basis demonstrating how design for all supports the general quality and usability of products without being recognized as Design for All (e.g., The Center for Universal Design, 2001; Darzentas and Miesenberger, 2005):

- Smooth ground surfaces of entranceways, avoiding stairs
- Wide interior doors and hallways
- Lever handles for opening doors rather than twisting knobs
- Light switches with large flat panels rather than small toggle switches
- Buttons on control panels that can be distinguished by touch
- Bright and appropriate lighting, particularly task lighting
- Auditory output redundant with information on visual displays
- Visual output redundant with information in auditory output
- Contrast controls on visual output
- Use of meaningful icons as well as text labels
- Clear lines of sight (to reduce dependence on sound)
- Volume controls on auditory output
- Speed controls on auditory output
- Choice of language on speech output
- Ramp access in swimming pools
- Closed captioning on television networks[8]

A very good and comprehensive example of the fundamental influence of accessibility, usability, and design for all to the quality, usability, and success of products is the graphical user interface (GUI) (Myers, 1998; Müller-Prove, 2002; Rheingold, 2000; Reimer, 2005). A lot of the usability-enhancing features that everybody enjoys and takes for granted today (e.g., adaptability of colors, font size, etc.) result from accessibility and usability requests of interest groups of people with disabilities.

---

[5] IDeA: Center for Inclusive Design & Environmental Access, School of Architecture & Planning, University at Buffalo: http://www.ap.buffalo.edu/IDEA/Home/index.asp.

[6] The Universal Design Toolkit: http://www.inclusivedesigntoolkit.com.

[7] The Universal Design Net: http://www.universaldesign.net.

[8] Wikipedia: Universal Design: http://en.wikipedia.org/wiki/Universal_design.

As soon as these qualities became standard usability features, and also in some cases legal requirements, the accessibility and design for all origin got lost. Such features are embedded into modern design approaches and everybody expects them to be present in many products without even thinking about design for all. This is one of the most convincing examples of the influence of design for all. People can't simply imagine a GUI which, for example, would not allow adjusting the visual display of information.

It could be said that the more successful, the less recognized design for all seems to be, as the qualities of design for all are absorbed as general qualities of products and services. It might also well be that a user-centered design process oriented toward the diversity of user needs leads to qualities that are in accordance with, and are examples of, design for all without having the concept explicitly in mind. Therefore, the qualities of design for all seem to be of a conceptual and process, and not product, nature. Design for all qualities are those that are expected as basic requirements, and not as convincing add-ons advertising for design for all itself. This should be borne in mind when discussing examples of best practice and also the role of design for all.

58.3 The Context of Design for All

The process-oriented and often invisible nature of design for all invites a closer look at the more general context of design, in which design for all is embedded, to find indicators for examples of best practice. The term *design* is of a very general nature; it is so familiar that people often do not reflect on its meaning in various contexts. Therefore, it may be useful to begin by asking: what is design? *Design* is defined as the support of creation, planning, and/or development of new products, or related activities in redesigning existing products to better fulfill their roles in supporting people to reach their goals and intentions (e.g., Heider et al., 2004; Lidwell et al., 2004).

Design has the role of providing better tools that ease or support the life of people in all its aspects. Therefore, the first key aspect of design is functionality. Design has to support the goal of an activity by providing tools to be successful. At a subsequent level, users expect that tools are efficient and practical. Design should not only support a goal but should also contribute to an efficient process toward reaching the goal. Further on, in particular when several tools are available to reach a goal, people also tend to take other expectations into account, for example, if and how tools make the user feel good, support individual style, and present or communicate this to others in an expected way (e.g., when choosing a car it is not only a tool to go from A to B in the most efficient and economic way but a process influenced by our actual or intended lifestyle and status). Design theory therefore refers to at least three functions of design (see, e.g., Steffen, 2000; Schneider, 2005):

- *Practical functionality (functionality toward an object)*: The designed artifact needs to be efficient in reaching a goal.

- *Aesthetic functionality (functionality toward a user)*: The designed artifact needs to be effective, practical, and ergonomic for the user.
- *Symbolic/semantic/persuasive functionality (psychological functionality)*: The designed artifact has an affective, emotional, identifying role.

There are several other definitions using a more detailed structure (e.g., Bonsiepe[9]). The role of design could vary among these broad aspects from guiding the whole process of development and application, over guaranteeing the practical and aesthetic functionality toward a standalone artificial supplement for persuasive effects.

Design for all seems to refer to the practical and aesthetic level that a designed artifact can be used in accordance with the needs of users. In general we should say that the more holistic the approach to design is in terms of involving all functions discussed previously, the smaller is the risk that important aspects are neglected, also in terms of design for all. Design for all by its nature asks for a holistic process starting from the practical and aesthetic level. Design for all outlines a set of criteria and guidelines on how to take practical and aesthetic functionalities for a broad diversity of potential users as the starting point of design.

It is also interesting to look at the differences of the meaning or the focus of design in different cultural contexts (Walker, 2005):

- In the Anglo-American context, the focus is on a practical, usability-oriented level.
- In French- and Spanish-speaking countries, *design* or *diseño* is much more related to terms like concept, draft, plan, outline.
- In German-speaking countries, *design* is first of all related to a creative process and arts.

This might also be a reason why design for all meets with different levels of awareness and understanding. In the Anglo-American context, where design has a clearer focus on the practical level, design for all seems to better meet with the general concept of design. It might also be worth analyzing what is subject to design and to look at different domains of design leading perhaps to different domains of design for all. Modern design concepts do not only see physical objects as their goal, but also (Spillers, 2003):

- Virtual objects
- Information
- Interaction/interface
- Emotion
- Attitudes, social situations

This defines a considerable number of design disciplines and application areas that address the mind, the heart, and the body of users. More and more aspects of daily life become subject to design. This outlines the difficulty for design for all to be recognized as a general principle of design. How does design in

---

[9] Bonsiepe, g.: Erziehung zur visuellen Gestaltung, in: Ulm 12/13, Zeitschrift der Hochschule für Gestaltung Ulm, HfG Ulm, Ulm 1964.

general relate to domains that are seen as similar or that include design as a methodological approach[10]?

- *Engineering*: Design in engineering should lead to practical and effective tools; design could be seen as the creative part of engineering.
- *Production, construction*: Design as production is a planning and executing activity; the design ("plan") should anticipate the final product and could be seen as a step before production.
- *Marketing*: Design as part of PR activities should be persuasive in competitive markets and information spaces that a product at a similar or at the same functional and aesthetic level is more successful on the market due to additional affective and emotional qualities of persuasive design.
- *Art*: Design as part of artwork should be creative and develop new concepts, but in relation to art it has clear goals in terms of producing a final result and most of the time commercial purposes. Design could be referred to as applied arts.

The role of design considerably varies in the different domains. Practical and aesthetic functionalities are of particular importance in engineering and production. They are also of particular importance for design for all as the needs of users with disabilities or users in specific situations have to be addressed at practical level. In competitive markets where similar products provide almost equal levels of practical and aesthetical functionalities, the persuasive functionality might become predominant. Design as part of art might also be related to its persuasive function, but as an independent discipline it is first of all seen as independent from any functional preconditions and relations to other domains.

Due to these different roles, design becomes subject to different interests. The increasing importance of persuasive design tends to bring the focus away from practical and also aesthetical aspects that might lead to reduced usability and accessibility. It might happen that products with lower practical and aesthetic functionality are produced and chosen due to overwhelming persuasive functionalities. Design could tend to become the playground of marketing and drift away into a very loosely or selective user focus, and might tend not to take into account fundamental needs of specific user groups, in particular those with disabilities and functional limitations.

An analysis of the workflow of web design projects with a focus on design for all in the public and private sector (Miesenberger et al., 2005) demonstrated that very often the public relation or marketing department first defines the design of a web page based on criteria primarily deriving from requirements related to the persuasive level. Functional and aesthetic design requirements are not recognized as problematic and worth being analysed. Functional and aesthetic features are seen as add-ons to the

persuasive design that should be implemented at a later stage of the design process. The functional and aesthetic requirements, which are a focus in accessibility, usability, and design for all, are at risk of becoming secondary or forgotten. This turns the design process around. Starting from persuasive interests risks drifting away from the process from user-centered design and design for all principles.

This also seems to be a reason for one of the most important critics or fears of marketing when being confronted with design for all requirements. Marketing is afraid that persuasive concepts are interfered by requirements of design for all. Practical and aesthetic considerations (e.g., information and interaction design) are often not seen as a source that could lead to better and more successful design. Accessibility and usability problems are discredited as minority problems, and the potential negative market impact due to exclusion of consumer groups is neglected. As design for all asks for a holistic process, starting from the practical and aesthetic functionalities, it is often seen as a barrier to modern, marketing-oriented (and therefore primarily persuasive) design.

Of course engineering and production are based on solid functional design concepts and theories that are open to design for all. But they tend to work on the basis of external, predefined requirements and have to put in place what is requested—often by marketing. When practical and aesthetic requirements are not taken into account seriously, accessibility and usability requirements are most probably not defined. When the workflow starts from the wrong end with pure persuasive requirements, accessibility and usability are at risk of not being taken into account. Design often seems to be valued too much against its persuasive impact as a criterion for economic impact (see Section 58.8) and the risk of reduced usability and accessibility, as well as long-lasting success based on practical and aesthetic quality, is neglected.

As an obvious consequence, redesign for accessibility and usability (e.g., due to legal requirements), tends to be complex and cost intensive (The Center for Universal Design, 2001; Stephanidis, 2001; Darzentas and Miesenberger, 2005). But the reason for this later on is often not seen in this problematic, inverted design workflow, but in the demand expressed by design for all. This tends to discredit design for all as a burden to an already established design process. This misunderstanding seems to be very much related to the dominant role of persuasive design.

Design for all, as outlined in Section 58.2, therefore asks for awareness and rethinking established workflows in the design process. Design for all (re)focuses the process to the needs of a diversity of users. Today more than ever such a diversity of user needs can be supported due to the flexibility of tools and technologies available for design. Also the multimedia power of information and communications technology (ICT) and assistive technologies, as will be discussed in more detail in Section 58.7 as an example of best practice, do play a major role in this.

Design for all does not neglect the need for attractive and persuasive design in competitive markets. Design for all demands

---

[10] American Psychological Association: Design, in *The American Heritage Dictionary of the English Language*, 4th ed.

starting from the practical and aesthetic needs of a diversity of user groups and using the potential of new tools and technologies to support these needs. The challenge of design for all is to integrate the user-centered approach into marketing-driven design practice. Having the diversity of users' needs in mind will be beneficial to all users and the success of design. This will also encourage, not hinder, creativity and art. In such a way, design for all might become a method to address new markets and user groups with new designs, as some examples show (e.g., oxo[11], Solutions Marketing Group[12]).

58.4 Best Practice: Principles and Criteria for Design for All

The wide range of domains of design leads to a variety of sets of indicators for the quality of design for all. This section aims at identifying such indicators—principles, criteria, and standards. Most of these indicators are discussed in more detail in other sections of this handbook (see, e.g., Chapter 3 of this handbook). The examples and the discussion of best practice in subsequent parts of this chapter are based on these indicators.

Design for all is closely related to usability or usability engineering, which have evolved as a discipline over the last decades. It provides a set of principles, guidelines, and methods to achieve better usability in terms of (Nielsen, 1994; Norman, 2002; Richter and Flückiger, 2007):

- Learnability
- Efficiency of use
- Memorability
- Few and noncatastrophic errors
- Subjective satisfaction

Usability as a discipline contributes to human-computer interaction (HCI) and software engineering (e.g., ISO 9241). Usability leads to the need for a user-centered design process that is applicable in, and adaptable to, different contexts and situations. Principles like those provided by the Userfit method (The Userfit Tool)[13] or IS4All[14] approach help in guiding design toward users' diversity.

Design for all emphasizes that usability should be extended to the broadest possible number of user groups, taking different contexts and situations into account and including groups like people with disabilities, migrating populations, and people with diverse social and economic backgrounds. The Center for Universal Design (1997) defines the following principles for universal design:

1. *Equitable use*: The design is useful and marketable to people with diverse abilities.
2. *Flexibility in use*: The design accommodates a wide range of individual preferences and abilities.
3. *Simple and intuitive*: Use of the design is easy to understand, regardless of the user's experience, knowledge, language skills, or current concentration level.
4. *Perceptible information*: The design communicates necessary information effectively to the user, regardless of ambient conditions or the user's sensory abilities.
5. *Tolerance for error*: The design minimizes hazards and the adverse consequences of accidental or unintended actions.
6. *Low physical effort*: The design can be used efficiently and comfortably and with a minimum of fatigue.
7. *Size and space for approach and use*: Appropriate size and space is provided for approach, reach, manipulation, and use regardless of user's body size, posture, or mobility.

The IDeA center, which in general provides an excellent resource for information on design for all, offers information on best practice regarding how these principles were used for the city of New York (Giuliani, 2001; Levine, 2003).

Such general indicators for design for all are further specified for different application domains, as well as for different end-user groups.

Principles for built environments and environmental design offer criteria on how to address the needs of different user groups, including those with disabilities. An example is (ISO/TR 9527): Building construction—needs of disabled people in buildings—design guidelines extended according to regional and national norms. Examples of guidelines are provided, for example, in the Build for All reference manual (Build for All, 2008), from the Center on Universal Design (The Universal Design Center, 2001) or in extended scientific discussions (Stephanidis, 2005). Guidelines for designing for usability of everyday products can be found in:

- (ISO 20282-1): Ease of operation of everyday products—Part 1: Context of use and user characteristics
- (ISO 20282-2): Ease of operation of everyday products—Part 2: Test method

Later, guidelines and principles support design for all of man-machine interactions such as:

- (ISO 9241): Ergonomics of human-system interaction
- (ISO 13407:1999): Human-centered design processes for interactive systems
- (ISO 9355-3:2006): Ergonomic requirements for the design of displays and control actuators—Part 3: Control actuators
- (ISO 9241-110:2006): Ergonomics of human-system interaction—Part 110: Dialogue principles

Guidelines of particular importance for user groups with disabilities in design for all are:

---

[11] Oxo: http://www.oxo.com.
[12] Solutions Marketing Group.
[13] The Userfit Tool: http://www.sc.ehu.es/acwusfit.
[14] IS4ALL: International Scientific Forum "Towards an Information Society for All": http://ui4all.ics.forth.gr/isf_is4all.

- Web Accessibility Guidelines of the W3C/WAI for
  - Content (W3C/WAI/WCAG)
  - Authoring tools (W3C/WAI/ATAG)
  - User agents (W3C/WAI/UAAG)
- The (Wab Cluster) with its Unified Web Evaluation Methodology, which provides a framework for evaluation and development of web design for all and which is also used for the development of web evaluation tools[15]
- (ISO 924)1 which includes since 2006 (ISO 9241-171) for software accessibility (based on (ISO/TS 16071)), which provides criteria for
  - Accessibility impact (core/primary/secondary)
  - Implementation responsibility (OS/application)
- (Section 508) Paragraph 1194.21 on "Software applications and operating systems"

Several other collections of guidelines exist that are often applications or derivations of the previously mentioned international norms and guidelines for national/regional laws and regulations. Besides that, there are numerous guidelines of software developing companies that take up these guidelines at a more practical level (see Section 58.10).

Design for all of consumer electronics is taken into account in working groups (WG) of standardization committees (SC) such as:

- (ISO/IEC JTC 1 SC 36): Learning technologies
- (ISO/IEC JTC 1 SC 25 WG 1): Home electronic systems

Guidelines and standards relevant to universal access are discussed in more detail in Chapters 54 and 55 of this handbook. Issues of legislation are discussed in Chapter 53.

Other candidates for indicators are criteria and guidelines for awards in design for all like the IFI Design for All Award (International Federation of Interior Architects/Designers),[16] the European DfA-Award,[17] or Biene Award,[18] which will be discussed in Section 58.9. Such criteria start to enter general design award schemes like the Spanish National Awards of Design of the Design for All Foundation.[19] It has to be mentioned that there are only few empirical studies that would allow analyzing what criteria form the basis of the evaluation of best practice. Awards, besides outlining general submission and evaluation criteria, tend to keep their decision-making process secret.

The process-oriented nature of design for all asks for indicators and methods supporting the process and management of design for all, like the British Norm BS7000 Part 6 (Keates, 2004)

and related advice documents (Clarkson et al., 2003; Keates and Clarkson, 2003). The process-oriented nature of design for all is also demonstrated by the work on meta guidelines, which are guidelines for the process of guideline development to incorporate design for all, like in CEN/CENELEC Guide 6: Guidelines for standards developers to address the needs of older persons and persons with disabilities (CEN/CENLEC, 2008).

Thimbleby (2007) outlines and demonstrates that there is a lack in applying formal and quantitative methods in usability engineering. This also seems to be applicable for design for all, and demonstrates that ongoing efforts are needed to provide better applicable and reliable indicators for usability engineering and evaluation.

A number of brochures and references exist on guidelines, criteria, and indicators for design for all. As an example of best practice, the TIRESIAS (RNIB)[20] system should be mentioned, which provides a frequently updated reference to related materials.

These criteria form the basis for guiding and evaluating practice. They form a set of very general indicators for design for all, which have to be further analyzed and exemplified for the regional or local context and the field of evaluation. More specific and in some aspects quantified indicators tend to focus on singular user groups, such as people with disabilities and their demand for accessibility (e.g., WAI guidelines). The broad field of design for all, the various contexts to be considered, and the different and perhaps contradicting interests that drive design for all make the process of selection difficult. So far, there is a lack of frameworks helping to guide the evaluation, both for quantitative as well as qualitative measurements, of best practice.

This overview seems to support the already outlined impression that indicators for best practice in design for all are much more process oriented than related to final products. Criteria and first standards and norms that guide the process of design toward taking the diversity of user needs into account in various situations and contexts are the core of best practice in design for all.

58.5 Best Practice: Awareness Raising

Design for all, as outlined, asks for taking different contexts and a diversity of needs of users into account in the process of design. Design for all advocates for employing the potential of modern ICT, HCI, and assistive technology (AT) to address the diversity of user needs. It therefore asks for awareness and understanding of the needs of the users, the importance of user involvement, and the potential of modern design methodologies. Due to this it is worth having a look at best practice in design for all related to awareness raising.

---

[15] BenToWeb home page: http://www.bentoweb.org/home.

[16] International Federation of Interior Architects/Designers: http://www.ifiworld.org/index.cfm?GPID=61.

[17] Forschungsinstitut Technologie-Behindertenhilfe: First European award for outstanding achievements in the areas of Assistive Technology and Universal Design: http://en.ftb-net.de/aktuell/dfa_at.html.

[18] BIENE Award: Barrierefreies Internet eröffnet neue Einsichten: http://www.biene-award.de/award.

[19] Design for All Foundation: The Spanish "National Awards of Design 2006" Demand a Design sustainable, ecologic and for all: http://www.designforall.org/en/novetats/noticia.php?id=378.

[20] RNIB (Royal National Institute for the Blind), Scientific Research Unit: Guidelines on accessibility issues for all types of disabilities, for designers of information and communication technology (ICT) systems: Tiresias home page: http://www.tiresias.org/index.htm.

New legal frameworks are a consequence of changes in attitudes and understanding of society. Legal rules are also tools to make this new understanding a consensus in society. But they are also key instruments to raise awareness and change practice. Both processes, legal and social change, are driven by individual people and interest groups. Raising awareness is a key factor in the whole process of changing attitudes, legal frameworks, and practice. Societies as a whole, decision makers, and designers in particular have to become aware of the need to change.

The Disability Rights and Independent Living Movement[21] starting from the U.S. civil rights and consumer movements spread around the world and led to numerous initiatives to put equal rights for access in place. Design for all, accessibility, and usability are of course key factors that support this movement. It would be an endless list of people and organizations that could be mentioned here, and it is a difficult task to find out examples of best practice as again their qualities very much depend on their local context. For a history of this movement refer to, for example, Welch and Palames (1995).

Looking at the recently published 7th Framework Programme of the EU (eInclusion@EU)[22] and the European Council's Resolution on eAccessibility (Council of the European Union, 2003), it is remarkable how deeply it is influenced by the idea of design for all and eInclusion. The 7th Framework Programme of the European Commission eInclusion outlines in its introductory note on 17 October 2000 concerning the "Issues in the fight against poverty and social exclusion" that "the emergence of new information and communication technologies constitutes an exceptional opportunity, provided that the risk of creating an ever-widening gap between those who have access to the new knowledge and those who do not is avoided" (eInclusion@EU). Only if design for all in this movement toward the information society is taken into account can a widening of the access gap be avoided.

An example of good practice in putting the political will of European resolutions and according directives in practice at national level is the European Design for All and eAccessibility Network (EDeAN). It is an example of a community devoted to promoting universal access and design for all in Europe. EDeAN was established in 2002, in accordance with the specific goals of the eEurope 2002 Action Plan.[23] Based on and related to the work of other groups like Cost219[24] and AAATE, the EDeAN network provokes and supports national programs and activities, and has been the driving force of important European activities, transferring the EU-driven ideas of eAccessibility and eInclusion as part of the eEurope and i2010[25] plans into place.

The network got support from different projects, such as D4AllNet, which provided a suitable and accessible platform for international cooperation, and IDCNet, which identified and made available an elaborated interdisciplinary body of knowledge and curricula in the field of design for all (D4AllNet). More recently, the D4All@eInclusion project[26] supports the network by, for example, establishing a resource center/knowledge base, providing online services, training, and consultation, strengthening the relations with industry for a better uptake of design for all, and supporting the mainstreaming of accessibility.

58.6 Best Practice: Political and Legal Development

The idea of universal access to society—and therefore in principle also design for all, or universal design—is imminent to democratic societies. For example, the Austrian Constitution Act of 1867 outlines that citizens are equal and have the right to access public buildings and information. Although not understood in the same way as today, such a democratic constitution already bears accessibility in mind. These are the points of reference for the upcoming equal rights movements of different groups in society (e.g., women, races). One of the most recent equal rights and opportunities movements is the disability rights movement.

The independent living movement in the United States is a starting point for an international movement for equal rights for people with disabilities and therefore also for legal and political frameworks in design for all (Welch and Palames, 1995). A series of standards for better accessibility was accepted in the more and more states that became uniformed in 1973 and published as the "Uniform Federal Accessibility Standard (UFAS)." These norms ad standards became the reference point and driving force for the upcoming anti-discrimination legislation regarding people with disabilities, for example, Architectural Barriers Act in 1966, Section 504 of the Rehabilitation Act of 1973, Education for Handicapped Children Act of 1975 (Individuals with Disabilities Education Act, IDEA), Americans with Disabilities Act of 1990 (ADA), Telecommunications Act of 1996, and Section 508 of the Rehabilitation Act of 1998.

It is worth mentioning these developments from a global perspective, as this practice initiated similar movements around the world. In the European context, different developments have taken place due to different legal and societal circumstances. In some countries, like the United Kingdom, the independent living movement brought forward related legislation. In other countries, such as Austria, where social care and services have been and still are at focus, it took a longer time until this movement took place. Design for all is primarily seen as an accessibility issue, and due to the social care approach, as an issue of the public sector. The awareness for design for all as a responsibility of mainstream design seems to be much harder to establish.

---

[21] Disability Rights and Independent Living Movement: http://bancroft.berkeley.edu/collections/drilm.

[22] eInclusion@EU: http://www.einclusion-eu.org.

[23] eEurope Action Plan: http://europa.eu.int/information_society/eeurope/2002/action_plan/index_en.htm.

[24] COST219ter: Accessibility for All to services and terminals for next generation mobile networks: http://www.tiresias.org/cost219ter/ (accessed March 30, 2007).

[25] i2010: A European Information Society for growth and employment: http://ec.europa.eu/information_society/eeurope/i2010/index_en.htm.

[26] D4All@eInclusion: http://www.dfaei.org.

The European Union is playing a special role, as it drives this movement toward according legislation with its directives (e.g., Council of the European Union, 2003), which forced the member states to implement antidiscrimination legislation at national levels. The directive is put in place in different ways in the different countries, but has led in any case to the implementation of antidiscrimination legislation in all EU member states. Due to this, eInclusion meets with increasing attention at the policy level.

In Austria, for example, the Anti-Discrimination Act (Bundesbehindertengleichstellungsgesetz) has been enacted following the EU Directive. Although enacted rather late after the deadline of the EU on 1 January 2006, it should be mentioned as it not only affects the public sector, but also offers possibilities to claim for anti-discrimination and accessibility in different private areas. It is also worth mentioning that it is based on reversal of the burden of proof, that is, that those offering goods and services have to demonstrate that they avoided discrimination. In case of discrimination it offers the right for compensation, which constitutes an important incentive for accessibility and therefore also design for all. An arbitration board has been set up that should help to solve discrimination before going to the courts. The costs for this arbitration board are covered by the government, which again supports equal access in society and design for all driven by the users.

Another important aspect should be mentioned related to ICT and in particular web accessibility. The act is very general in terms of guidelines and regulations. It asks for taking the up-to-date state of the art in ICT and AT into account. This leads to a very progressive orientation in accessibility. Regulations, guidelines, and standards may be slow in adapting to the fast technical developments; reference to the state of the art does not restrict the situation to standards or references that are not updated. For example, one might well refer already to WCAG2.0 (W3C/WCAG 2.0), which is not published yet, instead of WCAG1.0 (W3C/WCAG 1.0), which has been in force since 1999. This is positive for product/service providers as they can rely already now and also in the future on up-to-date technologies that WCAG refers to in its baseline for accessibility (W3C/WCAG 2.0). But this is also positive for people with disabilities and the field of design for all, as their claim for accessibility is not in danger of being rated as unattractive or out of date. Accessibility and usability is integrated into the context of most attractive, modern, and forward design. Of course, this openness asks for interpretations, which might lead to unintended discrimination or problems for the target group to keep up with the changes, a fact that is criticized by some nongovernmental organizations (Klagsverband).[27] But the implemented arbitration board at least supports an interpretation according to the meaning of the anti-discrimination act.

There are for sure a lot of other examples of best practice related to the political and legal development offering other approaches and aspects that are worthy of mention. The U.N. (United Nations, 1993) and several other organizations like standardization bodies (see Chapter 53 of this handbook) contribute to this political change. The fact that anti-discrimination legislation is now on the way to becoming a global standard makes best practices a driving force for design for all.

Once again, it is not only the final product that constitutes best practice; this might not be seen or presented as design for all. It is the process itself based on attitudes open for design for all that implements guidelines and principles leading to accessible and better usable products for all.

Standards and normative documents support the implementation of legal frameworks of design for all by providing manageable reference points for practice. Standards have become a major source driving design for all, and there are several examples of good practice. As they are addressed at a different place in this book, they are not discussed here. More detailed information is also available from the special-interest group (SIG) on standardization of the EDeAN network.

58.7 Best Practice: eAccessibility and ICT Design for All

Undoubtedly, inclusion and participation in society do not only depend on the availability of tools and ICT. Inclusion is, of course, a social concept, and depends on the individual and shared understandings and attitudes of society. But it is also obvious that the independent living movement has been provoked and strongly supported by the existence and availability of tools and devices enabling a better and easier design of the environment in accordance with the needs of people with disabilities.

Besides the political progress and awareness campaigning, equal access in society for people with disabilities depends on tools and techniques to enable independent interaction and communication with people, products, systems, and services. Modern ICT plays an outstanding role in this context, and HCI leads to a convergence of developments in AT and the general technological development. HCI enables AT to interact with standard ICT, which becomes used in almost any aspect of modern life. As AT also makes use of modern HCI-based ICT, it can seamlessly integrate into interaction and communication.

With ICT/HCI/AT, the independent living and antidiscrimination movement gets a powerful tool at hand to enable design for all and eInclusion. In particular, HCI enables (Miesenberger, 2004; Darzentas and Miesenberger, 2005; see Figure 58.1):

- The separation of the interaction from the parameters of the tools and services itself
- The presentation of information in multiple formats (multimedia)
- The interaction in different modes (multimodality)

---

[27] Klagsverabnd zur Durchsetzung der Rechte von Diskriminirungsopfern: http://www.klagsverband.at.

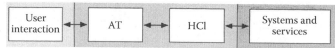

FIGURE 58.1 (From Miesenberger, K., "Equality = e-quality" 'design for all' und 'accessibility' als Grundlage für eine demokratische, offene und inklusive Gesellschaft, in *Qual-I-tät und Integration, Beiträge zum 8*, E. Feyerer and W. Pammer, eds., Linz, PraktikerInnenforum, Universitätsverlag Rudolf Trauner, 2004; and Darzentas, J. and Miesenberger, K., Design for all in information technology: A universal concern (keynote), in *Database and Expert Systems Applications, Proceedings of the 16th International Conference (DEXA 2005)*, K. V. Andersen, J. Debenham, and R. Wagner, eds., 22–26 August 2005, Copenhagen, Denmark, pp. 406–420, Berlin/Heidelberg, Springer-Verlag, 2005.)

These characteristics lead step by step to opportunities of adaptation to:

- A diversity of user needs
- Different interaction devices, including AT

As HCI enters almost any area of life, it has become a universal standard for interaction. Elements (windows, icons, menus, pointers, etc.), devices (screen, keyboard, mouse, etc.), and techniques (point and click, drag and drop, etc.) of HCI are limited but can be applied for an almost unlimited number of applications. This stability of interaction helps to manage the complexity of the information society as the same interaction skills can be applied to almost all domains where ICH/HCI is used.

This puts HCI at the center of usability and accessibility issues in the information society, and therefore in design for all. When looking at the increasing power of computers, one recognizes immediately that more and more of the resources available are consumed by features supporting HCI.

Moore's law states that the power of computers doubles every 2 to 3 years. Holzinger (2000) demonstrates (see Figure 58.2) that the power of the brain and therefore the skills of users have been more or less stable over thousands of years. Only due to the quality of HCI are we able to overcome this gap, by providing more usable and user-friendly interfaces. It is therefore with no

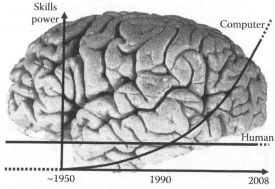

FIGURE 58.2 (From Holzinger, A., *Basiswissen Multimedia*, 3 Bände, Würzburg, Vogel, Band 2, Mensch, Human, 2000.)

surprise that most of the power of modern computers is consumed by the HCI to support the user in interacting with ICT-based systems and services. Due to the outlined convergence of ICT, HCI, and AT, accessibility, participation, and eInclusion depend on the quality of HCI: equality and equal opportunities in society become more and more dependant on eQuality, the quality of systems in terms of taking the diverse needs of users into account.

The application of modern ICT provides an increasing potential for people with disabilities to get access to this new way of handling systems and services. If people with disabilities can interface the HCI with AT, they get, in principle, access to all eSystems and eServices. Traditional human-human or human-machine interfaces did not provide the same level of flexibility and adaptability and tended to offer special solutions and exclusion. Therefore, the combination of ICT/HCI/AT forms a powerful basis for eInclusion—the participation of people with disabilities in the information society. Often for the first time in their life people with disabilities can handle processes independently without the support of other persons. People with disabilities can use the flexibility and the multimedia power of modern ICT and AT to independently interact with eSystems and eServices. Service providers can use more powerful and user-centered tools to support their clients. This of course is only possible if eSystems and eServices support accessibility.

However, the trend toward eSystems and eServices decreases the availability of traditional systems and services based on human-human, face-to-face interaction. If new eSystems and eServices cannot be used due to economic or educational reasons, or due to accessibility and usability problems, citizens are put at a disadvantage.

The more systems and services are based on ICT (e.g., eServices), the more eAccessibility and design for all enables users with or without assistive devices to interact with these systems. Equality, equal access, and equal opportunities in society are more and more dependent on eQuality, on taking eAccessibility and design for all into account in the information society (see Figure 58.3). Only if eSystems and eServices are developed under principles of eAccessibility and design for all can they be accessed by AT and deploy inclusive potential for people with disabilities.

Modern society and in particular ICT give tools at hand that show a growing potential for design for all. This potential has to be put in place in the upcoming information or knowledge society. Neglecting it or not putting it in place would violate the fundamental right of every citizen of not being discriminated against. If the 21st century forms the transition into the information or knowledge society, design for all has to be understood and taken up as a fundamental human right.

There is no other technology outlining in a clearer way that disability is not only an individual characteristic, but much more a pattern and an attribute of the design of the environment and society. ICT/HCI/AT make obvious that disability is a social construct and that tools are available to provide equal access and to put design for all in place.

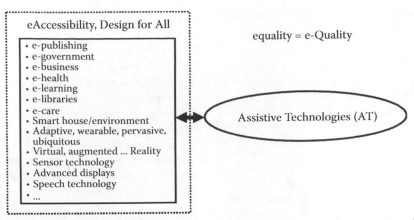

FIGURE 58.3 (From Miesenberger, K., "Equality = e-quality" 'design for all' und 'accessibility' als Grundlage für eine demokratische, offene und inklusive Gesellschaft, in *Qual-I-tät und Integration, Beiträge zum 8*, E. Feyerer and W. Pammer, eds., Linz, PraktikerInnenforum, Universitätsverlag Rudolf Trauner, 2004.)

This makes it worth mentioning that HCI is an example of best practice as it brings forward powerful tools for eInclusion. The development of modern HCI (Myers, 1998; Rheingold, 2000; Müller-Prove, 2002; Reimer, 2005), from punch cards over batch mode, command line to GUIs, WIMP (windows, icons, menus, pointers), SILK (speech, image, language, knowledge) and adaptive/learning interfaces brings forward more and more degrees of freedom in interaction that could be used for design for all. This best practice in and for design for all does not often receive the necessary attention.

58.8 Best Practice: Economic and Business Impact

In January 2006, the president of the European Commission outlined at the kick-off event of the project "Build for All: Public Procurement and Accessibility"[28] that an effective legal framework is important to reach accessibility for all, and that the public sector plays a central role, as its purchasing power consists of about 1500 billion €, nearly 16% of the EU cross-national product for goods and services. This provides a considerable market power to the public sector to change attitudes toward access for all. EU policies and directives intend to use this power.

Besides ethical and legal reasons, design for all is supported by economic reasons both in the welfare/public and in the business sector. In an aging society, more and more citizens depend on design for all to be able to use eSystems and eServices to participate independently in society. This is of growing importance for welfare systems to answer the exploding costs of support and care. This is also important for business, as people with specific needs form a growing consumer group.

There are several examples showing the impact of design for all on the social economy, as well as on business. However, as mentioned previously, these qualities are often not referred to as

qualities of design for all, but rather as general qualities of products and services. Many ideas and inventions have a design for all origin, as in the car industry or devices for daily use (e.g., TV remote control, typewriter, telephone, transistors, single lever tap, or mentioned features of adapting the GUI/HCI).

There is a good example showing the impact of design for all in the social economy: at the World Congress on Ergo Therapy it was reported that an investment of 8.8 million € in smart homes to enable 5100 citizens (on average 1725€ per person) to stay at home longer in life resulted in a positive return of investment on an average of 8.5 days due to the high costs of residential care (on average 200€ per day) (Zagler, 2004).

There are other good examples of how design for all products became a success in the market (e.g., Darzentas and Miesenberger, 2005) but, as already mentioned, these products are most of the time not recognized as design for all products.

Dong et al. (2004) in a survey on the opinion of manufacturers, retailers, and design consultants in the United Kingdom provide a more in-depth understanding of barriers and motivations for design for all in industry, as well as indicators for the quality of design for all in business. The U.K. framework is of course specific, due to a longer lasting and more advanced legal framework for accessibility and design for all, but it can be estimated that the gained criteria can be transferred to other contexts. They found that a good business case is key to industry and that design for all:

- Lacks in good business cases
- Lacks in perceived sacrifice of aesthetics
- Is seen as more expensive
- Is seen as more complex
- Lacks in clear requirements
- Suffers from a lack of budget for user research and orientation

It is critical to all business cases that the potential market based on consumer demand is shown, and that design for all shows related innovation. Based on their survey, and on a related

---

[28] Bauen für Alle: http://www.design-fuer-alle.de/download/Bauen_fuer_Alle_Leitfaden.pdf.

literature study, they outline business objectives that could be used as measurable indicators for the economic impact for design for all (Dong et al., 2004):

- Product/service-focused objectives contributing to
  - Performance
  - Safety
  - Desirability
  - Competitiveness
  - New markets
  - Leadership in technology
  - Costumer satisfaction
  - Better or new services
  - Customization
- Market-focused objectives contributing to
  - Market share
  - Market position and leadership
  - Repeat business
  - New customers
  - Competitiveness
  - Differentiation from competitors
- Revenue-focused objectives contributing to
  - Revenues
  - Profits
  - Earnings per share
  - Unit sales
- Image-focused objectives leading to increased recognition of
  - Technology leadership
  - Contribution to standards or industry cooperation
  - Quality and reliability
  - Outstanding customer services
  - Performance leadership

These criteria can be seen as a first basis for operable criteria to compare different design for all activities. So far, no studies on design for all examples can be found.

58.9 Best Practice: Web Design for All

Fortunately, there are numerous web pages that are examples of best practice today. The W3C/WAI guidelines and the legal frameworks, which are examples of best practice themselves, led to examples able to demonstrate design for all in this domain. There are several award programs supporting the awareness for good design based on design for all approaches. One of these awards is the German Biene Award, which supports raising awareness—an example for best practice itself. The Biene Award provides a set of indicators and criteria that are based on scientific studies (Pieper et al., 2004).

A winner of the Biene Gold in 2006 is the Internet portal of the Austrian government, help.gv.at (or, Help)[29] (Miesenberger

[29] Help.gv.at: Offizieller Amtshelfer: http://help.gv.at.

et al., 2005; see Figure 58.4). As it is again important for design for all not only to look at the final result, but also to have an insight into the process of design for all, this example is discussed in the following.

Of course, the legal framework in Austria, the Anti-Discrimination Act and related rules in the eGovernment legislation (eGovernment Law), made it necessary to work on accessibility of this system, but the work began before these legislations were in force, demonstrating an increasing awareness and willingness toward design for all.

It should be mentioned that help.gv.at is the central access point for citizens of Austria related to all public affairs, from federal down to municipal. This underlines the importance of access. It also underlines the complexity of the system and the impact of the design for all approach.

Public authorities plan to promote Help as a one-stop access point to administrative online services and information. Due to the importance of Help, the project HELP barrierefrei will make administrative services for people with disabilities more accessible, easy, and user friendly, thus promoting independent use. It is therefore hoped that Help will be a model example of a web portal and other initiatives will follow this example. Help is an Internet platform with links to a large number of public authorities. The portal provides information on all kinds of interactions with Austrian authorities required in the most frequent, so-called life situations. Help services describe the individual steps to be taken when interacting with authorities, describes background information, and indicates links for more detailed information. Whereas during the first years the main emphasis was on the provision of information to citizens in electronic format, today there is an increasing trend toward complete electronic processing of all administrative procedures. Help permits the electronic processing of an increasing number of standard administrative procedures. The life situations cover different subjects of the life cycle like pregnancy, childbirth, marriage, housing, business, disability, immigration, and so on, and offer information on administrative processes, the required documents, fees, and deadlines, and other interesting links. Specialized services for areas like business, immigration, and disability are put in place and optimized in cooperation with interest groups.

To date, Help offers access to more than 200 such life situations. Step-by-step additional life situations are added to the system, offering information and also electronic processing of administrative services. A growing number of partner authorities at national, regional/provincial, and local/municipal level use and integrate their services into Help. It is the political goal that Help should become a one-stop portal for all administrative services and information in Austria.

The project did not exclusively work toward accessibility for people with disabilities. It took accessibility and usability for people with disabilities as an integral part of user-centered design and design for all. To combine up-to-date design and technology with accessibility and usability at the highest level was the challenge of the project.

FIGURE 58.4   (Screenshot taken from help.gv.at.)

The design for all process adopted for help.gv.at is also remarkable in the sense that it had to include different stakeholders:

- The content owner and provider of the public sector
- The data-processing center of the public sector
- Private companies contracted for different aspects of the development
  - Design companies
  - Technical and programming support
- Accessibility and usability consultant

Such an approach using distributed responsibilities over different partners is typical of large-scale projects (see Figure 58.5). It had to be guaranteed that accessibility and usability issues are taken into consideration from the first steps of conceptual design toward implementation by all stakeholders. This could be seen as a good example of how awareness, understanding, and practical implementation of design for all could be put in place in the long run.

Accessibility experts became involved in the group as consultants from the beginning. Their role was to guide the process toward design for all, defining the requirements, developing prototypes and user testing. The challenge in this cooperation was to combine modern and up-to-date design from a design company with an existing technical framework based on newest Internet technology with the highest level of accessibility and usability possible. Not only should compliance with the WAI/WCAG guidelines (W3C/WAI/WCAG 1.0) at the AAA level be put in place, but also features like sign language integration, ease of reading, and ease of navigation should increase the usability for all citizens. The project defined the following user groups for operability reasons of the project:

- Standard users
- Blind people
- Visually impaired people
- Deaf people
- Hearing impaired people
- People with mobility and movement problems
- People with cognitive problems and
- Elderly people

First the help system was checked for accessibility and usability. Involvement of users with disabilities from all target groups had highest priority. Experts with disabilities were also involved. Requirements have been defined accordingly, and a plan for restructuring the interface has been specified. Fully accessible prototypes were provided by the accessibility consultants as a basis for the implementation of the help system by the involved companies and the ministry.

To guarantee a good accessibility for all users in all target groups, another important issue besides AAA conformance with the W3C/WAI/WCAG 1.0 guidelines (W3C/WAI/WCAG 1.0) was orientation toward usability in terms of elements such as (FTB)[30]:

- Perceptibility
- Understandability
- Operability
- Memorability
- Efficiency
- Technical robustness

---

[30] Forschungsinstitut Technologie-Behindertenhilfe (FTB): Web for All—Projekt Barrierefreiheit im Internet, Barrierefreies E-Government—Leitfaden für Entscheidungsträger, Grafiker und Programmierer, 2003.

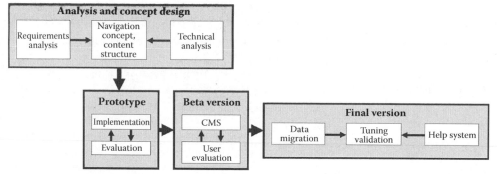

FIGURE 58.5 (From Miesenberger, K. and Pühretmair, F., Help.gv.at—Accessible e-Government in Austria, in *Assistive Technology: From Virtuality to Reality, Proceedings of AAATE 2005*, A. Pruski and H. Knops, eds., 6–9 September 2005, Lille, France. Amsterdam, IOS Press, 2005; and Miesenberger, K., Pühretmair, F., and Nussbaum, G., E-government Protal www.help.gv.at barrierefrei: Anforderungen–Herausforderungen–Erfahrungen, in *E-Government 2005: Knowledge Transfer and Status*, M. Wimmer, ed., Wien, OCG, 2005.)

These issues have been addressed by involving a competent partner in usability testing and engineering (CURE[31]).

Both automatic testing with appropriate tools (e.g., Watchfire, 2005) and manual/human usability testing methods were employed for all user groups, including those with disabilities.

In general, the involvement and the cooperation in setting up the Unified Web Evaluation Methodology (UWEM) developed in the Wab Cluster helped in developing and evaluating the quality of the redesigned system on a structural basis (Wab Cluster, 2006). UWEM defines a step-by-step procedure for analyzing web accessibility conformance of web pages. This methodology can also be easily integrated into the design and implementation process of new web pages. This has been the case for the work on help.gv.at. The UWEM methodology once again should be mentioned as a basis that could lead to a solid evaluation and development framework for web design for all.

Figure 58.6 summarizes several of the features implemented in favor of people with disabilities. Special attention has been given to accessibility for deaf people, as well as people with

cognitive problems, by supporting concepts like sign language, video integration, and easy to read language as equivalent alternatives. Easiness to read should be realized directly in the text, and when this is not possible due to legal reasons, for example, if the wording of a law has to be represented, an easy to read version can be provided as an alternative.

The system is based on a content management system (CMS) and several templates to guarantee that the content will be in accordance with the defined concepts and standards. To guarantee consistent quality of information, authoring guidelines tailored for this specific system have been defined, which put special attention on the preservation of accessibility and usability. Training has been provided to increase awareness and acceptance toward the new approach.

58.10 Best Practice: Software Design for All

As mentioned throughout this chapter, HCI plays a key role in design for all, and software accessibility should make use of the flexibility and adaptability of HCI to give access to systems and

---

[31] CURE: Center for usability research and engineering: http://www.cure.at.

- New navigation structure
- Sub-navigation structure
- Accessible wording
- Intuitive handlong
- Bread-crumb navigation
- Internal links for blind users
- Sign language videos
- Adapted content
- CSS instead of layout tables
- Device-browser-independent
- WCAG 1.0 - AAA

FIGURE 58.6 Screenshot of the website help.gv.at with a list of accessibility and usability features implemented.

services in the information society to the widest range of users. Leading software developers have put in place software accessibility features and frameworks that support accessibility of their own tools, and also support accessible software development using their software as a platform. Several examples are discussed in this book, and therefore this section lists some examples of good practice supporting design for all in alphabetical order without going into detail:

- *Adobe*: http://www.adobe.com/enterprise/accessibility/main. html
- *Apple*: http://www.apple.com/accessibility/, http://developer. apple.com/documentation/Accessibility
- *IBM*: http://www-306.ibm.com/able/guidelines
- *Linux/Unix/Gnome*: http://lars.atrc.utoronto.ca, http:// developer.gnome.org/projects/gap/ (Accessibility Toolkit, Gnopernikus)
- *Macromedia*: http://www.macromedia.com/resources/ accessibility
- *Microsoft Enable*: http://www.microsoft.com/enable, http://msdn.microsoft.com/library/default.asp?url=/ library/enus/dnanchor/html/accessibility.asp
- *Oracle*: http://www.oracle.com/accessibility/index.html
- *Sun Microsystems Accessibility Program*: http://www.sun. com/access/index.html

Together with criteria and methods for software usability already mentioned (Nielsen, 1994; Norman, 2002) and discussed in other chapters in this book, software and web accessibility form the revolutionary basis to adapt systems and services to the needs of a diversity of users. eAccessibility and HCI usability are the driving force for a more inclusive information society.

58.11 Best Practice: Games Accessibility

Games are important for both learning and entertainment. Usually, game interfaces ask for special interaction skills by users. Standard HCI in most aspects is ruled and determined by the user. Game interaction is in important ways determined by the game itself, and therefore games ask for very special accessibility features. The field of game accessibility is rather new. Most of the time games for people with disabilities were designed especially for this user group with therapeutic or educational purposes. Following the trend toward accessibility in the web and for software in general, it is time to make games entertaining for people with disabilities and to offer them possibilities to play the same games as their peers (Ossmann et al., 2006).

Games use challenging HCI concepts and ask for specific skills of users. Games are also receiving more attention as they are very important for learning, teaching, entertainment, and inclusion. But they are among the most challenging applications concerning accessibility and usability for people with disabilities. Game accessibility is discussed in detail in Chapter 17, "Designing Universally Accessible Games," of this handbook.

The International Games Developer Association's (IGDA)[32] Special Interest Group on Games Accessibility (GA-SIG) can be seen as an example of best practice that made available a first set of general guidelines.

Another approach is put forward by the Norwegian IT company MediaLT. MediaLT has developed a set of guidelines (MediaLT, 2008) that further elaborated on existing guidelines.

There are numerous activities working on examples of games that should be seen as examples of best practice in this domain (Grammenos et al., 2005, 2006; Ossmann and Miesenberger, 2006). Following these first attempts toward game accessibility, new initiatives have been started (Ossmann et al., 2006a) intending to work on:

- Guidelines helping game developers to design their products in a more accessible manner.
- A descriptive language (markup, XML scheme) based on the guidelines, which should help to better describe the interaction with the game by different I/O devices and to enable assistive technologies to interact with the game engine; this should help game developers during the game design process to fulfil the rules in the game development guidelines and show them which parts of the game needs additional information for the presentation on different output devices; a collection of different schemas for the different kinds of computer games in combination with the different kinds of disabilities and AT should be made available.
- A toolbox for developers based on the descriptive language (Active Games Accessibility, AGA).
- Collections of examples of games or scenarios of games.

A first set of guidelines has been made available for discussion (MediaLT). The other aspects are subject to running activities, which should lead to an extended network and cooperation, including AT users, AT experts, game developers (research and development), and industry.

58.12 Best Practice: Teaching Design for All

Initiatives like Telemate,[33] ATACP,[34] and the already mentioned project IDCNet addressed the need for education in the domain of D4All, as well as first attempts to identify the core curriculum. These initiatives led to first courses, which constitute the basis for more recent work in this field. Initiatives focus on two main areas:

- Integrating D4All in standard curricula
- Special courses in D4All for experts in the field of ICT/AT

---

[32] IGDA: International Games Developer Association: Accessibility in Games – Motivation and Approaches: http://www.igda.org/accessibility/ IGDA_Accessibility_WhitePaper.pdf.

[33] Telemate: Project home page: http://prt.fernuni-hagen.de/pro/telemate/ telemate.htm.

[34] ATACP: The Assistive Technology Applications Certificate Program: http://www.csun.edu/codtraining.

Another good example of resources in design for all is the U.K. National Contact Point for (EDeAN Education).[35] This web site is a resource tool provided by the European Design for All e-Accessibility Network (EDeAN). It was developed and donated by the RSA, a member organization in the UK-DeAN, from a prototype funded by the Engineering and Physical Sciences Research Council (EPSRC) as part of the EQUAL program.

This web site, originally named the RSA Inclusive Design Resource, was developed by The Helen Hamlyn Research Centre (HHRC), Royal College of Art, with the intention of providing an online resource for inclusive design to aid tutors, students, professional designers, and business/industry. The RSA handed over the resource to EDeAN. Of course there are many more resources for design for all like the already mentioned Center on Design for All (The Center for Universal Design, 1997), the IDeA center, the Inclusive Design Toolkit, the *Design for All Handbook* (Preiser and Ostroff, 2001) or the Universal Design Net, to mention a few.

Based on the experiences made in projects like IDCNet and D4AllNet and using the experiences and contacts of networks like AAATE and EDeAN, the University of Linz started to develop two university courses closely related to D4All. Both courses are based on the idea and the concepts of D4All and use them as a starting point for teaching in more detail in the two domains of accessible web design[36] and assistive technologies application.[37] These courses answer an increasing demand for education and certification due to changing understanding and attitudes toward people with disabilities and according to legal and economical changes.

Both areas, but in particular in related research and development activities, have established a considerable body of knowledge over recent years. The level of complexity of the technological knowledge and the complex situations of application ask for appropriate teaching and training possibilities, as it cannot be expected that practice is able to adapt to this changing situation by itself.

Accordingly, it is often stated that the application of the newest AT, ICT, and also the practice of implementing accessible web systems is lacking. By putting these two courses in place an appropriate answer to the challenging technical developments and opportunities should be given that the increasing potential of ICT and AT can be used in practice for the benefit of people with disabilities.

Both courses are special courses in the sense that they are not part of mainstream education. They are related to mainstream education as web accessibility and AT are mainstream by nature and in addition certain parts of courses necessary for special education can be integrated into standard courses (e.g., computer science).

The development of both courses has been supported by the European Social Fund (ESF),[38] the Federal Ministry for Education, Science and Culture (bm:bwk),[39] and the University of Linz.[40]

58.12.1 BFWD—University Course "Barrierefreies Web Design" (Accessible Web Design)

Awareness of web accessibility in the mainstream ICT sector and in society in general increases. This is driven by legal changes and a better understanding of the importance of web accessibility in ethical, socioeconomic, and business terms, both in the public sector and in industry. In particular, the idea and the concept of barrierefreies web design (BFWD) has been a proactive answer to expected legal changes in Austria (e-Government-Law)[41] and the previously discussed[42] anti-discrimination legislation. The development started before these laws had been enacted, and offered an early answer to the changing situation.

The course is eLearning-based and designed in an accessible way for students and teachers. This course is designed for postgraduate students and practitioners working in the field of web design. These practitioners or their employers responded very positively to the course layout, as they are invited to bring their real-world projects into the course and get guidance for the design of new or the redesign of existing web systems during the course.

People with disabilities themselves have been involved in the development of the eLearning system, as well as the development of the content (blind and visually impaired, mobility impaired, hearing impaired, and deaf). Intensive testing of eLearning platforms and related accessibility adaptations led to the implementation of a fully accessible system with a good level of usability for all user groups (Miesenberger and Ortner, 2006).

The most critical issue was obtaining the approval of the course by the university board and the involved bodies (politics, industry, administration), as the course leads to an official and statewide accepted university degree. The course was accepted in spring 2005 after almost 2 years of negotiation.

The curriculum encompasses a total of 44 semester hours embedded into a time frame of 4 semesters. It is divided into the following six modules:

- Technical Fundamentals
- Assistive Technologies
- Guidelines and Laws
- Accessibility
- Design and Usability
- Practical Experience

---

[35] EDeAN Education: Design for All Education and Training: http://www.education.edean.org.

[36] Bfwd: Barrierefreies Web Design: www.bfwd.at.

[37] Assistec: Asstistierende Technologien: www.assistec.at.

[38] ESF: Europäischer Sozialfond: http://www.esf.at.

[39] bm:bwk: Federal Ministry for Education, Science and Culture: http://www.bmbwk.gv.at.

[40] University of Linz: http://www.jku.at.

[41] E-Government-Law: http://www.parlament.gv.at/pls/portal/docs/page/PG/DE/XXII/BNR/BNR_00149/fname_014980.pdf.

[42] Behindertengleichstellungsgesetz: Regierungsvorlage (government bill): http://www.parlament.gv.at/portal/page?_pageid=908,848700&_dad=portal&_schema=PORTAL.

Each of the modules contains appropriate lectures. Mandatory attendance hours are reduced to a minimum during the whole course, as the main parts of the content are taught using an online eLearning system. This also allows for including participants who are already in a job to decide when and where to learn.

The preparation of the first parts of the course content was carried out over summer 2005, and the course started in October 2005 with 21 students from Austria and Germany. Most of the students work in web design companies that are confronted with the need to develop accessible web pages for their clients. Other target groups represented are unemployed people, people re-entering the labor market, and people with disabilities searching for new vocational challenges in this field with personal user experience in the background.

After successfully starting the course in Austria, first considerations and negotiations have been started to offer the course also in other countries and languages. This of course requires localizing the content. Such cooperations should guarantee the offer in the long run and an efficient and cost-effective organization and updating.

The curriculum is designed following a modular approach, which allows for single lectures to be modified and also used for other teaching purposes. Some lectures could, for example, be easily integrated into the regular lectures for computer science students, or in the training courses of companies and organizations.

The course is based on the open source eLearning platform called Moodle.[43] As this platform is developed by a big community, there is a constant progress in functionalities, as well as toward accessibility. This asks for ongoing updates, as well as for adaptations regarding features for accessibility and usability. Discussions with the community about integrating adaptations into their development, and therefore also into new releases, would not only be very useful for the purposes of the course, but also for other organizations providing eLearning courses via Moodle.

As soon as version 2.0 of the Web Content Accessibility Guidelines (W3C/WAI/WCAG 2.0) becomes an official recommendation of WAI, the current content of the course will have to be adapted according to the new specifications. The system itself has also to fulfill this new standard.

58.12.2 Assistec: University Course "Assistierende Technologien" (Assistive Technology)

AT and ICT have been recognized as playgrounds for specialists over the last decades. Although an increasing number of excellent tools and services have been developed, the end-users and the fields of support, services, and care for people with disabilities has not been using the potential of these new possibilities.

In addition, the last years led to increasing awareness of aging and disability concerns. The demographic situation shows an increasing need for new and efficient approaches to be able to meet the level of quality and quantity of services needed to answer this trend.

AT and ICT are increasingly recognized as major factors to overcome this situation. More research, development, and application projects underline this. But still a gap can be found between the technological progress and the social, pedagogical, and psychological field of application.

Of course the fast and revolutionary developments in ICT/AT often driven by technological needs caused problems in usability and applicability. ICT/AT research and development reacted to this and orients itself toward usability and applicability.

One particular reason that the gap still seems to become bigger lies in the fact that neither the aspects of AT/ICT have sufficiently entered established curricula of the support and care field, nor have the comprehensive courses or diplomas been established for the AT/ICT sector. In Austria, for example, none of the curricula for jobs in the social services and care sectors takes AT and ICT as a subject into account, and no course in AT/ICT exists.

This invites increasing efforts in transferring know-how into the social services and care sectors. At the University of Linz, a distance education course in AT has been developed and was approved by the university authorities following the model of the BFWD course. The organizational and administrative procedures, as well as the existing technical infrastructure of the BFWD course, as described above, have been used to implement this course.

The university course aims at educating people coming from different vocational backgrounds (e.g., welfare, social politics, health care, education, AT research/development/application). In particular people with disabilities themselves are invited to participate as they have unique experiences and knowledge in the field, and are confronted with increasing problems at the labor market (Matausch et al., 2006).

The curriculum offers in a time frame of four semesters four modules encompassing 18 seminars with a total of 42 semester hours:

- Module: Fundamentals
  - eLearning
  - Medical and Physiological Fundamentals
  - Legislative Framework and Funding
  - Design for All
  - Fundamentals in AT and Rehabilitation Technology
- Module 2: Assistive Technologies—Special Knowledge
  - Aids for Specific Learning Difficulties
  - Aids for Hearing Impairment
  - Aids for Vision Impairment
  - Mobility and Movement Aids
  - AAL
  - AT for Special Environments (work, home, education, leisure)
  - Practice
- Module 3: Process of Assortment and Provision of AT/ICT
  - Assessment

---

[43] Moodle: http://www.moodle.org.

- Analyzing the Environment: Technical, Social, and Economical Factors
- AT Management and Mediation
- AT in Practice: Examples and Application
  - Training on the Job
  - Thesis
  - Research AT and Future Trends

Graduates will be experts in the area of AT, especially concerning assortment of appropriate AT, usability of AT, funding, application, adaptation, management, and service and counseling. Moreover, the course stresses concentrated and goal-oriented transfer of knowledge according to the up-to-date state of the art in a multidisciplinary environment. One major key feature and goal is that the course substantially emphasizes practical training and application of the theoretically gained knowledge. Moreover, one major intention is to enhance quality in the practical treatment regarding the profession fields of health care and support and services of people with disabilities. In addition, the implementation of the university course aims to improve and foster product development of AT.

58.13 Conclusions

Design for all is recognized as (1) special design for people with disabilities or the aging population or as (2) an integral part of the design process itself. As the second is preferred as the most appropriate in design for all, its qualities are often absorbed by final products or services and no longer seen as design for all qualities. Design for all often does not obtain the reward for the influence it has in practice. The qualities of design for all are very much dependent on specific contexts. This leads to a complex situation where examples are not recognized as design for all or are not recognized as special design for users with special needs, what is seen as stigmatizing for general design.

Due to this, design for all is often criticized for contradicting modern design. We have outlined that this is more due to a design workflow that does not start from the needs of a diversity of users but has to work on predefined requirements that tend to forget functional requirements that are key to design for all.

This chapter presented design for all as a quality of the general design process. Design for all should reorient design toward functional and aesthetical needs for all users. The examples of good practice discussed should invite identifying related examples for one's own practice. The discussion also showed that there is a need for more empirical and comparative studies to define indicators and examples for design for all.

References

Build for All (2008). *Build for All Reference Manual.* http://www.build-for-all.net/en/home.

CEN/CENELEC (2008). *Guide 6: Guidelines for Standards Developers to Address the Needs of Older Persons and Persons with Disabilities.* http://www.cenelec.org.

The Center for Universal Design (1997). *The Principles of Universal Design (Version 2.0).* Raleigh, NC: North Carolina State University.

The Center for Universal Design (2001). *Cultural Facilities, Universal Design* [CD-ROM]. Raleigh, NC: North Carolina State University.

Clarkson, P. J., Coleman, R., and Keates, S. (2003). *Inclusive Design: Design for the Whole Population.* London: Springer-Verlag.

Council of the European Union (2003). *COUNCIL RESOLUTION on "eAccessibility": Improving the Access of People with Disabilities to the Knowledge Based Society.* http://register.consilium.eu.int/pdf/en/03/st05/st05165en03.pdf.

Darzentas, J. and Miesenberger, K. (2005). Design for all in Information Technology: A universal concern (keynote), in *Database and Expert Systems Applications, Proceedings of the 16th International Conference (DEXA 2005)* (K. V. Andersen, J. Debenham, and R. Wagner, eds.), 22–26 August 2005, Copenhagen, Denmark, pp. 406–420. Berlin/Heidelberg: Springer-Verlag.

Dong, H., Keates, S., and Clarkson, P. J. (2004). Inclusive design in industry: Barriers, drivers and the business case, in *User-Centred Interaction Paradigms for Universal Access in the Information Society, Proceedings of the 8th ERCIM International Workshop* (C. Stary and C. Stephanidis, eds.), 28–29 June 2004, Vienna, Austria, pp. 305–319. Berlin/Heidelberg: Springer-Verlag.

Giuliani, R. (2001). *Universal Design, The City of New York.* New York: Mayor's Office for People with Disabilities.

Grammenos, D., Savidis, A., Georgalis, Y., and Stephanidis, C. (2006). Access invaders: Developing a universally accessible action game, in *Computers Helping People with Special Needs, Proceedings of the 10th International Conference (ICCHP 2006)* (K. Miesenberger, J. Klaus, W. Zagler, and A. Karshmer, eds.), 10–11 July 2006, Linz, Austria, pp. 388–395. Berlin/Heidelberg: Springer-Verlag.

Grammenos, D., Savidis, A., and Stephanidis, C. (2005). UA-Chess: A universally accessible board game, in *Universal Access in HCI: Exploring New Interaction Environments, Volume 7 of the Proceedings of the 11th International Conference on Human-Computer Interaction (HCI International 2005)* (C. Stephanidis, ed.), Las Vegas, 22–27 July 2005 [CD-ROM]. Mahwah, NJ: Lawrence Erlbaum Associates.

Heider, T., Stegmann, M., and Zey, R. (2004). *Das Designlexikon* [The Design Lexicon] [CD-ROM]. Berlin: Directmedia Publishing.

Holzinger, A. (2000). *Basiswissen Multimedia* (3 Bände) [Basic knowledge in multimedia (3 volumes)]. Würzburg: Vogel, Band 2: Mensch (Human).

ISO 13407 (1999). *Human-Centred Design Processes for Interactive Systems.* ISO. http://www.iso.org.

ISO 20282-1 (2006). *Ease of Operation of Everyday Products—Part 1: Design Requirements for Context of Use and User Characteristics.* http://www.iso.org.

ISO 9241 (2006). *Ergonomics of Human-System Interaction.* http://www.iso.org.

ISO 9241-110 (2006). *Ergonomics of Human-System Interaction—Part 10: Dialogue Principles.* http://www.iso.org.

ISO 9355-3 (2006). *Ergonomic Requirements for the Design of Displays and Control Actuators—Part 3: Control Actuators.* http://www.iso.org.

ISO IEC: JTC1/SC36 (2008). *International Standards and Guidance in Information Technology for Learning, Education, and Training.* http://jtc1sc36.org/home.

ISO/IEC: JTC 1/SC 25/WG 1 (2008). *Information Technology: Home Electronic System.* http://hes-standards.org.

ISO/TR 9527 (2008). *Building Construction: Needs of Disabled People in Buildings—Design Guidelines.* http://www.iso.org.

ISO/TS 16071 (2003). *Ergonomics of Human-System Interaction: Guidance on Accessibility for Human-Computer Interfaces.* http://www.iso.org.

ISO/TS 20282-2:2006 (2006). *Ease of Operation of Everyday Products—Part 2: Test Method for Walk-Up-and-Use Products.* http://www.iso.org.

Keates, S. (2004). Developing BS700 Part 6—Guide to managing inclusive design, in *User-Centred Interaction Paradigms for Universal Access in the Information Society, Proceedings of the 8th ERCIM International Workshop* (C. Stary and C. Stephanidis, eds.), 28–29 June 2004, Vienna, Austria, pp. 332–339. Berlin/Heidelberg: Springer-Verlag.

Keates, S. and Clarkson, J. (eds.) (2003). *Countering Design Exclusion: An Introduction to Inclusive Design.* Berlin/Heidelberg: Springer-Verlag.

Levine, D. (ed.) (2003). *Universal Design New York 2.* Buffalo, NY: Center for Inclusive Design & Environmental Access, SUNY–Buffalo.

Lidwell, W., Holden, K., and Butler, J. (2004). *Design. Die 100 Prinzipien für erfolgreiche Gestaltung* [Design: 100 principles for successful designing]. Munich: Stiebner.

Matausch, K., Hengstberger, B., and Miesenberger, K. (2006). "Assistec": A university course on assistive technologies, in the *Proceedings of the 10th International Conference in Computers Helping People with Special Needs (ICCHP 2006)* (K. Miesenberger, A. Karshmer, J. Klaus, and W. Zagler, eds.), 11–13 July 2006, Linz, Austria, pp. 361–368. Berlin/Heidelberg: Springer-Verlag.

MediaLT (2008). *Guidelines for the Development of Entertaining Software for People with Multiple Learning Disabilities.* http://www.medialt.no/rapport/entertainment_guidelines/index.htm.

Miesenberger, K. (2004). "Equality = e-quality" 'design for all' und 'accessibility' als Grundlage für eine demokratische, offene und inklusive Gesellschaft ["Equality = e-quality" 'design for all' and 'accessibility' forming the basis of an open and inclusive society] in *Qual-I-tät und Integration, Beiträge zum 8* [Quality and integration, contributions to the 8th forum of practitioners] (E. Feyerer and W. Pammer, eds.). Linz: PraktikerInnenforum, Universitätsverlag Rudolf Trauner.

Miesenberger, K. and Ortner, D. (2006). Raising the expertise of web designers through training: The experience of B.F.W.D.—Accessible Web Design (Barrierefreies Web Design) in Austria, in *Computers Helping People with Special Needs, Proceedings of the 10th International Conference (ICCHP 2006)* (K. Miesenberger, J. Klaus, W. Zagler, and A. Karshmer, eds.), 11–13 July 2006, Linz, Austria, pp. 253–257. Berlin/Heidelberg: Springer-Verlag.

Miesenberger, K. and Pühretmair, F. (2005). Help.gv.at—Accessible e-Government in Austria, in *Assistive Technology: From Virtuality to Reality, Proceedings of AAATE 2005* (A. Pruski and H. Knops, eds.) 6–9 September 2005, Lille, France. Amsterdam: IOS Press.

Miesenberger, K., Pühretmair, F., and Nussbaum, G. (2005). E-government Protal www.help.gv.at barrierefrei: Anforderungen–Herausforderungen–Erfahrungen, in *E-Government 2005: Knowledge Transfer and Status* (M. Wimmer, ed.). Wien: OCG.

Müller-Prove, M. (2002). *Vision and Reality of Hypertext and Graphical User Interfaces.* PhD dissertation, University of Hamburg, Hamburg, Germany.

Myers, B. A. (1998). A brief history of human computer interaction technology. *ACM Interactions* 5: 44–54.

Nielsen, J. (1994). *Usability Engineering.* San Francisco: Morgan Kaufmann Publishers.

Norman, D. A. (2002). *The Design of Everyday Things.* New York: Basic Books.

Ossmann, R., Archambault, D., and Miesenberger, K. (2006). Computer game accessibility: From specific games to accessible games, in the *Proceedings of CGAMES'06, the 9th International Conference on Computer Games,* 22–24 November 2006, Dublin, Ireland. Wolverhampton, U.K.: University of Wolverhampton.

Ossmann, R. and Miesenberger, K. (2006). Guidelines for the development of accessible computer games, in *Computers Helping People with Special Needs, Proceedings of the 10th International Conference (ICCHP 2006)* (K. Miesenberger, J. Klaus, W. Zagler, and A. Karshmer, eds.), 11–13 July 2006, Linz, Austria, pp. 403–406. Berlin/Heidelberg: Springer-Verlag.

Pieper, M., Anderweit, R., Schulte, B., Peter, U., Croll, J., and Cornelssen, I. (2004). Methodological approaches to identify honorable best practice in barrier-free web design: Examples from Germany's 1st BIENE Award competition, in *User-Centred Interaction Paradigms for Universal Access in the Information Society, Proceedings of the 8th ERCIM International Workshop* (C. Stary and C. Stephanidis, eds.), 22–27 July 2004, Las Vegas, pp. 360–372. Berlin/Heidelberg: Springer-Verlag.

Preiser, W. and Ostroff, E. (eds.). (2001). *Universal Design Handbook.* New York: McGraw-Hill Professional.

Reimer, J. (2005). *A History of the GUI.* http://arstechnica.com/articles/paedia/gui.ars.

Rheingold, R. (2000). *Tools for Thought.* http://www.rheingold.com/texts/tft.

Richter, M. and Flückiger, M. (2007). *Usability Engeneering kompakt: Benutzbare Software gezielt entwickeln* [Usability engineering consolidated: Targeted design of usable software]. Munich: Elsevier.

Schneider, B. (2005). *Design: eine Einführung* [Design: An introduction]. Berlin: Birkhäuser Verlag.

Section 508. *Section 508 Standards – 1194.21 Software Applications and Operating Systems*. IT Accessibility & Workforce Division, U.S. General Services Administration, Washington, DC. http://www.section508.gov/index.cfm?FuseAction=content&ID=12#Software.

Spillers, F. (2003). *Web Usability Best Practice Handbook*. http://www.experiencedynamics.com.

Steffen, D. (2000). *Design als Produktsprache: Der »Offenbacher Ansatz« in Theorie und Praxis* [Design, the language of product: The "offenbach approach"]. Frankfurt/Main: Verlag.

Stephanidis, C. (ed.) (2001). *Universal Access in HCI: Towards an Information Society for All*. Mahwah, NJ: Lawrence Erlbaum Associates.

Stephanidis, C. (ed.) (2005). *Universal Access in Health Telematics: A Design Code of Practice*. Berlin/Heidelberg: Springer-Verlag.

Thimbleby, H. (2007). User-centred methods are insufficient for safety critical systems, in *HCI and Usability for Medicine and Health Care* (A. Holzinger, ed.). Berlin/Heidelberg: Springer-Verlag.

United Nations (1993). *The Standard Rules on the Equalization of Opportunities for Persons with Disabilities*. http://www.un.org/esa/socdev/enable/dissre00.htm.

W3C/WAI/ATAG (2008). *World Wide Web Consortium: Authoring Tool Accessibility Guidelines*. http://www.w3.org/WAI/intro/ATAG.php.

W3C/WAI/UAAG (2008). *World Wide Web Consortium: User Agent Accessibility Guidelines*. http://www.w3.org/WAI/intro/UAAG.php.

W3C/WAI/WCAG (2008). *World Wide Web Consortium: Web Content Accessibility Guidelines*. http://www.w3.org/WAI/intro/wcag.php.

W3C/WAI (2008). *World Wide Web Consortium: Web Accessibility Initiative*. http://www.w3.org/WAI.

W3C/WCAG 1.0 (2008). *The World Wide Web Consortium (W3C), Web Accessibility Initiative (WAI): Web Content Accessibility Guidelines 1.0*. http://www.w3.org/WAI/intro/wcag10docs.php.

W3C/WCAG 2.0 (2008). *The World Wide Web Consortium (W3C), Web Accessibility Initiative (WAI): Web Content Accessibility Guidelines 2.0*, draft. http://www.w3.org/WAI/intro/wcag20.php.

Wab Cluster (2006). *Unified Web Evaluation Methodology (UWEM)*. http://www.wabcluster.org/uwem05.

Walker, J. A. (2005). *Designgeschichte: Perspektiven einer wissenschaftlichen Disziplin* [History of design: Perspectives of a scientific discipline]. Munich: Springer-Verlag.

Watchfire (2005). *Bobby: Web Accessibility Testing Tool*. http://bobby.watchfire.com/bobby/html/en/index.jsp.

Welch, P. and Palames, C. (1995). A brief history of disability rights legislation in the United States, in *Strategies for Teaching Universal Design* (P. Welch, ed.). Boston: Adaptive Environments Center.

Zagler, W. (2004). *Rehabilitationstechnik (Habilitationsschrift), Teil A: Die Grundlagen* [Rehabilitation technology (habilitation), part A: fundamentals], p. A209. Vienna, Austria: Technische Universität Wien.

IX

Looking to the Future

59

Implicit Interaction

Alois Ferscha

59.1 The Environment Is the Interface

Computer science nowadays appears to be challenged (and driven) by technological progress and quantitative growth. Among the technological progress challenges are advances in submicron and system-on-a-chip designs, novel communication technologies, microelectromechanical systems, and nano and materials sciences. The vast pervasion of global networks over the past years, the growing availability of wireless communication technologies in the wide, local, and personal area, and the evolving ubiquitous use of mobile and embedded information and communication technologies are examples of challenges posed by quantitative growth. A shift is currently perceived from the "one person with one computer" paradigm, which is based on explicit human-computer interaction, toward a ubiquitous and pervasive computing landscape, in which implicit interaction and cooperation is the primary mode of computer-supported activity. This change—popularly referred to as *pervasive computing*—poses serious challenges to the conceptual architectures of computing, and the related engineering disciplines in computer science.

Historically, pervasive computing has its roots in ideas first coined by the term *ubiquitous computing*. "The most profound technologies are those that disappear. They weave themselves into the fabric of everyday life until they are indistinguishable from it" was Mark Weiser's central statement in his seminal paper in *Scientific American* in 1991 (Weiser 1991). The conjecture that "we are trying to conceive a new way of thinking about computers in the world, one that takes into account the natural human environment and allows the computers themselves to vanish into the background" has fertilized the embedding of ubiquitous computing technology into a physical environment that responds to people's needs and actions. Most of the services delivered through such a technology-rich environment are adapted to the context, particularly to the person, the time, and the place of use. Along Weiser's vision, it is expected that context-aware services will evolve, enabled by wirelessly ad-hoc networked, mobile, autonomous special-purpose computing devices (i.e., information appliances), providing largely invisible support for tasks performed by users. It is further expected that services with *explicit user input* will be replaced by a computing landscape sensing the physical world via a huge variety of sensors, and controlling it via a manifold of actuators in such a way that it becomes merged with the virtual world. This interaction principle is referred to in this chapter as *implicit interaction*, since input to such a system does not necessarily need to be given explicitly or attentively. Applications and services will have to be greatly based on the notion of context and knowledge, will have to cope with highly dynamic environments and changing resources, and will thus need to evolve toward a more *implicit* and *proactive interaction* with users.

A second historical vision impacting the evolution of pervasive computing claimed an intuitive, unobtrusive, and distraction-free interaction. In an attempt to bring interaction "back to the real world" after an era of keyboard and screen interaction, computers have started to be understood as secondary artifacts, embedded and operating in the background, whereas the set of all physical objects present in the environment have started to be understood as the primary artifacts (i.e., the interface). Instead of interacting via keyboard and screen, physical interaction with digital data (i.e., interaction by manipulating physical artifacts via graspable or tangible interfaces) was proposed. Inspired by the early approaches of coupling abstract data entities with everyday physical objects and surfaces, like Bishop's Marble Answering

Machine, Jeremijenko's Live Wire, and Wellner's Digital Desk, tangible interface research has evolved, where physical artifacts are considered as both representations and controls for digital information. A physical object thus represents information, while at the same time acting as a control for directly manipulating that information or underlying associations. With this seamless integration of representation and control into a physical artifact, input and output devices fall together. In this view, artifacts can exploit physical affordances suggesting and guiding user actions, while not compromising existing artifact use and habits of the user. Recent examples of embodied interaction, where input and output are fused into physical object manipulation, include architecture and landscape design and analysis, and object shape modeling interfaces using bricklike blocks or triangular tiles.

Although the first attempts toward realizing the ubiquitous and pervasive computing vision in the early 1990s fell short due to the lack of enabling hard- and software technologies, now, about 10 years later, new approaches are viable due to technological progress and quantitative growth. Pervasive computing initiatives and projects have emerged at major universities worldwide, and national and international research funding authorities (IST Future and Emerging Technologies program of the EU, DARPA, NSF, etc.) have accelerated the efforts of a rapidly growing, vibrant research community. Although originally suffering from a plethora of unspecific terms like *calm computing, hidden or invisible computing, ambient intelligence, sentient computing, post-PC computing, universal computing, autonomous computing, everyday computing,* and so on, the research field is now consolidating from its foundations in distributed systems and embedded systems, and is starting to codify its scientific concerns in technical journals, conferences, workshops, and textbooks. This process, however, is by far not settled today.

59.2 What Is Implicit Interaction?

To make computing part of everyday life, the interfacing with new computing devices should go beyond traditional explicit interaction. The vision of future information appliances drawn in Schmidt (1999) motivates the need for modalities of implicit interaction: "we will be able to create (mobile) devices that can see, hear and feel. Based on their perception, these devices will be able to act and react according to the situational context in which they are used." Accordingly, the following definition has been proposed (Schmidt, 1999):

> Implicit human computer interaction is an action performed by the user that is not primarily aimed to interact with a computerized system but which such a system understands as input.

Implicit interaction is based on two main concepts: perception and interpretation. Perception is information gathering about the environment and situations, usually involving (technological) sensors. This information is generally provided implicitly to the system and displayed naturally to the user. Interpretation

is the mechanism to understand the sensed data. Conceptually, perception and interpretation in combination are described as situational context.

59.2.1 Interacting with Landscapes of Digital Artifacts

As an increasing number of technology-enriched physical objects with embedded computing capabilities emerge, such as vehicles, computers, mobile phones, and portable music players (referred to here as *digital artifacts*, or just *artifacts*), the issue of their interaction becomes a dominant issue of human-computer interaction (HCI) research. Technology integrated into everyday objects like tools and appliances, and environments like offices, homes, or cars, turns these artifacts into entities subject to human-artifact interaction whenever humans use those appliances or become active in those environments. Moreover, built with networked embedded systems technology, they become increasingly interconnected, diverse, and heterogeneous, subject to artifact-artifact interaction, thus raising the challenge of an operative and semantically meaningful interplay among each other.

One approach to address this challenge is to design and implement systems able to manage themselves in a more or less autonomous way. While self-management stands for the ability of a single digital artifact to describe itself, to select and use adequate sensors to capture information describing its context, self-organizing stands for the ability of a group of possibly heterogeneous peers to establish a spontaneous network based on interest, purpose, or goal, and to negotiate and fulfill a group goal. A way of implementing digital artifacts is based on miniaturized stick-on embedded computing systems, integrating sensor, actuator, and wireless communication facilities. Such stick-on solutions can then be attached or built into everyday objects, and executing software stacks that implement self-organization in a totally distributed style. Interaction at the application level is invoked based on the analysis of self-describing profile data exchanged among nearby artifacts. Self-management builds a basis for the self-organization of artifact ensembles, their ability to establish a spontaneous network based on individual interest, purpose, or goal, and to negotiate and fulfill a group goal through cooperation. Research on self-managing and -organizing systems has attracted much interest in the computer science community (Kephart and Chess, 2003; Serugendo et al., 2003; Herrmann, Muhl, and Geihs, 2005; Jelasity et al., 2006; Mamei et al., 2006).

59.2.2 Context Awareness

Context awareness refers to the ability of the system to recognize and localize objects, as well as people and their intentions. The context of an application is understood as "any information that can be used to characterize the situation of an entity," an entity being "a person, place or object that is considered relevant to the interaction between a user and an application, including the user and applications themselves." A key architecture design principle for context-aware applications is to decouple mechanisms for collecting or sensing context information and its interpretation, from the provision and exploitation of this information to build and run context-aware applications. Sensing context must happen in an application-independent way, and context representation must be generic for all possible applications.

Technology trends in the domains of sensors and actuators, processing devices, embedded systems, and wireless communication have fertilized the evolution of software architectures for building context-aware systems (see Figure 59.1) (Ferscha et al., 2006). Usually, the adoption of a world model representing a set of objects and their state in the physical (or "real") world is suggested, with mechanisms to sense, track, manipulate, and trigger the real-world objects from within the world model. People living in the real world, acting, perceiving, and interacting with objects in their environment are represented in the world model by virtual objects or proxies. Proxies of persons, things, and places are linked to each other in the virtual world, such that this linkage is highly correlated with the linkage of physical persons, things, and places in the real world. A context-aware application then monitors the state and activity of the real-world objects via a set of sensors, coordinates the proxies according to the rules embodied in the application, and notifies, triggers, or modifies the physical world objects via a set of actuators. The extent to which context information can act as the trigger for actions that are either a direct consequence of intentional or implicit user input will be further discussed later in this chapter.

Understanding implicit interaction as a result of processing context within digital artifacts autonomously, and triggering appropriate actions accordingly, encourage the study of digital artifacts with respect to two important properties: autonomy and context awareness. As artifacts are by nature distributed throughout physical space, an obvious consideration for implicit interaction are their inherent spatial properties—in particular their position, direction, and shape—as well as spatial relationships among them. Awareness about spatial relationships among distributed entities is considered valuable for self-organizing systems, and most of the known phenomena of self-organization and -adaptation in nature are actually phenomena of self-organization in space (Mamei and Zambonelli, 2004).

The ability to describe itself is an important aspect of an autonomous digital artifact, which allows for expressing all kinds of context information. The use of a self-description is twofold: it provides local applications with an awareness of the artifact's context on the one hand, and it serves as a basis for achieving awareness about other artifacts by exchanging the self-descriptions on the other hand. The concept of self-describing artifacts is therefore considered a promising approach for implementing implicit interaction among autonomous systems, particularly with regard to an open-world assumption (i.e., interacting artifacts do not know each other in advance) where an ad hoc exchange of self-descriptions upon encountering other artifacts is required to obtain awareness of their context. Therefore, the approach is proposed of autonomous digital artifacts able to exchange self-descriptions upon becoming aware of the existence of other artifacts in a direct peer-to-peer manner (Ferscha et al., 2006). Further interaction can then be parameterized and contextualized considering the provided context information of the interaction counterpart. In this context, focus is specifically

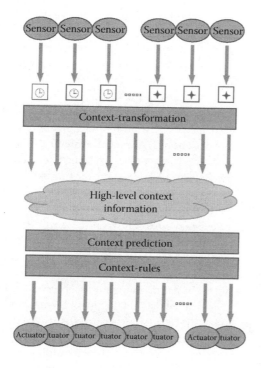

Context sensing
Time-triggered or event-triggered acquisition of low-level sensor data

Context transformation
Aggregation/interpretation toward high-level context information

Context representation
Data structures representing abstract context information

Context forecasting

Context rule base
Implicit or explicit triggering of control events

Actuators
Control the environment

FIGURE 59.1 Framework architecture for context-aware systems.

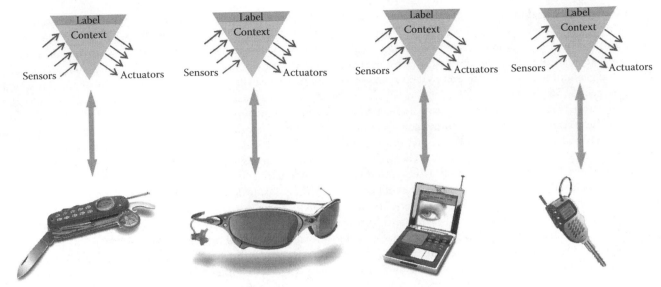

FIGURE 59.2 Context-aware autonomous digital artifacts.

on the spatial context of artifacts, as well as its quantitative and qualitative representation.

Explaining *implicit interaction* as an operational principle of self-describing, self-managed, self-organizing digital artifacts, this chapter first recalls essential enabling technologies, including location sensing, a foundation for spatiotemporal reasoning, and sensing of velocity, orientation, and general context. It then looks into categories of applications exploiting implicit interaction in (1) person-to-person interaction, (2) person-to-artifact interaction, and (3) artifact-to-artifact interaction. Above that, we structure the categories of implicit interaction along the type and amount of information that is used to steer implicit interaction, ranging from (1) pure *presence*, that is, "just being there," to (2) *identity*, or being there as a person or as an identified object, to (3) *proximity*, that is, being in a certain area of spatial proximity of someone or something, to (4) *profile*, which is describing preferences, abilities, goals, and so on, of someone or something, and finally (5) *context*, that is, any information collected via sensors that describes the situation of a person or an artifact. Potential fields of application are highlighted, and an outlook into open challenging research issues is offered.

59.3 Localization and Space

Starting from the idea of a digital artifact as an autonomic piece of technology, integrating context sensing and processing abilities, the provision of information related to an artifact's location (in physical space), orientation, velocity (if mobile), identification, or even profile is essential to implicitly invoke and steer interactions. Particularly the notions of space, and the methods for localizing objects in space, are of outstanding importance for almost any kind of investigation in implicit interaction. These are therefore summarized in some detail in the subsequent subsections.

59.3.1 Localization

Localization technologies and techniques aim at the location of persons, objects, artifacts, and so on, in the environment. Localization systems can be categorized as coarse- or fine-grained. Coarse-grained systems are designed for large spatial areas and are operating at coarse accuracy. Aside from GPS, systems include almost all radio-based systems like RFID, 802.11, Bluetooth, and cellular. Fine-grained systems were specifically designed for spatially constrained location sensing. Popular systems in this category are Active Bats, Cricket, Ubisense, and computer vision-based systems.

Methodologically, angulation, proximity, and scene analysis are three location-sensing techniques (Hightower and Borriello, 2001). Triangulation uses geometric properties of triangles, such as distances between vertices or angles between the vertices (angulation). Lateration computes the position of an object from its distance from multiple reference points. The distance calculation is done by (1) direct measurement of distance, (2) measuring distance from an object to some reference point using Time-of-Flight (ToF; i.e., measuring the time it takes to travel between the object and point P at a known velocity), or (3) estimating the distance from an object to some point P by measuring the strength of a signal emission when it reaches P (Time-of-Arrival, ToA, of a signal). The scene analysis location-sensing technique uses features of a scene observed from a particular viewpoint to draw conclusions about the location of the observer or of objects in the scene. In static scene analysis, observed features are looked up in a predefined data set that maps them to object locations. In contrast, differential scene analysis tracks the difference between successive scenes to estimate location. A proximity location-sensing technique entails determining when an object is "near" a known location. The object's presence is sensed using a physical phenomenon with limited range, like radio, noise, or light.

Figure 59.3 summarizes contemporary localization according to the following criteria: *physical* (e.g., GPS) versus *symbolic* (e.g., office, kitchen) and *absolute* versus *relative* delivery of location information, possibility of *local* computation of the location, and level of *accuracy* and *precision* a system can achieve.

Localization, after enjoying dominance by specialized location systems in the past, experiences a paradigm shift, after a popularity burst and extensive deployment of many RF systems, which are now being utilized for location sensing as well. There is, however, a trade-off between specialized and nonspecialized location systems. Since WiFi, mobile phones, TV, Bluetooth, and infrared-based location systems were not designed for location sensing, they lack the accuracy needed for many location-based applications. Location estimation with an accuracy of 5 to 100 meters is not sufficient, especially in indoor applications. Along with these systems, a specialized location system, GPS, has actually been deployed at a global level. Its accuracy of 5 to 10 meters is sufficient for

Technology	Technique	Accuracy & precision	Physical	Symbolic	Absolute	Relative	Local Comp.	Scale	Limitations	Recognition
GPS	Radio time-of-flight lateration	1-5 m (95-99%)	✓		✓		✓	24 satellites worldwide	Not indoors	
Active badges	Diffused infrared cellular proximity	Room size		✓	✓			1 base per room, badge per base per 10 second	Sunlight & fluorescent interference with infrared	✓
Active bats	Ultrasound time of flight lateration	9 cm (95%)	✓		✓			I base per 10 m square, 25 computations per room per second	Require ceiling sensor grid	✓
MotionStar	Scene analysis lateration	1 mm, 1 ms, 1.0 degree (nearly 100%)	✓		✓			Controller pre scene, 108 sensors per scene	Control unit tether, precise installation	✓
VHF omni directional ranging (VOR)	Angulation	I degree radial (nearly 100%)	✓		✓		✓	Several transmitters per metropolitan area	30-140 nautical miles line of sight	
Cricket	Proximity lateration	4*4 ft. regions, (nearly 100%)		✓	✓	✓	✓	1 beacon per 16 sq feet	No central management, receiver computations	
MSR RADAR	802.11 RF scene analysis & triangulation	3-4.3 m (50%)	✓		✓			3 bases per floor	Wireless NICs required	✓
PinPoint 3D-iD	RF lateration	1-3 m	✓		✓			Several bases per building	Proprietary, 802.11 interference	✓
Avalanche Transceivers	Radio signal strength proximity	Variable, 60-80 m range	✓			✓		I transceiver per person	Short radio range, unwanted signal attenuation	
Easy Living	Vision triangulation	Variable		✓	✓			3 cameras per single room	Ubiquitous public cameras	✓
Smart Floor	Physical contact proximity	Spacing of pressure sensors (100%)	✓		✓			Complete sensor grid per floor	Recognition may not scale to large population	✓
Wireless Andrews	802.11 cellular proximity	802.11 cell size (100 m indoors, 1 km free space)		✓	✓			Many bases per campus	Wireless NICs required, RF cell geometries	✓
E911	Triangulation	150-300 m (95%)	✓		✓			Density of cellular infrastructure	Only where cell coverage exists	✓
SpotON	Ad hoc lateration	Depends on cluster size	✓			✓		Cluster at least 2 tags	Attenuation less accurate than ToF	✓

FIGURE 59.3   Comparison of localization systems.

most applications outdoors. But for technological reasons it cannot be used indoors or in high-building sections in a populated city. Specialized location-sensing systems like UWB, ultrasound, and vision based, work with good accuracy, especially indoors.

In this sketched context, a system that seems to resolve the trade-off between deployment and accuracy, is RFID. There is optimism that RFID systems will be ubiquitous in the near future, replacing identification tags like barcodes. Since an RFID tag would be attached to every object of interest, it can be concluded that an environment would exist in which all objects are labeled. Another advantage that RFID systems have over many other systems is that in most cases their tags are passive and do not require power recharging. But on the other side, RFID systems require very specific settings for scanning the objects by the readers. Especially due to low power, the operating range of these systems is very small.

Most coarse-grained localization systems were developed for reasons other than localization, but are nowadays also used for localization (IrDA, WiFi, GSM, RFID). Fine-grained systems were specifically designed for location sensing. The most important of these systems are ultrasound-based systems, such as Cricket and Active Bats. In addition to ultrasound systems, systems based on computer vision are also popular. Among multi-robot localization and LuxTrace, EasyLiving is the most important system in this category. It uses two sets of color stereo cameras for tracking multiple people during live demonstrations in a living room. The stereo images are used for locating people, and the color images are used for maintaining their identities. The system performs well enough to make the room feel responsive, and it tracks multiple people standing, walking, sitting, occluding, and entering and leaving the space. Ubisense is specifically designed for location sensing and is based on UWB radio signals. UWB signals have very short pulse intervals to allow accurate ToA and AoA measurement, with an accuracy of about 15 cm. Another advantage of UWB is that the direct Line of Sight (LoS) between tags and readers is not required. On the other side, these waves are affected by environmental conditions. Electromagnetic systems face interference from metal. Other important fine-grained localization systems are Smart Floor and ubiTrack. See Figure 59.4 for a systematic comparison of related indoor and outdoor localizaion technologies.

59.3.2 Qualitative Space

Aside from the triggering of implicit interactions based on criteria of metric space (quantitative distance, absolute orientation), the relative spatial relations among objects in real space can be exploited (collocated, left, front, near, far, etc.). In other words, there are two ways in which spatiotemporal relations between objects can be represented and reasoned upon: quantitative and qualitative. Qualitative spatiotemporal reasoning (QSR) is reasoning about nonmetric properties

of space and time. QSR has been an area of research for quite some time (Gottfried et al., 2006), a prominent approach being Allen's (Allen, 1983) temporal logic based on time intervals (rather than points in time), and the region connected calculus (RCC) (Randell et al., 1992), defining topological relations between two regions of "space."

The RCC (see Figure 59.5) defines eight pragmatic relations between two spacially extended objects, X and Y. The left most relation represents two (spatial) regions, or objects being disconnected, followed by externally connected regions where two objects are just touching each other; two regions partially overlapping followed by regions that are exactly identical. The last four relations are those in which one region contains another region in its entirety. These relations allow the spatial context of, e.g., rooms in a building, to be characterized. Examples of such relations are:

{EC(living room, hallway), PO(passage, living room),
PO(passage, hallway),
TPP(window, living room), TPP(shelf, living room),
TPP(sofa, living room),
TPP(light-cone, living room), PO(light-cone, passage),
TPP(sofa, light-cone),
NTPP(chair1, light-cone), NTPP(chair2, light-cone),
NTPP(table, light-cone)}

Beyond topology, qualitative spatial reasoning is reasoning about space at a conceptual level. In natural language, for example, prepositions are used to express spatial relations between objects in space (Lehman and Bennardo, 2003). In the English language, for example, the sentence "The book is *on* the table" means that the book is positioned at the upper surface of the table, touching it from above. On the other hand, "near" in the sentence "We sat *near* the front" means close to, in close proximity to the front. Finally, the preposition *through* in "I went *through* the door" means from one side of an opening to the other, and so on. An abstract view of such propositions is systematically summarized in Figure 59.6.

Relationships expressed by prepositions in the English language are specified with respect to the following terminology (Figure 59.7). An area within effective range of a point is called its neighborhood. An object can have three kinds of relationships with its neighbors, *contact*, *in*, and *vicinity*. Two objects (primary and secondary) are in *contact* with each other if at least a bounding interior point of a primary object is at the limit of closeness (zero distance) from a bounding interior point of a secondary object. Two objects have an *in* relationship if at least some nonbounding interior points (with one of a limit) of one of the two objects must be in contact with some nonbounding ones (again with one as a limit) of the other object. For a 3D second object, its interior points will, by definition of three-dimensionality, include all points of space properly enclosed by that object. Any object not *in* or in *contact* with another object, but still in the neighborhood of the first object, is in the *vicinity* of that object.

Technology	Technique	Accuracy & Precision	Scale	Limitations	Form Factor	Battrey Life	Deploy-ments	Remarks
GPS	Radio, ToF, Lateration	1-5 m outdoors, 10 m indoors, 95-99%	World wide	Not suitable indoors, cost is still high	6.7*8.7 mm	75 mW, have to recharge	High	
A-GPS	GPS supported by cellular network	5-10 m with 50-90%^	World wide	Cost could be high, still unreliable for critical applications	13.6*17.5* 2.75 mm	125 mW, have to recharge	High	Work with the help of cellular network to respond in cases when GPS is not properly working
u-blox	GPS with high quality DSP	2 m	World wide	Cost is very high	100 mm square	50 mW, have to recharge		
GPS with IMU	GPS with inertial sensing	1 m	World wide	Cost is very high, bulky	23*23*23 mm			Accuracy is very good
GPS-Less (outdoors)	Radio, proximity	Avg: 1.87 m, <4.50 m, 90%	Limited	Needs beacons deployment		Have to recharge		Specific research project
INDOOR GPS	IR laser, ToA, AoA, triangula tion	1 m	Room size	Needs spcific infrastructure				Specific infrastructure of sensors, transmitters and a centralized hub is required
Ekahau (Wi-Fi)	Radio, fingerprin ting, RSSI	1 m	Availa ble where WiFi	Needs an offline calibration and learning mechanism	45*55*19 mm	Up to 5 years	High	Wirless NIC is required
RADAR (Wi-Fi)	Radio frequency, scene analysis	1-3 m	Availa ble where WiFi	Needs an offline calibration and learning mechanism, infrastructure installation			High	
Active badge	Diffused IR, cellular, proximity	Room size	Room	Sunlight & fluorescent interference with IR	Badge size		Limited	
Active bats	Ultrasound, ToF, lateration	3 cm (95%)	Indoors	Require ceiling sensor grid	7.5*3.5*1.5 cm	18 months	Limited	

FIGURE 59.4 Coarse-grain and fine-grain localization systems.

The directional and geometric relationship of one object (or group of objects) with another object (or group of objects) is called *orientation*. Orientation can be parameterized in four categories:

1. *Relative direction*: If two objects are approximately facing each other, they can be said to have *opposite* direction, relative to each other. The exact opposite is true for *same* direction. Two objects may have opposite direction to each other, still they may be *facing* or facing in opposite directions. The same is true for objects having the same direction. If an object has neither the same nor the opposite direction with reference to another, it can be assumed to be *at normal* to the other object.

2. *Gravity*: The gravity to which an object undergoes can be an indicator of a relation like *above* or *below*. It is an indicator of an object's height with respect to another object. If an object is neither above nor below the other object, it would be at the *same height*.

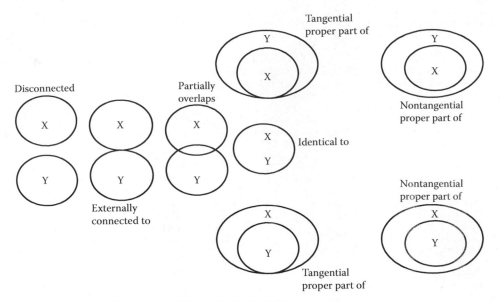

FIGURE 59.5 Spatial relations among "region" X and "region" Y in the RCC8 calculus.

3. *Distance*: Objects can be *near, far,* or at *intermediate* (if not indicated otherwise) distance from each other.
4. *Placement*: This is an abstract view of relative position. An object may *lag* in position from another object or may *lead* in position from another object (in the same line of action), when seen in this particular context.

Motion means that an object (over time) is changing its position in space (going from one place to another) in a specific direction for each successive pair of places (this direction can be the same for successive pairs, or different for specific groups of places). This change happens over specific ordered instances of time that are unique events. Motion can be *bound*, that is, assigned a boundary by indicating a *beginning* and an *end* to it. To do so, two places where the objects are not moving have to be considered as the end points of the motion. *Path* is a geometric (purely spatial) description of motion abstracted from motion itself. Motion takes place along a path.

Using these concepts, spatial relations can be described, distinguishing static and dynamic ones:

- *Placement*: Each object can be considered as a point in space. The first and foremost characteristic of an object with reference to another object is presence. Presence tells an object about another object near it. For a digital artifact, range defines the scope of presence, for example, the wireless communication range of the object. Presence can be equated both with present and not present, both equally important. The notion of nonpresence of an entity can be as useful as presence. When not considered for precise relations like in, contact, and vicinity, neighborhood can be equated to presence.
- *Proximity*: Proximity is knowledge of an object about its neighbors. It seems very similar to the presence property,

but it is more descriptive. Proximity of an object gives a combined view about its neighborhood. It is a specialized form of presence. An entity present would always be considered for proximity. Still, proximity does not provide any information, for example, about the formation of groups or crowds (Figure 59.6, part A).

- *Direction*: Direction is expressed based on an assumed frame of reference (FoR). In an *intrinsic* FoR, the location of an object is expressed in relation to a part of another object (binary), whereas in an *extrinsic* FoR, the location of an object is expressed in relation to arbitrary fixed bearings (like north, east, south, west), and in a *relative* FoR, location is expressed as a ternary relation involving the referred object, as well as the viewpoint of the perceiver and the position of another object.
- *Gravity*: The measure of gravitational force on two objects and their relative placement results in relations like *above* or *below*. Weight sensors and location of objects can help validate these relations.
- *Distance*: The distances among objects can be expressed quantitatively (in metric space, like, cm, m, km), as well as qualitatively (like *near, intermediate,* or *far*).
- *Stature*: An object can be tilted, inverted or sideways when seen with reference to its own normal position. Stature is thus expressed related to the geometry of the object itself. Acceleration sensors can be used to detect any sudden change in position of an object. A generalizing term of the relations *direction, gravity, distance,* and *stature* and their interplay could be *orientation* as in Figure 59.6, part B.
- *Shape, face, and size (geometry)*: The geometry of an object is a description of its appearance (in mathematical terms).

FIGURE 59.6 Prepositional (spatial) relations.

Part C

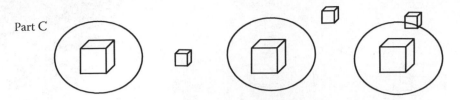

A moving to B
Need to assign a destination without specifying the exact direction; location-time pairs history can be used

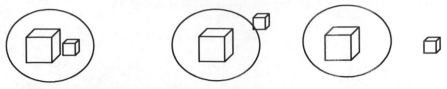

A moving from B
Need to assign a source without specifying the exact direction; location-time pairs history can be used

A moving via B
Need to assign an intermediate object only (direction) without specifying the source or destination, location-time pairs history can be used

Part D

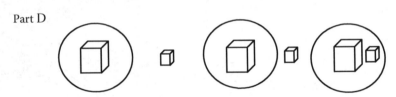

A moving towards B (specialized case of to)
Need to assign a destination specifying the exact direction; location-time pairs history can be used

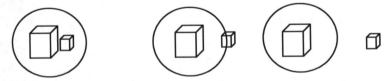

A moving away from B (specialized case of from)
Need to assign a source specifying the exact direction; location-time pairs history can be used

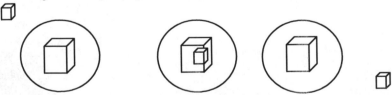

A moving through B (specialized case of via)
Need to assign an intermediate object only, without specifying the source destination; also, vicinity is not enough—object has to touch the intermediate object; location-time pairs history can be used

FIGURE 59.6   (*Continued*)

Part E

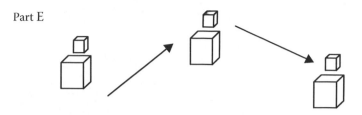

A moving by B (dynamic version of deside) - both objects moving
Need to assign a common path; object can be in vicinity or in contact with other object; location-time pairs
history can be used

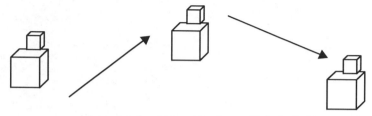

A moving along B (specialized case of by) - both objects moving
Need to assign a common path; object has to be in contact with other object; location-time pairs
history can be used

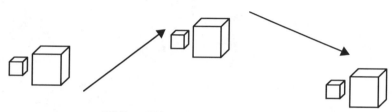

A follows B (dynamic version of behind) - both objects moving
Need to assign direction of object and placement relative to it; need to know direction of objects and
coordinates; location-time pairs history can be used

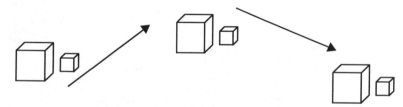

A follows B (dynamic version of front) - both objects moving
Need to assign direction of object and placement relative to it; need to know direction of objects and
coordinates; location-time pairs history can be used

FIGURE 59.6 *(Continued)*

- *Overlapping*: This is a classical problem of relating the range of influence between two objects. Identifying overlapping regions (or objects) helps in defining possible relations like inside, in, on, above, below, beside, and beneath.
- *Motion*: Movement is a sense of spatial activity related to an object. It should not be confused with more concrete movement patterns like velocity, acceleration, and rotation. Anything that is displaced from its previous position can be considered as having moved.

Looking at Figure 59.7, motion-based relations can be divided into four categories:

1. Simple motion relations like to, from, and via. As shown, there are no restrictions on these relations as far as

neighborhood is concerned. It can be of any type: in, contact, or vicinity. To describe a motion's path, only end is needed for to, only beginning is needed for from, and only body (route) is needed for via.

2. Implicit motion relations like by, along, and through. As shown there are restrictions on these relations as far as neighborhood is concerned. By can be defined when both objects have neighborhood relation of contact, or vicinity. Along can be defined when both objects have neighborhood relation of contact only. Through can be defined when both objects have neighborhood relation of in, or contact.

3. Directional motion relations like towards, and away from. As shown there are no restrictions on these

Preposition	Neighborhood			Orientation									Motion			
				Relative direction			Gravity		Distance		Placement		Path			
	In	Contact	Vicinity	Same	Opposite	Facing	Above	Below	Near	Far	Lag	Lead	Beginning	Body	End	Dir
On		X														
In	X															
Outside			X													
Inside																
At	X															
Across			X		X	X										
Against			X		X											
Over			X				X									
Below			X					X								
Near			X						X							
Far			X							X						
Behind			X	X							X					
Front			X	X								X				
Beside			X													
To	X														X	
From	X												X			
Via	X													X		
By		X	X										X			
Along		X											X			
Through	X	X											X			
Towards	X														X	X
Away from	X												X			X
Follow	X										X				X	X
Lead	X											X	X	X		X

FIGURE 59.7 Relationship among prepositions and neighborhood, orientation, and motion.

relations as far as neighborhood is concerned. On the other hand, for motion's path, end of it and dir is needed in towards, and beginning of it and dir is needed for away from.

4. Tracking motion relations like follow and lead. As shown there are no restrictions on these relations as far as neighborhood is concerned. On the other hand, for motion's path, end of it and dir are needed in follow, and beginning of it and dir are needed for lead. Both relations, in addition, also need placement descriptions as indicated.

59.4 Categories of Implicit Interaction

Depending on the type and number of involved entities and the kind of information used to trigger and steer interactions, a variety of different categories of implicit interaction appear possible.

- *Person-to-person* refers to the modes of interaction where technology acts on behalf of people, once a certain relation (e.g., spatial relation like "*close to each other*") has been verified. Example applications are friend finder systems, or lost and found systems.

- *Person-to-artifact* interaction provides functional autonomy to the artifact, but still the control remains with the user. In this mode of interaction, a physical device or a virtual object like a software agent acts as an entity in the system. The beneficiary of the system is still the user, but interacting with a digital object instead of a person on the other side. Example applications are variants of smart spaces like intelligent classrooms, interactive kitchens, or smart cabinets and smart refrigerators.
- *Artifact-to-artifact* is a very undeveloped, yet promising category of interaction mode. Here, autonomous digital objects interact with each other to achieve certain goal. The user does not necessarily need to be involved in the interactions among artifacts; this can happen in the background.

Furthermore, applications can be further categorized based on implicit interaction with respect to the kind and amount of information they use to trigger and steer interactions. The following categories can be identified according to their increasing complexity:

1. *Presence*: Systems exploiting information about the pure presence or absence of persons or things.
2. *Identity*: Systems exploiting identity information about the involved interaction entities, like the ID of a person, the serial number of a product, the MAC address of a network node, and so on.
3. *Proximity*: Systems that exploit information about the mutual spatial relationships among entities involved in the interaction, like the distance, the direction, or the topology.
4. *Profile*: Systems that exploit a predefined self-characterization (self-description) of the entities involved in the interaction, like preferences, taste, habit, goals, needs, capabilities, or even emotions.
5. *Context*: Systems that consider any available information that describes the situation of entities involved in the interaction.

The first consequence of interaction between things is the ability of things to record the history of their interaction. Applications based on things lacking this capability can be called *nonmemorizing*, in contrast to *memorizing* applications. Probing further, a direct consequence of logging the interaction and contextual information may lead to more advanced *memorizing and predicting* applications.

59.4.1 Presence

The presence of a thing perceived by another thing represents the simplest form of implicit input to a system. A thing in the *interaction range* of another thing indicates the presence of a sensed object in the vicinity of the sensing object. The sensed object, while being explored by a sensing object, may register the presence of the sensing object in its vicinity, though the exploration activity was not initiated by it. Further, it is important to note that the not-present status can be as useful as the present status in many applications.

Familiar technologies used for detecting presence are acoustic, electromagnetic, photoelectric, capacitive, pressure, and temperature sensing. All of these techniques are application oriented and specialized. For example, for acoustic sensing, the object to be sensed should have the capability to generate sound waves, whereas the sensing object should be able to listen to those similar to photoelectric sensing requiring light reflection and capture capability. Electromagnetic field sensing is primarily for detecting the presence of metal objects. Pressure and temperature sensing is basically connected with detecting human presence in an environment.

In addition to those specialized systems, wireless communication technologies based on radio provide presence sensing as a default functionality. Initial hand shaking and periodic beaconing in established protocols of cellular networks, radar systems, Bluetooth, wireless LANs, infrared and ultrasound communication, keep peer devices updated about their presence. Mobile ad hoc and sensor networks also implement communication protocols that enable peer devices to know about many parameters, presence being the default.

The simplest application of presence in use today is the *automatic door*, emphasizing ease of access. Similarly, in smart homes, *automatic lighting* is becoming popular, which operates

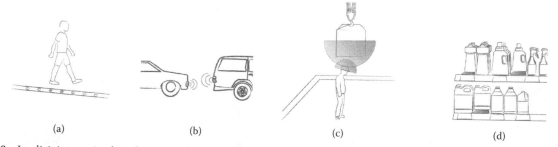

FIGURE 59.8 Implicit interaction based on presence recognition. (a) Automatic lighting as the user walks along a path: Presence sensors detect the user and lights along the way (turn on and off). (b) Automatic vehicle detection to help in parking: Both vehicles generate alarms as they feel presence nearby. (c) Workplace safety system: A worker's helmet recognizes being under a heavy load. (d) A smart shelf: Indicating in-stock and out-of-stock status of items on the shelf.

on the basis of instant presence (motion) detection. In the context of energy (power) saving, an application like automatic lighting, the sensing of the nonpresence of a person or an object is crucial to be able to switch lights off in case of absence. Similarly, *home and industrial alarm systems* can gain from advanced sensing technologies. In industrial environments or logistics, an object knowing about its safety requirements can generate alerts upon the detection of the presence of hazard. The simplest example can be a safety system for fragile goods: in a warehouse, this package can generate an alarm upon attempts to store another package on top of it.

All of these application scenarios reflect instantaneous action upon the detection of presence; i.e., objects as such are not memorizing. With memorization, one can imagine many application categories applicable in industry, marketplaces, and busy areas like airports. Memorizing time-stamped events of detected presence, and thus declaring the epochs of absence, can help to implement more sophisticated interactions, like readiness for the anticipated occurrence of recurring events as learned from the occurrence history, or the prediction of future events as a learning effect from patterns in the event history (future awareness). For memorization alone, the timestamp-based presence history enables applications broadly categorized as applications concentrating on *process speed*. In an automatic production environment, a sensor knowing the process speed coupling the presence with timestamps can alter the speed of, for example, a conveyor belt or assembly line to get the optimum performance (if speed is too slow) or to avoid a process deadlock (if speed is too fast). Extending the memorization capabilities to a whole group of individuals or objects, by, for example, counting and storing the number of objects detected to be simultaneously present, enables applications broadly categorized as *queuing* and *inventory control*. Sensing the increase or decrease of the population of objects in a queue waiting for service can help in taking alternative actions. It can be helpful in busy areas like airports, shopping malls, or during celebrations and special events. In manufacturing and logistics, knowing the length of queues can initiate alternative actions to increase performance and productivity. For inventory control, an object knowing that only a few things are left in the shelf can raise a low inventory alarm.

It is easy to extend any of the presented scenarios into the category of memorization and prediction. A prediction logic (based on the collected object history) embedded into an object can help implement proactive behavior. A *smart room* can start its heating system when it senses the presence of the user. It would be more convenient to predict on the future situation (user approaching), and to act in advance to accomplish the desired state in a timely manner.

59.4.2 Identity

This category is defined by the fact that not the mere presence, but the recognized identity of a person, object, or artifact defines the input to the system. For implicit interaction, the quest would only be limited to things identifying other things. Applications like access control, product inventories, or road toll systems are traditional representatives of this category of applications.

Identification of an entity by another entity can be done in three ways. First, an entity can identify another entity by verifying the identity an entity may have. Examples are identification of keys in a smart card, bar codes, visual markers, and RFID identities. Second, an entity can identify another entity by verifying the identity an entity may possess. Examples are verification of biometric identities like fingerprint detection, iris scan, human radiation detection, and voice recognition (see also Chapter 13 of this handbook). Last, an entity can identify another entity by verifying the identity an entity may know. Examples are password or security code matching. The second category is designed for human identification. The third category implies an explicit interaction, thus denying emphasis on implicit interaction. The technologies enabling the automatic exchange and authentication of identification keys are experimented with in many technological environments, RFID being the most prominent.

As discussed previously, the most common example of identity-based application is *access control*. For example, it is desirable that only employees of an office are allowed to enter an office building. It can easily be checked by a reader opening the door if the company identity of the employee matches with the identity printed on his company card. If enhanced with memorization, the system can perform *activity monitoring* by registering

         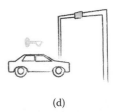

(a)   (b)   (c)   (d)

FIGURE 59.9 Implicit interaction based on identity sensing. (a) Access control: Only an authorized user is allowed to enter a passage. (b) Time management in sports: In track and field events, athletes are informed about their timing based on their identity. (c) Personalized controls: Car offers personalized settings based on the identity of the current driver. (d) Automatic toll payment: Car does not need to stop at the toll station. The identity of the vehicle is sensed and the toll is automatically debited from the driver's account.

the presence of a particular user in or out of the office. Activity monitoring is not restricted to human beings. In automatic production, logistics, and sales, the life cycle of the product can be monitored. *Disaster prediction* in a controlled environment can be achieved by both logging the activities of personnel and events. Similarly, identifying a thing at a place that can create a hazard can immediately be remedied.

In a manufacturing environment, automatic *component assembly* can be archived by applying identification mechanism. A subcomponent related to its compatible subcomponent can be ensured before assembling. Similarly, in logistics, *container* management could be supported, that is, loading and unloading based on the identity of containers, but also based on their collective self-organization while traveling to a certain destination. Extending the example with memorization capabilities leads to the category of *production line* applications. As an example, an RFID-enabled production could operate according to the following steps: (1) carts containing items move along the production line for processing, where each item is associated with a tag and each carrier has writable tag. An RFID reader is mounted at each station and writes the information about the performed steps onto the tags of the carts. (2) Readers at the end of the production line check for completeness of the production steps performed on the items. If complete, items are scanned and packed into packages. (3) Packages are packed into containers in a warehouse. (4) Containers are loaded onto a truck and transported. As illustrated with this trivial example, *identity management* solutions are among the most promising approaches of implicit interaction-based process automation.

59.4.3 Spatial Proximity

In this category of applications, facts like distance, orientation, and spatial relations are automatically recognized and used to control the system. Reasoning about space in the vicinity of another thing may be based on one of the two concepts of space: quantitative space and qualitative space. As was said, *quantitative* spatial reasoning is related with measurement of exact space metrics. *Qualitative* spatial reasoning is based on nonmetric abstractions of space related to topology, distance, and orientation.

Geographic information systems (GIS) based on global positioning technologies (e.g., GPS) are the main enabler for spatial proximity sensing, particularly concerning distance and orientation. On a smaller scale, a mobile ad hoc network, a sensor network, or an RFID system may be used to estimate distance and orientation between two objects based on relative signal strength, particularity in the qualitative domain. Coupling this qualitative measure with inference relative to a node having global coordinates can result in quantitative measures.

The categories of applications that build upon spatial proximity of objects among each other range from industrial production (e.g., congestion control in production lines), transport, logistics, storing, and so on. *Indication and placement* can be coined as a term representing a group of applications suitable for shelves in sales environments. By detecting proximity, an item in a shelf can indicate a gap on the right, left, top, or bottom of itself, taking a shelf containing books as an example. Knowing about objects in proximity has many applications in automotive systems, where future vehicles will behave more like intelligent agents traveling in intelligent spaces. Car-to-car interaction can guide the driver (or driver assistance system) to proceed or not, depending on traffic rules and speed. Car-to-roadside interaction can assist finding parking spaces, and so on. Large public displays on walls or mirrors are becoming popular in public places. Many of these displays use proximity sensing to select and control the content being displayed to the user. In many projects, the interaction is situation-based on three zones of interactions (see Figure 59.10). The closest zone is the space from where a user can touch the display to interact with the system (A). The second zone, outer to interaction zone, is the notification zone (B), where a user can be notified based on the information of the wall so that the user can start thinking about interaction and may decide to proceed further to interact. The last and outermost zone is the ambient zone (C), which indicates presence of a user near the wall. Different sensing technologies can help to scan these zones appropriately. Thinking about proximity sensing with the ability to memorize

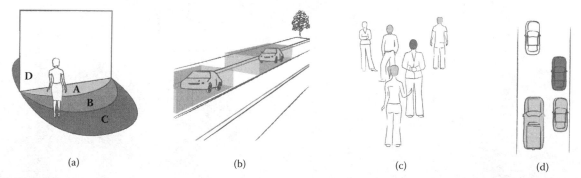

(a)   (b)   (c)   (d)

FIGURE 59.10 Implicit interaction based on spatial proximity sensing. (a) A public wall display: Based on the proximity (nearness) of the user, the display changes appearance and contents. (b) Drive-by-wire: A car adopts the right cruising speed with respect to the preceding car, or generates an alarm as it senses another car approaching too closely. (c) Spontaneous interaction in social settings: A user knows about other users near her. (d) Driver assistance system: Car knows about its proximity and can support decision for driving.

and predict, opens application possibilities like, for example, unmanned driving systems, within which autonomous vehicles can safely perform any activity like overtaking, turning, speeding up, slowing down, and stopping, and so on.

59.4.4 Profile

Exchanging the self-descriptions of objects encoded in metadata format (e.g., XML) to induce adequate (implicit) interaction is yet another increase in the quality of spontaneous interactions. The metadata related to an object may cover material properties, object structure, and general physical characteristics, whereas metadata referring to a person could indicate physiognomic properties of that person, personal preferences, likings or habits, intents, capabilities, goals, or even emotions. The exchange of profiles, automatically triggered among colocated objects, and the analysis of correspondences within them, enable interactions best adapted to the individual counterparts. In person-to-person interaction, for example, exchanged business profiles could help to check the level of "compatibility" between two persons. In person-to-artifact interaction, display systems can be imagined that present content that matches the interest of the person. In artifact-to-artifact interaction, an application may assist in alerting about a situation in which a container with certain contents must not go with a container with incompatible contents (see Figure 59.11).

In nonmemorizing implementations of profile-based implicit interactions, the most common and popular examples come in the form of *intelligent (or smart) appliances*. Examples range from web luggage, implementing containment recognition and inventory services accessible via the Internet, to magic wardrobes, knowing about clothes placed inside and suggesting combinations that fit together, to smart shelves, in which goods are monitored to enhance replenishment and identify misplaced products, to smart medicine cabinets, monitoring the medication taken by the patient while reminding, alerting, and displaying information about prescriptions, incompatibilities, or recalled medicines, and so on. In the memorizing category, an autonomic product solution appears viable, with technological support for every product to manage its own life cycle from

production to recycling autonomously. Implicit interaction principles deployed in manufacturing and production, transport and logistics, sales, customer support, and even separation and recycling, as well as at the points of handover from one life cycle phase to another, can make product life cycles self-managed, swift, and trustworthy. In *wildlife monitoring*, the tagging of animals with sensor beacons collecting environmental data to keep track of their position and movement are already common today. *Physical activity monitoring* is another potential application in this category. Wearable sensors can be imagined that sense the physical activity of a user (gym workout), and exercise machines communicating with the objects (sensors) worn by the user to provide guidelines for an optimal workout.

59.4.5 Context

Steering interactions based on any available information describing the situation within which the interaction occurs is the most versatile approach of implicit interaction. Basically, every circumstance, as identified from sensor data analysis, can be used to trigger or constrain interactions (see Figure 59.12). As already mentioned, a context-aware system can be designed to be reactive or proactive. A simple adaptive heating system was discussed in a previous section. Including context in the same example, the heating system can refine its function by knowing the number of persons present in the room, or it can start the ventilation function after sensing smoke (the system does not have to generate a smoke alarm in this case). This scenario is an example of a *reactive adaptive environment*. A *proactive adaptive environment* predicts and makes a decision well before time. A variety of assisted living examples implement services with features of being ready for the user just in time (proactive door openers, scene lighting, warm-up of coffee makers, etc.).

Finally, there is a coordination aspect of implicit interaction as soon as collections, groups, or ensembles of entities are concerned. This issue can be addressed through an abstract concept called *things that go together*. The criterion for things to belong to a group is to fulfill a certain relation that holds among all group members. There can be many types of relations between things like similarity, likeliness, compatibility, or suitability. In some

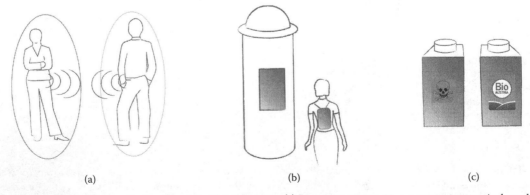

(a)   (b)   (c)

FIGURE 59.11 Implicit interaction based on exchanged self-descriptions. (a) Person-to-person: Two-persons unattentively exchanging profiles. (b) Person-to-thing: A person and a thing exchanging profiles spontaneously. (c) Thing-to-thing: Two colocated things exchanging profiles.

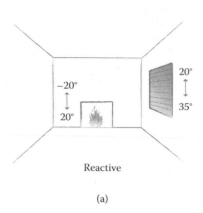

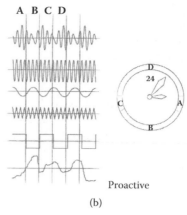

FIGURE 59.12 Implicit interaction based on information describing the situation of an entity (person, thing, artifact). (a) Reactive: As the temperature outdoors changes, the heating system indoors reacts and adjusts the thermostat. (b) Proactive: The system identifies recurring patterns of user activities from sensor data time series. Learning from this experience, the system is able to estimate the user preferences and to act accordingly.

sense, entities may be considered to belong to a group if they share the same context. Implicit interaction at the level of this group can then be used to implement coordinated, collective behavior.

59.5 Outlook

Technological advances, miniaturization of embedded computing technology and wireless communication, together with the evolution of global networks (like the Internet) has brought the vision of pervasive and ubiquitous computing to life: technology seamlessly woven into the fabric of everyday life. Along with this development go the need and challenge of interfaces supporting an intuitive, unobtrusive, and distraction-free interaction with such technology-rich environments. Considering the huge amount and ever-growing number of a vast manifold of heterogeneous, small, embedded, or mobile devices shaping this pervasive computing landscape makes traditional styles of interaction (explicit, e.g., via keyboard) appear hopeless.

Implicit interaction is based on the principle that a system does not have to be explicitly forced to act (by the user), but can autonomously trigger actions by collecting and interpreting information about its environment or situation. Input is not necessarily explicitly stated or intentionally given, but the system understands the information it collects as input. The active, driving role in the interaction is thus moved from the user to the system. Consequently, the user does not have to be attentive and responsive to a plethora of devices, but the devices as single entities or as ensembles of cooperative entities develop a sense for the user, and act accordingly. Implicit interaction, besides person-to-person interaction, is a promising approach also for person-to-artifact interaction, but most of all artifact-to-artifact interaction. Particularly for the latter, it can be seen as an operational principle of self-describing, self-managed, self-organizing digital artifacts.

This chapter has looked into categories of implicit interactions that make use of information at different levels of complexity. At the lowest level are systems that act based on the detection of the pure *presence* (or *absence*) of an entity, like a person or a thing. Systems that cannot only sense presence, but also recognize the *identity* of entities, like the ID of a person, the serial number of a tool, or the license plate number of a vehicle, can implement more specific implicit interactions based on that information. Space and relations of entities in (physical) space are very crucial concepts for implicit interaction. The *proximity* of an entity, that is, the recognized being of other entities in a certain area of spatial proximity of an entity, is a yet more powerful concept to steer interactions. *Profiles* describing attributes of objects, or preferences, abilities, or goals of users allows for the design of interactions that are particularly tailored for individual entities. Finally, the tailoring of interactions to particular situations is made possible when considering *context*, that is, any information collected via (technological) sensors that describes the situation of interacting entities.

Acknowledgments

The author wishes to thank Dr. Martin Scheurer and Dr. Birgit Otto (both from Siemens AG, Corporate Technology, User Interface Design) for inspiring me to elaborate on implicit interaction and for discussion and suggestions on how to structure the content. The valuable support of Kashif Zhia for developing a major part of the localization and space technologies in Section 59.3 is gratefully acknowledged. Thanks to Bernadette Emsenhuber for providing scribbles and drawings. My most sincere thanks go to Constantine Stephanidis, who—despite my (self-made) overload—has never lost confidence in me and this chapter. Costas—I thank you with my deepest respect for being on my side in good and in hard times!

Further Reading

Aalto, L., Göthlin, N., Korhonen, J., and Ojala, T. (2004). Bluetooth and WAP push based location-aware mobile advertising system, in the *Proceedings of the International Conference on Mobile Systems, Applications and Services*, 6–9 June 2004, Boston, pp. 49–58. New York: ACM Press.

Abdelmoty, A. I., El-Geresy, B. (1994). An intersection-based formalism for representing orientation relations in a geographic database, in the *2nd ACM Workshop on Advances in Geographic Information Systems, Workshop at CIKM 1995*, 1–2 December 1994, Baltimore, pp. 44–51. New York: ACM Press.

Addlesee, M., Curwen, R., Hodges, S., Newman, J., Steggles, P., Ward, A., and Hopper, A. (2001). Implementing a sentient computing system. *IEEE Computer Magazine* 34: 50–56.

Antifakos, S., Schiele, B., and Holmquist, L. E. (2003). Grouping mechanisms for smart objects based on implicit interaction and context proximity, in *UBICOMP 2003 Interactive Posters*, pp. 207–208. UbiComp 2003 Adjunct Proceedings. http://ubicomp.org/ubicomp2003/adjunct_proceedings.

Augusto, J. C. and Nugent, C. D. (eds.) (2006). *Designing Smart Homes: The Role of Artificial Intelligence*. Berlin/Heidelberg: Springer-Verlag.

Bahl, P. and Padmanabhan, V. N. (2000). RADAR: An in-building RF-based user location and tracking system, in the *Proceedings of the 9th Annual Joint Conference of the IEEE Computer and Communications Societies (INFOCOM 2000)*, Volume 2, 26–30 March 2000, Tel Aviv, Israel, pp. 775–784. Washington, DC: IEEE.

Baus, J. and Kray, C. (2002). Frames of reference, positional information and navigational assistance, in the *Proceedings of FLAIRS Conference 2002*, 14–16 May 2002, Pensacola Beach, FL, pp. 461–465. Menlo Park, CA: AAAI Press.

Borriello, G. (ed.) (2005). RFID: Tagging the world (special issue). *Communications of the ACM* 48: 34–37.

Boutilier, C., Das, R., Kephart, J. O., Tesauro, G., and Walsh, W. E. (2003). Cooperative negotiation in autonomic systems using incremental utility elicitation, in the *Proceedings of the 19th Conference on Uncertainty in Artificial Intelligence, UAI 2003*, 7–10 August 2003, Acapulco, Mexico, pp. 89–97. San Francisco: Morgan Kaufmann.

Bulusu, N., Heidemann, J., and Estrin, D. (2000). GPS-less low cost outdoor localization for very small devices. *IEEE Personal Communications Magazine* 7: 28–34.

Čapkun, S., Hamdi, M., and Hubaux, J.-P. (2002). GPS free positioning in ad hoc networks. *Cluster Computing* 5: 157–167.

Chang, A., Resner, B., Koerner, B., Wang, X., and Ishii, H. (2001). Lumitouch: An emotional communication device, in *CHI '01 Extended Abstracts on Human Factors in Computing Systems*, pp. 313–314. New York: ACM Press.

Chatzigiannakis, I., Nikoletseas, S., and Spirakis, P. (2005). Efficient and robust protocols for local detection and prorogation in smart dust networks. *Mobile Networks and Applications* 10: 133–149.

Chen, M., Haehnel, D., Hightower, J., Sohn, T., LaMarca, A., Smith, I., et al. (2006). Practical metropolitan-scale positioning for GSM phones, in the *Proceedings of the Eighth International Conference on Ubiquitous Computing (Ubicomp 2006)*, 17–21 September 2006, Orange County, CA, pp. 225–242. Berlin/Heidelberg: Springer-Verlag.

Clementini, E., Felice, P. D., and Hernández, D. (1997). Qualitative representation of positional information. *Artificial Intelligence* 95: 317–356.

Cohn, A. G., Bennett, B., Gooday, J., and Gotts, N. M. (1997). Representing and reasoning with qualitative spatial relations about regions, in *Spatial and Temporal Reasoning* (O. Stock, ed.), pp. 97–134. New York: Kluwer Academic Publishers.

Cohn, A. G. and Hazarika, S. M. (2001). Qualitative spatial representation and reasoning: An overview. *Fundamenta Informaticae* 46: 1–29.

Cox, S., Daisey, P., Lake, R., Portele, C., and Whiteside, A. (2007). *OpenGIS Geography Markup Language (GML) Encoding Standard v3.2.1*. http://www.opengeospatial.org/standards/gml.

DARPA (2007). *DARPA Urban Challenge*. http://www.darpa.mil/grandchallenge.

de Moraes, L.F.M. and Nunes, B. A. A. (2006). Calibration-free WLAN location system based on dynamic mapping of signal strength, in the *Proceedings of the 4th International Workshop on Mobility Management and Wireless Access*, 2 October 2006, Terromolinos, Spain, pp. 92–99. New York: ACM Press.

Dietz, P., Yerazunis, W. S., and Leigh, D. L. (2003). Very low-cost sensing and communication using bidirectional LEDs, in the *Proceedings of the International Conference on Ubiquitous Computing (UbiComp 2003)*, 12–15 October 2003, Seattle, pp. 175–191. Berlin/Heidelberg: Springer-Verlag.

Dodge, C. (1997). The bed: A medium for intimate communication, in the *Extended Abstracts of CHI'97*, pp. 371–372. New York: ACM Press.

Doherty, L., Pister, K. S. J., and El Ghaoui, L. (2001). Convex position estimation in wireless sensor networks, in the *Proceedings of the Twentieth Annual Joint Conference of the IEEE Computer and Communications Societies (INFOCOM 2001)*, Volume 3, 22–26 April 2001, Anchorage, AK, pp. 1655–1663. Washington, DC: IEEE Press.

Dourish, P. and Bellotti, V. (1992). Awareness and coordination in shared work spaces, in the *Proceedings of the Conference on Computer Supported Cooperative Work (CSCW 1992)*, 31 October–4 November 1992, Toronto, pp. 107–114. New York: ACM.

Egenhofer, M. J. (1989). A formal definition of binary topological relationships, in the *Proceedings of the 3rd International Conference on Foundations of Data Organization and Algorithms (FODO 1989)*, 21–23 June 1989, Paris, pp. 457–472. Berlin/Heidelberg: Springer-Verlag.

Ferscha, A., Hechinger, M., dos Santos Rocha, M., Mayrhofer, R., Zeidler, A., Riener, A., and Franz, M. (2008). Building flexible manufacturing systems based on peer-its. *EURASIP Journal on Embedded Systems.* Article ID 267560, doi:10.1155/2008/267560.

Forbus, K. D. (1997). Qualitative reasoning, in *The Computer Science and Engineering Handbook* (A. B. Tucker, ed.), pp. 715–733. Boca Raton, FL: CRC Press.

Fox, D., Burgard, W., Kruppa, H., and Thrun, S. (2000). A probabilistic approach to collaborative multi-robot localization. *Autonomous Robots* 8: 325–344.

Frank, A. U. (1998). Formal models for cognition: Taxonomy of spatial location description and frames of reference, in *Spatial Cognition: An Interdisciplinary Approach to Representing and Processing Spatial Knowledge* (C. Freksa, C. Habel, and K. F. Wender, eds.), pp. 293–312. Berlin/Heidelberg: Springer-Verlag.

Freksa, C. (1992). Using orientation information for qualitative spatial reasoning, in the *Proceedings of the International Conference GIS: From Space to Territory: Theories and Methods of Spatio-Temporal Reasoning,* 21–23 September 1992, Pisa, Italy, pp. 162–178. Berlin/Heidelberg: Springer-Verlag.

Freksa, C. and Röhrig, R. (1993). Dimensions of qualitative spatial reasoning, in the *Proceedings of the 3rd IMACS Workshop on Qualitative Reasoning and Decision Technologies (QUARDET 1993),* June 1993, Barcelona, Spain, pp. 483–492. Washington, DC: IEEE Press.

Hadjieleftheriou, M., Kollios, G., Bakalov, P., and Tsotras, V. J. (2005). Complex spatiotemporal pattern queries, in the *Proceedings of the 31st International Conference on Very Large Data Bases (VLDB),* 30 August–2 September 2005, Trondheim, Norway, pp. 877–888. New York: ACM Press.

Haeberlen, A., Flannery, E., Ladd, A. M., Rudys, A., Wallach, D. S., and Kavraki, L. E. (2004). Practical robust localization over large-scale 802.11 wireless networks, in the *Proceedings of the 10th ACM International Conference on Mobile Computing and Networking (MOBICOM 2004),* 26 September–1 October 2004, Philadelphia, pp. 70–84. New York: ACM Press.

Hazas, M., Kray, C., Gellersen, H. W., Agbota, H., Kortuem, G., and Krohn, A. (2005). A relative positioning system for co-located mobile devices, in the *Proceedings of the 3rd International Conference on Mobile Systems, Applications, and Services (MobiSys 2005),* 6–8 June 2005, Seattle, pp. 177–190. New York: ACM Press.

Hazas, M., Scott, J., and Krumm, J. (2004). Location-aware computing comes of age. *IEEE Computer Magazine* 37: 95–97.

He, T., Huang, C., Blum, B. M., Stankovic, J. A., and Abdelzaher, T. (2003). Range-free localization schemes for large scale sensor networks, in the *Proceedings of the 9th Annual International Conference on Mobile Computing and Networking (MOBICOM 2003),* 14–19 September 2003, San Diego, pp. 81–95. New York: ACM Press.

Hernández, D. (1994). *Qualitative Representation of Spatial Knowledge.* Berlin/Heidelberg: Springer-Verlag.

Hernández, D., Clementini, E., and Felice, P. D. (1995). Qualitative distances, in *Spatial Information Theory: A Theoretical Basis for GIS, Proceedings of the International Conference (COSIT 1995),* 21–23 September 1995, Semmering, Austria, pp. 45–57. Berlin/Heidelberg: Springer-Verlag.

Hightower, J., Want, R., and Borriello, G. (2000). *SpotON: An Indoor 3D Location Sensing Technology Based on RF Signal Strength.* UW CSE 2000-02-02, University of Washington, Seattle. http://seattle.intel-research.net/people/jhightower/pubs/hightower2000indoor/hightower2000indoor.pdf.

Hinske, S. (2007). Determining the position and orientation of multi-tagged objects using RFID technology, in the *Proceedings of the Fifth Annual IEEE International Conference on Pervasive Computing and Communications Workshops (PerCom Workshops '07),* March 2007, White Plains, NY, pp. 377–381. Washington, DC: IEEE Press.

Hobbs, J. R. and Narayanan, S. (2002). Spatial representation and reasoning, in *Encyclopedia of Cognitive Science.* London, U.K.: MacMillan.

Holzmann, C. (2007a). Inferring and distributing spatial context, in the *Proceedings of the 2nd European Conference on Smart Sensing and Context (EuroSSC 2007),* 23–25 October 2007, Kendal, U.K., pp. 77–92. Berlin/Heidelberg: Springer-Verlag.

Holzmann, C. (2007b). Rule-based reasoning about qualitative spatiotemporal relations, in the *Proceedings of the 5th International Workshop on Middleware for Pervasive and Ad-Hoc Computing (MPAC 2007),* 26–30 November 2007, Newport Beach, CA, pp. 49–54. New York: ACM Press.

Holzmann, C. and Ferscha, A. (2007). Towards collective spatial awareness using binary relations, in the *Proceedings of the 3rd International Conference on Autonomic and Autonomous Systems (ICAS 2007),* 19–25 June 2007, Athens, Greece, pp. 36–36. Washington, DC: IEEE CS Press.

Isli, A., Haarslev, V., and Möller, R. (2003). *Combining Cardinal Direction Relations and Relative Orientation Relations in Qualitative Spatial Reasoning.* CoRR cs.AI/0307048.

Jung, W. and Woo, W. (2006). ubiTrack location-based and user-centered situation-awareness method. *International Conference on Hybrid Information Technology,* 9–11 November 2006. Jeju Island, Korea. http://old.uvr.gist.ac.kr/papers/2006/ICHIT_2006_wjung.pdf.

Karalar, T. C. (2006). *Implementation of Localization Systems for Sensor Networks.* Technical Report, University of California, Berkeley.

Karimi, H. A. and Liu, X. (2003). A predictive location model for location-based services, in the *Proceedings of the 11th ACM International Symposium on Advances in Geographic Information Systems (GIS'03),* 7–8 November 2003, New Orleans, pp. 126–133. New York: ACM Press.

Kawsar, F., Fujinami, K., and Nakajima, T. (2007). A lightweight indoor location model for sentient artifacts using sentient artefacts, in the *Proceedings of the 2007 ACM Symposium on Applied Computing*, 11–15 March 2007, Seoul, Korea, pp. 1624–1631. New York: ACM Press.

Klatzky, R. L. (1998). Allocentric and egocentric spatial representations: Definitions, distinctions, and interconnections, in *Spatial Cognition, An Interdisciplinary Approach to Representing and Processing Spatial Knowledge*, pp. 1–18. Berlin/Heidelberg: Springer-Verlag.

Kortuem, G., Kray, C., and Gellersen, H. (2005). Sensing and visualizing spatial relations of mobile devices, in the *Proceedings of the 18th Annual ACM Symposium on User Interface Software and Technology (UIST 2005)*, 22–27 October 2005, Seattle, pp. 93–102. New York: ACM Press.

Krahnstoever, N., Rittscher, J., Tu, P., Chean, K., and Tomlinson, T. (2005). Activity recognition using visual tracking and RFID, in the *Proceedings of the Seventh IEEE Workshops on Application of Computer Vision, 2005 (WACV/MOTIONS '05), Volume 1*, 5–7 January 2005, pp. 494–500. Washington, DC: IEEE Press.

Krumm, J., Cermak, G., and Horvitz, E. (2003). RightSPOT: A novel sense of location for a smart personal object, in the *Proceedings of UBICOMP 2003*, 12–15 October 2003, Seattle, pp. 36–43. Berlin/Heidelberg: Springer-Verlag.

Krumm, J., Harris, S., Meyers, B., Brumitt, B., Hale, M., and Shafer, S. (2000). Multi-camera multi-person tracking for EasyLiving, in the *Proceedings of the 3rd IEEE Workshop on Visual Surveillance*, 1 July 2000. http://research.microsoft.com/easyliving/Documents/2000%2007%20Krumm.pdf.

Laasonen, K., Raento, M., and Toivonen, H. (2004). Adaptive on-device location recognition, in the *Proceedings of the Second International Conference on Pervasive Computing*, 21–23 April 2004, Vienna, Austria, pp. 287–304. Berlin/Heidelberg: Springer-Verlag.

Ladd, A. M., Bekris, K. E., Rudys, A., Marceau, G., Kavraki, L. E., and Wallach, D. S. (2002). Robotics-based location sensing using wireless Ethernet, in the *Proceedings of the 8th Annual International Conference on Mobile Computing and Networking*, 23–28 September 2002, Atlanta, pp. 227–238. New York: ACM Press.

LaMarca, A., Chawathe, Y., Consolvo, S., Hightower, J., Smith, I., Scott, J., et al. (2005). Place lab: Device positioning using radio beacons in the wild, in the *Proceedings of the Third International Conference on Pervasive Computing (PERVASIVE 2005)*, 8–13 May 2005, Munich, Germany, pp. 116–133. Berlin/Heidelberg: Springer-Verlag.

Li, B., Salter, J., Dempster, A. G., and Rizos, C. (2006). Indoor positioning techniques based on wireless LAN, in the *Proceedings of Auswireless 2006 Conference*, 13–16 March 2006, Sydney, Australia. http://epress.lib.uts.edu.au/dspace/bitstream/2100/170/1/113_Li.pdf.

Liu, X., Corner, M., and Shenoy, P. (2006). Ferret: RFID localization for passive multimedia, in the *Proceedings of the 8th Ubicomp Conference*, 17–21 September 2006, Orange County, CA, pp. 422–440. Berlin/Heidelberg: Springer-Verlag.

Liu, C., Wu, K., and He, T. (2004). Sensor localization with ring overlapping based on comparison of received signal strength indicator, in the *Proceedings of the 2004 IEEE International Conference on Mobile Ad-hoc and Sensor Systems*, 7–10 November 2004, Washington, DC, pp. 516–518. Washington, DC: IEEE Press.

Lorincz, K. and Welsh, M. (2006). MoteTrack: A robust, decentralized approach to RF-based location tracking. *Personal and Ubiquitous Computing* 11: 489–503.

Merrill, D., Kalanithi, J., and Maes, P. (2007). Siftables: Towards sensor network user interfaces, in the *Proceedings of the First International Conference on Tangible and Embedded Interaction (TEI'07)*, 15–17 February 2007, Baton Rouge, LA, pp. 75–78. New York: ACM Press.

Mezentsev, O. and Lachapelle, G. (2005). Pedestrian dead reckoning—A solution to navigation in GPS signal degraded areas? *Geomatica* 59: 175–182.

Miller, L. E. (2006). *Indoor Navigation for First Responders: A Feasibility Study*. Technical Report, NIST. http://www.antd.nist.gov/wctg/RFID/Report_indoornav_060210.pdf.

Moratz, R. (2004). *Qualitative Spatial Reasoning about Oriented Points*. Technical Report SFB/TR 8 Report No. 003-10/2004, University of Bremen, Bremen, Germany.

Moratz, R., Dylla, F., and Frommberger, L. (2005). A relative orientation algebra with adjustable granularity, in the *Proceedings of the Workshop on Agents in Real-Time and Dynamic Environments at IJCAI 2005*, July 2005, Edinburgh, Scotland. http://www.cosy.informatik.uni-bremen.de/staff/dylla/publications/Paper/cosy:dylla:2005:OPRA_RealRobotics.pdf.

Moratz, R. and Fischer, K. (2000). *Cognitively Adequate Modelling of Spatial Reference in Human-Robot Interaction*. Washington, DC: IEEE CS Press.

Moratz, R., Renz, J., and Wolter, D. (2000). Qualitative spatial reasoning about line segments, in the *Proceedings of the 14th European Conference on Artificial Intelligence (ECAI 2000)*, 20–25 August 2000, Berlin, Germany, pp. 234–238. Amsterdam: IOS Press.

Moratz, R., Tenbrink, T., Bateman, J. A., and Fischer, K. (2003). Spatial knowledge representation for human-robot interaction, in *Spatial Cognition III, Routes and Navigation, Human Memory and Learning, Spatial Representation and Spatial Learning*, pp. 263–286. Berlin/Heidelberg: Springer-Verlag.

Musto, A. (1999). *On Spatial Reference Frames in Qualitative Motion Representation*. Technical Report FKI-230-99, Institut für Informatik, Technische Universität München, Munich, Germany.

Muthukrishnan, K., Lijding, M., and Havinga, P. J. M. (2005). Towards smart surroundings: Enabling techniques and technologies for localization, in the *Proceedings of the First International Workshop on Location- and Context-Awareness (LoCA 2005)* (T. Strang and C. Linnhoff-Popien, eds.), 12–13 May 2005, Munich, Germany, pp. 350–362. Berlin/Heidelberg: Springer-Verlag.

Nagpal, R., Shrobe, H., and Bachrach, J. (2003). Organizing a global coordinate system from local information on an ad hoc sensor network, in the *Proceedings of Information Processing in Sensor Networks (IPSN 2003)*, April 2003, Palo Alto, CA, pp. 333–348. Berlin/Heidelberg: Springer-Verlag.

National Imagery and Mapping Agency (1984). *Department of Defense World Geodetic System 1984: Its Definition and Relationships with Local Geodetic Systems* (3rd ed.). National Geospatial-Intelligence Agency. http://earth-info.nga.mil/GandG/publications/tr8350.2/wgs84fin.pdf.

Ni, L. M., Liu, Y., Lau, Y. C., and Patil, A. P. (2003). LANDMARC: Indoor location sensing using active RFID, in the *Proceedings of the First IEEE International Conference on Pervasive Computing and Communications (PerCom'03)*, 23–26 March 2003, Fort Worth, TX, pp. 407–415. Washington, DC: IEEE Press.

Orr, R. J. and Abowd, G. D. (2000). The smart floor: A mechanism for natural user identification and tracking, in the *CHI '00 Extended Abstracts on Human Factors in Computing Systems*, pp. 275–276. New York: ACM Press.

Otsason, V., Varshavsky, A., LaMarca, A., and de Lara, E. (2005). Accurate GSM indoor localization, in the *Proceedings of the 7th International Conference on Ubiquitous Computing (UbiComp 2005)*, 11–14 September 2005, Tokyo, pp. 141–158. Berlin/Heidelberg: Springer-Verlag.

Philipose, M., Smith, J. R., Jiang, B., Mamishev, A., Sumit, R., and Sundara-Rajan, K. (2005). Battery-free wireless identification and sensing. *IEEE Pervasive Computing* 4: 37–45.

Randall, J., Arnft, O., Bohn, J., and Burri, M. (2006). LuxTrace: Indoor positioning using building illumination. *Personal and Ubiquitous Computing* 11: 417–428.

Raskar, R., Beardsley, P. A., van Baar, J., Wang, Y., Dietz, P. H., Lee, J. C., et al. (2004). RFIG lamps: Interacting with a self-describing world via photosensing wireless tags and projectors. *ACM Transactions on Graphics (TOG)* 23: 406–415.

Raskar, R., Beardsley, P. A., Dietz, P. H., and van Baar, J. (2005). Photosensing wireless tags for geometric procedures. *Communications of the ACM* 48: 46–51.

Ravi, N., Shankar, P., and Frankel, A. (2006). Indoor localization using camera phones, in the *Proceedings of the 7th IEE Workshop on Mobile Computing System & Applications: Supplement (WMCSA'06)*, 6–7 April 2006, Semiahmoo, Washington, pp. 1–7. Washington, DC: IEEE Press.

Renz, J. and Mitra, D. (2004). Qualitative direction calculi with arbitrary granularity, in the *Proceedings of the 8th Pacific Rim International Conference on Artificial Intelligence, (PRICAI 2004)*, 9–13 August 2004, Auckland, New Zealand, pp. 65–74. Berlin/Heidelberg: Springer-Verlag.

Sarma, S. (2004). Integrating RFID. *ACM Queue* 2: 50–57.

Satoh, I. (2005). A location model for pervasive computing environments, in the *Proceedings of the 3rd IEEE International Conference on Pervasive Computing and Communications (PerCom 2005)*, 8–12 March 2005, Kauai Island, HI, pp. 215–224. Washington, DC: IEEE Press.

Savvides, A., Park, H., and Srivastava, M. B. (2002). The bits and flops of the N-hop multilateration primitive for node localization problems, in the *Proceedings of the 1st ACM International Workshop on Wireless Sensor Networks and Applications*, 28 September 2002, Atlanta, pp. 112–121. New York: ACM Press.

Saxena, A., Ganguly, S., Bhatnagar, S., and Izmailov, R. (2007). RFInD: An RFID-based system to manage virtual spaces, in the *Proceedings of the Fifth Annual IEEE International Conference on Pervasive Computing and Communications Workshops (PerComW'07)*, 19 March 2007, New York, New York, pp. 382–387. Washington, DC: IEEE Press.

Scivos, A. and Nebel, B. (2004). The finest of its class: The natural point-based ternary calculus for qualitative spatial reasoning, in *Spatial Cognition IV: Reasoning, Action, Interaction, Proceedings of the 4th International Conference on Spatial Cognition 2004*, October 2004, Frauenchiemsee, Germany, pp. 283–303. Berlin/Heidelberg: Springer-Verlag.

Smith, A., Balakrishnan, H., Goraczko, M., and Priyantha, N. (2004). Tracking moving devices with the Cricket location system, in the *Proceedings of the 2nd USENIX/ACM MOBISYS Conference*, 6–9 June, Boston. Berkeley, CA: USENIX.

Smith, J., Fishkin, K. P., Jiang, B., Mamishev, A., Philipose, M., Rea, A. D., et al. (2005). RFID-based techniques for human activity detection. *Communications of the ACM* 48: 39–44.

Sohn, T., Varshavsky, A., LaMarca, A., Chen, M. Y., Choudhury, T., Smith, I., et al. (2006). Mobility detection using everyday GSM traces, in the *Proceedings of the 8th International Conference on Ubiquitous Computing (UbiComp 2006)*, 17–21 September 2006, Orange County, CA, pp. 212–224. Berlin/Heidelberg: Springer-Verlag.

Streitz, N. A., Röcker, C., Prante, Th., Stenzel, R., and van Alphen, D. (2003). Situated interaction with ambient information: Facilitating awareness and communication in ubiquitous work environments, in *Human-Centred Computing: Cognitive, Social, and Ergonomic Aspects, Vol. 3 of the Proceedings of HCI International 2003* (D. Harris, V. Duffy, M. Smith, and C. Stephanidis, eds.), 22–27 June 2003, Crete, Greece, pp. 133–137. Mahwah, NJ: Lawrence Erlbaum Publishers.

Tapia, E., Marmasse, N., Intille, S. S., and Larson, K. (2004). MITes: Wireless portable sensors for studying behavior, in the *Proceedings of Extended Abstracts of Ubicomp 2004*. http://web.media.mit.edu/~intille/papers-files/MunguiaTapiaETAL04.pdf.

Tenbrink, T. (2005). *Semantics and Application of Spatial Dimensional Terms in English and German.* Technical Report SFB/TR 8 Report No. 004-03/2005, University of Bremen, Bremen, Germany.

UPnP Forum (2006). *UPnP Device Architecture 1.0.* http://upnp.org/specs/arch/UPnP-DeviceArchitecture-v1.0-20060720.pdf.

Varshavsky, A., de Lara, E., Chen, M., Haehnel, D., Hightower, J., LaMarca, A., et al. (2006). Are GSM phones THE solution for localization?, in the *7th IEEE Workshop on Mobile Computing Systems and Applications (WMCSA'06)*, 6–7 April 2006, Semiahmoo Resort, WA. Washington, DC: IEEE Press.

Varshavsky, A., LaMarca, A., Hightower, J., and de Lara, E. (2007). The SkyLoc floor localization system, in the *Proceedings of the Fifth Annual IEEE International Conference on Pervasive Computing and Communications (PerCom'07)*, 19–23 March 2007, White Plains, NY, pp. 125–134. Washington, DC: IEEE Press.

Viqueira, J. R. R. and Lorentzos, N. A. (2007). SQL extension for spatio-temporal data. *International Journal on Very Large Data Bases* 16: 179–200.

Wallgrün, J. O., Frommberger, L., Dylla, F., and Wolter, D. (2007). *SparQ User Manual v0.7*. Technical Report, SFB/TR 8 Spatial Cognition—Project R3, University of Bremen, Bremen, Germany. http://www.sfbtr8.spatial-cognition.de/project/r3/material/SparQ-Manual-V0.7.pdf.

Want, R. (2004). RFID: A key to automating everything. *Scientific American* 290: 56–65.

Want, R., Hopper, A., Falcao, V., and Gibbons, J. (1992). The active badge location system. *ACM Transactions on Information Systems (TOIS)* 10: 91–102.

Welch, G. and Foxlin, E. (2002). Motion tracking: No silver bullet, but a respectable arsenal. *IEEE Computer Graphics and Applications* 22: 24–38.

Xu, B. and Gang, W. (2006). Random sampling algorithm in RFID indoor location system, in the *Proceedings of the 3rd IEEE International Workshop on Electronic Design, Test and Applications (DELTA 2006)*, 17–19 January 2006, Kuala Lumpur, Malaysia, p. 6. Washington, DC: IEEE Press.

References

Allen, J. (1983). Maintaining knowledge about temporal intervals. *Communications of the ACM* 26: 832–843.

Ferscha, A., Hechinger, M., Riener, A., Schmitzberger, H., Franz, M., dos Santos Rocha, M., and Zeidler, A. (2006). Context-aware profiles, in the *Proceedings of the 2nd International Conference on Autonomic and Autonomous Systems (ICAS 2006)*, 3–8 September 2006, Hamburg, Germany, pp. 48–48. Washington, DC: IEEE CS Press.

Gottfried, B., Guesgen, H. W., and Huebner, S. (2006). Spatiotemporal reasoning for smart homes, in *Designing Smart Homes: The Role of Artificial Intelligence* (C. Augusto and C. D. Nugent, eds.), pp. 16–34. Berlin/Heidelberg: Springer-Verlag.

Herrmann, K., Muhl, G., and Geihs, K. (2005). Self-management: The solution to complexity or just another problem? *IEEE Distributed Systems Online* 6. http://ieeexplore.ieee.org/xpls/abs_all.jsp?tp=&arnumber=1407752.

Hightower, J. and Borriello, G. (2001). Location sensing techniques. *IEEE Computer Magazine* 34: 57–66.

Jelasity, M., Babaoglu, Ö., Laddaga, R., Nagpal, R., Zambonelli, F., Sirer, E. G., et al. (2006). Interdisciplinary research: Roles for self-organization. *IEEE Intelligent Systems* 21: 50–58.

Kephart, J. O. and Chess, D. M. (2003). The vision of autonomic computing. *IEEE Computer* 36: 41–50.

Lehman, F. K. and Bennardo, G. (2003). A computational approach to the cognition of space and its linguistic expressions. *Mathematical Anthropology and Cultural Theory* 1: 1–83. http://www.mathematicalanthropology.org.

Mamei, M., Menezes, R., Tolksdorf, R., and Zambonelli, F. (2006). Case studies for self organization in computer science. *Journal of Systems Architecture: The EUROMICRO Journal* 52: 443–460.

Mamei, M. and Zambonelli, F. (2004). Spatial computing: The TOTA approach, in the *Proceedings of the 5th AI*IA/TABOO Joint Workshop "From Objects to Agents": Complex Systems and Rational Agents*, 30 November–1 December 2004, Torino, Italy, pp. 126–142. Bologna: Pitagora Editrice.

Randell, D. A., Cui, Z., and Cohn, A. G. (1992). A spatial logic based on regions and connection, in the *Proceedings of the 3rd International Conference on Principles of Knowledge Representation and Reasoning (KR-92)*, pp. 165–176. San Francisco: Morgan Kaufmann.

Schmidt, A. (1999). Implicit human-computer interaction through context, in the *Proceedings of the 2nd Workshop on Human Computer Interaction with Mobile Devices*, 31 August 1999, Edinburgh, Scotland. http://www.dcs.gla.ac.uk/mobile99/papers/schmidt.pdf.

Serugendo, G. D. M., Foukia, N., Hassas, S., Karageorgos, A., Most´efaoui, S. K., Rana, O. F., et al. (2003). Self-organisation: Paradigms and applications, in *Engineering Self-Organising Systems: Nature-Inspired Approaches to Software Engineering*, pp. 1–19. Berlin/Heidelberg: Springer-Verlag.

Weiser, M. (1991). Ubiquitous computing. *Scientific American*.

Zambonelli, F. and Mamei, M. (2004). Spatial computing: An emerging paradigm for autonomic computing and communication, in the *Proceedings of the 1st International IFIP Workshop on Autonomic Communication (WAC 2004)*, 18–19 October 2004, Berlin, Germany, pp. 44–57. Berlin/Heidelberg: Springer-Verlag.

60

Ambient Intelligence

Norbert A. Streitz
and Gilles Privat

60.1 Introduction

This chapter presents an overview of the basic concepts, trends, and perspectives of ambient intelligence. It starts with an overall technological frame of reference under which ambient intelligence may be understood (a "big picture"). Then, it presents a number of constituent approaches and technologies as the parallel and overlapping threads woven into the braid of ambient intelligence. Based on this, seven alternatives are proposed and discussed, characterizing the theoretical and practical challenges that have to be addressed in this field. We conclude with some final comments on current trends and perspectives for the future.

60.1.1 Microcosm, Telecosm, Ambicosm

A plethora of competing and often conflicting buzzwords and catchphrases surround and obfuscate the overall ambient intelligence evolution: *ubiquitous/pervasive/proactive computing, Internet of things, smart spaces, invisible/disappearing computer*, to cite only a few. This may obscure the fact that, beyond the hype, a singular and sweeping (r)evolution of information and communication technology is currently taking place.

The most frequently heard viewpoint on this evolution refers mainly to information processing and storage capabilities, highlighting the decentralization and distribution of computing into all kinds of devices beyond computers proper. This generalization and parallel miniaturization of embedded[1] computing are a direct consequence of Moore's law and the ceaselessly touted progress of microelectronics, but it actually predates ubiquitous computing/ambient intelligence by at least two decades. It is only one of the three forerunning evolutions that support ambient intelligence. This aspect of the evolution is implicitly what the most common expressions like *ubiquitous* and *pervasive computing* denote and connote.

George Gilder (2000) has eloquently articulated a broad vision of a further technology evolution that he baptized *telecosm*, succeeding the *microcosm*[2] described before. In Gilder's historical view of technology evolution, the defining limitation of one era becomes the defining abundance of the following one. The limitation of the microcosm (the era of overabundant silicon and processing power) was the *bandwidth* of both wired and wireless network, which became, jointly, the defining abundances of the telecosm.

---

[1] "Embedded" is taken throughout this chapter in a strict sense (advocated by, e.g., National Research Council, 2001) meaning information processing and storage integrated in devices that do not have information storage or communication as their primary function.

[2] *Microcosm* was actually the title of a previous book by Gilder (1989) describing the microelectronics revolution.

Besides fostering an overcapacity of optical pipes and the telecom bubble at the end of the 1990s, by merely increasing the capacity for existing applications, the telecosm fosters a less visible but more promising field of opportunity: as per Moore's law and the ensuing generalization of embedded computing, devices with embedded computing capabilities are due to become more numerous than human beings. Hence the grand idea of connecting everything to everything, corresponding at this stage to the networking of all kinds of physical devices (e.g., home appliances, industrial apparatuses) that are equipped with embedded processing capabilities but are primarily neither information-processing devices nor communication devices. This first extension of the reach of networks toward smart devices that become smart networked devices opens up a new range of entirely new, as yet unexploited, network-based services.

To complete this big picture of ambient intelligence, a third new wave of technology evolution, less obvious and more diverse, comes hard on the heels of the previous two, described previously as the microcosm and the telecosm. From this up-to-date viewpoint on ambient intelligence, the truly groundbreaking novelty is the potential miniaturization and dissemination of sensors and actuators that can be manufactured in novel technologies (MEMS[3] at first, later nanotechnologies) and cheaply integrated with smart networked devices. Just as the telecosm increased and leveraged the digital-to-digital bandwidth of classical networks, the new era increases and leverages the analogue to digital and digital to analogue bandwidth of interaction between the physical world and the digital world. This quantum leap results in a much tighter and richer coupling between the networked infospace and the environment that humans inhabit.[4] Among the many exciting and unforeseen possibilities that opened up, the grandest vision of generalized distributed sensing-actuating embedded in the physical environment has been touted as either (depending on the scale) "smart matter" or some kind of "nervous system of the earth."

Another, no less important aspect of this increased physical to digital interaction bandwidth is the enlargement and augmentation of user-to-information interfaces or human-information interaction (Streitz et al., 2001) in contrast to the traditional user/

human-computer interaction paradigm, as they may become distributed in physical devices throughout the environment. This facilitates overcoming the limitations of the single terminal window and its simulated worldview, be it a 2D (desktop) or 3D (VR) metaphor.

Balancing the two, it should be clear that the multiplication of environment- and network-aware devices is bound to shift the emphasis from human-centered, interactive computing to human-supervised proactive computing (Tennenhouse, 2000), if only because no human user will have the time and attention required to explicitly interact with all these devices.

Following Gilder's footsteps, the new era can be defined as the *ambicosm*. The defining limitation of the ambicosm is the time and attention of users, compounded by the exponentially growing amounts of content using the unlimited bandwidth of telecosmic networks and the corresponding bandwidth of interface devices that present this content to users. This limit, which follows on the previous bandwidth and storewidth[5] limitations, is called here *eyewidth*. Figure 60.1 presents an overview of the defining dimensions of the three eras, considered as fundamental for the current technology evolution.

60.1.2 Ambient Intelligence at the Center of the Ambicosm

The topic map in Figure 60.2 situates the concept of ambient intelligence in relation to other concepts that all pertain to the broad evolution outlined previously. Positions along the horizontal axis situate each of the concepts according to whether it results most directly from any of the three waves of ICT evolution described previously. Positions along the vertical axis represent the relative weight of physical interaction with the environment (by way of sensors and actuators in nonhuman modalities) versus interaction with users by way of classical interface devices and human sensory-motor modalities). Ambient intelligence is the most central concept, as it conjoins aspects of human and physical interaction, communication, and processing being implicit but necessarily present as resulting from the two previous technology waves outlined previously.

There are many characterizations of Ambient Intelligence (AmI), which will be elaborated upon in more detail in the following sections of this chapter. In any case, the description in Streitz and Savidis (2006) serves as a good starting point for the readers of this chapter:

> *Ambient Intelligence represents a vision of the (not too far) future where "intelligent" or "smart" environments and systems react in an attentive, adaptive, and active (sometimes even proactive) way to the presence and activities of humans and objects in order to provide intelligent/smart services to*

---

[3] MEMS, micro-electro-mechanical systems. The most widespread example of MEMS sensors are accelerometers and those of MEMS actuators are digital micro-mirror devices used in digital light processing (DLP) video projectors.

[4] It can be argued that brain-computer interfaces (BCI) could represent, in the long-term, the ultimate "tight coupling" between the user's own world and the information world, further increasing person-information bandwidth by several orders of magnitude. Yet the relationship of BCI to AmI is different from that of AmI to the previous stages of HCI evolution. Whereas AmI enlarges and enriches previous stages of ICT evolution, BCIs are certainly not destined to supersede or subsume AmI: one side of AmI is not concerned with HCI but with information-physical world interaction, and even for HCI, it is fairly obvious that not all interfaces will dispense with a physical intermediary: for all those cases where physical interaction is either mandatory or the most efficient, AmI and its evolution will remain central.

---

[5] Storewidth (Gilder, 2000) is a combination of latency and storage capacity against which the access to stored information is measured.

Era	Time scale	Defining technology	Defining abundance	Defining limitation
Microcosm	1980s onwards	Microelectronics	Silicon-based processing power	Bandwidth
Telecosm	1990s onwards	Optics	Wireless and wireline bandwidth	"Storewidth"
Ambicosm	2000 onwards	MEMS and nanotechnologies	Physical interaction capabilities	"Eyewidth"

FIGURE 60.1   Eras of technology evolution.

the inhabitants of these environments. *Ambient Intelligence technologies integrate sensing capabilities, processing power, reasoning mechanisms, networking facilities, applications and services, digital content, and actuating capabilities distributed in the surrounding environment. While a wide variety of different technologies is involved, the goal of Ambient Intelligence is to hide their presence from users, by providing implicit, unobtrusive interaction paradigms. People and their social situations, ranging from individuals to groups, be they work groups, families or friends and their corresponding environments (office buildings, homes, public spaces, cities, etc) are at the centre of the design considerations.*

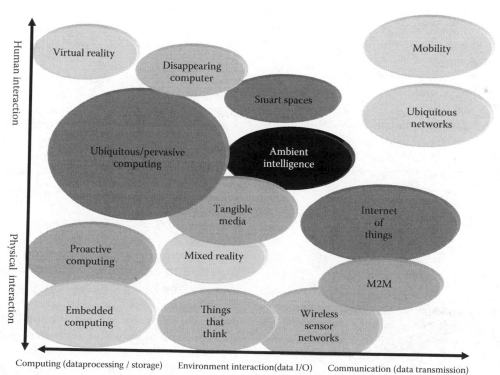

FIGURE 60.2   Ambient intelligence topic map.

60.2 Ambient Intelligence and Universal Access

According to the original Information Society Technology Advisory Group (ISTAG) vision statement (Ducatel, 2001)

> humans will, in an Ambient Intelligence (AmI) Environment, be surrounded by intelligent interfaces supported by computing and networking technology that is embedded in everyday objects such as furniture, clothes, vehicles, roads, and smart materials—even particles of decorative substances like paint. AmI implies a seamless environment of computing, advanced networking technology, and specific interfaces. This environment should be aware of the specific characteristics of human presence and personalities; adapt to the needs of users; be capable of responding intelligently to spoken or gestured indications of desire; and even result in systems that are capable of engaging in intelligent dialogue. Ambient Intelligence should also be unobtrusive—interaction should be relaxing and enjoyable for the citizen and not involve a steep learning curve.

Based on this characterization, one of the crossroads where universal access (UA) and AmI meet is the area of ambient assisted living (AAL), which is also an important strategic objective of the 7th Framework Program of the EU (FP 7, extending for the 2007–2013 period). The Ambient Assisted Living Joint Program addresses the needs of the aging population, while lowering future social security costs, and, at the same time, potentially reducing innovation barriers of forthcoming promising markets. AAL aims—by the use of intelligent products and the provision of remote services, including care services—at extending the time older people can live in their home environment by increasing their autonomy and assisting them in carrying out activities of daily living. This view is closely related to developments that are summarized under the notion of quality of life technology (QoLT).

But there is more to the relationship of UA and AmI. Both contribute to and converge in the general idea of assistive technologies that capture information about users, their spatial, temporal, and social situation, their planned activities, and the way they will be addressed and executed. The common concept is the notion of *context* taken here in a very broad sense. The central position of context will be further addressed in subsequent sections.

60.3 Threads of AmI

In terms of historical perspective, AmI was presented for the first time to a larger audience in 2001 via the often quoted report, "Scenarios for Ambient Intelligence in 2010" (Ducatel, 2001) published by ISTAG of the European Commission. While the term *ambient intelligence* originates in some early work that was carried out at Philips in 1999 (where Emile Aarts became later on its main propagator) one has to state that the underlying concept and ideas have been around several years before the ISTAG report was published. One milestone of this trend was the seminal paper by Mark Weiser (1991). Therefore, this section presents different approaches, at the same time discussing the origins and history of the overall theme that was propagated under different terminologies as, for example, ubiquitous, pervasive, proactive, ambient computing, and the disappearing computer.

An intriguing and thought-provoking (though maybe simplistic) viewpoint would state that three continents have spawned three different visions of the domain, characterized by three relevant key phrases:

- *Ubiquitous computing* in the United States
- *Ubiquitous networking* in Japan and Korea
- *Ambient intelligence* in Europe

These three terms may be interpreted further as placing the emphasis on, respectively, the computing/processing aspect, the transmission/bandwidth aspect, the human interface aspect, each corresponding to the three technology evolutions outlined in the introduction to this chapter. *Ambient intelligence* is arguably the most appropriate key phrase among these, as it implicitly builds upon and recapitulates the first two evolutions, yet places the emphasis on the third one, highlighting the role of physical interaction in the environment for human and social interfaces.

60.3.1 Ubiquitous Computing

As already mentioned, Mark Weiser at Xerox PARC can be considered a main originator of the ideas that are currently being discussed and investigated under different names and with different perspectives. He proposed the notion of ubiquitous computing in his seminal article, "The Computer for the 21st Century" (Weiser, 1991). Accordingly, computers and computer functionality, including network connections, are available in a ubiquitous fashion. It emphasizes the idea that information is available everywhere, any time, and so on. While the proposed concept seems strongly related to mobility, it is, however, different from mobile and wearable computing.

In his description of examples of ubiquitous computers developed at Xerox PARC, Weiser (1991) categorized them as tabs, pads, and boards, representing also three different scales of size (inches, feet, and yards). Augmenting these devices was the Active Badge system, developed originally for Olivetti by Roy Want who had joined PARC. A prominent example and starting point for these ideas was the so-called *live board* (designed primarily by Scott Elrod and Richard Bruce), a wooden cabinet with a large vertical interactive display, later known and marketed as the LiveBoard by the spin-off company LiveWorks. Although the wooden frame was intended to make it look different from a traditional computer, and rather appear as a piece of furniture, it was still a big box and not really becoming invisible. But it provided new forms of interacting with information while standing in front of it. This was especially due to the pen-based mode of interaction (Elrod et al., 1992) modeled after the real interaction with a felt pen on a traditional whiteboard or a piece of chalk on a blackboard. This, in combination with the large size of the display area of the rear projection unit using an internal hidden mirror, made it very different from traditional desktop computers, although it did not have a higher resolution than a workstation display at that time (see also Streitz, 2008).

Another important aspect of Weiser's ubiquitous computing was the number and availability of a range of different devices, all being able to exchange and communicate information among each other. He saw ubiquitous computing as the next trend where one person would have several computers at hand, compared to personal computing, where one person would have one computer, and also different from the previous mainframe computing paradigm where one large central computer was shared by many users using only terminals for accessing it.

60.3.2 The Disappearing Computer and Calm Technology

The notion of the disappearing computer is an implication of Mark Weiser's well-known statement: "The most profound technologies are those that disappear. They weave themselves into the fabric of everyday life until they are indistinguishable from it" (Weiser, 1991, p. 66). Furthermore, he argues for the development of a "calm technology" (Weiser, 1998; Weiser and Brown, 1996) by moving technology into the background, while the functionality is available in a ubiquitous fashion.

This constituted a starting point for current approaches (e.g., Streitz, 2001; Streitz et al., 2001), which argue that while in the past computers were primary objects of attention, resulting also in a well-developed research area called human-computer interaction, today the real goal is rather to interact with information, to communicate and collaborate with people, to enjoy and share experiences. Thus, shouldn't the computer move into the background and disappear?

This disappearance can take different forms. A typical distinction is between physical and mental disappearance (Streitz, 2001, 2008). *Physical disappearance* refers to the miniaturization of devices and their integration in everyday artifacts as, for example, clothes and other smart materials. In the case of *mental disappearance*, the artifacts can still be large, but they are not perceived as computers because people discern them as, for example, interactive walls or interactive tables. In this way, the computer disappears and is almost invisible, but its functionality is ubiquitously available and provides new forms of interacting with information and of making experiences. This leads to the core issue and questions: How can human-information interaction be designed and human-human communication and cooperation be supported by exploiting the affordances of existing objects in the environment? And, in doing so, how is the potential of computer-based support augmenting human activities exploited?

The notion of the disappearing computer also became the headline and the name of an EU-funded (IST-FET) proactive initiative (http://www.disappearing-computer.net), a cluster of 17 projects that were established under the following goal:

> To explore how everyday life can be supported and enhanced through the use of collections of interacting smart artefacts. Together, these artefacts will form new people-friendly environments in which the "computer-as-we-know-it" has no role.

An overview with in-depth articles of the concepts and the prototypes of the Disappearing Computer initiative can be found in the State-of-the-Art survey edited by Streitz et al. (2007). An example of reporting experiences of building disappearing computers in three different research contexts was provided by Russell et al. (2005).

> It seems like a paradox but it will soon become reality: The rate at which computers disappear will be matched by the rate at which information technology will increasingly permeate our environment and our lives. (Streitz and Nixon, 2005, p. 33)

This statement illustrates how computers are increasingly becoming an important part of day-to-day activities and will determine a wide range of physical and social contexts of future life. The availability and ubiquity of computers is the first step in this direction. It is to be followed by the integration of information, communication, and sensing technology into everyday objects, resulting in smart artifacts and making the computer disappear at the same time.

60.3.3 Pervasive and "Embedded, Everywhere" Computing

The phrases *pervasive/ubiquitous computing*, though the first to appear and still the most frequently used, do implicitly place a restrictive emphasis on the first of the three evolutions underlying AmI as defined here, at the expense of the communication and interaction aspects. The traditional computing industry has seen pervasive computing as an extension of its territory, and companies such as IBM have been at the forefront of this vision: where all kinds of information appliances diversify and decentralize the computing power that was previously concentrated, first in mainframe computers, then in personal computers (Hansmann et al., 2003).

Embedded Everywhere is the title of a book published by the U.S. National Research Council (2001), edited by Deborah Estrin. The book propounds a vision of "EmNets" (short for "networked systems of embedded computers"), as driving a new evolution of computing characterized by a tighter coupling with the physical world, a "digital nervous system to enable instrumentation of all sorts of spaces."

This report uses a stricter definition of *embedded* that is also used here (whereas a less precise use of the word refers to anything that is not general-purpose computing), and two important ideas: EmNets, operating in real-time without direct human intervention, "are likely to exhibit emergent or unintended behaviors," and that this safety issue has to be addressed upfront. Existing distributed computing models are inadequate for the new milieu of EmNets. New richer computing models need to be evolved that incorporate among other things, failures, resource constraints, new data models, and location.

60.3.4 Internet of Devices and Web of Things

The catchphrase *Internet of things* appeared first in a *Scientific American* paper about a bold new proposal for the evolution of the Internet (Gershenfeld et al., 2004). It was brought to a wider audience by an ITU report (2005),[6] though this document did not in itself propose a new vision of the field. Taken in a strict sense, this idea had been around ever since proponents of Ipv6 put forward a vision where the Internet would not only link computers' hosts in its traditional sense, nor even only mobile phones, but potentially all devices endowed with minimal communication capabilities. In this vision, every lightbulb would have its own IP address,[7] extending the reach of networks at the very level of the defining Internet protocol layer. The fact that this grand plan has been much slower to materialize than initially predicted[8] is for the time being attributable to the inertia of stakeholders and certainly does not invalidate it.

Even taking into account such new visions of networking as the Internet-0[9] (Gershenfeld and Cohen, 2006) that enlarge the reach of networks by several orders of magnitude toward not only new kinds of "things," but also new kinds of transmission media, the IP layer and IP addressing schemes (even extended to 128 bits by Ipv6) are definitely not suited to identify all the things that do potentially become attached to networks.

One step beyond, the Electronic Product Code (EPC) (http://www.epcglobalinc.org) and ubiquitous ID (http://www.uID-center.org) consortia are intent on identifying every manufactured item on earth, whenever barcodes will have been replaced with RFID tags, providing for each item, rather than for each class of items, a unique identification number stored in the tag. With such an extended addressing scheme, the reach of RFID networking may extend orders of magnitude beyond that of the already far-reaching Internet-0.

Taking stock of the most defining novelty of the new era, as outlined in the introduction, the range of things that are *indirectly*

attached to networks may be extended even much further than envisioned with RFID by EPC, to encompass all things that can be sensed through a sensor that is itself attached to a network. If a networked camera and an appropriate recognition software operating through this camera are available, one can consider that everything in the scene captured by this camera becomes ipso facto a networked thing, and it need not bear an RFID tag for this (an RFID reader may be considered a particular kind of sensor itself, but it can sense only things that were previously fitted with tags defined to the corresponding standard). From this vantage point, it should become clear that the Internet of things is a misnomer to convey a vision whereby the reach of networks extends to all things that can be sensed and/or actuated. Much as the early World Wide Web was a virtual network of hyperlinked HTML documents overlayed on top of the Internet, a web of things is currently emerging, a virtual network overlayed upon an Internet of devices and sensors, which is itself an extension by several orders of magnitude of the early Internet. As a virtual network (a graph in mathematical terms), the web of things does actually comprise a far larger number of nodes than an IP network ever will, and does also correspond to a different topology, as links between things correspond to their mutual sense-ability. Another key difference is that, whereas the graph representing an IP network is nondirected, the graphs representing either the classical document-centric web, or the web of things are *directed*.[10] A document hyperlinks to another, and the link target does not "know" directly that it is being linked to, and does not in general reciprocate, making up a link structure that is inherently unidirectional. Similarly, a sensed thing does not "know" directly that it is being sensed, and the corresponding link is also normally unidirectional. An interesting and important case arises when the thing being sensed has the capability to alter the information being captured by the sensor, rather than just being sensed passively. This happens when the sensed thing is directly an actuator, or when its state can be changed by something else (generally another actuator or a human being). In this case, the actuated and sensed thing links back to the sensor, adding a supplementary degree of complexity to the overall system that is modeled by this graph (Figure 60.3).

Things are thus hyperlinked physically through sensors, forming a graph of sensing links, but they also have digital shadows, their representations that are conveyed and made accessible through the network, whatever they are (symbolic IDs such as EPC numbers, direct or abstracted iconic representations, etc.).[11]

These representations may be linked in a way that is closer to the web of services, which itself grew out of the original web of documents, matching outgrowths of the physical web of things in

---

[6] This report was published on the occasion of the UN-ITU World Summit on the Information Society held in Tunis in November 2005.

[7] Ipv6 appeared in RFCs as early back as 1996 (and circulated as IPng before that), when the Internet itself only began to be known to the general public. The humungous address space provided by Ipv6 was never intended only to provide a flat fixed addressing plan, for which it is obviously a massive overkill, but also to facilitate dynamic allocation of addresses.

[8] As of the end of 2007, the deployment of IPv6 is still slow; only a tiny fraction of Internet hosts and routers actually implement the protocol, as overextended NAT and CIDR workarounds prolong the lifetime of the Ipv4 address space.

[9] This coinage was intended to be a counterpart to the touted high-speed Internet 2, oriented toward high performance.

---

[10] It should be noted here, that hypertext systems preceding the World Wide Web (WWW) had already all kinds of directed and typed node-link structures.

[11] This deep linking of the real and virtual worlds through sensors is related to the concept of dual reality (http://www.media.mit.edu/resenv/dual_reality_lab).

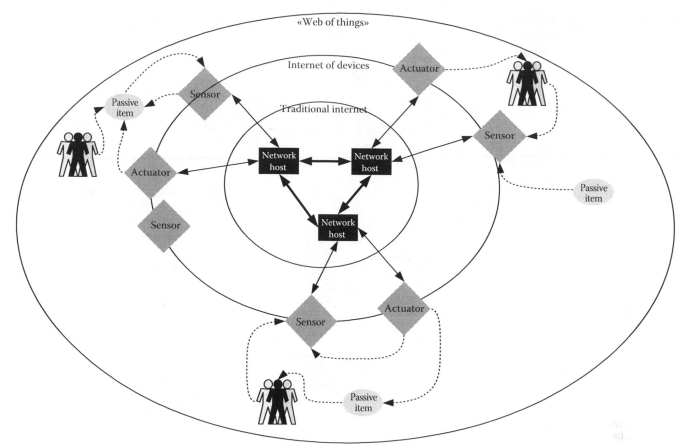

FIGURE 60.3 Web of things versus Internet of devices.

an extended digital web. These digital shadows may correspond at the very least, for passive items, to an entry in a database, or, in the case of more active devices like sensors or actuators, to a service registered in a networked service directory[12].

60.3.5 Ubiquitous Networking

Even though the original vision of the evolution that has become well known as ubiquitous computing is almost universally attributed to Marc Weiser at Xerox PARC in the United States, Ken Sakamura, professor at the University of Tokyo, claims to be the true inventor and supporter of the general idea, and, more importantly, to have contributed to support it in the real world through the spread of the TRON operating system that he created. Furthermore, he developed a real and comprehensive application in terms of the TRON concept building that was actually built in Tokyo in the late 1980s (Sakamura, 1990). Now director of the Ubiquitous Networking Laboratory (UNL) that he created, Ken Sakamura is a chief evangelist for a broad vision of *ubiquitous networking*, a phrase widely adopted in Japan and Korea at a political level. However, the take-up of this phrase by East Asian media and politicians has diluted its meaning beyond recognition, diverging as far as to include the whole evolution

of ICT, especially mobile and fixed broadband networking. Two fora, both chaired by Ken Sakamura, have been instrumental in supporting a more meaningful and concrete vision of ubiquitous networking: the T-engine forum (http://www.t-engine.org) that promotes an open platform (being the heir to the widely used TRON architecture) for embedded systems, and its offshoot, the ubiquitous ID center (http://www.uidcenter.org), that promotes a meta coding plan intended to subsume existing identification standards such as barcodes and, more controversially, the RFID numbering plan supported by the rival forum, EPCglobal. Ubiquitous ID "ucodes," as these unique identification numbers are called, may be attached to all material things, but also to places, and are intended to support all Internet of things applications on this basis, beyond the canonical supply chain management applications promoted by EPCglobal: mobility assistance, navigation, tourism guides, and so on.

60.3.6 Ambient Computing and Ambient Displays

There are many similarities between *ambient computing* and the other terms (ubiquitous, pervasive, etc.) discussed so far. The starting point is that computer functionality is not only available in the foreground (desktop PC, monitor), but moves more and

---

[12] UDDI is an example of such a directory, used by web services infrastructures.

more also in the background, in the periphery of human attention (=> ambient). The notion of ambient displays is especially central here. They go beyond the traditional notion of "display" encountered with conventional graphical user interfaces (GUI) found on PCs, notebooks, and PDAs. Taking a more general view, anything can be considered to act as a display—combined with multimodal presentation and interaction. The design of ambient displays is often based on observations in nature or employing corresponding metaphors. The ambient aspect is especially obvious when they are designed to display information without constantly demanding the user's full attention. Usually, this is achieved in a more implicit way by being available in the periphery compared to traditional explicit GUIs. Ambient displays are envisioned as being all around, and thereby moving information off the conventional screens into the physical environment. They present information via changes in light, sound, movement of objects, smell, and so on, or by using new smart materials. Early examples were described by Ishii et al. (1998) and Wisneski et al. (1998), and ways of evaluating them by Mankoff et al. (2003). A good example of using sound is realized as part of the AmbientROOM demonstrator by Ishii et al. (1998). It is based on sampling sounds from traffic in the street where the noise is an indicator for the number of vehicles passing by and mapping the sound level to the amount of invisible traffic on a computer network.

There is, of course, a close relationship between calm technology and ambient computing when it comes to its realization and when implementing the ideas of the disappearing computer described previously. A good example for this combined approach is the Hello.Wall ambient display based on light that was developed in the Ambient Agoras project (Streitz et al., 2003, 2005, 2007). In contrast to mechanisms like open video channels (Fish et al., 1992; Bly et al., 1993), dynamic light patterns were used for supporting informal social encounters and communication processes in a local context and especially between connected remote office environments. The Hello.Wall uses special light patterns to represent and communicate public awareness information in an ambient and unobtrusive way. Different types of light patterns were able to trigger the attention of team members in a subtle and peripheral way, communicating presence of people and atmosphere, thus providing a sense of a place.

60.3.7 Tangible Interfaces

Tangible interfaces were initially proposed as input interfaces (Ishii and Ullmer, 1997; Ullmer and Ishii, 2000) alternative to the classical controls associated with the WIMP-GUI interface model. These controls are multiplexed in the sense that they are used for different functions, depending on the current mode in which the interface they control is used. The representation of this functionality is distinct from the control device itself, and conveyed via a separate output interface (usually a display screen). Tangible user interfaces (TUI) overcome this separation, because tangible controls are dedicated (nonmultiplexed) and may be directly representational, in an iconic and concrete

rather than symbolic or abstract way, of the particular control functionality they support. Moreover, the fact that these interfaces represent a *physical embodiment* of this functionality is deeply significant with regard to all that has been mentioned previously: they provide affordances associated with this physical form that may be richer and more intuitive than those of classical controls (mouse, keypad, buttons) and may also be associated with nonvisual feedback (e.g., haptic force feedback) that markedly increases the bandwidth of interaction, especially by not requiring focused attention from the user.

Graspable interfaces (Fitzmaurice et al., 1995) were an earlier presentation of the TUI concept, placing emphasis on a one-to-one correspondence between input devices (the bricks) and the control functionality they embody. This is intended to eliminate the cognitive dissonance that supposedly results in the WIMP model from having the representations of these controls (usually GUI widgets) be space-multiplexed (i.e., separated in the 2D desktop space), whereas the corresponding physical manipulation is physically collapsed (time-multiplexed) into one device (the mouse).

In a broader and less specific view, the notion of tangible interface is related to the use of tactile/haptic/force feedback *output* interfaces, thus placing the emphasis on the role of the corresponding human sensory modalities (tactition, proprioception/kinesthesia) to complement the overused and often saturated visual or auditory modalities. Devices such as Braille displays could be considered typical tangible interfaces in this view, but their role is very similar to that of classical visual displays, substituting one sensory modality by another for the visually impaired. In an alternative view, tactile feedback is considered as a complementary means to convey information when the auditory or visual modalities are saturated or cannot be attended to by the intended recipient.

In both perspectives, whether used as input or output interfaces, tangible interfaces are congruent the general AmI objective of "broadening the bandwidth of interaction between people and digital information" (Ishii and Ullmer, 1997).

An example that the use of tangible physical objects in interaction does not necessarily require or involve electronic tagging as it is now often the case (e.g., using RFID) is the *Passage mechanism* (Konomi et al., 1999; Streitz et al., 2001). Passage is a mechanism for establishing relations between physical objects and virtual information structures, that is, bridging the border between the real world and the digital, virtual world. So-called "passengers" (passage-objects) enable people to have quick and direct access to a large amount of information and to carry them around from one location to another via physical representatives that are acting as physical bookmarks into the virtual world. Passage is a concept for ephemeral binding of content to an object. It provides an intuitive way for the transportation of information between computers and Roomware components (Streitz et al., 1998, 2001), for example, between offices or to and from meeting rooms. A passenger does not have to be a special physical object. Any uniquely detectable physical object may become a passenger. Since the information structures are not stored on the Passenger itself but only linked to it in a dedicated way, people can turn any object into a passenger: a watch, a ring, a pen, glasses, or other arbitrary objects. No electronic tagging is needed. Identification and recognition is facilitated via selected physical properties, for example, the weight of an object.

A strong connection exists here with the field of mixed and augmented reality, which has emerged in a completely independent way as an outgrowth of virtual reality (see also Chapter 12 of this handbook). Mixed and augmented reality interfaces cover a complete spectrum, from virtual-reality-like displays (usually head-mounted) that superimpose a few inputs from the real world on an immersive simulated environment, to ordinary tangible devices that get augmented with digital information, with all possible intermediates in-between. Of more interest here is the latter end of the spectrum, which resonates very much with the AmI idea of moving interfaces off the desktop, out of traditional displays and GUIs, back into the real world where people live, sense, and act. This kind of interface may be direct, where the augmented object (Holmquist et al., 2004) is directly used as a combined input/output interface: as an input interface as described previously; as an output interface, it may use a projected display (Pinhanez, 2001; Molyneaux et al., 2007) where a projector-camera combination is used to track the object or augmented surfaces as developed by Rekimoto and Saitoh (1999). The role of the physical interactive paradigm for mixed reality was described by Cheok et al. (2006), illustrating it via examples of entertainment applications. Much less radical but already demonstrated is an approach where an intermediary device, such as a smartphone, is used to convey information about locations or artifacts that it can sense, either directly or indirectly, in the real world. Optical technologies (such as QR codes) or radio-based technologies (such as NFC) may be used for this purpose.

60.4 Parallel or Anterior Evolutions with Strong Relationships to AmI

60.4.1 AmI and Artificial Intelligence

AmI has been envisioned and publicized initially as being mostly a technology-centric evolution, with some degree of confusion, as highlighted in the introduction, as to what aspect of this evolution was the most distinctive. As previously discussed, the distinctive and defining feature of the AmI evolution is the stronger and richer coupling between the information world and the physical world. From this starting point, the Ubicomp/AmI agenda branches off into two broad domains of applications corresponding to:

- (Human ⇔ Information) Coupling

Human interactions with information are no longer confined to impoverished representations of the physical world such as the desktop. They can use a richer and more diverse repertoire of interaction resources, as exemplified by such trends as tangible interfaces, interactive spaces, augmented reality, and so on, through which, ultimately, the physical world itself becomes its own representation as a support for interaction with information or information-mediated interaction.

- (Physical World ⇔ Information) Coupling

All kinds of physical things get sensed and actuated through networks, even passively or indirectly. They can be seen as the analogues of the sensory-motor terminations of a peripheral nervous system, extending monitoring and controlling qualitatively and quantitatively beyond present-day applications.

Interestingly, a parallel tendency has been emerging[13] at a very different level in cognitive science and cognitive systems technology, with different schools of thought (embodied cognition, behavioral robotics, enaction, developmental learning) conjoining in the idea of *situating* and *grounding* cognition in the *hic et nunc* physical world and questioning the dominantly representational/objectivist stance of orthodox cognitivism.

The embodied cognition/behavioral robotics trend of cognitive sciences questions the basic tenet of artificial intelligence, which is the pivotal role of representations, as used in the still dominant symbolist paradigm of AI.

AmI has implicitly tended to adopt this representational model, especially for dealing with context awareness. Yet, context awareness, approached this way, has made little headway in real-world applications beyond exploiting the lowest levels of context data (e.g., geographical coordinates in outdoor settings for mobile location-based services, user connection status in instant messaging, etc.). This AI-heavy vision of AmI tends to enforce *a priori* delineation and modeling of context, aggregation/abstraction of numeric data into symbolic/logic variables and rule-based inference. Requiring an ad hoc set of rules and extensive manual programming, this approach may result in brittle and unscalable systems, especially when applied to a high-level context that is intrinsically ambiguous. Besides practical problems, more fundamental arguments question this approach. Ex ante formalization of context is completely at variance with the most salient and characteristic feature of AmI, which should lead to recentering on situation in the physical world, embracing the highly dynamic nature of this environment. Also, separation of context from content/activity negates the emergent and tightly coupled nature of their relationship: context arises from activity as much as activity arises from context (Dourish, 2004). As yet, little work has appeared to take stock of this remarkable convergence of technological and scientific evolutions, and even less to build upon it.

60.4.2 Mobile Computing and Communication

The distinction between embedded computing, mobile computing, and pervasive computing is made clearer by the diagram in Figure 60.4.

---

[13] Both evolutions arose from the 1990s onwards, largely independently as the two communities have had little interaction.

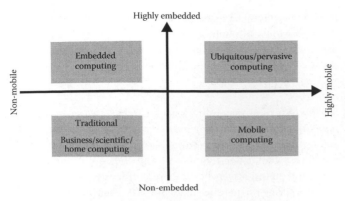

FIGURE 60.4 Topic map for embedded and mobile computing.

Mobile communication has been extensively successful in making space *irrelevant*. AmI is fundamentally different in that it takes a clear stance on *re-anchoring* communication in the here-and-now familiar space and time of users, and for this it draws upon both the infrastructure that supports mobile communication *and* the technologies of embedded computing; embedded computing being computing that interacts directly with the physical environment through sensors and actuators other than human interface devices.

Mobile communication has freed users from wire-line network attachments, but ties them to a handheld device that tends to look like a mobile avatar of the PC, a do-it-all device that just transposes the paradigm of personal computing/communication to mobile computing/communication, without fundamentally altering it. Contrary to a still dominant usage, mobile computing is *not* embedded computing. Today's mobile device is now a universal computing and communication platform, more powerful than a PC from 3 years ago. Only when it becomes equipped with special physical sensors and actuators (beyond its routine human interfaces) such as location sensors and biometric sensors, does it become a ubiquitous/pervasive computing device.

AmI aims at freeing users from mobile devices as exclusive intermediaries, by letting all these sensors and actuators come into play, grounding the locus of communication and the focus of interaction in the surrounding space, rather than confining them to a mobile device and a puny shrinkage of the 25-year-old GUI paradigm.

Location-based mobile communication services have been the very first step in going beyond the first stage of mobile computing, which might be seen as "making location irrelevant." This amounts to recognizing that location is a highly relevant piece of context that matters to the communication activity and may be taken into account as such by the communication system. Of course, it was necessary to untether communication first, so as to be able to re-anchor it in a different way than was previously the case, grounding it in space proper rather than tying it to a network termination.

There is much more to space and physical context than simply location, and ambient communication extends and enriches location-aware mobile communication in all these respects:

taking into account the presence and activity of persons rather than just whether they are hooked up to a network, adapting to their position or even posture in their environment. Ultimately, moving one's body around may become a way of interacting with a communication system, using that most natural of human capabilities (Borkowski and Privat, 2007).

60.5 Seven Alternatives about AmI

This section identifies and describes seven alternatives that are characteristic of the current situation of AmI, but at the same time are also pointing to future developments. Of course, as with most dichotomies, there is no either-or. Even though there will be a continuum of solutions between the two end points of each alternative, naming the alternatives helps to establish a more transparent view of the design options and opportunities that are available.

60.5.1 Alternative 1: It's All in the Environment versus Carrying Devices with You

The mobile computing approach is characterized by the fact that one can rely on a mobile infrastructure provided by mobile phones, PDAs, laptops, and so on, and the wireless networks to which they attach. This seems to be an advantage over the previous era of having stationary devices such as desktop PCs that were usually only available in dedicated places such as in offices, at home, or in an Internet cafe. One has to keep in mind that the mobile approach requires the user to carry the different devices always with him and to take care of power supplies, chargers, and so on. In some cases, weight and security plays an important role, too. Furthermore, compatibility with unknown infrastructures in new places can be an issue.

This contrasts with an alternative situation where it is envisioned that it will not be necessary to carry along all these different devices. In this vision, it is the environment that provides everything that is needed. This is similar to some situations today where, when entering a conference or meeting room, paper and pencils are provided on the table, or in a hotel room where one finds paper and a pen on the nightstand and a phone and a TV set as well. One would never think of carrying a TV set while traveling, yet laptops are being carried most of the time, especially on business travels. An important characteristic of the situations described is that the devices are abundantly available and ready to use. This is now generalized to all kinds of smart artifacts populating rooms, hallways, public and private spaces. Thus, *the world becomes the interface* to information and for communication and cooperation with other people.

The only requirement encountered in this environment is that the users identify themselves. This can be done in various device-less ways using only certain unique characteristics of the human body as, such as voice, eyes, fingers (see Chapter 13 of this handbook). For less security-critical applications, an implicit contextual pseudo-identification may even be sufficient. After such an identification process, that is also used to get the corresponding access rights to necessary information sources, one can appropriate the range of devices available. Which specific kinds of devices are available is not relevant in the context of the present argument. The devices are public in the first place. They are instances of the full range of devices possible. They can be pad-like mobile artifacts or—using another technology—electronic paper-like artifacts lying on the table. The table itself can be an interactive table or—using new smart materials—covered with interactive tablecloth; the wall can be an interactive wall or covered with interactive wallpaper. The devices can be integrated with other everyday objects being available in the local space. Depending on the characteristics and the affordances (Streitz et al., 2007) that are provided by the devices and the environment, the interaction with them can be explicit or implicit. Some devices or services might be even used without any identification and authentication. Once one identifies oneself, however, they can be turned into personalized tools that support each individual in a familiar way, independent of the specific instance of the device or the current location. Compatibility issues do not play a role, because the devices are connected to the local infrastructure that knows about them. The local infrastructure is connected with the global infrastructure, which provides the information resources, requested via the local infrastructure. Once the use of these personalized devices is finished, the service is quit, the devices are depersonalized and abandoned or left where they were picked up, so that they are available for the next person passing by.

Alternative 1 is closely related to the next alternative, especially with respect to the issue of investigating the continuum between the two end points.

60.5.2 Alternative 2: Swiss Army Knife versus Dedicated Devices

While the PC in its early phases and similarly the mobile phone in the initial stages were somehow dedicated devices providing specific functionality for defined purposes (word processing, painting and drawing, calculating in case of the PC; making and receiving phone calls in case of the phone), their developments over time showed an increasing amount and diversity of additional functionality being assigned to them. Of course, there are pros and cons for each approach and in the end one has to engage in trade-offs. Norman (1998) discussed some of these issues in his book, *The Invisible Computer*, and argues for dedicated *information appliances*, a term that was originally created by Jef Raskin who started the Macintosh project at Apple, and later founded his own company called Information Appliance. Norman uses the analogy of using a variety of dedicated tools when repairing something at home vs. having a Swiss Army knife with the "all-in-one" approach that people use while traveling. The advantage of a dedicated device is its simplicity and straightforward provision of the needed functionality, often in combination with a tangible interface. The complexity of the device does not exceed the complexity of the task. The disadvantage is the multitude of devices that one has to carry around, and—now especially for the case of information appliances—the difficulty of exchanging information between the different devices to use them in different contexts. Applying these ideas to AmI environments and combining them with Alternative 1 ("It's all in the environment") results again in two alternatives. On the one hand, one can consider, for example, an augmented room as a kind of Swiss Army knife where all the functionality is provided in a comprehensive way embedded and integrated in the walls, the furniture, and so on. This amounts to a virtualization of the variety of tools constituting a Swiss Army knife and an impersonation of the functionality. On the other hand, it can also be envisioned that there are dedicated devices lying around and available for whoever enters the space, an idea that corresponds to dedicated tangible interfaces of smart artifacts. And, of course, there is good reason to assume that a coexistence of both alternatives is to be foreseen.

60.5.3 Alternative 3: Physical versus Virtual (Physical Correspondences to Virtual Objects versus Virtual Reality and Immersion in a Simulated World)

Tangible interfaces (described previously) and virtual reality are the two opposites in a spectrum of potential evolutions of human interfaces. On the one side, proponents of AmI interfaces take a strong stance toward making interfaces stand on their own, "out there" in the real world, using all kinds of real-life material objects and various physically based interaction modalities (tactition, proprioception, spatiation). In the simplest possible viewpoint, real-world objects are mapped one-to-one with virtual objects such as interface widgets in a traditional GUI metaphor. A classical example is using traditional physical photo frames as representatives for the person whose photo they enclose (e.g., starting a communication with this person when the frame is swiped by a communication device).

Ultimately, the environment, through its spatial structure, is the interface, and traditional interface devices are mere foci of temporary fixation in a wide repertoire of interaction modalities afforded by this environment.

From within the virtual reality camp, a *representation* of the real world is the interface, a mere evolutionary step from the 2D desktop metaphor to a similarly impoverished 3D metaphor. Augmented objects are potentially a way to bridge these two opposite viewpoints. In this view, real-world objects do not merely stand for themselves, they may be enhanced with digital representation capabilities that make it possible for them to represent different virtual entities, either as output or input interfaces. This bears some similarity to augmented reality, but does potentially afford a richer interaction repertoire: augmented objects may be either input or output interfaces, whereas augmented reality deals mostly with output interfaces. The Passage mechanism (Konomi et al., 1999; Streitz et al., 2001) described in Section 60.3.7 is an example of the approach to use physical correspondences to virtual objects for interaction. A general comment on the different approaches could be formulated by contrasting them as "putting the real world into the computer" (e.g., a virtual reality simulation of an existing building) versus "putting the computer into the real world" (e.g., integrating computer technology into the walls, doors, or furniture of a building as it is done in the Roomware-approach by Streitz et al., 1998, 2001). The latter is to be favored when aiming at the future of human-*environment* interaction, although there is no doubt that there is also a justification for certain virtual reality applications.

60.5.4 Alternative 4: People-Oriented, Empowering versus System-Oriented, Importunate Smartness (Smart Spaces Make People Smarter)

Without entering into the philosophical discussion of when it is justified to call an artifact *smart* or what constitutes smart or intelligent behavior in general, the following distinction turns out to be useful (Streitz et al., 2005).

60.5.4.1 System-Oriented, Importunate Smartness

An environment is smart if it enables certain self-directed (re)actions of individual artifacts (or by the environment in case of an ensemble of artifacts) based on previously and continuously collected information. For example, a space or a place can be smart by having and exploiting knowledge about which people and artifacts are currently situated within its area, who and what was there before, when and how long, and what kind of activities took place. In this version of smartness, the space would be active (in many cases even proactive) and in control of the situation by making decisions on what to do next and actually take action and execute them automatically (without a human in the loop). For example, in a smart home, a person has access control to the house and other functions like heating, closing windows and blinds automatically. Some of these actions could be importunate. The almost classic example is a smart refrigerator in a home analyzing consumption patterns of the inhabitants and autonomously ordering depleting food. While one might appreciate that the fridge makes suggestions on recipes that are based on the food currently available (that would be still on the supportive side), it can be very upsetting if the fridge is autonomously ordering food that will not be consumed for reasons beyond its knowledge, such as a sudden vacation, sickness, or a temporal change in taste.

60.5.4.2 People-Oriented, Empowering Smartness

This view can be contrasted by another perspective where the empowering function is in the foreground, a view that can be summarized as *smart spaces make people smarter*. This is achieved by keeping *the human in the loop*, thus empowering people to make informed and mature decisions and take actions while being in control. In this case, the environment also collects data about what is going on and aggregates the data, but it provides and communicates the resulting information—appropriately in an intuitive way, so that ordinary people can comprehend it

easily—for guidance and subsequent actions determined by the people. In this case, a smart space might also make suggestions based on the information collected, but the people are still in the loop and in control of what to do next. Here, the place supports smart, intelligent behavior of the people present (or in remote interaction scenarios where people are away "on the road" but connected to the space). This view can be summarized as smart spaces make people smarter.

Of course, these two points of view will often not exist in their pure distinct forms. They rather represent the end points of a dimension where weighted combinations of both can be positioned somewhere in between. What kind of combination will be realized is different for different cases and depends very much on the application domain. It is also obvious that the system is not asking for user's feedback and confirmation for every single step in an action chain, because this would result in an information overload. The challenge is to find the right balance for feedback and requests for decisions to be made by the person. The position propagated here is that the overall design rationale should be guided and informed by the objective of having the human in the loop and in control as much as possible and feasible.

In this context, it is interesting to revisit the scenario at the end of Weiser's article (1991), where he describes parts of a day in the home, on the road, and in the office. One gets exposed to a world where comfort is provided; everything is calm and working smoothly. In case of some annoying occurrences (e.g., a traffic jam), they are remedied by being warned about it, following the suggestion of taking the next exit and having a cup of coffee. In the office, the existence of fresh coffee is indicated by a telltale odor at the door. Although not explicitly called smart environment or ambient intelligence, it is this view of the world that determines these ideas. The combination of extensive sets of sensor data and inference engines and other methods from artificial intelligence should allow taking care of the manifold situations to be encountered. Streitz (2008) provides a discussion putting Weiser's article in perspective of other developments.

60.5.5 Alternative 5: Calm Technology versus Cognitive Overload (More Services, More Interfaces, and More Devices Mean Cognitive Overload)

Weiser's vision (Weiser and Brown, 1996) of calm technology conjured up a compelling comparison between a traditional computer interface (or, for the sake of argument, a poorly designed AmI space) and a walk in the woods. Both provide a barrage of stimuli to the sensory system, but one may be bothersome and frustrating, while the other is deeply calming. The main difference between the two is in the way they *engage attention*: "Calm technology engages both the center and the periphery of our attention, and in fact moves back and forth between the two" (Weiser and Brown, 1996).

Moving the interface out of monopolistic interface devices, off the desktop, and into the environment, as advocated in the AmI vision, also means more devices (and more services offered on top of this wider interface bandwidth) requesting attention, and may in this sense be contradictory with the initial vision.

This is why these devices should not only be ubiquitous, connected, and smart: they should be *attentive* in the sense that, ideally, they should take into account both the relative importance of the information conveyed and the current activity and focus of attention of the user. Present-day synchronous communication (and its infamous ring tones) is the most primitive possible counterexample. It engages attention in the crudest possible way, inflicting an interruption and request for exclusive attention.

AmI services, however ubiquitously scattered throughout the environment they may become, will remain calmly unobtrusive if they can get weaved into the finest fabric of time by paying attention to human attention.

60.5.6 Alternative 6: Context versus Content

The distinction between context and content is central, at least implicitly, to a mainstream view of AmI. It is better illustrated by an example application of ambient remote communication. Audio or video content is the raw material of this kind of communication and is usually transmitted as such, with minimal coding/compression. The perimeter of this content is usually determined by the capabilities of the input interfaces, but is in general fixed (corresponding, e.g., to the sensitivity of the microphone or the field of view of the camera). Context, by contrast, has a more fluctuating perimeter: it may encompass any data that are relevant to the communication situation, such as location of either parties, their activities, whether someone is present in the environment, and so on. At a more technical level it is often thought of as comprising the status of their connection, its bandwidth, and so on. In classical communication, this context has to be transmitted "in-band" explicitly (e.g., by responding to "where do you call from" in mobile communication). In different forms of ambient communication, different types of context data may be transmitted "out-of-band," making it possible either for the system to adapt automatically to them, or to render them to the distant user. This context is not only transmitted separately, but also processed entirely differently, using context middleware (context management systems), which has been the subject of abundant research work (e.g., Ramparany et al., 2007), requiring complex and error-prone interpretation if it is to rise above basic raw data.

Processing and transmission of content is, by contrast, an extremely well-known process, extensively studied and incrementally refined since the beginning of telecommunications. As per the characterization of smart spaces articulated previously, physical interfaces distributed through the smart space that acquire context, and, if relevant, render it, will become more and more shared between context and content. Taking the previous example further, cameras intended to capture raw content to be transmitted as such for visual communication can also be used for vision-based tracking of people in their environments, thus providing context data for applications to exploit.

New modes of operation of visual communication make it possible to adaptively blend content with context: moving from narrowly focused visual communication to wide-angle communication, or, conversely, filtering out part of an image selectively, puts context back into content and vice versa. This is an example of how the two kinds of data may be processed and rendered similarly, rather than treated through two entirely separate channels, as is mostly the case at present. Escaping the context-content dichotomy makes it possible to better take into account the inherently dynamic and variable nature of the perimeter of context data. While mainstream context management in AmI enforces *a priori* delineation and modeling of context, accepting all inputs as potentially either context or content provides a more open viewpoint and makes it possible to reconfigure the context-content border dynamically in real time. The system that treats both types of data in this way is no longer merely context aware, it becomes a full-fledged cognitive system for which, just as for the human cognitive system, context is not explicit, even less formalized.

60.5.7 Alternative 7: Privacy as a Right versus Privacy as a Commodity

There is a controversy looming ahead between "privacy as a legal and moral right" versus "privacy as a commodity" that can be traded and that must be paid for; this is becoming of central importance for the development of AmI and its acceptance by people and society in general. AmI environments depend heavily on the deployment of sensors and the exploitation of the data that they are delivering. Actually, the intelligent or smart behavior is critically dependent on the sensor data obtained and processed. Looking at comprehensive ambient intelligent scenarios, it is obvious that large amounts of information about almost every aspect of life (past preferences, current activities, and future plans) are being collected to provide smart services. This leads to an unprecedented coverage of comprehensive activity monitoring and recording of personal data. Since a number of the scenarios derive their power mainly from the combination of a multitude of information sources, the interconnectivity implies an unprecedented level of data sharing and probably unwanted information flows. In combination with the notion of the disappearing computer, data collection in these sensor-based environments will happen in most cases in a hidden and unobtrusive way. Making technology invisible means that sensory borders disappear and common principles like "if I can see you, you can see me" no longer hold. Because collecting and processing personal information is a core function of smart environments, privacy and ubiquity seem to be in constant conflict. All this leads to a number of critical issues that are usually discussed under the term *privacy* (see Chapter 57, "Security and Privacy for Universal Access"). The fear of privacy infringements is calling for privacy enhancing technologies (PETs). Before discussing the implications, the following distinction is useful when addressing privacy issues:

- *Outgoing data* refers to the usual notion in combination with privacy (i.e., data logging, tracking, surveillance of people's behavior, etc.).

- *Incoming data* refers to intrusion, unsolicited communication, spam (be it e-mail, short text messages, phone calls, and postal mail). But also the loud and after some time unbearable conversation on the mobile phone of somebody (e.g., on a train) can be considered an intrusion into someone else's privacy. At the same time, these people are giving up their own privacy to some degree because some very personal information one might overhear they would rather hide (although it seems that some people are losing an appropriate and well-mannered reservation more and more).

In principle, everybody will start out with the notion that privacy is a legal and moral right and that this should be maintained by all means despite the increasing demand of collecting personal data. If more data are being collected, more legal measures should be taken to protect privacy. On the other hand, it can be observed that in a lot of situations, including outside the technology-based data collection, privacy is often a socially and commercially negotiated feature. People engage in trade-offs by compromising and balancing different interests they have. This is often the case in interpersonal and social interactions. In other situations, people are willing—sometimes not really being aware of the implications—to treat their privacy as a commodity they trade to obtain some (sometimes small) benefit. A good example is the increasing use of loyalty and customer cards where people are willing to allow that the store can collect personal shopping and consumption data and create customer profiles of personal preferences, and so on. Another example is the voluntarily provided exhibition of personal data on web sites of social communities as (e.g., FaceBook, My Space, etc.). It is amazing what kinds of information especially young people are providing there, in many cases probably without any awareness of the implications, especially in the long term. These developments and the lack of consciousness make it sometimes difficult to argue for more protection of privacy.

It seems that the privacy dilemma, that is, the trade-off between the amount of information necessary to provide personalized smart services and the right of people to be in control over which data are collected, by whom, how they are used, and so on, cannot be solved. Although there seems to be no escape from data collection (and data will be stored), it can be mitigated, for example, through new types of negotiation mechanisms that allow different decisions at different points in time and in different contexts. This amounts to proposing that people should be more involved in the conditions of the data collection processes. Similar to the discussion of Alternative 4 ("Smart spaces make people smarter"), it is necessary to keep the human in the loop, this time at the level of data collection. In correspondence with the previous discussion of how much intermediary level information (e.g., from the smart home) people want to have as feedback and are able to process, similar questions should be asked here. How much feedback on which data are being collected about an individual in a particular situation can be processed without being overloaded to engage in a negotiation process? Although the negotiation proposal tends to deviate from the original principle of the legal and moral right, it is an attempt to make the situation more transparent and to assure that the user has the final word. The implicit notion of trading and therefore treating privacy to some degree as a commodity (although under the full control of the user) will change future life in various ways. For example, it might result in a situation where privacy becomes a privilege and people who can afford to pay for it will have more privacy than those who cannot.

60.6 Summary and Conclusions

The preceding sections presented an overview of the basic concepts and trends in AmI currently being proposed and discussed in the scientific community but the chapter does not intend to provide a comprehensive account. Furthermore, a frame of reference was provided that allows for understanding that there might be an underlying (r)evolution in information and communication technology that goes beyond the variety of buzzwords and "new" research areas currently being witnessed; a variety that is also reflected in the range of different names of conferences, workshops, journals, and books.

The three eras identified in the framework are *microcosm*, *telecosm* (both based on Gilder's terminology), and *ambicosm* (a term created by the authors). The defining limitation of this ambicosm is the time and attention of users, compounded by the exponentially growing amounts of content using the unlimited bandwidth of telecosmic networks and the corresponding bandwidth of interface devices that present this content to users. The defining technologies of this era can be summarized as being mainly MEMS (micro-electro-mechanical systems) and nanotechnologies, the defining abundance is the physical interaction capability that is available.

Beyond the framework and presentation of various threads and characterizations of AmI, seven alternatives were proposed and elaborated to stimulate a discussion on design challenges and choices as well as on some fundamental conceptual issues. The *first alternative* suggests a vision that there will be no need anymore to carry all kinds of devices when mobile but that one can assume everything is already there, provided by the environment in which one is engaging. The only requirement is that users have to identify themselves, which can be done in a deviceless way using biometric methods. For less security-critical applications, an implicit contextual pseudo-identification may even be sufficient. In case of ambient environments, built-in appropriate affordances have to be provided allowing users to become aware of the potential interaction and communication possibilities available. The *second alternative* addresses the issue of providing a multitude of dedicated devices or information appliances versus the Swiss army knife approach. While the mainstream AmI emphasis would be leaning toward providing dedicated devices and services in a context-aware fashion, the proposal of considering the totality of an augmented smart room or a smart building as a Swiss army knife offers an additional design perspective. The *third alternative* discusses the always present design dimension physical versus virtual, and elaborates the existence of physical correspondences to virtual objects versus virtual reality and immersion in a simulated world. This can also be viewed as placing the real world into the computer versus integrating the computer into the real world. Again, a spectrum of possible combinations of both views exist; one is called ubiquitous virtual reality, emphasizing more the VR aspect, while others are mixed reality, augmented reality, and dual reality. The preferred notion here is the one of augmented objects and smart artefacts because they potentially afford a richer interaction repertoire. The *fourth alternative* contrasts the approach of people-oriented, empowering smartness with system-oriented, importunate smartness. The first one is favored with the implication that the human has to be in the loop, not giving up for a fully automated environment. Nevertheless, there will be exploitation of all kinds of sensor data and information collected by the system but they will be used to enable people to make more informed and mature decisions. This can be summarized as "smart spaces make people smarter."

The *fifth alternative* addresses the key limitation of the ambicosm, the perceptual and cognitive bandwidth of users. Contrary to a literal vision of ubiquitous devices that would bombard users with a multitude of stimuli, AmI assuages their cognitive overload by situating information in a continuum between the center and the periphery of their attention.

The *sixth alternative* contrasts a present-day view of AmI, where context is handled differently and separately from content proper, with a long-term perspective where the evolution toward more powerful cognitive systems would dissolve the distinction between content, context, and control.

The final *seventh alternative* addresses the issue of treating privacy as a legal and moral right versus a tendency that privacy might become more and more a commodity that one has to pay for and thus a privilege for a small group. It discusses privacy with respect to outgoing and ingoing data, which appear to be perceived very differently. It is also pointed out that there are alarming signs that people are willing to give up some of their privacy because they obtain small benefits (e.g., with loyalty cards) without actually being aware of it. Given this situation, it seems difficult to argue for a better protection of privacy in sensor-based environments although many individuals and organizations are demanding it. It shows that the general public has to be better informed about the implications for privacy infringements when providing data without their real intention and allowing data to be collected without their permission. Otherwise, the position is weakened and the argumentation for privacy will be difficult to uphold, which would have severe implications for our personal and social life and for society in general.

Looking at future trends, as is done for example in the EU-funded Coordinated Action InterLink (InterLink, 2008), one can observe that the evolution toward future information and knowledge societies is currently the theme of many efforts worldwide. Examples are the i2010-program of an Inclusive European Information Society, the plan for u-cities in Korea, the idea of U-Japan as a ubiquitous network society, and the iN-2015 master plan for the Intelligent Nation Singapore. They can be characterized by the development of personalized individual as well as collective services that exploit new qualities of infrastructures situated in smart environments and based on a range of ubiquitous and pervasive communication networks providing ambient computing at multiple levels. The underlying vision of pervasive and ambient computing assumes very large numbers of invisible, small computing devices embedded into our environment. They will interact with and be used by multiple users in a wide range of dynamically changing situations. In addition, this heterogeneous collection of devices will be supported by an infrastructure of intelligent sensors (and actuators) embedded in our homes, offices, hospitals, public spaces, leisure environments, providing the raw data and active responses via actuators needed for a wide range of smart services. Furthermore, new and innovative interaction techniques will move toward the integration of tangible and mixed reality interaction. Thus, the interaction experience of the user will be more integrated and intuitive than today. It is anticipated that economics will drive this technology to evolve from a large variety of specialized devices to a small number of universal, extremely small and low-cost devices that can be embedded in a variety of materials. It is anticipated that this technology will be part of all artifacts that will be built in the future or will be added to augment existing artifacts. Thus, we will be provided with a computing, communication, sensing, and interaction substrate for systems and services. We can characterize them as "smart ecosystems" (borrowing concepts from other disciplines) to emphasize the seamless integration of the components, their smooth interaction, the equilibrium achieved through this interaction and the emergent smartness of the overall environment. We assume that the degree of diffusion of smart devices and supporting sensors and actuators will result in smart ecosystems that might parallel other ecosystems in the not too far future and associated aggregated entities will show new kinds of emerging behavior.

Under all these sweeping plans, as well as under many of the research threads presented in this chapter, lurks a common potential pitfall: the over-engineered grand design that forces its own technology-centered vision upon users. These visions have all too often been presented as scenarios in the style of "A day in the life of the ambiently intelligent Ms. X in 2020." This chapter has not included any of these scenarios, which, replicated *ad infinitum,* have become a niche subgenre of technical literature. Beyond their lack of originality, most of these designs give insufficient credit to users for the serendipitous and paradoxical ways in which they can tweak, distort, and hijack the ready-made technology they are offered. This process ends up evolving something entirely new, a genuine cocreation of technology and usage. All the field studies and focus groups that have become a staple of research projects cannot match the "wisdom of crowds" effect that results from the evolutionary selection and recombination of features by large volumes of users, when they take control of technologies that are open-ended enough to support this. Such an AmI system that users can appropriate and play with may become the support of new media, a veritable engine for creation. In this view, the long-term goal of researchers, engineers, and designers in this field, and their ultimate vindication, should not be to impose on users their own vision of an ideal ambient intelligence world, but to open up and enrich the repertoire of interaction capabilities that their systems afford.

References

Bly, S., Harrison, S. R., and Irwin, S. (1993). Media spaces: Bringing people together in a video, audio, and computing environment. *Communications of the ACM* 36: 28–46.

Borkowski, S. and Privat, G. (2007). Spatial interaction in ambient communication, in the *Proceedings of ENACTIVE'07, 4th International Conference on Enactive Interfaces,* 19–24 November 2007, Grenoble, France. http://acroe.imag.fr/enactive07/proceedings.php.

Cheok, A. D., Teh, K. S., Nguyen, T. H., Qui, T. C., Lee, S. P., Liu, W., et al. (2006). Social and physical interactive paradigms for mixed-reality entertainment. *ACM Computers in Entertainment (CIE)* 4: article 5.

Dourish, P. (2004). What we talk about when we talk about context. *Personal and Ubiquitous Computing* 8: 19–30.

Ducatel, K., Bogdanowicz, M., Scapolo, F., Leijten, J., and Burgelman, J. C. (2001). *Scenarios for Ambient Intelligence in 2010.* ISTAG Final Report (IPTS-Seville, 2001).

Elrod, S., Bruce, R., Gold, R., Goldberg, D., Halasz, F., Janssen, W., et al. (1992). LiveBoard: A large interactive display supporting group meetings, presentations, and remote collaboration, in the *Proceedings of ACM CHI'92 Conference,* 3–7 May 1992, Monterey, CA, pp. 599–607. New York: ACM Press.

Fish, R. S., Kraut, R. E., Root, R. W., Rice, R. E. (1992). Evaluating video as a technology for informal communication, in *Proceedings of the ACM CHI '92 Conference,* 3–7 May 1992, Monterey, CA, pp. 37–48. New York: ACM Press.

Fitzmaurice, G. W., Ishii, H., and Buxton, W. (1995). Bricks: Laying the foundations for graspable user interfaces, in the *Proceedings of ACM CHI '95 Conference,* 7–11 May 1995, Denver, pp. 442–449. New York: ACM Press.

Gershenfeld, N. and Cohen, D. (2006). Internet 0: Interdevice internetworking. *IEEE Circuits and Devices Magazine,* pp. 48–55.

Gershenfeld, N., Krikorian, R., and Cohen, D. (2004). The Internet of things. *Scientific American* 291: 76–81.

Gilder, G. (1989). *Microcosm: The Quantum Revolution in Economics and Technology.* New York: Simon and Schuster.

Gilder, G. (2000). *Telecosm: How Infinite Bandwidth Will Revolutionize Our World.* New York: Free Press.

Hansmann, U., Merk, L., Nicklous, M., and Stober, T. (2003). *Pervasive Computing* (2nd ed.). New York: Springer-Verlag.

Holmquist, L-E., Gellersen, H-W., Kortuem, G., Schmidt, A., Strohbach, M., Antifakos, S., et al. (2004). Building intelligent environments with smart-its. *IEEE Computer Graphics and Applications* 24: 56–64.

InterLink (2008). *Report of the Working Group "Ambient Computing and Communication Environments" of the Coordinated Action InterLink.* http://interlink.ics.forth.gr.

Ishii, H. and Ullmer, B. (1997). Tangible bits: Towards seamless interfaces between people, bits and atoms, in the *Proceedings of ACM CHI'97 Conference,* 22–27 March 1997, Atlanta, pp. 234–241. New York: ACM Press.

Ishii, H., Wisneski, C., Brave, S., Dahley, A., Gobert, M., Ullmer, B., and Paul, Y. (1998). Ambient-ROOM: Integrating ambient media with architectural space, in *Proceedings of ACM CHI'98 Conference,* 18–23 April 1998, Los Angeles, pp. 173–174. New York: ACM Press.

ITU (2005). *The Internet of Things.* http://www.itu.int/internetofthings.

Konomi, S., Müller-Tomfelde, C., and Streitz, N. (1999). Passage: Physical transportation of digital information in cooperative buildings, in *Cooperative Buildings: Integrating Information, Organizations, and Architecture, Proceedings of the Second International Workshop (CoBuild'99)* (N. Streitz, J. Siegel, V. Hartkopf, and S. Konomi, eds.), 1–2 October 1999, Pittsburgh, pp. 45–54. Heidelberg: Springer-Verlag.

Mankoff, J., Dey, A., Hsieh, G., Kientz, J., Lederer, S., and Ames, M. (2003). Hueristic evaluation of ambient displays, in *Proceedings of ACM CHI'03 Conference,* 5–10 April 2003, Ft. Lauderdale, FL, pp. 169–176. New York: ACM Press.

Molyneaux, D., Gellersen, H-W., Kortuem, G. and Schiele, B. (2007). Cooperative augmentation of smart objects with projector-camera systems, in *Proceedings of Ubicomp 2007. 9th International Conference on Ubiquitous Computing,* September 2007, Innsbruck, Austria, pp. 501–518. Berlin/Heidelberg: Springer-Verlag.

National Research Council (2001). *Embedded Everywhere: A Research Agenda for Networked Systems of Embedded Computers.* Washington, DC: National Academies Press.

Norman, D. (1998). *The Invisible Computer.* Cambridge, MA: MIT Press.

Pinhanez, C. (2001). The everywhere displays projector: A device to create ubiquitous graphical interfaces, in *Proceedings of UbiComp 2001,* 30 September–2 October 2001, Atlanta, pp. 315–331. Berlin/Heidelberg: Springer-Verlag.

Ramparany, F., Poortinga, R., Stikic, M., Schmalenströer, J., and Prante, T. (2007). An open context information management infrastructure, in *Proceedings of the 3rd IET International Conference on Intelligent Environments (IE'07),* 24–25 September 2007, Ulm, Germany.

Rekimoto, J. and Saitoh, M. (1999). Augmented surfaces: A spatially continuous work space for hybrid computing environments, in *Proceedings of ACM CHI'99 Conference,* 15–20 May 1999, Pittsburgh, pp. 378–385. New York: ACM Press.

Russell, D., Streitz, N., and Winograd, T. (2005). Building disappearing computers. *Communications of the ACM* 48: 42–48.

Sakamura, K. (1990). The TRON intelligent house. *IEEE Micro* 10: 6–7.

Streitz, N. (2001). Augmented reality and the disappearing computer, in *Cognitive Engineering, Intelligent Agents and Virtual Reality, Volume 1 of the HCI International 2001 Proceedings* (M. Smith, G. Salvendy, G. Harris, and R. Koubek, eds.), pp. 738–742. Mahwah, NJ: Lawrence Erlbaum Associates.

Streitz, N. (2008). The disappearing computer, in *HCI Remixed: Reflections on Works That Have Influenced the HCI Community* (T. Erickson and D. McDonald, eds.), pp. 55–60. Cambridge, MA: The MIT Press.

Streitz, H., Geißler, J., and Holmer, T. (1998). Roomware for cooperative buildings: Integrated design of architectural spaces and information spaces, in *Cooperative Buildings, Integrating Information, Organization, and Architecture, Proceedings of CoBuild'98* (N. Streitz, S. Konomi, and H. Burkhardt, eds.), 25–26 February 1998, Darmstadt, Germany, pp. 4–21. Berlin/Heidelberg: Springer-Verlag.

Streitz, N., Kameas, A., and Mavrommati, I. (eds.) (2007). *The Disappearing Computer: Interaction Design, System Infrastructures and Applications for Smart Environments.* Springer "State-of-the-Art" Survey. Berlin/Heidelberg: Springer-Verlag.

Streitz, N. and Nixon, P. (2005). The disappearing computer. *Communications of the ACM* 48: 33–35.

Streitz, N., Prante, T., Röcker, C., van Alphen, D., Magerkurth, C., Stenzel, R., Plewe, D. (2003). Ambient displays and mobile devices for the creation of social architectural spaces: Supporting informal communication and social awareness in organizations, in *Public and Situated Displays: Social and Interactional Aspects of Shared Display Technologies* (K. O'Hara, M. Perry, E. Churchill, and D. Russell, eds.), pp. 387–409. Kluwer Academic Publisher.

Streitz, N., Prante, T., Röcker, C., van Alphen, D., Stenzel, R., Magerkurth, C., et al. (2007). Smart artefacts as affordances for awareness in distributed teams, in *The Disappearing Computer* (N. Streitz, A. Kameas, and I. Mavrommati, eds.), pp. 3–29. Springer-Verlag.

Streitz, N., Röcker, C., Prante, T., van Alphen, D., Stenzel, R., and Magerkurth, C. (2005, March). Designing smart artefacts for smart environments. *IEEE Computer*, pp. 41–49.

Streitz, N. and Savidis, A. (2006). *Mission Statement of the ERCIM Working Group SESAMI (Smart Environments and Systems for Ambient Intelligence).* http://www.ics.forth.gr/sesami/index.html.

Streitz, N., Tandler, P., Müller-Tomfelde, C., and Konomi, S. (2001). Roomware: Towards the next generation of human-computer interaction based on an integrated design of real and virtual worlds, in *Human-Computer Interaction in the New Millennium* (J. Carroll, ed.), pp. 55–578. Reading, MA: Addison-Wesley.

Tennenhouse, D. (2000). Proactive computing. *Communications of the ACM* 43: 43–50.

Ullmer, B. and Ishii, H. (2000). Emerging frameworks for tangible user interfaces. *IBM Systems Journal* 39: 915–931.

Weiser, M. (1991). The computer for the 21st century. *Scientific American*, pp. 66–75.

Weiser, M. (1998). The invisible interface: Increasing the power of the environment through calm technology. *Keynote at the First International Workshop on Cooperative Buildings (CoBuild'98),* in *Cooperative Buildings: Integrating Information, Organization, and Architecture, Proceedings of CoBuild'98* (N. Streitz, S. Konomi, and H-J. Burkhardt, eds.), 25–26 February 1998, Darmstadt, Germany, p. 1. Berlin/Heidelberg: Springer-Verlag.

Weiser, M. and Brown, J. (1996). Designing calm technology. *Power Grid Journal* 1: 75–85.

Wisneski, C., Ishii, H., Dahley, A., Gorbet, M., Brave, S., Ullmer, B., and Yarin, P. (1998). Ambient displays: Turning architectural space into an interface between people and digital information, in *Cooperative Buildings: Integrating Information, Organization, and Architecture, Proceedings of CoBuild'98* (N. Streitz, S. Konomi, and H-J. Burkhardt, eds.), 25–26 February 1998, Darmstadt, Germany, pp. 22–32. Berlin/Heidelberg: Springer-Verlag.

61

Emerging Challenges

Constantine Stephanidis

61.1 Introduction

Computing technology evolves rapidly, and each new generation of technology offers new opportunities to improve the quality of human life. The target user population addressed broadens, while the type, functionality, and content of new products and services expand and diversify, and access technologies proliferate. At the same time, however, each new generation of technology has the potential of introducing new difficulties and barriers in the use of products and services, and eventually, new risks for the health and safety of people, while new forms of social exclusion and discrimination emerge in everyday life.

As the information society develops and evolves, another technological paradigm shift is taking place toward the establishment of ambient intelligence (AmI) environments, characterized by invisible (i.e., embedded) computational power in everyday appliances and other surrounding physical objects, and populated by intelligent mobile and wearable devices.

A general description of the lines of anticipated technological development can be found in the report of the Information Society Technologies Advisory Group (2001), where a vision of the information society as ambient intelligence is offered:

> The concept of Ambient Intelligence (AmI) provides a vision of the Information Society where the emphasis is on greater user-friendliness, more efficient services support, user-empowerment, and support for human interactions. People are surrounded by intelligent intuitive interfaces that are embedded in all kinds of objects and an environment that is capable of recognising and responding to the presence of different individuals in a seamless, unobtrusive and often invisible way." (p. 1)

AmI is an emergting field of research and development that is rapidly gaining wide attention by an increasing number of researchers and practitioners worldwide, and in Europe in particular. The notion of AmI is also becoming a de facto key dimension of the emerging information society, since many of the new generation industrial digital products and services are clearly shifted toward an overall intelligent computing environment.

AmI will have profound consequences on the type, content, and functionality of the emerging products and services, as well as on the way people will interact with them, bringing about multiple new requirements for the development of the information society (e.g., Edwards and Grinter, 2001; Coroama et al., 2004; Emiliani and Stephanidis, 2005).

While a wide variety of different technologies is involved, the goal of AmI is to either hide the presence of technology from users, or smoothly integrate it within the surrounding context as enhanced environment artifacts. This way, the computing-oriented connotation of technology essentially fades out or disappears in the environment, providing seamless and unobtrusive interaction paradigms. Therefore, people and their social situations, ranging from individuals to groups, and their corresponding environments (office buildings, homes, public spaces, etc.), are at the center of the design considerations. In this context, accessibility and usability by users with different characteristics and requirements cannot be addressed through ad hoc technological solutions introduced after the main building components of the new environment are in place. The notions of universal access and design for all, therefore, are central to this vision (Emiliani and Stephanidis, 2005). Although the cost and availability of AmI technologies will be important factors for their wider spread and market success, the ultimate criterion for their public acceptance and use will be the extent to which people feel safe and comfortable using them.

This brings about the need for identifying a new role, contribution, and challenges of universal access. Although universal access by definition, and as demonstrated by the wide range of experiences and approaches collected in this handbook, aims toward the accessibility and usability of information society technologies by anyone, anywhere, and at any time, so far it has mainly addressed various interaction platforms (e.g., desktop computers, web, PDAs, mobile phones) working mostly independently from each other. In AmI environments, universal access will face new challenges, posed by pursuing proactive accessibility and usability in the context of ambient interaction. Therefore, universal access needs to build on achieved results and progress to broaden its scope and address such challenges.

Several chapters in this handbook address specific issues in an AmI perspective. Other chapters, on the other hand, discuss methodological and technological approaches that are likely to surface as critical in AmI. For example, Chapter 2 claims that the emergence of AmI environments, while asking for a more general application of the design for all approach, is also favoring its implementation, by making available in the environment the necessary interaction means and intelligence. Chapter 3 also describes a wide variety of novel and emerging technologies relevant to AmI, and their potential for built-in accessibility and better interaction for everybody. Development requirements for ambient interaction are discussed in Chapter 18.

Finally, the previous two chapters of this final part of the handbook, Chapter 59, "Implicit Interaction," and Chapter 60, "Ambient Intelligence," respectively, provide an overview of implicit interaction as one of the main interaction paradigms of AmI, and an overview of the basic concepts, trends, and perspectives of AmI.

This concluding chapter aims at both widening and deepening these perspectives in the context of a renewed account of universal access as an evolving field. For each main universal access dimension, new challenges emerging in AmI are identified, highlighting where possible current and anticipated directions and promising research approaches.

61.2 Diversity in the User Population

Undoubtedly, the target population of the envisaged technological environment includes every citizen in the information society, and the currently available body of knowledge on human characteristics and requirements, as addressed in Part II of this handbook, will need to be applied and further enriched with considerations related to environment interaction. For example, for blind, visually impaired, and motor-impaired users, requirements related to interaction need to be combined with requirements related to physical navigation in the interactive environment (see Chapter 49). Along the same lines, the complexity of the environment and the disappearing of technologies can become insurmountable obstacles for cognitively impaired users if not properly addressed. Age-related factors are also very important, particularly in light of the fact that a large part of AmI applications will be targeted to supporting independent living, and that current understanding of the needs and requirements of users of different ages in such a complex environment is limited (Kleinberger et al., 2007). Also, while AmI develops at a global level, it may be anticipated that cultural factors will become particularly relevant toward identifying and reasoning about users' goals and tasks, which may be highly influenced by different cultural backgrounds (Hartz Søraker and Brey, 2007).

Furthermore, computing is expected to become far more dependent on social factors than it has been in the past. Cooperation will be an important aspect, as communication and access to information will be concurrently used to solve common problems in a cooperative manner; moreover cooperation may be among human users themselves or among user representatives (agents and avatars), to whom variable degrees of trust can be assigned. Access to information and communications will no longer be the task of an individual, but will be extended to communities of users, who have at their disposal common (sometimes virtual) spaces in which they can interact. Finally, the dynamic reconfiguration of context will be highly dependent on social phenomena (users moving in the environment, meeting, communicating with each other, collaborating, etc.). Therefore, users in AmI environments can no longer be studied only at an individual level, and accounts of social behavior will become equally important. In this respect, work on virtual communities (see Chapter 42) and computer-supported cooperative work (see Chapter 43) is expected to provide useful insights.

61.3 Enabling Technologies for AmI

AmI is fostering rapid development in broad and diverse hardware and software technological areas that emerge as key enabling technologies (Information Society Technologies Advisory Group of the European Commission, 2003), and only some of them can be mentioned here.

Concerning hardware, ubiquitous, flexible, and reflective displays, extreme miniaturization, sensors and actuators, textile electronics, and solid-state lighting are considered as key emerging technologies (van Houten, 2006).

Complex heterogeneous networks need to function and communicate in a seamless and interoperable way in AmI. This implies the integration of mobile and fixed networks, which will have to be seamless and dynamically reconfigurable. Services are dynamic and can be reconfigured or recombined at run-time to accommodate different needs of different users in different contexts and environments. The web will play a critical role in such a context (Issarny et al., 2002), and content transformation technologies, as well as new and emerging web technologies (see Chapter 10) will be indispensable to ensure appropriate presentation of content throughout the distributed environment and the myriad of available fixed and mobile devices. For example, the Web 2.0 paradigm is considered to have a significant potential toward allowing the next wave of smart devices and digital objects to form ecosystems and societies of intelligent artifacts (Vazquez and López de Ipiña, 2008).

The AmI environment must also be safe, dependable, and secure. In this context, technologies such as biometrics, which have the potential to lead to intelligent and unobtrusive user and object identification, acquire particular importance (see Chapter 13).

Finally, AmI by definition needs to incorporate appropriate levels of intelligence, which are integrated in various devices and systems, but also distributed in the environment (Burzagli et al., 2007). Various artificial intelligence techniques are crucial to interpret the environment's state, represent the information and knowledge associated with the environment, model and represent entities in the environment, learn about the environment and its aspects, plan decisions and actions, interact with humans, and act on the environment (Ramos et al., 2008; see also Chapter 60). Agents (see Chapter 14) and natural dialogue interfaces (see Chapter 31) are key to enabling intelligence at the user interface. Another important advance is related to emotion-aware technologies, that is, technologies capable of taking into account the users' emotions (Zhou et al., 2007).

61.4 Development Life Cycle of User Interfaces

61.4.1 Human Needs in AmI Environments

A prerequisite for the successful development of any AmI environment is that user requirements are appropriately captured and analyzed. In particular, it is necessary to anticipate future computing needs in everyday life (Abascal and Civit, 2001), and acquire an in-depth understanding of the factors that determine the usefulness of diverse interactive artifacts in context. These requirements are likely to be more subjective, complex, and interrelated than in previous generations of technology. For example, traditional methods for capturing user requirements, due to their origin (business environments and application) are oriented toward highly structured tasks, with specific goals and steps. However, in everyday life, human activities do not always have a specific goal, and are characterized by a much looser structure, which may not easily decompose into discrete steps.

Under a universal access perspective, this implies incorporating the needs, requirements, and preferences of all individual users (see Chapter 15). As a starting point toward analyzing user requirements in AmI, studies have been conducted through the use of scenario techniques. Both positive (Information Society Technologies Advisory Group of the European Commission, 2003) and worst-case (Punie et al., 2006) scenarios have been proposed.

Extensions of scenarios to address various disabilities have also been elaborated (Antona et al., 2007). However, while scenarios provide a useful starting point, they cannot be considered as sufficient to fully capture future human needs and expectations in a context difficult to grasp. Current studies of human requirements in AmI environments mainly focus on homogeneous user groups and specific environments (Hellenschmidt and Wichert, 2005; Röcker et al., 2005). Some general nonfunctional requirements for AmI environments have also been identified and discussed (e.g., Bohn et al., 2005; Emiliani and Stephanidis, 2005; Kemppainen et al., 2007), and include accessibility, privacy, and safety. Accessibility in the context of AmI is usually intended as an inclusive mainstream product design (e.g., Kemppainen et al., 2007), although a definition is not yet available. However, a gradual transition from assistive technologies to mainstream accessible products is foreseeable (Antona et al., 2007).

Another relevant nonfunctional requirement is privacy, which concerns the effective protection of personal data that are collected through the continuous monitoring of people in AmI environments. New challenges arise concerning how a person will be able to know what information is recorded, when, by whom, and for what use (Friedewald et al., 2007). A thorough discussion of privacy issues in the context of AmI is provided in Chapter 60, "Ambient Intelligence."

In a situation in which technology may act on the physical environment and deal with critical situations without the direct intervention of humans, it is also likely that new hazards will emerge for people's health and safety. Possible malfunctions or wrong interpretations of monitored data can lead to unforeseeable consequences, especially for disabled users who will be more dependent on technology than others. Therefore, appropriate strategies for avoiding risks must be elaborated and validated. A related challenge is the consideration of interoperability among different technologies and devices, as the correct functioning of the intelligent environment as a whole needs to be ensured.

61.4.2 Design for All

In the context of AmI, the need of proactively embedding accessibility and usability in a variety of "hidden," interconnected, and multifunctional artifacts brings forward several implications for design for all (Emiliani and Stephanidis, 2005). There is a much wider range of individual and context-related requirements that must be taken into account, and, on the other hand, accessibility and usability of each device in isolation will be a necessary, but not sufficient condition for the accessibility and usability of the overall distributed environment. Design for all, therefore, is necessary to investigate, under a universal access perspective, the factors that are dynamically involved in the integration and cooperation of all elements in the technological environment.

So far, while design for all in the domain of interactive applications and services was inspired by previous approaches to design for all in the built environment, it departed from the latter in that it recognized the need of adapted and personalized interactive solutions (Stephanidis, 2001). Appropriate design methods and techniques will need to offer adequate means for the design of user interfaces capable of intelligent adaptation behavior (see Chapter 16). User-interface adaptation in AmI environments is likely to be affected by a larger number of factors than in current technological environments. For example, adaptation will also concern the selection of appropriate interaction devices at various interaction steps, according to the position of the user in the environment and the dynamic availability of personal and ambient devices (Savidis and Stephanidis, 2005).

At the level of interaction design, new design principles, which will cater to the newly introduced design parameters, will be required, stemming from the fusion and extension of design principles of everyday objects with software usability and accessibility principles. Furthermore, design approaches will need to take into account ways of dynamically combining the interaction capabilities of interconnected objects and devices (see Chapter 59). This also implies a more close study of the similarities and differences of design for all in the architectural and digital worlds, as these worlds will no longer be clearly separated, but, on the contrary, closely interrelated.

Therefore, it is necessary to elaborate design techniques and strategies that reflect and reason about dynamic change in the users' context and technologies diversity, and support complex forms of interrelated adaptations. Appropriate models need to be elaborated, empirically validated, and put to use.

61.4.3 Development Requirements

The development of AmI environments entails software engineering requirements addressing the dimensions of user, platform, and context diversity intrinsic in universal access (see Chapter 18, "Software Requirements for Inclusive User Interfaces"). Therefore, software engineering approaches elaborated in the context of universal access are emerging as crucial for AmI. Appropriate architectural frameworks and support tools for environment interaction are discussed in Section 61.5.

61.4.4 User Experience Evaluation

Evaluation of AmI technologies and environments needs to go beyond current accessibility and usability evaluation (see Chapter 20) in a number of dimensions, concerning assessment methods and tools, as well as metrics (Emiliani and Stephanidis, 2005). Ubiquitous AmI technologies and systems, such as personal digital assistants, wearable sensors, mobile phones, and so on, challenge traditional usability evaluation methods, because the context of use can be difficult to recreate. This suggests that the evaluation of a user's experience with AmI technologies should take place in real-world contexts. However, evaluation in real settings also presents difficulties, as there are limited possibilities of continuously monitoring users and their activities (Intille et al., 2006). In this respect, it is critical to combine user experience in context with the availability of the necessary technical infrastructure for studying the users' behavior over extended periods of time. Another relevant issue concerns the content of the evaluation. Performance-based approaches are not optimal for AmI environments, as they were developed for single user, desktop applications, and are usually applied in laboratory evaluations. Additionally, it is difficult to specify tasks that capture the complexity of everyday activities, and a more subjective view of the user experience is necessary. Qualities that may need to be taken into account in evaluating AmI technologies and environments include highly subjective factors such as attention, values, emotion, fun, privacy, and trust (Theofanos and Scholtz, 2004). Under a universal access perspective, accessibility of AmI environments is also an orthogonal concern, as AmI needs to be developed from the start as fully accessible and inclusive, and the accessibility of interactive technologies will be deeply interrelated with the accessibility of the environment.

It is expected that future testing and evaluation methods and activities will rely to a large extent upon the simulation of the interactive behavior of single devices and the overall technological environment.

61.5 User Interface Development: Architectures, Components, and Tools

AmI environments are intrinsically based on adaptation, and user and context awareness, as well as adaptation decision making becoming fundamental. Therefore, software architectures that foster and support adaptation (see Chapter 21, "A Unified Software Architecture for User Interface Adaptation") are required. An example of an architecture and development support tool for dynamic dialogue composition in ambient computing, based on abstract interaction objects along with a discussion of the involved issues and challenges, is provided by Savidis and Stephanidis (2005) (see also Chapter 27).

Other recent approaches to user interface architectures in AmI are mainly model-based (e.g., Clerckx et al., 2006), and therefore, in this respect, further advances are needed in present model-based design and development tools (see Chapter 25).

Appropriate user and context models for AmI need to be elaborated and present significant challenges, as they need to be dynamically updated through monitoring and highly influence each other. In AmI, all factors determining adaptation are highly dynamic and interrelated and vary over time (i.e., during interaction). For example, a user's abilities may change over short time intervals according to various conditions partly determined by external conditions (e.g., elements of the environment). Also, when a user moves in the environment, both the context and the set of available technologies change (e.g., a different room with different characteristics and technologies). Many users with different needs and requirements may be present at the same time in an environment, and interact concurrently, and this fact may introduce potential conflicts. Additionally, the context may change also due to factors not related to the user's position (e.g., according to different times of the day of different days). Technologies also vary depending on the user's position and context. Additionally, devices can be plugged in or removed from the environment at any time. The dynamic availability of services is also likely to lead to user tasks changing on the fly (e.g., Clerckx et al., 2006), that is, the content and functionality of interactive applications will be continuously adapted, bringing forward the need for user interface fusion (Bihler and Mügge, 2007), as well as dynamic adaptation strategies.

Therefore, current approaches to user modeling (see Chapter 24) need to integrate inference and reasoning capabilities toward capturing a user's intentions and dynamically changing tasks (González et al., 2005).

On the other hand, technological platforms cannot, any longer, be conceived in isolation, and a more articulated notion of context needs to be elaborated, including also the physical and the social environment (Schmidt, 2005; see also Chapter 59, "Implicit Interaction"). The number and potential impact of relevant factors increase dramatically with respect to conventional computing devices, particularly with respect to the (co)presence of people, computing devices and other elements in the surrounding environment. This implies the necessity of monitoring every relevant element in context, and raises the issue of identifying the elements that should be monitored by each device, and the conditions and parameters according to which monitoring should take place. Various proposals have been put forward in the literature to systematically analyze and address the notion of context in AmI environments (e.g., Dogac et al., 2003; Haya et al., 2004; Preuveneers et al., 2004). Available models of the AmI context distinguish among user-related context (i.e., the roles and actions of users, as individuals and as groups, within the environments), environment-related context (i.e., physical characteristics of the environment, such as temperature, light, noise, etc.), and platform-oriented context (i.e., the characteristics of the many available interactive devices). An important aspect of context concerns the location of the user, which can be identified through a variety of technologies (for an overview, see Chapter 59, "Implicit Interaction"). The user location is often considered to be constituting the center of context, as context needs to be defined around the user (Riva et al., 2003).

While sensor, monitoring, and recognition technologies offer the possibility of acquiring massive context data, one of the bigger challenges for AmI remains how to model, reason about, and exploit such data (Schmidt, 2005). Row data coming from sensors need to be appropriately interpreted (see Chapter 59, "Implicit Interaction").

Adaptation decision making (see Chapter 22) is also intrinsic in AmI. It needs to be realized at various levels, including the individual interacting device level, as well as the overall environment level.

Overall, the issues related to the context of use in AmI are in need of extensive experimentation to test and compare different context models, improve approaches to context inference, experiment with the usefulness and usability of context-based automated action of the AmI environment, and investigate the users' awareness and acceptance of such technologies.

61.6 Interaction Techniques and Devices

AmI environments integrate a wide variety of interactive devices, which will range from personal handheld and wearable microdevices, carrying individual and possibly private information (e.g., wristwatches, bracelets, personal mobile displays and notification systems, smart clothing), to public devices in the surrounding environment (e.g., embedded screens and speakers, ambient pixel or nonpixel displays). Interactivity and computing capacities will also be embedded in intelligent materials (e.g., Coelho, 2007).

Many of these devices will be equipped with built-in facilities for multimodal interaction and alternative input/output (e.g., voice recognition and synthesis, pen-based pointing devices, vibration alerting, touch screens, etc.; see Chapter 40), or with accessories that facilitate alternative ways of use (e.g., hands-free kits; see Chapter 11). Therefore, the design of such devices will address a wider range of user and context requirements than the traditional desktop computer.

In the context of AmI, interaction is no longer a one-to-one relationship between a human, a specific task, and a specific device in a static context. Rather, it becomes a many-to-many relationship between diverse users and a multitude of devices in a dynamically changing environment. AmI will bring about new interaction techniques, as well as novel uses and multimodal combinations of existing advanced techniques, such as, for example, speech and audio (see Chapters 30 and 32, respectively) gaze-based interaction (Gepner et al., 2007; see also Chapter 36), gesture (Ferscha et al., 2007; see also Chapter 34), and natural language (Zimmermann et al., 2004; Gárate et al., 2005; see also Chapter 31). Additionally, interaction will tangibly be embedded in everyday objects and smart artifacts (e.g., Streitz et al., 2005; see also Chapter 60).

It is likely that some of the built-in features of AmI environments, such as multimodality, will facilitate the provision of solutions that will be accessible by design (see Chapter 40). For example, blind users will benefit from the wider availability of voice input and output. Different modalities can be used concurrently, so as to increase the quantity of information made available or present the same information in different contexts, or redundantly, to address different interaction channels, both to reinforce a particular piece of information or to cater to the different abilities of users. Multimodality and recognition technologies are also claimed to have the potential to make human communication with the environment natural and much more similar to human communication than the traditional interaction paradigm (Nijholt et al., 2004), and to better exploit human perception abilities (Cai, 2007).

However, AmI is also anticipated to introduce increased complexity for its users. As technology disappears to humans both physically and mentally, devices will be no longer perceived as computers, but rather as augmented elements of the physical environment (Streitz, 2007). The nature of interaction in AmI environments will change radically, evolving from human-computer interaction to human-environment interaction (Streitz, 2007) and human-computer confluence (Ferscha et al., 2007). These concepts emphasize the fusion of the technology and the environment, as well as the inextricable role of interaction in all aspects of everyday life.

Therefore, interaction shifts from an explicit paradigm, in which the users' attention is on computing, toward an implicit paradigm, in which interfaces themselves drive human attention when required (see Chapter 59). Interaction in the emerging environment will no longer be based on a series of discrete steps, but on a continuous input/output exchange of information (Faconti and Massink, 2001). Continuous interaction differs from discrete interaction since it takes place over a relatively longer period of time, in which the exchange of information between the user and the system occurs at a relatively high rate, in real time. A first implication is that the system must be capable of dealing in real time with the distribution of input and output in the environment. This implies an understanding of the factors that influence the distribution and allocation of input and output resources in different situations for different individuals.

Due to the intrinsic characteristics of the new technological environment, interaction poses different perceptual and cognitive demands on humans compared to currently available technology (Gaggioli, 2005). It is, therefore, important to investigate how human perceptual and cognitive functions will be engaged in the emerging forms of interaction, and how this will affect an individual's perceptual and cognitive space (e.g., emotion, vigilance, information processing, and memory). The main challenge in this respect is to identify and avoid forms of interaction that may lead to negative consequences such as confusion, cognitive overload, frustration, and so on. This is particularly important given the pervasive impact of the new environment on all types of everyday activities and on the way of living. A thorough discussion of related challenges is provided in Chapter 60.

61.7 Application Domains

A variety of new products and services is made available by the emerging technological environment. Key application examples include intelligent home environments supporting home automation (Friedewald et al., 2005), communication and socialization (Romero et al., 2003), intelligent office environments (Hall et al., 2001), intelligent campuses (Sadeh et al., 2005), intelligent transportation systems (Brett Hall and Trivedi, 2002; see also Chapter 49), and intelligent edutainment (Ndiaye et al., 2005).

These types of applications are characterized by increasing ubiquity, nomadicity, and personalization, and pervade all human activities. They have the potential to enhance security in the physical environment, save human time, augment human memory, and support people in complex tasks, as well as in simple activities. Overall, AmI offers a range of new opportunities towards universal access for all citizens in the information society, through the possibility of integrating (or, in some cases, substituting) limited human abilities with the intelligence (electronic, tele-operated, and robotic devices) spread in the environment, in such a way as to support independent living, higher quality of health care, and easier interpersonal communication (e.g., Sevillano et al., 2004). AmI technologies can also be used for rehabilitation purposes (e.g., Morganti and Riva, 2005). The concept of ambient assisted living also fosters the utilization of AmI technologies toward enhancing the independent living of the disabled and the elderly and solving problems caused by the aging of the European population (Steg et al., 2005).

61.8 Nontechnological Issues

Although AmI is only taking its first steps, it already very clearly appears that its human-centered and inclusive development raises a variety of ethical and social issues that necessitate appropriate policy, standardization, and legislative intervention (Information Society Technologies Advisory Group of the European Commission, 2003; Bohn et al., 2005; Kemppainen et al., 2007). Relevant legislative areas are, for example, personal data protection, consumer protection, accessibility, and telecommunications regulation. In particular, privacy and security (see Chapter 57) are overwhelming discussion topics in relation to AmI. As discussed in Chapter 60, acceptability levels of data availability in the environment may vary depending on many factors, and may ultimately be a matter of trade-off between the need of privacy and the options offered by the available information about the users.

Another important horizontal issue is how research on universal access in AmI should be conducted. Both universal access and AmI are highly multidisciplinary fields, and it is, therefore, very important to bring together research teams and diverse user groups, so as to start a constructive dialogue and establish a common vocabulary.

Both fields are also deeply human-centered. It is, therefore, necessary to bring users in direct contact with AmI technologies and the possibilities they offer, and to make them aware of the technological possibilities and potential approaches to build up the new environment. Toward addressing this issue, facilities and laboratories have been set up where different AmI technologies are being developed, integrated, and tested in a real-life context, simulating AmI environments and supporting short-, medium-, and long-term experimentation with AmI technologies (e.g., Georgia Tech's Aware Home,[1] MIT's House_n, Philips'

---

[1] http://www.cc.gatech.edu/fce/ahri.

HomeLab,[2] Fraunhofer-Gesellschaft's inHaus[3]). The ongoing development of a fully accessible AmI research facility simulating various living environments is reported by Stephanidis et al. (2007).

61.9 Conclusions

This chapter has outlined some key research issues that emerge from the evolution of the information society toward AmI environments, focusing on the diversity of human needs, the dynamic evolution of context, and the multifaceted characteristics of interactive technologies. These challenges include:

- Investigation of human characteristics, abilities, and requirements in the context of AmI
- Suitable approaches to nonfunctional characteristics, such as accessibility, privacy and security, and safety
- Suitable models of the context of use
- Appropriate interaction devices and techniques for diverse users and contexts of use
- Interaction design for continuous and implicit interaction
- Elaboration of design methods suitable for very complex interactive environments
- Mechanisms for interaction adaptation
- A balance of policy, standardization, and legislation intervention

It is argued that only a global approach to systematically understanding these factors and their interplay can lead to universal access in AmI environments. Therefore, universal access methods and techniques need to be appropriately expanded and enhanced, as well as validated in practice in the AmI context. This requires multidisciplinary collaboration across multiple application domains, as well as the build up of appropriate research infrastructures to act as test beds and incubators of future technologies.

References

Abascal, J. and Civit, A. (2001). Mobile communication for older people: New opportunities for autonomous life, in the *Proceedings of the EC/NSF Workshop on "Universal Accessibility of Ubiquitous Computing: Providing for the Elderly,"* May 2001, Alcacer do Sal, Portugal. http://virtual. inesc.pt/wuauc01/procs/papers-list.html.

Antona, M., Burzagli, L., Emiliani, P.-L., and Stephanidis, C. (2007). The ISTAG scenarios: A case study, in *Towards an Inclusive Future: Impact and Wider Potential of Information and Communication Technologies* (P.R.W. Roe, ed.), section 4.1 of Chapter 4 "Ambient Intelligence and Implications

for People with Disabilities," pp. 158–187. Brussels: COST219ter.

Bihler, P. and Mügge, H. (2007). Supporting cross-application contexts with dynamic user interface fusion, in the *Proceedings of the MoBe Workshop at Informatik 2007*. http:// sam.iai.uni-bonn.de/people/PascalBihler/paper/mobe 07-bihler-muegge.pdf.

Bohn, J., Coroamà, V., Langheinrich, M., Mattern, F., and Rohs, M. (2005). Social, economic, and ethical implications of ambient intelligence and ubiquitous computing, in *Ambient Intelligence* (W. Weber, J. M. Rabaey, and E. H. L. Aarts, eds.), pp. 5–28. Berlin/Heidelberg: Springer-Verlag.

Brett Hall, T. and Trivedi, M. (2002). A novel interactivity environment for integrated intelligent transportation and telematic systems, in the *Proceedings of the 5th International IEEE Conference on Intelligent Transportation Systems*, 3–6 September 2002, Singapore, pp. 396–401. Washington, DC: IEEE Press.

Burzagli, L., Emiliani, P. L., and Gabbanini, F. (2007). Is the intelligent environment smart enough?, in *Universal Access in Human-Computer Interaction: Ambient Interaction, Volume 6 of the Proceedings of the 12th International Conference on Human-Computer Interaction (HCI International 2007)* (C. Stephanidis, ed.), 22–27 July 2007, Beijing, pp. 43–52. Berlin/Heidelberg: Springer-Verlag.

Cai, Y. (2007). Ambient intelligence: From interaction to insight. *International Journal of Human-Computer Studies* 65: 419–420.

Clerckx, T., Vandervelpen, C., Luyten, K., and Coninx, K. (2006). A task-driven user interface architecture for ambient intelligent environments, in the *Proceedings of the 11th International Conference on Intelligent User Interfaces (IUI '06)*, January 2006, Sydney, Australia, pp. 309–311. New York: ACM Press.

Coelho, M. (2007). Programming the material world: A proposition for the application and design of transitive materials, in the *Proceedings of the 9th International Conference on Ubiquitous Computing (Ubicomp 2007)*, 16–19 September 2007, Innsbruck, Austria. Berlin/Heidelberg: Springer-Verlag.

Coroama, V., Bohn, J., and Mattern, F. (2004). Living in a smart environment: Implications for the coming ubiquitous information society, in the *Proceedings of the International Conference on Systems, Man and Cybernetics 2004 (IEEE SMC 2004), Volume 6*, 10–13 October 2004, Amsterdam, The Netherlands, pp. 5633–5638. Washington, DC: IEEE Press.

Dogac, A., Laleci, G., and Kabak, Y. (2003). Context frameworks for ambient intelligence, in the *Proceedings of eChallenges 2003*, 22 October 2003, Bologna, Italy. Oxford: IOS Press. http://www.srdc.metu.edu.tr/webpage/publications/2003/ context.pdf.

Edwards, W. K. and Grinter, R. E. (2001). At home with ubiquitous computing: Seven challenges, in the *Proceedings of the 3rd International Conference on Ubiquitous Computing,*

---

30 September–2 October 2001, Atlanta, pp. 256–272. London: Springer-Verlag.

Emiliani, P. L. and Stephanidis, C. (2005). Universal access to ambient intelligence environments: Opportunities and challenges for people with disabilities. *IBM Systems Journal, Special Issue on Accessibility* 44: 605–619.

Faconti, G. and Massink, M. (2001). Continuous interaction with computers: Issues and requirements, in *Universal Access in HCI: Towards an Information Society for All, Volume 3 of the Proceedings of HCI International 2001* (C. Stephanidis, ed.), 5–10 August 2001, New Orleans, pp. 301–304. Mahwah, NJ: Lawrence Erlbaum Associates.

Ferscha, A., Resmerita, S., and Holzmann, C. (2007). Human computer confluence, in *Universal Access in Ambient Intelligence Environments, Proceedings of the 9th ERCIM Workshop on User Interfaces for All* (C. Stephanidis and M. Pieper, eds.), 27–28 September 2007, Bonn, Germany, pp. 14–27. Berlin/Heidelberg: Springer-Verlag.

Friedewald, M., Da Costab, O., Punieb, Y., Alahuhtac, P., and Heinonen, S. (2005). Perspectives of ambient intelligence in the home environment. *Telematics and Informatics* 22: 221–238.

Friedewald, M., Vildjiounaite, E., Punie, Y., and Wright, D. (2007). Privacy, identity and security in ambient intelligence: A scenario analysis. *Telematics and Informatics Archive* 24: 15–29.

Gaggioli, A. (2005). Optimal experience in ambient intelligence, in *Ambient Intelligence* (G. Riva, F. Vatalaro, F. Davide, and M. Alcañiz, eds.), pp. 35–43. Amsterdam, The Netherlands: IOS Press.

Gárate, A., Herrasti, N., and López, A. (2005). GENIO: An ambient intelligence application in home automation and entertainment environment, in the *Proceedings of the 2005 Joint Conference on Smart Objects and Ambient Intelligence: Innovative Context-Aware Services: Usages and Technologies (sOc-EUSAI '05), Volume 121*, 12–14 October 2005, Grenoble, France, pp. 241–245. New York: ACM Press.

Gepner, D., Simonin, J., and Carbonell, N. (2007). Gaze as a supplementary modality for interacting with ambient intelligence environments, in *Universal Access in Human-Computer Interaction—Ambient Interaction (Part II), Volume 6 of the Proceedings of 12th International Conference on Human-Computer Interaction (HCI International 2007)* (C. Stephanidis, ed.), 22–27 July 2007, Beijing, pp. 848–857. Berlin/Heidelberg: Springer-Verlag.

González, G., Angulo, C., López, B., and de la Rosa, J. L. (2005). Smart user models: Modelling the humans in ambient recommender systems, in the *Proceedings of the 10th International Conference on User Modeling (DASUM 2005)*, July 2005, Edinburgh, Scotland, pp. 11–20. Berlin/Heidelberg: Springer-Verlag.

Hall, D., Le Gal, C., Martin, J., Chomat, O., and Crowley, J. L. (2001). MagicBoard: A contribution to an intelligent office environment. *Robotics and Autonomous Systems* 35: 211–220.

Hartz Søraker, J. and Brey, P. (2007). Ambient intelligence and problems with inferring desires from behaviour. *International Review of Information Ethics* 8: 7–12.

Haya, P. A., Montoro1, G., and Alamán, X. (2004). A prototype of a context-based architecture for intelligent home environments, in *On the Move to Meaningful Internet Systems 2004: CoopIS, DOA, and ODBASE, Part I of the Proceedings of the OTM Confederated International Conferences*, October 2004, Cyprus, pp. 477–491. Berlin/Heidelberg: Springer-Verlag.

Hellenschmidt, M. and Wichert, R. (2005). *Goal-oriented Assistance in Ambient Intelligence.* Fraunhofer-IGD Technical Report, Darmstadt. http://www.igd.fhg.de/igd-a1/publications/publ/ERAmI_2005.pdf.

Information Society Technologies Advisory Group of the European Commission (2001). *Scenarios for Ambient Intelligence in 2010.* ftp://ftp.cordis.europa.eu/pub/ist/docs/istagscenarios2010.pdf.

Information Society Technologies Advisory Group of the European Commission (2003). *Ambient Intelligence: From Vision to Reality.* ftp://ftp.cordis.lu/pub/ist/docs/istag-ist2003_consolidated_report.pdf.

Intille, S. S., Larson, K., Munguia Tapia, E., Beaudin, J., Kaushik, P., Nawyn, J., and Rockinson, R. (2006). Using a live-in laboratory for ubiquitous computing research, in the *Proceedings of PERVASIVE 2006* (K. P. Fishkin, B. Schiele, P. Nixon, and A. Quigley, eds.), 7–10 May 2006, Dublin, Ireland, pp. 349–365. Berlin/Heidelberg: Springer-Verlag.

Issarny, V., Sacchetti, D., Tartanoglu, F., and Sailhan, F. (2002). Enabling ambient intelligence via the web, in the *Proceedings of DNAC 2002*, May 2002, San Francisco. http://www-rp.lip6.fr/dnac/6.4issarny-article.pdf.

Kemppainen, E., Abascal, J., Allen, B., Delaitre, S., Giovannini, C., and Soede, M. (2007). Ethical and legislative issues with regard to ambient intelligence, in *Impact and Wider Potential of Information and Communication Technologies* (P. R.W. Roe, ed.), pp. 188–205. Brussels: COST.

Kleinberger, T., Becker, M., Ras, E., Holzinger, A., and Müller, P. (2007). Ambient intelligence in assisted living: Enable elderly people to handle future interfaces, in *Universal Access in Human-Computer Interaction: Ambient Interaction, Volume 6 of the Proceedings of the 12th International Conference on Human-Computer Interaction (HCI International 2007)* (C. Stephanidis, ed.), 22–27 July 2007, Beijing, pp. 103–112. Berlin/Heidelberg: Springer-Verlag.

Morganti, F. and Riva, G. (2005). Ambient intelligence for rehabilitation, in *Ambient Intelligence* (G. Riva et al., eds.), pp. 283–295. Amsterdam: IOS Press. http://www.ambientintelligence.org.

Ndiaye, A., Gebhard, P., Kipp, M., Klesen, M., Schneider, M., and Wahlster, W. (2005). Ambient intelligence in edutainment: Tangible interaction with life-like exhibit guides, in *Intelligent Technologies for Interactive Entertainment* (J. G. Carbonell and J. Siekmann, eds.), pp. 104–113. London: Springer-Verlag.

Nijholt, A., Rist, T., and Tuijnenbreijer, K. (2004). Lost in ambient intelligence?, in the *Proceedings of CHI 2004*, 24–29 April 2004, Vienna, Austria, pp. 1725–1726. New York: ACM Press.

Preuveneers, D., Van den Bergh, J., Wagelaar, D., Georges, A., Rigole, P., Clerckx, T., et al. (2004). Towards an extensible context ontology for Ambient Intelligence, in the *Proceedings of EUSAI 2004*, 3–4 November 2004, Eindhoven, The Netherlands. http://research.edm.uhasselt.be/~tclerckx/eusai2004.pdf.

Punie, Y., Delaitre, S., and Maghiros, I. (2006). Dark scenarios in ambient intelligence: Highlighting risks and vulnerabilities. Deliverable D2. *Safeguards in a World of Ambient Intelligence (SWAMI)*. http://swami.jrc.es.pages/deliverables.htm.

Ramos, C., Augusto, J. C., and Shapiro, D. (2008). Ambient Intelligence: The next step for artificial intelligence. *IEEE Intelligent Systems* 23: 15–18.

Riva, G., Loreti, P., Lunghi, M., Vatalaro, F., and Davide, F. (2003). Presence 2010: The emergence of ambient intelligence, in *Being There: Concepts, Effects and Measurement of User Presence in Synthetic Environments* (G. Riva, F. Davide, and W. A. IJsselsteijn, eds.), pp. 59–82. Amsterdam: IOS Press.

Röcker, C., Janse, M. D., Portolan, N., and Streitz, N. (2005). User requirements for intelligent home environments: A scenario-driven approach and empirical cross-cultural study, in the *Proceedings of the 2005 Joint Conference on Smart Objects and Ambient intelligence: Innovative Context-Aware Services: Usages and Technologies*, 12–14 October 2005, Grenoble, France, pp. 111–116. New York: ACM Press.

Romero, N., v. Baren, J., Markopoulos, P., de Ruyter, B., and IJsselsteijn, W. (2003). Addressing interpersonal communication needs through ubiquitous connectivity: Home and away, in *Ambient Intelligence* (E. Aarts, R. Collier, E. van Loenen, and B. de Ruyter, eds.), pp. 419–430. Berlin/Heidelberg: Springer-Verlag.

Sadeh, N., Gandon, F., and Kwon, O. B. (2005, July). *Ambient Intelligence: The MyCampus Experience. School of Computer Science*. Technical Report CMU-ISRI-05-123, Carnegie Mellon University, Pittsburgh.

Savidis, A. and Stephanidis, C. (2005). Distributed interface bits: Dynamic dialogue composition from ambient computing resources. *Personal and Ubiquitous Computing* 9: 142–168.

Schmidt, A. (2005). Interactive context-aware systems interacting with ambient intelligence, in *Ambient Intelligence* (G. Riva, F. Vatalaro, F. Davide, and M. Alcañiz, eds.), pp. 159–178. Amsterdam, The Netherlands: IOS Press. http://www.ambientintelligence.org.

Sevillano, J. L., Falcó, J., Abascal, J., Civit-Balcells, A., Jiménez, G., Vicente, S., and Casas, R. (2004). On the design of ambient intelligent systems in the context of assistive technologies, in the *Proceedings of Computers Helping People with Special Needs (ICCHP 2004)* (K. Miesenberger et al., eds.),

7–9 July 2004, Paris, France, pp. 914–921. Berlin/Heidelberg: Springer-Verlag.

Steg, H., Strese, H., Hull, J., and Schmidt, S. (2005). *Europe Is Facing a Demographic Challenge. Ambient Assisted Living Offers Solutions*. Report of the AAL Project. http://www.aal169.org/Published/Final%20Version.pdf.

Stephanidis, C. (2001). Adaptive techniques for universal access. *User Modelling and User Adapted Interaction International Journal* 11: 159–179.

Stephanidis, C., Antona, M., and Grammenos, D. (2007). Universal access issues in an ambient intelligence research facility, in *Universal Access in Human-Computer Interaction: Ambient Interaction – Volume 6 of the Proceedings of the 12th International Conference on Human-Computer Interaction (HCI International 2007)* (C. Stephanidis, ed.), 22–27 July 2007, Beijing, pp. 208–217. Berlin/Heidelberg: Springer-Verlag.

Streitz, N. A. (2007). From human-computer interaction to human-environment interaction: Ambient intelligence and the disappearing computer, in *Universal Access in Ambient Intelligence Environments, Proceedings of the 9th ERCIM Workshop on User Interfaces for All* (C. Stephanidis and M. Pieper, eds.), 27–28 September 2006, Bonn, Germany, pp. 3–13. Berlin/Heidelberg: Springer-Verlag.

Streitz, N. A., Röcker, C., Prante, T., van Alphen, D., Stenzel, R., and Magerkurth, C. (2005). Designing smart artifacts for smart environments. *Computer* 38: 41–49.

Theofanos, M. and Scholtz, J. (2004). Towards a framework for evaluation of ubicomp applications. *IEEE Pervasive Computing* 3: 82–88.

van Houten, H. (2006). The physical basis of ambient intelligence, in *AmIware Hardware Technology Drivers of Ambient Intelligence, Volume 5* (S. Mukherjee et al., eds.), pp. 9–27. Amsterdam, The Netherlands: Springer-Verlag.

Vazquez, J. I. and López de Ipiña, D. (2008). Social devices: Autonomous artifacts that communicate on the Internet, in the *Proceedings of the First International Conference Internet of Things (IOT) 2008*, 26–28 March 2008, Zurich, Switzerland, pp. 308–324. Berlin/Heidelberg: Springer-Verlag.

Zhou, J., Yu, C., Riekki, J., and Kärkkäinen, E. (2007). AmE framework: A model for emotion-aware ambient intelligence, in the *Proceedings of the 2nd International Conference on Affective Computing and Intelligent Interaction (ACII2007): Doctoral Consortium*, 12–14 September 2007, Lisbon, Portugal, pp. 163–169. Berlin/Heidelberg: Springer-Verlag.

Zimmermann, G., Vanderheiden, G., Ma, M., Gandy, M., Trewin, S., Laskowski, S., and Walker, M. (2004). Universal remote console standard: Toward natural user interaction in ambient intelligence, in *CHI '04 Extended Abstracts on Human Factors in Computing Systems*, pp. 1608–1609. New York: ACM Press.

Index